BLE OF INTEGRALS

ic Forms

$$1. \int u^n\, du = \frac{1}{n+1} u^{n+1} + C, \ n \neq -1$$

$$2. \int \frac{du}{u} = \ln |u| + C$$

$$3. \int e^u\, du = e^u + C$$

$$4. \int a^u\, du = \frac{1}{\ln a} a^u + C$$

$$5. \int \sin u\, du = -\cos u + C$$

$$6. \int \cos u\, du = \sin u + C$$

$$7. \int \sec^2 u\, du = \tan u + C$$

$$8. \int \csc^2 u\, du = -\cot u + C$$

$$9. \int \sec u \tan u\, du = \sec u + C$$

$$10. \int \csc u \cot u\, du = -\csc u + C$$

$$11. \int \tan u\, du = \ln |\sec u| + C$$

$$12. \int \cot u\, du = \ln |\sin u| + C$$

$$13. \int \sec u\, du = \ln |\sec u + \tan u| + C$$

$$14. \int \csc u\, du = \ln |\csc u - \cot u| + C$$

$$15. \int \frac{du}{\sqrt{a^2 - u^2}} = \sin^{-1} \frac{u}{a} + C$$

$$16. \int \frac{du}{a^2 + u^2} = \frac{1}{a} \tan^{-1} \frac{u}{a} + C$$

$$17. \int \frac{du}{u\sqrt{u^2 - a^2}} = \frac{1}{a} \sec^{-1} \frac{u}{a} + C$$

onal Forms

$$18. \int \frac{du}{a + bu} = \frac{1}{b} \ln |a + bu| + C$$

$$19. \int \frac{u\, du}{a + bu} = \frac{1}{b^2} (a + bu - a \ln |a + bu|) + C$$

$$20. \int \frac{u\, du}{(a + bu)^2} = \frac{a}{b^2(a + bu)} + \frac{1}{b^2} \ln |a + bu| + C$$

$$21. \int \frac{u^2\, du}{(a + bu)^2} = \frac{1}{b^3} \left(a + bu - \frac{a^2}{a + bu} - 2a \ln |a + bu| \right) + C$$

$$22. \int \frac{du}{u(a + bu)^2} = \frac{1}{a(a + bu)} - \frac{1}{a^2} \ln \left| \frac{a + bu}{u} \right| + C$$

$$23. \int \frac{du}{u^2(a + bu)} = -\frac{1}{au} + \frac{b}{a^2} \ln \left| \frac{a + bu}{u} \right| + C$$

$$24. \int u(a + bu)^n\, du = \frac{(a + bu)^{n+1}}{b^2} \left(\frac{a + bu}{n + 2} - \frac{a}{n + 1} \right) + C, \ n \neq -1, -2$$

$$25. \int \frac{du}{a^2 - u^2} = \frac{1}{2a} \ln \left| \frac{u + a}{u - a} \right| + C$$

$$26. \int \frac{du}{(a + bu)(c + du)} = \frac{1}{ad - bc} \ln \left| \frac{c + du}{a + bu} \right| + C, \ ad - bc \neq 0$$

$$27. \int \frac{u\, du}{(a + bu)(c + du)} = \frac{1}{ad - bc} \left(\frac{a}{b} \ln |a + bu| - \frac{c}{d} \ln |c + du| \right) + C, \ ad - bc \neq 0$$

ns Containing $\sqrt{a + bu}$

$$28. \int u\sqrt{a + bu}\, du = \frac{2}{15b^2} (3bu - 2a)(a + bu)^{3/2} + C$$

$$29. \int u^n \sqrt{a + bu}\, du = \frac{2}{b(2n + 3)} \left[u^n(a + bu)^{3/2} - na \int u^{n-1}\sqrt{a + bu}\, du \right]$$

$$30. \int \frac{u\, du}{\sqrt{a + bu}} = \frac{2}{3b^2} (bu - 2a)\sqrt{a + bu} + C$$

$$31. \int \frac{u^2\, du}{\sqrt{a + bu}} = \frac{2}{15b^3} (8a^2 + 3b^2u^2 - 4abu)\sqrt{a + bu} + C$$

$$32. \int \frac{u^n\, du}{\sqrt{a + bu}} = \frac{2u^n \sqrt{a + bu}}{b(2n + 1)} - \frac{2na}{b(2n + 1)} \int \frac{u^{n-1}\, du}{\sqrt{a + bu}}$$

$$33. \int \frac{du}{u\sqrt{a + bu}} = \begin{cases} \frac{1}{\sqrt{a}} \ln \left| \frac{\sqrt{a + bu} - \sqrt{a}}{\sqrt{a + bu} + \sqrt{a}} \right| + C, \ a > 0 \\ \frac{2}{\sqrt{|a|}} \tan^{-1} \sqrt{\frac{a + bu}{|a|}} + C, \ a < 0 \end{cases}$$

$$34. \int \frac{du}{u^n \sqrt{a + bu}} = -\frac{\sqrt{a + bu}}{a(n - 1)u^{n-1}} - \frac{b(2n - 3)}{2a(n - 1)} \int \frac{du}{u^{n-1}\sqrt{a + bu}}$$

$$35. \int \frac{\sqrt{a + bu}}{u}\, du = 2\sqrt{a + bu} + a \int \frac{du}{u\sqrt{a + bu}}$$

$$36. \int \frac{\sqrt{a + bu}}{u^2}\, du = -\frac{\sqrt{a + bu}}{u} + \frac{b}{2} \int \frac{du}{u\sqrt{a + bu}}$$

ns Containing $\sqrt{u^2 \pm a^2}$

$$37. \int \sqrt{u^2 \pm a^2}\, du = \frac{u}{2} \sqrt{u^2 \pm a^2} \pm \frac{a^2}{2} \ln \left| u + \sqrt{u^2 \pm a^2} \right| + C$$

$$38. \int u\sqrt{u^2 \pm a^2}\, du = \frac{1}{3} (u^2 \pm a^2)^{3/2} + C$$

$$39. \int u^2 \sqrt{u^2 \pm a^2}\, du = \frac{u}{8} (2u^2 \pm a^2)\sqrt{u^2 \pm a^2} - \frac{a^4}{8} \ln \left| u + \sqrt{u^2 \pm a^2} \right| + C$$

$$40. \int \frac{\sqrt{u^2 + a^2}}{u}\, du = \sqrt{u^2 + a^2} - a \ln \left| \frac{a + \sqrt{u^2 + a^2}}{u} \right| + C$$

2

CALCULUS

EDITION

2

CALCULUS

Leonard I. Holder
Gettysburg College

James DeFranza
St. Lawrence University

Jay M. Pasachoff
Williams College

Brooks/Cole Publishing Company
Pacific Grove, California

I(T)P™ The trademark ITP is used under license.

To

Jean, Bill, and Steve
Regan, Sara, and David
Naomi, Eloise, and Deborah

for their love and support

Brooks/Cole Publishing Company
A Division of Wadsworth, Inc.

Printed in the United States of America

10 9 8 7 6 5 4 3 2

Library of Congress Cataloging in Publication Data
Holder, Leonard Irvin, [date]–
    Calculus / Leonard I. Holder, James DeFranza, Jay M. Pasachoff. — Ed. 2.
        p.   cm.
    Rev. ed. of: Single variable calculus. 1993.
    ISBN 0-534-23304-X
    1. Calculus.   I. DeFranza, James, [date]–   .   II. Pasachoff, Jay
M.   III. Holder, Leonard Irvin, [date]–   Single variable calculus.
IV. Title.
QA303.H578   1994
2nd ed.
515—dc20   94-6514
                                                                  CIP

Sponsoring Editor: *Jeremy Hayhurst*
Editorial Assistant: *Elizabeth Barelli Rammel*
Production Coordinator: *Marlene Thom*
Production Assistant: *Tessa A. McGlasson*
Manuscript Editor: *Janet Tilden*
Interior and Cover Design: *Sharon L. Kinghan*
Cover Photo: *Orrery, Adler Planetarium; Satellite, European Space Agency/Photo Researchers, Inc.*
Art Coordinator: *Lisa Torri*
Interior Illustration: *MacArt Design; Tech-Graphics*
Photo Coordinator: *Robert Western*
Photo Researchers: *Diana Mara Henry; Joan Meyers*
Typesetting: *Electronic Technical Publishing*
Cover Printing: *Phoenix Color Corporation*
Printing and Binding: *Arcata Graphics/Hawkins*

PREFACE

The cover depicts a working mechanical model of the solar system, called an *orrery* after the fourth Earl of Orrery, who commissioned the second such machine ever built, in about 1713. The earliest orreries showed only the movement of the Earth and Moon relative to the Sun, but later ones incorporated other planets such as Mercury and Venus, all of which were set in motion by a hand crank. This particular orrery includes Saturn and was built in the Netherlands in 1735. A well-crafted orrery was an elegant mechanical model used as both a calendar and a clock, and it furthered popular understanding of the physical and mathematical nature of the solar system.

The cover also portrays a currently deployed European Remote Sensing satellite (ERS1) because the artificial satellite is in large part an extension of the underlying physics and mathematics used in constructing the orrery. We have chosen the orrery and the satellite images to represent both the very beginning and the latest manifestation of the modern scientific age, which has been profoundly influenced by mathematics—especially calculus. They also serve to remind us of the remarkable duality of mathematics: its precision and truthfulness are unchanging; yet its application is amazingly flexible and adaptable to both the physical and mathematical universes.

The study of the orbits of artificial and natural satellites (planets) owes a great deal to the scientific insights made possible by calculus. Isaac Newton used calculus to show that the planets travel in elliptical orbits around the Sun. He thereby logically confirmed Johannes Kepler's earlier postulates based on years of astronomical observations by Kepler's mentor Tycho Brahe. Whereas mechanical orreries gave relatively crude approximations of the orbits of the planets, modern software-based digital orreries are vastly more accurate and are used to make predictions about the long-term (and perhaps ultimate) behavior of the solar system and the universe.

In the early eighteenth century, about the time that the first orrery was built, Bernard de Fontenelle said that the recent invention of the calculus was a revolution in mathematics. Fontenelle's enthusiasm was an understandable reaction to the astonishing success of calculus in enabling rapid progress on many previously intractable problems. Calculus was immediately seen to be effective and remarkably powerful in describing and predicting both change and motion. The complementary concepts of derivative and integral and their associated foundational problems—that of tangent (or velocity) and area—clearly set the stage for a profound reevaluation of the importance and very nature of mathematics. It was evident, even in Fontenelle's day, that calculus was a singular development in the history of mathematics, but in what sense was it a revolution? Many of the insights that calculus provided are worthy of that description, but many mathematical philosophers would deny that there have ever been revolutions in mathematics. After all, mathematics is unique among the sciences in that its truths, when based on common axioms, are unchanging once proven. Mathematics does not appear to be prone to revolutions in the sense of sudden and dramatic changes; but there is another connotation for the word *revolution*. It was first used as a technical term in the exact sciences, where it literally meant to return again, through cyclic succession, as in the changing seasons, or to ebb and flow, like the movement of tides. In this sense Fontenelle was on firmer ground. Perhaps it is natural to expect a kind of cyclic progression whereby calculus, and mathematics itself, is periodically reexamined and reexpressed.

CONTEMPORARY CALCULUS

One could argue that the teaching of calculus is currently undergoing a kind of revolution caused by widespread dissatisfaction with the current national calculus curriculum. Broadly stated, there is an awakened interest in teaching core concepts in a modern context as opposed to overwhelming students with interesting but secondary details. Although it is probably characteristic of revolutionary movements that—even among their supporters—they mean different things to different people, there are common themes in reform discussions. The principal goals seem to be: a focus on conceptual understanding rather than rote manipulation; increased relevance and applicability to other disciplines; an emphasis on visualization as a counterpoint to abstraction; and the use of relevant technology to streamline computation. The discussions seem to hold out the promise of a new and creative attitude to mathematics instruction.

We wholeheartedly approve of many of the goals of reform, but we have consciously not "thrown the baby out with the bathwater." For instance, despite calls for streamlining topics, we have retained a thorough treatment of limits and techniques of integration. We see attempts to all but eliminate their discussion as ill advised. After all, the limit is the single most important notion in calculus, which is the mathematics of limits; and techniques of integration have important theoretical implications in later mathematics courses. The same is true of computational skills. Though we want students to appreciate that mathematics is more than computation, nothing is gained by claiming that these skills are somehow suddenly less important in understanding calculus. We considered many options and weighed them carefully before revising this text. We have ultimately taken what might be called a modern but conservative approach. Our response to criticism of the calculus curriculum is evolutionary rather than revolutionary.

Some of the reform principles exhibited here are as follows: a thorough grounding in functions and their graphs; early introduction of differential equations; the proper emphasis on numerical procedures; the use of graphing calculators and computer algebra systems; discussion of broad-based and pedagogically realistic applications; a more streamlined discussion of integration techniques in favor of using CAS and tables; and the introduction of infinite series through the vehicle of polynomial approximation of functions. We emphasize the practicality and utility of calculus and its visualization through graphics and images, both computer-generated and photographic, whenever they can make an otherwise dry and abstract discussion more concrete. The occasional use of color photographs encourages visualization in relevant physical contexts and adds interest and visual appeal. A new design enables us to highlight important results and to distinguish important elements of graphs, logical connectives, and related concepts that need to be distinguished.

Computer algebra subsections also encourage efforts to visualize using appropriate technology. These subsections are intended to be useful reading for students even if they do not have access to a computer algebra system. When facilities are available, we recommend "computer-assisted algebra"—not thoughtless button pushing—and stress that using the technology is not a substitute for understanding the mathematics. Quite the contrary—one usually needs to understand the mathematics in order to verify and apply the results of the technology.

We have revised this edition with special attention to making the material clearer and more concrete as well as more readable and interesting, while retaining the appropriate level of rigor. The organization, the explanations and examples, the exercises, and the text layout have all been substantially improved. We have also added several new features to the text that we believe will enhance students' perception and understanding of the subject.

ORGANIZATION

- The material on functions and graphs has been expanded and now comprises the entire first chapter. Additional precalculus review material appears in the free Supplementary Appendices.

- Limits and continuity form a new second chapter. The last section on the theory of limits is optional and can be omitted.

- The transcendental functions have been consolidated in this edition—that is, exponential and logarithmic functions have been combined with inverse trigonometric and hyperbolic functions in one chapter. Trigonometric functions are reviewed in the first chapter. We have chosen to introduce the logarithm as an integral and present the exponential function as its inverse to reinforce the concept of the integral.

- The chapter on parametric equations and polar coordinates now precedes that on sequences and series. A brief review of conic sections is followed by the polar form of the conics. The former separate chapter on conic sections has been placed in the Supplementary Appendices so it can be presented with near-total flexibility.

- Discussion of numerical methods is treated where the material occurs naturally in the context of other topics, not in a separate chapter. Throughout the text numerical methods are given their proper due as an irreplaceable adjunct to symbolic and graphical techniques.

NEW FEATURES

- In addition to an ample number of drill exercises, applied problems, and more theoretical exercises, we included many problems calling for the use of graphing calculators or computer algebra systems. These latter problems are marked with a special icon so that instructors can easily skip over or emphasize them, depending on local preferences and resources. The number of examples and figures has more than doubled.

- Selected applied problems have been added to the end of each chapter in special exercise sets called Applying Calculus. Some of these problems will be appropriate for class projects and, at the request of some reviewers, answers to these exercises are not in the text. Solutions are provided in the Complete Solutions Manual, Volumes 1 and 2, for instructors.

- Preceding each chapter is a "Personal View" interview or essay "about calculus." People interviewed here include representatives of statistics, history, economics, physics, biology, engineering, and computer science, as well as mathematicians. The purposes are severalfold. We hope to give a glimpse of the diversity of opinion and varied personalities of scientists and mathematicians, to neutralize some myths about mathematics, and to show that mathematics is not a static discipline. We also want students to understand that calculus is widely useful in different professions, and that teaching and mathematical research are not without controversy. We hope these "personal views" will provide a little extra motivation and perhaps even inspiration for those students who ask "where will I ever use this stuff?" We hope more students can be led to perceive that the course has utility and significance beyond the immediate grade they achieve.

- Each chapter concludes with a Concept Quiz. Answers to these quizzes are provided in the Study Guides, but students should be encouraged to attempt these quizzes and verify the answers on their own.

PEDAGOGICAL CONSIDERATIONS

To succeed in any quantitative discipline, students must "get off on the right foot" in mathematics. A calculus course is critically important in this respect alone. Most students learn best by proceeding from the concrete to the more abstract, and our consistent approach is to discuss new concepts intuitively before stating formal definitions or theorems. Our style is conversational, speaking directly to the student. We have tried to explain the mathematics in simple language. We do not "talk down" to students and we do not avoid rigorous argument where it is needed. (A number of the more technical proofs are given in the Supplementary Appendices.)

We begin each chapter with a summary of the chapter goals so that students will have some idea of where the material is leading. We try to motivate new ideas via concrete problems and often ask students to make their own conjectures before we state the results. We hope that with the right tone in the exposition, many students can be persuaded to read the text before they come to class. We even dare to hope that more of them will truly enjoy calculus and decide to major in mathematics or in a mathematically related field.

ACKNOWLEDGMENTS

First, we give our sincere thanks to the contributors of the Personal Views: Paul R. Halmos, Owen Gingerich, Judith V. Grabiner, Mark J. T. Smith, Anna J. Schwartz, Phillip A. Griffiths, Keith Geddes, Deborah Tepper Haimo, Sheldon Glashow, Benoit Mandelbrot, Constance McMillan Elson, Allan Cormack, Joseph Newhouse, David Lieberman, N. Scott Urquhart, Alfred L. Goldberg, and Frank McGrath. The efforts of supplement authors and accuracy checkers Raymond Southworth, Ben Brown, John Banks, David Royster, Marian Hoyle, and Terri Bittner are also gratefully acknowledged. Special thanks are due to Raymond Southworth, who labored long and hard to ensure accurate answers and solutions to the exercises. Any remaining errors are our responsibility.

Jay Pasachoff thanks his daughters, Eloise and Deborah, for inspiring him through their studies of calculus to work on this book. He further thanks Deborah Pasachoff for her comments on the manuscript and proofs.

In addition, we thank the following reviewers and colleagues, who have also contributed a great deal to the text: Donna Bailey, Northeast Missouri State University; Nancy Blachman, Variable Symbols, Inc.; Barbara Bohannon, Hofstra University; Richard Bonanno, Deerfield Academy; Stephen Brady, Wichita State University; Fred Brauer, University of Wisconsin–Madison; Jon Breitenbucher, Ohio State University; Chris Caldwell, University of Tennessee–Martin; Mervin Childers; Roger Cooper, Canada College; Branko Curgus, Western Washington University; Deborah Frantz, Kutztown University; Anthony Ferzola, University of Scranton; Dante Giarusso, St. Lawrence University; James Hart, Middle Tennessee State University; Nabil Husni, Palm Beach Community College–North Campus; Nathaniel Martin, University of Virginia; Montie Monzingo, Southern Methodist University; Lance Nielsen, University of Nebraska–Omaha; John Randolph, West Virginia University; Janice Rech, University of Nebraska–Omaha; Robert Stanton, St. John's University; M. J. Still, Palm Beach Community College–North Campus; Richard Tucker, North Carolina A&T State University; Andrei Verona, California State University–Los Angeles; and Roger Waggoner, University of Southwestern Louisiana. We thank Joel Cohen of Rockefeller University for comments regarding the logistic equation.

We also thank the indexer, Nancy Kutner, for her expert work.

Thanks also to the Brooks/Cole staff: Elizabeth Barelli Rammel, editorial assistant, who coordinated volumes of handwritten, verbal, and electronic communication and resolved numerous details essential to the book's development; Audra Silverie, assistant editor; Sharon Kinghan, senior designer; Lisa Torri, senior art coordinator; Margaret Parks, promotions manager; Patrick Farrant, marketing manager; and Bob Western, photo coordinator. Special thanks go to Marlene Thom, production coordinator, for the superb job she did guiding the book through the complex all-electronic production process. She did so with skill, and somehow maintained her good humor, despite what was often a demanding schedule. It was a delight to work with her. Finally, we wish to thank our editor, Jeremy Hayhurst, for his support and guidance. His ideas, his vision, and his encouragement were critical in making this book possible.

Leonard I. Holder
James DeFranza
Jay M. Pasachoff

PROLOGUE FOR STUDENTS

The word *calculus* comes from the Latin term for a pebble used as a token in counting and calculating. The development of calculus has a long and rich history. Although rudiments of the subject can be traced back to the ancient Greek mathematicians, Archimedes in particular, the primary impetus came in the seventeenth century from mathematicians such as Caliveri in Italy, Fermat and Descartes in France, and Wallis and Barrow in England. The invention of analytic geometry by Descartes and Fermat was an especially important prerequisite for the development of calculus. However, it was Isaac Newton and Gottfried Wilhelm Leibniz who found the crucial mathematical relationships and provided the notational tools needed to bring the emerging ideas together into a coherent calculation system—that is, a calculus—and this remains an intellectual feat of enormous importance. (See the *Personal Views* before Chapters 2 and 3 for more discussion of the history of calculus.)

Calculus can be described as the mathematics of change and motion. Since change and motion are implicit in all aspects of the physical world, the methods of calculus are useful in all the physical, natural, and social sciences, including economics. Here are only a few of the things calculus enables us to do.

- forecast the size of a growing population after a given period of time
- maximize profit and minimize costs in a business
- model ocean and air currents and their effect on weather
- design turbines, boat hulls, and electric circuits
- model the spread of a disease, rumor, or fashion
- determine how to place a satellite in geosynchronous (stationary) orbit
- predict voting behavior

Given the incredible diversity of the applications of calculus, it is all the more surprising that it evolved from two seemingly unrelated geometric problems: finding the tangent line to a curve and finding the area bounded by a curve. In part (a) of the figure in the margin we show the tangent line to a curve at point P, but how do we know exactly which line it is? In other words, given the

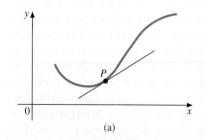
(a)

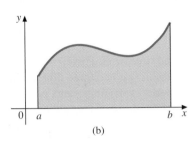

(b)

curve and the point P, how do we find the tangent line? In part (b) we show a region under a portion of a curve and above the x-axis. How do we find the area of the region?

Solving the tangent problem leads to the concept of derivative. That part of calculus that deals with derivatives is called *differential calculus*. Solving the area problem leads to the integral, which is the basis of *integral calculus*. How the derivative and the integral are related is one of the most fundamental insights in all of calculus. The key to solving both the tangent and area problems is the concept of limit. The limit is the single most important notion in calculus, and it is what separates calculus in a fundamental way from all of the mathematics that preceded it. Much of this course will deal with limits in one form or another and all of Chapter 2 is devoted to studying limits of functions.

Although the derivative has its origins in the solution of the tangent problem, it can be interpreted more broadly as a measure of the instantaneous rate of change of one quantity with respect to another. It is this rate-of-change interpretation that makes the derivative such a powerful tool. It is the derivative, for example, that enables us to find such things as the velocity and acceleration of a moving object, the rate at which blood is flowing through an artery at a given instant, the rate at which the profit on some item is increasing or decreasing, the rate at which a culture of bacteria is growing, and the rate of decay of a radioactive substance.

Although the integral is founded on a geometric problem involving an area, it can be more broadly interpreted as the total change in some quantity whose rate of change is known. For example, the integral can be used to determine the total distance traveled by an object even when its velocity changes, the work done by a variable force, the total force against a dam, the total length of a curve, and the total volume enclosed by a surface.

COMPUTERS, CALCULATORS, AND CALCULUS

During your course you will undoubtedly encounter some form of technology. Both graphing calculators and computer software can be valuable aids in studying calculus, if they are used properly; however, they are not substitutes for learning the basic mathematical concepts. In this text we discuss ways in which calculators and currently available computer software can be used to best advantage in learning calculus. We are trying to give you a general sense of where these systems are most helpful and appropriate, even if you don't have a computer or a calculator at hand. We have included discussion of software in the text and graphing calculators are discussed in the Supplementary Appendices.

With the software programs and mathematical languages known as *computer algebra systems* (abbreviated "CAS"), almost all calculus operations and computations can be done quickly and efficiently. The best-known of these systems are Maple, Mathematica, and DERIVE. The term *computer algebra system* is slightly misleading, because although these systems do algebra, they do much more, and they are forcing a reevaluation of what it is that we emphasize when teaching mathematics. CAS are designed to do symbolic manipulation, solve numerical problems, and graph functions, and they can be programmed as well. Other software programs, called *spreadsheets*, are optimized for numerical calculations. They are widely used in the business world and can be useful in calculus too, especially in work with numerical methods. Spreadsheets also have the ability to graph functions. Some widely used spreadsheets are Microsoft's Excel, Lotus's 1-2-3, and Novell's Quattro Pro.

The Graphing Calculator program distributed free with Power Macintosh computers, which use the PowerPC™ chip, graphs in two and three dimensions, carries out algebraic simplifications, does numeric evaluations, and performs simple differentiations. It does not include integration. Thus, although the program does not have as much capability as Maple, Mathematica, or DERIVE, it can be especially useful in Chapters 1 through 4 and Chapter 7 of this book as well as in graphing three-dimensional surfaces. Choosing "derivative" from a menu or "differentiate" from a keyboard that you select to have on the screen allows you to find the derivative of the function you type in. This Graphing Calculator program can be a helpful aid as you work problems to understand calculus.

Do not let yourself be intimidated by this type of technology. It is a servant in the course, not a master. No one is encouraging you to compete with or to try to outcompute an algebra system. It's important to understand that in many cases the software is using the same algorithm (correct mathematical procedure) that you use when you solve the problem "by hand," but in other cases it is using an algorithm that is only practical to use within the software. You will sometimes have to ask yourself how the computer or calculator is getting its results, and occasionally you will have to use the theory of calculus to verify that your results are correct. Technology changes rapidly, and there will undoubtedly be newer devices or releases of these programs available when you take this course. It is important to remember that the mathematics you are learning has not changed, and the problem-solving skills that you develop in this course will never be obsolete.

The text includes "computer algebra subsections" on a yellow screened background at the end of certain sections of the text. These subsections give guidance on carrying out calculus operations using one of the three common CAS. Not all the details are shown, but you should be able to follow the instructions as given. Don't panic if the commands don't work as indicated on your version of the software. You may have a different release that uses different commands. Also, your computer may have some other CAS, in which case you can adapt the instructions to your particular system. You may need to consult the program documentation carefully.

Basic scientific calculators have many uses in calculus just as they do in precalculus courses. The more advanced programmable calculators, especially those with graphing capabilities, are especially well suited to calculus. These advanced calculators share many features with computers. They have smaller viewing screens and do numerical calculations rather than the more general symbolic operations carried out by CAS. Although the calculators are more limited, they are more compact and portable and are cheaper than software and a computer. However, if you already have a computer, consider getting a CAS program. They rarely cost more than a good graphing calculator and will take you much further than a calculator will. Other advantages of the CAS are that you can program and update them easily as well as use them with a word processor.

Of course, it may not be necessary for you to choose between a computer and a calculator. If you have access to both, you will sometimes want to use one and sometimes the other; in most cases, we do not specify which device or program to use. We do not show instructions for using calculators in the text. (See the Supplementary Appendices for graphing calculator examples.) There are specially marked exercises for graphing calculators and CAS where technology is all but required, but we have deliberately not marked all the exercises where graphing calculators or CAS could be used.

APPROPRIATE USE OF TECHNOLOGY

What should you learn to do by hand and what should you have a computer or calculator do for you? This is analogous to asking whether grade-school students should learn fundamentals of multiplication and multiplication tables, or simply learn to push keys on calculators. Most mathematicians would argue that it is always better to understand the principles before you push the buttons. But perhaps there is more in common here than is first apparent. Using a multiplication or log table, memorized or not, and pushing buttons on a calculator both involve the use of an algorithm for carrying out a mathematical operation. In a sense, both you and the calculating device are simply employing an algorithm. The key is in understanding what the algorithm means and why it works. As your experience and confidence increase, you'll develop a sense of when the calculator or software will be most helpful. In many cases you will find that it is calculus itself, not the technology, that is the most useful tool.

In using technology to find derivatives, for example, it is essential that you first learn what a derivative is and what information it conveys. Only then is it meaningful to calculate derivatives using software, and only for more complicated derivatives (the simple ones can be found more easily and even faster by hand). If you rely solely on software to find all derivatives you encounter, you will have learned nothing about derivatives. Also, if you place too heavy a reliance on a computer or a calculator, you might draw incorrect conclusions from the output unless you have some idea of what to expect. When plotting a graph, for example, if you don't chose the x-scale and y-scale properly, you may not see all the important features of the graph. Without a knowledge of calculus, you would not know what the important features are and could easily come to incorrect conclusions. With a basic understanding of calculus, you can have total confidence in the correctness of your results.

Calculators and computers with CAS or spreadsheets are powerful devices and can be real assets in a calculus course, but they do not eliminate the need to learn the fundamental mathematical concepts. If your instructor prefers to overlook the CAS sections or the exercises marked for calculators or computers, you will not be disadvantaged in the course. We have tried to show how to take advantage of the availability of these devices, but the course is not dependent on them. If your instructor permits and encourages it, take advantage of what technology has to offer, but use it to learn your calculus. Once you truly understand what a derivative is, your instructor is unlikely to criticize you for using software to calculate one. If you don't understand what a derivative is, no degree of proficiency with software will compensate for this deficiency.

SOME SUGGESTIONS FOR STUDYING CALCULUS

Although a calculus course provides challenging goals, your success will have a great deal to do with how you study from your lecture notes and from this text. First, you will understand far more if you read the relevant section of the text before your lecture and before you attempt to work the exercises you are assigned. Don't treat this text as a novel. You need to read actively, with a pencil and paper at hand. Try to fill in any missing details you encounter in the algebraic steps. Second, do your assignments when they are assigned. Letting homework accumulate is a prescription for disaster in this or any other mathematics course. To learn mathematics, you must do mathematics. Watching your instructor do it is not enough. Don't give up too quickly if you have difficulty with a problem. If your instructor permits it, you might find it helpful to study with some of your classmates. Group study has proved to be more effective for some students than individual study.

As a student of calculus, you have much to look forward to as this rich and engaging subject unfolds before you. Read the chapter-opening Personal Views, not to learn calculus, but to learn about calculus and mathematics and their influence on the modern world. Recognize that calculus contains powerful ideas that are very general and applicable to a huge variety of problems, whatever your chosen discipline. Use whatever technology you have at your disposal to explore and investigate the mathematical ideas you encounter. This course should help you think about mathematics in a more logical fashion and to understand that definitions are not theorems, that theorems require proofs, and that abstract mathematical arguments are of great importance in efficiently describing the phenomena we see in the "real world." It is impossible to say what calculus is and to explain what all it can do in a brief introduction, because that, of course, is what the rest of the book is about. We hope that these preliminary remarks have served to whet your appetite and that you are ready now for the "main course." If you give calculus the attention it deserves, you will not regret the effort devoted to it in your future courses and career.

CONTENTS

Paul Halmos

What do you remember about your first mathematics course in college?

Most of what I learned in my first year of college was not taught in courses.... Calculus was considered too sophisticated for freshmen in those days. The mathematics I was taught in my freshman year at Illinois was algebra, trigonometry, and analytic geometry (a revelation to me). No one seemed to notice that virtually the only part of analytic geometry that calculus uses is the idea of a graph, and most of us knew about that before we entered college.

How did you do in your first calculus course?

It wasn't a big part of my life then. It was just a chore. I couldn't for the life of me understand it. B's were the best I could do. I could differentiate and integrate, but I had no understanding of things like the "four-step rule" in the text. I now know that it means "apply the definition to find the derivative."

Students sometimes have trouble in calculus because they don't know algebra and trigonometry. Why is this so important?

It's an old saying among teachers that students often trip up on the algebra—not the calculus itself. Mathematics is a hierarchical discipline. If you don't know enough algebra you're going to have trouble understanding the extensions of those ideas in calculus. My advice is to review as much precalculus as you can before you start calculus.

So what did you learn in your calculus class?

What I *did not* learn was how to study, or for that matter, what studying meant. I was just barely smart enough to know that it did not mean memorizing—but I did think that to have learned something really meant to remember it. Everything I had ever been taught was done by "here is how it goes—now you do it." That's how I was taught to count, to write, and to solve algebra problems. I didn't understand what it meant to understand something, and what one should do to get there.

Do you mean that much of what's taught in a mathematics course doesn't need to be memorized?

Not exactly. You should try to learn how to think about mathematics in a way that works for you. Don't rely on memory, but memory can't be ruled out as an asset in mathematical success. There's a story told about a mathematician named Tamarkin. On a Ph.D. oral exam, he asked the student about the convergence properties of certain series. "I don't remember," said the student, "but I can always look it up if I need it." Tamarkin was not pleased by this answer. "That doesn't seem to be true," he said, "because you sure need it now."

What would you say to a student just beginning a calculus course?

Here you sit, an undergraduate with a calculus book open in front of you.... How do you learn this, how do you study, how do you penetrate the darkness? All I can tell you is what I do, but I suspect that the same sort of thing works for everyone. It's been said before and often, but it cannot be overemphasized: study *actively*. Don't just read it; fight it! Ask your own questions, look for your own examples, discover your own proofs. Is the hypothesis necessary? Is the converse true? Why is this true? You'll learn much more this way.

What should a student concentrate on while attending a lecture?

Lectures are a standard way of learning something—one of the worst ways. Too passive, that's the trouble. *Standard recommendation*: take notes. *Counterargument*: yes, to be sure, taking notes is an activity. If you do it, you'll have something solid to refer back to afterward, but you are likely to miss the delicate details of the presentation as well as the big picture—you are too busy scribbling to pay attention. *Counter-counter argument*: if you don't take notes, you won't remember what happened, in what order it came, and chances are your attention will flag part of the time, you'll daydream, and you might even nod off. My own solution is a compromise: I take very skimpy notes, and then, whenever possible, I transcribe them in greater detail, as soon afterward as possible. By skimpy notes I mean one or two words per minute, plus, possibly a crucial formula or two and a crucial picture or two—just enough to fix the order of events, and incidentally, to keep me awake. The act of expanding your own notes is very helpful in learning what you did or didn't understand from the lecture.

We often hear that mathematics is in part a language. Is learning mathematics like learning a language?

Learning a language is different from learning a mathematical subject; in the first case the problem is to acquire a habit, in the other to understand a structure. The difference has some important implications. In learning a language from a textbook, you might as well go through the book as it stands and work the exercises in it; what matters is to keep practicing the use of the language. If, however, you want to learn calculus, it might not be a good idea to open the book on page 1 and read it, working all the problems in order.... The material is arranged so that its linear reading is logically defensible, to be sure, but we readers are human ... and each one of us is likely to find something difficult that is easy for someone else, like the author of the text. My advice is to read until you come to a definition that is new to you, and then stop and try to think of examples, and non-examples, or till you come to a theorem new to you, and then stop and try to understand it and maybe try to prove it for yourself. Read with a pencil in hand to sketch a solution or a diagram. Here's the key: when you come to an obstacle, a mysterious passage, an unsolvable problem, just skip it. Jump ahead, try the next problem, turn the page, go to the next chapter, or if it gets really bad, abandon the book and start another one.

Books may be linearly ordered, but our minds are not.

What mathematics is really all about is solving concrete problems. The biggest fault of even good students is that although they may be able to spout correct theorems and remember correct proofs they cannot give examples, or construct counterexamples, or solve special problems. So I try to start every class with a problem. A famous dictum of George Polya, the guru of problem solving, is: "if you can't solve a problem, then there is an easier problem that you can't solve. Find it!"

What about teaching calculus? I know you're an advocate of the Moore method.

Some people call it that. The "Moore method" is named for a famous professor, R. L. Moore, who had a lot of rules that he made students follow on their honor. He challenged his students to learn by asking themselves and each other questions, and getting them to present results to each other. He challenged students to learn and they did. Challenge is the best teaching tool there is, for arithmetic as well as for functional analysis. I'm convinced that it's the best way to teach anything and everything. There's an old Chinese proverb that I learned from Moore himself: "I hear, I forget; I see, I remember; I do, I understand." There's another saying I like: "The best way to learn is to do; the worst way to teach is to talk."

Do you have any advice for instructors who are new to teaching calculus?

Prepare. To really understand a subject, you must know more than the subject; to teach a course, you must know much more about the subject than you can put into the course. The first time I taught calculus at Chicago, I knew that I knew more calculus than I was expected to teach, but even so it was important to prepare for the course. Students are receptive and if you say something wrong, they'll remember it and it will do them damage. I try hard not to say something wrong on purpose and never, ever, write something deliberately wrong on the board. My teacher at Illinois once told me that he was making the point that "$0^0 = 1$" is a definition, not a theorem, and to underscore the point he said he might just as well have defined 0^0 to be 7, which he wrote on the board. Years later he got a letter from a student saying "I have in my notes that $0^0 = 7$, but I forgot how you proved it?" Obviously, this student missed the point, but at least was taking notes.

Mistakes are another matter. If I make a mistake, I admit it and if I can't answer a question I usually say: "Let me think about it." Although we try to avoid mistakes and don't like baffling questions, it's well known that they have educational value. A teacher who comes off as omniscient helps perpetuate the myth that mathematics is a rigid body of facts and perfect techniques. To see an expert admit ignorance and then reason the way to the truth can be a real eye-opener for students. The lesson is that mathematical thinking can be used to learn something that you didn't learn before. "Mathematics is not a spectator sport."

Does humor work in the classroom?

Sometimes. You can even be corny if it helps make a point. Here is a bit of innocent fun on the topic of integration. Partly as an integration drill and partly to make a point about the use of "dummy variables," I might call on several students one after another to tell me what is $\int \frac{dx}{x}$, $\int \frac{du}{u}$, $\int \frac{dz}{z}$, $\int \frac{da}{a}$, and then, as the clincher, I'll write $\int \frac{d(\text{cabin})}{(\text{cabin})}$. Some students shout out right away "log cabin." They're then surprised when I tell them that I can't agree, but I might accept the answer "houseboat" or "log cabin on the beach"—If they're still puzzled I emphasize that the answer is really "log cabin plus sea." [This riddle is explained in chapters 5 & 7.]

But let's get back to teaching by challenging. An intrinsic aspect of the method at all levels, elementary or advanced, is to concentrate attention on the definite, the concrete, the specific. Once a student understands, really and truly understands, why 3×5 is the same as 5×3, then he or she quickly gets the automatic and obvious but nevertheless exciting unshakable conviction that "it goes the same way" for all other numbers. We all have an innate ability to generalize; the teacher's function is to call attention to a concrete special case that hides (and, we hope, ultimately reveals) the germ of the student's conceptual difficulty.

You seem to have fun talking about mathematics. What would make teaching it more satisfying?

I dream of the ideal university, full of students who are brimming with intellectual curiosity. The subset of them who take a mathematics course do so because they want to learn mathematics. They may be future doctors or chemists, or executives in a shirt factory, or none of the above, but if, for whatever reason, they really want to find out what this mathematics stuff is all about, and they come to me free willing and ask me to teach it to them, I'd jump at the chance. If that happened more often, joy! But the appeal of mathematics is subtle. It takes some exposure and training to understand what makes it all so interesting. Be patient, and ask questions.

Paul R. Halmos is Professor of Mathematics at Santa Clara University. He has spent the greater part of his professional life at the University of Chicago and the University of Michigan, as well as five years at the Institute for Advanced Study in Princeton. Professor Halmos has held a Guggenheim Fellowship and was the recipient of  a Chauvenet Prize for mathematical exposition from the Mathematical Association of America. He is the author of numerous books and research articles in ergodic theory, algebraic logic, and operator theory.

1

FUNCTIONS AND GRAPHS

Functions play a central role in calculus as well as in other branches of mathematics. It is hard to overestimate how important functions are both in mathematics and the "real world." The concept of a function is one of the two most basic ideas on which calculus is built. (The second is the concept of a *limit*, which we will introduce in Chapter 2.) In this chapter we will first introduce the notion of a function in general. Then we will consider ways of creating new functions based on known ones. We then explore some specific types of functions that we will use throughout the text: polynomial, rational, rational power, and trigonometric functions. We defer our study of other basic functions (such as exponential and logarithmic functions) to Chapter 7, when we can use some of the tools of calculus to analyze them.

1.1 THE FUNCTION CONCEPT

Weight w in oz.	Cost in dollars
$0 < w \leq 1$	0.29
$1 < w \leq 2$	0.52
$2 < w \leq 3$	0.75
$3 < w \leq 4$	0.98
$4 < w \leq 5$	1.21
$5 < w \leq 6$	1.44
$6 < w \leq 7$	1.67
$7 < w \leq 8$	1.90
$8 < w \leq 9$	2.13
$9 < w \leq 10$	2.36
$10 < w \leq 11$	2.59
$11 < w \leq 12$	2.82

(For $w > 12$, priority rates apply.)

Stating that one quantity *is a function of* another means that the first quantity depends on the second in some specific way. For example, the cost of mailing a letter first-class is a function of its weight. The Postal Service has set the exact form of this functional relationship; after a letter has been weighed, a Postal Service employee knows what to charge for postage. As another example, the well-known formula $C = 2\pi r$ expresses the circumference C of a circle as a function of its radius r.

In the case of circumference, the functional relationship is expressed clearly and concisely by a formula. For first-class postage, a table such as the one shown in the margin expresses the postal cost as a function of weight in an easily understood way. A formula in this case is more difficult to determine. (See Exercise 23 of Exercise Set 1.1 for a formula of this function.)

Another way of specifying a function is to use a graph. For example, the graph in Figure 1.1 shows the concentration of carbon dioxide in the atmosphere as a function of time for the years 1958 through 1992.

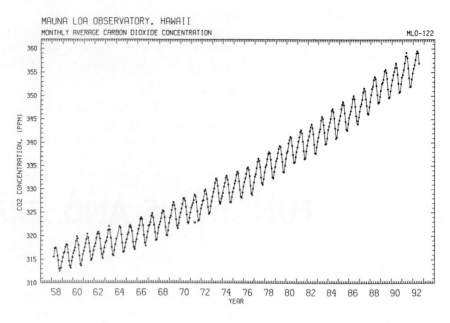

FIGURE 1.1

Atmospheric carbon dioxide concentrations as measured at the Mauna Loa observatory in Hawaii. This function is important for assessing the greenhouse effect and global warming.

Whether a function is specified by a formula (either in words or mathematical symbols), a table, or a graph, the basic idea is that *for each allowable value of one quantity, a value of another quantity is uniquely determined*. We make this idea more explicit in the following definition.

Definition 1.1 Function	A **function** from a set A to a set B is a rule that assigns to each element of A a unique element of B.

Note: Sometimes rules might appear to be different but yield identical results. For example, $x^2 - 1$ and $(x + 1)(x - 1)$ do not look alike, but for every value substituted for x, the two expressions yield the same value. When two rules assign the same value to each element of the set A in the definition, we will consider the rules to be the same.

The set A in Definition 1.1 is called the **domain** of the function, and the set B is called the **codomain**. If x is an element of the domain A, then the element y of B assigned by the function to x is called the **image** of x. The totality of images of points of A is called the **range** of the function (see Figure 1.2). So the range is a subset of B, but it may or may not include all of B. For example, in the case of first-class postage, the domain is the set A of all possible weights of letters between 0 and 12 oz. That is, A is the set of real numbers for which $0 < w \leq 12$. The set B could be taken as the set of all positive real numbers, but the range consists of just certain elements of B, namely, the cost for a letter weighing up to 1 oz., the cost for a weight between 1 and 2 oz., and so on, up to the cost for a weight between 11 and 12 oz.

Although the sets A and B in Definition 1.1 could consist of any sort of entities, we will assume they consist of real numbers unless otherwise specified.

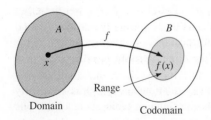

FIGURE 1.2

The set of all real numbers is denoted by $\mathbb{R}$. A function whose domain and range are subsets of $\mathbb{R}$ is called a **real-valued function of a real variable**. A function of this type is said to be *from $\mathbb{R}$ to $\mathbb{R}$*.

REMARK

■ You can think of the set $\mathbb{R}$ of real numbers as being the set of all decimal quantities. Decimals that terminate (such as $5/4 = 1.25$) and those that repeat (such as $5/3 = 1.666\ldots$) are *rational*, meaning that they can be expressed as the *ratio* of two integers. All other real numbers are *irrational*. Examples are $\sqrt{2} = 1.4142136237\ldots$ and $\pi = 3.1415926535\ldots$. Another way of characterizing real numbers is by associating them with points on a number line. Each real number corresponds to a unique point on the line, and each point on the line corresponds to a unique real number. We discuss real numbers in more detail in Appendix 1.

A function is usually designated by a single letter. The letter f is used most often, but other letters can be used. The image of an element x in A is then designated $f(x)$, read "f of x." Because an element x in A can be chosen arbitrarily, it is referred to as an **independent variable**. The image $f(x)$ is often designated by the letter y. Since its value depends on x and the function f, it is called the **dependent variable**.

It is important to distinguish between a single letter, such as f, that names a function, and the value of the function at x, symbolized by $f(x)$. So if we write, for example,

$$f(x) = x^2 + 1$$

we are in effect giving the defining rule of the function f. We often abbreviate a statement such as "Consider the function f defined by $f(x) = x^2 + 1$" by writing "Consider the function $f(x) = x^2 + 1$." Alternatively, for this same function we might write its rule by the equation

$$y = x^2 + 1$$

In this case the function is implied, but no letter such as f is given to name it.

Graphs of Functions

By the **graph of a function** f from $\mathbb{R}$ to $\mathbb{R}$, we mean the set of all points (x, y) in the plane such that x is in the domain of f and $y = f(x)$. It is usually impossible to plot all points of the graph, since there are usually infinitely many of them. In some cases we can plot a few well-chosen points and connect them with a smooth curve, but this procedure has some possible pitfalls. Calculus provides the necessary tools for removing uncertainty about the shape of the graph of a function. The simple function $f(x) = 2x$, whose graph is shown in Figure 1.3, can, however, be analyzed without calculus. Letting $y = f(x)$, we see that for each increase in x, y increases twice as much. Therefore, the graph is a straight line. Since the points $(0, 0)$ and $(1, 2)$ are on the line, we can draw the line through these two points.

Since the definition of a function requires that to each x in the domain a unique value of y in the range be assigned, we can use the following test to determine whether or not a curve is the graph of a function.

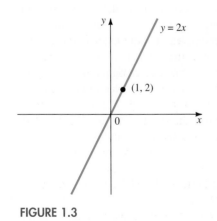

FIGURE 1.3

> ### Vertical Line Test
>
> A curve in the plane is the graph of a function if and only if each vertical line intersects the curve in at most one point.

Figure 1.4 illustrates this test.

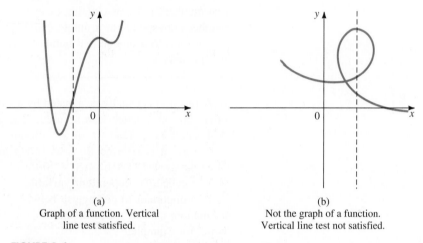

(a)
Graph of a function. Vertical line test satisfied.

(b)
Not the graph of a function. Vertical line test not satisfied.

FIGURE 1.4

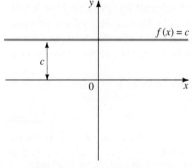

The constant function $f(x) = c$

FIGURE 1.5

A particularly simple type of function is one that gives the same value of the dependent variable for all values of the independent variable. Such a function is called a **constant function** and has the form $f(x) = c$. For example, $f(x) = 2$ is the constant function that has the value 2 for all values of x. The graph of a typical constant function is shown in Figure 1.5. It is a horizontal line and clearly passes the vertical line test. Note, however, that a vertical line does not pass this test and thus cannot be the graph of such a function.

It is sometimes helpful to think of a function as a machine. As illustrated in Figure 1.6, an element x is taken from the domain A and fed into the "f machine." The machine operates on x according to the rule for f, and out comes $f(x)$. This machine analogy suggests calling x the **input** value and $f(x)$ the **output** value. One of the best examples of such a function machine is a calculator. For example, you can input a number—say, 3—punch the $\boxed{x^2}$ key to activate the squaring function [$f(x) = x^2$], and the output 9 appears on the display screen. As a matter of fact, a calculator combines many different "function machines" into one unit.

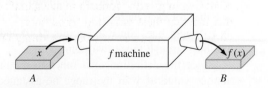

FIGURE 1.6

Different letters can be used to represent functions and variables. For example, in the first-class postage function we might let the independent variable (the weight) be denoted by w and the dependent variable (the cost of postage) be denoted by P. Then we could write

$$P = f(w)$$

with the rule for f determined by the Postal Service. Similarly, if we let g be the function giving the circumference of a circle in terms of its radius r, we have

$$C = g(r)$$

In this case we know the rule for g and can write

$$g(r) = 2\pi r$$

Then, for example, $g(2) = 4\pi$ expresses the fact that when $r = 2$, the circumference is 4π.

In many applications the independent variable is *time*, so it is natural to use the letter t to represent it. For example, in the function giving the concentration of carbon dioxide in the atmosphere (Figure 1.1), we might write $c = h(t)$, where t is the time in years. Then we could read from the graph to get values of $h(t)$. For example, $h(1970) \approx 325$, telling us that at the beginning of 1970 the concentration of carbon dioxide in the atmosphere was approximately 325 parts per million.

For a function to be fully defined, the rule as well as the domain and the codomain must be given. We will adopt the convention, however, that *when the domain of a function is not specified, we will understand it to be the largest subset of $\mathbb{R}$ for which the values of the function are real*. Also, *we will always assume the codomain is the set $\mathbb{R}$ of all real numbers* unless otherwise specified. This assumption means that the range of each of the functions we consider is a subset of $\mathbb{R}$. For example, the domain of

$$f(x) = \frac{1}{x - 2}$$

is the set of all real numbers except $x = 2$. We have to exclude $x = 2$, because division by 0 is not defined. Similarly, the domain of $f(x) = \sqrt{x - 2}$ is the set of all real numbers greater than or equal to 2. We can indicate this set symbolically by $\{x : x \geq 2\}$.* Values of x less than 2 result in square roots of negative numbers, which are not real.

As we have seen, functions are usually given in one of three ways: by a formula, by a table, or by a graph. Let us illustrate each way with the squaring function, that is, the function whose output is the square of the input value. We have just described the function in words, but it is more economical and less confusing to give the formula in mathematical symbols, such as $f(x) = x^2$. For this function there is no limitation on x, so we understand the domain of f to be all of $\mathbb{R}$. The three ways of describing this function are as follows:

---

*A symbol of the form $\{x : \ldots\}$ is read "the set of all x values (understood to be real) such that …"

Graph

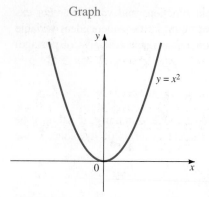

FIGURE 1.7

Formula	Table	
$f(x) = x^2$	0	0
	± 1	1
	± 2	4
	± 3	9

Of the three ways, the table is usually least satisfactory, since it is generally impossible to show all input and output values. Based on our table alone, we have no way of knowing that when $x = \frac{1}{2}$, $f(x) = \frac{1}{4}$, or even that when $x = 5$, $f(x) = 25$. Nevertheless, tables are useful, if for no other reason, to aid in drawing graphs. More important, data from experiments or from observations often are presented in tabular form, and the function may be difficult or impossible to describe by a formula. (Also, with spreadsheets on computers, tables can be used to represent functions with a high degree of accuracy.)

Graphing, too, has its limitations. For example, just looking at the graph in Figure 1.7 does not tell us that it is the graph of $f(x) = x^2$. Since it is impossible to show the entire domain on the portion of the x-axis we draw, we have to assume that the graph continues in the manner shown. This limitation is particularly important in using a graphing calculator (or a computer algebra system). When using one, you must select an x range and a y range in such a way that you can be confident you are seeing all the important features of the graph. (You really need calculus to be certain, however.) Figure 1.8 illustrates some inadequate graphs of a function resulting from poor choices of the y range, along with the graph as it should be when an appropriate y range is used.

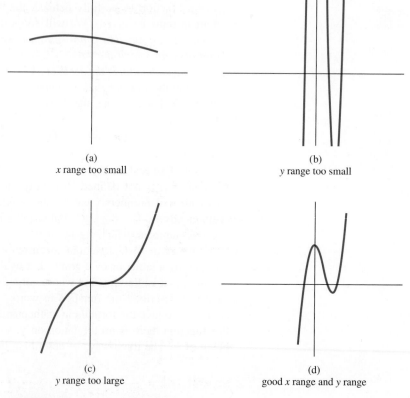

(a)
x range too small

(b)
y range too small

(c)
y range too large

(d)
good *x* range and *y* range

FIGURE 1.8

Graphs are helpful, despite their limitations. The graph of a function lets us see at a glance where the function is increasing or decreasing, whether it is bending up or bending down, where its high and low points are, and so on. All these behaviors are important in studying mathematical models of real-world phenomena, as we will see later.

A combination of ways of representing functions is usually best, especially a formula and a graph. It is possible, however, if enough tabular values are known, not only to draw the graph but also to find a formula that at least approximates the function. This "curve fitting" procedure is often used to construct mathematical models.

The following examples illustrate the main ideas introduced so far.

EXAMPLE 1.1   Let $f(x) = \sqrt{x - 1}$. What are the domain and range of f? Find $f(1)$, $f(2)$, $f(5)$, and $f(10)$. Draw the graph of f.

Solution   For the values of this function to be real numbers, the expression under the radical sign must be positive or zero. That is, we can choose only those x values for which $x - 1 \geq 0$, or, equivalently, $x \geq 1$. So the domain of f is the set of x values for which $x \geq 1$. By choosing x in this domain, we can make $\sqrt{x - 1}$ equal to any nonnegative real number, so the range of f is the set of all y values for which $y \geq 0$. By substituting, in turn, $x = 1$, $x = 2$, $x = 5$, and $x = 10$, we find

$$f(1) = 0, \qquad f(2) = 1, \qquad f(5) = 2, \qquad f(10) = 3$$

To draw the graph (Figure 1.9), we set $y = f(x)$ and plot the points (x, y) we have found, connecting them with a smooth curve.   ■

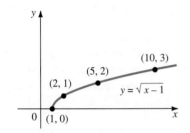

FIGURE 1.9

EXAMPLE 1.2   Let $f(t) = \dfrac{1}{2t - 1}$. Find the following:

(a) $f(1)$ (b) $f\left(\frac{1}{4}\right)$

(c) $f\left(\dfrac{1}{t}\right)$ (d) $\dfrac{1}{f(t)}$

(e) $f(t + h)$

Solution

(a) $f(1) = \dfrac{1}{2(1) - 1} = \dfrac{1}{1} = 1$

(b) $f\left(\dfrac{1}{4}\right) = \dfrac{1}{2\left(\frac{1}{4}\right) - 1} = \dfrac{1}{-\frac{1}{2}} = -2$

(c) We substitute $1/t$ for t and simplify:

$$f\left(\frac{1}{t}\right) = \frac{1}{2\left(\frac{1}{t}\right) - 1} = \frac{1}{\frac{2}{t} - 1} = \frac{1}{\frac{2}{t} - 1} \cdot \frac{t}{t} = \frac{t}{2 - t}$$

(d) $\dfrac{1}{f(t)} = \dfrac{1}{\frac{1}{2t-1}} = 2t - 1$ Note the difference between parts (c) and (d).

(e) $f(t + h) = \dfrac{1}{2(t + h) - 1} = \dfrac{1}{2t + 2h - 1}$ Replace t with $t + h$.   ■

As illustrated in Example 1.2, we evaluate a function at any admissible value by substituting that value for the independent variable wherever it occurs. For example, suppose $f(x) = x^3$. Then each of the following statements is valid:

$$f(2g) = (2g)^3 = 8g^3$$

$$f(3a + 4b) = (3a + 4b)^3$$

$$f(x + 1) = (x + 1)^3$$

$$f(x^2yz) = (x^2yz)^3 = x^6y^3z^3$$

When you substitute some expression for the independent variable, it is a good idea to place it in parentheses and then simplify. For example, with $f(x) = x^3$, we first wrote $f(2g) = (2g)^3$ and then simplified to get $8g^3$. This practice avoids the pitfall of failing to cube the 2 as well as the g.

EXAMPLE 1.3   Let $g(x) = \sqrt{x}$. Evaluate the expression

$$\frac{g(4 + h) - g(4)}{h}, \qquad h \neq 0$$

Then rationalize the numerator and simplify.

Note: We will see in Chapter 3 that the type of quotient appearing in this example forms the basis for defining what is called the *derivative*.

Solution   Since $g(4 + h) = \sqrt{4 + h}$ and $g(4) = \sqrt{4} = 2$, we have

$$\frac{g(4 + h) - g(4)}{h} = \frac{\sqrt{4 + h} - 2}{h}$$

To "rationalize the numerator" means to perform some valid operation on the function that results in a numerator that has only rational expressions—that is, no radicals (square roots, cube roots, or the like). Here, we accomplish this result by multiplying numerator *and* denominator by the factor $(\sqrt{4 + h} + 2)$. In doing this operation we rely on two facts from algebra: first, a fraction is unchanged if both numerator and denominator are multiplied by the same nonzero quantity (in effect, we are multiplying the fraction by 1); and second, the difference of two quantities times their sum is equal to the difference of their squares. Thus, we have

$$(a - b)(a + b)$$
$$= a^2 + ab - ab - b^2$$
$$= a^2 - b^2$$
Middle terms cancel.

$$\frac{\sqrt{4 + h} - 2}{h} = \frac{\sqrt{4 + h} - 2}{h} \cdot \frac{\sqrt{4 + h} + 2}{\sqrt{4 + h} + 2} = \frac{(\sqrt{4 + h})^2 - (2)^2}{h(\sqrt{4 + h} + 2)}$$

$$= \frac{(4 + h) - 4}{h(\sqrt{4 + h} + 2)} = \frac{h}{h(\sqrt{4 + h} + 2)} = \frac{1}{\sqrt{4 + h} + 2}$$

since $h \neq 0$.   ∎

The domain and range of a function frequently are subsets of $\mathbb{R}$ called **intervals**. The following notation and terminology are useful in describing types of intervals.

Symbol	Name	Meaning
$[a, b]$	closed interval	$\{x : a \leq x \leq b\}$
(a, b)	open interval	$\{x : a < x < b\}$

Intervals of these two types can be shown graphically as in Figure 1.10.

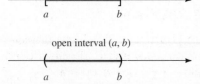

closed interval $[a, b]$

open interval (a, b)

FIGURE 1.10

A set of the form $\{x : a < x \leq b\}$ is designated $(a, b]$ and is called a **half-open** (or **half-closed**) interval, as is $[a, b)$, meaning $\{x : a \leq x < b\}$. It is also convenient to use the symbol ∞, read "infinity," in interval notation as follows:

$$[a, \infty) = \{x : x \geq a\}$$

$$(a, \infty) = \{x : x > a\}$$

$$(-\infty, a] = \{x : x \leq a\}$$

$$(-\infty, a) = \{x : x < a\}$$

$$(-\infty, \infty) = \text{the set of all real numbers}$$

EXAMPLE 1.4   Find the domain of the function F defined by the rule

$$F(x) = \sqrt{2 - x - x^2}$$

Solution   Because the square root of a negative number is not real, the domain consists of those x values for which $2 - x - x^2 \geq 0$. Factoring the left-hand side of this inequality gives

$$(1 - x)(2 + x) \geq 0$$

FIGURE 1.11

To see where the product is greater than zero, let us first see where it is equal to zero. The numbers $x = 1$ and $x = -2$ cause the product on the left to be 0. They divide the number line into three intervals, as shown in Figure 1.11. By testing the sign of the product in each interval, we obtain the results indicated in the figure. One way to test the signs is to use a *test value* in each interval. For example, we could use $x = 2$, $x = 0$, and $x = -3$ as test values. By substituting these values, in turn, for x in $(1 - x)(2 + x)$, we find the signs as indicated. Since we want the product to be positive *or* zero, we conclude that the domain of F is the closed interval $[-2, 1]$.   ■

REMARK

■ A figure such as Figure 1.11 showing the sign of an expression for various intervals on the real line is called a **sign graph** for the expression.

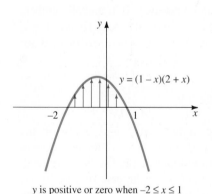

$y = (1 - x)(2 + x)$

y is positive or zero when $-2 \leq x \leq 1$

FIGURE 1.12

If the actual graph of the expression whose signs we want to determine is known, an alternative to a sign graph is to observe where it is positive (above the x-axis) and where it is negative (below the x-axis). For example, to find where $(1 - x)(2 + x) \geq 0$, we could sketch the graph of $y = (1 - x)(2 + x)$, as in Figure 1.12. (For more complicated graphs, a graphing calculator or computer algebra system is useful here.) We see from the graph that $y \geq 0$ when $-2 \leq x \leq 1$, just as we found in Example 1.4.

Applications often require finding the form of a functional relationship based on known information. The next example illustrates one such situation.

EXAMPLE 1.5   Given an isosceles triangle inscribed in a circle of radius 2, find the area A of the triangle as a function of its base b.

Solution   A typical triangle is pictured in Figure 1.13. We denote the perpendicular distance from the center of the circle to the base of the triangle by y, as shown. Then the altitude of the triangle is $y + 2$ and the area is

$$A = \frac{1}{2}b(y + 2)$$

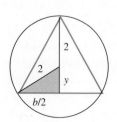

FIGURE 1.13

Pythagorean Theorem: In a right triangle the sum of the squares of the lengths of the legs equals the square of the length of the hypotenuse.

Now we must find y in terms of b. Using the Pythagorean Theorem on the shaded right triangle with legs $b/2$ and y and hypotenuse 2, we get

$$y^2 + \left(\frac{b}{2}\right)^2 = 4$$

Solving for y gives

$$y = \sqrt{4 - \frac{b^2}{4}} = \frac{\sqrt{16 - b^2}}{2}$$

Thus,

$$A = \frac{1}{2}b\left(\frac{\sqrt{16 - b^2}}{2} + 2\right) = \frac{b}{4}\left(\sqrt{16 - b^2} + 4\right) \qquad \blacksquare$$

Piecewise-Defined Functions

Some functions are defined by different rules on different parts of their domains. We call a function of this type a **piecewise-defined function**. The next example provides an illustration.

EXAMPLE 1.6   Draw the graph of the function h defined by

$$h(x) = \begin{cases} -1 & \text{if } x < 0 \\ 1 & \text{if } x = 0 \\ x + 2 & \text{if } x > 0 \end{cases}$$

Determine its domain and range.

Solution   The graph is shown in Figure 1.14. Notice that the point $(0, 1)$ is on the graph, as indicated by the solid circle there, but $(0, 2)$ and $(0, -1)$ are not, as shown by the open circles.

Although h is defined in a piecewise fashion, its domain is all of $\mathbb{R}$. Its range is the set of y values taken on by the function, namely, -1, 1, and all numbers greater than 2. We could indicate the range as $\{-1, 1\} \cup (2, \infty)$. The symbol $\cup$ is read "union," and $A \cup B$ represents the set of all elements in A or in B (or in both). $\qquad \blacksquare$

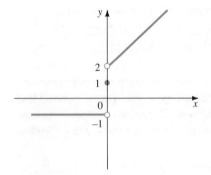

FIGURE 1.14

Even and Odd Functions; Symmetry

Suppose that for every x in the domain of a function f, $-x$ is also in the domain. It may happen that for each such x, $f(x)$ and $f(-x)$ are equal, as in the case of $f(x) = x^2$, whose graph is shown in Figure 1.15. On the other hand, $f(-x)$ may always be opposite in sign to $f(x)$, as in the case of $f(x) = x^3$, shown in Figure 1.16. These special types of functions are given names in the following definition.

Definition 1.2 Even and Odd Functions	A function f is said to be **even** if $f(-x) = f(x)$ for every x in the domain of f, and f is said to be **odd** if $f(-x) = -f(x)$ for every x in the domain of f.

For example, $f(x) = x^2$ is even and $f(x) = x^3$ is odd, since $(-x)^2 = x^2$ and $(-x)^3 = -x^3$.

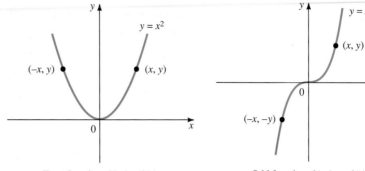

Even function: $f(-x) = f(x)$

FIGURE 1.15

Odd function: $f(-x) = -f(x)$

FIGURE 1.16

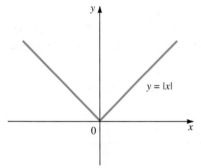

The absolute value function

FIGURE 1.17

Another important even function is the **absolute value function**, defined by $f(x) = |x|$, where

$$|x| = \begin{cases} x & \text{if } x \geq 0 \\ -x & \text{if } x < 0 \end{cases}$$

We can see that f is even, since the absolute value of any real number and of its negative are the same. For example, $|2| = 2$ and $|-2| = -(-2) = 2$. Thus, $f(2) = f(-2)$. The graph of f is shown in Figure 1.17.

REMARK ———————————————————————————————

■ It should be emphasized that even and odd functions are special types. Most functions are neither even nor odd. For example, if $f(x) = x^3 + x^2 - 1$, then $f(-x) = -x^3 + x^2 - 1$, which does not equal either $f(x)$ or $-f(x)$. So this function is neither even nor odd.

———————————————————————————————————————

For each point (x, y) on the graph of an even function, $(-x, y)$ is also on the graph, so the graph of an even function is **symmetric with respect to the y-axis**, as shown in Figure 1.18(a). When (x, y) is on the graph of an odd function, then $(-x, -y)$ is on the graph also, and the graph of an odd function is **symmetric with respect to the origin**, as shown in Figure 1.18(b).

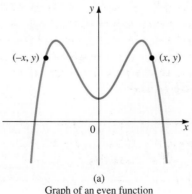

(a)
Graph of an even function
(symmetric with respect to the y-axis)

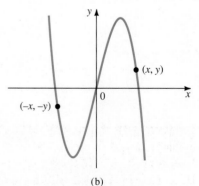

(b)
Graph of an odd function
(symmetric with respect to the origin)

FIGURE 1.18

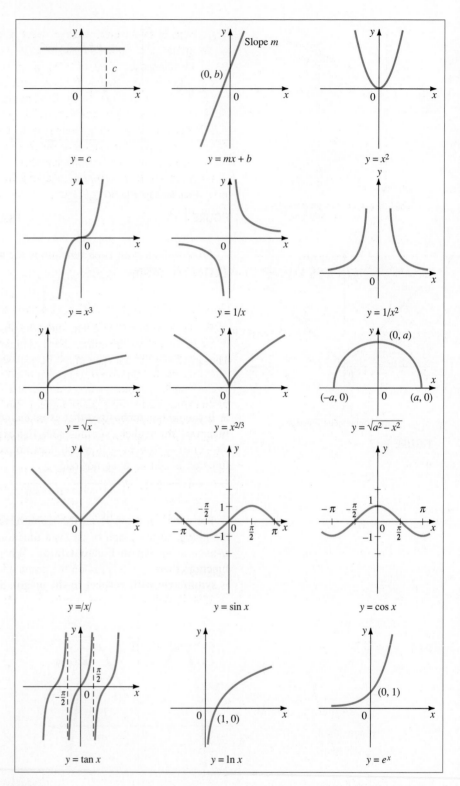

CHART 1.1

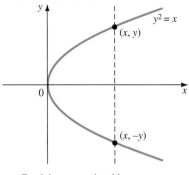

Graph is symmetric with respect to
the x-axis; not a function.

FIGURE 1.19

A graph is **symmetric with respect to the x-axis** if for each point (x, y) on the graph, the point $(x, -y)$ also is on the graph. Except for the x-axis itself, in which case $f(x) = 0$ for all x, such a graph could not be the graph of $y = f(x)$ for any function f, since the graph fails the vertical line test. The graph of $y^2 = x$ (Figure 1.19), for example, is symmetric with respect to the x-axis, but this equation does not define y as a function of x. In fact, for each $x > 0$ there are two corresponding y values, namely, $y = \sqrt{x}$ and $y = -\sqrt{x}$. (*Note:* The graph of an *equation*, such as $y^2 = x$, is the set of all points (x, y) whose coordinates satisfy the equation.)

We conclude this section by showing for reference the graphs of some commonly occurring functions (Chart 1.1). We will study each of these in succeeding sections.

GRAPHING CURVES USING COMPUTER ALGEBRA SYSTEMS

Using geometric or graphical representations of mathematical concepts or problems can greatly enhance our problem-solving abilities. Whether hand-drawn or computer-generated, visualization of important concepts plays a key role in our understanding of mathematics. In selected sections throughout the text, we will consider three computer algebra systems (CAS)—Maple, Mathematica, and DERIVE—as aids to computation and visualization. But in fact, as we will demonstrate, computer algebra systems can be used for much more.

You can follow these steps on your computer with Maple, Mathematica, or DERIVE. But you do not have to do so; you can just follow along in the text, since the syntax (command logic) of these computer algebra systems is close to what we would write in mathematical notation.

In this section, we will consider how Maple, Mathematica, and DERIVE can be used to sketch curves in the plane and to obtain information about functions. For consistency we will use Maple primarily and comment on the other CAS where necessary.

CAS 1

Plot the function $f(x) = x^2$ on the interval $[-2, 2]$.

In Maple, we type the command:

plot(x^2,x=-2..2);

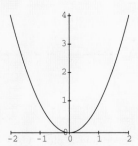

FIGURE 1.1.1
$y = f(x) = x^2$

Mathematica:[†]

Plot[x^2,{x,-2,2}]

DERIVE:

(At the □ symbol, go to the next step.)
a (author) □ x^2 □ p (plot window) □
p (plot) □ z (zoom out)

[†]In Mathematica, the braces { } surround information showing that we are plotting x on the horizontal axis from $x = -2$ to $x = 2$. In Maple this information is not inside braces, but is written x=−2..2, with two dots. Also note that all Maple commands end with a semicolon and all Mathematica commands must begin with a capital letter. In DERIVE, zoom out to change the scale to display the desired portion of the graph. In all three CAS, the symbol ^ (found over the 6 on the keyboard) means "raise to a power."

CAS 2

Plot the function $f(x) = x^2 - 2x - 3$ on the interval $[-4, 4]$.

Maple:

plot(x^2–2*x–3,x=–4..4);

Mathematica:

Plot[x^2–2*x – 3,{x,–4,4}]

DERIVE:

(At the □ symbol, go to the next step.) a (author) □ x^2–2x–3 □ p (plot window) □ p (plot) □ z (zoom out)

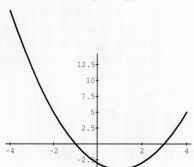

FIGURE 1.1.2
$y = f(x) = x^2 - 2x - 3$

CAS 3

Plot the function $f(x) = \frac{1}{3}x^3 + \frac{1}{2}x^2 - 2x + 3$ on the interval $[-4, 3]$. Find the point of intersection of the curve with the x-axis.

Maple:

plot(x^3/3+x^2/2–2*x+3, x=–4..3);

Mathematica:

Plot[x^3/3+x^2/2–2*x+3, {x,–4,3}]

DERIVE:

(At the □ symbol, go to the next step.) a (author) □ x^3/3+x^2/2–2x+3 □ p (plot window) □ p (plot) □ z (zoom out)

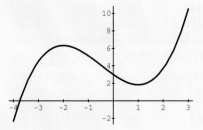

FIGURE 1.1.3
$y = f(x) = \frac{1}{3}x^3 + \frac{1}{2}x^2 - 2x + 3$
on $[-4, 3]$

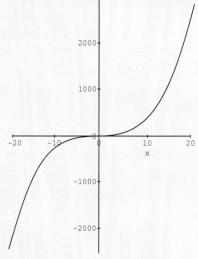

FIGURE 1.1.4
$y = f(x) = \frac{1}{3}x^3 + \frac{1}{2}x^2 - 2x + 3$
on $[-20, 20]$

The choice for the limits on x is important when generating a plot. CAS Figure 1.1.4 shows the Maple plot generated by the command

plot(x^3/3+x^2/2–2*x+3,x=–20 . . 20);

Although the domain contains the important features of the graph, the scale hides the true shape of the curve.

To find the point of intersection of the curve with the x-axis, we must find the roots of the equation $f(x) = \frac{1}{3}x^3 + \frac{1}{2}x^2 - 2x + 3$. That is, we need to solve $\frac{1}{3}x^3 + \frac{1}{2}x^2 - 2x + 3 = 0$. We can execute the following commands:

Maple:

f:=x^3/3+x^2/2–2*x+3;
fsolve(f=0,x);
Output: –3.744348422.

Mathematica:[†]

f=x^3/3+x^2/2–2*x+3
NRoots[f == 0,x]

DERIVE:

(At the □ symbol, go to the next step.)
a (author) □ x^3/3+x^2/2–2x+3 □ 1
(solve) □ [choose expression to solve]
□ x (approximate) □ [choose expression to approximate]

⌨ CAS 4

Plot the function $f(x) = \dfrac{(x+2)}{(x-1)}$ on the interval $[-10, 10]$.

Maple:

plot((x+2)/(x–1),x=–10..10,y=–5..5);

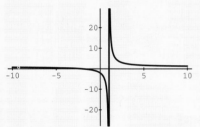

Mathematica:

Plot[(x+2)/(x–1),{x,–10,10}]

DERIVE:

(At the □ symbol, go to the next step.)
a (author) □ (x+2)/(x–1) □ p (plot window) □ p (plot) □ z (zoom out) □ z (zoom out) □

FIGURE 1.1.5
$y = f(x) = (x+2)/(x-1)$

The graph shows that f goes to infinity near $x = 1$; that is, it is asymptotic near $x = 1$. We will discuss this type of behavior in Section 1.5.

⌨ CAS 5

Use Maple to sketch the graph of $f(x) = x^2 \sin(1/x)$. This function involves the trigonometric sine function, which we will consider in Section 1.6. We show several sketches of the curve using Maple commands, starting with

    plot(x^2*sin(1/x),x=–5..5);

See Figure 1.1.6.

Something interesting appears to be happening near the origin. We zoom in closer.

    plot(x^2*sin(1/x),x=–1..1);   A little better
    plot(x^2*sin(1/x),x=–0.25..0.25);   Even better
    plot(x^2*sin(1/x),x=–0.025..0.025,numpoints=500);

[†] In Mathematica, the double equal sign is used to show that we are dealing with an equation. In both Maple and Mathematica, f is defined as the equation on the right of := and =, respectively.

This last plot is convincing as to the behavior of the function near the origin. The option, numpoints=500, plots additional points, adding to the accuracy of the plot. We see that care must be taken when interpreting computer-generated output.

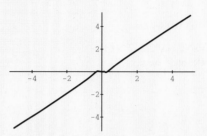

FIGURE 1.1.6
$y = f(x) = x^2 \sin(1/x)$

FIGURE 1.1.7
$y = f(x) = x^2 \sin(1/x)$

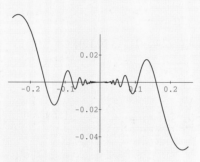

FIGURE 1.1.8
$y = f(x) = x^2 \sin(1/x)$

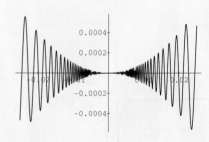

FIGURE 1.1.9
$y = f(x) = x^2 \sin(1/x)$

Maple:

plot(x^2*sin(1/x),x=–5..5);
plot(x^2*sin(1/x),x=–1..1);
plot(x^2*sin(1/x),x=–0.25..0.25);
plot(x^2*sin(1/x),x=–0.025..0.025);

Mathematica:[‡]

Plot[x^2*Sin[1/x],{x,–5,5}]
Plot[x^2*Sin[1/x],
    {x,–1,1}]
Plot[x^2*Sin[1/x],
    {x,–0.25,0.25}]
Plot[x^2*Sin[1/x],
    {x,–0.025,0.025}]

DERIVE

(At the □ symbol, go to the next step.)
a (author) □ x^2sin(1/x) □ p (plot win–
dow) □ p (plot) □ z (zoom in) □
z (zoom in) □ z (zoom in) □

[‡] In Mathematica, the sine function is Sin and must start with a capital letter. The argument for the function is in square braces, [ ]. In DERIVE, notice this time that we zoom in on the graph, rather than zoom out.

Exercise Set 1.1

1. If $f(x) = \dfrac{x^2 - 1}{x + 2}$, find
 (a) $f(1)$
 (b) $f(-1)$
 (c) $f(0)$
 (d) $f(a)$
 (e) $f(x + 1)$

2. If $g(x) = \dfrac{2x - 3}{3x + 4}$, find
 (a) $g(-2)$
 (b) $g\left(\frac{5}{6}\right)$
 (c) $g\left(-\frac{2}{3}\right)$
 (d) $g(2x)$
 (e) $g(1 - x)$

3. If $F(t) = |t - 3|$, find
 (a) $F(4)$
 (b) $F(1)$
 (c) $F(0)$
 (d) $F(t + 3)$
 (e) $F(3 - x^2)$

In Exercises 4 and 5, determine from the graph the domain and range of each function.

4.

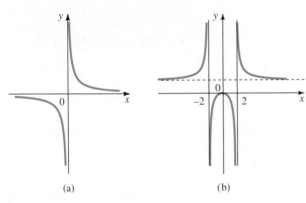

 (a)   (b)

5.

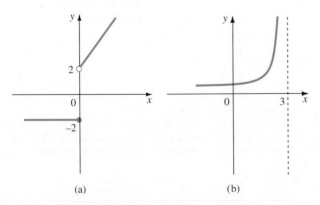

 (a)   (b)

In Exercises 6–11, find the domain and range of f.

6. $f(x) = 2x + 3$

7. $f(x) = x^2 - 4$

8. $f(x) = \dfrac{1}{x}$

9. $f(t) = \sqrt{1 - t}$

10. $f(u) = \dfrac{u + 1}{u - 2}$

11. $f(s) = s^3 + 1$

In Exercises 12–21, draw the graph of the given function.

12. $f(x) = 3x - 4$

13. $g(x) = x^2 + 1$

14. $h(x) = \dfrac{2}{x}$

15. $F(x) = 4x - x^2$

16. $f(x) = \sqrt{x + 9}$

17. $G(x) = |x + 1|$

18. $h(x) = \begin{cases} x^2 & \text{if } x \geq 1 \\ 1 - x & \text{if } x < 0 \end{cases}$

19. $f(x) = \begin{cases} 2x & \text{if } x \geq 1 \\ 2 & \text{if } 0 \leq x < 1 \\ x + 2 & \text{if } x < 0 \end{cases}$

20. $f(x) = \begin{cases} \sqrt{x} & \text{if } x > 0 \\ 1 & \text{if } x = 0 \\ -1 & \text{if } x < 0 \end{cases}$

21. $g(x) = |4 - x^2|$

22. Greatest Integer Function. The function whose value at x is the greatest integer that does not exceed x is called the *greatest integer function*. Its value is symbolized by $[\![x]\!]$. Thus, if $k \leq x < k + 1$, where k is an integer, then $[\![x]\!] = k$.
 (a) Give the value of each of the following: $[\![3.2]\!]$, $[\![5]\!]$, $[\![-3/2]\!]$, $[\![1.99]\!]$.
 (b) Draw the graph of $y = [\![x]\!]$.

23. Verify that the first-class postage function we discussed at the beginning of this section is defined by

$$P(w) = \begin{cases} 0.29 + 0.23[\![w]\!] & \text{if } w \text{ is not an integer} \\ 0.29 + 0.23(w - 1) & \text{if } w \text{ is an integer} \end{cases}$$

Here w is the weight in ounces ($0 < w \leq 12$) and $P(w)$ is the cost of postage in dollars. (See Exercise 22.)

24. You are driving along a straight road at a constant speed between cities A and B. Along the way you pass through city C. Draw a graph that indicates your distance from city C as a function of time, measured from the time you left city A.

25. If $f(x) = \dfrac{1}{x}$, find $\dfrac{f(x) - f(2)}{x - 2}$ for $x \neq 2$ and $x \neq 0$. Simplify the result.

26. If $g(x) = \sqrt{x}$, show that for $h \neq 0$,

$$\frac{g(x+h) - g(x)}{h} = \frac{1}{\sqrt{x+h} + \sqrt{x}}$$

27. Let $\phi(x) = \dfrac{x}{x-3}$. Find

$$\frac{\phi(4+h) - \phi(4)}{h} \qquad (h \neq 0)$$

and simplify the result.

28. Let $f(x) = \dfrac{x+2}{2x+1}$. Show that if $x \neq 0$,

$$f\left(\frac{1}{x}\right) = \frac{1}{f(x)}$$

(*Hint:* Simplify each side and show that the results are the same.)

29. For what positive integers n is $f(x) = x^n$ an even function? For what values of n is it odd?

30. Determine in each of the following whether the function is even, odd, or neither:
   (a) $f(x) = \sqrt{x^2 + 1}$ (b) $f(x) = x^3 - 2x$
   (c) $f(x) = x^2 + 2x$ (d) $g(x) = x - 1$
   (e) $h(x) = 2x^2 - 3$

31. Follow the instructions for Exercise 30 for the following:
   (a) $f(x) = \dfrac{x}{x^2 + 1}$ (b) $g(x) = \dfrac{1}{x^3 - 2x^5}$
   (c) $h(t) = \dfrac{t+2}{t-1}$ (d) $\phi(x) = \dfrac{x^2 - 1}{x^2 + 4}$
   (e) $F(t) = |t|$

32. (a) Is the sum of two even functions even? odd?
   (b) What can be said about the product of two even functions? two odd functions? an even function and an odd function?
   (c) What can be said about the sum of an even function and an odd function?
   (d) Can a function be both even and odd? Explain.

33. Classify each of the graphs shown as being graphs of functions that are even, odd, or neither.

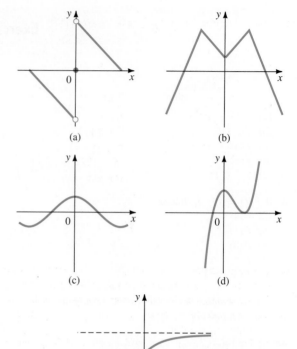

(a)   (b)

(c)   (d)

(e)

34. The graph of a function f is given for $x \geq 0$. Extend the graph for $x < 0$ if
   (a) f is even   (b) f is odd

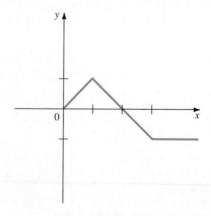

35. Repeat Exercise 34 for the graph shown here.

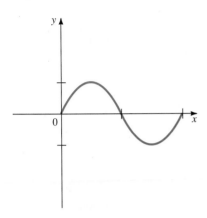

36. A piece of equipment is purchased for $200,000. For tax purposes, the value is decreased by $40,000 times the number of years x since the equipment was purchased. Let $f(x)$ = value of equipment x years after purchase.
   (a) Write the rule for the function f.
   (b) Find the domain of f if the value of the equipment cannot be negative.

37. If a price of x dollars per unit is charged for a certain product, the number of units sold will be $1000 - 25x$. Let $f(x)$ be the revenue when the price is x, where revenue is the price per unit times the number of units sold.
   (a) Write the rule for the function f.
   (b) Find the domain of f if the revenue is nonnegative.

38. The length of a rectangle is 3 units more than twice its width. Express each of the following as a function of the width: (a) perimeter and (b) area.

39. A can is in the form of a right circular cylinder with its height equal to twice the radius of the base. Express each of the following as a function of the radius of the base: (a) volume, (b) lateral surface area, and (c) total surface area.

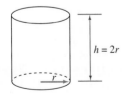

40. Express the area of an equilateral triangle as a function of the length of a side.

41. A rectangle is inscribed in an isosceles triangle of height 4 cm and base 6 cm, as shown in the figure. Express the area A of the rectangle as a function of its base b.

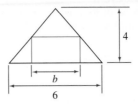

In Exercises 42–48, find the domain of the given function.

42. $f(x) = \sqrt{x^2 - 4}$ 　　　　　**43.** $f(x) = \sqrt{3 + 2x - x^2}$

44. $g(x) = \sqrt{\dfrac{x - 1}{x + 2}}$

45. $g(x) = \dfrac{\sqrt{x - 1}}{\sqrt{x + 2}}$. Compare with Exercise 44.

46. $g(x) = \dfrac{2}{\sqrt{1 - x^2}}$ 　　　**47.** $h(x) = \dfrac{x - 1}{\sqrt{6 - x - x^2}}$

48. $h(x) = \dfrac{\sqrt{x^2 - 2x - 8}}{x + 5}$

49. A right circular cylinder is inscribed in a sphere of radius a. Express the volume of the cylinder as a function of its base radius r.

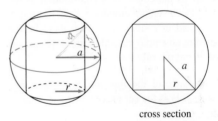

cross section

50. A right circular cone is inscribed in a sphere of radius a. Express the volume of the cone as a function of its base radius r.

51. A pyramid has a square base 4 m on a side, and its height is 10 m. Express the area of a cross section parallel to the base and h m above the base as a function of h.

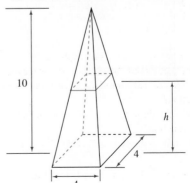

52. A storage tank is in the form of a cylinder x feet long and y feet in diameter, with hemispheres at both ends (see the figure).

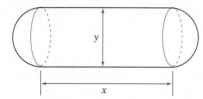

(a) Express the volume of the tank as a function of x and y.
(b) Express the surface area of the tank as a function of x and y.
(c) If $x = 4y$, express the volume and surface area as a function of y only.

53. An organization that issues credit cards charges cardholders interest of $1\frac{1}{2}\%$ per month for the first $1000 of unpaid balance after 30 days and 1% per month on the unpaid balance above $1000. Let $f(x)$ be the interest charged per month on an unpaid balance of x dollars, and find the rule for f.

54. A company deducts 2% of gross salary from each employee's pay as the employee's contribution to the pension fund. However, no employee pays more than $400 into the fund in a single year. Let $f(x)$ be the contribution from an employee with a gross annual salary of x dollars. Write the rule for f.

55. Prove that the range of the function f for which $f(x) = 2x/(x^2 + 1)$ is the set $\{y: |y| \le 1\}$.

56. Redraw the graph shown, and on the same set of axes draw the graph of
(a) $y = -f(x)$ (b) $y = f(-x)$
In each case describe in words the effect of introducing the minus sign.

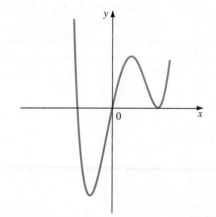

 In Exercises 57–59, use a graphing calculator or CAS.

57. Straight Lines. Let $y = ax + b$, where $a \in \mathbb{R}$ and $b \in \mathbb{R}$. Describe the effect of a and b on the graph. Use both positive and negative values.

58. Parabolas. Let $y = (x - a)^2 + b$, where $a \in \mathbb{R}$ and $b \in \mathbb{R}$. Describe the effect of a and b on the graph. Consider the separate cases

$a = 0, \ b \in \mathbb{R}$	Start with $b = 0$.
$a > 0, \ b \in \mathbb{R}$	Start with $b = 0$.
$a < 0, \ b \in \mathbb{R}$	Start with $b = 0$.

59. Let $f(x) = x(x - 2)^2$. Use a graphing calculator or CAS to obtain the graph of $y = f(x)$. Then find the graphs of the following, for $a = 1, 2, -1, -2$.
(a) $y = f(x) + a$ (b) $y = f(x + a)$
(c) $y = af(x)$ (d) $y = f(ax)$
Compare each graph with that of $y = f(x)$. How would you generalize your findings?

 In Exercises 60–63, first show that the given value of a is not in the domain of f, and then use a calculator to find $f(a + h)$ for each of the following values of h: ± 0.1, ± 0.01, ± 0.001, ± 0.0001. Conjecture how the function behaves as x comes arbitrarily close to a from each side.

60. $a = 3; \ f(x) = \dfrac{x^2 - 9}{x - 3}$

61. $a = -\dfrac{3}{2}; \ f(x) = \dfrac{6x^2 + 17x + 12}{2x^2 + x - 3}$

62. $a = 2; \ f(x) = \dfrac{3x^3 - 4x^2 - 8}{3x^2 + x - 14}$

63. $a = 8; \ f(x) = \dfrac{x - 8}{2 - \sqrt[3]{x}}$

64. Sketch the graph of the function $f(x) = |x - 1| + |2x + 1|$.

1.2 NEW FUNCTIONS FROM OLD

If we begin with one or more known functions, we can obtain new functions from them in a variety of ways, some of which we discuss in this section. After we have introduced some calculus concepts, we will see even more ways of creating new functions from old. One of the benefits of learning the techniques in this section is that we can frequently use them to go in reverse. That is, given a new function, we may be able to recognize it as having come from some known function or functions in a way we have studied. We will then have important knowledge of the new function as a result of known information about the original function(s).

We begin by adding a constant to the independent or the dependent variable. Then we multiply each of these variables by a constant and discuss the result.

Vertical and Horizontal Translations

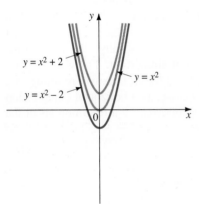

Suppose we know the graph of $y = f(x)$. (We will call f the *basic function* in the following discussion.) Let c denote a positive constant. We want to see how the graph of each of the following is related to the graph of the basic function:

$$y = f(x) + c \qquad y = f(x + c)$$
$$y = f(x) - c \qquad y = f(x - c)$$

These four new functions are special types of **transformations** of the basic function known as **translations**. We consider them in pairs.

FIGURE 1.20

1. $y = f(x) + c$ and $y = f(x) - c$
   Each y value of the graph of the basic function is shifted upward by c units in $y = f(x) + c$ and downward by c units in $y = f(x) - c$. These two transformations are called **vertical translations**. We illustrate them in Figure 1.20 with $y = x^2$ as the basic function and the vertical translations $y = x^2 + 2$ and $y = x^2 - 2$.

2. $y = f(x + c)$ and $y = f(x - c)$
   Suppose (a, b) is a point on the graph of the basic function; that is, $b = f(a)$. Then direct substitution shows that $(a - c, b)$ is on the graph of $y = f(x + c)$ and $(a + c, b)$ is on the graph of $y = f(x - c)$. So $y = f(x + c)$ shifts all points on the basic graph c units to the *left*, and $y = f(x - c)$ shifts all points on the basic graph c units to the *right*. These two transformations are called **horizontal translations**. We illustrate them in Figure 1.21, again using $y = x^2$ as the basic function, with horizontal translations $y = (x + 2)^2$ and $y = (x - 2)^2$. Note carefully the direction of each shift.

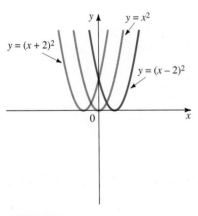

FIGURE 1.21

Stretching, Shrinking, and Reflecting

Next we consider the following four transformations of the basic function $y = f(x)$.

$$y = cf(x) \qquad y = f(cx)$$
$$y = -cf(x) \qquad y = f(-cx)$$

We again consider the transformations in pairs.

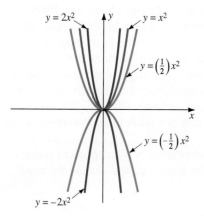

$y = 2x^2$ $y = x^2$

$y = \left(\frac{1}{2}\right)x^2$

$y = \left(-\frac{1}{2}\right)x^2$

$y = -2x^2$

FIGURE 1.22

3. $y = cf(x)$ and $y = -cf(x)$

In $y = cf(x)$, each y value of the basic function is multiplied by c, so the graph is stretched vertically if $c > 1$ and shrunk if $0 < c < 1$. For $c > 1$, we call the transformation a **vertical elongation** and for $0 < c < 1$, a **vertical compression**. In $y = -cf(x)$, each y value of $y = cf(x)$ is replaced by its negative, so the graph is **reflected about the x-axis**. Thus, $y = -cf(x)$ is a vertical elongation (if $c > 1$) or compression (if $0 < c < 1$), followed by a reflection about the x-axis. In Figure 1.22 we show the basic function $y = x^2$ along with $y = 2x^2$, $y = \frac{1}{2}x^2$, $y = -2x^2$, and $y = -\frac{1}{2}x^2$.

4. $y = f(cx)$ and $y = f(-cx)$

If the point (a, b) is on the graph of the basic function, so that $b = f(a)$, then direct substitution shows that $(a/c, b)$ is on the graph of $y = f(cx)$. Thus, if $c > 1$, the point (a, b) on the basic graph is replaced by $(a/c, b)$, which is closer to the y-axis. Thus, the transformation $y = f(cx)$ results in a shrinking, called a **horizontal compression**. If $0 < c < 1$, then $1/c > 1$, so that $(a/c, b)$ is farther from the y-axis than (a, b). The result is that $y = f(cx)$ stretches the basic curve horizontally. We refer to the transformation in this case as a **horizontal elongation**. For $y = f(-cx)$, each x-coordinate of $y = f(cx)$ is replaced by its negative, resulting in a **reflection about the y-axis**. So $y = f(-cx)$ is a horizontal compression (if $c > 1$) or elongation (if $0 < c < 1$), followed by a reflection about the y-axis. To illustrate these transformations we use $y = (x - 1)^2$ as the basic function and show graphs of $y = (2x - 1)^2$, $y = (\frac{1}{2}x - 1)^2$, $y = (-2x - 1)^2$, and $y = (-\frac{1}{2}x - 1)^2$ (Figure 1.23).

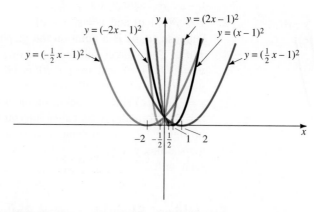

$y = (-2x - 1)^2$ $y = (2x - 1)^2$

$y = (-\frac{1}{2}x - 1)^2$ $y = (x - 1)^2$

$y = (\frac{1}{2}x - 1)^2$

$-2 \quad -\frac{1}{2} \quad \frac{1}{2} \quad 1 \quad 2$

FIGURE 1.23

In Table 1.1 we summarize the eight transformations we have discussed. The basic graph we use is meant to represent an arbitrary function. For simplicity, in $y = -cf(x)$ and $y = f(cx)$ we take $c = 1$.

TABLE 1.1

Transformation	Effect	Illustration
$y = f(x) + c$	Vertical translation c units upward	
$y = f(x) - c$	Vertical translation c units downward	
$y = f(x + c)$	Horizontal translation c units to the left	
$y = f(x - c)$	Horizontal translation c units to the right	
$y = cf(x)$	Vertical elongation if $c > 1$, compression if $0 < c < 1$	

TABLE 1.1 (*continued*)

Transformation	Effect	Illustration
$y = f(cx)$	Horizontal compression if $c > 1$, elongation if $0 < c < 1$	
$y = -f(x)$	Reflection about the x-axis	
$y = f(-x)$	Reflection about the y-axis	

As the next example shows, several of these transformations can be combined to create one new function.

EXAMPLE 1.7   Discuss the function defined by

$$y = 1 - 2(x - 3)^2$$

and draw its graph.

Solution   If we rewrite the equation as

$$y = -2(x - 3)^2 + 1$$

we can show the function as a sequence of transformations of $y = x^2$ in the following diagram. (Other sequences are also possible.)

$$x^2 \longrightarrow (x - 3)^2 \longrightarrow 2(x - 3)^2 \longrightarrow -2(x - 3)^2 \longrightarrow -2(x - 3)^2 + 1$$

shift 3 units to the right   stretch vertically by factor of 2   reflect about x-axis   shift 1 unit upward

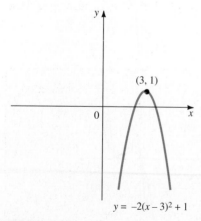

$y = -2(x - 3)^2 + 1$

FIGURE 1.24

The resulting graph is shown in Figure 1.24.

EXAMPLE 1.8   In each of the following, obtain the graph of $y = f(x)$ by considering transformations of an appropriate basic function.

(a) $f(x) = 2\sqrt{x-3}$ (b) $f(x) = |x+2| - 3$

Solution

(a) Starting with the basic function $F(x) = \sqrt{x}$, we first have the vertical stretching $y = 2F(x)$, followed by a horizontal translation 3 units to the right, $y = 2F(x-3)$. In Figure 1.25 we show both the graph of the basic function and the result of the transformations.

(b) Here we begin with the basic absolute value function $F(x) = |x|$, shift it 2 units to the left and then 3 units down, as shown in Figure 1.26.   ∎

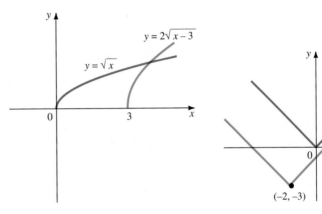

FIGURE 1.25   **FIGURE 1.26**

Sums, Differences, Products, and Quotients

Given two functions with a common domain, we can obtain new functions by adding them, subtracting them, multiplying them, or dividing them (provided the denominator is nonzero). The next definition gives precise meaning to these new functions.

Definition 1.3
Sums, Differences, Products, and Quotients of Functions

Let f and g be any two functions whose domains have nonempty intersection D. The functions $f + g$, $f - g$, $f \cdot g$, and f/g are defined as follows:

$$(f + g)(x) = f(x) + g(x) \qquad (f \cdot g)(x) = f(x)g(x)$$

$$(f - g)(x) = f(x) - g(x) \qquad \left(\frac{f}{g}\right)(x) = \frac{f(x)}{g(x)} \quad \text{if } g(x) \neq 0$$

The domain of each of the functions $f + g$, $f - g$, and $f \cdot g$ is all of D. The domain of f/g is the set of all points x in D for which $g(x) \neq 0$.

The next three examples illustrate how to form these new functions when f and g are given by formulas, tables, and graphs.

EXAMPLE 1.9   Let $f(x) = 2x - 3$ and $g(x) = x^2 - 1$. Find $(f + g)(x)$, $(f - g)(x)$, $(f \cdot g)(x)$, and $(f/g)(x)$, and give the domain of each.

Solution

$$(f + g)(x) = (2x - 3) + (x^2 - 1) = x^2 + 2x - 4; \text{ domain} = \mathbb{R}$$
$$(f - g)(x) = (2x - 3) - (x^2 - 1) = -x^2 + 2x - 2; \text{ domain} = \mathbb{R}$$
$$(f \cdot g)(x) = (2x - 3)(x^2 - 1) = 2x^3 - 3x^2 - 2x + 3; \text{ domain} = \mathbb{R}$$
$$\left(\frac{f}{g}\right)(x) = \frac{2x - 3}{x^2 - 1}; \text{ domain} = \{x : x \neq \pm 1\}$$

∎

EXAMPLE 1.10   Let f and g be defined by the following table.

x	$f(x)$	$g(x)$
-1	-5	4
0	-2	5
1	1	4
2	4	1
3	7	-4
4	10	-11
5	13	-20

Find $(f + g)(x)$, $(f - g)(x)$, $(f \cdot g)(x)$, and $(f/g)(x)$.

Solution   The required values are shown in the following table.

x	$(f + g)(x)$	$(f - g)(x)$	$(f \cdot g)(x)$	$(f/g)(x)$
-1	-1	-9	-20	$-5/4$
0	3	-7	-10	$-2/5$
1	5	-3	4	$1/4$
2	5	3	4	4
3	3	11	-28	$-7/4$
4	-1	21	-110	$-10/11$
5	-7	33	-260	$-13/20$

∎

EXAMPLE 1.11   The graphs of two functions f and g are shown in Figure 1.27. Find $f + g$ and $f \cdot g$ graphically.

Solution   At selected values of x, we estimate $f(x)$ and $g(x)$ from their graphs. Then we add the values to get the value of $(f + g)(x)$ and multiply them to get the value of $(f \cdot g)(x)$. Finally, we connect the points obtained to get the graphs of the functions $f + g$ and $f \cdot g$. The results are shown in Figure 1.28.

∎

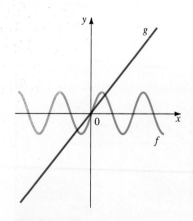

FIGURE 1.27

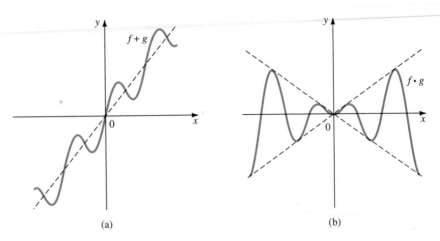

(a)

(b)

FIGURE 1.28

Exercise Set 1.2

In Exercises 1–6, the graph of a function $y = f(x)$ *is given.*
From it, obtain the graph of each of the following:
(a) $y = f(x) + 2$,   (b) $y = f(x - 3)$,   (c) $y = 2f(x)$,
(d) $y = f(2x)$,   (e) $y = -f(x)$,   (f) $y = f(-x)$.

1.

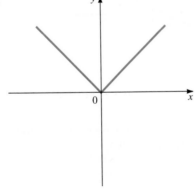

2.

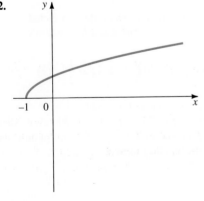

3.

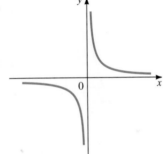

4.

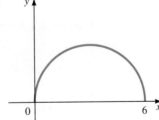

5.

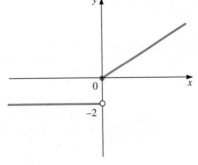

6.

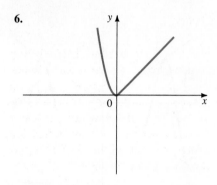

7. Draw the graph of $y = x^3$ and from it obtain the graph of $y = \frac{1}{2}(x-1)^3 + 2$.

8. Draw the graph of $y = \sqrt{x}$ and from it obtain the graph of $y = \sqrt{2-x}$.

9. The graph of a function f is given. From it obtain the graph of g, where $g(x) = 3 - 2f(\frac{1}{2}x - 1)$.

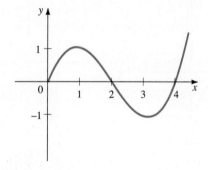

In Exercises 10–17, find $f + g$, $f - g$, $f \cdot g$, and f/g and give the domain of each.

10. $f(x) = 3x - 5$; $g(x) = 2x + 3$

11. $f(x) = 1 - x$; $g(x) = 1 + x$

12. $f(x) = \dfrac{1}{x-1}$; $g(x) = \dfrac{1}{x+1}$

13. $f(x) = \sqrt{x+4}$; $g(x) = \sqrt{4-x}$

14. $f(x) = x$; $g(x) = \sqrt{x-1}$

15. $f(x) = \dfrac{1}{x}$; $g(x) = \dfrac{x}{x-3}$

16. $f(x) = \begin{cases} -1 & \text{if } x < 0 \\ 1 & \text{if } x \geq 0 \end{cases}$; $g(x) = \begin{cases} 1 & \text{if } x < 0 \\ -1 & \text{if } x \geq 0 \end{cases}$

17. $f(x) = \begin{cases} x & \text{if } x \geq 0 \\ 0 & \text{if } x < 0 \end{cases}$; $g(x) = \begin{cases} 0 & \text{if } x \geq 0 \\ -x & \text{if } x < 0 \end{cases}$

In Exercises 18 and 19, the graphs of functions f and g are given. From these obtain the graphs of $f + g$ and $f \cdot g$.

18.

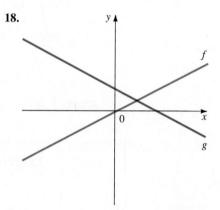

19.

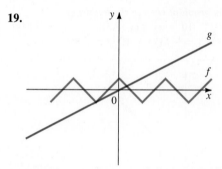

20. Refer to the greatest integer function, $f(x) = [\![x]\!]$, introduced in Exercise 22 of Exercise Set 1.1. Draw the graph of each of the following.
  (a) $y = [\![x + 1]\!]$ (b) $y = [\![x - 2]\!]$
  (c) $y = x - [\![x]\!]$ (d) $y = \sqrt{x - [\![x]\!]}$
  (e) $y = [\![x]\!] + \sqrt{x - [\![x]\!]}$
  (f) $y = (x - [\![x]\!])^2$

21. For an arbitrary function f whose domain includes $-x$ whenever it includes x, define g and h as follows:

$$g(x) = \frac{1}{2}\left[f(x) + f(-x)\right] \qquad h(x) = \frac{1}{2}\left[f(x) - f(-x)\right]$$

  (a) Show that g is even and h is odd.
  (b) Show that $f = g + h$. (This proves that every such function f can be expressed as the sum of an even function and an odd function.)

1.3 COMPOSITE FUNCTIONS AND INVERSE FUNCTIONS

In the last section we saw how two functions can be combined by addition, subtraction, multiplication, and division. Another very important way of combining two given functions to get a new function is called **composition**. In this method of combining, the output of one function becomes the input of another. For example, let $f(x) = x^3 + 1$ and $g(x) = 2x + 3$. If we input x into the g function, we get the output $2x + 3$. If we then input $2x + 3$ into the f function, we get $f(2x+3) = (2x+3)^3 + 1$. The result is the value of the function known as the **composition of f with g**, denoted by $f \circ g$. So for these functions we have

$$(f \circ g)(x) = f(g(x)) = f(2x + 3) = (2x + 3)^3 + 1$$

More generally, we have the following definition.

Definition 1.4 Composition	For two functions f and g, the **composition of f with g**, denoted by $f \circ g$, is defined by $$(f \circ g)(x) = f(g(x))$$ The domain of $f \circ g$ is the set of all x in the domain of g for which $g(x)$ is in the domain of f.

Figure 1.29 is a diagram of f, g, and $f \circ g$.

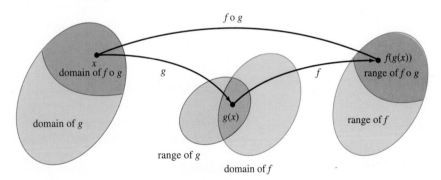

FIGURE 1.29

One reason for studying composition of functions is that if a complicated function can be recognized as the composition of two (or more) simpler functions, our knowledge of the simpler constituent functions might tell us much about the composite function. This viewpoint is similar to that taken in disciplines such as physics, where scientists study the basic constituents of matter (quarks, electrons, and other elementary particles) to understand the more complicated structures formed by grouping these constituents.

EXAMPLE 1.12   Let $f(x) = x^2$ and $g(x) = \sqrt{x - 4}$. Find $f \circ g$ and $g \circ f$, and give the domain of each.

Solution

$$(f \circ g)(x) = f(g(x)) = f(\sqrt{x-4}) = (\sqrt{x-4})^2 = x - 4$$

$$(g \circ f)(x) = g(f(x)) = g(x^2) = \sqrt{x^2 - 4}$$

The domain of f is all of $\mathbb{R}$, and the domain of g is the set $\{x : x \geq 4\}$. To determine the domain of $f \circ g$, we must find that subset of the domain of g for which $g(x)$ lies in the domain of f. But since the domain of f is all of $\mathbb{R}$, the entire range of g lies in the domain of f in this case. That is, the domain of $f \circ g$ is identical with the domain of g, namely, $\{x : x \geq 4\}$, which can also be written as the interval $[4, \infty)$. Note carefully that although $(f \circ g)(x) = x - 4$ and $x - 4$ has meaning for all of $\mathbb{R}$, the domain of $f \circ g$ is nevertheless limited to x values for which $g(x)$ is defined.

For $g \circ f$, the domain consists of all x values for which x^2 is in the domain of g, namely $\{x : x^2 \geq 4\}$. This set can be written in the equivalent form $\{x : |x| \geq 2\}$, or in interval notation as $(-\infty, -2] \cup [2, \infty)$. ∎

$\{x : x^2 \geq 4\} = \{x : x^2 - 4 \geq 0\}$

$x^2 - 4 = (x + 2)(x - 2)$

REMARK ───────────────────────────────

■ As the preceding example shows, the functions $f \circ g$ and $g \circ f$ are usually not the same.

Returning to our machine analogy, the function $f \circ g$ can be thought of as being composed of the g and f machines operating in sequence, as indicated in Figure 1.30.

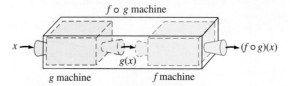

FIGURE 1.30

Sometimes it is important to express a function as the composition of two simpler functions, as illustrated in the next example.

EXAMPLE 1.13   Let $F(x) = (2x + 3)^5$. Find two functions f and g such that $F = f \circ g$.

Solution   To compute the value of F at x, we first compute $2x + 3$ and then raise the result to the fifth power. Thus, we have separated the evaluation of F into two operations that can be written as

$$g(x) = 2x + 3 \quad \text{and} \quad f(x) = x^5$$

So it is natural to express F as the composition $f \circ g$, since

$$F(x) = (2x + 3)^5 = [g(x)]^5 = f(g(x))$$

Note: There are other possible choices of the functions f and g, but the choice we made is the most natural. ∎

x	$p(x)$	$q(x)$
−2	1	5
−1	0	−8
0	−1	4
1	2	3
2	−2	−12

The next example illustrates how the values of a composite function can be found when the two functions in the composition are given by a table.

EXAMPLE 1.14   Values of $p(x)$ and $q(x)$ are shown in the table. Find the values of $(q \circ p)(x)$.

Solution   The solution is given in the table in the margin. To illustrate how the values are obtained, consider the first entry. When $x = -2$, we know that $p(x) = 1$. Also, $q(1) = 3$. Thus, $(q \circ p)(-2) = q(p(-2)) = q(1) = 3$.   ∎

x	$(q \circ p)(x)$
−2	3
−1	4
0	−8
1	−12
2	5

Inverses

If we are given a function, we can sometimes obtain a new function from it by reversing the roles of the dependent and independent variables. Consider the following example.

EXAMPLE 1.15   The formula

$$C = \frac{5}{9}(F - 32)$$

expresses the Celsius temperature C as a function of the Fahrenheit temperature F. For example, when $F = 32$, $C = 0$, and when $F = 212$, $C = 100$. Find F as a function of C.

Solution   We solve for F as follows.

$$C = \frac{5}{9}(F - 32)$$

$$\frac{9}{5}C = F - 32$$

$$F = \frac{9}{5}C + 32$$

This last equation expresses F as a function of C.   ∎

In Example 1.15 we began with a function of F. Suppose we call it $g(F)$. Then $g(F) = \frac{5}{9}(F - 32)$. After reversing the roles of the dependent and independent variables, we obtained a function of C, say $h(C)$, where $h(C) = \frac{9}{5}C + 32$. This function h is called the **inverse** of the function g.

Unfortunately, this procedure of reversing the roles of the dependent and independent variables does not always result in a function. Consider $f(x) = x^2$, for example. Let y be the dependent variable, so that $y = x^2$. When we solve for x in terms of y, we get $x = \pm\sqrt{y}$. But this equation does not define x as a function of y, since for a given positive value of y (now being treated as the independent variable) there are two values of x, which violates the requirement that to each admissible value of the independent variable of a function there must correspond a *unique* value of the dependent variable. Figure 1.31 clearly shows the problem here. The graph of $y = x^2$ passes the vertical line test, so the equation $y = x^2$ does define y as a function of x. But if we reverse the roles of dependent and independent variables, we see that x is not a function of y. For example, if $y = 4$ we get *two* values of x, namely $x = \sqrt{4} = 2$ and $x = -\sqrt{4} = -2$. So even though $2 \neq -2$, we have $f(2) = f(-2)$. That is, two

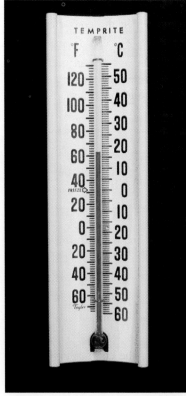

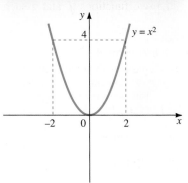

FIGURE 1.31

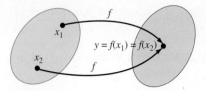

FIGURE 1.32

different x values correspond to the *same* y value. Figure 1.32 provides another way of viewing the problem. There we show two x values in A corresponding under the function to one y value in B. When we try to go in reverse, the value of y in B does not correspond to a unique x in A.

The example of $f(x) = x^2$ should make clear that only if we have a function for which no two different x values correspond to the same y value can we create a new function by interchanging independent and dependent variables. Such a function is said to be *one-to-one*.

Definition 1.5
One-to-One Function

A function f is said to be **one-to-one**, abbreviated 1–1, provided that when $x_1 \neq x_2$, then $f(x_1) \neq f(x_2)$.

Equivalently, f is 1–1 if whenever $f(x_1) = f(x_2)$, then $x_1 = x_2$. That is, each y value in the range of f corresponds to exactly one x value in the domain. We know that *all* functions must satisfy the vertical line test. By what we have just said, all 1–1 functions also must satisfy the horizontal line test.

Horizontal Line Test

A function is 1–1 if and only if every horizontal line intersects the graph of the function in at most one point.

Figure 1.33 illustrates a function that passes this test and one that does not.

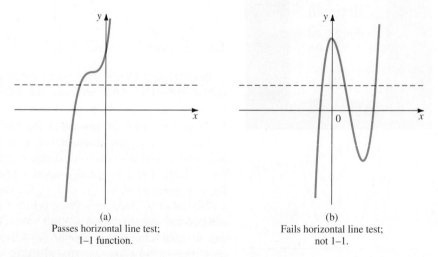

(a)
Passes horizontal line test;
1–1 function.

(b)
Fails horizontal line test;
not 1–1.

FIGURE 1.33

EXAMPLE 1.16   Let

$$f(x) = \sqrt{2x - 3} \qquad \text{and} \qquad g(x) = 2 + |x - 3|$$

Show by means of Definition 1.5 that f is 1–1 on its domain but g is not 1–1 on its domain. Confirm the results by the horizontal line test.

Solution   To show that f is 1–1 we suppose that $f(x_1) = f(x_2)$ and show that it necessarily follows that $x_1 = x_2$. Thus, we assume that

$$\sqrt{2x_1 - 3} = \sqrt{2x_2 - 3}$$

Squaring both sides gives

$$2x_1 - 3 = 2x_2 - 3$$
$$2x_1 = 2x_2$$
$$x_1 = x_2$$

So f is 1–1.

We can show that g is not 1–1 by exhibiting two different values of x that yield the same function value. For example, we can take $x_1 = 2$ and $x_2 = 4$. Then

$$g(x_1) = g(2) = 2 + |2 - 3| = 2 + |-1| = 2 + 1 = 3$$

and

$$g(x_2) = g(4) = 2 + |4 - 3| = 2 + |1| = 2 + 1 = 3$$

We show the graph of f in Figure 1.34 and the graph of g in Figure 1.35. Clearly f satisfies the horizontal line test and g does not.   ∎

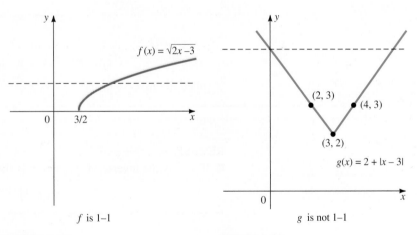

f is 1–1 g is not 1–1

FIGURE 1.34   **FIGURE 1.35**

When a function is 1–1, the new function obtained from it by interchanging independent and dependent variables is called the **inverse** of the original function. In effect, the inverse undoes what the original function does. If the given function operates on a to give b, the inverse function operates on b to give a.

Suppose y is given as a function of x by a table. For example, consider the following table.

x	0	1	2	−1	−2
y	2	5	9	0	3

To obtain the inverse, we use the y values as the values of the independent variable and the x values as the values of the dependent variable. Since it is customary to use x for independent and y for dependent variables, we can show the table for the inverse function by replacing x with y and y with x:

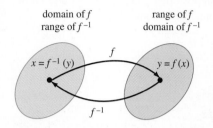

domain of f
range of f^{-1}

range of f
domain of f^{-1}

$x = f^{-1}(y)$ f $y = f(x)$

f^{-1}

FIGURE 1.36

x	2	5	9	0	3
y	0	1	2	−1	−2

If f is a 1–1 function, then it has a unique inverse. We customarily use the symbol f^{-1}, read "f inverse," for the inverse of f.

If f makes x correspond to y, then f^{-1} makes y correspond to x. (See Figure 1.36.) That is, if $y = f(x)$, then $f^{-1}(y) = x$, or equivalently

$$f^{-1}(f(x)) = x$$

The −1 in the notation f^{-1} is not to be treated as an exponent. So $f^{-1} \neq \frac{1}{f}$. This use of the superscript −1 is reserved for functions only and always indicates the inverse function.

We can now give the formal definition of the inverse of a 1–1 function.

Definition 1.6
Inverse Functions

Let f be a 1–1 function with domain A and range B. The function f^{-1} with domain B and range A satisfying

$$(f^{-1} \circ f)(x) = x \quad \text{for all } x \text{ in } A$$

and

$$(f \circ f^{-1})(y) = y \quad \text{for all } y \text{ in } B$$

is called the **inverse of f**.

REMARK ———————————————————————————
■ If f^{-1} is the inverse of f, then f is the inverse of f^{-1}. That is,

$$(f^{-1})^{-1} = f$$

The function that maps each element of a given set onto itself is called the **identity mapping function** and is often denoted by I. Thus, $I(x) = x$. Since $(f^{-1} \circ f)(x) = x$ for all x in the domain of f, we see that

$$f^{-1} \circ f = I$$

on this domain. This equation bears a striking similarity to the equation

$$a^{-1} \cdot a = 1$$

for a nonzero real number a. So the identity mapping function I plays a role similar to the multiplicative identity element 1 for real numbers.

As a consequence of Definition 1.6 we can conclude the following:

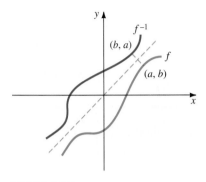

FIGURE 1.37

$$y = f(x) \quad \text{if and only if} \quad f^{-1}(y) = x$$

We have seen how to obtain a table for f^{-1} when f is given in tabular form. Graphically, it is also easy to obtain f^{-1} from f. When a point (a, b) is on the graph of f, so that $f(a) = b$, then, by Definition 1.6, $f^{-1}(b) = a$, so the point (b, a) is on the graph of f^{-1}. It follows that the graphs of f and f^{-1} are symmetrical with respect to the line $y = x$, as illustrated in Figure 1.37. So when you know the graph of a 1–1 function, to get the graph of f^{-1} simply reflect the graph of the given function about the line $y = x$.

Procedure for Finding the Inverse

Suppose now that a 1–1 function f is defined by a formula, and we set $y = f(x)$. To get f^{-1}, using x as the independent variable, we interchange x and y, getting $x = f(y)$. Then we solve this equation (if possible) for y in terms of x. To see that the result is $f^{-1}(x)$, apply f^{-1} to both sides of $x = f(y)$:

$$f^{-1}(x) = f^{-1}(f(y)) = y$$

To illustrate this procedure, consider the function f defined by

$$f(x) = \frac{2x - 3}{4}$$

Let $y = f(x)$, getting

$$y = \frac{2x - 3}{4}$$

Interchange x and y to get the inverse:

$$x = \frac{2y - 3}{4}$$

Solve for y:

$$2y - 3 = 4x$$

$$y = \frac{4x + 3}{2}$$

This last equation is in the form $y = f^{-1}(x)$; that is,

$$f^{-1}(x) = \frac{4x + 3}{2}$$

We summarize this procedure below.

Procedure for Finding the Inverse of a Function

If f is a 1–1 function, follow these steps to find f^{-1}:

1. Set $y = f(x)$.
2. Interchange x and y.
3. Solve for y in terms of x.

The resulting equation is in the form $y = f^{-1}(x)$.

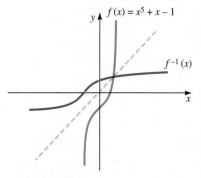

FIGURE 1.38

There is one possible problem with this procedure that comes in step 3. It is not always possible to solve for y. Suppose, for example, that after step 2 we have the equation

$$x = y^5 + y - 1$$

Now try to solve for y! The original function in this case is given by the formula $f(x) = x^5 + x - 1$. It is possible, by means you will learn later, to show that f^{-1} does exist, but trying to find a formula for it is hopeless. We could, however, find its graph, as we show in Figure 1.38. We used a computer algebra system to get the graph of f. Notice that f does appear to be 1–1, since by what we see of its graph, the horizontal line test is satisfied. We obtained the graph f^{-1} by reflecting the graph of f about the 45°-line $y = x$.

EXAMPLE 1.17   Show that $f(x) = \sqrt{x - 1}$ has an inverse on the domain $[1, \infty)$, and find the inverse. Draw the graphs of f and f^{-1}.

Solution   To show f is 1–1, suppose that x_1 and x_2 have the same image. That is, suppose

$$\sqrt{x_1 - 1} = \sqrt{x_2 - 1}$$

Squaring both sides gives

$$x_1 - 1 = x_2 - 1$$

$$x_1 = x_2$$

Thus, the only way that $f(x_1)$ and $f(x_2)$ can be the same is for x_1 to be the same as x_2. Therefore f is 1–1, and so f^{-1} exists. Following our three-step procedure, we write

$$y = \sqrt{x - 1}$$

and interchange x and y to get the inverse:

$$x = \sqrt{y - 1}$$

$$x^2 = y - 1$$

$$y = x^2 + 1$$

Thus, $f^{-1}(x) = x^2 + 1$. But we must state its domain. The domain of f^{-1} is the range of f, namely the set $[0, \infty)$, since $\sqrt{x - 1}$ ranges over the set of all nonnegative real numbers as x takes on all values greater than or equal to 1. So the domain of f^{-1} is the set $[0, \infty)$. We show the graphs of f and f^{-1} in Figure 1.39.   ■

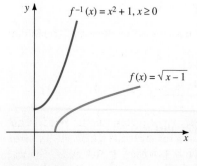

FIGURE 1.39

An Important Inverse Pair

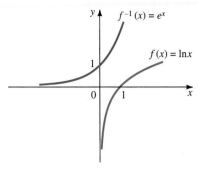

FIGURE 1.40

We conclude this section with a brief mention of the function $f(x) = \ln x$, called the **natural logarithm function**, and its inverse, $f^{-1}(x) = e^x$, called the **natural exponential function**. These two functions are especially important in applications; they may even be the most important function-inverse pair in mathematics and science applications. We will study them in detail in Chapter 7. The letter e is used to designate a certain irrational number approximately equal to 2.71828. Your calculator has a $\boxed{\ln x}$ key. Some calculators also have an $\boxed{e^x}$ key. On others, e^x is in the $\boxed{\text{2nd}}$ register above $\boxed{\ln x}$. By plotting points, or by using a graphing calculator or computer algebra system, you can obtain the graphs shown in Figure 1.40.

FINDING INVERSE FUNCTIONS USING A COMPUTER ALGEBRA SYSTEM

The procedure for finding an inverse of a one-to-one function can be carried out easily using a CAS such as Maple, Mathematica, or DERIVE. We describe several examples here, where we assume it is known that the functions are one-to-one.

CAS 6

Find the inverse for the function $f(x) = \sqrt{x - 1}$ defined on the interval $[1, \infty]$ (the domain of f). Draw the graphs of f, f^{-1}, and $y = x$.

Following the procedure described earlier, we interchange x and y, and then solve for y in terms of x.

In Maple:

```
solve(x=sqrt(y–1),y);
```

The output is: $1 + x^2$

To plot all three functions on the same set of axes, we enter the Maple plot command:

```
plot({[x,sqrt(x–1),x=1..3],1+x^2,x},x=0..3,y=0..3);
```

The plot is on the interval $[0, 3\}$ to give us a good view of the relationship between the three graphs. Notice that to plot more than one graph on the same set of axes, the expressions must be enclosed in braces, { }, and that individual expressions are separated by commas. The same is true in Mathematica.

Maple:

solve[x=sqrt(y–1),y);

Output: 1+x^2

plot({[x,sqrt(x–1),x=1..3],1+x^2,x},
   x=0..3,y=0..3);

Mathematica:

Solve[x == Sqrt[y–1],y]

Output: {{y → 1 + x^2}}

Plot[{Sqrt[x–1],x^2+1,x},
{x,0,3},PlotRange–>{0,3}]

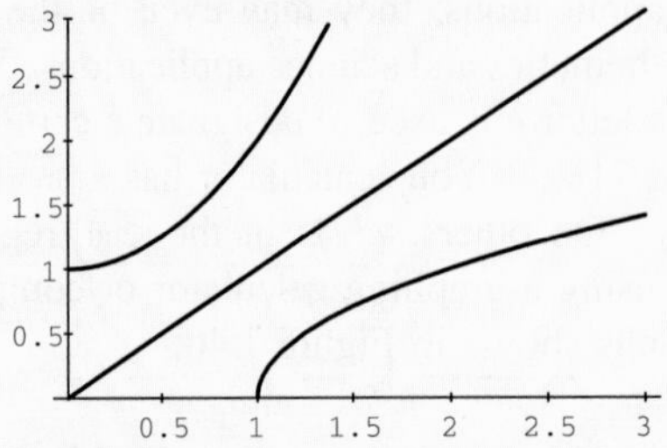

FIGURE 1.3.1
$y = f(x) = \sqrt{x - 1}$, $y = f^{-1}(x) = x^2 + 1$, $y = x$

DERIVE:

(At the □ symbol, go to the next step.)
a (author) □ x=sqrt(y–1) □ l (solve) □
[choose expression to solve for y] □ y
(solve for y)
Output: y=IF(x≥ 0,x^2+1)
a (author) □ sqrt(x–1) □ p (plot window) □ p (plot) □ a (algebra) □ a (author) □ x^2x+1 □ p (plot window) □
p (plot) □ a (algebra) □ a (author) □ x
□ p (plot window) □ p (plot)

CAS 7

Find the inverse for the function $f(x) = \dfrac{(1 + 5x)}{(5 - 3x)}$ defined on the interval $[1, \infty]$. Draw the graphs of f, f^{-1}, and $y = x$.

In Maple, we enter:

solve(x=(1+5*y)/(5–3*y),y);

The output is: $-\dfrac{-5x + 1}{3x + 5}$

To get the curve shown in Figure 1.3.2 we used the plot command:

plot({(1+5*x)/(5–3*x),–(–5*x+1)/(3*x+5),x},x=–5..5,y=–10..10);

The specification y=–10..10 restricts the size of the y-axis. If, for example, we use the Maple command

plot({(1+5*x)/(5–3*x),–(–5*x+1)/(3*x+5),x},x=–10..10);

we get the curve shown in Figure 1.3.3. Again we need to be careful of the output generated by computer algebra systems. Try several modifications in the specifications of the x- and y-axes to see what pictures are generated.

Maple:

solve (x=(1+5*y)/(5–3*y),y);

Output: $-\dfrac{-5x + 1}{3x + 5}$

plot({(1+5*x)/(5–3*x),–(–5*x+1)/
(3*x+5),x},x=–5..5,y=–10..10);

Mathematica:[†]

Solve[x == (1+5*y)/(5–3*y),y]

Output: $\left\{\left\{y \to -\left(\dfrac{1-5x}{5 + 3x}\right)\right\}\right\}$

Plot[{(1+5*x)/(5–3*x),
  –(1–5*x/(5+3*x),x},
{x,–5,5},
PlotRange–>{–10,10}]

DERIVE:

(At the □ symbol, go the next step.)
a (author) □ x=(1+5y)/(5–3y) □ l (solve)
□ [choose the expression to solve
for y]

Output: $y = \dfrac{5x - 1}{3x + 5}$

a (author) □ (1+5x)/(5–3x) □ p (plot
window) □ p (plot) □ a (algebra) □
a (author) □ (5x–1)/(3x+5) □ p (plot
window) □ p (plot) □ a (algebra) □ a
(author) □ x □ p (plot window) □ p
(plot) □ z (zoom-out)

---

† Using the * for multiplication is optional in Mathematica.

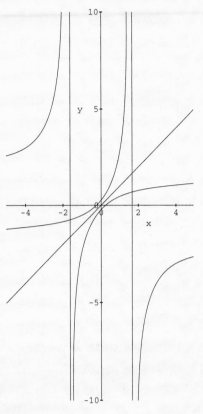

FIGURE 1.3.2
$y = f(x) = (1 + 5x)/(5 - 3x)$,
$y = f^{-1}(x) = (5x - 1)/(3x + 5)$, $y = x$

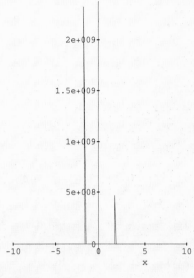

FIGURE 1.3.3

Exercise Set 1.3

In Exercises 1–4, find $f \circ g$ and $g \circ f$, and give the domain of each.

1. $f(x) = 2x - 1$; $g(x) = x + 2$

2. $f(x) = x^2 + 1$; $g(x) = 2x - 3$

3. $f(x) = \dfrac{x}{x - 1}$; $g(x) = \dfrac{3}{x}$

4. $f(x) = \sqrt{x - 1}$; $g(x) = x^2 - 3$

In Exercises 5–8, find functions f and g such that $F = f \circ g$.

5. $F(x) = (4 - 3x^2)^6$ **6.** $F(x) = \sqrt{2x + 3}$

7. $F(x) = \sqrt{(x^2 - 1)^3}$ **8.** $F(x) = 3(1 - x)^{-4}$

9. Complete the following table so that g is even and h is odd, and fill in the values for $f \circ g$ and $g \circ f$.

t	$g(t)$	$h(t)$	$(g \circ h)(t)$	$(h \circ g)(t)$
-3	2	2		
-2	-1	-1		
-1	3	3		
0	0	0		
1				
2				
3				

10. Is it ever true that $f \circ g = g \circ f$? Is it always true? Give examples or explanations to support your conclusions.

11. Show that $f \circ (g \circ h) = (f \circ g) \circ h$ whenever both sides are defined.

12. Which of the functions whose graphs are given have inverses, and which do not?

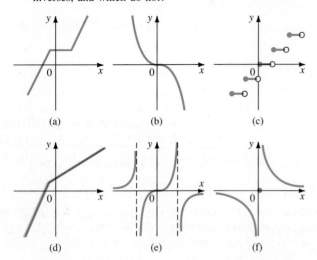

(a)   (b)   (c)

(d)   (e)   (f)

In Exercises 13–18, show that the given function has an inverse, and find it. Give the domain of the inverse.

13. $f(x) = \dfrac{5x - 2}{3}$ **14.** $g(x) = \dfrac{4 - 7x}{2}$

15. $F(x) = \sqrt{2x - 3}$ **16.** $G(x) = \dfrac{1}{\sqrt{x}}$

17. $\varphi(t) = t^2 - 1, t \geq 0$ **18.** $\psi(z) = \sqrt{z^2 - 1}, z \geq 1$

In Exercises 19 and 20, graphs of 1–1 functions are given. Copy each graph and on the same set of axes sketch the graph of its inverse.

19.

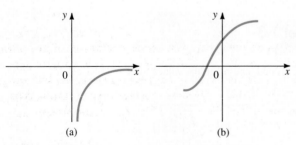

(a)   (b)

20.

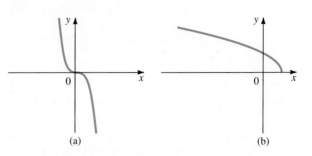

(a)   (b)

In Exercises 21–24, show that the given function is not 1–1. Then determine a subset of its domain on which the function is 1–1, and find its inverse on this restricted domain. Give the domain of the inverse.

21. $f(x) = \dfrac{1}{x^2 + 2}$ **22.** $g(x) = \dfrac{1}{\sqrt{4 - x^2}}$

23. $F(x) = |3 - 2x|$

24. $h(x) = 2x - x^2$ (*Hint:* Complete the square to obtain $h(x) = 1 - (x - 1)^2$.)

25. Let $f(x) = (x + a)/(x + b)$, where $a \neq b$. Show that f^{-1} exists, and find it. Give the domain and range of both f and f^{-1}. Verify that $(f^{-1} \circ f)(x) = x$ for all x in the domain of f and that $(f \circ f^{-1})(x) = x$ for all x in the domain of f^{-1}.

26. Let $f(x) = 1/x$. Prove that for functions g and h,

$$f \circ \left(\frac{g}{h}\right) = \frac{f \circ g}{f \circ h}$$

What restrictions must be placed on the images of g and h?

 In Exercises 27 and 28, use a graphing calculator or CAS to obtain the graph of the given function. Verify from the graph that the function is 1–1 and draw the graph of its inverse.

27. $f(x) = 2x^3 + x - 5$ **28.** $g(x) = 3 - 4x^3 - x^5$

1.4 POLYNOMIAL FUNCTIONS

The simplest functions are the **polynomial functions**, yet they are among the most important in applications. Almost all nonpolynomial functions encountered in engineering and science are evaluated numerically by polynomial approximations. Here are some examples of polynomials:

$$f(x) = 2x + 3 \qquad p(x) = 3x^2 - 4x + 1 \qquad q(x) = x^5 + x - 1$$

Every special case of a polynomial function can be written in the general form

$$p(x) = a_n x^n + a_{n-1} x^{n-1} + \cdots + a_0 \tag{1.1}$$

where n is a nonnegative integer and $a_0, a_1, \ldots, a_n$ are constants, which we will assume to be real unless otherwise specified. If $a_n \neq 0$, the expression on the right is called a **polynomial of degree** n. The coefficient $a_n \neq 0$ of the highest power of x is called the **leading coefficient**. The domain of every polynomial function is the set of all real numbers.

Functions defined by polynomials of degree 0, 1, 2, and 3, respectively, are given names as shown below.

Degree	Name	Example
0	Constant function	$p(x) = 2$
1	Linear function	$p(x) = 2x + 3$
2	Quadratic function	$p(x) = x^2 - 2x + 7$
3	Cubic function	$p(x) = 4x^3 + 2x - 8$

The names *quartic* (degree 4) and *quintic* (degree 5) are sometimes used, but polynomial functions of degree higher than 5 are seldom named.

Constant Functions and Linear Functions

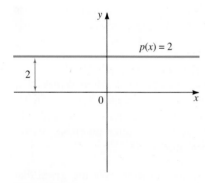

FIGURE 1.41

The graph of a constant function is a horizontal line. For example, the graph of the constant function $p(x) = 2$ consists of all points (x, y) with $y = 2$. So for every x, the value of y is 2. The graph of this function is shown in Figure 1.41.

Linear functions also have straight lines as graphs, as the name suggests. From Equation 1.1, a linear function is of the form $p(x) = a_1 x + a_0$, where $a_1 \neq 0$. It is customary to use the letters m and b for linear functions, rather than a_1 and a_0. So we write

$$p(x) = mx + b \qquad (m \neq 0)$$

for a typical linear function. By setting $x = 0$, we get $p(0) = b$. Thus, the graph crosses the y-axis at the point $(0, b)$. The number b is called the y-intercept. If (x_1, y_1) and (x_2, y_2) are any two different points on the line, we have $y_1 = mx_1 + b$ and $y_2 = mx_2 + b$. If we subtract the corresponding sides of the first equation from those of the second, we get $y_2 - y_1 = m(x_2 - x_1)$. Dividing by $(x_2 - x_1)$ gives

$$m = \frac{y_2 - y_1}{x_2 - x_1}$$

This number m is called the *slope* of the line. Its value is the same for *any* two distinct points (x_1, y_1) and (x_2, y_2) on the line. The numerator $y_2 - y_1$ is the change in y values corresponding to the change in x values $x_2 - x_1$ in the denominator. That is, m is the change in the vertical direction divided by the change in the horizontal direction, often described as "rise over run." Figure 1.42 illustrates a typical situation. When m is positive, the line slopes upward to the right, as in Figure 1.43(a), and when m is negative, the line slopes downward to the right, as in Figure 1.43(b).

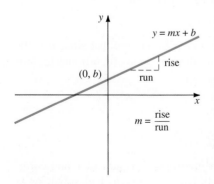

FIGURE 1.42

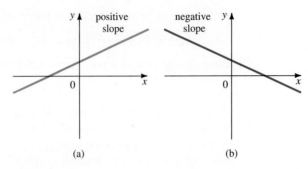

(a)   (b)

FIGURE 1.43

The next two examples illustrate how linear functions can arise in applications.

EXAMPLE 1.18   A car rental agency charges \$32 a day, plus 20¢ per mile, for renting a mid-size car. Write the formula for a cost function $C(x)$ for one day's rental, where x is the number of miles driven. Find the cost of renting a car of this type and driving a distance of 215 miles.

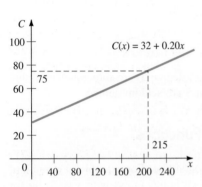

FIGURE 1.44

Solution   This situation is typical of many cost functions in that it involves both a **fixed cost** and a **variable cost**. In this example, the fixed cost is \$32 and the variable cost is $0.20x$ (the cost per mile, in dollars, times the number of miles driven). So the total cost $C(x)$, in dollars, is given by the formula

$$C(x) = 32 + 0.20x$$

Notice that $C(x)$ is a linear function with slope 0.20 and y-intercept 32. Its graph is shown in Figure 1.44. The cost for driving 215 miles is

$$C(215) = 32 + 0.20(215) = \$75 \qquad ∎$$

EXAMPLE 1.19   For income tax purposes, equipment is often *depreciated linearly* over a given period of time. Assume that a Mercedes-Benz originally valued at \$60,000 is depreciated linearly. After five years its value is \$40,000. Find a formula for the value $V(t)$ of the car t years after purchase, where $0 \le t \le 15$. What is its value eight years after purchase?

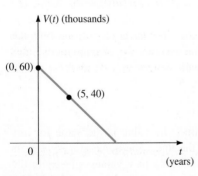

FIGURE 1.45

Solution   The graph in Figure 1.45 shows the situation. We are given that $V(0) = 60$ (in thousands) and $V(5) = 40$. So the points $(0, 60)$ and $(5, 40)$ are on the graph. Since $V(t)$ is linear, it must be of the form

$$V(t) = mt + b$$

There are two ways to find m and b. One way is to observe that since b is the y-intercept, $b = 60$. Then calculate m using the two points $(0, 60)$ and $(5, 40)$:

$$m = \frac{40 - 60}{5 - 0} = \frac{-20}{5} = -4$$

Thus,

$$V(t) = -4t + 60$$

The second method is to substitute the two known values of t and corresponding values of $V(t)$ into the equation $V(t) = mt + b$, and solve for m and b:

$$60 = m(0) + b \qquad \text{We substituted } t = 0, \ V(t) = 60.$$

$$40 = m(5) + b \qquad \text{We substituted } t = 5, \ V(t) = 40.$$

The first equation gives $b = 60$, and on substituting this value for b in the second, we get

$$40 = m(5) + 60$$

$$m = -\frac{20}{5} = -4$$

as before. The value after eight years is

$$V(8) = -4(8) + 60 = -32 + 60 = 28$$

That is, the car is worth \$28,000. ■

When data for a function are given in tabular form, we can test for linearity by seeing if for each pair (x_1, y_1) and (x_2, y_2) in the table, the ratio $(y_2 - y_1)/(x_2 - x_1)$ remains constant. If so, this constant value is the slope. Testing for a constant slope is particularly easy when the step between consecutive x values does not change. Then, all you have to do is to see if there is also a fixed change between consecutive y values. (We are using x to mean the independent variable and y to mean the dependent variable; clearly, other letters can be used.) The next example illustrates such a situation.

EXAMPLE 1.20   Show that the data in the accompanying table indicate a linear functional relationship, and find a formula for the function.

Solution   The change in x from one entry to the next is 2 in every case. The corresponding change in y is 6 in every case. So we expect that the data come from a linear function $f(x) = mx + b$ with

$$m = \frac{\text{change in } y}{\text{change in } x} = \frac{6}{2} = 3$$

x	y
-3	-14
-1	-8
1	-2
3	4
5	10

To find b we can substitute any pair of values (x, y) into $y = mx + b$. Using $(1, -2)$, we get

$$-2 = 3(1) + b$$

$$b = -5$$

So $f(x) = 3x - 5$.

It should be made clear that although the tabular values give strong evidence that the function we have found is correct, we have no way of knowing whether points would continue to fall on this line if the table were extended. ■

Quadratic Functions

We turn next to quadratic functions, that is, polynomial functions of degree 2. The simplest case is

$$f(x) = x^2$$

The range of f is the set $[0, \infty)$ of all nonnegative real numbers, and $f(x) = 0$ only when $x = 0$. Since $f(-x) = f(x)$, f is an even function, so its graph is symmetric with respect to the y-axis. The graph, shown in Figure 1.46, is a

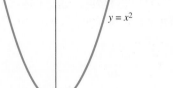

$y = x^2$

FIGURE 1.46

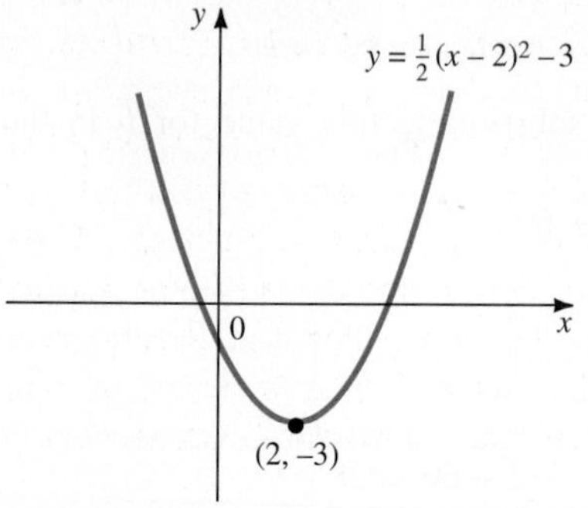

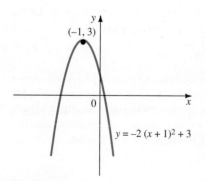

$y = \frac{1}{2}(x-2)^2 - 3$

$(2, -3)$

FIGURE 1.47

parabola with **vertex** at the origin and **axis** coinciding with the y-axis. From this basic parabola we can obtain the graph of every other quadratic function, using the transformations we considered in Section 1.2. For example, the graph of

$$y = a(x - h)^2 + k \qquad (1.2)$$

is obtained from that of $y = x^2$ by a vertical elongation (if $|a| > 1$) or compression (if $0 < |a| < 1$), as well as a reflection about the x-axis if $a < 0$, and then a translation so that the vertex is at the point (h, k). Equation 1.2 is called the **standard form** of the equation of a parabola. To illustrate, we show the graph of

$$y = \frac{1}{2}(x - 2)^2 - 3$$

in Figure 1.47.

The general quadratic function

$$f(x) = ax^2 + bx + c \qquad (a \neq 0)$$

can be put in the form (1.2), where $y = f(x)$, by *completing the square*, as illustrated in the next example.

EXAMPLE 1.21   Analyze and draw the graph of the function defined by

$$y = 1 - 4x - 2x^2$$

Solution   First, rearrange terms and factor out the coefficient of x^2 from the terms involving x:

$$y = -2(x^2 + 2x \qquad) + 1$$

We have left a space inside the parentheses for adding a term to complete the square. The term we need is 1 (the square of half the coefficient of x). Adding 1 inside the parentheses has the effect of adding -2 to the right-hand side, because of the coefficient preceding the parentheses. To compensate for this change, we must add $+2$. So we have

$$y = -2(x^2 + 2x + 1) + 1 + 2$$

or

$$y = -2(x + 1)^2 + 3$$

The equation is now in the form of Equation 1.2. Its graph is that of the basic parabola $y = x^2$, stretched so that each y value is doubled, reflected about the x-axis, and translated so that its vertex is at the point $(-1, 3)$. We show the graph in Figure 1.48. ∎

If we go through steps similar to those in Example 1.21 with the general quadratic function $f(x) = ax^2 + bx + c$, we get the result

$$y = a(x - h)^2 + k$$

where

$$h = -\frac{b}{2a} \qquad \text{and} \qquad k = \frac{4ac - b^2}{4a}$$

You will be asked to carry out the derivation in the exercises. Rather than memorizing these formulas for h and k, it is probably better to carry out the steps in each individual case, as we did in Example 1.21. Note, however, that $k = f(h)$, so it would not be too difficult to learn the formula $h = -b/2a$ and then find k by substituting h into the given function.

For completing the square the leading coefficient must be 1:

$$ax^2 + bx + c$$

$$= a\left(x^2 + \frac{b}{a}x\right) + c$$

Inside the parentheses, the coefficient of x^2 is 1.

$(-1, 3)$

$y = -2(x + 1)^2 + 3$

FIGURE 1.48

Applications often require finding the largest or smallest value of a given function, possibly subject to certain limitations on the independent variable. In business, we might want to maximize profit, for example, or minimize cost. In physics, we might want to determine how high an object will rise if projected at a given angle and initial velocity or how close to a planet a spacecraft can be safely sent. Problems of this type are called *optimization* problems. If the function in question happens to be quadratic, we can find the largest or smallest value by locating the vertex of the parabola. For most other functions we need the tools of calculus that we will introduce in Chapter 3. The following example illustrates the technique we can use in the case of a quadratic function.

EXAMPLE 1.22   The average midnight attendance at a "Rocky Horror Picture Show" is 100 when the ticket price is $5.00. The theater owner estimates that for each 10¢ decrease in the ticket price, the average attendance will increase by 10. What price should the owner charge to produce the maximum revenue, and what is the maximum revenue? What would the average attendance be with this new ticket price?

Solution   Let x denote the number of 10¢ reductions. Then

$$5 - 0.10x = \text{the price of each ticket (in dollars)}$$

and

$$100 + 10x = \text{the average nightly attendance}$$

The revenue, say $R(x)$, is found by multiplying the price per ticket by the number of tickets sold. That is,

$$R(x) = (5 - 0.1x)(100 + 10x)$$
$$= -x^2 + 40x + 500$$

To put this equation in the standard form of a parabola, we can complete the square, as in Example 1.21, obtaining

$$R(x) = -(x^2 - 40x + 400) + 500 + 400$$
$$= -(x - 20)^2 + 900$$

The graph of $R(x)$ is therefore a downward-opening parabola with vertex (20, 900). Thus, the maximum value of $R(x)$ is $900, occurring when $x = 20$. (See Figure 1.49.) The ticket price when there are 20 reductions of 10¢ each is

$$5 - 0.1(20) = \$3.00$$

and the average attendance is

$$100 + 10(20) = 300$$

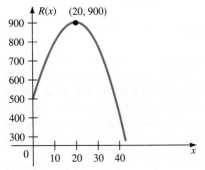

FIGURE 1.49

Polynomial Functions of Higher Degree

We will give only a brief discussion of polynomial functions of degree 3 or higher here. As we will see, calculus is needed to understand fully the nature of such functions. Figure 1.50 shows some computer-generated graphs of cubic

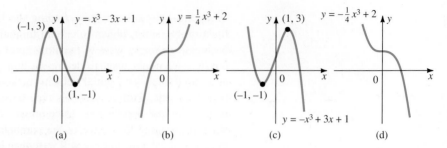

FIGURE 1.50

functions. These graphs are typical of cubics. If the leading coefficient is positive, as in parts (a) and (b), the function is positive for large positive values of x and negative for large negative values. When the leading coefficient is negative, as in parts (c) and (d), the opposite is true. The point $(-1, 3)$ that we have labeled in part (a) is called a **local maximum point**, because it is higher than all other points on the graph in its immediate vicinity. In part (c) the point $(1, 3)$ is a local maximum point. Similarly, the point $(1, -1)$ in (a) and $(-1, -1)$ in (c) are **local minimum points**. Parts (b) and (d) have no local maximum or minimum points.

Refer again to the graph in Figure 1.50(a). For $x < -1$, as we move from left to right on the x-axis, the curve rises, and we say the function is **increasing**. Between $x = -1$ and $x = 1$ as we move from left to right on the x-axis, the curve goes downward, and we say the function is **decreasing**. For $x > 1$, it again increases. By observing the graphs, you should verify the following results for the functions in parts (b), (c), and (d).

$$\begin{cases} \text{part (b): function is increasing for all } x \\ \text{part (c): function is decreasing for } x < -1 \text{ and for } x > 1 \\ \qquad\qquad \text{and increasing for } -1 < x < 1 \\ \text{part (d): function is decreasing for all } x \end{cases}$$

Besides knowing where a function is increasing or decreasing and knowing whether it has any local maxima (plural of maximum) or minima (plural of minimum), it is important to know if the function is bending up, as in Figure 1.51(a), or bending down, as in Figure 1.51(b). In the first instance we say the graph is **concave up** and in the second, **concave down**. For example, the graphs in Figure 1.50(a) and (b) are both concave down for $x < 0$ and concave

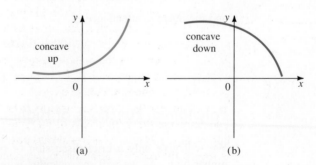

FIGURE 1.51

up for $x > 0$. Similarly, the graphs in parts (c) and (d) are both concave up for $x < 0$ and concave down for $x > 0$. Note that whether a function is increasing or decreasing on an interval has nothing to do with concavity.

We emphasize that the observations we have made about the graphs of the cubic functions in Figure 1.50 are based on computer-generated graphs. As we have seen, pictures can sometimes be deceiving. Maybe something unexpected happens beyond the range we are plotting. We need some means of determining exact locations of local maxima and minima, intervals over which a function is increasing and those over which it is decreasing, and intervals over which the graph is concave up and those over which it is concave down. As you might guess, it is calculus that provides the key, and we will return to these questions when we have introduced the appropriate calculus concepts.

Graphs of polynomial functions of degree 4 can have at most three local maxima and minima (two maxima and one minimum, or the reverse). There may be only one maximum or minimum, however. The graphs in Figure 1.52 illustrate typical possibilities in which the leading coefficient is positive. Notice that in each case the function is positive both for large positive and large negative values of x. After all, for these large values, the fourth power of x is larger than all lower powers combined. If the leading coefficient is negative, the graph always goes downward for large x (either positive or negative). Notice that the graph in part (c) looks almost like a parabola. So you cannot always tell the degree of a polynomial function just by looking at its graph.

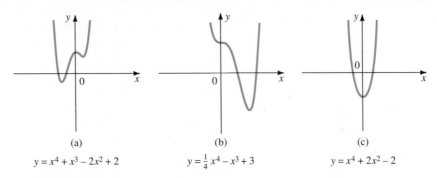

$$y = x^4 + x^3 - 2x^2 + 2 \qquad y = \tfrac{1}{4}x^4 - x^3 + 3 \qquad y = x^4 + 2x^2 - 2$$

$$\text{(a)} \qquad\qquad\qquad \text{(b)} \qquad\qquad\qquad \text{(c)}$$

FIGURE 1.52

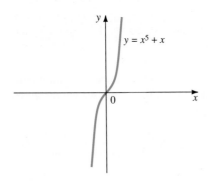

FIGURE 1.53

For each increase of one in the degree of a polynomial function, the graph *may* add one local maximum or minimum point. Thus, a polynomial function of degree n, with $n \geq 2$, has *at most* $n - 1$ local maxima or minima. It is possible, however, that an *odd*-degree polynomial function will have no maxima or minima. Consider, for example, the graph of $f(x) = x^5 + x$, which we show in Figure 1.53. Here again, you could not tell the degree just by looking at the graph. In fact, one might guess that this is the graph of a cubic polynomial. It is useful to note that the range of every odd-degree polynomial function is all of $\mathbb{R}$, since for large positive and large negative values of x, the function "goes to infinity" in opposite directions.

A value of x that causes a function to be zero is called a **zero** of the function. The zeros are the x-intercepts of the graph. Not all functions have real zeros. For example, $f(x) = x^2 + 1$ is always positive. If a polynomial can be factored, however, the zeros (if any) can be determined easily. For example, consider

$$f(x) = x^3 - 3x^2 - 4x + 12$$

$$= x^2(x - 3) - 4(x - 3)$$

$$= (x - 3)(x^2 - 4)$$

$$= (x - 3)(x + 2)(x - 2)$$

The only zeros are 3, 2, and −2, since if any of the factors is zero, the product is zero, and no other value of x can cause this product to be zero. If a polynomial cannot be factored, or is hard to factor, a graphing calculator or computer algebra system can be used to approximate the zeros. (We will give a method using calculus for approximating zeros in Chapter 4.)

Finding a Polynomial Function Given Its Zeros and Another Value

The next two examples illustrate how the degree of a polynomial function, its zeros, and one additional point uniquely determine the function.

EXAMPLE 1.23   The zeros of a cubic polynomial function f are −1, 2, and 4, and $f(0) = 3$. Find $f(x)$.

If $x = r$ is a zero of a polynomial function, then $x - r$ is a factor, and conversely.

Solution   For $f(x)$ to be 0 when $x = -1$, 2, or 4, the polynomial must contain the factors $(x + 1)$, $(x - 2)$, and $(x - 4)$. Since f is of degree 3, there can be no other nonconstant factors. Thus,

$$f(x) = a(x + 1)(x - 2)(x - 4)$$

where a is the unknown leading coefficient. We find a using the fact that $f(0) = 3$:

$$3 = a(1)(-2)(-4)$$

$$3 = 8a$$

$$a = \tfrac{3}{8}$$

So

$$f(x) = \tfrac{3}{8}(x + 1)(x - 2)(x - 4)$$

We show the graph of f in Figure 1.54.   ∎

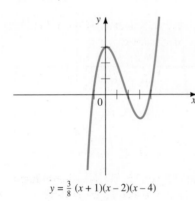

$y = \tfrac{3}{8}(x + 1)(x - 2)(x - 4)$

FIGURE 1.54

EXAMPLE 1.24   Find the formula for the function whose graph is shown in Figure 1.55 if it is known to be a polynomial function of lowest possible degree.

Solution   Since there are two local maximum points and one local minimum point, the lowest degree possible is 4. Reading from the graph, we get the zeros −4, −2, 1, and 5. So, as in Example 1.23, we have

$$f(x) = a(x + 4)(x + 2)(x - 1)(x - 5)$$

Since $(2, 3)$ is on the graph, it follows that $f(2) = 3$. So, on substituting $x = 2$ we get

$$3 = a(2 + 4)(2 + 2)(2 - 1)(2 - 5)$$

$$3 = a(6)(4)(1)(-3)$$

$$a = -\frac{1}{24}$$

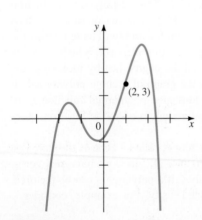

(2, 3)

FIGURE 1.55

Thus,

$$f(x) = -\frac{1}{24}(x+4)(x+2)(x-1)(x-5)$$

We could have anticipated that the leading coefficient would be negative, since the graph is negative both for large positive and large negative values of x. The fact that $(2, 3)$ is on the graph forces a to be negative. ∎

Dominant Term of a Polynomial

A technician inspects parabolic mirrors used in military optical systems.

The highest-degree term of a polynomial is the *dominant term*. For example, it has been found that the cost of building a large telescope increases approximately with the cube of the diameter of the main mirror. The cost may go up with the square of the number of instrument ducts and linearly with the number of staff. However, the diameter of the mirror determines the major cost of a large telescope. (Recent ideas of new ways of building telescopes have lowered the degree of the cost function below a cubic.) The dominant term determines the behavior of a polynomial function when x is large in absolute value. For example, consider the cubic function

$$f(x) = 2x^3 + 10x^2 - 5x + 4$$

We can demonstrate why the highest-degree term dominates by the following trick. By factoring out x^3, we can write $f(x)$ in the form

$$f(x) = x^3 \left(2 + \frac{10}{x} - \frac{5}{x^2} + \frac{4}{x^3}\right) \qquad (x \neq 0)$$

If x is large, either positively or negatively, each of the terms

$$\frac{10}{x}, \qquad -\frac{5}{x^2}, \qquad \text{and} \qquad \frac{4}{x^3}$$

is close to 0. Each can be made arbitrarily small by taking $|x|$ sufficiently large. Thus, $f(x)$ is approximately equal to $2x^3$, which is indicated by writing

$$f(x) \approx 2x^3 \qquad \text{for } |x| \text{ large}$$

More generally, if

$$f(x) = a_n x^n + a_{n-1} x^{n-1} + \cdots + a_0 \qquad (a_n \neq 0)$$

then we can factor out x^n to get

$$f(x) = x^n \left(a_n + \frac{a_{n-1}}{x} + \cdots + \frac{a_0}{x^n}\right) \qquad (x \neq 0)$$

Each term inside the parentheses except the first can be made as close to 0 as we please by taking $|x|$ sufficiently large. Thus,

$$f(x) \approx a_n x^n \qquad \text{for } |x| \text{ large}$$

So the highest-degree term, $a_n x^n$, dominates all other terms.

ESTIMATING ROOTS OF EQUATIONS USING A COMPUTER ALGEBRA SYSTEM

Finding the roots of an equation is typically a very hard problem, unless the function can be easily manipulated algebraically. There are techniques for approximating roots (see Section 4.5). Here we will use a computer algebra system to estimate roots of an equation.

CAS 8

Estimate the roots of the equation $\frac{2}{3} - \frac{x}{3} - \frac{5}{3}x^2 + x^3 = 0$.

Maple:[*]

Eq:=2/3–x/3–5/3*x^2+x^3;
solve(Eq=0,x);

Output:

$$\frac{2}{3}, \frac{1}{2} + \frac{1}{2}\sqrt{5}, \frac{1}{2} - \frac{1}{2}\sqrt{5}$$

fsolve(Eq=0,x);

Output:

–.6180339877, .6666666667,
1.618033989

Mathematica:

Eq = 2/3–x/3–5/3*x^2+x^3

[defines a name for the equation]

solve(Eq == 0,x);

Output:

$$\left\{ \left\{ x \to \frac{2}{3} \right\}, \left\{ x \to \frac{1 + \mathrm{Sqrt}[5]}{2} \right\}, \left\{ x \to \frac{1 - \mathrm{Sqrt}[5]}{2} \right\} \right\}$$

N[%]

[numerical roots of previous output]

Output:

$\{\{x \to 0.666667\}, \{x \to 1.61803\}, \{x \to -0.618034\}\}$

DERIVE:

(At the □ symbol, go to the next step.)
a (author) □ 2/3–x/3–5/3x^2+x^3 □ 1
(solve)

Output:

$$\frac{2}{3}$$

$$\frac{1}{2} - \frac{\sqrt{5}}{2}$$

$$\frac{\sqrt{5}}{2} + \frac{1}{2}$$

x (approximate) □
[choose expression]

Output: 0.666666

x (approximate) □
[choose expression]

Output: −0.618033

x (approximate) □
[choose expression]

Output: 1.61803

[*]Note that in Maple `fsolve` gives numerical answers instead of the radicals that result from `solve`. In Mathematica, $N[\%]$ simplifies the previous output. In general, $\%$ refers to the previous output in Mathematica. In Maple, the double quote "refers to the previous output.

Exercise Set 1.4

1. For a certain linear function f, it is known that $f(-1) = 3$ and $f(7) = -5$. Find $f(x)$. What is $f(10)$?

2. The graph of a function g is a straight line passing through the points $(2, -1)$ and $(5, 3)$. Find a formula for $g(x)$.

3. The daily car rental rate for agency A is $21, plus 21¢ per mile. For agency B the rate is $32, plus 18¢ per mile. A driver plans to rent from one of these agencies and take a 320-mile round trip. From which agency should the driver rent to get the better rate? What is the minimum number of miles beyond which agency B's cost is less than that of agency A?

4. In depreciating equipment it is reasonable to assume that certain items will not depreciate to 0 but will have some residual value, at least as junk. Suppose a car originally valued at $15,000 depreciates linearly to a residual value of $200 in 20 years. Find its value $V(t)$, t years after purchase, where $0 \leq t \leq 20$.

5. (a) Find a linear function f such that $(f \circ f)(x) = 9x + 16$.
   (b) For which linear functions $cx + d$ is there a function f such that $(f \circ f)(x) = cx + d$? Under what conditions is f unique?

6. One of the functions f or g in the accompanying table is linear and one is quadratic, of the form at^2. Determine which is which and find formulas for $f(t)$ and $g(t)$.

t	$f(t)$	$g(t)$
-2	2	2
2	2	14
6	18	26
10	50	38

7. Draw the graph of each of the following, using your knowledge of the basic quadratic function and the transformations studied in Section 1.2.
   (a) $f(x) = \frac{1}{4}x^2$ (b) $f(x) = -\frac{1}{2}x^2$
   (c) $f(x) = 2x^2 - 3$ (d) $f(x) = (x - 2)^2$
   (e) $f(x) = (x + 1)^2$
   (f) $f(x) = 2(x + 1)^2 - 3$

8. By completing the square, write the equation $y = 2x^2 - 4x + 5$ in the form $y = a(x - h)^2 + k$ and draw its graph.

9. Repeat Exercise 8 with the equation

$$y = 1 - 2x - \frac{x^2}{2}$$

10. For a small manufacturing firm the unit cost $C(x)$ in dollars of producing x items per day is given by

$$C(x) = x^2 - 120x + 4000$$

How many items should be produced per day to minimize the unit cost? What is the minimum unit cost?

11. A company that produces computer terminals analyzes production and finds that the profit function $P(x)$, in dollars, for selling x terminals per month is given by

$$P(x) = 160x - 0.1x^2 - 20,000$$

How many terminals should be sold per month to produce the maximum profit? What is the maximum profit?

12. In a contest, a baseball is thrown upward from the ground with an initial velocity of 57.6 meters per second. If air resistance is neglected, its distance $s(t)$, in meters, above the ground after t seconds is given by $s(t) = -4.8t^2 + 57.6t$. Find the maximum height the baseball will rise.

13. A restaurant has a fixed price of $18 for a complete dinner. The average number of customers per evening is 200. The owner estimates that for each 50¢ increase in the price of the dinner, there will be four fewer customers per evening on average. What price for the dinner will produce the maximum revenue?

14. From each of the data sets given, sketch the graph of f and determine a possible formula for $f(x)$.

(a)

x	$f(x)$
0	-26
1	-6
2	6
3	10
4	6
5	-6
6	-26

(b)

x	$f(x)$
-3	7
-2	5
-1	3
0	1
1	-1
2	-3
3	-5

15. (a) Is the composition of two linear functions a linear function? Prove your answer.
    (b) Is the composition of two quadratic functions quadratic? Prove your answer.
    (c) What is the degree of $p \circ q$ if p is a polynomial function of degree m and q is a polynomial function of degree n?

16. Let $f(x) = 100x^2$ and $g(x) = 0.1x^3$. For what values of x is $f(x) \geq g(x)$, and for what values is $g(x) \geq f(x)$?

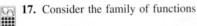 17. Consider the family of functions

$$f_n(x) = x^n$$

for n a positive integer. On one set of axes draw the graph of f_n for $n = 2$, 4, and 6. On another set of axes draw f_n for $n = 1$, 3, and 5. Discuss your results. How can you generalize them? For a given positive integer k, compare the sizes of $f_k(x)$ and $f_{k+1}(x)$ on each of the intervals $0 < x < 1$ and $x > 1$.

18. The zeros of a certain cubic function g are 1, 2, and -1, and $g(0) = -4$. Find $g(x)$, and draw its graph.

19. A cubic function f has zeros -1 and 2, with 2 a double zero, indicating that one of the factors of $f(x)$ is $(x-2)^2$. If $f(1) = 6$, find $f(x)$ and draw its graph.

20. The function h whose graph is shown is a polynomial function of degree 4. Find $h(x)$.

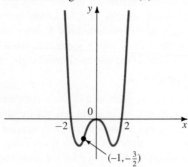

21. The function whose graph is shown is a polynomial function of degree n. What is the lowest possible value of n? Give the intervals over which the function is (a) increasing, (b) decreasing, and the intervals over which the graph is (c) concave up, and (d) concave down.

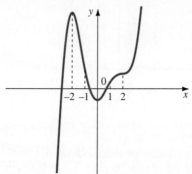

22. From the graph of the function f shown, draw the graphs of $y = |f(x)|$ and $y = f(|x|)$. Compare the graphs of $y = |f(x)|$ and $y = f(|x|)$ with the graph of f.

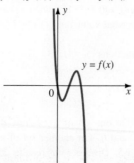

$y = f(x)$

23. Draw the graph of each of the following.

(a) $y = |2x - 3|$ (b) $y = |x^2 + x - 2|$

(c) $y = |2x - x^3|$

24. Let $f(x) = x^2 - \frac{3}{2}x$ and $g(x) = x^2 - 2$. Find the domains of $\sqrt{f(x)}$ and $\sqrt{g(x)}$, graphically and algebraically.

25. If $f(x) = ax^2 + bx + c$, what do you know about the values of a, b, and c if

(a) $(1, 1)$ is on the graph of f?

(b) $(1, 1)$ is the vertex?

(c) the y-intercept is 6?

(d) conditions (a), (b), and (c) are all satisfied?

What are the domain and range of the resulting function?

26. The rate R at which a population in a confined space increases is proportional to the product of the current population P and the difference between the maximum population the environment can sustain and the current population.

(a) Write R as a function of P. (Denote the proportionality constant by k.)

(b) For a fixed $k > 0$ sketch the graph of the function in part (a).

27. Cubics. In this exercise we show how the graph of every cubic function is obtainable from a basic cubic curve of the form $y = x^3 + kx$, through a series of transformations. We begin with the general cubic function

$$f(x) = ax^3 + bx^2 + cx + d \qquad (a \neq 0)$$

(a) Rewrite $f(x)$ as $af_1(x)$, where

$$f_1(x) = x^3 + b_1x^2 + c_1x + d_1$$

and specify b_1, c_1, and d_1 in terms of a, b, and c. How is the graph of f related to the graph of f_1?

(b) Let $f_2(x) = f_1\left(x - \dfrac{b_1}{3}\right)$. Show that

$$f_2(x) = x^3 + c_2x + d_2$$

and specify c_2 and d_2 in terms of b_1, c_1, and d_1. How is the graph of f_1 related to the graph of f_2?

(c) Let $f_3(x) = f_2(x) - d_2$. Then

$$f_3(x) = x^3 + c_2x$$

How is the graph of f_2 related to the graph of f_3?

(d) Let $k = c_2$, and write

$$g(x) = x^3 + kx$$

By collecting the results of parts (a), (b), and (c), explain how the graph of f can be obtained from the graph of g.

(e) Use a graphing calculator or CAS to graph g for various values of k. Try to describe the possible shapes of the graph.

In Exercises 28–30, obtain the graph from a graphing calculator or CAS and estimate each of the following: (a) local maximum and minimum points, (b) intervals where the function is increasing and those where it is decreasing, and (c) intervals where the graph is concave up and those where it is concave down.

28. $y = x^4 - 4x^3 + 5$

29. $y = \dfrac{x^5}{5} - \dfrac{10x^3}{3} + 9x$

30. $y = \dfrac{x^4}{2} - \dfrac{x^3}{3} - 4x^2 + 4x + 7$

1.5 RATIONAL FUNCTIONS AND RATIONAL POWER FUNCTIONS

A **rational function** is a function that can be expressed in the form

$$f(x) = \frac{p(x)}{q(x)}$$

where p and q are polynomial functions. The domain of f is the set of all real numbers x for which the denominator $q(x) \neq 0$. Here are some examples:

$$f(x) = \frac{x+1}{x-1} \qquad g(x) = \frac{x^2 - 2x + 3}{3x^3 + 4x - 1} \qquad h(x) = \frac{1}{x}$$

A polynomial function p is also a rational function, since we can take the denominator $q(x)$ to be 1 for all x; that is, $p(x) = p(x)/1$.

If $p(x)$ and $q(x)$ have a nonconstant factor in common, then we divide by that common factor, ruling out of the domain any value of x causing this factor to be 0. For example, let

$$f(x) = \frac{x^2 - 4}{x - 2}$$

The domain of f consists of all real numbers except $x = 2$. For all such x, we have

$$f(x) = \frac{(x-2)(x+2)}{x-2} = x+2 \qquad (x \neq 2)$$

The graph of f, shown in Figure 1.56, is the line $y = x + 2$ with the single point $(2, 4)$ missing (indicated by the open circle). *Throughout the remainder of this section, unless otherwise specified, we will assume that all common factors have been divided out, so that $p(x)/q(x)$ is in lowest terms (in which case p and q are said to be* **relatively prime***).*

The zeros of a rational function are the zeros of its numerator, $p(x)$, since a quotient is zero if and only if the numerator is 0 (when, as we are assuming, the denominator is nonzero). For example, the zeros of

$$f(x) = \frac{x^2 - 1}{x^2 - 4}$$

are 1 and −1.

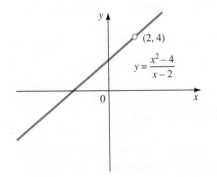

$(2, 4)$

$y = \dfrac{x^2 - 4}{x - 2}$

FIGURE 1.56

Vertical Asymptotes

The zeros of the denominator of a rational function are not in its domain, but investigating the behavior of f as x approaches each zero of the denominator is particularly useful in drawing its graph. Consider, for example, the function

(x > 1)		(x < 1)	
x	f(x)	x	f(x)
1.1	21	0.9	−19
1.01	201	0.99	−199
1.001	2001	0.999	−1999
1.0001	20001	0.9999	−19999

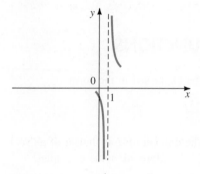

FIGURE 1.57

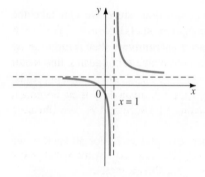

Graph of $f(x) = (x^2 - 1)/(x - 1)^2$. Same as $f(x) = (x + 1)/(x - 1)$. Line $x = 1$ is a vertical asymptote.

FIGURE 1.58

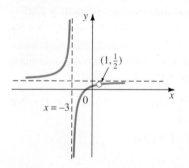

Graph of $f(x) = (x^2 - 1)/[(x + 3)(x - 1)]$. Same as $f(x) = (x + 1)/(x + 3)$, minus the point $(1, \frac{1}{2})$. Line $x = -3$ is a vertical asymptote, but $x = 1$ is not.

FIGURE 1.59

f defined by

$$f(x) = \frac{x + 1}{x - 1}$$

In the margin we show two tables, one for values of x greater than 1 but close to 1 and one for x less than 1 but close to 1. These tables provide strong evidence that when x approaches 1 through values greater than 1 (when x approaches 1 from the right), $f(x)$ gets arbitrarily large, and when x approaches 1 through values less than 1 (when x approaches 1 from the left), $f(x)$ gets arbitrarily large in absolute value but is negative. Symbolically, we can indicate this behavior as follows:

$$\text{As } x \to 1^+, \ f(x) \to \infty.$$

$$\text{As } x \to 1^-, \ f(x) \to -\infty.$$

Figure 1.57 illustrates how the graph of f looks in the vicinity of $x = 1$. The line $x = 1$ is called a **vertical asymptote** to the graph.

As this example illustrates, if $q(a) = 0$, then the line $x = a$ is a vertical asymptote to the graph of $f(x) = p(x)/q(x)$ when p and q have no common factor.

REMARK
■ If p and q have a factor $(x - a)$ in common, then $x = a$ *may* be a vertical asymptote. For example, if

$$f(x) = \frac{x^2 - 1}{(x - 1)^2}$$

we can rewrite $f(x)$ in the form

$$f(x) = \frac{(x + 1)(x - 1)}{(x - 1)(x - 1)} = \frac{x + 1}{x - 1} \qquad \text{if } x \neq 1$$

and conclude, as above, that $x = 1$ is a vertical asymptote (see Figure 1.58). However, if

$$f(x) = \frac{x^2 - 1}{x^2 + 2x - 3}$$

then, by factoring, we get

$$f(x) = \frac{(x + 1)(x - 1)}{(x + 3)(x - 1)} = \frac{x + 1}{x + 3} \qquad \text{if } x \neq 1$$

In this case, then, $x = 1$ is not a vertical asymptote even though the denominator of the original fraction is 0 when $x = 1$. In fact, as x approaches 1 from either side, $f(x) \to \frac{1}{2}$ (but note that the point $(1, \frac{1}{2})$ is not on the graph; see Figure 1.59). The point to remember is that when checking for vertical asymptotes, be sure to divide out any common factors first.

EXAMPLE 1.25   Find all vertical asymptotes to the graph of

$$f(x) = \frac{3x + 2}{x^2 - x - 2}$$

Solution   We factor the denominator to obtain

$$f(x) = \frac{3x + 2}{(x - 2)(x + 1)}$$

There is no common factor between numerator and denominator, so we conclude from the preceding discussion that the lines $x = 2$ and $x = -1$ are vertical asymptotes. By letting x approach 2 and -1, respectively, from the right and left, and examining the signs of all of the factors in each case, we can determine the following. (You should verify these results.)

$$\text{As } x \to 2^+, \ f(x) \to \infty$$

$$\text{As } x \to 2^-, \ f(x) \to -\infty$$

$$\text{As } x \to -1^+, \ f(x) \to \infty$$

$$\text{As } x \to -1^-, \ f(x) \to -\infty$$

We will show the graph of this function in Example 1.27, after we have discussed horizontal asymptotes.   ∎

Horizontal Asymptotes

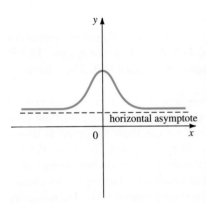

FIGURE 1.60

A line $y = b$ is a **horizontal asymptote** to the graph of a function f if $f(x) \to b$ as x becomes arbitrarily large, either positively or negatively, that is, as $x \to \infty$ or as $x \to -\infty$. (See Figure 1.60.) To determine if there is a horizontal asymptote to the graph of a rational function $f(x) = p(x)/q(x)$, we can use the fact that for $|x|$ large, both $p(x)$ and $q(x)$ are approximated by their dominant terms (their highest-degree terms), as we showed in Section 1.3. We illustrate this procedure in the next example.

EXAMPLE 1.26   For each of the following functions determine whether the graph has a horizontal asymptote, and if so, what the asymptote is.

(a) $f(x) = \dfrac{2x^2 - 3}{5x^3 + 4x - 7}$ (b) $g(x) = \dfrac{3x^2 - 1}{2x^2 + 5}$ (c) $h(x) = \dfrac{1 - x^3}{x^2 - 5}$

Solution

(a) For x large in absolute value, the numerator behaves approximately like its dominant term, $2x^2$, and the denominator like $5x^3$. So we can say that for $|x|$ large,

$$f(x) \approx \frac{2x^2}{5x^3} = \frac{2}{5x}$$

and since $2/5x \to 0$ as $|x| \to \infty$, we conclude that the line $y = 0$—that is, the x-axis—is a horizontal asymptote.

(b) Reasoning as in part (a), we see that for $|x|$ large,

$$g(x) \approx \frac{3x^2}{2x^2} = \frac{3}{2}$$

So $y = \frac{3}{2}$ is a horizontal asymptote.

(c) Again, for $|x|$ large,

$$h(x) \approx \frac{-x^3}{x^2} = -x$$

Since $-x$ does not approach a finite limiting value as x becomes arbitrarily large in absolute value, there is no horizontal asymptote.   ∎

The preceding example illustrates the following general result, which can be proved using the idea of a dominant term.

Let

$$f(x) = \frac{a_m x^m + a_{m-1} x^{m-1} + \cdots + a_0}{b_n x^n + b_{n-1} x^{n-1} + \cdots + b_0} \qquad (m, n \geq 1)$$

where $a_m \neq 0$ and $b_n \neq 0$. Then

(i)   If $m < n$, the line $y = 0$ is a horizontal asymptote.
(ii) If $m = n$, the line $y = a_m / b_n$ is a horizontal asymptote.
(iii) If $m > n$, there is no horizontal asymptote.

EXAMPLE 1.27   For the function in Example 1.25, find the horizontal asymptote. Also find the x- and y-intercepts and use this information, together with the results of Example 1.25, to draw the graph.

Solution   The function is

$$f(x) = \frac{3x + 2}{x^2 - x - 2}$$

Since the degree of the denominator is greater than the degree of the numerator (case (i) above), we conclude that $y = 0$ is a horizontal asymptote. The x-intercept is $-\frac{2}{3}$, found by setting the numerator $3x+2$ equal to 0. The y-intercept is -1, found by setting $x = 0$.

In Example 1.25 we found that the lines $x = 2$ and $x = -1$ are vertical asymptotes, and we saw how the graph approaches these asymptotes from each side. When we combine all this information, we can draw the graph as shown in Figure 1.61 with reasonable confidence. We should note, however, that we need the tools of calculus to be certain of the shape. In particular, using calculus we will be able to verify that the middle section of the graph does change from concave up to concave down and to determine exactly where the concavity changes.   ∎

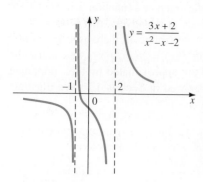

FIGURE 1.61

$$y = \frac{3x + 2}{x^2 - x - 2}$$

Rational Power Functions

We now consider functions of the form

$$f(x) = x^{m/n}$$

where m and n are positive integers. A function of this type is called a **rational power function**, since the exponent (power) is a rational number. For example, $f(x) = x^{2/3}$ is such a function. The meaning given to the expression $x^{m/n}$ is

$$x^{m/n} = (\sqrt[n]{x})^m$$

When m/n is in lowest terms, we also can write

$$x^{m/n} = \sqrt[n]{x^m}$$

If n is even, we must restrict x to be nonnegative when m/n is in lowest terms, since even roots ($\sqrt{N}$, $\sqrt[4]{N}$, ...) of negative numbers do not exist within the real number system.

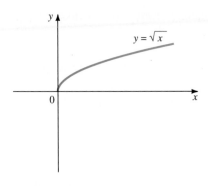

FIGURE 1.62

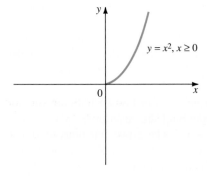

FIGURE 1.63

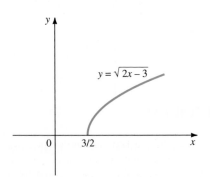

FIGURE 1.64

If the exponent is a negative rational number, say $-m/n$, then we define

$$x^{-m/n} = \frac{1}{x^{m/n}}$$

and in this case, we must also restrict x to be nonzero.

The simplest of the rational power functions is the **square root function**

$$f(x) = x^{1/2}$$

or equivalently, $f(x) = \sqrt{x}$. The domain, as well as the range, is the set of all nonnegative real numbers. The graph is shown in Figure 1.62. This curve looks as though it might be the upper half of a parabola with a horizontal axis. To show that this is the case, consider the function defined by $y = x^2$ with $x \geq 0$. This curve is the half-parabola shown in Figure 1.63. By the horizontal line test, this curve is the graph of a 1–1 function, so the function has an inverse. To find the inverse, we interchange x and y, then solve for y:

$$x = y^2$$
$$y = \sqrt{x}$$

(We used the positive square root only, since both x and y are nonnegative.) Thus, we have shown that the square root function $f(x) = \sqrt{x}$ is the inverse of the squaring function $y = x^2$ in which x is restricted to be nonnegative.

Using transformations of the types considered in Section 1.2, we can now analyze and graph all functions of the form

$$g(x) = \sqrt{ax + b}$$

The next example illustrates such a function.

EXAMPLE 1.28   Identify and draw the graph of the function

$$g(x) = \sqrt{2x - 3}$$

Solution   First we write $g(x)$ in the form

$$g(x) = \sqrt{2\left(x - \frac{3}{2}\right)}$$

We can view $g(x)$ as having come from $f(x) = \sqrt{x}$ by the horizontal compression $f(2x)$, followed by a horizontal translation $3/2$ units to the right. That is,

$$g(x) = f\left(2\left(x - \frac{3}{2}\right)\right)$$

So the graph of g is the half-parabola shown in Figure 1.64.   ∎

REMARK ────────────────────────────────
■ The horizontal compression $f(2x) = \sqrt{2x}$ also can be thought of as the vertical stretching $\sqrt{2}f(x) = \sqrt{2}\sqrt{x}$, since $\sqrt{2x} = \sqrt{2}\sqrt{x}$. The result is the same.

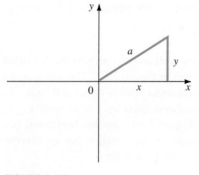

The square root function is one of the family of nth root functions

$$h_n(x) = x^{1/n}$$

You will be asked to discuss general characteristics of this family for n even and n odd in the exercises.

Another important root function is obtained from the composition of the square root function and the polynomial $a^2 - x^2$, with $a > 0$ and $-a \leq x \leq a$, giving

$$f(x) = \sqrt{a^2 - x^2}$$

If we set $y = f(x)$ and square both sides, we get $y^2 = a^2 - x^2$, or equivalently

$$x^2 + y^2 = a^2 \tag{1.3}$$

FIGURE 1.65
$x^2 + y^2 = a^2$

Points (x, y) whose coordinates satisfy this equation are all a units from the origin, as can be seen by the Pythagorean Theorem (see Figure 1.65). Thus, the graph of Equation 1.3 is a circle of radius a, centered at the origin. (See also Appendix 1.)

All points on the graph of $y = \sqrt{a^2 - x^2}$ lie on the circle given by Equation 1.3, but since $\sqrt{a^2 - x^2}$ is nonnegative, its graph is the upper half of the circle. That is, the graph of the function $f(x) = \sqrt{a^2 - x^2}$ is the upper semicircle shown in Figure 1.66.

The circle $x^2 + y^2 = a^2$ is not the graph of a function (it fails the vertical line test), but it can be thought of as combining the function $y = \sqrt{a^2 - x^2}$ and its negative, the function $y = -\sqrt{a^2 - x^2}$. These two functions represent the upper and lower semicircles, respectively (see Figure 1.67). We will have occasion in the next section to use the equation of a circle.

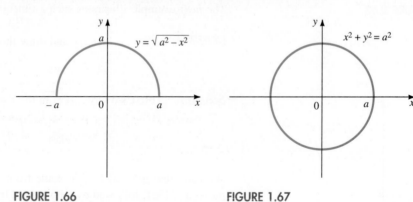

FIGURE 1.66   **FIGURE 1.67**

The next example illustrates a rational power function $y = x^{m/n}$ in which m is even and n is odd.

EXAMPLE 1.29   Discuss and sketch the graph of

$$y = x^{2/3}$$

Solution   Because cube roots of all real numbers exist, the domain is all of $\mathbb{R}$. Writing $x^{2/3} = \sqrt[3]{x^2}$, we see that replacing x with $-x$ leaves y unchanged, so that the graph is symmetric with respect to the y-axis. The range is contained in the set of all nonnegative real numbers, since $\sqrt[3]{x^2} \geq 0$ for all x. In fact,

x	y
1	1
2	1.59
3	2.08
4	2.52
5	2.92

for $k \geq 0$ we can take $x = k^{3/2}$ and get $(k^{3/2})^{2/3} = k$, which shows that the range is the set of all nonnegative real numbers. With the aid of the points shown in the table, and making use of symmetry, we can draw the graph in Figure 1.68.

After we have introduced some calculus concepts, we will be able to confirm that the graph is tangent to the y-axis at the origin (it is said to form a **cusp** there) and that the graph is always concave down. ∎

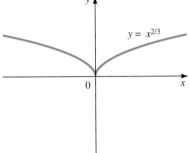

$y = x^{2/3}$

FIGURE 1.68

As our final example in this section, we consider the composition of the square root function and a rational function.

EXAMPLE 1.30   Discuss and sketch the graph of

$$y = \sqrt{\frac{x - 1}{x + 2}}$$

Solution   The domain consists of all real numbers x for which

$$\frac{x - 1}{x + 2} \geq 0$$

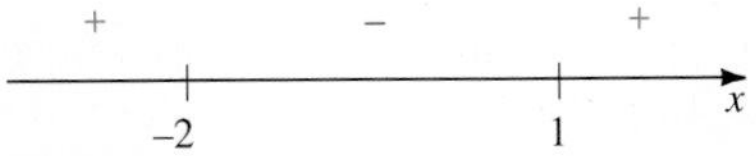

One way to determine the x values satisfying this inequality is to use the sign graph shown in Figure 1.69, where, after marking the zeros 1 and -2 of the numerator and denominator, respectively, we determined the sign of the fraction $(x - 1)/(x + 2)$ in each of the three intervals $(-\infty, -2)$, $(-2, 1)$, and $(1, \infty)$ into which these zeros divide the line. Using test values, such as $x = -3$, $x = 0$, and $x = 2$, shows that the signs are as given above the intervals in Figure 1.69. We conclude that $(x - 1)/(x + 2)$ is positive in $(-\infty, -2)$ and in $(1, \infty)$, and since the fraction is zero at $x = 1$ and undefined at $x = -2$, the domain of f is

$$(-\infty, -2) \cup [1, \infty)$$

Since $f(1) = 0$, the x-intercept of the graph is 1. There is no y-intercept because $x = 0$ is not in the domain.

To find asymptotes, we proceed as with rational functions, taking care to stay within the domain. The line $x = -2$ is a vertical asymptote, but to stay in the domain, x can approach -2 only from the left. More explicitly, we have,

$$\text{As } x \to -2^-, \ y \to \infty$$

Using the dominant term approximation for numerator and denominator, we see that for $|x|$ large,

$$y \approx \sqrt{\frac{x}{x}} = \sqrt{1} = 1$$

So $y = 1$ is a horizontal asymptote. In this case, the domain "extends to infinity" in both directions, so that the graph is asymptotic to the line $y = 1$ on the left and on the right. In general, you must check to see if points arbitrarily far out to the left and to the right are in the domain. Some graphs have a horizontal asymptote on one side only, and some have no horizontal asymptote at all.

Finally, note that $y \geq 0$ for all x in the domain. With the information we now have, we get the graph in Figure 1.70. ∎

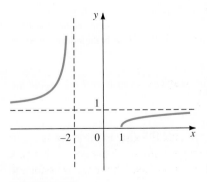

FIGURE 1.69

(Sign graph showing + over $(-\infty, -2)$, − over $(-2, 1)$, + over $(1, \infty)$, with -2 and 1 marked.)

FIGURE 1.70

Algebraic and Transcendental Functions

All the functions we have considered so far fall into a broad category known as **algebraic functions**. These are functions that can be obtained from polynomial functions by any finite combination of the following operations: addition, subtraction, multiplication, division, raising to integral powers, and extracting integral roots. Some examples that we have not yet discussed are

$$f(x) = (5x^4 - 2x^2 + 3)^{7/3} \quad \text{and} \quad g(x) = \frac{\sqrt{x - \sqrt[3]{x^2 - 7}}}{(x^3 - 1)^4}$$

You might well ask whether there are any functions that are not algebraic. The answer is that there are *many* others. In fact, there are more nonalgebraic functions than there are algebraic functions (a "higher order of infinity"). These nonalgebraic functions are called **transcendental**, and you have already encountered many of them. The most familiar ones are the trigonometric functions that we will discuss in the next section. Others include the exponential, logarithmic, inverse trigonometric, and hyperbolic functions, all of which we will study in some detail in Chapter 7, when we have calculus to aid in our analysis.

Exercise Set 1.5

In Exercises 1 and 2, find all vertical and horizontal asymptotes.

1. (a) $y = \dfrac{2x + 1}{x - 2}$ (b) $y = \dfrac{x}{x^2 - 4}$

2. (a) $y = \dfrac{2x^2 - 3x + 1}{3x^2 + 5x - 2}$ (b) $y = \dfrac{x^3 + 2x - 4}{x^2 + x - 2}$

In Exercises 3–6, find x- and y-intercepts, vertical and horizontal asymptotes, and sketch the graph. Make use of symmetry where applicable.

3. $f(x) = \dfrac{1}{x - 3}$ **4.** $f(x) = \dfrac{x - 1}{x + 2}$

5. $f(x) = \dfrac{x^2 - 4}{x^2 - 1}$ **6.** $f(x) = \dfrac{x}{x^2 + 1}$

7. Sketch a graph that satisfies each of the following conditions:
  (a) passes through $(0, 0)$
  (b) is concave down for all $x \leq 0$
  (c) is concave up for $0 < x < 2$
  (d) has a vertical asymptote at $x = 2$
  (e) is concave down for all $x \geq 2$
  (f) has a horizontal asymptote at $y = 4$

8. Consider the family of functions

$$g_n(x) = \frac{1}{x^n}$$

where $x \neq 0$ and n is a positive integer. On one set of axes draw the graph of g_n for $n = 2, 4,$ and 6. On another set of axes draw g_n for $n = 1, 3,$ and 5. Discuss your results. How can you generalize from them? For a given positive integer k, compare the sizes of $g_k(x)$ and $g_{k+1}(x)$ on each of the intervals $0 < x < 1$ and $x > 1$.

9. Show that if $f(x) = p(x)/q(x)$ is a rational function in lowest terms with the degree of p *exactly one greater* than the degree of q, then by long division $f(x)$ can be written in the form

$$f(x) = mx + b + \frac{r(x)}{q(x)}$$

where the degree of r is less than the degree of q. Explain why the graph of $y = f(x)$ approaches the line $y = mx + b$ as $x \to \infty$. (This line is called an **oblique asymptote**.)

In Exercises 10 and 11, use the idea of Exercise 9 to find the oblique asymptote.

10. (a) $f(x) = \dfrac{x^2 - 4}{x + 1}$ (b) $f(x) = \dfrac{x^2 - 3x + 4}{x - 2}$

11. (a) $f(x) = \dfrac{x^3 - x}{x^2 + 1}$ (b) $f(x) = \dfrac{x^3 - 2x^2 + 3x - 4}{x^2 + 2x - 1}$

12. Let $h_n(x) = x^{1/n}$ for n a positive integer. On one set of axes draw the graph of h_n for $n = 2, 4,$ and 6 and on another draw h_n for $n = 1, 3,$ and 5. Pay particular attention to the domain for n even and for n odd. How can you generalize your results? For a given positive integer k, compare $h_k(x)$ and $h_{k+1}(x)$ on each of the intervals $0 < x < 1$ and $x > 1$.

In Exercises 13–15, explain how the graph can be obtained from the graph of $y = \sqrt{x}$, and draw it.

13. (a) $y = \sqrt{x+2}$ (b) $y = \sqrt{x-1}$

14. (a) $y = \sqrt{1-x}$ (b) $y = \sqrt{2x+1}$

15. (a) $y = \sqrt{4-3x} - 2$ (b) $y = 1 + 3\sqrt{x}$

16. Identify and draw the graph of each of the following.
(a) $f(x) = \sqrt{4-x^2}$ (b) $g(x) = -\sqrt{1-x^2}$

17. Draw the graph of each of the following, using a transformation of the semicircle $y = \sqrt{a^2 - x^2}$ for the appropriate value of a. Identify the resulting curve.
(a) $y = \sqrt{9-4x^2}$ (b) $y = \frac{3}{2}\sqrt{4-x^2}$

18. Analyze and draw the graph of each of the following. Can you identify the curve in each case? Can you express in words why such a seemingly minor change from part (a) to part (b) has such a profound result?
(a) $y = \sqrt{x^2 + 4}$ (b) $y = \sqrt{x^2 - 4}$

19. Analyze and draw the graph of each of the following functions.
(a) $f(x) = x^{3/2}$ (b) $g(x) = (x-2)^{2/3}$

In Exercises 20–24, discuss the domain, intercepts, symmetry, and asymptotes, and draw the graph.

20. $y = \sqrt{\dfrac{x+2}{x-1}}$ **21.** $y = \sqrt{\dfrac{2-x}{2+x}}$

22. $y = \sqrt{\dfrac{x^2-1}{x^2+1}}$ **23.** $y = \dfrac{x}{\sqrt{4-x^2}}$

24. $y = \dfrac{2\sqrt{x^2-4}}{x-1}$

 25. In Exercises 25–29, use a graphing calculator or a CAS to obtain the graph of each of the functions in Exercises 20–24.

1.6 THE TRIGONOMETRIC FUNCTIONS

In this section we give a brief review of the trigonometric functions, emphasizing how they can be defined with real-number domains, just as is the case with the other functions we have studied. This approach is in contrast to the one in which these functions are defined in terms of *angles*, although the two are closely related, as we shall see.

You may have wondered why two definitions of trigonometric functions are needed. Whether we choose to use the definitions in terms of real numbers or in terms of angles often depends on how we intend to apply the trigonometric functions. For example, the angle definition is natural when we are considering applications that involve finding sides or angles of triangles. Architects, surveyors, and navigators, for example, use trigonometric functions of angles. The real-number definition is more natural when we work with harmonic motion and with more complicated functions. Thus, physicists and engineers working with electronics, optics, and so on, tend to use trigonometric functions of real numbers. In Appendix 2 we give a more extensive review of both approaches.

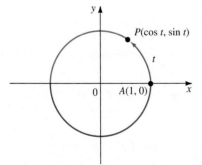

FIGURE 1.71

Sine and Cosine Functions

We begin with a circle of radius 1, called a **unit circle**, centered at the origin, as shown in Figure 1.71. Let A denote the point $(1, 0)$ and let t be any real number. If $t > 0$, measure t units of arc on the circle starting at A and going in a counterclockwise direction. If $t < 0$, measure $|t|$ units from A around the circle and in a clockwise direction. Let P denote the point on the circle arrived at in this way. In Figure 1.71 we have illustrated such a point P for $t > 0$. Since the circumference of the circle is $2\pi \approx 6.28$ ($C = 2\pi r$, and $r = 1$), if t exceeds 2π in absolute value, it is necessary to go around the circle more than once to arrive at P. The values of the **sine** and **cosine** functions at the real number t, abbreviated $\sin t$ and $\cos t$, are defined as follows:

$$\sin t = \text{the } y\text{-coordinate of } P$$

$$\cos t = \text{the } x\text{-coordinate of } P$$

Recall that in Section 1.5 we found the equation of a circle of radius a, centered at the origin, to be $x^2 + y^2 = a^2$. Since we have defined $\cos t$ and $\sin t$ as the x- and y-coordinates of a point P on the unit circle, we can conclude that for all real numbers t, $(\cos t)^2 + (\sin t)^2 = 1$. It is customary to write $\sin^2 t$ to mean $(\sin t)^2$, and similarly to write $\cos^2 t$ (we use the same type of notation for higher powers also). So we have

$$\sin^2 t + \cos^2 t = 1 \qquad (1.4)$$

Equation (1.4) is an example of a **trigonometric identity** since it is identically true for all t. Other identities are derived in Appendix 2.

Up to this point we have made no mention of angles. For example, $\sin 2$ means the sine function evaluated at the *number* 2. Theoretically, at least, we know how to find its value. We measure 2 units from A on the unit circle in a counterclockwise direction, arriving at the point P, as shown in Figure 1.72. Note that P is somewhere in the second quadrant since 2 is more than one-quarter but less than one-half of the entire circumference (approximately 6.28). The value of $\sin 2$ is the y-coordinate of P, as shown. Of course, this method is not a practical way of finding $\sin 2$, since we cannot measure distances around the circle very accurately. In practice you would find $\sin 2$ with high accuracy using a calculator by entering 2 and punching the $\boxed{\sin}$ key (or on some calculators doing this sequence in reverse). You might well ask how anyone is able to program a calculator to give values of the sine and cosine. At this stage we cannot answer this question, but in Chapter 10 you will see a way of evaluating $\sin t$ and $\cos t$ to any desired degree of accuracy, using what are known as infinite series. Calculators and computers do use infinite series (or approximations of them) in evaluating the trigonometric functions.

From Figure 1.71 we *can* obtain exact values of $\sin t$ and $\cos t$ when P falls on one of the coordinate axes. In particular, if $t = 0$, then P coincides with A and has coordinates $(1, 0)$. So we have

$$\sin 0 = 0 \qquad \text{and} \qquad \cos 0 = 1$$

One-quarter of the circumference is $\pi/2$, half is π, and three-quarters is $3\pi/2$. So we can construct Table 1.2, as follows.

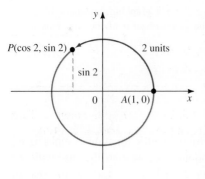

FIGURE 1.72

TABLE 1.2

t	P	$\sin t$	$\cos t$
0	$(1, 0)$	0	1
$\pi/2$	$(0, 1)$	1	0
π	$(-1, 0)$	0	-1
$3\pi/2$	$(0, -1)$	-1	0

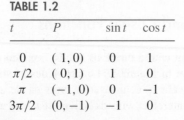

By using certain facts from geometry, we can also find exact values of $\sin t$ and $\cos t$ for $t = \pi/6$, $t = \pi/4$, and $t = \pi/3$. We carry out the details in Appendix 2. The results are shown in Table 1.3 for reference.

TABLE 1.3

t	$\sin t$	$\cos t$
$\pi/6$	$1/2$	$\sqrt{3}/2$
$\pi/4$	$\sqrt{2}/2$	$\sqrt{2}/2$
$\pi/3$	$\sqrt{3}/2$	$1/2$

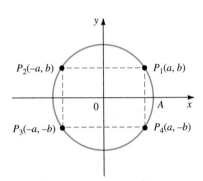

FIGURE 1.73

Now, we can get additional exact values, making use of the symmetries exhibited in Figure 1.73. There the point P_1 is t units from A, where $0 < t < \pi/2$. The points P_2, P_3, and P_4 are, respectively, $\pi - t$, $\pi + t$, and $2\pi - t$ units from A in a counterclockwise direction. As we have labeled P_1, we see that $\sin t = b$ and $\cos t = a$. Thus, from the relationship of the coordinates of P_2, P_3, and P_4 to those of P_1, we can conclude the following:

$$\sin(\pi - t) = \sin t \quad \sin(\pi + t) = -\sin t \quad \sin(2\pi - t) = -\sin t$$
$$\cos(\pi - t) = -\cos t \quad \cos(\pi + t) = -\cos t \quad \cos(2\pi - t) = \cos t \tag{1.5}$$

REMARK ——————————————————————————————

■ The requirement $0 < t < \pi/2$ was made only to simplify the picture in Figure 1.73. Equations 1.5 are true for all values of t.

To illustrate how we can use Equations 1.5, let $t = \pi/3$, for example. Then $\pi - t = \pi - \pi/3 = 2\pi/3$. So from the first and fourth equations in (1.5),

$$\sin \frac{2\pi}{3} = \sin \frac{\pi}{3}$$

$$\cos \frac{2\pi}{3} = -\cos \frac{\pi}{3}$$

and since we know $\sin \pi/3 = \sqrt{3}/2$ and $\cos \pi/3 = 1/2$, we now also know that $\sin 2\pi/3 = \sqrt{3}/2$ and $\cos 2\pi/3 = -1/2$. In the exercises you will be asked to extend Tables 1.2 and 1.3 of known exact values of the sine and cosine using each of Equations 1.5 with $t = \pi/6$, $\pi/4$, and $\pi/3$.

There is another bit of useful information we can get from Figure 1.73. To arrive at P_4, instead of going $2\pi - t$ units in the positive (counterclockwise) direction, we could go t units in the negative (clockwise) direction. That is, we can view the coordinates of P_4 as $(\cos(-t), \sin(-t))$. Comparing these with the coordinates of P_1, which are $(\cos t, \sin t)$, we conclude that

$$\sin(-t) = -\sin t$$

and

$$\cos(-t) = \cos t$$

That is, the **sine function is odd, and the cosine function is even**.

As we have seen, the exact values of $\sin t$ and $\cos t$ can be found for certain special values of t. But for most values of t there is no way to obtain exact values, and you will probably use a calculator to approximate them. (Before scientific calculators were invented, extensive printed tables were used to get values of the trigonometric functions. Such tables still exist, but they are seldom used now.) When using a calculator to get the sine or cosine of a number t, it is essential that the calculator be set in *radian mode* (as opposed to degree mode). To explain why, we will have to say a few words about angle measurement.

In our definition of $\sin t$ and $\cos t$ using Figure 1.71, think of P as initially coinciding with A and then moving around the circle until it comes to its final position t units from A. As it does so, the line segment OP turns through an angle at the center of the circle. The measure of this angle, in radians, is defined to be the number t, that is, the directed length of the arc from A to P. This definition is illustrated in Figure 1.74. For each 1 unit of arc on the circle, there is an angle of 1 radian at the center. In one complete revolution there are 2π units of arc. So there are 2π radians in a complete revolution. Since there also are $360°$ in one revolution, we see that 2π radians = 360 degrees, or

$$\pi \text{ radians} = 180°$$

Dividing by π, we find that 1 radian = $(180/\pi)° \approx 57.3°$.

Now suppose again that you want to find the sine of the real number 2 on a calculator. The calculator is programmed to give trigonometric functions of *angles*, either in degrees, radians, or grads (100 grad = $90°$). Since the sine of an angle of radian measure 2 is the same as the sine of the number 2, we find $\sin 2$ by setting the calculator in radian mode to get the answer:

$$\sin 2 \approx 0.9093$$

One of the most common mistakes students make in using a calculator with trigonometric functions is failing to use the right mode. For example, if your calculator is in degree mode and you ask for $\sin 2$, meaning the sine of 2 radians, the calculator interprets what you want as $\sin 2°$ and gives the answer as 0.034899, which is incorrect.

REMARK ───────────────────────────────────

■ To convert from degree measure to radian measure, or vice versa, you can use the relationships $1° = \pi/180$ radians and 1 radian = $(180/\pi)°$. So, for example,

$$2 \text{ radians} = 2\left(\frac{180}{\pi}\right)° = \left(\frac{360}{\pi}\right)°$$

and you could use this result to get $\sin 2$ even with the calculator in degree mode, but there is no advantage to doing so.

───────────────────────────────────

Let us now return to the definitions of $\sin t$ and $\cos t$ for an arbitrary real number t, using Figure 1.71. Since the circumference of the unit circle is 2π, for any value of t corresponding to the point P, increasing t by 2π units brings us back to the same point. Thus,

$$\sin(t + 2\pi) = \sin t$$

$$\cos(t + 2\pi) = \cos t$$

Any function f for which $f(t + k) = f(t)$ for some constant k and all t in the domain is said to be **periodic**, with **period** k. The smallest positive value of the period k is called the **fundamental period**. So both the sine and cosine functions are periodic with (fundamental) period 2π.

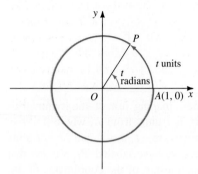

FIGURE 1.74

To obtain the graph of $f(t) = \sin t$, visualize how the y-coordinate of P in Figure 1.71 varies as P moves around the circle. You might imagine yourself at the point P on a ferris wheel of radius 1 unit, and visualize stationary horizontal and vertical axes through the center, in the plane of the ferris wheel. As the wheel rotates, your vertical displacement from the horizontal axis varies like a sine function, and your horizontal distance from the vertical axis varies like a cosine function. Visualizing these displacements, you should be able to get the general shape of the graphs of $f(t) = \sin t$ and $g(t) = \cos t$. Remember that t is a measure of the arc length along the circle. To draw the graphs with greater accuracy, you can use the values in Tables 1.2 and 1.3, and you can get additional points using a calculator. The graphs are shown in Figure 1.75. Notice that the two graphs have the same basic shape but are shifted horizontally from each other. In fact, if the cosine curve were shifted $\pi/2$ units to the right, it would coincide with the sine curve. That is,

$$\cos\left(t - \frac{\pi}{2}\right) = \sin t$$

Similarly,

$$\sin\left(t + \frac{\pi}{2}\right) = \cos t$$

as can be seen by shifting the sine curve $\pi/2$ units to the left.

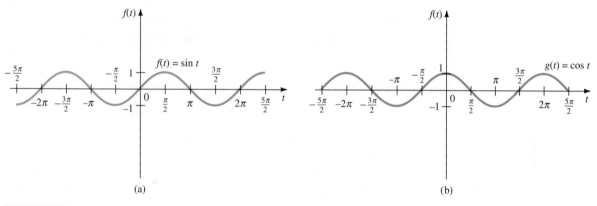

(a)   (b)

FIGURE 1.75

Since the radius of the unit circle is 1, the maximum height attained by either $\sin t$ or $\cos t$ is 1 and the minimum is -1. The maximum displacement from the t-axis is called the **amplitude**. So for both $f(t) = \sin t$ and $g(t) = \cos t$, the amplitude is 1. (See Figure 1.76.)

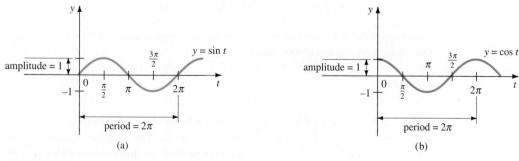

(a)   (b)

FIGURE 1.76

REMARK ──

■ Now that we have the graphs of the sine and cosine, there is no need to continue using the letter t to designate the independent variable. We can use the familiar x for this purpose (even though in the definition using the unit circle, x had a different meaning). In the future, then, we will often write $y = \sin x$ or $y = \cos x$.

──

The next two examples show the effects of various transformations (studied in Section 1.2) on the basic sine and cosine curves.

EXAMPLE 1.31   Discuss and sketch the graph of

$$y = 2 \sin 3x$$

Solution   If we let $f(x) = \sin x$, then the function we want is $2f(3x)$. We know from Section 1.2 that the factor 2 causes a vertical stretching, multiplying each y-coordinate of the basic function by 2. So the amplitude of the new function is 2. The factor 3 causes a horizontal compression by dividing each x-coordinate of the basic curve by 3. For the basic function $f(x) = \sin x$, as x goes through one period from 0 to 2π, the graph is said to complete one cycle. For the new function, one cycle is completed when x goes from 0 to $2\pi/3$. That is, the new period is $2\pi/3$. One cycle of the graph is shown in Figure 1.77.   ■

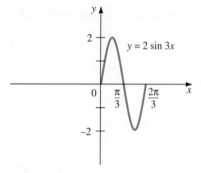

FIGURE 1.77

As Example 1.31 illustrates, for a function of the form

$$y = a \sin bx \qquad (a > 0, \ b > 0)$$

the amplitude is a and the period is $2\pi/b$. The same results hold true for the cosine function

$$y = a \cos bx \qquad (a > 0, \ b > 0)$$

EXAMPLE 1.32   Discuss and sketch the graph of

$$y = \cos(2x - 3) + 1$$

Solution   First we factor out the coefficient of x, getting

$$y = \cos 2 \left(x - \frac{3}{2} \right) + 1$$

This equation represents a transformation of the basic cosine function $g(x) = \cos x$ first by the horizontal compression $g(2x) = \cos 2x$, then by a horizontal translation 3/2 units to the right, and finally by a vertical translation 1 unit upward:

$$g \left[2 \left(x - \frac{3}{2} \right) \right] + 1 = \cos 2 \left(x - \frac{3}{2} \right) + 1$$

The period is $2\pi/2 = \pi \approx 3.14$, and the amplitude is 1. We show one cycle of the graph in Figure 1.78.   ■

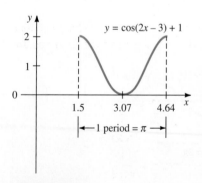

FIGURE 1.78

The sine and cosine functions are especially important in modeling periodic phenomena, such as ocean tides, brain waves, vibrations, and electric currents. A weighted spring set in motion vertically and moved horizontally (to take a multiflash exposure) exhibits a sinusoidal path. Often we are given data whose graph appears to be sinusoidal (i.e., in the shape of a sine curve) and we want to find the form of the sine (or cosine) function that seems to fit the graph. The next example illustrates how to work backward from the graph to the equation.

EXAMPLE 1.33   Find a function that seems to fit the graph in Figure 1.79.

Solution   There is no unique answer, since the curve could be the graph of either a shifted sine curve or a cosine curve. Let us attempt to find a sine function whose graph appears to match the given curve.

To determine the amplitude, think of a horizontal axis 3 units up from the x-axis. The maximum displacement from this horizontal axis is 2 units, since the point $(\frac{3}{2}, 1)$ is 2 units below this axis. So the amplitude is 2. The horizontal distance between the points $(\frac{1}{2}, 3)$ and $(\frac{3}{2}, 1)$ is 1, which is one-fourth of the period. So the period is 4. Before translation (and reflection), the equation would be of the form $y = a \sin bx$, with $a = 2$, and $2\pi/b = 4$. Thus, $b = \pi/2$, and therefore the equation would be

$$y = 2 \sin \frac{\pi}{2} x$$

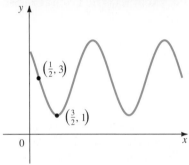

FIGURE 1.79

Next, we note that as we move to the right from $(\frac{1}{2}, 3)$, the curve goes downward, suggesting that our graph is reflected about the x-axis, giving

$$y = -2 \sin \frac{\pi}{2} x$$

Finally, the curve is shifted one-half unit to the right and 3 units upward, giving the result

$$y = -2 \sin \frac{\pi}{2} \left(x - \frac{1}{2} \right) + 3$$

You might wish to fit a cosine curve to the given graph. A correct answer (again, not unique) is

$$y = 2 \cos \frac{\pi}{2} \left(x + \frac{1}{2} \right) + 3 \qquad \blacksquare$$

The Tangent, Cotangent, Secant, and Cosecant Functions

The sine and cosine are the basic two trigonometric functions. The other four are the **tangent** (tan), **cotangent** (cot), **secant** (sec), and **cosecant** (csc) and are defined as follows:

$$\tan x = \frac{\sin x}{\cos x} \qquad \qquad \sec x = \frac{1}{\cos x}$$

$$\cot x = \frac{\cos x}{\sin x} \qquad \qquad \csc x = \frac{1}{\sin x}$$

For $\tan x$ and $\sec x$ the domain includes all real numbers x except those for which $\cos x = 0$, namely $x = \frac{\pi}{2} + n\pi$, $n = 0, \pm 1, \pm 2, \ldots$ For $\cot x$ and $\csc x$ the domain excludes x values for which $\sin x = 0$, namely $x = n\pi$, $n = 0, \pm 1, \pm 2, \ldots$ Each of the four functions has a vertical asymptote at each x value for which the denominator is 0. The graphs are shown in Figures 1.80–1.83.

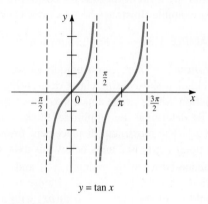

$y = \tan x$

FIGURE 1.80

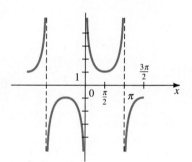

$y = \cot x$

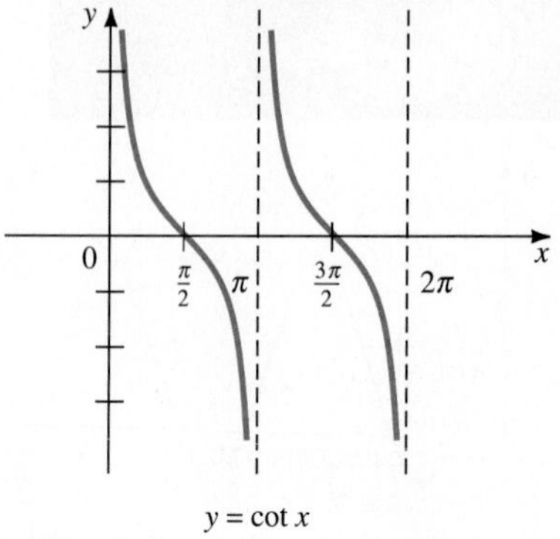

FIGURE 1.81

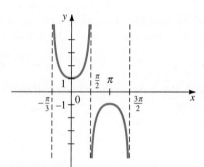

FIGURE 1.82

FIGURE 1.83

REMARK ———————————————————————————————

■ On a calculator you will not find keys for $\cot x$, $\sec x$, or $\csc x$. To obtain these values you can get reciprocals of $\tan x$, $\cos x$, and $\sin x$, respectively. For example, if you want $\sec 3$, which is the same as $1/\cos 3$, you can use the keystroke sequence (if the calculator is in radian mode) 3 $\boxed{\cos}$ $\boxed{1/x}$. The answer displayed will be approximately 1.00137.

INVESTIGATING TRIGONOMETRIC FUNCTIONS USING A COMPUTER ALGEBRA SYSTEM

The names used by Maple, Mathematica, and DERIVE for the trigonometric functions are the standard names given in the previous section.

CAS 9

Graph, on the same set of axes, the curves

$$y = \cos x$$
$$y = \cos 2x$$
$$y = 2\cos 3x + 2$$

for $-10 \le x \le 10$.

Maple:

plot({cos(x),cos(2*x),
  2*cos(3*x)+2},
  x=–10..10,numpoints=500);

Mathematica:*

Plot[{Cos[x],Cos[2*x],
  2*Cos[3*x]+3},
  {x,–10,10}]

DERIVE:

(At the □ symbol, go to the next step.)
a (author) □ cos(x) □ p (plot window)
□ p (plot) □ a (algebra) □ a (author) □
cos(2x) □ p (plot window) □ p (plot) □
a (algebra) □ a (author) □ 2cos(3x)+3
□ p (plot window) □ p (plot)

FIGURE 1.6.1
$y = \cos x$, $y = \cos(2x)$,
$y = 2\cos(3x) + 2$

CAS 10

(a) Graph $y = \tan(x)$, $-\frac{\pi}{2} < x < \frac{\pi}{2}$. (Figure 1.6.2)

(b) Graph $y = \sec(x)$ and $y = \cos(x)$ for $x = -10$ to 10 on the same set of axes. (Figure 1.6.3)

Maple:

(a) plot[tan(x),x=–Pi/2..Pi/2,
y=–5..5);
(b) plot ({sec(x), cos(x)},
x=–10..10,y=–5..5);

Mathematica:

(a) Plot[Tan[x],
  {x,–Pi/2,Pi/2},
  PlotRange–>{–5,5}]
(b) Plot[{Sec[x],Cos[x]},
  {x,–10,10},PlotRange–>{–5,5}]

DERIVE:

(At the □ symbol, go to the next step.)
(a) a (author) □ tan(x) □ p (plot window) □ p (plot)
(b) a (author) □ sec(x) □ p (plot window) □ p (plot) □ a (algebra) □ a (author) □ cos(x) □ p (plot window) □ p (plot) □ z (zoom-out)

---

*Recall that functions in Mathematica begin with capital letters.

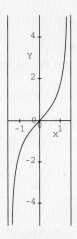

FIGURE 1.6.2

$y = \tan x, \; -\frac{\pi}{2} < x < \frac{\pi}{2}$

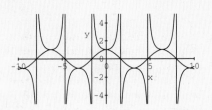

FIGURE 1.6.3

$y = \sec x, \; y = \cos x$

CAS 11

Solve the equation $x \sin x - x^2 = 0$.

Maple:

Eq := x*sin(x)–x^2;
solve(Eq=0,x);

Output:

0, 0

(Maple gives two solutions)
If we rewrite the equation as $x(\sin x - x) = 0$, we see one solution is at $x = 0$ and the other is where $\sin x = x$. A graph of $y = \sin x$ and $y = x$ on the same axes shows the second result is also 0. To see this graph enter

plot({x,sin(x)},x=–10..10,y=–4..4);

Mathematica:[†]

Solve[x*Sin[x]–x^2 == 0,x]

Output: None

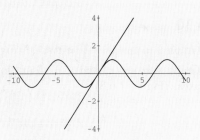

FIGURE 1.6.4

$y = \sin x, \; y = x$

DERIVE:

(At the □ symbol, go to the next step.)
a (author) □ xsin(x)–x^2 □ l (solve) □
[choose expression]

Output:

x=0
sin(x)–x=0

[†]Our version of Mathematica could not solve this equation. DERIVE was able to find $x = 0$ but also could not solve $x - \sin(x) = 0$.

Exercise Set 1.6

1. Construct a table giving values of $\sin x$, $\cos x$, $\tan x$, $\cot x$, $\sec x$, and $\csc x$ at each of the following values of x for which the function is defined: $x = 0$, $\pi/6$, $\pi/4$, $\pi/3$, $\pi/2$, $2\pi/3$, $3\pi/4$, $5\pi/6$, π, $7\pi/6$, $5\pi/4$, $4\pi/3$, $3\pi/2$, $5\pi/3$, $7\pi/4$, and $11\pi/6$. For a value of x for which a function is undefined, write DNE (for "does not exist").

In Exercises 2–13, give the amplitude and the period, and draw the graph.

2. $y = \cos 2x$

3. $y = 2\cos x$

4. $y = 2\sin 3x$

5. $y = 3\sin \frac{x}{2}$

6. $y = 1 - \cos x$

7. $y = 2 + \sin \pi x$

8. $y = 2\cos \pi x - 1$

9. $y = \frac{1}{2}\sin 2x + \frac{\pi}{2}$

10. $y = \sin(3x + \pi)$

11. $y = 3\cos(2x - \frac{\pi}{2})$

12. $y = 2 - \cos(x - 1)$

13. $y = -\frac{3}{2}\sin(2x + 1) + 2$

In Exercises 14–17, find a sine function or a cosine function whose graph matches the given curve.

14.

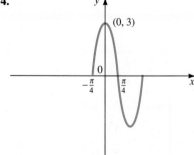

15.

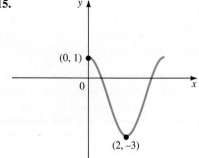

16.

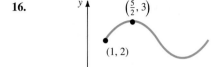

17.

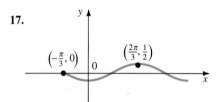

18. Fit a sine curve to the given data.

x	0	0.5	1	1.5	2	−0.5	−1	−1.5	−2
y	1	2.4	3	2.4	0	−0.4	−1	−0.4	0

19. Fit a cosine curve to the given data.

x	0	0.5	1	1.5	2	2.5	3	3.5	4
y	1.5	1.13	1	1.13	1.5	2	2.5	2.87	3

20. Let a, b, and c be positive constants. Describe what happens to the graph of $y = a\cos(bx + c)$ if
(a) a and b are fixed and c is doubled
(b) a and c are fixed and b is doubled
(c) b and c are fixed and a is doubled

21. Find the smallest positive value of x for which

$$f(x) = -2\sin\left(2x + \frac{\pi}{3}\right)$$

(a) has the value 0
(b) attains a maximum value
(c) attains a minimum value

22. Write each of the following as a composition of two functions f and g.
(a) $h(x) = 2\sin(3x + 2)$
(b) $h(x) = \sqrt{\cos x}$
(c) $h(x) = -5\tan^2 x$

Draw the graph of each function in Exercises 23–26.

23. $y = 2 \tan 3x$

24. $y = \frac{1}{2} \sec 2x$

25. $y = \cot \pi x + 1$

26. $y = -3 \csc(2x - \frac{\pi}{3})$

 27. Discuss the family of functions

$$y = a \tan(bx + c)$$

Where are the zeros? What are the asymptotes?

 28. Discuss the family of functions

$$y = a \sec(bx + c)$$

Where do the local minima and maxima occur? What are the asymptotes?

29. Draw the graph of $y = \sin x + \cos x$ by adding the ordinates (y-values) of $y_1 = \sin x$ and $y_2 = \cos x$ for selected values of x. How would you describe the resulting curve? Find its equation as a single sine function.

30. Repeat Exercise 29, with $y = \sqrt{3} \cos x - \sin x$ and $y_1 = \sqrt{3} \cos x$, $y_2 = -\sin x$.

 31. Use a graphing calculator or CAS to obtain the graph in each of the following. Discuss the main features of the graph in each case.

(a) $y = \dfrac{\sin x}{x}$ $(x \neq 0)$ (b) $y = \sin \dfrac{1}{x}$ $(x \neq 0)$

(c) $y = \sin^2 x$

(d) $y = |\sin x|$

(e) $y = \sin x^2$

(f) $y = x \sin \dfrac{1}{x}$ $(x \neq 0)$

 32. Use a graphing calculator or CAS to obtain graphs of

$$y = x \sin \frac{1}{x} \quad \text{and} \quad y = x^2 \sin \frac{1}{x}$$

By zooming in repeatedly on the curves near the origin, observe and discuss the differences in their behaviors near the origin.

33. (a) Find all points where $y = x \sin \dfrac{1}{x}$ and $y = x$ intersect.

(b) Find all points where $y = x \sin \dfrac{1}{x}$ and $y = -x$ intersect.

 34. Use a graphing calculator or CAS to graph $y = \sec x$ and $y = \cos x$ on the same axes, for $-2\pi \leq x \leq 2\pi$. Discuss the relationship between the two graphs.

 35. Repeat Exercise 34 with $y = \csc x$ and $y = \sin x$, for $-\dfrac{3\pi}{2} \leq x \leq \dfrac{5\pi}{2}$.

 36. Use a graphing calculator or CAS to find a third-degree polynomial that "closely" approximates $y = \sin x$ on $[\pi, \pi]$.

Chapter 1 Review Exercises

1. Give the domain of each of the following functions.

(a) $f(x) = \sqrt{3x^2 - 8x - 35}$

(b) $g(x) = \sqrt{\dfrac{x^2 - 1}{x^2 - 2x - 3}}$

2. A circle is inscribed in an equilateral triangle, each side of which has length s. Express the area of the circle as a function of s.

3. Let $f(t) = t^2 - 3$ and $g(t) = \sqrt{1 - t}$. Find $(f \circ g)(t)$ and $(g \circ f)(t)$ and give the domain of each.

4. At temperature x the thermal conductivity of a wall is given by $f(x) = k(1 + ax)$, where k and a are constants. Write the rule for the thermal resistance $g \circ f$, given that $g(u) = T/(Au)$, where T is the thickness of the wall and A is its cross-sectional area.

5. Let

$$f(x) = \begin{cases} 1 & \text{if } x > 2 \\ \dfrac{1}{|x - 2|} & \text{if } 0 \leq x < 2 \\ -x & \text{if } x < 0 \end{cases}$$

Find each of the following, and use your results to aid in drawing the graph of f.

(a) $f(-3)$ (b) $f(0)$

(c) $f(1)$ (d) $f(1.8)$

(e) all vertical and horizontal asymptotes

6. A manufacturer of solar energy cells finds that the cost, in dollars, of producing x cells per month is $30x + 1500$, and he sets the selling price per unit at $120 - 0.1x$ dollars. Find the profit function $P(x)$ and draw its graph. How many units should be sold in order for $P(x)$ to be greatest? What is the maximum profit?

7. Let $g(t) = 1/(t-1)$ and $h(t) = 1/(t+2)$. Find $g \circ h$ and $h \circ g$, and give the domain of each.

8. Let $f(x) = x^2 - 2x - 3$, $g(x) = x^2 + 5x + 4$, and $h = \dfrac{f}{g}$. Find all vertical and horizontal asymptotes to the graph of $y = h(x)$.

9. Let $f(x) = 1 + x^2$ and $g(x) = \tan \dfrac{\pi x}{4}$. Find each of the following:
   (a) $(f - g)(3)$ (b) $\left(\dfrac{f}{g}\right)(-1)$
   (c) $(f \circ g)(x)$

10. Find two functions f and g so that $F = f \circ g$, where
    (a) $F(x) = 1 + \sqrt{(2x-3)^3}$
    (b) $F(x) = \dfrac{3}{|x^2 - 4|}$
    (c) $F(x) = \sin^2 \left(\dfrac{1}{2x - 1}\right)$

11. For the function f whose graph is given, draw the graph of each of the following.
    (a) $y = f(x) + 1$ (b) $y = f(x + 1)$
    (c) $y = 2f(x)$ (d) $y = f(-x)$
    (e) $y = f\left(\dfrac{x}{2}\right)$

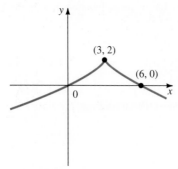

12. Draw the graphs of $f(x) = \sin 2x$ and $g(x) = \cos 3x$, and use your results to obtain the graph of $h(x) = \sin 2x + \cos 3x$. Give the period of each of the functions f, g, and h.

13. Show that f^{-1} exists and find it.
    (a) $f(x) = 1 - \dfrac{1}{\sqrt{x}}$ (b) $f(x) = \dfrac{x+2}{x-3}$

14. Discuss intercepts, symmetry, and asymptotes, and draw the graph of
    $$y = \frac{4x}{1 - x^2}$$

15. Find the zeros of
    $$f(x) = -\frac{1}{4}(x^4 - 2x^3 - 8x^2)$$

and draw its graph. Determine the intervals on which the function f is (a) increasing and (b) decreasing, and the intervals on which the graph is (c) concave up and (d) concave down.

16. Find the formula for the function f whose graph is shown if it is known that f is a polynomial function of degree 4 or less.

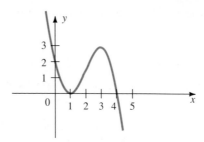

17. Discuss and draw the graph of each of the following:
    (a) $y = (2x - 1)^{2/3}$ (b) $y = \sqrt{\dfrac{1+x}{1-x}}$

18. Find the period and amplitude, and draw the graph.
    (a) $y = 1 - 2\sin(3x - \pi)$
    (b) $y = 3\cos\frac{\pi}{2}(x + 1) - 2$

19. Find a sine or cosine function whose graph matches the given graph.
    (a)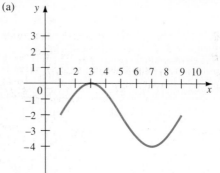

    (b)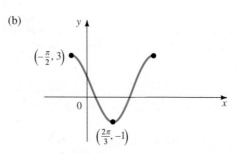

20. Prove the identity $\sec^2 t = 1 + \tan^2 t$, if $t \neq \frac{\pi}{2} + n\pi$, $n = 0, \pm 1, \pm 2, \ldots$ (Hint: Begin with the equation $x^2 + y^2 = 1$ of the unit circle.)

Chapter 1 Concept Quiz

1. Define each of the following. Try to do so in your own words before referring to the text.
   (a) a function from A to B
   (b) an even function
   (c) an odd function
   (d) the composite function $f \circ g$
   (e) a one-to-one function
   (f) the inverse of a function
   (g) a rational function
   (h) $\sin t$, where t is a real number
   (i) $\cos t$, where t is a real number
   (j) $\tan t$, where t is a real number

2. State the following in your own words:
   (a) the vertical line test
   (b) a necessary and sufficient condition for a function to have an inverse
   (c) a procedure for finding the inverse of a function
   (d) the horizontal line test
   (e) a way of determining vertical and horizontal asymptotes for a rational function

3. Describe the relationship between the graph of $y = f(x)$ and the graph of each of the following, where c is a positive constant:
   (a) $y = f(x + c)$ (b) $y = f(x) + c$
   (c) $y = f(cx)$ (d) $y = cf(x)$
   (e) $y = f(x) - c$ (f) $y = f(x - c)$
   (g) $y = -cf(x)$ (h) $y = f(-cx)$

4. Give the period and amplitude of each of the following, and explain how the graph is related to that of $y = \sin x$ or $y = \cos x$:
   (a) $y = 3 \sin 2x$ (b) $y = 2 \cos(\frac{\pi}{2}x)$
   (c) $y = -2 \sin(x + 1)$
   (d) $y = 1 - 3 \cos(2x - 3)$

5. Give the domain and range of $\sin x$, $\cos x$, $\tan x$, $\cot x$, $\sec x$, and $\csc x$.

6. Fill in the blanks:
   (a) The domain of $f \circ g$ is the largest subset of the domain of _____ such that _____ is in the domain of _____.
   (b) The sum of two odd functions is _____ and their product is _____ .
   (c) If f is a function from A to B and g is a function from B to A such that $(f \circ g)(x) = x$ for each x in _____ and $(g \circ f)(x) = x$ for each x in _____, then $g = $ _____.
   (d) The sine of a real number x is the same as the sine of x _____.
   (e) The graph of a polynomial function of degree n can have at most _____ maxima and minima.

7. Which of the following statements are true? Which are false?
   (a) The range of a function may be a single number.
   (b) The domain of the quotient function f/g is the intersection of the domains of f and g.
   (c) If f is 1–1, then $f^{-1}(x) = \dfrac{1}{f(x)}$.
   (d) If $f(x_1) < f(x_2)$ whenever $x_1 < x_2$, then f is an increasing function.
   (e) If $q(a) = 0$, the line $x = a$ is always a vertical asymptote for the rational function $p(x)/q(x)$.

APPLYING FUNCTIONS

1. *Queen Dido's Problem*   Legend tells us that a woman named Dido, from the ancient Phoenician city of Tyre, founded the African city-state of Carthage in 814 A.D. and became its queen. When Dido arrived as a refugee in North Africa she had very little, but bargained for some land by agreeing to pay a fixed price for all the property "lying within the bounds of the hide of an ox." Naturally she paid very little, but Queen Dido's ingenuity and knowledge of geometry made her purchase much better than it sounds. She cut the ox-hide into very narrow strips, sewed the ends together, and laid them out to obtain a much larger piece of property than those making the sale had in mind. Use her reasoning to solve the more general problem.

The importance of this type of problem stems from the need to choose a function $f(x)$ which meets a particular requirement, in this case A_{max}. It differs from those problems in which we analyze functions derived from particular situations or scientific laws. This type of problem is a simple example of what's now known as the calculus of variations. After Queen Dido there was little progress in this area until the work of the Bernoullis nearly 900 years later.

(a)   Given a fixed perimeter (the sewn together strips of ox-hide) determine whether she chose a circle, square, equilateral triangle, or a rectangle in which the length is twice the width. Express the area A of each of the four figures as a function of the perimeter P.

(b)   Suppose she were able to get 500 m of ox-hide strips. What would be the maximum area of her ox-hide territory?

(c)   If the intact ox-hide measured 2.5 m^2, by what factor had Queen Dido's ingenuity increased her intended real estate holdings? (This was probably the end of ox-hide measurements of real estate.) Express it as a percentage. Neglecting the length of material needed to fasten the strips, what was the average width of her original strips?

(d)   The legend goes that Dido's even better idea was to use the strips to form the boundary for a region along the coast (where no hide strips were needed). It was the land enclosed by this boundary that eventually became the city of Carthage. If she used her 500 m of strips to maximize her land this way, what would be the shape of the city limits of Carthage? How large would the original city be?

2.   *A baseball diamond is actually a square with sides 90 feet long.*
A (fast) player runs around the bases at a constant speed of 30 ft/sec, starting from home plate at time $t = 0$. Express his distance D from home plate as a function of time t.

3.   *The intensity of illumination (sound) E at a point P which is d units from a light (sound) source varies directly as the strength I of the source and inversely as the square of the distance from the source.*
(a)   Determine an equation that expresses the intensity of illumination (sound) in terms of the strength of the source and the distance from the source.

(b)   Suppose two light sources are 100 meters apart and one is 10 times brighter than the other. Find an equation for the set of all points $P(x, y)$ at which the amount of light from each will be the same.

(c)   Two sound speakers are in the center of a large room. They are 20 feet apart. Find an equation describing where the person should stand to receive an equal amount of sound from each speaker. Graph the equation.

 4.   The accompanying table gives statistics for marriages and divorces in the United States between the years 1980 and 1991 (*Information Please Almanac Atlas and Yearbook 1993*, Houghton Mifflin). The divorces include annulments and the rate is per 1,000 population.
(a)   Plot the marriage and divorce rates on a graph.

(b) Fit a straight line, as "best" as you can, through the data.

(c) Based on the function determined in part (b), predict the marriage and divorce rate in the year 2000.

(d) From the trends, do you expect the rate of divorce to exceed the rate of marriage? If so, in what year?

Marriages and divorces in the United States

Year	Marriages Number	Marriages Rate	Divorces Number	Divorces Rate
1980	2,406,708	10.6	1,182,000	5.2
1981	2,438,000	10.6	1,219,000	5.3
1982	2,495,000	10.8	1,180,000	5.1
1983	2,444,000	10.5	3,179,000	5.0
1984	2,487,000	10.5	1,155,000	4.9
1985	2,425,000	10.2	1,187,000	5.0
1986	2,400,000	10.0	1,159,000	4.8
1987	2,421,000	9.9	1,157,000	4.8
1988	2,389,000	9.7	1,183,000	4.8
1989	2,404,000	9.7	1,163,000	4.7
1990	2,448,000	9.8	1,175,000	4.7
1991	2,371,000	9.4	1,187,000	4.7

 5. The accompanying table shows the public debt in the United States for selected years from 1800 to 1991 (*Information Please Almanac Atlas and Yearbook 1993*, Houghton Mifflin).

(a) Plot the year versus debt and fit a sixth degree polynomial to the data. Plot the data points together with the polynomial determined.

(b) Estimate the error between the actual values and the values on the curve used to fit the data.

(c) Plot the error together with the data and the curve.

(d) Determine the years in which the public debt increased by more than 50% over the previously recorded amount or increased by more than 50 billion dollars. (Assume a constant rate of increase in years that are not listed in the table.)

The public debt

Year	Amount (in millions)	Year	Amount (in millions)
1800	83	1950	256,097
1860	65	1955	272,807
1865	2,678	1960	284,093
1900	1,263	1965	313,819
1920	24,299	1970	370,094
1925	20,516	1975	533,189
1930	16,185	1980	907,701
1935	28,701	1985	1,823,103
1940	42,968	1990	3,233,313
1945	258,682	1991	3,665,303

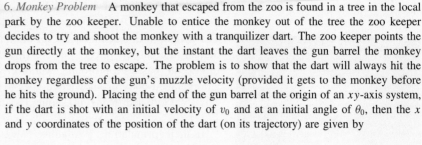

6. *Monkey Problem*   A monkey that escaped from the zoo is found in a tree in the local park by the zoo keeper. Unable to entice the monkey out of the tree the zoo keeper decides to try and shoot the monkey with a tranquilizer dart. The zoo keeper points the gun directly at the monkey, but the instant the dart leaves the gun barrel the monkey drops from the tree to escape. The problem is to show that the dart will always hit the monkey regardless of the gun's muzzle velocity (provided it gets to the monkey before he hits the ground). Placing the end of the gun barrel at the origin of an xy-axis system, if the dart is shot with an initial velocity of v_0 and at an initial angle of θ_0, then the x and y coordinates of the position of the dart (on its trajectory) are given by

$$x = (v_0 \cos \theta_0)t$$

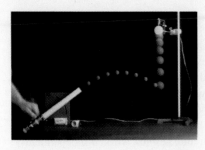

and

$$y = (v_0 \sin \theta_0)t - \frac{1}{2}gt^2$$

(Notice that x and y are both functions of time t.)

Since the monkey drops vertically downward, the x-coordinate of his position at time t is always the distance from the zoo keeper, d, and the y-coordinate is given by the equation for a freely falling body,

$$y = d \tan \theta_0 - \frac{1}{2}gt^2$$

where $d \tan \theta_0$ is the initial position of the monkey and g the constant acceleration due to gravity.

Show the monkey always gets hit with the tranquilizer dart.

7.   Agricultural research on a certain variety of peach has shown that if thirty trees are planted per acre, then each tree will yield an average of 300 pounds of peaches per season. For each additional tree planted per acre, the average yield per tree is reduced by 5 pounds.

(a)   Determine an equation expressing Y, the yield of a 1 acre plot, as a function of the number of trees planted, and draw its graph.

(b)   How many trees should be planted per acre to produce the maximum yield? What is the maximum yield?

The Founders of Calculus

Isaac Newton

A page from Newton's manuscript on optics and color and the prism he used in his experiments.

Gottfried Wilhelm Leibniz

The invention of calculus is one of the landmark achievements in the history of human thought. Its impact on scientific development, and therefore on society in general, has been profound. Major credit for the introduction of calculus as an organized body of knowledge goes to two great geniuses of the seventeenth century, Gottfried Wilhelm Leibniz (1646–1716) and Sir Isaac Newton (1642–1727).

In their formulation of calculus, Newton and Leibniz worked independently, and although the published work of Leibniz preceded that of Newton, it is now known that Newton obtained his results before Leibniz. At the time, their supporters engaged in an infamous and bitter argument over whose work came first. This unfortunate affair forced mathematicians to take sides, and the lack of communication significantly inhibited mathematical progress, especially in England, for close to a hundred years.

Much of the controversy resulted from Newton's reluctance to publish his results, and therefore it was Leibniz who first revealed calculus's general capabilities in 1684. Newton had only hinted at his work on calculus in 1687 and didn't discuss it thoroughly until he released his own papers on the subject in 1704 and, more completely, in 1711. Although they approached their ideas from different directions, Newton and Leibniz achieved essentially identical results. (See the following biographies and *A Personal View* on pages 82 and 137.)

Leibniz was known all over Europe and corresponded with many of the greatest figures of his day, including Newton. It was partly the known correspondence between the two that fanned the flames of controversy later on. Leibniz has been called a "universal genius" for his command of virtually all branches of knowledge at the time, something that would be impossible today. He made lifelong contributions to mathematics, history, general philosophy, as well as "natural philosophy," especially physics and geology. He developed his own versions of the laws of motion, although his reputation as a physicist was overshadowed by that of Newton.

Newton is among the greatest mathematicians in history and is certainly the greatest applied mathematician. Probably no other individual has had so great an impact on the development of science. He is equally famous as a physicist because of his remarkable accomplishments in physics and astronomy. In particular, he made huge contributions to the new science of optics and is credited with the invention of the reflecting telescope. It was his insight into gravitation, which he said was inspired by an apple falling from an undisturbed tree, that led him to the quantitative description of the laws of universal gravitation. Our understanding of the "Newtonian Universe" was not seriously improved on until the work of Einstein in the twentieth century. Newton described his goals in his masterwork, the *Principia Mathematica*: "For I have deigned only to give a mathematical notion of the forces, without considering their physical causes. . . ."

Newton and his supporters never accepted that Leibniz had done his work independently and persisted in believing that Leibniz had plagiarized Newton. Long after Leibniz's death, Newton was still attacking him in his written works. Leibniz was more charitable about Newton, saying: "From the beginning of the world until Newton, what he did is much the better half."

Isaac Newton (1642–1727)

Newton was born prematurely on Christmas day, 1642, into a farming family in Woolsthorpe, England. His father died before Newton was born and his mother despaired of making him into a farmer. At the recommendation of an uncle he was sent off to Trinity College, Cambridge, where his early interests focused on chemistry and alchemy. At Cambridge he also studied Euclid, Descartes, and Galileo, and became familiar with Copernican astronomy.

One of the most famous periods in the history of science and mathematics was the 18 months after June 1665, when Cambridge University was closed because of danger from the bubonic plague. Newton went home to Woolsthorpe, and during this period of sustained isolation and concentration he discovered his version of the calculus. However, it was nearly 20 years before he published any real account of his calculus methods. Newton's reluctance to publish even extended to the *Principia Mathematica*, and he had to be persuaded to do so by his friend Edmond Halley, who was supportive enough to pay for its publication. The development of the laws of motion and gravity stem from this monumental work. Prior to the *Principia*, Newton had written three separate accounts of his calculus methods, but his avoidance of controversy meant that only his friends knew of Newton's progress.

The Trinity library in Cambridge has Newton's copy of a book by Descartes, with "error" noted here and there in Newton's handwriting. He later wrote: "In the beginning of the year 1665, I found the Method of approximating series & the Rule for reducing any [power] of any Binomial into such a series," which we now call the Binomial Theorem. Newton continued his work on series and corresponded with many people about them, including Leibniz. In 1665 he began the study of rates of change, which he called *fluxions*, and this study led to the modern concept of the *derivative*, a term invented by Leibniz. Newton originally used the letters p, q, and r for the derivatives of x, y, and z with respect to time. Later he introduced a more cumbersome notation that was adopted by British mathematicians but was eventually displaced by that of Leibniz.

Newton's practical and political talent led to later public work as Master of the Royal Mint and as President of the Royal Society. He led English science for many years and on his death was buried with great fanfare in Westminster Abbey. On his tomb are engraved the words: "Mortals, congratulate yourselves that so great a man has lived for the honor of the human race." The famous poet Alexander Pope added in tribute:

Nature and Nature's law lay hid in night;
God said: 'Let Newton be', and all was light.

Gottfried Wilhelm Leibniz (1646–1716)

Leibniz was born in Hanover, Germany, in 1646. Although he loved mathematics from an early age, he had little opportunity to study it formally either in high school or at the University of Leipzig, which he entered at the age of 15. Here he studied law, philosophy, and logic and read the work of great scientists and philosophers such as Galileo, Hobbes, and Descartes. However, he was eventually denied his law degree from Leipzig simply because he was so young.

Leibniz not only made mathematical inventions such as the binary system of computation, but he also invented many practical devices. In the 1670s he invented a calculating machine, called the Leibniz Wheel, that could add, subtract, multiply, and divide. He offered it as a labor-saving device and wrote: "It is unworthy of excellent men to lose hours like slaves in the labor of computation." Leibniz exhibited his machine in Paris and specifically traveled to London to demonstrate it to the Royal Society, the most prestigious scientific society of the era. His visit came six years after the Royal Society had published Newton's *Principia*.

In Paris in 1672, Leibniz met the great Dutch physicist and mathematician Christiaan Huygens and persuaded Huygens to tutor him in mathematics. Within a year he was developing insights into the calculus. According to his own writings, it was by reading Pascal's work that he arrived at his first important insight, using an infinite series to find a circular quadrant, in 1673. We now know that Leibniz developed his version of calculus in the years 1673–1676 and published his research in 1684 and 1686. Between 1674 and 1676 he formulated the notation we now use for differential and integral calculus. In particular, the notations for derivative and integral were much better designed than those offered by Newton. By 1684, he had worked out much of differential calculus, which he published in a book entitled *New Method for the Greatest and the Least*. It was this book that brought calculus to the world's attention, before Isaac Newton's earlier, independent invention became widely known.

One of Leibniz's lifelong ambitions was the development of a kind of universal language—a *lingua chacteristica* that would not only incorporate mathematical computation but also serve as a medium for communicating all scientific and mathematical knowledge.

Sadly, and in dramatic contrast to the celebrity of Newton, Leibniz was ultimately overlooked by all and died a lonely man in near total obscurity. In the words of the eighteenth-century French philosopher and scientist Fontenelle, "He was glad to observe the flowering in other people's gardens of plants whose seeds he provided."

Owen Gingerich

Isaac Newton's Principia, published in 1687, has been called the greatest scientific book ever written. Is it true that Newton discovered many of his propositions using calculus, but then revised the proofs to disguise this?

For many years it was commonly believed that Newton systematically recast his proofs into geometrical form, but a detailed study of his manuscripts made in the past few decades shows that what he first got is what you see. In fact, the *Principia* was written very quickly, in just over two years, and part of the reason the book is so difficult to read is that Newton was working at the very frontier, and his initial attack was often clumsy or obscure by later standards. There were probably fewer than a dozen mathematicians in Europe who could really comprehend it.

Does this mean the Principia didn't use calculus?

Newton didn't explicitly use differentials or integrals, but he continuously used the basic concepts of letting quantities become very small to see what happens to a ratio. The book is teeming with theory and applications of the infinitesimal calculus.

What about integration?

There is a very famous proposition in which Newton proves that the gravitational attraction from a sphere acts as if the mass is all concentrated at the center. Sometimes it is said that Newton had to invent the integral calculus to solve this particular problem. He did it by dividing the sphere into very many tiny parts and devising an ingenious method to sum over their individual attractions. In concept it is very like integration.

But Newton had invented the calculus before he wrote the Principia, not so?

Yes and no. He had found the binomial theorem about 20 years earlier, and he saw the relation of that infinite series to differentiation and integration. But the evidence from the vast hoard of manuscripts remaining at (the English) Cambridge shows that the real development of the notation and its application to *Principia*-type problems came in the 1690s.

For centuries there has been a controversy over who invented calculus. Was it Newton or his German rival, Gottfried Wilhelm Leibniz?

Newton put together the basic ideas first, but Leibniz published first, in 1684. The priority battle was very much fueled by Newton later on. As president of the Royal Society, he appointed an impartial committee to investigate whether Leibniz had plagiarized the invention, and the committee handed the credit to Newton. That same hoard of manuscripts shows that Newton himself wrote the committee's report, making draft after draft of it. Today Leibniz shares the credit for the invention.

In fact, Leibniz had the last word, didn't he, because today we use his notation.

Yes, Leibniz was very attentive to questions of form and elegant presentation. The use of dx and dy is his, and also the integral sign, a stylized "s" for summation. The terms *differential* and *integral calculus* also arose from his vocabulary.

What did Newton call his system? Not the calculus?

No, he called it "The Method of Fluxions' and his book of that name was finally published in 1736. His first book on it, *De analysi* ("On analysis by equations with an infinite number of terms"), was composed in 1669 and circulated in manuscript among his friends, but wasn't published until 1711.

It's often said that mathematicians usually get their most creative ideas when they're young, but if Newton was working out his "fluxions" in the 1690s, he was already about 50 years old.

Yes, it's amazing how Newton's mathematical powers were so persistent, even when he wasn't especially interested anymore. In 1696 he became Warden of the Mint, and he was very serious about overseeing the new coinage, when mill marks were introduced for the first time to the edges of the English coins. The European mathematicians including Leibniz had begun to doubt whether this strange man who had abandoned science was really such a genius after all, so the Swiss mathematician Johann Bernoulli sent out a challenge problem. Although it was addressed "to all the mathematicians of Europe," the real quarry was Newton himself. Later [it was] called the brachistochrone problem (see Chapter 9). The Secretary of the Royal Society sent the problem to Newton, who pushed it aside saying, "I have not time for trifles while on the King's business." Newton took the problem home at about four in the afternoon, and though being very tired from the work at the Mint, he didn't sleep until he had it solved. The next morning he sent the full answer back to the Royal Society, showing that the required curve was a cycloid, and his solution was then printed anonymously. When Bernoulli saw the results, he exclaimed, "Ex unge leonem!"— "By the claw, the lion is revealed." There was no question as to the author's identity.

Owen Gingerich is a senior astronomer at the Smithsonian Astrophysical Observatory and Professor of Astronomy and of the History of Science at Harvard University. He was Chair of the Department of History of Science. Professor Gingerich is a leading authority on the astronomer Johannes Kepler and the cosmologist Nicholas Copernicus. In recognition of these studies, an asteroid has been named in Professor Gingerich's honor.

2

LIMITS AND CONTINUITY

Calculus is built on the two fundamental concepts of function and limit. We introduced the function concept in Chapter 1. In this chapter we take up limits of functions. We will see that the idea of limit underlies virtually all the concepts of calculus and that our understanding of the mathematics of change and motion depends on the limit.

We begin by seeing intuitively what it means for a function $f(x)$ to approach a limit L as x approaches some value a. Next, we give some of the important properties of limits of functions. We then introduce the closely related topic of continuity of a function, which is defined in terms of a limit. As will be evident later on, continuity is a key property that assures the validity of certain important calculus operations.

In our definition of limit, the numbers a and L are both understood to be (finite) real numbers. In Section 2.4 we extend the limit notion by permitting either a or L to be replaced by ∞ or $-\infty$. We refer to these extensions as "limits at infinity" and "infinite limits." As we will see, these ideas are closely related to horizontal and vertical asymptotes.

We conclude this chapter by giving precise formulations of the limit and continuity definitions. We will indicate how these formulations enable us to prove certain properties stated earlier. We defer most of the proofs, however, to Supplementary Appendix 5.

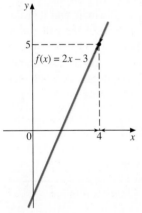

FIGURE 2.1
As $x \to 4$, $f(x) \to 5$

2.1 THE LIMIT OF A FUNCTION

Intuitively, the idea of the limit of a function is fairly simple. For example, let $f(x) = 2x - 3$, and consider what happens to $f(x)$ as x approaches 4. That is, if x is very close to 4, what can we say about $f(x)$? Clearly, the answer is that $f(x)$ is close to 5, as Figure 2.1 shows. We say that *the limit of $2x - 3$ as x approaches 4 is 5* and write this statement in symbols as

$$\lim_{x \to 4}(2x - 3) = 5$$

or equivalently, we may write $(2x - 3) \to 5$ as $x \to 4$. Not all limits are this obvious, as the next example shows.

x	$f(x)$
$(x < 0)$	
−0.01	0.12501954
−0.001	0.12500195
−0.0001	0.12500020
−0.00001	0.12500002

x	$f(x)$
$(x > 0)$	
0.01	0.12498047
0.001	0.12499805
0.0001	0.12499980
0.00001	0.12499998

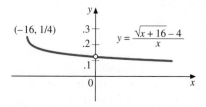

FIGURE 2.2

EXAMPLE 2.1   Use a calculator to evaluate the following function near $x = 0$. Estimate the limit as $x \to 0$.

$$f(x) = \frac{\sqrt{x + 16} - 4}{x}$$

Solution   Note first that 0 is not in the domain of the function, so we cannot substitute $x = 0$. We calculate and tabulate values of $f(x)$ for $x < 0$ and $x > 0$.

It appears that $f(x)$ approaches the limit 0.125, or 1/8, as $x \to 0$. Although our calculations give strong evidence that the limit is 1/8, we have not *proved* this result. We will soon see a way of proving it. The graph in Figure 2.2 seems to confirm our result. Notice the missing point for $x = 0$.   ∎

REMARK ─────────────────────────────

■ The fact that the function f in this example is undefined at $x = 0$ has no significance in terms of the limit of $f(x)$ as x approaches 0. More generally, in finding the limit of a function as $x \to a$, we consider x values close to a, *but not equal to a.*

─────────────────────────────────────

If you continued taking smaller values of x in Example 2.1, you would expect that the function values would come still closer to 0.125. But an unexpected thing happens. If you take $x = 0.00000001$, for example, you will find that the calculator gives the value 0! The problem is that $\sqrt{x + 16}$ with this value of x is so close to 4 that the calculator gives 0 for the numerator of $f(x)$. This result shows that you cannot always rely on numerical answers given by calculators or computer software that is not storing or carrying enough digits. A Computer Algebra System (CAS) can typically carry an arbitrary number of digits and does not have the limitations of a calculator.

We are now ready for the following informal definition.

Definition 2.1
Informal Definition of Limit

The statement

$$\lim_{x \to a} f(x) = L$$

means that $f(x)$ will be as close as we please to the number L for all x values sufficiently close to a, but not equal to a.

We call this definition informal because it contains the imprecise notions "as close as we please" and "sufficiently close." We will give a precise definition in Section 2.5. For many purposes, an intuitive understanding of the limit concept is sufficient, but for proving theorems about limits, we need the precise definition.

In Definition 2.1, how close to a we should make x depends, in general, on how close to L we want $f(x)$ to be. If we first say how close to L we want the $f(x)$ values to be, then we can find how close to a (that is, we can say what "sufficiently close" is) to choose the x values. Normally, the closer to L we require $f(x)$ to be, the closer to a we will need to take x. (We are assuming here that $f(x)$ does approach a limit L.) It should be clear that if the limit exists, that limit is unique. That is, a function cannot have more than one limit as x approaches a. The $f(x)$ values cannot all be arbitrarily close to two different limits for all x values near $x = a$.

The Tangent Problem

As another example of the limit of a function, we consider a special case of a type of problem (called the *tangent problem*) that will motivate our definition of the derivative in the next chapter. So this problem is a preview of things to come. In Figure 2.3 we show the graph of $y = x^2$, together with the fixed point $P(1, 1)$ and the variable point $Q(x, x^2)$, with $x \neq 1$. Our objective is to define the tangent line to the graph at P. First we will find the slope of the line PQ, called a *secant line*. (More generally, a secant line to a curve is a line joining any two points on the curve.) Since P and Q are distinct points, we can compute the slope of the secant line PQ. Then we will allow Q to approach P by letting x come arbitrarily close to 1, but not equal to 1. This latter qualification deserves emphasis, for if we permitted x to be equal to 1, then Q would coincide with P, so there would not be a well-defined line joining P and Q and we would not see how to compute a slope. If x is very close to 1, then Q is close to P, but as the magnified view in Figure 2.4 shows, the two points are distinct, and so a unique secant line through them exists.

In Figure 2.5 we show several secant lines PQ for points Q closer and closer to P. Notice how these secant lines seem to be approaching some limiting position (the tangent line). For any $x \neq 1$, the slope of the secant line PQ for our example is

$$m_{\text{sec}} = \frac{\text{change in } y}{\text{change in } x} = \frac{x^2 - 1}{x - 1}$$

It seems reasonable to *define* the tangent line at P to be that line through P whose slope is the limit of m_{sec} as x approaches 1, provided this limit exists. We show this tangent line in Figure 2.5. According to this definition, the slope of the tangent line is given by

$$m_{\text{tan}} = \lim_{x \to 1} m_{\text{sec}} = \lim_{x \to 1} \frac{x^2 - 1}{x - 1}$$

The problem, then, is to find this limiting value. We examine this limit in the next example.

EXAMPLE 2.2   The slope of the tangent line to the graph of $y = x^2$ at the point $(1, 1)$ is

$$m_{\text{tan}} = \lim_{x \to 1} \frac{x^2 - 1}{x - 1}$$

Use a calculator to estimate this limit, and then find the equation of the tangent line, based on the estimate.

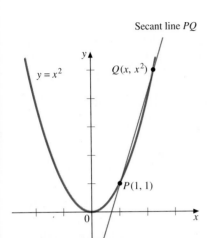

FIGURE 2.3

Secant line PQ

$y = x^2$

$Q(x, x^2)$

$P(1, 1)$

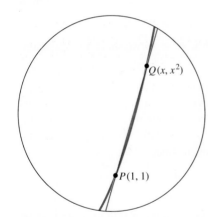

FIGURE 2.4

$Q(x, x^2)$

$P(1, 1)$

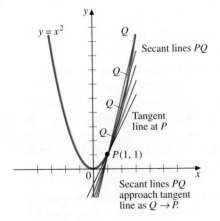

FIGURE 2.5

$y = x^2$

Q

Secant lines PQ

Q

Q

Tangent line at P

Q

$P(1, 1)$

Secant lines PQ approach tangent line as $Q \to P$.

$(x > 1)$	
x	m_{sec}
1.1	2.1
1.01	2.01
1.001	2.001
1.0001	2.0001

$(x < 1)$	
x	m_{sec}
0.9	1.9
0.99	1.99
0.999	1.999
0.9999	1.9999

Solution   As in the preceding discussion, we know that $m_{sec} = (x^2 - 1)/(x - 1)$, and we want to find what these slopes approach as x approaches 1. We cannot simply substitute $x = 1$, since we would get 0/0, which is undefined. In the margin we show tables giving values obtained from a calculator for m_{sec}, taking x slightly greater than 1 in one case and slightly less than 1 in the other. These tables give convincing evidence that the limit is 2. In the next section we will discuss an algebraic way of showing that the limit actually is 2. (Can you see a way?)

Let us assume that $m_{tan} = 2$ based on our evidence. Then we can use the *point-slope form* for the equation of a line (Supplementary Appendix 1):

$$y - y_1 = m(x - x_1)$$

Taking (x_1, y_1) as (1, 1) and m as $m_{tan} = 2$, we obtain $y - 1 = 2(x - 1)$, or

$$y = 2x - 1$$

as the equation of the tangent line.   ∎

The limit in the next example is an important one that we will have occasion to use later. Here again we rely on an estimate obtained by using a calculator. (A CAS or spreadsheet also could be used.) In Chapter 3 we will show a geometric way of verifying our result.

EXAMPLE 2.3   Use a calculator, a CAS, or a spreadsheet to conjecture whether

$$\lim_{x \to 0} \frac{\sin x}{x}$$

exists, and if so, what its value is.

Solution   Observe first that $(\sin x)/x$ is even, since

$$\frac{\sin(-x)}{-x} = \frac{-\sin x}{-x} = \frac{\sin x}{x}$$

So we need only consider $x > 0$, since the same value of $(\sin x)/x$ will occur for $-x$ as for x.

We show a table in the margin for selected values of x near 0. The table gives convincing evidence that

x	$(\sin x)/x$
1.0	0.84147
0.5	0.95885
0.1	0.99833
0.05	0.99958
0.01	0.99998

$$\lim_{x \to 0} \frac{\sin x}{x} = 1$$

The conjecture that the limit is 1 can be further supported by the graph of $f(x) = (\sin x)/x$ in Figure 2.6, generated by a computer algebra system. We have added a small circle at the point (0, 1) to emphasize that $x = 0$ is not in the domain of this function.   ∎

We emphasize that when we ask for $\lim_{x \to a} f(x)$, we want to know what value $f(x)$ *approaches* as x approaches a. We are not concerned with whether $f(x)$ actually reaches its limit, nor are we concerned with what happens when $x = a$. In fact, $f(a)$ may not even be defined, as the tangent problem illustrated. Even if $f(a)$ were defined, it may not equal the limit L that is approached by $f(x)$ as x approaches a. The next example illustrates this situation.

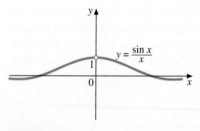

FIGURE 2.6

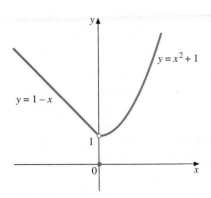

FIGURE 2.7

	$(x > 0)$		$(x < 0)$
x	$1 + x^2$	x	$1 - x$
0.5	1.25	−0.5	1.5
0.1	1.01	−0.1	1.1
0.01	1.0001	−0.01	1.01
0.001	1.000001	−0.001	1.001

EXAMPLE 2.4   Let

$$f(x) = \begin{cases} x^2 + 1 & \text{if } x > 0 \\ 0 & \text{if } x = 0 \\ 1 - x & \text{if } x < 0 \end{cases}$$

Determine whether $\lim_{x \to 0} f(x)$ exists, and if so, find its value.

Solution   We show the graph of f in Figure 2.7. When x is positive, $f(x) = x^2 + 1$, so if x is close enough to 0, we can make $f(x)$ as close to 1 as we please. Similarly, if x is negative, $f(x) = 1 - x$, which we can also make as close to 1 as we want by choosing x sufficiently small in absolute value. The tables in the margin give further evidence that $f(x)$ comes arbitrarily close to 1 as x approaches 0 through either positive or negative values. Although we have not formally proved this result, it is intuitively evident that

$$\lim_{x \to 0} f(x) = 1$$

We were given that $f(0) = 0$, but this fact has nothing to do with the limit. ∎

One-Sided Limits

In Example 2.4 we found the limit by considering first the value approached by $f(x)$ as x approached 0 through positive values and then the value approached by $f(x)$ as x approached 0 through negative values. In each case, we found the same value, 1, and so we concluded that the limit of $f(x)$ as x approached 0 was 1. We now give names and symbols to "one-sided" limits such as those in this example. If $f(x)$ is as close to a number L_1 as we please for all $x > a$ and sufficiently close to a, then we say that L_1 is the **right-hand limit** of $f(x)$ as x approaches a, and we indicate such a limit symbolically by

$$\lim_{x \to a^+} f(x) = L_1$$

The **left-hand limit** is defined analogously. If its value is L_2, we write

$$\lim_{x \to a^-} f(x) = L_2$$

In Example 2.4 both L_1 and L_2 were equal to the same value, 1, and so we concluded that $\lim_{x \to 0} f(x) = 1$. If $L_1 \neq L_2$, as is true for the function in Figure 2.8, then there is no limit, since $f(x)$ cannot be made arbitrarily close to any *one* value for all x values close to a. We can summarize this discussion as follows.

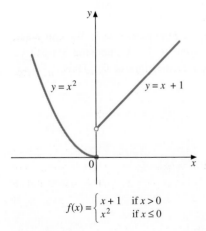

$$f(x) = \begin{cases} x + 1 & \text{if } x > 0 \\ x^2 & \text{if } x \le 0 \end{cases}$$

FIGURE 2.8
$\lim_{x \to 0^+} f(x) = 1$ and $\lim_{x \to 0^-} f(x) = 0$, so $\lim_{x \to 0} f(x)$ does not exist.

$$\lim_{x \to a} f(x) = L \text{ if and only if both } \lim_{x \to a^+} f(x) = L \text{ and } \lim_{x \to a^-} f(x) = L$$

That is, the (unrestricted) limit exists if and only if the right-hand and left-hand limits both exist and are equal.

REMARK
■ The expression $x \to a^+$ means that x approaches a from the right, and it has nothing to do with the sign of a. For example, $x \to -2^+$ means x is approaching -2 through values to the right of -2 (such as $-1.9, -1.99, -1.999, \ldots$). Similarly, the expression $x \to a^-$ means that x approaches a from the left.

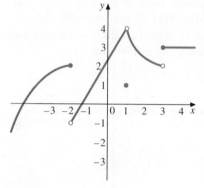

FIGURE 2.9

EXAMPLE 2.5   From the graph in Figure 2.9 determine $\lim_{x \to a^-} f(x)$ and $\lim_{x \to a^+} f(x)$, if they exist, for (a) $a = -2$, (b) $a = 1$, and (c) $a = 3$. In each case, state whether $\lim_{x \to a} f(x)$ exists.

Solution

(a) $\lim_{x \to -2^-} f(x) = 2$ and $\lim_{x \to -2^+} f(x) = -1$
   Thus, $\lim_{x \to -2} f(x)$ does not exist.

(b) $\lim_{x \to 1^-} f(x) = 4$ and $\lim_{x \to 1^+} f(x) = 4$
   Thus, $\lim_{x \to 1} f(x) = 4$.

(c) $\lim_{x \to 3^-} f(x) = 2$ and $\lim_{x \to 3^+} f(x) = 3$
   Thus, $\lim_{x \to 3} f(x)$ does not exist.   ■

EXAMPLE 2.6   Let

$$f(x) = \frac{|x|}{x}$$

Find $\lim_{x \to 0^+} f(x)$ and $\lim_{x \to 0^-} f(x)$. Does $\lim_{x \to 0} f(x)$ exist?

Solution   When $x > 0$, we have $|x| = x$, so $f(x) = x/x = 1$. We can make $f(x)$ as close to 1 as we please (namely, *equal* to 1) by choosing x positive and sufficiently close to 0 (namely, *anywhere* to the right of 0). Thus,

$$\lim_{x \to 0^+} \frac{|x|}{x} = 1$$

If $x < 0$, then $|x| = -x$, so

$$f(x) = \frac{|x|}{x} = \frac{-x}{x} = -1$$

Reasoning as above, we conclude that

$$\lim_{x \to 0^-} \frac{|x|}{x} = -1$$

Since the one-sided limits are not equal, we see that $\lim_{x \to 0} |x|/x$ does not exist. Note that $x = 0$ is not in the domain of f, but this fact has nothing to do with our investigation of *limits* as x approaches 0. We show the graph of f in Figure 2.10.   ■

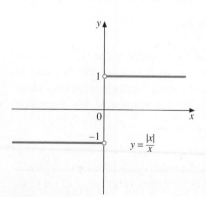

FIGURE 2.10

Other Functions Whose Limits Fail to Exist

In Examples 2.5 and 2.6 we saw some functions that failed to have a limit as x approached some number, because the one-sided limits were unequal. In the next example, we show some other ways that a function may fail to have a limit.

EXAMPLE 2.7   Show that the limit of each of the following functions fails to exist as $x \to 0$. Draw the graphs and explain in what way, or ways, Definition 2.1 fails.

(a) $f(x) = \dfrac{1}{x^2}$ (b) $g(x) = \sin\dfrac{1}{x}$

Solution   We show the graphs of $y = f(x)$ and $y = g(x)$ in Figure 2.11.

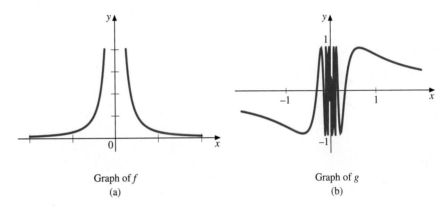

Graph of f Graph of g
   (a)   (b)

FIGURE 2.11

(a) The graph of f has a vertical asymptote at $x = 0$. That is, as $x \to 0$, $f(x) \to \infty$. But ∞ is not a real number, so $\lim_{x \to 0} f(x)$ does not exist. As we will show in Section 2.4, we often write in this case $\lim_{x \to 0} f(x) = \infty$. Saying the limit is "infinity" means that the function becomes arbitrarily large.

(b) As $x \to 0$, the function g takes on every value between -1 and 1 infinitely many times since $\sin(1/x)$ oscillates between -1 and 1 infinitely many times as $x \to 0$. In fact, $\sin(1/x) = 1$ if $1/x = \pi/2 + 2n\pi$, for any integer n, that is, if

$$x = \frac{2}{(4n+1)\pi}$$

Similarly, $\sin(1/x) = -1$ if

$$x = \frac{2}{(4n+3)\pi}$$

You can see that by choosing n large enough, each of these values of x can be made as close to 0 as we please. So the values of $g(x)$ cannot be made arbitrarily close to any fixed value for all x values sufficiently close to 0.   ∎

LIMITS USING COMPUTER ALGEBRA SYSTEMS

 ### CAS 12

Compute each of the following limits using Maple, Mathematica, or DERIVE.

1. (a) $\lim_{x \to -2} (2x - 1)$ **2.** (b) $\lim_{x \to \infty} \left(\dfrac{x^3 - 3x^2 + 1000x + 25000}{3x^3 + 25000x + 10} \right)$

Maple:

limit(2*x–1,x=–2);

Output: −5, as you would expect.

Mathematica:

Limit[2*x–1,x–>–2]

Output: −5

DERIVE:

(At the □ symbol, go to the next step.)
a (author) □ 2x−1 □ c (calculus) □ l
(limit) □ [choose expression] □ x (variable) □ −2 (limit point)

Output: $\lim_{x \to -2} 2x - 1$

s (simplify) □
[choose expression]

Output: −5

Maple:

h := x–>(x^3–3*x^2+1000*x+25000)/
   (3*x^3+25000*x+10);
limit(h(x), x=infinity);

Output: $\frac{1}{3}$

Mathematica:*

g[x_]=x^3–3*x^2+1000*x+25000)/
   (3*x^3+25000*x+10)
Limit[g[x],x–>Infinity]

Output: $\frac{1}{3}$

DERIVE:

(At the □ symbol, go to the next step.)
a (author) □ x^3−3x^2+1000x+25000)/
□ (3x^3+25000x+10) □ c (calculus) □
l (limit) □ [choose expression] □ x (variable) □ inf (limit point) □ s (simplify)
□ [choose expression]

Output: $\frac{1}{3}$

 ### CAS 13

Determine

$$\lim_{x \to 1} \frac{1}{(x - 1)}$$

*We have defined h and g as functions and not just equations in Maple and Mathematica, respectively. The functions h and g can then be evaluated at any expression for x. In Maple be sure to use the minus sign followed by the greater than sign, as in $h:=x->$ (x goes to the expression on the right). In Mathematica use the underscore character following the variable, as in $g[x_]$.

In Maple, we have:

g :=x -> 1/(x–1);
limit(g(x),x=1);

Output: undefined
This is the correct answer. Let's check the left- and right-hand limits. In Maple we have:

limit(g(x),x=1,right);

Output: infinity

limit(g(x),x=1,left);

Output: -infinity
The plot of the function makes this clear:

plot(g(x),x=–1..2);

In Mathematica, we proceed in a similar manner:

g[x_] = 1/(x–1)
Limit[g[x],x–>1]

Here we need to be cautious, since the output from our version of Mathematica is:

infinity

This is not the correct answer. If we check the left- and right-hand limits in Mathematica, we get the same answer as we did using Maple.

Limit[g[x],x–>1,
        Direction–> –1]
Limit[g[x],x–>1,Direction–>1]

The first command gives an output of infinity and the second, an output of − infinity. The −1 means from the right and 1 means from the left.

In DERIVE, we get a result similar to the one in Mathematica:

(At the □ symbol, go to the next step.)
a (author) □ 1/(x−1) □ c (calculus) □ 1 (limit) □ [choose expression] □ x (variable) □ 1 (limit point) □ s (simplify) □ [choose expression]

Output: $\frac{1}{0}$

Now proceed to compute the one-sided limits: c (calculus) □ 1 (limit) □ [choose expression] □ x (variable) □ 1 (limit point) □ left (left-hand limit) □ s (simplify) □ [choose expression]

Output: $-\infty$

c (calculus) □ 1 (limit) □ [choose expression] □ x (variable) □ 1 (limit point) □ right (right-hand point) □ s (simplify) □ [choose expression]

Output: ∞

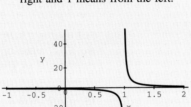

FIGURE 2.1.1
$$y = f(x) = \frac{1}{1-x}$$

Exercise Set 2.1

In Exercises 1–6, sketch the graph of the function and use it to determine the given limit.

1. $f(x) = 5 - x$; $\displaystyle\lim_{x \to 2} f(x)$

2. $g(x) = x^2 + 2x + 1$; $\displaystyle\lim_{x \to 1} g(x)$

3. $f(t) = \begin{cases} t^2 + 1 & \text{if } t \geq 0 \\ -2 & \text{if } t < 0 \end{cases}$; $\displaystyle\lim_{t \to 0} f(t)$

4. $h(x) = 3$; $\displaystyle\lim_{x \to 10} h(x)$

5. $f(x) = \begin{cases} 2x + 2 & \text{if } x \neq 1 \\ 0 & \text{if } x = 1 \end{cases}$; $\displaystyle\lim_{x \to 1} f(x)$

6. $g(t) = \dfrac{1}{t - 5}$; $\displaystyle\lim_{t \to 5} g(t)$

In Exercises 7–12, use the graph to determine $\lim_{x\to 2^+} f(x)$, $\lim_{x\to 2^-} f(x)$, *and* $\lim_{x\to 2} f(x)$, *or explain why the limit fails to exist.*

7.

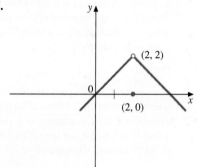

8.

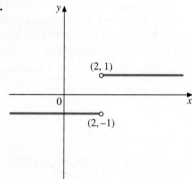

9.

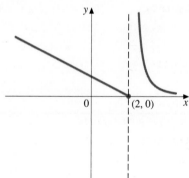

10.

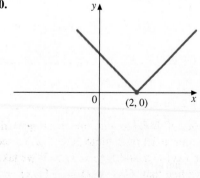

11.

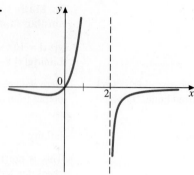

12.

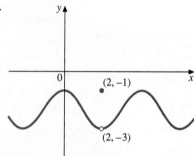

In Exercises 13–18, find $\lim_{x\to a^+} f(x)$ *and* $\lim_{x\to a^-} f(x)$ *for the given values of a and* $f(x)$. *State whether* $\lim_{x\to a} f(x)$ *exists, and if so, give its value.*

13. $a = 2;\ f(x) = \begin{cases} x^2 & \text{if } x \ge 2 \\ 6 - x & \text{if } x < 2 \end{cases}$

14. $a = 0;\ f(x) = \begin{cases} 1 & \text{if } x > 0 \\ 0 & \text{if } x = 0 \\ -1 & \text{if } x < 0 \end{cases}$

15. $a = 1;\ f(x) = \begin{cases} 2x & \text{if } x > 1 \\ 0 & \text{if } x = 1 \\ x^2 + 1 & \text{if } x < 1 \end{cases}$

16. $a = 0;\ f(x) = \dfrac{x}{\sqrt{x^2}}$

17. $a = 1;\ f(x) = \dfrac{x^2 - x}{|x - 1|}$

18. $a = 2;\ f(x) = \dfrac{|2 - x|}{x^2 - 4}$

In Exercises 19–28, use a calculator, CAS, or spreadsheet to make a table of values, and conjecture the limit based on the table.

19. $\displaystyle\lim_{x\to -2} \frac{x^3 + 8}{x + 2}$

20. $\displaystyle\lim_{x\to 2} \frac{x^2 - x - 6}{x^2 - 4}$

21. $\displaystyle\lim_{t\to 1} \frac{3 - 2t - t^2}{2t^2 - t - 1}$

22. $\displaystyle\lim_{x\to 4} \frac{\sqrt{x} - 2}{x - 4}$

23. $\lim\limits_{x \to 4} \dfrac{\sqrt{2x+1}-3}{x-4}$

24. $\lim\limits_{h \to 0} \dfrac{\sqrt{(4+h)^2+9}-5}{h}$

25. $\lim\limits_{x \to 0} \dfrac{1-\cos x}{x^2}$

26. $\lim\limits_{t \to 0} \dfrac{t^3}{t-\sin t}$

27. $\lim\limits_{x \to (\pi/2)^-} (\tan x)^{\cos x}$

28. $\lim\limits_{x \to 1} x^{1/(1-x)}$

In Exercises 29–32, use a calculator, CAS, or spreadsheet to estimate the limits. Then, after estimating that $\lim_{x \to a} f(x) = L$, find an interval centered at a such that for all x in this interval, $f(x)$ is less than 0.01 unit from L. Repeat, replacing 0.01 with 0.001.

29. $\lim\limits_{x \to 2} (3x-1)$

30. $\lim\limits_{x \to 2} \dfrac{x^2+1}{x-1}$

31. $\lim\limits_{x \to 0} x \sin \dfrac{1}{x}$

32. $\lim\limits_{x \to 0} \dfrac{\sin 5x}{x}$

2.2 LIMIT PROPERTIES AND TECHNIQUES FOR FINDING LIMITS

In Section 2.1, we made implicit use of certain properties of limits involving sums, differences, products, and quotients. We now state these properties explicitly. Each of the properties seems intuitively evident, and they can be proved formally. We will prove some of them in Section 2.5 and the others in Appendix 5.

Properties of Limits

If $\lim_{x \to a} f(x)$ and $\lim_{x \to a} g(x)$ both exist, then the following properties hold true:

1. $\lim\limits_{x \to a} cf(x) = c \lim\limits_{x \to a} f(x)$ for any constant c.
The limit of a constant times a function is the constant times the limit of the function.

2. $\lim\limits_{x \to a} [f(x) + g(x)] = \lim\limits_{x \to a} f(x) + \lim\limits_{x \to a} g(x)$
The limit of a sum is the sum of the limits.

3. $\lim\limits_{x \to a} [f(x) - g(x)] = \lim\limits_{x \to a} f(x) - \lim\limits_{x \to a} g(x)$
The limit of a difference is the difference of the limits.

4. $\lim\limits_{x \to a} [f(x) \cdot g(x)] = [\lim\limits_{x \to a} f(x)] \cdot [\lim\limits_{x \to a} g(x)]$
The limit of a product is the product of the limits.

5. $\lim\limits_{x \to a} \dfrac{f(x)}{g(x)} = \dfrac{\lim\limits_{x \to a} f(x)}{\lim\limits_{x \to a} g(x)}$, provided $\lim\limits_{x \to a} g(x) \neq 0$
The limit of a quotient is the quotient of the limits, provided the limit of the denominator is not 0.

REMARK

■ Let us emphasize that for these properties of limits to hold true, each individual limit on the right side of the equation must exist (as a finite number). To see a case where this condition fails, consider $f(x) = x$ and $g(x) = 1/x$. If we take $a = 0$ and attempt to apply Property 4, we find that $f(x) \cdot g(x) = x \cdot (1/x) = 1$

for all $x \neq 0$. So $\lim_{x \to 0}[f(x) \cdot g(x)] = 1$. But the right-hand side is

$$\left[\lim_{x \to 0} x\right] \cdot \left[\lim_{x \to 0} \frac{1}{x}\right]$$

and the second limit does not exist, so the product is meaningless. Thus, the equality in Property 4 does not hold true in this instance.

If we apply Property 4 to $[f(x)]^2 = [f(x)] \cdot [f(x)]$, we get

$$\lim_{x \to a}[f(x)]^2 = \left[\lim_{x \to a} f(x)\right] \cdot \left[\lim_{x \to a} f(x)\right] = \left[\lim_{x \to a} f(x)\right]^2$$

Similarly,

$$\lim_{x \to a}[f(x)]^3 = \left[\lim_{x \to a} f(x)\right]^2 \cdot \left[\lim_{x \to a} f(x)\right] = \left[\lim_{x \to a} f(x)\right]^3$$

and in general, for any positive integer n,

$$\lim_{x \to a}[f(x)]^n = \left[\lim_{x \to a} f(x)\right]^n$$

(The general result can be proved by the technique of mathematical induction, discussed in Supplementary Appendix 4.)

An analogous result holds for limits of roots. We list the following two results as further properties of limits.

Further Properties of Limits

If $\lim_{x \to a} f(x)$ exists and n is a positive integer, then the following properties hold true:

6. $\lim_{x \to a}[f(x)]^n = [\lim_{x \to a} f(x)]^n$
7. $\lim_{x \to a} \sqrt[n]{f(x)} = \sqrt[n]{\lim_{x \to a} f(x)}$, provided that when n is even, $\lim_{x \to a} f(x) > 0$.

In addition to these general properties, we have made use of the following special limits.

Two Special Limits

1. If c is a constant and $f(x) = c$ for all x, then $\lim_{x \to a} f(x) = c$ for every real number a. More briefly,

$$\lim_{x \to a} c = c \tag{2.1}$$

2. If $f(x) = x$, then $\lim_{x \to a} f(x) = a$ for every real number a. More briefly,

$$\lim_{x \to a} x = a \tag{2.2}$$

The graphs of these two special functions, shown in Figure 2.12, illustrate their limits as $x \to a$.

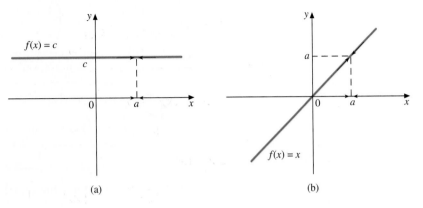

(a)   (b)

FIGURE 2.12

EXAMPLE 2.8   Show how the limit properties are used in finding each of the following limits:

(a) $\lim_{x \to 2}(3x^2 - 4x + 5) = 9$ (b) $\lim_{x \to -1}\dfrac{x^3 + 2x + 4}{3x^2 - 2} = 1$

Solution

(a) $\lim_{x \to 2}(3x^2 - 4x + 5) = \lim_{x \to 2}(3x^2) - \lim_{x \to 2}(4x) + \lim_{x \to 2}5$ Properties 2 and 3

$\qquad\qquad\qquad\quad = 3\lim_{x \to 2}x^2 - 4\lim_{x \to 2}x + 5$ Property 1 and Special Limit 2.1

$\qquad\qquad\qquad\quad = 3(2)^2 - 4(2) + 5$ Property 6 and Special Limit 2.2

$\qquad\qquad\qquad\quad = 12 - 8 + 5 = 9$

(b) $\lim_{x \to -1}\dfrac{x^3 + 2x + 4}{3x^2 - 2} = \dfrac{\lim_{x \to -1}(x^3 + 2x + 4)}{\lim_{x \to -1}(3x^2 - 2)}$ Property 5

$\qquad\qquad\qquad = \dfrac{\left(\lim_{x \to -1}x^3\right) + \left[\lim_{x \to -1}(2x)\right] + \left(\lim_{x \to -1}4\right)}{\lim_{x \to -1}(3x^2) - \lim_{x \to -1}(2)}$ Properties 2 and 3

$\qquad\qquad\qquad = \dfrac{(-1)^3 + 2(-1) + 4}{3(-1)^2 - 2}$ Properties 1 and 6 and Special Limits 2.1 and 2.2

$\qquad\qquad\qquad = \dfrac{-1 - 2 + 4}{3 - 2} = \dfrac{1}{1} = 1$ ■

EXAMPLE 2.9   Show how the limit properties are used in finding the limit

$$\lim_{x \to 0}\sqrt{\dfrac{1 + \sin^2 x}{\cos x}} = 1$$

Solution

$$\lim_{x \to 0} \sqrt{\frac{1 + \sin^2 x}{\cos x}} = \sqrt{\lim_{x \to 0}\left(\frac{1 + \sin^2 x}{\cos x}\right)} \qquad \text{Property 7}$$

$$= \sqrt{\frac{\lim\limits_{x \to 0}(1 + \sin^2 x)}{\lim\limits_{x \to 0}\cos x}} \qquad \text{Property 5}$$

$$= \sqrt{\frac{\lim\limits_{x \to 0}1 + \lim\limits_{x \to 0}\sin^2 x}{1}} \qquad \text{Property 2}$$

$$= \sqrt{1 + \left(\lim_{x \to 0}\sin x\right)^2} \qquad \text{Special Limit 2.1 and Property 6}$$

$$= \sqrt{1 + (0)^2} = 1 \qquad \blacksquare$$

Property 6, when applied to the function $f(x) = x$, gives

$$\lim_{x \to a} x^n = a^n$$

and this result, together with Property 1, gives

$$\lim_{x \to a} cx^n = ca^n$$

Since polynomials consist of sums or differences of terms of the form cx^n, we can apply Properties 2 and 3 repeatedly to obtain the following result, which we state as a theorem.

THEOREM 2.1

If p is any polynomial function, then for every real number a,

$$\lim_{x \to a} p(x) = p(a)$$

For example, if $p(x) = x^2 - 5x + 7$, then

$$\lim_{x \to 2} p(x) = 2^2 - 5(2) + 7 = 1$$

Theorem 2.1 says that you can find the limit of a polynomial function as x approaches a by direct substitution of $x = a$ into the function.

Now we can use Property 5 to get another theorem.

THEOREM 2.2

If f is a rational function

$$f(x) = \frac{p(x)}{q(x)}$$

where $p(x)$ and $q(x)$ are polynomials, then

$$\lim_{x \to a} f(x) = \frac{p(a)}{q(a)}$$

for all real numbers a for which $q(a) \neq 0$.

For example, if

$$f(x) = \frac{2x^2 - 3x + 4}{x^2 - 4}$$

then

$$\lim_{x \to 1} f(x) = \frac{2 - 3 + 4}{1 - 4} = \frac{3}{-3} = -1$$

Be careful, though, to observe the restriction that $q(a) \neq 0$ before applying Theorem 2.2. If $q(a)$ does equal 0, then, as we saw in Section 2.1, the limit of the quotient *may* exist if $p(a)$ also equals 0. In the next five examples we illustrate some algebraic techniques for finding limits when both numerator and denominator are zero at the point in question.

EXAMPLE 2.10   Show that

$$\lim_{x \to 1} \frac{x^2 - 1}{x - 1} = 2$$

Solution   Recall that this is the limit in Example 2.2 giving the slope of the tangent line to the graph of $y = x^2$ at the point $(1, 1)$. In that example, we used a calculator to estimate the limit.

By factoring the numerator, we can write

$$\lim_{x \to 1} \frac{x^2 - 1}{x - 1} = \lim_{x \to 1} \frac{(x + 1)(x - 1)}{x - 1} = \lim_{x \to 1} (x + 1) = 2$$

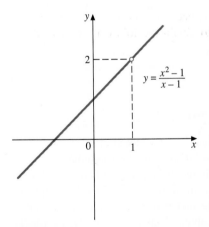

where in the last step we used Theorem 2.1. Note carefully that we can divide out the common factor $(x - 1)$ because in taking the limit we restrict x to be different from 1. Thus, when we write "$\lim_{x \to 1}$," we know that although we are letting x come arbitrarily close to 1, it is not equal to 1.

We show the graph of $y = (x^2 - 1)/(x - 1)$ in Figure 2.13. Notice that it differs from the graph of $y = x + 1$ only in that it has no value when $x = 1$. But since we exclude $x = 1$ in the limit calculation, the limit of the given function is identical to the limit of $y = x + 1$, namely, 2.   ■

FIGURE 2.13

REMARK ──────────────────────────────

■ Note that in the example we wrote the symbol "$\lim_{x \to 1}$" at each step until we found the limit at the last step. This symbol "$\lim_{x \to 1}$" indicates an operation to be performed and must be written until the limit is found. Similar remarks apply for other limits.

EXAMPLE 2.11   Determine whether the limit

$$\lim_{x \to -1} \frac{x^3 + 1}{x^2 - 1}$$

exists, and if it does, find its value.

x	$\dfrac{x^3 + 1}{x^2 - 1}$
−0.9	−1.43
−0.99	−1.493
−0.999	−1.4993
−1.1	−1.576
−1.01	−1.5075
−1.001	−1.50075

Solution   For this limit we see once again that we cannot simply substitute $x = -1$ into the numerator and denominator. The result of this substitution would be the undefined quantity 0/0.

To examine the behavior of $(x^3 + 1)/(x^2 - 1)$ as x approaches -1, we can first make a table of values, such as the one in the margin. The table suggests that the limit may actually exist and equal -1.5. Note that as x approaches -1, the numerator approaches 0, causing the fraction to get smaller, while the denominator also approaches 0 but causes the fraction to get larger. Apparently these two trends cancel each other.

Sum of two cubes:

$$x^3 + a^3 = (x + a)(x^2 - ax + a^2)$$

Difference of two cubes:

$$x^3 - a^3 = (x - a)(x^2 + ax + a^2)$$

To see this limit algebraically, we need to simplify the expression $(x^3 + 1)/(x^2 - 1)$. By factoring the numerator and denominator, we get

$$\lim_{x \to -1} \frac{x^3 + 1}{x^2 - 1} = \lim_{x \to -1} \frac{(x + 1)(x^2 - x + 1)}{(x + 1)(x - 1)}$$

$$= \lim_{x \to -1} \frac{x^2 - x + 1}{x - 1}$$

Note carefully that the equality of the last two fractions is valid because in taking the limit as $x \to -1$, we are excluding $x = -1$, so the common factor $(x + 1)$ in the numerator and denominator is not 0. Now we can apply Theorem 2.2, since the denominator $x - 1$ is not 0 when $x = -1$. Thus,

$$\lim_{x \to -1} \frac{x^3 + 1}{x^2 - 1} = \lim_{x \to -1} \frac{x^2 - x + 1}{x - 1} = -\frac{3}{2}$$

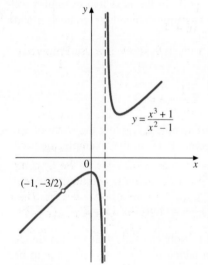

We show the graph of $y = (x^3 + 1)/(x^2 - 1)$ in Figure 2.14. Notice the missing point when $x = -1$. The limit approached, however, is the same as that approached by the function $(x^2 - x + 1)/(x - 1)$, which is identical to the original function except that the value for $x = -1$ is defined, namely, $-3/2$. ■

FIGURE 2.14

EXAMPLE 2.12   Evaluate the limit

$$\lim_{t \to 0} \frac{1 - \cos t}{\sin^2 t}$$

t	$f(t)$
±0.5	0.533
±0.1	0.501
±0.01	0.50001

Solution   Since $\cos 0 = 1$ and $\sin 0 = 0$, the function assumes the form 0/0 when $t = 0$ and so is not defined there. Using a calculator, CAS, or a spreadsheet we obtain the table in the margin, indicating that the limit appears to be 0.5. (Note that we made use of the fact that $f(t) = (1 - \cos t)/\sin^2 t$ is an even function in constructing the table. That is, $f(-t) = f(t)$.) To confirm the conclusion suggested by the table, we use the trigonometric identity $\sin^2 t + \cos^2 t = 1$ to replace $\sin^2 t$ with $1 - \cos^2 t$, getting

$$\lim_{x \to 0} \frac{1 - \cos t}{\sin^2 t} = \lim_{t \to 0} \frac{1 - \cos t}{1 - \cos^2 t} = \lim_{t \to 0} \frac{1 - \cos t}{(1 + \cos t)(1 - \cos t)}$$

$$= \lim_{t \to 0} \frac{1}{1 + \cos t} = \frac{1}{2}$$

since by the definition of $\cos t$ (Section 1.6), it is clear that $\cos t \to 1$ as $t \to 0$.

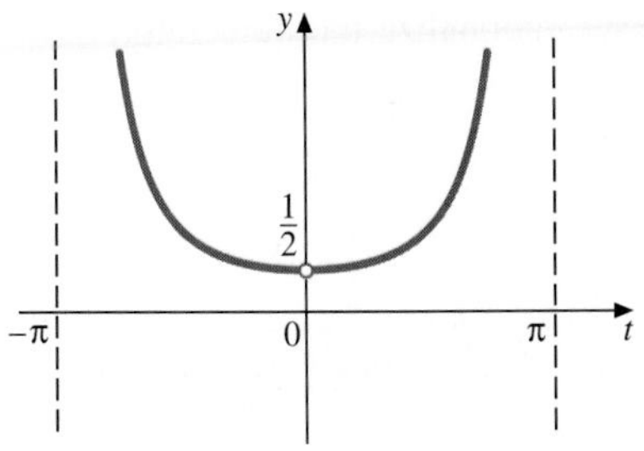

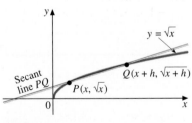

$$y = \frac{1 - \cos t}{\sin^2 t}$$

FIGURE 2.15

We show the graph of $y = (1 - \cos t)/\sin^2 t$ from $-\pi$ to π in Figure 2.15. As the graph shows, the function approaches the missing value, $1/2$, as x approaches 0.   ■

EXAMPLE 2.13   Verify the limit found in Example 2.1:

$$\lim_{x \to 0} \frac{\sqrt{x + 16} - 4}{x} = \frac{1}{8}$$

Solution   This time we change the fraction to a form in which we can use our limit properties by *rationalizing the numerator*. That is, we get rid of the radical on the numerator. To do so, we multiply both numerator and denominator by $(\sqrt{x + 16} + 4)$, using the fact that the difference of two numbers times their sum gives the difference of their squares. (Recall that we did a similar calculation in Example 1.3.)

$$\begin{aligned}
\lim_{x \to 0} \frac{\sqrt{x + 16} - 4}{x} &= \lim_{x \to 0} \frac{\sqrt{x + 16} - 4}{x} \cdot \frac{\sqrt{x + 16} + 4}{\sqrt{x + 16} + 4} \\
&= \lim_{x \to 0} \frac{(x + 16) - 16}{x(\sqrt{x + 16} + 4)} \\
&= \lim_{x \to 0} \frac{x}{x(\sqrt{x + 16} + 4)} \\
&= \lim_{x \to 0} \frac{1}{\sqrt{x + 16} + 4} \qquad \text{Since } x \neq 0
\end{aligned}$$

Now the denominator does not approach 0 as $x \to 0$, so we can use Properties 5 and 7, together with Theorem 2.1, to get

$$\lim_{x \to 0} \frac{1}{\sqrt{x + 16} + 4} = \frac{1}{\sqrt{16} + 4} = \frac{1}{8}$$

We showed the graph of $y = (\sqrt{x + 16} - 4)/x$ in Figure 2.2.   ■

The limit in the next example can be interpreted as the slope of the tangent line to a curve, much as in Example 2.2 but with different notation. To see why, consider Figure 2.16, where we have shown the graph of $y = \sqrt{x}$. The point $P(x, \sqrt{x})$ for $x > 0$ is arbitrary, but fixed. The point $Q(x + h, \sqrt{x + h})$ with $h \neq 0$ varies as h varies. In particular, Q approaches P as h approaches 0. The slope of the secant line PQ is

$$m_{\text{sec}} = \frac{\sqrt{x + h} - \sqrt{x}}{(x + h) - x} = \frac{\sqrt{x + h} - \sqrt{x}}{h}$$

The slope of the tangent line at P, if it exists, is then the limit of m_{sec} as $h \to 0$. It is this limit that we consider in the next example.

EXAMPLE 2.14   Evaluate the limit

$$\lim_{h \to 0} \frac{\sqrt{x + h} - \sqrt{x}}{h} \qquad (x > 0)$$

Solution   Note that in finding the limit as $h \to 0$, x is considered fixed and h is the variable. Again, if we allowed h to be 0, we would have $0/0$, so we must exclude $h = 0$. Since x is not specified, we cannot make a table of values

FIGURE 2.16

to give any indication of the limit. Again, we rationalize the numerator, as in the previous example.

$$\lim_{h \to 0} \frac{\sqrt{x+h} - \sqrt{x}}{h} = \lim_{h \to 0} \frac{\sqrt{x+h} - \sqrt{x}}{h} \cdot \frac{\sqrt{x+h} + \sqrt{x}}{\sqrt{x+h} + \sqrt{x}}$$

$$= \lim_{h \to 0} \frac{(x+h) - x}{h(\sqrt{x+h} + \sqrt{x})} = \lim_{h \to 0} \frac{h}{h(\sqrt{x+h} + \sqrt{x})}$$

$$= \lim_{h \to 0} \frac{1}{\sqrt{x+h} + \sqrt{x}}$$

Now the limit of the denominator is no longer 0 as $h \to 0$, so by our limit properties we have

$$\lim_{h \to 0} \frac{1}{\sqrt{x+h} + \sqrt{x}} = \frac{1}{\sqrt{x} + \sqrt{x}} = \frac{1}{2\sqrt{x}}$$

From the discussion preceding this example, we can conclude that for any $x > 0$, the slope of the tangent line to the graph of $y = \sqrt{x}$ at $(x, \sqrt{x})$ is $1/(2\sqrt{x})$. For example, at $(1, 1)$ the slope is $1/(2\sqrt{1}) = 1/2$, and at $(4, 2)$ the slope is $1/(2\sqrt{4}) = 1/4$. ∎

In each of Examples 2.10 through 2.14, we replaced the original function with a simpler one whose limit we could find. The two functions were equal except for the one point being approached by x. For instance, in Example 2.10 we replaced

$$\frac{x^2 - 1}{x - 1} \quad \text{with} \quad x + 1$$

since these two functions are equal except when $x = 1$. But in finding the limit as x approaches 1, we exclude $x = 1$ from consideration. So the limits are the same. We can state the general result as follows.

If $f(x) = g(x)$ except when $x = a$, and if $\lim_{x \to a} g(x) = L$, then $\lim_{x \to a} f(x) = L$ also.

The Squeeze Theorem

Sometimes it may be difficult to compute a limit directly, but it may be possible to compute it indirectly by "squeezing" the function between two known functions that have the same limit. For example, suppose we want to find

$$\lim_{x \to a} f(x)$$

but do not see a way of finding it directly. If we can find two functions g and h such that

$$g(x) \leq f(x) \leq h(x)$$

and for which $\lim_{x \to a} g(x) = \lim_{x \to a} h(x) = L$, then it is evident geometrically that $\lim_{x \to a} f(x) = L$ also (see Figure 2.17). We summarize this result in the following theorem.

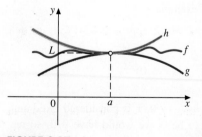

FIGURE 2.17

THEOREM 2.3

The Squeeze Theorem

If f, g, and h are functions satisfying

$$g(x) \le f(x) \le h(x)$$

for all x in some open interval containing a, except possibly at a itself, and if

$$\lim_{x \to a} g(x) = L \qquad \text{and} \qquad \lim_{x \to a} h(x) = L$$

then

$$\lim_{x \to a} f(x) = L$$

EXAMPLE 2.15   Prove that

$$\lim_{x \to 0} x \sin \frac{1}{x} = 0$$

We are using the following facts concerning absolute values: For all real numbers a and b,

$$|ab| = |a||b|$$

and

$$|a| \le |b|$$

is equivalent to

$$-|b| \le a \le |b|$$

Solution   Since the sine of a number never exceeds 1 in absolute value and since

$$\left| x \sin \frac{1}{x} \right| = |x| \left| \sin \frac{1}{x} \right|$$

we have

$$\left| x \sin \frac{1}{x} \right| \le |x|$$

That is,

$$-|x| \le x \sin \frac{1}{x} \le |x| \qquad (x \ne 0)$$

We can use limit properties and Special Limit 2.2 to show that both $|x|$ and $-|x|$ approach 0. For if $x > 0$, $|x| = x$, and we therefore have

$$\lim_{x \to 0^+} |x| = \lim_{x \to 0^+} x = 0 \qquad \text{Special Limit 2.2}$$

Whereas if $x < 0$, $|x| = -x = (-1)x$, so

$$\lim_{x \to 0^-} |x| = \lim_{x \to 0^-} (-1)x = (-1)\lim_{x \to 0^-} x = 0 \qquad \text{Property 1 and Special Limit 2.2}$$

Since these one-sided limits are equal, $\lim_{x \to 0} |x|$ must be 0. Now, using Property 1 again with $c = -1$, we get

$$\lim_{x \to 0} (-|x|) = (-1)\lim_{x \to 0} |x| = 0$$

Finally, we apply the Squeeze Theorem with $g(x) = -|x|$ and $h(x) = |x|$, both of which have limit 0 as $x \to 0$. Since

$$f(x) = x \sin \frac{1}{x}$$

is squeezed between $g(x)$ and $h(x)$, both of which approach 0, it follows that $f(x)$ also approaches 0 as $x \to 0$; that is,

$$\lim_{x \to 0} x \sin \frac{1}{x} = 0$$

The graphs of $g(x) = -|x|$, $h(x) = |x|$, and $f(x) = x \sin \frac{1}{x}$ are shown in Figure 2.18.

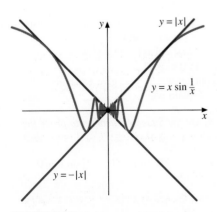

FIGURE 2.18

Exercise Set 2.2

In Exercises 1–16, evaluate the limits.

1. (a) $\lim_{x \to 1}(2x^2 - 3x + 1)$

   (b) $\lim_{x \to -2} \dfrac{2}{x+1}$

2. (a) $\lim_{x \to -3} \dfrac{2x^2 - 5x + 1}{x^2 - 2x}$ (b) $\lim_{x \to 3} \dfrac{x - 2}{\sqrt{x^2 - x - 2}}$

3. (a) $\lim_{x \to 0} \dfrac{1 - \sin x}{\cos x}$ (b) $\lim_{x \to \pi/2} \sqrt{2 \sin x - 3 \cos x}$

4. (a) $\lim_{x \to \pi} \dfrac{\sec x}{\tan x + 2}$ (b) $\lim_{x \to 1/2} \dfrac{\sin \pi x}{1 - \cos \pi x}$

5. $\lim_{x \to -2} \dfrac{x^2 - 4}{x + 2}$ **6.** $\lim_{x \to 3/2} \dfrac{4x^2 - 9}{2x - 3}$

7. $\lim_{x \to 3} \dfrac{x^2 - x - 6}{x - 3}$ **8.** $\lim_{x \to 7} \dfrac{x^2 - 5x - 14}{x - 7}$

9. $\lim_{x \to 1} \dfrac{x^2 + 3x - 5}{x^2 - 3x + 2}$ **10.** $\lim_{x \to -3} \dfrac{x^2 + 5x + 6}{x^2 + 6x + 9}$

11. $\lim_{x \to 1} \dfrac{\sqrt{x} - 1}{x - 1}$ **12.** $\lim_{x \to 2} \dfrac{\sqrt{2x - 3} - 1}{x - 2}$

13. $\lim_{x \to 5} \dfrac{\sqrt{3x + 1} - 4}{x - 5}$ **14.** $\lim_{x \to -2} \dfrac{\sqrt{2 - x} - 2}{x + 2}$

15. $\lim_{x \to \pi} \dfrac{1 + \cos x}{\sin^2 x}$ **16.** $\lim_{x \to -\pi/4} \dfrac{\sin x + \cos x}{\sin^2 x - \cos^2 x}$

In Exercises 17–24, determine the domain of f. Then write $f(x)$ in a simpler form for x in that domain and draw its graph, observing any restrictions on its domain. Then, for the given value of a, find $\lim_{x \to a^+} f(x)$, $\lim_{x \to a^-} f(x)$, and $\lim_{x \to a} f(x)$, or show they fail to exist.

17. $f(x) = \dfrac{x^2 - x - 2}{x + 1}$; $a = -1$

18. $f(x) = \dfrac{2x^2 + x - 1}{2x - 1}$; $a = \dfrac{1}{2}$

19. $f(x) = \dfrac{6x^2 + x - 2}{3x + 2}$; $a = -\dfrac{2}{3}$

20. $f(x) = \dfrac{6 + 7x - 3x^2}{x - 3}$; $a = 3$

21. $f(x) = \dfrac{x^3 + 2x^2}{x + 2}$; $a = -2$

22. $f(x) = \dfrac{x^3 - x^2 + x - 1}{x - 1}$; $a = 1$

23. $f(x) = \dfrac{x + 1}{x^2 - 1}$; $a = -1$

24. $f(x) = \dfrac{x - 2}{4 - x^2}$; $a = 2$

In Exercises 25–30, evaluate the limit

$$\lim_{h \to 0} \frac{f(a + h) - f(a)}{h}$$

for the given values of a and $f(x)$, or show that the limit does not exist. (We will see in Chapter 3 that this type of limit is the basis for defining the derivative of a function.)

25. $a = 2$; $f(x) = 2x - 1$

26. $a = 3$; $f(x) = x^2 - 1$

27. $a = 1$; $f(x) = \dfrac{1}{x}$

28. $a = -1$; $f(x) = \sqrt{3 - x}$

29. $a = 0$; $f(x) = \dfrac{x}{x + 1}$

30. $a = 2$; $f(x) = \dfrac{1}{\sqrt{x + 2}}$

31. Find functions f and g such that $\lim_{x \to a} f(x)$ and $\lim_{x \to a} g(x)$ both fail to exist but
   (a) $\lim_{x \to a}[f(x) + g(x)]$ exists.
   (b) $\lim_{x \to a}[f(x) \cdot g(x)]$ exists.

32. Do functions f and g exist such that $\lim_{x \to a} f(x)$ exists, and $\lim_{x \to a} g(x)$ does not exist, and
   (a) $\lim_{x \to a}[f(x) + g(x)]$ exists?
   (b) $\lim_{x \to a}[f(x) \cdot g(x)]$ exists?
   In each case, if your answer is yes, give examples. If your answer is no, explain why.

In Exercises 33 and 34, use the Squeeze Theorem to prove the given limit statement.

33. $\lim_{x \to 0} x^2 \cos 2x = 0$ **34.** $\lim_{x \to 1} |x - 1| \sin x = 0$

2.3 CONTINUITY

Informally stated, a function is *continuous* at a given point if its graph does not have a break there. Most functions describing physical phenomena are continuous. For example, the velocity of a moving object as a function of time is continuous. A car cannot suddenly jump from a speed of 20 mph to 40 mph. Rather, its speed varies continuously over all values betweeen 20 and 40. Similarly, air pressure is a continuous function of height above sea level. You can probably think of many other examples of continuous functions. On the other hand, the cost of domestic first-class postage as a function of weight is *discontinuous* at certain weights. A letter weighing 1 oz or less costs 29¢ in 1994, but if it exceeds 1 oz, even slightly, but is less than 2 oz, the cost jumps to 52¢.

Defining continuity of a function at a point as meaning that the graph does not break there is not totally satisfactory. How can we tell for sure whether the graph has a break? Even CAS-generated graphs might mask a break if the scale is poorly chosen. Let us be precise about what we mean by "no break." Suppose the domain of a function f includes an open interval containing the point $x = a$. Suppose further that $\lim_{x \to a} f(x)$ exists. If this limiting value is equal to the function value $f(a)$, then we say f is continuous at $x = a$. So continuity at $x = a$ means not only that the function *approaches* a limiting value as x approaches a, but also that this limit is *reached* when $x = a$ and is exactly the value of the function at that point.

Figure 2.19 illustrates this situation. The limit approached by $f(x)$ as x approaches a is the same as the value of f at a. Functions can be continuous at some points and not at others. When f is continuous at every point of some interval, we say it is continuous on that interval. The following definition summarizes this discussion.

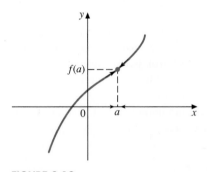

FIGURE 2.19

Definition 2.2
Continuity of a Function

Let f be defined in an open interval containing the point $x = a$. Then we say that **f is continuous at a**, provided that

$$\lim_{x \to a} f(x) = f(a)$$

If f is continuous at each point of an open interval I, then we say that **f is continuous on I**.

Continuity of f at a implies three conditions:

1. $\lim_{x \to a} f(x)$ exists.
2. $f(a)$ exists (i.e., a is in the domain of f).
3. The values in Conditions 1 and 2 are the same.

So a function can *fail* to be continuous—that is, it can be **discontinuous**—at $x = a$ by failing to satisfy one or more of these three conditions.

When a function is defined on a *closed* interval I, we extend the definition of continuity on I by additional requirements at the endpoints where only one-sided limits are used. In particular, if $I = [a, b]$, we require that f be **continuous from the right** at a, that is, $\lim_{x \to a^+} f(x) = f(a)$, and **continuous from the left** at b, that is, $\lim_{x \to b^-} f(x) = f(b)$.

It is instructive to look at various discontinuous functions. Examples are provided in Figures 2.20 through 2.23.

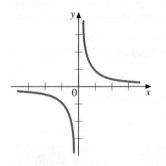

FIGURE 2.20
f is discontinuous at $x = 0$. $\lim_{x \to 0} f(x)$ does not exist and $f(0)$ does not exist.

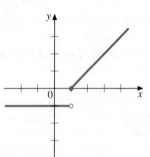

FIGURE 2.21
f is discontinuous at $x = 1$. $f(1) = 0$, but $\lim_{x \to 1^+} f(x) \neq \lim_{x \to 1^-} f(x)$, so $\lim_{x \to 1} f(x)$ does not exist.

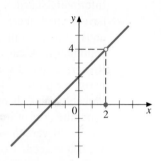

FIGURE 2.22
f is discontinuous at $x = 2$. $f(2) = 0$ but $\lim_{x \to 2} f(x) = 4$, so $\lim_{x \to 2} f(x) \neq f(2)$.

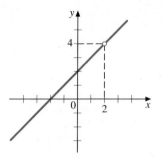

FIGURE 2.23
f is discontinuous at $x = 2$. $\lim_{x \to 2} f(x) = 4$, but $f(2)$ is not defined.

The type of discontinuity illustrated in Figure 2.20 is called an **infinite** discontinuity, that in Figure 2.21 is called a **simple** (or **jump**) discontinuity, and those in Figures 2.22 and 2.23 are called **removable** discontinuities (since by defining, or redefining, the function at one point only, the discontinuity can be removed). These examples illustrate the most common types of discontinuities. There are others, some of which we will see later.

The function

$$f(x) = \frac{\sin x}{x}$$

that we considered in Example 2.5 has a removable discontinuity at $x = 0$. Although $f(0)$ is not defined, $\lim_{x \to 0} f(x) = 1$. (We conjectured this limit in Example 2.5 and will give a proof later.) Thus, if we *define* $f(0)$ to be 1, in addition to $f(x) = (\sin x)/x$ for $x \neq 0$, we get a continuous function. We show again the graph of $(\sin x)/x$ given by a CAS in Figure 2.24. Although the point $(0, 1)$ is not on the graph, it is difficult to detect a missing point. So what we see appears to be the graph of the continuous function in which $f(0)$ has been defined as 1.

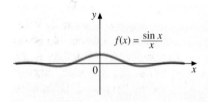

FIGURE 2.24

Based on results we have obtained, we can easily determine where polynomial and rational functions are continuous. Recall that, according to Theorem 2.1, if p is any polynomial function, then

$$\lim_{x \to a} p(x) = p(a)$$

But this is precisely the requirement for continuity at $x = a$. Thus, a polynomial function is continuous for every real number. Similarly, Theorem 2.2 states that if p and q are polynomial functions, and $f(x) = p(x)/q(x)$, then for all real numbers a for which $q(a) \neq 0$,

$$\lim_{x \to a} f(x) = f(a)$$

Thus, every rational function $p(x)/q(x)$ is continuous for all points in its domain. We state these results as a theorem.

THEOREM 2.4

Every polynomial function is continuous on all of $\mathbb{R}$, and every rational function is continuous wherever it is defined.

EXAMPLE 2.16 Determine where each of the following functions is continuous.

(a) $f(x) = \sqrt{3}x^5 + 2.7x^3 - \pi x + 3$

(b) $g(x) = \dfrac{2x^4 + 3x^2 - 7}{x^5 - 16x}$

Solution

(a) f is a polynomial function and so is continuous for all x. Its CAS-generated graph is shown in Figure 2.25.

(b) g is a rational function whose domain includes all real numbers x except those for which $x^5 - 16x = 0$. Solving for x, we get

$$x(x^4 - 16) = 0$$

$$x(x^2 - 4)(x^2 + 4) = 0$$

$$x(x - 2)(x + 2)(x^2 + 4) = 0$$

So the three points for which g is undefined are $x = 0$, $x = 2$, and $x = -2$. Thus, g is continuous for all other values of x. As the graph (Figure 2.26) shows, g has infinite discontinuities at $x = 0$, $x = 2$, and $x = -2$. ■

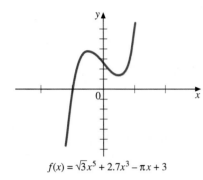

$f(x) = \sqrt{3}x^5 + 2.7x^3 - \pi x + 3$

FIGURE 2.25
f is continuous for all x

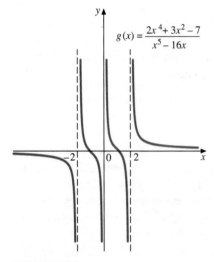

$g(x) = \dfrac{2x^4 + 3x^2 - 7}{x^5 - 16x}$

FIGURE 2.26
g has infinite discontinuities at $x = 0$, $x = 2$, and $x = -2$, and is continuous for all other x.

Combinations of Continuous Functions

Certain combinations of continuous functions result in a new continuous function. This fact enables us in many cases to recognize continuous functions without having to rely on Definition 2.2. For example, if a function is the sum of two continuous functions, it is also continuous. To see why, suppose $\lim_{x\to a} f(x) = f(a)$ and $\lim_{x\to a} g(x) = g(a)$. Then, by Property 2 of limits in Section 2.1,

$$\lim_{x\to a}[f(x) + g(x)] = \lim_{x\to a} f(x) + \lim_{x\to a} g(x) = f(a) + g(a)$$

Thus, if $h = f + g$, then $\lim_{x\to a} h(x) = h(a)$, so h is continuous at $x = a$. Similar results hold for differences, products, and quotients of continuous functions (where, as always for quotients, we must require that the denominator be nonzero at $x = a$). We therefore have the following theorem.

THEOREM 2.5

If f and g are continuous at $x = a$, then so are $f + g$, $f - g$, $f \cdot g$, and f/g, provided that in the case of f/g, $g(a) \neq 0$.

Thus, we are able to recognize a broad class of continuous functions. This class can be extended further by the next theorem, which we will prove in Section 2.5. In essence it says that *a continuous function of a continuous function is continuous.* Recall (Section 1.3) that the domain of the composite function $f \circ g$ consists of all x in the domain of g for which $g(x)$ is in the domain of f. In the following theorem we suppose $x = a$ to be such a point.

THEOREM 2.6

> Suppose the composite function $f \circ g$ is defined in some open interval containing the point $x = a$. If g is continuous at $x = a$ and f is continuous at $g(a)$, then $f \circ g$ is continuous at $x = a$.

EXAMPLE 2.17   Show that $F(x) = \sqrt{x^2 + 1}$ is continuous on all of $\mathbb{R}$.

Solution   We can express F as the composition $f \circ g$, where $f(x) = \sqrt{x}$ and $g(x) = x^2 + 1$. Since g is a polynomial function, it is continuous on all of $\mathbb{R}$. Furthermore, $g(x) > 0$ for all x, so $(f \circ g)(x) = \sqrt{x^2 + 1}$ is defined for all x. By Limit Property 7 in Section 2.2 and by Special Limit 2.2, we can see that for $a > 0$, $\lim_{x \to a} \sqrt{x} = \sqrt{a}$. Thus, f is continuous everywhere on its domain. By Theorem 2.6 it now follows that $f \circ g$ is continuous at every real number a. That is, F is continuous on all of $\mathbb{R}$.   ∎

Continuity of the Trigonometric Functions

From the definitions of $\cos t$ and $\sin t$ as the x- and y-coordinates of a point on the unit circle, we can see geometrically that both functions are continuous for all real values of t. Since the other four trigonometric functions are defined in terms of $\sin t$ and $\cos t$, we can make use of Theorem 2.5 to conclude that they are also continuous except where their denominators equal 0. We summarize these results in the next theorem.

THEOREM 2.7

> The functions $\sin t$ and $\cos t$ are continuous for all real t. The functions $\tan t$ and $\sec t$ are continuous except where $\cos t = 0$, namely, $t = \pi/2 + n\pi$, for $n = 0, \pm 1, \pm 2, \ldots$. The functions $\cot t$ and $\csc t$ are continuous except where $\sin t = 0$, namely, $t = n\pi$, for $n = 0, \pm 1, \pm 2, \ldots$.

EXAMPLE 2.18   Determine where each of the following functions is continuous.

(a) $\sin\left(\dfrac{1 + x}{1 - x}\right)$ (b) $\cot\left(\dfrac{1}{t}\right)$

Solution

(a) Let $f(x) = \sin x$ and $g(x) = (1 + x)/(1 - x)$. Then

$$(f \circ g)(x) = f(g(x)) = \sin\left(\frac{1 + x}{1 - x}\right)$$

According to Theorem 2.6, this composite function is continuous at $x = a$ if g is continuous at a and f is continuous at $g(a)$. In this instance, g is continuous everywhere except at $x = 1$, and f is continuous everywhere. So the given composite function is continuous for all x except possibly at $x = 1$. Now we check $x = 1$. At $x = 1$, the composite function is not defined, so it is discontinuous there.

(b) Reasoning as above, the composite function $\cot(1/t)$ is continuous except where $t = 0$ (where $1/t$ is discontinuous) and where $1/t = n\pi$ (where the cotangent is discontinuous). Thus, $\cot(1/t)$ is continuous for all real t except where $t = 0$ or $t = 1/n\pi$, for $n = \pm 1, \pm 2, \pm 3, \ldots$. ∎

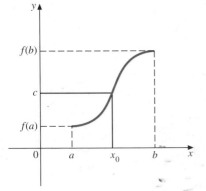

FIGURE 2.27

The Intermediate-Value Theorem

Functions that are continuous on an interval, especially on a closed and bounded interval, have many important properties. We will see some of these in later chapters. For now, we state one such property whose truth seems intuitively obvious, but which is surprisingly difficult to prove. The proof can be found in textbooks on advanced calculus.

THEOREM 2.8

The Intermediate-Value Theorem

Let f be continuous on the closed interval $[a, b]$ with $f(a) \neq f(b)$. If c is any number between $f(a)$ and $f(b)$, then there is at least one number x_0 in the open interval (a, b) such that $f(x_0) = c$.

Figure 2.27 illustrates the idea of this theorem. The graph crosses each horizontal line $y = c$, where c is between $f(a)$ and $f(b)$. One consequence of it is that if f is continuous on $[a, b]$ with $f(a)$ and $f(b)$ opposite in sign, then the equation $f(x) = 0$ has at least one real root. For example, every polynomial of odd degree must have at least one real zero. (You will be asked to prove this result in Exercise 45 of Exercise Set 2.3.)

EXAMPLE 2.19 Show that the equation

$$x^3 - 2x^2 + 4x - 7 = 0$$

has a root between $x = 1$ and $x = 2$.

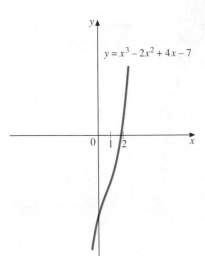

FIGURE 2.28

Solution Let $f(x) = x^3 - 2x^2 + 4x - 7$. We know that f is continuous for all x since it is a polynomial function. By substituting $x = 1$ and $x = 2$, we find that $f(1) = -4$ and $f(2) = 1$. The Intermediate-Value Theorem tells us that f assumes all values between -4 and 1 as x ranges from 1 to 2. Since 0 is between -4 and 1, it follows that there is a number x_0 in the open interval $(1, 2)$ such that $f(x_0) = 0$. That is, x_0 is a root of the equation $x^3 - 2x^2 + 4x - 7 = 0$ that lies between $x = 1$ and $x = 2$. We show the graph of f in Figure 2.28. ∎

Exercise Set 2.3

In Exercises 1–10, determine from the graph where the function is continuous and where it is discontinuous. At points of discontinuity, describe in what way or ways the definition of continuity fails.

1.

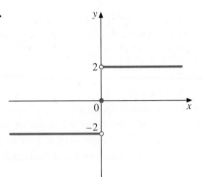

2.

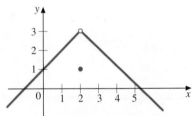

3.

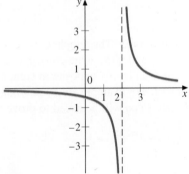

4.

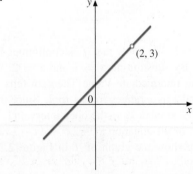

5.

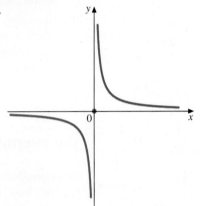

6.

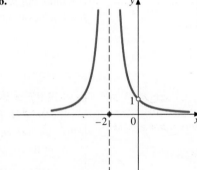

7.

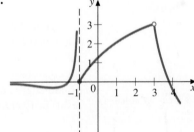

8.

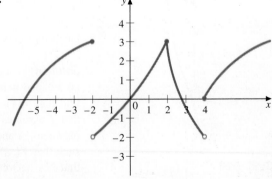

9.

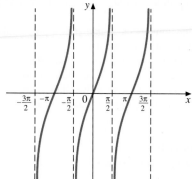

10.

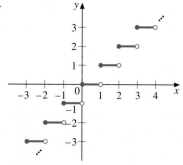

In Exercises 11–24, determine whether the function f is continuous at $x = a$. If it is discontinuous, state in what way the definition of continuity is not satisfied.

11. $f(x) = x^2 - 3x + 4;\ a = 2$

12. $f(x) = 2 - 3x + 4x^3;\ a = -1$

13. $f(x) = \dfrac{x}{x-1};\ a = 3$

14. $f(x) = \dfrac{2x-3}{x^2-1};\ a = 4$

15. $f(x) = \dfrac{x^2 - 2x + 1}{3x^2 - x - 3};\ a = -1$

16. $f(x) = \dfrac{x^3 - 1}{x^2 + 2x - 8};\ a = -4$

17. $f(x) = |x - 2|;\ a = 2$

18. $f(x) = \dfrac{|x|}{x};\ a = 0$

19. $f(x) = \dfrac{x-1}{x-2};\ a = 0$

20. $f(x) = \dfrac{x^2 - 4}{x - 2};\ a = 2$

21. $f(x) = \dfrac{x+3}{x^2 - 9};\ a = -3$

22. $f(x) = \dfrac{x^2 - 2x - 3}{x^2 - 5x + 6};\ a = 3$

23. $f(x) = \begin{cases} \sin \dfrac{1}{x} & \text{if } x \neq 0 \\ 0 & \text{if } x = 0 \end{cases};\ a = 0$

24. $f(x) = \begin{cases} x \sin \dfrac{1}{x} & \text{if } x \neq 0 \\ 0 & \text{if } x = 0 \end{cases};\ a = 0$

In Exercises 25–34, graph the given function and state on what intervals it is continuous. When an endpoint is included, indicate whether the function is continuous from the right or the left.

25. $f(x) = \begin{cases} x^2 + 1 & \text{if } x \geq 0 \\ 1 - x & \text{if } x < 0 \end{cases}$

26. $g(x) = \begin{cases} x + 2 & \text{if } x > 1 \\ 2x + 1 & \text{if } x \leq 1 \end{cases}$

27. $h(x) = \begin{cases} \dfrac{1}{x} & \text{if } x > 0 \\ -x & \text{if } x \leq 0 \end{cases}$

28. $F(x) = \begin{cases} \dfrac{1}{x^2} & \text{if } x \neq 0 \\ 0 & \text{if } x = 0 \end{cases}$

29. $G(x) = \begin{cases} \dfrac{|x|}{x} & \text{if } x \neq 0 \\ 1 & \text{if } x = 0 \end{cases}$

30. $H(x) = \begin{cases} x^2 - 4 & \text{if } x > 2 \\ 1 & \text{if } x = 2 \\ x - 1 & \text{if } x < 2 \end{cases}$

31. $f(x) = \begin{cases} \dfrac{x^2 - 9}{x - 3} & \text{if } x \neq 3 \\ 6 & \text{if } x = 3 \end{cases}$

32. $g(x) = \begin{cases} \dfrac{4 - x^2}{x - 2} & \text{if } x \neq 2 \\ -4 & \text{if } x = 2 \end{cases}$

33. $f(x) = \begin{cases} 2x + 1 & \text{if } x < 0 \\ 1 - x & \text{if } 0 \leq x < 2 \\ x - 2 & \text{if } x \geq 2 \end{cases}$

34. $f(x) = \begin{cases} \sqrt{x} & \text{if } x \geq 0 \\ -1 & \text{if } -2 \leq x < 0 \\ 2x + 5 & \text{if } x < -2 \end{cases}$

35. Let

$$f(x) = \frac{x^2 - 9}{x^2 - x - 6}$$

if $x \neq 3$. Define $f(3)$ so that f is continuous at $x = 3$.

36. Let

$$f(x) = \begin{cases} px & \text{if } x \le 2 \\ qx^2 + x + 3 & \text{if } x > 2 \end{cases}$$

Determine all choices of p and q such that f is continuous at $x = 2$.

37. Let $\phi(x) = \begin{cases} 1 & \text{if } x \text{ is rational} \\ 0 & \text{if } x \text{ is irrational} \end{cases}$.
Discuss the continuity of ϕ (phi). (*Hint*: The rationals and the irrationals are *dense* in $\mathbb{R}$. That is, between any two real numbers there is at least one rational number and one irrational number.)

38. Let $\psi(x) = \begin{cases} x & \text{if } x \text{ is rational} \\ 0 & \text{if } x \text{ is irrational} \end{cases}$.
Discuss the continuity of ψ (psi). (See the hint for Exercise 37.)

39. A function is said to be **bounded** if there is some positive number M such that the graph of f lies between the horizontal lines $y = M$ and $y = -M$. Give an example of a function that is continuous on an open interval but is not bounded there.

40. Give an example of a function that has a finite value at each point of a closed interval $[a, b]$ but is not bounded. (See Exercise 39.) (*Hint*: The function will have to be discontinuous.)

41. Does there exist a function that is discontinuous at every point? Does there exist a function that is continuous at

one and only one point? (*Hint*: See Exercises 37 and 38.)

42. Suppose f is continuous on $\mathbb{R}$ and p is a polynomial function for which it is known that $f(x) = p(x)$ for all $x \in \mathbb{R}$ except the single point $x = a$. Show that $f(a) = p(a)$; that is, $f(x)$ must be identical to $p(x)$ everywhere. What conclusion can we draw if we know only that $f(x) = p(x)$ except for finitely many values of x? Explain.

43. Prove or disprove: If f and g are discontinuous at $x = a$, then (a) $f + g$ is discontinuous at $x = a$; (b) $f \cdot g$ is discontinuous at $x = a$.

44. Prove that if f is continuous on an interval I, then so is $|f|$. Is the converse of this statement also true? Prove or disprove.

45. Prove that every polynomial function p_n of *odd* degree has at least one real zero; that is, $p_n(x_0) = 0$ for some x_0.

46. Refer to the first-class postage function described at the beginning of Chapter 1 (see also Exercise 23 in Exercise Set 1.1). Where is this function continuous? Where is it discontinuous?

47. Let $f(x)$ denote the greatest integer function $[\![x]\!]$ (see Exercise 22 in Exercise Set 1.1). Determine all real x for which f is continuous and all real x for which f is discontinuous.

2.4 INFINITE LIMITS AND LIMITS AT INFINITY

Throughout our discussion of limits, we have understood that when we write $\lim_{x \to a} f(x) = L$, both a and L represent finite real numbers. It is possible, however, to assign meaning to the limit statement when either a or L is replaced (or both are replaced) by one of the symbols ∞ or $-\infty$. In a sense, we have already discussed limits of this sort when we considered vertical and horizontal asymptotes in Section 1.6, although we did not use limit terminology. Consider, for example, the function

$$f(x) = \frac{1}{x^2}$$

x	$1/x^2$
± 1	1
± 0.1	100
± 0.01	10,000
± 0.001	1,000,000

As $x \to 0$, we see that $f(x)$ becomes arbitrarily large (see the table). In Section 1.6 we used the expression

$$\text{as } x \to 0, f(x) \to \infty$$

x	$1/x^2$
± 10	0.01
± 100	0.0001
± 1000	0.000001

to indicate this behavior. Equivalently, we can (and usually do) write

$$\lim_{x \to 0} f(x) = \infty$$

Graphically, we know then that the line $x = 0$ is a vertical asymptote.

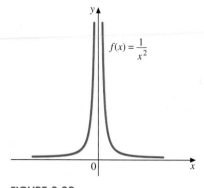

FIGURE 2.29
$$\lim_{x\to 0}\frac{1}{x^2}=\infty,\ \lim_{x\to\infty}\frac{1}{x^2}=0,$$
$$\lim_{x\to-\infty}\frac{1}{x^2}=0$$

When looking for horizontal asymptotes, we want to see if $f(x)$ approaches some finite value as x becomes arbitrarily large in absolute value. For $f(x) = 1/x^2$ the line $y = 0$ is a horizontal asymptote, since $1/x^2$ becomes arbitrarily close to 0 as $|x|$ becomes larger and larger (see the bottom table on p. 110). Previously, we indicated this behavior by saying

$$\text{as } x \to \infty,\ f(x) \to 0,\ \text{and as } x \to -\infty,\ f(x) \to 0$$

Again, we can use the alternative statements

$$\lim_{x\to\infty} f(x) = 0 \quad \text{and} \quad \lim_{x\to-\infty} f(x) = 0$$

The graph of f, given in Figure 2.29, shows the vertical and horizontal asymptotes.

We summarize this discussion in the following informal definition.

Definition 2.3
Infinite Limits and Limits at Infinity

(a) The statement

$$\lim_{x\to a} f(x) = \infty$$

means that $f(x)$ will be as large as we please for all x sufficiently close to a but not equal to a.

(b) The statement

$$\lim_{x\to\infty} f(x) = L$$

means that $f(x)$ will be as close to L as we please for all sufficiently large values of x.

Analogous definitions hold for

$$\lim_{x\to a} f(x) = -\infty \text{ and } \lim_{x\to-\infty} f(x) = L$$

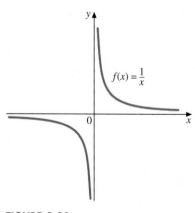

FIGURE 2.30
$$\lim_{x\to 0^+}\frac{1}{x}=\infty,\ \lim_{x\to 0^-}\frac{1}{x}=-\infty$$

REMARK
■ In part (a) when we say "as large as we please," we mean that $f(x)$ can be made *arbitrarily large*. That is, no matter how large a positive number M is chosen, $f(x)$ will be larger than M for all x values sufficiently close to a but different from a. In part (b) when we say "as close to L as we please," we mean that $f(x)$ comes *arbitrarily close* to L when x is large enough.

In Definition 2.3(a), by restricting x to one side of a, we get the right-hand and left-hand limits

$$\lim_{x\to a^+} f(x) = \infty \quad \text{and} \quad \lim_{x\to a^-} f(x) = \infty$$

and similarly when ∞ is replaced by $-\infty$.

For the function $f(x) = 1/x$, for example, the one-sided limits at 0 are (see Figure 2.30)

$$\lim_{x\to 0^+}\frac{1}{x}=\infty \quad \text{and} \quad \lim_{x\to 0^-}\frac{1}{x}=-\infty$$

REMARKS

■ We cannot write anything meaningful for $\lim_{x \to 0} 1/x$, since the left-hand and right-hand limits are not the same.

■ When we write $\lim_{x \to a} f(x) = \infty$ (or $-\infty$), we sometimes say "the limit is infinite." However, we do not mean that the limit exists but rather that it fails to exist in a particular way, namely, that the function gets large beyond all bound. When we say "the limit exists," we will always mean that it is finite.

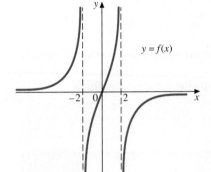

FIGURE 2.31

EXAMPLE 2.20   Refer to the function f whose graph is shown in Figure 2.31. Find each of the following:
(a) $\lim_{x \to -2^-} f(x)$ (b) $\lim_{x \to -2^+} f(x)$
(c) $\lim_{x \to 2^-} f(x)$ (d) $\lim_{x \to 2^+} f(x)$

Solution   From the graph we get the following results:
(a) $\lim_{x \to -2^-} f(x) = \infty$ (b) $\lim_{x \to -2^+} f(x) = -\infty$
(c) $\lim_{x \to 2^-} f(x) = \infty$ (d) $\lim_{x \to 2^+} f(x) = -\infty$ ■

EXAMPLE 2.21   Show that the following limits are infinite.
(a) $\displaystyle\lim_{x \to 1^+} \frac{2x + 3}{\sqrt{x - 1}}$ (b) $\displaystyle\lim_{x \to 0^-} \frac{1 + \cos x}{\sin x}$

Solution

(a) As $x \to 1^+$, $2x + 3$ comes arbitrarily close to 5, and $\sqrt{x - 1}$ comes arbitrarily close to 0 through positive numbers. Thus, the quotient $(2x + 3)/\sqrt{x - 1}$ is positive and becomes arbitrarily large. That is,

$$\lim_{x \to 1^+} \frac{2x + 3}{\sqrt{x - 1}} = \infty$$

We show the graph in Figure 2.32.

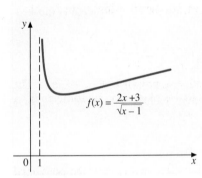

FIGURE 2.32
$\lim_{x \to 1^+} f(x) = \infty$

$f(x) = \dfrac{2x + 3}{\sqrt{x - 1}}$

(b) The numerator of the fraction

$$\frac{1 + \cos x}{\sin x}$$

approaches the positive number 2 as $x \to 0^-$, and the denominator approaches 0 through negative values. So we conclude that

$$\lim_{x \to 0^-} \frac{1 + \cos x}{\sin x} = -\infty$$

We show the graph in Figure 2.33.   ■

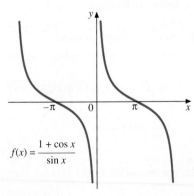

$f(x) = \dfrac{1 + \cos x}{\sin x}$

FIGURE 2.33
$\lim_{x \to 0^-} f(x) = -\infty$

EXAMPLE 2.22   Find the following limits.

(a) $\displaystyle\lim_{x \to \infty} \frac{3x + 5}{\sqrt{4x^2 - 9}}$ and $\displaystyle\lim_{x \to -\infty} \frac{3x + 5}{\sqrt{4x^2 - 9}}$

(b) $\displaystyle\lim_{x \to \infty} \frac{\cos x}{\sqrt{x + 2}}$

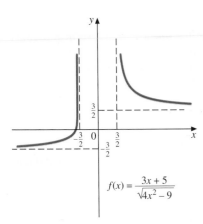

$$f(x) = \frac{3x + 5}{\sqrt{4x^2 - 9}}$$

FIGURE 2.34
$\lim\limits_{x \to \infty} f(x) = \frac{3}{2}$ and $\lim\limits_{x \to -\infty} f(x) = -\frac{3}{2}$

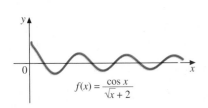

$$f(x) = \frac{\cos x}{\sqrt{x} + 2}$$

FIGURE 2.35
$\lim\limits_{x \to \infty} f(x) = 0$

Solution

(a) Using the dominant-term approach (see Section 1.3), for $|x|$ large, $3x + 5 \approx 3x$ and $4x^2 - 9 \approx 4x^2$. So

$$\frac{3x + 5}{\sqrt{4x^2 - 9}} \approx \frac{3x}{\sqrt{4x^2}} = \frac{3x}{2|x|} \qquad \text{For } |x| \text{ large}$$

Thus,

$$\lim_{x \to \infty} \frac{3x + 5}{\sqrt{4x^2 - 9}} = \lim_{x \to \infty} \frac{3x}{2x} = \frac{3}{2}$$

$$\lim_{x \to -\infty} \frac{3x + 5}{\sqrt{4x^2 - 9}} = \lim_{x \to -\infty} \frac{3x}{2|x|} = \lim_{x \to -\infty} \frac{3x}{2(-x)} = -\frac{3}{2}$$

For the limit as $x \to -\infty$, we made use of the fact that $\sqrt{x^2} = |x|$ and that $|x| = -x$ when x is negative. The graph in Figure 2.34 shows the asymptote $y = 3/2$ on the right and $y = -3/2$ on the left.

(b) We can see intuitively that

$$\lim_{x \to \infty} \frac{\cos x}{\sqrt{x} + 2} = 0$$

since the numerator varies between 1 and -1, while the denominator becomes arbitrarily large. So the fraction approaches 0. More formally, since $-1 \le \cos x \le 1$ for all x, we have

$$-\frac{1}{\sqrt{x} + 2} \le \frac{\cos x}{\sqrt{x} + 2} \le \frac{1}{\sqrt{x} + 2}$$

Clearly, both $1/(\sqrt{x} + 2)$ and its negative approach 0 as $x \to \infty$. The Squeeze Theorem (Theorem 2.3) continues to hold true when $x \to a$ is replaced by $x \to \infty$, so we conclude that

$$\lim_{x \to \infty} \frac{\cos x}{\sqrt{x} + 2} = 0$$

(See the graph in Figure 2.35.) Note that the graph crosses the horizontal asymptote $y = 0$ infinitely many times. ∎

Vertical and Horizontal Asymptotes

Let us restate the definition of vertical and horizontal asymptotes, using the limit terminology of this section.

Definition 2.4
Vertical and Horizontal Asymptotes

(a) The line $x = a$ is a vertical asymptote to the graph of $y = f(x)$ if either of the following holds true:
  (i) $\lim_{x \to a^+} f(x) = \infty$ or $-\infty$
  (ii) $\lim_{x \to a^-} f(x) = \infty$ or $-\infty$

(b) The line $y = b$ is a horizontal asymptote to the graph of $y = f(x)$ if either of the following holds true:
  (i) $\lim_{x \to \infty} f(x) = b$
  (ii) $\lim_{x \to -\infty} f(x) = b$

■ We can paraphrase part (a) by saying that the line $x = a$ is a vertical asymptote if $f(x)$ becomes infinite (either positively or negatively) as x approaches a from either side. Similarly, for part (b), the line $y = b$ is a horizontal asymptote if $f(x)$ comes arbitrarily close to $y = b$ as x becomes infinite (either positively or negatively).

In Figures 2.36 and 2.37 we show some examples of vertical and horizontal asymptotes. Notice that the graph of a function $y = f(x)$ can cross a horizontal asymptote but not a vertical one.

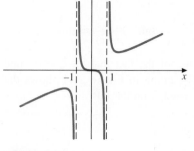

FIGURE 2.36
$x = -1$ and $x = 1$ are vertical asymptotes.

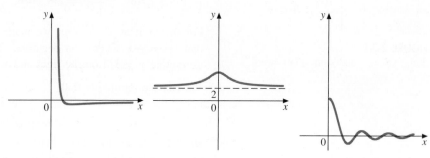

$y = 0$ is a horizontal asymptote.

(a)

$y = 2$ is a horizontal asymptote.

(b)

$y = 0$ is a horizontal asymptote.

(c)

FIGURE 2.37

EXAMPLE 2.23   Find all vertical and horizontal asymptotes.

(a) $f(x) = \dfrac{2x^2 + 3}{4 - x^2}$ (b) $f(x) = \dfrac{1 + \sin x}{\sqrt{x - 1}}$

Solution

(a) Note that f is even, so its graph is symmetric with respect to the y-axis. Thus, we need only consider $x \geq 0$. By factoring the denominator, we obtain

$$f(x) = \frac{2x^2 + 3}{(2 - x)(2 + x)}$$

Since

$$\lim_{x \to 2^+} \frac{2x^2 + 3}{(2 - x)(2 + x)} = -\infty \quad \text{and} \quad \lim_{x \to 2^-} \frac{2x^2 + 3}{(2 - x)(2 + x)} = \infty$$

the line $x = 2$ is a vertical asymptote. By symmetry, so is the line $x = -2$.
For horizontal asymptotes, we need to calculate

$$\lim_{x \to \infty} \frac{2x^2 + 3}{4 - x^2}$$

Using dominant terms on the numerator and denominator, we see that for large x,

$$\frac{2x^2 + 3}{4 - x^2} \approx \frac{2x^2}{-x^2}$$

so

$$\lim_{x \to \infty} \frac{2x^2 + 3}{4 - x^2} = \lim_{x \to \infty} \frac{2x^2}{-x^2} = -2$$

Thus, the line $y = -2$ is a horizontal asymptote on the right, and by symmetry, also on the left. We show the graph of f in Figure 2.38.

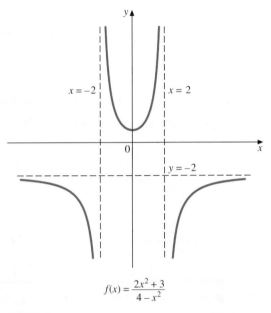

$$f(x) = \frac{2x^2 + 3}{4 - x^2}$$

FIGURE 2.38
Vertical asymptotes are $x = 2$ and $x = -2$; horizontal asymptote is $y = -2$.

(b) Note first that the domain of f is the interval $(1, \infty)$, since otherwise $\sqrt{x - 1}$ is either 0 (which is excluded since we cannot have 0 in the denominator) or not a real number. Since

$$\lim_{x \to 1^+} \frac{1 + \sin x}{\sqrt{x - 1}} = \infty$$

the line $x = 1$ is a vertical asymptote.

For horizontal asymptotes, we have

$$\lim_{x \to \infty} \frac{1 + \sin x}{\sqrt{x - 1}} = 0$$

since the numerator is always between 0 and 2 (inclusive) and the denominator becomes arbitrarily large. More explicitly, since $-1 \le \sin x \le 1$ for all x, we have

$$0 \le \frac{1 + \sin x}{\sqrt{x - 1}} \le \frac{2}{\sqrt{x - 1}}$$

Again, we can use the Squeeze Theorem to see that since

$$\lim_{x \to \infty} \frac{2}{\sqrt{x - 1}} = 0$$

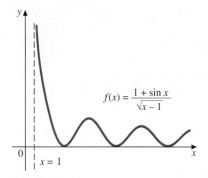

$$f(x) = \frac{1 + \sin x}{\sqrt{x-1}}$$

$x = 1$

FIGURE 2.39
Vertical asymptote is $x = 1$; horizontal asymptote is $y = 0$.

the function $(1 + \sin x)/\sqrt{x-1}$ is squeezed between 0 and something that approaches 0. So

$$\lim_{x \to \infty} \frac{1 + \sin x}{\sqrt{x-1}} = 0$$

We show the graph of f in Figure 2.39.   ∎

In Section 1.5 we showed how to find horizontal asymptotes to graphs of rational functions. We repeat those findings here, using limit terminology.

Limits of Rational Functions at Infinity

Let $p(x)$ and $q(x)$ be polynomials. Then

(a) $\displaystyle \lim_{x \to \pm\infty} \frac{p(x)}{q(x)} = 0$ if $\deg p < \deg q$

(b) $\displaystyle \lim_{x \to \pm\infty} \frac{p(x)}{q(x)} = \frac{\text{leading coefficient of } p(x)}{\text{leading coefficient of } q(x)}$ if $\deg p = \deg q$

(c) $\displaystyle \lim_{x \to \pm\infty} \frac{p(x)}{q(x)}$ is infinite if $\deg p > \deg q$

REMARK ─────────────────────────────────

∎ When we write $\lim_{x \to \pm\infty} f(x) = L$, we mean both $\lim_{x \to \infty} f(x) = L$ and $\lim_{x \to -\infty} f(x) = L$.

EXAMPLE 2.24   Find the following limits.

(a) $\displaystyle \lim_{x \to \infty} \frac{2x^2 - 3x + 4}{5 - 4x + x^4}$ (b) $\displaystyle \lim_{x \to -\infty} \frac{5x^3 - 4x + 7}{2x^3 + 6x^2 - 3}$ (c) $\displaystyle \lim_{x \to \infty} \frac{2x^3 + 1}{x^2 - 1}$

Solution

(a) $\displaystyle \lim_{x \to \infty} \frac{2x^2 - 3x + 4}{5 - 4x + x^4} = 0$ since the degree of the numerator is less than the degree of the denominator.

(b) $\displaystyle \lim_{x \to -\infty} \frac{5x^3 - 4x + 7}{2x^3 + 6x^2 - 3} = \frac{5}{2}$ since the numerator and denominator are of the same degree.

(c) $\displaystyle \lim_{x \to \infty} \frac{2x^3 + 1}{x^2 - 1} = \infty$ since the numerator is of higher degree. Note that for large x, the fraction is approximately $2x^3/x^2 = 2x$. So as $x \to \infty$, the fraction becomes infinite.   ∎

EXAMPLE 2.25   Find all vertical and horizontal asymptotes to the graph of

$$y = \frac{x^2 + 4}{x^2 - 4}$$

Make use of these asymptotes, along with intercepts and symmetry, to sketch the graph.

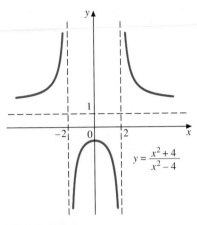

FIGURE 2.40

Solution   The vertical asymptotes occur when the denominator is 0, namely, when $x = 2$ and $x = -2$. In particular,

$$\lim_{x \to 2^+} \frac{x^2 + 4}{x^2 - 4} = \infty \quad \text{and} \quad \lim_{x \to 2^-} \frac{x^2 + 4}{x^2 - 4} = -\infty$$

The graph is symmetric with respect to the y-axis, since replacing x with $-x$ does not change the value of y. So we can limit our analysis to nonnegative values of x and then reflect the graph about the y-axis.

Since the degree of the numerator is the same as that of the denominator, we conclude that

$$\lim_{x \to \infty} \frac{x^2 + 4}{x^2 - 4} = 1$$

Thus, $y = 1$ is a horizontal asymptote.

The y-intercept is -1, found by setting $x = 0$. There is no x-intercept. We now have enough information to sketch the graph (Figure 2.40).   ∎

REMARK
■ For a rational function

$$y = \frac{p(x)}{q(x)}$$

in *lowest terms* (that is, $p(x)$ and $q(x)$ have no nonconstant factor in common), the vertical asymptotes occur at zeros of the denominator. So if $q(a) = 0$, then the line $x = a$ is a vertical asymptote. If $p(x)$ and $q(x)$ do have a common factor, that factor should first be divided out, bearing in mind that the x value that causes the factor to be 0 is not in the domain of the original function. Recall that we illustrated cases where $p(x)$ and $q(x)$ have a common factor in Examples 2.10 and 2.11.

Exercise Set 2.4

In Exercises 1–6, determine whether the functions approach ∞ *or* $-\infty$.

1. (a) $\displaystyle\lim_{x \to 2^-} \frac{1}{x - 2}$ (b) $\displaystyle\lim_{x \to 2^+} \frac{1}{x - 2}$

2. (a) $\displaystyle\lim_{x \to 3^-} \frac{x + 1}{x - 3}$ (b) $\displaystyle\lim_{x \to 3^+} \frac{x + 1}{x - 3}$

3. (a) $\displaystyle\lim_{x \to -1^-} \frac{2x - 3}{x + 1}$ (b) $\displaystyle\lim_{x \to -1^+} \frac{2x - 3}{x + 1}$

4. (a) $\displaystyle\lim_{x \to -2^-} \frac{1 - x}{x^2 - 4}$ (b) $\displaystyle\lim_{x \to -2^+} \frac{1 - x}{x^2 - 4}$

5. (a) $\displaystyle\lim_{x \to 5^-} \frac{3 - 4x}{(x - 5)^2}$ (b) $\displaystyle\lim_{x \to 5^+} \frac{3 - 4x}{(x - 5)^2}$

6. (a) $\displaystyle\lim_{x \to 4^-} \frac{x^2 + x - 1}{x^2 - 2x - 8}$ (b) $\displaystyle\lim_{x \to 4^+} \frac{x^2 + x - 1}{x^2 - 2x - 8}$

In Exercises 7–16, find the indicated limits or show that they fail to exist.

7. $\displaystyle\lim_{x \to \infty} \frac{1}{2x + 3}$

8. $\displaystyle\lim_{x \to -\infty} \frac{x}{x^2 - 4}$

9. $\displaystyle\lim_{x \to -\infty} \frac{x + 1}{x - 2}$

10. $\displaystyle\lim_{x \to \infty} \frac{x^2 + 3}{2x^2 - 5}$

11. $\displaystyle\lim_{x \to \infty} \frac{1 - x + 2x^2}{x^2 - 4}$

12. $\displaystyle\lim_{x \to -\infty} \frac{3x^3 - 2x - 7}{2 + 4x^2 - 5x^3}$

13. $\displaystyle\lim_{x \to -\infty} \frac{x^2 - 1}{x + 2}$

14. $\displaystyle\lim_{x \to \infty} \frac{4 - x - 2x^2}{3x - 7}$

15. (a) $\displaystyle\lim_{x \to \infty} \frac{2x + 3}{\sqrt{x^2 - 1}}$ (b) $\displaystyle\lim_{x \to -\infty} \frac{2x + 3}{\sqrt{x^2 - 1}}$

16. $\displaystyle\lim_{x \to -\infty} \frac{\sqrt{2 - x}}{x + 2}$

In Exercises 17–19, determine $\lim_{x \to \infty} f(x)$ *and* $\lim_{x \to -\infty} f(x)$
from the graph.

17.

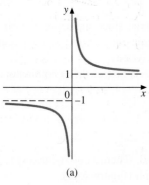

(a)

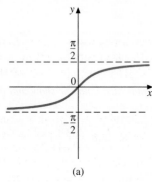

(b)

18.

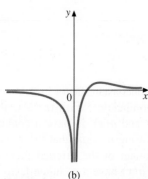

(a)

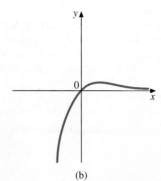

(b)

19.

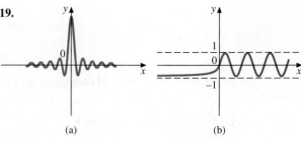

(a)   (b)

In Exercises 20–35, find all vertical and horizontal asymptotes.

20. $f(x) = \dfrac{x}{x - 3}$ **21.** $f(x) = \dfrac{x - 1}{x^2 - 4}$

22. $f(x) = \dfrac{x^2 - 1}{x^2 - 9}$ **23.** $f(x) = \dfrac{2x^2 + x - 1}{x^2 - x - 6}$

24. $f(x) = \dfrac{x - 3}{6x^2 + x - 2}$ **25.** $f(x) = \dfrac{x^2 - 4}{x + 1}$

26. $f(x) = \dfrac{x^2 - 9}{x + 3}$ **27.** $f(x) = \dfrac{x + 3}{x^2 - 9}$

28. $f(x) = \dfrac{x^2 - 5x + 4}{3x^2 - x - 2}$ **29.** $f(x) = \dfrac{2x^2 + x - 10}{x^2 - x - 2}$

30. $f(x) = 1 - \cos \dfrac{2}{x}$ **31.** $f(x) = \dfrac{1}{x - 1} \sin \dfrac{1}{x}$

32. $f(x) = \dfrac{\cos x}{x(1 + \sin^2 x)}$ **33.** $f(x) = \dfrac{\sin^2 x}{\sqrt{x} + \cos^2 x}$

34. $f(x) = \dfrac{2\cos^2 x - 3}{2\sin^2 x - \sin x - 1}$ (vertical only)

35. $f(x) = \dfrac{x^2 - 4}{4\cos^2 x - 3}$

36. Describe the manner in which the graph of

$$f(x) = 1 + \frac{\sin x}{x}$$

approaches its horizontal asymptote. Use a graphing
calculator or CAS to sketch the graph. Vary the x-scale
and y-scale so that the behavior for large x is clearly
displayed.

37. Use a calculator, CAS, or spreadsheet to conjecture
numerically the value of the limit

$$\lim_{x \to \infty} \left(1 + \frac{1}{x}\right)^x$$

Obtain the graph of this function using a graphing
calculator or CAS and see if the graph is consistent with
your conjecture.

38. Suppose $\lim_{x \to \infty} f(x) = 0$ and $\lim_{x \to \infty} g(x) = \infty$. Discuss the possibilities for $\lim_{x \to \infty} f(x) \cdot g(x)$. Give examples to support your answer.

39. Give examples to show that Limit Properties 2–5 in Section 2.2 need not hold true if either $\lim_{x \to a} f(x)$ or $\lim_{x \to a} g(x)$ is infinite.

40. Give examples of functions f and g such that

$$\lim_{x \to a} f(x) = \infty, \qquad \lim_{x \to a} g(x) = \infty$$

and

$$\lim_{x \to a} [f(x) - g(x)] = L$$

where L is a finite, nonzero number.

41. (a) If \$1,000 is invested at 10% interest compounded n times per year, show that the amount of money, $P(n)$, in the account at the end of one year is

$$P(n) = 1000 \left(1 + \frac{0.1}{n}\right)^n$$

(b) Complete the following table:

How compounded	Number of times (n) per year	$P(n)$ (to nearest cent)
Annually	1	\$1,100.00
Semiannually	2	\$1,102.50
Quarterly		
Monthly		
Weekly		
Daily		
Hourly		

(c) If interest is compounded *continuously*, the amount after one year is defined to be $\lim_{n \to \infty} P(n)$, Conjecture the result from the table you completed in part (b).

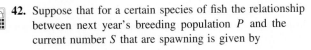

42. Suppose that for a certain species of fish the relationship between next year's breeding population P and the current number S that are spawning is given by

$$P(S) = \frac{aS^2}{S^2 + b}$$

where a and b are constants determined by the particular species.
(a) Graph $y = P(S)$ for several choices of a and b.
(b) What happens to $P(S)$ as the number of fish spawning increases indefinitely?

43. The number of bacteria of a certain type present in a culture at time t is given by

$$f(t) = 10,000 \left(\frac{2t^2 + 1}{t^2 + 1}\right) \qquad (t \geq 0)$$

As t increases, does the size of the bacteria colony stabilize? If so, what is the limiting number?

44. A rocket of mass m is projected upward from the earth at an initial velocity v_0. We wish to determine the smallest value of v_0 that will cause the rocket to escape from the earth's gravitational field. Denote the radius of the earth by R, and let v and r be the velocity of the rocket and its distance from the center of the earth at any time after is has been projected upward. Its kinetic energy is $\frac{1}{2}mv^2$ and its potential energy is $-GMm/r$, where M is the mass of the earth and G is the universal gravitational constant. By the law of conservation of energy, the sum of the kinetic energy and potential energy is constant. So this sum at time t is the same as when the rocket was initially projected:

$$\frac{1}{2}mv^2 - \frac{GMm}{r} = \frac{1}{2}mv_0^2 - \frac{GMm}{R} \qquad (2.3)$$

We can find the velocity v_0 for which the rocket just escapes by calculating v_0 for which $\lim_{v \to 0} r = \infty$. Carry out this calculation as follows:
(a) Solve Equation 2.3 for r.
(b) Find $\lim_{v \to 0} r$ from part (a).
(c) Find v_0 that causes the limit in part (b) to be ∞.
(d) Use $G = 6.67 \times 10^{-11}$ N $\cdot$ m^2/kg^2, $M = 5.97 \times 10^{24}$ kg, and $R = 6.37 \times 10^6$ m to find the escape velocity in m/s.

45. The radius of the moon is approximately 0.27 times the radius of the earth, and the mass is about 1/81 the mass of the earth. Find the escape velocity from the moon. (See Exercise 44.)

46. The number of stars in a spherical shell around our sun is the *number density*, N (the average number of stars per unit volume), times the volume of a shell. The volume of a shell of thickness Δr and radius r is $4\pi r^2 \Delta r$. The brightness of a star is I/r^2, where I is the intrinsic intensity.
(a) Find a formula for the brightness of a shell of radius R and find the limit as $r \to \infty$, holding Δr fixed.
(b) If we consider n shells with increasing radii and let B_n denote the sum of the brightnesses of these shells, find $\lim_{n \to \infty} B_n$.
Note: The discrepancy between the result in part (b) and the observed darkness of space is called Olbers's Paradox, after the nineteenth-century astronomer Wilhelm Olbers.

2.5 A RIGOROUS APPROACH TO LIMITS AND CONTINUITY

To motivate a precise formulation of the definition of the limit of a function, let us consider the linear function

$$f(x) = 2x - 1$$

and investigate its limit as x approaches 2. Of course, we know the answer is 3. That is,

$$\lim_{x \to 2} f(x) = 3$$

In the language of Definition 2.1, the fact that the limit is 3 means that $f(x)$ will be as close to 3 as we wish for all x sufficiently close to 2, but not equal to 2. (In this particular case the condition $x \neq 2$ is not necessary, because the function is continuous, but we have seen examples where such a restriction is necessary.) Our objective is to show precisely what we mean by "as close as we wish" and "sufficiently close."

Suppose as a first approximation we choose to make $f(x)$ within 0.5 units of 3. To accomplish this relationship, how close to 2 should we take x? Figure 2.41 helps answer this question. For $f(x)$ to be within 0.5 units of 3, it must lie between 2.5 and 3.5. That is,

$$2.5 < f(x) < 3.5$$

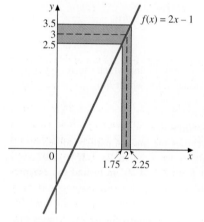

FIGURE 2.41

Graphically, we can determine corresponding x values by drawing the lines $y = 2.5$ and $y = 3.5$, determining their intersections with the graph of f, and projecting these points downward to the x-axis, as we have indicated in Figure 2.41. We should emphasize, however, that this graphical approach works in this case because f is an increasing function. It would also work with a decreasing function, but not necessarily with a function that is neither always increasing nor decreasing in the interval between the two x projections. Algebraically, we can find the x values from the inequality $2.5 < f(x) < 3.5$ by substituting the value of $f(x)$ and solving for x:

$$2.5 < 2x - 1 < 3.5$$
$$3.5 < 2x < 4.5$$
$$1.75 < x < 2.25$$

Since we are interested in knowing how far from 2 we may take x, we subtract 2 from each term of the last inequality to get

$$-0.25 < x - 2 < 0.25$$

This inequality can be written using absolute values as

$$|x - 2| < 0.25$$

Thus, if x is any number less than 0.25 units from 2, the corresponding $f(x)$ value will be less than 0.5 units from the asserted limit 3. That is,

$$|f(x) - 3| < 0.5$$

We have thus found a tolerance limit on x, given the desired tolerance for $f(x)$.

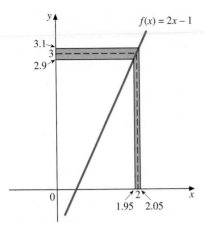

FIGURE 2.42

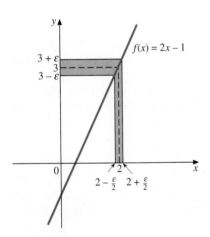

FIGURE 2.43

Suppose now that we specify a smaller tolerance for the difference between $f(x)$ and 3, say 0.1. Figure 2.42 illustrates how we can find graphically a suitable interval about $x = 2$. Algebraically, we proceed as before.

$$2.9 < f(x) < 3.1$$

$$2.9 < 2x - 1 < 3.1$$

$$3.9 < 2x < 4.1$$

$$1.95 < x < 2.05$$

If we choose any x such that

$$-0.05 < x - 2 < 0.05$$

or, equivalently,

$$|x - 2| < 0.05$$

then $f(x)$ will be within 0.1 unit of 3. That is, the inequality

$$|f(x) - 3| < 0.1$$

will be satisfied.

We could continue in this way, using smaller and smaller "degrees of closeness," or tolerance levels, but so long as we use specific values, such as 0.5 or 0.1, no matter how small we take them (greater than 0), we will not have shown that $f(x)$ can be made "as close as we please" to 3. That is, we must show that the difference between $f(x)$ and 3, in absolute value, can be made less than *any* arbitrarily small positive number by choosing x sufficiently close to 2. It is customary to use the Greek letter ε (epsilon) to designate an arbitrarily small positive number. Suppose we use ε to denote how close to 3 we want $f(x)$ to be. In our first two approximations we used $\varepsilon = 0.5$ and $\varepsilon = 0.1$, respectively. Now we are leaving ε unspecified. As before, we illustrate graphically (Figure 2.43) how to find an interval about $x = 2$ on the x-axis such that for any x in that interval, $f(x)$ will be between the lines $y = 3 - \varepsilon$ and $y = 3 + \varepsilon$. Algebraically we find this interval in the same way as before:

$$3 - \varepsilon < f(x) < 3 + \varepsilon$$

$$3 - \varepsilon < 2x - 1 < 3 + \varepsilon$$

$$4 - \varepsilon < 2x < 4 + \varepsilon$$

$$2 - \frac{\varepsilon}{2} < x < 2 + \frac{\varepsilon}{2}$$

Thus,

$$-\frac{\varepsilon}{2} < x - 2 < \frac{\varepsilon}{2}$$

or equivalently,

$$|x - 2| < \frac{\varepsilon}{2}$$

Working backward, we conclude that if x is any number less than $\varepsilon/2$ units from 2, then $f(x)$ is less than ε units from 3. That is, $|f(x) - 3| < \varepsilon$. Thus, finally and conclusively, we have shown that $f(x)$ can be made *arbitrarily* close to 3 (that is, within ε units) provided x is *sufficiently* close to 2 (that is, within $\varepsilon/2$ units). The degree of closeness of x to the number in question, in this case

2, is usually designated by the Greek letter δ (delta). So in this case we simply choose $\delta = \varepsilon/2$. Thus, whenever $|x - 2| < \delta$, we know that $|f(x) - 3| < \varepsilon$. Note that any smaller positive value of δ would also work, for if x is within a smaller interval about 2, it will also be within $\varepsilon/2$ units of 2, and so $f(x)$ will again be within ε units of 3. Note also that δ depends on ε. A smaller value of ε necessitates a smaller value of δ.

Let us use the example we just completed as a model for the precise meaning of

$$\lim_{x \to a} f(x) = L$$

For any $\varepsilon > 0$ we want to find how far from a we may allow x to be (we will call this distance δ) so that $f(x)$ will remain within ε units of L. Again, we show the situation graphically in Figure 2.44. We have shown a small circle about $x = a$, indicating that we are excluding this point from consideration, for as we have seen, there are cases in which the limit as $x \to a$ exists even though $f(a)$ does not exist, or if it exists, it may not equal L. In the figure, we show the lines $y = L - \varepsilon$ and $y = L + \varepsilon$ and their intersections with the graph of f. The x-coordinates of these points of intersection in general determine an interval about $x = a$, as shown. If f is either increasing or decreasing throughout this interval, we can determine δ graphically from this interval, as shown in Figure 2.44. The interval may not be symmetrical about $x = a$, as it was in the preceding example. If not, we choose a smaller interval that is symmetrical about a and denote it by $(a - \delta, a + \delta)$. Algebraically, we proceed as in our example, working from the inequality

$$L - \varepsilon < f(x) < L + \varepsilon$$

to solve for x in the form

$$a - \delta < x < a + \delta$$

and restricting x to be different from a. Equivalently, we can begin with the absolute value of the difference between $f(x)$ and $L, |f(x) - L|$. Then, by algebraic manipulation, we try to verify that this quantity can be made less than ε when $|x - a| < \delta$ for a suitable δ. We will illustrate this latter approach after stating the definition.

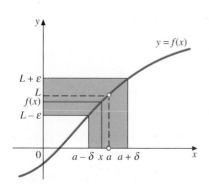

FIGURE 2.44

Definition 2.5
Formal Definition of the Limit

Let f be defined in some open interval containing the point $x = a$, except possibly at $x = a$ itself. Then the limit of $f(x)$ as x approaches a is the number L, written

$$\lim_{x \to a} f(x) = L$$

means that for each positive number ε there exists a positive number δ such that

$$\text{if } 0 < |x - a| < \delta, \text{ then } |f(x) - L| < \varepsilon$$

REMARK
■ The inequality $0 < |x - a|$ ensures that $x \neq a$.

EXAMPLE 2.26   Use Definition 2.5 to prove that

$$\lim_{x \to 2}(3x + 4) = 10$$

Solution   Let ε denote an arbitrary positive number. To find δ we will use the procedure suggested just before the statement of Definition 2.5. We let $f(x) = 3x + 4$ and begin by writing $|f(x) - 10|$ in equivalent forms, with the goal of obtaining an expression involving $|x - 2|$.

$$|f(x) - 10| = |(3x + 4) - 10| = |3x - 6| = 3|x - 2|$$

Now we can conclude that $|f(x) - 10|$ will be less than ε whenever $3|x-2| < \varepsilon$. That is, if

$$|x - 2| < \frac{\varepsilon}{3}$$

then

$$|f(x) - 10| < \varepsilon$$

So we have found a suitable value of δ, namely, $\delta = \varepsilon/3$, satisfying the requirements of the definition. Once again, in this case the restriction $x \neq 2$ is unnecessary, although imposing it does not alter the conclusion. ∎

In the next example, the restriction $x \neq a$ is necessary. Also, the function involved is nonlinear.

EXAMPLE 2.27   Determine the limit

$$\lim_{x \to 1} \frac{x^3 - x^2 + x - 1}{x - 1}$$

and then prove your result using Definition 2.5.
Note: This limit would arise, for example, in finding the slope of the tangent line to the graph of $y = x^3 - x^2 + x$ at $x = 1$.

Solution   Let $f(x)$ denote the given function. Clearly $f(1)$ is not defined. In fact, both numerator and denominator equal 0 at $x = 1$. We can factor the numerator by grouping terms two by two:

$$x^3 - x^2 + x - 1 = x^2(x - 1) + (x - 1) = (x^2 + 1)(x - 1)$$

Thus, if $x \neq 1$,

$$f(x) = \frac{(x^2 + 1)(x - 1)}{x - 1} = x^2 + 1$$

We can see now that

$$\lim_{x \to 1} f(x) = \lim_{x \to 1}(x^2 + 1) = 2$$

To prove that Definition 2.5 is satisfied, let ε denote any positive number, and consider the difference $|f(x) - 2|$, for $x \neq 1$.

$$|f(x) - 2| = |(x^2 + 1) - 2| = |x^2 - 1| = |(x + 1)(x - 1)| = |x + 1||x - 1|$$

Keep in mind that our goal is to find an inequality of the form $|x-1| < \delta$ that will guarantee that $|f(x) - 2| < \varepsilon$. We have shown that $|f(x) - 2| = |x + 1||x - 1|$. This product contains $|x - 1|$, which we want, but it also contains $|x + 1|$, which we do not want. A way out of our difficulty is to require at the outset that x

be, say, between 0 and 2. This restriction is equivalent to stipulating in advance that whatever δ we eventually choose, we will not let it exceed 1. For if $\delta \leq 1$ and $|x - 1| < \delta$, then $-1 < x - 1 < 1$, or $0 < x < 2$. Placing such a restriction on δ involves no loss of generality, since whenever a value of δ is found, any smaller value will do just as well.

If we make the restriction, then, that $0 < x < 2$, it follows from the triangle inequality (see marginal note) that $|x + 1| \leq |x| + 1 < 3$. So we can write

$$|f(x) - 2| = |x + 1||x - 1| < 3|x - 1|$$

Thus, if

$$|x - 1| < \frac{\varepsilon}{3} \quad (x \neq 1)$$

then

$$|f(x) - 2| < 3\left(\frac{\varepsilon}{3}\right) = \varepsilon$$

It appears, then, that we should choose $\delta = \varepsilon/3$. But remember our initial restriction that $\delta \leq 1$. We can take care of both restrictions by choosing δ to be the smaller of the numbers 1 and $\varepsilon/3$, which we write as

$$\delta = \min\left\{1, \frac{\varepsilon}{3}\right\}$$

("min" stands for "minimum"). Then, if

$$0 < |x - 1| < \delta$$

we have

$$|f(x) - 2| < \varepsilon$$

and the proof is complete.   ■

> **The triangle inequality**
>
> $$|a + b| \leq |a| + |b|$$
>
> holds true for all real numbers a and b (Appendix 1).

Most of the proofs of the limit properties given in Section 2.2 are relegated to Appendix 5, but we will prove the first property here, since the proof is particularly simple and it further illustrates Definition 2.5.

EXAMPLE 2.28   Prove that if $\lim_{x \to a} f(x)$ exists and c is any real number, then

$$\lim_{x \to a} cf(x) = c \lim_{x \to a} f(x)$$

Solution   Let L denote the limit of $f(x)$ as $x \to a$. Then we must show that

$$\lim_{x \to a} cf(x) = cL$$

Let ε denote any positive number. We consider two cases: $c = 0$ and $c \neq 0$. The proof for $c = 0$ is easy. As you will see, however, to prove the result for $c \neq 0$, we will need to divide by $|c|$, which is why we cannot include $c = 0$ in our argument.

Case (i): If $c = 0$, then

$$|cf(x) - cL| = |0 - 0| = 0$$

So $|cf(x) - cL| < \varepsilon$ for all $\varepsilon > 0$ whenever x is in the domain of f, which proves the result we want when $c = 0$.

Case (ii): Now suppose $c \neq 0$. Then

$$|cf(x) - cL| = |c(f(x) - L)| = |c||f(x) - L|$$

So the inequality $|cf(x) - cL| < \varepsilon$ will be satisfied if $|c||f(x) - L| < \varepsilon$, or equivalently, if

$$|f(x) - L| < \frac{\varepsilon}{|c|}$$

Now $\varepsilon/|c|$ is a positive number, and we can use it in place of ε in Definition 2.5. Because we are given that $\lim_{x \to a} f(x) = L$, we know that a number $\delta > 0$ exists such that

$$\text{if } 0 < |x - a| < \delta, \text{ then } |f(x) - L| < \frac{\varepsilon}{|c|}$$

or equivalently,

$$|cf(x) - cL| < \varepsilon$$

The proof is therefore complete.   ■

Infinite Limits

The next definition makes precise the meanings of the "infinite limits"

$$\lim_{x \to a} f(x) = \infty \quad \text{and} \quad \lim_{x \to a} f(x) = -\infty$$

Definition 2.6
Infinite Limits

Let f be defined in some open interval containing $x = a$, except possibly at $x = a$ itself.

(a) The limit statement

$$\lim_{x \to a} f(x) = \infty$$

means that for every positive number M (however large it might be) there exists a positive number δ such that

$$\text{if } 0 < |x - a| < \delta, \text{ then } f(x) > M$$

(See Figure 2.45.)

(b) The limit statement

$$\lim_{x \to a} f(x) = -\infty$$

means that for every positive number M (however large it might be) there exists a positive number δ such that

$$\text{if } 0 < |x - a| < \delta, \text{ then } f(x) < -M$$

(See Figure 2.46.)

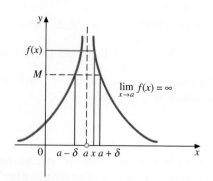

FIGURE 2.45

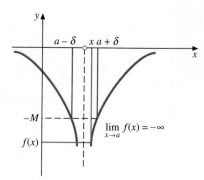

FIGURE 2.46

REMARK ───────────────────────────────────

■ If $\lim_{x \to a} f(x) = \infty$, we often say "$f(x)$ approaches infinity" or "$f(x)$ becomes arbitrarily large" as x approaches a. Similarly, if $\lim_{x \to a} f(x) = -\infty$, we often say "$f(x)$ approaches negative infinity" or, colloquially, "$f(x)$ approaches minus infinity." Of course, in each case, ∞ is not a number, so the limit does not exist, but it fails to exist by becoming arbitrarily large (positively in the first case and negatively in the second).

───────────────────────────────────

EXAMPLE 2.29   Find

$$\lim_{x \to -1} \frac{2}{(x+1)^2}$$

and verify the result using Definition 2.6.

Solution   Since $(x+1)^2 \to 0$ as $x \to -1$, and $(x+1)^2$ is always positive because it is a square, the quotient $2/(x+1)^2$ becomes arbitrarily large positively. That is,

$$\lim_{x \to -1} \frac{2}{(x+1)^2} = \infty$$

To show that Definition 2.6(a) is satisfied, let M denote any positive number. Then the inequality

$$\frac{2}{(x+1)^2} > M$$

will be true provided

$$(x+1)^2 < \frac{2}{M} \qquad (x \neq -1)$$

This latter inequality is satisfied if

$$|x+1| < \sqrt{\frac{2}{M}} \qquad \text{Since } \sqrt{a^2} = |a|$$

Let $\delta = \sqrt{2/M}$. We have shown that if

$$0 < |x - (-1)| < \delta$$

then

$$\frac{2}{(x+1)^2} > M$$

which, according to Definition 2.6(a), proves that

$$\lim_{x \to -1} \frac{2}{(x+1)^2} = \infty$$

Note that $x = -1$ is not in the domain of the function, so we use $0 < |x+1|$ to restrict x from being equal to -1.   ■

REMARK ──────────────────────────────────────

■ Definitions 2.5 and 2.6 can easily be modified to define the right-hand and left-hand limits. For example, if we want x to approach a from the right, then in both definitions the inequality $0 < |x - a| < \delta$ is replaced with

$$a < x < a + \delta$$

For the left-hand limit we use

$$a - \delta < x < a$$

For example,

$$\lim_{x \to 0^+} \left(\frac{1}{x} \right) = \infty$$

since for any $M > 0$, $1/x > M$ provided that $0 < x < 1/M$. Similarly,

$$\lim_{x \to 0^-} \left(\frac{1}{x} \right) = -\infty$$

since for $M > 0$, $1/x < -M$ provided that $-1/M < x < 0$. In each case we are using $\delta = 1/M$.

Limits at Infinity

We have seen limits that are infinite as x approaches a finite value. Now let us define the opposite: finite limits as x becomes positively or negatively infinite, symbolized by

$$\lim_{x \to \infty} f(x) = L \qquad \text{and} \qquad \lim_{x \to -\infty} f(x) = L$$

We sometimes refer to these limits as "limits at infinity." The precise definition is as follows.

Definition 2.7
Limits at Infinity

(a) Let the domain of f contain an interval of the form (a, ∞). The limit statement

$$\lim_{x \to \infty} f(x) = L$$

means that corresponding to each positive number ε there exists a positive number N such that

$$\text{if } x > N, \text{ then } |f(x) - L| < \varepsilon$$

(See Figure 2.47.)

(b) Let the domain of f include an interval of the form $(-\infty, a)$. The limit statement

$$\lim_{x \to -\infty} f(x) = L$$

means that corresponding to each positive number ε there is a positive number N such that

$$\text{if } x < -N, \text{ then } |f(x) - L| < \varepsilon$$

(See Figure 2.48.)

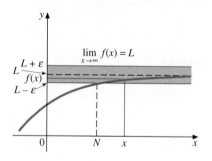

FIGURE 2.47   **FIGURE 2.48**

REMARK

■ If $\lim_{x \to \infty} f(x) = L$, we often say that $f(x)$ approaches L as x "approaches infinity," but the words "approaches infinity" mean merely that x becomes large without bound. Similar remarks apply to $\lim_{x \to -\infty} f(x) = L$.

EXAMPLE 2.30   Prove that

$$\lim_{x \to \infty} \frac{2x^2 + 1}{x^2} = 2$$

Solution   Let $f(x) = (2x^2 + 1)/x^2$. By dividing each term of the numerator by x^2, we can write $f(x)$ in the equivalent form

$$f(x) = 2 + \frac{1}{x^2} \qquad (x \neq 0)$$

Let ε denote any positive number. We can write $|f(x) - 2|$ as

$$\left| \left(2 + \frac{1}{x^2}\right) - 2 \right| = \frac{1}{x^2}$$

So we will have $|f(x) - 2| < \varepsilon$ whenever

$$\frac{1}{x^2} < \varepsilon$$

which is true provided $x > 1/\sqrt{\varepsilon}$. Thus, in Definition 2.7(a) if we take $N = 1/\sqrt{\varepsilon}$, we see that when $x > N, |f(x) - 2| < \varepsilon$, and we conclude that

$$\lim_{x \to \infty} f(x) = 2$$

$\blacksquare$

The $\varepsilon - \delta$ Definition of Continuity

Recall that for f to be continuous at $x = a$, the limit of $f(x)$ must exist as $x \to a$ and be equal not just to *some* number L, but specifically to the number $f(a)$. In Definition 2.5, then, we replace L by $f(a)$ and remove the restriction $0 < |x - a|$, since now we permit x to equal a.

Definition 2.8 The $\varepsilon - \delta$ Definition of Continuity	Let f be defined for all x in some interval containing the point $x = a$. Then **f is continuous at $x = a$** provided that for each positive number ε there exists a positive number δ such that $\qquad$ if $\|x - a\| < \delta$, then $\|f(x) - f(a)\| < \varepsilon$ If f is continuous at every point of some open interval I, then we say that **f is continuous on I.**

For continuity at the endpoints of a closed interval $[a, b]$, we use one-sided limits. In particular, we say that **f is continuous from the left** at b if for every $\varepsilon > 0$ there is a $\delta > 0$ such that

$$\text{if } b - \delta < x \le b, \text{ then } |f(x) - f(b)| < \varepsilon$$

Similarly, f **is continuous from the right** at a if for every $\varepsilon > 0$ there is a $\delta > 0$ such that

$$\text{if } a \le x < a + \delta, \text{ then } |f(x) - f(a)| < \varepsilon$$

REMARK ————

$\blacksquare$ In both cases, we must have $\delta \le b - a$ in order that x is in the interval $[a, b]$. In the future, when we say f is continuous on the closed interval $[a, b]$, we will mean f is continuous at each point between a and b and is continuous from the right at a and from the left at b.

EXAMPLE 2.31   Prove that $f(x) = \sqrt{x}$ is continuous on $[0, \infty)$.

Solution   First, we prove continuity from the right at $x = 0$. Let ε denote any positive number. Since

$$|f(x) - f(0)| = |\sqrt{x} - \sqrt{0}| = \sqrt{x}$$

the inequality $|f(x) - f(0)| < \varepsilon$ will be satisfied if $\sqrt{x} < \varepsilon$, or equivalently, if $0 \le x < \varepsilon^2$. So we let $\delta = \varepsilon^2$, and the requirement for right-hand continuity at $x = 0$ is satisfied.

Next, we prove continuity at all points in $(0, \infty)$. Let $x = a$ be in this open interval. Then $a > 0$. So we have, for $x > 0$,

$$|f(x) - f(a)| = |\sqrt{x} - \sqrt{a}| = \left| \frac{\sqrt{x} - \sqrt{a}}{1} \cdot \frac{\sqrt{x} + \sqrt{a}}{\sqrt{x} + \sqrt{a}} \right|$$

$$= \frac{|x - a|}{\sqrt{x} + \sqrt{a}} < \frac{|x - a|}{\sqrt{a}} \qquad \text{Since } \sqrt{x} > 0$$

Notice how we rationalized the expression $\sqrt{x} - \sqrt{a}$ in order to introduce the factor $|x - a|$. For any $\varepsilon > 0$ that we choose, the inequality

$$\frac{|x - a|}{\sqrt{a}} < \varepsilon$$

will be satisfied if $|x - a| < \varepsilon\sqrt{a}$. So it appears that we should take $\delta = \varepsilon\sqrt{a}$. We must also have $\delta \le a$, however, in order that x is not negative when

$$|x - a| < \delta$$

(otherwise $\sqrt{x}$ is not defined). Thus, we take $\delta = \min\{a, \varepsilon\sqrt{a}\}$. Then, whenever $|x-a| < \delta$, we have $|f(x) - f(a)| < \varepsilon$, as can be seen by reversing the preceding steps. So f is continuous at $x = a$. Since a represents *any* number in $(0, \infty)$, f is continuous for *all* points in this open interval. Finally, f is continuous from the right at 0, and we conclude that f is continuous on $[0, \infty)$.   ∎

Although the definitions we have given in this section are technical in nature, they are fundamental in proving the properties of limits and continuity on which much of calculus is based. We will see, for example, in the next chapter how the derivative is based on the limit concept, and its properties are based on those of the limit. So for a deep understanding of the derivative, the limit definitions and properties are crucial.

Continuity of Composite Functions

As indicated previously, we will defer most of the proofs of theorems concerning limits and continuity to Supplementary Appendix 5. We do choose to prove Theorem 2.6 here, however, as well as a corollary to it, because these results will be used repeatedly in subsequent chapters. We restate the theorem for completeness.

THEOREM 2.6

> Suppose the composite function $f \circ g$ is defined in some open interval containing the point $x = a$. If g is continuous at $x = a$ and f is continuous at $g(a)$, then $f \circ g$ is continuous at $x = a$.

Proof   Let ε denote any positive number. Write $y = g(x)$ and let $b = g(a)$. Since f is continuous at b, there exists a positive number η (eta) such that if $|y - b| < \eta$, then $|f(y) - f(b)| < \varepsilon$. But $|y - b| = |g(x) - g(a)|$ and since g

is continuous at a, there exists a $\delta > 0$ such that $|g(x) - g(a)| < \eta$ whenever $|x - a| < \delta$. Thus, if $|x - a| < \delta$, we have $|f(y) - f(b)| < \varepsilon$, or equivalently

$$|f(g(x)) - f(g(a))| < \varepsilon$$

So $f \circ g$ is continuous at $x = a$. ∎

COROLLARY 2.6

If $\lim_{x \to a} g(x) = b$ and f is continuous at b, then

$$\lim_{x \to a} f(g(x)) = f\left(\lim_{x \to a} g(x)\right)$$

Proof   We make g into a continuous function at $x = a$, if it is not already continuous, by defining (or redefining if necessary) $g(a)$ to equal the limiting value b. Then $f \circ g$ is continuous at $x = a$, by Theorem 2.6. Thus, by the definition of continuity,

$$\lim_{x \to a} f(g(x)) = f(g(a)) = f(b) = f\left(\lim_{x \to a} g(x)\right)$$ ∎

As a consequence of this corollary and Example 2.31, we can prove Property 7 for limits, stated in Section 2.2. We can restate the property as follows: if $\lim_{x \to a} g(x)$ exists and is positive,

$$\lim_{x \to a} \sqrt{g(x)} = \sqrt{\lim_{x \to a} g(x)}$$

To see that this result is true, recall that in Example 2.31 we proved that $f(x) = \sqrt{x}$ is continuous for all $x \geq 0$. So, by Corollary 2.6,

$$\lim_{x \to a} f(g(x)) = f\left(\lim_{x \to a} g(x)\right)$$

That is,

$$\lim_{x \to a} \sqrt{g(x)} = \sqrt{\lim_{x \to a} g(x)}$$

In the result just proved, we assumed that $\lim_{x \to a} g(x) > 0$. It follows that $g(x)$ itself is positive throughout some open interval $(a - \delta, a + \delta)$ except possibly at $x = a$. (See Exercise 38 in Exercise Set 2.5.) If we permitted $\lim_{x \to a} g(x)$ to be 0, then it is possible that there would be values of x arbitrarily close to a for which $g(x)$, although close to 0, would be negative, in which case $\sqrt{g(x)}$ would not be real. However, if we also know that $g(x) \geq 0$ for all x sufficiently close to a, then Property 7 continues to hold true even when $\lim_{x \to a} g(x) = 0$. For example, since for all x,

$$g(x) = 1 + \cos x \geq 0 \qquad \text{Why?}$$

and $\lim_{x \to \pi} g(x) = 0$, we can conclude that

$$\lim_{x \to \pi} \sqrt{1 + \cos x} = \sqrt{\lim_{x \to \pi} (1 + \cos x)} = \sqrt{0} = 0$$

Note carefully that passing to the limit under the radical sign was valid in this case since $g(x) = 1 + \cos x$ is never negative. In general, the equality of $\lim_{x \to a} \sqrt{g(x)}$ and $\sqrt{\lim_{x \to a} g(x)}$ holds true only when $\lim_{x \to a} g(x)$ is greater than zero.

Exercise Set 2.5

In Exercises 1–17, prove the statements by showing that Definition 2.5 is satisfied.

1. $\lim\limits_{x \to 1} (2x + 3) = 5$

2. $\lim\limits_{x \to 3} (4x - 5) = 7$

3. $\lim\limits_{x \to 2} \left(\dfrac{x + 1}{3} \right) = 1$

4. $\lim\limits_{x \to 4} \left(\dfrac{2x - 3}{5} \right) = 1$

5. $\lim\limits_{x \to 2} (3 - 4x) = -5$

6. $\lim\limits_{x \to 5} \left(\dfrac{1 - x}{2} \right) = -2$

7. $\lim\limits_{x \to -1} (2x + 5) = 3$

8. $\lim\limits_{x \to -2} (3 - 4x) = 11$

9. $\lim\limits_{x \to 0} \dfrac{5 - 2x}{10} = \dfrac{1}{2}$

10. $\lim\limits_{x \to -6} \dfrac{3 - x}{12} = \dfrac{3}{4}$

11. $\lim\limits_{x \to 2} \dfrac{x^2 - 4}{x - 2} = 4$

12. $\lim\limits_{x \to 3} \dfrac{9 - x^2}{x - 3} = -6$

13. $\lim\limits_{x \to -1/2} \dfrac{1 - 4x^2}{2x + 1} = 2$

14. $\lim\limits_{x \to 4/3} \dfrac{9x^2 - 16}{3x - 4} = 8$

15. $\lim\limits_{x \to a} x = a$

16. If $\lim\limits_{x \to a} f(x) = 0$ and $|g(x)| \le M$, then $\lim\limits_{x \to a} f(x) \cdot g(x) = 0$.

17. If $\lim\limits_{x \to a} f(x) = 0$, then $\lim\limits_{x \to a} [f(x)]^2 = 0$. (*Hint:* For x in a sufficiently small open interval containing a, excluding $x = a$ itself, show that $|f(x)| < 1$, and explain why it is true also that $[f(x)]^2 \le |f(x)|$ for x in this same open interval excluding a.)

18. Prove that if $\lim\limits_{x \to a} f(x) = L$, then $\lim\limits_{x \to a} |f(x)| = |L|$. Is the converse true? Prove your answer.

19. Prove that $f(x) = |x|$ is continuous on $\mathbb{R}$.

20. Use the result of Exercise 19 and Theorem 2.6 to prove that if f is continuous at $x = a$, then so is $|f|$.

In Exercises 21–24, use Definition 2.6 to prove that $\lim\limits_{x \to a} f(x) = \infty$ or $\lim\limits_{x \to a} f(x) = -\infty$.

21. $\lim\limits_{x \to 0} \dfrac{3x + 2}{x^2}$

22. $\lim\limits_{x \to 0} \left(1 - \dfrac{1}{x^2} \right)$

23. $\lim\limits_{x \to -1} \dfrac{-2}{|x + 1|}$

24. $\lim\limits_{x \to 1} \left(\dfrac{x + 1}{x - 1} \right)^2$ (*Hint:* Initially restrict x so that $0 \le x \le 2$.)

In Exercises 25–28, find the limit and then prove your result using Definition 2.7.

25. $\lim\limits_{x \to \infty} \dfrac{x + 2}{2x + 1}$

26. $\lim\limits_{x \to \infty} \dfrac{x^2}{x^2 + 3}$

27. $\lim\limits_{x \to -\infty} \dfrac{3 - x}{x^2 + 1}$

28. $\lim\limits_{x \to -\infty} \dfrac{x^2 - 1}{3 - x^2}$

29. Formulate a precise definition of each of the following.
   (a) $\lim\limits_{x \to \infty} f(x) = \infty$
   (b) $\lim\limits_{x \to \infty} f(x) = -\infty$
   (c) $\lim\limits_{x \to -\infty} f(x) = \infty$
   (d) $\lim\limits_{x \to -\infty} f(x) = -\infty$

In Exercises 30–35, use Definition 2.5 to prove the given limit statement.

30. $\lim\limits_{x \to 2} x^2 = 4$ (*Hint:* First restrict x to lie in the interval $(1, 3)$.)

31. For any real number a, $\lim\limits_{x \to a} x^2 = a^2$. (*Hint:* First restrict x to lie in the interval $(a - 1, a + 1)$.)

32. $\lim\limits_{x \to a} \dfrac{1}{\sqrt{x}} = \dfrac{1}{\sqrt{a}}$ if $a > 0$

33. $\lim\limits_{x \to 2} \dfrac{x}{x - 1} = 2$

34. $\lim\limits_{x \to 0} f(x) = 1$ where $f(x) = \begin{cases} x^2 + 1 & \text{if } x > 0 \\ 3 & \text{if } x = 0 \\ 1 - 2x & \text{if } x < 0 \end{cases}$

35. $\lim\limits_{x \to 1} g(x) = 0$ where $g(x) = \begin{cases} x^2 - 1 & \text{if } x > 1 \\ 1 & \text{if } x = 1 \\ \dfrac{x - 1}{3} & \text{if } x < 1 \end{cases}$

36. Let $\phi(x) = \begin{cases} 0 & \text{if } x \text{ is irrational} \\ 1 & \text{if } x \text{ is rational} \end{cases}$.
Prove that ϕ is discontinuous everywhere.

37. Let $f(x) = \begin{cases} 0 & \text{if } x \text{ is irrational} \\ x & \text{if } x \text{ is rational} \end{cases}$.
Prove that f is continuous at $x = 0$ but discontinuous at all other points.

38. Prove that if $\lim\limits_{x \to a} f(x) = L$ and $L > 0$, then there is a positive number δ such that $f(x) > 0$ for all x in the interval $(a - \delta, a + \delta)$, except possibly at $x = a$.

39. Use the result of Exercise 38 to show that if f is continuous at $x = a$ and $f(a) > 0$, then there is a positive number δ such that $f(x) > 0$ for all x in $(a - \delta, a + \delta)$. State and prove an analogous result for $f(x) < 0$.

40. Prove the Squeeze Theorem (Theorem 2.3). (*Hint:* First show that $g(x) - L \le f(x) - L \le h(x) - L$. Then show how it follows that $|f(x) - L| \le |g(x) - L| + |h(x) - L|$.)

Chapter 2 Review Exercises

1. Let $f(x) = \dfrac{x^2 - x}{x^2 - 4}$. Find each of the following limits, and then use the information from parts (a) through (g) to draw the graph of f.

  (a) $\lim\limits_{x \to 2^+} f(x)$ (b) $\lim\limits_{x \to 2^-} f(x)$

  (c) $\lim\limits_{x \to -2^+} f(x)$ (d) $\lim\limits_{x \to -2^-} f(x)$

  (e) $\lim\limits_{x \to \infty} f(x)$ (f) $\lim\limits_{x \to -\infty} f(x)$

  (g) The zeros of f

In Exercises 2–10, find the indicated limits or show they do not exist.

2. (a) $\lim\limits_{x \to 1} \dfrac{1 - 2x}{(x-1)^{2/3}}$ (b) $\lim\limits_{x \to \pi/3^-} \dfrac{3 + 2\tan x}{\sec x - 2}$

3. (a) $\lim\limits_{x \to 2} \dfrac{x^2 - 4x + 4}{x^2 + 2x - 8}$ (b) $\lim\limits_{x \to 3} \dfrac{\sqrt{x + 1} - 2}{x - 3}$

4. (a) $\lim\limits_{x \to 1^+} \dfrac{x^2 - 1}{\sqrt{x - 1}}$ (b) $\lim\limits_{x \to -1} \dfrac{x^{1/3} + 1}{x^{2/3} - 1}$

5. (a) $\lim\limits_{x \to 2} \dfrac{x - 2}{|x - 2|}$ (b) $\lim\limits_{x \to 0} \dfrac{x}{\sqrt{x^3 + x^2}}$

6. (a) $\lim\limits_{x \to -3^-} \dfrac{2x^2 - 1}{x^2 + 3x}$ (b) $\lim\limits_{x \to 2^-} \dfrac{\sqrt{4 - x}}{x - 2}$

7. (a) $\lim\limits_{x \to 2} \dfrac{\frac{1}{x} - \frac{1}{2}}{x - 2}$ (b) $\lim\limits_{x \to -1} \dfrac{|x - 1| - 2}{x + 1}$

8. (a) $\lim\limits_{x \to \pi/4} \dfrac{\sin x - \cos x}{\cos 2x}$ (*Hint:* Find an identity for $\cos 2x$ in Appendix 2.)

  (b) $\lim\limits_{x \to \pi/2} \dfrac{\sec x - \tan x}{\sin x - 1}$ (*Hint:* Write in terms of $\sin x$ and $\cos x$.)

9. (a) $\lim\limits_{x \to -\infty} \dfrac{\sin x}{\sqrt{1 - x}}$ (b) $\lim\limits_{x \to \infty} \dfrac{x^{2/3}}{\sqrt{4x^2 - 9}}$

10. (a) $\lim\limits_{x \to \pi^-} \dfrac{1 - 2\cos x}{\sin x}$ (b) $\lim\limits_{x \to 1} \dfrac{x^2 - 1}{x^3 - 1}$

11. Let $f(x) = \sqrt{x}$, $g(x) = 2x + 3$, $F = f \circ g$, and $G = g \circ f$. Find

$$\lim\limits_{h \to 0} \dfrac{F(x + h) - F(x)}{h} \quad \text{and} \quad \lim\limits_{h \to 0} \dfrac{G(x + h) - G(x)}{h}$$

In Exercises 12–14, determine all points of discontinuity of the given function, and at each such point tell in what ways the definition of continuity fails to be satisfied.

12. (a) $f(x) = \dfrac{x + 1}{x^2 - 3x - 10}$

  (b) $f(x) = \dfrac{1 - x^2}{x + 1}$

13. (a) $f(x) = \dfrac{|2 - x|}{x - 2}$ if $x \neq 2$; $f(2) = 1$

  (b) $f(x) = \dfrac{|x|}{x^2 - 2}$

14. (a) $f(x) = \begin{cases} x^2 + 1 & \text{if } x > 0 \\ 0 & \text{if } x = 0 \\ 1 - x & \text{if } x < 0 \end{cases}$

  (b) $f(x) = \begin{cases} \dfrac{2}{x} & \text{if } x > 0 \\ \dfrac{x}{2} & \text{if } x \leq 0 \end{cases}$

In Exercises 15–18, prove the statements using Definition 2.5.

15. $\lim\limits_{x \to 4}(2x - 5) = 3$ **16.** $\lim\limits_{x \to -1} \dfrac{1 - x}{2} = 1$

17. $\lim\limits_{x \to 4} |2 - x| = 2$ **18.** $\lim\limits_{x \to 0} \sqrt{2x + 1} = 1$

19. Using Definition 2.8, prove that the linear function $f(x) = mx + b$, where $m \neq 0$, is continuous everywhere.

20. Using Definition 2.6, prove that $\lim\limits_{x \to -2} \dfrac{3}{(x + 2)^2} = \infty$.

21. Using Definition 2.7, prove that $\lim\limits_{x \to \infty} \dfrac{2x^2 - 1}{x^2 + 4} = 2$.

Chapter 2 Concept Quiz

1. Define:
   (a) $\lim\limits_{x \to a} f(x) = L$
   (b) $\lim\limits_{x \to a^+} f(x) = L$
   (c) $\lim\limits_{x \to a^-} f(x) = L$
   (d) $\lim\limits_{x \to a} f(x) = \infty$
   (e) $\lim\limits_{x \to a} f(x) = -\infty$
   (f) $\lim\limits_{x \to \infty} f(x) = L$
   (g) $\lim\limits_{x \to -\infty} f(x) = L$
   (h) f is continuous at a.
   (i) f is continuous on $[a, b]$.

2. Suppose $\lim_{x \to a} f(x) = L$ and $\lim_{x \to a} g(x) = M$. What can be said about each of the following?
   (a) $\lim\limits_{x \to a} cf(x)$ (c a constant)
   (b) $\lim\limits_{x \to a}[f(x) + g(x)]$
   (c) $\lim\limits_{x \to a}[f(x) - g(x)]$
   (d) $\lim\limits_{x \to a} f(x) \cdot g(x)$
   (e) $\lim\limits_{x \to a} f(x)/g(x)$
   (f) $\lim\limits_{x \to a}[f(x)]^n$
   (g) $\lim\limits_{x \to a} \sqrt[n]{f(x)}$

3. (a) State three conditions that must be satisfied in order for a function f to be continuous at a.
   (b) For each of the conditions in part (a), give a graph of a function that fails to satify the condition.

4. Fill in the blanks.
   (a) If $\lim_{x \to a} f(x) = L$ and $L \neq f(a)$ or $f(a)$ does not exist, then f has a _____ discontinuity at a.
   (b) If $\lim_{x \to a^+} f(x)$ and $\lim_{x \to a^-} f(x)$ both exist but are unequal, then f has a _____ discontinuity at a.
   (c) If f is _____ on $[a, b]$ and c is any number between $f(a)$ and $f(b)$, then there is some x_0 between a and b such that _____.
   (d) If $f(x) \leq g(x) \leq h(x)$ for all x in an open interval containing a and $\lim_{x \to a} f(x) = \lim_{x \to a} h(x)$, then _____.

5. State which of the following are true and which are false.
   (a) If f has a finite value for each x in I, then f is continuous on I.
   (b) The sum of two discontinuous functions at a point is also discontinuous at the point.
   (c) If $f(a) \cdot f(b) < 0$ and f is continuous on $[a, b]$, then the equation $f(x) = 0$ is satisfied by at least one x in (a, b).
   (d) If $\lim_{x \to a} f(x) \cdot g(x)$ exists, then so do $\lim_{x \to a} f(x)$ and $\lim_{x \to a} g(x)$.
   (e) It is impossible for a function that is continuous on an interval I to be positive at just one point in I.

APPLYING CALCULUS

1. According to Einstein's special theory of relativity, if a body of initial mass m_0 is moving at a velocity v then its mass m is

$$m = \frac{m_0}{\sqrt{1 - \dfrac{v^2}{c^2}}}$$

where c is the speed of light (3×10^8 m/s). What happens to m as v approaches c from smaller values? (Note that when v is small relative to c, as is true of most moving bodies near the earth, the mass is approximately constant.)

2. Einstein's special theory of relativity explains how an object moving close to the speed of light can appear shortened. The following formula was earlier found by the Dutch physicist H. A. Lorentz and is known as the *Lorentz contraction*:

$$L = L_0 \sqrt{1 - \frac{v^2}{c^2}}$$

where L is the apparent length, L_0 the original length, v the velocity, and c the speed of light.

(a) What mathematical reason affects whether we can find $\lim_{v \to c^+} L$? Could we find $\lim_{v \to c^+} L$? Explain.

(b) Can we find $\lim_{v \to c^-} L$? Do so, and explain your reasoning.

(c) What would be the consequence of increasing the speed to the speed of light (if it were possible to do so)?

3. In economics, the profit function P gives the profit $P(x)$ to the manufacturer when x units of a commodity are produced and sold. The **marginal profit**, when x units are produced, is defined by

$$\lim_{h \to 0} \frac{P(x+h) - P(x)}{h}$$

Suppose P is given by

$$P(x) = 10,000 + 200x - x^2$$

where $P(x)$ is in dollars. Find the marginal profit.

4. In the beginning of Chapter 1, a table showing the cost of mailing a first-class letter (in 1994 in the United States) was given. In Exercise 23 of Exercise Set 1.1, you were asked to verify that this postage function p is given by

$$p(w) = \begin{cases} 0.29 + 0.23[\![w]\!] & \text{if } w \text{ is not an integer} \\ 0.29 + 0.23(w-1) & \text{if } w \text{ is an integer} \end{cases}$$

where w is in ounces and $0 < w \le 12$. (When $w > 12$ oz, Priority Mail Rates take over.)

(a) Resketch the curve. What are the domain and range of the function?

(b) Find $\lim_{w \to 2.5} p(w)$ and $p(2.5)$. Is p continuous at $w = 2.5$?

(c) Find $\lim_{w \to 4^+} p(w)$ and $\lim_{w \to 4^-} p(w)$. Is p continuous at $w = 4$?

(d) For what values of w is p continuous?

5. For a given (constant) velocity v, the distance s that an object travels in a straight line in time t is given by $s(t) = vt$. For a given (constant acceleration a, the velocity v and the distance traveled s at time t are given by $v(t) = at$ and $s(t) = \frac{1}{2}at^2$. An athlete runs a 100 meter race in 10 seconds, accelerating at a constant rate a for the first 4 seconds, attaining the velocity v, and then running at that velocity for 6 seconds.

(a) Determine the distance run s, velocity v, and acceleration a as functions of time. (Hint: s, v, a will be piecewise defined; determine s first.)

(b) Graph the function s, v and a determined in part (a). Which of these is a continuous function?

(c) Find $\lim_{t \to 4} a(t)$, $\lim_{t \to 4} v(t)$, and $\lim_{t \to 4} s(t)$.

(d) What is the runner's maximum speed?

6. In the spectrum of the sun and stars, a series of colors appear dark. This series converges towards the blue, that is, toward shorter wavelengths. In the last century, Johann Balmer, a Swiss schoolteacher, found that the wavelengths λ obey the following rule: $1/\lambda = $ (constant) $\left(\dfrac{1}{2^2} - \dfrac{1}{n^2} \right)$, $n = 3,\ 4,\ 5\ldots$, "the Balmer series."

(a) Given that the wavelength of the first line ($n = 3$), known as H-alpha, is 656.3 nm, find the constant.

(b) Using a calculator, spreadsheet, or CAS, calculate $\dfrac{1}{2^2} - \dfrac{1}{n^2}$ and, to four significant figures, the wavelengths of the lines $n = 4$ (H-beta), $n = 5$ (H-gamma), and $n = 6$ (H-delta).

(c) By choosing larger and larger values of $n(n = 50,\ n = 100,\ n = 200,\ n = 500)$, find the Balmer limit, the value of λ that the sequence converges.

(d) Compare this value with the limit you can find symbolically by letting $n \to \infty$ in $\dfrac{1}{2^2} - \dfrac{1}{n^2}$.

Judith V. Grabiner

We've heard that Isaac Newton and Gottfried Wilhelm Leibniz independently and almost simultaneously invented the calculus. What exactly did they invent and why was it different from what was known mathematically in that time?

By the mid-seventeenth century, methods already existed to find tangents, areas, volumes, and maxima and minima, mostly involving curves represented by polynomial functions. Newton and Leibniz saw that these separate problems could all be solved using the two general concepts that we now call derivative and integral.

Leibniz thought of the derivative of a function as the quotient of the infinitely small differences of the function and of the independent variable; he called it the *differential quotient*. His integral represents the area under a curve as the sum of rectangles of infinitely small width drawn under it. Newton thought of a function whose value changes as a "flowing quantity" or a quantity in flux; its rate of change or flow is what we call its derivative and what he called its *fluxion*. The changing function itself—that which flows—he called the *fluent*; this is the inverse of the fluxion and is what we now call the integral.

Newton and Leibniz each devised a notation for these concepts, which made calculating with them almost automatic. Because of the notation, the new methods were not only powerful, but also easy to learn and use.

Above all, both men realized—and demonstrated—that the basic processes of finding tangent and areas, that is, differentiating and integrating, are mutually inverse—a result we now call the Fundamental Theorem of Calculus. This was a revelation. Finally, their work made clear that the calculus could be used to solve a large number of problems. Mathematicians in the eighteenth century extended the calculus in many directions.

Is it true that in the eighteenth century, mathematicians didn't worry much about the underlying logic of the discipline?

Yes. Results were what counted. They could now solve the equations of motion for the whole solar system. They could analyze physical systems ranging from bridges to projectile motion to vibrating strings. Calculus is about rates of change and the results of those changes, that an area or volume is the sum of infinitely thin slices, and that the slope of the tangent is given by the ratio of an infinitely small change in a function to the corresponding change in the independent variable. They could then use these intuitions and the new notation to find distances and speeds, tangents and area, arc lengths and volumes. Calculus helped solve many important problems, so people paid less attention to its logical foundations. They certainly did not understand or care about foundations the way we do today.

Weren't eighteenth-century mathematicians making errors because they didn't have an explicitly formulated rigorous foundation for the calculus?

No. They were usually right. If one concentrates on functions of one variable, on functions that—because they arise from physical problems—are usually continuous, which take only real and not complex values, and if one considers only those infinite series that are power series with finite coefficients, the solution methods work very well without worrying about whether the derivative or integrals exist or whether the series converge. Also, mathematicians such as Euler and Laplace had a real insight into the ideas they used and were able to avoid logical pitfalls.

How did the concept of limit develop?

The term *limit* originally just meant "boundary." Think of the border of a county, or think of the circle as the "limit" of all inscribed polygons with arbitrarily many sides. Limits had been described since the time of Newton as bounds that could be approached closer and closer but not surpassed. Since average speed is defined as the ratio of the change in distance to the change in time, Newton understood an instantaneous speed to be that which is approached, closer and closer, by a ratio of finite changes in distance and time. In modern textbooks, we call this the limit of $[f(x + h) - f(x)]/h$ as h goes to zero. Thus, the instantaneous rate of change is the limit of the ratio of finite quantities. Newton called the instantaneous speed the "last value" that the ratio has right when the quantities h and $f(x + h) - f(x)$ vanish together. At first this sounds very modern but there are two differences between his concept and the modern one. First, he spoke of actually "reaching" the limit when the quantities vanish, but all that is required in proofs about limits is that the quantities "approach nearer than any given difference." Newton thought his formulation was more exact, but it later caused logical difficulties. Also, requiring that the variable "never go beyond" the limit means that it cannot oscillate about its limit. For these reasons, Newton's description of limit was not general enough to support proofs about all the concepts of the calculus. Finally, Newton and his colleagues did not translate this definition into algebraic terms and did not apply it algebraically to proofs; it was designed to make people feel more confident about the subject in general. The description is basically correct, but it needed refinement, which it got in the late eighteenth and early nineteenth centuries.

What about the derivative?

In a sense, the derivative was used before it was fully understood. In the 1630s Fermat developed a method for finding maxima and minima and a method for finding tangents. It worked, but he didn't fully understand why, or how the two problems were related. Fermat was implicitly using derivatives without realizing it. Newton and Leibniz effectively discovered the derivative by showing that this single concept unified the methods for finding maxima and minima, tangents, and velocities. But although they understood the derivative's most useful properties, even they didn't have the whole story. Mathematicians such as Taylor, Maclaurin, Euler, and Lagrange learned more about the derivative, especially through their work on series and on dif-on differential equations. Cauchy eventually gave a second algebraic definition of the concept of limit. He then used that definition to rigorously define the derivative as the limit, when it exists, of the ratio $[f(x+h)-f(x)]/h$ as h goes to zero. After Cauchy, the calculus was viewed differently. It was no longer seen just as a set of powerful methods, but a rigorous subject—like Euclidean geometry—with clear definitions and theorems whose proofs were based on those definitions. Cauchy's new rigor put the earlier results on a firm foundation, but it also provided a framework for new results, some of which could not even have been formulated before his work. However, Cauchy's original work had a subtle flaw—not distinguishing between convergence and uniform convergence—and it was left to later workers to produce a thoroughly rigorous treatment of calculus.

The historical development is the reverse of what we see in the textbooks. There, one starts with a definition, then explores some properties and results, and only then discusses applications. Historically, mathematicians began with the problems they wanted to solve and then developed the concepts they needed. When it was clear what the key properties were, they gave the subject its final, rigorous form, thus establishing the precise conditions under which the results were valid.

OK, but why did rigor suddenly become important?

Proofs have always been important to mathematicians but until the nineteenth century calculus wasn't ready to be made rigorous. Only then was there a well-developed algebra of inequalities. The key properties of calculus were better understood, and these properties could be expressed in the language of inequalities. Thus, Cauchy, Riemann, and Weierstrass gave the calculus a rigorous basis, using the already existing language of inequalities, and built a logically connected structure of theorems about the concepts of the calculus. Also, they started to think of themselves as mathematicians, not scientists, and believed that more abstract methods were appropriate to mathematics.

Another set of reasons was philosophical. In 1734 the philosopher Bishop Berkeley attacked calculus as being incorrect because it wasn't rigorous. One of his motives was to defend religion. Scientists, he said, attack religion for being unreasonable; well, let them improve their own reasoning first. He ridiculed Newton's treatment of vanishing increments, calling them "ghosts of departed quantities." He attacked the method of calculating even the simplest derivatives. Suppose, for instance, that $y = x^2$. Now take the ratio of the differences $[(x + h)^2 - x^2]/h$. When the ratio is simplified algebraically, one gets $2x + h$. To get the derivative $2x$, we have to let h "vanish." But what right did we have to get rid of the h? If h is zero, we cannot divide by it in setting up the ratio. But if it is not zero, we have no right to throw it away. A quantity is either zero or not; there is nothing in between! Berkeley described the mathematicians of his time as "rather accustomed to compute than to think." The answer to this apparent paradox requires recognizing that an equation involving limits is really a shorthand expression for a set of inequalities; the ratio of differences never really equals $2x$, but the ratio can be made as close

to $2x$ as one wishes, provided that h is taken small enough. This is a subtle and complex idea. No eighteenth-century mathematician managed to give a satisfactory response, though many tried. Maclaurin, d'Alembert, Lagrange, and Lazare Carnot, to name a few, all were concerned enough about these criticisms to think and write about the foundations of the calculus. Finally, it seemed to many late eighteenth century mathematicians that there was a limit to the results that could be obtained by their methods. The philosopher Diderot said that the mathematicians of the day had "erected the pillars of Hercules" beyond which it was impossible to go. Lagrange became interested in the foundations of calculus and influenced other mathematicians to work on the problem. Nobody solved it, though, so when Lagrange had to teach the calculus at the Ecole Polytechnique in Paris, he tried to solve it himself. And his work, in turn, decisively influenced Cauchy.

Teaching mathematics forces you to justify your concepts, and at the end of the eighteenth century there was an increased focus on teaching in Europe. It was obvious that mathematics was very useful, and therefore it was necessary to teach more of it to students of science and engineering. The need to teach calculus was the catalyst that produced the new rigor from the knowledge gained in the eighteenth century.

That's very interesting. One doesn't usually think of teaching being the impetus for research. It's usually the opposite, isn't it?

Explaining complex ideas to students makes you think deeply about what they really mean. Even in the nineteenth and twentieth centuries, important work on the foundations of analysis came from the need to teach. Weierstrass developed his ideas on foundations in his lectures at Berlin; Dedekind said he first realized the need for a deeper understanding of continuity while teaching at Zurich; Dini and Landau turned to foundations when teaching analysis. And—back to the first rigorous definition of limit and derivative—Cauchy's foundations of analysis originated as lectures at the Ecole Polytechnique.

Judith V. Grabiner teaches mathematics and the history of science at Pitzer College in Claremont, California. Among her publications are *The Origins of Cauchy's Rigorous Calculus* (MIT Press, 1981), *The Calculus as Algebra: J. L. Lagrange 1736–1813* (Garland, 1990) and "The Centrality of Mathematics in the History of Western Thought," *American Mathematical Monthly*, 1988. She is currently working on a history of the calculus of Colin Maclaurin. She received her B.S degree in Mathematics from the University of Chicago and her Ph.D. in the History of Science from Harvard University.

3

THE DERIVATIVE

In this chapter we begin the study of one of the two major branches of calculus, called **differential calculus**. It is based on the concept of the **derivative**, which is defined as a special type of limit. The second major branch of calculus, integral calculus, is also based on limits, and we will see that there is a very close connection between the two branches.

To understand the concept of the derivative we first consider two seemingly unrelated problems: (1) finding the tangent line to a given curve (a geometric problem) and (2) finding the velocity of a moving object (a physical problem). Imagine a line drawn between the points at which two consecutive wheels of a roller coaster contact the track. The slope of that line is approximately the slope of the tangent line to a curve (the track) at a given point (the position of the car at the instant the picture was taken). At that same instant the car has a specific "instantaneous" velocity. Finding the slope of the tangent line and the instantaneous velocity of an object are both problems that can be solved using the derivative. It is important to understand that although the position and speed of the roller coaster provide a useful physical model, the derivative is a general mathematical concept that is very powerful in developing applications for both the physical world and mathematics itself.

After we investigate the relationship between a function and its derivative, which is itself a function, we will develop some working rules for calculating derivatives of functions such as those introduced in Chapter 1. This development will provide the background needed for the study of applications of the derivative, which is the subject of Chapter 4.

3.1 THE TANGENT PROBLEM AND THE VELOCITY PROBLEM

We know that the tangent line to a circle at a point is perpendicular to the radius drawn to that point (see Figure 3.1), but how can we find tangent lines to other curves? More fundamentally, how is the tangent line defined? Students are

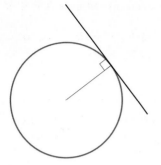

FIGURE 3.1

often tempted to say "the line that touches the curve in only one point" or "the line that touches the curve but doesn't cross it." The curves in Figure 3.2 show, however, that neither of these descriptions suffices.

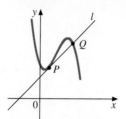

Line l is tangent to the curve at P but also crosses the curve at Q.

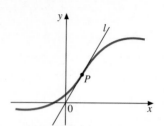

Line l is tangent to the curve at P and also crosses the curve there.

FIGURE 3.2

To see how to define the tangent line, consider the graph of $y = f(x)$. Let P be the point at which we want to draw the tangent line. As shown in Figure 3.3(a), let Q be another point on the graph, and consider the line that joins P and Q. This line is called a *secant line*. (The word *secant* comes from a Latin word meaning *to cut*.) Holding P fixed, we now let Q move along the curve toward P. In this way, we get a collection of secant lines through P, as shown in Figure 3.3(b). If these secant lines approach some limiting position as Q approaches P, then this limiting line is by definition the tangent line at P.

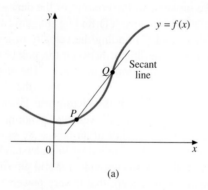

(a)

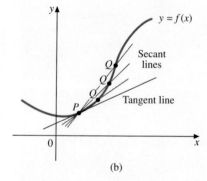

(b)

FIGURE 3.3

We can state the definition in a more useful way by considering slopes. Suppose P has coordinates $(a, f(a))$. The coordinates of Q can then be written in the form $(a+h, f(a+h))$ for some $h \neq 0$. The slope of the secant line PQ is then

$$m_{\text{sec}} = \frac{f(a+h) - f(a)}{h}$$

(See Figure 3.4.) Now we can make Q approach P simply by letting h approach 0. It is natural to expect that if the tangent line at P exists, then the slopes of these secant lines approach the slope of the tangent line at P, that is,

$$\lim_{h \to 0} m_{\text{sec}} = m_{\text{tan}}$$

FIGURE 3.4

Note: In Figure 3.4 we show $h > 0$. We could instead have $h < 0$, in which case Q would be on the other side of P.

This discussion leads to the following definition (made by the seventeenth-century French mathematician Pierre de Fermat).

Definition 3.1
The Tangent Line to the Graph of a Function

Let $P(a, f(a))$ be a point on the graph of the function f. If the limit

$$m_{\tan} = \lim_{h \to 0} \frac{f(a + h) - f(a)}{h} \tag{3.1}$$

exists (which implies that it is finite), the unique line through P with slope equal to $m_{\tan}$ is called the **tangent line** to the graph of f at P.

EXAMPLE 3.1   Show that a tangent line exists at $(1, 2)$ for the curve defined by $y = x^2 + 1$, and find its equation.

Solution   Let $(a, f(a)) = (1, 2)$ and investigate the limit in Equation 3.1:

$$\lim_{h \to 0} \frac{f(1 + h) - f(1)}{h} = \lim_{h \to 0} \frac{[(1 + h)^2 + 1] - 2}{h}$$

$$= \lim_{h \to 0} \frac{1 + 2h + h^2 + 1 - 2}{h}$$

$$= \lim_{h \to 0} \frac{h^2 + 2h}{h}$$

$$= \lim_{h \to 0} \frac{h(h + 2)}{h}$$

$$= \lim_{h \to 0} (h + 2) = 2 \quad \text{Remember that } h \neq 0.$$

Since this limit exists, the tangent line exists and has slope $m_{\tan} = 2$. Thus, from the point-slope form of a straight line $y - y_1 = m(x - x_1)$, we have

$$y - 2 = 2(x - 1)$$

or, after simplification,

$$y = 2x$$

In Figure 3.5 we show the graph of $y = x^2 + 1$ and its tangent line $y = 2x$ at the point $(1, 2)$. ∎

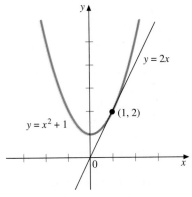

$y = 2x$

$(1, 2)$

$y = x^2 + 1$

FIGURE 3.5

EXAMPLE 3.2   Find the slope of the tangent line to the graph of $f(x) = 2\sqrt{x}$ at an arbitrary point $(x, f(x))$ on the curve, with $x > 0$.

Solution   By Equation 3.1, with a replaced by x,

$$m_{\tan} = \lim_{h \to 0} \frac{f(x + h) - f(x)}{h}$$

$$= \lim_{h \to 0} \frac{2\sqrt{x + h} - 2\sqrt{x}}{h}$$

We rationalized the numerator in order to change the form of the fraction to one whose limit we could find.

$$= \lim_{h \to 0} \frac{2(\sqrt{x+h} - \sqrt{x})}{h} \cdot \frac{\sqrt{x+h} + \sqrt{x}}{\sqrt{x+h} + \sqrt{x}} \qquad \text{Rationalizing the numerator}$$

$$= \lim_{h \to 0} \frac{2(x+h-x)}{h(\sqrt{x+h} + \sqrt{x})}$$

$$= \lim_{h \to 0} \frac{2h}{h(\sqrt{x+h} + \sqrt{x})}$$

$$= \lim_{h \to 0} \frac{2}{\sqrt{x+h} + \sqrt{x}} \qquad \text{Since } h \neq 0$$

$$= \frac{2}{2\sqrt{x}} = \frac{1}{\sqrt{x}}$$

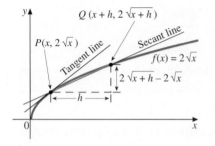

FIGURE 3.6

In Figure 3.6 we show the graph of f, along with a secant line joining $P(x, 2\sqrt{x})$ and $Q(x+h, 2\sqrt{x+h})$, as well as the tangent line at P. Note that we now have the formula for the slope of the tangent line to this curve at *any* point for which $x > 0$. For example, at the point $(1, 2)$ the slope is $1/\sqrt{1} = 1$, and at $(4, 4)$ it is $1/\sqrt{4} = 1/2$. ∎

The Velocity Problem

The problem of finding the tangent line to a curve at an arbitrary point was first studied systematically by the French mathematician Pierre de Fermat (1601–1665). His work preceded that of Newton and Leibniz and provided an important part of the background for their later formulation of the calculus. Fermat (along with René Descartes) was an originator of what is now known as analytical geometry. Fermat's work on tangent lines soon led to a way of finding the instantaneous velocity of a moving object. This instantaneous velocity can be interpreted as the slope of the tangent line to the position-versus-time curve. Thus, a purely geometric problem provided the key to solving a physical problem.

To see how the two problems are related, let us begin with an example. Suppose a skier moves downhill on a straight path for the first 3 seconds. The skier's distances $s(t)$ in meters from the starting point at half-second intervals are given in the table.

t	0	0.5	1.0	1.5	2.0	2.5	3.0	t in seconds,
$s(t)$	0	5.3	11.7	19.2	29.5	41.2	53.8	$s(t)$ in meters

Let us consider the problem of finding the skier's velocity at a particular instant, say $t = 2$. We can easily find the *average velocity* in the time interval from $t = 2.0$ to $t = 2.5$, using the familiar relationship "distance = rate × time." In this case, "rate" means average velocity. Thus, the average velocity over the time interval in question is

$$v_{\text{ave}} = \frac{\text{distance covered}}{\text{elapsed time}} = \frac{41.2 - 29.5}{2.5 - 2.0}$$

$$= \frac{11.7}{0.5} = 23.4 \text{ m/s}$$

t	$s(t)$
1.9	27.33
2.0	29.50
2.1	31.71

Since the skier's velocity is increasing with time, the velocity at $t = 2$ is somewhat less than 23.4. So v_{ave} in this case is an overestimate of the velocity at $t = 2$. Suppose now that distances are known at time intervals of 0.1 second, as shown, near $t = 2$. Then the average velocity from $t = 2.0$ to 2.1 is

$$v_{\text{ave}} = \frac{31.71 - 29.50}{2.1 - 2.0} = \frac{2.21}{0.1} = 22.1 \text{ m/s}$$

If we knew distances for still smaller time intervals, our average velocities would come closer and closer to the instantaneous velocity at $t = 2$. For example, if we know the distance is 29.720 when $t = 2.01$, then the average velocity from $t = 2.00$ to $t = 2.01$ is

$$v_{\text{ave}} = \frac{29.720 - 29.500}{2.01 - 2.00} = \frac{0.220}{0.01} = 22.0 \text{ m/s}$$

REMARK ───────────────────────────

■ We have used the time intervals [2.0, 2.5], [2.0, 2.1], and [2.00, 2.01] in this example. We could equally well have used intervals with 2 as the right endpoint: [1.5, 2.0], [1.9, 2.0], and [1.99, 2.00]. If we had used these intervals, we would have obtained average velocities that underestimate the instantaneous velocity at $t = 2$.

─────────────────────────────────

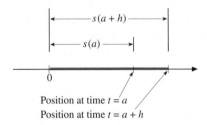

Position at time $t = a$
Position at time $t = a + h$

FIGURE 3.7

Let us try to generalize the procedure we have just described. Suppose an object (such as a person, a car, or a rocket) is moving along a straight path. We set up a coordinate system so that the object's position $s(t)$ on the path is known at time t. We wish to determine the velocity of the object at a particular instant, say $t = a$. As a first approximation, we calculate the average velocity over a small time interval from $t = a$ to $t = a + h$, where $h \neq 0$. Figure 3.7 shows the position at $t = a$ and at $t = a + h$. The average velocity over this time interval is

$$v_{\text{ave}} = \frac{\text{change in position}}{\text{change in time}} = \frac{s(a + h) - s(a)}{h}$$

If we now consider smaller and smaller values of h, the average velocities may approach some limiting value. If so, then this limiting value is by definition the **instantaneous velocity** at $t = a$:

$$v(a) = \lim_{h \to 0} \frac{s(a + h) - s(a)}{h} \qquad (3.2)$$

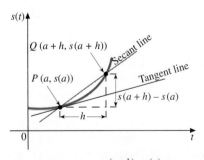

Slope of secant line $= \dfrac{s(a + h) - s(a)}{h}$

$=$ average velocity from P to Q

Slope of tangent line $= \lim\limits_{h \to 0} \dfrac{s(a + h) - s(a)}{h}$

$=$ instantaneous velocity at P

FIGURE 3.8

Compare Equation 3.2 for instantaneous velocity at $t = a$, with Equation 3.1 for the slope of the tangent line to a curve. Except for the letters used to name the functions, the equations are identical. To see more clearly why these should be the same, let us consider the graph of the position function $s(t)$, shown in Figure 3.8. Let P be the point $(a, s(a))$ and Q be the point $(a + h, s(a + h))$. Then the average velocity

$$v_{\text{ave}} = \frac{s(a + h) - s(a)}{h}$$

is the slope of the secant line through P and Q. As we let $h \to 0$, we see that $Q \to P$, and the secant lines approach the tangent line at P. So the instantaneous velocity at $t = a$, denoted by $v(a)$, is exactly the same as the slope of the tangent line at P.

EXAMPLE 3.3   A particle moves in a straight line in such a way that its position from the starting point, in meters, is given by

$$s(t) = 16 \left(1 - \frac{1}{t+1} \right)$$

where t is nonnegative and is given in seconds.

(a) Find the average velocity of the particle over the time interval from $t = 1$ to $t = 3$.

(b) Find the instantaneous velocity of the particle when $t = 3$.

Solution

(a) When $t = 1$, the particle is at a distance $s(1) = 16(1 - \frac{1}{2}) = 8$ m from the starting point. When $t = 3$, it is $s(3) = 16(1 - \frac{1}{4}) = 12$ m from the starting point. The average velocity is given by the distance traveled divided by the elapsed time:

$$v_{ave} = \frac{s(3) - s(1)}{3 - 1} = \frac{12 - 8}{2} = 2 \text{ m/s}$$

(b) In Equation 3.2 we let $a = 3$ and get

$$v(3) = \lim_{h \to 0} \frac{s(3 + h) - s(3)}{h}$$

$$= \lim_{h \to 0} \frac{16 \left(1 - \dfrac{1}{3 + h + 1} \right) - 12}{h}$$

$$= \lim_{h \to 0} \frac{16 - \dfrac{16}{4 + h} - 12}{h}$$

$$= \lim_{h \to 0} \frac{4 - \dfrac{16}{4 + h}}{h}$$

$$= \lim_{h \to 0} \frac{4(4 + h) - 16}{h(4 + h)} \qquad \text{We multiplied numerator and denominator by } 4 + h.$$

$$= \lim_{h \to 0} \frac{4h}{h(4 + h)}$$

$$= \lim_{h \to 0} \frac{4}{4 + h} = 1 \qquad \text{Dividing by } h \text{ is valid since } h \neq 0.$$

So at the instant when $t = 3$ seconds, the particle is traveling at the rate of 1 m/s.

In Figure 3.9 we have shown the graph of $s(t)$ and the solution to part (a) as the slope of the secant line joining $(1, 8)$ and $(3, 12)$. The solution to part (b) is the slope of the tangent line at $(3, 12)$. ∎

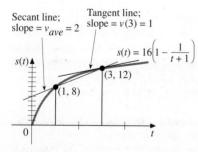

Secant line; slope = $v_{ave} = 2$ Tangent line; slope = $v(3) = 1$

$s(t) = 16\left(1 - \dfrac{1}{t+1}\right)$

$(3, 12)$

$(1, 8)$

FIGURE 3.9

From the preceding example, we note that the velocity is 1 m/s when $t = 3$, so at that instant the particle is moving in the positive direction on its path. We can see this conclusion intuitively by reasoning that if the velocity remained the same for 1 second, then the distance would be 1 meter greater after 1 second had elapsed. More generally, *when the instantaneous velocity is positive, the directed distance from the starting point is increasing*, and *when the instantaneous velocity is negative, the distance from the starting point is decreasing*. Figure 3.10 illustrates this conclusion. It should be emphasized that these are instantaneous conditions only. For example, at the instant when a particle changes direction, its velocity will be 0, but an observer would not be able to discern that it was not moving at that instant. For example, if a ball is thrown upward, gravity is pulling downward, so when the ball reaches its maximum height, $v = 0$. But v is always changing and does not remain at 0 for any finite time. It is an *instantaneous* velocity.

We will consider the motion of a particle in more detail after we have formalized the notion of a derivative.

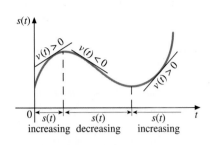

FIGURE 3.10

Exercise Set 3.1

1. Let $f(x) = x^2/2$, and let $P = (2, 2)$ and $Q = (2 + h, f(2 + h))$.
   (a) On a graph, show the curve $y = f(x)$ and the secant lines through P and Q for $h = 1$, $h = 0.5$, and $h = 0.1$, respectively.
   (b) Calculate the slopes of the secant lines in part (a).
   (c) Find the slope of the tangent line at P.
   (d) Find the equation of the tangent line at P.

2. Let $f(x) = 4/x$, and let $P = (2, 2)$ and $Q = (2 + h, f(2 + h))$. Repeat parts (a)–(d) of Exercise 1 for this function.

In Exercises 3–10, show that a tangent line to the graph of $y = f(x)$ exists at $(a, f(a))$, and find its equation. Draw the graph of the function and show its tangent line at $(a, f(a))$.

3. $f(x) = 1 - x^2$; $a = 1$ **4.** $f(x) = 4 - \dfrac{2}{x}$; $a = 1$

5. $f(x) = \dfrac{2}{x - 1}$; $a = -1$

6. $f(x) = x^2 - x$; $a = 1$

7. $f(x) = (x - 2)^2$; $a = 3$

8. $f(x) = 3x - x^2$; $a = 2$

9. $f(x) = \dfrac{1}{2}x(x - 2)$; $a = -1$

10. $f(x) = \dfrac{3}{x + 2}$; $a = -4$

11. A particle moves in a straight line in such a way that its directed distance from the origin at time t is $s(t) = 2t^2 - 3$.
   (a) Find the average velocity over each of the time intervals $[1, 2]$, $[1, 1.5]$, $[1, 1.1]$, and $[1, 1.01]$.
   (b) Find the average velocity over the time interval $[1, 1 + h]$ for h arbitrary but nonzero.
   (c) Find the instantaneous velocity at time $t = 1$.
   (d) Sketch the graph of $y = s(t)$ and show the instantaneous velocity at $t = 1$.

12. A driver sees traffic stopped ahead and applies the brakes. Assume that the table below gives the distance the car travels (in feet) in t seconds after the brakes are applied.

t	0	0.5	1.0	1.5	2.0	2.5	3.0
$s(t)$	0	38	74	106	131	149	160

   (a) Find the average velocity in the time interval $[2.0, 2.5]$.
   (b) Find the average velocity in the time interval $[1.5, 2.0]$.
   (c) Estimate the instantaneous velocity at $t = 2.0$ and explain the basis for your estimate.
   (d) What is the significance of the sign of the answers in parts (a)–(c)?

13. Let $s(t)$ be the position function of a particle moving in a straight line. Consider the graph of $s(t)$ versus t, and tell what you can infer about the velocity $v(t)$ in each of the following situations, assuming that $v(t)$ exists.
(a) $s(t)$ is increasing.
(b) $s(t)$ is decreasing.
(c) The graph is concave up.
(d) The graph is concave down.
(e) The point $(t_0, s(t_0))$ is a local maximum point.

14. The accompanying graph shows the position function $s(t)$ versus time t for a particle moving in a straight line. Give the time intervals during which the particle is doing the following things:
(a) moving in the positive direction;
(b) moving in the negative direction;
(c) speeding up;
(d) slowing down.
At what times is the particle's velocity 0?

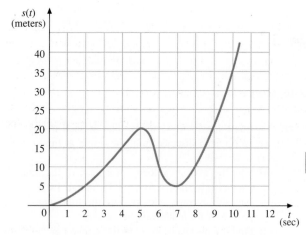

15. By estimating values from the graph in Exercise 14, obtain approximate values for the following.
(a) The average velocity during the time interval $[5, 6]$
(b) The instantaneous velocity at $t = 5$

16. A person on a bicycle leaves home, riding for several minutes on a straight, level road, gradually picking up speed. Then the road goes downhill and the cyclist picks up speed rapidly but slows down rapidly when the road goes uphill again. After reaching the top, the cyclist turns around and returns home over the same road. Sketch a graph of the position function $s(t)$ that would be consistent with the cyclist's ride.

In Exercises 17–26, use the given values of a and s(t) to find the instantaneous velocity v(a) of an object moving in a straight line in such a way that its directed distance from the origin is s(t) at time t.

17. $s(t) = 2t + 3$; $a = 2$ **18.** $s(t) = 3t - 4$; $a = 3$

19. $s(t) = \dfrac{t + 1}{2}$; $a = 5$ **20.** $s(t) = t^2$; $a = 1$

21. $s(t) = t^2 - t$; $a = 3$ **22.** $s(t) = 1 - \dfrac{1}{t}$; $a = 2$

23. $s(t) = \dfrac{t + 4}{t}$; $a = 3$ **24.** $s(t) = \dfrac{12}{t - 1}$; $a = 4$

25. $s(t) = t^2 - \dfrac{2}{t}$; $a = 2$ **26.** $s(t) = \dfrac{8}{t + 2}$; $a = 2$

In Exercises 27–30, find the slope of the tangent line to the graph of $y = f(x)$ at an arbitrary point $(x, f(x))$.

27. $f(x) = x^2 - 2x$ **28.** $f(x) = \dfrac{1}{x^2}$; $x \neq 0$

29. $f(x) = \dfrac{x}{x + 1}$; $x \neq -1$

30. $f(x) = x - \dfrac{1}{x}$; $x \neq 0$

31. Find the equation of the tangent line to the graph of $y = \sqrt{2x + 1}$ at $x = 4$.

32. Find the average velocity during the time interval from $t = 0$ to $t = 3$ of a particle moving in a straight line according to the formula $s(t) = 4/\sqrt{t + 1}$. What is the instantaneous velocity at $t = 3$? In what direction is the particle moving at the instant when $t = 3$?

 In Exercises 33–35, use a CAS or a graphing calculator.

33. The position function of an object moving in a straight line is given by

$$s(t) = \frac{\sin^2 t + t^2}{t^2 - 1}$$

where t is in seconds and $s(t)$ is in feet. Estimate the intervals on which the velocity is increasing and those on which it is decreasing.

34. The position function of an object moving along a line is given by

$$s(t) = \frac{\sin^2 t + t^2}{t^2 + 1}$$

where t is in seconds and $s(t)$ is in meters.
(a) Estimate the intervals on which $v(t)$ is increasing and those on which it is decreasing.
(b) Estimate the velocity at $t = 1$ using Equation 3.2 with $h = 0.1$, $h = 0.01$, and $h = 0.001$.

35. Repeat Exercise 34 with

$$s(t) = \frac{\cos^2 t + t^2}{t^2 + 1}$$

3.2 THE DERIVATIVE

We have seen that the geometric problem of finding the slope of the tangent line to the graph of a function involves the same type of limiting operation as the physical problem of finding the instantaneous velocity of a moving object. We now introduce a name and a symbol for the type of limit involved in each of these calculations. This limit is one of the fundamental concepts in calculus.

Definition 3.2 The Derivative of a Function at a Point	Let f be a function defined in some open interval containing the point $x = a$. The **derivative** of f at $x = a$, denoted by $f'(a)$, is defined by $$f'(a) = \lim_{h \to 0} \frac{f(a+h) - f(a)}{h} \qquad (3.3)$$ provided this limit exists.

If the graph of $y = f(x)$ has a tangent line at the point $(a, f(a))$, then its slope is $f'(a)$. Similarly, if $s(t)$ is the position function for a particle moving along a line, then its velocity at time $t = a$ is the derivative $s'(a)$. These are but two of the many interpretations of the derivative as the **instantaneous rate of change** of a function with respect to its independent variable. Velocity is the instantaneous rate of change of position with respect to time. The slope of the tangent line to a curve can also be thought of as a rate of change—namely, the rate at which y changes with respect to x.

Derivative as a Rate of Change

Here are some other examples that illustrate the derivative as an instantaneous rate of change.

1. Let $N(t)$ be the size of a bacteria culture at time t. Then $N'(t)$ is the rate at which the culture grows or diminishes.
2. Let $Q(t)$ be the quantity of a radioactive substance on hand at time t. Then $Q'(t)$ is the rate of radioactive decay.
3. Let $X(t)$ be the amount present at time t of a compound being formed by the chemical reaction of two other compounds. Then $X'(t)$ is the reaction rate.
4. Let $C(x)$ be the cost of manufacturing x items of a certain type. Then $C'(x)$ is the rate of change of cost with respect to the number of items produced. This rate is called **marginal cost** in economics. Similarly, there is **marginal revenue**, $R'(x)$, and **marginal profit**, $P'(x)$.
5. Let $A(t)$ be the accumulated amount of money after t years that results from a certain principal invested at continuously compounded interest. Then $A'(t)$ is the rate of growth of the account.
6. Let $p(h)$ be the atmospheric pressure at height h m above the earth's surface. Then $p'(h)$ is the rate at which pressure is changing with respect to altitude.

7. Let $L(t)$ be the current state of knowledge of a learner in an instructor/learner situation. Then $L'(t)$ is the rate at which the material is being learned.

We could give many other examples, but these should suffice to give you a feeling for the broad applications of the derivative as an instantaneous rate of change. In the next chapter we will return to applications. For the present, we concentrate on how to calculate the derivative and on the relationship between a function and its derivative.

EXAMPLE 3.4   Find $f'(3)$ if $f(x) = \sqrt{x+1}$.

Solution

$$f'(3) = \lim_{h \to 0} \frac{f(3+h) - f(3)}{h}$$

$$= \lim_{h \to 0} \frac{\sqrt{(3+h)+1} - \sqrt{3+1}}{h}$$

$$= \lim_{h \to 0} \frac{\sqrt{h+4} - 2}{h} \cdot \frac{\sqrt{h+4} + 2}{\sqrt{h+4} + 2} \qquad \text{Here, we are rationalizing the numerator.}$$

$$= \lim_{h \to 0} \frac{h+4-4}{h(\sqrt{h+4}+2)} = \lim_{h \to 0} \frac{h}{h(\sqrt{h+4}+2)}$$

$$= \lim_{h \to 0} \frac{1}{\sqrt{h+4}+2} = \frac{1}{4}$$

So the derivative of f at 3 is $\frac{1}{4}$.   ∎

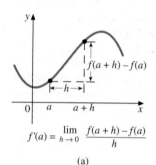

$$f'(a) = \lim_{h \to 0} \frac{f(a+h) - f(a)}{h}$$

(a)

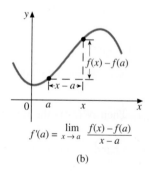

$$f'(a) = \lim_{x \to a} \frac{f(x) - f(a)}{x-a}$$

(b)

FIGURE 3.11
Two equivalent ways of writing a formula for $f'(a)$.

An alternative way of writing the limit in Definition 3.2 is sometimes useful. As we show in Figure 3.11, we let $x = a + h$, so that $h = x - a$, and observe that $h \to 0$ if and only if $x \to a$. So we can write $f'(a)$ as

$$f'(a) = \lim_{x \to a} \frac{f(x) - f(a)}{x - a} \qquad (3.4)$$

We can redo Example 3.4 using this alternative form as follows. Since $f(x) = \sqrt{x+1}$, we get $f(3) = \sqrt{4} = 2$. So, with $a = 3$, we use Equation 3.4 to obtain

$$f'(3) = \lim_{x \to 3} \frac{\sqrt{x+1} - 2}{x-3} = \lim_{x \to 3} \frac{\sqrt{x+1} - 2}{x-3} \cdot \frac{\sqrt{x+1} + 2}{\sqrt{x+1} + 2}$$

$$= \lim_{x \to 3} \frac{(x+1) - 4}{(x-3)(\sqrt{x+1}+2)} = \lim_{x \to 3} \frac{x-3}{(x-3)(\sqrt{x+1}+2)} = \frac{1}{4}$$

The alternative form of Equation 3.4 can also be readily used to prove the following important result.

THEOREM 3.1

If $f'(a)$ exists, then f is continuous at $x = a$.

Proof   According to Definition 2.9, we need to show that

$$\lim_{x \to a} f(x) = f(a)$$

or, equivalently, that $\lim_{x \to a}[f(x) - f(a)] = 0$. If $x \neq a$, we can write

$$f(x) - f(a) = [f(x) - f(a)] \cdot \left[\frac{x - a}{x - a} \right] \qquad \text{Since } \tfrac{x-a}{x-a} = 1 \text{ when } x \neq a$$

Thus, if $x \neq a$,

$$f(x) - f(a) = \frac{f(x) - f(a)}{x - a} \cdot (x - a)$$

Taking limits and using Limit Property 2 of Section 2.1, we get

$$\lim_{x \to a}[f(x) - f(a)] = \left[\lim_{x \to a} \frac{f(x) - f(a)}{x - a} \right] \left[\lim_{x \to a}(x - a) \right]$$

$$= [f'(a)] \cdot 0 = 0$$

So f is continuous at $x = a$. Notice that we made use of the fact that $f'(a)$ exists when we wrote the limit of the product as the product of the limits. ■

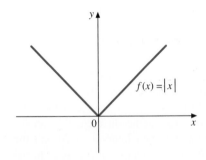

FIGURE 3.12
$f'(0)$ does not exist.

The converse of Theorem 3.1 is not true. That is, if f is continuous at $x = a$, we cannot necessarily conclude that $f'(a)$ exists. To see why, consider the example $f(x) = |x|$, whose graph is shown in Figure 3.12. If $x > 0$, then $f(x) = x$ and if $x < 0$, $f(x) = -x$, both of which are continuous linear functions. We have previously shown that $\lim_{x \to 0} f(x) = 0$; since $f(0) = 0$, this function is continuous on all of $\mathbb{R}$. Now let us investigate whether $f'(0)$ exists. First, consider $h > 0$.

$$\lim_{h \to 0^+} \frac{f(0 + h) - f(0)}{h} = \lim_{h \to 0^+} \frac{f(h) - 0}{h} = \lim_{h \to 0^+} \frac{|h|}{h}$$

$$= \lim_{h \to 0^+} \frac{h}{h} = 1$$

Whereas, if $h < 0$,

$$\lim_{h \to 0^-} \frac{f(0 + h) - f(0)}{h} = \lim_{h \to 0^-} \frac{f(h) - 0}{h} = \lim_{h \to 0^-} \frac{|h|}{h}$$

$$= \lim_{h \to 0^-} \frac{-h}{h} = -1$$

Thus, the right- and left-hand limits are not the same, and so

$$\lim_{h \to 0} \frac{f(0 + h) - f(0)}{h}$$

does not exist. That is, $f'(0)$ does not exist. Geometrically, we have shown that as we approach 0 from the right, the slope of each of the secant lines is 1, and when we approach 0 from the left, the slope is -1. Thus, *at 0* there is no well-defined tangent line. At all other points, however, the tangent lines do exist, so $f'(x)$ exists everywhere except at $x = 0$ for this function.

As another example, consider the function g whose graph is shown in Figure 3.13. The curve has a vertical tangent line at $x = -3$ and at $x = 1$. Since slope is not defined for vertical lines, it follows that for this function $g'(-3)$ and $g'(1)$ do not exist. Yet from the graph we see that g is continuous for all x.

In the case of $f(x) = |x|$, $f'(x)$ fails to exist at one point only. For $g(x)$ of Figure 3.13, $g'(x)$ fails to exist at two points. You can probably think of graphs of functions that are continuous everywhere but whose derivative fails to exist at infinitely many points. (See, for example, the "sawtooth" function

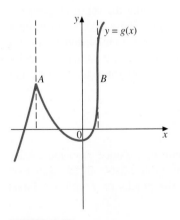

FIGURE 3.13
Tangent lines at points A and B are vertical, so slopes are undefined.

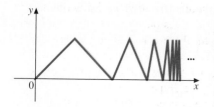

FIGURE 3.14
A "sawtooth" function that fails to have a derivative at infinitely many points.

in Figure 3.14.) It is harder to imagine a continuous function whose derivative fails to exist at *every* point. Yet such functions do exist! As you might imagine, their formulas are complicated, and they are impossible to draw, so we will not attempt to illustrate such a function. Examples can be found in advanced analysis textbooks. The first such example was given by the German mathematician Karl Weierstrass in 1861. The important point of this discussion is that the converse of Theorem 3.1 is not true. So while the existence of the derivative at a point necessarily implies continuity there, continuity at a point does not imply existence of the derivative there.

The process of calculating the derivative of a function is called **differentiation**, and to **differentiate** a function means to find its derivative. When the derivative of a function exists at a given point, we say the function is **differentiable** at that point. The quotient

$$\frac{f(a + h) - f(a)}{h} \qquad (h \neq 0)$$

that occurs in the definition of the derivative of f at a is called a **difference quotient** for f. So the derivative is a limit of a difference quotient. If a function is differentiable at all points of an open interval (a, b), we say it is **differentiable on (a, b)**.

In Definition 3.2, if we restrict h to be positive and consider the limit as $h \to 0^+$, we call the limit the **right-hand derivative at a**. Similarly, if we restrict h to be negative and consider the limit as $h \to 0^-$, we call the limit the **left-hand derivative at a**. A function is said to be differentiable on the closed interval $[a, b]$ if it is differentiable on the open interval (a, b) and the right-hand derivative at a exists and the left-hand derivative at b exists. In the future when we speak of differentiability on an interval, we will understand that if either endpoint is included, then the appropriate one-sided derivative must be used there.

The Derivative Function

So far, we have considered the derivative of a function at a fixed point, which we have designated by $x = a$. Now we want to consider the set of all points x at which the derivative of a given function f exists. These points constitute a subset of the domain of f. (The subset may even be the entire domain of f.) For each x in this subset, a value of $f'(x)$ exists. Thus, what we have is really a new function f', called the **derivative** of f. Its value at x is given by the limit that defines the derivative, namely,

$$f'(x) = \lim_{h \to 0} \frac{f(x + h) - f(x)}{h}$$

and its domain is the largest subset of the domain of f (since $f(x)$ must exist for the numerator to be defined) for which this limit exists. In the next two examples we explore the relationship between the graphs of f and f'. First, we find f', given f.

EXAMPLE 3.5   The graph of a function f is given in Figure 3.15. From it, obtain a sketch of the graph of f'. What is the domain of f'?

x	$f'(x)$
-2	2
-1	0
0	-1
1	0
2	2
3	Undefined
4	0
5	-3

Solution   At selected values of x, we estimate the values of $f'(x)$ by means of the slope of the tangent line. This method gives only a crude approximation. At the integer points on the interval $[-2, 5]$, these values appear to be approximately those in the table. Notice that at $x = 3$ the tangent line is vertical, so there is no slope. The corresponding point on the graph is an example of a **cusp**. As 3 is approached from the left, the slope gets arbitrarily large, and as 3 is approached from the right, the slope is negative but arbitrarily large in absolute value. This behavior results in a vertical asymptote for the graph of f'. We show the graph of f' in Figure 3.16. From the graph in Figure 3.15, it appears that the domain of f is all of $\mathbb{R}$. The domain of f' therefore is the set of all x in $\mathbb{R}$ except $x = 3$. ∎

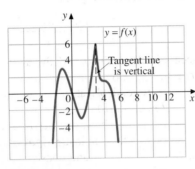

FIGURE 3.15

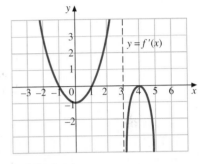

FIGURE 3.16

Next, we reverse the process and find f, given f'.

EXAMPLE 3.6   The graph of the derivative f' of a certain function f is given in Figure 3.17. It is known that $f(0) = 2$. Sketch the graph of f.

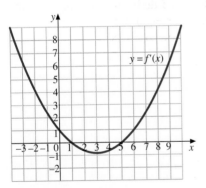

FIGURE 3.17

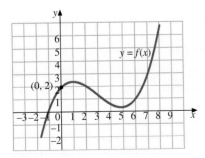

FIGURE 3.18

Solution   The y value that we read from the graph of f' corresponding to any x value represents the slope of the tangent line to the graph of f. We know that since $f(0) = 2$, the graph of f goes through the point $(0, 2)$. At $x = 0$, we read from Figure 3.17 that $f'(0) = 1$. So as the graph of f passes through $(0, 2)$, it does so with its tangent line having slope 1. As we move to the right, $f'(x)$ decreases but remains positive until $x = 1$. The graph of f therefore continues to rise on the interval $[0, 1]$ but is rising less and less rapidly. At $x = 1$, the tangent line is horizontal, since $f'(1) = 0$. From $x = 1$ to $x = 3$, the slope is more and more negative, so the graph of f goes downward more and more steeply. From $x = 3$ to $x = 5$, $f'(x)$ is still negative but approaches zero again as we move to the right, which means that f continues to decrease on this interval but is decreasing less and less rapidly. Since $f'(5) = 0$, the tangent line to the graph of f is horizontal at $x = 5$. Beyond $x = 5$, $f'(x)$ is positive and increases as we move to the right. So the graph of f rises more and more rapidly.

To the left of $x = 0$, $f'(x)$ is also positive, and as we move from left to right toward $x = 0$, the slope decreases, so the graph of f is rising on the interval $(-\infty, 0)$ but less and less rapidly.

When we put all this information together, we obtain the sketch of the graph of f in Figure 3.18. ∎

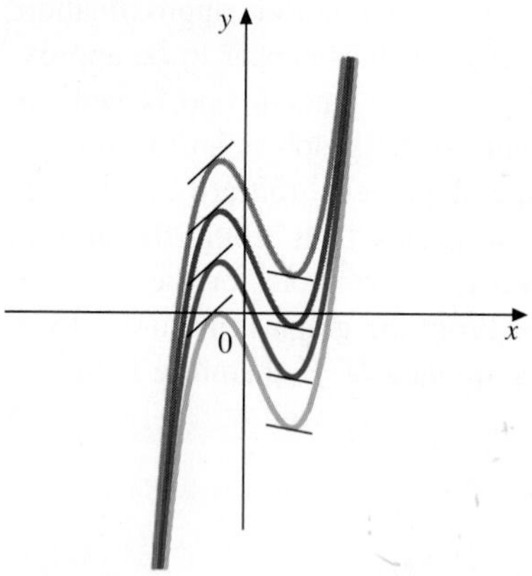

FIGURE 3.19
At each value of x these curves have the same slope.

REMARK ————————————

■ In this example if we had not been given a point on the graph of f, we could have found the slope of the graph but not its location vertically. After all, "parallel" curves all have the same slopes. That is, when two functions differ only by some constant value k—so that the graph of one of them is shifted k units vertically from the other—they clearly have the same derivative at every value of x. So when we know only the graph of f' and nothing more, we cannot say anything about the location of the graph of f in the vertical direction. The graph could be one of many curves, as illustrated in Figure 3.19.

The Δ-Notation

There is nothing special about using the letter h in the difference quotient $[f(x+h) - f(x)]/h$. In fact, the symbol Δx, read "delta x," has been used historically where we have used h, and it still is frequently used. The symbol Δ (capital Greek letter delta) is often used to denote a change, so Δx is understood to convey the idea of an *increment*, or *change*, in x—that is, the change from the point x to the point $x + \Delta x$. If we write $y = f(x)$, then the y increment that results from the increment Δx in x can be denoted by Δy; that is, $\Delta y = f(x + \Delta x) - f(x)$. In this notation, then,

$$f'(x) = \lim_{\Delta x \to 0} \frac{f(x + \Delta x) - f(x)}{\Delta x} = \lim_{\Delta x \to 0} \frac{\Delta y}{\Delta x}$$

Although in most cases we shall continue to use the h notation, at times the delta notation is more convenient.

Exercise Set 3.2

In Exercises 1–10, find $f'(a)$ using Definition 3.2. Also find the equation of the tangent line to the graph of f at the point where $x = a$.

1. $f(x) = 3x + 4$; $a = 1$

2. $f(x) = \dfrac{2x - 1}{3}$; $a = 6$

3. $f(x) = x^2 - 4$; $a = -2$

4. $f(x) = 4x - 3x^2$; $a = 2$

5. $f(x) = 1 - \dfrac{1}{x}$; $a = 1$

6. $f(x) = \dfrac{1}{x - 1}$; $a = 2$

7. $f(x) = \sqrt{x}$; $a = 9$

8. $f(x) = \sqrt{x - 1}$; $a = 5$

9. $f(x) = \dfrac{x + 2}{x}$; $a = -1$

10. $f(x) = \dfrac{x}{x - 1}$; $a = 2$

In Exercises 11–20, redo Exercises 1–10, using Equation 3.4.

In Exercises 21–26, find $f'(x)$ using Definition 3.2.

21. $f(x) = 2x + 3$

22. $f(x) = 3x^2 - 5$

23. $f(x) = \dfrac{2}{x}$

24. $f(x) = \sqrt{x + 2}$

25. $f(x) = \dfrac{x}{x + 1}$

26. $f(x) = \dfrac{1}{\sqrt{x}}$

In Exercises 27–29, sketch the graph of f' using the method suggested in Example 3.5.

27.

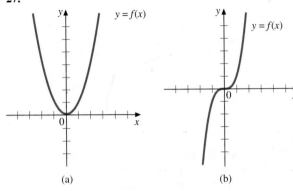

(a)   (b)

28.

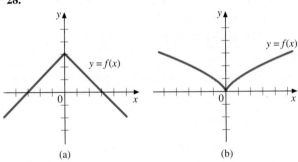

(a)   (b)

29.

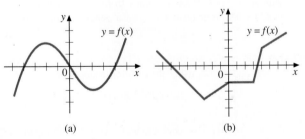

(a)   (b)

In Exercises 30–32, the graph of the derivative f' of a given function f is shown, and $f(a)$ is given for a certain value of a. Use the ideas of Example 3.6 to sketch a possible graph of f.

30.

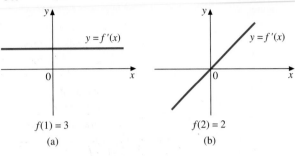

$f(1) = 3$ $f(2) = 2$
(a)   (b)

31.

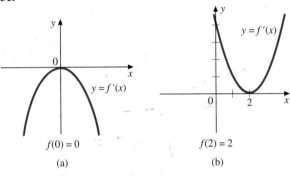

$f(0) = 0$ $f(2) = 2$
(a)   (b)

32.

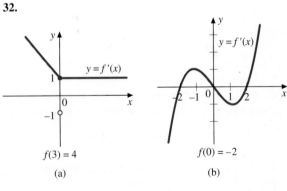

$f(3) = 4$ $f(0) = -2$
(a)   (b)

In Exercises 33–36, find $f'(x)$ and give its domain.

33. $f(x) = \dfrac{1}{\sqrt{2x - 3}}$ **34.** $f(x) = x^3 - 2x$

35. $f(x) = x^{3/2}$ **36.** $f(x) = \sqrt{4 - x^2}$

37. The graphs of two functions f and g are shown in the figure. What is the derivative of $h(x) = f(x) - g(x)$?

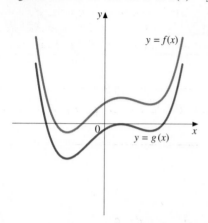

38. The graph of a function f is given. Sketch the graph of the function g that satisfies *both* of the following conditions.

(a) $g'(x) = f'(x)$ for all x

(b) $g(2) = 3$

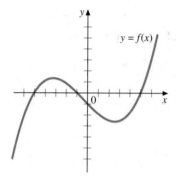

39. From the graph of $f(x) = |x|$ determine the graph of $f'(x)$. What is the domain of f'? Describe the relationship between f' and the function g defined by

$$g(x) = \frac{|x|}{x} \qquad (x \neq 0)$$

40. By changing variables in Equation 3.4, show that an alternative formula for $f'(x)$ is

$$f'(x) = \lim_{t \to x} \frac{f(t) - f(x)}{t - x}$$

and use this formula to find $f'(x)$ for $f(x) = x/(x-1)$.

41. A particle moves in a straight line so that its position function is

$$s(t) = \frac{t}{t+1}$$

Find its velocity function $v(t)$. [*Hint: $v(t) = s'(t)$.*]

42. The volume of water in a conical rain gauge, as the water rises, is given by

$$V(h) = \frac{1}{12}\pi h^3$$

where h is the height of the water level. Find the instantaneous rate of change of volume with respect to height when the height is 3 cm.

43. Newton's Universal Law of Gravitation states that two bodies of masses m_1 and m_2, respectively, at a distance r units apart, exert a force of attraction on each other given by the formula

$$F(r) = G\frac{m_1 m_2}{r^2}$$

where G is the universal gravitational constant. Find the instantaneous rate of change of F with respect to r. Explain the significance of the sign of your answer.

44. The revenue function $R(x)$ for selling x items of a certain type is given by

$$R(x) = 100x - 0.05x^2 \qquad (0 < x < 2000)$$

Find the marginal revenue function $R'(x)$. Show that $R'(1000) = 0$. Explain the significance of these results.

45. (a) Draw the graph of $f(x) = \sin x$ for $0 \leq x \leq 2\pi$. On the same set of axes sketch the graph of $f'(x)$, using the ideas of Example 3.5. Can you guess the formula for $f'(x)$, based on the graph you obtained?

(b) Repeat part (a), replacing $f(x)$ by $g(x) = \cos x$.

46. Let $f(x) = \ln x$. Approximate $f'(x)$ for each of the values $x = 0.5$, $x = 1$, $x = 2$, and $x = 3$, using the difference quotient

$$\frac{f(x+h) - f(x)}{h}$$

with $h = 0.001$. Obtain values from your calculator or CAS. Do your results suggest a general formula for $f'(x)$?

47. Let $g(x) = e^x$. Approximate $g'(x)$ for $x = -1$, $x = 0$, $x = 1$, and $x = 2$, using the difference quotient

$$\frac{g(x+h) - g(x)}{h}$$

for $h = 0.001$. Use your calculator or CAS. Compare your answers with $g(x)$ in each case. Make a conjecture about a general formula for $g'(x)$.

48. Use a graphing calculator or a CAS to find the approximate values of x at which the function $f(x) = |x^5 + x^4 - 2x^3 + x^2 + x - 2|$ is not differentiable.

3.3 DERIVATIVES OF SUMS, DIFFERENCES, AND POWERS

Calculating derivatives by means of the definition can become quite tedious (as you doubtless have discovered). Fortunately, in many cases the work can be greatly reduced by using results that we derive in the remainder of this chapter.

THEOREM 3.2

If $f(x) = c$, then $f'(x) = 0$ for all x; that is, *the derivative of a constant function is* 0.

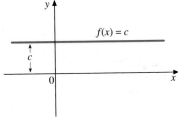

FIGURE 3.20

Proof

$$f'(x) = \lim_{h \to 0} \frac{f(x + h) - f(x)}{h} = \lim_{h \to 0} \frac{c - c}{h} = \lim_{h \to 0} 0 = 0$$

Figure 3.20 shows this result geometrically. The tangent line coincides with the graph of f and has slope 0. ∎

THEOREM 3.3

If c is any constant, then the derivative of the function cf is

$$(cf)'(x) = c \cdot f'(x)$$

for all x in the domain of f'.

Proof   Since

$$\frac{(cf)(x + h) - (cf)(x)}{h} = \frac{cf(x + h) - cf(x)}{h} = c\left[\frac{f(x + h) - f(x)}{h}\right]$$

it follows that

$$(cf)'(x) = c\left[\lim_{h \to 0} \frac{f(x + h) - f(x)}{h}\right] = c \cdot f'(x) \qquad \text{By Limit Property 2} \quad \blacksquare$$

In words, Theorem 3.3 states that *the derivative of a constant times a function is the constant times the derivative of the function.* So you can factor out constant multipliers when differentiating. Another way of looking at the result is to observe that multiplying the y-coordinates by a constant multiplies the slopes by the same constant.

THEOREM 3.4

For all x in the domain of f' and g',

$$(f + g)'(x) = f'(x) + g'(x)$$

and

$$(f - g)'(x) = f'(x) - g'(x)$$

Proof   The difference quotient for $f + g$ can be written as

$$\frac{(f+g)(x+h) - (f+g)(x)}{h} = \frac{[f(x+h) + g(x+h)] - [f(x) + g(x)]}{h}$$

$$= \frac{f(x+h) - f(x)}{h} + \frac{g(x+h) - g(x)}{h}$$

Taking limits of both sides as $h \to 0$ and using Limit Property 3, it follows that

$$(f+g)'(x) = f'(x) + g'(x)$$

A similar proof can be given for $(f - g)'$, or alternatively we may write $f - g = f + (-1)g$ and use Theorems 3.3 and 3.4.   ∎

Theorem 3.4 can be read as follows: *the derivative of a sum (or difference) is the sum (or difference) of the derivatives.* This result can be extended by associativity to any finite sum or difference. For example,

$$(f+g-h)'(x) = [(f+g) - h]'(x) = (f+g)'(x) - h'(x) = f'(x) + g'(x) - h'(x)$$

For the next theorem we need to use the *Binomial Theorem*, the proof of which is given in Appendix 4. It states that for any positive integer n,

$$(a+b)^n = a^n + na^{n-1}b + \frac{n(n-1)}{2!}a^{n-2}b^2 + \frac{n(n-1)(n-2)}{3!}a^{n-3}b^3 + \cdots + b^n$$

(Recall that $k!$, read "k factorial," means $1 \cdot 2 \cdot 3 \cdot \cdots \cdot k$.)

THEOREM 3.5

The Power Rule

Let $f(x) = x^n$, where n is any positive integer. Then for all real numbers x,

$$f'(x) = nx^{n-1} \tag{3.5}$$

Proof

$$f'(x) = \lim_{h \to 0} \frac{f(x+h) - f(x)}{h} = \lim_{h \to 0} \frac{(x+h)^n - x^n}{h}$$

$$= \lim_{h \to 0} \frac{\left[x^n + nx^{n-1}h + \frac{n(n-1)}{2!}x^{n-2}h^2 + \cdots + h^n \right] - x^n}{h}$$

$$= \lim_{h \to 0} \left[nx^{n-1} + \frac{n(n-1)}{2!}x^{n-2}h + \cdots + h^{n-1} \right]$$

$$= nx^{n-1}$$

Note that in the next-to-last step, all terms but the first contain h as a factor and so go to 0 in the limit.   ∎

EXAMPLE 3.7   Find $f'(x)$ if $f(x) = x^{10}$.

Solution   By Theorem 3.5 with $n = 10$, $f'(x) = 10x^9$. You might wish to compare this quick way with the work of finding $f'(x)$ directly from the definition.   ∎

By combining Theorems 3.3 and 3.5 we see that if $f(x) = cx^n$, then

$$f'(x) = c(nx^{n-1}) = (cn)x^{n-1}$$

For example, if $f(x) = 3x^4$, then $f'(x) = 12x^3$.

We are now ready to take the derivative of any polynomial function by using Theorems 3.2 through 3.5.

EXAMPLE 3.8   If $f(x) = 3x^4 + 5x^3 - 7x^2 + 2x - 9$, find $f'(x)$.

Solution

$$\begin{aligned} f'(x) &= 3(4x^3) + 5(3x^2) - 7(2x) + 2 \\ &= 12x^3 + 15x^2 - 14x + 2 \end{aligned}$$

You should check to see how we have made use of each of the four theorems. ∎

EXAMPLE 3.9   If $f(x) = 2x^4 - 6x^2 + 7$, find $f'(2)$.

Solution   First we find $f'(x)$ in general:

$$f'(x) = 8x^3 - 12x$$

Now we substitute $x = 2$:

$$f'(2) = 8(8) - 12(2) = 64 - 24 = 40 \qquad ∎$$

Do not conclude that because you have now learned the "easy way" to differentiate a polynomial function, the definition of the derivative is no longer useful. There are many important functions that are not polynomials, and each time we encounter one of these, we will have to return to Definition 3.2 to develop a differentiation formula. Further, we will see that an understanding of the definition is essential in finding new applications of the derivative. This definition also provides a key to the fundamental relationship between the derivative and the integral. So this definition is of fundamental importance and must be learned.

Exercise Set 3.3

In Exercises 1–20, find $f'(x)$, by using Theorems 3.1 through 3.5.

1. $f(x) = 3$

2. $f(x) = x$

3. $f(x) = x^4$

4. $f(x) = 3x^5$

5. $f(x) = 1 - 2x$

6. $f(x) = (3x + 4)/5$

7. $f(x) = ax + b$

8. $f(x) = 2x^2 - 3x + 4$

9. $f(x) = 4x^3 - 7x^2 + 8x - 6$

10. $f(x) = 3 - x - 7x^2 + 2x^3$

11. $f(x) = x^5 - 6x^4$

12. $f(x) = \frac{2}{3}x^6 - 5x^4 + \frac{1}{2}x^2 - 8$

13. $f(x) = ax^2 + bx + c$

14. $f(x) = (2x - 1)(x + 3)$

15. $f(x) = (2x - 1)(2 + x)$

16. $f(x) = x(1 - x)(2 + x)$

17. $f(x) = (3x + 4)^2$ **18.** $f(x) = (x + 2)^3$

19. $f(x) = 2x^3 - 3x(2 + x - x^2)$

20. $f(x) = x(2x - 3) - (3 - x)^2$

21. Find the equation of the tangent line to the graph of $y = 2x^2 - 3x + 4$ at the point $(2, 6)$.

22. Find the equation of the tangent line to the graph of $y = 3x^2 - x^4$ at the point for which $x = -2$.

23. Find k so that the slope of the tangent line to the graph of $y = 2x^2 - 3kx + 4$ is 4 at the point for which $x = 2$.

24. Find k so that the slope of the tangent line to the graph of $y = 2k^2x^2 - 2kx + 3$ is 2 at the point for which $x = 1$.

25. The line perpendicular to the tangent line to a curve at the point of tangency is called the **normal line**. Find the equations of the tangent and normal lines to $y = x^3 - 3x^2 - 7x + 18$ at the point for which $x = 3$.

26. Find all points on the graph of $y = x^4 - 4x^3 + 4x^2 + 3$ at which the tangent line is horizontal.

In Exercises 27–30, a particle moves in a straight line with position function $s(t)$. Find the instantaneous velocity of the particle at time t_1.

27. $s(t) = 3t^2 - 4t + 5$; $t_1 = 4$

28. $s(t) = t(t^2 - 3t)$; $t_1 = 3$

29. $s(t) = 1 - 2t(t + 1)$; $t_1 = 2$

30. $s(t) = (2t - 3)(3t + 4)$; $t_1 = 5$

31. Find a and b such that the graph of $y = ax^2 + bx$ passes through the point $(2, -3)$ and such that at this point the tangent line has slope 1.

32. Find all points on the graph of $y = 2x^3 - 3x^2 - 10x + 7$ at which the slope of the tangent line is 2. Find the equations of the tangent lines at these points.

33. Two tangent lines to the curve $y = x^2$ pass through the point $(0, -4)$. Find the points of tangency and the equations of the tangent lines.

34. Find the equations of all tangent lines to the graph of $f(x) = 3x - x^2$ that pass through the point $(4, 0)$. Draw the graph of f, showing these tangent lines along with their points of tangency.

35. If $s(t) = (2t^2 - 3)^3$ is the position function for a particle moving in a straight line, find the velocity function $v(t)$. Also find the acceleration $a(t)$, defined as $v'(t)$. (*Hint:* $(a - b)^3 = a^3 - 3a^2b + 3ab^2 - b^3$.)

36. Let $f(x) = (x^2 + 2x + 1)(x^3 + x)$.
   (a) Find $f'(x)$.
   (b) Let $g(x) = x^2 + 2x + 1$ and $h(x) = x^3 + x$. Compute

$$g'(x)h(x) + g(x)h'(x)$$

   and simplify. Is there any relationship between this answer and the answer to part (a)?

3.4 THE PRODUCT AND QUOTIENT RULES

Our main objective in this section is to develop formulas for differentiating products and quotients of functions. These formulas will prove to be indispensable in our continued study of derivatives.

THEOREM 3.6

> **The Product Rule**
>
> If f and g are both differentiable at x, then so is fg, and
>
> $$(fg)'(x) = f'(x)g(x) + f(x)g'(x) \qquad (3.6)$$

Proof   We write the difference quotient for fg in the following equivalent ways:

$$\frac{(fg)(x + h) - (fg)(x)}{h}$$

$$= \frac{f(x+h)g(x+h) - f(x)g(x)}{h} \qquad \text{Definition of } fg$$

$$= \frac{f(x+h)g(x+h) - f(x)g(x+h) + f(x)g(x+h) - f(x)g(x)}{h}$$

We subtracted $f(x)g(x+h)$
and then added it.

$$= \left[\frac{f(x+h) - f(x)}{h} \right] g(x+h) + f(x) \left[\frac{g(x+h) - g(x)}{h} \right] \qquad \begin{array}{l} \text{Algebraic} \\ \text{simplification} \end{array}$$

Since $g'(x)$ exists, we know by Theorem 3.1 that g is continuous at x. It follows that $\lim_{h \to 0} g(x+h) = g(x)$. (See Exercise 35 in Exercise Set 3.4.) Thus, on taking limits as $h \to 0$ and using Definition 3.2, as well as limit properties for sums and products, we get

$$(fg)'(x) = f'(x)g(x) + f(x)g'(x) \qquad \blacksquare$$

It is useful to learn this result in words. We call f the first function and g the second.

The derivative of a product of two functions is given by the derivative of the first times the second plus the first times the derivative of the second.

EXAMPLE 3.10   Let $F(x) = (2x^3 - x + 3)(x^4 + 3x^2 - 8)$. Find $F'(x)$.

Solution   One way to do the problem is to multiply the factors together. We leave it to you to show that the result is

$$F(x) = 2x^7 + 5x^5 + 3x^4 - 19x^3 + 9x^2 + 8x - 24$$

The derivative is therefore

$$F'(x) = 14x^6 + 25x^4 + 12x^3 - 57x^2 + 18x + 8$$

An easier way is to use the Product Rule without multiplying the factors. By Equation 3.6, we have

$$F'(x) = \overbrace{(6x^2 - 1)}^{\substack{\text{derivative} \\ \text{of} \\ \text{the first}}} \cdot \overbrace{(x^4 + 3x^2 - 8)}^{\text{second}} + \overbrace{(2x^3 - x + 3)}^{\text{first}} \cdot \overbrace{(4x^3 + 6x)}^{\substack{\text{derivative} \\ \text{of} \\ \text{the second}}}$$

The answer may be left in this form. You may wish to multiply and collect terms, however, to show that our two answers are the same. $\blacksquare$

In differentiating a product $f(x) \cdot g(x)$, it is tempting just to multiply the two derivatives and get $f'(x) \cdot g'(x)$, *but you must not yield to this temptation, because it is incorrect.* The correct result is given by Equation 3.6: $f'(x) \cdot g(x) + f(x) \cdot g'(x)$.

THEOREM 3.7

The Quotient Rule

If f and g are both differentiable at x and $g(x) \neq 0$, then f/g is differentiable at x, and

$$\left(\frac{f}{g}\right)'(x) = \frac{f'(x)g(x) - f(x)g'(x)}{[g(x)]^2} \tag{3.7}$$

Proof   We use a similar approach to the one we used in the proof of the Product Rule. We write the difference quotient for f/g in the following equivalent ways.

$$\frac{\left(\dfrac{f}{g}\right)(x+h) - \left(\dfrac{f}{g}\right)(x)}{h}$$

$$= \frac{\dfrac{f(x+h)}{g(x+h)} - \dfrac{f(x)}{g(x)}}{h} \qquad \text{Definition of } f/g$$

$$= \frac{f(x+h)g(x) - f(x)g(x+h)}{h[g(x)g(x+h)]} \qquad \text{We simplified the complex fraction.}$$

$$= \frac{f(x+h)g(x) - f(x)g(x) + f(x)g(x) - f(x)g(x+h)}{h[g(x)g(x+h)]} \qquad \begin{array}{l}\text{We subtracted} \\ f(x)g(x) \text{ and} \\ \text{then added it.}\end{array}$$

$$= \frac{\left[\dfrac{f(x+h) - f(x)}{h}\right] \cdot g(x) - f(x)\left[\dfrac{g(x+h) - g(x)}{h}\right]}{g(x)g(x+h)} \qquad \text{Factoring}$$

Now we take the limit as $h \to 0$, using limit properties and the definitions of $f'(x)$ and $g'(x)$, as well as the continuity of g (so that $g(x+h) \to g(x)$ as $h \to 0$):

$$\left(\frac{f}{g}\right)'(x) = \frac{f'(x)g(x) - f(x)g'(x)}{[g(x)]^2} \qquad \blacksquare$$

We can state the Quotient Rule as follows:

The derivative of a quotient of two functions is given by the derivative of the numerator times the denominator minus the numerator times the derivative of the denominator, all divided by the square of the denominator.

EXAMPLE 3.11   Find $F'(x)$, where

$$F(x) = \frac{x^2 - 1}{2x^2 + 3}$$

Solution

$$F'(x) = \frac{\overbrace{(2x)}^{\substack{\text{derivative} \\ \text{of} \\ \text{numerator}}} \cdot \overbrace{(2x^2 + 3)}^{\text{denominator}} - \overbrace{(x^2 - 1)}^{\text{numerator}} \cdot \overbrace{(4x)}^{\substack{\text{derivative} \\ \text{of} \\ \text{denominator}}}}{\underbrace{(2x^2 + 3)^2}_{\text{square of denominator}}}$$

$$= \frac{4x^3 + 6x - 4x^3 + 4x}{(2x^2 + 3)^2} = \frac{10x}{(2x^2 + 3)^2} \qquad \blacksquare$$

The following corollary to Theorem 3.7 will enable us to extend the Power Rule (Theorem 3.5) to negative exponents.

COROLLARY 3.7

If $g'(x)$ exists and $g(x) \neq 0$, then the derivative of $1/g$ exists at x and

$$\left(\frac{1}{g}\right)'(x) = -\frac{g'(x)}{[g(x)]^2}$$

Proof   In Equation 3.7 let $f(x)$ be the constant function $f(x) = 1$. Then $f'(x) = 0$, and we get the desired result. $\blacksquare$

Consider now a function of the form $f(x) = x^n$, with n a negative integer. Write $n = -m$, where m is a positive integer ($m = |n|$). Then

$$f(x) = x^{-m} = \frac{1}{x^m}$$

In Corollary 3.7, let $g(x) = x^m$. By Theorem 3.5,

$$g'(x) = mx^{m-1}$$

So the corollary gives

$$f'(x) = -\frac{mx^{m-1}}{(x^m)^2} = -m\frac{x^{m-1}}{x^{2m}} = (-m)x^{-m-1} = nx^{n-1}$$

Thus, Equation 3.5 continues to hold true when n is a negative integer. For example, if $f(x) = x^{-3}$, then $f'(x) = -3x^{-4}$.

EXAMPLE 3.12   If $f(x) = \dfrac{2x - 4}{x^2}$, find $f'(x)$.

Solution   We could use the Quotient Rule, but in cases such as this, in which the denominator is a monomial (a polynomial with only one term), it is usually easier to divide first and then use the Power Rule. Doing it this way, we first divide, getting

$$f(x) = 2x^{-1} - 4x^{-2}$$

Then we use the Power Rule, extended to negative exponents:

$$f'(x) = -2x^{-2} + 8x^{-3} \qquad \blacksquare$$

REMARK
■ Later in this chapter, we will see that the Power Rule holds true for fractional (rational) exponents, and in Chapter 7, after we have given meaning to irrational powers, we will prove it holds true for irrational exponents as well. So it actually is true for all real numbers as exponents.

Exercise Set 3.4

In Exercises 1–6, find $f'(x)$ in two ways: (a) by multiplying the factors first, and (b) by using the Product Rule. Show that the results are equal.

1. $f(x) = (2x - 3)(3x + 4)$

2. $f(x) = (x^2 - 2)(x^2 + 3)$

3. $f(x) = (1 - x^2)(2 + 5x^2)$

4. $f(x) = (x^3 - 3x^2)(2x^2 + 1)$

5. $f(x) = (2x^3 - 1)(x^4 + x)$

6. $f(x) = (x^2 + x - 1)(x^2 - x + 1)$

In Exercises 7–12, find $f'(x)$ by the Product Rule.

7. $f(x) = (x^2 - 2x + 3)(4 - 3x^2)$

8. $f(x) = (4x - x^3)(x^2 + 2x - 3)$

9. $f(x) = (x^4 - 2x^2 + 1)(x^2 - x + 3)$

10. $f(x) = (3x^4 - 2x^2)(2x^4 + 3x^2 + 5)$

11. $f(x) = (5x^4 + 6x^3 + 7)(2x^3 - 3x + 1)$

12. $f(x) = (4 - x - x^3)(3x + 2x^3 - x^5)$

In Exercises 13–22, find $f'(x)$ using the Quotient Rule.

13. $f(x) = \dfrac{x}{x - 2}$

14. $f(x) = \dfrac{3x + 2}{4x - 3}$

15. $f(x) = \dfrac{x^2 - 1}{x^2 + 1}$

16. $f(x) = \dfrac{2x^2 - 3x}{1 - x^2}$

17. $f(x) = \dfrac{x^2}{x^2 - 4}$

18. $f(x) = \dfrac{4 - 2x + 3x^2}{x^2 + 2}$

19. $f(x) = \dfrac{2x - 3}{x^2 + 2x}$

20. $f(x) = \dfrac{3x - x^2}{2x + 1}$

21. $f(x) = \dfrac{x^3 + 8}{x^3 - 1}$

22. $f(x) = \dfrac{4x^2 + 3x}{x^4 - 16}$

In Exercises 23–28, find $f'(x)$ using the Power Rule.

23. $f(x) = 2x^{-3}$

24. $f(x) = 3x^{-5}$

25. $f(x) = x^2 + x^{-2}$

26. $f(x) = 2 - 3x^{-1} + 4x^{-2}$

27. $f(x) = x + 3 + \dfrac{4}{x}$

28. $f(x) = 3 + 2x - \dfrac{4}{x^2}$

29. Let $g(x) = (x^2 - 4)/x^2$. Find $g'(x)$ in two ways: by the Quotient Rule and by writing $g(x)$ in the form $1 - 4x^{-2}$. Compare the two methods.

30. Let $h(x) = (3 - 4x + 5x^2)/2x$. Find $h'(x)$ without using the Quotient Rule.

31. Find the equation of the tangent line to the graph of

$$y = \frac{x + 3}{x - 1}$$

at the point $(2, 5)$.

32. Find the equation of the tangent line to the graph of

$$y = \frac{x^2}{2x - 3}$$

at the point for which $x = -3$.

33. A particle moves in a straight line so that its position function at time t is given by

$$s(t) = \frac{2t^2 + 3}{t + 1}$$

Find its velocity at the instant when $t = 3$.

34. The length of a rectangle is increasing at the rate of 5 cm/s, and the width is increasing at the rate of 3 cm/s. At what rate is the area increasing when the length is 8 cm and the width is 5 cm?

35. In the proof of the Product Rule, we used the fact that continuity of g at x implies that $\lim_{h \to 0} g(x + h) = g(x)$. Prove this result.

In Exercises 36 and 37, find $F'(x)$ by using the Quotient Rule and the Product Rule. Simplify the result.

36. $F(x) = \dfrac{(x^2 - 4)(x^2 + 1)}{x^2 - 9}$

37. $F(x) = \dfrac{2x - 1}{(3x^2 + 5)(x^2 - 2x)}$

38. Let $f(x) = (x^2 - 4)/(x^2 + 4)$. Find $f'(x)$ without using the Quotient Rule or Product Rule. (*Hint:* Divide first

and then make use of Corollary 3.7.)

39. Follow the instructions in Exercise 38 for

$$f(x) = \frac{(x^3 - x^2 + x)}{(x^2 + 1)}$$

40. Derive a formula for the derivative of the product fgh of three differentiable functions. (*Hint:* Write $(fgh)' = [(fg) \cdot h]'$ and apply the Product Rule twice.)

41. Using the result of Exercise 40, find $F'(x)$ if

$$F(x) = (2x + 1)(3x^2 - 4x)(x^2 + 4)$$

42. Using the result of Exercise 40, make a conjecture for a formula for the derivative $(f_1 \cdot f_2 \cdot \cdots \cdot f_n)'$ of the product of n differentiable functions. Prove your formula by mathematical induction. (See Supplementary Appendix 4.)

3.5 HIGHER-ORDER DERIVATIVES AND OTHER NOTATIONS

Since for a differentiable function f, the derivative f' is a function in its own right, there is no reason why it may not also have a derivative. When it does, we denote its derivative by f''. So f'' is the derivative of f'. We also call f'' the **second derivative** of f. For example, suppose $f(x) = 3x^4 - 2x^2 + 7$. Then

$$f'(x) = 12x^3 - 4x \qquad \text{First derivative}$$

and

$$f''(x) = 36x^2 - 4 \qquad \text{Second derivative}$$

There is no reason for stopping here, of course; we can take further derivatives, getting

$$f'''(x) = 72x \qquad \text{Third derivative}$$

$$f''''(x) = 72 \qquad \text{Fourth derivative}$$

Note that all derivatives higher than the fourth are identically zero in this case. It gets rather cumbersome to write so many primes, so we use the symbol

$$f^{(k)}(x)$$

to mean the *kth-order derivative*. (Usually we go to this notation for $k > 3$, though in practice the need for such higher-order derivatives rarely arises.) For example, $f^{(5)}(x)$ means the fifth derivative. The superscript k is put in parentheses to distinguish it from an exponent.

There are also other notations for derivatives. We introduce them not to make life more difficult for you, but rather because each notation has certain advantages, or historical status, and so it is useful to know them all. Moreover, in other books where calculus is used (higher math, physics, economics, statistics, epidemiology, and so on) you might encounter these other notations.

Suppose that the rule for a function is given by an equation of the form $y = f(x)$. Then the derivative, $f'(x)$, can be symbolized using any of the following alternative forms:

$$y' \qquad \frac{dy}{dx} \qquad \frac{df(x)}{dx} \qquad D_x y \qquad D_x f$$

The second notation, dy/dx (read "dee wye dee ex"), is widely used. This notation was introduced by Leibniz. It is suggestive of the limit

$$\lim_{\Delta x \to 0} \frac{\Delta y}{\Delta x}$$

that we introduced in Section 3.2. For the present you should not think of this form as a quotient but simply as a symbol for the derivative. Later we will see that we can give meaning to dy and dx separately. Even now we can, however, give meaning to the symbol d/dx. We refer to this symbol as a *derivative operator*, and when we follow it with an expression, as in

$$\frac{d}{dx}(\quad)$$

we mean to take the derivative with respect to x of what is inside the parentheses. For example,

$$\frac{d}{dx}(3x^2 + 2x) = 6x + 2$$

In this sense, then, dy/dx can be thought of as the operator d/dx operating on y. Alternatively, we can replace y with $f(x)$ and write $df(x)/dx$, or simply df/dx. The symbol D_x can be used in the same way as d/dx. Thus,

$$D_x(3x^2 + 2x) = 6x + 2$$

Suppose we are given that $f(x) = 3x^2 + 2x$. By setting $y = f(x)$, we can use any of the following to indicate the derivative:

$$f'(x) \qquad y' \qquad \frac{dy}{dx} \qquad \frac{df(x)}{dx} \qquad \frac{d}{dx}(3x^2 + 2x) \qquad D_x(3x^2 + 2x)$$

or even $(3x^2 + 2x)'$. In each case the value is $6x + 2$. If we want to evaluate the derivative at a specific point, say $x = 2$, the $f'(x)$ notation has a clear advantage; we simply write $f'(2)$. For the other notations we resort to the following device:

$$y'\big|_{x=2} \qquad \frac{dy}{dx}\bigg|_{x=2} \qquad \frac{d}{dx}(3x^2 + 2x)\bigg|_{x=2}$$

and so on. We will see advantages to some of the other notations as we progress.

Indicating higher-order derivatives with these new notations, we have

$$y', \ y'', \ y''', \ y^{(4)}, \dots$$

$$\frac{dy}{dx}, \ \frac{d^2y}{dx^2}, \ \frac{d^3y}{dx^3}, \ \frac{d^4y}{dx^4}, \dots$$

$$D_x y, \ D_x^2 y, \ D_x^3 y, \ D_x^4 y, \dots$$

The notation d^2y/dx^2 can be explained by observing that the second derivative is the derivative of dy/dx and so can be obtained from it by operating on dy/dx with the derivative operator d/dx:

$$\frac{d}{dx}\left(\frac{dy}{dx}\right)$$

In the next chapter we will see some applications of second derivatives, but we can point out one application now. We know that if a particle moves in a straight line so that its position function at time t is $s(t)$, then its instantaneous velocity $v(t)$ is the derivative $s'(t)$. Since acceleration is the rate of change of velocity, the average acceleration over the time interval from t to $t + h$ is

$$\frac{v(t + h) - v(t)}{h}$$

and the limit of this quotient as $h \to 0$ is the *instantaneous acceleration* at t. So we have

$$a(t) = v'(t) = s''(t)$$

The third derivative of the position function is the rate of change of acceleration and is sometimes referred to as the "jerk."

Exercise Set 3.5

In Exercises 1–6, find y', y'', and y'''.

1. $y = 2x^4$

2. $y = 3x^{-2}$

3. $y = 4x^3 - 2x^2 + 1$

4. $y = x^5 - 2x^3 + 3$

5. $y = x^2 - x^{-2}$

6. $y = \dfrac{2}{x} - 3x^2$

In Exercises 7–12, find dy/dx and d^2y/dx^2.

7. $y = (2x + 3)(3x - 1)$

8. $y = (x^2 - 1)(2x^3 + 5)$

9. $y = x^2(x^2 - 3) + \dfrac{5}{x}$

10. $y = \dfrac{2}{x^3} - \dfrac{3}{2x^2}$

11. $y = \dfrac{x^2 - x + 3}{x}$

12. $y = \dfrac{x^2 - 1}{x^4}$

In Exercises 13–16, find the indicated derivatives.

13. $\dfrac{d}{dx}(x^3 - 3x^2 + 1)$

14. $\dfrac{d}{dx}(x^2 - 2x^{-3})$

15. $D_x\left(\dfrac{x - 2}{x + 3}\right)$

16. $D_x\left(\dfrac{3x - 4}{2x + 3}\right)$

In Exercises 17–20, find $f''(x)$.

17. $f(x) = x^3 - x^{-2}$

18. $f(x) = \dfrac{3x - 2}{x^2}$

19. $f(x) = \left(x - \dfrac{1}{2x}\right)^2$

20. $f(x) = \sqrt{x^2 + 1 + \dfrac{1}{4x^2}}$, $x > 0$ (*Hint:* Compare with Exercise 19.)

21. Let $P(x) = a_0 + a_1 x + a_2 x^2 + \cdots + a_n x^n$. Find $P^{(n)}(x)$. What is $P^{(k)}(x)$ for $k > n$?

22. Let $g(x) = \dfrac{1}{x}$. Find $g^{(n)}(x)$.

23. Let $f(x) = \sqrt{x}$. Assuming the Power Rule holds true for fractional exponents, find the first four derivatives of f, and devise a formula for $f^{(n)}(x)$. (The formula for the derivative of $f^{(n)}(x)$ can be proved by mathematical induction, which is discussed in Appendix 4.)

24. Let $y = f(x) \cdot g(x)$, where f and g have derivatives of order 3 or more. Find a formula for the following:

(a) $\dfrac{d^2 y}{dx^2}$

(b) $\dfrac{d^3 y}{dx^3}$

25. A particle moves in a straight line so that its position function at time t is given by

$$s(t) = 2t^2 - \frac{t^2 - 1}{t + 3}$$

Find formulas for its velocity $v(t)$ and its acceleration $a(t)$ at any time $t \geq 0$. Also find the velocity and acceleration after 1 second, $v(1)$ and $a(1)$, respectively.

26. Show that $y = x^3 - 2x^2 + 4$ satisfies the equation $y'' - 2y' + y = x^3 - 8x^2 + 14x$.

27. Show that $y = (2x^2 + 1)/x$ satisfies the equation $x^3 y'' + 3x^2 y' - xy = 4x^2 - 2$.

28. The accompanying figure shows portions of graphs, generated by a CAS, of a certain polynomial function, $y = p(x)$, and its first two derivatives, $y = p'(x)$ and $y = p''(x)$. Each tick mark on the x-axis represents one unit, and on the y-axis, ten units. Can this graph be an accurate description of p? That is, can p be a quadratic polynomial? Explain. If your answer is "no," what is the least degree p can be?

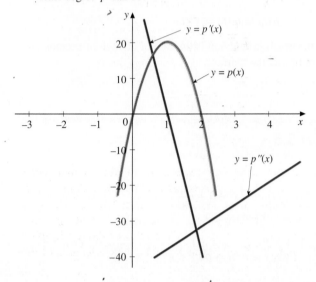

29. Compute dy/dx for each of the following by expanding and then using the Power Rule.
   (a) $y = (3x - 2)^2$
   (b) $y = (2x + 4)^2$
   (c) $y = (x^2 - 2x + 1)^2$
   (d) $y = (x^3 + 1)^2$

30. (a) Refer to Exercise 29. Notice that each equation can be written in the form $y = [f(x)]^2$. In each case find a function g so that the derivative can be written in the form

$$\frac{dy}{dx} = 2f(x)g(x)$$

   (b) From the result of part (a) conjecture a formula for

$$\frac{d}{dx}[f(x)]^2$$

3.6 DERIVATIVES OF THE TRIGONOMETRIC FUNCTIONS

Our main objective in this section is to obtain formulas for the derivatives of the sine and cosine functions. Since the other four trigonometric functions are defined in terms of these two basic ones, we can obtain formulas for their derivatives using results for derivatives of quotients and reciprocals from Section 3.4.

Before getting to the main theorem, we need a preliminary result. Such a preliminary result is called a *lemma* (a theorem used in proving another theorem). The first part of the lemma states that

$$\lim_{t \to 0} \frac{\sin t}{t} = 1$$

We have seen evidence earlier (Example 2.5) that this limit is correct. In the proof of the lemma we use a geometric argument based on the unit circle definition of $\sin t$ and $\cos t$.

LEMMA

(a) $\lim_{t \to 0} \dfrac{\sin t}{t} = 1$

(b) $\lim_{t \to 0} \dfrac{1 - \cos t}{t} = 0$

Proof

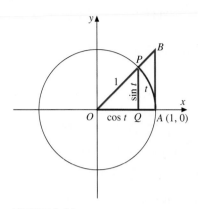

FIGURE 3.21

(a) In Figure 3.21 we show a unit circle, along with the point P that is t units along the circle from $A(1,0)$, where $0 < t < \pi/2$. The point Q is the projection of P on the x-axis, and B is the intersection of the tangent line $x = 1$ at A and the line through O and P. So the triangles OQP and OAB are similar right triangles.

By the definitions of $\sin t$ and $\cos t$, the coordinates of P are $(\cos t, \sin t)$. Thus, the length of the line segment PQ, which we symbolize by $\overline{PQ}$, is $\sin t$, and $\overline{OQ} = \cos t$. Also, $\overline{OA} = 1$, the radius of the circle. By similarity of triangles OQP and OAB, we have the proportion

$$\frac{\overline{AB}}{\overline{PQ}} = \frac{\overline{OA}}{\overline{OQ}}$$

That is,

$$\frac{\overline{AB}}{\sin t} = \frac{1}{\cos t}$$

So

$$\overline{AB} = \frac{\sin t}{\cos t}$$

We are going to make use of the fact that the area of the circular sector OAP (the "pie-shaped" region) lies between the area of the triangle OQP and the area of the triangle OAB. We know how to find the area of a triangle ($\frac{1}{2}$base × height). The area of the sector OAP is that fractional part of the area of the entire circle given by the ratio of the arc length t to the total circumference 2π ($C = 2\pi r$ and $r = 1$). Since the area of the circle is π (area $= \pi r^2$ and $r = 1$), we obtain

$$\text{area of sector } OAB = \left(\frac{t}{2\pi}\right)\pi = \frac{t}{2}$$

The area of the triangle OQP is $\frac{1}{2}(\overline{OQ})(\overline{QP}) = \frac{1}{2}(\cos t)(\sin t)$ and the area of the triangle OAB is

$$\frac{1}{2}(\overline{OA})(\overline{AB}) = \frac{1}{2}(1)\left(\frac{\sin t}{\cos t}\right) = \frac{1}{2}\left(\frac{\sin t}{\cos t}\right)$$

So we have

$$\frac{1}{2}(\cos t)(\sin t) < \frac{t}{2} < \frac{1}{2}\left(\frac{\sin t}{\cos t}\right)$$

Multiplying through by the positive quantity $2/\sin t$, and then inverting all members (thus reversing the sense of each inequality), we get

$$\cos t < \frac{\sin t}{t} < \frac{1}{\cos t}$$

As $t \to 0^+$, $\cos t \to 1$ and $1/\cos t \to 1$. By the Squeeze Theorem (Theorem 2.3), it follows that

$$\lim_{t \to 0^+} \frac{\sin t}{t} = 1$$

We complete the proof of part (a) by observing that $f(t) = (\sin t)/t$ is an even function. That is,

$$f(-t) = \frac{\sin(-t)}{(-t)} = \frac{-\sin t}{-t} = \frac{\sin t}{t} = f(t)$$

It follows from symmetry that

$$\lim_{t \to 0^-} \frac{\sin t}{t} = \lim_{t \to 0^+} \frac{\sin t}{t}$$

Since the limit on the right is 1, the limit on the left also is 1. Thus,

$$\lim_{t \to 0} \frac{\sin t}{t} = 1$$

(b) Since $\sin^2 t + \cos^2 t = 1$, we can write $1 - \cos^2 t = \sin^2 t$. Thus,

$$\frac{1 - \cos t}{t} = \frac{1 - \cos t}{t} \cdot \frac{1 + \cos t}{1 + \cos t} = \frac{1 - \cos^2 t}{t(1 + \cos t)} = \frac{\sin^2 t}{t(1 + \cos t)}$$

$$= \left(\frac{\sin t}{t}\right)\left(\frac{\sin t}{1 + \cos t}\right)$$

We know from part (a) that $\lim_{t \to 0}(\sin t)/t = 1$. Also,

$$\lim_{t \to 0} \frac{\sin t}{1 + \cos t} = \frac{0}{1 + 1} = \frac{0}{2} = 0$$

Finally, using the fact that the limit of a product is the product of the limits when they both exist (Limit Property 5), we have

$$\lim_{t \to 0} \frac{1 - \cos t}{t} = (1)(0) = 0 \qquad \blacksquare$$

REMARK ————————————————————————————

■ The limits given by the Lemma are based on the definitions of $\sin t$ and $\cos t$ as functions of the *real number* t. If t is an *angle*, then for these limits to be valid, t *must be expressed in radians*. Since the derivatives of the sine and cosine (that we will consider in the theorem that is to follow) are based on the Lemma, in calculus angles should always be expressed in radians.

————————————————————————————————————

Before going to the main theorem, we give an example to show how the Lemma can be used to obtain other related limits.

EXAMPLE 3.13   Evaluate the following limits.

(a) $\lim_{x \to 0} x \cot x$

(b) $\lim_{\theta \to 0} \dfrac{\sin 2\theta}{\sin 3\theta}$

Solution

(a) $x \cot x = x\left(\dfrac{\cos x}{\sin x}\right) = \left(\dfrac{x}{\sin x}\right)\cos x = \dfrac{1}{\left(\dfrac{\sin x}{x}\right)} \cdot \cos x \qquad (x \neq 0)$

So

$$\lim_{x \to 0} x \cot x = \lim_{x \to 0}\left(\frac{1}{\dfrac{\sin x}{x}}\right) \cdot \lim_{x \to 0} \cos x$$

$$= \frac{1}{\lim_{x \to 0}\left(\dfrac{\sin x}{x}\right)} \cdot 1$$

$$= \frac{1}{1} \qquad \text{Lemma, part (a)}$$

$$= 1$$

(b) We write $(\sin 2\theta)/(\sin 3\theta)$ in the form

$$\frac{\sin 2\theta}{\sin 3\theta} = \frac{\sin 2\theta}{2\theta} \cdot \frac{3\theta}{\sin 3\theta} \cdot \frac{2}{3} \qquad (\theta \neq 0 \text{ and } |\theta| < \pi/3)$$

By part (a) of the Lemma, replacing t by 2θ in one case and by 3θ in the other, we know that

$$\lim_{\theta \to 0} \frac{\sin 2\theta}{2\theta} = 1 \qquad \text{and} \qquad \lim_{\theta \to 0} \frac{\sin 3\theta}{3\theta} = 1$$

So

$$\lim_{\theta \to 0} \frac{\sin 2\theta}{\sin 3\theta} = \left(\lim_{\theta \to 0} \frac{\sin 2\theta}{2\theta} \right) \left(\lim_{\theta \to 0} \frac{3\theta}{\sin 3\theta} \right) \cdot \frac{2}{3}$$

$$= (1) \left(\frac{1}{1} \right) \cdot \frac{2}{3} = \frac{2}{3} \qquad \blacksquare$$

In the proof of the main theorem we also need the following two trigonometric identities, called the *addition formulas* for the sine and cosine. They are proved in Supplementary Appendix 2.

$$\sin(\alpha + \beta) = \sin \alpha \cos \beta + \cos \alpha \sin \beta \qquad (3.8)$$

$$\cos(\alpha + \beta) = \cos \alpha \cos \beta - \sin \alpha \sin \beta \qquad (3.9)$$

Both formulas are valid for all real numbers α and β.

Before stating the theorem, let us try to anticipate the result from the graphs of $y = \sin x$ and $y = \cos x$. In the top graph of Figure 3.22 we have shown the graph of $f(x) = \sin x$. Using slopes of tangent lines drawn at selected points on this curve, we have obtained points on the graph of $y = f'(x)$, shown below the sine curve. The graph of f' looks suspiciously like the graph of $y = \cos x$. So a good conjecture is that the derivative of $\sin x$ is $\cos x$. In the theorem we prove that this conjecture is correct. Refer again to Figure 3.22, and assume the lower curve is the graph of $y = \cos x$. By considering slopes of tangent lines to this curve, can you conjecture what the derivative of the cosine is? Try it before going on to the theorem.

THEOREM 3.8

(a) $\dfrac{d}{dx} \sin x = \cos x$

(b) $\dfrac{d}{dx} \cos x = -\sin x$

Proof

(a) Let $f(x) = \sin x$. Then, by Definition 3.2,

$$f'(x) = \lim_{h \to 0} \frac{f(x+h) - f(x)}{h} = \lim_{h \to 0} \frac{\sin(x+h) - \sin x}{h}$$

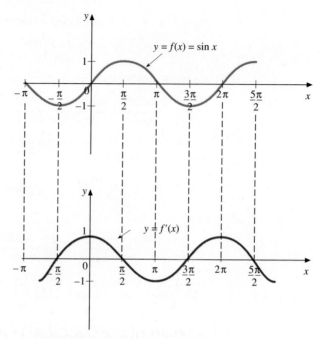

FIGURE 3.22

We use Identity 3.8 with $\alpha = x$ and $\beta = h$ to write the difference quotient on the right in the following equivalent ways.

$$\frac{\sin(x + h) - \sin x}{h} = \frac{\sin x \cos h + \cos x \sin h - \sin x}{h}$$

$$= \frac{\sin x(\cos h - 1) + \cos x \sin h}{h}$$

$$= -\sin x \cdot \left(\frac{1 - \cos h}{h}\right) + \cos x \left(\frac{\sin h}{h}\right)$$

Applying the Lemma we have, on taking the limit as $h \to 0$,

$$f'(x) = (-\sin x)(0) + (\cos x)(1) = \cos x$$

That is,

$$\frac{d}{dx}\sin x = \cos x$$

(b) Let $g(x) = \cos x$. Then

$$g'(x) = \lim_{h \to 0}\frac{g(x + h) - g(x)}{h} = \lim_{h \to 0}\frac{\cos(x + h) - \cos x}{h}$$

Using Identity 3.9, we write

$$\frac{\cos(x + h) - \cos x}{h} = \frac{\cos x \cos h - \sin x \sin h - \cos x}{h}$$

$$= \frac{\cos x(\cos h - 1) - \sin x \sin h}{h}$$

$$= -\cos x \left(\frac{1 - \cos h}{h}\right) - \sin x \left(\frac{\sin h}{h}\right)$$

Letting $h \to 0$ and applying the Lemma we get

$$g'(x) = (-\cos x)(0) - (\sin x)(1) = -\sin x$$

That is,

$$\frac{d}{dx}\cos x = -\sin x \qquad \blacksquare$$

EXAMPLE 3.14   Let $f(x) = x\sin x + \cos x$. Find $f'(x)$ and $f''(x)$.

Solution

$$f'(x) = \underbrace{(\sin x + x\cos x)}_{\substack{\text{derivative of}\\ x\sin x}} + (-\sin x) = x\cos x$$

$$f''(x) = \cos x + x(-\sin x) = \cos x - x\sin x \qquad \blacksquare$$

EXAMPLE 3.15   Find

$$\frac{d}{dt}\left(\frac{1-\cos t}{\sin t}\right)$$

Solution   By the Quotient Rule, we have

$$\frac{d}{dt}\left(\frac{1-\cos t}{\sin t}\right) = \frac{(\sin t)(\sin t) - (1-\cos t)\cos t}{\sin^2 t}$$

$$= \frac{\sin^2 t - \cos t + \cos^2 t}{\sin^2 t}$$

$$= \frac{1-\cos t}{\sin^2 t} \qquad \text{Since } \sin^2 t + \cos^2 t = 1$$

$$= \frac{1-\cos t}{1-\cos^2 t} \qquad \text{Since } \sin^2 t = 1 - \cos^2 t$$

$$= \frac{1-\cos t}{(1-\cos t)(1+\cos t)}$$

$$= \frac{1}{1+\cos t} \qquad \blacksquare$$

REMARK ───────────────────────────────────
■ In problems such as the one in Example 3.15 we will always assume any restrictions that are necessary on the independent variable so that denominators are not 0.
──

Derivatives of $\tan x$, $\cot x$, $\sec x$, and $\csc x$

Using the definitions

$$\tan x = \frac{\sin x}{\cos x} \qquad \cot x = \frac{\cos x}{\sin x} \qquad \sec x = \frac{1}{\cos x} \qquad \csc x = \frac{1}{\sin x}$$

we can get the derivatives of these other four trigonometric functions from those for the sine and cosine simply by using our formulas for differentiating quotients and reciprocals. We illustrate the calculation for the tangent and leave the others as exercises:

$$\frac{d}{dx}(\tan x) = \frac{d}{dx}\left(\frac{\sin x}{\cos x}\right)$$

$$= \frac{(\cos x)(\cos x) - (\sin x)(-\sin x)}{\cos^2 x} \qquad \text{Quotient Rule}$$

$$= \frac{\cos^2 x + \sin^2 x}{\cos^2 x}$$

$$= \frac{1}{\cos^2 x}$$

$$= \sec^2 x$$

We summarize here the derivatives of all six trigonometric functions.

Derivatives of the Trigonometric Functions

$$\frac{d}{dx}(\sin x) = \cos x \qquad\qquad \frac{d}{dx}(\cot x) = -\csc^2 x$$

$$\frac{d}{dx}(\cos x) = -\sin x \qquad\qquad \frac{d}{dx}(\sec x) = \sec x \tan x$$

$$\frac{d}{dx}(\tan x) = \sec^2 x \qquad\qquad \frac{d}{dx}(\csc x) = -\csc x \cot x$$

The formulas for derivatives of the sine and cosine are valid for all real x, those for the tangent and secant for all $x \neq \frac{\pi}{2} + n\pi$, and those for the cotangent and cosecant for all $x \neq n\pi$, with $n = 0, \pm 1, \pm 2, \ldots$.

EXAMPLE 3.16   Find $\dfrac{dy}{dx}$.

(a) $y = \sec x(1 + \tan x)$ (b) $y = \dfrac{\cot x}{1 - \csc x}$

Solution

(a) Using the Product Rule, we get

$$\frac{dy}{dx} = (\sec x \tan x)(1 + \tan x) + \sec x(\sec^2 x)$$

$$= \sec x \tan x + \sec x \tan^2 x + \sec^3 x$$

(b) By the Quotient Rule,

$$\frac{dy}{dx} = \frac{(-\csc^2 x)(1 - \csc x) - \cot x(\csc x \cot x)}{(1 - \csc x)^2}$$

$$= \frac{-\csc^2 x + \csc^3 x - \csc x \cot^2 x}{(1 - \csc x)^2}$$

We can simplify the answer further by making use of the identity

$$\cot^2 x = \csc^2 x - 1$$

given in Supplementary Appendix 2. (You can get it from $\sin^2 x + \cos^2 x = 1$ by dividing both sides by $\sin^2 x$.) Replacing $\cot^2 x$ by $\csc^2 x - 1$ gives

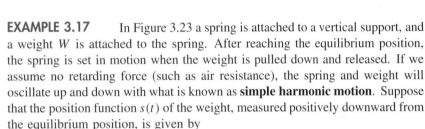

$$\frac{dy}{dx} = \frac{-\csc^2 x + \csc^3 x - \csc x(\csc^2 x - 1)}{(1 - \csc x)^2}$$

$$= \frac{-\csc^2 x + \csc x}{(1 - \csc x)^2}$$

$$= \frac{\csc x(1 - \csc x)}{(1 - \csc x)^2} = \frac{\csc x}{1 - \csc x} \quad \blacksquare$$

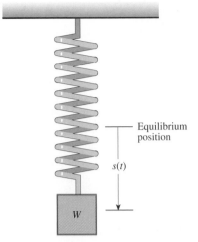

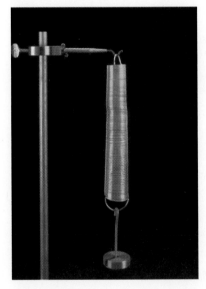

Equilibrium
position

$s(t)$

W

FIGURE 3.23

EXAMPLE 3.17   In Figure 3.23 a spring is attached to a vertical support, and a weight W is attached to the spring. After reaching the equilibrium position, the spring is set in motion when the weight is pulled down and released. If we assume no retarding force (such as air resistance), the spring and weight will oscillate up and down with what is known as **simple harmonic motion**. Suppose that the position function $s(t)$ of the weight, measured positively downward from the equilibrium position, is given by

$$s(t) = 8 \cos t$$

where t is in seconds and $s(t)$ is in centimeters. Discuss the motion.

Solution   We know that the first derivative of s gives velocity and the second derivative gives acceleration. So we have

$$v(t) = -8 \sin t$$

$$a(t) = -8 \cos t$$

At time $t = 0$, the position of the weight is $s(0) = 8$ cm below the equilibrium position, and the velocity $v(0) = 0$. So the weight was initially pulled down 8 cm from the equilibrium position and released from rest. The amplitude of the motion is 8, and the period is 2π. So the weight will oscillate from the lowest point 8 cm below equilibrium to its highest point 8 cm above equilibrium, and it will go through a complete cycle every 2π seconds.

From $t = 0$ to $t = \pi$, the velocity is negative, indicating the weight is moving upward (remember that downward is the positive direction). At $t = \pi$ it reaches its highest position. Then, from $t = \pi$ to $t = 2\pi$, the velocity is positive, and the weight is moving downward. It passes through the equilibrium position at times $t = \pi/2$, $3\pi/2$, $5\pi/2$, ... (See Figure 3.24.)

The acceleration is opposite to the position; that is, $a(t) = -s(t)$. Thus, when the weight is below equilibrium, the acceleration is positive, and when it is above equilibrium, the acceleration is negative. From Newton's Second Law of Motion, we know that force = mass × acceleration. So when the weight is below equilibrium, the force exerted by the spring is negative (upward), and when it is above equilibrium, the force is positive (downward). We summarize the nature of the motion in Figure 3.25. $\blacksquare$

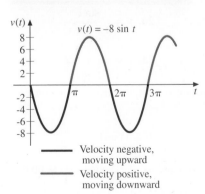

$v(t)$

$v(t) = -8 \sin t$

Velocity negative,
moving upward

Velocity positive,
moving downward

FIGURE 3.24

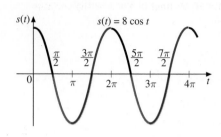

$s(t) = 8 \cos t$

— Moving up, speeding up
— Moving up, slowing down
— Moving down, speeding up
— Moving down, slowing down

FIGURE 3.25

REMARK ─────────────────────────────

■ All simple harmonic motion can be modeled by an equation of the form

$$s(t) = a \cos(\omega t) + b \sin(\omega t)$$

which can also be written in the form

$$s(t) = A \sin(\omega t + \alpha)$$

using Identity 3.8 (see Exercise 36 in this section), where a, b, ω, as well as A and α are constants that depend on the initial position and velocity, the mass of the object in motion, and a property of the spring known as the spring constant. In practice, pure simple harmonic motion is never achieved because of resisting forces, but the idealized model is nevertheless useful in approximating the motion.

Exercise Set 3.6

In Exercises 1–10, make use of the Lemma (on page 166) to find the limits.

1. $\displaystyle\lim_{t \to 0} \frac{\sin 2t}{t}$

2. $\displaystyle\lim_{x \to 0} \frac{\sin^2 x}{x}$

3. $\displaystyle\lim_{\theta \to 0} \frac{\tan \theta}{\theta}$

4. $\displaystyle\lim_{t \to 0} \frac{3t - \sin t}{t}$

5. $\displaystyle\lim_{x \to 0} \frac{\sec x - 1}{x}$

6. $\displaystyle\lim_{\theta \to 0} \frac{\sin \theta}{\theta + \sin \theta}$

7. $\displaystyle\lim_{t \to 0} \frac{\sin 3t}{\tan 2t}$

8. $\displaystyle\lim_{x \to 0} \frac{1 - \cos x}{\sin x}$

9. $\displaystyle\lim_{h \to 0} (\sin 2h)(\cot 5h)$

10. $\displaystyle\lim_{x \to \infty} x \sin \frac{1}{x}$ (*Hint:* Let $t = 1/x$.)

In Exercises 11–20, find dy/dx.

11. $y = x^2 \sin x$

12. $y = 3 \sin x - 2 \cos x$

13. $y = \dfrac{1 - \sin x}{1 + \sin x}$

14. $y = \dfrac{\sec x}{1 + \tan x}$

15. $y = \dfrac{\sin x}{x}$

16. $y = x \cos x - \sin x$

17. $y = \dfrac{2 \cot x + 1}{\csc x}$

18. $y = \sec x \tan x$

19. $y = \dfrac{\tan x}{1 - \tan x}$

20. $y = \dfrac{x \sin x}{1 - \cos x}$

21. (a) By writing $\sin^2 x = (\sin x)(\sin x)$, find a formula for $\dfrac{d}{dx}(\sin^2 x)$.
(b) Use the idea of part (a) to find a formula for $\dfrac{d}{dx}(\cos^2 x)$.

22. Use the results of Exercise 21 to find $f'(x)$ if $f(x) = \cos^2 x - \sin^2 x$.

23. Find the equation of the tangent line to the graph of $y = x \sin x + \cos x$ at the point for which $x = \pi/2$.

24. Find all points on the graph of $y = \sqrt{3} \sin x + \cos x$ between $x = 0$ and $x = 2\pi$ where the tangent line is horizontal.

In Exercises 25–28, find y''.

25. $y = x \cos x - \sin x$

26. $y = \dfrac{\sin x}{1 - \cos x}$

27. $y = \dfrac{x}{\cos x}$

28. $y = \sin^2 x + x$

29. The position function $s(t)$ for a particle moving on a line is given by

$$s(t) = (t + 1) \sin t$$

Find its velocity $v(t)$ and its acceleration $a(t)$ at time t.

30. Show that for all constants C_1 and C_2, the function $y = C_1 \cos x + C_2 \sin x$ satisfies the equation $y'' + y = 0$.

31. Show that $y = x(2\cos x + \sin x)$ satisfies the equation
$y'' + y = 2\cos x - 4\sin x$.

In Exercises 32–34, prove the given derivative formula.

32. $\dfrac{d}{dx}\cot x = -\csc^2 x$

33. $\dfrac{d}{dx}\sec x = \sec x \tan x$

34. $\dfrac{d}{dx}\csc x = -\csc x \cot x$

35. A spring with an attached weight, as in Example 3.17, moves in simple harmonic motion according to the formula

$$s(t) = 3\cos t + 2\sin t$$

(a) Where is the weight at time $t = 0$, and what is its velocity?

(b) At time $t = 2\pi/3$ give the location of the weight, its velocity, and its acceleration. In what direction is it moving? Is it speeding up or slowing down?

(c) When does the weight go through the equilibrium position for the first time?

36. (a) Show that the equation

$$s(t) = a\cos(\omega t) + b\sin(\omega t)$$

can be written in the equivalent form

$$s(t) = A\sin(\omega t + \alpha)$$

where $A = \sqrt{a^2 + b^2}$, $\sin\alpha = a/\sqrt{a^2 + b^2}$, and $\cos\alpha = b/\sqrt{a^2 + b^2}$. (*Hint:* Multiply and divide the right-hand side of the original equation by $\sqrt{a^2 + b^2}$. Draw a figure for the angle α. Then use Identity 3.8.)

(b) Rewrite the equation from Exercise 35 in the form $s(t) = A\sin(\omega t + \alpha)$. Sketch its graph, showing the amplitude and period.

37. If x is the *degree* measure of an angle, is the following statement still true?

$$\lim_{x\to 0}\frac{\sin x}{x} = 1$$

Prove your answer. If the limit is not 1, see if you can find what it is, and prove it.

38. Prove the following statement.

$$\frac{d}{dx}(\sin 2x) = 2\cos 2x$$

(*Hint:* $\sin 2x = 2\sin x \cos x$.)

39. Prove the following statement.

$$\frac{d}{dx}(\cos 2x) = -2\sin 2x$$

(*Hint:* $\cos 2x = 2\cos^2 x - 1$.)

40. A hot-air balloon lifts off from a point 2500 ft from an observer and rises vertically. At what rate (in feet per degree) is the distance from the observer to the balloon changing when the angle of elevation θ is $30°$? (See the figure.)

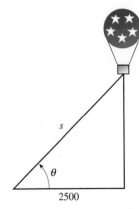

41. Let $f(x) = (x + \sin x)/\cos x$, $-\pi/2 < x < \pi/2$.

(a) Is f an even function, odd function, or neither? Prove your answer.

(b) Determine $f'(x)$.

(c) Find the equation of the tangent line to the curve $y = f(x)$ at the point for which $x = \pi/3$.

3.7 THE CHAIN RULE

In this section we will develop a formula, known as the **Chain Rule**, for the derivative of the composition of two functions. This is a powerful result that enables us to differentiate many otherwise intractable functions.

Consider the system of five gears shown in the accompanying photograph, where gear A is topmost and gears B, C, D, and E are clockwise from A. Gear A (12 teeth) drives B (14 teeth), B drives C (22 teeth), C drives D (36 teeth), and D drives E (17 teeth). Let's use the number of teeth on the gears as a measure of their relative circumference, assign a rate of rotation to each gear, and find the rate of rotation of gear E with respect to gear A.

Since the teeth on the gears are equally spaced, A is 12/14 the circumference of B, and therefore B turns 12/14 times every time A rotates once. Similarly, C rotates 14/22 times for every rotation of B and D rotates 22/36 times for every rotation of C. Gear E's rotation is 36/17 that of D. Visualize the gears turning. Do you see that the rate of rotation of gear E with respect to the drive gear, A, is the product of the rates of rotation?

$$\frac{\text{rate of rotation of E}}{\text{rate of rotation of A}} = \left(\frac{36}{17}\right)\left(\frac{11}{18}\right)\left(\frac{7}{11}\right)\left(\frac{6}{7}\right)$$

$$= \frac{12}{17} \approx .706$$

That is, gear E turns 12/17 times every time gear A turns once. We will see in this section that this example is a special case of the Chain Rule.

Before stating the rule, let us consider another example. Suppose we are given the function F defined by

$$F(x) = \sin(x^2 + 2)$$

and wish to find $F'(x)$. If we let $f(x) = \sin x$ and $g(x) = x^2 + 2$, then

$$F(x) = f(g(x))$$

That is, F is the composite function $f \circ g$. We already know how to differentiate f and g individually. The question is how to differentiate their composition. The Chain Rule provides the answer; namely, we *multiply* the individual derivatives of f and g, evaluated at appropriate points. We will return to this example after stating the Chain Rule.

The Chain Rule is one of the most important rules for differentiation. It is essential that you learn it and understand how to apply it.

THEOREM 3.9

The Chain Rule

Let g be differentiable at x and f be differentiable at $g(x)$. Then the composition $f \circ g$ is differentiable at x, and its derivative there is

$$(f \circ g)'(x) = f'(g(x)) \cdot g'(x) \qquad (3.10)$$

REMARK ———————————————

■ Since $(f \circ g)(x) = f(g(x))$, it is useful to refer to f as the *outer* function and g as the *inner* function. By the Chain Rule we first differentiate the outer function and evaluate it at the inner, getting $f'(g(x))$. Then we multiply this result by the derivative of the inner function at x, $g'(x)$.

Before we consider the proof of the Chain Rule, we give some more examples to show how it works.

EXAMPLE 3.18   Find $F'(x)$ if $F(x) = \sin(x^2 + 2)$.

Solution   As we showed in our preliminary discussion, $F = f \circ g$, where $f(x) = \sin x$ (the outer function) and $g(x) = x^2 + 2$ (the inner function). Thus, by the Chain Rule,

$$F'(x) = \overbrace{[\cos(x^2 + 2)]}^{f'(g(x))} \cdot \overbrace{(2x)}^{g'(x)} = 2x\cos(x^2 + 2) \qquad \blacksquare$$

EXAMPLE 3.19   Use the Chain Rule to find dy/dx if $y = (x^2 + 4)^{10}$.

Solution   We can write y in the form $y = (f \circ g)(x)$, where $f(x) = x^{10}$ and $g(x) = x^2 + 4$. Since $f'(x) = 10x^9$, we see that $f'(g(x)) = 10(x^2 + 4)^9$. Thus, by the Chain Rule,

$$\frac{dy}{dx} = f'(g(x)) \cdot g'(x) = 10(x^2 + 4)^9 \cdot 2x = 20x(x^2 + 4)^9 \qquad \blacksquare$$

EXAMPLE 3.20   Find y' if $y = \cos^3(2x)$.

Solution   Since $\cos^3(2x)$ means $[\cos(2x)]^3$, we can write y as the composition of the three functions $f(x) = x^3$, $g(x) = \cos x$, and $h(x) = 2x$. We can do this composition in stages, two at a time. For example, we can write y as

$$y = [f \circ (g \circ h)](x)$$

and first apply the Chain Rule separately to the two functions f and $g \circ h$. Then we apply the Chain Rule again to $g \circ h$. Thus,

$$\begin{aligned}
y' &= f'[(g \circ h)(x)] \cdot (g \circ h)'(x) & \text{First application} \\
&= f'[(g \circ h)(x)] \cdot g'(h(x)) \cdot h'(x) & \text{Second application} \\
&= 3[\cos(2x)]^2 \cdot [-\sin(2x)] \cdot 2 \\
&= -6\cos^2(2x)\sin(2x) \qquad \blacksquare
\end{aligned}$$

REMARK ———————————————————————————————

■ Although the analysis we went through in the preceding example is correct, in actual practice one would likely shortcut the procedure, as follows:

$$\begin{aligned}
y' &= 3[\cos(2x)]^2 \cdot \frac{d}{dx}[\cos(2x)] \\
&= 3[\cos(2x)]^2[-\sin(2x)] \cdot \frac{d}{dx}(2x) \\
&= 3[\cos(2x)]^2[-\sin(2x)]2 \\
&= -6\cos^2(2x)\sin(2x)
\end{aligned}$$

Note that we work from the outside in, continuing to take derivatives until we have differentiated the innermost function. With practice, you can shortcut the procedure even more, going from the original function to the final answer in one or two steps.

Partial Proof of the Chain Rule

Proof   Let $F(x) = (f \circ g)(x)$. That is, $F(x) = f(g(x))$. In applying the definition of the derivative (Definition 3.2), it is convenient here to use the delta notation, introduced in Section 3.2, writing Δx in place of h. Thus,

$$F'(x) = \lim_{\Delta x \to 0} \frac{F(x + \Delta x) - F(x)}{\Delta x} = \lim_{\Delta x \to 0} \frac{f(g(x + \Delta x)) - f(g(x))}{\Delta x}$$

Let $y = f(g(x))$ and $\Delta y = f(g(x + \Delta x)) - f(g(x))$. Then we have

$$F'(x) = \lim_{\Delta x \to 0} \frac{\Delta y}{\Delta x}$$

If we also write $u = g(x)$ and $\Delta u = g(x + \Delta x) - g(x)$, we obtain

$$f'(u) = \lim_{\Delta u \to 0} \frac{f(u + \Delta u) - f(u)}{\Delta u} = \lim_{\Delta u \to 0} \frac{\Delta y}{\Delta u}$$

and

$$g'(x) = \lim_{\Delta x \to 0} \frac{g(x + \Delta x) - g(x)}{\Delta x} = \lim_{\Delta x \to 0} \frac{\Delta u}{\Delta x}$$

By hypothesis, $g'(x)$ exists, so we know by Theorem 3.1 that g is continuous at x. Thus, $g(x + \Delta x) \to g(x)$ as $\Delta x \to 0$, and since $\Delta u = g(x + \Delta x) - g(x)$, it follows that $\Delta u \to 0$ as $\Delta x \to 0$. If $\Delta u \neq 0$ for all sufficiently small (nonzero) values of $|\Delta x|$, then we can write, for such values,

$$\frac{\Delta y}{\Delta x} = \frac{\Delta y}{\Delta u} \cdot \frac{\Delta u}{\Delta x}$$

and thus,

$$\begin{aligned} F'(x) = \lim_{\Delta x \to 0} \frac{\Delta y}{\Delta x} &= \lim_{\Delta x \to 0} \left(\frac{\Delta y}{\Delta u} \right) \left(\frac{\Delta u}{\Delta x} \right) \\ &= \left(\lim_{\Delta u \to 0} \frac{\Delta y}{\Delta u} \right) \left(\lim_{\Delta x \to 0} \frac{\Delta u}{\Delta x} \right) \qquad \text{Since } \Delta u \to 0 \text{ as } \Delta x \to 0 \\ &= f'(u)g'(x) \\ &= f'(g(x))g'(x) \end{aligned}$$

The theorem is therefore proved under the assumption $\Delta u \neq 0$ stated above.   ∎

Unfortunately, there are functions $u = g(x)$ for which $\Delta u = 0$ for some value of Δx in every interval about $\Delta x = 0$. In these cases, the proof we have given is not valid. A proof that works in all cases is given in Supplementary Appendix 5.

In the proof just given we introduced the notation $y = f(u)$ and $u = g(x)$. The result (Equation 3.10) of the Chain Rule can thus be written in the alternative form

---

The Chain Rule (Alternative Notation)

$$\frac{dy}{dx} = \frac{dy}{du} \cdot \frac{du}{dx} \tag{3.11}$$

For example, consider again the function $y = (x^2 + 4)^{10}$ of Example 3.19. We can let $u = x^2 + 4$. Then $y = u^{10}$. So, by Equation 3.11,

$$\frac{dy}{dx} = \underbrace{(10u^9)}_{\frac{dy}{du}} \underbrace{(2x)}_{\frac{du}{dx}} = 20x(x^2 + 4)^9$$

Powers of Differentiable Functions

Derivatives of functions of the form $[g(x)]^n$ occur with sufficient frequency to warrant special consideration. A function of this form can be regarded as the composition $(f \circ g)(x)$, where $f(x) = x^n$. Assuming g is differentiable and that $g(x) \neq 0$ if $n < 1$, we get from the Chain Rule

$$\frac{d}{dx}[g(x)]^n = n[g(x)]^{n-1} \cdot g'(x)$$

This special case of the Chain Rule is sometimes called the **Generalized Power Rule**. Our previous results show it to be valid for all integers n, and we will see later that it is true if n is any real number. By writing $u = g(x)$, the formula assumes the form given in Equation 3.12.

Generalized Power Rule

If $u = g(x)$ is differentiable at x, then for all integers n,

$$\frac{d}{dx}(u^n) = nu^{n-1}\frac{du}{dx} \tag{3.12}$$

provided that when $n < 1$, $u \neq 0$.

Using this result on the function $y = (x^2 + 4)^{10}$ of Example 3.19, we get the answer immediately:

$$\frac{dy}{dx} = 10(x^2 + 4)^9 \cdot 2x = 20x(x^2 + 4)^9$$

The next two examples further illustrate the Generalized Power Rule.

EXAMPLE 3.21   Find dy/dx if

$$y = \frac{2}{(3x - x^2)^4}$$

Solution   First write y in the form $y = 2(3x - x^2)^{-4}$. Then, by Equation 3.12,

$$\frac{dy}{dx} = -8(3x - x^2)^{-5}(3 - 2x)$$ ∎

EXAMPLE 3.22   Find dy/dx if $y = \sin^3 x \cos^2 x$.

Solution   Here y is a product of two functions, each of which is of the form u^n. By using the Product Rule together with Equation 3.12, we get

$$\frac{dy}{dx} = \left[3 \sin^2 x(\cos x)\right] \cos^2 x + \sin^3 x \left[2 \cos x(-\sin x)\right]$$

$$= 3 \sin^2 x \cos^3 x - 2 \sin^4 x \cos x.$$ ∎

One of the most frequent mistakes students make in differentiation is failure to apply the last step of the Chain Rule. For example, consider $\frac{d}{dx}(\sin 2x)$. All too frequently students give the answer as $\cos 2x$ and forget the final du/dx of Equation 3.11, which in this case is 2. The correct answer is $(\cos 2x) \cdot 2 = 2 \cos 2x$. So be alert to this danger. *Do not forget the final du/dx.*

The next example mixes algebraic and trigonometric functions.

EXAMPLE 3.23   Find $f'(x)$ for each of the following.

(a) $f(x) = x^2 \sin 2x$
(b) $f(x) = (\tan x - x)^2$

Solution

(a) The function is a product, so we have

We used the
Chain Rule here.
$$f'(x) = (2x)(\sin 2x) + x^2 \overbrace{(2 \cos 2x)}$$

$$= 2x(\sin 2x + x \cos 2x)$$

(b)

$$f'(x) = 2(\tan x - x)(\sec^2 x - 1)$$

$$= 2(\tan x - x)(\tan^2 x) \qquad \text{By the identity } 1 + \tan^2 x = \sec^2 x$$

$$= 2 \tan^2 x(\tan x - x)$$ ∎

Exercise Set 3.7

In Exercises 1–40, find dy/dx.

1. $y = (x^2 + 2)^4$

2. $y = (2x^3 - 3)^5$

3. $y = (1 - 2x^2)^5$

4. $y = (x^3 - x)^{-1}$

5. $y = (3x^4 - 2)^{10}$

6. $y = \left(\dfrac{x-1}{\sqrt{x}}\right)^{12}$

7. $y = (x^2 - 3x + 4)^3$

8. $y = 4(x^3 - 3x^2 + 5)^2$

9. $y = (3x^4 - 2x^2 + 1)^{-2}$

10. $y = (x^2 - 4x - 7)^{-3}$

11. $y = (2x + 1)^2(x^2 + 2)^3$

12. $y = x^5(x^3 - 2x^2 + 3)^4$

13. $y = \dfrac{2}{(x^2 + 1)^3}$

14. $y = \dfrac{3}{(2x^2 - 5)^4}$

15. $y = \dfrac{(3x - 4)^3}{x + 1}$

16. $y = \dfrac{(x^2 + 2)^2}{2x - 1}$

17. $y = \dfrac{x^2 + 1}{(x^2 - 1)^2}$

18. $y = \dfrac{2x^2 - x}{(3x - 4)^3}$

19. $y = \left(\dfrac{x-1}{x+1}\right)^3$

20. $y = \left(\dfrac{x^2}{x^2 - 1}\right)^{-2}$

21. $y = \sin 3x$

22. $y = \sin(2x + 3)$

23. $y = \cos(1 - 2x)$

24. $y = 2\cos(x^2 + 1)$

25. $y = \tan(3x^2)$

26. $y = 3\cot(1 - x)$

27. $y = 2\csc 3x$

28. $y = \sec\left(\dfrac{1}{x}\right)$

29. $y = 5\sin^2 x$

30. $y = 1 - 2\cos^2 x$

31. $y = \tan^2(2x + 3)$

32. $y = \csc 2x - \cot 2x$

33. $y = \tan^3 x^2$

34. $y = (1 + \sin 2x)^2$

35. $y = (\cos x - \sin x)^2$

36. $y = x^2 \cos 2x$

37. $y = (3x + 2)\csc 3x$

38. $y = (x + \tan x)^3$

39. $y = \sin(\cos x)$

40. $y = \tan(\sec^2 x)$

In Exercises 41–46, find $(f \circ g)'(x)$ for the given functions f and g.

41. $f(x) = x^3$; $g(x) = 2x + 3$

42. $f(x) = 2x^4$; $g(x) = 1 - 3x$

43. $f(x) = x^{-2}$; $g(x) = x^2 + \sin x$

44. $f(x) = \dfrac{1}{x}$; $g(x) = \cos x - 1$

45. $f(x) = 3x^2$; $g(x) = x^3 - 8$

46. $f(x) = x^{10}$; $g(x) = \tan x$

In Exercises 47–52, use Equation 3.11 to find dy/dx. Express the answer in terms of x.

47. $y = 2u^3$; $u = x^2 - 3$

48. $y = u^2 - 1$; $u = \cos 2x$

49. $y = 3/u^2$; $u = x^3 + 4$

50. $y = 4u^2 - 3$; $u = \sin 3x$

51. $y = u^2 - 2u$; $u = \cos x - \sin x$

52. $y = 1 - u^5$; $u = 1 - x^5$

In Exercises 53–60, find $f''(x)$.

53. $f(x) = (x^2 + 1)^4$

54. $f(x) = (2x^2 - 7)^5$

55. $f(x) = x/(1 - x)$

56. $f(x) = 1/(x^2 - 1)^2$

57. $f(x) = x^2 \cos 2x$

58. $f(x) = \tan x^2$

59. $f(x) = \sec^3 2x$

60. $f(x) = \cos x^3$

61. Let $h(t) = (t + 1)/(t - 1)$ and $g(t) = t^2$. Use the Chain Rule to find $(h \circ g)'(t)$ and $(g \circ h)'(t)$.

62. Let $\phi(u) = u^2 - 2u + 3$ and $\psi(u) = (2u - 1)/(3u + 1)$. Use the Chain Rule to find $(\phi \circ \psi)'(u)$ and $(\psi \circ \phi)'(u)$.

63. Let $z = t^3 - 3t^2 + 4$ and $t = y^2 + 3y - 1$. Use the Chain Rule to find dz/dy.

64. Find the equation of the tangent line and normal line (see Exercise 25 in Exercise Set 3.3) to the graph of

$$y = \dfrac{(3x - 4)^2}{2(x^2 - 2)^2}$$

at the point for which $x = 2$.

65. Find $f''(x)$ if $f(x) = (2x - 1)/(x^2 + 1)^2$.

66. Derive a formula for $(f \circ g \circ h)'(x)$, assuming appropriate differentiability properties by writing $f \circ g \circ h = (f \circ g) \circ h$.

67. Use the result of Exercise 66 to find y' if $y = [(2x^3 - 3)^2 + 4]^3$. What are f, g, and h in this case?

68. Show that $y = \cos 2x + \frac{1}{2}(1 - x \sin 2x)$ satisfies the equation $y'' + 4y = 4 \sin^2 x$.

69. A spring with a weight attached is set in motion, and its position $x(t)$ (in centimeters) from the equilibrium position at time t (in seconds) is given by

$$x(t) = \frac{1}{4}(1 - 8t) \cos 16t + \frac{1}{8} \sin 16t$$

Find its velocity and acceleration when $t = \dfrac{5\pi}{3}$.

70. Suppose f is differentiable on $\mathbb{R}$. Determine expressions for $\dfrac{d}{dx} f(\sin x)$ and $\dfrac{d}{dx} \sin(f(x))$.

71. Suppose f is differentiable on $\mathbb{R}$, and let $y = f(x)$. Determine an expression for $\dfrac{d}{dx} \cos(y^2 + 2xy)$.

72. Suppose f is differentiable on $\mathbb{R}$. If f is an even function, what can you say about f'? If f is an odd function, what can you say about f'? Prove your answers.

73. Let f be a differentiable function, and let

$$f_1(x) = f(x)$$
$$f_2(x) = (f \circ f)(x)$$
$$f_3(x) = (f \circ f \circ f)(x)$$
$$\vdots$$
$$f_n(x) = \underbrace{(f \circ f \circ \cdots \circ f)(x)}_{n \text{ times}}$$

Determine a formula for $f_n'(x)$ for $n \geq 2$.

74.(a) Use the fact that

$$\frac{d}{dx}|x| = \begin{cases} 1 & \text{if } x > 0 \\ -1 & \text{if } x < 0 \end{cases}$$

(see Exercise 39 in Exercise Set 3.2) and the Chain Rule to find $\dfrac{d}{dx}|\sin x|$. What is the domain of this derivative?

(b) On the same axes sketch the graphs of $y = |\sin x|$ and $y = \dfrac{d}{dx}|\sin x|$.

75. Let

$$f(x) = \begin{cases} x \sin \dfrac{1}{x} & \text{if } x \neq 0 \\ 0 & \text{if } x = 0 \end{cases}$$

(a) Find $f'(x)$ for $x \neq 0$.
(b) Use Definition 3.2 to prove that $f'(0)$ does not exist.
(c) Prove that f is continuous on all of $\mathbb{R}$.

76. Let

$$g(x) = \begin{cases} x^2 \sin \dfrac{1}{x} & \text{if } x \neq 0 \\ 0 & \text{if } x = 0 \end{cases}$$

(a) Find $g'(x)$ for $x \neq 0$.
(b) Use Definition 3.2 to prove that $g'(0)$ exists, and find its value.
(c) Prove that g' is discontinuous at $x = 0$.

3.8 IMPLICIT DIFFERENTIATION

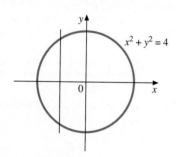

FIGURE 3.26

Certain equations in two variables x and y define y *implicitly* as a function of x. Consider, for example, the equation

$$2x^2 - 3y = 5$$

Although this equation is not in the form $y = f(x)$, we can easily solve for y to put it in this form:

$$y = \frac{2x^2 - 5}{3}$$

Now y is expressed *explicitly* as a function of x.
    Consider now the equation

$$x^2 + y^2 = 4$$

of a circle of radius 2, centered at the origin, as shown in Figure 3.26. Since the graph fails the vertical line test, the equation does not define y as a function of x. If we solve for y, we get

$$y = \pm\sqrt{4 - x^2}$$

Suppose we let $f_1(x) = \sqrt{4 - x^2}$ and $f_2(x) = -\sqrt{4 - x^2}$. Then, the graph of $y = f_1(x)$ is the upper semicircle in Figure 3.27(a) and the graph of $y = f_2(x)$ is the lower semicircle in Figure 3.27(b). Thus, we can say that the equation $x^2 + y^2 = 4$ defines y implicitly as two functions of x, namely, f_1 and f_2.

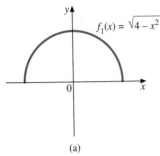

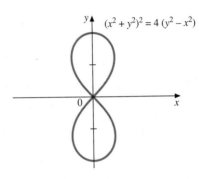

(a)   (b)

FIGURE 3.27

FIGURE 3.28

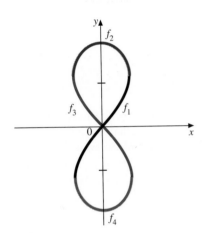

FIGURE 3.29

As another example of implicitly defined functions, consider the equation

$$(x^2 + y^2)^2 = 4(y^2 - x^2)$$

whose graph is shown in Figure 3.28. This curve is known as a **lemniscate**. (It is one of the curves we will consider in Chapter 9.) It would be difficult, to say the least, to solve the equation for y in terms of x. However, as we indicate in Figure 3.29 with the different colors, we can divide the graph into four parts, each of which passes the vertical line test and thus represents a function.

You might be thinking at this point that every equation in x and y defines y implicitly as one or more functions of x. This conclusion is not correct, as the next two examples illustrate.

EXAMPLE 3.24   Show that the equation $x^2 + y^2 + 1 = 0$ does not define y as a function of x.

Solution   Whatever real values x and y may be, the left-hand side of this equation is always positive. So it cannot equal 0. The equation is therefore not satisfied by any real values of x and y and hence cannot define y as a (real-valued) function of x.   ■

EXAMPLE 3.25   Show that the equation $(x + y)^2 = y^2 + 2xy + 4$ does not define y as a function of x.

Solution   If the left-hand side is expanded and terms are collected, the result is (verify it)

$$x^2 - 4 = 0$$

or equivalently,

$$x = 2 \quad \text{or} \quad x = -2$$

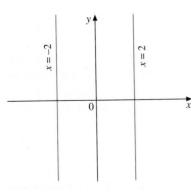

FIGURE 3.30

The graph therefore consists of the two vertical lines shown in Figure 3.30. This equation therefore does not define y as a function of x.   ■

Let us return now to the example of the circle $x^2 + y^2 = 4$, which defines y implicitly as $y = f_1(x)$ and $y = f_2(x)$, where $f_1(x) = \sqrt{4 - x^2}$ and $f_2(x) = -\sqrt{4 - x^2}$. We can find $f_1'(x)$ and $f_2'(x)$ using the Generalized Power Rule (it is true for nonintegral n, as remarked earlier). That is,

$$f_1'(x) = \frac{1}{2}(4 - x^2)^{-1/2}(-2x) = \frac{-x}{\sqrt{4 - x^2}}$$

Similarly,

$$f_2'(x) = \frac{x}{\sqrt{4 - x^2}}$$

We now illustrate an alternative method, called **implicit differentiation**, whereby we can find both derivatives at once, using the original equation.

We reason as follows. Let $f(x)$ be either of the functions defined implicitly by the equation $x^2 + y^2 = 4$. That is, $f(x)$ can be either $f_1(x)$ or $f_2(x)$. If we substitute $f(x)$ for y in the equation, we get

$$x^2 + [f(x)]^2 = 4$$

and this equation is an *identity* in x. That is, the equation is true for all values of x in the domain of f, namely, $[-2, 2]$. The left-hand side is therefore the constant function 4 and so its derivative is 0, wherever that derivative exists. Thus, by the Chain Rule,

$$2x + 2[f(x)]f'(x) = 0$$

Solving for $f'(x)$, we obtain the result

$$f'(x) = -\frac{x}{f(x)} = -\frac{x}{y}$$

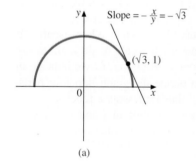

(a)

which is valid except where $y = 0$. (Note that when $y = 0$, the tangent lines are vertical.) Since $f(x)$ represents either $f_1(x)$ or $f_2(x)$, we have found a formula for the derivative of each of these functions. For example, consider the point $(\sqrt{3}, 1)$ on the circle $x^2 + y^2 = 4$. We can find the slope of the tangent line to the circle at this point simply by substituting the x- and y-coordinates of the point into the formula just derived. That is,

$$m_{\tan} = -\frac{x}{y} = -\frac{\sqrt{3}}{1} = -\sqrt{3}$$

Since we know the point $(\sqrt{3}, 1)$ is on the upper semicircle, we could have computed the slope of the tangent line using the formula for $f_1'(x)$. That is,

$$m_{\tan} = f_1'(\sqrt{3}) = \frac{-\sqrt{3}}{\sqrt{4 - (\sqrt{3})^2}} = -\sqrt{3}$$

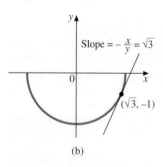

(b)

FIGURE 3.31

As suspected, the answers are the same. Similar computations can be done with the point $(\sqrt{3}, -1)$ on the lower semicircle, replacing f_1' by f_2'. The result in this case for the slope is $\sqrt{3}$. Both tangent lines are shown in Figure 3.31.

In practice, we usually shortcut the above procedure as follows.

$$x^2 + y^2 = 4$$

$$2x + 2yy' = 0 \qquad \text{We differentiated each term with respect to } x.$$

$$y' = -\frac{x}{y} \qquad \text{Solve for } y'.$$

Here, it is important to understand that y is a function of x, namely, one of the functions f_1 or f_2 determined implicitly by the original equation. Thus, when we differentiate y^2, we must use the Chain Rule

$$\frac{d}{dx}(y)^2 = 2(y)\frac{dy}{dx}$$

That is,

$$\frac{d}{dx}(y^2) = 2yy'$$

As we have seen, not all equations in x and y define y implicitly as one or more functions of x. Conditions that ensure the equation does define y as such a function, or functions, are usually given in advanced calculus courses. In our examples and exercises, we will *assume* that all the equations we use do define y as one or more functions that are differentiable on some suitable domain. The importance of implicit differentiation lies in the fact that we can find formulas for the derivatives when it is difficult, or even impossible, to solve for the explicit functions. In the case of $x^2 + y^2 = 4$, we could have found $f_1'(x)$ and $f_2'(x)$ directly, although doing so with implicit differentiation is easier. In the case of the equation $(x^2 + y^2)^2 = 4(y^2 - x^2)$ of the lemniscate we illustrated in Figures 3.28 and 3.29, solving for f_1, f_2, f_3, and f_4 is very difficult (try it). Yet we can still find y' implicitly. (See Exercise 17 in this section.) In the case of the equation $x^3 - 2x^2y + y^4 - 3y^5 = 0$, it is *impossible* to solve for y, but finding y' implicitly is straightforward. (See Exercise 18 in this section.)

EXAMPLE 3.26   Find y'.
(a) $x^3 - 2xy + y^3 = 4$ (b) $y = x \sin y$

Solution

(a) Proceeding as above, we have

$$3x^2 - 2(y + xy') + 3y^2 \cdot y' = 0 \qquad \text{The Product Rule is used on the } xy \text{ term.}$$

Now we rearrange terms and solve for y':

$$y'(3y^2 - 2x) = 2y - 3x^2$$

$$y' = \frac{2y - 3x^2}{3y^2 - 2x}$$

Of course, any points (x, y) for which the denominator is 0 must be excluded.

(b) At first, this looks like an explicit function, since it is of the form $y =$ something. But the "something" involves y, so it must be treated implicitly. Differentiating both sides, we get

$$y' = \sin y + x(\cos y)y'$$

$$y'(1 - x \cos y) = \sin y$$

$$y' = \frac{\sin y}{1 - x \cos y} \qquad (x \cos y \neq 1)$$

Note that the derivative of the term $x \sin y$ involved both the Product Rule and the Chain Rule. ∎

EXAMPLE 3.27   Find y' and y'' if $x^3 - y^3 = 8$.

Solution   We differentiate both sides with respect to x:

$$3x^2 - 3y^2 \cdot y' = 0$$

$$y' = \frac{x^2}{y^2} \qquad (y \neq 0)$$

Now we differentiate again, using the Quotient Rule on the right:

$$y'' = \frac{(2x)y^2 - x^2(2y \cdot y')}{y^4}$$

We next divide out the common factor y on the numerator and denominator, replace y' by its value we have just found, and simplify:

$$y'' = \frac{2xy - 2x^2\left(\dfrac{x^2}{y^2}\right)}{y^3} = \frac{2x(y^3 - x^3)}{y^5}$$

Finally, we observe that since the original equation was $x^3 - y^3 = 8$, it follows that $y^3 - x^3 = -8$, and the final answer can be written in the form

$$y'' = \frac{-16x}{y^5} \qquad (y \neq 0)$$

You should be alert to the possibility of simplifying final answers in this way. It frequently happens in finding second derivatives.   ■

Derivatives of Rational Powers

Implicit differentiation allows us to extend the Power Rule for differentiation to rational exponents, as the next theorem shows.

THEOREM 3.10

> **Power Rule for Rational Exponents**
>
> For any rational number r, if $y = x^r$, then for all x for which y is differentiable
>
> $$\frac{dy}{dx} = rx^{r-1}$$

Proof   Since r is rational, it can be written in the form $r = m/n$, where m and n are integers. So $y = x^{m/n}$. We raise both sides to the nth power and differentiate implicitly:

$$y^n = x^m$$

$$ny^{n-1} \cdot y' = mx^{m-1} \qquad \text{Since we know the Power Rule holds for integers}$$

$$y' = \frac{m}{n}\frac{x^{m-1}}{y^{n-1}} \qquad \text{If } y^{n-1} \neq 0$$

Replacing y with $x^{m/n}$ and simplifying, we get

$$y' = \frac{m}{n}\frac{x^{m-1}}{x^{m(n-1)/n}} = \frac{m}{n}\frac{x^{m-1}}{x^{m-(m/n)}} = \frac{m}{n}x^{(m-1)-m+(m/n)}$$
$$= \frac{m}{n}x^{(m/n)-1}$$

which is the desired result, since $r = \frac{m}{n}$.   ■

REMARK ───

■ In Chapter 7 we will show that $y = x^r$ is differentiable for all x, except that when $x = 0$, we must have $r \geq 1$.

───

By combining Theorem 3.10 with the Chain Rule, we can extend the Generalized Power Rule (Equation 3.12) for rational exponents. That is, if $u = g(x)$ is differentiable at x and r is any rational number,

$$\frac{d}{dx}u^r = ru^{r-1}\frac{du}{dx}$$

provided $u \neq 0$ if $r < 1$.

EXAMPLE 3.28   Find $f'(x)$:

(a) $f(x) = \dfrac{\sqrt{x^2+1}}{x}$ (b) $f(x) = \sqrt{\cos x}$

Solution

(a) We write $f(x)$ in the form

$$f(x) = \frac{(x^2+1)^{1/2}}{x}$$

and use the Quotient Rule, along with Theorem 3.10 and the Chain Rule, to get

$$f'(x) = \frac{\left[\frac{1}{2}(x^2+1)^{-1/2} \cdot 2x\right]x - (x^2+1)^{1/2}}{x^2}$$

This result can be simplified considerably by multiplying numerator and denominator by the factor $(x^2+1)^{1/2}$, since doing so eliminates the negative exponent and also brings the other fractional exponent up to 1. You should verify that the result is

$$f'(x) = \frac{x^2 - (x^2+1)}{x^2(x^2+1)^{1/2}} = \frac{-1}{x^2\sqrt{x^2+1}}$$

(b)

$$f(x) = \sqrt{\cos x} = (\cos x)^{1/2}$$
$$f'(x) = \frac{1}{2}(\cos x)^{-1/2}(-\sin x) = -\frac{\sin x}{2\sqrt{\cos x}}$$ ■

EXAMPLE 3.29   Find y' and y'' if

$$x^{2/3} + y^{2/3} = a^{2/3}$$

Solution   Note first of all that we *cannot* eliminate the fractional exponents by cubing each term (however tempting this might be). Differentiating implicitly, we obtain

$$\frac{2}{3}x^{-1/3} + \frac{2}{3}y^{-1/3} \cdot y' = 0 \quad \text{Since } a \text{ is a constant}$$

$$y' = -\frac{y^{1/3}}{x^{1/3}} \quad (x \neq 0)$$

We find y'' as in Example 3.27:

$$y'' = -\frac{\left(\frac{1}{3}y^{-2/3} \cdot y'\right) \cdot x^{1/3} - y^{1/3} \cdot \frac{1}{3}x^{-2/3}}{x^{2/3}}$$

$$= -\frac{x^{1/3}y^{-2/3}\left(\frac{-y^{1/3}}{x^{1/3}}\right) - y^{1/3}x^{-2/3}}{3x^{2/3}} \quad \text{Substituting for } y'$$

$$= -\frac{-y^{-1/3} - y^{1/3} \cdot x^{-2/3}}{3x^{2/3}} \cdot \frac{x^{2/3}y^{1/3}}{x^{2/3}y^{1/3}} \quad \begin{array}{l}\text{This multiplication eliminates}\\ \text{negative exponents.}\end{array}$$

$$= \frac{x^{2/3} + y^{2/3}}{3x^{4/3}y^{1/3}}$$

$$= \frac{a^{2/3}}{3x^{4/3}y^{1/3}} \quad \text{Since } x^{2/3} + y^{2/3} = a^{2/3}$$

Observe that y'' does not exist if $x = 0$ or $y = 0$.   ■

Exercise Set 3.8

Use implicit differentiation to find y' in Exercises 1–18.

1. $x^2 - y^2 = 1$

2. $x^2 + 2xy - 3y^2 = 0$

3. $x^2y^3 - 2x + y = 3$

4. $y^2 - 3xy - x^3 = 8$

5. $y = x^3 - y^3$

6. $x(y^2 - 3x) = 5$

7. $y(x^2 - 2y) = 3$

8. $\sqrt{x} + \sqrt{y} = 1$

9. $y = \sqrt{x + y}$

10. $x + 2\sqrt{xy} + y = 3$

11. $x \sin y + y \sin x = 1$

12. $y = \tan(x + y)$

13. $\sec(xy) = 3x$

14. $x = \cos\sqrt{xy}$

15. $\sin x \cos y = x - y$

16. $\sin^2 x + \cos^2 y = x^2 + y^2$

17. $(x^2 + y^2)^2 = 4(x^2 - y^2)$

18. $x^3 - 2x^2y + y^4 - 3y^5 = 0$

In Exercises 19–36, find dy/dx.

19. $y = \sqrt{x^2 + 1}$

20. $y = x\sqrt{1 - x}$

21. $y = \dfrac{1}{\sqrt{4 - x^2}}$

22. $y = \dfrac{x}{\sqrt{x + 4}}$

23. $y = \sqrt[3]{(2x^3 + 5)^2}$

24. $y = 1 + (1 - 3x^3)^{2/3}$

25. $y = \dfrac{x}{\sqrt{1 - x^2}}$

26. $y = \sin\sqrt{x + 1}$

27. $y = \sqrt{\sin x}$

28. $y = (\sqrt{\cos x})^3$

29. $y = \tan\left(\dfrac{1}{\sqrt{x}}\right)$

30. $y = \sin^2 x^2$

31. $y = \dfrac{\sec(2x+3)}{\tan(2x+3)}$

32. $y = (1 - \sin^2 x)^{3/2}$

33. $y = \sqrt{2(1 - \cos 2x)}$ $(0 \le x \le \pi)$

34. $y = \sin^4 x + \cos^4 x$

35. $y = \dfrac{\sqrt{4 - x^2}}{x}$

36. $y = \sqrt{\dfrac{1-x}{1+x}}$

In Exercises 37–44, find y''.

37. $x^2 + y^2 = 1$

38. $x^3 + y^3 = 8$

39. $x^{3/2} - y^{3/2} = 8$

40. $\sqrt{x} - \sqrt{y} = 1$

41. $y = \sqrt{1 - x^2}$

42. $y = 1/\sqrt{x^2 + 1}$

43. $y = \sin\sqrt{x}$

44. $y = \sqrt{\cos x}$

45. Find the equation of the tangent line to the ellipse defined by
$$\frac{x^2}{a^2} + \frac{y^2}{b^2} = 1$$
at an arbitrary point (x_0, y_0) on the ellipse for which $y_0 \ne 0$.

46. Find dy/dx and d^2y/dx^2 if $x^{4/3} - 2y^{4/3} = a^{4/3}$, where a is a constant.

47. Find $f'(x)$ and $f''(x)$ if $y = f(x)$ is a twice-differentiable solution of $x = \tan(x + y)$.

48. Assuming g is a differentiable function for which $x = g(t)$ is a solution of
$$x^2 t^3 - 4x^3 t^2 + 3x^2 - 2t^5 - 7 = 0$$
find $g'(t)$.

49. (a) Find two functions f_1 and f_2 that are differentiable on $(-2, 2)$ and such that $y = f_1(x)$ and $y = f_2(x)$ satisfy the equation $y^2 - 2xy + 2x^2 - 4 = 0$. (*Hint:* Use the Quadratic Formula.)
(b) Find $f_1'(x)$ and $f_2'(x)$.
(c) Using the equation from part (a), differentiate implicitly to find y'.
(d) Show that the answers in parts (b) and (c) are the same.

50. (a) Show that $y = x$ satisfies the equation $y^3 - xy^2 - x^4 y + x^5 = 0$, and find two other solutions. Differentiate each solution.
(b) Use implicit differentiation to find y' for the equation in part (a), and show that this answer agrees with the derivatives found in part (a).

51. A particle moves in a straight line so that its position function at time t is $s(t) = (t^2 - 1)^{2/3}$. Find its velocity $v(t)$ and its acceleration $a(t) = v'(t)$ at time t. What are $v(3)$ and $a(3)$?

52. Two curves are said to be **orthogonal** at a point of intersection if their tangent lines at that point are perpendicular to each other. Prove that the graphs of $x^2 = 4y$ and $2y^2 + x^2 = 6$ are orthogonal at each of their points of intersection. Draw both curves, showing their tangent lines at their points of intersection.

53. Prove that the *family* of curves $y^2 = c_1 x^3$ is orthogonal to the family $2x^2 + 3y^2 = c_2$ in the sense that for all values of the constants c_1 and c_2, each member of the first family is orthogonal to each member of the second at each point of intersection (see Exercise 52). Use a CAS or graphing calculator to draw several members of each family on the same set of axes. (*Hint:* Before differentiating, write the equation of the first family in the form $y^2/x^3 = c_1$.)

3.9 THE DIFFERENTIAL AND LINEAR APPROXIMATION

We have used the Leibniz notation dy/dx as one of several ways to designate the derivative $f'(x)$, where $y = f(x)$, but until now we have never given meaning to the symbols dx and dy individually. It turns out, though, that we can give them meanings that are consistent with the derivative notation, but which allow us to treat dx and dy as separate entities. These separate entities are called *differentials*.

Suppose f is differentiable at x. Let dx denote an independent variable that can take on any real value. Then the **differential** of y, designated dy, is defined as follows.

> ### The Differential of $y = f(x)$
>
> $$dy = f'(x)\, dx \qquad\qquad (3.13)$$

Note that dy is a dependent variable, dependent on dx for each fixed value of x. We call dx the differential of x.

EXAMPLE 3.30   Find dy if $y = \sin^2 x$.

Solution   Let $f(x) = \sin^2 x$. Then $f'(x) = 2\sin x \cos x = \sin 2x$. So by Equation 3.13,

$$dy = (\sin 2x)dx \qquad\qquad \blacksquare$$

If $dx \neq 0$ and we divide both sides of Equation 3.13 by dx, we get

$$\frac{dy}{dx} = f'(x)$$

Now this equation appears to be just a restatement of a familiar fact. But there really is something new here, since now the left-hand side is a quotient of two differentials rather than simply a symbol for the derivative, as we have previously used it. The two interpretations give the same result, however.

The notations dy/dx and $f'(x)$ for the derivative were devised by Leibniz and Newton, respectively, who took different approaches to the development of the ideas of calculus. It is interesting that the formal verification of the *correctness* of Leibniz's approach was not understood fully until the 1960s when Abraham Robinson introduced what is called *nonstandard analysis*.

All formulas for derivatives can now be phrased in terms of differentials as well. Suppose, for example, that $u = f(x)$ and $v = g(x)$, where f and g are differentiable in some common domain. Then, since $du = f'(x)dx$ and $dv = g'(x)dx$, we have the following formulas for differentials of sums, differences, products, and quotients:

(i) $d(u \pm v) = du \pm dv$

(ii) $d(uv) = u\, dv + v\, du$

(iii) $d\left(\dfrac{u}{v}\right) = \dfrac{v\, du - u\, dv}{v^2} \qquad (v \neq 0)$

To illustrate how to prove these differential forms, consider equation (ii). Since $uv = (fg)(x)$, by the definition of the differential,

$$
\begin{aligned}
d(uv) &= \big[f'(x)g(x) + f(x)g'(x)\big]dx \\
      &= \big[f(x)g'(x) + g(x)f'(x)\big]dx \\
      &= \underbrace{f(x)}_{u}\,\underbrace{g'(x)dx}_{dv} + \underbrace{g(x)}_{v}\,\underbrace{f'(x)dx}_{du}
\end{aligned}
$$

These results provide an alternative means of finding derivatives implicitly, as the next example shows.

EXAMPLE 3.31   Use differentials to find $\dfrac{dy}{dx}$ if

$$x^2 + xy - 3y^2 = 5$$

Solution   We take the differential of each term, getting

$$2x\,dx + (x\,dy + y\,dx) - 6y\,dy = 0$$

We collect the terms with dx on one side and those with dy on the other and then solve for the quotient dy/dx:

$$(2x + y)dx = (6y - x)dy$$

$$\frac{dy}{dx} = \frac{2x + y}{6y - x} \qquad (x \neq 6y) \qquad\blacksquare$$

Suppose now that $y = f(x)$ and $x = g(t)$, so that $y = f(g(t)) = (f \circ g)(t)$. Then, by Equation 3.13 and the Chain Rule, we have

$$dy = (f \circ g)'(t)dt$$

$$= f'(g(t)) \cdot g'(t)dt$$

But again by Equation 3.13 applied to $x = g(t)$, we see that $dx = g'(t)dt$. So

$$dy = f'(\underbrace{g(t)}_{x}) \cdot \underbrace{g'(t)dt}_{dx} = f'(x)dx$$

This calculation shows that when $y = f(x)$, the differential of y is given by Equation 3.13 even when x is dependent on some other variable (assuming appropriate differentiability conditions).

Geometric Interpretation

The differentials dx and dy have a useful geometric interpretation, as shown in Figure 3.32. Suppose $y = f(x)$. Let $P(x, f(x))$ be a point on the graph at which f' exists, and let l be the tangent line at P. Since l has slope dy/dx, it follows that the point $Q(x + dx, y + dy)$ always lies on l for all nonzero values of dx. Thus, for any horizontal displacement dx from P, dy is the vertical displacement RQ to the tangent line l. Recall that we have previously used the symbol Δx to indicate an increment, or change, in x, and correspondingly we have used Δy to be the change in the function:

$$\Delta y = f(x + \Delta x) - f(x)$$

Suppose such an increment Δx is given. Since dx may be chosen arbitrarily, let us choose $dx = \Delta x$. Now we want to examine the relationship between dy and Δy. As we have seen, dy gives the vertical displacement from $P(x, f(x))$ *to the tangent line* drawn at P. The increment Δy, on the other hand, gives the vertical displacement from P *to the curve itself*. The relationship between dy and Δy is shown in Figure 3.33. Observe carefully that, although we are free to choose the differential dx of the *independent* variable equal to the increment Δx, the differential dy of the *dependent* variable will not, in general, equal

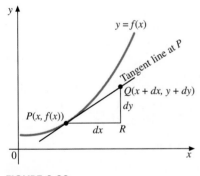

FIGURE 3.32

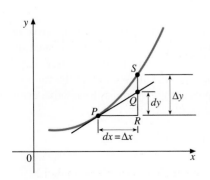

FIGURE 3.33

the increment Δy. However, if $\Delta x \ (= dx)$ is sufficiently small, it seems reasonable to suppose that $\Delta y \approx dy$. To illustrate this approximation, consider the following two examples.

EXAMPLE 3.32   Compare Δy and dy for $y = 2x^2 + 3x - 1$ when $x = 3$ and $\Delta x = 0.1$.

Solution   Denote the given function by f, and let $dx = \Delta x = 0.1$. Then

$$\Delta y = f(x + \Delta x) - f(x) = f(3.1) - f(3)$$
$$= [2(3.1)^2 + 3(3.1) - 1] - [2(3)^2 + 3(3) - 1]$$
$$= 27.52 - 26 = 1.52$$

Since $f'(x) = 4x + 3$, we get

$$dy = f'(3)dx = (15)(0.1) = 1.5$$

So in this case, the error in using dy to approximate Δy is only 0.02. ∎

EXAMPLE 3.33   A spherical natural gas storage tank is designed with a radius of 24 m. The allowable error in the actual radius is ± 10 cm. Use differentials to find the approximate excess capacity the tank could have over its designed volume. Compare this result with the volume of the maximum excess capacity as calculated from the given dimensions.

Solution   The volume $f(r)$ of a sphere is given by $V = \frac{4}{3}\pi r^3$. We take $r = 24$ and $\Delta r = dr = 0.10$, since 10 cm $= 0.10$ m. The approximate change in volume if the radius is 24.1 instead of 24 is

$$dV = f'(r)dr = 4\pi r^2 dr = 4\pi (24)^2 (0.10)$$
$$\approx 723.8 \text{ m}^3$$

The change calculated from the given measurements is

$$\Delta V = f(r + \Delta r) - f(r) = f(24.1) - f(24)$$
$$= \frac{4}{3}\pi (24.1)^3 - \frac{4}{3}\pi (24)^3$$
$$\approx 726.8 \text{ m}^3$$

So the error in using dV to approximate ΔV is about 3 m^3. ∎

Relative Error and Percentage Error

If $y = f(x)$ and x is in error by an amount Δx, then the corresponding error in y is $\Delta y = f(x + \Delta x) - f(x)$. As we have seen, Δy can be approximated by $dy = f'(x)\,\Delta x$ when Δx is small. Often it is more useful to know how large Δy, or its approximation dy, is in relation to y itself. We define the ratio of Δy to y as the **relative error** in y. The **percentage error** is the relative error multiplied by 100. So we have the following:

Absolute error	$\Delta y \approx dy$
Relative error	$\dfrac{\Delta y}{y} \approx \dfrac{dy}{y}$
Percentage error	$100\dfrac{\Delta y}{y} \approx 100\dfrac{dy}{y}$

EXAMPLE 3.34   Find the approximate relative errors and percentage errors in Examples 3.32 and 3.33.

Solution   In Example 3.32 we were given that $y = 2x^2 + 3x - 1$, $x = 3$, and $\Delta x = 0.1$. We found $dy = 1.5$. To find y, we substitute $x = 3$ in the given equation:

$$y = 2(3)^2 + 3(3) - 1 = 26$$

So the relative error and percentage error are approximately

$$\text{relative error} \approx \frac{dy}{y} = \frac{1.5}{26} \approx 0.0577$$

$$\text{percentage error} \approx 5.77\%$$

In Example 3.33, the volume when $r = 24$ is $\frac{4}{3}\pi(24)^3 \approx 57{,}905.84$. We found $dV \approx 723.8$. So we have

$$\text{relative error} \approx \frac{dV}{V} \approx \frac{723.8}{57{,}906} \approx 0.0125$$

$$\text{percentage error} \approx 1.25\% \blacksquare$$

Linear Approximation

If we know the value of a differentiable function f at a given point $x = a$, then we can get its value at a nearby point $a + \Delta x$ by adding Δy to the original value; that is,

$$f(a + \Delta x) = f(a) + \Delta y$$

Furthermore, since $f'(a)$ exists, we can approximate Δy by dy if we take $dx = \Delta x$. So we have

$$f(a + \Delta x) \approx f(a) + dy$$

or equivalently,

$$f(a + \Delta x) \approx f(a) + f'(a)\Delta x \qquad (3.14)$$

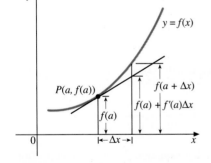

FIGURE 3.34
The distance from the x-axis to the tangent line at P approximates the distance to the curve if Δx is small.

This result is known as **linear approximation,** since we are, in effect, using values along the *tangent line* to approximate values on the *curve*. (See Figure 3.34.)

As we saw in Examples 3.33 and 3.34, it is frequently easier to compute dy than Δy, and so approximating $f(x)$ near a point where its value is known by the linear approximation formula (Equation 3.14) is easier than calculating its actual value.

EXAMPLE 3.35   Without using a calculator, find an approximation to $\sqrt[3]{65}$. Then check your result with a calculator.

Solution   Let $f(x) = \sqrt[3]{x} = x^{1/3}$. The point nearest 65 at which we know the exact value of f is $x = 64$, since $\sqrt[3]{64} = 4$. So in Equation 3.14 we take $x = 64$ and $\Delta x = 1$. Since $f'(x) = \frac{1}{3}x^{-2/3}$, we get

$$f(65) \approx f(64) + f'(64)\Delta x = 4 + \frac{1}{3}(64)^{-2/3}(1) = 4 + \frac{1}{3}\left(\frac{1}{16}\right) \approx 4.0208$$

With a calculator, we find $\sqrt[3]{65} = (65)^{1/3} = 4.02073$ (to five decimal places). So our linear approximation is quite good in this case.   ∎

Refer again to Figure 3.34. Since the tangent line at P has slope $f'(a)$, its equation is

$$y = f(a) + f'(a)(x - a)$$

(you can verify this result). Let us designate the function defined by this equation by L (for *linear*), so that

$$L(x) = f(a) + f'(a)(x - a) \qquad (3.15)$$

If in Equation 3.14 we replace Δx by $x - a$, we see that the equation becomes

$$f(x) \approx f(a) + f'(a)(x - a)$$

That is,

$$f(x) \approx L(x)$$

The closer x is to a (or equivalently, the smaller Δx is), the better the approximation. We call $L(x)$ the **linearization** of $f(x)$ at $x = a$.

In the next example we find the linearization of $f(x) = \sqrt{1 + x}$ at $x = 0$ and at $x = 8$, and use the results to approximate values of f near these two points.

EXAMPLE 3.36   Let $f(x) = \sqrt{1 + x}$.

(a) Find the linearization of $f(x)$ at $x = 0$ and use the result to approximate $\sqrt{1.1}$, $\sqrt{1.01}$, and $\sqrt{1.001}$.

(b) Find the linearization of $f(x)$ at $x = 8$ and use the result to approximate $\sqrt{8.9}$, $\sqrt{9.1}$, and $\sqrt{10}$.

Solution   First, we calculate $f'(x)$, since it will be used in both part (a) and part (b).

$$f'(x) = \frac{1}{2}(1 + x)^{-1/2}(1) = \frac{1}{2\sqrt{1 + x}}$$

(a) Taking $a = 0$, $f'(a) = \dfrac{1}{2\sqrt{1 + 0}} = \dfrac{1}{2}$. By Equation 3.15, the linearization at 0 is

$$L(x) = \sqrt{1 + 0} + \frac{1}{2}(x - 0) = 1 + \frac{x}{2}$$

Thus, for x near 0,

$$f(x) \approx 1 + \frac{x}{2}$$

Since $\sqrt{1.1} = \sqrt{1 + 0.1} = f(0.1)$, we have

$$\sqrt{1.1} \approx 1 + \frac{0.1}{2} = 1.05$$

Similarly,

$$\sqrt{1.01} \approx 1 + \frac{0.01}{2} = 1.005$$

and

$$\sqrt{1.001} \approx 1 + \frac{0.001}{2} = 1.0005$$

By calculator, $\sqrt{1.1} \approx 1.0488$, $\sqrt{1.01} \approx 1.004988$, and $\sqrt{1.001} \approx 1.0004999$. Our approximations are therefore correct to two, three, and four decimal places, respectively. The closer x is to 0, the more accurate is the approximation.

(b) Since

$$f'(8) = \frac{1}{2\sqrt{1+8}} = \frac{1}{6}$$

and $f(8) = \sqrt{1+8} = 3$, the linearization at $x = 8$ is

$$L(x) = 3 + \frac{1}{6}(x - 8)$$

So for x near 8,

$$\sqrt{1+x} \approx 3 + \frac{1}{6}(x - 8)$$

In particular, with $x = 7.9$,

$$\sqrt{8.9} \approx 3 + \frac{1}{6}(-0.1) \approx 2.98333 \qquad \text{By calculator, 2.98329}$$

Similarly, with $x = 8.1$,

$$\sqrt{9.1} \approx 3 + \frac{1}{6}(0.1) \approx 3.01666 \qquad \text{By calculator, 3.01662}$$

Finally, with $x = 9$,

$$\sqrt{10} \approx 3 + \frac{1}{6}(1) = 3.16666 \qquad \text{By calculator, 3.1623}$$

Notice that the first two results are correct to at least four places, but the last is correct only to one place. The error is greater in $\sqrt{10}$ because $x = 9$ is farther from 8 than either $x = 7.9$ or 8.1.

Figure 3.35 shows the graph of f along with the tangent lines $y = L(x)$ at the points $(0, 1)$ and $(8, 3)$. Notice that because the curve is flatter near $(8, 3)$ than it is near $(0, 1)$, the tangent line at $(8, 3)$ provides a better approximation to $f(x)$ than it does at $(0, 1)$. The figure clearly shows that it would be inappropriate to use the linearization at $(0, 1)$ to approximate values of $f(x)$ in part (b) of this example, since the tangent line at $(0, 1)$ is nowhere near the actual curve in the vicinity of $x = 8$.   ∎

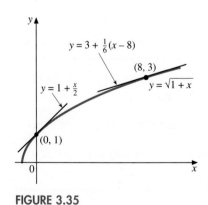

FIGURE 3.35

REMARK ─────────────────────────────────────

∎ You should bear in mind that values given by the calculator (or numerical software) are also approximations in most cases.

The Error in the Approximation of Δy by dy

We conclude this section by examining the size of the error in using the linear approximation given by Equation 3.14—that is, the error in using dy to approximate Δy. Assume $f'(x)$ exists and let $dx = \Delta x$. Then, $dy/dx = f'(x)$. Also,

$$\lim_{\Delta x \to 0} \frac{\Delta y}{\Delta x} = \lim_{\Delta x \to 0} \frac{f(x + \Delta x) - f(x)}{\Delta x} = f'(x)$$

and

$$\lim_{\Delta x \to 0} \frac{\Delta y - dy}{\Delta x} = \lim_{\Delta x \to 0} \left(\frac{\Delta y}{\Delta x} - \frac{dy}{\Delta x} \right)$$

Since $dx = \Delta x$, this limit is the same as

$$\lim_{\Delta x \to 0} \left(\frac{\Delta y}{\Delta x} - \frac{dy}{dx} \right)$$

which equals

$$\left(\lim_{\Delta x \to 0} \frac{\Delta y}{\Delta x} \right) - f'(x) = f'(x) - f'(x) = 0$$

Thus,

$$\lim_{\Delta x \to 0} \frac{\Delta y - dy}{\Delta x} = 0$$

This result tells us not only that the difference $\Delta y - dy$ approaches 0 as $\Delta x \to 0$, but also that it does so more rapidly than Δx itself. So the error $|\Delta y - dy|$ is small in comparison with the size of $|\Delta x|$. We now state a theorem that gives a more explicit estimate of the error. Its proof depends on results we will obtain in Chapter 10. The theorem refers to a *bounded second derivative* in an interval and an *upper bound* on $|f''(t)|$. To say that f has a bounded second derivative means that there is some positive constant M such that $|f''(t)| \le M$ for all t in the interval in question. Such a constant M is then called an upper bound for $|f''(t)|$.

THEOREM 3.11

If f has a bounded second derivative in some open interval $(x - r, x + r)$ centered on x, then for all Δx such that $|\Delta x| < r$,

$$|\Delta y - dy| \le \frac{M}{2}(\Delta x)^2 \qquad (3.16)$$

where M is an upper bound on $|f''(t)|$ for t in $[x, x + \Delta x]$ when $\Delta x > 0$ and in $[x - |\Delta x|, x]$ when $\Delta x < 0$.

To illustrate this inequality, recall Example 3.35. There we had $f(x) = x^{1/3}$ and $f'(x) = \frac{1}{3}x^{-2/3}$, and so

$$f''(x) = -\frac{2}{9}x^{-5/3} = -\frac{2}{9x^{5/3}}$$

In that example we set $x = 64$ and $\Delta x = 1$, so that we want to know a bound on $|f''(t)|$ for t in the interval $[64, 65]$. Since

$$|f''(t)| = \frac{2}{9t^{5/3}}$$

we obtain the largest value of $|f''(t)|$ by substituting the smallest value of t—namely, 64—since t is in the denominator. Thus, for t in the interval $[64, 65]$,

$$|f''(t)| \leq \frac{2}{9(64)^{5/3}} = \frac{2}{9(4)^5} = \frac{1}{4608} \approx 0.000217$$

So, by Inequality 3.16,

$$|\Delta y - dy| \leq \frac{0.000217}{2}(1)^2 \approx 0.0001$$

This result confirms that using dy to approximate Δy gives accuracy to at least three decimal places.

SYMBOLIC DIFFERENTIATION USING COMPUTER ALGEBRA SYSTEMS

Computer algebra systems have the capability for performing symbolic operations including differentiation. We have seen in this chapter that to differentiate a variety of functions, we need a collection of formulas. Even with the formulas available, many functions are difficult to differentiate because of the algebraic manipulation that is necessary. Computer algebra systems can be used to do the tedious work of finding derivatives, allowing us to investigate a greater number of problems.

CAS 14

Find the derivative of
$$f(x) = x^2 \cos\left(\sqrt{x^3 - x^{1/2}}\right)$$

This problem is easy to solve using Maple, Mathematica, or DERIVE.

Maple:

diff(x^2*cos(sqrt(x^3 − x^(1/2))),x);

where "diff" means "differentiate." The expression for the derivative given in Maple is:

$$2x\cos\left(\sqrt{x^3 - \sqrt{x}}\right) - \frac{1}{2}\frac{x^2\sin\left(\sqrt{x^3 - \sqrt{x}}\right)\left(3x^2 - \frac{1}{2}\frac{1}{\sqrt{x}}\right)}{\sqrt{x^3 - \sqrt{x}}}$$

Mathematica:*

f[x_]=x^2*Cos[Sqrt[x^3 − x^(1/2)]]
f'[x]

DERIVE:

(At the □ symbol, go to the next step.)
a (author) □ x^2cos(sqrt(x^3 − x^(1/2)))
□ c (calculus) □ d (differentiate) □ [choose expression] □ x (variable) □ 1 (order)
□ s (simplify) □ [choose expression]

---

*In Mathematica the derivative of f can then be computed by f'[x], as we did, or by using D[f[x],x].

▣ CAS 15

Plot the function $f(x) = \dfrac{x^4}{4} + \dfrac{x^3}{3} - x^2 + 1$ and the tangent line to the curve where $x = 2$. Find all points where the tangent line is horizontal.

Maple:[†]

f := x–>x^4/4+x^3/3–x^2+1;
m := x–>D(f)(x);

Output: $m := x \to x^3 + x^2 - 2x$

L := m(2)*(x–2)+f(2);

Output:

$$L := 8x - \frac{37}{3}$$

plot({f(x),L},x=–3..3,y=–5..5);

The third command defines the line from the point-slope formula $y = \text{slope} \cdot (x - x_0) + y_0$; we used m to stand for the slope.

Mathematica:

f[x_]=x^4/4+x^3/3–x^2+1
L[x_]=f'[2]*(x–2)+f[2]
Plot[{L[x],f{x}},{x,–3,3}]

DERIVE:

(At the □ symbol, go to the next step.)
a (author) □ f(x) := x^4/4+x^3/3–x^2+1
□ c (calculus) □ d (differentiate) □
[choose expression for f] □ x (variable) □ 1 (order) □ s (simplify) □
[choose expression] □ a (author) □ m(x)
:= [previous expression] □ a (author) □
m(2)(x–2)+f(2) □ s (simplify) □ [choose expression] □ p (plot window) □ p (plot)
□ a (algebra window) □ [highlight f]
□ p (plot window) □ p (plot)

FIGURE 3.9.1.
$y = f(x) = \dfrac{x^4}{4} + \dfrac{x^3}{3} - x^2 + 1,$
$y = 8x - \dfrac{37}{3}$

To find the points where the tangent line is horizontal, we need to find the x and y coordinates of the points where the tangent line has slope 0.

[†]In Maple the operator **D** computes derivatives of operators, whereas **diff** (see CAS 14) computes derivatives of expressions. That is, the argument and result of **D** are functions, whereas the argument and result of **diff** are expressions.

Maple:

solve(m(x)=0,x);

The output of the x-coordinates where the tangent has slope 0 is

0, 1, –2

Now find the y-coordinates of the three points:

f(0);
f(1);
f(–2);

The three points are given by:

$(0, 1)$, $(1, 7/12)$, and, $(-2, -5/3)$

Mathematica:

Solve[f′[x] == 0,x]

Output:

$\{\{x -> 0\}, \{x -> -2\}, \{x -> 1\}\}$

To find the y-coordinates:

f[0]
f[1]
f[–2]

DERIVE:

(At the □ symbol, go to the next step.)
1 (solve) □ [choose expression for $m(x)$]
Output:

$x = 0$

$x = 1$

$x = -2$

To find the y-coordinates: a (author) □
f(0) □ s (simplify) □ [choose expression] □ a (author) □ f(1) □ s (simplify)
□ [choose expression] □ a (author) □
f(−2) □ s (simplify) □ [choose expression]

CAS 16

Define the function f by

$$f(x) = \begin{cases} x^2 \sin(1/x) & \text{if } x \neq 0 \\ 0 & \text{if } x = 0 \end{cases}$$

Investigate the derivative of f at 0.

In this example the derivative at 0 cannot be computed by substituting 0 into the expression for the derivative. We need to use the definition of the derivative at a point as the limit of the difference quotient. Notice that since $f(0) = 0$, the numerator of the difference quotient is simply $f(x + h)$.

Maple:

f := x–>x^2*sin(1/x);
df := x–>f(x+h)/h;
limit(df(0),h=0);

The value of this last limit is 0, so the derivative is equal to 0 when $x = 0$.

Mathematica:

f[x_] = x^2*Sin[1/x]
f′[x]
df = (f[0+h])/h
Limit[df,h–>0]

DERIVE:

(At the □ symbol, go to the next step.)
a (author) □ x^2sin(1/x) □ m (manage)
□ s (substitute) □ [choose expression]
□ 0+h (substitute for x) □ c (calculus) □ 1 (limit) □ [choose expression]
□ h (limit with respect to h) □ 0 (limit point) □ s (simplify) □ [choose expression]

Exercise Set 3.9

In Exercises 1–16, find dy.

1. $y = x^2 + 2x - 3$

2. $y = 2x - x^3$

3. $y = \sqrt{4 - x^2}$

4. $y = 1/\sqrt{3x + 4}$

5. $y = x/\sqrt{1 - x}$

6. $y = x/\sqrt{1 - x^2}$

7. $y = x \sin x$

8. $y = \cot \sqrt{x}$

9. $y = \sec x + \tan x$

10. $y = \cos^2 3x$

11. $y = (x^6 + 1)^{2/3}$

12. $y = (x^{2/3} + a^{2/3})^{3/2}$

13. $y = (\tan x)/x$

14. $y = 2\csc^3 2x$

15. $y = \sin^2 x^2$

16. $y = (\sin 2x)/(1 - \cos 2x)$

In Exercises 17–24, find dy/dx implicitly using differentials.

17. $x^3 - 2xy^2 = 6$

18. $y^2 - x \sin y = \cos x$

19. $2x^2 - 3xy + 4y^2 - 10 = 0$

20. $\sqrt{x^2 + y^2} - 3 = 2xy$

21. $x^{2/3} + y^{2/3} = a^{2/3}$

22. $x^4 - 3x^2 y^3 + y^5 = 10$

23. $x \sin y - y \sin x = 1$

24. $x^{4/3} - y^{4/3} = a^{4/3}$

In Exercises 25–27, use differentials for the approximations.

25. The side of a square is measured to be 18 cm. If there is an error in the measurement of 0.05 cm, approximate the resulting error in (a) the area and (b) the perimeter.

26. Find the approximate error in the calculated area of a circle with a diameter measured as 20 in. when in fact it is 20.01 in. What is the error in the circumference?

27. Find the approximate relative error and percentage error in each part of Exercise 26.

In Exercises 28–31, estimate the value using linear approximation, and compare your result with the answer obtained by using a calculator. Find the percentage error.

28. $\sqrt{10}$

29. $\sqrt[3]{7.5}$

30. $\sqrt[3]{-25}$

31. $1/\sqrt{102}$

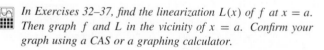

In Exercises 32–37, find the linearization $L(x)$ of f at $x = a$. Then graph f and L in the vicinity of $x = a$. Confirm your graph using a CAS or a graphing calculator.

32. $f(x) = \sin x; \ a = 0$

33. $f(x) = \tan x; \ a = \dfrac{\pi}{4}$

34. $f(x) = \sqrt{x}; \ a = 1$

35. $f(x) = x^{3/2}; \ a = 4$

36. $f(x) = 2\cos x - 1; \ a = \dfrac{\pi}{3}$

37. $f(x) = \dfrac{x^3}{3} + \dfrac{x^2}{2} - 2x - 1; \ a = 2$

38. Show that for x near 0, $(1 + x)^k \approx 1 + kx$ for all rational numbers k.

In Exercises 39–44, use the result of Exercise 38 to find a linear approximation of the given function near zero.

39. $(1 + x)^3$

40. $\sqrt[3]{1 + x}$

41. $\dfrac{1}{(1 + x)^2}$

42. $\dfrac{1}{\sqrt{1 + x}}$

43. $(1 - 2x)^{3/2}$ (*Hint:* In the formula in Exercise 38, replace x by $(-2x)$.)

44. $\dfrac{1}{\sqrt{4 - x}}$ (*Hint:* Write as $\dfrac{1}{2}\left(1 - \dfrac{x}{4}\right)^{-1/2}$ and use Exercise 38 with x replaced by $-x/4$.)

45. A cardboard box has a square base, and its height is double the length of a side of the base. The box is supposed to be 2 ft × 2 ft × 4 ft, but it is subject to an error of $\pm\frac{1}{8}$ inch in each dimension. Find the approximate maximum and minimum values for the actual volume. Then use a calculator to find the exact maximum and minimum values and compare the two sets of results.

46. The radius of the orbit (assumed circular) of communications satellite B is 1 mi larger than that of satellite A. Approximately how much farther does satellite B travel in completing one orbit than satellite A? Show that linear approximation gives the exact value in this case. Why doesn't the answer depend on whether the satellite is in a low earth orbit or in a synchronous orbit 22,600 mi high?

47. The cost $C(x)$ in dollars of manufacturing x television sets at a certain plant is given by

$$C(x) = 4000 - 15x + 0.015x^2$$

If 600 sets are typically manufactured each day, find the approximate cost of manufacturing one more set. (*Note:* Although x actually takes on only integer values, for purposes of calculation we can treat it as though it were a continuous variable that takes on all positive real values.)

48. The profit function $P(x)$ for manufacturing and selling x items of a certain type is given by

$$P(x) = 100\sqrt{x} - 0.75x - 600$$

Approximately what additional profit would result from manufacturing and selling 405 items instead of 400 items?

49. Suppose $y = x\sqrt{4 - x^2}$ and $x = 2\sin t$. Calculate dy in the following two ways and show that they are the same: (a) by replacing x in the first equation by its value from the second equation, and (b) by applying Equation 3.13.

50. Verify the differential formula

$$d\left(\frac{u}{v}\right) = \frac{v\,du - u\,dv}{v^2}$$

where u and v are differentiable functions of x, with $v \neq 0$.

51. Find du if

$$u = \frac{z \tan z}{\sqrt{1 + z^2}}$$

52. The inner radius of a section of steel pipe for an oil pipeline is 2 ft, and the thickness of the metal is $\frac{1}{4}$ in. If the section is 32 ft long, use differentials to approximate the number of cubic feet of metal in the pipe.

53. The height of Pike's Peak is calculated by measuring the angles of elevation α and β at points A and B, d meters apart, then using trigonometry. (See the figure.) Show that

$$h = \frac{d}{\cot \alpha - \cot \beta}$$

Suppose B is chosen so that $\beta = 2\alpha$. If $d = 2000$ (assumed exact for purposes of this problem) and $\alpha = 20°$, with a possible error of $\pm 0.5°$, find the approximate possible error in the calculation of h.

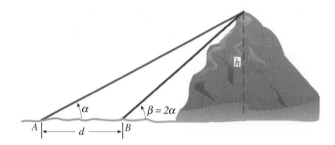

54. Use linear approximation to estimate each of the following, and find an upper bound on the error using Theorem 3.11:
(a) $\sqrt{4.2}$
(b) $(30)^{2/3}$

Chapter 3 Review Exercises

In Exercises 1–4, find $f'(a)$ using Definition 3.2.

1. $f(x) = 2x^2 - 3x$; $a = 3$

2. $f(x) = 2/\sqrt{x}$; $a = 1$

3. $f(x) = x/(x + 1)$; $a = 2$

4. $f(x) = \sqrt{2x + 3}$; $a = -1$

In Exercises 5–14, find $f'(x)$.

5. $f(x) = (3x^2 - 2)^{10}$

6. $f(x) = \dfrac{2x^2 - 3x}{x^2 + 1}$

7. $f(x) = \dfrac{x}{\sqrt{1 - 2x}}$

8. $f(x) = x^2 \sin x^3$

9. $f(x) = \dfrac{\sin x}{1 + \cos x}$

10. $f(x) = \sqrt{\dfrac{1 + x}{1 - x}}$

11. $f(x) = (a^{2/3} - x^{2/3})^{3/2}$

12. $f(x) = \sec^3 x + \tan^3 x$

13. $f(x) = \dfrac{\sqrt{1 - \cos 2x}}{\cos x}$, $0 \leq x < \dfrac{\pi}{2}$

14. $f(x) = \dfrac{x\sqrt{1 + x^2}}{1 - x^2}$

In Exercises 15 and 16, find the equation of the tangent line and normal line (line perpendicular to the tangent line) to the graph of f at the point $(a, f(a))$.

15. $f(x) = x^2/(x - 4)$; $a = 2$

16. $f(x) = (\sqrt{x} - 2)^3$; $a = 1$

In Exercises 17 and 18, the position function $s(t)$ of a particle moving on a straight line is given. Find its instantaneous velocity and acceleration when $t = a$.

17. $s(t) = 100\left(\dfrac{t^2 - 1}{t^2 + 1}\right)$; $a = 3$

18. $s(t) = 16\sqrt[3]{t^3 - 3t + 6}$; $a = 2$

In Exercises 19–22, find y' using implicit differentiation.

19. $2xy^2 + x^4 = y^3$

20. $y^3 + (x - y)^2 = 3$

21. $y \tan x - x \sec y = 1$

22. $\sqrt{x^2 + y^2}/(x + y) = 4$

23. Find y' and y'' if $x^{1/2} + y^{1/2} = a^{1/2}$, where $x > 0$, $y > 0$, and a is a positive constant.

24. Find the equations of the tangent line and normal line (line perpendicular to the tangent line) to the graph of $x^3 + x^2y^2 - y^3 = 1$ at the point $(-2, 3)$.

25. The height of a model of an ice cream cone is three times the radius of the top. The top is supposed to have a radius of 6 cm, but this measurement may be in error by as much as 1 mm. Use differentials to approximate the possible error in (a) the volume and (b) the lateral surface area. Also find the percentage error in each case. Assuming that the cone would have 50 calories if the radius measurement had no error, how uncertain is the calorie count?

26. Use differentials to approximate the area of the shaded circular segment in the figure. Leave your answer in exact form.

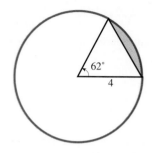

27. Verify that $y = x^{-1/2}(1 + \sin x)$ is a solution of the equation

$$x^2 y'' + xy' + \left(x^2 - \frac{1}{4} \right) y = x^{3/2}$$

Chapter 3 Concept Quiz

1. Let $y = f(x)$. Define each of the following:
(a) $f'(a)$
(b) the derivative function f'
(c) dy
(d) the linearization $L(x)$ of f at a

2. State each of the following:
(a) the Product Rule
(b) the Quotient Rule
(c) the Chain Rule
(d) the Power Rule
(e) the relationship between differentiability and continuity

3. Write the formulas for the derivatives of the six trigonometric functions.

4. (a) Write the equation of the tangent line to the graph of a differentiable function f at $x = a$.
(b) Show graphically the relationship between Δy and dy for $y = f(x)$, where f is differentiable at $x = a$.

5. Fill in the blanks.
(a) If f is differentiable at $x = a$, then $f'(a)$ is the _____ of the _____ _____ to the graph of f at $(a, f(a))$.
(b) If $s''(t)$ exists and s is the position function of a particle moving along a line, then $s'(t)$ is the _____ of the particle and $s''(t)$ is its _____.

(c) In words, the Product Rule states that the derivative of a product is the _____ of the _____ times the _____, plus the _____ times the _____ of the _____.

(d) In words, the Quotient Rule states that the derivative of a quotient is the _____ of the _____ times the _____, minus the _____ times the _____ of the _____, divided by the _____ of the _____.

(e) Linear approximation means that if f is differentiable at $x = a$, then $f(x)$ is approximated near a by _____.

6. State which of the following are true and which are false.
(a) If $y = f(x)g(x)$, then $y' = f'(x)g'(x)$.
(b) If f is not continuous at a, then f is not differentiable at a.
(c) If $y = f(x)/g(x)$, then $y' = f'g - fg'$.
(d) If $f(x) = \sin x$ and $g(x) = \cos x$, then $(f + g)'(x) = g(x) - f(x)$.
(e) If $f''(a)$ exists, then f' is continuous at a.

APPLYING CALCULUS

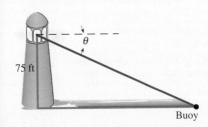

75 ft

Buoy

1. From the top of a 75-feet-high lighthouse, the angle of depression θ of a buoy that indicates the location of a sunken ship is measured to be $23°15'$, and the distance x from the lighthouse to the buoy is calculated using trigonometry (see the figure). Find the approximate error in x if the measurement of θ is in error by $5'$. (*Note:* Express θ and $d\theta$ in radians.)

2. A rocket fired at an angle θ from the horizontal with an initial velocity of v_0 meters per second will rise to a maximum height of

$$y = \frac{v_0^2 \sin \theta}{2g}$$

where $g = 9.80$ m/s^2 is the acceleration caused by gravity. Approximately how much higher will a rocket fired at an angle of $31°$ go than one that is fired at an angle of $30°$, if the initial velocity in each case is 80 m/s?

3. A new computer workstation costs $25,000. Its useful lifetime will be 5 years, at which time it will be worth an estimated $5,000. The company calculates its depreciation using the linear decline that is an option in the tax laws.
   (a) Find the linear equation that expresses the value V of the equipment as a function of time t, $0 \leq t \leq 5$.

   (b) How much will the equipment be worth after 2.5 years?

   (c) What is the average rate of change in the value of the equipment from 1 to 3 years?

   (d) What is the instantaneous rate of change in the value of the equipment at any time $t = a$?

4. According to Boyle's Law, if the temperature of a confined gas is held fixed, then the product of the pressure P (in lb./in.^{2}) and the volume V (in in.^{3}) is a constant. Suppose, for a certain gas, $PV = 8000$.
   (a) Determine the average rate of change of P as V increases from 200 in.^{3} to 250 in.^{3}

   (b) Express V as a function of P and show that the instantaneous rate of change of V with respect to P is proportional to the inverse square of P.

5. Consider the Earth as a sphere of radius $R = 6378$ km. Given the geometric formulas for the surface area and volume of a sphere and the formula for the circumference of a circle, use differentials to estimate:
   (a) The increase in the Earth's surface area if its radius were increased by 1 km.

   (b) The decrease in the Earth's volume if its radius were decreased by 1 km.

   (c) The difference in the circumference of the orbital path of the Hubble Space Telescope and that of the Earth, given that the height of the telescope's orbit is approximately 300 km above the Earth's surface. Express the answer both as a formula and in meters.

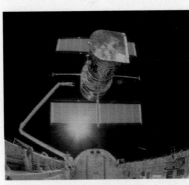

The Hubble Space Telescope being placed into an orbit 300 km above the Earth's surface by a space shuttle.

6. Let $P : (x_0, y_0)$ be a point on the ellipse $\frac{x^2}{a^2} + \frac{y^2}{b^2} = 1$.
   (a) Determine an equation for the tangent line to the ellipse at the point P.

   (b) Assume $a^2 > b^2$. Let $c = \sqrt{a^2 - b^2}$ and let F_1 and F_2 be the points $(c, 0)$ and $(-c, 0)$, respectively. Show that the angle between the line F_1P and the tangent line equals the angle between the line F_2P and the tangent line.
   Note: The points $(c, 0)$ and $(-c, 0)$ are the foci of the ellipse. The property suggested in (b) is the "reflection" property of ellipses—a line emanating from one

203

focus is "reflected" through the other focus. Elliptically shaped rooms are known as "whispering galleries."

7. The rate of change of electric charge with respect to time is called current. Thus, if $q = q(t)$ denotes the charge (in coulombs) at time t, then the current i is given by $i = dq/dt$.

(a) Suppose $q = 2t^2 - 6t + 3$ coulombs of charge flow through a conducting wire in t seconds.

   (i) Determine the current i (measured in amperes = coulombs per sec) after t seconds. What is the current after 4 seconds?

   (ii) When does the current reverse itself, i.e., when does i change sign?

   (iii) Suppose there is a 15-amp fuse in the line. How long will it last?

(b) Suppose $q = (1/3)t^3 - t^2 + t + 4$.

   (i) Determine the current i after t seconds.

   (ii) Does the current ever reverse itself?

   (iii) If there is a 25 amp fuse in the line, how long will it last?

8. Telescopes and solar collectors are often made in paraboloidal shapes (see the photo of parabolic solar collectors at left), because then on-axis parallel rays are reflected to the same point, called the *focus*. Consider the parabola $y^2 = 4px$.

(a) Show that the slope of the tangent to the curve is given by $dy/dx = \sqrt{p/x}$.

(b) For a horizontal ray of light approaching the parabola at an angle β with the tangent line, as shown, and with the reflected ray making an angle α with the tangent and an angle θ with the positive x-axis, show that $\theta = \alpha + \beta$.

(c) Show that the angle β of incidence = the angle α of reflection, using the trigonometric formula

$$\tan(a - b) = \frac{\tan a - \tan b}{1 + \tan a \tan b}$$

(d) Assuming only that $\beta = \alpha$, by finding the equation of the reflected ray, show that all incoming rays parallel to the axis of the parabola are reflected to the focus F.

9. National health care spending, in billions of dollars, has taken the shape of a parabola over the last few decades, increasing at an alarming rate. (See the table.)

(a) Using the 1965, 1980, and 1990 data only, fit a parabola of the form $y = ax^2 + bx + c$.

(b) Calculate the values for the other years given, and assess the error as a percentage of the value.

(c) Differentiating, calculate the slope of the parabola, and substitute values to get the numerical value of the derivative at each of the years listed.

(d) What percentage increase does the parabolic increase predict for 1993?

(e) A newspaper article reports that employers have budgeted for increases of 15% to 18% for 1993, whereas preliminary indications are that the costs have increased by only 8% to 12%. Compare these values and the slopes they imply with the predictions of part (d).

Year	Dollars (in billions)
1965	30
1970	80
1975	120
1980	250
1985	400
1990	690
1992	820 (estimated)

10. It has recently been discovered that "when fielders run backwards or forwards to catch a ball they run at a speed which keeps $d^2(\tan\alpha)/dt^2$ zero, where α is the angle of elevation of gaze from catcher to ball. This ensures that they intercept the ball before it reaches the ground (provided they can run fast enough to keep $d^2(\tan\alpha)/dt^2$ zero) whatever the effect of aerodynamic drag on the ball's trajectory."[†] Find the derivative of $\tan\alpha$ with respect to t, and make use of the given information about the second derivative to solve for $d\alpha/dt$. (Your answer will involve a constant.)

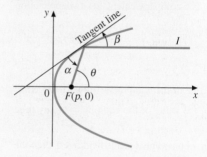

[†](Reference: Peter McLeod and Zoltan Dienes, *Nature*, vol. 362, p. 23, 1993.)

Mark J. T. Smith

What do you remember of your first calculus class?

I took precalculus in high school and calculus when I attended the Massachusetts Institute of Technology. I remember finding it more abstract than the mathematics courses I'd taken previously. There were aspects of the course that were interesting but overall I wouldn't call it exciting. I remember thinking that I would probably forget all the tricks for integration and was glad to learn that there were integral tables available. I had no real understanding of applications then. Calculus seemed to me to be about solving contrived problems, many of which were fun, but they still seemed contrived. I basically accepted the fact that calculus was an important tool and that I would use it later on.

Did you already intend to be an engineer at that point in your life?

I decided in junior high that I wanted to be either an engineer or a physicist. For a short period of time as a freshman, I thought I wanted to be a biomedical engineer, but by the end of my freshman year, I had chosen electrical engineering. I was drawn to it because I had a fascination with electronics and building electrical things. I started in electrical engineering because I had this naive idea that it was all about electronics. As I learned more I became more interested in aspects of electrical engineering that have nothing to do with electronics.

Are there particular mathematics courses that a freshman should pay special attention to if they intend to major in engineering?

That depends dramatically on what type of engineering you are going to be doing. The mathematics courses that I remember taking for which I needed calculus at the undergraduate level were differential equations and probability theory and at the graduate level, random processes and real analysis. I also took a course in special functions that used calculus a lot. Many electrical engineering students have trouble in electromagnetics. At M.I.T. it had a reputation of being a killer course, but I found all this to be exaggerated. It was actually quite an enjoyable course. Electromagnetics makes extensive use of differentiation and integration. It wasn't until I took this course that I felt I was really proficient in calculus.

Having said that, the use of calculus in electrical engineering is not evenly distributed throughout the discipline. Electromagnetics is very calculus intensive, and it's a required course, but digital logic design, for instance, requires other types of mathematics. My own area is digital signal processing. I rely on discrete mathematics, time series, convergence theorems, and numerical methods. The concept of approximating a continuous time function is one that seems essential in virtually any of the problems I do. In this context differentiation and integration are used heavily.

So you still use calculus in your current research?

My research area is digital signal processing, image processing, data compression and speech processing. We are always using mathematics to optimize our systems. This is typically done by computing or estimating derivatives.

Could you give us an example?

Engineers often first model systems with mathematical equations that they intend to build later on. Consider an automatic volume control device for a car radio as a hypothetical example. The purpose of the device is to keep the volume of the radio at a pleasing level, regardless of the background noise. It should gently raise the radio volume when you drive in noisy traffic and should lower the volume when you come to a stop light in a quiet hospital zone. This system has as its input a sensor input $x(t)$ that measures all the noise in the car for each instant in time t'. The output of the system is the radio signal you hear. The control parameter c is the volume control. Assume that the radio signal output $y(t)$ is represented by an equation where one of the variables is the volume control c. So $y(t)$ is a function in the variables c and t. The idea is to have the system automatically determine c for each instant in time t so that radio level is pleasing.

This optimal control value c can be found by minimizing an error function E, which measures performance. In particular, the error function value is large if the radio volume is too loud or too soft and is zero (or relatively small) if the volume is just right.

We can find the value of c by taking the derivative of the error function with respect to c and setting the result equal to zero. Solving this equation for c gives us the optimal value of c resulting in the perfect volume for the radio. In practice, the system would update the value of c frequently (say every second) thereby keeping the radio volume at a pleasing level for the duration of the car trip. Often, we have to introduce constraints in our problems. Let's assume now that our suggested system has problems. The value computed for c is sometimes negative. Clearly having a negative value for the volume does not make sense. This problem can be solved by introducing a constraint into the minimization equation. We can modify the formulation so that we minimize the error subject to the constraint that c is always positive. This is called *constrained optimization* and is used extensively in engineering.

To what extent is it important to understand the rigorous foundations of calculus if you intend to be an engineer?

Well, it's not as important as it would be to someone in a pure mathematics concentration. I find that proofs, per se, in calculus are not as important as having a functional understanding of the mathematics and how to apply it. However, engineers rely on proving theorems and understanding their properties all the time. This is because engineers are very often seeking optimal solutions to problems. We would like to first determine the best situation that is possible in theory. Even though we may not be able to implement this ideal level of performance in practice, knowing the optimal situation helps us set realistic goals and provides a measure we can use for comparison. Such optimal solutions are often derived using calculus.

To illustrate the importance of determining optimal upper bounds on performance, consider the design of a solar energy system for the home. We can model the atmosphere, earth position and rotation, and properties of the materials, and determine reasonable upper bounds on the amount of energy that can be extracted from such a solar collection system. This information can be extremely useful. For example, it may tell you immediately that it is impossible for this technology to provide enough supplementary energy to justify the cost of the system or that the design of the system cannot meet some target energy requirements. This kind of analysis is critical in effective engineering design.

You mentioned differentiation and integration, but what about sequences and series? How significant are they in understanding engineering principles?

The study of sequences and series is the foundation of digital signal processing and discrete systems. There is a rapidly growing trend toward digital implementations. Devices from music systems to hand-held personal communications systems are all starting to employ digital representations of signals. Sequences and series are indispensable in carrying out numerical computations because they can be used to approximate numbers like π and e or evalute functions like $\ln x$ or $\cos x$ to any desired degree of accuracy. That is, you can use an arbitrary number of terms to get as close as you like to the number or function's true value. Approximations to these constants and functions are repeatedly used in electrical engineering models. In other words, signals are digitally represented as sequences of numbers. These numbers are processed by computers to, perhaps, enhance the signal quality by filtering out background noise, or compress the data for efficient transmission. Digital signal processing is one of the most important areas of technology for the twenty-first century.

Some relatively new mathematical techniques such as wavelets are being applied to communications and coding problems. Is calculus necessary in understanding these new mathematical techniques?

In the areas of systems, telecommunications, controls, and signal processing, we often rely on transformations or transforms as they are called. These transforms convert a signal from its original form to another form that can be interpreted more easily. For example, the Fourier transform converts a signal into a frequency domain representation. In the Fourier domain we represent a signal in terms of its frequency components. Fourier analysis is a fundamental part of engineering. The Fourier transform as well as many other transforms are based on taking integrals. [See Section 10.10.]

Engineers always try to make the best use of new advances in mathematics, science, and technology. For example, the introduction of fractals and wavelets have created a tremendous amount of interest. Wavelets, for instance, are extremely important in data compression so that data can be transmitted far more efficiently than before. There are many engineers and scientists who believe that these techniques have the potential to vastly improve performance over conventional approaches.

Do you think it's a good idea for students to learn to use mathematical software such as computer algebra systems if they intend to go on in engineering?

I strongly recommend that students start to learn to use these tools when they are undergraduates. They will certainly use them later on and if they can help learn calculus, so much the better. Our engineering undergraduates and graduate students at Georgia Tech use Maple, Mathematica, and Matlab very frequently. In the future, I see mathematical software as being as important to calculus and engineering as the calculator is to arithmetic. It is a tool that makes our work easier, faster, and more reliable. It allows us to handle very large problems that could not be done easily by hand. These tools are by no means a substitute for learning the basics. Just because you know how to use a calculator does not mean that you should not know how to add or multiply. This is even more critical with calculus. Understanding the fundamentals of calculus provides insight into formulating the problems and their solutions. This is something that a computer program can't do for you.

I recall learning how to do matrix inversion when I was an undergraduate. It's a tedious process, particularly for matrix sizes 3×3 and larger. Computer tools now do matrix inversion for you instantaneously. By working with software tools like Matlab, Maple, and Mathematica, many of the more mundane aspects of the computation are removed. It's more than just a convenience. It gives you more time to concentrate on what the results of the computation mean and how they are applicable to your original problem.

Mark J. T. Smith is an Associate Professor in the School of Electrical Engineering at the Georgia Institute of Technology and a member of the faculty at Georgia Institute of Technology in Lorraine, France. He has received several awards for teaching and research and is the co-author of two introductory books in digital signal processing. Professor Smith has authored more than 80 papers in this area and is presently the Chairman of the IEEE Digital Signal Processing Technical Committee. Professor Smith's research interests are in the areas of speech, image, and video computer processing. He received the S.B. degree from MIT in 1978, and the M.S. and Ph.D. degrees from the Georgia Institute of Technology in 1979 and 1984.

4

APPLICATIONS OF THE DERIVATIVE

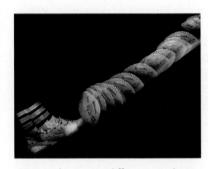

Among the many different applications of derivatives are projectile problems. For instance, the derivative can determine the optimal angle at which a football should be kicked to give maximum distance.

We have already seen two applications of the derivative. First, we found that it gave the slope of the tangent line to a curve. Second, we found that it gave the velocity of a body moving along a line. More generally, we have seen that the derivative of a function gives its instantaneous rate of change with respect to its independent variable, whatever the independent variable and the function may represent. In this chapter we explore further this interpretation of the derivative as a "rate of change."

We will show that if we know only where $f'(x)$ and $f''(x)$ are positive and where they are negative, we will then know where the function f is increasing and where it is decreasing, where its graph is concave up and concave down, and where f attains maximum and minimum values. We will make use of this knowledge of the behavior of functions to solve a wide variety of practical problems. We will emphasize those that involve maximizing or minimizing a function. For example, if you owned a pizza parlor, you would want to know what price per pizza to charge to maximize your profit.

Rates of change with respect to time (of which velocity is but one example) are especially important. We will show how knowing such a rate of change for one or more variables can often enable us to find the rate of change of another. As a further application, we will show how the derivative of a function can be used to approximate the zeros of the function. This process provides a way of approximating solutions to equations for which it is difficult or impossible to find exact solutions.

As another application of the derivative, we show a procedure, known as L'Hôpital's Rule, for finding limits (if they exist) of quotients in which the numerator and denominator both approach 0 or both become infinite. Limits of these types are examples of *indeterminate forms*. We will encounter other types of indeterminate forms in Chapter 7 and will see how L'Hôpital's Rule can again be used.

We conclude the chapter by showing how we can work backward from the derivative of a function to the function itself, provided we know the value of the function at one point. This process leads naturally to the subject of differential equations, which we introduce here and study in more detail later.

At the outset we proceed intuitively, relying mainly on geometrical reasoning. Toward the end of the chapter, however, we provide the theoretical basis for our conclusions in the form of a theorem with far-reaching consequences, called the Mean-Value Theorem. We will encounter the need for this theorem in other chapters as well, especially in Chapter 5 on the integral.

4.1 EXTREME VALUES

One of the most important applications of the derivative is in finding the largest or smallest value of a function. For example, we might wish to know the dimensions of a container that can hold a given volume yet requires the least amount of material. Or we might wish to know what price a theater owner should charge to obtain maximum revenue. As another example, the owner of an orchard may be interested in knowing how many apple trees to plant per acre to maximize the yield of apples. These problems and many others, when properly formulated, can be solved by making use of differentiation. Because we can find out a lot about a function by looking at its graph, we begin by considering how the derivative of a function can be used to determine certain information about its graph.

Rising curves; slopes are positive

Falling curves; slopes are negative

FIGURE 4.1

We know already that if $y = f(x)$, where f is a differentiable function, then $f'(x)$ gives the slope of the tangent line to the graph of the function at the point $(x, f(x))$. When this slope is positive at each point of some interval, we would expect the curve to be rising on that interval as we move from left to right. When the slope is negative, we expect the curve to be falling. We illustrate this relationship between the sign of the slope and direction of the curve in Figure 4.1. We have previously used the terms *increasing* and *decreasing* to describe a function whose graph is rising or falling, respectively. We now give a precise definition of these terms.

Definition 4.1
Increasing and Decreasing Functions; Monotone Functions

Let a function f be defined on an interval I. Then f is said to be **increasing** on I if $f(x_1) < f(x_2)$ for any two numbers x_1 and x_2 in I with $x_1 < x_2$. Similarly, f is said to be **decreasing** on I if $f(x_1) > f(x_2)$ for any two numbers x_1 and x_2 in I with $x_1 < x_2$. A function that is either increasing on I or decreasing on I is said to be **monotone** on I.

For example, if $f(x) = x^3$, then f is an increasing function on all of $\mathbb{R}$, since if $x_1 < x_2$, then $x_1^3 < x_2^3$. The function g for which $g(x) = x^2$ is decreasing on $(-\infty, 0)$ and increasing on $(0, \infty)$. The graphs of these two functions are shown in Figure 4.2.

We formalize our observation about the relationship between derivatives and monotone functions in the following theorem.

THEOREM 4.1

If $f'(x) > 0$ for all x in an open interval I, then f is increasing on I, and if $f'(x) < 0$ for all x in I, then f is decreasing on I.

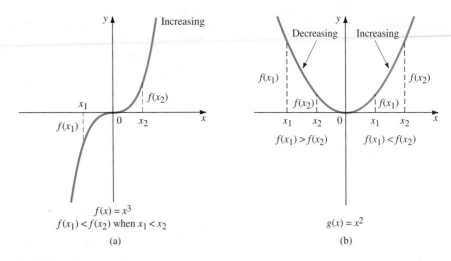

FIGURE 4.2

As we have seen, Theorem 4.1 is a plausible result, since $f'(x) > 0$ in I means that the tangent line is always sloping upward from left to right. This upward slope implies that the function is increasing. Similar remarks apply when $f'(x) < 0$. We will prove these statements in Section 4.7 after we have developed more background. For now, we will proceed intuitively. Let us emphasize that knowing where a function is increasing and where it is decreasing is not sufficient to describe the shape of the curve. For example, although the function $f(x) = x^3$ in Figure 4.2(a) is always increasing, its shape for $x < 0$ is quite different from its shape for $x > 0$. We will discuss the additional information we need in Section 4.2 when we take up the notion of concavity.

EXAMPLE 4.1   Determine the intervals on which the function f is increasing and the intervals on which it is decreasing:

$$f(x) = 2x^3 - 3x^2 - 12x + 4$$

Solution   First calculate $f'(x)$:

$$f'(x) = 6x^2 - 6x - 12 = 6(x + 1)(x - 2)$$

From the sign graph for f' (Figure 4.3), we see that $f'(x) > 0$ on $(-\infty, -1) \cup (2, \infty)$ and $f'(x) < 0$ on $(-1, 2)$. Thus,

$$f \text{ is increasing on } (-\infty, -1) \text{ and on } (2, \infty)$$

$$f \text{ is decreasing on } (-1, 2)$$

Figure 4.4 illustrates these results.   ■

High points and low points on the graph of a continuous function occur when the function changes from increasing to decreasing, or vice versa. They might also occur at endpoints of intervals. Figure 4.5 illustrates some possibilities. The points P_2, P_4, and P_7 are all high points in the sense that they are higher than other points in their immediate vicinity. Similarly, P_1, P_3, P_5, and P_8 are low points. If we ask for the *highest* point, it is P_7, and the *lowest* point is P_8. Note that except for P_1 and P_8, which are at the endpoints of the interval, the tangent lines at these high and low points appear to be either horizontal or

FIGURE 4.3
Sign graph for $f'(x) = 6(x + 1)(x - 2)$

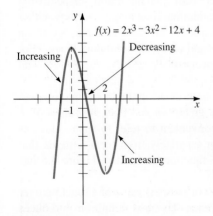

FIGURE 4.4

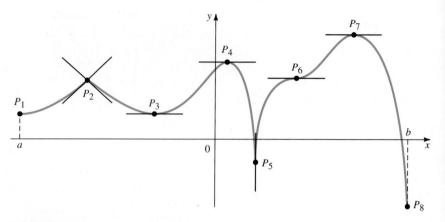

FIGURE 4.5

vertical, or else there is no tangent line. In particular, at P_2 there is no tangent line, since from the left the slopes approach one value and from the right they approach another. At P_3, P_4, and P_7 the tangent line is horizontal, and at P_5 it is vertical. Note also, however, that at P_6 there is a horizontal tangent line, even though P_6 is neither a high point nor a low point. The next definition makes these ideas more precise.

Definition 4.2
Local Maximum and Local Minimum

A function f is said to have a **local maximum** at the point $x = a$ if there is an open interval I containing a such that $f(a) \geq f(x)$ for all x in I. Similarly, f has a **local minimum** at $x = a$ if there is an open interval I containing a such that $f(a) \leq f(x)$ for all x in I.

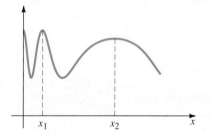

FIGURE 4.6
f has local maximum points when $x = x_1$ and when $x = x_2$. For x_1, the interval throughout which $f(x_1) \geq f(x)$ is small, whereas for x_2, the interval throughout which $f(x_2) \geq f(x)$ is large.

Note: There is no requirement on the size of the interval I in this definition. In some cases it can be very large, and in others it may be very small. (See Figure 4.6.)

If f has a local maximum at a, then we say the point $(a, f(a))$ is a **local maximum point** on the graph of f, and $f(a)$ is called a **local maximum value** of f. Similar terminology is used for the local minimum. The terms **maxima** and **minima** are the plural forms of maximum and minimum, respectively. When we wish to speak of maxima and minima collectively, we say either **extreme values** or **extrema**. Thus, if we ask for all local extrema, we mean the collection of all local maximum values and all local minimum values. (For example, the local extrema in Figure 4.5 occur at P_1, P_2, P_3, P_4, P_5, P_7, and P_8, but not at P_6.)

When the domain of f includes one or both of the endpoints of an interval, then the value of f at such a point may be greater or less than the values of f at all nearby points in the domain. It is convenient to refer to such values as **endpoint extrema** and to investigate them separately from local extrema that occur at interior points. For example, the function pictured in Figure 4.5 has endpoint minima at both $x = a$ and $x = b$.

Frequently we are interested in the largest or smallest value of a function over its entire domain, as opposed to local extrema. The next definition introduces terminology for such global extrema.

Definition 4.3
Absolute Maximum and
Absolute Minimum

A function f is said to have an **absolute maximum** (or **global maximum**) at the point a of its domain if $f(a) \geq f(x)$ for all x in its domain, and it is said to have an **absolute minimum** (or **global minimum**) at a if $f(a) \leq f(x)$ for all x in its domain.

If f has an absolute maximum at a, then $f(a)$ is called the **absolute maximum value of f** (or simply the maximum value of f), and the point $(a, f(a))$ is called the absolute maximum point on the graph of f. We use analogous terminology for the absolute minimum.

If a function has an absolute maximum value, this value is the largest among all local maximum values and endpoint values. But as the graphs in Figures 4.7 through 4.9 show, not all functions have absolute maximum and minimum values. There may not even be local maxima or minima.

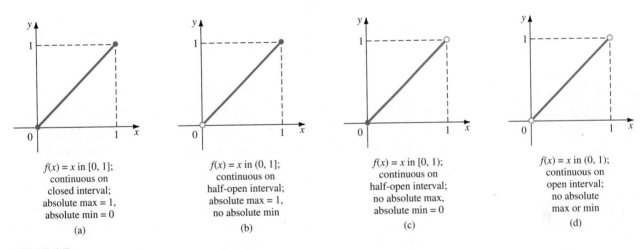

$f(x) = x$ in $[0, 1]$;
continuous on
closed interval;
absolute max = 1,
absolute min = 0

(a)

$f(x) = x$ in $(0, 1]$;
continuous on
half-open interval;
absolute max = 1,
no absolute min

(b)

$f(x) = x$ in $[0, 1)$;
continuous on
half-open interval;
no absolute max,
absolute min = 0

(c)

$f(x) = x$ in $(0, 1)$;
continuous on
open interval;
no absolute
max or min

(d)

FIGURE 4.7

In Figure 4.7 we show the function $f(x) = x$ on intervals from $x = 0$ to $x = 1$, including both endpoints in (a), one endpoint in (b) and (c), and neither endpoint in (d). Notice that only in (a) where f is defined on the closed interval $[0, 1]$ does f attain both an absolute maximum and an absolute minimum value. In (b) and (d) the function comes arbitrarily close to 0 but does not get there, so there is no local or absolute minimum. Similarly, in (c) and (d) there is no local or absolute maximum. In Figure 4.8 both f and g are defined on the closed interval $[0, 2]$ but f fails to have an absolute maximum and g fails to have an absolute minimum. Each of the functions in Figures 4.7 and 4.8 is bounded, that is, their graphs are contained between two horizontal lines. In Figure 4.9 the function h is unbounded both above and below and hence has no absolute maximum or minimum, although it is continuous on the domain $(-3, 3)$. There is a local minimum at P_1 and a local maximum at P_2.

The following theorem gives conditions that guarantee that a function will have an absolute maximum and minimum. The theorem is thus of paramount importance in the study of extreme values. Its proof requires concepts we have not studied, but it can be found in texts on advanced calculus.

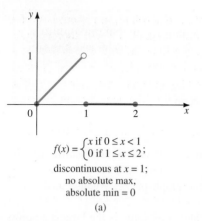

$$f(x) = \begin{cases} x \text{ if } 0 \le x < 1 \\ 0 \text{ if } 1 \le x \le 2 \end{cases};$$
discontinuous at $x = 1$;
no absolute max,
absolute min $= 0$
(a)

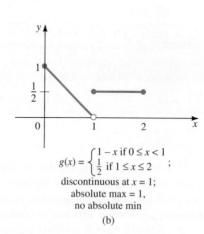

$$g(x) = \begin{cases} 1 - x \text{ if } 0 \le x < 1 \\ \frac{1}{2} \text{ if } 1 \le x \le 2 \end{cases};$$
discontinuous at $x = 1$;
absolute max $= 1$,
no absolute min
(b)

FIGURE 4.8

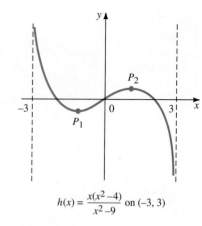

$$h(x) = \frac{x(x^2 - 4)}{x^2 - 9} \text{ on } (-3, 3)$$

FIGURE 4.9

THEOREM 4.2

Extreme-Value Theorem

If f is continuous on the closed interval I, then f attains both an absolute maximum value and an absolute minimum value on I.

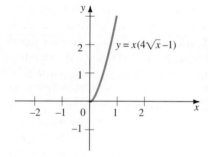

FIGURE 4.10

REMARK

■ Theorem 4.2 is an example of an *existence theorem*. When the conditions of the theorem are satisfied, we can conclude that the function does have an absolute maximum and an absolute minimum, but the theorem does not tell us what the extreme values are nor where they occur. The next theorem will provide a way for finding these values.

Re-examining the functions shown in Figures 4.7 through 4.9 reveals in what ways one of the conditions of this theorem fails to hold true in each case. The function f in Figure 4.7 is continuous on an interval, but the interval is not closed in parts (b), (c), and (d). The functions f and g in Figure 4.8 are defined on a closed interval but are not continuous. The function h in Figure 4.9 is continuous, but the interval is open.

Suppose now that we are given a function and we want to find its local maximum and minimum points, if any exist. One way of finding them is to obtain an accurate graph by using a graphing calculator or a CAS, then zooming in on the points that appear to be local maximum or minimum points and finding their coordinates to any desired degree of accuracy. For many functions this graphical approach is quite satisfactory. There are some possible pitfalls, however. Depending on the scale you use (determining the x range and y range), you may or may not correctly identify all the local extrema. Consider, for example, the graph of $f(x) = x(4\sqrt{x} - 1)$ shown in Figure 4.10. This graph was generated with a CAS using a normal scale in which each tick mark is 1 unit on each axis. Unless you look very carefully, you would probably conclude that the function reaches its lowest point at the origin. This conclusion is not correct, however, as the graph in Figure 4.11 shows. Here we used a scale of 0.1

FIGURE 4.11

for x and 0.1 for y. Clearly, there is a local (also absolute) minimum point for some value of x between 0 and 0.1. In fact, by methods we will soon discuss, we can find the exact minimum point. Its coordinates are $(1/36, -1/108)$.

The point of the preceding example is that care must be taken when drawing conclusions based solely on graphs generated by a computer or calculator. The analysis we provide here is one way to be certain of the nature of the function under consideration.

An important key to locating local extrema is given in the next theorem.

THEOREM 4.3

Let f be defined on an interval I. If f has a local maximum or minimum at an interior point a of I, and if $f'(a)$ exists, then $f'(a) = 0$.

Note carefully that the theorem does *not* say that if $f'(a) = 0$, then f has a local maximum or minimum at a. Figure 4.12 illustrates the various possibilities that can occur when $f'(a) = 0$. In each case $f'(0) = 0$, so the tangent line at $(0, 0)$ is horizontal. In parts (c) and (d), however, the point $(0, 0)$ is neither a local maximum nor a local minimum (it is called a *point of inflection*, as we will learn in the next section). The theorem also does not say anything about the nature of points where the derivative fails to exist. In Figure 4.13(a) (on the following page) we show the graph of $f(x) = |x|$, which we have previously shown to be nondifferentiable at $x = 0$. Still, the function does have a minimum value there. On the other hand, in Figure 4.13(b) we show the graph of $g(x) = x^{1/3}$. It also fails to have a derivative at $x = 0$ (the slope is "infinite") and there is no maximum or minimum at $x = 0$. Thus, when $f'(a)$ fails to exist, there may or may not be a local extreme point when $x = a$. Finally, if we omit the hypothesis that $x = a$ be an interior point of an interval, the function can attain a maximum or minimum value even though the derivative is not zero. The function $f(x) = x$ on the closed interval $[0, 1]$ in Figure 4.14 clearly has an endpoint (and absolute) maximum at $x = 1$ and an endpoint (and absolute) minimum at $x = 0$, yet at both points the (one-sided) derivative equals 1. So the conclusion of the theorem may not be true if $x = a$ is an endpoint of the interval.

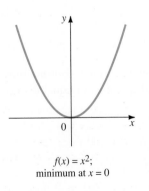

$f(x) = x^2$;
minimum at $x = 0$

(a)

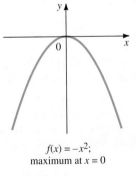

$f(x) = -x^2$;
maximum at $x = 0$

(b)

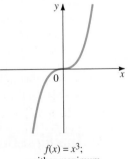

$f(x) = x^3$;
neither maximum
nor minimum at $x = 0$

(c)

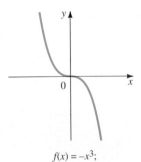

$f(x) = -x^3$;
neither maximum
nor minimum at $x = 0$

(d)

FIGURE 4.12

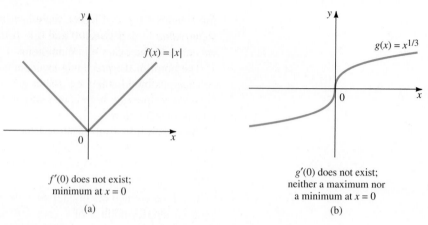

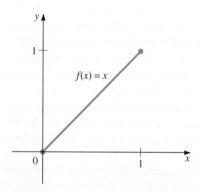

f'(0) does not exist;
minimum at $x = 0$

(a)

g'(0) does not exist;
neither a maximum nor
a minimum at $x = 0$

(b)

FIGURE 4.13

FIGURE 4.14
f has a maximum at $x = 1$ and a
minimum at $x = 0$, but $f'(0) = 1$ and
$f'(1) = 1$.

Proof of Theorem 4.3   We prove the theorem for a local maximum only.
The proof for a local minimum is similar. We suppose that the derivative at a
exists. Using the form for $f'(a)$ from Equation 3.4, we have

$$f'(a) = \lim_{x \to a} \frac{f(x) - f(a)}{x - a}$$

So the limit from the right and the limit from the left both exist and are the
same. We consider these one-sided limits separately. First, consider $x > a$.
By hypothesis, the point $(a, f(a))$ is a local maximum point, so if x is near a,
$f(a) \geq f(x)$. Thus, $f(x) - f(a) \leq 0$. Since we are taking $x > a$, it follows
that $x - a > 0$, and

$$\frac{f(x) - f(a)}{x - a} \leq 0 \qquad \text{Numerator negative, denominator positive}$$

So

$$\lim_{x \to a^+} \frac{f(x) - f(a)}{x - a} \leq 0 \qquad \text{See Exercise 69 in Exercise Set 4.1.}$$

Now consider $x < a$. Then $x - a < 0$, and

$$\frac{f(x) - f(a)}{x - a} \geq 0 \qquad \begin{array}{l}\text{Numerator negative,}\\\text{denominator negative}\end{array}$$

So

$$\lim_{x \to a^-} \frac{f(x) - f(a)}{x - a} \geq 0$$

Since $f'(a)$ exists, the right-hand and left-hand limits are equal. But the only
number that is both greater than or equal to zero and less than or equal to zero
is zero itself. Thus, $f'(a) = 0$. ∎

The First-Derivative Test

As we saw in Figures 4.12(c) and 4.12(d), the fact that $f'(a) = 0$ does not
necessarily imply that f has a maximum or minimum value at $x = a$. We
can, however, treat such a point as a *candidate* for an x-coordinate at which f
has an extreme value. The theorem does tell us that if a is an interior point of

the domain of f and $f'(a) \neq 0$, then $f(a)$ is definitely not a local maximum or minimum value. So as a first step in locating points at which f has local extrema, we set $f'(x) = 0$ and solve for x. Then we test each such value to determine the nature of f at that point. Another possibility, as we saw in Figure 4.13(a), is that f has a local extreme value at a point where $f'(x)$ fails to exist. So a second candidate to test is any point at which $f'(x)$ fails to exist. Finally, as we saw in Figure 4.14, when the domain of f includes one or both endpoints of an interval, we must also test these points. In summary, we conclude that maximum and minimum values for a function defined on an interval can occur only at the following types of points in the interval:

1. An interior point where $f'(x) = 0$
2. An interior point where $f'(x)$ does not exist
3. An endpoint of the interval

Types 1 and 2 are called **critical points** of f. Critical points are not themselves maximum or minimum points but are x-coordinates of points on the graph that *may* be maximum or minimum points. So if a is a critical point, we substitute into the given function, f, to get $f(a)$, which is then a candidate for a maximum or minimum value. The endpoint values, if they are in the domain, must also be checked.

We saw in Figure 4.12 the four possibilities that can exist at critical points of type 1. A critical point of type 2, for which f' does not exist, also may or may not be the x-coordinate of a maximum or minimum point. We saw such a situation in Figure 4.13. Some of the many other possibilities when f' does not exist are illustrated in Figure 4.15. In parts (a) and (c) the tangent line is vertical and so the slope is undefined. In (b) the slopes from the left and from the right are unequal. In (d), (e), (f), and (g) the functions are discontinuous and so the derivatives cannot exist. The function in part (h) is given by $f(x) = x \sin \frac{1}{x}$ for $x \neq 0$, with $f(0)$ defined as 0. You will be asked in the exercises to show that $f'(0)$ does not exist in this latter case.

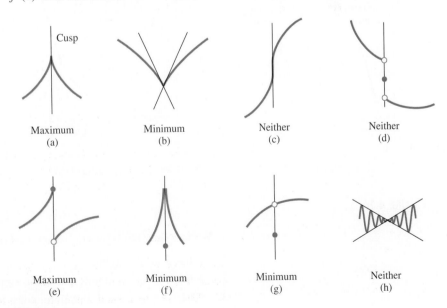

FIGURE 4.15

After finding the critical points, we must test in some way to determine the nature of the corresponding points on the graph. For a continuous function f there are two ways of doing this, one using the first derivative and one using the second derivative. We will discuss the latter in the next section. When f is discontinuous at a critical value, the possibilities are numerous, as Figures 4.15(d)–(g) show. The only feasible test is to apply Definition 4.2 directly, comparing $f(a)$ with $f(x)$ for x near a.

The basis for the First-Derivative Test is that for a continuous function, a maximum occurs when the function is increasing to the left of the critical value and decreasing to the right. The reverse is true at a minimum. Thus, by Theorem 4.1 we have the following test.

THEOREM 4.4

First-Derivative Test for Local Maxima and Minima

Let $x = a$ be a critical point of the function f, and suppose that f is continuous at $x = a$. Suppose further that $f'(x)$ exists for all x in some open interval I containing a, except possibly at $x = a$.

1. If for x in I, $f'(x) > 0$ when $x < a$ and $f'(x) < 0$ when $x > a$, then f has a local maximum at a.
2. If for x in I, $f'(x) < 0$ when $x < a$ and $f'(x) > 0$ when $x > a$, then f has a local minimum at a.
3. If for x in I, $f'(x)$ is of the same sign for $x < a$ as for $x > a$, then f has neither a local maximum nor a local minimum at a.

Figure 4.16 illustrates the situation when $x = a$ is a critical point for which $f'(a) = 0$ (type 1).

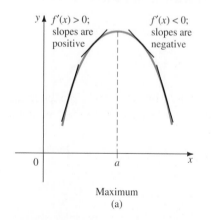

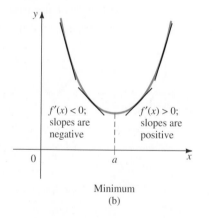

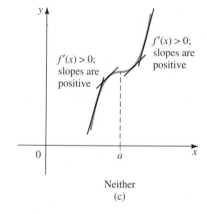

FIGURE 4.16

We can state the First-Derivative Test more briefly as follows:

1. f has a local maximum at a if the sign of f' goes from $+$ to $-$.
2. f has a local minimum at a if the sign of f' goes from $-$ to $+$.
3. f has neither a local maximum nor a local minimum if the sign of f' goes from $-$ to $-$ or from $+$ to $+$.

In applying this briefer version, we must understand that f is continuous at a and the signs indicated occur as we move from left to right through a.

We illustrate this test in the next four examples.

EXAMPLE 4.2   Find all local maxima and minima for the function f for which

$$f(x) = 2x^3 - 3x^2 - 12x + 4 \qquad (x \in \mathbb{R})$$

Solution   To find the critical values we first calculate $f'(x)$:

$$f'(x) = 6x^2 - 6x - 12 = 6(x+1)(x-2)$$

FIGURE 4.17
Sign graph for $f'(x) = 6(x+1)(x-2)$

Since $f'(x) = 0$ when $x = -1$ or $x = 2$, and $f'(x)$ exists everywhere, it follows that -1 and 2 are the only critical points. To test whether these are extrema, we use the First-Derivative Test. A useful device for doing this test is the sign graph for f' in Figure 4.17. Since f' changes from $+$ to $-$ as we pass through -1 and from $-$ to $+$ as we pass through 2, we conclude that f has a local maximum at -1 and a local minimum at 2. To determine these maximum and minimum values, we substitute into the original function:

$$f(-1) = 11 \text{ is a local maximum value.}$$

$$f(2) = -16 \text{ is a local minimum value.}$$

Alternatively, we can say $(-1, 11)$ is a local maximum point on the graph of f and $(2, -16)$ is a local minimum point on the graph.

If we take $|x|$ large enough, $f(x) \approx 2x^3$, so $f(x)$ can be made arbitrarily large for large positive values of x. By taking negative x values large in absolute value, $f(x)$ can be made arbitrarily large in absolute value and negative. So since the domain of f is unlimited, it follows that there is no absolute maximum or minimum value of f. We showed the graph of f in Example 4.1 (Figure 4.4). ∎

EXAMPLE 4.3   Find all local maxima and minima for the function

$$f(x) = \frac{x+1}{x^2 - 3x} \qquad (x \neq 0, 3)$$

Solution   The domain of f consists of the union of the three intervals $(-\infty, 0)$, $(0, 3)$, and $(3, \infty)$. So we can apply our theory to each of these intervals. First we find the critical points:

$$f'(x) = \frac{1(x^2 - 3x) - (x+1)(2x - 3)}{(x^2 - 3x)^2} = -\frac{x^2 + 2x - 3}{(x^2 - 3x)^2}$$

$$= -\frac{(x+3)(x-1)}{(x^2 - 3x)^2} = -\frac{(x+3)(x-1)}{[x(x-3)]^2}$$

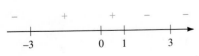

FIGURE 4.18
Sign graph for $f'(x) = -\dfrac{(x+3)(x-1)}{[x(x-3)]^2}$

Since $f'(x) = 0$ when $x = -3$ or when $x = 1$, these are critical points. There are no values of x in the domain of f for which $f'(x)$ is undefined, so there are no other critical points. (We have already omitted from the domain of f the points $x = 0$ and $x = 3$, where the denominator is zero.)

To test $x = -3$ and $x = 1$, we look at a sign graph for f' in Figure 4.18. Notice that we show the points 0 and 3 on the line, since they are endpoints of intervals on which f is defined.

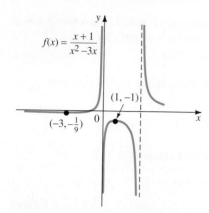

$f(x) = \dfrac{x+1}{x^2 - 3x}$

$(1, -1)$

$(-3, -\frac{1}{9})$

FIGURE 4.19

Since f' changes from $-$ to $+$ as we pass through -3 and from $+$ to $-$ as we pass through 1, we conclude that f has a local minimum at -3 and a local maximum at 1. The corresponding points on the graph are

$$\left(-3, -\frac{1}{9}\right), \text{ local minimum point}$$

$$(1, -1), \text{ local maximum point}$$

The graph of f is shown in Figure 4.19.   ■

EXAMPLE 4.4   Find all local and absolute extrema of the function f defined on the interval $[0, 2\pi]$ by $f(x) = \sin 2x + 2 \sin x$.

Solution $f'(x) = 2\cos 2x + 2\cos x$

To write $f'(x)$ in an equivalent form in which each trigonometric function is evaluated at x only, rather than at both $2x$ and x, we can use the identity $\cos 2x = 2\cos^2 x - 1$ (see Supplementary Appendix 2) to get

$$f'(x) = 2(2\cos^2 x - 1) + 2\cos x$$

$$= 2(2\cos^2 x + \cos x - 1)$$

$$= 2(\cos x + 1)(2\cos x - 1)$$

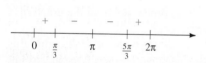

FIGURE 4.20
Sign graph for
$f'(x) = 2(\cos x + 1)(2\cos x - 1)$

The critical points are those in the closed interval $[0, 2\pi]$ for which $\cos x = -1$ or $\cos x = \frac{1}{2}$, namely, $x = \pi$, $\pi/3$, and $5\pi/3$. The sign graph for f' (Figure 4.20) shows that f has a local maximum at $\pi/3$ and a local minimum at $5\pi/3$. At $x = \pi$ there is neither a maximum nor a minimum. Substituting into the formula for $f(x)$, we find

$$f\left(\frac{\pi}{3}\right) = \frac{3\sqrt{3}}{2} \text{ is a local maximum value.}$$

$$f\left(\frac{5\pi}{3}\right) = -\frac{3\sqrt{3}}{2} \text{ is a local minimum value.}$$

The endpoint values are $f(0) = 0$ and $f(2\pi) = 0$. Thus, the absolute maximum value of f is $3\sqrt{3}/2$ and the absolute minimum value is $-3\sqrt{3}/2$. We show the graph of f in Figure 4.21.   ■

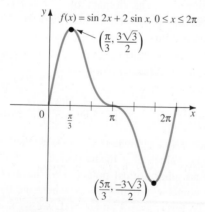

$f(x) = \sin 2x + 2\sin x, \ 0 \le x \le 2\pi$

$\left(\dfrac{\pi}{3}, \dfrac{3\sqrt{3}}{2}\right)$

$\left(\dfrac{5\pi}{3}, \dfrac{-3\sqrt{3}}{2}\right)$

FIGURE 4.21

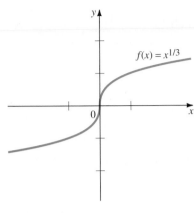

$f(x) = x^{1/3}$

FIGURE 4.22

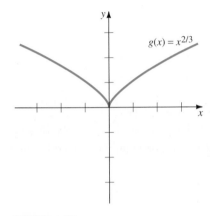

$g(x) = x^{2/3}$

FIGURE 4.23

EXAMPLE 4.5   **Critical points of type 2.** Find and test the critical points of the following functions:

(a) $f(x) = x^{1/3}$

(b) $g(x) = x^{2/3}$

Solution

(a) $f'(x) = \dfrac{1}{3}x^{-2/3} = \dfrac{1}{3x^{2/3}}$

Since $f'(0)$ does not exist, $x = 0$ is a critical point of type 2. There are no other critical points. Since $x^{2/3}$ is positive for all $x \neq 0$, $f'(x)$ does not change sign at 0. So f does not have a local maximum nor a local minimum there. The graph of f is shown in Figure 4.22.

(b) $g'(x) = \dfrac{2}{3}x^{-1/3} = \dfrac{2}{3x^{1/3}}$

Again, $x = 0$ is a critical point of type 2. If $x < 0$, $g'(x) < 0$ and if $x > 0$, $g'(x) > 0$. By the First-Derivative Test, we conclude that g has a local minimum at $x = 0$. Since $g(0) = 0$, and for $x \neq 0$, $g(x) > 0$, we see that $(0, 0)$ is also the absolute minimum point on the graph. The graph is shown in Figure 4.23. ■

If a function f is continuous on a closed interval $[a, b]$, we know by Theorem 4.2 that f has an absolute maximum and an absolute minimum on $[a, b]$. The absolute maximum is the largest of the local and endpoint maxima, and similarly for the absolute minimum. Thus, if the only objective is to find the absolute extrema, the following procedure is sufficient:

Procedure for Finding Absolute Extrema

If f is continuous on $[a, b]$:

1. Find all critical points.
2. Evaluate f at each critical point and at each endpoint.
3. Select the largest and smallest values in step 2.

Note that it is not necessary in this case to test the critical points to see if they give local maxima or minima.

We illustrate this procedure in the next example.

EXAMPLE 4.6   Find the absolute extrema of f, where

$$f(x) = (2x - 5)(x - 5)^{2/3} \quad \text{on } [3, 6]$$

Solution   By the Product Rule,

$$f'(x) = 2(x - 5)^{2/3} + (2x - 5) \cdot \frac{2}{3}(x - 5)^{-1/3}$$

$$= \frac{2}{3}(x - 5)^{-1/3}[(3(x - 5) + (2x - 5)]$$

$$= \frac{2}{3}(x-5)^{-1/3}(3x-15+2x-5)$$

$$= \frac{10(x-4)}{3(x-5)^{1/3}}$$

The critical points are $x = 4$ and $x = 5$.

$$f(4) = (3)(-1)^{2/3} = 3$$

$$f(5) = (5)(0) = 0$$

$$f(3) = (1)(-2)^{2/3} = \sqrt[3]{4} \quad \text{Left endpoint}$$

$$f(6) = (7)(1)^{2/3} = 7 \qquad \text{Right endpoint}$$

Conclusions:

Absolute maximum value $= 7$ when $x = 6$.

Absolute minimum value $= 0$ when $x = 5$.   ∎

EXAMPLE 4.7   Find the absolute extreme values of $f(x) = x^3 - 3x + 2$ on the interval $[0, 2]$.

Solution $f'(x) = 3x^2 - 3 = 3(x^2 - 1) = 3(x + 1)(x - 1)$
The critical points are $x = 1$ and $x = -1$, but $x = -1$ is not in the interval $[0, 2]$, so we ignore it. We calculate $f(x)$ at $x = 1$ and at the endpoints $x = 0$ and $x = 2$:

$$f(1) = 0$$

$$f(0) = 2$$

$$f(2) = 4$$

Thus, the absolute minimum value is 0, attained when $x = 1$. The absolute maximum value is 4, attained when $x = 2$. We show the graph of f in Figure 4.24.   ∎

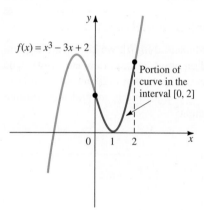

$f(x) = x^3 - 3x + 2$

Portion of curve in the interval $[0, 2]$

FIGURE 4.24

Exercise Set 4.1

In Exercises 1–12, find the intervals on which the function f is increasing and those on which it is decreasing.

1. $f(x) = 3 - 2x - x^2$ **2.** $f(x) = 2x^2 - x - 3$

3. $f(x) = x^3 - x^2 - 8x$

4. $f(x) = 2x^3 + 3x^2 - 36x + 5$

5. $f(x) = x^3 - 2x^2 - 4x + 7$

6. $f(x) = 12 + 40x + 7x^2 - 4x^3$

7. $f(x) = 3x^4 - 4x^3$ **8.** $f(x) = x^4 - 2x^3 - 5x^2$

9. $f(x) = \dfrac{1}{(x-2)^2}$ **10.** $f(x) = \dfrac{x}{x+3}$

11. $f(x) = \dfrac{1-x}{3+x^2}$ **12.** $f(x) = \dfrac{x+3}{(x-1)^2}$

In Exercises 13–24, find all local maximum and minimum points.

13. $f(x) = x^2 - 8x + 7$

14. $f(x) = 3 - x - 2x^2$

15. $f(x) = x^3 - 3x + 6$

16. $f(x) = 2x^3 + 3x^2 - 12x$

17. $f(x) = x^3 - 3x^2 - 24x + 4$

18. $f(x) = x^3 + 5x^2 - 8x + 3$

19. $f(x) = x^4 - 8x^2 + 12$

20. $f(x) = x^4 - 4x^3 - 8x^2 + 12$

21. $f(x) = x^4 - 4x^3 + 8$

22. $f(x) = 2x^5 + 5x^4 + 10$

23. $f(x) = 1 + 2(x - 3)^{2/3}$

24. $f(x) = (x^2 - 1)^{2/3}$

In Exercises 25–36, find local as well as absolute maximum and minimum values on the given interval.

25. $f(x) = x^2 - 4x + 3$ on $[0, 5]$

26. $f(x) = 5 - 3x - 2x^2$ on $[-6, 0]$

27. $f(x) = x^3 - 2x^2 - 4x + 8$ on $[-2, 3]$

28. $f(x) = 12x - x^3 - 9$ on $[-1, 3]$

29. $f(x) = x^4 - 2x^3 - 9x^2 + 27$ on $[-2, 4]$

30. $f(x) = 3x^4 - 16x^3 + 24x^2 - 21$ on $[-1, 3]$

31. $f(x) = \sin x + \cos x$ on $[0, \pi]$

32. $f(x) = \sin 2x - x$ on $[-\pi, \pi]$

33. $f(x) = \tan x - 2x + 1$ on $[-\frac{\pi}{3}, \frac{\pi}{3}]$

34. $f(x) = 2x + \sin 2x - 4\sin x$ on $[-\pi, \pi]$

35. $f(x) = \sin 2x + 4\cos x + x - 1$ on $[0, 2\pi]$

36. $f(x) = x\sin x + \cos x - \pi\sin x$ on $[0, 2\pi]$

Exercises 37–46 refer to earlier problems in this exercise set. In each case find the absolute maximum and absolute minimum values of f on the indicated interval. Make use of previously obtained results.

37. Exercise 15, on $[-3, 3]$

38. Exercise 16, on $[-3, 2]$

39. Exercise 17, on $[-3, 6]$

40. Exercise 18, on $[-6, 1]$

41. Exercise 19, on $[-3, 3]$

42. Exercise 20, on $[-2, 3]$

43. Exercise 21, on $[-4, 4]$

44. Exercise 22, on $[-3, 1]$

45. Exercise 23, on $[-5, 4]$

46. Exercise 24, on $[-3, 3]$

47. Show that $f(x) = x^7$ has a critical point at $x = 0$ but that f does not have a local extremum there. For which positive integers n does $f(x) = x^n$ satisfy this same property? If, in addition, a and b are real numbers, what can be said about $f(x) = (x - a)^n + b$ at the point $x = a$?

48. Find all critical points of the greatest integer function $f(x) = [\![x]\!]$.

49. If f has a maximum value at $x = a$, show that $g(x) = -f(x)$ has a minimum value at $x = a$.

50. The graph of the derivative, $y = f'(x)$, of a certain function f is shown in the figure. How many solutions can the equation $f(x) = 0$ have? Explain.

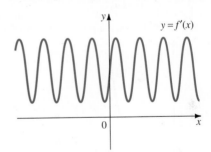

51. Repeat Exercise 50 for the graph of $y = f'(x)$ shown here.

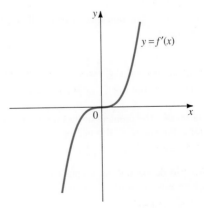

52. The accompanying figure shows the graphs of two functions f and g. If $x = c$ is the point where the vertical distance between the curves is greatest for x in the interval $[a, b]$, what can you say about the tangent lines to the curves $y = f(x)$ and $y = g(x)$ at $x = c$? Prove your answer. Assume that $f'(x)$ and $g'(x)$ exist for all x. (*Hint:* Consider $h(x) = f(x) - g(x)$.)

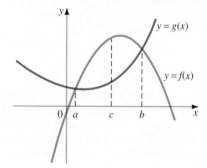

53. Let f be defined on $(0, 1)$, and suppose that $f'(x)$ exists for all x in $(0, 1)$. Define a function g on $(0, 1)$ by $g(x) = f(x^3)$. If f is increasing on $(0, 1)$, what can be said about g? Explain.

54. The graph of the derivative, $y = f'(x)$, of a function f is shown. From it, determine where f is increasing and where it is decreasing. Find the x-coordinates of all maximum and minimum points (indicate which) on the graph of f. If $f(0) = 0$, sketch a possible graph of f.

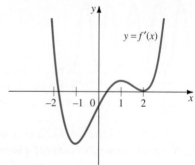

55. Sketch the graph of a function that satisfies all of the following conditions:
  (a) $f'(-2) = f'(4) = 0$
  (b) $f'(x) > 0$ for $x < -2$ or $x > 4$
  (c) $f'(x) < 0$ for $-2 < x < 4$
  (d) $\lim_{x \to \infty} f(x) = 5$ and $\lim_{x \to -\infty} f(x) = -\infty$

56. Cubic Polynomials. The problem of describing the graph of the general cubic polynomial function

$$f(x) = ax^3 + bx^2 + cx + d$$

can be reduced to describing the graphs of all curves $y = g(x)$, where g is given by

$$g(x) = x^3 + kx \qquad k \in \mathbb{R}$$

(See Exercise 27 in Exercise Set 1.4.)
  (a) Show that $g(x) = x^3 + kx$ can have no critical points, one critical point, or two critical points. Sketch graphs to illustrate each possibility.
  (b) How many local extreme values can the general cubic function f have?
  (c) Determine where the function g is increasing and where it is decreasing for each of the cases $k \geq 0$ and $k < 0$.
  (d) How do the slopes of the tangent lines to the curves $y = x^3$ and $y = x^3 + kx$, for $k > 0$, differ at $x = 0$?

In Exercises 57–68, find the local and absolute extrema and state where they occur.

57. $f(x) = x\sqrt{x^2 - 1}$ on $[1, \infty)$

58. $f(x) = (x - 1)^{2/3}(6 - x)$ on $[0, 6]$

59. $f(x) = \dfrac{x}{(x - 1)^2}$ on $\left(-\infty, \frac{1}{2}\right]$. Also, discuss absolute extrema on $(-\infty, 1)$ and on $(1, \infty)$.

60. $f(x) = x\sqrt[3]{x - 4}$ on $[0, 5]$

61. $f(x) = \dfrac{(x - 4)^{2/3}}{x}$ on $[3, 12]$

62. $f(x) = \sin^2 x - 2\cos x + \sin x + x$ on $\left[\dfrac{\pi}{2}, \dfrac{3\pi}{2}\right]$

63. $f(x) = x^4 - 4x^3 - 18x^2 + 108x + 120$ on $[-5, 4]$

64. $f(x) = 2x^4 + 4x^3 - 9x^2 - 27x + 5$ on $[-2, 2]$

65. $f(x) = x^4 - 4x^3 + 16x - 8$ on $[-2, 3]$

66. $f(x) = x^4 - 2x^3 - 23x^2 + 24x + 64$ on $[-4, 5]$

67. $f(x) = 2.13x^3 + 3.02x^2 - 11.9x + 32.4$ on $[-3, 2]$

68. $f(x) = x^3 - 3.256x^2 + 1.432x + 5.875$ on $[-1.256, 2.435]$

69. Suppose $f(x) \geq 0$ for all x in some open interval containing a, except possibly at $x = a$ itself, and suppose $\lim_{x \to a} f(x) = L$ exists. Prove that $L \geq 0$. Show by an example that L may be 0 even if $f(x) > 0$ in the interval. (*Hint:* For the first part, assume $L < 0$ and use Definition 2.5 to arrive at a contradiction.)

70. Complete the proof of Theorem 4.3 by showing that if f' exists at an interior point a at which f attains a local minimum, then $f'(a) = 0$.

71. Use the definition of the derivative to show that $f'(0)$ does not exist for the function defined by

$$f(x) = \begin{cases} x \sin \dfrac{1}{x} & \text{if } x \neq 0 \\ 0 & \text{if } x = 0 \end{cases}$$

72. Use the definition of the derivative to show that for the function defined by

$$f(x) = \begin{cases} x^2 \sin \dfrac{1}{x} & \text{if } x \neq 0 \\ 0 & \text{if } x = 0 \end{cases}$$

$f'(0)$ exists, and find its value.

73. A function is **bounded** on its domain D if there exists a positive real number M such that $|f(x)| \leq M$ for all $x \in D$. Give an example of a function, either by a graph or by a formula, that is bounded on a closed interval I but does not attain a maximum value on I.

74. Is it possible for a continuous function on a closed interval to be unbounded? Explain. (See Exercise 73.)

75. Suppose both f and g are defined on an interval I and each has an absolute maximum there. Let $\max_{(I)} f$ and $\max_{(I)} g$ denote these maximum values. State everything you can about $\max_{(I)}(f + g)$ and $\max_{(I)}(fg)$.

76. The accompanying figure shows a portion of the graph of

$$f(x) = \frac{1}{4}x^4 - 10x^3 + 100x^2 + 100$$

Use information from $f'(x)$ to explain why this is or is not an accurate depiction of the graph of f. If it is not accurate, obtain a graph that does give an accurate description. What scales did you use on the x- and y-axes?

$y = \frac{1}{4}x^4 - 10x^3 + 100x^2 + 100$

4.2 CONCAVITY

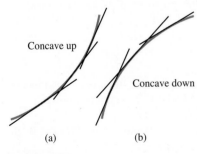

Concave up

Concave down

(a)   (b)

FIGURE 4.25

In Chapter 1 we used the terms *concave up* and *concave down* to describe curves that bend upward or downward, respectively. In this section we give a more precise definition of concavity and show how it is related to the second derivative of a function.

In Figure 4.25(a) we show a portion of the graph of a function that is concave up and in part (b) a portion of a graph that is concave down. Observe the slopes of the tangent lines as we move from left to right in each case. In the case of upward concavity these slopes increase, and in the case of downward concavity the slopes decrease. Since the slopes of the tangent lines are given by the derivative of the function, we can say that upward concavity corresponds to an increasing derivative and downward concavity to a decreasing derivative. In fact, we take this characterization as the definition of upward and downward concavity.

Definition 4.4
Concavity

Let f be differentiable on an open interval I. Then the graph of f is said to be **concave up** on I if f' is an increasing function on I, and the graph of f is said to be **concave down** on I if f' is a decreasing function on I.

Consider again the graphs in Figure 4.25. Notice that in part (a) the graph is always above the tangent line at any point, whereas in part (b) the graph is below the tangent line. These relationships hold true in general, although we will not give a proof.

Note also that concavity is independent of whether the function itself is increasing, decreasing, or neither. In Figure 4.26 each of the graphs is concave up, for example.

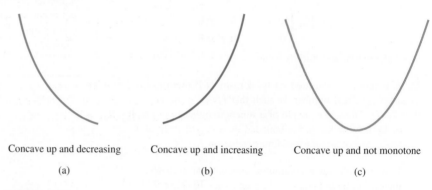

Concave up and decreasing   Concave up and increasing   Concave up and not monotone

(a)   (b)   (c)

FIGURE 4.26

There is a simple test for concavity when $f''(x)$ exists. According to Theorem 3.1 (with f' replacing f), f' is an increasing function if $f''(x) > 0$ and a decreasing function if $f''(x) < 0$. So we have the following theorem.

THEOREM 4.5

> Suppose $f''(x)$ exists for all x on an interval I.
>
> 1. If $f''(x) > 0$ for all x in I, then the graph of f is concave up on I.
> 2. If $f''(x) < 0$ for all x in I, then the graph of f is concave down on I.

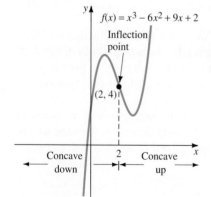

FIGURE 4.27

EXAMPLE 4.8   Determine where the graph of the function f defined by $f(x) = x^3 - 6x^2 + 9x + 2$ is concave up and where it is concave down.

Solution   We need to determine where $f''(x) > 0$ and where $f''(x) < 0$. So we first calculate $f''(x)$:

$$f'(x) = 3x^2 - 12x + 9$$
$$f''(x) = 6x - 12 = 6(x - 2)$$

Thus, $f''(x) > 0$ for all $x > 2$ and $f''(x) < 0$ for all $x < 2$. By Theorem 4.5 we conclude that the graph of f is concave up on $(2, \infty)$ and concave down on $(-\infty, 2)$. The graph in Figure 4.27 shows these results.   ■

Knowing where the graph of a function is concave up and where it is concave down is very useful in drawing the graph. Consider the graph of $f(x) = x^3$ in Figure 4.28, which we have encountered earlier. Up to now, we have simply

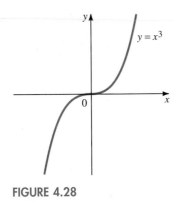

$y = x^3$

FIGURE 4.28

assumed that the curve bends upward to the right of $x = 0$ and downward to the left. Knowing only that it increases everywhere is not enough. From the second derivative $f''(x) = 6x$, though, we see that the graph is concave up for all $x > 0$ and concave down for all $x < 0$, confirming the shape shown in Figure 4.28.

Inflection Points

A point on the graph of a continuous function at which the concavity changes is called an **inflection point**. In Example 4.8 the point $(2, 4)$ is such a point. To the left of $x = 2$, the graph is concave down, and to the right it is concave up. So the point for which $x = 2$ is an inflection point. Since $f(2) = 2^3 - 6(2^2) + 9(2) + 2 = 4$, the point is $(2, 4)$.

If the graph of a twice-differentiable function has an inflection point at $x = a$, we know that the sign of $f''(x)$ for $x < a$ is opposite to that for $x > a$, since the concavity changes at $x = a$. There is an Intermediate-Value Theorem for Derivatives (see Exercise 34 of Section 4.7) that guarantees $f''(a) = 0$ in this case. So when $f''(x)$ exists for all x, we can find all possible x-coordinates of points of inflection simply by setting $f''(x) = 0$ and solving for x. After finding these values, we must test to see if the concavity changes.

Figure 4.29 illustrates a graph with several inflection points. Observe the change in concavity at each of the points P_1 through P_5; each of these is an inflection point. The points P_1 and P_4 are the most common types. At each of these both f' and f'' exist, with $f' \neq 0$ and $f'' = 0$. The points P_2, P_3, and P_5 all occur at critical points of f. At P_2 and P_5, f' is not defined, and at P_3, $f' = 0$. The tangent line is vertical at P_2 (and so its slope is not defined), and no tangent line exists at P_5. At P_3 both f' and f'' equal 0. This type of inflection point always has an appearance similar to one of those in Figure 4.30.

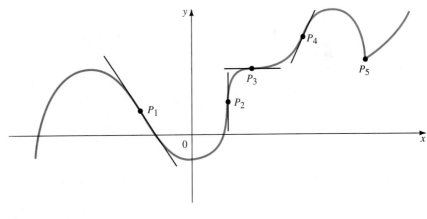

FIGURE 4.29

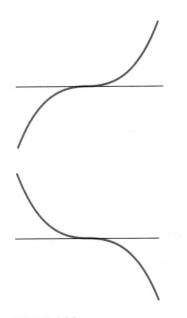

FIGURE 4.30
Inflection points at which both $f' = 0$ and $f'' = 0$

EXAMPLE 4.9

(a) Show that the graph of $y = x^4 - 4x^3 + 12x + 3$ has two inflection points where $y'' = 0$.

(b) Show that the graph of $y = (x - 2)^{5/3}$ has an inflection point but $y'' \neq 0$.

(c) Show that the graph of $y = x^4 - 2x + 3$ has no inflection points even though y'' can be 0.

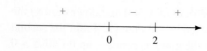

FIGURE 4.31

Sign graph for $y'' = 12x(x-2)$

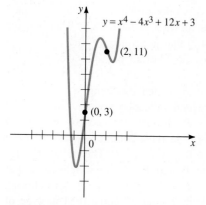

FIGURE 4.32

Inflection points $(0, 2)$ and $(2, 11)$

Solution

(a) Since y'' exists everywhere in this case, all inflection points can be found by setting $y'' = 0$. So we first calculate y''.

$$y' = 4x^3 - 12x^2 + 12$$

$$y'' = 12x^2 - 24x = 12x(x-2)$$

Setting $y'' = 0$ gives $x = 0$ or $x = 2$. These two values are x-coordinates of *possible* points of inflection. We must test to see if they correspond to actual points of inflection. As the sign graph for y'' in Figure 4.31 shows, y'' changes sign at $x = 0$ and at $x = 2$. Thus, there is a change in concavity both at $x = 0$ and at $x = 2$. We calculate the y values in each case from the original equation by substituting the x value and solving for y. We find that the points of inflection are $(0, 3)$ and $(2, 11)$. We show the graph in Figure 4.32.

(b) First, we compute y''.

$$y' = \frac{5}{3}(x-2)^{2/3}$$

$$y'' = \frac{10}{9}(x-2)^{-1/3}$$

In this case y'' is never zero. It is undefined at $x = 2$, but the original function $y = (x-2)^{5/3}$ is defined there and equals 0. Since $y'' > 0$ for $x > 2$ and $y'' < 0$ for $x < 2$, we conclude that $(2, 0)$ is a point of inflection. This example shows that a point of inflection *may* occur at $x = a$ even if $f''(a) \neq 0$, in particular, where $f''(a)$ is not defined. We show the graph in Figure 4.33.

(c) Again, we find y''.

$$y' = 4x^3 - 2$$

$$y'' = 12x^2$$

Thus, $y'' = 0$ when $x = 0$. But for $x > 0$, $y'' > 0$ and for $x < 0$, $y'' > 0$. So there is no change in concavity, and hence no inflection point, at $x = 0$. This example shows that having $f''(a) = 0$ does not guarantee that an inflection point occurs at $x = a$. We show the graph in Figure 4.34. ■

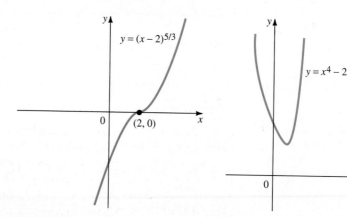

FIGURE 4.33

Inflection point $(2, 0)$

FIGURE 4.34

No inflection points

Be careful to note that *an inflection point must be on the graph of the function.* For example, consider $f(x) = 1/x$. Since $f'(x) = -1/x^2$, we see that $f''(x) = 2/x^3$. Thus, for $x > 0$, the graph is concave up, and for $x < 0$, the graph is concave down. Yet there is no inflection point at $x = 0$, because $f(0)$ is not defined. (See Figure 4.35.)

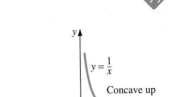

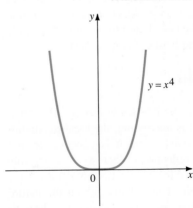

$y = \frac{1}{x}$

Concave up

0

Concave down

FIGURE 4.35
No inflection point since $x = 0$ is not in the domain

The Second-Derivative Test

Suppose that $x = a$ is a critical point for which $f'(a) = 0$ (type 1). In addition, suppose that $f''(a) > 0$. We can then show that the graph of f is concave up in some open interval containing a. So we know two things: since $f'(a) = 0$, the graph has a horizontal tangent line at the point $(a, f(a))$, and since $f''(a) > 0$, the graph is concave up in an interval containing a. Figure 4.36(a) illustrates this situation. Clearly, the point $(a, f(a))$ is a local minimum point on the graph. If $f''(a) < 0$, then the graph is concave down near a, and the situation is as pictured in Figure 4.36(b). The point $(a, f(a))$ is a local maximum point in this case. We therefore have an alternative to the First-Derivative Test for finding local extrema in the case of critical points of type 1, provided the second derivative exists at $(a, f(a))$.

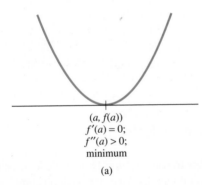

$(a, f(a))$
$f'(a) = 0;$
$f''(a) > 0;$
minimum

(a)

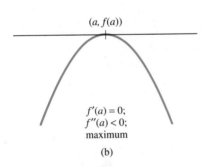

$(a, f(a))$

$f'(a) = 0;$
$f''(a) < 0;$
maximum

(b)

FIGURE 4.36

THEOREM 4.6

y

$y = x^4$

0   x

FIGURE 4.37

> **Second-Derivative Test for Maxima and Minima**
>
> Let $x = a$ be a critical point of f for which $f'(a) = 0$ and $f''(a)$ exists.
>
> 1. If $f''(a) > 0$, the point $(a, f(a))$ is a local minimum point.
> 2. If $f''(a) < 0$, the point $(a, f(a))$ is a local maximum point.

REMARK ────

■ If $f''(a) = 0$, this test is inconclusive. The point $(a, f(a))$ may be an inflection point of the type shown in Figure 4.30, or it may be a local maximum or minimum. For example, if $f(x) = x^3$, then $f''(0) = 0$, and as we have seen, the point $(0, 0)$ is an inflection point (see Figure 4.28). If, however, $f(x) = x^4$, again $f''(0) = 0$, but since $f'(x) = 4x^3$, the slope changes from $-$ to $+$ as we move through 0, so the point $(0, 0)$ is a local (and absolute) minimum point. (See Figure 4.37.)

EXAMPLE 4.10   Use the Second-Derivative Test to find local maxima and minima for the function given by

$$f(x) = x^3 + 3x^2 - 24x - 20$$

Solution   First, we find the critical points.

$$
\begin{aligned}
f'(x) &= 3x^2 + 6x - 24 \\
&= 3(x^2 + 2x - 8) \\
&= 3(x + 4)(x - 2)
\end{aligned}
$$

The critical points are $x = -4$ and $x = 2$.

Next, we find $f''(x)$ and substitute the critical points for x.

$$f''(x) = 6x + 6 = 6(x + 1)$$

$$f''(-4) = -18 < 0$$

$$f''(2) = 18 > 0$$

By Theorem 4.6 we conclude that f has a local maximum at $x = -4$ and a local minimum at $x = 2$.

Finally, we substitute -4 and 2, respectively, into the original function to find the points on the graph:

$(-4, 60)$ is a local maximum point.

$(2, -48)$ is a local minimum point.   ■

Choice of Tests

Whether to use the First-Derivative Test or the Second-Derivative Test depends on several factors. First, we have no choice if a is a critical point for which either $f''(a) = 0$ or $f'(a)$ is not defined; the Second-Derivative Test will not work. When the function is a polynomial, the Second-Derivative Test often is simpler to apply. For other functions, it depends on the difficulty of calculating f''. If you need f'' anyway (for example, to determine concavity and inflection points), then use the Second-Derivative Test. If the only objective is to determine maxima and minima, and if it appears that the calculation of f'' would be quite laborious, then it is probably easier to use the First-Derivative Test.

A Procedure for Sketching Graphs

From the first two derivatives of a function, we now know how to determine where the function is increasing and where it is decreasing, the local maximum and minimum points on its graph, the intervals where it is concave up and where it is concave down, and the points of inflection. When we combine this information with the aids to graphing we studied in Chapter 1, such as intercepts, symmetry, and asymptotes, we have a wealth of information about the nature of the function and its graph. In the next three examples we illustrate how to use all this information to draw the graph, showing all of its important features. In most cases we will need to plot few, if any, additional points.

EXAMPLE 4.11   For the function f, given by

$$f(x) = 2x^3 - 3x^2 - 12x + 8 \qquad (x \in \mathbb{R})$$

find the intervals on which f is increasing and on which f is decreasing, the local maximum and minimum points, the intervals where the graph of f is concave up and where concave down, and the points of inflection. Use the information obtained, along with intercepts, symmetry, and asymptotes, to draw the graph of f.

Solution   Since $f(x)$ is a cubic polynomial, the graph of f has no asymptotes. To find the x-intercepts, we would need to solve a cubic equation. We will see a way of approximating the roots in Section 4.5, but for now we will not attempt to find them. As we get other information about the graph, the approximate location of the places where it crosses the x-axis will become evident. The y-intercept is 8, found by setting $x = 0$.

To locate maximum and minimum points, to find where f is increasing and where it is decreasing, and to see where the graph is concave up and where it is concave down, we calculate $f'(x)$ and $f''(x)$.

$$f(x) = 2x^3 - 3x^2 - 12x + 8$$
$$f'(x) = 6x^2 - 6x - 12 = 6(x^2 - x - 2) = 6(x - 2)(x + 1)$$
$$f''(x) = 12x - 6 = 6(2x - 1)$$

FIGURE 4.38
Sign graph for $f'(x) = 6(x-2)(x+1)$

The critical points are $x = 2$ and $x = -1$, where $f'(x) = 0$. From the sign graph for f' (Figure 4.38), we see that f is increasing on $(-\infty, -1)$ and on $(2, \infty)$, and f is decreasing on $(-1, 2)$.

Since $f''(-1) = -18$ and $f''(2) = 18$, we see by the Second-Derivative Test that f has a local maximum at -1 and a local minimum at 2. We calculate $f(-1) = 15$ and $f(2) = -12$. Thus, $(-1, 15)$ is a local maximum point and $(2, -12)$ is a local minimum point.

For concavity, we consider $f''(x)$. When $x > 1/2$, $f''(x) > 0$, and when $x < 1/2$, $f''(x) < 0$. So the graph is concave up on $(1/2, \infty)$ and concave down on $(-\infty, 1/2)$. Furthermore, the point $(1/2, f(1/2)) = (1/2, 3/2)$ is a point of inflection.

This information enables us to draw the graph of f with reasonable accuracy (see Figure 4.39 on the following page). Plotting the additional points $(3, -1)$ and $(-2, 4)$ increases the accuracy.   ∎

EXAMPLE 4.12   Draw the graph of the function g given by $g(x) = x^4 - 4x^3 + 16$, making use of the information obtained from g' and g''.

Solution   Since $g(0) = 16$, the y-intercept is 16. Again, we will not attempt to find the x-intercepts (although you might observe that $g(2) = 0$). There are no asymptotes and no symmetry. We calculate $g'(x)$ and $g''(x)$.

$$g(x) = x^4 - 4x^3 + 16$$
$$g'(x) = 4x^3 - 12x^2 = 4x^2(x - 3)$$
$$g''(x) = 12x^2 - 24x = 12x(x - 2)$$

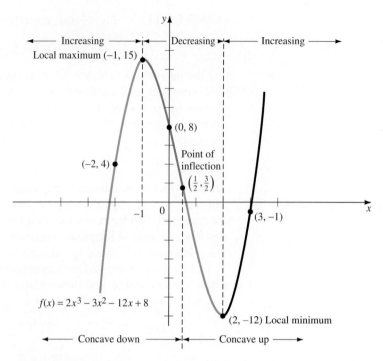

FIGURE 4.39

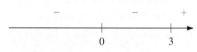

FIGURE 4.40
Sign graph for $g'(x) = 4x^2(x - 3)$

FIGURE 4.41
Sign graph for $g''(x) = 12x(x - 2)$

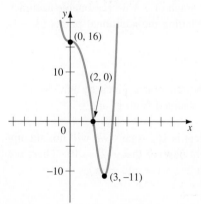

FIGURE 4.42

The critical points are $x = 0$ and $x = 3$. Since $g''(0) = 0$ and $g''(3) > 0$, we see that g has a local minimum at $x = 3$. We cannot test $x = 0$ by the Second-Derivative Test since $g''(0) = 0$, so we use the First-Derivative Test. The sign graph for g' (Figure 4.40) shows that the slope does not change at $x = 0$, so g does not have a local maximum or a local minimum there. It is, rather, decreasing both to the left and to the right of $x = 0$.

Next, look at the sign graph for g'' (Figure 4.41). We see that the concavity changes at $x = 0$ and $x = 2$, so we conclude that g has points of inflection at $x = 0$ and $x = 2$.

We calculate $g(x)$ at the points in question and summarize our findings:

g is increasing on $(3, \infty)$.

g is decreasing on $(-\infty, 0)$ and on $(0, 3)$.

g is concave up on $(-\infty, 0)$ and on $(2, \infty)$.

g is concave down on $(0, 2)$.

$(3, -11)$ is a local minimum point.

$(0, 16)$ is a point of inflection where the tangent line is horizontal.

$(2, 0)$ is a point of inflection.

The graph is shown in Figure 4.42.   ∎

Asymptotes

Before we go to the next example, let us review briefly horizontal and vertical asymptotes. If any one of the following holds true, then the line $x = a$ is a vertical asymptote:

$$\lim_{x \to a} f(x) = \infty \text{ or } -\infty,$$

$$\lim_{x \to a^+} f(x) = \infty \text{ or } -\infty,$$

$$\lim_{x \to a^-} f(x) = \infty \text{ or } -\infty.$$

In particular, if f is the rational function

$$f(x) = \frac{p(x)}{q(x)}$$

with $q(a) = 0$ and $p(a) \neq 0$, the line $x = a$ is an asymptote. For example, if

$$f(x) = \frac{x^2 - 4}{x^2 - 9}$$

then both $x = 3$ and $x = -3$ are vertical asymptotes.

A line $y = b$ is a horizontal asymptote to the graph of f if

$$\lim_{x \to \infty} f(x) = b \qquad \text{or} \qquad \lim_{x \to -\infty} f(x) = b$$

For a rational function f, such that

$$f(x) = \frac{p(x)}{q(x)}$$

we saw in Chapter 1 that by considering dominant terms on numerator and denominator, we can draw the following conclusions.

(i) If $\deg p < \deg q$, then $y = 0$ is a horizontal asymptote (we are using "deg" to mean "degree of").

(ii) If $\deg p = \deg q$, then $y = b$ is a horizontal asymptote, where b is the leading coefficient of p divided by the leading coefficient of q.

(iii) If $\deg p > \deg q$, there is no horizontal asymptote.

We also saw that if $\deg p = \deg q + 1$, there is an inclined asymptote, $y = mx + b$, where $mx + b$ is the quotient obtained by dividing $p(x)$ by $q(x)$. To illustrate, the graph of

$$f(x) = \frac{x^2 - 4}{x^2 - 9}$$

has the horizontal asymptote $y = 1$ [case (ii) above]. The graph of

$$g(x) = \frac{x^2 - 4}{x + 1} = x - 1 - \frac{3}{x + 1} \qquad \text{See the long division in the margin.}$$

has no horizontal asymptote [case (iii)], but since

$$\lim_{x \to \infty} [g(x) - (x - 1)] = \lim_{x \to \infty} \left(-\frac{3}{x + 1} \right) = 0$$

it has the inclined asymptote

$$y = x - 1$$

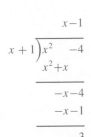

EXAMPLE 4.13   Discuss and sketch the graph of

$$f(x) = \frac{x^2 - 1}{x^2 - 4}$$

Solution   Let $y = f(x)$.

(a) *Intercepts.* Setting $y = 0$, we get $x = \pm 1$. Setting $x = 0$, we get $y = 1/4$.

(b) *Symmetry.* Since $f(x) = f(-x)$, the graph is symmetric with respect to the y-axis.

(c) *Asymptotes.* The vertical asymptotes are $x = 2$ and $x = -2$, and the horizontal asymptote is $y = 1$.

(d) *Intervals of Increase and Decrease.* The derivative is found by the Quotient Rule to be

$$f'(x) = \frac{-6x}{(x^2 - 4)^2}$$

so $x = 0$ is the only critical point. Note that although $f'(x)$ is not defined at $x = \pm 2$, these points are not critical points of the second kind. In fact, they are not in the domain of the function, since $f(x)$ also is not defined for $x = \pm 2$. If $x \neq 2$, and $x > 0$, $f'(x) < 0$, so f is decreasing for all positive x values in the domain of f, namely, $(0, 2) \cup (2, \infty)$. Similarly, when $x \neq -2$ and $x < 0$, $f'(x) > 0$, so f is decreasing on $(-\infty, -2) \cup (-2, 0)$.

(e) *Maxima and Minima.* We have seen that $x = 0$ is the only critical point. Since $f'(x)$ changes from $+$ to $-$ as x goes from left to right through 0, we conclude from the First-Derivative Test that f has a local maximum value at $x = 0$. Since $f(0) = 1/4$, the point $(0, 1/4)$ is a local maximum point.

(f) *Concavity and Inflection Points.* Calculating $f''(x)$, we get, after simplifying,

$$f''(x) = \frac{6(3x^2 + 4)}{(x^2 - 4)^3} = \frac{6(3x^2 + 4)}{(x - 2)^3(x + 2)^3}$$

We show a sign graph for f'' in Figure 4.43. Note that the numerator for $f''(x)$ is always positive, so it does not need to be considered. We see from the sign graph that when $x > 2$ or $x < -2$, $f''(x) > 0$, so the graph is concave up. For $-2 < x < 2$, $f''(x) < 0$, and the graph is therefore concave down. Since $x = 2$ and $x = -2$ are not in the domain of f, there are no points on the graph where the concavity changes, so there are no inflection points.

By plotting a few additional points, we can now draw the graph, as shown in Figure 4.44. ■

FIGURE 4.43

Sign graph for $f''(x) = \dfrac{6(3x^2 + 4)}{(x - 2)^3(x + 2)^3}$

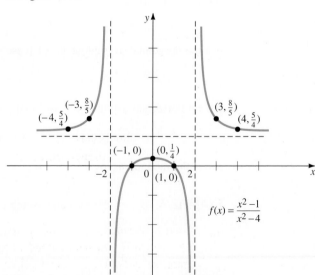

FIGURE 4.44

Infinite limits for f' can also provide useful information about the graph. Suppose that f is continuous at a and $\lim_{x \to a} f'(x) = \infty$. Since $f'(x)$ gives the slope of the tangent line, we know that the slope is positive near a, and the tangent line becomes steeper and steeper as x approaches a. At a itself, the tangent line is vertical. So the graph must be similar to Figure 4.45(a) in the vicinity of a. The other possibilities are shown in parts (b), (c), and (d).

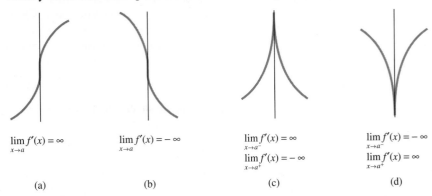

$$\lim_{x \to a} f'(x) = \infty$$

(a)

$$\lim_{x \to a} f'(x) = -\infty$$

(b)

$$\lim_{x \to a^-} f'(x) = \infty$$
$$\lim_{x \to a^+} f'(x) = -\infty$$

(c)

$$\lim_{x \to a^-} f'(x) = -\infty$$
$$\lim_{x \to a^+} f'(x) = \infty$$

(d)

FIGURE 4.45

In parts (a) and (b) of Figure 4.45, an inflection point occurs at $(a, f(a))$. The maxima and minima in parts (c) and (d) occur at cusps. The four possibilities shown assume that $\lim_{x \to a^+} f'(x)$ and $\lim_{x \to a^-} f'(x)$ are both infinite. If only one of these one-sided limits is infinite, or if f is discontinuous, various possibilities exist, some of which are shown in Figure 4.46.

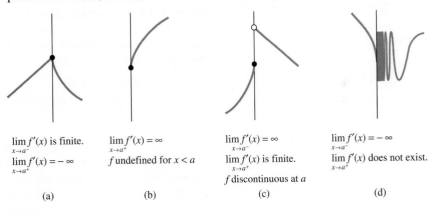

$$\lim_{x \to a^-} f'(x) \text{ is finite.}$$
$$\lim_{x \to a^+} f'(x) = -\infty$$

(a)

$$\lim_{x \to a^+} f'(x) = \infty$$
f undefined for $x < a$

(b)

$$\lim_{x \to a^-} f'(x) = \infty$$
$$\lim_{x \to a^+} f'(x) \text{ is finite.}$$
f discontinuous at a

(c)

$$\lim_{x \to a^-} f'(x) = -\infty$$
$$\lim_{x \to a^+} f'(x) \text{ does not exist.}$$

(d)

FIGURE 4.46

EXAMPLE 4.14   Discuss and sketch the graph of

$$f(x) = x + 3(1 - x)^{1/3}$$

Solution

(a) *Intercepts.* $f(0) = 3$, so the y-intercept is 3. The x-intercepts are not easily obtained, but with the aid of a calculator it can be shown that one x-intercept is between -5 and -6 and another is between 4 and 5.

(b) *Symmetry.* There is no symmetry, since $f(-x) = -x + 3(1 + x)^{1/3}$, so $f(-x)$ is not equal to $f(x)$ or $-f(x)$.

(c) *Asymptotes.* There are no asymptotes.

(d) *Intervals of Increase and Decrease.* We calculate $f'(x)$ to determine the critical points.

$$f'(x) = 1 - (1-x)^{-2/3}$$

The critical points are x values in the domain of f for which $f'(x) = 0$ or for which $f'(x)$ does not exist. Clearly, $f'(1)$ does not exist, and since $f(1)$ does exist (that is, $x = 1$ is in the domain of f), we conclude that $x = 1$ is a critical point. To see if there are critical points for which $f'(x) = 0$, we solve

$$1 - (1-x)^{-2/3} = 0$$

or, equivalently,

$$1 = \frac{1}{(1-x)^{2/3}}$$

Inverting, we get

$$(1-x)^{2/3} = 1 \qquad \text{Cube both sides.}$$

$$(1-x)^2 = 1 \qquad \text{Take square roots.}$$

$$1 - x = \pm 1$$

$$x = 0 \text{ or } 2$$

The critical points are therefore 0, 1, and 2. To determine the sign graph for f', we can test a value in each of the four regions into which the critical points divide the number line. For example, we can use $x = 3$ in the interval $(2, \infty)$, $x = 1.5$ in the interval $(1, 2)$, $x = 0.5$ in the interval $(0, 1)$, and $x = -1$ in the interval $(-\infty, 0)$. With the aid of a calculator, we obtain the signs as indicated in Figure 4.47. We conclude that

$$f \text{ is increasing in } (2, \infty) \text{ and in } (-\infty, 0)$$

$$f \text{ is decreasing in } (0, 1) \text{ and in } (1, 2)$$

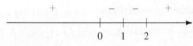

FIGURE 4.47

Sign graph for $f'(x) = 1 - \dfrac{1}{(1-x)^{2/3}}$

(e) *Maxima and Minima.* From the sign graph in Figure 4.47, we conclude that f has a local maximum at $x = 0$ and a local minimum at $x = 2$. There is no local maximum or minimum at $x = 1$. Substituting $x = 0$ and $x = 2$ into the original formula for $f(x)$, we obtain the following points:

$$(0, 3) \text{ is a local maximum point.}$$

$$(2, -1) \text{ is a local minimum point.}$$

We can gain further insight into the nature of the graph near the critical point $x = 1$ by examining the right-hand and left-hand limits of $f'(x)$ as x approaches 1:

$$\lim_{x \to 1^+} f'(x) = \lim_{x \to 1^+} \left[1 - \frac{1}{(1-x)^{2/3}} \right] = -\infty$$

$$\lim_{x \to 1^-} f'(x) = \lim_{x \to 1^-} \left[1 - \frac{1}{(1-x)^{2/3}} \right] = -\infty$$

Thus, the tangent line is vertical at $x = 1$.

(f) *Concavity and Inflection Points.* We first calculate $f''(x)$.

$$f''(x) = -\frac{2}{3}(1-x)^{-5/3} = -\frac{2}{3(1-x)^{5/3}}$$ Don't forget to use the Chain Rule.

For $x > 1$, $f''(x) > 0$, so the graph is concave up. For $x < 1$, $f''(x) < 0$, so the graph is concave down. Since $x = 1$ is in the domain of f, we conclude that there is an inflection point at $x = 1$. Since $f(1) = 1$, the coordinates of the inflection point are $(1, 1)$.

In summary, the point $(0, 3)$ is a local maximum point, $(2, -1)$ is a local minimum point, and $(1, 1)$ is an inflection point at which the tangent line is vertical. Furthermore, the graph is concave down to the left of 1 and concave up to the right of 1. We get a few additional points and draw the graph as shown in Figure 4.48. ∎

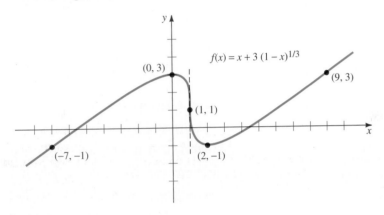

FIGURE 4.48

ANALYSIS OF FUNCTIONS USING A COMPUTER ALGEBRA SYSTEM

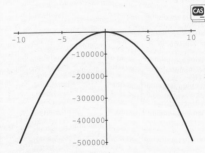

FIGURE 4.2.1

$y = f(x) = \dfrac{x^4}{4} - 5000x^2 + 5,$

$-10 \le x \le 10$

CAS 17

Let $f(x) = \frac{x^4}{4} - 5000x^2 + 5$. Give a complete analysis of the graph of f.

We can begin by simply graphing the curve, specifying an initial range for x (Figure 4.2.1). In Maple we could enter

```
f:=x->x^4/4–5000*x^2+5;
plot(f(x), x=–10..10);
```

The plot suggests the curve is a parabola. We should be wary of simply accepting this initial scale—all the important information may not be displayed. Rather than trying to guess a more appropriate scale, we can use the tools of calculus to sketch the curve more accurately. We start with finding the critical values of the function.

In Maple enter:

fprime := diff(f(x),x);   fprime is a name we give to the derivative of f with respect to x.
solve(fprime=0);

Maple gives the output 0, 100, -100 for the three critical values for f. We could see this set of critical values immediately if we could factor the derivative. Enter

factor(fprime);

and the output is

x(x − 100) (x + 100)

The location of the critical values suggest that we should sketch the curve over an interval such as $[-200, 200]$ (Figure 4.2.2). Enter

plot(f(x),x=–200..200);

The curve now shows more complicated behavior. The complete analysis of the curve can be done as easily using Maple, Mathematica, or DERIVE. We complete the analysis using Maple.

Increasing and Decreasing Intervals:

solve(fprime > 0,x);

Output: $\{x < 0, -100 < x\}, \{100 < x\}$

solve(fprime < 0,x);

Output: $\{x < -100\}, \{x < 100, 0 < x\}$

Local Extremes: They can be read off the graph or the First-Derivative Test can be invoked.

Local maxima: $(0, f(0)) = (0, 0)$

Local minima: $(-100, f(-100))$ and $(100, f(100))$

Use Maple to determine the y-coordinates.

f(–100); –24999995
f(100); –24999995

Concavity: First, compute the second derivative.

f2prime := diff(fprime,x);

Concave up:

solve(f2prime > 0,x);

Output:

$$\left\{x < -\frac{100}{3}\sqrt{3}\right\}, \left\{\frac{100}{3}\sqrt{3} < x\right\}$$

Concave down:

solve(f2prime < 0,x);

Output:

$$\left\{-\frac{100}{3}\sqrt{3} < x, x < \frac{100}{3}\sqrt{3}\right\}$$

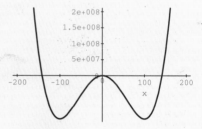

FIGURE 4.2.2

$y = f(x) = \dfrac{x^4}{4} - 5000x^2 + 5,$
$-200 \leq x \leq 200$

Inflection Points: They can now be read off as

$$\left(\frac{100}{3}\sqrt{3}, f\left(\frac{100}{3}\sqrt{3}\right)\right) \quad \text{and} \quad \left(-\frac{100}{3}\sqrt{3}, f\left(-\frac{100}{3}\sqrt{3}\right)\right)$$

Use the CAS to determine the y-coordinates:

f(100/3*sqrt(3)); $-\dfrac{124999955}{9}$

f(−100/3*sqrt(3)); $-\dfrac{124999955}{9}$

Asymptotes: There are none.

Exercise Set 4.2

In Exercises 1–10, determine the intervals on which the graph of the given function is concave up and the intervals on which it is concave down.

1. $f(x) = 2x^2 - 3x + 4$ **2.** $g(x) = 5 - 6x - x^2$

3. $f(x) = ax^2 + bx + c$ **4.** $f(x) = x^3 - 3x^2 + 4x$

5. $h(x) = x^3 + 6x^2 - 5x + 1$

6. $F(x) = 2x^3 - 3x^2 - x + 5$

7. $G(x) = x^4 - 2x^3$

8. $g(x) = x^4 + 2x^3 - 12x^2 + 10$

9. $f(x) = x^4 - 6x^2 + 8$

10. $F(x) = x^4 - 3x^3 - 15x^2 - 12x + 9$

11. Refer to the points A, B, C, and D on the accompanying graph of the function f.
(a) At which points is $f'(x) > 0$?
(b) At which points is $f''(x) > 0$?

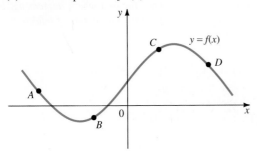

In Exercises 12–21, find all local maximum and minimum points, using the Second-Derivative Test when applicable and the First-Derivative Test otherwise.

12. $f(x) = 2x^3 - 6x + 4$

13. $g(x) = x^3 + 2x^2 - 4x + 5$

14. $h(x) = 2x^3 - 3x^2 + 7$

15. $\phi(x) = 2 - 3x + 5x^2 - x^3$

16. $F(x) = x^4 + 4x^3 - 8x^2 + 4$

17. $G(x) = 3x^5 - 25x^3 + 60x$

18. $f(x) = x + \dfrac{1}{x}$ **19.** $f(x) = \dfrac{1}{x^2} + \dfrac{2}{x} + 3$

20. $g(x) = 3x^4 - 8x^3 + 6x^2$

21. $h(x) = 3x^5 - 20x^3 + 7$

In Exercises 22–41, determine the intervals where the function is increasing and where decreasing, local maxima and minima, concavity, and all points of inflection. Use the information obtained to draw the graph of $y = f(x)$.

22. $f(x) = x^3 - 3x^2$ **23.** $f(x) = x^3 - 12x$

24. $f(x) = x^3 + 3x^2 - 9x - 15$

25. $f(x) = x^3 - x^2 - x + 1$

26. $f(x) = 4 + 3x - x^3$

27. $f(x) = 2x^3 + 4x^2 - 8x + 2$

28. $f(x) = x^4 - 4x^3 + 5$

29. $f(x) = x^3 - 6x^2 + 12x - 3$

30. $f(x) = x^4 - 6x^2 + 10$

31. $f(x) = 3 + 4x - x^4$

32. $f(x) = \dfrac{x^5}{5} - \dfrac{10x^3}{3} + 9x$

33. $f(x) = (x^2 - 4)(x - 2)$

34. $f(x) = (x - 3)^2(x + 3)$

35. $f(x) = (x - 2)^2(2x + 5)$

36. $f(x) = 2\sin\dfrac{x}{2} + 1$ on $[0, 4\pi]$

37. $f(x) = \sin x + \cos x$ on $[0, 2\pi]$

38. $f(x) = x + \sin x$ on $[0, 4\pi]$

39. $f(x) = \cos x - \sqrt{3}\sin x$ on $\left[-\dfrac{\pi}{3}, \dfrac{5\pi}{3}\right]$

40. $f(x) = \cos 2x - 4\cos x$ on $[-\pi, \pi]$

41. $f(x) = \sin 2x - 2\sin^2 x$ on $[-\pi, \pi]$

In Exercises 42–51, discuss intercepts, symmetry, vertical and horizontal asymptotes, intervals where f is increasing and where decreasing, local maxima and minima, concavity, and points of inflection; then sketch the graph.

42. $f(x) = \dfrac{x - 1}{x + 2}$

43. $f(x) = \dfrac{x}{x^2 - 1}$

44. $f(x) = \dfrac{2}{x^2 + 1}$

45. $f(x) = \dfrac{x}{x^2 + 1}$

46. $f(x) = \dfrac{x^2}{x^2 - 4}$

47. $f(x) = \dfrac{x^2 - 4}{x^2 - 1}$

48. $f(x) = \dfrac{x^2}{x^2 + 4}$

49. $f(x) = \dfrac{1 - x^2}{x^2 - 4}$

50. $f(x) = \dfrac{1}{x - 1} + \dfrac{1}{x + 1}$

51. $f(x) = \dfrac{1}{x} - \dfrac{1}{x + 1}$

52. Sketch a possible graph of a function f satisfying all of the following properties.
   (a) $f''(x) > 0$ for $-2 < x < 2$
   (b) $f'(0) = 0$, $f(0) = 1$, $f'(3) = 0$, $f'(5) = 0$
   (c) $f'(x) > 0$ for $2 < x < 3$ or $x > 5$
   (d) $f'(x) < 0$ for $3 < x < 5$
   (e) $\lim_{x \to -2} f(x) = \infty$
   (f) $\lim_{x \to 2^-} f(x) = \infty$ and $\lim_{x \to 2^+} f(x) = -\infty$
   (g) $\lim_{x \to -\infty} f(x) = 0$

In Exercises 53–60, show that the function has a vertical tangent line, and determine the nature of the graph at that point.

53. $f(x) = x^{2/3}$

54. $f(x) = 1 - x^{1/3}$

55. $f(x) = \sqrt[3]{1 - x}$

56. $f(x) = x^{3/5} - 2x^{2/5}$

57. $f(x) = \dfrac{x^{2/3}}{x - 1}$

58. $f(x) = \sqrt{1 - x^2}, \quad 0 \le x \le 1$

59. $f(x) = 1 - x^{1/2}, \quad x \ge 0$

60. $f(x) = \sqrt{\dfrac{x - 1}{x + 1}}, \quad x \ge 1$

In Exercises 61–66, find all local extrema, using the Second-Derivative Test when applicable and the First-Derivative Test otherwise.

61. $f(x) = 3 + 4x - 2x^2 - \dfrac{x^3}{3} + \dfrac{x^4}{4}$

62. $f(x) = \dfrac{x^2 - x + 2}{x + 1}$

63. $f(x) = \dfrac{x^2}{2} - \dfrac{1}{x} + 2$

64. $f(x) = x^4 - 6x^2 + 8x - 3$

65. $f(x) = 3x + 2(4 - x)^{3/2}$

66. $f(x) = x^{2/3}(2x - 5)$

In Exercises 67–82, discuss intercepts, symmetry, asymptotes, intervals where f is increasing and where decreasing, local maxima and minima, concavity, and points of inflection; then draw the graph.

67. $f(x) = \frac{1}{16}(3x^4 - 8x^3 - 24x^2 + 96x)$

68. $f(x) = \dfrac{x^4}{2} - \dfrac{x^3}{3} - 4x^2 + 4x + 7$

69. $f(x) = x^5 - 5x^3 + 10x - 3$

70. $f(x) = 2 - 8x + 6x^2 - x^4$

71. $f(x) = x^4 - 4x^3 + 16x - 4$

72. $f(x) = x^4 - 2x^3 + 2x + 3$

73. $f(x) = \dfrac{3\cos x}{2 - \sin x}$ on $[-\pi, \pi]$

74. $f(x) = \dfrac{\tan x}{2\sec x - 1}$ on $[0, 2\pi]$

75. $f(x) = \cos^4 x - 4\cos x + 3$ on $\left[-\dfrac{3\pi}{2}, \dfrac{3\pi}{2}\right]$

76. $f(x) = 2x + 3(1 - x)^{2/3}$

77. $f(x) = \dfrac{3 - x^2}{x^3}$

78. $f(x) = \dfrac{x^2 - 1}{x^3 - 3x}$

79. $f(x) = \dfrac{3x^{1/3}}{2 - x}$

80. $f(x) = \dfrac{1 - \sin x}{\cos x - x}$

81. $f(x) = \dfrac{x^3 + 1}{x^2}$

82. $f(x) = \dfrac{x^2 - 8}{x - 3}$

83. Cubic Polynomials. Use information from the first and second derivatives to give a complete description of all possible shapes of the graphs of the family of cubic functions

$$f(x) = x^3 + kx, \qquad k \in \mathbb{R}$$

Consider the cases $k = 0$, $k > 0$, and $k < 0$ separately. Use your results to discuss possible shapes of the general cubic function

$$g(x) = ax^3 + bx^2 + cx + d \qquad a, b, c, d \in \mathbb{R}, \ a \neq 0$$

(See Exercise 27 in Section 1.4 and Exercise 56 in Section 4.1.)

84. Let $f(x) = -x^3 - 2x^2 + x + 2$.
(a) Obtain the graphs of f, f', and f''.
(b) Using the graphs in part (a), approximate where $f'(x)$ has its maximum value.
(c) Find the exact value of x where $f'(x)$ has its maximum value. Justify your answer, and compare it with the approximation in part (b).

85. Investigate the family of curves given by

$$f(x) = \dfrac{1}{1 + x^2} + cx, \qquad c \in \mathbb{R}$$

(a) Conjecture for which values of c the function f is increasing for all $x \in \mathbb{R}$ and for which values of c the function f is decreasing for all $x \in \mathbb{R}$.
(b) Determine the inflection points for the graph of f.
(c) Is there any relationship between the location of the inflection points and the answer you found in part (a)?

86. Let

$$f(x) = \dfrac{ax + b}{cx + d}$$

where a, b, c, and d are positive constants. Use a graphing calculator or a computer algebra system to obtain several graphs by varying a, b, c, and d. Do the graphs have any common features? If so, what are they? What can be said about the graph of the given function f in general?

4.3 APPLIED MAXIMUM AND MINIMUM PROBLEMS

As we stated at the beginning of Section 4.1, one of the most important applications of the derivative is in finding maximum or minimum values of functions. Usually certain restrictions, known as "constraints," are imposed on the variable (or variables) involved. Such problems are referred to collectively as **optimization problems**, since we are trying to find the best (optimal) result under the given constraints. Some optimization problems can be formulated in terms of one *continuous* variable (one that can assume any real value on an interval), and some require more than one such variable. We will concentrate here on the first type and study the second type later. It should also be mentioned that certain problems involve *discrete* variables (for example, variables that can assume only integer values, such as the number of seats in a theater). Such problems often can be solved by supposing the variable to be continuous and interpreting the final result appropriately. Other problems require totally different approaches, as are studied in discrete mathematics courses.

Our previous studies have given us the capability of finding extreme values, *provided the quantity to be maximized or minimized is expressed as a function of one continuous variable*. This restriction is what we will concentrate on in this section. The main difficulty is that in real-life situations, problems are usually presented in words, not as neat "textbook-type" mathematical formulas. So our first step is to translate the word problem into an appropriate mathematical form. Then we solve the mathematical problem, and finally, we interpret the answer in terms of the original problem. This process is known as **mathematical modeling**, and it is one of the most important aspects of the work of applied mathematicians and others who use mathematics in their work. Although there are no guaranteed paths to success in setting up a model of an optimization problem for which calculus is appropriate, the following steps can serve as a general guide.

1. Read the problem through (more than once if necessary) and identify the function to be maximized or minimized. We refer to this function as the **objective function**. Give it a name. (Using initial letters such as C for cost, P for profit, V for volume, and so on, is a good idea.)

2. Determine a variable on which the objective function depends and give it a name, such as x or t. This variable is the *independent variable*. Note any limitations on its size. Sometimes there may be two (or more) such variables, but in the present chapter we will always give enough information to express any additional variables in terms of the first.

3. Express the objective function in terms of the independent variable. This is a key step and will be made easier if steps 1 and 2 have been carried out carefully. Frequently it is helpful to draw a figure. Sometimes well-known relationships such as "distance = rate × time" or "profit = revenue − cost" are used. Also, mensuration formulas, such as those for area or volume, may be useful.

4. Find the derivative of the objective function with respect to the independent variable identified in step 2, and determine the critical points.

5. Test critical points that lie in the allowable domain to see if they yield maximum points, minimum points, or neither. Often there is only one critical point in the allowable domain. The nature of the problem might ensure that there is an extreme value interior to the interval in question, so that this extreme value must occur at the critical point. In such cases it is not necessary to test the critical point, but it is still a good idea to test it as a check.

6. Find the value of the objective function at each critical point identified in step 5. If the endpoint values also are feasible solutions (that is, if they are in the domain of the objective function), find the value of the objective function at these points as well. The value that gives the desired extreme can now be determined by direct computation and comparison of the functional values at the critical points and endpoints (if any).

7. Mentally check to see that the answer you found is reasonable. Sometimes it is possible to see that you made a mistake because the answer you found is clearly incorrect. (If you were to find that the price of a one-family house comes out as 2¢ or 2 billion dollars, for example, or if you find that the length of a rectangle is −10 ft, you would clearly know you had made a mistake.)

The following examples illustrate this procedure.

EXAMPLE 4.15   An open-top box is to have a square base and a volume of 10 ft^3. If the bottom costs 15¢ per square foot and the sides 6¢ per square foot, find the dimensions of the most economical box, and find the minimum cost.

Solution   The most economical box is the one with the minimum cost. So the objective function is the cost C, and we want to find its minimum value. We introduce as the independent variable the side x of the square base (Figure 4.49). The height y is also a relevant variable, but it can be written in terms of x as follows. We know the volume of the box is 10 ft^3, and since the volume of a box is the area of the base times the height, we have

$$x^2 y = 10$$

So

$$y = \frac{10}{x^2}$$

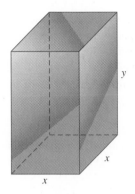

FIGURE 4.49

Clearly we must have $x > 0$, because we are discussing a real box.

Now we can find $C(x)$ from the given information. The base of the box has area $x^2 \text{ ft}^2$, and each square foot costs 15¢. There are four vertical sides, each having area $xy = x(10/x^2) = 10/x$, each costing 6¢ per square foot. Thus, the value $C(x)$ of the cost function, in cents, is

$$C(x) = 15x^2 + 4(6)\left(\frac{10}{x}\right) = 15x^2 + \frac{240}{x}$$

The computer-generated graph of $y = C(x)$, for $x > 0$ (Figure 4.50), clearly shows the existence of a minimum value. To determine exactly where this minimum occurs, we find the critical points by taking the derivative of $C(x)$.

$$C'(x) = 30x - \frac{240}{x^2}$$

Set $C'(x) = 0$:

$$30x - \frac{240}{x^2} = 0 \qquad \text{Multiply by } x^2 \text{ and divide by 30.}$$

$$x^3 - 8 = 0$$

$$x^3 = 8$$

$$x = 2$$

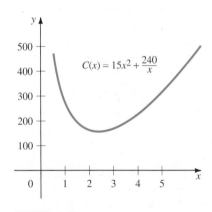

FIGURE 4.50

It seems reasonable that this value of x gives a minimum, since mathematically we can see from $C(x)$ that for large x or x close to 0, the cost is high. The graph of $C(x)$ in Figure 4.50 confirms this observation. We should check to see that the value $x = 2$ gives a minimum by substituting in $C''(x)$:

$$C''(x) = 30 + \frac{480}{x^3}$$

$$C''(2) = 30 + \frac{480}{8} = 90$$

Thus, $C''(2) > 0$, so the minimum value of C indeed occurs when $x = 2$. At this value of x we have

$$y = \frac{10}{2^2} = \frac{5}{2}$$

and

$$C(2) = 15(2^2) + \frac{240}{2} = 60 + 120 = 180$$

Thus, the most economical box has dimensions (in feet) of $2 \times 2 \times 2\frac{1}{2}$, and the minimum cost is 180¢, or \$1.80.   ∎

EXAMPLE 4.16   A rectangular flower bed is to be designed with one side adjacent to a building and the other three sides bordered by a walkway 3 ft wide. If the outer perimeter of the walkway is to be 172 ft (we have 172 linear feet of decorative edging available), what should be the dimensions of the flower bed if it is to have maximum area? What is this maximum area?

Solution   The objective function is the area A of the flower bed. As in Figure 4.51, let x be the width of the bed and y its length. Since the outer perimeter of the walkway is $2(x + 3) + (y + 6)$ (verify this), we must have

$$2x + 6 + y + 6 = 172$$

$$y = 160 - 2x$$

Since x and y are both positive, we must have $0 < x < 80$. The area of the bed is $xy = x(160 - 2x)$, so

$$A(x) = 160x - 2x^2$$

$$A'(x) = 160 - 4x$$

To find the critical points, we set $A'(x) = 0$, getting $x = 40$. To test whether this value gives a maximum or a minimum area, we calculate the second derivative and get $A''(x) = -4$. The only critical point is $x = 40$, and since $A''(40) < 0$, this value produces a maximum value of the area. Thus, the dimensions that produce the maximum area are

$$x = 40 \qquad \text{and} \qquad y = 160 - 2(40) = 80$$

and the maximum area is

$$A(40) = (40)(80) = 3200 \text{ ft}^2$$

Figure 4.52 illustrates our results.   ∎

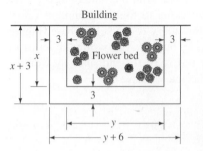

FIGURE 4.51

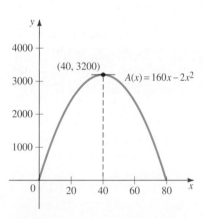

FIGURE 4.52

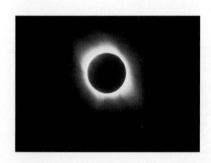

EXAMPLE 4.17   An airplane is chartered for a flight to Hawaii to view a total solar eclipse for 100 passengers at \$1,000 per passenger. The price per ticket will be reduced by \$5 per passenger for each person in excess of 100 who goes on the trip. Find the number of passengers that will provide maximum revenue, and find the maximum revenue. The plane will accommodate up to 200 passengers.

Solution   Let R denote the revenue, the function we want to maximize. The revenue is calculated by multiplying the number of passengers by the price of each ticket. Each of these quantities depends on the number of passengers *in excess* of 100. Call this number x. Then we have

$$100 + x = \text{total number of passengers}$$

$$1,000 - 5x = \text{price of each ticket, in dollars}$$

Because of the capacity of the airplane, we must have $0 \le x \le 100$. Thus, we have

$$R(x) = (100 + x)(1,000 - 5x) = 100,000 + 500x - 5x^2$$

$$R'(x) = 500 - 10x$$

Setting $R'(x) = 0$, we find the critical value $x = 50$. Since $R''(x) = -10$, we conclude from the Second-Derivative Test that $x = 50$ gives a local maximum value. The revenue when $x = 50$ is

$$R(50) = (150)(1,000 - 250) = (150)(750) = 112,500$$

We need to check the endpoint values also:

$$R(0) = (100)(1,000) = 100,000 \qquad R(100) = (200)(500) = 100,000$$

We conclude that a maximum revenue of \$112,500 is obtained when 50 passengers in excess of 100 go—that is, when 150 passengers are on the flight. ∎

REMARK ——————————————————————————————

■ We could have used x for the total number of passengers rather than the number in excess of 100. The calculations would have been different, but the final result would have been the same. You might try this alternative and compare the difficulty with the method we used.

EXAMPLE 4.18   Towns A and B are 5 km apart and are located on the same side of a straight river. Town A is 1 km from the river and town B is 4 km from the river. A pumping station is to be built at the river's edge and pipes are to go to the two towns. Where should the pumping station be located to minimize the amount of pipe used?

Solution   In Figure 4.53, C is the point on the shore nearest to A and D is the point on the shore nearest to B. Then the distance CD is the same as AE, which by the Pythagorean Theorem is 4. Let P be the point at which the pumping station is to be located, and use x to denote its distance from C. Then $4 - x$ is its distance from D. Clearly, $0 \le x \le 4$. If $L(x)$ denotes the total length of pipe, then $L(x) = \overline{PA} + \overline{PB}$. Using the Pythagorean Theorem on triangles ACP and BDP, we get

$$L(x) = \sqrt{x^2 + 1} + \sqrt{(4 - x)^2 + 16}$$

Differentiating,

$$L'(x) = \frac{x}{\sqrt{x^2 + 1}} - \frac{4 - x}{\sqrt{(4 - x)^2 + 16}}$$

Setting $L'(x) = 0$, we get

$$x\sqrt{(4 - x)^2 + 16} = (4 - x)\sqrt{x^2 + 1}$$

and on squaring and simplifying, this gives (check the details)

$$15x^2 + 8x - 16 = 0$$

$$(5x - 4)(3x + 4) = 0$$

$$x = \frac{4}{5} \text{ or } x = -\frac{4}{3}$$

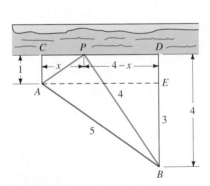

FIGURE 4.53

The answer $x = -4/3$ is ruled out on two counts. First, it is outside the admissible domain. Second, it fails to satisfy the equation we obtained by setting $L'(x) = 0$, since it would cause one side to be positive and the other side to be negative. The solution $x = 4/5$ does satisfy this equation (check this), so it is a critical point. It can be shown that $L''(4/5) > 0$, so a minimum value of L occurs at $x = 4/5$. However, calculating L'' is a little messy, and we can avoid finding it (and also avoid using the First-Derivative Test) by calculating $L(4/5)$ and comparing it with the endpoint values $L(0)$ and $L(4)$. After a bit of arithmetic, we get

$$L\left(\tfrac{4}{5}\right) = \sqrt{41} \approx 6.40 \qquad L(0) = 1 + 4\sqrt{2} \approx 6.66 \qquad L(4) = 4 + \sqrt{17} \approx 8.12$$

Since L is continuous on $[0, 4]$, it reaches both a maximum and a minimum value there, and the only candidates for these extreme values are the three given here. So $L(4/5)$ is the minimum value. Thus, to use the least amount of pipe, the pumping station should be located $4/5$ km from C and $16/5$ km from D. ∎

REMARK ──────────────────────────────

■ It is not unusual for algebraic solutions of the equation obtained by setting the derivative of the objective function equal to 0 (to get critical points) to be invalid for the applied problem. For example, in the pumping station problem, we had to rule out $x = -4/3$. You should be alert to such limitations.

────────────────────────────────────

EXAMPLE 4.19   Prove that the isosceles triangle of maximum area that can be inscribed in a circle of fixed radius a is equilateral.

Solution   Referring to Figure 4.54, we see by the Pythagorean Theorem that $y = \sqrt{a^2 - x^2}$, so the area $A(x)$ of the isosceles triangle is

$$A(x) = x(\sqrt{a^2 - x^2} + a), \qquad 0 \le x \le a$$

We calculate $A'(x)$ and set it equal to 0. (We omit some of the algebra.)

$$A'(x) = \frac{-x^2}{\sqrt{a^2 - x^2}} + \sqrt{a^2 - x^2} + a = \frac{a^2 - 2x^2 + a\sqrt{a^2 - x^2}}{\sqrt{a^2 - x^2}}$$

$$A'(x) = 0 \text{ if } 2x^2 - a^2 = a\sqrt{a^2 - x^2}$$

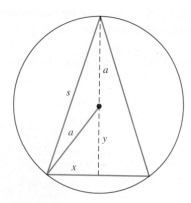

FIGURE 4.54

After squaring both sides and collecting terms, we get

$$x^2(4x^2 - 3a^2) = 0$$

so that $x = 0$ or $x = \pm a\sqrt{3}/2$. Since $x = 0$ gives an area of 0 and since x cannot be negative, the only critical point of interest is $a\sqrt{3}/2$. We calculate

$$A\left(\frac{a\sqrt{3}}{2}\right) = \frac{3a^2\sqrt{3}}{4}$$

which is greater than the endpoint value $A(a) = a^2$ because $3\sqrt{3}/4 > 1$, so $x = a\sqrt{3}/2$ does produce the maximum area.

To show that the triangle with this value of x is equilateral, we calculate the height,

$$y + a = \sqrt{a^2 - x^2} + a = \sqrt{a^2 - \frac{3a^2}{4}} + a = \sqrt{\frac{a^2}{4}} + a = \frac{a}{2} + a = \frac{3a}{2}$$

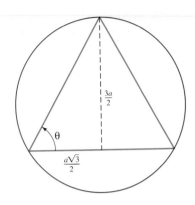

FIGURE 4.55

The distance formula:

$$d = \sqrt{(x_2 - x_1)^2 + (y_2 - y_1)^2}$$

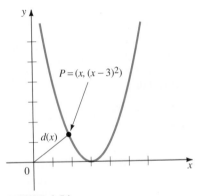

FIGURE 4.56

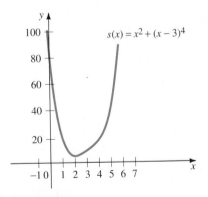

FIGURE 4.57

The tangent of the base angle θ (Figure 4.55) is

$$\tan \theta = \frac{\dfrac{3a}{2}}{\dfrac{a\sqrt{3}}{2}} = \sqrt{3}$$

so $\theta = 60°$. The base angles are equal, and since the three angles add to $180°$, all three angles are equal to $60°$. So the triangle is equiangular and hence equilateral. ■

EXAMPLE 4.20   Find the point on the parabola $y = (x - 3)^2$ that is nearest the origin.

Solution   An arbitrary point P on the parabola has coordinates $(x, (x-3)^2)$, and its distance $d(x)$ from the origin is

$$d(x) = \sqrt{x^2 + (x - 3)^4}$$

(Figure 4.56). This is the function we want to minimize. We can make the problem simpler, however, by observing that the distance will be minimum if and only if the square of the distance is a minimum. We denote the squared distance by $s(x)$:

$$s(x) = x^2 + (x - 3)^4$$

The computer-generated graph of $s(x)$ in Figure 4.57 shows clearly the existence of a minimum. To find where it occurs, we first calculate $s'(x)$:

$$s'(x) = 2x + 4(x - 3)^3 = 4x^3 - 36x^2 + 110x - 108$$

We set $s'(x) = 0$ and divide through by 2 to get

$$2x^3 - 18x^2 + 55x - 54 = 0$$

Various techniques exist for solving equations of degree 3 or higher, some of which you may have studied in your precalculus course. If there are any rational roots, we can find them by trial and error, with the trials limited as follows. If $x = p/q$ is a root, then p must be a factor of the constant term, in this case 54, and q must be a factor of the leading coefficient, in this case 2. Starting with integer possibilities, we eliminate $x = 1$ but find that $x = 2$ works. (A calculator is useful here.) After dividing out the factor $x - 2$, the other factor is found to be $2x^2 - 14x + 27$, which has no real zeros (you can verify that the zeros are complex by using the quadratic formula). So $x = 2$ is the only critical point. Since the geometry of the problem assures us that a minimum distance exists, we can conclude that $x = 2$ produces this minimum. (We could also check this conclusion by observing that $s''(2) > 0$.) The point on the graph nearest the origin is therefore $(2, 1)$, and its distance from the origin is $d(2) = \sqrt{5}$. ■

Concepts from Economics

Economists often use calculus, so before giving the last two examples of this section, we discuss briefly some basic concepts of economics. We will be concerned here with maximizing the profit from the manufacture and sale of some commodity. If we let $C(x)$ denote the cost of producing x items and $R(x)$ denote the gross revenue obtained from selling them, then the profit, $P(x)$, is

$$P(x) = R(x) - C(x)$$

Although x in this situation can actually take on only positive integer values, we can treat it as if it could take on any positive real value. By doing so, we can apply the techniques of calculus to the functions C, P, and R. Then when we obtain the critical point x that maximizes the profit, if x is not an integer, we test the two closest integers to see which gives a maximum.

The cost function C usually consists of a **fixed cost** and an additional **variable cost**. The fixed cost is a constant, independent of the number of items produced. Its value is determined by such things as rent, mortgage payments, taxes, maintenance, basic utility bills, and any salaries that are not based on the number of items produced. The variable cost is dependent on the number of items produced. For example, the cost of plastic for making pitchers is a variable cost, since it depends on how many pitchers you make. On the other hand, the salary of the company president is a fixed cost. The formula for the variable cost may be obtained by analyzing the factors that affect the cost, or it may be obtained empirically. In the latter case, records may be kept over a period of time and a graph plotted showing the cost versus the number of items produced. Then by a process known as "curve fitting," a function whose graph matches as closely as possible the one obtained experimentally is determined. The resulting function can take a variety of forms, but it is often a polynomial of first, second, or third degree.

To see how the variable cost might be calculated in one situation, consider the following example.

EXAMPLE 4.21   The admissions office of a college expects to need between 20,000 and 40,000 catalogs, and it receives the following bid from a printer: For the first 20,000 copies the cost will be $3.00 per catalog. For additional catalogs ordered, the cost per copy for these additional catalogs will be reduced by 5¢ per 1,000 catalogs ordered. For example, if the college orders 26,000 catalogs, the cost will be $3.00 per copy for the first 20,000; then for the next 6,000 the cost will be $2.70 per copy, a reduction of 30¢ (six times 5¢) per copy. Suppose x thousand catalogs are ordered, where $20 \leq x \leq 40$. Find the variable cost function $C_v(x)$.

Solution   The cost in thousands of dollars is

$$C_v(x) = x[\underbrace{3 - 0.05 \underbrace{(x - 20)}_{\substack{\text{number of} \\ \text{thousands in} \\ \text{excess of 20}}}}_{\substack{\text{cost per catalog} \\ \text{of } x \text{ thousand}}}] = -0.05x^2 + 4x$$

In this case, then, the variable cost function is quadratic over the interval in question. We show its graph in Figure 4.58.   ∎

Quadratic cost functions are common, as are linear ones. Note that in Figure 4.58, although the curve is rising, it is concave down, reflecting the fact that the unit cost decreases as more units are produced. In some similar situations it may happen that if x increases still further, the cost curve turns upward again,

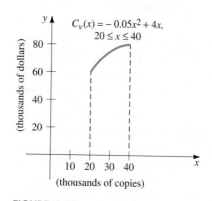

FIGURE 4.58

since more machines, more employees, more space, and so on, may be required. These considerations would probably lead to a cubic cost function.

The **revenue function** R is obtained by multiplying the number of items sold by the price p at which they are sold, so that

$$R(x) = xp(x)$$

We have indicated p as a function of x, since there is a definite functional relationship between price and number of items sold. The function $p(x)$ is sometimes called the **demand function**, since it gives the price per unit when there is demand for x units. Typically, when the price goes down, the number of items that can be sold at that price goes up.

The next example illustrates these ideas.

EXAMPLE 4.22   The Sun-Ray Company produces cells for solar collectors. It finds through experience that on the average it can sell x cells per day when the price $p(x) = 100 - 0.05x$, where $250 \le x \le 800$. The fixed cost is \$4000 per day and the variable cost *per unit* is $60 - 0.01x$. Find the number of cells that Sun-Ray should produce each day to maximize profit, and find the maximum profit.

Solution   The revenue is the number of units times the price per unit:

$$R(x) = xp(x) = x(100 - 0.05x) = 100x - 0.05x^2$$

The variable cost is the number of units times the cost per unit. So the total cost is

$$C(x) = 4000 + x(60 - 0.01x) = 4000 + 60x - 0.01x^2$$

Thus,

$$\begin{aligned}
P(x) &= R(x) - C(x) \\
&= 100x - 0.05x^2 - (4000 + 60x - 0.01x^2) \\
&= -0.04x^2 + 40x - 4000 \\
P'(x) &= -0.08x + 40 \\
P'(x) &= 0 \text{ if } x = 500
\end{aligned}$$

Since $P''(x) < 0$, the critical point $x = 500$ gives a maximum. We calculate $P(500) = 6000$ and the endpoint values $P(250) = 3500$ and $P(800) = 2400$. So a maximum profit of \$6000 per day is obtained when 500 cells are produced. ∎

Marginal Revenue and Marginal Cost

In general, as in the preceding example, the maximum profit will occur at a critical value where $P'(x) = 0$. Let x_0 be such a point. Then we have

$$P'(x_0) = R'(x_0) - C'(x_0) = 0$$

so that

$$R'(x_0) = C'(x_0)$$

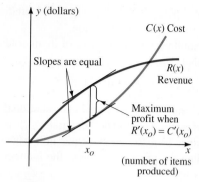

FIGURE 4.59

That is, maximum profit occurs where R' and C' are equal. We illustrate this result in Figure 4.59. As we mentioned in Section 3.2, these quantities, $R'(x)$ and $C'(x)$, are called, respectively, **marginal revenue** and **marginal cost**. If we let x change by a small amount, say dx, then the corresponding changes in $R(x)$ and $C(x)$ are approximated by their differentials:

$$\Delta R \approx dR = R'(x)\,dx \qquad \text{and} \qquad \Delta C \approx dC = C'(x)\,dx$$

In particular, if $dx = 1$, we have $\Delta R \approx R'(x)$ and $\Delta C \approx C'(x)$. Thus, the marginal revenue is approximately the change in revenue caused by selling one additional item, and the marginal cost is approximately the change in cost caused by producing one additional item. Also, we have

$$\Delta P \approx dP = [R'(x) - C'(x)]dx$$

As long as the marginal revenue $R'(x)$ exceeds the marginal cost $C'(x)$, there is an increase in profit by increasing the number of items. The value x_0 at which $R'(x_0) = C'(x_0)$ is the value where no further increase in profit would occur by increasing x, and so the profit is then a maximum.

Exercise Set 4.3

1. A rectangular plot of ground is to be fenced on three sides to form a garden, with the fourth side bounded by a retaining wall. If there are 50 ft of fencing available, what should be the dimensions of the garden to produce the maximum area?

2. A rectangular lot adjacent to a road is to be enclosed by a fence. The side along the road requires reinforced fencing, which costs $7 per foot. Fencing that costs $5 per foot can be used for the other three sides. What is the largest area that can be enclosed for a cost of $3600, and what are the dimensions that give this maximum area?

3. A farmer wishes to fence a rectangular pasture having a total area of 12,000 m², and wants to divide it into two parts with a fence across the middle. Fencing around the outside costs $7.50 per meter, but the farmer can use less expensive fencing at $3 per meter as the divider. What dimensions will result in the least cost?

4. A water trough is to be formed from a sheet of metal 2 ft wide and 10 ft long by bending up at a right angle equal amounts from each long side. (Pieces will then be welded onto the ends.) How many inches on each side should be bent up to give the maximum volume?

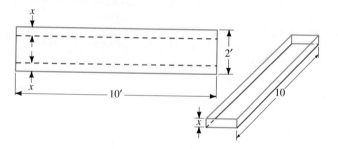

5. A cardboard box is to be constructed having a volume of 9 ft³ and such that the bottom is a rectangle twice as long as it is wide. What should be the dimensions so that the least amount of cardboard is used?

6. An open-top box is to be made from a piece of cardboard 4 ft long and $2\frac{1}{2}$ ft wide by cutting out squares of equal size at each corner and bending up the flaps. What size squares should be removed in order to produce the maximum volume?

7. Find the dimensions of the rectangle of maximum area that can be inscribed in an isosceles triangle of height 6 and base 4 (see the figure). (*Hint:* Use similar triangles to find y in terms of x.)

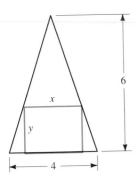

8. Find the dimensions of the rectangle of maximum area that can be inscribed as shown in a right triangle with legs of lengths 6 and 8, respectively.

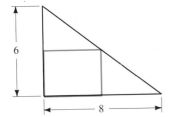

9. Show that the rectangle of maximum area that can be inscribed in a circle with fixed radius a is a square.

10. Find the dimensions of the rectangle of maximum area that can be inscribed in a semicircle of radius a as shown in the figure.

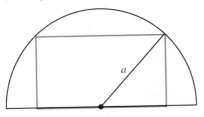

11. Find the dimensions of the right circular cylinder of maximum volume that can be inscribed in a sphere of fixed radius a. (*Hint:* Consider a cross section as shown in the figure to find the altitude h of the cylinder in terms of its base radius r.)

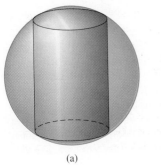

(a)

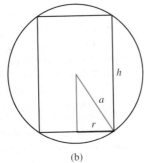

(b)

12. In Exercise 11, maximize the lateral surface area instead of the volume.

13. The unit price of a microwave oven is constant at $250. The fixed cost is $80, and the variable cost per unit is found by experience to be approximately $x + 90$, when the number x of ovens is between 10 and 100. Assuming x is in this range and that all ovens manufactured are sold, find the number that should be manufactured to produce the maximum profit.

14. The total cost function for producing x tables in a certain furniture factory is determined to be $C(x) = 0.02x^2 + 20x + 100$ for $10 \le x \le 100$, and the unit price function $p(x) = 25 + \frac{120}{x} - 0.03x$ for x in this range. Find the number of tables that should be produced to maximize profit. What is the maximum profit?

15. The owner of an orchard has a 10-acre plot planted with 200 apple trees, and the average yield is 500 apples per tree. The orchard owner plans to put apple trees of the same type on another 6-acre plot that has the same soil conditions. If the trees are planted closer together on the second plot than on the first plot, the yield per tree will be less, but it is possible that with more trees the total yield per acre will be greater. For each additional tree per acre planted, the average yield per tree will be reduced by 10 apples. How many trees should be planted on the 6-acre plot to give the maximum total yield?

16. An excursion train is to be run to the Super Bowl. Amtrak sets the fare at $100 per ticket if 200 people go but agrees to lower the cost of all tickets by 20¢ each for every passenger in excess of 200. How many passengers will produce the greatest revenue, and what is the maximum revenue? (The train has a capacity of 450 passengers.)

17. A yacht is rented for an excursion for 60 passengers at $20 per passenger. For each person in excess of 60 who goes, up to an additional 40 passengers, the fare of each passenger will be reduced by 25¢. Find the number of passengers that will produce the maximum revenue.

18. A restaurant has a fixed price of $24 for a complete dinner. The average number of customers per evening is 160. The owner estimates on the basis of experience that for each $1 increase in the cost of a dinner, there will be an average of 5 fewer customers per evening. What price should be charged to produce the maximum revenue, and what is the maximum revenue?

19. At a certain movie theater the price of admission is $5, and the average daily attendance is 200. As an experiment, the manager reduces the price by 10¢ and finds that the average attendance increases by 5 people per day. Assuming that for each additional 10¢ reduction the average attendance rises by 5, find the number of 10¢ reductions that would result in the maximum revenue.

20. A manufacturer makes aluminum cups in the form of right circular cylinders open at the top (no handle), having a volume of 16π in.^{3}. If the cost of the material for the bottom is twice that for the sides, find the dimensions that will give the lowest cost.

21. The strength of a rectangular beam of fixed length is jointly proportional to the width and the square of the depth. Find the dimensions of the strongest beam that can be cut from a log 3 ft in diameter.

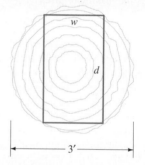

22. Find the point on the graph of $y = \sqrt{2x - 3}$ that is nearest the point $(3, 0)$.

23. A picture with an area of 120 in.^{2} is rectangular and is framed so that there is a $2\frac{1}{2}$-in. matting on each side and a 3-in. matting at the top and bottom. Find the dimensions of the picture if the total area of the framed picture is to be as small as possible. (Test to see that you have found the minimum.)

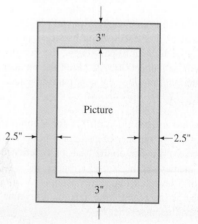

24. Find the point on the upper half of the ellipse $4x^2 + 9y^2 = 36$ that is nearest the point $(1, 0)$.

25. The printed matter on a page is to occupy an area of 80 in.^{2}, and there are to be margins of 1 in. at the top, bottom, and right side, and $1\frac{1}{2}$ in. at the left. Find the overall dimensions of the page that requires the least amount of paper.

26. A rain gutter is to be formed as shown in the figure by bending up at equal angles sides of 10 cm from a long metal sheet 30 cm wide. Find the angle θ that will enable the gutter to handle the greatest flow of water without overflowing.

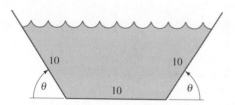

27. Find the dimensions of the right circular cone of maximum volume that can be inscribed in a sphere of fixed radius a.

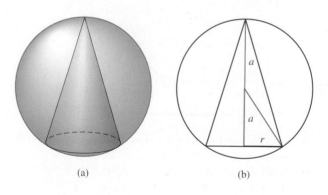

(a)   (b)

28. Change Exercise 27 so that the cone is to have maximum lateral surface area.

29. A telephone cable is to be laid under a river from point A to point C (see the figure at the top of page 251) and then underground to point B. The river is 20 m wide, and the distance from point B to the point directly across the river from A is 40 m. If it costs $1\frac{1}{2}$ times as much to lay the cable under the river as it does to put it underground, find where point C should be located to minimize the cost.

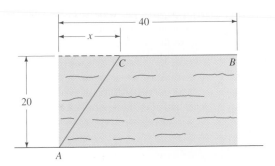

30. A window is to be designed in the shape of a rectangle surmounted by an equilateral triangle. The overall outside perimeter is to be 12 ft. Find the dimensions of the window if the area is to be a maximum.

31. Repeat Exercise 30 with the triangle replaced by a semicircle.

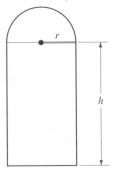

32. A silo is constructed in the form of a right circular cylinder surmounted by a hemisphere of the same radius as the cylinder. Find the dimensions that give the minimum cost if it is to have a volume of 600 m³, and the unit cost of the material for the hemisphere is twice that for the cylinder.

33. A car leaves town A traveling at 80 km/hr due east on a straight road toward town B, 50 km away. At the same time, a second car leaves town B traveling 60 km/hr due south on a straight road (see the figure). How much time must the cars travel until the distance between them is a minimum? At that instant where are the cars and how far apart are they?

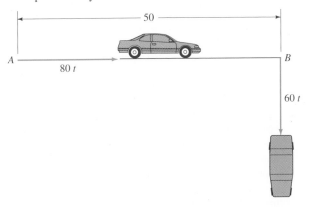

34. The cost of operating a certain truck is determined from experience to be $20 + x/12$ cents per kilometer when the truck is driven at a speed of x km/hr. The truck driver earns $9 per hour of actual driving time. What is the most economical speed at which to operate the truck on a 600-km trip?

35. Find the length of the longest piece of straight pipe that can be carried horizontally around the corner from a 5-ft-wide hallway into a 3-ft-wide hallway. Assume the pipe has negligible thickness. (*Hint:* Choose as independent variable the angle θ as shown on the following page.)

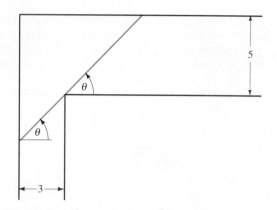

37. A long sheet of paper 6 in. wide is to be folded up as shown in the figure. Find where the point P of the fold should be so that (a) the area of the folded part is a minimum and (b) the length PQ of the fold is a minimum.

36. The illumination from a source of light varies directly as the intensity of the light source and inversely as the square of the distance from the source. One light is 1 m directly above one end of a table 3 m in length, and another light of twice the candlepower is directly opposite the first at the other end of the table and 2 m above it (see the figure). At what point on the table directly between the lights will the sum of the intensities of lights be the least?

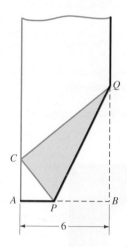

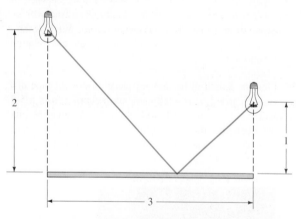

38. Reconsider the situation in Exercise 36 with two identical light sources s units apart, positioned at the same distance above the table. Show that the point at which the intensity of illumination on the table is a minimum is not necessarily $s/2$, but rather depends on the distance, d, of the light sources from the table. For definiteness take $s=2$, and introduce an x-axis on the table, with $x=-1$ under one light and $x=1$ under the other. Express the illumination at a point on the table in terms of x and d.

4.4 RELATED RATES

In this section we will apply the derivative to functions that have *time* as the independent variable. We already know that if a function represents the position of a particle moving in a straight line, then the derivative with respect to time is the velocity of the particle, and the derivative of the velocity is the acceleration. But many other time-dependent functions exist. The type of problem we wish to consider is one in which we know one or more rates of change and want to find another rate of change that is dependent on those we know. Such problems are called **related-rate problems**. The next two examples illustrate how one rate of change is influenced by another.

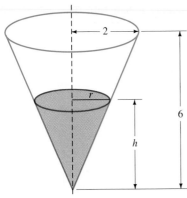

FIGURE 4.60

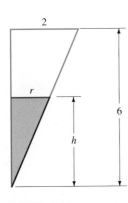

FIGURE 4.61

EXAMPLE 4.23   Suppose a cone-shaped container is 6 m tall and the radius of the top is 2 m. (See Figure 4.60.) Water is being pumped into the container at the constant rate of 500 L/min (L is the symbol for liters, and 1 L = 10^{-3} m^3). Find the rate at which the water level is rising (a) when the water level is 3 m and (b) when the container is half full.

Solution   We denote by V, h, and r the volume of water in the container, the height of the water, and the radius of the circular surface of the water, respectively, all at time t. It is important to note that all three variables are functions of time and that they are not constants. It would be incorrect at this stage, for example, in solving part (a) to put $h = 3$, because h is not always 3, and we must first find the relationship between the relevant rates of change before we consider what they are at a particular instant. *In short, before substituting a particular value for a variable quantity, we must differentiate.*

By the formula for the volume of a cone, the volume of water at any instant is given by

$$V = \frac{1}{3}\pi r^2 h \tag{4.1}$$

In both parts (a) and (b) we want to know dh/dt at a particular instant, so we will need to differentiate. First, however, it will simplify matters to replace r by its equivalent in terms of h, since h is the variable of interest in this problem. We can establish a relation between r and h using similar triangles. Figure 4.61 is the right-hand half of a vertical slice along the axis of the cone. From it, we get

$$\frac{r}{2} = \frac{h}{6}$$

so $r = \frac{1}{3}h$. Substituting this equivalent value for r in Equation 4.1 and simplifying, we get

$$V = \frac{1}{27}\pi h^3 \tag{4.2}$$

Now we can differentiate both sides with respect to time t:

$$\frac{dV}{dt} = \frac{1}{9}\pi h^2 \frac{dh}{dt}$$

(Note the use of the Chain Rule.) Since dV/dt is the rate of change of the volume of water, and since we are given that water is entering the container at the rate of 500 L/min, or $\frac{1}{2}$ m^3/min, we have

$$\frac{dV}{dt} = \frac{1}{2}$$

So

$$\frac{1}{2} = \frac{1}{9}\pi h^2 \frac{dh}{dt}$$

or

$$\frac{dh}{dt} = \frac{9}{2\pi h^2} \tag{4.3}$$

Now we can find dh/dt for any value of h. For part (a) we have

$$\left.\frac{dh}{dt}\right|_{h=3} = \frac{9}{2\pi(3^2)} = \frac{1}{2\pi} \approx 0.159 \text{ m/min}$$

For part (b) we must find h when the container is half full. The volume of the entire container is $\frac{1}{3}\pi(2)^2 \cdot 6 = 8\pi$, so when it is half full, the volume is 4π. Substituting in Equation 4.2, we get

$$4\pi = \frac{1}{27}\pi h^3$$

$$h^3 = 4(27)$$

$$h = 3\sqrt[3]{4}$$

Now we use Equation 4.3 again:

$$\left.\frac{dh}{dt}\right|_{h=3\sqrt[3]{4}} = \frac{9}{2\pi(3\sqrt[3]{4})^2} = \frac{1}{4\pi\sqrt[3]{2}} \approx 0.0632 \text{ m/min}$$ ∎

EXAMPLE 4.24   A car leaves town A and travels due north at an average speed of 40 km/hr. At the same instant, another car leaves town B, 50 km due east of town A, and travels due south at an average speed of 80 km/hr. Find how fast the distance between the cars is increasing 1 hr later.

Solution   As in Figure 4.62 let x denote the distance traveled by the first car, y the distance traveled by the second car, and z the distance between them, all after t hr. Note that x, y, and z are all variable quantities. The only constant is the distance 50 between A and B. From the given information, we know that

$$\frac{dx}{dt} = 40 \qquad \text{and} \qquad \frac{dy}{dt} = 80$$

and we want to find dz/dt at a particular time. The first thing we must do is to find an equation relating the three variables. Then we will differentiate with respect to t.

By means of the dashed lines in Figure 4.62 we have constructed a right triangle with legs of length 50 and $x+y$, and hypotenuse z. By the Pythagorean Theorem we have

$$z^2 = (50)^2 + (x + y)^2 \tag{4.4}$$

We differentiate both sides with respect to time, t:

$$2z\frac{dz}{dt} = 2(x + y)\left(\frac{dx}{dt} + \frac{dy}{dt}\right)$$

Substituting the known values of dx/dt and dy/dt and solving for dz/dt gives

$$\frac{dz}{dt} = 120\left(\frac{x + y}{z}\right) \tag{4.5}$$

which is the rate of change of the distance between the cars at any time. Now we wish to let $t = 1$. At that instant $x = 40$ and $y = 80$, and so from Equation 4.4,

$$z = \sqrt{(50)^2 + (120)^2} = \sqrt{16,900} = 130$$

Thus, by Equation 4.5,

$$\left.\frac{dz}{dt}\right|_{t=1} = (120)\left(\frac{40 + 80}{130}\right) = \frac{1440}{13} \approx 111 \text{ km/hr}$$ ∎

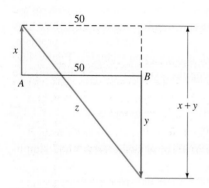

FIGURE 4.62

The General Procedure

Examples 4.23 and 4.24 are typical of "related-rate" problems. Let us examine what they have in common. In both cases we were given one or more rates of change, and we wished to find another at a given instant. We were able to find a relationship between the variables whose rates of change we knew and the variable whose rate of change we were seeking. In Example 4.23 we found a relationship by using the formula for the volume of a cone and then using similar triangles to get r in terms of h. In Example 4.24 we obtained a relationship by using the Pythagorean Theorem. In both cases drawing a figure was an important aid in finding the relationships. Next, we differentiated with respect to time and substituted the known rates of change. Finally, we substituted the values of all variables at the particular instant in question and solved for the unknown rate of change.

We summarize these steps here.

Procedure for Solving Related-Rate Problems

1.   If it is appropriate, draw a figure. On it label all variable parts with letters and all fixed parts with numbers.

2.   Find an equation relating the variables whose rates of change are known and the variable whose rate of change is desired. Look for such things as areas or volumes of geometric figures, similar triangles, the Pythagorean Theorem, or trigonometric relationships.

3.   Differentiate both sides of the equation found in step 2 with respect to time, and substitute the known rates of change. Don't forget to apply the Chain Rule.

4.   Calculate the values of all variables involved at the particular instant in question, and substitute them in the result of step 3.

5.   Solve for the unknown rate of change.

We cannot overemphasize the importance of waiting until step 4 to substitute instantaneous values of variable quantities. This substitution is to be done only *after* differentiation. Substituting such values too soon is the most frequent mistake students make.

We will give one further example of a related-rate problem.

EXAMPLE 4.25   A television camera is 30 m directly opposite the finish line at an automobile racetrack and is trained on the lead car as it approaches the finish. If the car is traveling at the rate of 150 km/hr, find the rate in radians per second at which the camera is turning when the car is 40 m from the finish line. What is the rate in degrees per second?

Solution   In Figure 4.63 we have labeled the relevant variables x and θ. We will measure x in meters and θ in radians. Note carefully that the distance from the car to the finish line is not shown as 40 because this distance is a *variable* quantity. We will put $x = 40$ only *after* we differentiate.

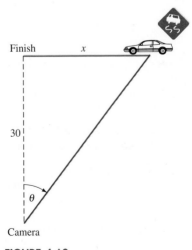

FIGURE 4.63

We are given that the car is approaching the finish line at 150 km/hr. Since other units are in meters and we want the answer to be in radians per second, we convert 150 km/hr to meters per second:

$$150 \text{ km/hr} = \frac{150 \text{ km} \times 1000 \text{ m/km}}{1 \text{ hr} \times 3600 \text{ s/hr}}$$

$$= \frac{125}{3} \text{ m/s}$$

Since x decreases with time, dx/dt must be negative. So

$$\frac{dx}{dt} = -\frac{125}{3}$$

An equation relating x and θ is

$$\tan \theta = \frac{x}{30}$$

We differentiate both sides with respect to time, recalling that $\frac{d}{d\theta} \tan \theta = \sec^2 \theta$:

$$\sec^2 \theta \frac{d\theta}{dt} = \frac{1}{30} \frac{dx}{dt} = \frac{1}{30}\left(-\frac{125}{3}\right) = -\frac{25}{18}$$

So

$$\frac{d\theta}{dt} = \frac{1}{\sec^2 \theta}\left(-\frac{25}{18}\right) = -\frac{25}{18}\cos^2 \theta$$

Now we are ready to consider the situation when $x = 40$. At that instant, the right triangle is as shown in Figure 4.64, so the hypotenuse is 50. Thus, $\cos \theta = 3/5$. Therefore, we have

$$\frac{d\theta}{dt}\bigg|_{x=40} = -\frac{25}{18}\left(\frac{3}{5}\right)^2 = -\frac{25}{18}\cdot\frac{9}{25} = -\frac{1}{2} \text{ radian/s}$$

The significance of the negative sign is that θ decreases with time. Since 1 radian $= (180/\pi)°$, or approximately $57.3°$, the camera is turning at about $-28.6°/s$, which would lead to one revolution in about $12\frac{1}{2}$ seconds, and this rate seems reasonable. (If the rate of turning had been several revolutions per second, we would have rejected it as being unreasonable and rechecked.) ∎

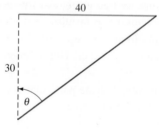

FIGURE 4.64

Exercise Set 4.4

1. Two joggers leave simultaneously from the same point, one going east at 3 m/s and the other going south at 4 m/s. Find how fast the distance between them is increasing 30 min later.

2. The commander of a Coast Guard cutter is notified from an aircraft that a boat suspected of carrying illegal cargo is headed north in the Gulf of Mexico at the rate of 15 knots (1 knot is 1 nautical mile per hour). By heading east and traveling at 36 knots, the commander estimates that the cutter can intercept the boat in 50 min. Find how fast the two boats are approaching each other after 30 min.

3. A child 5 ft tall is walking away from a lamppost at the rate of 3 ft/s. If the height of the lamppost is 20 ft, find the rate at which the child's shadow is lengthening.

4. In Exercise 3, find the rate at which the tip of the child's shadow is moving.

5. A child throws a rock into a pond, causing a circular ripple. If the radius of the circle increases at the constant rate of 1 m/s, find how fast the area is increasing when the radius is 5 m.

6. A V-shaped trough has a cross section in the form of an inverted isosceles triangle with base 40 cm and altitude 50 cm. If the trough is 3 m long and is being filled with water at the rate of 200 L/min ($1 \text{ L} = 10^{-3} \text{ m}^3$), find how fast the water level is rising when the water level is 30 cm.

7. A lighthouse is 500 m from a straight shore. If its light is revolving at the rate of 10 rev/min, find how fast the beam of light is moving along the shoreline when the angle between the light beam and the shoreline is 30°.

8. Helium is being pumped into a spherical balloon at the rate of 600 cm³/s. At what rate is the radius increasing when the radius is 20 cm? At what rate is the surface area of the balloon increasing at this same instant?

9. Two runners are 7 m apart at the starting line, and they run on parallel paths, one averaging 8 m/s and the other

7.6 m/s. At what rate is the distance between the runners increasing after 1 min?

10. Coal is being transported on a conveyor to conical piles each of whose height is one-fourth the diameter of the base. If the coal is being dropped at the rate of 3m³/min, find how fast the height of a pile is increasing when it is 10 m high. How fast is the circumference of the base increasing at this same instant?

11. A watering trough 10 ft long has a cross section in the form of an isosceles trapezoid with dimensions as shown in the figure. If water is being pumped into the tank at the rate of 8 ft³/min, find how fast the water level is rising when it is 1 ft deep.

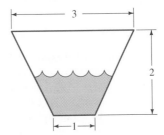

12. In Exercise 11, find how fast the water level is rising when the trough is half full.

13. A boat is being pulled in by means of a rope attached to a windlass on the dock. The windlass is 4.5 m above the level at which the rope is attached to the boat, and the rope is being pulled in at the rate of 2 m/s. Find how fast the boat is approaching the base of the dock when it is 6 m from the base.

14. A helicopter is 500 m directly above an observer on the ground at a given instant and is flying horizontally at the rate of 50 m/s. Find how fast the angle at the observer from the vertical to the line of sight of the helicopter is changing after 10 seconds (see the figure on the following page).

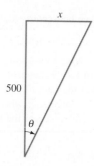

15. When a gas under pressure expands or contracts with no gain or loss of heat, the relationship between pressure p and volume v is given by $pv^k = C$, where k and C are constants. This relation is known as the *adiabatic law*. Suppose for a certain gas $k = 1.4$ and at a given instant the pressure is 2 N/m² (newtons per square meter), the volume is 0.064 m³, and the pressure is increasing at the rate of 0.5 N/m² each second. Find the rate of change of the volume at this instant.

16. An oil spill from a drilling platform in the Gulf of Mexico is approximately circular, and the diameter is observed to be increasing at the constant rate of 100 m per day. How fast is the area of the spill increasing when the diameter is 500 m?

17. Newton's Universal Law of Gravitation states that two bodies of masses m_1 and m_2, at a distance r units apart, exert a force of attraction on each other given by the formula

$$F = G\frac{m_1 m_2}{r^2}$$

where G is the universal gravitational constant. Suppose that for two bodies of fixed masses m_1 and m_2, respectively, the force of attraction is 20 N (newtons) when they are 10 m apart, and at that instant they are separating at the rate of 5 m/s. How fast is the force of attraction decreasing?

18. A small airplane is flying 120 mi/hr at an altitude of 5000 ft. An observer on the ground is keeping a transit (a surveyor's instrument that measures angles) trained on the plane. Through how many radians per second must the observer rotate the scope of the transit when the plane is flying away from the observer and the angle of elevation of the plane is 60°? Neglect the height of the transit. (Be careful with units.)

19. Some children are rolling a ball of snow to make a snowman. If the radius of the ball is increasing at the rate of 20 cm/min, how fast is the volume of the ball increasing when it is 60 cm in diameter?

20. A searchlight located 30 m from a straight road is kept trained on a car traveling on the road at 50 km/hr.

Through how many radians per second is the searchlight turning when the car is 20 m past the point on the road nearest the searchlight?

21. A child flying a kite is letting string out at the rate of 2 m/s, and the kite is 30 m high (from hand level) moving horizontally. Find the speed of the kite at the instant when 34 m of string are out.

22. The buildup of mineral deposits in a water pipe in an area of hard water gradually restricts the rate of flow. Assume that the buildup forms a uniform circular ring around the inside of a pipe (see the figure), and suppose that in a pipe of inside diameter 1.2 cm the area of the annular ring increases at the rate of 10π mm² per year. At what rate is the effective diameter of the pipe decreasing when the thickness of the deposit is 2 mm?

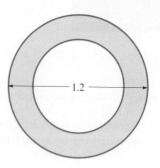

23. The larvae of a certain insect have the shape of a right circular cylinder with hemispheres at both ends and with length five times the diameter (see the figure). During the first stages of growth, the larvae increase in diameter at the rate of 0.2 mm per day. At what rate does the volume increase when the diameter is 3 mm?

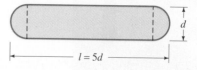

24. A water tank in the form of an inverted cone of height 6 m and base radius 3 m is being drained at the rate of 2 m³/min. How fast is the water level falling when there are $(2\pi)/3$ m³ of water in the tank?

25. In a certain mechanical system, parts A and B are connected by a cable 84 cm long that passes over a pulley at C 30 cm above the point D on a horizontal track on which parts A and B are constrained to move (see the figure). The parts move back and forth in such a way that the cable is always taut. At a certain instant, part B

is 16 cm from D and is moving to the right at 17 cm/s. How fast is part A moving at that instant?

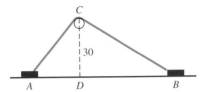

26. One end of an 85-ft rope is attached to a weight on the ground. The rope goes over a pulley 45 ft above the ground, with the other end held by a worker standing beside the weight. The end of the rope held by the worker is 5 ft above the ground, and continuing to hold it at this height, the worker begins to walk away at the constant rate of 5 ft/s, thereby raising the weight. How fast is the weight rising when the worker has walked 30 ft?

27. An oil tank is in the form of a right circular cylinder with a horizontal axis, having length 3 m and radius 1 m. If the tank is being filled at the rate of 60 L/min, find the rate at which the oil depth is increasing at the instant when the depth is 20 cm.

28. Cars A and B leave from the same point, traveling on roads that are $120°$ apart. Car A travels at an average speed of 60 km/hr. Car B leaves the intersection one-half hour after car A and travels at an average speed of 45 km/hr. How fast are the cars separating when car A has traveled 150 km?

29. A water tank is in the form of a frustum of a cone (portion of a cone between two parallel planes; see the figure) with upper radius 3 m, lower radius 1 m, and height 4 m. Water is being pumped into the tank at the rate of 5 m³/min and is being taken from the

tank at the rate of 2 m³/min. Find the rate at which the water level is rising when the tank contains 30 m³ of water.

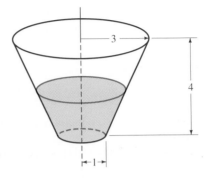

30. A train passes over the middle of a bridge at the same time a car goes under the middle on a path at right angles to the bridge. The bridge is 24 m above the road. If the train's speed is 10 m/s and the car's speed is 15 m/s, find the rate at which the distance between them is increasing 4 s later.

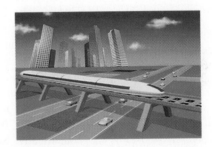

4.5 APPROXIMATING ROOTS OF EQUATIONS

In this section we show how to use derivatives to find approximate roots of equations. If an equation is of the form $f(x) = 0$, then $x = r$ is a **root** of the equation provided $f(r) = 0$. So a root of the equation is a zero of the function f, or equivalently, an x-intercept of its graph. One approach to solving the equation $f(x) = 0$, then, is to draw the graph of the function f and estimate the values where it crosses the x-axis. Obtaining the graph with the aid of a CAS or graphing calculator and zooming in on the x-intercepts repeatedly is one way of approximating roots of equations. The method we will describe is often more efficient and will yield answers to a high degree of accuracy in just a few steps in most cases.

Of course, there are equations for which exact solutions can be found. For example, if f is a polynomial function of degree 1 or 2, we can obtain exact roots. For the first-degree equation $ax + b = 0$, the solution is simply $x = -b/a$, and for the second-degree equation $ax^2 + bx + c = 0$, the solutions are given by the quadratic formula

$$x = \frac{-b \pm \sqrt{b^2 - 4ac}}{2a}$$

Just as the quadratic formula works on every second-degree polynomial equation, formulas exist for solving third- and fourth-degree equations. These formulas are seldom used, however, as they are rather complicated. Surprisingly, though, no such formulas exist for polynomial equations of degree 5 or higher. It is not just that no one has been clever enough to find such formulas; they do not exist! This amazing result was proved in 1824 by the brilliant young Norwegian mathematician Niels Henrik Abel (1802–1829). So, for example, if you wanted to solve the equation

$$x^5 - 3x^2 + 4x - 7 = 0$$

you would have to use some approximating technique. (It should be noted, however, that some fifth-degree, or higher-degree, equations have rational roots, which can be obtained by methods you may have studied. For example, you can verify that the roots of $x^5 + 3x^4 - 5x^3 - 15x^2 + 4x + 12 = 0$ are $x = \pm 1$, ± 2, and -3.) For nonpolynomial functions, the situation is likely to be even worse, as only very special equations can be solved in an exact way. Even the simple equation $\tan x - x = 0$ cannot be solved in an exact form for any root other than the obvious one, $x = 0$.

The basic idea of the method we describe originated with Newton, and it is generally referred to as **Newton's Method**. However, the form in which it is presently used was given by Joseph Raphson (1648–1715), so it is also known as the **Newton–Raphson Method**. The method is easy to describe geometrically. Suppose that by graphical means or otherwise, we find that the equation $f(x) = 0$ has at least one real root. Let r be the true value of one such root. We assume that f is differentiable on an interval I containing r. We make an initial guess at the value of r and denote this first guess by x_0. Sketching the graph of f or making a table of values is usually sufficient to provide a reasonable guess for x_0. Unless we are unusually lucky, x_0 will not equal r. So we locate the point $(x_0, f(x_0))$ and draw the tangent line to the graph of f there. As in Figure 4.65 we let x_1 be the x-intercept of this tangent line. This value, x_1, is our next approximation to r. In general, if x_0 is close enough to r, x_1 will be closer to r than x_0. (If this is not so, try another x_0 as the starting point.) Now we repeat the process. We draw the tangent line to the graph of f at $(x_1, f(x_1))$ and obtain the next approximation to r from its x-intercept, which we call x_2. We continue this process until we obtain the desired degree of accuracy.

To develop a formula for x_{n+1}, once we know x_n, we need to find the equation of the tangent line at $(x_n, f(x_n))$. Its slope is $f'(x_n)$. So by the point-slope form of the equation of a line, its equation is

$$y - f(x_n) = f'(x_n)(x - x_n)$$

To get the x-intercept, we set $y = 0$:

$$-f(x_n) = f'(x_n)(x - x_n)$$

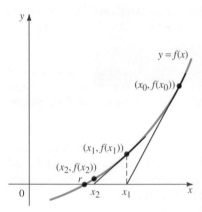

FIGURE 4.65

We solve for x and name the result x_{n+1}:

$$x_{n+1} = x_n - \frac{f(x_n)}{f'(x_n)} \qquad (4.6)$$

Of course, $f'(x_n)$ must be different from 0. The successive approximations $x_0, x_1, x_2, \ldots$ form what is called a **sequence**, and the approximations are called **terms** of the sequence. Equation 4.6, giving the term x_{n+1} when x_n is known, is called a **recursive formula** for the sequence.

Newton's Method (The Newton–Raphson Method)

1. Make an initial guess x_0 for a root r of the equation $f(x) = 0$.
2. Find $x_1, x_2, x_3, \ldots$ using

$$x_{n+1} = x_n - \frac{f(x_n)}{f'(x_n)} \qquad (f'(x_n) \neq 0)$$

by taking $n = 0, 1, 2, 3, \ldots$

If the terms of the sequence of approximations come arbitrarily close to the root r as n increases, we say the sequence **converges** to r. In general, the sequence of approximations will converge to the root r you are seeking if your initial choice x_0 is close to r. There are various tests to guarantee convergence, and we will mention one of these in a moment, but in practice it is usually sufficient to try a starting point x_0 and then to observe whether subsequent approximations seem to be approaching a constant value. We illustrate this approach in the next two examples.

EXAMPLE 4.26   Approximate the positive root of $x^2 - 2x - 2 = 0$ using Newton's Method. Continue the process until two consecutive approximations agree to the first five decimal places, and round the answer to four places. Compare with the exact answer.

Solution   We show the graph of $f(x) = x^2 - 2x - 2$ in Figure 4.66. Since $f(2) = -2$ and $f(3) = 1$, the root we are seeking is between 2 and 3, and from the graph it appears to be closer to 3. A good first approximation might therefore be $x_0 = 2.6$. To apply Newton's Method we need f':

$$f'(x) = 2x - 2 = 2(x - 1)$$

Now we apply the recursion formula (4.6), using a calculator:

$$n = 0: \quad x_1 = x_0 - \frac{f(x_0)}{f'(x_0)} = 2.6 - \frac{f(2.6)}{f'(2.6)} = 2.7375$$

$$n = 1: \quad x_2 = x_1 - \frac{f(x_1)}{f'(x_1)} = 2.72918$$

$$n = 2: \quad x_3 = x_2 - \frac{f(x_2)}{f'(x_2)} = 2.73205$$

$$n = 3: \quad x_4 = x_3 - \frac{f(x_3)}{f'(x_3)} = 2.7320508$$

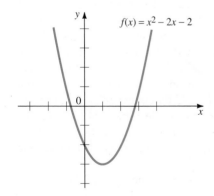

$f(x) = x^2 - 2x - 2$

FIGURE 4.66

Since x_3 and x_4 agree to the first five decimal places, we stop here and say that $r \approx 2.7321$. The fact that x_3 and x_4 agree to five places does not guarantee that the answer is correct to four places, but it strongly suggests that is true.

To compare with the exact answer, we solve for x by the quadratic formula, getting

$$x = \frac{2 \pm \sqrt{4 + 8}}{2} = 1 \pm \sqrt{3}$$

The positive root is $1 + \sqrt{3}$, which to seven places is 2.7320508. Note that x_4 agrees with this answer to all seven places, so four iterations of Newton's Method yielded very good results.   ∎

EXAMPLE 4.27   Use Newton's Method to approximate the smallest positive root of the equation $\tan x - x = 0$.

Solution   Let $f(x) = \tan x - x$. We will need f':

$$f'(x) = \sec^2 x - 1 = \tan^2 x$$

Rather than graphing f, it is easier to graph $y_1 = \tan x$ and $y_2 = x$ and estimate where $y_1 = y_2$, as we have done in Figure 4.67. It appears that the root r we are seeking is close to $3\pi/2 \approx 4.71$. We try $x_0 = 4.5$ as a first approximation of r, and proceed with Newton's Method. The recursion formula is

$$x_{n+1} = x_n - \frac{f(x_n)}{f'(x_n)} = x_n - \frac{\tan x_n - x_n}{\tan^2 x_n}$$

This time we use a table:

n	x_n	x_{n+1}
0	4.5	4.4936
1	4.4936	4.49341
2	4.49341	4.49341

So in just three iterations it appears that we have five-place accuracy and can say that $r \approx 4.49341$.   ∎

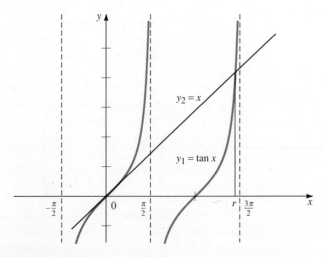

FIGURE 4.67

REMARK ———————————————————————————

■ It was essential in the preceding example that we have the calculator set in the *radian* mode.

——

Pitfalls of Newton's Method

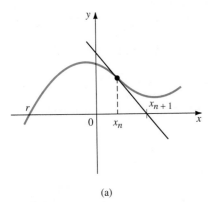

FIGURE 4.68

We turn our attention now to ways in which Newton's Method can fail. We know already that if $f'(x_n) = 0$ for some n, then the method fails. In such a case we try another starting point. As shown in Figure 4.68, if for some n, x_n is simply near a point where the derivative is 0, it can be sufficient to cause all subsequent approximations to be farther from r than x_n because of the flatness of the slope of the tangent line at $(x_n, f(x_n))$. Some other ways in which the process may go awry are shown in Figure 4.69. In part (a) the approximations again get farther and farther from r. In part (b) the process is in an endless loop, and in part (c) Newton's Method does converge, but to a different root from the one we expect.

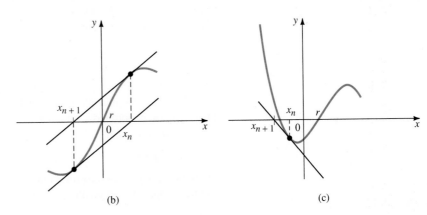

(a)   (b)   (c)

FIGURE 4.69

Criteria for Convergence

If there is an open interval containing the root r that we are trying to approximate throughout which f is monotone and the concavity is always in the same direction, then r is the only root in the interval. Furthermore, if we choose x_0 such that $f(x_0)$ is positive when the graph is concave up and negative when the graph is concave down, then it can be proved that the successive approximations in Newton's Method always converge to r. Figure 4.70 illustrates such a choice of x_0 for each type of concavity for both increasing and decreasing functions. Notice that in each case as you proceed along the curve from the point $(x_0, f(x_0))$ toward $(r, 0)$, the curve bends toward the x-axis. We emphasize that the conditions stated here for convergence are *sufficient* only. So even if these conditions are not met, the approximations may (and usually will) converge. That is, if the conditions are satisfied, then approximations produced by Newton's Method definitely will converge to the root, but it is possible they will converge to the root in the absence of one or more of the conditions stated.

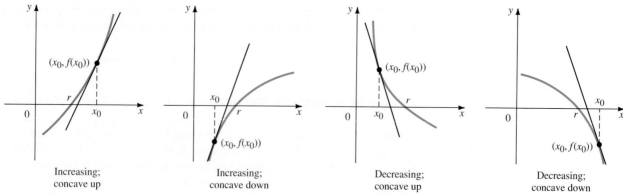

Increasing;
concave up

Increasing;
concave down

Decreasing;
concave up

Decreasing;
concave down

FIGURE 4.70

Suppose that the conditions stated in the preceding paragraph are met. If there are positive constants m and M such that $|f'(x)| > m$ in the interval (the slope is not too small) and $|f''(x)| \le M$ in the interval (the concavity is not too great), then the error in x_{n+1} is related to the error in x_n by the inequality

$$E_{n+1} \le \frac{M}{2m} E_n^2 \tag{4.7}$$

where $E_n = |x_n - r|$ and $E_{n+1} = |x_{n+1} - r|$. We will see how to prove this result in Chapter 10. One consequence of Inequality 4.7 is that when $M/2m < 2$, if x_n approximates r correctly to k decimal places, then x_{n+1} approximates r correctly to at least $2k$ decimal places. (See Exercise 29 in Exercise Set 4.5.) That is, the precision is doubled with each successive approximation, which explains the rapidity of convergence of Newton's Method.

EXAMPLE 4.28   Find a starting point x_0 for Newton's Method that will guarantee convergence to the largest positive root of $x^3 - 3x^2 - 24x + 48 = 0$. Find an upper bound on the error in using x_3 to approximate the root.

Solution   Let $f(x) = x^3 - 3x^2 - 24x + 48$. Then

$$f'(x) = 3x^2 - 6x - 24 = 3(x + 2)(x - 4)$$
$$f''(x) = 6x - 6 = 6(x - 1)$$

We analyze the signs of $f'(x)$ and determine that f is increasing on $(4, \infty)$ and on $(-\infty, -2)$, and f is decreasing on $(-2, 4)$. From the sign of $f''(x)$, we see that the graph is concave up for $x > 1$ and concave down for $x < 1$. A sketch of the graph of f is given in Figure 4.71. By calculator, we find that $f(5) = -22$ and $f(6) = 12$. So a root lies in the interval $I = (5, 6)$. No root can be to the right of this one since f is increasing and concave up everywhere to the right of $x = 5$. According to the convergence criteria we have stated, we should choose x_0 so that $f(x_0) > 0$, since the concavity is upward in the interval $(5, 6)$ containing the root. Checking $f(5.8)$, we find its value to be 2.992. So $x_0 = 5.8$ is a suitable starting point. The table shows the calculated values of x_1, x_2, and x_3.

Now we wish to estimate the error in the approximation to r given by x_3, which is the last entry in the table. To use Inequality 4.7, we need positive constants m and M such that for x in the original interval $I = (5, 6)$, $|f'(x)| \ge m$ and $|f''(x)| \le M$. Both f' and f'' are increasing and nonnegative in I. So the

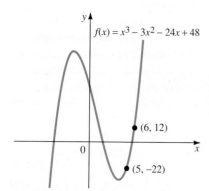

$f(x) = x^3 - 3x^2 - 24x + 48$

(6, 12)

(5, −22)

FIGURE 4.71

n	x_n	x_{n+1}
0	5.8	5.729
1	5.729	5.72716
2	5.72716	5.72716014

minimum value of $f'(x)$ is $f'(5) = 21$, and the maximum value of $f''(x)$ is $f''(6) = 30$. Thus, Inequality 4.7 becomes

$$E_{n+1} \leq \frac{30}{2(21)} E_n^2 = \frac{5}{7} E_n^2$$

We need E_0 to begin with. To get as small a bound on the error as we can, we should make E_0 small. We know that $f(5.8) > 0$. Checking 5.7, we find $f(5.7) < 0$. So the root lies between 5.7 and 5.8, which means that $x_0 = 5.8$ is less than 0.1 unit from the root. That is, $E_0 \leq 0.1$. Thus,

$$E_1 \leq \frac{5}{7}(0.1)^2 \approx 0.00714$$

$$E_2 \leq \frac{5}{7}(0.00714)^2 \approx 0.0000364$$

$$E_3 \leq \frac{5}{7}(0.0000364)^2 \approx 0.000000000946$$

Clearly, the answer we have found for x_3 is accurate to at least eight decimal places. ∎

We will mention briefly two other methods for approximating roots of equations. Although these methods do not involve the derivative, they are easily understood and simple to apply. Furthermore, the first method provides an alternative way of estimating the error in Newton's Method.

The Method of Secants

Suppose the interval I contains only one root r of the equation $f(x) = 0$. For example, this condition will be met, as we have previously seen, when f is monotone in I and the direction of its concavity does not change there. Choose as a starting point, x_0, any point in I such that $f(x_0) > 0$ when the concavity is upward and $f(x_0) < 0$ when the concavity is downward. In Figure 4.72, we illustrate such a choice for a graph that is concave up. Now choose any other point z_0 in I such that $f(z_0)$ is opposite in sign to $f(x_0)$. Then, by the Intermediate-Value Theorem we know that the function takes on the value 0 somewhere between z_0 and x_0; that is, the root r lies between z_0 and x_0. As shown in Figure 4.72 we now draw the *secant line* joining the points $(x_0, f(x_0))$ and $(z_0, f(z_0))$. Its x-intercept is the next approximation to the root. Call it z_1. Then, let z_2 be the x-intercept of the secant line joining $(x_0, f(x_0))$ and $(z_1, f(z_1))$. Continue this process in such a way that z_{n+1} is the x intercept of the secant line joining $(x_0, f(x_0))$ and $(z_n, f(z_n))$ for $n = 0, 1, 2, 3, \ldots$ In Exercise 30 of this section you will be asked to derive the formula

$$z_{n+1} = z_n - \frac{f(z_n)}{m_n} \tag{4.8}$$

where m_n is the slope of the secant line joining $(x_0, f(x_0))$ and $(z_n, f(z_n))$. The method we have described is called the **Method of Secants**, for obvious reasons.

It is evident geometrically (see Figure 4.73) that at the nth stage of the Method of Secants, the point z_n and the point x_n at the nth stage of Newton's Method are on opposite sides of the root r. It follows, then, that the error $E_n = |x_n - r|$ at the nth stage of Newton's Method does not exceed the distance between x_n and z_n: $E_n \leq |x_n - z_n|$.

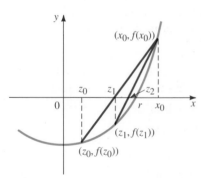

FIGURE 4.72

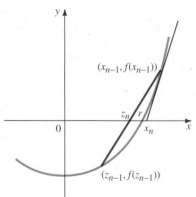

FIGURE 4.73

The Bisection Method

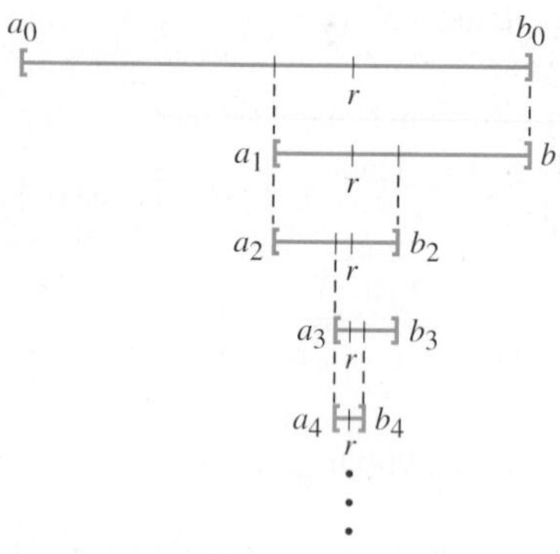

The other method we describe is called the **Bisection Method**. Assume again that f has only one root r in an interval I throughout which f is continuous. Let a_0 and b_0 (with $a_0 < b_0$) be two points in I such that $f(a_0)$ and $f(b_0)$ are opposite in sign so that the root r lies in the interval $[a_0, b_0]$. What we are going to do is bisect the interval $[a_0, b_0]$ and choose the half containing the root r. Then we will bisect that half and again choose the half containing r. By continuing in this manner, we get a sequence (called a *nest*) of intervals, each half as long as the preceding one, and each containing the root r. Since each interval is half as long as the previous one, the lengths become arbitrarily small as we continue the process indefinitely. Thus we can obtain an approximation to the root to whatever degree of accuracy we choose. In Figure 4.74 we show a possible sequence of intervals such as we have described. We designate the half of $[a_0, b_0]$ containing r by $[a_1, b_1]$, and we show it below $[a_0, b_0]$, although it actually coincides with the right half of $[a_0, b_0]$. Similarly, $[a_2, b_2]$ is the half of $[a_1, b_1]$ containing r, and so on. If we stop with the interval $[a_n, b_n]$, then we use its midpoint

$$M_n = \frac{a_n + b_n}{2}$$

as the approximation to r.

Since the width of each new interval in the Bisection Method is half of the preceding one, the width of the nth interval is $(b_0 - a_0)/2^n$. The root r lies in the nth interval. So the distance from the midpoint M_n to r is less than half the width of the interval (see Figure 4.75). Thus, the error, E_n, in stopping at the nth stage satisfies

$$E_n \le \frac{b_0 - a_0}{2^{n+1}} \tag{4.9}$$

FIGURE 4.74

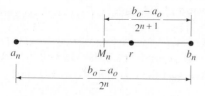

FIGURE 4.75

REMARK ——————————————————————————

■ It may happen at some stage that the midpoint of some interval is the exact root r. If so, we stop the process at that stage.

———————————————————————————————

The Bisection Method is easy to understand and is easy to program on a computer. The convergence is quite slow, however, compared with Newton's Method or the Method of Secants.

EXAMPLE 4.29   Approximate the largest positive root of $f(x) = 0$, where

$$f(x) = x^3 - 3x^2 - 24x + 48$$

Use (a) the Method of Secants and (b) the Bisection Method. Compare the results with the answer obtained by Newton's Method in Example 4.28.

Solution   In Example 4.28 we found that the largest positive root was in the interval $(5, 6)$ and that $f(5) = -22$ and $f(6) = 12$.

(a) By the Method of Secants we can take $x_0 = 6$ and $z_0 = 5$. The slope of the secant line joining the two points $(5, -22)$ and $(6, 12)$ is

$$m_0 = \frac{12 - (-22)}{6 - 5} = 34$$

So by the recursion formula (4.8)

$$z_1 = 5 - \frac{-22}{34} = 5.647$$

We calculate $f(z_1)$ to be

$$f(5.647) = -3.119$$

The slope of the secant line joining $(z_1, f(z_1))$ and $(x_0, f(x_0))$ is

$$m_1 = \frac{12 + 3.119}{6 - 5.647} = 43.376$$

By the recursion formula (4.8), we find that

$$z_2 = 5.647 - \frac{-3.119}{43.376} = 5.7198$$

Continuing in this way, we find (omitting the details) that

$$z_3 = 5.7264 \qquad z_4 = 5.727095 \qquad z_5 = 5.727154$$

(b) By the Bisection Method, we take the initial interval $[a_0, b_0]$ to be $[5, 6]$. Its midpoint is 5.5. We know already that $f(6)$ is positive, and we find that $f(5.5)$ is negative. So we take $[a_1, b_1]$ to be $[5.5, 6]$. The value of f at the midpoint, 5.75, is found to be positive, so the root lies in $[5.5, 5.75]$, which we call $[a_2, b_2]$. The midpoint of $[a_2, b_2]$ is 5.625, and $f(5.625)$ is negative, so we take $[a_3, b_3]$ to be $[5.625, 5.75]$. Continuing in this way, we find after 14 bisections that the root lies in the interval $[5.727155, 5.727185]$. If we stop here, we take the midpoint M_{14} as the approximation to r:

$$M_{14} = 5.72717$$

The error, according to the estimate (4.9), is

$$E_{14} \leq \frac{6 - 5}{2^{15}} = 0.0000305$$

Thus, M_{14} is accurate to at least four decimal places. We round off this estimate to 5.7272, to indicate this level of accuracy.   ■

In Example 4.28, we found that after only three iterations by Newton's Method we had accuracy to at least eight decimal places. In fact, the accuracy after only two iterations was to four places. By contrast, we had to carry the Method of Secants through five iterations to get four-place accuracy, and we had to go through 14 bisections in the Bisection Method to obtain the same accuracy. Newton's Method would therefore seem to be the one to use. Aside from the pitfalls of this method that we already discussed, however, it requires that the function be differentiable, whereas the other two do not. So there are times when Newton's Method simply will not work. The Bisection Method is the least restrictive of the three, so despite the relative slowness of its convergence, it may be the best of the three choices in certain cases.

REMARK ——————————————————————————————————

■ We should note that there are still other methods for approximating roots of equations, some of which are studied in numerical analysis courses.

APPROXIMATING ROOTS USING A COMPUTER ALGEBRA SYSTEM—NEWTON'S METHOD

🖳 CAS 18

Use Newton's Method to approximate a root of the function $f(x) = x^3 - x - 1$. If we first plot the function we can determine where a root lies and then choose an appropriate starting value for Newton's Method.

Maple:

```
f: = x->x^3–2*x–1;
plot(f(x),x=–3..3,y=–5..5);
```

The plot suggests there is a root in the interval [1, 2]. We will start Newton's Method with a value of $x = 2$. First we define a function we call newton.

```
newton: = x->x–f(x)/D(f)(x);
```

Now start the algorithm.

```
evalf(newton(2));
```

The output is 1.7. Now iterate by placing the previous value into newton, using the ditto mark.

```
evalf(newton("));    1.623088456
evalf(newton("));    1.618055040
evalf(newton("));    1.618033989
evalf(newton("));    1.618033989
```

When the values stabilize the computation is complete. Calculating these 6 terms could have been done quicker by entering:

```
x:=2;
for i from 1 by 1 to 6 do
evalf(newton(x));
x:=";
od;
```

The end of a loop that begins with "do" is marked by the reverse letters, "od."

Mathematica:

```
f[x_] = x^3–2*x–1
Plot [f[x],{x,–3, 3}]
newton[x_] = x – f[x]/f'[x]
newton[2]
```

Output:

$$\frac{17}{10}$$

For decimal output:

```
newton[N[2,20]]
```

Now iterate:

```
newton[%]
newton[%]
newton[%]
newton[%]
newton[%]
```

To calculate all six terms:

```
NestList[newton,N[2,20],7]
```

DERIVE:*

(At the □ symbol, go to the next step.)
a (author) □ f(x): =x^3–2x–1 □ p (plot window) □ p (plot) □ a (author) □ newton(x) := x–f(x)/dif(f(x),x) □ a (author) □ iterates(newton(x),x,2,6) □ x (approximate) □ [choose expression]

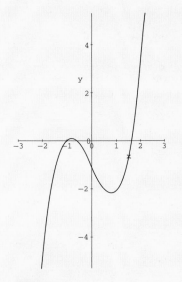

FIGURE 4.5.1
$y = f(x) = x^3 - 2x - 1$

---

*Note that in DERIVE, dif(f(x),x) can be used for the derivative of the function f with respect to x.

Exercise Set 4.5

In Exercises 1–8, approximate each root using Newton's Method, continuing until two successive approximations agree to the first five decimal places. Round the answer to four places.

1. The positive root of $2x^2 - 4x - 1 = 0$
Compare with the exact answer.

2. The real root of $x^3 - 4x^2 + 7x - 5 = 0$

3. The negative root of $x^4 - 2x^3 - 2 = 0$

4. The positive root of $2 \sin x - x = 0$

5. The real root of $e^x + x = 2$

6. The smallest positive root of $\ln x - x + 2 = 0$

7. Both real roots of $2x^4 - 3x^3 - 2x + 2 = 0$

8. The positive root of $x^2 - \cos x = 0$

In Exercises 9–12, approximate the root to four decimal places of accuracy by writing an equation that has the given root as a solution, and using Newton's Method.

9. $\sqrt{2}$ **10.** $\sqrt[3]{4}$

11. $\sqrt[3]{-9}$ **12.** $\sqrt[4]{6}$

In Exercises 13–16, without the aid of a calculator or CAS, find an interval I containing the indicated root r of $f(x) = 0$ such that f is monotone in I and the direction of the concavity of the graph of f does not change in I. Then find a starting point x_0 in I such that $f(x_0) > 0$ if the concavity is upward and $f(x_0) < 0$ if the concavity is downward. Sketch the graph of f for x in I.

13. $f(x) = x^3 - 5x^2 - 8x - 3; r > 0$

14. $f(x) = x^4 - 8x^2 - 3; r > 0$

15. $f(x) = 3(x - 1)^{2/3} - x; 1 < r < 9$

16. $f(x) = 2 \cos x + x - 2; 0 < r < \pi$

In Exercises 17–20, show that the interval I contains a root of $f(x) = 0$ and that I and the starting point x_0 satisfy the conditions stated in the instructions for Exercises 13–16. Find the approximation x_3 in Newton's Method, and use Inequality 4.7 to obtain a bound on the error.

17. $f(x) = 2x^3 - 3x^2 + 4x - 8; I = (1, 2), x_0 = 1.8$

18. $f(x) = x^4 - 2x^3 - 3x - 8; I = (-2, -1), x_0 = -1.2$

19. $f(x) = 2 \sin x - \dfrac{x^2}{2}; I = (\pi/2, \pi), x_0 = 2.1$

20. $f(x) = x - \cos x; I = (0, \pi/2), x_0 = 0.9$

In Exercises 21–24, approximate the root in each of Exercises 17–20 by the Method of Secants with $n = 3$. Begin by choosing z_0 in I such that $f(z_0)$ and $f(x_0)$ are opposite in sign. Compare your answer with x_3 as obtained by Newton's Method. Which answer is more accurate?

In Exercises 25 and 26, use the Method of Bisection to approximate the real root correct to two decimal places.

25. $x^3 - 3x^2 + 3x - 4 = 0$

26. $xe^{-x} + 1 = 0$ (Use your calculator to evaluate e^{-x}.)

27. Find all local maximum and minimum points correct to four decimal places, by Newton's Method, on the graph of the function f defined by

$$f(x) = 3x^4 - 8x^3 - 5x^2 + 4x - 20$$

28. How many real zeros does the function f of Exercise 27 have? Find them, using Newton's Method, correct to four decimal places. Draw the graph of f.

29. Prove that if f satisfies the conditions for Inequality 4.7 and $M < 4m$, then in Newton's Method whenever x_n is correct to k decimal places, x_{n+1} is correct to $2k$ decimal places. (*Hint:* The approximation x_n is correct to k places if $E_n \leq 5 \times 10^{-(k+1)}$.)

30. For the Method of Secants, derive the recursion relation

$$z_{n+1} = z_n - \frac{f(z_n)}{m_n}$$

where

$$m_n = \frac{f(z_n) - f(x_0)}{z_n - x_0}$$

4.6 L'HÔPITAL'S RULE

Sometimes the limit properties we discussed in Chapter 1 are not applicable. For example, in a quotient both numerator and denominator might approach 0, or they might both become infinite. In this section we show how derivatives can often be used to find limits of such functions. Actually, we have already encountered some limits of this type. For example,

$$\lim_{x \to 1} \frac{x^2 - 1}{x - 1} \quad \text{and} \quad \lim_{x \to 0} \frac{\sin x}{x}$$

cannot be evaluated by taking limits of the numerator and denominator individually and then dividing. If we did so, in both cases the division would give us the meaningless result $0/0$. Such limits are called "indeterminate" as they stand, since the limit may or may not exist. Some other procedure is needed. In the two examples just cited we were able to determine the limits. In the first case (Chapter 1) we factored the numerator to get

$$\lim_{x \to 1} \frac{x^2 - 1}{x - 1} = \lim_{x \to 1} \frac{(x + 1)(x - 1)}{(x - 1)} = \lim_{x \to 1}(x + 1) = 2$$

(remember that in the limiting process we do not let $x = 1$, so $x - 1 \neq 0$). In the second case we used a geometric argument (the Lemma on p. 166, part (a)) and found that

$$\lim_{x \to 0} \frac{\sin x}{x} = 1$$

These techniques will not work on many other indeterminate limits. For example, the limit

$$\lim_{x \to \pi/4} \frac{\tan x - 1}{4x - \pi}$$

also would assume the form $0/0$ if we tried to take limits of the numerator and denominator individually. Further, it is not clear that any geometric argument or factoring technique would enable us to find the limit or show that it fails to exist. The method we will introduce shortly will enable us to evaluate this limit, as well as many others.

Closely related to quotients in which both numerator and denominator approach 0 is the case where they both become infinite. Again, we have encountered some limits of this type in finding horizontal asymptotes to graphs of rational functions. For example, to find the horizontal asymptote to the graph of

$$f(x) = \frac{x^3 - 2x}{3x^3 + 10x^2 - 5}$$

we need to determine the limit

$$\lim_{x \to \infty} \frac{x^3 - 2x}{3x^3 + 10x^2 - 5}$$

As it stands, this limit is indeterminate because both numerator and denominator become infinite. Recall that we found the limit in problems of this type by using dominant terms in numerator and denominator, getting

$$\lim_{x \to \infty} \frac{x^3 - 2x}{3x^3 + 10x^2 - 5} = \lim_{x \to \infty} \frac{x^3}{3x^3} = \frac{1}{3}$$

There are many other cases, however, where this dominant-term approach will not work. An example is

$$\lim_{x \to (\pi/2)^-} \frac{\sec x}{1 + \tan x}$$

Here again, both numerator and denominator become infinite as $x \to (\pi/2)^-$. We need some new procedure, then, for evaluating the limit.

Indeterminate Forms of Types 0/0 and ∞/∞

In the following definition we make the notion of the two types of indeterminate limits we have discussed more precise.

Definition 4.5 Indeterminate Forms of Type 0/0 and ∞/∞	The limit $$\lim_{x \to a} \frac{f(x)}{g(x)}$$ is said to be an **indeterminate form of type 0/0** if $$\lim_{x \to a} f(x) = 0 \text{ and } \lim_{x \to a} g(x) = 0$$ It is **indeterminate of type ∞/∞** if $$\lim_{x \to a} f(x) = \pm\infty \quad \text{and} \quad \lim_{x \to a} g(x) = \pm\infty$$ In each case, $x \to a$ can be replaced by any of the following: $x \to a^+$, $x \to a^-$, $x \to \infty$, or $x \to -\infty$.

We are now ready to state the main result of this section. It appeared in the first calculus book ever published, written by the French mathematician G. F. L'Hôpital (the Marquis de L'Hôpital, 1661–1704). The result has been known ever since as **L'Hôpital's Rule**. (L'Hôpital is pronounced "lo-pee-tal'"). The result had actually been discovered earlier by his teacher, Johann Bernoulli (1667–1748), who communicated it to L'Hôpital in 1694.

The idea of the rule is very simple. When we have an indeterminate limit of the form 0/0 or ∞/∞ we simply replace the given limit with the limit of the *derivative* of the numerator over the *derivative* of the denominator. For example, we write

$$\lim_{x \to 0} \frac{\sin x}{x} = \lim_{x \to 0} \frac{\cos x}{1}$$

Now the second limit is no longer indeterminate. Since $\cos x \to 1$ as $x \to 0$, the answer is 1. It is not at all obvious why the rule works. We will prove it for a special case (which is a commonly occurring case) shortly, but first we state the rule formally and give some more examples.

THEOREM 4.7

L'Hôpital's Rule

Let

$$\lim_{x \to a} \frac{f(x)}{g(x)}$$

be an indeterminate form of type $\frac{0}{0}$ or $\frac{\infty}{\infty}$. If $f'(x)$ and $g'(x)$ exist and $g'(x) \neq 0$ at all points in some open interval I containing a, except possibly at a itself, then

$$\lim_{x \to a} \frac{f(x)}{g(x)} = \lim_{x \to a} \frac{f'(x)}{g'(x)} \qquad (4.10)$$

provided the limit on the right exists or is ∞ or $-\infty$.

L'Hôpital's Rule is also true when a is an endpoint of the interval I, in which case $x \to a$ is to be replaced by $x \to a^-$ or $x \to a^+$, as appropriate. The rule is also true when a is replaced by ∞ or $-\infty$, provided I is infinite.

Equation 4.10 does *not* say that $f(x)/g(x) = f'(x)/g'(x)$, but rather that their *limits* are the same, when the hypotheses are satisfied. Note also that $f'(x)/g'(x)$ is *not* the derivative of the quotient $f(x)/g(x)$. In other words, do not use the Quotient Rule for differentiation here.

We illustrate the rule with two of the indeterminate forms mentioned earlier.

EXAMPLE 4.30   Use L'Hôpital's Rule to evaluate the following limits.

(a) $\displaystyle \lim_{x \to \pi/4} \frac{\tan x - 1}{4x - \pi}$

(b) $\displaystyle \lim_{x \to (\pi/2)^-} \frac{\sec x}{1 + \tan x}$

Solution

(a) This limit is indeterminate of the form $0/0$, and the hypotheses of L'Hôpital's Rule are satisfied. So we have

$$\lim_{x \to \pi/4} \frac{\tan x - 1}{4x - \pi} = \lim_{x \to \pi/4} \frac{\sec^2 x}{4} = \frac{2}{4} = \frac{1}{2}$$

(b) The limit is indeterminate of the form ∞/∞, and again, the hypotheses of L'Hôpital's Rule are satisfied. Applying the rule, we get

$$\lim_{x \to (\pi/2)^-} \frac{\sec x}{1 + \tan x} = \lim_{x \to (\pi/2)^-} \frac{\sec x \tan x}{\sec^2 x} = \lim_{x \to (\pi/2)^-} \frac{\tan x}{\sec x}$$

$$= \lim_{x \to (\pi/2)^-} \frac{\dfrac{\sin x}{\cos x}}{\dfrac{1}{\cos x}} = \lim_{x \to (\pi/2)^-} (\sin x) = 1$$

Note that we simplified the fraction $\tan x / \sec x$ by changing to sines and cosines before taking the limit. In applying L'Hôpital's Rule you should simplify fractions as much as possible at each stage. ∎

Proof of L'Hôpital's Rule for a Special Case

We give here the proof of a special case of L'Hôpital's Rule for a 0/0 indeterminate form. A proof for the general case is given in Appendix 5. The proof for the form ∞/∞ is more difficult, and we omit it. For the special case, we assume $f(a) = g(a) = 0$ (so that we are dealing with a 0/0 form), and we assume further that f'/g' is a continuous function at $x = a$. This continuity hypothesis implies that $f'(x)$ and $g'(x)$ both exist in some open interval I containing a (including a itself) and that $g'(x) \neq 0$ in I. By the definition of continuity, it also implies that

$$\lim_{x \to a} \frac{f'(x)}{g'(x)} = \frac{f'(a)}{g'(a)}$$

Although the continuity hypothesis on f'/g' is more restrictive than the hypotheses stated in the rule, in actual application of the rule the continuity hypothesis often is satisfied. We proceed now to the proof.

$$\lim_{x \to a} \frac{f(x)}{g(x)} = \lim_{x \to a} \frac{f(x) - f(a)}{g(x) - g(a)} \qquad \text{Since } f(a) = g(a) = 0.$$

$$= \lim_{x \to a} \frac{\dfrac{f(x) - f(a)}{x - a}}{\dfrac{g(x) - g(a)}{x - a}} \qquad \begin{array}{l}\text{Since } x \neq a, \text{ we can divide} \\ \text{numerator and denominator by } x - a.\end{array}$$

$$= \frac{\displaystyle\lim_{x \to a} \frac{f(x) - f(a)}{x - a}}{\displaystyle\lim_{x \to a} \frac{g(x) - g(a)}{x - a}} \qquad \begin{array}{l}\text{Since the limit of the denominator is } g'(a) \neq 0, \\ \text{we can apply the limit property for quotients.}\end{array}$$

$$= \frac{f'(a)}{g'(a)} \qquad \text{By definition of the derivative}$$

$$= \lim_{x \to a} \frac{f'(x)}{g'(x)} \qquad \text{By continuity of } f'/g' \qquad\qquad \blacksquare$$

As the next example shows, we sometimes have to apply L'Hôpital's Rule more than once.

EXAMPLE 4.31   Find the limit

$$\lim_{x \to 0^-} \frac{\cos x - 1}{x^3}$$

or show it does not exist.

Solution   The given limit is indeterminate of the type 0/0, and the hypotheses of L'Hôpital's Rule are satisfied. So we have

$$\lim_{x \to 0^-} \frac{\cos x - 1}{x^3} = \lim_{x \to 0^-} \frac{-\sin x}{3x^2}$$

Now we have another 0/0 indeterminate form. So we apply the rule again.

$$\lim_{x \to 0^-} \frac{-\sin x}{3x^2} = \lim_{x \to 0^-} \frac{-\cos x}{6x}$$

The last limit is not indeterminate, since the numerator approaches 1 and the denominator approaches 0, both through negative values. Thus,

$$\lim_{x \to 0^-} \frac{\cos x - 1}{x^3} = \infty$$ ∎

If we had carelessly applied L'Hôpital's Rule a third time in the preceding example, we would have gotten

$$\lim_{x \to 0^-} \frac{\cos x}{6x} = \lim_{x \to 0^-} \frac{\sin x}{6} = 0$$

But this result is *incorrect*. The problem lies in the fact that the limit on the left is not of the form 0/0, since $\cos x \to 1$ as $x \to 0^-$. Thus, L'Hôpital's Rule does not apply. Remember that L'Hôpital's Rule works only on 0/0 or ∞/∞ indeterminate forms.

The Indeterminate Forms $0 \cdot \infty$ and $\infty - \infty$

Many times an indeterminate limit that does not fit L'Hôpital's Rule can be transformed to one that does. For example,

$$\lim_{x \to \infty} x^3 \sin\left(\frac{1}{x}\right)$$

involves the product of two factors, one of which goes to ∞ (the x^3 factor) and the other to 0 (the $\sin(1/x)$ factor, since $1/x$ goes to 0 and $\sin 0 = 0$). Because one factor becomes arbitrarily large and the other arbitrarily small, it is not clear what, if anything, the product approaches. The limit is thus indeterminate. By rewriting the limit as

$$\lim_{x \to \infty} \frac{\sin(1/x)}{1/x^3}$$

we change it to the form 0/0, and L'Hôpital's Rule can be applied.

More generally, if $\lim_{x \to a} f(x) = 0$ and $\lim_{x \to a} g(x) = \pm\infty$, we say that $\lim_{x \to a} f(x) \cdot g(x)$ is an **indeterminate form of the type** $0 \cdot \infty$. As we just illustrated, to handle this type of indeterminate form, we rewrite the product as a quotient, either

$$f(x) \cdot g(x) = \frac{f(x)}{1/g(x)} \qquad \text{or} \qquad f(x) \cdot g(x) = \frac{g(x)}{1/f(x)}$$

and then see if L'Hôpital's Rule can be applied.

EXAMPLE 4.32   Evaluate the following limits or show that they fail to exist.

(a) $\lim_{x \to 0} x \cot x$

(b) $\lim_{x \to \infty} x^3 \sin\left(\frac{1}{x}\right)$

Solution

(a) As it stands, the limit is indeterminate of the form $0 \cdot \infty$. Since $\cot x = 1/\tan x$, we can write

$$\lim_{x \to 0} x \cot x = \lim_{x \to 0} \frac{x}{\tan x}$$

The last limit is of the form 0/0, so we can apply L'Hôpital's Rule to get

$$\lim_{x \to 0} x \cot x = \lim_{x \to 0} \frac{x}{\tan x} = \lim_{x \to 0} \frac{1}{\sec^2 x} = 1$$

(b) This limit is the one we discussed earlier. We rewrite it as a quotient and apply L'Hôpital's Rule.

$$\lim_{x \to \infty} x^3 \sin\left(\frac{1}{x}\right) = \lim_{x \to \infty} \frac{\sin(1/x)}{1/x^3} = \lim_{x \to \infty} \frac{\cos(1/x) \cdot (-1/x^2)}{-3/x^4}$$

$$= \lim_{x \to \infty} \frac{1}{3} x^2 \cos\left(\frac{1}{x}\right)$$

As $x \to \infty$, $1/x \to 0$, so $\cos 1/x \to \cos 0 = 1$. Thus, the limit is ∞. That is,

$$\lim_{x \to \infty} x^3 \sin\left(\frac{1}{x}\right) = \infty \qquad \blacksquare$$

Another type of indeterminate form occurs when we have the limit of the difference of two functions, both of which become infinite. In particular, if

$$\lim_{x \to a} f(x) = \infty \qquad \text{and} \qquad \lim_{x \to a} g(x) = \infty$$

then we say that $\lim_{x \to a}[f(x) - g(x)]$ is an **indeterminate form of the type** $\infty -$ ∞. We use the same designation when $\lim_{x \to a} f(x) = -\infty$ and $\lim_{x \to a} g(x) = -\infty$.

REMARK
■ If $f(x)$ and $g(x)$ become infinite with opposite signs, then $\lim_{x \to a}[f(x) - g(x)]$ is not indeterminate (but $\lim_{x \to a}[f(x) + g(x)]$ is). For example, symbolically, $\infty - (-\infty) = \infty$ and $-\infty - (\infty) = -\infty$.

You might be tempted to say that $\infty - \infty = 0$, but remember that ∞ is not a number, so ordinary rules of arithmetic do not apply. We can often handle $\infty - \infty$ forms by expressing $f(x) - g(x)$ as a single fraction and then using L'Hôpital's Rule. We illustrate this approach in the next example.

EXAMPLE 4.33   Evaluate the limits

(a) $\displaystyle \lim_{x \to 1} \left(\frac{2x}{x^2 - 1} - \frac{1}{x - 1}\right)$

(b) $\displaystyle \lim_{x \to 0} \left(\frac{1}{x} - \csc x\right)$

Solution

(a) As it stands, the limit is of the form $\infty - \infty$, since both fractions approach ∞ as $x \to 1^+$ and both approach $-\infty$ as $x \to 1^-$. By getting a common denominator, we can write

$$\lim_{x \to 1} \left(\frac{2x}{x^2 - 1} - \frac{1}{x - 1} \right) = \lim_{x \to 1} \frac{2x - (x + 1)}{x^2 - 1} = \lim_{x \to 1} \frac{x - 1}{x^2 - 1}$$

We then find the last limit, which is of the form 0/0, either by factoring the denominator or by using L'Hôpital's Rule. By the latter method, we have

$$\lim_{x \to 1} \frac{x - 1}{x^2 - 1} = \lim_{x \to 1} \frac{1}{2x} = \frac{1}{2}$$

(b) Again, we have an $\infty - \infty$ form. To express the difference as a fraction, we write $\csc x = 1/\sin x$ and then get a common denominator.

$$\lim_{x \to 0} \left(\frac{1}{x} - \frac{1}{\sin x} \right) = \lim_{x \to 0} \left(\frac{\sin x - x}{x \sin x} \right) \qquad \text{0/0 form}$$

$$= \lim_{x \to 0} \left(\frac{\cos x - 1}{\sin x + x \cos x} \right) \qquad \text{Still 0/0}$$

$$= \lim_{x \to 0} \frac{-\sin x}{\cos x - x \sin x + \cos x} = 0 \qquad \blacksquare$$

In Chapter 7 we will discuss still other types of indeterminate forms. There, too, we will have to transform the limit to one of the forms 0/0 or ∞/∞ before we can apply L'Hôpital's Rule.

Exercise Set 4.6

In Exercises 1–25, use L'Hôpital's Rule to evaluate the limits or to show that they do not exist.

1. $\lim\limits_{x \to 3} \dfrac{x^2 - 9}{3 - x}$

2. $\lim\limits_{x \to 2} \dfrac{x^2 + x - 6}{x^2 - 4}$

3. $\lim\limits_{x \to 1} \dfrac{\sqrt{x} - 1}{x - 1}$

4. $\lim\limits_{x \to \infty} \dfrac{2x^4 - 3x}{1 - x - x^4}$

5. $\lim\limits_{x \to 0} \dfrac{\tan x}{x}$

6. $\lim\limits_{x \to 0} \dfrac{1 - \cos x}{x^2}$

7. $\lim\limits_{x \to 1} \dfrac{x - 5x^4 + 4x^6}{(x - 1)^2}$

8. $\lim\limits_{x \to (\pi/2)^-} \dfrac{\cot x}{\sin 2x}$

9. $\lim\limits_{x \to 0} \dfrac{x^3}{x - \sin x}$

10. $\lim\limits_{x \to 1} \dfrac{x^2 + x - 2}{\sin \pi x}$

11. $\lim\limits_{x \to 2} \dfrac{x - 2}{\sqrt{x + 2} - 2}$

12. $\lim\limits_{x \to 0} \dfrac{x - \sin x}{\cos 2x - 1}$

13. $\lim\limits_{x \to 0} \dfrac{\cos x - 1}{\cos 2x - 1}$

14. $\lim\limits_{x \to 0} \dfrac{\cot x + x}{\csc x}$

15. $\lim\limits_{x \to 0} \dfrac{1 - \cos x}{\sin x^2}$

16. $\lim\limits_{x \to 0} \dfrac{x \sin x}{\sec x - 1}$

17. $\lim\limits_{x \to 0} x \csc x$

18. $\lim\limits_{x \to 1/2} (2x - 1) \tan \pi x$

19. $\lim\limits_{x \to 0} (\sin 2x)(\cot x)$

20. $\lim\limits_{x \to \pi} (\cos x + 1)(\csc x)$

21. $\lim\limits_{x \to 0} \left(\cot x - \dfrac{1}{x} \right)$

22. $\lim\limits_{x \to 0} (\csc x - \cot x)$

23. $\lim\limits_{x \to 1} \left(\dfrac{3}{x^2 - 1} + \dfrac{\sqrt{4x + 5}}{1 - x^2} \right)$

24. $\lim\limits_{x \to 0^+} \left(\dfrac{x + 1}{x} - \dfrac{2x + 3}{x^2} \right)$

25. $\lim\limits_{x \to \infty} \left(\sqrt{x^2 + x} - x \right)$ (*Hint*: Let $x = 1/t$.)

26. In each of the following, find the limit without using L'Hôpital's Rule. Then see what happens when you try to apply L'Hôpital's Rule. Explain why you do not get the correct result by L'Hôpital's Rule.

(a) $\lim\limits_{x \to 0} \dfrac{x^2 \sin(1/x)}{\sin x}$ (*Hint:* Write as $\left(\dfrac{x}{\sin x}\right)\left(x \sin \frac{1}{x}\right)$.)

(b) $\lim\limits_{x \to \infty} \dfrac{x + \cos x}{x}$

 27. Use a calculator, CAS, or spreadsheet with $x = 0.1, 0.01,$ $0.001,$ and 0.0001 in part (a) and $x = 10, 100, 1,000,$ and $10,000$ in part (b) to give further evidence that the limits you found in Exercise 26 are correct.

 28. Let

$$f(x) = \frac{\cos x - 1}{x^3}$$

(a) Determine all horizontal and vertical asymptotes to the graph of f. (See Example 4.31.)

(b) Use a CAS or graphing calculator to obtain the graph of f. Pay particular attention to values of x near 0 and to the behavior of f for large values of x.

 29. Let

$$f(x) = \frac{2x^2 - 2x + 1}{x^3 + 2x - 1}$$

(a) Use L'Hôpital's Rule to determine the horizontal asymptote to the graph of f.

(b) Use Newton's Method to approximate any vertical asymptotes to the graph.

(c) Use a CAS or graphing calculator to obtain the graph of f.

(d) Approximate all local extrema of the graph.

4.7 THE MEAN-VALUE THEOREM

In this section we prove a theorem that has far-reaching consequences, known as the **Mean-Value Theorem** for differential calculus, or the **Law of the Mean**. As one application of this theorem, we will use it to prove some of the results of this chapter that we accepted earlier on intuitive grounds. We precede the main theorem with another result, which is used in the proof and so could be referred to as a lemma. This preliminary result is important in its own right, however, and it is known as **Rolle's Theorem**, after the seventeenth-century French mathematician Michel Rolle.

Before we state Rolle's Theorem, we will describe what it says informally. Consider the graph in Figure 4.76. There we have pictured a continuous function on the closed interval $[a, b]$ that equals 0 at each of the endpoints $x = a$ and $x = b$. We assume also that f has a derivative at all points between a and b. That is, the graph has a tangent line (nonvertical) at every point, except possibly at the endpoints a and b (where the tangent line could be vertical). Then, as we indicate, there must be some point on the curve (we have shown this point as having x-coordinate c) where the tangent line is horizontal. So when $x = c$, the slope is 0. This seemingly obvious result has some important consequences. As we have noted, one application of it is in proving the Mean-Value Theorem. There are many others.

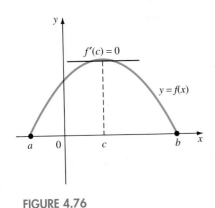

FIGURE 4.76

THEOREM 4.8	**Rolle's Theorem**

Suppose the function f is continuous at every point of $[a, b]$ and differentiable at every point of (a, b), with $f(a) = 0$ and $f(b) = 0$. Then there exists a number c in the open interval (a, b) such that $f'(c) = 0$.

Proof   Since f is continuous on $[a, b]$, we know by Theorem 4.2 that it attains both an absolute maximum and an absolute minimum there. If $f(x) = 0$ for all x on $[a, b]$, then any number c between a and b satisfies the conclusion. If f is not constant, then its graph goes above the x-axis or below it, or both. Thus, either its maximum is greater than 0 or its minimum is less than 0, or both, and hence at least one of these extremes must occur at a point c between a and b. But since $f'(c)$ exists, it must be 0 by Theorem 4.3. The proof is therefore complete.   ∎

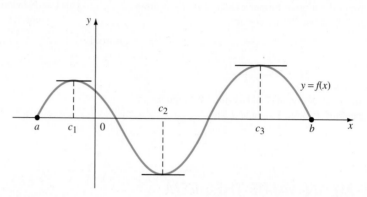

FIGURE 4.77

Rolle's Theorem guarantees the existence of *at least one* number c such that $f(c) = 0$, but there may be more than one. In Figure 4.77 we show a graph having three points satisfying this property.

If one or more of the hypotheses of Rolle's Theorem fail to hold true, then the conclusion may or may not be true. In Figure 4.78 we show three functions for which there is no number c in the interval in question such that $f'(c) = 0$. In part (a) the function is continuous in $[0, 2]$ and the derivative exists in

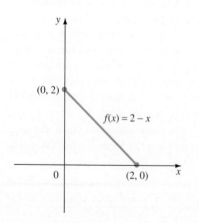

$f(0) \neq 0$, so $f(a)$ and $f(b)$ are not both 0, where $a = 0$ and $b = 2$.

(a)

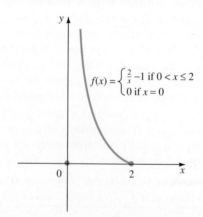

f is not continuous at $x = 0$, so f is not continuous on $[a, b]$, where $a = 0$ and $b = 2$.

(b)

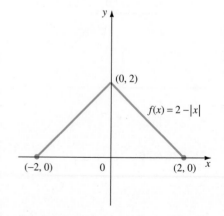

$f'(0)$ does not exist, so $f'(x)$ does not exist at all points of (a, b), where $a = -2$ and $b = 2$.

(c)

FIGURE 4.78

$(0, 2)$, but $f(0) \neq 0$. In part (b) the function is 0 at both endpoints and its derivative exists in $(0, 2)$, but the function is not continuous at the left endpoint, $x = 0$. In part (c) the function is continuous on $[-2, 2]$ and it is 0 at both endpoints, but the derivative does not exist at $x = 0$. These three examples show that all the hypotheses of Rolle's Theorem are needed to guarantee the conclusion.

Let us now consider what conclusion we can draw if we drop the hypothesis in Rolle's Theorem that $f(a)$ and $f(b)$ both equal 0. We have pictured such a function in Figure 4.79. The function is continuous on the closed interval $[a, b]$ and differentiable on the open interval (a, b), but no requirement is placed on the value of $f(a)$ or of $f(b)$. As we indicate in the figure, there appears to be some number c between a and b such that the tangent line drawn at the point $(c, f(c))$ is parallel to the secant line joining the endpoints $(a, f(a))$ and $(b, f(b))$. So their slopes are the same:

$$f'(c) = \frac{f(b) - f(a)}{b - a}$$

What we have observed is exactly the content of the Mean-Value Theorem, which we now state formally.

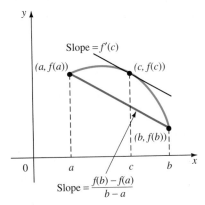

FIGURE 4.79

THEOREM 4.9

Mean-Value Theorem

Let f be continuous on the closed interval $[a, b]$ and differentiable on the open interval (a, b). Then there exists a point c in the open interval (a, b) such that

$$f'(c) = \frac{f(b) - f(a)}{b - a}$$

Proof   For convenience, let

$$m = \frac{f(b) - f(a)}{b - a}$$

Then the equation of the secant line that joins the endpoints is

$$y = f(a) + m(x - a) \qquad \text{Verify.}$$

We define a new function F on $[a, b]$ as follows:

$$F(x) = f(x) - [f(a) + m(x - a)] \qquad (4.11)$$

Note that $F(x)$ is the vertical (directed) distance from the secant line to the curve, which is positive when the curve is above the secant line and negative when it is below. Now F is the difference of two functions, each of which is continuous on $[a, b]$ and differentiable on (a, b), and so F also has these properties. Furthermore, $F(a) = 0$ and $F(b) = 0$, as can be seen by direct substitution. Thus, F satisfies all the hypotheses of Rolle's Theorem and hence its conclusion; that is, there exists a number c in (a, b) such that $F'(c) = 0$. But from Equation 4.11,

$$F'(x) = f'(x) - m$$

So $F'(c) = 0$ implies that $f'(c) - m = 0$, or equivalently,

$$f'(c) = m$$

That is,

$$f'(c) = \frac{f(b) - f(a)}{b - a}$$

■

EXAMPLE 4.34   Find a number c in the interval $(-1, 3)$ that satisfies the conclusion of the Mean-Value Theorem for the function

$$f(x) = x^3 - x^2 + 3$$

Solution   Note first that the hypotheses of Theorem 4.9 are satisfied, since f is continuous on $[-1, 3]$ and differentiable on $(-1, 3)$. In fact, f is differentiable and hence continuous everywhere, since it is a polynomial. We want to show the existence of a number c in $(-1, 3)$ for which

$$f'(c) = \frac{f(3) - f(-1)}{3 - (-1)} = \frac{21 - 1}{4} = 5$$

Since $f'(x) = 3x^2 - 2x$, we have $f'(c) = 3c^2 - 2c$, so we solve the equation

$$3c^2 - 2c = 5$$
$$3c^2 - 2c - 5 = 0$$
$$(3c - 5)(c + 1) = 0$$
$$c = \frac{5}{3} \quad \text{or} \quad c = -1$$

But the requirement is for c to be in the *open* interval $(-1, 3)$, so we must have $-1 < c < 3$. The solution $c = 5/3$ is therefore the only one that satisfies the theorem.

■

REMARK ───

■ If f fails to satisfy the hypotheses of the Mean-Value Theorem, then the conclusion may or may not be true. Consider, for example,

$$f(x) = 2 + (1 - x)^{2/3} \qquad \text{on } [0, 2]$$

Here, $f(0) = 3$ and $f(2) = 3$. So $[f(2) - f(0)]/2 = 0$. Since

$$f'(x) = -\frac{2}{2(1 - x)^{1/3}}$$

can never be 0, the conclusion of the Mean-Value Theorem cannot be satisfied. This failure does not contradict the theorem, however, because $f'(x)$ does not exist at $x = 1$, which is in the open interval $(0, 2)$.

Average versus Instantaneous Velocity

An interesting physical interpretation of the Mean-Value Theorem relates the instantaneous velocity of a moving object and its average velocity over some time interval. Suppose that $s(t)$ represents the position of the object along a line at time t, and that we are interested in the time interval from $t = a$ to $t = b$. Assume that s satisfies the hypotheses of the Mean-Value Theorem (replacing

$f(x)$ by $s(t)$). Then the conclusion of that theorem states that there is some value $t = c$ between $t = a$ and $t = b$ such that

$$s'(c) = \frac{s(b) - s(a)}{b - a}$$

But $s'(c)$ is the instantaneous velocity of the object at time $t = c$, and the right-hand side of the equation is the average velocity over the time interval from $t = a$ to $t = b$ (change in distance divided by change in time). Thus, *at some value of t between a and b, the instantaneous velocity equals the average velocity over the time interval from $t = a$ to $t = b$.* We illustrate this result in the next example.

EXAMPLE 4.35   A car starts from point A and accelerates, reaching point B 10 seconds later. Its position, measured from A, is given by

$$s(t) = 0.8t^2 \qquad (0 \le t \le 10)$$

where t is in seconds and $s(t)$ is in meters. Find (a) the car's average velocity in going from A to B, and (b) the time and location of the car when its instantaneous velocity equals its average velocity.

Solution   Note first that $s(t)$ is differentiable, and hence continuous, everywhere, so the hypotheses of the Mean-Value Theorem are satisfied.

(a) $v_{\text{ave}} = \dfrac{s(10) - s(0)}{10 - 0} = \dfrac{(0.8)(100)}{10} = 8$ m/s

(b) The instantaneous velocity $v(t)$ is given by

$$v(t) = s'(t) = 1.6t$$

To find where $v(t)$ equals v_{ave}, we set $1.6t$ equal to 8 and solve for t:

$$1.6t = 8$$

$$t = 5$$

So the instantaneous velocity after 5 seconds equals the average velocity. At that time the car is $(0.8)(5)^2 = 20$ m from A. ∎

Using the Mean-Value Theorem

We will now use the Mean-Value Theorem to prove Theorem 4.1, which we accepted on intuitive grounds earlier. We will also establish two important corollaries to the Mean-Value Theorem. First, we restate Theorem 4.1.

THEOREM 4.1

(restated)

If $f'(x) > 0$ for all x in an open interval I, then f is increasing on I, and if $f'(x) < 0$ for all x in I, then f is decreasing on I.

Proof   Suppose first that $f'(x) > 0$ on I. Choose x_1 and x_2 in I, with $x_1 < x_2$. We want to show that $f(x_2) > f(x_1)$. Since $f'(x)$ exists on the interval I, f is differentiable and therefore continuous on the interval $[x_1, x_2]$, so the hypotheses of the Mean-Value Theorem are definitely satisfied. In that theorem we replace a by x_1 and b by x_2 and conclude that a number c exists in the open interval (x_1, x_2) such that

$$f'(c) = \frac{f(x_2) - f(x_1)}{x_2 - x_1}$$

Since $f'(c) > 0$ by hypothesis, and since $x_2 > x_1$ by choice, it follows that the numerator on the right must be positive; that is, $f(x_2) > f(x_1)$. In summary, we have shown that if $x_1 < x_2$, then $f(x_1) < f(x_2)$. Thus, by definition, f is an increasing function. The proof of the first part of the theorem is therefore complete. We leave the second part as an exercise.   ■

We conclude this section with two important corollaries of the Mean-Value Theorem.

COROLLARY 4.9(a)

If $f'(x) = 0$ for all x in an interval I, then f is a constant function on I.

Proof   Let a denote any fixed point in I, and let x denote any other point in I. We will show that $f(x) = f(a)$, and since $f(a)$ is a constant, this will tell us that all functional values are equal to this constant. We prove the result for $x > a$ only. The proof for $x < a$ is similar. We can apply the Mean-Value Theorem to the interval $[a, x]$. The existence of f' throughout I guarantees that the hypotheses of the Mean-Value Theorem are satisfied. (Why?) Thus, there is a number c in (a, x) such that

$$f'(c) = \frac{f(x) - f(a)}{x - a}$$

But since c is in I, $f'(c) = 0$. Thus, since $x - a \neq 0$ it follows that $f(x) - f(a) = 0$, and $f(x) = f(a)$. Thus, $f(x)$ is constant for all x in I.   ■

REMARK ─────────────────────────────────

■ We already know that if $f(x)$ is a constant, then $f'(x) = 0$. Thus, we now know that $f(x)$ is a constant if and only if $f'(x) = 0$.

COROLLARY 4.9(b)

If $f'(x) = g'(x)$ for all x in an interval I, then for all x in I, $f(x) = g(x) + C$ for some constant C.

Proof   Define $h(x) = f(x) - g(x)$. Then

$$h'(x) = f'(x) - g'(x) = 0$$

for all x in I. So by Corollary 4.9(a), $h(x)$ is a constant, that is, $h(x) = C$. Thus,

$$f(x) - g(x) = C \qquad (x \in I)$$

which is equivalent to the desired conclusion.   ■

This corollary says that when two functions have the same slopes at all points, the curves are separated by a constant. In a certain sense we can think of the curves as being parallel. Figure 4.80 illustrates this relationship.

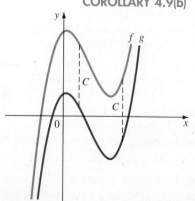

FIGURE 4.80

Exercise Set 4.7

In Exercises 1–8, show that the conditions of Rolle's Theorem are satisfied by the function on the given interval, and find a number c that satisfies the conclusion of the theorem.

1. $f(x) = x^2 - 3x - 4$ on $[-1, 4]$

2. $f(x) = x^2 - 4x + 3$ on $[1, 3]$

3. $f(x) = 2x^2 - 3x - 2$ on $[-\frac{1}{2}, 2]$

4. $f(x) = x^3 + x^2 - x - 1$ on $[-1, 1]$

5. $f(x) = x^3 - 9x$ on $[0, 3]$

6. $f(x) = 2x^3 - x^2 - 8x + 4$ on $[-2, 2]$

7. $f(x) = (x^2 - 9)\sqrt{x + 1}$ on $[-1, 3]$

8. $f(x) = (x - 1)\sqrt{3 - x^2}$ on $[1, \sqrt{3}]$

In Exercises 9–16, show that the conditions of the Mean-Value Theorem are satisfied by the function on the given interval, and find a number c that satisfies the conclusion of the theorem.

9. $f(x) = x^2 - 3x + 1$ on $[-1, 2]$

10. $f(x) = 4 - x - 3x^2$ on $[0, 3]$

11. $f(x) = x^3 - 2x^2 + 3$ on $[-1, 4]$

12. $f(x) = x^3 - 2x^2 + 5x + 8$ on $[-2, 2]$

13. $f(x) = x^3 + 5x^2 - 2x - 5$ on $[-1, 2]$

14. $f(x) = x - \frac{1}{x}$ on $[\frac{1}{2}, 2]$

15. $f(x) = \sqrt{2x + 1}$ on $[0, 4]$

16. $f(x) = \dfrac{x - 1}{x + 1}$ on $[0, 5]$

In Exercises 17–20, show that the conclusion of Rolle's Theorem is not satisfied by the function on the given interval, and explain why there is no contradiction with the theorem.

17. $f(x) = 1 - x^{2/3}$ on $[-1, 1]$

18. $f(x) = \dfrac{x^2 - 4}{x + 1}$ on $[-2, 2]$

19. $f(x) = \dfrac{2 \sin x - 1}{\cos x}$ on $\left[0, \dfrac{5\pi}{6}\right]$

20. $f(x) = \dfrac{1 - 2 \cos x}{\sin x}$ on $\left[\dfrac{-\pi}{3}, \dfrac{\pi}{3}\right]$

In Exercises 21–24, show that the conclusion of the Mean-Value Theorem is not satisfied by the function on the given interval, and explain why there is no contradiction with the theorem.

21. $f(x) = \dfrac{x^2 - 1}{x}$ on $[-1, 2]$

22. $f(x) = x - \dfrac{2}{x} + 3$ on $[-2, 2]$

23. $f(x) = (x - 1)^{2/3}$ on $[0, 2]$

24. $f(x) = x - 3x^{2/3}$ on $[-8, 8]$

25. Show that $f(x) = 1 - x + (3x - 1)^{1/3}$ fails to satisfy one or more of the hypotheses of Rolle's Theorem on $[0, 3]$, yet the conclusion of the theorem is true. Does this represent a contradiction? Explain.

26. Show that $f(x) = x + 3(1 - x)^{1/3}$ fails to satisfy one or more of the hypotheses of the Mean-Value Theorem on $[0, 2]$, yet the conclusion of the theorem is true. Does this represent a contradiction? Explain.

27. Find two points in the interval $(-1, 3)$ for which the tangent line to the graph of $f(x) = 2x^3 - 5x^2 - 3x + 1$ is parallel to the secant line that joins $(-1, f(-1))$ and $(3, f(3))$. Draw a figure showing the graph of f, the two tangent lines, and the secant line.

28. Let f satisfy the conditions of the Mean-Value Theorem on $[a, b]$, and let x be any number in $[a, b]$. Show that for some number c in (a, x),

$$f(x) = f(a) + f'(c)(x - a)$$

29. Let f satisfy the conditions of the Mean-Value Theorem on $[a, b]$, and let h be any number that satisfies $0 < h \le b - a$. Show that $f(a + h) = f(a) + h \cdot f'(a + \theta h)$ for some number θ such that $0 < \theta < 1$.

30. Use the Mean-Value Theorem to show that for any two numbers x_1 and x_2,

$$|\sin x_1 - \sin x_2| \le |x_1 - x_2|$$

31. Use the Mean-Value Theorem to show that $\tan x > x$ for all x that satisfy $0 < x < \frac{\pi}{2}$. (*Hint:* First pick an x in the given interval and then apply the Mean-Value Theorem to $f(x) = \tan x$ on $[0, x]$.)

32. Complete the proof of Theorem 4.1 by showing that if $f'(x) < 0$ for all x on an open interval I, then f is a decreasing function.

33. Prove that if f is differentiable on $[a, b]$ with $f'(a)$ and $f'(b)$ of opposite sign, then there is a number c between

a and b such that $f'(c) = 0$. [*Hint:* Show that f must attain its maximum or its minimum (or both) somewhere between a and b. Then use Theorem 4.2.]

34. Prove the following **Intermediate-Value Theorem for Derivatives**: If f is differentiable on $[a, b]$ and k is any number between $f'(a)$ and $f'(b)$, then there is a number c between a and b such that $f'(c) = k$. (Note that f' need not be a continuous function.) (*Hint:* Show that the result of Exercise 33 can be applied to $g(x) = f(x) - kx$.)

4.8 ANTIDERIVATIVES AND DIFFERENTIAL EQUATIONS

It is very important to be able to do the opposite of differentiating—that is, to find a function whose derivative we know. This process is called **antidifferentiation**, and a function arrived at in this way is called an **antiderivative** of the given function. More formally we have the following definition.

Definition 4.6 Antiderivative of a Function	A function F is called an **antiderivative** of f if $F'(x) = f(x)$ for all x in the domain of f.

For example, suppose we are given the function $3x^2$, and we are asked to find a function for which $3x^2$ is the derivative. We can do this calculation mentally and arrive at x^3 as an answer. So we can say that x^3 is an antiderivative of $3x^2$. We are being very careful in using the article *an* here rather than *the* because there will always be many (infinitely many, in fact) antiderivatives of a function if there are any at all. In the present case, for example, we can see that each of the following is an antiderivative of $3x^2$: $x^3 + 2$, $x^3 - 4$, and $x^3 + 100$. In fact, $x^3 + C$, where C is *any* real number, is an antiderivative of $3x^2$. Whether there can be any that are not of this form is answered by the next theorem, whose proof further illustrates the importance of the Mean-Value Theorem and its corollaries.

THEOREM 4.10

Let $F(x)$ be an antiderivative of $f(x)$ on an interval I. Then every other antiderivative of $f(x)$ on I is of the form $F(x) + C$ for some constant C.

Proof   We are given that $F'(x) = f(x)$ for all $x \in I$. Suppose G is any other antiderivative of $f(x)$ on I, so that $G'(x) = f(x)$ on I. Then for all x in I, $G'(x) = F'(x)$. Therefore, by Corollary 4.9(b),

$$G(x) = F(x) + C$$

for some constant C.   ■

If $F(x)$ is an antiderivative of $f(x)$, then we sometimes call $F(x) + C$ the **general antiderivative** of $f(x)$, so called because we leave the constant C unspecified. In the sense of Theorem 4.10, $F(x) + C$ represents the *family of all* antiderivatives of $f(x)$. We will soon learn certain techniques for finding antiderivatives that are less obvious than the one we have considered so far. For the present, we will concentrate on certain basic types.

REMARK ————————————————————————————
■ Strictly speaking, we should write the family of all antiderivatives of f as the set

$$\{F(x) + C : C \in \mathbb{R} \text{ and } F'(x) = f(x)\}$$

For brevity, though, and in accordance with customary usage, we will write simply $F(x) + C$.

————————————————————————————————

We know that to differentiate a rational power of x, say x^n, we multiply by the exponent and then subtract 1 from the exponent, getting nx^{n-1}. To find an antiderivative of x^n we do the reverse; that is, we *add* 1 to the exponent and then *divide* by the resulting exponent, getting

$$\frac{x^{n+1}}{n+1}$$

For this quotient to have meaning we must have $n \neq -1$. With this restriction, we can check the result by differentiating:

$$\frac{d}{dx}\left(\frac{x^{n+1}}{n+1}\right) = (n+1) \cdot \frac{x^n}{n+1} = x^n$$

For example, an antiderivative of x^3 is $x^4/4$, and an antiderivative of $x^{-1/2}$ is $2x^{1/2}$.

Because the derivative of a constant times a function is the constant times the derivative of the function, the same is true of antiderivatives. For example, an antiderivative of $5x^3$ is $5x^4/4$. Also, since the derivative of the sum or difference of functions is the sum or difference of the derivatives, the same is true for antiderivatives. These facts enable us to find antiderivatives of polynomials. For example, an antiderivative of $2x^2 - 3x + 4$ is

$$\frac{2x^3}{3} - \frac{3x^2}{2} + 4x$$

The following table summarizes some of the basic antiderivatives we already know.

Function	Antiderivative
x^n	$\dfrac{x^{n+1}}{n+1} + C \qquad (n \neq -1)$
$\sin x$	$-\cos x + C$
$\cos x$	$\sin x + C$
$\sec^2 x$	$\tan x + C$
$\csc^2 x$	$-\cot x + C$
$\sec x \tan x$	$\sec x + C$
$\csc x \cot x$	$-\csc x + C$

All these antiderivatives can be verified by differentiating the function in the right-hand column to get the function on the left. It should be clear that, whenever we find a formula for the derivative of a function, we also have found a formula for an antiderivative.

If we denote by F any antiderivative of f, then we also have the following:

Function	Antiderivative
$cf(x)$	$cF(x)$
$f_1(x) \pm f_2(x)$	$F_1(x) \pm F_2(x)$

EXAMPLE 4.36   Find the general antiderivative of the function

$$f(x) = 3x^2 - 2x - 2$$

and draw graphs for several different values of the constant C.

Solution   The general antiderivative of f is

$$x^3 - x^2 - 2x + C$$

We show the graphs for $C = -1$, 0, 1, 2, and 3 in Figure 4.81.   ∎

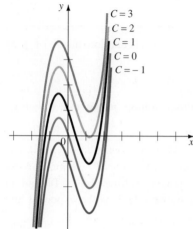

FIGURE 4.81
The family $y = x^3 - x^2 - 2x + C$ of antiderivatives of $f(x) = 3x^2 - 2x - 2$

The preceding example further illustrates the nature of the general antiderivative of a function as being not just one function but a *family* of functions. The members of the family differ from one another only by constant values. The graphs in Figure 4.81 clearly show this property. At each value of x, the slopes of all members of the family are the same.

Often, we will want to single out one particular member of the family of antiderivatives of a function. For example, we may want the member of the family whose graph goes through a particular point. The next example shows how to find such a particular antiderivative.

EXAMPLE 4.37   Find the particular antiderivative of the function

$$f(x) = 3x^2 - 2x - 2$$

whose graph passes through the point $(2, -1)$.

Solution   In Example 4.36 we found the general antiderivative of f to be

$$x^3 - x^2 - 2x + C$$

Let us designate by $F(x)$ the particular member we are seeking. Thus,

$$F(x) = x^3 - x^2 - 2x + C$$

where the value of C is yet to be determined. Since the point $(2, -1)$ is on the graph of F, we must have $F(2) = -1$. So to evaluate the constant C we substitute $x = 2$ and $F(x) = -1$:

$$-1 = (2)^3 - (2)^2 - 2(2) + C$$

$$-1 = 8 - 4 - 4 + C$$

$$C = -1$$

The particular antiderivative we are seeking is therefore

$$F(x) = x^3 - x^2 - 2x - 1$$ ∎

Differential Equations

An equivalent way of asking for the general antiderivative of a function f is to ask for all functions F such that $F'(x) = f(x)$. Thus, we could have rephrased the instructions in Example 4.36 as follows: Find all functions F such that

$$F'(x) = 3x^2 - 2x - 2 \qquad (4.12)$$

The answer is what we found in the example, namely, $F(x) + C$, where

$$F(x) = x^3 - x^2 - 2x$$

Equation 4.12 is called a **differential equation**. In general, differential equations involve one or more derivatives of some unknown function. To *solve* a differential equation means to find the unknown function or family of functions.

REMARK ————————————————————
■ Leibniz used the term *differential equation* to mean an equation relating differentials dx and dy. This terminology has stayed with us. A better name would be *derivative* equation.
————————————————————————

Equation 4.12 is the simplest type of differential equation and is the only type we will consider in this section. Here are some other examples of differential equations:

$$x\frac{dy}{dx} - 3y = 4$$

$$(y+1)dy = (x-2)dx$$

$$y'' - 2y' + y = \sin x$$

(We will show how to solve the first two of these in Chapter 7.)

Differential equations are of fundamental importance in virtually all the sciences. They usually arise as mathematical models involving rates of change. The subject of differential equations is very broad, and we can only touch on it here. Whole courses at both the undergraduate and graduate levels are devoted to the subject.

If we use the Leibniz notation dy/dx for the derivative, simple differential equations such as Equation 4.12 can be written in the form

$$\frac{dy}{dx} = f(x)$$

The solution is the general antiderivative

$$y = F(x) + C$$

where $F'(x) = f(x)$, and it is called the **general solution** of the differential equation. For example, Equation 4.12 can be written as

$$\frac{dy}{dx} = 3x^2 - 2x - 2$$

and its general solution is

$$y = x^3 - x^2 - 2x + C$$

We showed the graphs of several members of this family in Figure 4.81.

If we know a particular value of y for some x value, the differential equation, together with the known pair (x, y), is called an **initial-value problem**. In the next example we rephrase Example 4.37 as such an initial-value problem.

EXAMPLE 4.38   Solve the initial-value problem

$$\frac{dy}{dx} = 3x^2 - 2x - 2; \quad y = -1 \text{ when } x = 2$$

Solution   The general solution of the differential equation is

$$y = x^3 - x^2 - 2x + C$$

Now we substitute $x = 2$ and $y = -1$ to find C:

$$-1 = (2)^3 - (2)^2 - 2(2) + C$$

$$C = -1$$

Thus, the particular solution we are seeking is

$$y = x^3 - x^2 - 2x - 1 \qquad \blacksquare$$

Often differential equations involve derivatives with respect to *time*. For example, equations relating to an object moving along a line might involve its velocity or its acceleration. If $s(t)$ represents the position of the object from a designated origin at time t, then we know that $v(t) = s'(t)$ and $a(t) = v'(t)$, or equivalently $a(t) = s''(t)$. The next two examples illustrate differential equations of this type.

EXAMPLE 4.39   From just off the top edge of a 10-story building 96 ft high, a ball is thrown vertically upward with an initial velocity of 40 ft/s. Find the position $s(t)$ of the ball above the ground t seconds later. What is the velocity after 2 seconds? Find the maximum height of the ball and also the ball's velocity when it hits the ground. Assume that acceleration caused by gravity is 32 ft/s², acting downward, and that air resistance is negligible.

Solution   We take the distance as 0 at the ground level and choose the positive direction upward, as indicated in Figure 4.82. Since the only force on the ball is assumed to be gravity, and it acts downward, the acceleration $a(t)$ of the ball is $a(t) = -32$. Since $a(t) = v'(t)$, we can find $v(t)$ by solving the initial-value problem

$$v'(t) = -32; \qquad v(0) = 40$$

The antiderivative is

$$v(t) = -32t + C$$

and we find C by the initial condition $v(0) = 40$. That is, we let $t = 0$ and $v(t) = 40$:

$$40 = -32(0) + C$$

$$C = 40$$

Thus,

$$v(t) = -32t + 40 \qquad (4.13)$$

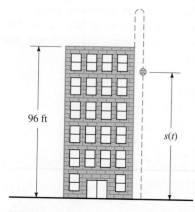

FIGURE 4.82

96 ft

$s(t)$

Also, we know that $v(t) = s'(t)$, so we have a second initial-value problem to solve:

$$s'(t) = -32t + 40; \qquad s(0) = 96$$

(since the initial height is 96 ft). Again taking an antiderivative, we get

$$s(t) = -16t^2 + 40t + C_1$$

We denote the constant by C_1, since it is not necessarily the same constant we had in the velocity problem. The initial condition $s(0) = 96$ gives

$$C_1 = 96$$

So the position of the ball at time t is

$$s(t) = -16t^2 + 40t + 96 \qquad (4.14)$$

To get the velocity after 2 seconds, we use Equation 4.13 with $t = 2$:

$$v(2) = -32(2) + 40 = -24$$

The negative sign indicates that $s(t)$ is decreasing when $t = 2$. So at that instant the ball is falling at the rate of 24 ft/s, which is one of the results requested.

Now let us find the maximum height the ball reaches. The distance $s(t)$ is maximum when its derivative, $v(t)$, is 0. So we set $v(t) = 0$ and solve for t:

$$-32t + 40 = 0$$

$$t = \frac{5}{4}$$

The velocity is positive for $t < 5/4$ and negative for $t > 5/4$, confirming that the critical value $5/4$ produces a maximum value of the position function. From Equation 4.14, we get

$$s_{\max} = s\left(\frac{5}{4}\right) = -16\left(\frac{25}{16}\right) + 40\left(\frac{5}{4}\right) + 96 = 121$$

So the ball reaches a maximum height above the ground of 121 ft (25 ft above the building). In Figure 4.83 we show the graph of the position function $s(t)$ versus time t. The graph clearly shows the maximum value of $s(t)$ when $t = 5/4$.

Finally, we want to know the velocity with which the ball hits the ground. First we find the time when the ball hits the ground by setting $s(t) = 0$ and solving for t:

$$-16t^2 + 40t + 96 = 0$$

$$2t^2 - 5t - 12 = 0$$

$$(2t + 3)(t - 4) = 0$$

$$t = -\frac{3}{2} \quad \text{or} \quad t = 4$$

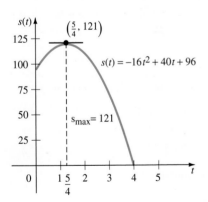

FIGURE 4.83

We reject $t = -\frac{3}{2}$ as being meaningless in the context of the problem. So the ball hits the ground when $t = 4$ seconds. From Equation 4.13 we therefore have

$$v(4) = -32(4) + 40 = -88$$

The ball is traveling (downward) at 88 ft/s at the instant it strikes the ground. (Note that 88 ft/s is 60 mph, so you wouldn't want to let the ball hit you on the head!) ∎

It is useful to note that the constant C in the solution for $v(t)$ is equal to the initial velocity $v(0)$. Similarly, the constant C_1 in the solution for $s(t)$ is equal to the initial position $s(0)$. These values for the constants are typical for problems involving motion along a line.

REMARK ───

■ An equivalent way of setting up the initial-value problem in Example 4.39 is

$$s''(t) = -32; \qquad s'(0) = 40, \; s(0) = 96$$

The differential equation now is of *second order* (it involves a second derivative), and two initial conditions are required to find the constants that arise in the solution. The solution in this case would proceed just as in the example, with $s'(t)$ replacing $v(t)$.

───

EXAMPLE 4.40   A particle moves on a line so that its acceleration at time $t \geq 0$ is given by $a(t) = 6(1 - t)$. The initial conditions are $v(0) = 9$ and $s(0) = 0$.

(a) Find $v(t)$ and $s(t)$.

(b) Find when and where the particle reaches its greatest positive distance from the origin.

(c) Determine the time intervals and corresponding distance intervals in which the particle is moving to the right (positive direction) and those in which it is moving to the left.

(d) Find the time and distance intervals when it is slowing down and those in which it is speeding up.

Solution

(a) Solving the initial-value problem

$$v'(t) = 6(1 - t); \qquad v(0) = 9$$

we obtain

$$v(t) = 6t - 3t^2 + C$$

By the comment above concerning constants in motion problems such as this one, we have $C = v(0)$. So, since we are given that $v(0) = 9$,

$$v(t) = -3t^2 + 6t + 9$$

Thus, we have the second initial-value problem

$$s'(t) = -3t^2 + 6t + 9; \qquad s(0) = 0$$

with the following solution:

$$s(t) = -t^3 + 3t^2 + 9t + s(0) \qquad \text{The constant } C_1 = s(0).$$

The answer to part (a) is therefore

$$v(t) = -3t^2 + 6t + 9$$

$$s(t) = -t^3 + 3t^2 + 9t$$

(b) To find the greatest positive distance from the origin, we look for the absolute maximum value of $s(t)$. First, we find critical points by setting $v(t) = s'(t) = 0$:

$$-3t^2 + 6t + 9 = 0$$
$$t^2 - 2t - 3 = 0$$
$$(t - 3)(t + 1) = 0$$
$$t = 3 \text{ or } t = -1$$

We reject $t = -1$, since we are given that $t \geq 0$. Since $s(3) = -27 + 3(9) + 9(3) = 27$, and $s''(3) = a(3) < 0$, it follows that 27 is a local maximum value of $s(t)$. The only endpoint value is $s(0) = 0$, so 27 is the absolute maximum value of $s(t)$.

(c) Since $v(t) = -3(t+1)(t-3)$, we see that for $t \geq 0$, the velocity is positive for $t < 3$ and negative for $t > 3$. So the particle moves to the right from $s(0) = 0$ to $s(3) = 27$. Then it turns around and moves to the left for all $t > 3$.

(d) Since acceleration measures the rate of change of velocity, we know that if $a(t) > 0$, the velocity is increasing. Thus, if both $v(t) > 0$ and $a(t) > 0$, the particle is moving to the right and is speeding up. On the other hand, if $v(t) < 0$ and $a(t) > 0$, the particle is moving to the left but slowing down, since the velocity is negative but is tending toward becoming positive. Suppose now that $a(t) < 0$. Then, if $v(t)$ also is negative, the particle is moving to the left, and its velocity is becoming more negative; that is, the particle is speeding up. On the other hand, if $v(t) > 0$ but $a(t) < 0$, the particle is moving to the right but with decreasing velocity. We can summarize this discussion as follows:

1. If $v(t)$ and $a(t)$ are *like* in sign, the particle is speeding up.

2. If $v(t)$ and $a(t)$ are *opposite* in sign, the particle is slowing down.

In either case the direction of motion is determined by $v(t)$: to the right if $v(t) > 0$ and to the left if $v(t) < 0$. A comparison of sign graphs (Figure 4.84) for v and a makes clear when each situation occurs. We see that from $t = 0$ to $t = 1$, the particle is speeding up and moving to the right ($v > 0$, $a > 0$); from $t = 1$ to $t = 3$ it continues to the right but is slowing down ($v > 0$, $a < 0$), and for $t > 3$ it moves to the left faster and faster ($v < 0$, $a < 0$). Figure 4.85 summarizes our findings. The motion actually takes place on a line, but in this diagram we show it on two levels to see more clearly the nature of the motion in each direction. ∎

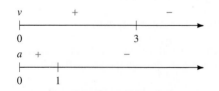

For $0 < t < 1$, $v(t) > 0$ and $a(t) > 0$;
for $1 < t < 3$, $v(t) < 0$ and $a(t) > 0$;
for $t > 3$, $v(t) < 0$ and $a(t) < 0$

FIGURE 4.84

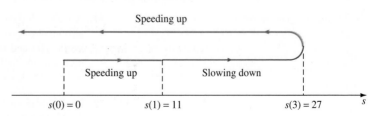

FIGURE 4.85

Exercise Set 4.8

In Exercises 1–20, find the general antiderivative of the given function.

1. $f(x) = x^5$

2. $f(x) = 3x^4$

3. $f(x) = 2x^2 - 4x + 5$

4. $f(x) = 6x^2 - 2x + 3$

5. $f(x) = 3x^3 - 5x^2 + 2x - 3$

6. $f(x) = 2 - x + 4x^2 - 6x^3$

7. $f(x) = 1/(2x^2) - x^2/2$

8. $f(x) = 2x^{-3} - 4x^{-2} + 1$

9. $f(x) = 2x^{1/2} - 3x^{-1/2}$

10. $f(x) = 1/\sqrt{x} - 2/x^2 + 1$

11. $f(x) = 3x - 4\sqrt{x} + 7$

12. $f(x) = x^{3/2} + 2x^{1/2}$

13. $f(x) = x(\sqrt{x} + 3)$

14. $f(x) = (x - 1)(1/\sqrt{x} + 1)$

15. $f(x) = 2\sin x - 3\cos x$

16. $f(x) = \sec x(\sec x + \tan x)$

17. $f(x) = x - 1/\sin^2 x$

18. $f(x) = \cos x + 1/\cos^2 x$

19. $f(x) = 1 - \cos x/\sin^2 x$

20. $f(x) = \tan x \cos x - \sin x \cot x$

In Exercises 21–28, find the general solution of the differential equation.

21. $f'(t) = \sqrt{t}(t - 1)$

22. $g'(x) = \cos x - x$

23. $\dfrac{dy}{dx} = \dfrac{1 - x}{x^3}$

24. $dy = (x^2 + 2)dx$

25. $(\cos^2 x)dy = (\sin x + 1)dx$

26. $\dfrac{dx}{dt} = 1 - t^{-0.2}$

27. $v'(t) = (t - 1)(t - 2)$

28. $s'(t) = t^{3/2} - 2t^{1/2} + 3$

In Exercises 29–38, solve the given initial-value problem.

29. $f'(x) = 2x - 3$; $f(0) = 5$

30. $f'(x) = 3x^2 - 2x + 1$; $f(-2) = 3$

31. $g'(x) = 1 - x - 4x^2$; $g(3) = 2$

32. $g'(x) = \dfrac{3}{x^2} - 2x + 3$; $g(1) = 5$

33. $y' = 2x - x^2$; when $x = 1$, $y = -2$

34. $dy = (\sqrt{x} - 1)dx$; when $x = 1$, $y = 3$

35. $f''(x) = 3x$; $f'(0) = 0$, $f(0) = 2$

36. $f''(x) = 1 - x$; $f'(1) = 2$, $f(1) = -1$

37. $f''(x) = x^2 - 1$; $f'(-1) = 0$; $f(0) = 2$

38. $f'''(x) = 2$; $f''(0) = 2$; $f'(0) = -1$; $f(0) = 4$

39. Find y as a function of x if $dy/dx = x + \sin x$ and the graph of the function goes through the point $(0, 3)$.

40. The graph of $y = f(x)$ goes through the point $(1, -2)$ and has a slope of $\frac{1}{2}$ at this point. If $y'' = 1 - x$, find $f(x)$.

In Exercises 41 and 42, assume that the acceleration due to gravity is 32 ft/s² downward, and assume there is no air resistance.

41. A ball is thrown vertically upward from the ground with an initial velocity of 64 ft/s. Find expressions for its height $s(t)$ and its velocity $v(t)$ after t seconds. How high will the ball rise? When will it strike the ground, and what will its velocity be at that instant?

42. From a balloon 1600 ft above the ground, a projectile is fired vertically downward at an initial velocity of 240 ft/s. How long will it take to reach the ground, and what will be its terminal velocity?

43. From just over the top edge of a sheer cliff 58.8 m above the canyon floor, a rock is thrown vertically upward at an initial velocity of 19.6 m/s. How high will the rock rise, and when will it strike the canyon floor? (Take acceleration caused by gravity to be 9.8 m/s^2.)

44. The accompanying figure shows the graphs of three functions f, g, and h. The function f is an antiderivative of one of the functions g or h. Which is it?

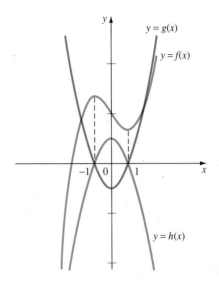

Chapter 4 Review Exercises

In Exercises 1 and 2, find the intervals on which f is increasing and those on which it is decreasing.

1. (a) $f(x) = x^4 - 4x^3 - 2x^2 + 12x$
   (b) $f(x) = (x + 2)^3(2x - 1)^2$

2. (a) $f(x) = (3x^2 - 8x)/(x^2 + 4)$
   (b) $f(x) = (x - 1)^{2/3}(x + 7)^2$

In Exercises 3–10, find all local maxima and minima. Also find absolute maximum and minimum values if they exist.

3. $f(x) = 5 + 4x + 2x^2 - x^3$

4. $f(x) = 3x^5 - 25x^3 + 60x$

5. $f(x) = x^2/(x - 1)$

6. $f(x) = (x^2 + 3)/(x^3 - 5x)$ on $[-2, 2]$

7. $f(x) = x^4 - 6x^2 - 8x + 5$ on $[-2, 3]$

8. $f(x) = \sqrt{x - 1}\,(x - 6)^2$ · on $[1, 10]$

9. $f(x) = \cos 2x + x$ on $\left[-\dfrac{\pi}{2}, \dfrac{\pi}{2}\right]$

10. $f(x) = 2\cos x + \sin 2x$ on $\left[-\pi, \dfrac{\pi}{2}\right]$

In Exercises 11 and 12, determine intervals on which the graph of f is concave up and those on which it is concave down.

11. (a) $f(x) = x^4 - x^3 - 3x^2 - 5x + 7$
   (b) $f(x) = 3x^5 - 15x^4 - 10x^3 + 90x^2 - 100$

12. (a) $f(x) = (x - 2)/(x + 1)^2$
   (b) $f(x) = (x - 2)^4(x + 1)^2$

In Exercises 13–30, discuss intercepts, symmetry, asymptotes, local maxima and minima, and points of inflection. Use this information to draw the graph of f.

13. $f(x) = (x^2 - 1)(x + 1)$

14. $f(x) = 2x^3 - 3x^2 - 12x + 4$

15. $f(x) = x^4 - 18x^2 + 60$

16. $f(x) = \dfrac{x^2}{x^2 - 4}$ **17.** $f(x) = \dfrac{8x}{x + 4}$

18. $f(x) = 3x^5 - 5x^3$ **19.** $f(x) = x^4 - 4x^3 + 15$

20. $f(x) = \frac{1}{3}(x - 1)^3(x + 3)$

21. $f(x) = x - \dfrac{2}{x-1}$

22. $f(x) = x^{1/3}(4+x)$

23. $f(x) = x - (1-3x)^{2/3}$

24. $f(x) = 3x^4 - 8x^3 + 6x^2 - 1$

25. $f(x) = \dfrac{4x}{(x-1)^2}$

26. $f(x) = \dfrac{4(1-x)}{x^2}$

27. $f(x) = 3x^{-1} - x^{-3}$

28. $f(x) = \dfrac{\sin x}{\cos x - 2}$

29. $f(x) = \dfrac{2\sin^2 x + 1}{2\sin^2 x - 1}$

30. $f(x) = \frac{1}{12}(3x^4 + 4x^3 - 30x^2 + 36x)$

31. A closed box with a square base is to be constructed so that its volume is 324 ft³. The material for the top and bottom costs $3 per square foot, and that for the sides costs $2 per square foot. Find the dimensions so that the cost will be a minimum.

32. Find the dimensions of the right circular cylinder of maximum volume that can be inscribed in a cone of base radius r and altitude h.

33. The largest box the United Parcel Service will accept is one for which the sum of the length and the girth (distance around) is 108 in. Find the dimensions of the box with square cross section having the greatest volume that can be sent via the United Parcel Service.

34. A car dealer sells an average of 100 cars of a certain type per month when the selling price is $15,000. The cost to the dealership is $11,000 per car. The dealer estimates that for every $200 in rebate, they will sell 10 more cars per month. What total rebate should the dealer offer to obtain the maximum profit?

35. An open-top cylindrical boiler pan is to have a volume of 250π in.³. The lateral surface is to be made of stainless steel and the bottom of copper. If copper costs twice as much as stainless steel, what should be the dimensions of the pan for minimum cost?

36. An oil drilling platform in the Gulf of Mexico is 9 km from point A, the nearest point on shore. A second oil drilling platform is 3 km from B, the nearest point on shore to it. The distance from A to B is 5 km. A supply depot is located at a point C on the shore between A and B in such a way that the sum of the distances from C to the two platforms is a minimum. What is the distance from A to C? (See the figure.)

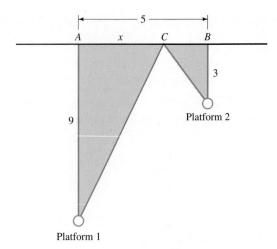

37. According to one model of population growth, the rate dx/dt at which a population grows is jointly proportional to the current size x of the population and the remaining capacity to grow, $m - x$, where m is the theoretical maximum size of the population. Find how large the population is when it is growing most rapidly.

38. To produce a certain manufactured item there is for each unit a fixed cost of $30 and a variable cost of $20 - 0.1x$, where the number x of items does not exceed 100. The unit price function is

$$p(x) = 700 + \frac{100}{x} - 0.3x$$

Find how many units should be produced and sold to maximize profit.

39. An 8-ft-high fence stands 1 ft from a building. A ladder is to be placed against the building and over the fence (see the figure). What is the length of the shortest ladder that can be used?

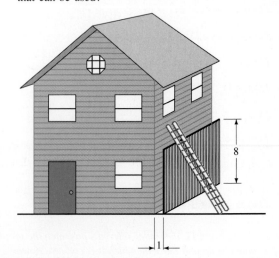

40. Car A and Car B are approaching a town on roads that are at right angles to each other. Car A is traveling at 80 km/hr and Car B at 55 km/hr. At the instant when Car A is 5 km from the town, Car B is 12 km from the town. How fast is the distance between the cars decreasing at that instant?

41. At the instant a hot-air balloon begins rising, a runner leaves from the takeoff point, running in a straight line. If the balloon rises at the constant rate of 22 ft/s and the runner goes 8 mph, find the rate at which the distance between the runner and the balloon is increasing after 10 min.

42. A yacht is cruising at 12 mph. A person standing in the stern 15 ft above the water is pulling in a rope (assumed taut) attached to a rowboat at the rate of 50 ft/min. How fast is the rowboat moving through the water when there are 39 ft of rope out?

43. A cone-shaped container is being filled with water at the rate of 4 ft^3/min, at the same time that water is being drained out through the bottom at the rate of 2 ft^3/min. The top radius of the container is 2 ft and its height is 8 ft. Find how fast the water level is rising when (a) the water level is 4 ft and (b) the container is half full.

44. At a track meet a TV camera operator is focusing the camera on the lead runner in a 100 m dash. The camera is positioned 9 m from the track, directly opposite the finish line. If the runner is going at the constant rate of 10 m/s, how many radians is the camera turning through per second at the instant the runner is 12 m from the finish line?

In Exercises 45–48, approximate the specified roots of the equation using Newton's Method, continuing until two successive approximations agree to the first five decimal places. Round the answer to four places.

45. The real root of $x^3 + 2x^2 + 2x - 3 = 0$

46. All real roots of $x^4 - 2x^2 + 5x - 8 = 0$

47. All real roots of $x^2 - \frac{1}{x} = 4$

48. All real roots of $x^{-2} = \sec x$

49. Use Newton's Method to find the maximum value of $f(x) = x \cos x$ on $(0, \pi/2)$ correct to six decimal places.

50. Show that if $a \neq 0$, $\sqrt[3]{a}$ can be approximated by the sequence of iterations

$$x_{n+1} = \frac{2x_n^3 + a}{3x_n^2}$$

and this will converge as long as x_0 has the same sign as a.

51. What value of n will ensure that in using Newton's Method, x_n will approximate the root of $x^3 - 3x^2 + 2x - 4 = 0$ that lies in $[2, 3]$ to an accuracy of ten decimal places, if x_0 is taken as 2.8? (*Hint:* First show that the root lies between 2.7 and 2.8.)

In Exercises 52 and 53, show that a root r of $f(x) = 0$ lies in the given interval and that f is monotone and its graph does not change direction of concavity there. Then find a suitable starting point x_0 that will ensure convergence of Newton's Method to the root r.

52. $f(x) = \dfrac{3}{x-1} - x$ on $(2, 3)$

53. $f(x) = \cos x - \sqrt{x}$ on $\left(\dfrac{\pi}{6}, \dfrac{\pi}{3}\right)$

54. Use Inequality 4.7 to find n such that the nth approximation by Newton's Method to the root of $\tan x - \sqrt{2x} = 0$ that lies in $(0, \pi/2)$ is correct to eight decimal places. What is x_n?

In Exercises 55–58, use L'Hôpital's Rule to find each limit.

55. (a) $\displaystyle\lim_{x \to 1} \frac{x^3 - 3x + 2}{2x^3 - 3x^2 + 1}$ (b) $\displaystyle\lim_{x \to 0} \frac{\sin x - \tan x}{x}$

56. (a) $\displaystyle\lim_{x \to \infty} \frac{1 - 3x + 2x^3}{3x^3 + 2x^2 + 5}$ (b) $\displaystyle\lim_{x \to 4} \frac{\sqrt[3]{2x} - 2}{\sqrt{x} - 2}$

57. (a) $\displaystyle\lim_{x \to 0} x^2 \cot 2x$ (b) $\displaystyle\lim_{x \to 0} (\sin 3x)(\csc 2x)$

58. (a) $\displaystyle\lim_{x \to -1} \left(\frac{x}{x+1} - \frac{4 - 3x^2}{x^2 + x} \right)$
(b) $\displaystyle\lim_{x \to \pi/4} (\sec 2x - \tan 2x)$

In Exercises 59 and 60, show that Rolle's Theorem is satisfied by f on the given interval, and find a number c that satisfies the conclusion of that theorem.

59. (a) $f(x) = x^3 - 3x - 2$ on $[-1, 2]$
(b) $f(x) = (x^2 - 1)/(x^2 + 2)$ on $[-1, 1]$

60. (a) $f(x) = x^{2/3}(1 - x)$ on $[0, 1]$
(b) $f(x) = \sqrt{1 - x}(x + 3)$ on $[-3, 1]$

In Exercises 61 and 62, show that the Mean-Value Theorem is satisfied by f on the given interval, and find a number c that satisfies the conclusion of that theorem.

61. (a) $f(x) = 2x^2 - 4x + 3$ on $[-1, 2]$

(b) $f(x) = x^3 - 2x^2 + 4x - 4$ on $[1, 3]$

62. (a) $f(x) = (x + 3)/(x - 2)$ on $[0, 1]$

(b) $f(x) = (3x - 4)^{4/3}$ on $[1, 4]$

63. (a) Suppose $f'(x) = 0$ for all x, and $f(2) = 7$. What is $f(5)$, and why?

(b) Suppose $f'(x) = g'(x)$ for all x, and $f(x) = x^2 + 4$. If $g(1) = 2$, what is $g(x)$, and why?

In Exercises 64–66, find the general antiderivative of f.

64. (a) $f(x) = 3x^2 - 2x + \dfrac{4}{x^2}$

(b) $f(x) = \dfrac{x^2 - 1}{\sqrt{x}}$

65. (a) $f(x) = 4 \sin x - 3 \cos x$

(b) $f(x) = \dfrac{\sec x + 2 \tan x}{\cos x}$

66. (a) $f(x) = \tan^2 x + \sec^2 x$

(b) $f(x) = \sin^2 x - \cos^2 x$

In Exercises 67–69, find the solution to the initial-value problem.

67. (a) $f'(x) = 3x^2 - 2/x^2;\ f(1) = 4$

(b) $\dfrac{dy}{dx} = 2 \sin x - x$; when $x = 0$, $y = 5$

68. (a) $f''(x) = 2x;\ f(1) = -3,\ f'(1) = 2$

(b) $y'' = 4$; when $x = 0$, $y = 7$, and $y' = 0$

69. (a) $f'''(x) = 1/x^4;\ f(-1) = 2,\ f'(-1) = 1,\ f''(-1) = 0$

(b) $\dfrac{d^4 y}{dx^4} = \cos x$; when $x = 0$, $y = 2$, $y' = -1$, $y'' = 1$, and $y''' = 0$

70. Find the function f such that $y = f(x)$ satisfies the differential equation

$$\frac{d^2 y}{dx^2} + x = 2$$

so that the graph of f passes through the point $(-1, 3)$ and has slope $\frac{1}{2}$ at that point.

71. A particle moves on a line directed positively to the right so that its acceleration at time t is $a(t) = 6t$. At time $t = 0$ it is at the origin and has initial velocity $v(0) = -12$. Consider $t \geq 0$.

(a) Find its velocity and position at time t.

(b) Find when and where it reaches its leftmost position.

(c) Find the intervals in which it is speeding up and those in which it is slowing down.

(d) Indicate its motion by means of a sketch.

72. From the edge of a building 128 ft high, a rock is thrown vertically upward at a speed of 32 ft/s. Find (a) how high the rock rises and (b) the speed with which it strikes the ground.

Chapter 4 Concept Quiz

1. Define each of the following:

(a)   an increasing function on I

(b)   a decreasing function on I

(c)   a monotone function on I

(d)   a local maximum value of f

(e)   a local minimum value of f

(f)   an absolute maximum value of f on I

(g)   an absolute minimum value of f on I

(h)   a critical point

(i)   concave up

(j)   concave down

(k)   an inflection point

(l)   an antiderivative of f

2. State each of the following:

(a)   The Extreme-Value Theorem

(b)   The First-Derivative Test for Extrema

(c)   The Second-Derivative Test for Extrema

(d)   L'Hôpital's Rule

(e)   Rolle's Theorem

(f)   The Mean-Value Theorem

(g)   Newton's Method

(h)   Two corollaries to the Mean-Value Theorem

3. Describe the graph of f in each case.

(a) $f'(x) > 0$, $f''(x) > 0$ on I

(b) $f'(x) > 0$, $f''(x) < 0$ on I

(c) $f'(x) < 0$, $f''(x) > 0$ on I

(d) $f'(x) < 0$, $f''(x) < 0$ on I

(e) $f'(x_0) = 0$, $f''(x_0) < 0$

(f) $f'(x_0) = 0$, $f''(x_0) > 0$

(g) $f(x_0)$ finite but $\lim_{x \to x_0} f'(x) = \infty$

(h) $f'(x_0) = 0$, $f'(x) < 0$ for $x \neq x_0$

4. Fill in the blanks.

(a)   If f is continuous at x_0 and _____ for $x < x_0$ and _____ for $x > x_0$, then f has a minimum value at x_0.

(b)   The limit $\lim_{x \to a}[f(x)/g(x)]$ is an indeterminate form of type $0/0$ if _____ and _____.

(c)   In Newton's Method, a root of $f(x) = 0$ is approximated using x-intercepts of the _____ line to the graph of f.

(d)   If $f(x_0)$ exists but $\lim_{x \to x_0^+} f'(x) = \infty$ and $\lim_{x \to x_0^-} f'(x) = -\infty$, then the tangent line to the graph of f at x_0 is _____ and the point $(x_0, f(x_0))$ is a _____ point on the graph.

(e)   According to the Mean-Value Theorem, if f is continuous on $[a, b]$ and differentiable on (a, b), there is a point c between a and b such that the _____ line at $x = c$ is parallel to the _____ line joining the points _____ and _____.

5. Which of the following are true and which are false?

(a)   If $f'(x) = g'(x)$ for all x, then $f(x) = g(x)$.

(b)   L'Hôpital's Rule states that when $\lim_{x \to a} f(x) = 0$ and $\lim_{x \to a} g(x) = 0$, then

$$\lim_{x \to a} \frac{f(x)}{g(x)} = \lim_{x \to a} \left(\frac{f}{g} \right)'(x)$$

(c)   If $f''(x_0) = 0$, then $(x_0, f(x_0))$ is an inflection point.

(d)   If $f'(x) \cdot f''(x) < 0$ for all x, then f is either increasing and concave down or decreasing and concave up.

(e)   If f is continuous and has a maximum at $x = x_0$, then $f'(x_0) = 0$.

APPLYING CALCULUS

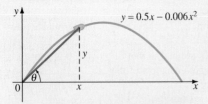

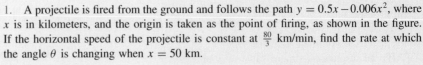

1. A projectile is fired from the ground and follows the path $y = 0.5x - 0.006x^2$, where x is in kilometers, and the origin is taken as the point of firing, as shown in the figure. If the horizontal speed of the projectile is constant at $\frac{80}{3}$ km/min, find the rate at which the angle θ is changing when $x = 50$ km.

2. The formula

$$\frac{1}{f} = \frac{1}{u} + \frac{1}{v}$$

expresses the relationship between the focal length f of a concave mirror or lens, the distance u of an object from the lens, and the distance v of its image from the lens. Suppose the focal length of a certain lens is 30 cm and an object is moving toward the lens at the rate of 3 cm/s. Find the rate at which the image is receding from the lens when the object is 90 cm away.

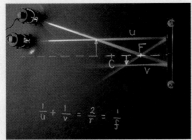

3. Campbell's and Progresso both sell soup in cans designed to hold an equivalent weight, approximately 19 ounces or 538 grams. The cans have the dimensions shown in the table. Assume that the soups are of equal density (weight per volume) and that both companies manufacture an equivalent amount of soup at equivalent costs. Thus, differences in production costs should be a result of differences in the costs of the cans, which include the tops, bottoms, and sides.

	Progresso	Campbell
Outside sizes		
diameter	3 3/8″	3 1/8″
height	4 5/16″	5″
calc. volume	38.6	38.4
On label		
net weight	19 oz; 538 g	19 oz; 539 g

(a) Which company should have the lower production costs?

(b) Suppose the materials used to make the cans are such that the tops and bottoms cost \$0.14/ft^2 and the sides cost \$0.08/ft^2. Which company has the lowest costs under these conditions?

(c) Suppose both companies require that the cans have a volume of 33.4 cu. in. Assuming that the same material is used for the tops, bottoms, and sides, what dimensions should the cans have to minimize the cost?

4. An engineer is designing a drainage canal that has a trapezoidal cross section. The bottom and sides of the channel are each L feet long, and the sides make an angle θ with the horizontal.

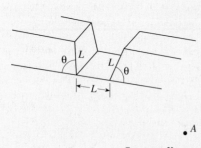

(a) Find an expression for the cross-sectional area of the channel.

(b) Determine the angle θ that will maximize the cross-sectional area.

5. A manufacturer has outlets at 3 locations A, B, and C (see the figure). The owner wishes to locate the plant at a point X so that total shipping costs to all outlets are as small as possible. Assuming that the number of shipments to all 3 outlets are equal, and that the cost of transportation is proportional to the distance, shipping costs will be minimized if x is chosen so that the sum $S(x)$ of the distances from X to A, B, and C is as small as possible. Using the simplifying assumption that the points A and B are symmetric with respect to C (i.e., the line connecting C to the midpoint M of segment AB is perpendicular to that segment as in the figure), you are asked to locate the point X by following the steps below.

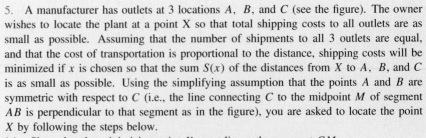

(a) Show that the minimizing point X must lie on the segment CM.

(b) Introduce rectangular coordinates so that C is the origin and the ray CM is the positive x-axis. If the coordinates of A are (a, b), what are the coordinates of B?

(c)   If we denote the coordinates of X by $(x, 0)$, what inequality(ies) must x satisfy? Show that S, which depends on x, is given by the formula

$$S(x) = x + 2\sqrt{(a-x)^2 + b^2}$$

(d)   Finally, show that if $a \leq b/\sqrt{3}$, $S(x)$ achieves its minimum at $x = 0$, and if $a > b/\sqrt{3}$, $S(x)$ achieves its minimum at $x = a - b/\sqrt{3}$.

6.   The Massachusetts Turnpike is 123 miles long. When entering the turnpike the driver of a passenger car is given a card indicating the date and time and listed are the toll charges at each exit off the turnpike. Suppose you enter the turnpike on the east end at 1:00 P.M. and travel the entire 123 miles of the turnpike. You arrive at the toll gate on the west end of the turnpike at 2:30 P.M. When the toll gate attendant examines your toll card he informs you that not only must you pay the toll but an additional $80 for speeding. He claims you have driven at least 75 mph at one point in your journey and the maximum turnpike speed limit is only 65 mph. When you protest, he insists your arguments are useless since he has calculus on his side. Why is the toll booth attendant correct in his claim that you were speeding?*

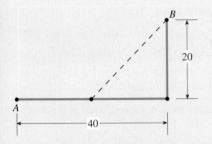

7.   A public-works department is to construct a new road between towns A and B. Town A lies on an abandoned road running east-west. Town B is 20 miles north of this road and 40 miles east of town A. An engineer proposes that the road be constructed by restoring a section of the old road leaving town A and joining it to an entirely new section of the old road at a point to be determined, connecting that point with town B (see the figure).

(a)   If the cost of restoring the old road is $200,000 per mile and the cost of the new road is $400,000 per mile, how much of the old road should be restored in order to minimize the department's cost?

(b)   If the cost of restoring the old road is $200,000 per mile, what would the maximum cost per mile of the new road have to be in order for the department to construct a direct route from A to B? [Hint: Treat the cost of the new road as an unknown parameter.]

(c)   Under what conditions would it make sense to construct an L-shaped route connecting A and B?

(d)   Extending the problem: Treat both the cost of restoring the old road and the construction of the new road as unknown parameters. Determine a relationship between the two costs that makes a direct link between A and B the most economical.

8.   A trucking company would like to determine the highway speed that it should require of its drivers. The decision is to be made purely on economical grounds, and the two primary factors to be considered are driver wages and fuel consumption. Wage information is easily obtained: drivers earn from $11.00 to $15.00 an hour, depending on experience. Incorporating the fuel consumption question is much more difficult and the company has hired you as consultants in order to solve the problem for them. Correctly assuming that fuel consumption is closely related to fuel economy at various highway speeds, they have provided you with the following statistics taken from a U.S. Department of Transportation study:[†]

---

*Adapted from J. Chover, *The Green Book of Calculus.*

[†]U.S. Department of Transportation. *The Effect of Speed on Truck Fuel Consumption Rates*, by E. M. Cope. [Washington]: U. S. Dept. of Transportation, Federal Highway Administration, Office of Highway Planning, Highway Statistics Division. 1974. (TD2.2:Sp3)

Miles per Gallon at Selected Speeds

Vehicle	50 mph	55 mph	60 mph	65 mph
Truck #1	5.12	5.06	4.71	*
Truck #2	5.41	5.02	4.59	4.08
Truck #3	5.45	4.97	4.52	*
Truck #4	5.21	4.90	4.88	4.47
Truck #5	4.49	4.40	4.14	3.72
Truck #6	4.97	4.51	4.42	*

*The truck could not go this fast.

Driver wage	Ideal speed
$11.00	
$12.00	
$13.00	
$14.00	
$15.00	

Compute the average mpg at each speed and fit a cubic polynomial through the points. This gives a function relation f between miles per gallon and miles per hour. Then let D represent distance traveled, F the cost of fuel per gallon, and W, the driver's hourly wage. Write the total cost in terms of W, D, $F(x)$, and x, where x represents speed. Apply extrema theory to the cost function obtained by letting $F = 1$ and $W = 11$, 12, 13, 14, and then 15. Complete the table in the margin.

9. The Babylonians discovered a rule for approximating the square root of a positive number k. Begin by making a reasonable estimate a of $\sqrt{k}$. Then

$$b = \tfrac{1}{2}\left(a + \frac{k}{a}\right)$$

will be a better estimate. Continue the process.
(a) Use this method to approximate $\sqrt{13}$ to five decimal places.

(b) Show that this method of approximating square roots is a special case of Newton's method.

10. Given a function F defined on an interval I a point $x \in I$ is a *fixed point* of F if $F(x) = x$. Many important applications of mathematics involve finding fixed points of functions. It can be shown using the Intermediate-Value Theorem that if F is continuous on the interval $[a, b]$ and satisfies

$$a \leq F(x) \leq b$$

for all $x \in [a, b]$, then F has at least one fixed point on the interval. Moreover, it can be shown that if F is differentiable and satisfies $|F'(x)| < r$, for $a < x < b$, where $0 < r < 1$, then F has exactly one fixed point on $[a, b]$.

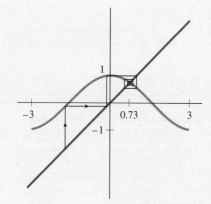

(a) One method for finding a fixed point of a given function F is by *iteration*. If F satisfies the second condition above, the sequence (x_n) defined by x_1, $x_2 = F(x_1)$, $x_3 = F(x_2), \ldots, x_{n+1} = F(x_n), \ldots$, where x_1 is an initial "guess," will converge to the unique fixed point. Show that $F(x) = \cos x$, $-3 \leq x \leq 3$, satisfies the two conditions stated above and find an approximate solution of $F(x) = x$, accurate to 4 decimal places.

(b) Since a fixed point of a given function F is a root of the equation $G(x) = 0$, where $G(x) = F(x) - x$, fixed points can be found by Newton's method. Indeed, in most instances, Newton's method is faster than iteration. Determine the recursion formula corresponding to Newton's method for finding the roots of $G(x) = F(x) - x = 0$. Use this method to approximate the fixed point of $F(x) = \cos x$ on $[0, 3]$.

Anna J. Schwartz

Is mathematics necessary or merely useful in economics? For instance, is calculus an essential course to take if you plan to become an economist?

It's definitely more than just useful. Mathematics helps explain economic phenomena. We express the problem of how one variable changes in relation to another by means of a functional relationship between variables. The derivative measures the instantaneous rate of growth of one quantity with respect to another and approximates the actual average rate of growth when the variation considered is small. In economics, derivatives are of special importance, especially in dealing with cost and profit functions. In econometrics, which is the main empirical [experimental] way in which economics is done, the basic mathematical tool is calculus. If you plan to major in economics, it makes sense to take both calculus and linear algebra, especially if you want to go on to graduate school.

What exactly is econometrics?

Econometrics is a combination of economic theory, mathematics, and statistics. All three disciplines are needed to understand the quantitative relationships that are inherent in modern economic life. Economic theory, which is derived from statistical data and other facts, suggests hypotheses to be tested by empirical analysis. Theory identifies the factors thought to be involved in the behavior of individual economic units and the markets in which they interact, as well as economic aggregates, such as aggregate consumption and aggregate investment. Econometric methods apply mathematics, such as calculus-based probability, to the analysis of economic data. The aim is to test the validity of the theory under consideration with appropriate methods and data. Empirical economic research deals with functions with variables expressed as levels or as rates of change. For this purpose, one needs familiarity with calculus.

Calculus deals in part with rates of growth. In what sense is the economy a result of many competing rates of growth?

The question is not so much rates of growth, per se, but the functional relationship among variables with different rates of growth. In the United States, until the end of World War II, the rate of growth of (per-capita) money holdings was greater than the rate of growth of (per-capita) income. Since then, on average, the rate of growth of money holdings has been lower than the rate of growth of income. It is an interesting economic question to explain this difference in behavior.

Do you remember a particular problem?

One such problem is hyperinflation, the radical devaluing of currencies, in the countries of post–World War II Europe. In the problem of hyperinflation you want to know how acceleration (of growth) in the money supply interacts with the demand for money. The empirical problem is to identify the variables that determine the amount of money people want to hold, rather than spend. Tests of various hyperinflations suggest that people have a well-defined demand for money in terms of the cost of holding it. Two expected costs of holding money are the expected yield on some alternative asset, and the rate at which the value of money is expected to decline due to rising prices. To estimate the second component of costs, we assume that individuals modify their expectations about future prices on the basis of the error in the period before. Therefore, previous price changes determine expectations of current prices. In Europe, hyperinflation was a really daunting problem that was eventually solved by replacing the money that had lost value, because of oversupply and the public's psychological rejection of that money, with substitute currency that had real purchasing power.

Could you give us some examples of the use of the derivative in economics?

In economics, two central concepts are *average* and *marginal* value. The average variation of total cost (C_t), with respect to total output (O_t), is the total cost to the firm (at any level of output) divided by the total output; and the marginal variation is the variation of C_t for small values of O_t. Marginal cost, however, is the change in C_t at a certain level of output brought about by a very small increase in the quantity of output from that level. The marginal value concept is basic to many economic ideas about supply and demand, costs and prices, production rates, and revenues. Of course it is the derivative of the function that relates C_t and O_t. The numerical value of the derivative measures how rapidly the function C_t increases or decreases with respect to O_t. The results indicate the nature of the function over a whole range of values of the variable O_t. Determination of maximum or minimum positions is important in economic theory. For example, a firm chooses output or price that will produce the largest net revenue; similarly, the consumer seeks the highest position on a scale of consumption possibilities. The variation of a function with more than one independent variable involves differentiation to obtain the differential of a function. For example, production of a certain item requires inputs of capital and labor. Via differentiation we can calculate the marginal rate of substitution of capital for labor in the production of the item. Similar calculations enable us to determine the marginal rate of substitution by the consumer of one item for another.

What about integration? Is it useful as well?

Very. Economists use integration to measure the area under a curve, just as other scientists do. We commonly plot graphs of functions using known or assumed values of variables. The market demand for an item can be represented by three functions and curves showing total, average, and marginal revenue, respectively, for various outputs of the item. Total revenue from any output is then the area (the integral) under either the average or marginal revenue curves. A similar result holds for any pair of average and marginal curves.

Integration is an important tool in both microeconomics, the study of the economic behavior of individuals and of firms, and macroeconomics, the study of the economy as a whole. In microeconomics you might want to study an integral that maximizes profit. In macroeconomics you might want to use integration to evaluate aggregate supply and demand. For example, supply and demand of a commodity may depend not only on present prices but also on prices in the recent past. Farmers might plant more wheat this year if last year's price was high and plant less this year if last year's price was low. This relationship can be expressed and computed in integral form, so calculus makes this kind of application possible. This kind of decision making is very important to industry and the economy as a whole.

Calculus is also useful in problems involving capital and interest. Logarithmic functions are particularly important for this purpose. Another application of the integral is the capital value at the present time of an income stream from some asset. The capital value depends on the size of the income stream, the number of years of the income flow, and the rate of interest. Yet another example is consumer surplus, which is found by the area under a demand curve that measures the consumer value of an item in excess of its price.

Is calculus used in developing theoretical economic models?

Yes. Calculus is useful both in an empirical way and in setting up theoretical structures or models. For instance, Arthur Laffer, the originator of supply-side economics, drew a curve that showed the revenue that a government obtained with various marginal rates of income tax. He used this curve to assert that governments could obtain more revenue within a certain range of marginal rates of taxation and that the revenue would diminish as the rate of marginal taxation rose. This "Laffer curve" was used as a theoretical basis and computed by measuring different points at different marginal rates of taxation using derivatives. These models may also involve the use of integration. For instance, you could set up a model that assumes people will try to maximize what turns out to be the integral of three elements that determine the behavior of economic time series: (1) maximizing quantities and prices subject to budget constraints as a function of time, (2) minimizing the effect of erratic shocks, and (3) minimizing the effect of excessive inventories. These abstract concepts have become part and parcel of the way economists think and work.

When did you realize how important the tools of calculus were to your profession?

Not soon enough. When I was a graduate student, it just wasn't part of prescribed graduate training. Now it's taken for granted. Every graduate student has to take a course in mathematics for economists, which includes at least some calculus. I learned by doing and by working with people who had better mathematical training. They introduced me to the kinds of exercises and applications that one would ordinarily encounter in class. My exposure was on the job where there were real problems to solve.

I wouldn't suggest that the use of calculus is the end of the story on how economists use mathematics in their work. There are economists who are working with chaos theory or fractals, for instance, which is a different mathematical subdiscipline. These methods don't seem to have the generality of calculus, however.

What if a student decided not to go to graduate school? Could this person still become a working economist?

Well, the calculus course they're taking now would give them the basic concepts—I don't want to alarm anyone about needing to have exceptional mathematical aptitude, but I think a graduate degree is very important in this day and age. You might get an entry-level position without a graduate degree but remain in that position throughout your career, while those with Ph.Ds were assigned the more interesting work. Your salary wouldn't grow at the same pace as someone with a graduate degree either.... There's an economic fact that you don't need calculus to understand.

Do you see these economic issues as being more understandable if you understand calculus? Would Isaac Newton, for instance, have understood the application of calculus to today's economic issues?

Well, I'd be speculating of course, but Newton was a genius and in his later years was Britain's Master of the Mint. As such he had a major influence over the money supply. He took his job very seriously, personally supervising the hanging of guilty counterfeiters. At one point he established a "Mint price" for guineas, which overvalued them and therefore inadvertently displaced the then-current silver standard by the gold standard. If the Mint price for guineas had been lower or the price of silver higher, no change in the standard would have occurred.

Anna J. Schwartz is a native New Yorker and a member of the research staff at the National Bureau of Economic Research. Recently named a Distinguished Fellow of the American Economic Association, she has collaborated with Milton Friedman on studies of monetary economics, including *A Monetary History of the United States, 1867–1960* (1963). She currently serves on the editorial boards of three economics journals. Dr. Schwartz holds a B.A. from Barnard and M.A. and Ph.D. degrees in Economics from Columbia University.

5

THE INTEGRAL

In Chapters 3 and 4 we explored **differential calculus**, based on the concept of the derivative. In this chapter we begin the study of the other major branch of calculus, **integral calculus**, based on the concept of the *definite integral*, which we will define in Section 5.2.

In differential calculus, we used the problem of defining tangent lines to curves (a geometric problem) and the problem of computing the velocity of a moving body from its change in position (a physical problem) to develop the definition of the derivative in terms of a limit. In a similar manner we will use the problem of determining the area under a curve (a geometric problem) and the problem of determining the total distance traveled by an object with a given velocity (a physical problem) to develop the definition of the integral of a function, also in terms of a limiting process. Table 5.1 shows the relationship between these fundamental problems and the two branches of calculus. It is critically important to understand the connection between the two branches of calculus, which we will make precise in Section 5.3.

TABLE 5.1

	Differential calculus	Integral calculus
Geometric problem	Tangent lines to curves	Area under a curve
Physical problem	Velocity of a moving body (from its distance traveled over time)	Total distance traveled by a moving body with a given velocity

The study of the area problem—computing the area under a curve—and the theory of integration that leads from it have intrinsic mathematical value. That is, they are worth studying in their own right. They become even more important when we realize we can use integrals to solve an incredible range of diverse problems. For example, we can find volumes of solids, lengths of curves, forces on dams, the work done by variable forces, and probabilities of certain events. Still more applications of the integral occur in such fields as economics, biology, and chemistry. It is testimony to the power of mathematics that a single abstract concept, such as the integral, applies to a huge variety of real-world problems.

5.1 THE AREA PROBLEM AND THE DISTANCE PROBLEM

Before beginning study of the integral, let us consider two geometric problems involving area and distance. Imagine that we want to estimate the *area* of the "two-point zone" at one end of the basketball court in Boston Garden, as shown in the photograph. The difficulty in estimating this area results from its shape, which appears to be a rectangle surmounted by the arc of a circle (of unknown diameter).

Instead of trying to estimate the area directly, we establish an upper and a lower bound for the area in question by counting the wooden floor tiles that enclose the area (A_{max}) and those that are entirely inside the area (A_{min}), as shown in Figure 5.1(a). Since the court measures 94 ft by 50 ft, and ten tiles span the width of the court, each tile is 5 ft square and has area 25 ft^2. There are 34 whole tiles inside the two-point zone; therefore, A_{min} is 34×25 ft$^2 = 850$ ft^2. Similarly, there are 48 whole tiles that include all or part of the two-point zone; therefore, A_{max} is 48×25 ft$^2 = 1200$ ft^2. From our crude visual estimate we know only that $850 < A < 1200$. How can we improve our estimate? What if we had smaller tiles? If we subdivide each tile to create 25 tiles of 1 ft^2 each, can we use the smaller tiles to approximate the area of the two-point zone more accurately? Why? What would happen if we made the tiles very small so that their areas approached zero? Could the limit of this process somehow be used to find the area we are trying to estimate? Before we address these questions in this section, let's look at a situation involving *distance*.

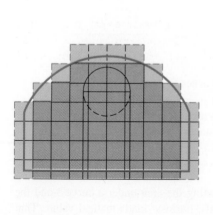

(a)

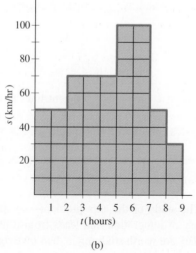

(b)

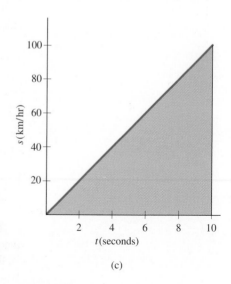

(c)

FIGURE 5.1

The graphs in Figures 5.1(b) and 5.1(c) represent the speed of a car in km/hr over the given period of time. In Figure 5.1(b) we see the average speed of the automobile over a several *hour* period and in 5.1(c) we see the automobile's speed over a several *second* period. By inspecting Figure 5.1(b), we know that the car was traveling at 100 km/hr five hours into the trip (we ignore the time spent accelerating to the average speed). What total distance did the car travel during the nine-hour trip? We can see from the graph that the area under the speed "curve" is the product of the speed of the car and the time elapsed (kilometers/hour × hours = kilometers). If we sum the individual areas of 10 km/hr increments ×1 hr periods, we get the total distance traveled, in this case 59 km. Note that the area under the speed curve represents distance, not distance squared. Similarly, from Figure 5.1(c), we see that the car accelerates from 0 to 100 km/hr in 10 seconds (1/360 hr). What distance did the car travel during this 10-second period? From the right-triangular area under the curve, $A = (1/2)(\text{base})(\text{height})$; that is, $(1/2)(1/360 \text{ hr}) \times (100 \text{ km/hr}) = 5/36$, or about 0.139 km. This distance is not as simple to conceive as that for the average speed situation in Figure 5.1(b), but the calculation is the same in that we find distance traveled by computing the area under a curve.

Both the area problem and the distance problem are solved using the general mathematical notion of the integral (although an integral does not have to be an area or a distance). In this chapter we will make these ideas more precise. Let's look at each type of problem in more detail.

The Area Problem—A Geometric Problem

We continue our discussion of the area problem. Specifically, we want to find the area, as shown in Figure 5.2, bounded above by the graph of a continuous nonnegative function f, below by the x-axis, and on the left and right by the vertical lines $x = a$ and $x = b$. We will refer to the area of this shaded region as *the area under the graph of f from x = a to x = b*. When there is no ambiguity about the curve and interval being discussed, we can simply say *the area under the curve*. The real question is: How do we *define* what we mean by the area of such a region?

Area formulas for many of the basic geometric figures date back to the earliest mathematical writings thousands of years ago. The area of a rectangle is "base times height," a triangle has area "one-half base times height," and the area of a polygon can be found by dividing it into triangles (Figure 5.3).

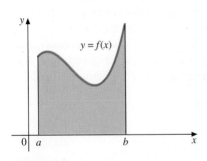

$y = f(x)$

FIGURE 5.2

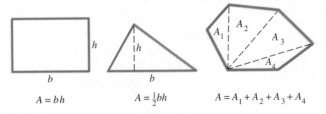

$A = bh$ $A = \frac{1}{2}bh$ $A = A_1 + A_2 + A_3 + A_4$

FIGURE 5.3
A parabolic region

FIGURE 5.4

FIGURE 5.5
Area approximated by triangles

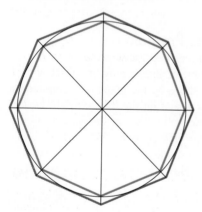

FIGURE 5.6

The problem of finding areas of regions with curved boundaries was of great interest to the Greeks, especially Archimedes (287–211 B.C.), who devised the "method of exhaustion" to solve special cases of this problem. For example, Archimedes was able to find the areas of certain parabolic regions (Figure 5.4) by inscribing an increasing number of triangles in the region (Figure 5.5), getting at each stage a better approximation to the actual area. In this manner the region is eventually exhausted, and the area is the limit of the approximations.

Archimedes also gave arguments for the areas of circles and spirals. In the case of circles, he extended his "method of exhaustion" to what has been called his "method of compression." Here the circle is inscribed (red lines) and circumscribed (blue lines) with polygons constructed from isosceles triangles (Figure 5.6). The area of the circle is then compressed between the larger and smaller approximations, and once again the area is the limit of these approximations.

The first solutions to the general area problem took almost another 2000 years when, independently, solutions were given in 1669 by Sir Isaac Newton, the great British mathematician and physicist, and in 1673 by the German mathematician Gottfried Wilhelm Leibniz. The technique we will describe and use was developed during the seventeenth and eighteenth centuries, culminating with the work of another German mathematician, Georg Riemann (1826–1866).

There are several reasonable ways of approximating the area of a region such as the one in Figure 5.2, but we concentrate on only two. The approach we will take is similar to Archimedes' "method of compression." The first method is called the method of **inscribed rectangles**. In this method we divide the interval $[a, b]$ into a number of subintervals. For simplicity, we take them to be of equal width, though this specification is not required. Next, we construct rectangles that have these subintervals as bases and such that the height of each rectangle is the minimum value of $f(x)$ for all x in the subinterval forming the base. We know by the theorem on extreme values for continuous functions (Theorem 4.2) that f does assume a minimum value (as well as a maximum value) on each of the closed subintervals. In Figure 5.7 we have shown two sets of inscribed rectangles. In part (b) we have used a finer subdivision than in part (a), causing the rectangles to be thinner and more numerous.

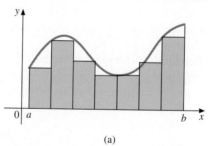

(a)

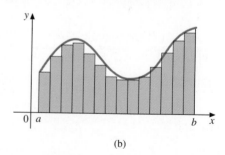

(b)

FIGURE 5.7

We refer to the sum of the areas of the inscribed rectangles for a given subdivision as a **lower sum**. Although we have not yet defined the area under the graph, intuitively it would appear that a given lower sum approximates that area from below. That is, each lower sum is an underestimate of the area. Furthermore, as we increase the number of rectangles, thereby making the widths smaller and smaller, the rectangles exhaust more and more of the area, so the approximations become increasingly better.

A second method is to use the maximum value of f on each subinterval, instead of the minimum value. Then we arrive at what we call **circumscribed rectangles**, as in Figure 5.8, and the sum of their areas, called an **upper sum**, approximates the area from above.

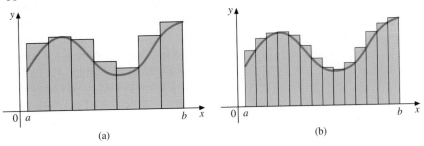

(a)   (b)

FIGURE 5.8

For any particular subdivision of $[a, b]$, we always have

$$\text{lower sum} \leq \text{upper sum}$$

It appears that as we allow the number n of subintervals to increase indefinitely (that is, let $n \to \infty$), the lower and upper sums approach each other, and their common limit is the area we are seeking.

Before we develop more general methods of approximating the area under a curve, we will illustrate the procedure we have described with an example of a region whose area we already know, to show that our method does give the correct answer in this case.

EXAMPLE 5.1   Approximate the area under the graph of the function $f(x) = x - 1$ from $x = 1$ to $x = 3$, using both inscribed and circumscribed rectangles.

Solution   The graph is shown in Figure 5.9. Notice that the region is triangular, so we know in advance that the area is

$$A = \frac{1}{2}bh = \frac{1}{2}(2)(2) = 2$$

Let us introduce the rectangles by partitioning the interval $[1, 3]$ into subintervals of equal lengths. Figure 5.10 shows the interval $[1, 3]$ divided into two such subdivisions, figure (a) with 8 equal parts and figure (b) with 16 equal parts; the resulting inscribed and circumscribed rectangles are displayed. We

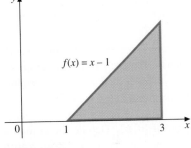

$f(x) = x - 1$

FIGURE 5.9

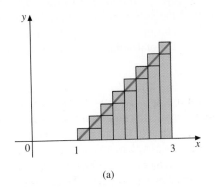

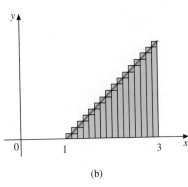

(a)   (b)

FIGURE 5.10

can now compute the lower sum by multiplying the base of each rectangle by its height, and then summing the areas of the rectangles. To find the base, we divide the width of the total interval from 1 to 3 by the number of rectangles. So for 8 rectangles, each base is $(3 - 1)/8 = 2/8 = 1/4$. The height of each lower rectangle is computed by substituting the left endpoint of each subinterval for x in $f(x) = x - 1$. For the upper sum, we use the right endpoints of the subintervals instead. Choosing left endpoints to get the heights of lower rectangles and right endpoints to get the heights of upper rectangles is typical for all *increasing* functions, such as the one in this example. In the case of *decreasing* functions these choices would be reversed. For our function, using 8 rectangles, the lower sum (L_8) and the upper sum (U_8) are then, respectively,

$$L_8 = \tfrac{1}{4}(1 - 1) + \tfrac{1}{4}(\tfrac{5}{4} - 1) + \tfrac{1}{4}(\tfrac{6}{4} - 1) + \tfrac{1}{4}(\tfrac{7}{4} - 1) + \tfrac{1}{4}(\tfrac{8}{4} - 1)$$
$$+ \tfrac{1}{4}(\tfrac{9}{4} - 1) + \tfrac{1}{4}(\tfrac{10}{4} - 1) + \tfrac{1}{4}(\tfrac{11}{4} - 1)$$
$$= \tfrac{1}{4}(0 + \tfrac{1}{4} + \tfrac{2}{4} + \tfrac{3}{4} + \tfrac{4}{4} + \tfrac{5}{4} + \tfrac{6}{4} + \tfrac{7}{4}) = 1.75$$

and

$$U_8 = \tfrac{1}{4}(\tfrac{5}{4} - 1) + \tfrac{1}{4}(\tfrac{6}{4} - 1) + \tfrac{1}{4}(\tfrac{7}{4} - 1) + \tfrac{1}{4}(\tfrac{8}{4} - 1) + \tfrac{1}{4}(\tfrac{9}{4} - 1)$$
$$+ \tfrac{1}{4}(\tfrac{10}{4} - 1) + \tfrac{1}{4}(\tfrac{11}{4} - 1) + \tfrac{1}{4}(3 - 1)$$
$$= \tfrac{1}{4}(\tfrac{1}{4} + \tfrac{2}{4} + \tfrac{3}{4} + \tfrac{4}{4} + \tfrac{5}{4} + \tfrac{6}{4} + \tfrac{7}{4} + \tfrac{8}{4}) = 2.25$$

The error involved is less than or equal to $U_8 - L_8 = 0.50$.

Table 5.2 shows the computation using the number of rectangles n doubling in sequence from 4 to 512. It is easy to use a spreadsheet or a CAS to compute the values in the table. Notice that as the number of subintervals increases, the upper sums decrease toward 2, the lower sums increase toward 2, and the differences between them decrease. The data suggest that the limit of these approximations is the actual area. ∎

TABLE 5.2

n	U_n	L_n	$U_n - L_n$
4	2.5	1.5	1
8	2.25	1.75	0.5
16	2.125	1.875	0.25
32	2.063	1.938	0.125
64	2.031	1.969	0.0625
128	2.016	1.984	0.0312
256	2.008	1.992	0.0156
512	2.004	1.996	0.0078

Formulas for Upper and Lower Sums

To arrive at some general computational formulas for upper and lower sums, we introduce the following notation. Let A denote the area under the graph of the nonnegative, continuous function f from $x = a$ to $x = b$. Let the points of the subdivision of $[a, b]$ into n subintervals be denoted by $x_0, x_1, x_2, \ldots, x_n$, with endpoints $a = x_0$ and $b = x_n$, such that $x_0 < x_1 < x_2 < \cdots < x_n$. Such a set of ordered points is said to form a **partition** of $[a, b]$. Denote the common length of the subintervals by Δx. Since the distance on the x-axis from a to b is $b - a$, and there are n subintervals, it follows that the length of each subinterval is the total length divided by the number of subintervals, which we can write as

$$\Delta x = \frac{b - a}{n}$$

Let m_k denote the minimum value of $f(x)$ on the kth subinterval, $[x_{k-1}, x_k]$, and let M_k denote the maximum value of $f(x)$ on this subinterval. So m_1 is the minimum height from the x-axis to the graph of f on the interval $[x_0, x_1]$, m_2 the minimum height on the interval $[x_1, x_2]$, and so on. Similarly, $M_1, M_2, \ldots, M_n$ are the maximum heights. These meanings are indicated in Figure 5.11, where we have shown a typical subinterval.

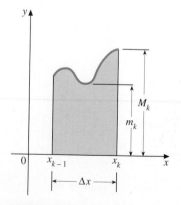

FIGURE 5.11

Denote the lower sum for this subdivision by L_n and the upper sum by U_n, where the subscript n indicates the number of subintervals. Then

$$L_n = m_1 \Delta x + m_2 \Delta x + m_3 \Delta x + \cdots + m_n \Delta x$$
$$= (m_1 + m_2 + m_3 + \cdots + m_n) \Delta x \tag{5.1}$$

Similarly,

$$U_n = M_1 \Delta x + M_2 \Delta x + M_3 \Delta x + \cdots + M_n \Delta x$$
$$= (M_1 + M_2 + M_3 + \cdots + M_n) \Delta x \tag{5.2}$$

EXAMPLE 5.2   Find L_n and U_n for the area under $y = f(x) = x^2 + 1$ from $x = 0$ to $x = 2$, for $n = 4, 10, 20, 100,$ and 500.

Solution   Figure 5.12 shows the inscribed and circumscribed rectangles for $n = 4$. In this case, $\Delta x = \frac{b-a}{4} = \frac{2}{4} = \frac{1}{2}$. Since the graph is increasing on the given interval, the minimum value of the function on each subinterval occurs at the left endpoint and the maximum value at the right endpoint. So we have

$$
\begin{array}{ll}
m_1 = f(0) = 1 & M_1 = f(0.5) = 1.25 \\
m_2 = f(0.5) = 1 + (0.5)^2 = 1.25 & M_2 = f(1) = 2 \\
m_3 = f(1) = 1 + (1)^2 = 2 & M_3 = f(1.5) = 3.25 \\
m_4 = f(1.5) = 1 + (1.5)^2 = 3.25 & M_4 = f(2) = 5
\end{array}
$$

Thus, by Equation 5.1,

$$L_4 = (1 + 1.25 + 2 + 3.25)(0.5) = 3.75$$

TABLE 5.3

n	L_n	U_n
4	3.75	5.75
10	4.28	5.08
20	4.47	4.87
100	4.6268	4.7068
500	4.6587	4.6747

and by Equation 5.2,

$$U_4 = (1.25 + 2 + 3.25 + 5)(0.5) = 5.75$$

The calculations for the other values of n are carried out in a similar way. The results are shown in Table 5.3. (We used a spreadsheet to compute these values, which could also be easily computed with a CAS or a programmable calculator.) We conclude that the area under the curve is between 4.6587 and 4.6747.   ∎

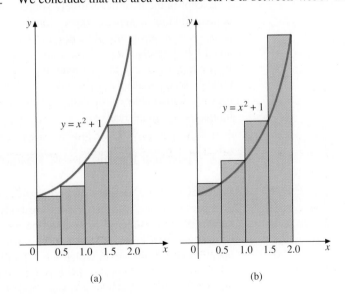

(a)   (b)

FIGURE 5.12

We have been discussing the area under the graph of a continuous, non-negative function without having *defined* the area, proceeding on an intuitive basis. Now we are ready to give a definition. The upper sums $U_1, U_2, U_3, \ldots$ form a *sequence*, which we indicate briefly by $\{U_n\}$. Similarly, we denote the sequence $L_1, L_2, L_3, \ldots$ of lower sums by $\{L_n\}$. When the terms of a sequence come arbitrarily close to some fixed value, say A, as $n \to \infty$, we say that the sequence *converges* to A. (Recall that we mentioned sequences in Chapter 4 in connection with Newton's Method. We will study sequences in more detail in Chapter 10.) As our examples illustrate, and as our intuition tells us, the two sequences $\{U_n\}$ and $\{L_n\}$, of upper sums and lower sums, respectively, both converge to a common limit. It is this common limit that we take as the definition of the area. The formal proof that $\{U_n\}$ and $\{L_n\}$ converge to a common limit for continuous functions involves a concept called *uniform continuity* that is studied in advanced calculus courses.

Definition 5.1 Area Under a Curve	Let f be a continuous, nonnegative function on the closed interval $[a, b]$. Then the area A of the region under the graph of f and above the x-axis from $x = a$ to $x = b$ is the common limit of the sequences $\{U_n\}$ and $\{L_n\}$ of upper and lower sums, as $n \to \infty$. That is, $$A = \lim_{n\to\infty} U_n, \text{ or equivalently, } A = \lim_{n\to\infty} L_n$$

REMARK ——————————————————————————————

■ Since the area A lies between U_n and L_n for every n, it follows that the **error** in using either U_n or L_n to approximate A cannot exceed $U_n - L_n$.

——————————————————————————————————————

Except in the case of relatively simple functions, such as polynomials of low degree, finding a general expression for L_n or U_n and determining the limit as $n \to \infty$ is prohibitively difficult. (See Example 5.5 and Exercises 10, 12, and 14 in Exercise Set 5.1 for some cases in which this *can* be done.) As in Example 5.1, we can approximate the limit (and hence the area) to any desired number of decimal places by taking n large enough. In Section 5.3, we will prove a remarkable result that enables us to find exact values of many such limits, using antiderivatives.

The Distance Problem—A Physical Problem

Consider next the problem of finding the distance covered by an object (such as a person, a car, a train, or a falling body) moving in a straight line, with a known positive velocity, for a fixed period of time. Of course, if the velocity is constant, you know the answer (distance = rate × time), so the problem of interest to us is the one in which the velocity (the rate) changes with time. Call this velocity $v(t)$. Suppose, for example, that the velocity of a relay runner in meters per second, measured at one-second intervals for a period of 5 seconds, is as shown in the table. We want to estimate the distance the runner covers.

t	0	1	2	3	4	5
$v(t)$	3.3	3.9	4.5	5.0	5.5	5.8

We assume that throughout the time interval, the velocity is increasing. Thus, during the first one-second interval, the runner went at least 3.3 m/s and at most 3.9 m/s. So the runner covered a distance of no less than $(3.3)(1) = 3.3$ m (using $d = rt$) and no more than $(3.9)(1) = 3.9$ m. Similarly, during the next one-second interval the runner went at least 3.9 m and at most 4.5 m. Continuing in this way, we see that the total distance is between

$$3.3 + 3.9 + 4.5 + 5.0 + 5.5 = 22.2 \text{ m}$$

and

$$3.9 + 4.5 + 5.0 + 5.5 + 5.8 = 24.7 \text{ m}$$

Suppose now that the runner's velocities are known at half-second intervals, as shown in the table below. Reasoning as above, in the first half-second interval the runner's speed was at least 3.3 m/s, so the distance covered was at least $(3.3)(0.5) = 1.65$ m. Similarly, the runner's speed during this time interval was at most 3.59 m/s, so the distance covered was at most $(3.59)(0.5) = 1.795$ m.

t	0	0.5	1	1.5	2	2.5	3	3.5	4	4.5	5
$v(t)$	3.3	3.59	3.9	4.26	4.5	4.90	5.0	5.26	5.5	5.66	5.8

Continuing in this way, using the velocity at the beginning of each half-second interval to get the underestimate of distance and the velocity at the end to get the overestimate, we obtain

$$22.93 \text{ m} \leq \text{distance} \leq 24.19 \text{ m}$$

If we knew velocities at even smaller time intervals, we could come closer with our low and high estimates.

If the underestimates for the distance in this example seem to resemble a lower sum for an area under a curve and the overestimates an upper sum, it is no accident. In fact, when we draw the graph of velocity versus time, as we have done in Figure 5.13, the reason becomes clear. We have shown inscribed rectangles for subintervals of width 1 (representing a time of 1 second). The area of the first rectangle is $(3.3)(1) = 3.3$, but this value is also the distance that we obtained previously when we were underestimating the distance covered in the first second. In a similar way, the area of each rectangle can be interpreted as a distance (its height represents velocity and its base, time, so the product

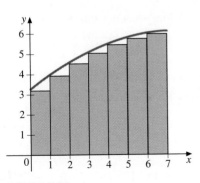

FIGURE 5.13

represents distance). So the lower sum L_5 for the area is precisely the underestimate, 22.2, that we obtained earlier. In a similar way, by using circumscribed rectangles, we find that the upper sum U_5 is the overestimate of distance, 24.7.

You should be convinced by now that the problem of finding area under a curve and the problem of finding distance when velocities are known are basically the same. In the next section we will look at the common aspects of these problems independently of their geometrical or physical interpretations.

In our examples for both area and distance, we were dealing with increasing functions. The calculations are basically the same for decreasing functions. However, if a function is not monotone, finding the minimum and maximum values on subintervals may be difficult. In the next section we will show that it is not essential to use inscribed or circumscribed rectangles. Instead, we can use any rectangles lying between these two extremes. Our examples thus far also consider only nonnegative functions. In the next section we will show that this requirement may also be relaxed. An example may help motivate how this is done.

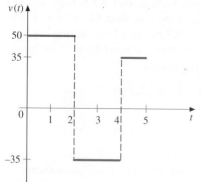

FIGURE 5.14

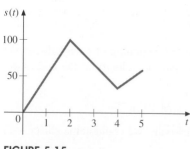

FIGURE 5.15

EXAMPLE 5.3   Suppose the velocity of an automobile on a straight road is given by Figure 5.14. Draw a graph of the distance, $s(t)$, of the car from its starting point as a function of time.

Solution   The graph in Figure 5.14 tells us that the driver travels at 50 mph for 2 hr, then goes 35 mph in the opposite direction for 2 hr, and then travels 35 mph for another hour in the starting direction. We cannot simply compute the area between the velocity curve and the t-axis as we did in the case of the runner, since this area would be the total distance traveled by the car, that is, the distance shown on the odometer, namely,

$$(50)(2) + (35)(2) + (35)(1) = 205 \text{ mi}$$

We know this result cannot be the correct answer for the distance of the car from its starting position, though, because we would be calculating the distance as though the car went only in one direction. However, the distance from the starting point can be computed as "net area" (that is, the area bounded by the t-axis and the portion of the curve above the t-axis, minus the area bounded by the t-axis and the portion of the curve below the t-axis):

$$\text{net area} = (50)(2) + (-35)(2) + (35)(1) = 65$$

Figure 5.15 is a graph of the distance from the starting point versus time.   ■

We shall see in the next section that the concept of area under a curve can be extended to "net area" under a curve. Before showing this fact, we will introduce a more compact notation that can be used for L_n and U_n as well as for sums lying between them.

A Useful Notation—Sigma Notation

The Greek capital letter sigma, $\sum$, is used to denote *sum*. To illustrate how sigma notation works, consider the sum of the cubes of the first n positive integers.

$$1^3 + 2^3 + 3^3 + \cdots + n^3$$

For example, if $n = 3$, the sum is $1^3 + 2^3 + 3^3$ and if $n = 6$, the sum is $1^3 + 2^3 + 3^3 + 4^3 + 5^3 + 6^3$. In general, we add up all terms of the form k^3, where k assumes all positive integer values from 1 to n. In sigma notation this sum becomes

$$\sum_{k=1}^{n} k^3$$

and is read "the summation of k^3, where k goes from 1 to n."

If $x_1, x_2, \ldots, x_n$ represents n numbers, then

$$x_1 + x_2 + \cdots + x_n$$

is written as

$$\sum_{k=1}^{n} x_k$$

and is read "the summation of x sub k as k goes from 1 to n." The letter k is called the *index of summation*. Other letters can be used as the index; it is most common to use i, j, k, m, n.

EXAMPLE 5.4   Evaluate each of the following expressions.
(a) $\sum_{i=1}^{5} i$ (b) $\sum_{k=1}^{4}(2k^2 - 1)$ (c) $\sum_{m=1}^{10} 5$

Solution

(a) $\displaystyle\sum_{i=1}^{5} i = 1 + 2 + 3 + 4 + 5 = 15$

(b) $\displaystyle\sum_{k=1}^{4}(2k^2 - 1) = (2(1) - 1) + (2(4) - 1) + (2(9) - 1) + (2(16) - 1) = 56$

(c) $\displaystyle\sum_{m=1}^{10} 5 = 5 + 5 + 5 + 5 + 5 + 5 + 5 + 5 + 5 + 5 = 5(10) = 50$ ■

Calculations involving summation symbols often can be simplified by using the following properties.

Properties of Sigma Notation

(i) $\displaystyle\sum_{k=1}^{n} cx_k = c \sum_{k=1}^{n} x_k$, where c is any constant

(ii) $\displaystyle\sum_{k=1}^{n}(x_k + y_k) = \sum_{k=1}^{n} x_k + \sum_{k=1}^{n} y_k$

(iii) $\displaystyle\sum_{k=1}^{n}(x_k - y_k) = \sum_{k=1}^{n} x_k - \sum_{k=1}^{n} y_k$

These properties are simple extensions of the distributive and associative properties of real numbers.

Carl Friedrich Gauss (1777–1855) came from a poor and uneducated family. Despite this background, he entered advanced study at the age of 11, and by 15 he was receiving a stipend (allowance) from a Duke.

Gauss's talent with calculations enabled him to make interesting conjectures. At college, he discovered that some of his independent inventions had been made previously by others. At 19, he found how to construct regular polygons using only ruler and compass, a problem dating from the ancient Greeks. His discovery that one could construct a regular polygon of 17 sides in this way was the first advance in this area since Euclid. Gauss proved the Fundamental Theorem of Algebra at age 22.

Gauss was unbelievably prolific. Some of his work systematized whole fields of mathematics. His discoveries of non-Euclidean geometry and of noncommutative algebra remain especially noteworthy.

In 1801, an object was discovered that was thought to be a new planet, but not enough observations had been made and the object could no longer be found in the sky. Gauss worked out new ways of calculating orbits, and within a few months, the object was located again, right where Gauss had predicted. We know it now as the first asteroid, Ceres, a so-called "minor planet." Because such a remarkable prediction had been made on the basis of such a small amount of information, Gauss's reputation was even more enhanced.

Gauss became an astronomer but continued his work on mathematics and on other subjects simultaneously. In middle age, he became a geologist, measuring the shape of the Earth, and later provided basic discoveries that are still at the heart of physics. Indeed, a unit of magnetic field—the gauss—is named for him. The average magnetic field of the Earth is about 1 gauss.

It is possible to find the sums of certain expressions in what is called *closed form*; that is, we can find a formula for the given sum. One of the simplest examples is the sum of the first n natural numbers, $\sum_{k=1}^{n} k$. It can be shown (see Exercise 9 in Exercise Set 5.1) that the sum is $\frac{1}{2}n(n+1)$. That is,

$$1 + 2 + 3 + \cdots + n = \frac{n(n+1)}{2} \tag{5.3}$$

The right-hand side of Equation 5.3 is a closed form for the sum on the left.

The finite sequence

$$1, 2, 3, \ldots, n$$

whose terms are added on the left-hand side of Equation 5.3 is an example of an *arithmetic progression*, in which the difference between any two consecutive terms remains constant (in this case, the constant difference is 1).

An interesting story about Carl Friedrich Gauss involves an arithmetic progression. In 1787, when Gauss was 10 years old, his teacher gave the class a list of 100 large numbers to add up, thinking it would keep them occupied for the entire period. The numbers formed an arithmetic progression, with a constant difference between consecutive terms. The teacher knew what the answer was, but the pupils had been told nothing of arithmetic progressions. The teacher's instructions were for the first pupil to get the answer to put his slate down on his desk. Then, as others finished, they were to place their slates, in order, on the stack. Imagine the teacher's surprise when Gauss placed his slate down only seconds after the problem was given. His surprise was even greater when he saw but one number written on Gauss's slate—the correct answer for the sum!

Gauss reasoned as follows: If I were to write the sum a second time, with the order reversed, then add, term-by-term, the second sum to the first sum, I would get 100 terms, each having the same value. So all I need to do is add the first term of the original sum to the last and multiply by 100. This will give twice what I want; so I will divide the result by 2.

Let us apply Gauss's reasoning to the sum on the left-hand side of Equation 5.3, with $n = 100$ (assuming we do not know the closed form on the right-hand side):

$$1 + 2 + 3 + \cdots + 98 + 99 + 100$$

In reverse we get

$$100 + 99 + 98 + \cdots + 3 + 2 + 1$$

Adding, term-by-term, we get

$$101 + 101 + 101 + \cdots + 101$$

where there are 100 terms, each equal to 101. So the sum is 10100. This is double what we want. The answer is thus 5050.

Gauss's teacher recognized his extraordinary ability and encouraged him to move on to higher mathematics. Gauss is generally recognized as one of the greatest mathematical geniuses who have ever lived.

We show the result (Equation 5.3) again here, along with two others that you will be asked to verify in the exercises (see Exercises 11 and 13 in Exercise Set 5.1).

Some Useful Summation Formulas

$$\sum_{k=1}^{n} k = \frac{n(n+1)}{2} \qquad (5.4)$$

$$\sum_{k=1}^{n} k^2 = \frac{n(n+1)(2n+1)}{6} \qquad (5.5)$$

$$\sum_{k=1}^{n} k^3 = \left[\frac{n(n+1)}{2}\right]^2 \qquad (5.6)$$

Upper and Lower Sums Using Sigma Notation

We can rewrite Equations 5.1 and 5.2 using sigma notation. For Equation 5.1 we have

$$L_n = m_1\Delta x + m_2\Delta x + m_3\Delta x + \cdots + m_n\Delta x$$
$$= \Delta x(m_1 + m_2 + m_3 + \cdots + m_n)$$
$$= \Delta x \sum_{k=1}^{n} m_k$$

and for Equation 5.2,

$$U_n = M_1\Delta x + M_2\Delta x + M_3\Delta x + \cdots + M_n\Delta x$$
$$= \Delta x(M_1 + M_2 + M_3 + \cdots + M_n)$$
$$= \Delta x \sum_{k=1}^{n} M_k$$

If an interval $[a, b]$ is partitioned into n subintervals of equal length $\Delta x = (b - a)/n$, then the partition points $x_1, x_2, x_3, \ldots, x_n$ are given by

$$x_0 = a$$
$$x_1 = a + \Delta x$$
$$x_2 = x_1 + \Delta x = a + 2\Delta x$$
$$x_3 = x_2 + \Delta x = a + 3\Delta x$$
$$\vdots$$
$$x_{n-1} = x_{n-2} + \Delta x = a + (n - 1)\Delta x$$
$$x_n = x_{n-1} + \Delta x = a + n\Delta x = b$$

For an increasing function, as in Example 5.2, the minimum height of the function occurs at the left endpoint of each subinterval, and the maximum height occurs at the right endpoint. For a typical subinterval $[x_{k-1}, x_k]$, therefore,

$$m_k = f(x_{k-1}) \qquad \text{and} \qquad M_k = f(x_k)$$

Thus, for such an increasing function the lower and upper sums are

$$L_n = \sum_{k=1}^{n} f(x_{k-1})\Delta x = f(x_0)\Delta x + f(x_1)\Delta x + f(x_2)\Delta x + \cdots + f(x_{n-1})\Delta x$$

$$U_n = \sum_{k=1}^{n} f(x_k)\Delta x = f(x_1)\Delta x + f(x_2)\Delta x + f(x_3)\Delta x + \cdots + f(x_n)\Delta x$$

If f is a decreasing function, then the sums for L_n and U_n are reversed.

We have seen that the error in approximating the area using either L_n or U_n is at most $U_n - L_n$. For an increasing function we can write $U_n - L_n$ in the form

$$U_n - L_n = \sum_{k=1}^{n} f(x_k)\Delta x - \sum_{k=1}^{n} f(x_{k-1})\Delta x$$

$$= \sum_{k=1}^{n} \left[f(x_k) - f(x_{k-1}) \right] \Delta x$$

In Exercise 15 (in Exercise Set 5.1) you will be asked to show that, for a decreasing function, the result is the negative of the sum on the right, and that the result in both cases (increasing or decreasing) can be written as follows:

$$\text{error} \leq |f(b) - f(a)|\Delta x \qquad \text{If } f \text{ is monotone}$$

The next example illustrates how with the aid of Equation 5.5 we can find the exact area under the graph of a portion of the parabola $y = x^2$ by calculating U_n and letting $n \to \infty$. (We could use L_n instead.)

EXAMPLE 5.5   Find the upper sum U_n for the area under the graph of the function $f(x) = x^2$ from $x = 0$ to $x = 2$. Then obtain the exact area by letting $n \to \infty$.

Solution   We show the graph, together with some circumscribed rectangles, in Figure 5.16. As we have shown, for a monotone increasing function such as this one,

$$U_n = \sum_{k=1}^{n} f(x_k)\,\Delta x$$

where $x_k = a + k\,\Delta x$ and $\Delta x = (b-a)/n$. In our case, $a = 0$ and $b = 2$. So $\Delta x = 2/n$ and $x_k = 0 + k \cdot 2/n = 2k/n$. Thus,

$$U_n = \sum_{k=1}^{n} f\left(\frac{2k}{n}\right) \cdot \frac{2}{n} = \sum_{k=1}^{n} \left(\frac{2k}{n}\right)^2 \cdot \frac{2}{n} = \sum_{k=1}^{n} \frac{8k^2}{n^3}$$

We can factor out $8/n^3$ to get

$$U_n = \frac{8}{n^3} \sum_{k=1}^{n} k^2$$

By Equation 5.6,

$$\sum_{k=1}^{n} k^2 = \frac{n(n+1)(2n+1)}{6}$$

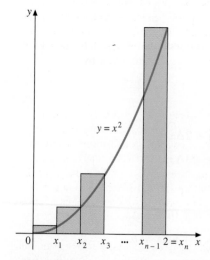

FIGURE 5.16

Thus,

$$U_n = \frac{8}{n^3} \left[\frac{n(n+1)(2n+1)}{6} \right] = \frac{4(n+1)(2n+1)}{3n^2} = \frac{4(2n^2+3n+1)}{3n^2}$$

To find the limit as $n \to \infty$, we can use the dominant-term approach or we can divide numerator and denominator by n^2 to get

$$U_n = \frac{4(2 + \frac{3}{n} + \frac{1}{n^2})}{3}$$

Thus,

$$\lim_{n \to \infty} U_n = \frac{8}{3}$$

since both $3/n$ and $1/n^2$ approach 0. The exact area is therefore 8/3. ∎

Exercise Set 5.1

 In Exercises 1–5, use a calculator, CAS, or spreadsheet to find L_n and U_n for the area under the graph of f on the specified interval for n = 5, 10, 20, and 100.

1. $f(x) = \dfrac{x+3}{2}$ on $[0, 5]$

2. $f(x) = \dfrac{x^2}{4}$ on $[0, 4]$

3. $f(x) = 4 - x^2$ on $[0, 2]$

4. $f(x) = 2x - 1$ on $[1, 6]$

5. $f(x) = x^2 + 1$ on $[-1, 2]$

6. By reading values from the graph shown, find upper and lower bounds, U_n and L_n, for the area under the curve from $x = -2$ to $x = 5$, using $n = 14$.

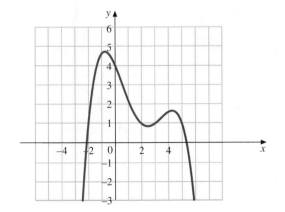

7. A driver traveling on an interstate highway sees traffic coming to a stop ahead and applies the brakes. The car's speed, in feet per second (60 mph = 88 ft/s), is given at half-second intervals in the accompanying table. Use upper sums and lower sums to give an upper bound and a lower bound on the distance covered.

t	0	0.5	1	1.5	2	2.5	3	3.5	4
$v(t)$	80	72	63	53	42	29	17	7	0

 8. A container of food is dropped from a helicopter high above a refugee camp. When air resistance is taken into consideration, the container's velocity, $v(t)$, t seconds after it is dropped, is approximately given by the formula

$$v(t) = 320(1 - e^{-0.1t})$$

(See Section 1.3, where e^x is introduced.) Use a calculator, a CAS, or a spreadsheet to make a table of values showing t and $v(t)$ at half-second intervals for $t = 0$ to $t = 5$. Find an underestimate and an overestimate of the distance the container falls in the first 5 seconds. Sketch a graph of $v(t)$ versus t and show these underestimates and overestimates as sums of areas of rectangles.

9. Using Gauss's reasoning, derive Equation 5.3 for the sum of the first n natural numbers:

$$1 + 2 + 3 + \cdots + n = \frac{n(n+1)}{2}$$

10. Make use of the formula derived in Exercise 9 to find the upper sum U_n for an unspecified n for the area under the graph of $f(x) = 2x + 3$ from $x = 0$ to $x = 4$. Then find the exact area by letting $n \to \infty$. Check your answer, using what you know about the area of a trapezoid. (*Hint:* Draw the figure.)

11. Prove Equation 5.5,

$$1^2 + 2^2 + 3^2 + \cdots + n^2 = \frac{n(n+1)(2n+1)}{6}$$

using mathematical induction. (See Appendix 4.)

12. Use the result of Exercise 11 to find the lower sum L_n for the area under the graph of $f(x) = 4 - x^2$ from $x = 0$ to $x = 2$. Then find the exact area by letting $n \to \infty$.

13. Prove Equation 5.6,

$$1^3 + 2^3 + 3^3 + \cdots + n^3 = \left[\frac{n(n+1)}{2} \right]^2$$

using mathematical induction. (See Appendix 4.)

14. Use the result of Exercise 13 to find the upper sum U_n for the area under the graph of $f(x) = x^3$ from $x = 0$ to $x = 2$. Then let $n \to \infty$ to get the exact area.

15. Let f be a decreasing, nonnegative function on $[a, b]$, and let $x_0, x_1, x_2, \ldots, x_n$ be a partition of $[a, b]$ with all subintervals of equal width $\Delta x = (b - a)/n$. Express the minimum values $m_1, m_2, \ldots, m_n$ and maximum values $M_1, M_2, \ldots, M_n$ of the function on the subintervals in terms of $f(x_k)$ for appropriate values of k. Then substitute these in Equations 5.1 and 5.2 for L_n and U_n, and calculate $U_n - L_n$. Using your result, together with the result of $U_n - L_n$ for an increasing function found in this section, show that the error in using L_n or U_n to approximate the area under a monotone function satisfies

$$\text{error} \le |f(b) - f(a)| \Delta x$$

16. Use the result of Exercise 15 to determine a value of n that will guarantee an error that is less than 0.05 when approximating the area under the graph of f on $[a, b]$ by L_n or U_n, and find the corresponding area, where
(a) $f(x) = x^2$ on $[0, 4]$
(b) $f(x) = \sqrt{x}$ on $[1, 9]$

17. Repeat Exercise 16 for the functions given. Recall that we introduced e^x and $\ln x$ in Section 1.3. Their values can be obtained from a calculator. (We will study these functions in Chapter 7.)
(a) $f(x) = \ln x$ on $[1, 7]$
(b) $f(x) = 6e^{-0.24x}$ on $[2, 10]$

Exercises 18 and 19 illustrate Archimedes' "method of compression" for calculating the area of a circle.

18. The accompanying figure shows an n-sided regular polygon (that is, all sides are of equal length) inscribed in a circle of radius r. The polygon has been divided into n congruent triangles, each with central angle $2\pi/n$ radians.
(a) Show that the area of each triangle is

$$\frac{1}{2} r^2 \sin \left(\frac{2\pi}{n} \right)$$

(*Hint:* Use the identity $\sin 2t = 2 \sin t \cos t$.)

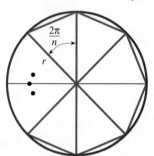

(b) Show that as $n \to \infty$ the area of the polygon approaches πr^2, which is the area of the circle. That is, show that

$$\lim_{n \to \infty} \frac{1}{2} n r^2 \sin \left(\frac{2\pi}{n} \right) = \pi r^2$$

(*Hint:* Use the fact that $\lim_{t \to 0} (\sin t / t) = 1$.)

19. Rather than using inscribed polygons as in Exercise 18, we can use circumscribed polygons, as in the accompanying figure, where we have again divided the polygon into n congruent triangles.

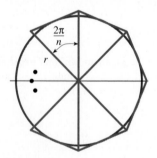

(a) Show that the area of each triangle is

$$r^2 \tan \left(\frac{\pi}{n} \right)$$

(b) Show that the areas of the circumscribed polygons approach πr^2 as $n \to \infty$. That is, show that

$$\lim_{n \to \infty} n r^2 \tan \left(\frac{\pi}{n} \right) = \pi r^2$$

(See the hint for Exercise 18(b).)

20. The Keck Telescope in Hawaii is made of 36 hexagons (see the figure), each 1.8 m in its largest diagonal. By dividing a typical hexagon into triangles, calculate its area. Then give the area of the total mirror. Find the diameter of a circular mirror of the same total area. Compare the result with that of the venerable Hale Telescope on Palomar Mountain in California, whose mirror is 5 m (200 in.) in diameter.

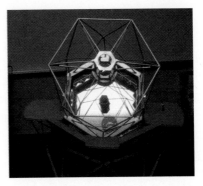

A model of the Keck Telescope, the largest telescope in the world

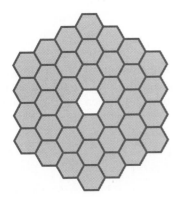

21. Use Equations 5.4, 5.5, and 5.6, along with a calculator (or CAS or spreadsheet), to find the smallest positive integer n such that

(a) $\displaystyle\sum_{k=1}^{n} k \geq 1{,}000$

(b) $\displaystyle\sum_{k=1}^{n} k^2 \geq 10{,}000$

(c) $\displaystyle\sum_{k=1}^{n} k^3 \geq 100{,}000$

5.2 THE DEFINITE INTEGRAL

Our objective in this section is to define the *definite integral* and to investigate its properties. This concept is the unifying idea that relates many diverse problems. In particular, we will see that the area problem we considered in Section 5.1 can be solved by a definite integral. In Section 5.4 we will define an *indefinite integral* and show how the two types of integrals are related.

As a first step in arriving at a definition of the definite integral, let us return to the problem of finding the area under the graph of a nonnegative, continuous function f from $x = a$ to $x = b$. We have seen that the area A is given by either

$$\lim_{n \to \infty} U_n \quad \text{or} \quad \lim_{n \to \infty} L_n$$

where

$$U_n = \sum_{k=1}^{n} M_k \, \Delta x \quad \text{and} \quad L_n = \sum_{k=1}^{n} m_k \, \Delta x$$

For monotone functions, finding the maximum values M_k and the minimum values m_k of $f(x)$ on the subintervals is straightforward, since these extreme values occur at the endpoints of the subintervals. For functions that are not monotone, it may be very difficult to find m_k and M_k, other than by estimating the values from the graph. (See Figure 5.17.) As the following argument shows, we do not need to use either m_k or M_k in approximating the area. Instead, we can use any value between these two extremes.

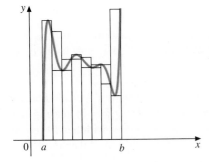

FIGURE 5.17
Maximum and minimum values of f do not necessarily occur at endpoints of subintervals.

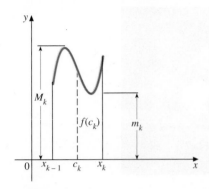

FIGURE 5.18

Select arbitrary points $c_1, c_2, \ldots, c_n$ such that for each k, c_k lies in the kth subinterval. Thus,

$$x_{k-1} \leq c_k \leq x_k$$

for $k = 1, 2, \ldots, n$. Such a choice is indicated for the kth subinterval in Figure 5.18. Since M_k is the maximum value of $f(x)$ and m_k the minimum value of $f(x)$ on the kth subinterval, $f(c_k)$ must be between these extremes. That is,

$$m_k \leq f(c_k) \leq M_k$$

Thus,

$$m_k \Delta x \leq f(c_k) \Delta x \leq M_k \Delta x$$

Geometrically, we are saying that the area of the rectangle with height $f(c_k)$ and base Δx lies between the area of the rectangle with height m_k and the one with height M_k (see Figure 5.18), each with base Δx. When we add up all the areas, we get

$$\sum_{k=1}^{n} m_k \Delta x \leq \sum_{k=1}^{n} f(c_k) \Delta x \leq \sum_{k=1}^{n} M_k \Delta x$$

and so

$$L_n \leq \sum_{k=1}^{n} f(c_k) \Delta x \leq U_n \tag{5.7}$$

We know that L_n and U_n approach the area under the curve A as $n \to \infty$. Thus, by Equation 5.7, the sum in the middle also approaches A, since it is squeezed between L_n and U_n (the Squeeze Theorem (2.3) is true also for sequences). That is,

$$\lim_{n \to \infty} \sum_{k=1}^{n} f(c_k) \Delta x = A \tag{5.8}$$

The importance of this result is that we do not have to know the maximum or minimum values of the function on each of the subintervals but can use any value in between. For example, we could take c_k as the left endpoint, the right endpoint, the midpoint, or any other point of the kth subinterval. The limiting value in Equation 5.8 is exactly the same as the limit of L_n or U_n, namely, the area A.

EXAMPLE 5.6   Find an expression in the form of Equation 5.8 for the area under the curve $y = f(x) = \sin(\pi x)/2 + 2$ from $x = 0$ to $x = 4$, using c_k as the right endpoint of the kth subinterval.

Solution   Figure 5.19 shows the graph of f, along with a typical subdivision of the interval $[0, 4]$. Note that since the curve is neither increasing nor decreasing throughout the whole interval, building rectangles with heights equal to values of the function at the right endpoint does not give either the lower sum L_n or the upper sum U_n, but something in between.

For an arbitrary choice of n, we have

$$\Delta x = \frac{4 - 0}{n} = \frac{4}{n}$$

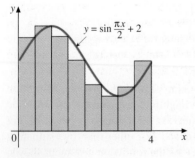

FIGURE 5.19

Thus,

$$x_1 = 0 + \Delta x = \frac{4}{n}$$

$$x_2 = 0 + 2\Delta x = \frac{8}{n}$$

$$x_3 = 0 + 3\Delta x = \frac{12}{n}$$

and so on. The right endpoint of the kth subinterval $[x_{k-1}, x_k]$ is x_k, so

$$x_k = 0 + k\Delta x = \frac{4k}{n}$$

This term is the c_k of Equation 5.8. That is, $c_1 = x_1$, $c_2 = x_2, \ldots, c_n = x_n$. Thus,

$$f(c_k) = f(x_k) = 2 + \sin \frac{\pi x_k}{2} = 2 + \sin \frac{2k\pi}{n}$$

and so by Equation 5.8,

$$A = \lim_{n \to \infty} \sum_{k=1}^{n} \underbrace{\left(2 + \sin \frac{2k\pi}{n}\right)}_{\text{This is } f(c_k)} \underbrace{\left(\frac{4}{n}\right)}_{\text{This is } \Delta x}$$

If we wanted to approximate the area, we could take some large value of n, say $n = 100$, and calculate the sum on the right using a CAS or a spreadsheet. At this stage we have no way of finding the exact value of the limit as $n \to \infty$. We will soon see a way, however, of getting this exact answer. ∎

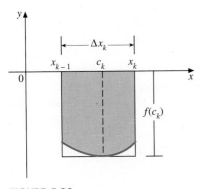

FIGURE 5.20

Up to this point we have considered only continuous functions that are non-negative. Now we want to relax this latter requirement. If the continuous function f is sometimes positive and sometimes negative, what geometric interpretation can be given to the sum $\sum_{k=1}^{n} f(c_k)\Delta x$? When $f(c_k)$ is negative, we can think of the rectangle with height $f(c_k)$ and base Δx as contributing "negative area," since it extends below the x-axis (Figure 5.20). So the limit in Equation 5.8 would give what we have called "net area," that is, the algebraic sum of positive and negative areas. In Figure 5.21, we show a typical situation. Regions I and III would be counted as positive area and region II as negative area. If we wanted the actual total area between the curve and the x-axis (to compute, for example, the weight of a sheet of metal cut to that shape), we would have to take the absolute value of the negative area of II and add it to the areas of I and III (area = area I + |area II| + area III).

To interpret the situation in Figure 5.21 in terms of distance, suppose x represents time and $f(x)$ represents velocity. Then, since the velocity is sometimes positive and sometimes negative, the object is sometimes moving in the direction we have designated as positive and sometimes moving in the opposite direction. The areas of I and III give the total distance covered in the positive direction, and the area of II gives the distance covered in the negative direction. So Equation 5.8 could be interpreted as the "net distance" covered—that is, the distance from the starting point of the object to its final position (see Example 5.3 in Section 5.1). For example, if you drive forward and then back to your starting point, the distance from the starting point (the "net distance") is 0. The area above the x-axis under the distance function and the area below would be equal. If you wanted to find the actual total distance covered (say, to calculate your gas mileage), you would need to add the absolute values of the areas.

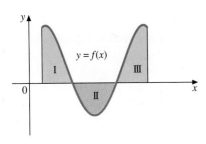

FIGURE 5.21

We indicated earlier that it is not necessary to use subintervals of equal width. We did so only for convenience. Choosing division points in this way gave the common width $\Delta x = (b - a)/n$ for all subintervals, and by letting $n \to \infty$, we were assured that the widths of all the rectangles became arbitrarily small. We can now generalize our process by allowing the points of division $x_0, x_1, x_2, \ldots, x_n$ to be spaced in any way whatsoever in $[a, b]$, subject only to the requirements that $x_0 = a$, $x_n = b$, and $x_0 < x_1 < x_2 < \cdots < x_n$. Denote the width of the kth subinterval by Δx_k, so that

$$\Delta x_k = x_k - x_{k-1} \qquad (k = 1, 2, \ldots, n)$$

We may no longer be guaranteed that as $n \to \infty$, the rectangles exhaust the area under the curve. For example, we can define points that divide only a portion of the interval into more and more subintervals and leave the remainder of the interval unchanged. As an example, half the interval could always be considered as one subinterval in each successive finer subdivision (Figure 5.22). In this case, the number of rectangles increases to infinity, but the sum of the areas does not approach the area under the curve. To remedy this problem, we also require that as $n \to \infty$, the largest of the widths, Δx_k, approaches 0. We indicate the largest Δx_k by max Δx_k. With this notation, the area is given by

$$A = \lim_{\max \Delta x_k \to 0} \sum_{k=1}^{n} f(c_k) \Delta x_k \qquad (5.9)$$

where, again, A is to be interpreted as "net area." A sum of the type appearing on the right,

$$\sum_{k=1}^{n} f(c_k) \Delta x_k$$

is called a **Riemann sum** for f in honor of the German mathematician G. F. B. Riemann, who made major contributions to the theory of the integral.

The limit $\lim_{n \to \infty} \sum_{k=1}^{n} f(c_k) \Delta x$ in Equation 5.8 is a special case of the limit in Equation 5.9, in which $\Delta x_k = \Delta x$ for all k. Since $\Delta x = (b - a)/n$, we see that $\Delta x \to 0$ as $n \to \infty$, so it follows that max $\Delta x_k \to 0$ as $n \to \infty$. (In fact, there is no need to speak of the maximum Δx_k in this case, since all subintervals are of the same width.)

EXAMPLE 5.7   Form the Riemann sum for the function $f(x) = x^2 - 2x$ on the interval $[-1, 3]$, using the points of division

$$x_0 = -1, \quad x_1 = -0.2, \quad x_2 = 0.6, \quad x_3 = 1.8, \quad x_4 = 2.4, \quad x_5 = 3$$

Choose c_k as the midpoint of the kth subinterval.

Solution   The points c_k are found by averaging x_{k-1} and x_k, as follows:

$$c_1 = \frac{-1 - 0.2}{2} = -0.6 \qquad c_2 = \frac{-0.2 + 0.6}{2} = 0.2$$

$$c_3 = \frac{0.6 + 1.8}{2} = 1.2 \qquad c_4 = \frac{1.8 + 2.4}{2} = 2.1$$

$$c_5 = \frac{2.4 + 3}{2} = 2.7$$

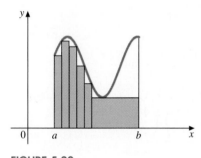

FIGURE 5.22

In Figure 5.23 we show the graph of f, the division points, the midpoints c_k, and the rectangles whose (signed) areas make up the Riemann sum. The Δx_k values are as follows:

$$\Delta x_1 = x_1 - x_0 = -0.2 - (-1) = 0.8$$

$$\Delta x_2 = x_2 - x_1 = 0.6 - (-0.2) = 0.8$$

$$\Delta x_3 = x_3 - x_2 = 1.8 - 0.6 = 1.2$$

$$\Delta x_4 = x_4 - x_3 = 2.4 - 1.8 = 0.6$$

$$\Delta x_5 = x_5 - x_4 = 3 - 2.4 = 0.6$$

In this case max $\Delta x_k = 1.2$, and

$$
\begin{aligned}
\sum_{k=1}^{5} f(c_k)\Delta x_k &= [f(-0.6)](0.8) + [f(0.2)](0.8) + [f(1.2)](1.2) \\
&\quad + [f(2.1)](0.6) + [f(2.7)](0.6) \\
&= (1.56)(0.8) + (-0.36)(0.8) + (-0.96)(1.2) \\
&\quad + (0.21)(0.6) + (1.89)(0.6) \\
&= 1.068
\end{aligned}
$$

This result is an approximation of the net area. Since our division points were not very close together, we would not expect this to be a good approximation. The exact value, as we will show in the next section, is 4/3.   ■

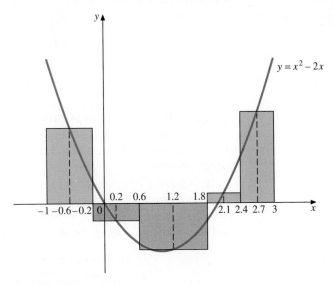

FIGURE 5.23

Definition of the Integral

To define the integral, we drop the assumption that f be continuous and assume only that f is *defined* and *bounded* on $[a, b]$. To say that f is "defined" on $[a, b]$ means that $f(x)$ has a finite value for every x on the closed interval $[a, b]$. That is, the closed interval $[a, b]$ is included in the domain of f. To say that f is "bounded" means that $f(x)$ does not get arbitrarily large either positively or negatively. More precisely, f is bounded on $[a, b]$ provided there is some positive constant M such that $|f(x)| \leq M$ for all x in the closed interval $[a, b]$. Geometrically, the graph of f lies between the horizontal lines $y = M$ and $y = -M$. Although we do not require that f be continuous, most of the functions we deal with will be continuous. We set aside for the moment the idea of area (or distance) and simply look at Riemann sums for f. For every subdivision of $[a, b]$ and for every choice of the intermediate points c_k, such a Riemann sum exists. The question is, as we take finer and finer subdivisions (that is, points closer and closer together), do these Riemann sums approach some limit? If the answer is yes, then this limit is called the *integral* of f over $[a, b]$. It is also called the *definite integral*, or the *Riemann integral*. We state this definition formally, with the symbols and terminology understood to be as we have previously defined them.

Definition 5.2
The Definite Integral

Let f be defined and bounded on $[a, b]$. If

$$\lim_{\max \Delta x_k \to 0} \sum_{k=1}^{n} f(c_k) \Delta x_k$$

exists, independently of the choices of the division points and the choices of the intermediate points c_k, then this limit is called the **definite integral of f over $[a, b]$**, and is denoted by

$$\int_a^b f(x)dx$$

It is read "the integral of f of x with respect to x from a to b."

The symbol $\int$ is called an **integral sign**. It has the appearance of an elongated S, the first letter of the Latin word *summa*, meaning sum. The letters a and b at the bottom and top of the integral sign $\int_a^b f(x)dx$ are called **limits of integration**, and $f(x)$ is called the **integrand**.

The symbol $\int_a^b f(x)dx$ looks something like a Riemann sum $\sum_{k=1}^{n} f(c_k)\Delta x_k$, with the integral sign replacing the summation symbol, $f(x)$ replacing $f(c_k)$, and dx replacing Δx_k. We can think of the integral as a sort of idealized sum.

The precise meaning of the limit in Definition 5.2 is as follows:

$$\lim_{\max \Delta x_k \to 0} \sum_{k=1}^{n} f(c_k)\Delta x_k = L$$

provided that corresponding to each positive number ε (however small it may be), there exists a (sufficiently small) positive number δ such that for every partition of $[a, b]$ for which $\max \Delta x_k < \delta$ (so that all subintervals are of length less than δ), then

$$\left| \sum_{k=1}^{n} f(c_k)\Delta x_k - L \right| < \varepsilon$$

regardless of how c_k is chosen in the interval $[x_{k-1}, x_k]$ for $k = 1, 2, 3, \ldots, n$. In simpler (but less precise) language, all Riemann sums for f on $[a, b]$ can be made as close as we please to L, provided only that all of the widths Δx_k are sufficiently small. This type of limit is more complex than those we studied in Chapter 2.

Because the definite integral is defined by a limit, the integral may or may not exist, depending on the behavior of the function f. When the integral exists on $[a, b]$, the function f is said to be **integrable** on $[a, b]$. Determining which functions are integrable is generally a very difficult problem. However, we state without proof that if f satisfies any of the following conditions, it is integrable on $[a, b]$:

1. f is continuous on $[a, b]$.
2. f is either only increasing or only decreasing (i.e., f is monotone) on $[a, b]$.
3. f is bounded on $[a, b]$ and has only finitely many points of discontinuity.

In the next two examples we show how the limit of a Riemann sum can be recognized as an integral.

EXAMPLE 5.8   Write the following limit as an integral.

$$\lim_{\max \Delta x_k \to 0} \sum_{k=1}^{n} (c_k^2 + 3c_k + 1)\Delta x_k, \text{ for the interval } [-1, 3]$$

Solution   This limit conforms exactly to the definition of the integral of $f(x) = x^2 + 3x + 1$ on $[-1, 3]$. So the limit is

$$\int_{-1}^{3} (x^2 + 3x + 1)dx \qquad\blacksquare$$

EXAMPLE 5.9   Write the following limit as an integral.

$$\lim_{n \to \infty} \sum_{k=1}^{n} \left[\cos\left(1 + \frac{3k}{n}\right) \right] \left(\frac{3}{n}\right)$$

Solution   The sum appears to be a Riemann sum in which the division points are equally spaced, with $\Delta x_k = 3/n$ for all k. Also, $c_k = 1 + 3k/n$. So

$$c_1 = 1 + \frac{3}{n}, \ c_2 = 1 + 2\left(\frac{3}{n}\right), \dots, \ c_n = 1 + n\left(\frac{3}{n}\right) = 4$$

The c_k's are the right endpoints of an equally spaced subdivision of $[1, 4]$. Since $\cos x$ is a continuous function everywhere, its integral from 1 to 4 exists, and according to the definition, no restriction is placed on subdivisions except that the maximum widths of the subintervals approach 0. In our case this maximum width is $3/n$, which does approach 0 as $n \to \infty$. We conclude that the limit is the integral

$$\int_{1}^{4} \cos x \, dx \qquad\blacksquare$$

In the next section we will see how to find the exact values of the integrals in Examples 5.8 and 5.9.

If $f(x) \geq 0$, we know that one way of interpreting the integral $\int_{a}^{b} f(x)dx$ is as the area under the graph of f from $x = a$ to $x = b$. In certain cases, we can calculate this area by elementary geometry, and thus the value of the integral

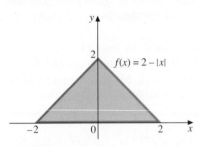

FIGURE 5.24

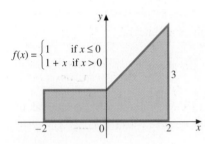

FIGURE 5.25

Recall that the area of a trapezoid is the average of the lengths of the two parallel sides times the distance between them.

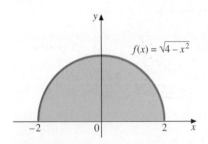

FIGURE 5.26

will have been found. In the next example we illustrate this "backdoor" method of evaluating certain integrals.

EXAMPLE 5.10 Evaluate the integral

$$\int_{-2}^{2} f(x)\,dx$$

for each of the following functions.

(a) $f(x) = 2 - |x|$

(b) $f(x) = \begin{cases} 1 & \text{if } x \le 0 \\ 1+x & \text{if } x > 0 \end{cases}$

(c) $f(x) = \sqrt{4 - x^2}$

Solution

(a) The graph of f is shown in Figure 5.24. The region below it is a triangle with base 4 and altitude 2. The area is therefore $\frac{1}{2}(4)(2) = 4$. Thus,

$$\int_{-2}^{2} (2 - |x|)\,dx = 4$$

(b) The graph is shown in Figure 5.25. The region under it consists of a rectangle and a trapezoid. The combined area is

$$(2)(1) + \left(\frac{1+3}{2}\right)(2) = 2 + 4 = 6$$

So,

$$\int_{-2}^{2} f(x)\,dx = 6$$

(c) The graph of $y = \sqrt{4 - x^2}$ is the upper half of $y^2 = 4 - x^2$, or equivalently of $x^2 + y^2 = 4$, which is a circle of radius 2. We know that the area of the entire circle is $\pi r^2 = \pi(2)^2 = 4\pi$. So the area we want (shown in Figure 5.26) is 2π. Thus,

$$\int_{-2}^{2} \sqrt{4 - x^2}\,dx = 2\pi \qquad \blacksquare$$

In part (c) of the preceding example, we assumed that you knew the formula $A = \pi r^2$ for the area enclosed by a circle of radius r. If you completed Exercise 18 or 19 in Exercise Set 5.1, you have seen a *proof* of this area formula. We will present another proof later, but for now we assume that the result is understood and we will make use of it from time to time.

By proceeding as in part (c) above, we can give the value of the more general integral of the form

$$\int_{-a}^{a} \sqrt{a^2 - x^2}\,dx$$

where a is an arbitrary positive constant. The graph of $y = \sqrt{a^2 - x^2}$ is the upper half of the circle $x^2 + y^2 = a^2$, whose total area is πa^2. Thus we have the following result.

Area Under the Semicircle with Radius a

$$\int_{-a}^{a} \sqrt{a^2 - x^2}\, dx = \frac{\pi a^2}{2} \qquad (5.10)$$

We will have several occasions later to use this result.

Unfortunately, most integrals cannot be evaluated by the method of Example 5.10, since the area under the curve cannot usually be found by elementary geometry. With a few exceptions (see Example 5.5 and Exercises 10, 12, and 14 in Exercise Set 5.1), the best we can do at this point is to approximate the integral using a Riemann sum made up of many small subintervals. For this purpose, when we know the integral exists, we might as well take subintervals of equal width, as we did in Section 5.1. For some functions it may be necessary to take a large number of subintervals to get much accuracy in approximating an integral. Of course, with a computer or programmable calculator, this requirement does not present much of a problem. Fortunately, we can often avoid any approximating procedure and get the exact value of the integral using one of the main results we will prove in the next section. There will still be some functions, however, whose integrals can only be approximated. In Section 5.5 we will give approximating techniques that are more efficient than using Riemann sums.

Using Different Variables of Integration

The letter x in the integral

$$\int_a^b f(x)dx$$

is called the **variable of integration**. The value of the integral is unchanged if we use a different letter. For example, for any integrable function f on $[a,b]$,

$$\int_a^b f(x)dx = \int_a^b f(t)dt$$

The variable of integration is called a "dummy variable," which means that changing to a different variable does not alter the value of the integral. You can see why the letter makes no difference when you think of the integral as representing area. The area is the same no matter which letter is used to name the axis of the independent variable.

In the definition of the integral $\int_a^b f(x)dx$, we have assumed that $a < b$. To take care of the other possibilities we adopt the following conventions:

$$\int_a^a f(x)dx = 0 \qquad \text{Provided } f(a) \text{ exists} \qquad (5.11)$$

(That is, the area of a vertical line segment is zero.)

If $a > b$ and f is integrable on $[b, a]$, we define

$$\int_a^b f(x)dx = -\int_b^a f(x)dx \qquad (5.12)$$

(That is, integrating from right to left is the negative of integrating from left to right.) For example, if f is integrable on $[0, 3]$,

$$\int_3^3 f(x)dx = 0 \qquad \text{and} \qquad \int_3^0 f(x)dx = -\int_0^3 f(x)dx$$

Some Properties of Integrals

We conclude this section by listing some of the properties of integrals. They can all be justified on the basis of the definition as a limit of Riemann sums, and in the exercises you will be asked to carry out the arguments for some of the simpler cases. We show by means of the figures the plausibility of most of the properties based on the area interpretation when the functions are nonnegative. We assume that all the functions are integrable on $[a, b]$.

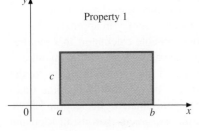

Property 1

Area $= c(b - a)$

FIGURE 5.27

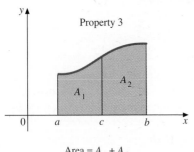

Property 3

Area $= A_1 + A_2$

FIGURE 5.28

Properties of the Integral

1. For any constant c, $\int_a^b c\,dx = c(b - a)$. (See Figure 5.27.)

2. For any constant c, $\int_a^b c f(x)dx = c \int_a^b f(x)dx$.

3. If $a < c < b$, $\int_a^b f(x)dx = \int_a^c f(x)dx + \int_c^b f(x)dx$. (See Figure 5.28.)

4. $\int_a^b [f(x) + g(x)]dx = \int_a^b f(x)dx + \int_a^b g(x)dx$

5. If $f(x) \geq 0$ on $[a, b]$, then $\int_a^b f(x)dx \geq 0$. (See Figure 5.29.)

6. If $f(x) \leq g(x)$ on $[a, b]$, then $\int_a^b f(x)dx \leq \int_a^b g(x)dx$. (See Figure 5.30.)

7. $\left| \int_a^b f(x)dx \right| \leq \int_a^b |f(x)|dx$

8. If $m \leq f(x) \leq M$ on $[a, b]$, then $m(b-a) \leq \int_a^b f(x)dx \leq M(b-a)$. (See Figure 5.31.)

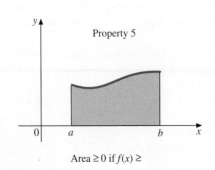

Property 5

Area ≥ 0 if $f(x) \geq$

FIGURE 5.29

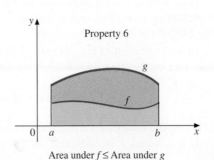

Property 6

Area under $f \leq$ Area under g

FIGURE 5.30

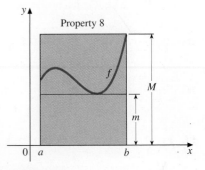

Property 8

FIGURE 5.31

Exercise Set 5.2

In Exercises 1–4, find the Riemann sum for the given function on the interval specified, corresponding to the given division points and choices of c_k. Simplify the result.

1. $f(x) = 2x - 3$ on $[-3, 2]$; $x_0 = -3$, $x_1 = -2$, $x_2 = -1$, $x_3 = 0$, $x_4 = 1$, $x_5 = 2$; $c_1 = -\frac{5}{2}$, $c_2 = -1$, $c_3 = -\frac{1}{4}$, $c_4 = 0$, $c_5 = \frac{7}{4}$

2. $f(x) = 5 - 2x$ on $[0, 4]$; $x_0 = 0$, $x_1 = 1$, $x_2 = 3$, $x_3 = 4$; $c_1 = \frac{1}{2}$, $c_2 = \frac{5}{2}$, $c_3 = 4$

3. $f(x) = 10(x - 2)$ on $[1, 2]$; $x_0 = 1$, $x_1 = 1.2$, $x_2 = 1.3$, $x_3 = 1.6$, $x_4 = 1.9$, $x_5 = 2$; $c_1 = 1.1$, $c_2 = 1.225$, $c_3 = 1.4$, $c_4 = 1.8$, $c_5 = 1.95$

4. $f(x) = x(3 - x)$ on $[-1, 3]$; $x_0 = -1$, $x_1 = 0$, $x_2 = \frac{3}{2}$, $x_3 = 3$; $c_1 = -\frac{1}{2}$, $c_2 = \frac{1}{2}$, $c_3 = 2$

In Exercises 5–8, write the Riemann sum for f on $[a, b]$ using n subintervals of equal width. Take c_k as the right endpoint of the kth subinterval.

5. $f(x) = \frac{1}{x}$ on $[1, 2]$

6. $f(x) = 1 + \cos x$ on $[0, \pi]$

7. $f(x) = \sqrt[3]{3x - 4}$ on $[1, 4]$

8. $f(x) = \sin(\pi x)$ on $[0, 2]$

In Exercises 9–12, repeat Exercises 5–8, using as c_k the left endpoint of the kth subinterval.

In Exercises 13–18, write the given limit as an integral.

13. $\displaystyle\lim_{\max \Delta x_k \to 0} \sum_{k=1}^{n} \sqrt{1 + c_k^3}\, \Delta x_k$, for the interval $[0, 2]$

14. $\displaystyle\lim_{\max \Delta x_k \to 0} \sum_{k=1}^{n} \left[\frac{\ln c_k}{c_k}\right] \Delta x_k$, for the interval $[1, 3]$

15. $\displaystyle\lim_{n \to \infty} \sum_{k=1}^{n} \left[\frac{2k}{n}\right]^2 \left[\frac{2}{n}\right]$

16. $\displaystyle\lim_{n \to \infty} \sum_{k=1}^{n} \left[\frac{1}{1 + \frac{4(k-1)}{n}}\right] \left[\frac{4}{n}\right]$

17. $\displaystyle\lim_{n \to \infty} \sum_{k=1}^{n} (\sin c_k) \left[\frac{5}{n}\right]$, where $-2 + \dfrac{5(k-1)}{n} \le c_k \le -2 + \dfrac{5k}{n}$

18. $\displaystyle\lim_{n \to \infty} \sum_{k=1}^{n} \left[\frac{c_k}{1 + c_k}\right] \left[\frac{1}{n}\right]$, where $1 + \dfrac{k - 1}{n} \le c_k \le 1 + \dfrac{k}{n}$

In Exercises 19–21 express the area of the shaded region as a definite integral.

19.

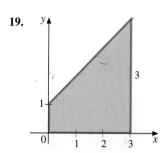

20.

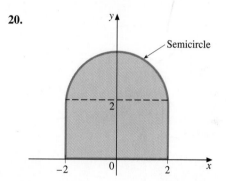

Semicircle

21.

$Z - |x - 2|$

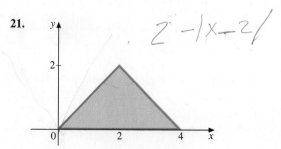

In Exercises 22–30, use geometric formulas for area to evaluate the integral.

22. $\int_{-1}^{3} (2x + 1)\,dx$

23. $\int_{-3}^{3} |x|\,dx$

24. $\int_{0}^{3} |x - 2|\,dx$

25. $\int_{0}^{\sqrt{8}} \sqrt{8 - x^2}\,dx$

26. $\int_{-2}^{4} (|2x - 3| - 3)\,dx$

27. $\int_{-2}^{0} (3 + \sqrt{4 - x^2})\,dx$

28. $\int_{-2}^{4} f(x)dx$, where $f(x) = \begin{cases} x+2 & \text{if } -2 \le x < 0 \\ 2 & \text{if } 0 \le x \le 2 \\ 4-x & \text{if } 2 < x \le 4 \end{cases}$

29. $\int_{-3}^{6} f(x)dx$, where $f(x) = \begin{cases} \sqrt{9-x^2} & \text{if } -3 \le x < 0 \\ \frac{1}{2}(6-x) & \text{if } 0 \le x \le 6 \end{cases}$

30. $\int_{0}^{7} g(t)dt$, where $g(t) = \begin{cases} 2(1+t) & \text{if } 0 \le x < 3 \\ 4(5-t) & \text{if } 3 \le x \le 7 \end{cases}$

In Exercises 31 and 32, the graph of f is given. Make use of areas to evaluate $\int_a^b f(x)dx$.

31.

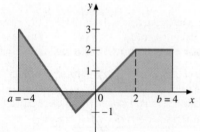

32.

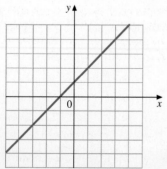

33. Evaluate $\int_a^b |f(x)|dx$ for each of the functions given in Exercises 31 and 32.

In Exercises 34–36, use the graph to estimate, by upper and lower sums, with $\Delta x = 1$, the integrals $\int_1^4 f(x)dx$ and $\int_{-3}^2 f(x)dx$. Each grid square has side 1.

34.

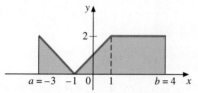

35.

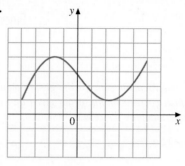

36.

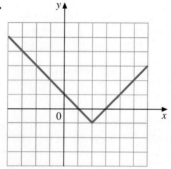

In Exercises 37–40, use a CAS, a spreadsheet, or a calculator to construct a table of left-hand and right-hand Riemann sums with n = 2, 8, 16, 32, 100, and 200, respectively. Estimate the value of the integral.

37. $\int_0^1 x^2 dx$ **38.** $\int_0^{\pi/2} \cos x \, dx$

39. $\int_{0.2}^3 \sin\left(\frac{1}{x}\right) dx$ **40.** $\int_0^3 \sqrt{x}\, dx$

In Exercises 41 and 42, use a CAS, a spreadsheet, or a calculator to compute the definite integral to one decimal place of accuracy. In each case determine the number of subdivisions that will guarantee this accuracy. Show that each function is increasing on the interval of integration.

41. $\int_0^2 x^2 dx$ **42.** $\int_0^{1/2}(x - x^2)dx$

43. Use the figure (shown at top of page 331) to give an approximate value of $\int_0^{2.2} \cos(x^2)dx$.

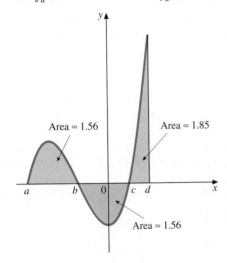

$y = \cos(x^2)$

Area ≈ 0.98

1.25   2.2

Area ≈ 0.57

44. From the areas shown in the figure, give the approximate value of each of the following.

(a) $\int_a^b f(x)dx$ (b) $\int_b^c f(x)dx$

(c) $\int_a^c f(x)dx$ (d) $\int_a^d f(x)dx$

Area ≈ 1.56 Area ≈ 1.85

a b 0 c d

Area ≈ 1.56

45. For the function f whose graph is given in Exercise 44, evaluate $\int_a^d |f(x)|dx$.

46. Obtain an approximation to π as follows. Use a CAS, a spreadsheet, or a calculator to compute $\int_{-1}^{1} \sqrt{1-x^2}dx$ using left-hand sums with 100 subdivisions. Explain why π is equal to twice the value of this integral. (*Hint:* Draw the graph of $y = \sqrt{1-x^2}$ on $[-1, 1]$.)

47. (a) Use a CAS or a graphing calculator to obtain graphs of

$$f(x) = \cos^5 x \quad \text{and} \quad g(x) = \sin^5 x$$

(b) Using the graphs obtained in part (a), what can you say about the value of the following integrals?

$$\int_0^{2\pi} \cos^5 x\, dx \quad \text{and} \quad \int_0^{2\pi} \sin^5 x\, dx$$

48. Use geometric reasoning to explain why

$$\int_0^1 x^n dx + \int_0^1 x^{1/n} dx = 1$$

for $n \geq 1$. (*Hint:* $y = x^n$ and $y = x^{1/n}$ are inverse functions.)

Exercises 49–56 refer to properties of the integral on page 328. In Exercises 49–52, use Riemann sums to prove the indicated property.

49. Property 1   **50.** Property 2

51. Property 4   **52.** Property 8

53. Assuming Property 5, prove Property 6. [*Hint:* Consider $f(x) - g(x)$.]

54. Prove Property 7 as follows. Explain why $f(x) \leq |f(x)|$, and then use Property 6 to conclude that $\int_a^b f(x)dx \leq \int_a^b |f(x)|dx$. Next, explain why $-f(x) \leq |f(x)|$, and then use Property 6 and Property 2 to conclude that $-\int_a^b f(x)dx \leq \int_a^b |f(x)|dx$. How does Property 7 follow?

55. Use mathematical induction to show that Property 4 can be extended to any finite number of functions. That is, show that if $f_1, f_2, \ldots, f_n$ are integrable on $[a, b]$,

$$\int_a^b \left(\sum_{k=1}^{n} f_k(x) \right) dx = \sum_{k=1}^{n} \left(\int_a^b f_k(x)dx \right)$$

56. Use Property 6 to show that $\int_0^1 \sin x\, dx \leq \frac{1}{2}$.

In Exercises 57–59, verify the result by making use of Equations 5.4, 5.5, or 5.6, as appropriate, together with the definition of the integral. In the Riemann sums use c_k as the right endpoint of the kth subinterval. (Hint: $\Delta t = x/n$.)

57. $\int_0^x t\, dt = \dfrac{x^2}{2}$ **58.** $\int_0^x t^2 dt = \dfrac{x^3}{3}$

59. $\int_0^x t^3 dt = \dfrac{x^4}{4}$

60. Let f be the function whose graph is shown. Define a new function A by

$$A(x) = \int_0^x f(t)\,dt \qquad (0 \le x \le 10)$$

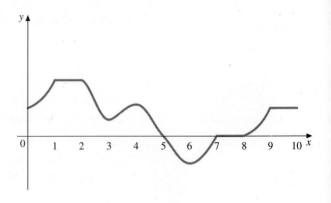

(a) Where is the function A increasing?
(b) Where is the function A decreasing?
(c) Is A constant on any interval? If so, where?
(d) Where do local maximum values of A occur?

5.3 THE FUNDAMENTAL THEOREMS OF CALCULUS

The concepts of the derivative, derived from the tangent line problem, and the integral, derived from the area problem, appear to be completely different from each other. There is, however, a very close relationship between the two concepts. Isaac Newton and Gottfried Leibniz are credited with its discovery—one of the most profound insights in the history of mathematics. We will show this relationship in two different ways. The results are two theorems, called the First and Second Fundamental Theorems of Calculus. These two theorems provide the vital links between differential and integral calculus, and we cannot overemphasize their importance. We will see a much simpler way of evaluating certain integrals than by using the definition as a limit of Riemann sums. Before we get to the main theorems, we need a preliminary result called the Mean-Value Theorem of Integral Calculus. (It states that the value of a definite integral is equal to some mean, or intermediate, value of the integrand times the length of the interval of integration.)

THEOREM 5.1

> **Mean-Value Theorem of Integral Calculus**
>
> If f is continuous on $[a, b]$, there exists a point c in $[a, b]$ such that
>
> $$\int_a^b f(x)\,dx = f(c)(b - a)$$

Proof   Since f is continuous on the interval $[a, b]$, we know by the Extreme-Value Theorem (Theorem 4.2) that f attains somewhere an absolute maximum value M and somewhere an absolute minimum value m on $[a, b]$.

By Property 8 of integrals (Section 5.2), since $m \le f(x) \le M$, we have

$$m(b - a) \le \int_a^b f(x)\,dx \le M(b - a)$$

As long as $b - a > 0$, we can divide by $b - a$ to get

$$m \le \frac{\int_a^b f(x)\,dx}{b - a} \le M$$

Thus, the number

$$\frac{\int_a^b f(x)dx}{b-a}$$

is between the smallest and largest values of f on the interval. We can therefore apply the Intermediate-Value Theorem (Theorem 2.8) and conclude that there exists a number c in $[a, b]$ such that

$$f(c) = \frac{\int_a^b f(x)dx}{b-a}$$

Multiplying both sides by $b - a$ yields the desired result. ∎

This theorem has a simple geometric interpretation. Suppose f is continuous and $f(x) \geq 0$ on $[a, b]$. We know, then, that $\int_a^b f(x)dx$ gives the area under the graph of f. The Mean-Value Theorem for integrals says this area is equal to $f(c)(b - a)$ for some number c in $[a, b]$. But $f(c)(b - a)$ is just the area of the rectangle with height $f(c)$ and base equal to the interval $[a, b]$, as shown in Figure 5.32. The height $f(c)$ is therefore a sort of average (mean) value of the ordinates $f(x)$ of the graph; that is, if we replace the curve by the horizontal line $f(c)$ units high, the area under the line is the same as the area under the curve. Although we have assumed f is nonnegative and continuous on $[a, b]$, we extend this notion of the average value of f to any integrable function by the following definition.

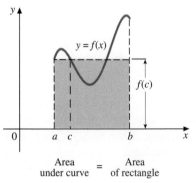

Area under curve $=$ Area of rectangle

FIGURE 5.32

Definition 5.3
Average Value of a
Function

Let f be integrable on $[a, b]$. The **average value of f** on $[a, b]$ is defined as

$$f_{\text{ave}} = \frac{1}{b-a} \int_a^b f(x)dx$$

EXAMPLE 5.11   Find the average value of the function

$$f(x) = \sqrt{4 - x^2}$$

on the interval $[-2, 2]$.

Solution   We found in Example 5.10(c) that

$$\int_{-2}^2 \sqrt{4 - x^2}\, dx = 2\pi$$

Thus, by Definition 5.3,

$$f_{\text{ave}} = \frac{1}{2 - (-2)} \int_{-2}^2 \sqrt{4 - x^2}\, dx$$

$$= \frac{1}{4}(2\pi)$$

$$= \frac{\pi}{2}$$

∎

The First Fundamental Theorem

Suppose now that f is continuous on the closed interval $[a, b]$ and that x is any point in this interval. We consider the integral from a to x, which we will call $A(x)$. So

$$A(x) = \int_a^x f(t)dt \qquad (5.13)$$

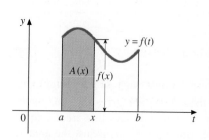

FIGURE 5.33

We are using t as a dummy variable of integration to avoid confusion with the independent variable x, which appears in the upper limit of integration. (We saw an integral of this type in Exercise 60 of Exercise Set 5.2.) If f is nonnegative on $[a, b]$, then we know that $A(x)$ is the area under the graph of f from a to x. (See Figure 5.33.) For this reason, we call $A(x)$ the *area function* for f on $[a, b]$. Since f is assumed to be continuous, we know that the integral exists on $[a, b]$ and is a unique number for each $a \leq x \leq b$. Consequently, we can see that $A(x)$ is a function of x. Even if f is of variable sign, we will still call $A(x)$ the area function, although we will then have to understand "area" to mean what we have called "net area." Before we state the first theorem, let us give an indication of what it will tell us. For instance, if we take $f(t) = t^2$ and $a = 0$, then by Exercise 58 in Exercise Set 5.2, we have

$$A(x) = \int_0^x t^2 dt = \frac{x^3}{3}$$

Now notice that $A'(x) = f(x)$ and hence we see $A' = f$, or equivalently, A is an antiderivative of f.

Now we can state the First Fundamental Theorem of Calculus. It says that for continuous functions, the derivative of this area function—that is, the derivative of the integral of f from a to x—is always $f(x)$. Or, in symbols,

$$A'(x) = f(x)$$

Referring to Figure 5.33, this theorem states that the area function is differentiable and the rate at which the area $A(x)$ is changing is exactly equal to the ordinate $f(x)$. In the statement of the theorem we replace $A(x)$ by the integral defining it in Equation 5.13.

THEOREM 5.2

> **The First Fundamental Theorem of Calculus**
>
> Let f be continuous on $[a, b]$. Then for x in $[a, b]$, the derivative of the integral $\int_a^x f(t)dt$ exists, and
>
> $$\frac{d}{dx} \int_a^x f(t)dt = f(x) \qquad (5.14)$$

This is a remarkable result and is worth studying carefully. It provides a crucial link between integration and differentiation. It tells us that every function f continuous on an interval $[a, b]$ has an antiderivative there, namely, $\int_a^x f(t)dt$.

Proof We show that the derivative of $A(x) = \int_a^x f(t)dt$ exists by showing the existence of the limit

$$\lim_{h \to 0} \frac{A(x+h) - A(x)}{h}$$

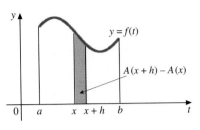

FIGURE 5.34

If $h > 0$, the numerator is the area under the graph of f from a to $x + h$ minus the area from a to x, and therefore is the shaded area from x to $x + h$, as indicated in Figure 5.34. As we have pictured it, f is nonnegative, but our result is valid without this restriction. That is,

$$A(x+h) - A(x) = \int_x^{x+h} f(t)dt$$

(This equality is justified without reference to the graph by Property 3 of integrals.) By the Mean-Value Theorem for integrals (Theorem 5.1), there is a number c between x and $x + h$ such that

$$\int_x^{x+h} f(t)dt = f(c) \cdot h$$

Thus, we obtain, on substituting $f(c) \cdot h$ for $A(x + h) - A(x)$,

$$\lim_{h \to 0} \frac{A(x+h) - A(x)}{h} = \lim_{h \to 0} \frac{f(c) \cdot h}{h} = \lim_{h \to 0} f(c)$$

Since c lies between x and $x+h$, it follows from the Squeeze Theorem (Theorem 2.3) that as $h \to 0$, $c \to x$, and so by continuity (Definition 2.2) $f(c) \to f(x)$. Thus,

$$\lim_{h \to 0^+} \frac{A(x+h) - A(x)}{h} = f(x)$$

The argument is similar for $h < 0$, and you will be asked to supply it in Exercise 69 of Exercise Set 5.3. We conclude from the definition of the derivative that $A'(x)$ exists and that $A'(x) = f(x)$. ∎

EXAMPLE 5.12 Find the derivative of $\int_2^x \sin t \, dt$.

Solution Since $\sin x$ is continuous everywhere, by Equation 5.14,

$$\frac{d}{dx} \int_2^x \sin t \, dt = \sin x$$ ∎

EXAMPLE 5.13 Find the derivative of $\int_0^x \frac{du}{\sqrt{1 + u^3}}$, $x > 0$.

Solution The function $1/\sqrt{1 + u^3}$ is continuous on $[0, \infty)$, so by Equation 5.14,

$$\frac{d}{dx} \int_0^x \frac{du}{\sqrt{1 + u^3}} = \frac{1}{\sqrt{1 + x^3}}$$

Note that u is merely a dummy variable of integration. The result would be the same no matter which letter we used, as long as we used the same letter for the function $1/\sqrt{1 + u^3}$ and the differential du. The integral is a function of x, not of u. ∎

The Second Fundamental Theorem

In a certain sense, Theorem 5.2 expresses an inverse relationship between integration and differentiation. If we begin with a continuous function f on $[a, b]$, integrate it from a to x, then differentiate the result, we come back to the original function, evaluated at x. The next theorem turns this procedure around and shows that under certain hypotheses, if we differentiate first and then integrate the result, we also return to the original function, evaluated at the endpoints of the interval.

THEOREM 5.3

> **The Second Fundamental Theorem of Calculus**
>
> Let f be continuous on $[a, b]$ and let F be any antiderivative of f on $[a, b]$; that is, $F'(x) = f(x)$. Then
>
> $$\int_a^b f(x)dx = F(b) - F(a) \qquad (5.15)$$

Proof   By Theorem 5.2 we know that $\int_a^x f(t)dt$ is one antiderivative of f on $[a, b]$, and we are given that F is also an antiderivative of f there. So by Corollary 4.9b, these two antiderivatives differ by a constant; that is,

$$\int_a^x f(t)dt = F(x) + C$$

for some constant C. In fact, we can determine C. If we put $x = a$, we get

$$\int_a^a f(t)dt = F(a) + C$$

But the integral on the left is 0, so $C = -F(a)$. Thus,

$$\int_a^x f(t)dt = F(x) - F(a) \qquad (5.16)$$

Finally, we put $x = b$ in Equation 5.16 to get

$$\int_a^b f(t)dt = F(b) - F(a)$$

which is what we wanted to prove.   ■

To give an idea of the significance of this theorem, recall the work involved in evaluating an integral such as $\int_0^2 x^2 dx$ by making use of the limit of Riemann sums, as we did in Example 5.5. Using the Second Fundamental Theorem, we can get the value of $\int_0^2 x^2 dx$ quickly and easily by first finding *any* antiderivative of x^2—for example, $F(x) = x^3/3$—and then calculating $F(2) - F(0)$:

$$\int_0^2 x^2 dx = F(2) - F(0) = \frac{8}{3} - 0 = \frac{8}{3}$$

It is convenient to introduce the notation $F(x)]_a^b$ to mean $F(b) - F(a)$. Then we can write

$$\int_a^b f(x)dx = F(x)]_a^b$$

for the result of the Second Fundamental Theorem. For example,

$$\int_0^2 x^2 dx = \frac{x^3}{3}\Big]_0^2 = \frac{8}{3} - 0 = \frac{8}{3}$$

EXAMPLE 5.14   Evaluate each of the following using the Second Fundamental Theorem:

(a) $\displaystyle\int_{-1}^2 (3x^2 - 2x + 4)dx$ (b) $\displaystyle\int_0^{\pi/2} \cos x \, dx$ (c) $\displaystyle\int_{-\pi/4}^{\pi/4} \sec^2 x \, dx$

Solution   Note first that all the integrands are continuous on the intervals involved, so the Second Fundamental Theorem applies.

(a) $\displaystyle\int_{-1}^2 (3x^2 - 2x + 4)dx = x^3 - x^2 + 4x\Big]_{-1}^2 = (8 - 4 + 8) - (-1 - 1 - 4)$

$$= 12 + 6 = 18$$

(b) $\displaystyle\int_0^{\pi/2} \cos x \, dx = \sin x]_0^{\pi/2} = \sin \frac{\pi}{2} - \sin 0 = 1$

(c) $\displaystyle\int_{-\pi/4}^{\pi/4} \sec^2 x \, dx = \tan x]_{-\pi/4}^{\pi/4} = \tan \frac{\pi}{4} - \tan \left(\frac{-\pi}{4}\right) = 1 - (-1) = 2$ ∎

REMARK ───────────────────────────────

■ Since in the Second Fundamental Theorem we are allowed to use *any* antiderivative of the integrand, we might as well choose the simplest one. Any other would give the same result. For example,

$$\int_{-1}^2 x^2 dx = \frac{x^3}{3}\Big]_{-1}^2 = \frac{8}{3} - \frac{(-1)}{3} = 3$$

If we had used another antiderivative, instead of the simplest, it would necessarily have been of the form $x^3/3 + C$, so

$$\int_{-1}^2 x^2 dx = \frac{x^3}{3} + C\Big]_{-1}^2 = \left[\frac{8}{3} + C\right] - \left[\frac{-1}{3} + C\right]$$

$$= \frac{8}{3} + \frac{1}{3} = 3$$

In other words, the constant C will always cancel out.

Do not infer from this result that it is no longer necessary to add the constant when forming an antiderivative. We are speaking here of one particular use of the antiderivative where the constant can be deleted, namely, in evaluating definite integrals by the Second Fundamental Theorem. In most cases the constant is essential.

The Integral of the Derivative as Net Change in the Function

In the Second Fundamental Theorem, F is an antiderivative of f; that is, $F'(x) = f(x)$. Therefore, we could write the result in the alternative form

$$\int_a^b F'(x)dx = F(b) - F(a) \qquad (5.17)$$

as long as F' is continuous on $[a, b]$. It can be proved, in fact, that Equation 5.17 holds true even if F' is not continuous as long as it is integrable on $[a, b]$. (See Exercise 70 in Exercise Set 5.3.) Going back to Equation 5.16 in the proof of the theorem, if we also replace f by F', we get

$$\int_a^x F'(t)dt = F(x) - F(a)$$

Since the only function involved here is F, we can also write the result with another letter to designate the function. In particular, we can use our usual letter f and get

$$\int_a^x f'(t)dt = f(x) - f(a) \qquad (5.18)$$

which is true for all $x \in [a, b]$ provided f' is integrable there. Let us compare this result with the result of the First Fundamental Theorem, Equation 5.14, which states that if f is continuous on $[a, b]$, then

$$\frac{d}{dx} \int_a^x f(t)dt = f(x), \qquad x \in [a, b]$$

Here we differentiate the integral of f from a to x and get $f(x)$ as the result. In Equation 5.18, we integrate from a to x the derivative of f and get $f(x)$ minus a constant, $f(a)$. It is in this sense that integration and differentiation can be considered inverse processes.

Equation 5.17 provides a useful way of interpreting the definite integral. Since $F(b) - F(a)$ is the change in the function values $F(x)$ from $x = a$ to $x = b$, and $F'(x)$ is the rate of change of the function, we reach the following conclusion.

> The integral from a to b of the *rate of change* of a function is the *net change* in the function from a to b.

We have seen this interpretation already in Example 5.3. The velocity function $v(t)$ is the rate of change of the position function $s(t)$ (that is, $v(t) = s'(t)$). So

$$\int_a^b v(t)dt = s(b) - s(a) = \text{net change in distance}$$

The next two examples further illustrate this interpretation of the integral of rate of change as the net change in the function.

EXAMPLE 5.15   A pump is pumping water into a tank at the rate of $2t$ liters per minute (L/min), where t is in minutes. Find the amount of water pumped into the tank during the first half-hour the pump is in operation.

Solution   Let $w(t)$ denote the number of liters of water pumped into the tank during the first t min of pumping. Then we can conclude from the given information that

$$w'(t) = 2t$$

An antiderivative of $2t$ is t^2. So the total amount of water pumped into the tank in the first half-hour (30 min) is

$$\int_0^{30} 2t\, dt = t^2 \Big]_0^{30} = 900 \text{ L} \qquad \blacksquare$$

EXAMPLE 5.16   In the fall of 1991, the poinsettia white fly began devastating vegetable crops in the Imperial Valley of California. Assume that the graph in Figure 5.35 gives the rate at which fields of broccoli were being destroyed, in acres per day, for the month of November. Approximate the total number of acres of broccoli destroyed during the first 10 days of the month and during the last 10 days.

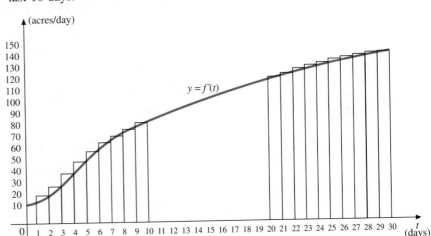

$f'(t)$ = rate of destruction of broccoli fields by
the poinsettia white fly in the Imperial Valley, in
acres per day, during November 1991

FIGURE 5.35

Solution   Let $f(t)$ denote the number of acres of broccoli destroyed in the first t days. Then the graph gives $f'(t)$. In the first 10 days, the total number of acres destroyed is given by

$$\int_0^{10} f'(t)\,dt$$

If we knew a formula for f, we could evaluate this integral using the Second Fundamental Theorem. Since we do not know f, we can approximate the integral with a Riemann sum

$$\sum_{k=1}^{n} f'(c_k)\Delta t_k$$

We might choose to use subintervals of equal length, with $\Delta t = 1$. A reasonable choice for c_k is the midpoint of the kth subinterval. When we do this, we get (reading approximate values of $f'(c_k)$ from the graph)

$$\int_0^{10} f'(t)dt \approx (11)(1) + (19)(1) + (27)(1) + (36)(1) + (48)(1) + (58)(1)$$
$$+ (65)(1) + (70)(1) + (75)(1) + (81)(1)$$
$$= 490$$

So approximately 490 acres were destroyed in the first 10 days of November.

For the last 10 days we approximate the value in a similar way, getting

$$\int_{20}^{30} f'(t)dt \approx (121)(1) + (122)(1) + (128)(1) + (130)(1) + (133)(1)$$
$$+ (135)(1) + (138)(1) + (140)(1) + (144)(1) + (142)(1)$$
$$= 1,333$$

Thus, approximately 1,333 acres of broccoli were destroyed during the last 10 days of November.   ∎

Let us conclude this section by summarizing the two fundamental theorems. It is important that you learn these, as we will use them many times throughout the remainder of this textbook. We use the abbreviations FTC1 and FTC2 for the First and Second Fundamental Theorems of Calculus. Below each statement we paraphrase what it says in simple language.

The Two Fundamental Theorems

FTC1   If f is continuous on $[a, b]$, then

$$\frac{d}{dx} \int_a^x f(t)dt = f(x)$$

for each x in $[a, b]$.

That is, when you differentiate an integral (with respect to a variable upper limit), you get the integrand (evaluated at the upper limit).

FTC2   If f is continuous on $[a, b]$ and F is any antiderivative of f, then

$$\int_a^b f(x)dx = F(b) - F(a)$$

That is, to evaluate a definite integral, find an antiderivative of the integrand and subtract its value at the lower limit from its value at the upper limit.

Since for an antiderivative F of f, we have $F' = f$, an alternative form of FTC2 is

$$\int_a^b F'(x)dx = F(b) - F(a) \qquad \text{If } F' \text{ is continuous on } [a, b]$$

In this form we can say that to integrate the derivative of a function, you subtract its value at the lower limit from its value at the upper limit.

Exercise Set 5.3

Use the Second Fundamental Theorem to evaluate the integrals in Exercises 1–30.

1. $\int_1^3 (x^2 + 2x - 1)dx$

2. $\int_{-1}^1 (3x - x^2)dx$

3. $\int_{-2}^1 (x - 1)^2 dx$

4. $\int_3^5 x(2x - 3)dx$

5. $\int_0^4 (x^3 - 3x^2 + 4)dx$

6. $\int_{-2}^0 (x + 2)^3 dx$

7. $\int_0^1 (x - \sqrt{x} + 3)dx$

8. $\int_{-1}^1 (x^5 - 2x^3 + 4x)dx$

9. $\int_{-2}^2 (x^2 - 3)^2 dx$

10. $\int_4^1 (x - 1)(x + 2)dx$

11. $\int_{-2}^{-1} \frac{x^3 - 1}{x^2}dx$

12. $\int_1^4 \frac{x + 1}{\sqrt{x}}dx$

13. $\int_0^\pi 2 \sin x \, dx$

14. $\int_{-\pi/2}^{\pi/2} 3 \cos x \, dx$

15. $\int_0^{\pi/3} \sec^2 x \, dx$

16. $\int_{-\pi/3}^{\pi/4} \sec x \tan x \, dx$

17. $\int_{\pi/6}^{\pi/2} 2 \csc x \cot x \, dx$

18. $\int_{\pi/4}^{3\pi/4} 4 \csc^2 x \, dx$

19. $\int_4^9 (t^{3/2} - 2t^{1/2} + 1)dt$

20. $\int_1^8 (t^{2/3} + 2t^{-1/3})dt$

21. $\int_1^8 \frac{u - 2}{\sqrt[3]{u}}du$

22. $\int_1^4 \frac{(v - 1)^3}{\sqrt{v}}dv$

23. $\int_{\pi/6}^{5\pi/6} (x + \sin x)dx$

24. $\int_{\pi/4}^{\pi/2} \frac{1 - \cos\theta}{\sin^2\theta}d\theta$

25. $\int_0^{\pi/4} (1 + \sin^2\phi \sec^2\phi)d\phi$

26. $\int_{\pi/6}^{\pi/3} \tan^2 x \, dx$

27. $\int_{2\pi/3}^{5\pi/6} \frac{\sec t - \csc t}{\sec t \csc t}dt$

28. $\int_{\pi/4}^{\pi/2} \cot^2 v \, dv$

29. $\int_1^3 \left(x - \frac{1}{x}\right)\left(2x - \frac{3}{x}\right)dx$

30. $\int_1^9 \frac{(\sqrt{t} - 2)(\sqrt{t} + 3)}{2\sqrt{t}}dt$

In Exercises 31–34, find the average value of the function on the given interval.

31. $f(x) = x^2 - 2x - 3$ on $[-2, 4]$

32. $f(x) = 1 + x - 2x^2$ on $[-1, 2]$

33. $f(x) = x^3 - 3x^2 - 4x + 2$ on $[-2, 3]$

34. $f(x) = 2x^3 - 3x - 4$ on $[-2, 2]$

35. Suppose an object moving along a line has distance function $s(t)$.
  (a) Give its average velocity from time $t = a$ to time $t = b$.
  (b) Find the average value of the velocity function $v(t) = s'(t)$ by using Definition 5.3 together with the Second Fundamental Theorem. Show that the result is the same as in part (a).

In Exercises 36–39, find a number c in $[a, b]$ that satisfies $\int_a^b f(x)dx = f(c) \cdot (b - a)$.

36. $f(x) = \sqrt{x}$ on $[0, 9]$

37. $f(x) = 4/x^2$ on $[1, 4]$

38. $f(x) = x^2 - 1$ on $[0, 3]$

39. $f(x) = x^2 - 2x$ on $[-2, 1]$

In Exercises 40–43, use the First Fundamental Theorem to find the derivative. State on what intervals the result is valid.

40. $\dfrac{d}{dx}\int_1^x \dfrac{2}{\sqrt{1 + t^2}}dt$

41. $\dfrac{d}{dx}\int_2^x \dfrac{t}{t - 1}dt$

42. $F'(x)$ if $F(x) = \int_0^x \sqrt{t^3 + 8}\,dt$

43. $F'(x)$ if $F(x) = \int_{-3}^x \dfrac{\sin t}{2 + \cos t}dt$

44. Find $\frac{d}{dx}\int_{-1}^x (t^3 - 3t^2 + 1)dt$ by two methods: (a) the First Fundamental Theorem and (b) first performing the integration using the Second Fundamental Theorem and then differentiating.

45. The accompanying figure shows the graph of a function f on the interval $[0, 3]$. Let $A(x) = \int_0^x f(t)dt$. Sketch, as well as you can, the graph of $y = A(x)$ for $0 \le x \le 3$.

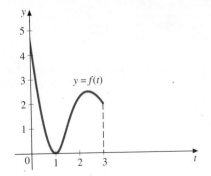

46. If a population of elephants is increasing at the rate of $500 + 2t$, where t is measured in years, find the amount by which the elephant population increases between the third and seventh years.

47. In recent years the Black Sea has become one of the most polluted bodies of water in the world. Assume that since 1980 the rate at which pollutants have been entering the sea is approximately $6300t^{5/2}$ kg/yr, where t is measured in years since 1980. Find approximately how many kilograms of pollutants were added from 1980 to 1990. If this rate continues, how many kilograms would be added from 1990 to 2000?

48. The following graph shows the inflation rate, $r(t)$, of a certain country in percent per year for the past 15 years. Estimate the total percentage increase in the cost of living during the same time period.

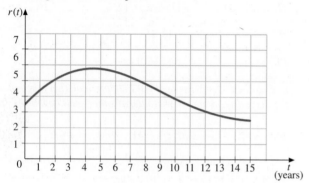

49. The accompanying graph shows the power consumption by a certain city on a given day in millions of kilowatts. Approximate the energy used on that day in kilowatt-hours. (The relationship between power, P, and energy, E, is $P = dE/dt$.)

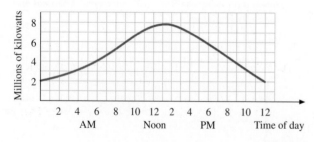

50. If $f(x) = x/(x + 1)$, find $\int_0^2 f'(x)dx$.

51. Find $\displaystyle\int_0^2 \frac{d}{dt}\left(1/\sqrt{1 + t^3}\right) dt$.

52. Derive a formula for

$$\frac{d}{dx}\int_x^b f(t)dt$$

where f is continuous on $[a, b]$ and $x \in [a, b]$.

53. Use the result of Exercise 52 to find $g'(x)$ if

$$g(x) = \int_x^4 \frac{\sin t}{t}dt$$

where $x > 0$.

54. Derive a formula for

$$\frac{d}{dx}\int_a^{g(x)} f(t)dt$$

where f is continuous and g is differentiable. [*Hint:* Let $u = g(x)$, and use the chain rule.]

55. Use the result of Exercise 54 to evaluate

$$\frac{d}{dx}\int_1^{x^2 - 2x} \sin^3 t\, dt$$

56. Find the number a for which the line $x = a$ divides the area under the graph of $y = (x - 3)^2$ from $x = -1$ to $x = 3$ in a $2 : 1$ ratio.

57. Evaluate the integral

$$\int_0^{3\pi/4} |\cos x|dx$$

58. Evaluate $\int_{-1}^2 |x^3 - 4x|dx$. (*Hint:* Draw the graph of $y = x^3 - 4x$.)

59. Find the area bounded by the curve $y = x^2 - 4x + 3$ and the x-axis.

60. In the accompanying figure, what is the ratio of the area of the shaded region to the area of the rectangle $ABCD$?

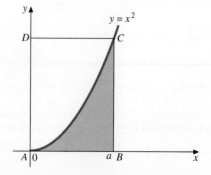

61. Find the area of the parabolic segment between the graph of $y = x^2$ and the line AB in the figure by subtracting the area under the graph of $y = x^2$ from the area of the triangle ABC.

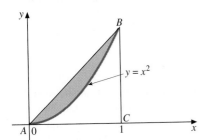

62. On the surface of the earth acceleration caused by gravity is approximately -10 m/s^2, whereas on the surface of the moon it is approximately -1.6 m/s^2.

(a) If an astronaut on the moon throws a baseball directly upward with an initial velocity of 15 m/s, what will be its velocity and height 2.5 seconds later? When will it reach its maximum height? What is the maximum height?

(b) If, on returning to the earth, the astronaut throws the ball directly upward with the same initial velocity, answer the questions of part (a).

In Exercises 63–68, simplify the integrands and then use the Second Fundamental Theorem to evaluate the integrals.

63. $\displaystyle \int_{2}^{3} \sqrt{\frac{1}{x^4} - 1 + \frac{x^4}{4}} \, dx$ **64.** $\displaystyle \int_{-1}^{2} \frac{2t^3 + 3t^2 + 4}{t + 2} \, dt$

65. $\displaystyle \int_{-\pi/6}^{\pi/3} \left(\frac{1}{1 + \sin \theta} + \frac{1}{1 - \sin \theta} \right) d\theta$

66. $\displaystyle \int_{\pi/2}^{5\pi/6} \frac{2}{1 - \cos 2x} \, dx$

67. $\displaystyle \int_{-\pi/4}^{\pi/3} \frac{1 - \sin \theta}{1 + \sin \theta} \, d\theta$ (*Hint:* Multiply numerator and denominator by $1 - \sin \theta$.)

68. $\displaystyle \int_{5\pi/6}^{\pi} \frac{1 - \sin 2x}{\sin x - \cos x} \, dx$ (*Hint:* Replace 1 by $\sin^2 x + \cos^2 x$ and use an identity for $\sin 2x$.)

69. Complete the proof of Theorem 5.2 by considering $h < 0$.

70. Prove that Equation 5.17 holds true if we assume only that F' is integrable on $[a, b]$. (*Hint:* Use the Mean-

Value Theorem of Differential Calculus to show that for any partition of $[a, b]$, there is a point $c_k \in (x_{k-1}, x_k)$, $k = 1, 2, \ldots, n$, such that

$$\sum_{k=1}^{n} [F(x_k) - F(x_{k-1})] = \sum_{k=1}^{n} F'(c_k) \Delta x_k$$

Show that the left-hand side reduces to $F(b) - F(a)$. Then let max $\Delta x_k \to 0$.)

71. Let

$$f(x) = \begin{cases} x^2 \sin \dfrac{1}{x^2} & \text{if } x \neq 0 \\ 0 & \text{if } x = 0 \end{cases}$$

Show that $f'(x)$ exists on $[0, 1]$ but that

$$\int_{0}^{1} f'(x)dx \neq f(1) - f(0)$$

Why does this not contradict Exercise 70? (*Hint:* Show that f' is unbounded on $[0, 1]$.)

72. Prove that if f is integrable on $[a, b]$, the function $F(x) = \int_{a}^{x} f(t)dt$ is continuous on $[a, b]$. [*Hint:* Let M be such that $|f(x)| \leq M$ on $[a, b]$. (Justify.) Then use the $\varepsilon - \delta$ definition of continuity (Definition 2.9) of F at x_0, making use of the properties of the integral.]

73. Show that the Mean-Value Theorems of Integral Calculus and of Differential Calculus are essentially different formulations of the same result. [*Hint:* In $\int_{a}^{b} f(x)dx = f(c)(b - a)$, let F be an antiderivative of f so that $f = F'$.]

74. Prove that if f and g are defined on $[a, b]$, with $f(x) = g(x)$ on $[a, b)$ and g continuous on $[a, b]$, then $\int_{a}^{b} f(x)dx$ exists and equals $\int_{a}^{b} g(x)dx$ regardless of how $f(b)$ is defined. (*Hint:* Use the result of Exercise 72.)

75. Prove that if f is integrable on $[a, b]$, then $F(x) = \int_{x}^{b} f(t)dt$ is continuous on $[a, b]$. (*Hint:* Use the hint for Exercise 72.)

76. Prove that if f and g are defined on $[a, b]$, with $f(x) = g(x)$ on $(a, b]$, and g is continuous on $[a, b]$, then $\int_{a}^{b} f(x)dx$ exists and equals $\int_{a}^{b} g(x)dx$ regardless of how $f(a)$ is defined. (*Hint:* Use the result of Exercise 75.)

77. Suppose f is defined on $[a, b]$ and is continuous on (a, b), and $\lim_{x \to a^+} f(x)$ and $\lim_{x \to b^-} f(x)$ both exist. Prove that $\int_{a}^{b} f(x)dx$ exists, regardless of how $f(a)$ and $f(b)$ are defined. (*Hint:* Use the results of Exercises 74 and 76.)

5.4 THE INDEFINITE INTEGRAL; THE TECHNIQUE OF SUBSTITUTION

According to the Second Fundamental Theorem of Calculus, we can evaluate the integral of a continuous function f over $[a, b]$ quickly and efficiently by the formula

$$\int_a^b f(x)dx = F(b) - F(a)$$

provided we can find F, where F is an antiderivative of f (that is, F is a function such that $F' = f$). To appreciate the importance of this result, recall the work involved in evaluating the integral as a limit of Riemann sums. There is only one hitch. To apply the theorem, you must be able to find a function F for which $F' = f$. Now we know by the First Fundamental Theorem that every continuous function on a closed interval has an antiderivative there, but finding such an antiderivative in the form of a recognizable elementary function is another matter. In some cases we can recognize the integrand as the derivative of a known function. In Section 4.7 we summarized a number of such functions along with their general antiderivatives, based on known derivatives of power functions and trigonometric functions. In this section we will learn a method that will enable us to greatly expand the list, and in Chapter 8 we will learn additional ways of finding antiderivatives. Unfortunately, no matter how many techniques we learn, there will be some continuous functions for which no one can find antiderivatives in the form of elementary functions. The Second Fundamental Theorem is no help in these cases, and we can only approximate the integral. In the next section we consider approximation techniques. Many of the functions used in describing scientific concepts fall into this category.

We need a symbol for the antiderivative:

Definition 5.4 The Indefinite Integral	The symbol $\int f(x)dx$ is called the **indefinite integral** of f and means the general antiderivative of f. That is, $$\int f(x)dx = F(x) + C$$ where F is any function whose derivative is f. The arbitrary constant C is called the **constant of integration**.

Although definite and indefinite integrals differ only slightly in appearance, their meanings differ significantly. Consider the following:

$$\text{Definite integral:} \quad \int_{-1}^2 x^2 dx = \frac{x^3}{3}\Bigg]_{-1}^2 = \frac{8}{3} - \left(-\frac{1}{3}\right) = 3$$

$$\text{Indefinite integral:} \quad \int x^2 dx = \frac{x^3}{3} + C$$

In the first case the answer is a *number*, whereas in the second it is a *family of functions*. The definite and indefinite integrals are related by the Second Fundamental Theorem. In fact, we could write the conclusion of that theorem in the following form.

Restatement of FTC2

$$\int_a^b f(x)dx = \int f(x)dx \bigg]_a^b$$

Just as with a definite integral, we refer to $f(x)$ in $\int f(x)dx$ as the *integrand*, and the process by which we obtain the answer is called *integration*. To *evaluate* an indefinite integral means to find the general antiderivative.

The following properties of indefinite integrals hold true on any interval I where the functions involved have antiderivatives.

Properties of Indefinite Integrals

1. $\dfrac{d}{dx} \int f(x)dx = f(x)$
2. $\int f'(x)dx = f(x) + C$, if f is differentiable on I
3. $\int cf(x)dx = c \int f(x)dx$, where c is any constant
4. $\int [f(x) \pm g(x)]dx = \int f(x)dx \pm \int g(x)dx$

The proofs of these properties follow from Definition 5.4 and are left as exercises.

All the antidifferentiation formulas summarized in Section 4.7 can now be restated using indefinite integral notation. In particular, we have

---

$$\int x^n dx = \frac{x^{n+1}}{n+1} + C, \text{ if } n \neq -1 \qquad \text{Power Rule}$$

$$\int \sin x \, dx = -\cos x + C$$

$$\int \cos x \, dx = \sin x + C$$

The Guess-and-Check Method

We now turn to the question of how to evaluate an indefinite integral when the integrand is not among those in our list in Section 4.7. We want to find a function F for which

$$\int f(x)dx = F(x) + C$$

We know that $F'(x)$ must equal $f(x)$. Sometimes we can recognize the integrand as being the derivative of a composite function. Consider the following examples.

EXAMPLE 5.17   Evaluate the indefinite integral $\int 2x \cos x^2 dx$.

Solution   We know that an antiderivative of $\cos x$ is $\sin x$, so we might guess that the antiderivative we are looking for involves the sine function. Also, since we know that $d(x^2)/dx = 2x$, we might further guess that $2x \cos(x^2)$ is the derivative of the composite function $\sin x^2$, by recalling how the Chain Rule works. Checking, we see that

$$\frac{d}{dx} \sin x^2 = (\cos x^2)(2x) = 2x \cos x^2$$

So our guess is correct, and we can write the answer as

$$\int 2x \cos x^2 dx = \sin x^2 + C$$ ∎

EXAMPLE 5.18   Evaluate the indefinite integral $\int (3x^2 + 1)(x^3 + x + 1)^5 dx$.

Solution   Because $3x^2 + 1$ is the derivative of $x^3 + x + 1$, it looks as though the integrand might be the derivative of the composite function $(x^3 + x + 1)^6$, since by the Generalized Power Rule the derivative of a function to the nth power equals n times the function to the $(n - 1)$st power multiplied by the derivative of the function. So we check and find that

$$\frac{d}{dx}(x^3 + x + 1)^6 = 6(x^3 + x + 1)^5(3x^2 + 1)$$

$$= 6(3x^2 + 1)(x^3 + x + 1)^5$$

We were almost right in our guess, but it is off because of the factor 6. We can easily remedy the situation by dividing our original guess by 6. That is, we use

$$\frac{1}{6}(x^3 + x + 1)^6$$

So the correct result is

$$\int (3x^2 + 1)(x^3 + x + 1)^5 dx = \frac{1}{6}(x^3 + x + 1)^6 + C$$ ∎

We refer to the reasoning used in Examples 5.17 and 5.18 as the *guess-and-check method*. When the integrand looks as though it might be the result of applying the Chain Rule to some composite function, we guess what the function is and check by differentiating it to see if we get the integrand. If so, we add an arbitrary constant C, and we have the answer. If the derivative of our guess is off only because it is k times what we want, where k is a constant (obviously nonzero), then the correct antiderivative is our original guess divided by k.

The Method of Substitution

The guess-and-check method relies on the fact that if $F' = f$, then by the Chain Rule,

$$\frac{d}{dx} F(g(x)) = F'(g(x))g'(x) = f(g(x))g'(x)$$

so that

$$\int f(g(x))g'(x)dx = F(g(x)) + C \qquad (5.19)$$

An alternative to guessing is to substitute a new variable such as u for $g(x)$ in Equation 5.19. Then we have $u = g(x)$, $du = g'(x)dx$, so that Equation 5.19 becomes

$$\int f(u)du = F(u) + C \quad \text{(where } F' = f)$$

Then, to get the final answer we replace u by $g(x)$ in the right-hand side of this equation. As you might expect, we call this technique the *method of substitution*.

REMARK ───────────────────────────────

■ The substitution method shows that the dx appearing in an indefinite integral can be treated as a differential, since $du = g'(x)dx$. The same is true in a definite integral, as we will see shortly.

───

In the substitution method the goal is to choose a substitution u that will make $f(u)$ the derivative of a known function $F(u)$. That is, we want $\int f(u)du$ to be among our list of known antiderivatives.

Let us redo the problems in Examples 5.17 and 5.18 using the method of substitution. In Example 5.17 we let $u = x^2$. Then $du = 2x \, dx$. Thus,

$$\int 2x \cos(x^2)dx = \int \cos u \, du = \sin u + C$$

Now all that remains is to replace u in the final answer with its value in terms of x, namely, $u = x^2$. So we get

$$\int 2x \cos(x^2)dx = \sin x^2 + C$$

In Example 5.18 we let $u = x^2 + x + 1$, so that $du = (3x^2 + 1)dx$. Then we have

$$\int (3x^2 + 1)(x^3 + x + 1)^5 dx = \int u^5 \, du = \frac{u^6}{6} + C = \frac{1}{6}\left(x^3 + x + 1\right)^6 + C$$

Here are some more examples.

EXAMPLE 5.19   Use the method of substitution to evaluate $\int (x^2 - 1)^3 \cdot 2x \, dx$.

Solution   When the integrand involves a function of x raised to a power, such as $(x^2 - 1)^3$, a general rule of thumb is to substitute u for the function on the inside. So in this case we let $u = x^2 - 1$. Then $du = 2x \, dx$. So we get

$$\int \underbrace{(x^2 - 1)^3}_{u^3} \cdot \underbrace{2x \, dx}_{du} = \int u^3 du = \frac{u^4}{4} + C$$

Now we replace u by $x^2 - 1$ to get the final answer

$$\int (x^2 - 1)^3 \cdot 2x \, dx = \frac{(x^2 - 1)^4}{4} + C \qquad ■$$

EXAMPLE 5.20   Use the method of substitution to evaluate $\int x\sqrt{1-x^2}\,dx$.

Solution   First, write the integral in the form

$$\int x(1-x^2)^{1/2}dx$$

Again, the integrand involves a function raised to a power, namely, $(1-x^2)^{1/2}$. So we try $u = 1 - x^2$. Then $du = -2x\,dx$. Note that the x and dx are present in the integral. The only thing missing is the factor (-2). We can therefore get the correct differential du if we multiply by (-2) inside the integral. To compensate, we multiply by its reciprocal, $(-\frac{1}{2})$, outside the integral. Then we have

$$-\frac{1}{2}\int \underbrace{(1-x^2)^{1/2}}_{u^{1/2}}\underbrace{(-2x\,dx)}_{du} = -\frac{1}{2}\int u^{1/2}du$$

$$= -\frac{1}{2}\frac{u^{3/2}}{\frac{3}{2}} + C$$

$$= -\frac{1}{2}\left(\frac{2}{3}\right)u^{3/2} + C$$

$$= -\frac{1}{3}u^{3/2} + C$$

$$= -\frac{1}{3}(1-x^2)^{3/2} + C \qquad\blacksquare$$

REMARK

■ By Property 2 of integrals in Section 5.2, for any constant c,

$$c\int_a^b f(x)dx = \int_a^b cf(x)dx$$

It is this property that we used in Example 5.20. We can show the steps in more detail by noting that since $(-\frac{1}{2})(-2) = 1$,

$$\int (1-x^2)^{1/2}x\,dx = \int (1-x^2)^{1/2}\left(-\frac{1}{2}\right)(-2x\,dx)$$

$$= -\frac{1}{2}\int (1-x^2)^{1/2}(-2x\,dx)$$

by Property 2. Thus, multiplying by (-2) inside the integral and by its reciprocal $(-1/2)$ outside the integral leaves the integral unchanged.

EXAMPLE 5.21   Evaluate the integral

$$\int \frac{\sin\sqrt{x}}{\sqrt{x}}\,dx$$

Solution   If we could make a substitution that would result in an integral of the form $\int \sin u \, du$, we would know the antiderivative. So we try the substitution $u = \sqrt{x}$. Then

$$du = \frac{1}{2\sqrt{x}} \, dx$$

All that is missing in the original integral is the factor $\frac{1}{2}$. So we multiply by $\frac{1}{2}$ on the inside of the integral and by 2 on the outside.

$$\int \frac{\sin \sqrt{x}}{x} \, dx = 2 \int \sin \sqrt{x} \left(\frac{1}{2\sqrt{x}} \, dx \right)$$

$$= 2 \int \sin u \, du$$

$$= -2 \cos u + C$$

$$= -2 \cos \sqrt{x} + C \qquad \blacksquare$$

EXAMPLE 5.22   Evaluate the integral $\int \cos^3(2x) \sin(2x) \, dx$.

Solution   Since $\cos^3(2x)$ means $(\cos 2x)^3$, we have a function raised to a power. So, as in Examples 5.19 and 5.20, we substitute u for the function. That is, we let $u = \cos 2x$. Then $du = -2 \sin 2x \, dx$. Multiplying and dividing by -2 gives

$$\int \cos^3(2x) \sin(2x) \, dx = -\frac{1}{2} \int (\cos 2x)^3 (-2 \sin 2x \, dx)$$

$$= -\frac{1}{2} \int u^3 \, du$$

$$= -\frac{1}{2} \frac{u^4}{4} + C$$

$$= -\frac{1}{8} \cos^4(2x) + C \qquad \blacksquare$$

It is important that the substitution be chosen in such a way that everything inside the integral sign is expressed in terms of the new variable u (or whatever letter is used), including the differential. If the only thing missing is a *constant* factor, then you can multiply by that factor on the inside and by its reciprocal on the outside. We will refer to this process as "adjusting the differential."

 Only constants can be factored out of an integral. So you *cannot* adjust the differential if it requires multiplying by an expression containing the variable of integration.

There is nothing special about using the letter u for a substitution, any more than there is in using x for the original variable, although both are traditional. In the following example we use different letters.

EXAMPLE 5.23   Evaluate the integral $\int \cos(3+4t)dt$.

Solution   Let $v = 3 + 4t$. Then $dv = 4\,dt$. So we adjust the differential to get

$$\int \cos(3+4t)dt = \frac{1}{4}\int \cos(3+4t)(4\,dt)$$

$$= \frac{1}{4}\int \cos v\,dv$$

$$= \frac{1}{4}\sin v + C$$

$$= \frac{1}{4}\sin(3+4t) + C \qquad \blacksquare$$

With a little practice, you will probably be able to do some substitutions mentally, without actually writing the substitution. Consider, for example, the following:

$$\int \frac{x}{\sqrt{1-x^2}}dx = -\frac{1}{2}\int (1-x^2)^{-1/2}(-2x\,dx)$$

$$= -\frac{1}{2}\frac{(1-x^2)^{1/2}}{1/2} + C$$

$$= -\sqrt{1-x^2} + C$$

Here we *mentally* let $u = 1-x^2$ and on observing that its differential is $-2x\,dx$, we saw that we could adjust the differential by multiplying and dividing by -2. Finally, we saw that since our integral was in the form $-\frac{1}{2}\int u^{-1/2}du$, we could use the Power Rule to get the answer, with u being $1-x^2$. After you have done several problems by writing the substitutions, you should try others mentally.

Evaluating Definite Integrals by Substitution

When evaluating an indefinite integral by the method of substitution, always express the final answer in terms of the original variable. Substituting a new variable is a means to an end only, and the new variable should not appear in the final answer. In the case of a *definite* integral, however, the answer can be obtained from the Second Fundamental Theorem using the antiderivative in terms of the new variable, *provided* the limits of integration are expressed in terms of the new variable. In the next example we illustrate this method of handling a substitution in a definite integral, as well as an alternative procedure.

EXAMPLE 5.24   Evaluate the integral

$$\int_0^2 x^2\sqrt{1+x^3}dx$$

Solution   *Method (i):* We begin, as with an indefinite integral, by letting $u = 1+x^3$. Then $du = 3x^2dx$. We can adjust the differential to get the proper form of du. The new thing to consider is that there are limits of integration. The lower and upper limits 0 and 2 are x values. We now find the corresponding u

values. Since $u = 1 + x^3$, we see that when $x = 0$, $u = 1$, and when $x = 2$, $u = 9$. Thus, we have

$$\int_0^2 x^2\sqrt{1 + x^3}\, dx = \frac{1}{3}\int_0^2 (1 + x^3)^{1/2}(3x^2\, dx)$$

$$= \frac{1}{3}\int_1^9 u^{1/2}du \qquad \text{Note the new limits.}$$

$$= \frac{1}{3}\cdot\frac{2}{3}u^{3/2}\Big]_1^9 = \frac{2}{9}(9^{3/2} - 1^{3/2}) = \frac{2}{9}(27 - 1) = \frac{52}{9}$$

The answer was obtained using the antiderivative in terms of u, because the limits of integration were changed to u limits. We did not return to the original x variable.

Method (ii): First find the indefinite integral, just as in Examples 5.19 through 5.23. That is, find $\int x^2\sqrt{1 + x^3}\, dx$. Use $u = 1 + x^3$ as before, with $du = 3x^2\, dx$. Adjusting the differential, we get

$$\int x^2\sqrt{1 + x^3}\, dx = \frac{1}{3}\int (1 + x^3)^{1/2}(3x^2\, dx) = \frac{1}{3}\int u^{1/2}du$$

$$= \frac{1}{3}\cdot\frac{2}{3}u^{3/2} + C = \frac{2}{9}(1 + x^3)^{3/2} + C$$

Note that here, as with all substitutions in indefinite integrals, we wrote the answer in terms of the original variable x.

Now to evaluate the given definite integral, we need an antiderivative of the integrand. But we have just found its general antiderivative. Remember that in applying the Second Fundamental Theorem, any antiderivative of the integrand will do. So we use the one with the constant $C = 0$ to simplify the calculation. (It would "drop out" anyway.) The result is

$$\int_0^2 x^2\sqrt{1 + x^3}\, dx = \frac{2}{9}(1 + x^3)^{3/2}\Big]_0^2 = \frac{2}{9}\left[(1 + 8)^{3/2} - 1^{3/2}\right]$$

$$= \frac{2}{9}(9^{3/2} - 1) = \frac{52}{9} \qquad\blacksquare$$

Of the two methods illustrated in the preceding example, the first is usually easier. However, in "mental" substitution the second method is in effect being used.

When you use Method (i) for evaluating a definite integral in which you have made a substitution, you *must* change the limits of integration to those for the new variable. Failure to do so is a mistake students often make. For example, after making the substitution $u = 1 + x^3$ in the integral of Example 5.24, note carefully that

$$\int_0^2 x^2\sqrt{1 + x^3}\, dx \neq \frac{1}{3}\int_0^2 u^{1/2}\, du$$

since the integrand on the right is in terms of u and the upper and lower limits on the integral are x limits.

Method (i) used in Example 5.24 is summarized in the formula

$$\int_a^b f(g(x))g'(x)dx = \int_{g(a)}^{g(b)} f(u)du = F(g(b)) - F(g(a)) \qquad (5.20)$$

where $u = g(x)$ and $F' = f$. We assume that both $f(g(x))$ and $g'(x)$ are continuous on $[a, b]$.

In the next example we show how to find the area inside an ellipse using a substitution that changes the problem to that of the area inside a circle, multiplied by an appropriate constant.

EXAMPLE 5.25   Find the area enclosed by the ellipse

$$\frac{x^2}{a^2} + \frac{y^2}{b^2} = 1$$

Solution   The graph of a typical ellipse is shown in Figure 5.36. By symmetry we can find the area above the x-axis and double it. Note that the entire ellipse is not the graph of a function, but the upper half (as well as the lower half) is the graph of a function. The equation of the upper semi-ellipse is found by solving for y and using the positive square root, giving

$$y = b\sqrt{1 - \frac{x^2}{a^2}}$$

The area under this curve from $-a$ to a is

$$b\int_{-a}^{a} \sqrt{1 - \frac{x^2}{a^2}}\, dx$$

We make the substitution $u = x/a$. Then $du = (1/a)dx$, so that $dx = a\,du$. When $x = -a$, $u = -1$, and when $x = a$, $u = 1$. So the integral becomes

$$b\int_{-1}^{1} \sqrt{1 - u^2}\cdot a\,du = ab\int_{-1}^{1}\sqrt{1 - u^2}\,du$$

Now the integral $\int_{-1}^{1}\sqrt{1 - u^2}\,du$ gives the area under the semicircle $y = \sqrt{1 - u^2}$ of radius 1. (See Figure 5.37.) But we know this area is $\pi/2$ (half of πr^2, with $r = 1$). Note that the integral is of the form expressed by Equation 5.10 in Section 5.2, with $a = 1$. Thus, the area under the semi-ellipse is $\pi ab/2$, and the area enclosed by the entire ellipse is πab. ∎

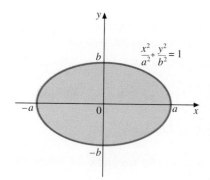

FIGURE 5.36

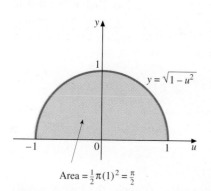

Area $= \frac{1}{2}\pi(1)^2 = \frac{\pi}{2}$

FIGURE 5.37

Integrals of Even and Odd Functions over Symmetric Intervals

We can use the method of substitution to prove an interesting and useful result concerning integrals of even and odd functions over symmetric intervals. Recall that f is an even function provided $f(-x) = f(x)$, and f is odd provided $f(-x) = -f(x)$. Even functions are symmetric about the y-axis, and odd functions are symmetric about the origin. We want to consider integrals of these two types of functions on an interval of the type $[-a, a]$. Figure 5.38 illustrates a typical situation.

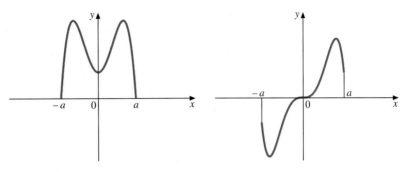

(a) Even function on symmetric
interval $[-a, a]$

$$\int_{-a}^{a} f(x)\, dx = 2 \int_{0}^{a} f(x)\, dx$$

(b) Odd function on symmetric
interval $[-a, a]$

$$\int_{-a}^{a} f(x)\, dx = 0$$

FIGURE 5.38

It is evident geometrically, using the area interpretation of the integral, that when f is even,

$$\int_{-a}^{a} f(x)\, dx = 2 \int_{0}^{a} f(x)\, dx$$

and when f is odd,

$$\int_{-a}^{a} f(x)\, dx = 0$$

The fact that the integral is 0 when f is odd follows because the area below the x-axis equals the area above, making the "net area" 0.

We can prove these results without reference to geometry as follows. By Property 3 in Section 5.2 we can write

$$\int_{-a}^{a} f(x)\, dx = \int_{-a}^{0} f(x)\, dx + \int_{0}^{a} f(x)\, dx \tag{5.21}$$

In the first integral on the right, make the substitution $u = -x$. Then $du = -dx$, or equivalently, $dx = -du$. Also, when $x = -a$, $u = a$, and when $x = 0$, $u = 0$. So we have

$$\int_{-a}^{0} f(x)\, dx = \int_{a}^{0} f(-u)(-du) = -\int_{a}^{0} f(-u)\, du$$

$$= \int_{0}^{a} f(-u)\, du$$

(Remember that interchanging upper and lower limits changes the sign of the integral.) If f is even, $f(-u) = f(u)$, so the last integral becomes

$$\int_{0}^{a} f(u)\, du$$

whereas, if f is odd, $f(-u) = -f(u)$, and the integral becomes

$$-\int_{0}^{a} f(u)\, du$$

A definite integral is unchanged if some other letter is used for the dummy variable. In particular, we can replace u by x (ignoring at this stage our earlier

substitution). Thus, for f even,

$$\int_{-a}^{0} f(x)\,dx = \int_{0}^{a} f(x)\,dx$$

and for f odd,

$$\int_{-a}^{0} f(x)\,dx = -\int_{0}^{a} f(x)\,dx$$

Returning now to Equation 5.21, we have the result that we summarize below.

If f is integrable on the symmetric interval $[-a, a]$, then

$$\int_{-a}^{a} f(x)\,dx = 2\int_{0}^{a} f(x)\,dx \qquad \text{if } f \text{ is even} \qquad (5.22)$$

$$\int_{-a}^{a} f(x)\,dx = 0 \qquad \text{if } f \text{ is odd} \qquad (5.23)$$

When using a CAS it is critical to remember these results. Most CAS occasionally fail to evaluate integrals correctly, especially when the integrand involves a square root.

EXAMPLE 5.26   Evaluate the following integrals.

(a) $\int_{-2}^{2}(x^2 + 1)\,dx$

(b) $\int_{-1}^{1} x^2 \sin x \,dx$

Solution

(a) Since $f(x) = x^2 + 1$ is even, we can write

$$\int_{-2}^{2}(x^2 + 1)\,dx = 2\int_{0}^{2}(x^2 + 1)\,dx = 2\left[\frac{x^3}{3} + x\right]_{0}^{2} = 2\left(\frac{8}{3} + 2\right) = \frac{28}{3}$$

One advantage of evaluating the integral from -2 to 2 by doubling its value from 0 to 2 is that it is easier to substitute 0 in the antiderivative than -2.

(b) Since $\sin x$ is odd, and x^2 is even, the product $x^2 \sin x$ is odd. So

$$\int_{-1}^{1} x^2 \sin x \, dx = 0 \qquad\blacksquare$$

If the interval of integration is not symmetric about 0, then integrals of even and odd functions cannot be simplified as we have shown in Equations 5.22 and 5.23. For example,

$$\int_{-2}^{3} x^2 \, dx \neq 2\int_{0}^{3} x^2 \, dx$$

even though the integrand is even. Similarly,

$$\int_{-2}^{3} x^3 \, dx \neq 0$$

even though the integrand is odd. Each of these problems would be solved by a direct application of the Second Fundamental Theorem. To apply Equations 5.22 and 5.23, *the interval of integration must be symmetric and of the form −a to a.*

Differential Equations Revisited

In Section 4.7 we saw how antiderivatives arise in the solutions of certain differential equations. Since an indefinite integral is a general antiderivative, we can use indefinite integrals in solving differential equations. In the next two examples we illustrate how this is done.

EXAMPLE 5.27   The slope of the tangent line to the graph of a certain function at an arbitrary point (x, y) is $2x^{-3}$, and the graph passes through the point $(1, 3)$. Find the function.

Solution   Let $y = f(x)$ denote the unknown function. Then we know that $f'(x) = 2x^{-3}$ and that $f(1) = 3$. So we have an initial-value problem to solve. The general solution of the differential equation is

$$y = \int f'(x)dx = \int 2x^{-3}dx = -x^{-2} + C = -\frac{1}{x^2} + C$$

To find C, we use the initial value $f(1) = 3$. That is, when $x = 1$, $y = 3$:

$$3 = -1 + C$$

$$C = 4$$

The particular solution we want is therefore

$$y = -\frac{1}{x^2} + 4$$

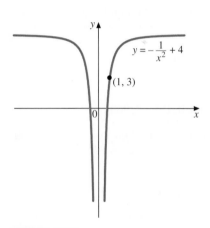

FIGURE 5.39

We show its graph in Figure 5.39.   ■

EXAMPLE 5.28   A projectile is fired vertically upward from the ground at an initial velocity of 196 m/s. Ignoring air resistance, find formulas for (a) the velocity, and (b) the distance above the ground at any time t before it strikes the ground. How high does the projectile rise?

Solution   The acceleration of the projectile at any time t is the acceleration of gravity, which is approximately 9.8 m/s^2 acting downward. Thus, since acceleration is the derivative of the velocity, $v(t)$, we have

$$v'(t) = -9.8$$

Also, we have the initial condition $v(0) = 196$. So

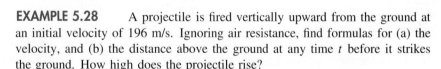

$$v(t) = \int (-9.8)dt = -9.8t + C_1$$

Using the initial condition, we get

$$196 = (-9.8)(0) + C_1$$

$$C_1 = 196$$

The velocity is therefore

$$v(t) = -9.8t + 196$$

which is the answer to part (a).

Let $s(t)$ denote the distance above the ground at time t. Then $s'(t) = v(t)$, and we have the differential equation

$$s'(t) = -9.8t + 196$$

together with the initial condition $s(0) = 0$. Solving, we get

$$s(t) = \int (-9.8t + 196)dt = -4.9t^2 + 196t + C_2$$

The initial condition gives $C_2 = 0$. Thus,

$$s(t) = -4.9t^2 + 196t$$

which is the answer for part (b).

At the instant the projectile reaches its maximum height, the velocity is 0. So we set $v(t) = 0$ and solve for t to see when the maximum height is attained:

$$-9.8t + 196 = 0$$

$$t = 20$$

Now we substitute $t = 20$ into the distance formula $s(t)$:

$$s(20) = (-4.9)(400) + (196)(20) = 1,960$$

The projectile therefore rises to a maximum height of 1,960 m.   ∎

Exercise Set 5.4

In Exercises 1–28, evaluate each indefinite integral by making an appropriate substitution.

1. $\int x(1 + x^2)^3 dx$

2. $\int x\sqrt{x^2 - 4}\, dx$

3. $\int x^2(x^3 - 1)^5 dx$

4. $\int \dfrac{x}{\sqrt{9 - x^2}}\, dx$

5. $\int x\sqrt{3x^2 + 4}\, dx$

6. $\int \dfrac{(\sqrt{x} + 1)^5}{\sqrt{x}}\, dx$

7. $\int \sqrt{1 + 3x}\, dx$

8. $\int (4x - 3)^{2/3}\, dx$

9. $\int (x - 1)\sqrt{x^2 - 2x + 3}\, dx$

10. $\int \dfrac{2x + 1}{\sqrt{x^2 + x - 4}}\, dx$

11. $\int \sin 2x\, dx$

12. $\int \cos 3x\, dx$

13. $\int \sin kx\, dx$, k a nonzero real constant

14. $\int \cos kx\, dx$, k a nonzero real constant

15. $\int 2\sin x \cos x\, dx$

16. $\int \sin x \cos^2 x\, dx$

17. $\int x\sec^2 x^2 dx$

18. $\int \sec 2x \tan 2x\, dx$

19. $\int \sqrt{\tan x}\, \sec^2 x\, dx$

20. $\int \sqrt{1 + \sin x}\, \cos x\, dx$

21. $\int \dfrac{\sin x}{(1 - \cos x)^2}\, dx$

22. $\int \dfrac{x + \cos 2x}{\sqrt{x^2 + \sin 2x}}\, dx$

23. $\int \dfrac{\csc^2 \sqrt{x}}{\sqrt{x}} dx$

24. $\int \dfrac{1}{x^2} \cos \dfrac{1}{x} dx$

25. $\int \left(\dfrac{x-1}{x}\right)^5 \cdot \dfrac{1}{x^2} dx$

26. $\int (x^2 - x)(2x^3 - 3x^2 + 1)^{3/2} dx$

27. $\int \sqrt{x}(x^{3/2} - 1)^{2/3} dx$

28. $\int \dfrac{(x^{2/3} + a^{2/3})^{3/2}}{\sqrt[3]{x}} dx$, a a nonzero real constant

In Exercises 29–44, evaluate each definite integral by making an appropriate substitution.

29. $\int_0^2 (1 + 4x)^{3/2} dx$

30. $\int_0^4 x\sqrt{x^2 + 9} dx$

31. $\int_{-1}^2 x(x^2 - 2)^4 dx$

32. $\int_3^4 \dfrac{x}{\sqrt{25 - x^2}} dx$

33. $\int_4^9 \dfrac{(\sqrt{x} - 4)^6}{\sqrt{x}} dx$

34. $\int_0^2 x(3x^2 + 4)^{3/2} dx$

35. $\int_1^5 (2x + 1)(x^2 + x - 3)^{1/3} dx$

36. $\int_1^2 \left(\dfrac{5x^2 - 8}{3x^2}\right)^{2/3} \dfrac{dx}{x^3}$

37. $\int_0^{\pi/6} \cos 3x\, dx$

38. $\int_0^{\pi/2} \sin \dfrac{x}{2} dx$

39. $\int_{\pi/6}^{2\pi/3} \sin^3 x \cos x\, dx$

40. $\int_0^{\pi/2} \sqrt{\cos x} \sin x\, dx$

41. $\int_{-\pi/6}^{\pi/8} \tan^3 2x \sec^2 2x\, dx$

42. $\int_{-\pi/3}^{\pi/2} \dfrac{\sin x}{(1 + \cos x)^2} dx$

43. $\int_0^{\pi/3} \sec^3 x \tan x\, dx$

44. $\int_{\pi^2/4}^{\pi^2} \dfrac{\sin \sqrt{x}}{\sqrt{x}} dx$

45. Find the area under the graph of
$$y = \dfrac{1}{(2x - 3)^3}$$
over the interval [2, 3].

46. Find $\int 9x^2(x^3 + 1)dx$ by the following two methods:
(a) Multiply the factors in the integrand and integrate term by term.
(b) Make the substitution $u = x^3 + 1$.

Explain the difference in the answers in parts (a) and (b). Are they both correct?

47. Find $\int \sec^2 x \tan x\, dx$ in the following two ways:
(a) Make the substitution $u = \tan x$.
(b) Make the substitution $u = \sec x$.
Explain the difference in the answers in parts (a) and (b). Are they both correct?

48. Find $f(x)$ if $\int f(x)dx = 6x^4 - 2x^2 + x + C$.

49. Find a function g such that $g'(x) = \sqrt{4x + 2}$ and $g(1/2) = 1$.

50. The marginal profit for a company is given by
$$P'(x) = \dfrac{x}{(100 + x^2)^{2/3}}$$
where x is the number of units sold and $P(x)$ is in hundreds of dollars. If the company breaks even by making 30 units, find the profit function $P(x)$.

51. The ABC Company has determined that its marginal cost, in hundreds of dollars, for producing x dishwashers per month is given by
$$C'(x) = \dfrac{x}{\sqrt{10{,}000 + x^2}}$$
If the fixed cost of production per month is \$50,000, find the cost of producing 500 dishwashers per month.

 52. The cost, in hundreds of dollars, of producing x units of a certain commodity is increasing at the rate of $x^2/(1 + x^2)$. Use a CAS, a spreadsheet, or a calculator to determine an upper and lower estimate, accurate to one decimal place, of the cost of increasing production from 10 to 50 units.

In Exercises 53–58, find the general solution of the differential equation.

53. $\dfrac{dy}{dx} = \dfrac{x}{\sqrt{x^2 + 1}}$

54. $\dfrac{dy}{dt} = (2t - 1)(t^2 - t + 4)^{2/3}$

55. $y' - \cos 2x = x$

56. $\sqrt{x}\, y' = (\sqrt{x} - 2)^4$

57. $f'(t) = \dfrac{(1 + \tan t)^{3/2}}{\cos^2 t}$

58. $g'(u) = \sqrt{u^4 - 9u^2}$ $(u > 0)$

In Exercises 59–62, solve the initial-value problem.

59. $\dfrac{dy}{dx} = \dfrac{x^2}{\sqrt{x^3 + 1}}$; when $x = 2$, $y = 5$

60. $dx/dt = \sin^3 t \cos t$; when $t = \pi/2$, $x = 1$

61. $y' = \sin 2x \sqrt{1 + \sin^2 x}$; when $x = 0$, $y = 2$

62. $f'(t) = t \cos t^2 \sin t^2$; $f(0) = -1$

63. The graph of a certain function passes through the point $(0, -1)$. Its slope at an arbitrary point (x, y) is $x/(x^2 + 1)^2$. Find the function.

64. A certain function has the following properties: $f''(x) = -2(3x + 2)^{-4/3}$, $f'(2) = 0$, $f(2) = 5$. Find $f(x)$.

65. (a) When the brakes of a car, initially traveling at 65 mph, are applied, they produce a constant deceleration of 11 ft/s^2. What is the shortest distance in which the car can be braked to a complete stop?
(b) Redo part (a) if the car is initially traveling at 100 km/h and the deceleration is 3 m/s^2.

66. If a car starts from a dead stop, what constant acceleration will allow the car to travel 300 m in 12.8 seconds?

67. From a balloon 1,176 m above the ground an object is thrown downward with a velocity of 39.2 m/s. Assuming no wind and no air resistance, find (a) the velocity, and (b) the distance of the object from the balloon at any time t before it strikes the ground. How long does it take for the object to reach the ground? (Take $g = 9.8$ m/s^2.)

In Exercises 68–73, evaluate the indefinite integral.

68. $\displaystyle\int x\sqrt{(x^4 + 12x^2 + 36)^3}\, dx$

69. $\displaystyle\int \frac{\sec^2 x}{\sec^2 x + 2\tan x}\, dx$

70. $\displaystyle\int \cos\frac{x}{2}\sin x\, dx$

71. $\displaystyle\int \frac{dx}{\sqrt{x - x\sqrt{x}}}$

72. $\displaystyle\int (\sin 2x - \cos 2x)^2\, dx$

73. $\displaystyle\int \cos 2x \cos x\, dx$

Verify the equations in Exercises 74 and 75.

74. $\displaystyle\int x \sin 2x\, dx = -\frac{x\cos 2x}{2} + \frac{\sin 2x}{4} + C$

75. $\displaystyle\int \frac{1}{x^2} \cos\frac{1}{\sqrt{x}}dx = -2\left[\frac{1}{\sqrt{x}}\sin\frac{1}{\sqrt{x}} + \cos\frac{1}{\sqrt{x}}\right] + C$

76. Let $f(x) = \begin{cases} 1 & \text{if } x \geq 0 \\ -1 & \text{if } x < 0 \end{cases}$. Show that $\int_{-1}^{1} f(x)dx$ exists, but $\int f(x)dx$ does not exist on $[-1, 1]$. (*Hint:* Consider the result of Exercise 34 in Exercise Set 4.7.)

77. Prove Properties 1–4 of the indefinite integral.

5.5 APPROXIMATING DEFINITE INTEGRALS

We have already noted that some continuous elementary functions have no antiderivative in the form of an elementary function. If we want to evaluate a definite integral of such a function, the Second Fundamental Theorem is of no help. The best we can do is to approximate its value. Here are three integrals of this type:

$$\int_1^2 \frac{\sin x}{x}dx \qquad \int_0^1 \frac{dx}{\sqrt{1 + x^3}} \qquad \int_0^4 \cos \sqrt{x}\, dx$$

In each of these cases the integrand is continuous on the interval in question, so each integral exists, but we cannot find its exact value. These are but three examples; there are many others for which approximation is also necessary.

Functions arising in applications frequently are known only graphically or by tables of values, rather than by formulas. If we need to find the integral of such a function, our only recourse is to approximate its value by some numerical procedure. Even when the exact value of an integral *can* be found, the work involved in obtaining it may not be justified, especially if accuracy to one or two decimal places is all that is needed. In such a case it may be more efficient, and equally useful, to approximate the integral numerically rather than to obtain the exact answer.

Left- and Right-Endpoint Rules

We have already seen that we can approximate a definite integral by a Riemann sum. Since, by definition, the integral is the limit of Riemann sums as the widths of the subintervals shrink to zero, any particular Riemann sum can be used to approximate the integral. That is,

$$\int_a^b f(x)dx \approx \sum_{k=1}^n f(c_k)\Delta x_k$$

where the Riemann sum on the right has the meaning we assigned in Section 5.2. Recall that the points $c_1, c_2, \ldots, c_n$ can be arbitrarily chosen in the n consecutive subintervals. For convenience, throughout this section we will consider equally spaced partition points only, so that all subintervals have the same width, $\Delta x = (b-a)/n$.

If for each integer k from 1 to n we take c_k to be the left endpoint of the kth subinterval $[x_{k-1}, x_k]$, the approximation to the integral is called the **Left-Endpoint Rule.**

$$\int_a^b f(x)dx \approx \Delta x\big[f(x_0) + f(x_1) + \cdots + f(x_{n-1})\big] \qquad (5.24)$$

Similarly, if for each integer k from 1 to n we take c_k to be the right endpoint of the kth subinterval, the approximation is called the **Right-Endpoint Rule.**

$$\int_a^b f(x)dx \approx \Delta x\big[f(x_1) + f(x_2) + \cdots + f(x_n)\big] \qquad (5.25)$$

Midpoint Rule

Instead of using the right or left endpoints of the subintervals, we might choose to use the midpoints. That is, for each k from 1 to n, we can take c_k to be $\frac{1}{2}(x_{k-1} + x_k)$. Let us denote this midpoint by $\bar{x}_k$. Approximating the integral by a Riemann sum with $c_k = \bar{x}_k$ results in what is frequently called the **Midpoint Rule** and is stated as follows.

The Midpoint Rule

$$\int_a^b f(x)dx \approx \Delta x\big[f(\bar{x}_1) + f(\bar{x}_2) + \cdots + f(\bar{x}_n)\big]$$

where $\Delta x = \dfrac{b-a}{n}$ and $\bar{x}_k = \dfrac{x_{k-1} + x_k}{2}$ for $k = 1, 2, \ldots, n$.

In Figure 5.40 we show a nonnegative continuous function f on an interval $[a, b]$ with its integral (which equals the area under the graph) approximated by left-endpoint, right-endpoint, and midpoint Riemann sums. Although it may not be obvious, the Midpoint Rule generally is more accurate than the right- and left-endpoint approximations.

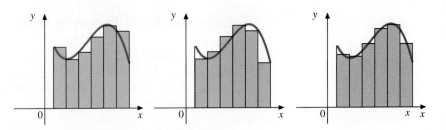

Left-Endpoint Approximation   Right-Endpoint Approximation   Midpoint Approximation

FIGURE 5.40

Trapezoidal Rule

Yet another method, which might have occurred to you already, is to average the approximations given by the Left-Endpoint and Right-Endpoint Rules (5.24 and 5.25). Doing so, we get

$$\int_a^b f(x)dx \approx \frac{1}{2}\left[\sum_{k=1}^n f(x_{k-1})\Delta x + \sum_{k=1}^n f(x_k)\Delta x\right]$$

$$= \frac{\Delta x}{2}\sum_{k=1}^n \left[f(x_{k-1}) + f(x_k)\right]$$

$$= \frac{\Delta x}{2}\left[\left(f(x_0) + f(x_1)\right) + \left(f(x_1) + f(x_2)\right) + \cdots \right.$$
$$\left. + \left(f(x_{k-1}) + f(x_k)\right)\right]$$

$$= \frac{\Delta x}{2}\left[f(x_0) + 2f(x_1) + 2f(x_2) + \cdots + 2f(x_{k-1}) + f(x_k)\right]$$

This result is known as the **Trapezoidal Rule**, for reasons we explain below. First we restate the rule for emphasis.

The Trapezoidal Rule

$$\int_a^b f(x)dx \approx \frac{\Delta x}{2}\left[f(x_0) + 2f(x_1) + 2f(x_2) + \cdots + 2f(x_{n-1}) + f(x_n)\right]$$

where $\Delta x = \dfrac{b-a}{n}$.

REMARK ———————————————————————————————

■ The Left-Endpoint, Right-Endpoint, and Midpoint Rules are all special cases of Riemann sums, but the Trapezoidal Rule is not.

To see why we use the name "Trapezoidal" Rule, consider a nonnegative continuous function f over an interval $[a, b]$, as shown in Figure 5.41. Partition the interval $[a, b]$ in the usual way with the equally spaced points $x_0 = a$, $x_1, x_2, \ldots, x_n = b$. Instead of approximating the integral $\int_a^b f(x)dx$ (that is,

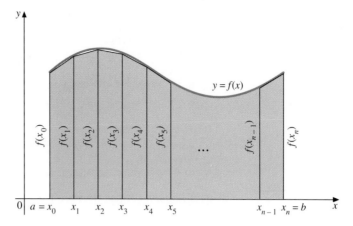

FIGURE 5.41

the area under the graph) with rectangular regions, as we do in Riemann sums, we do so with areas of trapezoids (quadrilaterals with two parallel sides), as shown. Now the area inside a trapezoid is the average of the lengths of the two parallel sides times the distance between them. For example, the leftmost trapezoid in Figure 5.41 has area

$$\frac{1}{2}\big[f(x_0) + f(x_1)\big]\Delta x$$

When we add the areas of all the trapezoids we get the approximate value of the integral:

$$\int_a^b f(x)dx \approx \frac{1}{2}\big[f(x_0) + f(x_1)\big]\Delta x + \frac{1}{2}\big[f(x_1) + f(x_2)\big]\Delta x + \cdots$$

$$+ \frac{1}{2}\big[f(x_{n-1}) + f(x_n)\big]\Delta x$$

$$= \frac{\Delta x}{2}\big[f(x_0) + 2f(x_1) + 2f(x_2) + \cdots + 2f(x_{n-1}) + f(x_n)\big]$$

which is exactly the value given by the Trapezoidal Rule. We have shown, then, that averaging the approximations given by the Left-Endpoint Rule and the Right-Endpoint Rule gives the same result as finding the sum of the areas of the trapezoids we have described.

REMARK —————————————————————————————

■ Although we used a nonnegative continuous function to explain the origin of the name "Trapezoidal" Rule, the rule is applicable as a means of approximating the definite integral of any integrable function. (See Figure 5.42.)

————————————————————————————————————

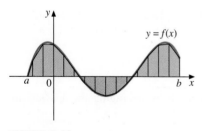

FIGURE 5.42
The Trapezoidal Rule approximates "net area" when f takes on both positive and negative values.

In the next example we use a function with a known antiderivative so that we can compare our approximations with the exact value.

EXAMPLE 5.29   Approximate the integral

$$\int_1^3 \cos x\, dx$$

using (a) the Left-Endpoint Rule, (b) the Right-Endpoint Rule, (c) the Midpoint Rule, and (d) the Trapezoidal Rule, with $n = 10$. Repeat, with $n = 20, 50$, and 100. Compare the results, and find the error in each case.

Solution   With $n = 10$, we find that

$$\Delta x = \frac{3 - 1}{10} = \frac{2}{10} = 0.2$$

So the points of division are (see Figure 5.43)

$$1.0, \ 1.2, \ 1.4, \ 1.6, \ 1.8, \ 2.0, \ 2.2, \ 2.4, \ 2.6, \ 2.8, \ 3.0$$

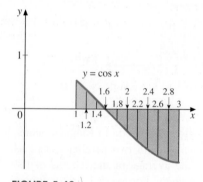

FIGURE 5.43

(a) *Left-Endpoint Rule.* By Formula 5.24,

$$\int_1^3 \cos x \, dx \approx 0.2 \big[\cos(1.0) + \cos(1.2) + \cos(1.4) + \cdots + \cos(2.8) \big]$$

$$\approx -0.5450$$

(b) *Right-Endpoint Rule.* By Formula 5.25,

$$\int_1^3 \cos x \, dx \approx 0.2 \big[\cos(1.2) + \cos(1.4) + \cos(1.6) + \cdots + \cos(3.0) \big]$$

$$\approx -0.8510$$

(c) *Midpoint Rule.* The midpoints of the intervals are

$$m_1 = 1.1, \quad m_2 = 1.3, \quad m_3 = 1.5, \quad m_4 = 1.7, \quad m_5 = 1.9,$$

$$m_6 = 2.1, \quad m_7 = 2.3, \quad m_8 = 2.5, \quad m_9 = 2.7, \quad m_{10} = 2.9$$

So the Midpoint Rule gives

$$\int_1^3 \cos x \, dx \approx 0.2 \big[\cos(1.1) + \cos(1.3) + \cos(1.5) + \cdots + \cos(2.9) \big]$$

$$\approx -0.7015$$

(d) *Trapezoidal Rule*

$$\int_1^3 \cos x \, dx \approx \frac{0.2}{2} \big[\cos(1.0) + 2\cos(1.2) + 2\cos(1.4) + \cdots + 2\cos(2.8)$$

$$+ \cos(3.0) \big]$$

$$\approx -0.6980$$

Note that we could have obtained this result by averaging the answers in parts (a) and (b).

We use the Second Fundamental Theorem to get the exact value:

$$\int_1^3 \cos x \, dx = \sin 3 - \sin 1$$

Although $\sin 3 - \sin 1$ is the exact value, it is not in a useful form to compare with our approximations. We can estimate the value, however, using a calculator. Of course, the calculator also gives only an approximate value, but it is accurate to more decimal places than any of our approximations and so gives a suitable answer for comparison purposes. We find that, to six decimal places of accuracy,

$$\int_1^3 \cos x \, dx \approx -0.700351$$

Method	Error, $n = 10$
LEFT	0.1554
RIGHT	−0.1506
MID	−0.00115
TRAP	0.00235

To get the error in each of our approximations, we subtract this six-place true answer from each of the four approximations we found in parts (a)–(d). We show the results in the table in the margin, where we have used the abbreviations LEFT, RIGHT, MID, and TRAP for the approximations in parts (a), (b), (c), and (d), respectively.

Now we repeat the process with $n = 20$, 50, and 100. A computer algebra system, spreadsheet, or programmable calculator is almost a necessity for working with these larger values of n. We summarize all the results in the following two tables.

	Approximations			
Method	$n = 10$	$n = 20$	$n = 50$	$n = 100$
LEFT	−0.5450	−0.6232	−0.6697	−0.6850
RIGHT	−0.8510	−0.7763	−0.7309	−0.7156
MID	−0.7015	−0.7006	−0.700397	−0.700362
TRAP	−0.6980	−0.6998	−0.700257	−0.700327

	Errors			
Method	$n = 10$	$n = 20$	$n = 50$	$n = 100$
LEFT	0.1554	0.0772	0.03065	0.15535
RIGHT	−0.1506	−0.0760	−0.03055	−0.15249
MID	−0.00115	−0.000249	−0.000046	−0.000011
TRAP	0.00235	0.000551	0.000094	0.000024

Several conclusions about the methods are immediately evident for this integral:

1. LEFT and RIGHT are much less accurate than MID or TRAP.
2. The magnitudes of the errors for LEFT and RIGHT are about the same, but the errors are opposite in sign.
3. The errors for MID and TRAP are opposite in sign.
4. The magnitude of the error for MID is about half that for TRAP. ∎

The Midpoint Rule uses areas of rectangles to approximate the area under a curve. These rectangles, however, can be replaced by certain trapezoids, as we indicate in Figure 5.44. There we show a typical subinterval and the rectangle used in the Midpoint Rule to approximate the area under the curve. By rotating the top edge of the rectangle about the point $\left(\bar{x}_k, f(\bar{x}_k)\right)$ until it coincides with the tangent line to the curve at that point, we form a trapezoid whose area is the same as the area of the rectangle, since the area added by the rotation equals that subtracted.

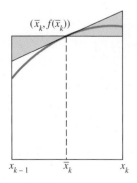

FIGURE 5.44

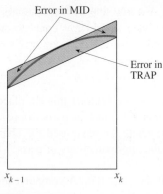

FIGURE 5.45

Now consider Figure 5.45. There again we show a typical subinterval and two trapezoids, the one described above, equivalent to the rectangle in the Midpoint Rule, and the one used in the Trapezoidal Rule. You can see now that the error in the Midpoint Rule is less than that in the Trapezoidal Rule, appearing from the figure to be about half as much. In the figure we show the curve as concave down, and the Midpoint Rule overestimates, whereas the Trapezoidal Rule underestimates the actual area. If the graph were concave up, the opposite would be true.

Of the four approximation methods we have considered so far, the Midpoint Rule is clearly the most accurate in most cases. (There are certain cases that we will discuss later when two or more of the methods give the exact answer.) So why do we bother with the others? One reason is that when the function is known only by means of a table, we might not know the function values at the midpoints of the subintervals. In this case we would choose the Trapezoidal Rule.

The methods we have considered so far are called *linear approximation methods*, because in each case the actual curve is replaced by a straight line on each of the subintervals. The next method is a *quadratic approximation method*. Here we replace segments of the curve by second-degree polynomials (parabolas).

Simpson's Rule

The quadratic approximation method we consider here is called **Simpson's Rule** (after the eighteenth-century English mathematician Thomas Simpson). In it, we use a series of parabolic arcs (rather than the straight line segments of our other methods) to approximate the curve. As with the other methods, we divide the interval $[a, b]$ into n subintervals with common width $\Delta x = (b - a)/n$. For Simpson's Rule, we require that n be an *even* number, since we are considering subintervals by pairs.

Let $y_k = f(x_k)$ for $k = 0, 1, 2, \ldots, n$. We are going to fit a second-degree polynomial (whose graph is a parabola) to the three points (x_0, y_0), (x_1, y_1), and (x_2, y_2). Then we will fit another second-degree polynomial to the next group of three, (x_2, y_2), (x_3, y_3), and (x_4, y_4), and so on. In Figure 5.46 we show the parabolic arcs for a nonnegative function f, where $P_0, P_1, \ldots, P_n$ designate the points $(x_0, y_0), (x_1, y_1), \ldots, (x_n, y_n)$.

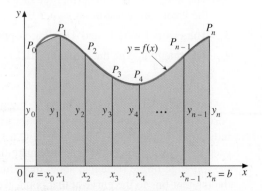

FIGURE 5.46

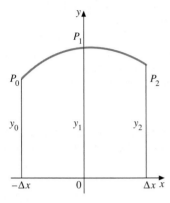

FIGURE 5.47

Consider first the problem of fitting a second-degree polynomial to the points P_0, P_1, and P_2. What we really want is the integral, which is equal to the area under the graph, and this area is unchanged if we translate the origin of our coordinate system to coincide with the point $(x_1, 0)$, as in Figure 5.47. With respect to the new axes, the coordinates of P_0, P_1, and P_2 are $(-\Delta x, y_0)$, $(0, y_1)$, and $(\Delta x, y_2)$. Let

$$y = ax^2 + bx + c$$

be the equation of the parabola in the new coordinate system that passes through P_0, P_1, and P_2. The coordinates of P_0, P_1, and P_2 each satisfy this equation, so we have

$$\begin{cases} y_0 = a(-\Delta x)^2 + b(-\Delta x) + c = a(\Delta x)^2 - b(\Delta x) + c \\ y_1 = a(0)^2 + b(0) + c = c \\ y_2 = a(\Delta x)^2 + b(\Delta x) + c \end{cases} \qquad (5.26)$$

Since it is the area we really want, we calculate

$$\int_{-\Delta x}^{\Delta x} (ax^2 + bx + c)\,dx$$

Observe that ax^2 is even, bx is odd, and c is even. Since the integration is over a symmetric interval, we know by Equations 5.22 and 5.23 that the integral can be written as

$$\int_{-\Delta x}^{\Delta x} ax^2\,dx + \int_{-\Delta x}^{\Delta x} bx\,dx + \int_{-\Delta x}^{\Delta x} c\,dx = 2\int_0^{\Delta x} ax^2\,dx + 0 + 2\int_0^{\Delta x} c\,dx$$

$$= 2\frac{ax^3}{3} + 2cx \Big]_0^{\Delta x}$$

$$= 2a\frac{(\Delta x)^3}{3} + 2c\Delta x$$

$$= \frac{\Delta x}{3}\left[2a(\Delta x)^2 + 6c\right]$$

From Equations 5.26 we see that $c = y_1$, and by adding the first and third equations, we get

$$y_0 + y_2 = 2a(\Delta x)^2 + 2c$$

or

$$2a(\Delta x)^2 = y_0 + y_2 - 2y_1$$

Thus,

$$\int_{-\Delta x}^{\Delta x} (ax^2 + bx + c)\,dx = \frac{\Delta x}{3}[(y_0 + y_2 - 2y_1) + 6y_1]$$

$$= \frac{\Delta x}{3}[(y_0 + 4y_1 + y_2)] \qquad (5.27)$$

This result is the area under the parabolic arc from P_0 to P_2, passing through P_1. Notice that it depends only on the width Δx and the ordinates of the three points.

Exactly the same reasoning could be used to get the area under the parabolic arc that passes through the next group of three points, P_2, P_3, and P_4. But there is no need to go through the derivation again because we already know the result. We simply change the subscripts in Equation 5.27, and for the next group of three, and so on. The sum of these areas is

$$\frac{\Delta x}{3}(y_0 + 4y_1 + y_2) + \frac{\Delta x}{3}(y_2 + 4y_3 + y_4) + \cdots + \frac{\Delta x}{3}(y_{n-2} + 4y_{n-1} + y_n)$$

When we factor out $\Delta x/3$ and combine terms, replacing each y_k by $f(x_k)$, we have the result known as Simpson's Rule.

Simpson's Rule

$$\int_a^b f(x)dx \approx \frac{\Delta x}{3}\big[f(x_0) + 4f(x_1) + 2f(x_2) + 4f(x_3) + \cdots$$

$$+ \ 2f(x_{n-2}) + 4f(x_{n-1}) + f(x_n)\big]$$

where $\Delta x = (b - a)/n$ and n is even.

REMARK
- Note the pattern of the coefficients here:

$$1, 4, 2, 4, 2, \ldots, 2, 4, 1$$

Contrast this pattern with the one for the Trapezoidal Rule:

$$1, 2, 2, 2, \ldots, 2, 1$$

Also, note that for Simpson's Rule the initial factor is $\Delta x/3$, whereas for the Trapezoidal Rule it is $\Delta x/2$.

Although our derivation of Simpson's Rule was based on area under a nonnegative function, it is valid for any integrable function, as is also true for the other methods we have studied.

REMARK
- Simpson's Rule can be shown to be a weighted average of the Midpoint and Trapezoidal Rules. In fact, if we denote by M_n the value given by the Midpoint Rule with n subintervals, T_n the value given by the Trapezoidal Rule, and S_{2n} the value given by Simpson's Rule for $2n$ subintervals, we can show (see Exercise 22 in this section) that

$$S_{2n} = \frac{2}{3}M_n + \frac{1}{3}T_n$$

For comparison with the other methods, in the next example we apply Simpson's Rule to the same integral we used in Example 5.29.

EXAMPLE 5.30   Use Simpson's Rule with $n = 10$ to approximate the integral

$$\int_1^3 \cos x \, dx$$

Solution   Since $\Delta x = (3-1)/10 = 0.2$, we have

$$\int_1^3 \cos x \, dx \approx \frac{0.2}{3}\big[\cos(1.0) + 4\cos(1.2) + 2\cos(1.4) + 4\cos(1.6)$$

$$+ \cdots + 2\cos(2.6) + 4\cos(2.8) + \cos(3.0)\big]$$

$$\approx -0.700357$$

In Example 5.29 we found that the value of the integral, to six decimal places, is -0.700351. So with the relatively small value of $n = 10$, we have an error of only -0.000006, which is considerably more accurate than any of the other methods, even with $n = 100$.   ∎

Error Analysis of the Approximation Methods

In Examples 5.29 and 5.30 we applied our methods to an integral we could evaluate in an exact form. Of course, the reason for studying approximation methods is primarily (but not solely) for integrals that cannot be evaluated exactly. For such integrals we need some way of determining the accuracy of our approximations. We state below, without proof, formulas for determining bounds on the errors for the Midpoint Rule, the Trapezoidal Rule, and Simpson's Rule. In each case E_n denotes the absolute value of the difference between the value given by the method and the exact value of the integral, for a fixed n. We again use MID and TRAP for Midpoint and Trapezoidal Rules, and we use SIMP for Simpson's Rule. By means of these formulas, we will also be able to determine how large an n will guarantee a given degree of accuracy.

Error Bounds for MID, TRAP, and SIMP

MID	TRAP

$$E_n \le \frac{M_2(b-a)(\Delta x)^2}{24} \qquad E_n \le \frac{M_2(b-a)(\Delta x)^2}{12}$$

SIMP

$$E_n \le \frac{M_4(b-a)(\Delta x)^4}{180}$$

where M_2 and M_4 are nonnegative constants satisfying

$$|f''(x)| \le M_2 \qquad \text{and} \qquad |f^{(4)}(x)| \le M_4$$

for all x in $[a, b]$, and $\Delta x = (b-a)/n$.

We illustrate how to find these error bounds in the next example.

EXAMPLE 5.31   Determine n so that the approximation of the integral

$$\int_1^2 \frac{1}{x} \, dx$$

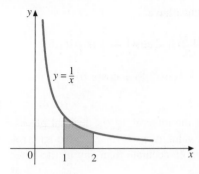

FIGURE 5.48

When the error is less than or equal to 0.00005, the answer is correct as rounded to the fourth decimal place. In general, to get accuracy to n decimal places, we want

$$E_n \leq 0.\underbrace{00\cdots05}_{n \text{ zeros}}$$

given by (a) the Midpoint Rule, (b) the Trapezoidal Rule, and (c) Simpson's Rule is accurate to four decimal places.

Solution   The integral represents the area under the graph of $f(x) = 1/x$ from $x = 1$ to $x = 2$. (See Figure 5.48.) For accuracy to four places, we want the error E_n in each approximation not to exceed 0.00005. To find numbers M_2 and M_4, we need the first four derivatives of $f(x) = 1/x$.

$$f'(x) = -\frac{1}{x^2}$$

$$f''(x) = \frac{2}{x^3}$$

$$f'''(x) = -\frac{6}{x^4}$$

$$f^{(4)}(x) = \frac{24}{x^5}$$

For x in the interval $[1, 2]$, each derivative takes on its largest absolute value when the denominator is smallest, that is, when $x = 1$ (since dividing by a smaller value yields a larger quotient). So we have,

$$|f''(x)| \leq 2 \qquad \text{and} \qquad |f^{(4)}(x)| \leq 24$$

for x in $[1, 2]$. Therefore, we can take $M_2 = 2$ and $M_4 = 24$.

(a)   Midpoint Rule   Since $\Delta x = (b - a)/n$, we have

$$E_n \leq \frac{M_2(b-a)(\Delta x)^2}{24} = \frac{M_2(b-a)^3}{24n^2} = \frac{2(1)^3}{24n^2} = \frac{1}{12n^2}$$

We want to find the smallest integer n for which

$$\frac{1}{12n^2} \leq 0.00005$$

or equivalently,

$$n^2 \geq \frac{1}{12(0.00005)}$$

$$n \geq 40.8$$

Since n must be an integer, we take $n = 41$.

(b)   Trapezoidal Rule   In this case, again replacing Δx by $(b-a)/n$, we have

$$E_n \leq \frac{M_2(b-a)(\Delta x)^2}{12} = \frac{M_2(b-a)^3}{12n^2} = \frac{2(1)^3}{12n^2} = \frac{1}{6n^2}$$

So we want to find n such that

$$\frac{1}{6n^2} \leq 0.00005$$

or,

$$n^2 \geq \frac{1}{6(0.00005)}$$

$$n \geq 57.7$$

Thus, we take $n = 58$.

(c) Simpson's Rule   The error bound in this case is

$$E_n \le \frac{M_4(b-a)(\Delta x)^4}{180} = \frac{M_4(b-a)^5}{180n^4} = \frac{24(1)^5}{180n^4} = \frac{2}{15n^4}$$

For the desired accuracy, we want

$$\frac{2}{15n^4} \le 0.00005$$

$$n^4 \ge \frac{2}{15(0.00005)}$$

$$n \ge 7.186$$

We take $n = 8$, which is even, as required. In general, for Simpson's Rule, we select n as the smallest *even* integer greater than or equal to the value calculated from the error bound.

In summary, we can be guaranteed of accuracy to four places by the Trapezoidal Rule with $n = 58$, by the Midpoint Rule with $n = 41$, and by Simpson's Rule with $n = 8$. Clearly, Simpson's Rule is much more efficient than either of the other methods in this case. ∎

EXAMPLE 5.32   Approximate the integral

$$\int_1^2 \frac{1}{x}\, dx$$

correct to four decimal places by Simpson's Rule.

Solution   In part (c) of the preceding example we found that $n = 8$ will guarantee accuracy to at least four places for this integral. Since the interval of integration is from $x = 1$ to $x = 2$, we have

$$\Delta x = \frac{1}{8} = 0.125$$

So the points of division are 1, 1.125, 1.250, 1.375, 1.500, 1.625, 1.750, 1.875, 2. Thus, by Simpson's Rule,

$$\int_1^2 \frac{1}{x}\, dx \approx \frac{0.125}{3}\left[\frac{1}{1} + 4\left(\frac{1}{1.125}\right) + 2\left(\frac{1}{1.250}\right) + 4\left(\frac{1}{1.375}\right) + 2\left(\frac{1}{1.500}\right)\right.$$

$$\left. + 4\left(\frac{1}{1.625}\right) + 2\left(\frac{1}{1.750}\right) + 4\left(\frac{1}{1.875}\right) + \frac{1}{2}\right]$$

$$\approx 0.69315$$

Rounding to four places, we get 0.6932.

In Chapter 7 we will find that the exact value of this integral is the natural logarithm of 2, denoted $\ln 2$. Using a calculator, we find $\ln 2 \approx 0.693147$. So our approximation by Simpson's Rule with $n = 8$ is actually correct to five decimal places. ∎

The constants M_2 and M_4 in the error bounds for our methods are determined from the second and fourth derivatives, respectively, of $f(x)$ on the interval $[a, b]$ in question. In Example 5.31, calculating these derivatives was straightforward and relatively simple. Also, it was easy to determine their largest absolute values on the interval in question. Unfortunately, for some functions calculation of the derivatives, especially the fourth, can be quite tedious. Then, determining upper bounds on their absolute values can again be difficult. One drawback of Simpson's Rule, in fact, is this difficulty in finding the number M_4 that is a bound for $|f^{(4)}(x)|$. The next example illustrates what we are talking about.

EXAMPLE 5.33   Find numbers M_2 and M_4 for the function

$$f(x) = \frac{1}{\sqrt{1+x^3}}$$

on $[0, 1]$ satisfying

$$|f''(x)| \leq M_2 \quad \text{and} \quad |f^{(4)}(x)| \leq M_4$$

Solution   Write $f(x) = (1 + x^3)^{-1/2}$. Using the Generalized Power Rule, we find

$$f'(x) = -\frac{1}{2}(1 + x^3)^{-3/2}(3x^2)$$

$$= \frac{-3x^2}{2(1 + x^3)^{3/2}}$$

Then applying the Quotient Rule, we get (omitting some details)

$$f''(x) = \frac{15x^4 - 12x}{4(1 + x^3)^{5/2}}$$

As you can see, the details of calculating $f'''(x)$ and then $f^{(4)}(x)$ are tedious. The final result is

$$f^{(4)}(x) = \frac{135x^2(7x^6 - 40x^3 + 16)}{16(x^3 + 1)^{9/2}}$$

We would like to know the absolute maximum values of $|f''(x)|$ and $|f^{(4)}(x)|$ on $[0, 1]$, but a good bit of work is involved in these computations. We might settle for finding upper bounds on these derivatives, that is, numbers at least as large as their maximum values. One way is to *maximize the numerator* and *minimize the denominator*. In each case the numerator cannot exceed the sum of the absolute values of the terms. That is, for $f''(x)$, $|15x^4 - 12x| \leq 15x^4 + 12|x|$. As x varies from 0 to 1, $15x^4 + 12|x|$ is at most $15 + 12 = 27$, since the maximum value occurs at $x = 1$. Similarly, the numerator of $f^{(4)}(x)$ does not exceed $135(7 + 40 + 16) = 8,505$ in absolute value. In each case the minimum value of the denominator occurs when $x = 0$. Thus, for all x in $[0, 1]$,

$$|f''(x)| \leq \frac{27}{4} = 6.75 \quad \text{and} \quad |f^{(4)}(x)| \leq \frac{8505}{16} \approx 531.6$$

so to three significant figures we can take $M_2 = 6.75$ and $M_4 = 532$.

An alternative procedure is to use a CAS or graphing calculator to obtain graphs of $|f''(x)|$ and $|f^{(4)}(x)|$ and to estimate from the graphs the maximum values for x in $[0, 1]$. We show these graphs in Figures 5.49 and 5.50. From the graphs we read the maximum values to be approximately 0.97 and 14.4, respectively. So we can take $M_2 = 1$ and $M_4 = 15$. These are much better values to use than the ones we previously obtained, since they give smaller bounds on the errors.   ∎

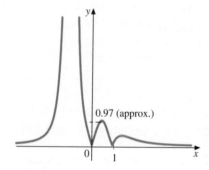

FIGURE 5.49

Graph of $|f''(x)|$ for $f(x) = \dfrac{1}{\sqrt{1+x^3}}$

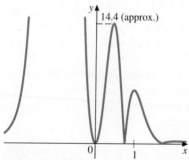

FIGURE 5.50

Graph of $|f_{(x)}^{(4)}|$ for $f(x) = \dfrac{1}{\sqrt{1+x^3}}$

REMARK ————————————————————————————————

■ With a computer algebra system (CAS) such as DERIVE, Maple, or Mathematica, derivatives such as $f^{(4)}(x)$ above can be calculated with ease. In fact, any of these systems could be used to approximate the value of the integral.

Cases Where Approximations Are Exact

If f is a polynomial function of degree 0 or 1, its second derivative is 0 for all x. For example, if $f(x) = 2x + 3$, then $f'(x) = 2$, and $f''(x) = 0$. Thus, in the error bounds for the Midpoint and Trapezoidal Rules, we may take $M_2 = 0$ (since $|f''(x)| = 0$ for all x). This choice of M_2 implies that $E_n = 0$ for every value of n. That is, both the Midpoint Rule and the Trapezoidal Rule give the *exact* value of the integral $\int_a^b f(x)dx$ when $f(x)$ is a polynomial of degree 0 or 1.

Similarly, if f is a polynomial function of degree 3 or less, its fourth derivative is identically 0, so in the formula for the error bound for Simpson's Rule we may take $M_4 = 0$, which implies that $E_n = 0$ for every choice of n. Thus, for such polynomial functions, Simpson's Rule gives the exact value of the integral. This result is not surprising if f is of second degree, since in this case, the graph of f is a parabola. After all, in Simpson's Rule we use parabolas to approximate the actual curve. So the approximation and the actual curve are the same. But it is not intuitively evident why fitting parabolas to cubic curves should result in the true value of the integral.

REMARK ————————————————————————————————

■ The fact that the approximation rules give exact answers for certain polynomial functions—those of degree 0 or 1 for the Midpoint and Trapezoidal Rules, and those of degree 0, 1, 2, or 3 for Simpson's Rule—is of more theoretical than practical interest. After all, if f is *any* polynomial, we can easily find an antiderivative of it and evaluate its integral by means of the Second Fundamental Theorem.

There are some cases in which the approximation methods yield exact answers for integrals even when the integrand is not a polynomial. The next example shows one such integral. The error analysis in this case is very misleading.

EXAMPLE 5.34   Consider the integral

$$\int_0^\pi \sin^2 x \, dx$$

(a) Determine values of n for both the Trapezoidal Rule and Simpson's Rule that will guarantee $E_n \leq 10^{-7}$ for the approximations of the integral by these rules.

(b) Show that both the Trapezoidal Rule and Simpson's Rule actually give the exact answer when $n = 6$.

Solution

(a) First, we find $f''(x)$ and $f^{(4)}(x)$ for the integrand $f(x) = \sin^2 x = (\sin x)^2$

$$f'(x) = 2(\sin x)\cos x = \sin 2x \qquad \text{By the trigonometric identity} \atop \sin 2x = 2\sin x \cos x$$

$$f''(x) = 2\cos 2x$$

$$f'''(x) = -4\sin 2x$$

$$f^{(4)}(x) = -8\cos 2x$$

Since $|\cos 2x| \le 1$ for all x, we have $|f''(x)| \le 2$ and $|f^{(4)}(x)| \le 8$. Thus, in the formulas for E_n we can take $M_2 = 2$ and $M_4 = 8$. Our interval of integration is from $x = 0$ to $x = \pi$, so in the formulas, $b - a = \pi$. We show the calculations below for $E_n \le 10^{-7} = 1/10^7$. Since $\Delta x = \pi/n$, we have

<table>
<tr><td align="center">TRAP</td><td align="center">SIMP</td></tr>
</table>

$$E_n \le \frac{M_2(b-a)(\Delta x)^2}{12} \qquad\qquad E_n \le \frac{M_4(b-a)(\Delta x)^4}{180}$$

$$\frac{2\pi^3}{12n^2} \le \frac{1}{10^7} \qquad\qquad \frac{8\pi^5}{180n^4} \le \frac{1}{10^7}$$

$$n^2 \ge \frac{10^7\pi^3}{6} \qquad\qquad n^4 \ge \frac{2(10^7)\pi^5}{45}$$

$$n \ge 7,189 \qquad\qquad n \ge 108$$

(b) To find the exact value of the integral, we use the trigonometric identity (see Appendix 2)

$$\sin^2 x = \frac{1}{2}(1 - \cos 2x)$$

With this identity we can write

$$\int_0^\pi \sin^2 x\, dx = \frac{1}{2}\int_0^\pi (1 - \cos 2x)\,dx = \frac{1}{2}\left[x - \frac{\sin 2x}{2}\right]_0^\pi = \frac{\pi}{2}$$

Using $n = 6$ in the Trapezoidal Rule and in Simpson's Rule, we have $\Delta x = \pi/6$ in each case. So the division points are

$$x_0 = 0, \ x_1 = \pi/6, \ x_2 = \pi/3, \ x_3 = \pi/2, \ x_4 = 2\pi/3, \ x_5 = 5\pi/6, \ x_6 = \pi$$

The approximations of the integral by the two methods are as follows.
TRAP:

$$\frac{\pi}{12}\left[\sin^2 0 + 2\sin^2 \frac{\pi}{6} + 2\sin^2 \frac{\pi}{3} + 2\sin^2 \frac{\pi}{2} + 2\sin^2 \frac{2\pi}{3}\right.$$

$$\left. + 2\sin^2 \frac{5\pi}{6} + \sin^2 \pi\right]$$

$$= \frac{\pi}{12}\left[0 + 2\left(\frac{1}{4}\right) + 2\left(\frac{3}{4}\right) + 2(1) + 2\left(\frac{3}{4}\right) + 2\left(\frac{1}{4}\right) + 0\right]$$

$$= \frac{\pi}{12}(6) = \frac{\pi}{2}$$

SIMP:

$$\frac{\pi}{18}\left[\sin^2 0 + 4\sin^2\frac{\pi}{6} + 2\sin^2\frac{\pi}{3} + 4\sin^2\frac{\pi}{2} + 2\sin^2\frac{2\pi}{3}\right.$$

$$\left. + 4\sin^2\frac{5\pi}{6} + \sin^2\pi\right]$$

$$= \frac{\pi}{18}\left[0 + 4\left(\frac{1}{4}\right) + 2\left(\frac{3}{4}\right) + 4(1) + 2\left(\frac{3}{4}\right) + 4\left(\frac{1}{4}\right) + 0\right]$$

$$= \frac{\pi}{18}(9) = \frac{\pi}{2}$$

Both approximations give the exact value of the integral! That is, the actual error with $n = 6$ is 0. Yet in part (a) we found by our formulas that to be assured of an error of less than 10^{-7}, we should take $n = 7,189$ for the Trapezoidal Rule and $n = 108$ for Simpson's Rule. The lesson to be learned from this example is that the value of n obtained from the formula for the bound on the error, E_n, for each method will indeed guarantee that the error does not exceed the specified amount, but it is possible that a smaller value of n will also work (and might even give the exact value, as in this example).

We should note that the value $n = 6$ in this example is special. A larger value of n need not yield the exact answer. You might find it interesting to try some larger values. In fact, even with the values of n obtained in part (a), the answers would not have been exact (but the errors would have been less than 10^{-7}). ∎

Approximating an Integral of a Function with No Known Formula

We conclude this section with an example to illustrate that Simpson's Rule, as well as the Trapezoidal Rule, can be used even if we do not know the function to be integrated, provided we know the function's values at equally spaced intervals. For the Midpoint Rule, we would also need the function's values at the midpoints of the intervals.

EXAMPLE 5.35   A flat iceberg (a sheet of ice) is measured across its width at 3 m intervals, as shown in Figure 5.51. A bore hole through the iceberg shows

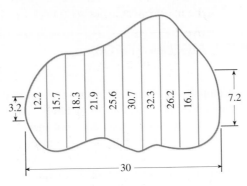

FIGURE 5.51

the thickness of the ice sheet to be 6 m. Use Simpson's Rule to approximate the volume of the ice sheet. How many ice cubes, 2 cm on an edge, would the iceberg make?

Solution   The volume is the area of the surface multiplied by the thickness. Let $w(x)$ denote the width of the ice sheet x m from the left end in Figure 5.51. The area of the ice sheet is then given by the integral $\int_0^{30} w(x)dx$. (You can see why, by thinking of Riemann sums composed of areas of narrower and narrower rectangles, so that in the limit you get the integral.) Although we do not know the formula for $w(x)$, we do know its values at the endpoints of 10 equally spaced intervals, 3 m wide. So we can use Simpson's Rule with $\Delta x = 3$.

$$\int_0^{30} w(x)dx \approx \frac{3}{3}\big[3.2 + 4(12.2) + 2(15.7) + 4(18.3) + 2(21.9)$$

$$+ 4(25.6) + 2(30.7) + 4(32.3) + 2(26.2) + 4(16.1) + 7.2\big]$$

$$= 617.4$$

The thickness of the ice is 6 m, so the volume is approximately

$$(617.4)(6) \approx 3700 \text{ m}^3$$

Since we do not know the formula for $w(x)$, there is no way to check the error in our approximation.

The volume of an ice cube 2 cm on an edge is 8 cm^3. The volume of the ice sheet in cubic centimeters is approximately $(3700)(100 \text{ cm/m})^3$. So the number of ice cubes is

$$\frac{3,700,000,000}{8} \approx 460,000,000$$

which is enough for more than 10 million soft drinks!   ■

Approximating Integrals by Computer

With a CAS such as DERIVE, Maple, or Mathematica, many definite integrals can be approximated rapidly and with a high degree of accuracy. So one might well question the need for the approximation techniques we have studied in this section. For one thing, it is worthwhile knowing how the computers do it. In fact, most of these programs use Simpson's Rule. Even if you do use a CAS, it is still important to understand the underlying theory that we have presented. It will serve you well when the software eventually makes a mistake.

INTEGRATION USING A COMPUTER ALGEBRA SYSTEM

Computer algebra systems can be used to perform both exact symbolic integration and approximations to integrals. The fact that we know that an antiderivative of a function exists does not mean it is always possible to determine the antiderivative explicitly. In some cases, the best that can be done is to approximate the integral. Even when it is possible to express the antiderivative in terms of simple functions, the algebra required may be very difficult. In both cases a computer algebra system can prove very useful.

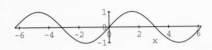

FIGURE 5.5.1
$y = f(x) = \sin x, -2\pi \leq x \leq 2\pi$

CAS 19

Approximate $\int_0^\pi \sin x \, dx$.

We start by using Maple, which allows each of the types of approximation discussed in Section 5.5. We need to load the package called "student" in order to access some additional Maple commands.

```
with (student):
f:=x->sin(x);
plot (f(x),x=-2*Pi..2*Pi);
leftsum (f(x),x=0..Pi,10);
```

The output from this command is the expression for the left sum, where the interval $[0, \pi]$ is partitioned into 10 subintervals of equal width.

$$\frac{1}{10}\pi \left(\sum_{i=0}^{9} \sin\left(\frac{1}{10}i\pi \right) \right)$$

To get a numerical value we need to evaluate this sum.

```
evalf(value(leftsum(f(x),x=0..Pi,10)));
```

Output: 1.983523538

If we had not used the evalf command, the expression output would be:

$$\frac{1}{10}\pi \left(\sqrt{5} + 1 + \frac{1}{2}\sqrt{2}\sqrt{5 - \sqrt{5}} + \frac{1}{2}\sqrt{2}\sqrt{5 + \sqrt{5}} \right)$$

To increase the number of subintervals over which the sum is taken, we change the last value specified in the command.

```
evalf(value(leftsum(f(x),x=0..Pi,100)));  1.999835643
```

Maple also allows graphic visualization of the process (see Figure 5.5.2).

```
leftbox (f(x),x=0..Pi,50);
```

To show the Right-Endpoint Rule, use:

```
rightbox(f(x),x = 0. .Pi,50);
```

The next output shows the approximations for $n = 100$ and 500 using the Left- and Right-Endpoint Rules, the Midpoint Rule, the Trapezoid Rule, and Simpson's Rule.

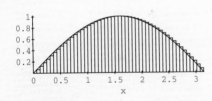

FIGURE 5.5.2
Left Riemann sum for $f(x) = \sin x$ on $[0, \pi]$ with $n = 50$

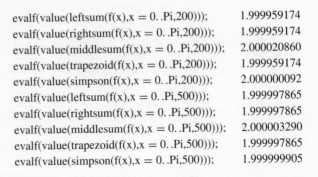

evalf(value(leftsum(f(x),x = 0. .Pi,200)));	1.999959174
evalf(value(rightsum(f(x),x = 0. .Pi,200)));	1.999959174
evalf(value(middlesum(f(x),x = 0. .Pi,200)));	2.000020860
evalf(value(trapezoid(f(x),x = 0. .Pi,200)));	1.999959174
evalf(value(simpson(f(x),x = 0. .Pi,200)));	2.000000092
evalf(value(leftsum(f(x),x = 0. .Pi,500)));	1.999997865
evalf(value(rightsum(f(x),x = 0. .Pi,500)));	1.999997865
evalf(value(middlesum(f(x),x = 0. .Pi,500)));	2.000003290
evalf(value(trapezoid(f(x),x = 0. .Pi,500)));	1.999997865
evalf(value(simpson(f(x),x = 0. .Pi,500)));	1.999999905

We can take this one step further and let Maple compute the integral, in this case the area under the sine curve on $[0,\pi]$ as the limit of Reimann sums.

Enter

    leftsum(f(x),x=0..Pi,n);

Output:

$$\frac{\pi\left(\sum_{i=0}^{n-1}\sin\left(\frac{i\pi}{n}\right)\right)}{n}$$

    limit(",n=infinity);

Output:

$$\lim_{n\to\infty}\frac{\pi\left(\sum_{i=0}^{n-1}\sin\left(\frac{i\pi}{n}\right)\right)}{n}$$

To compute the actual limit in Maple enter:

    leftsum(f(x),x=0..Pi,n);   (same output as above)
    value (");

Output:

$$-\frac{\pi\sin\left(\frac{\pi}{n}\right)}{n\left(\cos\left(\frac{\pi}{n}\right)-1\right)}$$

    limit(", n = infinity);

Output: 2

In Mathematica and DERIVE, similar analysis can be carried out by defining the methods to be used. For example, to use the Left-Endpoint Rule:

Mathematica:

```
f[x_] = Sin[x]
a = 0
b = Pi
n = 50
dx = (b–a)/n
x[k_] = a+k*dx
leftsum = N[dx*Sum[f[x[k–1]],
{k,1,n}],10]
```

Output: 1.999341983

DERIVE:

(At the □ symbol, go to the next step.)
a (author) □ f(x) := sin(x) □ a (author) □ dx := (b–a)/n □ a (author) □ dxf(a+(k–1)dx) □ a (author) □ a := 0 □ a (author) □ b := Pi □ a (author) □ n := 50 □ c (calculus) □ s (sum) □ [choose expression] □ k (variable) □ 1 (lower limit of sum] □ n (upper limit of sum] □ s (simplify) □ [choose expression] □ x (approximate) □ [choose expression]
Output: 1.98352
c (calculus) □ 1 (limit) □ [choose expression for Reimann sum] □ n (variable) □ inf (limit point ∞) □ s (simplify) □ [choose expression]

 CAS 20

Use the Fundamental Theorem of Calculus to find

$$\int_0^\pi \sin x\, dx \quad\text{and}\quad \int_0^4 \sin x\, dx$$

Computer algebra systems can perform symbolic integration as well as integral approximation. Both antiderivatives (indefinite integrals) and definite integrals can be computed in most cases.

To find the indefinite integral $\int \sin x\, dx$:

Maple:

f: = x–> sin(x);
int(f(x),x):

Output: $-\cos(x)$

Mathematica:

f[x_] = Sin[x]
Integrate[f[x],x]

DERIVE:

(At the □ symbol, go to the next step.)
a (author) □ sin(x) □ c (calculus) □
i (integrate) □ [choose expression] □
x (variable) □ [enter] (no limits) □ s
(simplify) □ [choose expression]

To find the definite integral $\int_0^\pi \sin x\, dx$:

Maple:

int(f(x),x = 0..Pi);

The value returned for this definite integral is 2. Compare this result with the results that were given in CAS 19.

Mathematica:

Integrate [f[x],{x,0,Pi}]

DERIVE:

(At the □ symbol, go to the next step.)
a (author) □ sin(x) □ c (calculus) □
i (integrate) □ [choose expression] □ x
(variable) □ 0 (lower limit) □ Pi (upper
limit) □ s (simplify) □ [choose expression]

To find the definite integral $\int_0^4 \sin x\, dx$:

Maple:

evalf (int(f(x),x = 0..4));

Mathematica:

N[Integrate[f[x],{x,0,4}],10]

DERIVE:

(At the □ symbol, go to the next step.)
a (author) □ sin(x) □ c (calculus) □ i
(integrate) □ [choose expression] □ x
(variable) □ 0 (lower limit) □ Pi (upper limit) □ s (simplify) □ [choose ex-
pression] □ x (approximate) □ [choose
expression]

The output from Maple and Mathematica is 1.653643621 whereas DERIVE gives the value 1.645364. Notice that the value of the integral is less than 2, because the graph of $\sin x$ between π and 4 lies below the x-axis and the contribution to the integral will be negative. We can see this clearly by executing the Maple command:

leftbox(f(x),x=0..4,50);

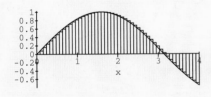

FIGURE 5.5.3
Left Riemann sum for $f(x) = \sin x$ on
[0,4] with $n = 50$

In Exercises 1–6, approximate the integral using (a) the Trapezoidal Rule, (b) the Midpoint Rule, and (c) Simpson's Rule, for the given value of n. Find the exact answer using the Second Fundamental Theorem, and determine the error in each approximation.

1. $\displaystyle\int_0^1 \frac{dx}{(x+1)^2}$; $n = 4$

2. $\displaystyle\int_2^4 \frac{dx}{\sqrt{2x+1}}$; $n = 10$

3. $\displaystyle\int_1^3 x\sin x^2 dx$; $n = 8$

4. $\displaystyle\int_{-1}^3 x\sqrt{x^2+1}\, dx$; $n = 8$

5. $\int_{-2}^{1} (1 + x^3)dx$; $n = 6$

6. $\int_{0}^{1} (\sin \pi x)\sqrt{1 - \cos \pi x}\, dx$; $n = 6$

In Exercises 7–12, determine a value of n for (a) the Trapezoidal Rule, (b) the Midpoint Rule, and (c) Simpson's Rule that will ensure that the error in approximating the integral does not exceed 10^{-4}.

7. $\int_{1}^{3} \frac{dx}{\sqrt{x+1}}$

8. $\int_{-1/2}^{2} \sqrt[3]{x+1}\, dx$

9. $\int_{3}^{5} \cos^2 x\, dx$

10. $\int_{1}^{3} \frac{1}{(4-x)^2}\, dx$

11. $\int_{0}^{2} \sin^3 x\, dx$

12. $\int_{0}^{2} (2x+1)^{-3/2}dx$

13. From the table given, approximate $\int_{0}^{5} f(x)dx$ using (a) the Trapezoidal Rule, and (b) Simpson's Rule.

x	1.0	1.4	1.8	2.2	2.6	3.0
$f(x)$	2.15	1.82	0.43	-0.15	-0.92	-1.27

x	3.4	3.8	4.2	4.6	5.0
$f(x)$	-0.84	0.16	1.18	2.35	3.76

14. By estimating values from the accompanying graph, approximate the area under the curve from $x = 2$ to $x = 6$ using Simpson's Rule with $n = 8$.

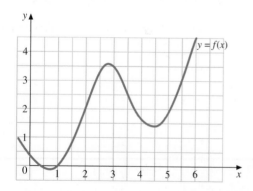

15. The average depth of a pond is known to be approximately 2 m. The length of the pond is 16 m, and the width of the pond at 2-m increments is measured and found to be as shown in the figure (in meters). Use Simpson's Rule to estimate the volume of water in the pond.

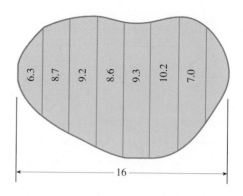

In Exercises 16 and 17, approximate the integral using (a) the Trapezoidal Rule, (b) the Midpoint Rule, and (c) Simpson's Rule, with $n = 10$. Give an upper bound on the error in each case.

16. $\int_{-1}^{1} \sin x^2 dx$

17. $\int_{1}^{2} \frac{\sin x}{x}\, dx$

18. Using the values of M_2 and M_4 found from the graphs of $|f''(x)|$ and $|f^{(4)}(x)|$ in Example 5.33, find a value of n for (a) the Midpoint Rule and (b) Simpson's Rule that will ensure an error of less than 0.00005 in approximating the integral

$$\int_{0}^{1} \frac{dx}{\sqrt{1+x^3}}$$

In each case use the value of n that you find to obtain the approximation to the integral.

19. Use the Midpoint Rule to approximate the integral

$$\int_{0}^{2} \sqrt{\sin^2 x + 3\cos^2 x}\, dx$$

with $n = 10$. Use a graphing calculator or a CAS to graph $|f''(x)|$, where $f(x)$ is the integrand in this integral. Estimate the smallest number M_2 for which $|f''(x)| \le M_2$ for x in $[0, 2]$, and use the value obtained to find a bound on the error in approximating the integral.

20. Consider the region bounded above by the graph of $y = \sin(x^2)$, below by the x-axis, and on each side by the vertical lines $x = -1$ and $x = 1$. (See the figure.)
(a) Use Simpson's Rule to approximate the area of the region using $n = 10$ and $n = 20$. In each case find an upper bound on the error.

(b) Find a value of n so that the approximation to the error using Simpson's Rule is correct to seven decimal places.

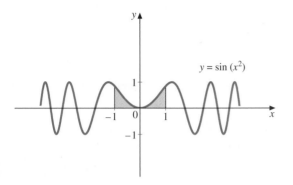

$y = \sin(x^2)$

21. Use Simpson's Rule to approximate

$$\int_0^1 \frac{dx}{1 + x^2}$$

with $n = 10$, 20, 100, and 1000. Conjecture the exact value of the integral. (*Hint:* Multiply your approximations by 4 and see if you recognize the value being approached.)

22. Prove that

$$S_{2n} = \frac{2}{3} M_n + \frac{1}{3} T_n$$

where S_{2n} is the value given by Simpson's Rule with $2n$ subdivisions, and M_n and T_n are the values given by the Midpoint and Trapezoidal Rules, respectively, for n subintervals.

Chapter 5 Review Exercises

1. Let $f(x) = x^2 + 1$ on $[0, 2]$. Find the upper sum U_n and the lower sum L_n for n arbitrary, and show that they have the same limit as $n \to \infty$. Interpret the result geometrically.

2. Let $f(x) = x(1+x)(3-x)$ on $[-1, 4]$. Find the Riemann sum for f using the partition points -1, 0, 2, 3, 4, with $c_1 = -0.5$, $c_2 = 0.6$, $c_3 = 2.3$, and $c_4 = 3.4$. Compare the result with the exact value of $\int_{-1}^{4} f(x)dx$.

3. Find the exact value of the integral $\int_{-1}^{3}(x^2 - 2x - 1)dx$ in two ways: (a) by taking the limit of the Riemann sum with equally spaced partition points, where $c_k = x_k$ for $k = 1, 2, \ldots, n$, and (b) by using the Second Fundamental Theorem of Calculus.

4. Let

$$f(x) = \begin{cases} 2x + 1 & \text{if } -1 \le x < 1 \\ 3 & \text{if } 1 \le x < 4 \\ 3(5 - x) & \text{if } 4 \le x \le 5 \end{cases}$$

Find $\int_{-1}^{5} f(x)dx$ making use of areas of known geometric figures.

5. A tank originally full of water was observed after one minute to be leaking at the rate of 12.6 L/min. Continued efforts to stop the leak were only partially successful, as the table shows. The table gives the rate $R(t)$ of leakage at half-minute intervals. Give an overestimate and an underestimate of the amount of water that leaked out from $t = 1$ to $t = 4$.

t	1.0	1.5	2.0	2.5	3.0	3.5	4.0
$R(t)$	12.6	8.4	6.3	5.2	4.2	3.6	3.2

6. Find the exact value of the limit

$$\lim_{n \to \infty} \sum_{k=1}^{n} x_k^3 \Delta x$$

where $\{x_0, x_1, x_2, \ldots, x_n\}$ is a partition of the interval $[0, 4]$ with equally spaced partition points, so that $\Delta x = 4/n$. (*Hint:* Write the limit as an integral.)

7. Use the values of the areas shown in the figure for the graph of f to find the following:

(a) $\displaystyle\int_{-2}^{3} f(x)dx$

(b) $\displaystyle\int_{-2}^{3} |f(x)|dx$

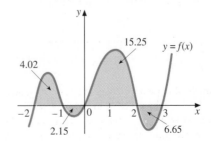

$y = f(x)$

In Exercises 8–19, use the Second Fundamental Theorem to evaluate the integral.

8. $\displaystyle\int_{2}^{4} x(1 - 2x)\,dx$

9. $\displaystyle\int_{-3}^{0} \left(x - \sqrt{1 - x}\right)dx$

10. $\displaystyle\int_{1}^{4} \left(x - \dfrac{1}{\sqrt{x}}\right)^{3} dx$

11. $\displaystyle\int_{0}^{\pi/3} \dfrac{1 - \sin x}{\cos^2 x}\,dx$

12. $\displaystyle\int_{-8}^{-1} (u^{1/3} - u^{-2/3})\,du$

13. $\displaystyle\int_{0}^{\pi/4} \sin^3 t \cos t\,dt$

14. $\displaystyle\int_{-\pi/3}^{\pi/3} (\theta + \cos\theta)\,d\theta$

15. $\displaystyle\int_{-1}^{2} (2x - 1)(x + 4)\,dx$

16. $\displaystyle\int_{0}^{2} \dfrac{x^2}{\sqrt{1 + x^3}}\,dx$

17. $\displaystyle\int_{9}^{16} \dfrac{(\sqrt{x} - 2)^5}{\sqrt{x}}\,dx$

18. $\displaystyle\int_{0}^{\pi/4} \dfrac{\sin 2x}{(1 + \cos 2x)^3}\,dx$

19. $\displaystyle\int_{27}^{125} t^{-1/3}(t^{2/3} - 9)^{3/2}\,dt$

In Exercises 20–27, find the indicated antiderivatives.

20. $\displaystyle\int \dfrac{1}{x^2}\sec^2\left(\dfrac{2}{x}\right)dx$

21. $\displaystyle\int \dfrac{\tan^3\theta - \sin\theta}{\cos^2\theta}\,d\theta$

22. $\displaystyle\int \dfrac{x - 1}{\sqrt{x^2 - 2x + 4}}\,dx$

23. $\displaystyle\int x(x^4 + 4x^2 + 4)^{5/2}\,dx$ (*Hint:* $x^4 + 4x^2 + 4$ is a perfect square.)

24. $\displaystyle\int \dfrac{\sqrt{1 + \sqrt{x}}}{\sqrt{x}}\,dx$

25. $\displaystyle\int \dfrac{x(1 + \cos^2 x^2)}{\sin^2 x^2}\,dx$

26. $\displaystyle\int \left[\dfrac{1}{1 + \cos\theta} - \dfrac{1}{1 - \cos\theta}\right]d\theta$ (*Hint:* Combine fractions.)

27. $\displaystyle\int \sin 2t(1 + \cos^2 t)^3\,dt$ (*Hint:* Use an identity.)

28. Verify that

$$\int x^2 \cos 3x\,dx = \dfrac{1}{27}\left[(9x^2 - 2)\sin 3x + 6x\cos 3x\right] + C$$

29. Evaluate the integral $\int_0^2 f'(t)\,dt$ if it is known that $f(x) = (3x^2 - 4)\sqrt{x^3 + 1}$.

30. Find $f'(t)$ if $f(t) = \displaystyle\int_0^t \dfrac{x}{\sin^2 x + 1}\,dx$.

31. (a) Find $F'(\pi)$ if $F(x) = \displaystyle\int_{\pi/6}^{x} \dfrac{\sin t}{t}\,dt$.

(b) Find the value of $\displaystyle\int_{\pi/6}^{\pi} \dfrac{d}{dx}\left(\dfrac{\sin x}{x}\right)dx$.

32. Find the average value of $f(x) = 2x^3 - 4x^2 - 5x + 2$ on $[-2, 2]$.

33. Find the average value of $g(x) = x/\sqrt{x^2 + 9}$ on $[0, 4]$.

34. Find c such that $\int_{-1}^{4} \sqrt{3x + 4}\,dx = 5\sqrt{3c + 4}$.

In Exercises 35–38, find the general solution of the differential equation.

35. $\dfrac{dy}{dx} = 2x - \sin 2x$

36. $(\cos^2 x)y' = 1 - \sin x$

37. $f'(t) = \sqrt{t^4 + 4t^2},\ (t > 0)$

38. $(x^2 + 1)^2\,dy = x\,dx$

In Exercises 39–41, solve the initial-value problem.

39. $g'(t) = t(t^2 + 4)^{3/2};\ g(0) = 2$

40. $\dfrac{dy}{dx} = \sin x \cos^2 x;\ y = 3$ when $x = \dfrac{\pi}{2}$

41. $\sqrt{x}\,y' = (\sqrt{x} + 2)^3;\ y = 5$ when $x = 1$

In Exercises 42–45, approximate the integral using (a) the Trapezoidal Rule, (b) the Midpoint Rule, and (c) Simpson's Rule, for the given value of n. Find an upper bound on the error in each case. Compare with the exact value.

42. $\displaystyle\int_0^2 \sin^2\dfrac{\pi x}{2}\,dx;\ n = 8$

43. $\displaystyle\int_0^1 \dfrac{x}{\sqrt{x^2 + 1}}\,dx;\ n = 10$

44. $\displaystyle\int_0^1 \tan^2 x\,dx;\ n = 10$ **45.** $\displaystyle\int_1^8 \dfrac{x^{2/3} + 1}{\sqrt[3]{x}}\,dx;\ n = 8$

In Exercises 46 and 47, find n such that the error in using (a) the Trapezoidal Rule, (b) the Midpoint Rule, and (c) Simpson's Rule to approximate the given integral does not exceed 10^{-4}.

46. $\displaystyle\int_1^4 \frac{dx}{\sqrt{5-x}}$ **47.** $\displaystyle\int_0^3 \sin^3 x \, dx$

48. Use the Trapezoidal Rule with $n = 20$ to approximate

$$\int_{-1}^1 \tan x \, dx$$

Give an upper bound on the error.

49. Repeat Exercise 48 using the Midpoint Rule.

 50. Repeat Exercise 48 using Simpson's Rule.

51. The accompanying table gives velocities of an object moving along a line, at half-second intervals. Use Simpson's Rule to approximate the total distance covered by the object. Assume t is in seconds and $v(t)$ is in meters per second.

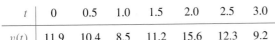

t	0	0.5	1.0	1.5	2.0	2.5	3.0
$v(t)$	11.9	10.4	8.5	11.2	15.6	12.3	9.2

52. Approximating Finite Sums with Integrals. Approximate the sum of the cube roots of the first n positive integers,

$$S_n = \sqrt[3]{1} + \sqrt[3]{2} + \sqrt[3]{3} + \cdots + \sqrt[3]{n}$$

by completing the following steps:
(a) Let $f(x) = \sqrt[3]{x}$. Partition the interval $[0, 1]$ into n subintervals of equal width. (See the figure.) Compute the upper sum U_n.

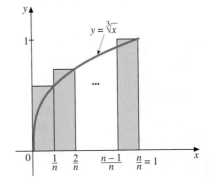

(b) Show that $S_n = n^{4/3} U_n$. Explain how it follows that for large n,

$$S_n \approx \frac{3}{4} n^{4/3}$$

 53. Use a CAS to approximate the following:

(a) $\displaystyle\int_3^4 \frac{\sin^3 x}{x^2} \, dx$

(b) $\displaystyle\int_{-1}^1 \cos^4 x \sin^2 x \tan^5 x \, dx$

(c) $\displaystyle\int_2^5 \frac{t^4 + t^3}{\sqrt{t^4 + 17}} \, dt$

Chapter 5 Concept Quiz

1. Define the following for a continuous function f on $[a, b]$:
   (a) The upper sum U_n
   (b) The lower sum L_n
   (c) The area under the graph of f from $x = a$ to $x = b$, if f is nonnegative
   (d) A Riemann sum for f
   (e) The definite integral $\int_a^b f(x) \, dx$
   (f) The average value of f on $[a, b]$
   (g) The indefinite integral $\int f(x) \, dx$

2. State the following:
   (a) The First Fundamental Theorem of Calculus
   (b) The Second Fundamental Theorem of Calculus
   (c) The Mean-Value Theorem of Integral Calculus
   (d) The Midpoint Rule

   (e) The Trapezoidal Rule
   (f) Simpson's Rule

3. Assuming appropriate continuity conditions, give the value of each of the following in a form that does not involve an integral.
   (a) $\int_{t_1}^{t_2} r(t) \, dt$, where R is a function for which $R'(t) = r(t)$
   (b) $\int_a^b f'(x) \, dx$
   (c) $\dfrac{d}{dt} \displaystyle\int_a^t h(x) \, dx$
   (d) $\int_a^b f(g(x)) g'(x) \, dx$, given that $F'(x) = f(x)$

4. Discuss in what sense integration and differentiation can be considered inverse processes.

5. Fill in the blanks. (More than one word may be needed.)

(a) If f is nonnegative and continuous on $[a, b]$, the area under the graph of f from $x = a$ to $x = b$ is given by _____ .

(b) If $A(x)$ is the area under the graph of the continuous, nonnegative function f from a to x, then $A'(x)$ exists and equals _____ .

(c) The integral $\int_a^b f(x)\, dx$ is the limit of _____ , provided this limit exists.

(d) The _____ Rule uses parabolic arcs to appoximate the integral.

(e) The _____ Rule is the average of the values given by the Left-Endpoint Rule and the Right-Endpoint Rule.

(f) The integral of the _____ of a function gives the net change in the function.

(g) In general, the Midpoint Rule is more accurate than _____ but less accurate than _____ .

6. Indicate which of the following statements are true and which are false.

(a) If $\int_a^b f(x)\, dx = \int_a^b g(x)\, dx$, then $f(x) = g(x)$.

(b) If $\int f(x)\, dx = \int g(x)\, dx$, then $f(x) = g(x)$.

(c) If $g(t) = \int_0^t \sin u^2\, du$, then $g'(t) = \sin t^2$.

(d) If $\int_a^b f(x)\, dx = 0$, then $f(x) = 0$.

(e) If $v(t)$ is the velocity of a particle moving along a line, and its acceleration $a(t)$ is continuous on $[a, b]$, then

$$\int_a^b a(t)\, dt = v(b) - v(a)$$

APPLYING CALCULUS

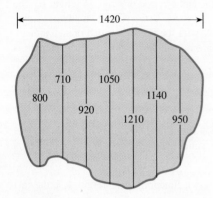

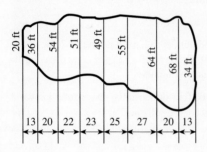

1. The Fish and Game Warden of a rural county is planning to stock a small pond with fish prior to the opening of the fishing season. However, before stocking the pond the warden has been advised to allow 1000 cubic feet of water for each fish and therefore needs to know the volume of water in the pond. The warden makes the measurements indicated in the figure and determines that the average depth of the pond is 20 feet.
   (a) Use the given information to estimate the volume of water in the pond.
   (b) How many fish will be needed to stock the pond?
   (c) If the warden wants to have at least 20% of the original fish population left in the pond at the end of the season, what is the maximum number of fishing licenses that can be sold, assuming that the average seasonal catch per license is 20 fish?

2. The town council wants to build a new parking lot using the area with the measurements shown in the figure, but only $11,000 is allocated in the budget for new parking. The cost to clear the land is $.15 per square foot and the cost to pave is $2.00 per square foot. It's your job as town engineer to report on whether it can be done within the budget. What would you say and why?

3. The rate at which a person's heart pumps blood is called the *cardiac output*. For a person at rest, this rate is normally 5 liters per minute. Strenuous exercise significantly raises this rate. Also, certain diseases of the heart, circulatory system or nervous system have a significant effect on the cardiac output of a person's heart. There is a technique for measuring cardiac output called *dye dilution* which can be described as follows: At time $t = 0$, a known amount D of dye is injected into a vein near the heart. The dyed blood circulates through the right side of the heart, to the lungs, then through the left side of the heart and into the arterial system. The concentration of dye is measured at fixed time intervals at some convenient point in the arterial system, e.g., the aorta. If the heart pumps blood at the constant rate of R liters per second and the concentration of dye at time t is $C(t)$ milligrams per liter, then the amount of dye flowing through the aorta is $RC(t)$ milligrams per second. It takes about 20 seconds for blood to circulate through the body, so

$$D = \int_0^{20} RC(t)dt = R\int_0^{20} C(t)dt$$

is the total amount of dye injected. Since D is known, this equation can be solved for R to obtain

$$R = \frac{D}{\int_0^{20} C(t)dt}$$

   (a) Suppose that at time t (in seconds), the dye concentration is given by

$$C(t) = -bt(t - 20)$$

   where b is a positive constant. Graph C. Determine R in terms of b and D, the total amount of dye injected.

   (b) Suppose $D = 5$ milligrams of dye is injected and $C = C(t)$ is measured at 1 second intervals to obtain the following data:

t	1	2	3	4	5	6	7	8	9	10	11	12	13	14	15	16	17	18	19	20
$C(t)$	0	0	.1	.6	.9	1.4	1.9	2.7	3.0	3.7	4.0	4.1	4.0	3.8	3.7	2.9	2.2	1.5	1.1	.9

   Approximate the integral $\int_0^{20} C(t)dt$ using a Riemann sum with right endpoints. Estimate the cardiac output R.

An erupting volcano

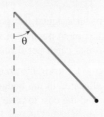

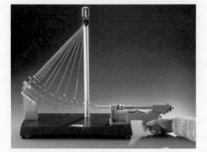

4. Geologists studying an eruption of a volcano want to measure the height of the curtain of fire without getting too close. They do so by using a stopwatch to time how long a bit of molten lava takes to fall at a constant acceleration of gravity 10 m/s². In the photograph lava at the top takes 5 seconds to fall. How high is the curtain of fire?

5. (CAS) A pendulum consists of a mass at the end of a rod or string of length L. The angle, θ, that the string makes with the vertical changes as the pendulum swings. Let β be the maximum angle that the pendulum makes with the vertical (called the amplitude). Then the time T (called the period) it takes for one complete swing of the pendulum is given by

$$T = 2\pi \sqrt{\frac{L}{g}} \int_0^{\pi/2} \frac{d\phi}{\sqrt{1 - k^2 \sin^2 \phi}}$$

where g is the acceleration due to gravity, $g = 32$ ft/s², and $k = \sin(\beta/2)$. This integral is an elliptic integral of the first kind whose antiderivative cannot be computed as a combination of simple functions.

(a) Show that $\lim_{\beta \to 0} T = 2\pi \sqrt{\frac{L}{g}}$. Note that T is a function of β. (Hint: Use the fact that T is continuous at $\beta = 0$, or approximate $\sin \phi$ by ϕ.)

(b) Approximate the period T corresponding to $\beta = 0.25$.

(c) Approximate the amplitude, β, if the pendulum is to have a period of 2.5 s. Find a function to fit T by plotting 10 points $(\beta, T(\beta))$ for $0 \le \beta \le \pi/2$ and having the computer algebra system connect the points by a curve.

6. The Universe is described by a measure of four-dimensional space-time known as the Robertson-Walker metric, which for flat space is:

$$(ds)^2 = c^2 (dt)^2 - a(t)^2 [(d\Theta)^2 + r^2 d\Omega]$$

where s is a measure of distance known as the "proper distance," c is the speed of light, t is the time, and $a(t)$ is a scale factor that describes the expansion of the universe. The variables Θ (capital theta) and Ω (capital omega) are radial and angular coordinates, respectively. We can derive that the equation of motion that relates the radial coordinate Θ to the time is:

$$d\Theta = \pm \frac{c \, dt}{a(t)}$$

For a set of galaxies expanding with the Universe, light emitted by a galaxy at position θ_g at time t_e will be observed at $\Theta = 0$ at a later time t_0 where Θ_g is the integral from t_e to t_0 of $d\Theta$.

(a) Write the integral and use one of the Fundamental Theorems of Calculus (specify which one) to differentiate it with respect to t_0.

(b) Since the galaxy's position is constant in its moving frame of reference, solve the equation of the previous part to find the relation between the time intervals dt_0 and dt_e in terms of the scale factors at the two times.

(c) Since the frequencies at which we observe the light depends on the time interval, derive the relation between the scale factors and the wavelengths, which are the inverses of the frequencies. Do the wavelengths become larger or smaller over time?

7. A beam of energy in a star, as it flows through a cubic meter of gas, loses a portion of its energy to absorption. This portion is proportional to the incoming intensity I, the density of the gas ρ, and a property of the gas known as absorptivity that is measured by the absorption coefficient κ. Also, the same volume of gas emits additional radiation in an amount proportional to both the density and to an emission coefficient ε.

(a) Express the differential of the optical depth τ in terms of its proportionality to κ, ρ, and the differential of radial distance x from the center of the star. Use a minus sign to show that the optical depth measured by us from Earth increases as the distance from the center of the star toward us decreases. Then integrate as x goes from a to b to find τ in terms of κ, ρ, a, and b.

(b) Derive the Equation of Transfer, which shows the change in intensity as a function of the absorption and emission coefficients and of the incoming intensity. To do so, express the differential of intensity through a small volume in terms of the radiation added to and subtracted from the radiation hitting that volume. Then express the differential equation in terms of optical depth to get the Equation of Transfer.

The Equation of Transfer has no simple solution, and is either solved with simplifying assumptions or using numerical methods.

Phillip A. Griffiths

Mathematics is still a "small science," an individual activity done with a lively imagination. In this sense it is highly interactive. The most productive environment for doing mathematics is one in which an individual can share ideas with others when appropriate, then retire into privacy for quiet reflection. This approach allows mathematicians to span many topics, remain flexible, and respond effectively and quickly to the needs of the rest of science.

But mathematics is fundamentally different from the sciences and the relationship between mathematics and other disciplines has changed substantially in recent years. Mathematics has been called both the "Queen of the Sciences" and the "Servant of the Sciences," but neither description seems adequate today. Mathematics is not so much above and below other disciplines, but within and around them. It has become a full and interactive partner. This is healthy not only for mathematics and associated fields like physics, but also for fields like business, psychology, and health policy analysis.

Mathematics is also a language—one on which the sciences depend when they need to quantify what they are doing. But it is more than that. Richard Feynman, a Nobel laureate in physics, said that the universe seems indescribable except in the language of mathematics. Sir Isaac Newton wanted a theoretical framework to express the motion of objects under the influence of gravity. This desire led to his law of universal gravitation as well as the calculus, one of the great achievements in the history of science. Albert Einstein spent years trying to formulate his insight that gravitation is really a reflection of the curvature of space–time, but couldn't express it mathematically until Marcell Grossman told him about Bernhard Riemann's work with curved space, following earlier work by Gauss, Bolyai, and Lobachevsky. This foundation enabled Einstein to develop the theory of relativity. Another Nobel laureate in physics, Steven Weinberg, has spoken of "spooky" coincidence, noting that physicists often find that mathematicians have been there before them. Group theory, for instance, was invented in the early nineteenth century to find the roots of polynomial equations. Evariste Galois made one of the most original leaps in the history of mathematics in introducing the concept of a group to describe symmetry. When physicists eventually discovered group theory in the twentieth century, they found it ideal for describing the conservation laws of energy, spin, momentum, and so on. The point is that even though Galois' motivation was purely internal to mathematics, his ideas eventually found application in the "real world."

Mathematicians have long been accused of losing themselves in the abstract beauty of their own conjectures. We all know that in geometry, for instance, we are studying infinitely small points, infinitely narrow lines, and perfectly round circles, all ideal objects. The concept of the ideal is as old as Plato, but in popular opinion is of little relevance to the real world. But abstraction is not a bad thing. Mathematicians concern themselves with internal consistency—that is, they are absolutely faithful to their own rules. They model the natural world on an idealized mathematical basis. They are criticized because they look at the world not as it is, but as it might be. This seems suspect until you realize that the world is not what it appears to be anyway. Who would imagine, for instance, by looking out a window, that the mass of an object becomes infinite when it is accelerated to near the speed of light? A mathematician might care about this phenomenon years before anyone else saw the need to care. Riemannian Geometry, for instance, was worked out 60 years before Einstein needed it to describe relativity and the algebra of Lie groups predated their application to particle physics by at least 30 years.

Another example of mathematics turning out to be "realer than real" was the discovery of the positron by the mathematical physicist Paul Dirac. While working on an equation describing the motion of the electron, Dirac realized his equation predicted the existence of a particle identical to the electron in every respect except charge. Experimental physicists then set out to look for this never-before-dreamed-of particle. So the discovery of the positron was a triumph of physics, but equally a triumph of mathematics.

The ulimate relevance of mathematical research is unpredictable. Armand Borel has likened the "shape" of mathematics to an iceberg. The portion above the surface is like applied mathematics, visible and apparent, but the bulk is below the surface, in the realm of pure mathematical research.

Around 1950, a mathematician named Herbert Hauptman became interested in the structure of crystals. Chemists knew about atoms in a crystal scattering X rays but couldn't pinpoint their position in the crystal itself. Hauptman was able to show that a special function called the *electron density function* could yield information about the phase of the X rays traveling through the crystal. From the phase information, one could then determine the atomic arrangement of the crystal. Even though he had taken only one chemistry course in his life, Hauptman won the 1985 Nobel Prize in Chemistry for this application of "classical" mathematics that had been available for over a century.

In another situation, a physicist, Allan Cormack, was trying to pinpoint the location and density of an object in the human body without surgery. At the time, only X rays were available and they could only give two-dimensional pictures. As it turned out, the mathematical solution of this problem had been around for many years, the work of a mathematician named Johann Radon. Using Radon's solutions, Cormack saw that by analyzing X rays from many different angles you can construct a three-dimensional image of an object in the body. This work led first to CAT scans (computer-assisted tomography) and later to MRI (magnetic resonance imaging), and then to the PET scan (positron emission tomography), an even more accurate technique.

Dirac's positron particle, itself predicted from mathematics, has become a tool used to measure both the metabolism and the anatomy of a living organism. Even more surprising, the analysis of this new emission tomography uses mathematics originally developed in the analysis of Soviet communication codes. The same techniques have now moved from medicine to paleoanthropology, where they have been used to determine that human ancestors walked around in an upright position. Radon's technique has also been applied to oceanography, where it is used to find ocean temperature in places where direct measurement is inconvenient or impossible, and has also been used to determine the distribution of stars near the sun, and the brightness of the moon. In 1979 Cormack won his own Nobel Prize—in Medicine.

Another instance of classical mathematics finding application in new technology comes from the mature subject of harmonic analysis, which has recently proved very useful in compressing data. Researchers at Yale have found that they can compress and restore almost any kind of image or sound by using mathematically generated shapes that resemble tiny waves—wavelets. These wavelets are so efficient at encapsulating information that it is estimated that the FBI's collection of 300 million fingerprints could be reduced by a factor of 20, and the time required to send the fingerprint information over the phone lines can be reduced from 20 minutes to one minute. This could save over $25 million that would otherwise be spent on optical storage disks alone. Scotland Yard in Britain plans to use wavelets for the same purpose.

Mathematical research is healthy today because internal barriers to its use are being broken down. After World War II, mathematics became more and more specialized to the point where a mathematician in one field often cannot communicate effectively with a mathematician in another field. This trend is partly offset by interesting problems that attract mathematicians in many different subfields. For example, Yang-Mills equations are differential equations that extend the famous equations of James Clerk Maxwell, who unified the phenomena of electricity and magnetism under the single theory of electromagnetism. Maxwell's equations and the Yang-Mills equations are so useful that they have been studied by mathematicians in many diverse fields, and this common interest has led to a rich exchange of mathematical ideas.

Mathematics is making numerous contributions to a whole range of other disciplines. These disciplines are challenging mathematicians with interesting problems that in turn lead to new mathematical applications. The more fundamental the mathematics involved, the wider the range of applications. A good example is the field of fluid dynamics. Underlying much of this field are the so-called Navier-Stokes equations. The "fluids" that they describe can be air, liquids, or even some solids. Mathematicians use these equations to study an incredible range of phenomena: blood flow in the heart; oil moving through porous ground; fuel mixing in carburetors; aircraft flying; crystals forming from liquids; plasma in fusion reactors; the motion of galaxies, clouds, winds, currents, hurricanes; and on and on. Even though a full description of the movement of fluids is too complex for complete theoretical understanding or even computer simulation, researchers make progress by combining theoretical

modeling, computer simulations, and experiments. These techniques are possible because of advances in computer hardware and software.

Computers are now far more than fast number crunchers. They allow mathematicians and scientists to do modeling in ways that were inconceivable a generation ago. Costly physical experiments can now be simulated more quickly, more cheaply, and with more confidence on a computer than by building a real model. In aircraft designs, for instance, wind tunnels have given way to shape optimization using software alone. These kinds of models have opened doors to techniques too complex to test in theory, such as the ways in which proteins fold and unfold, or how oil flows through deep rock. In recent years the hull design of yachts has been done by computer models. One specifies the characteristic to be optimized, like speed, and the computer generates the best design from that model. Other characteristics, like seaworthiness, are still best left to human designers. The point is that both techniques can now be combined to get optimum results.

New scientific problems are inspiring new mathematical work, and new discoveries in mathematics are finding application in science, and not just physical science. Mathematics is finally able to deal with some of the complexities of biological organisms. For instance, the unique capability of mathematics to discern patterns and organize information is starting to penetrate systems such as neural networks, the basic communication pathways of the brain. In yet another partnership, mathematicians and biologists are exploring the mechanisms of DNA replication. The mathematical subfields of knot theory, probability theory, and combinatorics, are all helping us understand the complex three-dimensional mechanics of DNA replication. Another important application for mathematics is epidemiology, which is taking advantage of high-speed computers to model epidemics. It's been found that AIDS, for instance, does not follow the usual pattern of infection found in other epidemics. The complexity of the models here is so great that even the fastest supercomputers are struggling, so mathematicians are engaged in using their skills to simplify the problems and models used so that the computers can handle the computations.

Computer modeling in industry has become so critical to economic progress that industries that do not use it will fall further and further behind. It is made possible by rapid advances in mathematical modeling as well as improvements in computer hardware and mathematical algorithms—that is, improved software. The design of microprocessor chips themselves is a highly mathematical process involving testing circuitry at hundreds or thousands of points. New algorithms can give quick solutions to problems involving thousands of points to within one percent of perfection. Minimizing the time involved in testing is crucial and advances in graph theory and computability have yielded better ways to do this.

One of the most interesting areas in which mathematics has formed a partnership with technology is the subject of *pattern theory*. The term was first introduced in the 1970s as a subfield of mathematics associated with computer vision, speech recognition, and artificial intelligence. Workers in this field are beginning to believe that this subject might encompass the basis of a universal theory of thought itself. From a practical point of view, pattern theorists are faced with the challenge of designing a machine that will translate speech into printed text. If everyone pronounced words in exactly the same way this would be an easy task, but we don't and it isn't.

In the past, theorists have tried to model the human brain as a kind of very complex computer, but this approach has failed in practice. A computer is designed to respond to logical input and computers are superb at this task. Some researchers now believe that the operation of the brain is fundamentally different from the design of a computer. What our brains perceive is not the raw sensory signal, which is usually fuzzy and ambiguous, but a remarkable reconstruction of that signal. The reconstruction makes use of memory, expectation, and logic. In this analysis, we do our thinking by use of pattern recognition and even though this is far from a full theory of cognition, it is already more successful than any competing theory. So we now find mathematics in the company of biology and behavioral psychology, propelling a new, broad, and intriguing theory that is only in its infancy.

I hope I have conveyed reasons for my belief that mathematics is extremely useful to our society. If this is true, one would assume that as a society we would vigorously support the research that leads to new uses of mathematics and that student interest would be at an all time high. Unfortunately this is not the case. The mathematics community has not effectively communicated to the public and their elected representatives that our subject is different, but no less valid, than the sciences. We do not design widgets or cure diseases, but our impact on science and engineering is tremendous and fundamental. It seems that mathematicians have dwelled so long in isolation that the public doesn't understand what it is we do.

In many cases mathematicians are disadvantaged by their training—it is too compressed to allow the development of both deep interests and broad knowledge, both of which are necessary for the kind of work I've described. Instead, there is much pressure for mathematicians to specialize early in careers in an area that has not been adequately worked, to apply for support to develop that research, and to get professional recognition for doing something of interest only to mathematics. This is true even though the intellectual trends within mathematics and related fields are very positive. There is an emerging balance between looking outward and looking inward that has not always been present, or even possible. Mathematicians are not accustomed to explaining to others what it is we do, much less promoting our skills as being useful to others.

One step toward better support from society is in producing better mathematics teachers. I can honestly say that the most important influence on my deciding to become a mathematician was my high school teacher, Lottie Wilson. Mrs. Wilson had the two essential qualities for getting her message across—she understood the majesty and the mystery of mathematics, and she knew there is no substitute for getting the right answer.

These are principles that came back to me when I found myself teaching calculus years later. Students expected "partial credit" on exams just because they got started on the solution or arrived at some "reasonable" but incorrect answer. I asked my class one day, "Suppose you end up being a doctor, are your patients going to be satisfied with a diagnosis that is partially correct?" This question was not well received, but my point was a serious one. Mathematics is one subject where there are definite answers and students can take real satisfaction in "getting it right." However, teachers have a responsibility in communicating the beauty as well as the utility of the subject.

It is a well-kept secret that mathematics is fun—at least to those who know some mathematics. I am still amazed at how often I hear the term "beautiful" to describe work satisfying to a mathematician. I am reminded of a story told by a mathematician named Jacques Tits. He was discussing the origin of human experiments with fire with a group of anthropologists. One said that he thought that humans were motived to understand fire by a need for cooking food. Another said that they were probably after a dependable source of heat. But Jacques said that he believed fire came under human control because of a fascination with the flame. I agree that the best mathematicians are fascinated by the flame. Fortunately for society, their fascination has in the end provided both the good cooking and the reliable heat that we all need.

Phillip A. Griffiths is a native of Raleigh, North Carolina, and is director of the Institute for Advanced Study in Princeton, New Jersey. Prior to joining the Institute he was professor of mathematics at both Harvard and Duke Universities. Dr. Griffiths is Chairman of the Committee on Science, Engineering, and Public Policy of the National Academy of Sciences, National Academy of Engineering, and Institute of Medicine. He is a member of the National Science Board and the American Philosophical Society. He received his doctorate in mathematics from Princeton University.

6

APPLICATIONS OF THE INTEGRAL

In the last chapter we saw a geometric interpretation of the definite integral: for a nonnegative continuous function over a specified finite interval, the integral gives the area under the graph of the function. Then we extended this geometric interpretation to what we referred to as "net area," for functions that took on both positive and negative values. We also gave a physical interpretation: when the integrand represents the velocity function of an object moving along a line, the integral represents the net distance covered by the object. More generally, we saw that the net change in any function is given by the integral of its rate of change with respect to its independent variable (that is, the integral of the derivative of the function).

In this chapter we give two further extensions of the area interpretation of the integral. We will also show how certain types of **volumes** can be calculated by integration. As some further geometric applications, we show how the integral can be used to find the **length of a curve**, and how we can use integration to find the areas of certain types of **surfaces**. We also discuss interpretations of the integral as the **work** done by a variable force acting along a line, the total **force** on an object submerged in a liquid (such as the force on a dam), and the **center of mass** of a continuously distributed one-dimensional mass (such as a wire) or two-dimensional mass (such as a thin plate). Finally, we show some applications of integration in **probability theory** and in **economics**. These applications, although by no means exhaustive, illustrate some of the many areas where integration is useful.

In all the applications mentioned and many more, the integrals involved arise in a natural way as the limit of Riemann sums of continuous functions on closed intervals. Once you master the techniques for arriving at the integrals in our applications, you should be able to apply them to many other types of problems. Scientists and engineers routinely use integrals in solving problems related to their work or areas of research.

6.1 MORE ON AREAS

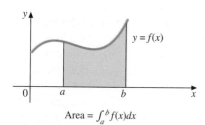

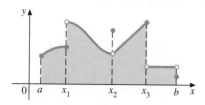

Area = $\int_a^b f(x)dx$

FIGURE 6.1

We know already that if f is a nonnegative continuous function on the closed interval $[a, b]$, then the area of the region bounded by the graph of f, the x-axis, and the lines $x = a$ and $x = b$ (the shaded region in Figure 6.1) is given by the definite integral

$$A = \int_a^b f(x)dx$$

In this section we give two extensions of this result.

Regions Bounded by Piecewise-Continuous Functions

First, suppose f is not continuous but is still nonnegative and bounded on $[a, b]$. Suppose further that it has a finite number of discontinuities, and that for each point of discontinuity c in the open interval (a, b), $\lim_{x \to c^+} f(x)$ and $\lim_{x \to c^-} f(x)$ both exist. We also assume that at the endpoints the one-sided limits, $\lim_{x \to a^+} f(x)$ and $\lim_{x \to b^-} f(x)$, exist. Such a function is said to be **piecewise-continuous**. An example is shown in Figure 6.2.

FIGURE 6.2
A piecewise-continuous function

Suppose f is such a piecewise-continuous function. Let us designate a by x_0, b by x_n, and the consecutive points of discontinuity between a and b by $x_1, x_2, \ldots, x_{n-1}$. Then we define the area of the region under the graph of f between a and b to be

$$A = \sum_{k=1}^n \int_{x_{k-1}}^{x_k} f(x)dx$$

The fact that each of the integrals $\int_{x_{k-1}}^{x_k} f(x)dx$ exists follows from the continuity of f on (x_{k-1}, x_k) and the existence of the right-hand and left-hand limits, $\lim_{x \to x_{k-1}^+} f(x)$ and $\lim_{x \to x_k^-} f(x)$. (See Exercise 77 in Exercise Set 5.3.)

EXAMPLE 6.1 Find the area under the graph of

$$f(x) = \begin{cases} 1 - x^2 & \text{if } -1 \le x < 0 \\ 2 - x & \text{if } 0 \le x < 1 \\ 0 & \text{if } x = 1 \\ x^2 - 2x + 3 & \text{if } 1 < x \le 2 \end{cases}$$

from $x = -1$ to $x = 2$.

Solution The graph of f is shown in Figure 6.3. Since f is nonnegative and piecewise-continuous, the area is given by

$$A = \int_{-1}^0 f(x)dx + \int_0^1 f(x)dx + \int_1^2 f(x)dx$$

$$= \int_{-1}^0 (1 - x^2)dx + \int_0^1 (2 - x)dx + \int_1^2 (x^2 - 2x + 3)dx$$

$$= \left[x - \frac{x^3}{3} \right]_{-1}^0 + \left[2x - \frac{x^2}{2} \right]_0^1 + \left[\frac{x^3}{3} - x^2 + 3x \right]_1^2$$

$$= \left[0 - \left(-1 + \frac{1}{3} \right) \right] + \left[\left(2 - \frac{1}{2} \right) - 0 \right] + \left[\left(\frac{8}{3} - 4 + 6 \right) - \left(\frac{1}{3} - 1 + 3 \right) \right]$$

$$= \frac{9}{2}$$

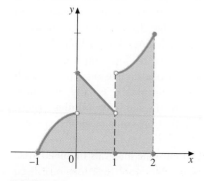

FIGURE 6.3

Areas of Regions Between Two Curves

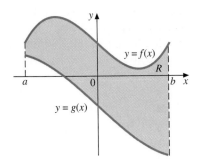

FIGURE 6.4

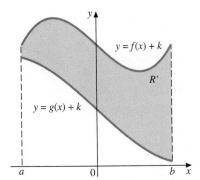

FIGURE 6.5

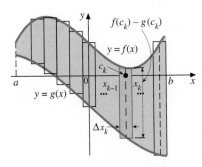

FIGURE 6.6

As a second extension of the use of integration to find areas, we consider the area of the region between the graphs of two continuous functions f and g from $x = a$ to $x = b$, where $f(x) \geq g(x)$ for all $x \in [a, b]$. Figure 6.4 illustrates a typical situation. Note that neither function needs to be nonnegative. To find the area, we will raise both curves by a constant amount k so that the entire region between them lies above the x-axis (Figure 6.5). The region R' between the graphs of $f+k$ and $g+k$ has the same area as the region R between the graphs of f and g.

To get the area of R', we subtract the area under $g + k$ from the area under $f + k$ to get

$$A = \int_a^b [f(x) + k]dx - \int_a^b [g(x) + k]dx$$

$$= \int_a^b [f(x) + k - g(x) - k]dx$$

or

$$A = \int_a^b [\overbrace{f(x)}^{\substack{\text{upper} \\ \text{curve}}} - \overbrace{g(x)}^{\substack{\text{lower} \\ \text{curve}}}]dx \tag{6.1}$$

Observe that the constant k has disappeared. We could have obtained the integrand simply by subtracting the ordinate (the y value) of the original lower curve from the ordinate of the original upper curve. This observation means that it is no longer necessary to raise the curves. We did so to arrive at Equation 6.1, but now that we have the result, we can forget about the shift upward.

It is instructive to see how we can also arrive at the result in Equation 6.1 directly, as a limit of Riemann sums, by partitioning the closed interval $[a, b]$ and forming rectangles as shown in Figure 6.6. The kth rectangle has width $\Delta x_k = x_k - x_{k-1}$ and height $f(c_k) - g(c_k)$, where c_k is any point in the interval $[x_{k-1}, x_k]$, and so its area is $[f(c_k) - g(c_k)]\Delta x_k$. The sum of all such areas,

$$\sum_{k=1}^n [f(c_k) - g(c_k)]\Delta x_k$$

is a Riemann sum for the continuous function $f - g$, and it approximates the area we want. Its limit, as the width of the largest subinterval approaches 0, is exactly this area; that is,

$$A = \lim_{\max \Delta x_k \to 0} \sum_{k=1}^n [f(c_k) - g(c_k)]\Delta x_k = \int_a^b [f(x) - g(x)]dx$$

We summarize our result as follows.

Area Between Two Curves

Let f and g be continuous, with $f(x) \geq g(x)$, for $a \leq x \leq b$. Then the area A of the region between the graphs of f and g from $x = a$ to $x = b$ is given by

$$A = \int_a^b [f(x) - g(x)]\,dx$$

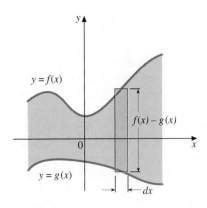

$y = f(x)$

$f(x) - g(x)$

$y = g(x)$

dx

FIGURE 6.7

A rectangle such as the one shown in blue in Figure 6.6 is called a *typical element of area*. The sum of such typical elements is a Riemann sum, and the limit of this sum as max $\Delta x_k \to 0$ is the integral. A useful informal device for setting up the integral is shown in Figure 6.7. There we indicate a typical element of area on the figure but label its width as the differential dx, rather than Δx_k, and its height as $f(x) - g(x)$, rather than $f(c_k) - g(c_k)$. The area of the typical element is then $[f(x) - g(x)] \, dx$. To get the total area, all we need to do is affix the integral sign $\int_a^b$. Since the integral is a limit of Riemann sums, we can think of the integral sign as a sort of idealized summation symbol. So we are thinking of adding up (in the limiting sense) all typical elements of area. We can summarize the procedure as follows.

Shorthand Technique for Finding the Area Between Two Curves

To find the area between $y = f(x)$ and $y = g(x)$ from $x = a$ to $x = b$, where $f(x) \geq g(x)$:

Step 1. Draw a figure.

Step 2. Find a and b if they are not given.

Step 3. Show a typical element of area with its width labeled dx and its height $f(x) - g(x)$.

Step 4. Write the area of the typical element as $[f(x) - g(x)] dx$.

Step 5. Affix the integral sign $\int_a^b$ to the area in step 3 and perform the integration.

REMARK ────────────────────────────

■ The procedure we have outlined is a useful memory device for setting up the correct integral for the area. Let us emphasize, however, that the mathematically precise way of arriving at this result is by Riemann sums, as we previously indicated.

EXAMPLE 6.2   Find the area of the region bounded by the curves $y = 4 - x^2$ and $y = x^3 - x^2 - 3x$ between $x = -1$ and $x = 2$.

Solution   Let $f(x) = 4 - x^2$ and $g(x) = x^3 - x^2 - 3x$. We show the graphs of f and g in Figure 6.8, along with a typical element of area. We can either use Equation 6.1 directly or follow the five-step shorthand procedure just given. Let us follow the latter procedure. We have already completed the first three steps. The typical element of area is

$$[f(x) - g(x)]dx = [(4 - x^2) - (x^3 - x^2 - 3x)]dx$$
$$= (-x^3 + 3x + 4)dx$$

So the area is

$$\int_{-1}^{2} (-x^3 + 3x + 4)dx = -\frac{x^4}{4} + \frac{3x^2}{2} + 4x \Big]_{-1}^{2}$$

$$= \left(-\frac{16}{4} + \frac{12}{2} + 8\right) - \left(-\frac{1}{4} + \frac{3}{2} - 4\right)$$

$$= \frac{51}{4}$$

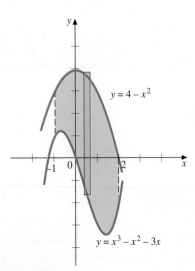

$y = 4 - x^2$

$y = x^3 - x^2 - 3x$

FIGURE 6.8

Sometimes the upper or lower boundaries of the region may consist of more than one curve. In such a case it may be necessary to divide the region into two or more subregions. The next example illustrates this situation.

EXAMPLE 6.3   Find the area of the region bounded by the curves $y = \sqrt{2x}$, $x + y = 4$, and $x = 3y$ that lies to the left of the line $x + y = 4$.

Solution   We find the points of intersection of the bounding curves to be $(0, 0)$, $(2, 2)$, and $(3, 1)$. As we see in Figure 6.9, the upper boundary of the region is $y = \sqrt{2x}$ if $0 \le x \le 2$, and it is $y = 4 - x$ if $2 \le x \le 3$. The lower boundary is $y = x/3$ ($x = 3y$, solved for y) for the entire region. We designate the two parts of the region by R_1 and R_2, as shown, and we find the areas of R_1 and R_2 and add the two results. This time, let us use Equation 6.1 directly.

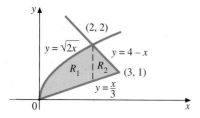

FIGURE 6.9

$$\text{area of } R_1 = \int_0^2 \left(\overbrace{\sqrt{2x}}^{\text{upper}} - \overbrace{\frac{x}{3}}^{\text{lower}} \right) dx = \int_0^2 \left(\sqrt{2}x^{1/2} - \frac{x}{3} \right) dx$$

$$= \sqrt{2}x^{3/2} \cdot \frac{2}{3} - \frac{x^2}{6} \Bigg]_0^2 = \frac{2\sqrt{2}}{3}(2^{3/2}) - \frac{4}{6} = \frac{8}{3} - \frac{2}{3} = 2$$

$$\text{area of } R_2 = \int_2^3 \left(\overbrace{4 - x}^{\text{upper}} - \overbrace{\frac{x}{3}}^{\text{lower}} \right) dx = \int_2^3 \left(4 - \frac{4x}{3} \right) dx$$

$$= 4x - \frac{2x^2}{3} \Bigg]_2^3 = (12 - 6) - \left(8 - \frac{8}{3} \right) = \frac{2}{3}$$

The total area is therefore $2 + \frac{2}{3} = \frac{8}{3}$. ∎

Horizontal Elements of Area

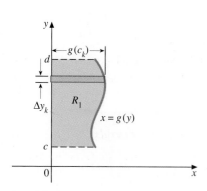

FIGURE 6.10

Sometimes when we want to find the area of a region, the roles of x and y are reversed; that is, the region whose area we want is between the y-axis and a curve whose equation is of the form $x = g(y)$ from $y = c$ to $y = d$, where $g(y) \ge 0$, as in Figure 6.10. Or it may lie between two curves, $x = g(y)$ and $x = h(y)$, with $g(y) \ge h(y)$, as in Figure 6.11. All the theory we have had so far on areas can be modified in obvious ways to get the areas

$$A = \int_c^d g(y)\,dy \tag{6.2}$$

in the first case, and

$$A = \int_c^d [\overbrace{g(y)}^{\substack{\text{rightmost}\\\text{curve}}} - \overbrace{h(y)}^{\substack{\text{leftmost}\\\text{curve}}}]\,dy \tag{6.3}$$

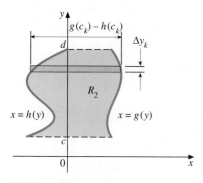

FIGURE 6.11

in the second.

The shorthand technique using typical elements of area, appropriately labeled, can be modified in obvious ways for the situations shown in Figures 6.10 and 6.11, interchanging the roles of x and y. When the typical rectangular element is vertical, then the integration is with respect to x (indicated by dx). When it is horizontal, the integration is with respect to y (indicated by dy).

EXAMPLE 6.4   Find the area of the region bounded by the curves $y = x^2$, $y = 6 - x$, and $y = 0$.

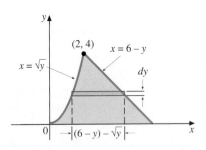

FIGURE 6.12

Solution   We show the region in Figure 6.12. We first find the point $(2, 4)$ of intersection of the curves $y = x^2$ and $y = 6 - x$ by solving the two equations simultaneously. One way of solving the problem is to use vertical elements (integration with respect to x) and divide the region into two parts, as we did in Example 6.3. Looking at the graph, however, reveals that if we use horizontal elements (integration with respect to y), the region does not have to be divided. Its right-hand boundary is always $y = 6 - x$ and its left-hand boundary is always $y = x^2$. These boundaries must be expressed as functions of y, namely, $x = 6 - y$ and $x = \sqrt{y}$, respectively. Using either Equation 6.3 with $g(y) = 6 - y$ and $h(y) = \sqrt{y}$ or the five-step shorthand procedure, we arrive at the result

$$A = \int_0^4 (6 - y - \sqrt{y})dy = 6y - \frac{y^2}{2} - \frac{2}{3}y^{3/2} \Big]_0^4$$

$$= 24 - 8 - \frac{16}{3} = \frac{32}{3}$$

Areas Between Curves That Cross Each Other

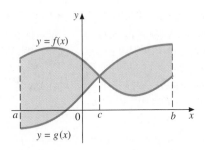

FIGURE 6.13

Finally, we note that if two curves cross each other at some point, then finding the area between them requires that we split the interval of integration at the crossing point, since the upper and lower (or rightmost and leftmost) functions are interchanged at this point. Only in this way is each area positive. For example, consider the region pictured in Figure 6.13. To get the area between f and g from $x = a$ to $x = b$, we use

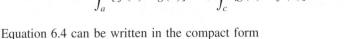

$$A = \int_a^c [\overset{\text{upper}}{f(x)} - \overset{\text{lower}}{g(x)}]dx + \int_c^b [\overset{\text{upper}}{g(x)} - \overset{\text{lower}}{f(x)}]dx \qquad (6.4)$$

Equation 6.4 can be written in the compact form

$$A = \int_a^b |f(x) - g(x)|dx \qquad (6.5)$$

since

$$|f(x) - g(x)| = \begin{cases} f(x) - g(x) & \text{if } a \le x \le c \\ g(x) - f(x) & \text{if } c \le x \le b \end{cases}$$

To apply Equation 6.5, we must split the integral at the crossing point, so in effect we are applying Equation 6.4.

EXAMPLE 6.5   Find the area between the curves $y = \cos x$ and $y = \sin x$ for $0 \le x \le \pi$.

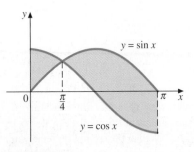

FIGURE 6.14

Solution   Let $f(x) = \cos x$ and $g(x) = \sin x$. We show their graphs, along with the region in question, in Figure 6.14. Since $\cos x \ge \sin x$ for $0 \le x \le \pi/4$ and $\sin x \ge \cos x$ for $\pi/4 \le x \le \pi$, we have

$$|f(x) - g(x)| = \begin{cases} \cos x - \sin x & \text{if } 0 \le x \le \pi/4 \\ \sin x - \cos x & \text{if } \pi/4 \le x \le \pi \end{cases}$$

So by Equation 6.5, since $\sin \frac{\pi}{4} = \cos \frac{\pi}{4} = \frac{\sqrt{2}}{2}$, $\sin 0 = \sin \pi = 0$, $\cos 0 = 1$, and $\cos \pi = -1$,

$$A = \int_0^\pi |\cos x - \sin x| dx = \int_0^{\pi/4} (\cos x - \sin x) dx + \int_{\pi/4}^\pi (\sin x - \cos x) dx$$

$$= [\sin x + \cos x]_0^{\pi/4} + [-\cos x - \sin x]_{\pi/4}^\pi$$

$$= \left(\frac{\sqrt{2}}{2} + \frac{\sqrt{2}}{2} - 0 - 1 \right) + \left(1 - 0 + \frac{\sqrt{2}}{2} + \frac{\sqrt{2}}{2} \right)$$

$$= 2\sqrt{2} \qquad \blacksquare$$

FINDING THE AREA BETWEEN CURVES USING COMPUTER ALGEBRA SYSTEMS

A computer algebra system can be used as an aid in exploring virtually all of the applications presented in this chapter. As an example, we show how a computer algebra system can be useful when finding the area bounded by several curves.

CAS 21

Estimate the area bounded by the curves $y = \sin x$ and $y = x^2$ on the interval $[0, 2]$.

A sketch of the two curves would be a good place to start.

Maple:

```
f:=x->sin(x);
g:=x->x^2;
plot({f(x),g(x)},x=-2..2);
```

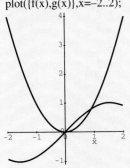

Mathematica:

```
f[x_] = Sin[x]
g[x_] = x^2
Plot[{f[x],g[x]},{x,-2,2}]
```

DERIVE:

(At the □ symbol, go to the next step.)
a (author) □ sin(x) □ p (plot window)
□ p (plot) □ a (algebra) □ a (author) □
x^2 □ p (plot window) □ p (plot)

FIGURE 6.1.1
$y = f(x) = \sin x$, $y = g(x) = x^2$

To estimate the area between the two curves, we see that the point of intersection near $x = 1$ must be found or estimated. Then the interval $[0, 2]$ can be split into two intervals, in which we integrate $f - g$ on the first part and $g - f$ on the second.

Maple:

P:=fsolve(f(x)=g(x),x);

Output: P := .8767262154

A:=int(f(x)–g(x),x=0..P)
   +int(g(x)–f(x),x=P..2);
evalf(A);

Output: 1.521914845

Mathematica:

P = FindRoot[f[x] == g[x],
      {x,1}]

Output: {x–> 0.876726}

P = 0.876726
A = Integrate[f[x]–g[x],
      {x,0,P}]+
   Integrate[g[x]–f[x],
      {x,P,2}]
N[%,10]

Output: 1.521914845

DERIVE:*

(At the □ symbol, go to the next step.)
a (author) □ x^2–sin(x) □ 1 (solve) □
s (simplify) □ [choose expression]
DERIVE does not seem to be able to
solve this. Let's try approximation.
x (approximate) □ [choose expression]
This also fails to work. Let's use Newton's Method as in CAS 18.
a (author) □ newton(x):=x–((x^2–
sin(x)))/dif(x^2–sin(x),x) □ iterates
(newton(x),x, 1,6) □ x (approximate)
□ [choose expression]

Output: 0.876726

a (author) □ P:=0.876726 □ a (author)
□ int(sin(x)–x^2,x,0,P)+int(x^2–sin(x),
x,P,2) □ x (approximate) □ [choose
expression]

Output: 1.52191

*Notice that in DERIVE an integral can be computed by selecting the "integrate" option from the calculus
menu or it can be authored using **int** as we did above.

Exercise Set 6.1

*In Exercises 1–10, find the area of the region between the x-
axis and the graph of f from $x = a$ to $x = b$, where $[a, b]$ is
the interval given. Sketch the graph in each case.*

1. $f(x) = \begin{cases} x^2 & \text{if } 0 \le x < 2 \\ 4 - x & \text{if } 2 \le x \le 4 \end{cases}$ on $[0, 4]$

2. $f(x) = \begin{cases} -x & \text{if } -1 \le x < 0 \\ 1 & \text{if } x = 0 \\ 2 - x & \text{if } 0 < x \le 1 \end{cases}$ on $[-1, 1]$

3. $f(x) = \begin{cases} 4 - x^2 & \text{if } 0 \le x < 2 \\ x - 2 & \text{if } 2 \le x \le 4 \end{cases}$ on $[0, 4]$

4. $f(x) = \begin{cases} x^2 & \text{if } 0 \le x < 1 \\ 2x - x^2 & \text{if } 1 \le x \le 2 \end{cases}$ on $[0, 2]$

5. $f(x) = \begin{cases} \sqrt{x + 4} & \text{if } -4 \le x < 0 \\ 1 & \text{if } x = 0 \\ \dfrac{x^2}{2} & \text{if } 0 < x \le 2 \end{cases}$ on $[-4, 2]$

6. $f(x) = \begin{cases} x^3 + 1 & \text{if } -1 \le x < 1 \\ x^2 - 4x + 5 & \text{if } 1 \le x \le 3 \end{cases}$ on $[-1, 3]$

7. $f(x) = \begin{cases} \cos x & \text{if } -\dfrac{\pi}{2} \le x < 0 \\ \sin x & \text{if } 0 \le x \le \pi \end{cases}$ on $\left[-\dfrac{\pi}{2}, \pi\right]$

8. $f(x) = \begin{cases} \cos \dfrac{x}{2} & \text{if } -\pi \le x < \dfrac{\pi}{2} \\ \sin \dfrac{x}{2} & \text{if } \dfrac{\pi}{2} \le x \le 2\pi \end{cases}$ on $[-\pi, 2\pi]$

9. $f(x) = \begin{cases} 2 \sin \dfrac{3x}{2} & \text{if } 0 \le x < \dfrac{2\pi}{3} \\ 4 + 2 \sin \dfrac{3x}{2} & \text{if } \dfrac{2\pi}{3} \le x \le \dfrac{4\pi}{3} \end{cases}$ on $\left[0, \dfrac{4\pi}{3}\right]$

10. $f(x) = \begin{cases} x^2 + 2x + 2 & \text{if } -1 \le x < 0 \\ 1 + \sin \dfrac{\pi x}{2} & \text{if } 0 \le x \le 2 \end{cases}$ on $[-1, 2]$

In Exercises 11–18, find the area of the shaded region.

11.

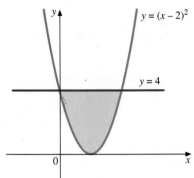

$y = (x - 2)^2$

$y = 4$

12.

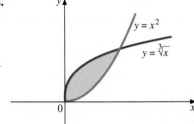

$y = x^2$

$y = \sqrt[3]{x}$

13.

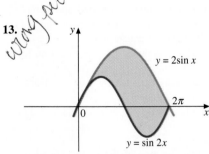

$y = 2\sin x$

$y = \sin 2x$

2π

14.

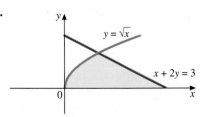

$y = \sqrt{x}$

$x + 2y = 3$

15.

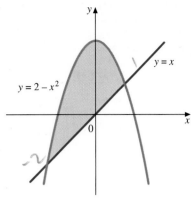

$y = x$

$y = 2 - x^2$

16.

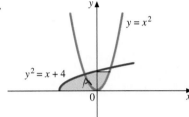

$y = x^2$

$y^2 = x + 4$

A

17.

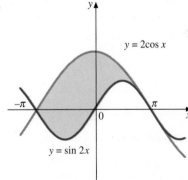

$y = 2\cos x$

$-\pi$ π

$y = \sin 2x$

18.

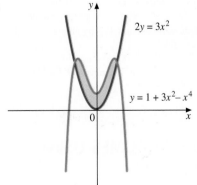

$2y = 3x^2$

$y = 1 + 3x^2 - x^4$

In Exercises 19–38, find the area bounded by the graphs of the given equations. Sketch the graphs.

19. $y = x^2 + 1$, $y = x$, $x = 0$, $x = 1$

20. $y = 4 - x^2$, $y = x^2 - 1$, $x = -1$, $x = 1$

21. $y = x^3 - 4x$, $y = 8 - x^2$, $x = -2$, $x = 2$

22. $x^2 + y = 0$, $x - y + 2 = 0$, $x = -1$, $x = 1$

23. $y = x^2$, $y = x + 2$

24. $x + y = 4$, $y = -\sqrt{x + 5}$, $x = -5$, $x = 4$

25. $x - y = 3$, $x^2 - 3x - y = 0$

26. $y = 8 - x^2$, $y = x^2$

27. $y = x^2$, $y^2 = x$

28. $y = x^2 - 3x + 4$, $y = -x^2 + 5x - 2$

29. $y = \sin \pi x$, $y = 4 + \cos 2\pi x$, $x = 0$, $x = 2$

30. $y = \sin \dfrac{\pi x}{2}$, $y = x^2 - 2x$

31. $y = \sin 2x$, $y = \cos x$, between $x = -\dfrac{\pi}{2}$ and $x = \dfrac{\pi}{6}$

32. $x = 4 - y^2$, $x = 0$

33. $x = y^2 - 2$, $x = y$

34. $y = \sqrt{x}$, $x - y = 2$, $y = 0$

35. $x = 2y + 1$, $x = y - 1$, $y = 2$

36. $x = y^3$, $y = x^2$ (Do this in two ways.)

37. $x = y^2 - 4y - 3$, $x = 1 - 2y^2$

38. $x = 2(y - 2)^2$, $x = y^2 - 4y + 8$

39. Find the area bounded by the graphs of the equations $y = 2x^2$, $x + y = 3$, and $y = 0$ (a) by using x as the independent variable and (b) by using y as the independent variable.

40. Follow the instructions for Exercise 39 for the area

bounded by $x = \sqrt{y + 1}$, $x = 2 + \sqrt{3 - y}$, and $y = -1$.

41. Use integration to find the area of the triangle with vertices $(-1, 1)$, $(3, 2)$, and $(5, 4)$.

 In Exercises 42–44, use a graphing calculator or CAS to obtain the graphs.

42. Find the total area bounded by the curves $y = x^3 - 3x^2 - 2x + 4$ and $y = 1 - x$.

43. Find the area between the graphs of $f(x) = 1 + 2\sin x$ and $g(x) = \cos 2x$ on $[0, 2\pi]$.

44. Find the area between the graphs of $x = \sin \dfrac{\pi y}{2}$ and $y = x$.

45. Find the area of the region bounded on the left by the curves $x^2 - 2x - 4y + 9 = 0$ and $x + 2y = 5$ and on the right by $y^2 - 4y + x = 0$.

46. Find the limit of the ratio of the area of triangle AOB to the area of the parabolic region (in darker red) as $\alpha \to 0$.

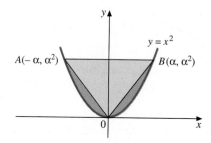

47. Suppose the area of the region bounded above by $y = f(x)$ and below by the x-axis, between the lines $x = 0$ and $x = 1$, equals A square units.
   (a) Show that the area bounded by $y = f(2x)$ on $[0, \frac{1}{2}]$ is $A/2$.
   (b) When are the areas bounded by $y = f(x)$ and $y = f(2x)$ on $[0, 1]$ equal?
   (c) Find the area bounded by $y = f(x)$ and $y = 2f(x)$ on $[0, 1]$.

48. Use a CAS to find the area of the region bounded by the two curves $f(x) = x^4 - 3x^2 + 2$ and $g(x) = x^3 + 7x^2 + x - 10$.

6.2 FINDING VOLUMES USING CROSS SECTIONS

We have seen how the integral is used in defining areas. In this section we will see that integrals also can be used to define volumes of certain solids. Consider the solid depicted in Figure 6.15. It is bounded on the left by a plane perpendicular to the x-axis at $x = a$ and bounded on the right by a plane perpendicular to the x-axis at $x = b$. If we pass a plane perpendicular to the x-axis at any point x between a and b, its intersection with the solid is a plane region called a **cross section** with respect to the x-axis. We denote the area of such a cross section by $A(x)$. *Throughout the remainder of this section we will consider only solids for which $A(x)$ exists and is a continuous function of x on the closed interval $[a, b]$.*

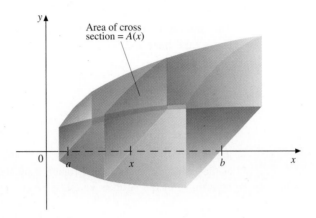

Area of cross section $= A(x)$

FIGURE 6.15

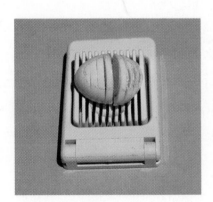

To find the volume of the solid, we partition the closed interval $[a, b]$ in the usual way. If we pass planes perpendicular to the x-axis at each of the partition points, we divide the solid into n parallel slices (like a sliced hard-boiled egg). In Figure 6.16 we have pictured the solid sliced in this way, highlighting a typical slice with faces that are the cross sections taken at x_{k-1} and x_k. The entire volume of the solid is the sum of the volumes of all the slices. In general, the cross section at x_{k-1} will be different from the one at x_k, but for Δx_k small, if we replace the kth slice with one that has constant cross-sectional area $A(c_k)$, as shown on the right in Figure 6.16, the volume will not differ much from that of the actual slice. We call this slice, which has width Δx_k and cross-sectional

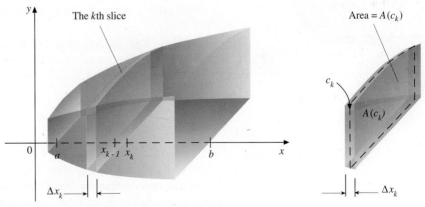

The kth slice

Area $= A(c_k)$

c_k

$A(c_k)$

FIGURE 6.16

area $A(c_k)$, a *typical element of volume* and denote its volume by ΔV_k. Its volume is defined as the *area of the cross section times the thickness*:

$$\Delta V_k = A(c_k)\Delta x_k$$

and the total volume is approximated by the sum of all the ΔV_k values:

$$V \approx \sum_{k=1}^{n} \Delta V_k = \sum_{k=1}^{n} A(c_k)\Delta x_k$$

The sum on the right is a Riemann sum that approaches the integral $\int_a^b A(x)\,dx$ as we take finer and finer partitions, which leads to the following definition.

Definition 6.1 Volumes Using Cross Sections	If a solid is bounded on the left by a plane perpendicular to the x-axis at $x = a$ and on the right by a plane perpendicular to the x-axis at $x = b$, and if the area $A(x)$ of the cross section with respect to the x-axis of the solid at the point x is continuous for all x in the closed interval $[a, b]$, then the **volume** V of the solid is given by $$V = \int_a^b A(x)\,dx \qquad (6.6)$$

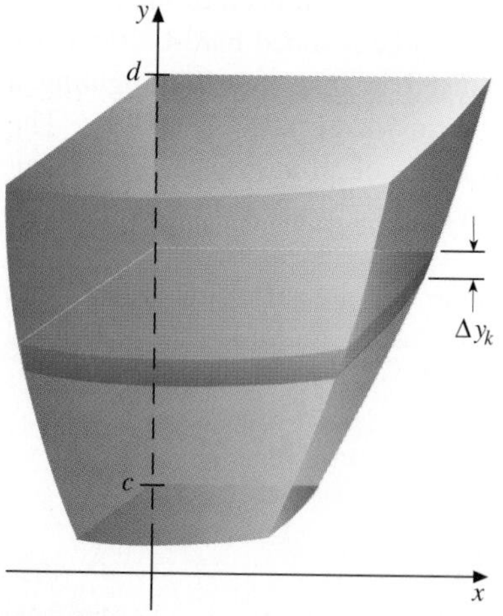

FIGURE 6.17

If the solid is bounded by planes perpendicular to the y-axis at $y = c$ and $y = d$, and if the cross-sectional area with respect to the y-axis is a continuous function $A(y)$ on $[c, d]$, then similar reasoning leads to the equation

$$V = \int_c^d A(y)\,dy \qquad (6.7)$$

(See Figure 6.17.)

EXAMPLE 6.6   The base of a solid is the region in the xy-plane bounded by the graphs of $y = \sqrt{x}$, $y = 0$, and $x = 4$, and each cross section of the solid perpendicular to the x-axis is a square. Find the volume of the solid.

Solution   The solid is depicted in Figure 6.18. The cross section at an arbitrary point x between 0 and 4 is a square whose base has length $y = \sqrt{x}$. Thus, the area $A(x) = y^2 = (\sqrt{x})^2 = x$. So by Equation 6.6, the volume is

$$V = \int_0^4 A(x)\,dx = \int_0^4 x\,dx = \left.\frac{x^2}{2}\right]_0^4 = 8 \qquad \blacksquare$$

As an alternative to Equation 6.6, we can use a shorthand procedure similar to the one outlined in Section 6.1 for finding areas. As before, we must view the procedure as an aid to memory only and not a mathematically precise formulation. The final result, however, is correct.

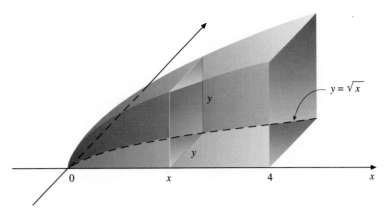

$y = \sqrt{x}$

FIGURE 6.18

Shorthand Technique for Finding Volumes Using Cross Sections

To find the volume of a solid with known cross-sectional area $A(x)$, between $x = a$ and $x = b$:

Step 1. Draw a figure.

Step 2. Find the integration limits a and b if they are not given.

Step 3. Show a typical element of volume and label its thickness dx and the area of its face $A(x)$.

Step 4. Write the volume of the typical element as $A(x)dx$.

Step 5. Affix the integral sign $\int_a^b$ to the volume in step 3 and perform the integration.

Obvious modifications in the procedure are needed when the cross-sectional areas are known with respect to the y-axis instead of the x-axis. We will illustrate the procedure after introducing a class of solids with simple cross sections.

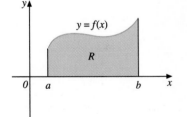

$y = f(x)$

R

(a)

Volumes of Revolution: Methods of "Disks" and "Washers"

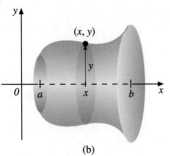

(x, y)

(b)

FIGURE 6.19

The "Disk" Method

When a plane area is revolved about a line, the resulting solid is known as a **solid of revolution**. Cross sections with respect to the line are circular. For example, suppose the region R under the graph of a nonnegative continuous function f in Figure 6.19(a) is rotated about the x-axis. The result is the solid of revolution in part (b) of the figure. At any point x between a and b, the height of the curve is $y = f(x)$. Since the area of a circle of radius r is πr^2, the area of the circular cross section after rotation is

$$A(x) = \pi y^2 = \pi[f(x)]^2$$

The volume, by Equation 6.6, is therefore given by the following equation.

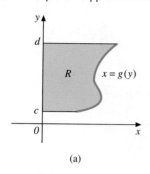

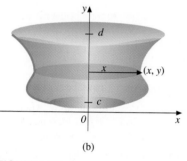

(a)

(b)

FIGURE 6.20

Volumes by the Disk Method: Rotation About the x-axis

$$V = \pi \int_a^b [f(x)]^2 dx \qquad (6.8)$$

For revolving about the y-axis instead of the x-axis, the analogous formula for a region bounded by $x = g(y)$, for $g(y) \geq 0$, between $y = c$ and $y = d$, is as follows.

Volumes by the Disk Method: Rotation About the y-axis

$$V = \pi \int_c^d [g(y)]^2 dy \qquad (6.9)$$

(See Figure 6.20.)

For a solid of revolution such as that pictured in Figure 6.19, the slicing technique produces elements of volume in the form of circular disks (like coins). We have shown a typical element in Figure 6.21. Because of the shape of such a typical element of volume, we often refer to the method described as the **disk method**.

FIGURE 6.21
Typical
element of volume
is a disk.

EXAMPLE 6.7   Find the volume of a cone of altitude h and base radius r.

Solution   For convenience, we orient the cone so that its vertex is at the origin and its axis is along the positive x-axis. The cone can be thought of as the solid of revolution formed by revolving the triangular region OAB in Figure 6.22(a) about the x-axis. The resulting cone is shown in Figure 6.22(b).

To apply Formula 6.8, we need to know the function $f(x)$ whose graph is the upper boundary of the region being rotated. In this case the graph is the

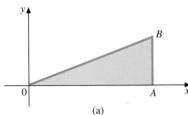

(a)

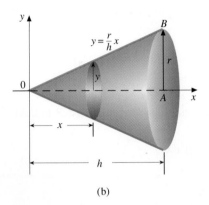

(b)

FIGURE 6.22

straight line through the origin, with slope $m = r/h$. Its equation is

$$y = \frac{r}{h}x$$

Thus, $f(x) = rx/h$, and from Formula 6.8 we obtain the volume as follows:

$$V = \pi \int_0^h [f(x)]^2 \, dx = \pi \int_0^h \frac{r^2 x^2}{h^2} \, dx = \frac{\pi r^2}{h^2} \int_0^h x^2 \, dx$$

$$= \frac{\pi r^2}{h^2} \left[\frac{x^3}{3}\right]_0^h = \frac{\pi r^2}{h^2} \left(\frac{h^3}{3}\right) = \frac{1}{3}\pi r^2 h$$

Since πr^2 is the area of the base of the cone, we can restate our result in the following words: *The volume of a right circular cone is one-third the area of the base times the altitude.* ■

EXAMPLE 6.8   Find the volume of the solid obtained by revolving the region under the graph of $y = \sqrt{x}$ from $x = 1$ to $x = 4$ about the x-axis.

Solution   This time we will illustrate the five-step shorthand technique. In Figure 6.23 we have shown in part (a) the region to be revolved, in part (b) a typical element of volume after rotation, and in part (c) the rotated solid. The typical volume element is a circular disk with radius $\sqrt{x}$, and we have labeled its thickness dx. Its volume is therefore $\pi(\sqrt{x})^2 \, dx = \pi x \, dx$. Thus, the volume of the solid is

$$V = \int_1^4 \pi x \, dx = \pi \left[\frac{x^2}{2}\right]_1^4 = \pi \left[\frac{16}{2} - \frac{1}{2}\right] = \frac{15\pi}{2} \qquad ■$$

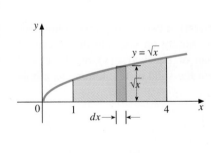

(a)

(b)

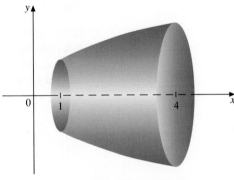

The solid of rotation
(c)

FIGURE 6.23

REMARK
■ Although up to now we have considered f (or g) to be nonnegative, this requirement is unnecessary, as the following reasoning shows. Suppose f is continuous on the closed interval $[a, b]$. Then the absolute value of f, $|f|$, is also continuous there (see Exercise 20 in Exercise Set 2.4) and is nonnegative. Figure 6.24 illustrates the graphs of f and $|f|$ for a typical function. When we rotate each of the shaded regions about the x-axis, we obtain identical solids,

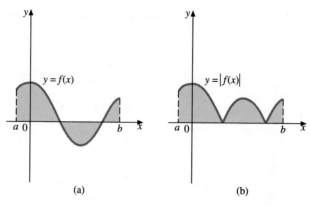

(a)   (b)

FIGURE 6.24

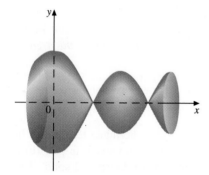

FIGURE 6.25
The same solid of revolution re-
sults from rotating either region in
Figure 6.24.

shown in Figure 6.25. But $|f|$ satisfies the conditions for Equation 6.8 to hold.
Thus,

$$V = \pi \int_a^b |f(x)|^2 dx$$

and since $|f|^2 = f^2$, we get

$$V = \pi \int_a^b [f(x)]^2 dx$$

That is, Equation 6.8 (and also 6.9) holds true even when we drop the requirement that f (or g) be nonnegative.

The "Washer" Method

If the region to be rotated lies between the graphs of two nonnegative continuous
functions $y_1 = g(x)$ and $y_2 = f(x)$ and is entirely on one side of the axis of
rotation, where $f(x)$ is always greater than or equal to $g(x)$ on the closed
interval $[a, b]$, as shown in Figure 6.26, then the solid of revolution has a hole

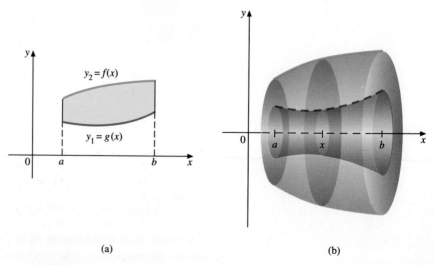

(a)   (b)

FIGURE 6.26

parallel to the axis in it. A cross section is the region between two concentric circles of radii y_1 and y_2, respectively. Thus, its area is

$$A(x) = \pi y_2^2 - \pi y_1^2 = \pi(y_2^2 - y_1^2) = \pi\{[f(x)]^2 - [g(x)]^2\}$$

The volume, by Equation 6.8, is therefore given by the following equation.

Volumes by the Washer Method: Rotation About the x-axis

$$V = \pi \int_a^b \{[f(x)]^2 - [g(x)]^2\}dx \qquad (6.10)$$

The analogous formula for a region bounded by the graphs of $x_1 = h(y)$ and $x_2 = g(y)$, where $g(y) \geq h(y) \geq 0$, between $y = c$ and $y = d$, revolved about the y-axis, is as follows.

Volumes by the Washer Method: Rotation About the y-axis

$$V = \pi \int_c^d \{[g(y)]^2 - [h(y)]^2\}dy \qquad (6.11)$$

The use of the dependent variable notations $y_1 = g(x)$ and $y_2 = f(x)$ for Equation 6.10 and $x_1 = h(y)$ and $x_2 = g(y)$ for Equation 6.11 simplifies the appearance of these formulas to

$$V = \pi \int_a^b (y_2^2 - y_1^2)dx \quad \text{about the x-axis}$$

and

$$V = \pi \int_c^d (x_2^2 - x_1^2)dy \quad \text{about the y-axis}$$

but you must remember that before integration can take place, the integrand must be expressed as a function of the variable of integration in each case.

Note that we can think of the volume in question as the volume within the outer surface minus the volume of the hole, just as though we drilled out a hole from the full volume.

Do not confuse $y_2^2 - y_1^2$ with $(y_2 - y_1)^2$. These are *not* the same, since $(y_2 - y_1)^2 = y_2^2 - 2y_2 y_1 + y_1^2$.

A typical element of volume in the slicing technique for solids of the type described is in the shape of a washer such as that used by carpenters or machinists, as shown in Figure 6.27 (for rotation about the x-axis). For this reason, the method described here is often referred to as the **washer method.** You can think of a washer as a disk with a hole drilled through its center.

EXAMPLE 6.9   Find the volume of the solid obtained by revolving the region between the graphs of $y = x^2 + 1$ and $y = \sqrt{x}$ from $x = 0$ to $x = 1$ about the x-axis.

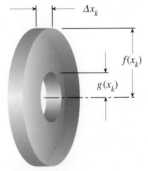

FIGURE 6.27

Typical element of volume is washer-shaped.

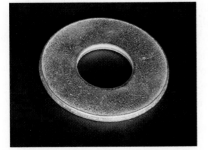

Solution   Let $f(x) = x^2 + 1$ and $g(x) = \sqrt{x}$. The region to be revolved is shown in Figure 6.28(a). By Equation 6.10, the volume is

$$V = \pi \int_0^1 [(x^2 + 1)^2 - (\sqrt{x})^2] dx$$

$$= \pi \int_0^1 (x^4 + 2x^2 + 1 - x) dx = \pi \left[\frac{x^5}{5} + \frac{2x^3}{3} + x - \frac{x^2}{2} \right]_0^1$$

$$= \pi \left[\frac{1}{5} + \frac{2}{3} + 1 - \frac{1}{2} \right] = \frac{41\pi}{30}$$

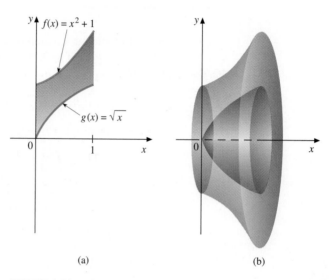

(a)   (b)

FIGURE 6.28

As an aid in remembering Equations 6.10 and 6.11, observe that the area of a cross section for the washer method is

$$\pi[(\text{outer radius})^2 - (\text{inner radius})^2]$$

and this is the form of the integrand in each case. (See Figure 6.29.) As the next example shows, this formula for the cross-sectional area enables us to rotate areas about lines other than the x- and y-axes.

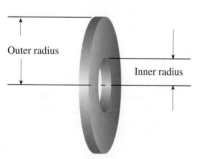

FIGURE 6.29
Area of cross section = $\pi [(\text{outer radius})^2 - (\text{inner radius})^2]$

EXAMPLE 6.10   The region bounded by the graphs of $x = 2\sqrt{y}$ and $x = y$ is to be revolved about the line $x = 4$. Find the volume of the resulting solid of revolution.

Solution   Figure 6.30 shows the region in question, together with a typical element of volume of the revolved solid. Since the axis of revolution is vertical, we will use y as the variable of integration. By solving the two equations simultaneously, we find their points of intersection to be $(0, 0)$ and $(4, 4)$. (Verify.) Thus, the limits on y are 0 and 4. As shown in part (a) of the figure, we let x_1 and x_2 denote the distances from the y-axis to the leftmost and rightmost boundaries, respectively. Thus, for each y in $[0, 4]$,

$$x_1 = y \quad \text{and} \quad x_2 = 2\sqrt{y}$$

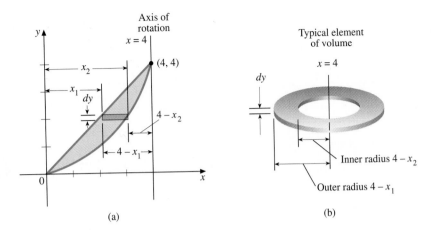

FIGURE 6.30

So the corresponding distances from the line $x = 4$ (the axis of rotation) are $4 - x_1$ and $4 - x_2$. As shown in Figure 6.30(b), these are the outer and inner radii, respectively, of the cross section of the solid of revolution taken at y:

$$\text{outer radius} = 4 - x_1 = 4 - y$$

$$\text{inner radius} = 4 - x_2 = 4 - 2\sqrt{y}$$

Thus, the area $A(y)$ of the cross section, by Equation 6.11, is

$$A(y) = \pi[(4 - y)^2 - (4 - 2\sqrt{y})^2]$$

and the volume of the solid is

$$V = \pi \int_0^4 A(y)dy = \pi \int_0^4 [(4 - y)^2 - (4 - 2\sqrt{y})^2]dy$$

$$= \pi \int_0^4 [16 - 8y + y^2 - 16 + 16\sqrt{y} - 4y]dy$$

$$= \pi \int_0^4 (y^2 - 12y + 16y^{1/2})dy = \pi \left[\frac{y^3}{3} - 6y^2 + \frac{32}{3}y^{3/2}\right]_0^4$$

$$= \pi \left[\frac{64}{3} - 96 + \frac{32}{3}(8)\right] = \frac{32\pi}{3} \qquad \blacksquare$$

EXAMPLE 6.11   A **torus**—a solid shaped like a doughnut—is generated by revolving a circle about an axis that does not intersect the circle. Figure 6.31(a) (on the following page) shows a region enclosed by a circle of radius r centered at $(0, R)$, with $R > r$. A torus is generated by revolving the region about the x-axis (Figure 6.31(b)). Use the washer method to find the volume of the torus. From your result, determine how many doughnuts can be made from a cube of dough 10 cm on a side, if the radius r of a cross section of the doughnut is 1 cm and the radius R is 4 cm.

Solution   The area of the circular shaded region in Figure 6.31(a) is πr^2, and since the center of the region is rotated a distance of $2\pi R$ (the circumference of a circle of radius R), intuitively we may guess that the volume of the torus

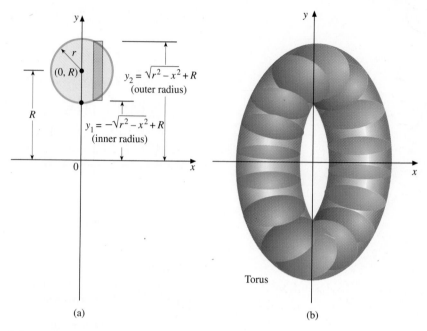

FIGURE 6.31

is

$$(\pi r^2)(2\pi R) = 2\pi^2 r^2 R$$

In Section 6.8 we will give a justification for this intuitive approach in the theorem called the First Theorem of Pappus. For now, we can verify the result using the washer method.

The circle in Figure 6.31 is simply the circle $x^2 + y^2 = r^2$ shifted vertically upward R units. The upper semicircle therefore has the equation

$$y_2 = \sqrt{r^2 - x^2} + R \qquad \text{Outer radius}$$

and the lower semicircle has the equation

$$y_1 = -\sqrt{r^2 - x^2} + R \qquad \text{Inner radius}$$

By Equation 6.10

$$V = \pi \int_{-r}^{r} (y_2^2 - y_1^2)\, dx = \pi \int_{-r}^{r} \left[\left(\sqrt{r^2 - x^2} + R \right)^2 - \left(-\sqrt{r^2 - x^2} + R \right)^2 \right] dx$$

$$= \pi \int_{-r}^{r} \left\{ \left[(r^2 - x^2) + 2R\sqrt{r^2 - x^2} + R^2 \right] \right.$$

$$\left. - \left[(r^2 - x^2) - 2R\sqrt{r^2 - x^2} + R^2 \right] \right\} dx$$

$$= \pi \int_{-r}^{r} 4R\sqrt{r^2 - x^2}\, dx$$

$$= 4\pi R \int_{-r}^{r} \sqrt{r^2 - x^2}\, dx$$

We have noted previously (see Example 5.8) that $\int_{-r}^{r} \sqrt{r^2 - x^2}\, dx$ is one-half the area of the circle $x^2 + y^2 = r^2$. Thus,

$$V = 4\pi R \left(\frac{\pi r^2}{2}\right) = 2\pi^2 r^2 R$$

With $r = 1$ cm and $R = 4$ cm, the volume of a doughnut is

$$V = 2\pi^2 \cdot 1^2 \cdot 4 = 8\pi^2 \approx 79 \text{ cm}^3$$

A cube of dough 10 cm on a side has a volume of $10^3 = 1,000$ cm^3. Since $1,000 \div 79 \approx 12.66$, we could make a dozen doughnuts—almost a baker's dozen—and have a little more than 50 cm^3 of dough left over. ■

Exercise Set 6.2

In Exercises 1–10, find the volume of the solid obtained by revolving the region under the given curve between the specified limits about the x-axis.

1. $y = (x + 3)/2$ from $x = 0$ to $x = 1$

2. $y = x^2 + 1$ from $x = 0$ to $x = 2$

3. $y = 4 - x^2$ from $x = 0$ to $x = 2$

4. $y = 2x - x^2$ from $x = 0$ to $x = 2$

5. $y = \sqrt{x + 1}$ from $x = -1$ to $x = 3$

6. $y = \sqrt{2 - x}$ from $x = -2$ to $x = 2$

7. $y = x^3 + 1$ from $x = -1$ to $x = 1$

8. $y = \sqrt[3]{x}$ from $x = 1$ to $x = 8$

9. $y = \sqrt{25 - x^2}$ from $x = -4$ to $x = 3$

10. $y = 4x^2 - x^4$ from $x = -1$ to $x = 1$

In Exercises 11–20, find the volume of the solid obtained by rotating the region bounded by the given curves about the x-axis.

11. The x-axis and $y = x^2 - 1$ from $x = -2$ to $x = 2$

12. The x-axis and $y = \sqrt[3]{3x + 1}$ from $x = -3$ to $x = 0$

13. $y = x + 4$ and $y = 3 - \dfrac{x}{2}$ from $x = 1$ to $x = 4$

14. $y = x$ and $y = \sqrt{x}$

15. $y = x + 2$ and $y = x^2$

16. $y = \sqrt{2x - 3}$ and $y = \dfrac{x}{2}$

17. $y = x^2$ and $y = 2 - x^2$

18. $y = \sqrt{x}$ and $y = \sqrt{3x + 4}$ from $x = 0$ to $x = 4$

19. $y = 1$ and $y = \sqrt{5 - x^2}$

20. $y = 2/\sqrt[3]{x}$ and $y = \sqrt[3]{9 - x}$

In Exercises 21–30, find the volume of the solid obtained by rotating the region bounded by the given curves about the y-axis.

21. The y-axis and $x + 2y = 4$ from $y = 0$ to $y = 2$

22. The y-axis and $x = \sqrt{y}$ from $y = 4$ to $y = 9$

23. The y-axis and $x = y^{1/3}$ from $y = -1$ to $y = 8$

24. The y-axis and $x = y^2 - y - 2$

25. $y = 2x$, $x = 2y$, and $y = 2$

26. $y = 2x$, $y = \sqrt{3 - 4x}$, and the x-axis

27. $y = x$, $x = 6 - y^2$, and $x = 0$, for $x > 0$

28. $x = \sqrt{16 - y^2}$, $x = 10 - 2y$, $y = 0$, and $y = 4$

29. $y^3 + x = 0$, $y + 2 = 0$, and $x - 1 = 0$

30. $y^2 = x$ and $y^2 = 2x - 4$

31. Derive the formula for the volume of a sphere created by revolving the region under the semicircle $y = \sqrt{a^2 - x^2}$ about the x-axis.

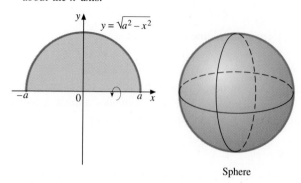

Sphere

32. If the region under the semiellipse $y = \frac{b}{a}\sqrt{a^2 - x^2}$ is revolved about the x-axis, the resulting solid is called an *ellipsoid of revolution*. Find its volume.

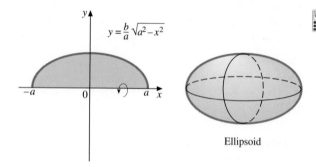

Ellipsoid

33. The region bounded by $y = x$, $x = 2$, and $y = 0$ is revolved about the line $y = 3$. Find the volume of the resulting solid.

34. Find the volume of the solid obtained by revolving the region under $y = \sqrt{x}$ from $x = 1$ to $x = 4$ about the line $y + 1 = 0$.

35. Find the volume of the solid obtained by revolving the region bounded by $x = \sqrt{y}$, $y = 8 - 2x$, and the x-axis about (a) the y-axis and (b) the line $x = 4$.

36. Find the volume of the solid obtained by revolving the region in the first quadrant bounded by $x^3 = 2y$ and $y = 2x$ about (a) the line $x = 2$ and (b) the line $y = 4$.

In Exercises 37–48, refer to the regions R_1, R_2, and R_3 in the figure. Find the volume of the solid of revolution formed by revolving the specified region about the line indicated.

37. R_1 about OA

38. R_1 about OC

39. R_1 about CB

40. R_1 about AB

41. R_2 about OA

42. R_2 about OC

43. R_2 about CB

44. R_2 about AB

45. R_3 about OA

46. R_3 about OC

47. R_3 about CB

48. R_3 about AB

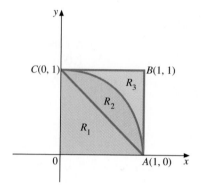

In Exercises 49–52, find the volume of the solid of revolution obtained by revolving the region between the graph of f and the x-axis from $x = a$ to $x = b$ about the x-axis. Use a graphing calculator or a CAS to sketch the regions.

49. $f(x) = \sin x + \cos x$; $a = -\dfrac{\pi}{3}$, $b = \dfrac{\pi}{6}$

50. $f(x) = \sec x$; $a = 0$, $b = \dfrac{\pi}{4}$

51. $f(x) = 2 \csc x$; $a = \dfrac{\pi}{4}$, $b = \dfrac{3\pi}{4}$

52. $f(x) = \sqrt{2(1 - \sin x)}$; $a = -\dfrac{\pi}{2}$, $b = \dfrac{\pi}{6}$

In Exercises 53–56, find the volume using Definition 6.1.

53. The base of a solid is the triangular region bounded by $y = \frac{x}{2}$, $x = 4$, and $y = 0$, and each cross section perpendicular to the x-axis is a square. Find the volume.

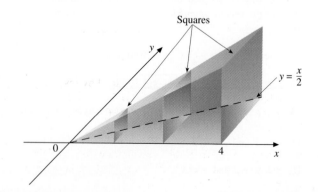

54. Find the volume of the pyramid shown in the figure, whose base is an isosceles right triangle with each leg 3 and altitude 5.

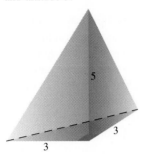

55. The base of a solid is the region enclosed by the parabola $y^2 = 2x$ and the line $x = 2$, and each cross section perpendicular to the x-axis is semicircular. Find the volume.

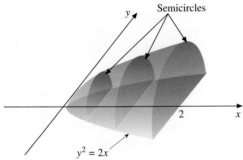

56. The base of a solid is the region enclosed by the graph of $y = 4 - x^2$ and the x-axis, and each cross section perpendicular to the y-axis is an isosceles triangle whose altitude is one-half its base. Find its volume.

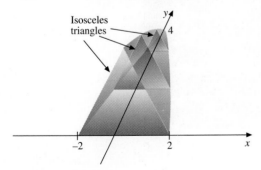

57. Find the volume of the solid obtained by revolving the leftmost region bounded by $y = 3x$, $y = x^2/2$, and $x + y = 4$ about the x-axis. (*Hint:* Divide the region into two parts.)

58. Find the volume of the solid obtained by revolving the region bounded by the x-axis and the curve $y = x^3 - x^2 - 2x$ about the x-axis.

59. An oil tank is in the shape of a sphere with radius 2 m,

and the oil in it is 1 m deep (see the figure). Find how many liters of oil are in the tank ($1 \text{ m}^3 = 1000$ L).

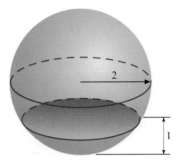

60. In a sphere of radius a, at a point b units above the center on a vertical axis ($b < a$) a horizontal plane is passed. The portion of the sphere above the plane is called a *spherical segment*. Find its volume.

61. The base of a solid is the region inside the ellipse $4x^2 + 9y^2 = 36$, and each cross section perpendicular to the x-axis is an equilateral triangle. Find the volume of the solid.

62. The upper half of the region inside the *four-cusp hypocycloid* $x^{2/3} + y^{2/3} = a^{2/3}$ (see the figure) is revolved around the x-axis. Find the volume generated.

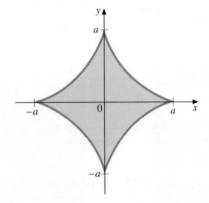

63. The region between the graphs of $y = \sqrt{x}$ and $y = 4 - \sqrt{x}$ from $x = 0$ to $x = 9$ is rotated about the line $y = -2$. Find the volume.

64. The region between the graphs of $y = \sin x$ and $y = \cos x$ from $x = \frac{\pi}{4}$ to $x = \pi$ is rotated about the line $y = 1$. Find the volume. (*Hint:* To perform the integration, make use of a trigonometric identity.)

65. The region bounded above by the graph of $f(x) = \sqrt[3]{x^3 + 2}$, below by the x-axis, and on each side by the vertical lines $x = 0$ and $x = 2$ is rotated about the x-axis. Approximate the volume of the solid of revolution using
(a) a right-hand Riemann sum with $n = 100$;
(b) Simpson's Rule with $n = 16$.

6.3 FINDING VOLUMES OF SOLIDS OF REVOLUTION USING CYLINDRICAL SHELLS

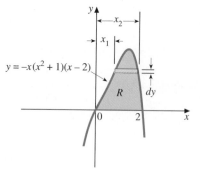

$y = -x(x^2 + 1)(x - 2)$

FIGURE 6.32

Using the method of disks or of washers to find the volume of a solid of revolution is sometimes difficult, or even impossible, because the resulting integral cannot always be evaluated easily. In other cases, the process is tedious because the shape of the region requires that it be divided into two or more parts in order to apply the method. To illustrate one such difficulty, consider the problem of rotating the region shown in Figure 6.32 about the y-axis. To use the washer method we would have to find the inner radius x_1 and the outer radius x_2 as functions of y. Thus, we would have to solve the equation

$$y = -x(x^2 + 1)(x - 2)$$

for x. (Try it!) We would also have to think of the graph in two parts—the part to the left of the maximum point and the part to the right. For each part x would have to be expressed as a function of y. Clearly, using the washer method for this problem is extremely difficult.

These difficulties can sometimes be eliminated by an alternative procedure for finding volumes of solids of revolution that we present in this section. The method involves the calculation of volumes of thin-walled hollow cylinders, called **cylindrical shells**. Pieces of pipe are such shells. More precisely, a cylindrical shell is the solid between two concentric right circular cylinders of fixed height. We show such a cylindrical shell in Figure 6.33. Our approach to finding the volume of a solid of revolution will be to use a collection of concentric cylindrical shells as an approximation to the volume. We will see that the sum of the volumes of the shells is a Riemann sum. Its limit is an integral that gives the total volume we are seeking.

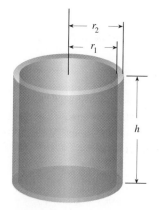

FIGURE 6.33
A cylindrical shell

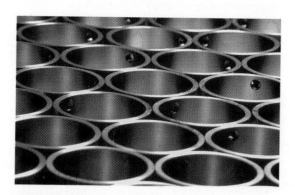

For a given cylindrical shell, we denote the inner radius by r_1, the outer radius by r_2, and the height by h. Then the volume V of the shell is the volume of the outer cylinder minus that of the inner cylinder:

$$V = \pi r_2^2 h - \pi r_1^2 h = \pi (r_2^2 - r_1^2) h$$

To write the sum of volumes of such cylindrical shells as a Riemann sum, we will find it convenient to write this formula in the following equivalent way:

$$V = \pi (r_2 + r_1)(r_2 - r_1)h = 2\pi \left(\frac{r_2 + r_1}{2} \right) (r_2 - r_1)h$$

Since $\frac{1}{2}(r_1 + r_2)$ is the average radius, multiplying it by 2π gives the average circumference. Also, $r_2 - r_1$ is the thickness of the shell. Thus, we can state the formula for its volume in words:

$$\text{volume of a shell} = (\text{average circumference}) \times (\text{height}) \times (\text{thickness})\quad (6.12)$$

Now let us consider the region R between the x-axis and the graph of a continuous nonnegative function f between $x = a$ and $x = b$. Our objective is to rotate R about the y-axis and to obtain its volume as a limit of Riemann sums of volumes of shells. Let $P = \{x_0, x_1, x_2, \ldots, x_n\}$ be a partition of the closed interval $[a, b]$. Let c_k be the midpoint of the kth subinterval:

$$c_k = \frac{x_{k-1} + x_k}{2}\qquad k = 1, 2, \ldots, n$$

which matches the form we saw earlier for average radius.

Next, we construct a rectangle of height $f(c_k)$, with base equal to the subinterval $[x_{k-1}, x_k]$, for $k = 1, 2, 3, \ldots, n$. In Figure 6.34 we have shown a typical rectangle. If we rotate the kth rectangle about the y-axis, we get a cylindrical shell of average radius c_k, height $f(c_k)$, and thickness Δx_k, as shown in Figure 6.35. If this is done for $k = 1, 2, 3, \ldots, n$, we get n such concentric shells. Taken altogether, they approximate the solid of revolution that results from rotating the region R about the y-axis. By Equation 6.12 the volume of the kth cylindrical shell is

$$\Delta V_k = \underbrace{2\pi c_k}_{\substack{\text{average}\\ \text{circumference}}} \cdot \underbrace{f(c_k)}_{\text{height}} \cdot \underbrace{\Delta x_k}_{\text{thickness}}$$

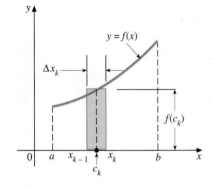

FIGURE 6.34

Thus, the volume V of the solid of revolution is approximated by

$$V \approx \sum_{k=1}^{n} \Delta V_k = \sum_{k=1}^{n} 2\pi c_k f(c_k) \Delta x_k$$

The sum on the right is the Riemann sum for the continuous function $2\pi x f(x)$ on $[a, b]$. We can see now the reason for choosing c_k as we did and for rewriting the original volume formula $V = \pi(r_2^2 - r_1^2)h$ in the form of Equation 6.12. By

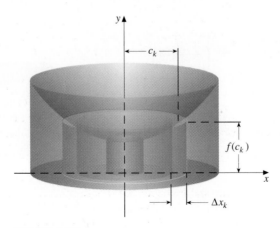

FIGURE 6.35

doing so we obtained a Riemann sum. By letting max $\Delta x_k \to 0$, we obtain the following result:

Volume by Method of Cylindrical Shells

The volume of the solid obtained by revolving about the y-axis the area of the region under the graph of the continuous nonnegative function f from $x = a$ to $x = b$ is

$$V = 2\pi \int_a^b x f(x)\,dx \qquad (6.13)$$

Shorthand Technique for Finding Volumes Using Cylindrical Shells

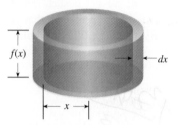

FIGURE 6.36
Volume of shell $= 2\pi x f(x)\,dx$

As an aid to setting up the integral in Equation 6.13, we can label a typical element of volume (a cylindrical shell) as in Figure 6.36. Its volume is, by Equation 6.12,

$$\underbrace{2\pi x}_{\substack{\text{average} \\ \text{circumference}}} \cdot \underbrace{f(x)}_{\text{height}} \cdot \underbrace{dx}_{\text{thickness}}$$

Affixing the integral sign and the limits of integration is all that is needed to get the entire volume.

Analogous formulas hold for rotation about the x-axis and for regions between two graphs. Regions to be revolved along with typical cylindrical shells, are shown in Figures 6.37, 6.38, and 6.39. You can convince yourself of the validity of these formulas by finding the volume of a typical cylindrical shell. All functions involved are assumed to be continuous.

$$V = 2\pi \int_c^d y g(y)\,dy \qquad (6.14)$$

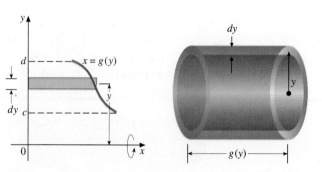

FIGURE 6.37
Rotation about the x-axis

$$V = 2\pi \int_a^b x[f(x) - g(x)]dx \qquad (6.15)$$

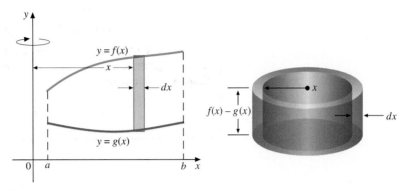

FIGURE 6.38
Rotation about the y-axis

$$V = 2\pi \int_c^d y[g(y) - h(y)]dy \qquad (6.16)$$

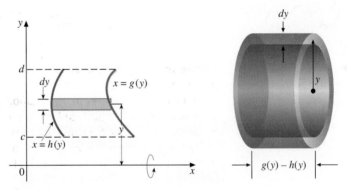

FIGURE 6.39
Rotation about the x-axis

REMARK

■ Rather than memorize Equations 6.14–6.16, it is easier in each case to draw a typical rectangle and the resulting cylindrical shell. Calculating its volume leads to the appropriate integrand.

For obvious reasons, calculating volumes in the way we have described is called the **method of cylindrical shells.** We illustrate the technique with several examples.

EXAMPLE 6.12   Use the method of cylindrical shells to find the volume of the solid obtained by revolving the region under the graph of $y = x^2$ from $x = 0$ to $x = 1$ about the y-axis.

Solution   We show the region to be rotated, together with the resulting solid of revolution, in Figure 6.40. Writing $f(x) = x^2$ and using Equation 6.13, we get

$$V = 2\pi \int_0^1 xf(x)dx = 2\pi \int_0^1 x(x^2)dx = 2\pi \int_0^1 x^3\, dx$$

$$= 2\pi \left[\frac{x^4}{4}\right]_0^1 = \frac{\pi}{2} \qquad\blacksquare$$

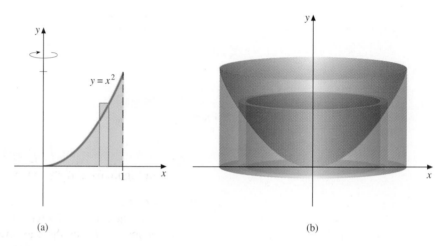

(a)   (b)

FIGURE 6.40

The volume in Example 6.12 can also be calculated by the method of washers. You should do it by the washer method, using Equation 6.11, with $g(y) = 1$ and $h(y) = \sqrt{y}$, and compare the difficulty with the cylindrical shell method.

EXAMPLE 6.13   Find the volume of the solid obtained by revolving about the x-axis the region bounded by $g(y) = 2y - y^2$ and the y-axis.

Solution   We show the region in Figure 6.41, along with the solid of revolution. We choose the method of cylindrical shells, since cross-sectional areas perpendicular to the x-axis are not easily calculated in this case. (It would be instructive for you to try to find the volume by the washer method to see the difficulties involved.) By Equation 6.14, we obtain

$$V = 2\pi \int_0^2 yg(y)dy = 2\pi \int_0^2 y(2y - y^2)dy$$

$$= 2\pi \int_0^2 (2y^2 - y^3)dy = 2\pi \left[\frac{2y^3}{3} - \frac{y^4}{4}\right]_0^2$$

$$= 2\pi \left[\frac{16}{3} - 4\right] = \frac{8\pi}{3} \qquad\blacksquare$$

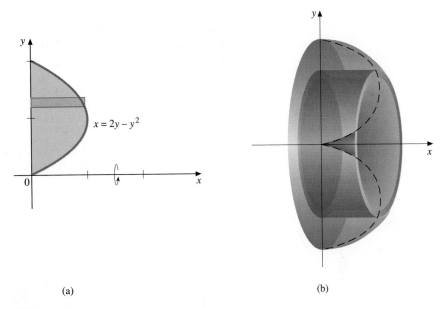

(a)

(b)

FIGURE 6.41

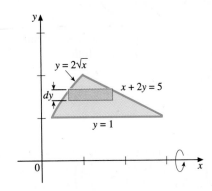

(a)

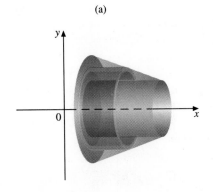

(b)

FIGURE 6.42

EXAMPLE 6.14   Find the volume of the solid obtained by revolving the region shown in Figure 6.42(a) about the x-axis.

Solution   We choose the method of cylindrical shells, since in this case the method avoids having to divide the region into two parts, as would be necessary with the washer method (since there are two different upper boundaries). The typical rectangle to be rotated is horizontal, so we need to express the left and right boundaries as functions of y:

$$x = \frac{y^2}{4} \qquad \text{Left boundary}$$

$$x = 5 - 2y \qquad \text{Right boundary}$$

We solve these two equations simultaneously to get their point of intersection. The y-coordinate gives the upper limit of integration. By setting the x values equal to each other, we obtain

$$\frac{y^2}{4} = 5 - 2y$$

$$y^2 + 8y - 20 = 0$$

$$(y - 2)(y + 10) = 0$$

$$y = 2 \text{ or } y = -10$$

but $y = -10$ is outside the domain of definition, so the upper limit is $y = 2$. Using Equation 6.16 with $g(y) = 5 - 2y$ and $h(y) = y^2/4$, we obtain the volume of the solid shown in Figure 6.42(b):

$$V = 2\pi \int_1^2 y \left(5 - 2y - \frac{y^2}{4}\right) dy = 2\pi \int_1^2 \left(5y - 2y^2 - \frac{y^3}{4}\right) dy$$

$$= 2\pi \left[\frac{5y^2}{2} - \frac{2y^3}{3} - \frac{y^4}{16}\right]_1^2$$

$$= 2\pi \left[\left(10 - \frac{16}{3} - 1 \right) - \left(\frac{5}{2} - \frac{2}{3} - \frac{1}{16} \right) \right] = \frac{91\pi}{24}$$ ∎

EXAMPLE 6.15   Find the volume of the solid of revolution obtained by revolving the region bounded by the graphs of $y = x^2/2$ and $y = x + 4$ about the line $x = 6$.

Solution   In Figure 6.43(a), we show a typical area element. Figure 6.43(b) shows the cylindrical shell that is generated when this element is revolved about the line $x = 6$. In Figure 6.43(c), we show the generated solid. We have used our simplified labeling technique, so the average radius of the shell is $6 - x$, the height is $f(x) - g(x)$, where $f(x) = x + 4$ and $g(x) = x^2/2$, and the thickness is dx. Thus, the volume of the shell is

$$\underbrace{2\pi(6 - x)}_{\substack{\text{average} \\ \text{circumference}}} \cdot \underbrace{\left[(x + 4) - \frac{x^2}{2} \right]}_{\text{height}} \cdot \underbrace{dx}_{\text{thickness}}$$

The total volume is therefore

$$V = 2\pi \int_{-2}^{4} (6 - x) \left(x + 4 - \frac{x^2}{2} \right) dx$$

The limits of integration were found by solving the two equations simultaneously. We simplify the integrand and perform the integration as follows:

$$V = 2\pi \int_{-2}^{4} \left(\frac{x^3}{2} - 4x^2 + 2x + 24 \right) dx$$

$$= 2\pi \left[\frac{x^4}{8} - \frac{4x^3}{3} + x^2 + 24x \right]_{-2}^{4}$$

$$= 2\pi \left[\left(32 - \frac{4 \cdot 64}{3} + 16 + 96 \right) - \left(2 + \frac{4 \cdot 8}{3} + 4 - 48 \right) \right]$$

$$= 2\pi(90) = 180\pi$$ ∎

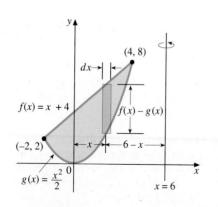

(a)

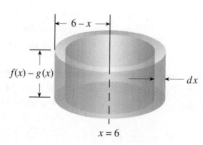

(b)

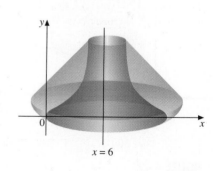

(c)

FIGURE 6.43

Exercise Set 6.3

In Exercises 1–10, find the volume obtained by revolving the region bounded by the graphs of the given equations about the axis specified. Use the method of cylindrical shells.

1. $y = x$, $y = 0$, $x = 2$; y-axis

2. $y = \sqrt{x}$, $y = 0$, $x = 4$; y-axis

3. $y = \sqrt{4 - x}$, $x = 0$, $y = 0$; x-axis

4. $x = 1/\sqrt{y}$, $y = 1$, $y = 4$, $x = 0$; x-axis

5. $x = \sqrt{y}$, $x = 0$, $x + y = 6$; y-axis

6. $y = \sqrt{6 - x}$, $y = x$, $y = 0$; x-axis

7. $y = (16 - x^2)/4$, $x - 2y + 6 = 0$, $x = 2$, $x = 4$; y-axis

8. $y = \sqrt{x + 5}$, $y = x - 1$, $y = 1$; x-axis

9. $y = x^3/2$, $y = 1/2$, $y = 4$, $x = 0$; x-axis

10. $y^2 = x^3$, $x = 1$; y-axis

In Exercises 11–16, find by two methods the volume of revolution obtained by revolving the region bounded by the graphs of the given equations about the axis specified.

11. $y = x^2$, $y = 2x$; y-axis

12. $y = \sqrt{2x - x^2}$, $y = x$; x-axis

13. $y^2 = 2x$, $x^2 = 2y$; x-axis

14. $x = 2\sqrt{4 - y^2}$, $x + 2y = 4$; y-axis

15. $y^2 = x^2 - 1$, $2x - y = 2$; x-axis

16. $y = \sqrt{8x}$, $y = x$, y-axis

In Exercises 17–20, use the method of cylindrical shells to find the volume of the solid generated by revolving the shaded region about the axis indicated.

17. y-axis

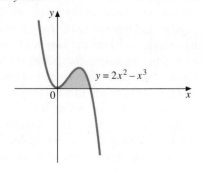

18. x-axis

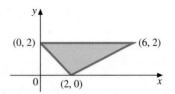

19. x-axis

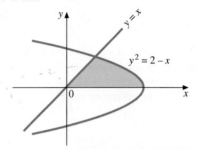

20. y-axis

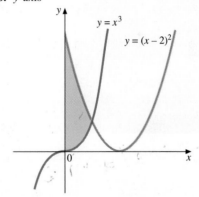

21. Find the volume of the solid obtained by revolving the region between the x-axis and the graph of $f(x) = 1 - x^2$ about the line $x = 2$.

22. Find the volume of the solid obtained by revolving the region between the graphs of $h(y) = y$ and $g(y) = 2 - y^2$ about the line $y = 1$.

23. Use the method of cylindrical shells to find the volume of a torus (see Example 6.11) generated by revolving a circle of radius a about an axis b units from the center of the circle, where $b > a$. (*Hint:* Orient the axes so that the center of the circle is at the origin.)

24. Find the volume of the solid obtained by revolving the region between the graphs of $y = 4x - x^2$ and $y = x^2 - 2x + 4$ about (a) the y-axis and (b) the x-axis.

25. Find the volume of the solid obtained by revolving the region between the graphs of $y = \frac{1}{3}(x - 4)(x^2 - 8x - 3)$ and $x + y = 4$ about the y-axis.

26. Let

$$f(x) = \begin{cases} x^2 & \text{if } 0 \le x < 2 \\ 6 - x & \text{if } 2 \le x \le 6 \end{cases}$$

Find the volume of the solid obtained by revolving the region bounded by the graph of f and the x-axis about the line $x = -1$. Use (a) the method of washers and (b) the method of cylindrical shells.

27. Find the volume generated by revolving the region bounded by the graphs of $y = 3x - x^2$ and $y = x^2 - 2$ about (a) the line $x = -1$ and (b) the line $y = -2$.

28. Maccabee the goldfish lived in a bowl that is in the shape of a sphere with its top and bottom cut off. (See the photo.) The radius of the sphere is 12 cm, and the top and bottom 3.5 cm of the sphere are cut off.

(a) When the bowl was half-filled with water, how much water was there, and how much did it weigh? (Water weighs 1 gm/cm^3.)

(b) If water evaporates in 30 days so that the level is 2.5 cm below the middle, what volume of water evaporates per day?

 In Exercises 29 and 30, use a CAS to obtain the graph. Set up the integral for the volume of the solid described, and evaluate it using the CAS.

29. The solid obtained by revolving the region bounded by $y = \sin^2 x$ and the x-axis from $x = 0$ to $x = \pi$
(a) about the y-axis
(b) about the line $x + \pi = 0$

30. The solid obtained by revolving the region bounded by $y = x^2$ and $y = 2\cos\left(\dfrac{\pi x}{3}\right)$
(a) about the x-axis
(b) about the line $y = 2$

6.4 ARC LENGTH

FIGURE 6.44
The length of the polygonal path approximates the length of the curve from a to b.

In this section we show how integration can be used to find the length of a curve. The basic idea is very simple. As we show in Figure 6.44, we approximate the length from A to B by a series of straight line segments joining points on the curve. This collection of line segments is called a **polygonal path** from A to B. The length of the polygonal path is easily calculated, since the length of each segment is just the distance between two points. By taking the points on the curve closer and closer together, we get polygonal paths that more and more closely approximate the curve. We will show that the length of each polygonal path can be expressed as a Riemann sum, so in the limit we get an integral.

To carry out the process just described, let us assume that f is a continuous function on the closed interval $[a, b]$. For each k from 0 to n, let P_k be the point $(x_k, f(x_k))$ on the graph of f. Note that the initial point A is P_0 and

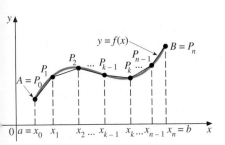

FIGURE 6.45

the final point B is P_n. (See Figure 6.45.) The straight line segments joining the consecutive points $P_0, P_1, P_2, \ldots, P_n$ form a polygonal path from A to B corresponding to our partition. To find its length, consider the distance between two typical consecutive points P_{k-1} and P_k. As we indicate in Figure 6.46, we denote the horizontal distance between them by Δx_k and the vertical distance by Δy_k. So by the Pythagorean Theorem, the length of the segment joining P_{k-1} and P_k is

$$\sqrt{(\Delta x_k)^2 + (\Delta y_k)^2}$$

The length of the polygonal path is therefore

$$\sum_{k=1}^{n} \sqrt{(\Delta x_k)^2 + (\Delta y_k)^2} \tag{6.17}$$

To write the sum in Formula 6.17 in a form that looks like a Riemann sum, we apply the Mean-Value Theorem (Theorem 4.6) to f on each of the closed subintervals $[x_{k-1}, x_k]$. To do so, we require that $f'(x)$ exists, at least on each open interval (x_{k-1}, x_k). Assuming f satisfies this condition, the Mean-Value Theorem tells us that there is a point c_k in the open interval (x_{k-1}, x_k) such that

$$\frac{f(x_k) - f(x_{k-1})}{x_k - x_{k-1}} = f'(c_k)$$

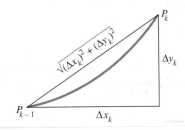

FIGURE 6.46

or, equivalently,

$$\frac{\Delta y_k}{\Delta x_k} = f'(c_k) \qquad k = 1, 2, \ldots, n$$

Figure 6.47 illustrates this relationship. We solve the last equation for Δy_k to get

$$\Delta y_k = f'(c_k) \Delta x_k$$

When we substitute this value for Δy_k in the formula for the distance between P_{k-1} and P_k, we have

$$\sqrt{(\Delta x_k)^2 + (\Delta y_k)^2} = \sqrt{(\Delta x_k)^2 + [f'(c_k)\Delta x_k]^2} = \sqrt{1 + [f'(c_k)]^2}\, \Delta x_k$$

The length formula (6.17) for the polygonal path can now be written as

$$\sum_{k=1}^{n} \sqrt{1 + [f'(c_k)]^2}\, \Delta x_k$$

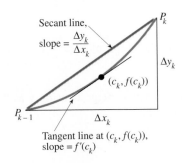

FIGURE 6.47
The slope of the tangent line at $(c_k, f(c_k))$ equals the slope of the secant line $P_{k-1}P_k$.

which we recognize as a Riemann sum for the function $\sqrt{1 + [f'(x)]^2}$. If this latter function is integrable on $[a, b]$, we know that its Riemann sums will approach its integral. To assure this integrability condition, we make the further assumption that $f'(x)$ be continuous on $[a, b]$. (See the marginal comment for a function that is differentiable everywhere but whose derivative is not continuous.) It follows that $\sqrt{1 + [f'(x)]^2}$ is also continuous there. A function whose derivative exists and is continuous on an interval is said to be a **smooth function**. Its graph has a continuously turning tangent line, with no sharp corners. We now have set the stage for the following definition.

The derivative of the function

$$f(x) = \begin{cases} x^2 \sin\left(\dfrac{1}{x}\right) & \text{if } x \neq 0 \\ 0 & \text{if } x = 0 \end{cases}$$

exists everywhere, but the derivative is discontinuous at $x = 0$. (See Exercise 76 in Exercise Set 3.7.)

Definition 6.2	Let f be a smooth function on $[a, b]$. Then the length L of its graph from $A = (a, f(a))$ to $B = (b, f(b))$ is given by

Definition 6.2
The Length of a Curve (Arc Length)

Let f be a smooth function on $[a, b]$. Then the length L of its graph from $A = (a, f(a))$ to $B = (b, f(b))$ is given by

$$L = \int_a^b \sqrt{1 + [f'(x)]^2}\, dx \qquad (6.18)$$

Similarly, when g is a smooth function of y on the closed interval $[c, d]$, the length of the graph of g from $A = (g(c), c)$ to $B = (g(d), d)$ is given by

$$L = \int_c^d \sqrt{1 + [g'(y)]^2}\, dy \qquad (6.19)$$

EXAMPLE 6.16   Find the length of the curve $y = 2(x + 1)^{3/2}$ from $x = -1$ to $x = 6$.

Solution   Let $f(x) = 2(x + 1)^{3/2}$. Then $f'(x) = 3(x + 1)^{1/2}$, so by Equation 6.18,

$$L = \int_{-1}^6 \sqrt{1 + 9(x + 1)}\, dx = \int_{-1}^6 \sqrt{9x + 10}\, dx \qquad \text{Mentally, let } u = 9x + 10,$$
$$\text{so that } du = 9\, dx.$$

$$= \frac{1}{9} \int_{-1}^6 (9x + 10)^{1/2} \cdot 9\, dx = \frac{1}{9} \cdot \frac{2}{3} (9x + 10)^{3/2} \Big]_{-1}^6$$

$$= \frac{2}{27} [(64)^{3/2} - (1)^{3/2}] = \frac{2}{27} (8^3 - 1) = \frac{2}{27} (511) = \frac{1022}{27} \qquad \blacksquare$$

EXAMPLE 6.17   Find the length of arc on the graph of $x = \frac{2}{3}(y^2 - 1)^{3/2}$ from $y = 1$ to $y = 3$.

Solution   The form in which the equation is written suggests using y as the independent variable, so Equation 6.19 is applicable:

$$g'(y) = \frac{dx}{dy} = \frac{2}{3} \cdot \frac{3}{2} (y^2 - 1)^{1/2} \cdot 2y = 2y(y^2 - 1)^{1/2}$$

Thus,

$$L = \int_1^3 \sqrt{1 + [2y(y^2 - 1)^{1/2}]^2}\, dy = \int_1^3 \sqrt{1 + 4y^4 - 4y^2}\, dy$$

$$= \int_1^3 \sqrt{(2y^2 - 1)^2}\, dy = \int_1^3 (2y^2 - 1)\, dy \qquad \text{Since } \sqrt{(2y^2 - 1)^2} = |2y^2 - 1|$$
$$= 2y^2 - 1 \text{ for } 1 \le y \le 3$$

$$= \frac{2y^3}{3} - y \Big]_1^3 = (18 - 3) - \left(\frac{2}{3} - 1\right)$$

$$= \frac{46}{3} \qquad \blacksquare$$

Approximating Arc Length

We chose the functions in the two preceding examples carefully so that the integrals involved could be evaluated in exact form by means we have studied. When the integral does not turn out so nicely (which is usually the case), we can use one of our approximation methods. The next example illustrates such a case.

EXAMPLE 6.18   Use Simpson's Rule with $n = 10$ to approximate the arc of the parabola $y = x^2$ from $x = 0$ to $x = 2$.

Solution   Since $y = x^2$, we have $y' = 2x$, so

$$L = \int_0^2 \sqrt{1 + 4x^2}\, dx$$

In Chapter 8 we will show how to evaluate an integral of this type by means of a substitution. Here, however, the best we can do is to approximate its value. With $n = 10$, we have

$$\Delta x = \frac{2}{10} = 0.2$$

Writing $f(x) = \sqrt{1 + 4x^2}$, we have, by Simpson's Rule,

$$\int_0^2 \sqrt{1 + 4x^2}\, dx \approx \tfrac{0.2}{3}[f(0) + 4f(0.2) + 2f(0.4) + 4f(0.6) + 2f(0.8)$$

$$+ 4f(1.0) + 2f(1.2) + 4f(1.4) + 2f(1.6) + 4f(1.8)$$

$$+ f(2.0)]$$

By calculator, we obtain the answer

$$L \approx 4.65 \qquad \blacksquare$$

In the next example we consider the circumference of an ellipse. As in the previous example, we arrive at an integral of a function for which no antiderivative in the form of an elementary function is immediately apparent. In contrast to the previous example, however, there is no substitution, nor any other technique, by which the integral can be brought around to one we can evaluate in an exact form. We could use Simpson's Rule, or some other approximation technique. An alternative, however, is to consult a table. Not surprisingly, the integral in question is called an **elliptic integral**, and approximate values of such integrals have been widely tabulated. They also can be approximated using a CAS. Elliptic integrals occur in a variety of applications, particularly in physics.

EXAMPLE 6.19   Find an integral whose value gives the length of the ellipse

$$\frac{x^2}{a^2} + \frac{y^2}{b^2} = 1$$

Solution   We show a typical ellipse in Figure 6.48. The upper half of the ellipse has the equation

$$y = \frac{b}{a}\sqrt{a^2 - x^2}$$

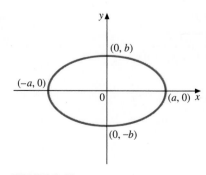

FIGURE 6.48
The ellipse $\dfrac{x^2}{a^2} + \dfrac{y^2}{b^2} = 1$

(obtained by solving the original equation for y and taking the positive square root). By symmetry, the entire length L of the ellipse is four times the arc in the first quadrant. Thus, by Equation 6.18,

$$L = 4 \int_0^a \sqrt{1 + (y')^2} \, dx$$

To find y', we can differentiate the equation

$$\frac{x^2}{a^2} + \frac{y^2}{b^2} = 1$$

implicitly to get

$$\frac{2x}{a^2} + \frac{2yy'}{b^2} = 0$$

Solving for y', we get

$$y' = -\frac{b^2 x}{a^2 y}$$

Thus,

$$L = 4 \int_0^a \sqrt{1 + \frac{b^4 x^2}{a^4 y^2}} \, dx$$

Since

$$y^2 = \frac{b^2}{a^2}(a^2 - x^2)$$

we finally obtain the result

$$L = 4 \int_0^a \sqrt{1 + \frac{b^2 x^2}{a^2(a^2 - x^2)}} \, dx$$

If we knew particular values of a and b, we could approximate the length by Simpson's Rule, or we could use a CAS or consult a table of elliptic integrals.

The special case in which $a = b$ is of interest. In this case the ellipse is a circle, and we know from previous experience (but have not yet proved) that the length (circumference) is $2\pi a$. So we have

$$2\pi a = 4 \int_0^a \sqrt{1 + \frac{x^2}{a^2 - x^2}} \, dx$$

$$= 4 \int_0^a \sqrt{\frac{a^2 - x^2 + x^2}{a^2 - x^2}} \, dx$$

$$= 4a \int_0^a \frac{1}{\sqrt{a^2 - x^2}} \, dx$$

Solving for the integral, we obtain

$$\int_0^a \frac{1}{\sqrt{a^2 - x^2}} \, dx = \frac{\pi}{2}$$

In Chapter 7 we will see another way of obtaining this result by making use of the Second Fundamental Theorem of Calculus. ∎

REMARK ─────────────────────────────────────

■ You might have noticed that each of the integrals we arrived at in Example 6.19 has an unbounded integrand on the interval $[0, a]$. The integrals are said to be *improper* in this case. We will study improper integrals in Chapter 7. We will also reconsider both integrals in Chapter 8 after we have studied trigonometric substitutions.

─────────────────────────────────────

Piecewise-Smooth Functions

Suppose f is continuous but f' is not continuous on the closed interval $[a, b]$ (so that f is not smooth), but we can divide the interval into a finite number of closed subintervals, on each of which f is smooth. Then we say that f is **piecewise smooth**. The length of its graph on $[a, b]$ is defined as the sum of the lengths on each of the subintervals. So if L_k denotes the arc length on the kth subinterval and there are m subintervals, then

$$L = \sum_{k=1}^{m} L_k$$

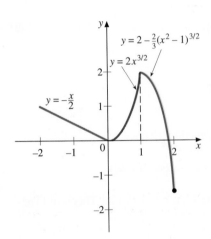

FIGURE 6.49

EXAMPLE 6.20   Find the length of the piecewise-smooth curve shown in Figure 6.49.

Solution   Let L_1 denote the length of the segment from $x = -2$ to $x = 0$, L_2 the length from $x = 0$ to $x = 1$, and L_3 the length from $x = 1$ to $x = 2$. We do not need calculus to find L_1, since it is the length of a straight line segment, but we will apply our theory anyway and show that the result agrees with what we would get from the Pythagorean Theorem.
Curve 1:

$$y = -\frac{x}{2} \qquad \text{on } [-2, 0]$$

$$y' = -\frac{1}{2}$$

$$L_1 = \int_{-2}^{0} \sqrt{1 + \frac{1}{4}} \, dx = \frac{\sqrt{5}}{2} x \bigg]_{-2}^{0} = -\frac{\sqrt{5}}{2}(-2) = \sqrt{5}$$

Using the Pythagorean Theorem, we also get $L_1 = \sqrt{2^2 + 1^2} = \sqrt{5}$.
Curve 2:

$$y = 2x^{3/2} \qquad \text{on } [0, 1]$$

$$y' = 2 \cdot \frac{3}{2} x^{1/2} = 3x^{1/2}$$

$$L_2 = \int_0^1 \sqrt{1 + 9x} \, dx = \frac{1}{9} \int_0^1 (1 + 9x)^{1/2} (9 \, dx) \qquad \text{\small Mentally, let } u = 1 + 9x, \text{\small so that } du = 9 \, dx.$$

$$= \frac{1}{9} \cdot \frac{2}{3} (1 + 9x)^{3/2} \bigg]_0^1 = \frac{2}{27} \left[(10)^{3/2} - 1 \right]$$

Curve 3:

$$y = 2 - \frac{2}{3}(x^2 - 1)^{3/2} \qquad \text{on } [1, 2]$$

$$y' = -\frac{2}{3} \cdot \frac{3}{2}(x^2 - 1)^{1/2}(2x) = -2x(x^2 - 1)^{1/2}$$

$$L_3 = \int_1^2 \sqrt{1 + 4x^2(x^2 - 1)}\, dx$$

$$= \int_1^2 \sqrt{4x^4 - 4x^2 + 1}\, dx$$

$$= \int_1^2 \sqrt{(2x^2 - 1)^2}\, dx$$

$$= \int_1^2 (2x^2 - 1)\, dx \qquad \sqrt{(2x^2-1)^2} = |2x^2 - 1| = 2x^2 - 1 \text{ for } 1 \le x \le 2$$

$$= \frac{2x^3}{3} - x \Big]_1^2$$

$$= \left(\frac{16}{3} - 2\right) - \left(\frac{2}{3} - 1\right) = \frac{10}{3} - \left(-\frac{1}{3}\right) = \frac{11}{3}$$

The total length is therefore

$$L = L_1 + L_2 + L_3 = \sqrt{5} + \frac{2}{27}\left[(10)^{3/2} - 1\right] + \frac{11}{3} \approx 8.17 \qquad \blacksquare$$

The Arc-Length Function

It is instructive to consider the *arc-length function*, defined by

$$s(x) = \int_a^x \sqrt{1 + [f'(t)]^2}\, dt \qquad (x \in [a, b])$$

for the smooth function f on $[a, b]$. By the First Fundamental Theorem (Theorem 5.2),

$$s'(x) = \sqrt{1 + [f'(x)]^2}$$

Thus, the *differential of arc length*, $ds = s'(x)\, dx$, is given by

$$ds = \sqrt{1 + [f'(x)]^2}\, dx = \sqrt{1 + \left(\frac{dy}{dx}\right)^2}\, dx \tag{6.20}$$

If $x = g(y)$, where g is smooth on $[c, d]$, then ds is given by

$$ds = \sqrt{1 + [g'(y)]^2}\, dy = \sqrt{1 + \left(\frac{dx}{dy}\right)^2}\, dy \tag{6.21}$$

We now can express both Equations 6.18 and 6.19 in the briefer form

$$L = \int_{l_1}^{l_2} ds$$

where the correct limits l_1 and l_2 must be supplied. They will be x limits if ds is given by Equation 6.20, and they will be y limits if ds is given by Equation 6.21.

In the equation

$$ds = \sqrt{1 + \left(\frac{dy}{dx}\right)^2}\, dx$$

if we bring the final dx inside the radical and simplify, we get

$$ds = \sqrt{(dx)^2 + (dy)^2}$$

The same result can be obtained by bringing dy inside the radical in Equation 6.21. Squaring both sides gives

$$(ds)^2 = (dx)^2 + (dy)^2 \tag{6.22}$$

Equation 6.22 has an interesting geometric interpretation, as shown by Figure 6.50. The differential ds can be interpreted as the length of the hypotenuse of a right triangle with legs dx and dy. We let $dx = \Delta x$ be an arbitrary increment on x, which causes a corresponding increment Δy on y. From Section 3.9, we know that the distance dy to the tangent line approximates the distance Δy to the curve when Δx is small. Using Δs to denote the arc length from the point (x, y) to the point $(x + \Delta x, y + \Delta y)$, we see that $ds \approx \Delta s$. So we can think of the differential of arc as almost equal to a true section of arc when the change in x is small.

The use of this differential notation is illustrated in the following example.

FIGURE 6.50

EXAMPLE 6.21   An engineer plans to shape a length of fiber-optic cable according to the curve defined by $x^4 - 6xy + 3 = 0$ from $(1, \frac{2}{3})$ to $(3, \frac{14}{3})$, with distances in meters. The cable costs \$1 per meter. What is the total cost?

Solution   We show the computer-generated graph of the curve in Figure 6.51. Solving the equation for y gives

$$y = \frac{x^3}{6} + \frac{1}{2x}$$

For x in the interval $[1, 3]$ this equation defines a smooth function of x. (Why?)

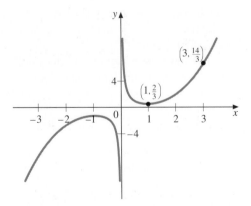

FIGURE 6.51

So we use Equation 6.20 to get ds:

$$ds = \sqrt{1 + \left(\frac{dy}{dx}\right)^2}\, dx = \sqrt{1 + \left(\frac{x^2}{2} - \frac{1}{2x^2}\right)^2}\, dx$$

$$= \sqrt{1 + \frac{x^4}{4} - \frac{1}{2} + \frac{1}{4x^4}}\, dx = \sqrt{\frac{x^4}{4} + \frac{1}{2} + \frac{1}{4x^4}}\, dx$$

$$= \sqrt{\left(\frac{x^2}{2} + \frac{1}{2x^2}\right)^2}\, dx = \left(\frac{x^2}{2} + \frac{1}{2x^2}\right) dx$$

Thus,

$$L = \int_{x=1}^{x=3} ds = \int_1^3 \left(\frac{x^2}{2} + \frac{1}{2x^2}\right) dx$$

$$= \frac{x^3}{6} - \frac{1}{2x}\Bigg]_1^3 = \left(\frac{27}{6} - \frac{1}{6}\right) - \left(\frac{1}{6} - \frac{1}{2}\right)$$

$$= \frac{14}{3}$$

The cable in question is therefore $4\frac{2}{3}$ m long, so its cost is \$4.67. ∎

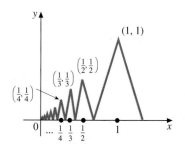

FIGURE 6.52
A curve with infinite length

It may surprise you to learn that there are curves defined by continuous functions on finite intervals that are infinite in length. One such example is the "sawtooth" function shown in Figure 6.52. The peaks occur at the points $\left(\frac{1}{n}, \frac{1}{n}\right)$ for $n = 1, 2, 3, \ldots$, with the function defined as 0 at the origin. The sum of the lengths of the two rightmost segments that come together at the point $(1, 1)$ exceeds the vertical distance 1 from the x-axis to the peak. Similarly, the sum of the lengths of the two segments meeting at the point $\left(\frac{1}{2}, \frac{1}{2}\right)$ is greater than the vertical distance $\frac{1}{2}$ from the x-axis to the peak. In general, the sum of the lengths of the two segments meeting at $\left(\frac{1}{n}, \frac{1}{n}\right)$ is greater than $\frac{1}{n}$. Thus, if the total length L of the curve were finite, we would have

$$L > 1 + \frac{1}{2} + \frac{1}{3} + \frac{1}{4} + \cdots$$

In Chapter 10 we will show that by continuing to add terms on the right-hand side, the sum can be made arbitrarily large. That is, the sum is infinite. So L cannot be finite.

The function in Figure 6.52 does not have a simple formula. An example of a curve with infinite length that does have a simple formula is the graph of

$$f(x) = \begin{cases} x \sin \frac{1}{x} & \text{if } x \neq 0 \\ 0 & \text{if } x = 0 \end{cases}$$

The length over any interval containing the origin is infinite. We will ask you to find this result for the interval $0 \le x \le 1$ in Exercise 24 in Exercise Set 6.4.

Curves that have finite length are said to be **rectifiable**. The smoothness condition of a function f is sufficient (but not necessary) to guarantee that its graph is rectifiable.

Exercise Set 6.4

In Exercises 1–14, find the arc length of the graph of the given equation between the specified limits.

1. $y = 3x - 4$ from $x = -1$ to $x = 2$

2. $y = 3 - 2x$ from $x = 0$ to $x = 2$

3. $2x - 3y = 4$ from $(-1, -2)$ to $(2, 0)$

4. $4y - 5x = 7$ from $(-3, -2)$ to $(1, 3)$

5. $y = \dfrac{2}{3}x^{3/2}$ from $x = 0$ to $x = 3$

6. $x = (y - 1)^{3/2}$ from $y = 1$ to $y = 6$

7. $y^2 = x^3$ from $(0, 0)$ to $(5, 5\sqrt{5})$

8. $y = x^{2/3}$ from $x = 1$ to $x = 8$

9. $y = \dfrac{2}{3}(x^2 + 1)^{3/2}$ from $x = 0$ to $x = 2$

10. $y = 1 + \left(\dfrac{3x}{2}\right)^{2/3}$ from $(0, 1)$ to $(\tfrac{2}{3}, 2)$

11. $y = \left(\dfrac{1 - x}{2}\right)^{3/2} + 4$ from $x = -4$ to $x = 1$

12. $y = \dfrac{x^4}{8} + \dfrac{1}{4x^2}$ from $x = 1$ to $x = 2$

13. $4y^4 - 12xy + 3 = 0$ from $\left(\tfrac{7}{12}, 1\right)$ to $\left(\tfrac{109}{12}, 3\right)$

14. $9y^2 = (x^2 - 2)^3$ from $(\sqrt{2}, 0)$ to $\left(\sqrt{6}, \tfrac{8}{3}\right)$

15. Find the total length of the four-cusp hypocycloid $x^{2/3} + y^{2/3} = a^{2/3}$.

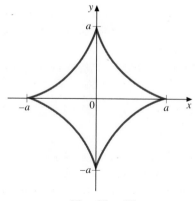

$$x^{2/3} + y^{2/3} = a^{2/3}$$

16. Find the length of the arc on the graph of $x = \tfrac{1}{3}\sqrt{y}(y - 3)$ from $y = 1$ to $y = 4$.

17. Find the fallacy in the following attempt to find the length of the curve $2y = 3(x + 1)^{2/3}$ from $(-2, \tfrac{3}{2})$ to $(7, 6)$, and correct it.

$$x = \left(\dfrac{2y}{3}\right)^{3/2} - 1, \qquad \text{so} \quad \dfrac{dx}{dy} = \left(\dfrac{2y}{3}\right)^{1/2}$$

$$L = \int_{3/2}^{6} \sqrt{1 + \dfrac{2y}{3}}\, dy = \left. \left(1 + \dfrac{2y}{3}\right)^{3/2} \right]_{3/2}^{6}$$

$$= 5\sqrt{5} - 2\sqrt{2}$$

18. Let

$$f(x) = \begin{cases} 1 - x & \text{if } 0 \le x < 1 \\ (x - 1)^{3/2} & \text{if } 1 \le x < 2 \\ 2 - (x - 1)^{2/3} & \text{if } 2 \le x \le 9 \end{cases}$$

(a) Draw the graph of f.

(b) Find the length of the graph of f from $x = 0$ to $x = 9$.

In Exercises 19 and 20, use ds to approximate the arc length in question.

19. Arc of $y = x^2$ from $x = 1.0$ to $x = 1.1$

20. Arc of $y = x/(x - 1)$ from $x = 2.05$ to $x = 2.07$

In Exercises 21–23, use Simpson's Rule with $n = 10$ to approximate the arc length.

21. $y = x^2$ from $x = 0$ to $x = 2$

22. $y = \sin x$ from $x = 0$ to $x = \pi$

23. $y = 4 - x^3$ from $x = -2$ to $x = 2$

24. Let

$$f(x) = \begin{cases} x \sin \dfrac{1}{x} & \text{if } x \ne 0 \\ 0 & \text{if } x = 0 \end{cases}, \qquad 0 \le x \le 1$$

Calculate $f(x)$ for $x = \dfrac{2}{\pi}, \dfrac{2}{3\pi}, \dfrac{2}{5\pi}, \ldots$. Show how it follows that if the length of the graph of f is L, then

$$L > \dfrac{2}{\pi}\left(1 + \dfrac{1}{3} + \dfrac{1}{5} + \dfrac{1}{7} + \cdots\right)$$

In Chapter 10 we will show that the sum inside the parentheses is infinite. Thus the graph has infinite length.

6.5 FINDING AREAS OF SURFACES OF REVOLUTION

If an arc of a curve is revolved about an axis, a **surface of revolution** is generated. Our purpose in this section is to develop a means of finding the areas of such surfaces. First, we must define precisely what we mean by the area in this situation.

Our basic approach is to approximate the curve by a polygonal path and then rotate the polygonal path about the axis, to obtain an approximation to the actual surface area. We illustrate this idea in Figure 6.53. When a typical segment of the polygonal path is rotated about the axis, a band is formed. We highlight one such band in the figure. The band can be viewed as a section of a cone cut off by two parallel planes. The name given to such a portion of a cone is "frustum." So each segment of the polygonal path, when rotated, generates a frustum of a cone. As we indicate below, there is a simple formula for the area of a frustum of a cone. When we add up the areas of all the frustums (the bands), we obtain an approximation to the area of the surface formed by rotating the curve itself about the axis. We will show that our approximation is similar to a Riemann sum. As we take polygonal paths with points closer and closer together on the curve, we are led to an integral as the limit of Riemann sums.

As a starting point, we need to know the area of a frustum of a cone. Such a frustum is shown in Figure 6.54. An outline of how to find its area is presented in Exercise 22 at the end of this section. The result is that the surface area is given by

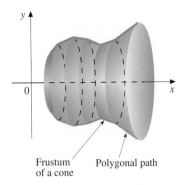

FIGURE 6.53

Each segment of the polygonal path generates a band, called a frustum of a cone, when rotated.

$$\text{area} = (\text{average circumference}) \times (\text{slant height})$$

or

$$\text{Area} = 2\pi r l \qquad (6.23)$$

where r is the average radius $= \frac{1}{2}(r_1 + r_2)$ and l is the slant height.

Now consider the arc of the graph of $y = f(x)$ between $x = a$ and $x = b$, where f is a smooth nonnegative function. Just as in defining arc length, we partition the closed interval $[a, b]$ and consider the chords that join successive points $P_k = (x_k, f(x_k))$ on the curve. (See Figure 6.55.) For each $k = 1, 2, \ldots, n$, we rotate the chord $P_{k-1}P_k$ about the x-axis. This process generates a band that is a frustum of a cone. In Figure 6.55 we show a typical band. We designate its area as ΔS_k. In Section 6.4 we found the slant height (the length of the line

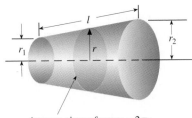

Average circumference $= 2\pi r$, where $r = \frac{1}{2}(r_1 + r_2)$;
Area $= 2\pi r l$

FIGURE 6.54

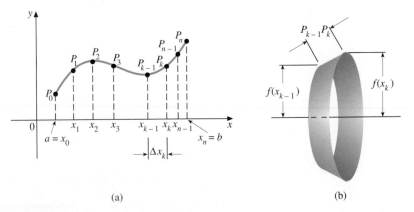

(a)

(b)

FIGURE 6.55

segment joining P_{k-1} and P_k) to be

$$\sqrt{1 + [f'(c_k)]^2}\, \Delta x_k$$

where c_k is between x_{k-1} and x_k. The average circumference is

$$2\pi \cdot \overbrace{\frac{1}{2}[f(x_{k-1}) + f(x_k)]}^{\text{average radius}}$$

and since $\frac{1}{2}[f(x_{k-1}) + f(x_k)]$ lies between $f(x_{k-1})$ and $f(x_k)$, we know by the Intermediate-Value Theorem (Theorem 2.8) that there is a number d_k in the interval $[x_{k-1}, x_k]$ such that

$$f(d_k) = \frac{1}{2}[f(x_{k-1}) + f(x_k)]$$

Thus, the surface area ΔS_k of the kth band is

$$\Delta S_k = \overbrace{2\pi f(d_k)}^{\substack{\text{average}\\\text{circumference}}} \cdot \overbrace{\sqrt{1 + [f'(c_k)]^2}\,\Delta x_k}^{\text{slant height}}$$

Summing all such areas, we arrive at what we would expect to be an approximation to the total surface area S:

$$S \approx \sum_{k=1}^{n} 2\pi f(d_k)\sqrt{1 + [f'(c_k)]^2}\, \Delta x_k \tag{6.24}$$

The sum on the right looks almost like a Riemann sum for the function $2\pi f(x)\sqrt{1 + [f'(x)]^2}$. The problem is that the points c_k and d_k are not necessarily the same. Fortunately, in the limit, it can be shown that this discrepancy does not matter, and the sum approaches the integral of this function. We therefore have the following definition.

Definition 6.3
Area of a Surface of Revolution

Let f be a nonnegative smooth function on the closed interval $[a, b]$. The area S of the surface of revolution generated by revolving the graph of f from $x = a$ to $x = b$ about the x-axis is given by

$$S = 2\pi \int_a^b f(x)\sqrt{1 + [f'(x)]^2}\, dx \tag{6.25}$$

The corresponding result for revolving the graph of the nonnegative smooth function g from $y = c$ to $y = d$ about the y-axis is

$$S = 2\pi \int_c^d g(y)\sqrt{1 + [g'(y)]^2}\, dy \tag{6.26}$$

In Section 6.4 we showed that the differential of arc length ds can be written as either

$$ds = \sqrt{1 + [f'(x)]^2}\, dx \quad \text{or} \quad ds = \sqrt{1 + [g'(y)]^2}\, dy$$

Using these equations, we can rewrite Equations 6.25 and 6.26 more simply as follows:

Area of a Surface of Revolution

rotation about x-axis: $S = 2\pi \displaystyle\int_{l_1}^{l_2} y \, ds$ (6.27)

rotation about y-axis: $S = 2\pi \displaystyle\int_{l_1}^{l_2} x \, ds$ (6.28)

The limits l_1 and l_2 of integration depend on the form of ds being used. When $ds = \sqrt{1 + [f'(x)]^2}\, dx$ is used, then x is the independent variable, and l_1 and l_2 are x-limits. When $ds = \sqrt{1 + [g'(y)]^2}\, dy$ is used, y is independent, so l_1 and l_2 are y-limits.

When the function f has an inverse (see Section 1.2), either Equation 6.27 or 6.28 may be used. In particular, when f is monotone increasing or monotone decreasing (still assuming f is smooth and nonnegative), either of the forms may be used. To be more explicit, suppose f is monotone increasing and that g is the inverse of f where both f and g are nonnegative. Then if $y = f(x)$, we have $x = g(y)$. If x satisfies $a \le x \le b$, then y satisfies $c \le y \le d$, where $c = f(a)$ and $d = f(b)$. Thus, Equation 6.27 can be written as

$$S = 2\pi \int_a^b f(x)\sqrt{1 + [f'(x)]^2}\, dx$$

or as

$$S = 2\pi \int_c^d y\sqrt{1 + [g'(y)]^2}\, dy$$

Similar remarks apply to Equation 6.28. We will illustrate this equivalence with the next example.

EXAMPLE 6.22   Find in two ways the area of the surface generated by revolving about the x-axis the arc of the curve $y = \sqrt{x}$ from $x = 1$ to $x = 4$.

Solution   Write $f(x) = \sqrt{x}$. Then for $1 \le x \le 4$ f is smooth, nonnegative, and monotone increasing (Figure 6.56). So we have a choice on how to write ds in Equation 6.27.

First, let x be the independent variable:

$$ds = \sqrt{1 + [f'(x)]^2}\, dx = \sqrt{1 + \left(\frac{1}{2\sqrt{x}}\right)^2}\, dx$$

$$= \sqrt{1 + \frac{1}{4x}}\, dx = \frac{\sqrt{4x + 1}}{2\sqrt{x}}\, dx$$

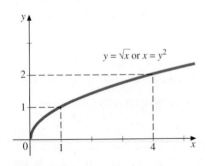

FIGURE 6.56

By Equation 6.27, with l_1 and l_2 as x-limits,

$$S = 2\pi \int_{l_1}^{l_2} y \, ds = 2\pi \int_1^4 \sqrt{x}\, \frac{\sqrt{4x + 1}}{2\sqrt{x}}\, dx$$

$$= \pi \int_1^4 \sqrt{4x + 1}\, dx$$

Now we mentally substitute $u = 4x + 1$, so that $du = 4\,dx$. We need a 4 to go with the dx, so we multiply by 4 inside the integral and by 1/4 outside. Thus,

$$S = \frac{\pi}{4}\int_1^4 \overbrace{(4x+1)^{1/2}}^{u^{1/2}}\ \overbrace{(4\,dx)}^{du} = \frac{\pi}{4}\cdot\frac{2}{3}(4x+1)^{3/2}\Big]_1^4$$

$$= \frac{\pi}{6}\left[(17)^{3/2} - (5)^{3/2}\right] \approx 30.85$$

Second, let us take y as the independent variable. We solve $y = \sqrt{x}$ for x to get $x = y^2$. When $x = 1$, $y = 1$, and when $x = 4$, $y = 2$, so the limits are from $y = 1$ to $y = 2$. Defining g by $g(y) = y^2$, we have

$$ds = \sqrt{1 + [g'(y)]^2}\,dy = \sqrt{1 + (2y)^2}\,dy = \sqrt{1 + 4y^2}\,dy$$

So by Equation 6.27, with l_1 and l_2 now given as y-limits,

$$S = 2\pi\int_{l_1}^{l_2} y\,ds = 2\pi\int_1^2 y\sqrt{1+4y^2}\,dy \quad u = 1+4y^2,\ du = 8y\,dy$$

$$= \frac{2\pi}{8}\int_1^2 y(1+4y^2)^{1/2}(8\,dy)$$

$$= \frac{\pi}{4}\cdot\frac{2}{3}(1+4y^2)^{3/2}\Big]_1^2$$

$$= \frac{\pi}{6}[(17)^{3/2} - (5)^{3/2}] \approx 30.85 \qquad \blacksquare$$

Shorthand Technique for Finding Areas of Surfaces of Revolution

Another advantage of Equations 6.27 and 6.28 is that they can be arrived at quite easily using the following technique. Referring to Figure 6.57, we treat ds as if it were a small arc on the curve. It may be thought of as being so small that it can be considered straight. Let (x, y) be its center. Then, when this arc is rotated about the x-axis, we get a frustum of a cone whose slant height is almost equal to the arc length. Its surface area is

(average circumference) $\times$ ("slant height") $= 2\pi y\,ds$

The total surface area is then the "sum" $2\pi\int_{l_1}^{l_2} y\,ds$. The same sort of reasoning leads to $2\pi\int_{l_1}^{l_2} x\,ds$ for the area of the surface of rotation about the y-axis.

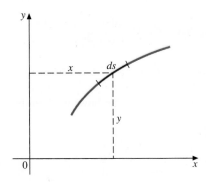

FIGURE 6.57

REMARK ——————

■ Just as with our other shorthand techniques, what we have described is simply a convenient device for arriving at a correct result. The reasoning is not mathematically rigorous.

————————————————

The next two examples further illustrate the use of Equations 6.27 and 6.28.

EXAMPLE 6.23   Find the surface area of the segment of a sphere of radius r cut off by a plane at a point a units from the center, where $0 < a < r$.

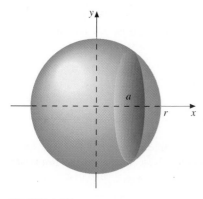

FIGURE 6.58

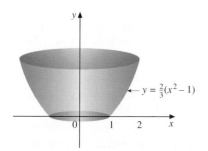

FIGURE 6.59

Solution   In Figure 6.58 we show the sphere with center at the origin and the segment cut off by a plane through $(a, 0)$ perpendicular to the x-axis. In Figure 6.59 we show an arc of the circle $x^2 + y^2 = r^2$ that generates the spherical segment when it is revolved about the x-axis. Treating y as a function of x, we have

$$x^2 + y^2 = r^2$$

Differentiating implicitly with respect to x, we obtain

$$2x + 2yy' = 0$$

Solving for y', we obtain

$$y' = -\frac{x}{y} \qquad y \neq 0$$

So

$$ds = \sqrt{1 + (y')^2}\, dx = \sqrt{1 + \frac{x^2}{y^2}}\, dx = \frac{\sqrt{x^2 + y^2}}{y}\, dx = \frac{r}{y}\, dx$$

Here, we have used the fact that $\sqrt{x^2 + y^2} = \sqrt{r^2} = r$ and also the fact that since $y \geq 0$ on the arc in question, $\sqrt{y^2} = y$. Thus,

$$S = 2\pi \int_{l_1}^{l_2} y\, ds = 2\pi \int_{a}^{r} y\left(\frac{r}{y}\right) dx = 2\pi r [x]_{a}^{r}$$
$$= 2\pi r (r - a) \qquad \blacksquare$$

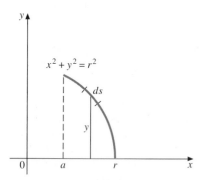

FIGURE 6.60

EXAMPLE 6.24   A ceramic bowl is turned on a potter's wheel to approximate the surface obtained by revolving the curve $y = \frac{2}{3}(x^2 - 1)$ from $x = 1$ to $x = 2$ about the y-axis. (See Figure 6.60.) What is the approximate outside area of the bowl? Assume that each unit on the x- and y-axes represents 10 cm.

Solution   Letting $f(x) = \frac{2}{3}(x^2 - 1)$, we have $f'(x) = \frac{4}{3}x$, so

$$ds = \sqrt{1 + \frac{16}{9}x^2}\, dx = \frac{\sqrt{9 + 16x^2}}{3}\, dx$$

By Equation 6.28 the surface area S of the curved part of the bowl is given by

$$S = 2\pi \int_{l_1}^{l_2} x\, ds = 2\pi \int_{1}^{2} x \frac{\sqrt{9 + 16x^2}}{3}\, dx$$

If we think of u as the quantity $9 + 16x^2$ under the radical, we see that its differential is $32x\, dx$. So we multiply and divide by 32 to get

$$S = \frac{2\pi}{3} \cdot \frac{1}{32} \int_{1}^{2} (9 + 16x^2)^{1/2} (32x\, dx)$$
$$= \frac{\pi}{48} \cdot \frac{2}{3}(9 + 16x^2)^{3/2} \Big]_{1}^{2} = \frac{\pi}{72}\left[(73)^{3/2} - (25)^{3/2}\right] \approx 21.76$$

The circular bottom of the bowl has area $\pi(1^2) \approx 3.14$. Thus, the total outside area of the bowl is approximately 24.9 square units, or $(24.9)(10)^2 = 2,490$ cm². $\blacksquare$

Exercise Set 6.5

In Exercises 1–10, find the area of the surface generated by revolving about the specified axis the arc of the curve whose equation is given, between the specified limits.

1. $y = 2x + 1$ from $x = 0$ to $x = 3$; x-axis

2. $2x + 3y = 6$ from $(0, 2)$ to $(3, 0)$; y-axis

3. $y = \sqrt{2x}$ from $x = 0$ to $x = 2$; x-axis

4. $y = x^2$ from $(1, 1)$ to $(2, 4)$; y-axis

5. $y = \sqrt{1 - x^2}$ from $(0, 1)$ to $(1, 0)$; y-axis

6. $y = x^3$ from $x = 0$ to $x = 1$; x-axis

7. $y = x^{1/3} + 1$ from $(1, 2)$ to $(8, 3)$; y-axis

8. $x^2 + y^2 = 9$ from $(0, 3)$ to $(2, \sqrt{5})$; x-axis

9. $x = \sqrt{5 - y}$ from $y = 1$ to $y = 4$; y-axis

10. $y^2 - 2x + 3 = 0$ from $(2, 1)$ to $(6, 3)$; x-axis

11. Derive the formula $S = 4\pi r^2$ for the surface area of a sphere of radius r.

12. Show that the surface area of a cone of height h and base radius r is $\pi r \sqrt{r^2 + h^2}$ by finding the area of the surface generated when the line segment $y = \dfrac{r}{h}x$ from $x = 0$ to $x = h$ is revolved about the x-axis.

13. Find the area of the surface that results from rotating the upper half of the four-cusp hypocycloid $x^{2/3} + y^{2/3} = a^{2/3}$ about the x-axis. (See Exercise 15 in Exercise Set 6.4.) Make use of implicit differentiation.

14. Find the area of the surface generated by revolving the segment of the line $2x + 3y = 6$ cut off by the x- and y-axes about (a) the line $x = -1$ and (b) the line $y = -2$.

15. Show that if f is smooth on $[a, b]$ but not necessarily nonnegative, the area of the surface generated by revolving the graph of f between $x = a$ and $x = b$ about the x-axis is

$$S = 2\pi \int_a^b |f(x)| \sqrt{1 + [f'(x)]^2}\, dx$$

16. Use the result of Exercise 15 to find the surface area generated by revolving the arc of the curve $y = x^3/6$ from $x = -1$ to $x = 2$ about the x-axis.

17. Find the area of the surface generated by revolving the arc of the curve

$$y = \frac{x}{2}\sqrt{\frac{1 - x^2}{2}}$$

from $x = -1$ to $x = 1$ about the x-axis. (See Exercise 15 and use the fact that $y = f(x)$ is odd in this case.)

18. Find the area of the surface generated by revolving the arc of the curve $y^4 + 3 = 6xy$ from $y = 1$ to $y = 2$ about the y-axis.

19. Find the area of the surface generated by revolving the arc of the curve $y = \sqrt{x}(x - 3)/3$ from $(1, -\frac{2}{3})$ to $(9, 6)$ about the x-axis.

20. A *zone* of a sphere is the band cut off by two parallel planes that intersect the sphere. Show that the surface area of a zone of a sphere of radius a cut off by two parallel planes h units apart ($h \leq 2a$) is independent of the location of the planes (as long as both intersect the sphere). What is the area of such a zone?

21. (a) The earth is approximately a sphere of radius 6,374 km. The south-temperate zone extends from $-23\frac{1}{2}^\circ$ (the Tropic of Capricorn) to $-67\frac{1}{2}^\circ$ (the Antarctic Circle). What is the surface area of this south-temperate zone? (See Exercise 20.)

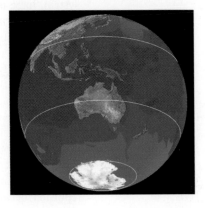

(b) A rain forest in Borneo covers a region 100 km by 200 km. What fraction of the area of the equatorial zone, from the Tropic of Cancer to the Tropic of Capricorn at $-23\frac{1}{2}^\circ$, does the rain forest cover?

22. (a) Find a formula for the lateral surface area of a cone
as follows. Imagine cutting a cone of base radius r
along a line AB as shown in part (a) of the figure.
Then flatten the cone so that it becomes a circular
sector as in part (b). Calculate the area of this sector.

(b) Use the result of part (a) to find the area of a frustum
of a cone with radii r_1 and r_2 and slant height l.

23. In each of the following use Simpson's Rule to approximate to two decimal places the area of the surface
generated by revolving the curve about the x-axis.
(a) $y = \sin x$, $0 \le x \le \pi$
(b) $y = x^2$, $0 \le x \le 2$

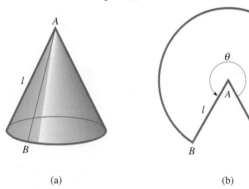

(a)   (b)

6.6 WORK

When a constant force F acts on an object, causing it to move a distance d in
the direction of the force, physicists define the **work** W done by the force to
be

$$W = Fd = (\text{force}) \times (\text{distance}) \tag{6.29}$$

Scientists now use units based on combinations of meters, kilograms, and seconds. In this *Système International d'Unités* (SI), force is measured in *newtons*
(symbol: N), and the corresponding unit of work, a *newton-meter*, is called a
joule (symbol: J). Energy is also measured in joules. Another set of metric
units, now less used, involves centimeters, grams, and seconds (cgs). In this
system, force is measured in *dynes* (1 newton = 10^5 dyne), so the unit of work
is the *dyne-centimeter*, also called an *erg* (1 joule = 10^7 erg). In the English
system of units, force is typically measured in pounds and distance in feet or
inches, so the unit of work is the *foot-pound* or the *inch-pound*. These English
units are hardly used anymore in science or internationally, though the British
thermal unit (Btu), a unit of energy, is in common use in the United States.

Equation 6.29 applies only to constant forces that act in the direction of
motion. There are many situations in which the acting force is variable. Either
the magnitude or the direction of the force, or both, may change over time. For
now, let us consider a variable force acting in a straight line. An example of
a variable force is the force necessary to stretch a spring. The force is higher
when the spring has been stretched more. As another example, the force of
repulsion between two like electrostatic charges increases as the charges get
closer to each other.

Let us now define the work done by a variable force F acting in a straight
line over a fixed distance. For convenience, let us orient the x-axis so that the
direction of force is along the axis, and let $F(x)$ denote the amount of the force
at x. Suppose further that F is a continuous function on the closed interval
from $x = a$ to $x = b$. As in Figure 6.61, we partition this closed interval

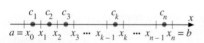

FIGURE 6.61

$[a, b]$, and let c_k denote an arbitrary point in the kth subinterval. If the width Δx_k is sufficiently small, the error in replacing the variable force $F(x)$ with the constant force $F(c_k)$ on the kth subinterval will not be great. Using this constant force enables us to apply Equation 6.29. For this constant force, the work ΔW_k done across the kth subinterval is

$$\Delta W_k = \underbrace{F(c_k)}_{\text{force}} \cdot \underbrace{\Delta x_k}_{\text{distance}}$$

We can reasonably suppose that the total work W done by the force on the closed interval $[a, b]$ is approximated by the sum of the ΔW_k values:

$$W \approx \sum_{k=1}^{n} \Delta W_k = \sum_{k=1}^{n} F(c_k) \Delta x_k$$

As the widths of the subintervals all approach 0, the approximation becomes better and better. Since the Riemann sum on the right approaches the integral $\int_a^b F(x)\, dx$, we are led to the following definition.

Definition 6.4 Work	The **work** W done by a continuous force F acting along the x-axis from $x = a$ to $x = b$ is given by $$W = \int_a^b F(x)\,dx \qquad (6.30)$$

EXAMPLE 6.25   A 100-kg mass is to be hoisted from the ground to a point 10 m high by a uniform chain passing over a pulley. If the chain has mass 3 kg/m, find the work done in raising the mass.

Solution   As shown in Figure 6.62, let x denote the distance from the pulley to the mass at any stage. Then x varies between 0 and 10. By Newton's Second Law the total force $F(x)$ is the total mass times the acceleration of gravity g (approximately 9.81 m/s^2 near the earth's surface). The total mass is the 100-kg mass plus the mass of the chain still to be raised. When a section of chain measuring x m is out, its mass is $3x$ kg. So $F(x) = (100 + 3x)g$ is the variable force. Thus, by Equation 6.30, the work W in joules is

$$W = \int_0^{10} (100 + 3x)g\, dx = g\left[100x + \frac{3x^2}{2} \right]_0^{10}$$

$$= (1{,}000 + 150)g = 1{,}150g$$

Using $g = 9.81$ m/s^2, we get $W \approx 11{,}280$ J.   ■

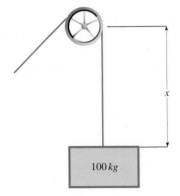

FIGURE 6.62

Hooke's Law for Springs

The force necessary to stretch a spring x units beyond its natural length is given by $F(x) = kx$ (provided the force does not exceed what is called the *elastic limit* of the material from which the spring is made). The factor k is a constant, called the *spring constant*. This result is known as **Hooke's Law**, after the scientist Robert Hooke, a seventeenth-century contemporary (and rival) of Isaac Newton. It also holds true if the spring is compressed x units from its natural length.

EXAMPLE 6.26   A certain spring has a natural length of 50 cm, and a 5-N force is required to stretch it 0.15 m. Find the work done in stretching it from its natural length to a length of 60 cm. Find the additional work done in stretching it 10 cm more (from 60 cm to 70 cm).

Solution   By Hooke's Law, $F(x) = kx$, and we know that $F(x) = 5$ when $x = 0.15$. So we can find k:

$$5 \text{ kg} = k(0.15 \text{ m})$$

$$k = \frac{5}{0.15} = \frac{100}{3} \text{ N/m}$$

Thus, for the spring,

$$F(x) = \frac{100}{3}x$$

where $F(x)$ is in newtons and x is in meters. In stretching the spring from its natural length of 50 cm to a length of 60 cm, x varies from 0 to 0.1, since the spring is stretched 10 cm $= 0.1$ m. (Remember that x is the distance in meters that the spring is stretched *beyond its natural length*.) So we have

$$W = \int_0^{0.1} F(x)dx = \int_0^{0.1} \frac{100}{3}x \, dx = \frac{100}{3}\left[\frac{x^2}{2}\right]_0^{0.1}$$

$$= \frac{0.5}{3} \approx 0.17 \text{ J}$$

For the spring to be stretched an additional 10 cm, x will vary from 0.1 m to 0.2 m, so the additional work done is

$$W = \int_{0.1}^{0.2} \frac{100}{3}x \, dx = \frac{100}{3}\left[\frac{x^2}{2}\right]_{0.1}^{0.2} = \frac{50}{3}(0.04 - 0.01) = 0.5 \text{ J}$$

Note that the work required to stretch the spring the second 10 cm was almost three times that required to stretch it the first 10 cm.   ∎

Pumping Liquid from Tanks

The next example is typical of a class of problems. In preparation, let us consider a tank of height b units filled with a liquid to a depth a units from the top. Figure 6.63 shows this situation. We want to find the work done in pumping the liquid over the top rim of the tank. We can imagine pumping (lifting) the liquid out of the tank one thin slab at a time and applying Equation 6.29 for the work done by a constant force (the weight of the slab) to each slab. The total work done is approximately equal to the sum that we get by adding up the work of lifting all the slabs. We will see that this sum is a Riemann sum whose limit, as the slab thicknesses approach zero, is an integral. The exact form of the integral will depend on the weight of the liquid and the cross-sectional area of the container.

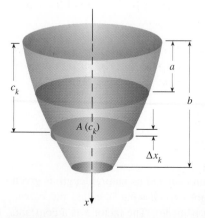

FIGURE 6.63

Suppose the density of the liquid is ρ, defined as mass per unit volume (for example, kilograms per cubic meter (kg/m^3)). As previously suggested, we can think of the liquid as being divided into thin layers by planes passed perpendicular to the vertical axis at points $\{x_0, x_1, x_2, \ldots, x_n\}$ that form a partition of the closed interval $[a, b]$. Suppose at the distance x units down from the top of the tank, the cross-sectional area $A(x)$ is known and is a continuous function of x on $[a, b]$. As in our study of volumes by the cross-sectional method, the kth layer is approximated by a disk of constant cross-sectional area $A(c_k)$ and thickness $\Delta x_k = x_k - x_{k-1}$, where c_k is any point in $[x_{k-1}, x_k]$. We have shown a typical disk in Figure 6.63. We imagine this "liquid disk" to be lifted to the top. The work required to do this lifting is the weight of the disk (a force) multiplied by the distance c_k through which it is lifted. The weight of the kth disk is its volume $A(c_k)\Delta x_k$ times its weight per unit volume, ρg (ρ is mass per unit volume and weight is mass times the acceleration g of gravity). So the work ΔW_k of lifting this weight to the top is

$$\Delta W_k = \overbrace{\rho g A(c_k)\Delta x_k}^{\text{force}} \cdot \overbrace{c_k}^{\text{distance}}$$

The total work is approximated by the sum

$$W \approx \sum_{k=1}^{n} \Delta W_k = \sum_{k=1}^{n} \rho g c_k A(c_k)\Delta x_k$$

On taking the limit as the thicknesses Δx_k all approach 0, we arrive at the result

$$W = \int_a^b \rho g x A(x) dx \tag{6.31}$$

REMARK

■ As we have noted, the product ρg is weight per unit volume. For this reason it is sometimes called *weight density*. For water, in SI units, $\rho = 1,000$ kg/m^3. So, near the earth's surface, $\rho g \approx (1,000)(9.81) = 9,810$ N/m^3. In the English system, the weight density is given directly. For example, for water near the earth's surface, $\rho g \approx 62.4$ lb/ft^3.

EXAMPLE 6.27   A water tank 8 m high is in the form of a frustum of a cone, as shown in Figure 6.64. It is filled to a depth of 5 m. Find the work done in pumping all the water over the top of the tank.

Solution   As in the preceding discussion, we let x be the distance from the top of the tank to a typical element. To find the cross-sectional area $A(x)$, we need to find the radius r. By similar triangles (see Figure 6.65), we have

$$\frac{r - 2}{4} = \frac{8 - x}{8}$$

So

$$r - 2 = \frac{4}{8}(8 - x)$$

or

$$r = 2 + \frac{1}{2}(8 - x) = \frac{1}{2}(12 - x)$$

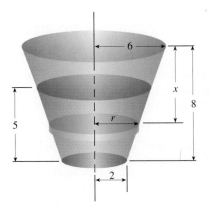

FIGURE 6.64

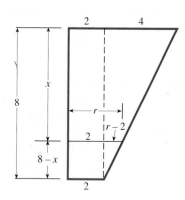

FIGURE 6.65

Thus,

$$A(x) = \pi r^2 = \frac{\pi}{4}(12 - x)^2$$

The density of water is 1000 kg/m³. So, by Equation 6.31, we have (taking g = 9.81 m/s²),

$$W = \int_a^b \rho g x A(x)\, dx = \int_3^8 9810x \left(\frac{\pi}{4}\right)(12 - x)^2\, dx$$

$$= \frac{9810\pi}{4} \int_3^8 x(144 - 24x + x^2)\, dx$$

$$= \frac{9810\pi}{4}\left[72x^2 - 8x^3 + \frac{x^4}{4}\right]_3^8$$

$$\approx 8.350 \times 10^6 \text{ J} \quad \blacksquare$$

Exercise Set 6.6

1. A variable force of $2x - x^2$ N acts in the direction of the positive x-axis from $x = 0$ to $x = 2$ (in meters). Find the work done by the force.

2. A weight of 200 lb is to be pushed a distance of 12 ft up a smooth ramp that makes an angle of 30° with the horizontal. Neglecting any friction, find the work done.

3. A force of 4 lb is required to stretch a certain spring 3 in. beyond its natural length of 10 in. Find the work done in stretching it from its natural length to a length of 16 in.

4. For the spring in Exercise 3, find the work done in stretching it from a length of 15 in. to a length of 18 in.

5. A spring of natural length 20 cm is stretched to a length of 28 cm by a force of 5 N. Find the work done in compressing the spring 6 cm from its natural length. What is the work done in stretching the spring from a length of 22 cm to a length of 30 cm?

6. A heavy-duty spring with spring constant 10,000 N/m is used to cushion the shock of a subway car at the end of the line in the event it does not come to a complete stop before bumping the end. Find the work done by the force of the subway car if in bumping the end, the spring is compressed a distance of 1 m.

7. The work done in stretching a spring from its natural length of 25 cm to a length of 30 cm is 0.5 J. Find the spring constant.

8. A spring of natural length 14 in. requires 30 ft-lb of work

to be compressed from a length of 12 in. to a length of 8 in. Find the spring constant. (Be careful with units.)

9. A uniform chain weighing 3 lb/ft is hanging over the edge of a building and is to be pulled up to the top. If a 30-ft length of chain initially hangs down, find the work required to raise it to the top.

10. A crane at a construction site is used to lift a bucket of cement from the ground to a point 10 m above the ground. The distance from the end of the crane arm to the ground is 25 m. If the bucket of cement has mass 250 kg and the cable has mass 7 kg/m, find the work required to lift the cement.

11. Water is to be raised from a well 15 m deep by means of a bucket attached to a rope. When the bucket is full of water its mass is 15 kg, but the bucket has a leak that causes it to lose water at the constant rate of $\frac{1}{4}$ kg for each meter that the bucket is raised. Neglecting the mass of the rope, find the work done in raising the bucket to the top.

12. A water tank is in the form of a vertical right circular cylinder with radius 2 m and height 8 m. If the tank is half full of water, find the work required to pump all the water over the top rim.

13. In Exercise 12 suppose the tank is initially full of water. Find the work required to pump all of it to a point 3 m above the top of the tank.

14. A storage tank for liquid ammonia is in the form of a sphere with radius 3 m. If the tank is full, find the approximate work required to pump half of the ammonia to a point 5 m above the top of the tank. The density of ammonia is $\rho = 891$ kg/m^3.

15. According to Coulomb's Law, the repulsive force between two like electrostatic charges Q_1 and Q_2 is given by

$$F = \frac{kq_1q_2}{x^2}$$

where q_1 and q_2 are the magnitudes of Q_1 and Q_2, respectively, x is the distance between them, and k is a constant of proportionality. Suppose $q_1 = 1$ C (coulomb) and $q_2 = 2$ C, and the charges are initially 6 cm apart. If Q_1 is stationary, find the work done in moving Q_2 on a line toward Q_1 to a point 3 cm from Q_1. (Leave the answer in terms of k.)

16. According to Newton's Universal Law of Gravitation, the gravitational force exerted by the earth on a body of mass m at a distance x above the earth's surface is

$$F(x) = G\frac{mM}{(x+R)^2}$$

where R is the radius of the earth, M is the earth's mass, and G is a constant called the universal gravitational constant. It can be shown that $G = gR^2/M$. Taking $g = 9.8$ m/s^2 and $R = 6.37 \times 10^6$ m, find the work done in lifting a satellite of mass 3.2×10^4 kg a distance of 300 km (300 km $= 3 \times 10^5$ m) above the earth's surface.

17. Hooke's Law can be extended to metal bars under tension or compression. The force necessary to elongate the bar, or compress it, by x units is given by

$$F(x) = \frac{EA}{L}x$$

where A is the cross-sectional area of the bar and L is its length. E is a constant called the *modulus of elasticity*

that is a property of the metal. Suppose a cylindrical aluminum bar of natural length 15 cm and cross-sectional radius 3 cm is stretched to a length of 15.5 cm. Find the work done. For aluminum, $E = 7.30 \times 10^{10}$ N/m^2.

18. A water tank is in the form of a frustum of a cone that has bottom radius 5 ft, top radius 8 ft, and height 12 ft. If the tank is initially full of water, find the work done in pumping half the water to a point 10 ft above the top. (Use $\rho g = 62.4$ lb/ft^3.)

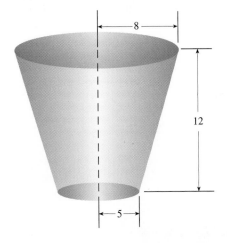

19. A swimming pool slopes from a depth of 1 m at the shallow end to a depth of 4 m at the deep end. It is 20 m long and 10 m wide. If it is full of water, find the work required to pump all the water over the top. (Use $\rho = 1000$ kg/m^3 and $g = 9.81$ m/s^2.)

20. A crane raises a bucket of sand weighing 500 lb from the ground to a height of 40 ft in 10 seconds. Sand spills out of the bucket at the rate of 10 lb/s as it is being raised. If the cable weighs 5 lb/ft, find the work done in raising the sand.

21. A tank has the shape of the surface generated by revolving the right half of the parabola $y = x^2/2$ from $x = 0$ to $x = 2$ about the y-axis. If x and y are measured in meters and the tank is initially half full of water (half of the volume), find the work done in pumping all the water over the top.

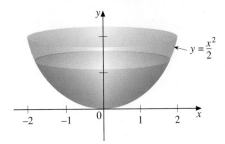

22. The force on a piston caused by the expansion of a gas in a cylinder (see the figure) is $F = pA$, where A is the area of the face of the piston and p is the pressure (force per unit area) of the gas. The volume v occupied by the gas, when the distance from the piston to the end of the cylinder is x, is given by $v = Ax$. For an ideal gas under an adiabatic process (in which no heat is transferred into or out of the cylinder), the relationship between pressure and volume is $pv^{1.4} = c$, where c is a constant. If when $x = a$ the volume is v_1 and when $x = b$ the volume is v_2, show that the work done by the gas in expanding from v_1 to v_2 is

$$W = \int_a^b F(x)dx = \int_{v_1}^{v_2} cv^{-1.4}\,dv$$

If the radius of the piston is 3 in., find the work done by the gas in moving the piston from $x = 5$ in. to $x = 8$ in. Express the answer in terms of c.

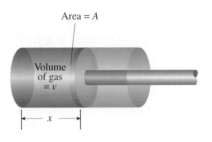

Area = A

Volume of gas = v

x

6.7 FLUID PRESSURE

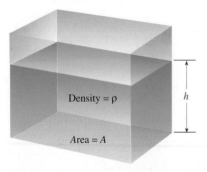

The Hoover Dam

Density = ρ

h

Area = A

FIGURE 6.66
Force on plate = $\rho g A h$

Our objective in this section is to show how integration can be used to calculate the force exerted by a liquid pressing against one side of a flat surface. For example, we can find the force of the water in a lake against the side of a dam. (We ignore the curvature of the dam's face.) We will consider only static forces rather than dynamic ones. This limitation means that we will suppose the liquid is not in motion.

Suppose the liquid in question has density ρ (mass per unit volume). If a flat plate of area A is submerged horizontally at a depth h (see Figure 6.66), then the total force on the plate is just the weight of the liquid directly above the plate. The volume occupied by this liquid is the area times the height, Ah. We multiply this volume by the weight per unit volume, ρg, to get the total weight of the liquid. So

$$F = \rho g A h$$

For example, the force exerted by the oil in an oil storage tank on the bottom of the tank is the area of the bottom times the depth of the oil times its density. Note that the force does not depend on the shape of the submerged plate, but only on its area and the depth.

The **pressure** exerted by the liquid is defined as the force on each square unit of area. So, taking $A = 1$, we get for the pressure p at depth h,

$$p = \rho g h \tag{6.32}$$

As the depth increases, the pressure also increases, which explains why dams must be thicker at the bottom than at the top. It is an interesting fact, known as *Pascal's Principle*, that the pressure p at a given depth is the same in all directions. Thus, if a plate is submerged horizontally, vertically, or on a slant, the pressure at a fixed depth h units below the surface is given in all cases by Equation 6.32.

When a plate of known area is submerged horizontally, calculating the force on it is straightforward, as we have seen. When it is submerged vertically, however, finding the force is more involved. After all, the pressure varies continuously from the top of the plate to the bottom. We introduce a vertical axis that we show in Figure 6.67 as being directed positively downward. (In certain applications it will be more convenient to choose the positive direction upward.) The origin may be taken at any point, usually determined by the shape of the plate. In the examples we will see how certain choices can often simplify calculations. We assume that the vertical extremities of the plate are at $x = a$ and $x = b$, where $a < b$. We further assume that enough information is known about the shape of the plate that its width, $w(x)$, can be determined at each value of x in the interval. Finally, we assume that w is a continuous function of x on the closed interval $[a, b]$. We denote by $h(x)$ the depth at x.

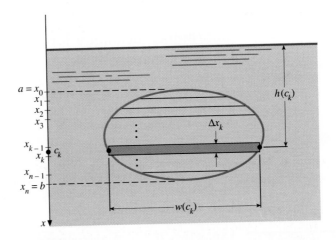

FIGURE 6.67

We partition the interval $[a, b]$ in the usual way. Horizontal lines drawn at the partition points divide the plate into thin strips. The kth strip can be approximated by a rectangle that has dimensions $w(c_k)$ by Δx_k, where c_k is an intermediate point in the kth subinterval. A typical such rectangle is shown in Figure 6.67. Although the pressure varies from the top to the bottom of this rectangle, if Δx_k is very small, the error introduced by assuming the entire rectangle is at depth $h(c_k)$ will not be great. The pressure at this depth, by Equation 6.32, is

$$p = \rho g h(c_k)$$

We approximate the total force on the kth rectangle, which we designate by ΔF_k, by its area times this pressure:

$$\Delta F_k \approx \overbrace{\rho g}^{\substack{\text{weight} \\ \text{density}}} \cdot \overbrace{h(c_k)}^{\text{depth}} \cdot \overbrace{w(c_k)\Delta x_k}^{\text{area}}$$

It is useful to learn this result in words:

$$\text{force on a rectangular strip} \approx \text{weight density} \times \text{depth} \times \text{area}$$

We approximate the total force as the sum of all ΔF_k values:

$$F \approx \sum_{k=1}^{n} \Delta F_k \approx \sum_{k=1}^{n} \rho g h(c_k) w(c_k) \Delta x_k$$

As we take finer and finer partitions, the Riemann sum on the right approaches the integral $\int_a^b \rho g h(x) w(x) dx$. This discussion leads to the following definition.

| Definition 6.5 | The total **force** F on one side of a flat plate submerged vertically in a liquid of density ρ is given by |
| Fluid Force | |

$$F = \int_a^b \rho g h(x) w(x)\, dx \tag{6.33}$$

where the vertical bounds of the plate are $x = a$ and $x = b$, respectively, $h(x)$ is the depth at x, and $w(x)$ is the width of the plate at x.

Shorthand Technique for Finding Force on a Submerged Plate

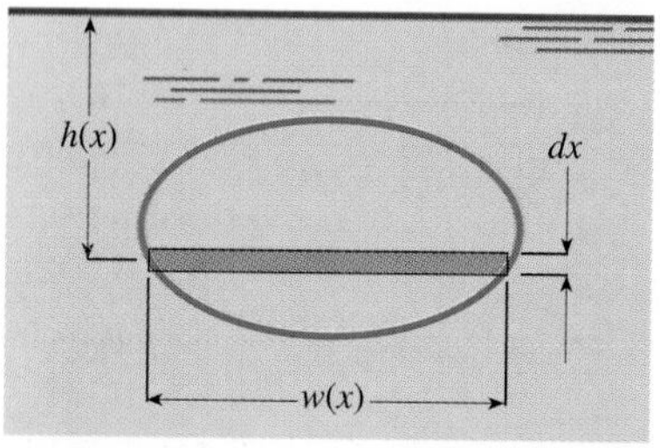

FIGURE 6.68

As a memory device, we can indicate the width of a typical rectangular element by dx, its length by $w(x)$, and its depth by $h(x)$, as shown in Figure 6.68. So the force on this element is

$$\text{weight density} \times \text{depth} \times \text{area} = \rho g h(x) w(x)\, dx$$

"Summing up" all such elements of force (in the sense of integration) we get Equation 6.33. Again, this technique is only a memory device, not a mathematically rigorous derivation.

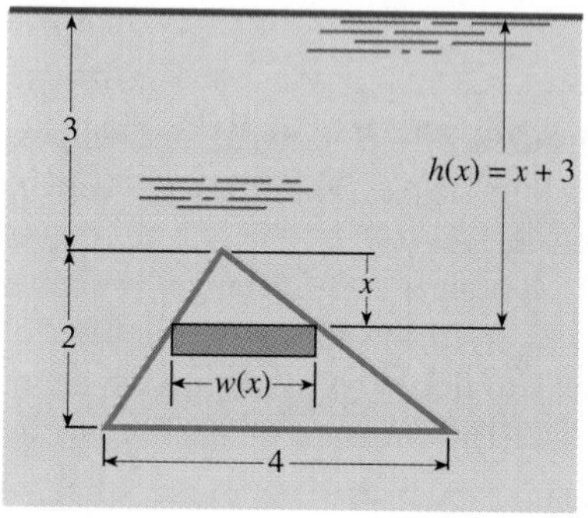

EXAMPLE 6.28   A triangular plate with base 4 m and altitude 2 m is submerged vertically in a liquid of density ρ kg/m^3. It is oriented as in Figure 6.69, with its upper vertex 3 m below the surface of the liquid. Find the total force on one side of the plate.

Solution   We choose the origin for x as the upper vertex and direct x positively downward, as shown. The limits on x are then 0 and 2. To find the width $w(x)$, we use similar triangles:

FIGURE 6.69

$$\frac{w(x)}{4} = \frac{x}{2}$$

$$w(x) = 2x$$

The depth $h(x)$ is given by $h(x) = x + 3$. Thus, by Equation 6.33,

$$F = \int_0^2 \rho g (x + 3)(2x)\, dx$$

$$= 2\rho g \int_0^2 (x^2 + 3x)\, dx$$

$$= 2\rho g \left[\frac{x^3}{3} + \frac{3x^2}{2} \right]_0^2$$

$$= \frac{52 \rho g}{3} \text{ N}$$

In the exercises you will be asked to rework this example, choosing the origin for x as the level of the fluid so that you can see the advantages of the choice we made.   ∎

EXAMPLE 6.29   A dam across a river gorge is in the form of an inverted isosceles trapezoid, with upper base 100 ft, lower base 50 ft, and altitude 80 ft (Figure 6.70). Find the force of the water on the face of the dam (assumed vertical) when the water is within 10 ft of the top of the dam.

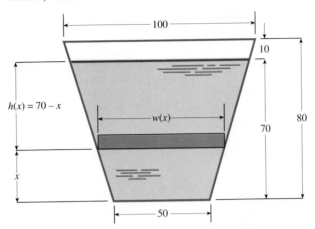

FIGURE 6.70

Solution   Because the trapezoid is inverted, it is convenient to choose as the origin for x the bottom of the dam and take the positive direction as upward. Doing so simplifies the calculation of the width $w(x)$, as we will show. The limits on x are then 0 and 70, since it is only this portion of the dam that has water against it. The depth from the water level to a point with coordinate x is $h(x) = 70 - x$. To find the width $w(x)$, we again make use of similar triangles, as shown in Figure 6.71. From this figure we see that $w(x) = 50 + 2d$, and using similar triangles, we have

$$\frac{d}{x} = \frac{25}{80}$$

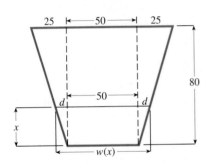

FIGURE 6.71

So $d = 5x/16$ and

$$w(x) = 50 + 2\left(\frac{5x}{16}\right)$$

$$= 50 + \frac{5x}{8}$$

$$= \frac{5}{8}(80 + x)$$

Now, using Equation 6.33, we have

$$F = \int_0^{70} \rho g (70 - x) \cdot \frac{5}{8}(80 + x)\,dx$$

$$= \frac{5\rho g}{8} \int_0^{70} (5{,}600 - 10x - x^2)\,dx$$

$$= \frac{5\rho g}{8}\left[5{,}600x - 5x^2 - \frac{x^3}{3}\right]_0^{70}$$

$$= \frac{5\rho g}{8}\left(\frac{759{,}500}{3}\right)$$

Taking $\rho g = 62.4$ lb/ft^3 as the approximate density of water, we finally get $F = 9{,}873{,}500$ lb. At 2,000 lb/ton, this force is nearly 5,000 tons.   ∎

Exercise Set 6.7

In Exercises 1–6, a plate is submerged vertically in a liquid of density ρ, as shown. Find the force of the liquid on one side of the plate.

1.

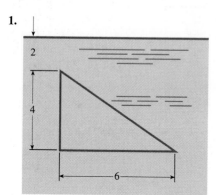

2.

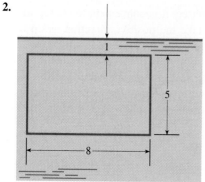

3.

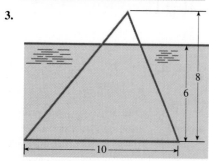

4.

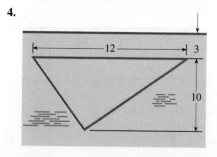

5.

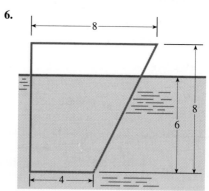

6.

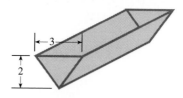

7. The ends of a watering trough are in the form of inverted isosceles triangles, having base 3 ft and altitude 2 ft. Find the force of the water in the trough on one of the ends when the trough is full of water. What is the force when the water level is 1 ft?

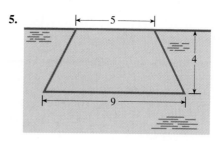

8. If instead of a triangle, each end of the watering trough in Exercise 7 forms the lower half of a circle of radius 2 ft, find the force on one of the ends when the trough is full of water.

9. A swimming pool is 10 m wide and 15 m long. At the shallow end its depth is 1 m, and at the deep end it is 4 m; the depth increases linearly from the shallow to the deep end. When the pool is full of water find the force on (a) each of the ends and (b) each of the sides.

10. An oil drum is cylindrical, with diameter 1 m. If it is half full of oil that has density 920 kg/m³, and it is lying on its side, find the force on each end.

11. Rework Example 6.28, choosing the level of the fluid as the origin for x.

12. Rework Example 6.29, choosing the water level as the origin for x.

13. The plate in the accompanying figure is parabolic and is submerged as shown in a fluid of density ρ. Find the force on one side of the plate. (*Hint:* If the origin is taken at the vertex of the parabola, its equation is of the form $y = ax^2$. Find a, and use y instead of x as the variable of integration.)

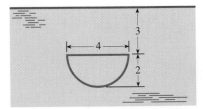

14. Each end of a gasoline tank is elliptical, with equation $x^2 + 4y^2 = 1$ when referred to a coordinate system with origin at the center of the ellipse and y-axis vertical, where x and y are in feet. Find the force on one end of the tank when it is half full. Take the weight density of gasoline as 45 lb/ft^3. (*Hint:* Use y as the variable of integration.)

15. Show that if a plate is submerged at an angle θ with the vertical, where $0 < \theta < \frac{\pi}{2}$, the force on it is given by

$$F = \int_a^b \rho g h(x) w(x) \sec \theta \, dx$$

with $h(x)$ and $w(x)$ having the same meaning as in Definition 6.5, with the x-axis directed vertically downward.

16. Find the force on the bottom of the swimming pool in Exercise 9. (Use the result of Exercise 15.)

17. A porthole in an undersea observation laboratory is circular, with a 0.3 m diameter. Assume the porthole lies in a vertical plane. If its center is at a depth of 400 m, find the force on the porthole. (The density of seawater is approximately 1040 kg/m^3.) (*Hint:* In carrying out the integration, use your knowledge of the area of the circle.)

18. Use Simpson's Rule with $n = 10$ to approximate the force on a vertical plate bounded by the x-axis and the top half of the hypocycloid $x^{2/3} + y^{2/3} = 4^{2/3}$, if the plate is submerged in water with its base 10 m beneath the surface.

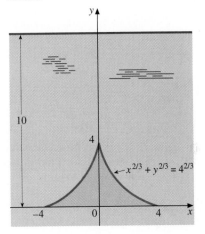

6.8 MOMENTS, CENTROIDS, AND CENTERS OF MASS

In this section we present some applications of the integral that are especially important in physics and engineering. The ideas are also used in statistics.

Children playing on a seesaw quickly learn that the heavier child must sit closer to the pivot point, called the *fulcrum*, to bring the seesaw into balance. Denoting the weights of the children by w_1 and w_2 and the corresponding distances from the fulcrum by d_1 and d_2, as in Figure 6.72, we must have $w_1 d_1 = w_2 d_2$ for the seesaw to be balanced. This relation is called the *law of the lever* and was known to Archimedes.

Now consider a finite number of weights, $w_1, w_2, \ldots, w_n$, placed at various points on an axis, with w_i at x-coordinate x_i, as in Figure 6.73. We wish to find the location of the balance point, also called the *center of gravity*, of the system. Denote its x-coordinate by $\bar{x}$. The product of a weight (or force) and its distance from a point is called the **moment** produced by that weight (or force)

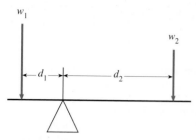

FIGURE 6.72

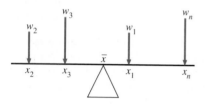

FIGURE 6.73

about that point. The moment thus is a measure of the tendency for the weight to cause a turning, or rotation, about the point. The directed distance of the weight from the point is called the **moment arm** of the weight. If the moment arm is positive, then the weight tends to produce a clockwise rotation, and if negative, a counterclockwise rotation. In Figure 6.73, for example, the moment arm of w_1 from the center of gravity is the positive distance $x_1 - \bar{x}$, and the moment about this point is $w_1(x_1 - \bar{x})$. The moment arm of w_3 is $x_3 - \bar{x}$, which is a negative number, and so w_3 produces the negative moment $w_3(x_3 - \bar{x})$. For balance, it must be true that the algebraic sum of all the moments is 0:

$$\sum_{i=1}^{n} w_i(x_i - \bar{x}) = 0$$

We can now solve for $\bar{x}$ from this equation:

$$\sum_{i=1}^{n} w_i x_i - \sum_{i=1}^{n} \bar{x} w_i = 0$$

Since $\bar{x}$ is a constant, we can factor it out of the second summation and then solve, getting

$$\bar{x} = \frac{\displaystyle\sum_{i=1}^{n} w_i x_i}{\displaystyle\sum_{i=1}^{n} w_i}$$

Since by Newton's Second Law the weight w of an object is its mass m times the acceleration of gravity g, we have for each i, $w_i = m_i g$, so that $\bar{x}$ can be rewritten in terms of masses as

$$\bar{x} = \frac{\displaystyle\sum_{i=1}^{n} m_i g x_i}{\displaystyle\sum_{i=1}^{n} m_i g} = \frac{\displaystyle\sum_{i=1}^{n} m_i x_i}{\displaystyle\sum_{i=1}^{n} m_i} \tag{6.34}$$

(assuming g is a constant and can therefore be divided out). Because of the shift in emphasis from weight to mass as given by Equation 6.34, it is customary to use the term **center of mass** instead of center of gravity for the balance point of the system. (Only if the force of gravity varies over the distance being studied is the center of gravity different from the center of mass.) The product $m_i x_i$ is the moment of mass m_i about the origin, since x_i is the directed distance of m_i from the origin. Thus, we can restate the result of Equation 6.34 in words as

$$\bar{x} = \frac{\text{sum of moments about the origin}}{\text{sum of masses}}$$

If we let $m = \sum_{i=1}^{n} m_i$ (the total mass m equals the sum of the individual masses m_i), then on clearing Equation 6.34 of fractions, we see that

$$m\bar{x} = \sum_{i=1}^{n} m_i x_i$$

From this result, we conclude that if the masses were all concentrated at $\bar{x}$—that is, at the center of mass—the moment about the origin of the total mass m would equal the sum of the moments about the origin of the individual masses.

EXAMPLE 6.30   Three masses, $m_1 = 35$ g, $m_2 = 56$ g, and $m_3 = 47$ g, are located on the x-axis at distances $x_1 = 15$, $x_2 = 36$, and $x_3 = -14$, respectively. Distances are in cm. Find the x-coordinate of the center of mass.

Solution   By Equation 6.34,

$$\bar{x} = \frac{(35)(15) + 56(36) + (47)(-14)}{35 + 56 + 47} = \frac{1883}{138} \approx 13.6 \text{ cm}$$ ∎

Continuous One-Dimensional Mass

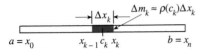

FIGURE 6.74

So far, we have been talking about a discrete system of masses. Now we consider a continuously distributed mass along a line, such as a thin rod or a wire. For definiteness, let us say it is a rod, and that we have introduced an x-axis as in Figure 6.74 such that the rod reaches from $x = a$ to $x = b$. Imagine that we cut the rod into n small segments by means of the partition points $a = x_0 < x_1 < x_2 < \cdots < x_n = b$. In Figure 6.74 we show a typical segment. Denote its center coordinate by c_k.

We will allow for the possibility that the density of the rod may vary with respect to distance along the axis, where density is defined as mass per unit length. (In general, density is mass per unit volume, but we are considering our rod to be essentially one-dimensional.) We denote the density by $\rho(x)$, and assume that ρ is continuous on the closed interval $[a, b]$. Let Δm_k denote the mass of the typical segment of the rod we have shown in Figure 6.74. If its length, Δx_k, is small, then the density over this short distance can be considered approximately the constant value $\rho(c_k)$. Thus,

$$\Delta m_k \approx \rho(c_k)\Delta x_k$$

We can consider our n small segments, $\Delta m_1, \Delta m_2, \ldots, \Delta m_n$, as point masses concentrated at the points $c_1, c_2, \ldots, c_n$, respectively. Thus, by Equation 6.34, the center of mass of the rod is approximately

$$\bar{x} \approx \frac{\sum_{k=1}^{n}(\Delta m_k)c_k}{\sum_{k=1}^{n}\Delta m_k} \approx \frac{\sum_{k=1}^{n}\rho(c_k)c_k\Delta x_k}{\sum_{k=1}^{n}\rho(c_k)\Delta x_k}$$

The numerator is a Riemann sum for the function $\rho(x)x$, and the denominator is a Riemann sum for the function $\rho(x)$. As we take finer and finer partitions, we get, from the limits on the numerator and denominator, the following formula for $\bar{x}$:

Center of Mass of a One-Dimensional Continuous Mass

$$\bar{x} = \frac{\int_a^b \rho(x)x\,dx}{\int_a^b \rho(x)dx} \qquad (6.35)$$

EXAMPLE 6.31   A rod 20 cm long has density given by $\rho(x) = (x+1)/2$. Find its center of mass.

Solution   By Equation 6.35,

$$\bar{x} = \frac{\displaystyle\int_0^{20} \left(\frac{x+1}{2}\right) x\, dx}{\displaystyle\int_0^{20} \left(\frac{x+1}{2}\right) dx} = \frac{\displaystyle\frac{1}{2}\int_0^{20} (x^2+x)\,dx}{\displaystyle\frac{1}{2}\int_0^{20} (x+1)\,dx} = \frac{\left.\dfrac{x^3}{3}+\dfrac{x^2}{2}\right]_0^{20}}{\left.\dfrac{x^2}{2}+x\right]_0^{20}}$$

$$= \frac{430}{33} \approx 13.03 \text{ cm}$$

Note that the center of mass is to the right of the geometric center, which makes sense because the density increases as we move from left to right. ∎

Point Masses in the Plane

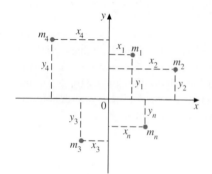

FIGURE 6.75

Now we extend these ideas to a system of masses in the xy-plane, as shown in Figure 6.75. We assume each mass to be concentrated at a point and refer to it as a *point mass*. The concept of a point mass is, of course, idealized. Let the point mass m_i be located at (x_i, y_i). If l is any line in the plane and d_i is the directed distance from l to (x_i, y_i), then the moment of m_i about l is $m_i d_i$, and the sum of the moments of all the masses about l is $\sum_{i=1}^{n} m_i d_i$. In particular, the sum of the moments about the x-axis is $\sum_{i=1}^{n} m_i y_i$ and about the y-axis is $\sum_{i=1}^{n} m_i x_i$. The point $(\bar{x}, \bar{y})$ for which the sum of the moments about the lines $x = \bar{x}$ and $y = \bar{y}$ each equals 0 is the center of mass of the system. By proceeding as in the one-dimensional case, we find that

$$\bar{x} = \frac{\displaystyle\sum_{i=1}^{n} m_i x_i}{\displaystyle\sum_{i=1}^{n} m_i} = \frac{\text{sum of moments about } y\text{-axis}}{\text{sum of masses}}$$

$$\bar{y} = \frac{\displaystyle\sum_{i=1}^{n} m_i y_i}{\displaystyle\sum_{i=1}^{n} m_i} = \frac{\text{sum of moments about } x\text{-axis}}{\text{sum of masses}}$$

(6.36)

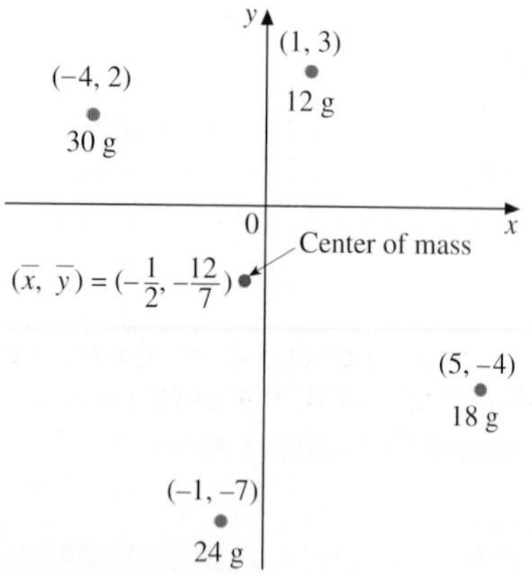

EXAMPLE 6.32   Four point masses, of 12 g, 18 g, 24 g, and 30 g, are located in the plane as shown in Figure 6.76. Find the center of mass.

Solution   By Equations 6.36, we have

$$\bar{x} = \frac{(1)(12) + (5)(18) + (-1)(24) + (-4)(30)}{12 + 18 + 24 + 30} = -\frac{42}{84} = -\frac{1}{2}$$

$$\bar{y} = \frac{(3)(12) + (-4)(18) + (-7)(24) + 2(30)}{84} = -\frac{144}{84} = -\frac{12}{7}$$

So the center of mass is located at the point $(-\frac{1}{2}, -\frac{12}{7})$, as indicated in Figure 6.76. ∎

FIGURE 6.76

Continuous Mass in the Plane—A Lamina

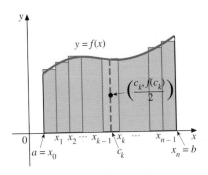

FIGURE 6.77

If, instead of a system of discrete point masses in the plane, we have a continuously distributed mass, called a **lamina** (such as a thin sheet of material), we divide the lamina into vertical or horizontal strips (depending on its shape) and treat each strip as a point mass. For the present we will suppose the lamina has constant density ρ. When the density is constant, the lamina is said to be *homogeneous*. In multivariable calculus, using the concept of the double integral, we could consider a variable density. With the density constant, we can assume that each strip is a point mass concentrated at its geometric center.

In Figure 6.77 we show a lamina that occupies an area below the graph of a continuous nonnegative function, $y = f(x)$, between $x = a$ and $x = b$. We partition $[a, b]$ in the usual way, obtaining n rectangles. In the figure we show a typical rectangle, where we have taken c_k as the midpoint of the subinterval $[x_{k-1}, x_k]$. The "mass" of this rectangle is its area times the density ρ. Letting Δm_k be this mass, we have

$$\Delta m_k = \rho f(c_k)\Delta x_k$$

Since the density is constant, we can consider the rectangular element as a point mass concentrated at its geometric center,

$$\left(c_k, \frac{f(c_k)}{2}\right)$$

By the first of Equations 6.36, then, we have

$$\bar{x} \approx \frac{\displaystyle\sum_{k=1}^{n} c_k \Delta m_k}{\displaystyle\sum_{k=1}^{n} \Delta m_k}$$

and by the second,

$$\bar{y} \approx \frac{\displaystyle\sum_{k=1}^{n} \frac{f(c_k)}{2}\Delta m_k}{\displaystyle\sum_{k=1}^{n} \Delta m_k}$$

where $\Delta m_k = \rho f(c_k)\Delta x_k$. In each case, the numerator and denominator are Riemann sums. So as we let $n \to \infty$, with all $\Delta x_i \to 0$, the sums approach integrals. To simplify notation, we write

$$dm = \rho f(x)dx$$

and call dm the *differential of mass*. Then we can write $\bar{x}$ and $\bar{y}$ as follows.

Center of Mass of a Homogeneous Lamina
Under the Graph of $y = f(x)$ from $x = a$ to $x = b$

$$\bar{x} = \frac{\displaystyle\int_a^b x\,dm}{\displaystyle\int_a^b dm} \qquad \text{and} \qquad \bar{y} = \frac{\displaystyle\int_a^b \frac{f(x)}{2}\,dm}{\displaystyle\int_a^b dm} \qquad (6.37)$$

By replacing dm by $\rho f(x)\,dx$ and noting that since ρ is a constant, we can divide numerator and denominator by ρ to get the equivalent equations

$$\bar{x} = \frac{\displaystyle\int_a^b xf(x)\,dx}{\displaystyle\int_a^b f(x)\,dx} \qquad \text{and} \qquad \bar{y} = \frac{\displaystyle\int_a^b \frac{[f(x)]^2}{2}\,dx}{\displaystyle\int_a^b f(x)\,dx}$$

REMARK
■ If in the second of Equations 6.37 we replace $f(x)$ by y, we obtain

$$\bar{y} = \frac{\dfrac{1}{2}\displaystyle\int_a^b y\,dm}{\displaystyle\int_a^b dm}$$

(Of course, to apply this formula, we do have to replace y with its equivalent value in terms of x.) Since

$$\bar{x} = \frac{\displaystyle\int_a^b x\,dm}{\displaystyle\int_a^b dm}$$

there is a lack of symmetry in the equations for $\bar{x}$ and $\bar{y}$ because the factor $\frac{1}{2}$ appears in the formula for $\bar{y}$. From the preceding derivation, we can see why the factor $\frac{1}{2}$ is used for $\bar{y}$ but for $\bar{x}$. The center of the typical element in Figure 6.77, thought of as a point mass, is $\left(c_k, \frac{f(c_k)}{2}\right)$. The width of this rectangular element shrinks to 0 in the limit, but the height does not. So the y-coordinate continues to require the factor $\frac{1}{2}$.

EXAMPLE 6.33   Find the center of mass of a lamina of constant density ρ in the shape of a semicircular region of radius 2, shown in Figure 6.78.

Solution   Since the lamina is of constant density and the y-axis is a line of symmetry, it follows that $\bar{x} = 0$. That is, the center of mass lies on the y-axis. To find $\bar{y}$, let us first calculate the mass, m, given by

$$m = \int_{-2}^{2} dm = \int_{-2}^{2} \rho f(x)\,dx = \int_{-2}^{2} \rho\sqrt{4-x^2}\,dx = \rho\int_{-2}^{2}\sqrt{4-x^2}\,dx = \rho A$$

where A is the area of the region. Since the region is semicircular, we know its area A is $\frac{1}{2}\pi(2)^2 = 2\pi$. Thus, $m = 2\pi\rho$.

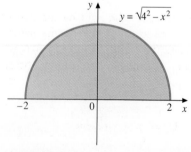

$y = \sqrt{4^2 - x^2}$

FIGURE 6.78

The numerator for $\bar{y}$ is

$$\int_{-2}^{2} \frac{f(x)}{2} dm = \frac{1}{2} \int_{-2}^{2} \sqrt{4 - x^2} \left(\rho\sqrt{4 - x^2}\, dx\right)$$

$$= \frac{\rho}{2} \int_{-2}^{2} (4 - x^2) dx$$

$$= \rho \int_{0}^{2} (4 - x^2) dx \quad \text{Since the integrand is even}$$

$$= \rho \left[4x - \frac{x^3}{3} \right]_{0}^{2} = \frac{16\rho}{3}$$

Thus,

$$\bar{y} = \frac{16\rho/3}{2\pi\rho} = \frac{8}{3\pi}$$

The center of mass, therefore, is the point

$$(\bar{x}, \bar{y}) = \left(0, \frac{8}{3\pi} \right)$$

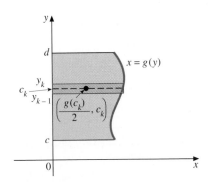

FIGURE 6.79

If the lamina occupies the region bounded by the nonnegative function $x = g(y)$, the lines $y = c$ and $y = d$, and the y-axis (see Figure 6.79), then the formulas for $\bar{x}$ and $\bar{y}$ become

$$\bar{x} = \frac{\displaystyle\int_{c}^{d} \frac{g(y)}{2} dm}{\displaystyle\int_{c}^{d} dm} \quad \text{and} \quad \bar{y} = \frac{\displaystyle\int_{c}^{d} y\, dm}{\displaystyle\int_{c}^{d} dm} \tag{6.38}$$

where $dm = \rho g(y) dy$.

Shorthand Technique for Finding the Center of Mass of a Homogeneous Lamina

A useful device for obtaining the integrals in Equations 6.37 and 6.38, as well as those for the more general cases in which the region in question is bounded by two curves, is as follows. In Figure 6.80(a) we show the width of a typical vertical element as dx and in part (b) the width of a typical horizontal strip as dy. In either case let us call the area of the strip dA and its "mass"

$$dm = \rho\, dA$$

Call the moment arm from the y-axis to the center of the strip by the name x-*arm*, and call the moment arm from the x-axis to the center by the name y-*arm*. Then the moment of this elementary mass (the strip) about the y-axis is $(x\text{-arm})\, dm$ and the moment about the x-axis is $(y\text{-arm})\, dm$. "Summing" these moments (in the sense of integration), we get the total moments M_y and M_x of the lamina,

$$M_y = \int (x\text{-arm}) dm \quad \text{and} \quad M_x = \int (y\text{-arm}) dm$$

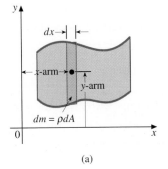

(a)

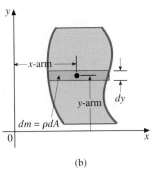

(b)

FIGURE 6.80

The appropriate limits of integration must be applied, depending on whether vertical strips (x independent) or horizontal strips (y independent) are being used. Also, x-arm and y-arm have to be calculated in each individual case and expressed in terms of the independent variable. We will illustrate how to do this shortly.

The total mass, m, of the lamina is

$$m = \int dm$$

where, again, the appropriate limits must be supplied. Now, using the earlier results, we have

$$\bar{x} = \frac{M_y}{m} \quad \text{and} \quad \bar{y} = \frac{M_x}{m}$$

We summarize our results as follows.

Center of Mass of a Homogeneous Lamina

$$\bar{x} = \frac{\int (x\text{-arm})dm}{\int dm} \quad \text{and} \quad \bar{y} = \frac{\int (y\text{-arm})dm}{\int dm} \qquad (6.39)$$

where $dm = \rho\, dA$. The appropriate limits of integration must be supplied.

EXAMPLE 6.34   Find the center of mass of the homogeneous lamina bounded by the curves $y = 2 - x^2$, $y = x$, and $x = 0$.

Solution   We show the region in Figure 6.81. Solving $y = 2 - x^2$ and $y = x$ simultaneously, we find the point of intersection to be $(1, 1)$. Denote the upper curve by y_2 and the lower curve by y_1. The center of the vertical element shown is then

$$\left(x, \frac{y_1 + y_2}{2}\right)$$

so

$$x\text{-arm} = x \quad \text{and} \quad y\text{-arm} = \frac{y_1 + y_2}{2}$$

The area of the element is

$$dA = (y_2 - y_1)dx$$

and its mass is $dm = \rho\, dA$. The total mass is

$$m = \int dm = \int_0^1 \rho(y_2 - y_1)dx$$

$$= \int_0^1 \rho[(2 - x^2) - x]dx$$

$$= \rho\left[2x - \frac{x^3}{3} - \frac{x^2}{2}\right]_0^1$$

$$= \frac{7\rho}{6}$$

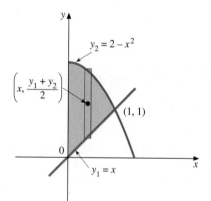

FIGURE 6.81

The denominator in each of Equations 6.39 is therefore $7\rho/6$. We calculate the numerators separately.

$$
\begin{aligned}
M_y = \int (x\text{-arm})dm &= \int_0^1 x\rho(y_2 - y_1)dx \\
&= \int_0^1 x\rho[(2 - x^2) - x]dx \\
&= \rho \int_0^1 (2x - x^3 - x^2)dx \\
&= \rho \left[x^2 - \frac{x^4}{4} - \frac{x^3}{3} \right]_0^1 \\
&= \frac{5\rho}{12}
\end{aligned}
$$

Thus,

$$
\bar{x} = \frac{M_y}{m} = \frac{5\rho/12}{7\rho/6} = \frac{5}{14}
$$

For the moment about the x-axis, we have

$$
\begin{aligned}
M_x = \int (y\text{-arm})dm &= \int_0^1 \left(\frac{y_1 + y_2}{2} \right) \rho(y_2 - y_1)dx \\
&= \frac{\rho}{2} \int_0^1 (y_2^2 - y_1^2)dx \\
&= \frac{\rho}{2} \int_0^1 [(2 - x^2)^2 - x^2]dx \\
&= \frac{\rho}{2} \int_0^1 (4 - 5x^2 + x^4)dx \\
&= \frac{\rho}{2} \left[4x - \frac{5x^3}{3} + \frac{x^5}{5} \right]_0^1 \\
&= \frac{19\rho}{15}
\end{aligned}
$$

Thus,

$$
\bar{y} = \frac{19\rho/15}{7\rho/6} = \frac{38}{35}
$$

So the center of mass is $\left(\frac{5}{14}, \frac{38}{35} \right)$.

In Equations 6.39 if we replace dm by $\rho\,dA$ we get

$$
\begin{aligned}
\bar{x} &= \frac{\int (x\text{-arm})\rho\,dA}{\int \rho\,dA} = \frac{\rho \int (x\text{-arm})dA}{\rho \int dA} = \frac{\int (x\text{-arm})dA}{\int dA} \\
\bar{y} &= \frac{\int (y\text{-arm})\rho\,dA}{\int \rho\,dA} = \frac{\rho \int (y\text{-arm})dA}{\rho \int dA} = \frac{\int (y\text{-arm})dA}{\int dA}
\end{aligned}
\tag{6.40}
$$

That is, the density ρ cancels, since it is a constant factor. The results show that $\bar{x}$ and $\bar{y}$ are determined solely by the geometry of the region. If the density were variable, this conclusion would not hold true. To distinguish between the geometric center of a region and the center of mass of a physical lamina, we refer to the former as the **centroid** of the region. For a homogeneous lamina, then, the center of mass is the same as the centroid of the region occupied by the lamina. In summary, we have the following equations.

Centroid of a Plane Region

$$\bar{x} = \frac{\int (x\text{-arm})\, dA}{\int dA} = \frac{M_y}{A}$$

$$\bar{y} = \frac{\int (y\text{-arm})\, dA}{\int dA} = \frac{M_x}{A} \qquad (6.41)$$

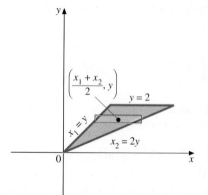

FIGURE 6.82

EXAMPLE 6.35   Find the centroid of the triangular region shown in Figure 6.82.

Solution   We choose to use horizontal elements this time to avoid having to divide the region into two parts. Denote the left boundary by x_1 and the right boundary by x_2. The center of a typical element of area is then

$$\left(\frac{x_1 + x_2}{2}, y \right)$$

so

$$x\text{-arm} = \frac{x_1 + x_2}{2} \qquad \text{and} \qquad y\text{-arm} = y$$

The area of the region is

$$A = \int dA = \int_0^2 (x_2 - x_1)\, dy = \int_0^2 (2y - y)\, dy = \left. \frac{y^2}{2} \right]_0^2 = 2$$

The moment with respect to the y-axis is

$$M_y = \int (x\text{-arm})\, dA = \int_0^2 \left(\frac{x_1 + x_2}{2} \right)(x_2 - x_1)\, dy$$

$$= \frac{1}{2} \int_0^2 (x_2^2 - x_1^2)\, dy$$

$$= \frac{1}{2} \int_0^2 [(2y)^2 - y^2]\, dy$$

$$= \frac{1}{2} [y^3]_0^2 = 4$$

and the moment with respect to the x-axis is

$$M_x = \int (y\text{-arm})\, dA = \int_0^2 y(x_2 - x_1)\, dy = \int_0^2 y[2y - y]\, dy = \left. \frac{y^3}{3} \right]_0^2 = \frac{8}{3}$$

Thus, by Equations 6.41

$$\bar{x} = \frac{M_y}{A} = \frac{4}{2} = 2 \quad \text{and} \quad \bar{y} = \frac{M_x}{A} = \frac{8/3}{2} = \frac{4}{3}$$

So the centroid is

$$\left(2, \frac{4}{3}\right)$$

$\blacksquare$

REMARK
$\blacksquare$ In the preceding example, if we had been seeking the center of mass of a homogeneous lamina occupying the same region, we would have replaced dA by dm, but since $dm = \rho \, dA$, the constant ρ would cancel, as we showed above. So the final result would be the same as the centroid.

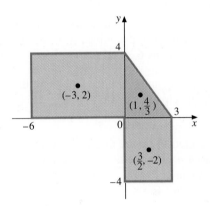

FIGURE 6.83

EXAMPLE 6.36   Find the centroid of the region shown in Figure 6.83.

Solution   Here we can make use of the fact that if the centroid $(\bar{x}, \bar{y})$ of a region is known, its moments with respect to the x- and y-axes, respectively, are

$$M_x = A\bar{y} \quad \text{and} \quad M_y = A\bar{x}$$

(In the case of a lamina of constant density ρ, the area A would be replaced by the mass $m = \rho A$.) Furthermore, we know that if a region consists of several parts, its total moment with respect to either axis is the sum of the moments of each individual part with respect to that axis.

In Exercise 22 of Exercise Set 6.8 you will be asked to show that the centroid of any triangular region is located one-third of the perpendicular distance from any side to the opposite vertex (at the intersection of the medians). Thus, the centroid of the triangular region in Figure 6.83 has coordinates $(1, \frac{4}{3})$. The area of the triangle is $\frac{1}{2}(3)(4) = 6$. The rectangular region to the left of the y-axis has centroid $(-3, 2)$ and area 24, and the rectangle below the x-axis has centroid $(\frac{3}{2}, -2)$ and area 12.

The total moment M_y is

$$M_y = (-3)(24) + (1)(6) + \left(\frac{3}{2}\right)(12) = -72 + 6 + 18 = -48$$

Similarly, the moment M_x is

$$M_x = 2(24) + \left(\frac{4}{3}\right)(6) + (-2)(12) = 48 + 8 - 24 = 32$$

The total area is $24 + 6 + 12 = 42$. Thus, the coordinates of the centroid of the total region are

$$\bar{x} = \frac{M_y}{A} = \frac{-48}{42} = -\frac{8}{7}$$

$$\bar{y} = \frac{M_x}{A} = \frac{32}{42} = \frac{16}{21}$$

$\blacksquare$

Centroids and Volumes of Solids of Revolution

We conclude this section by stating a theorem relating centroids and volumes of solids of revolution. You will be asked to supply the proof in Exercise 28 in Exercise Set 6.8. The theorem is attributed to Pappus of Alexandria, who flourished about A.D. 320. He was the last great Greek geometer.

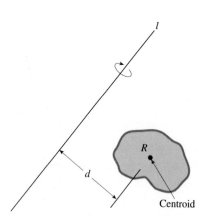

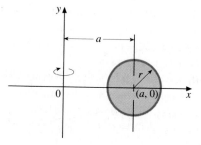

FIGURE 6.84
When R is rotated about l, the volume is $(2\pi d) \cdot$ (area of R).

The First Theorem of Pappus

If R is a region in a plane and l is a line in the plane that does not intersect R, then the volume of the solid formed by revolving R about l is the area of R times the distance traveled by the centroid of R.

Figure 6.84 illustrates the idea of this theorem.

In the next example we show how Pappus's Theorem enables us to find the volume of a torus quickly and easily. (Recall the work involved in using the methods of washers and cylindrical shells in Sections 6.2 and 6.3.)

EXAMPLE 6.37   Use the First Theorem of Pappus to find the volume of the torus formed by revolving the region inside the circle $(x - a)^2 + y^2 = r^2$ about the y-axis, where $a > r$.

Solution   In Figure 6.85, we show the region to be revolved. The centroid of the circle is its center, located at a distance a from the axis of rotation. The area of the circular region is πr^2. So, by Pappus's Theorem, the volume V of the torus generated after rotation (see Figure 6.86) is

$$V = 2\pi a(\pi r^2) = 2\pi^2 a r^2$$

In Example 6.11, we gave an intuitive argument for arriving at the volume of a torus as we have indicated in this example. The First Theorem of Pappus provides the needed justification for our earlier intuitive argument. ∎

FIGURE 6.85

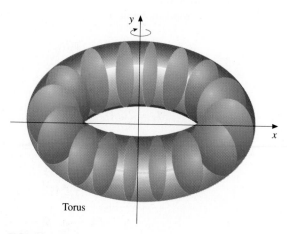

Torus

FIGURE 6.86

Exercise Set 6.8

In Exercises 1 and 2, point masses m_i are located on the x-axis, with x-coordinates x_i, for $i = 1, 2, 3, 4$. Find the center of mass.

1. $m_1 = 10$; $x_1 = 3$
   $m_2 = 12$; $x_2 = -4$
   $m_3 = 17$; $x_3 = 5$
   $m_4 = 20$; $x_4 = -2$

2. $m_1 = 13$; $x_1 = -5$
   $m_2 = 7$; $x_2 = -1$
   $m_3 = 11$; $x_3 = 6$
   $m_4 = 9$; $x_4 = 4$

In Exercises 3 and 4, point masses m_i are located in the plane, with coordinates as shown, for $i = 1, 2, 3, 4$. Find the center of mass.

3. $m_1 = 5$ at $(3, -2)$
   $m_2 = 13$ at $(-4, 5)$
   $m_3 = 6$ at $(2, 4)$
   $m_4 = 8$ at $(-1, -2)$

4. $m_1 = 15$ at $(-4, 3)$
   $m_2 = 18$ at $(2, 5)$
   $m_3 = 11$ at $(-3, -4)$
   $m_4 = 14$ at $(5, -6)$

In Exercises 5–8, a bar having density $\rho(x)$ is located on the x-axis between $x = a$ and $x = b$. Find the center of mass of the bar.

5. $\rho(x) = 2$; $a = 0$, $b = 3$

6. $\rho(x) = \frac{x}{2}$; $a = 1$, $b = 5$

7. $\rho(x) = \sqrt{x}$; $a = 4$, $b = 9$

8. $\rho(x) = 1 + x$; $a = 0$, $b = 7$

In Exercises 9–12, find M_x, M_y, and the center of mass of the lamina pictured, having the given density ρ.

9.

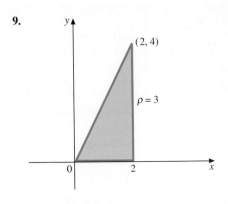

10.

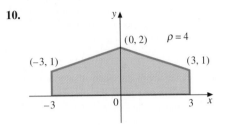

11.

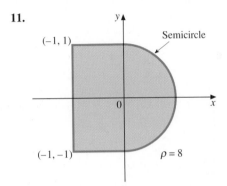

12.

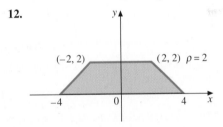

In Exercises 13–20, find the centroid of the region described.

13. Triangular region with vertices $(0, 0)$, $(4, 2)$, $(4, 8)$

14. Below $y = 4$ and above $y = |x|$

15. First-quadrant portion of circular region centered at the origin, with radius 2

16. Bounded by $y = \sqrt{x}$, $x = 4$, and the x-axis

17. Between $y = x$ and $y = 2 - x^2$

18. Under $y = 2x - x^2$ and above the x-axis

19. Between the y-axis and $y^2 = 4 - x$

20. Under $y = \cos x$ and above the x-axis, between $x = -\pi/2$ and $x = \pi/2$. (*Hint:* In one of the integrals you will need to use the identity $\cos^2 x = (1 + \cos 2x)/2$.)

21. Prove that the centroid of the parabolic region bounded by $y = ax^2$ and $y = b$, where $a > 0$ and $b > 0$, is independent of a. What is the value of $\bar{y}$?

22. Prove that the distance from any side of a triangular region to its centroid is one-third the perpendicular distance from that side to the opposite vertex. (*Hint:* Orient the axes so that one side is on the x-axis and the opposite vertex is on the y-axis. See the figure. Since any side can be placed on the x-axis in this way, you need only find $\bar{y}$ for the triangle shown.)

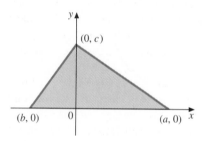

23. Prove that the distance from the base of a semicircular region of radius a to its centroid is $4a/3\pi$.

In Exercises 24–27, use the method of Example 6.36 to find the centroid of the region shown. Use the results of Exercises 22 and 23 where needed.

24.

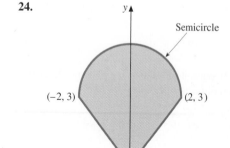

25.

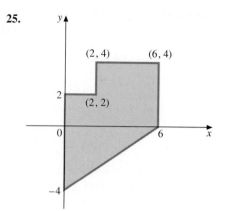

26.

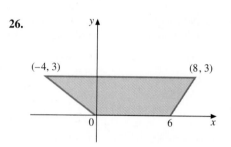

27. (*Hint:* Treat the hole as having negative area.)

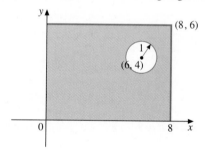

28. Prove the First Theorem of Pappus. (*Hint:* Orient the axes so that l coincides with the y-axis and take R to the right of the y-axis.)

In Exercises 29–31, use the First Theorem of Pappus to find the volume of the indicated solid.

29. Sphere of radius a (*Hint:* Use the result of Exercise 23.)

30. Cone of altitude h and base radius r

31. Solid obtained by revolving the region enclosed by the triangle with vertices $(2, 1)$, $(8, 1)$, and $(5, 4)$ about (a) the y-axis, (b) the x-axis, (c) the line $y = 6$

6.9 INTEGRATION IN PROBABILITY

Probability Distributions

If you were to flip a balanced coin 100 times, you would expect it to come up heads approximately 50 times. On any one toss of the coin there are two equally likely outcomes (heads or tails), so the likelihood it will be heads is 1 out of 2. We say that the **probability** of getting heads is 1/2. Similarly, if you toss a die a large number of times, you would expect that the number 3, say, would show up on the upper face about 1/6th of the time. That is, the probability of getting a 3 (or any other particular number from 1 to 6) is 1/6. In both of these cases we could reason in advance what the probability should be. In many cases, however, probabilities cannot be determined in advance. Suppose, for example, that as a quality-control measure, a manufacturer of light bulbs samples 100 of them periodically to see how many are defective. If, say, two of the 100 are found to be defective, then the manufacturer would assume that among the entire batch, approximately 2% are defective. That is, the probability that a bulb selected at random is defective is about 0.02.

In each of the cases cited, the number of possible outcomes is finite. In many cases, an investigator is interested in some characteristic, or attribute, that can take on any one of infinitely many possibilities. In particular, it may in some cases assume any real number on a given interval. Typically, these cases involve a measurement of some type, such as weight, height, or distance. For example, suppose the weights (in kilograms) of 100 college students were taken.* We could depict the results by means of a diagram such as that shown in Figure 6.87. In the diagram, points stacked vertically represent students who have approximately the same weight. The regions of approximately the same weight are called *bins*, and the procedure is called *binning*. The *density* of weights is greatest in the 50–100 kg range and diminishes to the right and left of this interval. As another example, a researcher might be interested in the cholesterol levels in the blood of persons in a certain age group. These values, too, can be any real numbers within some interval, say from 100 to 400.

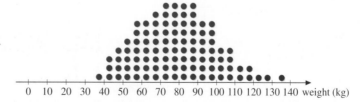

FIGURE 6.87

---

*Strictly speaking, we should refer to the *masses* of the students in kilograms, but it is standard practice in this context to call these masses weights.

A Galton machine, on display in Boston's Museum of Science. Balls drop through posts and distribute themselves randomly in the approximate shape of the red curve, called a *normal curve* in probability.

It is interesting to note the similarity between the binning of students' weights and an experiment first introduced by the English scholar Sir Francis Galton (1822–1911). He devised an apparatus, subsequently known as a **Galton machine** like the one shown in the margin. In his experiment, many identical balls are dropped, one at a time, onto a triangular array of posts, as shown. Each ball hits the first post, bounces one way or the other and onto a post in the second row, then to the third row, and so on. In a random fashion the balls make their way into the bins at the bottom of the machine. Note the similarity between the typical distribution of balls shown in the Galton machine photograph and the distribution of students' weights in Figure 6.87.

Weighing a group of students, counting the number of defective light bulbs in a sample, or dropping balls in a Galton machine can be thought of as an *experiment* in which the outcomes are not predictable in advance. Such an experiment is sometimes called a *random* experiment, and the attribute or characteristic being measured is called a **random variable**. Some random variables can take on discrete values only—such as the number of defective light bulbs in a sample. We call these *discrete* random variables. Others can take on real values on some interval—examples include weights or cholesterol levels of people in a given population. These are called *continuous* random variables. We will be concerned with continuous random variables, since they lend themselves to the use of calculus.

It is often useful to be able to determine in advance the likelihood, or **probability**, that the value of such a continuous random variable will be in some specified interval. For example, we might want to know the probability that a college student chosen at random will weigh between 70 and 90 kg, or that the cholesterol level of a person in the age group under study will exceed 250. In general, these probabilities cannot be determined with absolute certainty, but they often can be approximated by the use of integration.

Suppose we indicate bins by rectangles as in Figure 6.88. We let the area of each rectangle represent the *proportion* of the total population falling within that bin. For example, if we are dealing with weights of 100 college students, and the area of the bin from 50 to 60 is 0.15, then 15% of the students weigh between 50 kg and 60 kg. The sum of the areas of all the rectangles must equal 1, since this sum is the proportion of students weighing between the lowest and highest weights, which includes all of them.

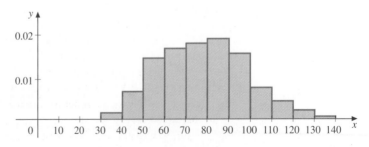

FIGURE 6.88

Figure 6.88 is an example of what is called a **histogram**. It looks suspiciously like the rectangles whose areas form a Riemann sum. In fact, if we were to consider narrower and narrower bins, the areas of the rectangles would approach more and more closely the area under a curve such as that shown in Figure 6.89.

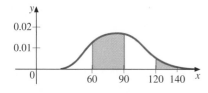

FIGURE 6.89

We can think of Figure 6.89 as an idealization of Figure 6.88. The shaded area between $x = 60$ and $x = 90$ represents the proportion of students weighing between 60 kg and 90 kg. Equivalently, we can say this area is the *probability* that the weight of a student selected at random will fall somewhere in this range. Similarly, the shaded area between $x = 120$ and $x = 140$ is the probability that the weight of a student selected at random is between 120 kg and 140 kg. The total area under the graph must be 1, since this area is the probability that a student's weight is between the lowest weight and the highest weight.

We can generalize the preceding discussion by considering any nonnegative integrable function such that the area under its graph on some specified interval is 1. (At present we consider only finite intervals. Later, we will remove this restriction.) In the following definition we follow the standard practice of using a capital letter, such as X, to designate a random variable, with the corresponding lowercase letter, such as x, designating one of its values.

Definition 6.6 Probability Density Function	Let f be a nonnegative integrable function on an interval I, such that the area under the graph of f on I is 1. If there exists a random variable X having I as the set of all of its possible values and such that for any two numbers a and b in I with $a < b$, the probability that X takes on a value between a and b is given by $$P(a \leq X \leq b) = \int_a^b f(x)dx \qquad (6.42)$$ then f is called a **probability density function** for X.

REMARK ────────────────────────

■ Since omitting one or both endpoints from the integral in Equation 6.42 will not change its value,

$$P(a \leq X \leq b) = P(a < X \leq b) = P(a \leq X < b) = P(a < X < b)$$

In actual practice the probabilities given by Equation 6.42 are usually only approximate, and the function f is an idealized mathematical model of the true probability density function for X. Many random variables have an apparent bell-shaped density function, similar to that shown in Figure 6.89. This type of graph is called a **normal curve** and the corresponding function a **normal density function**. We will come back to this type of function later. The choice of the particular density function to use depends on the nature of the random variable in question. Sometimes it can reasonably be predicted that the values of a random variable will be distributed in a certain way, and in other cases the choice of the model is based on experimental evidence.

A particularly simple probability density function is the **uniform probability density function**, defined by

$$f(x) = \begin{cases} \dfrac{1}{b - a} & \text{if } a \leq x \leq b \\ 0 & \text{otherwise} \end{cases} \qquad (6.43)$$

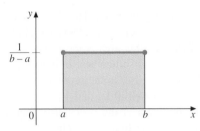

FIGURE 6.90

Uniform probability density function

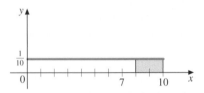

FIGURE 6.91

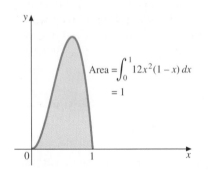

FIGURE 6.92

The beta density function with $\alpha = 3, \beta = 2$

The graph is shown in Figure 6.90. Observe that the function f is constant over the specified interval. The next example shows one case for which such a density function is an appropriate model.

EXAMPLE 6.38   Suppose a bus is known to arrive at a certain bus stop between 9:00 and 9:10 each weekday morning and that from observation over a period of time, it can be assumed that its arrival time during the 10-min interval is uniformly distributed; that is, the bus is just as likely to arrive at one time in the interval as at any other time. Find the probability that a person who arrives at 9:00 will have to wait more than 7 min for a bus.

Solution   The random variable X of interest in this case is the number of minutes after 9:00 that the bus arrives, and from the given information we know that the values have a uniform distribution on the interval [0, 10], as shown in Figure 6.91. Although we can find the area in this case without integration, we choose to illustrate Equation 6.42. We want $P(7 < X \le 10)$:

$$P(7 < X \le 10) = \int_{7}^{10} f(x)dx = \int_{7}^{10} \frac{1}{10}dx$$

$$= \frac{1}{10}[x]_{7}^{10} = \frac{3}{10}$$

So if this person arrives at 9:00 each morning, we would expect that on average he or she will have to wait at least 7 min about 30% of the time.   ∎

The model for the next example is a special case of what is known as a **beta distribution**, with the general form

$$f(x) = Bx^{\alpha-1}(1-x)^{\beta-1}, \qquad 0 \le x \le 1 \qquad (6.44)$$

The Greek letters α (alpha) and β (beta) can be any positive numbers. The constant B is chosen so that the area under the graph is 1. In Figure 6.92 we show the graph of the beta distribution in which $\alpha = 3$ and $\beta = 2$. You can verify that $B = 12$ in this case. (See Exercise 28 in Exercise Set 6.9.)

EXAMPLE 6.39   A certain grass seed is advertised as a mixture containing at least 70% Kentucky bluegrass, with the remainder composed of red fescue, rye, and other extraneous seeds. As a means of quality control, a sample of seeds is tested periodically, and if the proportion of non–Kentucky-bluegrass seeds exceeds 30%, the production is halted so that corrective measures can be taken. By experiment, the probability density function for the proportion of undesirable seeds is found to be approximated by

$$f(x) = 10(1-x)^9, \qquad 0 \le x \le 1$$

Find the probability that on testing a random sample the production process will be halted.

Solution   We define the random variable X to be the proportion of undesirable grass seeds in the sample. From the given information, we see that this variable has a beta density function, Equation 6.44, with $\alpha = 1$ and $\beta = 10$ (so $B = 10$). Production will be halted if the value of X exceeds 0.3. So the probability of halting production is

$$P(0.3 < X \leq 1) = \int_{0.3}^{1} 10(1-x)^9 dx$$

$$= -10 \left[\frac{(1-x)^{10}}{10} \right]_{0.3}^{1}$$

$$= -[0 - (0.7)^{10}] \approx 0.028$$

There is about a 3% chance of having to halt the production process.   ∎

The Mean, Expected Value, Variance, and Standard Deviation

An important characteristic of a probability density function is its **mean**, usually denoted by the Greek letter μ (mu), and defined by

$$\mu = \int_a^b xf(x)dx \tag{6.45}$$

It can be shown that if values of the random variable X, with probability density function $f(x)$, are repeatedly sampled, there is a high probability they will be concentrated near μ. In a sense, μ is the average value of X. Looked at geometrically, μ is the x-coordinate of the centroid of the region under the graph of f. For, as we know from the previous section,

$$\bar{x} = \frac{\displaystyle\int_a^b xf(x)dx}{\displaystyle\int_a^b f(x)dx}$$

But the denominator in this case is 1. Thus, $\bar{x} = \mu$.

Another way of viewing the mean is to compare it with the center of mass of a one-dimensional bar having variable density. We can take the density as $f(x)$ (it is even called a probability *density*), so according to Equation 6.35,

$$\bar{x} = \frac{\displaystyle\int_a^b xf(x)dx}{\displaystyle\int_a^b f(x)dx}$$

Again, the denominator is 1, so $\bar{x} = \mu$. Thus, the mean μ can be thought of as the center of mass, with "mass" (i.e., probabilities) distributed on a line from a to b, with density $f(x)$.

The mean of a distribution is a special case of what is called an **expected value**. In general, if X is a random variable with probability density $f(x)$ on the interval $[a, b]$, the **expected value of $g(X)$**, symbolized by $E[g(X)]$, is defined by

$$E[g(X)] = \int_a^b g(x)f(x)dx \tag{6.46}$$

We assume $g(x)f(x)$ is integrable, so that $E[g(X)]$ is defined. In the same sense that μ is an average value of X, the expected value of $g(X)$ is an average of the values taken on by $g(X)$. In repeated experiments in which $g(X)$ is measured, these measurements would be concentrated near $E[g(X)]$. By taking $g(X) = X$, we see that

$$\mu = E[X]$$

That is, the mean is the expected value of X.

Another expected value that is important in describing a distribution is obtained by taking $g(X) = (X - \mu)^2$ in Equation 6.46. We call this expected value the **variance** of X and symbolize it by σ^2 (σ is the lowercase Greek letter *sigma*). Thus,

$$\sigma^2 = E[(X - \mu)^2]$$

That is,

$$\sigma^2 = \int_a^b (x - \mu)^2 f(x)dx \qquad (6.47)$$

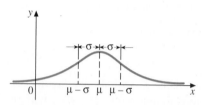

FIGURE 6.93

The mean μ and standard deviation σ for the normal curve

The number $x - \mu$ is called the *deviation* of x from the mean. By squaring the deviations, we eliminate the effect of their variations in sign (some positive, some negative). The variance is therefore a measure of how closely the values of X are clustered about the mean. The smaller the variance, the more closely they are clustered. The square root of the variance, denoted by σ, is called the **standard deviation** of X. In Figure 6.93 we show a normal probability density function with mean μ and standard deviation σ. The equation of the function is

$$f(x) = \frac{1}{\sqrt{2\pi}\,\sigma} e^{-(x-\mu)^2/2\sigma^2}$$

We will study functions of this type (exponential functions) in Chapter 7. There you will be asked to show that for this normal curve, σ is the distance from the mean μ to each of the points of inflection to the right and left of the mean.

EXAMPLE 6.40   A certain random variable X has probability density function $f(x) = kx^2$ on $[0, 2]$. Find the value of k and calculate μ, σ^2, and σ. Also find the approximate probability that X is no more than one standard deviation from the mean.

Solution   The constant k must be chosen so that the total area under the graph of f is 1. Since

$$\int_0^2 kx^2dx = k\left[\frac{x^3}{3}\right]_0^2 = \frac{8k}{3}$$

we must have $8k/3 = 1$ or $k = 3/8$. Thus, $f(x) = 3x^2/8$.

The mean, μ, is

$$\mu = E[X] = \int_0^2 xf(x)dx = \int_0^2 x\left(\frac{3}{8}x^2\right)dx = \frac{3}{8}\int_0^2 x^3dx$$

$$= \frac{3}{8}\left[\frac{x^4}{4}\right]_0^2 = \frac{3}{2}$$

The variance, σ^2, is

$$\sigma^2 = E[(X - \mu)^2] = E\left[\left(X - \frac{3}{2}\right)^2\right] = \int_0^2 \left(x - \frac{3}{2}\right)^2 f(x)dx$$

$$= \int_0^2 \left(x^2 - 3x + \frac{9}{4} \right) \cdot \frac{3}{8} x^2 dx = \frac{3}{8} \int_0^2 \left(x^4 - 3x^3 + \frac{9x^2}{4} \right) dx$$

$$= \frac{3}{8} \left[\frac{x^5}{5} - \frac{3x^4}{4} + \frac{3x^3}{4} \right]_0^2 = \frac{3}{8} \left[\frac{32}{5} - 12 + 6 \right]$$

$$= \frac{3}{20}$$

Thus, the standard deviation, σ, is

$$\sigma = \sqrt{\frac{3}{20}} = \sqrt{\frac{3}{20} \cdot \frac{5}{5}} = \frac{\sqrt{15}}{10} \approx 0.3873$$

Finally, we want $P(\mu - \sigma \le X \le \mu + \sigma)$ where

$$\mu - \sigma \approx 1.5 - 0.3873 = 1.1127$$

and

$$\mu + \sigma \approx 1.5 + 0.3873 = 1.8873$$

So we have

$$P(\mu - \sigma \le X \le \mu + \sigma) \approx \int_{1.1127}^{1.8873} \frac{3}{8} x^2 \, dx = \frac{1}{8} x^3 \Big]_{1.1127}^{1.8873}$$

$$\approx 0.668$$

Notice that about two-thirds of the time we would expect X to fall within one standard deviation of the mean. ∎

Exercise Set 6.9

In Exercises 1–10, find the constant k that will make f a probability density function on the given interval.

1. $f(x) = k(2x + 3)$ on $[0, 4]$

2. $f(x) = k(4 - 2x)$ on $[0, 2]$

3. $f(x) = k(x^2 + x)$ on $[1, 3]$

4. $f(x) = kx(1 - x)$ on $[0, 1]$

5. $f(x) = k/x^2$ on $[1, 4]$

6. $f(x) = k\sqrt{x - 1}$ on $[1, 5]$

7. $f(x) = kx^3(1 - x)^2$ on $[0, 1]$

8. $f(x) = kx/(1 + x^2)^2$ on $[0, 1]$

9. $f(x) = kx/\sqrt{1 + 2x^2}$ on $[0, 2]$

10. $f(x) = kx^2/\sqrt{1 + x^3}$ on $[0, 2]$

In Exercises 11–16, verify that the given function is a probability density function. Also find the probability specified for a random variable X that has f as its probability density function.

11. $f(x) = \frac{1}{2}x$ on $[0, 2]$; $P(\frac{1}{2} \le X \le \frac{3}{2})$

12. $f(x) = \frac{1}{6}(2x + 3)$ on $[-1, 1]$; $P(0 \le X < 1)$

13. $f(x) = \frac{1}{24}(x^2 + 1)$ on $[-3, 3]$; $P(-1 < X \le 2)$

14. $f(x) = 12x^2(1 - x)$ on $[0, 1]$; $P(\frac{1}{2} < X < 1)$

15. $f(x) = \frac{3}{125} x\sqrt{25 - x^2}$ on $[0, 5]$; $P(3 \le X \le 4)$

16. $f(x) = x/\left(2\sqrt{x^2 + 9}\right)$ on $[0, 4]$; $P(0 < X \le 1.25)$

In Exercises 17–22, find the mean and variance of the random variable that has the given probability density function.

17. $f(x) = \frac{2}{9}(x + 1)$ on $[-1, 2]$

18. $f(x) = \frac{1}{6}(3 - 2x)$ on $[-1, 1]$

19. $f(x) = 6x(1 - x)$ on $[0, 1]$

20. $f(x) = \frac{3}{16}\sqrt{x}$ on $[0, 4]$

21. $f(x) = 4x^3$ on $[0, 1]$

22. $f(x) = 3[1 - (x - 3)^2]/4$ on $[2, 4]$

23. Find the mean and variance of a random variable that has the uniform density function of Equation 6.43.

24. The number X of parts per million of a certain pollutant emitted from the smokestack of a cement plant is known to range from a low of 1 to a high of 27. Measurements of samples collected over a period of time suggest that an appropriate probability density model for X is $f(x) = cx^{-2/3}$, where $1 \le x \le 27$.
(a) Find the value of c.
(b) Find the expected value of X.
(c) Find the standard deviation of X.

25. The income level of persons in a certain city whose incomes exceed a certain amount $x_0 > 0$ is approximately modeled by a **Pareto distribution**:

$$f(x) = \begin{cases} \dfrac{\alpha x_0^{\alpha}}{x^{\alpha+1}} & \text{if } x \ge x_0 \\ 0 & \text{otherwise} \end{cases}$$

where α is a positive constant. If x is measured in thousands of dollars, $\alpha = 2$, and $x_0 = 10$, find the probability that a person with an income that exceeds \$10,000 makes less than \$20,000.

26. A **truncated Pareto distribution** (see Exercise 25) is of the form

$$f(x) = \begin{cases} c\left(\dfrac{x_0}{x}\right)^{\alpha+1} & \text{if } x_0 \le x \le x_1 \\ 0 & \text{otherwise} \end{cases}$$

where c and α are positive constants. Find the constant c in terms of x_0, x_1, and α so that f will be a probability density function.

27. The truncated Pareto function of Exercise 26 is sometimes used to model the distribution of the size of oil fields in a certain region. Suppose that x is measured in hundreds of millions of barrels, $\alpha = 2$, $x_0 = 2$, and $x_1 = 10$ for a certain region.
(a) Find the probability that a randomly selected oil field

in that region would produce more than 500 million barrels of oil.
(b) Find the average (mean) number of barrels for oil fields in that region.

28. The proportion X of impurities in a product resulting from a certain chemical process has the beta distribution of Equation 6.44 with $\alpha = 3$ and $\beta = 2$.
(a) Find the constant B.
(b) Find the expected value of X.

29. The lifetime, in hundreds of hours, of a certain type of light bulb has been found empirically to have a probability density function approximated by

$$f(x) = \frac{5\sqrt{17}}{12(1 + x^2)^{3/2}}, \qquad 0 \le x \le 12$$

Find the expected lifetime of a bulb of this type chosen at random.

30. Let X be a random variable with probability density function f on $[a, b]$. Prove the following properties of expected value:
(a) $E[c] = c$, where c is a constant.
(b) $E[cg(x)] = cE[g(x)]$, where c is a constant.
(c) $E[g_1(x) + g_2(x)] = E[g_1(x)] + E[g_2(x)]$

31. Prove that $\sigma^2 = E[X^2] - \mu^2$. (*Hint:* Use the results of Exercise 30, along with the definitions of μ and σ^2.)

32. Use the result of Exercise 31 to find σ^2 for a random variable with the probability density function given by

$$f(x) = \frac{3}{20}\left(\frac{x+1}{\sqrt{x}}\right), \qquad 1 \le x \le 4$$

33. The random variable X with probability density function

$$f(x) = \frac{8}{(\pi + 2)(1 + x^2)^2} \qquad \text{on } [0, 1]$$

is known to have variance

$$\sigma^2 = \frac{\pi^2 - 8}{(\pi + 2)^2}$$

Use this together with the result of Exercise 31 to evaluate the integral

$$\int_0^1 \frac{x^2 dx}{(1 + x^2)^2}$$

34. Show that if X is a random variable with finite mean and t can assume any real value, then $E[(x - t)^2]$ takes on its minimum value when $t = \mu$.

6.10 INTEGRATION IN ECONOMICS

Total Cost and Total Revenue

In Section 4.3 we introduced the notion of the **marginal cost** of a product, defined as the derivative of the cost function. We saw that the marginal cost evaluated at $x = n$ approximates the cost of producing the $(n + 1)$st unit. For example, the product in question might be pairs of shoes of a certain type. The marginal cost evaluated at $x = 100$, say, approximates the cost of producing the 101st pair, after having already produced 100 pairs. This marginal-cost concept is a key tool in the decision-making process.

In this section we are interested in the reverse problem to that of finding marginal cost when we know the cost function. That is, we now want to find the cost function when we know the marginal cost. We will refer to the cost function as the **total-cost function**, to emphasize that it gives the total cost (fixed cost plus variable cost) of producing a given number of units of the product in question.

Let $C(x)$ denote this total-cost function, and let $M_C(x)$ denote the marginal-cost function. Then

$$C'(x) = M_C(x)$$

We assume that $C(0) = C_0$, where C_0 is the fixed cost. We want to find $C(x)$, so we have an initial-value problem to solve. We can write the solution as

$$C(x) = \int_0^x M_C(t)dt + C_0 \tag{6.48}$$

You can check this solution by applying the First Fundamental Theorem of Calculus to get $C'(x) = M_C(x)$. It is also easily seen that $C(0) = C_0$.

REMARK ─────────────────────────────

■ We are assuming here that $M_C(x)$ is continuous for $x > 0$. Even though x might actually take on integer values only, we assume it can be any positive real number so that we can make use of calculus techniques.

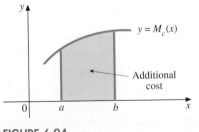

FIGURE 6.94

Suppose now that a units of the product in question are already being produced, and we want to know the *additional* cost of increasing the production level to b units, where $b > a$. This additional cost is

$$C(b) - C(a) = \int_a^b M_C(t)dt \tag{6.49}$$

as illustrated in Figure 6.94.

A similar analysis can be applied to the **total-revenue function**, $R(x)$. The **marginal revenue**, $M_R(x)$, is by definition the derivative of $R(x)$:

$$R'(x) = M_R(x)$$

We assume that $R(0) = 0$. (There is no income if no items are sold.) The solution of this initial-value problem is

$$R(x) = \int_0^x M_R(t)dt \tag{6.50}$$

As with the total cost, the change in revenue when production is increased from a units to b units is

$$R(b) - R(a) = \int_a^b M_R(t)dt \qquad (6.51)$$

Whether to increase production from one level to another clearly depends both on the change in cost and the change in revenue. So long as the increased revenue is greater than the increased cost, it is advantageous to increase production.

EXAMPLE 6.41   In the production of a new type of tennis racquet, the King Company determines that the fixed initial cost is \$65,000 and the marginal cost is

$$M_C(x) = -0.4x + 25 \qquad (0 \le x \le 60)$$

where x is in hundreds of racquets and $M_C(x)$ is in hundreds of dollars. Find

(a) the total-cost function

(b) the increased cost of producing 3,000 racquets instead of 2,000 racquets.

Solution

(a) By Equation 6.49, the total-cost function (in hundreds of dollars) is

$$C(x) = \int_0^x (-0.4t + 25)dt + 65,000 = -0.2x^2 + 25x + 65,000$$

(b) By Equation 6.48, the increased cost is

$$C(30) - C(20) = \int_{20}^{30} (-0.4t + 25)dt$$

$$= -0.2t^2 + 25t \big]_{20}^{30}$$

$$= [-0.2(900) + 25(30)] - [-0.2(400) + 25(20)]$$

$$= (-180 + 750) - (-80 + 500)$$

$$= 150$$

That is, the additional cost of producing an extra 1,000 racquets for a total of 3,000, is \$15,000.   ■

Consumers' Surplus and Producers' Surplus

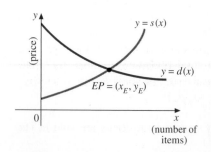

FIGURE 6.95

Two important functions in economics are the **demand function**, $d(x)$, giving the price per unit when there is a consumer demand of x units, and the **supply function**, $s(x)$, giving the price per unit at which the producer is willing to supply x units. In general, from a consumer's point of view, a lower price corresponds to a higher demand for the product in question. So the demand function decreases as x increases. Similarly, from a producer's point of view, a higher price will correspond to an increased supply; that is, a producer is willing to produce more items when the price goes up. Thus, the function s increases as x increases. In Figure 6.95 we show typical supply and demand curves. The point where the graphs of $y = s(x)$ and $y = d(x)$ intersect is called the **equilibrium point**, $EP = (x_E, y_E)$. Clearly, consumers want to buy as many items as possible at the lowest possible price, and producers want to sell as many items as possible at the highest possible price. The equilibrium point therefore represents the optimum quantity and price, taking into consideration the competing goals of consumers and producers.

Whatever the price, it is generally the case that some consumers would have been willing to pay more. So they saved money they were willing to spend by getting the item at a lower price. The total amount saved in this way is called the **consumers' surplus**. That is,

consumers' surplus = (total amount consumers were willing to pay)

− (total amount actually paid)

Similarly, whatever the actual price, some producers would likely have been willing to sell at a lower price. So these producers gained by selling at the higher price. This total amount gained is called the **producers' surplus**. That is,

producers' surplus = (total amount producers actually received)

− (total amount they were willing to accept)

To illustrate the concept of consumers' surplus, consider the following example.

EXAMPLE 6.42   Suppose that to attract new customers, a computer retailer slashes the price on its newest microcomputer system to $2,200. Further, assume that the sale attracts 50 customers who, before the sale, would have been willing to pay up to the amounts shown in the table. Find the consumers' surplus.

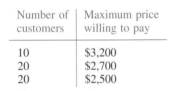

Number of customers	Maximum price willing to pay
10	$3,200
20	$2,700
20	$2,500

Solution   If the cost of the system had been $3,200, only 10 customers would have bought it; they would have spent a total of $10(\$3,200) = \$32,000$. If the price were $2,700, an additional 20 customers would have purchased the system, at an additional cost of $54,000. If the price had been $2,500, another 20 customers would have purchased the system, at an additional total cost of $50,000. By comparison, at the sale price of $2,200 we assume that all 50 customers purchased one of the systems, at a total cost of $50(\$2,200) = \$110,000$. Consequently, the total savings to the customers (the consumers' surplus) is

$$(\$32,000 + \$54,000 + \$50,000) - (\$110,000) = \$26,000$$

amount customers were willing to pay   amount paid

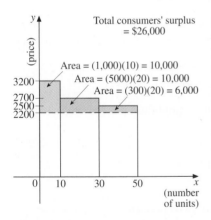

FIGURE 6.96

We can indicate the result graphically as shown in Figure 6.96.   ■

If we assume continuous supply and demand functions for some product and that the product is actually sold at the equilibrium price, then the areas shown in Figure 6.97 represent the consumers' surplus and producers' surplus, as indicated. These areas are given by the integrals

$$\text{consumers' surplus} = \int_0^{x_E} [d(x) - y_E]\,dx \qquad (6.52)$$

$$\text{producers' surplus} = \int_0^{x_E} [y_E - s(x)]\,dx \qquad (6.53)$$

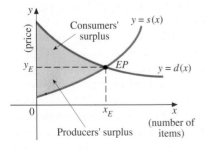

FIGURE 6.97

EXAMPLE 6.43   For a certain commodity, the demand function is

$$d(x) = 12 - 0.25x^2$$

and the supply function is

$$s(x) = 0.25x^2 + 0.50x + 2$$

where x is in thousands of units and both $d(x)$ and $s(x)$ are in hundreds of dollars. Find

(a) the equilibrium price, which is the y-coordinate of the equilibrium point;

(b) the consumers' surplus and the producers' surplus.

Show the results graphically.

Solution

(a) To find the equilibrium price we set $d(x) = s(x)$ and solve for x:

$$12 - 0.25x^2 = 0.25x^2 + 0.50x + 2$$

$$0.50x^2 + 0.50x - 10 = 0 \quad \text{Multiply by 2.}$$

$$x^2 + x - 20 = 0$$

$$(x + 5)(x - 4) = 0$$

$$x = -5 \quad \text{or} \quad x = 4$$

We reject $x = -5$, since x must be positive in the context of the problem. Thus, $x_E = 4$. To find y_E we can substitute 4 for x in $d(x)$ or in $s(x)$. Choosing $d(x)$, we get

$$y_E = d(4) = 12 - 0.25(4)^2 = 12 - 4 = 8$$

Since $d(x)$ is in hundreds of dollars, the equilibrium price is \$800.

(b) By Equation 6.52,

$$\text{consumers' surplus} = \int_0^4 [(12 - 0.25x^2) - 8]dx$$

$$= 4x - 0.25\frac{x^3}{3}\Bigg]_0^4 = 16 - \frac{16}{3} = \frac{32}{3}$$

By Equation 6.53,

$$\text{producers' surplus} = \int_0^4 [8 - (0.25x^2 + 0.50x + 2)]dx$$

$$= 6x - 0.25\frac{x^3}{3} - 0.50\frac{x^2}{2}\Bigg]_0^4 = 24 - \frac{16}{3} - 4 = \frac{44}{3}$$

In dollars, the consumers' surplus is approximately \$1,067 and the producers' surplus is approximately \$1,467. The results are shown graphically in Figure 6.98. ∎

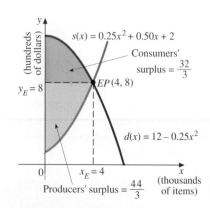

FIGURE 6.98

Exercise Set 6.10

1. A company determines that the marginal-cost function for a certain product is $M_C(x) = x^3 - 2x$. Find the total-cost function, assuming a fixed cost of $200.

2. The assembly line at the True-View Company produces 300 VCRs per day. The manager knows that the marginal-cost function (in dollars) is given by

   $$M_C(x) = 600 - 1.5x \qquad (0 \le x \le 400)$$

   where x is the number of VCRs made each day. Find the increase in total cost if the manager decides to have the assembly line increase production to 350 units daily.

3. The price per bushel of apples that a small apple grower receives depends on the size of the crop. From experience, the grower knows that if the size of the crop is x bushels, the marginal revenue is

   $$M_R(x) = 10 - 0.004x + \frac{1}{\sqrt{x}} \qquad (500 \le x \le 2000)$$

   Last year, the grower harvested 1,000 bushels. How much more revenue will the grower receive if the crop is 1,400 bushels? Sketch the marginal-revenue function, and shade the region whose area corresponds to the increased revenue.

4. Assume that in a market under pure competition (in which the actual price of an item is the equilibrium price, where supply equals demand), the supply function for a certain item is given by $s(x) = 2x + 1$, and the demand function for the same item is given by $d(x) = 12 - x^2$. Determine (a) the price at which the item will be sold, (b) the consumers' surplus, and (c) the producers' surplus.

5. The demand for a certain commodity in a local market is given by

   $$d(x) = \frac{30x + 6000}{x + 100} \qquad (0 \le x \le 500)$$

   where $d(x)$ is in hundreds of dollars. If the supply function is given by

   $$s(x) = 8\sqrt{x + 1} \qquad (0 \le x \le 500)$$

   use a graphing calculator or CAS to approximate the equilibrium point, and then find the consumers' surplus and producers' surplus at market equilibrium.

6. The publisher of a successful magazine wants to sell more copies of each issue. Current circulation is 60,000 copies each month. When x is in thousands of copies circulated per month, the marginal revenue (in thousands of dollars) is given by

   $$M_R(x) = 3.5 - 0.04x - \frac{50}{x}$$

where $40 \le x \le 80$. How much additional revenue would be produced by increasing the circulation to 70,000 copies?

7. The demand and supply functions for a certain commodity are, respectively, $d(x) = -x^2 - 20x + 1,020$ and $s(x) = x^2 + 60x + 20$, where $0 \le x \le 30$. The price is in dollars, and the quantity x is in thousands of units.
   (a) Find the equilibrium price and quantity.
   (b) Find the consumers' surplus.
   (c) Find the producers' surplus.
   Show your results graphically.

8. **Consumers' surplus when the supplier has a monopoly.** When a supplier has a monopoly on a commodity, market equilibrium will not exist (the product is not sold at the equilibrium price). In this situation the price will be set to maximize the supplier's profit. Suppose that a supplier has a monopoly on a certain product, with cost function $c(x) = x^2 + 920x + 3,000$ and demand function $d(x) = -x^2 - 2x + 2,000$ where $0 \le x \le 50$. It is assumed that $c(x)$ is in thousands of dollars, x is in thousands of units, and $d(x)$ is in dollars.
   (a) Find the number of units and the price per unit that will maximize profit. (*Hint:* profit = revenue − cost, and revenue is the number of units sold times the price per unit.)
   (b) Find the consumers' surplus if the product is sold at the price that maximizes the supplier's profit.

Chapter 6 Review Exercises

In Exercises 1–10, find the area of the region indicated. Sketch the region.

1. Under the graph of
$$f(x) = \begin{cases} 1 + \sqrt{2x} & \text{if } 0 \le x < 2 \\ x^2 - 4 & \text{if } 2 \le x \le 4 \end{cases} \quad \text{on } [0, 4]$$

2. Under the graph of
$$f(x) = \begin{cases} 1 - (x+2)^{-2} & \text{if } -1 \le x < 0 \\ 0 & \text{if } x = 0 \\ 4 - 2x & \text{if } 0 < x \le 2 \end{cases} \quad \text{on } [-1, 2]$$

3. Bounded by $y = 3x + 4$, $y = 2 - x$, $x = 0$, and $x = 3$

4. Between $y = x^3 - x$ and $y = 1 - x^2$

5. Between $y = x^2 - 2x$ and $y = 4\sqrt{x}$

6. Bounded by $y = \sqrt{x + 5}$, $4y = 3x$, and $y = 0$

7. Between $y = x^2 - x$ and $y = \sin \pi x$

8. Between $y^2 = x + 2$ and $y^2 + x - 2y - 2 = 0$

9. One of the regions bounded above by $y = \sqrt{3} \cos x$ and below by $y = \sin x$

10. Bounded by $y = 1/x^2$, $y = 1$, $2x - 4y = 3$, and $x = 0$

In Exercises 11–20, find the volume obtained by revolving the region about the specified axis. Sketch the region.

11. Bounded by $y = \frac{1}{x}$, $x = 1$, $x = 2$, and the x-axis; x-axis

12. Bounded by $y = \sqrt{2x + 3}$, $x = -1$, $x = 3$, and the x-axis; x-axis

13. Bounded by $x = 8 - 2y$, $x = 2$, $x = 6$, and the x-axis; y-axis

14. Bounded by $y = x$, $y = 2x$, $x = 1$, and $x = 4$; y-axis

15. Between $y = x^2$ and $y = x + 6$; x-axis

16. Between $y = x^2 + 1$ and $y = 3x - x^2$; y-axis

17. Bounded by $y = \sqrt{2x + 1}$, $y = x - 1$, and $y = 1$; x-axis

18. Between $x + \sqrt{2y} = 0$ and $x = y - y^2$; x-axis

19. Region of Exercise 15 revolved about $x + 2 = 0$

20. Region of Exercise 18 revolved about $x = 1$

21. The base of a wedge is the region enclosed by the ellipse $x^2 + 2y^2 = 4$, and each cross section perpendicular to the x-axis is an isosceles right triangle with right angle on the upper half of the ellipse. Find the volume of the wedge.

22. In the accompanying figure, ABC and DEF are equilateral triangles of sides 4 and 1, respectively, in planes perpendicular to edge AD, which is of length 6, as shown. Find the volume of the solid.

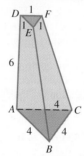

In Exercises 23–25, find the arc length of the curve on the specified interval.

23. $x = 2 + (y - 1)^{3/2}$ from $(2, 1)$ to $(10, 5)$

24. $y = \frac{1}{20}x^5 + \frac{1}{3}x^{-3}$ from $x = \frac{1}{2}$ to $x = 2$

25. $x = y^{3/2} - \frac{1}{3}y^{1/2}$ from $(\frac{2}{3}, 1)$ to $(\frac{22}{3}, 4)$

In Exercises 26–29, find the area of the surface formed by revolving the given arc about the specified axis.

26. $y = 2\sqrt{3x - 2}$ from $x = 3$ to $x = 6$, about x-axis

27. $x^2 - 3y = 3$ from $(0, -1)$ to $(2, \frac{1}{3})$, about y-axis

28. $x = \left(\dfrac{y + 5}{4}\right)^3$ from $(1, -1)$ to $(8, 3)$, about y-axis

29. The arc of Exercise 25 about (a) the x-axis and (b) the y-axis

30. A watering trough 20 ft long has ends in the shape of inverted isosceles trapezoids 4 ft across the top, 2 ft across the bottom, and 3 ft high. If the trough has water in it to a depth of 2 ft, find the work required to pump all the water over the top of the trough.

31. A ship's anchor, weighing 1000 lb, is being raised. The anchor chain weighs 10 lb/ft, and it is being hoisted over a pulley that is 40 ft above the water. How much work is required to raise the anchor from the water level to a position 10 ft below the pulley?

32. A force of 12 lb is required to stretch a certain spring from its natural length of 20 in. to a length of 24 in. Find the work required to (a) stretch it from a length of 24 in. to a length of 30 in. and (b) compress it 4 in. from its natural length.

33. The ends of a long trough are parabolic segments with dimensions as shown in the figure (in feet). Find the total force on one end of the trough when it (a) is full of water and (b) has water in it to a depth of 1 ft.

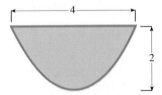

34. A plate as shown in the figure is submerged vertically in water so that the top edge of the plate is at a depth of 9 ft. Find the total force on one side of the plate.

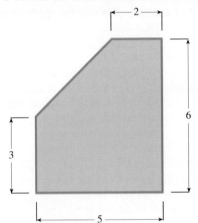

35. A bar with density $\rho(x) = \sqrt{x+1}$ is located on the x-axis between $x = -1$ and $x = 8$. Find its center of mass.

36. A homogeneous lamina with density $\rho = 3$ occupies the region in the plane bounded by the curves $y = 2 - x^2$, $y = x$, and $x = 0$. Find its center of mass.

In Exercises 37–40, find the centroid of the region described.

37. Between the graphs of $y = x^2$ and $y = x + 2$

38. The triangular region bounded by $y = x/2$, $y = 2x$, and $x = 4$

39. Between the y-axis and the parabola $y^2 - x - 4y = 0$

40. Bounded by $y = \sqrt{x}$, $x + y = 6$, and the x-axis

In Exercises 41 and 42, use the results of Exercises 22 and 23 of Exercise Set 6.8 to find the centroid.

41.

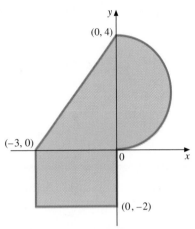

42.

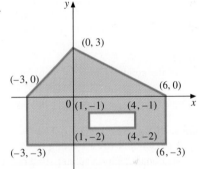

43. The probability density function for the random variable X, denoting the time for students in a certain class to complete a 3-hr final exam, is of the form $f(x) = \frac{x}{9}(1 + kx)$, where $0 \le x \le 3$. Find the following:
(a) The constant k
(b) The mean μ
(c) The variance σ^2
(d) $P(X < 1)$
(e) $P(2 \le X \le 3)$

44. The proportion X of defective items in a large batch has a probability density function that is approximated by a beta function with $\alpha = 1$ and $\beta = 8$. Find (a) the constant B and (b) the probability that more than 20% are defective.

45. Suppose that the marginal-cost and marginal-revenue functions for production of a certain product are

$$M_C(x) = 4 - \frac{x}{1000}$$

and

$$M_R(x) = 40 - \frac{x}{100}$$

The fixed cost is $5,400. Find the total-cost function $C(x)$ and the total-revenue function $R(x)$. Also find the profit function $P(x)$, and show that the maximum profit occurs when $M_C(x) = M_R(x)$. Find the value of x that yields the maximum profit. What is the maximum profit?

46. The supply and demand functions for a certain product are given by

$$s(x) = 5 - \frac{(x-8)^2}{16}$$

and

$$d(x) = 5 - \frac{x^2}{8}$$

respectively, where $0 \leq x \leq 6$. The quantity x is in thousands of units, and both $s(x)$ and $d(x)$ are in hundreds of dollars. Find

(a) the equilibrium price for the product;
(b) the consumers' surplus;
(c) the producers' surplus.

Show the results graphically.

Chapter 6 Concept Quiz

1. Let f be a nonnegative continuous function of x on $[a, b]$, where $b > a > 0$. Denote the graph of $y = f(x)$ by C, and let R be the region under C and above the x-axis from $x = a$ to $x = b$. Write integrals for the following:

(a) The area of R
(b) The volume of the solid obtained by revolving R about the x-axis
(c) The volume of the solid obtained by revolving R about the y-axis
(d) The coordinates $\bar{x}$ and $\bar{y}$ of the centroid of R
(e) The length of C (assuming f is smooth)
(f) The surface area obtained by revolving C about the x-axis (assuming f is smooth)
(g) The surface area obtained by revolving C about the y-axis (assuming f is smooth)

2. Let g be a nonnegative continuous function of y on $[c, d]$ where $d > c > 0$. Denote the graph of $x = g(y)$ by C, and let R be the region to the left of C and to the right of the y-axis from $y = c$ to $y = d$. Write integrals for parts (a)–(g) of Exercise 1, replacing f with g.

3. Repeat parts (a)–(d) of Exercise 1 for the region R

between the graphs of $y = f(x)$ and $y = g(x)$ for x on $[a, b]$, where $f(x) \geq g(x) \geq 0$ and $b > a > 0$.

4. Repeat parts (a)–(d) of Exercise 1 for the region R between the graphs of $x = g(y)$ and $x = h(y)$ for y on $[c, d]$, where $g(y) \geq h(y) \geq 0$ and $d > c > 0$.

5. Write integrals for the following.

(a) The center of mass of a thin rod that reaches from $x = a$ to $x = b$ if the density is $\rho(x)$
(b) The work done by the continuous force function $F(x)$ acting from $x = a$ to $x = b$
(c) The total force on a plate submerged vertically in a liquid of density $\rho(x)$ if the top of the plate is a units below the surface and the bottom is b units below the surface. Assume that the width of the plate at depth x ($a \leq x \leq b$) is $w(x)$.
(d) The probability that $a < X < b$ if the probability density function on $[a, b]$ is $f(x)$
(e) The mean μ of a random variable whose probability density function is $f(x)$ on $[a, b]$
(f) The variance σ^2 of a random variable with mean μ and probability density function $f(x)$ on $[a, b]$

APPLYING CALCULUS

1. *Income distribution—Lorenz curve and Gini coefficient.* When a high percentage of the population earns a low percentage of the total national income, there are income inequities. To measure the degree of inequity, economists use a Lorenz curve, which is the graph of a function showing the cumulative percentage, $I(x)$, of total income received, versus the cumulative percentage, x, of income recipients. The function I would be given by $y = x$ if all income recipients had equal shares (when $x = 0.2$, $y = 0.2$, so the poorest 20% earn 20% of the total income). When income is not equally distributed, $I(x)$ has some other form, giving a Lorenz curve that deviates from the 45° line $y = x$, *called the line of absolute income equality.* The accompanying figure illustrates such a situation. The Gini coefficient is defined as the area between the Lorenz curve and the line $y = x$ divided by the area of the right triangle with legs one unit long on the x and y axes and hypotenuse along the line $y = x$. Since the area of the right triangle is $(\frac{1}{2})(1)(1) = \frac{1}{2}$, the Gini coefficient is

$$G = 2 \int_0^1 [x - I(x)]\,dx$$

Lorenz curve

$y = x$

Unequally distributed income

Suppose that

$$I(x) = \frac{x^2 + 5x}{7} \qquad 0 \le x \le 1$$

(a) Sketch the Lorenz curve.

(b) What portion of the total income is earned by the lowest 15% of the population?

(c) Calculate the Gini coefficient.

2. *(Refer to Problem 1 above)* In 1990 the percent of aggregate income received by each fifth of families in the U.S. was as follows:

lowest fifth	5.1
second fifth	11.6
third fifth	17.5
fourth fifth	24.3
highest fifth	41.5

(a) Find a function I so that the Lorenz curve $y = I(x)$ closely fits the given data.

(b) Using the function you found in part (a), calculate the Gini coefficient.

3. *Marginal propensity to consume and to save.* The rate at which total national consumption changes with respect to change in total national disposable income is called the **marginal propensity to consume**. So if $C = f(I)$, where I is the total national disposable income and C is the total national consumption, the derivative dC/dI is the marginal propensity to consume. Similarly, if $S = g(I)$ represents the total national savings, then dS/dI represents the **marginal propensity to save**. Since the amount spent on consumption plus the amount saved equals the total income, we have $C + S = I$, and thus

$$\frac{dC}{dI} + \frac{dS}{dI} = 1$$

Suppose that the marginal propensity to save (in billions of dollars) is given by

$$\frac{dS}{dI} = 0.9 - \frac{0.4}{\sqrt{I+1}}$$

and consumption is $7.8 billion when disposable income is $15 billion. Find the consumption function. what is the consumption when disposable income is $20 billion?

4. Coulomb's Law states that the force of attraction or repulsion between two point charges is directly proportional to the product of the charges and inversely proportional to the square of the distance between them:

$$F = k\frac{q_1 q_2}{r^2}$$

where q_1 and q_2 are the magnitudes of the charges at the points P_1 and P_2, respectively, r is the distance between P_1 and P_2 and k is the constant of proportionality.

(a) Determine the rate of change of the force with respect to the distance between the points. Suppose the distance between the charges is changing with respect to time according to the equation

$$\frac{dr}{dt} = r^2 + 2r$$

What is the change in the force with respect to time?
Suppose the point charges are both negative and have equal magnitudes, $q_1 = q_2 = q$.

(b) Determine the constant k if the points P_1 and P_2 are 4 centimeters apart and the repulsion force is 10 dynes.

(c) How much work is required to bring the charge at P_2 to a distance of 2 centimeters from the charge at P_1?

5. Poiseuille discovered that the velocity v (centimeters per second) of a blood particle flowing through a blood vessel having a circular cross section of radius R is related to its distance r from the central axis by the equation

$$v(r) = \frac{P(R^2 - r^2)}{4\mu L}$$

where P is the blood pressure, μ (mu) is the coefficient of viscosity of the blood and L is the length of the vessel. Flow characterized by this equation is called laminar flow, since the blood particles are assumed to flow in cylindrical shells, or laminae, parallel to the walls of the vessel at a distance r from the center, with all the blood particles in a particular shell moving with the same velocity.

If the midpoint of the ith shell is at a distance r_i^* from the center of the vessel, then the volume of blood flowing through that shell is approximately

$$V_i \approx 2\pi r_i^* v(r_i^*) \Delta r_i$$

Thus, the total volume V of blood per second flowing through the cross section is given approximately by

$$\sum_{i=1}^{n} 2\pi r_i^* \mu(r_i^*) \Delta r_i$$

from which it follows that

$$V = \int_0^R 2\pi r v(r) dr$$

(a) Show that the volume of blood flowing through the blood vessel is proportional to the 4th power of the radius by evaluating the integral derived above. This result is known as Poiseuille's Law.

(b) Suppose that blood pressure is constant. Calculate the percentage decrease in the volume of blood that flows through the vessel per unit of time when the radius is constricted from R to $R/2$.

(c) The average velocity of blood flow through a vessel at a circular cross section of radius R and area A is defined to be

$$\bar{v} = \frac{1}{A} \int_0^R 2\pi r v(r)\, dr$$

Show that $V = PR^2/8\mu L$ and that $\bar{v}$ is 1/2 the maximum velocity of the blood flow.

6. *Volumes of barrels.*

(a) A barrel is formed by rotating an ellipse around its major axis and then cutting off equal caps from the top and bottom. Determine the volume of the barrel if the (maximum) radius at the center is R, its height is H and the top and bottom have radius r. (Problems of this type were considered by Kepler in his studies of the volumes of barrels.)

(b) Suppose a barrel has the shape of the solid of revolution formed by rotating the region between the y-axis and the parabola $x = R - ky^2$, $-h/2 \le y \le h/2$, R, $k > 0$, and $h < R/k$, around the y-axis. Show that the volume of the barrel is

$$V = \frac{1}{3}\pi h \left(2R^2 + r^2 - \frac{6}{15}\delta^2 \right)$$

where h is the height of the barrel, R is the radius at the center, r is the radius of the top and bottom, and $\delta = R - r$.

7. It can be shown that the shape of a flexible and inelastic cable supporting a load that is uniformly distributed horizontally is a parabola. As an application, consider the main cable in a suspension bridge. The dimensions L and h shown in the figure are called the span and the sag, respectively, of the bridge.

(a) Let the origin of an xy-coordinate system be at the lowest point of a parabolic cable of a suspension bridge whose span is L and whose sag is h. Determine the equation for the cable.

(b) Determine the length of the cable in part (a).

(c) The George Washington Bridge across the Hudson River connecting New York City to Fort Lee, New Jersey, has a span of 3500 feet and a sag of 316 feet. Determine the length of the main cables of the bridge.

8. According to *Newton's Law of Universal Gravitation*, the gravitational force on an object of mass m that has been projected vertically upward from the Earth's surface is

$$F = \frac{-mgR^2}{(x + R)^2}$$

where $x = x(t)$ is the object's distance above the surface at time t, R is the radius of the Earth and g is the acceleration due to gravity. (Assume all the force is exerted at the instant of launch.) Also, by *Newton's second law*, $F = ma = m\,dv/dt$ and so

$$m\frac{dv}{dt} = -\frac{mgR^2}{(x + R)^2}$$

(a) Suppose a space shuttle is fired vertically upward with an initial velocity v_0. Let h be the maximum height above the surface reached by the object. Show that

$$v_0 = \sqrt{\frac{2gRh}{R + h}}$$

(Hint: By the Chain Rule, $m\,dv/dt = mv\,dv/dx$.)

(b) Calculate $v_e = \lim_{h \to \infty} v_0$. This limit is called the *escape velocity* for the Earth.

(c) Use $R = 3960$ mi and $g = 32$ ft/s^2 to calculate v_e in ft/s and mi/s.

9. A space shuttle of mass m is fired vertically upward from the surface of the Earth. The force F of the earth's gravitational field acting on the space shuttle is given by Newton's Law of Universal Gravitation.

(a) Determine the work that must be done against F to lift the space shuttle to a height h above the surface of the Earth.

(b) Determine the work that must be done to lift the space shuttle out of the Earth's gravitational field, that is, to lift the space shuttle to an infinite height.

(c) Suppose that the space shuttle is to be projected upward by imposing an initial kinetic energy $\frac{1}{2}mv_0^2$. Determine the initial velocity required to lift the space shuttle infinitely high. Compare with the result in Problem 8.

10. The sun's rate of travel across the sky is uneven, sometimes ahead and sometimes behind the average solar time shown on our watches. Part (a) below shows a photo on a single piece of film that shows the position of the Sun in the sky at 8:00 a.m. every 10 days for a year. The figure-8 is known as an analemma. Part (b) shows a sundial with an analemma rotated around its axis as the shadow stick (known as the *gnomon*). As the Sun rises higher in the sky as summer approaches, the bulge in the analemma compensates for the sun's divergence from its average rate of travel. We can thus read exact time as the position where the shadow of one side of the analemma crosses a circular bar at the bottom of the sundial. This analemma is a solid of revolution. If we assume that it is 30 cm long and 4 cm in its greatest diameter, and is approximated by $y = 2\sin(2\pi t/182.6)$ rotated around the axis $y = -0.2$, calculate its volume. It is suspended by thin wires, so we have to know its weight. If it is made of aluminum, whose density is 2 g/cm^3, how much does this solid analemma weigh?

(a)

(b)

Keith Geddes

You're a professor of computer science in the faculty of mathematics at the University of Waterloo. Are you a computer scientist or a mathematician?

I see myself as a computer scientist and a mathematician. Computer science spans topics from mathematics through physics and engineering and more, but mathematics is even broader. In high school I found mathematics interesting but it never occurred to me that you could specialize in mathematics. I thought it was only a subject that supported "real" disciplines like engineering and physics. In second year, I switched my major from physics to mathematics.

What exactly is "computer algebra"?

The subject of computer algebra deals with mathematical computation in a general sense. This includes the manipulation of unevaluated symbols and unevaluated functions according to the rules of algebra and other branches of mathematics. In computer algebra programs, variables and expressions don't have to be evaluated to numerical values prior to being manipulated, but computer algebra programs can also work with arbitrary precision arithmetic. Computer algebra programs are like calculators that do algebra as well as arithmetic. They can also display your results graphically in a fast and sometimes spectacular way.

You and Gaston Gonnet are the original architects of the computer algebra program Maple. What motivated you to start the Maple project?

When we conceived of the project in November of 1980, the goal was to design a computer algebra system that would be useful to large numbers of researchers in mathematics, engineering, and the sciences—and also to assist students in their mathematical education. Existing computer algebra systems required large computers that were only usable by specialists. The other important motivation was to provide a focus for research into algorithms for algebraic computation. Maple has grown considerably since we started. There are now more than 2000 mathematical functions in Maple's libraries. But designers can't possibly anticipate all the needs of the users of mathematics. If Maple doesn't already know how to carry out a particular computation, you can program it to do so.

Maple's libraries are programs written in Maple's own language. How is the Maple language different from, say, Pascal or FORTRAN?

The fundamental difference is that the Maple language was designed specifically for algebraic computation. It incorporates commands for many high-level mathematical operations that are not built into other languages. Unlike FORTRAN or Pascal, the Maple language is designed to be used interactively. Specific directions such as "integrate $\cos x$ with respect to x" are given with one-line commands, which are easy to learn. Also, if you need to, it's not hard to program in Maple.

What types of research is Maple useful for?

Maple has proved useful in research that uses mathematics at almost any level. This includes pure mathematics, applied mathematics, mechanical engineering, economics, computer science, and so on. A few areas that come to mind are ongoing research in robotics, econometric modeling, general relativity, number theory, and computer-aided design. I recently heard about some models to improve voting systems and to study the AIDS epidemic that were developed and are taught using Maple. In the case of the AIDS model, students are exploring changes in parameters and using Maple to numerically integrate the differential equations involved.

How did you come to apply your own research in algebra to other areas of mathematics like those of calculus?

I came to the topic from the point of view of algebraic algorithms. I was attracted by the precision and beauty of algebraic computation and the very powerful algorithms that could be developed to factor multivariate polynomials, for instance. Many of the capabilities of computer algebra are straightforward, but indefinite integration is an exception. Many people view integration not as an algorithmic process, but as a collection of tricks that can solve a limited number of problems. Finding that the Risch algorithm applied algebraic algorithms to the indefinite integration problem of calculus is what led me deeper into computer algebra.

What is the Risch algorithm and why is it important?

In 1969 Robert Risch presented a decision procedure for the indefinite integration of the elementary functions. These are the functions typically encountered in calculus: algebraic, trigonometric, inverse trigonometric, exponential, logarithmic, and hyperbolic. By performing a finite number of computational steps, the algorithm can determine whether or not a given elementary function has an antiderivative that is also an elementary function. The Risch algorithm either determines a closed formula expressing the integral as an elementary function, or else it proves that expressing the integral as an elementary function is impossible. It has a clear advantage over standard integration techniques. If students learn traditional integration techniques only, and then fail to find an answer, they are sometimes left wondering if they simply did not find the right trick.

It's clear that Maple is a powerful research tool, but does it actually help students learn mathematics?

Yes. The greatest benefit of Maple to students is in making it possible to carry out "experimental mathematics." It allows mathematical topics that might seem distant and abstract to be played with on a level that is meaningful to students. It's been said that systems like Maple allow students and professors to devote more time to the ideas: the theory and the applications of the mathematics, and less to the routine computations. In the future I expect to see more emphasis on the problem-solving process with a significant decrease in the emphasis on teaching "tricks of the trade."

How would you contrast this approach with your own mathematical education?

My mathematical education is continuing, but it seems that today's students have more opportunity to explore "what if" scenarios related to core mathematical concepts. They can now learn more by "doing mathematics." Of course, they can also do many more problems than they can by hand. By removing some mathematical drudgery, there is greater potential for bringing some mathematical excitement to the learning process. It shouldn't just be drudgery. We should be teaching students how to take advantage of computers, not to compete with them in algorithmic computations.

O.K., but if you can now solve integrals in closed form with Maple, why should you learn to integrate by hand with pencil and paper?

People once worried that pocket calculators would make finding square roots by hand, and even knowing multiplication and division, less important. Of course, calculators didn't do this. The calculators made the concepts more important since more people were using mathematics because of them. In the same way, a small computer running a powerful computer algebra system is bound to change some aspects of using higher mathematics. I believe problem solving, as a way of mathematical thinking, will become more important. The real goal of a mathematical education is to be able to take a problem, stated in natural language, and express that problem in mathematical terms (a new language, if you will), to solve that mathematical problem, and finally to reinterpret the results in terms of the originally stated problem. The intermediate step [to] "solve the mathematical problem" will continue to require many skills and creativity, but the aspects that are pure calculation like "factor a polynomial" or "find an indefinite integral" can be relegated to the computer algebra calculator. The important thing, it seems to me, is to teach the understanding of mathematical concepts such as indefinite integration, or the meaning of a solution to a differential equation, rather than teaching how to "turn the crank." Sadly, there are generations of students who think, because of their education, that learning mathematics is learning how to turn the crank. Much of mathematics is really in trying to "find the crank" or, at a higher level, "is there a crank?" or "prove that there isn't a crank." Mathematics involves much more creativity than most people are aware.

Students seem to take well to Maple, after they learn the syntax and the computer keyboard. Is it true that students have been active in testing and developing Maple code?

Yes. Students have been very active with Maple from the beginning. At Waterloo most of our students are in the co-operative work/study program where they alternate four-month study terms with four-month work terms. We're lucky to have some good students who want to work on Maple itself. Many students have started with Maple in their calculus courses and have continued throughout their careers. There have always been several graduate students working on Maple at any given time. Waterloo has five separate departments in the Faculty of Mathematics. Faculty and students from all of these subdisciplines have had a hand in developing Maple. Maple is now an international research effort coordinated from Waterloo and other locations throughout the world. Many talented and hardworking people deserve credit for its success.

What are the limits to the type of mathematics Maple can help with? Will we ever be able to use Maple to set up a word problem or prove a theorem?

Well, I'm pretty sure that Maple can solve all of the limit problems in this text, for instance, but it's important to remember that Maple is just an algebraic calculator that does what it is programmed to do. While it is extremely useful to have such a calculator, calculation is only one aspect of doing mathematics. Maple will sometimes make mistakes because there are trade-offs between accuracy and efficiency in its design. Generally speaking, Maple doesn't "know calculus"; rather, it knows how to carry out the algebraic manipulations associated with calculus. I do not see Maple evolving into a system that could set up a word problem and solve it, or prove a theorem, although there is computer science research being done on these subjects. These aspects of mathematics require creative human mathematicians. Mathematics is very much a human endeavor.

Keith Geddes is Professor of Computer Science and Associate Dean for Graduate Studies and Research in the Faculty of Mathematics at the University of Waterloo. His research interests lie in the areas of algebraic algorithms, computer algebra systems, and scientific computation. Professor Geddes has coauthored a graduate-level textbook, *Algorithms for Computer Algebra* and continues to be involved in the research development of the Maple computer algebra system. He holds a B.A. in Mathematics and a Ph.D. in Computer Science from the University of Toronto.

7

TRANSCENDENTAL FUNCTIONS

In Chapter 1 we characterized the algebraic functions as those that can be obtained from polynomial functions by any finite combination of the operations of addition, subtraction, multiplication, raising to an integral power, and extracting integral roots. Stated precisely, an algebraic function is any function f such that $y = f(x)$ satisfies an equation of the form

$$P_n(x)y^n + P_{n-1}(x)y^{n-1} + \cdots + P_1(x)y + P_0(x) = 0$$

where each $P_i(x)$ is a polynomial with rational coefficients. For example, $f(x) = 1/\sqrt{x^2 + 1}$ is algebraic since $y = 1/\sqrt{x^2 + 1}$ satisfies the polynomial equation $(x^2 + 1)y^2 - 1 = 0$. The class of algebraic functions is very broad, but it by no means includes all possible functions.

In fact, we have already studied one class of nonalgebraic (or *transcendental*) functions, namely, the trigonometric functions. In this chapter we introduce the following additional transcendental functions: logarithmic, exponential, inverse trigonometric, hyperbolic, and inverse hyperbolic functions. These, together with the trigonometric and algebraic functions, constitute what are called the **elementary functions**. After introducing our new functions and studying their properties, we will concentrate on their derivatives and integrals. We will also show some important applications.

We begin with the natural logarithm function, $\ln(x)$, which we define as a special area function. This approach enables us to use properties of area functions that we studied in Chapter 5. In particular, we will see that the First Fundamental Theorem of Calculus plays an important role. Next, we define the natural exponential function, $\exp(x)$, as the inverse of $\ln(x)$. The natural logarithm and exponential functions comprise one of the most important function–inverse function pairs in mathematics and science. From these two basic functions we will obtain more general logarithm and exponential functions. (We will see, for example, how irrational powers such as $\pi^{\sqrt{2}}$ are defined.) We conclude the chapter by studying the inverses of each of the trigonometric functions (with suitably restricted domains) and then the so-called hyperbolic functions and their inverses.

7.1 THE NATURAL LOGARITHM FUNCTION

Logarithms were invented in about 1590 by John Napier (1550–1617), a Scottish mathematician who was interested in methods of rapid calculation for practical use in such fields as astronomy, navigation, and surveying. He developed his idea and calculated tables to 7 places of accuracy. Such "log tables" were in use well into the twentieth century. Though logarithms are easy to use, the idea had apparently not been thought of before Napier. A Swiss watchmaker, Jobst Bürgi, independently invented logarithms and published his results in 1620.

Word of Napier's discovery reached the Danish astronomer Tycho Brahe in the late 1590s, but Napier's tables were not ready in time for Brahe to use. But Brahe's successor, Johannes Kepler, did correspond with Napier and found logarithms important for astronomical use. Kepler's astronomical tables of 1620 made use of Napier's discovery.

You are probably familiar with what is known as the *common* logarithm of a number, usually designated $\log x$. This logarithm is by definition the power to which 10 must be raised to give the number x. For example, $\log 100 = 2$, since $10^2 = 100$. On your calculator you will find a $\boxed{\log x}$ key. Before scientific calculators were available, tables of common logarithms were used to obtain their values. These common logarithms were formerly used in simplifying arithmetical computations involving products, quotients, powers, and roots, based on the properties

$$\log ab = \log a + \log b$$

$$\log \frac{a}{b} = \log a - \log b$$

$$\log a^p = p \log a$$

Their use in computations is of little importance now, however, because of the calculator and computer.

Although common logarithms still have some uses, it is the *natural* logarithm that is of fundamental importance in calculus and its applications. The natural logarithm of x is denoted by $\ln(x)$, or more often by $\ln x$ (read "el en x"). It is possible to define $\ln x$ in a manner similar to the way $\log x$ is defined, with 10 being replaced by a number e whose value is approximately 2.71828. You may have seen this definition before in previous mathematics courses. Now that we have calculus at our disposal we will take a different approach, but we will see that this characterization of $\ln x$ as a power of e is valid.

We use the symbol ln to denote the **natural logarithm function**. We have previously (in Chapter 1) referred to the fact that your calculator has an $\boxed{\ln x}$ key. With it, you can obtain values of natural logarithms to a high degree of accuracy. But where these values come from is something of a mystery. One way of obtaining them is based on the following definition. (Another way will be given in Chapter 10.)

For any positive real number x, the value of $\ln x$ is defined as the integral of the function $f(t) = 1/t$ from $t = 1$ to $t = x$ (Figure 7.1). That is,

Definition 7.1 The Natural Logarithm Function	$$\ln x = \int_1^x \frac{1}{t}\,dt \qquad (x > 0) \qquad\qquad (7.1)$$

If $x > 1$, we can interpret $\ln x$ as the area under the graph of $f(t) = 1/t$ from 1 to x. (See Figure 7.1(a).) If $0 < x < 1$, then

$$\ln x = \int_1^x \frac{1}{t}\,dt = -\int_x^1 \frac{1}{t}\,dt$$

which is the negative of the area under the graph of f from x to 1. (See Figure 7.1(b).)

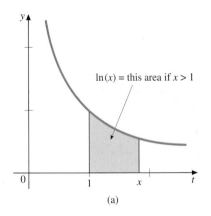

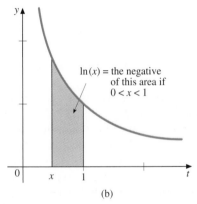

FIGURE 7.1

If $x = 1$, we see by Equation 7.1 that $\ln x = 0$. We therefore have

$$\ln(x) > 0 \qquad \text{if } x > 1$$
$$\ln 1 = 0$$
$$\ln x < 0 \qquad \text{if } 0 < x < 1$$

Note that $\ln x$ is not defined for $x \leq 0$. That is, the domain of $\ln x$ is the set of all *positive* real numbers. We will see what the range is after investigating some of the properties of $\ln x$.

Properties of $\ln x$

Recall that by the First Fundamental Theorem of Calculus, when a function f is continuous on a closed interval, for a fixed number a and any x in this interval

$$\frac{d}{dx} \int_a^x f(t)\,dt = f(x)$$

Applying this theorem to the integral in Equation 7.1, we obtain the important result

$$\frac{d}{dx}(\ln x) = \frac{1}{x} \qquad (x > 0) \qquad\qquad (7.2)$$

Since $x > 0$, we see that

$$\frac{d}{dx}(\ln x) > 0$$

for all x in the domain of $\ln x$. So $\ln x$ is an increasing function, and hence is one-to-one, throughout its domain. Differentiating again, we get

$$\frac{d^2}{dx^2}(\ln x) = -\frac{1}{x^2}$$

Since this second derivative is negative, we conclude that the graph of $y = \ln x$ is concave down throughout its domain.

We now know quite a bit about the graph of $y = \ln x$. Its x-intercept is 1, it is always rising as x increases, and it is always concave down. We are not sure, on the basis of what we have shown so far, whether there are any asymptotes. To investigate this question, we consider the two limits

$$\lim_{x \to \infty} (\ln x) \qquad \text{and} \qquad \lim_{x \to 0^+} (\ln x)$$

Before doing so, we need to establish some additional properties of the natural logarithm function.

Arithmetic Properties

By means of the following derivation, we will show that the three properties of the common logarithm, involving products, quotients, and powers, also hold true for the natural logarithm.

First, consider $\ln(ax)$, where a is a positive constant and x is any positive number. By Equation 7.2 and the Chain Rule, we have

$$\frac{d}{dx}\ln(ax) = \frac{1}{ax} \cdot a = \frac{1}{x}$$

But by Equation 7.2 we know that $1/x$ is also the derivative of $\ln x$. So

$$\frac{d}{dx}\ln(ax) = \frac{d}{dx}\ln x$$

By Corollary 4.9(b) we know that if two functions have the same derivative, they differ by a constant. Thus,

$$\ln(ax) - \ln x = C$$

for some constant C. We can determine C by substituting $x = 1$ and using the fact that $\ln 1 = 0$:

$$\ln a - \ln 1 = C$$

or, $C = \ln a$. So we have, for any $x > 0$ and $a > 0$,

$$\ln(ax) - \ln x = \ln a$$

In this equation we substitute $x = b$, where b is any positive number. After rearranging terms, we get

$$\ln(ab) = \ln a + \ln b \tag{7.3}$$

In words, we can say *the natural logarithm of the product of two positive real numbers is the sum of their natural logarithms*; or more briefly, we might say "the log of a product is the sum of the logs," with the understanding that "log" here means "natural logarithm." (Later, we will see that the same result holds for other types of logarithms as well.)

We can obtain a formula for the natural logarithm of a quotient by the following trick. Let a and b again denote any two positive numbers. Then we can write a as the product of b and a/b, so

$$\ln a = \ln\left(b \cdot \frac{a}{b}\right) = \ln b + \ln\left(\frac{a}{b}\right) \qquad \text{By Equation 7.3}$$

Solving for $\ln(a/b)$ gives

$$\ln\left(\frac{a}{b}\right) = \ln a - \ln b \tag{7.4}$$

So "the log of a quotient is the log of the numerator minus the log of the denominator."

A third property, involving the log of a power, can be derived as follows. Let r denote any rational number, and let x be positive. Then, by Equation 7.2 and the Chain Rule,

$$\frac{d}{dx}\ln(x^r) = \frac{1}{x^r} \cdot rx^{r-1} = \frac{r}{x}$$

Since

$$\frac{r}{x} = r \cdot \frac{1}{x} = r\frac{d}{dx}(\ln x) = \frac{d}{dx}(r\ln x) \quad \text{Since } r \text{ is a constant}$$

we see that $\ln x^r$ and $r \ln x$ have the same derivative for all x in $(0, \infty)$. So, again using Corollary 4.9(b), we conclude that these two functions differ only by some constant value C:

$$\ln(x^r) = r\ln x + C$$

Putting $x = 1$ gives $\ln 1 = r \ln 1 + C$, and since $\ln 1 = 0$, we find that $C = 0$. So we have the result

$$\ln(x^r) = r\ln x \qquad\qquad (7.5)$$

In words we can say "the log of x to a power is the exponent times the log of x."

REMARK —————————————————————————————————

■ At this stage the exponent in Equation 7.5 is limited to being a rational number, since in our derivation we used the Power Rule for differentiating x^r, and we have proved the Power Rule only for rational r. In Section 7.3 we will show that this restriction can be removed and that r can be any real number, rational or irrational. (See Theorem 7.5 and the Remark following it.)

Formulas 7.3, 7.4, and 7.5 account for much of the utility of logarithms. We pull them together here for emphasis, referring to them as arithmetic properties.

Arithmetic Properties of the Natural Logarithm Function

For $a > 0$, $b > 0$, and r rational,

$$\ln(ab) = \ln a + \ln b$$

$$\ln\left(\frac{a}{b}\right) = \ln a - \ln b$$

$$\ln a^r = r\ln a$$

EXAMPLE 7.1   Use the arithmetic properties of $\ln x$ to write each of the following in terms of $\ln 2$ or $\ln 3$, or both.

(a) $\ln 6$ (b) $\ln 1.5$ (c) $\ln 8$

Solution

(a) Since $6 = 2 \cdot 3$, by Formula 7.3 we have $\ln 6 = \ln 2 + \ln 3$
(b) Since $1.5 = 3/2$, by Formula 7.4 we have $\ln 1.5 = \ln \frac{3}{2} = \ln 3 - \ln 2$
(c) Since $8 = 2^3$, by Formula 7.5 we have $\ln 8 = \ln 2^3 = 3 \ln 2$ ∎

The Graph of $y = \ln x$

We return now to our investigation of the graph of $y = \ln x$. First, let us show that $\ln x \to \infty$ as $x \to \infty$. Choose any number greater than 1, so that its logarithm is positive. For definiteness, let us choose 2. Our plan is to show that $\ln(2^n)$ can be made arbitrarily large by choosing n sufficiently large. By Equation 7.5 we know that for any positive integer n,

$$\ln(2^n) = n \ln 2$$

If M is any positive number whatsoever, no matter how large, we can make $n \ln 2$ even larger than M by choosing

$$n > \frac{M}{\ln 2}$$

(multiplying by the positive number $\ln 2$ gives $n \ln 2 > M$). For such an n, therefore,

$$\ln(2^n) > M$$

Since $\ln$ is an increasing function, it follows that for all $x > 2^n$, $\ln x > \ln(2^n)$, so

$$\ln x > M$$

What we have shown is that $\ln x$ can be made as large as we please (greater than any number M) provided x is sufficiently large (greater than 2^n, where $n > M/\ln 2$). It follows that

$$\lim_{x \to \infty} \ln x = \infty$$

So while the graph of $y = \ln x$ is concave down, it has no horizontal asymptote. It continues to rise as x increases (but ever so slowly, as we will soon see).

Now we use a similar argument to show that $\ln x \to -\infty$ as $x \to 0^+$ (i.e., as x approaches 0 from the right). For n a positive integer, we have, by Equation 7.5,

$$\ln(2^{-n}) = -n \ln 2$$

since $-n$ is rational. So if M is an arbitrarily large positive number, $-n \ln 2$ will be less than $-M$, provided

$$n > \frac{M}{\ln 2}$$

Remember that dividing an inequality by a negative number reverses the direction of the inequality.

For any such n, therefore,

$$\ln(2^{-n}) < -M$$

Now choose any $x < 2^{-n}$ and positive. Then

$$\ln x < -M$$

since ln is an increasing function on its domain. That is, by choosing x sufficiently close to zero (but positive), we can make its logarithm as large negatively as we wish. In other words,

$$\lim_{x \to 0^+} \ln x = -\infty$$

The graph of $\ln x$ is therefore asymptotic to the negative y-axis as x approaches 0 from the right.

Because $\ln x$ is continuous for all x in its domain (its derivative exists everywhere in the domain, which implies it is continuous for all x in the domain), and since $\ln x$ takes on arbitrarily large positive and negative values, as we have just shown, we can conclude by the Intermediate-Value Theorem (Theorem 2.7) that $\ln x$ takes on every real value. That is, the range of the natural logarithm function is $(-\infty, \infty)$.

We can now indicate the general shape of the graph of $y = \ln x$, as shown in Figure 7.2. Note that we arrived at this graph using the properties we have shown but knowing only one specific value of $\ln x$, namely, $\ln 1 = 0$. We will discuss how to approximate other values shortly. (Of course, the easiest way is to use your calculator.)

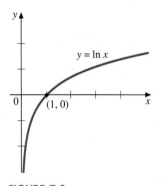

FIGURE 7.2

EXAMPLE 7.2   Use the graph of $\ln x$ shown in Figure 7.2, together with the arithmetic properties, to estimate the value of each of the following.

(a) $\ln 2$ (b) $\ln 4096$ (c) $\ln \dfrac{1}{64}$

Solution

(a) From the graph we find that $\ln 2 \approx 0.7$.

(b) Since $4096 = 2^{12}$ (you should verify this result) we have, by Formula 7.5

$$\ln 4096 = \ln 2^{12} = 12 \ln 2$$

Using our estimate for $\ln 2$ from part (a), we see that $\ln 4096 \approx 12(0.7) = 8.4$. The fact that 4096 is a rather large number but its natural logarithm is only about 8.3 illustrates that the graph in Figure 7.2 rises very slowly.

(c) Since $1/64 = 1/2^6 = 2^{-6}$, we have, again by Formula 7.5 and the result of part (a),

$$\ln \frac{1}{64} = \ln 2^{-6} = -6 \ln 2 \approx -6(0.7) = -4.2$$

■

Let us summarize what we know so far about the natural logarithm function.

The Natural Logarithm Function

1. $\ln x = \int_1^x \frac{1}{t}\, dt,\ x > 0$

2. $\text{domain} = (0, \infty),\ \text{range} = (-\infty, \infty)$

3. $\dfrac{d}{dx} \ln x = \dfrac{1}{x}$

4. $\ln x$ is an increasing function on its domain and its graph is concave down.

5. $\ln x > 0$ if $x > 1$

   $\ln x < 0$ if $0 < x < 1$

   $\ln 1 = 0$

6. $\lim\limits_{x \to \infty} \ln x = \infty$ and $\lim\limits_{x \to 0^+} \ln x = -\infty$

7. For $a > 0$, $b > 0$, and r rational,

$$\ln(ab) = \ln a + \ln b$$

$$\ln \frac{a}{b} = \ln a - \ln b$$

$$\ln a^r = r \ln a$$

EXAMPLE 7.3   Make use of the facts summarized above to do the following.

(a) Estimate the value of $\int_1^2 \frac{du}{u}$.

(b) Arrange in increasing order of magnitude:

  $\ln 22/7$, $\ln 2/3$, $\ln \pi$, $\ln 10^4$, $\ln 1 + \ln 0.02$, $\ln 2^{13}$, $\ln 4 - \ln 5$

(c) Show that $y = \ln x$ is a solution of the differential equation

$$x^2 y'' + 2xy' = 1$$

  for all $x > 0$.

Solution

(a) By the first item in the summary, we see that

$$\int_1^2 \frac{du}{u} = \ln 2$$

(The fact that the variable of integration here is u instead of t is of no consequence, since it is a "dummy" variable.) We saw in Example 7.2 that $\ln 2 \approx 0.7$. So

$$\int_1^2 \frac{du}{u} \approx 0.7$$

(b) First note that $\ln 1 + \ln 0.02 = \ln (0.02)$, since $\ln 1 = 0$. Also, $\ln 4 - \ln 5 = \ln (4/5)$ by the second arithmetic property in item 7. Since the natural logarithm function is monotone increasing, it suffices to arrange the numbers themselves in increasing order. That is, if we can show $x_1 < x_2$, it will follow that $\ln x_1 < \ln x_2$.

Clearly, $0.02 < 2/3 < 4/5$. In decimal form $22/7 = 3.1428\ldots$ and $\pi = 3.14159\ldots$, so $\pi < 22/7$. (It is interesting to note that $22/7$ is sometimes used as a rational approximation to π; in fact, a state legislature once passed a law declaring that π was *equal* to $22/7$!) We saw in Example 7.2 that $2^{12} = 4096$, so $2^{13} = 8192$. Since $10^4 = 10{,}000$, then, we see that $2^{13} < 10^4$.

We can now arrange the given logarithms in increasing order, as follows:

$$\ln 1 + \ln 0.02 < \ln \frac{2}{3} < \ln 4 - \ln 5 < \ln \pi < \ln \frac{22}{7} < \ln 2^{13} < \ln 10^4$$

(c) If $y = \ln x$, then we know by item 3 that $y' = 1/x$. Thus $y'' = -1/x^2$. So we have for $x > 0$

$$x^2 y'' + 2xy' = x^2 \left(-\frac{1}{x^2} \right) + 2x \left(\frac{1}{x} \right)$$

$$= -1 + 2 = 1$$

Thus, $y = \ln x$ is a solution to the given differential equation. ■

The Slow Growth Rate of $\ln x$

Although we have shown that $\ln x$ becomes arbitrarily large as $x \to \infty$, it is interesting to note how slowly the graph rises. For $x = 1$ billion, $\ln x$ is just somewhat more than 20. How large do you suppose x would need to be for $\ln x$ to equal 100? The answer is about $(2.7)^{100}$. As a way of getting some idea of the size of this number, we know by Definition 7.1 that for $\ln x$ to equal 100, we must have

$$\int_1^x \frac{1}{t}\, dt = 100$$

If we were to approximate the integral on the left by using a left-hand sum with $\Delta x = 1$, we would need to find n such that (see Figure 7.3)

$$\sum_{k=1}^n \frac{1}{k} = 100$$

(See Exercise 57 in Exercise Set 7.1.) That is, we would need to form the sum

$$1 + \frac{1}{2} + \frac{1}{3} + \frac{1}{4} + \cdots$$

and keep going until this sum first exceeds 100. How many terms are required? Well, as we said, it is about $(2.7)^{100}$. But if the fastest computer available today had started adding at the time of the Big Bang, about 20 billion years ago, it still would not have added enough terms!

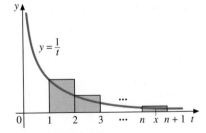

FIGURE 7.3
$\ln x = \int_1^x \frac{1}{t}\, dt \approx \sum_{k=1}^n \frac{1}{k}$, where $n \le x < n + 1$

Do not confuse the graph of $f(t) = 1/t$, which we used to define $\ln x$, with the graph of $y = \ln x$ itself. The graph of $f(t) = 1/t$ was a means to an end.

The fact that $\ln x$ is defined in terms of the area under $f(t) = 1/t$ does enable us to approximate $\ln x$ for particular values of x, using an approximation technique such as Simpson's Rule. For example, it will be useful to obtain approximate values of ln 2 and ln 3. By Equation 7.1,

$$\ln 2 = \int_1^2 \frac{1}{t} dt$$

Let us use Simpson's Rule with $n = 10$ to approximate this value. We have $\Delta x = (2 - 1)/10 = 0.1$, so

$$\ln 2 \approx \frac{0.1}{3} \left[1 + 4 \left(\frac{1}{1.1} \right) + 2 \left(\frac{1}{1.2} \right) + 4 \left(\frac{1}{1.3} \right) + 2 \left(\frac{1}{1.4} \right) + \cdots \right.$$

$$\left. + 2 \left(\frac{1}{1.8} \right) + 4 \left(\frac{1}{1.9} \right) + \frac{1}{2} \right]$$

$$\approx 0.6932$$

Similarly, using $n = 20$, we find that $\ln 3 \approx 1.0986$. Using the error bounds we gave in Chapter 5, you can verify that these answers are correct to the number of decimal places given (see Exercise 72 in Exercise Set 7.1). In the next section we will have occasion to use the fact that $\ln 2 < 1$ and $\ln 3 > 1$.

Derivatives of Natural Logarithm Functions

If g is any positive differentiable function of x, then by Equation 7.2 and the Chain Rule,

$$\frac{d}{dx} \ln(g(x)) = \frac{1}{g(x)} g'(x)$$

Letting $u = g(x)$, we can write this result using Leibniz notation as

$$\frac{d}{dx} \ln u = \frac{1}{u} \frac{du}{dx} \qquad (u > 0) \qquad (7.6)$$

We illustrate the use of this formula in the next example.

EXAMPLE 7.4   State the domain of each of the following, and find $f'(x)$.

(a) $f(x) = \ln(x\sqrt{1 + x^2})$ (b) $f(x) = \dfrac{\ln x}{x}$

(c) $f(x) = (\ln \sqrt{2x - 1})^3$ (d) $f(x) = \ln(\cos x)$

Solution

(a) The domain consists of all x for which $x\sqrt{1 + x^2} > 0$, which is the set of all $x > 0$. To find the derivative, we first simplify, making use of Equations 7.3 and 7.5.

$$\ln(x\sqrt{1 + x^2}) = \ln x + \ln(1 + x^2)^{1/2} \qquad \text{By Equation 7.3}$$

$$= \ln x + \frac{1}{2} \ln(1 + x^2) \qquad \text{By Equation 7.5}$$

Now we can differentiate each term, using Equation 7.6 on the second term, to get

$$f'(x) = \frac{1}{x} + \frac{1}{2} \frac{1}{1 + x^2} \cdot 2x = \frac{1}{x} + \frac{x}{1 + x^2}$$

Using the arithmetic properties of the natural logarithm function before taking the derivative greatly simplified our work. To see why, suppose that we had differentiated without applying these properties. Then, using Equation 7.6, we would have obtained

$$f'(x) = \frac{1}{x\sqrt{1+x^2}} \frac{d}{dx}\left(x\sqrt{1+x^2}\right)$$

$$= \frac{1}{x\sqrt{1+x^2}}\left[\sqrt{1+x^2} + x \cdot \frac{1}{2}\left(1+x^2\right)^{-1/2}(2x)\right]$$

$$= \frac{1}{x\sqrt{1+x^2}}\left[\sqrt{1+x^2} + \frac{x^2}{\sqrt{1+x^2}}\right]$$

$$= \frac{x}{1+x^2} + \frac{1}{x}$$

The result is the same as before, but getting there is harder. As a general rule, when taking the derivative of $\ln(g(x))$, where $g(x)$ involves products, quotients, or powers, make use of the arithmetic properties of the ln function *before* taking the derivative.

(b) The domain of the numerator is the set of all $x > 0$, so the requirement on the denominator that $x \neq 0$ is automatically satisfied. Thus, the domain is the set of all $x > 0$.

   In this case, we can make use of the Quotient Rule to find the derivative.

$$f'(x) = \frac{\frac{1}{x} \cdot x - \ln x}{x^2} = \frac{1 - \ln x}{x^2}$$

(c) For $\ln\sqrt{2x-1}$ to be defined, we must have $2x - 1 > 0$, or equivalently, $x > \frac{1}{2}$. The domain is therefore the set of all $x > \frac{1}{2}$.

   By Equation 7.5 we can write $f(x)$ as $[\frac{1}{2}\ln(2x-1)]^3$, or equivalently, $\frac{1}{8}[\ln(2x-1)]^3$. Now we use the Power Rule, together with Equation 7.6.

$$f'(x) = \frac{1}{8} \cdot 3[\ln(2x-1)]^2 \cdot \frac{1}{2x-1} \cdot 2 = \frac{3[\ln(2x-1)]^2}{4(2x-1)}$$

(d) The domain consists of all x values for which $\cos x > 0$ (the shaded regions in Figure 7.4), namely,

$$\left\{x : -\frac{\pi}{2} + 2n\pi < x < \frac{\pi}{2} + 2n\pi, \; n = 0, \pm1, \pm2, \ldots\right\}$$

Then, by Equation 7.6, we have

$$f'(x) = \frac{1}{\cos x}(-\sin x) = -\tan x \qquad \blacksquare$$

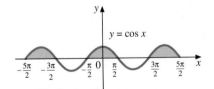

FIGURE 7.4

$\cos x$ is positive for
$-\frac{\pi}{2} + 2n\pi < x < \frac{\pi}{2} + 2n\pi$

We can generalize Equation 7.6 to include cases where $g(x) < 0$ by making use of absolute values. When $g(x) < 0$, $|g(x)| = -g(x)$, so

$$\frac{d}{dx}\ln|g(x)| = \frac{d}{dx}\ln[-g(x)] = \frac{1}{-g(x)} \cdot [-g'(x)] = \frac{g'(x)}{g(x)}$$

In terms of u, where $u = g(x)$, this result is

$$\frac{d}{dx}\ln|u| = \frac{1}{u}\frac{du}{dx} \qquad (u \neq 0) \qquad (7.7)$$

When $u > 0$, Equation 7.7 is equivalent to Equation 7.6. Thus, Equation 7.7 is true if $u > 0$ or if $u < 0$. The next example shows how we can use this result.

EXAMPLE 7.5   Find $f'(x)$ if

$$f(x) = \ln\left[\frac{x(x-1)}{x+2}\right]$$

Solution   First, we find the domain of f. A sign graph (Figure 7.5) shows that the quotient

$$\frac{x(x-1)}{x+2}$$

is positive if x is in $(-2, 0) \cup (1, \infty)$. This set is therefore the domain of f. For x in this domain, we can write $f(x)$ as

$$f(x) = \ln\left[\frac{x(x-1)}{x+2}\right] = \ln\left|\frac{x(x-1)}{x+2}\right| \qquad \text{Since for } N > 0, \\ N = |N|$$

$$= \ln\frac{|x||x-1|}{|x+2|} \qquad \text{Using properties of absolute value}$$

$$= \ln|x| + \ln|x-1| - \ln|x+2| \qquad \text{By Equations 7.3 and 7.4}$$

Thus, by Equation 7.7,

$$f'(x) = \frac{1}{x} + \frac{1}{x-1} - \frac{1}{x+2} \qquad \blacksquare$$

FIGURE 7.5
Sign graph for $\dfrac{x(x-1)}{x+2}$

REMARK ――――――――――――――――――――――――――――――――――――

■ In the preceding example, if we had ignored questions of the domain and forgotten about absolute values, we could have written, formally,

$$f(x) = \ln\left[\frac{x(x-1)}{x+2}\right] = \ln x + \ln(x-1) - \ln(x+2)$$

and on differentiation we would have obtained

$$f'(x) = \frac{1}{x} + \frac{1}{x-1} - \frac{1}{x+2}$$

which is exactly the answer we got before. So it is tempting to do the problem in this way. Unfortunately, it is implicit in this shortened procedure that each of the factors x, $x-1$, and $x+2$ is positive (otherwise their individual logarithms would not be defined). So the result would appear to be valid only on the interval $(1, \infty)$, whereas, as we showed in the example, it is actually valid on $(-2, 0)$ as well.

Before attempting to differentiate a function consisting of the natural logarithm of products, quotients, powers, or roots, *always* simplify it using the arithmetic properties 7.3, 7.4, and 7.5. Failure to do so is almost a guarantee that the problem will be more difficult algebraically.

Logarithmic Differentiation

In the next example we show how we can sometimes simplify the calculation of derivatives of complicated expressions involving products, quotients, and powers by first finding the logarithm of the expression and then differentiating. This technique is called **logarithmic differentiation**.

EXAMPLE 7.6   Find y' if

$$y = \frac{(x-1)^3}{(2x+3)\sqrt{1-2x}}$$

Solution   Of course, we could differentiate in a straightforward manner, using the Quotient Rule and Product Rule, along with the Chain Rule, but it would be messy. (Try it!) Instead, we will use logarithmic differentiation.

First, let us find the domain of the function. For $\sqrt{1-2x}$ to be a real number, we must have $1 - 2x \geq 0$, or equivalently, $x \leq \frac{1}{2}$. But we cannot allow $x = \frac{1}{2}$, since it causes the denominator to be 0. Similarly, we rule out $x = -\frac{3}{2}$, since we cannot allow $2x + 3$ to be 0. Thus, the domain is the set $D = \{x : x < \frac{1}{2}$ and $x \neq -\frac{3}{2}\}$.

To take the logarithm, we must have a positive function, since the logarithm is defined only for positive values. We can make the function positive by taking its absolute value:

$$|y| = \frac{|x-1|^3}{|2x+3||1-2x|^{1/2}} \qquad (x \in D)$$

Note that if $x = 1$ were in the domain D, we would have to restrict x to be different from 1 before taking the logarithm in order not to have ln of 0. However, $x = 1$ is not in the domain. So for all x in D, $|y| > 0$. Thus, we can take the logarithm and get

$$\ln|y| = \ln \frac{|x-1|^3}{|2x+3||1-2x|^{1/2}} = 3\ln|x-1| - \left[\ln|2x+3| + \frac{1}{2}\ln|1-2x|\right]$$

where we have made use of the arithmetic properties of the ln function. Now we differentiate both sides with respect to x, using Equation 7.7, bearing in mind that on the left-hand side y is a function of x:

$$\frac{1}{y}y' = \frac{3}{x-1} - \left[\frac{1}{2x+3}\cdot 2 + \frac{1}{2}\frac{1}{1-2x}(-2)\right]$$

$$= \frac{3}{x-1} - \frac{2}{2x+3} + \frac{1}{1-2x}$$

Finally, we solve for y' by multiplying both sides by y, and substitute its value in terms of x on the right:

$$y' = \frac{(x-1)^3}{(2x+3)\sqrt{1-2x}}\left[\frac{3}{x-1} - \frac{2}{2x+3} + \frac{1}{1-2x}\right]$$

Just as in the preceding example, we could have arrived at the same answer even if we had ignored absolute values. Had we done so, however, we would have been tacitly assuming $y > 0$, whereas our derivation is valid also if $y < 0$. ∎

Note the difference between Examples 7.5 and 7.6. In the former we were *given* the function in the form of a logarithm. No initiative on our part was required. We simply took the derivative. In the second, there was no logarithm. We began with the initial steps of *taking* the absolute value and then the logarithm of both sides, and then we differentiated.

An Integral Whose Value Is a Natural Logarithm

Each time we find a new derivative formula, we also have a new antiderivative formula. In particular, from Formula 7.7, we obtain

$$\int \frac{1}{u}\,du = \ln |u| + C \qquad (7.8)$$

This formula fills a gap in the Power Rule for antiderivatives. That is, the Power Rule, given by

$$\int u^n\,du = \frac{u^{n+1}}{n+1} + C$$

is valid *except* when $n = -1$. Formula 7.8 tells what the result is when $n = -1$. This new integration formula enables us to integrate many functions that we have been unable to integrate until now.

EXAMPLE 7.7   Evaluate the following indefinite integrals:

(a) $\displaystyle\int \frac{x}{x^2+1}\,dx$ (b) $\displaystyle\int \frac{\cos x}{1+2\sin x}\,dx$ (c) $\displaystyle\int \tan 3x\,dx$

Solution   In each case our strategy will be to make a substitution that brings the given integral around to the form of Equation 7.8.

(a) Let $u = x^2 + 1$. Then $du = 2x\,dx$. So we have

$$\int \frac{x}{x^2+1}\,dx = \frac{1}{2}\int \frac{2x\,dx}{x^2+1} = \frac{1}{2}\int \frac{du}{u} = \frac{1}{2}\ln|u| + C$$

$$= \frac{1}{2}\ln|x^2+1| + C$$

$$= \frac{1}{2}\ln(x^2+1) + C$$

In the last step we dropped the absolute values, since $x^2 + 1$ is always positive. We could also write the answer, using Equation 7.5, as $\ln \sqrt{x^2+1} + C$.

(b) Here we will do the substitution mentally and write

$$\int \frac{\cos x}{1 + 2\sin x}dx = \frac{1}{2}\int \frac{\overbrace{2\cos x\,dx}^{du}}{\underbrace{1 + 2\sin x}_{u}} = \frac{1}{2}\ln|1 + 2\sin x| + C$$

(c) $\displaystyle\int \tan 3x\,dx = \int \frac{\sin 3x}{\cos 3x}dx = -\frac{1}{3}\int \frac{(-3\sin 3x)dx}{\cos 3x}$

$$= -\tfrac{1}{3}\ln|\cos 3x| + C$$

(What mental substitution did we use?)   ■

EXAMPLE 7.8   Find a formula for the general antiderivative of $\sec x$.

Solution   The solution involves a trick. We multiply the numerator and denominator by $\sec x + \tan x$:

$$\int \sec x\,dx = \int \frac{\sec x}{1}\cdot\frac{\sec x + \tan x}{\sec x + \tan x}dx = \int \frac{\sec^2 x + \sec x\tan x}{\sec x + \tan x}dx$$

Now recall from Section 3.6 that the derivative of $\tan x$ is $\sec^2 x$ and the derivative of $\sec x$ is $\sec x \tan x$. So if we let $u = \sec x + \tan x$, we have $du = (\sec x \tan x + \sec^2 x)dx$. Thus,

$$\int \sec x = \int \frac{du}{u} = \ln|u| = \ln|\sec x + \tan x| + C$$ ■

Because we will have occasions later to refer to the integration formula from Example 7.8, we set it off here for emphasis.

$$\int \sec x\,dx = \ln|\sec x + \tan x| + C \qquad\qquad (7.9)$$

Graphs Involving Logarithms

We conclude this section with an example showing how to analyze and draw the graph of a function involving a natural logarithm.

EXAMPLE 7.9   Discuss and sketch the graph of $y = \ln|x^2 - 1|$.

Solution   Since $|x^2 - 1| > 0$ except when $x^2 = 1$, the domain consists of all real numbers except $x = \pm 1$.

(a) *Intercepts.* Setting $x = 0$, we get $y = \ln|-1| = \ln 1 = 0$. So the y-intercept is 0. Setting $y = 0$, we get $\ln|x^2 - 1| = 0$, which is true only when $|x^2 - 1| = 1$. (Why?) So $x^2 - 1 = \pm 1$. If $x^2 - 1 = 1$, then $x^2 = 2$, and $x = \pm\sqrt{2}$. If $x^2 - 1 = -1$, then $x^2 = 0$, and so $x = 0$ also. Thus, the x-intercepts are 0, $\sqrt{2}$, and $-\sqrt{2}$.

(b) *Symmetry.* Since no change results when x is replaced by $-x$, the graph is symmetric with respect to the y-axis.

(c) *Maxima and Minima.* By Equation 7.6,

$$y' = \frac{1}{x^2 - 1} \cdot 2x = \frac{2x}{x^2 - 1}$$

The only critical value is $x = 0$ (since $x = 1$ and $x = -1$ are outside the domain). By the sign graph for y' in Figure 7.6 we see that $(0, 0)$ is a local maximum point. We also see that the graph of our function is decreasing on $(-\infty, -1) \cup (0, 1)$ and is increasing on $(-1, 0) \cup (1, \infty)$.

(d) *Concavity and Inflection Points.* For the second derivative, we obtain

$$y'' = \frac{2 \cdot (x^2 - 1) - 2x(2x)}{(x^2 - 1)^2} = \frac{-2(x^2 + 1)}{(x^2 - 1)^2}$$

For all $x \neq \pm 1$, $y'' < 0$, so the graph is concave down at all points of the domain. There are no inflection points.

(e) *Asymptotes.* As $x \to 1$, $|x^2 - 1| \to 0^+$, so $\ln|x^2 - 1| \to -\infty$. Thus, $x = 1$ is a vertical asymptote. By symmetry, so is $x = -1$. As $|x| \to \infty$, $|x^2 - 1| \to \infty$, so $\ln|x^2 - 1| \to \infty$. Thus, there are no horizontal asymptotes. Combining all our information, we are able to draw the graph shown in Figure 7.7. ∎

FIGURE 7.6

Sign graph for $y' = \dfrac{2x}{x^2 - 1}$

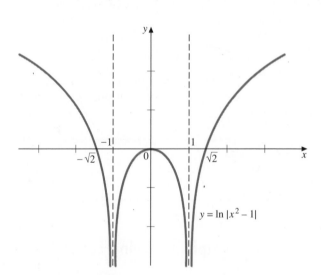

FIGURE 7.7

Exercise Set 7.1

In Exercises 1–20, give the domain of f and find $f'(x)$.

1. $f(x) = \ln \dfrac{1}{x}$

2. $f(x) = \ln \sqrt{x}$

3. $f(x) = \ln x^3$

4. $f(x) = (\ln x)^3$

5. $f(x) = \ln(x\sqrt{1-x})$

6. $f(x) = \ln\left(\dfrac{x^2}{x-1}\right)$

7. $f(x) = x \ln x$

8. $f(x) = \ln x^2$

9. $f(x) = \ln \sqrt{\dfrac{1-x}{1+x}}$

10. $f(x) = \sqrt{\ln x}$

11. $f(x) = \ln(\ln x)$

12. $f(x) = \ln \sin x$

13. $f(x) = \ln \sec x$

14. $f(x) = \sin(\ln x)$

15. $f(x) = \dfrac{1 - \ln x}{x}$

16. $f(x) = \ln\left(\dfrac{x-2}{x+1}\right)$

17. $f(x) = \ln \dfrac{\sqrt{x^2 - 1}}{2 - x}$

18. $f(x) = \ln \sqrt[3]{2x - x^2}$

19. $f(x) = \ln\left(\dfrac{x+3}{x^2 - 2x}\right)$

20. $f(x) = \ln\left(\dfrac{x\sqrt{1-x}}{1+x}\right)$

In Exercises 21–26, use L'Hôpital's Rule to find the limits.

21. $\lim\limits_{x \to 1} \dfrac{\ln x}{x - 1}$

22. $\lim\limits_{x \to \infty} \dfrac{\ln x}{x}$

23. $\lim\limits_{x \to 0^+} \dfrac{x^{-1/2}}{\ln x}$

24. $\lim\limits_{x \to 0^+} \dfrac{\cot x}{\ln x}$

25. $\lim\limits_{x \to \infty} \dfrac{1 + (\ln x)^3}{x \ln x}$

26. $\lim\limits_{x \to -1^-} \dfrac{x \ln |x|}{(x+1)^2}$

27. (a) Show that $\ln x < \sqrt{x}$ for $x > 0$.
(b) Show that $\ln x \le x - 1$ for $x > 0$.

In Exercises 28–37, find the antiderivatives.

28. $\displaystyle\int \dfrac{2x\, dx}{x^2 + 4}$

29. $\displaystyle\int \dfrac{\sin x\, dx}{1 - \cos x}$

30. $\displaystyle\int \dfrac{x\, dx}{2 - 3x^2}$

31. $\displaystyle\int \dfrac{x^2 dx}{x^3 + 1}$

32. $\displaystyle\int \dfrac{(x - 1)dx}{x^2 - 2x + 3}$

33. $\displaystyle\int \cot x\, dx$

34. $\displaystyle\int \dfrac{dx}{\sqrt{x}(\sqrt{x} + 1)}$

35. $\displaystyle\int \dfrac{dx}{x \ln x}$

36. $\displaystyle\int \dfrac{x \cos x}{x \sin x + \cos x}dx$

37. $\displaystyle\int \dfrac{\sqrt{x} - 1}{x(2\sqrt{x} - 3)}dx$

In Exercises 38–41, use logarithmic differentiation to find y'.

38. $y = \dfrac{(x - 1)^2\sqrt{3x + 4}}{x + 2}$

39. $y = \dfrac{x\sqrt{1 - x}}{(x + 2)(x - 3)}$

40. $y = \sqrt{\dfrac{(x - 1)(x + 2)}{(x - 3)^3}}$

41. $y = \dfrac{\sqrt[3]{(x + 1)^2}}{x^4(3 - 4x)^2}$

In Exercises 42–47, find all local maxima and minima.

42. $y = x - \ln x$

43. $y = \ln \sqrt{x} - x^2$

44. $y = x^3 - \ln x^3$

45. $y = x + \ln \dfrac{x}{2x - 1}$

46. $y = \ln \dfrac{\sqrt{1 - x}}{3 - 2x}$

47. $y = \ln \dfrac{x(2 - x)}{(x - 3)^2}$

In Exercises 48–55, discuss domain, intercepts, symmetry, asymptotes, maxima and minima, concavity, and points of inflection and sketch the graph.

48. $y = \ln 2x$

49. $y = \ln x^2$

50. $y = \ln 1/x$

51. $y = \ln(1 - x^2)$

52. $y = \ln \sqrt{x^2 - 1}$

53. $y = x - \ln x$

54. $y = \ln |x|$

55. $y = \ln(1 + x^2)$

56. Using the approximations $\ln 2 \approx 0.70$ and $\ln 3 \approx 1.1$, together with the arithmetic properties 7.3, 7.4, and 7.5, find approximations for the following:
(a) $\ln 6$ (b) $\ln 8$ (c) $\ln 81$ (d) $\ln 1.5$
(e) $\ln 36$

57. If $x > 0$ and n is an integer such that $n \le x < n + 1$, show that

$$\int_1^x \dfrac{1}{t}dt \le \sum_{k=1}^n \dfrac{1}{k}$$

Hence, conclude that for $\ln x$ to equal 100, x must be greater than the smallest integer n for which

$$\sum_{k=1}^n \dfrac{1}{k} \ge 100$$

(*Hint:* Use a left-hand Riemann sum on the interval $[1, n + 1]$, taking $\Delta t = 1$.)

In Exercises 58–61, find $f'(x)$ and $f''(x)$.

58. $f(x) = \dfrac{\ln(x + 1)}{x + 1}$

59. $f(x) = x(\ln x)^2$

60. $f(x) = \dfrac{x}{\ln x}$

61. $f(x) = \sqrt{\ln x^2}$

In Exercises 62 and 63, use implicit differentiation to find y'.

62. $x \ln y - y \ln x = 1$　　**63.** $\ln x^2 y = (\ln y)/x$

In Exercises 64–67, find the antiderivatives.

64. $\displaystyle\int \frac{\tan 2x}{\ln \cos 2x}\,dx$　　**65.** $\displaystyle\int \frac{\sin 2x}{1 + \cos^2 x}\,dx$

66. $\displaystyle\int \frac{dx}{x + x^{1/3}}$　(*Hint:* Factor the denominator.)

67. $\displaystyle\int \left(\frac{x-2}{x-1}\right)^2 dx$　(*Hint:* First divide.)

68. Derive the formula

$$\int \csc x\,dx = \ln|\csc x - \cot x| + C$$

(*Hint:* Use a trick similar to the one used in Example 7.8 to get the answer $-\ln|\csc x + \cot x|$. Then show that this answer can be put in the form given.)

In Exercises 69–71, discuss domain, intercepts, symmetry, asymptotes, maxima and minima, concavity, and points of inflection and sketch the graph.

69. $y = \ln\left|\dfrac{x-1}{x+1}\right|$　　**70.** $y = \ln \dfrac{x^2}{|x^2 - 2|}$

71. $y = \ln(x^2 - 3x) - \ln(x^2 + 3)$

 72. Verify the result stated in the text that $\ln 3 \approx 1.0986$, using Simpson's Rule with $n = 20$. Prove that this result is correct to the four decimal places given. Similarly, using Simpson's Rule with $n = 10$, verify that the result $\ln 2 \approx 0.6932$ obtained in the text is correct to four decimal places.

73. Show that for $n \geq 2$,

$$\frac{1}{2} + \frac{1}{3} + \cdots + \frac{1}{n} < \ln n < 1 + \frac{1}{2} + \frac{1}{3} + \cdots + \frac{1}{n-1}$$

 74. Let $f(x) = \ln x$ and $g(x) = \sqrt[n]{x}$, where n is a positive integer. Both $f(x)$ and $g(x)$ become infinite as $x \to \infty$. Use a graphing calculator or a CAS to graph f and g for several values of n. Conjecture which function grows more rapidly as $x \to \infty$. Prove your conjecture.

75. Consider the family of curves $y = a \ln(bx)$, where a is any real number and $b > 0$. Use a CAS or a graphing calculator to graph this function for several different values of a and b. How does the parameter b affect the growth of the function? How does the parameter a affect the shape of the graph?

76. Use a CAS to calculate each of the following to 12 decimal places.

　(a) $\ln 2$　　(b) $\ln 3$　　(c) $\ln 100$　　(d) $\ln 10^9$

7.2 THE NATURAL EXPONENTIAL FUNCTION

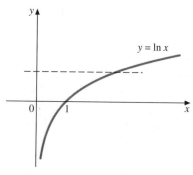

FIGURE 7.8
$f(x) = \ln x$ passes the Horizontal Line Test and thus is one-to-one.

We have seen that $f(x) = \ln x$ has domain $(0, \infty)$ and range $(-\infty, \infty)$. Furthermore, f is increasing for all x in its domain. Its graph (Figure 7.8) clearly shows it passes the horizontal line test. Hence, f is one-to-one and therefore has an inverse, which we designate by $f^{-1}(x) = \exp(x)$ and call the **exponential function** (with "natural" understood). The reason for the name "exponential" will soon become clear. Recall from Chapter 1 that the graph of the inverse of a function can be obtained by reflecting the graph of the function about the line $y = x$. So from the graph of $f(x) = \ln x$, we immediately obtain the graph of $f^{-1}(x) = \exp(x)$, shown in Figure 7.9. We know also that the domain and range of f^{-1} are the range and domain, respectively, of f. So the domain of exp is $(-\infty, \infty)$ and the range is $(0, \infty)$.

The Number e

Before analyzing the exponential function further, we single out for special emphasis the number whose natural logarithm is 1. We know that such a number exists, since $\ln x$ assumes every real value. This special number is designated by the letter e (after Leonhard Euler, the greatest mathematician of

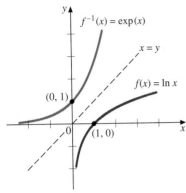

FIGURE 7.9

the eighteenth century and one of the most prolific mathematicians of all time). Thus, e is defined by the equation

$$\ln e = 1$$

We show e graphically in Figure 7.10. Since, as we have seen, $\ln 2 < 1$ and $\ln 3 > 1$, it follows from the fact that $\ln$ is an increasing function that $2 < e < 3$. It can be shown that e is irrational and that its approximate value is

$$e \approx 2.71828$$

With a computer algebra system, you can easily obtain e to a large number of decimal places. For example, in Maple, to get e to 100 places, simply type

evalf(E,100);

and you get the result

2.718281828459045235360287471352662497757247093699959574966961
7627724076630353547594571382178525166427

We will see a way of approximating e to any desired degree of accuracy in Chapter 10.

Properties of e^x

We know from arithmetic property 7.5 that if r is any rational number

$$\ln(e^r) = r \ln e$$

and since $\ln e = 1$ (by definition of e),

$$\ln(e^r) = r$$

Let us apply the exponential function to each side of this equation:

$$\exp[\ln(e^r)] = \exp(r)$$

Since exp and ln are inverses, the left-hand side is just e^r. (Remember that $f^{-1}(f(x)) = x$.) So we have

$$e^r = \exp(r)$$

for all rational numbers r. Up to now we have not assigned any meaning to an irrational power of e (or of any other number). Since $\exp(x)$ is defined for *all* real numbers x (rational or irrational), we define e^x so as to be consistent with its meaning when x is rational. That is, if x is *any* real number, we define e^x to be $\exp(x)$.

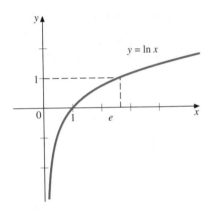

FIGURE 7.10

$$e^x = \exp(x) \qquad\qquad (7.10)$$

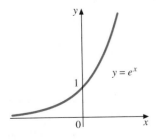

FIGURE 7.11

The graph of $y = e^x$ is therefore identical to the graph of $y = \exp(x)$. We show this important graph again in Figure 7.11. In the future when we speak of the exponential function, we can write it as either $f(x) = e^x$ or $f(x) = \exp(x)$. They are the same. Just as $\ln x$ grows extremely slowly as x increases, e^x

grows extremely rapidly. The term "exponential growth," as applied, say, to world population growth, comes from this fact.

If we apply ln to both sides of Equation 7.10, and again use the fact that ln and exp are inverse functions, we get, for all x,

$$\ln(e^x) = x \tag{7.11}$$

It is useful to learn the meaning of Equation 7.11 in words:

e^x is that number whose natural logarithm is x.

In Equation 7.10 if we replace x by $\ln x$ and again use the inverse relationship between exp and ln, so that $\exp(\ln(x)) = x$ for $x > 0$, we get

$$e^{\ln x} = x \tag{7.12}$$

In particular, since $\ln 1 = 0$, we have $e^0 = 1$.

Equations 7.11 and 7.12 are of sufficient importance to warrant special attention. First, let us reiterate the fact that since e^x is the inverse of $\ln x$, its domain and range are the range and domain, respectively, of $\ln x$. Thus, since the range of $\ln x$ is the set of all real numbers, $(-\infty, \infty)$, e^x has domain equal to this set. Also, the domain of $\ln x$ is the set of all positive real numbers, $(0, \infty)$, so e^x has this set as its range. Thus, $e^x > 0$ for all x.

Inverse Properties of $\ln x$ and e^x

$$\ln(e^x) = x \quad \text{for all real } x$$

$$e^{\ln x} = x \quad \text{for all } x > 0$$

We may also express these relationships in the following alternative form.

$$y = e^x \text{ if and only if } x = \ln y$$

EXAMPLE 7.10   Evaluate each of the following.

(a) $\ln e^2$ (b) $\ln\left(\dfrac{1}{\sqrt{e}}\right)$ (c) $e^{\ln 3}$ (d) $e^{2\ln 3}$

Solution

(a) By Equation 7.11, $\ln e^2 = 2$.

(b) By Equation 7.11, $\ln\left(\dfrac{1}{\sqrt{e}}\right) = \ln(e^{-1/2}) = -\dfrac{1}{2}$.

(c) By Equation 7.12, $e^{\ln 3} = 3$.

(d) By Equations 7.5 and 7.12, $e^{2\ln 3} = e^{\ln(3^2)} = 3^2 = 9$. ∎

EXAMPLE 7.11   Solve for x:

(a) $e^x = 3$ (b) $e^{(1-2x)} = 7$ (c) $\ln x = 5$ (d) $\ln(x-1) = -2$

Solution

(a) We take the natural logarithm of both sides, making use of Equation 7.11.

$$\ln(e^x) = \ln 3$$
$$x = \ln 3$$

(b) Proceeding as in part (a) we get

$$\ln(e^{1-2x}) = \ln 7$$
$$1 - 2x = \ln 7$$
$$2x = 1 - \ln 7$$
$$x = \frac{1 - \ln 7}{2}$$

(c) Here we "exponentiate." That is, we use the fact that when $a = b$, then $e^a = e^b$. So we have

$$e^{\ln x} = e^5$$

But by Equation 7.12, $e^{\ln x} = x$, so

$$x = e^5$$

(d) Again, we exponentiate.

$$e^{\ln(x-1)} = e^{-2}$$
$$x - 1 = \frac{1}{e^2}$$
$$x = 1 + \frac{1}{e^2}$$ ∎

Laws of Exponents

There is nothing in the definition of e^x (Equation 7.10) to suggest that x really behaves like an exponent, as we understand exponents. The next theorem, however, confirms that it does, and that we are therefore justified in thinking of e^x as "e raised to the power x." Also, since $\exp(x)$ and e^x are equal, the reason for the name "exponential function" will now be clear.

THEOREM 7.1

If x and y are any real numbers and r is any rational number,

1. $e^{x+y} = e^x \cdot e^y$
2. $e^{x-y} = \dfrac{e^x}{e^y}$
3. $(e^x)^r = e^{rx}$

Proof

1. Using Equation 7.11, we have

$$\ln e^{x+y} = x + y$$

Also, by Equation 7.3,

$$\ln(e^x \cdot e^y) = \ln e^x + \ln e^y = x + y$$

where the last equality again follows by Equation 7.11. Since the natural logarithm function is one-to-one, the equality of $\ln e^{x+y}$ and $\ln(e^x \cdot e^y)$ guarantees the equality of e^{x+y} and $e^x \cdot e^y$.

2. This formula is proved in a similar way, and we leave it as an exercise (Exercise 58 in Exercise Set 7.2).

3. Using the fact that $\ln a^r = r \ln a$, we have

$$\ln(e^x)^r = r \ln e^x = rx$$

Also,

$$\ln e^{rx} = rx$$

Again, the one-to-one nature of ln gives the desired result.   ■

REMARK ─────────────────────────────

■ At this stage, in part 3 of the preceding theorem we must limit r to be rational since we have so far proved $\ln a^r = r \ln a$ only for rational values of r. We will remove this restriction in the next section.

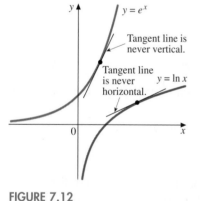

FIGURE 7.12

Derivatives and Integrals of Exponential Functions

The fact that $f(x) = \ln x$ has a nonzero derivative at each point of its domain suggests that the inverse function $f^{-1}(x) = \exp(x)$ (or equivalently, $f^{-1}(x) = e^x$) also has a derivative at each point in its domain. We can see this conclusion geometrically by reasoning that the graph of $y = \ln x$ has a nonhorizontal tangent line at all points on the graph, so the graph of $y = e^x$ has a nonvertical tangent line at all points on its graph. (See Figure 7.12.) There is a theorem, which we will not prove here, that confirms our conclusion without reference to geometry. It can be stated as follows. (See Appendix 5 for the proof.)

THEOREM 7.2

If f is differentiable on an open interval I, with range J, and if $f'(x) \neq 0$ for all x in I, then f^{-1} exists and is differentiable in J. Furthermore,

$$\frac{d}{dx} f^{-1}(x) = \frac{1}{f'(f^{-1}(x))} \qquad (7.13)$$

To find the derivative of e^x, we apply the result (Equation 7.13) to

$$y = e^x = f^{-1}(x)$$

with $f(x) = \ln x$. Since $f'(x) = 1/x$, we have $f'(f^{-1}(x)) = 1/e^x$. So by Equation 7.13,

$$\frac{d}{dx}e^x = \frac{1}{\dfrac{1}{e^x}} = e^x$$

That is,

$$\frac{d}{dx}e^x = e^x$$

We set off this important result for emphasis.

$$\frac{d}{dx}e^x = e^x \qquad\qquad (7.14)$$

This result is quite remarkable. The derivative of the exponential function is the function itself! So the second, third, fourth, ... derivatives all equal the original function. No other function has this self-replicating derivative property (except for constant multiples of e^x). Graphically, Equation 7.14 says that the slope of the tangent line at any point on the graph of $y = e^x$ is equal to the y-coordinate of that point.

If $u = g(x)$ is any differentiable function of x, we can use Equation 7.14 and the Chain Rule to obtain the formula

$$\frac{d}{dx}e^u = e^u \frac{du}{dx} \qquad\qquad (7.15)$$

Consequently, we also have the antidifferentiation formula

$$\int e^u \, du = e^u + C \qquad\qquad (7.16)$$

EXAMPLE 7.12

(a) Find dy/dx if $y = e^{\sin x}$.
(b) Evaluate the integral $\int x e^{x^2} \, dx$.

Solution

(a) By Equation 7.15,

$$\frac{dy}{dx} = e^{\sin x} \cos x$$

(b) Let $u = x^2$. Then $du/dx = 2x$. So we have

$$\int xe^{x^2}dx = \frac{1}{2}\int 2xe^{x^2}dx = \frac{1}{2}\int e^u du = \frac{1}{2}e^u + C = \frac{1}{2}e^{x^2} + C \qquad \blacksquare$$

EXAMPLE 7.13   Evaluate the following integrals:

(a) $\displaystyle\int \frac{e^x}{1+e^x}dx$ (b) $\displaystyle\int_0^{\ln 3} e^{-2x}dx$

Solution

(a) Let $u = 1 + e^x$. Then $du = e^x dx$. So we have

$$\int \frac{e^x}{1+e^x}dx = \int \frac{du}{u} = \ln|u| + C = \ln(1+e^x) + C$$

Note that the absolute value symbol can be deleted, since $1 + e^x$ is always positive.

(b) $\displaystyle\int_0^{\ln 3} e^{-2x}dx = -\frac{1}{2}\int_0^{\ln 3} \underbrace{e^{-2x}}_{e^u}\underbrace{(-2\,dx)}_{du} = -\frac{1}{2}\left[e^{-2x}\right]_0^{\ln 3}$

$$= -\frac{1}{2}(e^{-2\ln 3} - e^0) = -\frac{1}{2}(e^{\ln(3^{-2})} - 1)$$

$$= -\frac{1}{2}[(3)^{-2} - 1] = -\frac{1}{2}\left[\frac{1}{3^2} - 1\right] = -\frac{1}{2}\left(\frac{1}{9} - 1\right) = \frac{4}{9}$$

Observe how we used each of the following facts:

$$-2\ln 3 = \ln(3)^{-2} \qquad \text{By Equation 7.5}$$

$$e^{\ln(3^{-2})} = 3^{-2} \qquad \text{By Equation 7.12}$$

$$e^0 = 1 \qquad \text{Since } \ln 1 = 0 \qquad \blacksquare$$

Graphs Involving the Exponential Function

The next example illustrates how to analyze and draw the graph of a function involving a power of e.

EXAMPLE 7.14   Analyze and draw the graph of $y = xe^{-x}$.

Solution   Since xe^{-x} is defined for all real values of x, the domain is all of $\mathbb{R}$.

(a) *Intercepts.* When $x = 0$, $y = 0$, and conversely. So both the x-intercept and y-intercept are 0.

(b) *Symmetry.* None, since replacing x by $-x$ does not leave y unchanged, nor does it give its negative.

(c) *Maxima and Minima.*

$$y' = e^{-x} + x(-e^{-x}) = e^{-x}(1 - x) \qquad \text{By the Product Rule}$$

$$y'' = (-e^{-x})(1 - x) + e^{-x}(-1) = e^{-x}(x - 2)$$

Since $e^{-x} \neq 0$ for all x, $y' = 0$ only when $x = 1$, so the only critical point is $x = 1$. Substituting $x = 1$ in y'', we get

$$y''\big|_{x=1} = e^{-1}(-1) = -\frac{1}{e} < 0$$

So by the Second-Derivative Test, the function has a local maximum at $x = 1$. The corresponding y value is $1(e^{-1}) = 1/e$. The maximum point is therefore $(1, 1/e)$.

(d) *Concavity.* From $y'' = e^{-x}(x - 2)$ we see that $y'' > 0$ for $x > 2$ and $y'' < 0$ for $x < 2$. Thus, the graph is concave up on $(2, \infty)$ and concave down on $(-\infty, 2)$. The concavity changes at $x = 2$, so the point $(2, 2/e^2)$ is a point of inflection.

(e) *Asymptotes.* There is no vertical asymptote, since y is finite for all x values. For horizontal asymptotes, we consider the limits

$$\lim_{x \to -\infty} (xe^{-x}) \qquad \text{and} \qquad \lim_{x \to \infty} (xe^{-x})$$

For large negative values of x, e^{-x} is a large positive value (e.g., when $x = -100$, $e^{-x} = e^{100}$, which is a *huge* number). So the product of x and e^{-x} is a very large negative number. That is,

$$\lim_{x \to -\infty} (xe^{-x}) = -\infty$$

For the limit as $x \to \infty$, we have an indeterminate form of the type $0 \cdot \infty$, so, by L'Hôpital's Rule

$$\lim_{x \to \infty} xe^{-x} = \lim_{x \to \infty} \frac{x}{e^x} = \lim_{x \to \infty} \frac{1}{e^x} = 0$$

(It is also intuitively evident that the limit is 0, since e^x grows much faster than x, so that the fraction x/e^x becomes arbitrarily small.) We conclude that the line $y = 0$ (the x-axis) is an asymptote on the right.

When we combine our results, we get the graph in Figure 7.13. ∎

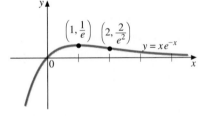

FIGURE 7.13

Relative Growth Rates of $\ln x$, e^x, and x^r

We conclude this section by considering the relative growth rates of the three functions $\ln x$, e^x, and x^r, where r is a positive rational number. (See also Exercise 74 in Exercise Set 7.1.) First, let us state precisely what we mean by one function growing faster than another.

Suppose

$$\lim_{x \to \infty} f(x) = \infty \qquad \text{and} \qquad \lim_{x \to \infty} g(x) = \infty$$

Then, we say that f **grows faster than** g as $x \to \infty$ if

$$\lim_{x \to \infty} \frac{f(x)}{g(x)} = \infty$$

or equivalently, if

$$\lim_{x \to \infty} \frac{g(x)}{f(x)} = 0$$

We also say in this case that g **grows more slowly than** f.

EXAMPLE 7.15   Let $f(x) = e^x$, $g(x) = x^r$, and $h(x) = \ln x$, where r is a positive rational number. Prove that as $x \to \infty$, f grows faster than g, and g grows faster than h.

Solution   Suppose n is an integer such that $n \le r < n + 1$ (i.e., $n = [\![r]\!]$ the greatest integer in r). Then, by repeated (if necessary) applications of L'Hôpital's Rule,

$$\lim_{x \to \infty} \frac{g(x)}{f(x)} = \lim_{x \to \infty} \frac{x^r}{e^x} = \lim_{x \to \infty} \frac{rx^{r-1}}{e^x} = \lim_{x \to \infty} \frac{r(r-1)x^{r-2}}{e^x}$$

$$= \cdots = \lim_{x \to \infty} \frac{r(r-1)\cdots(r-n+1)x^{r-n}}{e^x}$$

If $r = n$, we are through, since the numerator is a constant. In this case, the limit is 0. If $r > n$, we apply L'Hôpital's Rule one more time, and the exponent on x becomes $r - n - 1$, which is negative. Thus,

$$\lim_{x \to \infty} \frac{g(x)}{f(x)} = \frac{r(r-1)\cdots(r-n)}{e^x x^{n+1-r}} = 0$$

Therefore, e^x grows more rapidly than x^r, regardless of how large r is. For example, e^x grows more rapidly than x^{100}, or x^{1000}, or $x^{10,000}$.

Now consider the comparison of g and h.

$$\lim_{x \to \infty} \frac{h(x)}{g(x)} = \lim_{x \to \infty} \frac{\ln x}{x^r} = \lim_{x \to \infty} \frac{\dfrac{1}{x}}{rx^{r-1}} = \lim_{x \to \infty} \frac{1}{rx^r} = 0$$

Thus, x^r grows more rapidly than $\ln x$ regardless of how small the positive exponent r is. For example, even $x^{0.0001}$ grows more rapidly than $\ln x$.   ∎

Exercise Set 7.2

In Exercises 1–8, solve for x.

1. (a) $e^{2\ln x} = 3$
   (b) $e^{-(1/2)\ln x} = 4$

2. (a) $\ln(e^{-2}) = x$
   (b) $\ln(e^{2x+3}) = 1$

3. (a) $e^{x+1} = 5$
   (b) $e^{1-x} = 2$

4. (a) $\dfrac{1}{e^{2x}} = 3$
   (b) $e^{x^2} = 4$

5. (a) $\ln x = -1$
   (b) $\ln(x+1) = 2$

6. (a) $\ln\left(\dfrac{x}{x+1}\right) = 3$
   (b) $\ln(2x - 7) = 0$

7. (a) $3\ln x + 1 = 0$
   (b) $2 - \ln(x - 1) = 5$

8. (a) $\ln x + \ln(2x - 1) = 0$
   (b) $2\ln x - \ln(3x + 4) = 0$

In Exercises 9–28, find $f'(x)$.

9. $f(x) = e^{x^2}$

10. $f(x) = e^{\cos x}$

11. $f(x) = e^{3\ln x}$

12. $f(x) = \ln e^{2x}$

13. $f(x) = e^{\sqrt{x}}$

14. $f(x) = \sqrt{e^x}$

15. $f(x) = e^{e^x}$

16. $f(x) = e^{1/x}$

17. $f(x) = e^{x+\ln x}$

18. $f(x) = e^{2x-3\ln x}$

19. $f(x) = e^{2x} \sin 3x$

20. $f(x) = e^{\sqrt{1-x^2}}$

21. $f(x) = \dfrac{e^x - e^{-x}}{2}$

22. $f(x) = \dfrac{e^x + e^{-x}}{2}$

23. $f(x) = \dfrac{e^x - e^{-x}}{e^x + e^{-x}}$

24. $f(x) = e^{2x} \ln x$

25. $f(x) = \sin^2 e^x$

26. $f(x) = e^{-x} \cos 2x$

27. $f(x) = \dfrac{1}{1 - e^{-x}}$

28. $f(x) = e^{-x}(\sin x + \cos x)$

In Exercises 29–32, find y' using implicit differentiation.

29. $e^{x+y} = xy$

30. $e^{xy} = x + y$

31. $e^{-x} \sin y = e^{-y} \cos x$

32. $y = 1 - e^{-(x+y)}$

In Exercises 33–48, evaluate the integrals.

33. $\displaystyle\int x e^{(1-x^2)} dx$

34. $\displaystyle\int \frac{x\,dx}{e^{x^2}+1}$

35. $\displaystyle\int \frac{\sin x}{e^{\cos x}} dx$

36. $\displaystyle\int \frac{e^x}{1-e^x} dx$

37. $\displaystyle\int \frac{e^x - e^{-x}}{e^x + e^{-x}} dx$

38. $\displaystyle\int \frac{e^x}{\sqrt{e^x - 1}} dx$

39. $\displaystyle\int_0^{\ln 2} e^{3x} dx$

40. $\displaystyle\int_{\ln 2}^{\ln 5} \frac{e^x}{(1+e^x)^2} dx$

41. $\displaystyle\int_0^3 e^{2x} \sqrt[3]{1 - e^{2x}}\, dx$

42. $\displaystyle\int_{2+1/e}^3 \frac{dx}{x - 2}$

43. $\displaystyle\int_0^{\ln 2} \frac{dx}{1 + e^{-x}}$

44. $\displaystyle\int \frac{dx}{1 + e^x}$ (Hint: Multiply numerator and denominator by e^{-x}.)

45. $\displaystyle\int_{1/\ln 3}^{2/\ln 3} \frac{e^{1/x}}{x^2} dx$

46. $\displaystyle\int_0^{\ln 2} \frac{e^{-x}}{4 - 3e^{-x}} dx$

47. $\displaystyle\int \frac{dx}{e^x - 2 + e^{-x}}$

48. $\displaystyle\int_{\ln \pi}^{\ln 4\pi/3} e^x \tan e^x dx$

In Exercises 49–57, analyze the function and draw its graph.

49. $y = e^{-x^2/4}$

50. $y = e^{1-|x|}$

51. $y = x e^{-x/2}$

52. $y = \dfrac{e^x + e^{-x}}{2}$

53. $y = \dfrac{1}{1 + e^{-x}}$

54. $y = \dfrac{e^x}{1 - e^x}$

55. $y = x^3 e^{-x}$

56. $y = \dfrac{\ln x}{x^2}$

57. $f(x) = \begin{cases} e^{-1/x^2} & \text{if } x \neq 0 \\ 0 & \text{if } x = 0 \end{cases}$

58. Prove part 2 of Theorem 7.1.

59. Prove that $e^{-x} = 1/e^x$ for all real numbers x.

60. Prove the following, based on the definition of e^x:

(a) $e^0 = 1$ (b) $e^1 = e$.

61. Find the area of the region under the graph of $y = e^{-x}$ from x_1 to x_2, where $e^{x_1} = 1$ and $e^{x_2} = 2$.

62. Find the volume of the solid obtained by revolving the region under the graph of $y = \sqrt{1 + e^x}$ from $x = 0$ to $x = 1$ about the x-axis.

63. Find the length of the curve $y = (e^x + e^{-x})/2$ from $x = 0$ to $x = \ln 2$.

64. Find the volume of the solid obtained by revolving the region under the graph of $y = \sqrt{x} e^{(1-x^2)/2}$ from $x = 1$ to $x = 2$ about the x-axis.

65. Show that the function f defined by

$$f(x) = \frac{e^x - e^{-x}}{2}$$

is 1–1 on $\mathbb{R}$, and find f^{-1}.

66. Find $(f^{-1})'(x)$ for the function in Exercise 65.

In Exercises 67–69, analyze the function and draw its graph.

67. $f(x) = e^{1/x}$

68. $f(x) = \dfrac{e^{2x}}{x^2}$

69. $f(x) = x e^{(1-x^2)/2}$

70. A number of types of oscillatory motion, such as that of an oscillating spring, when the oscillation is occurring in a resisting medium, can be described by an equation of the type

$$y(t) = e^{-kt}(C_1 \cos \omega t + C_2 \sin \omega t)$$

where $y(t)$ is the displacement from the equilibrium position at time t. Suppose $k = 1$, $C_1 = 0$, $C_2 = 1$, and $\omega = 2$. Find the velocity and acceleration when $t = \pi$. Draw the graph of $y(t)$.

71. When limitations on size are considered for the growth of a colony of bacteria (because of physical limitations or limitations on nutrients, for example), the number $Q(t)$ present at time t is given with reasonable accuracy by

$$Q(t) = \frac{m}{1 + \left(\dfrac{m - Q_0}{Q_0} \right) e^{-kmt}}$$

where $Q_0 = Q(0)$, m is the maximum size, and k is a positive constant. If $Q_0 = 200$, $m = 1000$, and $Q(\frac{1}{2}) = 400$, find the constant k. Using this value of k, draw the graph of $Q(t)$. When does the maximum rate of growth occur?

72. The normal probability density function with mean μ and standard deviation σ is given by

$$f(x) = \frac{1}{\sqrt{2\pi}\,\sigma} e^{-(x-\mu)^2/2\sigma^2}$$

(a) Draw the graph of f.

(b) Show that inflection points occur at $x = \mu \pm \sigma$.

73. Let $f(x) = e^{ax}$, $g(x) = (\ln x)^b$, and $h(x) = x^c$ where a, b, and c are positive constants with b and c rational. Prove that as $x \to \infty$,

(a) f grows faster than h;

(b) h grows faster than g.

(Refer to Example 7.15.)

7.3 EXPONENTIAL AND LOGARITHMIC FUNCTIONS WITH OTHER BASES

Irrational Powers

We know that for rational numbers r and any real number $a > 0$,

$$a^r = e^{\ln a^r} = e^{r \ln a} \tag{7.17}$$

We still have not assigned a meaning to a raised to an irrational power. We now do so. Let x denote any real number, rational or irrational. We define a^x as follows:

**Definition 7.2
The General Exponential
Function**

$$a^x = e^{x \ln a}, \qquad a > 0 \tag{7.18}$$

This definition agrees with Equation 7.17 when x is rational and extends the meaning in the natural way to irrational values. Note that this definition is meaningful, since $\ln a$ is a well-defined real number for each $a > 0$, so that $x \ln a$ is also a real number, and since the exponential function is defined on all of $\mathbb{R}$, $e^{x \ln a}$ is also well defined. For example, according to our definition,

$$\pi^{\sqrt{2}} = e^{\sqrt{2}\ln \pi}$$

With a calculator we obtain the approximation

$$\pi^{\sqrt{2}} \approx 5.0475$$

Observe that since $e^{g(x)}$ is always positive, it follows that $a^x > 0$ for all x.

Taking the natural logarithm of both sides of Equation 7.18 gives an extension of Property 7.5 of the natural logarithm function:

$$\ln a^x = x \ln a, \qquad a > 0 \tag{7.19}$$

Laws of Exponents

The next theorem shows that the laws of exponents for rational exponents extend to all real exponents.

THEOREM 7.3

> **Laws of Exponents**
>
> Let a and b denote positive real numbers. If x and y are any real numbers, then
>
> 1. $a^{x+y} = a^x \cdot a^y$
> 2. $a^{x-y} = \dfrac{a^x}{a^y}$
> 3. $(a^x)^y = a^{xy}$
> 4. $(ab)^x = a^x b^x$
> 5. $\left(\dfrac{a}{b}\right)^x = \dfrac{a^x}{b^x}$

Proof   We prove the first and third laws of exponents and leave the others as exercises.

$$
\begin{array}{lll}
\textbf{1.} \quad a^{x+y} = e^{(x+y)\ln a} & & \text{By Equation 7.18}\\[4pt]
\phantom{\textbf{1.} \quad a^{x+y}} = e^{x\ln a + y\ln a} & & \text{Distributive Law}\\[4pt]
\phantom{\textbf{1.} \quad a^{x+y}} = e^{x\ln a}\cdot e^{y\ln a} & & \text{Theorem 7.1, part 1}\\[4pt]
\phantom{\textbf{1.} \quad a^{x+y}} = a^x \cdot a^y & & \text{By Equation 7.18}\\[8pt]
\textbf{3.} \quad (a^x)^y = e^{y\ln a^x} & & \text{By Equation 7.18}\\[4pt]
\phantom{\textbf{3.} \quad (a^x)^y} = e^{y(x\ln a)} & & \text{By Equation 7.19}\\[4pt]
\phantom{\textbf{3.} \quad (a^x)^y} = e^{(xy)\ln a} & & \text{Associative and Commutative Laws}\\[4pt]
\phantom{\textbf{3.} \quad (a^x)^y} = a^{xy} & & \text{By Equation 7.18} \qquad\blacksquare
\end{array}
$$

The next example illustrates how the natural logarithm can be used to solve certain types of equations for a variable that appears as an exponent.

EXAMPLE 7.16   Solve for x:

(a) $2^x = 3^{1-x}$ (b) $3 \cdot 4^{2x} = \dfrac{2}{\sqrt{5^x}}$

Solution   In each case we take the natural logarithm of both sides, using the fact that the natural logarithm is a one-to-one function.

(a)

$$
\begin{aligned}
\ln 2^x &= \ln 3^{1-x}\\[4pt]
x \ln 2 &= (1-x)\ln 3\\[4pt]
x \ln 2 + x \ln 3 &= \ln 3\\[4pt]
x(\ln 2 + \ln 3) &= \ln 3\\[4pt]
x &= \frac{\ln 3}{\ln 2 + \ln 3} = \frac{\ln 3}{\ln 6} \approx 0.6131
\end{aligned}
$$

(b) Rewrite the equation in the form

$$
3 \cdot 4^{2x} = \frac{2}{5^{x/2}}
$$

Then

$$\ln(3 \cdot 4^{2x}) = \ln\left(\frac{2}{5^{x/2}}\right)$$

$$\ln 3 + 2x \ln 4 = \ln 2 - \frac{x}{2} \ln 5$$

$$2 \ln 3 + 4x \ln 4 = 2 \ln 2 - x \ln 5$$

$$x(4 \ln 4 + \ln 5) = 2 \ln 2 - 2 \ln 3$$

$$x = \frac{2 \ln 2 - 2 \ln 3}{4 \ln 4 + \ln 5} = \frac{\ln 4 - \ln 9}{4 \ln 4 + \ln 5} \approx -0.1133$$

Note that any further attempts to simplify the answer before using your calculator are not worthwhile in this case.   ■

The Power Rule Extended to Real Exponents

We have often used the Power Rule,

$$\frac{d}{dx}x^r = rx^{r-1} \qquad (x > 0)$$

for *rational* numbers r. Now we can extend this rule to all real exponents.

THEOREM 7.4

> **Power Rule for Real Exponents**
>
> If $x > 0$ and α is any real number, then
>
> $$\frac{d}{dx}x^{\alpha} = \alpha x^{\alpha-1}$$

Proof   By Equation 7.18 and the derivatives of e^u and $\ln u$,

$$\frac{d}{dx}x^{\alpha} = \frac{d}{dx}e^{\alpha \ln x} = \left(\frac{\alpha}{x}\right)e^{\alpha \ln x}$$

$$= \left(\frac{\alpha}{x}\right)x^{\alpha} = \alpha x^{\alpha-1}$$ ■

For example, if $y = x^{\pi}$, then $dy/dx = \pi x^{\pi-1}$.

The Exponential Function a^x

If we take a as a fixed positive real number, then for each real number x, a^x is a well-defined real number, by Equation 7.18. Thus, $f(x) = a^x$ is a function from $\mathbb{R}$ to $\mathbb{R}$ (but its range is the set of positive real numbers). We call f the **exponential function with base** a. If the base $a = 1$, we have

$$1^x = e^{x \ln 1} = e^{x \cdot 0} = e^0 = 1$$

Thus, when $a = 1$, f is just the constant function $f(x) = 1$ for all x. Since the properties of constant functions are well known, *we will henceforth consider only those cases where $a > 0$ and $a \neq 1$.*

Derivatives of a^x

Since e^u is differentiable when u is differentiable and

$$\frac{d}{dx}e^u = e^u \frac{du}{dx}$$

it follows from Equation 7.18 that $a^x = e^{x \ln a}$ is differentiable, and the derivative is given by

$$\frac{d}{dx}a^x = \frac{d}{dx}e^{x \ln a} = e^{x \ln a} \cdot \ln a = a^x \ln a$$

Thus,

$$\frac{d}{dx}a^x = a^x \ln a \qquad (7.20)$$

Note that when $a = e$, this formula reduces to

$$\frac{d}{dx}e^x = e^x$$

since $\ln e = 1$. This derivative formula for e^x shows one reason why e is singled out as a special base. The derivative formula is simplified when e is the base.

The Graph of a^x

If $a > 1$, then $\ln a > 0$, and since a^x is always positive, it follows from Equation 7.20 that

$$\frac{d}{dx}a^x = a^x \ln a > 0 \qquad (a > 1)$$

If $0 < a < 1$, then $\ln a < 0$, so that

$$\frac{d}{dx}a^x < 0 \qquad (0 < a < 1)$$

Thus, the graph of $f(x) = a^x$ is increasing everywhere if $a > 1$ and decreasing everywhere if $0 < a < 1$. Consequently, there are no local (or absolute) maxima or minima. For f'' we have

$$f''(x) = \frac{d}{dx}(a^x \ln a) = \left(\frac{d}{dx}a^x\right)\ln a = a^x (\ln a)^2$$

which is always positive. So the graph is always concave up and there are no inflection points. Using the properties of e^x, we can now graph $f(x) = a^x$ as in Figure 7.14.

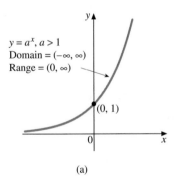

$y = a^x, a > 1$
Domain $= (-\infty, \infty)$
Range $= (0, \infty)$

$(0, 1)$

(a)

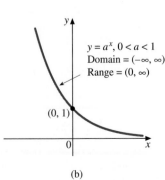

$y = a^x, 0 < a < 1$
Domain $= (-\infty, \infty)$
Range $= (0, \infty)$

$(0, 1)$

(b)

FIGURE 7.14

Derivative of $a^{g(x)}$

Using the Chain Rule together with Equation 7.20, we obtain

$$\frac{d}{dx}a^u = a^u \ln a \cdot \frac{du}{dx} \qquad (7.21)$$

where $u = g(x)$ is a differentiable function of x.

It is important to distinguish between a^x and x^a, where a is fixed (see Figure 7.15). We can contrast their derivatives as follows:

$$\begin{array}{cc} y = a^x & y = x^a \\[2mm] \dfrac{dy}{dx} = a^x \ln a & \dfrac{dy}{dx} = ax^{a-1} \end{array}$$

For example,

$$\frac{d}{dx}(3^x) = 3^x \ln 3, \quad \text{whereas} \quad \frac{d}{dx}x^3 = 3x^2$$

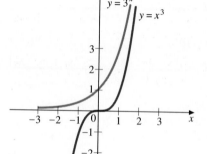

FIGURE 7.15

EXAMPLE 7.17   Find the derivative of each of the following:

(a) $y = 2^{x^2}$ (b) $y = x \cdot 3^{\sqrt{1-x}}$

Solution

(a) $\dfrac{dy}{dx} = 2^{x^2} \ln 2 \cdot 2x = (2 \ln 2)x \cdot 2^{x^2}$

(b) From the Product Rule and Equation 7.21,

$$\frac{dy}{dx} = 3^{\sqrt{1-x}} + x \cdot 3^{\sqrt{1-x}} \ln 3 \left(-\frac{1}{2\sqrt{1-x}} \right)$$

$$= 3^{\sqrt{1-x}} \left[1 - \frac{x \ln 3}{2\sqrt{1-x}} \right]$$

Integral of a^u

From Equation 7.21, we also obtain the antidifferentiation formula

$$\int a^u \, du = \frac{a^u}{\ln a} + C \qquad (7.22)$$

for which we are continuing to require $a > 0$ and $a \neq 1$.

EXAMPLE 7.18   Evaluate the integral

$$\int \frac{3^{\sqrt{x}}}{\sqrt{x}}\, dx$$

Solution   Let $u = \sqrt{x}$. Then $du = dx/(2\sqrt{x})$. So by Equation 7.22,

$$\int \frac{3^{\sqrt{x}}}{\sqrt{x}}\, dx = 2\int \underbrace{3^{\sqrt{x}}}_{3^u}\, \underbrace{\left(\frac{1}{2\sqrt{x}}\, dx\right)}_{du} = 2\left(\frac{3^{\sqrt{x}}}{\ln 3}\right) + C$$ ∎

Differentiating $[f(x)]^{g(x)}$ (Variables to Variable Powers)

We now know how to differentiate constants to variable powers (such as 3^x) and variables to constant powers (such as x^3). The next example shows how to differentiate variables to variable powers.

EXAMPLE 7.19   Find y' if $y = (\sin x)^x$, with $0 < x < \pi$.
Note: We restrict x to be in the open interval $(0, \pi)$ so that $\sin x > 0$. If we allowed $\sin x$ to be negative, $(\sin x)^x$ would not be well defined.

Solution   We will find the derivative by two methods. In Method 1 we use Equation 7.18 to get $y = (\sin x)^x = e^{x \ln \sin x}$. So, by Equation 7.15,

$$y' = \frac{d}{dx} e^{x \ln \sin x} = e^{x \ln \sin x}\left(\ln \sin x + x \cdot \frac{1}{\sin x} \cdot \cos x\right) \quad \begin{array}{l}\text{Using the Product Rule}\\\text{on } x \ln \sin x\end{array}$$

$$= (\sin x)^x (\ln \sin x + x \cot x)$$

In Method 2 we use logarithmic differentiation:

$$\ln y = x \ln \sin x$$

$$\frac{y'}{y} = \ln \sin x + x \cdot \frac{1}{\sin x} \cdot \cos x \quad \text{Using the Product Rule}$$

$$= \ln \sin x + x \cot x$$

$$y' = y(\ln \sin x + x \cot x)$$

$$= (\sin x)^x (\ln \sin x + x \cot x)$$ ∎

The Indeterminate Forms 0^0, 1^∞, and ∞^0

By using logarithms and exponentials, we can often evaluate limits of the form

$$\lim_{x \to a} f(x)^{g(x)}$$

where either

(i) $\lim\limits_{x \to a} f(x) = 0$ and $\lim\limits_{x \to a} g(x) = 0$ (the form 0^0)

(ii) $\lim\limits_{x \to a} f(x) = 1$ and $\lim\limits_{x \to a} g(x) = \pm\infty$ (the form 1^∞)

or

(iii) $\lim\limits_{x \to a} f(x) = \pm\infty$ and $\lim\limits_{x \to a} g(x) = 0$ (the form ∞^0)

The basic approach is to set $y = f(x)^{g(x)}$ and then take the natural logarithm of both sides:

$$\ln y = \ln f(x)^{g(x)} = g(x) \ln f(x)$$

Then we write the product on the right as a quotient (in most cases writing $\ln f(x)$ as the numerator), and use L'Hôpital's Rule, if it is applicable (either $0/0$ or ∞/∞). If we determine that

$$\lim_{x \to a} g(x) \ln f(x) = L$$

then we can conclude that

$$\lim_{x \to a} f(x)^{g(x)} = \lim_{x \to a} e^{g(x) \ln f(x)} = e^L$$

since $y = e^{\ln y}$. We used the continuity of the exponential function to get

$$\lim_{x \to a} e^{g(x) \ln f(x)} = e^L$$

We illustrate this approach in the next example.

EXAMPLE 7.20   Evaluate the following limits.

(a) $\displaystyle\lim_{x \to 0^+} (\sin x)^x$ (b) $\displaystyle\lim_{x \to 0} (1 - x)^{1/x}$

Solution

(a) As it stands, the limit is indeterminate, of the form 0^0. Proceeding as above, we set $y = (\sin x)^x$ and take logs and then write the product as a quotient.

$$\ln y = x \ln \sin x = \frac{\ln \sin x}{\dfrac{1}{x}}$$

As $x \to 0^+$, the fraction on the right is indeterminate, of the form ∞/∞, and since the conditions of L'Hôpital's Rule are met, we have

$$\lim_{x \to 0^+} \ln y = \lim_{x \to 0^+} \frac{\ln \sin x}{\dfrac{1}{x}}$$

$$= \lim_{x \to 0^+} \frac{\dfrac{1}{\sin x}(\cos x)}{-\dfrac{1}{x^2}}$$

$$= \lim_{x \to 0^+} \frac{-x^2 \cos x}{\sin x} \qquad \text{0/0 form; apply L'Hôpital's Rule again}$$

$$= \lim_{x \to 0^+} \frac{-2x \cos x + x^2 \sin x}{\cos x}$$

$$= \frac{0}{1} = 0$$

Thus,

$$\lim_{x \to 0^+} (\sin x)^x = \lim_{x \to 0^+} y = \lim_{x \to 0^+} e^{\ln y} = e^0 = 1$$

(b) As given, the limit is indeterminate, of the form 1^∞. We proceed as in part (a). Let

$$y = (1 - x)^{1/x}$$

Then

$$\ln y = \frac{1}{x} \ln(1 - x) = \frac{\ln(1 - x)}{x}$$

So

$$\lim_{x \to 0} \ln y = \lim_{x \to 0} \frac{\ln(1 - x)}{x}$$

Since we have a 0/0 indeterminate form, we apply L'Hôpital's Rule to get

$$\lim_{x \to 0} \ln y = \lim_{x \to 0} \frac{\dfrac{-1}{1 - x}}{1} = -1$$

Thus,

$$\lim_{x \to 0} (1 - x)^{1/x} = \lim_{x \to 0} y = \lim_{x \to 0} e^{\ln y} = e^{-1} = \frac{1}{e} \qquad \blacksquare$$

In limits of the type illustrated in the preceding example, it is tempting to stop after having found the limit of $\ln y$ by L'Hôpital's Rule. But it is essential that you find the limit of y itself, using

$$\lim_{x \to a} y = \lim_{x \to a} e^{\ln y} = e^{\lim_{x \to a} \ln y}$$

Otherwise, you haven't solved the problem that was posed.

In the next example we make use of L'Hôpital's Rule to aid in drawing the graph of a function.

EXAMPLE 7.21   Let $f(x) = x^x$ for $x > 0$.

(a) Find $\lim_{x \to 0^+} f(x)$.

(b) Find all maximum and minimum points on the graph of f.

(c) Draw the graph of f.

Solution

(a) Let $y = x^x$. Then $\ln y = x \ln x$. So

$$\lim_{x \to 0^+} \ln y = \lim_{x \to 0^+} \frac{\ln x}{1/x} \qquad \infty/\infty \text{ form}$$

$$= \lim_{x \to 0^+} \frac{1/x}{-1/x^2} = \lim_{x \to 0^+} (-x) = 0$$

So $\lim_{x \to 0^+} x^x = e^0 = 1$.

(b) From $\ln y = x \ln x$ we obtain by implicit differentiation

$$\frac{y'}{y} = \ln x + x \left(\frac{1}{x} \right)$$

$$y' = y(\ln x + 1) = x^x (\ln x + 1)$$

The only critical point occurs when $1 + \ln x = 0$ (since $x^x \neq 0$), or equivalently when $\ln x = -1$. Thus, $x = e^{-1} = 1/e \approx 0.37$. The corresponding

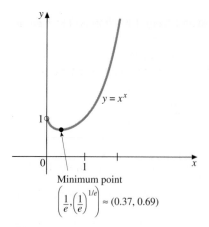

$y = x^x$

Minimum point
$$\left(\frac{1}{e}, \left(\frac{1}{e}\right)^{1/e}\right) \approx (0.37, 0.69)$$

FIGURE 7.16

y value is $f(1/e) = (1/e)^{1/e} \approx 0.69$. Since $y' > 0$ when $x > 1/e$ and $y' < 0$ when $x < 1/e$ (you should verify these inequalities), the point $\left(1/e, (1/e)^{1/e}\right)$ is a minimum point.

(c) To check concavity we calculate y'':

$$y'' = \left[\frac{d}{dx}(x^x)\right] \cdot (\ln x + 1) + x^x \left(\frac{1}{x}\right) = x^x(\ln x + 1)^2 + x^{x-1}$$

(using the derivative of x^x found in part (b)). Thus, $y'' > 0$ for all $x > 0$, which tells us that the graph is always concave up. Clearly,

$$\lim_{x \to \infty} x^x = \infty$$

We can now draw the graph as in Figure 7.16. ∎

The Logarithm Function with Base a

Since for $a > 0$ and $a \neq 1$, $f(x) = a^x$ is monotone (increasing if $a > 1$ and decreasing if $0 < a < 1$), f is one-to-one and therefore has an inverse. We denote its inverse by

$$f^{-1}(x) = \log_a x \qquad (a > 0, \ a \neq 1, \ x > 0)$$

and call f^{-1} the **logarithm function with base** a. Note that the domain of $f^{-1}(x)$ is the set of all $x > 0$, since the range of $f(x) = a^x$ is $(0, \infty)$. Remember that the domain of f^{-1} is the range of f. We call $\log_a x$ the **logarithm to the base** a **of** x. Thus, for $a > 0$, $a \neq 1$, we have

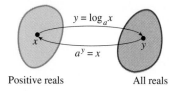

$y = \log_a x$

$a^y = x$

Positive reals   All reals

FIGURE 7.17

$$y = \log_a x \text{ if and only if } a^y = x \qquad (x > 0) \qquad (7.23)$$

This inverse relationship between a^y and $\log_a x$ is shown schematically in Figure 7.17. We can also state the same result as follows.

Inverse Properties

$$\log_a a^x = x \qquad \text{for all } x$$
$$a^{\log_a x} = x \qquad \text{for } x > 0 \qquad\qquad (7.24)$$

If $a = e$, then $\log_e x$ is the inverse of e^x. But $\ln x$ is the inverse of e^x, so we have

$$\log_e x = \ln x$$

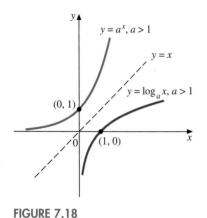

FIGURE 7.18

That is, the natural logarithm of x is the same as the logarithm to the base e of x.

We obtain the graph of $y = \log_a x$ by reflecting the graph of $y = a^x$ about the line $y = x$, with the result shown in Figure 7.18. We show the case $a > 1$ only.

Arithmetic Properties of $\log_a x$

The next theorem shows that the arithmetic properties for $\ln x$ also hold true for $\log_a x$. We leave the proof for the exercises.

THEOREM 7.5

Assume x and y are in the domain of the logarithm function of base a, where $a > 0$ and $a \neq 1$. Then

1. $\log_a xy = \log_a x + \log_a y$
2. $\log_a \dfrac{x}{y} = \log_a x - \log_a y$
3. $\log_a x^\alpha = \alpha \log_a x$ for any real number α

REMARK

■ In particular, when $a = e$, Property 3 can be stated as

$$\ln x^\alpha = \alpha \ln x$$

This property extends Equation 7.5 for $\ln x$ to all real numbers. We previously had proved this property for rational α only.

The Derivative of $\log_a x$

To find the derivative of $y = \log_a x$, we use Equation 7.23 to convert to the equivalent form

$$a^y = x$$

and then differentiate implicitly to get

$$(a^y \ln a) y' = 1 \qquad \text{By Equation 7.21}$$

Thus,

$$y' = \frac{1}{a^y \ln a}$$

But $a^y = x$, so we have the result

$$\frac{d}{dx} \log_a x = \frac{1}{x \ln a} \qquad\qquad (7.25)$$

Here we see why e is the "natural" base for logarithms, for if $a = e$, the derivative becomes simply $1/x$. By the Chain Rule, from Equation 7.25 we obtain for $u = g(x)$ a differentiable function with $u > 0$,

$$\frac{d}{dx} \log_a u = \frac{1}{u \ln a} \cdot \frac{du}{dx} \tag{7.26}$$

We could use Equation 7.26 to obtain an antiderivative of $1/u$ in terms of $\log_a u$, but this is unnecessary, since we already know that an antiderivative is $\ln|u| + C$. That is, there is no need to write the integral formula coming from Equation 7.26.

Change of Base

Sometimes it is useful to express a logarithm to one base in terms of a logarithm to another base. Suppose a and b are positive and different from 1. For $x > 0$, let

$$y = \log_b x \qquad \text{so that } x = b^y$$

Now take the logarithm to the base a of both sides of the last equation:

$$\log_a x = \log_a b^y = y \log_a b$$

Thus, $y = (\log_a x)/(\log_a b)$, and since $y = \log_b x$, we have the **Change-of-Base Formula**:

$$\log_b x = \frac{\log_a x}{\log_a b} \tag{7.27}$$

Since in calculus the most convenient base to work with is e, we usually want to use the following special case of Equation 7.27:

$$\log_b x = \frac{\ln x}{\ln b} \tag{7.28}$$

For example, by Equation 7.28 we have

$$\log_2 7 = \frac{\ln 7}{\ln 2} \approx 2.8074 \qquad \text{By calculator}$$

Integrals Involving $\log_a x$

The base e is by far the most commonly used base for logarithms and exponentials in calculus and its applications. The only other bases of any widespread use are 2 and 10. The base 2 is the foundation of binary arithmetic, which is very important in computer science. As we have previously noted, logarithms with base 10 are called common logarithms. In the next example we use the base 5, and in the exercises we will use a variety of bases to give practice in the theory of this section, but in the rest of the book we will concentrate on natural logarithms and exponentials.

EXAMPLE 7.22   Evaluate the indefinite integral

$$\int \frac{\log_5 x}{x}\, dx$$

Solution   By Formula 7.28,

$$\log_5 x = \frac{\ln x}{\ln 5}$$

So we can rewrite the integral as

$$\int \frac{\ln x}{x \ln 5}\, dx = \frac{1}{\ln 5} \int \frac{\ln x}{x}\, dx$$

Now we make the substitution $u = \ln x$. Then $du = \dfrac{1}{x}\, dx$. So we have

$$\frac{1}{\ln 5} \int \frac{\ln x}{x}\, dx = \frac{1}{\ln 5} \int u\, du = \frac{1}{\ln 5}\left(\frac{u^2}{2} + C\right) = \frac{1}{\ln 5}\left[\frac{(\ln x)^2}{2} + C\right]$$

If we let $C_1 = \dfrac{1}{\ln 5} C$, we can write the answer as

$$\int \frac{\log_5 x}{x}\, dx = \frac{(\ln x)^2}{2 \ln 5} + C_1 \qquad\blacksquare$$

The Number e as a Limit

We conclude this section with a theorem that provides a means of estimating e, along with an application of the result.

THEOREM 7.6

The number e can be expressed in either of the following ways:

1. $e = \lim\limits_{x \to 0} (1 + x)^{1/x}$

2. $e = \lim\limits_{n \to \infty} \left(1 + \dfrac{1}{n}\right)^n$

Proof

1. Let $F(x) = \ln x$. We know that $F'(x) = 1/x$, so $F'(1) = 1$. From the definition of the derivative, we have

$$F'(1) = \lim_{h \to 0} \frac{F(1+h) - F(1)}{h}$$

$$= \lim_{h \to 0} \frac{\ln(1+h) - \ln 1}{h}$$

$$= \lim_{h \to 0} \frac{1}{h} \ln(1+h) \qquad \text{Since } \ln 1 = 0$$

$$= \lim_{h \to 0} \ln(1+h)^{1/h} \qquad \text{By Equation 7.5}$$

$$= \ln \left[\lim_{h \to 0} (1+h)^{1/h} \right] \qquad \text{By continuity of the ln function}$$

Thus,

$$\ln \left[\lim_{h \to 0} (1+h)^{1/h} \right] = 1 \qquad \text{Since } F'(1) = 1$$

and since $\ln a = b$ implies $a = e^b$,

$$\lim_{h \to 0} (1+h)^{1/h} = e$$

By replacing h with x, we have proved part 1.

2. Let $1/n = x$. As $n \to \infty$, $x \to 0$, so we have

$$\lim_{n \to \infty} \left(1 + \frac{1}{n} \right)^n = \lim_{x \to 0} (1+x)^{1/x} = e$$

by part 1.   ■

n	$(1 + \frac{1}{n})^n$
1	2.000000
10	2.59374
100	2.704814
1,000	2.716924
10,000	2.718146
100,000	2.718268
1,000,000	2.718280

With the aid of a calculator we get the approximate values for $(1 + 1/n)^n$ shown in the margin. This table gives evidence that $e \approx 2.71828$, and it can be proved that this approximation is correct to five decimal places.

Compound Interest

The limit in part 1 of Theorem 7.6 arises naturally in compound-interest problems. When an amount of money P (called the principal) is invested at an annual interest rate r (for example, if the interest rate is 8%, $r = 0.08$), and interest is compounded annually, the total amount A_1 in the account after t yr is

$$A_1(t) = P(1+r)^t$$

(In the exercises, you will be asked to show why this formula is correct.) If interest is compounded semiannually, the amount after t yr becomes

$$A_2(t) = P \left(1 + \frac{r}{2} \right)^{2t}$$

and if it is compounded k times per year,

$$A_k(t) = P \left(1 + \frac{r}{k} \right)^{kt} \qquad (7.29)$$

Some banks compound interest daily, using the formula

$$A_{365}(t) = P\left(1 + \frac{r}{365}\right)^{365t}$$

What we want to do now is to let $k \to \infty$ in Equation 7.29. The result will give the amount that results from *continuous* compounding of interest. If we write $x = r/k$, then $x \to 0$ as $k \to \infty$, since r is fixed. So we have

$$A(t) = \lim_{k\to\infty} A_k(t) = \lim_{k\to\infty} P\left(1 + \frac{r}{k}\right)^{kt}$$

$$= \lim_{k\to\infty} P\left(1 + \frac{r}{k}\right)^{(k/r)rt}$$

$$= \lim_{x\to 0} P(1 + x)^{(1/x)rt}$$

$$= P\left[\lim_{x\to 0}(1 + x)^{1/x}\right]^{rt} \qquad \text{Using continuity of the exponential function}$$

$$= Pe^{rt} \qquad \text{By Part 1 of Theorem 7.6}$$

So we have the following formula for continuously compounded interest.

Continuously Compounded Interest

The amount $A(t)$ after t yr is

$$A(t) = Pe^{rt}$$

where P is the principal and r is the annual interest rate.

Exercise Set 7.3

In Exercises 1 and 2, express each number in terms of the base e (natural exponentials or logarithms) and then approximate its value correct to five decimal places using a calculator or CAS.

1. (a) $\pi^{\sqrt{3}}$
   (b) π^{π}
   (c) π^{e}
   (d) $(\sqrt{2})^{\sqrt{2}}$
   (e) $(\sin 2)^{\pi}$

2. (a) $\log_3 4$
   (b) $\log_{1/2} 5$
   (c) $\log_6 4$
   (d) $\log_7 3$
   (e) $\log_2 100$

In Exercises 3–18, find the derivative.

3. $y = 2^x$

4. $y = 3^{1-x^2}$

5. $y = x^2(4^{-x})$

6. $y = \log_2 x$

7. $y = \log_3 \dfrac{x}{1+x^3}$

8. $y = 3^x \cdot \log_3 x$

9. $y = \log_{10} \dfrac{\sqrt{x^2 - 1}}{x}$

10. $y = (\log_3 \sqrt{4 - x})^2$

11. $y = \log_2(\log_3 x)$

12. $y = (3^x)(x^3)$

13. $y = x^x$

14. $y = x^{\sin x}$

15. $y = x^{1/(1-x)}$

16. $y = (\cos x)^{\sin x}$

17. $y = (1 - x^2)^{1/x}$

18. $y = (\ln x)^x$

In Exercises 19–25, evaluate the limits.

19. (a) $\displaystyle\lim_{x\to 0} \frac{x^2}{e^x - x - 1}$ (b) $\displaystyle\lim_{x\to\infty} \frac{e^{\sqrt{x}}}{x^2}$

20. (a) $\displaystyle\lim_{x\to\infty} x^2 e^{-x}$ (b) $\displaystyle\lim_{x\to 0^+} x^2 e^{1/x}$

21. (a) $\displaystyle\lim_{x\to 0}\left(\frac{1}{e^x - 1} - \frac{1}{x}\right)$ (b) $\displaystyle\lim_{x\to 0}\left(\frac{1}{xe^x} - \frac{1}{x}\right)$

22. (a) $\lim_{x \to 0^+} x^x$ (b) $\lim_{x \to \pi/2^-} (\tan x)^{\cos x}$

23. (a) $\lim_{x \to 0} (1 + \sin x)^{1/x}$ (b) $\lim_{x \to 1} x^{1/(1-x)}$

24. (a) $\lim_{x \to 0^-} \left(1 - \dfrac{3}{x}\right)^x$ (b) $\lim_{x \to \infty} \left(\dfrac{x+1}{x}\right)^{2x}$

25. (a) $\lim_{x \to 1^+} (\ln x)^{1-x}$ (b) $\lim_{x \to 0} (e^x + x)^{1/x}$

In Exercises 26–35, evaluate the integrals.

26. $\displaystyle\int 5^x \, dx$

27. $\displaystyle\int \dfrac{x}{2^{x^2}} \, dx$

28. $\displaystyle\int \dfrac{3^x}{1 + 3^x} \, dx$

29. $\displaystyle\int (\cos x) \cdot 2^{\sin x} \, dx$

30. $\displaystyle\int x^2 \cdot 2^{x^3 - 1} \, dx$

31. $\displaystyle\int \dfrac{dt}{1 + 2^{-t}}$ (*Hint:* Multiply by $2^t/2^t$.)

32. $\displaystyle\int_1^4 \dfrac{2^{\sqrt{x}}}{\sqrt{x}} \, dx$

33. $\displaystyle\int_1^3 \dfrac{3^t}{\sqrt{3^t - 2}} \, dt$

34. $\displaystyle\int \dfrac{dx}{3^x(1 - 3^{-x})^2}$

35. $\displaystyle\int \dfrac{dt}{(2^t - 2^{-t})^2}$ (*Hint:* Multiply by $2^{2t}/2^{2t}$.)

In Exercises 36 and 37, draw the graphs of the functions in parts (a) through (d) on the same set of axes.

36. (a) $y = 2^x$
    (b) $y = e^x$
    (c) $y = 10^x$
    (d) $y = (\frac{1}{2})^x$

37. (a) $y = \log_2 x$
    (b) $y = \log_e x$
    (c) $y = \log_{10} x$
    (d) $y = \log_{1/2} x$

38. Prove that for $x \neq 0$, $a > 0$, and $a \neq 1$,
$$\dfrac{d}{dx} \log_a |x| = \dfrac{1}{x \ln a}$$

39. Using the result of Exercise 38 along with the Chain Rule, differentiate each of the following:

(a) $y = \log_{10} \left| \dfrac{x-1}{x+2} \right|$ (b) $y = \log_2 \left| \dfrac{x\sqrt{1-x}}{(2x-3)^3} \right|$

In Exercises 40 and 41, analyze the function and sketch its graph.

40. $y = 2^{1-x^2}$

41. $y = \log_{10} \sqrt{x^2 - 1}$

In Exercises 42–45, solve for x in terms of natural logarithms and then approximate the answer using a calculator or CAS.

42. $3^{2x-1} = 4^x$

43. $2 \cdot 3^{x+1} = 3 \cdot 2^{2-x}$

44. $3e^{x+2} = 4 \cdot 5^x$

45. $2^x \cdot 3^{1-2x} = 5 \cdot 6^{-x}$

46. Complete the proof of Theorem 7.3 by proving Properties 2, 4, and 5.

47. Prove Theorem 7.5.

48. Prove the formula $A_1(t) = P(1 + r)^t$ for the total amount of money $A_1(t)$ after t yr when a principal P is invested at a rate of interest r, compounded annually.

49. Suppose you invest $10,000. Find the total amount in your account after 10 yr if the interest rate is 5% and is compounded (a) annually, (b) semiannually, (c) daily, (d) continuously.

50. Using the properties of logarithms, solve for x and check your answers:
    (a) $\log_4(x + 2) - \log_4(x - 2) = \frac{1}{2}$
    (b) $2 \log_2 |x - 1| = 2 - \log_2(x + 2)$

51. Find the volume of the solid obtained by revolving the region under the graph of $y = 2^{1-x^2}$ from $x = 0$ to $x = 1$ about the y-axis.

52. Find the volume of the solid obtained by revolving the region under the graph of $y = \log_3 x$ from $x = 1$ to $x = 3$ about the y-axis.

53. Obtain an alternative proof of part 2 of Theorem 7.6 by the following steps:
    (a) For $t \in [1, 1 + \frac{1}{n}]$ show that
$$\int_1^{1+1/n} \dfrac{1}{1 + 1/n} \, dt \leq \int_1^{1+1/n} \dfrac{1}{t} \, dt \leq \int_1^{1+1/n} 1 \, dt$$
    (b) From part (a) show that
$$\dfrac{1}{n+1} \leq \ln\left(1 + \dfrac{1}{n}\right) \leq \dfrac{1}{n}$$
    (c) In part (b), multiply through by n and show that the result can be put in the form
$$\dfrac{1}{1 + \dfrac{1}{n}} \leq \ln\left(1 + \dfrac{1}{n}\right)^n \leq 1$$
    (d) In part (c), let $n \to \infty$ and show how it follows that
$$\lim_{n \to \infty} \left(1 + \dfrac{1}{n}\right)^n = e$$

54. (a) By graphing $f(x) = a^x$ for different values of $a > 0$, discuss the effect a has on the growth rate of a^x as $x \to \infty$.

(b) Repeat part (a) with $f(x) = \log_a x$.

55. Compare the growth rates of $f(x) = x^n$ and $g(x) = a^x$ as $x \to \infty$. Consider various values of n and a.

56. Use a graphing calculator or a CAS to graph $f(x) = x^n$ and $g(x) = a^x$ by pairs for various values of n and a. In each case determine approximate points of intersection (e.g., graph x^2 and 2^x). (*Hint:* Choose your scales carefully.)

57. Carry out the following steps (supplying reasons) to determine the set of all positive x values for which $x^e < e^x$.

(a) Show that $x^e < e^x$ if and only if $(\ln x)/x < 1/e$. (*Hint:* Take logs.)

(b) Let $f(x) = (\ln x)/x$. Show that $f(e) = 1/e$ is the absolute maximum value of f on $(0, \infty)$.

(c) For what positive x values is $f(x) < 1/e$?

58. Follow steps similar to those in Exercise 57 to determine the set of all positive x values for which $x^a < a^x$, where $a > 0$.

59. Investigate how the number of intersections of the graphs of $y = a^x$ and $y = x^b$ is affected by the values of a and b. (Consider $a > 0$ only.)

60. Use a CAS to compare the growth rates of the following functions as $x \to \infty$. Arrange in order according to increasing growth rates:

$$\ln x, \quad x^x, \quad e^{2x}, \quad x^{150}, \quad x^{-4}, \quad x^{10}e^{-x}, \quad x^{1/200}, \quad \frac{e^{4x}}{x^8}$$

61. (a) Redo Example 7.15 with r replaced by the positive real number α.

(b) Redo Exercise 73 in Exercise Set 7.2 with b and c replaced by positive real numbers β and γ.

7.4 EXPONENTIAL GROWTH AND DECAY

The exponential function is used to model certain phenomena of nature. Often, the rate of change of some quantity at any given instant is approximately proportional to the amount of the quantity present at that instant. For example, under certain conditions, a culture of bacteria will grow at a rate proportional to the number present (the more bacteria there are, the faster the rate of growth of the culture).

Even in human populations, a crude model is that the rate of population increase is proportional to the size of the population at any given time. This model was used by the English demographer Thomas Malthus, in the late eighteenth century, to predict future world populations. It is thus referred to as the **Malthus model**. The Malthus model is reasonably accurate for relatively short time intervals but because it does not take into account such limiting factors as war, famine, epidemics, and limits on available food supply and available space, its growth predictions are much too large when applied to longer time intervals. We will indicate in Exercise 24 (in Exercise Set 7.4) a more realistic model.

Bacteria cultures and human (as well as other) populations are quantities that increase with time. Indeed, the observation that the rate of oxygen given off in a sample of Martian soil examined by NASA's *Viking* spacecraft started to grow exponentially gave hope that there might be life on Mars. But when the rate of growth dropped below the exponential rate, that hope was dashed.

For some quantities, rate of change is proportional to the amount present, but with a negative growth rate; that is, the quantity decreases with time. An example is the rate of change of a radioactive substance, called *radioactive decay*. One interesting application of radioactive decay is in determining the

A Viking spacecraft on the surface of Mars

age of certain plant or animal remains. Archaeologists in particular have used this technique. It is called *radiocarbon dating*, and we will describe how it works in Exercise 25 in Exercise Set 7.4.

We could cite many other examples, but for now we want to show how this class of problems can be solved using calculus. We let $Q(t)$ denote the quantity in question at time t, where t is measured in appropriate units starting from a convenient time origin. Thus, $Q(t)$ might denote the number of bacteria in a culture t days after the culture was prepared, or $Q(t)$ might be the population of the world t yr after 1900, or $Q(t)$ might be the amount of strontium-90 (a radioactive substance) that will be left t yr from now out of 200 g currently on hand. In some cases the actual quantity in question might change by only discrete amounts; population, for example, is always an integer. Nevertheless, in our model we will assume Q is a differentiable function, so that it varies in a continuous fashion over some interval (remember that differentiability implies continuity). In cases where the answer must be an integer, we round our answer to the nearest integer. We assume in all cases that $Q(t) > 0$.

The basic assumption is that *the rate of change of Q is proportional to Q itself*. Thus, we have the differential equation

$$Q'(t) = kQ(t) \tag{7.30}$$

where k is the constant of proportionality, sometimes called the *growth constant*. We want to solve Equation 7.30 to determine the explicit form of $Q(t)$. Since $Q(t) \neq 0$, we can write the equation in the equivalent form

$$\frac{Q'(t)}{Q(t)} = k$$

and since

$$\frac{d}{dt} \ln Q(t) = \frac{Q'(t)}{Q(t)}$$

we have

$$\frac{d}{dt} \ln Q(t) = k$$

Taking the antiderivative, we get

$$\int \frac{d}{dt} \ln Q(t)\, dt = \int k\, dt$$

$$\ln Q(t) = kt + C$$

Hence,

$$Q(t) = e^{kt+C} \qquad \text{Since } e^{\ln a} = a$$

Since $e^{kt+C} = e^{kt} \cdot e^C$, we can write this result in the form

$$Q(t) = C_1 e^{kt}$$

where $C_1 = e^C$. Now suppose that when $t = 0$, $Q = Q_0$; that is, $Q(0) = Q_0$ (the initial amount of the quantity is Q_0 units). Then we get, on setting $t = 0$,

$$Q(0) = C_1 e^{k \cdot 0} \quad \text{or} \quad Q_0 = C_1$$

Thus, we obtain the formula for $Q(t)$:

Exponential Growth $(k > 0)$ and Decay $(k < 0)$ Formula

$$Q(t) = Q_0 e^{kt} \qquad (7.31)$$

where $Q_0 = Q(0)$.

It is easy to show that the value of Q given by Equation 7.31 does actually satisfy the differential equation (Equation 7.30) and has the correct value when $t = 0$. So this value is the solution we were seeking. The growth constant k is determined in each individual instance from known data, usually based on observed values of Q at certain times. When $k > 0$ we say that Q **grows exponentially**, and when $k < 0$ we say that Q **decays exponentially**.

The equation $Q'(t) = kQ(t)$ is an example of what is called a *separable* first-order differential equation, and the method we used to solve the equation for $Q(t)$ is known as *separation of variables*. The reason for this terminology becomes clearer if we write Q for $Q(t)$ and dQ/dt for $Q'(t)$. Then Equation 7.30 becomes $dQ/dt = kQ$ or, equivalently,

$$dQ = kQ\,dt$$

On dividing by Q, we get

$$\frac{dQ}{Q} = k\,dt$$

Now the variables Q and t are *separated* on opposite sides of the equation. Taking antiderivatives, we get, as before,

$$\ln Q = kt + C$$

and then we solve for $Q = Q(t)$ as we did earlier.

EXAMPLE 7.23   A culture of HeLa cells is observed to triple in size in 2 days. Assuming it grows exponentially, how large will the culture be in 5 days?

Solution   We are not told the original size of the culture, so we let this initial value be denoted by Q_0. We let $Q(t)$ be the size of the culture t days after the initial observation. When $t = 2$ we know that the size has tripled, so $Q(2) = 3Q_0$. By Equation 7.31, we have

$$Q(t) = Q_0 e^{kt}$$

On setting $t = 2$, we get

$$3Q_0 = Q_0 e^{k \cdot 2}$$
$$e^{2k} = 3$$

We write this equation in the equivalent logarithmic form to find k:

$$2k = \ln 3$$
$$k = \frac{\ln 3}{2}$$

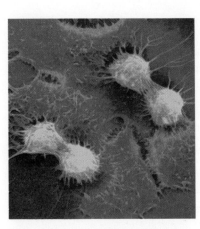

HeLa cells originated in 1951 in a tumor from a woman named Henrietta Lacks. They are now found worldwide in laboratories that do cell biology research. Cultures of HeLa cells grow exponentially and have been known to double in a single day.

Using a calculator, we could get a decimal approximation for k. However, it is often an advantage to leave k in an exact form at this stage. Substituting this value for k gives

$$Q(t) = Q_0 e^{t(\ln 3)/2} = Q_0 \left(e^{\ln 3}\right)^{t/2}$$

$$Q(t) = Q_0(3)^{t/2} \quad \text{Since } e^{\ln 3} = 3$$

(You can see now the advantage of leaving k in an exact form.) When $t = 5$ we have

$$Q(5) = Q_0(3)^{5/2} \approx 15.6 Q_0$$

So after 5 days the culture will be approximately 15.6 times its original size.

∎

REMARK ─────────────────────────────────

■ An alternative to solving for k in the preceding example is to solve for e^k, as follows. Since $e^{2k} = 3$, we have $(e^{2k})^{1/2} = (3)^{1/2}$ or $e^k = 3^{1/2}$. Now write

$$Q(t) = Q_0 e^{kt} = Q_0(e^k)^t = Q_0(3^{1/2})^t$$

or

$$Q(t) = Q_0(3)^{t/2}$$

as we had before. A similar procedure can be used in other problems of this type.

───

A useful measure of the rate of decomposition of a radioactive substance is its **half-life**, defined to be the time required for half the original amount to decompose. Thus, if 100 g of a radioactive substance with a half-life of 5 yr is on hand, then after 5 yr, 50 g will remain; after 10 yr, 25 g will remain; and so on. The next example makes use of this concept.

EXAMPLE 7.24   The radioactive isotope strontium-90 has a half-life of 25 yr. If 200 kg of the substance is on hand, how much will remain after 10 yr? How long will it be until 90% of the original amount has decomposed? Strontium-90 has not been added to the atmosphere since the nuclear test ban of 1963. What percentage of the amount present in 1963 remained in 1993?

Solution   Let $Q(t)$ be the amount, in kilograms, that remains after t yr. Then we know that $Q(0) = 200$, and so by Equation 7.31,

$$Q(t) = Q_0 e^{kt} = 200 e^{kt}$$

Since the half-life is 25 yr, one-half of the original 200 kg will remain after 25 yr. That is, $Q(25) = 100$. Thus, setting $t = 25$ gives

$$100 = 200 e^{k \cdot 25}$$

$$e^{25k} = \frac{1}{2}$$

$$25k = \ln \frac{1}{2} = -\ln 2$$

$$k = -\frac{1}{25} \ln 2$$

So we obtain

$$Q(t) = 200e^{-(\ln 2/25)t} = 200(e^{\ln 2})^{-t/25} = 200(2)^{-t/25}$$

When $t = 10$, we get

$$Q(10) = 200(2)^{-10/25} = 200(2)^{-2/5} \approx 151.6 \text{ kg}$$

To determine how long it will be until 90% has decomposed, we first observe that at this time 10%, or 20 kg, will remain. Thus, we want to find t for which $Q(t) = 20$; that is,

$$200(2)^{-t/25} = 20$$

$$2^{-t/25} = \frac{1}{10}$$

We take the natural logarithm of both sides to find t:

$$-\frac{t}{25} \ln 2 = \ln \frac{1}{10}$$

or

$$-\frac{t}{25} \ln 2 = -\ln 10$$

$$t = \frac{25 \ln 10}{\ln 2} \approx 83 \text{ yr}$$

To determine the percentage of strontium-90 remaining in 1993 of the amount present in 1963, we let Q_0 be the unspecified amount present in 1963, and we let $t = 30$ (30 yr after 1963). In our equation for $Q(t)$, we replace the initial amount 200 by the unspecified amount Q_0, giving

$$Q(t) = Q_0(2)^{-t/25}$$

Thus,

$$Q(30) = Q_0(2)^{-30/25} = Q_0(2)^{-1.2} \approx 0.435Q_0$$

As a fractional part of Q_0, we have

$$\frac{Q(30)}{Q_0} \approx \frac{0.435Q_0}{Q_0} = 0.435$$

That is, about 43.5% of the amount in 1963 remained in 1993. ∎

EXAMPLE 7.25   In 1930 the world population was approximately 2 billion and in 1960 approximately 3 billion. Using the Malthus model, estimate the population in the year 2000.

Solution   Let $Q(t)$ be the population in billions, with t in years measured from 1930. Then $Q(0) = 2$, $Q(30) = 3$, and, according to the Malthus model,

$$Q(t) = Q_0 e^{kt} = 2e^{kt}$$

Setting $t = 30$ and proceeding in the usual way, we get

$$3 = 2e^{30k}$$

$$e^{30k} = \frac{3}{2}$$

$$30k = \ln \frac{3}{2}$$

$$k = \frac{1}{30} \ln \frac{3}{2}$$

So

$$Q(t) = 2(e^{\ln 3/2})^{t/30} = 2 \cdot \left(\frac{3}{2}\right)^{t/30}$$

In the year 2000, $t = 70$. So, according to the model, the population should be

$$Q(70) = 2 \cdot \left(\frac{3}{2}\right)^{70/30} = 2 \cdot \left(\frac{3}{2}\right)^{7/3} \approx 5.15 \text{ billion} \quad \blacksquare$$

EXAMPLE 7.26 **Newton's Law of Cooling** states that when a body at initial temperature T_0 is introduced into a medium of temperature T_m, where $T_m < T_0$, the rate at which the body cools at any given instant is proportional to the difference between its temperature at that instant and the temperature T_m of the surrounding medium.

(a) Find a formula for the temperature of the body at time t units after being introduced into the medium.

(b) Suppose a body at 120°C is placed in air at 20°C and it has cooled to a temperature of 80°C after $\frac{1}{2}$ hr. Find its temperature after 1 hr.

Solution

(a) Let $T(t)$ be the temperature of the body t units of time after it is introduced into the medium. Then, by Newton's Law of Cooling,

$$T'(t) = k[T(t) - T_m]$$

Writing $T = T(t)$ and $dT/dt = T'(t)$, we obtain the equivalent differential equation

$$dT = k(T - T_m)dt$$

Now we separate variables and take antiderivatives:

$$\int \frac{dT}{T - T_m} = \int k \, dt$$

$$\ln(T - T_m) = kt + C \quad \text{Remember that } T - T_m > 0.$$

$$T - T_m = e^{kt+C} = e^{kt} \cdot e^C = C_1 e^{kt}$$

So, on solving for T, we get

$$T(t) = T_m + C_1 e^{kt}$$

To find the constant C_1 we set $t = 0$, observing that $T(0) = T_0$:

$$T_0 = T_m + C_1 e^0$$

$$C_1 = T_0 - T_m$$

Finally,

$$T(t) = T_m + C_1 e^{kt}$$

$$= T_m + (T_0 - T_m)e^{kt} \quad (7.32)$$

(b) We are given that $T_0 = 120$, $T_m = 20$, and $T(\frac{1}{2}) = 80$. So from Equation 7.32,

$$T(t) = 20 + 100e^{kt} \quad (7.33)$$

$$80 = 20 + 100e^{k(1/2)}$$

$$60 = 100e^{k/2}$$

$$e^{k/2} = 0.6$$

$$\frac{k}{2} = \ln(0.6)$$

$$k = 2\ln(0.6) = \ln(0.36)$$

Substituting for k in Equation 7.33 gives

$$T(t) = 20 + 100e^{(\ln 0.36)t}$$

$$= 20 + (100)(e^{\ln 0.36})^t$$

$$= 20 + 100(0.36)^t$$

Finally, we want $T(1)$:

$$T(1) = 20 + 100(0.36) = 56$$

So after 1 hr the temperature should be approximately 56°C.   ■

REMARK
■ Equation 7.32 is valid also if the body is introduced into a warmer medium (so that $T_m > T_0$), and you will be asked to show this result in Exercise 17 in Exercise Set 7.4.

Exercise Set 7.4

1. A culture of HeLa cells originally numbers 500. After 2 hr there are 1500 cells in the culture. Assuming exponential growth, find how many are present after 6 hr.

2. A radioactive substance has a half-life of 64 hr. If 200 g of the substance is present initially, how much will remain after 4 days?

3. A culture of 100 bacteria doubles after 2 hr. How long will it take for the number of bacteria to reach 3200?

4. How long will it take 100 g of a radioactive substance that has a half-life of 40 yr to be reduced to 12.5 g?

5. One hundred kilograms of a certain radioactive substance decays to 40 kg after 10 yr. Find how much will remain after 20 yr.

6. In the initial stages the growth of a HeLa cell culture is approximately exponential. If the number doubles after 3 hr, what proportion of the original number will be present after 6 hr? After 12 hr?

7. If a culture of bacteria doubles in size after 2 hr, how long will it take for it to triple in size? Assume exponential growth.

8. If 100 g of a radioactive material diminishes to 80 g in 2 yr, find the half-life of the substance.

9. Find the half-life of a radioactive substance that is reduced by 30% in 20 hr.

10. A radioactive substance has a half-life of 6 yr. If 20 lb of the substance is present initially, how much will remain after 2 yr?

11. A culture of bacteria increases from 1000 to 5000 in 8 hr. Assuming the growth is exponential, find how much will be present after 20 hr.

12. The population of a certain Florida city increased by 40% between 1970 and 1978. If the population in 1978 was 210,000, what is the expected population in the year 2000, according to the Malthus model?

13. A certain radioactive substance has a half-life of 8 yr. If 200 g of the substance is present initially, how much will remain at the end of 12 yr? How long will it be until 90% of the original amount has decayed?

14. A body heated to 120°F is brought into a room in which the temperature is 70°F. After 15 min the temperature of the body is 100°F. According to Newton's Law of Cooling, how long will it take for the temperature of the body to drop to 80°F?

15. A thermometer registering 20°C is placed in a freezer in which the temperature is −10°C. After 10 min the thermometer registers 5°C. According to Newton's Law of Cooling, what will it register after 30 min?

16. According to Newton's Law of Cooling, will the temperature of the body ever reach the temperature T_m of the surrounding medium? Show why or why not. Does this result agree with reality? If not, how do you explain the discrepancy?

17. Prove that the result in Equation 7.32 remains valid if $T_m > T_0$. Use this to find how long it will take a thermometer that registers 5°C when on the outside to rise to 18°C after being brought to the inside where the temperature is 20°C, if it takes 10 min to rise to 12°C.

18. The rate of change of air pressure p with respect to altitude h is approximately proportional to the pressure. If at sea level the pressure is p_0, find a formula for p as a function of h. If $p_0 = 1.01 \times 10^5$ pascals (Pa), and at altitude $h = 2$ km (6,560 ft) the pressure is 8.08×10^4 Pa, find the approximate pressure at the 4 km (13,800 ft) altitude of the Mauna Kea Observatory in Hawaii.

19. The atmospheric pressure at sea level is approximately 15 lb/in.^{2}, and at Denver, which is 1 mi above sea level, the pressure is approximately 12 lb/in.^{2}. What is the approximate pressure at the top of Vail Pass, which is 2 mi above sea level? (See Exercise 18.)

20. The rate of change of the intensity I of light passing through a translucent material, with respect to the depth x of penetration of the light ray, is approximately proportional to the intensity at depth x. If the sunlight that strikes water is reduced to half of its original intensity at a depth of 3 m, what will be the intensity (as a fractional part of the original intensity) at a depth of 9 m?

21. Use the data of Exercise 20 to find at what depth the intensity of illumination will be 10% of the original.

22. In a certain type of chemical reaction a substance is converted in such a way that the rate of conversion is always proportional to the amount of unconverted substance. If 20 g of the substance is present originally, and half of it has been converted in 2 min, find how long it will be until only 4 g remains unconverted.

23. The rate at which the number of items of a certain type of merchandise are sold is proportional to the number still on hand. Let $N(t)$ be the number of items still on hand t days after the items were first introduced, and let $N_1 = N(1)$.
 (a) Derive a formula for $N(t)$.
 (b) If 20 items are on hand on the first day and 10 on the fifth day, how many can be expected to be on hand on the ninth day?

24. A more realistic model than the Malthus model for population growth is the **Verhulst model**, which assumes that the rate of growth of the population is jointly proportional to the size of the population and the further growth capacity. Thus, if $Q(t)$ denotes the size of the population at time t and m is the ultimate limit on population, then

$$\frac{dQ}{dt} = kQ(m - Q)$$

 (a) Prove, by direct substitution into the differential equation, that

$$Q(t) = \frac{mQ_0}{Q_0 + (m - Q_0)e^{-mkt}}$$

  is a solution of the Verhulst model, where $Q_0 = Q(0)$. (A typical graph of $Q(t)$, called a **logistic curve**, is shown in the accompanying figure.)
 (b) Prove that for the solution in part (a), $\lim_{t \to \infty} Q(t) = m$.
 (c) Redo Example 7.25 using the Verhulst model with the assumption that the maximum supportable world population is 10 billion.

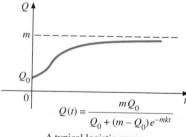

$$Q(t) = \frac{mQ_0}{Q_0 + (m - Q_0)e^{-mkt}}$$

A typical logistic curve for population growth

25. A radioactive isotope of carbon, called carbon-14, is found in all living organisms. During the lifetime of an organism the amount that decays is replenished through the atmosphere, so that the amount in the organism remains constant, but after the death of the organism the decayed carbon-14 is no longer replenished. Because the half-life of carbon-14 is very long (approximately 5730 yr), the remains of organisms that died thousands of years ago may still contain measurable amounts of carbon-14. By comparing the amount remaining with the amount known to be present in the particular organism when it was alive, the age of the remains can be determined. Suppose archaeologists find a human skull and determine that 10% of the original amount of carbon-14 remains. (The original amount is determined by measuring the ratio of carbon-14 to a stable isotope of carbon in a present-day object.) Find the approximate age of the skull.

26. The radioactive substance einsteinium-253 decays at a rate given approximately by $dQ/dt = -0.0939Q$, where t is measured in days, and $Q(t)$ is the amount present at time t.
(a) Find the half-life.
(b) Determine the proportion of an original amount that will remain after 8 days.
(c) Find the number of days required for 80% of an original amount to decay.

27. Certain radioactive substances decay into a second radioactive substance. If $Q(t)$ denotes the amount of such a radioactive substance at time t and $R(t)$ denotes the amount of the second substance, then

$$Q(t) = Q_0 e^{kt}, \qquad Q_0 = Q(0)$$

and since the amount of substance that has decomposed is the original amount Q_0 minus the amount $Q(t)$ still present,

$$R(t) = Q_0 - Q(t) = Q_0(1 - e^{kt})$$

(a) Find a formula for the time required for the ratio of R to Q to equal a given value λ.
(b) Rubidium-87, with a half-life of 5×10^{11} yr, decays exponentially into strontium-87. Find the time required for there to be 1% as much strontium-87 as rubidium-87.

28. Refer to Exercise 27, part (b). By measuring the ratio of strontium-87 to rubidium-87 in fossils, scientists can estimate the age of the fossil. Use this method to estimate the age of a fossil that contains 237 parts per million of rubidium and 1.85 parts per million of strontium.

7.5 INVERSE TRIGONOMETRIC FUNCTIONS

We have seen that the natural logarithm function and its inverse, the exponential function, are important in modeling many physical phenomena, especially those involving growth and decay. Other things in nature are modeled using trigonometric functions, especially the sine and cosine. The wave generated by the apparatus in the photo traces a sine and a cosine curve. Such things as ocean tides, electric impulses, human heartbeats, and many other periodic phenomena can be described using trigonometric functions. A natural question to ask is whether inverses of the trigonometric functions might also be useful. But there is a problem here. Recall that for a function to have an inverse, its graph must pass the horizontal line test. Clearly, none of the six trigonometric functions passes this test. For example, the graph of the sine function is shown in Figure 7.19, and it is clear that any horizontal line drawn between $y = -1$ and $y = 1$ intersects the graph in infinitely many points.

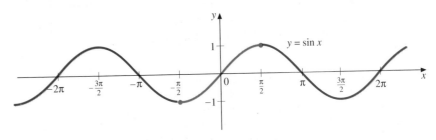

FIGURE 7.19
The part of the sine curve in red is used for the inverse.

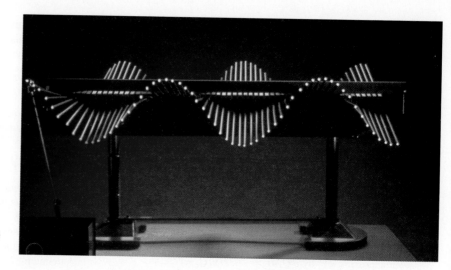

A mechanical way of showing a sine
and a cosine

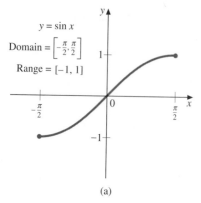

$y = \sin x$

Domain $= \left[-\frac{\pi}{2}, \frac{\pi}{2}\right]$

Range $= [-1, 1]$

(a)

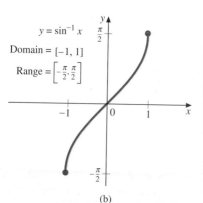

$y = \sin^{-1} x$

Domain $= [-1, 1]$

Range $= \left[-\frac{\pi}{2}, \frac{\pi}{2}\right]$

(b)

FIGURE 7.20

The Inverse Sine Function

We can remedy the problem of the failure of the sine function to have an inverse by considering a restricted domain on which the sine function is one-to-one. A convenient choice of such a restriction is the interval $[-\pi/2, \pi/2]$. We show the portion of the graph on this restricted domain in red in Figure 7.19. On that interval, the sine is increasing (since its derivative, $\cos x$, is positive between $-\pi/2$ and $\pi/2$), and it assumes all values in its range $[-1, 1]$. We denote the inverse of the sine restricted in this way by $\sin^{-1}$ or by **arcsin**. We will normally use $\sin^{-1}$, since it is consistent with the notation f^{-1} that we have been using for inverses in general. However, since both notations are in widespread use, you should be familiar with both. Relationships between the restricted sine function and its inverse are shown in Figure 7.20.

In summary, we have the following.

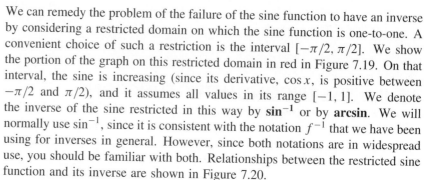

The Inverse Sine Function

$$y = \sin^{-1} x \text{ if and only if } \sin y = x$$

where $-1 \leq x \leq 1$ and $-\frac{\pi}{2} \leq y \leq \frac{\pi}{2}$

For example, we have

$$\sin^{-1} 1 = \frac{\pi}{2} \qquad\qquad \sin^{-1} \frac{1}{2} = \frac{\pi}{6}$$

$$\sin^{-1}(-1) = -\frac{\pi}{2} \qquad \sin^{-1}\left(-\frac{1}{2}\right) = -\frac{\pi}{6}$$

and so on.

Note carefully that $\sin^{-1} x$ *must* be a number in the interval $[-\pi/2, \pi/2]$.

It would be *incorrect*, for example, to say that $\sin^{-1}(-1/2) = (11\pi)/6$ even though $\sin(11\pi)/6 = -1/2$, because $(11\pi)/6$ does not lie in the interval $[-\pi/2, \pi/2]$.

The value of $\sin^{-1} x$, then, can be stated in words:

$\sin^{-1} x$ is that number in $[-\pi/2, \pi/2]$ whose sine is x.

For any function f for which an inverse f^{-1} exists, we know that

$$f^{-1}(f(x)) = x \qquad \text{for all } x \text{ in the domain of } f$$

and

$$f(f^{-1}(x)) = x \qquad \text{for all } x \text{ in the domain of } f^{-1}$$

Thus, we have the *cancellation laws*

$$\sin^{-1}(\sin x) = x \qquad \text{for } -\tfrac{\pi}{2} \le x \le \tfrac{\pi}{2}$$

and

$$\sin(\sin^{-1} x) = x \qquad \text{for } -1 \le x \le 1$$

 Be very careful in applying the first of these equations. For example,

$$\sin^{-1}\left(\sin \frac{2\pi}{3}\right) \ne \frac{2\pi}{3}$$

since $(2\pi)/3$ is not in the interval $[-\pi/2, \pi/2]$. Instead, we should write

$$\sin^{-1}\left(\sin \frac{2\pi}{3}\right) = \sin^{-1}\left(\frac{\sqrt{3}}{2}\right) = \frac{\pi}{3}$$

since $\pi/3$ is between $-\pi/2$ and $\pi/2$, and $\sin(\pi/3) = \sqrt{3}/2$.

The Derivative of $\sin^{-1} x$

To find the derivative of the inverse sine, we let $y = \sin^{-1} x$, so that $\sin y = x$. Then, differentiating the latter equation implicitly, we get

$$(\cos y)y' = 1$$

Thus, if $\cos y \ne 0$,

$$y' = \frac{1}{\cos y}$$

Since $\cos y > 0$ for $-\pi/2 < y < \pi/2$, we conclude that the derivative of $\sin^{-1} x$ exists on the open interval $(-1, 1)$. To obtain the answer in terms of x, observe that $\sin^2 y + \cos^2 y = 1$, so $\cos y = \sqrt{1 - \sin^2 y} = \sqrt{1 - x^2}$ (we choose the positive sign when taking the square root, since we know that $\cos y > 0$ when $-\pi/2 < y < \pi/2$).

An alternative way of finding $\cos y$ in terms of x is by the right triangle in Figure 7.21. There we label an acute angle as y, the side opposite as x, and the hypotenuse as 1. Labeling the triangle in this way expresses the fact that

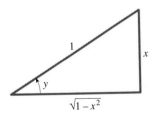

FIGURE 7.21
$y = \sin^{-1} x$

$y = \sin^{-1} x$; that is, $\sin y = x$. By the Pythagorean Theorem, the side adjacent to y is $\sqrt{1 - x^2}$. Thus, $\cos y = \sqrt{1 - x^2}$. We therefore have

$$\frac{d}{dx} \sin^{-1} x = \frac{1}{\sqrt{1 - x^2}} \qquad \text{if } x \in (-1, 1) \qquad (7.34)$$

Combining this formula with the Chain Rule in the usual way, we have the following formula for the derivative of $\sin^{-1} u$, where $u = g(x)$ is a differentiable function such that $g(x) \in (-1, 1)$:

$$\frac{d}{dx} \sin^{-1} u = \frac{1}{\sqrt{1 - u^2}} \frac{du}{dx} \qquad (7.35)$$

EXAMPLE 7.27   Find the derivative of each of the following:

(a) $y = \sin^{-1} 1/x$ (b) $y = x \sin^{-1} 2x$

Solution

(a) To apply Equation 7.35, we need to have $g(x) = 1/x$ in the interval $(-1, 1)$, which will be true if $x < -1$ or $x > 1$. With such values of x, we have, by Equation 7.35,

$$\frac{dy}{dx} = \frac{1}{\sqrt{1 - \dfrac{1}{x^2}}} \left(-\frac{1}{x^2} \right)$$

$$= \frac{1}{\dfrac{1}{|x|}\sqrt{x^2 - 1}} \left(-\frac{1}{x^2} \right) \qquad \text{Remember that } \sqrt{x^2} = |x|.$$

$$= -\frac{|x|}{x^2} \cdot \frac{1}{\sqrt{x^2 - 1}}$$

$$= \begin{cases} -\dfrac{1}{x\sqrt{x^2 - 1}} & \text{if } x > 1 \\[3mm] \dfrac{1}{x\sqrt{x^2 - 1}} & \text{if } x < -1 \end{cases}$$

(b) Using the Product Rule and Equation 7.35, we get

$$\frac{dy}{dx} = \sin^{-1} 2x + x \cdot \frac{1}{\sqrt{1 - 4x^2}} \cdot 2 = \sin^{-1} 2x + \frac{2x}{\sqrt{1 - 4x^2}}$$

This result is valid if $2x \in (-1, 1)$—that is, if $x \in (-\frac{1}{2}, \frac{1}{2})$.   ∎

Integral Formulas Involving the Inverse Sine

Formula 7.35 for the derivative of $\sin^{-1} u$ gives rise to the following antidifferentiation formula:

$$\int \frac{du}{\sqrt{1 - u^2}} = \sin^{-1} u + C, \qquad u \in (-1, 1) \qquad (7.36)$$

In Exercise 45 in Exercise Set 7.5 we will ask you to prove the more general formula given below, where $u = g(x)$ is a differentiable function on the open interval $(-a, a)$, for any positive constant a.

$$\int \frac{du}{\sqrt{a^2 - u^2}} = \sin^{-1} \frac{u}{a} + C, \qquad u \in (-a, a) \qquad (7.37)$$

EXAMPLE 7.28   Evaluate the following integrals:

(a) $\displaystyle\int_0^1 \frac{dx}{\sqrt{4 - x^2}}$ (b) $\displaystyle\int \frac{dx}{\sqrt{2x - x^2}}$

Solution

(a) By Equation 7.37,

$$\int_0^1 \frac{dx}{\sqrt{4 - x^2}} = \sin^{-1} \frac{x}{2} \Big]_0^1 = \sin^{-1} \frac{1}{2} - \sin^{-1} 0 = \frac{\pi}{6} - 0 = \frac{\pi}{6}$$

(b) We can make the integral conform to the pattern of Equation 7.36 by completing the square on $2x - x^2$:

$$\int \frac{dx}{\sqrt{2x - x^2}} = \int \frac{dx}{\sqrt{1 - (x^2 - 2x + 1)}} = \int \frac{dx}{\sqrt{1 - (x - 1)^2}}$$

Now let $u = x - 1$; then $du = dx$. So we have

$$\int \frac{dx}{\sqrt{1 - (x - 1)^2}} = \int \frac{du}{\sqrt{1 - u^2}} = \sin^{-1} u + C = \sin^{-1}(x - 1) + C$$

■

EXAMPLE 7.29   **The Circumference of a Circle** Verify the formula $C = 2\pi r$ for the circumference of a circle of radius r.

Solution   Place the center of the circle at the origin, so that its equation is $x^2 + y^2 = r^2$. By symmetry, we can find the length of the upper semicircle (see Figure 7.22) and double it. The equation of the upper semicircle is

$$y = \sqrt{r^2 - x^2}$$

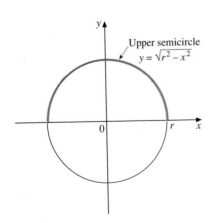

FIGURE 7.22

By Equation 6.18 for arc length, we know that the length of this semicircle is

$$\int_{-r}^{r} \sqrt{1 + \left(\frac{dy}{dx}\right)^2}\, dx = \int_{-r}^{r} \sqrt{1 + \left(\frac{-x}{\sqrt{r^2 - x^2}}\right)^2}\, dx$$

$$= \int_{-r}^{r} \sqrt{\frac{r^2 - x^2 + x^2}{r^2 - x^2}}\, dx$$

$$= \int_{-r}^{r} \frac{r}{\sqrt{r^2 - x^2}}\, dx \qquad \sqrt{r^2} = r \text{ since } r > 0$$

$$= r \int_{-r}^{r} \frac{dx}{\sqrt{r^2 - x^2}}$$

$$= r \left[\sin^{-1} \frac{x}{r}\right]_{-r}^{r} \qquad \text{By Equation 7.37}$$

$$= r[\sin^{-1} 1 - \sin^{-1}(-1)]$$

$$= r\left[\frac{\pi}{2} - \left(-\frac{\pi}{2}\right)\right] = \pi r$$

Therefore, the circumference is $2\pi r$.   ■

The Other Inverse Trigonometric Functions

We can obtain inverses of the remaining trigonometric functions also by restricting their domains so that they become one-to-one functions. We show them (along with $\sin^{-1} x$ for completeness) in the following box.

The Inverse Trigonometric Functions

1. $y = \sin^{-1} x$ if and only if $\sin y = x$,
    where $-1 \leq x \leq 1$ and $-\pi/2 \leq y \leq \pi/2$

2. $y = \cos^{-1} x$ if and only if $\cos y = x$,
    where $-1 \leq x \leq 1$ and $0 \leq y \leq \pi$

3. $y = \tan^{-1} x$ if and only if $\tan y = x$,
    where $-\infty < x < \infty$ and $-\pi/2 < y < \pi/2$

4. $y = \cot^{-1} x$ if and only if $\cot y = x$,
    where $-\infty < x < \infty$ and $0 < y < \pi$

5. $y = \sec^{-1} x$ if and only if $\sec y = x$, where $1 \leq x < \infty$ or
    $-\infty < x \leq -1$ and $0 \leq y < \pi/2$ or $\pi \leq y < 3\pi/2$

6. $y = \csc^{-1} x$ if and only if $\csc y = x$, where $1 \leq x < \infty$ or
    $-\infty < x \leq -1$ and $0 < y \leq \pi/2$ or $\pi < y \leq 3\pi/2$

REMARK ────────────────────────────────

■ There are other equally valid ways to restrict the domains of the secant and cosecant functions for forming the inverse. The choices we have made are motivated by the fact that the derivative formulas with these choices are simpler than with some of the other possible choices.

In Figures 7.23–7.27 we show the graphs of the restricted functions along with the graphs of their inverses.

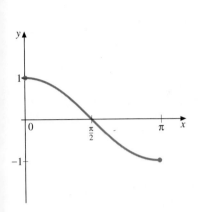

$y = \cos x$
Domain $= [0, \pi]$
Range $= [-1, 1]$

FIGURE 7.23

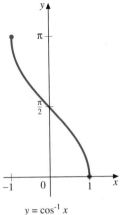

$y = \cos^{-1} x$
Domain $= [-1, 1]$
Range $= [0, \pi]$

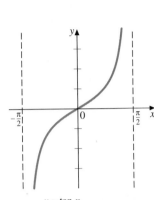

$y = \tan x$
Domain $= \left(-\frac{\pi}{2}, \frac{\pi}{2}\right)$
Range $= (-\infty, \infty)$

FIGURE 7.24

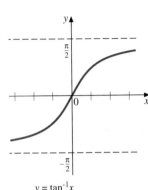

$y = \tan^{-1} x$
Domain $= (-\infty, \infty)$
Range $= \left(-\frac{\pi}{2}, \frac{\pi}{2}\right)$

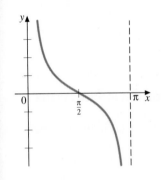

$y = \cot x$
Domain $= (0, \pi)$
Range $= (-\infty, \infty)$

FIGURE 7.25

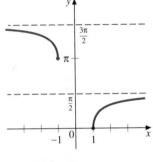

$y = \cot^{-1} x$
Domain $= (-\infty, \infty)$
Range $= (0, \pi)$

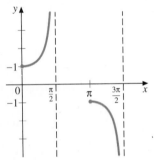

$y = \sec x$
Domain $= \left[0, \frac{\pi}{2}\right) \cup \left[\pi, \frac{3\pi}{2}\right)$
Range $= (-\infty, -1] \cup [1, \infty)$

FIGURE 7.26

$y = \sec^{-1} x$
Domain $= (-\infty, -1] \cup [1, \infty)$
Range $= \left[0, \frac{\pi}{2}\right) \cup \left[\pi, \frac{3\pi}{2}\right)$

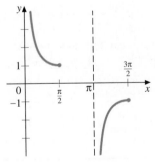

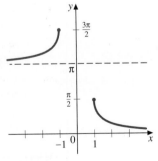

$y = \csc x$
Domain $= \left(0, \frac{\pi}{2}\right] \cup \left(\pi, \frac{3\pi}{2}\right]$
Range $= (-\infty, -1] \cup [1, \infty)$

$y = \csc^{-1} x$
Domain $= (-\infty, -1] \cup [1, \infty)$
Range $= \left(0, \frac{\pi}{2}\right] \cup \left(\pi, \frac{3\pi}{2}\right]$

FIGURE 7.27

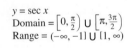

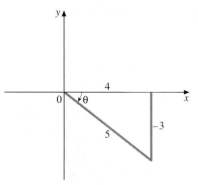

FIGURE 7.28

EXAMPLE 7.30   Evaluate each of the following:

(a) $\cos\left[\tan^{-1}\left(-\dfrac{3}{4}\right)\right]$ (b) $\sec^{-1}\left(\tan\dfrac{3\pi}{4}\right)$

Solution

(a) Let $\theta = \tan^{-1}(-3/4)$. Then $\tan\theta = -3/4$ and $-\pi/2 < \theta < \pi/2$. Since $\tan\theta$ is negative, θ is between $-\pi/2$ and 0. Viewed as an angle, θ is in the fourth quadrant, measured clockwise, as we show in Figure 7.28. By selecting the point $(4, -3)$ on the terminal side of θ we form the right triangle shown and see that the hypotenuse is 5. Thus,

$$\cos\left[\tan^{-1}\left(-\frac{3}{4}\right)\right] = \cos\theta = \frac{4}{5}$$

(b) Since $\tan 3\pi/4 = -1$, we want to find $\sec^{-1}(-1)$. Viewed as an angle, we want an angle that lies between π and $3\pi/2$ and whose secant is -1. The answer is π. Thus,

$$\sec^{-1}\left(\tan\frac{3\pi}{4}\right) = \pi$$ ■

Derivatives of the Other Inverse Trigonometric Functions

We will ask you to derive the following differentiation formulas in the exercises.

$$\frac{d}{dx}\cos^{-1}x = -\frac{1}{\sqrt{1-x^2}}, \qquad x \in (-1, 1) \qquad (7.38)$$

$$\frac{d}{dx}\tan^{-1}x = \frac{1}{1+x^2}, \qquad x \in (-\infty, \infty) \qquad (7.39)$$

$$\frac{d}{dx}\cot^{-1}x = -\frac{1}{1+x^2}, \qquad x \in (-\infty, \infty) \qquad (7.40)$$

$$\frac{d}{dx}\sec^{-1}x = \frac{1}{x\sqrt{x^2-1}}, \qquad |x| > 1 \qquad (7.41)$$

$$\frac{d}{dx}\csc^{-1}x = -\frac{1}{x\sqrt{x^2-1}}, \qquad |x| > 1 \qquad (7.42)$$

Formulas for the derivatives of $\cos^{-1}u$, $\tan^{-1}u$, $\cot^{-1}u$, $\sec^{-1}u$, and $\csc^{-1}u$, where u is a differentiable function of x that satisfies suitable restrictions, follow from these by the Chain Rule.

Integral Formulas Involving the Inverse Tangent and the Inverse Secant

From Formulas 7.39 and 7.41, we obtain

$$\int \frac{du}{1+u^2} = \tan^{-1} u + C \tag{7.43}$$

$$\int \frac{du}{u\sqrt{u^2-1}} = \sec^{-1} u + C, \qquad |u| > 1 \tag{7.44}$$

Observe that there is no need to write the integration formulas that arise from Equations 7.38, 7.40, and 7.42, since the negative sign can be factored out, leaving integrals for $\sin^{-1} x$, $\tan^{-1} x$, or $\sec^{-1} x$. For example, rather than write

$$\int -\frac{dx}{\sqrt{1-x^2}} = \cos^{-1} x + C$$

(which is correct), we write

$$\int -\frac{dx}{\sqrt{1-x^2}} = -\int \frac{dx}{\sqrt{1-x^2}} = -\sin^{-1} x + C$$

The following generalizations of Equations 7.43 and 7.44 are often useful.

For $a > 0$,

$$\int \frac{du}{a^2+u^2} = \frac{1}{a} \tan^{-1} \frac{u}{a} + C \tag{7.45}$$

$$\int \frac{du}{u\sqrt{u^2-a^2}} = \frac{1}{a} \sec^{-1} \frac{u}{a} + C, \qquad |u| > a \tag{7.46}$$

We will verify the first of these and leave the second for the exercises. The goal is to change the integrand so that it conforms to the form in Equation 7.43. To replace a^2 by 1, we divide numerator and denominator by a^2, getting

$$\int \frac{\dfrac{1}{a^2}}{1+\dfrac{u^2}{a^2}}\, du = \int \frac{\dfrac{1}{a^2}}{1+\left(\dfrac{u}{a}\right)^2}\, du$$

Now we make the substitution $v = u/a$. Then $dv = 1/a\, du$. Thus, we can write the integral as

$$\int \frac{\dfrac{1}{a}\left(\dfrac{1}{a}\right)}{1+\left(\dfrac{u}{a}\right)^2}\, du = \int \frac{\dfrac{1}{a}}{1+v^2}\, dv = \frac{1}{a} \int \frac{dv}{1+v^2}$$

By Equation 7.43,

$$\frac{1}{a} \int \frac{dv}{1+v^2} = \frac{1}{a} \left[\tan^{-1} v + C \right]$$

We therefore have, on replacing v by u/a,

$$\int \frac{du}{a^2 + u^2} = \frac{1}{a} \tan^{-1} \left(\frac{u}{a} \right) + C_1$$

where $C_1 = \frac{1}{a} C$. (Since C is an arbitrary constant, so is C_1.)

EXAMPLE 7.31   Differentiate each of the following:

(a) $y = \cos^{-1} \sqrt{x}$ $(0 < x < 1)$
(b) $y = \tan^{-1} e^x$
(c) $y = x \sec^{-1} x$, $|x| > 1$
(d) $y = \cot^{-1}(\tan x)$

Solution

(a) $\dfrac{dy}{dx} = \dfrac{-1}{\sqrt{1-x}} \cdot \dfrac{1}{2\sqrt{x}} = \dfrac{-1}{2\sqrt{x - x^2}}$

(b) $\dfrac{dy}{dx} = \dfrac{1}{1+e^{2x}} \cdot e^x = \dfrac{e^x}{1 + e^{2x}}$

(c) $\dfrac{dy}{dx} = \sec^{-1} x + x \cdot \dfrac{1}{x\sqrt{x^2 - 1}} = \sec^{-1} x + \dfrac{1}{\sqrt{x^2 - 1}}$

(d) $\dfrac{dy}{dx} = -\dfrac{1}{1 + \tan^2 x} \cdot \sec^2 x = -\dfrac{\sec^2 x}{\sec^2 x} = -1$

■

EXAMPLE 7.32   Evaluate the following integrals:

(a) $\displaystyle\int_{-\sqrt{3}}^{3\sqrt{3}} \frac{dx}{9 + x^2}$ (b) $\displaystyle\int \frac{dx}{x\sqrt{4x^2 - 9}}$

(c) $\displaystyle\int \frac{dx}{x^2 + 2x + 5}$ (d) $\displaystyle\int_0^{(\ln 3)/2} \frac{e^x}{e^{2x} + 1} dx$

Solution

(a) By Equation 7.45, with $a = 3$, we obtain

$$\int_{-\sqrt{3}}^{3\sqrt{3}} \frac{dx}{9 + x^2} = \frac{1}{3} \tan^{-1} \frac{x}{3} \Bigg]_{-\sqrt{3}}^{3\sqrt{3}} = \frac{1}{3} \left[\tan^{-1} \sqrt{3} - \tan^{-1} \left(-\frac{\sqrt{3}}{3} \right) \right]$$

$$= \frac{1}{3} \left[\frac{\pi}{3} - \left(-\frac{\pi}{6} \right) \right] = \frac{1}{3} \left(\frac{\pi}{2} \right) = \frac{\pi}{6}$$

As an alternative, we could divide the numerator and denominator by 9 to write the integrand in the form

$$\frac{1}{9} \left[\frac{1}{1 + \dfrac{x^2}{9}} \right] = \frac{1}{9} \left[\frac{1}{1 + \left(\dfrac{x}{3} \right)^2} \right]$$

Then we could substitute $u = x/3$ and proceed as in the derivation of Formula 7.45.

(b) To change this integral to the form of Equation 7.46, we let $u = 2x$. Then $du = 2\,dx$. So on multiplying numerator and denominator by 2, we have

$$\int \frac{dx}{x\sqrt{4x^2 - 9}} = \int \frac{2\,dx}{2x\sqrt{4x^2 - 9}} = \int \frac{du}{u\sqrt{u^2 - 9}} = \frac{1}{3}\sec^{-1}\frac{u}{3} + C$$

$$= \frac{1}{3}\sec^{-1}\frac{2x}{3} + C$$

(c) First, we complete the square in the denominator:

$$\int \frac{dx}{x^2 + 2x + 5} = \int \frac{dx}{(x^2 + 2x + 1) + 4} = \int \frac{dx}{(x + 1)^2 + 4}$$

Now if we mentally substitute $u = x + 1$, we see that this integral is in the form that occurs in Equation 7.45. So the answer is

$$\int \frac{dx}{x^2 + 2x + 5} = \frac{1}{2}\tan^{-1}\frac{x + 1}{2} + C$$

(d) This integral is in the form of Equation 7.45, with $u = e^x$ and $a = 1$. So we have

$$\int_0^{(\ln 3)/2} \frac{e^x}{e^{2x} + 1}dx = \tan^{-1} e^x\Big]_0^{(\ln 3)/2} = \tan^{-1}(e^{(\ln 3)/2}) - \tan^{-1} e^0$$

$$= \tan^{-1}\sqrt{3} - \tan^{-1} 1$$

$$= \frac{\pi}{3} - \frac{\pi}{4} = \frac{\pi}{12} \qquad \blacksquare$$

EXAMPLE 7.33   Find the area under the graph of

$$f(x) = \frac{1}{1 + x^2}$$

from $x = -1$ to $x = 1$.

Solution   The area A of the region (see Figure 7.29) is given by

$$A = \int_{-1}^{1} \frac{dx}{1 + x^2} = \tan^{-1} x\Big]_{-1}^{1}$$

$$= \tan^{-1} 1 - \tan^{-1}(-1)$$

$$= \frac{\pi}{4} - \left(-\frac{\pi}{4}\right) = \frac{\pi}{2}$$

Note that we therefore have the following representation of π:

$$\pi = 2\int_{-1}^{1} \frac{dx}{1 + x^2} \qquad \blacksquare$$

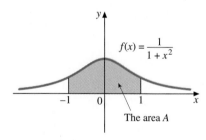

$f(x) = \dfrac{1}{1 + x^2}$

The area A

FIGURE 7.29

EXAMPLE 7.34   The TV broadcasting antennas for New York City are 10 m high and are on top of one of the towers of the World Trade Center, which is 330 m high. How far should a portable-television viewer at street level stand from the base of the building to get the best reception from the TV antennas, if the portable TV's antenna is about 2 m above the ground? (The reception is assumed to be best where the angle of the observer's eye subtended by the antenna is a maximum.)

Solution   The problem is to find x as shown in Figure 7.30 so as to maximize θ. We see that $\theta = \alpha - \beta$ and since $\cot \alpha = x/338$ and $\cot \beta = x/328$, we have

$$\alpha = \cot^{-1} \frac{x}{338} \quad \text{and} \quad \beta = \cot^{-1} \frac{x}{328}$$

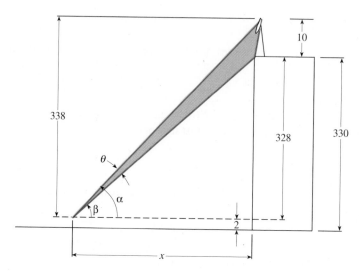

FIGURE 7.30

(We could have used tangents instead, but the derivatives involved are somewhat simpler using cotangents.) So

$$\theta = \cot^{-1} \frac{x}{338} - \cot^{-1} \frac{x}{328} \qquad (x \geq 0)$$

Thus,

$$\frac{d\theta}{dx} = -\frac{1}{1 + \left(\dfrac{x}{338}\right)^2} \cdot \frac{1}{338} + \frac{1}{1 + \left(\dfrac{x}{328}\right)^2} \cdot \frac{1}{328}$$

$$= -\frac{338}{(338)^2 + x^2} + \frac{328}{(328)^2 + x^2}$$

Setting $d\theta/dx$ equal to 0 gives the critical points. After simplification we get

$$10x^2 = 328(338)^2 - 338(328)^2$$

$$= (328)(338)(338 - 328)$$

$$= (328)(338)(10)$$

$$x^2 = (328)(338)$$

$$x = \sqrt{110864} \approx 333 \text{ m} \qquad \text{\small Since } x > 0, \text{ ignore the negative square root.}$$

It is intuitively evident that as x varies from 0 to arbitrarily large values, θ attains a maximum value, and since we found only one critical point, it is the one we are seeking. We could confirm that this value of θ gives a maximum by using either the First- or Second-Derivative Test, but we will not do so. (See also the CAS-generated graph of $\theta(x)$ in Figure 7.31, showing the maximum value occurring at about $x = 333$.)   ∎

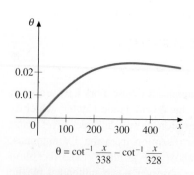

$$\theta = \cot^{-1} \frac{x}{338} - \cot^{-1} \frac{x}{328}$$

FIGURE 7.31

Exercise Set 7.5

InExercises 1–12, give the exact value. (If answers involve π, leave them in terms of π.)

1. (a) $\sin^{-1} \frac{1}{2}$

   (b) $\cos^{-1} \frac{1}{2}$

   (c) $\arctan 1$

   (d) $\text{arcsec } 2$

   (e) $\cot^{-1} 0$

2. (a) $\arccos(-\frac{1}{\sqrt{2}})$

   (b) $\tan^{-1}(-1)$

   (c) $\sin^{-1}(-\frac{\sqrt{3}}{2})$

   (d) $\text{arccsc}(-2)$

   (e) $\sec^{-1}(-\frac{2}{\sqrt{3}})$

3. (a) $\arcsin(-1)$

   (b) $\cos^{-1}(-1)$

   (c) $\arctan 0$

   (d) $\cos^{-1}(-\frac{1}{2})$

   (e) $\cot^{-1}(-1)$

4. (a) $\sin[\cos^{-1}(-\frac{1}{3})]$

   (b) $\sec[\sin^{-1}(-\frac{1}{4})]$

5. (a) $\tan[\arccos \frac{3}{5}]$

   (b) $\cos[\tan^{-1}(-\frac{4}{3})]$

6. (a) $\sin[\arctan \frac{5}{12}]$

   (b) $\csc[\arccos(-\frac{2}{3})]$

7. (a) $\cos[2\sin^{-1}(-\frac{1}{3})]$

   (b) $\sin[2\cos^{-1}(-\frac{3}{5})]$

8. (a) $\sin^{-1}(\cos \frac{\pi}{3})$

   (b) $\cos^{-1}[\sin(-\frac{\pi}{6})]$

9. (a) $\cos^{-1}[\cos(-\frac{\pi}{5})]$

   (b) $\sin^{-1}(\sin \frac{9\pi}{8})$

10. (a) $\tan^{-1}(\tan \frac{3\pi}{4})$

   (b) $\cot^{-1}(\cot \frac{5\pi}{3})$

11. $\sin[\sin^{-1} \frac{3}{5} + \cos^{-1}(-\frac{5}{13})]$ *(Hint: Use the identity for $\sin(\alpha + \beta)$.)*

12. $\cos[\tan^{-1}(-\frac{1}{2}) - \sec^{-1} \frac{5}{3}]$ *(Hint: Use the identity for $\cos(\alpha - \beta)$.)*

In Exercises 13 and 14, use a calculator or CAS to find the approximate radian measure of θ, where $0 \le \theta < 2\pi$.

13. (a) $\sin \theta = \frac{2}{3}$, $\cos \theta < 0$

   (b) $\cos \theta = \frac{1}{4}$, $\sin \theta < 0$

14. (a) $\tan \theta = 2$, $\sec \theta < 0$

   (b) $\cos \theta = -\frac{2}{5}$, $\cot \theta > 0$

In Exercises 15–24, find y'.

15. $y = \tan^{-1} \frac{2x}{3}$

16. $y = \sin^{-1}(2x - 1)$

17. $y = \arccos \sqrt{1 - x}$

18. $y = \sec^{-1} \frac{1}{x}$ $(0 < x < 1)$

19. $y = (1 + x^2) \tan^{-1} x$

20. $y = \sin^{-1} \sqrt{1 - x^2}$ $(0 < x < 1)$

21. $y = \dfrac{\cos^{-1} x}{\sqrt{1 - x^2}}$

22. $y = x^2 \sec^{-1} x^2$

23. $y - \sin^{-1} y = x^2$

24. $y = \tan^{-1} \dfrac{y}{x}$

In Exercises 25–40, evaluate the integrals.

25. $\displaystyle\int_{-1/2}^{\sqrt{3}/2} \frac{dx}{\sqrt{1 - x^2}}$

26. $\displaystyle\int_{-1}^{\sqrt{3}} \frac{dx}{1 + x^2}$

27. $\displaystyle\int_{\sqrt{2}}^{2} \frac{dx}{x\sqrt{x^2 - 1}}$

28. $\displaystyle\int_{-1}^{1} \frac{dx}{\sqrt{4 - x^2}}$

29. $\displaystyle\int_{0}^{4} \frac{dx}{16 + x^2}$

30. $\displaystyle\int_{4/\sqrt{3}}^{4} \frac{dx}{x\sqrt{x^2 - 4}}$

31. $\displaystyle\int \frac{dx}{\sqrt{16 - 9x^2}}$

32. $\displaystyle\int \frac{dx}{4x^2 + 25}$

33. $\displaystyle\int \frac{dx}{x\sqrt{16x^2 - 9}}$

34. $\displaystyle\int \frac{dx}{\sqrt{4x - 4x^2}}$

35. $\displaystyle\int \frac{dx}{9x^2 - 12x + 5}$

36. $\displaystyle\int_{0}^{4} \frac{dx}{x^2 - 2x + 4}$

37. $\displaystyle\int \frac{x\,dx}{\sqrt{1 - x^4}}$

38. $\displaystyle\int \frac{dx}{e^x + e^{-x}}$ *(Hint: Multiply numerator and denominator by e^x.)*

39. $\displaystyle\int_{-\sqrt{3}}^{-1} \frac{dx}{x\sqrt{4x^2 - 3}}$

40. $\displaystyle\int_{1}^{2} \frac{dx}{\sqrt{12x - 3x^2 - 8}}$

41. Find the area of the region under the graph of

$$y = \frac{1}{\sqrt{4 - x^2}}$$

from $x = -1$ to $x = 1$.

42. Find the area under the graph of

$$y = \frac{4}{4 + x^2}$$

from $x = -2$ to $x = 2$.

43. Find the volume of the solid generated by revolving the region under the graph of

$$y = \frac{1}{\sqrt{x^2 + 3}}$$

from $x = -1$ to $x = 3$ about the x-axis.

44. Find the volume of the solid generated by revolving the region under the graph of

$$y = \frac{1}{x^2\sqrt{x^2 - 9}}$$

from $x = 2\sqrt{3}$ to $x = 6$ about the y-axis.

In Exercises 45–52, prove Formulas 7.37–7.44, and in Exercise 53, prove Formula 7.46.

54. Give the exact value of $\tan^{-1}\frac{1}{3} + \tan^{-1}(-2)$. (*Hint:* Use $\tan(\alpha + \beta)$.)

55. Prove that if $0 < x < 1$,

$$\sin^{-1} x + \sin^{-1}\sqrt{1 - x^2} = \frac{\pi}{2}$$

(*Hint:* Use differentiation to prove that the left-hand side is a constant, and then substitute a particular value of x to show that the constant is $\pi/2$.)

56. Prove that if $x > 0$,

$$\tan^{-1} x + \tan^{-1}\frac{1}{x} = \frac{\pi}{2}$$

(See the hint for Exercise 55.)

In Exercises 57–60, evaluate the integrals.

57. $\displaystyle\int_1^3 \frac{dx}{\sqrt{x}(1 + x)}$

58. $\displaystyle\int \frac{dx}{\sqrt{e^{2x} - 1}}$

59. $\displaystyle\int_{\pi/2}^{\pi} \frac{\sin x}{1 + \cos^2 x}\, dx$

60. $\displaystyle\int \frac{dx}{x\sqrt{x^6 - 1}}$

61. Find the volume of the solid generated by rotating the region under the graph of $y = (1 + x^4)^{-1}$ from $x = 0$ to $x = 1$ about the y-axis.

62. A lighthouse is located on an island 2 km from the nearest point P on a straight shore. At a point on the shore 1 km from P the beam of light from the lighthouse beacon is moving at 40 km/min. Through how many radians per minute is the beacon turning? Through how many revolutions per minute is it turning?

63. A helicopter is initially 2 km above an observer on the ground, and it then moves horizontally at a speed of 30 m/s. What is the rate of change of the angle of elevation of the helicopter from the observer when the helicopter has gone 0.5 km?

 64. Use a calculator or CAS to approximate

$$\int_{-n}^{n} \frac{1}{1 + x^2}\, dx$$

for large values of n. As n increases, does the value of the integral get close to a familiar number? If so, which?

7.6 THE HYPERBOLIC FUNCTIONS

The combinations of exponential functions

$$\frac{e^x + e^{-x}}{2} \quad \text{and} \quad \frac{e^x - e^{-x}}{2}$$

have many interesting and useful properties. Because of their importance, they are given special names, as stated in the following definition.

Definition 7.3
The Hyperbolic Sine and Hyperbolic Cosine

The **hyperbolic sine**, abbreviated **sinh**, and the **hyperbolic cosine**, abbreviated **cosh**, are defined by

$$\sinh x = \frac{e^x - e^{-x}}{2}$$

$$\cosh x = \frac{e^x + e^{-x}}{2}$$

The domain of each of these functions is the set of all real numbers.

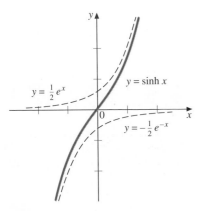

FIGURE 7.32

REMARK ────────────────────────────────

■ It is common practice to pronounce sinh like "cinch" and cosh to rhyme with "gosh."

──

We can obtain the graphs of the hyperbolic sine and cosine from their definitions in terms of the exponential function. Since

$$\sinh x = \frac{e^x - e^{-x}}{2} = \frac{1}{2}e^x + \left(-\frac{1}{2}e^{-x}\right)$$

we can draw the graphs of $y = \frac{1}{2}e^x$ and $y = -\frac{1}{2}e^{-x}$ and add the y values at different x values to find points on the graph $y = \sinh x$. In Figure 7.32 we show $y = \frac{1}{2}e^x$ and $y = -\frac{1}{2}e^{-x}$ (in dashed curves) and the graph of $y = \sinh x$ that results from this addition.

Similarly, for $\cosh x$, we write

$$\cosh x = \frac{e^x + e^{-x}}{2} = \frac{1}{2}e^x + \frac{1}{2}e^{-x}$$

So we can add the y values on the graphs of $y = \frac{1}{2}e^x$ and $y = \frac{1}{2}e^{-x}$ to get the graph of $y = \cosh x$, shown in Figure 7.33.

Since

$$\sinh(-x) = \frac{e^{-x} - e^x}{2} = -\sinh x$$

and

$$\cosh(-x) = \frac{e^{-x} + e^x}{2} = \cosh x$$

we see that $\sinh x$ is odd and $\cosh x$ is even, confirming the symmetries our graphs indicate. Also note that

$$\sinh 0 = \frac{e^0 - e^{-0}}{2} = \frac{1 - 1}{2} = 0$$

and

$$\cosh 0 = \frac{e^0 + e^{-0}}{2} = \frac{1 + 1}{2} = 1$$

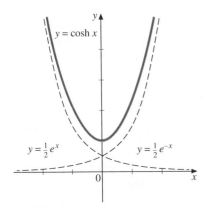

FIGURE 7.33

Recall that for the trigonometric functions $\sin x$ and $\cos x$, we also know that $\sin x$ is odd, $\cos x$ is even, $\sin 0 = 0$, and $\cos 0 = 1$. So while the graphs of the hyperbolic sine and cosine do not resemble those of the trigonometric sine and cosine, they do have these even–odd characteristics in common, and their values at $x = 0$ are the same. These common features give the first hint of why the names *sine* and *cosine* are used with the hyperbolic functions. We will soon see that the analogy between the hyperbolic and trigonometric functions goes much further. We will also see why the name *hyperbolic* is used.

The hyperbolic cosine curve has an interesting application. When a uniform flexible cable hangs between two supports, if the only force acting on it is that caused by gravity, then it assumes a shape known as a **catenary**. For example, an electric power line hangs in this shape (see Figure 7.34). Any flexible object suspended between two points hangs in the shape of a catenary, even a spider web. If the axes are suitably chosen, the equation of the catenary can be shown to be of the form

$$y = a \cosh \frac{x}{a}$$

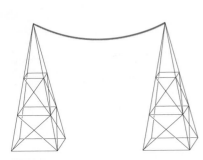

FIGURE 7.34

where a is the height of the catenary at its lowest point.

Another instance of the hyperbolic cosine is the Gateway Arch in St. Louis, Missouri. It is in the shape of an inverted hyperbolic cosine curve in which its height and span are the same: 630 feet.

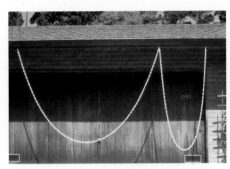

The Gateway Arch in St Louis is in the form of a catenary.

Flexible objects suspended between two supports assume the shape of catenaries.

We define four other hyperbolic functions by analogy with the corresponding trigonometric functions. They are the **hyperbolic tangent** (tanh), **hyperbolic cotangent** (coth), **hyperbolic secant** (sech), and **hyperbolic cosecant** (csch).

Definition 7.4
Other Hyperbolic Functions

$$\tanh x = \frac{\sinh x}{\cosh x} \qquad\qquad \text{sech}\, x = \frac{1}{\cosh x}$$

$$\coth x = \frac{\cosh x}{\sinh x}, \;\; x \neq 0 \qquad \text{csch}\, x = \frac{1}{\sinh x}, \;\; x \neq 0$$

We show the graphs of these four functions in Figures 7.35, 7.36, 7.37, and 7.38.

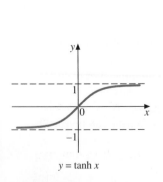

$y = \tanh x$

FIGURE 7.35

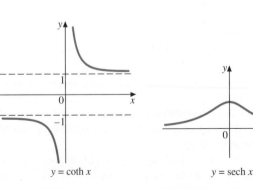

$y = \coth x$

FIGURE 7.36

$y = \text{sech}\, x$

FIGURE 7.37

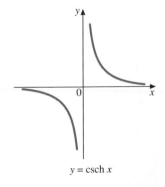

$y = \text{csch}\, x$

FIGURE 7.38

Hyperbolic Identities

There are many relationships among the hyperbolic functions, called hyperbolic identities, that bear a striking resemblance to trigonometric identities. We list the most important ones here.

Basic Hyperbolic Identities

1. $\cosh^2 x - \sinh^2 x = 1$

2. $\sinh(x + y) = \sinh x \cosh y + \cosh x \sinh y$

3. $\cosh(x + y) = \cosh x \cosh y + \sinh x \sinh y$

4. $\sinh 2x = 2 \sinh x \cosh x$

5. $\cosh 2x = \cosh^2 x + \sinh^2 x = 2\cosh^2 x - 1 = 1 + 2\sinh^2 x$

6. $\sinh \dfrac{x}{2} = \pm\sqrt{\dfrac{\cosh x - 1}{2}}$

7. $\cosh \dfrac{x}{2} = \sqrt{\dfrac{\cosh x + 1}{2}}$

8. $1 - \tanh^2 x = \operatorname{sech}^2 x$

9. $\coth^2 x - 1 = \operatorname{csch}^2 x$

10. $\tanh 2x = \dfrac{2 \tanh x}{1 + \tanh^2 x}$

11. $\tanh \dfrac{x}{2} = \dfrac{\sinh x}{1 + \cosh x} = \dfrac{\cosh x - 1}{\sinh x}$

To verify Identity 1, we use the definitions of $\sinh x$ and $\cosh x$ to get

$$\cosh^2 x - \sinh^2 x = \left(\frac{e^x + e^{-x}}{2}\right)^2 - \left(\frac{e^x - e^{-x}}{2}\right)^2$$

$$= \frac{e^{2x} + 2 + e^{-2x}}{4} - \frac{e^{2x} - 2 + e^{-2x}}{4} = \frac{4}{4} = 1$$

For Identity 2, it is easier to show that the right-hand side equals the left:

$$\sinh x \cosh y + \cosh x \sinh y = \frac{e^x - e^{-x}}{2} \cdot \frac{e^y + e^{-y}}{2} + \frac{e^x + e^{-x}}{2} \cdot \frac{e^y - e^{-y}}{2}$$

After multiplying terms and simplifying, using the laws of exponents, we get

$$\frac{e^{x+y} - e^{-(x+y)}}{2}$$

which, by Definition 7.3, is $\sinh(x + y)$. Identity 3 can be verified similarly. The remaining identities can be verified using Identities 1, 2, and 3, along with Definition 7.4.

As we have noted, the hyperbolic identities are very similar to trigonometric identities. Some have exactly the same form; for example,

$$\sinh 2x = 2 \sinh x \cosh x \qquad \text{and} \qquad \sin 2x = 2 \sin x \cos x$$

Others differ only by a sign; for example,

$$\cosh^2 x - \sinh^2 x = 1 \quad \text{and} \quad \cos^2 x + \sin^2 x = 1$$

These similarities provide further explanation for the fact that the names of the hyperbolic functions involve the names of the trigonometric functions.

Derivatives of Hyperbolic Functions

To find the derivatives of $\sinh x$ and $\cosh x$, we use Definition 7.3 again. We have

$$\frac{d}{dx}(\sinh x) = \frac{d}{dx}\left(\frac{e^x - e^{-x}}{2}\right) = \frac{e^x + e^{-x}}{2} = \cosh x$$

and

$$\frac{d}{dx}(\cosh x) = \frac{d}{dx}\left(\frac{e^x + e^{-x}}{2}\right) = \frac{e^x - e^{-x}}{2} = \sinh x$$

Again, observe the similarities with the derivatives of $\sin x$ and $\cos x$. The derivatives of the other hyperbolic functions can now be obtained from the derivatives of $\sinh x$ and $\cosh x$, along with Definition 7.4. We ask you to derive these in the exercises. We list the results in the following box, where we have given their generalized forms, using the Chain Rule, with u a differentiable function of x.

Derivatives of Hyperbolic Functions

$$\frac{d}{dx}\sinh u = \cosh u \,\frac{du}{dx} \qquad\qquad (7.47)$$

$$\frac{d}{dx}\cosh u = \sinh u \,\frac{du}{dx} \qquad\qquad (7.48)$$

$$\frac{d}{dx}\tanh u = \operatorname{sech}^2 u \,\frac{du}{dx} \qquad\qquad (7.49)$$

$$\frac{d}{dx}\coth u = -\operatorname{csch}^2 u \,\frac{du}{dx} \qquad\qquad (7.50)$$

$$\frac{d}{dx}\operatorname{sech} u = -\operatorname{sech} u \tanh u \,\frac{du}{dx} \qquad\qquad (7.51)$$

$$\frac{d}{dx}\operatorname{csch} u = -\operatorname{csch} u \coth u \,\frac{du}{dx} \qquad\qquad (7.52)$$

EXAMPLE 7.35   Find dy/dx for the following:

(a) $y = \sinh(\ln x)$ (b) $y = x - \tanh x$

Solution

(a) By Formula 7.47,

$$\frac{dy}{dx} = \cosh(\ln x) \cdot \frac{d}{dx}(\ln x) = \frac{1}{x}\cosh(\ln x)$$

(b) By Formula 7.49 and Identity 8,

$$\frac{dy}{dx} = 1 - \text{sech}^2 x = \tanh^2 x$$

∎

Integrals of Hyperbolic Functions

From the derivative formulas, we immediately have the following antiderivative formulas (indefinite integrals):

Integrals of Hyperbolic Functions

$$\int \sinh u \, du = \cosh u + C \qquad (7.53)$$

$$\int \cosh u \, du = \sinh u + C \qquad (7.54)$$

$$\int \text{sech}^2 u \, du = \tanh u + C \qquad (7.55)$$

$$\int \text{csch}^2 u \, du = -\coth u + C \qquad (7.56)$$

$$\int \text{sech}\, u \tanh u \, du = -\text{sech}\, u + C \qquad (7.57)$$

$$\int \text{csch}\, u \coth u \, du = -\text{csch}\, u + C \qquad (7.58)$$

EXAMPLE 7.36   Evaluate the following integrals:

(a) $\displaystyle\int \frac{\sinh x}{1 + \cosh x}\, dx$ (b) $\displaystyle\int_0^{\ln 2} \frac{\text{sech}^2 x}{1 + \tanh^2 x}\, dx$ (c) $\displaystyle\int \cosh^2 x \, dx$

Solution

(a) If we let $u = 1 + \cosh x$, then $du = \sinh x \, dx$. So we have

$$\int \frac{\sinh x}{1 + \cosh x}\, dx = \int \frac{du}{u} = \ln|u| + C = \ln(1 + \cosh x) + C$$

Absolute values are not necessary since $1 + \cosh x$ is always positive (remember that $\cosh x \geq 1$).

(b) Since $\dfrac{d}{dx}(\tanh x) = \operatorname{sech}^2 x$, we observe that the integral is in the form

$$\int \frac{du}{1 + u^2}$$

with $u = \tanh x$. The antiderivative is therefore $\tan^{-1} u$ (by Equation 7.43). So we have

$$\int_0^{\ln 2} \frac{\operatorname{sech}^2 x\, dx}{1 + \tanh^2 x} = \tan^{-1}(\tanh x)\big]_0^{\ln 2}$$

$$= \tan^{-1}[\tanh(\ln 2) - \tanh 0]$$

Since

$$\tanh x = \frac{\sinh x}{\cosh x} = \frac{e^x - e^{-x}}{e^x + e^{-x}}$$

we have $\tanh 0 = 0$, and

$$\tanh(\ln 2) = \frac{e^{\ln 2} - e^{-\ln 2}}{e^{\ln 2} + e^{-\ln 2}} = \frac{2 - \frac{1}{2}}{2 + \frac{1}{2}} = \frac{3}{5}$$

Thus,

$$\int_0^{\ln 2} \frac{\operatorname{sech}^2 x}{1 + \tanh^2 x}\, dx = \tan^{-1}\frac{3}{5} \approx 0.5404$$

(c) From Identity 5, we can write

$$\cosh^2 x = \frac{\cosh 2x + 1}{2}$$

So

$$\int \cosh^2 x\, dx = \frac{1}{2}\int (\cosh 2x + 1)dx$$

$$= \frac{1}{2}\cdot\frac{1}{2}\int \cosh 2x(2\,dx) + \frac{1}{2}\int dx$$

$$= \frac{1}{4}\sinh 2x + \frac{1}{2}x + C \qquad\blacksquare$$

EXAMPLE 7.37   High-tension wires hang in the form of a catenary between poles that are 20 m apart. With the axes chosen as in Figure 7.39 (x-axis on the ground), the equation of the catenary is $y = 10\cosh(x/10)$.

(a) Find the height of the wire at its lowest and highest points.

(b) Find the length of the wire between the two poles.

Solution

(a) The lowest point occurs at $x = 0$ and is given by

$$y_{\min} = 10\cosh\frac{0}{10} = 10 \quad \text{Since } \cosh 0 = 1$$

The curve assumes its maximum height at each end. So

$$y_{\max} = 10\cosh\frac{10}{10} = 10\cosh 1 = 10\cdot\frac{e + e^{-1}}{2} = 5\left(e + \frac{1}{e}\right) \approx 15.43$$

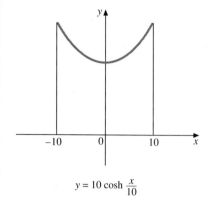

$y = 10 \cosh \dfrac{x}{10}$

FIGURE 7.39

(b) The total length is twice the length from $x = 0$ to $x = 10$. So we have

$$L = 2 \int_{x=0}^{x=10} ds = 2 \int_0^{10} \sqrt{1 + \left(\frac{dy}{dx}\right)^2}\, dx \quad \text{By Formula 6.18}$$

$$= 2 \int_0^{10} \sqrt{1 + \sinh^2 \frac{x}{10}}\, dx$$

$$= 2 \int_0^{10} \cosh \frac{x}{10}\, dx \qquad \text{By Identity 1}$$

$$= 20 \int_0^{10} \cosh \frac{x}{10} \left(\frac{1}{10}\, dx\right)$$

$$= 20 \sinh \frac{x}{10} \Big]_0^{10} = 20(\sinh 1 - \sinh 0)$$

$$= 20 \sinh 1 = 20 \left(\frac{e - e^{-1}}{2}\right) = 10 \left(e - \frac{1}{e}\right) \approx 23.50$$

$\blacksquare$

Origin of the Name "Hyperbolic"

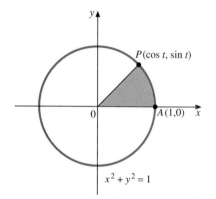

FIGURE 7.40

To understand why the functions we have been studying in this section are called *hyperbolic* functions, let us first recall how the trigonometric functions $\sin t$ and $\cos t$ are defined in terms of the unit circle $x^2 + y^2 = 1$. In Section 1.6 we saw that the point P that is t units along the circle from $A(1, 0)$ has coordinates $(\cos t, \sin t)$, as we indicate in Figure 7.40. These coordinates therefore satisfy the equation of the circle, giving the well-known identity

$$\cos^2 t + \sin^2 t = 1$$

Because of this relationship of $\sin t$ and $\cos t$ to the unit circle, these functions (as well as the other trigonometric functions) are sometimes called *circular* functions.

Since the hyperbolic functions $\sinh t$ and $\cosh t$ satisfy the identity

$$\cosh^2 t - \sinh^2 t = 1$$

it follows that the point P with coordinates $(\cosh t, \sinh t)$ lies on the "unit" hyperbola

$$x^2 - y^2 = 1$$

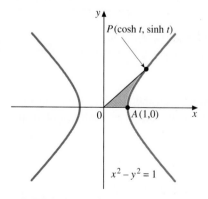

FIGURE 7.41

whose graph is shown in Figure 7.41. The point P is on the right-hand branch, since $\cosh t$ is always positive. It is above the x-axis when $t > 0$ and below when $t < 0$.

The geometric analogy between the relationship of trigonometric functions to the unit circle and hyperbolic functions to the unit hyperbola goes even further. The shaded areas in Figures 7.40 and 7.41 are called a *circular sector* and a *hyperbolic sector*, respectively. For $0 < t < 2\pi$, the area of each sector can be shown to be $t/2$; that is, in each case

$$\text{area of sector } OAP = \frac{t}{2}$$

(See Exercises 54 and 55 in Exercise Set 7.6.)

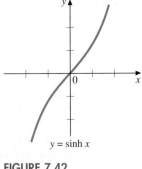

$y = \sinh x$

FIGURE 7.42

Inverse Hyperbolic Functions

Just as we defined inverses of the trigonometric functions, after suitable restrictions on their domains, we can also define inverses of the hyperbolic functions. Restrictions on the domain are not necessary in all cases. Consider first $\sinh x$. Since $\frac{d}{dx}(\sinh x) = \cosh x$ and $\cosh x > 0$ for all x, it follows that $\sinh x$ is always increasing, so it is one-to-one. (See Figure 7.42.) It therefore has an inverse, which we designate by $\sinh^{-1}$. Thus,

$$y = \sinh^{-1} x \quad \text{if and only if} \quad x = \sinh y$$

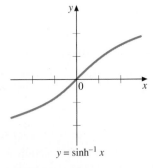

$y = \sinh^{-1} x$

FIGURE 7.43

The graph of $y = \sinh^{-1} x$ is shown in Figure 7.43.

To find the derivative of $\sinh^{-1} x$, we set $y = \sinh^{-1} x$ and solve for x to get

$$x = \sinh y$$

Now we differentiate implicitly.

$$1 = (\cosh y)y'$$

$$y' = \frac{1}{\cosh y}$$

By Identity 1, $\cosh^2 y = 1 + \sinh^2 y$. Also, $\sinh y = x$, so $\cosh^2 y = 1 + x^2$. We therefore have

$$\frac{d}{dx}(\sinh^{-1} x) = \frac{1}{\sqrt{1 + x^2}}$$

The hyperbolic cosine is not 1–1 (apply the horizontal line test to its graph in Figure 7.33). But when we restrict its domain to the interval $[0, \infty)$, it is then 1–1. So it has an inverse, designated by $\cosh^{-1}$. Thus,

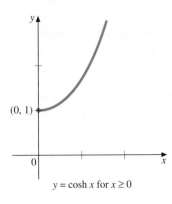

$y = \cosh x$ for $x \geq 0$

$$y = \cosh^{-1} x \quad \text{if and only if} \quad x = \cosh y, \text{ with } y \geq 0$$

We show the graphs of the restricted hyperbolic cosine and its inverse in Figure 7.44. Differentiating both sides of the equation $x = \cosh y$ and using Identity 1 again, we obtain

$$1 = (\sinh y)y'$$

$$y' = \frac{1}{\sinh y} = \frac{1}{\sqrt{\cosh^2 y - 1}} \qquad y > 0$$

$(1, 0)$

$y = \cosh^{-1} x$

FIGURE 7.44

Since $\cosh y = x$, we have

$$\frac{d}{dx}\cosh^{-1}x = \frac{1}{\sqrt{x^2-1}} \qquad x > 1$$

Derivatives of the other inverse hyperbolic functions can be obtained in a similar way. We will ask you to draw their graphs and to derive formulas for their derivatives in the exercises. Just as with $\cosh x$, we restrict the domain of $\operatorname{sech} x$ to $[0,\infty)$ in order to define its inverse. Except for these two, the hyperbolic functions are 1–1 on their entire domains. We summarize all the formulas for derivatives of the inverse hyperbolic functions in the following box, replacing x by u, where u is a differentiable function of x.

Derivatives of Inverse Hyperbolic Functions

$$\frac{d}{dx}\sinh^{-1}u = \frac{1}{\sqrt{1+u^2}}\frac{du}{dx} \qquad\qquad (7.59)$$

$$\frac{d}{dx}\cosh^{-1}u = \frac{1}{\sqrt{u^2-1}}\frac{du}{dx}, \qquad u > 1 \qquad (7.60)$$

$$\frac{d}{dx}\tanh^{-1}u = \frac{1}{1-u^2}\frac{du}{dx}, \qquad |u| < 1 \qquad (7.61)$$

$$\frac{d}{dx}\coth^{-1}u = \frac{1}{1-u^2}\frac{du}{dx}, \qquad |u| > 1 \qquad (7.62)$$

$$\frac{d}{dx}\operatorname{sech}^{-1}u = -\frac{1}{u\sqrt{1-u^2}}\frac{du}{dx}, \qquad 0 < u < 1 \qquad (7.63)$$

$$\frac{d}{dx}\operatorname{csch}^{-1}u = -\frac{1}{|u|\sqrt{1+u^2}}\frac{du}{dx}, \qquad u \neq 0 \qquad (7.64)$$

EXAMPLE 7.38   Find the derivative of each of the following:

(a) $y = \sinh^{-1}(\tan x)$, $\quad -\pi/2 < x < \pi/2$

(b) $y = \tanh^{-1}\dfrac{1+x}{1-x}$, $\quad x \neq 1$

Solution

(a) By Formula 7.59,

$$\frac{dy}{dx} = \frac{1}{\sqrt{1+\tan^2 x}}\cdot\sec^2 x = \frac{1}{\sec x}\cdot\sec^2 x = \sec x$$

Here we used the fact that $\sqrt{1+\tan^2 x} = \sqrt{\sec^2 x} = \sec x$, since $\sec x > 0$ for x in $(-\pi/2, \pi/2)$.

(b) By Formula 7.61,

$$\frac{dy}{dx} = \frac{1}{1 - \left(\dfrac{1+x}{1-x}\right)^2} \cdot \frac{(1-x) - (1+x)(-1)}{(1-x)^2}$$

$$= \frac{(1-x)^2}{(1-x)^2 - (1+x)^2} \cdot \frac{2}{(1-x)^2}$$

$$= \frac{2}{(1-x)^2 - (1+x)^2}$$

$$= \frac{2}{[(1-x) + (1+x)][(1-x) - (1+x)]}$$

$$= -\frac{1}{2x}$$

Note how we used the algebraic identity $a^2 - b^2 = (a+b)(a-b)$ to simplify the denominator in the next-to-last step.   ∎

Integral Formulas Involving Inverse Hyperbolic Functions

One of the primary uses of inverse hyperbolic functions is in integration. From the derivative formulas we have the following antiderivative formulas.

Integral Formulas

$$\int \frac{du}{\sqrt{1+u^2}} = \sinh^{-1} u + C \tag{7.65}$$

$$\int \frac{du}{\sqrt{u^2-1}} = \cosh^{-1} u + C, \qquad u > 1 \tag{7.66}$$

$$\int \frac{du}{1-u^2} = \begin{cases} \tanh^{-1} u + C, & \text{if } |u| < 1 \\ \coth^{-1} u + C, & \text{if } |u| > 1 \end{cases} \tag{7.67}$$

$$\int \frac{du}{u\sqrt{1-u^2}} = -\operatorname{sech}^{-1} u + C, \qquad 0 < u < 1 \tag{7.68}$$

$$\int \frac{du}{u\sqrt{1+u^2}} = -\operatorname{csch}^{-1} u + C, \qquad u > 0 \tag{7.69}$$

Before giving examples involving integration, we give alternative ways of writing the inverse hyperbolic functions, involving natural logarithms. These formulas will be useful in integration problems. Since the hyperbolic functions can be expressed in terms of exponentials, it is not surprising that their inverses can be expressed in terms of logarithms. The next theorem makes this representation explicit.

THEOREM 7.7

The inverse hyperbolic functions have the following explicit representations:

$$\sinh^{-1} x = \ln(x + \sqrt{x^2 + 1}) \qquad -\infty < x < \infty \qquad (7.70)$$

$$\cosh^{-1} x = \ln(x + \sqrt{x^2 - 1}) \qquad x \geq 1 \qquad (7.71)$$

$$\tanh^{-1} x = \frac{1}{2} \ln \frac{1 + x}{1 - x} \qquad |x| < 1 \qquad (7.72)$$

$$\coth^{-1} x = \frac{1}{2} \ln \frac{x + 1}{x - 1} \qquad |x| > 1 \qquad (7.73)$$

$$\text{sech}^{-1} x = \ln \left(\frac{1 + \sqrt{1 - x^2}}{x} \right) \qquad 0 < x \leq 1 \qquad (7.74)$$

$$\text{csch}^{-1} x = \ln \left(\frac{1}{x} + \frac{\sqrt{x^2 + 1}}{|x|} \right) \qquad x \neq 0 \qquad (7.75)$$

We will prove Equation 7.70 and leave the others as exercises.

Proof of Equation 7.70   Let $y = \sinh^{-1} x$. Then $x = \sinh y$, so by the definition of $\sinh y$,

$$x = \frac{e^y - e^{-y}}{2}$$

Clearing the equation of fractions and rearranging, we get

$$e^y - 2x - e^{-y} = 0$$

Now we multiply both sides by e^y:

$$e^{2y} - 2xe^y - 1 = 0$$

This equation is quadratic in e^y (with coefficients $a = 1$, $b = -2x$, and $c = -1$), so by the Quadratic Formula,

$$e^y = \frac{2x \pm \sqrt{4x^2 + 4}}{2} = x \pm \sqrt{x^2 + 1}$$

But $e^y > 0$ and $\sqrt{x^2 + 1} > x$, so the negative sign cannot hold. Thus,

$$e^y = x + \sqrt{x^2 + 1}$$

In the equivalent logarithmic form,

$$y = \ln(x + \sqrt{x^2 + 1}) \qquad \blacksquare$$

By Theorem 7.7, each of the integral formulas (7.65–7.69) can be written in an equivalent form involving logarithms. We call attention in particular to the fact that the two parts of Formula 7.67 can be written in the single form

$$\int \frac{du}{1 - u^2} = \frac{1}{2} \ln \left| \frac{1 + u}{1 - u} \right| + C, \qquad u \neq 1 \qquad (7.76)$$

You will be asked to verify this result in the exercises. We should note that because the integral formulas can be given in different ways, the answers given by computer algebra systems will vary, though they are equivalent.

EXAMPLE 7.39   Evaluate the following integrals:

(a) $\displaystyle\int \frac{x}{\sqrt{x^4 + 1}} dx$ (b) $\displaystyle\int_1^{7/2} \frac{dx}{\sqrt{4x^2 - 1}}$

Solution

(a) Let $u = x^2$. Then $du = 2x\,dx$, and we have, by Equation 7.65,

$$\int \frac{x}{\sqrt{x^4 + 1}} dx = \frac{1}{2} \int \frac{2x\,dx}{\sqrt{x^4 + 1}} = \frac{1}{2} \int \frac{du}{\sqrt{u^2 + 1}} = \frac{1}{2} \sinh^{-1} u + C$$

$$= \frac{1}{2} \sinh^{-1} x^2 + C$$

(b) Let $u = 2x$; then $du = 2\,dx$. When $x = 1$, $u = 2$, and when $x = 7/2$, $u = 7$. So we have

$$\int_1^{7/2} \frac{dx}{\sqrt{4x^2 - 1}} = \frac{1}{2} \int_2^7 \frac{du}{\sqrt{u^2 - 1}} = \frac{1}{2} \cosh^{-1} u \Big]_2^7 \qquad \text{By Equation 7.66}$$

$$= \frac{1}{2} [\cosh^{-1} 7 - \cosh^{-1} 2]$$

$$= \frac{1}{2} [\ln(7 + \sqrt{48}) - \ln(2 + \sqrt{3})] \qquad \text{By Equation 7.71}$$

$$= \frac{1}{2} [\ln(7 + 4\sqrt{3}) - \ln(2 + \sqrt{3})] = \frac{1}{2} \ln \frac{7 + 4\sqrt{3}}{2 + \sqrt{3}}$$

$$= \frac{1}{2} \ln \left[\frac{7 + 4\sqrt{3}}{2 + \sqrt{3}} \cdot \frac{2 - \sqrt{3}}{2 - \sqrt{3}} \right] = \frac{1}{2} \ln \frac{14 + \sqrt{3} - 12}{4 - 3}$$

$$= \frac{1}{2} \ln(2 + \sqrt{3}) \qquad\blacksquare$$

Exercise Set 7.6

1. Give the domain and range of each of the hyperbolic functions.

2. Express $\tanh x$, $\coth x$, $\operatorname{sech} x$, and $\operatorname{csch} x$ in terms of exponential functions, and evaluate the following limits:
   (a) $\lim_{x \to \infty} \tanh x$ (b) $\lim_{x \to 0^+} \coth x$
   (c) $\lim_{x \to \infty} \coth x$ (d) $\lim_{x \to \infty} \operatorname{sech} x$
   (e) $\lim_{x \to 0^+} \operatorname{csch} x$ (f) $\lim_{x \to \infty} \operatorname{csch} x$

3. Use a calculator or a CAS to approximate the value of each of the following for $x = 0.5$, 1, 2, and 4:
   (a) $\tanh x$ (b) $\coth x$
   (c) $\operatorname{sech} x$ (d) $\operatorname{csch} x$

4. Show that sech is an even function and that tanh, coth, and csch are odd functions.

In Exercises 5–12, verify the hyperbolic identities.

5. $\cosh(x+y) = \cosh x \cosh y + \sinh x \sinh y$

6. $\sinh 2x = 2 \sinh x \cosh x$

7. $\cosh 2x = \cosh^2 x + \sinh^2 x$

$$= 2\cosh^2 x - 1$$

$$= 1 + 2\sinh^2 x$$

8. $\sinh \dfrac{x}{2} = \pm \sqrt{\dfrac{\cosh x - 1}{2}}$

9. $\cosh \dfrac{x}{2} = \sqrt{\dfrac{\cosh x + 1}{2}}$

10. $1 - \tanh^2 x = \operatorname{sech}^2 x$

11. $\coth^2 x - 1 = \operatorname{csch}^2 x$

12. $\tanh 2x = \dfrac{2\tanh x}{1 + \tanh^2 x}$

In Exercises 13–24, find dy/dx.

13. $y = \tanh^3 2x$

14. $y = \dfrac{\cosh x}{1 - \sinh x}$

15. $y = \ln(\sinh x)$

16. $y = \tan^{-1}(\sinh x)$

17. $y = \sec^{-1}(\cosh x)$

18. $y = x - \coth x$

19. $y = \sqrt{\operatorname{sech} x^2}$

20. $y = \dfrac{\tanh x}{1 + \operatorname{sech} x}$

21. $x \cosh y - y \sinh x = 1$

22. $y^2 = \ln(x \cosh y)$

23. $y = (\sinh x)^x$

24. $y = (\sinh x)^{\operatorname{sech} x}$

In Exercises 25–42, evaluate the integrals.

25. $\displaystyle \int \sinh x \cosh x \, dx$

26. $\displaystyle \int \dfrac{1 + \tanh x}{\cosh^2 x} \, dx$

27. $\displaystyle \int \dfrac{\sinh x \, dx}{\sqrt{1 + \sinh^2 x}}$

28. $\displaystyle \int \sinh^2 x \, dx$

29. $\displaystyle \int \dfrac{1 + \cosh x}{x + \sinh x} \, dx$

30. $\displaystyle \int e^{\sinh^2 x} \sinh 2x \, dx$

31. $\displaystyle \int \dfrac{\tanh x}{\ln \cosh x} \, dx$

32. $\displaystyle \int x \sinh x^2 \cosh^2 x^2 \, dx$

33. $\displaystyle \int \sinh^3 x \, dx$

34. $\displaystyle \int \tanh^2 x \, dx$

35. $\displaystyle \int \cosh^3 x \, dx$

36. $\displaystyle \int \operatorname{sech}^3 x \tanh x \, dx$

37. $\displaystyle \int \operatorname{sech}^4 x \tanh x \, dx$

38. $\displaystyle \int \cosh^2 x \sinh^3 x \, dx$

39. $\displaystyle \int_{\ln 2}^{\ln 3} \coth x \, dx$

40. $\displaystyle \int_{-1}^{1} (\sinh x + \cosh x)^2 \, dx$

41. $\displaystyle \int_{0}^{\ln 2} \dfrac{\cosh x}{\sqrt{9 - 4\sinh^2 x}} \, dx$

42. $\displaystyle \int_{-\ln 3}^{\ln 3} \dfrac{\sinh x}{1 + \cosh^2 x} \, dx$

In Exercises 43–45, evaluate the limits.

43. (a) $\displaystyle \lim_{x \to 0} \dfrac{\cosh x - x^2 - 1}{\sinh^2 x}$ (b) $\displaystyle \lim_{x \to 1^-} \dfrac{\tanh^{-1} x}{\ln(1 - x)}$

44. (a) $\displaystyle \lim_{x \to \infty} x^2 \operatorname{sech} x$

(b) $\displaystyle \lim_{x \to 0} (\operatorname{csch} x - \coth x)$

45. (a) $\displaystyle \lim_{x \to 0^+} (\sinh x)^x$

(b) $\displaystyle \lim_{x \to \infty} (\tanh x)^{\cosh x}$

46. Find the volume of the solid obtained by revolving the region under the graph of $y = \operatorname{sech} x$ between $x = -1$ and $x = 1$ about the x-axis.

47. Find the volume of the solid obtained by revolving the region under the graph of $y = \tanh x$ between $x = 0$ and $x = 2$ about the x-axis.

48. Find the surface area obtained by revolving the arc of the curve $y = \cosh x$ from $x = 0$ to $x = 2$ about the x-axis.

In Exercises 49–52, verify that the equation is an identity for all admissible values of x by transforming the left-hand side into the right-hand side.

49. $\sinh x (\coth x - \tanh x) = \operatorname{sech} x$

50. $\dfrac{\sinh x}{\coth x - \operatorname{csch} x} = \cosh x + 1$

51. $\dfrac{2}{\coth x - \tanh x} = \sinh 2x$

52. $\dfrac{\coth x + \tanh x}{\coth x - \tanh x} = \cosh 2x$

53. Verify the identities

$$\tanh \frac{x}{2} = \frac{\sinh x}{1 + \cosh x}$$

and

$$\tanh \frac{x}{2} = \frac{\cosh x - 1}{\sinh x}$$

54. Prove that the area of the circular sector OAP shown in the figure is $t/2$, where $0 < t < \pi/2$. (*Hint:* Make use of your knowledge of the area of the circle.)

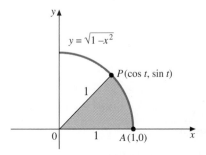

55. Prove that the area of the hyperbolic sector OAP shown in the figure is $t/2$, where $t > 0$, by completing the following steps.

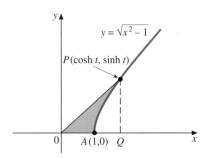

(a) Find the area of the triangle OPQ.
(b) Find the area under the arc AP by using Area $= \int_{1}^{\cosh t} y \, dx$. To evaluate the integral, substitute $x = \cosh u$.
(c) Subtract the area in part (b) from the area in part (a).

In Exercises 56–59, evaluate the integrals.

56. $\displaystyle \int \sinh^4 x \, dx$ **57.** $\displaystyle \int \tanh^5 x \, dx$

58. $\displaystyle \int \frac{dx}{\sech^2 x - \sech^4 x}$

59. $\displaystyle \int \frac{dx}{\cosh x - 1}$ (*Hint:* Multiply numerator and denominator by $\cosh x + 1$.)

In Exercises 60 and 61, derive the formulas.

60. $\int \sech x \, dx = \tan^{-1} |\sinh x| + C$
(*Hint:* Write $\sech x = \cosh x / \cosh^2 x$.)

61. $\int \csch x \, dx = \ln |\tanh \frac{x}{2}| + C$.   (*Hint:* Write $\csch x = \sinh x / \sinh^2 x$.)

In Exercises 62–64, verify that the given equation is an identity for all admissible values of x.

62. $\cosh^4 x = \dfrac{3}{8} + \dfrac{\cosh 2x}{2} + \dfrac{\cosh 4x}{8}$

63. $\dfrac{\sinh 3x}{\sinh x} - \dfrac{\cosh 3x}{\cosh x} = 2$

64. $2 \cosh^2 \dfrac{x}{2} \csch x = \coth \dfrac{x}{2}$

65. A uniform flexible cable suspended between two supporting towers 60 m apart hangs in the form of a catenary with the equation $y = 20 \cosh \frac{x}{20} + 10$, where y is the distance above the ground.
(a) Find the height of the cable at its lowest point.
(b) Find the height at each end.
(c) Determine the length of the cable.

66. (a) Show that $y = A \cosh \omega t + B \sinh \omega t$ satisfies the differential equation

$$\frac{d^2 y}{dt^2} - \omega^2 y = 0$$

(b) In part (a), find A and B such that when $t = 0$, $y = 3$ and $y' = 0$.

In Exercises 67–70, give the domain and range of f and draw its graph.

67. $f(x) = \tanh^{-1} x$ **68.** $f(x) = \coth^{-1} x$

69. $f(x) = \sech^{-1} x$ **70.** $f(x) = \csch^{-1} x$

71. (a) Prove $\sech^{-1} x = \cosh^{-1} \dfrac{1}{x}$, $x > 0$.
(b) Prove $\csch^{-1} x = \sinh^{-1} \dfrac{1}{x}$, $x \neq 0$.

In Exercises 72–75, derive the formula for the derivative.

72. $\dfrac{d}{dx} \tanh^{-1} x$ **73.** $\dfrac{d}{dx} \coth^{-1} x$

74. $\dfrac{d}{dx} \sech^{-1} x$ **75.** $\dfrac{d}{dx} \csch^{-1} x$

In Exercises 76–85, find y′.

76. $y = \sinh^{-1}\sqrt{x}$

77. $y = \cosh^{-1}x^2$

78. $y = \tanh^{-1}\dfrac{1}{x}$

79. $y = \mathrm{sech}^{-1}(\cos x), \quad 0 < x < \dfrac{\pi}{2}$

80. $y = \coth^{-1}\dfrac{x-1}{x+1}$

81. $y = x\,\mathrm{csch}^{-1}x, \quad x > 0$

82. $y = \cosh^{-1}(\ln x)$

83. $y = \mathrm{sech}^{-1}\sqrt{x+1}, \quad -1 < x < 0$

84. $y = (\sinh^{-1}2x)^3$

85. $y = \tanh^{-1}\dfrac{e^x}{1+e^x}$

In Exercises 86–95, evaluate the integrals.

86. $\displaystyle\int \dfrac{dx}{\sqrt{1+9x^2}}$

87. $\displaystyle\int \dfrac{dx}{\sqrt{4x^2-1}}$

88. $\displaystyle\int_{\ln 1/4}^{\ln 1/2} \dfrac{e^x}{1-e^{2x}}\,dx$

89. $\displaystyle\int \dfrac{dx}{x\sqrt{1-4x^2}}$

90. $\displaystyle\int \dfrac{\sin x\,dx}{\sqrt{1+\cos^2 x}}$

91. $\displaystyle\int \dfrac{\sec^2 x\,dx}{\sqrt{\tan^2 x-1}} \quad \left(\dfrac{\pi}{4} < x < \dfrac{\pi}{2}\right)$

92. $\displaystyle\int \dfrac{\sec^2 x}{\tan^2 x-1}\,dx \quad \left(-\dfrac{\pi}{4} < x < \dfrac{\pi}{4}\right)$

93. $\displaystyle\int_0^{1/2} \dfrac{x\,dx}{1-x^4}$

94. $\displaystyle\int_0^5 \dfrac{dx}{\sqrt{x^2+4x+3}}$

95. $\displaystyle\int \dfrac{dx}{\sqrt{x^2+2x+2}}$

In Exercises 96–100, verify the given antidifferentiation formula, where a > 0.

96. $\displaystyle\int \dfrac{du}{\sqrt{a^2+u^2}} = \sinh^{-1}\dfrac{u}{a} + C$

97. $\displaystyle\int \dfrac{du}{\sqrt{u^2-a^2}} = \cosh^{-1}\dfrac{u}{a} + C, \ u > a$

98. $\displaystyle\int \dfrac{du}{a^2-u^2} = \begin{cases} \dfrac{1}{a}\tanh^{-1}\dfrac{u}{a} + C & \text{if } |u| < a \\[2mm] \dfrac{1}{a}\coth^{-1}\dfrac{u}{a} + C & \text{if } |u| > a \end{cases}$

99. $\displaystyle\int \dfrac{du}{u\sqrt{a^2-u^2}} = -\dfrac{1}{a}\,\mathrm{sech}^{-1}\dfrac{u}{a} + C, \ 0 < u < a$

100. $\displaystyle\int \dfrac{du}{u\sqrt{a^2+u^2}} = -\dfrac{1}{a}\,\mathrm{csch}^{-1}\dfrac{u}{a} + C, \ u > 0$

In Exercises 101–106, use the results of Exercises 96–100 to evaluate the integrals.

101. $\displaystyle\int \dfrac{dx}{\sqrt{4+9x^2}}$

102. $\displaystyle\int \dfrac{dx}{\sqrt{4x^2-9}}$

103. $\displaystyle\int_0^1 \dfrac{dx}{9-4x^2}$

104. $\displaystyle\int_3^4 \dfrac{dx}{x\sqrt{25-x^2}}$

105. $\displaystyle\int \dfrac{dx}{\sqrt{4+e^{2x}}}$

106. $\displaystyle\int \dfrac{dx}{x^2-6x+5}$

107. Prove Equations 7.71 and 7.72.

108. Prove Equations 7.73 and 7.74.

109. Prove Equations 7.75 and 7.76.

110. Prove $\dfrac{d}{dx}\,\mathrm{sech}^{-1}|x| = -\dfrac{1}{x\sqrt{1-x^2}}$ if $|x| < 1$ and $x \neq 0$. Using this result, show that the formula of Exercise 99 can be generalized to

$$\int \dfrac{du}{u\sqrt{a^2-u^2}} = -\dfrac{1}{a}\,\mathrm{sech}^{-1}\dfrac{|u|}{a} + C, \qquad 0 < |u| < a$$

111. Following the idea of Exercise 110, show that

$$\int \dfrac{du}{u\sqrt{a^2+u^2}} = -\dfrac{1}{a}\,\mathrm{csch}^{-1}\dfrac{|u|}{a} + C, \qquad u \neq 0$$

112. Make use of inverse hyperbolic functions to obtain the result

$$\int \sec u\,du = \ln|\sec u + \tan u| + C$$

(*Hint:* $\sec u = \dfrac{1}{\cos u} = \dfrac{\cos u}{\cos^2 u} = \dfrac{\cos u}{1-\sin^2 u}$. Now use Equations 7.67 and 7.73.)

113. Consider a body falling under the influence of gravity, taking air resistance into consideration. The downward force is the weight of the object (mg, where m is its mass) and the resisting force is known experimentally to be approximately proportional to the square of the velocity. According to Newton's Second Law of Motion, the sum of forces acting on the body is its mass times its acceleration ($F = ma$). So we have

$$ma = mg - kv^2$$

Suppose that when $t = 0$, we have $v = 0$ and $s = 0$. (Measure distance positively downward from the initial position.) Using the fact that $a = dv/dt$ and $v = ds/dt$, find the following: (a) $v(t)$, (b) $s(t)$, and (c) the "terminal velocity" $\lim_{t \to \infty} v(t)$. (*Note:* We assume $mg > kv^2$.)

TRANSCENDENTAL FUNCTIONS USING COMPUTER ALGEBRA SYSTEMS

All the standard transcendental functions defined so far have definitions in every computer algebra system. The systems typically contain long lists of mathematical functions that can be accessed directly.

CAS 22

Let $f(x) = x^2 e^{1-x^2}$. Plot the graph and approximate $\displaystyle\int_0^4 x^2 e^{1-x^2}\, dx$.

To get a reasonable plot of the function, sketch the curve on the interval $[-4, 4]$ (see Figure 7.6.1).

Maple:

f:= x–>x^2*exp(1–x^2);
plot (f(x),x=–4..4,numpoints=500);

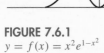

FIGURE 7.6.1
$y = f(x) = x^2 e^{1-x^2}$

Mathematica:

f[x_]=x^2*Exp[1–x^2]
Plot[f[x],{x,–4,4}]

DERIVE:

(At the □ symbol, go to the next step.)
a (author) □ x^2exp(1–x^2) □ p (plot window) □ p (plot)

The graph is shown in the margin. We see that the graph always lies above the x-axis, but that it approaches 0 very rapidly as x approaches plus or minus infinity. We can easily check this behavior using the computer algebra system.

Maple:

limit(f(x),x=infinity);   0
limit(f(x),x=–infinity);   0

Mathematica:

Limit[f[x], x–> Infinity]   0
Limit[f[x],x –> –Infinity]   0

DERIVE:

(At the □ symbol, go to the next step.)
c (calculus) □ l (limit) □ [choose expression] □ x (variable) □ inf (limit point) □ s (simplify) □ [choose expression]

Output: 0
c (calculus) □ l (limit) □ [choose expression] □ x (variable) □ [choose expression] □ –inf (limit point) □ s (simplify) □ [choose expression]
Output: 0

The computer algebra systems Maple, Mathematica, or DERIVE *cannot* find an antiderivative for the function f. In fact, the antiderivative cannot be expressed in terms of simple functions.

Maple:

int(f(x),x);

Output:

$$\left(-\frac{1}{2}\frac{x}{e^{x^2}} + \frac{1}{4}\sqrt{\pi}\,\mathrm{erf}(x) \right) e$$

Mathematica:

Integrate[f[x],x]

DERIVE:

int(x^2exp(1–x^2),x)
s (simplify) □
[choose expression]

In each case we see something of the form erf(x), which is called the error function and is another integral that cannot be evaluated.

The integral can be approximated by Maple, Mathematica, or DERIVE.

Maple:

evalf(int(f(x),x=0..4),6);

Mathematica:*

NIntegrate[f[x],{x,0,4}]

DERIVE:

int(x^2exp(1–x^2),x,0,4)
x (approximate)
[choose expression]

The approximate value for the integral is given as 1.20451 by Maple and Mathematica and as 1.20450 by DERIVE.

CAS 23

Approximate the volume of the solid of revolution obtained by rotating the region under $y = f(x) = \cosh(x)$ from $x = 0$ to $x = 2$ about the x axis.

Maple:

f := x–> cosh(x);
plot(f(x),x=–5..5,y=0..5);
evalf(int(Pi*f(x)^2,x=0..2));

The approximate volume is

24.57504350

Mathematica:

g[x_] = Cosh[x]
Plot[g[x],{x,–5,5},Plotrange–>{0,5}]
NIntegrate[Pi*(g[x])^2,
   {x,0,2}]

Output: 24.575

DERIVE:

(At the □ symbol, go to the next step.)
a (author) □ g(x):=cosh(x) □ p (plot window) □ p (plot) □ int(Pi(g(x))^2,x,0,2) □ x (approximate) □ [choose expression]

Output: 24.5750

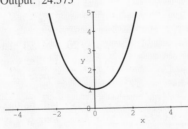

FIGURE 7.6.2
$y = f(x) = \cosh x$

---

*Mathematica's NIntegrate gives a numerical approximation of the definite integral.

Chapter 7 Review Exercises

In Exercises 1–7, find $f'(x)$ and give its domain.

1. (a) $f(x) = \ln(x\sqrt{x-2})$

(b) $f(x) = \ln\left(\dfrac{x^2 - x}{x^2 - 2x - 8}\right)$

2. (a) $f(x) = \sqrt{\ln(\sec^2 x)}$

(b) $f(x) = x \ln \dfrac{1}{x^2}$

3. (a) $f(x) = x^2 e^{-x}$

(b) $f(x) = e^{\ln(e^{-x})}$

4. (a) $f(x) = e^{\sin x + \ln x}$

(b) $f(x) = \ln\left(\dfrac{e^x - 1}{e^x + 1}\right)$

5. (a) $f(x) = \cos e^{-x} - \sin e^{-x}$

(b) $f(x) = (\ln \sqrt{e^x + e^{-x}})^2$

6. (a) $f(x) = x \log_3 x$

(b) $f(x) = \dfrac{2^x}{1 - 2^x}$

7. (a) $f(x) = x^{\ln x}$

(b) $f(x) = (\sec x)^{\tan x}$

8. Use logarithmic differentiation to find y':

(a) $y = \dfrac{x^3 \sqrt{2x + 1}}{(x - 3)^2}$

(b) $y = \dfrac{(x - 1)^5 (x^2 + 1)^{2/3}}{\sqrt{x^3 - 5x^2 + 6x}}$

9. Find y' using implicit differentiation:

(a) $e^y - \ln \dfrac{x}{y} = 1$

(b) $x = \ln\left(\dfrac{x + y}{x - y}\right)$

In Exercises 10–15, evaluate the given definite or indefinite integral.

10. (a) $\displaystyle\int \dfrac{\cos 2x}{1 + \sin 2x}\,dx$

(b) $\displaystyle\int \dfrac{x + 1}{x^2 + 2x + 4}\,dx$

11. (a) $\displaystyle\int e^{-2\ln x}\,dx$

(b) $\displaystyle\int \dfrac{e^x - 1}{e^{2x} - 2xe^x + x^2}\,dx$

12. (a) $\displaystyle\int \dfrac{\sqrt{x}}{\sqrt{x^3} + 4}\,dx$

(b) $\displaystyle\int \dfrac{dx}{x(\ln 2x + 3)}$

13. (a) $\displaystyle\int_{\ln 2}^{\ln 3} \dfrac{e^{-2x}}{1 + e^{-2x}}\,dx$

(b) $\displaystyle\int_{e^{-2}}^{e^2} \dfrac{dx}{x + \sqrt{x}}$

14. (a) $\displaystyle\int_{(1/2)\ln 7}^{2\ln 2} e^{2x}\sqrt{9 + e^{2x}}\,dx$

(b) $\displaystyle\int_{\ln 2}^{\ln 3} \dfrac{e^{2x} - 1}{e^{2x} + 1}\,dx$ (*Hint:* Multiply the integrand by e^{-x}/e^{-x}.)

15. (a) $\displaystyle\int \dfrac{2^{2x}}{4^x + 1}\,dx$

(b) $\displaystyle\int_0^2 \dfrac{\sqrt{1 - 2^{-t}}}{2^t}\,dt$

In Exercises 16–19, analyze the function and draw its graph.

16. $y = \dfrac{e^x - e^{-x}}{e^x + e^{-x}}$

17. $y = \ln|1 - x|$

18. $y = x \ln x^2$

19. $y = x^3 e^{-x^2/2}$

20. Find the volume of the solid obtained by revolving the region under the curve $y = e^{x^2/2}$ from $x = 0$ to $x = \sqrt{\ln 4}$ about the y-axis.

21. Find the surface area obtained by revolving the arc of the curve $y = e^{x/2} + e^{-x/2}$ from $x = -\ln 3$ to $x = \ln 3$ about the x-axis.

22. Solve for x in terms of natural logarithms and approximate the answer using a calculator or CAS:

(a) $2 \cdot 3^{-x} = 5 \cdot 4^{2x}$

(b) $6e^{2x-1} - 3^x \cdot 2^{1-x} = 0$

23. Solve for x and check your answers:

(a) $\ln x + \ln(x - 1) = \ln 2$

(b) $2 \log_3 x - \log_3(2x - 3) = 1$

24. The radioactive isotope cobalt-60 has a half-life of approximately 5.26 yr. Of a given initial amount, how much will remain after 3 yr? How long will it take for 80% of the original amount to decay?

25. A culture of bacteria is observed to double in size in 13 hr. How long will it take to triple in size? Assume exponential growth.

26. A steel plate that has been heated to 40°C by exposure to direct sunlight is brought into an air-conditioned room where the temperature is 20°C. After 2 min it has cooled to 38°C. According to Newton's Law of Cooling, what will be the temperature of the plate after 5 min? How long will it take for the temperature to drop to 21°C?

27. A modification of the Malthus model for population growth is given by

$$Q'(t) = kQ \ln \dfrac{m}{Q}, \qquad Q < m$$

where k and m are positive constants. (This differential equation is known as the **Gompertz model**.) Find $Q(t)$ for this model. What is $\lim_{t \to \infty} Q(t)$?

In Exercises 28–31, give the exact value of the expression.

28. (a) $\sin^{-1} 1$ (b) $\cos^{-1}\left(-\dfrac{\sqrt{3}}{2}\right)$

     (c) $\tan^{-1}\left(-\dfrac{1}{\sqrt{3}}\right)$ (d) $\operatorname{arcsec}(-\sqrt{2})$

     (e) $\operatorname{arccot}(0)$ (f) $\operatorname{arccsc}\left(-\dfrac{2}{\sqrt{3}}\right)$

29. (a) $\tan\left[\cos^{-1}\left(-\dfrac{3}{\sqrt{13}}\right)\right]$ (b) $\sin\left[2\tan^{-1}\left(-\dfrac{1}{2}\right)\right]$

30. (a) $\cos\left[2\sin^{-1}\left(-\dfrac{3}{5}\right)\right]$ (b) $\sin^{-1}\left(\sin\dfrac{8\pi}{9}\right)$

31. (a) $\sin\left[\tan^{-1}\dfrac{4}{3} + \sec^{-1}\left(-\dfrac{13}{5}\right)\right]$

     (b) $\cos\left[\cos^{-1}\left(-\dfrac{1}{3}\right) - \tan^{-1}(-2)\right]$

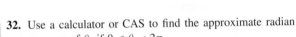 **32.** Use a calculator or CAS to find the approximate radian measure of θ, if $0 \le \theta < 2\pi$.

     (a) $\cos\theta = \dfrac{1}{3}$, $\tan\theta < 0$

     (b) $\sin\theta = -\dfrac{3}{4}$, $\tan\theta > 0$

In Exercises 33–39, find y'.

33. (a) $y = \dfrac{1}{x}\sin^{-1}x^2$, $0 < |x| < 1$

     (b) $y = \tan^{-1}\sqrt{x^2 - 1}$, $|x| > 1$

34. (a) $y = \cos^{-1}(e^{-x})$, $x \ge 0$

     (b) $y = x^2 \sec^{-1}\dfrac{1}{x} + \sin^{-1}x$, $0 < x < 1$

35. (a) $y = (\sin^{-1}\sqrt{x})^2 + (\cos^{-1}\sqrt{x})^2$

     (b) $\tan^{-1}\dfrac{y}{x} + \ln\sqrt{x^2 + y^2} = 1$

36. (a) $y = \ln(\cosh^2 x)$

     (b) $y = \cos^{-1}(\operatorname{sech} x)$, $x > 0$

37. (a) $y = \dfrac{\sinh x}{1 + \cosh x}$

     (b) $\tan^{-1}(\sinh y) + \cot^{-1}(\sinh x) = 1$

38. (a) $y = \sinh^{-1}\sqrt{x^2 - 1}$, $x > 1$

     (b) $y = \tanh^{-1}(\sin x)$, $|x| < \dfrac{\pi}{2}$

39. (a) $y = \coth^{-1}\dfrac{1 + x}{1 - x}$, $0 < x < 1$

     (b) $y = \sqrt{x}\cosh^{-1}(x + 1)$, $x > 0$

In Exercises 40–55, evaluate the given definite or indefinite integrals.

40. $\displaystyle\int \dfrac{dx}{\sqrt{9 - 4x^2}}$ **41.** $\displaystyle\int_{-3/2}^{3/2} \dfrac{dx}{4x^2 + 3}$

42. $\displaystyle\int_{-2\sqrt{2}}^{-2} \dfrac{dx}{x\sqrt{x^2 - 2}}$ **43.** $\displaystyle\int_{-\sqrt{2}/4}^{0} \dfrac{dx}{\sqrt{1 - 4x^2}}$

44. $\displaystyle\int_{1}^{e} \dfrac{dx}{x[1 + (\ln x)^2]}$ **45.** $\displaystyle\int_{1/\sqrt{3}}^{1} \dfrac{dx}{x\sqrt{4x^2 - 1}}$

46. $\displaystyle\int \dfrac{dx}{\sqrt{e^{2x} + 1}}$ **47.** $\displaystyle\int \dfrac{x}{1 + x^4}dx$

48. $\displaystyle\int \dfrac{\sinh 2x}{1 + \cosh 2x}dx$ **49.** $\displaystyle\int \dfrac{\sinh 2x}{e^{\sinh^2 x}}dx$

50. $\displaystyle\int x\tanh x^2 dx$ **51.** $\displaystyle\int_{0}^{\ln(1 + \sqrt{2})} \dfrac{\cosh x}{1 + \sinh^2 x}dx$

52. $\displaystyle\int_{0}^{\ln 2} \dfrac{\tanh x}{\cosh^2 x}dx$ **53.** $\displaystyle\int \dfrac{e^x dx}{\sqrt{e^{2x} - 1}}$

54. $\displaystyle\int_{-1}^{1} \dfrac{dx}{4 - x^2}$ **55.** $\displaystyle\int_{0}^{1} \dfrac{dx}{x^2 + 4x + 3}$

In Exercises 56–61, evaluate the limits.

56. (a) $\displaystyle\lim_{x \to 1^+} \dfrac{\ln(\ln x)}{\tan\frac{\pi x}{2}}$ (b) $\displaystyle\lim_{x \to 0^-} e^{1/x}\ln|x|$

57. (a) $\displaystyle\lim_{x \to 0} x\coth 2x$ (b) $\displaystyle\lim_{x \to 0^+} x\ln(\tanh x)$

58. (a) $\displaystyle\lim_{x \to 1^+} (\ln x)[\ln(\ln x)]$

     (b) $\displaystyle\lim_{x \to \infty} [\ln x - \ln(x + 2)]$

59. (a) $\displaystyle\lim_{x \to 0}\left[\dfrac{1}{e^x - 1} - \operatorname{csch} x\right]$

     (b) $\displaystyle\lim_{x \to 1}\left[\dfrac{1}{x - 1} - \dfrac{1}{\ln x}\right]$

60. (a) $\displaystyle\lim_{x\to 1^+} (x-1)^{\ln x}$ (b) $\displaystyle\lim_{x\to 0^+} [\ln(x+1)]^x$

61. (a) $\displaystyle\lim_{x\to 0^+} \left(\frac{1}{e^x+1}\right)^{\sqrt{x}}$ (b) $\displaystyle\lim_{x\to 0}(\cosh x)^{\coth^2 x}$

62. Find the volume of the solid obtained by revolving the region under the graph of

$$y = \frac{4}{x\sqrt{4-x^2}}$$

from $x=1$ to $x=\sqrt{3}$ about the y-axis.

63. Find the area between the graphs of

$$y = \frac{2e^x}{1+e^x} \quad\text{and}\quad y = \frac{2e^x}{1+e^{2x}}$$

from $x=0$ to $x=\frac{1}{2}\ln 3$.

64. The bottom of a 5 m-high billboard is 2 m above an observer's eye level. At what distance from the billboard should the observer stand to get the best view of the billboard?

65. A helicopter is initially 120 m directly above an observer on the ground. Then the helicopter begins to ascend at a speed of 20 m/s on a 45° path with the vertical. How fast is the angle of elevation of the helicopter changing, as measured from the observer, after 4 seconds?

66. Derive the formula for $\coth(x+y)$ that involves only $\coth x$ and $\coth y$ and no other hyperbolic functions.

In Exercises 67–70, verify the identities.

67. $\dfrac{1}{1-\tanh x} - \dfrac{1}{1+\tanh x} = \sinh 2x$

68. $\dfrac{\coth^2 x + \cosh x\,\operatorname{csch}^2 x}{1+\cosh x} = \operatorname{csch} x \coth x$

69. $\dfrac{\cosh x - \sinh x}{\cosh 2x - \sinh 2x} = \cosh x + \sinh x$

70. $\dfrac{1}{\operatorname{csch}^2 x + \operatorname{csch}^4 x} = \sinh^2 x - \tanh^2 x$

71. Let $f(x) = 2\cosh \dfrac{x}{2}$ on $[-1, 1]$.
(a) Find the area under the graph of f.
(b) Find the volume of the solid obtained by revolving the region under the graph of f about the x-axis.
(c) Determine the length of the graph of f.
(d) Find the surface area obtained by revolving the graph of f about the x-axis.

72. Show that the function

$$y = e^{-kt}(A\cosh \omega t + B\sinh \omega t)$$

satisfies the differential equation

$$\frac{d^2 y}{dt^2} + 2k\frac{dy}{dt} + (k^2 - \omega^2)y = 0$$

where k and ω are positive constants. For $k=2$ and $\omega=1$, find A and B such that when $t=0$, $y=-1$ and $dy/dt = 2$.

Chapter 7 Concept Quiz

1. Define each of the following:
(a) $\ln x$
(b) e^x
(c) $\log_a x$
(d) a^x
(e) $\sin^{-1} x$
(f) $\cos^{-1} x$
(g) $\tan^{-1} x$
(h) $\sinh x$
(i) $\cosh x$
(j) $\tanh x$

2. (a) State the three arithmetic properties of the natural logarithm.
(b) Write an equivalent equation to $\ln a = b$ involving the exponential function.
(c) Express $e^{a\ln b}$ in an equivalent form that does not involve e or $\ln$.
(d) What restriction must be placed on x for the equation $\sin^{-1}(\sin x) = x$ to be true?

3. Give the derivative formula for each of the following, where u is a differentiable function of x:
(a) $\ln u$
(b) e^u
(c) $\log_a u$
(d) a^u
(e) $\sin^{-1} u$
(f) $\tan^{-1} u$
(g) $\sec^{-1} u$
(h) $\sinh u$
(i) $\cosh u$
(j) $\tanh u$

4. Give the following indefinite integral formulas:

(a) $\displaystyle\int \frac{du}{u}$

(b) $\displaystyle\int e^u \, du$

(c) $\displaystyle\int \frac{du}{\sqrt{a^2 - u^2}}$

(d) $\displaystyle\int \frac{du}{a^2 + u^2}$

(e) $\displaystyle\int \frac{du}{u\sqrt{u^2 - a^2}}$

(f) $\displaystyle\int \sinh u \, du$

(g) $\displaystyle\int \cosh u \, du$

(h) $\displaystyle\int \tan u \, du$

5. Fill in the blanks:

(a) e is the number whose _____ is _____.

(b) The inverse of $f(x) = \ln x$ is $f^{-1}(x) = $

_____.

(c) The domain of $f(x) = \ln x$ is _____ and its range is _____.

(d) The domain of $f(x) = e^x$ is _____ and its range is _____.

(e) In order to define $\sin^{-1} x$, the domain of $f(x) = \sin x$ is restricted to the interval _____.

(f) In order to define $\cos^{-1} x$, the domain of $f(x) = \cos x$ is restricted to the interval _____.

(g) The domain of $f(x) = \tan^{-1} x$ is _____ and its range is _____.

6. Which of the following statements are true, and which are false?

(a) If $\ln a = \ln b$, then $a = b$.

(b) $\displaystyle\int \frac{dx}{x^2} = \ln x^2 + C$

(c) If $Q'(t) = kQ(t)$, then $Q(t) = Q(0) \cdot e^{kt}$.

(d) $\ln(a + b) = \ln a + \ln b$

(e) $e^{\ln a + \ln b} = ab$

(f) $\tan^{-1} x = \dfrac{1}{\tan x}$

APPLYING CALCULUS

1. Psychologists attempt to measure the reaction R of a subject to an amount of stimulus. For example, one might have a change of one unit in the reaction correspond to the smallest change in the stimulus that most individuals can distinguish. This is the idea behind the decibel scale—1 decibel is intended to be the smallest change in loudness that the human ear can detect. The *Weber-Fechner Law* states that if the stimulus is proportional to the fraction ds/s by which the stimulus increased by an amount ds, then the corresponding change in the reaction dR is increased. That is: $dR = k(ds/s)$, where k is a positive constant. This equation can be written in the form of a differential equation:

$$\frac{dR}{ds} = \frac{k}{s}$$

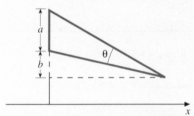

(a) Determine R as a function of s where s_0 is the smallest stimulus detectable by the subject.

(b) Suppose that a person can distinguish between two weights only if the heavier one weighs at least 5% more than the lighter one. Use the Weber-Fechner Law to devise a scale $R(s)$ of apparent weights such that a difference of 1 unit in R corresponds to the smallest detectable change in actual weight s. Let $R = 1$ correspond to a weight of $s = 1$ pound.

(c) A variation of the Weber-Fechner Law is the *Brentano-Stevens Law*, which states that

$$\frac{dR}{ds} = k\frac{R}{s}$$

Determine R as a function of s in this case.

2. A large picture measures a ft from top to bottom. Its bottom is b ft above the eye level of an observer who is standing x ft from the wall on which the picture is hung.

(a) Show that the angle θ subtended by the picture at the eye of the observer is given by

$$\theta = \tan^{-1}\left(\frac{a+b}{x}\right) - \tan^{-1}\left(\frac{b}{x}\right)$$

(b) A picture hanging in an art gallery measures 5 ft from top to bottom and its bottom is 7 ft above the floor. Suppose that an observer's eye level is 5 ft above the floor. How far should the observer stand from the wall to maximize her angle of vision?

3. The concentration of a drug in the blood after a single injection decreases with time as the drug is either absorbed or eliminated from the body. Clinical evidence supports the assumption that the rate of decrease of the concentration at time t is proportional to the concentration at time t. Thus, if $C = C(t)$ is the concentration at time t, then

$$C'(t) = -kC(t)$$

where k is a positive constant, called the *elimination constant*.

(a) Suppose that the initial concentration of a drug in the bloodstream is 20 milligrams per liter and 3 hours later the concentration is 12 milligrams per liter. Determine an expression for the concentration $C = C(t)$ at any time t. What is the half-life of the drug?

(b) Sodium phenobarbitol, which has a half-life of 5 hours, is to be used to anesthetize a dog for an operation. The dog is anesthetized when its bloodstream contains at least 30 milligrams of the drug for each kilogram of body weight. Assuming that the concentration of sodium phenobarbitol decreases at a rate proportional to the amount present, what single dose of the drug should be administered in order to anesthetize a 25 kilogram dog for 1 hour?

4. Another model for the concentration $C = C(t)$ of a drug in the bloodstream at time t after an injection is

$$C(t) = C_0(e^{-\alpha t} - e^{-\beta t}), \qquad t > 0$$

where C_0 is the initial concentration, and α and β are constants with $\beta > \alpha$.

(a) Determine the relative extrema of this function and determine the intervals on which it is increasing and decreasing.

(b) Determine where the graph is concave up and concave down; find the points of inflection.

(c) Sketch the graph, including $\lim_{t \to \infty} C(t)$.

5. Let $A = A(t)$ be the amount of a given natural resource that is consumed from time $t = 0$ to time $t > 0$. Let $c = c(t)$ be the rate at which the resource is being consumed. If $c(t) = C$, a constant, then the amount consumed from $t = 0$ to $t = T$ is $A(T) = CT$. On the other hand, since population and industry tend to grow exponentially, it is reasonable to assume that the rate of consumption will increase exponentially; that is $c(t) = Ce^{kt}$, where C and k are positive constants.

(a) Suppose that the rate of consumption is growing exponentially. Show that

$$A(t) = \frac{C(e^{kt} - 1)}{k}$$

(b) Let R be the amount of the resource that exists at time $t = 0$. Show that if the consumption rate is a constant, C, then the resource will last R/C years. This value is called the *static index* of the resource, which is denoted by s. Show that if the consumption rate increases exponentially, then the resource will last

$$r = \frac{\ln(ks + 1)}{k} \qquad \text{years.}$$

This value is called the *exponential index* of the resource.

(c) Suppose the known global reserves of oil at this time are 500×10^8 barrels. Determine the exponential index of this resource if the static index is 31 years and the rate of consumption increases 3.9% per year. What is the exponential index if the known reserves are 5 times as large?

6. Suppose income is to be received at the rate of $f(t)$ dollars per year from time $t = 0$ to $t = T$. For the purposes of analysis, it is convenient to assume that the income is generated and received continuously. In this case, the continuous function f is called the *rate of flow* for a *continuous income stream*. The total income received from time $t = 0$ to time $t = T$ is

$$I(T) = \int_0^T f(t)dt$$

If f is the rate of flow of a continuous income stream, then the *present value*, V_P, at an interest rate r (expressed as a decimal) compounded continuously for T years is

$$V_P(T, r) = \int_0^T e^{-rt} f(t)dt$$

If the income is to be received for all future time T, then the present value of all future income depends only on the interest rate r and is given by

$$V_P(r) = \int_0^\infty e^{-rt} f(t)dt$$

(Note: This equation can be viewed as "transforming" the function f into the function V_P. This particular relation has many important applications; it is called the *Laplace transform.*)

569

(a) Suppose that the rate of flow f is a constant, that is $f(t) = c$, $c > 0$. Calculate the total income I and the present value V_P at the interest rate r over the time interval [0,T]. Also, calculate $\lim_{T \to \infty} I(T)$ and $\lim_{T \to \infty} V_P(T, r)$.

(b) In a "cyclical" company, the rate of flow fluctuates with time. Suppose that the rate of flow of a cyclical company is given by

$$f(t) = 2500 + 1000 \cos \pi t \text{ dollars per year.}$$

Calculate the total income and the present value at the interest rate $r = 0.10$ for $0 < t < 10$ years.

(c) Determine the present value of all future income for the company in part (b).

7. In the investigation of a homicide or accidental death, it is usually important to estimate the actual time of death. Newton's Law of Cooling (see Section 7.4), which states that rate of change of temperature of an object is proportional to the difference in temperature between the object and the surrounding medium, can be used for this purpose. The mathematical formulation of this law is

$$\frac{dy}{dt} = -k(y - T)$$

where $y = y(t)$ is the temperature of the object at time t, T is the (constant) temperature of the surrounding medium, and k is a positive constant.

(a) Suppose a human body has temperature y_0 at time $t = 0$, and temperature y_1 at time $t = t_1 > 0$. Use Newton's Law of Cooling to derive an expression for the temperature of the body at any time t.

(b) A corpse with a body temperature of 77°F has been found in a ditch at 8:00 a.m. One hour later its temperature is 68°F. The (constant) air temperature is $T = 60$°F. Assuming that the person had a normal body temperature of 98.6°F at the moment of death, determine the time of death.

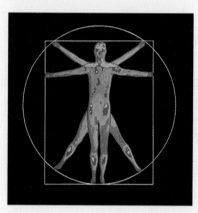

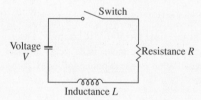

Switch

Voltage V

Resistance R

Inductance L

8. An electric circuit consists of a resistor whose resistance is R (ohms) connected in series to an inductor with inductance L (henries) and a battery which supplies a constant voltage V. The switch is closed at time $t = 0$.

Let $i = i(t)$ denote the current in the circuit at time $t > 0$. The voltage across the resistor is Ri and the voltage across the inductor is $L \, di/dt$. By *Kirchhoff's Law*, the sum of these two voltages must equal the applied voltage V. Thus,

$$L\frac{di}{dt} + Ri = V$$

The initial condition is $i(0) = 0$, since there is no current in the circuit until the switch is closed at time $t = 0$.

(a) Show that the solution of the initial value problem is

$$i = \frac{V}{R}\left(1 - e^{-Rt/L}\right) = \frac{V}{R} - \frac{V}{R}e^{-Rt/L}$$

(b) Sketch the graph of this solution, including $\lim_{t \to \infty} i(t)$.

(c) The term $i_0 = V/R$ is called the *steady-state current* and the term $i_{tr} = \frac{V}{R}e^{-Rt/L}$ is the *transient current*. The time $T = L/R$ is called the *time constant* of the circuit. Calculate the current at the time $T = L/R$. How long does it take for the current to reach 99% of its steady state current?

9. On a certain day, snow began to fall at a heavy and constant rate. A snowplow started out at noon, going 2 miles the first hour and 1 mile the second hour. What time did it start snowing? Assume the plow clears snow at a constant rate.

Deborah Tepper Haimo

When do you remember first getting interested in mathematics?

I did not find mathematics particularly interesting until after high school algebra. In sophomore year I studied geometry and a new and exciting world opened up for me. Here we weren't required to carry out calculations involving specialized formulas. Instead, we were given a set of axioms, self-evident truths, that we used to justify every step we took in establishing a result. Mathematical rules now made much more sense to me. We could add reasonable assumptions to the axioms that we accepted at the start, and only these could be used to advance and develop a geometric result. Until then I had no idea how fascinating mathematics could be. I spent much time trying to derive geometric properties beyond the routine classroom problems. I was thrilled to observe the dramatic consequences of our few hypotheses.

One particular class made an impression on me. We were learning about indirect proofs (as opposed to direct proofs) and were assigned a problem using this method. I tried to imagine circumstances in which only an indirect proof would work and I thought I'd found one involving geometric figures positioned in a certain way. I had forgotten an axiom that we'd never had occasion to use previously. It stated that geometric figures could be moved in space without affecting their properties. Either a direct or indirect proof could be used and both would be sound. This was wondrous to me. Mathematics was so plausible. You could get the same correct result in many ways.

How and when did you decide to go into mathematics professionally?

At Radcliffe I thought a lot about what I wanted to do. I knew that I loved mathematics, but I had no idea what I could do with it. I knew I didn't have the temperament to be a high school teacher. I finally decided that there might be opportunities in the physical sciences that use mathematics, so I enrolled in Astronomy and Physics courses. I obtained part-time employment at the Harvard Observatory, which had some appeal but the women astronomers there seemed to have really tedious and laborious jobs. Physics, on the other hand, seemed more promising initially, but another classroom situation helped me change my mind. We were doing a routine experiment using magnets to record the patterns made by iron filings around the magnetic poles. Everyone else did the experiment easily but my iron filings would not conform to the expected pattern. Substituting magnets didn't help. Everyone was mystified, including the lab instructor, until a student opened the drawer under the table. Of course, we found it full of magnets!

Although a trivial experience, it really did have an influence on my early attitudes to mathematics and science. Unlike my earlier revelation when studying geometry in which I failed to notice an assumption that was already there, in this case I had visions of developing a theory in physics that might be distorted by a drawerful of magnets! This thought helped reconfirm my earlier inclination toward mathematics. I was drawn to the control of hypotheses, and the intrinsic elegance of the mathematics when these were well chosen.

Did your calculus course further develop your interest in mathematics?

My calculus class at Harvard was taught by Hassler Whitney, a creative and imaginative mathematician who was acclaimed for his ground-breaking work in differential geometry. His lectures were hard to follow since he became very excited over results, stuttered a bit, and wrote illegibly. But I credit him with transmitting an air of excitement about mathematics that was inescapable. He recommended Courant's two volumes on Differential and Integral Calculus that became my "bible" for calculus.

Professor Whitney was the assigned advisor for our class, and when I consulted him I expected advice on what opportunities I would have if I decided to continue in mathematics. I was surprised to find that he was totally uninterested in this issue. Instead, he asked me why I bothered to come to class when I could clearly learn much more rapidly on my own. This question totally startled me. In my earlier academic experience, there was never any question that one was expected to attend class. When I explained that I was concerned about my grades and had to maintain my scholarship, he encouraged me to stop coming to class, skip the next required course, and take advanced calculus in my sophomore year. He then said that in my junior year he would arrange to enroll me in graduate courses at Harvard that were open to undergraduates and that he would be my tutor to direct me in my area of concentration and to supervise me in writing the required undergraduate thesis. All this was very unusual at the time, especially for Radcliffe students.

You're the only person I've heard of who was told to cut class by her own professor.

Yes. At the time I was both confused and also elated—because he confirmed what I already knew—that mathematics was my first love as a discipline. I had been holding back because of my concerns about the future. There and then I made up my mind to follow my interests whatever the ultimate outcome. I followed Whitney's advice even though my original Radcliffe tutor was appalled at what she considered totally unrealistic advice. She was sure I would fail and that it would reflect badly on Radcliffe. Nonetheless, in my junior year, as promised, Whitney became my tutor, and I enrolled in graduate-level courses.

How did you eventually decide on what research to pursue?

Whitney was known for his uncanny intuition, which led him to important conjectures, and he assigned me an undergraduate thesis topic involving what he thought was a simple version of a major theory he had just developed. As I discovered, the problem was far more intractable than he had surmised. I was not able to do more than raise doubts, through an example, of the validity of his initial conjecture.

During the war years, job vacancies were being created that had never before been available to women. I began my first teaching position and was appointed for one semester as head of both mathematics and physics, with a 22-contact-hour teaching schedule! I enjoyed the experience, but I was eager to return to Cambridge to marry the fellow student to whom I was engaged and to study for my Ph.D. My husband, Frank, was initially surprised to learn that I wanted to pursue a Ph.D., since we had both come from traditional family backgrounds where the women had devoted all their time to their homes. However, he was very happy with my decision and became my most ardent professional supporter. He continued with his doctoral work and I remained intrigued with the problem that I had failed to solve as an undergraduate. As time passed, and my efforts did not yield substantive results, Frank urged me to abandon the problem and move into analysis, the field for which calculus is a foundation, and in which I had shown early interest and aptitude. I was stubborn, though, and Whitney encouraged me to continue. He kept saying that he was sure that I was on the verge of a major result.

When we moved to St. Louis, I had done all my course work but had not written a thesis, nor did I seem close to doing so. At Washington University I was given an irregular position with a very heavy teaching load. I had to develop the lower division college algebra course, transform the course to the large lecture format, and plan help sessions and multiple-choice examinations. I was also assigned to teach the graduate course in linear programming to engineering students.

After 12 years at Washington University, I accepted an offer at Southern Illinois University. All that time, our family was growing, I continued teaching, and I worked on my thesis whenever I could. Whitney had moved to the Institute for Advanced Study and I communicated with him. I could not make headway on the problem, but I refused to give up. One summer I determined that I would devote all my time to the problem, but when fall came and I hadn't made substantial progress, I reluctantly decided to give up. I followed the advice I'd been given for several years and decided on a problem that seemed tractable. Professor Widder at Harvard told me that, despite the passage of two decades, I would be granted my Ph.D. if I did an acceptable thesis and passed the general oral doctoral examination. With the availability for consultation of an expert in my new field of interest, Professor Hirschman at Washington U., I succeeded in finishing my thesis in a few months. Defending my thesis at Harvard was unquestionably one of the most traumatic experiences of my life. Except for Widder, I faced a roomful of strangers who could ask me any question in mathematics that they liked, and they did. After what seemed an eternity, Widder left, indicating that he felt the session had continued long enough, but this only invited more questions in other areas from the other examiners. At long last, the session ended. The results were almost immediate as the office filled with smiling faces and, at long last, I had my Ph.D.

It was after my thesis adventures that I learned that, after I had abandoned my original problem, new mathematical tools had been developed that showed that the problem I had worked on for so long, the classification of certain singular points, was proven to be impossible!

Isn't it frustrating to know now that you spent all that time on an impossible problem? Is this a risk that all research mathematicians face?

Well, as far as research goes, the important thing is to know that a solution to a problem is impossible. My real frustration occurred when I realized that I should not have devoted so much time exclusively to a problem that didn't yield any useful material within a reasonable period of time. Indeed, when I finally decided to change, I found that there were plenty of sound and interesting problems that lent themselves to serious investigation and that could be expected to produce worthwhile results. I was surprised to find writing a thesis much less overwhelming than I had expected. After completing Ph.D. requirements, my research activities centered on generalizations of the classical heat equation. My field now is classical harmonic analysis.

Being a research mathematician is very satisfying and interesting. Useful work can be done wherever one happens to be because neither a laboratory nor a library is essential for mathematical research. The major requirement is to have good ideas; and good thinking can be carried out anywhere. In addition, there is the challenge of educating students and arousing their appreciation of the intrinsic beauty of mathematics. Opportunities exist to travel both in the States and abroad. Invitations often are issued to lecture on one's results at various institutions and to participate in speciality conferences and in regional, national, and international meetings.

In recent years, I've become increasingly concerned about our educational system. I have become directly involved at the national level in trying to address some of the major issues in mathematics education.

8

TECHNIQUES OF INTEGRATION

By now you have learned formulas for antiderivatives of quite a few functions. That is, you know how to evaluate indefinite integrals for these functions. You have also seen how other integrals can be converted to one of the standard forms, using the technique of substitution. Although substitution is a powerful technique, there are simple-looking integrals that defy any substitution. For example, consider

$$\int x \cos x \, dx$$

Try as you may, you will not be able to find a substitution that will change this integral to a form you can integrate. Some other technique is needed.

Finding antiderivatives is important for two primary reasons: (1) we can use them to evaluate definite integrals by means of the Second Fundamental Theorem, and (2) we can use them to solve differential equations. Since definite integrals and differential equations arise in a wide variety of applications, it is worthwhile to learn further techniques for finding antiderivatives. In this chapter we introduce the most important of these techniques.

Differentiation is usually a straightforward process once you have learned the basic derivatives and a few rules. On the other hand, antidifferentiation (indefinite integration) often requires a good deal of ingenuity. You now know how to differentiate many of the elementary functions, but you don't know how to integrate all of them. In fact, as we have previously noted, there are elementary functions for which no antiderivative exists in the form of an elementary function. The function $f(x) = e^{-x^2}$ (see Figure 8.1) is one of these (a very important one, especially in probability and statistics). Although the First Fundamental Theorem of Calculus tells us that this function does have

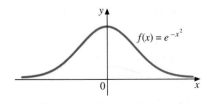

$f(x) = e^{-x^2}$

FIGURE 8.1
This simple function, important especially in probability and statistics, cannot be integrated in terms of elementary functions.

an antiderivative, there is no way of expressing it in terms of functions you know. In Chapter 10 we will see a way of approximating its antiderivative. We can, however, find antiderivatives of a great many functions without resorting to approximations. Learning the techniques in this chapter will greatly extend the class of functions you can integrate.

There are really only two significant techniques of integration: (1) **integration by parts** and (2) **substitution.** We will study the first of these in Section 8.1. It is based on the Product Rule for derivatives and is a powerful technique that enables us to integrate many functions we could not integrate by other means. We have used the method of substitution extensively already. In this chapter we show two further applications, both involving trigonometric functions. We also introduce a procedure called *partial-fraction decomposition*, which is not an integration technique per se but is an algebraic process by which complicated rational functions (quotients of polynomials) can be broken down into simpler ones. These simpler functions can either be integrated directly or by one of our other techniques.

The techniques of this chapter, together with those you have already learned, will provide you with the means of integrating most of the functions you are likely to encounter. Integration is an art, though, that can only be learned by practice. It would be nice if there were some definite step-by-step procedure (some *algorithm*) that you could apply to every integration problem and be sure of arriving at the answer. Unfortunately, there is no such procedure—the antiderivative may not even exist. With practice, you will learn which technique is appropriate for a given problem.

It was only in the late 1960s that it became possible to determine whether a function necessarily has an elementary antiderivative. The algorithm discovered by Robert Risch determines whether the elementary antiderivative exists and, if so, finds it. Modern computer algebra systems, such as Maple, use the same algorithm to accomplish this task.

Even when you know what technique to use on a particular integral, actually carrying out the steps can be tedious and time-consuming. Furthermore, if you need to evaluate a large number of integrals, each requiring a special technique, the time and effort needed may be more than you can afford to spend. In such situations, tables of integrals can be very helpful. Also, computer algebra systems can handle with ease many integrals that would be prohibitively difficult by other methods. We devote a section of this chapter, following the study of the various techniques, to an explanation of how to use integral tables and computer algebra systems to evaluate integrals.

We conclude the chapter by showing how the definition of the definite integral can be extended to two previously excluded cases: (1) the case in which the interval of integration is infinite, and (2) the case in which the integrand becomes infinite. Integrals of either type are referred to as *improper integrals*. Such integrals have both theoretical and practical applications. We have seen an example of an improper integral already, in Example 6.19, when we were deriving a formula for the length of an ellipse. We will consider this integral again in this chapter.

Before introducing the techniques, we summarize in Table 8.1 basic indefinite integral formulas that we have derived or have asked you to derive in the exercises. Our list is by no means exhaustive. A more extensive list is given on the front and back endpapers of this book. Tables of integrals that give hundreds of formulas are also available.

TABLE 8.1 Basic list of integrals

$$\int u^\alpha du = \frac{u^{\alpha+1}}{\alpha + 1} + C \quad (\alpha \neq -1) \qquad \int \frac{du}{u} = \ln|u| + C$$

$$\int e^u du = e^u + C \qquad \int a^u du = \frac{a^u}{\ln a} + C \quad (a \neq 1)$$

$$\int \sin u \, du = -\cos u + C \qquad \int \cos u \, du = \sin u + C$$

$$\int \sec^2 u \, du = \tan u + C \qquad \int \csc^2 u \, du = -\cot u + C$$

$$\int \sec u \tan u \, du = \sec u + C \qquad \int \csc u \cot u \, du = -\csc u + C$$

$$\int \tan u \, du = -\ln|\cos u| + C \qquad \int \cot u \, du = \ln|\sin u| + C$$

$$\int \sec u \, du = \ln|\sec u + \tan u| + C$$

$$\int \csc u \, du = \ln|\csc u - \cot u| + C$$

$$\int \frac{du}{\sqrt{a^2 - u^2}} = \sin^{-1}\frac{u}{a} + C \qquad \int \frac{du}{a^2 + u^2} = \frac{1}{a}\tan^{-1}\frac{u}{a} + C$$

$$\int \frac{du}{u\sqrt{u^2 - a^2}} = \frac{1}{a}\sec^{-1}\frac{u}{a} + C$$

$$\int \sinh u \, du = \cosh u + C \qquad \int \cosh u \, du = \sinh u + C$$

$$\int \operatorname{sech}^2 u \, du = \tanh u + C \qquad \int \operatorname{csch}^2 u \, du = -\coth u + C$$

$$\int \operatorname{sech} u \tanh u \, du = -\operatorname{sech} u + C \qquad \int \tanh u \, du = \ln(\cosh u) + C$$

$$\int \operatorname{csch} u \coth u \, du = -\operatorname{csch} u + C \qquad \int \coth u \, du = \ln|\sinh u| + C$$

$$\int \operatorname{sech} u \, du = \tan^{-1}|\sinh u| + C$$

$$\int \operatorname{csch} u \, du = \ln\left|\tanh\frac{u}{2}\right| + C$$

$$\int \frac{du}{\sqrt{a^2 + u^2}} = \sinh^{-1}\frac{u}{a} + C = \ln(u + \sqrt{u^2 + a^2}) + C$$

$$\int \frac{du}{\sqrt{u^2 - a^2}} = \cosh^{-1}\frac{u}{a} + C = \ln(u + \sqrt{u^2 - a^2}) + C$$

$$\int \frac{du}{a^2 - u^2} = \begin{cases} \dfrac{1}{a}\tanh^{-1}\dfrac{u}{a} + C, & |u| < a \\[2mm] \dfrac{1}{a}\coth^{-1}\dfrac{u}{a} + C, & |u| > a \end{cases} = \frac{1}{2a}\ln\left|\frac{a+u}{a-u}\right| + C$$

$$\int \frac{du}{u\sqrt{a^2 - u^2}} = -\frac{1}{a}\operatorname{sech}^{-1}\frac{|u|}{a} + C = -\frac{1}{a}\ln\left(\frac{a + \sqrt{a^2 - u^2}}{|u|}\right) + C$$

$$\int \frac{du}{u\sqrt{a^2 + u^2}} = -\frac{1}{a}\operatorname{csch}^{-1}\frac{|u|}{a} + C = -\frac{1}{a}\ln\left(\frac{a + \sqrt{a^2 + u^2}}{|u|}\right) + C$$

8.1 INTEGRATION BY PARTS

One of the most important techniques of integration is derived from the formula for the derivative of a product of two functions. If $u = f(x)$ and $v = g(x)$ are differentiable functions on some common domain, then we know that

$$\frac{d}{dx}[f(x)g(x)] = f'(x)g(x) + f(x)g'(x)$$

Taking antiderivatives of both sides, we get

$$f(x)g(x) = \int f'(x)g(x)\,dx + \int f(x)g'(x)\,dx$$

which can also be written in the form

$$\int f(x)g'(x)\,dx = f(x)g(x) - \int g(x)f'(x)\,dx \qquad (8.1)$$

Formula 8.1 is called the **integration by parts** formula. It often enables us to replace an integral that we *can't* integrate directly (the integral on the left) with one we *can* integrate (the one on the right). In our examples we will illustrate how, by appropriate choices of $f(x)$ and $g(x)$, the integral on the right is often simpler to work with than the one on the left.

It is customary in Formula 8.1 to write $u = f(x)$ and $v = g(x)$. Then $du = f'(x)\,dx$ and $dv = g'(x)\,dx$, giving the following formula.

Integration by Parts Formula

$$\int u\,dv = uv - \int v\,du \qquad (8.2)$$

Integration by parts can be applied to a wide variety of integrals. In particular, it often works when the integrand is a product of a polynomial and a transcendental function, such as

$$\int x^3 e^x, \quad \int x \ln x\,dx, \quad \text{and} \quad \int (x^2 + 2x + 3)\cos x\,dx$$

We illustrate how to apply the technique in the following example.

EXAMPLE 8.1   Evaluate the integral $\int x e^x\,dx$.

Solution   We want to express the integral $\int x e^x\,dx$ in the form $\int u\,dv$. The secret lies in choosing the "parts" u and dv. Usually, we want to choose u so that du/dx is simpler than u, thereby obtaining a simpler integral on the right-hand side of Equation 8.2 than the one on the left. Also, dv *must* always be chosen such that $v = \int dv$ can be found; otherwise, we could not evaluate either term on the right. In the integral in this problem, these criteria suggest that we choose $u = x$ and $dv = e^x\,dx$. (Note that when a choice is made for u, the choice for dv is determined.) With these choices, we have $du/dx = 1$ (which is certainly simpler than u), and v can be evaluated as

$$v = \int dv = \int e^x\,dx = e^x$$

It is not necessary to add an arbitrary constant at this stage. (See the discussion following this example.) We can organize our work as follows:

$$u = x \qquad dv = e^x\,dx$$
$$du = dx \qquad v = \int e^x\,dx = e^x$$

Now we apply Equation 8.2:

$$\int \overset{u}{\overbrace{x}}\ \overset{dv}{\overbrace{e^x\,dx}} = \overset{u}{\overbrace{x}}\ \overset{v}{\overbrace{e^x}} - \int \overset{v}{\overbrace{e^x}}\ \overset{du}{\overbrace{dx}}$$

The integral on the right is one we can integrate: $\int e^x \, dx = e^x$. So we have

$$\int x e^x \, dx = x e^x - e^x + C$$

∎

REMARK

■ In the preceding example, when we calculated v by integrating dv, we did not include an arbitrary constant of integration. Suppose we had included it. Then we would have had

$$v = \int e^x \, dx = e^x + C$$

By Equation 8.2 we then would have had

$$\int x e^x \, dx = x(e^x + C) - \int (e^x + C) \, dx$$
$$= x e^x + Cx - (e^x + Cx) + C_1$$
$$= x e^x - e^x + C_1$$

which is equivalent to our first answer. In general, as in this example, you may omit the arbitrary constant in calculating v from dv. That is, v can be taken as the simplest antiderivative.

Wrong Choice of Parts

What if we had tried a different choice of u and dv in the preceding example? If we had chosen $u = e^x$ and $dv = x \, dx$, then we would have had

$$du = e^x \, dx \qquad \text{and} \qquad v = \frac{x^2}{2}$$

Thus, by Equation 8.2,

$$\int x e^x \, dx = \frac{x^2}{2} e^x - \int \frac{x^2}{2} e^x \, dx$$

Even though this is a correct equality, we would be worse off than when we started, since the integral on the right is more complicated than the original integral. The lesson to be learned here is that if you make a bad choice of parts, abandon the choice when it becomes evident you are not getting anywhere, and try another.

EXAMPLE 8.2   Evaluate the integral $\int x^2 \cos x \, dx$.

Solution   Guided by the criteria stated in Example 8.1, we make the following choice of parts:

$$u = x^2 \qquad dv = \cos x \, dx$$

Then

$$du = 2x \, dx \qquad v = \int \cos x \, dx = \sin x$$

So, by Equation 8.2,

$$\int x^2 \cos x\, dx = x^2 \sin x - \int \sin x\, (2x\, dx)$$

$$= x^2 \sin x - 2 \int x \sin x\, dx$$

We have made progress (since the power of x in the last integral is less than in the first) but we still cannot evaluate the final integral directly. To evaluate it, we use integration by parts again, with

$$u_1 = x \qquad dv_1 = \sin x\, dx$$

$$du_1 = dx \qquad v_1 = \int \sin x\, dx = -\cos x$$

So

$$\int x \sin x\, dx = -x \cos x + \int \cos x\, dx$$

$$= -x \cos x + \sin x + C_1$$

Thus,

$$\int x^2 \cos x\, dx = x^2 \sin x - 2 \int x \sin x\, dx$$

$$= x^2 \sin x - 2[-x \cos x + \sin x + C_1]$$

$$= x^2 \sin x + 2x \cos x - 2 \sin x - 2C_1$$

But since C_1 is arbitrary, so is $-2C_1$. Calling this constant C, we get the final result:

$$\int x^2 \cos x\, dx = x^2 \sin x + 2x \cos x - 2 \sin x + C \qquad \blacksquare$$

Repeated Applications of Integration by Parts

A useful formula can be derived to give the result of repeated integration by parts in problems such as Example 8.2. Suppose u is a polynomial of degree n, with $n \geq 2$. We assume also that v and its successive integrals can be obtained. Let u_0 and v_0 be the initial choices of u and v, and for $k = 1, 2, \ldots, n$, let

$$u_k = u'_{k-1} \quad \text{and} \quad v_k = \int v_{k-1}\, dx$$

Then the result of integrating by parts n times is as follows:

$$\int u\, dv = u_0 v_0 - u_1 v_1 + u_2 v_2 - \cdots + (-1)^n u_n v_n + C$$

$$= \sum_{k=0}^{n} (-1)^k u_k v_k + C \qquad (8.3)$$

This formula was first given by the German mathematician Leopold Kronecker (1823–1891); you will be asked to prove it in Exercise 52 of Exercise Set 8.1. (See also Exercise 42 in Exercise Set 8.2 for an unusual function that is named after Kronecker.)

To illustrate Equation 8.3, consider again the integral in Example 8.2, where we initially took $u = x^2$ and $dv = \cos x\, dx$. Thus, our first u and v are $u_0 = x^2$ and $v_0 = \int \cos x\, dx = \sin x$. The accompanying table is a convenient way to organize the calculations. The first entries are u_0 and v_0. Each subsequent u_k is obtained by differentiating the preceding one, whereas each subsequent v_k is obtained by integrating the previous one. This sequence of steps is continued until a constant occurs in the u_k column. (This constant is u_n.) We now multiply each u_k by the corresponding v_k, assign $+$ signs and $-$ signs alternately, and add the results, supplying a constant of integration at the end:

u_k	v_k	
x^2	$\sin x$	$(k = 0)$
$2x$	$-\cos x$	$(k = 1)$
2	$-\sin x$	$(k = 2)$

$$\int x^2 \cos x\, dx = x^2 \sin x - (-2x \cos x) + (-2 \sin x) + C$$

$$= x^2 \sin x + 2x \cos x - 2 \sin x + C$$

EXAMPLE 8.3   Evaluate the integral $\int x^3 e^{2x}\, dx$ using Equation 8.3.

Solution   Let $u_0 = x^3$ and $dv_0 = e^{2x}\, dx$, so that $v_0 = e^{2x}/2$. From the accompanying table we get

u_k	v_k
x^3	$\frac{1}{2}e^{2x}$
$3x^2$	$\frac{1}{4}e^{2x}$
$6x$	$\frac{1}{8}e^{2x}$
6	$\frac{1}{16}e^{2x}$

$$\int x^3 e^{2x}\, dx = \frac{x^3 e^{2x}}{2} - \frac{3x^2 e^{2x}}{4} + \frac{6x e^{2x}}{8} - \frac{6 e^{2x}}{16} + C$$

$$= \frac{e^{2x}}{8}(4x^3 - 6x^2 + 6x - 3) + C \qquad \blacksquare$$

EXAMPLE 8.4   Evaluate the integral $\int x \ln x\, dx$.

Solution   In view of the choices of u and dv in the preceding examples, it might be tempting to try $u = x$ and $dv = \ln x\, dx$. But this choice would not help, since we do not know an integral of $\ln x$. So we could not find v. Instead, we take

$$u = \ln x \qquad dv = x\, dx$$
$$du = \frac{1}{x}\, dx \qquad v = \int x\, dx = \frac{x^2}{2}$$

Then we have

$$\int x \ln x\, dx = (\ln x)\left(\frac{x^2}{2}\right) - \int \frac{x^2}{2} \cdot \frac{1}{x}\, dx$$

$$= \frac{x^2}{2} \ln x - \frac{1}{2} \int x\, dx = \frac{x^2}{2} \ln x - \frac{x^2}{4} + C \qquad \blacksquare$$

Definite Integrals

Integration by parts can be applied to a definite integral in two ways.

Method 1: We include limits of integration in Equation 8.1, getting

$$\int_a^b f(x) g'(x)\, dx = f(x) g(x) \Big]_a^b - \int_a^b g(x) f'(x)\, dx$$

or, in the briefer notation of Equation 8.2,

$$\int_a^b u\, dv = uv \Big]_a^b - \int_a^b v\, du$$

It is understood in this last formula that a and b are limits for x and, as before, $u = f(x)$ and $v = g(x)$.

Method 2: First find the indefinite integral

$$\int u\, dv = uv - \int v\, du$$

and after the result is obtained, substitute upper and lower limits.

We illustrate both methods in the next example.

EXAMPLE 8.5   Evaluate the integral $\displaystyle\int_0^1 \tan^{-1} x\, dx$.

Solution   Note that we can interpret this integral geometrically as the area under the graph of $y = \tan^{-1} x$ from $x = 0$ to $x = 1$. (See Figure 8.2.)

Method 1: At first glance this integral does not seem suited to integration by parts, since there appears to be only one "part." But we can take dv as dx itself. (We wouldn't want to take $dv = \tan^{-1} x\, dx$. Why not?) Then we have

$$u = \tan^{-1} x \qquad dv = dx$$
$$du = \frac{1}{1+x^2}\, dx \qquad v = \int dx = x$$

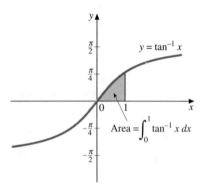

$y = \tan^{-1} x$

Area $= \displaystyle\int_0^1 \tan^{-1} x\, dx$

FIGURE 8.2

Thus,

$$\int_0^1 \tan^{-1} x\, dx = x \tan^{-1} x \Big]_0^1 - \int_0^1 \frac{x}{1+x^2}\, dx \qquad \text{Let } w = 1 + x^2. \text{ Then } dw = 2x\, dx.$$

$$= 1 \tan^{-1} 1 - 0 \tan^{-1} 0 - \frac{1}{2} \int_0^1 \frac{2x\, dx}{1+x^2}$$

$$= \frac{\pi}{4} - \frac{1}{2} \left[\ln(1 + x^2)\right]_0^1$$

$$= \frac{\pi}{4} - \frac{1}{2} \ln 2$$

Method 2: Taking u and dv as above, we have

$$\int \tan^{-1} x\, dx = x \tan^{-1} x - \int \frac{x}{1+x^2}\, dx$$

$$= x \tan^{-1} x - \frac{1}{2} \int \frac{2x}{1+x^2}\, dx$$

$$= x \tan^{-1} x - \frac{1}{2} \ln(1 + x^2) + C$$

Thus, the simplest antiderivative of $\tan^{-1} x$ is the one with $C = 0$:

$$x \tan^{-1} x - \frac{1}{2} \ln(1 + x^2)$$

So, by the Second Fundamental Theorem,

$$\int_0^1 \tan^{-1} x \, dx = \left[x \tan^{-1} x - \frac{1}{2} \ln(1 + x^2) \right]_0^1$$

$$= \tan^{-1} 1 - \frac{1}{2} \ln 2$$

$$= \frac{\pi}{4} - \frac{1}{2} \ln 2$$

In general, the first method is preferable, since limits are substituted in the uv term as the work progresses, which sometimes simplifies the expression.

■

REMARK ──────────────────────────────────────

■ Examples 8.4 and 8.5 suggest that when an integration involving a logarithm or an inverse trigonometric function (also an inverse hyperbolic function) is to be done by parts, u should be taken as the logarithm or the inverse function.

──

Recurrence of the Original Integral

The next two examples show how repeated integration by parts can result in the recurrence of the original integral and how it can then be evaluated.

EXAMPLE 8.6   Evaluate the integral $\int e^{-x} \cos 2x \, dx$.

Solution   There is no clear choice for u and dv in this case. Let us try

$$u = e^{-x} \qquad dv = \cos 2x \, dx$$

Then

$$du = -e^{-x} \, dx \qquad v = \int \cos 2x \, dx = \frac{1}{2} \sin 2x$$

Note that Equation 8.3 for repeated integration by parts is not applicable, since u is not a polynomial. We have

$$\int e^{-x} \cos 2x \, dx = \frac{1}{2} e^{-x} \sin 2x - \int \left(\frac{1}{2} \sin 2x \right) (-e^{-x} \, dx)$$

$$= \frac{1}{2} e^{-x} \sin 2x + \frac{1}{2} \int e^{-x} \sin 2x \, dx \qquad (8.4)$$

Although we did not achieve any simplification in this case, we can use integration by parts again on the last integral, with

$$u_1 = e^{-x} \qquad dv_1 = \sin 2x \, dx$$
$$du_1 = -e^{-x} \, dx \qquad v_1 = \int \sin 2x \, dx = -\tfrac{1}{2} \cos 2x$$

Thus,

$$\int e^{-x} \sin 2x \, dx = -\frac{1}{2} e^{-x} \cos 2x - \int \left(-\frac{1}{2} \cos 2x \right) (-e^{-x} \, dx)$$

$$= -\frac{1}{2} e^{-x} \cos 2x - \frac{1}{2} \int e^{-x} \cos 2x \, dx$$

Substituting this result in Equation 8.4 gives

$$\int e^{-x} \cos 2x \, dx = \frac{1}{2} e^{-x} \sin 2x + \frac{1}{2} \left[-\frac{1}{2} e^{-x} \cos 2x - \frac{1}{2} \int e^{-x} \cos 2x \, dx \right]$$

or

$$\int e^{-x} \cos 2x \, dx = \frac{1}{2} e^{-x} \sin 2x - \frac{1}{4} e^{-x} \cos 2x - \frac{1}{4} \int e^{-x} \cos 2x \, dx \quad (8.5)$$

It might seem that we are going in circles, since the integral we are attempting to evaluate occurs on both sides of Equation 8.5. But the way out of this dilemma is simple—we just solve the equation algebraically for the unknown integral. It is helpful to introduce a single letter—say, I—for the unknown integral:

$$I = \int e^{-x} \cos 2x \, dx$$

Then Equation 8.5 becomes

$$I = \frac{1}{2} e^{-x} \sin 2x - \frac{1}{4} e^{-x} \cos 2x - \frac{1}{4} I$$

To solve for I, we add $\frac{1}{4} I$ to both sides:

$$\frac{5I}{4} = \frac{e^{-x}}{4} (2 \sin 2x - \cos 2x)$$

$$I = \frac{e^{-x}}{5} (2 \sin 2x - \cos 2x)$$

In this process we did not introduce a constant of integration. To get the most general antiderivative, we must do so now:

$$\int e^{-x} \cos 2x \, dx = \frac{e^{-x}}{5} (2 \sin 2x - \cos 2x) + C \qquad \blacksquare$$

REMARK ───

■ Example 8.6 is typical of integrals of the form

$$\int e^{ax} \cos bx \, dx \quad \text{or} \quad \int e^{ax} \sin bx \, dx$$

In such integrals it is immaterial whether we take u as the exponential or the trigonometric function. We will ask you to redo Example 8.6 in the exercises (Exercise 23), letting $u = \cos 2x$ and $dv = e^{-x} dx$ to show that this choice works just as well. We must emphasize, however, that whatever choice you make initially, you *must* make the same type of choice when you integrate by parts the second time. Otherwise, you really will go in circles, winding up with the true but unhelpful statement that the original integral equals itself!

───

EXAMPLE 8.7   Evaluate the integral $\int \sec^3 x \, dx$.

Solution   We write the integral in the form $\int \sec x (\sec^2 x \, dx)$. The reason for writing it in this way is to get the factor $\sec^2 x$ with the dx, since we know its antiderivative is $\tan x$. Now we choose

$$u = \sec x \qquad dv = \sec^2 x \, dx$$

Then

$$du = \sec x \tan x \, dx \qquad v = \int \sec^2 x \, dx = \tan x$$

Then we have

$$\int \sec^3 x \, dx = \sec x \tan x - \int \tan x (\sec x \tan x \, dx)$$

$$= \sec x \tan x - \int \sec x \tan^2 x \, dx$$

Now we use the trigonometric identity $\tan^2 x = \sec^2 x - 1$ in the last integral to obtain

$$\int \sec^3 x \, dx = \sec x \tan x - \int \sec x (\sec^2 x - 1) \, dx$$

$$= \sec x \tan x - \int \sec^3 x \, dx + \int \sec x \, dx$$

$$= \sec x \tan x - \int \sec^3 x \, dx + \ln|\sec x + \tan x| \qquad \text{From Equation 7.9}$$

Here again, the unknown integral appears on both sides. Calling this integral I, we get

$$I = \sec x \tan x - I + \ln|\sec x + \tan x|$$

$$2I = \sec x \tan x + \ln|\sec x + \tan x|$$

Finally, we divide by 2 and add a constant of integration:

$$\int \sec^3 x \, dx = \frac{1}{2}(\sec x \tan x + \ln|\sec x + \tan x|) + C \qquad \blacksquare$$

Exercise Set 8.1

Use integration by parts to evaluate the integrals in Exercises 1–29.

1. $\displaystyle\int xe^{-x} \, dx$

2. $\displaystyle\int x \sin x \, dx$

3. $\displaystyle\int_0^{\pi/4} x \cos 2x \, dx$

4. $\displaystyle\int x \sec^2 x \, dx$

5. $\displaystyle\int x^4 e^x \, dx$

6. $\displaystyle\int x^2 \cos^3 x \, dx$

7. $\displaystyle\int x \sec x \tan x \, dx$

8. $\displaystyle\int x \sinh 2x \, dx$

9. $\displaystyle\int_0^{\ln 2} x^2 \cosh x \, dx$

10. $\displaystyle\int \ln x \, dx$

11. $\displaystyle\int x \csc^2 x \, dx$

12. $\displaystyle\int_0^{\sqrt{3}/2} \sin^{-1} x \, dx$

13. $\displaystyle\int x \ln \frac{1}{x} \, dx$

14. $\displaystyle\int_1^4 \frac{\ln x}{\sqrt{x}} \, dx$

15. $\displaystyle\int \cos x \ln \sin x \, dx$

16. $\displaystyle\int x^3 (1 - x)^9 \, dx$

17. $\displaystyle\int_1^2 \frac{x}{(2x - 1)^2} \, dx$

18. $\displaystyle\int xe^{1-2x} \, dx$

19. $\displaystyle\int x^3 e^{-2x} \, dx$

20. $\displaystyle\int x^3 e^{x^2} \, dx$ (*Hint:* Take $u = x^2$.)

21. $\displaystyle\int x \tan^2 x \, dx$

22. $\displaystyle\int \sqrt{x} \ln x^3 \, dx$

23. Redo Example 8.6, taking $u = \cos 2x$ and $dv = e^{-x} \, dx$.

24. $\displaystyle\int e^x \sin x\, dx$ **25.** $\displaystyle\int e^x \cosh x\, dx$

26. $\displaystyle\int \csc^3 x\, dx$ **27.** $\displaystyle\int \mathrm{sech}^3 x\, dx$

28. $\int x \tan^{-1} x\, dx$ (*Hint:* In evaluating the integral that results from integration by parts, use long division.)

29. $\int x^2 \cot^{-1} x\, dx$ (See the hint for Exercise 28.)

30. Find the area under the graph of $y = \ln x^2$ from $x = 1$ to $x = 3$. Does $\ln x^2 = 2\ln x$ for x between 1 and 3? Does $\ln x^2 = 2\ln x$ for all $x \neq 0$? Explain.

31. Find the area of the region bounded by the graphs of $y = x \sin x$ and $y = x \cos x$ from $x = \pi/2$ to $x = \pi$.

32. Find the volume of the solid obtained by revolving the region under the graph of $y = e^x$ from $x = 0$ to $x = \ln 3$ about the y-axis.

33. Find the volume of the solid obtained by revolving the region under the graph of $y = \cos(x/2)$ from $x = 0$ to $x = \pi$ about the y-axis.

34. Find the surface area generated by revolving the arc of $y = \cosh x$ from $(0, 1)$ to $(2, \cosh 2)$ about the y-axis.

35. (a) Find the x-coordinate of the centroid of the region bounded by $y = \sin x$ and the x-axis from $x = 0$ to $x = \pi/2$.
(b) Redo part (a) with $y = \cos x$ replacing $y = \sin x$.

36. Find the centroid of the region bounded by $y = \ln x$ and the x-axis from $x = 1$ to $x = e$.

37. Evaluate the integral $\int \sec^2 x \tan x\, dx$ by parts in two ways:
(a) choosing $u = \sec x$
(b) choosing $u = \tan x$

Show that the two answers are equivalent.

Use integration by parts to evaluate the integrals in Exercises 38–45.

38. $\displaystyle\int \sin\sqrt{x}\, dx$ (*Hint:* First make a substitution.)

39. $\displaystyle\int \cos(\ln x)\, dx$ (*Hint:* See the hint for Exercise 38.)

40. $\displaystyle\int x \sin^2 x\, dx$ **41.** $\displaystyle\int x \cos^2 x\, dx$

42. $\displaystyle\int \frac{x^3\, dx}{\sqrt{1 - x^2}}$ **43.** $\displaystyle\int x^3 (x^2 - 4)^{3/2}\, dx$

44. $\displaystyle\int e^{ax} \sin bx\, dx$ **45.** $\displaystyle\int e^{ax} \cos bx\, dx$

In Exercises 46–48, derive the given reduction formulas, where n is a positive integer greater than 1.

46. $\displaystyle\int (\ln x)^n\, dx = x(\ln x)^n - n \int (\ln x)^{n-1}\, dx$

47. (a) $\displaystyle\int x^n \sin x\, dx = -x^n \cos x + n \int x^{n-1} \cos x\, dx$

(b) $\displaystyle\int x^n \cos x\, dx = x^n \sin x - n \int x^{n-1} \sin x\, dx$

48. (a) $\displaystyle\int \sin^n x\, dx$

$\displaystyle = -\frac{\sin^{n-1} x \cos x}{n} + \frac{n-1}{n} \int \sin^{n-2} x\, dx$

(b) $\displaystyle\int \cos^n x\, dx$

$\displaystyle = \frac{\cos^{n-1} x \sin x}{n} + \frac{n-1}{n} \int \cos^{n-2} x\, dx$

49. Prove **Wallis's Formula** for $n > 1$:

$$\int_0^{\pi/2} \sin^n x\, dx = \int_0^{\pi/2} \cos^n x\, dx$$

$$= \begin{cases} \dfrac{2 \cdot 4 \cdot 6 \cdots (n-1)}{1 \cdot 3 \cdot 5 \cdots n} & \text{if } n \text{ is odd} \\[2ex] \dfrac{1 \cdot 3 \cdot 5 \cdots (n-1)}{2 \cdot 4 \cdot 6 \cdots n} \cdot \dfrac{\pi}{2} & \text{if } n \text{ is even} \end{cases}$$

(*Hint:* Use the result of Exercise 48.)

50. Use Wallis's Formula from Exercise 49 to evaluate each of the following:

(a) $\displaystyle\int_0^{\pi/2} \sin^5 x\, dx$

(b) $\displaystyle\int_0^{\pi/2} \cos^6 x\, dx$

(c) $\displaystyle\int_0^{\pi/2} \sin^{10} x\, dx$

(d) $\displaystyle\int_0^{\pi/2} \cos^7 x\, dx$

51. Prove the reduction formulas
(a) $\displaystyle\int \sec^n x\, dx$

$\displaystyle = \frac{\sec^{n-2} x \tan x}{n-1} + \frac{n-2}{n-1} \int \sec^{n-2} x\, dx$

(b) $\displaystyle\int \csc^n x\, dx$

$\displaystyle = -\frac{\csc^{n-2} x \cot x}{n-1} + \frac{n-2}{n-1} \int \csc^{n-2} x\, dx$

where n is a positive integer greater than 1.

52. Prove Equation 8.3 for repeated integration by parts.

8.2 INTEGRATION OF TRIGONOMETRIC FORMS

In this section we consider integrals of the following types:

I. $\displaystyle\int \sin^m x \cos^n x \, dx$

II. $\displaystyle\int \sec^m x \tan^n x \, dx$

III. $\displaystyle\int \sin mx \cos nx \, dx, \quad \int \sin mx \sin nx \, dx, \quad \text{and} \quad \int \cos mx \cos nx \, dx$

Unless otherwise specified, we will assume that m and n are nonnegative integers.

Type I: $\displaystyle\int \sin^m x \cos^n x \, dx$

The approach to take with integrals of the form $\int \sin^m x \cos^n x \, dx$ depends on whether at least one of the exponents is odd or both are even. We begin with the case in which at least one of these exponents is odd. For definiteness, suppose n is odd. Our objective is to replace the given integral by one or more integrals of the form

$$\int \sin^k x \cos x \, dx$$

We can then make the substitution $u = \sin x$, so that $du = \cos x \, dx$, to obtain

$$\int \sin^k x \cos x \, dx = \int u^k \, du = \frac{u^{k+1}}{k+1} + C = \frac{\sin^{k+1} x}{k+1} + C$$

We can accomplish our objective by using the trigonometric identity $\cos^2 x = 1 - \sin^2 x$. We illustrate this procedure in the next example.

m or n Odd

EXAMPLE 8.8   Evaluate the integral

$$\int \sin^2 x \cos^3 x \, dx$$

Solution   We make use of the fact that the differential of $\sin x$ is $\cos x \, dx$. First we rewrite the integral in the form

$$\int (\sin^2 x)(\cos^2 x)(\cos x \, dx)$$

putting one of the $\cos x$ factors with the differential dx. We use the trigonometric identity $\cos^2 x = 1 - \sin^2 x$ to obtain

$$\int \sin^2 x (1 - \sin^2 x)(\cos x \, dx) = \int (\sin^2 x - \sin^4 x)(\cos x \, dx)$$

Now we see that the substitution $u = \sin x$, $du = \cos x \, dx$, gives

$$\int (\sin^2 x - \sin^4 x)(\cos x \, dx) = \int (u^2 - u^4) \, du = \frac{u^3}{3} - \frac{u^5}{5} + C$$

All that remains is to replace u by $\sin x$ to get the final result

$$\int \sin^2 x \cos^3 x \, dx = \frac{\sin^3 x}{3} - \frac{\sin^5 x}{5} + C$$

By now, you can probably handle substitutions of this sort mentally. The important step is to write the integrand in powers of $\sin x$, with the differential $\cos x \, dx$. ∎

If the exponent m in the integral $\int \sin^m x \cos^n x \, dx$ is odd, we follow a similar procedure to that in Example 8.8, using the fact that the differential of $\cos x$ is $(-\sin x) \, dx$ and that $\sin^2 x = 1 - \cos^2 x$.

m and n Both Even

The next example shows the approach to take when both m and n are even. We will need the following trigonometric identities that are obtained from the *double-angle formulas* for $\cos 2x$. (See Appendix 2.)

$$\sin^2 x = \frac{1 - \cos 2x}{2} \tag{8.6}$$

$$\cos^2 x = \frac{1 + \cos 2x}{2} \tag{8.7}$$

When we use these identities, we can replace our integral with integrals involving powers of $\cos 2x$. For odd powers, we can use the technique of Example 8.8. For even powers, we apply Identity 8.7 again (with $2x$ in place of x), obtaining integrals involving powers of $\cos 4x$. We continue in this way until we are able to evaluate each of the integrals. We illustrate this technique in the next example.

EXAMPLE 8.9   Evaluate the integral

$$\int \sin^4 x \cos^2 x \, dx$$

Solution   Since $\sin^4 x = (\sin^2 x)^2$, we can use Identities 8.6 and 8.7 to get

$$\int \sin^4 x \cos^2 x \, dx = \int \left(\frac{1 - \cos 2x}{2}\right)^2 \left(\frac{1 + \cos 2x}{2}\right) dx$$

Now we expand the first factor on the right and then multiply the result by the second factor. The algebra involved is a bit tedious, and we omit the details. The result (which you should check) is

$$\frac{1}{8} \int (1 - \cos 2x - \cos^2 2x + \cos^3 2x) \, dx$$

which can be written as four integrals:

$$\frac{1}{8} \left[\int 1 \, dx - \int \cos 2x \, dx - \int \cos^2 2x \, dx + \int \cos^3 2x \, dx \right] \tag{8.8}$$

The first two of these integrals can easily be evaluated. The third integral again involves an even power, and we can evaluate it by applying Identity 8.7 again (with $2x$ replacing x), as follows.

$$\int \cos^2 2x\, dx = \int \frac{1 + \cos 4x}{2}\, dx = \frac{1}{2}\int (1 + \cos 4x)\, dx$$

$$= \frac{x}{2} + \frac{\sin 4x}{8}$$

(We will supply the constant of integration at the end.) The fourth integral in Formula 8.8 involves an odd power of $\cos 2x$, so we proceed as in Example 8.8. That is, we write

$$\int \cos^3 2x\, dx = \int \cos^2 2x (\cos 2x\, dx)$$

$$= \int (1 - \sin^2 2x)(\cos 2x\, dx)$$

Let $u = \sin 2x$. Then $du = 2\cos 2x\, dx$. So we have

$$\int \cos^3 2x\, dx = \frac{1}{2}\int (1 - \sin^2 2x)(2\cos 2x) = \frac{1}{2}\int (1 - u^2)\, du$$

$$= \frac{1}{2}\left(u - \frac{u^3}{3}\right) = \frac{1}{2}\left(\sin 2x - \frac{\sin^3 2x}{3}\right)$$

Now we can put everything together in Formula 8.8 to write the final answer (adding a constant of integration),

$$\int \sin^4 x \cos^2 x\, dx$$

$$= \frac{1}{8}\left[x - \frac{\sin 2x}{2} - \left(\frac{x}{2} + \frac{\sin 4x}{8}\right) + \frac{1}{2}\left(\sin 2x - \frac{\sin^3 2x}{3}\right)\right] + C$$

$$= \frac{x}{16} - \frac{\sin^3 2x}{48} - \frac{\sin 4x}{64} + C \qquad\blacksquare$$

We can summarize our approaches to integrals of Type I as follows:

Type I: $\displaystyle\int \sin^m x \cos^n x\, dx$

Case (i): n is odd. Factor off $\cos x$ to go with dx. Then use the identity $\cos^2 x = 1 - \sin^2 x$ to write the rest of the integrand in powers of $\sin x$. Make the substitution $u = \sin x$.

Case (ii): m is odd. Factor off $\sin x$ to go with dx. Then use the identity $\sin^2 x = 1 - \cos^2 x$ to write the rest of the integrand in powers of $\cos x$. Make the substitution $u = \cos x$ and note that $du = -\sin x\, dx$. You will have to adjust the differential to obtain the minus sign.

Case (iii): m and n are both even. Use Identities 8.6 and 8.7 to reduce the powers. In the resulting integrals if an odd power of $\cos 2x$ occurs, use Case (ii). If an even power occurs, use Case (iii) again.

REMARKS ———————————————————————————————————

■ In Case (i), m need not be a nonnegative integer. It may be negative, or it may not be an integer. For example,

$$\int \frac{\cos^3 x}{\sin^2 x} dx \qquad \text{and} \qquad \int \frac{\cos^5 x}{\sqrt{\sin x}} dx$$

both can be handled as in Case (i). Similarly, in Case (ii) n need not be a nonnegative integer.

■ In Case (iii) m and n must both be nonnegative integers (so one or the other could be 0). For example, both

$$\int \sin^2 x \, dx \qquad \text{and} \qquad \int \cos^4 x \, dx$$

fall under Case (iii).

■ An alternative way to handle integrals of the form

$$\int \sin^m x \, dx \qquad \text{and} \qquad \int \cos^n x \, dx$$

is to use the reduction formulas in Exercise 48 of Exercise Set 8.1.

Type II: $\int \sec^m x \tan^n x \, dx$

For integrals of the form $\int \sec^m x \tan^n x \, dx$, the approach again depends on whether the exponents m and n are even or odd. The approach in each case is similar to that used in Type I. The trigonometric identities we will need are

$$\sec^2 x = 1 + \tan^2 x$$

$$\tan^2 x = \sec^2 x - 1$$

(You can obtain these identities from $\sin^2 x + \cos^2 x = 1$ by dividing both sides by $\cos^2 x$ and then rearranging terms.) We also need the differentials

$$d(\tan x) = \sec^2 x \, dx$$

$$d(\sec x) = \sec x \tan x \, dx$$

We illustrate the various cases with examples.

m Even

EXAMPLE 8.10   Evaluate the integral

$$\int \sec^4 x \tan^6 x \, dx$$

Solution   Since the differential of $\tan x$ is $\sec^2 x \, dx$, we split off the factor $\sec^2 x$ to go with the differential and write the rest of the integrand in terms of $\tan x$, using the identity $\sec^2 x = 1 + \tan^2 x$. We *must* have an even power of $\sec x$. Otherwise, when we factor off $\sec^2 x$, we will be left with an odd power of $\sec x$, and the substitution $\sec^2 x = 1 + \tan^2 x$ will lead to an irrational expression.

$$\int \sec^4 x \tan^6 x \, dx = \int \sec^2 x \tan^6 x (\sec^2 x \, dx)$$

$$= \int (1 + \tan^2 x) \tan^6 x (\sec^2 x \, dx)$$

$$= \int (\tan^6 x + \tan^8 x)(\sec^2 x \, dx)$$

Now we let $u = \tan x$. Then $du = \sec^2 x \, dx$. So our integral becomes

$$\int (u^6 + u^8) \, du = \frac{u^7}{7} + \frac{u^9}{9} + C$$

Replacing u by $\tan x$ gives the final result

$$\int \sec^4 x \tan^6 x \, dx = \frac{\tan^7 x}{7} + \frac{\tan^9 x}{9} + C$$

The key to this approach is that $\sec x$ is raised to a positive *even* power. That is, in the integral $\int \sec^m x \tan^n x \, dx$, the exponent m is even and positive.   ∎

n Odd

EXAMPLE 8.11   Evaluate the integral

$$\int \sec^5 x \tan^3 x \, dx$$

Solution   This time we use the fact that the differential of $\sec x$ is $\sec x \tan x \, dx$, and split off the factor $\sec x \tan x$ to go with the differential. Then we are left with an even power of $\tan x$ (since the exponent on $\tan x$ was odd and we factored off one $\tan x$), so we can use the identity $\tan^2 x = \sec^2 x - 1$ to write the rest of the integrand in terms of $\sec x$:

$$\int \sec^5 x \tan^3 x \, dx = \int \sec^4 x \tan^2 x (\sec x \tan x \, dx)$$

$$= \int \sec^4 x (\sec^2 x - 1)(\sec x \tan x \, dx)$$

$$= \int (\sec^6 x - \sec^4 x)(\sec x \tan x \, dx)$$

Letting $u = \sec x$, we see that $du = \sec x \tan x \, dx$. Thus,

$$\int \sec^5 x \tan^3 x \, dx = \int (u^6 - u^4) \, du = \frac{u^7}{7} - \frac{u^5}{5} + C = \frac{\sec^7 x}{7} - \frac{\sec^5 x}{5} + C$$

The key to assuring success in this approach is that $\tan x$ is raised to a positive *odd* power. That is, in the integral $\int \sec^m x \tan^n x \, dx$, the exponent n is an odd integer greater than or equal to 1.   ∎

REMARK ───────────────────────────────────

■ When $\tan x$ is raised to an odd power (and $\sec x$ is raised to *any* power, not necessarily an integral power), an alternative approach is to substitute

$$\sec x = \frac{1}{\cos x} \quad \text{and} \quad \tan x = \frac{\sin x}{\cos x}$$

The integral then is of Type I. In Exercise 41 of Exercise Set 8.2 we will ask you to redo Example 8.11, using this approach.

───

m **Odd,** n **Even**

There are four possible combinations for m and n:

m even, n even m even, n odd

m odd, n even m odd, n odd

We have already handled the first two cases by considering m even (n could be either even or odd) and the fourth case by considering n odd (m could be even or odd). The remaining case, in which m is odd and n is even (both nonnegative), can be changed to one or more integrals of the form $\int \sec^k x \, dx$ by replacing $\tan^2 x$ with $\sec^2 x - 1$. Then we can use integration by parts, as in Example 8.7, or alternatively, the reduction formula given in Exercise 51(a) of Exercise Set 8.1. For example, we can write

$$\int \sec^3 x \tan^2 x \, dx = \int \sec^3 x (\sec^2 x - 1) \, dx = \int (\sec^5 x - \sec^3 x) \, dx$$

$$= \int \sec^5 x \, dx - \int \sec^3 x \, dx$$

To complete the integration, we could use integration by parts, along with the result of Example 8.7 (see Exercise 38 in Exercise Set 8.2).

We can summarize the three cases for integrating $\int \sec^m x \tan^n x \, dx$ as follows:

───

Type II: $\int \sec^m x \tan^n x \, dx$

Case (i): m is even. Factor off $\sec^2 x$ to go with the differential. Then use the identity $\sec^2 x = 1 + \tan^2 x$ to write the rest of the integrand in terms of $\tan x$. Make the substitution $u = \tan x$.

Case (ii): n is odd. Factor off $\sec x \tan x$ to go with the differential. Then use the identity $\tan^2 x = \sec^2 x - 1$ to write the rest of the integrand in terms of $\sec x$. Make the substitution $u = \sec x$.

Case (iii): m is odd and n is even. Use the identity $\tan^2 x = \sec^2 x - 1$ to change the integral to one or more integrals of the form $\int \sec^k x \, dx$. Then use integration by parts, with $dv = \sec^2 x \, dx$, or use the reduction formula of Exercise 51(a) in Exercise Set 8.1.

───

REMARK ——————————————————————————————

■ Integrals of the form $\int \csc^m x \cot^n x \, dx$ are handled as in Type II, using $\csc^2 x = 1 + \cot^2 x$, together with the differentials $d(\cot x) = -\csc^2 x \, dx$ and $d(\csc x) = -\csc x \cot x \, dx$.

———————————————————————————————————————

Type III $\int \sin mx \cos nx \, dx,$ $\int \sin mx \sin nx \, dx,$ **or** $\int \cos mx \cos nx \, dx$ **(m and n Positive Integers)**

For integrals of one of these forms where $m \neq n$, we use one of the following trigonometric identities, known as *product formulas* (see Appendix 2):

$$\sin A \cos B = \frac{1}{2}[\sin(A + B) + \sin(A - B)] \tag{8.9}$$

$$\sin A \sin B = \frac{1}{2}[\cos(A - B) - \cos(A + B)] \tag{8.10}$$

$$\cos A \cos B = \frac{1}{2}[\cos(A + B) + \cos(A - B)] \tag{8.11}$$

We illustrate the technique with the following example, which is typical of Type III problems.

EXAMPLE 8.12   Evaluate the integral

$$\int \sin 5x \cos 2x \, dx$$

Solution   By Identity 8.9, we can write

$$\int \sin 5x \cos 2x \, dx = \frac{1}{2} \int (\sin 7x + \sin 3x) \, dx$$

$$= \frac{1}{2}\left[-\frac{\cos 7x}{7} - \frac{\cos 3x}{3}\right] + C$$

$$= -\frac{\cos 7x}{14} - \frac{\cos 3x}{6} + C$$

An alternative approach to this problem is to integrate by parts twice, solving the resulting equation for the original integral. We would obtain a different form of the answer, but the two results would be equivalent.   ■

Exercise Set 8.2

In Exercises 1–40, evaluate the integrals.

1. $\int \cos^3 x \, dx$

2. $\int \sin^5 x \, dx$

3. $\int \sin^4 x \, dx$

4. $\int \cos^2 x \, dx$

5. $\int_0^{\pi/4} \sin^3 x \cos^2 x \, dx$

6. $\int \sin^2 x \cos^5 x \, dx$

7. $\int \sin^2 x \cos^2 x \, dx$

8. $\int \sin^2 x \cos^4 x \, dx$

9. $\displaystyle\int_0^{\pi/4} \sec^2 x \tan^3 x \, dx$

10. $\displaystyle\int \sec^4 x \tan^2 x \, dx$

11. $\displaystyle\int \sec x \tan^3 x \, dx$

12. $\displaystyle\int_0^{\pi/3} \sec^5 x \tan x \, dx$

13. $\displaystyle\int \sec x \tan^2 x \, dx$

14. $\displaystyle\int \sec^4 x \tan^3 x \, dx$ (Integrate by two methods.)

15. $\displaystyle\int_0^{\pi/3} \frac{\sin^3 x \, dx}{\cos x}$

16. $\displaystyle\int_{\pi/2}^{\pi} \sqrt{\sin x}\, \cos^3 x \, dx$

17. $\displaystyle\int \frac{\sec^4 x \, dx}{\tan^2 x}$

18. $\displaystyle\int \sec^{3/2} x \tan x \, dx$

19. $\displaystyle\int_0^{\pi/3} \tan^3 x \, dx$

20. $\displaystyle\int \tan^4 x \, dx$

21. $\displaystyle\int \sin^2 3x \cos^3 3x \, dx$

22. $\displaystyle\int_{\pi/12}^{\pi/4} \frac{\cos^5 2x}{\sin^4 2x} \, dx$

23. $\displaystyle\int_0^{\pi/8} \sqrt{\tan 2x}\, \sec^4 2x \, dx$

24. $\displaystyle\int \csc x \cot^2 x \, dx$

25. $\displaystyle\int x \sin^2(x^2) \cos^5(x^2) \, dx$ (*Hint:* Make the substitution $u = x^2$.)

26. $\displaystyle\int \frac{\tan^3 \sqrt{x}\, \sec^4 \sqrt{x}\, dx}{\sqrt{x}}$ (*Hint:* Make the substitution $u = \sqrt{x}$.)

27. $\displaystyle\int \csc^2 \frac{x}{2} \cot^2 \frac{x}{2} \, dx$

28. $\displaystyle\int \sin^4 x \cos^4 x \, dx$

29. $\displaystyle\int \sinh x \cosh^3 x \, dx$

30. $\displaystyle\int \sinh^2 x \cosh^2 x \, dx$

31. $\displaystyle\int \tanh^3 x \operatorname{sech}^4 x \, dx$

32. $\displaystyle\int \tanh^3 x \operatorname{sech}^3 x \, dx$

33. $\displaystyle\int \sin 3x \cos 2x \, dx$

34. $\displaystyle\int \sin 3x \sin 2x \, dx$

35. $\displaystyle\int \cos 3x \cos 2x \, dx$

36. $\displaystyle\int \sin x \cos 2x \, dx$

37. $\displaystyle\int \sin^6 x \, dx$

38. $\displaystyle\int \sec^3 x \tan^2 x \, dx$

39. $\displaystyle\int \frac{\tan^5 x \, dx}{\sec x}$

40. $\displaystyle\int (\sinh^3 x)(1 + \sinh^2 x)^{3/2} \, dx$

41. Redo Example 8.11 by writing the integrand in terms of $\sin x$ and $\cos x$.

42. For m and n positive integers, let

$$\delta_{mn} = \begin{cases} 1 & \text{if } m = n \\ 0 & \text{if } m \neq n \end{cases}$$

(δ_{mn} is known as the *Kronecker delta*.) Prove the following:

(a) $\displaystyle\int_{-\pi}^{\pi} \sin mx \sin nx \, dx = \int_{-\pi}^{\pi} \cos mx \cos nx \, dx = \pi \delta_{mn}$

(b) $\displaystyle\int_{-\pi}^{\pi} \sin mx \cos nx \, dx = 0$

43. A set $\{\phi_1(x), \phi_2(x), \ldots\}$ of continuous functions is said to be **orthogonal** on $[a, b]$ if

$$\int_a^b \phi_m(x)\phi_n(x) \, dx = 0 \qquad (m \neq n)$$

Prove that the set

$$\{1, \sin x, \cos x, \sin 2x, \cos 2x, \sin 3x, \cos 3x, \ldots\}$$

is orthogonal on $[-\pi, \pi]$. (*Hint:* See Exercise 42.)

44. Prove the following:
(a) $\int \sin^m x \cos^{2n+1} x \, dx = \int u^m (1 - u^2)^n \, du$,   where $u = \sin x$ and n is a nonnegative integer.
(b) $\int \sin^{2m+1} x \cos^n x \, dx = -\int u^n (1 - u^2)^m \, du$,   where $u = \cos x$ and m is a nonnegative integer.
(c) $\int \sec^{2m+2} x \tan^n x \, dx = \int u^n (1 + u^2)^m \, du$,   where $u = \tan x$ and m is a nonnegative integer.
(d) $\int \sec^m x \tan^{2n+1} x \, dx = \int u^{m-1} (u^2 - 1)^n \, du$,   where $u = \sec x$ and n is a nonnegative integer.

8.3 TRIGONOMETRIC SUBSTITUTIONS

We have previously encountered integrals of the form $\int_{-a}^{a} \sqrt{a^2 - x^2}\, dx$ and have used the fact that the graph of $y = \sqrt{a^2 - x^2}$ is the upper half of the circle $x^2 + y^2 = a^2$. Then, with our knowledge that the area of the entire circle is πa^2, we were able to arrive at the answer $\pi a^2/2$. In this section we will see another way of obtaining this result without reference to geometry.

More generally, we describe a method called *trigonometric substitution* that often works on integrals involving any one of the following expressions:

$$\sqrt{a^2 - u^2}, \quad \sqrt{a^2 + u^2}, \quad \text{or} \quad \sqrt{u^2 - a^2}$$

where $a > 0$. The appropriate substitutions are as follows:

If integral contains	Substitute
$\sqrt{a^2 - u^2}$	$u = a\sin\theta, \quad -\dfrac{\pi}{2} \le \theta \le \dfrac{\pi}{2}$
$\sqrt{a^2 + u^2}$	$u = a\tan\theta, \quad -\dfrac{\pi}{2} < \theta < \dfrac{\pi}{2}$
$\sqrt{u^2 - a^2}$	$u = a\sec\theta, \quad \begin{cases} 0 \le \theta < \dfrac{\pi}{2} & \text{if } u \ge a \\ \pi \le \theta < \dfrac{3\pi}{2} & \text{if } u \le -a \end{cases}$

(We are using the lowercase Greek letter θ as the substitution variable because it is suggestive of an angle, but other letters could be used.) These choices are based on the trigonometric identities

$$1 - \sin^2\theta = \cos^2\theta$$
$$1 + \tan^2\theta = \sec^2\theta$$
$$\sec^2\theta - 1 = \tan^2\theta$$

Using these identities, called the Pythagorean Identities, and the substitutions shown above, we eliminate the radicals:

$$\sqrt{a^2 - u^2} = \sqrt{a^2 - a^2\sin^2\theta} = \sqrt{a^2(1 - \sin^2\theta)} = a\sqrt{\cos^2\theta} = a\cos\theta$$
$$\sqrt{a^2 + u^2} = \sqrt{a^2 + a^2\tan^2\theta} = \sqrt{a^2(1 + \tan^2\theta)} = a\sqrt{\sec^2\theta} = a\sec\theta$$
$$\sqrt{u^2 - a^2} = \sqrt{a^2\sec^2\theta - a^2} = \sqrt{a^2(\sec^2\theta - 1)} = a\sqrt{\tan^2\theta} = a\tan\theta$$

The final step in each case is justified as follows, by the restrictions we placed on θ. In the first case, since $\cos\theta \ge 0$ for $-\pi/2 \le \theta \le \pi/2$, we have $\sqrt{\cos^2\theta} = |\cos\theta| = \cos\theta$. Similarly, $\sec\theta > 0$ for $-\pi/2 < \theta < \pi/2$, so that $\sqrt{\sec^2\theta} = |\sec\theta| = \sec\theta$, and $\tan\theta > 0$ for θ satisfying either $0 \le \theta \le \pi/2$ or $\pi \le \theta < 3\pi/2$, so that $\sqrt{\tan^2\theta} = |\tan\theta| = \tan\theta$.

REMARK ——————————————————————

■ In the past when we have made a substitution in an integral, we have expressed the new variable in terms of the original one. For example, in the integral $\int \sin^3 x \cos x\, dx$, we would let $u = \sin x$. The new variable u is expressed in terms of the original variable x. Now we are going in reverse. We are saying, for example, that when an integral contains an expression of the form $\sqrt{a^2 - u^2}$

we let $u = a \sin\theta$, with $-\pi/2 \le \theta \le \pi/2$. That is, the original variable u is expressed in terms of the new variable θ. With the restrictions we have placed on θ, however, we can rewrite $u = a \sin\theta$ as $\theta = \sin^{-1}(u/a)$. Similarly, when we substitute $u = a \tan\theta$ with the specified restrictions, the substitution is equivalent to $\theta = \tan^{-1}(u/a)$, and when $u = a \sec\theta$, we have $\theta = \sec^{-1}(u/a)$. From now on, when we make the trigonometric substitutions $u = a \sin\theta$, $u = a \tan\theta$, or $u = a \sec\theta$ in the cases we have described, we will understand that these mean $\theta = \sin^{-1}(u/a)$, $\theta = \tan^{-1}(u/a)$, or $\theta = \sec^{-1}(u/a)$, respectively.

EXAMPLE 8.13   Evaluate the integral

$$\int \frac{x^3}{\sqrt{4 - x^2}}\, dx$$

Solution   We make the substitution $x = 2 \sin\theta$. Then $dx = 2 \cos\theta\, d\theta$ and $\theta = \sin^{-1}(x/2)$. So we have

$$\int \frac{x^3}{\sqrt{4 - x^2}}\, dx = \int \frac{(2 \sin\theta)^3 (2 \cos\theta\, d\theta)}{\sqrt{4 - 4\sin^2\theta}} = \int \frac{(8 \sin^3\theta)(2 \cos\theta\, d\theta)}{2\sqrt{1 - \sin^2\theta}}$$

$$= 8 \int \frac{\sin^3\theta \cos\theta\, d\theta}{\sqrt{\cos^2\theta}} = 8 \int \frac{\sin^3\theta \, \cos\theta\, d\theta}{\cos\theta} = 8 \int \sin^3\theta\, d\theta$$

Now we use ideas of the preceding section:

$$8 \int \sin^3\theta\, d\theta = 8 \int \sin^2\theta(\sin\theta\, d\theta) = 8 \int (1 - \cos^2\theta)\sin\theta\, d\theta \qquad \text{Let } u = \cos\theta.$$

$$= 8 \left(-\cos\theta + \frac{\cos^3\theta}{3} \right) + C$$

But we are not through, since this result must be expressed in terms of x. An easy way is to draw a right triangle as in Figure 8.3, showing θ as an acute angle with its opposite side equal to x and hypotenuse 2. Then the adjacent side is $\sqrt{4 - x^2}$, so that $\cos\theta = \sqrt{4 - x^2}/2$. (We could also find $\cos\theta$ without reference to the figure, using $\cos\theta = \sqrt{1 - \sin^2\theta} = \sqrt{1 - (\frac{x}{2})^2} = \sqrt{4 - x^2}/2$.) Thus,

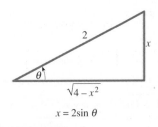

$x = 2\sin\theta$

FIGURE 8.3

$$\int \frac{x^3}{\sqrt{4 - x^2}}\, dx = 8 \left[-\frac{\sqrt{4 - x^2}}{2} + \frac{1}{3} \left(\frac{\sqrt{4 - x^2}}{2} \right)^3 \right] + C$$

$$= -4\sqrt{4 - x^2} + \frac{(4 - x^2)^{3/2}}{3} + C \qquad \blacksquare$$

 When making a substitution, do not forget to substitute for dx.

EXAMPLE 8.14   Evaluate the integral

$$\int_0^4 \frac{dx}{(9 + x^2)^{3/2}}$$

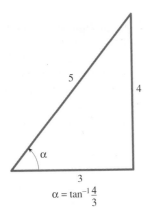

FIGURE 8.4

Solution   We let $x = 3\tan\theta$. Then $dx = 3\sec^2\theta\,d\theta$, and $\theta = \tan^{-1}(x/3)$. We also change the limits of integration. When $x = 0$, we have $\theta = \tan^{-1} 0 = 0$, and when $x = 4$, $\theta = \tan^{-1}(4/3)$. Let us denote $\tan^{-1}(4/3)$ by α (that is, α is the angle whose tangent is $4/3$). Figure 8.4 shows a right triangle with α as one of its acute angles. Our integral now becomes

$$\int_0^\alpha \frac{3\sec^2\theta\,d\theta}{(9 + 9\tan^2\theta)^{3/2}} = \int_0^\alpha \frac{3\sec^2\theta\,d\theta}{27(1 + \tan^2\theta)^{3/2}}$$

$$= \frac{1}{9}\int_0^\alpha \frac{\sec^2\theta}{\sec^3\theta}\,d\theta$$

$$= \frac{1}{9}\int_0^\alpha \cos\theta\,d\theta \quad \text{Since } \frac{1}{\sec\theta} = \cos\theta$$

$$= \frac{1}{9}[\sin\theta]_0^\alpha$$

$$= \frac{1}{9}\sin\alpha$$

From Figure 8.4 we see that $\sin\alpha = 4/5$. Thus, we have the final result

$$\int_0^4 \frac{dx}{(9 + x^2)^{3/2}} = \frac{1}{9}\left(\frac{4}{5}\right) = \frac{4}{45} \qquad \blacksquare$$

EXAMPLE 8.15   Evaluate the integral

$$\int \frac{\sqrt{4x^2 - 9}}{x}\,dx$$

Solution   The radical is of the form $\sqrt{u^2 - a^2}$, with $u = 2x$ and $a = 3$. Using $u = a\sec\theta$, we have $2x = 3\sec\theta$, or

$$x = \frac{3\sec\theta}{2}$$

Then $dx = \frac{3}{2}\sec\theta\tan\theta\,d\theta$ and $\theta = \sec^{-1}\frac{2x}{3}$. So the integral becomes

$$\int \frac{\sqrt{4x^2 - 9}}{x}\,dx = \int \frac{\sqrt{9\sec^2\theta - 9}}{\frac{3}{2}\sec\theta}\cdot\left(\frac{3}{2}\sec\theta\tan\theta\,d\theta\right)$$

$$= \int 3\sqrt{\sec^2\theta - 1}\,\tan\theta\,d\theta$$

$$= 3\int \tan^2\theta\,d\theta$$

$$= 3\int (\sec^2\theta - 1)\,d\theta$$

$$= 3(\tan\theta - \theta) + C$$

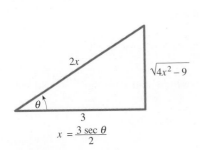

$$x = \frac{3\sec\theta}{2}$$

FIGURE 8.5

From the definition of θ, we get Figure 8.5, so

$$\int \frac{\sqrt{4x^2 - 9}}{x}\,dx = 3\left(\frac{\sqrt{4x^2 - 9}}{3} - \sec^{-1}\frac{2x}{3}\right) + C$$

$$= \sqrt{4x^2 - 9} - 3\sec^{-1}\frac{2x}{3} + C \qquad \blacksquare$$

The next example shows that a trigonometric substitution is sometimes helpful even when the integrand does not contain a radical.

EXAMPLE 8.16   Evaluate the integral

$$\int \frac{x^3\, dx}{1 + 2x^2 + x^4}$$

Solution   The integral can be written in the form

$$\int \frac{x^3\, dx}{(1 + x^2)^2}$$

Now let $x = \tan\theta$, so that $dx = \sec^2\theta\, d\theta$ and $\theta = \tan^{-1} x$. This substitution gives

$$\int \frac{\tan^3\theta \sec^2\theta\, d\theta}{(1 + \tan^2\theta)^2} = \int \frac{\tan^3\theta \sec^2\theta\, d\theta}{\sec^4\theta}$$

$$= \int \frac{\tan^3\theta\, d\theta}{\sec^2\theta} = \int \frac{\sin^3\theta}{\cos^3\theta} \cdot \cos^2\theta\, d\theta$$

$$= \int \frac{\sin^3\theta}{\cos\theta}\, d\theta = \int \frac{1 - \cos^2\theta}{\cos\theta} \sin\theta\, d\theta \qquad \text{Mentally, let } u = \cos\theta.$$

$$= \int \left(\frac{1}{\cos\theta} - \cos\theta\right) \sin\theta\, d\theta = -\ln|\cos\theta| + \frac{\cos^2\theta}{2} + C$$

From Figure 8.6, we therefore have

$$\int \frac{x^3\, dx}{1 + 2x^2 + x^4} = -\ln\frac{1}{\sqrt{1 + x^2}} + \frac{1}{2}\left(\frac{1}{1 + x^2}\right) + C$$

$$= -\ln(1 + x^2)^{-1/2} + \frac{1}{2}\left(\frac{1}{1 + x^2}\right) + C$$

$$= -\left(-\frac{1}{2}\right)\ln(1 + x^2) + \frac{1}{2}\left(\frac{1}{1 + x^2}\right) + C$$

$$= \frac{1}{2}\left[\ln(1 + x^2) + \frac{1}{1 + x^2}\right] + C \qquad\blacksquare$$

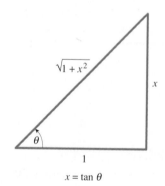

$x = \tan\theta$

FIGURE 8.6

EXAMPLE 8.17   Derive the integration formula

$$\int \sqrt{a^2 - u^2}\, du = \frac{1}{2}\left(a^2 \sin^{-1}\frac{u}{a} + u\sqrt{a^2 - u^2}\right) + C$$

Solution   Let $u = a\sin\theta$; then $du = a\cos\theta\, d\theta$ and $\theta = \sin^{-1}\dfrac{u}{a}$. So we have

$$\int \sqrt{a^2 - u^2}\, du = \int \sqrt{a^2 - a^2\sin^2\theta}\, (a\cos\theta\, d\theta)$$

$$= a^2 \int \sqrt{1 - \sin^2\theta}\, (\cos\theta\, d\theta)$$

$$= a^2 \int \cos^2\theta\, d\theta = \frac{a^2}{2}\int (1 + \cos 2\theta)\, d\theta$$

$$= \frac{a^2}{2}\left(\theta + \frac{\sin 2\theta}{2}\right) + C$$

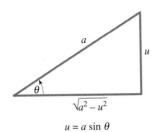

FIGURE 8.7

To express the answer in terms of u, use the trigonometric identity $\sin 2\theta = 2\sin\theta\cos\theta$ and refer to Figure 8.7. This gives

$$\int \sqrt{a^2 - u^2}\, du = \frac{a^2}{2}\left(\sin^{-1}\frac{u}{a} + \frac{u}{a}\cdot\frac{\sqrt{a^2 - u^2}}{a}\right) + C$$

$$= \frac{1}{2}\left(a^2\sin^{-1}\frac{u}{a} + u\sqrt{a^2 - u^2}\right) + C \qquad \blacksquare$$

REMARK

■ It is unlikely that you would want to commit the formula we just derived to memory. The important thing is to see how to derive it. In the future, when an integral of this type arises, you can use a trigonometric substitution, as we did in the example, or you can refer to the formula we derived. (It is listed in the tables on the endpapers.) You could also use a computer algebra system. We show one application of the formula in Example 8.18.

Area Enclosed by an Ellipse

In Example 5.25 we showed how to find the area enclosed by an ellipse by making use of our knowledge of the area of a circle. In the next example we calculate the area inside an ellipse again, without reference to the area of a circle. In fact, we will see that the area formula for a circle can be obtained as a special case of our new result.

EXAMPLE 8.18   Find the area inside the ellipse

$$\frac{x^2}{a^2} + \frac{y^2}{b^2} = 1 \qquad (a > 0, b > 0)$$

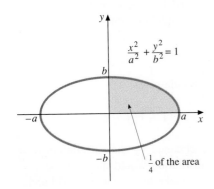

FIGURE 8.8

Solution   We show the ellipse in Figure 8.8. Taking symmetry into account, we see that it is sufficient to find the area in the first quadrant and multiply the result by 4.

The equation of the upper semiellipse is found by solving the given equation for y and taking the positive square root, since $y > 0$ for the upper half of the ellipse.

$$\frac{y^2}{b^2} = 1 - \frac{x^2}{a^2}$$

$$y^2 = b^2\left(1 - \frac{x^2}{a^2}\right) = \frac{b^2}{a^2}(a^2 - x^2)$$

$$y = \frac{b}{a}\sqrt{a^2 - x^2} \qquad\qquad \text{For upper semiellipse}$$

The total area inside the ellipse is

$$A = 4\int_0^a \frac{b}{a}\sqrt{a^2 - x^2}\, dx = \frac{4b}{a}\int_0^a \sqrt{a^2 - x^2}\, dx$$

To evaluate the integral, we could make the substitution $x = a\sin\theta$. But in the previous example we already evaluated the indefinite integral $\int \sqrt{a^2 - u^2}\, du$.

So we use that result and the Second Fundamental Theorem to get

$$A = \frac{4b}{a} \left[\frac{1}{2} \left(a^2 \sin^{-1} \frac{x}{a} + x\sqrt{a^2 - x^2} \right) \right]_0^a = \frac{4b}{a} \left(\frac{a^2}{2} \sin^{-1} 1 \right)$$

$$= 2ab \left(\frac{\pi}{2} \right) = \pi ab$$

In words, we can say that *the area of an ellipse is pi times the product of the semiaxes.*   ∎

Area of a Circle

Since a circle is an ellipse in which $a = b$, we also have, from the formula πab for the area of an ellipse, that the area of a circle is πa^2, where a is the radius. Although this latter result is well known, and we have used it on several occasions, we now have verified it.

Trigonometric Versus Hyperbolic Functions

The next example illustrates further similarities between trigonometric and hyperbolic functions. Recall that we have already made some comparisons in Section 7.6.

EXAMPLE 8.19   Compare the area under the upper half of the unit circle $x^2 + y^2 = 1$ with the area under the upper half of the unit hyperbola $y^2 - x^2 = 1$ from $x = -1$ to $x = 1$.

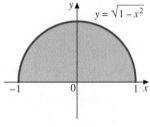

Solution   We show the two regions in question in Figure 8.9. The graph of $y = \sqrt{1 - x^2}$ is the upper half of the unit circle $x^2 + y^2 = 1$, and the graph of $y = \sqrt{1 + x^2}$ is the upper half of the unit hyperbola $y^2 - x^2 = 1$.
   In Example 8.17 we found a formula for the indefinite integral

$$\int \sqrt{a^2 - u^2} \, du$$

Taking $a = 1$ and replacing u by x in that formula gives

$$\int \sqrt{1 - x^2} \, dx = \frac{1}{2} \left(\sin^{-1} x + x\sqrt{1 - x^2} \right) + C$$

Recall that the result in Example 8.17 was obtained by the substitution $u = a \sin\theta$. In the special case we have, the substitution would therefore be $x = \sin\theta$. To get the area under the semicircle, we make use of the Second Fundamental Theorem:

$$A = \frac{1}{2} \left[\sin^{-1} x + x\sqrt{1 - x^2} \right]_{-1}^1 = \left[\frac{1}{2} \sin^{-1} 1 + 0 \right] - \left[\frac{1}{2} \sin^{-1}(-1) + 0 \right]$$

$$= \frac{1}{2} \left(\frac{\pi}{2} \right) - \frac{1}{2} \left(-\frac{\pi}{2} \right) = \frac{\pi}{2}$$

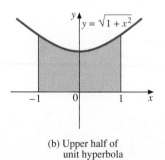

(a) Upper half of unit circle

(b) Upper half of unit hyperbola

FIGURE 8.9

Of course, we know this result already, since the area of a circle of radius 1 is $\pi (1)^2$, making half the area $\pi/2$. The emphasis we want to make in this example, however, is not on the numerical answer, but on the antiderivative used to get it, namely,

$$A = \int_{-1}^{1} \sqrt{1 - x^2}\, dx = \frac{1}{2}\left[\sin^{-1} x + x\sqrt{1 - x^2}\right]_{-1}^{1} \qquad (8.12)$$

Let us now consider the second curve, $y = \sqrt{1 + x^2}$. The area is given by

$$A = \int_{-1}^{1} \sqrt{1 + x^2}\, dx$$

To evaluate the integral, we will again use the method of finding the indefinite integral (antiderivative)

$$\int \sqrt{1 + x^2}\, dx$$

and then use the Second Fundamental Theorem.

A natural way to evaluate this indefinite integral is to make the substitution $x = \tan\theta$. However, to keep the parallelism with the method used for the area under the semicircle, we choose instead to use the hyperbolic sine (Section 7.6), letting

$$x = \sinh\theta$$

Then

$$dx = \cosh\theta\, d\theta$$

and

$$\sqrt{1 + x^2} = \sqrt{1 + \sinh^2\theta} = \sqrt{\cosh^2\theta} = |\cosh\theta| = \cosh\theta$$

Here we used the hyperbolic identity (derived in Section 7.6)

$$\cosh^2\theta - \sinh^2\theta = 1$$

giving $1 + \sinh^2\theta = \cosh^2\theta$. Thus, we have

$$\int \sqrt{1 + x^2}\, dx = \int (\cosh\theta)(\cosh\theta\, d\theta) = \int \cosh^2\theta\, d\theta$$

Now from Hyperbolic Identity 5 (in Section 7.6) we find that $\cosh^2\theta$ can be written as

$$\cosh^2\theta = \frac{1 + \cosh 2\theta}{2}$$

Thus,

$$\int \sqrt{1 + x^2}\, dx = \frac{1}{2}\int (1 + \cosh 2\theta)\, d\theta = \frac{1}{2}\left[\theta + \frac{\sinh 2\theta}{2}\right] + C$$

By Hyperbolic Identity 4,

$$\sinh 2\theta = 2\sinh\theta \cosh\theta$$

Thus,

$$\int \sqrt{1 + x^2}\, dx = \frac{1}{2}[\theta + \sinh\theta \cosh\theta] + C$$

Finally, since $x = \sinh\theta$, we have

$$\theta = \sinh^{-1} x$$

Also, $\cosh\theta = \sqrt{1 + \sinh^2\theta} = \sqrt{1 + x^2}$. Thus,

$$\int \sqrt{1 + x^2}\, dx = \frac{1}{2}\left[\sinh^{-1} x + x\sqrt{1 + x^2}\right] + C$$

By the Second Fundamental Theorem, the area is

$$A = \int_{-1}^{1} \sqrt{1 + x^2}\, dx = \frac{1}{2}\left[\sinh^{-1} x + x\sqrt{1 + x^2}\right]_{-1}^{1} \qquad (8.13)$$

To get the final answer, we can use Formula 7.70 to write

$$\sinh^{-1} x = \ln\left(x + \sqrt{1 + x^2}\right)$$

Thus,

$$A = \frac{1}{2}\left[\left(\sinh^{-1} 1 + 1\sqrt{2}\right) - \left(\sinh^{-1}(-1) - 1\sqrt{2}\right)\right]$$

$$= \frac{1}{2}\left[\sinh^{-1} 1 - \sinh^{-1}(-1) + 2\sqrt{2}\right]$$

$$= \frac{1}{2}\left[\ln\left(1 + \sqrt{2}\right) - \ln\left(-1 + \sqrt{2}\right) + 2\sqrt{2}\right]$$

$$= \sqrt{2} + \frac{1}{2}\ln\frac{\sqrt{2} + 1}{\sqrt{2} - 1}$$

Rationalizing, we can write

$$\frac{\sqrt{2} + 1}{\sqrt{2} - 1} \cdot \frac{\sqrt{2} + 1}{\sqrt{2} + 1} = \frac{2 + 2\sqrt{2} + 1}{2 - 1} = 3 + 2\sqrt{2}$$

Thus,

$$A = \sqrt{2} + \frac{1}{2}\ln(3 + 2\sqrt{2}) \approx 3.18$$

Again, it is not the final answer we want to emphasize but the result in Equation 8.13. The analogy with Equation 8.12 is striking. Let us summarize what we have found:

Area Under Semicircle	**Area Under Hyperbola**
$A = \displaystyle\int_{-1}^{1} \sqrt{1 - x^2}\, dx$	$A = \displaystyle\int_{-1}^{1} \sqrt{1 + x^2}\, dx$
Trigonometric Substitution	Hyperbolic Substitution
$x = \sin\theta$	$x = \sinh\theta$
$A = \dfrac{1}{2}\left[\sin^{-1} x + x\sqrt{1 - x^2}\right]_{-1}^{1}$	$A = \dfrac{1}{2}\left[\sinh^{-1} x + x\sqrt{1 + x^2}\right]_{-1}^{1}$

Beyond this point, the analogy breaks down, since the values of $\sin^{-1} x$ and $\sinh^{-1} x$ are quite different. Nevertheless, this example provides further reinforcement of the similarities between trigonometric and hyperbolic functions.

Exercise Set 8.3

In Exercises 1–9, evaluate the integral by making a trigono-metric substitution.

1. $\displaystyle\int \frac{dx}{\sqrt{x^2-4}}$

2. $\displaystyle\int \frac{\sqrt{9-x^2}}{x}\,dx$

3. $\displaystyle\int_{-1}^{1} \frac{dx}{(2-x^2)^{3/2}}$

4. $\displaystyle\int_{1}^{3} \frac{\sqrt{x^2+3}\,dx}{x^4}$

5. $\displaystyle\int_{\sqrt{3}}^{3} x^3\sqrt{4x^2-9}\,dx$

6. $\displaystyle\int_{0}^{2} \frac{x^2\,dx}{(3x^2+4)^{3/2}}$

7. $\displaystyle\int_{-1/\sqrt{3}}^{1} \frac{x^3\,dx}{(4-3x^2)^{5/2}}$

8. $\displaystyle\int \frac{dx}{(x^2-1)^2}$

9. $\displaystyle\int \frac{x^5}{(1-x^2)^3}\,dx$

In Exercises 10–16, evaluate the integrals by two methods: (a) trigonometric substitution and (b) algebraic substitution.

10. $\displaystyle\int \frac{x\,dx}{\sqrt{9-4x^2}}$

11. $\displaystyle\int_{0}^{1} x(3x^2+1)^{3/2}\,dx$

12. $\displaystyle\int_{1}^{\sqrt{3}} \frac{x}{\sqrt{4x^2-3}}\,dx$

13. $\displaystyle\int \frac{x}{1+2x^2+x^4}\,dx$

14. $\displaystyle\int x^3\sqrt{1-x^2}\,dx$ (*Hint for (b):* Let $u = 1 - x^2$. Then $du = -2x\,dx$ and $x^2 = 1 - u$.)

15. $\displaystyle\int \frac{x^3}{\sqrt{1+x^2}}\,dx$ (See the hint for Exercise 14.)

16. $\displaystyle\int \frac{x^3}{\sqrt{x^2-1}}\,dx$ (See the hint for Exercise 14.)

In Exercises 17–23, verify the given integration formula by us-ing a trigonometric substitution to evaluate the integral.

17. $\displaystyle\int \frac{du}{\sqrt{a^2-u^2}} = \sin^{-1}\frac{u}{a} + C$

18. $\displaystyle\int \frac{du}{a^2+u^2} = \frac{1}{a}\tan^{-1}\frac{u}{a} + C$

19. $\displaystyle\int \frac{du}{u\sqrt{u^2-a^2}} = \frac{1}{a}\sec^{-1}\frac{u}{a} + C$

20. $\displaystyle\int \frac{du}{\sqrt{a^2+u^2}} = \ln\left|u+\sqrt{a^2+u^2}\right| + C$

21. $\displaystyle\int \frac{du}{u^2-a^2} = \frac{1}{2a}\ln\left|\frac{u-a}{u+a}\right| + C$

22. $\displaystyle\int \frac{du}{u\sqrt{a^2-u^2}} = -\frac{1}{a}\ln\left|\frac{a+\sqrt{a^2-u^2}}{u}\right| + C$

23. $\displaystyle\int \frac{du}{u\sqrt{a^2+u^2}} = -\frac{1}{a}\ln\left|\frac{a+\sqrt{a^2+u^2}}{u}\right| + C$

24. Determine an appropriate rationalizing hyperbolic substitution for integrands involving the following:
(a) $\sqrt{a^2-u^2}$
(b) $\sqrt{a^2+u^2}$
(c) $\sqrt{u^2-a^2}$

In Exercises 25–27, redo Exercises 2, 3, and 5, respectively, using the appropriate hyperbolic substitution determined in Ex-ercise 24.

28. Find the length of the parabola $y = x^2/3$ from $x = 0$ to $x = 2$.

29. Find the area of the region under the curve $y = \sqrt{1+x^2}$ from $x = 0$ to $x = 2\sqrt{2}$.

In Exercises 30–33, after completing the square, use a trigono-metric substitution to evaluate the integral.

30. $\displaystyle\int \frac{dx}{\sqrt{x^2+2x+2}}$

31. $\displaystyle\int (2x-x^2)^{3/2}\,dx$

32. $\displaystyle\int \frac{dx}{x^2+4x-5}$

33. $\displaystyle\int \frac{(3x+2)^2\,dx}{(5-12x-9x^2)^{3/2}}$

Evaluate the integrals in Exercises 34–37.

34. $\displaystyle\int \frac{\cos t\,dt}{(1+\sin^2 t)^{3/2}}$ (*Hint:* First let $u = \sin t$.)

35. $\displaystyle\int \frac{\sec^2 t\tan^3 t\,dt}{\sqrt{\tan^2 t - 1}}$

36. $\displaystyle\int \frac{\sin\theta\,d\theta}{(1+\cos^2\theta)^3}$

37. $\displaystyle\int \frac{e^{3x}\,dx}{\sqrt{1-e^{2x}}}$

38. Derive the formula for the length of the circumference of a circle of radius a.

39. The torus revisited. Derive the formula for the volume of a torus by revolving the region under the upper half of the circle $x^2 + y^2 = r^2$ about the line $x = R$, where $R > r$, and doubling the result. Use the method of cylindrical shells, and use an appropriate trigonometric substitution to evaluate the integral.

40. The length of an ellipse revisited. In Example 6.19 of Section 6.4 we derived the formula

$$L = 4 \int_0^a \sqrt{1 + \frac{b^2 x^2}{a^2(a^2 - x^2)}} \, dx$$

for the length of the ellipse

$$\frac{x^2}{a^2} + \frac{y^2}{b^2} = 1$$

We noted that the integrand was unbounded at $x = a$ (the integral is improper).

(a) Use the trigonometric substitution $x = a \sin t$ to

convert the integral to

$$L = 4a \int_0^{\pi/2} \sqrt{1 - e^2 \sin^2 t} \, dt$$

where $e = \sqrt{a^2 - b^2}/a$ (e is called the *eccentricity* of the ellipse).

(b) The integral in part (a) is no longer improper, but there is no elementary antiderivative of the integrand, so the Second Fundamental Theorem of Calculus cannot be used. Use Simpson's Rule with $n = 20$ to approximate its value for $a = 4$ and $b = 3$.

8.4 PARTIAL FRACTIONS

We now turn to techniques for integrating rational functions—that is, functions of the form $P(x)/Q(x)$, where $P(x)$ and $Q(x)$ are polynomials. Actually, what we study is an algebraic technique for expressing $P(x)/Q(x)$ as a sum of simpler rational functions when the degree of Q is 2 or greater. The simpler functions can then be integrated by means already familiar to us. As a way of introducing the ideas, consider the following sum of two fractions:

$$\frac{2}{x - 1} + \frac{3}{x + 2} = \frac{5x + 1}{x^2 + x - 2}$$

Now suppose we are asked to perform the integration

$$\int \frac{5x + 1}{x^2 + x - 2} \, dx$$

By replacing the integrand with the sum of the two simpler fractions, whose integrals we can easily find, we get

$$\int \frac{5x + 1}{x^2 + x - 2} \, dx = \int \left(\frac{2}{x - 1} + \frac{3}{x + 2} \right) dx$$

$$= 2 \ln |x - 1| + 3 \ln |x + 2| + C$$

$$= \ln |(x - 1)^2 (x + 2)^3| + C$$

This procedure works, *provided* we know the two original fractions. The objective of this section is to introduce ways of determining the component fractions when we are given their sum; that is, we will introduce a procedure allowing us to do the reverse of adding fractions: given a rational function, we want to find the simplest fractions that add together to give that function. The process is called **decomposition into partial fractions** (or partial-fraction decomposition).

We begin with the assumption that $P(x)/Q(x)$ is in lowest terms and that the *degree of P is less than the degree of Q*—that is, that $P(x)/Q(x)$ is a *proper* fraction. If the fraction is not proper, then we divide to obtain

$$\frac{P(x)}{Q(x)} = P_1(x) + \frac{R(x)}{Q(x)}$$

$$\begin{array}{r} x - 3 \\ x^3 + x \overline{\smash{\big)}\ x^4 - 3x^3 + x^2 \quad\;\; - 5} \\ \underline{x^4 \qquad\;\; + x^2} \\ -3x^3 \qquad\;\; - 5 \\ \underline{-3x^3 \qquad - 3x} \\ 3x - 5 \end{array}$$

where the degree of R is less than the degree of Q. We then apply our theory to $R(x)/Q(x)$. For example, the long division in the margin shows that

$$\frac{x^4 - 3x^3 + x^2 - 5}{x^3 + x} = x - 3 + \frac{3x - 5}{x^3 + x}$$

Since the quotient $x - 3$ is easily integrated, we need only work with the fraction on the right.

It is proved in algebra courses that every polynomial with real coefficients can (in theory, at least) be factored into real linear and/or quadratic factors that are **irreducible** over the reals. A quadratic factor $ax^2 + bx + c$ is irreducible over the reals if it cannot be factored into linear factors with real coefficients, which is true when the *discriminant*, $b^2 - 4ac$, is negative. The factors $x^2 + 1$ and $x^2 + 2x + 4$ are examples of such irreducible quadratic factors. So we need to concern ourselves with only linear and irreducible quadratic factors for $Q(x)$. In particular, if $Q(x)$ is a cubic, or quartic, or a polynomial of higher degree, we first factor it into linear and/or quadratic factors. For example,

$$x^3 - 2x^2 + x - 2 = (x - 2)(x^2 + 1)$$

Computer algebra systems can do such factoring.

We illustrate in the next two examples how to handle situations in which $Q(x)$ has only linear factors. If a factor occurs to a power greater than 1, we say it is *repeated*. For example, we would say $x - 2$ is a repeated factor if it occurs as $(x - 2)^2$ or $(x - 2)^3$. The simplest case is the one in which no factor is repeated, as in the following example.

EXAMPLE 8.20   **Linear factors, none repeated**   Evaluate the integral

$$\int \frac{5x + 1}{x^2 + x - 2}\, dx$$

Solution   This integral is the one we considered at the beginning of this section. We know the answer, but we will do it again, this time showing how to decompose the integrand into its component fractions. First we factor the denominator:

$$\frac{5x + 1}{x^2 + x - 2} = \frac{5x + 1}{(x - 1)(x + 2)}$$

Now we would expect that the denominator $(x - 1)(x + 2)$ would arise from adding a fraction with $x - 1$ as its denominator to one with $x + 2$ as its denominator; that is, we expect that constants A and B exist such that

$$\frac{5x + 1}{(x - 1)(x + 2)} = \frac{A}{x - 1} + \frac{B}{x + 2} \tag{8.14}$$

Assume for the moment that this equation is true, and multiply both sides by the least common denominator $(x - 1)(x + 2)$ to obtain

$$5x + 1 = A(x + 2) + B(x - 1) \tag{8.15}$$

When we multiplied Equation 8.14 by the least common denominator, we were assuming that this common denominator was not 0. That is, we were assuming that $x \neq 1$ and $x \neq -2$. It would appear, then, that in Equation 8.15 we have to retain these restrictions. However, it can be proved that if two polynomials are equal for all but a finite number of values of the variable, they are equal for *all* values of the variable. Thus, whereas in Equation 8.14 we must restrict x to be different from 1 or -2, Equation 8.15 must be true for all values of x, including $x = 1$ and $x = -2$.

Equation 8.15 is particularly easy to solve for A or B if one of the terms on the right-hand side is 0. This situation occurs when $x = 1$ or when $x = -2$, respectively. By substituting these values for x, in turn, we will be able to solve for both A and B.

So we substitute first $x = 1$ and then $x = -2$:

$$\underline{x = 1}: \quad 5 + 1 = A(3) + B(0) \qquad \underline{x = -2}: \quad 5(-2) + 1 = A(0) + B(-3)$$
$$3A = 6 \qquad\qquad\qquad\qquad\qquad -3B = -9$$
$$A = 2 \qquad\qquad\qquad\qquad\qquad\quad B = 3$$

It is easy now to verify that these values of A and B do work—that is, that

$$\frac{5x + 1}{(x - 1)(x + 2)} = \frac{2}{x - 1} + \frac{3}{x + 2}$$

The integration is now straightforward and we get, as we have already seen,

$$\int \frac{5x + 1}{x^2 + x - 2} \, dx = \ln |(x - 1)^2 (x + 2)^3| + C \qquad\qquad \blacksquare$$

This example is typical of those integrands $P(x)/Q(x)$ (where the fraction is proper, that is, the degree of P is less than the degree of Q) in which $Q(x)$ factors into distinct linear factors, each appearing to the first power. Each such linear factor gives rise to a fraction in the decomposition that has a constant in the numerator. The constants can be determined by clearing of fractions and substituting, in turn, values of x that cause each factor to be 0. This procedure is called the **method of substitution.**

In the next example, one of the factors is repeated.

EXAMPLE 8.21   **Repeated linear factors**   Evaluate the integral

$$\int \frac{x^2 - 13x + 20}{(x - 1)(x - 3)^2} \, dx$$

Solution   The denominator again involves only linear factors, but $x - 3$ is repeated. The factor $x - 1$ is handled as in the preceding example, but for $(x - 3)^2$ we must allow for the denominator $x - 3$ as well as $(x - 3)^2$:

$$\frac{x^2 - 13x + 20}{(x - 1)(x - 3)^2} = \frac{A}{x - 1} + \frac{B}{x - 3} + \frac{C}{(x - 3)^2}$$

Note carefully the presence of the term $B/(x - 3)$ on the right-hand side. If the factor $(x - 3)^3$ had appeared, then we would need to allow for fractions with each of the denominators $x - 3$, $(x - 3)^2$, and $(x - 3)^3$.

As before, we clear the equation of fractions:

$$x^2 - 13x + 20 = A(x - 3)^2 + B(x - 1)(x - 3) + C(x - 1) \qquad (8.16)$$

Again we use the method of substitution, first setting $x = 1$ and then $x = 3$:

$$\underline{x = 1}: \quad 8 = A(-2)^2 \qquad \underline{x = 3}: \quad -10 = 2C$$
$$4A = 8 \qquad\qquad\qquad\qquad C = -5$$
$$A = 2$$

But now we have exhausted the substitutions that cause factors to be 0 and we still do not know B. One procedure is to substitute any other value for x and make use of the known values of A and C. An easy value to substitute is $x = 0$. When we do so, we get

$$x = 0: \qquad 20 = 9A + 3B - C$$

$$20 = 9(2) + 3B - (-5)$$

$$3B = -3$$

$$B = -1$$

An alternative method for finding B is known as **comparison of coefficients**. We will find this method is also useful when the denominator contains irreducible quadratic factors. With it, returning to Equation 8.16, we rewrite the right-hand side, expanding and collecting like powers of x. The result is (verify it)

$$x^2 - 13x + 20 = (A + B)x^2 + (-6A - 4B + C)x + (9A + 3B - C)$$

Now this equation is to be an *identity* in x; that is, the polynomial on the right must be the same polynomial as the one on the left. So coefficients of corresponding powers of x must be equal:

$$\text{Coefficients of } x^2: \qquad 1 = A + B$$

$$\text{Coefficients of } x: \qquad -13 = -6A - 4B + C$$

$$\text{Constant terms:} \qquad 20 = 9A + 3B - C$$

This system of three equations in three unknowns can be solved algebraically. However, we already know A and C, so we can get B using any one of the equations. From the first one, we have

$$1 = 2 + B$$

$$B = -1$$

We now know A, B, and C, so we can write

$$\int \frac{x^2 - 13x + 20}{(x - 1)(x - 3)^2}\, dx = \int \left[\frac{2}{x - 1} - \frac{1}{x - 3} - \frac{5}{(x - 3)^2} \right] dx$$

$$= 2 \int \frac{dx}{x - 1} - \int \frac{dx}{x - 3} - 5 \int \frac{dx}{(x - 3)^2}\, dx$$

$$= 2 \ln |x - 1| - \ln |x - 3| + 5(x - 3)^{-1} + C$$

$$= \ln \frac{(x - 1)^2}{|x - 3|} + \frac{5}{x - 3} + C \qquad \blacksquare$$

REMARK ———

■ The method of comparison of coefficients always works and can be used on all problems without first using substitution. However, it is usually easier to determine as many of the coefficients as possible by substitution (especially substitutions that cause a factor to be 0) and then go to comparison of coefficients to determine the remaining coefficients, as we did in the preceding example.

The next two examples show how to handle irreducible quadratic factors in the denominator.

EXAMPLE 8.22   **Nonrepeated, irreducible quadratic factor**

Evaluate the integral

$$\int \frac{7x - 4}{x^3 - 2x^2 + x - 2} \, dx$$

Solution   We can factor the denominator by grouping as follows:

$$x^3 - 2x^2 + x - 2 = (x^3 - 2x^2) + (x - 2)$$

$$= x^2(x - 2) + (x - 2)$$

$$= (x - 2)(x^2 + 1)$$

We therefore want to find the partial fraction decomposition of

$$\frac{7x - 4}{(x - 2)(x^2 + 1)}$$

We expect two fractions, one with denominator $x - 2$ and one with denominator $x^2 + 1$. The linear factor is handled as before, but for the irreducible quadratic factor $x^2 + 1$, we must assume a numerator that is a first-degree polynomial, rather than a constant. By allowing for a first-degree polynomial in the numerator, we are allowing for the worst case, in the sense that the fraction remains proper. Any higher degree in the numerator would cause an improper fraction. We therefore assume a decomposition of the form

$$\frac{7x - 4}{(x - 2)(x^2 + 1)} = \frac{A}{x - 2} + \frac{Bx + C}{x^2 + 1}$$

When we clear of fractions, we get

$$7x - 4 = A(x^2 + 1) + (Bx + C)(x - 2)$$

Now we substitute $x = 2$.

$$x = 2: \qquad 10 = A(5)$$

$$A = 2$$

To find B and C, we compare coefficients of like powers of x. It is not actually necessary to multiply things out and collect terms on the right, since we can easily identify the terms of the same degree. It is particularly easy to read off the coefficients of x^2 (the highest degree) and the constant term. Note that there is no x^2 term on the left, so its coefficient is 0.

$$\text{Coefficient of } x^2: \qquad 0 = A + B$$

$$B = -A = -2$$

$$\text{Constant term:} \qquad -4 = A - 2C$$

$$2C = 6$$

$$C = 3$$

So we have

$$\int \frac{7x - 4}{(x - 2)(x^2 + 1)} \, dx$$

$$= \int \frac{2}{x - 2} \, dx + \int \frac{-2x + 3}{x^2 + 1} \, dx \qquad \text{Write the second integral as two integrals.}$$

$$= 2 \ln |x - 2| - \int \frac{2x}{x^2 + 1} \, dx + 3 \int \frac{dx}{x^2 + 1}$$

$$= 2 \ln |x - 2| - \ln(x^2 + 1) + 3 \tan^{-1} x + C \qquad \blacksquare$$

 Do not confuse a repeated linear factor in the denominator with an irreducible quadratic factor. For example,

$$x^2 - 2x + 1 = (x - 1)^2$$

is a repeated linear factor, but

$$x^2 - 2x + 2$$

is an irreducible quadratic factor, since $b^2 - 4ac = 4 - 8 = -4 < 0$. The first gives rise to the two fractions

$$\frac{A}{x - 1} + \frac{B}{(x - 1)^2}$$

whereas the second gives rise to the single fraction

$$\frac{Ax + B}{x^2 - 2x + 2}$$

The following example shows how to handle a repeated irreducible quadratic factor in the denominator.

EXAMPLE 8.23   **Repeated irreducible quadratic factor**   Evaluate the integral

$$\int \frac{2x - 3}{x^5 + 2x^3 + x} \, dx$$

Solution   The integrand can be written as

$$\frac{2x - 3}{x(x^4 + 2x^2 + 1)} = \frac{2x - 3}{x(x^2 + 1)^2}$$

Here the denominator has the linear factor x and the repeated irreducible quadratic factor $x^2 + 1$. The decomposition follows a pattern similar to that for repeated linear factors:

$$\frac{2x - 3}{x(x^2 + 1)^2} = \frac{A}{x} + \frac{Bx + C}{x^2 + 1} + \frac{Dx + E}{(x^2 + 1)^2}$$

As usual, we clear fractions and, because we will rely extensively on comparison of coefficients, we group the right-hand side in powers of x. We leave it to you to verify that the result is

$$2x - 3 = (A + B)x^4 + Cx^3 + (2A + B + D)x^2 + (C + E)x + A$$

Setting $x = 0$ gives $A = -3$. Comparing coefficients of like powers yields the following system. Note that powers of x on the left greater than 1 have a coefficient of 0.

$$A + B = 0$$

$$C = 0$$

$$2A + B + D = 0$$

$$C + E = 2$$

Since we have already found that $A = -3$, we get $B = 3$, $C = 0$, $D = 3$, and $E = 2$. Thus, the integral becomes

$$\int \frac{2x - 3}{x^5 + 2x^3 + x} \, dx$$

$$= \int \left[\frac{-3}{x} + \frac{3x}{x^2 + 1} + \frac{3x + 2}{(x^2 + 1)^2} \right] dx$$

$$= -3 \int \frac{dx}{x} + \frac{3}{2} \int \frac{2x}{x^2 + 1} \, dx + \int \frac{3x + 2}{(x^2 + 1)^2} \, dx \qquad \text{We mentally let } u = x^2 + 1 \text{ in the second integral.}$$

$$= -3 \ln |x| + \frac{3}{2} \ln(x^2 + 1) + \int \frac{3x + 2}{(x^2 + 1)^2} \, dx$$

To evaluate the last integral we make the substitution $x = \tan \theta$, so that $du = \sec^2 \theta \, d\theta$. Then we get

$$\int \frac{3x + 2}{(x^2 + 1)^2} \, dx = \int \frac{3 \tan \theta + 2}{\sec^4 \theta} \sec^2 \theta \, d\theta$$

$$= 3 \int \frac{\tan \theta}{\sec^2 \theta} \, d\theta + 2 \int \frac{d\theta}{\sec^2 \theta}$$

$$= 3 \int \frac{\sin \theta}{\cos \theta} \cdot \cos^2 \theta \, d\theta + 2 \int \cos^2 \theta \, d\theta \qquad \text{Since } \sec^2 \theta = \frac{1}{\cos^2 \theta}$$

$$= 3 \int \sin \theta \cos \theta \, d\theta + \int (1 + \cos 2\theta) \, d\theta \qquad \begin{aligned} &\text{Since } \cos^2 \theta \\ &= \frac{1 + \cos 2\theta}{2} \end{aligned}$$

$$= \frac{3}{2} \sin^2 \theta + \theta + \frac{\sin 2\theta}{2} + C \qquad \begin{aligned} &\text{Since } d(\sin \theta) \\ &= \cos \theta \, d\theta \end{aligned}$$

$$= \frac{3}{2} \sin^2 \theta + \theta + \sin \theta \cos \theta + C \qquad \begin{aligned} &\text{Since } \sin 2\theta \\ &= 2 \sin \theta \cos \theta \end{aligned}$$

$$= \frac{3}{2} \cdot \frac{x^2}{1 + x^2} + \tan^{-1} x + \frac{x}{1 + x^2} + C$$

$$= \frac{3x^2 + 2x}{2(1 + x^2)} + \tan^{-1} x + C$$

Observe how we made use of Figure 8.10 to express the answer in terms of x. So we have the final result:

$$\int \frac{2x - 3}{x^5 + 2x^3 + x} \, dx = \ln \frac{(x^2 + 1)^{3/2}}{|x|^3} + \frac{3x^2 + 2x}{2(1 + x^2)} + \tan^{-1} x + C \qquad \blacksquare$$

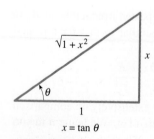

$$\sqrt{1 + x^2}$$

$$x$$

$$\theta$$

$$1$$

$$x = \tan \theta$$

FIGURE 8.10

We now summarize the procedures we have illustrated.

Partial-Fraction Decomposition

1. If $P(x)/Q(x)$ is improper (degree of $P \geq$ degree of Q), divide to obtain a quotient plus a proper fraction. The next steps apply to proper fractions only.

2. Factor the denominator $Q(x)$ into linear and irreducible quadratic factors.

3. For each factor of the form $(ax + b)^n$, write fractions

$$\frac{A_1}{ax + b} + \frac{A_2}{(ax + b)^2} + \cdots + \frac{A_n}{(ax + b)^n}$$

(If $n = 1$, only one fraction is present.)

4. For each factor of the form $(ax^2 + bx + c)^m$ where $b^2 - 4ac < 0$, write fractions of the form

$$\frac{B_1 x + C_1}{ax^2 + bx + c} + \frac{B_2 x + C_2}{(ax^2 + bx + c)^2} + \cdots + \frac{B_m x + C_m}{(ax^2 + bx + c)^m}$$

(If $m = 1$ only one fraction is present.)

5. Set $P(x)/Q(x)$ equal to the sum of all fractions obtained in steps 3 and 4 and clear of fractions.

6. Use substitution together with comparison of coefficients to find all unknown constants.

We will give one more example to illustrate how certain common complications can be handled.

EXAMPLE 8.24   Evaluate the integral

$$\int_0^1 \frac{x^3 - x^2 - 11x + 10}{x^3 - 2x + 4} \, dx$$

Solution   The first thing to observe is that the integrand is an improper fraction, so we divide:

$$
\begin{array}{r}
1 \\
x^3 - 2x + 4 \overline{\smash{\big)}\, x^3 - x^2 - 11x + 10} \\
\underline{x^3 - 2x + 4} \\
-x^2 - 9x + 6
\end{array}
$$

We can therefore write

$$\frac{x^3 - x^2 - 11x + 10}{x^3 - 2x + 4} = 1 + \frac{-x^2 - 9x + 6}{x^3 - 2x + 4} = 1 - \frac{x^2 + 9x - 6}{x^3 - 2x + 4}$$

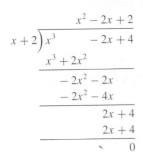

We concentrate on the last fraction, which is proper. To factor the denominator, we first try to determine any rational zeros. By trial and error, we find -2 to be a zero. Thus, $x + 2$ is a factor. By division (shown in the margin), we find the other factor to be the irreducible quadratic $x^2 - 2x + 2$. So we have

$$\frac{x^2 + 9x - 6}{x^3 - 2x + 4} = \frac{x^2 + 9x - 6}{(x + 2)(x^2 - 2x + 2)} = \frac{A}{x + 2} + \frac{Bx + C}{x^2 - 2x + 2}$$

$$x^2 + 9x - 6 = A(x^2 - 2x + 2) + (Bx + C)(x + 2) \qquad \text{Clear of fractions.}$$

$$\underline{x = -2}: \quad 4 - 18 - 6 = A(4 + 4 + 2) + 0 \qquad \text{Substitute } x = -2.$$

$$10A = -20$$

$$A = -2$$

$$\text{Coefficients of } x^2: \quad 1 = A + B \qquad \text{Equate coefficients.}$$

$$1 = -2 + B$$

$$B = 3$$

$$\text{Constants:} \quad -6 = 2A + 2C$$

$$-6 = -4 + 2C$$

$$2C = -2$$

$$C = -1$$

We therefore have the decomposition

$$\frac{x^2 + 9x - 6}{x^3 - 2x + 4} = \frac{-2}{x + 2} + \frac{3x - 1}{x^2 - 2x + 2}$$

Returning to the original fraction and the result of the first division,

$$\int_0^1 \frac{x^3 - x^2 - 11x + 10}{x^3 - 2x + 4}\, dx = \int_0^1 \left[1 - \frac{x^2 + 9x - 6}{x^3 - 2x + 4} \right] dx$$

$$= \int_0^1 \left[1 - \left(\frac{-2}{x + 2} + \frac{3x - 1}{x^2 - 2x + 2} \right) \right] dx$$

$$= \int_0^1 dx + 2 \int_0^1 \frac{dx}{x + 2} - \int_0^1 \frac{3x - 1}{x^2 - 2x + 2}\, dx$$

$$= \Big[x + 2 \ln(x + 2) \Big]_0^1 - \int_0^1 \frac{3x - 1}{x^2 - 2x + 2}\, dx$$

$$= 1 + 2 \ln 3 - 2 \ln 2 - \int_0^1 \frac{3x - 1}{x^2 - 2x + 2}\, dx$$

To evaluate the last integral we force the derivative of the denominator, $2x - 2$, into the numerator as follows (see the algebra in the margin):

$$\int \frac{3x - 1}{x^2 - 2x + 2}\, dx = \int \frac{\frac{3}{2}(2x - 2) + 2}{x^2 - 2x + 2}\, dx$$

$$= \frac{3}{2} \int \frac{2x - 2}{x^2 - 2x + 2}\, dx + 2 \int \frac{dx}{(x - 1)^2 + 1}$$

In the last integral we completed the square in the denominator to write

$$x^2 - 2x + 2 = (x^2 - 2x + 1) + 1$$

$$= (x - 1)^2 + 1$$

Margin algebra:

$$3x - 1 = 3\left(x - \frac{1}{3} \right)$$

$$= \frac{3}{2}\left(2x - \frac{2}{3} \right)$$

$$= \frac{3}{2}\left(2x - 2 + 2 - \frac{2}{3} \right)$$

$$= \frac{3}{2}(2x - 2) + \frac{3}{2}\left(2 - \frac{2}{3} \right)$$

$$= \frac{3}{2}(2x - 2) + \frac{3}{2}\left(\frac{4}{3} \right)$$

$$= \frac{3}{2}(2x - 2) + 2$$

The first integral is now in the form $\int du/u$ with $u = x^2 - 2x + 2$, and the second $\int du/(u^2 + 1)$ with $u = x - 1$. So we have

$$\int_0^1 \frac{3x - 1}{x^2 - 2x + 2}\, dx = \left[\frac{3}{2} \ln |x^2 - 2x + 2| + 2 \tan^{-1}(x - 1) \right]_0^1$$

$$= \frac{3}{2} \ln 1 + 2 \tan^{-1} 0 - \frac{3}{2} \ln 2 - 2 \tan^{-1}(-1)$$

$$= -\frac{3}{2} \ln 2 - 2\left(-\frac{\pi}{4}\right) = -\frac{3}{2} \ln 2 + \frac{\pi}{2}$$

Combining this answer with the results already obtained, we have

$$\int_0^1 \frac{x^3 - x^2 - 11x + 10}{x^3 - 2x + 4}\, dx = 1 + 2 \ln 3 - 2 \ln 2 - \left(-\frac{3}{2} \ln 2 + \frac{\pi}{2}\right)$$

$$= 1 + 2 \ln 3 - \frac{1}{2} \ln 2 - \frac{\pi}{2} \qquad \blacksquare$$

Exercise Set 8.4

In Exercises 1–26, make use of partial fractions to evaluate the integrals.

1. $\displaystyle\int \frac{dx}{x^2 + 2x}$

2. $\displaystyle\int \frac{dx}{x^2 + 5x + 6}$

3. $\displaystyle\int \frac{x\, dx}{x^2 - x - 6}$

4. $\displaystyle\int_{-2}^0 \frac{5x + 7}{x^2 + 2x - 3}\, dx$

5. $\displaystyle\int_{-3}^2 \frac{x - 17}{x^2 + x - 12}\, dx$

6. $\displaystyle\int \frac{x + 9}{2x^2 + x - 6}\, dx$

7. $\displaystyle\int \frac{3x + 10}{x^2 + 5x + 6}\, dx$

8. $\displaystyle\int \frac{2}{x^2 - 9x + 20}\, dx$

9. $\displaystyle\int \frac{dx}{x^3 - x}$

10. $\displaystyle\int \frac{10x + 4}{4x - x^3}\, dx$

11. $\displaystyle\int \frac{x^2 + 4x - 2}{x^2 + x - 2}\, dx$

12. $\displaystyle\int_0^2 \frac{2x^2 - x - 20}{x^2 - x - 6}\, dx$

13. $\displaystyle\int \frac{3x^2 + 32x + 44}{(x + 3)(x^2 + 2x - 8)}\, dx$

14. $\displaystyle\int \frac{2x^2 + 19x - 45}{(x - 1)(x^2 - x - 6)}\, dx$

15. $\displaystyle\int \frac{2x^3 - x^2 - 4x + 5}{x^2 - 1}\, dx$

16. $\displaystyle\int \frac{5x + 1}{(x + 2)(x^2 - 2x + 1)}\, dx$

17. $\displaystyle\int_{-1}^1 \frac{x^2 - 6x + 23}{(x + 3)(x^2 - 4x + 4)}\, dx$

18. $\displaystyle\int_1^2 \frac{4 + 4x - x^2}{x^3 + 2x^2}\, dx$

19. $\displaystyle\int \frac{2x^2 - 6x + 12}{x^3 - 4x^2 + 4x}\, dx$

20. $\displaystyle\int \frac{6x - 4}{(x - 2)(x^2 - 4)}\, dx$

21. $\displaystyle\int \frac{5x + 6}{(x - 2)(x^2 + 4)}\, dx$

22. $\displaystyle\int_0^1 \frac{x^2 + 3}{(x + 1)(x^2 + 1)}\, dx$

23. $\displaystyle\int_1^3 \frac{3x - 6}{x^4 + 3x^2}\, dx$

24. $\displaystyle\int \frac{dx}{x^4 - 1}$

25. $\displaystyle\int \frac{dx}{x^3 - 1}$

26. $\displaystyle\int \frac{12}{x^3 + 8}\, dx$

27. (a) Find the area of the region under the graph of

$$y = \frac{1}{x^2 - x}$$

from $x = 2$ to $x = 3$.

(b) Find the volume of the solid obtained by revolving the region of part (a) about the x-axis.

28. Find the volume of the solid obtained by revolving the region under the graph of

$$y = \frac{1}{x\sqrt{x^2 + 4}}$$

from $x = 1$ to $x = 2$ about the x-axis.

In Exercises 29–33, derive the integration formulas for the given integrals, using partial fractions. For Exercises 29–32, compare your results with those given in the tables on the front and back endpapers.

29. $\displaystyle\int \frac{du}{u(au + b)}$ **30.** $\displaystyle\int \frac{du}{u^2(au + b)}$

31. $\displaystyle\int \frac{du}{u(au + b)^2}$ **32.** $\displaystyle\int \frac{du}{u^2 - a^2}$

33. $\displaystyle\int \frac{du}{u^3 - a^3}$

In Exercises 34–38, use partial fractions to evaluate the integrals.

34. $\displaystyle\int \frac{x^4 + 5x^3 + 3x}{(x - 1)(x^2 + 2)^2}\, dx$ **35.** $\displaystyle\int \frac{4x}{x^3 - x^2 - x + 1}\, dx$

36. $\displaystyle\int \frac{x^2 + 3x + 6}{x^3 + x - 2}\, dx$ **37.** $\displaystyle\int \frac{x^3 - 3x^2 - 5x}{x^4 + 5x^2 + 4}\, dx$

38. $\displaystyle\int \frac{20 - 5x^2}{4x^4 + 9x^2 - 11x + 3}\, dx$

39. In Exercise 24 of Exercise Set 7.4 we gave the Verhulst model for population growth in the form of the differential equation

$$\frac{dQ}{dt} = kQ(m - Q)$$

where $Q(t)$ is the size of the population at time t, k and

m are positive constants, and $m > Q$. Let $Q_0 = Q(0)$, and find Q as a function of t by separating variables and using partial fractions.

40. A second-order chemical reaction involves the interaction of molecules of a substance A with those of a substance B to form molecules of a new substance X. If the initial concentrations of substances A and B are a and b, respectively, and $x(t)$ is the concentration of X at time t, then the rate at which the reaction occurs is given by

$$\frac{dx}{dt} = k(a - x)(b - x)$$

where k is a positive constant. If $x(0) = 0$ and $a \neq b$, find $x(t)$ at any time t by separating variables and using partial fractions.

8.5 INTEGRAL TABLES AND COMPUTER ALGEBRA SYSTEMS

Tables containing long lists of integration formulas have been used for many years, often eliminating the need for evaluating a particular integral by pencil and paper. These tables are very useful but cannot possibly contain every integral you may encounter. Instead, many times an integral must be made to conform to one in the table, by using a substitution, integration by parts, resolving the integrand into partial fractions, or some other technique. Trying to fit an integral into the pattern of one in the table in this way can be a difficult task in itself.

An alternative is to use a computer algebra system (CAS). These powerful systems can do symbolic integration and so eliminate many of the algebraic difficulties encountered in using tables or in evaluating integrals by hand. However, not every CAS can integrate every function, even when the function is known to have an antiderivative that is a combination of simple functions. So, just as tables are not the final answer, computer algebra systems—although a significant improvement—also have limitations. Both tables and computer algebra systems are powerful aids in evaluating integrals. However, as we will see in the examples that follow, they cannot eliminate the need for understanding the underlying concepts.

Using Tables of Integrals

We illustrate with examples some of the techniques used to match integrals with forms found in the tables on the front and back endpapers of this book. Far more extensive tables are available, such as Burrington's *Handbook of Mathematical Tables and Formulas* (New York: McGraw-Hill) and the Chemical Rubber Company's *Standard Mathematical Tables* (Cleveland: Chemical Rubber Publishing Company).

EXAMPLE 8.25   Evaluate the following, using the integral tables on the endpapers.

(a) $\displaystyle\int \sqrt{2x^2 - 3}\,dx$ (b) $\displaystyle\int \frac{x\,dx}{\sqrt{3x - 4x^2}}$

Solution

(a) Formula 37 in the table, for $\int \sqrt{u^2 \pm a^2}\,du$, seems to be the appropriate one to use (with the negative sign). We take $u = \sqrt{2}\,x$ and $a = \sqrt{3}$. To have the proper form of du—namely, $du = \sqrt{2}\,dx$—we must multiply and divide by $\sqrt{2}$. So we have

$$\int \sqrt{2x^2 - 3}\,dx = \frac{1}{\sqrt{2}} \int \sqrt{\left(\sqrt{2}\,x\right)^2 - \left(\sqrt{3}\right)^2}\,\left(\sqrt{2}\,dx\right)$$

$$= \frac{1}{\sqrt{2}} \int \sqrt{u^2 - a^2}\,du$$

$$= \frac{1}{\sqrt{2}} \left[\frac{u}{2}\sqrt{u^2 - a^2} - \frac{a^2}{2}\ln\left|u + \sqrt{u^2 - a^2}\right|\right] + C$$

By Formula 37

$$= \frac{1}{\sqrt{2}} \left[\frac{\sqrt{2}\,x}{2}\sqrt{2x^2 - 3} - \frac{3}{2}\ln\left|\sqrt{2}\,x + \sqrt{2x^2 - 3}\right|\right] + C$$

(b) Formula 64 seems to be the one to use in this case. We let $u = 2x$ and $a = 3/4$. Then, adjusting the differential by multiplying and dividing by 2, we have

$$\int \frac{x\,dx}{\sqrt{3x-4x^2}} = \frac{1}{2}\int \frac{2x\,dx}{\sqrt{3x-4x^2}} = \frac{1}{2}\int \frac{u\,du}{\sqrt{2au-u^2}}$$

$$= \frac{1}{2}\left[-\sqrt{2au-u^2} + a\cos^{-1}\left(\frac{a-u}{a}\right)\right] + C$$

By Formula 64

$$= \frac{1}{2}\left[-\sqrt{3x-4x^2} + \frac{3}{4}\cos^{-1}\left(\frac{\frac{3}{4}-2x}{\frac{3}{4}}\right)\right] + C$$

$$= -\frac{\sqrt{3x-4x^2}}{2} + \frac{3}{8}\cos^{-1}\left(\frac{3-8x}{3}\right) + C \qquad ■$$

EXAMPLE 8.26   Evaluate the integrals

(a) $\displaystyle\int \frac{x^2\,dx}{2x^2-7x+6}$ and   (b) $\displaystyle\int x\sqrt{x^2-4x+5}\,dx$

$$2x^2 - 7x + 6\,\overline{\smash{)}\,x^2}\;\overset{\displaystyle \frac{1}{2}}{}$$

$$\underline{x^2 - \frac{7}{2}x + 3}$$

$$\frac{7}{2}x - 3$$

using the tables on the endpapers.

Solution

(a) As the integral stands, there is no form in the table to match it. Since the integrand is an improper rational fraction, we divide (as shown in the margin) to get

$$\frac{x^2}{2x^2-7x+6} = \frac{1}{2} + \frac{\frac{7}{2}x - 3}{2x^2-7x+6} = \frac{1}{2} + \frac{\frac{7}{2}x - 3}{(2x-3)(x-2)}$$

We split the integral into three parts:

$$\int \frac{x^2\,dx}{2x^2-7x+6} = \int \frac{1}{2}\,dx + \frac{7}{2}\int \frac{x\,dx}{(2x-3)(x-2)} - 3\int \frac{dx}{(2x-3)(x-2)}$$

The second integral matches Formula 27 and the third matches Formula 26, both with $u = x$, $a = -3$, $b = 2$, $c = -2$, and $d = 1$. So $ad - bc = -3 + 4 = 1 \neq 0$. Thus,

$$\int \frac{x\,dx}{(2x-3)(x-2)} = \int \frac{u\,du}{(a+bu)(c+du)}$$

$$= \frac{1}{ad-bc}\left(\frac{a}{b}\ln|a+bu| - \frac{c}{d}\ln|c+du|\right) + C$$

Formula 27

$$= -\frac{3}{2}\ln|2x-3| + 2\ln|x-2| + C$$

and

$$\int \frac{dx}{(2x-3)(x-2)} = \int \frac{du}{(a+bu)(c+du)} = \frac{1}{ad-bc} \ln\left|\frac{c+du}{a+bu}\right| + C$$

Formula 26

$$= \ln\left|\frac{x-2}{2x-3}\right| + C$$

Finally, then, from our original long division, we have (writing only one arbitrary constant)

$$\int \frac{x^2\,dx}{2x^2-7x+6} = \frac{x}{2} + \frac{7}{2}\left[-\frac{3}{2}\ln|2x-3| + 2\ln|x-2|\right] - 3\ln\left|\frac{x-2}{2x-3}\right| + C$$

Using properties of logarithms and some algebra, you can show that the answer can be written in the simpler form

$$\frac{x}{2} - \frac{9}{4}\ln|2x-3| + 4\ln|x-2| + C$$

In all probability we could have gotten the answer faster without using tables. All we would need to do, after the long division, is to resolve the proper fraction into partial fractions.

(b) Again, we do not find a form in the table that matches the integral, but by completing the square under the radical, we convert it to a standard form. We have

$$\int x\sqrt{x^2-4x+5}\,dx = \int x\sqrt{(x^2-4x+4)+1}\,dx$$

$$= \int x\sqrt{(x-2)^2+1}\,dx$$

Now if we let $u = x-2$, we have $x = u+2$, and $dx = du$. So our integral becomes

$$\int (u+2)\sqrt{u^2+1}\,du = \int u\sqrt{u^2+1}\,du + 2\int \sqrt{u^2+1}\,du$$

We can easily evaluate the first integral on the right without tables (using a substitution for u^2+1), after adjusting the differential. For the second integral we use Formula 37 again, with $a=1$ (and with the plus sign). So we have

$$\int (u+2)\sqrt{u^2+1}\,du = \frac{1}{2}\int (u^2+1)^{1/2}(2u\,du) + 2\int \sqrt{u^2+1}\,du$$

$$= \frac{1}{2}(u^2+1)^{3/2}\cdot\frac{2}{3}$$

$$+ 2\left[\frac{u}{2}\sqrt{u^2+1} + \frac{1}{2}\ln\left|u+\sqrt{u^2+1}\right|\right] + C$$

or, on replacing u by $x-2$ and simplifying,

$$\int x\sqrt{x^2-4x+5}\,dx = \frac{1}{3}(x^2-4x+5)^{3/2} + (x-2)\sqrt{x^2-4x+5}$$

$$+ \ln\left|(x-2)+\sqrt{x^2-4x+5}\right| + C \qquad \blacksquare$$

INTEGRALS USING COMPUTER ALGEBRA SYSTEMS—TECHNIQUES OF INTEGRATION

We will illustrate how Maple, Mathematica, and DERIVE can be used to compute a wide range of integrals. Each of the examples duplicates an example that we previously have done by hand. In each case the actual input and output as they appear on the screen of the computer terminal are shown.

 CAS 24

Redo Example 8.3 using Maple, Mathematica, and DERIVE.

Maple:

int(x^3*exp(2*x),x);

Output:

$$\frac{1}{2}x^3e^{2x} - \frac{3}{4}x^2e^{2x} + \frac{3}{4}xe^{2x} - \frac{3}{8}e^{2x}$$

which can easily be shown to be equivalent to the answer obtained by hand. Note that the computer does not supply the constant of integration, which is true of other CAS as well.

Mathematica:

Integrate[x^3Exp[2x],x]

Note that in Mathematica, functions such as the exponential, as well as procedures such as integration, start with a capital letter. The x after the comma means that we are integrating with respect to x.

Output:

$$E^{2x}\left(-\left(\frac{3}{8}\right) + \frac{3x}{4} - \frac{3x^2}{4} + \frac{x^3}{2}\right)$$

Here E^{2x} means e^{2x}. We see, then, that the output is equivalent to the answer we found in Example 8.3.

DERIVE:

(At the □ symbol, go to the next step.)
a (author) □ x^3exp(2x) □ c (calculus)
□ i (integrate) □ [choose expression]
□ x (variable) □ enter (no limits) □ s
(simplify) □ [choose expression]
Output:
$$\frac{e^{2x}(4x^3 - 6x^2 + 6x - 3)}{8}$$

 CAS 25

Redo Example 8.7 using Maple, Mathematica, and DERIVE.

Maple:

int(sec(x)^3,x);

Output:
$$\frac{1}{2}\frac{\sin(x)}{\cos(x)^2} + \frac{1}{2}\ln(\sec(x) + \tan(x))$$

Mathematica:

Integrate[(Sec[x])^3,x]

Output: $-\dfrac{\text{Log}[\text{Cos}[\frac{x}{2}] - \text{Sin}[\frac{x}{2}]]}{2}$ +

$\dfrac{\text{Log}[\text{Cos}[\frac{x}{2}] + \text{Sin}[\frac{x}{2}]]}{2}$ +

$\dfrac{[\text{Sec}[x]\ \text{Tan}[x]]}{2}$

DERIVE:

(At the □ symbol, go to the next step.)
a (author) □ sec^3(x) □ c (calculus) □ i
(integrate) □ [choose expression] □ enter (no limits) □ s (simplify) □ [choose expression]

Output: $\dfrac{\text{LN}\left[\dfrac{\text{SIN}(x) + 1}{\text{COS}(x)}\right]}{2} + \dfrac{\text{SIN}(x)}{2\text{COS}(x)^2}$

In the preceding example the answers given by Maple and DERIVE are the same. It is not immediately evident that the answer given by Maple and DERIVE is equivalent to the answer given by Mathematica, nor that either answer is equivalent to the answer we got by hand in Example 8.7, namely $\frac{1}{2}(\sec x \tan x + \ln|\sec x + \tan x|) + C$. The answer given by Maple and DERIVE is essentially the same, however, since

$$\frac{1}{2}\frac{\sin x}{\cos^2 x} = \frac{1}{2}\left(\frac{\sin x}{\cos x}\right)\left(\frac{1}{\cos x}\right) = \frac{1}{2}\tan x \sec x$$

Note, however, that the absolute value signs on the logarithm terms are missing in Maple, Mathematica, and DERIVE. (In Mathematica "Log" means "ln.") You should be on the lookout for slight differences like this; otherwise, answers obtained could be interpreted incorrectly. The absolute value signs are essential in the solution of this problem.

In Exercise 25 of Section 8.5 we will ask you to show that the answer given by Mathematica in CAS 25 is equivalent to the one given by Maple and DERIVE.

CAS 26

Redo Example 8.9 using Maple, Mathematica, and DERIVE.

Maple:

int(sin(x)^4*cos(x)^2,x);

Output: $-\frac{1}{6}\sin(x)^3\cos(x)^3$
$-\frac{1}{8}\sin(x)\cos(x)^3$
$+\frac{1}{16}\cos(x)\sin(x) + \frac{1}{16}x$

Mathematica:

Integrate[Sin[x]^4Cos[x]^2,x]

Output: $\dfrac{x}{16} - \dfrac{\text{Cos}[x]\,\text{Sin}[x]^3}{24}$

$+ \dfrac{\text{Cos}[x]\,\text{Sin}[x]^5}{6} - \dfrac{\text{Sin}[2\,x]}{32}$

DERIVE:

(At the □ symbol, go to the next step.)
a (author) □ sin^4(x)cos^2(x) □ c (calculus) □ i (integrate) □ [choose expression] □ x (variable) □ enter (no limits) □ s (simplify) □ [choose expression]

Output:

$$-\text{COS}(x)^3 \cdot \left[\dfrac{\text{SIN}(x)^3}{6} + \dfrac{\text{SIN}(x)}{8}\right]$$

$$+ \dfrac{\text{SIN}(x)\text{COS}(x)}{16} + \dfrac{x}{16}$$

Again, in this example Maple and DERIVE give answers essentially in the same form. Mathematica gives something quite different. Neither answer conforms to what we obtained in Example 8.9. Again, we will ask you to show the equivalence of the three answers in Exercise 26 of Section 8.5.

CAS 27

Redo Example 8.15 using Maple, Mathematica, and DERIVE.

Maple:

int(sqrt(4*x^2-9)/x,x);

Output:

$$\sqrt{x^2 - \frac{9}{4}} - 3\arccos\left(\frac{2x}{3}\right)$$

Mathematica:

Integrate[(Sqrt[4x^2-9])/x,x]

Output: $\text{Sqrt}[-9 + 4\,x^2]$

$+ 3\,\text{ArcTan}\left[\dfrac{3}{\text{Sqrt}[-9 + 4\,x^2]}\right]$

DERIVE:

(At the □ symbol, go to the next step.)
a (author) □ sqrt(4x^2-9)/x □ c (calculus) □ i (integrate) □ [choose expression] □ x (variable) □ enter (no limits) □ s (simplify) □ [choose expression]
Output:

$$\sqrt{(4x^2 - 9)} - 3\text{ATAN}\left[\dfrac{\sqrt{(4x^2 - 9)}}{3}\right]$$

Observe that this time the result given by Maple is the same as the answer we got by hand in Example 8.15. This time Mathematica and DERIVE give essentially the same form for the answer. In Exercise 27 of Section 8.5 we will ask you to show that the result given by Mathematica is also equal to the result we got by hand.

⊡ CAS 28

Redo Example 8.23 using Maple, Mathematica, and DERIVE.

Maple:

int((2*x–3)/(x^5–2*x^3+x),x);

Output: $-3\ln(x) + \dfrac{3}{2}\ln(x^2 + 1)$
$+\dfrac{1}{4}\dfrac{4x - 6}{x^2 + 1} + \arctan(x)$

Mathematica:

Integrate[(2x–3)/(x^5+2x^3+x),x]

Output: $\dfrac{-3 + 2\,x}{2\,(1 + x^2)} + \text{ArcTan}[x]$

$-3\,\text{Log}[x] + \dfrac{3\,\text{Log}[1 + x^2]}{2}$

DERIVE:

(At the ⊡ symbol, go to the next step.)
a (author) ⊡ (2x-3)/(x^5+2x^3+x) ⊡ c (calculus) ⊡ i (integrate) ⊡ [choose expression] ⊡ x (variable) ⊡ enter (no limits) ⊡ s (simplify) ⊡ [choose expression]

Output: $\text{ATAN}\left[\dfrac{3\text{LN}(x^2 + 1)}{2}\right]$

$-3\text{LN}(x) + \dfrac{2x - 3}{2(x^2 + 1)}$

This time the three answers can easily be seen to be equivalent, but none of the three matches the answer we obtained by hand in Example 8.23. The answers do, however, differ by a constant, as we will ask you to show in Exercise 28 of Section 8.5. Again, notice the missing absolute values in $\ln|x|$.

Maple, Mathematica, and DERIVE certainly save work on this problem (we needed more than a page to do it by hand). Using a CAS, though, cannot replace an understanding of the techniques.

We give one final example to show the computational power of a CAS. To do this problem by hand would be prohibitively difficult.

⊡ CAS 29

Evaluate the integral

$$\int \frac{1}{1 + x^5}dx$$

using Maple, Mathematica, and DERIVE.

Maple:

int(1/(1+x^5),x);

Output:

$$\frac{1}{5}\ln(x+1)-$$

$$\frac{1}{20}\ln(-2x^2+x+\sqrt{5}x-2)\sqrt{5}-$$

$$\frac{1}{20}\ln(-2x^2+x+\sqrt{5}x-2)$$

$$-\frac{\arctan\left(\frac{-4x+1+\sqrt{5}}{\sqrt{\%2}}\right)}{\sqrt{\%2}}$$

$$+\frac{1}{5}\frac{\arctan\left(\frac{-4x+1+\sqrt{5}}{\sqrt{\%2}}\right)\sqrt{5}}{\sqrt{\%2}}$$

$$+\frac{1}{20}\ln(2x^2-x+\sqrt{5}x+2)\sqrt{5}$$

$$-\frac{1}{20}\ln(2x^2-x+\sqrt{5}x+2)+$$

$$\frac{\arctan\left(\frac{4x-1+\sqrt{5}}{\sqrt{\%1}}\right)}{\sqrt{\%1}}+$$

$$\frac{1}{5}\frac{\arctan\left(\frac{4x-1+\sqrt{5}}{\sqrt{\%1}}\right)\sqrt{5}}{\sqrt{\%1}}$$

$$\%1:=10+2\sqrt{5}$$

$$\%2:=10-2\sqrt{5}$$

Mathematica:

Integrate[1/(1+x^5),x]

Output:

$$\frac{\mathrm{Log}[1+x]}{5}-$$

$$\frac{\mathrm{Cos}[\frac{Pi}{5}]\,\mathrm{Log}[1+x^2-2\,x\,\mathrm{Cos}[\frac{Pi}{5}]]}{5}-$$

$$\frac{\mathrm{Cos}[\frac{3\,Pi}{5}]\,\mathrm{Log}[1+x^2-2\,x\,\mathrm{Cos}[\frac{3\,Pi}{5}]]}{5}+$$

$$\frac{2\,\mathrm{ArcTan}[(x-\mathrm{Cos}[\frac{Pi}{5}])\,\mathrm{Csc}[\frac{Pi}{5}]]\,\mathrm{Sin}[\frac{Pi}{5}]}{5}+$$

$$\frac{2\,\mathrm{ArcTan}[(x-\mathrm{Cos}[\frac{3\,Pi}{5}])\,\mathrm{Csc}[\frac{3\,Pi}{5}]]\,\mathrm{Sin}[\frac{3\,Pi}{5}]}{5}$$

DERIVE:

(At the □ symbol, go to the next step.)
a (author) □ 1/(1+x^5) □ c (calculus)
□ i (integrate) □ [choose expression]
□ x (variable) □ enter (no limits) □ s
(simplify) □ [choose expression]
Output:

$$\sqrt{\frac{\sqrt{5}}{50}+\frac{1}{10}}\,\mathrm{ATAN}\left[\sqrt{\frac{1}{8}-\frac{\sqrt{5}}{40}}(4x+\sqrt{5}-1)\right]$$

$$+\sqrt{\frac{1}{10}-\frac{\sqrt{5}}{50}}\,\mathrm{ATAN}\left[(4x-\sqrt{5}-1)\sqrt{\frac{\sqrt{5}}{40}+\frac{1}{8}}\right]$$

$$+\left[\frac{\sqrt{5}}{20}-\frac{1}{20}\right]\mathrm{LN}(2x^2+x(\sqrt{5}-1)+2)$$

$$-\left[\frac{\sqrt{5}}{20}+\frac{1}{20}\right]\mathrm{LN}(2x^2-x(\sqrt{5}+1)+2)$$

$$+\frac{\mathrm{LN}(x+1)}{2}$$

The three answers certainly do not seem to bear much relation to one another. Trying to show they are equivalent would be a formidable task, so we will spare you time. (One easy way, however, to gain confidence that they are equivalent is to use Maple, Mathematica, or DERIVE to evaluate each answer at several values of x. The results should be the same, or else they should differ by a constant value.) It is a nontrivial problem.

Exercise Set 8.5

In Exercises 1–24, use the integral tables on the front and back endpapers to evaluate the integrals.

1. $\displaystyle\int\frac{dx}{x(2+3x)^2}$

2. $\displaystyle\int\frac{\sqrt{x^2-4}}{x}\,dx$

3. $\displaystyle\int x\tan^{-1}2x\,dx$

4. $\displaystyle\int x^2\sin 3x\,dx$

5. $\displaystyle\int\frac{dx}{x\sqrt{4x-x^2}}$

6. $\displaystyle\int\frac{x\,dx}{2x-3}$

7. $\displaystyle\int \frac{dx}{x^2(x-2)}$

8. $\displaystyle\int \frac{\sqrt{x^2-1}}{x^2}\,dx$

9. $\displaystyle\int \frac{dx}{(x^2+9)^{3/2}}$

10. $\displaystyle\int x^2\sqrt{9-4x^2}\,dx$

11. $\displaystyle\int \sqrt{2x-x^2}\,dx$

12. $\displaystyle\int (\ln x)^3\,dx$

13. $\displaystyle\int x^4 e^{3x}\,dx$

14. $\displaystyle\int \sin^6 x\,dx$

15. $\displaystyle\int \frac{dx}{x(3x+5)}$

16. $\displaystyle\int \frac{dx}{x^2\sqrt{3-2x^2}}$

17. $\displaystyle\int \frac{x\,dx}{(4x-3)^2}$

18. $\displaystyle\int \sec^5 x\,dx$

19. $\displaystyle\int e^{-2x}\cos 3x\,dx$

20. $\displaystyle\int \sqrt{x^2+2x+2}\,dx$

21. $\displaystyle\int \frac{x^2+1}{4+3x-x^2}\,dx$

22. $\displaystyle\int \frac{2x-3}{\sqrt{5+4x-x^2}}\,dx$

23. $\displaystyle\int x^2\sin^{-1}x\,dx$

24. $\displaystyle\int \frac{\sqrt{3x^2-6x+5}}{x-1}\,dx$

25. Show that the answer given by Mathematica in CAS 25 is equivalent to the one given by Maple.

26. Show that the answer found in Example 8.9 can be changed to the form in CAS 26 given by (a) Mathematica and (b) Maple. (*Hint:* Use the identities $\sin 2\theta = 2\sin\theta\cos\theta$, $\cos 2\theta = 1-2\sin^2\theta$, and $\cos^2\theta = 1-\sin^2\theta$.)

27. Show that the answer given by Mathematica in CAS 27 is equivalent to the one given by Maple.

28. Show that the answer obtained in Example 8.23 can be put in a form that agrees with the output given by Mathematica and Maple in CAS 28, except that it differs by a constant (which can be incorporated into the constant of integration). (*Hint:* Write $3x^2+2x$ as $3(1+x^2)-3+2x$.)

Do Exercises 29–32 only if you have access to a CAS other than Maple, Mathematica, or DERIVE. In each case redo the specified example using this CAS and show the answer is equivalent to the one obtained by either Mathematica, Maple, or DERIVE.

29. CAS 26

30. CAS 27

31. CAS 28

32. CAS 29

In Exercises 33–56, redo Exercises 1–24 using a CAS. Show that the answer is equivalent to the one found by using the integral table.

8.6 IMPROPER INTEGRALS

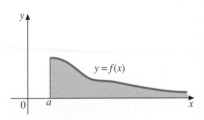

FIGURE 8.11
f is defined on the unbounded interval $[a, \infty)$.

In the definition of the definite integral $\int_a^b f(x)\,dx$, we understood that the interval of integration $[a, b]$ was *finite*; that is, a and b were finite real numbers. In this section we extend the definition of the integral to *unbounded* intervals of integration of the forms $[a, \infty)$, $(-\infty, a]$, and $(-\infty, \infty)$. In Figure 8.11 we illustrate the graph of a function defined on an unbounded interval of the form $[a, \infty)$.

Another requirement for the integral $\int_a^b f(x)\,dx$ was that the function f had to be *bounded*; that is, some positive constant M existed for which f had to satisfy an inequality of the form $-M \le f(x) \le M$. Now we extend the definition of the integral so as to remove this restriction also. Our extended definition will allow for f to have one or more infinite discontinuities (its graph

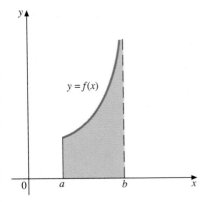

FIGURE 8.12
f is unbounded as $x \to b^-$.

will have vertical asymptotes) in $[a, b]$. In Figure 8.12 we show such a function with a vertical asymptote at $x = b$. Recall that we saw an integral with an unbounded integrand in calculating the length of an ellipse (Example 6.19).

Each of the two new types of integrals is called an **improper integral**. We consider the two types separately.

Unbounded Intervals of Integration

To motivate a plausible definition for the integral of a function on an infinite interval of integration, consider the function $f(x) = 1/x^2$. Suppose we wish to give some meaning to the area under the graph of f, to the right of $x = 1$. (See Figure 8.13.) On any finite interval $[1, t]$ with $t > 1$, the integral

$$\int_1^t \frac{1}{x^2}\, dx$$

is well defined (since f is continuous on any interval of this form). The value of this integral can be interpreted geometrically (Figure 8.14) as the area under the graph of $y = 1/x^2$ from $x = 1$ to $x = t$, namely,

$$\text{Area} = \int_1^t \frac{1}{x^2}\, dx = \int_1^t x^{-2}\, dx = -\frac{1}{x}\Big]_1^t = -\frac{1}{t} + 1$$

t	$\int_1^t \frac{1}{x^2}dx$
2	0.5
10	0.9
100	0.99
1000	0.999

Suppose we now assign larger and larger values to t. Some specific values of t and the corresponding values of the integral (the area) are shown in the table in the margin. Clearly, the areas seem to approach 1 as $t \to \infty$. We therefore can say that

$$\int_1^\infty \frac{1}{x^2}\, dx = \lim_{t \to \infty} \int_1^t \frac{1}{x^2}\, dx = 1$$

The preceding example suggests the following definition.

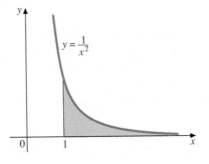

FIGURE 8.13
Unbounded region under the graph on $[1, \infty)$

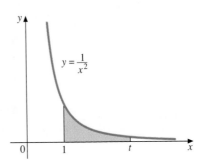

FIGURE 8.14

Definition 8.1
Improper Integrals with
Unbounded Intervals of
Integration

(a) If $\int_a^t f(x)dx$ exists for all $t \geq a$, then

$$\int_a^\infty f(x)dx = \lim_{t \to \infty} \int_a^t f(x)dx$$

provided the limit exists. (See Figure 8.15.)

(b) If $\int_t^a f(x)dx$ exists for all $t \leq a$, then

$$\int_{-\infty}^a f(x)dx = \lim_{t \to -\infty} \int_t^a f(x)dx$$

provided the limit exists. (See Figure 8.16.)

(c) If both $\int_{-\infty}^a f(x)dx$ and $\int_a^\infty f(x)dx$ exist for some real number a, then

$$\int_{-\infty}^\infty f(x)dx = \int_{-\infty}^a f(x)dx + \int_a^\infty f(x)dx$$

(See Figure 8.17).

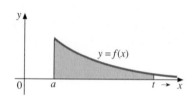

FIGURE 8.15
$\int_a^\infty f(x)\,dx = \lim_{t \to \infty} \int_a^t f(x)dx$

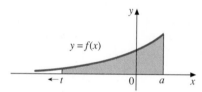

FIGURE 8.16
$\int_{-\infty}^a f(x)\,dx = \lim_{t \to -\infty} \int_t^a f(x)\,dx$

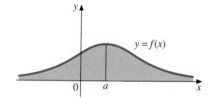

FIGURE 8.17
$\int_{-\infty}^\infty f(x)dx =$
$\int_{-\infty}^a f(x)\,dx + \int_a^\infty f(x)\,dx$

REMARKS ─────────────────────

■ When the limit in (a) or (b) exists, the corresponding integral is said to **converge**. Otherwise it is said to **diverge**. The integral in (c) converges if both integrals on the right converge, and it diverges if either integral on the right diverges.

■ In (c), any real number a can be used. It can be shown (see Exercise 54 in Exercise Set 8.6) that if both integrals on the right converge for any given real number a, they will converge for any other real number, and the sum will be the same.

■ If the function f is nonnegative on the specified infinite interval, then the improper integral can be interpreted as the area under the graph of f on that infinite interval, if the integral converges. If the integral diverges, we say the area does not exist.

■ The requirement in (c) that $\int_a^\infty f(x)dx$ and $\int_{-\infty}^a f(x)dx$ both exist means that the limits

$$\lim_{t_1 \to \infty} \int_a^{t_1} f(x)dx \qquad \text{and} \qquad \lim_{t_2 \to -\infty} \int_{t_2}^\infty f(x)dx$$

must exist (be a finite number) independently of one another.

EXAMPLE 8.27   Determine whether the integral $\int_1^\infty (1/x)\,dx$ converges or diverges.

Solution   Since

$$\lim_{t\to\infty}\int_1^t \frac{1}{x}\,dx = \lim_{t\to\infty}[\ln x]_1^t = \lim_{t\to\infty}[\ln t - \ln 1]$$

$$= \lim_{t\to\infty}(\ln t) = \infty$$

it follows that the integral diverges.   ■

It is interesting to compare the integrals

$$\int_1^\infty \frac{1}{x^2}\,dx \qquad \text{and} \qquad \int_1^\infty \frac{1}{x}\,dx$$

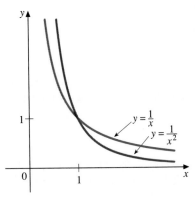

When we were preparing for Definition 8.1, we showed that the first integral converges to the value 1, and in Example 8.27 we have shown that the second integral diverges to ∞. In Figure 8.18 we show the graphs of $y = 1/x^2$ and $y = 1/x$. Interpreted in terms of area, our results show that the area under $y = 1/x^2$ to the right of $x = 1$ is finite (it equals 1), whereas the area under $y = 1/x$ to the right of $x = 1$ is infinite. Although the graphs look very much alike, the key difference is that $1/x^2$ goes to 0 faster than $1/x$ as x increases. So although the graphs of $y = 1/x^2$ and $y = 1/x$ both have the horizontal asymptote $y = 0$, the first curve approaches the asymptote more rapidly than the second does, as x increases.

Both $\int_1^\infty (1/x^2)\,dx$ and $\int_1^\infty (1/x)\,dx$ are special cases of the family of integrals

$$\int_1^\infty \frac{1}{x^p}\,dx \qquad (p > 0)$$

FIGURE 8.18

We have seen that when $p = 2$ the integral converges and when $p = 1$ it diverges. As the next example shows, $p = 1$ is the dividing point between convergence and divergence, in the sense that the integral converges for any value of p greater than 1 and diverges for any value of p less than (or equal to) 1.

EXAMPLE 8.28   Show that the integral

$$\int_1^\infty \frac{1}{x^p}\,dx$$

converges if $p > 1$ and diverges if $p \le 1$.

Solution   For $p \ne 1$, we have

$$\int_1^t \frac{1}{x^p}\,dx = \int_1^t x^{-p}\,dx = \frac{x^{-p+1}}{-p+1}\Bigg]_1^t = \frac{1}{1-p}\left[\frac{1}{t^{p-1}} - 1\right]$$

Thus,

$$\int_1^\infty \frac{1}{x^p}\,dx = \lim_{t\to\infty}\int_1^t \frac{1}{x^p}\,dx = \lim_{t\to\infty}\frac{1}{1-p}\left[\frac{1}{t^{p-1}} - 1\right]$$

If $p > 1$ the exponent $p - 1$ is positive, so as $t \to \infty$, $1/t^{p-1} \to 0$, and the limit is finite, namely,

$$\int_1^\infty \frac{1}{x^p}\,dx = \frac{1}{p-1} \qquad (\text{if } p > 1)$$

If $p < 1$, the exponent on t is negative, so

$$\frac{1}{t^{p-1}} = t^{1-p}$$

with $1 - p > 0$. Thus, as $t \to \infty$, $t^{1-p} \to \infty$, and the integral diverges. We saw in Example 8.27 that the integral diverges when $p = 1$. So we have completed the proof. ∎

Because of its importance, we summarize what we have just proved:

$$\int_1^\infty \frac{1}{x^p}\, dx \quad \begin{cases} \text{converges if } p > 1 \\ \text{diverges \ \ if } p \le 1 \end{cases}$$

REMARK ─────────────────────────────────

■ In determining convergence or divergence, the lower limit does not need to be 1; it can be any positive number. So our conclusions continue to hold true for

$$\int_a^\infty \frac{1}{x^p}\, dx$$

where $a > 0$. Of course, when the integral converges, the *value* is affected by what a is, but whether the integral converges or diverges depends only on p.

EXAMPLE 8.29   **An interesting representation of π**   Show that

$$\int_{-\infty}^\infty \frac{1}{1+x^2}\, dx = \pi$$

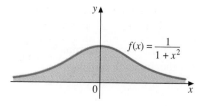

$f(x) = \dfrac{1}{1+x^2}$

FIGURE 8.19

Solution   We show the graph of $f(x) = 1/(1 + x^2)$ in Figure 8.19. By Definition 8.1(c), we must show that both of the integrals

$$\int_a^\infty \frac{1}{1+x^2}\, dx \qquad \text{and} \qquad \int_{-\infty}^a \frac{1}{1+x^2}\, dx$$

converge and that their sum is π. We are free to choose the number a. Let us take $a = 0$. Then we have

$$\int_0^\infty \frac{1}{1+x^2}\, dx = \lim_{t \to \infty} \int_0^t \frac{1}{1+x^2}\, dx = \lim_{t \to \infty} \left[\tan^{-1} x\right]_0^t$$

$$= \lim_{t \to \infty} \left[\tan^{-1} t - \tan^{-1} 0\right] = \frac{\pi}{2}$$

since $\tan^{-1} t \to \pi/2$ as $t \to \infty$, and $\tan^{-1} 0 = 0$.

Similarly,

$$\int_{-\infty}^0 \frac{1}{1+x^2}\, dx = \lim_{t \to -\infty} \int_t^0 \frac{1}{1+x^2}\, dx = \lim_{t \to -\infty} \left[\tan^{-1} x\right]_t^0$$

$$= \lim_{t \to -\infty} \left[\tan^{-1} 0 - \tan^{-1} t\right] = -\left(-\frac{\pi}{2}\right) = \frac{\pi}{2}$$

since $\tan^{-1} t \to -\pi/2$ as $t \to -\infty$.

Thus,

$$\int_{-\infty}^{\infty} \frac{1}{1+x^2}\, dx = \int_0^{\infty} \frac{1}{1+x^2}\, dx + \int_{-\infty}^0 \frac{1}{1+x^2}\, dx = \frac{\pi}{2} + \frac{\pi}{2} = \pi \quad \blacksquare$$

REMARK ──────────────────────

■ Since $f(x) = 1/(1 + x^2)$ is even, we could have used symmetry in Example 8.29 to conclude from the fact that the integral on $[0, \infty)$ is $\pi/2$ that the integral on $(-\infty, 0]$ also has the value $\pi/2$.

──────────────────────

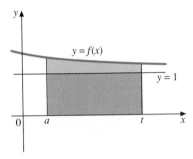

FIGURE 8.20
$\int_a^t f(x)\, dx > (t - a)$ and $t - a \to \infty$ as $t \to \infty$

In our examples the x-axis ($y = 0$) was a horizontal asymptote in each case. Suppose that the graph of f has some other horizontal asymptote, say $y = k$, where $k \neq 0$. Then the improper integral of f, whether on $[0, \infty)$ or on $(-\infty, a]$, diverges. Figure 8.20 illustrates this result for the case where the graph of f approaches the horizontal asymptote $y = 1$ from above, as $x \to \infty$. The area under the graph from a to t is greater than the area of the rectangle $(t - a)$ units long and 1 unit high. Thus,

$$\int_a^t f(x)\, dx > t - a \qquad (t > a)$$

and since $t - a \to \infty$ as $t \to \infty$, it follows that the integral $\int_a^{\infty} f(x)\, dx$ diverges. Although it is less evident in the case where the graph approaches a nonzero horizontal asymptote from below, the integral can again be shown to be divergent.

It is tempting to conclude from the preceding discussion that the only functions that have a chance of having convergent improper integrals over an unbounded interval of integration are those with the horizontal asymptote $y = 0$. Examples exist, however, to contradict this conclusion. We show the graph of one such function in Figure 8.21. The x-axis is not an asymptote, since $f(x) = 1$ throughout some small rectangle centered at each integer value. In Chapter 10 we will show that the areas, added together, give the value 1.

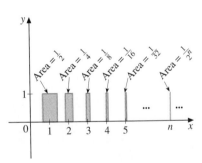

$$\int_0^{\infty} f(x)\, dx = \tfrac{1}{2} + \tfrac{1}{4} + \tfrac{1}{8} + \tfrac{1}{16} + \tfrac{1}{32} + \cdots = 1$$

FIGURE 8.21

A Comparison Test for Improper Integrals

Sometimes it may be difficult or impossible to find the exact value of an improper integral, but by comparing it with an improper integral whose value is known, we may be able to determine whether the integral in question converges or diverges. The next theorem shows how this comparison works.

THEOREM 8.1

> **A Comparison Test for Improper Integrals**
>
> Suppose that for all $x \geq a$, f and g are continuous and $0 \leq f(x) \leq g(x)$.
>
> (a) If $\int_a^{\infty} g(x)\, dx$ converges, then $\int_a^{\infty} f(x)\, dx$ converges.
> (b) If $\int_a^{\infty} f(x)\, dx$ diverges, then $\int_a^{\infty} g(x)\, dx$ diverges.

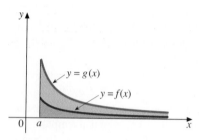

FIGURE 8.22

Intuitively, this result seems plausible. From Figure 8.22, the area under the graph of f is smaller than the area under the graph of g. So if the area under g is finite, we would expect the area under f to be finite also. On the other hand, if the area under f is infinite, the area under g, being greater, must also be infinite.

EXAMPLE 8.30   Test each of the following for convergence or divergence:

(a) $\displaystyle\int_1^\infty \frac{dx}{\sqrt{x^3 + 4}}$ (b) $\displaystyle\int_1^\infty \frac{dx}{\sqrt[3]{2x^2 - 1}}$

Solution

(a) Since for all $x \geq 1$,

$$0 \leq \frac{1}{\sqrt{x^3 + 4}} < \frac{1}{\sqrt{x^3}} = \frac{1}{x^{3/2}}$$

(see Figure 8.23), and since by Example 8.28,

$$\int_1^\infty \frac{dx}{x^{3/2}} \qquad p = \frac{3}{2} > 1$$

converges, it follows by the comparison test (Theorem 8.1) that

$$\int_1^\infty \frac{dx}{\sqrt{x^3 + 4}}$$

also converges.

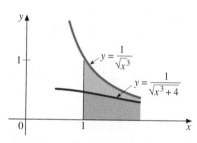

FIGURE 8.23

(b) We observe that for all $x \geq 1$, $2x^2 - 1 < 2x^2$. So $\sqrt[3]{2x^2 - 1} < \sqrt[3]{2x^2}$. Thus, taking reciprocals, we get the reverse inequality:

$$\frac{1}{\sqrt[3]{2x^2 - 1}} > \frac{1}{\sqrt[3]{2x^2}} = \frac{1}{\sqrt[3]{2}} \cdot \frac{1}{x^{2/3}} > 0$$

(see Figure 8.24.) Also, by Example 8.28,

$$\int_1^\infty \frac{dx}{x^{2/3}}\, dx \qquad p = \frac{2}{3} < 1$$

diverges, and multiplying by the constant $1/\sqrt[3]{2}$ does not alter this divergence. It follows by the comparison test (Theorem 8.1) that

$$\int_1^\infty \frac{dx}{\sqrt[3]{2x^2 - 1}}$$

also diverges. ■

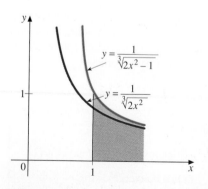

FIGURE 8.24

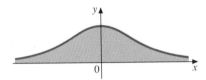

FIGURE 8.25
The standardized normal probability
density function $\phi(x) = \dfrac{1}{\sqrt{2\pi}}e^{-x^2/2}$

EXAMPLE 8.31   Show that the area under the standardized normal probability density function (see Figure 8.25)

$$\phi(x) = \frac{1}{\sqrt{2\pi}}\,e^{-x^2/2} \qquad (-\infty < x < \infty)$$

is finite. (*Note:* For $\phi(x)$ to be a probability density function, not only must the area be finite, but also its value must be 1. Here we are concerned only with showing convergence. It is possible to show its value is 1 using methods studied in multivariable calculus.)

Solution   Since ϕ is an even function, it suffices to show that the integral from 0 to ∞ is finite. First, we divide the interval of integration into the two intervals $[0, 1]$ and $[1, \infty)$, getting

$$\int_0^\infty e^{-x^2/2}dx = \int_0^1 e^{-x^2/2}dx + \int_1^\infty e^{-x^2/2}dx$$

The first integral on the right is finite, since $e^{-x^2/2}$ is continuous on $[0, 1]$. For the second integral, we use the comparison test, since we cannot integrate $e^{-x^2/2}$ directly. If, however, we could replace x^2 by x, then we could perform the integration. Because $x^2 \geq x$ for $x \geq 1$, we see that $e^{x^2/2} \geq e^{x/2}$ since the exponential function is increasing. Taking reciprocals gives the reverse inequality:

$$\frac{1}{e^{x^2/2}} \leq \frac{1}{e^{x/2}} \qquad (\text{for } x \geq 1)$$

or, equivalently, $e^{-x^2/2} \leq e^{-x/2}$. So if we can show that $\int_1^\infty e^{-x/2}dx$ converges, it will follow that $\int_1^\infty e^{-x^2/2}$ converges. For $t > 1$,

$$\int_1^t e^{-x/2}dx = -2\int_1^t e^{-x/2}\left(-\frac{1}{2}\,dt\right) = -2\big[e^{-x/2}\big]_1^t$$

$$= -2\big[e^{-t/2} - e^{-1/2}\big]$$

So

$$\int_1^\infty e^{-x/2}dx = \lim_{t\to\infty}\big[-2e^{-t/2} + 2e^{-1/2}\big] = \frac{2}{\sqrt{e}}$$

Since this limit is finite, it follows by the comparison test (Theorem 8.1) that $\int_1^\infty e^{-x^2/2}dx$ converges.

We can now conclude that

$$\int_0^\infty e^{-x^2/2}dx = \int_0^1 e^{-x^2/2}dx + \int_1^\infty e^{-x^2/2}dx$$

exists. By symmetry, then, we see that

$$\int_{-\infty}^\infty e^{-x^2/2}dx$$

also exists. Finally, the convergence is unaltered if we multiply by the constant $1/\sqrt{2\pi}$. (The factor $1/\sqrt{2\pi}$ provides the "normalization" to make the integral have the value 1.)   ∎

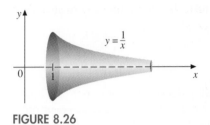

FIGURE 8.26

EXAMPLE 8.32   **Gabriel's Horn**   The infinite surface formed by revolving the graph of $f(x) = 1/x$ for $1 \leq x < \infty$ about the x-axis is often referred to as *Gabriel's Horn*. (See Figure 8.26.) Show that (a) the volume enclosed by the surface is finite, but (b) the area of the surface is infinite. (Someone has described this phenomenon by saying that a vessel in the shape of Gabriel's Horn would hold a finite amount of paint, but it would take an infinite amount of paint to cover its surface!)

Solution

(a) Using the disk method of Chapter 6, the volume V is

$$V = \pi \int_1^\infty \left(\frac{1}{x}\right)^2 dx = \pi \int_1^\infty \frac{1}{x^2} dx$$

From the result of Example 8.28, with $p = 2$, we know that this integral converges.

(b) Again, from Chapter 6 we know that the surface area S is given by

$$S = 2\pi \int_1^\infty y \, ds$$

where $y = \dfrac{1}{x}$ and

$$ds = \sqrt{1 + \left(\frac{dy}{dx}\right)^2} \, dx = \sqrt{1 + \left(-\frac{1}{x^2}\right)^2} \, dx = \frac{\sqrt{x^4 + 1}}{x^2} \, dx$$

Thus,

$$S = 2\pi \int_1^\infty \left(\frac{1}{x}\right) \frac{\sqrt{x^4 + 1}}{x^2} \, dx = 2\pi \int_1^\infty \frac{\sqrt{x^4 + 1}}{x^3} \, dx$$

Now, for $x \geq 1$,

$$\frac{\sqrt{x^4 + 1}}{x^3} > \frac{\sqrt{x^4}}{x^3} = \frac{x^2}{x^3} = \frac{1}{x}$$

and since $\int_1^\infty \frac{1}{x} dx$ diverges ($p = 1$), it follows by the comparison test that the integral for S also diverges.   ∎

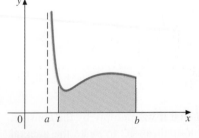

FIGURE 8.27

Unbounded Integrands

We consider first the case in which f is continuous for $a < x \leq b$ and $\lim_{x \to a^+} f(x) = \pm\infty$ (so the graph of f has a vertical asymptote at $x = a$). We illustrate such a function in Figure 8.27. As we indicate in that figure, we choose a number t in the open interval (a, b). Then, since f is continuous on the closed interval $[t, b]$, the integral $\int_t^b f(x)dx$ exists. After evaluating this

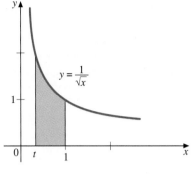

FIGURE 8.28

integral, we let $t \to a^+$ and see whether there is a finite limit. If so, this limit is the value we assign to the improper integral $\int_a^b f(x)dx$.

To illustrate what we have described, consider the integral

$$\int_0^1 \frac{dx}{\sqrt{x}}$$

We show the graph of $f(x) = 1/\sqrt{x}$ in Figure 8.28. Because the function f is unbounded in $(0, 1]$, the integral is improper. Following the procedure we outlined above, we let t be in the open interval $(0, 1)$ and evaluate the integral

$$\int_t^1 \frac{dx}{\sqrt{x}} = \int_t^1 x^{-1/2}dx = 2x^{1/2}\Big]_t^1 = 2 - 2\sqrt{t}$$

Now we let $t \to 0^+$ to get

$$\int_0^1 \frac{dx}{\sqrt{x}} = \lim_{t \to 0^+}(2 - 2\sqrt{t}) = 2$$

We can say that the area under the curve is 2.

If f is continuous for $a \le x < b$ and is unbounded as x approaches b from the left, we integrate from a to t and then let $t \to b^-$. (See Figure 8.29.)

A third possibility is that f becomes unbounded as x approaches some number c between the endpoints a and b of the interval of integration. We illustrate one such possibility in Figure 8.30. In this case we divide the interval into the two parts $[a, c]$ and $[c, b]$. For the integral over the entire interval $[a, b]$ to exist, the integral over each of these two subintervals (with c as an endpoint) must exist, as we have described above.

We are ready now for the formal definition of improper integrals with unbounded integrands.

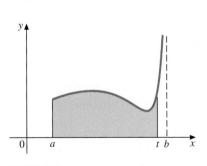

FIGURE 8.29

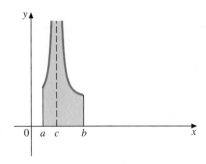

FIGURE 8.30
$\int_a^b f(x)dx =$
$\int_a^c f(x)dx + \int_c^b f(x)\,dx$

Definition 8.2
Improper Integrals with
Unbounded Integrands

(a) If f is continuous for $a < x \leq b$ and $\lim_{x \to a^+} f(x) = \pm\infty$, then

$$\int_a^b f(x)dx = \lim_{t \to a^+} \int_t^b f(x)dx$$

provided the limit on the right exists.

(b) If f is continuous for $a \leq x < b$ and $\lim_{x \to b^-} f(x) = \pm\infty$, then

$$\int_a^b f(x)dx = \lim_{t \to b^-} \int_a^t f(x)dx$$

provided the limit on the right exists.

(c) If f is continuous for $a \leq x < c$ and $c < x \leq b$ but has an infinite discontinuity at $x = c$, then

$$\int_a^b f(x)dx = \int_a^c f(x)dx + \int_c^b f(x)dx$$

provided both integrals on the right exist.

REMARKS ———————————————————————————————

■ As with improper integrals over unbounded intervals of integration, when the limit in (a) or (b) exists, we say the integral *converges*, and when both limits exist, we say the improper integral in (c) *converges*. Otherwise, these integrals *diverge*.

■ Definition 8.2 concerns integrals with unbounded integrands; that is, the integrands have infinite discontinuities. If f is continuous and *bounded* on the open interval (a, b), then it cannot have an infinite discontinuity at either $x = a$ or $x = b$, and even if $f(a)$ or $f(b)$ is undefined, it can be shown that $\int_a^b f(x)dx$ always exists. It is not an improper integral. For example, even though $(\sin x)/x$ is discontinuous at $x = 0$, it is bounded on $(0, 1)$, since

$$\lim_{x \to 0} \frac{\sin x}{x} = 1$$

Its integral,

$$\int_0^1 \frac{\sin x}{x} dx$$

therefore exists. The graph of $y = (\sin x)/x$ (Figure 8.31) clearly shows the boundedness property.

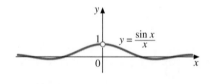

FIGURE 8.31

EXAMPLE 8.33   Evaluate the following integrals, or show that they are divergent:

(a) $\int_0^1 \frac{dx}{x}$ (b) $\int_0^1 \frac{dx}{\sqrt{1 - x^2}}$

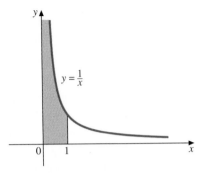

FIGURE 8.32

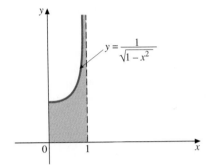

FIGURE 8.33

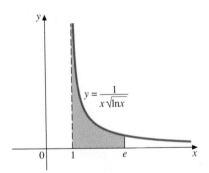

FIGURE 8.34

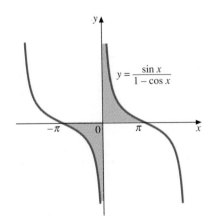

FIGURE 8.35

Solution

(a) The integrand is unbounded as $x \to 0^+$ (see Figure 8.32), so by Definition 8.2 we have

$$\int_0^1 \frac{dx}{x} = \lim_{t \to 0^+} \int_t^1 \frac{dx}{x} = \lim_{t \to 0^+} [\ln x]_t^1 = \lim_{t \to 0^+} (-\ln t) = \infty$$

So the integral diverges.

(b) Since the denominator is 0 when $x = 1$, the point of discontinuity is $x = 1$. We show the graph of $y = 1/\sqrt{1 - x^2}$ in Figure 8.33. Thus,

$$\int_0^1 \frac{dx}{\sqrt{1 - x^2}} = \lim_{t \to 1^-} \int_0^t \frac{dx}{\sqrt{1 - x^2}} = \lim_{t \to 1^-} [\sin^{-1} x]_0^t$$

$$= \lim_{t \to 1^-} (\sin^{-1} t - 0) = \sin^{-1} 1 = \frac{\pi}{2}$$

The integral converges to $\pi/2$. ∎

EXAMPLE 8.34   Evaluate the following integrals, or show that they are divergent:

(a) $\displaystyle\int_1^e \frac{dx}{x\sqrt{\ln x}}$ (b) $\displaystyle\int_{-\pi}^{\pi} \frac{\sin x}{1 - \cos x}\, dx$

Solution

(a) Since $\ln 1 = 0$, the integrand is unbounded near 1. We show its graph in Figure 8.34. Since $d(\ln x) = (1/x)\, dx$, we make the (mental) substitution $u = \ln x$ to get

$$\int_1^e \frac{dx}{x\sqrt{\ln x}} = \lim_{t \to 1^+} \int_t^e \frac{dx}{x\sqrt{\ln x}} = \lim_{t \to 1^+} \left[2\sqrt{\ln x}\right]_t^e$$

$$= \lim_{t \to 1^+} (2 - 2\sqrt{\ln t}) = 2$$

(b) To determine whether $\sin x/(1 - \cos x)$ has a vertical asymptote on $[-\pi, \pi]$, we set the denominator equal to 0 and see that $x = 0$. However, the numerator is also 0 at this point. Thus, we have an indeterminate form of the type $0/0$. So we use L'Hôpital's Rule.

$$\lim_{x \to 0} \frac{\sin x}{1 - \cos x} = \lim_{x \to 0} \frac{\cos x}{\sin x}$$

When $x \to 0^+$ the limit is ∞, and when $x \to 0^-$ the limit is $-\infty$. In either case, the integrand is unbounded. We show the graph in Figure 8.35. Using Definition 8.2(c), we write

$$\int_{-\pi}^{\pi} \frac{\sin x}{1 - \cos x}\, dx = \int_{-\pi}^0 \frac{\sin x}{1 - \cos x}\, dx + \int_0^{\pi} \frac{\sin x}{1 - \cos x}\, dx$$

For the last integral we have

$$\int_0^{\pi} \frac{\sin x}{1 - \cos x}\, dx = \lim_{t \to 0^+} \int_t^{\pi} \frac{\sin x}{1 - \cos x}\, dx \quad \text{Mentally, let } u = 1 - \cos x.$$

$$= \lim_{t \to 0^+} [\ln(1 - \cos x)]_t^{\pi}$$

$$= \lim_{t \to 0^+} [\ln 2 - \ln(1 - \cos t)] = \infty$$

(Note that since $1 - \cos x \geq 0$, we did not need absolute values for $\ln(1 - \cos x)$.) So this integral diverges, and hence the original integral also diverges. Observe that

$$f(x) = \frac{\sin x}{1 - \cos x}$$

is an odd function, so its graph is symmetric with respect to the origin. It is tempting, therefore, to say that the "net area" from $-\pi$ to π is 0. But remember that by Definition 8.2(c), each of the integrals $\int_0^\pi f(x)\,dx$ and $\int_{-\pi}^0 f(x)\,dx$ must exist independently of the other. In other words, we cannot say that ∞ and $-\infty$ add to give 0.   ■

REMARK ──

■ In contemporary physics, scientists doing quantum electrodynamics deal with equations in which "infinite terms" seem to cancel out. They use a process called "renormalization" to do so. The development of renormalization techniques helped lead to the Nobel Prize in physics for Richard Feynman, Julian Schwinger, and Sin-Itiro Tomonaga.

──

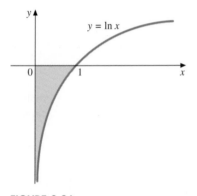

FIGURE 8.36

EXAMPLE 8.35   Determine whether the integral

$$\int_0^1 \ln x\,dx$$

converges or diverges.

Solution   We show the graph of $y = \ln x$ in Figure 8.36. It is unbounded as $x \to 0^+$. Thus, by Definition 8.2(a),

$$\int_0^1 \ln x\,dx = \lim_{t \to 0^+} \int_t^1 \ln x\,dx$$

We need to find an antiderivative of $\ln x$ to complete the problem. We use integration by parts, with $u = \ln x$ and $dv = dx$. Then $du = (1/x)\,dx$ and $v = x$. So we have

$$\int \ln x\,dx = x \ln x - \int x \cdot \frac{1}{x}\,dx = x \ln x - x$$

Now we return to the evaluation of the improper integral.

$$\int_0^1 \ln x\,dx = \lim_{t \to 0^+} \int_t^1 \ln x\,dx = \lim_{t \to 0^+} [x \ln x - x]_t^1$$

$$= \lim_{t \to 0^+} [(1 \ln 1 - 1) - (t \ln t - t)]$$

$$= \lim_{t \to 0^+} [-1 - t \ln t + t] \quad \text{Since } \ln 1 = 0$$

$$= -1 - \lim_{t \to 0^+} (t \ln t) \quad \text{Since } \lim_{t \to 0^+} t = 0$$

provided this last limit exists. Now, we are confronted in $\lim_{t \to 0^+} t \ln t$ with an indeterminate form of the type $0 \cdot \infty$, since $\ln t \to -\infty$ as $t \to 0^+$. So we rewrite the product as a quotient and apply L'Hôpital's Rule.

$$\lim_{t \to 0^+} t \ln t = \lim_{t \to 0^+} \frac{\ln t}{\dfrac{1}{t}} = \lim_{t \to 0^+} \frac{\dfrac{1}{t}}{-\dfrac{1}{t^2}} = \lim_{t \to 0^+} (-t) = 0$$

Finally, we have the result

$$\int_0^1 \ln x \, dx = -1 \qquad \blacksquare$$

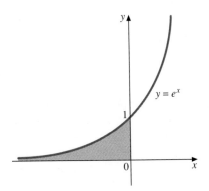

FIGURE 8.37

REMARK ————————————————————————————

■ It is interesting to note that the shaded region in Figure 8.36 is equivalent to the shaded region in Figure 8.37, obtained by reflecting through the line $y = x$. Thus, the integral $\int_0^1 \ln x \, dx$ in Example 8.35 can be replaced with

$$\int_{-\infty}^0 e^x \, dx$$

which is easier to analyze. (See Exercise 3 in Exercise Set 8.6.)

Special care must be taken when the integrand becomes unbounded at an interior point of the interval of integration, since failure to recognize the bad point can lead to incorrect results. For example, consider the integral

$$\int_0^2 \frac{dx}{(x-1)^2}$$

If we carelessly applied the Second Fundamental Theorem, we would get

$$\int_0^2 \frac{dx}{(x-1)^2} = -\frac{1}{x-1} \Big]_0^2 = -1 - 1 = -2$$

But this result cannot be right, since the integrand is always positive, so that if the integral exists, it must be nonnegative. The trouble, of course, is that the integrand becomes unbounded in a neighborhood of $x = 1$, as the graph in Figure 8.38 shows. The correct procedure is to apply Definition 8.2(c) and write

$$\int_0^2 \frac{dx}{(x-1)^2} = \int_0^1 \frac{dx}{(x-1)^2} + \int_1^2 \frac{dx}{(x-1)^2}$$

For the first integral on the right, we have

$$\int_0^1 \frac{dx}{(x-1)^2} = \lim_{t \to 1^-} \int_0^t \frac{dx}{(x-1)^2} = \lim_{t \to 1^-} \left[-\frac{1}{x-1} \right]_0^t$$

$$= \lim_{t \to 1^-} \left[-\frac{1}{t-1} - 1 \right] = \infty$$

There is no need to evaluate the second integral, since the divergence of the first one implies the divergence of the original integral.

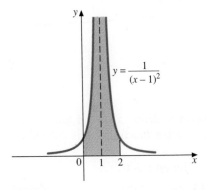

FIGURE 8.38

EXAMPLE 8.36   Evaluate the integral

$$\int_0^\infty \frac{dx}{x^2 + x}$$

or show that it diverges.

Solution   This integral is improper both because it has an unbounded interval of integration and because the integrand is unbounded in a neighborhood of $x = 0$. So that we can treat each problem separately, we write

$$\int_0^\infty \frac{dx}{x^2 + x} = \int_0^1 \frac{dx}{x^2 + x} + \int_1^\infty \frac{dx}{x^2 + x}$$

Without evaluating it, we can see by the comparison test that the second integral converges. (Do you see why?) So now we need consider only the first integral. By partial fractions, we find that

$$\frac{1}{x^2 + x} = \frac{1}{x(x + 1)} = \frac{1}{x} - \frac{1}{x + 1} \qquad \text{Verify.}$$

So

$$\int \frac{dx}{x^2 + x} = \int \left(\frac{1}{x} - \frac{1}{x + 1} \right) dx = \ln |x| - \ln |x + 1| = \ln \left| \frac{x}{x + 1} \right|$$

Thus,

$$\int_0^1 \frac{dx}{x^2 + x} = \lim_{t \to 0^+} \int_t^1 \frac{dx}{x^2 + x} = \lim_{t \to 0^+} \left[\ln \frac{x}{x + 1} \right]_t^1$$

$$= \lim_{t \to 0^+} \left[\ln \frac{1}{2} - \ln \frac{t}{t + 1} \right]$$

$$= \infty$$

since as $t \to 0^+$, $\ln[t/(t + 1)] \to -\infty$. Thus, this integral diverges, and hence the original integral also diverges.   ■

EVALUATING IMPROPER INTEGRALS USING A COMPUTER ALGEBRA SYSTEM

CAS 30

The **Gamma Function:** Frequently it is useful to define a special function as an improper integral. The gamma function is such a function and is defined as

$$\Gamma(t) = \int_0^\infty x^{t-1} e^{-x} \, dx \quad \text{for all } t > 0$$

Investigate $\Gamma(n)$ for positive integers n.

In Maple we can enter:

G:= t–>int(x^(t–1)*exp(–x),x=0..a);

The output is:

$$G := t \to \int_0^a x^{(t-1)} \exp^{-x} dx$$

Enter: Output:

  limit (G(1),a=infinity);   1
  limit (G(2),a=infinity);   1
  limit (G(3),a=infinity);   2
  limit (G(4),a=infinity);   6
  limit (G(5),a=infinity);   24
  limit (G(6),a=infinity);   120

Do you recognize these values? They
are factorials! That is

$$\Gamma(n + 1) = n!$$

We can check this in Maple:

  factorial(1);   1
  factorial(2);   2
  factorial(3);   6
  factorial(4);   24
  factorial(5);   120

The Gamma Function is such an important function in mathematics that Maple, Mathematica, and DERIVE have it built in. In Maple the function is called GAMMA (the function name must be specified in all caps), in Mathematica it is called Gamma, and it can be written as gamma in DERIVE. The Gamma Function is defined for all positive real numbers. For example,

Maple:

evalf(GAMMA(Pi));   2.288037796

Mathematica:

N[Gamma[Pi],10] 2.288037795

DERIVE:

a (author)
gamma(Pi)
x (approximate)
Output: 2.28803

Exercise Set 8.6

In Exercises 1–47, evaluate the integral or show that it diverges.

1. $\int_1^\infty \frac{dx}{x^3}$

2. $\int_1^\infty \frac{dx}{\sqrt{x^3}}$

3. $\int_{-\infty}^0 e^x \, dx$

4. $\int_0^\infty e^{-2x} dx$

5. $\int_{-\infty}^\infty \frac{dx}{x^2 + 4}$

6. $\int_2^\infty \frac{dx}{x\sqrt{x^2 - 1}}$

7. $\int_0^\infty \frac{e^{-x} dx}{1 + e^{-x}}$

8. $\int_0^\infty \frac{x \, dx}{(1 + x^2)^2}$

9. $\int_2^\infty \frac{x \, dx}{\sqrt{x^2 + 1}}$

10. $\int_1^\infty x^2 e^{-x} dx$

11. $\displaystyle\int_2^\infty \frac{dx}{x(\ln x)}$

12. $\displaystyle\int_2^\infty \frac{dx}{x(\ln x)^2}$

13. $\displaystyle\int_{-\infty}^\infty \frac{(x-1)dx}{(x^2-2x+4)^2}$

14. $\displaystyle\int_1^\infty \frac{dx}{x^2+x}$

15. $\displaystyle\int_2^\infty \frac{2\,dx}{x^2-1}$

16. $\displaystyle\int_3^\infty \frac{3}{x^2-x-2}\,dx$

17. $\displaystyle\int_0^\infty \tan^{-1} x\,dx$

18. $\displaystyle\int_1^\infty \frac{dx}{x\sqrt{1+x^2}}$

19. $\displaystyle\int_{-\infty}^{-3} \frac{5x\,dx}{x^2-x-6}$

20. $\displaystyle\int_1^\infty \frac{dx}{\sqrt{x}(1+\sqrt{x})^2}$

21. $\displaystyle\int_{-\infty}^\infty \frac{dx}{(x^2+1)^{3/2}}$

22. $\displaystyle\int_0^1 \frac{dx}{1-x}$

23. $\displaystyle\int_0^1 \frac{dx}{\sqrt{1-x}}$

24. $\displaystyle\int_0^1 \frac{x\,dx}{\sqrt[3]{1-x^2}}$

25. $\displaystyle\int_1^2 \frac{dx}{x\sqrt{x^2-1}}$

26. $\displaystyle\int_0^2 \frac{dx}{\sqrt{4-x^2}}$

27. $\displaystyle\int_2^3 \frac{dx}{(x-2)^{3/2}}$

28. $\displaystyle\int_2^3 \frac{dx}{(x-2)^{2/3}}$

29. $\displaystyle\int_0^3 \frac{dx}{(x-1)^2}$

30. $\displaystyle\int_{-2}^1 \frac{dx}{x+1}$

31. $\displaystyle\int_1^2 \frac{dx}{x\ln x}$

32. $\displaystyle\int_{-2}^{-1} \frac{dx}{x\sqrt{\ln|x|}}$

33. $\displaystyle\int_0^{\ln 2} \frac{e^x\,dx}{\sqrt{e^x-1}}$

34. $\displaystyle\int_0^{\pi/2} \frac{\cos x}{1-\sin x}\,dx$

35. $\displaystyle\int_0^\pi \frac{\sin x}{\sqrt[3]{\cos x}}\,dx$

36. $\displaystyle\int_0^\pi \tan^2 x \sec^2 x\,dx$

37. $\displaystyle\int_0^{\pi/3} \frac{\sec^2 x\,dx}{(\tan x-1)^2}$

38. $\displaystyle\int_{-2}^2 \frac{x^3 dx}{\sqrt{4-x^2}}$

39. $\displaystyle\int_0^2 \frac{dx}{x\sqrt{x^2+1}}$

40. $\displaystyle\int_0^2 \frac{dx}{x^2-2x}$

41. $\displaystyle\int_1^3 \frac{dx}{x^2-4x+3}$

42. $\displaystyle\int_{-1}^3 \frac{dx}{x^2+x-2}$

43. $\displaystyle\int_{-1}^1 \frac{dx}{x^3+x^2}$

44. $\displaystyle\int_{-1}^8 \frac{x^{1/3}dx}{x^{2/3}-4}$

45. $\displaystyle\int_0^\pi \frac{dx}{1-\sin x}$

46. $\displaystyle\int_0^2 \frac{dx}{\sqrt{2x-x^2}}$

47. $\displaystyle\int_a^{2a} \frac{dx}{\sqrt{x^2-a^2}}$

In Exercises 48–53, determine whether the integral is convergent or divergent, using Theorem 8.1.

48. $\displaystyle\int_1^\infty \frac{dx}{x^3+x+1}$

49. $\displaystyle\int_2^\infty \frac{dx}{x^{2/3}-1}$

50. $\displaystyle\int_3^\infty \frac{x\,dx}{\sqrt{x^3-1}}$

51. $\displaystyle\int_0^\infty \frac{x\,dx}{x^4+4}$

52. $\displaystyle\int_3^\infty \frac{dx}{x^2\ln x}$

53. $\displaystyle\int_1^\infty \frac{dx}{x^2+\sin^2 x}$

54. Prove that if $\int_a^\infty f(x)dx$ and $\int_{-\infty}^a f(x)dx$ both exist, then for any number $b \neq a$, $\int_b^\infty f(x)dx$ and $\int_{-\infty}^b f(x)dx$ also exist.

In Exercises 55–58, find the value of k so that f will be a probability density function for a random variable X on the indicated interval, and then find the specified probability.

55. $f(x) = ke^{-x/3}$ on $[0, \infty)$; $P(X \geq 1)$

56. $f(x) = kxe^{-x/2}$ on $[0, \infty)$; $P(X > 2)$

57. $f(x) = kxe^{-x^2}$ on $[0, \infty)$; $P(1 < X \leq 2)$

58. $f(x) = k[e^x/(1+e^x)^2]$ on $[0, \infty)$; $P(X \geq 1)$

59. Find the area of the region under the graph of $y = 1/(1+x^2)$ on $[1, \infty)$.

60. Find the volume of the solid obtained by revolving the region under the graph of $y = 1/(1+x)$ on $[0, \infty)$ about the x-axis.

61. Find the volume of the solid obtained by revolving the region under the graph of

$$y = \frac{1}{\sqrt{x}\ln x} \qquad (2 \leq x < \infty)$$

about the x-axis.

62. Show that

$$\lim_{t\to\infty} \int_{-t}^t \frac{x\,dx}{x^2+1} = 0$$

but that

$$\int_{-\infty}^\infty \frac{x\,dx}{x^2+1}$$

does not exist. Explain this result.

63. Find the area of the region under the graph of $y = 1/\sqrt{1-x}$ from 0 to 1, or show that it fails to exist.

64. Find the volume of the solid (if it exists) generated by rotating the region above $y = x$ and below $y = x^{-3/2}$ between $x = 0$ and $x = 1$ about the y-axis.

65. Prove each of the following.
(a) If f is even and $\int_0^\infty f(x)dx$ converges, then
$$\int_{-\infty}^{\infty} f(x)dx = 2\int_0^\infty f(x)dx$$
(b) If f is odd and $\int_0^\infty f(x)dx$ converges, then
$$\int_{-\infty}^{\infty} f(x)dx = 0$$

66. A random variable X has the **exponential** distribution if its probability density function is
$$f(x) = \frac{1}{\alpha} e^{-x/\alpha} \qquad (x \geq 0)$$
where α is a positive constant. Find the mean and variance of X. (See Section 6.9.) (This function is frequently used to model the distribution of times between successive events such as customers arriving at some facility.)

67. The lifetime X, in hours, of a certain type of light bulb is a random variable with an exponential probability density function having $\alpha = 1000$. (See Exercise 66.)
(a) Find the probability that a light bulb of this type selected at random will last at least 1500 hr.
(b) Of 10,000 light bulbs installed in an illuminated sign, how many will likely fail in less than 500 hr? in less than 100 hr?

68. In an experiment with mice in a maze, it is found that a reasonable model of the probability density function of the time X, in minutes, for a mouse to go through the maze is given by
$$f(x) = \frac{a}{x^2} \qquad (x \geq a)$$
where a is the minimum time required.
(a) Verify that f is a probability density function on $[a, \infty)$.
(b) Find the probability that a mouse selected at random will take longer than $2a$ minutes to go through the maze.

In Exercises 69–74, evaluate the integral or show that it diverges.

69. $\displaystyle\int_0^\infty \frac{dx}{\sqrt{x}(1+x)}$

70. $\displaystyle\int_1^\infty \frac{dx}{x\sqrt{x-1}}$

71. $\displaystyle\int_0^\infty \frac{dx}{e^x - 1}$

72. $\displaystyle\int_0^\infty \frac{dx}{\sqrt{x + x^2}}$

73. $\displaystyle\int_{-\infty}^\infty \frac{dx}{x(\ln |x|)^2}$

74. $\displaystyle\int_0^\pi \frac{\cot x\, dx}{\ln \sin x}$

75. Show that $\displaystyle\int_0^1 \cos\frac{1}{x}\, dx$ exists. (*Hint:* Let $u = \dfrac{1}{x}$.)

76. Evaluate the integral
$$\int_0^e \frac{dx}{x \ln x(\ln |\ln x|)^2}$$
or show that it diverges. (*Hint:* Look for all zeros of the denominator.)

77. Prove that the integral
$$\int_0^1 \frac{dx}{x^p}$$
converges if $p < 1$ and diverges if $p \geq 1$.

78. There are comparison tests for improper integrals with unbounded integrands analogous to those that have an unbounded interval of integration. For $f(x)$ unbounded in a neighborhood of a, one such test is as follows: If $0 \leq f(x) \leq g(x)$ for all x on $(a, b]$, and f and g are continuous on $(a, b]$, then
$$\int_a^b f(x)dx \text{ converges if } \int_a^b g(x)dx \text{ converges}$$
and
$$\int_a^b g(x)dx \text{ diverges if } \int_a^b f(x)dx \text{ diverges}$$
A similar test applies if f is unbounded near b. Use this test along with the result of Exercise 77 to determine the convergence or divergence of the following integrals.
(a) $\displaystyle\int_0^\pi \frac{\cos^2 x}{\sqrt{x}}\, dx$
(b) $\displaystyle\int_0^{\pi/3} \frac{\sec x}{x^{3/2}}\, dx$

79. Follow the instructions for Exercise 78 to determine convergence or divergence of the following integrals.
(a) $\displaystyle\int_0^{\pi/2} \frac{1 - \sin x}{x^{2/3} + x}\, dx$
(b) $\displaystyle\int_0^1 \frac{1 + e^{-x}}{x - x^2}\, dx$

80. Show that the formula for the length of an arc of the four-cusp hypocycloid $x^{2/3} + y^{2/3} = a^{2/3}$ (see the figure) results in an improper integral. Find the total length.

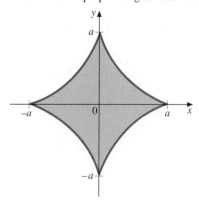

81. (a) Find the area of the region under the graph of $f(x) = 1/\sqrt{4 - x^2}$ from $x = 0$ to $x = 2$.
  (b) Find the volume of the solid generated by revolving the region in part (a) about the line $x = 2$.
  (c) Without finding its value, show that the surface obtained by revolving the graph of f in part (a) about the line $x = 2$ has a finite area.

82. Determine the values of p for which the integral

$$\int_2^\infty \frac{dx}{x(\ln x)^p}$$

converges and those for which it diverges.

83. Use the following outline to prove that

if $\displaystyle\int_a^\infty |f(x)|dx$ converges, then $\displaystyle\int_a^\infty f(x)dx$ converges.

  (a) Explain why $0 \le f(x) + |f(x)| \le 2|f(x)|$.
  (b) Why does it follow that $\int_a^\infty [f(x) + |f(x)|]dx$ converges?
  (c) Write $f(x) = [f(x) + |f(x)|] - |f(x)|$, and explain why $\int_a^\infty f(x)dx$ must therefore converge.

84. Use the result to Exercise 83, together with the comparison test, to show that each of the following converges:

  (a) $\displaystyle\int_0^\infty \frac{\sin x}{1 + x^2}\,dx$ (b) $\displaystyle\int_0^\infty e^{-x}\cos x\,dx$

85. Let $f(t)$ be continuous on $0 \le t < \infty$. The **Laplace transform** of $f(t)$, denoted by $F(s)$, is defined by

$$F(s) = \int_0^\infty f(t)e^{-st}dt$$

provided the integral converges. Laplace transforms are useful in the study of differential equations. They also occur in various parts of physics, including, for example, the signal processing in compact discs and digital audiotapes. Find the Laplace transform of each of the following.
  (a) $f(t) = 1$ (b) $f(t) = t$
  (c) $f(t) = e^t$ (d) $f(t) = \sin t$

86. Refer to Exercise 85. Prove that if there are positive constants M and k such that

$$|f(t)| \le Me^{kt} \quad \text{for } 0 \le t < \infty$$

then $F(s)$ exists for $s > k$.

87. The normal probability density function is of the form

$$f(x) = \frac{1}{\beta\sqrt{2\pi}}e^{-(x-\alpha)^2/2\beta^2} \quad (-\infty < x < \infty)$$

Take it as given that $\int_{-\infty}^\infty f(x)dx = 1$, and prove that the mean $\mu = \alpha$ and the standard deviation $\sigma = \beta$. (Refer to Section 6.9.)

88. The probability density function for the random variable X representing the length of life of certain plants and animals is often modeled by the **Weibull** distribution:

$$f(x) = \frac{m}{\alpha}x^{m-1}e^{-x^m/\alpha} \quad (x > 0)$$

This function was first introduced by the Swedish physicist Waloddi Weibull in 1939. Let $m = 2$ and $\alpha = 4$ and find the following:
  (a) $P(X \ge 2)$ (b) $E[X^2]$

Chapter 8 Review Exercises

In Exercises 1–30, evaluate the given definite or indefinite integral.

1. $\displaystyle\int x^2 e^{-2x}dx$

2. $\displaystyle\int \sqrt{x}\ln\sqrt{x}\,dx$

3. $\displaystyle\int_0^1 x\tan^{-1}x^2\,dx$

4. $\displaystyle\int e^{-x}\cosh 2x\,dx$

5. $\displaystyle\int \sec^2\sqrt{x}\,dx$ (*Hint:* Let $t = \sqrt{x}$.)

6. $\displaystyle\int_{-\pi/6}^{\pi/3} \sin^2 x(\sin x - 1)\,dx$

7. $\displaystyle\int \sin^5 x \cos^3 x\,dx$

8. $\displaystyle\int \sec^3 x \tan^5 x\,dx$

9. $\displaystyle\int_{-\pi/3}^{\pi/4} \frac{\sin^2 x}{\cos^4 x}\,dx$

10. $\displaystyle\int \frac{\sin^2 x}{\cos^3 x}\,dx$

11. $\displaystyle\int x \sin x \cos^3 x\,dx$

12. $\displaystyle\int \sec^2 \sqrt{x}\, \tan \sqrt{x}\,dx$ (*Hint:* Let $t = \sqrt{x}$.)

13. $\displaystyle\int \sin 2x \cos 4x\,dx$

14. $\displaystyle\int \frac{\ln\cos x}{\sin^2 x}\,dx$ (*Hint:* Integrate by parts.)

15. $\displaystyle\int_{2}^{2\sqrt{3}} \frac{x^2}{(16 - x^2)^{3/2}}\,dx$

16. $\displaystyle\int_{1}^{2} x^3\sqrt{x^2 + 1}\,dx$ (Do by two methods: integration by parts and trigonometric substitution.)

17. $\displaystyle\int_{1/\sqrt{3}}^{1} \frac{\sqrt{4x^2 - 1}}{x^3}\,dx$

18. $\displaystyle\int \frac{x\,dx}{\sqrt{x^2 - 2x + 5}}$

19. $\displaystyle\int \frac{e^{4x}\,dx}{\sqrt{1 + e^{2x}}}$

20. $\displaystyle\int \frac{\sinh x \cosh^3 x}{\sqrt{\cosh^2 x - 4}}\,dx$

21. $\displaystyle\int \frac{x^2\,dx}{\sqrt{4x - x^2}}$

22. $\displaystyle\int \frac{4x - 9}{2x^2 + 5x - 3}\,dx$

23. $\displaystyle\int_{1}^{2} \frac{x^2 - 3x + 2}{x^2 - 2x - 8}\,dx$

24. $\displaystyle\int \frac{5x + 1}{x^3 - x^2 - x + 1}\,dx$

25. $\displaystyle\int \frac{x^2 + x + 1}{x(x^2 + 1)^2}\,dx$

26. $\displaystyle\int \frac{x^4 + 20}{x^3 - 3x + 2}\,dx$

27. $\displaystyle\int_{-1}^{0} \frac{12\,dx}{x^3 - 8}$

28. $\displaystyle\int_{0}^{4} \frac{dx}{\sqrt{x} + 1}$ (*Hint:* Let $u = \sqrt{x}$.)

29. $\displaystyle\int \frac{x^3}{\sqrt{x^2 - 4}}\,dx$ (Do by two methods: trigonometric substitution and algebraic substitution.)

30. $\displaystyle\int_{0}^{\pi/2} \frac{1 + \cos x}{1 + \sin x}\,dx$ (*Hint:* Multiply numerator and denominator by $1 - \sin x$.)

31. Find the volume of the solid obtained by revolving the region under the graph of $y = \sin x$ from $x = 0$ to $x = \pi$ about (a) the x-axis and (b) the y-axis.

32. Find the length of the curve $y = x^2/4$ from $x = 0$ to $x = 2$.

33. Find the volume of the solid generated by revolving the region under the graph of

$$y = \frac{1}{\sqrt{5x - x^2}}$$

from $x = 1$ to $x = 4$ about (a) the x-axis and (b) the y-axis.

34. Find the area of the surface formed by revolving the arc of the curve $y = \cosh x$ from $x = 0$ to $x = \ln 3$ about (a) the x-axis and (b) the y-axis.

35. A model for a so-called *learning curve* in psychology is given by

$$\frac{dy}{dt} = k\sqrt{y^3(1 - y)^3} \qquad (0 \le y \le 1)$$

where k is a positive constant. Here $y = y(t)$ is the proportion of a task or of a body of knowledge learned by the subject at time t. For $0 < y < 1$, find y as a function of t by writing the equation in the form

$$\frac{dy}{y^{3/2}(1 - y)^{3/2}} = k\,dt$$

and integrating both sides. (*Hint:* To integrate the left-hand side, substitute $y = \sin^2 \theta$.)

In Exercises 36–51, evaluate the integral or show that it diverges.

36. $\displaystyle\int_{1}^{\infty} \frac{dx}{x^4 + x^2}$

37. $\displaystyle\int_{1}^{\infty} \frac{dx}{x^4 - x^2}$

38. $\displaystyle\int_{2}^{\infty} \frac{dx}{x\sqrt{x^2 - 4}}$

39. $\displaystyle\int_{-1}^{8} \frac{x + 1}{\sqrt[3]{x}}\,dx$

40. $\displaystyle\int_{2}^{4} \frac{x^3\,dx}{\sqrt{x^2 - 4}}$

41. $\displaystyle\int_{1}^{\infty} \frac{x\,dx}{\sqrt{x^4 + 1}}$

42. $\displaystyle\int_{0}^{1} \sqrt{\frac{1 - x}{x}}\,dx$

43. $\displaystyle\int_{3}^{\infty} \frac{3\,dx}{x^2 - x - 2}$

44. $\displaystyle\int_{-1}^{1} \frac{dx}{x^2\sqrt{1-x^2}}$ **45.** $\displaystyle\int_{0}^{1} x\ln x\,dx$

46. $\displaystyle\int_{-\infty}^{\infty} \frac{dx}{1+x^3}$ **47.** $\displaystyle\int_{1}^{e} \frac{dx}{x\sqrt{\ln x}}$

48. $\displaystyle\int_{-\ln 2}^{\ln 2} \coth x\,dx$ **49.** $\displaystyle\int_{0}^{\infty} x^3 e^{-x^2}\,dx$

50. $\displaystyle\int_{0}^{\infty} \frac{e^x\,dx}{e^{2x}+1}$ **51.** $\displaystyle\int_{\pi/3}^{\pi/2} \frac{\tan x}{(\ln\cos x)^2}\,dx$

52. Show that the volume of the solid generated by revolving the region under the graph of $y = xe^{-x}$ on $[0, \infty)$ about the x-axis is finite, and find its value.

53. Show that $\int_0^1 \sin(1/x^\alpha)\,dx$ converges for $\alpha > 0$.

54. Let

$$f(x) = \frac{1}{x\ln x[\ln(\ln x)]^p} \qquad (x > e)$$

Show that $\int_e^{e^2} f(x)\,dx$ converges if and only if $p < 1$, whereas $\int_{e^2}^{\infty} f(x)\,dx$ converges if and only if $p > 1$.

55. The **Pareto probability density function** is of the form

$$f(x) = \frac{k\theta^k}{x^{k+1}}$$

if $x \geq \theta$, and $f(x) = 0$ otherwise, where k and θ are positive constants.
  (a) Show that f is a probability density function.
  (b) For what values of k is the mean μ finite? Find μ for these values.
  (c) For what values of k is the variance σ^2 finite? Find σ^2 for these values. (See Equation 6.47.)

56. The **gamma function,** defined by

$$\Gamma(\alpha) = \int_0^{\infty} x^{\alpha-1}e^{-x}\,dx \qquad (\alpha > 0)$$

has many applications in both pure and applied mathematics. Show the following:
  (a) For $x \in (0, 1]$, $x^{\alpha-1}e^{-x} \leq x^{\alpha-1}$.
  (b) For x sufficiently large, $x^{\alpha-1}e^{-x} \leq 1/x^2$.
  (c) The integral that defines $\Gamma(\alpha)$ converges if $\alpha > 0$.
    (*Hint:* Use the results of parts (a) and (b).)

57. Refer to Exercise 56. Prove the following:
  (a) $\Gamma(\alpha + 1) = \alpha\Gamma(\alpha)$
  (b) For all natural numbers n,

$$\Gamma(\alpha + n) = (\alpha + n - 1)(\alpha + n - 2)\cdots(\alpha + 1)\Gamma(\alpha)$$

  (c) For all natural numbers n, $\Gamma(n + 1) = n!$. Show why it is natural to define $0! = 1$.

58. The **gamma probability density function** is defined by

$$f(x) = \frac{1}{\beta^\alpha\Gamma(\alpha)} x^{\alpha-1}e^{-x/\beta} \qquad (x > 0)$$

and $f(x) = 0$ if $x \leq 0$, where α and β are positive constants. (See Exercise 57.) Prove that f is a probability density function, and find μ and σ^2. (The gamma distribution is often useful in modeling probability distributions that are skewed, or nonsymmetric.)

59. Analyze the function

$$f(x) = \begin{cases} e^{1-1/x} & \text{if } x \neq 0 \\ 0 & \text{if } x = 0 \end{cases}$$

and draw its graph. Pay particular attention to the right-hand derivative at 0.

Chapter 8 Concept Quiz

1. State the Integration by Parts Formula in terms of u and v. In each of the following apply the formula, indicating appropriate choices for u and dv. Do not complete the integration.
  (a) $\int x^3 \sin x\,dx$
  (b) $\int(2x^2 - 3x + 1)e^{-2x}\,dx$
  (c) $\int x^2 \ln x\,dx$
  (d) $\int \sin^{-1} x\,dx$
  (e) $\int e^{-3x}\cos 2x\,dx$

2. Explain how you would evaluate the integral

$$\int \sin^m x \cos^n x\,dx \qquad (m, n \text{ nonnegative integers})$$

in each of the following cases:
  (a) m is odd
  (b) n is odd
  (c) m and n are even

3. Explain how you would evaluate the integral

$$\int \sec^m x \tan^n x\,dx \qquad (m, n \text{ nonnegative integers})$$

in each of the following cases:
  (a) m is even
  (b) n is odd
  (c) m is odd and n is even

4. Tell what substitution you would try in an integral involving the following:

(a) $\sqrt{u^2 - a^2}$

(b) $\sqrt{u^2 + a^2}$

(c) $\sqrt{a^2 - u^2}$

In each case indicate restrictions on the new variable, and sketch a right triangle that shows the relationship between u and the new variable.

5. Consider an integral of the form

$$\int \frac{P(x)}{Q(x)} \, dx$$

where $P(x)$ and $Q(x)$ are polynomials. If $\deg P \geq \deg Q$, what do you do first? Suppose that $\deg P < \deg Q$ in each of the following cases. Show the appropriate form of the partial-fraction decomposition.

(a) $Q(x) = x^2 - 4x - 5$

(b) $Q(x) = (2x + 3)(x^2 - 6x + 9)$

(c) $Q(x) = (x - 6)(x^2 + x + 1)$

(d) $Q(x) = (x + 2)^3(x^2 + 1)$

(e) $Q(x) = x^2(x^2 + 4)^2$

6. (a) How is $\int_a^\infty f(x) \, dx$ defined?

(b) If f is continuous on $[a, b)$ and $\lim_{x \to b^-} f(x) = \pm\infty$, how is $\int_a^b f(x) \, dx$ defined?

(c) If f is continuous on $(a, b]$ and $\lim_{x \to a^+} f(x) = \pm\infty$, how is $\int_a^b f(x) \, dx$ defined?

(d) If f is continuous on $[a, c) \cup (c, b]$ and f is unbounded in a neighborhood of $x = c$, how is $\int_a^b f(x) \, dx$ defined?

(e) For what values of p does the integral

$$\int_1^\infty \frac{1}{x^p} \, dx$$

converge, and for what values does it diverge?

APPLYING CALCULUS

1. A catalyst for a chemical reaction is a substance that increases the speed of the reaction without itself undergoing change. An *autocatalytic reaction* is a reaction whose product acts as a catalyst for its own formation. Such a reaction proceeds slowly at first when the amount of catalyst present is small, and slowly at the end when most of the original substance is used up. In between, when both the product and the substance are present in some quantity, the reaction proceeds at a faster rate. One mathematical model for an autocatalytic reaction is based on the assumption that the rate $v = dx/dt$ of the reaction is proportional to both the amount of the original substance and the amount of the product; that is

$$v = \frac{dx}{dt} = kx(A - x)$$

where x is the amount of the product, A is the amount of the substance at the beginning of the reaction, and k, $k > 0$, is the constant of proportionality.

(a) At what value x is the rate of the reaction a maximum? What is the maximum value of v?

The differential equation above can be solved by rewriting it in the form

$$\frac{dx}{x(A - x)} = kdt$$

and integrating both sides

$$\int \frac{dx}{x(A - x)} = \int kdt = kt + C.$$

(b) Determine x as a function of t by calculating the integral derived above. Assume that $x = x_0$ when $t = 0$.

(c) Sketch the graph of the solution $x = x(t)$. Where does the curve have a point of inflection? Compare with the result in part (a).

2. A student who has just heard a new joke returns to his college campus whose enrollment is 10,000 students. Four hours later, 10 students have heard the joke. Let $x = x(t)$ be the number of students who have heard the joke at time t. A common model used by social scientists for the spread of a rumor through a population is that the rumor spreads at a rate $v = dx/dt$ which is proportional to the product of the number of people who have heard the rumor and the number of people who have not yet heard it. Applying the model in this case,

$$v = \frac{dx}{dt} = kx(10,000 - x)$$

where k is a positive constant. To solve this differential equation, rewrite it in the form

$$\int \frac{dx}{x(10,000 - x)} = \int kdt = kt + C$$

(a) Find x as a function of t by evaluating the integral and using $x(0) = 1$ and $x(4) = 10$ to determine C and k.

(b) Sketch the graph of $x = x(t)$, including $\lim_{t \to \infty} x(t)$, relative extrema (if any) and points of inflection.

(c) When is the joke spreading at its most rapid rate? How long will it take for 75% of the students to have heard the joke?

3. Consider the logistic differential equation

$$\frac{dy}{dt} = ky(M - y)$$

with initial condition $y(0) = y_0$. Show that the solution of the initial-value problem is

$$y(t) = \frac{y_0 M}{y_0 + (M - y_0)e^{-kMt}}$$

(a) Suppose $y_0 < M$. Show that the solution is increasing on $[0, \infty)$. Determine the point(s) of inflection and the intervals on which the graph is concave up and the intervals on which it is concave down. Use a CAS to draw an accurate sketch of the graph of the solution.

(b) Suppose $y_0 = M$. Draw the graph of the solution in this case.

(c) Suppose $y_0 > M$. Show that the solution is decreasing and concave up on $[0, \infty)$. Draw the graph of the solution, including $\lim_{t \to \infty} y(t)$. Note that, in all three cases, $\lim_{t \to \infty} y(t) = M$. This value is called *carrying capacity*.

4. In the logistic model of population growth

$$\frac{dy}{dt} = ky(M - y) = my - ky^2 \qquad (m = kM)$$

it is assumed that there is a "birth" rate that is proportional to the population size (the term "my") and a "death" rate which is proportional to the square of the population (the term "$-ky^2$"). In contrast, there are species in which the birth rate depends on chance encounters between males and females, and is therefore proportional to the product of the number of males and the number of females. If males and females are equally distributed throughout the population, then the birth rate is proportional to the square of the population. Suppose it is further assumed that the death rate is proportional to the population size. The differential equation corresponding to this model for population growth is

$$\frac{dy}{dt} = my^2 - ky = my(y - M) \qquad (M = k/m)$$

(a) Determine the solution of the differential equation which satisfies $y(0) = y_0$.

(b) Graph the solution in the three cases (i) $y_0 < M$, (ii) $y_0 = M$, and (iii) $y_0 > M$. How does this model compare with the logistic model?

5. A simplified model for the spread of a disease assumes that the rate at which the disease spreads is proportional to the product of the number of infected people and the number of susceptible people. Suppose that an influenza epidemic is spreading through a population of 10,000 people and that 500 people have it initially.

(a) Let $y = y(t)$ denote the number of people who have the flu at time t. Determine the differential equation which models the spread of the disease.

(b) Suppose that 1000 people have the disease after 10 days. Determine an expression for the number of infected people at any time t.

(c) How long will it take for 75% of the people to have the flu?

6. A spacecraft at a distance r above a planet of radius R can see less than half of the planet.

(a) For a rectangular coordinate system with origin at A, with $(g(y), y)$ giving the coordinates of a point on arc EC, explain why the surface area A_z is

$$A_z = 2\pi \int_{y_B}^{y_E} g(y)\sqrt{1 + [g'(y)]^2}\, dy$$

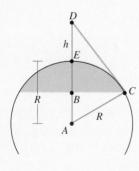

An astronaut in orbit

(Hint: Find $g(y)$, y_B, and y_E geometrically in terms of y and R.)

(b) Find A_z/A_p, where A_p is the planet's entire area.

(c) A space shuttle orbits at an altitude of about 300 km. What fraction of the Earth's surface can those astronauts see at one instant? The Earth's radius is 6378 km. What is the radius of a circle on the Earth's surface with this area? Give an example of a region of Earth of this size.

(d) How high would an astronaut have to go to see one-fourth of the Earth's surface?

(e) What fraction of the Earth could Neil Armstrong see when he became the first human to set foot on the Moon on July 20, 1969?

(f) Find $\displaystyle\lim_{h \to \infty} \frac{A_z}{A_{\text{Earth}}}$ and compare with the answer for part (e).

7. On September 19, 1991, the frozen remains of a man were found encased in ice in the mountains on the Swiss-Italian border. (See the photo.) Using the technique of radiocarbon-dating (see Exercise 25 in Exercise Set 7.4), researchers determined that the man had been dead for 5,000 years (*Nature*, vol. 362, 4 March 1993). The half-life of carbon-14 is approximately 5,730 years. Assuming exponential decay, write and solve the differential equation for the amount of carbon-14 remaining in his body t years after his death. How much carbon-14 did the researchers find in his body, as a percentage of the amount present when he died?

Sheldon Glashow

How did you first get interested in mathematics and physics?

I got interested in science as a youngster. I remember my older brother explaining the kinematics of airplanes flying and dropping bombs and how the bombs would explode directly underneath the airplane unless the airplane took some evasive action. After a brief infatuation with biology, I realized that I had to know some chemistry. I learned some chemistry on my own and realized that it didn't make sense without physics. So I figured out by the time I graduated from high school that I pretty much had to know physics. I went to the Bronx High School of Science, where they didn't have a calculus course at that time. There formal mathematics ended with solid geometry, so I was forced to learn calculus in the lunchroom from a friend who had somehow picked some of it up. It was clear from the beginning that physics deals with things that change and that you can't deal with things that change unless you know calculus.

In view of new developments in mathematics, and discussions of discrete mathematics and computers, is it still useful to learn calculus?

I'm a computer illiterate but I'm pretty good at calculus, and physics is a continuous subject. It deals with things that change continuously, most of the time, and discrete mathematics is not usually useful for things that change like that. My father was a plumber, and despite all his knowledge and wisdom about plumbing, he couldn't work without a blowtorch and a wrench and the other tools of his trade. In exactly the same sense, the practicing physicist—the licensed quantum mechanic—has to have a set of tools, and primary among them is calculus.

How necessary is calculus for doing physics?

It is hard to explain, because you cannot isolate mathematics from physics. It is so much ingrained into all of physics so that you can't even think physics without knowing calculus. But I didn't win any prizes for doing integrals. Again and again, physicists had to invent the mathematics they needed to deal with the problems they encountered. Newton and Leibniz were forced to invent calculus. Einstein was forced to invent another form of mathematics. And today, string theorists [those working on a theory used to try to explain the basic structure of matter] are using fields of mathematics that have not yet been developed, but which depend at the very beginning on calculus.

In the discussion of polar coordinates, we use concepts of symmetry. How has symmetry proved of interest to you?

When we study mechanics, we learn that it doesn't matter where you put the origin of your coordinate system. That is, you can translate from one point to another. That is a symmetry of nature. You can also rotate the coordinate system—you can do physics while standing on your head and the laws come out the same—since the laws of physics are invariant under rotations in space. But from the point of view of calculus, you can show that the first symmetry—translational symmetry—is equivalent to the conservation of momentum and that the second symmetry—rotational symmetry—is equivalent to the rotation law for angular momentum. These are remarkable implications of symmetry but they can only be deduced within the framework of variational principles and calculus. Most of the progress in physics in the last few years, especially in the construction of the standard model of quarks and leptons and their interactions, involves a mathematical marriage between calculus and group theory (the theory of continuous groups). All of the forces of nature are described mathematically by things called Lie groups, which are the offspring of this marriage. A Lie group is a complicated set like the set of all possible rotations of a rigid body—that is, the rotation group. There are many other such groups. But the miracle is that the properties of finite rotations, which are very complicated things, can almost all be described in terms of very tiny rotations applied in sequence. Once again, you are building up finite objects by putting together infinitesimals, thereby describing symmetry groups in terms of calculus.

Your high school didn't teach calculus. How does that compare with the preparation of today's students?

There are two kinds of calculus students—those who know calculus and those who think they know calculus. The numbers who think they know calculus is very high, but it turns out that many don't know algebra and therefore cannot know calculus. Calculus is not a stand-alone subject. It depends crucially on a knowledge of geometry, trigonometry, and algebra. Without that foundation, there is no possibility of learning calculus. Many people want to learn about quarks and their behavior. But they don't know anything about calculus. They have no hope of understanding what they want to know without knowing some calculus.

Sheldon Glashow is a theoretical physicist at Harvard University. He shared the Nobel Prize for Physics in 1979 for his theory that unified two of the fundamental forces of nature—the weak force and the electromagnetic force. He is also known for predicting the existence of what is known as the charmed quark, one of the six kinds of quarks that make up the fundamental particles of nature.

Pierre de Fermat shares credit with René Descartes for the independent invention of analytic geometry, which is essentially coordinate geometry. The Cartesian coordinate system is so named to acknowledge Descartes's contribution. We have used aspects of analytic geometry without explicitly mentioning it throughout the text, every time we give graphs of equations. In the present chapter, we will use it even more, as we study graphs of parametric equations and of polar equations. An especially important use of analytic geometry is in the study of the conic sections, which we take up in relation to polar coordinates.

Pierre de Fermat (1601–1665) was a jurist in Toulouse, France. Mathematically, he contributed to many aspects of number theory, analytic geometry, and calculus. As the discoverer of methods for finding tangents to curves and the curves' maxima and minima, ideas later taken up by Isaac Newton, Fermat is considered one of the developers of differential calculus, although he used it without fully understanding it. He is also known to physicists for "Fermat's principle," which states that light traveling between two points seeks the path that is either a minimum or a maximum of time.

Fermat often wrote his mathematical results in letters to friends. Other results were kept to himself, and he sometimes wrote them as marginal comments in his books, such as the *Arithmetica of Diophantus* (c. A.D. 250), which was a then-recent translation. Fermat's notes were such important additions to this text that within 10 years after his death, a second edition of the translation had been published with Fermat's marginal notes included.

Fermat's most famous result, "Fermat's last theorem," can be stated as follows:

No positive integers x, y, z, and n exist for which

$$x^n + y^n = z^n \text{ for } n > 2$$

In the margin of his copy of *Arithmetica*, Fermat wrote, "It is impossible to divide a cube into two cubes, or a fourth power into two fourth powers, or, in general, any power from then on past the square into two like powers; of this I have discovered a very wonderful demonstration. This margin is too narrow to contain it."

"Fermat's last theorem" has tantalized mathematicians and been the subject of detailed investigation and debate for years. Euler (1707–1783) proved it for $n = 3$ and $n = 4$, and various other "proofs" have been advanced over the years, but all have turned out to contain flaws. Did Fermat himself really have a valid proof? We shall probably never know, but probably he did not. The theorem wasn't really his "last"; it may even have been written down 30 years before he died, and he was known to elaborate on other theorems.

The English mathematician Louis Mordell found in 1922 that equations such as Fermat's are linked to the numbers of holes in surfaces. Thus Fermat's ideas in number theory are important for topology, which deals with such surfaces.

Computers have been used to evaluate Fermat's conjecture, which has been tested for all $n < 3$ million! But it would take only one example to disprove it.

In 1993, the mathematical world was astounded by the proof of Fermat's last theorem by Andrew Wiles of Princeton University, using ideas of elliptic integrals and geometry. His proof is hundreds of pages long and is still being checked by his mathematical colleagues. If his proof is found to be correct, then a 250-year-old mystery will finally have been laid to rest.

9

PARAMETRIC EQUATIONS AND POLAR COORDINATES

Most of the curves we have considered so far have been graphs of functions—that is, graphs of equations of the form $y = f(x)$. One characteristic feature of such a curve is that each vertical line intersects it in at most one point. Many curves do not have this property, however. For example, none of the curves in Figure 9.1 passes the vertical line test, so none of these curves is the graph of a function of x.

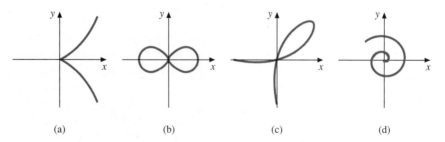

| (a) | (b) | (c) | (d) |

FIGURE 9.1

In this chapter we introduce two new methods of representing curves in the plane. We will see that each of the curves in Figure 9.1, and many others, can be described by one of these methods. The first method makes use of an auxiliary variable, called a **parameter**. The coordinates x and y are each expressed in terms of the parameter, yielding what are called **parametric equations**. In the second method, we make use of an alternative coordinate system for locating points in the plane, called the **polar coordinate system**. We will see that certain curves that are difficult to describe by an equation in x and y (*rectangular* coordinates) can be conveniently represented by an equation in polar coordinates.

We conclude the chapter with a discussion of the curves known as conic sections—the circle, ellipse, parabola, and hyperbola—and show how their equations assume a particularly simple form using polar coordinates.

9.1 CURVES DEFINED PARAMETRICALLY

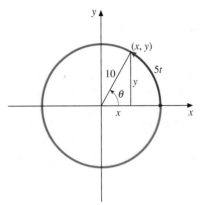

FIGURE 9.2

To introduce parametric equations, we consider the following problem. Suppose an object is moving at a constant speed of 5 m/s counterclockwise around a circle of radius 10 m, centered at the origin, as indicated in Figure 9.2. The path of the object is not the graph of a function. So we cannot express y as a (single) function of x (nor x as a (single) function of y). Instead, we describe the position of the object by expressing x and y separately as functions of time t. If at time $t = 0$ the object is at the point $(10, 0)$, what are the x- and y-coordinates of the object t seconds later?

As indicated in Figure 9.2, after t seconds the object has covered a distance of $5t$ meters on the circle. The angle θ, measured in radians, is therefore

$$\theta = \frac{5t}{10} = \frac{t}{2}$$

Here we used the fact that $\theta = s/r$, where s is the length of arc on a circle of radius r, subtended by a central angle of θ radians. (See Appendix 2.) Thus, $x = 10\cos\theta = 10\cos(t/2)$ and $y = 10\sin\theta = 10\sin(t/2)$. The two equations

$$\begin{cases} x = 10\cos\left(\dfrac{t}{2}\right) \\ y = 10\sin\left(\dfrac{t}{2}\right) \end{cases}$$

are called **parametric equations** for the path of the object. The variable t is called the **parameter**.

More generally, parametric equations are of the form

$$\begin{cases} x = f(t) \\ y = g(t) \end{cases}$$

where f and g are functions of the independent variable t. If no domain for t is specified, we will assume it to be the largest subset of the real numbers $\mathbb{R}$ for which both x and y are real numbers. As t varies over the domain, values of x and y are determined by the given parametric equations. In general, we will restrict the domain so that the curve is traced out only one time. The set of all pairs (x, y) determined in this way is the graph of the pair of parametric equations. In the example of the object moving on a circular path, we knew the graph at the outset, and our objective was to determine parametric equations for x and y. In the next example, we are given parametric equations and asked to determine the graph.

EXAMPLE 9.1   Make a table of values and draw the graph of the curve defined by the parametric equations

$$\begin{cases} x = t^2 \\ y = 2t - 1 \end{cases}$$

Solution   Since both x and y are defined for all real t, the domain is all of $\mathbb{R}$. We construct a table by assigning arbitrary values to t, thus determining the corresponding x and y values.

t	0	1	2	3	4	-1	-2	-3	-4
x	0	1	4	9	16	1	4	9	16
y	-1	1	3	5	7	-3	-5	-7	-9

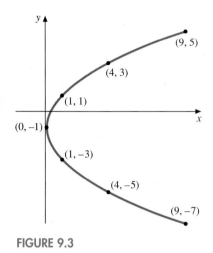

FIGURE 9.3

Now we plot the points (x, y) and connect them with a smooth curve, with the result shown in Figure 9.3. The curve appears to be a parabola with a horizontal axis. We will confirm this observation shortly.

Note that the parameter t does not appear on the graph. It is a means to an end only. The objective is to find x and y. ∎

The letter t is customarily used as a parameter, because, as in our introductory example, parametric equations often result from a consideration of the location of a moving object at time t. Other letters can be used for the parameter, however. For example, parametric equations of the circle given by $x^2 + y^2 = a^2$ can be obtained as shown in Figure 9.4 by using the central angle θ as a parameter. Since the x-coordinate of a point on the circle is $a \cos \theta$ and the y-coordinate is $a \sin \theta$, the parametric equations are

$$\begin{cases} x = a \cos \theta \\ y = a \sin \theta \end{cases} \quad 0 \le \theta \le 2\pi$$

The circle is traced out in a counterclockwise direction (as indicated by the arrows), starting at $(a, 0)$, as θ varies from 0 to 2π. (If θ is allowed to vary from 0 to 4π, the circle is traced two complete times.)

A curve defined parametrically is assigned a direction, or **orientation**, that is determined by the way points on the curve are traced out as the parameter increases.

Until now, we have relied on our intuitive understanding of what we mean by a curve. Now we give a precise definition.

Definition 9.1
A Plane Curve

A **curve** in the plane is a set of points (x, y) such that $x = f(t)$ and $y = g(t)$, where t is in an interval I and f and g are continuous on I.

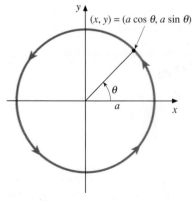

FIGURE 9.4
The circle given by $x^2 + y^2 = a^2$ has parametric equations
$$\begin{cases} x = a \cos \theta \\ y = a \sin \theta \end{cases} \quad 0 \le \theta \le 2\pi$$

The key word in this definition is *continuous*. Its significance is that small changes in t produce small changes in x and y, thus assuring that there are no breaks in the graph.

Consider again the parametric equations

$$\begin{cases} x = t^2 \\ y = 2t - 1 \end{cases} \quad -\infty < t < \infty$$

These equations define a curve, since both $f(t) = t^2$ and $g(t) = 2t - 1$ are continuous for all values of t.

Eliminating the Parameter

Sometimes we can eliminate the parameter between two parametric equations to find how x and y are related. We refer to the result as the **rectangular equation** of the curve. In the next two examples we illustrate some typical ways of eliminating the parameter.

EXAMPLE 9.2   Find the rectangular equation of the curve defined parametrically by the given equations:

(a) $\begin{cases} x = t^2 \\ y = 2t - 1 \end{cases}$ $-\infty < t < \infty$

(b) $\begin{cases} x = 2\cos\theta \\ y = 3\sin\theta \end{cases}$ $0 \le \theta \le 2\pi$

Solution

(a) We solve the second equation for t, getting

$$t = \frac{y+1}{2}$$

Now we substitute this value of t in the first equation to get

$$x = \frac{(y+1)^2}{4} \qquad \text{or} \qquad (y+1)^2 = 4x$$

which is the equation of a parabola with vertex $(0, -1)$, opening to the right, confirming our observation about the graph in Figure 9.3.

(b) To eliminate θ, we make use of the identity $\sin^2\theta + \cos^2\theta = 1$. We divide the first equation by 2 and the second by 3, and then square and add:

$$\left(\frac{x}{2}\right)^2 = \cos^2\theta$$
$$\left(\frac{y}{3}\right)^2 = \sin^2\theta$$
$$\overline{\frac{x^2}{4} + \frac{y^2}{9} = 1}$$

The graph of this equation is the ellipse shown in Figure 9.5, with x-intercepts ± 2 and y-intercepts ± 3. The curve is traced out in a counterclockwise direction, starting at $(2, 0)$ when $\theta = 0$ and terminating at the same point when $\theta = 2\pi$.   ∎

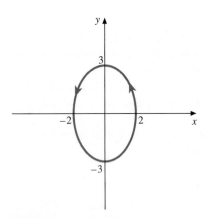

FIGURE 9.5
The graph of
$$\begin{cases} x = 2\ \cos\theta \\ y = 3\ \sin\theta \end{cases} \quad 0 \le \theta \le 2\pi$$
is the ellipse given by $x^2/4 + y^2/9 = 1$.

EXAMPLE 9.3   Find the rectangular equation of each of the following, and draw the graph:

(a) $\begin{cases} x = e^t \\ y = e^{-t} \end{cases}$

(b) $\begin{cases} x = t^2 \\ y = t^4 + 2 \end{cases}$

Solution

(a) Since $y = 1/e^t$ and $x = e^t$, we have

$$y = \frac{1}{x}$$

as the rectangular equation. Its graph is a hyperbola with asymptotes $x = 0$ and $y = 0$.

Note that the graph of this rectangular equation consists of two distinct branches, one in the first quadrant and one in the third (see Figure 9.6). However, the parametric equations require that both x and y be positive, since $e^t > 0$ and $e^{-t} > 0$ for all t. So the graph determined by the parametric equations is the right-hand branch only, shown in red in Figure 9.6. Since e^t is an increasing function, we see that x increases and y decreases with t, giving the orientation shown in the figure.

(b) Since $x = t^2$, we have $t^4 = x^2$, so on substituting x^2 for t^4 in the second equation, we get $y = x^2 + 2$ as the rectangular equation. Its graph is an upward-opening parabola with vertex $(0, 2)$. But here, again, the parametric equations impose a limitation not present in the rectangular equation, namely, that x be nonnegative. (After all, $x = t^2$, which is nonnegative.) We therefore have only the right half of the parabola $y = x^2 + 2$, as shown in red in Figure 9.7.   ∎

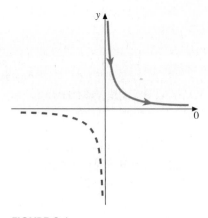

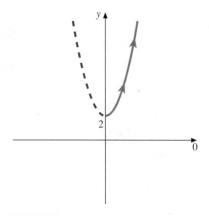

FIGURE 9.6
The graph of

$$\begin{cases} x = e^t \\ y = e^{-t} \end{cases}$$

is the right-hand branch only of the hyperbola $y = 1/x$.

FIGURE 9.7
The graph of

$$\begin{cases} x = t^2 \\ y = t^4 + 2 \end{cases}$$

is the right half of the parabola $y = x^2 + 2$.

As the preceding example makes clear, when you eliminate the parameter between two parametric equations, it is essential that you check the original equations for any restrictions imposed on x and y, and that you take these into account when you draw the graph.

The Cycloid

We conclude this section by deriving parametric equations of a curve known as a **cycloid**. A cycloid can be described as the path traced out by a point on a circle as the circle rolls along a line. For example, if a point is marked on the outer edge of a bicycle tire and the bicycle moves along a straight path, the point traces out a cycloid. (See Figure 9.8.) The cycloid has many interesting properties, two of which we will mention after we derive the equations.

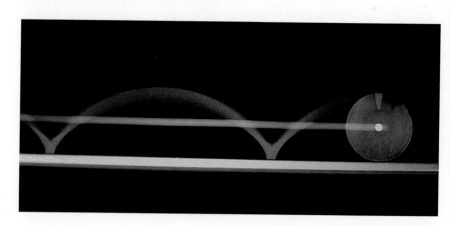

FIGURE 9.8

A cycloid generated by a circle rolling on a line.

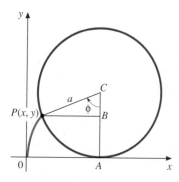

FIGURE 9.9

In Figure 9.9 we show the moving point P, originally at the origin, after the circle has rolled to the right. The angle ϕ (in radians) shown will be our parameter. Our objective is to express both x and y in terms of ϕ. Let the radius of the circle be a. From the figure we see that the x- and y-coordinates of P can be expressed as

$$x = \overline{OA} - \overline{PB} \qquad \text{and} \qquad y = \overline{AC} - \overline{BC}$$

Using the right triangle PBC, we have

$$\overline{PB} = a\sin\phi \qquad \text{and} \qquad \overline{BC} = a\cos\phi$$

Also, $\overline{AC} = a$. The distance $\overline{OA}$ is equal to the arc length $\overparen{AP}$. (Think of rolling the circle back to its starting position; the arc $\overparen{AP}$ would "unroll" as the distance $\overline{OA}$.) Since the arc length $\overparen{AP}$ equals $a\phi$ (because arc length equals the central angle in radians times the radius), we therefore have

$$\overline{OA} - \overline{PB} = a\phi - a\sin\phi \qquad \text{and} \qquad \overline{AC} - \overline{BC} = a - a\cos\phi$$

Thus, the parametric equations are

$$\begin{cases} x = a(\phi - \sin\phi) \\ y = a(1 - \cos\phi) \end{cases} \tag{9.1}$$

where ϕ can be any real number.

We can obtain the graph of the cycloid from Equations 9.1 by assigning values to ϕ. Finding the rectangular equation in this case is difficult (see Exercise 29 in Exercise Set 9.1), and it is not very helpful in drawing the graph. We show the graph in Figure 9.10. Note that the graph is periodic, with period equal to $2\pi a$, which is the circumference of the generating circle. Its maximum height is the diameter $2a$ of the circle.

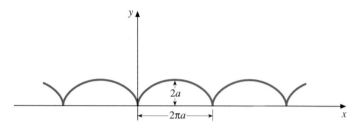

FIGURE 9.10
The cycloid

$$\begin{cases} x = a(\phi - \sin \phi) \\ y = a(1 - \cos \phi) \end{cases}$$

EXAMPLE 9.4   Find the area under one arch of the cycloid.

Solution   From Figure 9.10 we see that one arch is traced out as x goes from 0 to $2\pi a$. Thus, the area A is given by

$$A = \int_0^{2\pi a} y \, dx$$

To evaluate the integral, we use Equations 9.1, replacing y with $a(1 - \cos \phi)$ and dx with the differential of $a(\phi - \sin \phi)$, namely, $dx = a(1 - \cos \phi) \, d\phi$. The limits on ϕ are from $\phi = 0$ to $\phi = 2\pi$. So we have

$$A = \int_0^{2\pi} \overbrace{a(1 - \cos \phi)}^{y} \cdot \overbrace{a(1 - \cos \phi) \, d\phi}^{dx}$$

$$= a^2 \int_0^{2\pi} (1 - \cos \phi)^2 \, d\phi$$

$$= a^2 \int_0^{2\pi} (1 - 2\cos \phi + \cos^2 \phi) \, d\phi$$

$$= a^2 \int_0^{2\pi} [1 - 2\cos \phi + \tfrac{1}{2}(1 + \cos 2\phi)] \, d\phi$$

$$= a^2 \left[\frac{3\phi}{2} - 2\sin \phi + \frac{\sin 2\phi}{4} \right]_0^{2\pi} = 3\pi a^2$$

Thus, the area is three times the area inside the generating circle. Galileo guessed this result. It was first proved by Roberval around 1634 and was also proved independently by Torricelli in 1644.   ■

Two famous problems in mechanics are known as the **Brachistochrone Problem** and the **Tautochrone Problem**. The brachistochrone is the "curve of quickest descent." Suppose, for example, a bead is allowed to slide (without

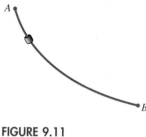

FIGURE 9.11

FIGURE 9.12

friction) along a wire from a point A to a lower point B, as in Figure 9.11. In what shape should the wire be bent so that the time required for the bead to go from A to B is a minimum? Surprisingly, the answer is not a straight line (the shortest distance) but is an inverted cycloid in which A is on the x-axis at a cusp of the cycloid and B is at the bottom of the arch. In effect, the bead picks up more speed by descending relatively quickly at first. This discovery was made independently by a number of seventeenth-century mathematicians (see the historical note).

The tautochrone is the "curve of equal time." Suppose beads are initially placed on a wire joining A and B, as in the Brachistochrone Problem, and then released. What should be the shape of the wire so that all the beads arrive at B at the same time, no matter how far down the wire they start? (See Figure 9.12.) Again, the answer is the inverted cycloid. This result is quite remarkable, for even if one of the beads starts very near to B and another starts at A, when the curve is a cycloid, they will arrive at the same instant. Dutch physicist and astronomer Christian Huygens (1629–1695) made this discovery in his effort to construct a pendulum clock in which the period of the pendulum is independent of the amplitude of its swing.

The brachistochrone problem represents the historical first of a new type of problem. For previous minimization problems, the object being minimized depended on only a few variables. In the brachistochrone problem, time is being minimized and it depends on the length of the whole curve. This approach introduced a whole new field of study that came to be known as the *calculus of variations*. This technique has many applications in physical science.

Historical Note: The Brachistochrone Problem

During the seventeenth and eighteenth centuries, mathematicians sometimes issued challenges to one another. In 1697, Johann Bernoulli had solved and published the question of finding the path by which a heavy body would descend most quickly from one position to another, given that the second position was not directly below the first. After the six-month deadline had passed, no answer had been given. Leibniz had written that he had solved the problem but suggested that the time be extended by a few months. Bernoulli added a second problem, sent the two problems for publication, and sent copies directly to both Newton and to one other person. Bernoulli and Leibniz apparently thought that the problem had stumped Newton and wanted to show their superiority over him.

Newton's recent biographer Richard Westfall concludes that Newton thought—probably correctly—that a direct challenge was being issued to his status as a mathematician. Newton wrote, "I do not love...to be...teezed by forreigners about Mathematical things...." But he had solved the problem by the next day. He wrote, "When the problem in 1697 was sent by Bernoulli-Sr I. N. was in the midst of the hurry of the great recoinage did not come home till four from the Tower very much tired, but did not sleep till he had solved it s^{ch} was by 4 in the morning."

Bernoulli received three solutions: one from Leibniz, one from L'Hôpital, and an anonymous one from England. Bernoulli immediately realized that this last solution must be from Newton. He said: "I recognize the lion from his claw" (*"tanquam ex ungue leonem"*).

Drawn from *Never at Rest*, by Richard S. Westfall, Cambridge University Press, 1980.

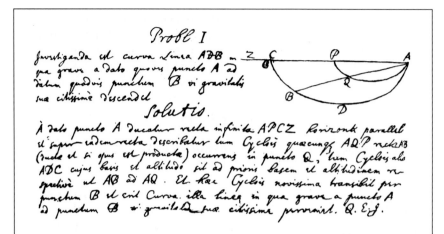

The Brachistochrone Problem and its solution in Newton's handwriting.

SKETCHING PARAMETRIC CURVES USING COMPUTER ALGEBRA SYSTEMS

Parametric representation of curves provides a method for describing a great variety of curves in the plane. The computer algebra systems Maple, Mathematica, and DERIVE have specific notations for easily plotting parametric curves.

 CAS 31

The Cycloid: Sketch the cycloid given by the parametric equations

$$x = 2(t - \sin t), \quad y = 2(1 - \cos t)$$

for $-10 \le t \le 10$.

FIGURE 9.1.1
$x = 2(t - \sin t)$, $y = 2(1 - \cos t)$,
$-10 \le t \le 10$

Maple:

```
plot([2*(t-sin(t)),
    2*(1-cos(t)),t=-10..10],
    numpoints=500);
```

Mathematica:

```
ParametricPlot[{2*(t-Sin[t]),
    2*(1-Cos[t])},
    {t,-10,10}]
```

DERIVE:

(At the □ symbol, go to the next step.)
a (author) □ [2(t−sin(t)),2(1−cos(t))]
□ p (plot window) □ s (scale) □ 3 (x-scale) □ 2 (y-scale) □ −10 (min) □ 10 (max)

Notice that in Maple and DERIVE, the parametric equations are enclosed in square brackets and that in Mathematica they are enclosed in curly brackets. Also, in Maple, the specification for the range of the parameter is given inside the square brackets.

CAS 32

Sketch the curve given by the parametric equations

$$x = \cos t(1 - 2\sin t), \quad y = \sin t(1 - 2\sin t)$$

for $0 \le t \le 2\pi$.

Maple:

```
x := t-> cos(t)*(1-2*sin(t));
y := t-> sin(t)*(1-2*sin(t));
plot([x(t),y(t),t=0..2*Pi]);
```

Mathematica:

```
x = Cos[t]*(1-2*Sin[t])
y = Sin[t]*(1-2*Sin[t])
ParametricPlot[{x,y},
    {t,0,2*Pi}]
```

DERIVE:

(At the □ symbol, go to the next step.)
(author) □ [cos(t)(1−2sin(t)), sin(t)
(1−2sin(t))] □ p (plot window) □ p
(plot) □ 0 (min) □ 2Pi (max)

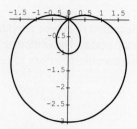

FIGURE 9.1.2
$x = \cos t(1 - 2\sin t),$
$y = \sin t(1 - 2\sin\ t), \ 0 \le t \le \pi$

Exercise Set 9.1

In Exercises 1–6, draw the curve defined by the parametric equations by substituting values for the parameter. Indicate the orientation on the graph by the use of arrows.

1. $\begin{cases} x = 2t + 3 \\ y = 3t - 1 \end{cases}$

2. $\begin{cases} x = t^2 - 1 \\ y = t^2 + 1 \end{cases} \quad -3 \le t \le 0$

3. $\begin{cases} x = t - 2 \\ y = t^2 \end{cases} \quad -2 \le t \le 3$

4. $\begin{cases} x = u^2 \\ y = u^3 \end{cases}$

5. $\begin{cases} x = \dfrac{1}{v} \\ y = v \end{cases} \quad v > 0$

6. $\begin{cases} x = \sqrt{t} \\ y = t \end{cases} \quad t > 0$

In Exercises 7–20, eliminate the parameter and make use of the rectangular equation to draw the given curve. Indicate the orientation.

7. $\begin{cases} x = 3t - 2 \\ y = 1 - t \end{cases} \quad -1 \le t \le 4$

8. $\begin{cases} x = 4 - t^2 \\ y = t^2 + 2 \end{cases} \quad 0 \le t \le 3$

9. $\begin{cases} x = \cos t \\ y = -\sin t \end{cases} \quad 0 \le t \le \pi$

10. $\begin{cases} x = 1 - \cos \theta \\ y = \sin \theta \end{cases} \quad 0 \le \theta \le 2\pi$

11. $\begin{cases} x = \cosh t \\ y = \sinh t \end{cases}$

12. $\begin{cases} x = 2t - 1 \\ y = 1 - t^2 \end{cases} \quad t \ge -1$

13. $\begin{cases} x = \sin^2 \theta + 1 \\ y = \cos \theta \end{cases} \quad 0 \le \theta \le \pi$

14. $\begin{cases} x = 3 \sin \theta \\ y = 4 \cos \theta \end{cases} \quad 0 \le \theta \le 2\pi$

15. $\begin{cases} x = \sec \theta \\ y = \tan \theta \end{cases} \quad -\dfrac{\pi}{2} < \theta < \dfrac{\pi}{2}$

16. $\begin{cases} x = 2 \cosh t \\ y = 3 \sinh t \end{cases}$

17. $\begin{cases} x = 2 - \cos t \\ y = 1 + 3 \sin t \end{cases} \quad 0 \le t \le 2\pi$

18. $\begin{cases} x = \sqrt{4 - t} \\ y = \sqrt{t} \end{cases} \quad 0 \le t \le 4$

19. $\begin{cases} x = \sin \theta \\ y = \cos 2\theta \end{cases} \quad 0 \le \theta \le \dfrac{\pi}{2}$

20. $\begin{cases} x = 2 \sec^2 \theta - 3 \\ y = \tan \theta + 1 \end{cases} \quad 0 \le \theta < \dfrac{\pi}{2}$

21. Compare the curves C_1, C_2, C_3, and C_4. Include a discussion of orientation.

C_1: $\begin{cases} x = 1 + 2t \\ y = -2 + 4t \end{cases} \quad 0 \le t \le 1$

C_2: $\begin{cases} x = 3 - 2t \\ y = 2 - 4t \end{cases} \quad 0 \le t \le 1$

C_3: $\begin{cases} x = t \\ y = 2t - 4 \end{cases} \quad 1 \le t \le 3$

C_4: $\begin{cases} x = \tan^2 t \\ y = 2 \sec^2 t - 6 \end{cases} \quad \dfrac{\pi}{4} \le t \le \dfrac{\pi}{3}$

22. If a curve crosses itself, the point where this occurs is called a **double point**. Show that the curve C defined by

$$\begin{cases} x = t^3 - 3t + 1 \\ y = t^2 + t - 1 \end{cases} \quad -\infty < t < \infty$$

has a double point, and find this point. What are the two distinct values of t that produce the double point?

23. If a curve has no double points (see Exercise 22), it is said to be a **simple curve.** Show that the curve C defined by

$$\begin{cases} x = t^3 + 1 \\ y = t^2 - 1 \end{cases} \quad 0 \le t \le 2$$

is simple. Draw its graph.

24. A curve C defined by $x = f(t)$ and $y = g(t)$, for $a \le t \le b$, is said to be a **simple closed curve** if it is simple for $a \le t < b$ (see Exercise 23) and $(f(a), g(a))$ is the same point as $(f(b), g(b))$. Show that the curve defined by

$$\begin{cases} x = t^2 - 3 \\ y = t^3 - 4t - 1 \end{cases} \quad -2 \le t \le 2$$

is a simple closed curve.

In Exercises 25–28, draw each curve, showing its orientation.

25. $\begin{cases} x = t^2 - 1 \\ y = t^3 - t \end{cases} \quad -\infty < t < \infty$

26. $\begin{cases} x = \cos^3 \theta \\ y = \sin^3 \theta \end{cases} \quad 0 \le \theta \le 2\pi$

27. $\begin{cases} x = \cos^2 \theta \\ y = \frac{1}{2} \sin 2\theta \end{cases} \quad 0 \le \theta \le \pi$

28. $\begin{cases} x = (t^2 - 2)(t + 1) \\ y = (t^2 - 2)(t - 1) \end{cases} \quad -\infty < t < \infty$

29. Find the rectangular equation of the cycloid.

30. A **hypocycloid** is a curve traced out by a fixed point on a circle rolling on the inside of a fixed larger circle. (See the accompanying figure.) If the radius of the larger circle is a and the radius of the smaller circle is b, derive the parametric equations

$$\begin{cases} x = (a - b)\cos\theta + b\cos\dfrac{a - b}{b}\theta \\ y = (a - b)\sin\theta - b\sin\dfrac{a - b}{b}\theta \end{cases} \quad 0 \le \theta \le 2\pi$$

(*Hint:* Observe that $\overset{\frown}{AB} = \overset{\frown}{BP}$, so that $a\theta = b\phi$.)

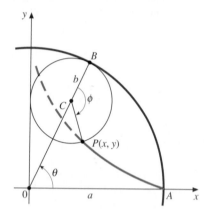

 31. Use a CAS or a graphing calculator to obtain the graph of the hypocycloid (see Exercise 30) for each of the following values of a and b.
(a) $a = 4$, $b = 1$
(b) $a = 6$, $b = 2$
(c) $a = 8$, $b = 1$

 32. Refer to Exercise 30 and show that if $a = 4b$ the equations can be put in the form

$$\begin{cases} x = a\cos^3\theta \\ y = a\sin^3\theta \end{cases}$$

Also find the rectangular equation. (This is the four-cusp hypocycloid.) Use a CAS or graphing calculator to obtain its graph. How does changing a affect the graph?

 33. If the smaller circle in Exercise 30 rolls on the outer circumference of the larger circle, the point traces out an **epicycloid**. (See the accompanying figure.) Derive the following parametric equations of the epicycloid:

$$\begin{cases} x = (a + b)\cos\theta - b\cos\left(\dfrac{a + b}{b}\right)\theta \\ y = (a + b)\sin\theta - b\sin\left(\dfrac{a + b}{b}\right)\theta \end{cases} \quad 0 \le \theta \le 2\pi$$

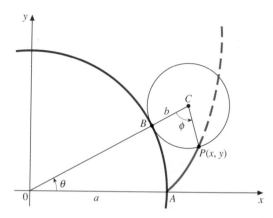

Use a CAS or graphing calculator to sketch the epicycloid for $a = 4$ and $b = 1$.

34. Suppose a point P is attached to an extended spoke of a wheel as in the figure and the wheel rolls along the x-axis. If the distance from the center of the wheel to P is b and the radius of the wheel is a (with $a < b$), find parametric equations of the path of P. This curve is called a **trochoid**. (See the accompanying figure.)

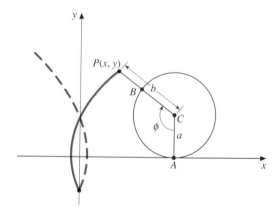

35. Find parametric equations for the x- and y-coordinates of a point that traces out the circle $x^2 + y^2 = r^2$ in the following manner:
(a) once around, counterclockwise, starting at $(r, 0)$;
(b) twice around, counterclockwise, starting at $(r, 0)$;
(c) twice around, clockwise, starting at $(r, 0)$;
(d) three times around, clockwise, starting at $(0, r)$;
(e) three times around, counterclockwise, starting at $(-r, 0)$.

36. Find parametric equations for the x- and y-coordinates of a point that traces out the ellipse $x^2/a^2 + y^2/b^2 = 1$ in the following manner:
(a) once around, counterclockwise, starting at $(a, 0)$;
(b) twice around, counterclockwise, starting at $(a, 0)$;

(c) twice around, clockwise, starting at $(a, 0)$;

(d) three times around, clockwise, starting at $(0, b)$;

(e) three times around, counterclockwise, starting at $(-a, 0)$.

37. Verify that

$$\begin{cases} x = x_1 + (x_2 - x_1)t \\ y = y_1 + (y_2 - y_1)t \end{cases} \quad 0 \le t \le 1$$

are parametric equations for the straight line segment from (x_1, y_1) to (x_2, y_2).

38. Find parametric equations for the line passing through the points (x_1, y_1) and (x_2, y_2).

39. Describe the difference between the curves defined by $x = \cos t$, $y = \sin t$, where $-\pi \le t \le \pi$, and $x = \sin t$, $y = \cos t$, where $-\pi \le t \le \pi$.

40. Use a CAS or graphing calculator to investigate the family of curves

$$\begin{cases} x = \cos t(a - b \sin t) \\ y = \sin t(a - b \sin t) \end{cases} \quad 0 \le t \le 2\pi$$

where a and b are real numbers. Consider cases where $a < b$, $a = b$, and $a > b$.

41. Repeat Exercise 40 with

$$\begin{cases} x = \cos t(a - b \cos t) \\ y = \sin t(a - b \cos t) \end{cases} \quad 0 \le t \le 2\pi$$

42. Use a CAS or graphing calculator to investigate the family of curves given by

$$\begin{cases} x = \cos t(1 - \sin(at)) \\ y = \sin t(1 - \sin(at)) \end{cases} \quad 0 \le t \le 2\pi$$

where a is a real number.

43. Find parametric equations for a circle of radius r with a center at (h, k).

9.2 DERIVATIVES OF FUNCTIONS DEFINED PARAMETRICALLY

We begin this section by considering the question of how to find the slope (when it exists) of the tangent line to a curve given parametrically by

$$x = f(t), \quad y = g(t) \quad (a \le t \le b)$$

We saw in the previous section that it is sometimes possible, by eliminating the parameter, to write y as a function of x. Doing so explicitly is often difficult, however, and sometimes it is impossible. In many instances, it can be concluded that y is a differentiable function of x, even when the explicit function is unknown. In the remark following this discussion, we outline some general conditions that guarantee y is a differentiable function of x.

If we assume for the present discussion that y is a differentiable function of x, so that dy/dx exists, and if both dx/dt and dy/dt exist, then the Chain Rule gives

$$\frac{dy}{dt} = \frac{dy}{dx}\frac{dx}{dt}$$

If $dx/dt \ne 0$, we can solve for dy/dx to get

$$\frac{dy}{dx} = \frac{\dfrac{dy}{dt}}{\dfrac{dx}{dt}} \quad \left(\frac{dx}{dt} \ne 0\right) \tag{9.2}$$

Note that formally if we treat dy and dx as differentials, with $dx \neq 0$, we could get Formula 9.2 by dividing the numerator and denominator of dy/dx by dt.

If $dy/dt = 0$ and $dx/dt \neq 0$, then by Formula 9.2 the slope of the tangent line is 0, so the curve has a horizontal tangent. Reversing the roles of x and y, when $dx/dt = 0$ and $dy/dt \neq 0$, the curve has a vertical tangent. When both $dx/dt = 0$ and $dy/dt = 0$, the curve is said to have a **singular point**. The curve may or may not have a tangent line at a singular point. A curve is said to be **smooth** if dx/dt and dy/dt are continuous and there are no singular points. A smooth curve has a continuously turning tangent line. There are no sharp corners.

REMARK ───

■ It can be shown that if the curve defined by the parametric equations $x = f(t)$, $y = g(t)$, for $a \leq t \leq b$, is smooth and $f'(t) \neq 0$ for $a \leq t \leq b$, then y can be expressed as a differentiable function of x for $f(a) \leq x \leq f(b)$, say $y = F(x)$. Then, since $y = g(t)$ and $x = f(t)$, we have $g(t) = F(f(t))$. So by the Chain Rule,

$$g'(t) = F'(f(t))\, f'(t) = F'(x)\, f'(t)$$

Thus, since $f'(t) \neq 0$,

$$F'(x) = \frac{g'(t)}{f'(t)}$$

or, equivalently,

$$\frac{dy}{dx} = \frac{\dfrac{dy}{dt}}{\dfrac{dx}{dt}}$$

That is, Formula 9.2 is true under the hypotheses stated.

───

EXAMPLE 9.5   Find dy/dx when $t = 2$ for the curve C defined parametrically by

$$\begin{cases} x = 2t^2 - 1 \\ y = t^3 - 3t \end{cases} \qquad 1 \leq t \leq 4$$

Solution   Since $dx/dt = 4t$ and $dy/dt = 3t^2 - 3$, we see that these derivatives are continuous and not simultaneously 0 in the interval $[1, 4]$. So C is smooth. Also $dx/dt \neq 0$ for t in the interval $[1, 4]$. Thus, by Equation 9.2,

$$\frac{dy}{dx} = \frac{\dfrac{dy}{dt}}{\dfrac{dx}{dt}} = \frac{3t^2 - 3}{4t}$$

When $t = 2$, we obtain

$$\left. \frac{dy}{dx} \right|_{t=2} = \frac{9}{8}$$

■

EXAMPLE 9.6   Find the equation of the tangent line to the curve C defined by

$$\begin{cases} x = \dfrac{4}{t} \\[2mm] y = \sqrt{t} \end{cases} \quad 1 \le t \le 9$$

at the point $(1, 2)$.

Solution   We have, by Equation 9.2,

$$\frac{dy}{dx} = \frac{\dfrac{dy}{dt}}{\dfrac{dx}{dt}} = \frac{\dfrac{1}{2\sqrt{t}}}{-\dfrac{4}{t^2}} = -\frac{t^{3/2}}{8}$$

The value of t that produces the point $(1, 2)$ is $t = 4$. Thus, at this point,

$$\left.\frac{dy}{dx}\right|_{t=4} = -\frac{(4)^{3/2}}{8} = -1$$

Using the point-slope form of the equation of a line, we obtain the tangent line

$$y - 2 = -(x - 1)$$

Point-slope form of the equation of a line:

$$y - y_1 = m(x - x_1)$$

or

$$x + y = 3 \qquad \blacksquare$$

The Second Derivative from Parametric Equations

As we have previously seen, second derivatives are useful in a variety of ways, such as in determining the concavity of a curve or the acceleration of a moving particle. Since $y'' = dy'/dx$, we can find the second derivative of y with respect to x, if it exists, by replacing y with y' in Equation 9.2. That is,

$$y'' = \frac{dy'}{dx} = \frac{\dfrac{dy'}{dt}}{\dfrac{dx}{dt}} \qquad \left(\frac{dx}{dt} \ne 0\right) \qquad (9.3)$$

We illustrate this way of calculating y'' in the next two examples.

EXAMPLE 9.7   Find y'' for the function defined in Example 9.6.

Solution   In Example 9.6 we found that

$$y' = \frac{dy}{dx} = -\frac{t^{3/2}}{8}$$

Then, since

$$\frac{dx}{dt} = -\frac{4}{t^2}$$

we have, by Equation 9.3,

$$y'' = \frac{dy'}{dx} = \frac{\dfrac{dy'}{dt}}{\dfrac{dx}{dt}} = \frac{-\dfrac{3t^{1/2}}{16}}{-\dfrac{4}{t^2}} = \frac{3t^{5/2}}{64}$$ ∎

EXAMPLE 9.8   For the cycloid

$$\begin{cases} x = a(t - \sin t) \\ y = a(1 - \cos t) \end{cases} \qquad -\infty < t < \infty$$

do the following:

(a) Find $\dfrac{dy}{dx}$ and $\dfrac{d^2y}{dx^2}$.

(b) Find all points where the tangent line is horizontal.

(c) Discuss concavity.

(d) Show that the origin is a singular point but that the curve has a vertical tangent line there.

Solution

(a) Since

$$\frac{dx}{dt} = a(1 - \cos t) \qquad \text{and} \qquad \frac{dy}{dt} = a \sin t$$

we have, by Equation 9.2,

$$y' = \frac{dy}{dx} = \frac{\dfrac{dy}{dt}}{\dfrac{dx}{dt}} = \frac{a \sin t}{a(1 - \cos t)} = \frac{\sin t}{1 - \cos t}$$

provided $\cos t \neq 1$. Similarly, by Equation 9.3,

$$\frac{d^2y}{dx^2} = \frac{dy'}{dx} = \frac{\dfrac{dy'}{dt}}{\dfrac{dx}{dt}} = \frac{\dfrac{\cos t(1 - \cos t) - \sin t(\sin t)}{(1 - \cos t)^2}}{a(1 - \cos t)}$$

$$= \frac{\cos t - \cos^2 t - \sin^2 t}{a(1 - \cos t)^3}$$

$$= \frac{\cos t - 1}{a(1 - \cos t)^3} \qquad \text{Since } \cos^2 t + \sin^2 t = 1$$

$$= \frac{-1}{a(1 - \cos t)^2}$$

provided again that $\cos t \neq 1$.

(b) The tangent line is horizontal where $dy/dx = 0$, or equivalently where $dy/dt = 0$ and $dx/dt \neq 0$. Setting $dy/dt = 0$ gives

$$a \sin t = 0$$

$$\sin t = 0$$

This equation is true when t is any integral multiple of π. But we must also have $dx/dt \neq 0$, that is,

$$a(1 - \cos t) \neq 0$$

$$\cos t \neq 1$$

Thus, t cannot be an *even* multiple of π (i.e., 0, $\pm 2\pi$, $\pm 4\pi, \ldots$). We conclude that the tangent line is horizontal when t is an odd multiple of π (i.e., $\pm \pi$, $\pm 3\pi$, $\pm 5\pi, \ldots$), which we can write as

$$t = (2n + 1)\pi \qquad n = 0, \pm 1, \pm 2, \ldots$$

At each such value of t, $\sin t = 0$ and $\cos t = -1$. Thus, the points on the curve are

$$(a(2n + 1)\pi, 2a) \qquad n = 0, \pm 1, \pm 2, \ldots$$

(c) Since

$$\frac{d^2 y}{dx^2} = \frac{-1}{a(1 - \cos t)^2}$$

is negative for all t except where $\cos t = 1$, the graph is concave down everywhere except for points corresponding to these values of t.

(d) When $t = 0$, we have $x = a(0 - \sin 0) = 0$ and $y = a(1 - \cos 0) = 0$. So the origin corresponds to $t = 0$. Also, at this point

$$\frac{dx}{dt} = a(1 - \cos 0) = 0 \qquad \text{and} \qquad \frac{dy}{dt} = a \sin 0 = 0$$

So the origin is a singular point. For t near 0 but not equal to 0, we have, by part (a),

$$\frac{dy}{dx} = \frac{\sin t}{1 - \cos t}$$

Using L'Hôpital's Rule, we have

$$\lim_{t \to 0^+} \frac{\sin t}{1 - \cos t} = \lim_{t \to 0^+} \frac{\cos t}{\sin t} = \infty$$

Similarly,

$$\lim_{t \to 0^-} \frac{\sin t}{1 - \cos t} = \lim_{t \to 0^-} \frac{\cos t}{\sin t} = -\infty$$

Thus, as t approaches 0 from the right, the slopes become larger and larger. That is, the tangent lines become more and more nearly vertical. As we approach from the left, the slopes are negative but large in absolute value, so again the tangent lines are nearly vertical. We illustrate this situation in Figure 9.13. The curve has a cusp at the origin. By periodicity it also has cusps at the points $(2n\pi a, 0)$ for $n = \pm 1, \pm 2, \ldots$.

It is easy to see that our conclusions are in agreement with the graph of the cycloid in Figure 9.10. ∎

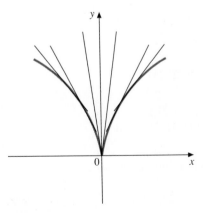

FIGURE 9.13
As $x \to 0^+$, slopes $\to \infty$. As $x \to 0^-$, slopes $\to -\infty$. When $x = 0$, tangent line is vertical.

Higher-Order Derivatives

To compute higher-order derivatives, we follow a procedure similar to the one used for y''. For example,

$$y''' = \frac{dy''}{dx} = \frac{\dfrac{dy''}{dt}}{\dfrac{dx}{dt}} \qquad \left(\frac{dx}{dt} \neq 0\right)$$

and in general, for $n \geq 1$,

$$y^{(n+1)} = \frac{dy^{(n)}}{dx} = \frac{\dfrac{dy^{(n)}}{dt}}{\dfrac{dx}{dt}} \qquad \left(\frac{dx}{dt} \neq 0\right)$$

provided the derivative exists. Note that in every case we divide by dx/dt and so must have this derivative nonzero.

Exercise Set 9.2

In Exercises 1–10, find dy/dx at the indicated point without eliminating the parameter.

1. $\begin{cases} x = t^2 - 1 \\ y = 2t + 3 \end{cases}$ at $t = 1$

2. $\begin{cases} x = \sqrt{t - 1} \\ y = t^2 \end{cases}$ at $t = 2$

3. $\begin{cases} x = t^3 \\ y = t^2 \end{cases}$ at $t = -2$

4. $\begin{cases} x = 2\sin t \\ y = 3\cos t \end{cases}$ at $t = \dfrac{\pi}{3}$

5. $\begin{cases} x = 4\sin^2 t - 3 \\ y = 2\cos t \end{cases}$ at $t = \dfrac{\pi}{3}$

6. $\begin{cases} x = 2\sinh t \\ y = \tanh t \end{cases}$ at $t = 0$

7. $\begin{cases} x = e^{-t} \\ y = 1 + e^t \end{cases}$ at $t = \ln 2$

8. $\begin{cases} x = \ln t \\ y = 2 - \ln \sqrt{t} \end{cases}$ at $t = e^2$

9. $\begin{cases} x = t^3 - 2t \\ y = \sqrt{t - 1} \end{cases}$ at $(4, 1)$

10. $\begin{cases} x = \dfrac{1}{t} \\ y = 4(1 - t^2) \end{cases}$ at $(-2, 3)$

In Exercises 11–14, find the equation of the tangent line to the curve defined by the parametric equations at the given point.

11. $\begin{cases} x = 3t^2 - 2t \\ y = t^2 + t - 1 \end{cases}$ at $t = -1$

12. $\begin{cases} x = \dfrac{1}{t - 1} \\ y = \dfrac{1}{t + 1} \end{cases}$ at $t = \frac{1}{2}$

13. $\begin{cases} x = t^3 + 3t^2 \\ y = 2t^2 - 3 \end{cases}$ at $(2, -1)$

14. $\begin{cases} x = t - \dfrac{2}{t} \\ y = t^2 \end{cases}$ at $(1, 4)$

15. Find all points on the curve C defined by

$$\begin{cases} x = t^3 - 4t^2 - 3t \\ y = 2t^2 + 3t - 5 \end{cases} \quad -\infty < t < \infty$$

where the tangent line is vertical and all points where the tangent line is horizontal.

16. (a) Show that the curve C defined by

$$\begin{cases} x = t^2 - 2t - 8 \\ y = t^2 - t - 2 \end{cases} \quad -\infty < t < \infty$$

is smooth.

(b) In what intervals do the parametric equations in part (a) define y as a differentiable function of x? Find dy/dx in terms of t for the intervals in which it exists.

17. Follow the instructions in Exercise 16 for the curve C defined by

$$\begin{cases} x = \dfrac{t^2 - 5}{t - 2} \\ y = \dfrac{1}{t - 2} \end{cases} \quad t \neq 2$$

In Exercises 18–21, find d^2y/dx^2 in terms of t without eliminating the parameter.

18. $\begin{cases} x = t^2 \\ y = 2t^3 \end{cases} \quad t > 0$

19. $\begin{cases} x = 2\cos t \\ y = 3\sin t \end{cases} \quad 0 < t < \pi$

20. $\begin{cases} x = \sinh t \\ y = \cosh^2 t \end{cases} \quad -\infty < t < \infty$

21. $\begin{cases} x = \sqrt{1 - t^2} \\ y = 1 + t^2 \end{cases} \quad 0 < |t| < 1$

22. A **double point** on a curve defined by $x = f(t)$ and $y = g(t)$ is a point for which $(f(t_1), g(t_1)) = (f(t_2), g(t_2))$ and $t_1 \neq t_2$. Show that the curve defined by

$$\begin{cases} x = t^2 + 1 \\ y = t^3 - t - 1 \end{cases} \quad -\infty < t < \infty$$

has a double point, and find equations of the two tangent lines at this point. Use a CAS or graphing calculator to obtain the graph of the curve.

23. When a smooth curve C is defined parametrically by $x = f(t)$ and $y = g(t)$ for t in some interval I, the notation $\dot{x} = dx/dt$, $\dot{y} = dy/dt$, $\ddot{x} = d^2x/dt^2$, and $\ddot{y} = d^2y/dt^2$ is sometimes used (it was introduced by Newton). Assuming the equations define y as a twice differentiable function of x, show that

$$\frac{d^2y}{dx^2} = \frac{\dot{x}\ddot{y} - \dot{y}\ddot{x}}{\dot{x}^3}$$

Similarly, find an expression for d^3y/dx^3.

24. Find the equation of the normal line (line perpendicular to the tangent line) to the cycloid at an arbitrary point $P(x, y)$ on the cycloid for which the parameter ϕ is in the open interval $(0, 2\pi)$. Show that the x-intercept of this line is the x-coordinate of the point of contact of the generating circle with the x-axis for the given value of ϕ.

9.3 ARC LENGTH AND SURFACE AREA

In Section 6.4, we showed that the differential of arc length, ds, can be written in the form

$$ds = \sqrt{(dx)^2 + (dy)^2} \tag{9.4}$$

If x and y are given parametrically by

$$\begin{cases} x = f(t) \\ y = g(t) \end{cases} \quad a \leq t \leq b$$

and f' and g' are continuous functions, we have

$$dx = f'(t)\,dt \quad \text{and} \quad dy = g'(t)\,dt$$

Thus, by Equation 9.4,

$$ds = \sqrt{[f'(t)\,dt]^2 + [g'(t)\,dt]^2} = \sqrt{[f'(t)]^2 + [g'(t)]^2}\,dt$$

or, equivalently,

$$ds = \sqrt{\left(\frac{dx}{dt}\right)^2 + \left(\frac{dy}{dt}\right)^2}\,dt \tag{9.5}$$

Thus, the length L of the curve is

Length of a Curve Defined Parametrically

$$L = \int_{t=a}^{t=b} ds = \int_a^b \sqrt{\left(\frac{dx}{dt}\right)^2 + \left(\frac{dy}{dt}\right)^2}\, dt \qquad (9.6)$$

In deriving Equation 9.4 for ds in Chapter 6, we assumed that y was a differentiable function of x. For Formula 9.6 to be valid, we no longer need this assumption. This formula remains true provided that f' and g' are continuous for $a \le t \le b$ and that the curve defined by $x = f(t)$, $y = g(t)$ is traced out exactly once as t increases from a to b. A formal proof can be constructed along the same lines as in the derivation of Formula 6.18 in Chapter 6, where we approximated the curve more and more closely with a series of smaller and smaller line segments (that is, by polygonal paths). We will omit the details.

It is easy to see that Formula 9.6 is equivalent to Formula 6.18 when y is a continuously differentiable function of x, say $y = F(x)$, with $a \le x \le b$. Here, we can parameterize the graph of F by the parametric equations

$$\begin{cases} x = t \\ y = F(t) \end{cases} \qquad a \le t \le b$$

So $dx/dt = 1$ and $dy/dt = F'(t)$. Thus, Equation 9.6 becomes

$$L = \int_a^b \sqrt{1 + [F'(t)]^2}\, dt$$

which is equivalent to Equation 6.18. A similar argument can be given when $x = G(y)$ to obtain Equation 6.19 as a special case of Formula 9.6.

EXAMPLE 9.9   A portion of a metal sculpture follows the curve C defined by

$$\begin{cases} x = t^3 - 1 \\ y = 2t^2 + 3 \end{cases} \qquad 0 \le t \le 1$$

where x and y are measured in meters. If the metal costs \$100 per meter, find the cost of the material.

Solution   We show the graph of C in Figure 9.14. The derivatives

$$\frac{dx}{dt} = 3t^2 \qquad \text{and} \qquad \frac{dy}{dt} = 4t$$

are both continuous for $0 \le t \le 1$, so we can apply Equation 9.6.

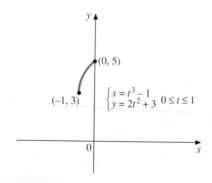

$$L = \int_0^1 \sqrt{(3t^2)^2 + (4t)^2}\, dt = \int_0^1 \sqrt{9t^4 + 16t^2}\, dt$$

$$= \int_0^1 t\sqrt{9t^2 + 16}\, dt = \frac{1}{18}\left[(9t^2 + 16)^{3/2} \cdot \frac{2}{3}\right]_0^1 \qquad \text{Substitute } u = 9t^2 + 16.$$

$$= \frac{1}{27}[125 - 64] = \frac{61}{27}$$

FIGURE 9.14

So the length is $\dfrac{61}{27}$ m. At \$100/m, the cost is $\$100\left(\dfrac{61}{27}\right) \approx \$225.93$.   ∎

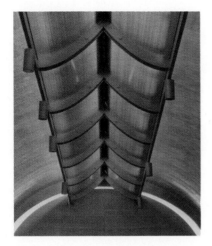

Ceiling of the Kimbell Art Museum

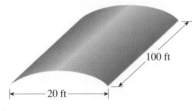

100 ft

20 ft

FIGURE 9.15

EXAMPLE 9.10   The ceiling of the Kimbell Art Museum in Forth Worth, Texas, consists of a series of arched sections, each in the shape of a cycloid, as shown in Figure 9.15. Find the parametric equations of the cycloid, and find the area of each section.

Solution   In Section 9.1 we showed that the period of the cycloid is $2\pi a$ and that the parametric equations are

$$\begin{cases} x = a(t - \sin t) \\ y = a(1 - \cos t) \end{cases}$$

The distance 20 ft in Figure 9.15 represents one period. So we have $2\pi a = 20$, or

$$a = \frac{10}{\pi}$$

For one arch we take t in the interval $[0, 2\pi]$. Thus, the equations are

$$\begin{cases} x = \dfrac{10}{\pi}(t - \sin t) \\ y = \dfrac{10}{\pi}(1 - \cos t) \end{cases} \quad 0 \le t \le 2\pi$$

To find the area, we will first find the length L of the arch of the cycloid. Then, by imagining that we flatten the curved surface, we have a rectangle whose area is the length times the width, namely, $100L$ ft^2.

To find the length of the arch, we use Equation 9.6. To simplify calculations, we use $x = a(t - \sin t)$ and $y = a(1 - \cos t)$ and replace a by $10/\pi$ at the end. Thus,

$$\frac{dx}{dt} = a(1 - \cos t) \qquad \text{and} \qquad \frac{dy}{dt} = a \sin t$$

By Equation 9.6,

$$L = \int_0^{2\pi} \sqrt{a^2(1 - \cos t)^2 + a^2 \sin^2 t}\, dt$$

$$= a \int_0^{2\pi} \sqrt{1 - 2 \cos t + \cos^2 t + \sin^2 t}\, dt$$

$$= a \int_0^{2\pi} \sqrt{2 - 2 \cos t}\, dt$$

Now we use the half-angle formula (see Appendix 2)

$$\sin \tfrac{1}{2} t = \sqrt{\frac{1 - \cos t}{2}} \qquad (0 \le t \le 2\pi)$$

Thus,

$$\sqrt{2(1 - \cos t)} = \sqrt{4 \left(\frac{1 - \cos t}{2} \right)} = 2 \sin \tfrac{1}{2} t$$

Making this substitution, we have

$$L = 2a \int_0^{2\pi} \sin \tfrac{1}{2} t\, dt = 2a(2) \int_0^{2\pi} (\sin \tfrac{1}{2} t)(\tfrac{1}{2}\, dt)$$

$$= 4a \left[-\cos \tfrac{1}{2} t \right]_0^{2\pi} = 8a$$

Thus, since $a = 10/\pi$ ft,

$$L = \frac{80}{\pi} \text{ ft}$$

The area is therefore

$$A = 100 \left(\frac{80}{\pi} \right) = \frac{800}{\pi} \approx 254.65 \text{ ft}^2 \qquad \blacksquare$$

In using Equation 9.6 to find arc lengths, it is important that the parameter range be such that the curve is traversed only once. Particular care must be exercised when x and y are expressed as periodic functions of t. For example, the equations $x = \cos 2t$ and $y = \sin 2t$ are parametric equations of the unit circle $x^2 + y^2 = 1$. The entire circle is traced out as t ranges from 0 to π. If we carelessly took t from 0 to 2π, then Equation 9.6 would give double the correct length, since the circle would have been traversed twice.

Surface Area

If C is defined parametrically by

$$\begin{cases} x = f(t) \\ y = g(t) \end{cases} \qquad a \leq t \leq b$$

we can derive formulas for the surface area formed by revolving C about the x-axis or the y-axis, analogous to Equations 6.25 and 6.26. We again assume f' and g' are continuous on $[a, b]$. For rotation about the x-axis, we require that $g(t)$ be nonnegative, so that C lies above the x-axis. Similarly, for rotation about the y-axis, we require that $f(t)$ be nonnegative. We omit the details of the derivation since it is similar to that in Section 6.5. For rotation about the x-axis, the result is

$$S = 2\pi \int_a^b g(t) \sqrt{[f'(t)]^2 + [g'(t)]^2} \, dt$$

or, equivalently,

$$S = 2\pi \int_a^b y \sqrt{\left(\frac{dx}{dt} \right)^2 + \left(\frac{dy}{dt} \right)^2} \, dt$$

We can put the result in a simpler form by using Equation 9.5:

$$S = 2\pi \int_{t=a}^{t=b} y \, ds \qquad \text{Rotation about } x\text{-axis} \qquad (9.7)$$

Similarly, for rotation about the y-axis, we have

$$S = 2\pi \int_{t=a}^{t=b} x \, ds \qquad \text{Rotation about } y\text{-axis} \qquad (9.8)$$

To carry out the integration in Equation 9.7, we must replace y with its value $g(t)$. In Equation 9.8, x must be replaced with $f(t)$.

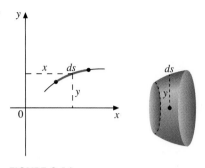

FIGURE 9.16
The rotated arc is an approximate
frustum of a cone with area $2\pi y\,ds$.

Just as we indicated in Section 6.5, an easy way to remember the integrals in Equations 9.7 and 9.8 is to think of ds as a small element of arc, as in Figure 9.16. When this element is rotated about the x-axis we get (approximately) a frustum of a cone, with surface area $2\pi y\,ds$ (average circumference $2\pi y$ times "slant height" ds). Similarly, we get $2\pi x\,ds$ on rotation about the y-axis. When we "sum" each of these areas by integration, we get Equations 9.7 and 9.8, respectively.

EXAMPLE 9.11   Find the area of the surface generated by revolving the curve C defined by

$$\begin{cases} x = t^2 - 1 \\ y = 3t \end{cases} \quad 0 \le t \le 2$$

about the x-axis. (See Figure 9.17.)

Solution   We have

$$\frac{dx}{dt} = 2t \quad \text{and} \quad \frac{dy}{dt} = 3$$

So, by Equation 9.5,

$$ds = \sqrt{\left(\frac{dx}{dt}\right)^2 + \left(\frac{dy}{dt}\right)^2}\,dt = \sqrt{4t^2 + 9}\,dt$$

and by Equation 9.7, the surface area is

$$S = 2\pi \int_{t=0}^{t=2} y\,ds = 2\pi \int_0^2 3t\sqrt{4t^2 + 9}\,dt$$

$$= \frac{6\pi}{8} \int_0^2 8t\sqrt{4t^2 + 9}\,dt \quad \text{Substitute } u = 4t^2 + 9.$$

$$= \frac{3\pi}{4}(4t^2 + 9)^{3/2} \cdot \frac{2}{3}\Big]_0^2$$

$$= \frac{\pi}{2}\left[(25)^{3/2} - 9^{3/2}\right]$$

$$= \frac{\pi}{2}[125 - 27] = \frac{\pi}{2}(98) = 49\pi \qquad \blacksquare$$

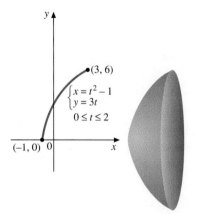

The rotated surface

FIGURE 9.17

Exercise Set 9.3

In Exercises 1–16, find the length of the curve defined by the given parametric equations over the specified interval.

1. $\begin{cases} x = 2t - 1 \\ y = 3t + 4 \end{cases} \quad 0 \le t \le 4$

2. $\begin{cases} x = 2 - t \\ y = 2t - 1 \end{cases} \quad -2 \le t \le 2$

3. $\begin{cases} x = t \\ y = t^{3/2} \end{cases} \quad 0 \le t \le 5$

4. $\begin{cases} x = \sin t \\ y = \cos t \end{cases} \quad 0 \le t \le \pi$

5. $\begin{cases} x = 4t^2 \\ y = t^3 \end{cases} \quad 0 \le t \le 2$

6. $\begin{cases} x = t + \dfrac{1}{t} \\ y = 2\ln t \end{cases}$ $1 \le t \le 3$

7. $\begin{cases} x = 3e^t \\ y = 2e^{3t/2} \end{cases}$ $\ln 3 \le t \le 2\ln 2$

8. $\begin{cases} x = \sqrt{1 - t^2} \\ y = 1 - t \end{cases}$ $0 \le t \le \dfrac{1}{2}$

9. $\begin{cases} x = e^t \sin t \\ y = e^t \cos t \end{cases}$ $0 \le t \le \ln 3$

10. $\begin{cases} x = \cos^3 t \\ y = \sin^3 t \end{cases}$ $0 \le t \le \dfrac{\pi}{2}$

11. $\begin{cases} x = \sin t - t\cos t \\ y = \cos t + t\sin t \end{cases}$ $1 \le t \le 2$

12. $\begin{cases} x = \ln\cos t \\ y = t - \tan t \end{cases}$ $0 \le t \le \dfrac{\pi}{3}$

13. $\begin{cases} x = 4\sqrt{t} \\ y = t^2 + \dfrac{1}{2t} \end{cases}$ $\dfrac{1}{2} \le t \le 2$

14. $\begin{cases} x = \tanh t \\ y = \ln(\cosh^2 t) \end{cases}$ $-2 \le t \le 3$

15. $\begin{cases} x = 2\sin t \\ y = \cos 2t \end{cases}$ $0 \le t \le \dfrac{\pi}{3}$

16. $\begin{cases} x = 3\cos t \\ y = \cos 2t \end{cases}$ $0 \le t \le \dfrac{\pi}{2}$

17. Use Formula 9.6 to show that the circumference of a circle of radius r is $2\pi r$.

18. Use Formula 9.6 to find the length of the curve with the specified parametric equations.

(a) $\begin{cases} x = \cos t \\ y = \sin t \end{cases}$ $0 \le t \le 2\pi$

(b) $\begin{cases} x = \cos 3t \\ y = \sin 3t \end{cases}$ $0 \le t \le 2\pi$

Explain the difference in the results for (a) and (b).

In Exercises 19–26, find the area of the surface formed by revolving the given curve about the specified axis.

19. $\begin{cases} x = 3t^2 - 1 \\ y = 2t \end{cases}$ $0 \le t \le \sqrt{7}$; x-axis

20. $\begin{cases} x = 6t^2 \\ y = t^4 \end{cases}$ $0 \le t \le 2$; y-axis

21. $\begin{cases} x = \sin t + 2 \\ y = \cos t - 3 \end{cases}$ $0 \le t \le \pi/2$; y-axis

22. $\begin{cases} x = \cos 2t \\ y = 4\sin t \end{cases}$ $0 \le t \le \pi/2$; x-axis

23. $\begin{cases} x = \tan^2 t \\ y = 2\sec t \end{cases}$ $0 \le t \le \pi/3$; x-axis

24. $\begin{cases} x = 6t^2 \\ y = t^4 - 2 \end{cases}$ $0 \le t \le 2$; y-axis

25. $\begin{cases} x = 4e^t - 1 \\ y = 3e^t - 1 \end{cases}$ $0 \le t \le \ln 2$; the line $x + 2 = 0$

26. $\begin{cases} x = \cos^3 t \\ y = \sin^3 t \end{cases}$ $0 \le t \le \pi/2$; the line $y + 1 = 0$

27. Find the surface area generated by revolving the arch of the cycloid

$$\begin{cases} x = a(\theta - \sin\theta) \\ y = a(1 - \cos\theta) \end{cases} \quad 0 \le \theta \le 2\pi$$

about the x-axis.

28. Find the length of the epicycloid

$$\begin{cases} x = 2a\cos t - a\cos 2t \\ y = 2a\sin t - a\sin 2t \end{cases} \quad 0 \le t \le 2\pi$$

29. For the curve defined by

$$\begin{cases} x = 1/t \\ y = \ln t \end{cases} \quad 1 \le t \le \sqrt{3}$$

find (a) its length and (b) the surface area formed by revolving the curve about the y-axis.

30. Show that the graph of

$$\begin{cases} x = t^2 - 3 \\ y = \frac{1}{3}t^3 - t \end{cases} \quad -\infty < t < \infty$$

has a loop, and find the length of the loop. Draw the graph.

31. Elliptic integrals
(a) Show that the length of the ellipse given parametrically by $x = a\cos t$, $y = b\sin t$ for $0 \le t \le 2\pi$ and $a > b > 0$ is

$$4a \int_0^{\pi/2} \sqrt{1 - e^2\sin^2 t}\, dt$$

where e is the eccentricity of the ellipse $e = \sqrt{a^2 - b^2}/a$. (See Example 6.19 and Exercise 40 in Exercise Set 8.3.)
(b) Approximate the length of the ellipse given by $x = 3\cos t$, $y = 2\sin t$ for $0 \le t \le 2\pi$ using Simpson's Rule with the number of subintervals $n = 10$.

32. Use a calculator or CAS to find the length of the **spiral of Archimedes** defined by

$$\begin{cases} x = t \cos t \\ y = t \sin t \end{cases} \quad 0 \le t \le 30$$

Sketch the spiral using a graphing calculator or a CAS.

In Exercises 33–48, sketch the graphs of the curves in Exercises 1–16 using a CAS or a graphing calculator.

9.4 THE POLAR COORDINATE SYSTEM

Although the rectangular system of coordinates we have used in all of our work so far is suitable for most graphing needs, other systems are sometimes more convenient to use. The most important alternative system is called the **polar coordinate system**. In it we take as references a fixed point O, called the **pole**, and a ray emanating from O, called the **polar axis**. It is customary to direct the polar axis horizontally to the right. If P is any point in the plane other than the pole, we indicate its location by means of the radial distance r from O to P and the measure θ of the angle from the polar axis to the ray from O through P, as shown in Figure 9.18. As usual, measures of angles resulting from a counterclockwise rotation are positive. The numbers r and θ are called the **polar coordinates** of P, and we indicate them as the ordered pair (r, θ). Although it is possible to take θ as either the degree or radian measure of the angle, we will always use radian measure. Coordinates of the pole are taken to be $(0, \theta)$ for any value of θ.

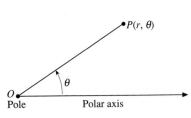

FIGURE 9.18

If a ray from the pole makes an angle α (alpha) with the polar axis, we refer to it as the **α ray**. For example, the polar axis is the 0 ray, and the $\pi/2$ ray is directed vertically upward. These correspond to the positive x- and y-axes, respectively, in a rectangular system. On a given ray, distances are measured positively in the direction of the ray. Figure 9.19 shows the points $A(3, \pi/4)$, $B(1, 2\pi/3)$, and $C(2, 3\pi/2)$. Also, we define the negative direction on a ray to be the opposite of the positive direction. More precisely, the negative direction of an α ray is in the direction of the $(\alpha + \pi)$ ray. With this understanding we can permit r to be negative, and to plot (r, θ) for r negative, we measure $|r|$ units in the negative direction of the θ ray. In Figure 9.20 we have plotted $P(-2, 0)$ and $Q(-3, 4\pi/3)$.

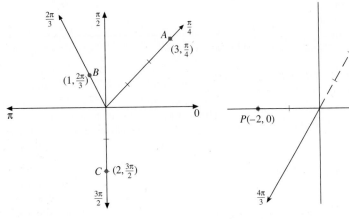

FIGURE 9.19   **FIGURE 9.20**

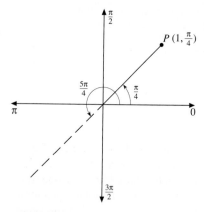

FIGURE 9.21

Point P has polar coordinates $(1, \frac{\pi}{4} + 2n\pi)$ and also $(-1, \frac{5\pi}{4} + 2n\pi)$ for $n = 0, \pm 1, \pm 2, \ldots$.

When we are given polar coordinates (r, θ), the point P represented by them is uniquely determined. Unfortunately, the converse is not true. If we are given a point P, there are infinitely many sets of polar coordinates for it, in contrast with the rectangular system, where there is a one-to-one correspondence between points in the plane and the set of all ordered pairs (x, y). You should convince yourself that $(1, \pi/4)$, $(1, 9\pi/4)$, $(1, -7\pi/4)$, $(-1, 5\pi/4)$, and $(-1, -3\pi/4)$ are all polar coordinates of the same point. In fact, $(1, \pi/4 + 2n\pi)$ and $(-1, 5\pi/4 + 2n\pi)$ for $n = 0, \pm 1, \pm 2, \ldots$ represent this point. (See Figure 9.21.) This lack of uniqueness of representation is not a serious drawback to the polar system, but it does mean that special care must be exercised when analyzing polar graphs or finding points of intersection of two curves, for example. Although two different ordered pairs may represent the same point, they are not equal as ordered pairs. For example, $(1, \pi/4)$ and $(-1, 5\pi/4)$ are coordinates of the same point, but $(1, \pi/4) \neq (-1, 5\pi/4)$, since by equality of ordered pairs we mean that first coordinates are equal and second coordinates are equal.

Change of Coordinates

It is frequently useful to change from the polar coordinate system to the rectangular system, or vice versa. To see how to make these changes, superimpose a rectangular system onto a polar system so that the origin coincides with the pole, and the positive x-axis coincides with the polar axis. Let P be any point other than the origin. Then P has unique rectangular coordinates (x, y). As in Figure 9.22, let (r, θ) be polar coordinates of P for which $r > 0$ and θ is between 0 and 2π. Then from the definitions of $\sin\theta$ and $\cos\theta$, we have $\cos\theta = x/r$ and $\sin\theta = y/r$, or, equivalently, the following equations.

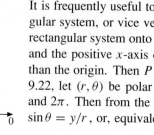

FIGURE 9.22

> ### Polar Coordinates to Rectangular Coordinates
>
> $$x = r\cos\theta$$
> $$y = r\sin\theta$$
>
> (9.9)

Adding a multiple of 2π to the angle θ does not alter $\sin\theta$ or $\cos\theta$, so when $r > 0$, any equivalent polar representation of the point P does not change x or y. The fact that we have shown P in the first quadrant is not significant. Equations 9.9 are true wherever P is located. It can also be shown that Equations 9.9 hold true when $r < 0$. (See Exercise 32 in Exercise Set 9.4.) They are true as well when $r = 0$, since for any value of θ, the point $(0, \theta)$ is the pole, or equivalently, the origin.

If we square both sides of Equations 9.9 and add, we get $r^2 = x^2 + y^2$, and if $x \neq 0$, on dividing the second equation by the first, we get $\tan\theta = y/x$. Thus, equations for r and θ in terms of x and y are as follows.

Rectangular Coordinates to Polar Coordinates

$$r^2 = x^2 + y^2$$

$$\tan\theta = \frac{y}{x} \qquad (x \neq 0)$$

(9.10)

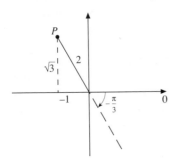

(a) If we choose $\theta = \frac{2\pi}{3}$ then $r = 2$.
Polar coordinates of P are $(2, \frac{2\pi}{3})$.

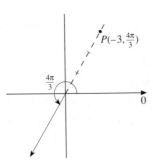

(b) If we choose $\theta = -\frac{\pi}{3}$ then $r = -2$.
Polar coordinates of P are $(-2, -\frac{\pi}{3})$.

FIGURE 9.23

To change from polar coordinates to rectangular coordinates, we use Equations 9.9 and get an unambiguous answer. To change from rectangular coordinates to polar coordinates, we use Equations 9.10, taking account of the quadrant in which the point lies and choosing $r = \sqrt{x^2 + y^2}$ or $-\sqrt{x^2 + y^2}$, depending on which value of θ is chosen. For example, if the rectangular coordinates of P are $(-1, \sqrt{3})$, then by Equations 9.10,

$$r^2 = (-1)^2 + (\sqrt{3})^2 = 4$$

and $\tan\theta = -\sqrt{3}$. If we take $\theta = 2\pi/3$, then $r = \sqrt{4} = 2$, whereas if we take $\theta = -\pi/3$, then we must choose $r = -\sqrt{4} = -2$. We illustrate this example in Figure 9.23. It will always work to take $r > 0$ if we choose θ as a positive angle from the polar axis to the ray from O through P, where P is the point with rectangular coordinates (x, y). There are times, however, when we might prefer some other polar representation of P, consistent with Equations 9.10. We illustrate these ideas in the next example.

EXAMPLE 9.12

(a) Change the polar coordinates $(-3, 4\pi/3)$ to rectangular coordinates. Plot the point.

(b) Give three sets of polar coordinates for the point P whose rectangular coordinates are $(\sqrt{3}, -1)$. For one set choose $r > 0$ and $\theta > 0$, and for the other two choose r and θ opposite in sign.

Solution

(a) The point P with polar coordinates $(-3, 4\pi/3)$ is plotted in Figure 9.24. Note that a simpler set of coordinates for P is $(3, \pi/3)$, but it is not necessary to make this change. By Equations 9.9, we have

$$x = r\cos\theta = (-3)\cos\frac{4\pi}{3} = (-3)\left(-\frac{1}{2}\right) = \frac{3}{2}$$

$$y = r\sin\theta = (-3)\sin\frac{4\pi}{3} = (-3)\left(-\frac{\sqrt{3}}{2}\right) = \frac{3\sqrt{3}}{2}$$

(b) The point P is plotted in Figure 9.25. From Equations 9.10 we have

$$r^2 = x^2 + y^2 = 3 + 1 = 4$$

From our discussion following Equations 9.10, we will choose $r = 2$ or $r = -2$, depending on which value of θ we choose. Since

$$\tan\theta = \frac{y}{x} = \frac{-1}{\sqrt{3}}$$

FIGURE 9.24

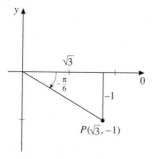

FIGURE 9.25

from our knowledge of special angles in trigonometry, we see that θ has a reference angle of $30°$, or $\pi/6$ radians. Since x is positive and y is negative, P is in the fourth quadrant. By choosing $\theta = 11\pi/6$ or $\theta = -\pi/6$, we take $r = 2$. By choosing $\theta = 5\pi/6$ (or any angle coterminal with it), we take $r = -2$. So three sets of polar coordinates for P can be taken as follows:

$$
\begin{array}{ccc}
(2, 11\pi/6) & (2, -\pi/6) & (-2, 5\pi/6) \\
r > 0,\ \theta > 0 & r > 0,\ \theta < 0 & r < 0,\ \theta > 0
\end{array}
$$

The second of these is probably the simplest form in this case. ■

Equations in Polar Coordinates

A **polar equation** is an equation whose only variables are the polar variables r and θ. For example, each of the following is a polar equation.

$$r = 1 - 2\cos\theta \qquad r^2 = \sin 2\theta \qquad r = 2 \qquad \theta = \pi/4$$

By the **graph** of a polar equation, we mean the set of all points P that have *at least one* set of coordinates (r, θ) that satisfy the equation. We will consider general graphing techniques in the next section. For now, we observe that graphs of polar equations can sometimes be more easily obtained by finding a rectangular equation that has the same graph. On the other hand, complicated rectangular equations can sometimes be graphed more easily using an equivalent polar equation. Thus, it is desirable to be able to change an equation in one system to an equation in the other system that has the same graph. Equations 9.9 and 9.10 enable us to make such changes. (We should note, however, that a CAS or graphing calculator can handle both rectangular and polar equations in most cases.)

In the examples and exercises that follow, when we ask for a polar equation to be transformed to an equivalent rectangular equation, or vice versa, we mean that the resulting equation will have the same graph as the given equation.

EXAMPLE 9.13

(a) Transform $r = 2\sin\theta$ to an equivalent rectangular equation. Identify and draw the graph.

(b) Transform the equation $(x^2+y^2)^2 = x^2 - y^2$ to an equivalent polar equation.

Solution

(a) Since $r = \sqrt{x^2 + y^2}$, we could substitute this value for r, but we can avoid the square root by multiplying both sides of the original equation by r. We get $r^2 = 2r\sin\theta$. Then since $r^2 = x^2 + y^2$ and $r\sin\theta = y$, we have the rectangular equation

$$x^2 + y^2 = 2y$$

The graph of this equation is a circle. By completing the square in y, we can put it in the standard form $x^2 + (y-1)^2 = 1$. This form is the equation of a circle with center $(0, 1)$ and radius 1. (See Figure 9.26.)

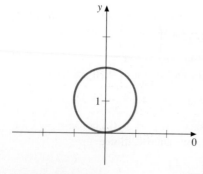

FIGURE 9.26

We should note that multiplication by r in this procedure did not change the graph, since the only possible new point it could have introduced would have been the pole, $r = 0$. We can see the reason for this result by writing $r^2 = 2r \sin \theta$ in the form

$$r^2 - 2r \sin \theta = 0$$

So

$$r(r - 2 \sin \theta) = 0$$

That is, either $r = 0$ or $r = 2 \sin \theta$. But the graph of $r = 0$ is just the pole, and the pole is on the graph of the original equation, $r = 2 \sin \theta$, since when $\theta = 0$, $r = 0$.

(b) Substitute for x and y from Equations 9.9 and use $x^2 + y^2 = r^2$ to get

$$r^4 = r^2 \cos^2 \theta - r^2 \sin^2 \theta$$

or, since $\cos 2\theta = \cos^2 \theta - \sin^2 \theta$ (see Appendix 2),

$$r^4 = r^2 \cos 2\theta$$

Rewriting this equation as $r^4 - r^2 \cos 2\theta = 0$, we can factor to get $r^2(r^2 - \cos 2\theta) = 0$. So the graph consists of all points with coordinates that satisfy $r^2 = 0$ (the pole) or $r^2 = \cos 2\theta$. But the pole is on the graph of $r^2 = \cos 2\theta$, since when $\theta = \pi/4$, $r = 0$. Thus, the equation $r^2 = \cos 2\theta$ is equivalent to the original equation. We will see in the next section how to draw the graph using its polar equation. ∎

Exercise Set 9.4

In Exercises 1 and 2, plot the points in a polar coordinate system.

1. (a) $(2, \pi)$

(b) $\left(3, -\dfrac{\pi}{2}\right)$

(c) $\left(-1, \dfrac{\pi}{4}\right)$

(d) $\left(0, \dfrac{5\pi}{9}\right)$

2. (a) $\left(-2, -\dfrac{\pi}{3}\right)$

(b) $\left(4, \dfrac{7\pi}{2}\right)$

(c) $(2, -5\pi)$

(d) $\left(-3, -\dfrac{5\pi}{6}\right)$

3. Show the approximate location of the points that have the following polar coordinates (r, θ), where θ is in radians.

(a) $(1, 2)$

(b) $(2, 1)$

(c) $(-5, 8)$

(d) $(-3, -4)$

4. Let P be the point determined by the given polar coordinates. Determine two other pairs of polar coordinates for P with $r > 0$ and two with $r < 0$.

(a) $\left(2, \dfrac{\pi}{4}\right)$

(b) $\left(-1, -\dfrac{\pi}{2}\right)$

(c) $\left(-3, \dfrac{4\pi}{5}\right)$

(d) $\left(4, -\dfrac{2\pi}{9}\right)$

In Exercises 5 and 6, change the given polar coordinates to equivalent rectangular coordinates.

5. (a) $\left(2, \dfrac{5\pi}{6}\right)$

(b) $\left(-3, \dfrac{3\pi}{2}\right)$

(c) $\left(0, \dfrac{17\pi}{13}\right)$

(d) $\left(-2, -\dfrac{3\pi}{4}\right)$

6. (a) $(3, 0)$

(b) $\left(4, -\dfrac{11\pi}{6}\right)$

(c) $(-2, 5\pi)$

(d) $\left(7, \dfrac{4\pi}{3}\right)$

In Exercises 7 and 8, change the given rectangular coordinates to an equivalent pair of polar coordinates in which r > 0 and $0 \le \theta < 2\pi$.

7. (a) $(4, 0)$
(b) $(0, 4)$
(c) $(-4, 0)$
(d) $(0, -4)$

8. (a) $(1, -\sqrt{3})$
(b) $(-5, 5)$
(c) $(-4\sqrt{3}, 4)$
(d) $(-2\sqrt{2}, -2\sqrt{2})$

In Exercises 9–14, find an equivalent rectangular equation. Also identify and draw the graph.

9. (a) $r = 2$
(b) $\theta = \dfrac{\pi}{4}$

10. (a) $r = -2$
(b) $\theta = -\dfrac{\pi}{4}$

11. (a) $r \cos \theta = 3$
(b) $r \sin \theta = -1$

12. (a) $r = -2 \csc \theta$
(b) $r = 3 \sec \theta$

13. (a) $2r \cos \theta - 3r \sin \theta = 4$
(b) $r(3 \cos \theta + 2 \sin \theta) = 5$

14. (a) $r = 2 \cos \theta$
(b) $r = -2 \sin \theta$

In Exercises 15–22, find an equivalent rectangular equation.

15. $r = a(1 + \cos \theta)$

16. $r^2 = 2a \cos 2\theta$

17. $r = 1 - 2 \cos \theta$

18. $r = 2 + \sin \theta$

19. $r = a \cos 2\theta$

20. $r = a \sin 2\theta$

21. $r = \dfrac{2}{1 - \cos \theta}$

22. $r = \dfrac{1}{1 + 2 \sin \theta}$

In Exercises 23–31, find an equivalent polar equation.

23. (a) $x^2 + y^2 = a^2$
(b) $y = \sqrt{3}\, x$

24. (a) $x + y = 0$
(b) $2x - 3y = 4$

25. (a) $y + 1 = 0$
(b) $x - 3 = 0$

26. $x^2 - y^2 = 4$

27. $y = x^2$

28. $x^2 + y^2 = 2x$

29. $x^2 - y^2 - 4x = 0$

30. $x^2 y^2 = 4(x^2 + y^2)^{3/2}$

31. $x^2 - y^2 = \dfrac{2xy}{x^2 + y^2}$

32. Prove that Equations 9.9 hold true when $r < 0$.

33. Prove that if P has rectangular coordinates (x, y) with $x \ne 0$, then as one set of coordinates (r, θ) for P we may take $\theta = \tan^{-1}(y/x)$ where $r = \sqrt{x^2 + y^2}$ if $x > 0$, and $r = -\sqrt{x^2 + y^2}$ if $x < 0$. (Recall that $-\pi/2 < \tan^{-1}(y/x) < \pi/2$.)

In Exercises 34–36, show that an equivalent polar equation is of the form $r = ed/(1 + e \cos \theta)$ for suitable values of d and e.

34. $y^2 = 1 - 2x$

35. $3x^2 + 4y^2 + 4x = 4$

36. $3x^2 - y^2 - 4x + 1 = 0$

37. By means of an example, show that a point P can lie on the graph of a polar equation even though a particular set of coordinates for P does not satisfy the equation.

9.5 GRAPHING POLAR EQUATIONS

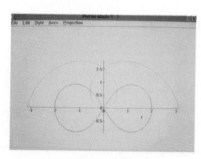

Maple plot of two polar graphs

As we have seen, the graph of a polar equation can sometimes be obtained by changing to an equivalent rectangular equation. Many interesting curves, however, have rather complicated rectangular equations, so it is important to be able to draw the graph directly from the polar equation. By substituting convenient values of θ and solving the equation for r, we can obtain a table of values from which to draw a reasonable approximation of the graph. This "point plotting" method is not entirely satisfactory, though, since it is not always clear how to connect the points unless a great many points are used (whether in polar or in rectangular coordinates). Some additional analysis is usually needed. Computer algebra systems and graphing calculators can draw graphs of equations in polar form. At the end of this section we will describe how Maple, Mathematica, and DERIVE do such plotting.

The simplest polar equations to graph are of the form

$$\text{(a) } r = a \quad \text{and} \quad \text{(b) } \theta = \alpha$$

where a and α are constants. (a) The first equation, $r = a$, puts no restriction on θ, so its graph consists of points (a, θ) where θ takes on all real values.

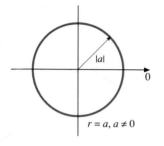

$r = a, a \neq 0$

(a)

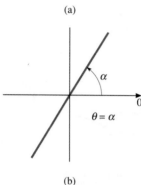

$\theta = \alpha$

(b)

FIGURE 9.27

Since a is a constant, the graph is a circle of radius $|a|$ if $a \neq 0$ and is the pole if $a = 0$. (b) The equation $\theta = \alpha$ places no restriction on r, and so its graph consists of all points of the form (r, α) where r takes on all real values. Thus, its graph is the line consisting of the α ray together with its extension in the opposite direction. Graphs of these equations are shown in Figure 9.27. For more complicated equations we want to consider several aids to graphing that can reduce reliance on plotting a large number of points.

Symmetry

We consider three types of symmetry: (a) symmetry with respect to the line $\theta = 0$ (x-axis, in rectangular coordinates), (b) symmetry with respect to the line $\theta = \pi/2$ (y-axis, in rectangular coordinates), and (c) symmetry with respect to the pole (origin). As Figure 9.28 indicates, in each case symmetry will exist if, whenever $P(r, \theta)$ is on the graph, the point P' as indicated is also on the graph. Since each point P' has many equivalent representations, devising a satisfactory test for symmetry is complicated. The following is perhaps the simplest test, but it should be emphasized that it provides *sufficient* conditions only (that is, if the conditions are satisfied, the curve will have the given type of symmetry, but the curve *may* have one or more of the symmetries even if the conditions are not satisfied). In Exercise 35 of Exercise Set 9.5 you will be asked to give a more general test, providing conditions that are both necessary and sufficient (by considering all possible polar representations of the points P' in Figure 9.28).

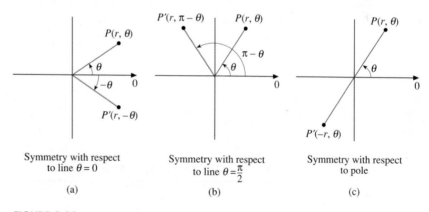

Symmetry with respect to line $\theta = 0$

(a)

Symmetry with respect to line $\theta = \dfrac{\pi}{2}$

(b)

Symmetry with respect to pole

(c)

FIGURE 9.28

Tests for Symmetry

(a) If replacing θ with $-\theta$ results in an equivalent equation, the graph is symmetric with respect to the line $\theta = 0$ (the x-axis).

(b) If replacing θ with $\pi - \theta$ results in an equivalent equation, the graph is symmetric with respect to the line $\theta = \pi/2$ (the y-axis).

(c) If replacing r with $-r$ results in an equivalent equation, the graph is symmetric with respect to the pole (the origin).

■ If a curve has any two of the three types of symmetry in (a), (b), or (c), it also has the third. You can convince yourself of this fact geometrically by combining any two of the types.

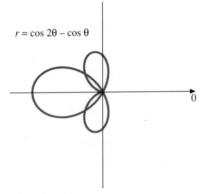

$r = \cos 2\theta - \cos \theta$

FIGURE 9.29
Symmetry with respect to the line $\theta = 0$

To illustrate Test (a), consider the equation

$$r = \cos 2\theta - \cos \theta$$

If we replace θ with $-\theta$, we get

$$r = \cos(-2\theta) - \cos(-\theta)$$

But the cosine is an even function, so $\cos(-2\theta) = \cos 2\theta$ and $\cos(-\theta) = \cos \theta$. Thus, we obtain an equivalent equation, and we conclude that the graph is symmetric with respect to the line $\theta = 0$. The CAS-generated graph in Figure 9.29 illustrates this symmetry.

To illustrate Test (b), consider the equation

$$r = \sin^3 \theta$$

Replacing θ with $\pi - \theta$ gives $r = \sin^3(\pi - \theta)$. But $\sin(\pi - \theta) = \sin \theta$. So we get an equivalent equation and conclude that the graph is symmetric with respect to the line $\theta = \pi/2$. We show the CAS-generated graph in Figure 9.30.

To illustrate Test (c), consider the equation

$$r^2 = e^\theta$$

Clearly, replacing r with $-r$ yields an equivalent equation, so the graph (shown in Figure 9.31) is symmetric with respect to the pole.

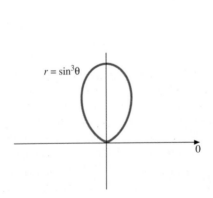

$r = \sin^3\theta$

FIGURE 9.30
Symmetry with respect to the line $\theta = \frac{\pi}{2}$

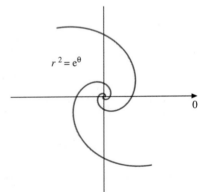

$r^2 = e^\theta$

FIGURE 9.31
Symmetry with respect to the pole

To illustrate that the conditions stated in the test are sufficient only, and not necessary for symmetry, consider the equation

$$r = \sqrt{\sin \frac{\theta}{2}}$$

Since $\sin(-\theta/2) = -\sin(\theta/2)$, Test (a) fails. Also, $\sin((\pi - \theta)/2) = \sin((\pi/2) - (\theta/2)) = \cos(\theta/2)$, so Test (b) fails. Finally, replacing r with $-r$ gives $-r = \sqrt{\sin(\theta/2)}$, or $r = -\sqrt{\sin(\theta/2)}$. Thus, Test (c) fails. Yet the

CAS-generated graph in Figure 9.32 clearly shows symmetry with respect to the line $\theta = 0$. To see that we do, in fact, have this type of symmetry, we can use the coordinates $(r, 2\pi - \theta)$ for P' in Figure 9.28(a). Thus, if we can show that replacing θ with $2\pi - \theta$ leads to an equivalent equation, we will be able to conclude that we have symmetry with respect to the line $\theta = 0$. Since

$$\sin\left(\frac{2\pi - \theta}{2}\right) = \sin\left(\pi - \frac{\theta}{2}\right) = \sin\frac{\theta}{2}$$

we do get an equivalent equation.

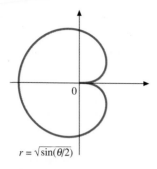

$r = \sqrt{\sin(\theta/2)}$

FIGURE 9.32
Symmetry with respect to the line $\theta = 0$.

Intercepts

The intercepts on what would correspond in a rectangular system to the x- and y-axes are almost always key points in drawing the graph. These are found by setting $\theta = 0$, $\pi/2$, π, and $3\pi/2$ and then solving for r.

Periodicity

Since many polar equations involve $\sin\theta$ or $\cos\theta$, the fact that each of these has period 2π can limit the domain for θ that needs to be considered. For example, in graphing $r = 1 + \cos\theta$ it is sufficient to consider values of θ that satisfy $-\pi < \theta \le \pi$. In fact, since the graph is symmetric with respect to the line $\theta = 0$, we need only consider θ in the range $0 \le \theta \le \pi$. The graph for $-\pi < \theta \le 0$ is then found by reflecting through the line $\theta = 0$.

Tangents at the Origin

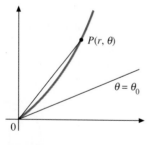

FIGURE 9.33

If the graph has one or more tangent lines at the origin, these can be found by setting $r = 0$ and solving for θ. We can reason as follows to see why. In Figure 9.33, the line OP that joins the pole O and a point $P(r, \theta)$ on the graph is a secant line. Now suppose that when $\theta = \theta_0$, $r = 0$. If we let $\theta \to \theta_0$, the secant line OP approaches the tangent line at O. The slope of OP is $\tan\theta$, which approaches $\tan\theta_0$ as $\theta \to \theta_0$. So the tangent line at O is the line with slope $\tan\theta_0$—namely, $\theta = \theta_0$. This reasoning will be made more precise in the next section, but we will use the result now, since it is often helpful in graphing.

To illustrate, consider the equation $r = 2 - 4\sin\theta$. Setting $r = 0$, we get $\sin\theta = 1/2$, so the lines $\theta = \pi/6$ and $\theta = 5\pi/6$ are tangent to the graph at the origin.

The next example uses these graphing aids.

EXAMPLE 9.14   Draw the graph of the equation $r = 2\cos\theta$.

Solution   We will consider the four aids to graphing discussed above.

Symmetry. Since $\cos(-\theta) = \cos\theta$, the equation is unchanged when θ is replaced with $-\theta$. So we have symmetry with respect to the line $\theta = 0$. Because of this symmetry and the periodicity of $\cos\theta$, we need only consider values of θ in the interval $[0, \pi]$. When we get the graph on this interval, we can flip it over (i.e., reflect it through) the line $\theta = 0$ (x-axis) to get the entire graph.

Intercepts. The intercepts on the rays $\theta = 0$, $\theta = \pi/2$, and $\theta = \pi$ are $r = 2$, $r = 0$, and $r = -2$, respectively. So the corresponding points are $(2, 0)$, $(0, \pi/2)$, and $(-2, \pi)$. Note that the points $(-2, \pi)$ and $(2, 0)$ are the same.

Periodicity. As already noted, $\cos\theta$ has period 2π, so taking values of θ between 0 and 2π is sufficient to give the entire graph. (We saw that by symmetry we can actually limit θ to be between 0 and π.)

Tangents at the Origin. Setting $r = 0$ yields $\cos\theta = 0$, so $\theta = \pi/2$. The curve is therefore tangent to the line $\theta = \pi/2$ (the y-axis) at the origin.

Finally, we make a table, choosing some convenient values of θ between 0 and π, and we plot the corresponding points.

θ	$\dfrac{\pi}{6}$	$\dfrac{\pi}{4}$	$\dfrac{\pi}{3}$	$\dfrac{2\pi}{3}$	$\dfrac{3\pi}{4}$	$\dfrac{5\pi}{6}$
r	$\sqrt{3}$	$\sqrt{2}$	1	-1	$-\sqrt{2}$	$-\sqrt{3}$

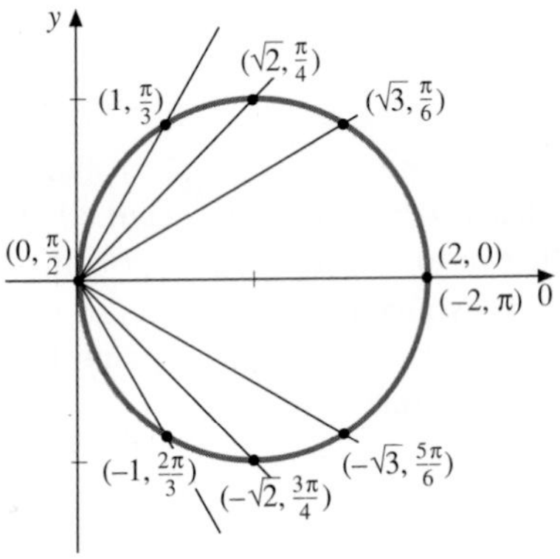

We show the graph in Figure 9.34. Notice that the entire graph is traced out for θ varying from 0 to π. The graph appears to be a circle. In fact, we can show that it is a circle by changing to rectangular coordinates as we did in Example 9.13(a). We first multiply by r, getting $r^2 = 2r\cos\theta$. Since $r^2 = x^2 + y^2$ and $r\cos\theta = x$, we have

$$x^2 + y^2 = 2x$$

and on completing the square in x,

$$(x - 1)^2 + y^2 = 1$$

We recognize the graph of this equation as a circle of radius 1 centered at $(1, 0)$. ∎

FIGURE 9.34
The circle $r = 2\cos\theta$

The equations in Examples 9.13(a) and 9.14 are special cases of the circles illustrated in Figure 9.35.

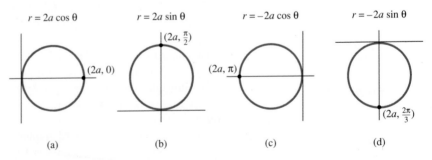

FIGURE 9.35
Family of circles $r = \pm 2a\cos\theta$ and $r = \pm 2a\sin\theta$

In the next four examples we illustrate particular cases of some well-known polar graphs that have special names. These curves are more easily analyzed from their polar equations than from their rectangular equations.

EXAMPLE 9.15 **A cardioid** Draw the graph of the equation $r = 1 + \sin\theta$.

Solution Since $\sin(\pi - \theta) = \sin\theta$, the graph is symmetric with respect to the line $\theta = \pi/2$. Because of this symmetry and the periodicity of $\sin\theta$, it is sufficient to consider θ in the interval $[-\pi/2, \pi/2]$. The key intercepts in this interval are $(0, -\pi/2)$, $(1, 0)$, and $(2, \pi/2)$. Setting $r = 0$ yields $\sin\theta = -1$, so the tangent line at the origin is $\theta = -\pi/2$. Additional selected points are found as shown in the table.

θ	$-\pi/3$	$-\pi/4$	$-\pi/6$	$\pi/6$	$\pi/4$	$\pi/3$
r	$1 - \frac{\sqrt{3}}{2} \approx 0.13$	$1 - \frac{\sqrt{2}}{2} \approx 0.29$	$1 - \frac{1}{2} = 0.50$	$1 + \frac{1}{2} = 1.50$	$1 + \frac{\sqrt{2}}{2} \approx 1.71$	$1 + \frac{\sqrt{3}}{2} \approx 1.87$

The graph can now be drawn with reasonable accuracy, as shown in Figure 9.36.

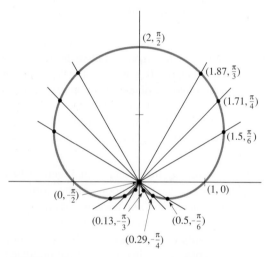

FIGURE 9.36
The cardioid $r = 1 + \sin\theta$

The graph in Figure 9.36 is called a **cardioid** because of its heartlike shape. It is a member of a family of curves called **limaçons** (pronounced lim-a-sons). The three types of limaçons are shown in Figure 9.37, along with their standard

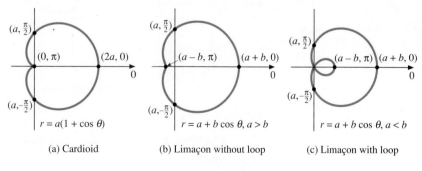

(a) Cardioid   (b) Limaçon without loop   (c) Limaçon with loop

(If $a > 2b$, the "dimple" on the left is absent.)

FIGURE 9.37
Family of limaçons $r = a + b\cos\theta$

equations when the axis of symmetry is horizontal. When the axis of symmetry is vertical, $\cos\theta$ is replaced with $\sin\theta$. The curves shown are for positive a and b. If b is negative and a is positive, the graph is rotated $180°$. If a is negative and b positive, the graph is unchanged, since when a point P with coordinates (r, θ) is on the graph of $r = a + b\cos\theta$, the same point with coordinates $(-r, \theta + \pi)$ is on the graph of $r = -a + b\cos\theta$. (See Exercise 25 in Exercise Set 9.5.)

EXAMPLE 9.16   **A lemniscate**   Draw the graph of the equation $r^2 = 4\cos 2\theta$.

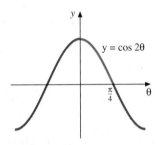

FIGURE 9.38
$\cos 2\theta \geq 0$ when $0 \leq \theta \leq \pi/4$.

Solution   The graph is symmetric with respect to the line $\theta = 0$ and with respect to the pole, as can be seen by replacing, in turn, θ with $-\theta$ and r with $-r$. From these two symmetries, it follows that the graph is also symmetric with respect to the vertical line $\theta = \pi/2$. Since the period of $\cos 2\theta$ is π, it is sufficient to consider θ in the interval $[0, \pi]$. Solving for r yields $r = \pm 2\sqrt{\cos 2\theta}$, and we see that θ must be restricted to satisfy $\cos 2\theta \geq 0$ in order for the square root to be real. Thus, we must have $0 \leq 2\theta \leq \pi/2$, or equivalently, $0 \leq \theta \leq \pi/4$. (See Figure 9.38.) The tangent lines at the origin are solutions of $\cos 2\theta = 0$—namely, $\theta = \pi/4$ and $\theta = -\pi/4$. Using the intercepts $(2, 0)$ and $(0, \pi/4)$, together with the one additional point $(\sqrt{2}, \pi/6)$, we can now draw the portion of the graph shown in Figure 9.39(a). Then, by symmetry, we get the entire graph, shown in Figure 9.39(b). ∎

The "figure 8" curve in Figure 9.39 is called a **lemniscate**. These curves typically have equations of the form $r^2 = a^2 \cos 2\theta$ or $r^2 = a^2 \sin 2\theta$. The graphs of these two basic forms are shown in Figure 9.40. Note that when $\cos 2\theta$ is replaced by $\sin 2\theta$, the graph is rotated through an angle of $\pi/4$ radians. In Exercise 39 of Exercise Set 9.5 you will be asked to show in general that if $\cos n\theta$ is replaced by $\sin n\theta$, where the original equation is of the form $r = f(\cos n\theta)$, then the graph is rotated through an angle of $\pi/(2n)$ radians.

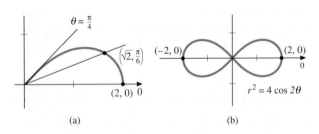

FIGURE 9.39
The lemniscate $r^2 = 4\cos 2\theta$

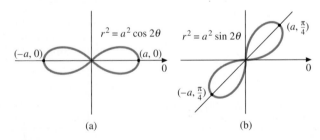

FIGURE 9.40
Lemniscates

EXAMPLE 9.17   **A four-leaf rose**   Draw the graph of the equation $r = \sin 2\theta$.

Solution   To test for symmetry, we use alternative tests given in Exercise 36 of Exercise Set 9.5. If we replace r with $-r$ and θ with $-\theta$, we get $-r = \sin(-2\theta)$, or $-r = -\sin 2\theta$, which is equivalent to $r = \sin 2\theta$. Since the

θ	r
0	0
$\pi/6$	$\sqrt{3}/2$
$\pi/4$	1
$\pi/3$	$\sqrt{3}/2$
$\pi/2$	0

points $P(r,\theta)$ and $P'(-r,-\theta)$ are symmetrically placed with respect to the line $\theta = \pi/2$, it follows that the graph is symmetric with respect to this line. Also, $\sin 2(\pi + \theta) = \sin(2\pi + 2\theta) = \sin 2\theta$, and since the points $P(r,\theta)$ and $P'(r, \pi + \theta)$ are symmetrically placed with respect to the pole, it follows that the graph is symmetric with respect to the pole. Thus, it is also symmetric with respect to the line $\theta = 0$. The period of $\sin 2\theta$ is π and so, because of symmetry, we can limit our consideration of θ to values between 0 and $\pi/2$. Tangent lines at the origin are $\theta = 0$ and $\theta = \pi/2$, found by solving $\sin 2\theta = 0$. We list some points in the table and show the graph of the portion of the curve for θ in $[0, \pi/2]$ in Figure 9.41(a). We then use symmetry to get the complete graph, shown in Figure 9.41(b). ∎

The graph in Figure 9.41(b) is called a **four-leaf rose**. Standard equations of rose curves are $r = a \cos n\theta$ and $r = a \sin n\theta$. When n is even, there are $2n$ petals, as in Example 9.17, whereas when n is odd, there are n petals. These two cases are illustrated in Figure 9.42, using $r = a \cos 2\theta$ and $r = a \cos 3\theta$.

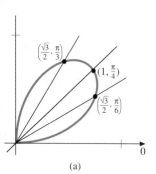
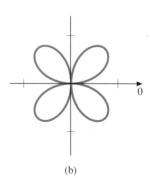

(a)   (b)

FIGURE 9.41
The four-leaf rose $r = \sin 2\theta$

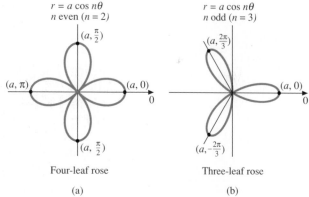

Four-leaf rose   Three-leaf rose

(a)   (b)

FIGURE 9.42
Rose curves

A nautilus shell has the shape of a logarithmic spiral

EXAMPLE 9.18   A logarithmic spiral Draw the graph of the equation $r = e^{\theta/4}$.

Solution   The graph has no symmetry, and $e^{\theta/4}$ is not periodic. Nor can r ever be 0, so there are no tangent lines at the origin. We see, in fact, from the properties of the exponential function, that $r > 0$ for all values of θ, and that r increases as θ increases. Furthermore,

$$\lim_{\theta \to -\infty} r = 0 \quad \text{and} \quad \lim_{\theta \to \infty} r = \infty$$

There are infinitely many intercepts on every ray since when θ increases by 2π, the point is on the same ray, but the value of r is increased. We show the graph in Figure 9.43. Points can be found using a calculator.   ■

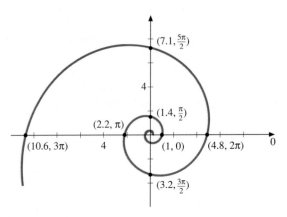

FIGURE 9.43
The logarithmic spiral $r = e^{\theta/4}$

The graph in Figure 9.43 is called a **logarithmic spiral**. There are other types of spirals as well. You will be asked to draw some of these in the exercises.

Polar Equations in Parametric Form

For the purpose of graphing a polar equation on some graphing calculators it is necessary to convert the polar equation to a pair of parametric equations. We will also see in Section 9.7 how the parametric representation is useful in deriving a formula for the length of a curve in polar coordinates.

Formulas 9.9, which involve polar to rectangular conversion,

$$x = r \cos\theta$$
$$y = r \sin\theta$$

provide a means of changing a polar equation

$$r = f(\theta), \qquad \alpha \le \theta \le \beta$$

to parametric form. We simply replace r with $f(\theta)$ to get

$$\begin{cases} x = f(\theta)\cos\theta \\ y = f(\theta)\sin\theta \end{cases} \qquad \alpha \le \theta \le \beta$$

To illustrate this conversion, parametric equations of the cardioid

$$r = 1 + \cos\theta, \qquad 0 \le \theta \le 2\pi$$

are

$$\begin{cases} x = (1 + \cos\theta)\cos\theta \\ y = (1 + \cos\theta)\sin\theta \end{cases} \qquad 0 \le \theta \le 2\pi$$

Intersections of Graphs in Polar Coordinates

In the next two examples we show how to find points of intersection of two polar graphs. The second example illustrates a problem that can arise because the coordinates of a given point are not unique.

EXAMPLE 9.19   Find the points of intersection of the curves

$$r = 2 + \sin\theta \qquad \text{and} \qquad r = 5\sin\theta$$

Show the graphs and label the points of intersection.

Solution   We set the two r values equal to one another and solve for θ.

$$2 + \sin\theta = 5\sin\theta$$

$$4\sin\theta = 2$$

$$\sin\theta = \frac{1}{2}$$

So $\theta = \pi/6$ or $5\pi/6$. Substituting these values into either of the original equations gives $r = 5/2$. The points of intersection are therefore $(5/2, \pi/6)$ and $(5/2, 5\pi/6)$.

We show the graphs in Figure 9.44. The first curve is a limaçon without a loop, and the second is a circle.   ■

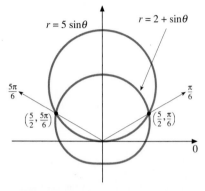

FIGURE 9.44

EXAMPLE 9.20   Find the points of intersection of the curves $r = 2 + 2\cos\theta$ and $r = -2\cos\theta$. Show the curves graphically.

Solution   Proceeding as in Example 9.19, we set the r values equal to one another and solve for θ:

$$2 + 2\cos\theta = -2\cos\theta$$

$$4\cos\theta = -2$$

$$\cos\theta = -\frac{1}{2}$$

Thus, $\theta = \pm 2\pi/3$ and $r = 1$. So we find the points of intersection $(1, 2\pi/3)$ and $(1, -2\pi/3)$. But as the graphs (Figure 9.45) clearly show, the pole is also a point of intersection.

Why did we fail to find this third point of intersection when we solved the equations simultaneously? The answer lies in the fact that the cardioid $r = 2 + 2\cos\theta$ goes through the pole when $\cos\theta = -1$, namely, when $\theta = \pi$. But the circle $r = -2\cos\theta$ goes through the pole when $\cos\theta = 0$, namely, when $\theta = \pi/2$. So the first curve goes through the pole because $(0, \pi)$ is on its graph, whereas the second curve goes through the pole because $(0, \pi/2)$ is on the graph. The points are the same (the pole) but the coordinates are different.   ■

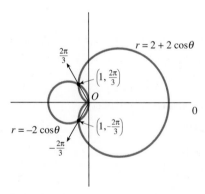

FIGURE 9.45
The pole is a point of intersection.

As the preceding example shows, there can be points of intersection of polar graphs that do not show up when we solve the equations simultaneously. One set of coordinates for such a point satisfies one of the equations and a different set of coordinates for the same point satisfies the other equation. Often, you can discover such "nonsimultaneous" solutions by drawing the graphs, as we did in Example 9.20. For a more general approach, see Exercise 54 in Exercise Set 9.5.

SKETCHING POLAR GRAPHS USING COMPUTER ALGEBRA SYSTEMS

As we have seen in this section, polar graphs can be very hard to sketch. Polar equations can also provide a wide range of beautiful and interesting curves. With the aid of a computer algebra system, many of these beautiful shapes that we could not otherwise envision become accessible to us.

In this section, we will illustrate how to use Maple, Mathematica, and DERIVE to sketch polar graphs. In the exercises, you will be asked to further experiment with graphing polar equations.

 CAS 33

A Cardioid: Sketch the polar curve $r = 2 + 2\cos t$.

Maple:

plot([2+2*cos(t),t,t=0..2*Pi],
    coords=polar);

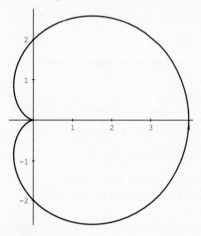

Mathematica:

PolarPlot[2+2 Cos[t],
    {t,0,2 Pi}]

DERIVE:

(At the □ symbol, go to the next step.)
a (author) □ [2+2cos(t),t] □ p (plot window) □ o (options) □ p (polar) □ p (plot) □ 0 (min) □ 2Pi (max)

FIGURE 9.5.1
$r = 2 + 2\cos t,\ 0 \le t \le 2\pi$

In Mathematica, you may first need to load a graphics library by executing the command <<Graphics'Graphics'.

In the examples that follow, assume that the plot mode in DERIVE has been changed to polar as we did in this example.

Notice the difference between the orientation of this graph and that of the cardioid in Example 9.15.

 CAS 34

A Limaçon with a Loop: Sketch the polar curve $r = 1 + 2\sin(t)$.

Maple:

```
plot([1+2*sin(t),t,t=0..2*Pi],
    coords=polar);
```

Mathematica:

```
PolarPlot[1+2 Sin[t],
    {t,0,2 Pi}]
```

DERIVE:

(At the □ symbol, go to the next step.)
a (author) □ [1+2sin(t),t] □ p (plot window) □ p (plot) □ 0 (min) □ 2Pi (max)

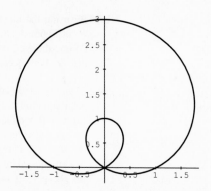

FIGURE 9.5.2
$r = 1 + 2\sin\ t,\ 0 \le t \le 2\pi$

🖳 **CAS 35**

A Rose: Sketch the polar curve $r = \cos 7t$.

Maple:

```
plot([cos(7*t),t,t=0..2*Pi],
    coords=polar,numpoints=200);
```

Mathematica:

```
PolarPlot[Cos[7 t],
    {t,0,2 Pi}]
```

DERIVE:

(At the □ symbol, go to the next step.)
a (author) □ [cos(7t),t] □ p (plot window) □ p (plot) □ 0 (min) □ 2Pi (max)

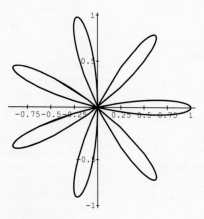

FIGURE 9.5.3
$r = \cos\ (7t),\ 0 \le t \le 2\pi$

Exercise Set 9.5

In Exercises 1–24, identify the curve and draw its graph.

1. (a) $r = 3$
  (b) $r = -3$

2. (a) $\theta = 3\pi/4$
  (b) $\theta = \pi/6$

3. (a) $r\cos\theta = 2$
  (b) $r\sin\theta = -3$

4. (a) $r = -3\csc\theta$
  (b) $r = 2\sec\theta$

5. $r = 3\cos\theta$

6. $r = -2\sin\theta$

7. $r = 2(1 - \cos\theta)$

8. $r = 1 - \sin\theta$

9. $r = -2(1 + \sin\theta)$

10. $r = \cos\theta - 1$

11. $r = 5 - 4\cos\theta$

12. $r = 4 - 5\sin\theta$

13. $r = 2 + 3\sin\theta$

14. $r = 3 + 2\cos\theta$

15. $r = 4 - 2\sin\theta$

16. $r = -2 + 4\cos\theta$

17. $r^2 = 4\sin 2\theta$

18. $r^2 = -2\cos 2\theta$

19. $r^2 = 6\sin\theta\cos\theta$

20. $r^2 = \cos^2\theta - \sin^2\theta$

21. $r = 2\cos 3\theta$

22. $r = 3\cos 2\theta$

23. $r = -2\sin 2\theta$

24. $r = \sin 3\theta$

25. Verify that the graphs of $r = a + b\cos\theta$ and $r = -a + b\cos\theta$, where $a > 0$ and $b > 0$, are the same.

In Exercises 26–29, draw the spirals.

26. $r = a\theta$, with $a = 1$ **(spiral of Archimedes)**

27. $r = a/\theta$, with $a = 2$ **(hyperbolic spiral)**

28. $r^2 = a\theta$, with $a = 1$ **(parabolic spiral)**

29. $r = e^{a\theta}$, with $a = \frac{1}{2}$ **(logarithmic spiral)**

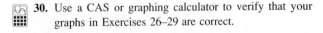

30. Use a CAS or graphing calculator to verify that your graphs in Exercises 26–29 are correct.

In Exercises 31–34, draw the graph by first changing to rectangular coordinates.

31. $r = \dfrac{6}{2\cos\theta - 3\sin\theta}$

32. $r^2 - r(2\cos\theta + 4\sin\theta) + 1 = 0$

33. $r = \dfrac{2}{1 - \cos\theta}$

34. $r = \dfrac{3}{1 + \cos\theta}$

35. By using the most general representations of the points P' in Figure 9.28, devise tests for symmetry that cover all possibilities; that is, devise tests that give conditions that are both necessary and sufficient.

36. Using the result of Exercise 35, show that the graph of a polar equation is symmetric with respect to the following:
  (a) $\theta = 0$ if replacing r by $-r$ and θ by $\pi - \theta$ results in an equivalent equation;
  (b) $\theta = \pi/2$ if replacing r by $-r$ and θ by $-\theta$ results in an equivalent equation;
  (c) the pole if replacing θ by $\pi + \theta$ results in an equivalent equation.

37. Use the results of Exercise 36 to show the specified type of symmetry for the following. Also show that the test as stated in this section fails in each case.
  (a) $r = \tan\theta$; symmetric with respect to $\theta = 0$, $\theta = \pi/2$, and the pole.
  (b) $r = \sin 2\theta\cos\theta$; symmetric with respect to $\theta = \pi/2$.
  (c) $r = 2\csc 2\theta - 1$; symmetric with respect to the pole.
  (d) $r = \sin 2\theta + \cos\theta$; symmetric with respect to $\theta = 0$.

38. Prove that if the polar axis is rotated through an angle α radians, a point P that has coordinates (r, θ) with respect to the original system has coordinates (r', θ') with respect to the new system, where $r' = r$ and $\theta' = \theta - \alpha$. Conclude from this that if the graph of a polar equation is known, the graph of the equation obtained by replacing θ by $\theta - \alpha$ is the original graph rotated through the angle α.

39. Use Exercise 38 to prove that if an equation can be expressed in the form $r = f(\cos\theta)$, then
  (a) when $\cos\theta$ is replaced by $\sin\theta$, the graph is rotated $\pi/2$ radians;
  (b) when $\cos\theta$ is replaced by $-\cos\theta$, the graph is rotated π radians;
  (c) when $\cos\theta$ is replaced by $-\sin\theta$, the graph is rotated $3\pi/2$ radians.

40. Prove that the graph of an equation of the form $r = f(\sin n\theta)$ is the same as that of $r = f(\cos n\theta)$ after rotating through an angle of $\pi/(2n)$ radians. (See Exercise 38.)

41. By multiplying and dividing the right-hand side of the equation $r = a\sin\theta + b\cos\theta$ by $\sqrt{a^2 + b^2}$, show that its graph is the circle $r = \sqrt{a^2 + b^2}\cos\theta$ rotated through the angle α for which $\sin\alpha = a/\sqrt{a^2 + b^2}$ and $\cos\alpha = b/\sqrt{a^2 + b^2}$. (See Exercise 38.)

42. Use the result of Exercise 41 to draw the graph of each of the following:
(a) $r = \sin\theta + \cos\theta$ (b) $r = 3\cos\theta - 4\sin\theta$

In Exercises 43–47, draw the graph. Use a CAS or a graphing calculator to confirm your results.

43. $r = 2\cos\dfrac{\theta}{2}$ **44.** $r = \sqrt{4\sin\theta}$

45. $r = a\sin 2\theta\cos\theta$, for $a = 1$ **(bifolium)**

46. $r = a\sin\theta\tan\theta$, for $a = 2$ **(cissoid)**

47. $r = a\sec\theta + b$, for $a = 1, b = -2$ **(conchoid)**

In Exercises 48–53, find all points of intersection. Draw the graphs.

48. $r = 1 + \cos\theta, r = 1$

49. $r = 2(1 - \cos\theta), r = 1$

50. $r^2 = 2\cos 2\theta, r = 1$

51. $r = 2\cos\theta, r = 2(1 - \cos\theta)$

52. $r = 2\sqrt{2} + \sin\theta, r = 5\sin\theta$

53. $r = 1 + 2\cos\theta, r = 1$

54. Show that a point P_0 is a point of intersection of the graphs of $r = f(\theta)$ and $r = g(\theta)$ if and only if (a) there exist numbers θ_1 and θ_2 such that $f(\theta_1) = 0$ and $g(\theta_2) = 0$, in which case P_0 is the pole, or (b) there exists a number θ_0 such that

$$f(\theta_0) = (-1)^k g(\theta_0 + k\pi)$$

for some integer k. In (b), P_0 has coordinates $(f(\theta_0), \theta_0)$ or, equivalently, $(g(\theta_0 + k\pi), \theta_0 + k\pi)$. (*Hint:* If P_0 has coordinates (r_0, θ_0) for $r_0 \neq 0$, then all other coordinates of P_0 have one of the forms $(r_0, \theta_0 + 2n\pi)$ or $(-r_0, \theta_0 + (2n + 1)\pi)$, where $n = 0, \pm 1, \pm 2, \ldots$.)

In Exercises 55 and 56, find all points of intersection, making use of Exercise 54.

55. $r = 2, r = 4\sin 2\theta$

56. $r = 25\cos\theta, r = 3 + 2\cos\theta$

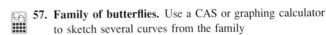

57. Family of butterflies. Use a CAS or graphing calculator to sketch several curves from the family

$$r = 1 + \sin(n\theta) + \cos^2(2n\theta), \qquad n \geq 1$$

58. Family of hands. Repeat Exercise 57 with the family

$$r = 1 + \sin(2\theta) + \cos^2(n\theta), \qquad n \geq 5$$

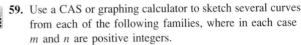

59. Use a CAS or graphing calculator to sketch several curves from each of the following families, where in each case m and n are positive integers.
(a) $r = \sin(m\theta) + \sin^2(n\theta)$
(b) $r = \cos(m\theta) + \cos^2(n\theta)$
(c) $r = \sin(m\theta) + \cos^2(n\theta)$
(d) $r = \cos(m\theta) + \sin^2(n\theta)$

60. Use a CAS or a graphing calculator to sketch several curves from the family

$$r = [1 + \sin(m\theta) + \cos^2(n\theta)]^k$$

where m, n, and k are positive integers.

9.6 AREAS IN POLAR COORDINATES

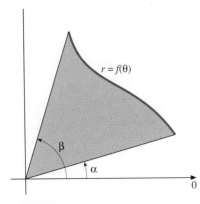

FIGURE 9.46

In Chapter 6 we derived formulas for areas between graphs whose equations were given in rectangular form. We cannot apply these formulas to regions between graphs of polar equations, however. In this section we develop formulas applicable to such regions. We begin with the simplest case, as shown in Figure 9.46. The region in question is bounded by the graph of $r = f(\theta)$ and the rays $\theta = \alpha$ and $\theta = \beta$, with $\beta \leq \alpha + 2\pi$. We assume f is continuous and nonnegative for $\alpha \leq \theta \leq \beta$. Our objective is to find a formula for the area of the region.

We partition the interval $[\alpha, \beta]$ by means of the numbers $\alpha = \theta_0 < \theta_1 < \theta_2 < \cdots < \theta_n = \beta$. For each k from 1 to n, let $\Delta\theta_k = \theta_k - \theta_{k-1}$. The θ_k rays divide the region in question into n subregions, as we show in Figure 9.47(a). Each of these subregions is shaped somewhat like a sector of a circle. In Figure

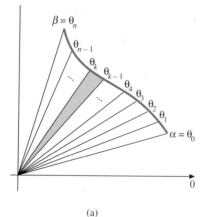

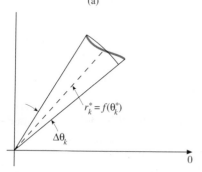

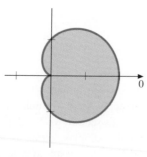

FIGURE 9.47

9.47(b) we show a typical subregion and a circular sector that approximates it. We find the radius r_k^* of the approximating sector by evaluating the function f at an arbitrary value of θ between θ_{k-1} and θ_k. (We have labeled this value θ_k^*.) If $\Delta\theta_k$ is small, we would expect that the area of the circular sector would differ very little from the actual area of the subregion. The area ΔA_k of the circular sector is (see Exercise 32 in Exercise Set 9.6)

$$\Delta A_k = \frac{1}{2}(r_k^*)^2 \Delta\theta_k = \frac{1}{2}\left[f(\theta_k^*)\right]^2 \Delta\theta_k$$

By summing all such ΔA_k values for $k = 1, 2, \ldots, n$, we obtain an approximation to the desired area A of the total region:

$$A \approx \sum_{k=1}^{n} \Delta A_k = \frac{1}{2}\sum_{k=1}^{n}\left[f(\theta_k^*)\right]^2 \Delta\theta_k$$

The sum on the right is a Riemann sum. Thus, by taking finer and finer partitions, such that each of the angles $\Delta\theta_k$ approaches 0, we get the exact value of A as the limit:

$$A = \lim_{\max \Delta\theta_k \to 0} \sum_{k=1}^{n} \Delta A_k = \lim_{\max \Delta\theta_k \to 0} \sum_{k=1}^{n} \frac{1}{2}\left[f(\theta_k^*)\right]^2 \Delta\theta_k = \int_{\alpha}^{\beta} \frac{1}{2}\left[f(\theta)\right]^2 d\theta$$

That is,

$$A = \int_{\alpha}^{\beta} \frac{1}{2}\left[f(\theta)\right]^2 d\theta$$

Since $r = f(\theta)$, we can write this result in the following briefer form:

Area Bounded by a Polar Curve $r = f(\theta)$

$$A = \frac{1}{2}\int_{\alpha}^{\beta} r^2\, d\theta \qquad (9.11)$$

EXAMPLE 9.21   Find the area inside the cardioid $r = 1 + \cos\theta$.

Solution   The graph is shown in Figure 9.48. By symmetry, it is sufficient to find the area between $\theta = 0$ and $\theta = \pi$ and then double the result. So we have, by Equation 9.11,

$$A = 2 \cdot \frac{1}{2}\int_0^{\pi} r^2 d\theta = \int_0^{\pi}(1 + \cos\theta)^2 d\theta$$

$$= \int_0^{\pi}(1 + 2\cos\theta + \cos^2\theta)d\theta$$

$$= \int_0^{\pi}\left(1 + 2\cos\theta + \frac{1 + \cos 2\theta}{2}\right)d\theta$$

$$= \int_0^{\pi}\left(\frac{3}{2} + 2\cos\theta + \frac{\cos 2\theta}{2}\right)d\theta$$

$$= \frac{3\theta}{2} + 2\sin\theta + \frac{\sin 2\theta}{4}\Big]_0^{\pi} = \frac{3\pi}{2}$$

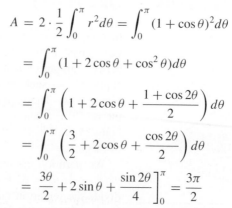

FIGURE 9.48

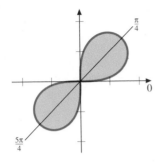

FIGURE 9.49

EXAMPLE 9.22   Find the area inside the lemniscate $r^2 = 4\sin 2\theta$.

Solution   We show the graph in Figure 9.49. We again use symmetry (to the pole) by finding the area between $\theta = 0$ and $\theta = \frac{\pi}{2}$ and doubling the result:

$$A = 2 \cdot \frac{1}{2} \int_0^{\pi/2} r^2 d\theta = \int_0^{\pi/2} 4\sin 2\theta \, d\theta = -2\cos 2\theta \big]_0^{\pi/2}$$
$$= -2[-1 - 1] = 4 \qquad \blacksquare$$

EXAMPLE 9.23   **Family of "Butterflies"**   Consider the family

$$r = 1 + \sin(n\theta) + \cos^2(2n\theta), \qquad 0 \le \theta \le 2\pi$$

where n is a positive integer. Show that the area enclosed by the graph is the same for all positive integers n.

Solution   In Exercise 57 of Exercise Set 9.5, we asked you to use a CAS or graphing calculator to graph several members of this family. In Figure 9.50 we show the curves corresponding to $n = 2$ and $n = 3$.

For any positive integer n, the area enclosed is

$$A = \frac{1}{2}\int_0^{2\pi} r^2 d\theta = \frac{1}{2}\int_0^{2\pi} \left[1 + \sin(n\theta) + \cos^2(2n\theta)\right]^2 d\theta$$

$$= \frac{1}{2}\int_0^{2\pi} \left[1 + \sin^2(n\theta) + \cos^4(2n\theta) + 2\sin(n\theta) + 2\cos^2(2n\theta)\right.$$

$$\left. + 2\sin(n\theta)\cos^2(2n\theta)\right] d\theta$$

$$= \frac{1}{2}\int_0^{2\pi} \left[1 + \frac{1 - \cos(2n\theta)}{2} + \left(\frac{1 + \cos(4n\theta)}{2}\right)^2 + 2\sin(n\theta)\right.$$

$$\left. + (1 + \cos(4n\theta)) + \sin(n\theta)\,(1 + \cos(4n\theta))\right] d\theta$$

Since

$$\left(\frac{1 + \cos(4n\theta)}{2}\right)^2 = \frac{1}{4}\left(1 + 2\cos(4n\theta) + \cos^2(4n\theta)\right)$$

$$= \frac{1}{4}\left[1 + 2\cos(4n\theta) + \frac{1 + \cos(8n\theta)}{2}\right]$$

$$= \frac{3}{8} + \frac{\cos(4n\theta)}{2} + \frac{\cos(8n\theta)}{8}$$

we have, on combining like terms in the last integral,

$$A = \frac{1}{2}\int_0^{2\pi} \left[\frac{23}{8} - \frac{\cos(2n\theta)}{2} + \frac{3\cos(4n\theta)}{2} + \frac{\cos(8n\theta)}{8} + 3\sin(n\theta)\right.$$

$$\left. + \sin(n\theta)\cos(4n\theta)\right] d\theta$$

It is not difficult to show that the only term of the integrand that does not give 0 when integrated over the interval $[0, 2\pi]$ is the constant term, $23/8$. Thus,

$$A = \frac{1}{2}\int_0^{2\pi} \frac{23}{8} d\theta = \frac{1}{2} \cdot \frac{23}{8} \cdot 2\pi = \frac{23\pi}{8}$$

Since the area is the same for all positive integers n, each of the butterfly curves encloses the same area. $\blacksquare$

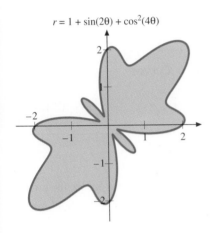

$r = 1 + \sin(2\theta) + \cos^2(4\theta)$

$r = 1 + \sin(3\theta) + \cos^2(6\theta)$

FIGURE 9.50
Two "butterflies"

Area Between Two Polar Graphs

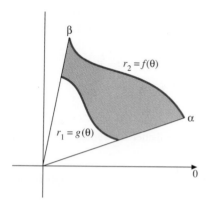

FIGURE 9.51

Suppose now that we want the area between the graphs of two continuous functions f and g for which $0 \leq g(\theta) \leq f(\theta)$, for θ between α and β, where again we have $\beta \leq \alpha + 2\pi$. We illustrate a typical situation in Figure 9.51. Then we have

$$\begin{aligned} A &= \frac{1}{2}\int_{\alpha}^{\beta}[f(\theta)]^2\,d\theta - \frac{1}{2}\int_{\alpha}^{\beta}[g(\theta)]^2\,d\theta \\ &= \frac{1}{2}\int_{\alpha}^{\beta}\left\{[f(\theta)]^2 - [g(\theta)]^2\right\}d\theta \end{aligned}$$

Writing $r_1 = g(\theta)$ and $r_2 = f(\theta)$, we can write the result in the following briefer form:

Area Between Two Polar Curves

$$A = \frac{1}{2}\int_{\alpha}^{\beta}(r_2^2 - r_1^2)d\theta \qquad (r_2 \geq r_1) \qquad (9.12)$$

EXAMPLE 9.24 Find the area outside the circle $r = 1$ and inside the circle $r = 2\cos\theta$.

Solution We show the two circles and the area in question in Figure 9.52. Our first task is to find the points of intersection so that we will know the limits on θ. Solving the equations simultaneously, we get

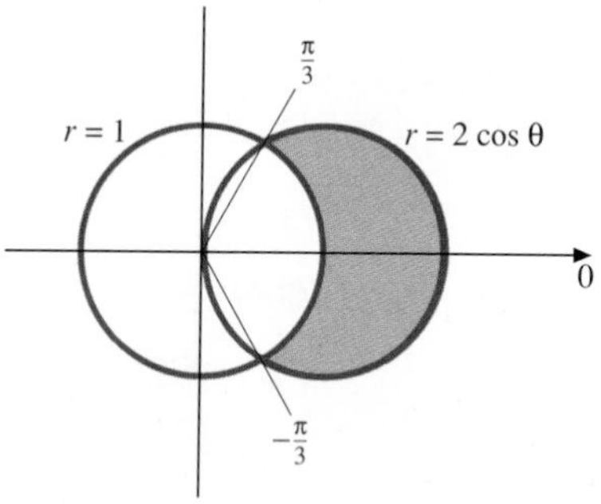

FIGURE 9.52

$$\begin{aligned} 1 &= 2\cos\theta \\ \cos\theta &= \frac{1}{2} \end{aligned}$$

So $\theta = \pm\pi/3$. Because of symmetry we will find the area from 0 to $\pi/3$ and double the result. By Equation 9.12, we have

$$\begin{aligned} A &= 2\cdot\frac{1}{2}\int_0^{\pi/3}\left[(2\cos\theta)^2 - 1^2\right]d\theta \\ &= \int_0^{\pi/3}(4\cos^2\theta - 1)d\theta = \int_0^{\pi/3}[2(1+\cos 2\theta) - 1]\,d\theta \\ &= \int_0^{\pi/3}(2\cos 2\theta + 1)d\theta = \sin 2\theta + \theta\Big]_0^{\pi/3} = \frac{\sqrt{3}}{2} + \frac{\pi}{3} \end{aligned}$$ ■

EXAMPLE 9.25 Find the area that is both inside the circle $r = 3\sin\theta$ and outside the cardioid $r = 1 + \sin\theta$.

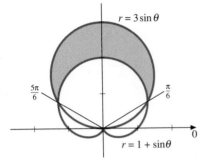

FIGURE 9.53

Solution   We show the graphs in Figure 9.53. Setting the two r values equal to each other gives

$$3 \sin \theta = 1 + \sin \theta$$

$$2 \sin \theta = 1$$

$$\sin \theta = \frac{1}{2}$$

$$\theta = \frac{\pi}{6}, \frac{5\pi}{6}$$

So we have, using symmetry,

$$A = 2 \cdot \frac{1}{2} \int_{\pi/6}^{\pi/2} \left[(3 \sin \theta)^2 - (1 + \sin \theta)^2 \right] d\theta$$

$$= \int_{\pi/6}^{\pi/2} (8 \sin^2 \theta - 2 \sin \theta - 1) d\theta$$

$$= \int_{\pi/6}^{\pi/2} \left[4(1 - \cos 2\theta) - 2 \sin \theta - 1 \right] d\theta$$

$$= \int_{\pi/6}^{\pi/2} (-4 \cos 2\theta - 2 \sin \theta + 3) d\theta$$

$$= \left[-2 \sin 2\theta + 2 \cos \theta + 3\theta \right]_{\pi/6}^{\pi/2} = \frac{3\pi}{2} - \left(-\sqrt{3} + \sqrt{3} + \frac{\pi}{2} \right) = \pi \quad \blacksquare$$

Exercise Set 9.6

In Exercises 1–10, find the area enclosed by the graph of the given equation. Sketch the graph.

1. $r = 2(1 - \cos \theta)$ **2.** $r = 3 + 2 \sin \theta$

3. $r^2 = 9 \cos 2\theta$ **4.** $r = \cos 2\theta$

5. $r = \sin \theta - 1$ **6.** $r = 4 + 2 \cos \theta$

7. $r = 2 \sin 3\theta$ **8.** $r^2 = 2 \sin \theta \cos \theta$

9. $r = -4 \cos \theta$ **10.** $r = \sin \theta + \cos \theta$

In Exercises 11–20, find the area outside the graph of the first equation and inside the graph of the second. Sketch the graphs, showing the area in question.

11. $r = 2$; $r = 2(1 + \cos \theta)$

12. $r = \sqrt{3}$; $r = 2 \sin \theta$

13. $r = \sqrt{2} - \cos \theta$; $r = \cos \theta$

14. $r = \sqrt{2}$; $r^2 = 4 \cos 2\theta$

15. $r = 2 \cos \theta$; $r^2 = 2\sqrt{3} \sin 2\theta$

16. $r = 1$; $r = 2 \cos 2\theta$

17. $r = 3 + 2 \cos \theta$; $r = 1 - 2 \cos \theta$

18. $r = 2 - \sin \theta$; $r = 2(1 + \sin \theta)$

19. $r = 3(1 + \cos \theta)$; $r = 1 - \cos \theta$

20. $r = \sin \theta$; $r = \sqrt{\sin 2\theta}$

21. Find the area enclosed by the small loop of the limaçon $r = 1 - 2 \cos \theta$.

22. Find the area inside the limaçon $r = 1 + 2 \cos \theta$ that is to the right of the line $r = \sec \theta$.

23. Find the area inside the limaçon $r = 3 - 2 \sin \theta$ that lies below the line $r = -2 \csc \theta$.

24. Find the area inside both the circle $r = 3 \sin \theta$ and the cardioid $r = 1 + \sin \theta$.

25. Find the area inside the cardioid $r = 2(1 + \sin\theta)$ and also inside the circle $r = 2\cos\theta$.

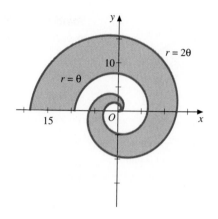

26. Find the smallest area bounded by $r = \ln\theta$, $\theta = 1$, and $\theta = e$. Sketch the graph.

27. Find the area inside the limaçon $r = 1 - \sqrt{2}\sin\theta$ that lies outside its small loop.

28. Find the area inside the graph of $r = 2\cos 3\theta$ and outside the graph of $r = -2\cos\theta$.

29. Find the area inside the circle $r = -2\sin\theta$ that lies outside the small loop of the limaçon $r = 1 - 2\sin\theta$.

30. (a) Find all points of intersection of the graphs of $r = 2\cos\theta/2$ and $r = -2\cos\theta$. Draw the graphs.
(b) Find the area inside the graph of $r = -2\cos\theta$ that lies outside the two small loops of $r = 2\cos\theta/2$.

31. Find the shaded area shown in the figure at the top of the next column.

32. Derive the formula

$$A = \tfrac{1}{2}r^2\theta$$

for the area of a circular sector of radius r and central angle θ (in radians).

33. Show that for $n \geq 1$ the rose curves $r = \cos n\theta$ and $r = \sin n\theta$ all enclose the same area when
(a) n is even
(b) n is odd

34. Use a CAS or graphing calculator to show that

$$\int_0^{2\pi} \left[1 + \sin(m\theta) + \cos^2(n\theta)\right]^k \, d\theta = c_k$$

where for each positive integer k, c_k is a constant, independent of m and n ($m, n \geq 1$). In particular, show that $c_5 = \dfrac{8829}{128}\pi$.

9.7 TANGENT LINES TO POLAR GRAPHS; ARC LENGTH

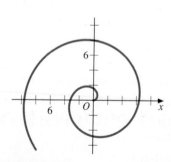

FIGURE 9.54

In rectangular coordinates, if $y = f(x)$ and f is a differentiable function, then $f'(x)$ gives the slope of the tangent line to the graph. In polar coordinates, if $r = f(\theta)$ and f is a differentiable function, it might be tempting to assume that $f'(\theta)$ also gives the slope of the tangent line to the graph. But this apparent result clearly is not so, as the example $r = \theta$ shows. Here, $f(\theta) = \theta$, and so $f'(\theta) = 1$. But the graph (Figure 9.54) is the Spiral of Archimedes, which clearly does *not* always have slope 1. So we have to approach the slope of a polar graph in a different way.

The secret of success lies in changing the equation of a polar curve to parametric equations, as we did in Section 9.5. Recall that the equation $r = f(\theta)$ has the parametric representation

$$\begin{cases} x = f(\theta)\cos\theta \\ y = f(\theta)\sin\theta \end{cases} \qquad \alpha \leq \theta \leq \beta \qquad (9.13)$$

Now, by Equation 9.2 (with t replaced by θ),

$$\frac{dy}{dx} = \frac{\dfrac{dy}{d\theta}}{\dfrac{dx}{d\theta}} = \frac{f'(\theta)\sin\theta + f(\theta)\cos\theta}{f'(\theta)\cos\theta - f(\theta)\sin\theta} \qquad (9.14)$$

provided the denominator is nonzero. This result—though true—is not very satisfactory, since the formula is rather complicated. There is a more suitable measure than slope for describing the behavior of the tangent line in polar coordinates that we discuss presently. First, though, we can use Equation 9.14 to confirm our observation in Section 9.5 about tangent lines at the pole.

Tangent Lines at the Pole

Suppose that $r = f(\theta)$ and f is differentiable at the pole. Set $r = 0$ and let $\theta = \theta_0$ be a root of the equation $f(\theta) = 0$ (that is, $f(\theta_0) = 0$). Assume further that $f'(\theta_0) \neq 0$. Then, since $f(\theta_0) = 0$, we have, from Equation 9.14, the slope at the origin

$$\left.\frac{dy}{dx}\right|_{0,\theta_0} = \frac{f'(\theta_0)\sin\theta_0}{f'(\theta_0)\cos\theta_0} = \tan\theta_0$$

if $\cos\theta_0 \neq 0$. Since the slope of the tangent line is the tangent of its angle of inclination (see Figure 9.55), it follows that its angle of inclination is θ_0, or that it differs from this by a multiple of π. In any case, the line $\theta = \theta_0$ is tangent to the curve at the origin. If $\cos\theta_0 = 0$, then the tangent line is vertical, and again $\theta = \theta_0$ is this tangent line, since $\theta_0 = \pi/2$ or an odd multiple of $\pi/2$. Note that if $f'(\theta_0) = 0$, then both $dx/d\theta$ and $dy/d\theta$ are 0 at the origin, so the origin is a singular point.

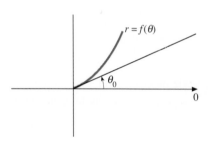

FIGURE 9.55
Slope of tangent line $= \tan\theta_0$

An Alternative to Slope of the Tangent Line

To find a way to describe the tangent line to a polar graph that is more convenient than its slope, consider the graph of the differentiable function $r = f(\theta)$. At an arbitrary point P on the curve, other than the pole, let ϕ (phi) be the angle of inclination of the tangent line at P. Then $\tan\phi$ is the slope of the tangent line. Its value must then be given by Equation 9.14, provided $\phi \neq \pi/2$ (if $\phi = \pi/2$, $\tan\phi$ is not defined):

$$\tan\phi = \frac{f'(\theta)\sin\theta + f(\theta)\cos\theta}{f'(\theta)\cos\theta - f(\theta)\sin\theta} \qquad (9.15)$$

As we show in Figure 9.56, we let ψ (psi) denote the smallest positive angle from the ray OP to the tangent line. Since the sum of the angles in a triangle is π radians (i.e., 180°), we must have

$$\theta + \psi + (\pi - \phi) = \pi$$

or $\psi = \phi - \theta$. Although this figure illustrates the special case in which $\phi > \theta$ and $0 \leq \theta < 2\pi$, it can be shown that in all cases

$$\psi = \phi - \theta + n\pi \qquad (9.16)$$

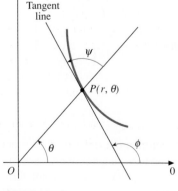

FIGURE 9.56

where n is an integer. Because the period of the tangent function is π, it follows that if $\psi \neq \pi/2$, $\tan \psi = \tan(\phi - \theta + n\pi) = \tan(\phi - \theta)$. By the trigonometric identity for $\tan(\phi - \theta)$ (see Appendix 2), therefore,

$$\tan \psi = \frac{\tan \phi - \tan \theta}{1 + \tan \phi \tan \theta}$$

In the exercises we ask you to show that when $\tan \phi$ is replaced by the right-hand side of Equation 9.15 and $\tan \theta$ is replaced by $\sin \theta / \cos \theta$, we get the simple formula

$$\tan \psi = \frac{f(\theta)}{f'(\theta)}$$

provided both $f(\theta)$ and $f'(\theta)$ are nonzero. Since $r = f(\theta)$, this equation can also be written in the form

$$\tan \psi = \frac{r}{dr/d\theta} \qquad (9.17)$$

EXAMPLE 9.26   Find the angle ψ at the point $(3, \pi/3)$ on the graph of $r = 2(1 + \cos \theta)$. Show the result graphically.

Solution   By Equation 9.17 for $\theta = \pi/3$,

$$\tan \psi = \frac{r}{dr/d\theta} = \frac{2(1 + \cos \theta)}{-2 \sin \theta} = -\frac{1 + \cos \theta}{\sin \theta} = -\frac{\dfrac{3}{2}}{\dfrac{\sqrt{3}}{2}} = -\sqrt{3}$$

So $\psi = 2\pi/3$. We show the graph in Figure 9.57. Note that at the point in question the tangent line is horizontal. ∎

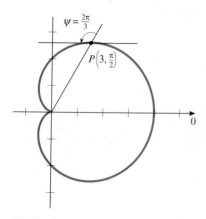

FIGURE 9.57

From Equation 9.16, it follows that the angle of inclination ϕ of the tangent line is $\theta + \psi$ or differs by a multiple of π. In any case, if the tangent line is nonvertical,

$$\tan \phi = \tan(\theta + \psi) \qquad (9.18)$$

This formula is an alternative to Equation 9.14 for calculating the slope, should it be needed. For example, suppose we want the slope of $r = \cos^2 \theta$ at $\theta = \pi/4$. First we calculate $\tan \psi$:

$$\tan \psi = \frac{r}{dr/d\theta} = \frac{\cos^2 \theta}{-2 \sin \theta \cos \theta} = -\frac{1}{2} \cot \theta$$

At $\theta = \pi/4$, $\tan \psi = -\frac{1}{2}$ and $\tan \theta = 1$. So the slope is

$$\tan \phi = \tan(\theta + \psi) = \frac{\tan \theta + \tan \psi}{1 - \tan \theta \tan \psi} = \frac{1 + \left(-\dfrac{1}{2}\right)}{1 - (1)\left(-\dfrac{1}{2}\right)} = \frac{1}{3}$$

Angle Between Two Curves

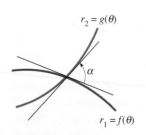

FIGURE 9.58
The angle α from $r_1 = f(\theta)$ to $r_2 = g(\theta)$ is the smallest positive angle from the tangent line of r_1 to the tangent line of r_2 at their point of intersection.

When two curves intersect, the angle α between them at a point of intersection is defined as the smallest positive angle between their tangent lines. (See Figure 9.58.) If the equations are $r_1 = f(\theta)$ and $r_2 = g(\theta)$, then it is not difficult to show that the smallest positive angle α from the first curve to the second is

$\psi_2 - \psi_1$, or differs from that result by an integral multiple of π (see Exercise 47 in Exercise Set 9.7), where ψ_1 and ψ_2 are the angles defined above, for $r_1 = f(\theta)$ and $r_2 = g(\theta)$, respectively. So

$$\tan \alpha = \tan(\psi_2 - \psi_1) \tag{9.19}$$

This result assumes that $\alpha \neq \pi/2$. When $\alpha = \pi/2$, the tangent lines are perpendicular, and the curves are said to be *orthogonal*. For such orthogonal curves, $\tan \alpha$ is undefined, but $\cot \alpha = 0$. You will be asked to show in Exercise 31 of Exercise Set 9.7 that setting $\cot \alpha = 0$ leads to the condition

$$\tan \psi_1 \tan \psi_2 = -1 \tag{9.20}$$

for orthogonality, provided neither ψ_1 nor ψ_2 is $\pi/2$. Note the similarity to the perpendicularity condition $m_1 m_2 = -1$ for slopes m_1 and m_2.

EXAMPLE 9.27   Find the angle from $r_1 = 2 - \cos \theta$ to $r_2 = 2(1 + \cos \theta)$ at each of the points of intersection.

Solution   First we find the points of intersection. Setting $r_1 = r_2$ gives

$$2 - \cos \theta = 2 + 2 \cos \theta$$

$$3 \cos \theta = 0$$

$$\cos \theta = 0$$

$$\theta = \pm \frac{\pi}{2}$$

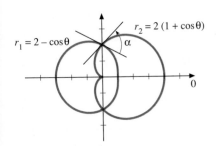

FIGURE 9.59

So the curves intersect at $(2, \pm \pi/2)$. An examination of the graphs (Figure 9.59) shows that there are no other points of intersection. By symmetry we need only consider the angle between the curves when $\theta = \pi/2$. To use Equation 9.19 we need $\tan \psi_1$ and $\tan \psi_2$:

$$\tan \psi_1 = \frac{r_1}{dr_1/d\theta} = \frac{2 - \cos \theta}{\sin \theta}$$

$$\tan \psi_2 = \frac{r_2}{dr_2/d\theta} = \frac{2(1 + \cos \theta)}{-2 \sin \theta} = \frac{1 + \cos \theta}{-\sin \theta}$$

At $\theta = \pi/2$, $\tan \psi_1 = 2$ and $\tan \psi_2 = -1$. Thus,

$$\tan \alpha = \tan(\psi_2 - \psi_1) = \frac{\tan \psi_2 - \tan \psi_1}{1 + \tan \psi_1 \tan \psi_2}$$

$$= \frac{-1 - 2}{1 + 2(-1)} = 3$$

Using a calculator and the inverse tangent, we find that $\alpha \approx 1.25$ radians $= (1.25)(180/\pi)°$, or approximately $71.6°$.   ∎

Arc Length in Polar Coordinates

Equations 9.13 enable us to use the results from Section 9.3 to find the length of the curve $r = f(\theta)$, where f' is continuous for $\alpha \le \theta \le \beta$. We calculate $dx/d\theta$ and $dy/d\theta$ from Equations 9.13:

$$\frac{dx}{d\theta} = f'(\theta) \cos \theta - f(\theta) \sin \theta$$

and

$$\frac{dy}{d\theta} = f'(\theta)\sin\theta + f(\theta)\cos\theta$$

So, by Equation 9.4, with t replaced by θ,

$$ds = \sqrt{\left(\frac{dx}{d\theta}\right)^2 + \left(\frac{dy}{d\theta}\right)^2}\, d\theta$$

$$= \sqrt{[f'(\theta)\cos\theta - f(\theta)\sin\theta]^2 + [f'(\theta)\sin\theta + f(\theta)\cos\theta]^2}\, d\theta$$

We leave it for you in Exercise 48 of Exercise Set 9.7 to show that after simplification, this equation becomes

$$ds = \sqrt{[f(\theta)]^2 + [f'(\theta)]^2}\, d\theta \tag{9.21}$$

We also note that if $f'(\theta)$ is continuous for θ in $[\alpha, \beta]$, both $dx/d\theta$ and $dy/d\theta$ are continuous there. It therefore follows from Equation 9.6 that arc length in polar coordinates is given by

$$L = \int_\alpha^\beta \sqrt{[f(\theta)]^2 + [f'(\theta)]^2}\, d\theta$$

provided f' is continuous in $[\alpha, \beta]$. Since $r = f(\theta)$, we can also write the equation in the following form:

Arc Length in Polar Coordinates

$$L = \int_\alpha^\beta \sqrt{r^2 + \left(\frac{dr}{d\theta}\right)^2}\, d\theta \tag{9.22}$$

EXAMPLE 9.28   Find the length of the cardioid $r = 1 + \cos\theta$.

Solution   Because of symmetry it is sufficient to find the length for θ in $[0, \pi]$ and double the result. Since $dr/d\theta = -\sin\theta$, we have

$$L = 2\int_0^\pi \sqrt{r^2 + \left(\frac{dr}{d\theta}\right)^2}\, d\theta$$

$$= 2\int_0^\pi \sqrt{1 + 2\cos\theta + \cos^2\theta + \sin^2\theta}\, d\theta$$

$$= 2\int_0^\pi \sqrt{2(1 + \cos\theta)}\, d\theta$$

To complete the integration, we need to eliminate the square root. To do so, we make use of the identity (see Appendix 2)

$$\cos\frac{\theta}{2} = \pm\sqrt{\frac{1 + \cos\theta}{2}}$$

choosing the positive sign, since $\cos(\theta/2)$ is positive for θ in $[0, \pi]$. If we multiply both sides of this equation by 2, we get $\sqrt{2(1 + \cos\theta)} = 2\cos(\theta/2)$. So we have

$$L = 2\int_0^\pi 2\cos\frac{\theta}{2}\,d\theta = 8\sin\frac{\theta}{2}\bigg]_0^\pi = 8 \qquad \blacksquare$$

Exercise Set 9.7

In Exercises 1–10, find $\tan\psi$ at the indicated value of θ.

1. $r = 2(1 - \sin\theta)$; $\theta = \pi/6$

2. $r = 3\cos 2\theta$; $\theta = \pi/6$

3. $r = \cos^2\theta$; $\theta = \pi/4$

4. $r = 4\sin\theta$; $\theta = \pi/3$

5. $r = 4 - 2\sin\theta$; $\theta = 5\pi/6$

6. $r = \sqrt{\cos 2\theta}$; $\theta = -\pi/6$

7. $r = \dfrac{1}{1 - \cos\theta}$; $\theta = 2\pi/3$

8. $r = \dfrac{2}{2 - \sin\theta}$; $\theta = -\pi/6$

9. $r = \sin\theta\cos\theta$; $\theta = \pi/3$

10. $r = 1 - 2\sin\dfrac{\theta}{2}$; $\theta = \pi/2$

In Exercises 11–20, make use of the fact that $\tan\phi = \tan(\theta + \psi)$, where ϕ is the angle of inclination of the tangent line, to find the slope of the curve at the given value of θ.

11. $r = 1 - \cos\theta$; $\theta = 2\pi/3$

12. $r = 2 + \sin\theta$; $\theta = -\pi/6$

13. $r = \dfrac{1}{1 - \cos\theta}$; $\theta = \pi/3$

14. $r = 2\cos\theta$; $\theta = \pi/4$

15. $r = 2\cos(\theta/2) - 1$; $\theta = \pi$

16. $r = \sqrt{\sin 2\theta}$; $\theta = \pi/3$

17. $r = 1 - 2\sin\theta$; $\theta = \pi$

18. $r = \cos^2\theta$; $\theta = \pi/4$

19. $r = \dfrac{2}{3 - 2\sin\theta}$; $\theta = \pi$

20. $r = \sec\theta + \tan\theta$; $\theta = 2\pi/3$

In Exercises 21 and 22, show that the curve has a horizontal tangent line at the indicated value of θ, using $\tan\phi = \tan(\theta + \psi)$.

21. $r = 4 + 4\cos\theta$; $\theta = -\pi/3$

22. $r = 2\sqrt{\cos 2\theta}$; $\theta = 5\pi/6$

In Exercises 23 and 24, show that the curve has a vertical tangent line at the indicated point, using the fact that $\cot\phi = \cot(\theta + \psi)$.

23. $r = 1 - \cos\theta$; $\theta = \pi/3$

24. $r = 3 + \tan^2\theta$; $\theta = 3\pi/4$

25. Show that for the logarithmic spiral $r = ae^{b\theta}$, the angle ψ is constant. What is ψ when $b = 1$?

In Exercises 26–30, find the tangent of the angle from the first curve to the second at each of their points of intersection.

26. $r_1 = 2 - \cos\theta$; $r_2 = 3\cos\theta$

27. $r_1 = \dfrac{1}{1 - \cos\theta}$; $r_2 = \dfrac{1}{2 + \cos\theta}$

28. $r_1 = \dfrac{3}{1 + \sin\theta}$; $r_2 = 4\sin\theta$

29. $r_1 = \sqrt{3}\cos\theta$; $r_2 = \sin 2\theta$

30. $r_1 = \sqrt{2} + \sin\theta$; $r_2 = 3\sin\theta$

31. Prove the orthogonality condition, Equation 9.20.

In Exercises 32–34, show that the graphs are orthogonal at all points of intersection.

32. $r_1 = \sin\theta$; $r_2 = \cos\theta$

33. $r_1 = 3\cos\theta$; $r_2 = 2 - \cos\theta$

34. $r_1 = \dfrac{1}{1 - \sin\theta}$; $r_2 = \dfrac{1}{1 + \sin\theta}$

In Exercises 35–37, show that the curves intersect orthogonally for all nonzero values of a and b.

35. $r_1 = a \cos \theta$; $r_2 = b \sin \theta$

36. $r_1 = a\theta$; $r_2 = be^{-\theta^2/2}$

37. $r_1 = a(1 - \sin \theta)$; $r_2 = b(1 + \sin \theta)$, except at the pole

In Exercises 38–44, find the length of the curve on the given interval.

38. $r = 2a \cos \theta$; $0 \leq \theta \leq 2\pi$

39. $r = e^{3\theta}$; $0 \leq \theta \leq \ln 2$

40. $r = \cos \theta - 1$; $0 \leq \theta \leq \pi$

41. $r = 2 \cos^2(\theta/2)$; $0 \leq \theta \leq \pi$

42. $r = \sin^3(\theta/3)$; $0 \leq \theta \leq 3\pi$

43. $r = \theta^2 - 1$; $0 \leq \theta \leq 3$

44. $r = \sin \theta - \cos \theta$; $-\pi/2 \leq \theta \leq \pi$

45. Prove Equation 9.17.

46. Prove that the condition for orthogonality of the graphs of $r = f(\theta)$ and $r = g(\theta)$ can be put in the form $f(\theta)g(\theta) + f'(\theta)g'(\theta) = 0$.

47. Verify Equation 9.19. (*Hint:* Express $\phi_2 - \phi_1$ in terms of ψ_2 and ψ_1.)

48. Verify Equation 9.21.

49. Find the length of the curve $r = \sin^2 \theta$ from $\theta = 0$ to $\theta = \pi/2$.

50. Find the length of that portion of the curve $r = 2(1 - \sin \theta)$ that lies inside $r = -6 \sin \theta$.

51. (a) Suppose that f' is continuous and that $f(\theta) \sin \theta \geq 0$ for $\alpha \leq \theta \leq \beta$. Derive the following formula for the surface area obtained by rotating the graph of $r = f(\theta)$ about the line $\theta = 0$:

$$S = 2\pi \int_\alpha^\beta f(\theta) \sin \theta \sqrt{[f(\theta)]^2 + [f'(\theta)]^2} \, d\theta$$

(b) State appropriate conditions and derive an analogous formula for the surface area formed by rotation about the line $\theta = \pi/2$.

In Exercises 52–55, use the results of Exercise 51 to find the surface area formed when the graph of $r = f(\theta)$ is rotated about the specified axis.

52. $r = 1 - \cos \theta$, $0 \leq \theta \leq \pi$; about $\theta = 0$

53. $r = \sqrt{\sin 2\theta}$, $0 \leq \theta \leq \pi/2$; about $\theta = \pi/2$

54. $r = e^{2\theta}$, $0 \leq \theta \leq \pi$; about $\theta = 0$

55. $r = 2 \sin^2(\theta/2)$, $0 \leq \theta \leq \pi/2$; about (a) $\theta = 0$ and (b) $\theta = \pi/2$

9.8 POLAR EQUATIONS OF CONIC SECTIONS

Our goal in this section is to obtain polar equations of the curves known as **conic sections**—the circle, parabola, ellipse, and hyperbola. The name *conic sections* (also called **conics**) comes from the fact that when a plane passes through a right-circular cone at various angles, the curves of intersection are the conic sections, as we illustrate in the following photos.

Slicing a cone perpendicular to its axis gives a circle, and slicing it parallel to its axis gives a branch of a hyperbola.

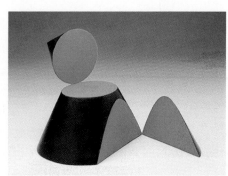

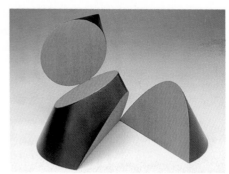

Slicing through a cone at an angle gives an ellipse, and slicing it parallel to an edge gives a parabola.

We concentrate on the parabola, ellipse, and hyperbola, since we already know that a circle of radius a centered at the origin has the polar equation $r = a$.

Let us begin by reviewing some facts concerning the three types of curves in question, including their rectangular equations. (See also Supplementary Appendix 3.)

The Parabola

A parabola is defined as the set of points in the plane that are equidistant from a fixed point, called the **focus**, and a fixed line, called the **directrix**. If we take the focus as the point $(p, 0)$, in a rectangular coordinate system, and the directrix as the line $x = -p$ (see Figure 9.60), the definition requires that $\overline{PF} = \overline{PP'}$. That is,

$$\sqrt{(x - p)^2 + y^2} = x + p$$

If we square both sides and collect terms, we obtain the equation

$$y^2 = 4px \tag{9.23}$$

In an analogous way, by choosing the focus on the y-axis at $(0, p)$ and the directrix as the horizontal line $y = -p$, we would arrive at the equation

$$x^2 = 4py \tag{9.24}$$

with the graph shown in Figure 9.61.

If we replace x with $-x$ in Equation 9.23 we get $y^2 = -4px$, and the resulting graph is the reflection through the y-axis of the first graph (Figure 9.62). Similarly, the graph in Figure 9.63 is the reflection through the x-axis of the parabola $x^2 = 4py$, and its equation is $x^2 = -4py$.

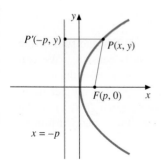

FIGURE 9.60
The parabola $y^2 = 4px$

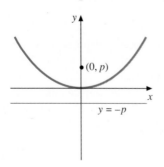

FIGURE 9.61
The parabola $x^2 = 4py$

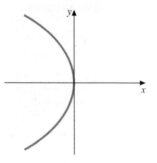

FIGURE 9.62
The parabola $y^2 = -4px$

FIGURE 9.63
The parabola $x^2 = -4py$

Each of the parabolas $y^2 = \pm 4px$ and $x^2 = \pm 4py$ has its vertex at the origin. If we translate the curves h units horizontally and k units vertically, the new equations are obtained from the old with replacing x with $x - h$ and y by $y - k$. So we have the following equations.

Standard Forms for Equations of Parabolas

$(y - k)^2 = \pm 4p(x - h)$, vertex at (h, k), axis horizontal
$(x - h)^2 = \pm 4p(y - k)$, vertex at (h, k), axis vertical

Parabolas are important in optics, since light rays parallel to the axis of a parabola are reflected to the focus. (See Figure 9.64.) Thus, many telescope and searchlight mirrors are parabolic in shape. In the case of a searchlight, light emitted from the focus is reflected in parallel beams. Similarly, radio telescopes have parabolic shapes, since radio frequencies are reflected to the focus.

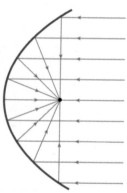

FIGURE 9.64
Light rays parallel to the axis of a parabola are reflected to the focus.

The Ellipse

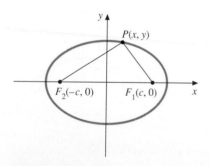

FIGURE 9.65

An ellipse is defined as the set of points in the plane such that the sum of the distances from each point in the set to two fixed points is constant. We call the two fixed points the **foci** (pronounced "fo sye"). Let us take the foci as $F_1(c, 0)$ and $F_2(-c, 0)$, and the constant as $2a$ (the final result is simpler when we use $2a$ rather than a). Then the definition requires of a point $P(x, y)$ on the ellipse that $\overline{PF_1} + \overline{PF_2} = 2a$ (see Figure 9.65). Using the distance formula, this requirement becomes

$$\sqrt{(x - c)^2 + y^2} + \sqrt{(x + c)^2 + y^2} = 2a$$

To simplify this equation, we can isolate one of the radicals and square, then

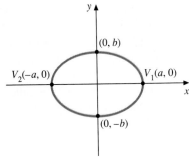

(a) Position I $\dfrac{x^2}{a^2} + \dfrac{y^2}{b^2} = 1$

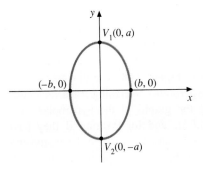

(b) Position II $\dfrac{x^2}{b^2} + \dfrac{y^2}{a^2} = 1$

FIGURE 9.66

isolate the remaining radical and square again. The result is

$$\frac{x^2}{a^2} + \frac{y^2}{b^2} = 1 \qquad \text{Position I} \qquad (9.25)$$

where $b^2 = a^2 - c^2$ (see Appendix 3 for the details).

If the foci are on the y-axis at $(0, c)$ and $(0, -c)$, the resulting equation is

$$\frac{x^2}{b^2} + \frac{y^2}{a^2} = 1 \qquad \text{Position II} \qquad (9.26)$$

Note that in both Equations 9.25 and 9.26, $a > b$.

In Figure 9.66 we show typical graphs of ellipses conforming to Equations 9.25 and 9.26. The points labeled V_1 and V_2 are called **vertices**. The line segment joining the vertices is called the **major axis** of the ellipse. Its length is $2a$. The **minor axis** is the line segment of length $2b$, through the center, perpendicular to the major axis and terminating on the ellipse.

When the center is translated from the origin to the point (h, k), the equations are obtained from Equations 9.25 and 9.26 by replacing x with $x - h$ and y with $y - k$.

Standard Forms for Equations of Ellipses

$$\frac{(x - h)^2}{a^2} + \frac{(y - k)^2}{b^2} = 1, \text{ center } (h, k), \text{ major axis horizontal}$$

$$\frac{(x - h)^2}{b^2} + \frac{(y - k)^2}{a^2} = 1, \text{ center } (h, k), \text{ major axis vertical}$$

Halley's Comet

Ellipses are particularly important because, as Johannes Kepler showed in 1609, the planets and other objects in our solar system all orbit the sun in ellipses. Figure 9.67 shows the orbit of Halley's Comet, which came closest to the sun most recently in 1986. It won't return to that point in its orbit (perihelion) until 2061.

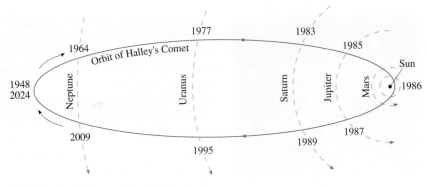

FIGURE 9.67
Orbits of Halley's Comet and the planets, showing the years when the comet crosses each planetary orbit.

The Hyperbola

A hyperbola is defined as the set of points in the plane for which the absolute value of the *difference* in their distances from two fixed points (the foci) is constant. As with the ellipse, the simplest equations result when the foci are either $(\pm c, 0)$ or $(0, \pm c)$, and the constant referred to in the definition is called $2a$. We leave it as an exercise for you to show that the two resulting equations can be put in the forms

$$\frac{x^2}{a^2} - \frac{y^2}{b^2} = 1 \qquad \text{Position I} \tag{9.27}$$

and

$$\frac{y^2}{a^2} - \frac{x^2}{b^2} = 1 \qquad \text{Position II} \tag{9.28}$$

where $b^2 = c^2 - a^2$. We show the graphs in Figure 9.68. The diagonals of the rectangles shown are asymptotes to the hyperbolas. Sketching these rectangles and their diagonals is an aid to drawing the graphs of the hyperbolas. For Position I the asymptotes have slopes $\pm b/a$, and for Position II they have slopes $\pm a/b$. The line through the vertices V_1 and V_2 is called the **transverse axis** and the line perpendicular to it through the center is called the **conjugate axis**.

When the center is translated to the point (h, k), we obtain the following:

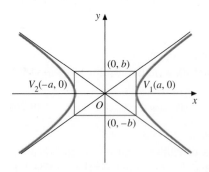

(a) Position I $\dfrac{x^2}{a^2} - \dfrac{y^2}{b^2} = 1$

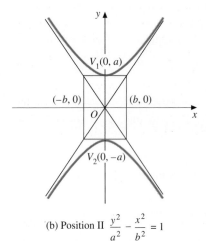

(b) Position II $\dfrac{y^2}{a^2} - \dfrac{x^2}{b^2} = 1$

FIGURE 9.68

Standard Forms of Equations of Hyperbolas

$$\frac{(x - h)^2}{a^2} - \frac{(y - k)^2}{b^2} = 1, \text{ center } (h, k), \text{ transverse axis horizontal}$$

$$\frac{(y - k)^2}{a^2} - \frac{(x - h)^2}{b^2} = 1, \text{ center } (h, k), \text{ transverse axis vertical}$$

Hyperbolas, too, have uses in optics. Mirrors for X-ray telescopes often use grazing rays (rays incident at very low angles) to bounce the X rays. The mirrors of NASA's Advanced X-Ray Facility will use combinations of hyperbolas and parabolas. In the photo at left we see one such mirror being constructed. The curvature of the inside of the cylinder, from top to bottom as we see it, is first a parabola and then a hyperbola. The combination is wrapped around to make a cylinder, and four such mirrors of different sizes will be nested.

Alternative Definition of Conic Sections

The following unites the parabola, ellipse, and hyperbola in a single definition:

> ### Definition 9.2
> ### Conics Defined by Eccentricity
>
> Let F be a fixed point and l a fixed line in the plane. The set of all points P in the plane such that
>
> $$\frac{\text{distance between } P \text{ and } F}{\text{distance between } P \text{ and } l} = e \qquad e \text{ is a positive constant}$$
>
> is a parabola if $e = 1$, an ellipse if $e < 1$, and a hyperbola if $e > 1$. The constant e is called the **eccentricity** of the conic, F is a **focus**, and l is a **directrix**.

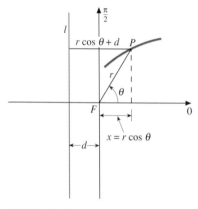

FIGURE 9.69

For the parabola this definition is equivalent to the one previously given. It is also equivalent in the case of the ellipse and the hyperbola, but we will not carry out the proof. We will show, however, that the equation that results when $e < 1$ is that of an ellipse, and the one for $e > 1$ is a hyperbola, as previously defined. The parabola has only one focus and one directrix, whereas the ellipse and the hyperbola have two foci and two directrices (plural of directrix).

Definition 9.2 provides a means of writing equations of the conic sections (other than the circle) in polar coordinates. To start with, we take the focus F at the pole and the directrix l perpendicular to the line $\theta = 0$, d units to the left of the pole, as shown in Figure 9.69. Then for a point P on the curve, the distance between P and F is r, and the distance between P and l is $r \cos \theta + d$, since $r \cos \theta$ is the x-coordinate of P. By Definition 9.2, we have

$$\frac{r}{r \cos \theta + d} = e$$

or, on solving for r,

$$r = \frac{ed}{1 - e \cos \theta} \qquad (9.29)$$

To verify that the graph of Equation 9.29 is one of the conic sections specified in Definition 9.2, let us obtain the corresponding rectangular equation. On clearing fractions and observing that $r \cos \theta = x$, we get

$$r = e(d + x)$$

If we square both sides and use the fact that $r^2 = x^2 + y^2$, we get

$$x^2 + y^2 = e^2(d^2 + 2xd + x^2) \qquad (9.30)$$

Now we consider three cases:

Case i: $e = 1$

Equation 9.30 becomes

$$y^2 = 2d\left(x + \frac{d}{2}\right)$$

which we recognize as a parabola with horizontal axis, vertex $(-d/2, 0)$, opening to the right. In fact, if we write $d = 2p$, the equation becomes

$$y^2 = 4p(x + p)$$

which is in the form $(y - k)^2 = 4p(x - h)$, with $h = -p$ and $k = 0$.

Case ii: $e < 1$

Equation 9.30 can be rearranged to give

$$x^2(1 - e^2) - 2e^2 xd + y^2 = e^2 d^2$$

If we divide by $1 - e^2$ (we know that $1 - e^2 > 0$ since $e < 1$) and complete the square in x, we get (you should verify it)

$$\left(x - \frac{e^2 d}{1 - e^2} \right)^2 + \frac{y^2}{1 - e^2} = \frac{e^2 d^2}{(1 - e^2)^2}$$

Finally, when we divide by the right-hand side, we get the standard form of the ellipse with center (h, k),

$$\frac{(x - h)^2}{a^2} + \frac{(y - k)^2}{b^2} = 1$$

in which

$$h = \frac{e^2 d}{1 - e^2}, \quad k = 0, \quad a = \frac{ed}{1 - e^2}, \quad \text{and} \quad b = \frac{ed}{\sqrt{1 - e^2}}$$

(Note that $\sqrt{1 - e^2}$ is a real number, since $1 - e^2 > 0$.)

Case iii: $e > 1$

We leave it for you in Exercise 26 of Exercise Set 9.8 to show that the rectangular equation in this case can be put into the standard form of the hyperbola with center (h, k),

$$\frac{(x - h)^2}{a^2} - \frac{(y - k)^2}{b^2} = 1$$

in which

$$h = -\frac{e^2 d}{e^2 - 1}, \quad k = 0, \quad a = \frac{ed}{e^2 - 1}, \quad \text{and} \quad b = \frac{ed}{\sqrt{e^2 - 1}}$$

From the three cases we have considered, we see that the polar equation

$$r = \frac{ed}{1 - e \cos \theta}$$

does have a parabola, an ellipse, or a hyperbola as its graph, according to whether $e = 1$, $e < 1$, or $e > 1$, respectively. To obtain this polar equation, we chose the directrix d units to the left of the focus (taken as the pole). If, instead, we take the directrix d units to the right, the resulting equation is

$$r = \frac{ed}{1 + e \cos \theta}$$

As a memory device, when the denominator $1 \pm e \cos \theta$ has a minus sign, the directrix in question is to the *left* of the pole, and when it has a plus sign, the directrix is to the *right* of the pole.

If we take the directrix as a horizontal line d units above or below the pole, we get

$$r = \frac{ed}{1 + e \sin \theta} \qquad \text{or} \qquad r = \frac{ed}{1 - e \sin \theta}$$

We will ask you to verify these forms in Exercise 27 of Exercise Set 9.8. Can you see how to determine from the sign in the denominator $1 \pm e \sin \theta$ when the directrix in question is above, and when it is below, the pole?

We summarize now our discussion of polar equations of conics having a focus at the pole and with directrix d units from the pole, directed vertically (when $\cos \theta$ is involved) or horizontally (when $\sin \theta$ is involved).

Polar Equations of Conic Sections

$$r = \frac{ed}{1 \pm e \cos \theta} \qquad \text{Directrix vertical} \qquad (9.31)$$

$$r = \frac{ed}{1 \pm e \sin \theta} \qquad \text{Directrix horizontal} \qquad (9.32)$$

The graph is a parabola if $e = 1$, an ellipse if $e < 1$, and a hyperbola if $e > 1$. In each case a focus is at the pole and the corresponding directrix is d units from the pole.

As the next two examples show, we can sketch the conic directly from its polar equation. It is not necessary to change to rectangular coordinates.

EXAMPLE 9.29   Identify the conic defined by

(a) $r = \dfrac{2}{1 - \cos \theta}$ (b) $r = \dfrac{3}{2 + \sin \theta}$

and draw its graph.

Solution

(a) The equation of this conic conforms to Equation 9.31, with $e = 1$, so the graph is a parabola with focus at the pole. The directrix is vertical and is $d = 2$ units to the left of the pole. (Its rectangular equation is $x = -2$.) When $\theta = \pm \pi/2$, $r = 2$, and when $\theta = \pi$, $r = 1$. The point $(1, \pi)$ is the vertex. Note that it is midway between the focus (the pole) and the directrix. Another easy pair of points to find are those corresponding to $\theta = \pi/3$ and $\theta = -\pi/3$. We find the corresponding r to be 4 in each case. With these points we can draw the graph as shown in Figure 9.70. We show the directrix, although it is not necessary for the graph.

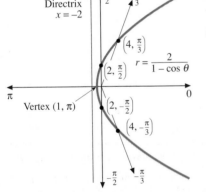

FIGURE 9.70

REMARK ───────────────

■ If you had not remembered which way the parabola opened by looking at its equation, you could see that since r is undefined only when $\theta = 2n\pi$ ($n = 0$, ± 1, ± 2, ...), the parabola must open to the right.

(b) To make the equation conform to Equation 9.32, we divide numerator and denominator by 2. The key is to get the 1 in the denominator. Thus,

$$r = \frac{3/2}{1 + \frac{1}{2} \sin \theta}$$

θ	r
0	$3/2$
$\pi/2$	1
π	$3/2$
$3\pi/2$	3

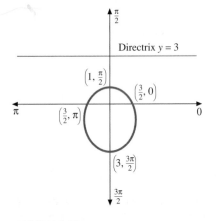

FIGURE 9.71

So we have $e = 1/2$. The curve is therefore an ellipse with major axis vertical. Since $ed = 3/2$ and $e = 1/2$, it follows that $d = 3$. So the directrix corresponding to the focus at the pole is 3 units from the pole, and is horizontal. Four key points on the ellipse can be obtained by taking $\theta = 0, \pi/2, \pi$, and $3\pi/2$ (corresponding to intercepts on the x- and y-axes). We show the corresponding values of r in the table in the margin. Using these four points and our knowledge of the symmetry of the ellipse, we obtain the graph in Figure 9.71.

Again, we did not need the directrix to draw the graph. Notice, however, that since the directrix corresponding to the focus at the pole is 3 units above the pole, if we consider the point $P(1, \pi/2)$, its distance from the pole (a focus) is 1, and its distance from the corresponding directrix is 2. Thus, the ratio of the first of these distances to the second is $\frac{1}{2}$, confirming that for this one point, the definition of e (Definition 9.2) is correct. (Of course, the ratio of the corresponding distances for any other point on the ellipse would also equal $\frac{1}{2}$.)   ∎

EXAMPLE 9.30   Identify and sketch the graph of

$$r + 2r\cos\theta - 4 = 0$$

Solution   Solving for r, we obtain

$$r = \frac{4}{1 + 2\cos\theta} \tag{9.33}$$

θ	r
0	$4/3$
$\pm\pi/2$	4
π	-4

This equation conforms to Equation 9.31, with $e = 2$. Hence, the graph is a hyperbola with a vertical directrix. Consequently, its transverse axis is horizontal. Again, we make a table of values, using the four key angles $\theta = 0, \pm\pi/2$, and π. With these four points and our knowledge of symmetry of the hyperbola, we can make a sketch as in Figure 9.72. The accuracy can be improved by noting that

$$\lim_{\theta \to (2\pi/3)^+} \frac{4}{1 + 2\cos\theta} = -\infty$$

and

$$\lim_{\theta \to (2\pi/3)^-} \frac{4}{1 + 2\cos\theta} = \infty$$

These limits tell us that the asymptotes to the hyperbola, which go through its center, are parallel to the extended $2\pi/3$ ray and, by symmetry, the $-2\pi/3$ ray, shown in blue (see Exercise 28 in Exercise Set 9.8). We show these asymptotes in black in Figure 9.72.

If we superimpose a rectangular coordinate system on the polar system in the usual way, we can make further observations. We see, in particular, that the center is at $(\frac{8}{3}, 0)$, halfway between the two vertices. In rectangular coordinates the vertices are $(\frac{4}{3}, 0)$ and $(4, 0)$. The other focus (besides the pole) has rectangular coordinates $(\frac{16}{3}, 0)$, since it is $\frac{4}{3}$ units to the right of its corresponding vertex.

We can verify that $d = 2$, so the directrix corresponding to the focus at the pole is $x = 2$ (in rectangular coordinates). Using the point $P(\frac{4}{3}, 0)$, we can verify that the ratio defining e in Definition 9.32 is correct, namely,

$$\frac{4}{3} \div \frac{2}{3} = \frac{4}{3} \cdot \frac{3}{2} = 2$$ ∎

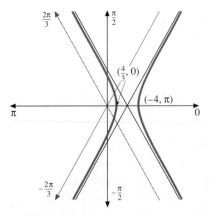

FIGURE 9.72
The graph of $r = 4/(1 + 2\cos\theta)$.

Exercise Set 9.8

In Exercises 1–10, identify the curve and draw its graph.

1. $r = \dfrac{1}{1 + \cos\theta}$

2. $r = \dfrac{2}{1 - \frac{1}{2}\sin\theta}$

3. $r = \dfrac{3}{1 - 2\sin\theta}$

4. $r = \dfrac{3}{2(1 - \sin\theta)}$

5. $r = \dfrac{6}{3 - 2\cos\theta}$

6. $r = \dfrac{3}{2 - \cos\theta}$

7. $r = \dfrac{9}{3 + 4\sin\theta}$

8. $r(4 + 3\cos\theta) = 8$

9. $2r = 12 - 3r\sin\theta$

10. $2r\sin\theta = 5 - 3r$

In Exercises 11–14, find the equation of the conic with the given eccentricity, having a focus at the pole, and with the corresponding directrix as specified.

11. $e = \frac{1}{3}$; directrix horizontal, 5 units above the pole.

12. $e = 1$; directrix vertical, 3 units to the right of the pole.

13. $e = \frac{3}{2}$; directrix vertical, 1 unit to the left of the pole.

14. $e = 0.3$; directrix horizontal, 1.2 units below the pole.

In Exercises 15–18, find the rectangular equation of the conic. If it is a parabola, give its vertex and its focus; if it is an ellipse, give its center, its vertices, and its foci; if it is a hyperbola, give its center, its vertices, its foci, and its asymptotes.

15. $r = \dfrac{2}{1 - \cos\theta}$

16. $r = \dfrac{4}{3 + \cos\theta}$

17. $r = \dfrac{6}{3 - 4\sin\theta}$

18. $r = \dfrac{3}{3 + \sin\theta}$

19. Denote the length of a semimajor axis of an ellipse by a. If its eccentricity is e and a focus is at the pole, verify the distances shown in the accompanying figure for the distances from the center to a focus and to a directrix. If c is the distance from a focus to the center, notice that $c = ae$, which implies that $e = c/a$ and also that $c < a$. (The same equations hold true for the hyperbola, but since $e > 1$, we see that $c > a$.)

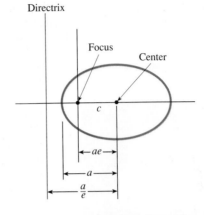

Directrix

Focus

Center

c

$\leftarrow ae \rightarrow$

$\leftarrow a \rightarrow$

$\dfrac{a}{e}$

20. Use the result of Exercise 19 to show that the equation of the ellipse shown there can be written in the form

$$r = \frac{a(1 - e^2)}{1 - e\cos\theta}$$

Show that when $e = 0$, the ellipse becomes a circle.

21. Show that the polar form of the ellipse

$$\frac{x^2}{a^2} + \frac{y^2}{b^2} = 1$$

can be written as

$$r^2 = \frac{a^2(1 - e^2)}{1 - e^2\cos^2\theta}$$

22. The planets travel in elliptical orbits around the sun, with the sun at a focus. Let a polar coordinate system be introduced so that the sun is at the pole. Let r_a denote the greatest distance from the planet to the sun (called the *aphelion* distance) and r_p denote the minimum distance (called the *perihelion* distance).

(a) Prove that the eccentricity e of the orbit can be written in the form

$$e = \frac{r_a - r_p}{r_a + r_p}$$

(b) Prove that $r_a = a(1 + e)$ and $r_p = a(1 - e)$, where a is the semimajor axis of the ellipse.

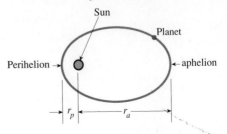

Sun

Planet

Perihelion →

← aphelion

r_p r_a

23. The planet Jupiter's semimajor axis is approximately 9.503 A.U. (astronomical units), and the eccentricity is 0.0484. Find the equation of its orbit (see Exercise 20). Find the aphelion and perihelion distances (see Exercise 22). *Note:* One A.U. is the semimajor axis of the earth's orbit, or approximately 1.496×10^{11} m. (Since the eccentricity of the earth's orbit is only 0.0167, its distance from the sun is always close to its semimajor axis, which we can think of as an average orbital radius.)

Jupiter, photographed with the Hubble Space Telescope

24. Repeat Exercise 23 for Halley's Comet, whose semi-major axis is approximately 18 A.U. and whose orbit has eccentricity 0.9.

25. Repeat Exercise 23 for Asteroid (5100) Pasachoff, whose semimajor axis is approximately 2.47 A.U. and whose eccentricity is approximately 0.135.

26. Prove that the rectangular form of Equation 9.29 for $e > 1$ is

$$\frac{(x-h)^2}{a^2} - \frac{(y-k)^2}{b^2} = 1$$

where $h = -\dfrac{e^2 d}{e^2 - 1}$, $k = 0$, $a = \dfrac{ed}{e^2 - 1}$, and $b = \dfrac{ed}{\sqrt{e^2 - 1}}$.

27. Prove that if a focus of a conic section having eccentricity e is at the pole and its corresponding directrix is d units above the pole and is horizontal, then the equation of the conic is

$$r = \frac{ed}{1 + e \sin \theta}$$

Similarly, show that the equation is

$$r = \frac{ed}{1 - e \sin \theta}$$

if the directrix is d units below the pole.

28. Show that if $e > 1$, the asymptotes of the hyperbolas

$$r = \frac{ed}{1 \pm e \cos \theta} \qquad \text{and} \qquad r = \frac{ed}{1 \pm e \sin \theta}$$

are parallel to the lines obtained by setting the denominator equal to zero and solving for θ. (*Hint:* Use the formulas $a = ed/(e^2 - 1)$ and $b = ed/\sqrt{e^2 - 1}$, together with the slopes of the asymptotes $\pm b/a$ for Position I and $\pm a/b$ for Position II.)

Chapter 9 Review Exercises

In Exercises 1–6, eliminate the parameter and use the result to sketch the curve defined by the parametric equations.

1. $\begin{cases} x = 2t - 1 \\ y = \sqrt{1 - t} \end{cases} \quad -\infty < t \le 1$

2. $\begin{cases} x = \sin^2 t \\ y = \cos t \end{cases} \quad -\infty < t < \infty$

3. $\begin{cases} x = \ln t \\ y = 1 - \dfrac{1}{t} \end{cases} \quad 0 < t < \infty$

4. $\begin{cases} x = 1 + e^t \\ y = 1 - e^{-t} \end{cases} \quad -\infty < t < \infty$

5. $\begin{cases} x = \tan^2 \theta - 1 \\ y = \sec \theta + 1 \end{cases} \quad 0 \le \theta < \dfrac{\pi}{2}$

6. $\begin{cases} x = 2 \sinh t - 1 \\ y = 3 \cosh t + 2 \end{cases} \quad -\infty < t < \infty$

In Exercises 7–10, find dy/dx and d^2y/dx^2 without eliminating the parameter.

7. $\begin{cases} x = t^2 - 1 \\ y = 2t^3 \end{cases}$

8. $\begin{cases} x = \sqrt{t} \\ y = t^2 + 2 \end{cases}$

9. $\begin{cases} x = \sin^2 t \\ y = \ln \cos t \end{cases}$

10. $\begin{cases} x = (t - 1)/t \\ y = \ln \sqrt{t} \end{cases}$

11. Find the equations of the tangent line and normal line to the curve defined by $x = \ln(t^2 - 3)$ and $y = t/\sqrt{t^2 - 3}$ at $(0, 2)$.

12. Without eliminating the parameter, find all maximum and minimum points on the curve defined by

$$\begin{cases} x = \sqrt{t^2 - 1} \\ y = 10t^3 - 3t^5 \end{cases} \quad t \geq 1$$

Use the Second Derivative Test for testing critical values.

13. Find the length of the curve defined by

$$\begin{cases} x = \cosh^2 t \\ y = 2\sinh t \end{cases} \quad 0 \leq t \leq \ln 3$$

14. Find the area of the surface formed by revolving the curve of Exercise 13 about (a) the x-axis and (b) the y-axis.

15. For the curve defined by

$$\begin{cases} x = \ln t - 1 \\ y = 2\sqrt{t} \end{cases} \quad 1 \leq t \leq 3$$

find (a) its length and (b) the area of the surface formed by revolving the curve about the x-axis.

16. Let R denote the region enclosed by the x-axis and one arch of the cycloid

$$\begin{cases} x = a(t - \sin t) \\ y = a(1 - \cos t) \end{cases}$$

Find (a) the area of R and (b) the volume of the solid formed by revolving R about the x-axis.

In Exercises 17 and 18, change to an equivalent rectangular equation, identify the curve, and draw its graph.

17. (a) $r = 2(\sin\theta + \cos\theta)$
(b) $r(2 + \sin\theta) = 1$

18. (a) $r = \dfrac{\sec\theta}{1 - \tan\theta}$
(b) $r = \sec^2(\theta/2)$

In Exercises 19–21, discuss and graph the given equation.

19. (a) $r = 2\sin^2(\theta/2)$
(b) $r^2 = 2\sin\theta\cos\theta$

20. (a) $r = -1 - 2\cos\theta$
(b) $r = 5 + 3\sin\theta$

21. (a) $r = \sin^2\theta - \cos^2\theta$
(b) $r\theta = 1, \ \theta > 0$

In Exercises 22–24, find the area of the specified region.

22. Inside both $r = \sqrt{3}\cos\theta$ and $r = \sin\theta$

23. Inside both $r = 1 - \cos\theta$ and $r = 2 + \cos\theta$

24. Outside $r = 2\cos 2\theta$ and inside $r^2 = 6\sin 2\theta$

25. Let C_1 be the curve $r = 2$ and C_2 the curve $r = 4\cos 3\theta$. Find the areas of the following regions:
(a) outside C_1 and inside C_2
(b) inside C_2 and inside C_1
(c) outside C_2 and inside C_1

In Exercises 26 and 27, find the angle between the two curves at each of their points of intersection.

26. $r = 5\sin\theta, r = 2 + \sin\theta$

27. $r = \dfrac{1}{1 + \sin\theta}, r = 2(1 - \sin\theta)$

In Exercises 28 and 29, show that the curves intersect orthogonally.

28. $r = \tan(\theta/2), r = e^{\cos\theta}; 0 < \theta < \pi$

29. $r(1 + \cos\theta) = 3, r = 4\cos\theta; -\pi/2 < \theta < \pi/2$

In Exercises 30 and 31, find the length of the given curve.

30. $r = \sqrt{1 - \sin 2\theta}; -\pi/4 \leq \theta \leq \pi/4$

31. $r = 7/\theta; 1 \leq \theta \leq 7$

32. Find the area of the surface formed by revolving the arc of the curve $r = \sec^2(\theta/2)$ from $\theta = 0$ to $\theta = 2\pi/3$ about the polar axis.

33. The **witch of Agnesi** is a curve generated by the point P (as in the figure) moving so that QPR is always a right triangle with right angle at P. The point Q is on the circle of radius a centered at $(0, a)$ and R is on the line $y = 2a$. Find parametric equations for this curve, using θ as a parameter, and draw its graph. (*Hint:* Make use of the polar equation of the circle.)

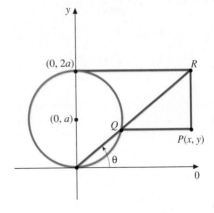

In Exercises 34–39, identify the curve defined by the given equation, and draw its graph.

34. $r = \dfrac{2}{\cos\theta - 1}$

35. $r = \dfrac{-2}{1 + \sin\theta}$

36. $r = \dfrac{1}{-1 + 2\cos\theta}$

37. $r = \dfrac{4}{5 - 3\cos\theta}$

38. $r(4\sin\theta - 3) = 8$

39. $r(\cos\theta - 2) = 5$

40. Show that the polar form of the hyperbola

$$\frac{x^2}{a^2} - \frac{y^2}{b^2} = 1$$

can be written as

$$r^2 = \frac{a^2(e^2 - 1)}{e^2\cos^2\theta - 1}$$

41. Show that the parabolas

$$r = \frac{ed}{1 + \cos\theta} \quad \text{and} \quad r = \frac{ed}{1 - \cos\theta}$$

are orthogonal (intersect at right angles).

42. The planet Pluto's semimajor axis is approximately 39.529 A.U. and its eccentricity is 0.2481. Find each of the following:
(a) the equation of its orbit;
(b) its longest and shortest distances from the sun. (See Exercise 22 in Exercise Set 9.8.)

Chapter 9 Concept Quiz

1. Let a curve C be given parametrically by $x = f(t)$ and $y = g(t)$ for $a \le t \le b$. In the following, assume that f and g satisfy relevant differentiability conditions. Give formulas for each of the following:
(a) dy/dx and d^2y/dx^2 in terms of t;
(b) the length of C;
(c) the area of the surface formed by revolving C about the x-axis and about the y-axis. (Assume that C lies in the first quadrant.)

2. (a) Write equations expressing rectangular coordinates x and y in terms of polar coordinates r and θ.
(b) Suppose a point P_1 has polar coordinates (r_1, θ_1). Give three other sets of polar coordinates for P_1, one of which involves a negative angle.

3. Identify the graph of each of the following polar equations.
(a) $r = a$
(b) $\theta = \alpha$
(c) $r = 2a\cos\theta$
(d) $r = a(1 + \cos\theta)$
(e) $r = a + b\cos\theta,\ a > b > 0$
(f) $r = a + b\cos\theta,\ 0 < a < b$
(g) $r^2 = a^2\cos 2\theta$
(h) $r = a\cos n\theta$
(i) $r = a\theta$
(j) $r\sec\theta = a$

4. (a) Let $r = f(\theta)$, where f is a nonnegative, continuous function for $\alpha \le \theta \le \beta$. Write the formula for the area of the region bounded by the graph of $r = f(\theta)$ and the rays $\theta = \alpha$ and $\theta = \beta$.
(b) Let $r = f(\theta)$ and $r = g(\theta)$ be continuous, nonnegative functions for $\alpha \le \theta \le \beta$, with $g(\theta) \le f(\theta)$. Write the formula for the area of the region bounded by the graphs of $r = f(\theta)$, $r = g(\theta)$, and the rays $\theta = \alpha$ and $\theta = \beta$.

5. In each of the following, identify the conic and give the equation (you may use the rectangular equation) of the directrix corresponding to the focus at the pole:
(a) $r = \dfrac{3}{2 + \cos\theta}$
(b) $r = \dfrac{2}{1 - \sin\theta}$
(c) $r = \dfrac{4}{1 - 2\cos\theta}$
(d) $r = \dfrac{6}{3\sin\theta - 2}$
(e) $r = \dfrac{-4}{4 - 2\cos\theta}$

6. Indicate which of the following statements are true and which are false.
(a) If x and y are given parametrically and you eliminate the parameter, the graph of the resulting rectangular equation is the same as the graph defined by the parametric equations.
(b) All points of intersection of the graphs of $r = f(\theta)$ and $r = g(\theta)$ are found by solving the equations simultaneously.
(c) The graph of the equation $r = f(\theta)$ is symmetric with respect to the pole if $f(\theta - \pi) = f(\theta)$.
(d) The slope of the tangent line to the graph of $r = f(\theta)$ at a point (r, θ) on the graph is $f'(\theta)$.
(e) If $r = f(\theta)$, $\alpha \le \theta \le \beta$, then parametric equations for the graph are $x = f(t)\cos t$, $y = f(t)\sin t$, $\alpha \le t \le \beta$.

APPLYING CALCULUS

1. The electrostatic charge distribution consisting of a charge q at the point with polar coordinates, $(s, 0)$ and a charge $-q$ at the point (s, π) is called a *dipole*.

   (a) The lines of force (force field) are given by the family of curves $r = C \sin^2 \theta$, where each value of $C > 0$ determines a specific line of force. Sketch the lines of force corresponding to $C = 1$, $C = 2$ and $C = 3$.

   (b) The equipotential lines for the dipole are given by the family of curves $r^2 = B \cos \theta$, $B > 0$. Sketch the equipotential lines corresponding to $B = 1$, $B = 2$ and $B = 3$.

   (c) Show that the lines of force and the equipotential lines are orthogonal trajectories.

2. Consider orbits, which follow Kepler's Laws.

   (a) The orbit of the earth is approximately an ellipse with the sun at one focus. The length of the semimajor axis is 9.26×10^7 miles and the eccentricity of the orbit is 0.02. Determine the minimum and maximum distances between the sun and the earth (the earth's *perihelion* and *aphelion* distances, respectively).

   (b) Halley's comet travels in an elliptic orbit with the sun at one focus. In terms of astronomical units (1 AU = 9.26×10^7 miles = the semimajor axis of the earth's orbit), the major and minor axes of the comet's orbit are 36.18 AU and 5.12 AU, respectively. What is the eccentricity of the orbit? What are the perihelion and aphelion distances? Express your answers in terms of astronomical units.

3. If a point P is on a fixed radius and at a distance h from the center of a circle of radius r which is rolling without slipping around a fixed circle of radius R, then the path traced out by P is called an *epitrochoid*. The geometrical design of the Wankel rotary engine (used in the Mazda RX-7) involves an epitrochoid in which $R = 2r$ and $0 < h < r$. The orbit of the planet Mercury relative to the Earth (i.e., in a coordinate system with the Earth at the origin) is an epitrochoic. Consider the epitrochoid given parametrically by

$$x = h \cos 3\theta + 3r \cos \theta, \qquad y = h \sin 3\theta + 3r \sin \theta$$

This particular curve is generated by having a circle of radius r roll on a circle of radius $2r$.

   (a) Sketch the graph of this epitrochoid in the two cases: $0 < h < r$ and $r < h < 3r$.

   (b) Suppose that $0 < h < r$ and let $R = 3r$. Show that the arc length of the curve is given by

$$L = \int_0^{2\pi} \sqrt{9h^2 + R^2 + 6hR \cos 2\theta}\, d\theta$$

which can be written in the form

$$L = \int_0^{2\pi} (3h + R)\sqrt{1 - k^2 \sin^2 \theta}\, d\theta$$

where $k^2 = \dfrac{12hR}{(3h + R)^2} \leq 1$. There is no elementary antiderivative for the integrand. The integral belongs to a class of integrals known as *elliptic integrals* and tables of values are given in mathematical handbooks.

   (c) Let $h = 1/2$ and $r = 1$ in the integral in (b). Use Simpson's rule with $n = 8$ to find an approximation for the length of the corresponding epitrochoid.

4. Kepler's laws of planetary orbits can be derived from Newton's law of gravity. In particular, Kepler's First Law is that the planets orbit the Sun in ellipses with the Sun at one focus. From the law of conservation of energy, given Newton's law of gravity, we have

$$\left[\left(\frac{dr}{d\theta}\right)^2 + r^2\right]\frac{J^2}{2\mu r^4} - \frac{k}{r} = W$$

where r is the distance between the Sun and a planet, J is the total angular momentum, μ is the mass function $m_1 m_2/(m_1 + m_2)$, k is the gravitational constant Gm_1m_2 where G is the universal constant of gravitation, and W is the total energy. W and J are constant.

(a) Solve the equation of conservation of momentum for θ as an indefinite integral.

(b) Show that the integral can be evaluated as an inverse trigonometric function:

$$\theta = \sin^{-1}\left(\frac{\mu k r - J^2}{r(\mu^2 k^2 + 2J^2\mu W)^{1/2}}\right) - \alpha$$

(c) Set the constant $\alpha = \pi/2$ and use $\sin(\theta + (\pi/2)) = \cos\theta$, and then solve for r as a function of θ.

(d) Show that this equation can be put into the form of the general polar equation of an ellipse

$$r = \frac{a(1 - e^2)}{1 \pm e\cos\theta}$$

where a is the semimajor axis of the ellipse and e is the eccentricity.

(e) From the equations relating the numerators and the denominators, respectively, solve for W, the total energy, and show that it depends only on the semimajor axis of the orbit.

5. Kepler's third law of planetary motion, the harmonic law, states that the period squared of a planet is proportional to the cube of its semimajor axis. Derive this law from the law of conservation of angular momentum

$$r^2(d\theta/dt) = J/\mu$$

where r is the distance between the planet and the Sun, J is the constant angular momentum, k is Gm_1m_2, and μ is a mass function $m_1m_2/(m_1 + m_2)$.

(a) Putting terms with r and θ on the left side of the equation and the constants and t on the right side, and integrating around the ellipse, evaluate the integral of the variables in terms of the area of an ellipse using its major and minor axes. Remember that the integral of time over the orbit is the period P and that the area element is $\frac{1}{2}(r^2)d\theta$. Show that $JP/2\mu = \pi ab$.

(b) Using the formula $b = a\sqrt{(1 - e^2)}$ for the semiminor axis of the ellipse in terms of the semimajor axis and eccentricity, and $J^2/\mu k = a(1 - e^2)$ for the angular momentum (see Problem 4(d)), solve for P in terms of a.

(c) Show that this result can be put in the usual form of Kepler's third law:

$$\frac{a^3}{P^2} = \frac{G}{4\pi^2}(m_1 + m_2)$$

6. A rope is anchored at a point on the ground at the base of a round silo that is 8 meters in diameter. The rope's length is half the circumference of the silo. A girl takes the free end of the rope and wraps the rope tightly around the silo until she is diametrically opposite the point where the rope is anchored. Then, while continuing to hold the rope taut, she starts walking away from the silo, so that the rope is always tangent to the circle. She continues until the rope stretches its full length in a straight line from the

anchor point. She then continues walking, holding the rope taut, now going in a circular path, until the rope again comes in contact with the silo on the opposite side from before. Finally, she comes back to the starting point with the rope wrapped all the way around the other half of the silo.

(a) Find parametric equations of the path of the girl. (*Hint:* The path will have to be divided into two parts.)

(b) Find the area of the region between the silo and the path of the girl.

Benoit B. Mandelbrot

Calculus is required as the first real mathematics course taught in most colleges. Is that still a reasonable requirement in view of the mathematics that is most applicable today?

It remains perfectly reasonable for many, or even most, students, but I would prefer to see the coexistence of several distinct introductory courses. At Yale, I work actively at promoting one that is based on fractal geometry. Calculus can be taught so as to leave little room for subjective grading, so it is tempting to use it as a kind of filter to select students among otherwise qualified candidates for professional schools. This may increase the number of jobs for math teachers, but mathematics as a profession is left the poorer. On the other hand, mathematicians find it hard to claim that their field must be supported simply because it is a noble and great adventure of humanity, like great music and drama. This is because music and drama are widely understood and appreciated in the general public, whereas mathematics is even less understood than the most extreme avant-garde in art.

Do fractals have a place in the teaching of calculus?

Yes, I think they have a very important place. The popular appeal of fractals shows, in my opinion, that almost everybody has some interest for some part of mathematics. Even when the overall presentation of calculus is very traditional and basic, fractals can help with a critically important lesson early on, when derivatives and tangents are introduced. My claim is that to understand tangents, it is best to start with curves that have no tangents, such as fractal curves. To the contrary, around 1977, an eminent international mathematical education committee recommended that teachers not confuse students by explaining that a curve may fail to have a tangent at any point. The only exceptions they allowed were those courses reserved for future professional mathematicians. Obviously, I think that this recommendation was badly timed and terribly out of touch. First of all, it happened to come out the same year my book *Fractals, Form, Chance and Dimension* was published. This book treated fractal geometry as a mathematical subspecialty. It effectively destroyed the well-entrenched belief that nondifferentiability was an abstract nonphysical notion, a kind of mathematical pathology. On the contrary, it is a very useful and broad notion in describing nature. In fact, the very same features that had been called "mathematical pathologies" are the very stuff of nature. Secondly, this committee seemed to be convinced that before starting calculus students' minds are blank slates. I take issue with this perception. I am convinced that every person, at a very early age, has a kind of vague but valid intuition based on their everyday experience that denies the physical existence of tangent lines or planes. Even a child knows that one cannot draw a tangent line to a coastline or a tangent plane to the bark of an oak tree. Even those who have never heard of fractals are at ease with the basic ideas of fractal geometry; yet textbooks throughout history have argued as if those intuitive ideas simply did not exist. Today, many high school students are exposed to fractal geometry through popular science books. They already know examples of theoretical curves without tangents. Therefore, a calculus course would benefit from including some form of the following argument. "Let's talk about something that you probably already know, either intuitively or from reading or class work. Most curves you either see or can draw are best viewed as having no tangents. On the other hand, a circle has a tangent and so do other simple abstract shapes. You may feel that this is a difficult notion; but practice makes it intuitive." Calculus can serve to model natural phenomena as well as abstract phenomena. It is indispensable in science as well as higher mathematics, and if you pursue calculus far enough, you'll find it extraordinarily beautiful.

Your life has been very interesting and unconventional. How do you think your experiences influenced your decision to become a mathematician?

One of my uncles was a successful mathematician. I could have followed directly in his footsteps, but despite this family connection, my path into mathematics was convoluted. My schooling was always chaotic and thoroughly disrupted by the Second World War. My mother was a doctor and terrified of epidemics, so she made me skip early school grades and I studied privately with another of my uncles. He despised rote learning, including the alphabet and multiplication tables. This definitely made an impression; I'm still not good with either. Instead, he encouraged me to read and train my memory. Mostly we played chess. My father was a map nut, and I could read maps as early as I can remember. I've since thought that maybe the constant daily practice of playing chess and reading maps in my childhood helped develop my geometric intuition, which became my most important scientific skill.

Before World War II started we moved from Poland to Paris, then to the country outside Paris. From mid–1942 to the end of 1943, the German occupation of France was tightening, and my family decided that it was not safe for me to attend school. Periods of intense danger and fear—several times I narrowly escaped being executed or deported—were separated by periods when not much was happening. I had plenty of time for study on my own. Finally I landed in a large lycée (high school) that included two postgraduate grades designed to prepare good students for Ecole Normale and Ecole Polytechnique. You had to pass a series of notoriously tough "killer exams." Many students, some of whom later became great scientists, had to repeat the second grade. During the second week after I joined this postgraduate high school in Lyons, I found out that I had a talent I had never known or suspected.

Your geometric intuition?

Yes. Our math professor would describe a problem that he enjoyed, always stating it in terms of algebra or analytic geometry. But I didn't seem to hear the analytic problem; instead I heard a geometric problem. I soon realized that during my self-schooling, I had become intimately familiar with a large "zoo" of geometric shapes that I could instantly recognize and call upon, even when they were dressed up in analytic clothes.

The five months I spent in Lyons were among the most important of my life. I hardly ever left the school grounds because I couldn't afford to buy a meal elsewhere and we feared and avoided the Nazi boss in Lyons (I later found out his name was Klaus Barbie). I had a burning desire to do well. I worked at a rate that I could not have sustained indefinitely, learning the basics for the exams and polishing my geometric skills. But becoming an algebra whiz was not my goal. Anyhow, I was supposed to take the exams "for the practice." To my amazement and that of nearly everyone else, not only did I pass, I finished a very close second in all of France.

Did these exams contain calculus topics?

Yes, I remember a conversation with a professor who complained about a problem involving a triple integral that he couldn't solve in the time allowed on the exam. I told him that I had recognized that the integral was simply the volume of the sphere in certain coordinates suggested by the geometry. After I had explained my reasoning he went away repeating "of course, of course."

So you were mathematically precocious because of your gift?

Yes. My university schooling continued and I became certified in several disciplines but only because of my geometric gift. But I missed the main goal of certification, which is to train students for a profession by providing guidance and role models. This never happened to me, in any profession. Unlike most mathematicians I did my best known work later in life. I didn't devote substantial time to mathematical research until I was 40 and I discovered the Mandelbrot Set in my mid-fifties.

Your schooling may have adversely affected your early academic career, but it obviously gave you other opportunities.

There are many reasons why my particular story isn't a good model for academic success. Even though established fields find it hard to deal with diversity in training, diversity is just as indispensable to science as it is to society. There are those who believe that a given discipline at a given instant should have a single "best way" of training new workers in that field, and many believe that this best way is determined by some kind of inevitable tide of history. Were this true, those who depart from the one true way would deserve to be shunned by the mainstream. That is indeed how things happen, but this is an absolutely deplorable situation. It may not be corrected, but it must be mitigated. To mitigate it, society must intervene to allow at least a few exceptions to survive.

Could you say very briefly how you came to the ideas of fractal geometry?

It would be nice to say that fractal geometry was born in some kind of "Eureka moment," like when Archimedes ran from his overflowing bathtub. It would also be nice to respond by identifying the source in some early application, like records of prices, errors in telephone lines, or the shape of coastlines. But this would be misleading. Fractal geometry didn't start with an overreaching idea centered on an already formulated problem, much less a problem already recognized as significant enough to justify the effort needed to answer it. Let me point out that *problem solving* and *problem setting* seem to involve different talents. In the popular view—and in the restricted views of many mature disciplines—science is viewed as problem solving. I regret this, perhaps because my talent lies mainly in problem setting.

The study of fractals proceeded from bottom to top. It went from a mess of disregarded and disconnected odds and ends, on to ill-organized parts, and then toward increasing organization. Of the disconnected odds and ends with which I started, some were theoretical. In that sense, fractal geometry has incorporated many geometric shapes that were known long before I was born, but remained scattered with no common idea behind them. This is also why they were given no name, other than "mathematical monsters." My present name for them is *protofractals*. They were part of the zoo of shapes that played a central role in my early career. A second source of disconnected odds and ends were miscellaneous facts about nature. Most people don't register facts that do not yet fit into any established theory, but I was blessed with a sponge-like memory and could recall without conscious effort all these odd shapes and all these odd facts. I also read about mathematics a lot. On many occasions I was reading about some new mathematical shape and thought instantly and spontaneously of some real empirical phenomenon that somehow had the same "taste." On other occasions, I went from fact to formula. Each time, I dropped whatever I had been doing and followed up this new link between form and substance. So the "flavor" of my scientific activity has been oscillating back and forth. It was, at different times, dominated either by the search for more facts—to be identified in the literature or discovered in the laboratory, or by the search for new mathematics—again, to be identified in the literature or developed from scratch. My skills can best be described as "having a nose" for the subject.

Benoit B. Mandelbrot is best known for his *Fractal Geometry of Nature* (1982). "The father of fractal geometry," Professor Mandelbrot continues to work in this new field, with its concrete applications, and its growing impact on art and on high-school and college science teaching. Professor Mandelbrot worked for 35 years at the IBM T. J. Watson Research Center, where he is now IBM Fellow Emeritus. His contributions, first viewed as isolated and peripheral, are now widely recognized as central to the new ideas concerning chaos and complexity. Latest among his numerous awards is the 1993 Wolf Prize for Physics.

10

INFINITE SERIES

Leonhard Euler (1707–1783) began his research in mathematics and physics at age 18. He joined the two Bernoulli brothers in St. Petersburg, Russia, in 1725 at the newly organized St. Petersburg Academy of Sciences and stayed for 14 years.

Euler's mathematical work ranged from the foundations of mathematical physics to the theory of lunar and planetary motion to differential geometry of surfaces. He devoted much attention to the problem of logarithms of negative numbers, the equation of a vibrating string, and problems in optics.

Euler investigated many functions and extensively studied power series. He discovered, for example, that $\frac{\pi}{2} - \frac{x}{2} = \sin x + \sin \frac{2x}{2} + \sin \frac{3x}{3} + \ldots$, a type of Fourier series.

Euler provided many notations we use today: the symbol e for the base of natural logarithms, the use of f and of parentheses for a function, the modern signs for trigonometric functions, the Greek Sigma (Σ) for sum, and the letter i for the square root of -1. The use of the letters $a, b,$ and c for the sides of a triangle is also due to Euler. One of the most prolific mathematicians in history, Euler published more than 500 books and articles.

In 1736 the great Swiss mathematician Leonhard Euler (pronounced "oiler") discovered the remarkable formula

$$1 + \frac{1}{2^2} + \frac{1}{3^2} + \frac{1}{4^2} + \cdots = \frac{\pi^2}{6} \tag{10.1}$$

The sum on the left-hand side of the equation is an example of what is called an *infinite series* (or simply a *series*). The three dots mean that we are to continue adding terms indefinitely. The question immediately arises as to how we can find the sum of infinitely many terms. We will give a formal definition soon, but for now you can think of the sum as the value that is *approached* (if such a value exists) as we add more and more terms. The discovery of Formula 10.1 is a fascinating story in the history of mathematics.

We will not try to show how Euler found the sum $\pi^2/6$ for the series in Equation 10.1 but will indicate another way of obtaining the result in Exercise 8 of Exercise Set 10.10. To illustrate the idea of the sum of an infinite series, we consider the simpler series

$$1 + \frac{1}{2} + \frac{1}{4} + \frac{1}{8} + \frac{1}{16} + \cdots \tag{10.2}$$

in which each term after the first is one-half the preceding one. By starting with the first term, and listing the finite sums obtained by successively adding on one more term, we get the following:

1	$1\frac{1}{2}$	$1\frac{3}{4}$	$1\frac{7}{8}$	$1\frac{15}{16}$
first term	sum of first two terms	sum of first three terms	sum of first four terms	sum of first five terms

This list of finite sums is called the *sequence of partial sums* of the series. It appears that as we add more and more terms, the finite sums come arbitrarily close to 2. For this reason we say that the *infinite* sum *is* 2, and write

$$1 + \frac{1}{2} + \frac{1}{4} + \frac{1}{8} + \frac{1}{16} + \cdots = 2$$

We say that the series *converges* to 2.

Euler was one of the most prolific writers of mathematics in history.

Not all series converge. For example, the sum of the series $1+2+3+4+\cdots$ clearly becomes arbitrarily large as we add on more and more terms. As another example, the series $1-1+1-1+1-\cdots$ does not approach any one value as we add more and more terms, since the finite sums alternate between 1 and 0, depending on where we stop. When a series fails to converge to a unique finite number, we say that it *diverges*. So both of the series $1+2+3+4+\cdots$ and $1-1+1-1+\cdots$ diverge.

As one final example, consider the series

$$1 + \frac{1}{2} + \frac{1}{3} + \frac{1}{4} + \cdots$$

composed of the reciprocals of consecutive positive integers. Does this series converge? If so, can you find its sum? Or does it diverge? You should make a conjecture by considering the sums of the first 10, 20, or even 100 terms using a CAS, spreadsheet, or calculator. We will return to this series in Section 10.2 and answer the questions raised here.

Infinite series can also be made up of variable terms. The simplest ones involve integral powers of a variable. An example is

$$1 + x + x^2 + x^3 + \cdots \tag{10.3}$$

We call such a series a *power series* in x, since each term is a power of x. We would expect the sum, if it exists, to depend on x; that is, the sum is a function of x. So, for the series 10.3, we can define a function f by

$$f(x) = 1 + x + x^2 + x^3 + \cdots$$

In general, a power series will converge for some values of x and diverge for other values (although it may converge for all values). For example, if we set $x = \frac{1}{2}$ in the series 10.3, we get the series 10.2, which converges to 2, as we have seen; but if we set $x = 1$, we get $1 + 1 + 1 + \cdots$, which clearly diverges.

If we consider only a finite number of terms of the series 10.3, we have a polynomial of the form

$$1 + x + x^2 + \cdots + x^n \qquad (n \geq 1)$$

which is an approximation to the infinite series and consequently to the function determined by the series. As we take polynomials of higher and higher degree (that is, as $n \to \infty$), the approximations become better and better, and in the limit we obtain the exact value of the function (assuming x is restricted to those values for which the series converges). Suppose now, instead of beginning with a series such as 10.3, we begin with a function $f(x)$. An important question is whether we can find an nth degree polynomial approximating $f(x)$ in such a way that as $n \to \infty$, the resulting "infinite polynomial" converges to $f(x)$ exactly.

For example, we will show in the next section that

$$e^x \approx 1 + x + \frac{x^2}{2!} + \frac{x^3}{3!} + \frac{x^4}{4!} + \cdots + \frac{x^n}{n!} \tag{10.4}$$

and that the approximation becomes better and better as we let n increase. So we might expect that we can let $n \to \infty$ and obtain the exact equality

$$e^x = 1 + x + \frac{x^2}{2!} + \frac{x^3}{3!} + \frac{x^4}{4!} + \cdots \tag{10.5}$$

Passing from the finite sum in Equation 10.4 to the infinite sum in Equation 10.5 and concluding that this latter sum converges to e^x requires justification. In this case the result is valid, as we will see later.

Our procedure in this chapter will be to begin with approximation of functions by polynomials. Next, we will discuss sequences and their limits, since, as we have indicated, the sum of an infinite series is determined from its sequence of partial sums. Then we will consider the general question of when an infinite series converges or diverges. In particular, we will study various convergence criteria for series of constants. Afterward, we will study power series and their properties. Then we will return to the crucial question of whether we can extend finite polynomial approximations of a function to an infinite power series that represents the function. We conclude the chapter by showing another way to represent functions by infinite series, one that involves sines and cosines instead of powers of x. This last kind of series, known as a *Fourier series,* is especially important in physics.

10.1 APPROXIMATION OF FUNCTIONS BY POLYNOMIALS; TAYLOR'S THEOREM

Polynomial functions are in many ways the simplest functions of all. First, they are the easiest to evaluate, requiring only addition and multiplication (since raising to positive integral powers is just repeated multiplication). Second, they possess derivatives of all orders, which are easily computed, for all real numbers. Third, they can be integrated with ease over any interval. In contrast, nonpolynomial functions can be difficult or impossible to evaluate, differentiate, or integrate. Even the exponential function $f(x) = e^x$ does not lend itself to easy evaluation. For example, try finding $e^{0.1}$ without the aid of a calculator, a computer, or tables. Similarly, $\ln 2$, $\sin 0.5$, and $\sqrt[3]{9.2}$ cannot readily be approximated as decimal quantities. We encounter the same problem in evaluating definite integrals. Although we can sometimes readily get the "answer," as in

$$\int_{\pi/6}^{\pi/2} \cot x\, dx = \ln(\sin x)\big]_{\pi/6}^{\pi/2} = \ln\left(\sin\frac{\pi}{2}\right) - \ln\left(\sin\frac{\pi}{6}\right)$$

$$= \ln 1 - \ln\frac{1}{2} = \ln 2$$

we are again confronted with the difficulty of giving this result as a decimal quantity. The situation is even worse with an integral such as

$$\int_0^1 e^{-x^2} dx$$

where no elementary antiderivative of the integrand exists. We could use Simpson's Rule (see Section 5.5) to approximate the integral in this case. If, however, we could find a polynomial approximation to e^{-x^2}, we could easily perform the integration on the polynomial, and the result would be an approximation to the given integral. We have seen that the function $f(x) = e^{-x^2}$ is especially important in probability theory.

Of course, calculators, computers, and tables *are* available, and with these a very large number of functions can be evaluated to a high degree of accuracy.

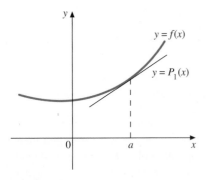

FIGURE 10.1
$P_1(x) = f(a) + f'(a)(x - a)$ is the linear approximation to $f(x)$ near $x = a$.

It is instructive, however, to learn how calculators and computers might arrive at the answers and how tables are constructed. Also, it is important to be able to determine just how accurate the approximations are. We will see answers to these questions in the procedures we develop in this section.

Linear and Quadratic Approximation

In Section 3.9 we discussed linear approximation of a function f near a point $x = a$. We found that if $f'(a)$ exists, then for x near a,

$$f(x) \approx f(a) + f'(a)(x - a)$$

If we let

$$P_1(x) = f(a) + f'(a)(x - a) \tag{10.6}$$

then P_1 is the first-degree polynomial that best approximates f near a. The graph of P_1 is just the tangent line to the curve $y = f(x)$ at the point for which $x = a$. (See Figure 10.1.) The graph of P_1 has two things in common with the graph of f: both graphs go through the point $(a, f(a))$, and they both have the slope $f'(a)$ at that point.

If we also know the concavity of the curve $y = f(x)$ at $x = a$, it would make sense to approximate f with a polynomial that not only goes through the point $(a, f(a))$ and has slope $f'(a)$ there, but also has the same concavity as f at $x = a$. (See Figure 10.2.) That is, we want a quadratic polynomial $P_2(x)$ that satisfies the following three conditions:

$$P_2(a) = f(a) \quad \text{Same value at } x = a$$
$$P_2'(a) = f'(a) \quad \text{Same slope at } x = a$$
$$P_2''(a) = f''(a) \quad \text{Same concavity at } x = a$$

(We use the subscript 2 to indicate that P_2 is a polynomial of degree 2.) Let us write such a polynomial as

$$P_2(x) = c_0 + c_1 x + c_2 x^2$$

and determine what values the coefficients c_0, c_1, and c_2 must have so that our three conditions are satisfied. By differentiating $P_2(x)$ and substituting $x = a$ in $P_2(x)$, $P_2'(x)$, and $P_2''(x)$, our three conditions can be written as

$$P_2(a) = c_0 + c_1 a + c_2 a^2 = f(a)$$
$$P_2'(a) = c_1 + 2c_2 a \qquad = f'(a)$$
$$P_2''(a) = 2c_2 \qquad\qquad = f''(a)$$

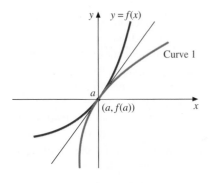

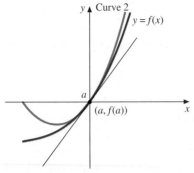

FIGURE 10.2
Curves 1 and 2 both go through the point $(a, f(a))$ with slope $f'(a)$, but Curve 2 better approximates the graph of f near that point, because its concavity agrees with that of f.

From the third equation, we see that

$$c_2 = \frac{f''(a)}{2}$$

Thus, on substituting this value of c_2 in the second equation, we have

$$c_1 = f'(a) - 2\left(\frac{f''(a)}{2}\right)a = f'(a) - af''(a)$$

and from the first equation,

$$c_0 = f(a) - \frac{f''(a)}{2}a^2 - \big[f'(a) - af''(a)\big]a$$

$$= f(a) - af'(a) + \frac{a^2}{2}f''(a)$$

Our polynomial can now be written in the form

$$P_2(x) = \overbrace{f(a) - af'(a) + \frac{a^2}{2}f''(a)}^{c_0} + \overbrace{\big[f'(a) - af''(a)\big]}^{c_1}x + \overbrace{\frac{f''(a)}{2}}^{c_2}x^2$$

By rearranging the right-hand side and factoring out the derivatives $f'(a)$ and $f''(a)$, we obtain

$$P_2(x) = f(a) + f'(a)(x - a) + \frac{f''(a)}{2}(x^2 - 2ax + a^2)$$

or, finally,

$$P_2(x) = f(a) + f'(a)(x - a) + \frac{f''(a)}{2}(x - a)^2 \tag{10.7}$$

EXAMPLE 10.1   Find the linear and quadratic polynomial approximations $P_1(x)$ and $P_2(x)$ for the function $f(x) = e^x$ near $x = 0$. Approximate $e^{0.1}$ using both $P_1(0.1)$ and $P_2(0.1)$.

Solution   Taking $a = 0$ in Equations 10.6 and 10.7, we get

$$P_1(x) = f(0) + f'(0)x$$

and

$$P_2(x) = f(0) + f'(0)x + \frac{f''(0)}{2}x^2$$

Since $f(x) = e^x$, both $f'(x)$ and $f''(x)$ also equal e^x. So $f(0) = 1$, $f'(0) = 1$, and $f''(0) = 1$. We therefore have

$$P_1(x) = 1 + x \qquad \text{and} \qquad P_2(x) = 1 + x + \frac{x^2}{2}$$

In Figure 10.3 we show the graph of $y = e^x$ along with the graphs of $y = P_1(x)$ and $y = P_2(x)$.

When $x = 0.1$, we have

$$P_1(0.1) = 1.1 \qquad \text{and} \qquad P_2(0.1) = 1.105$$

Using a calculator, we can find that

$$e^{0.1} \approx 1.10517092$$

to eight places. So $P_2(0.1)$ approximates $e^{0.1}$ to three places of accuracy.   ■

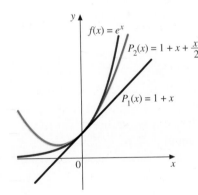

FIGURE 10.3

Taylor Polynomials of Degree n

Suppose now that f possesses derivatives up through the nth order at $x = a$. Our objective now is to find the polynomial of the nth degree, $P_n(x)$, whose first n derivatives at $x = a$ agree with the corresponding first n derivatives of

f at $x = a$ and whose value agrees with f at $x = a$. We know the result for $n = 1$ and $n = 2$. The pattern of the polynomials in these two cases suggests that we write $P_n(x)$ in the form

$$P_n(x) = c_0 + c_1(x - a) + c_2(x - a)^2 + \cdots + c_n(x - a)^n$$

Computing the successive derivatives, we obtain

$$P_n'(x) = c_1 + 2c_2(x - a) + 3c_3(x - a)^2 + \cdots + nc_n(x - a)^{n-1}$$
$$P_n''(x) = \qquad 2c_2 \qquad + 3 \cdot 2c_3(x - a) + \cdots + n(n-1)c_n(x - a)^{n-2}$$
$$P_n'''(x) = \qquad\qquad\qquad 3 \cdot 2c_3 \qquad + \cdots + n(n-1)(n-2)c_n(x - a)^{n-3}$$
$$\vdots \qquad\qquad\qquad\qquad\qquad \vdots$$
$$P_n^{(n)}(x) = \qquad\qquad\qquad\qquad\qquad\qquad n!\,c_n$$

Setting $x = a$ and equating $P_n(a)$ with $f(a)$ and $P_n^{(k)}(a)$ with $f^{(k)}(a)$ for $k = 1, \ldots, n$, gives

$$P_n(a) = c_0 \quad = f(a)$$
$$P_n'(a) = c_1 \quad = f'(a)$$
$$P_n''(a) = 2c_2 \quad = f''(a), \quad \text{so } c_2 = \frac{f''(a)}{2!}$$
$$P_n'''(a) = 3 \cdot 2c_3 = f'''(a), \quad \text{so } c_3 = \frac{f'''(a)}{3!}$$
$$\vdots \qquad \vdots \qquad\qquad \vdots$$
$$P_n^{(n)}(a) = n!\,c_n \quad = f^{(n)}(a), \quad \text{so } c_n = \frac{f^{(n)}(a)}{n!}$$

We can write the kth coefficient c_k as

$$c_k = \frac{f^{(k)}(a)}{k!}, \qquad k = 0, 1, 2, \ldots, n$$

where $f^{(0)}(a)$ means $f(a)$. The polynomial $P_n(x)$ with these coefficients is named after the English mathematician Brook Taylor (1685–1731).

Definition 10.1
The nth Taylor Polynomial

Let f possess derivatives up through the nth order at $x = a$. The polynomial

$$P_n(x) = f(a) + f'(a)(x - a) + \frac{f''(a)}{2!}(x - a)^2 + \cdots + \frac{f^{(n)}(a)}{n!}(x - a)^n$$

(10.8)

is called the **nth Taylor polynomial for f about $x = a$**.

REMARK ────────────

■ As our derivation of $P_n(x)$ shows, $P_n(x)$ is the polynomial of degree n in powers of $x - a$ that agrees with $f(x)$ at $x = a$ and whose first n derivatives agree with the corresponding derivatives of $f(x)$ at $x = a$.

────────────

EXAMPLE 10.2   Find the Taylor polynomials $P_3(x)$, $P_4(x)$, and $P_5(x)$ about $x = 0$ for the function $f(x) = e^x$. Discuss the approximations to $e^{0.1}$ and $e^{1.0}$ by each of these polynomials.

Solution   Recall that in Example 10.1 we found $P_1(x)$ and $P_2(x)$ for this same function. To find $P_3(x)$, $P_4(x)$, and $P_5(x)$ we use Equation 10.8 with $a = 0$. Since all derivatives of $f(x) = e^x$ also equal e^x, their common value at $x = 0$ is 1. So we have

$$P_3(x) = 1 + x + \frac{x^2}{2!} + \frac{x^3}{3!}$$

$$P_4(x) = 1 + x + \frac{x^2}{2!} + \frac{x^3}{3!} + \frac{x^4}{4!}$$

$$P_5(x) = 1 + x + \frac{x^2}{2!} + \frac{x^3}{3!} + \frac{x^4}{4!} + \frac{x^5}{5!}$$

where $k! = 1 \cdot 2 \cdot 3 \cdot \cdots \cdot k$. The computer-generated graphs in Figure 10.4 show that P_3, P_4, and P_5 approximate f more and more closely as the degree of the Taylor polynomial increases. For x near 0, the polynomial graphs are almost indistinguishable from the graph of f. As we move farther from 0 the approximation is not as good. To confirm this observation, we compute values of each of the polynomials at $x = 0.1$ and at $x = 1.0$.

$$P_3(0.1) \approx 1.105167 \qquad P_3(1.0) = 2.6167$$
$$P_4(0.1) \approx 1.1051708 \qquad P_4(1.0) = 2.7083$$
$$P_5(0.1) \approx 1.105170917 \qquad P_5(1.0) = 2.71667$$

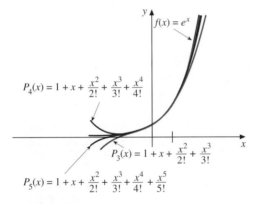

FIGURE 10.4

As we found with a calculator in Example 10.1, $f(0.1) = e^{0.1} \approx 1.10517092$ to eight places. Also, $f(1.0) = e^{1.0} \approx 2.71828183$ to eight places. We see, then, that for $x = 0.1$ each of the polynomials gives an answer close to the true value. In fact, $P_3(0.1)$, if rounded, is correct to five places, $P_4(0.1)$ is correct to six places, and $P_5(0.1)$ is correct to eight places.

In contrast, $P_3(1.0)$ is not even correct to one decimal place, $P_4(1.0)$ is correct to one place only, and $P_5(1.0)$ is correct only to two places. If we wanted more accuracy at $x = 1$, we would have to use a higher-degree Taylor polynomial. Also, to approximate e^x for values of x near $x = 1$, it would be better to use Taylor polynomials about $x = 1$ than about $x = 0$. ∎

As this example shows, the farther x is from the point $x = a$ (in our case $a = 0$), the less accurate is the Taylor polynomial approximation $P_n(x)$ for any fixed n. Also, for a fixed n we get better and better approximations by $P_n(x)$ as

n increases. Although these observations are valid for the function $f(x) = e^x$, it is not clear that they are true for other functions. We will have more to say about errors in approximating functions by their Taylor polynomials shortly. First, let us look at another example.

EXAMPLE 10.3　　Find the Taylor polynomial $P_n(x)$ about $x = 0$ for the function $f(x) = \cos x$.

Solution　　To apply Formula 10.8, we need to calculate successive derivatives of $f(x) = \cos x$ and evaluate them at $x = 0$.

$$
\begin{array}{llll}
f(x) & = \cos x & f(0) & = 1 \\
f'(x) & = -\sin x & f'(0) & = 0 \\
f''(x) & = -\cos x & f''(0) & = -1 \\
f'''(x) & = \sin x & f'''(0) & = 0 \\
f^{(4)}(x) & = \cos x & f^{(4)}(0) & = 1 \\
\;\;\;\;\vdots & & \;\;\;\;\vdots &
\end{array}
$$

The pattern is now clear: $f^{(n)}(0) = \pm 1$ when n is even, and $f^{(n)}(0) = 0$ when n is odd. Let $n = 2m$, where $m = 0, 1, 2, 3, \ldots$. Then n is even. Since the signs of $f^{(2m)}(0)$ alternate from $+1$ to -1, we can write

$$f^{(2m)}(0) = (-1)^m, \qquad m = 0, 1, 2, \ldots$$

Thus, by Formula 10.8,

$$P_{2m}(x) = 1 - \frac{x^2}{2!} + \frac{x^4}{4!} - \frac{x^6}{6!} + \cdots + \frac{(-1)^m x^{2m}}{(2m)!} = \sum_{k=0}^{m} \frac{(-1)^k x^{2k}}{(2k)!}$$

Note that since $f^{(2m+1)}(0) = 0$, it follows that $P_{2m+1}(x) = P_{2m}(x)$ for $f(x) = \cos x$.

In Figure 10.5 we show a computer-generated graph of $f(x)$ along with $P_0(x)$, $P_2(x)$, $P_4(x)$, $P_6(x)$, $P_8(x)$, and $P_{10}(x)$.　■

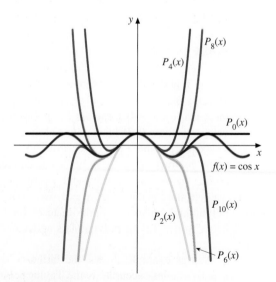

FIGURE 10.5

The Error in the Taylor Polynomial Approximation

We know that the nth Taylor polynomial $P_n(x)$ and its first n derivatives agree with $f(x)$ at $x = a$, and that for x near a, $P_n(x) \approx f(x)$. But just how good is this approximation? The following theorem provides the answer. We give its proof in Supplementary Appendix 5.

THEOREM 10.1

Taylor's Formula with Remainder

Let f have derivatives up through the $(n+1)$st order in an open interval I centered at $x = a$. Then for each x in I there is a number c between a and x such that

$$f(x) = f(a)+f'(a)(x-a)+\frac{f''(a)}{2!}(x-a)^2+\cdots+\frac{f^{(n)}(a)}{n!}(x-a)^n+R_n(x)$$

$$\text{(10.9)}$$

where

$$R_n(x) = \frac{f^{(n+1)}(c)}{(n+1)!}(x-a)^{n+1} \qquad \text{(10.10)}$$

REMARKS

■ Formula 10.9 (Taylor's Formula) can be written more briefly as $f(x) = P_n(x) + R_n(x)$.

■ $R_n(x)$ is called the **Lagrange form of the remainder**, after Joseph Louis Lagrange (1736–1813), one of the greatest mathematicians of the eighteenth century. The term *remainder* is used here to mean what is left when the approximation $P_n(x)$ is subtracted from $f(x)$. It is the error in approximating $f(x)$ by $P_n(x)$.

■ The number c as stated in the theorem is dependent on x. It is between a and x, whether x is greater than a or less than a.

■ The interval I may be the entire real line, $(-\infty, \infty)$, or it may be finite, in which case it will be of the form $I = (a - r, a + r)$ for some $r > 0$. The size of the interval is determined by the point closest to a for which $f^{(n+1)}(x)$ fails to exist. If this point is r units from a, then $I = (a - r, a + r)$. If $f^{(n+1)}(x)$ exists everywhere, then $I = (-\infty, \infty)$.

■ Note that the remainder term $R_n(x)$ follows the same pattern as the terms in $P_n(x)$ *except* that the derivative is evaluated at the intermediate value c between a and x, rather than at a.

When $n = 0$ in Taylor's Formula and we replace x with b, where b is in the interval I and $b \neq a$, the result is

$$f(b) = f(a) + f'(c)(b - a)$$

or equivalently,

$$f'(c) = \frac{f(b) - f(a)}{b - a}$$

where c is between a and b. But this formula is just the statement of the conclusion in the Mean-Value Theorem (Theorem 4.9). For this reason Theorem 10.1 is sometimes called the *Extended Mean-Value Theorem*. (We should note that the Mean-Value Theorem is used in the proof of Theorem 10.1, so we cannot use this theorem as an independent way of getting the Mean-Value Theorem.)

EXAMPLE 10.4   Find $P_n(x)$ and $R_n(x)$ about $x = 0$ for $f(x) = e^x$. Use the result to find a bound on the error in estimating $e^{0.1}$ using $P_3(0.1)$.

Solution   Note first that $f(x) = e^x$ possesses derivatives of all orders every-where, all equal to e^x, so that in this case Theorem 10.1 is valid on $(-\infty, \infty)$. Taking $a = 0$ in Theorem 10.1, we have

$$P_n(x) = 1 + x + \frac{x^2}{2!} + \frac{x^3}{3!} + \cdots + \frac{x^n}{n!}$$

and

$$R_n(x) = \frac{f^{(n+1)}(c)}{(n+1)!} x^{n+1} = \frac{e^c}{(n+1)!} x^{n+1}$$

where c is between 0 and x. Thus, the error in using $P_3(0.1)$ to estimate $e^{0.1}$ is

$$R_3(0.1) = \frac{e^c}{4!}(0.1)^4, \qquad 0 < c < 0.1$$

Although we do not know c, we can get an upper bound on e^c by the fact that

$$e^c < e^{0.1} < e^1 < 3$$

This is a very generous upper bound, since $e^{0.1}$ is much less than e, but since we presumably do not know $e^{0.1}$, we can take something larger, with a value that we do know. Thus,

$$R_3(0.1) < \frac{3}{4!}(0.1)^4 = \frac{3}{4 \cdot 3 \cdot 2 \cdot 1}(0.0001) = 0.0000125$$

Even using the generous upper bound on e^c, we see that the error is small, resulting in accuracy to at least four decimal places. Recall that in Example 10.2 we found that $P_3(0.1) \approx 1.10517$, which is correct to five decimal places.   ■

EXAMPLE 10.5   Find Taylor's Formula with remainder for arbitrary n for the function $f(x) = \ln x$ about $x = 1$, and state the interval in which it is valid.

Solution   In abbreviated form, Taylor's Formula is $f(x) = P_n(x) + R_n(x)$. For P_n we need the first n derivatives of f, and for R_n we need $f^{(n+1)}$. So we begin by calculating the successive derivatives:

$$
\begin{aligned}
f(x) &= \ln x & f(1) &= 0 \\
f'(x) &= \tfrac{1}{x} = x^{-1} & f'(1) &= 1 \\
f''(x) &= -x^{-2} & f''(1) &= -1 \\
f'''(x) &= 2x^{-3} & f'''(1) &= 2 \\
f^{(4)}(x) &= -3 \cdot 2x^{-4} = -3!\,x^{-4} & f^{(4)}(1) &= -3! \\
f^{(5)}(x) &= 4 \cdot 3!\,x^{-5} = 4!\,x^{-5} & f^{(5)}(1) &= 4! \\
\vdots \qquad & \qquad\qquad \vdots & \vdots \qquad & \\
f^{(n)}(x) &= (-1)^{n-1}(n-1)!x^{-n} & f^{(n)}(1) &= (-1)^{n-1}(n-1)!
\end{aligned}
$$

For $R_n(x)$ we need $f^{(n+1)}(x)$ evaluated at $x = c$:

$$f^{(n+1)}(x) = (-1)^n n! \, x^{-(n+1)} \qquad f^{(n+1)}(c) = (-1)^n n! \, c^{-(n+1)}$$

Observe carefully how we handled the coefficients of the successive derivatives. First, they alternate in sign—they are positive when the order is odd and negative when the order is even. The factor $(-1)^{n-1}$ in the nth derivative accomplishes the alternation of the signs. Second, we did not multiply out the factors as they accumulated but rather indicated the product as a factorial. Doing so enabled us to recognize the general pattern. We also used the fact that $k! = k(k-1)!$ for $k \geq 1$. You should convince yourself of this equality.

Since

$$f^{(n+1)}(x) = \frac{(-1)^n n!}{x^{n+1}}$$

is not defined when $x = 0$, the largest interval about $x = 1$ for which Taylor's Formula is valid is $(0, 2)$. (Later we will see that the formula remains valid at the right endpoint $x = 2$, but for now we will consider only the open interval.)
Now we can find $P_n(x)$ and $R_n(x)$. For $x \in (0, 2)$,

$$f(x) = P_n(x) + R_n(x)$$

$$= f(1) + f'(1)(x-1) + \frac{f''(1)}{2!}(x-1)^2 + \frac{f'''(1)}{3!}(x-1)^3$$

$$+ \frac{f^{(4)}(1)}{4!}(x-1)^4 + \cdots + \frac{f^{(n)}(1)}{n!}(x-1)^n + \frac{f^{(n+1)}(c)}{(n+1)!}(x-1)^{n+1}$$

$$= 0 + 1(x-1) + \frac{-1}{2!}(x-1)^2 + \frac{2}{3!}(x-1)^3 + \frac{-3!}{4!}(x-1)^4 + \cdots$$

$$+ \frac{(-1)^{n-1}(n-1)!}{n!}(x-1)^n + \frac{(-1)^n n!}{(n+1)! \, c^{n+1}}(x-1)^{n+1}$$

or

$$f(x) = (x-1) - \frac{(x-1)^2}{2} + \frac{(x-1)^3}{3} - \frac{(x-1)^4}{4} + \cdots$$

$$+ \frac{(-1)^{n-1}(x-1)^n}{n} + R_n(x)$$

where

$$R_n(x) = \frac{(-1)^n (x-1)^{n+1}}{(n+1)c^{n+1}}, \qquad c \text{ between } 1 \text{ and } x$$

In Figure 10.6 we show the graphs of $f(x) = \ln x$, along with the graphs of $P_n(x)$ for $n = 2$, 5, and 10. Notice how all the graphs of $P_n(x)$ differ significantly from that of $\ln x$ for $x > 2$. ■

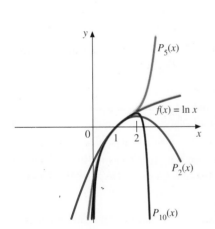

FIGURE 10.6

EXAMPLE 10.6   Using the results of Example 10.5,

(a) find $P_5(1.2)$ and an upper bound on the error in using $P_5(1.2)$ to approximate $\ln 1.2$, and

(b) find the smallest integer n that will guarantee that $P_n(1.2)$ estimates $\ln 1.2$ with eight-decimal-place accuracy.

Solution

(a) From Example 10.5,

$$P_5(x) = (x-1) - \frac{(x-1)^2}{2} + \frac{(x-1)^3}{3} - \frac{(x-1)^4}{4} + \frac{(x-1)^5}{5}$$

and
$$R_5(x) = \frac{(-1)^5(x-1)^6}{6c^6}, \qquad c \text{ between } 1 \text{ and } x$$

So
$$P_5(1.2) = 0.2 - \frac{(0.2)^2}{2} + \frac{(0.2)^3}{3} - \frac{(0.2)^4}{4} + \frac{(0.2)^5}{5}$$
$$\approx 0.182331$$

The error is given by
$$|R_5(1.2)| = \left| \frac{(-1)^5(0.2)^6}{6c^6} \right| = \frac{0.000064}{6c^6}$$

Since $1 < c < 1.2$, we get a larger fraction by replacing c by 1. Thus,
$$|R_5(1.2)| \leq \frac{0.000064}{6} \approx 0.000011$$

So the answer is correct to at least four decimal places.

(b) From Example 10.5,
$$R_n(x) = \frac{(-1)^n(x-1)^{n+1}}{(n+1)c^{n+1}}, \qquad c \text{ between } 1 \text{ and } x$$

With $x = 1.2$, an upper bound on $|R_n(1.2)|$ can be obtained by replacing c with 1. Then $c^{n+1} = 1^{n+1} = 1$, so we have
$$|R_n(1.2)| \leq \frac{(0.2)^{n+1}}{n+1}$$

To obtain eight decimal places of accuracy, we find the smallest integer n such that
$$\frac{(0.2)^{n+1}}{n+1} \leq 5 \times 10^{-9}$$

By trial and error, using a calculator, a CAS, or a spreadsheet, we find that $n + 1 = 11$, or $n = 10$, is the smallest value that works. So $P_{10}(1.2)$ estimates $\ln 1.2$ to at least eight decimal places of accuracy. ∎

EXAMPLE 10.7   Find an upper bound on the error in using the nth Taylor polynomial $P_n(x)$ about $x = 0$ to approximate $f(x) = \cos x$. Use $P_6(x)$ to approximate $\cos 10°$ and estimate the error.

Solution   In Example 10.3 we found that
$$P_{2n}(x) = 1 - \frac{x^2}{2!} + \frac{x^4}{4!} + \cdots + \frac{(-1)^n}{(2n)!}x^{2n}$$

and that $P_{2n+1}(x) = P_{2n}(x)$. The remainder term in Taylor's Formula (Equation 10.9) can therefore be written
$$R_{2n}(x) = R_{2n+1}(x) = \frac{f^{(2n+2)}(c)}{(2n+2)!}x^{2n+2}$$

where c is between 0 and x. Since an even-order derivative of $\cos x$ is either $\cos x$ or $-\cos x$, we therefore have
$$|R_{2n}(x)| = \frac{|\cos c|}{(2n+2)!}|x|^{2n+2} \leq \frac{|x|^{2n+2}}{(2n+2)!}$$

since $|\cos c| \leq 1$. Thus, the error in using $P_{2n}(x)$ to approximate $\cos x$ for any real number x is at most $|x|^{2n+2}/(2n+2)!$.

To use $P_6(x)$ to approximate $\cos 10°$, we first express $10°$ in radians:

$$10° = 10° \left(\frac{\pi \text{ radians}}{180°} \right) = \frac{\pi}{18} \text{ radians}$$

Then we have

$$P_6 \left(\frac{\pi}{18} \right) = 1 - \frac{1}{2!} \left(\frac{\pi}{18} \right)^2 + \frac{1}{4!} \left(\frac{\pi}{18} \right)^4 - \frac{1}{6!} \left(\frac{\pi}{18} \right)^6 \approx 0.9848077530$$

The upper bound we found above on the error is

$$\left| R_6 \left(\frac{\pi}{18} \right) \right| \le \frac{\left(\frac{\pi}{18} \right)^8}{8!} \approx 2.135 \times 10^{-11}$$

So our answer is correct to at least ten decimal places. ∎

A Preview of Taylor Series

In Example 10.4 we found the nth Taylor polynomial about $x = 0$ for $f(x) = e^x$ to be

$$P_n(x) = 1 + x + \frac{x^2}{2!} + \frac{x^3}{3!} + \cdots + \frac{x^n}{n!}$$

So we have the approximation

$$e^x \approx 1 + x + \frac{x^2}{2!} + \frac{x^3}{3!} + \cdots + \frac{x^n}{n!}$$

and the approximation gets better and better as n increases. Later we will show that the approximation is valid for all real values of x. A natural question to ask is whether we can let n "go to infinity" and get the exact value of e^x. That is, can we say that

$$e^x = 1 + x + \frac{x^2}{2!} + \frac{x^3}{3!} + \frac{x^4}{4!} + \cdots \tag{10.11}$$

for all real values of x? Recall from this chapter's introduction that we call a series such as this one a *power series in x*. If we set $x = 1$, for example, our question is whether it is valid to write

$$e = 1 + 1 + \frac{1}{2!} + \frac{1}{3!} + \frac{1}{4!} + \cdots$$

We are immediately confronted with another question: What do we mean by the sum of infinitely many numbers? Clearly, we must have some new definition of addition in such a case. We will give a definition in the next section and will explore its consequences in the sections that follow. We will find that Equation 10.11 is valid for all real values of x.

For now, we will say that the expression on the right-hand side of Equation 10.11 is the **Taylor Series of e^x about $x = 0$** and that, under the meaning yet to be given, the series **converges** to e^x for all real x. More generally, by letting $n \to \infty$ in Taylor's Formula with Remainder, Equation 10.9, we obtain the **Taylor Series for the function f about $x = a$:**

$$f(a) + f'(a)(x - a) + \frac{f''(a)}{2!}(x - a)^2 + \frac{f'''(a)}{3!}(x - a)^3 + \cdots$$

As you might guess, the condition under which this series converges to $f(x)$ is that the error in the approximation of $f(x)$ by its nth Taylor polynomial gets smaller and smaller as $n \to \infty$; that is,

$$\lim_{n \to \infty} R_n(x) = 0$$

We will return to this important question of the representation of a function by its Taylor series after our discussion of infinite series in general.

COMPUTING TAYLOR POLYNOMIALS USING A COMPUTER ALGEBRA SYSTEM

Taylor's Theorem presents a method for using polynomials to approximate functions that cannot be explicitly evaluated. The theorem also provides a useful means of analyzing the error in such approximations. The degree of accuracy in using Taylor polynomials to approximate functions generally increases as the degree of the polynomial increases. The higher the degree of the polynomial desired, the greater the amount of computation necessary. For this reason, a computer algebra system becomes a valuable aid.

CAS CAS 36

Use Taylor polynomials to approximate the function $f(x) = e^{-x^2}$ close to 0.

We will begin by using Maple, Mathematica, and DERIVE to compute the Taylor polynomial of degree 8 for f near $x = 0$.

Maple:

f := x–>exp(–x^2);
taylor(f(x),x=0,9);

The output is:

$$1 - x^2 + \frac{1}{2}x^4 - \frac{1}{6}x^6 + \frac{1}{24}x^8 + O(x^9)$$

Notice the big-O term on the end of the output from Maple. This indicates that all higher degree terms have an exponent of at least 9. Because of this form of the output in Maple, if the Taylor polynomial of degree 8 is desired, the last parameter in the **taylor** command should be specified as 9.

Mathematica:

f[x_] = Exp[–x^2]
Normal[Series[f[x],{x,0,8}]]

DERIVE:

(At the □ symbol, go to the next step.)
a (author) □ exp(–x^2) □ c (calculus)
□ t (taylor) □ [choose expression] □
x (variable) □ 8 (degree) □ 0 (point) □
s (simplify)

Notice that Mathematica and DERIVE do not include a big-O term in the output. In Mathematica, the Normal command removes this big-O term.

To see the relative accuracy of the approximation near 0 a plot would be helpful. Maple does not directly plot the result of the command   taylor(f(x),x=0,9);. We could enter a plot and retype the 8th degree polynomial as the function to plot. We can, however, avoid the problem by using a more general command in Maple, called mtaylor. To access this command, it must be read from a Maple library via readlib(`mtaylor`):

Maple:

readlib('mtaylor'):
P := mtaylor(f(x),[x],9);

The output is:

$$P := 1 - x^2 + \frac{1}{2}x^4 - \frac{1}{6}x^6 + \frac{1}{24}x^8$$

plot({f(x),P},x=–2..2);

We see from Figure 10.1.1 that near 0 the approximation looks very good.

Mathematica:

P[x_] = Normal[Series[f[x], {x,0,8}]]
Plot[{f[x],P[x]},{x,–2,2}]

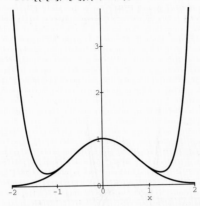

DERIVE:

(At the □ symbol, go to the next step.)
[choose exp(−x^2)] □ p (plot window)
□ p (plot) □ a (algebra) □ [choose Taylor polynomial] □ p (plot window)
□ p (plot)

FIGURE 10.1.1

To see the relative accuracy obtained in approximating f by Taylor polynomials of different degrees, we could repeat what we have already done with different polynomials and compare the pictures obtained. This would be tedious. However, the computer algebra systems allow us to make this kind of comparison very quickly by using a loop structure.

Maple:

for n from 1 by 1 to 10 do
  P := mtaylor(f(x),[x],n);
od;

The output gives Taylor polynomials of degrees 0 to 9.

$$P := 1$$

$$P := 1$$

$$P := 1 - x^2$$

$$P := 1 - x^2$$

$$P := 1 - x^2 + \frac{1}{2}x^4$$

$$P := 1 - x^2 + \frac{1}{2}x^4$$

$$P := 1 - x^2 + \frac{1}{2}x^4 - \frac{1}{6}x^6$$

$$P := 1 - x^2 + \frac{1}{2}x^4 - \frac{1}{6}x^6$$

$$P := 1 - x^2 + \frac{1}{2}x^4 - \frac{1}{6}x^6 + \frac{1}{24}x^8$$

$$P := 1 - x^2 + \frac{1}{2}x^4 - \frac{1}{6}x^6 + \frac{1}{24}x^8$$

Notice the 1st and 2nd, 3rd and 4th, 5th and 6th, etc., Taylor polynomials are identical.

To get a series of plots in Maple, enter:

for n from 1 by 1 to 10 do
P[n] := mtaylor(f(x),[x],n);
plot({f(x),P[n]},x=–2..2);
od;

To plot f and the Taylor polynomials for f about $x = 0$, of orders 2, 4, 6, 8, and 10 on the same axes, use the Maple command:

plot({f(x),P[2],P[4],P[6],P[8],P[10]},
x=–5..5,y=–5..5);

Mathematica:

Do[p[x_,k] = Normal[Series[f[x],{x,0,k}]],
    {k,0,8}]
Do[Plot[{f[x],p[x,k]},
    {x,–2,2}],{k,0,8}]

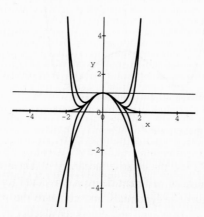

FIGURE 10.1.2
$f(x)$, $P_2(x)$, $P_4(x)$, $P_6(x)$, $P_8(x)$, and $P_{10}(x)$

The graphical analysis above gives an intuitive feel for where the Taylor polynomial is a good approximation to the function. For more precise numerical information, the error term should be analyzed.

Exercise Set 10.1

In Exercises 1–6, find the Taylor polynomial of the specified degree for f about $x = 0$.

1. $f(x) = e^{-x}$; $n = 5$

2. $f(x) = \sin x$; $n = 7$

3. $f(x) = 1/(x + 1)$; $n = 6$

4. $f(x) = \sqrt{x + 1}$; $n = 5$

5. $f(x) = \cosh x$; $n = 8$

6. $f(x) = \tan x$; $n = 5$

In Exercises 7–12, find the Taylor polynomial of degree n for f about the specified point a.

7. $f(x) = 1/x$; $a = 1$ **8.** $f(x) = e^{x-2}$; $a = 2$

9. $f(x) = \sin x$; $a = 0$

10. $f(x) = 1/(1 - x)^2$; $a = -1$

11. $f(x) = \ln(2 - x)$; $a = 1$

12. $f(x) = \sinh x$; $a = 0$

In Exercises 13–18, find $R_n(x)$ for the functions specified in Exercises 7–12 and give the interval about which $f(x) = P_n(x) + R_n(x)$.

In Exercises 19–32, estimate the specified function value using P_n for the given value of n and an appropriate value of a. Find an upper bound on the error, and determine the number of decimal places of accuracy. (When appropriate, you may use the results of preceding exercises or examples.)

19. e^{-1}; $n = 5$

20. $\sqrt{1.2}$; $n = 5$

21. $\sin 1.5$; $n = 7$

22. $\ln 1.5$; $n = 6$

23. $\cos 2$; $n = 8$

24. $\sinh 2$; $n = 7$

25. $\sqrt{5}$; $n = 5$ (*Hint:* Take $a = 4$.)

26. $\ln 4/5$; $n = 8$

27. e^3; $n = 12$

28. $\sqrt{e}$; $n = 4$

29. $\cosh 1$; $n = 6$

30. $\sin 32°$; $n = 5$ (*Hint:* Change to radians and use $a = \pi/6$.)

31. $\cos 58°$; $n = 4$ (*Hint:* Take $a = \pi/3$.)

32. $\sqrt[3]{7}$; $n = 4$

33. Use the result of Example 10.7 to find the degree of the Taylor polynomial for $\cos x$ about $x = 0$ that approximates $\cos 72°$ to six decimal places of accuracy. Find this approximation and compare it with the value given by a calculator.

34. Find $\sin 80°$ correct to five decimal places of accuracy using the appropriate Taylor polynomial about $x = 0$. (See Exercises 9 and 15.)

35. Find $\sinh 2$ to five decimal places of accuracy using the appropriate Taylor polynomial about $x = 0$. (See Exercises 12 and 18.)

36. Find $P_n(x)$ and $R_n(x)$ for $f(x) = \sqrt{x}$ about $x = 4$. In what interval is the result valid? Use the result to find $\sqrt{3}$ correct to four decimal places.

37. Let $f(x) = (1 + x)^\alpha$, where α is real.
(a) Find $P_n(x)$ and $R_n(x)$ for f about $x = 0$.
(b) Show that if $\alpha = m$, where m is a positive integer, then $f(x) = P_m(x)$ for all x.
(c) If α is not a positive integer, show that $f(x) = P_n(x) + R_n(x)$ for $-1 < x < 1$.
(d) Use the result of part (a) to estimate $\sqrt[3]{1.5}$, taking $n = 4$. Find an upper bound on the error.

38. (a) Show that if $f(x)$ is a polynomial of degree n, then the Taylor polynomial $P_n(x)$ for f about $x = a$ is identical to $f(x)$.

(b) Express the polynomial
$$f(x) = 2x^5 - 3x^4 + 7x^3 - 8x^2 + 2$$
in powers of $x - 1$.

39. For $-1 < x < 1$, let
$$f(x) = \ln \frac{1 + x}{1 - x}$$
(a) Show that if u is any positive number, there is an x in $(-1, 1)$ such that $f(x) = \ln u$.
(b) Find $P_n(x)$ and $R_n(x)$ for f about $x = 0$. (*Hint:* Rewrite $f(x)$ using properties of logarithms before taking derivatives.)
(c) Use parts (a) and (b) to find $\ln 2$ correct to three decimal places.

40. (a) In Exercise 39(a), let $u = (N + 1)/N$, where N is a positive integer. What is x so that $f(x) = \ln u$?
(b) Use the result of part (a), together with Exercise 39, to show that
$$\ln(N + 1) = \ln N + 2 \sum_{k=1}^{n} \frac{1}{(2k - 1)(2N + 1)^{2k-1}}$$
$$+ R_{2n}\left(\frac{1}{2N + 1}\right)$$
and that
$$\left| R_{2n}\left(\frac{1}{2N + 1}\right) \right| \le \frac{2}{(2n + 1)(2N)^{2n+1}}$$
(Note that $R_{2n-1}(x) = R_{2n}(x)$, since the Taylor polynomial consists of odd powers only.)
(c) Find $\ln 3$ correct to three decimal places, making use of part (b) and Exercise 39(c).

41. Let $f(x) = \ln(1 + x)$.
(a) Find the nth Taylor polynomial $P_n(x)$ for f about $x = 0$.
(b) Use a CAS or graphing calculator to obtain the graphs of $f(x)$, $P_5(x)$, $P_{10}(x)$, and $P_{20}(x)$ on the same set of axes.
(c) Describe the behavior of the graphs obtained in part (b). Pay particular attention to values of x in the interval $(-1, 1)$ as compared with values outside this interval.
(d) Conjecture, based on your graphs, the values of x for which the remainder term $R_n(x)$ approaches 0 as $n \to \infty$.

42. Use the remainder term $R_n(x)$ in Taylor's Formula to determine the values of x for which the approximation
$$\sin x \approx x - \frac{x^3}{6}$$
is accurate to within 0.001 unit.

43. Repeat Exercise 42 with

$$\cos x \approx 1 - \frac{x^2}{2}$$

44. Repeat Exercise 42 with

$$\sqrt{1 + x^2} \approx 1 + \frac{x^2}{2}$$

45. Letting $f(x) = \sin x$,
   (a) find the nth Taylor polynomial for f about $x = 0$;
   (b) generate graphs of f and P_n on the same axes for $n = 1, 3, 5, 10$;
   (c) use the graphs to estimate on which interval $P_3(x)$ approximates $f(x)$ to within about 0.01 unit. (Zoom in.) Repeat for $P_5(x)$ and for $P_{10}(x)$.

46. Repeat Exercise 45 with $f(x) = e^{-x}$.

47. Repeat Exercise 45 with $f(x) = (1 - x)^{1/3}$.

48. Repeat Exercise 47 using the nth Taylor polynomial about $x = 2$.

49. Repeat Exercise 45 with $f(x) = \sqrt{x + 1}$, first using $P_n(x)$ about $x = 0$, and then using $P_n(x)$ about $x = 3$.

50. Let $f(x) = e^x$. Find a value of n so that $P(0.1)$ is within 10^{-6} units of $e^{0.1}$.

10.2 SEQUENCES

In the introduction to this chapter we indicated how the sum of an infinite *series* can be defined. The key lies in computing what we called the *sequence of partial sums* and determining if these partial sums approach some finite value as more and more terms are added. To illustrate this relationship between a series and its sequence of partial sums, we used the series

$$1 + \frac{1}{2} + \frac{1}{4} + \frac{1}{8} + \frac{1}{16} + \cdots \tag{10.12}$$

Its sequence of partial sums is

$$1, \quad 1\frac{1}{2}, \quad 1\frac{3}{4}, \quad 1\frac{7}{8}, \quad 1\frac{15}{16}, \ldots \tag{10.13}$$

which comes arbitrarily close to 2 as we add more and more terms. Thus, we concluded that the sum of series 10.12 is 2. Recall that we obtain the sequence 10.13 by writing, in order, the first term of the series 10.12, the sum of the first two terms, the sum of the first three terms, and so on.

Note carefully that in a series, such as 10.12, the terms are added together. In contrast, in a sequence, such as 10.13, the members (also called terms) are separated by commas. Though the words *sequence* and *series* are often used interchangeably in nonmathematical usage, in mathematics we are careful to distinguish between them.

We will explore more fully the relationship between a series and its sequence of partial sums in the next section. In this section we investigate general properties of sequences. Later, we will apply these properties to sequences of partial sums of a series. Since the convergence or divergence of a series is of central importance, one of our main goals is to define the notion of the *limit* of a sequence and to learn some methods of finding this limit when it exists.

Let us be more precise about the meaning of a sequence. One way of defining a sequence is to say it is an ordered collection of elements, called *terms*, formed according to some rule that gives a unique term for each positive integer. Since a rule that assigns a unique element to each positive integer is a function having the positive integers as its domain, we can state the definition more succinctly as follows:

Definition 10.2 A Sequence	A **sequence** is a function whose domain is the set of all positive integers.

REMARK
■ According to the definition, a sequence is a function, but it is customary to refer to the ordered list of function values (the range) as the sequence itself.

For most of the sequences we will be dealing with, the terms will be real numbers, but they could be other mathematical entities. Note that by our definition, there will always be infinitely many terms, one for each positive integer (although the terms may not be distinct—there can be repetitions). So when we speak of a sequence, we will always mean an *infinite sequence*. In Definition 10.2, if the domain is replaced by the set of the first n positive integers, the resulting function is called a *finite sequence*.

Notation

It is customary, in giving the rule for a sequence, to use subscript notation, such as a_n, rather than the usual functional notation $f(n)$. For example, if we write

$$a_n = \frac{1}{n}$$

then by substituting, in turn, $n = 1, 2, 3, \ldots$, we get the sequence

$$\frac{1}{1}, \frac{1}{2}, \frac{1}{3}, \frac{1}{4}, \ldots \tag{10.14}$$

We refer to a_n as the **nth term**, or **general term**, of the sequence. Giving a formula for the nth term effectively defines the entire sequence.

When a formula for the nth term a_n of a sequence is known, we can designate the sequence with braces around a_n, such as $\{a_n\}$ or, more explicitly, $\{a_n\}_{n=1}^{\infty}$. The latter notation makes it explicit that n takes on the positive integers starting with 1. If the simple notation $\{a_n\}$ is used (as will usually be the case), we will understand it to mean the same as $\{a_n\}_{n=1}^{\infty}$. (There may, however, be times when we want to start with $n = 0$ or some other positive integer n_0 with $n_0 \neq 1$. If so, we will indicate the limits.) Using this notation, we can designate the sequence 10.14 by $\{1/n\}$.

In the next example we show several sequences using this notation.

EXAMPLE 10.8   Show the first five terms of each of the following sequences, indicating its continuation by three dots.

(a) $\{2n - 1\}$ (b) $\{(-1)^{n-1}\}$ (c) $\left\{\dfrac{n}{n+1}\right\}$

(d) $\left\{\dfrac{2^n}{n!}\right\}$ (e) $\{\sin n\}$

Solution

(a) $1, 3, 5, 7, 9, \ldots$ (the odd positive integers)

(b) $(-1)^0, (-1)^1, (-1)^2, (-1)^3, (-1)^4, \ldots$, or on simplifying, $1, -1, 1, -1, 1, \ldots$

(c) $\dfrac{1}{2}, \dfrac{2}{3}, \dfrac{3}{4}, \dfrac{4}{5}, \dfrac{5}{6}, \ldots$

(d) $\dfrac{2^1}{1!}, \dfrac{2^2}{2!}, \dfrac{2^3}{3!}, \dfrac{2^4}{4!}, \dfrac{2^5}{5!}, \ldots$ Since $1! = 1$, $2! = 2 \cdot 1 = 2$, $3! = 3 \cdot 2 \cdot 1 = 6$, $4! = 4 \cdot 3 \cdot 2 \cdot 1 = 24$, and $5! = 5 \cdot 4 \cdot 3 \cdot 2 \cdot 1 = 120$, we can rewrite the sequence as

$$\frac{2}{1}, \frac{4}{2}, \frac{8}{6}, \frac{16}{24}, \frac{32}{120}, \ldots$$

which becomes, after reducing the fractions,

$$2, 2, \frac{4}{3}, \frac{2}{3}, \frac{4}{15}, \ldots$$

(e) $\sin 1, \sin 2, \sin 3, \sin 4, \sin 5, \ldots$. Using a calculator (in radian mode), we get (rounding to four decimal places) approximately $0.8415, 0.9093, 0.1411, -0.7568, -0.9589, \ldots$. ■

If the pattern of the terms is clear, so that we can infer the formula for the nth term, it is permissible to designate the sequence by merely showing a few of its terms. For example, the sequence 10.14 could be given simply by writing

$$1, \frac{1}{2}, \frac{1}{3}, \frac{1}{4}, \ldots$$

We could also show the sequence in parts (a), (b), and (c) of Example 10.8 in this way. However, showing the sequence in part (d) in its simplified form,

$$2, 2, \frac{4}{3}, \frac{2}{3}, \frac{4}{15}, \ldots$$

is *not* in itself a satisfactory way of designating the sequence, since it is not clear from this list alone how to form succeeding terms.

EXAMPLE 10.9   Assume that each of the following sequences continues in the pattern of the terms shown. Write the general term.

(a) $\dfrac{1}{2}, -\dfrac{1}{4}, \dfrac{1}{6}, -\dfrac{1}{8}, \ldots$ (b) $\dfrac{1}{2}, \dfrac{3}{4}, \dfrac{5}{8}, \dfrac{7}{16}, \dfrac{9}{32}, \ldots$

Solution

(a) The alternating signs can be indicated by means of the factor $(-1)^{n-1}$, as we saw in Example 10.8(b). Each denominator is twice the number of the term. So we can write

$$a_n = \frac{(-1)^{n-1}}{2n}$$

(b) The numerators are the odd positive integers, which are indicated by the factor $2n - 1$, as we saw in Example 10.8(a). The denominators appear to be powers of 2, with the nth term denominator equal to 2^n. So we can write

$$a_n = \frac{2n - 1}{2^n}$$ ∎

It is not always possible to give an explicit formula for the nth term. For example, consider the sequence $\{a_n\}$ where a_n is the nth prime number. Recall that a prime number is a positive integer greater than 1 whose only positive divisors are 1 and itself. Several terms of this sequence are

$$2, 3, 5, 7, 11, 13, 17, 19, 23, \ldots$$

It is known that there are infinitely many primes, but there is no known formula for the nth prime.

As another example, let a_n denote the digit in the nth decimal place of the decimal representation of π. The first few terms are

$$1, 4, 1, 5, 9, 2, 6, 5, \ldots$$

Incidentally, π has been calculated to more than two billion places! Again, no formula has been found for the nth decimal digit.

Recursive Definitions

Sometimes sequences are defined *recursively*. For example, suppose we are given that $a_1 = \sqrt{2}$ and that for all $n \geq 2$,

$$a_n = \sqrt{2 + a_{n-1}}$$

This formula giving a_n in terms of a_{n-1} is called a **recursion formula**. Since we know that $a_1 = 1$, we can find a_2 by the recursion formula as follows:

$$a_2 = \sqrt{2 + a_1} = \sqrt{2 + \sqrt{2}}$$

Now we use the recursion formula again to get a_3:

$$a_3 = \sqrt{2 + a_2} = \sqrt{2 + \sqrt{2 + \sqrt{2}}}$$

and again to get a_4:

$$a_4 = \sqrt{2 + a_3} = \sqrt{2 + \sqrt{2 + \sqrt{2 + \sqrt{2}}}}$$

and so on. We could approximate each term by using a calculator, but the pattern would no longer be evident.

Another example of a recursive definition is the sequence of approximations by Newton's Method to a root of an equation $f(x) = 0$ (see Section 4.5). After the initial approximation x_0 is made, we find subsequent approximations by the recursion formula:

$$x_{n+1} = x_n - \frac{f(x_n)}{f'(x_n)} \qquad (n = 0, 1, 2, \ldots)$$

Note that in this case we begin with $n = 0$.

The Fibonacci sequence is named after the Italian businessman and mathematician Leonardo Fibonacci (who lived about 1170–1250). He posed the following problem:

> How many pairs of rabbits can be produced from a single pair in a year if every month each pair begets a new pair that from the second month onward becomes productive?

If we assume the rabbits in the initial pair are newborn (so that they produce the first new pair after two months), then the number of pairs present in the nth month is f_n, the nth term of the Fibonacci sequence.

Perhaps the best-known sequence defined recursively is the **Fibonacci sequence** (see the note in the margin), defined by

$$f_1 = 1, \ f_2 = 1, \ \text{and} \ f_n = f_{n-1} + f_{n-2} \qquad \text{for } n \geq 2$$

In this case two initial terms are given, and the recursion formula gives the nth term as the sum of the two preceding terms. The first few terms are

$$1, 1, 2, 3, 5, 8, 13, \ldots$$

The numbers f_n are called *Fibonacci numbers*, and they occur in nature in various ways. For example, if you count the number of spirals on a pine cone or a pineapple, you will get a Fibonacci number. The same is true for the number of spirals formed by the seeds of a sunflower. Also, the number of petals on a daisy is normally equal to a Fibonacci number.

Graphical Representation of Sequences

There are two useful ways of showing sequences graphically. In the first we use a two-dimensional coordinate system just as with other functions. We plot the points (x, y), where $x = n$ and $y = a_n$ for $n = 1, 2, 3, \ldots$. The graph consists of isolated points that we do not connect. In the second method we show the numbers a_n, where $n = 1, 2, 3, \ldots$, as points on a number line. We illustrate both methods in the next example.

EXAMPLE 10.10   Show each of the following sequences graphically by two methods.

(a) $\left\{ \dfrac{1}{n} \right\}$ (b) $\left\{ \dfrac{n}{n+1} \right\}$

Solution

(a) By the first method we plot the points $(1, 1)$, $(2, \frac{1}{2})$, $(3, \frac{1}{3})$, $(4, \frac{1}{4})$, ... We show the result in Figure 10.7(a). By the second method, we show the terms

$$1, \frac{1}{2}, \frac{1}{3}, \frac{1}{4}, \ldots$$

as points on a number line in Figure 10.7(b).

(b) In the first method we plot the points $(1, \frac{1}{2})$, $(2, \frac{2}{3})$, $(3, \frac{3}{4})$, $(4, \frac{4}{5})$, ..., as shown in Figure 10.8(a) at the top of page 741. In the second method we show the points

$$\frac{1}{2}, \frac{2}{3}, \frac{3}{4}, \frac{4}{5}, \ldots$$

as points on a number line in Figure 10.8(b). ∎

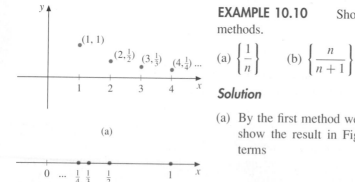

(a)

(b)

FIGURE 10.7

Limits of Sequences

In Figure 10.7(a) it appears that if we were to plot more and more points, they would come arbitrarily close to the x-axis, that is, to the line $y = 0$. Similarly, in Figure 10.7(b) the points seem to be coming arbitrarily close to 0. In fact, we can see from the nth term

$$a_n = \frac{1}{n}$$

that as $n \to \infty$, a_n comes arbitrarily close to 0.

In Figure 10.8(a) the points seem to be getting arbitrarily close to the horizontal line $y = 1$, and in Figure 10.8(b) the points on the number line seem to be getting close to 1. If we consider the nth term in this case,

$$a_n = \frac{n}{n + 1}$$

we can see intuitively that for large n, a_n is almost 1, since the numerator and denominator differ by only one unit. For example, with $n = 1{,}000$,

$$a_n = \frac{1000}{1001}$$

which is only slightly less than 1.

When the terms of a sequence come arbitrarily close to some number L as $n \to \infty$, we say that the sequence **converges** and that its **limit** is L. From our discussion in Example 10.10, we strongly suspect that both sequences $\{1/n\}$ and $\{n/(n + 1)\}$ converge, the first to the limit 0 and the second to the limit 1. In the case of the second sequence, we can use the "dominant term" approach, just as we did with rational functions of a continuous variable in Chapters 1 and 2. We reason that for large n, both numerator and denominator are approximately the same as their terms of highest degree. In this case, then, we could say that

$$\frac{n}{n + 1} \approx \frac{n}{n} = 1$$

for large n.

Another way to see this same result is to divide numerator and denominator by n:

$$\frac{n}{n + 1} = \frac{1}{1 + \frac{1}{n}}$$

As $n \to \infty$, $1/n \to 0$, so we conclude that the limit is 1.

The precise definition of the limit of a sequence is given below.

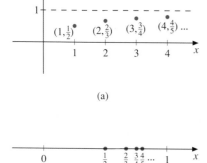

(a)

(b)

FIGURE 10.8

Definition 10.3
The Limit of a Sequence

We say that the **limit** of the sequence $\{a_n\}$ is L and write

$$\lim_{n \to \infty} a_n = L$$

provided that corresponding to each positive number ε, there is a positive integer N such that

$$|a_n - L| < \varepsilon$$

for all $n > N$. If the limit of $\{a_n\}$ is L, we say that the sequence **converges** to L. If the limit of $\{a_n\}$ does not exist, we say that the sequence **diverges**.

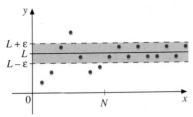

For $n > N$, the points (n, a_n) lie within an ε-band about L.

(a)

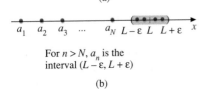

For $n > N$, a_n is the interval $(L - \varepsilon, L + \varepsilon)$

(b)

FIGURE 10.9

REMARKS ─────────────────────────

■ Since ε can be any positive number, by saying that $|a_n - L| < \varepsilon$ for all $n > N$, we are saying that the term a_n of the sequence will be *arbitrarily* close to L (closer than ε, however small ε may be) if we go far enough out in the sequence (beyond the Nth term).

■ A sequence can have at most one limit, since the terms cannot simultaneously all be arbitrarily close to two different values.

■ Definition 10.3 is the same as Definition 2.3 for the limit of a function $f(x)$ as $x \to \infty$, except that the continuous variable x is replaced by the integer variable n.

─────────────────────────────────

In Figure 10.9 we illustrate Definition 10.3 graphically by both methods.

EXAMPLE 10.11   Use Definition 10.3 to show that

$$\lim_{n \to \infty} \frac{n}{n + 1} = 1$$

Solution   Consider the difference $|a_n - L|$, where $a_n = n/(n+1)$ and $L = 1$:

$$\left| \frac{n}{n + 1} - 1 \right| = \left| \frac{n - n - 1}{n + 1} \right| = \left| \frac{-1}{n + 1} \right| = \frac{1}{n + 1}$$

Now let ε denote any positive number. We want to see how large to make n so that $1/(n + 1) < \varepsilon$. Since

$$\frac{1}{n + 1} < \frac{1}{n}$$

we can simplify our choice by making $1/n < \varepsilon$. Solving this inequality for n, we get $n > 1/\varepsilon$. Thus, if we let N denote any integer larger than $1/\varepsilon$, we will have, for all $n > N$,

$$\left| \frac{n}{n + 1} - 1 \right| < \varepsilon$$

So by Definition 10.3,

$$\lim_{n \to \infty} \frac{n}{n + 1} = 1 \qquad \blacksquare$$

EXAMPLE 10.12   Show that each of the following sequences is divergent.
(a) $\{(-1)^{n-1}\}$ (b) $\{2n - 1\}$ (c) $\{\sin n\}$

Solution

(a) The sequence is $1, -1, 1, -1, 1, \ldots$. There is no *single* number L that all terms come arbitrarily close to, no matter how far out in the sequence we go. More formally, if we choose $\varepsilon = \frac{1}{2}$, say, then regardless of the value of L, and regardless of how large N is, there are infinitely many terms a_n with $n > N$ such that $|a_n - L| \geq \frac{1}{2}$, contrary to the requirement on L given in Definition 10.3. The graph of $\{(-1)^{n-1}\}$ in Figure 10.10 clearly shows that the terms do not approach a limit.

(b) The sequence is $1, 3, 5, 7, 9, \ldots$, whose terms become arbitrarily large as $n \to \infty$, so the sequence diverges. We show its graph in Figure 10.11.

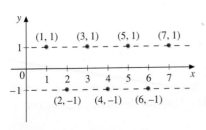

FIGURE 10.10
Graph of $\{(-1)^{n-1}\}$

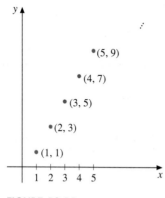

FIGURE 10.11
Graph of $\{2n - 1\}$

(c) In Figure 10.12 we show by means of the dashed line the graph of $f(x) = \sin x$. The points $(n, \sin n)$ lie on this graph. It is evident that as $n \to \infty$ these points do not come arbitrarily close to any horizontal line $y = L$. That is, $\lim_{n \to \infty} \sin n$ does not exist, so the sequence $\{\sin n\}$ diverges. ■

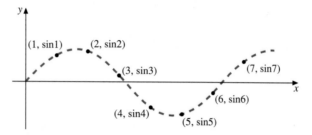

FIGURE 10.12
Graph of $\{\sin n\}$

EXAMPLE 10.13   Use Definition 10.3 to prove that if $|r| < 1$, then $\lim_{n \to \infty} r^n = 0$.

Solution   If $|r| < 1$, then its reciprocal $1/|r|$ is greater than 1, so we can write

$$\frac{1}{|r|} = 1 + p$$

for some positive number p. By the Binomial Theorem (see Appendix 4),

$$(1 + p)^n = 1 + np + \frac{n(n - 1)}{2!}p^2 + \cdots + p^n > 1 + np > np$$

Thus, inverting, we get

$$|r|^n = \frac{1}{(1 + p)^n} < \frac{1}{np}$$

If ε denotes any positive integer, then we can make $1/np < \varepsilon$ by choosing $n > 1/\varepsilon p$. So if we choose N any integer greater than $1/\varepsilon p$, we have, for all $n > N$,

$$|r^n - 0| = |r^n| = |r|^n < \frac{1}{np} < \varepsilon$$

By Definition 10.3, it follows that

$$\lim_{n \to \infty} r^n = 0$$ ■

n	$(0.9999)^n$
10	0.9990
100	0.9900
1000	0.9048
10,000	0.3679
100,000	0.000045

Even if we choose r only slightly less than 1—say, $r = 0.9999$—it is still true that r^n approaches 0 as $n \to \infty$. The table in the margin gives pretty convincing evidence for $r = 0.9999$.

If $r = 1$, then $r^n = 1$ for all n, and the sequence $\{r^n\}$ is the *constant* sequence

$$1, 1, 1, \ldots$$

whose limit is clearly 1. (After all, $|a_n - 1| = |1 - 1| = 0$, which is less than ε for any positive ε and for all n.)

If $r = -1$, the sequence $\{r^n\}$ is

$$-1, 1, -1, 1, -1, \ldots$$

which diverges, similar to the sequence in Example 10.12(a).

If $|r| > 1$, then, as we showed in Example 10.13, $|r| = 1 + p$ for some $p > 0$, and $|r|^n = (1 + p)^n > np$. Since $p > 0$, $np \to \infty$ as $n \to \infty$. Thus, $\{r^n\}$ diverges. For $r > 1$, r^n becomes arbitrarily large as $n \to \infty$. For $r < -1$, r^n oscillates in sign but becomes large in absolute value.

We now summarize our findings about the sequence $\{r^n\}$. (See also Figure 10.13.)

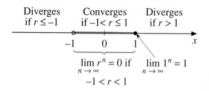

| Diverges | Converges | Diverges |
| if $r \le -1$ | if $-1 < r \le 1$ | if $r > 1$ |

$\lim\limits_{n \to \infty} r^n = 0$ if $\lim\limits_{n \to \infty} 1^n = 1$

$-1 < r < 1$

FIGURE 10.13
Convergence of $\{r^n\}$

The Sequence $\{r^n\}$

(a) If $|r| < 1$, $\lim\limits_{n \to \infty} r^n = 0$.

(b) If $r = 1$, $\lim\limits_{n \to \infty} r^n = 1$.

(c) For all other values of r, the sequence diverges.

Infinite Limits

As we saw with the sequence $\{2n - 1\}$ in Example 10.12(b) and again with the sequence $\{r^n\}$ when $r > 1$, some sequences diverge because their terms become arbitrarily large as $n \to \infty$. We indicate this type of divergence of a sequence $\{a_n\}$ by writing

$$\lim_{n \to \infty} a_n = \infty$$

Similarly, if the terms a_n are negative (at least for all sufficiently large n) but become arbitrarily large in absolute value as $n \to \infty$, we write

$$\lim_{n \to \infty} a_n = -\infty$$

In the first case we may say that "a_n approaches infinity" and in the second that "a_n approaches minus infinity," but in neither case does the sequence $\{a_n\}$ converge. We are simply saying it diverges in a particular way.

REMARKS

■ A sequence can diverge without approaching ∞ or $-\infty$, as we saw in Example 10.12, parts (a) and (c).

■ When we write $\lim_{n \to \infty} a_n = L$, we will always mean that L is finite.

Limit Properties of Sequences

The limit properties for functions given in Section 2.2 also hold true for sequences. After all, sequences are particular kinds of functions. We state these properties below in terms of sequences.

Limit Properties for Sequences

If $\lim\limits_{n \to \infty} a_n$ and $\lim\limits_{n \to \infty} b_n$ both exist, then the following properties hold true:

1. $\lim\limits_{n \to \infty} ca_n = c \left(\lim\limits_{n \to \infty} a_n \right)$ for any constant c

2. $\lim\limits_{n \to \infty} (a_n + b_n) = \lim\limits_{n \to \infty} a_n + \lim\limits_{n \to \infty} b_n$

3. $\lim\limits_{n \to \infty} (a_n - b_n) = \lim\limits_{n \to \infty} a_n - \lim\limits_{n \to \infty} b_n$

4. $\lim\limits_{n \to \infty} a_n b_n = \left(\lim\limits_{n \to \infty} a_n \right) \left(\lim\limits_{n \to \infty} b_n \right)$

5. $\lim\limits_{n \to \infty} \dfrac{a_n}{b_n} = \dfrac{\lim\limits_{n \to \infty} a_n}{\lim\limits_{n \to \infty} b_n}$ if $b_n \neq 0$ for all n and $\lim\limits_{n \to \infty} b_n \neq 0$

The next two theorems are often helpful in finding limits of sequences. We omit the proofs.

THEOREM 10.2

If $\lim\limits_{n \to \infty} a_n = L$, and f is a function whose domain includes L and a_n for $n \geq N$, and if f is continuous at $x = L$, then

$$\lim_{n \to \infty} f(a_n) = f(L)$$

In particular, since $f(x) = x^k$, for k a positive integer, is continuous for all x, we have

$$\lim_{n \to \infty} (a_n)^k = L^k$$

provided the sequence $\{a_n\}$ converges to L. Similarly,

$$\lim_{n \to \infty} \sqrt[k]{a_n} = \sqrt[k]{L}$$

provided $a_n > 0$ and $L > 0$ for even-ordered kth roots.

THEOREM 10.3

Let $\{a_n\}$ be a sequence and f a function such that

$$f(n) = a_n, \qquad n = 1, 2, 3, \ldots$$

If

$$\lim_{x \to \infty} f(x) = L$$

then also

$$\lim_{n \to \infty} a_n = L$$

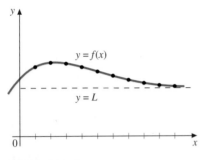

FIGURE 10.14
If $\lim\limits_{x \to \infty} f(x) = L$, then $\lim\limits_{n \to \infty} a_n = L$, where $a_n = f(n)$.

You will be asked to give a proof of Theorem 10.3 in Exercise 48 of Exercise Set 10.2, but the result is obvious intuitively by looking at the graph in Figure 10.14. The points (n, a_n) lie on the graph of $y = f(x)$, since $a_n = f(n)$. So if the graph of f is asymptotic to the line $y = L$, the a_n values come arbitrarily close to L.

REMARK

■ The converse of Theorem 10.3 is not true, as can be seen from the sequence $\{\sin \pi n\}$, which is constantly 0 and hence is convergent, but the corresponding function $f(x) = \sin \pi x$ does not approach a limit as $x \to \infty$. (See Figure 10.15.) So in this case $\lim\limits_{n \to \infty} f(n) = 0$, but $\lim\limits_{x \to \infty} f(x)$ does not exist.

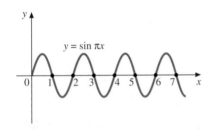

FIGURE 10.15
The sequence $\{\sin \pi n\} = \{0\}$, but $\lim\limits_{x \to \infty} \sin \pi x$ does not exist.

In the next example we illustrate how we can make use of Theorem 10.3 to find limits of sequences.

EXAMPLE 10.14   Find the limit of each of the following sequences.

(a) $\left\{\dfrac{\ln n}{n}\right\}$ (b) $\{\sqrt[n]{n}\}$

Solution

(a) Let
$$f(x) = \frac{\ln x}{x}$$

As $x \to \infty$, the function is indeterminant of the type ∞/∞. Applying L'Hôpital's Rule, we get
$$\lim_{x \to \infty} \frac{\ln x}{x} = \lim_{x \to \infty} \frac{1/x}{1} = 0$$

Since $f(n) = (\ln n)/n$, we conclude by Theorem 10.3 that
$$\lim_{n \to \infty} \frac{\ln n}{n} = 0$$

(b) Let $f(x) = x^{1/x}$ for $x > 0$. To find the limit, write $y = x^{1/x}$ and take the natural logarithm of both sides:
$$\ln y = \frac{1}{x} \ln x = \frac{\ln x}{x}$$

As we have seen in part (a), $(\ln x)/x \to 0$ as $x \to \infty$. Thus, since $\ln y \to 0$, it follows that $y \to e^0 = 1$. By Theorem 10.3 we conclude that
$$\lim_{n \to \infty} n^{1/n} = \lim_{n \to \infty} \sqrt[n]{n} = 1 \qquad ■$$

The Squeeze Theorem (Theorem 2.3) also holds true for sequences. We restate it here in terms of sequences.

THEOREM 10.4

The Squeeze Theorem for Sequences

If $\lim\limits_{n \to \infty} a_n = L$ and $\lim\limits_{n \to \infty} b_n = L$ and if for all sufficiently large n the inequality
$$a_n \le c_n \le b_n$$
holds true, then $\lim\limits_{n \to \infty} c_n = L$ also.

EXAMPLE 10.15   Find the limit of the sequence

$$\left\{ \frac{n \sin n}{1 + n^2} \right\}$$

or show that it diverges.

Solution   Since $|\sin n| \leq 1$ for all n, we have

$$\left| \frac{n \sin n}{1 + n^2} \right| \leq \frac{n}{1 + n^2} < \frac{n}{n^2} = \frac{1}{n}$$

Thus,

$$-\frac{1}{n} < \frac{n \sin n}{1 + n^2} < \frac{1}{n}$$

for all positive integers n. Since both $\{1/n\}$ and $\{-1/n\}$ converge to 0, it follows by the Squeeze Theorem that $\{n \sin n/(1 + n^2)\}$ converges to 0 also. That is,

$$\lim_{n \to \infty} \frac{n \sin n}{1 + n^2} = 0$$

■

Monotonicity and Boundedness

Sequences whose terms either always increase or always decrease (or else remain the same) play an especially important role in the study of infinite series. We use the following terminology to describe such sequences.

Definition 10.4
Monotone Sequences

A sequence $\{a_n\}$ is said to be

(a) **increasing** if $a_n \leq a_{n+1}$, or

(b) **decreasing** if $a_n \geq a_{n+1}$

for all positive integers n. A sequence that is either always increasing or always decreasing is said to be **monotone**.

REMARKS ——————————————————————

■ Frequently we will say *monotone increasing* or *monotone decreasing* to emphasize the monotonicity.

■ The terms *nondecreasing* and *nonincreasing* are sometimes used where we have used increasing and decreasing, respectively.

If it is necessary to distinguish between a sequence for which $a_n < a_{n+1}$ for all n and one for which $a_n \leq a_{n+1}$, we will refer to the former as *strictly* increasing. By our definition, each of the following is increasing:

$$1, 2, 3, 4, 5, \ldots$$

$$1, 2, 2, 3, 4, 4, 5, \ldots$$

Only the first one is strictly increasing. Similarly, we can say a sequence in which $a_n > a_{n+1}$ is *strictly* decreasing.

Definition 10.5 Bounded Sequences	A sequence $\{a_n\}$ is said to be **bounded** if there is some positive constant M such that $$|a_n| \leq M$$ for all positive integers n.

The sequences $\{(-1)^{n-1}\}$ and $\{\sin n\}$ considered earlier are bounded, since in each case $|a_n| \leq 1$. On the other hand, the sequence $\{2n - 1\}$ is unbounded.

Alone, neither monotonicity nor boundedness guarantees convergence. The sequence $\{(-1)^{n-1}\}$ is bounded but not convergent (also not monotone). The sequence $\{2n - 1\}$ is monotone but not convergent (also not bounded). We will shortly prove the important result that if a sequence is *both* monotone and bounded, then it must be convergent.

Before proving this result we need several more notions related to boundedness. A set S of real numbers is said to be **bounded above** if there is a real number M such that all elements of S are less than or equal to M. The number M is then said to be an **upper bound** of the set S. Similarly, S is **bounded below** if there is a real number m such that all elements of S are greater than or equal to m, in which case m is a **lower bound** of S. Clearly, if a sequence is bounded both above and below, then it is bounded, and the converse is also true. A fundamental property of the real-number system, called the *completeness property*, is that every nonempty set of real numbers that is bounded above has a **least upper bound**. To say L is the least upper bound of S means not only that L is an upper bound of S, but also that no number less than L is an upper bound of S. Similarly, if S is bounded below, it has a **greatest lower bound**. (These concepts are discussed further in Appendix 1.)

We are ready now for the main theorem on monotone bounded sequences.

THEOREM 10.5

> **Monotone Bounded Sequence Theorem**
>
> If $\{a_n\}$ is a sequence of real numbers that is both monotone and bounded, then it converges.

Proof   We will prove the theorem for a monotone increasing sequence only. The proof for the decreasing case is similar.

Let $\{a_n\}$ be monotone increasing and bounded. The terms of the sequence constitute a set of real numbers that is bounded above. By the completeness property of real numbers just discussed, there is a least upper bound, say L, for this set. That is, $a_n \leq L$ for all n, where L is the smallest number having this property.

Let ε denote any positive number. Then, since $L - \varepsilon < L$, the number $L - \varepsilon$ is *not* an upper bound to the sequence $\{a_n\}$. (Remember that L is the *least* upper bound, so $L - \varepsilon$ cannot be an upper bound.) Thus, there is some member

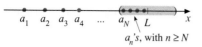

FIGURE 10.16

of the sequence, say a_N, such that $a_N > L - \varepsilon$. Since the sequence is increasing, it follows that $a_n \geq a_N$ for all $n > N$ (see Figure 10.16). Thus, for all $n > N$,

$$L - \varepsilon < a_N \leq a_n \leq L < L + \varepsilon$$

In particular, for $n > N$,

$$L - \varepsilon < a_n < L + \varepsilon$$

So, subtracting L from each member of this inequality,

$$-\varepsilon < a_n - L < \varepsilon$$

or equivalently,

$$|a_n - L| < \varepsilon \qquad \text{if } n > N$$

By Definition 10.3, we see that

$$\lim_{n \to \infty} a_n = L \qquad \blacksquare$$

Note that we have proved not only that the sequence converges but also that its limit is the least upper bound of the points in the sequence. (In many cases, however, finding this least upper bound is difficult.)

In applying Theorem 10.5 it is sufficient that the sequence be monotone *from some point onward*, since convergence or divergence depends only on the behavior of the sequence as n gets arbitrarily large. For example, the sequence

$$20, \quad 33, \quad -75, \quad 1, \quad \frac{1}{2}, \quad \frac{1}{4}, \quad \frac{1}{8}, \quad \frac{1}{16}, \ldots$$

converges since it is monotone decreasing starting with the fourth term and is bounded. In fact, in this case we see that the limit is 0, which is the greatest lower bound of the terms starting with the fourth term.

Showing Monotonicity

To apply Theorem 10.5, we need to know that the sequence is monotone. Sometimes monotonicity is obvious. For example, the sequence $\{1/n\}$ is clearly decreasing. Some sequences, however, we must analyze further to determine if they are monotone. The next example illustrates three ways to test for monotonicity.

EXAMPLE 10.16 Show that each of the following sequences is monotone.

(a) $\left\{\dfrac{2n + 3}{n}\right\}$ (b) $\left\{\dfrac{n}{\sqrt{1 + n^2}}\right\}$ (c) $\left\{\dfrac{n!}{n^n}\right\}$

Solution

(a) The first technique to test for monotonicity is to look at the difference $a_{n+1} - a_n$ between successive terms. If this difference is always greater than or equal to 0, the sequence is increasing; if it is always less than or equal to 0, the sequence is decreasing. For the sequence $\{(2n + 3)/n\}$, we have

$$a_{n+1} - a_n = \frac{2(n + 1) + 3}{n + 1} - \frac{2n + 3}{n} = \frac{(2n + 5)n - (2n + 3)(n + 1)}{n(n + 1)}$$

$$= \frac{2n^2 + 5n - (2n^2 + 5n + 3)}{n(n + 1)} = \frac{-3}{n(n + 1)} < 0$$

Thus, for all n, $a_{n+1} - a_n < 0$, so $a_{n+1} < a_n$, and we conclude that the sequence is (strictly) decreasing.

(b) As a second technique for testing for monotonicity, we consider a function f for which $f(n) = a_n$. If we can show monotonicity of f on the interval $[1, \infty)$, then we can conclude that $\{a_n\}$ is also monotone for $n \geq 1$. If f is differentiable, we can test it for monotonicity by considering $f'(x)$.

For the sequence $\{a_n\} = \{n/\sqrt{1+n^2}\}$, the natural choice for f is

$$f(x) = \frac{x}{\sqrt{1+x^2}}$$

Taking its derivative, we have

$$f'(x) = \frac{\sqrt{1+x^2} - x(\frac{1}{2})(1+x^2)^{-1/2}(2x)}{1+x^2}$$

$$= \frac{1+x^2-x^2}{(1+x^2)^{3/2}} = \frac{1}{(1+x^2)^{3/2}} > 0$$

(You should supply the missing algebra here.) Since $f'(x) > 0$ for all x, we know by Theorem 4.1 that f is an increasing function. Thus, since $a_n = f(n)$, we see that the sequence $\{a_n\}$ is also increasing.

(c) In the third approach to testing for monotonicity, we consider the ratio a_{n+1}/a_n of a given term (after the first) to the preceding term. If this ratio is always greater than or equal to 1, we conclude that $a_{n+1} \geq a_n$, and the sequence is increasing. If the ratio is less than or equal to 1, then $a_{n+1} \leq a_n$, and the sequence is decreasing. (In order to apply this technique, we must have $a_n > 0$ for all n.)

With $a_n = n!/n^n$, we have

$$\frac{a_{n+1}}{a_n} = \frac{\dfrac{(n+1)!}{(n+1)^{n+1}}}{\dfrac{n!}{n^n}} = \frac{(n+1)!}{(n+1)^{n+1}} \cdot \frac{n^n}{n!} = \frac{(n+1)!}{n!} \cdot \frac{n^n}{(n+1)^{n+1}}$$

$$= (n+1) \cdot \frac{n^n}{(n+1)^{n+1}} \qquad \text{Since } (n+1)! = (n+1)n!$$

$$= \frac{n^n}{(n+1)^n} < \frac{n^n}{n^n} = 1 \qquad \text{Since } (n+1)^n > n^n$$

Thus,

$$\frac{a_{n+1}}{a_n} < 1 \qquad \text{for all } n$$

Hence, $a_{n+1} < a_n$, and we conclude that $\{a_n\}$ is a decreasing sequence. ∎

Let us summarize the three ways of showing monotonicity illustrated in the preceding example.

Tests for Monotonicity

1. Calculate the difference $a_{n+1} - a_n$.

    If $\begin{cases} a_{n+1} - a_n \geq 0 & \text{for all } n, \text{ then } \{a_n\} \text{ is increasing;} \\ a_{n+1} - a_n \leq 0 & \text{for all } n, \text{ then } \{a_n\} \text{ is decreasing.} \end{cases}$

2. Let $f(x)$ be such that $f(n) = a_n$. Calculate $f'(x)$ if it exists.

    If $\begin{cases} f'(x) \geq 0 & \text{on } [1, \infty), \text{ then } \{a_n\} \text{ is increasing;} \\ f'(x) \leq 0 & \text{on } [1, \infty), \text{ then } \{a_n\} \text{ is decreasing.} \end{cases}$

3. If $a_n > 0$ for all n, calculate the ratio a_{n+1}/a_n.

    If $\begin{cases} \dfrac{a_{n+1}}{a_n} \geq 1 & \text{for all } n, \text{ then } \{a_n\} \text{ is increasing;} \\[2ex] \dfrac{a_{n+1}}{a_n} \leq 1 & \text{for all } n, \text{ then } \{a_n\} \text{ is decreasing.} \end{cases}$

Again, for purposes of applying Theorem 10.5, it is sufficient that the sequence be monotone from some point onward, say, for $n \geq n_0$. Thus, in applying Tests 1 and 3 listed above, we can modify "for all n" by requiring only that the conditions be true "for all $n \geq n_0$." For Test 2, $f'(x)$ need be either positive or negative on $[a, \infty)$ for some $a > 0$.

If a sequence of nonnegative terms is found to be decreasing, then it must converge, since it is bounded (bounded below by 0 and above by its first term). Thus, we can conclude that the sequences in parts (a) and (c) of Example 10.16 both converge. In fact, for part (a) we can find the limit directly:

$$\lim_{n \to \infty} \frac{2n + 3}{n} = \lim_{n \to \infty} \left(2 + \frac{3}{n} \right) = 2$$

The limit in part (c) is 0, as you will be asked to show in the exercises (Exercise 50 of this section). In part (b) we showed that the sequence is increasing. It is bounded above by 1, since

$$\frac{n}{\sqrt{1 + n^2}} < \frac{n}{\sqrt{n^2}} = \frac{n}{n} = 1$$

(Of course, it is also bounded below by its first term, since subsequent terms are larger.) Thus, the sequence converges. The limit is actually 1, as we see from the dominant term approach:

$$\frac{n}{\sqrt{1 + n^2}} \approx \frac{n}{\sqrt{n^2}} = \frac{n}{n} = 1 \qquad \text{for large } n$$

So

$$\lim_{n \to \infty} \frac{n}{\sqrt{1 + n^2}} = 1$$

Exercise Set 10.2

In Exercises 1–3, write out the first five terms of the sequence $\{a_n\}$.

1. (a) $a_n = \dfrac{(-1)^{n-1}n}{2n-1}$ (b) $a_n = \dfrac{2^n}{n!}$

2. (a) $a_n = \dfrac{2 \cdot 4 \cdot 6 \cdot \cdots \cdot (2n)}{1 \cdot 3 \cdot 5 \cdot \cdots \cdot (2n-1)}$

   (b) $a_n = \dfrac{\sin[(2n+1)\pi/2]}{n(n+1)}$

3. (a) $a_1 = 1,\ a_n = -na_{n-1}$ if $n \geq 2$
   (b) $a_1 = 1,\ a_2 = 2,\ a_n = \dfrac{a_{n-1}}{a_{n-2}}$ if $n \geq 3$

In Exercises 4–6, determine a formula for the nth term of the sequence.

4. (a) $1, -\dfrac{1}{3}, \dfrac{1}{5}, -\dfrac{1}{7}, \dfrac{1}{9}, \ldots$

   (b) $\dfrac{1}{2}, \dfrac{2}{5}, \dfrac{3}{10}, \dfrac{4}{17}, \dfrac{5}{26}, \ldots$

5. (a) $\dfrac{1}{2}, \dfrac{3}{4}, \dfrac{7}{8}, \dfrac{15}{16}, \dfrac{31}{32}, \ldots$

   (b) $\dfrac{2}{5}, -\dfrac{4}{7}, \dfrac{6}{9}, -\dfrac{8}{11}, \dfrac{10}{13}, \ldots$

6. (a) $\dfrac{1}{7}, -\dfrac{3}{10}, \dfrac{5}{13}, -\dfrac{7}{16}, \dfrac{9}{19}, \ldots$

   (b) $-1, 3, \dfrac{5}{3}, \dfrac{7}{5}, \dfrac{9}{7}, \ldots$

In Exercises 7–30, find $\lim_{n \to \infty} a_n$ *or show that the sequence diverges.*

7. $a_n = \dfrac{n^2 - 2n}{3n^2 + n - 1}$ **8.** $a_n = \dfrac{n+1}{2n^2 - n + 1}$

9. $a_n = \dfrac{2n - n^3}{(n+1)(n+3)}$

10. $a_n = \dfrac{n}{\sqrt{1+n^2}}$ **11.** $a_n = \dfrac{\cos n\pi}{\sqrt{n}}$

12. $a_n = \dfrac{1 + \sin n}{\ln(n+1)}$ **13.** $a_n = \dfrac{1 - (-1)^n}{2}$

14. $a_n = n^3 e^{-n}$ **15.** $a_n = \tanh n$

16. $a_n = \dfrac{\ln n}{\sqrt{n}}$ **17.** $a_n = (-0.999)^n$

18. $a_n = \dfrac{1 - \cosh n}{\sinh n}$ **19.** $a_n = \ln\left(1 + \dfrac{1}{n}\right)$

20. $a_n = \dfrac{\ln n}{\ln(\ln n)}$ **21.** $a_n = \dfrac{2^n}{e^{n-1}}$

22. $a_n = \dfrac{(-1)^{n-1}(2\sin n + \cos n)}{n \sec n}$

23. $a_n = \sin n + \cos n$ **24.** $a_n = (1.00001)^n$

25. $a_n = \dfrac{n+1}{n}\left(\dfrac{2}{3}\right)^n$ **26.** $a_n = \dfrac{(-1)^n e^{2n}}{1 + 9^n}$

27. $a_n = \dfrac{n \ln n}{1 + n^2}$ **28.** $a_n = \dfrac{\sqrt{1+n^3}}{\sqrt[3]{1+n^4}}$

29. $a_n = n^2 \cdot 2^{-n}$ **30.** $a_n = n \sin \dfrac{1}{n}$

In Exercises 31–40, show that the sequence $\{a_n\}$ *is monotone for* $n \geq n_0$. *If* $n_0 \neq 1$, *give the smallest value of* n_0.

31. $a_n = \dfrac{2n-1}{n+1}$ **32.** $a_n = \dfrac{n}{2n-3}$

33. $a_n = n^2 e^{-n}$ **34.** $a_n = \dfrac{(\ln n)^2}{n}$

35. $a_n = \dfrac{2^n}{n!}$ **36.** $a_n = 2^{1/n}$

37. $a_n = \dfrac{n!}{1 \cdot 3 \cdot 5 \cdot \cdots \cdot (2n-1)}$

38. $a_n = \dfrac{1 \cdot 3 \cdot 5 \cdot \cdots \cdot (2n-1)}{2^n \cdot n!}$

39. $a_n = \ln n - \ln(n+1)$

40. $a_n = \tan^{-1} n$

In Exercises 41–46, prove that the sequence $\{a_n\}$ *converges.*

41. $a_n = \dfrac{e^n}{n!}$ **42.** $a_n = \dfrac{2^n n!}{(2n)!}$

43. $a_n = n^{1/n}$

44. $a_n = \dfrac{1 \cdot 3 \cdot 5 \cdot \cdots \cdot (2n-1)}{2 \cdot 4 \cdot 6 \cdot \cdots \cdot 2n}$

45. $a_1 = 1,\ a_n = 1 + \dfrac{a_{n-1}}{2}$ for $n \geq 2$

46. $a_1 = 2,\ a_n = \sqrt{a_{n-1}}$ for $n \geq 2$

47. Prove that a sequence can have at most one limit.

48. Prove Theorem 10.3.

49. Prove that $\lim_{n \to \infty} a_n = 0$ if and only if $\lim_{n \to \infty} |a_n| = 0$.

50. Prove that $\lim_{n \to \infty} (n!/n^n) = 0$. (*Hint:* Write

$$a_n = \frac{n!}{n^n} = \frac{1 \cdot 2 \cdot 3 \cdots n}{n \cdot n \cdot n \cdots n}$$

and show that $a_n < 1/n$.)

51. Prove that if

$$a_n = \frac{1}{n+1} + \frac{1}{n+2} + \cdots + \frac{1}{2n}$$

then $\{a_n\}$ is convergent. *Hint:* Use Theorem 10.4. To show boundedness, observe that

$$a_n \le n \left(\frac{1}{n+1} \right)$$

52. Let $a_1 = \sqrt{3}$ and for $n \ge 2$, $a_n = \sqrt{3a_{n-1}}$. Prove that $\{a_n\}$ converges, and find its limit. (*Hint:* Use mathematical induction (see Supplementary Appendix 4) to prove that $a_{n+1} > a_n$ and $a_n < 3$ for all n.)

53. For the Fibonacci sequence $\{f_n\}$, define

$$a_n = \frac{f_{n+1}}{f_n}, \qquad n = 1, 2, 3, \ldots$$

Use a spreadsheet, CAS, or calculator to find the first twenty terms of the sequence $\{a_n\}$. Assuming that $\lim_{n \to \infty} a_n = L$, find the exact value of L. (*Hint:* Show that $a_n = 1 + (1/a_{n-1})$ for $n \ge 2$ and take the limit as $n \to \infty$ to get $L = 1 + (1/L)$. Now solve for L.)

Note: The number L is the *golden mean*, defined as shown in the figure.

The golden mean L is the mean proportional between 1 and $1 - L$:
$$\frac{1}{L} = \frac{L}{1 - L}$$

54. Use a spreadsheet, CAS, or calculator to do the following:
  (a) Find the first twenty terms of the Fibonacci sequence $\{f_n\}$.
  (b) Find the first twenty terms of the "Fibonacci-like" sequence $\{g_n\}$ in which $g_1 = 1$, $g_2 = 3$, and $g_n = g_{n-1} + g_{n-2}$ for $n \ge 2$.
  (c) Find the ratio g_n/f_n for $n = 1, 2, \ldots, 20$ for f_n and g_n in parts (a) and (b). Conjecture the limit

$$\lim_{n \to \infty} \frac{g_n}{f_n}$$

55. (a) Let x_1 be given and $x_{n+1} = \sqrt{x_n}$ for $n \ge 1$. Use a spreadsheet, CAS, or calculator to find the first

twenty terms of the sequence $\{x_n\}$ for four different values of x_1. Conjecture the limit in each case. Prove your conjecture. Does the answer depend on x_1?
  (b) Repeat part (a) if $x_{n+1} = x_n^2 - 2$.

56. (1) Choose an arbitrary integer $A > 0$.
  (2) If $A = 1$, stop.
  (3) If A is even, replace A with $A/2$ and go to step 2.
  (4) If A is odd, replace A with $3A + 1$ and go to step 2.
  Let $A = 3$, $A = 34$, and $A = 75$, and in each case compute the terms generated. Is it true that the larger the value of A, the more steps are required before the algorithm stops?

57. Let $a_1 = 2$ and $a_n = \sqrt{2 + a_{n-1}}$ for $n \ge 2$.
  (a) Plot several points of the sequence $\{a_n\}$ on a number line. Does the sequence appear to be bounded above? If so, what appears to be an upper bound?
  (b) Use mathematical induction to prove that $a_n \le 2$ for all n.
  (c) Show that $a_{n+1}^2 - a_n^2 = (2 - a_n)(1 + a_n)$. Use this result to show that $\{a_n\}$ is monotone increasing.
  (d) Let $\lim_{n \to \infty} a_n = L$. By taking the limit of both sides of the equation $a_n = \sqrt{2 + a_{n-1}}$ as $n \to \infty$, find the value of L. (*Hint:* Note that $a_{n-1} \to L$ as $n \to \infty$.)

58. (a) Is it possible for a term of a convergent sequence to exceed the limit of the sequence? Is it possible for some terms to exceed and some terms to be less than the limit? Is it possible for some terms to equal the limit? Use examples in your answers to these questions.
  (b) If the terms of a sequence alternate in sign, can the sequence converge? If you think it is possible, give an example.

59. In each of the following give an example, whenever possible, satisfying the given condition. If it is not possible, explain why.
  (a) A sequence with limit 12 and no term equal to 12
  (b) An increasing sequence that is not bounded
  (c) A bounded sequence that does not converge
  (d) A bounded, convergent sequence that is not monotone
  (e) An increasing sequence that does not converge
  (f) A convergent sequence with infinitely many terms equal to 7, infinitely many terms not equal to 7, that converges to 7
  (g) A convergent sequence with infinitely many terms equal to 7 and infinitely many terms equal to 8
  (h) A convergent sequence that is not bounded
  (i) Two sequences with the same limit but with no common terms
  (j) A sequence with limit 2 that has infinitely many negative terms

10.3 INFINITE SERIES OF CONSTANTS

In the introduction to this chapter we discussed the question of how to define the sum of infinitely many terms. Now we want to explore this question in more detail.

Consider first the familiar decimal representation of the fraction 1/3. By repeated division, we get

$$\frac{1}{3} = 0.3333\ldots$$

We can write the repeating decimal on the right as the infinite sum

$$0.3 + 0.03 + 0.003 + 0.0003 + \cdots \tag{10.15}$$

We know, then, that the sum of this infinite series is 1/3.

As a second example, consider again the infinite series

$$1 + \frac{1}{2} + \frac{1}{4} + \frac{1}{8} + \cdots \tag{10.16}$$

in which each term after the first is half of the preceding term. We proceed as we did in the introduction to the chapter by recording its **partial sums**—that is, the successive finite sums:

$$\text{First term} \qquad\qquad = 1$$
$$\text{Sum of first two terms} \quad = 1\frac{1}{2}$$
$$\text{Sum of first three terms} \; = 1\frac{3}{4}$$
$$\text{Sum of first four terms} \;\; = 1\frac{7}{8}$$
$$\vdots$$

FIGURE 10.17

In Figure 10.17 we show these partial sums as points on a number line. Notice that each successive point after the first is halfway between the previous point and 2. Clearly, as we add more and more terms, the partial sums come arbitrarily close to 2, so we say that the sum of the series is 2 and write

$$1 + \frac{1}{2} + \frac{1}{4} + \frac{1}{8} + \cdots = 2$$

Let us now generalize what we have illustrated with the two particular infinite series 10.15 and 10.16. Suppose that $a_1, a_2, a_3, \ldots$ are numbers (not necessarily all distinct). The sum

$$a_1 + a_2 + a_3 + \cdots \tag{10.17}$$

is called an **infinite series**, or simply a **series**. Writing the three dots at the end is important as an indication that the series continues indefinitely. For brevity, we will often write series 10.17 in summation notation as

$$\sum_{n=1}^{\infty} a_n$$

The particular letter used as the index of summation is not important, since the index is replaced by $1, 2, 3, \ldots$ when we write out the series. For example, we could write $\sum_{k=1}^{\infty} a_k$ for the series 10.17. Sometimes we will omit the range of the index and simply write $\sum a_n$, with the understanding that the index goes from 1 to ∞. There will be times when we want the index to begin with 0. Note, for example, that

$$\sum_{n=0}^{\infty} \frac{1}{2^n} \quad \text{and} \quad \sum_{n=1}^{\infty} \frac{1}{2^{n-1}}$$

are two ways of representing the same series. (Try writing the first few terms in each case.)

Just as we did with the series 10.16, we list the partial sums of the series 10.17:

First partial sum $= a_1$
Second partial sum $= a_1 + a_2$
Third partial sum $= a_1 + a_2 + a_3$
$\vdots$
nth partial sum $= a_1 + a_2 + a_3 + \cdots + a_n$
$\vdots$

The question is: Do these partial sums approach more and more closely some limiting value as we let n increase indefinitely? If so, this limiting value is what we define as the sum of the series 10.17. In order to state this definition more succinctly, we introduce the notation

$$S_n = a_1 + a_2 + a_3 + \cdots + a_n$$

That is, $S_1 = a_1$, $S_2 = a_1 + a_2$, $S_3 = a_1 + a_2 + a_3$, and so on. For each positive integer n, S_n is the **nth partial sum** of the series 10.17. These partial sums, in order, $S_1, S_2, S_3, \ldots$, constitute a sequence $\{S_n\}$, called the **sequence of partial sums** of the series.

Note carefully that when we list the terms of a sequence, we separate them with commas, whereas the terms of a series are added together. For example, the *series* 10.16 is

$$1 + \frac{1}{2} + \frac{1}{4} + \frac{1}{8} + \cdots$$

and its *sequence* of partial sums is

$$1, 1\frac{1}{2}, 1\frac{3}{4}, 1\frac{7}{8}, \ldots$$

For this example, we can find a simple formula for S_n, namely,

$$S_n = 2 - \frac{1}{2^{n-1}}$$

since S_1 is 1 less than 2, S_2 is $\frac{1}{2}$ less than 2, S_3 is $\frac{1}{4}$ less than 2, and so on. As $n \to \infty$, we get the limit

$$\lim_{n \to \infty} S_n = 2$$

since $2 - (1/2^{n-1})$ can be made arbitrarily close to 2 by taking n sufficiently large. This result confirms again that series 10.16 converges to 2.

We now give the formal definition of the sum of an infinite series.

Definition 10.6
The Sum of an Infinite Series

The infinite series

$$\sum_{n=1}^{\infty} a_n = a_1 + a_2 + a_3 + \cdots$$

is said to **converge to the sum S** provided

$$\lim_{n \to \infty} S_n = S$$

where $S_n = a_1 + a_2 + \cdots + a_n$. In this case we write $\sum_{n=1}^{\infty} a_n = S$. If $\lim_{n \to \infty} S_n$ does not exist, we say that the series **diverges**.

Briefly, we can write

$$\sum_{n=1}^{\infty} a_n = \lim_{n \to \infty} (a_1 + a_2 + \cdots + a_n)$$

provided the limit on the right exists (that is, the limit is finite).

Recall from Section 10.2 that the precise meaning of

$$\lim_{n \to \infty} S_n = S$$

is that, given any positive number ε, there exists a positive number N such that for all $n > N$, $|S_n - S| < \varepsilon$.

In most cases it is not feasible to apply Definition 10.6 directly to determine whether a given series converges and if so, the value of its sum. In the next two sections we will develop means of testing for convergence or divergence without having to apply the definition directly. There are, however, certain series for which the definition can be used. The most important of these is called a *geometric series*, which we now describe.

Geometric Series

A **geometric series** is an infinite series of the form

$$\sum_{n=1}^{\infty} ar^{n-1} = a + ar + ar^2 + ar^3 + \cdots \tag{10.18}$$

in which a and r are fixed real numbers with $a \neq 0$. The number a is its first term, and the number r is called the **common ratio**. Note that the ratio of any term (after the first) to the preceding term is r if $r \neq 0$ (for example, $ar^3 / ar^2 = r$), hence the name common ratio. Geometric series are particularly easy to work with not only because we can completely describe when they converge and diverge, but also because when they converge, we can easily compute the sum. To verify these facts, we investigate the sequence $\{S_n\}$ of partial sums, where

$$S_n = a + ar + ar^2 + \cdots + ar^{n-1}$$

First, suppose $r = 1$. Then

$$S_n = a + a + a + \cdots + a = na$$

Clearly, $\lim_{n\to\infty} S_n$ does not exist as a finite number in this case, since na will be arbitrarily large in absolute value if n is sufficiently large. That is, $\lim_{n\to\infty} na = \infty$ if $a > 0$ and $\lim_{n\to\infty} na = -\infty$ if $a < 0$. Thus, when $r = 1$, the geometric series 10.18 diverges.

If $r \neq 1$, we can write S_n in what is called a *closed form* by the following trick. We note that if we multiply S_n by r, we get many terms just like those in S_n. If we subtract $r S_n$ from S_n, we get

$$S_n - r S_n = (a + ar + ar^2 + ar^3 + \cdots + ar^{n-1}) - (ar + ar^2 + ar^3 + \cdots + ar^n)$$

All the terms except a and ar^n cancel, so we have

$$S_n - r S_n = a - ar^n$$

or, equivalently,

$$S_n(1 - r) = a(1 - r^n)$$

Since with $r \neq 1$, the factor $1 - r$ is nonzero, we can solve for S_n:

$$S_n = \frac{a(1 - r^n)}{1 - r}$$

We know from Example 10.13 and the discussion following it that when $|r| < 1$,

$$\lim_{n\to\infty} r^n = 0$$

and when $r = -1$ or $|r| > 1$,

$$\lim_{n\to\infty} r^n \text{ does not exist}$$

Thus, when $|r| < 1$,

$$\lim_{n\to\infty} S_n = \lim_{n\to\infty} \frac{a(1 - r^n)}{1 - r}$$

$$= \frac{a}{1 - r}$$

and the geometric series converges to $a/(1 - r)$. If $|r| \geq 1$, $\{S_n\}$ does not converge, so the geometric series diverges.

Because of the importance of geometric series, we state our findings as a theorem.

THEOREM 10.6

> If $|r| < 1$, the geometric series $\sum_{n=1}^{\infty} ar^{n-1}$ converges to the sum $a/(1-r)$. That is,
>
> $$a + ar + ar^2 + ar^3 + \cdots = \frac{a}{1 - r} \qquad \text{if } |r| < 1 \qquad (10.19)$$
>
> If $|r| \geq 1$, the series diverges.

Both of the series 10.15 and 10.16 are geometric. For 10.15 we see from the series

$$0.3 + 0.03 + 0.003 + \cdots$$

that $a = 0.3$ and $r = 0.1$. Since $|r| = 0.1 < 1$, we have, by Equation 10.19,

$$0.3 + 0.03 + 0.003 + \cdots = \frac{0.3}{1 - 0.1} = \frac{0.3}{0.9} = \frac{3}{9} = \frac{1}{3}$$

confirming what we already knew.

For the series 10.16,

$$1 + \frac{1}{2} + \frac{1}{4} + \frac{1}{8} + \cdots$$

$a = 1$ and $r = \frac{1}{2}$. Again, by Equation 10.19, the sum is

$$\frac{a}{1 - r} = \frac{1}{1 - \frac{1}{2}} = \frac{1}{\frac{1}{2}} = 2$$

The next four examples further illustrate Theorem 10.6.

EXAMPLE 10.17   Determine whether each of the following series is convergent or divergent. If convergent, find the sum.

(a) $1 - \dfrac{2}{3} + \dfrac{4}{9} - \dfrac{8}{27} + \cdots$

(b) $3 + 4 + \dfrac{16}{3} + \dfrac{64}{9} + \cdots$

Solution

(a) The series is geometric, with $a = 1$ and $r = -2/3$. One way to determine r is to divide any term after the first by the one before it. Another way is to ask by what factor you would multiply a given term to get the next term. In either case, be sure to check that the ratio (or the multiplicative factor) remains the same for all pairs of consecutive terms. Confirming that $r = -2/3$, we see that its absolute value is less than 1, so by Equation 10.19 the series converges to

$$\frac{a}{1 - r} = \frac{1}{1 - \left(-\frac{2}{3}\right)} = \frac{1}{\frac{5}{3}} = \frac{3}{5}$$

We can therefore write

$$1 - \frac{2}{3} + \frac{4}{9} - \frac{8}{27} + \cdots = \frac{3}{5}$$

(b) The series is geometric, with $a = 3$ and $r = 4/3$. Since $r > 1$, the series diverges. ∎

EXAMPLE 10.18   Show that each of the following series is geometric and determine whether it is convergent or divergent. If convergent, find the sum.

(a) $\displaystyle\sum_{n=0}^{\infty} e^{-n}$ (b) $\displaystyle\sum_{k=1}^{\infty} \frac{(-1)^{k-1}}{2^k}$

Solution

(a) When a series is written in summation notation, it is sometimes helpful to write out the first few terms. In this case, we have

$$\sum_{n=0}^{\infty} e^{-n} = 1 + e^{-1} + e^{-2} + e^{-3} + \cdots = 1 + \frac{1}{e} + \frac{1}{e^2} + \frac{1}{e^3} + \cdots$$

Each term is obtained by multiplying the preceding one by $1/e$. So $r = 1/e$, and the series is geometric. Since $e \approx 2.7$, $|1/e| < 1$. The sum of the series is therefore

$$\frac{a}{1-r} = \frac{1}{1 - \frac{1}{e}} = \frac{e}{e-1}$$

(b) By writing out the first few terms,

$$\sum_{k=1}^{\infty} \frac{(-1)^{k-1}}{2^k} = \frac{1}{2} - \frac{1}{4} + \frac{1}{8} - \frac{1}{16} + \cdots$$

we see that the series is geometric, with $a = \frac{1}{2}$ and $r = -\frac{1}{2}$. Since $|-\frac{1}{2}| < 1$, the sum is

$$\frac{a}{1-r} = \frac{\frac{1}{2}}{1 - (-\frac{1}{2})} = \frac{\frac{1}{2}}{\frac{3}{2}} = \frac{1}{3} \qquad \blacksquare$$

EXAMPLE 10.19   Use a geometric series to express the repeating decimal $1.272727\ldots$ as the ratio m/n of two positive integers.

Solution   First, write the repeating decimal as the infinite series

$$1 + (0.27 + 0.0027 + 0.000027 + \cdots)$$

The series in parentheses is geometric, with $a = 0.27$ and $r = 0.01$. Thus, its sum is

$$\frac{a}{1-r} = \frac{0.27}{1 - 0.01} = \frac{0.27}{0.99} = \frac{27}{99} = \frac{3}{11}$$

The entire sum is therefore $1 + \frac{3}{11}$. That is,

$$1.272727\ldots = 1 + \frac{3}{11} = \frac{14}{11} \qquad \blacksquare$$

EXAMPLE 10.20   A ball is dropped from a height of 50 cm, and on each bounce it goes three-fourths as high as before. Approximate the total distance traveled by the ball in coming to rest.

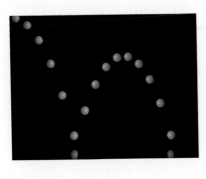

Solution   Although the ball moves only along a vertical line, we can get a better picture by showing the path as in Figure 10.18 (on page 760). First it drops 50 cm, then it rises $37\frac{1}{2}$ cm and falls the same distance. Thereafter it rises and falls three-fourths of the up-and-down total for the previous bounce. To make the first motion analogous to the others, we can pretend it started from the ground, rising 50 cm and then falling 50 cm. We will subtract this imaginary initial 50-cm rise after getting the total. With this understanding, the total distance covered is

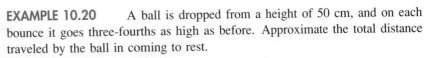

$$\left[100 + \frac{3}{4}(100) + \left(\frac{3}{4}\right)^2 (100) + \left(\frac{3}{4}\right)^3 (100) + \cdots \right] - 50$$

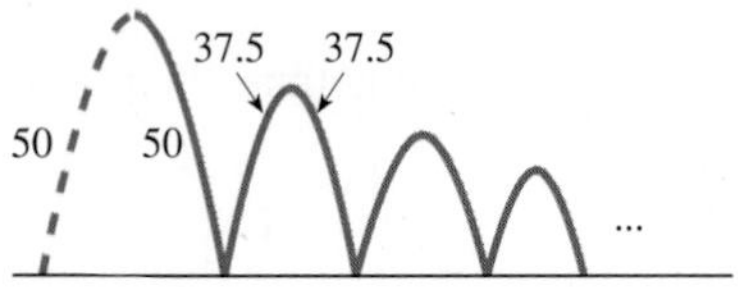

FIGURE 10.18

The expression in brackets is a geometric series, with $a = 100$ and $r = \frac{3}{4}$. So its sum is

$$\frac{a}{1-r} = \frac{100}{1 - \frac{3}{4}} = 400$$

Thus, on subtracting the initial imaginary 50-cm rise, we find the total distance the ball covers to be 350 cm. ∎

Telescoping Series

The next example illustrates another type of series in which the limit of the nth partial sum can be found. For reasons that will become apparent, the series is called a **telescoping series**.

EXAMPLE 10.21   Show that the series

$$\sum_{k=1}^{\infty} \frac{1}{k^2 + k}$$

converges, and find its sum.

Solution   Using partial fractions, we can rewrite the general term in the form

$$\frac{1}{k^2 + k} = \frac{1}{k(k+1)} = \frac{1}{k} - \frac{1}{k+1}$$

Thus, the nth partial sum of the series is

$$S_n = \sum_{k=1}^{n} \left(\frac{1}{k} - \frac{1}{k+1} \right)$$

$$= \left(\frac{1}{1} - \frac{1}{2} \right) + \left(\frac{1}{2} - \frac{1}{3} \right) + \left(\frac{1}{3} - \frac{1}{4} \right) + \cdots + \left(\frac{1}{n} - \frac{1}{n+1} \right)$$

By regrouping the parentheses, we see that all terms except the first and last cancel out (this expression "telescopes," as in an old type of telescope that is collapsible), and so

$$S_n = 1 - \frac{1}{n+1}$$

As $n \to \infty$, $S_n \to 1$. This calculation shows that the original series converges, since the sequence $\{S_n\}$ converges, and also that its sum is 1. That is,

$$\sum_{k=1}^{\infty} \frac{1}{k^2 + k} = 1$$

∎

A Necessary Condition for Convergence

Another very useful result can be obtained from Definition 10.6, which we state as a theorem.

THEOREM 10.7

If the series $\sum_{n=1}^{\infty} a_n$ converges, then $\lim_{n \to \infty} a_n = 0$.

Proof   Let S_n be the nth partial sum of $\sum_{n=1}^{\infty} a_n$; that is,

$$S_n = a_1 + a_2 + a_3 + \cdots + a_n$$

Then, if $n > 1$, we also have

$$S_{n-1} = a_1 + a_2 + a_3 + \cdots + a_{n-1}$$

On subtracting S_{n-1} from S_n, all terms drop out except a_n.

$$a_n = S_n - S_{n-1}$$

Now, since the series converges, we know that $\lim_{n \to \infty} S_n$ exists. Call its value S. But as $n \to \infty$, we also have $(n-1) \to \infty$. So both $\lim_{n \to \infty} S_n = S$ and $\lim_{n \to \infty} S_{n-1} = S$. Thus,

$$\lim_{n \to \infty} a_n = \lim_{n \to \infty} (S_n - S_{n-1})$$

$$= \lim_{n \to \infty} S_n - \lim_{n \to \infty} S_{n-1}$$

$$= S - S = 0 \qquad \blacksquare$$

REMARK

■ Here we have made use of the limit property from Section 10.2 that says if $\lim_{n \to \infty} a_n$ and $\lim_{n \to \infty} b_n$ both exist, then $\lim_{n \to \infty}(a_n - b_n)$ exists and equals $\lim_{n \to \infty} a_n - \lim_{n \to \infty} b_n$.

An equivalent way of stating Theorem 10.7 (a logical equivalent called its *contrapositive*) is the following:

The nth-Term Test for Divergence

If $\lim_{n \to \infty} a_n \neq 0$, then $\sum_{n=1}^{\infty} a_n$ diverges.

Thus, if for a given series we can show that $\lim_{n \to \infty} a_n \neq 0$, we can conclude that the series $\sum a_n$ definitely diverges. For example, the series

$$\sum_{n=1}^{\infty} \frac{n}{2n+1}$$

diverges, since

$$\lim_{n \to \infty} a_n = \lim_{n \to \infty} \frac{n}{2n+1} = \frac{1}{2} \neq 0$$

 If you find merely that $\lim_{n \to \infty} a_n = 0$, you cannot draw any definite conclusion regarding the convergence or divergence of $\sum a_n$ except that the series at least has a chance to converge. It is a common mistake to assume that the converse of Theorem 10.7 is true; that is, that $\lim_{n \to \infty} a_n = 0$ implies the convergence of

$\sum a_n$. That this converse is *not true* can be seen by finding a divergent series whose terms approach 0, as we do now.

The Harmonic Series

The best-known example of a divergent series whose nth term tends to 0 is the so-called **harmonic series**—namely,

$$\sum_{n=1}^{\infty} \frac{1}{n} = 1 + \frac{1}{2} + \frac{1}{3} + \cdots + \frac{1}{n} + \cdots$$

Clearly, $\lim_{n\to\infty} 1/n = 0$, yet the series diverges, as the following analysis of certain of its partial sums shows:

$$S_1 = 1$$

$$S_2 = 1 + \frac{1}{2}$$

$$S_4 = 1 + \frac{1}{2} + \frac{1}{3} + \frac{1}{4} > 1 + \frac{1}{2} + \left(\frac{1}{4} + \frac{1}{4}\right) = 1 + \frac{1}{2} + \frac{1}{2}$$

$$S_8 = 1 + \frac{1}{2} + \frac{1}{3} + \frac{1}{4} + \frac{1}{5} + \frac{1}{6} + \frac{1}{7} + \frac{1}{8}$$

$$> 1 + \frac{1}{2} + \left(\frac{1}{4} + \frac{1}{4}\right) + \left(\frac{1}{8} + \frac{1}{8} + \frac{1}{8} + \frac{1}{8}\right) = 1 + \frac{1}{2} + \frac{1}{2} + \frac{1}{2}$$

Notice that the relationship between these partial sums and the number of times $\frac{1}{2}$ is added is as follows:

$$S_1 = S_{2^0} = 1 + 0\left(\frac{1}{2}\right)$$

$$S_2 = S_{2^1} > 1 + 1\left(\frac{1}{2}\right)$$

$$S_4 = S_{2^2} > 1 + 2\left(\frac{1}{2}\right)$$

$$S_8 = S_{2^3} > 1 + 3\left(\frac{1}{2}\right)$$

$$\vdots \qquad\qquad \vdots$$

In general, we have

$$S_{2^n} \geq 1 + n\left(\frac{1}{2}\right), \qquad n = 0, 1, 2, \ldots$$

(Mathematical induction can be used to give a formal proof.) Thus, $\lim_{n\to\infty} S_{2^n} = \infty$ and it follows that $\lim_{n\to\infty} S_n$ does not exist. Thus, the series diverges. Because we will have frequent occasions to refer to this result, we set it off for emphasis.

The Harmonic Series

The harmonic series

$$\sum_{n=1}^{\infty} \frac{1}{n} = 1 + \frac{1}{2} + \frac{1}{3} + \cdots + \frac{1}{n} + \cdots$$

diverges.

Since for the harmonic series $\lim_{n \to \infty} S_n = \infty$, we sometimes write $\sum_{n=1}^{\infty} 1/n = \infty$. It is interesting to note that the sum of the first billion terms of the harmonic series is only about 21. That is,

$$S_{10^9} = 1 + \frac{1}{2} + \frac{1}{3} + \cdots + \frac{1}{10^9} \approx 21$$

and each succeeding term is *very* small. Yet if we continue adding terms indefinitely, the cumulative effect is to cause the sum to become infinite! Sometimes infinite processes defy one's intuition.

Properties of Convergent Series

The next theorem is a direct consequence of Definition 10.6. We call for a proof in Exercise 38 of Exercise Set 10.3.

THEOREM 10.8

If $\sum_{n=1}^{\infty} a_n$ and $\sum_{n=1}^{\infty} b_n$ are convergent series, and if c is any real number, then each of the series $\sum_{n=1}^{\infty} ca_n$, $\sum_{n=1}^{\infty} (a_n + b_n)$, and $\sum_{n=1}^{\infty} (a_n - b_n)$ converges, and

1. $\displaystyle\sum_{n=1}^{\infty} ca_n = c \sum_{n=1}^{\infty} a_n$

2. $\displaystyle\sum_{n=1}^{\infty} (a_n + b_n) = \sum_{n=1}^{\infty} a_n + \sum_{n=1}^{\infty} b_n$

3. $\displaystyle\sum_{n=1}^{\infty} (a_n - b_n) = \sum_{n=1}^{\infty} a_n - \sum_{n=1}^{\infty} b_n$

EXAMPLE 10.22   Show that the series

$$\sum_{n=1}^{\infty} \left(\frac{1}{2^{n-1}} + \frac{1}{n^2 + n} \right)$$

converges, and find its sum.

Solution   The series

$$\sum_{n=1}^{\infty} \frac{1}{2^{n-1}}$$

is the geometric series with $a = 1$ and $r = \frac{1}{2}$, whose sum is

$$\frac{a}{1-r} = \frac{1}{1 - \dfrac{1}{2}} = 2$$

The series

$$\sum_{n=1}^{\infty} \frac{1}{n^2 + n}$$

is the telescoping series we considered in Example 10.21, where we found that it converges to 1. It follows by Property 2 of Theorem 10.8 that

$$\sum_{n=1}^{\infty} \left(\frac{1}{2^{n-1}} + \frac{1}{n^2 + n} \right) = 2 + 1 = 3 \qquad \blacksquare$$

If $\sum a_n$ converges and $\sum b_n$ diverges, then we can see from Theorem 10.8 that $\sum (a_n + b_n)$ and $\sum (a_n - b_n)$ both diverge. For suppose $\sum (a_n + b_n)$ converges. Then we can write $b_n = (a_n + b_n) - a_n$. Property 3 of Theorem 10.8 then gives

$$\sum b_n = \sum [(a_n + b_n) - a_n] = \sum (a_n + b_n) - \sum a_n$$

since by our assumption both series on the right converge. But we are given that $\sum b_n$ diverges. So $\sum (a_n + b_n)$ cannot converge. We can prove that $\sum (a_n - b_n)$ diverges in a similar way. If $\sum a_n$ and $\sum b_n$ both diverge, no definite conclusion can be drawn about convergence or divergence of $\sum (a_n + b_n)$ or $\sum (a_n - b_n)$. They may converge or diverge, depending on the particular series in question. (See Exercise 39 in Exercise Set 10.3.)

Another consequence of Theorem 10.8 is that if $\sum a_n$ diverges, so does $\sum c a_n$ for any $c \neq 0$. This result follows from the fact that we can write

$$a_n = \frac{1}{c}(c a_n)$$

So by Property 1 of Theorem 10.8 (with c replaced by $1/c$ and a_n replaced by $c a_n$), if $\sum c a_n$ converges, it would follow that $\sum a_n = \sum 1/c(c a_n)$ also would converge, contrary to what we are given.

We summarize these properties as a corollary to Theorem 10.8.

COROLLARY 10.8

1. If $\sum a_n$ converges and $\sum b_n$ diverges, then both $\sum (a_n + b_n)$ and $\sum (a_n - b_n)$ diverge.

2. If $\sum a_n$ diverges and $c \neq 0$, then $\sum c a_n$ diverges.

We mention one other property that will sometimes be useful.

The convergence or divergence of an infinite series is unaltered if we delete or add any finite number of terms.

Suppose, for example, we delete some or all of the first N terms of the series $\sum a_n$. For convenience of notation, let us replace each term we deleted with 0 and call the new series $\sum b_n$. Then some or all of the terms $b_1, b_2, \ldots, b_N$ equal 0, but for $n > N$, $b_n = a_n$. Let $\{A_n\}$ denote the sequence of partial sums of the series $\sum a_n$ and $\{B_n\}$ denote the sequence of partial sums of the series $\sum b_n$. Then for all $n > N$,

$$A_n - A_N = a_{N+1} + a_{N+2} + \cdots + a_n = b_{N+1} + b_{N+2} + \cdots + b_n = B_n - B_N$$

Since A_N and B_N are fixed constants (each being the sum of the first N terms of a series, where N is fixed), we see that if either A_n or B_n approaches a finite limit as $n \to \infty$, so does the other. It follows that $\sum a_n$ and $\sum b_n$ either both converge or both diverge. A similar argument can be given if a finite number of terms are added.

Note carefully that while altering a finite number of terms does not alter convergence or divergence, it *does* alter the sum when the series is convergent.

Fractals

In 1904 the Swedish mathematician Helge von Koch gave an example of a curve without a tangent anywhere. To his contemporaries this was a "pathological" example, whereas now mathematicians are finding that such examples occur frequently in pure and applied mathematics. They are called **fractals**, geometric figures in which a set pattern repeats on an ever-decreasing scale. The Polish-born mathematician Benoit Mandelbrot, while working in this country at IBM during the late 1970s, did pioneering work in the theory of fractals. His findings have played a key role in the rapidly expanding field of fractal geometry. To construct the Koch curve, we start with the unit interval, remove the middle third and replace it with two pieces of equal length, forming the sides of an equilateral triangle. Then we repeat the process over and over. The first three stages of the construction are shown in Figure 10.19. A good approximation is given in Figure 10.20. The Koch curve is the "limit" of the sets shown. This limiting curve can never be drawn, existing only as a mathematical concept. In Figure 10.21, we can see that magnification of one-third of the edge of the curve by a factor of 3 results in the entire curve. If this property is preserved at every stage of any iterative construction, the curve is called self-similar. Fractals possess this property of self-similarity.

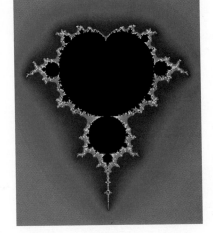

The Mandelbrot Set

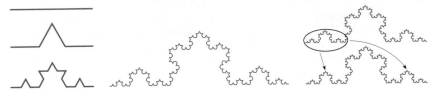

FIGURES 10.19–10.21

The construction of the **Koch island** is similar to the construction of the Koch curve described above. In the case of the island, we start with an equilateral triangle (Figure 10.22(a)). Now we remove the middle third of each of the three sides and replace each with two pieces of equal length, resulting in the star-shaped region in Figure 10.22(b). Now we repeat the process over and over. The next two approximations to the Koch island are given in Figure 10.23.

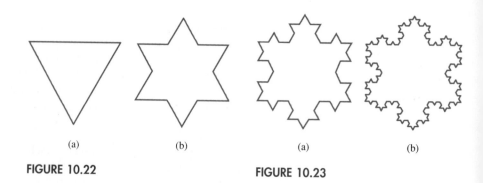

FIGURE 10.22   FIGURE 10.23

In Exercise 49 of Exercise Set 10.3 we will ask you to show that the perimeter of the Koch island is infinite and that its area is finite.

INFINITE SERIES USING COMPUTER ALGEBRA SYSTEMS

Computer algebra systems have commands for finding partial sums of infinite series and, in some cases, the sum of a convergent series.

CAS 37

Find several partial sums for the series

$$\sum_{n=1}^{\infty} \frac{1}{n^2}$$

If possible, determine the sum of the series.

Maple:

sum(1/k^2, k=1..4);
$$\frac{205}{144}$$
evalf(sum(1/k^2, k=1..100));
    1.634983901
evalf(sum(1/k^2,
    k=1..1000));
    1.643934568

Mathematica:

Sum[1/n^2,{n,1,4}]
$$\frac{205}{144}$$
N[Sum[1/n^2,{n,1,100}],10]
    1.6349839
N[Sum[1/n^2,{n,1,1000}],10]
    1.643934567

DERIVE:

(At the □ symbol, go to the next step.)
a (author) □ 1/k^2 □ c (calculus) □ s (sum) □ [choose expression] □ k (variable) □ 1 (lower limit) □ 4 (upper limit)

Output: $\dfrac{205}{144}$

Another approach:
a (author) □ 1/k^2 □ c (calculus) □ s (sum) □ [choose expression] □ k (variable) □ 1 (lower limit) □ n (upper limit) □ m (manage) □ s (substitute) □ k (variable) □ 100 (value for *n*) □ x (approximate) □ [choose expression]
Output: 1.63498

Maple can take an example like this one step further. In fact, the sum of this infinite series can be computed.

In Maple enter:

sum(1/k^2,k=1..infinity);

The output is the constant given by Euler, $\pi^2/6$.

Maple can also return formal infinite sums by using the Sum command (note the capital S). In Maple, sum and Sum are different commands. The command sum computes a sum, and Sum returns the formal sum in summation notation.

For example, in Maple enter:

Sum(1/k^2,k=1..infinity);

The output is:

$$\sum_{k=1}^{\infty} \frac{1}{k^2}$$

As one final example, in Maple enter:

sum(k^2, k=1..n);

The output is:

$$\frac{1}{3}(n+1)^3 - \frac{1}{2}(n+1)^2 + \frac{1}{6}n + \frac{1}{6}$$

If the output is factored, we see the formula given in Equation 5.5.

factor(");

The output is:

$$\frac{1}{6}n(n+1)(2n+1)$$

Exercise Set 10.3

In Exercises 1–4, write out several terms of the series and give the first four partial sums.

1. $\displaystyle\sum_{k=1}^{\infty} \frac{k}{k^2+1}$

2. $\displaystyle\sum_{n=1}^{\infty} \frac{(-1)^{n-1}}{\sqrt{2n-1}}$

3. $\displaystyle\sum_{m=1}^{\infty} \frac{\ln(m+1)}{m!}$

4. $\displaystyle\sum_{n=0}^{\infty} \frac{(-1)^n(n+1)}{2^n}$

In Exercises 5–8, write the series using summation notation. Also, give the nth partial sum.

5. $1 - \dfrac{1}{3} + \dfrac{1}{9} - \dfrac{1}{27} + \dfrac{1}{81} - \cdots$

6. $1 - \dfrac{1}{2!} + \dfrac{1}{4!} - \dfrac{1}{6!} + \dfrac{1}{8!} - \cdots$

7. $\dfrac{1}{2\ln 2} + \dfrac{1}{3\ln 3} + \dfrac{1}{4\ln 4} + \cdots$

8. $1 - \dfrac{2}{3^2} + \dfrac{3}{5^2} - \dfrac{4}{7^2} + \dfrac{5}{9^2} - \cdots$

In Exercises 9–12, the nth partial sum, S_n, of a series $\sum a_n$ is given. Determine whether the series converges and, if it does, give its sum.

9. $S_n = \dfrac{n}{n+1}$

10. $S_n = 1 - (-1)^n$

11. $S_n = 2 - \dfrac{\ln(n+1)}{n+1}$

12. $S_n = \dfrac{n^3 - 2n + 1}{\sqrt{n^4 + n^2 + 4}}$

In Exercises 13–20, show that the given series is geometric and determine whether it converges. If it converges, find its sum.

13. $2 - 1 + \dfrac{1}{2} - \dfrac{1}{4} + \dfrac{1}{8} - \cdots$

14. $\displaystyle\sum_{k=1}^{\infty} \dfrac{2}{3^{k-1}}$

15. $\displaystyle\sum_{n=0}^{\infty} 3 \cdot 2^{-n}$

16. $\dfrac{1}{\ln 3} + \dfrac{1}{(\ln 3)^2} + \dfrac{1}{(\ln 3)^3} + \cdots$

17. $\dfrac{2}{e} - \dfrac{4}{e^2} + \dfrac{8}{e^3} - \dfrac{16}{e^4} + \cdots$

18. $\displaystyle\sum_{n=1}^{\infty} \left(\dfrac{5}{4}\right)^n$

19. $\displaystyle\sum_{n=0}^{\infty} (0.99)^n$

20. $\displaystyle\sum_{k=1}^{\infty} \dfrac{(-1)^{k-1} 3^k}{4^{k-1}}$

In Exercises 21–24, express the repeating decimal in the form m/n, in lowest terms, where m and n are integers, making use of geometric series.

21. $0.151515\ldots$

22. $2.181818\ldots$

23. $0.148148148\ldots$

24. $1.135135135\ldots$

In Exercises 25–28, show that the series is telescoping, and find its sum.

25. $\displaystyle\sum_{k=1}^{\infty} \dfrac{1}{(k+1)(k+2)}$

26. $\displaystyle\sum_{n=1}^{\infty} \dfrac{2}{n(n+2)}$

27. $\displaystyle\sum_{n=1}^{\infty} \dfrac{2}{4n^2 - 1}$

28. $\displaystyle\sum_{k=1}^{\infty} \dfrac{1}{k^2 + 4k + 3}$

In Exercises 29–32, show that the series converges, and find its sum.

29. $\displaystyle\sum_{n=1}^{\infty} \left(\dfrac{1}{2^n} + \dfrac{2}{3^n}\right)$

30. $\displaystyle\sum_{k=1}^{\infty} \left[\left(\dfrac{2}{3}\right)^k - \left(\dfrac{3}{4}\right)^{k-1}\right]$

31. $\displaystyle\sum_{n=1}^{\infty} \left(\dfrac{1}{n^2 + n} - \dfrac{1}{3^{n-1}}\right)$

32. $\displaystyle\sum_{n=1}^{\infty} \left[\dfrac{(-1)^{n-1} 5}{2^n} + \dfrac{2}{n^2 + 2n}\right]$

33. A ball is dropped from a height of 10 m, and on each

successive bounce it rises two-thirds as high as on the preceding bounce. Find the total distance the ball travels.

34. Determine whether the series $\sum_{k=100}^{\infty} 1/k$ converges or diverges. Justify your answer.

In Exercises 35 and 36, show that each series diverges.

35. (a) $\displaystyle\sum_{n=1}^{\infty} \dfrac{n}{100n + 1}$ (b) $\displaystyle\sum_{n=2}^{\infty} \dfrac{n}{(\ln n)^2}$

36. (a) $\displaystyle\sum_{n=1}^{\infty} \dfrac{2n^2 - 3n + 4}{3n^2 + n + 5}$

(b) $\displaystyle\sum_{n=1}^{\infty} \dfrac{(-1)^{n-1} n}{\sqrt{1 + n^2}}$

37. Indicate which of the following statements are true and which are false.
(a) If $\sum a_n$ diverges, then $\lim_{n \to \infty} a_n = 0$.
(b) If $\lim_{n \to \infty} a_n = 0$, then $\sum a_n$ converges.
(c) If $a_1 + a_2 + \cdots + a_n = 1/n$, then $\sum a_n$ converges to 0.
(d) If $\sum (a_n + b_n)$ converges, so do $\sum a_n$ and $\sum b_n$.
(e) If $c \neq 0$ and $\sum c a_n$ converges, then $\sum a_n$ also converges.

38. Prove Theorem 10.8, making use of the properties of limits of sequences from Section 10.2.

39. (a) Prove that if $\sum a_n$ converges and $\sum b_n$ diverges, then $\sum (a_n - b_n)$ diverges.
(b) Give examples to show that if $\sum a_n$ and $\sum b_n$ both diverge, then $\sum (a_n + b_n)$ and $\sum (a_n - b_n)$ may converge or may diverge.

40. Show that the series

$$\sum_{n=1}^{\infty} \ln \dfrac{n}{n+1}$$

diverges. (*Hint:* Show that it is a telescoping series and find S_n.)

41. Prove that if $a_n \geq 0$ and $a_1 + a_2 + a_3 + \cdots + a_n \leq k$ for all n, where k is a constant, then $\sum a_n$ converges.

In Exercises 42 and 43, show that the series is geometric. Find the values of x for which the series converges, and give the sum as a function of x.

42. (a) $\displaystyle\sum_{n=0}^{\infty} \left(\dfrac{x}{2}\right)^n$ (b) $\displaystyle\sum_{n=0}^{\infty} \dfrac{(x-1)^n}{3^{n+1}}$

43. (a) $\displaystyle\sum_{n=0}^{\infty} \frac{(-1)^n 3^n}{(x+2)^{n+1}}$ (b) $\displaystyle\sum_{n=0}^{\infty} (\ln x)^n$

44. A pendulum 1 m long is released from a position in which its angle with the vertical is 60°. On each swing after the first, it reaches a maximum angle with the vertical that is 0.9 times as large as the angle reached on the previous swing. Find the total distance covered by the bob of the pendulum in coming to rest.

45. Regrouping of terms. Consider the divergent series $\displaystyle\sum_{n=1}^{\infty} (-1)^{n+1}$. If we group the terms by pairs starting with the first two terms, we get

$$(1-1) + (1-1) + (1-1) + \cdots = 0 + 0 + 0 + \cdots = 0$$

If, instead, we group by pairs starting with the second and third terms, we can write the result as

$$1 - (1-1) - (1-1) - (1-1) - \cdots = 1 - 0 - 0 - 0 - \cdots = 1$$

Prove that this phenomenon cannot occur in a convergent series. That is, prove that the terms of a convergent series

can be regrouped, preserving order, in any way and the resulting series has the same sum as the original one.

 46. Use a CAS or a spreadsheet to approximate the value of $\displaystyle\sum_{n=1}^{\infty} 1/n^2$ by the partial sum S_{100}. Compare your answer with the known result $\pi^2/6$.

 47. Use a CAS or a spreadsheet to approximate S_{100} for the harmonic series $\displaystyle\sum_{n=1}^{\infty} 1/n$.

 48. Find the smallest integer n for which the partial sum S_n of the series $\displaystyle\sum_{n=0}^{\infty} 1/3^n$ agrees with the exact sum to 10 decimal places.

49. For the Koch island, let the initial equilateral triangle have sides each of length a.
   (a) Find formulas for the number of sides, the length of each side, and the perimeter of the nth approximation to the Koch island. Denote these by s_n, l_n, and P_n, respectively.
   (b) Show that the perimeter of the Koch island is infinite. That is,

$$\lim_{n \to \infty} P_n = \infty$$

   (c) Express the area of the Koch island as an infinite geometric series and show that the area equals

$$\left(\frac{2\sqrt{3}}{5}\right) a^2$$

(*Hint:* The initial triangle has area $(\sqrt{3}/4)a^2$, and three triangles are added at the second stage, each with area $(\sqrt{3}/4)(a/3)^2$. So the combined area at the second stage is $(\sqrt{3}/4)a^2 + 3(\sqrt{3}/4)(a/3)^2$. Continue in this manner.)

10.4 SERIES OF POSITIVE TERMS; THE INTEGRAL TEST AND COMPARISON TESTS

Given an infinite series of constants, we usually want to answer two primary questions: (1) Does the series converge? and (2) If it does converge, what is its sum? The second question generally is much harder to answer than the first. (Geometric series and telescoping series are exceptions.) However, if we know that a series converges, we can at least approximate its sum by using a partial sum S_n for sufficiently large n. So it is very useful just to be able to answer the first question. In this section and the next we give certain tests for convergence that are applicable when the terms of the series are all positive or zero.

The basis for our tests is the following theorem.

THEOREM 10.9

If $a_n \geq 0$ for all n, then $\sum a_n$ converges if and only if its sequence of partial sums is bounded.

Proof   Let $\{S_n\}$ be the sequence of partial sums. Then, since $S_1 = a_1$, $S_2 = a_1 + a_2$, $S_3 = a_1 + a_2 + a_3, \ldots$, and all of the a_n's are nonnegative, we see that $S_1 \leq S_2 \leq S_3 \leq \cdots$. That is, $\{S_n\}$ is a monotone increasing sequence. If $\{S_n\}$ is bounded, we know from Theorem 10.5 that the sequence converges. Thus, the series $\sum a_n$ converges.

Suppose now we are given that $\sum a_n$ converges. Since $\{S_n\}$ is an increasing sequence, it is either bounded or it diverges to ∞. But if $\lim_{n \to \infty} S_n = \infty$, the series would diverge, contrary to our hypothesis. Thus $\{S_n\}$ must be bounded. Our proof is therefore complete.   ∎

The Integral Test

As our first application of Theorem 10.9, we prove a test for convergence of a series based on our knowledge of improper integrals (see Section 8.7).

THEOREM 10.10

The Integral Test

Let f be a continuous, positive, monotone decreasing function for $x \geq 1$, and let $a_n = f(n)$ for $n = 1, 2, 3, \ldots$. Then $\sum_{n=1}^{\infty} a_n$ is convergent if and only if the improper integral $\int_1^{\infty} f(x)dx$ is convergent. That is:

(a) If $\int_1^{\infty} f(x)dx$ converges, then $\sum_{n=1}^{\infty} a_n$ converges.

(b) If $\int_1^{\infty} f(x)dx$ diverges, then $\sum_{n=1}^{\infty} a_n$ diverges.

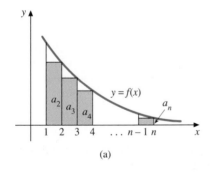

(a)

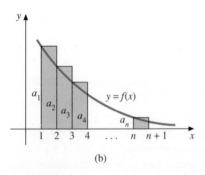

(b)

FIGURE 10.24

Proof   Suppose first that $\int_1^{\infty} f(x)\,dx$ converges. As Figure 10.24(a) shows, if we partition the interval $[1, n]$ at the integer points, the sum of the areas of the inscribed rectangles is

$$f(2) \cdot 1 + f(3) \cdot 1 + f(4) \cdot 1 + \cdots + f(n) \cdot 1$$

Since $a_n = f(n)$, the sum can be written as

$$a_2 + a_3 + a_4 + \cdots + a_n = S_n - a_1$$

where S_n is the nth partial sum of the series. The areas of the inscribed rectangles cannot exceed the area under the graph of f. So we have

$$S_n - a_1 \leq \int_1^n f(x)dx \leq \int_1^{\infty} f(x)dx$$

Thus,

$$S_n \leq a_1 + \int_1^{\infty} f(x)dx$$

By our hypothesis, the integral on the right converges. Therefore, the sequence $\{S_n\}$ is bounded. We also know that $a_n \geq 0$, since $a_n = f(n)$ and f is a positive function. It now follows from Theorem 10.9 that $\sum a_n$ converges.

Suppose now that $\int_1^\infty f(x)dx$ diverges. From Figure 10.24(b), by partition-ing the interval $[1, n+1]$ and using circumscribed rectangles, we see that

$$S_n \geq \int_1^{n+1} f(x)dx$$

As $n \to \infty$, the integral on the right becomes arbitrarily large (otherwise $\int_1^\infty f(x)dx$ would be finite). Thus, $\lim_{n\to\infty} S_n = \infty$, and $\sum a_n$ diverges. ∎

REMARK
■ Since deleting or adding any finite number of terms of a series does not affect its convergence or divergence, it is sufficient that the conditions of the Integral Test hold for $x \geq m$, where m is some positive integer.

EXAMPLE 10.23   Test each of the following series for convergence or di-vergence. (a) $\displaystyle\sum_{n=1}^{\infty} \frac{1}{1+n^2}$ (b) $\displaystyle\sum_{n=2}^{\infty} \frac{1}{n \ln n}$

Solution

(a) Let $f(x) = 1/(1+x^2)$ (see Figure 10.25). Then f satisfies the hypotheses of Theorem 10.10. So we consider the improper integral

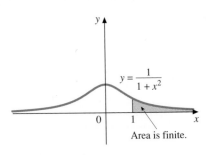

$$\int_1^\infty \frac{dx}{1+x^2} = \lim_{t\to\infty} \int_1^t \frac{dx}{1+x^2} = \lim_{t\to\infty} \left[\tan^{-1} t - \tan^{-1} 1\right] = \frac{\pi}{2} - \frac{\pi}{4} = \frac{\pi}{4}$$

Since the integral converges, we conclude from Theorem 10.10 that the series

$$\sum_{n=1}^{\infty} \frac{1}{1+n^2}$$

FIGURE 10.25

also converges.

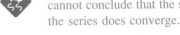

Note that while we know the value of the improper integral is $\pi/4$, we cannot conclude that the series converges to this value. All we know is that the series does converge.

(b) The function $f(x) = 1/(x \ln x)$ (see Figure 10.26) satisfies the hypotheses of Theorem 10.10 for $x \geq 2$. So we consider

$$\int_2^\infty \frac{1}{x \ln x}dx = \lim_{t\to\infty} \int_2^t \frac{1}{x \ln x}dx$$

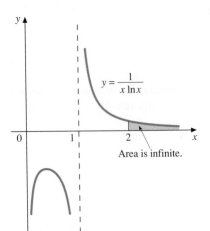

Since $d(\ln x) = (1/x)dx$, the integrand is in the form du/u with $u = \ln x$. So an antiderivative is $\ln|u| = \ln|\ln x|$. Thus,

$$\lim_{t\to\infty} \int_2^t \frac{1}{x \ln x}dx = \lim_{t\to\infty} \left[\ln|\ln t| - \ln|\ln 2|\right] = \infty$$

since $\ln t \to \infty$, and hence, $\ln|\ln t| \to \infty$ as $t \to \infty$. Thus, by Theo-rem 10.10, the series

$$\sum_{n=2}^{\infty} \frac{1}{n \ln n}$$

diverges. ∎

FIGURE 10.26

EXAMPLE 10.24   Determine all values of p for which the series $\sum 1/n^p$ converges.

Solution   If $p > 0$, let $f(x) = 1/x^p$. Then f is a positive, continuous, monotone decreasing function for $x \geq 1$. Furthermore, $f(n) = 1/n^p$. So the conditions of Theorem 10.10 are met. We found in Example 8.28 that

$$\int_1^\infty \frac{1}{x^p} dx$$

converges if $p > 1$ and diverges if $p \leq 1$. Thus, by Theorem 10.10, $\sum 1/n^p$ also converges if $p > 1$ and diverges if $p \leq 1$.

If $p \leq 0$, write $p = -\alpha$, where $\alpha \geq 0$. Then $1/n^p = n^\alpha \geq 1$. That is, each term of the series is at least 1, so the partial sums become arbitrarily large, and the series diverges.   ∎

The series in the preceding example is referred to as the **p-series**. We give special prominence to what we have found.

The p-Series

The p-series

$$\sum_{n=1}^\infty \frac{1}{n^p}$$

converges if $p > 1$ and diverges if $p \leq 1$.

Note that by taking $p = 1$, we see again that the harmonic series diverges. Also, we confirm the fact that the series $\sum_{n=1}^\infty 1/n^2$ converges. This series is the one Euler proved converges to $\pi^2/6$. In fact, Euler also proved that $\sum_{n=1}^\infty 1/n^4$ converges to $\pi^4/90$, and he found a general formula for the sum of $\sum_{n=1}^\infty 1/n^{2k}$, but he was unable to find the sum of the reciprocals of odd powers (greater than or equal to 3) of the positive integers. In particular, he did not find the sum of the series

$$\sum_{n=1}^\infty \frac{1}{n^3}$$

Nor has anyone else ever been able to find this sum. It is one of the famous unsolved problems in mathematics. You will become famous overnight if you find the sum.

Comparison Tests

We can frequently determine whether a series converges or diverges by comparing its terms with those of some known convergent series or known divergent series. There are two ways to make such comparisons. We state the first way in the following theorem.

THEOREM 10.11

The Comparison Test

Let $\sum a_n$ and $\sum b_n$ be series of nonnegative terms, with $a_n \le b_n$ for all n.

(a) If $\sum b_n$ converges, $\sum a_n$ converges.
(b) If $\sum a_n$ diverges, $\sum b_n$ diverges.

In other words, if each term of a positive-term series is smaller than the corresponding term of a known convergent series, the original series also converges, and if each term is larger than the corresponding term of a known divergent series of positive terms, the original series also diverges.

Proof   Let A_n and B_n denote the nth partial sums of $\sum a_n$ and $\sum b_n$, respectively. Suppose first that $\sum b_n$ converges. Then, by Theorem 10.9, $B_n \le M$ for some positive constant M. Since $a_n \le b_n$, it follows that

$$A_n = a_1 + a_2 + \cdots + a_n \le b_1 + b_2 + \cdots + b_n = B_n$$

So $A_n \le B_n \le M$. Using Theorem 10.9 again, we see that $\sum a_n$ converges, since $\{A_n\}$ is bounded.

Suppose now that $\sum a_n$ diverges. Then the sequence $\{A_n\}$ is unbounded (if it were bounded, $\sum a_n$ would be convergent). Since $B_n \ge A_n$, it follows that $\{B_n\}$ also is unbounded. Thus, by Theorem 10.9, $\sum b_n$ diverges.   ∎

REMARKS

■ It is sufficient that $a_n \le b_n$ from some n onward, say for $n \ge N$, since convergence or divergence is unaffected by neglecting any finite number of terms.

■ When a_n and b_n are nonnegative and $a_n \le b_n$ for all $n \ge N$, we say that the series $\sum b_n$ *dominates* the series $\sum a_n$. Equivalently, we may say that $\sum a_n$ *is dominated by* $\sum b_n$. Theorem 10.11 can therefore be rephrased by saying that for series of nonnegative terms, when a series to be tested is dominated by a known convergent series, the series converges. Similarly, when a series to be tested dominates a known divergent series, the series diverges.

If $a_n \le b_n$ and $\sum a_n$ converges, we can draw no conclusion from the Comparison Test about convergence or divergence of $\sum b_n$. Similarly, if $\sum b_n$ diverges, we can draw no conclusion about $\sum a_n$. For example,

$$\frac{1}{n^2} \le \frac{1}{n}$$

but $\sum 1/n^2$ converges and $\sum 1/n$ diverges. Figure 10.27 shows the situation graphically. In the figure, C represents a known convergent series and D a known divergent series. Any series with terms less than those of C converges, and any series with terms greater than those of D diverges. No conclusion can be drawn about series between C and D, however.

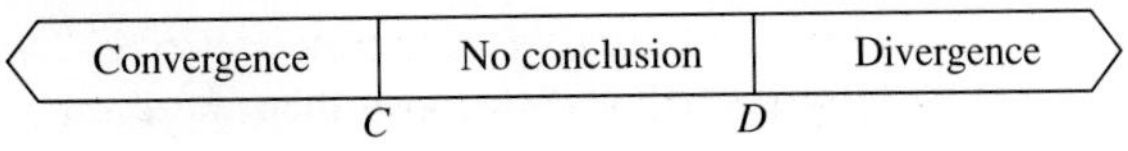

Convergence	No conclusion	Divergence
C		D

FIGURE 10.27

EXAMPLE 10.25   Test each of the following series for convergence or divergence.

(a) $\displaystyle\sum_{n=1}^{\infty} \frac{1}{\sqrt{n^3 + 3}}$ (b) $\displaystyle\sum_{n=1}^{\infty} \frac{1}{2n - 1}$

Solution

(a) By ignoring the 3 under the radical in the denominator, we see that the general term behaves like $1/n^{3/2}$. But $\sum 1/n^{3/2}$ is a convergent *p*-series, and we might suspect, by the Comparison Test, that the given series also converges. Indeed, since for $n \geq 1$,

$$\frac{1}{\sqrt{n^3 + 3}} < \frac{1}{\sqrt{n^3}} = \frac{1}{n^{3/2}}$$

it follows by the Comparison Test that

$$\sum_{n=1}^{\infty} \frac{1}{\sqrt{n^3 + 3}}$$

also converges.

(b) Here we might suspect that the terms $1/(2n - 1)$ behave similarly to $1/n$. In fact,

$$\frac{1}{2n - 1} > \frac{1}{2n} \qquad \text{for } n \geq 1$$

since a smaller denominator results in a larger fraction. Now, we know that the harmonic series $\sum 1/n$ diverges. It follows that

$$\sum \frac{1}{2n}$$

also diverges (see Corollary 10.8). Thus, by the Comparison Test,

$$\sum \frac{1}{2n - 1}$$

diverges.   ∎

The next test is a variation on the Comparison Test that often is easier to apply than the Comparison Test itself. In it we consider the *ratio* of the terms of a series we are testing to those of a series that we know to be either convergent or divergent. If this ratio approaches some positive (finite) limit as $n \to \infty$, we can conclude that the two series are similar with regard to convergence or divergence. That is, both series converge or else both series diverge.

THEOREM 10.12

> **The Limit Comparison Test**
>
> If $\sum a_n$ and $\sum b_n$ are series of positive terms and
>
> $$\lim_{n \to \infty} \frac{a_n}{b_n} = L$$
>
> where $0 < L < \infty$, then $\sum a_n$ and $\sum b_n$ both converge or else both diverge.

Proof   Since a_n/b_n approaches L as a limit, it follows that a_n/b_n will lie in the interval $(L/2, 3L/2)$ for all sufficiently large n, say for $n \geq N$. Thus, for $n \geq N$, since $a_n/b_n < 3L/2$, we have

$$a_n < \frac{3L}{2}b_n$$

If we know that $\sum b_n$ converges, then so does $\sum (3L/2)b_n$ (Theorem 10.8). Thus, by the Comparison Test, $\sum a_n$ also converges.

Since for $n \geq N$, the ratio a_n/b_n is in the interval $(L/2, 3L/2)$, we also have $a_n/b_n > L/2$, or

$$a_n > \frac{L}{2}b_n$$

when $n \geq N$. Thus, if we know that $\sum b_n$ diverges, so does $\sum (L/2)b_n$, by Corollary 10.8. Thus, again by the Comparison Test, $\sum a_n$ diverges.   ∎

REMARK

■ When the conditions of Theorem 10.12 are met (i.e., $\lim_{n \to \infty} a_n/b_n = L$, with $0 < L < \infty$), we say that a_n and b_n are *of the same order of magnitude* as $n \to \infty$. You can often tell the order of magnitude of the nth term of a series by neglecting all but the highest powers in the numerator and denominator. You can also ignore any coefficients. (See also Exercise 50 in Exercise Set 10.4.)

EXAMPLE 10.26   Test each of the following for convergence or divergence.

(a) $\displaystyle\sum_{n=1}^{\infty} \frac{n^2 - n}{2n^3 + 3n - 4}$ (b) $\displaystyle\sum_{n=1}^{\infty} \frac{1}{\sqrt{3n^4 - 2n}}$

Solution   Observe that in each case it would be difficult to apply the Integral Test. Also, finding a suitable series for comparison and showing that the appropriate inequality in Theorem 10.11 is satisfied would be difficult. So we try the Limit Comparison Test in each case.

(a) Neglecting the coefficients and all but the highest powers, we see that

$$\frac{n^2 - n}{2n^3 + 3n - 4} \sim \frac{n^2}{n^3} = \frac{1}{n}$$

where we are using "$\sim$" to mean "is of the same order of magnitude." To confirm that the orders of magnitude are the same, we take the limit of the quotient:

$$\lim_{n \to \infty} \left[\frac{n^2 - n}{2n^3 + 3n - 4} \div \frac{1}{n} \right] = \lim_{n \to \infty} \frac{n^3 - n^2}{2n^3 + 3n - 4} = \frac{1}{2}$$

Since we know that the harmonic series $\sum 1/n$ diverges, we conclude by Theorem 10.12 that the given series also diverges.

(b) Again using the dominant-term approach, we have

$$\frac{1}{\sqrt{3n^4 - 2n}} \sim \frac{1}{\sqrt{n^4}} = \frac{1}{n^2}$$

To confirm that the orders of magnitude are the same, we divide and take the limit as $n \to \infty$:

$$\lim_{n\to\infty} \left[\frac{1}{\sqrt{3n^4 - 2n}} \div \frac{1}{n^2} \right] = \lim_{n\to\infty} \frac{n^2}{\sqrt{3n^4 - 2n}}$$

$$= \lim_{n\to\infty} \sqrt{\frac{n^4}{3n^4 - 2n}} = \frac{1}{\sqrt{3}}$$

We know that the p-series $\sum 1/n^2$ converges (since $p = 2 > 1$), so by Theorem 10.12,

$$\sum_{n=1}^{\infty} \frac{1}{\sqrt{3n^4 - 2n}}$$

also converges.   ∎

Exercise Set 10.4

In Exercises 1–10, use the Integral Test to determine convergence or divergence.

1. $\displaystyle\sum_{n=1}^{\infty} \frac{1}{\sqrt{2n-1}}$

2. $\displaystyle\sum_{n=2}^{\infty} \frac{1}{n(\ln n)^2}$

3. $\displaystyle\sum_{n=1}^{\infty} \frac{1}{n^2 + 4}$

4. $\displaystyle\sum_{n=1}^{\infty} ne^{-n^2}$

5. $\displaystyle\sum_{n=3}^{\infty} \frac{1}{n\sqrt{\ln n}}$

6. $\displaystyle\sum_{n=1}^{\infty} \frac{n}{(n^2+1)^{3/2}}$

7. $\displaystyle\sum_{n=1}^{\infty} \frac{1}{3n+2}$

8. $\displaystyle\sum_{n=1}^{\infty} \frac{\ln n}{n}$

9. $\displaystyle\sum_{n=1}^{\infty} \frac{n}{n^2 + 1}$

10. $\displaystyle\sum_{n=1}^{\infty} ne^{-n}$

In Exercises 11–20, use the Comparison Test to determine convergence or divergence.

11. $\displaystyle\sum_{n=1}^{\infty} \frac{1}{n^2 + 1}$

12. $\displaystyle\sum_{n=1}^{\infty} \frac{1}{\sqrt{n^3 + 2n}}$

13. $\displaystyle\sum_{n=1}^{\infty} \frac{2}{2n-1}$

14. $\displaystyle\sum_{n=1}^{\infty} \frac{n+1}{n^2}$

15. $\displaystyle\sum_{n=1}^{\infty} \sqrt{\frac{n+1}{n}}$

16. $\displaystyle\sum_{n=1}^{\infty} \frac{n}{2n^3 + 1}$

17. $\displaystyle\sum_{n=1}^{\infty} \frac{2}{3^n + 1}$

18. $\displaystyle\sum_{n=1}^{\infty} \frac{3^n}{2^n - 1}$

19. $\displaystyle\sum_{n=1}^{\infty} \frac{1 + \sin^2 n}{n\sqrt{n}}$

20. $\displaystyle\sum_{n=1}^{\infty} \frac{n+1}{\sqrt{2n^3 - 1}}$

In Exercises 21–30, use the Limit Comparison Test to determine convergence or divergence.

21. $\displaystyle\sum_{n=1}^{\infty} \frac{2n-1}{3n^2 + 4n - 2}$

22. $\displaystyle\sum_{n=1}^{\infty} \frac{3n}{n^4 - 2}$

23. $\displaystyle\sum_{n=1}^{\infty} \frac{\sqrt{2n+3}}{n^3}$

24. $\displaystyle\sum_{n=1}^{\infty} \frac{n}{\sqrt{n^3 - 2n + 4}}$

25. $\displaystyle\sum_{n=1}^{\infty} \frac{1 + 2n}{(n^2+1)^{3/2}}$

26. $\displaystyle\sum_{n=1}^{\infty} \frac{n^2 - 2}{\sqrt{3n^5 + 1}}$

27. $\displaystyle\sum_{n=1}^{\infty} \frac{2n^3 - 3n^2 + 4}{5n^4 + 2n^3 - 1}$

28. $\displaystyle\sum_{n=1}^{\infty} \sqrt{\frac{n}{n^5 + 2}}$

29. $\displaystyle\sum_{n=1}^{\infty} \frac{\sqrt{n^2 + 3n}}{n^3 + 1}$

30. $\displaystyle\sum_{n=2}^{\infty} \frac{1}{\sqrt{n^3 - n}}$

In Exercises 31–48, test for convergence or divergence by any appropriate means.

31. $\displaystyle\sum_{n=1}^{\infty} \frac{n\cos^2 n}{1 + n^3}$

32. $\displaystyle\sum_{k=2}^{\infty} \frac{\sqrt{\ln k}}{k}$

33. $\displaystyle\sum_{n=1}^{\infty} \frac{1}{n \cdot 2^n}$

34. $\displaystyle\sum_{i=1}^{\infty} \frac{2i^2}{3i^2 - 2}$

35. $\displaystyle\sum_{n=0}^{\infty} \frac{\sqrt{n+1}}{n^2+1}$

36. $\displaystyle\sum_{j=0}^{\infty} \frac{j+1}{j^3+2}$

37. $\displaystyle\sum_{m=0}^{\infty} \frac{3^m}{4^m+3}$

38. $\displaystyle\sum_{n=0}^{\infty} \frac{1}{\cosh n}$

39. $\displaystyle\sum_{k=0}^{\infty} \frac{e^k}{1+e^{2k}}$

40. $\displaystyle\sum_{k=2}^{\infty} \frac{2k^2-3k+1}{k^5+2k^2+1}$

41. $\displaystyle\sum_{n=1}^{\infty} \frac{3n-2}{(n+1)(n+2)}$

42. $\displaystyle\sum_{m=1}^{\infty} \frac{m\sec^2 m}{1+m^2}$

43. $\displaystyle\sum_{n=1}^{\infty} \frac{1+\cos^2 n}{n^{3/2}}$

44. $\displaystyle\sum_{k=1}^{\infty} \frac{k}{\sqrt{100k^2+1}}$

45. $\displaystyle\sum_{n=3}^{\infty} \frac{1}{n(\ln n)[(\ln(\ln n)]}$

46. $\displaystyle\sum_{n=1}^{\infty} \frac{\sinh n}{\cosh 2n}$

47. $\displaystyle\sum_{n=1}^{\infty} \frac{n\ln n}{1+n^3}$

48. $\displaystyle\sum_{n=1}^{\infty} \frac{\sqrt{n+1}-\sqrt{n}}{n}$

49. Determine all values of p for which the series
$$\sum_{n=2}^{\infty} \frac{1}{n(\ln n)^p} \text{ converges.}$$

50. Prove the following extensions of the Limit Comparison Test for series $\sum a_n$ and $\sum b_n$ of positive terms:
(a) If $\lim_{n\to\infty} a_n/b_n = 0$ and $\sum b_n$ converges, then $\sum a_n$ also converges.
(b) If $\lim_{n\to\infty} a_n/b_n = \infty$ and $\sum b_n$ diverges, then $\sum a_n$ also diverges.

51. Use the results of Exercise 50 to test each of the following series for convergence or divergence.
(a) $\displaystyle\sum_{n=2}^{\infty} \frac{1}{\ln n}$ (b) $\displaystyle\sum_{n=2}^{\infty} \frac{\ln n}{n}$ (c) $\displaystyle\sum_{n=2}^{\infty} \frac{\ln n}{n^2}$

52. Prove that if $a_n \geq 0$ and $\sum a_n$ converges, then $\sum a_n^2$ also converges.

53. Prove that if $a_n \geq 0$ and $\lim_{n\to\infty} na_n$ exists and is positive, then $\sum a_n$ diverges.

54. Prove that if $\sum_{k=1}^{\infty} a_k$ converges, then $\lim_{n\to\infty} \sum_{k=n}^{\infty} a_k = 0$.

55. Use a CAS or a spreadsheet to find the first 100 terms of each of the series in Example 10.26. In part (a) compare your result with the first 100 terms of the harmonic series, and in part (b) compare your result with the first 100 terms of the series $\sum 1/n^2$. Interpret your findings.

10.5 SERIES OF POSITIVE TERMS; THE RATIO TEST AND THE ROOT TEST

The next test is particularly well suited to series whose nth terms involve powers or products, especially factorials. In it, we consider the ratio of each term (after the first) to the preceding one.

THEOREM 10.13

The Ratio Test

Let $\sum a_n$ be a series of positive terms such that
$$\lim_{n\to\infty} \frac{a_{n+1}}{a_n} = L.$$
(a) If $L < 1$, then $\sum a_n$ converges.
(b) If $L > 1$, then $\sum a_n$ diverges.
(c) If $L = 1$, then the test is inconclusive.

For $n \geq N$, the ratios $\dfrac{a_{n+1}}{a_n}$ are all to the left of r.

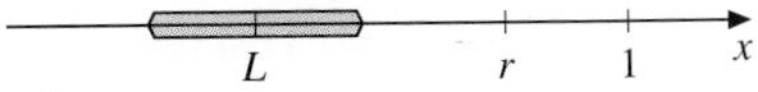

$\dfrac{a_{n+1}}{a_n}$ is in the shaded interval for $n \geq N$.

FIGURE 10.28

Proof   Suppose first that $L < 1$. Let r denote any number such that $L < r < 1$. Since the ratios a_{n+1}/a_n eventually come arbitrarily close to L, they all lie to the left of r for n sufficiently large, say for $n \geq N$. (See Figure 10.28.)

Then we have

$$\frac{a_{n+1}}{a_n} < r \qquad \text{for all } n \geq N$$

or equivalently, $a_{n+1} < ra_n$ for all $n \geq N$. So

$$a_{N+1} < ra_N$$
$$a_{N+2} < ra_{N+1} < r^2 a_N$$
$$a_{N+3} < ra_{N+2} < r^3 a_N$$
$$\vdots$$

Thus, the terms of the series $\sum_{n=N+1}^{\infty} a_n$ are dominated by the terms of the series $\sum_{n=1}^{\infty} a_N r^n$. This latter series is a convergent geometric series, since the common ratio r is positive and less than 1. It follows by the Comparison Test that $\sum_{n=N+1}^{\infty} a_n$ converges. When we add the first N terms, the resulting series is also convergent; that is, the entire series $\sum_{n=1}^{\infty} a_n$ also converges.

If $L > 1$, then for all sufficiently large n,

$$\frac{a_{n+1}}{a_n} > 1, \qquad \text{or equivalently,} \qquad a_{n+1} > a_n$$

so the terms increase and thus cannot approach 0. Thus, $\sum a_n$ diverges (the nth term test for divergence).

Finally, the series $\sum 1/n$ and $\sum 1/n^2$ both satisfy

$$\lim_{n \to \infty} \frac{a_{n+1}}{a_n} = 1$$

(see Exercise 27 in Exercise Set 10.5), yet the first series diverges and the second converges. So when $L = 1$, the Ratio Test does not give us any conclusive information, and some other test is required. ∎

In applying the Ratio Test to show convergence, it is not enough to show $a_{n+1}/a_n < 1$ for all n. For example, consider the harmonic series

$$\sum \frac{1}{n} = 1 + \frac{1}{2} + \frac{1}{3} + \frac{1}{4} + \cdots$$

which we know to be divergent. Yet

$$\frac{a_{n+1}}{a_n} = \frac{1}{n+1} \div \frac{1}{n} = \frac{n}{n+1}$$

which is less than 1 for all n. It is the *limit* of this ratio as $n \to \infty$ that must be less than 1 to ensure convergence. In the case of the harmonic series, the limit of the ratio is 1. So the Ratio Test is not applicable.

EXAMPLE 10.27   Test each of the following series for convergence.

(a) $\displaystyle\sum_{n=1}^{\infty} \frac{2^n}{n!}$ (b) $\displaystyle\sum_{n=1}^{\infty} \frac{3^{n-1}}{n^2 \cdot 2^n}$

Solution   Since products and powers are involved, the Ratio Test appears to be the test to use.

(a) $\displaystyle\lim_{n \to \infty} \frac{a_{n+1}}{a_n} = \lim_{n \to \infty} \frac{2^{n+1}}{(n+1)!} \cdot \frac{n!}{2^n} = \lim_{n \to \infty} \frac{2^{n+1}}{2^n} \cdot \frac{n!}{(n+1)!} = \lim_{n \to \infty} \frac{2}{n+1} = 0$

Note that instead of writing the quotient of a_{n+1} over a_n, we inverted the denominator and multiplied. Note also that $(n + 1)! = (n + 1)n!$, so we canceled the $n!$ that appeared in both numerator and denominator. Since $L = 0$ and $0 < 1$, the series converges.

(b)

$$\lim_{n\to\infty} \frac{a_{n+1}}{a_n} = \lim_{n\to\infty} \frac{3^n}{(n+1)^2 \, 2^{n+1}} \cdot \frac{n^2 2^n}{3^{n-1}}$$

$$= \lim_{n\to\infty} \frac{n^2}{(n+1)^2} \cdot \frac{3^n}{3^{n-1}} \cdot \frac{2^n}{2^{n+1}}$$

$$= \lim_{n\to\infty} \left(\frac{n}{n+1}\right)^2 \cdot \frac{3}{2} = \frac{3}{2}$$

Since $L = \frac{3}{2} > 1$, the series diverges.   ∎

EXAMPLE 10.28   Test the following series for convergence or divergence.

$$\sum_{n=1}^{\infty} \frac{1 \cdot 3 \cdot 5 \cdot \cdots \cdot (2n - 1)}{3^n \cdot n!}$$

Solution   Before applying the Ratio Test, we simplify the quotient a_{n+1}/a_n. The numerator of the nth term is the product of the first n odd positive integers, so the numerator of the $(n + 1)$st term is the product of the first $n + 1$ odd positive integers, namely,

$$1 \cdot 3 \cdot 5 \cdot \cdots \cdot [2(n + 1) - 1] = 1 \cdot 3 \cdot 5 \cdot \cdots \cdot (2n + 1)$$

This product has all of the factors of the numerator of the nth term, together with the additional factor $2n + 1$. Thus,

$$\frac{a_{n+1}}{a_n} = \frac{1 \cdot 3 \cdot 5 \cdot \cdots \cdot (2n + 1)}{3^{n+1}(n + 1)!} \cdot \frac{3^n n!}{1 \cdot 3 \cdot 5 \cdot \cdots \cdot (2n - 1)} = \frac{2n + 1}{3(n + 1)}$$

Now we take the limit:

$$\lim_{n\to\infty} \frac{a_{n+1}}{a_n} = \lim_{n\to\infty} \frac{2n + 1}{3n + 3} = \frac{2}{3}$$

Since $\frac{2}{3} < 1$, we conclude that the series converges.   ∎

We conclude the tests of positive term series with the **Root Test**. Its proof is similar to that of the Ratio Test, and we leave it for the exercises (Exercise 29 in Exercise Set 10.5).

THEOREM 10.14

The Root Test

Let $\sum a_n$ be a series of nonnegative terms such that

$$\lim_{n\to\infty} \sqrt[n]{a_n} = L$$

(a) If $L < 1$, then $\sum a_n$ converges.
(b) If $L > 1$, then $\sum a_n$ diverges.
(c) If $L = 1$, then the test is inconclusive.

The Root Test is more powerful than the Ratio Test in the sense that whenever the Ratio Test gives a definite conclusion concerning convergence or divergence, the same will be true of the Root Test. But there are series for which the Ratio Test is inconclusive and the Root Test gives definite information. (See Exercise 31 in Exercise Set 10.5.) However, the Ratio Test is usually easier to apply.

EXAMPLE 10.29   Test the series

$$\sum_{n=1}^{\infty} \left(\frac{n}{2n+1} \right)^n$$

Solution   Here, the Root Test seems appropriate, since the general term involves the nth power. When we take the nth root, we get a simple expression whose limit we can find:

$$\lim_{n \to \infty} \sqrt[n]{a_n} = \lim_{n \to \infty} \frac{n}{2n+1} = \frac{1}{2} < 1$$

So the series converges.   ∎

REMARK ——————————————————————————————

■ In applying the Root Test you may sometimes need the following limits:

$$\lim_{n \to \infty} \sqrt[n]{a} = 1 \quad \text{for } a > 0$$

$$\lim_{n \to \infty} \sqrt[n]{n} = 1$$

We proved the second of these limits in Example 10.14. You will be asked to prove the first in Exercise 28 of this section. You can also convince yourself that the results are reasonable with a CAS or a spreadsheet (Exercise 32 of this section).

You may have wondered why there are so many tests for convergence (there are still more, but we have given the main ones). The answer is that no single test works on all positive-term series. Even when more than one test *could* be used, one may be much easier to apply than the others. With a little practice you will probably develop a feel for which test to apply in a given situation.

Exercise Set 10.5

In Exercises 1–26, use the Ratio Test or Root Test to determine convergence or divergence.

1. $\displaystyle\sum_{n=1}^{\infty} ne^{-n}$

2. $\displaystyle\sum_{n=1}^{\infty} \frac{n^2 \cdot 2^n}{3^{n-1}}$

3. $\displaystyle\sum_{n=1}^{\infty} \frac{n!}{10^n}$

4. $\displaystyle\sum_{n=1}^{\infty} \left(\frac{n-1}{2n+1} \right)^n$

5. $\displaystyle\sum_{n=1}^{\infty} \frac{2^{3n}}{3^{2n}}$

6. $\displaystyle\sum_{n=1}^{\infty} \frac{n^2}{2^n}$

7. $\displaystyle\sum_{n=1}^{\infty} \frac{2^{n^2}}{n!}$

8. $\displaystyle\sum_{n=1}^{\infty} n^3 3^{-n}$

9. $\displaystyle\sum_{n=1}^{\infty} \frac{n}{\sqrt{e^n}}$

10. $\displaystyle\sum_{k=2}^{\infty} \left(\frac{4k^2+3}{3k^2-4} \right)^k$

11. $\displaystyle\sum_{m=1}^{\infty} \frac{(m+1)^{2m}}{3^{m^2}}$

12. $\displaystyle\sum_{k=1}^{\infty} k^2 \left(\frac{2}{3}\right)^k$

13. $\displaystyle\sum_{n=0}^{\infty} \frac{(100)^{2n+1}}{(2n+1)!}$

14. $\displaystyle\sum_{n=1}^{\infty} \frac{(n+1)^{n/2}}{2 \cdot 2^{n+1}}$

15. $\displaystyle\sum_{n=1}^{\infty} \frac{\sqrt{3^n}}{2^n}$

16. $\displaystyle\sum_{n=1}^{\infty} \frac{2^{3n+1}}{e^{2n}}$

17. $\displaystyle\sum_{n=1}^{\infty} \frac{n^n}{2^{n^2}}$

18. $\displaystyle\sum_{n=1}^{\infty} \frac{4^n}{n \cdot 3^n}$

19. $\displaystyle\sum_{n=1}^{\infty} \frac{2^n \cdot n!}{(2n)!}$

20. $\displaystyle\sum_{n=1}^{\infty} \frac{(n!)^2}{(2n)!}$

21. $\displaystyle\sum_{n=1}^{\infty} \frac{n!\,(2n+1)!}{(2n+3)!}$

22. $\displaystyle\sum_{n=1}^{\infty} \frac{n^n}{n!}$

23. $\displaystyle\sum_{n=1}^{\infty} \frac{n!}{1 \cdot 3 \cdot 5 \cdot \,\cdots\, \cdot (2n-1)}$

24. $\displaystyle\sum_{n=1}^{\infty} \frac{n!}{2 \cdot 4 \cdot 6 \cdot \,\cdots\, \cdot (2n)}$

25. $\displaystyle\sum_{n=1}^{\infty} \frac{1 \cdot 4 \cdot 7 \cdot \,\cdots\, \cdot (3n-2)}{3 \cdot 5 \cdot 7 \cdot \,\cdots\, \cdot (2n+1)}$

26. $\displaystyle\sum_{n=1}^{\infty} \frac{2^n \cdot n!}{1 \cdot 3 \cdot 5 \cdot \,\cdots\, \cdot (2n-1)}$

27. Prove that

$$\lim_{n\to\infty} \frac{a_{n+1}}{a_n} = 1$$

where (a) $a_n = 1/n$ and (b) $a_n = 1/n^2$. What conclusion can you draw?

28. Prove that

$$\lim_{n\to\infty} a^{1/n} = 1 \qquad \text{for } a > 0$$

(*Hint:* Let $y = a^{1/x}$. Take logs and use L'Hôpital's Rule. Then apply Theorem 10.2.)

29. Prove the Root Test (Theorem 10.14).

30. Prove the following version of the Ratio Test for the positive-term series $\sum a_n$.
(a) If there is a number r such that $0 < r < 1$ for which

$$\frac{a_{n+1}}{a_n} \le r \qquad \text{for all } n \ge N$$

where N is some positive integer, then $\sum a_n$ converges.
(b) If

$$\frac{a_{n+1}}{a_n} \ge 1 \qquad \text{for all } n \ge N$$

where N is some positive integer, then $\sum a_n$ diverges.

31. Consider the series

$$\sum_{n=1}^{\infty} 2^{(-1)^n - n} = \frac{1}{2^2} + \frac{1}{2} + \frac{1}{2^4} + \frac{1}{2^3} + \frac{1}{2^6} + \frac{1}{2^5} + \cdots$$

(a) Use the Root Test to show that the series converges.
(b) Show that

$$\frac{a_{n+1}}{a_n} = 2^{2(-1)^{n+1} - 1}$$

and $\lim_{n\to\infty}(a_{n+1}/a_n)$ does not exist, so that the Ratio Test does not apply.

32. Use a CAS or a spreadsheet to find
(a) $\sqrt[n]{2}$ for $n = 1$ to $n = 20$
(b) $\sqrt[n]{100}$ for $n = 1$ to $n = 20$
What do you observe in each case?

33. Use a CAS or a spreadsheet to find $\sqrt[n]{n}$ for $n = 1$ to $n = 20$. What do you observe?

10.6 SERIES WITH TERMS OF VARIABLE SIGNS; ABSOLUTE CONVERGENCE

In general, when a series consists of both positive and negative terms, tests for convergence are more complex than those for series consisting only of positive terms. There is one important test, however, that is relatively simple, and that is applicable to series in which the terms are alternately positive and negative. Such series are called **alternating series**. Though alternating series in general could begin with either a positive or a negative first term, it is sufficient to consider such series in which the first term is positive, writing, for example,

$$\sum_{n=1}^{\infty} (-1)^{n-1} a_n = a_1 - a_2 + a_3 - a_4 + \cdots$$

where for all n, $a_n > 0$. If the first term were negative, we could write

$$-a_1 + a_2 - a_3 + a_4 - \cdots = -(a_1 - a_2 + a_3 - a_4 + \cdots)$$

and determine convergence or divergence by studying the series in parentheses.
    An example of an alternating series is

$$\sum_{n=1}^{\infty} \frac{(-1)^{n-1}}{n} = 1 - \frac{1}{2} + \frac{1}{3} - \frac{1}{4} + \cdots$$

This series is called the **alternating harmonic series**. Though the harmonic series itself, with all terms positive, diverges, we will soon see (Example 10.30) that the alternating harmonic series converges.
    The following test shows that if the absolute values of the terms of an alternating series decrease monotonically and approach 0 as $n \to \infty$, then the series converges.

THEOREM 10.15

Alternating Series Test

If $a_n > 0$ for all n and the following two conditions are satisfied for all n,

(i) $a_{n+1} \leq a_n$

(ii) $\lim_{n\to\infty} a_n = 0$

then the alternating series

$$\sum_{n=1}^{\infty} (-1)^{n-1} a_n$$

converges.

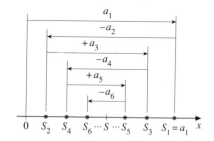

FIGURE 10.29

Proof   The idea of the proof is indicated in Figure 10.29. We start with the first partial sum S_1, which is just a_1. For the purpose of this illustration, let us suppose that the strict inequality $a_{n+1} < a_n$ holds true. Then $S_2 = a_1 - a_2$ is to the left of S_1, and $S_3 = a_1 - a_2 + a_3 = S_2 + a_3$ is to the right of S_2. Also, since $a_3 < a_2$, S_3 is less than S_1. Similarly, $S_4 = S_3 - a_4$ is to the left of S_3 but to the right of S_2, and so on. We see that the even-ordered partial sums $S_2, S_4, S_6, \ldots$ increase and are all to the left of the odd-ordered partial sums $S_1, S_3, S_5, \ldots$, which decrease. It seems reasonable to suppose that both even-ordered and odd-ordered partial sums converge to the same limit S, from which we would conclude that S is the sum of the alternating series. We now proceed to the proof.
    First, consider the partial sums of even order:

$$S_2 = a_1 - a_2$$

$$S_4 = (a_1 - a_2) + (a_3 - a_4)$$

$$\vdots$$

$$S_{2n} = (a_1 - a_2) + (a_3 - a_4) + \cdots + (a_{2n-1} - a_{2n})$$

By condition (i), the pairs in parentheses are all nonnegative. So $\{S_{2n}\}$ is a sequence of nonnegative terms that is monotone increasing. Furthermore, this sequence is bounded above, since we can also write S_{2n} in the form

$$S_{2n} = a_1 - (a_2 - a_3) - (a_4 - a_5) - \cdots - (a_{2n-2} - a_{2n-1}) - a_{2n}$$

The pairs in parentheses are all positive or zero, and a_{2n} is positive. So $S_{2n} \le a_1$. As we saw in the proof of Theorem 10.5, such a bounded, monotone increasing sequence must converge. Let the limit be S. That is,

$$\lim_{n \to \infty} S_{2n} = S$$

Now let us consider an odd-ordered partial sum. Except for S_1, we can write a typical odd-ordered partial sum as

$$S_{2n+1} = S_{2n} + a_{2n+1}$$

For example,

$$S_3 = (a_1 - a_2) + a_3 = S_2 + a_3$$

$$S_5 = (a_1 - a_2 + a_3 - a_4) + a_5 = S_4 + a_5$$

and so on. By condition (ii), $\lim_{n \to \infty} a_{2n+1} = 0$. So we have

$$\lim_{n \to \infty} S_{2n+1} = \lim_{n \to \infty} S_{2n} + \lim_{n \to \infty} a_{2n+1} = S + 0 = S$$

That is, both even- and odd-ordered partial sums approach the same limit S.
Thus,

$$\lim_{n \to \infty} S_n = S$$

So the series $\sum_{n=1}^{\infty} (-1)^{n-1} a_n$ converges to S.   ■

REMARK ————————————————————————————————

■ As with our other tests for convergence, it is sufficient that the conditions of Theorem 10.15 hold true from some point onward. That is, the series converges if the signs alternate and $a_{n+1} \le a_n$ from some n onward (not necessarily starting with $n = 1$), provided $\lim_{n \to \infty} a_n = 0$.

Error in Using S_n to Approximate S for Alternating Series

When a series satisfies the hypotheses of the Alternating Series Test, there is an easy way of getting an upper bound on the error when using any given partial sum, S_n, to approximate the sum S. The following corollary shows how.

COROLLARY 10.15

Under the conditions of Theorem 10.15,

$$|S_n - S| \le a_{n+1}$$

That is, the error in approximating the sum S by the nth partial sum S_n does not exceed the absolute value of the $(n + 1)$st term of the series.

Proof   In the proof of Theorem 10.15 we showed that the even-ordered partial sums were bounded above by a_1, the first term of the series. It follows that $\lim_{n \to \infty} S_{2n} = S$ must also not exceed a_1. Figure 10.29 clearly shows the fact that $S \leq a_1$.

Thus, for any alternating series whose terms in absolute value decrease monotonically and approach 0 as $n \to \infty$, the sum does not exceed the first term of the series (which we assumed to be positive). But

$$|S_n - S| = a_{n+1} - a_{n+2} + a_{n+3} - a_{n+4} + \cdots$$

is itself an alternating series whose terms in absolute value decrease monotonically and have 0 as a limit. By what we have just shown, the sum of the alternating series on the right does not exceed its first term, a_{n+1}. That is,

$$|S_n - S| \leq a_{n+1} \qquad \blacksquare$$

EXAMPLE 10.30   **The Alternating Harmonic Series**   Show that the alternating harmonic series

$$\sum_{n=1}^{\infty} \frac{(-1)^{n-1}}{n} = 1 - \frac{1}{2} + \frac{1}{3} - \frac{1}{4} + \cdots$$

converges. Determine an upper bound on the error in estimating the sum using S_{100}.

Solution   The conditions of Theorem 10.15 are clearly satisfied. That is, the series is an alternating series in which (i) $1/(n+1) < 1/n$, and (ii) $\lim_{n \to \infty} 1/n = 0$. The series therefore converges. Call its sum S. Then, by Corollary 10.15,

$$|S_{100} - S| \leq \frac{1}{101} \approx 0.009$$

So if we take the first 100 terms of the series to estimate the sum, we can be assured of accuracy to only one decimal place. This series converges very slowly. $\blacksquare$

EXAMPLE 10.31   Determine the convergence or divergence of each of the following series: (a) $\displaystyle\sum_{n=1}^{\infty}(-1)^{n-1} \ln\left(\frac{n+1}{n}\right)$ (b) $\displaystyle\sum_{n=1}^{\infty}(-1)^{n-1}\frac{n}{2n+1}$

Solution

(a) To apply Theorem 10.15, we must show both that $a_{n+1} \leq a_n$ and that $a_n \to 0$ as $n \to \infty$. To show that the terms $a_n = \ln((n+1)/n)$ decrease, we will use Method 2 of the tests for monotonicity given in Section 10.2. Replacing n with x in the formula for a_n and differentiating, we have

$$\frac{d}{dx} \ln\left(\frac{x+1}{x}\right) = \frac{d}{dx}[\ln(x+1) - \ln x]$$

$$= \frac{1}{x+1} - \frac{1}{x} = \frac{x - (x+1)}{x(x+1)} = \frac{-1}{x(x+1)}$$

Since the derivative is negative for all positive values of x, the function $\ln((x+1)/x)$ decreases. It therefore follows that for $n \geq 1$

$$\ln\left(\frac{n+2}{n+1}\right) \leq \ln\left(\frac{n+1}{n}\right)$$

That is, $a_{n+1} \leq a_n$ for all $n \geq 1$.

To see if $a_n \to 0$ as $n \to \infty$, we calculate the limit

$$\lim_{n \to \infty} \ln \left(\frac{n+1}{n} \right) = \ln \left[\lim_{n \to \infty} \left(\frac{n+1}{n} \right) \right] = \ln 1 = 0$$

(Note that we used the continuity of $\ln x$ here.)

Both conditions of the Alternating Series Test are met, so the series converges.

(b) Since

$$\lim_{n \to \infty} \frac{n}{2n+1} = \frac{1}{2} \neq 0$$

condition (ii) of the Alternating Series Test is not satisfied. We know, in fact, by the nth-Term Test for Divergence given in Section 10.3 that the series diverges. ∎

Absolute Convergence

Some series with variable signs have the property that when the signs are changed to be all positive, the resulting series converges. In this case we say that the original series is *absolutely convergent*. We can state the definition more succinctly as follows.

Definition 10.7 Absolute Convergence	The series $\sum a_n$ is said to be **absolutely convergent** if $\sum \lvert a_n \rvert$ converges.

REMARKS

■ Here we are allowing a_n to be either positive, negative, or zero, and we are not explicitly showing the signs, as we did with alternating series. No restriction is placed on the signs of the terms. They need not alternate but may do so.

■ If $a_n \geq 0$ for all n, then absolute convergence is equivalent to ordinary convergence. The concept of absolute convergence is useful only when the original series has variable signs (which may or may not alternate).

EXAMPLE 10.32   Test each of the following for absolute convergence.

(a) $\displaystyle \sum_{n=1}^{\infty} \frac{(-1)^{n-1}}{n}$ (b) $\displaystyle \sum_{n=1}^{\infty} \frac{(-1)^{n-1}}{n^2}$

Solution

(a) This series is the alternating harmonic series, which we have shown to be convergent, using the Alternating Series Test. However, when we take absolute values, we get the harmonic series $\sum 1/n$ itself, which we know to be divergent. So the alternating harmonic series is not absolutely convergent.

(b) Taking absolute values, we get the p-series $\sum 1/n^2$, with $p = 2$. Since $p > 1$, the series converges. Thus, the original series converges absolutely.

∎

Conditional Convergence

The alternating harmonic series in part (a) of Example 10.32 converges, but it is not absolutely convergent. This type of convergence is called *conditional convergence*.

Definition 10.8 Conditional Convergence	If $\sum a_n$ converges but $\sum |a_n|$ diverges, then $\sum a_n$ is said to be **conditionally convergent**.

Thus, of the two series in Example 10.32, the first is conditionally convergent and the second is absolutely convergent.

We now know that a series may be convergent without being absolutely convergent (in which case it is conditionally convergent). However, if a series is absolutely convergent, then it must also be convergent, as the next theorem shows.

THEOREM 10.16	If $\sum a_n$ is absolutely convergent, then it is convergent.

Proof   Since $|a_n|$ is either a_n or $-a_n$, it follows that

$$0 \le a_n + |a_n| \le 2|a_n|$$

(For example, if $a_n = -2$, then $|a_n| = 2$, and $0 = -2 + 2 < 2(2)$. Similarly, if $a_n = 4$, then $|a_n| = 4$, and $0 < 4 + 4 = 2(4)$.) Thus, since $\sum |a_n|$ (and hence also $\sum 2|a_n|$) converges, it follows by the Comparison Test that $\sum (a_n + |a_n|)$ converges. Finally, $a_n = (a_n + |a_n|) - |a_n|$, so by Theorem 10.8, part 3,

$$\sum a_n = \sum (a_n + |a_n|) - \sum |a_n|$$

Since each series on the right converges, we conclude that $\sum a_n$ converges.

∎

The next theorem, which we state without proof, shows that absolute convergence is a much stronger condition than conditional convergence. It deals with *rearrangements* of terms—that is, altering the order in which the terms occur. For example,

$$1 + \frac{1}{3} - \frac{1}{2} + \frac{1}{5} - \frac{1}{4} + \frac{1}{7} - \frac{1}{6} + \cdots$$

is a rearrangement of the alternating harmonic series.

THEOREM 10.17

If $\sum a_n$ is absolutely convergent, then every rearrangement of it converges to the same sum. If $\sum a_n$ is conditionally convergent, then if S is any real number, there exists a rearrangement of the series that converges to S. Also, there are rearrangements that diverge to ∞ or to $-\infty$.

(See Exercises 39–43 in Exercise Set 10.6 for the ideas involved in the proof.)

REMARK

■ According to this theorem, no amount of rearranging of terms disturbs the convergence or the value of the sum of an absolutely convergent series. On the other hand, a conditionally convergent series is delicately balanced. Any change in the order may cause it to diverge or to converge to another sum. In fact, by a suitable rearrangement, the series can be made to converge to any sum we want. (See Exercise 43 in Exercise Set 10.6.) The sum is therefore *conditional* on the particular way the terms are written.

A Strategy for Testing Series with Variable Signs

When you are confronted with a series of variable signs, a good way to proceed is to test first to see that $a_n \to 0$ (so that the series has a chance to converge). If $a_n \to 0$, then test for absolute convergence, using one of the tests for positive-term series. If the given series is absolutely convergent, then we know by Theorem 10.16 that it is convergent. If it is not absolutely convergent and the signs alternate, then the Alternating Series Test can be applied. If the signs do not alternate, there are more delicate tests that can be used, but we will not study them here.

Exercise Set 10.6

In Exercises 1–10, use the Alternating Series Test to determine convergence or divergence.

1. $\displaystyle\sum_{n=1}^{\infty} \frac{(-1)^{n-1}}{\sqrt{n}}$

2. $\displaystyle\sum_{n=1}^{\infty} \frac{(-1)^{n-1}}{\ln(n+1)}$

3. $\displaystyle\sum_{n=1}^{\infty} \frac{(-1)^{n-1} \ln n}{n}$

4. $\displaystyle\sum_{n=1}^{\infty} \frac{(-1)^n n}{\sqrt{1+n^2}}$

5. $\displaystyle\sum_{n=2}^{\infty} \frac{(-1)^n}{n \ln n}$

6. $\displaystyle\sum_{n=1}^{\infty} (-1)^n \ln\left(1 + \frac{1}{n}\right)$

7. $\displaystyle\sum_{n=1}^{\infty} \frac{(-1)^n n}{100n + 1}$

8. $\displaystyle\sum_{n=1}^{\infty} \frac{\cos n\pi}{n}$

9. $\displaystyle\sum_{n=1}^{\infty} \frac{(-1)^n (n+1)}{n^2 - 2}$

10. $\displaystyle\sum_{n=1}^{\infty} (-1)^n \ln \frac{2n+3}{2n+1}$

In Exercises 11–14, show that the given series converges, and determine an upper bound on the error in using S_n to approximate the sum, for the specified value of n.

11. $\displaystyle\sum_{k=1}^{\infty} \frac{(-1)^{k-1}}{k^3}$; $n = 9$

12. $\displaystyle\sum_{k=1}^{\infty} \frac{(-1)^{k-1}}{k!}$; $n = 7$

13. $\displaystyle\sum_{k=1}^{\infty} \frac{(-1)^{k-1}}{k^k}$; $n = 5$

14. $\displaystyle\sum_{k=1}^{\infty} \frac{(-1)^k \sqrt{k}}{k+1}$; $n = 9999$

In Exercises 15–18, show that the series converges, and determine the smallest value of n for which the error in using S_n to approximate the sum does not exceed 0.005.

15. $\displaystyle\sum_{k=1}^{\infty} \frac{(-1)^k}{\sqrt{k+1}}$

16. $\displaystyle\sum_{k=2}^{\infty} \frac{(-1)^k}{k(\ln k)^2}$

17. $\displaystyle\sum_{k=1}^{\infty} \frac{(-1)^k(k+1)}{2k^2-3}$

18. $\displaystyle\sum_{k=1}^{\infty} \frac{(-1)^{k-1}(\ln k)^3}{k^2}$

19. Find an estimate, with an error of at most 0.0005, of the sum of the series

$$\sum_{n=1}^{\infty} \frac{(-1)^{n-1}(0.7)^n}{n}$$

In Exercises 20–31, test for absolute convergence, conditional convergence, or divergence.

20. $\displaystyle\sum_{n=1}^{\infty} \frac{(-1)^{n-1}}{\sqrt{n^3+1}}$

21. $\displaystyle\sum_{n=1}^{\infty} \frac{(-1)^{n-1}\sqrt{n+1}}{n}$

22. $\displaystyle\sum_{n=2}^{\infty} \frac{(-1)^n}{n \ln \sqrt{n}}$

23. $\displaystyle\sum_{n=1}^{\infty} \frac{(-1)^{n-1}n}{2^n}$

24. $\displaystyle\sum_{n=2}^{\infty} \frac{(-1)^n}{\ln n}$

25. $\displaystyle\sum_{n=0}^{\infty} (-1)^n \frac{e^n}{e^{2n}+1}$

26. $\displaystyle\sum_{n=1}^{\infty} \frac{(-1)^{n-1}n}{\sqrt{2n^2+3}}$

27. $\displaystyle\sum_{n=1}^{\infty} \frac{(-1)^{n-1}}{n(n+2)}$

28. $\displaystyle\sum_{n=1}^{\infty} \frac{\cos n}{n^2}$

29. $\displaystyle\sum_{n=1}^{\infty} \frac{n \sin n}{1+n^3}$

30. $\displaystyle\sum_{n=0}^{\infty} ne^{-n}\cos n$

31. $\displaystyle\sum_{n=0}^{\infty} \frac{(1-\pi)^n}{2^{n+1}}$

32. Determine the fallacy in the following "proof" that $2=1$. Let S denote the sum of the alternating harmonic series:

$$S = 1 - \frac{1}{2} + \frac{1}{3} - \frac{1}{4} + \frac{1}{5} - \frac{1}{6} + \frac{1}{7} - \frac{1}{8}\cdots$$

Then, on multiplying both sides by 2, we get

$$2S = 2 - 1 + \frac{2}{3} - \frac{1}{2} + \frac{2}{5} - \frac{1}{3} + \frac{2}{7} - \frac{1}{4} + \cdots$$

$$= 1 - \frac{1}{2} + \left(\frac{2}{3} - \frac{1}{3}\right) - \frac{1}{4} + \left(\frac{2}{5} - \frac{1}{5}\right) - \frac{1}{6} + \cdots$$

$$= 1 - \frac{1}{2} + \frac{1}{3} - \frac{1}{4} + \frac{1}{5} - \frac{1}{6} + \cdots = S$$

So $2S = S$, and since $S \neq 0$, $2 = 1$.

33. Let

$$a_n = \frac{2^n n!}{3 \cdot 5 \cdot 7 \cdot \cdots \cdot (2n+1)}$$

Prove that $\sum_{n=0}^{\infty}(-1)^n a_n$ is conditionally convergent by carrying out the following steps: Show that a_n can be written in the form

$$a_n = \frac{2 \cdot 4 \cdot 6 \cdot \cdots \cdot (2n)}{3 \cdot 5 \cdot 7 \cdot \cdots \cdot (2n+1)}$$

To show the series is not absolutely convergent, write

$$a_n = \frac{2}{1} \cdot \frac{4}{3} \cdot \frac{6}{5} \cdots \frac{2n}{2n-1} \cdot \frac{1}{2n+1}$$

Explain how it follows that

$$a_n > \frac{1}{2n+1}$$

To show $\{a_n\}$ is monotone decreasing, consider a_{n+1}/a_n. To show that $a_n \to 0$, verify the following:

$$a_n < \frac{3 \cdot 5 \cdot 7 \cdot \cdots \cdot (2n+1)}{4 \cdot 6 \cdot 8 \cdot \cdots \cdot (2n+2)} = \frac{1}{a_n(n+1)}$$

so

$$a_n^2 < \frac{1}{n+1}, \quad \text{or equivalently,} \quad a_n < \frac{1}{\sqrt{n+1}}$$

Now show that $a_n \to 0$ as $n \to \infty$.

34. Prove that

$$\sum_{n=1}^{\infty}(-1)^{n-1}\frac{1 \cdot 3 \cdot 5 \cdot \cdots \cdot (2n-1)}{2 \cdot 4 \cdot 6 \cdot \cdots \cdot (2n)}$$

is conditionally convergent. (*Hint:* Use steps similar to those in Exercise 33.)

35. Under the conditions of Theorem 10.15 show that the partial sums S_{2n+1} of odd order form a monotone decreasing, bounded sequence.

36. It can be shown (see Exercise 43 in Exercise Set 10.3) that in any convergent series the associative law holds unrestrictedly; that is, we may group terms by introducing parentheses (but not rearranging the order of the terms). Use this fact to show that

$$\sum_{n=1}^{\infty} \frac{1}{(2n-1)2n} = \frac{1}{1 \cdot 2} + \frac{1}{3 \cdot 4} + \frac{1}{5 \cdot 6} + \cdots$$

converges to the same sum as the alternating harmonic series. (*Hint:* In the alternating harmonic series, group terms by pairs.)

37. Let $\sum a_n$ be conditionally convergent and write

$$p_n = \begin{cases} a_n & \text{if } a_n \geq 0 \\ 0 & \text{if } a_n < 0 \end{cases} \quad \text{and} \quad q_n = \begin{cases} 0 & \text{if } a_n \geq 0 \\ -a_n & \text{if } a_n < 0 \end{cases}$$

(a) Show that $\sum a_n = \sum(p_n - q_n)$ and $\sum |a_n| = \sum(p_n + q_n)$.

(b) Show that $\sum p_n$ and $\sum q_n$ both diverge.

38. Show that in an absolutely convergent series the series $\sum p_n$ and $\sum q_n$ both converge, where p_n and q_n are defined as in Exercise 37. (*Hint:* Use part (a) of Exercise 37.)

39. Let p_n and q_n be defined as in Exercise 37 for the alternating harmonic series. Use a CAS or a spreadsheet to do the following. Add terms of $\sum p_n$ until the sum first exceeds 2. Then add one or more terms of $\sum(-q_n)$ until the sum is less than 2. Then continue with terms of $\sum p_n$ until the sum again exceeds 2. Continue in

this way until you have added a total of 50 nonzero terms. If this process were continued, what could you conclude about the convergence of this rearrangement of the alternating harmonic series? Justify your conclusion.

40. Use the idea of Exercise 39 with the roles of $\sum p_n$ and $\sum(-q_n)$ reversed in order to obtain a rearrangement of the alternating harmonic series with partial sums that can be made arbitrarily close to -1.

41. Rearrange the alternating harmonic series so that the resulting series diverges.

42. Rearrange the alternating harmonic series so that the resulting series converges to 4.

43. Use the result of Exercise 37 to explain how a suitable rearrangement of any conditionally convergent series can be made to converge to any sum S we choose.

10.7 POWER SERIES

In Section 10.1 we indicated that we would be considering "infinite polynomials" that result from letting $n \to \infty$ in the Taylor polynomial representation of a function. Such infinite series differ from those we have been studying in the previous two sections in that the terms involve powers of a variable x, or more generally of $(x - a)$. In fact, our motivation for studying infinite series of constants was that such series arise when we substitute a particular value of x into a series involving variables.

In the next section we will return to the question of representing a function by its extended Taylor polynomial. In this section we consider properties of series of powers of x, or of $(x - a)$. Our emphasis will be on the series itself, rather than on what function may have given rise to the series.

Definition 10.9 Power Series	A series of the form

$$\sum_{n=0}^{\infty} a_n x^n = a_0 + a_1 x + a_2 x^2 + \cdots \qquad (10.20)$$

is called a **power series in x**, and a series of the form

$$\sum_{n=0}^{\infty} a_n (x - a)^n = a_0 + a_1(x - a) + a_2(x - a)^2 + \cdots \qquad (10.21)$$

is called a **power series in $(x - a)$**.

REMARK ———————————————————————————————

■ If we let $x = 0$ on the left-hand side of Equation 10.20, the first term of the summation is $a_0(0)^0$, which is not defined. We will adopt the convention, however, that the first term in this case is a_0, in agreement with the right-hand side. Similar remarks apply to the summation in Equation 10.21 when $x = a$ and $n = 0$.

———————————————————————————————————————

The power series 10.20 is clearly the special case of 10.21, in which $a = 0$. Although we could develop our theory for the general case of power series in $(x - a)$, we will concentrate instead primarily on the simpler case of power series in x. All our results can be extended in obvious ways to power series in $(x - a)$.

An appropriate question to ask for a power series is not "Does the series converge?" but rather "For what values of x does the series converge?" This change arises because convergence is defined in terms of series of constants, and a power series becomes a constant series only when x is given a value. All power series in x converge in a trivial way for $x = 0$, as can be seen by putting $x = 0$ in Equation 10.20. Similarly, the series in Equation 10.21 converges for $x = a$. Some series converge for all real values of x, and others converge on an interval and diverge outside that interval. We will see examples of each of these types shortly. The nature of the set of x values where convergence occurs can be determined from the following theorem.

THEOREM 10.18

> If the power series $\sum a_n x^n$ converges at $x_0 \neq 0$, then it converges absolutely for all x such that $|x| < |x_0|$. If the series diverges at $x_1 \neq 0$, then it diverges for all x such that $|x| > |x_1|$.

Simply stated, if a power series in x converges for any nonzero value of x, it converges absolutely for all x that are smaller in absolute value. If it diverges for a given value of x, it diverges for all x that are greater in absolute value.

REMARK ———————————————————————————————

■ For simplicity, we will sometimes drop the range of the index n (from 0 to ∞) in $\sum_{n=0}^{\infty} a_n x^n$ and write simply $\sum a_n x^n$.

———————————————————————————————————————

Proof   Since $\sum a_n x_0^n$ converges, we know by Theorem 10.7 that $\lim_{n \to \infty} a_n x_0^n = 0$. So for n sufficiently large, say $n > N$, $|a_n x_0^n| < 1$. Now let x be any number for which $|x| < |x_0|$, and denote the ratio $|x/x_0|$ by r, so that $0 \leq r < 1$. Then we have for $n > N$

$$|a_n x^n| = |a_n x_0^n| \cdot \left| \frac{x}{x_0} \right|^n = |a_n x_0^n| \cdot r^n < r^n$$

Since $\sum r^n$ is a convergent geometric series, it follows by the Comparison Test that $\sum |a_n x^n|$ converges; that is, $\sum a_n x^n$ converges absolutely.

For the second part, if $\sum a_n x_1^n$ diverges and $|x| > |x_1|$, the series could not converge at x. If it did, since $|x_1| < |x|$, it would converge absolutely at x_1

(and thus would converge), by what we have just proved. But this conclusion contradicts the given fact that the series is assumed to diverge at x_1. So $\sum a_n x^n$ diverges when $|x| > |x_1|$. ∎

Radius of Convergence and Interval of Convergence

As a consequence of Theorem 10.18 the following result can be proved. (See Exercise 40 in Exercise Set 10.7.)

THEOREM 10.19

> For the power series $\sum a_n x^n$, one and only one of the following cases holds true:
>
> 1. The series converges only for $x = 0$.
> 2. The series converges absolutely for all x.
> 3. There is a positive number R such that the series converges absolutely for $|x| < R$ and diverges for $|x| > R$.

The number R in Case 3 is called the **radius of convergence** of the series. For convenience, if Case 1 holds, we agree to call the radius of convergence 0, and if Case 2 holds, we say the radius of convergence is ∞. For Case 3, in which $0 < R < \infty$, nothing is said about convergence at $x = R$ or at $x = -R$. The series may or may not converge at these points, and if it does converge, the convergence may be absolute or conditional. These endpoints of the interval $(-R, R)$ must be individually tested using means studied in the previous three sections. Depending on whether the series converges or diverges at these endpoint values, the series will converge in an interval of one of the types $(-R, R)$, $[-R, R]$, $[-R, R)$, or $(-R, R]$, and it will diverge elsewhere. Do you see that if only one endpoint is included, the convergence is necessarily conditional there? The appropriate interval for a given series is called its **interval of convergence**. When $R = 0$, the interval of convergence degenerates to the single point $x = 0$, and if $R = \infty$, it is the entire real line $(-\infty, \infty)$.

Theorem 10.19 as well as the notions of radius of convergence and interval of convergence extend, with obvious modifications, to power series of the form $\sum a_n (x - a)^n$. If the radius of convergence is R, for example, where $0 < R < \infty$, the interval of convergence is of the form $(a - R, a + R)$, or this interval together with one or both of its endpoints.

Using the Ratio Test to Find the Radius of Convergence

When $\lim_{n \to \infty} |a_{n+1}/a_n|$ exists, the radius of convergence can be found using the Ratio Test. We illustrate this procedure in the next three examples.

EXAMPLE 10.33   Find the interval of convergence of the series

$$\sum_{n=0}^{\infty} \frac{x^n}{2n + 1}$$

Solution   We use the Ratio Test on the series of absolute values. (Remember that the Ratio Test is applicable only to series of positive terms, and since x can be either positive or negative, we must consider absolute values.)

$$\lim_{n\to\infty} \left| \frac{x^{n+1}}{2(n+1)+1} \cdot \frac{2n+1}{x^n} \right| = \lim_{n\to\infty} \frac{2n+1}{2n+3}|x| = |x|$$

(Note that for this limit x is held fixed and $n \to \infty$.) The series therefore converges absolutely when $|x| < 1$ and diverges when $|x| > 1$. The radius of convergence is then $R = 1$. Now we must test the endpoint values $x = \pm 1$. We test them by substituting in the original series:

$$x = -1: \qquad \sum_{n=0}^{\infty} \frac{(-1)^n}{2n+1} = 1 - \frac{1}{3} + \frac{1}{5} - \frac{1}{7} + \cdots$$

This series is an alternating series, and $1/(2n+1)$ decreases monotonically to 0. Thus, the series converges.

$$x = 1: \qquad \sum_{n=0}^{\infty} \frac{1}{2n+1} = 1 + \frac{1}{3} + \frac{1}{5} + \frac{1}{7} + \cdots$$

The term $1/(2n+1)$ appears to be of the same order of magnitude as $1/n$, so we can try the Limit Comparison Test:

$$\lim_{n\to\infty} \frac{1}{2n+1} \div \frac{1}{n} = \lim_{n\to\infty} \frac{n}{2n+1} = \frac{1}{2}$$

This limit confirms that our series and the harmonic series $\sum 1/n$ behave in the same way. Thus, at $x = 1$ our series diverges.

The complete interval of convergence of the original series is therefore $-1 \le x < 1$. Note that the convergence is conditional at $x = -1$.   ∎

EXAMPLE 10.34   Find the interval of convergence of the series

$$\sum_{n=0}^{\infty} \frac{(-1)^n x^{2n}}{(2n)!}$$

Solution   We will again apply the Ratio Test to the series of absolute values. Since $|(-1)^n| = 1$ for any n, we can ignore the factor $(-1)^n$. We then have

$$\lim_{n\to\infty} \left| \frac{x^{2(n+1)}}{[2(n+1)]!} \div \frac{x^{2n}}{(2n)!} \right| = \lim_{n\to\infty} \left| \frac{x^{2n+2}}{(2n+2)!} \cdot \frac{(2n)!}{x^{2n}} \right|$$

$$= \lim_{n\to\infty} \frac{x^2}{(2n+2)(2n+1)} = 0$$

for all fixed values of x. Since $0 < 1$, we conclude that the series converges absolutely for all real x; that is, $R = \infty$ and the interval of convergence is $(-\infty, \infty)$.   ∎

EXAMPLE 10.35   Find the interval of convergence of the series $\sum_{n=0}^{\infty} n!\, x^n$.

Solution   To apply the Ratio Test we consider the following limit for $x \ne 0$,

$$\lim_{n\to\infty} \left| \frac{(n+1)!\, x^{n+1}}{n!\, x^n} \right| = \lim_{n\to\infty} (n+1)|x|$$

Thus, if $x \ne 0$, the limit is ∞, and so by the Ratio Test the series diverges. If $x = 0$, only the first term of the series is different from 0, so the series is finite, hence convergent. Thus, $R = 0$ and the interval of convergence degenerates to the single point $x = 0$. ∎

We will now see how the same technique can be applied to power series in $(x - a)$ to find the interval of convergence.

EXAMPLE 10.36   Find the interval of convergence of the series

$$\sum_{n=0}^{\infty} \frac{(-1)^n (x - 2)^n}{(n + 1)^2 \cdot 3^n}$$

Solution   Consider the limit

$$\lim_{n \to \infty} \left| \frac{(x - 2)^{n+1}}{(n + 2)^2 3^{n+1}} \cdot \frac{(n + 1)^2 3^n}{(x - 2)^n} \right| = \lim_{n \to \infty} \frac{1}{3} \left(\frac{n + 1}{n + 2} \right)^2 |x - 2| = \frac{|x - 2|}{3}$$

Thus, by the Ratio Test the series converges absolutely if $|x - 2|/3 < 1$ or, equivalently, if $|x - 2| < 3$, and it diverges if $|x - 2| > 3$. Now we test the values $(x - 2) = \pm 3$, which correspond to the endpoint values $x = 5$ and $x = -1$.

$$\underline{(x - 2) = 3}: \qquad \sum_{n=0}^{\infty} \frac{(-1)^n 3^n}{(n + 1)^2 \cdot 3^n} = \sum_{n=0}^{\infty} \frac{(-1)^n}{(n + 1)^2}$$

and this series converges absolutely, since the series of absolute values is of the same order of magnitude as a p-series with $p = 2$.

$$\underline{(x - 2) = -3}: \qquad \sum_{n=0}^{\infty} \frac{(-1)^n (-3)^n}{(n + 1)^2 \cdot 3^n} = \sum_{n=0}^{\infty} \frac{(-1)^n \cdot (-1)^n \cdot 3^n}{(n + 1)^2 \cdot 3^n}$$

$$= \sum_{n=0}^{\infty} \frac{1}{(n + 1)^2}$$

and again we have a convergent p-series. So the complete interval of convergence is defined by $|x - 2| \le 3$, and the convergence is absolute for all values of x satisfying this inequality. We can determine the interval of convergence as follows:

$$|x - 2| \le 3$$
$$-3 \le x - 2 \le 3$$
$$-1 \le x \le 5$$

Thus, the interval of convergence is $[-1, 5]$. Notice that the center of the interval is $x = 2$ and the radius is $R = 3$. ∎

Exercise Set 10.7

In Exercises 1–37, find the interval of convergence.

1. $\displaystyle\sum_{n=0}^{\infty} x^n$

2. $\displaystyle\sum_{n=1}^{\infty} \frac{x^n}{n}$

3. $\displaystyle\sum_{n=0}^{\infty} \frac{x^n}{n!}$

4. $\displaystyle\sum_{n=0}^{\infty} \frac{(-1)^n n x^n}{n + 1}$

5. $\displaystyle\sum_{n=0}^{\infty} \frac{(-1)^n x^n}{1 + n^2}$

6. $\displaystyle\sum_{n=0}^{\infty} \frac{x^{2n+1}}{(2n + 1)!}$

7. $\displaystyle\sum_{n=0}^{\infty} \frac{(x-1)^n}{\sqrt{n^2+1}}$

8. $\displaystyle\sum_{n=0}^{\infty} \frac{(x+1)^n}{2^n}$

9. $\displaystyle\sum_{n=0}^{\infty} \frac{(x+2)^n}{3^n(n+1)}$

10. $\displaystyle\sum_{n=0}^{\infty} \frac{(-1)^n 2^{n+1} x^n}{3^n}$

11. $\displaystyle\sum_{n=2}^{\infty} \frac{(-1)^n x^{n-2}}{n \ln n}$

12. $\displaystyle\sum_{n=2}^{\infty} \frac{\ln n}{n} x^{n-2}$

13. $\displaystyle\sum_{n=1}^{\infty} \frac{2^n x^{n-1}}{n^2}$

14. $\displaystyle\sum_{n=0}^{\infty} (-1)^n \sqrt{\frac{n+1}{n+2}}\, x^n$

15. $\displaystyle\sum_{n=0}^{\infty} \frac{n+1}{n^2+1} x^n$

16. $\displaystyle\sum_{n=0}^{\infty} \frac{n!\,(x-1)^n}{2^n}$

17. $\displaystyle\sum_{n=0}^{\infty} \frac{n!}{(2n)!} x^n$

18. $\displaystyle\sum_{n=0}^{\infty} n e^{-n} x^n$

19. $\displaystyle\sum_{n=1}^{\infty} \frac{(-1)^{n-1} 2^n (x-3)^{n-1}}{n^2}$

20. $\displaystyle\sum_{n=0}^{\infty} \frac{(-1)^n \sqrt{n}}{n+1} (x+2)^n$

21. $\displaystyle\sum_{n=2}^{\infty} \left(\frac{3}{4}\right)^{n-2} (x+1)^n$

22. $\displaystyle\sum_{n=2}^{\infty} \frac{(-2x)^{n-2}}{\ln n}$

23. $\displaystyle\sum_{n=0}^{\infty} \frac{n(-3x)^n}{2n+1}$

24. $\displaystyle\sum_{n=0}^{\infty} \frac{(-1)^n n^4 x^n}{e^n}$

25. $\displaystyle\sum_{n=0}^{\infty} \frac{\sqrt{n}(x-5)^n}{1+n^2}$

26. $\displaystyle\sum_{n=0}^{\infty} \frac{n^2-1}{n^2+1} x^{2n}$

27. $\displaystyle\sum_{n=0}^{\infty} \frac{(-1)^n x^{2n+1}}{(2n+1)!}$

28. $\displaystyle\sum_{n=1}^{\infty} \frac{(2x-1)^n}{n\sqrt{n+1}}$

29. $\displaystyle\sum_{n=0}^{\infty} \frac{(3x-2)^n}{(n+1)^{2/3}}$

30. $\displaystyle\sum_{n=0}^{\infty} \frac{(1-x)^n}{2^{n+1}}$

31. $\displaystyle\sum_{n=0}^{\infty} \frac{n!\,(x+4)^{2n}}{3^n}$

32. $\displaystyle\sum_{n=0}^{\infty} \frac{n!\,x^n}{1\cdot 3\cdot 5\cdot\,\cdots\,\cdot(2n+1)}$ (See Exercise 33 in Exercise Set 10.6.)

33. $\displaystyle\sum_{n=1}^{\infty} \frac{1\cdot 3\cdot 5\cdot\,\cdots\,\cdot(2n-1)}{2\cdot 4\cdot 6\cdot\,\cdots\,\cdot(2n)} x^{n-1}$ (See Exercise 34 in Exercise Set 10.6.)

34. $\displaystyle\sum_{n=1}^{\infty} \frac{n^n x^n}{2^{n+1}}$

35. $\displaystyle\sum_{n=0}^{\infty} \frac{(n!)^2}{(2n)!} x^{2n}$

36. $\displaystyle\sum_{n=0}^{\infty} \frac{2^n+n}{3^n+2} x^n$

37. $\displaystyle\sum_{n=1}^{\infty} \frac{x^n}{(\sqrt{n})^n}$

38. If $\sum a_n x^n$ has radius of convergence R, where $0 < R < \infty$, prove that the radius of convergence of $\sum a_n x^{2n}$ is $\sqrt{R}$.

39. Find the radius of convergence of the series $\sum_{n=0}^{\infty} \binom{\alpha}{n} x^n$, where α is a fixed real number and
$$\binom{\alpha}{n} = \frac{\alpha(\alpha-1)(\alpha-2)\cdots(\alpha-n+1)}{n!}$$

40. Prove Case 3 of Theorem 10.19 as follows.
(a) Let R denote the least upper bound of the set of x values for which $\sum a_n x^n$ is absolutely convergent. If $x_0 \in (-R, R)$, choose x_1 such that $|x_0| < |x_1| \le R$, and explain why $\sum a_n x_1^n$ is absolutely convergent. Based on the definition of R, why can such an x_1 be chosen? (*Hint:* Otherwise $|x_0|$ would be an upper bound, smaller than R, of the set of points where $\sum a_n x^n$ is absolutely convergent.)
(b) Now use Theorem 10.18. How do you conclude that $\sum a_n x^n$ converges absolutely for *all* x in $(-R, R)$?
(c) To prove divergence for $|x| > R$, assume to the contrary that $\sum a_n x^n$ converges for some x such that $|x| > R$, and arrive at a contradiction.

10.8 DIFFERENTIATION AND INTEGRATION OF POWER SERIES

Within its interval of convergence, a power series defines a function. So it is appropriate to write, for example,

$$f(x) = \sum_{n=0}^{\infty} a_n x^n$$

where x is in the interval I of convergence. For any x_0 in I, $f(x_0)$ is equal to the sum of the convergent series of constants, $\sum a_n x_0^n$. It is appropriate to ask, then, what properties such a function has. For example, we may want to know about continuity, differentiability, or integrability. The following rather remarkable theorem answers these questions. We do not give its proof, which is quite technical. (A proof can be found in most textbooks on advanced calculus.) The theorem says that a power series can essentially be treated as a polynomial within its interval of convergence.

THEOREM 10.20

Let $\sum a_n x^n$ have nonzero radius of convergence R, and for $-R < x < R$ write

$$f(x) = \sum_{n=0}^{\infty} a_n x^n$$

Then

1. f is continuous on the interval $(-R, R)$.
2. f is differentiable on $(-R, R)$ and

$$f'(x) = \sum_{n=0}^{\infty} \frac{d}{dx}(a_n x^n) = \sum_{n=1}^{\infty} n a_n x^{n-1}$$

The series on the right also has radius of convergence R.

3. f is integrable over any interval $[a, b]$ contained in $(-R, R)$, and

$$\int_a^b f(x)dx = \sum_{n=0}^{\infty} \int_a^b a_n x^n dx$$

Furthermore, f has an antiderivative in $(-R, R)$ given by

$$\int f(x)dx = \sum_{n=0}^{\infty} \int a_n x^n dx = \sum_{n=0}^{\infty} \frac{a_n x^{n+1}}{n+1} + C$$

The series on the right also has radius of convergence R.

REMARKS

■ Property 2 says that a power series may be differentiated term by term, and the resulting power series converges to the derivative of the function and has the same radius of convergence as the original series. Since the differentiated series is itself a power series with radius of convergence R, it too can be differentiated term by term to give $f''(x)$. Continuing in this way gives the very powerful result that within its interval of convergence (excluding endpoints) *a power series has derivatives of all orders.*

■ We can paraphrase Properties 2 and 3 by saying that a power series can be differentiated or integrated term by term within $(-R, R)$. If the integral is a definite integral over $[a, b]$, then we must have $-R < a < b < R$. If it is an indefinite integral (an antiderivative), the integration is valid for any x in $(-R, R)$.

■ None of the properties can be assumed to be true at the endpoints $x = \pm R$, even if the original series converges at one or both of these points. Actually, it can be proved that continuity *does* extend to an endpoint if the series converges there. So a power series is continuous on its entire interval of convergence.

■ The theorem extends in an obvious way to $f(x) = \sum a_n (x - a)^n$.

The following examples illustrate some of the consequences of Theorem 10.20. In the examples we will make repeated use of the geometric series

$$a + ar + ar^2 + \cdots = \frac{a}{1-r}, \qquad |r| < 1$$

By setting $r = x$ and $a = 1$ and reversing the sides of the equation, we obtain the following result.

$$\frac{1}{1-x} = 1 + x + x^2 + x^3 + \cdots = \sum_{n=0}^{\infty} x^n, \qquad |x| < 1 \qquad (10.22)$$

EXAMPLE 10.37   Find the sum of the series

$$\sum_{n=1}^{\infty} nx^n = x + 2x^2 + 3x^3 + 4x^4 + \cdots$$

and state the domain of validity. Use the result to find the sum of the series $\sum_{n=1}^{\infty} n(1/2)^n$.

Solution   On factoring out an x we get

$$x(1 + 2x + 3x^2 + 4x^3 + \cdots)$$

and we observe that the series in parentheses is the derivative of

$$x + x^2 + x^3 + x^4 + \cdots$$

Factoring out an x, we can write this series as $x(1 + x + x^2 + x^3 + \cdots)$. The series in parentheses is the geometric series in Equation 10.22. So if $|x| < 1$,

$$x + x^2 + x^3 + x^4 + \cdots = x(1 + x + x^2 + x^3 + \cdots) = x \left(\frac{1}{1-x} \right) = \frac{x}{1-x}$$

Thus, by Property 2 of Theorem 10.20, we have

$$\sum_{n=1}^{\infty} nx^n = x \frac{d}{dx} \left(x + x^2 + x^3 + x^4 + \cdots \right)$$

$$= x \left(\frac{d}{dx} \frac{x}{1-x} \right)$$

$$= \frac{x}{(1-x)^2}$$

and this result is valid for $|x| < 1$. In particular, for $x = 1/2$, we have

$$\sum_{n=1}^{\infty} n \left(\frac{1}{2} \right)^n = \frac{\frac{1}{2}}{\left(1 - \frac{1}{2} \right)^2} = 2$$

∎

EXAMPLE 10.38   Find a power series whose sum is $\ln(1 + x)$. Use the result to find a series that converges to $\ln(1/2)$.

Solution   We begin with the fact that
$$\ln(1 + x) = \int \frac{dx}{1 + x}$$

If in Equation 10.22 we replace x with $-x$, we obtain

$$\frac{1}{1 + x} = \sum_{n=0}^{\infty}(-1)^n x^n, \qquad |x| < 1$$

Thus, by Theorem 10.20,

$$\int \frac{dx}{1 + x} = \int \sum_{n=0}^{\infty}(-1)^n x^n \, dx$$

$$= \sum_{n=0}^{\infty}\int (-1)^n x^n dx$$

We therefore have
$$\ln(1 + x) = \sum_{n=0}^{\infty}\frac{(-1)^n x^{n+1}}{n + 1} + C, \qquad |x| < 1$$

To find C, we can substitute $x = 0$:

$$\ln 1 = 0 + C$$

Thus,
$$C = 0$$

So we have
$$\ln(1 + x) = \sum_{n=0}^{\infty}\frac{(-1)^n x^{n+1}}{n + 1} = x - \frac{x^2}{2} + \frac{x^3}{3} - \frac{x^4}{4} + \cdots$$

and this result is valid for $|x| < 1$. In particular, setting $x = -1/2$, we have

$$\ln\left(\frac{1}{2}\right) = -\frac{1}{2} - \frac{1}{2 \cdot 2^2} - \frac{1}{3 \cdot 2^3} - \frac{1}{4 \cdot 2^4} - \cdots$$

since $\ln(1/2) = -\ln 2$, we can also conclude that

$$\ln 2 = \frac{1}{2} + \frac{1}{2 \cdot 2^2} + \frac{1}{3 \cdot 2^3} + \frac{1}{4 \cdot 2^4} + \cdots$$

$$= \sum_{n=1}^{\infty}\frac{1}{n \cdot 2^n} \qquad \blacksquare$$

Note that when $x = 1$, the series for $\ln(1 + x)$ that we obtained in Example 10.38 is convergent (verify), and as stated in the third remark after Theorem 10.20, the continuity of the series thus extends to include the endpoint $x = 1$. We can therefore say that

$$\lim_{x \to 1}\ln(1 + x) = \lim_{x \to 1}\sum_{n=0}^{\infty}\frac{(-1)^n x^{n+1}}{n + 1}$$

and since the natural logarithm function also is continuous, we get the following result.

The Sum of the Alternating Harmonic Series

$$\ln 2 = 1 - \frac{1}{2} + \frac{1}{3} - \frac{1}{4} + \cdots$$

So we now know that the sum of the alternating harmonic series is $\ln 2$.

EXAMPLE 10.39   Find a power series representation of $\tan^{-1} x$ valid near $x = 0$.

Solution   We use the idea of the previous example. Since

$$\tan^{-1} x = \int \frac{1}{1 + x^2} dx$$

we can solve the problem provided we can determine a power series representation of $f(x) = 1/(1 + x^2)$. All we need to do is to replace x with $-x^2$ in Equation 10.22, giving

$$\frac{1}{1 + x^2} = 1 - x^2 + x^4 - x^6 + \cdots = \sum_{n=0}^{\infty} (-1)^n x^{2n}$$

where $x^2 < 1$, or equivalently, $|x| < 1$. So

$$\int \frac{1}{1 + x^2} dx = \int \sum_{n=0}^{\infty} (-1)^n x^{2n} dx$$

$$\tan^{-1} x = \sum_{n=0}^{\infty} \int (-1)^n x^{2n} dx \qquad \text{By Theorem 10.20}$$

$$= \sum_{n=0}^{\infty} \frac{(-1)^n x^{2n+1}}{2n + 1} + C$$

For $x = 0$ we get $\tan^{-1} 0 = 0 + C$. Thus, $C = 0$. So

$$\tan^{-1} x = \sum_{n=0}^{\infty} \frac{(-1)^n x^{2n+1}}{2n + 1}$$

$$= x - \frac{x^3}{3} + \frac{x^5}{5} - \frac{x^7}{7} + \cdots$$

valid so long as we have $|x| < 1$.

We can readily see that the series on the right converges at both endpoints. (Check this.) So, as we have indicated, the series represents a continuous function on the closed interval $[-1, 1]$. Since $\tan^{-1} x$ is also continuous at $x = \pm 1$, it follows as in the preceding example that

$$\tan^{-1} 1 = 1 - \frac{1}{3} + \frac{1}{5} - \frac{1}{7} + \cdots$$

and

$$\tan^{-1}(-1) = -1 + \frac{1}{3} - \frac{1}{5} + \frac{1}{7} - \cdots$$

Using the first of these equations we find that

$$\frac{\pi}{4} = 1 - \frac{1}{3} + \frac{1}{5} - \frac{1}{7} + \cdots$$

or

$$\pi = 4\left[1 - \frac{1}{3} + \frac{1}{5} - \frac{1}{7} + \cdots\right]$$

We can view this result in two ways: first, it gives a means of calculating π to any degree of accuracy (although it is not efficient, since the series converges very slowly), and second, it is a formula for the sum of the series on the right. ∎

In the next example we give one way to verify that the nth Taylor polynomial for e^x about $x = 0$ can be extended to an infinite power series (the Taylor series) that converges to e^x.

EXAMPLE 10.40   Show that the series

$$\sum_{n=0}^{\infty} \frac{x^n}{n!} = 1 + x + \frac{x^2}{2!} + \frac{x^3}{3!} + \cdots + \frac{x^n}{n!} + \cdots$$

converges to e^x for all real x.

Solution   Let

$$f(x) = \sum_{n=0}^{\infty} \frac{x^n}{n!}$$

We first show that this series converges for all real x. Then we will show that $f'(x) = f(x)$, suggesting the possibility that $f(x) = e^x$. Applying the Ratio Test, we have

$$\lim_{n \to \infty} \left| \frac{x^{n+1}}{(n+1)!} \cdot \frac{n!}{x^n} \right| = \lim_{n \to \infty} \frac{|x|}{n+1} = 0$$

Since this limit is always less than 1, the given series converges for all values of x. Its derivative is, by Theorem 10.20,

$$f'(x) = \sum_{n=1}^{\infty} \frac{nx^{n-1}}{n!} = \sum_{n=1}^{\infty} \frac{x^{n-1}}{(n-1)!} \qquad \text{Since } n!/n = (n-1)!$$

$$= 1 + x + \frac{x^2}{2!} + \frac{x^3}{3!} + \cdots = f(x)$$

That is, $f'(x) = f(x)$ for all values of x. We know that e^x is a function with the property that its derivative equals itself. To show that our $f(x)$ and e^x are identical, consider the derivative

$$\frac{d}{dx}\left(\frac{f(x)}{e^x}\right) = \frac{f(x) \cdot e^x - e^x \cdot f'(x)}{e^{2x}}$$

Since $f'(x) = f(x)$, the numerator is 0 and it follows that

$$\frac{d}{dx}\left(\frac{f(x)}{e^x}\right) = 0$$

Thus,

$$\frac{f(x)}{e^x} = C$$

for some constant C. Setting $x = 0$, we get $C = 1$, since both $f(0)$ and e^0 equal 1. Thus, $f(x) = e^x$. ∎

The result of the preceding example deserves special emphasis.

A Power Series for e^x

$$e^x = \sum_{n=0}^{\infty} \frac{x^n}{n!} = 1 + x + \frac{x^2}{2!} + \frac{x^3}{3!} + \cdots + \frac{x^n}{n!} + \cdots \qquad (10.23)$$

valid for $-\infty < x < \infty$.

A consequence of the convergence of the series in Equation 10.23 is that the nth term goes to 0 as $n \to \infty$, and this result, too, is one we will need later.

A Special Limit

$$\lim_{n \to \infty} \frac{x^n}{n!} = 0 \qquad \text{for all real } x \qquad (10.24)$$

Exercise Set 10.8

In Exercises 1–6, differentiate the given series term by term, and determine in what interval the resulting series is the derivative of the function defined by the given series.

1. $1 + x + \dfrac{x^2}{2} + \dfrac{x^3}{3} + \cdots$

2. $x - \dfrac{x^3}{3} + \dfrac{x^5}{5} - \dfrac{x^7}{7} + \cdots$

3. $\displaystyle\sum_{n=0}^{\infty} \frac{(-1)^n x^{2n}}{(2n)!}$

4. $\displaystyle\sum_{n=0}^{\infty} \frac{(-1)^n x^{2n+1}}{(2n+1)!}$

5. $\displaystyle\sum_{n=0}^{\infty} \frac{x^n}{2^{n+1}}$

6. $\displaystyle\sum_{n=1}^{\infty} \frac{x^n}{n^2 + n}$

In Exercises 7–12, find the indicated antiderivatives and give the domain of validity.

7. $\displaystyle\int \sum_{n=0}^{\infty} \frac{x^n}{n+1} \, dx$

8. $\displaystyle\int \sum_{n=1}^{\infty} (-1)^{n-1} n x^n \, dx$

9. $\displaystyle\int \sum_{n=0}^{\infty} \frac{(-1)^n x^{2n}}{(2n)!} \, dx$

10. $\displaystyle\int \sum_{n=1}^{\infty} \frac{x^{2n}}{1 \cdot 3 \cdot 5 \cdot \, \cdots \, \cdot (2n-1)} \, dx$

11. $\displaystyle\int \sum_{n=0}^{\infty} \frac{(-1)^n (2n+1) x^{2n}}{2^n} \, dx$

12. $\displaystyle\int \sum_{n=1}^{\infty} \frac{(-1)^{n-1} x^{2n-1}}{(2n-1)!} \, dx$

In Exercises 13–18, show that term-by-term integration is valid over the given interval, and carry out the integration. Where possible, express the answer in closed form.

13. $\displaystyle\int_0^{1/2} \sum_{n=0}^{\infty} (n+1) x^n \, dx$

14. $\displaystyle\int_{-1}^{1} \sum_{n=0}^{\infty} \frac{x^n}{2^n} \, dx$

15. $\displaystyle\int_1^2 \sum_{n=0}^{\infty} \frac{x^n}{n!} \, dx$

16. $\displaystyle\int_{-1}^{2}\sum_{n=0}^{\infty}\frac{(-1)^n x^n}{3^{n+1}}\,dx$

17. $\displaystyle\int_{-1/2}^{1/2}\sum_{n=0}^{\infty}\frac{(-1)^n x^{2n}}{2n+1}\,dx$

18. $\displaystyle\int_{-1}^{3}\sum_{n=0}^{\infty}\frac{(-1)^n x^{2n}}{(2n)!}\,dx$

In Exercises 19–31, use differentiation or integration of an appropriate geometric series to find a series representation in powers of x of the given function, and give its radius of convergence.

19. $\dfrac{1}{(1+x)^2}$

20. $\dfrac{2}{(2-x)^2}$

21. $\ln(1-x)$

22. $\tanh^{-1} x$

23. $\dfrac{2x}{(1-x^2)^2}$

24. $\dfrac{2x}{(1-2x)^2}$

25. $\ln\sqrt{1+x}$

26. $\ln\dfrac{1+x}{1-x}$

27. $\tan^{-1}\dfrac{x}{2}$

28. $6\left(\dfrac{x}{3-2x}\right)^2$

29. $\dfrac{1}{(1-x)^3}$

30. $\ln(3+2x)$

31. $\dfrac{x^2}{(1-x)^2}$

32. Show, by integrating the series in Equation 10.23, that $\int e^x\,dx = e^x + C$.

33. Let $f(x) = e^{2x}$. In Equation 10.23 replace x with $2x$ to find a series representation of f. Use this result to calculate $\int_0^1 f(x)\,dx$. What conclusion can you draw concerning the sum of the following series?
$$1 + \frac{2}{2!} + \frac{2^2}{3!} + \frac{2^3}{4!} + \cdots + \frac{2^n}{(n+1)!} + \cdots$$

34. Find the sum of the series $\sum_{n=1}^{\infty}(-1)^{n-1}nx^n$ for $|x| < 1$. (*Hint:* Differentiate an appropriate geometric series and then multiply by x.)

35. Find the sum of the series $\sum_{n=0}^{\infty} n^2 x^n$. (*Hint:* Write $n^2 = n + n(n-1)$ and use the idea of Example 10.37.)

36. Find the sum of the series
$$\sum_{n=1}^{\infty}\frac{n}{2^n}$$
(*Hint:* See Example 10.37.)

37. Find the sum of the series
$$\sum_{n=1}^{\infty}\frac{n^2}{3^n}$$
(*Hint:* See Exercise 35.)

38. Let $f(x) = xe^x$. Make use of Equation 10.23 and $f'(1)$ to find the sum of the series
$$\sum_{n=0}^{\infty}\frac{n+1}{n!}$$

39. Make use of Equation 10.23 to find a series for $(e^x - 1)/x$. By differentiating the result, show that
$$\frac{1}{2!} + \frac{2}{3!} + \frac{3}{4!} + \cdots = 1$$

40. (a) Use partial fractions and geometric series to find a power series in x for the function
$$f(x) = \frac{1}{2 + x - x^2}$$
and give its radius of convergence.

(b) Using part (a), find a power series representation of
$$\frac{2x - 1}{(2 + x - x^2)^2}$$
What is the radius of convergence?

41. Use Equation 10.23 to evaluate
$$\int_0^1 e^{-x^2}\,dx$$
correct to eight decimal places.

42. Define f and g on $(-\infty, \infty)$ by
$$f(x) = \sum_{n=0}^{\infty}\frac{(-1)^n x^{2n}}{(2n)!} \qquad g(x) = \sum_{n=0}^{\infty}\frac{(-1)^n x^{2n+1}}{(2n+1)!}$$

Show the following:
(a) $f''(x) + f(x) = 0$ and $g''(x) + g(x) = 0$
(b) $f(0) = 1$, $f'(0) = 0$, $g(0) = 0$, and $g'(0) = 1$
(c) $f'(x) = -g(x)$ and $g'(x) = f(x)$
Do the results in part (b) and part (c) remind you of functions you have seen before? If so, what are they?

43. Use a CAS, a spreadsheet, or a calculator to verify the plausibility of Equation 10.24 with the following values of x and n.
(a) $x = 10$, $n = 50$ (b) $x = -20$, $n = 100$
(c) $x = 100$, $n = 1000$

10.9 TAYLOR SERIES

Recall (from Section 10.1) Taylor's Formula, which states that if f has derivatives up through the $(n + 1)$st order in an open interval I centered at $x = a$, then

$$f(x) = P_n(x) + R_n(x) \qquad (10.25)$$

for all x in I. Here $P_n(x)$ is the Taylor polynomial

$$P_n(x) = f(a) + f'(a)(x - a) + \frac{f''(a)}{2!}(x - a)^2 + \cdots + \frac{f^{(n)}(a)}{n!}(x - a)^n \quad (10.26)$$

and $R_n(x)$ is the remainder term, given by

$$R_n(x) = \frac{f^{(n+1)}(c)}{(n + 1)!}(x - a)^{n+1} \qquad (10.27)$$

where c is some number between a and x.

We want to consider now what happens when we let $n \to \infty$. First of all, we must require that f have derivatives of *all* orders in the interval I. By letting $n \to \infty$ in Equation 10.26 we get the infinite series shown below.

Taylor Series for f about $x = a$

$$\sum_{n=0}^{\infty} \frac{f^{(n)}(a)}{n!}(x - a)^n = f(a) + f'(a)(x - a) + \frac{f''(a)}{2!}(x - a)^2$$

$$+ \frac{f'''(a)}{3!}(x - a)^3 + \cdots \qquad (10.28)$$

This series is called the **Taylor series for f about $x = a$**. In particular, if $a = 0$, we get the **Taylor series for f about $x = 0$**:

Taylor Series for f about $x = 0$: Maclaurin Series

$$\sum_{n=0}^{\infty} \frac{f^{(n)}(0)}{n!}x^n = f(0) + f'(0)x + \frac{f''(0)}{2!}x^2 + \frac{f'''(0)}{3!}x^3 + \cdots \quad (10.29)$$

The Taylor series for f about $x = 0$ in Equation 10.29 is also called the **Maclaurin series for f**, after the Scottish mathematician Colin Maclaurin (1698–1746).

The crucial question is whether the Taylor series for a function f converges to that function in some interval. If so, then we say that f is *represented* by its Taylor series in the interval where the convergence is valid. The answer to this question of representation hinges on the remainder term $R_n(x)$ given by Equation 10.27. Note that the Taylor polynomial $P_n(x)$ is the nth partial sum (starting with $n = 0$) of the Taylor series in Equation 10.28. By Equation 10.25, we have

$$|f(x) - P_n(x)| = |R_n(x)|$$

It follows that $P_n(x)$ will come arbitrarily close to $f(x)$ if and only if $|R_n(x)|$ comes arbitrarily close to 0 as $n \to \infty$. That is, $\lim_{n \to \infty} P_n(x) = f(x)$ if and only if $\lim_{n \to \infty} R_n(x) = 0$. We summarize our findings as a theorem.

THEOREM 10.21

Taylor's Theorem

Let f have derivatives of all orders in an open interval I centered at $x = a$. Then the Taylor series for f about $x = a$ converges to $f(x)$ for x in I if and only if for all x in I,

$$\lim_{n \to \infty} R_n(x) = 0$$

where $R_n(x)$ is the remainder term in Taylor's Formula, given in Equation 10.27.

EXAMPLE 10.41   Show that the Taylor series for $f(x) = e^x$ about $x = 0$ converges to e^x for all real x.

Solution   In Example 10.4 we found that the nth Taylor polynomial for e^x is

$$P_n(x) = 1 + x + \frac{x^2}{2!} + \frac{x^3}{3!} + \cdots + \frac{x^n}{n!}$$

and that

$$R_n(x) = \frac{e^c}{(n+1)!} x^{n+1}$$

where c is between 0 and x. If $0 < c < x$, then $e^c < e^x$ since $f(x) = e^x$ is an increasing function. Thus, for $x > 0$,

$$|R_n(x)| \le \frac{e^x}{(n+1)!} x^{n+1}$$

From the special limit given in Equation 10.24 we know that $x^{n+1}/(n+1)! \to 0$ as $n \to \infty$. So for fixed $x > 0$, e^x is constant and it follows that

$$\lim_{n \to \infty} R_n(x) = 0$$

If $x < c < 0$, then $e^c < e^0 = 1$. Thus,

$$|R_n(x)| \le \left| \frac{x^{n+1}}{(n+1)!} \right|$$

Again, by the special limit in Equation 10.24 the right-hand side goes to 0 as $n \to \infty$. Thus, for all $x < 0$,

$$\lim_{n \to \infty} R_n(x) = 0$$

By Theorem 10.21, we conclude that the Taylor series for e^x about $x = 0$ converges to e^x for all real x. That is,

$$e^x = 1 + x + \frac{x^2}{2!} + \frac{x^3}{3!} + \cdots = \sum_{n=0}^{\infty} \frac{x^n}{n!}$$

In the preceding section (see Example 10.40) we arrived at the result of Example 10.41 in a different way. We began with the power series $\sum x^n/n!$ and proved that its radius of convergence was infinite. Then we proved that the function represented by the series was, in fact, identical to e^x. As you might suspect, it is no accident that we got the same series for e^x by the two different approaches. The next theorem explains why.

THEOREM 10.22

> If f can be represented by a power series in $x - a$ in an open interval I centered at $x = a$, then that power series is the Taylor series for f about $x = a$.

Proof   For simplicity, we prove the theorem for $a = 0$. The proof for $a \neq 0$ is similar. Suppose, then, that $f(x)$ is represented by the power series

$$f(x) = a_0 + a_1 x + a_2 x^2 + a_3 x^3 + a_4 x^4 + a_5 x^5 + \cdots$$

for all x in an interval I centered at $x = 0$. Then we have, on successive differentiation,

$$f'(x) = a_1 + 2a_2 x + 3a_3 x^2 + 4a_4 x^3 + 5a_5 x^4 + \cdots$$
$$f''(x) = 2a_2 + (3 \cdot 2)a_3 x + (4 \cdot 3)a_4 x^2 + (5 \cdot 4)a_5 x^3 + \cdots$$
$$f'''(x) = (3 \cdot 2)a_3 + (4 \cdot 3 \cdot 2)a_4 x + (5 \cdot 4 \cdot 3)a_5 x^2 + \cdots$$
$$f^{(4)}(x) = (4 \cdot 3 \cdot 2)a_4 + (5 \cdot 4 \cdot 3 \cdot 2)a_5 x + \cdots$$
$$\vdots$$

On setting $x = 0$ in each equation, we get

$$f(0) = a_0, \quad f'(0) = a_1, \quad f''(0) = 2a_2, \quad f'''(0) = 3!\, a_3,$$

$$f^{(4)}(0) = 4!\, a_4, \quad \ldots$$

and in general,

$$f^{(n)}(0) = n!\, a_n$$

If we solve for a_n, we get

$$a_n = \frac{f^{(n)}(0)}{n!}, \qquad n = 0, 1, 2, \ldots$$

Thus, the coefficients in the given power series representation of f are precisely those of the Taylor series for f about $x = 0$. The given series therefore is the Taylor series. ∎

An important consequence of Theorem 10.22 is that by whatever valid means we arrive at a power series representation of a function (such as by differentiation or integration of a geometric series, for example), the series will be the Taylor series. In particular, then, the series for $\ln(1 + x)$ and for $\tan^{-1} x$ that we found in Examples 10.37 and 10.38 are in fact the Taylor series about $x = 0$ for these functions.

It is important to emphasize in Theorem 10.22 the hypothesis that f *can* be represented by a power series. That is, we begin with the assumption that there is a power series in $x - a$ that *does converge to* f throughout some interval centered at $x = a$. Then we can conclude that f has derivatives of all orders at a, and that the given power series is, in fact, the Taylor series for f about $x = a$. If, on the other hand, we knew only that f had derivatives of all orders at $x = a$, we would know that it has a Taylor series about $x = a$, *but we would not know that f is represented by its series.* That is, we would not know that the series converges to $f(x)$. Theorem 10.21 gives the necessary and sufficient condition for the function to be represented by its Taylor series, namely, that $\lim_{n \to \infty} R_n(x) = 0$. In Exercise 29 of Exercise Set 10.9 we will ask you to show that the function f defined by

$$f(x) = \begin{cases} e^{-1/x^2} & \text{if } x \neq 0 \\ 0 & \text{if } x = 0 \end{cases}$$

has a Taylor series about $x = 0$ but that, except for $x = 0$, the series does not converge to $f(x)$.

In the next example we show two ways of finding the Taylor series about $x = 0$ of $f(x) = (1 - x)^{-2}$.

EXAMPLE 10.42   Find the Maclaurin series for $f(x) = 1/(1 - x)^2$ by two methods. (*Note:* Recall that the Maclaurin series is the Taylor series about $x = 0$.)

Solution   We first use the direct method to calculate the coefficients in the Taylor series about $x = 0$. We calculate the successive derivatives of f and evaluate them at $x = 0$.

$$f(x) = \frac{1}{(1 - x)^2} = (1 - x)^{-2} \qquad\qquad f(0) = 1!$$
$$f'(x) = -2(1 - x)^{-3}(-1) = 2!(1 - x)^{-3} \qquad f'(0) = 2!$$
$$f''(x) = -3 \cdot 2(1 - x)^{-4}(-1) = 3!\,(1 - x)^{-4} \qquad f''(0) = 3!$$
$$f'''(x) = -4 \cdot 3!\,(1 - x)^{-5}(-1) = 4!\,(1 - x)^{-5} \qquad f'''(0) = 4!$$
$$\vdots \qquad\qquad\qquad\qquad\qquad \vdots$$
$$f^{(n)}(x) = (n + 1)!\,(1 - x)^{-(n+2)} \qquad\qquad f^{(n)}(0) = (n + 1)!$$

The general term of the Maclaurin series in Equation 10.29 is therefore

$$\frac{f^{(n)}(0)}{n!}x^n = \frac{(n + 1)!}{n!}x^n = (n + 1)x^n$$

So the Maclaurin series for f is

$$\sum_{n=0}^{\infty}(n + 1)x^n = 1 + 2x + 3x^2 + 4x^3 + \cdots$$

Note that we have not shown that this series represents f in any interval. To do so, we would have to show that the remainder term in Taylor's Formula, $R_n(x)$, approaches 0 as $n \to \infty$. Rather than show this limit, we defer the question of whether the series represents f to our second method, where the representation is an automatic consequence of the method.

For the second method, we note that

$$\frac{1}{(1 - x)^2} = \frac{d}{dx}\left(\frac{1}{1 - x}\right)$$

From Equation 10.22 we know that

$$\frac{1}{1-x} = 1 + x + x^2 + x^3 + \cdots = \sum_{n=0}^{\infty} x^n, \qquad |x| < 1$$

By Theorem 10.20, therefore,

$$\frac{1}{(1-x)^2} = \frac{d}{dx}(1 + x + x^2 + x^3 + \cdots)$$

$$= 1 + 2x + 3x^2 + \cdots = \sum_{n=0}^{\infty}(n+1)x^n, \qquad |x| < 1$$

This result is the same as we found by the direct method, but we actually have the added information that the series does represent the function in the interval $-1 < x < 1$. ■

EXAMPLE 10.43

(a) Find the Maclaurin series for $f(x) = \cos x$.

(b) Use the result of part (a) to approximate the integral

$$\int_0^1 \frac{\sin^2 x}{x^2}\,dx$$

correct to three decimal places.

Solution

(a) In Example 10.3 we found the nth Taylor polynomial for $\cos x$ about $x = 0$ to be

$$P_n(x) = 1 - \frac{x^2}{2!} + \frac{x^4}{4!} - \frac{x^6}{6!} + \cdots + \frac{(-1)^n}{(2n)!}x^{2n}$$

and in Example 10.7, we found the remainder term

$$R_{2n}(x) = \frac{(-1)^{n+1}\cos c}{(2n+2)!}x^{2n+2}, \qquad c \text{ between } 0 \text{ and } x$$

To show where the Taylor series converges to $\cos x$, we need to find the values of x for which

$$\lim_{n\to\infty} R_{2n}(x) = 0$$

For any real value of x,

$$|R_{2n}(x)| = \frac{|\cos c|}{(2n+2)!}|x|^{2n+2} \le \frac{|x|^{2n+2}}{(2n+2)!}$$

since $|\cos c| \le 1$. By the special limit in Equation 10.24, the last expression approaches 0 as $n \to \infty$. Therefore, $\lim_{n\to\infty} R_{2n}(x) = 0$ for all real x. By Theorem 10.21 it follows that

$$\cos x = 1 - \frac{x^2}{2!} + \frac{x^4}{4!} - \frac{x^6}{6!} + \cdots = \sum_{n=0}^{\infty} \frac{(-1)^n}{(2n)!}x^{2n}$$

for all real x.

(b) To use the result of part (a) to evaluate the integral, we need to relate $\sin^2 x$ to the first power of the cosine. Such a relationship is provided by the trigonometric identity

$$\sin^2 x = \frac{1}{2}(1 - \cos 2x) \quad \text{See Supplementary Appendix 2.}$$

Now we replace x with $2x$ in the result of part (a) to get

$$\sin^2 x = \frac{1}{2}\left[1 - \left(1 - \frac{(2x)^2}{2!} + \frac{(2x)^4}{4!} - \frac{(2x)^6}{6!} + \cdots\right)\right]$$

$$= \frac{2x^2}{2!} - \frac{2^3 x^4}{4!} + \frac{2^5 x^6}{6!} - \cdots, \qquad -\infty < x < \infty$$

Dividing by x^2 gives

$$\frac{\sin^2 x}{x^2} = 1 - \frac{2^3 x^2}{4!} + \frac{2^5 x^4}{6!} - \cdots \qquad (x \neq 0)$$

Thus, by Theorem 10.20,

$$\int_0^1 \frac{\sin^2 x}{x^2}\,dx = \int_0^1 \left(1 - \frac{2^3 x^2}{4!} + \frac{2^5 x^4}{6!} - \frac{2^7 x^6}{8!} + \cdots\right) dx$$

$$= x - \frac{(2x)^3}{3 \cdot 4!} + \frac{(2x)^5}{5 \cdot 6!} - \frac{(2x)^7}{7 \cdot 8!} + \cdots\Bigg]_0^1$$

$$= 1 - \frac{2^3}{3 \cdot 4!} + \frac{2^5}{5 \cdot 6!} - \frac{2^7}{7 \cdot 8!} + \cdots$$

Using the first three terms, we get

$$\int_0^1 \frac{\sin^2 x}{x^2}\,dx \approx 0.898$$

Since the series alternates, we know from Corollary 10.15 that the error does not exceed the fourth term:

$$\text{Error} \leq \frac{2^7}{7 \cdot 8!} \approx 0.000454$$

So our approximation is correct to three places. ∎

REMARK ────────────

■ Although $(\sin^2 x)/x^2$ is not defined when $x = 0$, it has a removable discontinuity there. In fact, we see from its series representation that its limit is 1 as $x \to 0$.

Binomial Series

We wish to consider finally an important series known as the **binomial series**, which is the Maclaurin series for $f(x) = (1 + x)^\alpha$ for an arbitrary real number α. If $\alpha = 0$, then $f(x) = 1$. If α is a positive integer, say $\alpha = n$, we know from the **Binomial Formula** that

$$(1 + x)^\alpha = (1 + x)^n = 1 + nx + \frac{n(n-1)}{2!}x^2 + \frac{n(n-1)(n-2)}{3!}x^3 + \cdots + x^n$$

$$(10.30)$$

which is a finite series—that is, a polynomial—so there is no question of convergence. For all other values of α we proceed as usual to find the Maclaurin series:

$$
\begin{aligned}
f(x) &= (1+x)^\alpha & f(0) &= 1 \\
f'(x) &= \alpha(1+x)^{\alpha-1} & f'(0) &= \alpha \\
f''(x) &= \alpha(\alpha-1)(1+x)^{\alpha-2} & f''(0) &= \alpha(\alpha-1) \\
f'''(x) &= \alpha(\alpha-1)(\alpha-2)(1+x)^{\alpha-3} & f'''(0) &= \alpha(\alpha-1)(\alpha-2) \\
&\;\;\vdots & &\;\;\vdots \\
f^{(n)}(x) &= \alpha(\alpha-1)\cdots & f^{(n)}(0) &= \alpha(\alpha-1)\cdots \\
&\quad\cdots(\alpha-n+1)(1+x)^{\alpha-n} & &\quad\cdots(\alpha-n+1)
\end{aligned}
$$

So the binomial series is

$$
1 + \alpha x + \frac{\alpha(\alpha-1)}{2!}x^2 + \frac{\alpha(\alpha-1)(\alpha-2)}{3!}x^3 + \cdots
$$
$$
+ \frac{\alpha(\alpha-1)(\alpha-2)\cdots(\alpha-n+1)}{n!}x^n + \cdots \tag{10.31}
$$

Note that if $\alpha = n$, then the series terminates with the term of nth degree since the factor $(\alpha - n)$ appears in all successive terms and is 0 when $\alpha = n$. Thus, series 10.31 reduces to the finite series 10.30, that is, to the Binomial Formula. It is convenient to introduce the notation

$$
\binom{\alpha}{n} = \frac{\alpha(\alpha-1)(\alpha-2)\cdots(\alpha-n+1)}{n!}
$$

with $\binom{\alpha}{0}$ defined as 1. Then we can write the series in the compact form

$$
\sum_{n=0}^{\infty} \binom{\alpha}{n} x^n
$$

It is not difficult to show that the radius of convergence for this series is $R = 1$ (see Exercise 27 in Exercise Set 10.9), so that the series converges in $(-1, 1)$, with only convergence at the endpoints remaining in question. This convergence in itself, however, does not guarantee that the function represented by the series in this interval is what we hope it will be—namely, $(1 + x)^\alpha$. We can show that the series does converge to $(1 + x)^\alpha$ by proving that $\lim_{n\to\infty} R_n(x) = 0$. The details are tedious, however. We simply state the result in the following theorem.

THEOREM 10.23

If α is any real number other than a nonnegative integer, the binomial series

$$
\sum_{n=0}^{\infty} \binom{\alpha}{n} x^n = 1 + \alpha x + \frac{\alpha(\alpha-1)}{2!}x^2 + \frac{\alpha(\alpha-1)(\alpha-2)}{3!}x^3 + \cdots
$$

converges to $(1 + x)^\alpha$ in the interval $(-1, 1)$ and diverges if $|x| > 1$. If α is a nonnegative integer, the series is finite and represents $(1 + x)^\alpha$ for all real values of x.

REMARK ——

■ When endpoint values are taken into consideration, it can be shown that the series represents the function precisely in the following intervals, depending on the size of α:

$$\begin{cases} -1 \le x \le 1 & \text{if } \alpha > 0 \\ -1 < x < 1 & \text{if } \alpha \le -1 \\ -1 < x \le 1 & \text{if } -1 < \alpha < 0 \end{cases}$$

——

EXAMPLE 10.44   Find the Maclaurin series for $f(x) = 1/\sqrt{1 - x^2}$.

Solution   Since $f(x) = (1 - x^2)^{-1/2}$, we can use the binomial series in which $\alpha = -\frac{1}{2}$ and x is replaced by $-x^2$. So we have

$$\frac{1}{\sqrt{1-x^2}} = \sum_{n=0}^{\infty} \binom{-\frac{1}{2}}{n} (-x^2)^n = \sum_{n=0}^{\infty} (-1)^n \binom{-\frac{1}{2}}{n} x^{2n}$$

$$= 1 - \left(-\frac{1}{2}\right) x^2 + \frac{\left(-\frac{1}{2}\right)\left(-\frac{3}{2}\right)}{2!} x^4$$

$$- \frac{\left(-\frac{1}{2}\right)\left(-\frac{3}{2}\right)\left(-\frac{5}{2}\right)}{3!} x^6 + \cdots$$

$$= 1 + \frac{1}{2} x^2 + \frac{1 \cdot 3}{2^2 \cdot 2!} x^4 + \frac{1 \cdot 3 \cdot 5}{2^3 \cdot 3!} x^6 + \cdots$$

$$= 1 + \sum_{n=1}^{\infty} \frac{1 \cdot 3 \cdot 5 \cdot \cdots \cdot (2n - 1)}{2^n \cdot n!} x^{2n}$$

The series converges to the given function for $|-x^2| < 1$ or, equivalently, $-1 < x < 1$.   ■

EXAMPLE 10.45   Find the Maclaurin series for $\sin^{-1} x$ and give its radius of convergence.

Solution   Since

$$\sin^{-1} x = \int \frac{dx}{\sqrt{1 - x^2}} + C$$

we can integrate the series obtained in the preceding example to get

$$\sin^{-1} x = \int \left[1 + \sum_{n=1}^{\infty} \frac{1 \cdot 3 \cdot 5 \cdot \cdots \cdot (2n - 1)}{2^n \cdot n!} x^{2n} \right] dx + C$$

$$= x + \sum_{n=1}^{\infty} \int \frac{1 \cdot 3 \cdot 5 \cdot \cdots \cdot (2n - 1)}{2^n \cdot n!} x^{2n} dx + C$$

$$= x + \sum_{n=1}^{\infty} \frac{1 \cdot 3 \cdot 5 \cdot \cdots \cdot (2n - 1)}{2^n \cdot n!} \cdot \frac{x^{2n+1}}{2n + 1} + C$$

Putting $x = 0$ on each side, we see that $C = 0$. So we can write

$$\sin^{-1} x = x + \frac{1}{2} \cdot \frac{x^3}{3} + \frac{1 \cdot 3}{2^2 \cdot 2!} \cdot \frac{x^5}{5} + \frac{1 \cdot 3 \cdot 5}{2^3 \cdot 3!} \cdot \frac{x^7}{7} + \cdots$$

Since the radius of convergence of the series for $(1 - x^2)^{-1/2}$ is $R = 1$, it follows from Theorem 10.20 that the series for $\sin^{-1} x$ has the same radius of convergence. ∎

Summary of Some Important Maclaurin Series

For reference, we list some of the more frequently used Maclaurin series. The first three, in particular, are especially useful to know.

Frequently Used Maclaurin Series

$$e^x = \sum_{n=0}^{\infty} \frac{x^n}{n!} = 1 + x + \frac{x^2}{2!} + \frac{x^3}{3!} + \cdots, \quad -\infty < x < \infty$$

$$\sin x = \sum_{n=0}^{\infty} \frac{(-1)^n x^{2n+1}}{(2n+1)!} = x - \frac{x^3}{3!} + \frac{x^5}{5!} - \frac{x^7}{7!} + \cdots, \quad -\infty < x < \infty$$

$$\cos x = \sum_{n=0}^{\infty} \frac{(-1)^n x^{2n}}{(2n)!} = 1 - \frac{x^2}{2!} + \frac{x^4}{4!} - \frac{x^6}{6!} + \cdots, \quad -\infty < x < \infty$$

$$\ln(1 + x) = \sum_{n=1}^{\infty} \frac{(-1)^{n-1} x^n}{n} = x - \frac{x^2}{2} + \frac{x^3}{3} - \frac{x^4}{4} + \cdots, \quad -1 < x \le 1$$

$$\tan^{-1} x = \sum_{n=1}^{\infty} \frac{(-1)^{n-1} x^{2n-1}}{2n-1} = x - \frac{x^3}{3} + \frac{x^5}{5} - \frac{x^7}{7} + \cdots, \quad -1 \le x \le 1$$

$$\sinh x = \sum_{n=0}^{\infty} \frac{x^{2n+1}}{(2n+1)!} = x + \frac{x^3}{3!} + \frac{x^5}{5!} + \frac{x^7}{7!} + \cdots, \quad -\infty < x < \infty$$

$$\cosh x = \sum_{n=0}^{\infty} \frac{x^{2n}}{(2n)!} = 1 + \frac{x^2}{2!} + \frac{x^4}{4!} + \frac{x^6}{6!} + \cdots, \quad -\infty < x < \infty$$

$$(1 + x)^\alpha = \sum_{n=0}^{\infty} \binom{\alpha}{n} x^n = 1 + \binom{\alpha}{1} x + \binom{\alpha}{2} x^2 + \binom{\alpha}{3} x^3 + \cdots, \quad -1 < x < 1,$$

$$\text{where } \binom{\alpha}{k} = \frac{\alpha(\alpha - 1) \cdots (\alpha - k + 1)}{k!}$$

COMPUTATIONS WITH POWER SERIES AND COMPUTER ALGEBRA SYSTEMS

When a function can be approximated via the Taylor series representation, then the function can be treated essentially as an infinite polynomial. In particular, differentiation and integration can be performed on the series representation term by term.

CAS 38

Approximate $\int_0^1 e^{-x^2}\, dx$.

We will use the representation

$$e^{-x^2} = \sum_{n=1}^{\infty} (-1)^n \frac{x^{2n}}{n!}$$

and to approximate the integral, we replace the integrand with a 10th degree Taylor polynomial, which is a partial sum of the series representation for the function.

Maple:

f := x−>exp(−x^2);
readlib('mtaylor'):   Access mtaylor
P := mtaylor(f(x),[x],12);

Output:

$$P := 1 - x^2 + \frac{1}{2}x^4 - \frac{1}{6}x^6 + \frac{1}{24}x^8 - \frac{1}{120}x^{10}$$

int(P,x);

Output:

$$x - \frac{1}{3}x^3 + \frac{1}{10}x^5 - \frac{1}{42}x^7 + \frac{1}{216}x^9 - \frac{1}{1320}x^{11}$$

int(P,x=0..1); $\dfrac{31049}{41580}$

evalf(");   .7467291967

Mathematica:

f[x_] = Exp[−x^2]
P[x_] = Normal[Series[f[x],
   {x,0,10}]]
Integrate[P[x],x]
Integrate[P[x],{x,0,1}]

DERIVE:

(At the □ symbol, go to the next step.)
a (author) □ exp(−x^2) □ c (calculus)
□ t (taylor) □ [choose expression] □ x
(variable) □ 10 (degree) □ 0 (point) □
s (simplify) □ [choose expression] □
c (calculus) □ i (integrate) □ [choose
expression] □ x (variable) □ enter (no
limits) □ s (simplify) □ [choose expres-
sion] □ c (calculus) □ i (integrate) □
[choose expression] □ x (variable) □
0 (lower limit) □ 1 (upper limit) □ x
(approximate) □ [choose expression]

Finally, we compare this result with that obtained from Simpson's Rule.

Maple:

with(student):
evalf(value(simpson(f(x),
   x=0..1,100)));

Output: .7468241331

Exercise Set 10.9

In Exercises 1–18, find the Maclaurin series for the given function and determine its interval of convergence.

1. $\sin x$ **2.** $\sinh x$

3. $\cosh x$ **4.** $\ln(1 - x)$

5. $\tanh^{-1} x$

6. $\sqrt{1 - x}$ (Do not test endpoints.)

7. $\sin^2 x$ (*Hint:* $\cos 2x = 1 - 2\sin^2 x$.)

8. $\cos^2 x$ (*Hint:* $\cos 2x = 2\cos^2 x - 1$.)

9. $\dfrac{1}{(1 - x)^3}$

10. $\sin x \cos x$ (*Hint:* $\sin 2x = 2\sin x \cos x$.)

11. 2^x **12.** xe^{-x}

13. $\dfrac{1 - \cos x}{x^2}$ (defined as its limiting value at 0)

14. $\sin x^2$

15. $\sqrt[3]{8 - x}$ (Do not test endpoints.)

16. $\dfrac{\sin x}{x}$ (defined at 0 to be its limiting value as $x \to 0$)

17. $\cos \sqrt{x}$ **18.** $x \ln \sqrt{1 + x}$

In Exercises 19–24, find the Taylor series for the given function about $x = a$ for the specified value of a, and determine the interval of convergence of the series.

19. e^x; $a = 1$ **20.** $\cos x$; $a = \dfrac{\pi}{3}$

21. $\dfrac{1}{x}$; $a = -1$ **22.** $\sqrt{x + 3}$; $a = 1$

23. $\sin \dfrac{\pi x}{6}$; $a = 1$ **24.** $1/(x - 1)^2$; $a = 2$

25. For each of the following, show that the Maclaurin series of the given function represents the function for all values of x.
(a) $\sin x$ (b) $\sinh x$ (c) $\cosh x$

26. Find the Taylor series about $x = 1$ for $f(x) = \ln x$ by replacing x with $x - 1$ in the Maclaurin series for $\ln(1 + x)$. In what interval does the series represent $\ln x$?

27. Show that the radius of convergence of the binomial series is 1.

28. Prove that if $f^{(n)}(x)$ exists for all n and x and there exists a constant M such that $|f^{(n)}(x)| \le M$ for all n and x, then the Taylor series of f about $x = a$ exists and converges to $f(x)$ everywhere.

29. For $x \ne 0$, define $f(x) = e^{-1/x^2}$ and let $f(0) = 0$. Prove that the Maclaurin series for f exists and converges everywhere but that it represents f only at $x = 0$.

CAS *In Exercises 30–33, approximate the integrals, making use of three terms of the Maclaurin series of the integrand. Estimate the error.*

30. $\displaystyle\int_{-1}^{1} \dfrac{1 - \cos x}{x^2}\,dx$ **31.** $\displaystyle\int_{0}^{0.5} \sqrt{1 + x^3}\,dx$

32. $\displaystyle\int_{0}^{0.4} \dfrac{\tan^{-1} x}{x}\,dx$ **33.** $\displaystyle\int_{-0.2}^{0} e^{-x^2}\,dx$

34. Find the first three nonzero terms of the Maclaurin series for $\tan x$.

35. Find the first three nonzero terms of the Maclaurin series for $\sec^2 x$. (*Hint:* Use the results of Exercise 34.)

CAS **36.** Use a CAS to find the first eight terms of the Maclaurin series for
(a) $\tan x$;   (b) $\sec^2 x$.
(Compare Exercises 34 and 35.)

37. The **error function** is defined by

$$\mathrm{erf}(x) = \dfrac{2}{\sqrt{x}} \int_{0}^{x} e^{-t^2}\,dt, \qquad -\infty < x < \infty$$

Find the Maclaurin series for $\mathrm{erf}(x)$, and use it to approximate $\mathrm{erf}(1)$ correct to three decimal places.

CAS **38.** Approximate the integral

$$\int_{0}^{2} xe^x\,dx$$

(a) using Simpson's Rule with 10 subintervals;
(b) using the first 10 nonzero terms of the Taylor series about $x = 0$.
Compare the results.

39. (a) Use a CAS to obtain the graph of

$$f(x) = |\sin^{-1}(\sin x)|$$

Explain the appearance of the graph.

(b) Define the function

$$g(x) = \sum_{n=0}^{\infty} \left(\frac{3}{4}\right)^n f(4^n x)$$

where f is the function of part (a). Approximate the graph of $g(x)$ by several partial sums of the series. *Note:* $g(x)$ is a nowhere-differentiable function first given by Karl Weierstrass (1815–1897).

10.10 FOURIER SERIES

The representation of functions by Taylor series is of fundamental importance. Yet there are many functions that cannot be expanded in Taylor series. For one thing, only functions that are infinitely differentiable have such expansions. Another type of series, whose terms consist of sines and cosines, can be used to represent a much broader class of functions. Series of this type are called Fourier (pronounced "foor-ee-yay") series, after the French mathematician and physicist Jean Baptiste Joseph Fourier (1768–1830), who studied these series in connection with his investigation of heat conduction.

Fourier series are especially important in studying all sorts of wave phenomena, such as light waves, sound waves, radio waves, and signals in electronic instruments. A concept closely related to a Fourier series, called a Fourier integral, has been used in such diverse ways as analyzing the sunspot cycle and investigating X rays in computer-assisted tomography (CAT) scanners.

We will give only a brief introduction here to Fourier series. There is a vast literature on this subject, and research related to Fourier series and Fourier integrals is ongoing. In this discussion we can only scratch the surface.

Let us begin by looking at the series

$$\cos x - \frac{\cos 3x}{3} + \frac{\cos 5x}{5} - \cdots = \sum_{k=1}^{\infty} \frac{(-1)^{k-1}\cos(2k-1)x}{2k-1} \qquad (10.32)$$

This series is an example of a Fourier series involving cosine terms only. (Others involve only sine terms or combine sines and cosines.) In Figure 10.30 we show graphs of the first three partial sums,

$$S_1(x) = \cos x, \quad S_2(x) = \cos x - \frac{\cos 3x}{3}, \quad S_3(x) = \cos x - \frac{\cos 3x}{3} + \frac{\cos 5x}{5}$$

Notice how adding the second term modifies the first by adding more "ripples" with smaller amplitude. The tops and bottoms seem to be flattening out. Adding the third term (getting $S_3(x)$) further flattens the tops and bottoms and adds "ripples" of still smaller amplitude. To see that this trend continues, we show the graph of the first ten terms,

$$S_{10}(x) = \sum_{k=1}^{10} \frac{(-1)^{k-1}\cos(2k-1)x}{2k-1}$$

in Figure 10.31. It appears that the partial sums are converging to the "square wave" shown in Figure 10.32. In Exercise 9 of Exercise Set 10.10 we will ask you to verify that series 10.32 is the Fourier series for the function whose graph is given in Figure 10.32.

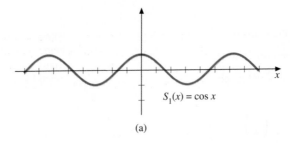

$$S_1(x) = \cos x$$

(a)

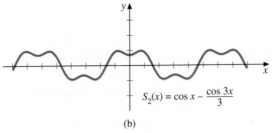

$$S_2(x) = \cos x - \frac{\cos 3x}{3}$$

(b)

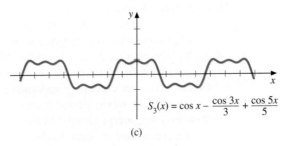

$$S_3(x) = \cos x - \frac{\cos 3x}{3} + \frac{\cos 5x}{5}$$

(c)

FIGURE 10.30

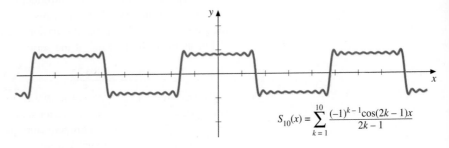

$$S_{10}(x) = \sum_{k=1}^{10} \frac{(-1)^{k-1}\cos(2k-1)x}{2k-1}$$

FIGURE 10.31

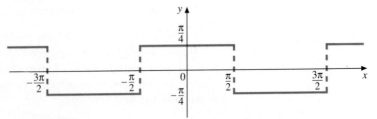

FIGURE 10.32
The square wave
$$f(x) = \begin{cases} \frac{\pi}{4} & \text{if } -\frac{\pi}{2} < x < \frac{\pi}{2} \\ -\frac{\pi}{4} & \text{if } \frac{\pi}{2} < x < \frac{3\pi}{2} \end{cases}$$
$$f(x + 2\pi) = f(x).$$

In our example above, we began with a series and conjectured the function to which it converged. Now we want to reverse this procedure. We will begin with a function and attempt to find a series of sines and cosines converging to the function. This process will lead us to the definition of a Fourier series.

Deriving the Fourier Series

We begin with a function f that is integrable on the interval $[-\pi, \pi]$. (Later, we will indicate how we can use a more general interval of the form $[-L, L]$ for any $L > 0$.) At present we place no other restriction on f. It may not even be continuous. Let us assume for the moment that $f(x)$ can be represented by the following series of sines and cosines:

$$f(x) = \frac{a_0}{2} + \sum_{n=1}^{\infty}(a_n \cos nx + b_n \sin nx) \tag{10.33}$$

We will see later why it is convenient to write the constant term as $a_0/2$ rather than a_0. (By doing so, it will turn out that the formula we develop for a_n when $n \geq 1$ will be true for $n = 0$ as well.)

To find the coefficients in Equation 10.33, we will need the following results, which can be obtained using the integration techniques of Section 8.3 (see Exercise 42 in Exercise Set 8.3).

For m and n positive integers,

$$\int_{-\pi}^{\pi} \cos mx \cos nx \, dx = \begin{cases} 0 & \text{if } m \neq n \\ \pi & \text{if } m = n \end{cases} \tag{10.34}$$

$$\int_{-\pi}^{\pi} \sin mx \sin nx \, dx = \begin{cases} 0 & \text{if } m \neq n \\ \pi & \text{if } m = n \end{cases} \tag{10.35}$$

$$\int_{-\pi}^{\pi} \sin mx \cos nx \, dx = 0 \tag{10.36}$$

By direct integration, you can also show that

$$\int_{-\pi}^{\pi} \cos nx \, dx = \int_{-\pi}^{\pi} \sin nx \, dx = 0 \tag{10.37}$$

Let us return now to Equation 10.33 and integrate both sides from $-\pi$ to π. We will assume at this stage that term-by-term integration on the right-hand side is valid.

$$\int_{-\pi}^{\pi} f(x) \, dx = \int_{-\pi}^{\pi} \frac{a_0}{2} \, dx + \sum_{n=1}^{\infty}\left[\int_{-\pi}^{\pi} a_n \cos nx \, dx + \int_{-\pi}^{\pi} b_n \sin nx \, dx\right]$$

By Equation 10.37 the integrals in the summation both equal 0 for all n. Thus, we get

$$\int_{-\pi}^{\pi} f(x) \, dx = \frac{a_0}{2}(2\pi)$$

Solving for a_0 gives

$$a_0 = \frac{1}{\pi}\int_{-\pi}^{\pi} f(x) \, dx \tag{10.38}$$

We now have found the value of a_0. To find a_m for $m \geq 1$, we multiply both sides of Equation 10.33 by $\cos mx$, where m is a fixed, but unspecified, positive integer. Then, we again integrate over the interval $[-\pi, \pi]$:

$$\int_{-\pi}^{\pi} f(x) \cos mx \, dx = \frac{a_0}{2} \int_{-\pi}^{\pi} \cos mx \, dx + \sum_{n=1}^{\infty} \left[a_n \int_{-\pi}^{\pi} \cos mx \cos nx \, dx \right.$$
$$\left. + b_n \int_{-\pi}^{\pi} \cos mx \sin nx \, dx \right]$$

By Equation 10.37 the first integral on the right-hand side is 0 if $m \geq 1$. By Equation 10.34 the only nonzero term in the summation is the term for which $n = m$. Thus,

$$\int_{-\pi}^{\pi} f(x) \cos mx \, dx = a_m \int_{-\pi}^{\pi} \cos^2 mx \, dx$$

By Equation 10.34 the integral on the right equals π. So, on solving for a_m, we have

$$a_m = \frac{1}{\pi} \int_{-\pi}^{\pi} f(x) \cos mx \, dx, \quad m = 1, 2, \ldots \tag{10.39}$$

Notice that if we let $m = 0$ in this equation, the result agrees with the value of a_0 in Equation 10.38.

In order to find b_m, we proceed in a similar way, multiplying both sides of Equation 10.33 by $\sin mx$ and integrating. The result is (see Exercise 3 in Exercise Set 10.10)

$$b_m = \frac{1}{\pi} \int_{-\pi}^{\pi} f(x) \sin mx \, dx, \quad m = 1, 2, \ldots \tag{10.40}$$

REMARK

■ We used the letter m as a subscript in deriving Equations 10.39 and 10.40, since n was used as the index of summation. Now that we have the formulas, we can replace m with n.

Although we have made a number of assumptions in deriving the formulas for the coefficients, the formulas themselves have meaning provided only that $f(x)$ is integrable on $[-\pi, \pi]$. We therefore can make the following definition.

Definition 10.10
The Fourier Series for f

Let f be integrable on $[-\pi, \pi]$. Then the series

$$\frac{a_0}{2} + \sum_{n=1}^{\infty} (a_n \cos nx + b_n \sin nx)$$

where

$$a_n = \frac{1}{\pi} \int_{-\pi}^{\pi} f(x) \cos nx \, dx, \quad n = 0, 1, 2, \ldots$$

and

$$b_n = \frac{1}{\pi} \int_{-\pi}^{\pi} f(x) \sin nx \, dx, \quad n = 1, 2, 3, \ldots$$

is called the **Fourier series for f** on $[-\pi, \pi]$.

EXAMPLE 10.46   Find the Fourier series for the function f defined by $f(x) = x$ on $[-\pi, \pi]$.

Solution   By Equation 10.39,

$$a_n = \frac{1}{\pi} \int_{-\pi}^{\pi} x \cos nx \, dx$$

Since $x \cos nx$ is the product of an odd function and an even function, the result is odd. Thus, as we showed in Section 5.4, the integral over a symmetric interval is 0. That is,

$$a_n = 0 \qquad \text{for } n = 0, 1, 2, \ldots$$

Also, by Equation 10.40 and by what we showed in Section 5.4,

$$b_n = \frac{1}{\pi} \int_{-\pi}^{\pi} x \sin nx \, dx = \frac{2}{\pi} \int_0^{\pi} x \sin nx \, dx$$

since $x \sin nx$ is an even function (the product of two odd functions is even). Now we integrate by parts, with $u = x$ and $dv = \sin nx$.

$$b_n = \frac{2}{\pi} \left[-\frac{x \cos nx}{n} \Big|_0^{\pi} + \frac{1}{n} \int_0^{\pi} \cos nx \, dx \right]$$

$$= \frac{2}{\pi} \left[-\frac{\pi \cos n\pi}{n} \right] + \frac{2}{\pi n^2} \left[\sin nx \ \right]_0^{\pi}$$

$$= -\frac{2}{n} \cos n\pi$$

Since $\cos n\pi = 1$ when n is even and $\cos n\pi = -1$ when n is odd, we can write $\cos n\pi = (-1)^n$. Thus,

$$b_n = \frac{2(-1)^{n+1}}{n} \qquad n = 1, 2, 3, \ldots$$

By Definition 10.10 the Fourier series for f is

$$2 \left[\sin x - \frac{\sin 2x}{2} + \frac{\sin 3x}{3} - \cdots \right] \qquad (10.41)$$

We will graph some of the partial sums shortly.   ∎

A Convergence Theorem

The question of convergence of the Fourier series for a function is a deep one and is better left to more advanced courses. We will state below one set of criteria that will ensure that the series converges to the function, at least for most values of x. First, though, observe that since $\sin nx$ and $\cos nx$ have period 2π for all values of n, if the Fourier series converges in the interval $[-\pi, \pi]$, then it also converges outside that interval in a periodic manner. Thus, if $f(x)$ is to be represented by its Fourier series for all x, it must be periodic, of period 2π. That is, $f(x)$ must satisfy

$$f(x + 2\pi) = f(x)$$

If f is not periodic, its Fourier series still may converge to $f(x)$ on $[-\pi, \pi]$. Outside that interval it would then converge to the *periodic extension* of f.

The following theorem was first stated by the German mathematician P. G. Lejeune Dirichlet (1805–1859).

THEOREM 10.24

Dirichlet's Theorem

Let f be bounded, piecewise-continuous, and have a finite number of maxima and minima on the interval $-\pi < x \leq \pi$. For x outside this interval let f be defined periodically by $f(x + 2\pi) = f(x)$. Then the Fourier series for f converges to $f(x)$ for all points of continuity of f. At each point of discontinuity the series converges to the average of the right-hand and left-hand limits of f at that point.

REMARK

■ A function is piecewise continuous on an interval if it has at most finitely many points of discontinuity, and each such discontinuity is a simple (jump) discontinuity. (See Section 2.3.)

Graphing the Partial Sums

Let us illustrate the manner in which the partial sums of a Fourier series converge by looking again at some graphs. We will use series 10.41 for the function $f(x) = x$. In Figure 10.33 we show the graph of f on $(-\pi, \pi]$ and extended periodically outside that interval. Because of the appearance of its graph, this function is often referred to as a "sawtooth" function.

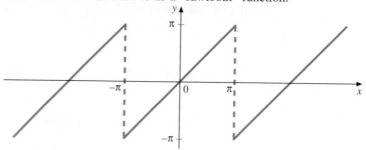

FIGURE 10.33
The "sawtooth" function $f(x) = x$ on $(-\pi, \pi]$, $f(x + 2\pi) = f(x)$

The hypotheses of Theorem 10.24 are satisfied for our function in Figure 10.33, so its Fourier series converges in the manner specified by the theorem. The points of discontinuity are the integral multiples of π. At each such point the average of the right-hand and left-hand limits of f is 0. If we redefine $f(x)$ to be 0 at these points, then we can write

$$f(x) = 2\left[\sin x - \frac{\sin 2x}{2} + \frac{\sin 3x}{3} - \cdots\right] = 2\sum_{k=1}^{\infty} \frac{(-1)^{k+1} \sin kx}{k}$$

In Figure 10.34 we show the graphs of the three partial sums $S_3(x)$, $S_6(x)$, and $S_9(x)$. Notice how they increasingly approach the sawtooth shape of the graph of f as n increases. Also, observe that each partial sum graph crosses the x-axis (the value of the partial sum is 0) at each integral multiple of π.

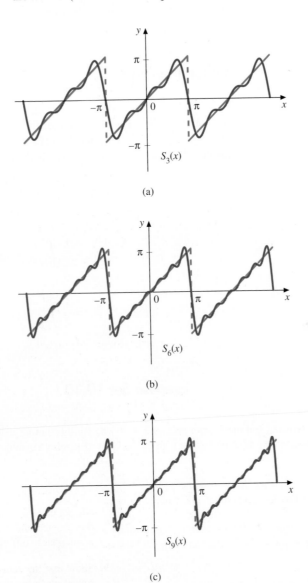

(a)

(b)

(c)

FIGURE 10.34

Harmonics

In the Fourier series for a function f,

$$\frac{a_0}{2} + \sum_{n=1}^{\infty}(a_n \cos nx + b_n \sin nx)$$

the constant term $a_0/2$ is known as *the fundamental.* In Exercise 1 of this section you will be asked to show that it is the average value of f over $[-\pi, \pi]$. The term $a_1 \cos x + b_1 \sin x$ is called the *first harmonic,* the term $a_2 \cos 2x + b_2 \sin 2x$ the *second harmonic,* and so on. When we listen to music, we are hearing the sum of many harmonics. The human ear and brain make a wonderful Fourier analyzer of these harmonics.

Other Intervals

By means of a horizontal stretching or shrinking of the type we studied in Section 1.2, it is possible to obtain the Fourier series for a function f defined on $[-L, L]$ for an arbitrary positive number L instead of on $[-\pi, \pi]$. The result, which we ask you to show in Exercise 16 of Exercise Set 10.10, is

$$f(x) = \frac{a_0}{2} + \sum_{n=1}^{\infty} \left(a_n \cos \frac{n\pi x}{L} + b_n \sin \frac{n\pi x}{L} \right) \tag{10.42}$$

where

$$a_n = \frac{1}{L} \int_{-L}^{L} f(x) \cos \frac{n\pi x}{L} \, dx \tag{10.43}$$

and

$$b_n = \frac{1}{L} \int_{-L}^{L} f(x) \sin \frac{n\pi x}{L} \, dx \tag{10.44}$$

Exercise Set 10.10

1. Show that the constant term $a_0/2$ in the Fourier series for a function f is the average value of f on $[-\pi, \pi]$. (See Definition 5.3.)

2. (a) Show that if f is even, then $b_n = 0$ for $n = 1, 2, 3, \ldots$, and the Fourier series for f consists of cosine terms only.
   (b) Show that if f is odd, then $a_n = 0$ for $n = 0, 1, 2, \ldots$, and the Fourier series for f consists of sine terms only.

3. Carry out the details of deriving Formula 10.40 for b_m.

In Exercises 4–7, find the Fourier series for f.

4. $f(x) = |x|$ on $[-\pi, \pi]$

5. $f(x) = \begin{cases} 0 & \text{if } -\pi < x < 0 \\ x & \text{if } 0 \le x \le \pi \end{cases}$ and $f(x + 2\pi) = f(x)$

6. $f(x) = x^2$ on $[-\pi, \pi]$

7. $f(x) = \cos 2x$ on $[-\pi, \pi]$ (How could you have determined the answer without calculation?)

8. By substituting $x = \pi$ in the Fourier series for $f(x) = x^2$ (Exercise 6), obtain the sum

$$\sum_{n=1}^{\infty} \frac{1}{n^2} = \frac{\pi^2}{6}$$

(Recall that this sum was referred to in the introduction to this chapter, and it was first discovered by Euler. He used a completely different approach from the one used here.)

9. Show that the series 10.32 is the Fourier series for the square wave function shown in Figure 10.32.

10. Find the Fourier series for $f(x) = \cos^2 x$ for $-\pi \le x \le \pi$.

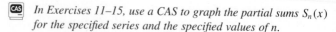 *In Exercises 11–15, use a CAS to graph the partial sums $S_n(x)$ for the specified series and the specified values of n.*

11. The series of Example 10.46; $n = 2, 5, 7, 10$

12. The series of Exercise 4; $n = 2, 5, 8, 11$

13. The series of Exercise 5; $n = 1, 3, 5, 7$

14. The series of Exercise 6; $n = 2, 4, 6, 8$

15. The series of Exercise 10; $n = 2, 5, 8, 11$

16. Derive Equations 10.42, 10.43, and 10.44 as follows:
(a) Let f satisfy the conditions given in Theorem 10.24 on the interval $[-L, L]$. Make the change of

variables $u = \pi x/L$, and let

$$g(u) = f\left(\frac{Lu}{\pi}\right)$$

(b) Show that $g(u)$ is defined for $-\pi \le u \le \pi$.
(c) Write the Fourier series for $g(u)$ and the formulas for a_n and b_n in terms of u.
(d) In the results of part (c) replace u with $\pi x/L$. Simplify the formulas for a_n and b_n.

Chapter 10 Review Exercises

1. Find the Taylor polynomial of degree 5 about $x = 0$ for $f(x) = \ln(1 - x)$, and determine how accurately it approximates $f(x)$ for $|x| \le 0.2$.

2. Compute $\cos 62°$ with six decimal places of accuracy using Taylor's Formula with an appropriate value of n and a.

3. Find the Taylor polynomial of degree 5 about $x = \pi/4$ for $f(x) = \tan x$.

4. Find the nth Taylor polynomial for $f(x) = \cos^2 x$ about $x = 0$. What is $R_n(x)$ for this function? (*Hint:* Use a trigonometric identity.)

5. Find the nth Taylor polynomial about $x = \pi/4$ for $f(x) = \sin x - \cos x$. (*Hint:* Show that $f(x) = \sqrt{2}\sin(x - \pi/4)$.)

6. Use a cubic polynomial to approximate $\sqrt[3]{25}$. Estimate the error. (*Hint:* $\sqrt[3]{25} = (27 - 2)^{1/3} = (27)^{1/3}(1 - \frac{2}{27})^{1/3} = 3(1 - \frac{2}{27})^{1/3}$.)

7. Find $P_3(x)$ and $R_3(x)$ about $x = 1$ for $f(x) = \tan^{-1} x$.

8. Show that the Taylor polynomial

$$1 - \frac{x^2}{2!} + \frac{x^4}{4!} - \frac{x^6}{6!} + \frac{x^8}{8!}$$

for $\cos x$ gives accuracy to at least seven decimal places for x in $[-\pi/4, \pi/4]$.

9. Let $P_n(x)$ be the nth Taylor polynomial about $x = 0$ for $f(x) = e^x$. Use a calculator to make a table showing each of the following for $x = \pm 0.25$, $x = \pm 0.50$, $x = \pm 0.75$, $x = \pm 1$: e^x, $P_1(x)$, $P_2(x)$, $P_3(x)$, and $P_4(x)$.

10. Write out Taylor's Formula with Remainder with $a = 0$ for $f(x) = xe^{-x}$.

In Exercises 11 and 12, find $\lim_{n \to \infty} a_n$, or show the limit fails

to exist.

11. (a) $a_n = \dfrac{2n^3 - 3n + 1}{n^3 + 2n + 3}$ (b) $a_n = \dfrac{\sqrt{n + 10}}{n + 2}$

12. (a) $a_n = \dfrac{\sqrt{n^5 + 1}}{10n^2 + 3n}$ (b) $a_n = \dfrac{n \cos n\pi}{n + 1}$

In Exercises 13 and 14, show that each sequence is monotone. Then determine whether it converges by investigating boundedness.

13. (a) $\{e^{1-(1/n)}\}$ (b) $\left\{\dfrac{2n - 3}{n + 3}\right\}$

14. (a) $\left\{\dfrac{e^{2n}n!}{(2n - 1)!}\right\}$
(b) $\{n \ln(n + 1) - (n + 1)\ln n\}$

15. Prove that if $\lim_{n \to \infty} a_n = L$, then $\lim_{n \to \infty} a_n^2 = L^2$. Is the converse also true? Prove or disprove.

16. Let

$$S_n = \frac{1}{2n + 1} + \frac{1}{2n + 2} + \frac{1}{2n + 3} + \cdots + \frac{1}{3n}$$

Prove that $\{S_n\}$ is a convergent sequence.

17. A sequence $\{S_n\}$ is defined recursively by $S_1 = 1$, and for $n \ge 2$, $S_n = S_{n-1} + 2/(n^2 + n)$. Find an explicit formula for S_n. Then show that $\{S_n\}$ converges, and find its limit.

18. What can you conclude about convergence or divergence of the series $\sum_{k=1}^{\infty} a_k$ under the following conditions?
(a) $a_k = \dfrac{2k^2 - 1}{k^2 + 1}$

(b) $a_1 + a_2 + \cdots + a_n = \dfrac{2n^2 - 1}{n^2 + 1}$

Give reasons.

19. Find the sum of each of the following series:

(a) $\displaystyle\sum_{n=0}^{\infty} \frac{(-1)^n 2^{n-1}}{3^n}$ (b) $\displaystyle\sum_{k=2}^{\infty} \frac{1}{k^2 - 1}$

20. Use geometric series to express each of the following repeating decimals as the ratio of two integers:

(a) $1.297297297\ldots$ (b) $3.2454545\ldots$

In Exercises 21–44, test for convergence or divergence. If the series has variable signs, test also for absolute convergence.

21. $\displaystyle\sum_{n=1}^{\infty} \frac{n}{\sqrt{n^3 + 2}}$

22. $\displaystyle\sum_{n=1}^{\infty} \frac{\sin n + \cos n}{n\sqrt{n+1}}$

23. $\displaystyle\sum_{n=0}^{\infty} \frac{e^{-n}}{1 + e^{-n}}$

24. $\displaystyle\sum_{n=2}^{\infty} \frac{\cos n\pi}{\ln n}$

25. $\displaystyle\sum_{n=1}^{\infty} \frac{\tanh n}{n}$

26. $\displaystyle\sum_{n=1}^{\infty} n e^{-2n}$

27. $\displaystyle\sum_{k=1}^{\infty} \frac{\ln k}{k^2}$

28. $\displaystyle\sum_{n=1}^{\infty} \left(\frac{n}{3n+4}\right)^n$

29. $\displaystyle\sum_{k=1}^{\infty} \frac{(-1)^{k-1} \ln k}{\sqrt{k}}$

30. $\displaystyle\sum_{k=1}^{\infty} \frac{3k^2 + 2k - 1}{k^5 + 2}$

31. $\displaystyle\sum_{k=2}^{\infty} \frac{\sin k}{k(\ln k)^2}$

32. $\displaystyle\sum_{n=1}^{\infty} \frac{n-1}{(n+1)^2}$

33. $\displaystyle\sum_{n=0}^{\infty} \frac{2^n n!}{(2n)!}$

34. $\displaystyle\sum_{k=0}^{\infty} (-1)^k k^2 e^{-k}$

35. $\displaystyle\sum_{k=1}^{\infty} \frac{k^k}{3^k k!}$

36. $\displaystyle\sum_{n=2}^{\infty} \frac{\ln n}{n^3 + 4}$

37. $\displaystyle\sum_{n=2}^{\infty} (-1)^n \ln\left(\frac{n+1}{n-1}\right)$

38. $\displaystyle\sum_{k=0}^{\infty} \frac{\sin(k + \frac{1}{2})\pi}{k + \frac{1}{2}}$

39. $\displaystyle\sum_{n=1}^{\infty} \frac{n^{2n}}{(n^3 + 1)^n}$

40. $\displaystyle\sum_{n=1}^{\infty} \left(\frac{1}{n} - \frac{1}{n+2}\right)$

41. $\displaystyle\sum_{n=1}^{\infty} \frac{2 + \cos n}{\sqrt{2n^2 - 1}}$

42. $\displaystyle\sum_{n=0}^{\infty} \frac{\operatorname{sech}^2 n}{1 + \tanh n}$

43. $\displaystyle\sum_{n=0}^{\infty} (-1)^n (\sqrt{n+1} - \sqrt{n})$

44. $\displaystyle\sum_{n=0}^{\infty} \frac{(-1)^n (n-1)}{2n + 1}$

In Exercises 45–52, find the interval of convergence.

45. $\displaystyle\sum_{n=0}^{\infty} \frac{2^n x^n}{3^{n+1}}$

46. $\displaystyle\sum_{n=0}^{\infty} \frac{(-1)^n n x^n}{2n + 1}$

47. $\displaystyle\sum_{k=0}^{\infty} \frac{(x-2)^k}{\sqrt{2k + 1}}$

48. $\displaystyle\sum_{k=0}^{\infty} \frac{(-1)^k (x+1)^{2k}}{3^k}$

49. $\displaystyle\sum_{k=1}^{\infty} \frac{k x^k}{\ln(k + 1)}$

50. $\displaystyle\sum_{n=1}^{\infty} \frac{(nx)^n}{n! e^n}$ (Do not check endpoints.)

51. $\displaystyle\sum_{n=0}^{\infty} \frac{(-1)^n (2x - 1)^n n!}{1 \cdot 3 \cdot 5 \cdot \cdots \cdot (2n + 1)}$ (*Hint:* In testing endpoints, see Exercise 33 in Exercise Set 10.6.)

52. $\displaystyle\sum_{n=1}^{\infty} \frac{\cosh n}{n^2} (x - 1)^n$

In Exercises 53–56, use integration or differentiation of an appropriate series to find the Maclaurin series of f. Give the radius of convergence.

53. $f(x) = \ln(1 - x^2)$

54. $f(x) = x/(2 + x)^2$

55. $f(x) = \sinh^{-1} x$

56. $f(x) = \dfrac{1}{a} \tan^{-1} \dfrac{x}{a}$

57. Use infinite series to verify the following limits:

(a) $\displaystyle\lim_{x \to 0} \frac{\sin x}{x} = 1$

(b) $\displaystyle\lim_{x \to 0} \frac{1 - \cos x}{x^2} = \frac{1}{2}$

(c) $\displaystyle\lim_{x \to 0} \frac{e^x - 1}{x} = 1$

In Exercises 58–60, find the Taylor series about $x = a$.

58. $f(x) = 1/(2 - x)$; $a = 1$

59. $f(x) = \ln \sqrt{x + 2}$; $a = -1$

60. $f(x) = \sin x$; $a = \dfrac{\pi}{4}$. Show that the series represents the function for all values of x.

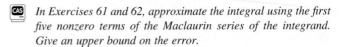

In Exercises 61 and 62, approximate the integral using the first five nonzero terms of the Maclaurin series of the integrand. Give an upper bound on the error.

61. $\displaystyle\int_0^{1/2} \frac{1 - e^{-x}}{x} \, dx$

62. $\displaystyle\int_0^{1/2} \frac{dx}{\sqrt{1 + x^3}}$

63. Find the Taylor series for $f(x) = \cos^2 x - \sin^2 x$ about $x = \pi/4$, and show that it converges to f for all x. (*Hint:* First use a trigonometric identity.)

64. Find the Fourier series for each of the functions
(a) $f(x) = \begin{cases} 0 & \text{if } -\pi < x < 0 \\ 1 & \text{if } 0 \le x \le \pi \end{cases}$
(b) $f(x) = |\sin x|$ on $[-\pi, \pi]$

Chapter 10 Concept Quiz

1. Define each of the following:
(a) the limit of a sequence $\{a_n\}$
(b) the sum of a series $\sum_{n=1}^{\infty} a_n$
(c) a geometric series
(d) an absolutely convergent series
(e) the nth Taylor polynomial for a function f about $x = a$
(f) the Fourier series for a function f on $[-\pi, \pi]$

2. State each of the following:
(a) the nth-Term Divergence Test
(b) the Integral Test
(c) the Comparison Test
(d) the Limit Comparison Test
(e) the Ratio Test
(f) the Alternating Series Test
(g) Taylor's Formula with Remainder
(h) a necessary and sufficient condition for the Taylor series for a function f about $x = a$ to converge to $f(x)$
(i) conditions that guarantee the Fourier series for a function will converge to the function

3. (a) Suppose that $a_n \le a_{n+1} \le M$ for all n, where M is a constant. What can you conclude about convergence or divergence of the sequence $\{a_n\}$, and why?
(b) Suppose the power series $\sum a_n x^n$ converges when $x = -1$ and diverges when $x = 2$. State everything you can about where else the series converges and where else it diverges.
(c) Suppose that $a_n \ge 0$, $b_n \ge 0$, and $c_n \ge 0$ for all n, and that $\sum a_n$ converges and $\sum b_n$ diverges. State everything you can about convergence or divergence of $\sum c_n$ if for all n:
    (i) $c_n \le a_n$
    (ii) $a_n \le c_n \le b_n$
    (iii) $c_n \ge b_n$
(d) Let $f^{(n)}(x)$ exist for all n for $-R < x < R$, and

suppose $\sum_{n=0}^{\infty} a_n x^n$ converges to $f(x)$ for all x in $(-R, R)$. What can you conclude about a_n?
(e) Explain under what conditions on a and b the equation

$$\int_a^b \left(\sum_{n=0}^{\infty} a_n x^n \right) dx = \sum_{n=0}^{\infty} \int_a^b a_n x^n \, dx$$

is valid.

4. Fill in the blanks.
(a) The series $\sum_{n=1}^{\infty} a_n$ converges to the sum S if and only if its _____ converges to S.
(b) The p-series $\sum_{n=1}^{\infty} 1/n^p$ converges if _____ and diverges if _____.
(c) The geometric series $\sum_{n=1}^{\infty} ar^{n-1}$ converges to _____ if _____ and diverges if _____.
(d) A sequence is a function whose domain is _____.
(e) If the power series $\sum_{n=0}^{\infty} a_n x^n$ has positive radius of convergence R and its sum is $f(x)$, then $\sum_{n=1}^{\infty} na_n x^{n-1}$ converges to _____ for each x in the interval _____.

5. Indicate which of the following statements are true and which are false.
(a) If $\lim_{n \to \infty} a_n = 0$, then $\sum_{n=1}^{\infty} a_n$ converges.
(b) If $\sum_{n=1}^{\infty} a_n$ diverges, then $\lim_{n \to \infty} a_n \ne 0$.
(c) If $\lim_{n \to \infty} a_n \ne 0$, then $\sum_{n=1}^{\infty} a_n$ does not converge.
(d) If $\sum_{n=1}^{\infty} a_n$ is a series of nonnegative terms, and $a_1 + a_2 + a_3 + \cdots + a_n \le 10$ for all n, then $\sum_{n=1}^{\infty} a_n$ converges.
(e) If $\{a_n\}$ is a monotone decreasing sequence of positive numbers, then the sequence converges.
(f) If $\sum_{n=1}^{\infty} a_n$ is a series of positive terms and $a_{n+1}/a_n < 1$ for all n, then $\sum_{n=1}^{\infty} a_n$ converges.

APPLYING CALCULUS

1. When money is spent on goods and services, those that receive money also spend some of it. The people receiving some of the twice-spent money will spend some of that, and so on. Economists call this chain reaction the *multiplier effect*. In a hypothetical isolated community, the local government begins the process by spending $D. Suppose that each recipient of spent money spends 100 c% and saves 100 s% of the money that he or she receives. The values c and s are called the *marginal propensity to consume* and the *marginal propensity to save*, and, of course, $c + s = 1$.

   (a) Let S be the total spending that has been generated after n transactions. Determine an equation for S.

   (b) Show that $\lim_{n \to \infty} S = kD$, where $k = 1/s$. The number k is called the *multiplier*. What is the multiplier if the marginal propensity to consume is 0.8?
   Note: The Federal Government uses this principle to justify deficit spending. Banks use this principle to justify lending out a large percentage of the money that they receive in deposits.

2. A certain ball has the property that each time it falls from height h onto a hard level surface, it rebounds to a height rh, where $0 < r < 1$. Suppose the ball is dropped from an initial height of H meters.

   (a) Assuming that the ball continues to bounce indefinitely, determine the total distance that it travels.

   (b) Calculate the total time that the ball is traveling. (Hint: If the ball is dropped from a height h, then the distance s it travels in t seconds is $s = 4.9t^2$).

   (c) Suppose that each time the ball strikes the surface with velocity v, it rebounds with velocity kv, where $0 < k < 1$. How long will it take for the ball to come to rest?

3. Let $S(x)$ and $C(x)$ be defined by the power series

$$S(x) = \sum_{k=0}^{\infty} \frac{(-1)^k (\omega x)^{2k+1}}{(2k+1)!}$$

$$C(x) = \sum_{k=0}^{\infty} \frac{(-1)^k (\omega x)^{2k}}{(2k)!}$$

where ω is a constant.

   (a) Determine the domain of S and of C.

   (b) Show that

$$\int_0^x C(t)\,dt = \frac{1}{\omega} S(x) \qquad \text{and} \qquad C'(x) = -\omega S(x)$$

   (c) Show that S and C are solutions of the differential equation

$$y'' + \omega^2 y = 0$$

satisfying the initial conditions: $y(0) = 0, y'(0) = 1$ and $y(0) = 1, y'(0) = 0$, respectively.

   (d) Show that $S^2(x) + C^2(x) = 1$. (Hint: Put $U(x) = S^2(x) + C^2(x)$ and calculate U'.)

4. Consider the initial value problem

$$y' - 2y = e^x; \qquad y(0) = 0$$

Assume that the problem has a solution $y = y(x)$ that can be expressed as a power series expansion

$$y(x) = \sum_{n=0}^{\infty} a_n x^n = a_0 + a_1 x + a_2 x^2 \ldots$$

with radius of convergence $R > 0$.

(a) Express y' as a power series and calculate the power series expansion for $y' - 2y$.

(b) Since two power series are equal if and only if the coefficients of the corresponding powers of x are equal, equate the power series found in part (a) to the MacLaurin series for e^x and solve for the coefficients. The result will be a recursion formula that gives a_{n+1} in terms of a_n, $n = 0, 1, 2, \ldots$.

(c) Show that the initial condition $y(0) = 0$ implies $a_0 = 0$ and determine the solution of the initial value problem.

(d) Show that the power series found in (c) is equal to $y(x) = e^{2x} - e^x$.

5. What if you watch a TV showing a TV camera watching it?

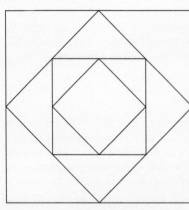

(a) The "nested" sequence of squares shown in the margin is formed as follows: join the midpoints of the sides of the outermost square S_1 to form the square S_2; join the midpoints of the sides of S_2 to form S_3; and so on.
Determine the sum of the areas of all the squares if the first square has area A.

(b) The "nested" sequence of circles and squares shown in the margin is formed as follows: Inscribe a square in the circle of radius r; inscribe a circle in the square; inscribe a square in the circle; and so on.
Determine the area of the shaded region.

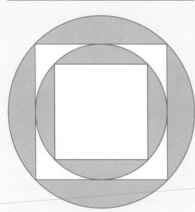

6. The curve defined parametrically by the equations

$$x = a \cos t, \qquad y = b \sin t, \qquad 0 < t < 2\pi$$

where a and b are positive numbers, is an ellipse whose center is at the origin and whose axes are on the coordinate axes. Assuming that $a > b$, the major axis is on the x-axis. The arc length of this ellipse is

$$L = \int_0^{2\pi} \sqrt{a^2 \sin^2 t + b^2 \cos^2 t} \, dt \qquad (*)$$

(a) Show that the formula $(*)$ can be simplified to

$$L = 4a \int_0^{\pi/2} \sqrt{1 - e^2 \cos^2 t} \, dt$$

where e is the eccentricity of the ellipse. This form is an elliptic integral; the integrand does not have an elementary antiderivative.

(b) Use the binomial series for $\sqrt{1 - x}$ to show that

$$L = 4a \int_0^{\pi/2} \left[1 - \frac{1}{2} e^2 \cos^2 t - \frac{1}{8} e^4 \cos^4 t - \frac{1}{16} e^6 \cos^6 t - \ldots \right] dt$$

(c) Show that the arc length of the ellipse is given by

$$L = 2\pi a \left[1 - \frac{1}{4} e^2 - \frac{3}{64} e^4 - \frac{5}{256} e^6 - \ldots \right]$$

by evaluating the integral in part (b). Hint: Integrate termwise using the identity

$$\int_0^{\pi/2} \cos^{2n} t \, dt = \frac{1 \cdot 3 \cdot 5 \cdots (2n-1)}{2 \cdot 4 \cdot 6 \cdots (2n)} \cdot \frac{\pi}{2}$$

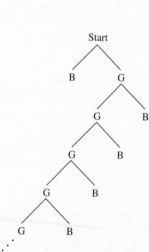

See Exercise 49 in Exercise Set 8.1. This formula gives the arc length of an ellipse in terms of its semimajor axis and its eccentricity. Note that this formula reduces to $2\pi a$, the circumference of a circle of radius a when $e = 0$.

(d) The orbit of the moon about the Earth is (almost) a perfect ellipse with the Earth at one focus. Assume that $a = 400,000$ km and $e = 0.055$. Use the first three terms of the series in (c) to determine the length of the moon's path around the earth.

7. In the accompanying figure is a region bounded by two circles of radius 1 that are tangent to one another and by a straight line which is tangent in both circles. A sequence of smaller circles, each having the largest possible radius, is inscribed in the region as shown. Show geometrically that the lengths of the diameters of these smaller circles are the terms of a convergent series whose sum is one. Then show that this series is

$$\sum_{n=1}^{\infty} \frac{1}{n(n+1)}.$$

8. Construct a sequence of circles with radii $r_1, r_2, r_3, \ldots,$ as follows. Inscribe a circle of radius $r_1 = 1$ in an equilateral triangle. Inscribe a circle of radius r_2, passing through the vertices of the triangle, in a square. Inscribe a circle of radius r_3, passing through the vertices of the square, in a regular pentagon. Continue in this way: inscribe a circle of radius r_{n-1}, passing through the vertices of a regular polygon of n sides, in a regular polygon of $n + 1$ sides.

(a) Sketch a diagram of what is happening.

(b) Show that $r_2 = r_1 \sec(\pi/3)$ and $r_3 = r_2 \sec(\pi/4)$. Find a general formula for r_n.

(c) Show that

$$\ln r_n = \ln r_1 + \ln \sec \frac{\pi}{3} + \ln \sec \frac{\pi}{4} + \cdots + \ln \sec \frac{\pi}{n+1}$$

(d) Does r_n have a limit as n tends to infinity? (Hint: Use the Limit Comparison Test with $\sum[\ln \sec(\pi/n)]$ and $\sum(1/n^2)$.)

9. An indicator of future population growth is the fertility rate, which is the average number of children a woman has during her childbearing years. If this rate exceeds the theoretical replacement level of 2 (it is slightly greater than 2 in the more developed countries and even higher in the less developed countries), the size of the population will continue to increase. One factor that affects a country's fertility rate is the type of family considered desirable. In each of the following assume that the probability of bearing male and female children is the same.

(a) Suppose each family will have children until they have a son. What is the fertility rate for the culture? (See the tree diagram.)

(b) Suppose each family will have children until they have at least one male and one female. Draw a tree diagram showing the possible outcomes. What is the fertility rate for this culture?

(c) Suppose each family has children until they have at least two males. Draw a tree diagram showing the possible outcomes. What is the fertility rate for this culture?

(d) A more general case is to suppose each family has children until they have at least m males and n females. Draw a tree diagram showing the possible outcomes. What is the fertility rate for this culture?

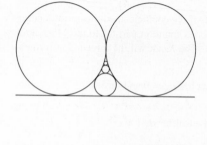

Constance McMillan Elson

What's the basic diference between multivariable calculus and vector calculus?

Some people use the terms interchangeably, but one could argue that multivariable calculus is the study of points, called vectors, in higher-dimensional spaces (dimension 2, 3, ..., n) and the study of functions whose inputs or outputs are vectors. Multivariable calculus deals with differentiating such functions and integrating them over flat regions in Euclidean space. It has applications in almost any area of human inquiry that is carried out quantitatively.

Vector calculus can include multivariable calculus, but it specifically refers to the extension of this calculus to curved spaces. Its principal mathematical objects are vector fields (functions whose inputs and outputs are vectors), special types of derivatives called the curl and divergence of these vector fields, and integrals of vector fields over curves and surfaces. In particular, vector calculus includes the study of Green's, Stokes's, and Gauss's Theorems, which are higher-dimensional analogs of the Fundamental Theorem of Calculus and relate various types of integrals. Vector calculus is the mathematical tool of greatest significance in the study of the basic forces of nature: electricity, magnetism, the nuclear forces, and gravity.

Are there different representations of the term vector?

There are three common ways to represent vectors: as algebraic objects called n-tuples, as points in n-dimensional Euclidean space, and, in the two- and three-dimensional cases, as arrows. Each representation has its advantages in specific situations, and it is important to be able to move conceptually among all three representations. Since an n-tuple is just an ordered collection of n real numbers, it is particularly easy to define the basic algebra of vectors in terms of n-tuples and to perform computations in this setting.

Equating vectors with points in space allows us to explore the geometry of multidimensional spaces, even in spaces of dimension 4 or higher. We do this by using vectors to describe the geometry of spaces we can visualize, i.e., regions in two- and three-dimensional Euclidean space. Then we use vector algebra to extend this to higher dimensions.

Finally, vectors can be interpreted as quantities having magnitude and direction; in $\mathbb{R}^2$ and $\mathbb{R}^3$ this means that we can represent them as arrows. These arows can be shifted at will (preserving magnitude and direction) without being altered as a vector; this property is one of the main reasons that vectors were developed and are used in physics.

The concepts of motion and force are fundamental in understanding the natural world. Since they are characterized by direction and magnitude, they are intrinsically vector quantities. When an object moves under the influence of two or more forces, its motion is related to the vector sum of the various forces. For instance, a canoe being paddled straight across a moving river will be carried downstream by the current. The actual course of the canoe is the vector sum of its still-water velocity and the velocity of the current. Similarly, an airplane flight path requires a pilot to set her course in an ocean of air. All the fancy navigational aids available to pilots and ship captains depend on the basic principal of vector addition.

Juggling the different representations of a vector can be tricky. It is important to be aware of what representation we are actually using at any given moment, but the interplay of these three points of view provides rich results.

Do vectors play a role in disciplines other than physics and engineering?

Definitely. Vectors can be used to organize information in any quantitative area. An investment officer of a bank or corporation, for instance, might manage a number of different stock portfolios, each of which can be represented as a vector. The components of the vector are the number of shares of the different stocks in the portfolio. If the prices of the stocks are summarized in a vector $\mathbf{P}$, the current value of the portfolio represented by the vector $\mathbf{X}$ can be computed using the dot product from vector algebra. One can track the value of the portfolio over time by simply updating the price vector and recomputing the dot product. Modern spreadsheet software depends heavily on this kind of vectorization of information.

Vectors have important applications in social sciences. You could use vectors to try to find connections between several different socioeconomic variables. Often statistical techniques are used to test the strength of the various relationships between variables. These multivariate statistical methods are based on the algebra and geometry of vectors.

For example, say an archeologist is excavating a site and finds a huge number of different pottery fragments all jumbled together. He might conjecture that at least two different cultures have occupied the site and hopes to demonstrate this by means of the pottery fragments (shards). By assigning three numbers to each fragment, giving its thickness, granularity, and color, the archaeologist maps each shard into a vector. If these vectors are then interpreted as points in space and are found to separate into two (or more) distinct clusters, the archaeologist has found evidence supporting his conjecture. If there were several clusters of vectors, the geometric relationship between clusters might indicate a migration of skills or materials from one culture to another.

Are you concerned that students sometimes view calculus as just a set of rules for differentiating and integrating functions?

Describing calculus as a set of rules for differentiation and integration is about as informative as saying that a rock concert is loud. Both descriptions miss the essence of the event. It's better to develop the idea that a certain mathematical process is repeated indefinitely, with some controlled change taking place, and the resulting numerical or algebraic values converge to a limiting value. This idea underlies all of calculus. What gives calculus its practical utility is that this involved procedure can usually be reduced to certain computational rules for differentiation and integration. However, what gives calculus its power is that the process itself can be applied in very general circumstances. To make full use of this power requires a deep understanding of the idea of limit. For most students this kind of understanding is achieved only in an advanced analysis course, and so to many students in engineering and the sciences, limits seem abstruse and irrelevant because "you can get the answers without them." Using graphics and visualization, students can develop an intuitive and basic grasp of the idea of a limit in several dimensions, together with some algebraic results for computing limits. This gives a foundation that allows a student to understand other concepts in vector calculus quite rigorously. This is important because the interesting problems that lead to new results require thinking about things from first principles, rather than just applying an existing integration formula.

How well do most students make the transition from derivatives to partial derivatives and eventually to partial differential equations?

For functions of several variables, students usually understand partial derivatives very well, but they are often foggy about what a derivative is in this context. This is because a partial derivative is a number that represents a slope in a particular direction; thus it seems similar to the derivative that students are already familiar with from one-variable calculus. One way to visualize a partial derivative is to use the analogy between a function of several variables and a machine with several input hoppers and one output hopper. What happens to the output when we tweak one of the inputs by changing the quantity that we add to that particular hopper? Since that is the only input we are changing, the partial derivative with respect to that variable contains the answer to that question. It tells us the instantaneous rate at which the function output changes as one input variable changes.

A derivative of a multivariable function is a more abstract entity, but it's not hard to understand why it is the correct generalization of the single-variable derivative. We define the derivative at a point as a vector, or as a matrix, which makes a certain difference quotient go to zero in a controlled way. The derivative is composed of partial derivatives that makes it easy to compute, but the derivative is much richer than any single partial derivative and it generates a wealth of information about the function. If the derivative is not zero, we can find the rate of change of the function for any combination of changes in the inputs, the maximum rate of change of the function at that point, the direction (i.e., the combination of inputs) that produces the maximum rate of change, and the direction that produces no change in the function. Functions of several variables are mathematically rich objects and their derivatives reflect this.

A partial differential equation is an equation involving one or more partial derivatives and solving it involves finding a function whose partials satisfy the equation. For instance, three of the most famous partial differential equations—Laplace's equation, the heat equation, and the wave equation—involve only second partials. These equations are elegantly simple to write down, but efforts to understand their solutions have led to the development of an amazingly large part of modern mathematics.

The problems that a student encounters may take a few hours to solve or at most, for a project, a few weeks. These three equations have engaged some of the best mathematical minds for the past two centuries and they're still yielding interesting results. The creation of mathematics is a continuing endeavor.

How do you expect students' attitudes and understanding of mathematics to develop while taking this course?

Ideally I hope three modes of learning will happen simultaneously. First, a student should develop proficiency in three-dimensional thinking and see vector functions as natural extensions of scalar functions. The goal is to make a student as comfortable and intuitively capable of solving problems in several dimensions as in one dimension. Second, the multidimensional calculus should reinforce and enrich a student's understanding of one-dimensional calculus. I try to stress the differences in moving from one to several variables, while revealing the unity of the concepts underlying the calculus. Good notation can facilitate this, but it is even more productive to state explicitly that the concepts of function, limit, derivative, and integral are at the core of calculus, and to explore in what way they are independent of dimension. The third mode is less tangible, and often neglected, but is the hallmark of a first-rate course. It's really about mathematical thinking. A student should progress from the perception that "mathematics is just a way to solve problems" to a true appreciation of the power of mathematics to find similarity and unity in seemingly different objects. Multivariable calculus has a dual geometric and algebraic approach that develops an appreciation for the uses and power of mathematical abstraction, without a heavy emphasis on the idea of "proof." I think multivariable calculus is the best possible foundation for a subsequent study of modern mathematics. It provides wonderful motivation for the study of linear algebra, differential geometry, topology, integration theory, and partial differential equations. And for a student in any other field that uses mathematics, the subject provides an opportunity to demonstrate how mathematics builds on known structures to explore unknown areas.

Constance McMillan Elson is professor of mathematics at Ithaca College. She has conducted mathematical research at NASA; M.I.T.; University of Wisconsin, Madison; and the National Center for Atmospheric Research. Professor Elson has a strong interest in teaching and promotes the use of computer algebra systems and other interactive media, particularly in undergraduate courses beyond calculus. She received her B.S. in mathematics from Stanford University and her Ph.D. in mathematics from the University of California at San Diego.

11

VECTORS IN TWO AND THREE DIMENSIONS

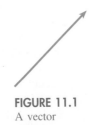

FIGURE 11.1
A vector

Certain physical quantities have both *magnitude* and *direction*; examples are force, velocity, acceleration, and displacement of a moving particle. A convenient way to represent such quantities is with a directed line segment, such as the one in Figure 11.1. The length of the segment, to some scale, represents the magnitude of the quantity in question, and the direction is indicated by the inclination of the segment and by the arrowhead. Such a directed line segment is called a **vector**. For example, the vector in Figure 11.1 might represent a wind velocity of 20 mph blowing in a northeasterly direction. The length would then be taken as 20 units, to some convenient scale. Any quantity that has both magnitude and direction can be represented in this way and is therefore called a *vector quantity*.

Vector quantities are different from measures of area, mass, time, and distance, which can be adequately described by a single number. These are called *scalar quantities* (since they are measured according to some scale), and the numbers used to measure them are called **scalars**. For our purposes, then, scalars are just real numbers.

The subject now called vector analysis was developed in the latter part of the nineteenth century by the American physicist and mathematician Josiah Willard Gibbs (1839–1903) and the English engineer Oliver Heaviside (1850–1925), working independently. Many of the ideas came earlier, however, especially from the Irish mathematician William Rowan Hamilton (1805–1865) in his work on *quaternions*, and the Scottish physicist James Clerk Maxwell (1831–1879), who used some of Hamilton's ideas in his study of electromagnetic field theory. So the subject is strongly grounded in the physical sciences and engineering and is still an important tool in these fields. The applications have expanded greatly now, even to economics and some of the other social sciences. Moreover, vectors have contributed significantly to the continuing development of mathematics itself.

In early work with vectors, their geometric properties were dominant. As we shall see, though, the advantages of vectors can be realized fully only when their algebraic properties are used in conjunction with their geometric ones. In the next section we explore the geometric nature of vectors in the plane and then formulate vectors and their properties in algebraic terms.

Distance is a scalar quantity, but a quantity with both magnitude (distance) *and* direction is a vector quantity.

11.1 VECTORS IN THE PLANE

FIGURE 11.2
Equivalent vectors

FIGURE 11.3

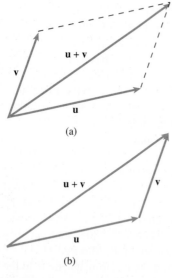

FIGURE 11.4

Geometric Vectors

Typically, boldface letters such as **u**, **v**, and **w** are used to designate vectors. Throughout this section we consider vectors that lie in a plane, although most of the results have natural extensions to three (or more) dimensions. Two vectors **u** and **v** are said to be **equivalent** if they have the same magnitude and direction, and in this case we write **u** = **v**. Three equivalent vectors are illustrated in Figure 11.2. We do not distinguish between equivalent vectors, so that in effect we can shift a vector from one location to another as long as its original magnitude and direction are retained. Because of this freedom of movement, we say that we are working within a system of *free* vectors.

Suppose, as in Figure 11.3, a vector extends from a point P to a point Q. When we wish to emphasize this fact we use the notation $\overrightarrow{PQ}$ to designate the vector. The point P is called the **initial point** and Q the **terminal point**. Sometimes we also use "tail" and "tip" instead of initial point and terminal point, respectively.

Vector Addition

Two nonparallel vectors are **added** according to the **parallelogram law**, illustrated in Figure 11.4(a). The vectors are drawn with a common initial point, and a parallelogram is constructed with **u** and **v** as adjacent sides. The vector **u** + **v** is then defined as the vector along the diagonal from the common initial point to the opposite vertex. An alternative method is to place the initial point of **v** at the terminal point of **u**. Then **u** + **v** is the vector shown in Figure 11.4(b), drawn from the initial point of **u** to the terminal point of **v**. You should convince yourself that the triangle in part (b) is just the lower half of the parallelogram in part (a). This second method is sometimes called the "tail to tip" method of adding. If **u** and **v** are parallel vectors, then the parallelogram of part (a) is degenerate. The tail to tip method still works, however.

From our definition of addition it is easy to see that

$$\mathbf{u} + \mathbf{v} = \mathbf{v} + \mathbf{u}$$

In other words, vector addition is commutative. It is also associative; that is,

$$\mathbf{u} + (\mathbf{v} + \mathbf{w}) = (\mathbf{u} + \mathbf{v}) + \mathbf{w}$$

You will be asked in the exercises to give a geometric argument for this property.

Vector addition is consistent with observed results. For example, if **u** and **v** represent forces acting on an object, then the net effect is **u** + **v**; that is, the two individual forces **u** and **v** could be replaced by the force **u** + **v**, and the effect would be the same. In this case we call **u** + **v** the **resultant** of **u** and **v**. Similarly, if **u** is a vector representing the indicated velocity of an airplane and **v** is the wind velocity vector, then the true velocity of the airplane relative to the ground is **u** + **v**.

It is convenient to introduce the notion of the **zero vector**, denoted by **0**, with a magnitude of 0 and assigned no direction. We may think of the zero vector as a single point. If **v** is a nonzero vector, then −**v** is the vector that has the same length as **v** but with a direction opposite to that of **v**. We now define

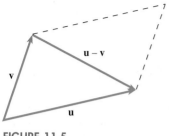

subtraction of vectors by

$$\mathbf{u} - \mathbf{v} = \mathbf{u} + (-\mathbf{v})$$

Thus, $\mathbf{u} - \mathbf{v}$ is the vector that when added to $\mathbf{v}$ gives $\mathbf{u}$. This definition is illustrated in Figure 11.5. Notice that when $\mathbf{u}$ and $\mathbf{v}$ are drawn with the same initial point, $\mathbf{u} - \mathbf{v}$ is the vector *from the tip of* $\mathbf{v}$ *to the tip of* $\mathbf{u}$. Notice also that when we construct the parallelogram with $\mathbf{u}$ and $\mathbf{v}$ as adjacent sides, $\mathbf{u} - \mathbf{v}$ is directed along the diagonal from the tip of $\mathbf{v}$ to the tip of $\mathbf{u}$, in contrast to $\mathbf{u} + \mathbf{v}$, which is directed along the other diagonal. From this definition we see that, as we would expect,

FIGURE 11.5

$$\mathbf{v} - \mathbf{v} = \mathbf{0}$$

Scalar Multiplication

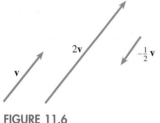

FIGURE 11.6

Vectors can be multiplied by scalars as follows. If $k > 0$ and $\mathbf{v}$ is a nonzero vector, then $k\mathbf{v}$ is a vector that has the same direction as $\mathbf{v}$ and magnitude k times the magnitude of $\mathbf{v}$. If $k < 0$, then $k\mathbf{v}$ has direction opposite to that of $\mathbf{v}$ and magnitude $|k|$ times the magnitude of $\mathbf{v}$. If $k = 0$, we define $k\mathbf{v}$ as the zero vector, and if $\mathbf{v} = \mathbf{0}$, then $k\mathbf{v} = \mathbf{0}$ for all scalars k. In Figure 11.6, we depict a vector $\mathbf{v}$, along with the vectors $2\mathbf{v}$ and $-\frac{1}{2}\mathbf{v}$.

Algebraic Vectors

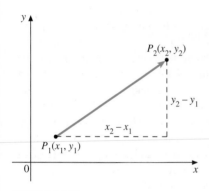

FIGURE 11.7

We can gain further insight into the properties of vectors by introducing a rectangular coordinate system. Suppose that a vector $\mathbf{v} = \overrightarrow{P_1 P_2}$, where the coordinates of P_1 and P_2 are (x_1, y_1) and (x_2, y_2), respectively. As shown in Figure 11.7, the horizontal displacement from P_1 to P_2 is $x_2 - x_1$ and the vertical displacement is $y_2 - y_1$. We call $x_2 - x_1$ the **horizontal component** (or x component) and $y_2 - y_1$ the **vertical component** (or y component) of $\mathbf{v}$. For example, if the coordinates of P_1 are $(3, 2)$ and those of P_2 are $(7, 5)$, then the horizontal component of $\mathbf{v}$ is $7 - 3 = 4$ and the vertical component is $5 - 2 = 3$. So every vector has a unique pair of components. Conversely, if we are given a pair of components, then these uniquely determine the collection of equivalent vectors that have these components. For example, given the x component 3 and y component 2, we can determine all vectors that have these components. The simplest of these is the one with initial point at the origin and terminal point at $(3, 2)$. Since we do not distinguish between equivalent vectors, we can in effect say that a vector is uniquely determined by its components.

This identification of a vector with its components enables us to look at vectors in a new way. We use the symbol $\langle a, b \rangle$ to indicate a vector with x component a and y component b, and we refer to this ordered pair of numbers as a vector. When we wish to distinguish between vectors as directed line segments and vectors as ordered pairs, we say the former is a *geometric vector* and the latter an *algebraic vector*. By the preceding discussion, given a geometric vector $\mathbf{v}$, we can determine the corresponding algebraic vector $\langle a, b \rangle$ and conversely. Because of this correspondence we write $\mathbf{v} = \langle a, b \rangle$. Any geometric vector corresponding to $\langle a, b \rangle$ is called a **geometric representative** of $\langle a, b \rangle$. The simplest geometric representative of a vector $\langle a, b \rangle$ is the vector from the origin to the point (a, b). We call this geometric representative the **position vector** of (a, b). (See Figure 11.8.)

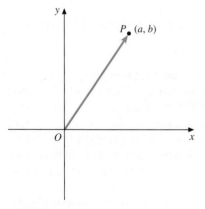

FIGURE 11.8

Position vector of P is $\overrightarrow{OP}$, O the origin $(0, 0)$

In summary, we have the following:

---

1. If $P_1 = (x_1, y_1)$ and $P_2 = (x_2, y_2)$, then $\overrightarrow{P_1 P_2} = \langle x_2 - x_1, y_2 - y_1 \rangle$.
2. If $\mathbf{v} = \langle a, b \rangle$ and the initial point of $\mathbf{v}$ is $P_1 = (x_1, y_1)$, then the terminal point is $P_2 = (x_1 + a, y_1 + b)$. In particular, if P_1 is the origin, then $P_2 = (a, b)$, and $\overrightarrow{P_1 P_2}$ is the position vector of P_2.

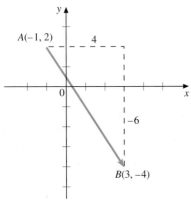

FIGURE 11.9

EXAMPLE 11.1

(a) Express the vector $\overrightarrow{AB}$ in algebraic form, where $A = (-1, 2)$ and $B = (3, -4)$. Draw the geometric vector.

(b) Draw the geometric representative of the vector $\langle -2, 3 \rangle$ whose initial point is $(4, -1)$. What is its terminal point?

Solution

(a) $\overrightarrow{AB} = \langle 3 - (-1), -4 - 2 \rangle = \langle 4, -6 \rangle$
    We show the vector geometrically in Figure 11.9.

(b) Beginning at $(4, -1)$, we go 2 units to the left and 3 units up, giving the terminal point $(2, 2)$, as shown in Figure 11.10. ∎

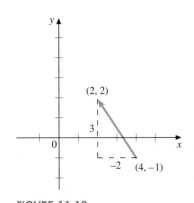

FIGURE 11.10

Properties of Vectors

We can now state properties of vectors in terms of their algebraic representations. First, we give a definition of addition and multiplication by a scalar. You will be asked in the exercises to show that these are consistent with the corresponding geometric definitions.

Definition 11.1
Two-Dimensional Vector Space

For ordered pairs $\langle a, b \rangle$ and $\langle c, d \rangle$ of real numbers,

1. $\langle a, b \rangle = \langle c, d \rangle$ if and only if $a = c$ and $b = d$
2. $\langle a, b \rangle + \langle c, d \rangle = \langle a + c, b + d \rangle$
3. $k \langle a, b \rangle = \langle ka, kb \rangle$ for any scalar k

The set of all such ordered pairs of real numbers with the definitions of equality, addition, and multiplication by a scalar given by Equations 1, 2, and 3 is called a **vector space of dimension two**, and each ordered pair in this set is called a **two-dimensional vector**.

THEOREM 11.1

If $\mathbf{u} = \langle a, b \rangle$, $\mathbf{v} = \langle c, d \rangle$, and $\mathbf{w} = \langle e, f \rangle$ are arbitrary vectors, then

1. $\mathbf{u} + \mathbf{v} = \mathbf{v} + \mathbf{u}$
2. $\mathbf{u} + (\mathbf{v} + \mathbf{w}) = (\mathbf{u} + \mathbf{v}) + \mathbf{w}$

and for any scalars k and l,

3. $k(\mathbf{u} + \mathbf{v}) = k\mathbf{u} + k\mathbf{v}$
4. $(k + l)\mathbf{u} = k\mathbf{u} + l\mathbf{u}$
5. $k(l\mathbf{u}) = (kl)\mathbf{u}$

The proof will be called for in the exercises.

Definition 11.2
The Zero Vector, the
Negative of a Vector, and
Subtraction

1. The element $\langle 0, 0 \rangle$ is called the **zero vector** and is denoted by $\mathbf{0}$.
2. If $\mathbf{u} = \langle a, b \rangle$ is any vector, then $-\mathbf{u} = \langle -a, -b \rangle$.
3. If $\mathbf{u} = \langle a, b \rangle$ and $\mathbf{v} = \langle c, d \rangle$ are arbitrary vectors, then

$$\mathbf{u} - \mathbf{v} = \mathbf{u} + (-\mathbf{v})$$

THEOREM 11.2

For any vector $\mathbf{u} = \langle a, b \rangle$,

1. $\mathbf{u} + \mathbf{0} = \mathbf{u}$
2. $\mathbf{u} + (-\mathbf{u}) = \mathbf{0}$
3. $1\mathbf{u} = \mathbf{u}$
4. $(-1)\mathbf{u} = -\mathbf{u}$
5. $0\mathbf{u} = \mathbf{0}$
6. $k\mathbf{0} = \mathbf{0}$ for all scalars k.

We again call for the proof in the exercises.

REMARK ——————
■ Because the addition of $\mathbf{0}$ to a vector leaves that vector unchanged, $\mathbf{0}$ is the *additive identity*. Also, because $-\mathbf{u}$ added to $\mathbf{u}$ gives $\mathbf{0}$, $-\mathbf{u}$ is the *additive inverse* of $\mathbf{u}$.

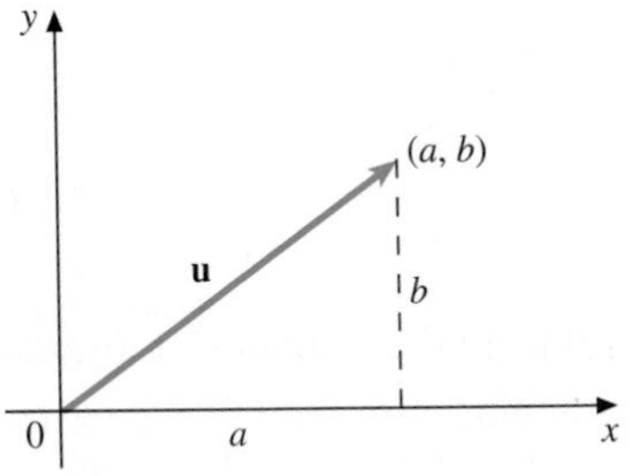

The Magnitude of a Vector

FIGURE 11.11
Length of $\mathbf{u} = \sqrt{a^2 + b^2}$

For a vector $\mathbf{u} = \langle a, b \rangle$, its geometric counterpart can be represented by the position vector of the point (a, b), as in Figure 11.11. The length of this geometric vector, by the Pythagorean Theorem, is $\sqrt{a^2 + b^2}$, which leads to the following definition.

Definition 11.3
The Magnitude of a Vector

Let $\mathbf{u} = \langle a, b \rangle$ be any vector. The **magnitude** (or *length*) of $\mathbf{u}$, denoted by $|\mathbf{u}|$, is defined by

$$|\mathbf{u}| = \sqrt{a^2 + b^2} \tag{11.1}$$

REMARK ———

■ It is reasonable to use the same symbol to designate the magnitude of a vector as the one used to denote the absolute value of a real number. If x is a real number, then $|x|$ can be interpreted geometrically as the distance from 0 to x on a number line. Similarly, when a vector $\mathbf{u}$ is interpreted geometrically with initial point at the origin, $|\mathbf{u}|$ is the distance from the origin to the terminal point of $\mathbf{u}$.

The components of a vector determine its magnitude, as Definition 11.3 shows. They also determine the *direction* of the vector. For if $\mathbf{u} = \langle a, b \rangle$, the direction of $\mathbf{u}$ is the same as that of the position vector shown in Figure 11.11, directed from $(0, 0)$ toward the point (a, b).

THEOREM 11.3

Let $\mathbf{u} = \langle a, b \rangle$ and $\mathbf{v} = \langle c, d \rangle$ be any two vectors, and let k be any scalar. Then

1. $|\mathbf{u}| \geq 0$ and $|\mathbf{u}| = 0$ if and only if $\mathbf{u} = \mathbf{0}$
2. $|-\mathbf{u}| = |\mathbf{u}|$
3. $|k\mathbf{u}| = |k|\,|\mathbf{u}|$
4. $|\mathbf{u} + \mathbf{v}| \leq |\mathbf{u}| + |\mathbf{v}|$ Triangle Inequality

Proofs of Properties 1, 2, and 3 are called for in the exercises. We will give an algebraic proof of Property 4 in the next section, but its validity is evident geometrically, as an examination of Figure 11.12 shows. The inequality simply reflects the fact that the length of one side of a triangle is less than the sum of the lengths of the other two sides (which explains the name "triangle inequality"). This inequality also reflects the fact that the shortest distance between two points is a straight line. You should think about the circumstances in which equality occurs.

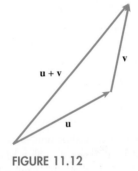

FIGURE 11.12

Unit Vectors and Basis Vectors

A **unit vector** is a vector with magnitude equal to 1. For example, $\langle 1, 0 \rangle$, $\langle 0, 1 \rangle$, $\langle \frac{3}{5}, \frac{4}{5} \rangle$, and $\left\langle \frac{\sqrt{2}}{2}, -\frac{\sqrt{2}}{2} \right\rangle$ are unit vectors. For any nonzero vector $\mathbf{v}$, the vector

$$\left(\frac{1}{|\mathbf{v}|} \right) \mathbf{v} = \frac{\mathbf{v}}{|\mathbf{v}|}$$

is a unit vector in the same direction as **v**. The fact that its length is 1 can be seen by

$$\left|\frac{\mathbf{v}}{|\mathbf{v}|}\right| = \left(\frac{1}{|\mathbf{v}|}\right)|\mathbf{v}| = 1$$

In words, to make a nonzero vector **v** into a unit vector, divide **v** by its own length.

EXAMPLE 11.2   Find a unit vector in the direction from $P_1(3, 2)$ toward $P_2(5, -2)$.

Solution   Let $\mathbf{v} = \overrightarrow{P_1 P_2}$. Then $\mathbf{v} = \langle 5-3, -2-2 \rangle = \langle 2, -4 \rangle$. So the desired unit vector is

$$\frac{\mathbf{v}}{|\mathbf{v}|} = \frac{\langle 2, -4 \rangle}{\sqrt{20}} = \frac{\langle 2, -4 \rangle}{2\sqrt{5}} = \left\langle \frac{1}{\sqrt{5}}, -\frac{2}{\sqrt{5}} \right\rangle \qquad \blacksquare$$

If **v** is a nonzero vector, we know that $\mathbf{v}/|\mathbf{v}|$ is a unit vector, so for any scalar k, $k\mathbf{v}/|\mathbf{v}|$ is a vector of magnitude $|k|$ that has the same direction as **v** when $k > 0$ and the opposite direction when $k < 0$. This fact enables us to find a vector of a specified length in the direction of a given vector or in the opposite direction.

EXAMPLE 11.3   Find a vector in the same direction as $\langle -1, 2 \rangle$ that has magnitude 10.

Solution   Let $\mathbf{v} = \langle -1, 2 \rangle$. The desired vector is

$$10\frac{\mathbf{v}}{|\mathbf{v}|} = 10\left\langle -\frac{1}{\sqrt{5}}, \frac{2}{\sqrt{5}} \right\rangle = \langle -2\sqrt{5}, 4\sqrt{5} \rangle \qquad \blacksquare$$

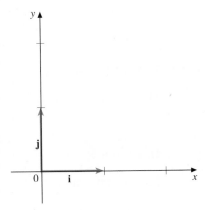

FIGURE 11.13

The two unit vectors $\langle 1, 0 \rangle$ and $\langle 0, 1 \rangle$ are of particular importance. They are given the special names

$$\mathbf{i} = \langle 1, 0 \rangle \qquad \text{and} \qquad \mathbf{j} = \langle 0, 1 \rangle$$

So **i** is a unit vector in the positive x direction and **j** is a unit vector in the positive y direction. (See Figure 11.13.)

Suppose $\mathbf{u} = \langle a, b \rangle$ is an arbitrary vector. Then we can write

$$\mathbf{u} = \langle a, 0 \rangle + \langle 0, b \rangle = a\langle 1, 0 \rangle + b\langle 0, 1 \rangle$$

or

$$\langle a, b \rangle = a\mathbf{i} + b\mathbf{j} \qquad (11.2)$$

The expression $a\mathbf{i} + b\mathbf{j}$ is called a **linear combination** of **i** and **j**. So every two-dimensional vector is uniquely expressible as a linear combination of **i** and **j**. Because of this property the vectors **i** and **j** constitute what is called a **basis** for two-dimensional vector space. As Exercise 49 in Exercise Set 11.1 shows, any two nonzero vectors that are not parallel also form a basis for this vector space, but **i** and **j** provide the simplest basis.

EXAMPLE 11.4   Let $P_1 = (7, -4)$ and $P_2 = (-3, 1)$. Express $\overrightarrow{P_1 P_2}$ as a linear combination of **i** and **j**.

Solution $\overrightarrow{P_1 P_2} = \langle -3-7, 1+4 \rangle = \langle -10, 5 \rangle = -10\mathbf{i} + 5\mathbf{j} \qquad \blacksquare$

Because of Equation 11.2, we can use the representation $a\mathbf{i} + b\mathbf{j}$ as an alternative to $\langle a, b \rangle$. In the future both representations will be used. Thus, for example, we may speak of the vector $2\mathbf{i} - 3\mathbf{j}$, which we will understand to mean the same thing as the vector $\langle 2, -3 \rangle$. Using this alternative representation, we have, in particular,

$$|a\mathbf{i} + b\mathbf{j}| = \sqrt{a^2 + b^2}$$

Parallel Vectors

In keeping with the geometric relationship between $\mathbf{u}$ and $k\mathbf{u}$, we have the following definition.

Definition 11.4 Parallel Vectors	Two nonzero vectors $\mathbf{u}$ and $\mathbf{v}$ are said to be **parallel** if there exists a nonzero scalar k such that $\mathbf{v} = k\mathbf{u}$. We also say that $\mathbf{0}$ is parallel to every vector.

Exercise Set 11.1

Exercises 1–12 refer to the vectors $\mathbf{u} = \langle -2, 3 \rangle$, $\mathbf{v} = \langle 4, 2 \rangle$, and $\mathbf{w} = \langle -1, -2 \rangle$. In each case, give the result as an algebraic vector. Also give a geometric construction illustrating the given operations.

1. $\mathbf{u} + \mathbf{v}$ **2.** $\mathbf{v} + \mathbf{w}$

3. $\mathbf{u} + \mathbf{w}$ **4.** $\mathbf{u} + \mathbf{v} + \mathbf{w}$

5. $\mathbf{u} - \mathbf{v}$ **6.** $\mathbf{v} - \mathbf{w}$

7. $\mathbf{w} - \mathbf{u}$ **8.** $2\mathbf{u} + \frac{1}{2}\mathbf{v}$

9. $\mathbf{u} - 2\mathbf{w}$ **10.** $\mathbf{u} + \frac{3}{2}\mathbf{v} - \mathbf{w}$

11. $-2\mathbf{u} + \mathbf{v} - 3\mathbf{w}$ **12.** $2\mathbf{u} - \frac{1}{2}\mathbf{v} + 3\mathbf{w}$

In Exercises 13–16, find the algebraic vector corresponding to $\overrightarrow{P_1 P_2}$.

13. $P_1 = (3, 4)$, $P_2 = (-1, 2)$

14. $P_1 = (-4, -2)$, $P_2 = (3, -1)$

15. $P_1 = (0, 4)$, $P_2 = (-3, 0)$

16. $P_1 = (7, -3)$, $P_2 = (-1, -8)$

In Exercises 17–20, draw the vector $\overrightarrow{P_1 P_2}$ that corresponds to the given algebraic vector and the given initial point P_1. Determine the coordinates of P_2.

17. $\langle 3, -2 \rangle$; $P_1 = (0, 0)$ **18.** $\langle -2, 4 \rangle$; $P_1 = (1, 2)$

19. $\langle 0, 3 \rangle$; $P_1 = (-2, -3)$ **20.** $\langle -3, -4 \rangle$; $P_1 = (4, 2)$

In Exercises 21–26, find $|\mathbf{v}|$.

21. $\mathbf{v} = \langle 3, 4 \rangle$ **22.** $\mathbf{v} = \langle -8, 6 \rangle$

23. $\mathbf{v} = 8\mathbf{i} + 15\mathbf{j}$ **24.** $\mathbf{v} = -12\mathbf{i} - 5\mathbf{j}$

25. $\mathbf{v} = 2\mathbf{i} + \mathbf{j}$ **26.** $\mathbf{v} = 4\mathbf{i} - 6\mathbf{j}$

27. If $\mathbf{u} = -3\mathbf{i} + \mathbf{j}$ and $\mathbf{v} = 2\mathbf{i} - 3\mathbf{j}$, find each of the following:
   (a) $|\mathbf{u} + \mathbf{v}|$ (b) $|\mathbf{u} - \mathbf{v}|$
   (c) $|2\mathbf{u} + 3\mathbf{v}|$ (d) $|3\mathbf{u} - 2\mathbf{v}|$

In Exercises 28–31, find a unit vector in the direction of $\mathbf{v}$.

28. $\mathbf{v} = 3\mathbf{i} - 4\mathbf{j}$ **29.** $\mathbf{v} = 5\mathbf{i} + 12\mathbf{j}$

30. $\mathbf{v} = \langle -4, 8 \rangle$ **31.** $\mathbf{v} = \langle -2, -3 \rangle$

In Exercises 32–37, find a vector $\mathbf{w}$ that is in the direction of the given vector $\mathbf{v}$, with the specified magnitude.

32. $\mathbf{v} = \langle -4, 3 \rangle$; $|\mathbf{w}| = 10$ **33.** $\mathbf{v} = \langle 2, -4 \rangle$; $|\mathbf{w}| = 10$

34. $\mathbf{v} = \mathbf{i} + \mathbf{j}$; $|\mathbf{w}| = 2$ **35.** $\mathbf{v} = 6\mathbf{i} + 8\mathbf{j}$; $|\mathbf{w}| = 4$

36. $\mathbf{v} = 7\mathbf{i} - 24\mathbf{j}$; $|\mathbf{w}| = 5$ **37.** $\mathbf{v} = 3\mathbf{i} - 6\mathbf{j}$; $|\mathbf{w}| = 15$

38. In the accompanying figure, $\mathbf{F}_1$ and $\mathbf{F}_2$ are forces acting on the object as shown. If $|\mathbf{F}_1| = 80$ lb, $|\mathbf{F}_2| = 60$ lb, $\alpha = 25°$, and $\beta = 115°$, find the magnitude and direction of the resultant both geometrically and by using trigonometry. (*Hint:* Use the Pythagorean Theorem.)

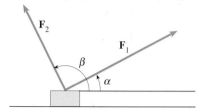

39. Repeat Exercise 38 if $|\mathbf{F}_1| = 20$ N, $|\mathbf{F}_2| = 30$ N, $\alpha = 10°$, and $\beta = 70°$. (*Hint:* For the trigonometric solution use the law of cosines.)

40. In air navigation, direction is given by the angle measured clockwise from north (this angle is called the *heading*). If an airplane is flying at an indicated heading of $120°$ at a speed of 300 kph, and a wind of 50 kph is blowing from $210°$, find the actual speed and direction of the airplane (relative to the ground). Do this geometrically and also by using trigonometry.

41. Repeat Exercise 40 for an airplane flying at an indicated heading of $230°$ at 260 kph with a 60 kph wind blowing from $110°$.

42. Show by a geometric argument that vector addition is associative; that is, $\mathbf{u} + (\mathbf{v} + \mathbf{w}) = (\mathbf{u} + \mathbf{v}) + \mathbf{w}$.

43. Show that Definition 11.1 is consistent with the corresponding geometric definitions of equivalence of vectors, addition of vectors, and multiplication of a vector by a scalar.

44. Prove Theorem 11.1.

45. Prove Theorem 11.2.

46. Prove Parts 1, 2, and 3 of Theorem 11.3.

47. Let $\mathbf{u} = \langle 1, 2 \rangle$, $\mathbf{v} = \langle -2, 1 \rangle$, and $\mathbf{w} = \langle 4, 3 \rangle$. Find scalars a and b such that $\mathbf{w} = a\mathbf{u} + b\mathbf{v}$.

48. Prove that if $\mathbf{u} = \langle u_1, u_2 \rangle$ and $\mathbf{v} = \langle v_1, v_2 \rangle$ are nonzero vectors, then they are parallel if and only if $u_1 v_2 - u_2 v_1 = 0$.

49. Prove that if $\mathbf{u} = \langle u_1, u_2 \rangle$ and $\mathbf{v} = \langle v_1, v_2 \rangle$ are nonzero vectors with $\mathbf{u} \neq k\mathbf{v}$ for all scalars k, and $\mathbf{w} = \langle w_1, w_2 \rangle$ is any other vector, then there exist scalars a and b such that $\mathbf{w} = a\mathbf{u} + b\mathbf{v}$.

50. Prove that if $\mathbf{u} = k\mathbf{v}$, with $k \geq 0$, then

$$|\mathbf{u} + \mathbf{v}| = |\mathbf{u}| + |\mathbf{v}|$$

Give a geometric argument to show that if $\mathbf{u}$ and $\mathbf{v}$ are not related in this way, $|\mathbf{u} + \mathbf{v}| < |\mathbf{u}| + |\mathbf{v}|$.

51. Use vectors to show that the diagonals of a parallelogram bisect each other.

52. Let P_1, P_2, P_3, and P_4 be any four points in the plane. Show both geometrically and algebraically that $\overrightarrow{P_1 P_2} + \overrightarrow{P_2 P_3} + \overrightarrow{P_3 P_4} + \overrightarrow{P_4 P_1} = \mathbf{0}$.

53. Use vectors to show that the line segments joining consecutive midpoints of the sides of an arbitrary quadrilateral form a parallelogram.

54. Three forces of magnitude $|\mathbf{F}_1| = 30$ lb, $|\mathbf{F}_2| = 45$ lb, and $|\mathbf{F}_3| = 56$ lb are acting on an object as shown in the figure. Find the magnitude of the resultant and its angle from $\mathbf{F}_1$ both geometrically and by using trigonometry. (See the hint for Exercise 39.)

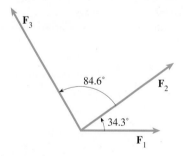

55. A pilot is to fly from A to B, 350 km due north of A, and then return to A. There is a wind blowing from $310°$ at 55 kph. If the average air speed of the plane is 210 kph, find the heading (see Exercise 40) the pilot should take on each part of the trip. What will be the total flying time?

11.2 THE DOT PRODUCT

In the previous section we considered addition and subtraction of vectors and multiplication by a scalar, but we have not yet considered the product of two vectors. We introduce one type of product now and later consider a second type, applicable to three-dimensional vectors only.

Definition 11.5 The Dot Product	Let $\mathbf{u} = \langle u_1, u_2 \rangle$ and $\mathbf{v} = \langle v_1, v_2 \rangle$ be any two vectors. Then the **dot product** of $\mathbf{u}$ and $\mathbf{v}$, written $\mathbf{u} \cdot \mathbf{v}$, is defined as

$$\mathbf{u} \cdot \mathbf{v} = u_1 v_1 + u_2 v_2 \qquad (11.3)$$

REMARKS

■ Observe that the dot product of two vectors is not a vector but is a scalar. For this reason the dot product is sometimes called the **scalar product**.
■ If $\mathbf{u}$ and $\mathbf{v}$ are written in the form $\mathbf{u} = u_1 \mathbf{i} + u_2 \mathbf{j}$ and $\mathbf{v} = v_1 \mathbf{i} + v_2 \mathbf{j}$, then we have

$$\mathbf{u} \cdot \mathbf{v} = (u_1 \mathbf{i} + u_2 \mathbf{j}) \cdot (v_1 \mathbf{i} + v_2 \mathbf{j}) = u_1 v_1 + u_2 v_2$$

EXAMPLE 11.5   Let $\mathbf{u} = \langle 3, -2 \rangle$ and $\mathbf{v} = \langle -4, -5 \rangle$. Find $\mathbf{u} \cdot \mathbf{v}$.

Solution

$$\mathbf{u} \cdot \mathbf{v} = 3(-4) + (-2)(-5) = -12 + 10 = -2$$

Using the alternative way of writing $\mathbf{u}$ and $\mathbf{v}$ in terms of the basis vectors $\mathbf{i}$ and $\mathbf{j}$, we can also write

$$\mathbf{u} \cdot \mathbf{v} = (3\mathbf{i} - 2\mathbf{j}) \cdot (-4\mathbf{i} - 5\mathbf{j}) = -12 + 10 = -2 \qquad \blacksquare$$

Properties of the Dot Product

The dot product of vectors shares several properties with products of real numbers, as the following theorem shows.

THEOREM 11.4

If $\mathbf{u}$, $\mathbf{v}$, and $\mathbf{w}$ are vectors and k is a scalar, then

1. $\mathbf{u} \cdot \mathbf{v} = \mathbf{v} \cdot \mathbf{u}$ Commutative law
2. $\mathbf{u} \cdot (\mathbf{v} + \mathbf{w}) = \mathbf{u} \cdot \mathbf{v} + \mathbf{u} \cdot \mathbf{w}$ Distributive law
3. $k(\mathbf{u} \cdot \mathbf{v}) = (k\mathbf{u}) \cdot \mathbf{v} = \mathbf{u} \cdot (k\mathbf{v})$
4. $\mathbf{u} \cdot \mathbf{0} = 0$
5. $\mathbf{u} \cdot \mathbf{u} = |\mathbf{u}|^2$

Proof   We will verify Property 2 and leave the other properties for the exercises. Let $\mathbf{u} = \langle u_1, u_2 \rangle$, $\mathbf{v} = \langle v_1, v_2 \rangle$, and $\mathbf{w} = \langle w_1, w_2 \rangle$. We verify Property 2 by calculating the value of each side independently and showing that we get the same result. For the left-hand side, we have

$$
\begin{aligned}
\mathbf{u} \cdot (\mathbf{v} + \mathbf{w}) &= \langle u_1, u_2 \rangle \cdot [\langle v_1, v_2 \rangle + \langle w_1, w_2 \rangle] \\
&= \langle u_1, u_2 \rangle \cdot \langle v_1 + w_1, v_2 + w_2 \rangle && \text{By Definition 11.1} \\
&= u_1(v_1 + w_1) + u_2(v_2 + w_2) && \text{By Definition 11.5} \\
&= (u_1 v_1 + u_1 w_1) + (u_2 v_2 + u_2 w_2) && \text{By the distributive law of real numbers}
\end{aligned}
$$

For the right-hand side, we have

$$\mathbf{u} \cdot \mathbf{v} + \mathbf{u} \cdot \mathbf{w} = \langle u_1, u_2 \rangle \cdot \langle v_1, v_2 \rangle + \langle u_1, u_2 \rangle \cdot \langle w_1, w_2 \rangle$$

$$= (u_1 v_1 + u_2 v_2) + (u_1 w_1 + u_2 w_2) \qquad \text{By Definition 11.5}$$

$$= (u_1 v_1 + u_1 w_1) + (u_2 v_2 + u_2 w_2) \qquad \text{By commutativity and associativity of real numbers}$$

The equality of the results proves Property 2. ∎

The Angle Between Two Vectors

An important property of the dot product has to do with the angle between two vectors, which we now define.

Definition 11.6
Angle Between Vectors

The **angle** between two nonzero vectors $\mathbf{u}$ and $\mathbf{v}$ is the smallest positive angle between geometric representatives of $\mathbf{u}$ and $\mathbf{v}$ that have the same initial point.

If we denote the angle between $\mathbf{u}$ and $\mathbf{v}$ as θ, it follows that $0 \leq \theta \leq \pi$. The angle is 0 if $\mathbf{u}$ and $\mathbf{v}$ are in the same direction, and it is π if they are in opposite directions. Figure 11.14 illustrates various possibilities for θ.

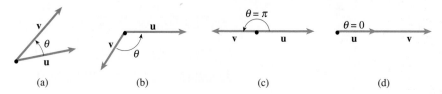

(a)   (b)   (c)   (d)

FIGURE 11.14

THEOREM 11.5

If θ is the angle between nonzero vectors $\mathbf{u}$ and $\mathbf{v}$, then

$$\mathbf{u} \cdot \mathbf{v} = |\mathbf{u}||\mathbf{v}| \cos \theta \qquad (11.4)$$

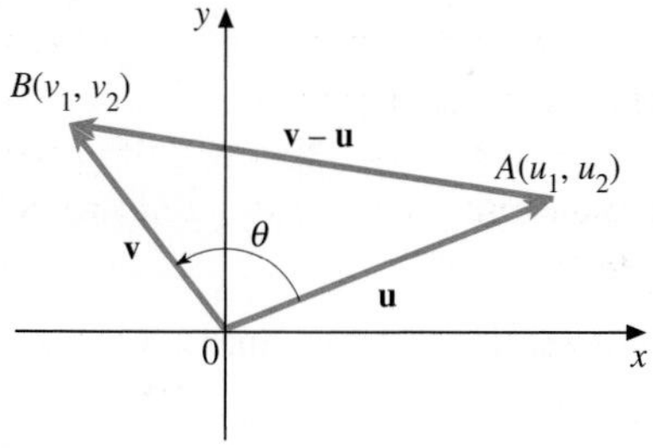

FIGURE 11.15

Proof   Assume first that $\mathbf{u}$ and $\mathbf{v}$ are not parallel, and choose geometric representatives of these vectors such that each has its initial point at the origin, as in Figure 11.15. If $\mathbf{u}$ and $\mathbf{v}$ have components given by $\langle u_1, u_2 \rangle$ and $\langle v_1, v_2 \rangle$, respectively, then it follows that the terminal point A of $\mathbf{u}$ is (u_1, u_2) and the terminal point B of $\mathbf{v}$ is (v_1, v_2). The vector $\overrightarrow{AB} = \mathbf{v} - \mathbf{u}$, and so by the Law of Cosines (see the margin note on next page) we have

$$|\mathbf{v} - \mathbf{u}|^2 = |\mathbf{u}|^2 + |\mathbf{v}|^2 - 2|\mathbf{u}||\mathbf{v}| \cos \theta$$

or,

$$(v_1 - u_1)^2 + (v_2 - u_2)^2 = u_1^2 + u_2^2 + v_1^2 + v_2^2 - 2|\mathbf{u}||\mathbf{v}| \cos \theta$$

The Law of Cosines states that in a triangle with sides a, b, and c, if C is the angle opposite side c, then

$$c^2 = a^2 + b^2 - 2ab \cos C$$

After expanding and collecting terms, we get

$$|\mathbf{u}| \, |\mathbf{v}| \cos \theta = u_1 v_1 + u_2 v_2$$

$$= \mathbf{u} \cdot \mathbf{v}$$

If $\mathbf{u}$ and $\mathbf{v}$ are parallel, then $\mathbf{v} = k\mathbf{u}$, and $\theta = 0$ or $\theta = \pi$, according to whether $k > 0$ or $k < 0$.

For $k > 0$, we have $\cos \theta = \cos 0 = 1$, and

$$|\mathbf{u}||\mathbf{v}| \cos \theta = |\mathbf{u}||k\mathbf{u}|(1) = k|\mathbf{u}|^2 = k(\mathbf{u} \cdot \mathbf{u}) = \mathbf{u} \cdot (k\mathbf{u}) = \mathbf{u} \cdot \mathbf{v}$$

so the result is true in this case. For $k < 0$, $\cos \theta = \cos \pi = -1$, and

$$|\mathbf{u}||\mathbf{v}| \cos \theta = |\mathbf{u}||k\mathbf{u}|(-1) = -|k||\mathbf{u}|^2 = k(\mathbf{u} \cdot \mathbf{u}) = \mathbf{u} \cdot (k\mathbf{u}) = \mathbf{u} \cdot \mathbf{v}$$

Here we used the fact that since $k < 0$, $-|k| = -(-k) = k$. The result is therefore true in all cases. ∎

From Equation 11.4, if $\mathbf{u}$ and $\mathbf{v}$ are nonzero, we have

$$\cos \theta = \frac{\mathbf{u} \cdot \mathbf{v}}{|\mathbf{u}||\mathbf{v}|} \tag{11.5}$$

In the next example, we use Equation 11.5 to find the angle between two vectors.

EXAMPLE 11.6   Find the angle between the vectors $\mathbf{u} = \langle 3, -4 \rangle$ and $\mathbf{v} = \langle 1, 7 \rangle$. Draw the vectors with initial point at the origin.

Solution   By Equation 11.5,

$$\cos \theta = \frac{\mathbf{u} \cdot \mathbf{v}}{|\mathbf{u}| \, |\mathbf{v}|} = \frac{3(1) + (-4)(7)}{\sqrt{9 + 16} \sqrt{1 + 49}} = \frac{-25}{\sqrt{25} \sqrt{50}} = \frac{-25}{5(5\sqrt{2})} = -\frac{1}{\sqrt{2}}$$

So $\theta = \frac{3\pi}{4}$. We show the vectors in Figure 11.16. ∎

EXAMPLE 11.7   Find the angle between $2\mathbf{i} - 4\mathbf{j}$ and $\mathbf{i} + \mathbf{j}$.

Solution

$$\cos \theta = \frac{(2\mathbf{i} - 4\mathbf{j}) \cdot (\mathbf{i} + \mathbf{j})}{|2\mathbf{i} - 4\mathbf{j}| \, |\mathbf{i} + \mathbf{j}|} = \frac{2 - 4}{\sqrt{20} \sqrt{2}} = \frac{-2}{2\sqrt{10}} = \frac{-1}{\sqrt{10}}$$

Using a calculator, we find $\theta \approx 1.65$ radians. ∎

The following corollary is an immediate consequence of Equation 11.5.

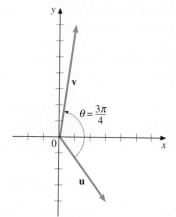

FIGURE 11.16

COROLLARY 11.5a

> **The Cauchy-Schwarz Inequality**
>
> For any two vectors $\mathbf{u}$ and $\mathbf{v}$,
>
> $$|\mathbf{u} \cdot \mathbf{v}| \leq |\mathbf{u}| \, |\mathbf{v}| \tag{11.6}$$

Proof   The result is trivial if either $\mathbf{u}$ or $\mathbf{v}$ is the zero vector, since both sides of Equation 11.6 are 0. For nonzero vectors $\mathbf{u}$ and $\mathbf{v}$, with angle θ between them, we have, by Equation 11.5,

$$\frac{|\mathbf{u} \cdot \mathbf{v}|}{|\mathbf{u}| \, |\mathbf{v}|} = |\cos \theta| \leq 1$$

Thus, $|\mathbf{u} \cdot \mathbf{v}| \leq |\mathbf{u}| \, |\mathbf{v}|$. ■

This corollary enables us to give an algebraic proof of the Triangle Inequality, as follows.

$$|\mathbf{u} + \mathbf{v}|^2 = (\mathbf{u} + \mathbf{v}) \cdot (\mathbf{u} + \mathbf{v})$$

$$= \mathbf{u} \cdot \mathbf{u} + 2\mathbf{u} \cdot \mathbf{v} + \mathbf{v} \cdot \mathbf{v} = |\mathbf{u}|^2 + 2\mathbf{u} \cdot \mathbf{v} + |\mathbf{v}|^2$$

$$\leq |\mathbf{u}|^2 + 2 \, |\mathbf{u} \cdot \mathbf{v}| + |\mathbf{v}|^2 \leq |\mathbf{u}|^2 + 2 \, |\mathbf{u}| \, |\mathbf{v}| + |\mathbf{v}|^2 \qquad \text{By the Cauchy-Schwarz Inequality}$$

$$= (|\mathbf{u}| + |\mathbf{v}|)^2$$

Now we take square roots to get the desired result:

$$|\mathbf{u} + \mathbf{v}| \leq |\mathbf{u}| + |\mathbf{v}|$$

Orthogonal Vectors

If the angle between two nonzero vectors is $\frac{\pi}{2}$, the vectors are said to be **orthogonal**. So geometric representatives of orthogonal vectors are perpendicular to each other. Since $\cos \frac{\pi}{2} = 0$, it follows from Theorem 11.5 that if $\mathbf{u}$ and $\mathbf{v}$ are orthogonal, then $\mathbf{u} \cdot \mathbf{v} = 0$. Conversely, if $\mathbf{u} \cdot \mathbf{v} = 0$, then $\cos \theta = 0$ by Equation 11.5, and so $\theta = \frac{\pi}{2}$. Thus, $\mathbf{u}$ and $\mathbf{v}$ are orthogonal. If either $\mathbf{u}$ or $\mathbf{v}$ is the zero vector, then $\mathbf{u} \cdot \mathbf{v} = 0$, and it is convenient in this case, too, to call $\mathbf{u}$ and $\mathbf{v}$ orthogonal; that is, we agree to say that $\mathbf{0}$ is orthogonal to every vector. We therefore have the following additional corollary to Theorem 11.5.

COROLLARY 11.5b

> Two vectors $\mathbf{u}$ and $\mathbf{v}$ are orthogonal if and only if $\mathbf{u} \cdot \mathbf{v} = 0$.

EXAMPLE 11.8   Show that the vectors $\mathbf{u} = \langle 6, -4 \rangle$ and $\mathbf{v} = \langle -2, -3 \rangle$ are orthogonal.

Solution   Since $\mathbf{u} \cdot \mathbf{v} = 6(-2) + (-4)(-3) = -12 + 12 = 0$, by Corollary 11.5b, $\mathbf{u}$ and $\mathbf{v}$ are orthogonal. We show the vectors $\mathbf{u}$ and $\mathbf{v}$ as position vectors of the points $(6, -4)$ and $(-2, -3)$ in Figure 11.17. ■

EXAMPLE 11.9   Find x so that the vectors $\langle x, 2 \rangle$ and $\langle 1 - x, 3 \rangle$ are orthogonal.

Solution   Let $\mathbf{u} = \langle x, 2 \rangle$ and $\mathbf{v} = \langle 1 - x, 3 \rangle$. Then

$$\mathbf{u} \cdot \mathbf{v} = \langle x, 2 \rangle \cdot \langle 1 - x, 3 \rangle = x - x^2 + 6 = 6 + x - x^2$$

$$= (3 - x)(2 + x)$$

and so $\mathbf{u} \cdot \mathbf{v} = 0$ if $x = 3$ or $x = -2$. Either value of x causes $\mathbf{u}$ and $\mathbf{v}$ to be orthogonal. For $x = 3$ we get $\mathbf{u} = \langle 3, 2 \rangle$ and $\mathbf{v} = \langle -2, 3 \rangle$, and for $x = -2$ we get $\mathbf{u} = \langle -2, 2 \rangle$ and $\mathbf{v} = \langle 3, 3 \rangle$. ■

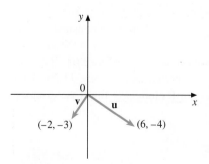

FIGURE 11.17

The Component of a Vector Along Another Vector

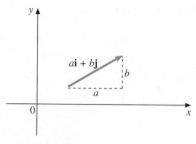

FIGURE 11.18

When we write $\mathbf{u} = a\mathbf{i} + b\mathbf{j}$, the numbers a and b are the components of $\mathbf{u}$ in the directions of $\mathbf{i}$ (horizontal) and $\mathbf{j}$ (vertical), respectively. (See Figure 11.18.) Sometimes it is useful to find the component of a vector in a direction other than horizontal and vertical. To understand what we mean, let $\mathbf{u}$ be any vector and suppose we want to find the displacement of $\mathbf{u}$ in the direction of some nonzero vector $\mathbf{v}$, as we show in Figure 11.19. We use geometric representatives $\overrightarrow{OP}$ and $\overrightarrow{OQ}$ of $\mathbf{u}$ and $\mathbf{v}$, respectively, and designate by P' the foot of the perpendicular from P to the line joining O and Q. Let θ be the angle between $\mathbf{u}$ and $\mathbf{v}$. Then we define the **component of u along v**, designated Comp$_{\mathbf{v}}\mathbf{u}$, by

$$\text{Comp}_{\mathbf{v}}\mathbf{u} = |\mathbf{u}| \cos \theta \qquad (11.7)$$

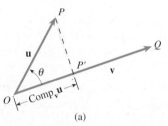

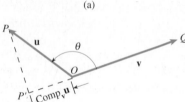

FIGURE 11.19

If $0 \leq \theta \leq \frac{\pi}{2}$, then Comp$_{\mathbf{v}}\,\mathbf{u} \geq 0$, as in Figure 11.19(a), whereas if $\frac{\pi}{2} < \theta \leq \pi$, Comp$_{\mathbf{v}}\,\mathbf{u} < 0$, as in Figure 11.19(b). If $\mathbf{u} \neq \mathbf{0}$, by Equation 11.5, we have $\cos \theta = \frac{\mathbf{u}\cdot\mathbf{v}}{|\mathbf{u}|\,|\mathbf{v}|}$, so that Equation 11.7 becomes

$$\text{Comp}_{\mathbf{v}}\,\mathbf{u} = \frac{\mathbf{u} \cdot \mathbf{v}}{|\mathbf{v}|} \qquad (11.8)$$

This result is valid also for $\mathbf{u} = \mathbf{0}$. It is easy to show (see Exercise 22 in Exercise Set 11.2) that if $\mathbf{u} = a\mathbf{i} + b\mathbf{j}$ and $\mathbf{v}$ is directed along the positive x-axis, then Comp$_{\mathbf{v}}\mathbf{u} = a$. Similarly, if $\mathbf{v}$ is directed along the positive y-axis, then Comp$_{\mathbf{v}}\mathbf{u} = b$. So our definition generalizes horizontal and vertical components.

Work

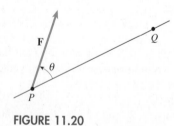

FIGURE 11.20

As an application of this concept, consider the work done by a constant force $\mathbf{F}$ in moving a particle along a straight line from P to Q. If $\mathbf{F}$ acts in the direction of $\overrightarrow{PQ}$, then since work equals force times distance, work $= |\mathbf{F}|\,|\overrightarrow{PQ}|$. Suppose, however, that $\mathbf{F}$ acts at some fixed angle θ with $\overrightarrow{PQ}$, as in Figure 11.20. Then it is natural to define work as the component of $\mathbf{F}$ along $\overrightarrow{PQ}$ times the distance; that is,

$$W = (\text{Comp}_{\overrightarrow{PQ}}\mathbf{F})|\overrightarrow{PQ}|$$

Substituting from Equation 11.8, we get

$$W = \mathbf{F} \cdot \overrightarrow{PQ} \qquad (11.9)$$

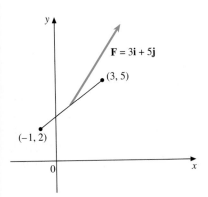

FIGURE 11.21

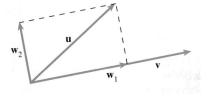

FIGURE 11.22

EXAMPLE 11.10   The force $\mathbf{F} = 3\mathbf{i} + 5\mathbf{j}$ moves an object along the line segment from $(-1, 2)$ to $(3, 5)$. If the magnitude of $\mathbf{F}$ is in newtons and distance is measured in meters, find the work done by $\mathbf{F}$. (See Figure 11.21.)

Solution   Let $P = (-1, 2)$ and $Q = (3, 5)$. Then $\overrightarrow{PQ} = 4\mathbf{i} + 3\mathbf{j}$, and

$$W = \mathbf{F} \cdot \overrightarrow{PQ} = (3\mathbf{i} + 5\mathbf{j}) \cdot (4\mathbf{i} + 3\mathbf{j})$$
$$= 12 + 15 = 27 \text{ joules} \qquad \blacksquare$$

Vector Projections

When we write a vector $\mathbf{u} = \langle a, b \rangle$ in the form $\mathbf{u} = a\mathbf{i} + b\mathbf{j}$, we are in effect expressing $\mathbf{u}$ as the sum of two mutually perpendicular vectors, one acting horizontally and the other vertically. Sometimes it is desirable to express $\mathbf{u}$ as the sum of two mutually perpendicular vectors, one in a prescribed nonhorizontal direction and the other perpendicular to this direction. We show how to find these vectors geometrically in Figure 11.22. There we show the given vector $\mathbf{u}$ and a direction as determined by the vector $\mathbf{v}$. We construct a rectangle with sides $\mathbf{w}_1$ and $\mathbf{w}_2$ having $\mathbf{u}$ as a diagonal and one side along $\mathbf{v}$. Then $\mathbf{u} = \mathbf{w}_1 + \mathbf{w}_2$, as required. We call the vector $\mathbf{w}_1$ the **projection of u on v** and designate it by $\text{Proj}_{\mathbf{v}}\mathbf{u}$. The vector $\mathbf{w}_2$ is called the **projection of u orthogonal to v** and is designated by $\text{Proj}_{\mathbf{v}}^{\perp}\mathbf{u}$. So we always have, for any vector $\mathbf{u}$ and $\mathbf{v} \neq 0$,

$$\mathbf{u} = \text{Proj}_{\mathbf{v}}\mathbf{u} + \text{Proj}_{\mathbf{v}}^{\perp}\mathbf{u} \qquad (11.10)$$

To find algebraic representatives of these projections, observe that $\mathbf{w}_1$ can be obtained by multiplying the component of $\mathbf{u}$ along $\mathbf{v}$ by a unit vector in the direction of $\mathbf{v}$. Thus,

$$\text{Proj}_{\mathbf{v}}\mathbf{u} = (\text{Comp}_{\mathbf{v}}\mathbf{u})\frac{\mathbf{v}}{|\mathbf{v}|} \qquad (11.11)$$

If we replace $\text{Comp}_{\mathbf{v}}\mathbf{u}$ by its value from Equation 11.8, we obtain

$$\text{Proj}_{\mathbf{v}}\mathbf{u} = \left(\frac{\mathbf{u} \cdot \mathbf{v}}{|\mathbf{v}|}\right)\frac{\mathbf{v}}{|\mathbf{v}|} = \left(\frac{\mathbf{u} \cdot \mathbf{v}}{|\mathbf{v}|^2}\right)\mathbf{v} \qquad (11.12)$$

and from Equation 11.9, we get

$$\text{Proj}_{\mathbf{v}}^{\perp}\mathbf{u} = \mathbf{u} - \text{Proj}_{\mathbf{v}}\mathbf{u} = \mathbf{u} - \left(\frac{\mathbf{u} \cdot \mathbf{v}}{|\mathbf{v}|^2}\right)\mathbf{v} \qquad (11.13)$$

EXAMPLE 11.11   Let $\mathbf{u} = \langle 3, 4 \rangle$ and $\mathbf{v} = \langle -4, 8 \rangle$. Express $\mathbf{u}$ as the sum of two vectors, one parallel to $\mathbf{v}$ and the other perpendicular to $\mathbf{v}$. Show the results geometrically.

Solution   Note that the desired vectors are $\text{Proj}_{\mathbf{v}}\mathbf{u}$ and $\text{Proj}_{\mathbf{v}}^{\perp}\mathbf{u}$. By Equation 11.12,

$$\begin{aligned}
\text{Proj}_{\mathbf{v}}\mathbf{u} &= \left(\frac{\mathbf{u} \cdot \mathbf{v}}{|\mathbf{v}|^2}\right)\mathbf{v} = \frac{\langle 3, 4 \rangle \cdot \langle -4, 8 \rangle}{|\langle -4, 8 \rangle|^2}\langle -4, 8 \rangle \\
&= \frac{-12 + 32}{80}\langle -4, 8 \rangle = \frac{1}{4}\langle -4, 8 \rangle \\
&= \langle -1, 2 \rangle
\end{aligned}$$

By Equation 11.13,

$$\text{Proj}_{\mathbf{v}}^{\perp}\mathbf{u} = \mathbf{u} - \text{Proj}_{\mathbf{v}}\mathbf{u} = \langle 3, 4 \rangle - \langle -1, 2 \rangle = \langle 4, 2 \rangle$$

We illustrate the results in Figure 11.23. ∎

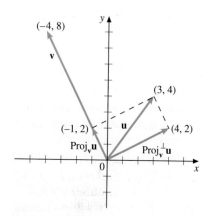

FIGURE 11.23

REMARK
■ $\text{Proj}_{\mathbf{v}}\mathbf{u}$ and $\text{Proj}_{\mathbf{v}}^{\perp}\mathbf{u}$ are sometimes called **vector components** of $\mathbf{u}$ in the direction of $\mathbf{v}$ and orthogonal to $\mathbf{v}$, respectively, and when these are found, the vector $\mathbf{u}$ is said to have been *resolved into vector components* in these directions.

Exercise Set 11.2

In Exercises 1–4, find the dot product of $\mathbf{u}$ and $\mathbf{v}$.

1. $\mathbf{u} = \langle 4, 7 \rangle$, $\mathbf{v} = \langle -5, 2 \rangle$

2. $\mathbf{u} = \langle -3, -6 \rangle$, $\mathbf{v} = \langle 5, -2 \rangle$

3. $\mathbf{u} = 2\mathbf{i} - 3\mathbf{j}$, $\mathbf{v} = \mathbf{i} + 2\mathbf{j}$

4. $\mathbf{u} = -\mathbf{i} + 4\mathbf{j}$, $\mathbf{v} = 3\mathbf{i} + \mathbf{j}$

In Exercises 5–8, show that $\mathbf{u}$ and $\mathbf{v}$ are orthogonal.

5. $\mathbf{u} = \langle 4, -8 \rangle$, $\mathbf{v} = \langle -4, -2 \rangle$

6. $\mathbf{u} = \langle -10, 6 \rangle$, $\mathbf{v} = \langle 12, 20 \rangle$

7. $\mathbf{u} = 2\mathbf{i} - 3\mathbf{j}$, $\mathbf{v} = 9\mathbf{i} + 6\mathbf{j}$

8. $\mathbf{u} = -3\mathbf{i} + 4\mathbf{j}$, $\mathbf{v} = -12\mathbf{i} - 9\mathbf{j}$

In Exercises 9–15, find the cosine of the angle θ between $\mathbf{u}$ and $\mathbf{v}$.

9. $\mathbf{u} = \langle 4, -4 \rangle$, $\mathbf{v} = \langle 1, 7 \rangle$

10. $\mathbf{u} = \langle 1, -2 \rangle$, $\mathbf{v} = \langle -1, 1 \rangle$

11. $\mathbf{u} = -\mathbf{i} + 3\mathbf{j}$, $\mathbf{v} = -2\mathbf{i} - \mathbf{j}$

12. $\mathbf{u} = 4\mathbf{i} + 6\mathbf{j}$, $\mathbf{v} = 4\mathbf{i} - 2\mathbf{j}$

13. $\mathbf{u} = \langle 6, 8 \rangle$, $\mathbf{v} = \langle -3, 4 \rangle$

14. $\mathbf{u} = \langle 4, 8 \rangle$, $\mathbf{v} = \langle -1, 3 \rangle$; also find θ.

15. $\mathbf{u} = \mathbf{i} + \sqrt{3}\,\mathbf{j}$, $\mathbf{v} = 2\mathbf{i}$; also find θ.

16. Find all values of x so that $\mathbf{u} = \langle 3x, 1 - x \rangle$ and $\mathbf{v} = \langle x, -4 \rangle$ will be orthogonal.

17. Find all values of x so that the angle between $\mathbf{u} = \langle 4, -3 \rangle$ and $\mathbf{v} = \langle x, 1 \rangle$ will be $\frac{\pi}{4}$.

In Exercises 18–21, find $\text{Comp}_\mathbf{v}\mathbf{u}$.

18. $\mathbf{u} = \langle 7, -4 \rangle$, $\mathbf{v} = \langle -3, 4 \rangle$

19. $\mathbf{u} = \langle -2, -3 \rangle$, $\mathbf{v} = \langle 1, 1 \rangle$

20. $\mathbf{u} = \mathbf{i} - 2\mathbf{j}$, $\mathbf{v} = 2\mathbf{i} - \mathbf{j}$

21. $\mathbf{u} = 3\mathbf{i} - 4\mathbf{j}$, $\mathbf{v} = \mathbf{i} + 7\mathbf{j}$

22. Show that if $\mathbf{u} = \langle a, b \rangle$ and $\mathbf{v} = k\mathbf{i}$ for $k > 0$, then $\text{Comp}_\mathbf{v}\mathbf{u} = a$. Also show that if $\mathbf{v} = k\mathbf{j}$ for $k > 0$, then $\text{Comp}_\mathbf{v}\mathbf{u} = b$.

In Exercises 23–26, find the work done by the force $\mathbf{F}$ acting on a particle along a line segment from the first point to the second. Assume $|\mathbf{F}|$ is in newtons and distance is in meters.

23. $\mathbf{F} = 2\mathbf{i} + 3\mathbf{j}$; $(1, 2)$ to $(6, 8)$

24. $\mathbf{F} = -\mathbf{i} + 4\mathbf{j}$; $(-2, 3)$ to $(3, 5)$

25. $\mathbf{F} = 10\mathbf{i} + 20\mathbf{j}$; $(2, 3)$ to $(1, 5)$

26. $\mathbf{F} = 5\mathbf{i} - 7\mathbf{j}$; $(-4, -1)$ to $(6, -6)$

In Exercises 27–30, find $\text{Proj}_\mathbf{v}\mathbf{u}$ and $\text{Proj}_\mathbf{v}^\perp\mathbf{u}$.

27. $\mathbf{u} = \langle 3, -2 \rangle$, $\mathbf{v} = \langle 2, 4 \rangle$

28. $\mathbf{u} = \langle -2, -1 \rangle$, $\mathbf{v} = \langle -3, 4 \rangle$

29. $\mathbf{u} = 6\mathbf{i} + 2\mathbf{j}$, $\mathbf{v} = 3\mathbf{i} - 4\mathbf{j}$

30. $\mathbf{u} = -2\mathbf{i} + 3\mathbf{j}$, $\mathbf{v} = 7\mathbf{i} + \mathbf{j}$

31. A block that weighs 1000 lb and is on an inclined plane that makes a 30° angle with the horizontal is being held in place by a person pulling on a rope attached to the block and passing over a pulley, as shown in the figure. Assuming no friction, what is the magnitude of the force $\mathbf{F}$ that must be exerted?

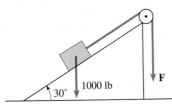

32. Resolve the vector $\mathbf{w} = 6\mathbf{i} - 4\mathbf{j}$ into vector components parallel and perpendicular, respectively, to the line that joins $(-1, -2)$ and $(2, 2)$.

33. Use vector methods to show that the points $(2, 1)$, $(6, 9)$, and $(-2, 3)$ are vertices of a right triangle. What is the area of the triangle?

34. Use vector methods to show that the points $(3, -1)$, $(5, 4)$, $(-5, 8)$, and $(-7, 3)$ are vertices of a rectangle. What is the area of the rectangle?

35. Use vector methods to show that the points $(-5, 2)$, $(-3, -2)$, $(6, 1)$, and $(4, 5)$ are vertices of a parallelogram. Find the interior angles of the parallelogram.

Prove the identities in Exercises 36 and 37.

36. (a) $(\mathbf{u} + \mathbf{v}) \cdot (\mathbf{u} - \mathbf{v}) = |\mathbf{u}|^2 - |\mathbf{v}|^2$
   (b) $(\mathbf{u} + \mathbf{v}) \cdot (\mathbf{u} + \mathbf{v}) = |\mathbf{u}|^2 + 2\mathbf{u} \cdot \mathbf{v} + |\mathbf{v}|^2$

37. (a) $|\mathbf{u} + \mathbf{v}|^2 + |\mathbf{u} - \mathbf{v}|^2 = 2(|\mathbf{u}|^2 + |\mathbf{v}|^2)$
   (b) $|\mathbf{u} + \mathbf{v}|^2 - |\mathbf{u} - \mathbf{v}|^2 = 4\mathbf{u} \cdot \mathbf{v}$

38. Prove that $\mathbf{u}$ and $\mathbf{v}$ are orthogonal if and only if

$$|\mathbf{u} + \mathbf{v}| = |\mathbf{u} - \mathbf{v}|$$

39. Give an algebraic proof that $\text{Proj}_\mathbf{v}^\perp\mathbf{u}$ is orthogonal to $\text{Proj}_\mathbf{v}\mathbf{u}$.

40. Prove that the vector $\mathbf{n} = a\mathbf{i} + b\mathbf{j}$ is perpendicular to the line $ax + by + c = 0$. (*Hint:* Consider two points on the line.)

41. Let $P_0(x_0, y_0)$ be any point on the line $ax + by + c = 0$ and let $P_1(x_1, y_1)$ be any point not on the line. Show that the distance d from the line to the point P_1 is $d = |\text{Comp}_\mathbf{n} \overrightarrow{P_0 P_1}|$, where $\mathbf{n} = a\mathbf{i} + b\mathbf{j}$. From this result, verify the Distance Formula

$$d = \frac{|ax_1 + by_1 + c|}{\sqrt{a^2 + b^2}}$$

(*Hint:* Use the result of Exercise 40.)

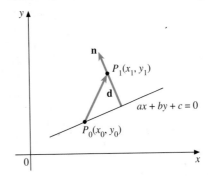

42. Use vector methods to prove that any triangle inscribed in a semicircle, with one side coinciding with the diameter, is a right triangle. (*Hint:* In the figure, find $\overrightarrow{AB}$ and $\overrightarrow{BC}$ in terms of $\mathbf{u}$ and $\mathbf{v}$, and use $|\mathbf{u}| = |\mathbf{v}|$ together with the result of Exercise 36(a).)

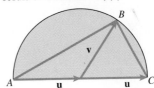

11.3 VECTORS IN SPACE

FIGURE 11.24

FIGURE 11.25

Three-Dimensional Coordinate Systems

In this section we extend the vector concept to three-dimensional space. It is necessary first to introduce a rectangular coordinate system. We begin with a horizontal plane that has a two-dimensional rectangular coordinate system with x- and y-axes in their usual orientation, the positive y-axis being $90°$ counterclockwise from the positive x-axis. Through the origin we introduce a vertical axis, called the z-axis, directed positively upward, with its origin coinciding with that of the x- and y-axes, as illustrated in Figure 11.24. We now have three mutually perpendicular axes oriented according to what is called the *right-hand rule*: if you point the index finger of your right hand in the positive x direction and the middle finger in the positive y direction, as in Figure 11.25, then your thumb will point in the positive z direction.

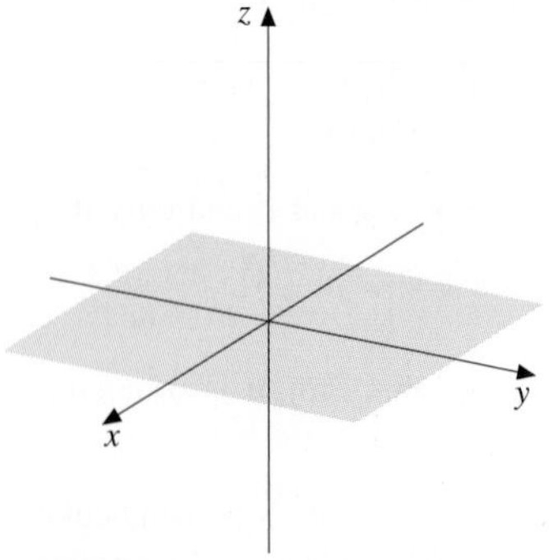

Each pair of axes determines a plane. We call these the **xy-plane**, the **xz-plane**, and the **yz-plane**. Frequently we will refer to the xy-plane as the **horizontal plane**. These three planes are the **coordinate planes.** Now let P denote any point in space. Through P pass planes parallel to each of the coordinate planes. If these cut the x-axis, y-axis, and z-axis at x_0, y_0, and z_0, respectively, then these three numbers are called the *coordinates* of P, and we write them as the ordered triple (x_0, y_0, z_0). We illustrate a typical such point in Figure 11.26. If we begin with the ordered triple (x_0, y_0, z_0), we locate P by proceeding x_0 units from the origin along the x-axis, then y_0 units parallel to the y-axis, and then z_0 units parallel to the z-axis, in each case using directed distances. In this way we establish a one-to-one correspondence between all points in three-dimensional space and all ordered triples of real numbers. We often will not distinguish between a point and its coordinates, saying, for example, "the point $(2, 3, -4)$" rather than "the point whose coordinates are $(2, 3, -4)$."

The three coordinate planes divide space into eight regions, called **octants**. The octant in which all coordinates are positive is called the first octant. There is no need to number the others. In plotting points it is useful to show lines, as we have done in plotting $P(3, 4, 6)$ in Figure 11.27. These help to make it appear that P is not in the plane of the paper. In this case we have shown the positive axes only, since the point is in the first octant.

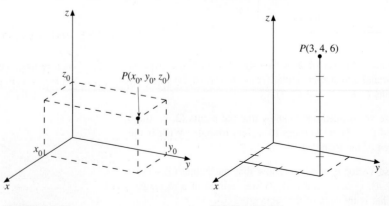

FIGURE 11.26

FIGURE 11.27

The Distance Formula

To determine a formula for the length of a vector, we need to know the distance between two points in space. Let $P_1(x_1, y_1, z_1)$ and $P_2(x_2, y_2, z_2)$ be any two such points. Construct a rectangular box with sides parallel to the coordinate planes so that P_1 and P_2 are at opposite corners of the box, as in Figure 11.28.

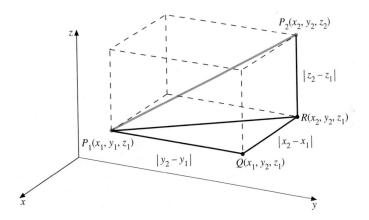

FIGURE 11.28

With vertices Q and R as shown, triangle P_1QR is a right triangle in a horizontal plane, and triangle P_1RP_2 is a right triangle in a vertical plane. Using $d(P_1, P_2)$ to mean the distance from P_1 to P_2, we have, from the first triangle, by the Pythagorean Theorem,

$$[d(P_1, R)]^2 = [d(P_1, Q)]^2 + [d(Q, R)]^2$$

and from the second,

$$[d(P_1, P_2)]^2 = [d(P_1, R)]^2 + [d(R, P_2)]^2$$

So

$$[d(P_1, P_2)]^2 = [d(P_1, Q)]^2 + [d(Q, R)]^2 + [d(R, P_2)]^2$$

But $d(P_1, Q) = |y_2 - y_1|$, $d(Q, R) = |x_2 - x_1|$, and $d(R, P_2) = |z_2 - z_1|$. Making these substitutions, we get the Distance Formula:

Distance Formula in Three Dimensions

$$d(P_1, P_2) = \sqrt{(x_2 - x_1)^2 + (y_2 - y_1)^2 + (z_2 - z_1)^2} \qquad (11.14)$$

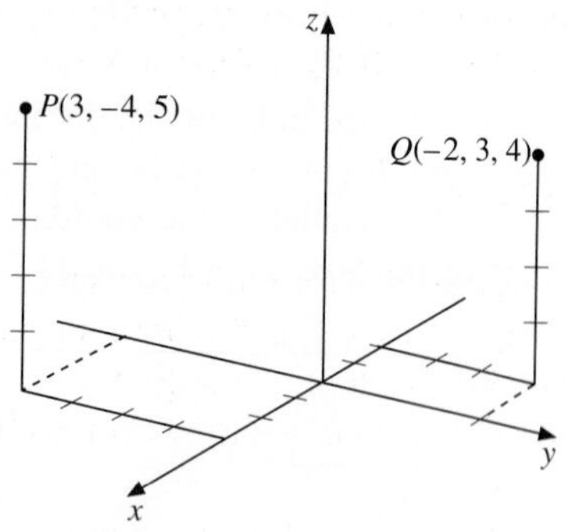

FIGURE 11.29

EXAMPLE 11.12   Plot the points $P(3, -4, 5)$ and $Q(-2, 3, 4)$, and find the distance between them.

Solution   We show the points in Figure 11.29. By Equation 11.14,

$$
\begin{aligned}
d(P, Q) &= \sqrt{(3+2)^2 + (-4-3)^2 + (5-4)^2} \\
&= \sqrt{25 + 49 + 1} \\
&= \sqrt{75} \\
&= 5\sqrt{3}
\end{aligned}
$$
∎

Vectors in Three Dimensions

With this background we are ready to extend vectors to three dimensions. The notion of a geometric vector as a directed line segment is exactly as it was for two dimensions, with operations on vectors done in exactly the same way. Suppose a geometric vector has its initial point at $P_1(x_1, y_1, z_1)$ and its terminal point at $P_2(x_2, y_2, z_2)$. Then, analogous to the two-dimensional case, we identify the ordered triple $\langle x_2 - x_1, y_2 - y_1, z_2 - z_1 \rangle$ with the vector $\overrightarrow{P_1 P_2}$. Conversely, if we are given an ordered triple $\langle a, b, c \rangle$, we identify with this triple any geometric vector that has x displacement a, y displacement b, and z displacement c. The simplest such vector is the one with initial point at the origin and terminal point at (a, b, c), called the **position vector** of (a, b, c). In keeping with the geometric definitions of equivalence, addition, and multiplication by a scalar, we have the following:

1. $\langle a_1, a_2, a_3 \rangle = \langle b_1, b_2, b_3 \rangle$ if and only if $a_1 = b_1$, $a_2 = b_2$, and $a_3 = b_3$
2. $\langle a_1, a_2, a_3 \rangle + \langle b_1, b_2, b_3 \rangle = \langle a_1 + b_1, a_2 + b_2, a_3 + b_3 \rangle$
3. $k\langle a_1, a_2, a_3 \rangle = \langle ka_1, ka_2, ka_3 \rangle$ for any scalar k

With these definitions the set of all such ordered triples is called a **vector space of dimension 3**, and each element of this space is called a **three-dimensional vector**. If $\mathbf{u} = \langle u_1, u_2, u_3 \rangle$, then u_1, u_2, and u_3 are called the **components** of $\mathbf{u}$. The negative of $\mathbf{u}$ is $-\mathbf{u} = \langle -u_1, -u_2, -u_3 \rangle$, and the zero vector is $\langle 0, 0, 0 \rangle$. Subtraction is defined by $\mathbf{u} - \mathbf{v} = \mathbf{u} + (-\mathbf{v})$.

The **magnitude** of a vector $\mathbf{u} = \langle u_1, u_2, u_3 \rangle$ is defined by

$$|\mathbf{u}| = \sqrt{u_1^2 + u_2^2 + u_3^2} \tag{11.15}$$

Magnitude is, by Equation 11.14, the length of a geometric representative of $\mathbf{u}$. The **dot product** of two vectors $\mathbf{u} = \langle u_1, u_2, u_3 \rangle$ and $\mathbf{v} = \langle v_1, v_2, v_3 \rangle$ is defined by

$$\mathbf{u} \cdot \mathbf{v} = u_1 v_1 + u_2 v_2 + u_3 v_3 \tag{11.16}$$

The angle θ between two nonzero vectors $\mathbf{u}$ and $\mathbf{v}$ is defined as for two-dimensional vectors, and $\cos \theta$ is again given by

$$\cos \theta = \frac{\mathbf{u} \cdot \mathbf{v}}{|\mathbf{u}|\,|\mathbf{v}|} \tag{11.17}$$

The proof is the same as before. With the agreement again that $\mathbf{0}$ is orthogonal to every vector, we have that $\mathbf{u}$ *and* $\mathbf{v}$ *are orthogonal if and only if* $\mathbf{u} \cdot \mathbf{v} = 0$.

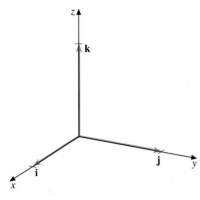

FIGURE 11.30

For any nonzero vector $\mathbf{u}$, $\mathbf{u}/|\mathbf{u}|$ is a *unit* vector, since its magnitude is 1. The unit vectors $\mathbf{i} = \langle 1, 0, 0 \rangle$, $\mathbf{j} = \langle 0, 1, 0 \rangle$, and $\mathbf{k} = \langle 0, 0, 1 \rangle$ form a *basis* for three-dimensional vector space, since

$$\langle a, b, c \rangle = a\mathbf{i} + b\mathbf{j} + c\mathbf{k} \tag{11.18}$$

means that every three-dimensional vector is expressible as a linear combination of $\mathbf{i}$, $\mathbf{j}$, and $\mathbf{k}$. Note that $\mathbf{i} \cdot \mathbf{j} = \mathbf{i} \cdot \mathbf{k} = \mathbf{j} \cdot \mathbf{k} = 0$, so that $\mathbf{i}$, $\mathbf{j}$, and $\mathbf{k}$ are mutually orthogonal. Geometrically, when placed with initial points at the origin, they are unit vectors directed along the positive x-axis, y-axis, and z-axis, respectively, as shown in Figure 11.30.

All other definitions and theorems in Sections 11.1 and 11.2 have natural extensions to three-dimensional vectors, and we will not repeat them. Proofs of the theorems are in many cases identical with proofs for the two-dimensional case, and at most require obvious modifications. Some of the results are illustrated in the examples that follow.

EXAMPLE 11.13   Let $\mathbf{u} = \langle 1, -2, 2 \rangle$ and $\mathbf{v} = \langle -3, -4, 5 \rangle$. Find each of the following.

(a) $|3\mathbf{u} - 2\mathbf{v}|$

(b) The angle between $\mathbf{u}$ and $\mathbf{v}$

Solution

(a) $3\mathbf{u} - 2\mathbf{v} = 3\langle 1, -2, 2 \rangle - 2\langle -3, -4, 5 \rangle$
$= \langle 3, -6, 6 \rangle - \langle -6, -8, 10 \rangle$
$= \langle 3, -6, 6 \rangle + \langle 6, 8, -10 \rangle$
$= \langle 9, 2, -4 \rangle$

So

$$|3\mathbf{u} - 2\mathbf{v}| = \sqrt{81 + 4 + 16} = \sqrt{101}$$

(b) $\cos\theta = \dfrac{\mathbf{u} \cdot \mathbf{v}}{|\mathbf{u}|\,|\mathbf{v}|} = \dfrac{\langle 1, -2, 2 \rangle \cdot \langle -3, -4, 5 \rangle}{\sqrt{1+4+4}\,\sqrt{9+16+25}}$

$= \dfrac{-3+8+10}{3\sqrt{50}} = \dfrac{15}{3(5\sqrt{2})} = \dfrac{1}{\sqrt{2}}$

So $\theta = \pi/4$.   ■

EXAMPLE 11.14   Find the work done by the force $\mathbf{F} = 4\mathbf{i} + 5\mathbf{j} - 8\mathbf{k}$ in moving a particle from $P(-1, 2, 4)$ to $Q(3, 6, -8)$. Assume $|\mathbf{F}|$ is in newtons and distance is in meters.

Solution

$$\overrightarrow{PQ} = \langle 4, 4, -12 \rangle = 4\mathbf{i} + 4\mathbf{j} - 12\mathbf{k}$$

so by Equation 11.9,

$$W = \mathbf{F} \cdot (\overrightarrow{PQ}) = (4\mathbf{i} + 5\mathbf{j} - 8\mathbf{k}) \cdot (4\mathbf{i} + 4\mathbf{j} - 12\mathbf{k})$$
$$= 16 + 20 + 96 = 132 \text{ joules}$$ ■

EXAMPLE 11.15   Find $\text{Comp}_v\mathbf{u}$ and $\text{Proj}_v\mathbf{u}$ if $\mathbf{u} = 4\mathbf{i} - 6\mathbf{j} + \mathbf{k}$ and $\mathbf{v} = -3\mathbf{i} - 2\mathbf{j} + 5\mathbf{k}$.

Solution   From Equation 11.8,

$$\text{Comp}_v\mathbf{u} = \frac{\mathbf{u}\cdot\mathbf{v}}{|\mathbf{v}|} = \frac{4(-3) + (-6)(-2) + (1)(5)}{\sqrt{9 + 4 + 25}} = \frac{5}{\sqrt{38}}$$

By Equation 11.11,

$$\text{Proj}_v\mathbf{u} = (\text{Comp}_v\mathbf{u})\frac{\mathbf{v}}{|\mathbf{v}|} = \frac{5}{\sqrt{38}}\,\frac{-3\mathbf{i} - 2\mathbf{j} + 5\mathbf{k}}{\sqrt{38}}$$

$$= -\frac{15}{38}\mathbf{i} - \frac{5}{19}\mathbf{j} + \frac{25}{38}\mathbf{k} \qquad\blacksquare$$

Direction Angles and Direction Cosines

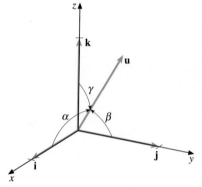

FIGURE 11.31

The angles a nonzero vector $\mathbf{u}$ makes with $\mathbf{i}$, $\mathbf{j}$, and $\mathbf{k}$ are called **direction angles** of $\mathbf{u}$ and are designated by α, β, and γ, respectively. We illustrate these angles in Figure 11.31. The cosines of these angles are called **direction cosines** of $\mathbf{u}$. If $\mathbf{u} = \langle u_1, u_2, u_3 \rangle$, we have

$$\cos\alpha = \frac{\mathbf{u}\cdot\mathbf{i}}{|\mathbf{u}|\,|\mathbf{i}|} = \frac{u_1}{|\mathbf{u}|} \quad \cos\beta = \frac{\mathbf{u}\cdot\mathbf{j}}{|\mathbf{u}|\,|\mathbf{j}|} = \frac{u_2}{|\mathbf{u}|} \quad \cos\gamma = \frac{\mathbf{u}\cdot\mathbf{k}}{|\mathbf{u}|\,|\mathbf{k}|} = \frac{u_3}{|\mathbf{u}|} \quad (11.19)$$

If we square and add, we get

$$\cos^2\alpha + \cos^2\beta + \cos^2\gamma = \frac{u_1^2}{|\mathbf{u}|^2} + \frac{u_2^2}{|\mathbf{u}|^2} + \frac{u_3^3}{|\mathbf{u}|^2}$$

or

$$\cos^2\alpha + \cos^2\beta + \cos^2\gamma = 1 \qquad (11.20)$$

If $\mathbf{u}$ is a unit vector, then by Equation 11.19 its components are precisely its direction cosines:

$$\mathbf{u} = \langle \cos\theta, \cos\beta, \cos\gamma \rangle \qquad \text{if } |\mathbf{u}| = 1$$

EXAMPLE 11.16   Find the direction cosines of the vector with initial point $P(7, -2, 4)$ and terminal point $Q(5, 3, 0)$.

Solution   Let $\mathbf{u} = \overrightarrow{PQ} = \langle -2, 5, -4 \rangle$. Then $|\mathbf{u}| = \sqrt{4 + 25 + 16} = \sqrt{45} = 3\sqrt{5}$. So by Equation 11.19,

$$\cos\alpha = \frac{-2}{3\sqrt{5}} \qquad \cos\beta = \frac{5}{3\sqrt{5}} \qquad \cos\gamma = \frac{-4}{3\sqrt{5}} \qquad\blacksquare$$

EXAMPLE 11.17   A unit vector $\mathbf{u}$ makes an angle of $60°$ with the positive x-axis and with the positive y-axis. What angle does it make with the positive z-axis? What are the components of $\mathbf{u}$?

Solution   By the given information, $\alpha = \beta = \frac{\pi}{3}$, and by Equation 11.20,

$$\cos^2\gamma = 1 - \cos^2\alpha - \cos^2\beta = 1 - \frac{1}{4} - \frac{1}{4} = \frac{1}{2}$$

So $\cos \gamma = \pm 1/\sqrt{2}$. Thus, $\gamma = \frac{\pi}{4}$ or $\frac{3\pi}{4}$. There are therefore two possibilities for **u**:

$$\mathbf{u} = \left\langle \frac{1}{2}, \frac{1}{2}, \frac{1}{\sqrt{2}} \right\rangle \qquad \text{or} \qquad \mathbf{u} = \left\langle \frac{1}{2}, \frac{1}{2}, -\frac{1}{\sqrt{2}} \right\rangle \qquad \blacksquare$$

Exercise Set 11.3

1. Plot each of the following points.
   (a) $(3, 2, 4)$
   (b) $(4, -2, 1)$
   (c) $(-3, 2, 4)$
   (d) $(0, -5, -2)$
   (e) $(-4, -3, -6)$

2. Find the distance between P and Q.
   (a) $P(2, 0, -1)$, $Q(3, 5, 7)$
   (b) $P(-3, 5, 2)$, $Q(-1, -1, 4)$
   (c) $P(8, 2, 0)$, $Q(7, 6, -3)$
   (d) $P(4, -2, -3)$, $Q(-1, -3, 2)$

3. Identify each of the following three-dimensional point sets.
   (a) The set of all points for which $z = 0$
   (b) The set of all points for which $x = 0$
   (c) The set of all points for which $y = 0$
   (d) All points of the form $(x, 0, 0)$
   (e) All points for which $xyz = 0$

4. Let $P(x, y, z)$ be an arbitrary point in space and Q be the fixed point (h, k, l). Write an equation expressing the fact that $d(P, Q) = a$, where a is a positive constant. Clear the equation of radicals. How would you describe the set of all points P that satisfy this equation?

5. Express the vector $\overrightarrow{PQ}$ in terms of its components.
   (a) $P(7, 3, -1)$, $Q(5, -1, 2)$
   (b) $P(2, -3, -4)$, $Q(-1, -2, 0)$
   (c) $P(0, -2, 7)$, $Q(3, 0, 5)$
   (d) $P(-4, 6, 10)$, $Q(2, 4, 8)$

6. Find the magnitude of $\overrightarrow{PQ}$ for each part of Exercise 5.

7. Let $\mathbf{u} = \langle 3, 1, -2 \rangle$, $\mathbf{v} = \langle -1, 0, 4 \rangle$, and $\mathbf{w} = \langle 4, 1, 5 \rangle$. Find the following.
   (a) $\mathbf{u} \cdot \mathbf{w} - |\mathbf{v}|^2$
   (b) $\mathbf{v} \cdot (\mathbf{u} - \mathbf{w})$
   (c) $|\mathbf{u} - 2\mathbf{v}|$
   (d) $3\mathbf{u} + 2\mathbf{v} - \mathbf{w}$

8. Let $\mathbf{u} = 3\mathbf{i} - 2\mathbf{j} - \mathbf{k}$, $\mathbf{v} = 5\mathbf{i} - 4\mathbf{k}$, and $\mathbf{w} = -4\mathbf{i} + 6\mathbf{j} + 2\mathbf{k}$. Find the following.
   (a) $\mathbf{u} \cdot (\mathbf{v} - \mathbf{w})$

   (b) $|3\mathbf{u} + 2\mathbf{w}|$
   (c) $|\mathbf{u}| \, |\mathbf{w}| - |\mathbf{u} \cdot \mathbf{w}|$
   (d) $(\mathbf{v} + \mathbf{w}) \cdot (\mathbf{v} - \mathbf{w})$

9. Find a unit vector in the direction of **u**.
   (a) $\mathbf{u} = \langle 2, -1, 2 \rangle$
   (b) $\mathbf{u} = \langle 4, 3, -5 \rangle$
   (c) $\mathbf{u} = \mathbf{i} - \mathbf{j} + \mathbf{k}$
   (d) $\mathbf{u} = 2\mathbf{i} - 4\mathbf{j} + 5\mathbf{k}$

10. Find the cosine of the angle between **u** and **v**.
    (a) $\mathbf{u} = \langle 3, -2, 6 \rangle$, $\mathbf{v} = \langle 1, 1, 1 \rangle$
    (b) $\mathbf{u} = 4\mathbf{i} + 2\mathbf{j} - 2\mathbf{k}$, $\mathbf{v} = -7\mathbf{i} + 4\mathbf{j} + 5\mathbf{k}$

11. (a) Show that $\mathbf{u} = 2\mathbf{i} - 3\mathbf{j} + \mathbf{k}$ and $\mathbf{v} = 4\mathbf{i} + 2\mathbf{j} - 2\mathbf{k}$ are orthogonal.
    (b) Find x so that $\mathbf{u} = \langle x, -1, 2 \rangle$ and $\mathbf{v} = \langle 6, 4, x \rangle$ will be orthogonal.

12. Use vector methods to show that the points $A(-1, 2, -3)$, $B(1, -1, 2)$, and $C(0, 5, 6)$ are vertices of a right triangle.

13. Find the direction cosines of **u**.
    (a) $\mathbf{u} = 2\mathbf{i} - \mathbf{j} + 2\mathbf{k}$
    (b) $\mathbf{u} = \langle 3, -5, 4 \rangle$

14. Find the vector **u** with magnitude 3 whose z component is positive, if $\cos \alpha = \frac{1}{3}$ and $\cos \beta = -\sqrt{2}/3$.

15. Find the unit vector for which $\beta = \pi/3$ and $\gamma = \pi/4$ and whose x component is negative.

16. If a vector makes equal acute angles with **i**, **j**, and **k**, what is this angle?

17. Show that if the direction angles α, β, and γ of a vector are all acute and $\alpha \geq \pi/3$ and $\beta \geq \pi/3$, then $\gamma \leq \pi/4$.

In Exercises 18–21, find the component of **u** *along* **v**.

18. $\mathbf{u} = \langle -3, 1, 4 \rangle$, $\mathbf{v} = \langle 2, -1, -2 \rangle$

19. $\mathbf{u} = \langle 5, 4, -4 \rangle$, $\mathbf{v} = \langle 3, 0, -4 \rangle$

20. $u = 7i - 2k$, $v = 3i - 5j + 4k$

21. $u = i + 2j - k$, $v = i - j + k$

*In Exercises 22–24, find the work done by the force **F** in moving a particle along the line segment from P to Q.*

22. $F = 10i + 12j - 8k$; $P(2, -1, 4)$, $Q(3, 5, 2)$; $|F|$ in newtons, distance in meters

23. $F = 20i - 12j + 6k$; $P(3, 4, 6)$, $Q(8, -1, 10)$; $|F|$ in dynes, distance in centimeters

24. $F = 6i + 2j + 8k$; $P(-1, 3, 5)$, $Q(4, -1, 9)$; $|F|$ in pounds, distance in feet

In Exercises 25–28, find $\text{Proj}_v u$ and $\text{Proj}_v^\perp u$.

25. $u = \langle 5, -1, 3 \rangle$, $v = \langle 2, 6, -4 \rangle$

26. $u = \langle 2, -3, 0 \rangle$, $v = \langle -5, 1, -2 \rangle$

27. $u = 2i - 3j - 5k$, $v = i + 2j - 3k$

28. $u = 4j - 5k$, $v = 3i - 5j + 4k$

In Exercises 29–33, prove that the indicated theorem continues to hold true for three-dimensional vectors.

29. Theorem 11.1

30. Theorem 11.2

31. Theorem 11.3

32. Theorem 11.4

33. Theorem 11.5

34. Find a nonzero vector $x = \langle x_1, x_2, x_3 \rangle$ that is perpendicular to each of the vectors $u = \langle 2, 1, -3 \rangle$ and $v = \langle -1, 1, 2 \rangle$. (*Hint:* Obtain two equations with three unknowns. Choose one of the unknowns arbitrarily.)

35. Find a unit vector orthogonal to each of the vectors $u = i - j + 2k$ and $v = 3i + 2j - 2k$. (See the hint in Exercise 34.)

36. Find scalars a and b such that $w = au + bv$, where $u = \langle 3, -2, 4 \rangle$, $v = \langle 1, 1, -2 \rangle$, and $w = \langle 6, 1, -2 \rangle$. Interpret the result geometrically.

37. Let $u_1 = \langle 1, -1, 0 \rangle$, $u_2 = \langle 0, 1, -1 \rangle$, and $u_3 = \langle 1, 1, 1 \rangle$. Find scalars a, b, and c such that $v = au_1 + bu_2 + cu_3$, where $v = \langle 3, 2, -1 \rangle$.

38. With u_1, u_2, and u_3 as in Exercise 37, show that *every* three-dimensional vector v can be expressed as a linear combination of u_1, u_2, and u_3. (*Note:* This shows that u_1, u_2, and u_3 constitute a basis for three-dimensional vector space.)

39. Prove that the sum of any two of the three direction angles α, β, and γ must be greater than or equal to $\pi/2$. (*Hint:* Suppose, for example, that $\alpha + \beta < \pi/2$, so that $\alpha < \pi/2 - \beta$. Show that $\cos^2 \alpha + \cos^2 \beta > 1$.)

CAS *Use a CAS in Exercises 40–51. In Exercises 40–43, find the direction cosines and the direction angles, in degrees and radians, of the given vectors. Give the direction angles correct to three decimal places.*

40. $\langle 2, 3, 5 \rangle$ **41.** $\langle 2.1, 3.2, -2 \rangle$

42. $-4i + 3j - 5k$ **43.** $i - 8j + 10k$

44. Verify that the points $(4, 6, 8)$, $(2, 3, 9)$, and $(7, 2, 2)$ form the vertices of a right triangle.

45. Use the dot product to find two unit vectors that are perpendicular to the vectors $\langle 1, 2, 3 \rangle$ and $\langle 2, 4, 3 \rangle$.

46. Let $u = 3i + 4j - 2k$ and $v = i + 2j + ak$.
   (a) Find a so that u and v are perpendicular.
   (b) Is it possible to find a so that u and v are parallel? Explain.
   (c) Approximate a so that the angle between u and v is $\pi/6$.

47. Let u and v be vectors in three space. Use a CAS to verify the identities:
   (a) $(u + v) \cdot (u - v) = |u|^2 - |v|^2$
   (b) $(u + v) \cdot (u + v) = |u|^2 + 2u \cdot v + |v|^2$

48. Let u and v be vectors in three space. Use a CAS to verify the identities:
   (a) $|u + v|^2 + |u - v|^2 = 2(|u|^2 + |v|^2)$
   (b) $|u + v|^2 - |u - v|^2 = 4u \cdot v$

49. Let u and v be any two distinct nonzero vectors in three space. Use a CAS to show that the vector $w = u - (u \cdot v)v/|v|^2)$ is orthogonal to v.

50. Two tugboats are pulling a barge through a channel.
   (a) One tugboat is pulling with a force of magnitude 350 N at an angle of 25° NE with the horizontal and the second is pulling with a force of magnitude 500 N at an angle of 35° SW with the horizontal. If the barge is being moved 2 km, estimate the work done by each tugboat.
   (b) One tugboat is pulling with a force of magnitude 250 N at an angle of 30° NE with the horizontal and the second is pulling with a force of magnitude w N at an angle of 25° SW with the horizontal. If the barge is moving horizontally, find the magnitude w of the force at which the second tugboat is pulling the barge.

51. Define two vectors u and v as $u = \langle 2, -1, 5 \rangle$ and $v = \langle -4, 2, 7 \rangle$. Find
   (a) $u + v$, $u - v$, and $-2u$
   (b) the length of v
   (c) a unit vector in the direction of v
   (d) $u \cdot v$
   (e) the angle between u and v
   (f) the component of u in the direction of v
   (g) the projection of u on v

11.4 THE CROSS PRODUCT

For three-dimensional vectors $\mathbf{u}$ and $\mathbf{v}$ there is a second type of product, called the **cross product**, written $\mathbf{u} \times \mathbf{v}$, that results in another vector, rather than a scalar as with the dot product. For this reason the cross product is sometimes called the **vector product**.

Definition 11.7 The Cross Product	The **cross product** of $\mathbf{u} = \langle u_1, u_2, u_3 \rangle$ and $\mathbf{v} = \langle v_1, v_2, v_3 \rangle$ is the vector $$\mathbf{u} \times \mathbf{v} = \langle u_2 v_3 - u_3 v_2, u_3 v_1 - u_1 v_3, u_1 v_2 - u_2 v_1 \rangle \qquad (11.21)$$

Determinant Notation for the Cross Product

Definition 11.7 can be remembered more easily using determinant notation, which we review briefly. A second-order determinant is defined by

$$\begin{vmatrix} a_1 & a_2 \\ b_1 & b_2 \end{vmatrix} = a_1 b_2 - a_2 b_1$$

For example,

$$\begin{vmatrix} 3 & 2 \\ -1 & 4 \end{vmatrix} = 3(4) - (2)(-1) = 12 + 2 = 14$$

A third-order determinant can be evaluated as follows:

$$\begin{vmatrix} a_1 & a_2 & a_3 \\ b_1 & b_2 & b_3 \\ c_1 & c_2 & c_3 \end{vmatrix} = a_1 \begin{vmatrix} b_2 & b_3 \\ c_2 & c_3 \end{vmatrix} - a_2 \begin{vmatrix} b_1 & b_3 \\ c_1 & c_3 \end{vmatrix} + a_3 \begin{vmatrix} b_1 & b_2 \\ c_1 & c_2 \end{vmatrix}$$

Each second-order determinant is then evaluated as above. The formula we have given is sometimes referred to as *expansion by the first row*. It is possible to expand by any row or column, but for our purposes the first row is the most convenient. To illustrate, consider the following:

$$\begin{vmatrix} 2 & -1 & -3 \\ 4 & 2 & 1 \\ 0 & 5 & -4 \end{vmatrix} = 2 \begin{vmatrix} 2 & 1 \\ 5 & -4 \end{vmatrix} - (-1) \begin{vmatrix} 4 & 1 \\ 0 & -4 \end{vmatrix} + (-3) \begin{vmatrix} 4 & 2 \\ 0 & 5 \end{vmatrix}$$

$$= 2(-8 - 5) + (-16) - 3(20) = -26 - 16 - 60 = -102$$

Now observe that Equation 11.21 can be written in the form

$$\mathbf{u} \times \mathbf{v} = \left\langle \begin{vmatrix} u_2 & u_3 \\ v_2 & v_3 \end{vmatrix}, -\begin{vmatrix} u_1 & u_3 \\ v_1 & v_3 \end{vmatrix}, \begin{vmatrix} u_1 & u_2 \\ v_1 & v_2 \end{vmatrix} \right\rangle$$

as you can verify by evaluating the second-order determinants and comparing the result with Equation 11.21. Equivalently, we can write $\mathbf{u} \times \mathbf{v}$ as

$$\mathbf{u} \times \mathbf{v} = \begin{vmatrix} u_2 & u_3 \\ v_2 & v_3 \end{vmatrix} \mathbf{i} - \begin{vmatrix} u_1 & u_3 \\ v_1 & v_3 \end{vmatrix} \mathbf{j} + \begin{vmatrix} u_1 & u_2 \\ v_1 & v_2 \end{vmatrix} \mathbf{k} \qquad (11.22)$$

A convenient way of remembering this formula is to write the third-order determinant

$$\begin{vmatrix} \mathbf{i} & \mathbf{j} & \mathbf{k} \\ u_1 & u_2 & u_3 \\ v_1 & v_2 & v_3 \end{vmatrix}$$

Since the first row consists of vectors instead of numbers, this is not a proper determinant. Nevertheless, if we formally expand it by the first row, we get (writing the scalars times the vectors instead of the reverse)

$$\begin{vmatrix} u_2 & u_3 \\ v_2 & v_3 \end{vmatrix} \mathbf{i} - \begin{vmatrix} u_1 & u_3 \\ v_1 & v_3 \end{vmatrix} \mathbf{j} + \begin{vmatrix} u_1 & u_2 \\ v_1 & v_2 \end{vmatrix} \mathbf{k}$$

and the result is seen, by comparison with Equation 11.22, to be the $\mathbf{i}$, $\mathbf{j}$, $\mathbf{k}$ notation for $\mathbf{u} \times \mathbf{v}$. Thus, with this understanding of what is meant by the determinant with vector entries in the first row, we have

$$\mathbf{u} \times \mathbf{v} = \begin{vmatrix} \mathbf{i} & \mathbf{j} & \mathbf{k} \\ u_1 & u_2 & u_3 \\ v_1 & v_2 & v_2 \end{vmatrix} \tag{11.23}$$

EXAMPLE 11.18   Find $\mathbf{u} \times \mathbf{v}$, where $\mathbf{u} = \langle 3, -1, 4 \rangle$ and $\mathbf{v} = \langle -2, 2, 5 \rangle$.

Solution   By Equation 11.23,

$$\mathbf{u} \times \mathbf{v} = \begin{vmatrix} \mathbf{i} & \mathbf{j} & \mathbf{k} \\ 3 & -1 & 4 \\ -2 & 2 & 5 \end{vmatrix} = \begin{vmatrix} -1 & 4 \\ 2 & 5 \end{vmatrix} \mathbf{i} - \begin{vmatrix} 3 & 4 \\ -2 & 5 \end{vmatrix} \mathbf{j} + \begin{vmatrix} 3 & -1 \\ -2 & 2 \end{vmatrix} \mathbf{k}$$

$$= (-5 - 8)\mathbf{i} - (15 + 8)\mathbf{j} + (6 - 2)\mathbf{k}$$

$$= -13\mathbf{i} - 23\mathbf{j} + 4\mathbf{k}$$

Equivalently, $\mathbf{u} \times \mathbf{v} = \langle -13, -23, 4 \rangle$.   ∎

Geometric Interpretation of the Cross Product

One of the most important properties of the cross product is given by the following theorem.

THEOREM 11.6

The vector $\mathbf{u} \times \mathbf{v}$ is orthogonal to both $\mathbf{u}$ and $\mathbf{v}$.

Proof   We will show that $\mathbf{u} \cdot (\mathbf{u} \times \mathbf{v}) = 0$ and leave it as an exercise to show that $\mathbf{v} \cdot (\mathbf{u} \times \mathbf{v}) = 0$. By Equation 11.21,

$$\mathbf{u} \cdot (\mathbf{u} \times \mathbf{v}) = \langle u_1, u_2, u_3 \rangle \cdot \langle (u_2 v_3 - u_3 v_2), (u_3 v_1 - u_1 v_3), (u_1 v_2 - u_2 v_1) \rangle$$

$$= u_1(u_2 v_3 - u_3 v_2) + u_2(u_3 v_1 - u_1 v_3) + u_3(u_1 v_2 - u_2 v_1)$$

$$= u_1 u_2 v_3 - u_1 u_3 v_2 + u_2 u_3 v_1 - u_2 u_1 v_3 + u_3 u_1 v_2 - u_3 u_2 v_1$$

$$= 0$$

So $\mathbf{u}$ and $\mathbf{u} \times \mathbf{v}$ are orthogonal.   ∎

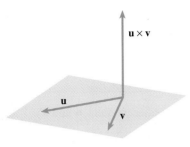

FIGURE 11.32

If **u** and **v** are nonzero and are not parallel, then we know from the preceding theorem that **u** × **v** is orthogonal to both **u** and **v**. Suppose geometric representatives of **u**, **v**, and **u** × **v** are drawn with the same initial point. Then **u** × **v** is perpendicular to the plane containing **u** and **v**, as shown in Figure 11.32. The direction of **u** × **v** is determined according to the following right-hand rule: if you curl the fingers of your right hand in the direction that would rotate **u** into **v** (through an angle of less than π), then your extended thumb will point in the direction of **u** × **v**.

The magnitude of **u** × **v** is related to the magnitudes of **u** and **v** as given in the following theorem.

THEOREM 11.7

If θ is the angle between the nonzero vectors **u** and **v**, then

$$|\mathbf{u} \times \mathbf{v}| = |\mathbf{u}|\,|\mathbf{v}|\sin\theta \tag{11.24}$$

Proof   By Equation 11.21 we have

$$|\mathbf{u} \times \mathbf{v}|^2 = (u_2 v_3 - u_3 v_2)^2 + (u_3 v_1 - u_1 v_3)^2 + (u_1 v_2 - u_2 v_1)^2$$

We leave it as an exercise for you to show that if the right-hand side is expanded and terms are appropriately grouped, it can be written in the form

$$(u_1^2 + u_2^2 + u_3^2)(v_1^2 + v_2^2 + v_3^2) - (u_1 v_1 + u_2 v_2 + u_3 v_3)^2$$

Thus,

$$|\mathbf{u} \times \mathbf{v}|^2 = |\mathbf{u}|^2\,|\mathbf{v}|^2 - (\mathbf{u} \cdot \mathbf{v})^2$$

Since $\mathbf{u} \cdot \mathbf{v} = |\mathbf{u}|\,|\mathbf{v}|\cos\theta$, we have

$$\begin{aligned}
|\mathbf{u} \times \mathbf{v}|^2 &= |\mathbf{u}|^2\,|\mathbf{v}|^2 - |\mathbf{u}|^2\,|\mathbf{v}|^2 \cos^2\theta \\
&= |\mathbf{u}|^2\,|\mathbf{v}|^2 (1 - \cos^2\theta) \\
&= |\mathbf{u}|^2\,|\mathbf{v}|^2 \sin^2\theta
\end{aligned}$$

Taking square roots, we get the desired result.   ∎

COROLLARY 11.7

Two three-dimensional vectors **u** and **v** are parallel if and only if $\mathbf{u} \times \mathbf{v} = \mathbf{0}$.

Proof   If either **u** or **v** is **0**, the result is trivial. If they are both nonzero, they are parallel if and only if $\theta = 0$ or $\theta = \pi$ or, equivalently, $\sin\theta = 0$. Thus, by Equation 11.24, they are parallel if and only if $|\mathbf{u} \times \mathbf{v}| = 0$, and hence, if and only if $\mathbf{u} \times \mathbf{v} = \mathbf{0}$.   ∎

Area of a Parallelogram

Equation 11.24 has an interesting geometric interpretation. Let **u** and **v** be nonzero, with angle θ between them. Choose geometric representatives of **u** and **v** that have the same initial point. Complete the parallelogram with **u**

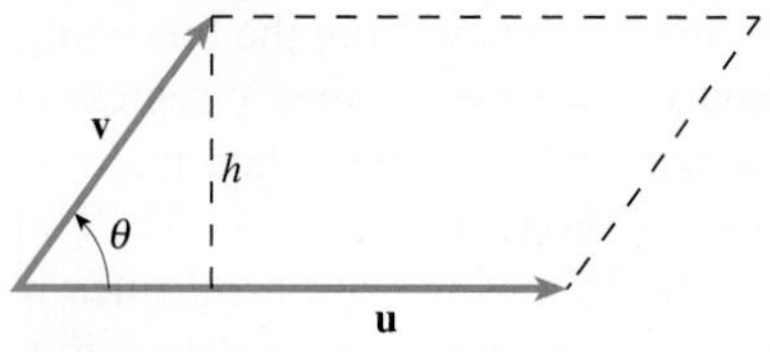

FIGURE 11.33

and **v** as adjacent sides, as in Figure 11.33. From that figure we see that the height h of the parallelogram from the base **u** is $h = |\mathbf{v}| \sin \theta$. Thus, its area is $|\mathbf{u}|h = |\mathbf{u}| |\mathbf{v}| \sin \theta$. So by Theorem 11.7,

$$|\mathbf{u} \times \mathbf{v}| = \text{the area of the parallelogram with adjacent sides } \mathbf{u} \text{ and } \mathbf{v}$$

EXAMPLE 11.19   Find the area of the triangle with vertices $A(2, 1, 4)$, $B(3, -1, 7)$, and $C(-1, 2, 5)$.

Solution   Let $\mathbf{u} = \overrightarrow{AB}$ and $\mathbf{v} = \overrightarrow{AC}$. Then the area of the triangle is one-half the area of the parallelogram determined by **u** and **v**, or

$$\text{Area} = \frac{1}{2} |\mathbf{u} \times \mathbf{v}|$$

We first find $\mathbf{u} \times \mathbf{v}$:

$$\mathbf{u} = \overrightarrow{AB} = \langle 1, -2, 3 \rangle$$

$$\mathbf{v} = \overrightarrow{AC} = \langle -3, 1, 1 \rangle$$

$$\mathbf{u} \times \mathbf{v} = \begin{vmatrix} \mathbf{i} & \mathbf{j} & \mathbf{k} \\ 1 & -2 & 3 \\ -3 & 1 & 1 \end{vmatrix} = \begin{vmatrix} -2 & 3 \\ 1 & 1 \end{vmatrix} \mathbf{i} - \begin{vmatrix} 1 & 3 \\ -3 & 1 \end{vmatrix} \mathbf{j} + \begin{vmatrix} 1 & -2 \\ -3 & 1 \end{vmatrix} \mathbf{k}$$

$$= -5\mathbf{i} - 10\mathbf{j} - 5\mathbf{k}$$

Thus, the area of the triangle is

$$\text{Area} = \frac{1}{2} |\mathbf{u} \times \mathbf{v}| = \frac{1}{2} \sqrt{25 + 100 + 25} = \frac{1}{2} \sqrt{150} = \frac{5}{2} \sqrt{6} \qquad \blacksquare$$

Algebraic Properties of the Cross Product

The next theorem provides some other properties of the cross product. The proof of each part can be shown by direct application of Definition 11.7. In some cases the proof can be facilitated, however, using Equation 11.23 and the following two properties of determinants:

1. If two rows in a determinant are identical, the value of the determinant is 0.
2. If two rows in a determinant are interchanged, the result is the negative of the original determinant.

In the exercises you will be asked to verify these properties for third-order determinants, and you will also be asked to verify the theorem.

THEOREM 11.8

For three-dimensional vectors $\mathbf{u}$, $\mathbf{v}$, and $\mathbf{w}$,

1. $\mathbf{u} \times \mathbf{v} = -(\mathbf{v} \times \mathbf{u})$ Anticommutative property
2. $k(\mathbf{u} \times \mathbf{v}) = (k\mathbf{u}) \times \mathbf{v} = \mathbf{u} \times (k\mathbf{v})$ for any scalar k
3. $\mathbf{u} \times \mathbf{0} = \mathbf{0}$
4. $\mathbf{u} \times \mathbf{u} = \mathbf{0}$
5. $\mathbf{u} \times (\mathbf{v} + \mathbf{w}) = \mathbf{u} \times \mathbf{v} + \mathbf{u} \times \mathbf{w}$ Left distributive property
6. $(\mathbf{u} + \mathbf{v}) \times \mathbf{w} = \mathbf{u} \times \mathbf{w} + \mathbf{v} \times \mathbf{w}$ Right distributive property
7. $\mathbf{u} \times (\mathbf{v} \times \mathbf{w}) = (\mathbf{u} \cdot \mathbf{w})\mathbf{v} - (\mathbf{u} \cdot \mathbf{v})\mathbf{w}$
8. $\mathbf{u} \cdot (\mathbf{v} \times \mathbf{w}) = (\mathbf{u} \times \mathbf{v}) \cdot \mathbf{w}$

It is useful to learn the various cross products involving pairs of the basis vectors $\mathbf{i}$, $\mathbf{j}$, and $\mathbf{k}$. Direct application of Definition 11.7 gives

$$\mathbf{i} \times \mathbf{j} = \mathbf{k} \qquad \mathbf{j} \times \mathbf{k} = \mathbf{i} \qquad \mathbf{k} \times \mathbf{i} = \mathbf{j}$$

By Property 1 in Theorem 11.8, if the factors on the left are reversed, the sign on the right becomes negative. That is,

$$\mathbf{j} \times \mathbf{i} = -\mathbf{k} \qquad \mathbf{k} \times \mathbf{j} = -\mathbf{i} \qquad \mathbf{i} \times \mathbf{k} = -\mathbf{j}$$

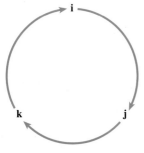

One way to remember these relationships is by the diagram in Figure 11.34. Going clockwise, crossing a vector with the following one produces the next one. Going counterclockwise produces the negative of the next one.

The cross product is neither commutative nor associative in general. Non-commutativity follows from Property 1 of Theorem 11.8, and nonassociativity can be seen, for example, by the following calculations:

$$(\mathbf{i} \times \mathbf{j}) \times \mathbf{j} = \mathbf{k} \times \mathbf{j} = -\mathbf{i} \qquad \text{but} \qquad \mathbf{i} \times (\mathbf{j} \times \mathbf{j}) = \mathbf{i} \times \mathbf{0} = \mathbf{0}$$

FIGURE 11.34

Triple Scalar Product

The product $\mathbf{u} \cdot (\mathbf{v} \times \mathbf{w})$ is called the **triple scalar product** of $\mathbf{u}$, $\mathbf{v}$, and $\mathbf{w}$. Applying Definition 11.7 to $\mathbf{v} \times \mathbf{w}$, we get

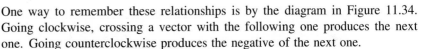

$$\mathbf{u} \cdot (\mathbf{v} \times \mathbf{w}) = u_1(v_2 w_3 - v_3 w_2) + u_2(v_3 w_1 - v_1 w_3) + u_3(v_1 w_2 - v_2 w_1)$$

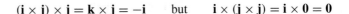

$$= u_1 \begin{vmatrix} v_2 & v_3 \\ w_2 & w_3 \end{vmatrix} - u_2 \begin{vmatrix} v_1 & v_3 \\ w_1 & w_3 \end{vmatrix} + u_3 \begin{vmatrix} v_1 & v_2 \\ w_1 & w_2 \end{vmatrix}$$

The right-hand side is the result of expanding by the first row the determinant whose rows, in order, are the components of $\mathbf{u}$, $\mathbf{v}$, and $\mathbf{w}$, respectively. Thus,

$$\mathbf{u} \cdot (\mathbf{v} \times \mathbf{w}) = \begin{vmatrix} u_1 & u_2 & u_3 \\ v_1 & v_2 & v_3 \\ w_1 & w_2 & w_3 \end{vmatrix} \tag{11.25}$$

Volume of a Parallelepiped

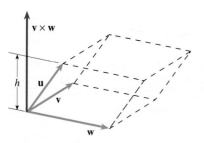

FIGURE 11.35

The triple scalar product has an interesting geometric interpretation. Let **u**, **v**, and **w** be nonzero vectors that do not lie in the same plane. Take geometric representatives of **u**, **v**, and **w** that have the same initial point and construct a parallelepiped, with these vectors as edges, as in Figure 11.35. Using as a base the parallelogram determined by **v** and **w**, the altitude h is the absolute value of the component of **u** perpendicular to this base; that is,

$$h = |\text{Comp}_{\mathbf{v} \times \mathbf{w}} \mathbf{u}|$$

As we have seen, the area of the base is $|\mathbf{v} \times \mathbf{w}|$, and the volume of the parallelepiped is the area of the base times the altitude. So we have

$$\text{vol} = h|\mathbf{v} \times \mathbf{w}| = |\text{Comp}_{\mathbf{v} \times \mathbf{w}} \mathbf{u}| \, |\mathbf{v} \times \mathbf{w}|$$

$$= \frac{|\mathbf{u} \cdot (\mathbf{v} \times \mathbf{w})|}{|\mathbf{v} \times \mathbf{w}|} |\mathbf{v} \times \mathbf{w}|$$

$$= |\mathbf{u} \cdot (\mathbf{v} \times \mathbf{w})|$$

We have shown, then, that

The volume of the parallelepiped with adjacent edges **u**, **v**, and **w** is

$$\text{vol} = |\mathbf{u} \cdot (\mathbf{v} \times \mathbf{w})|$$

EXAMPLE 11.20   Given the points $A(3, -1, 1)$, $B(2, 3, -2)$, $C(0, 1, 3)$, and $D(-1, 2, 4)$, find the volume of the parallelepiped determined by the vectors $\overrightarrow{AB}$, $\overrightarrow{AC}$, and $\overrightarrow{AD}$.

Solution   Let

$$\mathbf{u} = \overrightarrow{AB} = \langle -1, 4, -3 \rangle$$

$$\mathbf{v} = \overrightarrow{AC} = \langle -3, 2, 2 \rangle$$

$$\mathbf{w} = \overrightarrow{AD} = \langle -4, 3, 3 \rangle$$

By Equation 11.25,

$$\mathbf{u} \cdot (\mathbf{v} \times \mathbf{w}) = \begin{vmatrix} -1 & 4 & -3 \\ -3 & 2 & 2 \\ -4 & 3 & 3 \end{vmatrix} = -1 \begin{vmatrix} 2 & 2 \\ 3 & 3 \end{vmatrix} - 4 \begin{vmatrix} -3 & 2 \\ -4 & 3 \end{vmatrix} - 3 \begin{vmatrix} -3 & 2 \\ -4 & 3 \end{vmatrix}$$

$$= 0 - 4(-1) - 3(-1) = 7$$

The volume is $|\mathbf{u} \cdot (\mathbf{v} \times \mathbf{w})| = |7| = 7$.   ∎

Coplanar Vectors

In arriving at the volume of the parallelepiped determined by **u**, **v**, and **w** as $|\mathbf{u} \cdot (\mathbf{v} \times \mathbf{w})|$, we assumed the vectors were not coplanar. However, an analysis of the computations we made will show that for any nonzero vectors **u**, **v**, and **w**,

$$|\mathbf{u} \cdot (\mathbf{v} \times \mathbf{w})| = |\text{Comp}_{\mathbf{v} \times \mathbf{w}}\mathbf{u}| \, |\mathbf{v} \times \mathbf{w}|$$

and if **u** is in the same plane with **v** and **w**, $\text{Comp}_{\mathbf{v} \times \mathbf{w}}\mathbf{u} = 0$, since $\mathbf{v} \times \mathbf{w}$ is orthogonal to **u**. Conversely, if $\text{Comp}_{\mathbf{v} \times \mathbf{w}}\mathbf{u} = 0$, then **u** is orthogonal to $\mathbf{v} \times \mathbf{w}$ and hence in the plane of **v** and **w**. We therefore conclude that if **u**, **v**, and **w** have the same initial point,

---

u, **v**, and **w** are coplanar if and only if $\mathbf{u} \cdot (\mathbf{v} \times \mathbf{w}) = 0$.

VECTOR OPERATIONS USING COMPUTER ALGEBRA SYSTEMS

The examples considered in this section use Maple and Mathematica* to examine most of the vector operations that have been discussed in the first four sections of this chapter.

CAS 39

Define the vectors **u** and **v** as $\mathbf{u} = \langle 3, 2, 4 \rangle$ and $\mathbf{v} = \langle -1, 4, 2 \rangle$. Using Maple and Mathematica, calculate each of the following vector and/or scalar quantities.

(a) $\mathbf{u} + \mathbf{v}$, $\mathbf{u} - \mathbf{v}$ and $-3\mathbf{u}$.
(b) the length of **v**
(c) A unit vector in the direction of **v**
(d) the dot product of **u** and **v**
(e) the angle between **u** and **v**
(f) the component of **u** in the direction of **v**
(g) the projection of **u** on **v**
(h) the cross product of **u** and **v**

---

*From this chapter on, we no longer show the DERIVE commands for solving these examples. They are lengthy, and space considerations do not allow their inclusion.

Maple:[†]

In Maple, you will need to load the libraries plots and linalg using

with(plots):with(linalg):
u:=[3,2,4];
v:=[-1,4,2];
(a)
add(u,v);

Output: [2 6 6]

add(u,-v);

Output: [4 -2 2]

scalarmul(u,-3);

Output: [-9 -6 -12]

(b)
mag_v:=sqrt(v[1]^2+v[2]^2+v[3]^2);

Output: $mag\_v := \sqrt{21}$

(c)
unit_v:=scalarmul(v,l/mag_v);

Output: $unit\_v := \left[-\frac{1}{21}\sqrt{21} \quad \frac{4}{21}\sqrt{21} \quad \frac{2}{21}\sqrt{21} \right]$

(d)
dot_uv:=dotprod(u,v);

Output: $dot\_uv := 13$

(e)
mag_u:=sqrt(u[1]^2+u[2]^2+u[3]^2);

Output: $mag\_u := \sqrt{29}$

angle_uv:=arccos(dot_uv/(mag_u*mag_v));

Output: $angle\_uv := \arccos\left(\frac{13}{609}\sqrt{29}\sqrt{21} \right)$

evalf(");

Output: 1.015980710

(f)
comp_uv:=mag_u*cos(");

Output: $comp\_uv := \frac{13}{21}\sqrt{21}$

(g)
Proj_uv:=scalarmul(v,dot_uv/(mag_v)^2);

Output: $Proj\_uv := \left[\frac{-13}{21} \quad \frac{52}{21} \quad \frac{26}{21} \right]$

(h)
cross_uv:=crossprod(u,v);

Output: $cross\_uv:=[-12 \ -10 \ 14]$

Mathematica:

(a)
u={3,2,4}
v={-1,4,2}
u+v
u-v
-3u

(b)
mv=Sqrt[(v[[1]])^2+(v[[2]])^2+(v[[3]])^2]

(c)
unitv = (1/mv)v

(d)
dotuv = u.v

(e)
mu = Sqrt[(u[[1]])^2+(u[[2]])^2+(u[[3]])^2]
angleuv = ArcCos[dotuv/mu*mv]
N[%]

(f)
compuv = mu*Cos[angleuv]

(g)
Projuv = dotuv/mv^2*v

(h)
cuv = CrossProduct[u,v]

[†]To define vectors in Maple, enclose the coordinates in square brackets [ ], separated by commas; and in Mathematica, enclose the coordinates in curly braces { }, also separated by commas.

Maple:

In Maple, we continue and give the commands necessary to plot **u**, **v**, and the cross product **u** × **v**. Recall the vector **u** × **v** is perpendicular to both vectors **u** and **v**.

u_1:=convert(u,list);

Output: $u\_1 := [3, 2, 4]$

p1:=polygonplot3d([[0,0,0],u_1]);
v_1:=convert(v,list);

Output: $v\_1 := [-1, 4, 2]$

p2:=polygonplot3d([[0,0,0],v_1]);
cross_uv1:=convert(cross_uv,list);

Output: $cross\_uv1 := [-12, -10, 14]$

p3:=polygonplot3d([[0,0,0],cross_uvl]);
display3d({p1,p2,p3},axes=boxed,orientation=[−54,49]);

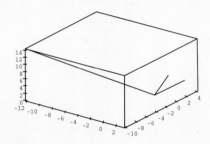

FIGURE 11.4.1

⊞ CAS 40

(a) Given the three points in space $A(-1, 2, 3)$, $B(2, 3, -1)$, and $C(3, -2, 4)$, find the area of the triangle that has these three points as vertices.

(b) Given the points $A(3, 2, -1)$, $B(2, 3, -2)$, $C(0, -2, 4)$, and $D(-1, 2, 3)$, find the volume of the parallelepiped determined by the vectors $\overrightarrow{AB}$, $\overrightarrow{AC}$, and $\overrightarrow{AD}$.

To compute the specified area and volume we will use the equations

$$\text{Area} = |\mathbf{u} \times \mathbf{v}|$$

$$\text{Volume} = |\mathbf{u} \cdot (\mathbf{v} \times \mathbf{w})|$$

where **u**, **v**, and **w** define the sides of the objects.

Maple:

(a)
A:=[–1,2,3];
B:=[2,3,–1];
C:=[3,–2,4];
A_B:=add(A,–B);
A_C:=add(A,–C);
Cross:=crossprod(A_B,A_C);
Area:=1/2*sqrt(Cross[1]^2+Cross[2]^2+Cross[3]^2);

Output: $Area := \frac{1}{2}\sqrt{842}$

(b)
A:=[3,2,–1];
B:=[2,3,–2];
C:=[0,–2,4];
D:=[–1,2,3];
u:=add(A,–B);
v:=add(A,–C);
w:=add(A,–D);
vol:=abs(dotprod(u,crossprod(v,w)));

Output: $vol := 24$

Mathematica:

(a)
A = {–1,2,3}
B = {2,3,–1}
C1 = {3,–2,4}
AB = A–B
AC = A–C1
CP = CrossProduct[AB,AC]
Area = 1/2*Sqrt[(CP[[1]])^2+(CP[[2]])^2+(CP[[3]])^2]

(b)
A = {3,2,–1}
B = {2,3,–2}
C1 = {0,–2,4}
D1 = {–1,2,3}
u = A–B
v = A–C1
w = A–D1
vol = Abs[u.CrossProduct[v,w]]

Exercise Set 11.4

In Exercises 1–8, find $\mathbf{u} \times \mathbf{v}$.

1. $\mathbf{u} = \langle 3, 1, -2 \rangle$, $\mathbf{v} = \langle -1, 1, 1 \rangle$

2. $\mathbf{u} = \langle 2, 0, -1 \rangle$, $\mathbf{v} = \langle 0, 2, 1 \rangle$

3. $\mathbf{u} = 4\mathbf{i} - 2\mathbf{j} + \mathbf{k}$, $\mathbf{v} = \mathbf{i} + \mathbf{j} - 2\mathbf{k}$

4. $\mathbf{u} = 3\mathbf{i} - 2\mathbf{j}$, $\mathbf{v} = 2\mathbf{i} + 3\mathbf{k}$

5. $\mathbf{u} = \langle 5, -3, -2 \rangle$, $\mathbf{v} = \langle -2, -3, 1 \rangle$

6. $\mathbf{u} = \langle 2, -1, 2 \rangle$, $\mathbf{v} = \langle -3, 4, -1 \rangle$

7. $\mathbf{u} = \mathbf{i} - 3\mathbf{j} + 4\mathbf{k}$, $\mathbf{v} = 2\mathbf{i} - \mathbf{j} - 5\mathbf{k}$

8. $\mathbf{u} = 6\mathbf{i} - 5\mathbf{j} + 4\mathbf{k}$, $\mathbf{v} = 4\mathbf{i} - 3\mathbf{j} - \mathbf{k}$

In Exercises 9–12, find a vector orthogonal to each of the given vectors.

9. $\mathbf{u} = \langle 0, 1, -3 \rangle$, $\mathbf{v} = \langle 2, 4, -1 \rangle$

10. $\mathbf{u} = \langle 3, 2, -3 \rangle$, $\mathbf{v} = \langle 2, 1, -4 \rangle$

11. $\mathbf{u} = 3\mathbf{i} - 2\mathbf{j} - 5\mathbf{k}$, $\mathbf{v} = \mathbf{i} + 4\mathbf{j} + 3\mathbf{k}$

12. $\mathbf{u} = 2\mathbf{i} - 3\mathbf{j} - \mathbf{k}$, $\mathbf{v} = 3\mathbf{i} - 2\mathbf{j} - \mathbf{k}$

In Exercises 13–19, $\mathbf{u} = 2\mathbf{i} - \mathbf{j} + \mathbf{k}$, $\mathbf{v} = \mathbf{i} + 2\mathbf{j} - 3\mathbf{k}$, *and* $\mathbf{w} = 3\mathbf{i} + 2\mathbf{j} - \mathbf{k}$. *Compute the value of the given expressions in Exercises 13–18.*

13. $\mathbf{u} \cdot (\mathbf{v} \times \mathbf{w})$

14. $(\mathbf{u} \times \mathbf{v}) \cdot \mathbf{w}$

15. $\mathbf{u} \times (\mathbf{v} \times \mathbf{w})$

16. $(\mathbf{u} \times \mathbf{v}) \times \mathbf{w}$

17. $(\mathbf{u} \times \mathbf{v}) \cdot (\mathbf{u} \times \mathbf{w})$

18. $(\mathbf{u} \times \mathbf{v}) \times (\mathbf{u} \times \mathbf{w})$

19. Show that $\mathbf{v}$ and $\mathbf{u} \times \mathbf{v}$ are orthogonal.

In Exercises 20–23, find the area of the parallelogram that has $\overrightarrow{AB}$ and $\overrightarrow{AC}$ as adjacent sides.

20. $A(3, 1, 0)$, $B(2, 2, -1)$, $C(4, 0, 2)$

21. $A(-1, 1, 3)$, $B(1, 3, 2)$, $C(-2, 2, -1)$

22. $A(4, -2, -7)$, $B(3, 1, -5)$, $C(-1, 2, 0)$

23. $A(0, 2, -1)$, $B(4, 0, 2)$, $C(3, -1, -4)$

24. Find the area of the triangle with vertices $A(4, -2, 3)$, $B(6, 1, -1)$, and $C(5, 2, 3)$.

25. Find the area of the triangle with vertices $A(1, 0, -2)$, $B(-3, 2, 1)$, and $C(4, -2, -3)$.

26. Find a vector perpendicular to the plane that contains the points $A(3, 4, 5)$, $B(-1, 2, 4)$, and $C(2, 3, 1)$.

27. Find a unit vector perpendicular to the plane that contains the points $P(0, -1, 3)$, $Q(1, 3, 2)$, and $R(2, -1, 4)$.

In Exercises 28 and 29 find the volume of the parallelepiped that has $\overrightarrow{AB}$, $\overrightarrow{AC}$, and $\overrightarrow{AD}$ as edges.

28. $A(3, 2, -5)$, $B(1, 4, -2)$, $C(-2, 3, 0)$, $D(4, 3, -8)$

29. $A(-2, 0, 4)$, $B(1, 1, 2)$, $C(0, 3, -1)$, $D(-3, -2, 4)$

30. Show that the vectors $\mathbf{u} = 2\mathbf{i} - 3\mathbf{j} + 4\mathbf{k}$, $\mathbf{v} = \mathbf{i} + 2\mathbf{j} - \mathbf{k}$, and $\mathbf{w} = 7\mathbf{i} + 5\mathbf{k}$ are coplanar.

31. Show that the points $A(1, -1, 2)$, $B(3, -4, 1)$, $C(0, 1, 2)$, and $D(1, 0, 1)$ all lie in the same plane.

32. Supply the missing steps in the proof of Theorem 11.7.

33. Prove that for a third-order determinant if two rows are identical, the value of the determinant is 0.

34. Prove that for a third-order determinant if two rows are interchanged, the resulting determinant is the negative of the original.

35. Prove Properties 1, 2, and 3 in Theorem 11.8.

36. Prove Properties 4, 5, and 6 in Theorem 11.8.

37. Prove Properties 7 and 8 in Theorem 11.8.

In Exercises 38–42, prove the given identities based on the properties given in Theorem 11.8, where $\mathbf{u}$, $\mathbf{v}$, $\mathbf{w}$, and $\mathbf{z}$ are three-dimensional vectors.

38. $\mathbf{u} \cdot (\mathbf{u} \times \mathbf{v}) = 0$

39. $(\mathbf{u} + \mathbf{v}) \times (\mathbf{u} - \mathbf{v}) = 2(\mathbf{v} \times \mathbf{u})$

40. $(\mathbf{u} \times \mathbf{v}) \times \mathbf{w} = (\mathbf{u} \cdot \mathbf{w})\mathbf{v} - (\mathbf{v} \cdot \mathbf{w})\mathbf{u}$

41. $\mathbf{u} \times (\mathbf{v} + \mathbf{w}) + \mathbf{v} \times (\mathbf{w} + \mathbf{u}) + \mathbf{w} \times (\mathbf{u} + \mathbf{v}) = \mathbf{0}$

42. $(\mathbf{u} \times \mathbf{v}) \cdot (\mathbf{w} \times \mathbf{z}) = \begin{vmatrix} \mathbf{u} \cdot \mathbf{w} & \mathbf{u} \cdot \mathbf{z} \\ \mathbf{v} \cdot \mathbf{w} & \mathbf{v} \cdot \mathbf{z} \end{vmatrix}$ (*Hint:* First apply Property 8 in Theorem 11.8 to the left-hand side, then Property 7, and use properties of the dot product.)

43. (a) Let $P_1(x_1, y_1, z_1)$, $P_2(x_2, y_2, z_2)$, and $P_3(x_3, y_3, z_3)$ be any three noncollinear points in space. Show that the area of the triangle that has these points as vertices is

$$A = \frac{1}{2} \left| \overrightarrow{P_1 P_2} \times \overrightarrow{P_1 P_3} \right|$$

(b) By treating points in two dimensions as points in three dimensions with the z-coordinate equal to 0, use the result of part (a) to show that the area of a triangle in a two-dimensional coordinate system that has $P_1(x_1, y_1)$, $P_2(x_2, y_2)$, and $P_3(x_3, y_3)$ as vertices can be put in the form

$$A = \pm \frac{1}{2} \begin{vmatrix} x_1 & y_1 & 1 \\ x_2 & y_2 & 1 \\ x_3 & y_3 & 1 \end{vmatrix}$$

where the sign is chosen so that the result is nonnegative.

44. Use a CAS to verify each of the following identities for vectors $\mathbf{u}$, $\mathbf{v}$, and $\mathbf{w}$ in three space.
   (a) $|\mathbf{u} \times \mathbf{v}|^2 = |\mathbf{u}|^2 |\mathbf{v}|^2 - (\mathbf{u} \cdot \mathbf{v})^2$
   (b) $\mathbf{u} \times (\mathbf{v} \times \mathbf{w}) = (\mathbf{u} \cdot \mathbf{w})\mathbf{v} - (\mathbf{u} \cdot \mathbf{v})\mathbf{w}$

45. Define the vectors $\mathbf{u}$ and $\mathbf{v}$ as $\mathbf{u} = \langle 2, 1, 4 \rangle$ and $\mathbf{v} = \langle -1, -3, 5 \rangle$. Find the cross product of $\mathbf{u}$ and $\mathbf{v}$ and show the three vectors in space.

46. Given the three points $(1, -2, 3)$, $(2, 1, 4)$, and $(1, -2, 1)$, find the area of the triangle that has these three points as vertices.

47. Given the four points $A(2, 4, -2)$, $B(1, -3, 6)$, $C(-1, 2, 4)$, and $D(1, 3, -2)$, find the volume of the parallelepiped determined by the vectors $\overrightarrow{AB}$, $\overrightarrow{AC}$, and $\overrightarrow{AD}$.

48. Find all vectors $\mathbf{v}$ for which $\langle 1, 1, 2 \rangle \times \mathbf{v} = \mathbf{0}$.

11.5 LINES IN SPACE

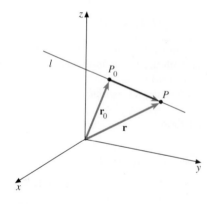

FIGURE 11.36

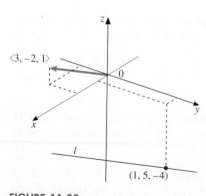

FIGURE 11.37

Vector Equation

A line in space can be described by a point on the line and a direction for the line. The direction is specified by means of a vector, called a **direction vector**, that is parallel to the line. This way of characterizing a line is similar to the two-dimensional case where a point and a slope are given. Suppose l is a line that passes through $P_0(x_0, y_0, z_0)$ and has direction vector $\mathbf{v} = \langle a, b, c \rangle$. Position $\mathbf{v}$ so that its initial point is at P_0, as in Figure 11.36. A point $P(x, y, z)$ will be on l if and only if

$$\overrightarrow{P_0P} = t\mathbf{v} \qquad (11.26)$$

for some scalar t. As we allow t to range over all real numbers, P traces out the entire line.

Equation 11.26 can be put in another form using position vectors. Recall that if $P(x, y, z)$ is a point in space, then its position vector is the vector $\overrightarrow{OP}$ that has initial point at the origin and terminal point at P. The components of the position vector for P are precisely the coordinates of P, namely, $\langle x, y, z \rangle$. Now let $\mathbf{r}_0$ and $\mathbf{r}$ be the position vectors of the points P_0 and P, respectively, on the line l. Then, as illustrated in Figure 11.37, $\overrightarrow{P_0P} = \mathbf{r} - \mathbf{r}_0$. So Equation 11.26 can be written in the form

$$\mathbf{r} = \mathbf{r}_0 + t\mathbf{v}, \qquad -\infty < t < \infty \qquad (11.27)$$

and we call this equation a **vector equation for the line l**.

EXAMPLE 11.21

(a) Find a vector equation of the line l passing through the point $(1, 5, -4)$ and parallel to the vector $\langle 3, -2, 1 \rangle$.

(b) Given the equation

$$\mathbf{r} = (2 - t)\mathbf{i} + (4 + 3t)\mathbf{j} + (-2 + 5t)\mathbf{k}, \qquad -\infty < t < \infty$$

describe its graph.

Solution

(a) Let $\mathbf{r}_0$ be the position vector $\langle 1, 5, -4 \rangle$ of the given point. The vector $\langle 3, -2, 1 \rangle$ is a direction vector for l. So by Equation 11.27, the position vector $\mathbf{r}$ of any point on l is

$$\mathbf{r} = \langle 1, 5, -4 \rangle + t\langle 3, -2, 1 \rangle = \langle 1 + 3t, 5 - 2t, -4 + t \rangle$$

We show the line l along with the direction vector $\langle 3, -2, 1 \rangle$ in Figure 11.38.

(b) We can rewrite the equation as

$$\mathbf{r} = (2\mathbf{i} + 4\mathbf{j} - 2\mathbf{k}) + t(-\mathbf{i} + 3\mathbf{j} + 5\mathbf{k})$$

By comparison with Equation 11.27, we see that this equation is a vector equation for the line passing through $(2, 4, -2)$ and parallel to the vector $-\mathbf{i} + 3\mathbf{j} + 5\mathbf{k}$. (See Figure 11.39.) ∎

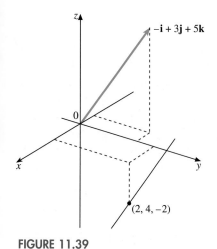

FIGURE 11.39

REMARK ───────────────────

■ Notice that we call Equation 11.27 *a* vector equation for l rather than *the* vector equation, because neither $\mathbf{r}_0$ nor $\mathbf{v}$ is unique. We are free to use *any* point on l as P_0 and *any* nonzero vector parallel to l as its direction vector $\mathbf{v}$.

───────────────────

EXAMPLE 11.22   Find a vector equation of the line l passing through the points $P(2, -1, 4)$ and $Q(3, 2, -1)$.

Solution   We can use either point P or Q as the fixed point with position vector $\mathbf{r}_0$ of Equation 11.27. Letting this point be P, we have $\mathbf{r}_0 = \langle 2, -1, 4 \rangle$. The direction vector $\mathbf{v}$ can be taken as $\overrightarrow{PQ} = \langle 1, 3, -5 \rangle$. Thus, a vector equation for l is

$$\mathbf{r} = \langle 2, -1, 4 \rangle + t\langle 1, 3, -5 \rangle = \langle 2 + t, -1 + 3t, 4 - 5t \rangle$$

Notice that when $t = 0$, $\mathbf{r}$ is the position vector of P and when $t = 1$, $\mathbf{r}$ is the position vector of Q.   ■

Parametric Equations of a Line

If we write Equation 11.27 in component form, we get

$$\langle x, y, z \rangle = \langle x_0, y_0, z_0 \rangle + t\langle a, b, c \rangle$$
$$= \langle x_0 + at, y_0 + bt, z_0 + ct \rangle$$

and equating components yields

$$\begin{cases} x = x_0 + at \\ y = y_0 + bt \\ z = z_0 + ct \end{cases} \quad -\infty < t < \infty \qquad (11.28)$$

Equations 11.28 are **parametric equations for the line** l. If we know a vector equation for l, we can write parametric equations, and conversely.

EXAMPLE 11.23   Find parametric equations of the line in Example 11.22.

Solution   In Example 11.22, we found that

$$\mathbf{r} = \langle 2 + t, -1 + 3t, 4 - 5t \rangle$$

So parametric equations are

$$\begin{cases} x = 2 + t \\ y = -1 + 3t \\ z = 4 - 5t \end{cases} \quad -\infty < t < \infty \qquad ■$$

EXAMPLE 11.24   Give two points and a direction vector for the line l that has parametric equations $x = 3 - 5t$, $y = -4 + 7t$, and $z = 10 - 8t$.

Solution   One point on the line is $(3, -4, 10)$ corresponding to $t = 0$. We get another point using a different value of t, say $t = 1$, giving $(-2, 3, 2)$. A direction vector is $\langle -5, 7, -8 \rangle$.   ■

Direction Numbers

The numbers a, b, and c in Equations 11.28 that are the components of a direction vector $\mathbf{v}$ for l are also called **direction numbers** for l. Since $k\mathbf{v} = \langle ka, kb, kc \rangle$ is also a direction vector for l for any nonzero scalar k, it follows that ka, kb, and kc also are direction numbers for l. Knowing a set of direction numbers for l is equivalent to knowing a direction vector for l.

EXAMPLE 11.25   Find a set of direction numbers for the line

$$\mathbf{r} = \langle 3 - 6t, 2 + 4t, -5 - 8t \rangle$$

Solution   One set of direction numbers is $-6, 4, -8$. Another set is $3, -2, 4$, obtained by multiplying the first set by $-\frac{1}{2}$.   ∎

The Angle Between Two Lines

By the *angle between two lines* l_1 and l_2, we mean the angle between a direction vector for l_1 and a direction vector for l_2, or its supplement, whichever does not exceed $\pi/2$. If $\mathbf{u}$ is any direction vector for l_1 and $\mathbf{v}$ is any direction vector for l_2, then by our definition and by Equation 11.5, the angle θ between l_1 and l_2 satisfies

$$\cos\theta = \frac{|\mathbf{u} \cdot \mathbf{v}|}{|\mathbf{u}|\,|\mathbf{v}|}, \qquad 0 \le \theta \le \frac{\pi}{2}$$

The absolute value of $\mathbf{u} \cdot \mathbf{v}$ is needed to ensure that $\cos\theta \ge 0$. The lines are *parallel* if $\mathbf{u}$ and $\mathbf{v}$ are parallel, and they are *orthogonal* if $\mathbf{u}$ and $\mathbf{v}$ are orthogonal. Lines that do not intersect and are not parallel are called **skew lines**.

EXAMPLE 11.26   Find the angle between the lines l_1 and l_2, defined by $\mathbf{r} = \langle 1 - 2t, 3 + t, -2 + 3t \rangle$ and $\mathbf{r} = \langle -2 + t, 4, 3 - t \rangle$, respectively.

Solution   A direction vector for l_1 is $\mathbf{u} = \langle -2, 1, 3 \rangle$, and a direction vector for l_2 is $\mathbf{v} = \langle 1, 0, -1 \rangle$. So for the angle θ between l_1 and l_2,

$$\cos\theta = \frac{|\mathbf{u} \cdot \mathbf{v}|}{|\mathbf{u}|\,|\mathbf{v}|} = \frac{|\langle -2, 1, 3 \rangle \cdot \langle 1, 0, -1 \rangle|}{\sqrt{4 + 1 + 9}\,\sqrt{1 + 1}} = \frac{5}{2\sqrt{7}}$$

$$\theta = \cos^{-1}\frac{5}{2\sqrt{7}} \approx 0.3335 \text{ radian} \approx 19.11° \qquad ∎$$

EXAMPLE 11.27   Let l_1 be the line with vector equation $\mathbf{r} = \langle 3 + 3t, 5 - t, -1 + 2t \rangle$ and l_2 be the line with vector equation $\mathbf{r} = \langle -1 + 4t, 7 + 2t, 3 - 5t \rangle$.

(a) Show that l_1 and l_2 are orthogonal.

(b) Find a vector equation of a line passing through $(4, -2, 0)$ that is parallel to l_1.

Solution

(a) Direction vectors for l_1 and l_2 are $\mathbf{u} = \langle 3, -1, 2 \rangle$ and $\mathbf{v} = \langle 4, 2, -5 \rangle$, respectively. Since $\mathbf{u} \cdot \mathbf{v} = 12 - 2 - 10 = 0$, it follows that $\mathbf{u}$ and $\mathbf{v}$, and hence l_1 and l_2, are orthogonal.

(b) We may use the same direction vector $\mathbf{u} = \langle 3, -1, 2 \rangle$ as for l_1. Only the fixed point need be changed. So the equation is

$$\mathbf{r} = \langle 4, -2, 0 \rangle + t \langle 3, -1, 2 \rangle = \langle 4 + 3t, -2 - t, 2t \rangle$$ ∎

EXAMPLE 11.28   Find a vector equation of the line passing through $(7, -1, 2)$ that intersects the line $\mathbf{r} = \langle 2 - t, 4 + 3t, 5 - 2t \rangle$ orthogonally.

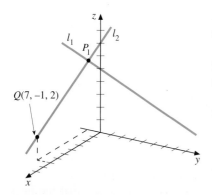

FIGURE 11.40

Solution   Let l_1 denote the given line and Q the given point. We need to determine the point on l_1, call it P_1, so that the line l_2 passing through P_1 and Q intersects l_1 in a right angle. (See Figure 11.40.) Equivalently, we need to find P_1, so that the dot product of the direction vector $\overrightarrow{P_1 Q}$ for the line l_2 and a direction vector for l_1 is zero.

Since P_1 lies on l_1, its coordinates can be written as $(2 - t_1, 4 + 3t_1, 5 - 2t_1)$ for some value of t_1, yet to be determined. Then

$$\overrightarrow{P_1 Q} = \langle 5 + t_1, -5 - 3t_1, -3 + 2t_1 \rangle$$

A direction vector for l_1 is

$$\mathbf{v} = \langle -1, 3, -2 \rangle$$

For orthogonality we must have

$$\mathbf{v} \cdot \overrightarrow{P_1 Q} = 0$$

or,

$$-5 - t_1 - 15 - 9t_1 + 6 - 4t_1 = 0$$
$$-14t_1 = 14$$
$$t_1 = -1$$

So the lines intersect at $(3, 1, 7)$, and the direction vector $\overrightarrow{P_1 Q}$ for l_2 is $\langle 4, -2, -5 \rangle$. An equation for l_2 is therefore

$$\mathbf{r} = \langle 3, 1, 7 \rangle + t \langle 4, -2, -5 \rangle$$

or

$$\mathbf{r} = \langle 3 + 4t, 1 - 2t, 7 - 5t \rangle$$ ∎

EXAMPLE 11.29   Let l_1 and l_2 be defined parametrically by

$$l_1 : \begin{cases} x = 3 - 2t \\ y = -1 + t \\ z = 2 + 3t \end{cases} \qquad l_2 : \begin{cases} x = 1 - s \\ y = 2 + 3s \\ z = 7 + 4s \end{cases}$$

Determine whether l_1 and l_2 intersect, are parallel, or are skew. If they intersect, find the point of intersection.

Solution   Notice that different letters are used to designate the parameters. This distinction is important, since otherwise we would disguise the fact that, if the lines intersect, the point of intersection might occur for different values of the parameters for l_1 or l_2.

Direction vectors for l_1 and l_2 are $\mathbf{u} = \langle -2, 1, 3 \rangle$ and $\mathbf{v} = \langle -1, 3, 4 \rangle$, and since $\mathbf{u} \neq k\mathbf{v}$, the lines are not parallel. To see whether they intersect, our approach will be to determine values of s and t that give the same values of x and y and then test to see whether the z values also are the same.

Setting the x and y values equal to each other gives

$$\begin{cases} 3 - 2t = 1 - s \\ -1 + t = 2 + 3s \end{cases} \quad \text{or} \quad \begin{cases} -2t + s = -2 \\ t - 3s = 3 \end{cases}$$

The simultaneous solution is easily found to be $t = \frac{3}{5}$, $s = -\frac{4}{5}$. The corresponding values of x and y are $x = \frac{9}{5}$, $y = -\frac{2}{5}$. The critical test is for z. For l_1, we have

$$z = 2 + 3t = 2 + 3\left(\frac{3}{5}\right) = \frac{19}{5}$$

and for l_2,

$$z = 7 + 4s = 7 + 4\left(-\frac{4}{5}\right) = \frac{19}{5}$$

The fact that we get the same value of z tells us that the lines do intersect. The point of intersection is $\left(\frac{9}{5}, -\frac{2}{5}, \frac{19}{5}\right)$.

If the z values had been different, we would have concluded that the lines were skew. ∎

Exercise Set 11.5

In Exercises 1–4, a point P_0 and a vector $\mathbf{v}$ are given. Find (a) a vector equation and (b) parametric equations for the line through P_0 that has direction vector $\mathbf{v}$.

1. $P_0(2, 5, -1)$, $\mathbf{v} = \langle -3, 1, 2 \rangle$

2. $P_0(-1, -2, 4)$, $\mathbf{v} = \langle 2, 5, -7 \rangle$

3. $P_0(5, 8, -6)$, $\mathbf{v} = 2\mathbf{i} - 3\mathbf{j} + 4\mathbf{k}$

4. $P_0(3, -9, 4)$, $\mathbf{v} = 3\mathbf{i} + 2\mathbf{j} - 5\mathbf{k}$

In Exercises 5–8, find a vector equation for the line through P and Q.

5. $P(4, -1, 8)$, $Q(3, 2, 5)$

6. $P(-1, 5, -6)$, $Q(2, 3, -1)$

7. $P(7, -2, -4)$, $Q(3, 1, -2)$

8. $P(4, 6, 9)$, $Q(1, -1, 5)$

In Exercises 9–12, find the angle between l_1 and l_2.

9. l_1: $\mathbf{r} = \langle 1 - 2t, 3 + t, 4 - 5t \rangle$
    l_2: $\mathbf{r} = \langle 2 - t, 1 - 2t, 3 + 2t \rangle$

10. l_1: $\mathbf{r} = (3 + 4t)\mathbf{i} + (2 - t)\mathbf{j} + (2 + 3t)\mathbf{k}$
     l_2: $\mathbf{r} = (1 - 3t)\mathbf{i} + (4 + t)\mathbf{j} + (7 - 2t)\mathbf{k}$

11. l_1: $x = 5 + 3t$, $y = 7 + 4t$, $z = 11 - 2t$
     l_2: $x = 4 - t$, $y = 5 + 2t$, $z = -1 + 3t$

12. l_1 has direction numbers $-1, 2, 3$, and l_2 has direction numbers $3, 5, -2$.

In Exercises 13 and 14, show that l_1 and l_2 are orthogonal.

13. l_1: $\mathbf{r} = \langle 8 + 3t, -6 - 2t, 7 + t \rangle$
     l_2: $\mathbf{r} = \langle 11 + 7t, 9 + 8t, 3 - 5t \rangle$

14. l_1: $x = 13 - 4t$, $y = -7 - 3t$, $z = 4 + 3t$
     l_2: $x = 6 + 6t$, $y = 8 - 5t$, $z = 12 + 3t$

15. Find parametric equations of the line through $(3, -1, 2)$ that is parallel to the line $\mathbf{r} = \langle 2 - 3t, 7 + t, 8 + 5t \rangle$.

16. Find a vector equation of the line through $P_1(4, -1, 3)$ that is parallel to the line through $P_2(-1, 0, 4)$ and $P_3(1, 3, 2)$.

In Exercises 17 and 18, find a vector equation of the line that passes through P_0 and has a direction vector that is orthogonal to both lines whose equations are given.

$n_1 = \langle 1, -2, -5 \rangle$

17. $P_0 = (5, 2, -3)$; $\mathbf{r} = \langle 2+t, 3-2t, 4-5t \rangle$,
  $\mathbf{r} = \langle 1-t, 2t, 3+4t \rangle$
  $n_2 = \langle -1, 2, 4 \rangle$

18. $P_0 = (0, -1, 2)$; $\mathbf{r} = (3+2t)\mathbf{i} + (4-3t)\mathbf{j} + (-2-t)\mathbf{k}$,
  $\mathbf{r} = (2-4t)\mathbf{i} + (-1+t)\mathbf{j} + 2\mathbf{k}$

19. Show that l_1 and l_2 intersect, and find parametric equations of a line orthogonal to both l_1 and l_2 at their point of intersection.

$$l_1 : \begin{cases} x = 2 - 3t \\ y = 1 + t \\ z = 5 - 4t \end{cases} \qquad l_2 : \begin{cases} x = 5 + 3s \\ y = -2 - 2s \\ z = 3 + s \end{cases}$$

In Exercises 20–24, determine whether l_1 and l_2 are parallel, intersecting, or skew. If they intersect, find their point of intersection.

20. l_1: $\mathbf{r} = \langle 11 - t, 7 + 2t, 8 - 3t \rangle$
  l_2: $\mathbf{r} = \langle 4 + 3s, 2 - 6s, 5 + 9s \rangle$

21. l_1: $x = 4 - t,\ y = 2t,\ z = 3 + 4t$
  l_2: $x = 2 + 3s,\ y = 1 - s,\ z = 4 + s$

22. l_1: $\mathbf{r} = (3 - 4t)\mathbf{i} + (2 + 3t)\mathbf{j} + (1 - t)\mathbf{k}$
  l_2: $\mathbf{r} = (2 + 2s)\mathbf{i} + (5 - 3s)\mathbf{j} + s\mathbf{k}$

23. l_1: $\mathbf{r} = \langle 3 - 4t, 2 + t, 2t \rangle$
  l_2: $\mathbf{r} = \langle 3 + 2s, 1 - s, 8 + 3s \rangle$

24. l_1: $\mathbf{r} = \langle 3 - 4t, 7 + 2t, -8 - 3t \rangle$
  l_2: $\mathbf{r} = \left\langle 11 + 2s, 3 - s, \dfrac{-4 + 3s}{2} \right\rangle$

25. If a, b, and c are all nonzero, show that the line $\mathbf{r} = \langle x_0 + at, y_0 + bt, z_0 + ct \rangle$ can be described by the equations

$$\frac{x - x_0}{a} = \frac{y - y_0}{b} = \frac{z - z_0}{c} \qquad (11.29)$$

These are called **symmetric equations** for the line. (*Hint:* Write parametric equations and eliminate the parameter.)

26. Referring to Exercise 25, suppose a line l has the symmetric equations

$$\frac{x - 2}{3} = \frac{y + 1}{2} = \frac{z - 3}{-5}$$

(a) Give two points on l.
(b) Find a unit direction vector for l.
(c) Give parametric equations for l.
(d) Give a vector equation for l.

27. A line l has direction numbers -2, 4, 3 and it contains the point $(3, -1, 4)$.
(a) Find symmetric equations for l. (See Exercise 25.)
(b) Find parametric equations for l.
(c) Determine a vector equation for l.
(d) Find the points where l pierces each of the coordinate planes.

28. Find a vector equation of the line passing through $(4, 0, -5)$ that intersects $\mathbf{r} = \langle 3 - t, 2t, 4 + 3t \rangle$ orthogonally.

29. Let l_1 be the line $\mathbf{r} = (3 + t)\mathbf{i} + (4 - 2t)\mathbf{j} + (5 + 2t)\mathbf{k}$ and let l_2 be a line passing through $(-1, 2, 8)$ that intersects l_1 so that the angle between l_1 and l_2 is $\frac{\pi}{4}$. Find a vector equation for l_2. (There are two solutions.)

30. Let l_1 and l_2 be the lines with vector equations $\mathbf{r} = \langle 3 - 2t, 4 + 3t, 1 - t \rangle$ and $\mathbf{r} = \langle 5 + 4t, 2 - 6t, 3 + 2t \rangle$, respectively.
(a) Show there is a line l_3 passing through $(2, 5, 0)$ that intersects l_1 and l_2 orthogonally, and find a vector equation for l_3.
(b) Find a vector equation of a line l_4 passing through the point $(5, -1, 7)$ that is perpendicular to the plane containing l_1 and l_2.

31. Let l be a line with direction vector $\mathbf{v}$. Show that the distance d from l to a point P not on l is

$$d = \frac{|\overrightarrow{PQ} \times \mathbf{v}|}{|\mathbf{v}|}$$

where Q is any point on l.

32. Find a vector equation of the line passing through the point $(1, -3, 5)$ and parallel to the vector $\langle 2, -3, 1 \rangle$. Draw the curve in space.

33. Find the angle between the lines defined by $\mathbf{r} = \langle 2 + t, 1 - t, -1 + 2t \rangle$ and $\mathbf{r} = \langle 2 - t, 1 + t, -2 - 3t \rangle$. Sketch the two lines in space and find their point of intersection.

34. Find the distance from the point $\left(1, \sqrt{2}\right)$ to the line $2x - 3y = 2$.

35. Use the formula $\mathbf{r} = (1 - t)\left(\overrightarrow{OP}\right) + t\left(\overrightarrow{OQ}\right)$, $0 \le t \le 1$ to find the parametric equations of the line segment from $P(1, -2, 4)$ to $Q(-2, 5, 6)$. Plot the line segment in space.

11.6 PLANES

Equations of Planes

Given a point $P_0(x_0, y_0, z_0)$ and a nonzero vector $\mathbf{n} = \langle a, b, c \rangle$, the set of all points P such that $\overrightarrow{P_0 P}$ and $\mathbf{n}$ are orthogonal is a plane (see Figure 11.41). The vector $\mathbf{n}$ is called a **normal vector** (or simply a **normal**) to the plane. The condition for orthogonality can be written as

$$\mathbf{n} \cdot \overrightarrow{P_0 P} = 0 \qquad (11.30)$$

As with equations of lines, if we let $\mathbf{r}_0$ and $\mathbf{r}$ be the position vectors of P_0 and P, respectively, then $\overrightarrow{P_0 P} = \mathbf{r} - \mathbf{r}_0$, so Equation 11.30 becomes

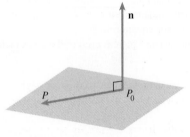

FIGURE 11.41

$$\mathbf{n} \cdot (\mathbf{r} - \mathbf{r}_0) = 0 \qquad (11.31)$$

We call Equation 11.31 a **vector equation of the plane**. In terms of components, $\mathbf{r} - \mathbf{r}_0 = \langle x - x_0, y - y_0, z - z_0 \rangle$, so Equation 11.31 becomes

$$a(x - x_0) + b(y - y_0) + c(z - z_0) = 0 \qquad (11.32)$$

Equation 11.32 is known as a **standard form** of the equation of the plane with normal $\langle a, b, c \rangle$ and containing the point (x_0, y_0, z_0). Neither the vector equation nor the standard form is unique, since we may use any point P_0 on the plane and any normal vector $\mathbf{n}$. Note, however, that all normal vectors to a given plane are parallel.

EXAMPLE 11.30   Find the equation of the plane passing through the point $(3, -1, 4)$ and having normal vector $\langle 2, 5, -3 \rangle$.

Solution   From Equation 11.32, we obtain

$$2(x - 3) + 5(y + 1) - 3(z - 4) = 0$$

Simplifying, we get

$$2x + 5y - 3z + 11 = 0 \qquad \blacksquare$$

If we carry out the indicated multiplications in Equation 11.32 and simplify, as we did in Example 11.30, we get an equation of the form

$$ax + by + cz + d = 0 \qquad (11.33)$$

where $d = -(ax_0 + by_0 + cz_0)$. We can also reverse this procedure. Suppose we are given an equation in the form of Equation 11.33. We find a point (x_0, y_0, z_0) that satisfies the equation (for example, by choosing x_0 and y_0 arbitrarily and solving for z_0). Then, $ax_0 + by_0 + cz_0 + d = 0$, so that $d = -ax_0 - by_0 - cz_0$. Substituting this value of d into Equation 11.33, we get

$$ax + by + cz + (-ax_0 - by_0 - cz_0) = 0$$

or

$$a(x - x_0) + b(y - y_0) + c(z - z_0) = 0$$

This equation is the standard form (11.32), so we know it represents a plane. An equation of the form of Equation 11.33 is called **linear** (meaning first degree) in x, y, and z. So what we have is that *every linear equation in x, y, and z represents a plane in space*. Furthermore, the coefficients of x, y, and z, in order, are components of a normal vector to the plane. We also refer to Equation 11.33 as a **general form** of the equation of a plane.

Finding Normal Vectors

A normal vector is not always given directly but can be found from the given information. Frequently, finding a normal involves finding a cross product, as the next two examples show.

EXAMPLE 11.31   Find an equation for the plane that contains the points $P(1, 0, -3)$, $Q(2, -5, -6)$, and $R(6, 3, -4)$.

Solution   As we show in Figure 11.42, the vectors $\overrightarrow{PQ}$ and $\overrightarrow{PR}$ lie in the plane, so a normal can be found by taking their cross product:

$$\overrightarrow{PQ} = \langle 1, -5, -3 \rangle \qquad \text{and} \qquad \overrightarrow{PR} = \langle 5, 3, -1 \rangle$$

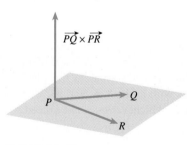

FIGURE 11.42

$$\overrightarrow{PQ} \times \overrightarrow{PR} = \begin{vmatrix} \mathbf{i} & \mathbf{j} & \mathbf{k} \\ 1 & -5 & -3 \\ 5 & 3 & -1 \end{vmatrix} = \begin{vmatrix} -5 & -3 \\ 3 & -1 \end{vmatrix} \mathbf{i} - \begin{vmatrix} 1 & -3 \\ 5 & -1 \end{vmatrix} \mathbf{j} + \begin{vmatrix} 1 & -5 \\ 5 & 3 \end{vmatrix} \mathbf{k}$$

$$= 14\mathbf{i} - 14\mathbf{j} + 28\mathbf{k}$$

Since any nonzero vector perpendicular to the plane is a suitable normal, we choose $\mathbf{n} = \frac{1}{14}(14\mathbf{i} - 14\mathbf{j} + 28\mathbf{k}) = \langle 1, -1, 2 \rangle$. We can use any one of P, Q, or R as the point (x_0, y_0, z_0). Choosing P, we get for the equation in standard form

$$(x - 1) - (y - 0) + 2(z + 3) = 0$$

or in general form

$$x - y + 2z + 5 = 0 \qquad \blacksquare$$

EXAMPLE 11.32   Find an equation of the plane that contains the line $\mathbf{r} = \langle 2 - t, 3 + 4t, -1 - 2t \rangle$ and the point $(5, -2, 7)$.

Solution   To use Equation 11.32 we need a normal vector to the plane and a point in the plane. To find a normal, we can take the cross product of two vectors in the plane. One vector in the plane is a direction vector of the given line. Since the equation of the line is $\mathbf{r} = \langle 2 - t, 3 + 4t, -1 - 2t \rangle$, a direction vector for it is

$$\mathbf{v} = \langle -1, 4, -2 \rangle$$

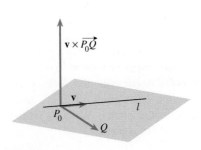

To find another vector in the plane, we can select any point P_0 on the given line and then use the vector $\overrightarrow{P_0Q}$, where Q is the given point $(5, -2, 7)$. (See Figure 11.43.) A convenient point P_0 on the line is found by taking $t = 0$, giving $P_0 = (2, 3, -1)$. Thus,

$$\overrightarrow{P_0Q} = \langle 3, -5, 8 \rangle$$

So a normal $\mathbf{n}$ is

$$\mathbf{v} \times \overrightarrow{P_0Q} = \langle -1, 4, -2 \rangle \times \langle 3, -5, 8 \rangle = \langle 22, 2, -7 \rangle$$

FIGURE 11.43

(You should supply the missing steps.) We can now write the equation of the plane as

$$22(x - 2) + 2(y - 3) - 7(z + 1) = 0$$

or, equivalently,

$$22x + 2y - 7z - 57 = 0 \qquad \blacksquare$$

Parallel and Perpendicular Planes; The Angle Between Two Planes

Two planes are parallel if their respective normals are parallel, and *two planes are perpendicular* if their respective normals are orthogonal. A line and a plane are parallel if a direction vector for the line and a normal to the plane are orthogonal. By the angle between two planes with normals $\mathbf{n}_1$ and $\mathbf{n}_2$, we mean the angle between $\mathbf{n}_1$ and $\mathbf{n}_2$ or its supplement, whichever does not exceed $\pi/2$. If we call this angle θ, then

$$\cos\theta = \frac{|\mathbf{n}_1 \cdot \mathbf{n}_2|}{|\mathbf{n}_1|\,|\mathbf{n}_2|}, \qquad 0 \le \theta \le \frac{\pi}{2} \qquad (11.34)$$

In Figure 11.44 we show planes that are parallel, planes that are perpendicular, and planes intersecting at an angle θ.

EXAMPLE 11.33   Find the angle between the planes

$$2x - 3y + 4z - 7 = 0 \qquad \text{and} \qquad x + y - 2z + 9 = 0$$

Solution   A normal vector to a plane of the form $ax + by + cz + d = 0$ is $\langle a, b, c \rangle$. So normals to the given planes are $\langle 2, -3, 4 \rangle$ and $\langle 1, 1, -2 \rangle$,

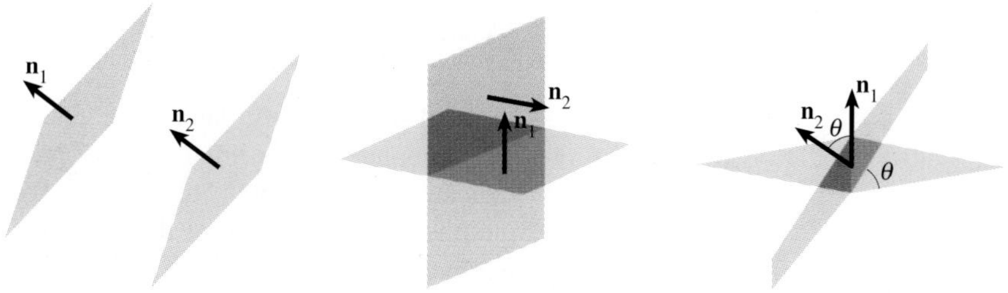

(a)

Parallel planes
$\mathbf{n}_1$ and $\mathbf{n}_2$ are parallel

(b)

Perpendicular planes
$\mathbf{n}_1$ and $\mathbf{n}_2$ are orthogonal

(c)

The angle between the
planes is θ if the acute
angle between their
normals is θ

FIGURE 11.44

respectively. So by Equation 11.34,

$$\cos \theta = \frac{|\langle 2, -3, 4 \rangle \cdot \langle 1, 1, -2 \rangle|}{|\langle 2, -3, 4 \rangle| \, |\langle 1, 1, -2 \rangle|}$$

$$= \frac{|2 - 3 - 8|}{\sqrt{29}\sqrt{6}} = \frac{9}{\sqrt{174}}$$

Using a calculator, we find that $\theta \approx 46.98°$. ∎

EXAMPLE 11.34   Find an equation of the plane that contains the line

$$\mathbf{r} = (2 + t)\mathbf{i} + (-3 + 4t)\mathbf{j} + (1 - t)\mathbf{k}$$

and that is perpendicular to the plane $3x - 4y + 5z + 7 = 0$.

Solution   Since the plane we want contains the line

$$\mathbf{r} = (2 + t)\mathbf{i} + (-3 + 4t)\mathbf{j} + (1 - t)\mathbf{k}$$

$$= \langle 2, -3, 1 \rangle + t\langle 1, 4, -1 \rangle$$

its normal must be orthogonal to the direction vector of the line,

$$\langle 1, 4, -1 \rangle = \mathbf{i} + 4\mathbf{j} - \mathbf{k}$$

The plane we want is also perpendicular to the plane $3x - 4y + 5z + 7 = 0$, with normal vector $\langle 3, -4, 5 \rangle = 3\mathbf{i} - 4\mathbf{j} + 5\mathbf{k}$. So the normal to the desired plane is orthogonal to this normal vector.

We therefore can obtain a normal to the plane we want by taking the cross product

$$(\mathbf{i} + 4\mathbf{j} - \mathbf{k}) \times (3\mathbf{i} - 4\mathbf{j} + 5\mathbf{k}) = 16\mathbf{i} - 8\mathbf{j} + 8\mathbf{k}$$

We use for $\mathbf{n}$ the simpler normal $2\mathbf{i} - \mathbf{j} + \mathbf{k}$. A point in the plane can be taken as any point on the given line. The one corresponding to $t = 0$ is $(2, -3, 1)$. Thus, the desired equation is

$$2(x - 2) - (y + 3) + (z - 1) = 0$$

or, in general form,

$$2x - y + z - 8 = 0$$ ∎

Sketching Planes

We will find it helpful later to be able to show planes graphically. Since planes are unbounded, it is impossible to graph an entire plane, but we can give an indication of its graph by showing its **traces** on the coordinate planes. These are the lines of intersection of the given plane with the coordinate planes. Finding *intercepts* on the coordinate axes is helpful in getting the traces.

For example, the plane

$$2x + 3y + 4z - 12 = 0$$

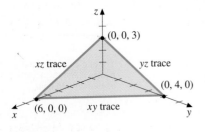

FIGURE 11.45

has x-intercept 6 (set $y = 0$ and $z = 0$), y-intercept 4 (set $x = 0$ and $z = 0$), and z-intercept 3 (set $x = 0$ and $y = 0$). Connecting these points gives the traces, as shown in Figure 11.45. Two-dimensional equations of the traces are found by setting one of the variables equal to 0. For example, if we set $z = 0$ in the equation of this plane, we get

$$2x + 3y = 12$$

which is the equation of the trace in the xy-plane.

Not all planes have intercepts on each axis. In the next example we illustrate some of these planes.

EXAMPLE 11.35   Describe the planes that have the following equations, and sketch their graphs.

(a) $x = 3$

(b) $z = 4$

(c) $2x + 3y = 6$

Solution   Recall that a normal vector to the plane $ax + by + cz + d = 0$ is $\langle a, b, c \rangle$.

(a) The equation $x = 3$ is of the form $ax + by + cz + d = 0$ with $a = 1$, $b = 0$, $c = 0$, and $d = -3$, so a normal vector is $\langle 1, 0, 0 \rangle = \mathbf{i}$. The plane is perpendicular to the x-axis, and hence parallel to the yz-plane. The x-intercept is 3. A portion of its graph is shown in Figure 11.46(a). The plane consists of all points (x, y, z) where x is always equal to 3, and y and z can have all possible values.

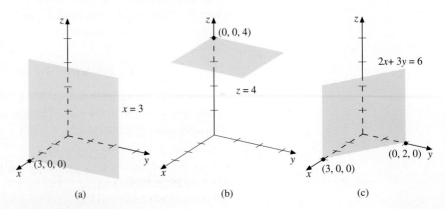

FIGURE 11.46

(b) A normal vector is $\langle 0, 0, 1 \rangle = \mathbf{k}$. So this plane is perpendicular to the z-axis, crossing it at $z = 4$. It is parallel to the xy-plane and 4 units above it. A portion of the graph is shown in Figure 11.46(b).

(c) The x-intercept is 3, and the y-intercept is 2. The xy trace has the same equation as that of the plane, since letting $z = 0$ does not alter the equation. A normal vector is $\langle 2, 3, 0 \rangle$, which is a vector in a horizontal plane, so the plane itself is vertical—that is, parallel to the z-axis. A portion of its graph is shown in Figure 11.46(c). ∎

Line of Intersection of Two Planes

Two nonparallel planes intersect in a line (see Figure 11.47), and we can find an equation of the line from the equations of the planes by the procedure shown in the next example.

EXAMPLE 11.36   Find a vector equation of the line of intersection of the planes $3x - 4y + 2z = 7$ and $x + 2y - 3z = 4$.

Solution   To find the equation of the line of intersection, we need a direction vector for the line and a point on it.

Note first that the planes do intersect, since their normal vectors $\mathbf{n}_1 = \langle 3, -4, 2 \rangle$ and $\mathbf{n}_2 = \langle 1, 2, -3 \rangle$ are not parallel (neither is a multiple of the other). So the planes themselves are not parallel. The line of intersection lies in both planes, hence is orthogonal to both of their normals, $\mathbf{n}_1$ and $\mathbf{n}_2$. A direction vector $\mathbf{v}$ of the line is therefore given by the cross product of these normals:

$$\mathbf{v} = \mathbf{n}_1 \times \mathbf{n}_2 = \langle 3, -4, 2 \rangle \times \langle 1, 2, -3 \rangle = \langle 8, 11, 10 \rangle$$

We can find a point on the line by solving simultaneously the equations of the planes. Since there are three variables and two equations, we can select one variable arbitrarily and solve for the other two. Letting $z = 0$, we have

$$\begin{cases} 3x - 4y = 7 \\ x + 2y = 4 \end{cases}$$

The solution is found to be $x = 3$, $y = \frac{1}{2}$. So a point on the line is $\left(3, \frac{1}{2}, 0\right)$. A vector equation of the line is therefore

$$\mathbf{r} = \left\langle 3, \frac{1}{2}, 0 \right\rangle + t \langle 8, 11, 10 \rangle$$

or

$$\mathbf{r} = \left\langle 3 + 8t, \frac{1}{2} + 11t, 10t \right\rangle$$ ∎

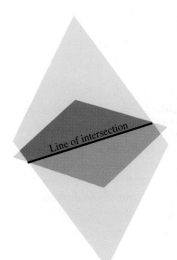

FIGURE 11.47

FIGURE 11.48

Distance Between a Plane and a Point

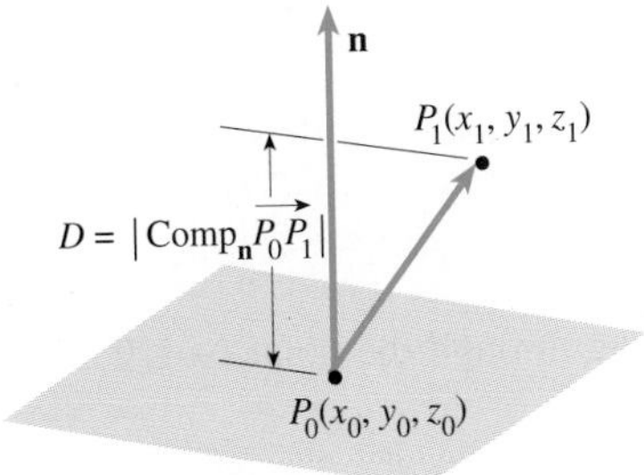

We conclude this section by deriving a formula for the distance between a plane and a point not on the plane. Let $ax + by + cz + d = 0$ be the equation of the plane, and let the point be $P_1(x_1, y_1, z_1)$. Let $P_0(x_0, y_0, z_0)$ be any point on the plane. Then, as Figure 11.48 shows, the distance D between the plane and the point P_1 is the length of the projection $\overrightarrow{P_0 P_1}$ along the normal $\mathbf{n} = \langle a, b, c \rangle$. The length of this projection is $|\text{Comp}_{\mathbf{n}} \overrightarrow{P_0 P_1}|$. So by Equation 11.8 we have

$$D = |\text{Comp}_{\mathbf{n}} \overrightarrow{P_0 P_1}| = \frac{|\mathbf{n} \cdot \overrightarrow{P_0 P_1}|}{|\mathbf{n}|} = \frac{|\langle a, b, c \rangle \cdot \langle x_1 - x_0, y_1 - y_0, z_1 - z_0 \rangle|}{\sqrt{a^2 + b^2 + c^2}}$$

$$= \frac{|a(x_1 - x_0) + b(y_1 - y_0) + c(z_1 - z_0)|}{\sqrt{a^2 + b^2 + c^2}}$$

$$= \frac{|ax_1 + by_1 + cz_1 - (ax_0 + by_0 + cz_0)|}{\sqrt{a^2 + b^2 + c^2}}$$

Since P_0 is on the plane, its coordinates satisfy the equation of the plane. So $ax_0 + by_0 + cz_0 = -d$. Thus, we obtain the following formula.

The distance D between the plane $ax + by + cz + d = 0$ and the point (x_1, y_1, z_1) is

$$D = \frac{|ax_1 + by_1 + cz_1 + d|}{\sqrt{a^2 + b^2 + c^2}} \tag{11.35}$$

REMARK

■ Compare this result with the two-dimensional case of the distance between a line and a point in Exercise 41 of Exercise Set 11.2.

EXAMPLE 11.37   Find the distance between the plane $3x - 4y + 5z - 8 = 0$ and the point $(2, 1, -1)$.

Solution   By Equation 11.35,

$$D = \frac{|3(2) - 4(1) + 5(-1) - 8|}{\sqrt{9 + 16 + 25}} = \frac{11}{5\sqrt{2}}$$ ■

INVESTIGATING LINES AND PLANES IN SPACE USING COMPUTER ALGEBRA SYSTEMS

CAS 41

(a) Find a vector equation for the line passing through the point $(-1, 2, 5)$ and parallel to the vector $\langle 2, -2, 1 \rangle$.

(b) Find the angle between the lines defined by $\mathbf{r} = \langle 1 + t, 2 + t, -1 + 2t \rangle$ and $\mathbf{r} = \langle 1 - t, 2 + t, -1 + 3t \rangle$.

(a) From Equation 11.27 the vector equation for the line is given by
$\langle -1, -2, 5 \rangle + t \langle 2, -2, 1 \rangle$.

Maple:

```
A:=[-1,2,5];
B:=[2,-2,1];
L:=add(A,scalarmul(B,t));
L:=convert(L,list);
spacecurve(L,t=-10..10,scaling=unconstrained,
   orientation=[-7,66],axes=framed);
```

Mathematica:

```
A = {-1,2,5}
B = {2,-2,1}
L = A+t*B
ParametricPlot3D[{-1+2*t,2-2*t,5+t},{t,-10,10}]
```

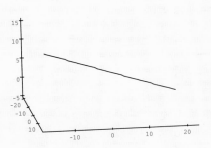

FIGURE 11.6.1

(b) To find the angle θ between the two lines we use the formula $\cos \theta = \dfrac{|\mathbf{u} \cdot \mathbf{v}|}{|\mathbf{u}||\mathbf{v}|}$ where $\mathbf{u}$ and $\mathbf{v}$ are direction vectors for the two lines respectively.

Maple:

```
L1:=add([1,2,-1],scalarmul([1,1,2],t));
L1:=convert(L1,list);
L2:=add([1,2,-1],scalarmul([-1,1,3],t));
L2:=convert(L2,list);
spacecurve({L1,L2},t=-10..10,scaling=unconstrained,
   orientation=[-35,65],axes=framed);
```

Mathematica:

```
L1 = {1,2,-1}+t*{1,1,2}
L2 = {1,2,-1}+t*{-1,1,3}
P1 = ParametricPlot3D[L1,{t,-10,10},
   DisplayFunction->Identity]
P2 = ParametricPlot3D[L2,{t,-10,10},
   DisplayFunction->Identity]
Show[{P1,P2},DisplayFunction->$DisplayFunction]
Solve[L1 == L2 , t]
x[t_] = L1
x[0]
u = {1,1,2}
v = {-1,1,3}
mu = Sqrt[(u[[1]])^2+(u[[2]])^2+(u[[3]])^2]
mv = Sqrt[(v[[1]])^2+(v[[2]])^2+(v[[3]])^2]
angle = ArcCos[Abs[u.v/(mu*mv)]]
anglerad = N[%]
angledeg = N[anglerad*180/Pi]
```

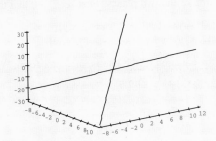

FIGURE 11.6.2

Now find the point of intersection of the two lines.

```
solve({L1[1]=L2[1],L1[2]=L2[2],L1[3]=L2[3]},t);
```

Output: $\{t = 0\}$

```
subs(t=0,L1);
```

Output: $[1, 2, -1]$

Direction vectors for the two lines are, respectively,

u:=[1,1,2];
v:=[-1,1,3];
mag_u:=sqrt(u[1]^2+u[2]^2+u[3]^2);

Output: $mag\_u := \sqrt{6}$

mag_v:=sqrt(v[1]^2+v[2]^2+v[3]^2);

Output: $mag\_v := \sqrt{11}$

angle:=arccos(abs(dotprod(u,v)/(mag_u*mag_u)));

Output: $angle := \arccos\left(\frac{1}{11}\sqrt{6}\sqrt{11}\right)$

anglerad:=evalf(angle);

Output: $anglerad := .7398807745$

angledeg:=evalf(anglerad*180/Pi);

Output: $angledeg := 42.39204571$

CAS 42

Make a plot displaying the planes $2x - y + 3z = 2$ and $3x - 2y + z = 6$ and find the vector equation for the line of intersection.

Maple:

plot3d({(2-2*x+y)/3,6-3*x+2*y},x=-10..10,y=-10..10,
    axes=boxed,orientation=[15,62]);

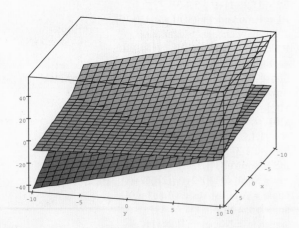

Mathematica:

P1 = Plot3D[(2-2*x+y)/3,{x,-10,10},{y,-10,10},
    DisplayFunction->Identity]
P2 = Plot3D[6-3*x+2*y,{x,-10,10},{y,-10,10},
    DisplayFunction->Identity]
Show [{P1,P2},DisplayFunction->$DisplayFunction]

To find the equation of the line of intersection of the two plane find two points on the line. That is two solutions of the system consisting of the two planes. First set $z = 0$ and then $z = 1$.

Solve[{2*x-y==2 , 3*x-2*y==6},{x,y}]
Solve[{2*x-y==-1 , 3*x-2*y==5},{x,y}]

Then the equation of the line is, in vector form,
$\{-2 - 7t, -6 - 13t, t\}$

FIGURE 11.6.3

We can let Maple find the equation of the line of intersection.

plane(p1,[2*x-y+3*z=2]),plane(p2,[3*x-2*y+z=6]):
inter(p1,p2,l);
l[equation];

Output:$[-2 + 5\_t, -6 + 7\_t, -\_t]$

CAS 43

Planes that are perpendicular to one of the axes can only be plotted using parametric form. Plot each of the following planes on the same coordinate axes:

$$x = 2,\ y = -2 \text{ and } z = 4$$

Maple:

plot3d({[2,t,s],[s,–2,t],[s,t,4]},t=–10..10,s=–10..10,axes=boxed);

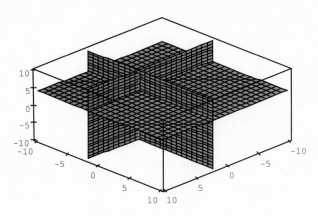

FIGURE 11.6.4

Mathematica:

P1 = ParametricPlot3D[{2,s,t},{s,–10,10},{t,–10,10}, DisplayFunction–>Identity]

P2 = ParametricPlot3D[{s,–2,t},{s,–10,10},{t,–10,10}, DisplayFunction–>Identity]

P3 = ParametricPlot3D[{s,t,4},{s,–10,10},{t,–10,10}, DisplayFunction–>Identity]
Show[{P1,P2,P3},DisplayFunction–>$DisplayFunction]

Exercise Set 11.6

In Exercises 1–4, find an equation of the plane through the given point perpendicular to the given vector in (a) vector form, (b) standard form, and (c) general form.

1. $(4, 2, 6),\ \langle 3, 2, -1 \rangle$

2. $(1, 0, -3),\ \langle -1, 2, 4 \rangle$

3. $(5, -3, -4),\ 2\mathbf{i} + 3\mathbf{j} - 4\mathbf{k}$

4. $(6, 1, 2),\ \mathbf{i} - 2\mathbf{j} + 2\mathbf{k}$

In Exercises 5–12, find a general form of the equation of the plane that satisfies the given conditions.

5. Perpendicular to the line $\mathbf{r} = \langle 2 - t, 3 + 2t, 1 + 4t \rangle$ at the point $(2, 3, 1)$

6. Perpendicular to the line $\mathbf{r} = (4 + 5t)\mathbf{i} + 3t\mathbf{j} + (1 - 2t)\mathbf{k}$ at the point $(4, 0, 1)$

7. Containing the point $(1, 4, -5)$ and parallel to the plane $3x - 2y + 4z = 7$

8. Containing the point $(-2, 3, 4)$ and parallel to the plane $2x - 3z = 4$

9. Containing the point $(3, -2, 4)$ and perpendicular to the z-axis

10. Containing the point $(-1, 8, 11)$ and perpendicular to the y-axis

11. Parallel to the plane $3x - 4y + 5z + 8 = 0$ and passing through the origin

12. Perpendicular to the line $\mathbf{r} = \langle 2 + t, 3 - 2t, 4 + 5t \rangle$ and passing through the origin

In Exercises 13–16, find a general equation of the plane that contains the three given points.

13. $(2, 4, -5),\ (1, -3, 4),\ (3, -1, 2)$

14. $(0, 3, -1)$, $(2, 4, 2)$, $(-1, 2, -3)$

15. $(1, 0, -1)$, $(2, 3, 1)$, $(4, -3, 2)$

16. $(5, 4, -3)$, $(2, -1, -2)$, $(4, 2, 3)$

In Exercises 17 and 18, find the angle between the two planes.

17. $2x - 3y - 4z = 8$
     $3x + 2y - z = 4$

18. $3x + 4y - 2z = 3$
     $2x - y - 3z = 5$

In Exercises 19 and 20, show that the two planes are perpendicular.

19. $3x - 4y + 2z = 5$
     $2x + 3y + 3z = 7$

20. $5x + 3y - 4z = 8$
     $2x - 6y - 2z = 15$

In Exercises 21–24, find a general equation of the plane that satisfies the given conditions.

21. Containing the line $\mathbf{r} = \langle 3 - 2t, 2 + t, 4 + 3t \rangle$ and the point $(-1, 2, 4)$

22. Containing the points $(4, -2, 1)$ and $(3, 1, 2)$ and perpendicular to the plane $3x + 2y - 4z = 5$

23. Perpendicular to each of the planes $3x + 5y - 4z = 4$ and $2x - 3y - z = 2$ and containing the point $(3, -3, 1)$

24. Containing the line $\mathbf{r} = 2\mathbf{i} + (3 - t)\mathbf{j} + (4 + 2t)\mathbf{k}$ and parallel to the line $\mathbf{r} = (3 + 2t)\mathbf{i} + (1 - t)\mathbf{j} + (-2 + 3t)\mathbf{k}$

In Exercises 25–34, use intercepts and traces on the coordinate planes to sketch the given plane.

25. $3x + 2y + z - 6 = 0$ **26.** $9x + 2y + 6z = 18$

27. $2x - y + z = 4$ **28.** $x + y - z + 4 = 0$

29. $3x + 4y = 12$ **30.** $2x + z = 8$

31. $y = 2z$ **32.** $x + 1 = 0$

33. $y = 4$ **34.** $x + y - 2z = 0$

In Exercises 35 and 36, find parametric equations of the line of intersection of the two planes.

35. $3x - y - 2z = 4$
     $5x + y + z = -2$

36. $x + 4y + 3z = 3$
     $2x - 7y + z = 11$

In Exercises 37–38, find the distance between the plane and the point.

37. $3x - 4y + 10z = 5$; $(1, -1, 2)$

38. $2x - y - 2z + 3 = 0$; $(6, -1, -4)$

39. Show that planes $x - 2y + 2z = 3$ and $3x - 6y + 6z + 5 = 0$ are parallel, and find the distance between them. (*Hint:*

Find the distance from one of the planes to a point on the other.)

40. Show that l_1 and l_2 are parallel, and find an equation of the plane that contains them.
 l_1: $\mathbf{r} = \langle 2 - 3t, 4 + 2t, -3 + t \rangle$
 l_2: $\mathbf{r} = \langle 6t, 3 - 4t, 5 - 2t \rangle$

41. Show that l_1 and l_2 intersect, and find an equation of the plane that contains them.
 l_1: $\mathbf{r} = \langle 1 + 2t, -1 + 3t, 2 - t \rangle$
 l_2: $\mathbf{r} = \langle 5 + 3t, 6 + 5t, -1 - 2t \rangle$

42. Find an equation of the plane perpendicular to the line of intersection of the planes $3x - y - 4z = 2$ and $x + 2y - z = 3$ at the point where this line pierces the xy-plane.

43. Find the point where the line $\mathbf{r} = (2 + t)\mathbf{i} + (3 - 2t)\mathbf{j} + (1 - t)\mathbf{k}$ pierces the plane $3x - 2y + 4z + 5 = 0$.

44. Find the minimum distance between the two skew lines
 l_1: $x = 3 + t$, $y = 2 - t$, $z = 4 + 3t$
 l_2: $x = -1 - 2t$, $y = 5 + 4t$, $z = -3t$
 (*Hint:* First find a plane that contains one of the lines and is parallel to the other.)

45. Make a plot displaying the two planes $x - 2y + 3z = 3$ and $3x - y - z = 4$ and find the equation of the line of intersection.

46. Find the equation of the plane that passes through the three points $A(-1, 2, 3)$, $B(2, 4, 5)$, and $C(1, 2, -3)$. Show the plane.

47. Consider the plane $2x + 3y + z = 3$ and the line L given parametrically by $x = t - 1$, $y = -t + 2$, and $z = t + 3$. By plotting several views of the line and the plane, show convincingly that they are parallel. Why are they parallel?

48. Do the plane $x + 2y + z = 1$ and the line given parametrically by $x = -3t - 1$, $y = 2t - 2$, and $z = t - 1$ intersect? Plot the line and plane first to answer the question; then give an argument for your answer.

49. Find the distance between the point $(-1, 2, 5)$ and the plane $-x + 2y + z = 7$.

50. Find the distance between the parallel planes $x + y - z = 5$ and $x + y - z = 10$. Draw the two planes.

51. A surveyor wishes to measure the height of a tall landmark. Using a rectangular coordinate system, the surveyor selects two points $P(81, 1, 0)$ and $Q(54, 131, 0)$ at ground level and sights from the points to the point T at the top of the landmark. Given that the direction vector for PT is $\langle -48/61, 11/61, 36/61 \rangle$ and the direction vector of QT is $\langle -23/49, -36/49, 24/49 \rangle$, determine the coordinates of the point T and hence deduce the height of the landmark.*

---

*From Chi-Keung Cheung and John Harer, *A Guidebook to Multivariable Calculus with Maple V* (John Wiley & Sons).

Chapter 11 Review Exercises

Exercises 1 and 2 refer to the vectors $\mathbf{u} = 3\mathbf{i} - 4\mathbf{j}$, $\mathbf{v} = \mathbf{i} + 2\mathbf{j}$, *and* $\mathbf{w} = 5\mathbf{i} - 7\mathbf{j}$. *Find the specified quantities.*

1. (a) $(2\mathbf{u} - 3\mathbf{v}) \cdot (\mathbf{v} + 2\mathbf{w})$
   (b) A vector in the direction of $\mathbf{v}$ with length $|3\mathbf{u} - 2\mathbf{w}|^2$

2. (a) $\text{Proj}_{\mathbf{v}}\mathbf{u}$ and $\text{Proj}_{\mathbf{v}}^{\perp}\mathbf{u}$
   (b) Scalars a and b such that $\mathbf{w} = a\mathbf{u} + b\mathbf{v}$

Exercises 3–6 refer to the vectors $\mathbf{u} = \langle 1, 1, -4 \rangle$, $\mathbf{v} = \langle -2, 0, 3 \rangle$, *and* $\mathbf{w} = \langle -2, 1, 2 \rangle$. *Find the specified quantities.*

3. (a) $|\mathbf{u}|^2 - (\mathbf{v} \cdot \mathbf{w})^2$
   (b) The angle between $\mathbf{u}$ and $\mathbf{w}$

4. (a) $(\mathbf{u} \times \mathbf{v}) \cdot \mathbf{w}$
   (b) $\mathbf{u} \times (\mathbf{v} \times \mathbf{w})$

5. (a) $|2\mathbf{u} - 3\mathbf{v} + \mathbf{w}|$
   (b) $\text{Proj}_{\mathbf{w}}\mathbf{v}$

6. (a) The component of $\mathbf{v}$ in the direction of $\mathbf{w} \times \mathbf{u}$
   (b) The volume of the parallelepiped with adjacent sides $\mathbf{u}$, $\mathbf{v}$, and $\mathbf{w}$

7. Using vector methods, prove that the line segment joining the midpoints of two sides of a triangle is parallel to the third side and one-half its length.

8. Forces $\mathbf{F}_1$, $\mathbf{F}_2$, and $\mathbf{F}_3$ of magnitudes 25.3 N, 14.8 N, and 19.6 N, respectively, are acting on an object as shown in the figure. Find the magnitude and direction of the resultant force.

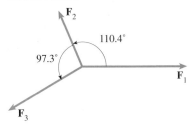

9. A pilot wants to fly from town A to town B, 400 mi due south of town A. A 60-mph wind is blowing from $210°$. To make the trip in 2 hr, at what heading and at what average speed should the pilot fly?

10. Prove that for arbitrary three-dimensional vectors $\mathbf{u}$, $\mathbf{v}$, and $\mathbf{w}$, $(\mathbf{u} \times \mathbf{v}) \times \mathbf{w} + (\mathbf{v} \times \mathbf{w}) \times \mathbf{u} + (\mathbf{w} \times \mathbf{u}) \times \mathbf{v} = \mathbf{0}$. (This is known as *Jacobi's identity*.)

In Exercises 11–13, find both vector and parametric equations of the line that satisfies the given conditions.

11. Through $(3, 1, -2)$, parallel to $\mathbf{r} = \langle 2 - t, 4 + 2t, 3 + 5t \rangle$

12. Formed by the intersection of the planes $2x - 3y - z = 4$ and $x + y + 2z + 3 = 0$

13. Through $(4, 2, -1)$, perpendicular to the plane $5x - 3y + 4z = 7$

14. Find the point where the line $\mathbf{r} = \langle 1 - 3t, 2t, 3 + t \rangle$ pierces the plane $3x - 5y + 4z = 5$. Also find the angle between the line and the plane (defined as the complement of the angle between the line and a normal to the plane).

15. Determine whether the lines $\mathbf{r}_1 = \langle 3 - t, 4 + 2t, -1 + t \rangle$ and $\mathbf{r}_2 = \langle 2s, 1 - s, 3 + 4s \rangle$ intersect. If so, find their point of intersection.

16. Show that the lines $\mathbf{r}_1 = \langle 2 - t, 4 + 2t, -3 + 4t \rangle$ and $\mathbf{r}_2 = \langle 1 + t, 5 - 3t, 3 - 2t \rangle$ intersect, and find an equation of the plane that contains them.

In Exercises 17 and 18, find an equation of the plane that satisfies the given conditions.

17. Containing the points $(2, 0, -1)$, $(3, 2, 0)$, and $(-4, -2, 3)$

18. Perpendicular to the line of intersection of the planes $3x - y + z + 3 = 0$ and $x + 2y - z = 9$ at the point where this line pierces the xy-plane

19. (a) Find the angles the vector $\mathbf{u} = -2\mathbf{i} + 4\mathbf{j} + 5\mathbf{k}$ makes with the coordinate axes.
   (b) For a certain vector $\mathbf{v}$ of length 12, $\cos \beta = \frac{7}{9}$ and $\cos \gamma = \frac{4}{9}$. Find $\mathbf{v}$, given that it has a negative x component.

20. A force $\mathbf{F}$ of magnitude 30 N acts in a direction perpendicular to the plane $3x - 4y + 5z = 10$, and its z component is positive. Find the work done by $\mathbf{F}$ in moving an object from $A(4, 1, -2)$ to $B(2, -5, -1)$, if distance is in meters.

21. (a) Find the distance from the line $3x - 4y = 7$ to the point $(5, -2)$.
   (b) Find the distance from the plane $x - 2y + 2z = 7$ to the point $(2, -1, 4)$.

22. Find x, y, and z so that the vectors $x\mathbf{i} + \mathbf{j} - 3\mathbf{k}$, $3\mathbf{i} - y\mathbf{j} - 7\mathbf{k}$, and $8\mathbf{i} + 5\mathbf{j} + z\mathbf{k}$ will be mutually orthogonal.

23. Prove that if $\mathbf{u}$, $\mathbf{v}$, and $\mathbf{w}$ are mutually orthogonal nonzero vectors, then they are *linearly independent*; that is, the equation

$$a\mathbf{u} + b\mathbf{v} + c\mathbf{w} = \mathbf{0}$$

is satisfied only if $a = b = c = 0$. (*Hint:* In turn, find the dot product of $\mathbf{u}$, $\mathbf{v}$, and $\mathbf{w}$ with both sides.)

24. Unit vectors that are mutually orthogonal are said to form an **orthonormal** set. Prove that if $\mathbf{u}$, $\mathbf{v}$, and $\mathbf{w}$ are any three noncoplanar three-dimensional vectors, then the vectors $\mathbf{e}_1$, $\mathbf{e}_2$, and $\mathbf{e}_3$, defined by

$$\mathbf{e}_1 = \frac{\mathbf{u}}{|\mathbf{u}|} \qquad \mathbf{e}_2 = \frac{\mathbf{v} - (\mathbf{v} \cdot \mathbf{e}_1)\mathbf{e}_1}{|\mathbf{v} - (\mathbf{v} \cdot \mathbf{e}_1)\mathbf{e}_1|}$$

$$\mathbf{e}_3 = \frac{\mathbf{w} - (\mathbf{w} \cdot \mathbf{e}_1)\mathbf{e}_1 - (\mathbf{w} \cdot \mathbf{e}_2)\mathbf{e}_2}{|\mathbf{w} - (\mathbf{w} \cdot \mathbf{e}_1)\mathbf{e}_1 - (\mathbf{w} \cdot \mathbf{e}_2)\mathbf{e}_2|}$$

form an orthonormal set.

Chapter 11 Concept Quiz

1. Define each of the following:
   (a) The magnitude of a vector
   (b) The dot product of two vectors
   (c) The cross product of two vectors
   (d) A unit vector
   (e) The unit vectors $\mathbf{i}$, $\mathbf{j}$, and $\mathbf{k}$
   (f) The vectors $\mathbf{u}$ and $\mathbf{v}$ are orthogonal
   (g) $\text{Comp}_\mathbf{v}\mathbf{u}$
   (h) $\text{Proj}_\mathbf{v}\mathbf{u}$

2. Let $P_0 = (x_0, y_0, z_0)$ and $\mathbf{v} = \langle a, b, c \rangle$. Write the following:
   (a) A vector equation of the line through P_0 and parallel to $\mathbf{v}$
   (b) Parametric equations of the line in part (a)
   (c) An equation of the plane containing P_0 and perpendicular to $\mathbf{v}$

3. Let $\mathbf{u}$, $\mathbf{v}$, and $\mathbf{w}$ be three two-dimensional or three-dimensional vectors. State five properties, each dealing with addition or scalar multiplication involving one or more of these vectors.

4. Draw two vectors $\mathbf{u}$ and $\mathbf{v}$ having the same initial point. Show each of the following vectors:
   (a) $\mathbf{u} + \mathbf{v}$ (b) $\mathbf{u} - \mathbf{v}$

   (c) $\frac{1}{2}\mathbf{u}$ (d) $2\mathbf{v}$
   (e) $2\mathbf{v} - \frac{1}{2}\mathbf{u}$

5. State each of the following:
   (a) A necessary and sufficient condition for two vectors to be orthogonal
   (b) The triangle inequality for vectors
   (c) The Cauchy-Schwarz inequality
   (d) A formula for the volume of the parallelepiped having the vectors $\mathbf{u}$, $\mathbf{v}$, and $\mathbf{w}$ as adjacent edges
   (e) A formula for the work done by a constant force $\mathbf{F}$ in moving a particle in a line from P to Q

6. Explain in each of the following cases how you would find an equation of a plane satisfying the given conditions:
   (a) Contains three noncollinear points
   (b) Passes through a given point and is parallel to a given plane
   (c) Contains a line and a point not on the line
   (d) Is the perpendicular bisector of the line segment joining two points

7. State which of the following are valid operations on three-dimensional vectors and which are not.
   (a) $(\mathbf{u} \times \mathbf{v}) \times \mathbf{w}$ (b) $(\mathbf{u} \times \mathbf{v}) \cdot \mathbf{w}$
   (c) $(\mathbf{u} \cdot \mathbf{v}) \times \mathbf{w}$ (d) $(\mathbf{u} \cdot \mathbf{v}) \cdot \mathbf{w}$
   (e) $(\mathbf{u} \cdot \mathbf{v})\mathbf{w}$ (f) $(\mathbf{u} \times \mathbf{v})\mathbf{w}$

APPLYING CALCULUS

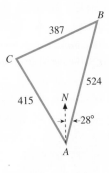

1. The pilot of a small airplane flies in the triangular path shown in the figure from A to B to C and back to A. The distances shown are in kilometers. If there had been no wind, the pilot would have flown at a heading of $28°$ (measured clockwise from north) to go on the straight path from A to B. Throughout the trip the pilot maintained an average airspeed of 332 kph. There was a constant wind of average velocity 113 kph blowing from $223°$.

(a) Use trigonometry to find the angles in the triangle at A, B, and C.

(b) Find the pilot's heading on each leg of the trip.

(c) What was the groundspeed (speed relative to the ground) on each leg?

(d) How long did the trip take?

2. In crystallography a linear combination of vectors $\mathbf{v}_1$, $\mathbf{v}_2$, and $\mathbf{v}_3$ of the form

$$m_1\mathbf{v}_1 + m_2\mathbf{v}_2 + m_3\mathbf{v}_3$$

where m_1, m_2, and m_3 are positive integers, is called a **lattice** for a crystal. The **reciprocal lattice** is

$$m_1\mathbf{w}_1 + m_2\mathbf{w}_2 + m_3\mathbf{w}_3$$

where m_1, m_2, and m_3 are again positive integers, and where

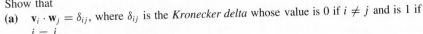

$$\mathbf{w}_1 = \frac{\mathbf{v}_2 \times \mathbf{v}_3}{\mathbf{v}_1 \cdot (\mathbf{v}_2 \times \mathbf{v}_3)}, \quad \mathbf{w}_2 = \frac{\mathbf{v}_3 \times \mathbf{v}_1}{\mathbf{v}_1 \cdot (\mathbf{v}_2 \times \mathbf{v}_3)}, \quad \text{and} \quad \mathbf{w}_3 = \frac{\mathbf{v}_1 \times \mathbf{v}_2}{\mathbf{v}_1 \cdot (\mathbf{v}_2 \times \mathbf{v}_3)}$$

Show that

(a) $\mathbf{v}_i \cdot \mathbf{w}_j = \delta_{ij}$, where δ_{ij} is the *Kronecker delta* whose value is 0 if $i \neq j$ and is 1 if $i = j$.

(b) $\mathbf{w}_1 \cdot (\mathbf{w}_2 \times \mathbf{w}_3) = \dfrac{1}{\mathbf{v}_1 \cdot (\mathbf{v}_2 \times \mathbf{v}_3)}$

3. If a force $\mathbf{F}$ acts on a rigid body at a point with position vector $\mathbf{r}$, then the vector

$$\boldsymbol{\tau} = \mathbf{r} \times \mathbf{F}$$

is called the **torque** produced by $\mathbf{F}$.

(a) A rod as shown in the figure is free to pivot about one end, and a force of magnitude 6.0 N is applied at the other end in the direction shown. Find the magnitude of the torque about the pivot point.

(b) In the accompanying figure, a continuously varying force is applied, with magnitude at the distance x from the left end given by

$$F(x) = F_0\left[1 - \left(\frac{x}{L}\right)^2\right]$$

Find the magnitude of the torque about $x = L$.

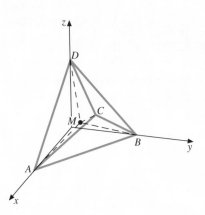

4. In chemistry the *bond angle* is the angle between the vectors from an atom to two atoms to which it is bonded. A *tetrahedral bond angle* is the angle from an atom at the centroid of a tetrahedron to two atoms at vertices of the tetrahedron. (See the figure.) This arrangement of atoms occurs in the methane molecule and in diamond crystals, for example. In methane, CH_4, the carbon atom is at the centroid and is bonded to four hydrogen atoms at the vertices. Find the bond angle AMD in the figure. The points A, B, C, and D have coordinates $(a, 0, 0)$, $(0, a, 0)$, (a, a, a), and $(0, 0, a)$. The centroid M has coordinates $\left(\frac{a}{2}, \frac{a}{2}, \frac{a}{2}\right)$.

5.

(a) Let l_1 and l_2 be two skew lines with vector equations $\mathbf{r} = \mathbf{a}_1 + t\mathbf{v}_1$, and $\mathbf{r} = \mathbf{a}_2 + t\mathbf{v}_2$, respectively. Derive the following formula for the minimum distance d between the lines:

$$d = \frac{|(\mathbf{a}_2 - \mathbf{a}_1) \cdot (\mathbf{v}_1 \times \mathbf{v}_2)|}{|\mathbf{v}_1 \times \mathbf{v}_2|}$$

(b) Suppose two airplanes are flying on the paths l_1 and l_2 of part (a), where $\mathbf{a}_1 = 5\mathbf{i} + 2\mathbf{j} + 8\mathbf{k}$, $\mathbf{v}_1 = 3\mathbf{i} - 2\mathbf{j} + 2\mathbf{k}$, $\mathbf{a}_2 = -\mathbf{i} + 2\mathbf{j} + 5\mathbf{k}$, and $\mathbf{v}_2 = -2\mathbf{i} + 3\mathbf{j} + \mathbf{k}$ (with coordinates in kilometers). What is the closest the airplanes could come to each other?

6. A light ray emanates from the point $(2, 3, 1)$ and follows a path in the direction of the vector $\mathbf{i} - \mathbf{j} - 2\mathbf{k}$.

(a) Find the point at which the light ray strikes the plane $2x + 3y + z = 8$ and the angle it makes with this plane.

(b) What is the minimum distance between the light ray and the origin?

(c) If the light ray strikes a vertical reflector that is perpendicular to the x-axis, what is the path of the reflected ray? (Note that the angle of incidence equals the angle of reflection.)

(d) Suppose that after hitting the vertical reflector in part (c), the reflected ray strikes a horizontal plane. What is the path of the reflected ray?

7. An incoming light ray hitting the inside of a cube reflects off three faces of the cube in such a way that it winds up heading in the opposite direction from which it came. The technique is used to reflect laser beams from earth off corner-cube reflectors placed on the moon by Apollo astronauts. By timing the light-travel time from earth to moon and back, we can now measure the distance to the moon to within a few centimeters. Further, we find out accurately where the telescope is on the earth; the technique is used to measure continental drift.

Using unit-vector notation, prove that the internal reflection off three adjacent faces of a cube indeed returns the light ray in the opposite direction.

This laser beam sent to the Moon reflects off corner reflectors left there by astronauts so that its time of travel to the Moon and back can be measured.

8. Points P and Q are directly opposite each other on the shores of a straight river that flows at a uniform speed of v miles per hour. A boat whose speed is u miles per hour must cross the river from P to Q.

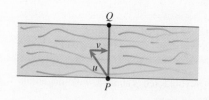

(a) In what direction should the boat head from P and what is the boat's actual speed; that is, what is the component of the velocity in the direction of the line from P to Q?

(b) Is the trip across the river from P to Q always possible? If not, give conditions under which the trip is possible.

(c) Suppose that it is only necessary for the boat to leave from P and to reach the other shore, it does not matter where. If the boat leaves P on a heading that makes an angle α with the line connecting P and Q, will the boat reach the opposite shore, and if so, at what point will it land?

Allan Cormack

How did mathematics play a role in the work that led to your Nobel Prize?

It was vital. The whole problem is a mathematical one.

Please tell us about the tomography project.

The problem is to infer the absorption coefficient of tissue from X rays from absorption measurements taken outside the object being X-rayed. In a single absorption measurement, what you are getting is a measurement of the line integral of the absorption coefficient along the line of the X-ray beam, or, if you like, the average X-ray absorption coefficient along that line. Now, if you do that for a number of lines in different directions, can you then infer the variation of the absorption coefficient from point to point? The answer is yes, and there are many different algorithms for doing that.

How did you come to that problem?

Well, it was just by an accident. The hospital physicist—and there was only one in those days—at the University hospital for the University of Cape Town Medical School quit. And incidentally, this was the Groot Schuur Hospital, where Christiaan Barnard subsequently did his heart transplants. This hospital physicist quit and went to Canada, and I was the only person in South Africa at the time who knew anything about handling of radioactive isotopes. There was no medical physics section of the hospital at the time (I'm talking about 1956) and so I was put in the radiology department. And there I couldn't avoid seeing the way they did radiotherapy-treatment planning. I was appalled by what I saw, even though it was a place where you got as good treatment as you would anywhere in the world. So my thought was that with a map of the point-by-point absorption coefficient in the body, one would be able to design radiotherapy treatments much more accurately. That was the motivation. And it occurred to me that such a map would be useful of itself, but I didn't realize just how useful it would be. Without taking anything away from physicians, the layman doesn't realize how almost impossible it is to see a tumor in the soft tissues of the head, where the tumor is obscured by all the images of the bones in the head that you get in an ordinary X ray. And so this is one of the reasons why it was the radiologists in the soft tissues who were the ones who were so delighted when the first commercial CT scanners came out, because they were seeing things with a clarity that they had never come close to before. But there are other people, like a distinguished physician at Mass General Hospital in Boston, who said "oh yes, this might be very well for the head, but it will never work in the chest." And, of course, he was dead wrong.

How did your training make you the person who was chosen to work at the hospital?

Well, I was the only person in Cape Town, certainly, and probably in all of South Africa who had been trained in handling radioactive isotopes, measuring absorption coefficients, and so on. I learned those skills when I was a graduate student in nuclear physics at Cambridge. I had done three years of engineering in Cape Town, and then I quit and changed to physics and went to Cambridge in England for a Ph.D.

Did you have any special mathematical training that led you to this problem?

No, and unfortunately, I took the mathematics sequence intended for engineers as an undergraduate and that leaves out a lot of the more interesting and fundamental parts of mathematics in the interest of getting to usable results quickly. For example, mathematicians in those days would spend a long time on the discussion of convergence of series and they probably do still. The engineering syllabus kind of brushed over those things, and I've regretted all my life not having had a more fundamental training in mathematics. So there's a plug for mathematics.

How did the mathematics get introduced to the CT problem?

Well, if you simply ask what you measure when you pass a beam of X rays through an inhomogeneous material, it turns out that the pertinent quantity is the line integral of the absorption coefficient along the line of the beam. It is Beer's Law, which has been known for a long time. In the early days of nuclear physics, for example, you got some information about a gamma-ray energy by measuring its absorption—not an accurate way by modern standards at all. So I knew about measuring gamma-ray absorption coefficients in a homogeneous medium, so the next question is what happens if the medium is inhomogeneous, as in the head.

Was this a three-dimensional problem?

Ideally, yes, three dimensions, but you can simplify by cutting into a series of two-dimensional slices. And tomography gets its name from that fact. A *tomogram* is a picture of a slice and it comes from the Greek *tomon*, which is the same root that occurs in *atom*, which means "that which cannot be cut [or sliced]."

You said CT, but wasn't it called CAT at one time?

People were trying to do tomography by different means as early as 1920. What they did was CAT, which originally meant "computed axial tomography," from an old-fashioned meaning, and then it became just CT, computer tomography. Now CAT stands for "computer-assisted tomography."

What relation did the CT work have to the development of MRI, medical-research imaging?

In the original development of MRI, people used CT algorithms to interpret their data, because they were also averaging over an NMR [nuclear magnetic resonance] signal over lines, or the three-dimensional version of this. The problem is, "Can you determine a function in an area knowing the line integral everywhere"—a two-dimensional problem. The three-dimensional problem is, "If you know the averages over planes, can you determine the three-dimensional function?" It turns out that the three-dimensional version is, in fact, simpler than the two-dimensional version. This is true for all odd-dimensional spaces as opposed to even-dimensional spaces. In the case of the even-dimensional spaces, such as two dimensions, you have to know the line integrals through all lines intersecting the plane in order to find the value of the function at a point, whereas in the three-dimensional case, you need to know only the averages over the planes that intersect a neighborhood of the point. All this was known in 1917 by Johann Radon, an Austrian mathematician. Radon had himself been anticipated in the three-dimensional case by Lorentz, in about 1905, who hadn't bothered to publish his result. Now Lorentz was interested in it, I think, because of the propagation of waves in crystals. One of the nice features of the so-called Radon transform is that if you take the wave equation and average over planes parallel to a given direction, you reduce the three-dimensional problem to a one-dimensional problem in that direction. So the Radon transformation is a very useful technique in the study of partial differential equations. And if I had known that, I would have been saved a great deal of work.

In view of the developments in mathematics and science over the last decades, is calculus still a good way to introduce students to mathematics?

I think it is, but there is a strong movement that says the contrary, why bother with classical analysis at all when you can do it all on computers. I think you would lose a great deal, but there are people who maintain this. Hounsfield, who won the Prize jointly with me, simply saw the problem as a successive approximation problem. His algorithm consisted of backprojecting an average density along every line and progressively correcting, so you get a better fit to the data. That is a perfectly good way to do it. It works.

Allan Cormack is University Professor at Tufts University. He was born and raised in South Africa, went to Cambridge, England, for graduate study, and then returned to South Africa. He was on a sabbatical at Harvard in 1956–57 when he was offered a position at Tufts, where he became professor of physics. He shared in the 1979 Nobel Prize in Physiology or Medicine with Godfrey Hounsfield for the development of "computer-assisted tomography."

12

VECTOR-VALUED FUNCTIONS

When we are describing the motion of an object such as a space vehicle or a planet moving in its orbit, the position of the object, its velocity, and its acceleration are usually given as vectors. These vectors are dependent on time, and as we will see, they provide a useful means of describing motion. The position of the object at any time t can be given by its position vector $\mathbf{r}(t)$ in reference to some coordinate system. Since values of $\mathbf{r}(t)$ are vectors, $\mathbf{r}$ is called a **vector-valued function,** or more briefly, a **vector function**. As we shall see, derivatives of $\mathbf{r}(t)$ give us information about the velocity and acceleration of the moving object. Similarly, to find the distance covered by the object in a given time interval, we will see that integration is needed.

We begin by discussing the nature of vector-valued functions in general, not just in relation to moving objects. While our emphasis will be on three-dimensional vector functions, most of our theory (except where cross products are involved) applies also to the two-dimensional case. Vector functions in two dimensions provide an alternative way of looking at the parametric representation of curves that we studied in Chapter 9.

Next, we discuss the calculus of vector functions and show how the length of a curve can be given as an integral involving a vector function. We introduce the concepts of the **unit tangent vector**, **unit normal vector**, and **curvature**, which are useful in describing a curve.

We then return to the important topic of motion along a curve and show how to apply the ideas we have introduced. We conclude the chapter by considering Kepler's Laws of Planetary Motion.

12.1 VECTOR-VALUED FUNCTIONS

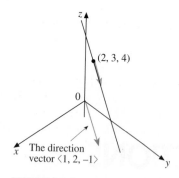

FIGURE 12.1
The graph of $\mathbf{r}(t) = \langle 2 + t,$ $3 + 2t, 4 - t \rangle$ is the line through $(2, 3, 4)$, parallel to the direction vector $\langle 1, 2, -1 \rangle$

We have already encountered one special type of vector function in studying straight lines in space. For example, we know that for each real t,

$$\mathbf{r}(t) = \langle 2 + t, 3 + 2t, 4 - t \rangle$$

is the position vector of a point on the line that passes through the point $(2, 3, 4)$ and is parallel to the direction vector $\langle 1, 2, -1 \rangle$. As t varies over all real numbers, the tip of the vector $\mathbf{r}(t)$ traces out the entire line. We show the line in Figure 12.1.

As another example, consider the vector function $\mathbf{r}$ for which

$$\mathbf{r}(t) = \langle 4 \cos t, 4 \sin t, t \rangle$$

As t varies, the tip of the position vector $\mathbf{r}(t)$ traces out a curve in space. In this case, the graph is not a straight line. We will analyze its graph shortly. More generally, we have the following definition.

Definition 12.1
Vector-Valued Function

A **vector-valued function,** or, more briefly, a **vector function,** is a function whose domain is a set of real numbers and whose range is a set of vectors.

If $\mathbf{r}$ is a vector function with values given by

$$\mathbf{r}(t) = \langle f(t), g(t) \rangle$$

then we call $\mathbf{r}$ a two-dimensional vector function with *component functions f* and g. Similarly, if

$$\mathbf{r}(t) = \langle f(t), g(t), h(t) \rangle$$

then $\mathbf{r}$ is a three-dimensional vector function having component functions f, g, and h. We will concentrate on the three-dimensional case, since we can think of a two-dimensional vector function as having three component functions, with the third component being the zero function.

REMARK ——————————————————————————
■ The letter $\mathbf{r}$ is frequently used to name a vector-valued function, because it suggests *radius vector,* which is another name for position vector. When we need to discuss more than one vector function, we will usually use letters such as $\mathbf{u}$, $\mathbf{v}$, $\mathbf{w}$.

——————————————————————————————————————

The domain of a vector function, unless otherwise specified, will be understood to mean the common domain of its component functions.

To distinguish vector functions from ordinary real-valued functions, we call the latter **scalar functions.** So the component functions of a vector function are scalar functions. In $\mathbf{r}(t) = \langle 4 \cos t, 4 \sin t, t \rangle$, the component functions are the scalar functions $f(t) = 4 \cos t$, $g(t) = 4 \sin t$, and $h(t) = t$.

Addition, subtraction, and multiplication by a scalar (or by a scalar function) are all carried out component by component, just as with constant vectors. The dot product of two vector functions is a scalar function, whereas the cross product is a vector function.

EXAMPLE 12.1   Find the domain of the vector function **r** defined by

$$\mathbf{r}(t) = \langle \ln(1 - t), \sqrt{t}, t^2 \rangle$$

Solution   Write $f(t) = \ln(1 - t)$, $g(t) = \sqrt{t}$, and $h(t) = t^2$. The domain of f is the set of t values for which $1 - t > 0$, or equivalently, $t < 1$. The domain of g is the set of all t values such that $t \geq 0$. The domain of h is all of $\mathbb{R}$. The domain of **r** is therefore the set of t values for which $0 \leq t < 1$. We can indicate this set more briefly as the half-open interval $[0, 1)$.   ∎

Definition 12.2
Graph of a Vector Function

The **graph** of a vector function **r** is the set of points taken on by the tip of the position vector $\mathbf{r}(t)$ as t varies over the domain of **r**.

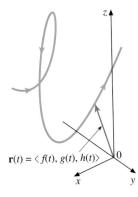

FIGURE 12.2
Graph of the vector function **r**

$\mathbf{r}(t) = \langle f(t), g(t), h(t) \rangle$

If

$$\mathbf{r}(t) = \langle f(t), g(t), h(t) \rangle$$

and f, g, and h are *continuous* on an interval I, then the graph of **r** is called a **space curve**. The curve consists of all points (x, y, z) of the form $(f(t), g(t), h(t))$ for t in I. In Figure 12.2, we show such a space curve. We assign a *direction* to the curve according to the manner in which it is traced out by the tip of $\mathbf{r}(t)$ as t increases through the domain values. We sometimes indicate the direction by arrows, as in Figure 12.2.

Since the tip of the vector $\mathbf{r}(t) = \langle f(t), g(t), h(t) \rangle$ lies on the graph of **r** for each value of t, we can give the coordinates (x, y, z) of such a point by the equations

$$\begin{cases} x = f(t) \\ y = g(t) \\ z = h(t) \end{cases} \quad t \in I$$

We call these three equations **parametric equations** of the curve C that is the graph of **r**. If we know the vector equation for C, we know the parametric equations, and conversely.

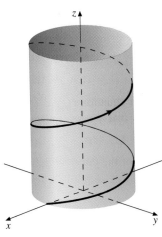

FIGURE 12.3
The circular helix
$\mathbf{r}(t) = \langle 4\cos t, 4\sin t, t \rangle$ for $t \geq 0$.

EXAMPLE 12.2   **The Circular Helix**   Sketch the graph of the vector function

$$\mathbf{r}(t) = \langle 4\cos t, 4\sin t, t \rangle$$

for $t \geq 0$.

Solution   The parametric equations of the graph are $x = 4\cos t$, $y = 4\sin t$, and $z = t$. Since $x^2 + y^2 = 16\cos^2 t + 16\sin^2 t = 16$, we see that the projection of the point (x, y, z) on the xy-plane lies on the circle $x^2 + y^2 = 16$ of radius 4. As t increases, the z-coordinate increases. The result is the climbing circular curve shown in Figure 12.3. Its direction is indicated by the arrows. The initial point, corresponding to $t = 0$, is $(4, 0, 0)$.   ∎

The curve in Figure 12.3 is called a **circular helix**. The general equation for a helix is of the form $\mathbf{r}(t) = \langle a\cos\omega t, b\sin\omega t, ct\rangle$. The helix is circular if $a = b$ and elliptical if $a \neq b$.

EXAMPLE 12.3   Discuss and sketch the graph of

$$\mathbf{r}(t) = (2t)\mathbf{i} + (3t^2)\mathbf{j} + (t^3)\mathbf{k}$$

for $t \geq 0$.

Solution   The parametric equations of the curve are $x = 2t$, $y = 3t^2$, and $z = t^3$. If we eliminate the parameter t between x and y, we get $y = (3x^2)/4$. Thus, the xy-projection of a point on the curve lies on this parabola. When $t = 0$, the point is at the origin. As t increases, the point moves upward, always in such a way that its xy-projection is on the parabola $y = (3x^2)/4$. We therefore have the graph shown in Figure 12.4.   ∎

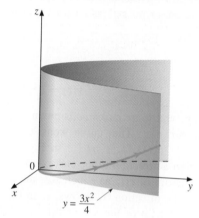

FIGURE 12.4
The twisted cubic $\mathbf{r}(t) = \langle 2t, 3t^2, t^3\rangle, t \geq 0$

The curve in Figure 12.4 is called a **twisted cubic**. The general equation is of the form $\mathbf{r}(t) = \langle at, bt^2, ct^3\rangle$.

If $\mathbf{r}(t) = \langle f(t), g(t)\rangle$ is a two-dimensional vector, with f and g continuous on an interval I, then the graph of $\mathbf{r}$ is a **plane curve**. Its parametric equations are

$$\begin{cases} x = f(t) \\ y = g(t) \end{cases} \quad t \in I$$

We studied such curves in Chapter 9. So the vector equation $\mathbf{r}(t) = \langle f(t), g(t)\rangle$ is simply an alternative way of describing curves that we represented by parametric equations in Chapter 9.

EXAMPLE 12.4   Identify and draw the graph of the curve C defined by the vector equation

$$\mathbf{r}(t) = \langle 3\cos t, 2\sin t\rangle$$

for $0 \leq t \leq 2\pi$.

Solution   The parametric equations of C are $x = 3\cos t$ and $y = 2\sin t$. Thus,

$$\frac{x}{3} = \cos t \quad \text{and} \quad \frac{y}{2} = \sin t$$

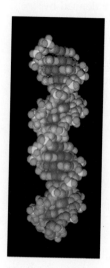

The DNA molecule is a double helix with a radius of 1 nanometer, 1×10^{-9} m.

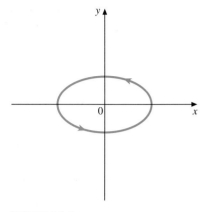

By squaring and adding, we get the rectangular equation

$$\frac{x^2}{9} + \frac{y^2}{4} = 1$$

The graph is the ellipse shown in Figure 12.5. Note that when $t = 0$, the tip of the vector $\mathbf{r}(t)$ is at the point $(3, 0)$, and as t increases, the curve is traced out in a counterclockwise direction, as shown. When $t = 2\pi$, the tip of $\mathbf{r}(t)$ is again at $(3, 0)$, so the ellipse has been traced out once in the interval $0 \leq t \leq 2\pi$. ∎

FIGURE 12.5
The graph of $r(t) = \langle 3\cos t, 2\sin t \rangle$ for $0 \leq t \leq 2\pi$

GRAPHING VECTOR-VALUED FUNCTIONS USING COMPUTER ALGEBRA SYSTEMS

Computer algebra systems are very useful for viewing curves in three space that are otherwise either very difficult or impossible to sketch and analyze. In this section, we define and sketch several curves in three space using Maple and Mathematica.

CAS 44

Make a sketch of the helix given in Example 12.2.

The helix is given by the vector-valued function $\mathbf{r}(t) = \langle 4\cos t, 4\sin t, t \rangle$ for $t \geq 0$. We will also display the cylinder on which the helix travels.

Maple:

```
with(plots):

r:=t->[4*cos(t),4*sin(t),t];
p1 :=plot3d([4*cos(t),4*sin(t),x],t=0..2*Pi,
   x=0..15,color=grey,grid=[20,5]);
p2 :=spacecurve(r(t),t=0..4*Pi,
   color=black,numpoints=200);
display3d(p1,p2,style=wireframe);
```

Mathematica:

```
x[t_] = {4*Cos[t],4*Sin[t],t}
ParametricPlot3D[{4*Cos[t],4*Sin[t],t},{t,0,2*Pi}]
```

FIGURE 12.1.1

The general helix can be written in the form $\mathbf{r}(t) = \langle a\cos\omega t, a\sin\omega t, bt\rangle$. If you have access to a CAS, explore the effect the constants a, b, and ω have on the graph of the helix. For example, try the Maple command

spacecurve(r(t),t=0..4*Pi,color=black,numpoints=200);

CAS 45

(a) Sketch the curve defined by $\mathbf{r}(t) = \langle e^{-t^2}\cos t, e^{-t^2}\sin t, t\rangle$.

(b) Sketch the curve defined by

$$\mathbf{r}(t) = \left\langle \left(3 + \cos\left(\frac{3t}{2}\right)\right)\cos t, \left(3 + \cos\left(\frac{3t}{2}\right)\right)\sin t, \sin\left(\frac{3t}{2}\right)\right\rangle$$

Maple:

(a)
r:=t–>[exp(–t^2)*cos(t),exp(–t^2)*sin(t),t];
spacecurve(r(t),t=–3..3,axes=boxed);

See Figure 12.1.2.

(b)
r:=t–>[(3+cos(3*t/2))*cos(t),(3+cos(3*t/2))*sin(t),sin(3*t/2)];
spacecurve(r(t),t=0..4*Pi,color=black,numpoints=200,
   axes=boxed);

See Figure 12.1.3.

Mathematica:

(a)
ParametricPlot3D[{Exp[–t^2]*Cos[t],Exp[–t^2]*Sin[t],t},
   {t,–3,3}]

(b)
ParametricPlot3D[{(3+Cos[3*t/2])*Cos[t],
   (3+Cos[3*t/2])*Sin[t],Sin[3*t/2]},{t,0,4*Pi}]

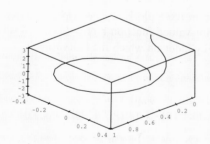

FIGURE 12.1.2

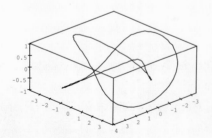

FIGURE 12.1.3

Exercise Set 12.1

In Exercises 1–6, give the domain of each of the vector functions.

1. $\mathbf{r}(t) = \left\langle \dfrac{t}{t-1}, \sqrt{1-t} \right\rangle$

2. $\mathbf{r}(t) = (\ln t)\mathbf{i} + \sqrt{1-t}\,\mathbf{j}$

3. $\mathbf{r}(t) = \left\langle 2t, \dfrac{1}{t}, \sqrt{t-1} \right\rangle$

4. $\mathbf{r}(t) = \langle \ln(t+2), e^{-t}, \ln(1-t) \rangle$

5. $\mathbf{r}(t) = (\tan t)\mathbf{i} + (\cot t)\mathbf{j} + \left(\sin \sqrt{t} \right)\mathbf{k}$

6. $\mathbf{r}(t) = \left(\sqrt{1-t^2} \right)\mathbf{i} + (\ln t)\mathbf{j} + e^{\sin t}\mathbf{k}$

In Exercises 7–18, sketch the graph of **r**. *Indicate the direction for increasing t values with arrows.*

7. $\mathbf{r}(t) = \langle t^2, t^3 \rangle$

8. $\mathbf{r}(t) = (t-2)\mathbf{i} + t^2\mathbf{j}$

9. $\mathbf{r}(t) = \left(\dfrac{1}{t} \right)\mathbf{i} + t\mathbf{j}, \quad t > 0$

10. $\mathbf{r}(t) = \langle \sin^2 t, \cos t - 1 \rangle, \quad 0 \le t \le 2\pi$

11. $\mathbf{r}(t) = \langle 1+t, 2t, 3t+2 \rangle$

12. $\mathbf{r}(t) = \langle 1 - t^2, t, 2 \rangle$

13. $\mathbf{r}(t) = (\sin t)\mathbf{i} + (\cos t)\mathbf{j} + t\mathbf{k}$

14. $\mathbf{r}(t) = \langle \cos t, t, \sin t \rangle$

15. $\mathbf{r}(t) = \langle 3\cos t, 2\sin t, t \rangle$

16. $\mathbf{r}(t) = t\mathbf{i} + t^2\mathbf{j} + t^3\mathbf{k}$

17. $\mathbf{r}(t) = \langle t, t, \sin t \rangle$

18. $\mathbf{r}(t) = \langle e^t, e^{-t}, t \rangle$

CAS **19.** Plot a sample of curves from the family of curves defined by $\mathbf{r}(t) = \langle \cos(nt), \sin(nt), \cos(pt) \rangle$, where n and p are positive integers.

CAS **20.** Repeat Exercise 19 with $\mathbf{r}(t) = \langle \cos(mt), \sin(nt), \sin(pt) \rangle$ for different values of m, n, and p.

CAS **21.** Plot a sample of curves from the family of curves given by $\mathbf{r}(t) = \langle (a + \cos(bt/2))\cos(t), (c + \sin(dt/2))\sin(t), \cos(et/2) \rangle$.

12.2 THE CALCULUS OF VECTOR FUNCTIONS

The calculus concepts of limit, continuity, differentiation, and integration are all defined for vector functions in terms of the component functions. In particular, we have the following.

Definition 12.3
The Limit of a Vector Function

Let $\mathbf{r}(t) = \langle f(t), g(t), h(t) \rangle$. The **limit** of $\mathbf{r}(t)$ as t approaches a exists if and only if the limits of the component functions,

$$\lim_{t \to a} f(t) = l_1, \qquad \lim_{t \to a} g(t) = l_2, \qquad \text{and} \qquad \lim_{t \to a} h(t) = l_3$$

all exist. In this case,

$$\lim_{t \to a} \mathbf{r}(t) = \mathbf{L} \qquad \text{where } \mathbf{L} = \langle l_1, l_2, l_3 \rangle$$

In particular, if $\lim_{t \to a} \mathbf{r}(t) = \mathbf{r}(a)$, then $\mathbf{r}$ is **continuous** at $t = a$, and conversely.

REMARK ──────────

■ To say that $\lim_{t \to a} \mathbf{r}(t) = \mathbf{L}$ means that the vectors $\mathbf{r}(t)$ come arbitrarily close to the constant vector $\mathbf{L}$ in *both magnitude and direction* for values of t sufficiently close to a (but not equal to a). From the definition, we can conclude that $\mathbf{r}$ *is continuous at $t = a$ if and only if its component functions are continuous there.*

───────────

EXAMPLE 12.5

(a) Let $\mathbf{u}(t) = \left\langle t^2, \cos t, \dfrac{\sin t}{t} \right\rangle$. Find $\lim_{t \to 0} \mathbf{u}(t)$.

(b) Let $\mathbf{v}(t) = \langle e^t, \ln t, \sinh t \rangle$. Show that $\mathbf{v}$ is continuous at all points of its domain.

Solution

(a) By Definition 12.3,

$$\lim_{t \to 0} \mathbf{u}(t) = \left\langle \lim_{t \to 0} t^2, \lim_{t \to 0} \cos t, \lim_{t \to 0} \frac{\sin t}{t} \right\rangle$$

$$= \langle 0, 1, 1 \rangle$$

(b) The only limitation on the domain of $\mathbf{v}$ is that t be positive, so that $\ln t$ is defined. If $t = a$, where a is any positive number, then each of the component functions is continuous there. So we have

$$\lim_{t \to a} \mathbf{v}(t) = \left\langle \lim_{t \to a} e^t, \lim_{t \to a} \ln t, \lim_{t \to a} \sinh t \right\rangle$$

$$= \langle e^a, \ln a, \sinh a \rangle = \mathbf{v}(a)$$

Thus, $\mathbf{v}$ is continuous at $t = a$.   ■

The Derivative of a Vector Function

Definition 12.4 The Derivative of a Vector Function	The **derivative** $\mathbf{r}'$ of a vector function $\mathbf{r}$ is defined by $$\mathbf{r}'(t) = \lim_{h \to 0} \frac{\mathbf{r}(t + h) - \mathbf{r}(t)}{h}$$ provided this limit exists.

Combining Definition 12.4 with the limit of a vector-valued function given in Definition 12.3, we get an easy method of computing the derivative of a vector-valued function. We simply differentiate each of the component functions. We set off this important result for emphasis.

> **The Derivative of a Vector Function**
>
> If $\mathbf{r}(t) = \langle f(t), g(t), h(t) \rangle$, and if $f'(t)$, $g'(t)$, and $h'(t)$ exist, then
>
> $$\mathbf{r}'(t) = \langle f'(t), g'(t), h'(t) \rangle$$

EXAMPLE 12.6   Find $\mathbf{r}'(\frac{\pi}{3})$ if

$$\mathbf{r}(t) = (\sin 2t)\mathbf{i} + (2\cos t)\mathbf{j} + (\tan t)\mathbf{k}$$

Solution   Each of the component functions is differentiable except where $\tan t$ is undefined (at odd multiples of $\frac{\pi}{2}$). At any value of t where they are differentiable, we have

$$\mathbf{r}'(t) = (2\cos 2t)\mathbf{i} + (-2\sin t)\mathbf{j} + (\sec^2 t)\mathbf{k}$$

For $t = \frac{\pi}{3}$, we have

$$\mathbf{r}'\left(\frac{\pi}{3}\right) = \left(2\cos\frac{2\pi}{3}\right)\mathbf{i} + \left(-2\sin\frac{\pi}{3}\right)\mathbf{j} + \left(\sec^2\frac{\pi}{3}\right)\mathbf{k}$$

$$= 2\left(-\frac{1}{2}\right)\mathbf{i} + (-2)\left(\frac{\sqrt{3}}{2}\right)\mathbf{j} + (2)^2\mathbf{k}$$

$$= -\mathbf{i} - \sqrt{3}\,\mathbf{j} + 4\mathbf{k} \qquad \blacksquare$$

Geometric Interpretation of the Derivative: Tangent Vectors and Tangent Lines

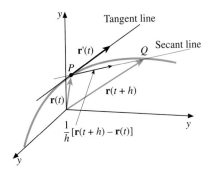

FIGURE 12.6

It should come as no surprise that the derivative of a vector function $\mathbf{r}$ has some relationship to the tangent line to the graph of $\mathbf{r}$. We show this relationship in Figure 12.6. The position vectors of the points P and Q are $\mathbf{r}(t)$ and $\mathbf{r}(t + h)$. The vector $\overrightarrow{PQ} = \mathbf{r}(t + h) - \mathbf{r}(t)$ is a direction vector for the secant line through P and Q (we are supposing that $h \neq 0$). The vector

$$\frac{\mathbf{r}(t + h) - \mathbf{r}(t)}{h} = \frac{1}{h}[\mathbf{r}(t + h) - \mathbf{r}(t)]$$

is just a scalar multiple of $\overrightarrow{PQ}$, so it is also a direction vector for this secant line (in the same direction if $h > 0$). It appears that as $h \to 0$, this direction vector for the secant line approaches a vector lying on the tangent line at P. It is natural, then, to define the tangent line at P as the line through P with direction vector equal to the limit

$$\lim_{h \to 0} \frac{\mathbf{r}(t + h) - \mathbf{r}(t)}{h}$$

But by Definition 12.4 this limit is $\mathbf{r}'(t)$. We must restrict $\mathbf{r}'(t)$ to be different from the zero vector, since otherwise there would be no well-defined direction for the tangent line. We therefore have the following definition.

Definition 12.5
Tangent Vector and Tangent Line to a Space Curve

Let C be the curve with vector equation $\mathbf{r}(t) = \langle f(t), g(t), h(t) \rangle$ for $t \in I$, and suppose that $\mathbf{r}'(t)$ exists and is not the zero vector. Let $P = (f(t), g(t), h(t))$. Then the vector $\mathbf{r}'(t)$ is called a **tangent vector** to C at P, and the **tangent line** to C at P is the line through P with direction vector $\mathbf{r}'(t)$.

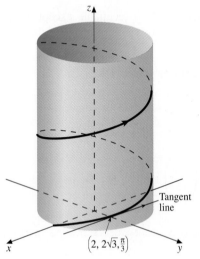

FIGURE 12.7

EXAMPLE 12.7   Find a tangent vector and a vector equation of the tangent line to the circular helix $\mathbf{r}(t) = \langle 4 \cos t, 4 \sin t, t \rangle$ at $t = \frac{\pi}{3}$.

Solution   Since $\mathbf{r}'(t) = \langle -4 \sin t, 4 \cos t, 1 \rangle$, a tangent vector at $t = \frac{\pi}{3}$ is

$$\mathbf{r}'\left(\frac{\pi}{3}\right) = \left\langle -4 \sin \frac{\pi}{3}, 4 \cos \frac{\pi}{3}, 1 \right\rangle = \left\langle -2\sqrt{3}, 2, 1 \right\rangle$$

The point on the curve at which $t = \frac{\pi}{3}$ is the tip of the position vector $\mathbf{r}\left(\frac{\pi}{3}\right)$, namely,

$$\left(4 \cos \frac{\pi}{3}, 4 \sin \frac{\pi}{3}, \frac{\pi}{3} \right) = \left(2, 2\sqrt{3}, \frac{\pi}{3} \right)$$

The tangent line through this point has the direction vector $\mathbf{r}'\left(\frac{\pi}{3}\right)$, so its equation can be written as

$$\mathbf{r}(t) = \left\langle 2, 2\sqrt{3}, \frac{\pi}{3} \right\rangle + t \left\langle -2\sqrt{3}, 2, 1 \right\rangle$$

$$= \left\langle 2 - 2\sqrt{3}t, 2\sqrt{3} + 2t, \frac{\pi}{3} + t \right\rangle$$

We show the graph in Figure 12.7, along with the tangent vector and tangent line at the given point.   ∎

Properties of Derivatives of Vector-Valued Functions

The following properties of differentiation follow from the analogous results for derivatives of the component functions.

Properties of the Derivative

If $\mathbf{u}$ and $\mathbf{v}$ are differentiable vector functions and f is a differentiable scalar function, then the following properties hold true.

1. $\frac{d}{dt}[\mathbf{u}(t) \pm \mathbf{v}(t)] = \mathbf{u}'(t) \pm \mathbf{v}'(t)$

2. $\frac{d}{dt}[c\mathbf{u}(t)] = c\mathbf{u}'(t)$ for any constant c

3. $\frac{d}{dt}[f(t)\mathbf{u}(t)] = f'(t)\mathbf{u}(t) + f(t)\mathbf{u}'(t)$

4. $\frac{d}{dt}[\mathbf{u}(t) \cdot \mathbf{v}(t)] = \mathbf{u}'(t) \cdot \mathbf{v}(t) + \mathbf{u}(t) \cdot \mathbf{v}'(t)$

5. $\frac{d}{dt}[\mathbf{u}(t) \times \mathbf{v}(t)] = \mathbf{u}'(t) \times \mathbf{v}(t) + \mathbf{u}(t) \times \mathbf{v}'(t)$

6. $\frac{d}{dt}[\mathbf{u}(f(t))] = \mathbf{u}'(f(t))f'(t)$ Chain Rule

---

In Property 5 it is essential that the order of the factors not be reversed, since the cross product is noncommutative.

EXAMPLE 12.8   Let $\mathbf{u}(t) = \langle t, 2 - t^3, 2t \rangle$ and $\mathbf{v}(t) = \langle 1, t^2, t^3 \rangle$. Find each of the following.

(a) $(\mathbf{u} \cdot \mathbf{v})'(1)$ (b) $(\mathbf{u} \times \mathbf{v})'(1)$

Solution

(a) By Property 4 of the derivative,

$$(\mathbf{u} \cdot \mathbf{v})'(t) = \mathbf{u}'(t) \cdot \mathbf{v}(t) + \mathbf{u}(t) \cdot \mathbf{v}'(t)$$

So for $t = 1$,

$$(\mathbf{u} \cdot \mathbf{v})'(1) = \mathbf{u}'(1) \cdot \mathbf{v}(1) + \mathbf{u}(1) \cdot \mathbf{v}'(1)$$

To evaluate the right-hand side, we first compute each of the vectors involved.

$$
\begin{array}{ll}
\mathbf{u}(t) = \langle t, 2 - t^3, 2t \rangle & \mathbf{u}(1) = \langle 1, 1, 2 \rangle \\
\mathbf{u}'(t) = \langle 1, -3t^2, 2 \rangle & \mathbf{u}'(1) = \langle 1, -3, 2 \rangle \\
\mathbf{v}(t) = \langle 1, t^2, t^3 \rangle & \mathbf{v}(1) = \langle 1, 1, 1 \rangle \\
\mathbf{v}'(t) = \langle 0, 2t, 3t^2 \rangle & \mathbf{v}'(1) = \langle 0, 2, 3 \rangle
\end{array}
$$

Thus,

$$(\mathbf{u} \cdot \mathbf{v})'(1) = \langle 1, -3, 2 \rangle \cdot \langle 1, 1, 1 \rangle + \langle 1, 1, 2 \rangle \cdot \langle 0, 2, 3 \rangle$$

$$= (1 - 3 + 2) + (0 + 2 + 6) = 8$$

(b) We make use of Derivative Property 5 and the values computed in part (a) to get

$$(\mathbf{u} \times \mathbf{v})'(1) = \mathbf{u}'(1) \times \mathbf{v}(1) + \mathbf{u}(1) \times \mathbf{v}'(1)$$

$$= \langle 1, -3, 2 \rangle \times \langle 1, 1, 1 \rangle + \langle 1, 1, 2 \rangle \times \langle 0, 2, 3 \rangle$$

$$= \langle -5, 1, 4 \rangle + \langle -1, -3, 2 \rangle = \langle -6, -2, 6 \rangle \qquad \blacksquare$$

EXAMPLE 12.9   Prove that if $\mathbf{r}(t)$ is a vector function such that for all t in the domain, $\mathbf{r}'(t)$ exists and $|\mathbf{r}(t)|$ is constant, then $\mathbf{r}(t)$ and $\mathbf{r}'(t)$ are orthogonal vectors for every t in the domain.

Solution   To show that the vectors $\mathbf{r}(t)$ and $\mathbf{r}'(t)$ are perpendicular, we will show that their dot product is zero. Denote the constant value of $|\mathbf{r}(t)|$ by c. That is, $|\mathbf{r}(t)| = c$. Recall that for any vector $\mathbf{v}$, we have $|\mathbf{v}|^2 = \mathbf{v} \cdot \mathbf{v}$. Thus,

$$\mathbf{r}(t) \cdot \mathbf{r}(t) = c^2$$

Now we take the derivative of each side, applying Property 4.

$$\mathbf{r}'(t) \cdot \mathbf{r}(t) + \mathbf{r}(t) \cdot \mathbf{r}'(t) = 0$$

But dot products are commutative. So we have

$$2\mathbf{r}(t) \cdot \mathbf{r}'(t) = 0$$

Thus, $\mathbf{r}(t) \cdot \mathbf{r}'(t) = 0$, which shows that $\mathbf{r}(t)$ and $\mathbf{r}'(t)$ are always orthogonal.

$\blacksquare$

Integrals of Vector Functions

Antiderivatives and definite integrals of vector functions can also be defined in terms of the component functions.

Definition 12.6
Integrals of Vector Functions

Let $\mathbf{r}(t) = \langle f(t), g(t), h(t) \rangle$. Then

(a) $\displaystyle \int \mathbf{r}(t)dt = \left\langle \int f(t)dt, \int g(t)dt, \int h(t)dt \right\rangle$

provided each of the indefinite integrals on the right exists.

(b) $\displaystyle \int_a^b \mathbf{r}(t)dt = \left\langle \int_a^b f(t)dt, \int_a^b g(t)dt, \int_a^b h(t)dt \right\rangle$

provided each of the definite integrals on the right exists.

Note that an antiderivative (indefinite integral) of a vector function is again a vector function, whereas a definite integral of a vector function is a constant vector.

If $\mathbf{R}'(t) = \mathbf{r}(t)$, we may write the result of part (a) of Definition 12.6 in the form

$$\int \mathbf{r}(t)dt = \mathbf{R}(t) + \mathbf{C}$$

where $\mathbf{C}$ is any constant vector.

The First and Second Fundamental Theorems of Calculus hold true for vector functions, as can be seen by a consideration of components. We state the two results in the following theorem.

THEOREM 12.1

(a) If $\mathbf{r}$ is continuous on the closed interval $[a, b]$, then for any t in $[a, b]$,

$$\frac{d}{dt}\int_a^t \mathbf{r}(u)du = \mathbf{r}(t) \qquad \text{First Fundamental Theorem}$$

(b) If $\mathbf{r}$ is continuous on $[a, b]$ and $\mathbf{R}$ is any antiderivative of $\mathbf{r}$, then

$$\int_a^b \mathbf{r}(t)dt = \mathbf{R}(b) - \mathbf{R}(a) \qquad \text{Second Fundamental Theorem}$$

EXAMPLE 12.10   Let $\mathbf{r}(t) = \langle 2t, 3, t^2 \rangle$. Find each of the following.

(a) $\displaystyle \int \mathbf{r}(t)dt$ (b) $\displaystyle \int_1^4 \mathbf{r}(t)dt$

Solution

(a) By Definition 12.6(a),

$$\int \mathbf{r}(t)dt = \left\langle \int 2t\,dt, \int 3\,dt, \int t^2 dt \right\rangle$$

$$= \left\langle t^2 + C_1, 3t + C_2, \frac{t^3}{3} + C_3 \right\rangle$$

$$= \left\langle t^2, 3t, \frac{t^3}{3} \right\rangle + \langle C_1, C_2, C_3 \rangle$$

We can write the answer as $\mathbf{R}(t) + \mathbf{C}$, where $\mathbf{R}(t) = \langle t^2, 3t, t^3/3 \rangle$ and $\mathbf{C} = \langle C_1, C_2, C_3 \rangle$.

(b) Using Definition 12.6(b) and Theorem 12.1(b), with the antiderivative $\mathbf{R}$ found in part (a), we have

$$\int_1^4 \mathbf{r}(t)dt = \mathbf{R}(4) - \mathbf{R}(1) = \left\langle t^2, 3t, \frac{t^3}{3} \right\rangle \Big]_1^4$$

$$= \left\langle 16, 12, \frac{64}{3} \right\rangle - \left\langle 1, 3, \frac{1}{3} \right\rangle$$

$$= \langle 15, 9, 21 \rangle \qquad \blacksquare$$

We list below the most important properties of the definite integral of a vector function. In most cases the proofs are a direct consequence of the analogous properties of the component functions. An exception is Property 6, which we will prove. You will be asked to prove the others in the exercises.

Properties of the Integral

If $\mathbf{u}$ and $\mathbf{v}$ are integrable vector functions on $[a, b]$, the following properties hold true.

1. $\int_a^b [\mathbf{u}(t) \pm \mathbf{v}(t)]\,dt = \int_a^b \mathbf{u}(t)dt \pm \int_a^b \mathbf{v}(t)dt$

2. $c \int_a^b \mathbf{u}(t)dt = \int_a^b c\mathbf{u}(t)dt$ \qquad for any scalar c

3. $\int_a^b \mathbf{u}(t)dt = \int_a^c \mathbf{u}(t)dt + \int_c^b \mathbf{u}(t)dt$ \qquad for $a < c < b$

4. $\mathbf{K} \cdot \int_a^b \mathbf{u}(t)dt = \int_a^b \mathbf{K} \cdot \mathbf{u}(t)dt$ \qquad for any constant vector $\mathbf{K}$

5. $\mathbf{K} \times \int_a^b \mathbf{u}(t)dt = \int_a^b \mathbf{K} \times \mathbf{u}(t)dt$ \qquad for any constant vector $\mathbf{K}$

6. $\left| \int_a^b \mathbf{u}(t)dt \right| \le \int_a^b |\mathbf{u}(t)|dt$ \qquad if $|\mathbf{u}(t)|$ is integrable on $[a, b]$

Proof of Property 6   Denote the integral $\int_a^b \mathbf{u}(t)dt$ by $\mathbf{K}$. Note that $\mathbf{K}$ is a constant vector. If $\mathbf{K} = \mathbf{0}$, then $|\mathbf{K}| = 0$, so the inequality of Property 6 is satisfied in a trivial way. Assume, then, that $\mathbf{K} \ne \mathbf{0}$. Then we have

$$|\mathbf{K}|^2 = \mathbf{K} \cdot \mathbf{K} = \mathbf{K} \cdot \int_a^b \mathbf{u}(t)dt = \int_a^b \mathbf{K} \cdot \mathbf{u}(t)dt \qquad \text{Property 4}$$

Since any real number cannot exceed its own absolute value, we have $\mathbf{K} \cdot \mathbf{u}(t) \leq |\mathbf{K} \cdot \mathbf{u}(t)|$. Thus,

$$\int_a^b \mathbf{K} \cdot \mathbf{u}(t)dt \leq \int_a^b |\mathbf{K} \cdot \mathbf{u}(t)|dt \leq \int_a^b |\mathbf{K}||\mathbf{u}(t)| \, dt$$

since $|\mathbf{K} \cdot \mathbf{u}(t)| \leq |\mathbf{K}||\mathbf{u}(t)|$ by the Cauchy-Schwarz Inequality (see Equation 11.4). By Property 2 we can factor out the constant $|\mathbf{K}|$ in the last integral. Combining our results, we have

$$|\mathbf{K}|^2 \leq |\mathbf{K}| \int_a^b |\mathbf{u}(t)| \, dt$$

Since $|\mathbf{K}| > 0$, we can divide both sides by $|\mathbf{K}|$ to get

$$|\mathbf{K}| \leq \int_a^b |\mathbf{u}(t)| \, dt$$

That is, $\left| \int_a^b \mathbf{u}(t)dt \right| \leq \int_a^b |\mathbf{u}(t)| \, dt$, which is what we wanted to prove.   ∎

Exercise Set 12.2

In Exercises 1–4, evaluate the limits.

1. $\lim\limits_{t \to 0} \langle e^{-t}, 2e^{3t}, 3e^t \rangle$

2. $\lim\limits_{t \to 1} \left\langle \dfrac{t-1}{t^2-1}, \dfrac{1-t}{1+t} \right\rangle$

3. $\lim\limits_{t \to 0} \left(\dfrac{\sin t}{t}\mathbf{i} + \dfrac{t^2}{1 - \cos t}\mathbf{j} \right)$

4. $\lim\limits_{t \to 0^+} \left[(t \ln t)\mathbf{i} + t^2 \left(1 - \dfrac{1}{t} \right)\mathbf{j} + 3t\mathbf{k} \right]$

In Exercises 5–8, determine the set of t values for which $\mathbf{r}$ is continuous.

5. $\mathbf{r}(t) = \left\langle t, \dfrac{1}{t-1}, \sqrt{1-t} \right\rangle$

6. $\mathbf{r}(t) = \langle t^{3/2}, e^{-t}, \ln t \rangle$

7. $\mathbf{r}(t) = \sqrt{\dfrac{1-t}{1+t}}\mathbf{i} + \dfrac{1}{(t-1)^2}\mathbf{j}$

8. $\mathbf{r}(t) = (\ln \cosh t)\mathbf{i} + (\ln \sinh t)\mathbf{j}$

In Exercises 9–11, evaluate the given expressions for $\mathbf{u}(t) = \langle 2t, t, 3 \rangle$, $\mathbf{v}(t) = \langle 1-t, 2, t \rangle$, and $f(t) = 1 - t$.

9. (a) $\mathbf{u}(t) - \mathbf{v}(t)$
   (b) $2\mathbf{u}(t) + 3\mathbf{v}(t)$

10. (a) $f(t)\mathbf{u}(t)$
   (b) $\mathbf{u}(t) \cdot \mathbf{v}(t)$

11. (a) $\mathbf{u}(t) \times \mathbf{v}(t)$
   (b) $\mathbf{v}(f(t))$

In Exercises 12–14, repeat Exercises 9–11 for $\mathbf{u}(t) = (\cos t)\mathbf{i} + (\sin t)\mathbf{j} + (\sin t)\mathbf{k}$, $\mathbf{v}(t) = (\sin t)\mathbf{i} + (\cos t)\mathbf{j} + \mathbf{k}$, and $f(t) = 2t$.

In Exercises 15–22, find $\mathbf{r}'(t)$.

15. $\mathbf{r}(t) = \left\langle t^2, e^{2t}, \dfrac{1}{t} \right\rangle$

16. $\mathbf{r}(t) = \langle \ln t, \sin t, \cos 2t \rangle$

17. $\mathbf{r}(t) = \langle t \sin t, t \ln t \rangle$

18. $\mathbf{r}(t) = \langle \sin^{-1} t, \tan^{-1} t, \ln(1+t) \rangle$

19. $\mathbf{r}(t) = (1 - e^{-t})\mathbf{i} + te^{2t}\mathbf{j}$

20. $\mathbf{r}(t) = (\sinh t)\mathbf{i} + (\cosh t)\mathbf{j} + (\tanh t)\mathbf{k}$

21. $\mathbf{r}(t) = \dfrac{t}{\sqrt{1-t^2}}\mathbf{i} + (\sin^{-1} t)\mathbf{j} + \sqrt{1-t^2}\,\mathbf{k}$

22. $\mathbf{r}(t) = \dfrac{1-t}{1+t}\mathbf{i} + \dfrac{t}{1-t^2}\mathbf{j}$

In Exercises 23–28, find parametric equations of the tangent line at t_0.

23. $\mathbf{r}(t) = \langle 2t, 3t^2, t^3 \rangle$; $t_0 = 1$

24. $\mathbf{r}(t) = \langle 2 \sin t, 3 \cos t, 4t \rangle$; $t_0 = \dfrac{\pi}{2}$

25. $\mathbf{r}(t) = \left(\dfrac{1}{t} \right) \mathbf{i} + \sqrt{t}\mathbf{j} + (4t^2)\mathbf{k}$; $t_0 = \dfrac{1}{4}$

26. $\mathbf{r}(t) = e^t \mathbf{i} + e^{-t}\mathbf{j} + e^{2t}\mathbf{k}$; $t_0 = \ln 2$

27. $\mathbf{r}(t) = \left\langle \dfrac{1}{1-t}, t\sqrt{t+2}, \dfrac{1}{t-1} \right\rangle$; $t_0 = 2$

28. $\mathbf{r}(t) = \left\langle t \ln 2, 2 \ln t, \dfrac{\ln t}{t} \right\rangle$; $t_0 = 1$

In Exercises 29–32, draw the graph of $\mathbf{r}$. *Find the tangent vector* $\mathbf{r}'(t)$ *at the specified point, and show it on the graph.*

29. $\mathbf{r}(t) = \langle 4 \cos t, 2 \sin t \rangle$; $t = \dfrac{\pi}{3}$

30. $\mathbf{r}(t) = \langle 1 - t, t^2 \rangle$; $t = 2$

31. $\mathbf{r}(t) = t^2 \mathbf{i} + t^3 \mathbf{j}$; $t = -1$

32. $\mathbf{r}(t) = (\sec t)\mathbf{i} + (\tan t)\mathbf{j}$; $t = \dfrac{3\pi}{4}$

In Exercise 33 and 34, find a tangent vector to the graph of $\mathbf{r}$ *at the specified point.*

33. $\mathbf{r}(t) = \langle t^2, 1 - t^3, 3t \rangle$; $(1, 0, 3)$

34. $\mathbf{r}(t) = \left\langle 2 + \ln t, 1 - t \ln t, \dfrac{2 \ln t}{t} \right\rangle$; $(2, 1, 0)$

In Exercises 35–42, evaluate the integrals.

35. $\displaystyle\int \langle \sin t, 1 - \cos t, t \rangle \, dt$

36. $\displaystyle\int \left\langle te^{-t^2}, \dfrac{\ln t}{t}, \dfrac{1}{t \ln t} \right\rangle dt$

37. $\displaystyle\int_0^1 \left\langle \dfrac{1}{1+t^2}, \sqrt{1-t}, t\sqrt{1-t^2} \right\rangle dt$

38. $\displaystyle\int_0^{\ln 2} \langle e^{-t}, e^{2t}, 6e^{-3t} \rangle \, dt$

39. $\displaystyle\int_0^{\pi/2} \left[(\sin^2 t)\mathbf{i} + (\sin t \cos^2 t)\mathbf{j} \right] dt$

40. $\displaystyle\int_1^{\sqrt{3}} \left(\dfrac{1}{1+t^2}\mathbf{i} + \dfrac{1}{\sqrt{4-t^2}}\mathbf{j} \right) dt$

41. $\displaystyle\int_{2\pi/3}^{\pi} \left[(\cos^3 t)\mathbf{i} + (\tan^2 t)\mathbf{j} + \mathbf{k} \right] dt$

42. $\displaystyle\int_0^1 \left[\dfrac{t-1}{t+1}\mathbf{i} + \dfrac{1}{\sqrt{t}(\sqrt{t}+1)}\mathbf{j} + \dfrac{t}{(t+1)^2}\mathbf{k} \right] dt$

43. Evaluate the integral

$$\int_1^{2\sqrt{2}} |\mathbf{r}(t)| \, dt$$

where $\mathbf{r}(t) = \langle t^2 \sin t, t^2 \cos t, t^2 \rangle$.

44. Prove that if $\mathbf{r}(t) = \mathbf{C}$, where $\mathbf{C} = \langle C_1, C_2, C_3 \rangle$ is a constant vector, then $\mathbf{r}'(t) = \mathbf{0}$ for all t.

45. Verify the inequality $\left| \int_a^b \mathbf{r}(t)dt \right| \leq \int_a^b |\mathbf{r}(t)| \, dt$ for $\mathbf{r}(t) = \langle \sin t, \cos t, 1 \rangle$.

In Exercises 46–56, prove the specified property.

46. Derivative Property 1

47. Derivative Property 2

48. Derivative Property 3

49. Derivative Property 4

50. Derivative Property 5

51. Derivative Property 6

52. Integral Property 1

53. Integral Property 2

54. Integral Property 3

55. Integral Property 4

56. Integral Property 5

57. Prove that if $\mathbf{u}'(t) = \mathbf{v}'(t)$ for all t on an interval I, then $\mathbf{u}(t) = \mathbf{v}(t) + \mathbf{C}$ on I.

58. Suppose the position of a particle moving in space is given by

$$\mathbf{r}(t) = \left\langle e^{(t-1)^2} \cos(t-1), e^{(t-1)^2} \sin(t-1), t \right\rangle$$

(a) Plot the curve in space for $0 \le t \le 2$.

(b) Calculate the velocity and speed of the particle at any time t.

(c) Calculate the acceleration at time t.

59. Suppose the position of a particle moving in space is given by $\mathbf{r}(t) = \left\langle t, t^2, t^3 \right\rangle$. Find the tangent line to the curve at the point $(2, 4, 8)$ and plot the curve and the tangent line.

60. Find the tangent line to the space curve $\mathbf{r}(t) = \langle \sin 2t, 2 \sin t, \cos t \rangle$ at the point where $t = 0$ and the point where $t = \pi/6$. Sketch the curve and the two tangent lines and find the point of intersection.

61. A stable camera is mounted on the nose of a spy airplane, so it can point directly ahead of the airplane. The path of the airplane is described by the curve $\mathbf{r}(t) = \langle 3 \cos t, 4 \sin t, 6 + \sin 3t \rangle$. Ground level is the plane $z = 0$. At time $t = \pi/3$, determine the coordinates of the ground-level point seen by the camera.*

---

*From Chi-Keung Cheung and John Harer, *A Guidebook to Multivariable Calculus with Maple V.*

12.3 ARC LENGTH

The distance traveled around one leaf of a freeway intersection can be found by parameterization of the appropriate function.

The **length** of a space curve C on an interval $I = [a, b]$ is defined analogously to the length of a plane curve given in Chapter 9. When C has a finite length, it is said to be **rectifiable**. For a plane curve C, with parameterization $x = f(t)$ and $y = g(t)$, we showed in Chapter 9 (Equation 9.6) that if f' and g' are continuous on $[a, b]$, then C is rectifiable and its length L is given by

$$L = \int_a^b \sqrt{[f'(t)]^2 + [g'(t)]^2}\, dt = \int_a^b \sqrt{\left(\frac{dx}{dt}\right)^2 + \left(\frac{dy}{dt}\right)^2}\, dt$$

Letting $\mathbf{r}(t) = \langle f(t), g(t) \rangle$, we see that the integrand is the length $|\mathbf{r}'(t)|$ of the tangent vector $\mathbf{r}'(t)$, so we can write L in the succinct form

$$L = \int_a^b |\mathbf{r}'(t)|\, dt$$

The following theorem generalizes this result.

THEOREM 12.2

> Let C be the graph of the continuous vector function $\mathbf{r}$ on $[a, b]$, and suppose $\mathbf{r}'$ also is continuous on $[a, b]$. Then C is rectifiable, and if C is traversed exactly once as t increases from a to b, its length L is given by
>
> $$L = \int_a^b |\mathbf{r}'(t)|\, dt \qquad\qquad (12.1)$$

REMARK ———————————————————————

■ Just as in Definition 9.2, we say that the graph of $\mathbf{r}$ is **smooth** if $\mathbf{r}'$ is continuous and $\mathbf{r}'(t) \ne 0$. So by Theorem 12.2, we see that every smooth curve is rectifiable.

———————————————————————

When C is a plane curve, Equation 12.1 is merely a restatement of Equation 9.6, but it holds true for space curves as well. If $\mathbf{r}(t) = \langle f(t), g(t), h(t) \rangle$, then Equation 12.1 can be written in the following form.

The Length of a Space Curve

$$L = \int_a^b \sqrt{[f'(t)]^2 + [g'(t)]^2 + [h'(t)]^2} \, dt$$

$$= \int_a^b \sqrt{\left(\frac{dx}{dt}\right)^2 + \left(\frac{dy}{dt}\right)^2 + \left(\frac{dz}{dt}\right)^2} \, dt$$

EXAMPLE 12.11   Find the length of the arc of the circular helix $\mathbf{r}(t) = \langle \cos t, \sin t, t \rangle$ for t varying from $t = 0$ to $t = 2\pi$.

Solution

$$\mathbf{r}'(t) = \langle -\sin t, \cos t, 1 \rangle$$

So by Equation 12.1,

$$L = \int_0^{2\pi} |\mathbf{r}'(t)| \, dt = \int_0^{2\pi} \sqrt{\sin^2 t + \cos^2 t + 1} \, dt$$

$$= \int_0^{2\pi} \sqrt{1 + 1} \, dt = 2\sqrt{2}\,\pi \qquad \blacksquare$$

EXAMPLE 12.12   Find the length of the curve
$$\mathbf{r}(t) = 3t^2\mathbf{i} + (1 - 4t^2)\mathbf{j} + 2t^3\mathbf{k}$$
from the point given by $t = 0$ to the point given by $t = 4$.

Solution   We have $\mathbf{r}'(t) = 6t\mathbf{i} - 8t\mathbf{j} + 6t^2\mathbf{k}$, so
$$|\mathbf{r}'(t)| = \sqrt{36t^2 + 64t^2 + 36t^4} = \sqrt{100t^2 + 36t^4} = 2t\sqrt{25 + 9t^2}$$
since $t \geq 0$. Thus, by Equation 12.1,

$$L = \int_0^4 2t\sqrt{25 + 9t^2} \, dt = \frac{1}{9} \cdot \frac{2}{3}\left[(25 + 9t^2)^{3/2}\right]_0^4 \qquad \text{Let } u = 25 + 9t^2.$$

$$= \frac{2}{27}\left[(169)^{3/2} - (25)^{3/2}\right] = \frac{4144}{27} \qquad \blacksquare$$

The Arc Length Function

If C has the vector equation $\mathbf{r}(t) = \langle f(t), g(t), h(t) \rangle$ for t in $[a, b]$, and if $\mathbf{r}'$ is continuous on $[a, b]$, we know that C is rectifiable. For an arbitrary t in $[a, b]$ we designate the length of C from $\mathbf{r}(a)$ to $\mathbf{r}(t)$ by $s(t)$; that is,

$$s(t) = \int_a^t |\mathbf{r}'(u)| \, du, \qquad a \leq t \leq b \tag{12.2}$$

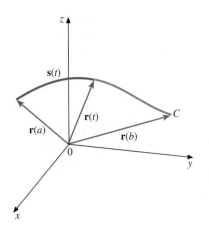

FIGURE 12.8

(where u is used as the variable of integration to avoid confusion with the upper limit t). We show the arc length $s(t)$ from $\mathbf{r}(a)$ to $\mathbf{r}(t)$ in Figure 12.8. In this way we are defining a scalar function s, called the **arc length function** for C. Note that $s(a) = 0$ and $s(b) = L$, the total length of C on $[a, b]$. Since $|\mathbf{r}'|$ is continuous on $[a, b]$, it follows by the First Fundamental Theorem of Calculus (Theorem 5.2) that

$$s'(t) = |\mathbf{r}'(t)| \tag{12.3}$$

Equivalently,

$$\frac{ds}{dt} = \sqrt{\left(\frac{dx}{dt}\right)^2 + \left(\frac{dy}{dt}\right)^2 + \left(\frac{dz}{dt}\right)^2} \tag{12.4}$$

or, in terms of differentials,

$$ds = \sqrt{\left(\frac{dx}{dt}\right)^2 + \left(\frac{dy}{dt}\right)^2 + \left(\frac{dz}{dt}\right)^2}\, dt \tag{12.5}$$

EXAMPLE 12.13   Find $s(t)$ for the circular helix

$$\mathbf{r}(t) = \langle \cos t, \sin t, t \rangle, \qquad t \geq 0$$

Solution   First, we calculate $|\mathbf{r}'(t)|$.

$$|\mathbf{r}'(t)| = \sqrt{\sin^2 t + \cos^2 t + 1} = \sqrt{2}$$

By Equation 12.2,

$$s(t) = \int_0^t |\mathbf{r}'(u)|\, du = \int_0^t \sqrt{2}\, du = \sqrt{2}\, t \qquad \blacksquare$$

Curves Parameterized by Arc Length

It is sometimes useful to make a change of parameter in the equation of a curve C. Under appropriate conditions, the length of the curve as found from the new parameterization will be the same as that found from the original equation (see Exercise 27 in Exercise Set 12.3). For example, if C is the graph of $\mathbf{r}(t) = \langle t^2, 1 - t^2, t^3 \rangle$ from $t = 1$ to $t = 2$, and we let $t = e^u$, so that $u = \ln t$, then the equation of C with parameter u is $\mathbf{R}(u) = \langle e^{2u}, 1 - e^{2u}, e^{3u} \rangle$ from $u = 0$ to $u = \ln 2$. In Exercise 21 of Exercise Set 12.3, you will be asked to verify that the same length L is given by the two integrals

$$\int_1^2 |\mathbf{r}'(t)|\, dt \qquad \text{and} \qquad \int_0^{\ln 2} |\mathbf{R}'(u)|\, du$$

It is particularly useful to consider a curve parameterized by arc length s, measured along the curve. If C has length L, then its vector equation with s as a parameter has the form

$$\mathbf{r}(s) = \langle f(s), g(s), h(s) \rangle, \qquad 0 \le s \le L$$

Equation 12.2 then gives

$$s = \int_0^s |\mathbf{r}'(u)| \, du$$

where we are assuming $\mathbf{r}'$ is continuous for s on $[0, L]$. If we differentiate both sides of this equation with respect to s, we get

$$1 = |\mathbf{r}'(s)|$$

by the First Fundamental Theorem of Calculus. Since $\mathbf{r}'(s)$ is a tangent vector to the curve, we have shown the following:

When a smooth curve C is parameterized by arc length, the tangent vector $\mathbf{r}'(s)$ is a unit vector at every point on C.

We will make use of this result in Section 12.4.

EXAMPLE 12.14   Obtain the parameterization by arc length of the circular helix in Example 12.13. Show that the tangent vector obtained by differentiation with respect to s is a unit vector.

Solution   In Example 12.13 we found that

$$s = \sqrt{2}\,t$$

Thus, $t = s/\sqrt{2}$. On substituting this value of t in the equation $\mathbf{r}(t) = \langle \cos t, \sin t, t \rangle$, we get

$$\mathbf{r}\left(\frac{s}{\sqrt{2}}\right) = \left\langle \cos\left(\frac{s}{\sqrt{2}}\right), \sin\left(\frac{s}{\sqrt{2}}\right), \frac{s}{\sqrt{2}} \right\rangle$$

as the parameterization with respect to arc length. Let $\mathbf{R}(s) = \mathbf{r}(s/\sqrt{2})$. Then

$$\mathbf{R}'(s) = \left\langle -\frac{1}{\sqrt{2}} \sin\left(\frac{s}{\sqrt{2}}\right), \frac{1}{\sqrt{2}} \cos\left(\frac{s}{\sqrt{2}}\right), \frac{1}{\sqrt{2}} \right\rangle$$

and

$$|\mathbf{R}'(s)| = \sqrt{\frac{1}{2} \sin^2\left(\frac{s}{\sqrt{2}}\right) + \frac{1}{2} \cos^2\left(\frac{s}{\sqrt{2}}\right) + \frac{1}{2}} = \sqrt{\frac{1}{2} + \frac{1}{2}} = 1 \qquad \blacksquare$$

Exercise Set 12.3

In Exercises 1–6, find the length of the curve on the specified interval.

1. $\mathbf{r}(t) = \langle 2 - t, 3 + 2t, 5 - 3t \rangle$, $1 \leq t \leq 3$

2. $\mathbf{r}(t) = \langle 3t, 2t^{3/2}, 4 \rangle$, $0 \leq t \leq 8$

3. $\mathbf{r}(t) = (3t^2 + 1)\mathbf{i} + 3t^2\mathbf{j} + 2t^3\mathbf{k}$, $0 \leq t \leq 2$

4. $\mathbf{r}(t) = 2 \sin^2 t\mathbf{i} + \cos^3 t\mathbf{j} + \sin^3 t\mathbf{k}$, $0 \leq t \leq \dfrac{\pi}{2}$

5. $x = 3t$, $y = 2 \cos 3t$, $z = 2 \sin 3t$, $0 \leq t \leq \dfrac{\pi}{3}$

6. $x = 2e^t$, $y = e^{-t}$, $z = 2t$, $-1 \leq t \leq 1$

In Exercises 7–12, find $s(t)$.

7. $\mathbf{r}(t) = \langle 3 - 2t, 4 + 6t, 5t \rangle$, $-1 \leq t \leq 10$

8. $\mathbf{r}(t) = \langle 2 \sin t, 4t, 2 \cos t \rangle$, $t \geq 0$

9. $x = 3t \sin t$, $y = 3t \cos t$, $z = (2t)^{3/2}$, $t \geq 0$

10. $x = 5e^{-t}$, $y = (3e^{-t} + 1)$, $z = -4e^{-t}$, $t \geq 0$

11. $\mathbf{r}(t) = 7\mathbf{i} + t^3\mathbf{j} - t^2\mathbf{k}$, $t \geq 0$

12. $\mathbf{r}(t) = \ln t^2\mathbf{i} + \dfrac{1}{t}\mathbf{j} + 2t\mathbf{k}$, $t \geq 1$

In Exercises 13–16, find an equivalent vector equation for the curve under the specified change in parameter. What is the parameter interval for u? State whether the direction is unchanged or reversed.

13. $\mathbf{r}(t) = \left\langle t, 2t, \dfrac{1}{t} \right\rangle$, $t > 0$; $u = t^2$

14. $\mathbf{r}(t) = \langle e^{-2t}, 1 + e^t, 2e^{2t} \rangle$, $-1 \leq t \leq 1$; $u = e^t$

15. $\mathbf{r}(t) = \dfrac{t}{t+1}\mathbf{i} + \dfrac{1}{t+1}\mathbf{j} + \dfrac{t+2}{t+1}\mathbf{k}, 0 \leq t \leq 1$;

$u = \dfrac{1}{t+1}$

16. $\mathbf{r}(t) = \sqrt{4 - t^2}\,\mathbf{i} + \dfrac{1}{\sqrt{4 - t^2}}\mathbf{j} + \dfrac{t}{\sqrt{4 - t^2}}\mathbf{k}$,

$0 \leq t \leq 1$; $u = \sqrt{4 - t^2}$

In Exercises 17–20, re-parameterize the given curve with respect to arc length.

17. $\mathbf{r}(t) = \langle a \cos t, a \sin t \rangle, 0 \leq t \leq 2\pi$, where a is a positive constant

18. $\mathbf{r}(t) = \langle 2t - 1, 3t, 1 - t \rangle, t \geq 0$

19. $\mathbf{r}(t) = (4 \cos t)\mathbf{i} + 3t\mathbf{j} + (4 \sin t)\mathbf{k}, t \geq 0$

20. $\mathbf{r}(t) = t^2\mathbf{i} + (1 - t^2)\mathbf{j} + \left(\dfrac{t^2 + 3}{2} \right)\mathbf{k}, t \geq 0$

21. Let C be the graph of $\mathbf{r}(t) = \langle t^2, 1 - t^2, t^3 \rangle$ for $1 \leq t \leq 2$. Find the length of C. Make the change of parameter $u = \ln t$, and let $\mathbf{R}(u)$ be the resulting function. What is the domain of $\mathbf{R}$ so that its graph is also C? Find the length of C using $\mathbf{R}(u)$, and verify that the answer is the same as that found using $\mathbf{r}(t)$.

22. Let $\mathbf{r}(t) = (2t - 1)\mathbf{i} + (2e^t + 3)\mathbf{j} + e^{-t}\mathbf{k}$ for t in the interval $0 \leq t \leq \ln 2$. Follow the instructions for Exercise 21 with the change of parameter $u = e^{-t}$.

23. Find the length of the curve $\mathbf{r}(t) = \langle t^2, t^3, -3t^2 \rangle$ from $(1, -1, -3)$ to $(1, 1, -3)$.

24. Let C be the curve defined parametrically by $x = t$, $y = t^2/2$, and $z = 2t$.
    (a) Find a unit vector tangent to C at the point $(2, 2, 4)$.
    (b) Find the length of C from $(-2, 2, -4)$ to $(2, 2, 4)$.
    (c) Sketch C over the range from $t = 0$ to $t = 4$, and show the unit tangent vector found in part (a).

25. Let C be the curve with position vector $\mathbf{r}(t) = \langle 3t^2/2, 4 + 2t^2, 3 \rangle$ on $[1, \infty)$. Make the change of parameter $u = s(t)$, where s is the arc length function, and let $\mathbf{R}(u)$ be the resulting position vector for C. Verify that for all $u \geq 0$, $|\mathbf{R}'(u)| = 1$.

26. Let C be the graph of the differentiable vector function $\mathbf{r}$ on the interval I. Let α be a differentiable scalar function on an interval J that has range I, suppose $\alpha'(u) \neq 0$ for all u in J, and let $\mathbf{R} = \mathbf{r} \circ \alpha$. If $t_0 = \alpha(u_0)$, prove that $\mathbf{r}'(t_0) = k\mathbf{R}'(u_0)$ for some scalar k. What is the geometric significance of this result?

27. Under the hypotheses of Exercise 26, prove that the length of the graph of $\mathbf{r}$ over I equals the length of the graph of $\mathbf{R}$ over J. (This result shows that the length of a curve is invariant under a parameter change of the type described.)

28. Let C be the graph of the continuously differentiable function $\mathbf{r}(t)$ for t in $[a, b]$, and let the length of C be L. Change the parameter to arc length $s = s(t)$ for s in $[0, L]$, and denote the result by $\mathbf{R}(s)$. Show that for every s in $[0, L]$, $|\mathbf{R}'(s)| = 1$. (This is an alternative proof to the one at the end of this section that when a smooth curve is parameterized by arc length, the tangent vector is always a unit vector.) (Hint: Let $t = t(s)$ be the inverse of $s = s(t)$, and use the Chain Rule, along with the fact that $dt/ds = 1/(ds/dt)$.)

29. Estimate the arc length of the space curve $\mathbf{r}(t) = \langle 2t, t^2, 3t^3 \rangle$, for t in $[-1, 1]$. Use Simpson's Rule with $n = 10$.

 30. Estimate the arc length of the space curve $\mathbf{r}(t) = \langle e^t \cos t, e^t \sin t, t \rangle$, for t in $[-1, 1]$. Use Simpson's Rule with $n = 10$.

12.4 UNIT TANGENT AND NORMAL VECTORS; CURVATURE

Unit Tangent and Normal Vectors

We saw in Section 12.3 that when a smooth curve C is parameterized by arc length, the tangent vector $\mathbf{r}'(s)$ always has length 1. We call this vector the **unit tangent vector** for C and designate it by $\mathbf{T}$.

The Unit Tangent Vector

$$\mathbf{T} = \frac{d\mathbf{r}}{ds} \qquad (12.6)$$

Since $|\mathbf{T}| = 1$, we have $\mathbf{T} \cdot \mathbf{T} = |\mathbf{T}|^2 = 1$, and if $d\mathbf{T}/ds$ exists, we can differentiate both sides of $\mathbf{T} \cdot \mathbf{T} = 1$ and use Property 4 for derivatives to get

$$\frac{d\mathbf{T}}{ds} \cdot \mathbf{T} + \mathbf{T} \cdot \frac{d\mathbf{T}}{ds} = 0$$

$$2\mathbf{T} \cdot \frac{d\mathbf{T}}{ds} = 0$$

Thus, $\mathbf{T} \cdot d\mathbf{T}/ds = 0$, so $\mathbf{T}$ and $d\mathbf{T}/ds$ are orthogonal. If $d\mathbf{T}/ds$ is nonzero and is not a unit vector, we make it into a unit vector by dividing it by its length. We designate the result by $\mathbf{N}$ and call it the **principal unit normal vector** (or **unit normal** for short). The unit tangent vector $\mathbf{T}$ and unit normal $\mathbf{N}$ to a representative curve $\mathbf{C}$ are shown in Figure 12.9.

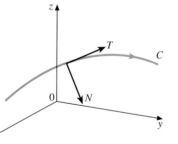

FIGURE 12.9

The Unit Normal Vector

$$\mathbf{N} = \frac{\dfrac{d\mathbf{T}}{ds}}{\left| \dfrac{d\mathbf{T}}{ds} \right|} \qquad (12.7)$$

Curvature

Since **T** has constant length 1, the derivative $d\mathbf{T}/ds$ reflects the change in *direction* of **T** only. Its magnitude $|d\mathbf{T}/ds|$ thus provides a measure of how rapidly the unit tangent vector **T** is turning as arc length is traversed. We call this magnitude the **curvature** of C and designate it by the Greek letter kappa, κ.

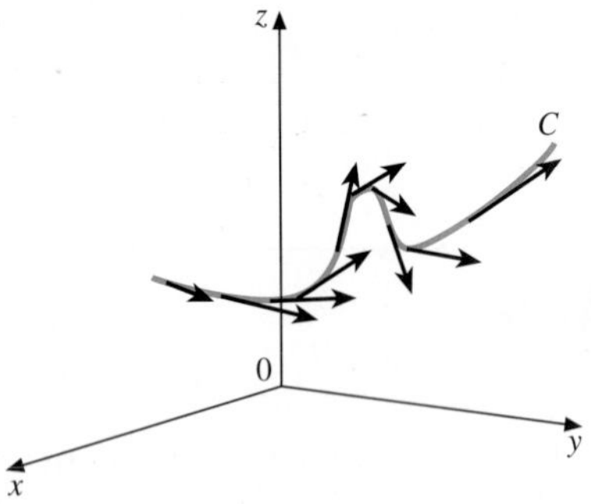

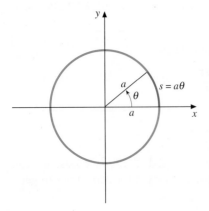

FIGURE 12.10
Unit tangent vector changes direction slowly when the curve is relatively flat, and it changes direction rapidly when the curve bends sharply.

Curvature

$$\kappa = \left| \frac{d\mathbf{T}}{ds} \right| \qquad\qquad (12.8)$$

In Figure 12.10 we show how the unit tangent vector changes direction slowly when the curve is fairly straight (curvature is small) but changes direction rapidly when the curve bends more sharply (curvature is large).

We can now write Equation 12.7 in the form

$$\mathbf{N} = \frac{1}{\kappa} \frac{d\mathbf{T}}{ds}$$

so that

$$\frac{d\mathbf{T}}{ds} = \kappa \mathbf{N} \qquad\qquad (12.9)$$

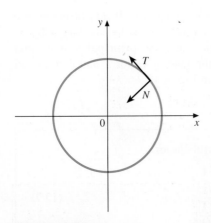

FIGURE 12.11

EXAMPLE 12.15   Find **T**, **N**, and κ at an arbitrary point on a circle of radius a in the xy-plane.

Solution   For convenience we take the center of the circle at the origin. From Figure 12.11 we see that parametric equations for the circle are $x = a\cos\theta$ and $y = a\sin\theta$, where $0 \le \theta \le 2\pi$. Equivalently, the circle, parameterized in terms of θ, has the vector equation $\mathbf{r}(\theta) = (a\cos\theta)\mathbf{i} + (a\sin\theta)\mathbf{j}$. Since the arc length s is given by $s = a\theta$, we have $\theta = s/a$. So the parameterization of the circle with respect to arc length s is given by

$$\mathbf{r}\left(\frac{s}{a}\right) = a\cos\left(\frac{s}{a}\right)\mathbf{i} + a\sin\left(\frac{s}{a}\right)\mathbf{j}, \qquad 0 \le s \le 2\pi a$$

Let $\mathbf{R}(s) = \mathbf{r}(s/a)$. Then, from Equation 12.6,

$$\mathbf{T} = \frac{d\mathbf{R}}{ds} = -\sin\left(\frac{s}{a}\right)\mathbf{i} + \cos\left(\frac{s}{a}\right)\mathbf{j}$$

To find **N** and κ we calculate $d\mathbf{T}/ds$:

$$\frac{d\mathbf{T}}{ds} = -\frac{1}{a}\cos\left(\frac{s}{a}\right)\mathbf{i} - \frac{1}{a}\sin\left(\frac{s}{a}\right)\mathbf{j}$$

$$\kappa = \left|\frac{d\mathbf{T}}{ds}\right| = \sqrt{\frac{1}{a^2}\cos^2\left(\frac{s}{a}\right) + \frac{1}{a^2}\sin^2\left(\frac{s}{a}\right)} = \frac{1}{a}$$

$$\mathbf{N} = -\left[\cos\left(\frac{s}{a}\right)\mathbf{i} + \sin\left(\frac{s}{a}\right)\mathbf{j}\right] = -\frac{1}{a}\mathbf{R}(s)$$

FIGURE 12.12

Observe that the direction of **N** is opposite that of $\mathbf{R}(s)$, so **N** points toward the center of the circle (Figure 12.12). ∎

REMARK ───────────────────────────────────

■ The idea that the curvature of a circle is the reciprocal of the radius makes sense on intuitive grounds. When the radius is small, the unit tangent vector changes direction rapidly as we progress around the circle and so the curvature is large, whereas for a circle with a large radius, this change is slow and the curvature is small.

───────────────────────────────────

T, N, and κ in Terms of Other Parameters

As we saw in Example 12.15, it is easy to parameterize the circle in terms of arc length. Unfortunately, for most curves doing so is difficult. So we need computational formulas for **T**, **N**, and κ in terms of other parameters. If t is any other parameter and $\mathbf{r}(t)$ is the position vector for C, we know that in theory at least, we can introduce as the parameter $s = s(t)$ given by Equation 12.2, provided $\mathbf{r}'$ is continuous on the t interval $[a, b]$ and $\mathbf{r}'(t) \neq 0$ there (that is, C is smooth). It follows that $s(t) = |\mathbf{r}'(t)| > 0$, so that s is an increasing function of t and hence has an inverse, say $t(s)$, that is also increasing. Under these conditions,

$$\frac{dt}{ds} = \frac{1}{\dfrac{ds}{dt}}$$

So, using the Chain Rule, we have

$$\mathbf{T} = \frac{d\mathbf{r}}{ds} = \frac{d\mathbf{r}}{dt}\frac{dt}{ds} = \frac{\dfrac{d\mathbf{r}}{dt}}{\dfrac{ds}{dt}}$$

and since $ds/dt = |\mathbf{r}'(t)|$, we can write

$$\mathbf{T} = \frac{\mathbf{r}'(t)}{|\mathbf{r}'(t)|} \tag{12.10}$$

To find **N** also as a function of t, we assume $\mathbf{r}''(t)$ exists and get, again by the Chain Rule,

$$\mathbf{N} = \frac{\dfrac{d\mathbf{T}}{ds}}{\left|\dfrac{d\mathbf{T}}{ds}\right|} = \frac{\dfrac{d\mathbf{T}}{dt}\dfrac{dt}{ds}}{\left|\dfrac{d\mathbf{T}}{dt}\dfrac{dt}{ds}\right|} = \frac{\dfrac{d\mathbf{T}}{dt}\dfrac{dt}{ds}}{\left|\dfrac{d\mathbf{T}}{dt}\right|\dfrac{dt}{ds}} = \frac{\dfrac{d\mathbf{T}}{dt}}{\left|\dfrac{d\mathbf{T}}{dt}\right|}$$

since $dt/ds > 0$. Thus,

$$\mathbf{N} = \frac{\mathbf{T}'(t)}{|\mathbf{T}'(t)|} \tag{12.11}$$

Equations 12.10 and 12.11 are valid both for plane curves and for space curves. For space curves, the cross product $\mathbf{T} \times \mathbf{N}$ is a vector orthogonal to both $\mathbf{T}$ and $\mathbf{N}$. It is called the **binormal** for C, and we designate it by $\mathbf{B}$:

$$\mathbf{B} = \mathbf{T} \times \mathbf{N} \tag{12.12}$$

The binormal is also a unit vector, as we see by

$$|\mathbf{B}| = |\mathbf{T} \times \mathbf{N}| = |\mathbf{T}||\mathbf{N}| \sin \frac{\pi}{2} = 1 \qquad \text{Since } \mathbf{T} \text{ and } \mathbf{N} \text{ are unit vectors}$$

The triple $\mathbf{T}, \mathbf{N}, \mathbf{B}$ thus is a set of mutually orthogonal unit vectors, much like the triple $\mathbf{i}, \mathbf{j}, \mathbf{k}$. But whereas the latter triple is fixed in direction, $\mathbf{T}, \mathbf{N}, \mathbf{B}$ can vary at different points on the curve. This triple is often called the *moving trihedral* for C (see Figure 12.13).

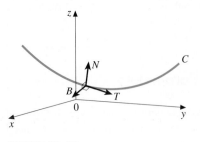

FIGURE 12.13

To find a more useful computational formula for κ than Equation 12.8, we derive some preliminary results. From Equation 12.10, we can write $\mathbf{r}'(t)$ in the form

$$\frac{d\mathbf{r}}{dt} = \left| \frac{d\mathbf{r}}{dt} \right| \mathbf{T} = \frac{ds}{dt} \mathbf{T}$$

so that

$$\frac{d^2\mathbf{r}}{dt^2} = \frac{d^2s}{dt^2} \mathbf{T} + \frac{ds}{dt} \frac{d\mathbf{T}}{dt}$$

By the Chain Rule and Equation 12.9,

$$\frac{d\mathbf{T}}{dt} = \frac{d\mathbf{T}}{ds} \frac{ds}{dt} = \kappa \frac{ds}{dt} \mathbf{N}$$

Thus,

$$\frac{d^2\mathbf{r}}{dt^2} = \frac{d^2s}{dt^2} \mathbf{T} + \kappa \left(\frac{ds}{dt} \right)^2 \mathbf{N} \tag{12.13}$$

We assume now that C is a space curve so that all vectors are three-dimensional. The calculations are valid as well, however, for curves in the xy-plane if we treat the vectors as being three-dimensional with third component 0. We take the cross product of $\mathbf{r}'(t)$ and $\mathbf{r}''(t)$, getting

$$\frac{d\mathbf{r}}{dt} \times \frac{d^2\mathbf{r}}{dt^2} = \frac{ds}{dt} \mathbf{T} \times \left[\frac{d^2s}{dt^2} \mathbf{T} + \kappa \left(\frac{ds}{dt} \right)^2 \mathbf{N} \right]$$

$$= \left(\frac{ds}{dt} \frac{d^2s}{dt^2} \right) (\mathbf{T} \times \mathbf{T}) + \kappa \left(\frac{ds}{dt} \right)^3 (\mathbf{T} \times \mathbf{N})$$

$$= \kappa \left(\frac{ds}{dt} \right)^3 \mathbf{B}$$

since $\mathbf{T} \times \mathbf{T} = \mathbf{0}$. Since $|\mathbf{B}| = 1$ and both κ and ds/dt are nonnegative,

$$\left| \frac{d\mathbf{r}}{dt} \times \frac{d^2\mathbf{r}}{dt^2} \right| = \kappa \left(\frac{ds}{dt} \right)^3$$

Finally, we solve for κ, writing $\mathbf{r}'$ and $\mathbf{r}''$ for $d\mathbf{r}/dt$ and $d^2\mathbf{r}/dt^2$, respectively, and replacing ds/dt by $|\mathbf{r}'|$:

Curvature in Terms of $\mathbf{r}(t)$

$$\kappa = \frac{|\mathbf{r}' \times \mathbf{r}''|}{|\mathbf{r}'|^3} \qquad (12.14)$$

EXAMPLE 12.16 Find $\mathbf{T}$, $\mathbf{N}$, and κ at an arbitrary point on the circular helix $\mathbf{r}(t) = \langle 2\cos 3t, 2\sin 3t, 8t \rangle$.

Solution We will need $\mathbf{r}'$, $\mathbf{r}''$, and $|\mathbf{r}'|$:

$$\mathbf{r}'(t) = \langle -6\sin 3t, 6\cos 3t, 8 \rangle$$

$$\mathbf{r}''(t) = \langle -18\cos 3t, -18\sin 3t, 0 \rangle$$

$$|\mathbf{r}'(t)| = \sqrt{36\sin^2 3t + 36\cos^2 3t + 64} = \sqrt{36 + 64} = 10$$

By Equation 12.10,

$$\mathbf{T} = \frac{\mathbf{r}'(t)}{|\mathbf{r}'(t)|} = \frac{\langle -6\sin 3t, 6\cos 3t, 8 \rangle}{10} = \left\langle -\frac{3}{5}\sin 3t, \frac{3}{5}\cos 3t, \frac{4}{5} \right\rangle$$

To find $\mathbf{N}$, we first calculate $\mathbf{T}'(t)$:

$$\mathbf{T}'(t) = \left\langle -\frac{9}{5}\cos 3t, -\frac{9}{5}\sin 3t, 0 \right\rangle$$

Then, by Equation 12.11,

$$\mathbf{N} = \frac{\mathbf{T}'(t)}{|\mathbf{T}'(t)|} = \frac{\left\langle -\frac{9}{5}\cos 3t, -\frac{9}{5}\sin 3t, 0 \right\rangle}{\sqrt{\frac{81}{25}\cos^2 3t + \frac{81}{25}\sin^2 3t}} = \frac{\left\langle -\frac{9}{5}\cos 3t, -\frac{9}{5}\sin 3t, 0 \right\rangle}{\frac{9}{5}}$$

$$= \langle -\cos 3t, -\sin 3t, 0 \rangle$$

Finally, we calculate κ, using Equation 12.14:

$$\mathbf{r}' \times \mathbf{r}'' = \begin{vmatrix} \mathbf{i} & \mathbf{j} & \mathbf{k} \\ -6\sin 3t & 6\cos 3t & 8 \\ -18\cos 3t & -18\sin 3t & 0 \end{vmatrix}$$

$$= 144\sin 3t\,\mathbf{i} - 144\cos 3t\,\mathbf{j} + (108\sin^2 3t + 108\cos^2 3t)\mathbf{k}$$

$$= 36(4\sin 3t\,\mathbf{i} - 4\cos 3t\,\mathbf{j} + 3\mathbf{k})$$

$$|\mathbf{r}' \times \mathbf{r}''| = 36\sqrt{16\sin^2 3t + 16\cos^2 3t + 9} = 36\sqrt{25} = 180$$

Thus,

$$\kappa = \frac{|\mathbf{r}' \times \mathbf{r}''|}{|\mathbf{r}'|^3} = \frac{180}{1000} = \frac{9}{50}$$

Notice that κ is constant in this case. ■

For reference, we summarize here Equations 12.10, 12.11, and 12.14, for **T**, **N**, and κ.

$$\mathbf{T} = \frac{\mathbf{r}'(t)}{|\mathbf{r}'(t)|} \qquad \text{Unit tangent vector}$$

$$\mathbf{N} = \frac{\mathbf{T}'(t)}{|\mathbf{T}'(t)|} \qquad \text{Unit normal vector}$$

$$\kappa = \frac{|\mathbf{r}' \times \mathbf{r}''|}{|\mathbf{r}'|^3} \qquad \text{Curvature}$$

Curvature in Two Dimensions; Radius of Curvature

For a curve C in the xy-plane, we can find the unit tangent and normal vectors by using Equations 12.10 and 12.11, respectively. In this case, however, there is no binormal vector. Equation 12.14 is valid, as remarked earlier, if we write two-dimensional vectors as three-dimensional vectors with third component 0. When we write the vectors in this way, we arrive at a formula for κ that is applicable to plane curves only, as follows. Suppose C has position vector $\mathbf{r}(t) = \langle f(t), g(t) \rangle$. Then we treat C as a space curve lying in the xy-plane, so that $\mathbf{r}(t) = \langle f(t), g(t), 0 \rangle$. For brevity, we can write $x = f(t)$ and $y = g(t)$, so $\mathbf{r}' = \langle x', y', 0 \rangle$ and $\mathbf{r}'' = \langle x'', y'', 0 \rangle$. Then $\mathbf{r}' \times \mathbf{r}'' = \langle 0, 0, x'y'' - x''y' \rangle$. (You should verify this result.) Finally, $|\mathbf{r}' \times \mathbf{r}''| = |x'y'' - x''y'|$, and we have the following equation.

Curvature for the Plane Curve $\mathbf{r}(t) = \langle x(t), \ y(t) \rangle$

$$\kappa = \frac{|x'y'' - x''y'|}{[(x')^2 + (y')^2]^{3/2}} \tag{12.15}$$

In case the equation of C is in the form $y = f(x)$, we can use the parameterization $x = t$ and $y = f(t)$, and since $x' = 1$ and $x'' = 0$, Equation 12.15 becomes

Curvature for the Plane Curve $y = f(x)$

$$\kappa = \frac{|y''|}{[1 + (y')^2]^{3/2}} \tag{12.16}$$

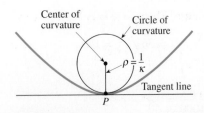

Center of curvature

Circle of curvature

$\rho = \dfrac{1}{\kappa}$

Tangent line

P

FIGURE 12.14

At a point P on a curve C where $\kappa \neq 0$, we define $\rho = 1/\kappa$ as the **radius of curvature** of C at P. If C is a plane curve, the circle of radius ρ that is tangent to C at P on its concave side (the direction of **N**) is called the **circle of curvature**, or **osculating circle**. Its center is called the **center of curvature**. (See Figure 12.14.) If C itself is a circle, Example 12.15 shows that its curvature is the reciprocal of its radius, $\kappa = 1/r$. Hence, $r = 1/\kappa$. That is, the radius of the circle is the radius of curvature. So C is its own circle of curvature. This result shows that for any plane curve C at a point P where $\kappa \neq 0$, the circle of curvature has the same curvature as C at P. In the sense that C and its circle of curvature have the same tangent line and the same curvature at P, the circle of curvature is the circle that best "fits" C in a neighborhood of P.

EXAMPLE 12.17   Find the curvature and center of curvature at the point $P(2, 1)$ on the curve $\mathbf{r}(t) = \langle 2t^2, 2 - t^3 \rangle$, where $t > 0$.

Solution   The parametric equations corresponding to $\mathbf{r}(t)$ are $x = 2t^2$ and $y = 2 - t^3$. So we have

$$x' = 4t \qquad y' = -3t^2$$
$$x'' = 4 \qquad y'' = -6t$$

The point P is given by $t = 1$. At this point, $x' = 4$, $x'' = 4$, $y' = -3$, and $y'' = -6$. So by Equation 12.15,

$$\kappa = \frac{|x'y'' - x''y'|}{[(x')^2 + (y')^2]^{3/2}} = \frac{|4(-6) - (4)(-3)|}{(16 + 9)^{3/2}} = \frac{12}{125}$$

The radius of curvature ρ is therefore $\frac{125}{12}$. The center of curvature is ρ units in the direction of the normal from P. So we need $\mathbf{N}$. Since $t > 0$, we have

$$\mathbf{T} = \frac{\mathbf{r}'(t)}{|\mathbf{r}'(t)|} = \frac{\langle 4t, -3t^2 \rangle}{\sqrt{16t^2 + 9t^4}}$$

$$= \frac{t\langle 4, -3t^2 \rangle}{t\sqrt{16 + 9t^2}} = \left\langle \frac{4}{\sqrt{16 + 9t^2}}, \frac{-3t}{\sqrt{16 + 9t^2}} \right\rangle$$

Thus, we find (omitting some details)

$$\mathbf{T}'(t) = \left\langle \frac{-36}{(16 + 9t^2)^{3/2}}, \frac{-48}{(16 + 9t^2)^{3/2}} \right\rangle$$

At $t = 1$,

$$\mathbf{T}'(1) = \left\langle \frac{-36}{125}, \frac{-48}{125} \right\rangle = \frac{-12}{125} \langle 3, 4 \rangle$$

and

$$|\mathbf{T}'(1)| = \frac{12}{25}$$

Thus, by Equation 12.14, the unit normal at P is

$$\mathbf{N}(1) = \frac{\mathbf{T}'(1)}{|\mathbf{T}'(1)|} = \left\langle -\frac{3}{5}, -\frac{4}{5} \right\rangle$$

Now let $Q(h, k)$ denote the center of curvature. Then, as seen in Figure 12.15, $\overrightarrow{PQ} = \rho\mathbf{N}$, or

$$\langle h - 2, k - 1 \rangle = \frac{125}{12} \left\langle -\frac{3}{5}, -\frac{4}{5} \right\rangle = \left\langle -\frac{25}{4}, -\frac{25}{3} \right\rangle$$

So $h - 2 = -\frac{25}{4}$ and $k - 1 = -\frac{25}{3}$, from which we get the center,

$$(h, k) = \left(\frac{-17}{4}, \frac{-22}{3} \right)$$

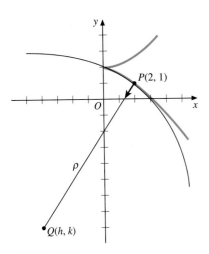

FIGURE 12.15

THE CALCULUS OF VECTOR-VALUED FUNCTIONS USING COMPUTER ALGEBRA SYSTEMS

All the calculus operations for vector-valued functions can be easily carried out by a CAS, including symbolic operations such as differentiation and integration. Analyzing the standard features of space curves, such as curvature, is generally very complicated, and, as we will see, they can be more easily examined using a computer algebra system such as Maple or Mathematica.

CAS 46

Suppose the position of a particle moving in space at time t is given by $\mathbf{r}(t) = \langle e^t \cos t, e^t \sin t, t \rangle$.

(a) Plot the path of the particle.

(b) Calculate the velocity, speed, and acceleration of the particle at time t.

(c) Calculate the curvature, κ, of the path and make a sketch of the curvature.

(d) Calculate the unit tangent vector and unit normal vector of the position.

(e) Calculate the tangential and normal components of the acceleration.

Maple:

(a) with(plots):with(linalg):
r:=t–>[exp(t)*cos(t),exp(t)*sin(t),t];
spacecurve(r(t),t=0..4*Pi,numpoints=200),axes=boxed;

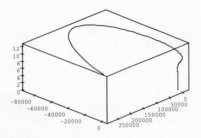

FIGURE 12.4.1

(b) v:=diff(r(t),t);

Output: $v := \left[e^t \cos(t) - e^t \sin(t), \right.$
$$\left. e^t \sin(t) + e^t \cos(t), 1 \right]$$

a:=diff(v,t);

Output: $a = \left[-2e^t \sin(t), 2e^t \cos(t), 0 \right]$

Now write a function to compute the length of a vector in three space. The length of a vector $\mathbf{w}$ can be written as $|\mathbf{w}| = \sqrt{\mathbf{w} \cdot \mathbf{w}}$.

Mathematica:

```
<<Calculus`VectorAnalysis`
ParametricPlot3D[{Exp[t]*Cos[t],Exp[t]*Sin[t],t},
    {t,0,4*Pi}]
r[t_] = {Exp[t]*Cos[t],Exp[t]*Sin[t],t}
v = Simplify[r'[t]]
START
a = Simplify[D[v,t]]
m[w_List] = Simplify[Sqrt[w.w],Trig–>True]
speed = Simplify[m[v]]
k = Simplify[m[CrossProduct[v,a]]/(m[v])^3]
Plot[k,{t,–10,10},PlotRange–>{0,1}]
T = Simplify[v/m[v]]
DT = Simplify[D[T,t]]
N = Simplify[DT/m[DT]]
Ta = Simplify[v.a/m[a]]
Na = Simplify[m[CrossProduct[v,a]]/m[v]]
```

Maple:

magnitude:=w–>sqrt(dotprod(w,w));
speed := simplify (magnitude(v));

Output: $speed: = \sqrt{2e^{2t} + 1}$

(c) For the curvature we use Equation 12.14
k:=simplify(magnitude(crossprod(v,a))/(magnitude(v))^3);

Output:

$$k := 2\frac{\sqrt{e^{2t} + e^{4t}}}{\left(2e^{2t} + 1\right)^{3/2}}$$

plot(k,t=–10..10,y=0..1,numpoints=200);

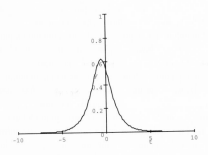

FIGURE 12.4.2

From the plot, we can easily see where the curvature is a maximum.

(d) T:=simplify(v/magnitude(v));

Output:

$$T = \frac{\left[e^t \cos(t) - e^t \sin(t), e^t \sin(t) + e^t \cos(t), 1\right]}{\sqrt{2e^{2t} + 1}}$$

Check that this is a unit vector:
simplify(magnitude(T));
Output: 1
DT:=simplify(diff(T,t));
N:=Simplify(DT/magnitude(DT));
(e)
T_a:=simplify(dotprod(v,a)/magnitude(a));

Output:

$$T\_a := e^t$$

N_a:=simplify(magnitude(crossprod(v,a))/magnitude(v));

Output:

$$N\_a := 2\frac{\sqrt{e^{2t} + e^{4t}}}{\sqrt{2e^{2t} + 1}}$$

Exercise Set 12.4

In Exercises 1–10, find **T** *and* **N** *at the prescribed point.*

1. $\mathbf{r}(t) = \langle t, t^2 \rangle$; $t = 2$

2. $\mathbf{r}(t) = \langle 1 - t^2, 2 + 3t \rangle$; $t = 1$

3. $\mathbf{r}(t) = 2 \cos t \mathbf{i} + 2 \sin t \mathbf{j}$; $t = \dfrac{\pi}{4}$

4. $\mathbf{r}(t) = 3 \cos t \mathbf{i} + 4 \sin t \mathbf{j}$; $t = \dfrac{\pi}{2}$

5. $\mathbf{r}(t) = \langle 2 \cos t, 2 \sin t, 4t \rangle$; $t = \dfrac{3\pi}{4}$

6. $\mathbf{r}(t) = \langle t, t^2, 3 - t^2 \rangle$; at $(1, 1, 2)$

7. $\mathbf{r}(t) = 2t^2 \mathbf{i} + t^3 \mathbf{j} + 3\mathbf{k}$; at $(2, 1, 3)$

8. $\mathbf{r}(t) = 2t^3 \mathbf{i} + 3t^2 \mathbf{j} + 3t\mathbf{k}$; $t = -1$

9. $\mathbf{r}(t) = \langle e^t, 2e^{-t}, 2t \rangle$; at $(1, 2, 0)$

10. $\mathbf{r}(t) = \left(2t, \dfrac{1}{t}, 2 \ln t \right)$; at $(2, 1, 0)$

In Exercises 11–25, find κ *at the prescribed point.*

11. $\mathbf{r}(t) = \langle t^2, t^3 \rangle$; $t = 1$

12. $\mathbf{r}(t) = \langle 1 + t, 1 - t^2 \rangle$; $t = 0$

13. $\mathbf{r}(t) = (t \ln t) \mathbf{i} + \dfrac{1}{t} \mathbf{j}$; $t = 1$

14. $\mathbf{r}(t) = \sqrt{t^2 - 3} \,\mathbf{i} + \dfrac{t}{\sqrt{t^2 - 3}} \mathbf{j}$; $t = 2$

15. $x = \sin 2t$, $y = 4 \cos 2t$; $t = \dfrac{\pi}{3}$

16. $x = t - \cos t$, $y = 1 - \sin t$; $t = \dfrac{\pi}{6}$

17. $y = 2x - x^2$; at $(0, 0)$

18. $y = x^3 - 2x^2 + 3$; at $(1, 2)$

19. $y = \dfrac{2}{x}$; at $(2, 1)$

20. $y = \sec x$; at $x = \dfrac{3\pi}{4}$

21. $\mathbf{r}(t) = \langle t^2, t^3, 1 - 2t \rangle$; $t = -1$

22. $\mathbf{r}(t) = \langle 4 \cos t, 4 \sin t, 3t \rangle$; $t = \dfrac{\pi}{3}$

23. $\mathbf{r}(t) = \langle 2e^t, e^{-t}, 2t \rangle$; at $(2, 1, 0)$

24. $\mathbf{r}(t) = t^2 \mathbf{i} + t^3 \mathbf{j} - 3t^2 \mathbf{k}$; $t = \dfrac{1}{2}$

25. $\mathbf{r}(t) = (e^t \sin t) \mathbf{i} + (e^t \cos t) \mathbf{j} + e^t \mathbf{k}$; $t = 0$

In Exercises 26–30, find the center of the circle of curvature at P. Sketch the graph in a neighborhood of P, showing the circle of curvature.

26. $\mathbf{r}(t) = \langle t, 1 + t^2 \rangle$; $P(0, 1)$

27. $\mathbf{r}(t) = \langle 4 \cos t, 3 \sin t \rangle$; $P(4, 0)$

28. $\mathbf{r}(t) = t \mathbf{i} + \sqrt{2t} \,\mathbf{j}$; $P(\frac{1}{2}, 1)$

29. $\mathbf{r}(t) = t \mathbf{i} + e^t \mathbf{j}$; $P(0, 1)$

30. $\mathbf{r}(t) = \langle t - \sin t, 1 - \cos t \rangle$; $t = \frac{\pi}{3}$

31. Let C be a plane curve that is the graph of $\mathbf{r}(s)$, where s is the arc length parameter, and suppose $\mathbf{r}''(s)$ exists on $[0, L]$, where L is the length of C. Let $\theta = \theta(s)$ be the angle between the vectors $\mathbf{i}$ and $\mathbf{T}$ at an arbitrary point on C. Show the following:
   (a) $\mathbf{T} = \langle \cos \theta, \sin \theta \rangle$
   (b) $\kappa = \left| \dfrac{d\theta}{ds} \right|$
   (c) $\mathbf{N} = \langle - \sin \theta, \cos \theta \rangle$ if $d\theta/ds > 0$
       $\mathbf{N} = \langle \sin \theta, - \cos \theta \rangle$ if $d\theta/ds < 0$

   Use the result of part (c) to show that $\mathbf{N}$ is always directed toward the concave side of C.

32. Suppose $\mathbf{N} = \langle n_1, n_2 \rangle$ is the unit normal vector at a point $P(x, y)$ on the plane curve C. Prove that the center of curvature at P is the point $(x + \rho n_1, y + \rho n_2)$, where ρ is the radius of curvature.

33. Find $\mathbf{T}$, $\mathbf{N}$, $\mathbf{B}$, and κ at an arbitrary point on the circular helix $\mathbf{r}(t) = \langle a \cos bt, a \sin bt, ct \rangle$.

34. Find $\mathbf{T}$, $\mathbf{N}$, $\mathbf{B}$, and κ at an arbitrary point on the curve $\mathbf{r}(t) = \langle t, t^2, 2t^3/3 \rangle$.

35. Use Equation 12.13 to show that $(\mathbf{r}' \times \mathbf{r}'') \times \mathbf{r}' = |\mathbf{r}'|^2 (ds/dt)^2 \kappa \mathbf{N}$, and hence obtain the formula

$$\mathbf{N} = \frac{(\mathbf{r}' \times \mathbf{r}'') \times \mathbf{r}'}{\kappa \, |\mathbf{r}'|^4}$$

*In Exercises 36 and 37, assume C is a smooth curve parameterized by arc length, and d**T**/ds, d**N**/ds, and d**B**/ds all exist.*

36. By differentiating both sides of $\mathbf{B} \cdot \mathbf{B} = 1$, show that $\mathbf{B}$ and $d\mathbf{B}/ds$ are orthogonal. Explain why it follows that

$$\frac{d\mathbf{B}}{ds} = \alpha \mathbf{T} + \beta \mathbf{N}$$

for some scalars α and β. By differentiating both sides of $\mathbf{B} \cdot \mathbf{T} = 0$, show that $\alpha = 0$.

37. (Continuation of Exercise 36) Write $\tau = -\beta$, so that $d\mathbf{B}/ds = -\tau \mathbf{N}$. (The scalar τ is called the **torsion**.) Show that $\mathbf{N} = \mathbf{B} \times \mathbf{T}$. Prove that

$$\frac{d\mathbf{N}}{ds} = \tau \mathbf{B} - \kappa \mathbf{T}$$

(*Note:* The formulas

$$\frac{d\mathbf{T}}{ds} = \kappa \mathbf{N} \qquad \frac{d\mathbf{N}}{ds} = \tau \mathbf{B} - \kappa \mathbf{T} \qquad \frac{d\mathbf{B}}{ds} = -\tau \mathbf{N}$$

are called the **Frenet Formulas**.)

38. Find the torsion τ as a function of t for the curve of Exercise 34. Verify the Frenet Formulas for this curve. (See Exercise 37.)

 39. Suppose the position of a particle moving in space at time t is given by $\mathbf{r}(t) = \langle t \cos t, t \sin t, t \rangle$.
  (a) Plot the space curve.
  (b) Calculate the velocity, speed, and acceleration of the particle at time t.
  (c) Calculate the curvature, κ, of the path and make a sketch of the curvature. From the sketch, estimate the point at which the curvature is a maximum.
  (d) Calculate the unit tangent vector and unit normal vector of the position.
  (e) Calculate the tangential and normal components of the acceleration.

 40. Calculate the unit tangent and unit normal vectors to the curve $\mathbf{r}(t) = \cos(t)\mathbf{i} + \sin(t)\mathbf{j}$ at $t = \pi/4$. Plot the curve, the unit tangent, and the unit normal. In which direction does the unit normal point?

 41. Calculate the unit tangent, unit normal, and curvature at $t = 2$, to the curve $\mathbf{r}(t) = (t^3/2 - t)\mathbf{i} - t^2\mathbf{j}$. Plot the curve, the unit tangent, and the unit normal vectors for $t = 2$.

12.5 MOTION ALONG A CURVE

In the preceding section, we considered curves as sets of terminal points of position vectors $\mathbf{r}(t)$ for continuous vector functions $\mathbf{r}$. No concept of movement was involved. We were looking at curves from the *static* point of view. Now we want to look at them from the *dynamic* point of view; that is, we want to consider a curve as the path traced out by a particle moving in space (or in a plane).

We let t denote time, measured in whatever units we choose, from some convenient time origin, and we let $\mathbf{r}(t) = \langle f(t), g(t), h(t) \rangle$ be the position vector of the moving particle at time t. Since a moving particle always goes in a continuous path, it follows that $\mathbf{r}$ is continuous, so that the path of the particle is a curve. Suppose the time interval of interest is I. We assume further that $\mathbf{r}'$ and $\mathbf{r}''$ both exist for all t in I. We know that

$$\mathbf{r}'(t) = \lim_{h \to 0} \frac{\mathbf{r}(t+h) - \mathbf{r}(t)}{h}$$

The vector $\mathbf{r}(t+h) - \mathbf{r}(t)$ is the displacement vector from the particle's position at time t to its position at time $t + h$. When we divide by the elapsed time h, we obtain the *average velocity*. Its limit as $h \to 0$ is the *velocity* $\mathbf{v}(t)$ at time t. The direction of the velocity vector indicates the (instantaneous) direction of movement of the particle, and its magnitude is the speed of the particle. A similar analysis applies to the derivative,

$$\mathbf{v}'(t) = \lim_{h \to 0} \frac{\mathbf{v}(t+h) - \mathbf{v}(t)}{h}$$

which we call the *acceleration* at time t.

Runners rounding a curve

In summary, we have the following definition.

Definition 12.7 Velocity, Acceleration, and Speed	Let $\mathbf{r}(t)$ be the position vector of a moving particle at time t, where t varies over the interval I, and suppose $\mathbf{r}'$ and $\mathbf{r}''$ both exist in I. Then the **velocity** $\mathbf{v}(t)$ and **acceleration** $\mathbf{a}(t)$ are defined as $$\mathbf{v}(t) = \mathbf{r}'(t)$$ $$\mathbf{a}(t) = \mathbf{v}'(t) = \mathbf{r}''(t)$$ The **speed** of the particle is defined as the magnitude of the velocity vector: $$\text{Speed} = |\mathbf{v}(t)| = |\mathbf{r}'(t)|$$

REMARK

■ Observe that both velocity and acceleration are vectors, whereas speed is a scalar. Velocity is the instantaneous rate of change of position, and acceleration is the instantaneous rate of change of velocity. These descriptions agree with our intuitive understanding of the concepts of velocity and acceleration. We know from Equation 12.3 that the arc length $s(t)$ satisfies $s'(t) = |\mathbf{r}'(t)|$; that is, speed $= ds/dt$. Thus, speed can be interpreted as the rate at which the arc length s is changing with time or, equivalently, the rate at which the distance along the path covered by the particle is changing with time. We observe further that the velocity vector is directed along the tangent line and points in the direction of motion.

EXAMPLE 12.18   A particle moves on the curve $\mathbf{r}(t) = \langle t, \frac{1}{t}, 2\sqrt{t} \rangle$ for $t > 0$. Find its velocity, acceleration, and speed when $t = 1$. Show the results graphically.

Solution

$$\mathbf{v}(t) = \mathbf{r}'(t) = \left\langle 1, -\frac{1}{t^2}, \frac{1}{\sqrt{t}} \right\rangle$$

$$\mathbf{a}(t) = \mathbf{v}'(t) = \left\langle 0, \frac{2}{t^3}, -\frac{1}{2t^{3/2}} \right\rangle$$

So when $t = 1$, $\mathbf{v}(1) = \langle 1, -1, 1 \rangle$ and $\mathbf{a}(1) = \langle 0, 2, -\frac{1}{2} \rangle$. The speed at this instant is therefore $|\mathbf{v}(1)| = \sqrt{3}$. From the parametric equations $x = t$, $y = \frac{1}{t}$, and $z = 2\sqrt{t}$, we see that for $t > 0$, all three coordinates are positive. Also, since $xy = 1$, the curve lies above the first-quadrant branch of this equilateral hyperbola. By plotting points corresponding to $t = \frac{1}{9}, \frac{1}{4}, 1, 4$, and 9, we get a reasonably accurate sketch, as shown in Figure 12.16. ■

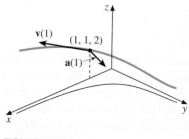

FIGURE 12.16

EXAMPLE 12.19   Find the force acting on a particle that moves in a circular path of radius r with constant speed v_0.

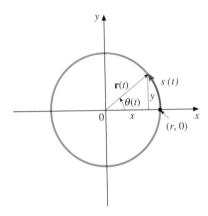

FIGURE 12.17

Solution   We introduce a coordinate system so that the center of the circle is at the origin and when $t = 0$ the particle is at $(r, 0)$. We assume the motion is in the positive (counterclockwise) direction. Let $\theta(t)$ be the polar angle to the position vector $\mathbf{r}(t)$ at time t, as shown in Figure 12.17, and let $s(t)$ be the arc length covered by the particle, measured from $(r, 0)$. Then, $x = r \cos \theta(t)$ and $y = r \sin \theta(t)$ are the parametric equations of the circle. In vector form, the equation is

$$\mathbf{r}(t) = (r \cos \theta(t))\mathbf{i} + (r \sin \theta(t))\mathbf{j}$$

To find an explicit expression for $\theta(t)$, we use the relationship $s = r\theta$ to get $ds/dt = r\, d\theta/dt$. But ds/dt is the speed of the particle, which by hypothesis is equal to the constant v_0. So $d\theta/dt = v_0/r$, which is also a constant. We designate this constant by ω (the Greek letter omega) and call it the **angular speed** because it is the rate of change of θ with respect to time. Thus,

$$\frac{d\theta}{dt} = \omega$$

and integrating both sides with respect to t, we obtain

$$\int \left(\frac{d\theta}{dt}\right) dt = \int \omega\, dt$$

$$\theta(t) = \omega t + C$$

Since $\theta(0) = 0$, it follows that $C = 0$. So $\theta(t) = \omega t$. The equations of motion now become

$$\mathbf{r}(t) = (r \cos \omega t)\mathbf{i} + (r \sin \omega t)\mathbf{j}$$

$$\mathbf{v}(t) = (-r\omega \sin \omega t)\mathbf{i} + (r\omega \cos \omega t)\mathbf{j}$$

$$\mathbf{a}(t) = (-r\omega^2 \cos \omega t)\mathbf{i} + (-r\omega^2 \sin \omega t)\mathbf{j}$$

$$= -\omega^2 [(r \cos \omega t)\mathbf{i} + (r \sin \omega t)\mathbf{j}]$$

$$= -\omega^2 \mathbf{r}(t)$$

We can calculate the speed as the magnitude of the velocity $\mathbf{v}(t)$:

$$|\mathbf{v}(t)| = \sqrt{r^2\omega^2 \sin^2 \omega t + r^2\omega^2 \cos^2 \omega t} = r\omega$$

This result confirms that the constant speed v_0 is $r\omega$: $v_0 = r\omega$.

Since $|\mathbf{r}(t)| = \sqrt{r^2 \cos^2 \omega t + r^2 \sin^2 \omega t} = r$, we see that $|\mathbf{a}(t)| = |-\omega^2| |\mathbf{r}(t)| = r\omega^2$. So each of the vectors $\mathbf{r}(t)$, $\mathbf{v}(t)$, and $\mathbf{a}(t)$ has constant magnitude.

According to Newton's Second Law of Motion, the force $\mathbf{F}$ acting on the body satisfies $\mathbf{F} = m\mathbf{a}$, where m is the mass of the body. Thus,

$$\mathbf{F} = -m\omega^2 \mathbf{r}(t)$$

This force, which is of constant magnitude $mr\omega^2$, is directed toward the center of the circle since it is opposite in sign to the position vector $\mathbf{r}(t)$. It is called **centripetal force.** ∎

EXAMPLE 12.20

Communication satellite in geostationary orbit

(a) Find a formula for the speed necessary to maintain a satellite of mass m in a fixed orbit h km above the earth's surface.

(b) Find a formula for the time required for the satellite of part (a) to complete one revolution around the earth.

(c) Taking the radius of the earth as approximately 6,370 km, find the speed of a satellite in orbit 1500 km above the earth's surface. Find the number of hours required to complete one revolution.

Solution   We make the simplifying assumptions that the satellite's orbit is circular, that the acceleration caused by gravity at the earth's surface is constant at $g \approx 9.81$ m/sec^2, and that the gravitational attraction of other bodies is negligible.

(a) Let R denote the radius of the earth. Then the radius of the orbit of the satellite is $r = R + h$. By what we found in Example 12.19, the centripetal force necessary to keep the satellite in orbit has magnitude $mr\omega^2$, and since $v = r\omega$, we get

$$|\mathbf{F}| = mr \left(\frac{v}{r}\right)^2 = \frac{mv^2}{r}$$

This force is produced by the earth's gravitational pull, which according to Newton's Law of Universal Gravitation is given by

$$|\mathbf{F}| = \frac{GMm}{r^2}$$

where M is the mass of the earth and G is a constant, called the universal gravitational constant. Equating the two expressions for $|\mathbf{F}|$ gives

$$v^2 = \frac{GM}{r}$$

A more useful form is obtained by observing that when the satellite is on the earth's surface, so that $r = R$, the attractive force is just the weight, mg, of the satellite. So by Newton's Gravitational Law with $r = R$,

$$mg = \frac{GMm}{R^2}$$

and thus $GM = R^2 g$. Making this substitution in the formula for v^2 gives $v^2 = R^2 g / r = R^2 g / (R + h)$. So

$$v = R\sqrt{\frac{g}{R + h}}$$

(b) The angular speed ω is the number of radians through which the position vector turns per unit of time. Since one revolution is equivalent to 2π radians, the time T required for the satellite to complete one revolution is

$$T = \frac{2\pi}{\omega} = \frac{2\pi r}{v} = \frac{2\pi (R + h)}{v} \qquad v = r\omega, \text{ so } \omega = v/r$$

From the result of part (a), we can substitute for v to obtain T in the alternative form

$$T = \frac{2\pi (R + h)^{3/2}}{R\sqrt{g}}$$

(c) We will use distances in kilometers and time in hours. So $R = 6370$ and $h = 1500$. To convert $g \approx 9.81$ m/sec^2 to km/hr^2, we divide by 1000 and multiply by $(3600)^2$, getting $g \approx 127,140$ km/hr^2. From part (a), the velocity is

$$v = R\sqrt{\frac{g}{R + h}} \approx 6370 \sqrt{\frac{127,140}{7870}} \approx 25,600 \text{ km/hr}$$

and from part (b), the time T for one revolution is

$$T = \frac{2\pi(R+h)}{v} \approx \frac{2\pi(7870)}{25,600} \approx 1.93 \text{ hr}$$ ∎

Projectile Motion

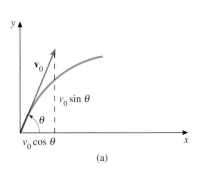

(a)

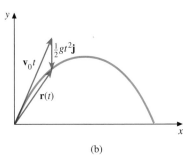

(b)

FIGURE 12.18

As a further application of motion on a curve, consider a projectile fired from the ground at an angle θ from the horizontal, with an initial velocity $\mathbf{v}_0$. We introduce x- and y-axes with the origin coinciding with the point from which the projectile is fired, as in Figure 12.18(a). We make the assumptions that the curvature of the earth is negligible in the interval in question, that air resistance and wind can be neglected, and that the acceleration caused by gravity, g, is constant. (It should be noted that the mathematical model we chose is an idealization, since none of these assumptions is, strictly speaking, valid, but for moderate distances the errors introduced by making them are small.) For simplicity we also assume that the projectile is fired from ground level. The questions of interest are the following:

1. What is the path of the projectile?
2. How high does it rise?
3. What is its range—that is, how far does it go in a horizontal direction?
4. What is its velocity at impact?

We consider each of these questions in order.

1.   If $\mathbf{r}(t)$ is the position vector of the projectile (considered as a point mass) at time t, then $\mathbf{r}(0) = \mathbf{0}$ and $\mathbf{v}(0) = \mathbf{v}_0$, since the initial position is at ground level and the initial velocity is $\mathbf{v}_0$. By our assumptions, the only force that acts on the projectile is the pull of gravity toward the earth. The magnitude of this force is the weight of the projectile—namely, mg, where m is its mass. Since the force is downward, we have $\mathbf{F} = -mg\mathbf{j}$. From Newton's Second Law of Motion we know that $\mathbf{F} = m\mathbf{a}$, where $\mathbf{a}$ is the acceleration of the projectile. Thus, $m\mathbf{a} = -mg\mathbf{j}$, or

$$\mathbf{a}(t) = -g\mathbf{j}$$

To find the velocity $\mathbf{v}(t)$ at time t, we integrate $\mathbf{a}(t)$, since acceleration and velocity are related by the differential equation $\mathbf{a}(t) = \mathbf{v}'(t)$:

$$\mathbf{v}(t) = \int \mathbf{a}(t)dt = -gt\mathbf{j} + \mathbf{C}_1$$

where $\mathbf{C}_1$ is a constant vector. Substituting $t = 0$, we get $\mathbf{v}(0) = \mathbf{C}_1$, and since $\mathbf{v}(0) = \mathbf{v}_0$,

$$\mathbf{v}(t) = -gt\mathbf{j} + \mathbf{v}_0$$

Now we integrate again to find $\mathbf{r}(t)$:

$$\mathbf{r}(t) = \int \mathbf{v}(t)dt = -\frac{1}{2}gt^2\mathbf{j} + \mathbf{v}_0t + \mathbf{C}_2$$

Again letting $t = 0$, we find that $\mathbf{C}_2 = \mathbf{r}(0) = \mathbf{0}$. Thus,

$$\mathbf{r}(t) = -\frac{1}{2}gt^2\mathbf{j} + \mathbf{v}_0t$$

If we denote the initial speed $|\mathbf{v}_0|$ by v_0, then we see from Figure 12.18 that $\mathbf{v}_0 = (v_0 \cos \theta)\mathbf{i} + (v_0 \sin \theta)\mathbf{j}$. So for $\mathbf{r}(t)$ we have

$$\mathbf{r}(t) = -\frac{1}{2}gt^2\mathbf{j} + [(v_0 \cos \theta)\mathbf{i} + (v_0 \sin \theta)\mathbf{j}]t$$

which can be rewritten as

$$\mathbf{r}(t) = (v_0 t \cos \theta)\mathbf{i} + \left(v_0 t \sin \theta - \frac{1}{2}gt^2\right)\mathbf{j} \qquad (12.17)$$

Similarly, $\mathbf{v}(t)$ becomes

$$\mathbf{v}(t) = (v_0 \cos \theta)\mathbf{i} + (v_0 \sin \theta - gt)\mathbf{j} \qquad (12.18)$$

Equation 12.17 is a vector equation of the path of the projectile. To analyze it further, we use the parametric equations for the path

$$\begin{cases} x = v_0 t \cos \theta \\ y = v_0 t \sin \theta - \frac{1}{2}gt^2 \end{cases} \qquad (12.19)$$

By eliminating the parameter t, we obtain the rectangular equation (verify)

$$y = x \tan \theta - \frac{gx^2}{2v_0^2 \cos^2 \theta}, \qquad 0 \le \theta < \frac{\pi}{2}$$

whose graph is the parabola shown in Figure 12.18(b).

2. Since $\mathbf{v}(t)$ is tangent to the path, the maximum height occurs when $\mathbf{v}(t)$ is horizontal, that is, when the y component of $\mathbf{v}(t)$ is 0. From Equation 12.18, this situation occurs when $v_0 \sin \theta - gt = 0$, or

$$t = \frac{v_0 \sin \theta}{g}$$

To find the maximum height, we substitute this value of t into the second of Equations 12.19:

$$y_{\max} = v_0 \left(\frac{v_0 \sin \theta}{g}\right) \sin \theta - \frac{1}{2}g \left(\frac{v_0 \sin \theta}{g}\right)^2 = \frac{1}{2}\frac{v_0^2 \sin^2 \theta}{g}$$

3. The projectile's maximum horizontal distance occurs when it strikes the ground, that is, when $y = 0$. So to find the range, we set $y = 0$ in the second of Equations 12.19 and solve for t:

$$t\left(v_0 \sin \theta - \frac{1}{2}gt\right) = 0$$

$$t = 0 \quad \text{or} \quad t = \frac{2v_0 \sin \theta}{g}$$

Clearly, the second value is the one we want, since $t = 0$ is when the projectile was fired. The range is found by substituting this value of t in the first of Equations 12.19:

$$\text{Range} = x_{\max} = v_0 \left(\frac{2v_0 \sin \theta}{g}\right) \cos \theta = \frac{v_0^2 \sin 2\theta}{g}$$

4. The velocity at impact with the ground occurs when $t = (2v_0 \sin \theta)/g$. Putting this value of t in Equation 12.18 and writing $\mathbf{v}_I$ for impact velocity, we get

$$\mathbf{v}_I = (\mathbf{v}_0 \cos \theta)\mathbf{i} + (v_0 \sin \theta - 2v_0 \sin \theta)\mathbf{j}$$

$$= (v_0 \cos \theta)\mathbf{i} - (v_0 \sin \theta)\mathbf{j}$$

Also, if $v_I = |\mathbf{v}_I|$, then

$$v_I = \sqrt{(v_0 \cos\theta)^2 + (-v_0 \sin\theta)^2} = v_0$$

So the speed at impact is the same as the initial speed.

Tangential and Normal Components of Acceleration

It is sometimes useful to express the acceleration of a moving particle as the sum of two orthogonal vectors, one parallel to the unit tangent vector $\mathbf{T}$ and the other parallel to the unit normal vector $\mathbf{N}$. The lengths of these two vectors are called the tangential and normal components, respectively, of the acceleration. In this section we derive formulas for these components.

Our starting point is Equation 12.13 that we obtained in Section 12.4:

$$\frac{d^2\mathbf{r}}{dt^2} = \frac{d^2 s}{dt^2}\mathbf{T} + \kappa\left(\frac{ds}{dt}\right)^2\mathbf{N}$$

But $d^2\mathbf{r}/dt^2 = \mathbf{a}$, the acceleration of the particle moving along the curve with position vector $\mathbf{r}(t)$. So we have

$$\mathbf{a} = \frac{d^2 s}{dt^2}\mathbf{T} + \kappa\left(\frac{ds}{dt}\right)^2\mathbf{N} \qquad (12.20)$$

The vectors $\mathbf{T}$ and $\mathbf{N}$ at a point on the path determine a plane, and since the right-hand side of Equation 12.20 is a linear combination of $\mathbf{T}$ and $\mathbf{N}$, it is a vector in that plane. That is, the acceleration vector at any point lies in the same plane as $\mathbf{T}$ and $\mathbf{N}$ at that point.

If we take the dot product of each side of Equation 12.20, first with $\mathbf{T}$ and then with $\mathbf{N}$, we get

$$\mathbf{a} \cdot \mathbf{T} = \frac{d^2 s}{dt^2}\mathbf{T} \cdot \mathbf{T} + \kappa\left(\frac{ds}{dt}\right)^2\mathbf{N} \cdot \mathbf{T} = \frac{d^2 s}{dt^2} \qquad (12.21)$$

$$\mathbf{a} \cdot \mathbf{N} = \frac{d^2 s}{dt^2}\mathbf{T} \cdot \mathbf{N} + \kappa\left(\frac{ds}{dt}\right)^2\mathbf{N} \cdot \mathbf{N} = \kappa\left(\frac{ds}{dt}\right)^2 \qquad (12.22)$$

where we have used the facts that $\mathbf{T}$ and $\mathbf{N}$ are orthogonal unit vectors so that $\mathbf{T} \cdot \mathbf{T} = |\mathbf{T}|^2 = 1$, $\mathbf{N} \cdot \mathbf{N} = |\mathbf{N}|^2 = 1$, and $\mathbf{T} \cdot \mathbf{N} = 0$. Now from Section 11.3 we know that the components of $\mathbf{a}$ along $\mathbf{T}$ and $\mathbf{N}$, respectively, are

$$\text{Comp}_\mathbf{T}\mathbf{a} = \frac{\mathbf{a} \cdot \mathbf{T}}{|\mathbf{T}|}$$

and

$$\text{Comp}_\mathbf{N}\mathbf{a} = \frac{\mathbf{a} \cdot \mathbf{N}}{|\mathbf{N}|}$$

But $|\mathbf{T}| = 1$ and $|\mathbf{N}| = 1$, since $\mathbf{T}$ and $\mathbf{N}$ are unit vectors. Thus, $\text{Comp}_\mathbf{T}\mathbf{a} = \mathbf{a} \cdot \mathbf{T}$ and $\text{Comp}_\mathbf{N}\mathbf{a} = \mathbf{a} \cdot \mathbf{N}$.

So from Equations 12.21 and 12.22, the components of $\mathbf{a}$ along $\mathbf{T}$ and $\mathbf{N}$, respectively, are

$$\text{Comp}_\mathbf{T}\mathbf{a} = \frac{d^2 s}{dt^2}$$

$$\text{Comp}_\mathbf{N}\mathbf{a} = \kappa\left(\frac{ds}{dt}\right)^2$$

We designate these components by the symbols a_T and a_N, respectively, and call them the **tangential** and **normal components of acceleration**.

Tangential and Normal Components of Acceleration

$$a_T = \frac{d^2 s}{dt^2} \qquad a_N = \kappa \left(\frac{ds}{dt} \right)^2 \tag{12.23}$$

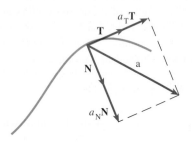

FIGURE 12.19

Equation 12.20 can now be written

$$\mathbf{a} = a_T \mathbf{T} + a_N \mathbf{N} \tag{12.24}$$

By Equation 12.11, we can write $a_T \mathbf{T} = \text{Proj}_{\mathbf{T}} \mathbf{a}$ and $a_N \mathbf{N} = \text{Proj}_{\mathbf{N}} \mathbf{a}$. We show a typical acceleration vector and its tangential and normal projections in Figure 12.19.

From Equation 12.24, we also have

$$\mathbf{a} \cdot \mathbf{a} = (a_T \mathbf{T} + a_N \mathbf{N}) \cdot (a_T \mathbf{T} + a_N \mathbf{N})$$

$$= a_T^2 \mathbf{T} \cdot \mathbf{T} + 2 a_T a_N \mathbf{T} \cdot \mathbf{N} + a_N^2 \mathbf{N} \cdot \mathbf{N}$$

$$= a_T^2 + a_N^2$$

since $\mathbf{T} \cdot \mathbf{T} = 1$, $\mathbf{T} \cdot \mathbf{N} = 0$, and $\mathbf{N} \cdot \mathbf{N} = 1$. That is,

$$|\mathbf{a}|^2 = a_T^2 + a_N^2 \tag{12.25}$$

(This result can also be seen from Figure 12.19, using the Pythagorean Theorem.) In practice it is usually easier to find a_T than a_N from Equations 12.23. Then Equation 12.25 can be used to get a_N, as shown in the next example.

EXAMPLE 12.21   Find the tangential and normal components of acceleration of a particle whose position vector is $\mathbf{r}(t) = 3t\mathbf{i} + 2t^3\mathbf{j} + 3t^2\mathbf{k}$. Also find the curvature κ of the path at an arbitrary time t.

Solution   We have

$$\mathbf{v} = \mathbf{r}'(t) = 3\mathbf{i} + 6t^2\mathbf{j} + 6t\mathbf{k} = 3(\mathbf{i} + 2t^2\mathbf{j} + 2t\mathbf{k})$$

$$\mathbf{a} = \mathbf{r}''(t) = 12t\mathbf{j} + 6\mathbf{k} = 6(2t\mathbf{j} + \mathbf{k})$$

$$\frac{ds}{dt} = |\mathbf{v}| = 3\sqrt{1 + 4t^4 + 4t^2} = 3\sqrt{(2t^2 + 1)^2} = 3(2t^2 + 1)$$

$$\frac{d^2 s}{dt^2} = 12t$$

$$|\mathbf{a}| = 6\sqrt{4t^2 + 1}$$

So by the first of Equations 12.23, $a_T = 12t$. Then, from Equation 12.25,

$$a_N = \sqrt{|\mathbf{a}|^2 - a_T^2} = \sqrt{36(4t^2 + 1) - 144t^2} = 6$$

We can now obtain κ using the formula for a_N in Equations 12.23. Since $a_N = \kappa (ds/dt)^2$, we have

$$\kappa = \frac{a_N}{\left(\dfrac{ds}{dt} \right)^2} = \frac{6}{9(2t^2 + 1)^2} = \frac{2}{3(2t^2 + 1)^2}$$

(You may wish to compare this "backdoor" method of finding κ with the direct method.) ■

The tangential component of acceleration is $a_{\mathbf{T}} = d^2 s / dt^2$. So $a_{\mathbf{T}}$ is the rate of change of speed, which agrees with our intuitive idea of acceleration. To help understand the normal component, $a_{\mathbf{N}} = \kappa (ds/dt)^2$, consider an automobile going around a curve. Since force is mass times acceleration, the normal component of the force necessary to hold the car on the road (the centripetal force) is $ma_{\mathbf{N}}$, where m is the mass of the car. This normal component is the magnitude of the force of friction between the tires and the road. If the curve is sharp, so that the curvature κ is large, this frictional force has a large magnitude. Similarly, if the speed ds/dt is large, the magnitude of the force is large. In fact, it increases as the square of the speed. Of course, if the curve is sharp *and* the speed is great, it is unlikely the car will stay on the road.

We conclude this section by giving alternative forms for $a_{\mathbf{T}}$ and $a_{\mathbf{N}}$ that are sometimes easier to use than those given in Equations 12.23. Note first that since by Equation 12.10,

$$\mathbf{T} = \frac{\mathbf{r}'(t)}{|\mathbf{r}'(t)|}$$

and $\mathbf{v}(t) = \mathbf{r}'(t)$, we have

$$\mathbf{T} = \frac{\mathbf{v}(t)}{|\mathbf{v}(t)|} \qquad (12.26)$$

or, $\mathbf{v}(t) = |\mathbf{v}(t)|\mathbf{T}$. For simplicity, we write this equation as $\mathbf{v} = |\mathbf{v}|\mathbf{T}$.

Thus, from Equation 12.24,

$$\mathbf{v} \cdot \mathbf{a} = |\mathbf{v}|\, \mathbf{T} \cdot (a_{\mathbf{T}}\mathbf{T} + a_{\mathbf{N}}\mathbf{N}) = |\mathbf{v}|\, a_{\mathbf{T}}$$

and

$$\mathbf{v} \times \mathbf{a} = |\mathbf{v}|\, \mathbf{T} \times (a_{\mathbf{T}}\mathbf{T} + a_{\mathbf{N}}\mathbf{N}) = |\mathbf{v}|\, a_{\mathbf{N}}\mathbf{B}$$

where we have used $\mathbf{T} \cdot \mathbf{T} = 1$, $\mathbf{T} \cdot \mathbf{N} = 0$, $\mathbf{T} \times \mathbf{T} = \mathbf{0}$, and $\mathbf{T} \times \mathbf{N} = \mathbf{B}$. Thus, since $|\mathbf{B}| = 1$ and $a_{\mathbf{N}} \geq 0$,

---

Alternative Formulas for Tangential and Normal Components of Acceleration

$$a_{\mathbf{T}} = \frac{\mathbf{v} \cdot \mathbf{a}}{|\mathbf{v}|}$$

$$a_{\mathbf{N}} = \frac{|\mathbf{v} \times \mathbf{a}|}{|\mathbf{v}|} \qquad (12.27)$$

EXAMPLE 12.22 Use Equations 12.27 to find $a_{\mathbf{T}}$ and $a_{\mathbf{N}}$ at time $t = 1$ for a particle with position vector

$$\mathbf{r}(t) = \left\langle 4\sqrt{t},\, 1 - 2t^2,\, \frac{8(t-1)}{\sqrt{t+3}} \right\rangle$$

Solution   First we calculate $\mathbf{v}(t)$ and $\mathbf{a}(t)$:

$$\mathbf{v}(t) = \left\langle \frac{2}{\sqrt{t}}, -4t, \frac{4(t+7)}{(t+3)^{3/2}} \right\rangle$$

$$\mathbf{a}(t) = \left\langle -\frac{1}{t^{3/2}}, -4, \frac{-2(t+15)}{(t+3)^{5/2}} \right\rangle$$

So

$$\mathbf{v}(1) = \langle 2, -4, -4 \rangle, \quad |\mathbf{v}(1)| = \sqrt{4 + 16 + 16} = 6,$$

and

$$\mathbf{a}(1) = \langle -1, -4, -1 \rangle$$

Then, by Equations 12.27,

$$a_T = \frac{|\mathbf{v} \cdot \mathbf{a}|}{|\mathbf{v}|} = \frac{\langle 2, -4, 4 \rangle \cdot \langle -1, -4, -1 \rangle}{6} = \frac{-2 + 16 - 4}{6} = \frac{5}{3}$$

$$a_N = \frac{|\mathbf{v} \times \mathbf{a}|}{|\mathbf{v}|} = \frac{|\langle 2, -4, 4 \rangle \times \langle -1, -4, -1 \rangle|}{6} = \frac{|\langle 20, -2, -12 \rangle|}{6}$$

$$= \frac{\sqrt{137}}{3} \qquad\blacksquare$$

Exercise Set 12.5

In Exercises 1–10, assume a particle moves so that its position vector at time t is $\mathbf{r}(t)$. Find its velocity, acceleration, and speed at t_0. Draw the graph of the curve followed by the particle, showing $\mathbf{v}(t_0)$ and $\mathbf{a}(t_0)$, drawn from the tip of $\mathbf{r}(t_0)$.

1. $\mathbf{r}(t) = \langle 2 \cos t, 3 \sin t \rangle; \ t_0 = \dfrac{\pi}{4}$

2. $\mathbf{r}(t) = t^2\mathbf{i} + t^3\mathbf{j}; \ t_0 = -1$

3. $\mathbf{r}(t) = 2e^t\mathbf{i} + 3e^{-t}\mathbf{j}; \ t_0 = 0$

4. $\mathbf{r}(t) = \langle \cosh t, \sinh t \rangle; \ t_0 = 0$

5. $\mathbf{r}(t) = \langle 2t - 1, t^2 + 3 \rangle; \ t_0 = 2$

6. $\mathbf{r}(t) = 2 \cos t\mathbf{i} + 2 \sin t\mathbf{j} + 3t\mathbf{k}; \ t_0 = \dfrac{\pi}{3}$

7. $\mathbf{r}(t) = 2t\mathbf{i} + t^2\mathbf{j} + t^3\mathbf{k}; \ t_0 = 1$

8. $\mathbf{r}(t) = \langle t + 1, 3t, t^2 \rangle; \ t_0 = 1$

9. $\mathbf{r}(t) = \langle e^t, 2t, e^{-t} \rangle; \ t_0 = 0$

10. $\mathbf{r}(t) = \cos^2 t\mathbf{i} + 2 \sin t\mathbf{j} + 2t\mathbf{k}; \ t_0 = \dfrac{\pi}{4}$

In Exercises 11–15, find $\mathbf{v}(t)$, $\mathbf{a}(t)$, and the speed at an arbitrary t in the given domain.

11. $\mathbf{r}(t) = (5 - 2t)^{3/2}\mathbf{i} + \dfrac{1}{2}(t^2 + 4t)\mathbf{j}; \ t < \dfrac{5}{2}$

12. $\mathbf{r}(t) = (\ln t^2)\mathbf{i} + \dfrac{1}{t}\mathbf{j} + 2t\mathbf{k}; \ t > 0$

13. $\mathbf{r}(t) = \left\langle t \cos t \sin t, \dfrac{(2t)^{3/2}}{3} \right\rangle; \ t > 0$

14. $\mathbf{r}(t) = \langle \cos^3 t, \sin^3 t, \cos 2t \rangle; \ 0 \leq t \leq \dfrac{\pi}{2}$

15. $\mathbf{r}(t) = \langle e^t \cos t, e^t \sin t, e^t \rangle; \ -\infty < t < \infty$

In Exercises 16–20, use the results of Examples 12.19 and 12.20.

16. A 2-kg mass attached to one end of a rope 3 m long is being whirled around horizontally in a circular path by a child holding the other end of the rope. If the speed of the mass is 4 m/s, find the force exerted by the child on the rope. Through how many revolutions per minute is the mass turning?

17. Using Exercise 16, find the effect on the force exerted by the child if (a) the speed is doubled and (b) the rope is half as long.

18. A satellite moves in a circular orbit 400 km above the earth. What is its speed? How long does it take to complete one orbit?

19. A satellite is in a circular orbit h km above the earth. If its speed is 28,000 km/hr, find h.

20. If a satellite circles the earth once every 90 min, find its height above the earth and its velocity.

21. A projectile is fired from the earth's surface with an initial speed of 600 m/s at an angle of 30° with the horizontal. Find the maximum height and range of the projectile.

22. At what angle θ should the projectile of Exercise 21 be fired for the range to be 30 km?

23. If a projectile is fired from the ground at an angle of 42° and attains a maximum height of 1500 m, what is its initial speed?

24. In Exercise 23 if, instead of the known height, we are given that the range is 36 km, what is the initial speed?

25. Prove that if a particle moves along a curve with constant speed, then its velocity and acceleration vectors are always orthogonal. (*Hint:* Use the fact that $|\mathbf{v}|^2 = \mathbf{v} \cdot \mathbf{v} = C$, and differentiate.)

26. Prove that if the position vector of a particle moving in space is of the form

$$\mathbf{r}(t) = (\cos \omega t)\mathbf{A} + (\sin \omega t)\mathbf{B}$$

where $\mathbf{A}$ and $\mathbf{B}$ are arbitrary constant vectors, then $\mathbf{a}(t) = -\omega^2 \mathbf{r}(t)$.

27. Prove that the position vector of a moving particle is of the form $\mathbf{r}(t) = t\mathbf{A} + \mathbf{B}$, where $\mathbf{A}$ and $\mathbf{B}$ are constant vectors if and only if $\mathbf{a}(t) = \mathbf{0}$ for all t. Describe the motion in this case.

28. A communications satellite is located above the equator, and its speed and altitude are such that it remains stationary relative to the earth. What are its speed and altitude?

29. A projectile is fired at an angle of 25° with the horizontal from the top of a hill 1000 m above the plain below. If the initial speed of the projectile is 500 m/s, find the range and the speed at impact.

30. In order to feed cattle in winter, a rancher drops bales of hay from a light airplane. If a bale is dropped from a height of 200 m while the airplane is flying horizontally at 50 m/s, how far is it from the point on the ground below the airplane when the bale is dropped to the point where it hits the ground? What is its speed at impact? (Neglect air resistance and assume the ground is flat.)

31. For a certain particle moving in space, it is known that $\mathbf{a}(t) = -t\mathbf{k}$ and that $\mathbf{v}(0) = 2\mathbf{i} - 3\mathbf{j} + \mathbf{k}$ and $\mathbf{r}(0) = 4\mathbf{i} + 2\mathbf{j}$. Find $\mathbf{v}(t)$ and $\mathbf{r}(t)$.

32. Redo Exercise 31 if $\mathbf{a}(t) = e^{-t}\mathbf{i} + 2e^{t}\mathbf{j} + te^{t}\mathbf{k}$, $\mathbf{v}(0) = 2\mathbf{i} + 6\mathbf{j} - \mathbf{k}$, and $\mathbf{r}(0) = \mathbf{i} + 2\mathbf{j} - 2\mathbf{k}$.

33. In Example 12.19, imbed the problem in a three-dimensional coordinate system, with the circle lying in the xy-plane, and write $\mathbf{r}(t)$ as a three-dimensional vector with third component 0. Let $\omega = \omega\mathbf{k}$. Prove that $\mathbf{v}(t) = \omega \times \mathbf{r}(t)$. Show that the same result holds true if the circle is in any plane parallel to the xy-plane, with the center on the z-axis.

In Exercises 34–39, find the tangential and normal components of acceleration at the indicated time. Draw the graph of $\mathbf{r}$, showing the acceleration, together with its tangential and normal projections at the point in question.

34. $\mathbf{r}(t) = 2t\mathbf{i} - t^2\mathbf{j};\ t = 2$

35. $\mathbf{r}(t) = \langle t^2, t^3 \rangle;\ t = 1$

36. $\mathbf{r}(t) = \langle 2\cos t, 4\sin t \rangle;\ t = \dfrac{\pi}{3}$

37. $\mathbf{r}(t) = \cosh t\mathbf{i} + \sinh t\mathbf{j};\ t = \ln 2$

38. $\mathbf{r}(t) = (t-1)\mathbf{i} + 4\sqrt{t}\,\mathbf{j};\ t = 1$

39. $\mathbf{r}(t) = \langle e^{2t} - 1, e^t \rangle;\ t = 0$

In Exercises 40–45, use the results of the specified problem to find the curvature at the indicated point in Exercises 34–39, respectively.

In Exercises 46–51, find $a_{\mathbf{T}}$ and $a_{\mathbf{N}}$ at an arbitrary value of t in the domain.

46. $\mathbf{r}(t) = \langle a\cos \omega t, a\sin \omega t, bt \rangle;\ t \geq 0, a > 0$

47. $\mathbf{r}(t) = \left\langle 2\ln t, \dfrac{t-1}{t}, 2t \right\rangle;\ t > 0$

48. $\mathbf{r}(t) = t\mathbf{i} + t^2\mathbf{j} + \frac{2}{3}t^3\mathbf{k};\ -\infty < t < \infty$

49. $\mathbf{r}(t) = (3t\sin t)\mathbf{i} + (3t\cos t)\mathbf{j} + (2t)^{3/2}\mathbf{k};\ t > 0$

50. $\mathbf{r}(t) = \langle 2t, 2e^t, e^{-t} \rangle;\ -\infty < t < \infty$

51. $\mathbf{r}(t) = \langle \sin t - t\cos t, \cos t + t\sin t, t^2 \rangle;\ t \geq 0$

52. Show that if $a_{\mathbf{N}} \neq 0$,

$$\mathbf{N} = \frac{\mathbf{a} - a_{\mathbf{T}}\mathbf{T}}{a_{\mathbf{N}}}$$

In Exercises 53–57, find $\mathbf{T}$, $\mathbf{N}$, and κ for the curve in Exercises 47–51, using $a_{\mathbf{T}}$ and $a_{\mathbf{N}}$ as previously found, together with the result of Exercise 52.

58. Let $v = |\mathbf{v}|$. Show that

$$\mathbf{a} = \frac{dv}{dt}\mathbf{T} + \frac{v^2}{\rho}\mathbf{N}$$

where ρ is the radius of curvature.

59. Use the formulas for a_N from Equations 12.23 and 12.27 to obtain the following formula for the curvature κ:

$$\kappa = \frac{|\mathbf{v} \times \mathbf{a}|}{|\mathbf{v}|^3}$$

60. Prove that if the normal component of acceleration of a particle is constantly 0, the particle moves in a straight line.

61. Prove that if the force on a particle is always centripetal (directed along the normal to the path), its speed is constant.

Exercises 62–66 form a sequential unit.

62. Let a particle move in the xy-plane in which a polar coordinate system is superimposed, with the polar axis coinciding with the positive x-axis. Let $\mathbf{r} = \mathbf{r}(t)$ be its position vector at time t, and let $r = |\mathbf{r}|$. If $\theta = \theta(t)$ is the polar angle to the vector $\mathbf{r}$ at time t, define $\mathbf{u}_r = \langle \cos\theta, \sin\theta \rangle$ and $\mathbf{u}_\theta = \langle -\sin\theta, \cos\theta \rangle$. Show each of the following.

(a) $\mathbf{u}_r$ and $\mathbf{u}_\theta$ are orthogonal unit vectors, and $\mathbf{u}_\theta$ is rotated 90° counterclockwise from $\mathbf{u}_r$.

(b) $\mathbf{r} = r\mathbf{u}_r$

(c) $\dfrac{d\mathbf{u}_r}{dt} = \mathbf{u}_\theta \dfrac{d\theta}{dt}$ and $\dfrac{d\mathbf{u}_\theta}{dt} = -\mathbf{u}_r \dfrac{d\theta}{dt}$

(*Note:* $\mathbf{u}_r$ and $\mathbf{u}_\theta$ are called the **radial** and **transverse** unit vectors, respectively.)

63. (a) Show that $\mathbf{v} = \dfrac{dr}{dt}\mathbf{u}_r + r\dfrac{d\theta}{dt}\mathbf{u}_\theta$.

(b) Show that speed $= \sqrt{\left(\dfrac{dr}{dt}\right)^2 + r^2\left(\dfrac{d\theta}{dt}\right)^2}$.

64. Show that

$$\mathbf{a} = \left[\frac{d^2r}{dt^2} - r\left(\frac{d\theta}{dt}\right)^2\right]\mathbf{u}_r + \left[r\frac{d^2\theta}{dt^2} + 2\frac{dr}{dt}\frac{d\theta}{dt}\right]\mathbf{u}_\theta$$

65. Show the following:

(a) $\text{Comp}_{\mathbf{u}_r}\,\mathbf{a} = \dfrac{d^2r}{dt^2} - r\left(\dfrac{d\theta}{dt}\right)^2$

(b) $\text{Comp}_{\mathbf{u}_\theta}\,\mathbf{a} = r\dfrac{d^2\theta}{dt^2} + 2\dfrac{dr}{dt}\dfrac{d\theta}{dt}$

66. Let $a_r = \text{Comp}_{\mathbf{u}_r}\,\mathbf{a}$ and $a_\theta = \text{Comp}_{\mathbf{u}_\theta}\,\mathbf{a}$. Show that $|\mathbf{a}|^2 = a_r^2 + a_\theta^2$.

67. A particle moves in the horizontal plane so that its polar coordinates at time t are $r = 1 + \cos t^2$ and $\theta = t^2$. Use the results of Exercises 63 and 64 to resolve $\mathbf{v}$ and $\mathbf{a}$ into radial and transverse vector components.

68. A particle moves in the xy-plane with position vector $\mathbf{r}(t) = (t^2 \cos t)\mathbf{i} + (t^2 \sin t)\mathbf{j}$ at time t. Find the polar coordinates of a point on the path at time t. Find the radial component a_r of acceleration and the transverse component a_θ, at time t. (See Exercise 65.)

69. A projectile is fired with an initial speed of 400 m/s and an angle of elevation of 30°. Find the range of the projectile, the maximum height the projectile attains, and the speed at impact. Plot the path of the projectile.

70. Repeat Exercise 69 with an initial speed of 600 m/s.

71. Repeat Exercise 69 with the initial speed 500 m/s and the angle of elevation 50°.

12.6 KEPLER'S LAWS

In the early seventeenth century, the German mathematician and astronomer Johannes Kepler (1571–1630) postulated the following three laws governing the orbits of planets around the sun.

Kepler's Laws of Planetary Motion

1. The orbit of each planet is an ellipse with the sun at one focus.

2. The radius vector from the sun to a planet sweeps out area at a constant rate.

3. The square of the time for a planet to complete one revolution around its elliptical orbit is proportional to the cube of the length of the semimajor axis of the ellipse.

Johannes Kepler, ca. 1627

Artist's rendition of the orbits of the planets

Kepler deduced these laws based on the astronomical observations of his mentor Tycho Brahe. His deductions required years of analyzing massive amounts of data with laborious calculations. Kepler's discoveries rank as one of the outstanding achievements in the history of science. Based as they were on empirical evidence, however, they lacked a sound theoretical basis until Newton, some fifty years later, deduced Kepler's Laws using his newly invented calculus. Newton published his findings in his book *Principia Mathematica* (1687), generally considered to be the most important scientific book ever written. Newton based his proofs on the following two principles:

1. *Newton's Second Law of Motion.* For a body of constant mass m moving under the action of a force $\mathbf{F}$,

$$\mathbf{F} = m\mathbf{a} \tag{12.28}$$

where $\mathbf{a}$ is the acceleration of the body.

2. *Newton's Law of Gravitation.* The force $\mathbf{F}$ of attraction between two bodies of masses M and m, respectively, is proportional to the product of the masses and inversely proportional to the square of the distance, r, between them:

$$\mathbf{F} = -\frac{GMm}{r^2}\mathbf{u}_r \tag{12.29}$$

where $\mathbf{u}_r$ is a unit vector directed from one mass toward the other.

The constant G in Equation 12.29 is called the *universal gravitational constant* and has the approximate value

$$G = 6.672 \times 10^{-11} \frac{\text{N} \cdot \text{m}^2}{\text{kg}^2}$$

To demonstrate the power of the vector calculus we have studied, we will prove the first of Kepler's Laws. In the exercises we outline proofs of the other two laws. You should follow the proof below using pencil and paper, verifying all steps. Note carefully the use of properties of the dot and cross products.

We take the origin as the center of mass of the sun and let M be its mass. Let $\mathbf{r}(t)$ be the position vector at time t of the center of mass of a planet that has mass m. Let $r = |\mathbf{r}|$, and let $\mathbf{u}_r$ be the unit vector $\mathbf{r}/r$, so that $\mathbf{r} = r\mathbf{u}_r$. We assume that the forces of attraction between this planet and all other bodies are negligible compared with the gravitational attraction between it and the sun.

We first show that the planet moves in one plane. Equating the right-hand sides of Equations 12.28 and 12.29, we get the acceleration in the form

$$\mathbf{a} = -\frac{GM}{r^2}\mathbf{u}_r \tag{12.30}$$

Then

$$\mathbf{r} \times \mathbf{a} = r\mathbf{u}_r \times \left(-\frac{GM}{r^2}\mathbf{u}_r\right) = -\frac{GM}{r}(\mathbf{u}_r \times \mathbf{u}_r) = \mathbf{0}$$

and since $\mathbf{v} \times \mathbf{v} = \mathbf{0}$, we also have

$$\frac{d}{dt}(\mathbf{r} \times \mathbf{v}) = \mathbf{r} \times \mathbf{a} + \mathbf{v} \times \mathbf{v} = \mathbf{0}$$

Thus,

$$\mathbf{r} \times \mathbf{v} = \mathbf{c} \tag{12.31}$$

where **c** is a constant vector. Since **r** is orthogonal to $\mathbf{r} \times \mathbf{v}$ and hence to **c**, this result says that at all times, the vector $\mathbf{r}(t)$ is perpendicular to the fixed vector **c**. So the path of the planet lies in a plane. There is no loss of generality in assuming that this plane is the xy-plane and that **c** is directed along the positive z-axis. We illustrate the path of the planet in Figure 12.20.

We will show that the planet moves in an elliptical orbit with the sun as a focus (Kepler's First Law) by showing that an equation for the graph of $\mathbf{r}(t)$ is the polar form of the equation of an ellipse with a focus at the pole. Let $\theta = \theta(t)$ be the polar angle from the positive x-axis to $\mathbf{r}(t)$. Then, $\mathbf{u}_r = \langle \cos\theta, \sin\theta, 0 \rangle$, and if we define $\mathbf{u}_\theta = \langle -\sin\theta, \cos\theta, 0 \rangle$, $\mathbf{u}_\theta$ is a unit vector such that $\mathbf{u}_r \cdot \mathbf{u}_\theta = 0$ and $\mathbf{u}_r \times \mathbf{u}_\theta = \langle 0, 0, 1 \rangle = \mathbf{k}$. As in Exercises 62 and 63 of Exercise Set 12.5, we obtain, from differentiating both sides of $\mathbf{r} = r\mathbf{u}_r$,

$$\mathbf{v} = \frac{dr}{dt}\mathbf{u}_r + r\frac{d\theta}{dt}\mathbf{u}_\theta$$

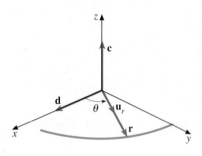

FIGURE 12.20

Thus, using Equation 12.31,

$$\mathbf{c} = \mathbf{r} \times \mathbf{v} = (r\mathbf{u}_r) \times \left(\frac{dr}{dt}\mathbf{u}_r + r\frac{d\theta}{dt}\mathbf{u}_\theta \right)$$

$$= r^2\frac{d\theta}{dt}(\mathbf{u}_r \times \mathbf{u}_\theta) = r^2\frac{d\theta}{dt}\mathbf{k} \qquad (12.32)$$

From Equations 12.30 and 12.32, we get

$$\frac{d}{dt}(\mathbf{v} \times \mathbf{c}) = \mathbf{a} \times \mathbf{c} = \left(-\frac{GM}{r^2}\mathbf{u}_r \right) \times \left[r^2\frac{d\theta}{dt}\mathbf{k} \right]$$

$$= -GM\frac{d\theta}{dt}[\mathbf{u}_r \times \mathbf{k}]$$

$$= GM\frac{d\theta}{dt}\mathbf{u}_\theta$$

$$= \frac{d}{dt}(GM\mathbf{u}_r)$$

The equality of these two derivatives implies that

$$\mathbf{v} \times \mathbf{c} = GM\mathbf{u}_r + \mathbf{d} \qquad (12.33)$$

The constant vector **d** is a linear combination of $\mathbf{u}_r$ and $\mathbf{v} \times \mathbf{c}$, each of which lies in the xy-plane, so **d** also lies in the xy-plane. We may assume, again without loss of generality, that **d** is directed along the positive x-axis. The relationship among the vectors **c**, **d**, $\mathbf{u}_r$, and **r** is shown in Figure 12.20.

Let $c = |\mathbf{c}|$ and $d = |\mathbf{d}|$. Then from Equations 12.31 and 12.33,

$$c^2 = \mathbf{c} \cdot \mathbf{c} = (\mathbf{r} \times \mathbf{v}) \cdot \mathbf{c}$$

$$= \mathbf{r} \cdot (\mathbf{v} \times \mathbf{c})$$

$$= (r\mathbf{u}_r) \cdot (GM\mathbf{u}_r + \mathbf{d})$$

$$= rGM + r\mathbf{u}_r \cdot \mathbf{d}$$

$$= r(GM + d\cos\theta)$$

Solving for r gives

$$r = \frac{c^2}{GM + d\cos\theta}$$

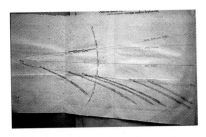

If we let $e = d/GM$ and $p = c^2/GMe$, this equation becomes, on dividing numerator and denominator by GM,

$$r = \frac{ep}{1 + e \cos \theta} \tag{12.34}$$

We know from Section 9.8 that Equation 12.34 is the polar equation of an ellipse if $0 < e < 1$, a hyperbola if $e > 1$, and a parabola if $e = 1$, each having a focus at the pole. Since it is known that planets travel in closed orbits, it follows that the equation is that of an ellipse. Kepler's First Law is therefore proved.

The comet of 1680 as drawn in Newton's *Principia*.

Exercise Set 12.6

1. Write the polar equation (12.34) for $0 < e < 1$ in the standard rectangular form

$$\frac{(x - h)^2}{a^2} + \frac{(y - k)^2}{b^2} = 1$$

2. From the result of Exercise 1, show the following:
   (a) $b = a\sqrt{1 - e^2}$
   (b) The center of the ellipse is at $\left(\dfrac{-e^2 p}{1 - e^2}, 0 \right)$.
   (c) The distance from the center to each focus is $\dfrac{e^2 p}{1 - e^2}$.
   (d) One focus is at $(0, 0)$ and the other is at $\left(\dfrac{-2e^2 p}{1 - e^2}, 0 \right)$.

In Exercises 3 and 4, each equation was used in the proof of Kepler's First Law. Verify each one.

3. (a) $\mathbf{u}_r \cdot \mathbf{u}_\theta = 0$
   (b) $\mathbf{u}_r \times \mathbf{u}_\theta = \mathbf{k}$
   (c) $\mathbf{v} = \dfrac{dr}{dt}\mathbf{u}_r + r\dfrac{d\theta}{dt}\mathbf{u}_\theta$

4. (a) $(r\mathbf{u}_r) \times \left(\dfrac{dr}{dt}\mathbf{u}_r + r\dfrac{d\theta}{dt}\mathbf{u}_\theta \right) = r^2 \dfrac{d\theta}{dt}(\mathbf{u}_r \times \mathbf{u}_\theta)$
   (b) $-GM\dfrac{d\theta}{dt}(\mathbf{u}_r \times \mathbf{k}) = GM\dfrac{d\theta}{dt}\mathbf{u}_\theta$
   (c) $GM\dfrac{d\theta}{dt}\mathbf{u}_\theta = \dfrac{d}{dt}(GM\mathbf{u}_r)$

5. Referring to the proof of Kepler's First Law, explain fully the justification for the following assertions:
   (a) $\mathbf{v} \times \mathbf{c}$ lies in the xy-plane.
   (b) $\mathbf{d}$ lies in the xy-plane.

In Exercises 6 and 7, fill in the details of the outlines given of proofs of Kepler's Second and Third Laws.

6. (a) Using the formula $A = \frac{1}{2}\int_\alpha^\beta r^2 \, d\theta$ for area in polar coordinates, show that the area between any fixed

angle $\theta_0 = \theta(t_0)$ and the angle $\theta = \theta(t)$, bounded by the ellipse of Equation 12.34, is

$$A(t) = \frac{1}{2}\int_{t_0}^{t} r^2 \frac{d\theta}{d\tau} d\tau$$

   (b) By using part (a) together with Equation 12.32, show that

$$\frac{dA}{dt} = \frac{c}{2}$$

   (c) Conclude that area is swept out by $\mathbf{r}(t)$ at a constant rate, proving Kepler's Second Law.

7. Let T be the time required for a planet to complete one revolution around the sun (the *period*).
   (a) Using Exercise 6, show that the total area enclosed by the ellipse is

$$A = \frac{1}{2}cT$$

   (b) Recall from Example 8.18 that the area also is given by $A = \pi ab$, where a and b are the lengths of the semimajor and semiminor axes of the ellipse. Combining this with part (a) and Exercise 2, part (a), show that

$$T = \frac{2\pi a^2}{c}\sqrt{1 - e^2}$$

   (c) From the result of Exercise 1, show that $1 - e^2 = \dfrac{ep}{a}$, and hence obtain

$$T^2 = \frac{4\pi^2 ep}{c^2}a^3$$

   thus proving Kepler's Third Law. By replacing p and e with the values assigned to them, rewrite the result in the form

$$T^2 = \frac{4\pi^2}{GM}a^3 \tag{12.35}$$

8. A reasonable approximation to the period of a planet is obtained by replacing the semimajor axis a in Equation 12.35 by the mean distance of the planet from the Sun.
   (a) The mean distance of Mars from the Sun is approximately $1\frac{1}{2}$ times that of the Earth. Find the approximate time (in "Earth days") it takes Mars to complete one revolution around the Sun.
   (b) It takes Mercury approximately 88 Earth days to orbit the Sun. If the mean distance of the Earth from the Sun is approximately 93 million miles, find the approximate mean distance of Mercury from the Sun.

9. Let r_0 denote the minimum value of the distance r of a planet from the sun and let v_0 be its speed when $r = r_0$.
   (a) Show that $r_0 = ep/(1 + e)$ when $\theta = 0$.

(b) Use Equation 12.33 to show that
$$v_0 = \frac{(GM + d)}{c}$$
   (*Hint:* Use $|\mathbf{v} \times \mathbf{c}| = |\mathbf{v}||\mathbf{c}|$. Why is this equation true?)
   (c) Show that
$$v_0 = \frac{c}{r_0}$$

10. (a) Use the result of Exercise 7, part (b), to express v_0 in terms of a, e, and T.
    (b) The earth takes approximately 365.26 days to complete its orbit around the sun. Its semimajor axis a is approximately 1.4959×10^8 km, and the eccentricity of the orbit is approximately 0.016732. Find r_0 and v_0 for the earth. (Express v_0 in kilometers per hour.)

Chapter 12 Review Exercises

1. Let $\mathbf{u}(t) = \langle t^2, t^3, 1 - t \rangle$, $\mathbf{v}(t) = \langle t \ln t, -2, t + 3 \rangle$, and $\alpha(t) = e^t$. Find the following:
   (a) $(\mathbf{u} \cdot \mathbf{v})(t)$
   (b) $\mathbf{u}(\alpha(t))$
   (c) $(\mathbf{u} - \mathbf{v})(t)$
   (d) $(\mathbf{v} \circ \alpha)(0)$
   (e) $(\mathbf{u}' \times \mathbf{v}')(1)$

2. Find $\mathbf{r}'(t)$ if:
   (a) $\mathbf{r}(t) = \langle t \ln \sqrt{t}, t^2 e^{-t} \rangle$, $t > 0$
   (b) $\mathbf{r}(t) = \left(\dfrac{t}{\sqrt{t^2 - 1}} \right)\mathbf{i} + \left(\sin^{-1}\dfrac{1}{t} \right)\mathbf{j} + \left(t\sqrt{t^2 - 1} \right)\mathbf{k}$

3. Evaluate the integrals:
   (a) $\displaystyle\int \left\langle \dfrac{1}{\sqrt{1 - t^2}}, \dfrac{t}{\sqrt{1 - t^2}}, \dfrac{1}{1 - t^2} \right\rangle dt$
   (b) $\displaystyle\int_0^{\pi/3} \left[(\cos^2 t)\mathbf{i} - (\sin^3 t)\mathbf{j} + (\tan^2 t)\mathbf{k} \right] dt$

4. Let $\mathbf{r}(t) = t^2\mathbf{i} + (3t - 2)\mathbf{j} + (1 - t^2)\mathbf{k}$ and $\alpha(t) = \cos t$. Find:
   (a) $\displaystyle\int (\alpha\mathbf{r})(t)dt$
   (b) $\displaystyle\int (\mathbf{r} \circ \alpha)(t)dt$

5. Let $\mathbf{r}(t) = (\ln t^2)\mathbf{i} - \dfrac{2}{\sqrt{t}}\mathbf{j} + 4\sqrt{t}\,\mathbf{k}$. Find:
   (a) $\displaystyle\int_1^4 \mathbf{r}'(t)dt$

(b) $\displaystyle\int_1^4 |\mathbf{r}'(t)|dt$

6. Verify Property 6 for integrals for $\mathbf{r}(t) = \langle 20t, 9t^2, 12t^2 \rangle$ on $[0, 1]$.

7. Let C be the graph of $\mathbf{r}(t) = (2t^3/3)\mathbf{i} + (1 - 2t^2)\mathbf{j} + 4t\mathbf{k}$ for $0 \le t \le 3$. Find the length of C.

8. Sketch the graph of $\mathbf{r}(t) = \langle \cos t, t, \sin t \rangle$ for $0 \le t \le 4\pi$. Find parametric equations of the tangent line to the graph at $t = 4\pi/3$.

9. Let C be the graph of $\langle \ln(\cosh t), 2 \tan^{-1} e^t, \sqrt{3}\,t \rangle$ on $[-1, 2]$.
   (a) Find the length of C.
   (b) Make the change of variables $u = e^{-t}$, and let $\mathbf{R}(u)$ be the new position vector for C. Find $\mathbf{R}(u)$ and the u interval. Is the orientation of C preserved or reversed?
   (c) Find the length of C using $\mathbf{R}(u)$ and show it is the same as that found in part (a).

10. Let C be the graph of $\mathbf{r}(t) = \langle 2e^t \sin t, 2e^t \cos t, e^t \rangle$ on $[0, \infty)$. Introduce arc length s as a parameter and let $\mathbf{R}(s)$ be the resulting position vector for C. Show that for all $s > 0$, $|\mathbf{R}'(s)| = 1$.

11. Use the arc length parameterization of C in Exercise 10 to find $\mathbf{T}$, $\mathbf{N}$, and κ at an arbitrary $s \ge 0$. Evaluate each of these at the point $(0, 2, 1)$.

12. A particle moves in the xy-plane so that its position vector at time t is $\mathbf{r}(t) = \langle 3 - 2\sqrt{t}, t + 1 \rangle$. Find its velocity, acceleration, and speed when $t = 4$. Identify the curve.

13. A particle moves so that its position vector at time t is $\mathbf{r}(t) = \langle e^{-t} \sin 2t, e^{-t} \cos 2t, 2e^{-t} \rangle$. Find $\mathbf{v}(t)$, $\mathbf{a}(t)$, and ds/dt at an arbitrary t.

14. A projectile is fired from ground level with an initial velocity $\mathbf{v}_0 = 0.4\mathbf{i} + 0.3\mathbf{j}$, with the magnitude in kilometers per second. Find the range of the projectile (x_{max}) and the maximum height it attains (y_{max}).

15. (a) A satellite is in orbit 240 km above the earth's surface. What is its speed?
(b) A satellite completes one orbit around the earth every 2 hr. Find its altitude and speed.

16. Let $\mathbf{r}(t) = t^3\mathbf{i} - 4t^2\mathbf{j}$ for $t \in [1, 4]$. At the point $(8, -16)$, find $\mathbf{T}$, $\mathbf{N}$, κ, and the center of curvature.

17. (a) Find the curvature of $y = \ln|\csc x|$ for any $x \neq n\pi$.
(b) Find the curvature of $x^3 + 3xy - y^3 = 3$ at the point $(2, -1)$.

18. Let C be defined by
$$\mathbf{r}(t) = \langle -1 + 5\sin t, 3\cos t, 1 - 4\cos t \rangle$$
for t in $[0, 2\pi]$. Find $\mathbf{T}$, $\mathbf{N}$, and $\mathbf{B}$ at an arbitrary point on C.

19. A particle moves so that its position vector at time $t \geq 0$ is $\mathbf{r}(t) = \langle \ln(t + 1), 2t, t^2 + 2t \rangle$. Find $\mathbf{a}$, $a_{\mathbf{T}}$, and $a_{\mathbf{N}}$.

20. A particle moves in the horizontal plane so that its polar coordinates at time t are $r = 1 + 2\sin e^t$ and $\theta = e^t$. Describe its path for $t \geq 0$. Find the radial and transverse components of acceleration. (See Exercise 65 in Exercise Set 12.5.)

21. If a planet has the elliptical orbit $r = ep/(1 + e\cos\theta)$, show that its speed v_m at the point on its orbit farthest from the sun is
$$v_m = \frac{2\pi a}{T} \sqrt{\frac{1 - e}{1 + e}}$$
where a is the length of the semimajor axis and e is the eccentricity of the ellipse and T is the period. Use the data given for the earth in Exercise 10, part (b), of Exercise Set 12.6 to find v_m for the earth.

Chapter 12 Concept Quiz

1. Define each of the following:
(a) the position vector of a point
(b) the derivative of the vector function $\mathbf{F}$ defined by $\mathbf{F}(t) = \langle f(t), g(t), h(t) \rangle$
(c) a space curve
(d) a simple closed curve C that is the graph of a vector function $\mathbf{r}(t)$ on $[a, b]$
(e) a smooth curve

2. Let C be the curve that is the graph of the twice-differentiable vector function $\mathbf{r}(t)$ for $a \leq t \leq b$. Give formulas for each of the following:
(a) the length of C
(b) the unit tangent vectors $\mathbf{T}$ and $\mathbf{N}$ at an arbitrary point on C
(c) the curvature of C at an arbitrary point

3. If the position vector of a moving point is $\mathbf{r}(t)$ at time t, give formulas for each of the following:

(a) its velocity
(b) its speed
(c) its acceleration
(d) the tangential and normal components of its acceleration

4. (a) If a curve C that is the graph of a vector function $\mathbf{r}$ is parameterized by arc length s, what can you say about $\mathbf{r}'(s)$?
(b) If the plane curve C has the equation $\mathbf{r}(t) = \langle x(t), y(t) \rangle$, express its curvature in terms of x', x'', y', and y''.
(c) If C is the graph of $y = f(x)$, express its curvature in terms of y' and y''.
(d) State a sufficient condition for the graph of a vector function $\mathbf{r}(t)$ on $[a, b]$ to be rectifiable (have finite length).

5. State Kepler's three laws of planetary motion.

APPLYING CALCULUS

1. In an oil refinery, one of the cylindrical storage tanks is 40 m in diameter and 15 m high. A staircase in the form of a helix goes around the outside of the tank and reaches the top after one revolution. Find the equation of the helix. What is the approximate length of handrail required along the outer edge of the staircase if the staircase is 60 cm wide?

2. A curve in a railroad track follows the parabola $25y = x^2$ from (0, 0) to (5, 1) and then follows the circle having the same curvature as the parabola at (5, 1). The x- and y-coordinates are measured in kilometers. The circular part of the track is 4 kilometers long, and then the track becomes straight in the direction tangent to the circle. Find:

(a) The center and radius of the circular arc;

(b) The direction of the straight track, where the positive y-axis points north.

3.

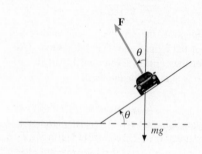

(a) Use the result of Example 12.19 to show that the magnitude of the centripetal force **F** required to keep an object of mass m on a circular path of radius r, with constant speed, can be written in the form

$$|\mathbf{F}| = \frac{m|\mathbf{v}|^2}{r}$$

where **v** is the velocity of the object.

(b) Use part (a) to show that the maximum safe speed v a car can travel around a curve of radius r without skidding, if the highway is banked at an angle θ (see the figure), is given by

$$v = \sqrt{rg \tan \theta}$$

(*Hint*: Show that $|\mathbf{F}| \cos \theta = m g$ and $|\mathbf{F}| \sin \theta = m v^2/r$. Then find $\tan \theta$).

(c) If a curve of radius 120 m is banked at an angle of 15°, what is the maximum safe speed for traversing the curve?

4. A golfer hits a golf ball so that the initial angle of elevation of its path is θ and its initial velocity is v_0. The fairway is sloping upward at an angle α, as shown in the figure.

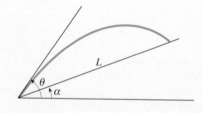

(a) If $\alpha = 10°$, $\theta = 45°$, and $v_0 = 120$ ft/s, find the distance L the ball will go along the fairway.

(b) For constant values of α and v_0, show that the angle θ that maximizes the distance L satisfies

$$\tan 2\theta = -\cot \alpha$$

5. A particle in space moves under the influence of a force that is always directed toward the origin. Let **r**, **v**, and **a** be the position, velocity, and acceleration vectors of the particle at time t.

(a) Show that $\mathbf{a}(t) = f(t)\mathbf{r}(t)$ for some scalar function f.

(b) Show that $(\mathbf{r} \times \mathbf{v})' = \mathbf{0}$ from which it follows that $\mathbf{r} \times \mathbf{v} = \mathbf{C}$ for some constant vector **C**. Conclude that the path of the particle lies in a plane. What is the plane if $\mathbf{C} \neq \mathbf{0}$? What is the path of the particle if $\mathbf{C} = \mathbf{0}$?

(c) Suppose that $\mathbf{a}(t) = -k^2\mathbf{r}(t)$ where k is a constant. From part (b), the path of the particle lies in a plane. Assume that the plane is the xy-plane and that when $t = 0$, the particle is on the x-axis with velocity vector parallel to the y-axis. Show that the path of the particle is an ellipse centered at the origin.

6. A new swimming pool was recently installed in the basement of a large urban high school, with the bottom resting on a floor below. Through a viewing window, it was observed that water was leaking out of the side of the pool (see the figure). It can be shown that the speed v at which water goes out of the hole is given by the formula

$$v = \sqrt{2g(h - y)}$$

(a) If $h = 2$ m and $y = 1.3$ m, find x.

(b) Assume h is constant (water is added at the same rate it is leaking out). If the hole is at the height y for which x is a maximum, find both x and y.

7. A disk of radius 1 is rotating in the counterclockwise direction at a constant angular speed ω. A particle starts at the center of the disk and moves toward the edge along a fixed radius so that its position at time t, $t > 0$, is given by $\mathbf{r}(t) = t\mathbf{R}(t)$ where

$$\mathbf{R}(t) = (\cos \omega t)\,\mathbf{i} + (\sin \omega t)\,\mathbf{j}$$

(a) Show that the velocity $\mathbf{v}$ of the particle is

$$\mathbf{v} = (\cos \omega t)\,\mathbf{i} + (\sin \omega t)\,\mathbf{j} + t\mathbf{v}_d$$

where $\mathbf{v}_d = \mathbf{R}'(t)$ is the velocity of a point on the edge of the disk.

(b) Show that the acceleration $\mathbf{a}$ of the particle is

$$\mathbf{a} = 2\mathbf{v}_d + t\mathbf{a}_d$$

where $\mathbf{a}_d = \mathbf{R}''(t)$ is the acceleration of a point on the edge of the disk. The extra term $2\mathbf{v}_d$ is called the *Coriolis acceleration*; it is the result of the interaction of the rotation of the disk and the motion of the particle. One can obtain a physical demonstration of this acceleration by walking toward the edge of a moving merry-go-round.

(c) Determine the Coriolis acceleration of a particle that moves on a rotating disk according to

$$\mathbf{r}(t) = e^{-t}\,(\cos \omega t)\,\mathbf{i} + e^{-t}\,(\sin \omega t)\,\mathbf{j}$$

Joseph Newhouse

At what point in your studies did you realize that calculus would be useful to you?

When I took my first course in microeconomic theory, I was a junior in college and I already had a year of calculus at that point. I discovered that it was much easier to write an equation than it was to write out long strings of words. There are some classic books in economics from the 1930s that are exceedingly difficult to understand because they try to write equations in words when it is much simpler just to write out the equations in mathematical notation.

For example, in economics, probably the most famous condition for the theory of the firm is *marginal revenue = marginal cost*, which means that the additional revenue you get from the last unit you sell equals the cost of producing it. Students taking their first economics course will understand this at about the level I just described, because it cannot be assumed that people who are studying economics know the calculus. However, what this equation really says is that profit is the total revenue minus the total cost. If I just differentiate both with respect to quantity, which are both functions of the quantity of goods the firm produces, I get the first derivative of total revenue minus the first derivative of total cost. There might be other things in those functions, in which case I have to take the cost function. Those derivatives are the marginal revenue and the marginal cost. So this condition that every first-year economics student learns is just a simple first-order condition. Although that condition is pretty easy to understand verbally, as one progresses, it gets harder and harder to verbalize and easier and easier to use the math. In fact, the things that economists get interested in, such as what happens when something changes, what happens at the optimum, frequently can only be done with the mathematics. It would be almost impossible to verbalize one's way through to a solution.

That is really why anybody pursuing economics on a more advanced level must learn calculus. Some people think that economists only use calculus because they want to obfuscate the evidence, but in fact it is impossible to derive answers for anything but the most simple problems without it (in continuous cases, where calculus applies; there is also a whole economics of discrete things in which calculus is not used).

You are an expert in health-care policy and economics. How is calculus useful in your field?

Broadly speaking it is useful in two ways. First of all, much of microeconomic theory is built on the calculus because it has an assumption of maximization or minimization. For example, for profit firms try to maximize profits or consumers try to maximize their utility or their well being, all of these subject to some constraints. So, traditional theories of consumer behavior and firm behavior are really problems in constrained maximization and how firms and consumers act as the constraints they face change. For example, how household incomes change behavior

will be analyzed by perturbing some kind of first-order condition and solving for the result. That is one way in which all of economics uses the calculus. What I do is no exception to that.

The second way in which I use the calculus is in statistical analysis of data to estimate certain population values (parameters). A useful technique is maximum-likelihood estimation, which involves finding the parameters that maximize the probability of observing the sample of data that one has.

Let's talk about maximum-likelihood estimation.

A simple idea is that I might want to estimate a person's demand for medical care as a function of the price of medical care the person faced and the person's income. In the simplest form, I might want to estimate use or demand as a linear function of price and income. So I would write an equation that says Use = constant + another constant times price plus another constant times income plus random error (written as $\beta_0 + \beta_1$ Price + β_2 Income + error). The estimation problem would be to estimate the three constants. There are various ways to do this, but if the random error were normally distributed, the least-squares estimator, which is the most popular estimator, is also the maximum-likelihood estimator. Maximum likelihood means that for each person I have, each observation, I write down the probability-density function for the error term. (See p. 463.) The likelihood of the error term will be a function of the three constants. I choose the three constants, in effect, so as to maximize that likelihood function. I multiply the probability of each observation to get a giant product, which I call the likelihood that I observe a particular sample. That likelihood is a function of the three constants, and it is convenient to take the logarithm of that product so that the result is a sum rather than a product. Then I will simply take the partial derivatives of the log likelihood function, which is the sum of the logs, with respect to each of the constants, and set that equal to zero and solve those equations for the constants.

It turns out that, under fairly general conditions, this method of going about estimating the constants gives me a best—in a sense that can be made precise—estimator. "Best" in this context means that it is best among a class of estimates that are asymptotically normally distributed. These will have the minimum variance. If I repeat this calculation in different samples, I will get a normal distribution of my estimate.

Those are the two main domains in which I personally use the calculus. There are certainly other applications of the calculus in economics, particularly in analysis of population dynamics or dynamic systems where control theory is used.

When we're dealing with multiple variables, do we need to use partial differential equations?

Yes. Because economists tend to be interested in partial effects—what happens if something changes and all else is constant—that is analogous to a partial derivative that holds everything else constant. Occasionally one is interested in a total derivative but most often in economics one is interested in a partial derivative.

You are a professor in the Medical School, the Kennedy School of Government, the School of Public Health, and the Faculty of Arts and Sciences. Do you see a difference in the mathematical training of students in those schools?

Students pursuing a Ph.D. in economics probably won't be admitted without calculus, including multivariable calculus. They may take a kind of refresher course to brush up, but right away in the first year the curriculum makes so many demands on the calculus that they simply wouldn't get in. So they tend to be well prepared. Masters-level students are a much more heterogeneous group, but in both the faculty of government and the faculty of public health, students can progress faster and go further who know the calculus in both economics and in statistics courses, which are used heavily by both faculties. Indeed, a masters degree in public policy at the school of government requires one year of economics and one year of statistics. Advanced courses are built on the calculus, and those courses go further with students knowing more at the end of the year, simply because you can go faster with the calculus than without it.

Joseph Newhouse is the John D. MacArthur Professor of Health Policy and Management at Harvard University and Director of the Division of Health Policy Research and Education. He is a member of the faculties of the John F. Kennedy School of Government, the Harvard Medical School, the Harvard School of Public Health, and the Faculty of Arts and Sciences. He is the editor of the *Journal of Health Economics* and an associate editor of the *Journal of Economic Perspectives*. He serves on the Council of the Institute of Medicine and is the President of the Association for Health Services Research.

13

FUNCTIONS OF MORE THAN ONE VARIABLE

In this chapter, we begin the study of real-valued functions of two or more independent variables. You already are familiar with some functions of this type. For example, the formula

$$V = \pi r^2 h$$

expresses the volume of a right circular cylinder (such as a can) as a function of the base radius r and the height h. So we can say that V is a function of the two independent variables r and h. Or consider the formula

$$A = P(1 + r)^t$$

where A is the amount of money that has accumulated after t years from an initial investment of P dollars at an interest rate r (expressed as a decimal), compounded annually. In this case A is a function of the three independent variables P, r, and t.

As another example, consider the Earth's surface temperature T as a function of position. If an xy-coordinate system is set up on a map, then at any fixed time, the temperature is a function of the two variables x and y. In this case, finding a formula for T in terms of x and y would be difficult, but by obtaining satellite data, we can show the temperature graphically. By connecting points where the temperature is the same, we get what are called **isotherms** ("equal temperature" curves). Figure 13.1 is a photograph that shows such isotherms for the Earth. This isothermic image is an example of what is called a **contour map** of a function. Contour maps provide one means of describing a function of two variables graphically.

Our emphasis in this chapter will be on functions of two or three variables, especially the former, since they can be pictured geometrically. We should mention, however, that functions involving large numbers of variables are often used in applications. For example, cosmologists are now considering the universe to be an eleven-dimensional space, and in economics as many as 100 independent variables are sometimes needed. Much of our theory can be extended to these higher dimensions.

FIGURE 13.1
Isotherms (equal-temperature curves) for the Earth

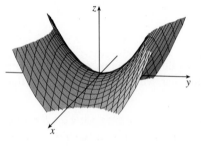

An example of a quadric surface

In the first section we give definitions of functions of two variables and of three variables, and then we discuss how to find their domains.

We discuss how to represent functions graphically, and we show a variety of graphs generated by a computer algebra system. These graphs are three-dimensional *surfaces*, which are hard to draw without the aid of a CAS, except in certain special cases. These special cases are included in a broad class called *quadric surfaces*, which we examine in Section 13.2.

We then define the notions of limit and continuity for functions of two or three variables. We conclude the chapter by considering what are called *partial derivatives*, and we indicate some of their applications.

13.1 FUNCTIONS OF TWO AND THREE VARIABLES

In Chapter 1, we defined a function from a set A to a set B as a rule that assigns to each element of A one, and only one, element of B. By specifying the set A (the domain) to be a set of ordered pairs of real numbers and the set B (the codomain) to be a subset of the set $\mathbb{R}$ of real numbers, we have the following definition of a real-valued function of two real variables.

Definition 13.1
A Real-Valued Function of Two Real Variables

Let A denote a set of ordered pairs of real numbers. A function f that assigns a unique real number z to each ordered pair (x, y) in A is called a **real-valued function of two real variables**, and we write

$$z = f(x, y)$$

The set A is the domain of f, and the subset B of $\mathbb{R}$ consisting of all numbers $f(x, y)$ for (x, y) in A is the range of f.

REMARK ─────────────────────────

■ For brevity, we will often refer to such a function simply as a function of two variables.

In Figure 13.2 we indicate how such a function f maps a point (x, y) in the domain A to a point z on the real line.

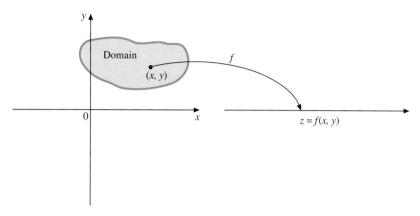

FIGURE 13.2
The function f maps the point (x, y) in the domain to a real number z in the range.

The set of all ordered pairs of real numbers (geometrically, the xy-plane) is frequently denoted by $\mathbb{R}^2$. So we can say that a function of two variables is from $\mathbb{R}^2$ to $\mathbb{R}$ and indicate this relationship in symbols by

$$f : \mathbb{R}^2 \to \mathbb{R}$$

If the domain A consists of ordered triples (x, y, z), then we call f a **real-valued function of three real variables** (or more simply, a function of three variables) and we write

$$f : \mathbb{R}^3 \to \mathbb{R}$$

The extension to a higher number of variables should be clear. To give one simple example of a function of many variables, the function f defined by

$$f(x_1, x_2, \ldots, x_n) = \frac{x_1 + x_2 + \cdots + x_n}{n}$$

is called the *mean* of the variables $x_1, x_2, \ldots, x_n$ and is commonly denoted by $\bar{x}$.

We might note that the vector-valued functions we studied in Chapter 12 assigned two-dimensional or three-dimensional vectors to each real number t in the domain. Since the coordinates of the tip of a position vector uniquely determine the vector, vector functions can be thought of as functions from $\mathbb{R}$ to $\mathbb{R}^2$ or $\mathbb{R}^3$. In this sense, then, the functions we are studying now reverse the roles of the domain and range of vector-valued functions.

The domain of a function of two variables is a subset of $\mathbb{R}^2$—that is, it is a subset of the xy-plane. In the next two examples we illustrate how to find domains and how to show them graphically.

EXAMPLE 13.1   Let $f(x, y) = \sqrt{x - y}$. Find the domain of f and show it graphically. Show that the point $(3, -1)$ lies in this domain, and find the value of $f(3, -1)$.

Solution   In order that $\sqrt{x - y}$ be a real number, we must have $x - y \geq 0$, or equivalently, $y \leq x$. The set of points satisfying this inequality are those that

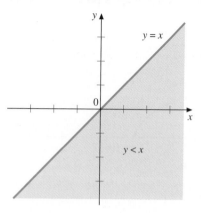

FIGURE 13.3
The domain of $f(x, y) = \sqrt{x - y}$

lie either *on* the line $y = x$ or *below* this line. This set is the shaded region in Figure 13.3.

Since $-1 < 3$, the point $(3, -1)$ satisfies the inequality $y < x$ and so lies in the domain of f. Furthermore,

$$f(3, -1) = \sqrt{3 - (-1)} = \sqrt{4} = 2$$ ∎

EXAMPLE 13.2   Find the domain of each of the following functions and show it graphically.

(a) $f(x, y) = \dfrac{1}{\sqrt{1 - x^2 - y^2}}$ (b) $f(x, y) = \ln(y - x^2)$

Solution

(a) For $\sqrt{1 - x^2 - y^2}$ to be real, x and y must satisfy $1 - x^2 - y^2 \geq 0$. All values of x and y for which $1 - x^2 - y^2 = 0$ must be ruled out, since they would make the denominator zero. Thus, the domain of f is the set of all points (x, y) for which $1 - x^2 - y^2 > 0$, or equivalently, $x^2 + y^2 < 1$. A point (x, y) satisfies this inequality if its distance from the origin is less than 1. So the domain consists of all points (x, y) *inside* the unit circle $x^2 + y^2 = 1$. We show this set in Figure 13.4. We indicate the bounding unit circle with a broken line to signify that it is not a part of the domain.

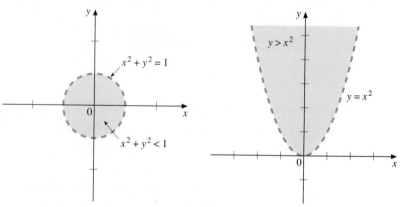

FIGURE 13.4
The domain of $f(x, y) = \dfrac{1}{\sqrt{1 - x^2 - y^2}}$

FIGURE 13.5
The domain of $f(x, y) = \ln(y - x^2)$

(b) Since the natural logarithm is defined only for positive values, x and y must satisfy $y - x^2 > 0$, or equivalently, $y > x^2$. In order to show the set of points satisfying this inequality, we first draw the parabola $y = x^2$, again with a broken line since it is not included, and then shade the region *above* this parabola. We show the resulting region in Figure 13.5. ∎

Graphs of Functions

In Figure 13.2 we indicated by an arrow from the xy-plane to the real line how a function of two variables associates a point (x, y) with a real number

$z = f(x, y)$. A more useful way to picture a function is by its graph, defined as follows:

Definition 13.2 The Graph of a Function of Two Variables	The **graph of a function f of two variables** is the set of all points (x, y, z) in $\mathbb{R}^3$ for which $z = f(x, y)$.

In general, the graph of a function of two variables is called a **surface**. We show such a surface in Figure 13.6.

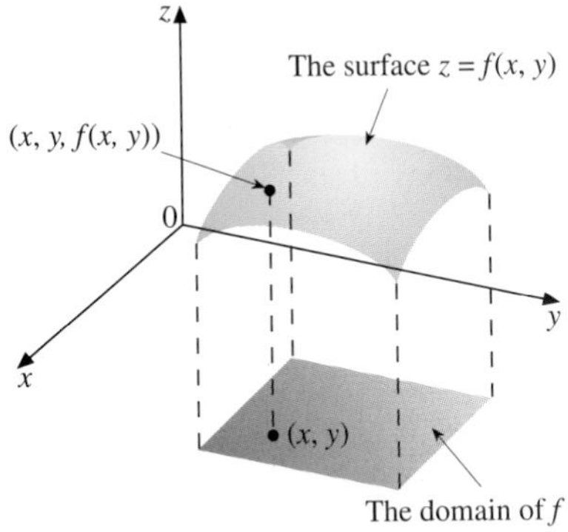

The surface $z = f(x, y)$

$(x, y, f(x, y))$

(x, y)

The domain of f

FIGURE 13.6

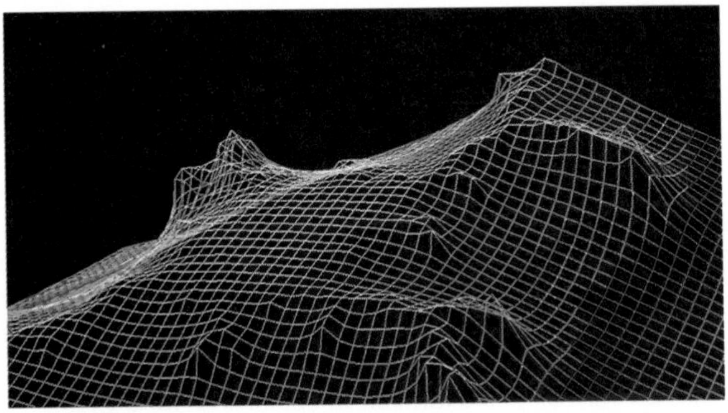

By plotting many points, a computer can generate a reasonably accurate description of a surface.

EXAMPLE 13.3   Let
$$f(x, y) = 6 - x - 2y$$
Show the portion of the graph of f that lies in the first octant.

Solution   Let $z = f(x, y)$. Then by Definition 13.2, the graph of f is the set of all points (x, y, z) satisfying $z = 6 - x - 2y$. From Chapter 11, we recognize this equation as being the equation of a plane. Its intercepts on the x-axis, y-axis, and z-axis are $x = 6$, $y = 3$, and $z = 6$, respectively. (We find the intercepts by setting two of the variables equal to 0 and solving for the other.) The traces of the plane on the coordinate planes are the lines connecting the intercepts. We show the graph in Figure 13.7.   ■

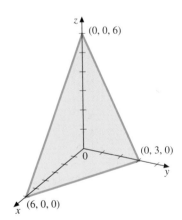

FIGURE 13.7
First-octant portion of the graph of $f(x, y) = 6 - x - 2y$

EXAMPLE 13.4   Identify and draw the graph of the function f defined by
$$f(x, y) = \sqrt{4 - x^2 - y^2}$$

Solution   The domain of f is the set of all points (x, y) satisfying $4 - x^2 - y^2 \geq 0$, or equivalently,
$$x^2 + y^2 \leq 4$$
We recognize this set as the circle $x^2 + y^2 = 4$, of radius 2, centered at the origin, together with all points inside the circle. To find the graph of f, let $z = f(x, y)$. Then $z = \sqrt{4 - x^2 - y^2}$. Note that $z \geq 0$ for all (x, y) in the domain. By squaring the equation and rearranging, we get
$$x^2 + y^2 + z^2 = 4$$

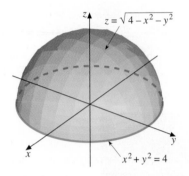

FIGURE 13.8
Graph of $f(x, y) = \sqrt{4 - x^2 - y^2}$ is a hemisphere of radius 3.

The left-hand side of the equation is the square of the distance of the point (x, y, z) from the origin. Hence, the equation is satisfied by points (x, y, z) if, and only if, the distance from the origin is 2. In other words, the graph is given by points on the *sphere* of radius 2 centered at the origin. For our function, $z \geq 0$. Hence, its graph is the upper half of the sphere, as we show in Figure 13.8.   ■

Most surfaces are difficult to draw by hand. In the next section we will discuss certain special surfaces that *can* be drawn by hand, and we will show how to obtain graphs of functions based on these special surfaces.

In the next example we show some graphs generated by a computer algebra system.

EXAMPLE 13.5   Use a CAS to obtain the graphs of each of the following functions.

(a) $f(x, y) = x^2 + \dfrac{y^2}{2}$ (b) $f(x, y) = \dfrac{1}{\sqrt{1 - x^2 - y^2}}$

(c) $f(x, y) = \ln(y - x^2)$ (d) $f(x, y) = \dfrac{\sin xy}{x^2 + y^2}$

Solution   We show the graphs in Figure 13.9. The surface in part (a) is called an *elliptical paraboloid* and is one of the special types we will consider in the next section. The functions in parts (b) and (c) are those for which we found the domains in Example 13.2. The graphs in parts (b), (c), and (d) would be difficult to obtain without a CAS.   ■

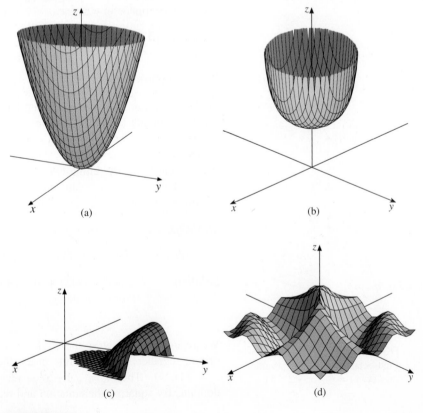

FIGURE 13.9

Level Curves and Contour Maps

Another way to gain insight into the nature of a function of two variables is by means of its *level curves,* defined as follows:

Definition 13.3 Level Curves and Contour Maps	A **level curve** for a function f of two variables is the graph of the equation $f(x, y) = c$ in the xy-plane, where c is any constant value in the range of f. A collection of level curves for different values of c is called a **contour map** for f.

FIGURE 13.10

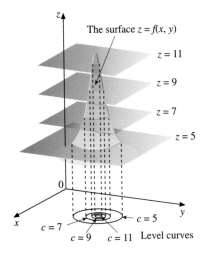

FIGURE 13.11

An isotherm on a weather map is a level curve for the temperature function. Along each such curve the temperature is constant. The collection of isotherms on the weather map is a contour map of the temperature function.

Another example of a contour map is a *topographical map* such as the one shown in Figure 13.10. The level curves are the curves of constant elevation. This topographical map is of Mount Shasta and the surrounding region, in California. To help understand the meaning of a level curve, consider the curve marked 13,800. Imagine a horizontal plane 13,800 feet above sea level, slicing through the mountain. The curve of intersection of the plane with the surface of the mountain is called a *trace* of the mountain surface on the plane. If you walked around the mountain on this trace, you would always be at the 13,800-foot level. The corresponding level curve on the topographical map is the *projection* of this trace on the xy-plane.

We illustrate these ideas further in Figure 13.11. There we show the planes $z = 5$, $z = 7$, $z = 9$, and $z = 11$, cutting the surface $z = f(x, y)$, along with the traces on these planes. The contour map in the xy-plane is the collection of projections of these traces. Note that with a constant difference in values of c (in this case a difference of 2), the closer together the level curves are, the steeper is the graph of the function.

Now suppose we were given only the contour map in Figure 13.11 and wanted to try to determine the corresponding surface. We could raise each curve $f(x, y) = c$ up to the height $z = c$ (or if c were negative, lower it to $z = c$). Then we could sketch the surface with these traces as guides. An experienced reader of topographical maps, for example, can visualize the general shape of the landscape in the manner described.

EXAMPLE 13.6   Let $f(x, y) = x^2 + (y^2/2)$. Draw a contour map showing the level curves $f(x, y) = c$ for $c = 2$, 4, 6, and 8.

Solution   We showed the CAS-generated graph of f in Figure 13.9(a). The equations of the level curves are

$$x^2 + \frac{y^2}{2} = 2, \quad \text{or} \quad \frac{x^2}{2} + \frac{y^2}{4} = 1$$

$$x^2 + \frac{y^2}{2} = 4, \quad \text{or} \quad \frac{x^2}{4} + \frac{y^2}{8} = 1$$

$$x^2 + \frac{y^2}{2} = 6, \quad \text{or} \quad \frac{x^2}{6} + \frac{y^2}{12} = 1$$

$$x^2 + \frac{y^2}{2} = 8, \quad \text{or} \quad \frac{x^2}{8} + \frac{y^2}{16} = 1$$

Each curve is an ellipse with major axis along the y-axis. In Figure 13.12(a) we repeat the graph of f (a paraboloid), and in Figure 13.12(b) we show the level curves. ∎

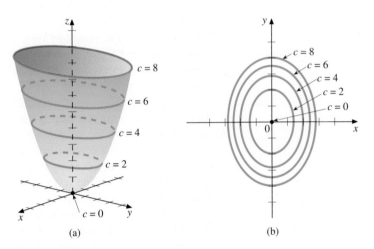

(a)

(b)

FIGURE 13.12

EXAMPLE 13.7   Draw the level curves $f(x, y) = c$ for $c = \pm 1, \pm 3$, and ± 5, for the function $f(x, y) = y^2 - x^2$. From the contour map obtained, sketch the graph of f.

Solution   For $c > 0$, the curves $y^2 - x^2 = c$ are hyperbolas with transverse axis along the y-axis. As c increases, the vertices move away from the origin. For $c < 0$, the level curves $y^2 - x^2 = c$ are hyperbolas with transverse axis along the x-axis. For example, with $c = -1$, we have

$$y^2 - x^2 = -1$$

or equivalently,

$$x^2 - y^2 = 1$$

We show the contour map in Figure 13.13(a).

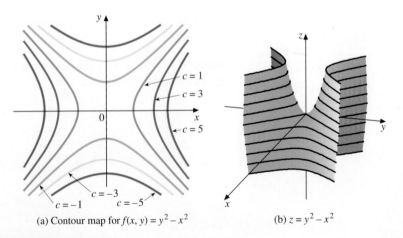

(a) Contour map for $f(x, y) = y^2 - x^2$

(b) $z = y^2 - x^2$

FIGURE 13.13

If we imagine the horizontal planes $z = c$ cutting the surface $z = f(x, y)$ at each of the given values of c, we can form a mental image of the surface. Above the xy-plane, the traces are hyperbolas with transverse axes parallel to the y-axis, with vertices moving away from the z-axis as c increases. Below the xy-plane the traces are hyperbolas with transverse axes parallel to the x-axis, with vertices again moving away from the z-axis for larger absolute values of c. The resulting surface is shown in Figure 13.13(b). It is called a **hyperbolic paraboloid** and is one of the special surfaces we will consider in the next section. ∎

Functions of Three Variables

When we go from functions of two variables to functions of three variables, the notion of a graph becomes more complicated. Although we can say that the graph of $w = f(x, y, z)$ is the set of "points" (x, y, z, w) in four-dimensional space for which w satisfies the given equation, it is impossible to draw such a graph. However, the extension of level curves to three dimensions is possible. In three dimensions, instead of level curves, we have the **level surfaces** $f(x, y, z) = c$ for constants c in the range of f. As an example, consider the function defined by

$$f(x, y, z) = \sqrt{16 - x^2 - y^2 - z^2}$$

The level surfaces $f(x, y, z) = c$ for $0 \le c \le 4$ are the spheres

$$x^2 + y^2 + z^2 = 16 - c^2$$

The function assumes its smallest value, 0, on the sphere of radius 4, and as c increases, the spheres on which $f(x, y, z) = c$ shrink in size, finally contracting to the single point $(0, 0, 0)$ when $c = 4$, which is the maximum value the function assumes.

SKETCHING SURFACES USING COMPUTER ALGEBRA SYSTEMS

Being able to make a mental image of a surface can make all the difference in understanding the essential nature of a problem. As we have seen, sketching surfaces, even in the simplest cases, can be very difficult. Computer algebra systems have the capability of quickly rendering accurate sketches of most surfaces we will encounter and many that we could not possibly envision without the aid of technology. Plotting two-dimensional projections of three-dimensional surfaces is one of the most interesting capabilities of computer algebra systems.

In this section, we will demonstrate how Maple and Mathematica sketch surfaces. The output displayed will be from the Maple system.

CAS **CAS 47**

Sketch the surface $z = f(x, y) = e^{-x^2 - y^2}$.

Maple:

plot3d(exp(–x^2–y^2),x=–2..2,y=–2..2,style=patch,
  axes=boxed,orientation=[45,57]);

Mathematica:

Plot3D[Exp[–x^2–y^2],{x,–2,2},{y,–2,2}]

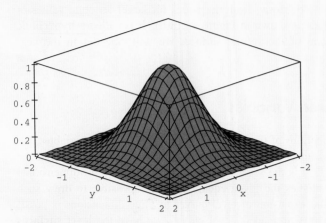

FIGURE 13.1.1

CAS 48

Sketch the surface $z = f(x, y) = \dfrac{-4x}{x^2 + y^2 + 2}$.

Maple:

plot3d(–4*x/(x^2+y^2+2),x=–10..10,y=–10..10,
  grid=[50,50],style=patch,axes=normal,
  tickmarks=[0,0,0],orientation=[14,54]);

Mathematica:

Plot3D[–4*x/(x^2+y^2+2),{x,–10,10},{y,–10,10}]

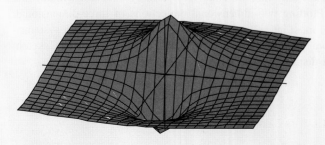

FIGURE 13.1.2

Maple:

Some insight into the nature of a surface can be obtained from a contour map that is a collection of level curves of the form $f(x, y) = c$. In Maple enter:

```
contourplot(-4*x/(x^2+y^2+2),x=-10..10,y=-10..10,
    numpoints=2000,axes=normal,tickmarks=[0,0,0],
    orientation=[90,0],scaling=constrained);
```

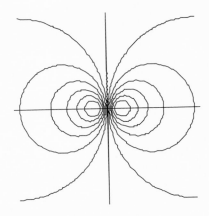

FIGURE 13.1.3

 CAS 49

Sketch the surface $z = f(x, y) = \sin x + \sin y$.

Maple:

```
plot3d(sin(x)+sin(y),x=-4*Pi..4*Pi,y=-4*Pi..4*Pi,
    numpoints=1500,style=patch,axes=none,
    scaling=constrained);
```

Mathematica:

```
Plot3D[Sin[x]+Sin[y],{x,-4*Pi,4*Pi},
    {y,-4*Pi,4*Pi},PlotPoints->40]
```

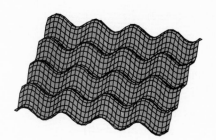

FIGURE 13.1.4

 CAS 50

Sketch the sphere of radius 1 and center $(1, 2, 1)$ given by

$$x^2 - 2x + y^2 - 4y + z^2 - 2z = -5$$

The surface this time is given implicitly by the equation involving the three variables x, y, and z. It is often easiest to represent such a surface in parametric form when using a CAS to generate the plot. However, many CAS include the capability of generating implicit plots.

Maple:

implicitplot3d(x^2–2*x+y^2–4*y+z^2–2*z=–5,
  x=0..2,y=1..3,z=0..3,view=[–1..3,0..4,0..3],
  scaling=constrained);

Mathematica:

ParametricPlot3D[{1+(Cos[t])*Cos[u],2+(Sin[t])*Cos[u],
  1+Sin[u]},{t,0,2*Pi},{u,–Pi/2,Pi/2}]

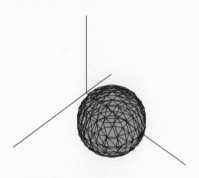

FIGURE 13.1.5

CAS 51

In Chapter 6 we considered finding volumes of solids of revolution. Computer algebra systems can be used to generate the solids of revolution using 3D plot capabilities.

(a) Generate the solid of revolution when the region bounded by $y = x + 4$ and $y = x^2/2$ is rotated about the y-axis.

(b) Generate the solid of revolution when the region bounded by $y = x^2 + 1$ and $\sqrt{x}$ is rotated about the x-axis.

Maple:

(a)

f:=x–>(x+4)–(x^2/2);
First plot the region in the plane:
plot({x+4,x^2/2},x=–5..5,y=0..8,scaling=constrained);

Now rotate the region about the y-axis:

plot3d([x*cos(t),x*sin(t),f(x)],x=–2..4,t=0..2*Pi,
  axes=normal,tickmarks=[0,0,0],labels=['x','y','z'],
  scaling=constrained);

Now change the orientation of the surface to view the inside:*

plot3d([x*cos(t),x*sin(t),f(x)],x=–2..4,t=0..2*Pi,orientation=
  [62,131],axes=normal,tickmarks=[0,0,0],
  scaling=constrained);

Mathematica:

(a)

f[x_] = (x+4)–(x^2/2)

Plot[{x+4,x^2/2},{x,–5,5},PlotRange–>{0,8},
  AspectRatio–>Automatic]

ParametricPlot3D[{x*Cos[t],x*Sin[t],f[x]},{x,–2,4},
  {t,0,2*Pi}]

ParametricPlot3D[{x*Cos[t],x*Sin[t],f[x]},{x,–2,4},
  {t,0,2*Pi},ViewPoint–>{–2,2,–2}]

---

*In Maple, the **orientation** command changes the view of the plot, and specifying the scaling to be constrained makes the aspect ratios of the axes the same.

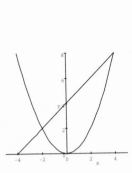

FIGURE 13.1.6a

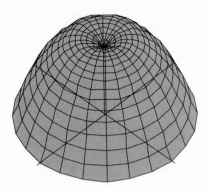

FIGURE 13.1.6b

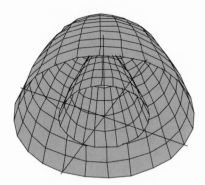

FIGURE 13.1.6c

Maple:

(b)

```
f:=x–>x^2+1–sqrt(x);
plot({x^2+1,sqrt(x),[1,x,x=0..2]},x=0..1,y=0..2,
    xtickmarks=2,ytickmarks=2,scaling=constrained);
plot3d([x,f(x)*cos(t),f(x)*sin(t)],x=0..1,t=0..2*Pi,
    axes=normal,tickmarks=[0,0,0],orientation=[69,81],
    scaling=constrained);
```

Mathematica:

(b)

```
f[x_] = x^2+1–Sqrt[x]

Plot[{x^2+1,Sqrt[x]},{x,0,1},PlotRange–>{0,2},
    AspectRatio–>Automatic]

ParametricPlot3D[{x,f[x]*Cos[t],f[x]*Sin[t]},{x,0,1},
    {t,0,2*Pi}]
```

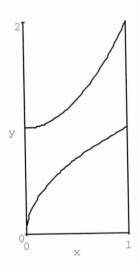

FIGURE 13.1.7a

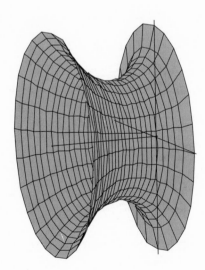

FIGURE 13.1.7b

Exercise Set 13.1

1. If $f(x, y) = \dfrac{2xy - y^2}{x^2 + 3xy}$, find

  (a) $f(-1, 1)$ (b) $f(1, 0)$

  (c) $f(2, 3)$ (d) $f(a, a); a \neq 0$

2. If $g(x, y) = 2 \tan^{-1} \dfrac{y}{x} - \sin^{-1} \dfrac{x}{\sqrt{x^2 + y^2}}$, find

  (a) $g(-1, 0)$ (b) $g(1, -1)$

  (c) $g(-1, \sqrt{3})$ (d) $g(-3, \sqrt{3})$

3. If $f(x, y, z) = \sqrt{12 - x^2 - y^2 - z^2}$, find

  (a) $f(1, -3, 1)$ (b) $f(-1, 1, -1)$

  (c) $f(2, 0, -2)$ (d) $f(-3, 1, \sqrt{2})$

4. If $f(x, y) = \ln(2x + y)$, find

  (a) $f(-1, 3)$ (b) $f(0, e^2)$

  (c) $f(x + h, y)$ (d) $f(x, y + k)$

5. If $g(x, y) = \dfrac{x^2 - y}{x - y^2}$, find

  (a) $g(-1, 1)$ (b) $g\left(a, \dfrac{1}{a}\right); a \neq 0$

  (c) $g(x + \Delta x, y)$ (d) $g(x, y + \Delta y)$

6. If $f(x, y) = x^2 - 2xy + 3y^2$, find each of the following and simplify:

  (a) $\dfrac{f(x + h, y) - f(x, y)}{h}; h \neq 0$

  (b) $\dfrac{f(x, y + k) - f(x, y)}{k}; k \neq 0$

7. If $f(x, y) = \dfrac{x + y}{x - y}$, find each of the following and simplify:

  (a) $\dfrac{f(-2 + h, 1) - f(-2, 1)}{h}; h \neq 0$

  (b) $\dfrac{f(-2, 1 + k) - f(-2, 1)}{k}; k \neq 0$

8. Let $f(x, y) = e^{x-y} \sin(x + y)$, $g(t) = 3t$, and $h(t) = t$. Find $f(g(t), h(t))$.

9. Let $f(x, y) = \dfrac{2xy}{x^2 - y^2} + \dfrac{1}{x^2 + y^2}$, $g(t) = \cos t$, and $h(t) = \sin t$. Find $f(g(t), h(t))$.

10. Let $f(x, y) = xy - \dfrac{x}{y}$, $g(u, v) = uv$, and $h(u, v) = \dfrac{u}{v}$. Find $f(g(u, v), h(u, v))$.

In Exercises 11–21, find the domain of f and show it graphically.

11. (a) $f(x, y) = \dfrac{1}{x^2 - y}$ (b) $f(x, y) = \dfrac{1}{x^2 - y^2}$

12. $f(x, y) = \sqrt{x + y}$ **13.** $f(x, y) = \sqrt{xy}$

14. $f(x, y) = \ln(2x - y)$ **15.** $f(x, y) = \dfrac{xy}{x^2 + y^2}$

16. $f(x, y) = \sqrt{x + y} - \sqrt{x - y}$

17. $f(x, y) = \ln(x^2 - y^2)$ **18.** $f(x, y) = \dfrac{x + y}{\sqrt{x^2 + y^2 - 1}}$

19. $f(x, y) = \ln \sinh(x^2 - 2y)$

20. $f(x, y) = \ln \left(\dfrac{2x + y}{2x - y} \right)$ **21.** $f(x, y) = \ln \sqrt{xy}$

In Exercises 22–25, give the domain of f and describe it geometrically.

22. $f(x, y, z) = \dfrac{\sin(xyz)}{\sqrt{x^2 + y^2 + z^2}}$

23. $f(x, y, z) = \dfrac{2x - 3y + z}{(x - 1)(y + 2)(z - 3)}$

24. $f(x, y, z) = \sqrt{xyz}$

25. $f(x, y, z) = e^{-(x^2+y^2)} \ln(x + y - z)$

26. Let $f(x, y) = \sqrt{4x + 3y}$. Show that if $h \neq 0$,

$$\frac{f(1 + h, 4) - f(1, 4)}{h} = \frac{2}{2 + \sqrt{h + 4}}$$

27. Let $f(x, y) = x^2 - 3xy + 2y^2$. Show that

$$f(x + \Delta x, y + \Delta y) = f(x, y) + (2x - 3y)\Delta x$$
$$+ (4y - 3x)\Delta y + \varepsilon_1 \Delta x + \varepsilon_2 \Delta y$$

where $\varepsilon_1 = g(\Delta x, \Delta y)$ and $\varepsilon_2 = h(\Delta x, \Delta y)$. Give the explicit forms of $g(x, y)$ and $h(x, y)$.

In Exercises 28 and 29, find the domain of f and show it graphically.

28. $f(x, y) = \sqrt{\dfrac{x - 3y}{x - y}}$

29. $f(x, y) = \ln(1 - |x| - |y|)$

30. (a) Express the volume of a right circular cone as a function of its base radius r and its altitude h.

(b) Express the surface area of an open-top box as a function of its length l, width w, and depth d.

31. A water tank is to be constructed in the form of a right circular cylinder of radius r and altitude h, with the top in the form of a hemisphere. The hemispherical top costs twice as much per unit area as the lateral surface and bottom. Express the total cost as a function of r, h, and the price p per square unit for the lateral surface and bottom.

32. A company manufactures two types of washing machines: deluxe model A and standard model B. When the price of each A model is p dollars and the price of each B model is q dollars, x model-A machines and y model-B machines can be sold each week. These price functions (called *demand functions*) are found by experience to be approximated by the equations

$$p(x, y) = 600 - 0.4x - 0.2y$$

$$q(x, y) = 400 - 0.3x - 0.5y$$

(a) Find the weekly revenue $R(x, y)$ from producing x model-A machines and y model-B machines.

(b) If the weekly cost of producing x model-A machines and y model-B machines is $C(x, y) = 120x + 90y + 600$, find the profit function $P(x, y)$ for this weekly production.

CAS *In Exercises 33–36, use a CAS to obtain the graph of f.*

33. $f(x, y) = x^2 + y^2$

34. $f(x, y) = \sqrt{4 - x^2 - y^2}$

35. $f(x, y) = \sqrt{x^2 + y^2}$

36. $f(x, y) = x^2 - y^2$

In Exercises 37–47, draw a contour map with at least six level curves. Use both positive and negative values of c where appropriate.

37. $f(x, y) = 3x - 5y$

38. $f(x, y) = \dfrac{x}{y}$

39. $f(x, y) = x^2 - 2y$

40. $f(x, y) = ye^{-x}$

41. $f(x, y) = x^2 y$

42. $f(x, y) = y - \cos x$

43. $f(x, y) = (x^2 - 1)/y$

44. $f(x, y) = \sqrt{4 - x^2 - y^2}$

45. $f(x, y) = x^2 - y^2$

46. $f(x, y) = y - \ln x$

47. $f(x, y) = y - \sin \pi x$

48. Below are six contour maps labeled (a)–(f) and on the next page are six surfaces labeled (1)–(6). Match up each contour map with the correct surface.

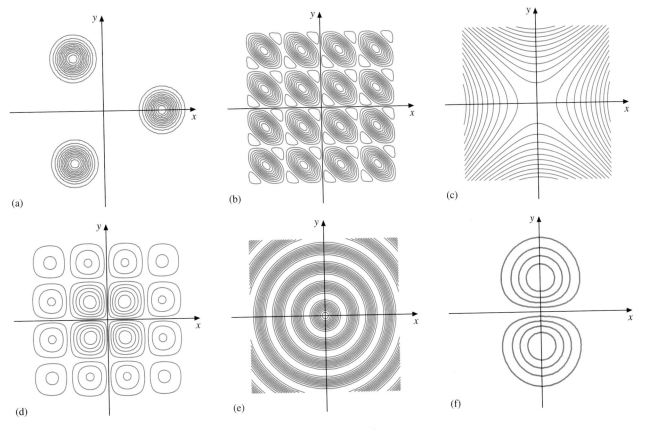

(a)   (b)   (c)   (d)   (e)   (f)

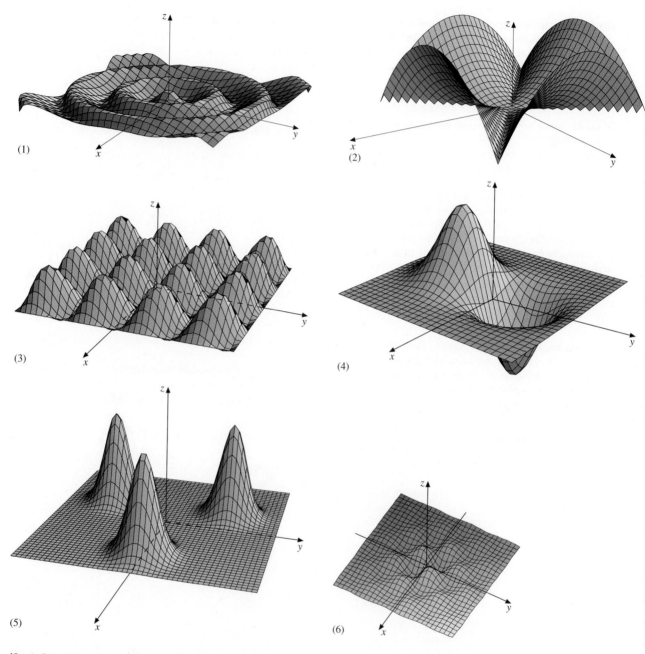

(1)

(2)

(3)

(4)

(5)

(6)

49. A flat plate in the xy-plane is heated from a point source at the origin. The temperature T in degrees Celsius at a point on the plate varies inversely as the square of its distance from the origin. Describe the isotherms. Suppose $T = 50$ at the point $(3, 4)$. Find all points at which $T = 30$.

50. The *ideal gas law* states that the temperature T of a gas, the volume V it occupies, and its pressure P are related by the equation $PV = kT$, where k is a constant. Draw several isothermal curves in the PV-plane and interpret the results. Express P as a function of T and V, and draw several isobars (lines of equal pressure) in the TV-plane. Interpret the results.

51. The speed of sound in an ideal gas is given by

$$v = \sqrt{\frac{kp}{d}}$$

where p is the pressure of the gas, d is its density, and k is a positive constant. Draw some level curves for v in the pd-plane. Solve the equation for p as a function of v and d, and draw some isobars in the vd-plane. (See Exercise 50.)

52. The accompanying figure shows a cross section of a circular cylinder lying on a horizontal plane. The cylinder and the plane are held at two different electric potentials. The electric potential in the shaded region is given by

$$V(x, y) = \frac{ky}{x^2 + y^2}$$

Draw several equipotentials (lines of equal potential) for this function.

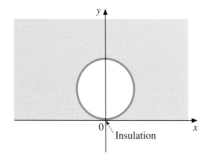

53. In hydrodynamics the level curves for the *stream function* ψ for two-dimensional fluid flow are called *streamlines*. A streamline is the path along which a given particle of the fluid moves. In the accompanying figure a fluid (such as water) flows from the negative x-axis toward the positive x-axis, with $y > 0$. There is a semicircular obstruction, as shown, centered at the origin. (You can think of the hump as a half-buried pipe at the bottom of a stream.) The stream function in this case is

$$\psi(x, y) = y - \frac{y}{x^2 + y^2}$$

Draw several streamlines. (*Hint:* Write the equations of the level curves in polar coordinates.)

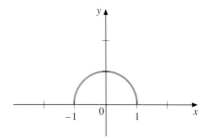

54. Plot the surface for the function $f(x, y) = \dfrac{x^2}{3} - \dfrac{y^2}{4}$ along with the horizontal planes $z = 0.5, z = 1, z = 2,$ and $z = 4$. Then generate a contour plot for the surface.

55. For each of the following surfaces, make a plot of the surface and make contour plots. In plotting the surfaces, be sure to determine a scale and orientation that shows the important information.

(a) $f(x, y) = \dfrac{x^2 y}{x^4 + y^2}$ (b) $f(x, y) = x^2 y e^{-x^2 - y^2}$

(c) $f(x, y) = \dfrac{-x}{x^2 + y^2 + 2}$ (d) $f(x, y) = \dfrac{100}{\sqrt{x^2 - y^2}}$

(e) $f(x, y) = \sin(x) + \cos(y)$

56. Plot level surfaces for each of the following functions of three variables.

(a) $f(x, y, z) = \dfrac{x^2 + y^2}{z}$

(b) $f(x, y, z) = z - x - \sin\sqrt{x^2 + y^2}$

13.2 SKETCHING SURFACES

In this section, we will illustrate some commonly occurring surfaces and give some guidelines for sketching them. Each of the surfaces is the *graph of an equation* in three variables, by which we mean the set of all points (x, y, z) whose coordinates satisfy the equation. For example, we have seen already (in Chapter 11) that the graph of every equation of the form $ax + by + cz + d = 0$ is a plane.

Not all the surfaces we consider are graphs of functions. In fact, we can identify which ones are functions by the **Vertical Line Test**: *for a surface to be the graph of a function, each vertical line must intersect the surface in at most one point.* This test follows from the fact that a function assigns one and only one value to each point in its domain.

One of the chief aids in sketching surfaces are **traces**. A trace of a surface on a plane is the curve of intersection of the surface with the plane. Of primary interest are traces on the coordinate planes and on planes parallel to the coordinate planes. To obtain the trace on the xy-plane, we set $z = 0$ in the equation of the surface. Similarly, setting $x = 0$ gives the yz-trace, and setting $y = 0$ gives the xz-trace. Traces on planes parallel to the coordinate planes are obtained by setting one of the variables equal to a constant. For example, if we set $z = c$, we get the trace on the plane $z = c$, parallel to the xy-plane.

Planes

First, we show planes parallel to the coordinate planes. For definiteness, we use $x = 3$, $y = 2$, and $z = 4$, as shown in Figure 13.14. For the plane $x = 3$ in Figure 13.14(a), we first mark the point three units out on the positive x-axis. Then from this point we draw a vertical line (the xz-trace), and a horizontal line (the xy-trace). In this case, there is no yz-trace because the plane is parallel to the yz-plane. Now we form a rectangle with portions of the two traces we have drawn as adjacent sides. The length and width of the rectangle are arbitrary. The region enclosed by the rectangle, of course, is only a part of the plane, but it is impossible to show the entire plane.

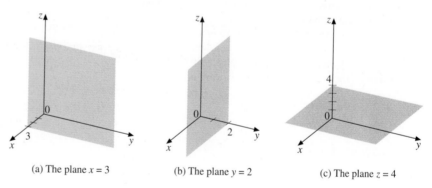

(a) The plane $x = 3$ (b) The plane $y = 2$ (c) The plane $z = 4$

FIGURE 13.14

The planes $y = 2$ and $z = 4$ are obtained in a similar way, and are shown in Figures 13.14(b) and (c), respectively.

For planes not parallel to any of the coordinate planes, we typically show a triangular portion of the plane bounded by the traces. The equations of the traces are found by setting just one variable at a time equal to 0. For example, consider the plane $2x + 3y + 4z = 12$. The equation of the xy-trace is the line $2x + 3y = 12$, that of the yz-trace is $3y + 4z = 12$, and that of the xz-trace is $2x + 4z = 12$, or equivalently, $x + 2y = 6$. A quick way to graph the traces is to connect the *intercepts* of the plane on the coordinate axes. The intercepts of the plane $2x + 3y + 4z = 12$ are found by setting two variables at a time equal to zero, giving the x-intercept 6, y-intercept 4, and z-intercept 3. The traces are the lines joining the intercepts. We show the portion of this plane bounded by the traces in Figure 13.15.

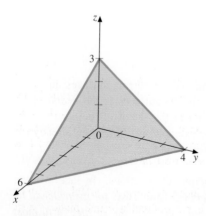

FIGURE 13.15
The plane $2x + 3y + 4z = 12$

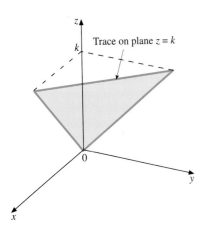

FIGURE 13.16
The plane $x + y - z = 0$

EXAMPLE 13.8   Sketch the following planes.
(a) $z = x + y$ (b) $4x + 3y = 12$

Solution

(a) When all three variables are set equal to zero, the equation is satisfied. Therefore, the plane passes through the origin. The xz-trace is the line $z = x$, and the yz-trace is the line $z = y$. Since the xy-trace is the single point at the origin, we find a trace on a plane parallel to the xy-plane to aid in sketching a part of the plane. To do so, we set $z = k$, for an arbitrary $k > 0$, giving $x + y = k$. Together, the traces we have found outline the triangular portion of the plane shown in Figure 13.16.

(b) The xy-trace is the line $4x + 3y = 12$, since setting $z = 0$ does not alter the equation. If the point (x, y) is on this line, then (x, y, z) is on the plane for any value of z. That is, the three-dimensional graph of $4x + 3y = 12$ is the plane formed by extending the line $4x + 3y = 12$ vertically. So the plane is parallel to the z-axis. The xz-trace is the vertical line $x = 4$, and the yz-trace is the vertical line $y = 4$. We again show a trace k units up on the plane $z = k$, with $k > 0$. The trace is simply a replica of the trace in the xy-plane in this case. We show the portion of the plane formed by these traces in Figure 13.17. ■

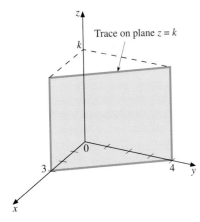

FIGURE 13.17
The plane $4x + 3y = 12$

REMARK
■ When you are sketching surfaces by hand, it is probably easiest to show the yz-plane in the plane of the paper, so that the yz-trace can be shown full-size, as with any other two-dimensional drawing. The positive x-axis should be shown at an angle of about 135° from the positive y-axis, and a unit of distance on it should be about two-thirds of a unit in the y and z directions. Following these conventions gives the illusion that the positive x-axis projects outward from the plane of the paper. For surfaces that are obtained using a CAS, the axes are oriented somewhat differently in order to show the surfaces from the most advantageous perspective.

Cylinders

A surface that is generated by a line moving along a plane curve C so that the line always remains parallel to some fixed line l not in or parallel to the plane of C is called a **cylindrical surface**, or more briefly, a **cylinder**. In Figure 13.18 we show such a surface, where the curve C is in the xy-plane and the line l is parallel to the z-axis. The use of the term *cylinder* here is more general than the usual notion of a cylinder in which the curve C is a circle and l is perpendicular to the plane of the circle (the correct name in this case is *right circular cylinder*).

We will restrict our consideration of cylinders to those for which the curve C is in one of the coordinate planes and l is perpendicular to that plane. Thus, l may be taken as one of the coordinate axes. In all such cases, the equation of the cylinder will simply be the equation of the curve C. As an example,

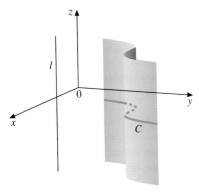

FIGURE 13.18
A cylinder generated by a line parallel to l moving along the curve C.

consider the equation $y = x^2$. Its graph in the xy-plane is a parabola, but as a three-dimensional surface, it is the **parabolic cylinder** shown in Figure 13.19, generated by a line parallel to the z-axis. To understand why, consider any point on the parabola in the xy-plane, say, $(2, 4, 0)$. When viewed as an equation in three variables, all points of the form $(2, 4, z)$ satisfy the equation, regardless of the value of z. Thus, the entire vertical line through the point $(2, 4, 0)$ lies on the surface. A similar result holds for all other points on the parabola. More generally, we have the following result.

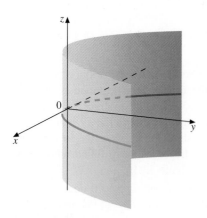

FIGURE 13.19
The parabolic cylinder $y = x^2$

Equations of Cylinders Parallel to a Coordinate Axis

If one of the variables x, y, or z is missing from an equation, then the graph of the equation is a cylinder generated by a line parallel to the axis of the missing variable.

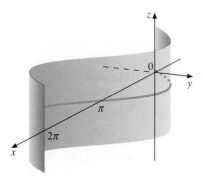

FIGURE 13.20
The cylinder $y = \sin x$

To sketch a cylinder of this type, first draw the two-dimensional curve in the plane of the variables present in the equation. Then extend the curve parallel to the axis of the missing variable, using traces where useful. We illustrate this technique in the next three examples.

EXAMPLE 13.9   Identify and sketch the graph of the equation $y = \sin x$ for $0 \le x \le \pi$.

Solution   Since z is missing from the equation, the graph is a cylinder with generating line parallel to the z-axis. To sketch it, we draw one cycle of the sine curve in the xy-plane and then extend it vertically. To give a better picture, we then show traces on two horizontal planes, one above and one below the xy-plane. These traces duplicate the original curve. We show the result in Figure 13.20. ∎

EXAMPLE 13.10   Identify and sketch the graph of the equation $9y^2 + 4z^2 = 36$.

Solution   In the yz-plane, the curve is an ellipse with standard form

$$\frac{y^2}{4} + \frac{z^2}{9} = 1$$

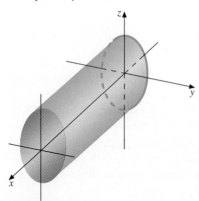

FIGURE 13.21
The elliptic cylinder
$9y^2 + 4z^2 = 36$

Since x is missing, the graph is a cylinder, called an **elliptical cylinder**, with generating line parallel to the x-axis. To sketch it, we first sketch the ellipse in the yz-plane. The y-intercepts are ± 2, and the z-intercepts are ± 3. Then we extend the ellipse in the x direction, showing a trace in a plane $x = k$ (where k is an arbitrary positive number). Again, the trace is a replica of the original curve. We show the sketch in Figure 13.21. ∎

EXAMPLE 13.11   Identify and sketch the graph of the equation $z^2 - x^2 = 4$.

Solution   In two dimensions, the graph is a hyperbola in the xz-plane. In three dimensions it is the cylinder sketched in Figure 13.22, generated by a line parallel to the y-axis. We use a trace in a plane perpendicular to the y-axis to aid in visualizing the surface. This cylinder is called a **hyperbolic cylinder**. ∎

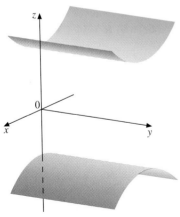

FIGURE 13.22
The hyperbolic cylinder $z^2 - x^2 = 4$

REMARK
■ Suppose you are given an equation such as $y = x^2$, involving only two variables. How do you know if the graph is a parabola or a parabolic cylinder? The answer is that without further information, there is no way to tell. The context is essential. The situation is even more uncertain with an equation such as $x = 4$. In one dimension the graph is a point, in two dimensions it is a vertical line, and in three dimensions it is a plane parallel to the yz-plane. *Throughout this chapter and the two that follow, all equations should be viewed as having graphs in three-dimensional space, unless otherwise specified.*

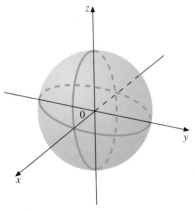

FIGURE 13.23
The sphere $x^2 + y^2 + z^2 = a^2$

Spheres

A sphere is the set of all points in $\mathbb{R}^3$ equidistant from a fixed point (the center). If the center is (h, k, l) and the common distance is a (that is, the radius is a), then a point (x, y, z) lies on the sphere if and only if

$$\sqrt{(x - h)^2 + (y - k)^2 + (z - l)^2} = a$$

or equivalently,

$$(x - h)^2 + (y - k)^2 + (z - l)^2 = a^2 \tag{13.1}$$

If the center is at the origin, the equation simplifies to

$$x^2 + y^2 + z^2 = a^2$$

We show the graph of a sphere with center at the origin in Figure 13.23. Each of the traces with the coordinate planes is a circle of radius a. By sketching these traces, using broken lines for the parts hidden from view, we get the figure. It is best to draw the yz-trace first, since it provides the outline of the sphere.

The next example shows how an equation of the form

$$x^2 + y^2 + z^2 + ax + by + cz + d = 0 \tag{13.2}$$

can, under certain conditions on the coefficients, be put in the form of Equation 13.1.

EXAMPLE 13.12   Discuss the graph of the equation

$$x^2 + y^2 + z^2 - 4x - 10y - 6z + k = 0$$

where
(a) $k = 34$ (b) $k = 38$ (c) $k = 42$

Solution

(a) We complete the squares on x, y, and z by adding appropriate constants to both sides:

$$(x^2 - 4x + 4) + (y^2 - 10y + 25) + (z^2 - 6z + 9) = -34 + 4 + 25 + 9$$

or

$$(x - 2)^2 + (y - 5)^2 + (z - 3)^2 = 4$$

This equation is in the form of Equation 13.1, so it represents a sphere with center $(2, 5, 3)$ and radius 2. We show a sketch of its graph in Figure 13.24.

(b) The left-hand side is the same as in part (a), but the right-hand side is 0:

$$(x - 2)^2 + (y - 5)^2 + (z - 3)^2 = 0$$

The only point satisfying this equation is $(2, 5, 3)$. This is an example of a *degenerate sphere*.

(c) Again, the only change is on the right-hand side. The result is

$$(x - 2)^2 + (y - 5)^2 + (z - 3)^2 = -4$$

Since the left-hand side cannot be negative, there is no graph. ∎

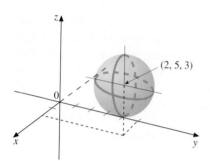

FIGURE 13.24
The sphere
$(x - 2)^2 + (y - 5)^2 + (z - 3)^2 = 4$

Quadric Surfaces

Equation 13.2 is a special case of the general second-degree equation in three variables,

$$Ax^2 + By^2 + Cz^2 + Dxy + Exz + Fyz + Gx + Hy + Iz + J = 0 \quad (13.3)$$

When the coefficients in Equation 13.3 are appropriately restricted, the graph is a sphere. Parabolic, elliptic, and hyperbolic cylinders also occur as special cases of Equation 13.3. For example, if $A = 1$, $H = -1$, and all other coefficients are 0, the equation reduces to $x^2 - y = 0$, or $y = x^2$, whose graph is the parabolic cylinder in Figure 13.19. There are six other possibilities (excluding degenerate cases) for graphs of Equation 13.3. (We are assuming that at least one of the coefficients A through F is nonzero, since otherwise the equation would not be of the second degree.) We will consider these other six types below. They, together with parabolic, elliptic, and hyperbolic cylinders, are known as **quadric surfaces**. Quadric surfaces can be thought of as three-dimensional analogues of the conic sections.

To describe the six quadric surfaces mentioned above, we write each equation in *standard form*. By rearranging, if necessary, you can verify that each equation is a special case of Equation 13.3. For simplicity, we take the coefficients D, E, and F to be 0. Later, we will show a computer-generated graph in which one of these coefficients is nonzero. It can be proved that when one or more of D, E, and F is nonzero, the graph, if it exists and is nondegenerate, is a rotation of one of the types we illustrate.

The equations we give are for surfaces conveniently placed with respect to the origin and the coordinate axes. When interchanges are made among the variables x, y, and z, the resulting surface is of the same type as illustrated but the orientation is changed. If x, y, and z are replaced by $x - h$, $y - k$, and $z - l$, respectively, the given surface is translated h units in the x direction, k units in the y direction, and l units in the z direction.

Ellipsoid

$$\frac{x^2}{a^2} + \frac{y^2}{b^2} + \frac{z^2}{c^2} = 1 \qquad (13.4)$$

xy-trace: ellipse $\dfrac{x^2}{a^2} + \dfrac{y^2}{b^2} = 1$

xz-trace: ellipse $\dfrac{x^2}{a^2} + \dfrac{z^2}{c^2} = 1$

yz-trace: ellipse $\dfrac{y^2}{b^2} + \dfrac{z^2}{c^2} = 1$

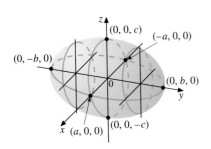

FIGURE 13.25
The ellipsoid $\dfrac{x^2}{a^2} + \dfrac{y^2}{b^2} + \dfrac{z^2}{c^2} = 1$

Traces on planes perpendicular to each coordinate axis between intercepts are ellipses. A special case of the ellipsoid is the sphere, in which $a = b = c$. The graph of Equation 13.4 is shown in Figure 13.25.

The key to recognizing an ellipsoid is that its standard equation involves the sum of the squares of all three variables.

Elliptic Paraboloid

$$\frac{x^2}{a^2} + \frac{y^2}{b^2} = cz \qquad (13.5)$$

We illustrate the case $c > 0$.

xy-trace: the origin

xz-trace: parabola $\dfrac{x^2}{a^2} = cz$

yz-trace: parabola $\dfrac{y^2}{b^2} = cz$

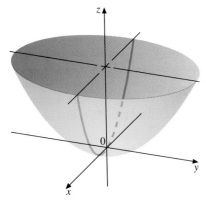

FIGURE 13.26
The paraboloid $\dfrac{x^2}{a^2} + \dfrac{y^2}{b^2} = cz$ $(c > 0)$

Traces on planes perpendicular to the positive z-axis are ellipses. For the given standard form, if $c > 0$, the paraboloid opens upward as in Figure 13.26, and if $c < 0$, it opens downward.

The key to recognizing an elliptic paraboloid is that its equation can be written so that one side involves the sum of the squares of two of the variables and the other side involves the first power of the third variable. The axis corresponds to the first-degree variable.

Elliptic Hyperboloid of One Sheet

$$\frac{x^2}{a^2} + \frac{y^2}{b^2} - \frac{z^2}{c^2} = 1 \qquad (13.6)$$

xy-trace: ellipse $\dfrac{x^2}{a^2} + \dfrac{y^2}{b^2} = 1$

xz-trace: hyperbola $\dfrac{x^2}{a^2} - \dfrac{z^2}{c^2} = 1$

yz-trace: hyperbola $\dfrac{y^2}{b^2} - \dfrac{z^2}{c^2} = 1$

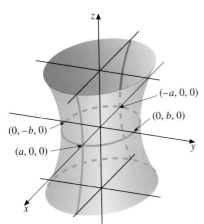

FIGURE 13.27
The hyperboloid of one sheet
$\dfrac{x^2}{a^2} + \dfrac{y^2}{b^2} - \dfrac{z^2}{c^2} = 1$

Traces on planes perpendicular to the z-axis are ellipses. We show the graph in Figure 13.27.

The key to recognizing the equation of a hyperboloid of one sheet is that the standard form involves the squares of all three variables, two with positive signs and one with a negative sign. The axis is that of the variable in the negative term.

Elliptic Hyperboloid of Two Sheets

$$\frac{z^2}{c^2} - \frac{x^2}{a^2} - \frac{y^2}{b^2} = 1 \tag{13.7}$$

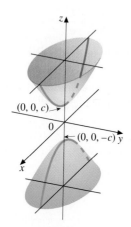

xy-trace:	none
xz-trace:	hyperbola $\dfrac{z^2}{c^2} - \dfrac{x^2}{a^2} = 1$
yz-trace:	hyperbola $\dfrac{z^2}{c^2} - \dfrac{y^2}{b^2} = 1$

For $|k| > c$, traces on planes $z = k$ are ellipses. We show the graph in Figure 13.28.

The key to recognizing the equation of a hyperboloid of two sheets is that the standard form involves the squares of all three variables, one with a positive sign and two with negative signs. The axis is that of the variable with a positive sign.

FIGURE 13.28
The hyperboloid of two sheets
$$\frac{z^2}{c^2} - \frac{x^2}{a^2} - \frac{y^2}{b^2} = 1$$

Elliptic Cone

$$\frac{x^2}{a^2} + \frac{y^2}{b^2} = \frac{z^2}{c^2} \tag{13.8}$$

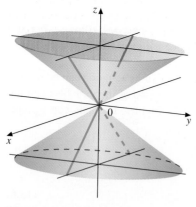

xy-trace:	origin
xz-trace:	two lines $\dfrac{x}{a} = \pm\dfrac{z}{c}$
yz-trace:	two lines $\dfrac{y}{b} = \pm\dfrac{z}{c}$

Traces on planes perpendicular to the z-axis are ellipses. We show the graph in Figure 13.29.

For the cone the equation can be written so that one side involves the sum of the squares of two of the variables and the other side involves the square of the third variable, with the axis that of the latter variable.

FIGURE 13.29
The cone $\dfrac{x^2}{a^2} + \dfrac{y^2}{b^2} = \dfrac{z^2}{c^2}$

REMARK

■ The upper and lower parts of the cone are called *nappes*. In customary usage when one refers to a cone (or to a right circular cone) only one nappe of the total cone is intended.

Hyperbolic Paraboloid

$$\frac{y^2}{b^2} - \frac{x^2}{a^2} = cz \qquad (13.9)$$

xy-trace: two lines $\dfrac{y}{b} = \pm\dfrac{x}{a}$

xz-trace: parabola $-\dfrac{x^2}{a^2} = cz$, opens downward

yz-trace: parabola $\dfrac{y^2}{b^2} = cz$, opens upward

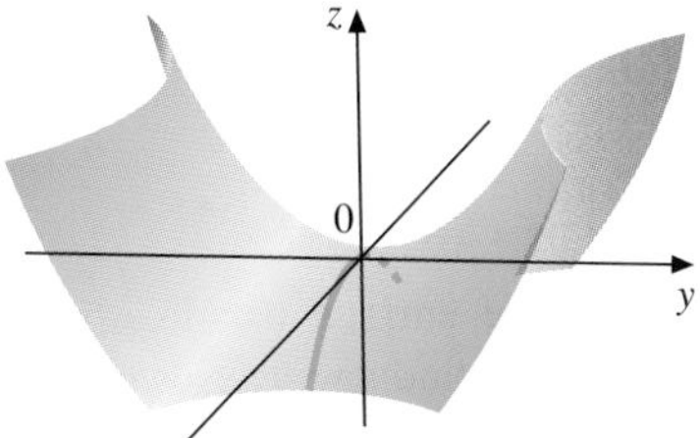

Traces on planes perpendicular to the z-axis are hyperbolas, and traces on planes perpendicular to the x-axis or y-axis are parabolas. We show a hyperbolic paraboloid with $c > 0$ in Figure 13.30.

The hyperbolic paraboloid has the appearance of a saddle near the origin. For this reason, the origin is called a **saddle point** of the surface.

The key to recognizing the hyperbolic paraboloid is that the first power of one of the variables in the standard equation equals the *difference* of the squares of the other two variables. In contrast, in the standard equation of the elliptic paraboloid (Figure 13.26), the first power of one of the variables equals the *sum* of the squares of the other two variables.

FIGURE 13.30
The hyperbolic paraboloid
$\dfrac{y^2}{b^2} - \dfrac{x^2}{a^2} = cz \quad (c > 0)$

REMARK

■ When traces on planes perpendicular to an axis of a surface are circular, the surface is called a *surface of revolution*, since it could be formed by revolving a plane curve about the axis. An ellipsoid of revolution can be recognized from its standard equation when any two of the numbers a, b, or c are equal (if all three are equal, the ellipsoid is a sphere). For the paraboloid, the hyperboloids, and the cone in the orientations shown in Figures 13.26 through 13.29, a surface of revolution results when $a = b$. (Obvious modifications are required when the axis is not the z-axis.) A cone of revolution is also called a *circular cone*.

EXAMPLE 13.13   Identify and sketch the graph of

$$9x^2 + 9y^2 + 4z^2 = 36$$

Solution   We divide both sides by 36 to obtain the standard form

$$\frac{x^2}{4} + \frac{y^2}{4} + \frac{z^2}{9} = 1$$

xy-trace: $\dfrac{x^2}{4} + \dfrac{y^2}{4} = 1$ or $x^2 + y^2 = 4$, circle

yz-trace: $\dfrac{y^2}{4} + \dfrac{z^2}{9} = 1$, ellipse

xz-trace: $\dfrac{x^2}{4} + \dfrac{z^2}{9} = 1$, ellipse

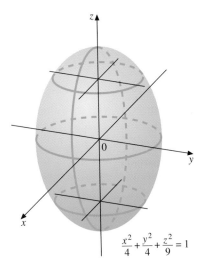

$\dfrac{x^2}{4} + \dfrac{y^2}{4} + \dfrac{z^2}{9} = 1$

FIGURE 13.31
Ellipsoid of revolution

The graph is the ellipsoid of revolution in Figure 13.31. Since the z-intercepts are ±3 and the x- and y-intercepts are ±2, the major axis of the ellipsoid is the z-axis.

EXAMPLE 13.14   Identify and sketch the surface whose equation is $4x^2 + z^2 = 4y$.

Solution   First, we divide by 4 to obtain the standard form

$$\frac{x^2}{1} + \frac{z^2}{4} = y$$

By comparison with Equation 13.5, we recognize the surface as an elliptic paraboloid with axis on the y-axis. To sketch it, we find the traces on the coordinate planes.

$$xy\text{-trace:} \quad x^2 = y, \text{ a parabola}$$

$$yz\text{-trace:} \quad z^2 = 4y, \text{ a parabola}$$

$$xz\text{-trace:} \quad \frac{x^2}{1} + \frac{z^2}{4} = 0, \text{ the origin}$$

After drawing the xy-trace and yz-trace, it appears that a trace on a plane perpendicular to the y-axis to the right of $y = 0$ would be helpful. If we let $y = k$, for an arbitrary positive k, we get

$$\frac{x^2}{1} + \frac{z^2}{4} = k$$

which we recognize as an ellipse. We can now sketch the graph as in Figure 13.32.   ∎

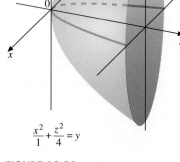

$$\frac{x^2}{1} + \frac{z^2}{4} = y$$

FIGURE 13.32
Elliptic paraboloid

EXAMPLE 13.15   Identify and sketch the graph of the equation

$$x^2 - 2y^2 - 3z^2 - 6 = 0$$

Solution   When we write the equation as

$$\frac{x^2}{6} - \frac{y^2}{3} - \frac{z^2}{2} = 1$$

we recognize it as representing a hyperboloid of two sheets with axis along the x-axis (compare with Equation 13.7).

$$xy\text{-trace:} \quad \frac{x^2}{6} - \frac{y^2}{3} = 1, \text{ hyperbola}$$

$$yz\text{-trace:} \quad \text{none}$$

$$xz\text{-trace:} \quad \frac{x^2}{6} - \frac{z^2}{2} = 1, \text{ hyperbola}$$

For $k > \sqrt{6}$, the trace on each of the planes $x = \pm k$ is the ellipse

$$\frac{y^2}{3} + \frac{z^2}{2} = \frac{k^2}{6} - 1$$

With these traces we obtain the sketch in Figure 13.33.   ∎

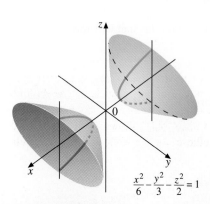

$$\frac{x^2}{6} - \frac{y^2}{3} - \frac{z^2}{2} = 1$$

FIGURE 13.33
Elliptic hyperboloid of two sheets

In the next two examples, we show how we can sometimes use knowledge of quadric surfaces to obtain graphs of functions of two variables.

EXAMPLE 13.16   Identify and sketch the graph of the function

$$f(x, y) = \sqrt{x^2 + y^2}$$

Solution   Let $z = f(x, y)$. If we square both sides, we get

$$z^2 = x^2 + y^2$$

which, by comparison with Equation 13.8, is a cone with axis on the z-axis. It is circular, since the coefficients of x^2 and y^2 are equal. The equation $z = f(x, y)$ represents the upper nappe of the cone only, since z is nonnegative. The traces are

xy-trace: $x^2 + y^2 = 0$, the origin

yz-trace: $z = \pm y$, two lines

xz-trace: $z = \pm x$, two lines

Traces on planes $z = \pm k$ are the circles $x^2 + y^2 = k^2$.
We can now sketch the graph (Figure 13.34).   ∎

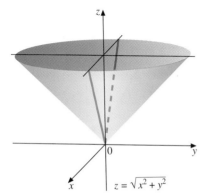

FIGURE 13.34
Upper nappe of circular cone

EXAMPLE 13.17   Identify and sketch the graph of the function

$$f(x, y) = \sqrt{1 + x^2 + y^2}$$

Give the domain and range of f.

Solution   Let $z = \sqrt{1 + x^2 + y^2}$. If we square both sides and rearrange, we get $z^2 - x^2 - y^2 = 1$. The graph of this latter equation is a hyperboloid of revolution of two sheets, with axis along the z-axis. Since $z > 0$, the graph of the original function is the upper sheet only, shown in Figure 13.35. The domain of f is all of $\mathbb{R}^2$, and the range is the set of z values for which $z \geq 1$.   ∎

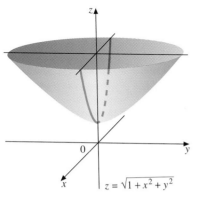

FIGURE 13.35
Upper sheet of hyperboloid of revolution of two sheets.

For reference, we summarize in Chart 13.1 the six quadric surfaces given by Equations 13.4 through 13.9. The figures shown were generated by a CAS and show many traces to help in visualizing the shapes.

CHART 13.1

Ellipsoid 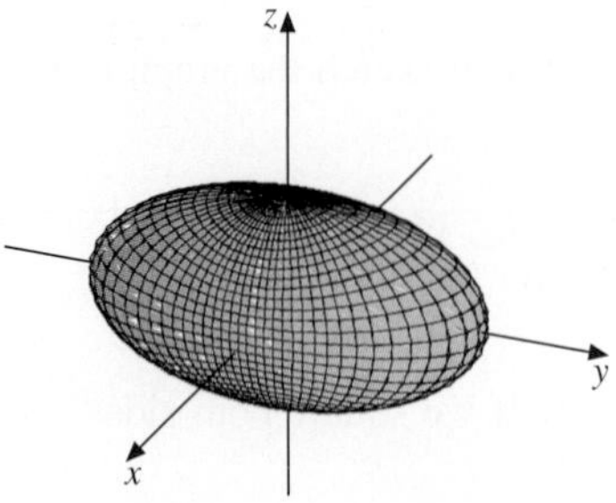	Elliptic Paraboloid 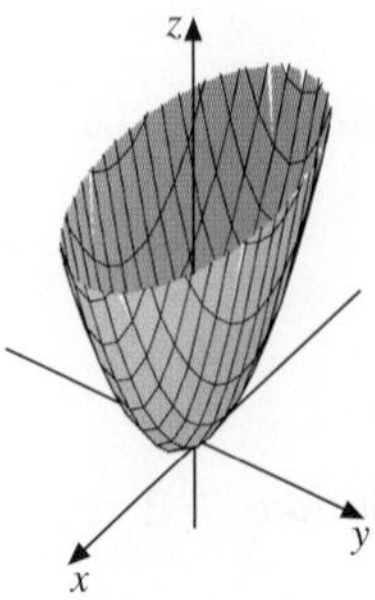	Elliptic Hyperboloid of One Sheet 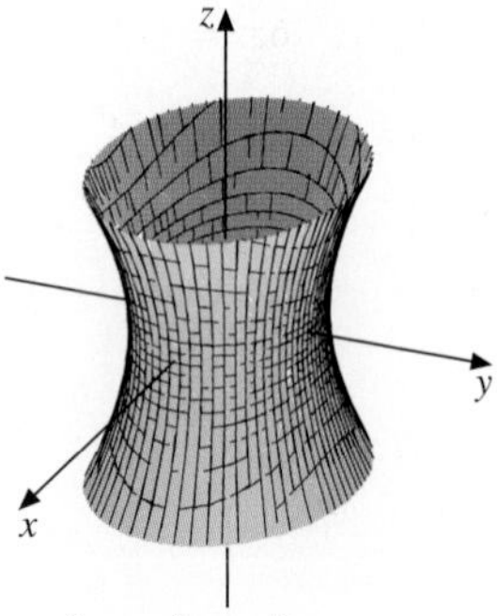

$$\frac{x^2}{a^2} + \frac{y^2}{b^2} + \frac{z^2}{c^2} = 1$$

$$\frac{x^2}{a^2} + \frac{y^2}{b^2} = cz$$

$$\frac{x^2}{a^2} + \frac{y^2}{b^2} - \frac{z^2}{c^2} = 1$$

Elliptic Hyperboloid of Two Sheets 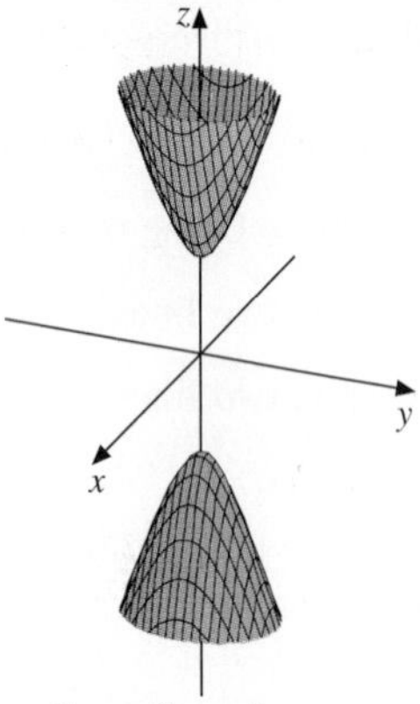	Elliptic Cone 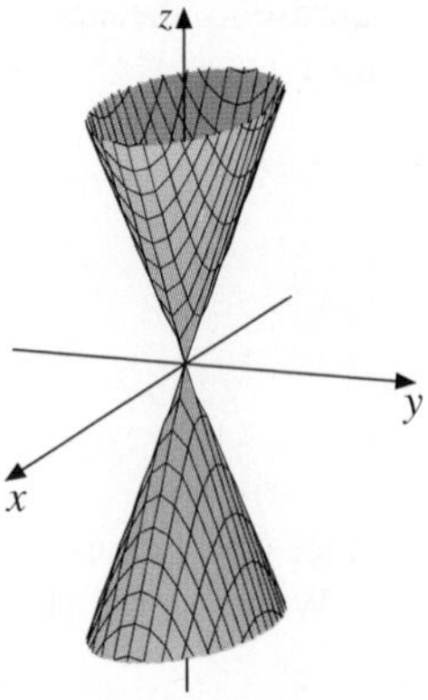	Hyperbolic Paraboloid 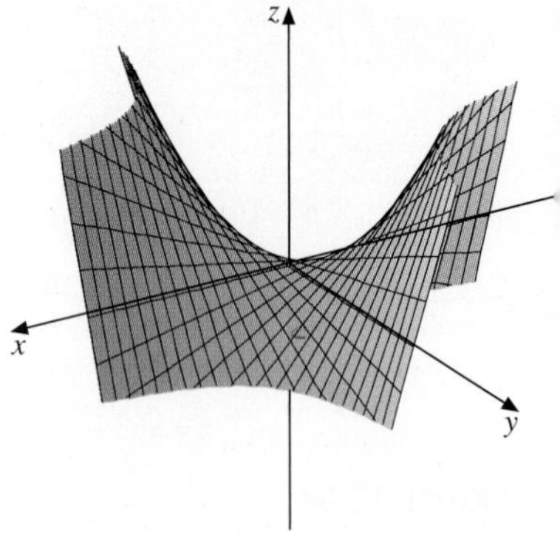

$$\frac{z^2}{c^2} - \frac{x^2}{a^2} - \frac{y^2}{b^2} = 1$$

$$\frac{x^2}{a^2} + \frac{y^2}{b^2} = \frac{z^2}{c^2} \ .$$

$$\frac{y^2}{b^2} - \frac{x^2}{a^2} = cz$$

Exercise Set 13.2

In Exercises 1–10, draw the graph of the given equation in three dimensions.

1. $y^2 = 2x$

2. $x^2 + y^2 = 4$

3. $y = 4 - x^2$

4. $z = \dfrac{x^2}{4}$

5. $z = \sqrt{4 - y^2}$

6. $4x^2 + 9y^2 = 36$

7. $x^2 - y^2 = 1$

8. $xy = 2$

9. $z = e^{-x}$

10. $z = \ln y$

In Exercises 11–15, determine whether the graph is a sphere or a degenerate sphere, or if there is no graph. If the graph is a sphere, give its center and radius and draw the graph.

11. $x^2 + y^2 + z^2 - 2x - 6y - 8z + 10 = 0$

12. $x^2 + y^2 + z^2 - 8y - 4z + 11 = 0$

13. $x^2 + y^2 + z^2 + 10x - 2y + 6z + 35 = 0$

14. $x^2 + y^2 + z^2 - 10x + 4y + 8z + 47 = 0$

15. $x^2 + y^2 + z^2 - 6x + 8y - 8z + 33 = 0$

In Exercises 16–24, identify the quadric surface and draw its graph, showing traces where useful.

16. $36x^2 + 9y^2 + 16z^2 = 144$

17. $36x^2 - 9y^2 + 16z^2 = 144$

18. $36x^2 - 9y^2 + 16z^2 + 144 = 0$

19. $36x^2 - 9y^2 + 16z^2 = 0$

20. $36z = 9x^2 + 4y^2$

21. $16x - 4y^2 - 9z^2 = 0$

22. $9x^2 - 4y^2 + 36z = 0$

23. $y = \sqrt{4x^2 + z^2}$

24. $3x^2 + 2z^2 - 6y = 0$

In Exercises 25–43, identify and draw the graph of f. State the domain and range.

25. $f(x, y) = 4 - x - y$ **26.** $f(x, y) = 4 - x$

27. $f(x, y) = x + 2y$ **28.** $f(x, y) = y^2$

29. $f(x, y) = \sqrt{x}$ **30.** $f(x, y) = \sqrt{4 - x^2}$

31. $f(x, y) = 9 - y^2$ **32.** $f(x, y) = \sqrt{1 + x^2}$

33. $f(x, y) = 2x^2 + y^2$ **34.** $f(x, y) = 4 - x^2 - y^2$

35. $f(x, y) = \sqrt{12 - 4x^2 + 3y^2}$

36. $f(x, y) = \sqrt{36 - 4x^2 - 9y^2}$

37. $f(x, y) = 1 + x^2 + y^2$ **38.** $f(x, y) = \sqrt{x^2 + y^2}$

39. $f(x, y) = \sqrt{x^2 + y^2 - 4}$ **40.** $f(x, y) = \sqrt{4 - x^2 + y^2}$

41. $f(x, y) = \sqrt{4y - x^2}$ **42.** $f(x, y) = \sqrt{4 + 4x^2 + y^2}$

43. $f(x, y) = y^2 - x^2$

In Exercises 44–49, make an appropriate translation to identify and draw the graph.

44. $25x^2 + 9y^2 + 15x^2 - 100x - 54y - 60z + 16 = 0$

45. $4x^2 + y^2 - 2z^2 - 8x - 4y - 8z + 8 = 0$

46. $9x^2 - 4y^2 + 9z^2 - 54x - 16y - 18z + 38 = 0$

47. $4x^2 + y^2 - 24x - 4y - 4z + 20 = 0$

48. $4x^2 + 3y^2 - z^2 - 32x - 12y + 2z + 75 = 0$

49. $2x^2 - 3y^2 - 8x - 12y + 12z - 52 = 0$

In Exercises 50–55, show the volume in the first octant bounded by the given surfaces.

50. $x^2 + z^2 = 4$, $y = x$, $y = 0$, $z = 0$

51. $z = x^2 + y^2$, $x + y + z = 4$

52. $x^2 + z^2 = 4$, $y^2 + z^2 = 4$, $x = 0$, $y = 0$, $z = 0$

53. $z = 4 - y^2$, $x^2 = 2y$, $x = 0$, $z = 0$

54. $z = 4 - x^2 - y^2$, $z^2 = x^2 + y^2$

55. $4x^2 + 2y^2 + 3z^2 = 48$, $y = 2x$, $y = 4$, $x = 0$, $z = 0$

13.3 LIMITS AND CONTINUITY

The concept of limit is as important in the context of functions of two, three, or even more variables as it is in the case of functions of a single variable. In particular, once we define the limit of a function of two variables, we will be able to discuss the slope of the tangent line to a curve on the surface, as well as rates of change in the function in different directions. The definition of the limit in this context is a natural extension of the limit concept for a function of one variable.

The Limit of a Function

In order to motivate the definition, let us begin with an example. Let

$$f(x, y) = \sqrt{1 - x^2 - y^2}$$

Setting $z = \sqrt{1 - x^2 - y^2}$ and squaring gives $z^2 = 1 - x^2 - y^2$, or equivalently, $x^2 + y^2 + z^2 = 1$. We recognize this equation as that of a sphere of radius 1, centered at the origin. For our original function, $z \geq 0$, so the graph is the upper hemisphere, shown in Figure 13.36. Suppose we now want to know what value $f(x, y)$ approaches as (x, y) approaches $(0, 0)$. That is, we want to know the limit

$$\lim_{(x,y) \to (0,0)} f(x, y)$$

We have not yet defined what such a limit means, but we can proceed intuitively.

Suppose we consider points in the domain of f that are close to $(0, 0)$. That is, we consider points (x, y) inside some small circle centered at the origin, as indicated in Figure 13.37(a). Then, as we see from Figure 13.37(b), the corresponding z values are close to 1. That is, $f(x, y)$ is close to 1.

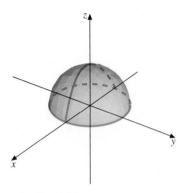

FIGURE 13.36
The graph of
$f(x, y) = \sqrt{1 - x^2 - y^2}$ is a hemisphere.

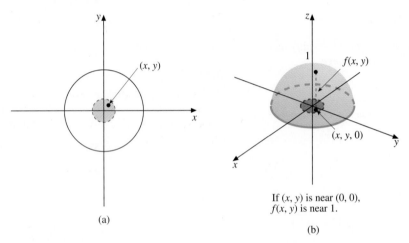

If (x, y) is near $(0, 0)$, $f(x, y)$ is near 1.

(a)

(b)

FIGURE 13.37

(Along $y = 0$)	
(x, y)	$f(x, y)$
$(\pm 0.5, 0)$	0.86603
$(\pm 0.1, 0)$	0.99499
$(\pm 0.01, 0)$	0.99995

(Along $x = 0$)	
(x, y)	$f(x, y)$
$(0, \pm 0.5)$	0.86603
$(0, \pm 0.1)$	0.99499
$(0, \pm 0.01)$	0.99995

(Along $y = x$)	
(x, y)	$f(x, y)$
$(\pm 0.5, \pm 0.5)$	0.70711
$(\pm 0.1, \pm 0.1)$	0.98995
$(\pm 0.01, \pm 0.01)$	0.99990

(Along $y = -x$)	
(x, y)	$f(x, y)$
$(\pm 0.5, \mp 0.5)$	0.70711
$(\pm 0.1, \mp 0.1)$	0.98995
$(\pm 0.01, \mp 0.01)$	0.9990

One major difference between the limit idea for a function of two (or more) variables and that for a function of one variable has to do with the manner of approach. In the case of

$$\lim_{x \to a} f(x) = L$$

there are only two possible ways for x to approach a—from the right and from the left. For the limit to exist, both one-sided limits have to exist and be the same value. In the case of the limit of a function of two variables, there are infinitely many ways (x, y) can approach the point in question. In the margin we show tables in which (x, y) approaches $(0, 0)$ along the paths $y = 0$ (the x-axis), $x = 0$ (the y-axis), the line $y = x$, and the line $y = -x$, respectively. In each case we show the corresponding values of $f(x, y) = \sqrt{1 - x^2 - y^2}$ (rounded to five places). Although these tables strongly suggest that the function is approaching 1 as (x, y) approaches $(0, 0)$, they do not provide a proof of this fact. We could never *prove* this limit is 1 by considering different paths of approach, because we could never consider all such paths.

The important thing is that for *all* points (x, y) sufficiently close to the origin, other than the origin itself, the function values must be as close to 1 as we please.

REMARK

■ Although we cannot prove that the limit of $f(x, y)$ exists as (x, y) approaches a point (x_0, y_0), by considering different paths of approach, we can prove the limit *does not* exist if we get different limiting values along two different paths. We will say more later about this method of showing the failure of the limit to exist.

Definition 13.4
Informal Limit Definition

Let f be a function of two variables whose domain includes all points inside some circle centered at (x_0, y_0), except possibly (x_0, y_0) itself. Then the **limit of $f(x, y)$ as (x, y) approaches (x_0, y_0) is L**, written

$$\lim_{(x,y)\to(x_0,y_0)} f(x, y) = L$$

means that $f(x, y)$ will be as close as we please to L for all points (x, y) sufficiently close to (x_0, y_0), but not equal to (x_0, y_0).

REMARK

■ The corresponding definition for the limit of a function of three (or more) variables should be clear. If f is a function of three variables defined inside some sphere centered at (x_0, y_0, z_0) except possibly at that point itself, then in Definition 13.4, we simply replace (x, y) by (x, y, z) and (x_0, y_0) by (x_0, y_0, z_0).

EXAMPLE 13.18   Find

$$\lim_{(x,y)\to(1,-2)} (8 - 2x^2 - y^2)$$

Solution   It seems intuitively evident that for x close to 1 and y close to -2, $f(x, y) = 8 - 2x^2 - y^2$ is close to $8 - 2(1)^2 - (-2)^2 = 2$. That is, by taking (x, y) close enough to $(1, -2)$, we can force $f(x, y)$ to be as close to 2 as we choose. Thus, by Definition 13.4, we conclude that

$$\lim_{(x,y)\to(1,-2)} (8 - 2x^2 - y^2) = 2$$

We show the graph of f in Figure 13.38. The surface is an elliptic paraboloid with vertex $(0, 0, 8)$. The trace on the xy-plane is the ellipse $2x^2 + y^2 = 8$, and the traces on the xz- and yz-planes are the parabolas $z = 8 - 2x^2$ and $z = 8 - y^2$, respectively.   ■

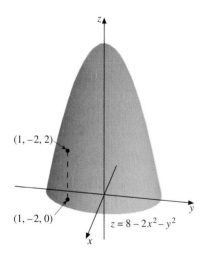

(1, -2, 2)

(1, -2, 0)

$z = 8 - 2x^2 - y^2$

FIGURE 13.38

EXAMPLE 13.19   Find the limit

$$\lim_{(x,y)\to(0,0)} \frac{x^4 - y^4}{x^2 + y^2}$$

or show that it fails to exist.

Solution   At $(0, 0)$ the function

$$f(x, y) = \frac{x^4 - y^4}{x^2 + y^2}$$

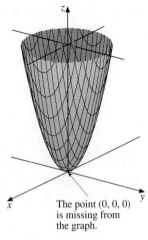

The point $(0, 0, 0)$ is missing from the graph.

FIGURE 13.39

$f(x, y) = \dfrac{x^4 - y^4}{x^2 + y^2} = x^2 - y^2$

if $(x, y) \neq (0, 0)$

is not defined. But it is defined for all other points. If we restrict (x, y) to be different from $(0, 0)$, we can factor the numerator and divide out the common factor to get

$$f(x, y) = \frac{(x^2 - y^2)(x^2 + y^2)}{x^2 + y^2} = x^2 - y^2 \qquad \text{if } (x, y) \neq (0, 0)$$

Since we are not concerned with the point $(0, 0)$ in finding the limit, we have

$$\lim_{(x,y)\to(0,0)} f(x, y) = \lim_{(x,y)\to(0,0)} (x^2 - y^2) = 0$$

We show the graph of f in Figure 13.39. It is identical to the paraboloid $z = x^2 + y^2$ except that the point $(0, 0, 0)$ is missing. ∎

Showing the Limit Fails to Exist

We remarked earlier that if we can find two different paths in the domain of f along which (x, y) approaches (x_0, y_0), for which $f(x, y)$ approaches two different limits, then we can conclude that

$$\lim_{(x,y)\to(x_0, y_0)} f(x, y)$$

does not exist. This conclusion follows from the fact that there are some points (x, y) arbitrarily close to (x_0, y_0) for which $f(x, y)$ will be close to one value and other points for which $f(x, y)$ will be close to a different value. So it is not the case that for *all* points sufficiently close (but not equal) to (x_0, y_0), the corresponding function values are as close as we please to a single limiting value. In the next two examples, we illustrate this method of showing the limit fails to exist.

EXAMPLE 13.20   Show that

$$\lim_{(x,y)\to(0,0)} \frac{x^2 - y^2}{x^2 + y^2}$$

does not exist.

Solution   Again, we restrict (x, y) to be different from $(0, 0)$. If (x, y) lies on the x-axis, with $x \neq 0$, then

$$f(x, y) = f(x, 0) = \frac{x^2}{x^2} = 1$$

whereas if (x, y) is on the y-axis, with $y \neq 0$, then

$$f(x, y) = f(0, y) = \frac{-y^2}{y^2} = -1$$

We can now see that however close to the origin (x, y) is, if the point approaches $(0, 0)$ on the x-axis, $f(x, y) = 1$, and if it approaches $(0, 0)$ on the y-axis, $f(x, y) = -1$. We conclude that there is no *single* value L (neither 1 nor -1) that $f(x, y)$ approaches as (x, y) approaches $(0, 0)$, without restriction as to the manner of approach. Thus, $\lim_{(x,y)\to(0,0)} f(x, y)$ does not exist.

We show the graph of f (generated by a CAS) in Figure 13.40. ∎

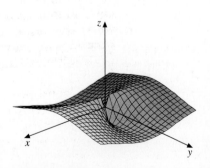

FIGURE 13.40

EXAMPLE 13.21   Show that $\lim_{(x,y)\to(0,0)} f(x, y)$ fails to exist, where

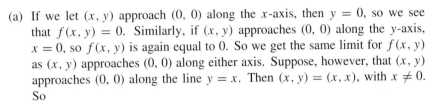

(a) $f(x, y) = \dfrac{2xy}{x^2 + y^2}$ (b) $f(x, y) = \dfrac{x^2 y}{x^4 + y^2}$

Solution

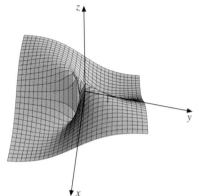

(a) If we let (x, y) approach $(0, 0)$ along the x-axis, then $y = 0$, so we see that $f(x, y) = 0$. Similarly, if (x, y) approaches $(0, 0)$ along the y-axis, $x = 0$, so $f(x, y)$ is again equal to 0. So we get the same limit for $f(x, y)$ as (x, y) approaches $(0, 0)$ along either axis. Suppose, however, that (x, y) approaches $(0, 0)$ along the line $y = x$. Then $(x, y) = (x, x)$, with $x \neq 0$. So

$$f(x, y) = \frac{2x^2}{x^2 + x^2} = \frac{2x^2}{2x^2} = 1$$

Now we see that $\lim_{(x,y)\to(0,0)} f(x, y)$ does not exist, since $f(x, y)$ approaches 0 along one path (either the x-axis or y-axis) and $f(x, y)$ approaches 1 along another path (the line $y = x$).

FIGURE 13.41

We show the CAS-generated graph of f in Figure 13.41.

(b) As in part (a), $f(x, y)$ approaches 0 along either axis. It also approaches 0 along the line $y = x$. In fact, (x, y) approaches $(0, 0)$ along any line $y = mx$ through the origin with $m \neq 0$. To see why, when $y = mx$, then $(x, y) = (x, mx)$, so that with $x \neq 0$,

$$f(x, y) = \frac{x^2(mx)}{x^4 + (mx)^2} = \frac{mx^3}{x^4 + m^2 x^2} = \frac{x^2(mx)}{x^2(x^2 + m^2)} \to \frac{0}{m^2} = 0$$

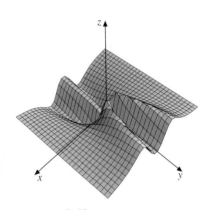

Now consider the situation where (x, y) approaches the origin along the parabola $y = x^2$. Then $(x, y) = (x, x^2)$. So with $x \neq 0$,

$$f(x, y) = \frac{x^2(x^2)}{x^4 + (x^2)^2} = \frac{x^4}{x^4 + x^4} = \frac{x^4}{2x^4} = \frac{1}{2}$$

We can now conclude that $\lim_{(x,y)\to(0,0)} f(x, y)$ does not exist, since as (x, y) approaches the origin along the x-axis (or the y-axis or any other line through the origin), $f(x, y)$ approaches 0, whereas along the parabola $y = x^2$, $f(x, y)$ approaches $\frac{1}{2}$.

FIGURE 13.42

We show the CAS-generated surface in Figure 13.42. ∎

For emphasis, we restate the technique we have illustrated for showing a limit fails to exist.

If $f(x, y)$ approaches different limits as (x, y) approaches (x_0, y_0) along two different paths, then $\lim_{(x,y)\to(x_0, y_0)} f(x, y)$ does not exist.

If you get the *same* limit for $f(x, y)$ as (x, y) approaches (x_0, y_0) along two or more different paths, you cannot conclude that the limit exists. In Example 13.21(b), we showed that $f(x, y)$ approached 0 along any straight-line path through the origin, suggesting perhaps that $\lim_{(x,y)\to(0,0)} f(x, y) = 0$. Yet we saw that $f(x, y)$ had the value $\frac{1}{2}$ when (x, y) approached $(0, 0)$ along the parabolic path $y = x^2$. Consequently, we concluded that the limit did not exist. Getting different values on different paths shows that the limit *fails* to exist. But we must use some other technique to show a limit *does* exist.

Limit Properties

We can often find limits by making use of the following properties, analogous to those for functions of one variable, given in Section 2.2.

Properties of Limits

If $\lim_{(x,y)\to(x_0,y_0)} f(x, y)$ and $\lim_{(x,y)\to(x_0,y_0)} g(x, y)$ both exist, then the following properties hold true.

1. $\displaystyle\lim_{(x,y)\to(x_0,y_0)} cf(x, y) = c\left[\lim_{(x,y)\to(x_0,y_0)} f(x, y)\right]$ for any constant c

2. $\displaystyle\lim_{(x,y)\to(x_0,y_0)} [f(x, y) \pm g(x, y)]$

    $\displaystyle= \lim_{(x,y)\to(x_0,y_0)} f(x, y) \pm \lim_{(x,y)\to(x_0,y_0)} g(x, y)$

3. $\displaystyle\lim_{(x,y)\to(x_0,y_0)} [f(x, y) \cdot g(x, y)]$

    $\displaystyle= \left[\lim_{(x,y)\to(x_0,y_0)} f(x, y)\right] \cdot \left[\lim_{(x,y)\to(x_0,y_0)} g(x, y)\right]$

4. $\displaystyle\lim_{(x,y)\to(x_0,y_0)} \frac{f(x, y)}{g(x, y)} = \frac{\lim_{(x,y)\to(x_0,y_0)} f(x, y)}{\lim_{(x,y)\to(x_0,y_0)} g(x, y)},$

    provided $\displaystyle\lim_{(x,y)\to(x_0,y_0)} g(x, y) \neq 0$

From Properties 1, 2, and 3, it follows that for $P(x, y)$ a polynomial in x and y,

$$\lim_{(x,y)\to(x_0,y_0)} P(x, y) = P(x_0, y_0)$$

We make use of this result in the next example.

EXAMPLE 13.22   Evaluate the limit

$$\lim_{(x,y)\to(2,-3)} \frac{x^2y - 3y^3}{x^3 + 2xy^2}$$

Solution   We use Property 4, taking limits on numerator and denominator separately and dividing:

$$\lim_{(x,y)\to(2,-3)} \frac{x^2y - 3y^3}{x^3 + 2xy^2} = \frac{\lim_{(x,y)\to(2,-3)} (x^2y - 3y^3)}{\lim_{(x,y)\to(2,-3)} (x^3 + 2xy^2)} = \frac{(2)^2(-3) - 3(-3)^3}{2^3 + 2(2)(-3)^2}$$

$$= \frac{-12 + 81}{8 + 36} = \frac{69}{44} \qquad \blacksquare$$

The Formal Limit Definition

For some purposes, such as for proving the limit properties and for proving other theorems concerning limits as well as for making the concept precise and unambiguous, we need a more precise statement of the definition of the limit of a function than that given in Definition 13.4. The following definition is analogous to Definition 2.5 for the one-variable case.

Definition 13.5 Formal Limit Definition	Let f be a function of two variables defined at all points inside some circle centered at (x_0, y_0), except possibly at (x_0, y_0) itself. Then we say that **the limit of $f(x, y)$ as (x, y) approaches (x_0, y_0) is L** and write $$\lim_{(x,y)\to(x_0,y_0)} f(x, y) = L$$ provided that corresponding to any positive number ε there exists a positive number δ such that if $0 < \sqrt{(x - x_0)^2 + (y - y_0)^2} < \delta$, then $|f(x, y) - L| < \varepsilon$ (13.10)

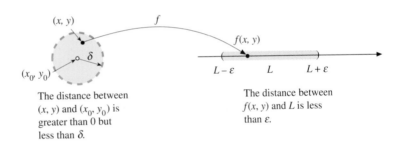

The distance between (x, y) and (x_0, y_0) is greater than 0 but less than δ.

The distance between $f(x, y)$ and L is less than ε.

FIGURE 13.43

In Figure 13.43 we show a diagram to help you understand this definition. The inequality $0 < \sqrt{(x - x_0)^2 + (y - y_0)^2} < \delta$ says that the distance between the point (x, y) and the fixed point (x_0, y_0) is greater than 0 and less than δ. So (x, y) is not equal to (x_0, y_0) but can be any other point inside the circle of radius δ centered at (x_0, y_0). Its image $f(x, y)$ will then lie within ε units of L. Since ε can be made as small as we want, the definition says that $f(x, y)$ comes arbitrarily close to L when (x, y) is close enough to (x_0, y_0) (within δ units of it) but is different from (x_0, y_0). The number δ usually depends on the value of ε.

REMARK ———

■ It *may* happen that the inequality $|f(x, y) - L| < \varepsilon$ is true even when $(x, y) = (x_0, y_0)$, but the definition does not *require* the inequality to hold true in this case. In fact, $f(x_0, y_0)$ need not be defined, as was the case in Example 13.19.

To extend Definition 13.5 to the three-variable case, we require that $f(x, y, z)$ be defined at all points inside a sphere centered at (x_0, y_0, z_0), except possibly at (x_0, y_0, z_0) itself. Then

$$\lim_{(x,y,z)\to(x_0,y_0,z_0)} f(x, y, z) = L$$

means that given any $\varepsilon > 0$, there is a $\delta > 0$ such that

$$\text{if } 0 < \sqrt{(x - x_0)^2 + (y - y_0)^2 + (z - z_0)^2} < \delta,$$

$$\text{then } |f(x, y, z) - L| < \varepsilon \tag{13.11}$$

Continuity

The definition of continuity for a function of two variables is the same as for a function of one variable. We require not only that the limit of $f(x, y)$ exist as $(x, y) \to (x_0, y_0)$ but also that it be equal to $f(x_0, y_0)$.

Definition 13.6
Continuity of a Function of Two Variables

Let f be a function of two variables that is defined at all points inside a circle centered at (x_0, y_0). Then f is **continuous** at (x_0, y_0) provided that

$$\lim_{(x,y)\to(x_0,y_0)} f(x, y) = f(x_0, y_0) \tag{13.12}$$

REMARK ————————————————————————————

■ Continuity of a function of three variables is defined in an analogous way, with (x_0, y_0) replaced by (x_0, y_0, z_0) and (x, y) replaced by (x, y, z). We assume that f is defined inside some sphere centered at (x_0, y_0, z_0).

Equation 13.12 implies three things:

1. $\lim_{(x,y)\to(x_0,y_0)} f(x, y)$ exists.
2. $f(x_0, y_0)$ exists.
3. The values in 1 and 2 are the same.

If any one of these conditions fails, the function is discontinuous at (x_0, y_0).

If f is continuous at all points of some set S, we say that f **is continuous on** S. When all points inside some circle centered at (x_0, y_0) lie in a set, the point (x_0, y_0) is called an **interior point** of the set. So in Definition 13.6 the point (x_0, y_0) is an interior point of the domain of f. A point belonging to a set that is not an interior point is called a **boundary point** of the set. More generally, a boundary point of a set (which may or may not be in the set) is a point such that every circle centered on it contains at least one point in the set and one point not in the set. The set of all boundary points of a set is called its **boundary**. We can alter Definition 13.6 to permit (x_0, y_0) to be a boundary point of the domain D of f (see Figure 13.44), provided (x_0, y_0) is in D, by adding the requirement that $(x, y) \in D$ to the definition. That is, we require only that points *of the domain* that are closer to (x_0, y_0) than δ units have function values closer to L than ε units.

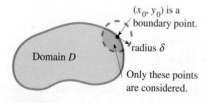

(x_0, y_0) is a boundary point.

radius δ

Domain D

Only these points are considered.

FIGURE 13.44

EXAMPLE 13.23   Show that the function

$$f(x, y) = 2x^2 + y^2$$

is continuous at all points of $\mathbb{R}^2$.

Solution   Since f is a polynomial function, for any point (x_0, y_0) in $\mathbb{R}^2$, we have

$$\lim_{(x,y)\to(x_0,y_0)} (2x^2 + y^2) = 2x_0^2 + y_0^2$$

Furthermore, $f(x_0, y_0) = 2x_0^2 + y_0^2$. So by Definition 13.6, f is continuous at (x_0, y_0). ∎

REMARK

■ We have seen that if P is any polynomial function of two variables,

$$\lim_{(x,y)\to(x_0,y_0)} P(x, y) = P(x_0, y_0)$$

It follows by Definition 13.6 that P is continuous at (x_0, y_0). Thus, every polynomial function of two variables is continuous on all of $\mathbb{R}^2$. Using Limit Property 4, we also then see that *every rational function of two variables is continuous at all points of $\mathbb{R}^2$ in its domain*, that is, wherever the denominator is nonzero.

EXAMPLE 13.24   Let

$$f(x, y) = \begin{cases} \dfrac{2xy}{x^2 + y^2} & \text{if } (x, y) \neq (0, 0) \\ 0 & \text{if } (x, y) = (0, 0) \end{cases}$$

Determine all points at which f is continuous.

Solution   At all points other than the origin, f is a rational function with nonzero denominator. So by the remark above, f is continuous everywhere except at the origin.

To check continuity at the origin, we check to see if $\lim_{(x,y)\to(0,0)} f(x, y) = f(0, 0)$. We are given that $f(0, 0) = 0$. But in Example 13.21(a) we showed that $f(x, y)$ does not approach a unique value as $(x, y) \to (0, 0)$. (We got different limits along two different paths.) Thus,

$$\lim_{(x,y)\to(0,0)} f(x, y)$$

does not exist, so f is discontinuous at $(0, 0)$. ∎

Based on the properties of limits, we can prove that the *sum*, *difference*, *product*, and *quotient* of two continuous functions are also continuous, provided that in the case of the quotient the denominator is nonzero. Furthermore, *a continuous function of a continuous function is continuous*. For example, the composite function $\sin(x^2 + y^2)$ is continuous at all points of $\mathbb{R}^2$, since $u = x^2 + y^2$ is continuous on all of $\mathbb{R}^2$, and $z = \sin u$ is continuous on all of $\mathbb{R}$.

Exercise Set 13.3

In Exercises 1–12, find the limit.

1. $\lim\limits_{(x,y)\to(2,-1)} (3x - 2y + 5)$ **2.** $\lim\limits_{(x,y)\to(-1,-2)} (2x^2 y - 3xy^3)$

3. $\lim\limits_{(x,y)\to(-2,3)} \dfrac{x - 2y}{x + y}$ **4.** $\lim\limits_{(x,y)\to(1,3)} \dfrac{x^2 - y^2}{x^2 + y^2}$

5. $\lim\limits_{(x,y,z)\to(1,-1,2)} (xz - yz + xy)$

6. $\lim\limits_{(x,y,z)\to(-4,5,3)} \dfrac{x^2 - 2yz + z^2}{x^2 + y^2 + z^2}$

7. $\lim\limits_{(x,y)\to(0,0)} e^{x^2 + y^2}$ **8.** $\lim\limits_{(x,y)\to(\pi/2,1)} y \cos(xy)$

9. $\lim\limits_{(x,y,z)\to(1,0,0)} \ln(x^2 + y^2 + z^2)$

10. $\lim\limits_{(x,y)\to(0,0)} \dfrac{\sin(x^2 + y^2)}{x^2 + y^2}$ (*Hint:* Let $u = x^2 + y^2$.)

11. $\lim\limits_{(x,y)\to(0,0)} f(x, y)$, where

$$f(x, y) = \begin{cases} \sqrt{x^2 + y^2} & \text{if } (x, y) \neq (0, 0) \\ 1 & \text{if } (x, y) = (0, 0) \end{cases}$$

12. $\lim\limits_{(x,y)\to(1,-1)} f(x, y)$, where

$$f(x, y) = \begin{cases} \dfrac{x^2 - 2xy}{2x + 3y} & \text{if } (x, y) \neq (1, -1) \\ 3 & \text{if } (x, y) = (1, -1) \end{cases}$$

In Exercises 13–18, show that $\lim_{(x,y)\to(0,0)} f(x, y)$ does not exist.

13. $f(x, y) = \dfrac{2x^2 - 3y^2}{x^2 + y^2}$ **14.** $f(x, y) = \dfrac{3xy}{2x^2 + 5y^2}$

15. $f(x, y) = \dfrac{x + y}{x^2 + y^2}$ **16.** $f(x, y) = \dfrac{x^2 - xy + y^2}{x^2 + y^2}$

17. $f(x, y) = \dfrac{xy^2}{x^2 + y^4}$ **18.** $f(x, y) = \dfrac{x^3 y}{x^6 + y^2}$

In Exercises 19–30, determine the largest subset of $\mathbb{R}^2$ on which f is continuous.

19. $f(x, y) = 2x^3 - 3xy + 7y^2$

20. $f(x, y) = \dfrac{x^2 - y^2}{x^2 + y^2}$ **21.** $f(x, y) = \dfrac{x^2 - y^2}{x - y}$

22. $f(x, y) = \dfrac{x^2 - 4y^2}{x - 2y}$ **23.** $f(x, y) = \dfrac{x - y}{x + y}$

24. $f(x, y) = \dfrac{2x - 3y}{x^2 + y^2 - 1}$ **25.** $f(x, y) = \dfrac{x^2 + y^2}{x^2 - y^2}$

26. $f(x, y) = \dfrac{2x^2 - 3xy + y^2}{x^2 - y + 1}$

27. $f(x, y) = e^{2xy}$ **28.** $f(x, y) = e^{-x/y}$

29. $f(x, y) = \ln \sqrt{x^2 + y^2}$ **30.** $f(x, y) = \tan^{-1} \dfrac{y}{x}$

In Exercises 31–35, determine if f is continuous at $(0, 0)$.

31. $f(x, y) = \begin{cases} \dfrac{x^4 - y^4}{x^2 + y^2} & \text{if } (x, y) \neq (0, 0) \\ 0 & \text{if } (x, y) = (0, 0) \end{cases}$

32. $f(x, y) = \begin{cases} \dfrac{x - y}{x^2 + y^2} & \text{if } (x, y) \neq (0, 0) \\ 0 & \text{if } (x, y) = (0, 0) \end{cases}$

33. $f(x, y) = \begin{cases} \dfrac{\sin \sqrt{x^2 + y^2}}{\sqrt{x^2 + y^2}} & \text{if } (x, y) \neq (0, 0) \\ 1 & \text{if } (x, y) = (0, 0) \end{cases}$

34. $f(x, y) = \begin{cases} \dfrac{x^2 + 2xy + y^2}{x^2 + y^2} & \text{if } (x, y) \neq (0, 0) \\ 1 & \text{if } (x, y) = (0, 0) \end{cases}$

35. $f(x, y) = \begin{cases} \dfrac{x^3 - y^3}{x^2 + xy + y^2} & \text{if } (x, y) \neq (0, 0) \\ 0 & \text{if } (x, y) = (0, 0) \end{cases}$

36. Let f and g be functions of two variables, each of which is continuous at (x_0, y_0). Use Definition 13.6 and the properties of limits to prove that each of the following functions is also continuous at (x_0, y_0).
   (a) cf, where c is a constant
   (b) $f \pm g$
   (c) fg
   (d) $\dfrac{f}{g}$, provided $g(x_0, y_0) \neq 0$

37. Use Definition 13.5 to prove that if

$$\lim\limits_{(x,y)\to(x_0,y_0)} f(x, y) = L \text{ and } \lim\limits_{(x,y)\to(x_0,y_0)} g(x, y) = M,$$

then $\lim_{(x,y)\to(x_0,y_0)} [f(x, y) + g(x, y)] = L + M$

38. Let f be a function of two variables with domain D. A point (a, b) is said to be an **isolated point** of D if (a, b) is in D and there is a circle with center (a, b) with no other point of D inside the circle.

(a) Show that an isolated point of D is always a boundary point of D.

(b) Show that any function is continuous at an isolated point of its domain.

(*Hint:* Choose δ so small that the circle of radius δ centered at (a, b) has no point of the domain D except (a, b) inside it. Then explain why the inequality $|f(x, y) - f(a, b)| < \varepsilon$ is satisfied regardless of what the value of the positive number ε is, provided (x, y) is in D and $\sqrt{(x - a)^2 + (y - b)^2} < \delta$.)

 39. Let $z = f(x, y) = \dfrac{x^2 + y}{3x}$.

(a) Plot the surface.

(b) Plot each of the following curves: $z = f(x, x)$, $z = f(x, 0)$, $z = f(x, 1)$, and $z = f(x, x^2)$. Does the limit as (x, y) approaches $(0, 0)$ exist?

 40. For each of the following functions, conjecture whether the function has a limit as (x, y) approaches $(0, 0)$.

(a) $f(x, y) = \dfrac{x^2 + 3xy + y^2}{x^2 + y^2}$

(b) $f(x, y) = \dfrac{x^3 + y^2}{x^3 + y^3}$

(c) $f(x, y) = |x|^y$

(d) $f(x, y) = \dfrac{\sin(x^2 + y^2)}{x^2 + y^2}$

13.4 PARTIAL DERIVATIVES

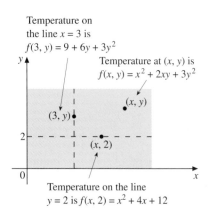

Temperature on the line $x = 3$ is $f(3, y) = 9 + 6y + 3y^2$

Temperature at (x, y) is $f(x, y) = x^2 + 2xy + 3y^2$

(x, y)

$(3, y)$

$(x, 2)$

Temperature on the line $y = 2$ is $f(x, 2) = x^2 + 4x + 12$

FIGURE 13.45

If f is a function of two variables x and y and we hold the variable y fixed, then f becomes a function of the single variable x. Similarly, if we hold x fixed but allow y to vary, then f is a function of the single variable y. For example, suppose that

$$f(x, y) = x^2 + 2xy + 3y^2$$

represents the temperature at a point (x, y) on a flat metal plate (see Figure 13.45). If we hold y fixed, say, $y = 2$, we get

$$f(x, 2) = x^2 + 4x + 12$$

This function of x gives the temperature at points along the horizontal line $y = 2$. If, instead, we hold x fixed, say, $x = 3$, we get

$$f(3, y) = 9 + 6y + 3y^2$$

which gives the temperature along the vertical line $x = 3$.

Since $f(x, 2) = x^2 + 4x + 12$ is a function of the single variable x, we can find its derivative, $2x + 4$, and interpret it as giving the rate of change of temperature in the x direction along the line $y = 2$. Similarly, the derivative of $f(3, y)$ is $6 + 6y$, which gives the rate of change of temperature in the y direction along the line $x = 3$.

More generally, for any function $f(x, y)$, when we hold y fixed, the resulting function of x may be differentiable. If so, we call the derivative the **partial derivative of f with respect to x** and denote it by $f_x(x, y)$. Similarly, if we hold x fixed and the resulting function of y is differentiable, we call its derivative the **partial derivative of f with respect to y** and denote it by $f_y(x, y)$.

To illustrate these partial derivatives, let us consider again the function

$$f(x, y) = x^2 + 2xy + 3y^2$$

Rather than assign a specific value to y, we treat it as a constant and differentiate with respect to x, to get

$$f_x(x, y) = 2x + 2y$$

Similarly,

$$f_y(x, y) = 2x + 6y$$

In the first case, since y is a constant, the derivative of $2xy$ with respect to x is $2y$, and the derivative of $3y^2$ is 0 (since $3y^2$ is a constant when differentiating with respect to x). Also, when we differentiate with respect to y, the term x^2 is constant, so its derivative is 0.

When you see f_x you will know to do two things: hold y fixed, and differentiate with respect to x. Similar remarks apply to the symbol f_y.

Before stating the formal definition of partial derivatives, we remind you of the definition of the derivative of a function of one variable. If f is a function of the single variable x, then its derivative at x is

$$f'(x) = \lim_{h \to 0} \frac{f(x+h) - f(x)}{h} \tag{13.13}$$

provided the limit on the right exists.

The partial derivatives $f_x(x, y)$ and $f_y(x, y)$ are defined exactly as in Equation 13.13.

Definition 13.7
The Partial Derivatives
f_x and f_y

Let f be a function of two variables x and y, with domain D. The functions f_x and f_y defined by

$$f_x(x, y) = \lim_{h \to 0} \frac{f(x+h, y) - f(x, y)}{h} \tag{13.14}$$

and

$$f_y(x, y) = \lim_{k \to 0} \frac{f(x, y+k) - f(x, y)}{k} \tag{13.15}$$

at all points D where these limits exist are called, respectively, the **partial derivative of f with respect to x** and **the partial derivative of f with respect to y**.

Other Notations

Just as with ordinary derivatives, different notations for partial derivatives are commonly used. Analogous to the Leibniz notation df/dx for the derivative of a function of one variable, the symbols $\partial f/\partial x$ and $\partial f/\partial y$ are alternative notations for f_x and f_y, respectively, when f is a function of two variables x and y. The symbol ∂ (sometimes referred to as "curly d") replaces the d in ordinary differentiation to signify that more than one variable is involved. The symbols $\partial/\partial x$ and $\partial/\partial y$ can be regarded as *partial derivative operators*, which instruct you to take the partial derivative of whatever follows. For example,

$$\frac{\partial}{\partial x}(2x^2y + x^3) = 4xy + 3x^2 \quad \text{and} \quad \frac{\partial}{\partial y}(2x^2y + x^3) = 2x^2$$

When a dependent variable is introduced, say, $z = f(x, y)$, the partial derivatives with respect to x and y, respectively, can be written as z_x or $\partial z/\partial x$ and z_y or $\partial z/\partial y$.

EXAMPLE 13.25 Let $f(x, y) = \tan^{-1} y/x$. Find $f_x(x, y)$, $f_y(x, y)$, $f_x(4, -3)$, and $f_y(4, -3)$.

Recall that $\dfrac{d}{dx} \tan^{-1} u = \dfrac{1}{1 + u^2} \dfrac{du}{dx}$.

Solution   Holding y fixed, we get

$$\frac{\partial}{\partial x}\left(\frac{y}{x}\right)$$

$$f_x(x, y) = \frac{1}{1 + \left(\dfrac{y}{x}\right)^2}\left(-\frac{y}{x^2}\right) = \frac{-y}{x^2 + y^2}$$

Holding x fixed, we get

$$f_y(x, y) = \frac{1}{1 + \left(\dfrac{y}{x}\right)^2}\left(\frac{1}{x}\right) = \frac{x}{x^2 + y^2}$$

Substituting $x = 4$ and $y = -3$, we have

$$f_x(4, -3) = \frac{3}{25} \quad \text{and} \quad f_y(4, -3) = \frac{4}{25} \qquad \blacksquare$$

EXAMPLE 13.26   Find $\dfrac{\partial}{\partial x}(e^{-xy}\cos y)$ and $\dfrac{\partial}{\partial y}(e^{-xy}\cos y)$.

Solution

$$\frac{\partial}{\partial x}(e^{-xy}\cos y) = -ye^{-xy}\cos y$$

$$\frac{\partial}{\partial y}(e^{-xy}\cos y) = -xe^{-xy}\cos y + e^{-xy}(-\sin y) \quad \text{Product Rule}$$

$$= -e^{-xy}(x\cos y + \sin y) \qquad \blacksquare$$

EXAMPLE 13.27   If $z = xy/(x + y)$, show that

$$x\frac{\partial z}{\partial x} + y\frac{\partial z}{\partial y} = z$$

Solution

$$\frac{\partial z}{\partial x} = \frac{y(x + y) - xy}{(x + y)^2} = \frac{y^2}{(x + y)^2} \qquad \text{Quotient rule}$$

$$\frac{\partial z}{\partial y} = \frac{x(x + y) - xy}{(x + y)^2} = \frac{x^2}{(x + y)^2}$$

$$x\frac{\partial z}{\partial x} + y\frac{\partial z}{\partial y} = \frac{xy^2}{(x + y)^2} + \frac{x^2 y}{(x + y)^2} = \frac{xy(y + x)}{(x + y)^2} = \frac{xy}{x + y} = z \qquad \blacksquare$$

Geometric Interpretation

If f is a function of two variables x and y, we have already seen that the partial derivatives f_x and f_y are really derivatives of functions of one variable. For example, we can write $f_x(x, y) = g'(x)$, where $g(x) = f(x, y)$ with y held fixed. So the geometric interpretation of a partial derivative is the same as in

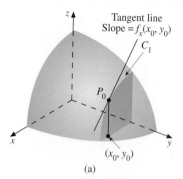

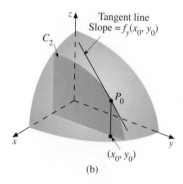

FIGURE 13.46

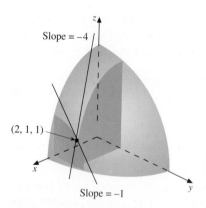

FIGURE 13.47

the single variable case, namely, the slope of the tangent line to a plane curve. To illustrate, let $P_0 = (x_0, y_0, z_0)$ be a point of the surface $z = f(x, y)$. Let C_1 denote the plane curve passing through P_0 that is the trace of the surface on the plane $y = y_0$. Also, let C_2 be the plane curve passing through P_0 that is the trace of the surface on the plane $x = x_0$. (See Figure 13.46.) Then $f_x(x_0, y_0)$ is the slope of the tangent line to C_1 at P_0, and $f_y(x_0, y_0)$ is the slope of the tangent line to C_2 at P_0, as we illustrate in Figure 13.46.

EXAMPLE 13.28   Find the slope of the tangent line to the curve of intersection of the surface $z = \sqrt{10 - 2x^2 - y^2}$ and the plane (a) $y = 1$ and (b) $x = 2$ at the point $(2, 1, 1)$. Show the results graphically.

Solution   By the preceding discussion, the slope of the tangent line in part (a) is $\partial z / \partial x$ and in part (b) is $\partial z / \partial y$, each evaluated at $(2, 1)$:

$$\frac{\partial z}{\partial x} = \frac{-2x}{\sqrt{10 - 2x^2 - y^2}} \quad \text{and} \quad \frac{\partial z}{\partial y} = \frac{-y}{\sqrt{10 - 2x^2 - y^2}}$$

So for the curve of part (a),

$$\text{slope} = \frac{\partial z}{\partial x}\bigg|_{(2, 1)} = \frac{-4}{1} = -4$$

and for the curve of part (b),

$$\text{slope} = \frac{\partial z}{\partial y}\bigg|_{(2, 1)} = \frac{-1}{1} = -1$$

We illustrate the results in Figure 13.47. Notice that the given surface is the upper half of the ellipsoid c

$$\frac{x^2}{5} + \frac{y^2}{10} + \frac{z^2}{10} = 1$$

We show the first-octant portion only.   ∎

Three or More Variables

If f is a function of three or more variables, we define partial derivatives in a similar way to their definitions for the two-variable case. All variables except the one in question are held fixed, and the derivative is taken with respect to the one that is not held fixed (which is the single independent variable). We illustrate how to find the first-order partials for a function of three variables in the next two examples.

EXAMPLE 13.29   Let

$$f(x, y, z) = z^2 \sin(xy)$$

Find f_x, f_y, and f_z.

Solution

$f_x(x, y, z) = yz^2 \cos(xy)$ *y* and *z* are held fixed.

$f_y(x, y, z) = xz^2 \cos(xy)$ *x* and *z* are held fixed.

$f_z(x, y, z) = 2z \sin(xy)$ *x* and *y* are held fixed.   ∎

EXAMPLE 13.30   Under certain conditions, the temperature at a point (x, y) on a flat metal plate at time $t > 0$ is given by

$$T(x, y, t) = \frac{k}{t^{3/2}} e^{-(x^2+y^2)/ct}$$

where k and c are constants. Find $\partial T/\partial x$, $\partial T/\partial y$, and $\partial T/\partial t$.

Solution   Holding y and t fixed, we get

$$\frac{\partial T}{\partial x} = \frac{k}{t^{3/2}} e^{-(x^2+y^2)/ct} \left[-\frac{2x}{ct} \right] = -\frac{2kx}{ct^{5/2}} e^{-(x^2+y^2)/ct}$$

Since the given function is symmetric in x and y, we can immediately write

$$\frac{\partial T}{\partial y} = -\frac{2ky}{ct^{5/2}} e^{-(x^2+y^2)/ct}$$

If we hold x and y fixed, we have a function of t alone, whose derivative can be found by the Product Rule:

$$\frac{\partial T}{\partial t} = -\frac{3k}{2t^{5/2}} e^{-(x^2+y^2)/ct} + \frac{k}{t^{3/2}} e^{-(x^2+y^2)/ct} \left[\frac{x^2 + y^2}{ct^2} \right]$$

$$= e^{-(x^2+y^2)/ct} \left[-\frac{3k}{2t^{5/2}} + \frac{k(x^2 + y^2)}{ct^{7/2}} \right]$$

$$= \frac{k}{2ct^{7/2}} e^{-(x^2+y^2)/ct} \left[2(x^2 + y^2) - 3ct \right] \qquad \blacksquare$$

Higher-Order Partials

Since f_x and f_y are themselves functions of the two variables x and y, it is possible that they too have partial derivatives, called *second-order partial derivatives of f* (or just "second partials" for short). These are denoted as follows:

1. f_{xx}, second partial of f with respect to x
2. f_{xy}, second mixed partial of f, first with respect to x and then with respect to y
3. f_{yy}, second partial of f with respect to y
4. f_{yx}, second mixed partial of f, first with respect to y and then with respect to x

Note carefully the order of differentiation in the mixed partials. For example, f_{xy} means $(f_x)_y$, so we first differentiate with respect to x and then differentiate the result with respect to y.

Using the Leibniz notation, we write

$$\frac{\partial^2 f}{\partial x^2} \qquad \text{Same as } f_{xx}$$

$$\frac{\partial^2 f}{\partial y \, \partial x} \qquad \text{Same as } f_{xy}$$

$$\frac{\partial^2 f}{\partial y^2} \qquad \text{Same as } f_{yy}$$

$$\frac{\partial^2 f}{\partial x \, \partial y} \qquad \text{Same as } f_{yx}$$

These notations are suggested by applying the partial differential operators twice—for example, as in

$$\frac{\partial}{\partial x}\left(\frac{\partial f}{\partial x}\right) = \frac{\partial^2 f}{\partial x^2}$$

and

$$\frac{\partial}{\partial y}\left(\frac{\partial f}{\partial x}\right) = \frac{\partial^2 f}{\partial y\,\partial x}$$

Again, observe carefully the order of differentiation. Compare the following, for example:

$$\frac{\partial^2 f}{\partial y\,\partial x} = f_{xy}$$

$$\begin{array}{cc}\uparrow\ \uparrow & \uparrow\ \nwarrow \\ \text{2nd 1st} & \text{1st 2nd}\end{array}$$

We could continue to higher-order partials, using notations such as

$$f_{xxy} \quad \text{or} \quad \frac{\partial^3 f}{\partial y\,\partial x^2}$$

You should verify that there are eight such third-order partials, and in general 2^n partials of nth order. In applications, partials of orders higher than 2 are seldom used.

EXAMPLE 13.31   Let $f(x, y) = 2x^3 y^2 - 3xy^4$. Find $f_{xx}(x, y)$, $f_{xy}(x, y)$, $f_{yy}(x, y)$, and $f_{yx}(x, y)$.

Solution

$$\begin{aligned} f_x(x, y) &= 6x^2 y^2 - 3y^4 & f_y(x, y) &= 4x^3 y - 12xy^3 \\ f_{xx}(x, y) &= 12xy^2 & f_{yy}(x, y) &= 4x^3 - 36xy^2 \\ f_{xy}(x, y) &= 12x^2 y - 12y^3 & f_{yx}(x, y) &= 12x^2 y - 12y^3 \end{aligned}$$ ∎

Note that in this example $f_{xy} = f_{yx}$. The equality of these mixed partials is no accident. Although it is not always true, it is true for "well-behaved" functions. (See Exercise 55 in Exercise Set 13.4 for a function where $f_{xy} \neq f_{yx}$.) The following theorem, whose proof is given in Appendix 5, gives sufficient conditions for the equality of f_{xy} and f_{yx}.

THEOREM 13.1

Let f be a function of x and y. If f_{xy} and f_{yx} both exist at all points inside some circle centered at (x_0, y_0) and they are continuous at (x_0, y_0), then

$$f_{xy}(x_0, y_0) = f_{yx}(x_0, y_0)$$

One advantage of Theorem 13.1 is that differentiating in one order can sometimes be much easier than in the other. For example, consider

$$f(x, y) = x^3 y + \frac{e^y}{y^2 \sin y}$$

Since $f_x(x, y) = 3x^2 y$, we have

$$f_{xy}(x, y) = 3x^2$$

If we were to first calculate $f_y(x, y)$ and then $f_{yx}(x, y)$, we would get the same result, but it would be much harder. (Try it.)

Partial Differential Equations

An equation such as

$$k\frac{\partial^2 u}{\partial x^2} = \frac{\partial u}{\partial t} \qquad (k > 0) \tag{13.16}$$

is an example of a **partial differential equation** because it involves partial derivatives. Equation 13.16 is called the one-dimensional **heat equation** and occurs in the theory of heat flow in a rod or thin wire. The function $u(x, t)$ is the temperature in the rod at time t at a distance x from one end. The constant k is called the *thermal diffusivity* of the rod or wire. The same equation also arises in the study of the flow of electricity in a transmission line. For this reason it is sometimes called a *telegraph equation*.

Two other important partial differential equations of physics are the **wave equation**,

$$\frac{\partial^2 y}{\partial t^2} = a^2 \frac{\partial^2 y}{\partial x^2} \tag{13.17}$$

and **Laplace's Equation**,

$$\frac{\partial^2 u}{\partial x^2} + \frac{\partial^2 u}{\partial y^2} + \frac{\partial^2 u}{\partial z^2} = 0 \tag{13.18}$$

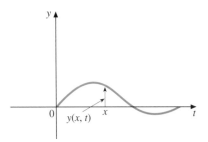

FIGURE 13.48
$y(x, t)$ is the vertical displacement at x at time t.

The wave equation relates the vertical displacements $y(x, t)$ in a vibrating string, where t represents time and x the horizontal distance from one end of the string (see Figure 13.48). Laplace's Equation holds true for the steady-state (time-independent) temperature $u(x, y, z)$ at points interior to a solid in which no heat is generated. Any function $u(x, y, z)$ that satisfies Laplace's Equation is said to be **harmonic**.

In the next three examples, we verify that certain functions are solutions to Equations 13.16, 13.17, and 13.18, respectively.

EXAMPLE 13.32   Show that the function

$$u(x, t) = e^{-kt} \cos x$$

satisfies the heat equation (Equation 13.16).

Solution   First, we compute the partials with respect to x.

$$\frac{\partial u}{\partial x} = -e^{-kt} \sin x$$

$$\frac{\partial^2 u}{\partial x^2} = -e^{-kt} \cos x$$

Now we obtain the partial of u with respect to t.

$$\frac{\partial u}{\partial t} = -ke^{-kt} \cos x$$

Comparing $\partial^2 u / \partial x^2$ and $\partial u / \partial t$, we see that

$$k \frac{\partial^2 u}{\partial x^2} = \frac{\partial u}{\partial t}$$

So the given function does satisfy the heat equation.   ∎

EXAMPLE 13.33   Show that if f is any twice-differentiable function of one variable, then

$$y(x, t) = \frac{1}{2} \left[f(x + at) + f(x - at) \right]$$

satisfies the wave equation (Equation 13.17).

Solution   Let $u = x + at$ and $v = x - at$. If we hold t fixed, then we can apply the Chain Rule to $f(u)$, with $u = x + at$, and $f(v)$, with $v = x - at$, to get

$$\frac{\partial y}{\partial x} = \frac{1}{2} \left[f'(u) \cdot 1 + f'(v) \cdot 1 \right] = \frac{1}{2} \left[f'(u) + f'(v) \right]$$

Similarly,

$$\frac{\partial^2 y}{\partial x^2} = \frac{1}{2} \left[f''(u) + f''(v) \right]$$

Now we hold x fixed and apply the Chain Rule to get

$$\frac{\partial y}{\partial t} = \frac{1}{2} \left[f'(u) \cdot a + f'(v)(-a) \right]$$

Applying the Chain Rule another time gives

$$\frac{\partial^2 y}{\partial t^2} = \frac{1}{2} \left[f''(u)a^2 + f''(v)(-a)^2 \right]$$

$$= \frac{a^2}{2} \left[f''(u) + f''(v) \right]$$

Comparing this result with $\partial^2 y / \partial x^2$, we see that

$$\frac{\partial^2 y}{\partial t^2} = a^2 \frac{\partial^2 y}{\partial x^2}$$ ∎

REMARK ——————————————————————

∎ The solution $y(x, t)$ to the wave equation given in Example 13.33 is known as *d'Alembert's Solution*, after the French mathematician, Jean Le Rond d'Alembert (1717–1783).

——————————————————————————————

EXAMPLE 13.34   The function

$$\psi(x, y, z) = \frac{-GMm}{\sqrt{x^2 + y^2 + z^2}}$$

is called the gravitational potential function for the force exerted by a body of mass M on a body of mass m, where the vector $\mathbf{r}$ from the first mass to the second is $\mathbf{r} = x\mathbf{i} + y\mathbf{j} + z\mathbf{k}$. The constant G is the universal gravitational constant. Show that ψ is a harmonic function.

Solution   Recall that a harmonic function is a function that satisfies Laplace's Equation (Equation 13.18). We first calculate $\partial\psi/\partial x$ and $\partial^2\psi/\partial x^2$.

$$\frac{\partial\psi}{\partial x} = -GMm\left(-\frac{1}{2}\right)(x^2+y^2+z^2)^{-3/2}(2x) = \frac{GMmx}{(x^2+y^2+z^2)^{3/2}}$$

$$\frac{\partial^2\psi}{\partial x^2} = GMm\frac{(x^2+y^2+z^2)^{3/2} - x\cdot\frac{3}{2}(x^2+y^2+z^2)^{1/2}\cdot 2x}{(x^2+y^2+z^2)^3} \qquad \text{Quotient Rule}$$

$$= GMm\frac{(x^2+y^2+z^2) - 3x^2}{(x^2+y^2+z^2)^{5/2}} \qquad \begin{array}{l}\text{We divided by}\\ (x^2+y^2+z^2)^{1/2}.\end{array}$$

$$= GMm\frac{-2x^2+y^2+z^2}{(x^2+y^2+z^2)^{5/2}}$$

Since $\psi(x,y,z)$ is symmetric in x, y, and z, we can immediately conclude that

$$\frac{\partial^2\psi}{\partial y^2} = GMm\frac{-2y^2+x^2+z^2}{(x^2+y^2+z^2)^{5/2}}$$

and

$$\frac{\partial^2\psi}{\partial z^2} = GMm\frac{-2z^2+x^2+y^2}{(x^2+y^2+z^2)^{5/2}}$$

Thus,

$$\frac{\partial^2\psi}{\partial x^2} + \frac{\partial^2\psi}{\partial y^2} + \frac{\partial^2\psi}{\partial z^2}$$

$$= GMm\frac{(-2x^2+y^2+z^2)+(-2y^2+x^2+z^2)+(-2z^2+x^2+y^2)}{(x^2+y^2+z^2)^{5/2}}$$

All the terms in the numerator cancel out, so we have

$$\frac{\partial^2\psi}{\partial x^2} + \frac{\partial^2\psi}{\partial y^2} + \frac{\partial^2\psi}{\partial z^2} = 0$$

which shows that ψ is harmonic. ∎

Applications in Economics

The cost C of producing a given commodity typically is a function of unit cost of raw materials, say x, and the average hourly wage rate, say y, of the work force. (It may depend on other variables as well, but we will consider this simplified model.) Then the partial derivatives $\partial C/\partial x$ and $\partial C/\partial y$ are called the **marginal cost of raw materials** and **marginal cost of labor**, respectively. For given values of x and y, $\partial C/\partial x$ is approximately the added cost if the unit price of raw materials were increased by \$1. Similarly, $\partial C/\partial y$ is approximately the added cost if the hourly wage rate were increased by \$1. We illustrate these marginal functions in the next example.

EXAMPLE 13.35   Suppose that the cost function (in dollars) for producing some commodity is given by

$$C(x,y) = 180 + x^2 + 2y^2 - xy$$

where the fixed cost is \$180, x is the cost per pound of raw materials, and y is the hourly wage rate. The current cost of materials is \$8 per pound, and the current hourly wage rate is \$12 per hour. Find each of the following:

(a) the approximate change in C if the cost of raw materials goes up to \$9 per pound;

(b) the approximate change in C if the hourly wage rate goes up to \$13 per hour.

Solution   We first calculate the partial derivatives C_x and C_y:

$$C_x(x, y) = 2x - y \quad \text{and} \quad C_y(x, y) = 4y - x$$

Thus, with $x = 8$ and $y = 12$, we have

$$C_x(8, 12) = 16 - 12 = 4 = \text{Marginal Cost of Raw Materials}$$

$$C_y(8, 12) = 48 - 8 = 40 = \text{Marginal Cost of Labor}$$

An increase of \$1 in the cost per pound of raw materials would result in approximately a \$4 increase in the cost of production, whereas an increase of \$1 in the hourly wage rate would result in approximately a \$40 increase in the cost of production. Clearly, in this case C is much more sensitive to a change in labor costs than to a change in the cost of raw materials.   ■

REMARK ──

■ Marginal revenue functions, marginal profit functions, marginal production functions, and the like are defined similarly to marginal cost functions.

──

The function

$$Q(K, L) = AK^\alpha L^{1-\alpha} \qquad (0 < \alpha < 1) \tag{13.19}$$

is known as the **Cobb-Douglas Production Function**. Here Q represents the number of units of some commodity that are produced, A is a positive constant, K is the amount of capital investment, and L is the size of the labor force. This function is used widely in economics. The variables can also have different meanings. For example, $Q(K, L)$ might represent the demand function for a commodity, K the disposable income, and L the amount spent on advertising.

EXAMPLE 13.36   For the Cobb-Douglas Production Function given by Equation 13.19, evaluate the ratio Q_K/Q_L when $K = 250$ (in thousands of dollars), $L = 500$, and $\alpha = \frac{2}{3}$. Interpret the result.

Solution   With $\alpha = \frac{2}{3}$, Equation 13.19 becomes

$$Q(K, L) = AK^{2/3}L^{1/3}$$

So

$$Q_K(K, L) = \frac{2}{3}AK^{-1/3}L^{1/3}$$

and

$$Q_L(K, L) = \frac{1}{3}AK^{2/3}L^{-2/3}$$

The ratio of Q_K to Q_L is

$$\frac{Q_K(K,L)}{Q_L(K,L)} = \frac{\frac{2}{3}AK^{-1/3}L^{1/3}}{\frac{1}{3}AK^{2/3}L^{-2/3}} = \frac{2L}{K}$$

Thus, when $K = 250$ and $L = 500$,

$$\frac{Q_K(250,500)}{Q_L(250,500)} = \frac{2(500)}{250} = 4$$

The partial derivative Q_K is the *marginal productivity of capital*, and Q_L is the *marginal productivity of labor*. At $K = 250$ and $L = 500$, we have found that $Q_K = 4Q_L$. Thus, an increase of 1 (thousand) in capital, holding the labor force fixed, will result in about four times as much additional production as increasing the labor force by 1, keeping the capital fixed. ∎

LIMITS AND PARTIAL DERIVATIVES OF FUNCTIONS OF SEVERAL VARIABLES USING COMPUTER ALGEBRA SYSTEMS

Determining whether limits exist for functions of several variables is typically much more difficult than for functions of one variable. The reason lies in the fact that there are infinitely many paths of approach in the multivariable case. When showing that a limit does not exist, one typically tries to select different paths that yield different limits. This algebraic process is perfectly suited for computer algebra systems. In this section we investigate limits and partial derivatives for functions of several variables.

CAS 52

Investigate the limits of each of the following:

(a) $\displaystyle\lim_{(x,y)\to(0,0)} \frac{x^2 - y^2}{x^2 + y^2}$ (b) $\displaystyle\lim_{(x,y)\to(0,0)} \frac{x^2 y}{x^4 + y^2}$

Maple:

(a)

First, define f as a function of the two variables x and y:

f:=(x,y)–>(x^2–y^2)/(x^2+y^2);

Now compute the limit:

limit(f(x,y), {x=0,y=0});

Output: *undefined*

Mathematica:

Our version of Mathematica does not have a limit command for functions of several variables.

(a)
f[x_,y_] = (x^2–y^2)/(x^2+y^2)
Plot 3D[f[x,y],{x,–1,1},{y,–1,1}]
t=m*x
g=f[x,t]
Limit[g,{x–>0}]

Maple:

A plot indicates that as the origin is approached along different lines, the value of the function approaches different values.

plot3d(f(x,y),x=−1..1,y=−1..1,orientation=[45, 45],axes=boxed);

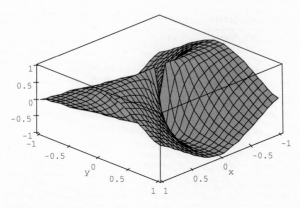

FIGURE 13.4.1

To verify this, we let (x, y) approach the origin along straight lines of varying slopes, all of which pass through the origin.

t:=m*x;
g:=f(x,t);

Output: $g := \dfrac{x^2 - m^2 x^2}{x^2 + m^2 x^2}$

limit(g, {x=0});

Output: $-\dfrac{-1 + m^2}{1 + m^2}$

(b)
f:=(x,y)–>x^2*y/(x^4+y^2);
limit(f(x,y),{x=0,y=0});

Output: $\text{limit}\left(\dfrac{x^2 y}{x^4 + y^2}, \{x = 0, y = 0\}\right)$

A limit is not computed in this case, so we proceed as in part (a).

t:=m*x;
g:=f(x,t);

Output: $g := \dfrac{x^3 m}{x^4 + m^2 x^2}$

limit(g,{x=0});

Output: 0

So if (x, y) approaches along lines through the origin, $f(x, y)$ does approach 0. But this does not say the limit exists. This time we need to consider other paths.

t:=x^2;
g:=f(x,t);

Mathematica:

(b)
f[x_,y_] = (x^2*y)/(x^4+y^2)
t = m*x
g = f[x,t]
Limit[g,{x–>0}]
t = x^2
g = f[x,t]
Limit[g,{x–>0}]

Maple:

Output: $g := \dfrac{1}{2}$

limit(g,{x=0});

Output: $\dfrac{1}{2}$

In this case, as (x, y) approaches $(0, 0)$ along the parabolic path $y = x^2$, the limit is 1/2, which does not agree with the limit along straight line paths through the origin, and again the limit does not exist.

CAS 53

Find the slope of the curve of intersection of the upper nappe of the cone $\dfrac{z^2}{4} = \dfrac{x^2}{4} + \dfrac{y^2}{9}$ and the plane $x = -2$ at the point $\left(-2, 1, \dfrac{\sqrt{40}}{3}\right)$.

Maple:

We first make a plot of the cone and the intersecting plane.

with(plots):
cone :=solve((x/2)^2+(y/3)^2=(z/2)^2,z);

Output: $cone := -\dfrac{1}{3}\sqrt{9x^2 + 4y^2},\ \dfrac{1}{3}\sqrt{9x^2 + 4y^2}$

The upper nappe is given by the second expression with the positive square root.

pl:=plot3d({cone[2]},x=-6..6,y=-8..8,view=-5..5,
    scaling=constrained,axes=boxed,tickmarks=[3,3,3]);
p2:=plot3d({[-2,s,t]},s=-10..10,t=-10..10);
display3d({pl,p2},orientation=[20,77]);

Mathematica:

cone = Solve[(x/2)^2+(y/3)^2==(z/2)^2,z]
P1 = Plot3D[Sqrt[9*x^2+4*y^2]/3,{x,-6,6},{y,-8,8},
    PlotRange->{-5,5},DisplayFunction->Identity]
P2 = ParametricPlot3D[{-2,s,t},{t,-10,10},{s,-10,10},
    DisplayFunction->Identity]
Show[{P1,P2},DisplayFunction->$DisplayFunction]
f[x_,y_] = Sqrt[9*x^2+4*y^2]/3
fp[x_,y_] = D[f[x,y],y]
fp[-2,1]

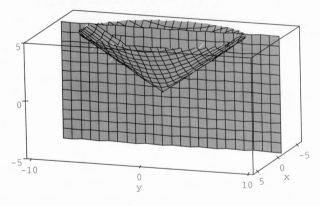

FIGURE 13.4.2

Maple:

Now define a function of two variables f as the upper nappe of the cone.

f:=(x,y)–>cone[2];

Find the partial of f with respect to y and substitute the x- and y-coordinates of the point $\left(-2, 1, \dfrac{\sqrt{40}}{3}\right)$ to find the slope of the curve.

m:=diff(f(x,y),y);

Output: $m := \dfrac{4}{3} \dfrac{y}{\sqrt{9x^2 + 4y^2}}$

subs(x=–2,y=1,m);

Output: $\dfrac{1}{30}\sqrt{40}$

Exercise Set 13.4

In Exercises 1–10, find $f_x(x, y)$ and $f_y(x, y)$.

1. $f(x, y) = x^2 + y^2$

2. $f(x, y) = \sqrt{x^2 + y^2}$

3. $f(x, y) = \sin xy$

4. $f(x, y) = e^x \cos y$

5. $f(x, y) = \dfrac{x + y}{x - y}$

6. $f(x, y) = \ln \dfrac{x}{y}$

7. $f(x, y) = x \ln y - y \ln x$

8. $f(x, y) = \tan^{-1} \dfrac{x}{y}$

9. $f(x, y) = \dfrac{xy}{\sqrt{x^2 - y^2}}$

10. $f(x, y) = e^{x-y} \sin(x - y)$

In Exercises 11–16, find $\partial z/\partial x$ and $\partial z/\partial y$.

11. $z = (1 - x^2 - y^2)^{-1/2}$

12. $z = \sin x \cosh y$

13. $z = \ln \cos(x - y)$

14. $z = \ln \left(\dfrac{x^2 - 2xy}{3xy - y^2}\right)$

15. $z = \sin^{-1} \sqrt{1 - x^2 y^2}$, $x > 0$, $y > 0$, $xy < 1$

16. $z = \sqrt{\dfrac{x - 2y}{x + 2y}}$

17. Find $f_x(3, -2)$ and $f_y(3, -2)$ if $f(x, y) = x^2 y - 2y^2$.

18. If $g(x, y) = 2xy/(x - y)$, find $g_x(-1, 1)$ and $g_y(-1, 1)$.

19. If $f(r, \theta) = e^r \cos \theta$, find $f_r\left(0, \dfrac{\pi}{3}\right)$ and $f_\theta\left(0, \dfrac{\pi}{3}\right)$.

20. If $w = \dfrac{u}{v} - \dfrac{v}{u}$, find $\partial w/\partial u$ and $\partial w/\partial v$ when $u = -2$ and $v = 2$.

21. If $w = 1/(2s - t^2)^2$, find $\partial w/\partial s$ and $\partial w/\partial t$ when $s = 3$ and $t = -2$.

22. Let $w = (u + v)/(u - v)$. Show that

$$v \dfrac{\partial w}{\partial u} + u \dfrac{\partial w}{\partial v} = 2w$$

23. Let $z = \tan^{-1} \dfrac{y}{x}$. Show that $x(\partial z/\partial y) - y(\partial z/\partial x) = 1$.

24. Find the slope of the curve of intersection of the cone $z = \sqrt{4x^2 + 3y^2}$ and
(a) the plane $y = 4$
(b) the plane $x = -2$ at the point $(-2, 4, 8)$.

25. Let $f(x, y) = 2x^2 - 3xy^2 + 3y^3$, and let C_1 and C_2 be the curves of intersection of the graph of f and the planes $y = 2$ and $x = 3$, respectively. Show that both C_1 and C_2 have horizontal tangent lines at the point $(3, 2, 6)$.

In Exercises 26–31, find all second-order partial derivatives of f.

26. $f(x, y) = \sqrt{x - 2y}$

27. $f(x, y) = \ln(x^2 + 3y^2)$

28. $f(x, y) = \sin xy$

29. $f(x, y) = e^{x^2 y}$

30. $f(x, y) = \dfrac{x - y}{x + y}$

31. $f(x, y) = e^{-x} \sin y + e^{-y} \cos x$

32. Let $f(x, y, z) = \dfrac{x - y}{y - z}$. Find f_x, f_y, and f_z.

In Exercises 33 and 34, verify that the given function satisfies the heat equation $ku_{xx} = u_t$.

33. $u(x, t) = 2(\cos 3x)e^{-9kt}$

34. $u(x, t) = (A \cos nx + B \sin nx)e^{-n^2 kt}$, where A, B, and n are constants.

In Exercises 35 and 36, verify that the given function satisfies the wave equation $y_{tt} = a^2 y_{xx}$.

35. $y(x, t) = \sin x \cos at$

36. $y(x, t) = e^x(A \cosh at + B \sinh at)$, where A and B are constants

*A function $u(x, y)$ is called **harmonic** if it satisfies Laplace's Equation in two variables, $u_{xx} + u_{yy} = 0$. In Exercises 37–40, show that u is harmonic at all points where u_{xx} and u_{yy} are defined.*

37. $u(x, y) = \cos x \cosh y$

38. $u(x, y) = \ln \sqrt{x^2 + y^2}$

39. $u(x, y) = \tan^{-1} \dfrac{y}{x}$

40. $u(x, y) = \dfrac{x}{x^2 + y^2}$

*If $u(x, y)$ is harmonic and $v(x, y)$ is a harmonic function such that $u_x = v_y$ and $u_y = -v_x$, then v is said to be a **harmonic conjugate** of u. In Exercises 41–44, show that v is a harmonic conjugate of u.*

41. $v(x, y) = -\sin x \sinh y$; $u(x, y)$ in Exercise 37

42. $v(x, y) = \tan^{-1} \dfrac{y}{x}$; $u(x, y)$ in Exercise 38

43. $v(x, y) = -\ln \sqrt{x^2 + y^2}$; $u(x, y)$ in Exercise 39

44. $v(x, y) = -\dfrac{y}{x^2 + y^2}$; $u(x, y)$ in Exercise 40

45. Let $u(x, y) = x^2 - y^2 + 2x + 1$ and $v(x, y) = 2xy + 2y$.
  (a) Show that u and v are harmonic.
  (b) Show that v is a harmonic conjugate of u. (See the instructions for Exercises 41–44.)
  (c) Show that the level curves of u and the level curves of v intersect at right angles, except for $u(x, y) = 0$ and $v(x, y) = 0$.
  (d) Draw several level curves for u and v on the same graph.

46. Verify that the function

$$f(x, y, z) = x^4 + y^4 + z^4 - 3(x^2 y^2 + x^2 z^2 + y^2 z^2)$$

satisfies Laplace's Equation in three variables.

47. Let $f(x, y, z) = xy \ln z + xz \ln y + yz \ln x$. Show the following:
  (a) $f_{xxy} = f_{xyx} = f_{yxx}$
  (b) $f_{xyy} = f_{yxy} = f_{yyx}$
  (c) $f_{xzz} = f_{zxz} = f_{zzx}$
  (d) The equality of all third-order partials of f with respect to x, y, and z, taken in any order

48. Suppose the cost function $C(x, y)$ for producing a certain commodity is given by

$$C(x, y) = 72 + x^2 + 2y^2 - xy + 3x + 4y$$

where x is the unit price of raw materials and y is the unit cost of labor. Find
  (a) the marginal cost of raw materials;
  (b) the marginal cost of labor.

49. Suppose the profit function $P(x, y, z)$ from the sale of a certain product is

$$P(x, y, z) = 30x + 50y + 80z - x^2 - 2yz - 3z^2$$

where x is the unit cost of raw materials, y is the unit cost of labor, and z is the unit cost of shipping and handling. If the current values of x, y, and z are $x = \$5$ per pound, $y = \$9$ per hour, and $z = \$10$ per item, find the approximate change in profit if
  (a) x is increased by \$1;
  (b) y is increased by \$1;
  (c) z is increased by \$1.

50. Suppose the production function $Q(K, L)$ where K represents available capital and L the size of the labor force, is given by

$$Q(K, L) = 325K^{0.72}L^{0.28}$$

Find the marginal productivity of capital and of labor.

51. In the Cobb-Douglas Production Function $Q(K, L) = AK^\alpha L^{1-\alpha}$, show that

$$K^2 Q_{KK} = L^2 Q_{LL}$$

52. Let $f(x, y) = (x^2 + y^2)^{1/3}$. Show that f is continuous at $(0, 0)$ but $f_x(0, 0)$ and $f_y(0, 0)$ do not exist.

53. Let

$$f(x, y) = \begin{cases} \dfrac{x^3 - y^3}{x^2 + y^2} & \text{if } (x, y) \neq (0, 0) \\ 0 & \text{if } (x, y) = (0, 0) \end{cases}$$

Use Definition 13.7 to show that $f_x(0, 0)$ and $f_y(0, 0)$ both exist. What are their values?

54. A Discontinuous Function Whose Partial Derivatives Exist

Let

$$f(x, y) = \begin{cases} \dfrac{xy}{x^2 + y^2} & \text{if } (x, y) \neq (0, 0) \\ 0 & \text{if } (x, y) = (0, 0) \end{cases}$$

Prove that $f_x(0, 0)$ and $f_y(0, 0)$ both exist but that f is discontinuous at $(0, 0)$.

55. A Function for Which $f_{xy} \neq f_{yx}$

Let

$$f(x, y) = \begin{cases} \dfrac{xy(x^2 - y^2)}{x^2 + y^2} & (x, y) \neq (0, 0) \\ 0 & (x, y) = (0, 0) \end{cases}$$

Show that $f_{xy}(0, 0) = -1$ but $f_{yx}(0, 0) = 1$.

56. Find the slope of the tangent line at $\left(1, 2, 1/e^3\right)$ to the curve of intersection of the surface $f(x, y) = e^{-x^2 - y^2}$ and the planes
(a) $x = 1$ and   (b) $y = 2$
In each case, make a plot of the surface and the intersecting plane.

57. Find the slope of the tangent line at $(0, 1, 1)$ to the curve of intersection of the surface $f(x, y) = x^2 + 2xy + y^2$ and the planes
(a) $x = 0$ and   (b) $y = 1$
In each case, make a plot of the surface and the intersecting plane.

58. Suppose $u(x, y) = f(x)g(y)$. Show that u satisfies the partial differential equation $uu_{xy} - u_x u_y = 0$.

Chapter 13 Review Exercises

1. Give the domain of f, and show it graphically.
(a) $f(x, y) = \ln(x^2 + y^2 - 1)$
(b) $f(x, y) = xy/\sqrt{x^2 - y^2}$

In Exercises 2–5, identify the surface and draw its graph.

2. (a) $4x^2 + y^2 + 4z = 8$
(b) $y = \sqrt{x^2 + y^2}$

3. (a) $y^2 - 2x - 4y + 4 = 0$
(b) $x^2 + y^2 + z^2 - 6x - 4y - 2z + 10 = 0$

4. (a) $x^2 + 4z^2 = 4$
(b) $9(x^2 + z^2) = 4(y^2 + 9)$

5. (a) $x^2 - 4y^2 + 4z = 0$
(b) $z = 1 + \tan^{-1} y$

6. Show the region in the first octant bounded by the given surfaces.
(a) $4x^2 + 3y^2 + 6z^2 = 48$, $x = 3$, $y = 2$, $z = 0$
(b) $4x^2 + y^2 = 4$, $z = 4 - y^2$, and the coordinate planes

7. Give the domain and range of f, identify its graph, and draw it.
(a) $f(x, y) = \sqrt{4 + x^2 + 4y^2}$

(b) $f(x, y) = 4 - x^2 - y^2$

8. Draw several level curves in part (a) and level surfaces in part (b) for both positive and negative values of the constant c.
(a) $f(x, y) = y^2 - x^2$
(b) $f(x, y, z) = \dfrac{x^2 + y^2}{z}$

9. A coordinate system is set up on a round metal plate with the origin at its center. A point source of heat is at the origin, and the temperature $T(x, y)$ in degrees Celsius at any point on the plate is given by

$$T(x, y) = 100e^{-(x^2 + y^2)/2}$$

Draw the isotherms $T(x, y) = C$ for $C = 80, 60, 40,$ and 20. How rapidly is the temperature changing in the direction of the positive y-axis at the point $(1, 3)$? Is it increasing or decreasing?

In Exercises 10 and 11, find f_x and f_y.

10. (a) $f(x, y) = e^{\sin xy}$
(b) $f(x, y) = \ln \dfrac{x^2 + 2y^2}{\sqrt{3x + 4y}}$

11. (a) $f(x, y) = \dfrac{y}{x} \cosh^2 \dfrac{x}{y}$

(b) $f(x, y) = \sin^{-1}\left(\dfrac{\sqrt{y^2 - 2x}}{y}\right), \ y > 0$

12. Let $w = (2u - v)/(u + 2v)$. Show that

$$u \frac{\partial w}{\partial u} + v \frac{\partial w}{\partial v} = 0$$

13. Show that $u(x, y) = \sin x \cosh y$ and $v(x, y) = \cos x \sinh y$ are harmonic conjugates. That is, show that each function satisfies Laplace's Equation in two variables and that $u_x = v_y$ and $u_y = -v_x$.

14. Let $f(x, y) = xy \ln \dfrac{x}{y}$. What is the domain of f? Find all second-order partials and show that $f_{xy} = f_{yx}$.

15. Let $w = \sqrt{z^2 - x^2 - y^2}$. Find each of the following:

(a) $\dfrac{\partial^2 w}{\partial z^2}$ (b) $\dfrac{\partial^2 w}{\partial x \, \partial z}$

(c) $\dfrac{\partial^2 w}{\partial y \, \partial z}$ (d) $\dfrac{\partial^2 w}{\partial y \, \partial x}$

16. Show that

$$\lim_{(x,y) \to (0,0)} \frac{\sin xy}{\sqrt{x^2 + y^2}} = 0$$

(*Hint:* $|\sin xy| \le |xy| \le x^2 + y^2$.)

17. Let $f(x, y) = (x^3 - 3x^2 y)/(x^2 + 2y^2)$ if $(x, y) \ne (0, 0)$. What value should be assigned to $f(0, 0)$ to make f continuous at the origin? Prove your result.

18. For $(x, y) \ne (0, 0)$ let

$$f(x, y) = (x^2 + y^2)\left(\sin \frac{1}{x^2 + y^2} + \cos \frac{1}{x^2 + y^2}\right)$$

and let $f(0, 0) = 0$. Use Definition 13.7 to show that $f_x(0, 0)$ and $f_y(0, 0)$ both exist and equal 0.

19. Prove that $\lim_{(x,y)\to(0,0)} f(x, y)$ does not exist for the following functions.

(a) $f(x, y) = \dfrac{x^3 - 2y^3}{3x^3 + 4y^3}$

(b) $f(x, y) = \dfrac{3xy^2}{2x^2 + 5y^4}$

20. When three resistors, with resistances of R_1 ohm, R_2 ohm, and R_3 ohm, respectively, are connected in parallel, their combined resistance R satisfies the equation

$$\frac{1}{R} = \frac{1}{R_1} + \frac{1}{R_2} + \frac{1}{R_3}$$

Find $\partial R / \partial R_1$. What can you conclude about $\partial R / \partial R_2$ and $\partial R / \partial R_3$?

21. The Ideal Gas Law relates the pressure P, the volume V, and the temperature T of a gas by the equation $PV = nRT$, where n and R are constant. Show that

$$\frac{\partial P}{\partial V} \frac{\partial V}{\partial T} \frac{\partial T}{\partial P} = -1$$

22. Show that each of the following functions satisfies the heat equation $ku_{xx} = u_t$.

(a) $u(x, t) = e^{-n^2 \pi^2 kt} \sin n\pi x$

(b) $u(x, t) = \sin x (\cosh kt - \sinh kt)$

23. Show that each of the following functions satisfies the wave equation $y_{tt} = a^2 y_{xx}$.

(a) $y(x, t) = \ln(x^2 - a^2 t^2)$

(b) $y(x, t) = Ae^{x+at} + Be^{x-at}$

24. Show that each of the following functions satisfies Laplace's Equation in three variables.

(a) $u(x, y, z) = \sin 3x \sin 4y \sinh 5z$

(b) $u(x, y, z) = \cos 5x \cos 12y \cosh 13z$

25. For the Cobb-Douglas Production Function $Q(K, L) = 30K^{0.6} L^{0.4}$, find and interpret $Q_K(240, 150)$ and $Q_L(240, 150)$.

26. Suppose the revenue $R(x, y)$ of a company as a function of advertising expenditures in printed form and on TV is given by

$$R(x, y) = 20x^{3/2} + 48y^{1/2} + 2xy$$

where x is the amount spent on TV and y is the amount spent on print advertising (both in hundreds of thousands of dollars). If the current expenditures are $x = 25$ and $y = 16$, use marginal analysis to approximate the additional revenue that would result from

(a) changing x to 26, leaving y alone;

(b) changing y to 17, leaving x alone.

27. Let

$$f(x, y) = \begin{cases} \dfrac{2x^3 - y^3}{x^2 + 2y^2} & \text{if } (x, y) \ne (0, 0) \\ 0 & \text{if } (x, y) = (0, 0) \end{cases}$$

(a) Use Definition 13.7 to find $f_x(0, 0)$.

(b) Find $f_x(x, y)$ for $(x, y) \ne (0, 0)$ using the Quotient Rule.

(c) Show that f_x is not continuous at the origin.

(d) Carry out steps (a), (b), and (c) for f_y.

28. Let

$$f(x, y) = \begin{cases} \dfrac{x^3 y - x y^3}{x^2 + y^2} & \text{if } (x, y) \neq (0, 0) \\ 0 & \text{if } (x, y) = (0, 0) \end{cases}$$

(a) Prove that f is continuous at $(0, 0)$.
(b) Find $f_x(0, y)$ and $f_y(x, 0)$ using Definition 13.7.
(c) From part (b), find $f_{xy}(0, 0)$ and $f_{yx}(0, 0)$.
(d) What conclusion about the continuity of f_{xy} and f_{yx} can you draw from part (c)?

29. A set is said to be **open** if it does not contain any of its boundary points. Prove that a nonempty set S in $\mathbb{R}^2$ or $\mathbb{R}^3$ is open if and only if, for every point P in S, all points inside some circle (in $\mathbb{R}^2$) or sphere (in $\mathbb{R}^3$) centered at P lie in S.

30. If S is a set, its **complement** is the set of all points not in S. A set is said to be **closed** if it contains all of its boundary points. Prove that a set S in $\mathbb{R}^2$ or $\mathbb{R}^3$ is closed if and only if its complement is open.

Chapter 13 Concept Quiz

1. Define each of the following:
  (a) The graph of a function f of two variables
  (b) A level curve of f
  (c) The partial derivatives $f_x(x, y)$ and $f_y(x, y)$
  (d) $\lim_{(x,y) \to (x_0, y_0)} f(x, y) = L$
  (e) Continuity of a function f of two variables at (x_0, y_0)

2. Identify each of the following surfaces, where a, b, c, and d are positive constants.
  (a) $x = ax^2 + by^2$
  (b) $z = ax^2$
  (c) $x^2 + y^2 + z^2 = a^2$
  (d) $ax^2 - by^2 + cz^2 = d$
  (e) $ax^2 + by^2 + cz^2 = d$
  (f) $ax^2 - by^2 - cz^2 = d$
  (g) $ax^2 - by^2 - cz^2 = 0$
  (h) $z = ax^2 - by^2$

3. (a) State a sufficient condition for $f_{xy}(x_0, y_0)$ to equal $f_{yx}(x_0, y_0)$.
  (b) Explain a way of showing that $\lim_{(x,y) \to (x_0, y_0)} f(x, y)$ does not exist.

4. (a) Explain the geometric significance of the partial derivative $f_x(x_0, y_0)$ and $f_y(x_0, y_0)$.
  (b) Explain the geometric nature of a contour map of a function of three variables.

5. Determine which of the following statements are true and which are false.
  (a) If the limit of $f(x, y)$ is L as (x, y) approaches (x_0, y_0) along every straight line through (x_0, y_0), then

$$\lim_{(x,y) \to (x_0, y_0)} f(x, y) = L$$

  (b) If $f_x(x_0, y_0)$ and $f_y(x_0, y_0)$ both exist, then f is continuous at (x_0, y_0).
  (c) If f and g are both continuous at (x_0, y_0), then so is f/g.
  (d) A three-dimensional surface is the graph of a function $z = f(x, y)$ if and only if every line parallel to the z-axis intersects the surface in at most one point.
  (e) The trace of the surface $z = f(x, y)$ in the plane $y = y_0$ has slope $f_x(x_0, y_0)$.

APPLYING CALCULUS

1. To estimate the value of a tree for lumber, the probable number of board feet in various sections of the tree are calculated. (A board 1 ft long with a cross-section of 12 in.² is one board-foot.) The following formula, called the *Doyle Log Rule*, is used for this calculation:

$$N(d, L) = \left(\frac{d-4}{4}\right)^2 L$$

where d is the diameter of the log in inches, L is its length in feet, and $N(d, L)$ is the number of board feet of lumber that can be obtained from the log.

CAS **(a)** Use a CAS to obtain the graph of $N(d, L)$ for $5 \le d \le 50$ and $1 \le L \le 40$.

(b) The lower part of a tree, before the first branch, is the most valuable, since it will have fewer knots and its diameter is greatest. Suppose that the first 17 ft of a tree has minimum diameter 24 in. and is valued at $1 per board foot, the next 17 ft has minimum diameter 18 in. and is valued at 30¢ per board foot, and the next 25 ft has minimum diameter 12 in. and is valued at 20¢ per board foot. Determine how many board feet there are in each section. What is the estimated value of the tree?

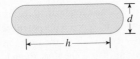

2. The larvae of certain insects have approximately the shape of a circular cylinder capped at each end by a hemisphere (see the figure). Let the length of the cylinder be h and its diameter d.

(a) Express the volume V and the surface area S as functions of h and d.

(b) The rate R of absorption of a chemical substance into the larvae is directly proportional to S and inversely proportional to V. Show that $\dfrac{\partial R}{\partial h} < 0$ and $\dfrac{\partial R}{\partial d} < 0$.

3. The temperature T in degrees Celsius at the point (x, y) on a flat plate in the form of the disk $\{(x, y): x^2 + y^2 \le 100\}$ is given by

$$T(x, y) = 100e^{-\sqrt{x^2+y^2}}$$

Suppose a bug is on the plate at the point $(4, -3)$.

(a) If the bug moves upward on the line $x = 4$, does the temperature increase or decrease? What is the instantaneous rate of change?

(b) If, instead, the bug moves to the left along the line $y = -3$, answer the questions in part (a).

(c) If the bug wants to travel in a path where the temperature doesn't change, what path should it take?

(d) If the bug heads on a straight path toward the origin, find the instantaneous rate of change of temperature in that direction. (*Hint:* Find a unit vector pointing toward the origin and use it to find a vector equation of the line. Then write T in terms of the parameter.)

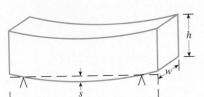

4. A rectangular beam of height h, width w, and length L is supported at each end and is subjected to a uniform load p. The deflection of the beam at its midpoint is called the sag, s, and is given by

$$s = C\frac{pL^3}{wh^3}$$

where C is a constant depending on the type of material and the units of measurement.

(a) Find $\partial s/\partial L$, $\partial s/\partial w$, and $\partial s/\partial h$. How would you interpret the results?

(b) Verify each of the following relationships:

$$L\frac{\partial s}{\partial L} + h\frac{\partial s}{\partial h} = 0, \qquad L\frac{\partial s}{\partial L} + 3w\frac{\partial s}{\partial w} = 0, \qquad h\frac{\partial s}{\partial h} = 3w\frac{\partial s}{\partial w}$$

(c) Suppose the length and the load are fixed. If one of the dimensions w or h can be increased by one unit, which one would you choose in order to reduce the sag the most? Justify your answer.

5. In the wave equation (Equation 13.17), giving the vertical displacement in a stretched string at time t at a point x units from the origin, it can be shown that the constant a^2 is F_0/μ, where F_0 is the tension in the string and μ is its mass per unit length. Show that

$$y(x, t) = y_0 \cos(kx - \omega t)$$

satisfies the equation provided ω and k are related to μ and F_0 by the equation

$$\frac{\omega}{k} = \sqrt{\frac{F_0}{\mu}}$$

(*Note:* The constants w and k are called *angular frequency* and *wave number*, respectively. The ratio ω/k is the wave speed.)

6. The equation

$$-\frac{h^2}{8\pi^2 m}\left(\frac{\partial^2 \psi}{\partial x^2} + \frac{\partial^2 \psi}{\partial y^2} + \frac{\partial^2 \psi}{\partial z^2}\right) + V\psi = E\psi$$

is called the three-dimensional, time-independent Schrödinger equation and is the basis for quantum mechanics. It was discovered by Erwin Schrödinger in 1926. The equation describes the motion of a particle of mass m moving in space with potential energy V and kinetic energy E. The function ψ has the property that ψ^2 is the probability density function for the position of the particle. The constant h is known as *Planck's constant* and is equal to 6.63×10^{-34} J $\cdot$ s. If the particle is confined to the cube $\{(x, y, z): 0 \le x \le L, 0 \le y \le L, 0 \le z \le L\}$ and the potential energy is zero, show that a solution to Schrödinger's equation is

$$\psi(x, y, z) = A \sin\left(\frac{n_x \pi x}{L}\right) \sin\left(\frac{n_y \pi y}{L}\right) \sin\left(\frac{n_z \pi z}{L}\right)$$

where the *quantum numbers* n_x, n_y, and n_z satisfy

$$E = \frac{h^2}{8mL^2}\left(n_x^2 + n_y^2 + n_z^2\right)$$

7. In an economy in competitive equilibrium, the real wage = the marginal product of labor and the real interest rate = the marginal product of capital. If Y is the aggregate production function, K is the capital factor of production, and L is the labor factor of production, then $Y = Y(K, L)$. (Aggregate means that all kinds of goods are combined as a single "good.")

(a) Since the real wage is the marginal product of labor, write the formula for the real wage W in terms of the production function Y.

(b) For the Cobb-Douglas Production Function

$$Y = K^{1-\alpha} L^\alpha, \quad 0 < \alpha < 1,$$

find W in terms of α, K, and L.

(c) Show that the labor share of the national product, WL/Y, is constant. (*Note:* This constant labor share is now about 3/4.)

(d) Show that the same calculation for the capital share of the national market gives a value that, when added to the constant in part (c), gives 1.

8. The electric field in a region can be expressed as the plane wave

$$\mathbf{E}(x, t) = E_0 \sin(kx - \omega t)\mathbf{j}$$

and the magnetic field as the plane wave

$$\mathbf{B}(x, t) = B_0 \sin(kx - \omega t)\mathbf{k}$$

where E_0 and B_0 are constants.

How the magnitude E of the electric field changes with position is related to the way the magnitude B of the magnetic field changes with time by **Faraday's Law**, named after the nineteenth-century British physicist Michael Faraday:

$$\frac{\partial E}{\partial x} = -\frac{\partial B}{\partial t}$$

The electric and magnetic fields are also related by **Ampère's Law**, named after the nineteenth-century French physicist André Marie Ampère:

$$\frac{\partial B}{\partial x} = -\mu_0 \varepsilon_0 \frac{\partial E}{\partial t}$$

where ε_0 is an electric constant, the *permittivity constant*, and μ_0 is a magnetic constant, the *permeability*.

(a) The speed of a wave is ω/k. Express the speed of electromagnetic waves—waves of simultaneously changing electric and magnetic fields—in terms of ε_0 and μ_0.

(b) Given that $\varepsilon_0 = 8.85 \times 10^{-12}$ F/m and $\mu_0 = 4\pi \times 10^{-7}$ H/m, where F/m is farads/meter and H/m is henries/meter, solve for the wave speed in part (a) numerically. You may use the fact that $1 \text{ F} \cdot \text{H} = 1 \text{ s}^2$.

(c) Relate the result of part (c) to the observed value of the speed of light, $c = 3.00 \times 10^8$ m/s. James Clerk Maxwell, in the 1860s, used this result to show that light is an electromagnetic wave.

(d) The angular velocity is $2\pi f$, where f is the frequency, and the wave constant is $k = 2\pi/\lambda$, where λ is the wave length. How are f, λ, and c related?

David Lieberman

Is calculus still the best way to start studying mathematics in college?

I think calculus is fundamental to understanding all of the theoretical and computational mathematical developments that go on today. A thorough knowledge of calculus is required to understand algorithms that computers are implementing and to know how one might use those algorithms to adapt to particular problems one faces. Also, to check whether computer results are reasonable, the techniques of calculus are valuable for approximating the answers that one expects. Just on the side, I think that more and more of the calculations—even symbolic calculations—are going to be done by computer, and the computer will replace the purely calculational aspect of calculus; but I think to understand how to use the computer, it is still going to require a deep appreciation and understanding of the methodology of the calculus.

Just what is it that you do?

I work on the mathematical foundations of cryptology and related subjects, such as processing speech by computer. I also do statistical analyses, trying to find statistical structure in data.

What kind of mathematics is involved with cryptology and speech processing?

Many modern cryptographic schemes attempt to use intractable mathematical problems as the basis for their security. One well-known scheme builds its security on the intractability of the problem of factoring large integers into prime factors. (Currently, to find the primes dividing a 500-digit number would require centuries running on the world's fastest computers.) Consequently, many areas of pure mathematics, particularly number theory and algebraic geometry, are playing an increasing role in the design and analysis of communications systems.

Mathematics also continues to play a central role in statistical analysis of data sets. A standard problem is that one has a set of data and needs to identify which of a large number of possible hypotheses best explain the properties of the data.

A simple case of this kind of problem is that in which the data are a collection of numbers, normally distributed, and the different hypotheses are distinguished by the expected values of the mean and standard deviation of the observed data set. Such problems are analyzed at length in any elementary statistics book. This analysis already requires interesting techniques from calculus.

A more complex problem concerns multivariate data, i.e., the case in which one observes n-tuples of numbers. You ask how the ith number varies when the jth number changes. And if you just make those measurements—that is, the correlation between the ith number and the jth number—you get a matrix. Those statistics—that is, the mean vector of the data and the covariance matrix—are all the parameters one needs to find the best fit—gaussian or multivariate gaussian—to the data. Then,

when other data come in, you can ask how well do they fit. You can calculate the probability that the data look like the old data by measuring them and using the covariance matrix to define a distance measure. That kind of simple analysis you will find in most statistics books. It already involves multiple variables, fitting gaussian distributions, and so on—so it involves a lot of multivariable calculus. But if one tries to build more complicated models for the data that take into account higher-order effects on the data, then one needs to use multivariable calculus to derive all the corresponding theorems—properties of the new sets of models. In particular, it is frequently much more difficult to estimate the parameters of the model directly from the data. The problem becomes an implicit problem in which one has to find the parameters that give the highest probability of an observed data set—the maximum-likelihood fitting of the parameters of the data. That typically involves a great deal of multivariable calculus, particularly maximization by iterative techniques. One studies how quickly such techniques converge, how often they converge, and so on. Another area has to do with formulation of hypotheses about data. That is, if you observe trends in data, you may build a model that takes account of those trends. And then you want to know whether you have taken account of all the information—whether the data exhibit roughness not predicted by the model you had so far. And you need a measure of how much additional information might be present in the data. That method is known as "maximal-entropy modeling." Fitting maximal-entropy models to data and analyzing residual information involves difficult analytic calculations. Calculus provides the theoretical framework but the ultimate calculations have to be done using a computer. Frequently you need to develop the computer algorithm yourself based on theoretical calculations made using multivariable calculus.

Another area of mathematics that is very hot—that is, very active—today is closely related to hard questions of combinatorial optimization. Finding, say, the shortest path through graphs or the best order in which to put objects in order to maximize some quantity. What one has is a large number of discrete variables, rather than continuously varying quantities. They take on a large but finite number of values. You must try all the values to find which values give the best answer. But the general ideas that one has for trying to solve such problems are based on continuous models that look like the discrete problem. And then you try to use the techniques of optimization, searching for improvement along gradient-like directions in order to find the answers—good approximate answers—to the intractable problems.

What uses does cryptography have today?

More and more messages and contracts are sent over computer; contract information and financial information are being sent digitally over telephone lines. To protect that information from eavesdroppers or tampering, banks and financial institutions are eager to encode and encrypt the information to prevent unauthorized listeners from hearing the information or being able to falsely identify themselves as people with access to that information.

Cryptography could be used to permit one to sign documents digitally so that you can prove who you are and prove that you signed the documents over a computer line without actually being there. Thus, one could provide authorization for financial transactions by computer, while protecting oneself from forgery and fraud. Fundamentally, cryptography is used as a lock and a key for protecting information.

How did you become interested in cryptography?

Originally, I applied for a summer job when I was an undergraduate without knowing what the work was going to be, but only knowing that they were looking for mathematicians. When I got there, the subject was extremely mathematical, and used a great deal of multivariate calculus and linear algebra, and I found the subject extremely intriguing. I served as a consultant for many years while a professor at a university, helping develop new tools for the government to use in its cryptologic program.

Is this a growing field with opportunities for new graduates?

Very much so. As more and more information is transmitted electronically, the need to protect that information and to permit long-distance contractual transactions will replace all kinds of written communications and paper checks and contracts. I believe it will all be done in the future electronically with greater assuredness of authenticity and greater safeguards than can be provided by traditional means. This type of remote contractual transaction is increasing at a great rate today. It is becoming ever more important, as the world is being wired up.

Do you have hope for speech recognition improving? Will I be able to stop typing at my computer and start talking to it?

A lot of the work we did here, early on, concerned the development of statistical models of speech in the hopes of allowing the computer to recognize and automatically transcribe human speech. Using mathematical models we have created here, a lot of theoretical and developmental work has been carried out by IBM, Bell Laboratories, BBN, and several universities. There now exist several speech-recognition systems that work with a 20,000 word vocabulary—and without the necessity for the speaker to pause in between words. But there still remains a lot to do to improve the performance of those systems, both in terms of vocabulary and in terms of making them independent of the speaker, to eliminate the necessity for each person to train the computer to recognize the way he pronounces words. Another area where multivariate calculus is used is in the design of codes to correct and detect errors in transmitted or stored data—the subject is called "the theory of error-correcting codes." And mathematics plays two roles there—major roles. One is the theory of how you actually find the best reconstruction of an original message given a corrupted message, and secondly, how do you design the set of messages so that they are not likely to be corrupted into confusion with one another. And those ques-

tions involve again combinatorial optimization, searching over large finite collections of things for ways of organizing them, with searches guided by analogous continuous questions.

Do you still find that mathematics is fun?

Very much. I enjoy thinking about mathematical problems. I get a tremendous satisfaction in seeing how mathematics will solve what appears to be a very difficult problem by simply organizing it and exposing its essential elements. I think that we continue to be surprised by the connections between the many fields of mathematics—geometry, analysis (including calculus), algebra. Most remarkable is the use of calculus techniques to obtain profound information about the theory of numbers. For example, the recent work of Wiles on Fermat's Last Theorem, which is apparently a purely number-theoretic question, might be thought to require discrete rather than continuous methods. But the method of solution grows out of work done 200 years ago on evaluating integrals for calculating the arc length of ellipses. These integrals led to the study of elliptic curves in algebraic geometry. This work was done long ago by Abel and Gauss and was then developed by the analysts and number theorists of the last two centuries.

David Lieberman is a Research Mathematician on the staff of the Center for Communications Research in Princeton, NJ. He majored in mathematics at Harvard and obtained his Ph.D. in 1966 from M.I.T., writing his thesis in the field of algebraic geometry. After two years at the Institute for Advanced Study at Princeton, he taught mathematics at Brandeis University for 10 years. He served as deputy director and director of the Institute for Defense Analyses at Princeton.

14

MULTIVARIABLE DIFFERENTIAL CALCULUS

We begin this chapter by defining what it means for a function of two variables to be *differentiable*. Most of the remainder of the chapter is devoted to analyzing some of the important consequences of differentiability. In particular, we will define the *differential* of a differentiable function, and we will see that near a given point the differential can be used to approximate the change in the function produced by a small change in x and a small change in y. We will also consider, for compositions of differentiable functions, various forms of chain rules for calculating derivatives, or partial derivatives, depending on the nature of the functions involved.

As we have seen in Section 13.4, the partial derivatives f_x and f_y give the rate of change of the function f at a given point in the x direction and y direction, respectively. These two partials are special cases of a more general derivative, called the *directional derivative*, that measures the rate of change of f in any prescribed direction.

We know that a function of one variable has a tangent line at each point on its graph at which the function is differentiable. Similarly, we will see that the graph of a function of two variables has a *tangent plane* at each point where the function is differentiable.

We conclude the chapter by analyzing how to find maximum and minimum values of a function of two variables. Here, again, we will see parallels with the one-variable case.

14.1 DIFFERENTIABILITY

In order to define the notion of differentiability for a function of two variables, let us first restate its meaning for a function of one variable. Let f be such a function, and suppose its derivative $f'(x_0)$ at x_0 exists. Then, by the definition of the derivative,

$$f'(x_0) = \lim_{\Delta x \to 0} \frac{f(x_0 + \Delta x) - f(x)}{\Delta x}$$

For brevity, we denote the numerator of the difference quotient on the right by Δf:

$$\Delta f = f(x_0 + \Delta x) - f(x_0)$$

We call Δf the *increment* in the function that corresponds to the increment Δx in x. So we can write

$$f'(x_0) = \lim_{\Delta x \to 0} \frac{\Delta f}{\Delta x}$$

We now denote the difference between $\Delta f / \Delta x$ and its limiting value $f'(x_0)$ by ε:

$$\varepsilon = \frac{\Delta f}{\Delta x} - f'(x_0)$$

Then $\varepsilon \to 0$ as $\Delta x \to 0$. Solving for Δf gives

$$\Delta f = f'(x_0)\Delta x + \varepsilon \Delta x \qquad (14.1)$$

where $\varepsilon \to 0$ as $\Delta x \to 0$. Equation 14.1, along with the stated condition on ε, can be taken as an alternative definition of differentiability for the function f at x_0, equivalent to the condition that $f'(x_0)$ exists.

Recall that we defined the *differential* of a differentiable function f of one variable by

$$df = f'(x)dx$$

where dx is an independent variable. (We can also denote the differential of $y = f(x)$ by dy.) If we let $dx = \Delta x$ and take $x = x_0$, then we see that the first term on the right-hand side of Equation 14.1 is df, so we can write that equation as

$$\Delta f = df + \varepsilon \Delta x$$

Since $\varepsilon \to 0$ as $\Delta x \to 0$, the differential df approximates the increment Δf for small values of Δx.

In Figure 14.1, we illustrate the relationships among the quantities involved in Equation 14.1. Observe that whereas Δf is the change in the y value of the *curve* $y = f(x)$ caused by changing x from x_0 to $x_0 + \Delta x$, df is the change in the y value of the *tangent line* at $(x_0, f(x_0))$ caused by this change in x. In the equation $\Delta f = df + \varepsilon \Delta x$, when we substitute the values of Δf and df and omit the "error term" $\varepsilon \Delta x$, we can solve for $f(x_0 + \Delta x)$ to obtain the *linear approximation* formula,

$$f(x_0 + \Delta x) \approx f(x_0) + f'(x_0)\Delta x$$

We can also write this result in the form

$$f(x) \approx f(x_0) + f'(x_0)(x - x_0)$$

where $x = x_0 + \Delta x$. The equation $y = f(x_0) + f'(x_0)(x - x_0)$ is the equation of the tangent line at $(x_0, f(x_0))$, which accounts for the name *linear* approximation.

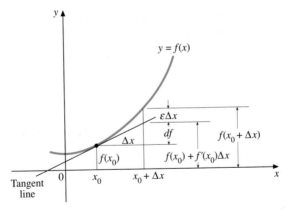

FIGURE 14.1

Now we extend these ideas to two dimensions. We begin by stating a theorem establishing an increment formula analogous to Equation 14.1. We give a proof in Appendix 5.

THEOREM 14.1

Fundamental Increment Formula

If the partial derivatives f_x and f_y of a function of two variables both exist inside some circle centered at (x_0, y_0) and are continuous at (x_0, y_0), then for all sufficiently small values of Δx and Δy, the increment $\Delta f = f(x_0 + \Delta x, y_0 + \Delta y) - f(x_0, y_0)$ can be written in the form

$$\Delta f = f_x(x_0, y_0)\Delta x + f_y(x_0, y_0)\Delta y + \varepsilon_1 \Delta x + \varepsilon_2 \Delta y \qquad (14.2)$$

where $\varepsilon_1 \to 0$ and $\varepsilon_2 \to 0$ as $(\Delta x, \Delta y) \to (0, 0)$.

By analogy with the one-variable case, we take Equation 14.2 as the definition of differentiability.

Definition 14.1
Differentiability of a
Function of Two Variables

A function f of two variables is **differentiable** at (x_0, y_0) provided the increment $\Delta f = f(x_0 + \Delta x, y_0 + \Delta y) - f(x_0, y_0)$ can be written in the form

$$\Delta f = f_x(x_0, y_0)\Delta x + f_y(x_0, y_0)\Delta y + \varepsilon_1 \Delta x + \varepsilon_2 \Delta y$$

where $\varepsilon_1 \to 0$ and $\varepsilon_2 \to 0$ as $(\Delta x, \Delta y) \to (0, 0)$.

One immediate consequence of differentiability of f at a point is that f is continuous there. In order to see why, write the condition for continuity (Definition 13.6),

$$\lim_{(x,y)\to(x_0,y_0)} f(x, y) = f(x_0, y_0)$$

in the equivalent form

$$\lim_{(x,y)\to(x_0,y_0)} [f(x, y) - f(x_0, y_0)] = 0$$

If we let $\Delta x = x - x_0$ and $\Delta y = y - y_0$, so that $f(x, y) = f(x_0 + \Delta x, y_0 + \Delta y)$, we see that the condition for continuity becomes

$$\lim_{(\Delta x, \Delta y) \to (0,0)} \Delta f = 0$$

which is true if f is differentiable, since each term on the right-hand side of Equation 14.2 approaches 0 as Δx and Δy approach 0.

We state this result again for emphasis.

Differentiability Implies Continuity

If a function of two variables is differentiable at a point, then it is continuous there.

REMARKS ─────────────────────────────────────

■ The converse of this result is not true, since there are continuous functions that are not differentiable (see, for example, Exercise 37 in Exercise Set 14.1).

■ Although differentiability at a point requires that f_x and f_y exist there, mere existence of these partial derivatives does not imply differentiability. In fact, the function in Exercise 54 of Exercise Set 13.4 is discontinuous, hence not differentiable, at $(0, 0)$, even though $f_x(0, 0)$ and $f_y(0, 0)$ both exist.

It can be difficult to test for differentiability directly from Definition 14.1. Theorem 14.1, however, provides sufficient conditions that often enable us to conclude that a function is differentiable without having to apply the definition. We state the result as Theorem 14.2.

THEOREM 14.2

If f is a function of two variables such that f_x and f_y exist at all points inside some circle centered at (x_0, y_0) and are continuous at (x_0, y_0), then f is differentiable at (x_0, y_0).

The key word here is *continuous*. It is not enough that f_x and f_y merely exist at (x_0, y_0) and at all nearby points, but if these partials are also continuous at (x_0, y_0), differentiability is assured.

By Theorem 14.2, we can conclude immediately that all *polynomial functions* in two variables are differentiable everywhere in the xy-plane, since the partial derivatives are also polynomials and hence are continuous everywhere. Similarly, all *rational functions* of two variables are differentiable except where the denominator is zero. Furthermore, if $f(x, y)$ is differentiable and g is a differentiable function of one variable whose domain includes the range of f, then the composite function $g(f(x, y))$ is a differentiable function of two variables. For example, e^{xy}, $\sin(2x - 3y)$, and $\ln(x^2 + y^2)$ are all differentiable (in the case of $\ln(x^2 + y^2)$ we must exclude $(0, 0)$).

Geometric Interpretation

When Δx and Δy are sufficiently small, we can conclude by Equation 14.2 that if f is differentiable at (x_0, y_0), then

$$f(x_0 + \Delta x, y_0 + \Delta y) \approx f(x_0, y_0) + f_x(x_0, y_0)\Delta x + f_y(x_0, y_0)\Delta y \quad (14.3)$$

as can be seen by writing $\Delta f = f(x_0 + \Delta x, y_0 + \Delta y) - f(x_0, y_0)$ and deleting the terms $\varepsilon_1 \Delta x$ and $\varepsilon_2 \Delta y$. In order to understand the geometric significance of this approximation, let us again write $x = x_0 + \Delta x$ and $y = y_0 + \Delta y$. Then Equation 14.3 becomes

$$f(x, y) \approx f(x_0, y_0) + f_x(x_0, y_0)(x - x_0) + f_y(x_0, y_0)(y - y_0)$$

The function on the right is linear (first-degree) in x and y, so its graph is a plane. If we set z equal to this function, and $z_0 = f(x_0, y_0)$, we can write the equation of this plane as

$$z - z_0 = f_x(x_0, y_0)(x - x_0) + f_y(x_0, y_0)(y - y_0) \quad (14.4)$$

Note that this plane passes through the point $P_0 = (x_0, y_0, z_0)$ and so coincides with the surface $z = f(x, y)$ there. In Section 14.4, we will define what is meant by a *tangent plane* to a surface and will show that Equation 14.4 is in fact the tangent plane to the surface $z = f(x, y)$ at the point (x_0, y_0, z_0). Now we can interpret the linear approximation given by Equation 14.3, as shown in Figure 14.2. Observe the analogy with Figure 14.1, in which the tangent plane takes the place of the tangent line. Since the right-hand side of Equation 14.3 approximates the value of f at a point (x, y) near (x_0, y_0) by the value of z on the tangent plane (the graph of a linear function), Equation 14.3 is called a *linear approximation* formula.

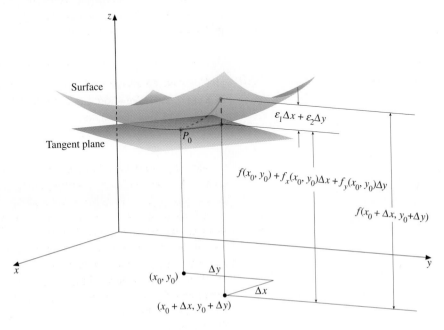

FIGURE 14.2

The Differential

The differential df of a function of one variable was defined in such a way that the increment formula (Equation 14.1) became

$$\Delta f = df + \varepsilon \Delta x$$

so that for Δx small, $df \approx \Delta f$. We define the differential of a function of two variables analogously, so that Equation 14.2 can be written as

$$\Delta f = df + \varepsilon_1 \Delta x + \varepsilon_2 \Delta y$$

Then it will follow that $df \approx \Delta f$ for Δx and Δy small.

Definition 14.2 The Differential of a Function of Two Variables	If f is differentiable at (x, y), then its **differential** df is given by $$df = f_x(x, y)dx + f_y(x, y)dy \qquad (14.5)$$ where dx and dy are independent variables.

REMARK ────

■ If we set $z = f(x, y)$ then we can use dz instead of df for the differential. We can also use the Leibniz notation for the partial derivatives, writing

$$dz = \frac{\partial z}{\partial x}dx + \frac{\partial z}{\partial y}dy$$

If we evaluate the differential at (x_0, y_0) with $dx = \Delta x$ and $dy = \Delta y$, then we see that df is the same as the right-hand side of Equation 14.2 except for the terms $\varepsilon_1 \Delta x$ and $\varepsilon_2 \Delta y$. Thus, as we anticipated,

$$\Delta f = df + \varepsilon_1 \Delta x + \varepsilon_2 \Delta y$$

and $df \approx \Delta f$ if Δx and Δy are small. The error in this approximation is $\varepsilon_1 \Delta x + \varepsilon_2 \Delta y$. We show the relationship between Δf and df in Figure 14.3.

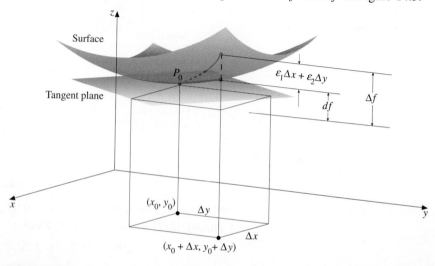

FIGURE 14.3

Applications of the Differential

As one application of the differential, suppose x and y are measured quantities, subject to possible errors of dx and dy, respectively. Suppose also that $z = f(x, y)$. Then the error in calculating z is approximately dz. The **relative error** is the true error divided by the calculated value. Thus, the relative error is approximated by dz/z. The relative error is often more important than the error itself. For example, an error of 0.1 cm in a calculated value of 2 cm gives a relative error of $\frac{0.1}{2} = 0.05$, whereas an error of 0.1 cm in a measured value of 2 m = 200 cm is $\frac{0.1}{200} = 0.0005$. Clearly, in the second case the relative error is much less serious than in the first case, even though the absolute error of 0.1 cm is the same in both cases. We summarize this discussion of errors below.

Suppose $z = f(x, y)$ is differentiable and that x and y are in error by amounts dx and dy, respectively. Then the approximate absolute error, relative error, and percentage error in z are as follows:

Approximate Absolute, Relative, and Percentage Errors

Absolute error in $z \approx dz$.

Relative error in $z \approx \dfrac{dz}{z}$.

Percentage error in $z \approx 100\dfrac{dz}{z}$.

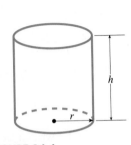

Estimating error tolerances is very important in manufacturing barrels to keep hazardous waste contained.

EXAMPLE 14.1   A can is in the shape of a right circular cylinder. The radius and height are measured as 0.40 m and 1.20 m, respectively. If the error in the radius is at most 0.005 m and the error in the height is at most 0.002 m, find the approximate maximum values of (a) the absolute error, (b) the relative error, and (c) the percentage error in the calculated value of the volume. Also find the actual absolute error in the volume, assuming the maximum errors in the radius and the height.

Solution

(a)   The volume of the can (see Figure 14.4) is given by

$$V = \pi r^2 h$$

We will use the differential dV to approximate the actual error ΔV. By Equation 14.5,

$$dV = \frac{\partial V}{\partial r}dr + \frac{\partial V}{\partial h}dh$$

$$= (2\pi rh)dr + (\pi r^2)dh$$

FIGURE 14.4
Volume =
(Area of Base) × (Altitude)
$= \pi r^2 h$

We take $dr = 0.005$ and $dh = 0.002$. Then, with the measured values of $r = 0.40$ and $h = 1.20$, we obtain

$$dV = 2\pi(0.40)(1.20)(0.005) + \pi(0.40)^2(0.002)$$

$$= 0.00512\pi \approx 0.01608$$

Thus, with the given measurements and errors in those measurements, the approximate absolute error in measuring the volume is 0.0161 m³.

(b) With $r = 0.40$ and $h = 1.20$, the volume is

$$V = \pi(0.40)^2(1.20) = 0.192\pi \approx 0.603 \text{ m}^3$$

The approximate relative error is therefore

$$\frac{dV}{V} = \frac{0.00512\pi}{0.192\pi} \approx 0.0267$$

(c) The approximate percentage error is

$$100\frac{dV}{V} = 2.67\%$$

The actual absolute error is ΔV, found by calculating V using $r = 0.405$ and $h = 1.202$ and subtracting the value of V using $r = 0.40$ and $h = 1.20$.

$$\Delta V = \pi(0.405)^2(1.202) - \pi(0.40)^2(1.20)$$

$$= 0.005158\pi \approx 0.0162044$$

In part (a) we found $dV \approx 0.01608$. The approximation of ΔV by dV does not appear to be as close as we might hope for. However, when we calculate the actual percentage error,

$$100\frac{\Delta V}{V} = 100\frac{0.005158\pi}{0.192\pi} \approx 2.69\%$$

we see that the difference between this percentage error and the approximation in part (c) is only about 0.02%. ∎

EXAMPLE 14.2   Let $f(x, y) = \sqrt{9 - x^2 - y^2}$. Use the differential of f to approximate $f(0.92, 2.12)$. Show the result graphically.

Solution   We choose (x_0, y_0) as $(1, 2)$, since this is the point nearest $(0.92, 2.12)$ at which we easily can obtain the exact value of f. We have

$$f(x_0, y_0) = f(1, 2) = \sqrt{9 - (1)^2 - (2)^2} = 2$$

Also, we take $dx (= \Delta x)$ as -0.08 and $dy (= \Delta y)$ as 0.12, in order that $x_0 + \Delta x = 0.92$ and $y_0 + \Delta y = 2.12$.

The differential of f is

$$df = \frac{\partial f}{\partial x}dx + \frac{\partial f}{\partial y}dy = \frac{-x}{\sqrt{9 - x^2 - y^2}}dx + \frac{-y}{\sqrt{9 - x^2 - y^2}}dy$$

Evaluating df at $(1, 2)$ with $dx = -0.08$ and $dy = 0.12$, we obtain

$$df = -\left(\frac{1}{2}\right)(-0.08) - \left(\frac{2}{2}\right)(0.12) = -0.08$$

The linear approximation given by Equation 14.3 can be written as

$$f(x_0 + \Delta x, y_0 + \Delta y) \approx f(x_0, y_0) + df$$

Thus,

$$f(0.92, 2.12) \approx 2 - 0.08 = 1.92$$

By calculator, the value (correct to four places) is 1.9129.

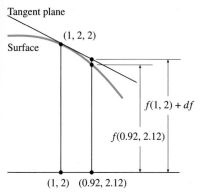

Tangent plane

Surface

(1, 2, 2)

$f(1, 2) + df$

$f(0.92, 2.12)$

(1, 2)   (0.92, 2.12)

FIGURE 14.5

The graph of f is the upper half of the sphere $x^2 + y^2 + z^2 = 9$. In Figure 14.5 we show a portion of the trace of this surface with the vertical plane containing the points $(1, 2)$ and $(0.92, 2.12)$ in the xy-plane, and the line of intersection of this plane and the tangent plane to the surface drawn at $(1, 2, 2)$. Since we approximated the z value on the hemisphere by the z value to the tangent plane, it is clear that our estimate is somewhat too high.   ∎

If $z = f(x, y)$, then we have seen that a change of dx in x and dy in y produces an approximate change of dz in z (assuming f is differentiable). We can use the notions of relative error and percentage error in this situation, where we replace *error* by *change*. That is,

$$\frac{dz}{z} = \text{approximate relative change in } z$$

$$100\,\frac{dz}{z} = \text{approximate percentage change in } z$$

We use these ideas in the next example, which is taken from economics.

EXAMPLE 14.3   The Cobb-Douglas Production Function, Equation 13.19, can be written in the form

$$Q(K, L) = AK^\alpha L^\beta$$

where $\beta = 1 - \alpha$ and $0 < \alpha < 1$. Here $Q(K, L)$ represents the number of units of some commodity that a company produces, A is a positive constant, K is the amount of capital investment, and L is the size of the labor force.

(a) By treating Q as a continuous function for $K > 0$ and $L > 0$ (it actually only takes on integer values, as does L), show that the relative change in Q caused by changes dK in K and dL in L is approximated by

$$\frac{dQ}{Q} = \alpha\left(\frac{dK}{K}\right) + \beta\left(\frac{dL}{L}\right)$$

(b) If $\alpha = 2/3$, find the approximate percentage change in Q caused by a 3% increase in K and a 2% decrease in L.

Solution

(a) First, we calculate dQ, using Equation 14.5.

$$dQ = \frac{\partial Q}{\partial K}dK + \frac{\partial Q}{\partial L}dL = \left(A\alpha K^{\alpha-1}L^\beta\right)dK + \left(A\beta K^\alpha L^{\beta-1}\right)dL$$

Thus,

$$\frac{dQ}{Q} = \frac{\left(A\alpha K^{\alpha-1}L^\beta\right)dK + \left(A\beta K^\alpha L^{\beta-1}\right)dL}{AK^\alpha L^\beta}$$

$$= \frac{A\alpha K^{\alpha-1}L^\beta}{AK^\alpha L^\beta}dK + \frac{A\beta K^\alpha L^{\beta-1}}{AK^\alpha L^\beta}dL = \frac{\alpha}{K}dK + \frac{\beta}{L}dL$$

$$= \alpha\left(\frac{dK}{K}\right) + \beta\left(\frac{dL}{L}\right)$$

(b) If $\alpha = 2/3$, then $\beta = 1/3$. A percentage increase of 3% in K means that

$$100\,\frac{dK}{K} = 3$$

so

$$\frac{dK}{K} = 0.03$$

Similarly, a percentage decrease of 2% in L means that

$$\frac{dL}{L} = -0.02$$

Substituting these values in the formula for dQ/Q from part (a) gives

$$\frac{dQ}{Q} = \frac{2}{3}(0.03) + \frac{1}{3}(-0.02) \approx 0.013$$

Thus, there is an approximate increase in production of 1.3%.   ∎

Functions of Three or More Variables

The notions of differentiability and the differential extend in natural ways to functions of more than two variables. For three variables, for example, a function $f(x, y, z)$ is differentiable at the point (x, y, z) provided that the change Δf in the function caused by a change Δx in x, Δy in y, and Δz in z can be written in the form

$$\Delta f = f_x(x, y, z)\Delta x + f_y(x, y, z)\Delta y + f_z(x, y, z)\Delta z + \varepsilon_1\Delta x + \varepsilon_2\Delta y + \varepsilon_3\Delta z$$

where $\varepsilon_1, \varepsilon_2$, and ε_3 all approach 0 as $(\Delta x, \Delta y, \Delta z) \to (0, 0, 0)$. If f is differentiable at a point (x, y, z), we define the differential of f as

$$df = f_x(x, y, z)dx + f_y(x, y, z)dy + f_z(x, y, z)dz$$

Just as in the two-variable case, we can set $dx = \Delta x$, $dy = \Delta y$, and $dz = \Delta z$, so that Δf becomes

$$\Delta f = df + \varepsilon_1\Delta x + \varepsilon_2\Delta y + \varepsilon_3\Delta z$$

Thus, $\Delta f \approx df$ when Δx, Δy, and Δz are small in absolute value.

Exercise Set 14.1

In Exercises 1–6, use Theorem 14.2 to show that the given function is differentiable on the specified domain.

1. $f(x, y) = x^3 + 3x^2y - 2y^4$, on $\mathbb{R}^2$

2. $f(x, y) = \dfrac{x + y}{x - y}$, $x \neq y$

3. $f(x, y) = e^x \sin y$, on $\mathbb{R}^2$

4. $f(x, y) = \sin^{-1}\dfrac{y}{x}$, $x^2 > y^2$

5. $f(x, y) = \ln\dfrac{xy}{x^2 + y^2}$, $xy > 0$

6. $f(x, y) = x \ln(\cos y)$, $-\pi/2 < y < \pi/2$

In Exercises 7–16, find df.

7. $f(x, y) = 2x^2 - 3xy + y^3$

8. $f(x, y) = x^4y^2 - 2xy^3$

9. $f(x, y) = e^{xy}$

10. $f(x, y) = \ln\sqrt{x^2 + y^2}$

11. $f(x, y) = \dfrac{x - y}{x + y}$

12. $f(x, y) = x \sin y + y \cos x$

13. $f(x, y) = \tan^{-1} \dfrac{x}{y}$

14. $f(x, y) = \ln\left(\dfrac{2x^2 y}{x + y}\right)$

15. $f(x, y, z) = \sqrt{x^2 + y^2 + z^2}$

16. $f(x, y, z) = \ln(x^2 y^3 z^4)$

In Exercises 17–22, use df to approximate the change Δf in f from P_0 to P_1.

17. $f(x, y) = 2x^3 - 3x^2 y + y^2; P_0 = (2, -1)$,
   $P_1 = (2.02, -0.99)$

18. $f(x, y) = \sqrt{\dfrac{2x - y}{x + 3y}}$;
   $P_0 = (4, -1)$, $P_1 = (3.97, -0.95)$

19. $f(x, y) = \ln(x - y)^2$; $P_0 = (3, -2)$,
   $P_1 = (3.04, -1.99)$

20. $f(x, y) = \tan^{-1} \dfrac{y}{x}$; $P_0 = (-3, 4)$, $P_1 = (-3.02, 3.98)$

21. $f(x, y) = \ln \sqrt{x^2 + y^2}$; $P_0 = (-4, 3)$,
   $P_1 = (-3.98, 2.99)$

22. $f(x, y, z) = (x^2 + y^2 + z^2)^{-1}$; $P_0 = (1, -1, 2)$,
   $P_1 = (0.97, -1.01, 2.05)$

23. The dimensions of a room are measured to be as follows: length 21 ft, width 12 ft, and height 8 ft. If each measurement is accurate only to the nearest tenth of a foot, find the approximate maximum error in volume calculated from the measured values, making use of differentials. What is the approximate percentage error?

24. In Example 14.1, find the approximate maximum error in total surface area, making use of differentials. What is the approximate relative error?

25. Suppose the temperature at a point in a thin, rectangular metal sheet is given by

$$T(x, y) = \dfrac{x^2 y^2}{x^2 + y^2}$$

where the origin is at the lower left corner. Using differentials, find the approximate difference in temperature at the points $(1.15, 2.05)$ and $(1, 2)$. Compare your answer with the actual change. What is the approximate percentage change?

26. A company manufactures and sells two models of automatic ice cream makers, the standard model, A, and the deluxe model, B. Through an analysis of sales over a period of time, it is found that the weekly profit from producing and selling x Model-A and y Model-B machines is approximately

$$P(x, y) = 200x + 300y - 0.2x^2 - 0.3y^2 - 0.4xy - 50$$

where P is in dollars. On average, 50 Model-A and 20 Model-B machines are sold each week. Use differentials to approximate the effect on profit of selling one fewer Model-A and two more Model-B machines per week than the average. What is the approximate percentage change in profit?

27. The cost of producing one unit of a certain manufactured item is given by

$$C(x, y) = 2x^2 + 3xy + y^2 + 15x + 6y + 50$$

where x is the hourly wage rate for the workers and y is the cost per pound of raw materials. Currently, the hourly wage is $9.50, and raw materials cost $6.00 per pound. Use differentials to estimate the increase in cost if the labor cost goes up by $0.50 per hour and the cost of raw materials increases by $0.40 per pound. By approximately what percentage does the cost go up?

28. At a certain factory, the weekly production is given by the Cobb-Douglas Production Function as

$$Q(K, L) = 120K^{0.6}L^{0.4}$$

units, where K represents the capital investment (in thousands of dollars) and L represents the size of the labor force. The current values of K and L are 450 and 260, respectively. Estimate, by differentials, the change in production that will result if K is increased by 2 and L is increased by 3. What is the approximate relative change in Q?

29. For the Cobb-Douglas Production Function in Exercise 28, estimate the percentage change in Q if K increases by 2% and L decreases by 1%.

In Exercises 30–33, find an expression for

$$\Delta f = f(x + \Delta x, y + \Delta y) - f(x, y)$$

and show that it can be written in the form of Equation 14.2. Identify ε_1 and ε_2 and show they both approach 0 as Δx and Δy approach 0.

30. $f(x, y) = 2x + 3y$

31. $f(x, y) = xy$

32. $f(x, y) = 2x^2 y$

33. $f(x, y) = (x - 3y)^2$

34. Prove that $f(x, y) = \sqrt{x^2 + y^2}$ is not differentiable at the origin. (*Hint:* Show that f_x does not exist at the origin.)

35. Let

$$f(x, y) = \begin{cases} \dfrac{x^2 y^2}{x^2 + y^2} & \text{if } (x, y) \neq (0, 0) \\ 0 & \text{if } (x, y) = (0, 0) \end{cases}$$

Prove that f is differentiable on all of $\mathbb{R}^2$. (*Hint:* Apply Theorem 14.2, paying particular attention to the origin.)

36. Let

$$f(x, y) = \begin{cases} \dfrac{xy}{x^2 + y^2} & \text{if } (x, y) \neq (0, 0) \\ 0 & \text{if } (x, y) = (0, 0) \end{cases}$$

(a) Show that f is discontinuous at $(0, 0)$.
(b) Show that $f_x(0, 0)$ and $f_y(0, 0)$ both exist.
(c) What can you conclude from part (a) about differentiability of f at the origin? Explain.
(d) What can you conclude from part (c) about continuity of f_x and f_y at the origin? Explain.

37. Let $f(x, y) = \sqrt{|xy|}$. Prove the following:
(a) f is continuous at $(0, 0)$
(b) $f_x(0, 0) = f_y(0, 0) = 0$
(c) f is not differentiable at $(0, 0)$
(*Hint:* Show that at $(0, 0)$ Equation 14.2 becomes $\sqrt{|\Delta x \Delta y|} = \varepsilon_1 \Delta x + \varepsilon_2 \Delta y$, and by taking $\Delta x = \Delta y$, conclude that ε_1 and ε_2 do not approach 0 as $(\Delta x, \Delta y) \to (0, 0)$.)

14.2 CHAIN RULES

For a function f of one variable x, if x in turn is a function of t, say $x = g(t)$, then the Chain Rule for differentiation (Equation 3.13) can be written as

$$(f \circ g)'(t) = f'(g(t))g'(t) \tag{14.6}$$

This formula holds true at all points t in the domain of g at which g' exists and for which $f'(g(t))$ exists. When we introduce the dependent variable $y = f(x)$, Equation 14.6 becomes, in Leibniz notation,

$$\frac{dy}{dt} = \frac{dy}{dx}\frac{dx}{dt} \tag{14.7}$$

Equation 14.7 is the familiar form of the Chain Rule. Care should be taken to distinguish between dy/dt and dy/dx. On the left, it is understood that x is first replaced by $g(t)$ and then the derivative is taken, so that

$$\frac{dy}{dt} \qquad \text{means} \qquad \frac{d}{dt}f(g(t))$$

On the right, dy/dx means that we first calculate $f'(x)$ and then replace x by $g(t)$, so that

$$\frac{dy}{dx} \qquad \text{means} \qquad f'(x)\Big|_{x=g(t)} \qquad \text{or } f'(g(t))$$

It is convenient in this context to call x the *intermediate variable* and t the *final (independent) variable.*

The form of the extension of Equation 14.6 or 14.7 to two or more variables depends on the number of intermediate variables and final variables. We begin with the simplest case, in which $z = f(x, y)$, $x = g(t)$, and $y = h(t)$. Here x and y are intermediate variables and t is the single, final, independent variable. If we first substitute for x and y, we get $z = f(g(t), h(t))$, so that z is finally a function of the single independent variable t. We want to find a formula for dz/dt. The secret lies in the definition of differentiability. We assume that $g'(t)$ and $h'(t)$ both exist and that f is differentiable at $(x, y) = (g(t), h(t))$. Let

$$\Delta x = g(t + \Delta t) - g(t)$$

$$\Delta y = h(t + \Delta t) - h(t)$$

Then, because g and h are continuous at t (why?), Δx and Δy both approach 0 as $\Delta t \to 0$. Now, by Definition 14.1, we can write

$$\Delta z = f_x(x, y)\Delta x + f_y(x, y)\Delta y + \varepsilon_1 \Delta x + \varepsilon_2 \Delta y \qquad (14.8)$$

where $\varepsilon_1 \to 0$ and $\varepsilon_2 \to 0$ as $(\Delta x, \Delta y) \to (0, 0)$ and hence also as $\Delta t \to 0$. Here Δz means

$$\Delta z = \Delta f = f(x + \Delta x, y + \Delta y) - f(x, y)$$

$$= f(g(t + \Delta t), h(t + \Delta t)) - f(g(t), h(t))$$

Dividing Equation 14.8 by Δt on both sides gives

$$\frac{\Delta z}{\Delta t} = f_x(x, y)\frac{\Delta y}{\Delta t} + f_y(x, y)\frac{\Delta y}{\Delta t} + \varepsilon_1\frac{\Delta x}{\Delta t} + \varepsilon_2\frac{\Delta y}{\Delta t}$$

Now we let $\Delta t \to 0$ to obtain the result

$$\frac{d}{dt}f(x, y) = f_x(x, y)g'(t) + f_y(x, y)h'(t)$$

or, in Leibniz notation, since $z = f(x, y)$,

$$\frac{dz}{dt} = \frac{\partial f}{\partial x}\frac{dx}{dt} + \frac{\partial f}{\partial y}\frac{dy}{dt} \qquad (14.9)$$

It is understood that $\partial f/\partial x$ and $\partial f/\partial y$ are to be evaluated at the point $(g(t), h(t))$. We can also use $\partial z/\partial x$ in place of $\partial f/\partial x$ and $\partial z/\partial y$ in place of $\partial f/\partial y$ to obtain the equivalent equation

$$\frac{dz}{dt} = \frac{\partial z}{\partial x}\frac{dx}{dt} + \frac{\partial z}{\partial y}\frac{dy}{dt}$$

We summarize our result as follows.

If $z = f(x, y)$ is a differentiable function of x and y, where $x = g(t)$ and $y = h(t)$ are differentiable functions of the single variable t, then z is a differentiable function of t, and

$$\frac{\partial z}{\partial t} = \frac{\partial z}{\partial x}\frac{dx}{dt} + \frac{\partial z}{\partial y}\frac{dy}{dt} \qquad (14.10)$$

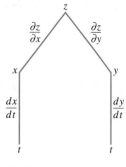

FIGURE 14.6

As an aid in obtaining Equation 14.10, refer to Figure 14.6. This figure is an example of a **tree diagram**. Starting from the dependent variable z, we move on the branch leading to the intermediate variable x, writing $\partial z/\partial x$. Then we move from x on the (only) branch from it leading to t, writing dx/dt. Note that we write dx/dt rather than $\partial x/\partial t$, since x is a function of just one variable, t. When we move from any one variable, if there are two (or more) branches, we use *partial* derivatives along each branch, whereas if there is only one branch, we use the *total* derivative (that is, the ordinary derivative of a function of one variable).

Now we multiply together the derivatives $\partial z/\partial x$ and dx/dt that we have obtained:

$$\frac{\partial z}{\partial x}\frac{dx}{dt}$$

Then we repeat the process, branching from z to y and from y to t, obtaining the product

$$\frac{\partial z}{\partial y}\frac{dy}{dt}$$

Finally, we add the results we have obtained, getting the right-hand side of Equation 14.10.

We will use tree diagrams for other combinations of intermediate and final variables, but for simplicity we will in the future not show the derivatives on the diagram.

EXAMPLE 14.4   Let

$$z = \frac{x - 2y}{2x + 3y} \quad \text{and} \quad \begin{cases} x = 2t - 3 \\ y = t^2 + 1 \end{cases}$$

Use the Chain Rule to find dz/dt when $t = -1$.

Solution   By Equation 14.10,

$$\frac{dz}{dt} = \frac{\partial z}{\partial x}\frac{dx}{dt} + \frac{\partial z}{\partial y}\frac{dy}{dt}$$

$$= \frac{7y}{(2x + 3y)^2}(2) + \frac{-7}{(2x + 3y)^2}(2t) \quad \text{Verify.}$$

When $t = -1$, we find that $x = -5$ and $y = 2$, so

$$\frac{dz}{dt}\bigg|_{t=-1} = \frac{14}{(-4)^2}(2) + \frac{(35)}{(-4)^2}(-2) = -\frac{21}{8} \quad \blacksquare$$

The next case we consider is that in which $z = f(x, y)$, and each of the variables x and y is a function of two other variables, say u and v (there are two intermediate variables, x and y, and two final variables u and v). The derivation goes in much the same way as for the previous case, but we obtain two equations, one for $\partial z/\partial u$ and one for $\partial z/\partial v$. We summarize the results as follows.

> If $z = f(x, y)$ is a differentiable function of x and y, where $x = g(u, v)$ and $y = h(u, v)$, and the partial derivatives $\partial x/\partial u$, $\partial x/\partial v$, $\partial y/\partial u$, and $\partial y/\partial v$ all exist, then
>
> $$\frac{\partial z}{\partial u} = \frac{\partial z}{\partial x}\frac{\partial x}{\partial u} + \frac{\partial z}{\partial y}\frac{\partial y}{\partial u}$$
>
> $$\frac{\partial z}{\partial v} = \frac{\partial z}{\partial x}\frac{\partial x}{\partial v} + \frac{\partial z}{\partial y}\frac{\partial y}{\partial v}$$
>
> (14.11)

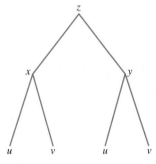

FIGURE 14.7

The tree diagram in Figure 14.7 illustrates a way to arrive at each of Equations 14.11. Since there are two branches from each of the variables z, x, and y, all derivatives are partial derivatives. To get the formula for $\partial z/\partial u$, we go along each branch leading from z to u. Remember that as we go from one variable to another, we take the partial derivative of the first with respect to the second. So, going from z to x to u, we get

$$\frac{\partial z}{\partial x}\frac{\partial x}{\partial u}$$

Similarly, going from z to y to u, we get

$$\frac{\partial z}{\partial y}\frac{\partial y}{\partial u}$$

Then we add these results to get $\partial z/\partial u$, as given by the first of Equations 14.11. When we repeat the process, going along all branches leading to v, we get $\partial z/\partial v$, as in the second of Equations 14.11.

EXAMPLE 14.5   If $z = x^2 - 2xy + 3y^2$, $x = uv$, and $y = u^2 - v^2$, find $\partial z/\partial u$ and $\partial z/\partial v$.

Solution   By Equations 14.11, or from the tree diagram in Figure 14.7,

$$\frac{\partial z}{\partial u} = \frac{\partial z}{\partial x}\frac{\partial x}{\partial u} + \frac{\partial z}{\partial y}\frac{\partial y}{\partial u} = (2x - 2y)(v) + (-2x + 6y)(2u)$$

$$= 2(uv - u^2 + v^2)v + 4(-uv + 3u^2 - 3v^2)u$$

$$= 12u^3 - 6u^2v - 10uv^2 + 2v^3$$

$$\frac{\partial z}{\partial v} = \frac{\partial z}{\partial x}\frac{\partial x}{\partial v} + \frac{\partial z}{\partial y}\frac{\partial y}{\partial v} = (2x - 2y)(u) + (-2x + 6y)(-2v)$$

$$= 2(uv - u^2 + v^2)u - 4(-uv + 3u^3 - 3v^2)v$$

$$= -2u^3 - 10u^2v + 6uv^2 + 12v^3$$

∎

REMARK ───
■ Note that in the preceding example we first obtained $\partial z/\partial u$ and $\partial z/\partial v$ as a mixed expression involving both intermediate and final variables. Then we replaced the intermediate variables x and y by their values in terms of the final variables u and v. In Example 14.4, we avoided the last step, since we were evaluating the derivative at a specific point, and we substituted the values of dx/dt and dy/dt at that point.

───

EXAMPLE 14.6   If $z = f(u/v, v/u)$, show that

$$u\frac{\partial z}{\partial u} + v\frac{\partial z}{\partial v} = 0$$

Solution   We can consider the variable z as the composition of $z = f(x, y)$ and $x = u/v$, $y = v/u$. Equations 14.11 then give

$$\frac{\partial z}{\partial u} = \frac{\partial z}{\partial x}\frac{1}{v} + \frac{\partial z}{\partial y}\left(-\frac{v}{u^2}\right)$$

$$\frac{\partial z}{\partial v} = \frac{\partial z}{\partial x}\left(-\frac{u}{v^2}\right) + \frac{\partial z}{\partial y}\left(\frac{1}{u}\right)$$

Even though we do not know the values of $\partial z/\partial x$ and $\partial z/\partial y$, we see that multiplying the first equation by u and the second by v and adding gives

$$u\frac{\partial z}{\partial u} + v\frac{\partial z}{\partial v} = 0 \qquad ■$$

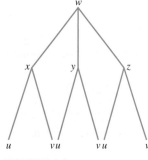

FIGURE 14.8

It should be clear now that the number of intermediate and final variables can be any number greater than or equal to one. For example, if $w = f(x, y, z)$ and, in turn, $x = x(u, v)$, $y = y(u, v)$, and $z = z(u, v)$, then the tree diagram in Figure 14.8 can be used to write the appropriate chain rule. In this case, each of the equations for $\partial w/\partial u$ and $\partial w/\partial v$ would contain three terms. For example,

$$\frac{\partial w}{\partial u} = \frac{\partial w}{\partial x}\frac{\partial x}{\partial u} + \frac{\partial w}{\partial y}\frac{\partial y}{\partial u} + \frac{\partial w}{\partial z}\frac{\partial z}{\partial u}$$

Rather than list formulas for various other combinations of intermediate and final variables, we state the following Generalized Chain Rule, where f is a function of n variables, each of which is a function of m variables. We assume differentiability of all functions involved and that the composite function is defined in the domain under consideration.

Generalized Chain Rule

Let $z = f(x_1, x_2, \ldots, x_n)$, and for each i from 1 to n, let $x_i = g_i(u_1, u_2, \ldots, u_m)$. Then

$$\frac{\partial z}{\partial u_j} = \frac{\partial z}{\partial x_1}\frac{\partial x_1}{\partial u_j} + \frac{\partial z}{\partial x_2}\frac{\partial x_2}{\partial u_j} + \cdots + \frac{\partial z}{\partial x_m}\frac{\partial x_m}{\partial u_j} \qquad (14.12)$$

for $j = 1, 2, \ldots, m$.

Here there are n intermediate variables $x_1, x_2, \ldots, x_n$ and m final variables $u_1, u_2, \ldots, u_m$. So there are m equations, each with n terms on the right-hand side.

There is nothing special about the particular letters used to designate the dependent variable or the intermediate and final variables. Consider the following example.

EXAMPLE 14.7   Let $w = r^2 + s^2 - 2t^2$ and $r = e^{-y} \cos z$, $s = e^{-y} \sin z$, and $t = e^{-y}$. Find $\partial w/\partial y$ and $\partial w/\partial z$.

Solution   We can use Equation 14.12, with appropriate changes in variable names, or we can use the tree diagram in Figure 14.9.

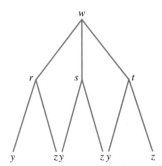

FIGURE 14.9

$$\frac{\partial w}{\partial y} = \frac{\partial w}{\partial r}\frac{\partial r}{\partial y} + \frac{\partial w}{\partial s}\frac{\partial s}{\partial y} + \frac{\partial w}{\partial t}\frac{\partial t}{\partial y}$$

$$= (2r)(-e^{-y}\cos z) + (2s)(-e^{-y}\sin z) + (-4t)(-e^{-y})$$

$$= (2e^{-y}\cos z)(-e^{-y}\cos z) + (2e^{-y}\sin z)(-e^{-y}\sin z) + (-4e^{-y})(-e^{-y})$$

$$= 2e^{-2y}(-\cos^2 z - \sin^2 z + 2)$$

$$= 2e^{-2y}$$

$$\frac{\partial w}{\partial z} = \frac{\partial w}{\partial r}\frac{\partial r}{\partial z} + \frac{\partial w}{\partial s}\frac{\partial s}{\partial z} + \frac{\partial w}{\partial t}\frac{\partial t}{\partial z}$$

$$= (2r)(-e^{-y}\sin z) + (2s)(e^{-y}\cos z) + (-4t)(0)$$

$$= (2e^{-y}\cos z)(-e^{-y}\sin z) + (2e^{-y}\sin z)(-e^{-y}\cos z)$$

$$= 0 \qquad \blacksquare$$

Second Derivatives by the Chain Rule

The next example shows how the Chain Rule given by Equation 14.10 can be used to compute the second derivative d^2z/dt^2 when $z = f(x, y)$ and $x = g(t)$, $y = h(t)$. In the exercises you will be asked to derive similar formulas for second-order partial derivatives where there are two final variables.

EXAMPLE 14.8   Let $z = f(x, y)$ and $x = g(t)$, $y = h(t)$. Assuming suitable differentiability conditions, derive a formula for d^2z/dt^2.

Solution   For the first derivative, we have from Equation 14.10 or from the tree diagram in Figure 14.10,

$$\frac{dz}{dt} = \frac{\partial z}{\partial x}\frac{dx}{dt} + \frac{\partial z}{\partial y}\frac{dy}{dt}$$

Using the Product Rule, we obtain

$$\frac{d^2z}{dt^2} = \frac{d}{dt}\left(\frac{\partial z}{\partial x}\right)\frac{dx}{dt} + \frac{\partial z}{\partial x}\frac{d^2x}{dt^2} + \frac{d}{dt}\left(\frac{\partial z}{\partial y}\right)\frac{dy}{dt} + \frac{\partial z}{\partial y}\frac{d^2y}{dt^2} \qquad (14.13)$$

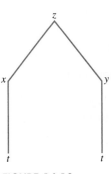

FIGURE 14.10

Now $\partial z/\partial x$ and $\partial z/\partial y$ are initially functions of the intermediate variables x and y. So to find their derivatives with respect to the final variable t, we use the Chain Rule given by Equation 14.10 again, with z replaced by $\partial z/\partial x$ in the

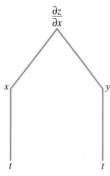

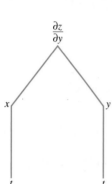

FIGURE 14.11

one case and by $\partial z/\partial y$ in the other (see also Figure 14.11).

$$\frac{d}{dt}\left(\frac{\partial z}{\partial x}\right) = \frac{\partial^2 z}{\partial x^2}\frac{dx}{dt} + \frac{\partial^2 z}{\partial y \partial x}\frac{dy}{dt}$$

$$\frac{d}{dt}\left(\frac{\partial z}{\partial y}\right) = \frac{\partial^2 z}{\partial x \partial y}\frac{dx}{dt} + \frac{\partial^2 z}{\partial y^2}\frac{dy}{dt}$$

If we assume equality of the second-order mixed partials, then on substitution of these expressions into Equation 14.13, we obtain

$$\frac{d^2 z}{dt^2} = \frac{\partial z}{\partial x}\frac{d^2 x}{dt^2} + \frac{\partial z}{\partial y}\frac{d^2 y}{dt^2} + \frac{\partial^2 z}{\partial x^2}\left(\frac{dx}{dt}\right)^2 + 2\frac{\partial^2 z}{\partial y \partial x}\left(\frac{dx}{dt}\frac{dy}{dt}\right) + \frac{\partial^2 z}{\partial y^2}\left(\frac{dy}{dt}\right)^2$$

∎

EXAMPLE 14.9   Let $z = f(x, y)$ and $x = t^2$, $y = \ln t$. Find $d^2 z/dt^2$.

Solution   We could use the formula found in Example 14.8, but it is probably more instructive to go through all the steps again.

Using the Chain Rule, we find

$$\frac{dz}{dt} = (f_x)(2t) + (f_y)\left(\frac{1}{t}\right)$$

To obtain the second derivative, we differentiate each term on the right-hand side as a product of two functions of t. For example, to differentiate the product $(f_x)(2t)$ with respect to t, we use the Product Rule to get

$$\frac{d}{dt}\left[(f_x)(2t)\right] = \left(\frac{df_x}{dt}\right)(2t) + (f_x)(2)$$

But to find df_x/dt, we must use the Chain Rule:

$$\frac{df_x}{dt} = f_{xx}\frac{dx}{dt} + f_{xy}\frac{dy}{dt} = (f_{xx})(2t) + (f_{xy})\left(\frac{1}{t}\right)$$

Thus,

$$\frac{d}{dt}\left[(f_x)(2t)\right] = (f_{xx})(4t^2) + 2f_{xy} + 2f_x$$

Similarly, for the product $(f_y)\left(\frac{1}{t}\right)$, we find that

$$\frac{d}{dt}\left[(f_y)\left(\frac{1}{t}\right)\right] = (f_{yx})(2t)\cdot\frac{1}{t} + (f_{yy})\left(\frac{1}{t}\right)^2 - \frac{1}{t^2}f_y$$

Combining these results gives

$$\frac{d^2 z}{dt^2} = 4t^2 f_{xx} + 4f_{xy} + \frac{1}{t^2}f_{yy} + 2f_x - \frac{1}{t^2}f_y$$

You can verify that this result agrees with the formula found in Example 14.8.

∎

In the next example, we illustrate how to find second-order partials when there are two final variables. In the exercises (Exercise 37), you will be asked to develop general formulas for this case.

EXAMPLE 14.10   Let $z = f(x, y)$ and let x and y be expressed in polar coordinates. Find $\partial^2 z/\partial r^2$ and $\partial^2 z/\partial \theta^2$.

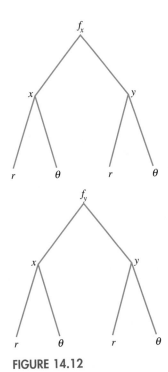

FIGURE 14.12

Solution   In terms of polar coordinates, $x = r \cos \theta$ and $y = r \sin \theta$. Applying the Chain Rule, we get

$$\frac{\partial z}{\partial r} = f_x \frac{\partial x}{\partial r} + f_y \frac{\partial y}{\partial r} = f_x \cos \theta + f_y \sin \theta$$

and

$$\frac{\partial z}{\partial \theta} = f_x \frac{\partial x}{\partial \theta} + f_y \frac{\partial y}{\partial \theta} = f_x(-r \sin \theta) + f_y(r \cos \theta)$$

To find the second-order partials, we will need to apply the Chain Rule again to f_x and f_y. The tree diagrams in Figure 14.12 are helpful in this regard. Since $\cos \theta$ and $\sin \theta$ are independent of r, we find that

$$\frac{\partial^2 z}{\partial r^2} = (f_{xx} \cos \theta + f_{xy} \sin \theta) \cos \theta + (f_{yx} \cos \theta + f_{yy} \sin \theta) \sin \theta$$

$$= f_{xx} \cos^2 \theta + 2 f_{xy} \sin \theta \cos \theta + f_{yy} \sin^2 \theta \quad \text{Assume } f_{xy} = f_{yx}.$$

For $\partial^2 z / \partial \theta^2$, we must use the Product Rule as well as the Chain Rule:

$$\frac{\partial^2 z}{\partial \theta^2} = \left[f_{xx}(-r \sin \theta) + f_{xy}(r \cos \theta) \right](-r \sin \theta) - r f_x \cos \theta$$

$$+ \left[f_{yx}(-r \sin \theta) + f_{yy}(r \cos \theta) \right](r \cos \theta) + r f_y(-\sin \theta)$$

$$= r^2 \sin^2 \theta f_{xx} - 2 r^2 \sin \theta \cos \theta f_{xy} + r^2 \cos^2 \theta f_{yy}$$

$$- r f_x \cos \theta - r f_y \sin \theta \qquad \blacksquare$$

Derivative Formulas for Implicit Functions

The Chain Rule can be used to derive a very convenient formula for the derivative of a function defined implicitly. As in Section 2.8, suppose that $F(x, y) = 0$ defines y as a differentiable function of x, say $y = f(x)$, on some domain D. We want to find a formula for dy/dx. We can think of this problem as a chain rule situation with intermediate variables x and y and final variable x. To distinguish the two roles played by x, however, we will use the letter t (temporarily) as the final variable. So we have

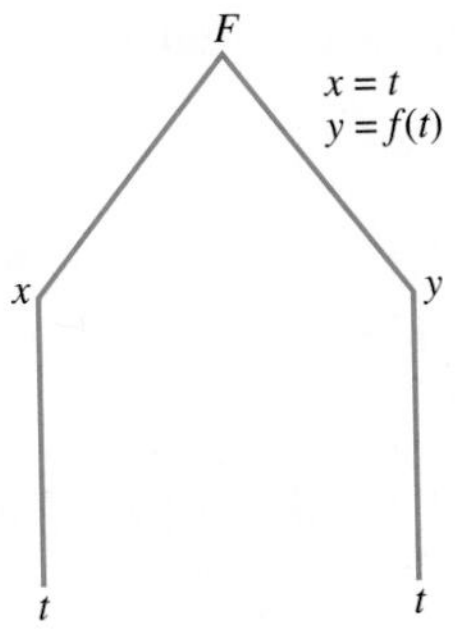

$$F(x, y) = 0 \qquad \text{and} \qquad \begin{cases} x = t \\ y = f(t) \end{cases}$$

Since $F(t, f(t)) = 0$ for all t in D, it follows that

$$\frac{d}{dt} F(t, f(t)) = 0$$

there also. By Equation 14.10 or by Figure 14.13,

$$\frac{d}{dt} F(t, f(t)) = \frac{\partial F}{\partial x} \frac{dx}{dt} + \frac{\partial F}{\partial y} \frac{dy}{dt}$$

$$= \frac{\partial F}{\partial x}(1) + \frac{\partial F}{\partial y} \frac{dy}{dt} = 0$$

FIGURE 14.13

Thus, if $\partial F/\partial y \neq 0$, we can solve for dy/dt to get

$$\frac{dy}{dt} = -\frac{\dfrac{\partial F}{\partial x}}{\dfrac{\partial F}{\partial y}}$$

Now we can replace t by x and write the answer in the following alternative form:

$$\frac{dy}{dx} = -\frac{F_x}{F_y} \qquad \text{if } F_y \neq 0 \tag{14.14}$$

EXAMPLE 14.11   Find dy/dx if

$$x^3 - 2xy + y^3 - 4 = 0$$

Solution   Let $F(x, y) = x^3 - 2xy + y^3 - 4$. Then the given equation becomes $F(x, y) = 0$. Assuming this equation defines y as a differentiable function of x on some domain D, we can use Equation 14.14 to get the result

$$\frac{dy}{dx} = -\frac{F_x}{F_y} = -\frac{3x^2 - 2y}{-2x + 3y^2} = \frac{2y - 3x^2}{3y^2 - 2x}$$

at all points of D for which $3y^2 - 2x \neq 0$. This same problem was solved in Example 3.26, part (a). You might be interested in comparing the new method with the one used in Chapter 3. ∎

A similar method can be applied when an equation $F(x, y, z) = 0$ defines z implicitly as a differentiable function of x and y, say $z = f(x, y)$, on some domain D of $\mathbb{R}^2$. We can obtain formulas for $\partial z/\partial x$ and $\partial z/\partial y$, as follows.

Treat $x, y,$ and z as intermediate variables and u and v as final variables, where

$$\begin{cases} x = u \\ y = v \\ z = f(u, v) \end{cases}$$

Then for (u, v) in D, $F(u, v, f(u, v)) = 0$. So $\partial F/\partial u = 0$ and $\partial F/\partial v = 0$. By Equations 14.11, or by the tree diagram in Figure 14.14, we have

$$\frac{\partial F}{\partial u} = \frac{\partial F}{\partial x}\frac{\partial x}{\partial u} + \frac{\partial F}{\partial y}\frac{\partial y}{\partial u} + \frac{\partial F}{\partial z}\frac{\partial z}{\partial u} = \frac{\partial F}{\partial x}(1) + \frac{\partial F}{\partial y}(0) + \frac{\partial F}{\partial z}\frac{\partial z}{\partial u}$$

and

$$\frac{\partial F}{\partial v} = \frac{\partial F}{\partial x}\frac{\partial x}{\partial v} + \frac{\partial F}{\partial y}\frac{\partial y}{\partial v} + \frac{\partial F}{\partial z}\frac{\partial z}{\partial v} = \frac{\partial F}{\partial x}(0) + \frac{\partial F}{\partial y}(1) + \frac{\partial F}{\partial z}\frac{\partial z}{\partial v}$$

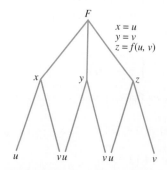

F

$x = u$
$y = v$
$z = f(u, v)$

$x \quad y \quad z$

$u \quad vu \quad vu \quad v$

FIGURE 14.14

Since $\partial F/\partial u$ and $\partial F/\partial v$ both equal 0, we can solve for $\partial z/\partial u$ and $\partial z/\partial v$ provided $\partial F/\partial z \neq 0$, getting

$$\frac{\partial z}{\partial u} = -\frac{\dfrac{\partial F}{\partial x}}{\dfrac{\partial F}{\partial z}} \quad \text{and} \quad \frac{\partial z}{\partial v} = -\frac{\dfrac{\partial F}{\partial y}}{\dfrac{\partial F}{\partial z}}$$

Finally, since $u = x$ and $v = y$,

$$\frac{\partial z}{\partial x} = -\frac{F_x}{F_z} \quad \text{and} \quad \frac{\partial z}{\partial y} = -\frac{F_y}{F_z} \quad \text{if } F_z \neq 0 \qquad (14.15)$$

Note the similarity between these equations and Equation 14.14.

EXAMPLE 14.12   If z is defined implicitly as a differentiable function of x and y by the equation

$$z^3 - 2xz + y^2 - x^3 - 13 = 0$$

find $\partial z/\partial x$ and $\partial z/\partial y$ at the point $(-1, 3, 1)$.

Solution   Here $F(x, y, z)$ is the left-hand side of the given equation. So we have, by Equations 14.15,

$$\frac{\partial z}{\partial x} = -\frac{F_x}{F_z} = -\frac{-2z - 3x^2}{3z^2 - 2x}$$

$$\frac{\partial z}{\partial y} = -\frac{F_y}{F_z} = -\frac{2y}{3z^2 - 2x}$$

On substituting $(-1, 3, 1)$ and simplifying, we get

$$\left.\frac{\partial z}{\partial x}\right|_{(-1,3,1)} = 1 \quad \text{and} \quad \left.\frac{\partial z}{\partial y}\right|_{(-1,3,1)} = -\frac{6}{5} \qquad \blacksquare$$

Exercise Set 14.2

In Exercises 1–16, make use of a chain rule. In Exercises 1–4, find dz/dt at the specified value of t.

1. $z = x^2 + y^2$, $x = t^2 + 1$, $y = 2 - t^2$; at $t = 2$

2. $z = 2x^2 - 3xy$, $x = e^t$, $y = te^t$; at $t = 0$

3. $z = xe^y - ye^x$, $x = \ln t$, $y = \ln\dfrac{1}{t}$; at $t = 1$

4. $z = x \cos xy$, $x = 2t^2$, $y = \dfrac{1}{t}$; at $t = \dfrac{\pi}{4}$

In Exercises 5–8, find dz/dt. Express answers in terms of t.

5. $z = \ln\dfrac{y}{x}$, $x = \sin t$, $y = \cos t$

6. $z = \tan^{-1}\dfrac{y}{x}$, $x = \sin 2t$, $y = \cos 2t$

7. $z = r^2(1 - \cos\theta)$, $r = \sqrt{t}$, $\theta = t^2$

8. $z = uv^2w^3$, $u = e^t$, $v = e^{-t}$, $w = te^t$

In Exercises 9–12, find $\partial z/\partial u$ and $\partial z/\partial v$ at the specified point.

9. $z = x^2 - 2xy$, $x = \dfrac{u}{v}$, $y = uv$; at $(u, v) = (2, -1)$

10. $z = \ln\sqrt{x^2 + y^2}$, $x = u + v$, $y = 2u - 3v$;
    at $(u, v) = (3, 1)$

11. $z = e^{xy}$, $x = \dfrac{u}{v}$, $y = 2v$; at $(u, v) = \left(1, \dfrac{1}{2}\right)$

12. $z = \sin(x + y)$, $x = u^2 - v^2$, $y = 2uv$;
    at $(u, v) = (\sqrt{\pi}, \sqrt{\pi}/2)$

In Exercises 13–16, find $\partial z/\partial u$ and $\partial z/\partial v$ at an arbitrary point (u, v).

13. $z = \ln\dfrac{x + y}{x - y}$; $x = \cos^2 uv$, $y = \sin^2 uv$

14. $z = x^2 + 2xy - y^2$; $x = \cosh u + \sinh v$,
    $y = \cosh u - \sinh v$

15. $z = r^2(3\sin^2\theta - 2\cos^2\theta)$; $r = \sqrt{u^2 + v^2}$, $\theta = \tan^{-1}(v/u)(u \neq 0)$

16. $z = \sqrt{rst}$; $r = u^2 v$, $s = u/v$, $t = v^2/u$ $(uv > 0)$

In Exercises 17–22, assume the given equation defines y as a differentiable function of x on some domain, and use Equation 14.14 to find dy/dx there.

17. $x^3 - 2xy^2 + y^4 = 0$

18. $2x^2 y^2 - 3xy^3 + 4y - 5 = 0$

19. $x\sin y - y\sin x = 1$

20. $x\ln y + y\ln x = xy$

21. $e^{-xy}(x^2 - y^2) = 4$

22. $xy\tan xy + 1 = 0$

In Exercises 23–28, assume the given equation defines z as a differentiable function of x and y on some domain, and use Equations 14.15 to find $\partial z/\partial x$ and $\partial z/\partial y$ there.

23. $x^2 yz - 2y^2 - 3z^2 = 0$

24. $x^2 + y^2 + z^2 - 2xy + 3xz - 4yz = 1$

25. $\ln\left[(x + y)/\sqrt{z}\right] - 2xyz = 4$

26. $\tan^{-1}\dfrac{x}{y} - \tan^{-1}\dfrac{y}{z} = 2$

27. $\sin xz + \ln\cos yz = 0$

28. $z - xy\ln z = 1$

29. If $w = (u + v)/u$ and $u = \sqrt{x^2 + y^2 + z^2}$, $v = 2xyz$, find $\partial w/\partial x$, $\partial w/\partial y$, and $\partial w/\partial z$ as functions of x, y, and z.

30. Let $f(x, y, z) = xyz$ and $F(s, t) = f(s + t, s - t, st)$. Using an appropriate chain rule, find $F_s(2, -1)$ and $F_t(2, -1)$.

31. Let $z = f(x, y, z)$, and let the xy coordinates be transformed to polar coordinates by the equations $x = r\cos\theta$, $y = r\sin\theta$. Show that

$$\left(\frac{\partial z}{\partial r}\right)^2 + \frac{1}{r^2}\left(\frac{\partial z}{\partial\theta}\right)^2 = \left(\frac{\partial z}{\partial x}\right)^2 + \left(\frac{\partial z}{\partial y}\right)^2$$

32. The relationship between the pressure P, volume V, and temperature T of a certain gas is given by $PV = 12T$. If the pressure is decreasing at the constant rate of 3 psi/min and the temperature is constantly increasing at $4°K$ per minute, find the rate at which the volume is changing when $P = 10$ psi and $T = 298°K$.

33. An oil slick in the Gulf of Mexico from a ruptured oil tanker is approximately triangular in shape. When the height of the triangle is 2 km, the base is 3 km, and at that instant the height and base are increasing at the rates of 200 m/hr and 320 m/hr, respectively. Find the rate at which the area is increasing at that instant.

34. Let $z = f(x - y, y - x)$. Show that

$$\frac{\partial z}{\partial x} + \frac{\partial z}{\partial y} = 0$$

(*Hint:* Let $u = x - y$ and $v = y - x$.)

35. Let $z = f(x/y)$. Show that

$$x\frac{\partial z}{\partial x} + y\frac{\partial z}{\partial y} = 0$$

36. Prove that $y = f(x + ct) + g(x - ct)$ is a solution of the wave equation

$$\frac{\partial^2 y}{\partial t^2} = c^2 \frac{\partial^2 y}{\partial x^2}$$

for any twice-differentiable functions f and g.

37. Let $z = f(x, y)$, $x = g(u, v)$, and $y = h(u, v)$. Derive formulas for $\partial^2 z / \partial u^2$ and $\partial^2 z / \partial v^2$.

38. Assume that the equation $F(x, y) = 0$ defines y as a twice-differentiable function, $y = f(x)$, on a domain D. Derive the following formula.

$$f''(x) = -\frac{F_{xx} F_y^2 - 2F_x F_y F_{xy} + F_{yy} F_x^2}{F_y^3} \qquad (F_y \neq 0)$$

39. For the function in Example 14.10, find $\partial^2 z / (\partial \theta \, \partial r)$.

40. If $z = f(x, y)$, the expression $\partial^2 z / \partial x^2 + \partial^2 z / \partial y^2$ is called the *Laplacian* of z. Show that on changing the xy coordinates to polar coordinates by means of the transformation $x = r \cos \theta$, $y = r \sin \theta$, the Laplacian of z becomes

$$\frac{\partial^2 z}{\partial r^2} + \frac{1}{r} \frac{\partial z}{\partial r} + \frac{1}{r^2} \frac{\partial^2 z}{\partial \theta^2}$$

41. In a grain elevator, grain deposited through a chute at the rate of 60 ft³/min assumes the form of a cone as it accumulates on the floor. After 5 min the base radius of the cone is 10 ft, and at that instant it is increasing at the rate of 0.5 ft/min. How fast is the height of the cone changing at that instant?

42. Part of a certain hydraulic lift has a triangular shape that varies in size as the lift is actuated. At the instant when two adjacent sides of the triangle are 3.1 m and 1.7 m long and the angle between them is 30°, these sides and this angle are increasing at the respective rates of 0.5 m/s, 0.8 m/s, and 2°/s. Find the rate at which the area of the triangle is changing at that instant.

CAS *Use a CAS for Exercises 43 through 46.*

43. Express a formula for the Chain Rule for a differentiable function $w = f(x, y, z)$, where x, y, and z are differentiable functions of t.

44. Use the formula obtained in Exercise 43 to find dw/dt in each of the following:

(a) $w = f(x, y, z) = xy + \sin\left(\dfrac{x^2}{z}\right) + x^3 z$, $x(t) = t$,

$y(t) = t^2$, $z(t) = t + 1$

(b) $w = f(x, y, z) = x \sin(y + z)$, $x(t) = 3t$, $y(t) = t^3$, $z(t) = e^t$

45. If $z = f(x, y)$ and x and y are both functions of two variables u and v, establish formulas for $\partial f / \partial u$ and $\partial f / \partial v$. Use the formula to compute $\partial f / \partial u$ and $\partial f / \partial v$ when $f(x, y) = e^{xy}$ and $x = u - 3uv + \sin v$, $y = u + e^{u^2 v}$.

46. Let $z = f(x, y)$ and show that $x \partial z / \partial x - y \partial z / \partial y = 0$.

14.3 DIRECTIONAL DERIVATIVES

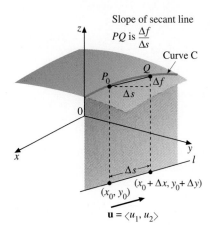

Slope of secant line PQ is $\dfrac{\Delta f}{\Delta s}$

Curve C

$\mathbf{u} = \langle u_1, u_2 \rangle$

FIGURE 14.15

In Section 13.4 we defined the partial derivatives f_x and f_y for a function of two variables $z = f(x, y)$ as

$$f_x(x_0, y_0) = \lim_{\Delta x \to 0} \frac{f(x_0 + \Delta x, y_0) - f(x_0, y_0)}{\Delta x}$$

and

$$f_y(x_0, y_0) = \lim_{\Delta y \to 0} \frac{f(x_0, y_0 + \Delta y) - f(x_0, y_0)}{\Delta y}$$

The first partials describe the rate of change of z in the x direction and y direction, respectively—that is, in the direction of the unit vectors $\mathbf{i} = \langle 1, 0 \rangle$ and $\mathbf{j} = \langle 0, 1 \rangle$. We now want to investigate the rate of change of z at a point (x_0, y_0) in the direction of an arbitrary unit vector. We will see that this rate of change can be expressed in a way that involves both partials f_x and f_y.

Suppose, then, that we want to determine the rate of change of $z = f(x, y)$ at (x_0, y_0) in the direction of the unit vector $\mathbf{u} = \langle u_1, u_2 \rangle$. The vertical plane that passes through the point $P_0 = (x_0, y_0, z_0)$ on the surface $z = f(x, y)$ in the direction $\mathbf{u}$ intersects the surface in some curve C, as we show in Figure 14.15.

The slope of the tangent line to the curve C at P_0 is the rate of change of z in the direction $\mathbf{u}$.

We can write a vector equation of the line l through (x_0, y_0) in the direction $\mathbf{u}$ as

$$\mathbf{r}(s) = \langle x_0 + su_1, y_0 + su_2 \rangle$$

where the parameter s is the directed distance along l, starting from (x_0, y_0), with positive direction given by $\mathbf{u}$.

A point on the line l that is Δs units from (x_0, y_0) has the position vector

$$r(\Delta s) = \langle x_0 + (\Delta s)u_1, y_0 + (\Delta s)u_2 \rangle$$

If we let $\Delta x = (\Delta s)u_1$, and $\Delta y = (\Delta s)u_2$, then the coordinates of this point are $x_0 + \Delta x$ and $y_0 + \Delta y$. Let Q be the point on the surface directly above $(x_0 + \Delta x, y_0 + \Delta y)$, as in Figure 14.15. As a point (x, y) moves from (x_0, y_0) to $(x_0 + \Delta x, y_0 + \Delta y)$ in the xy-plane, $f(x, y)$ changes by the amount

$$\Delta f = f(x_0 + \Delta x, y_0 + \Delta y) - f(x_0, y_0)$$
$$= f(x_0 + u_1 \Delta s, y_0 + u_2 \Delta s) - f(x_0, y_0)$$

The quotient

$$\frac{\Delta f}{\Delta s}$$

is the slope of the secant line through P and Q. (See Figure 14.15.) If this quotient approaches a limit as Δs approaches 0, we obtain the rate of change of z in the direction $\mathbf{u}$ (the slope of the tangent line to the curve C at P_0). We call this limit the *directional derivative* of f at (x_0, y_0) in the direction $\mathbf{u}$, and we denote it either by $D_{\mathbf{u}} f(x_0, y_0)$ or by df/ds. In summary, we have the following definition.

Definition 14.3
The Directional Derivative

The **directional derivative** of $f(x, y)$ at (x_0, y_0) in the direction of the unit vector $\mathbf{u} = \langle u_1, u_2 \rangle$ is

$$D_{\mathbf{u}} f(x_0, y_0) = \frac{df}{ds} = \lim_{\Delta s \to 0} \frac{\Delta f}{\Delta s} \qquad (14.16)$$

provided this limit exists, where

$$\Delta f = f(x_0 + u_1 \Delta s, y_0 + u_2 \Delta s) - f(x_0, y_0)$$

If we consider the special case of the directional derivative in which $\mathbf{u} = \langle 1, 0 \rangle$, then $\Delta x = \Delta s$ and $\Delta y = 0$. So, by Equation 14.16,

$$D_{\mathbf{u}} f(x_0, y_0) = \lim_{\Delta x \to 0} \frac{f(x_0 + \Delta x, y_0) - f(x_0, y_0)}{\Delta x} = f_x(x_0, y_0)$$

That is, the directional derivative in the direction of the x-axis is the same as the partial derivative with respect to x. Similarly, we can show that when $\mathbf{u} = \langle 0, 1 \rangle$, $D_{\mathbf{u}} f(x_0, y_0) = f_y(x_0, y_0)$. Thus, the partial derivatives of f with respect to x and y are special cases of the directional derivative.

Calculating the Directional Derivative

When f is *differentiable* at (x_0, y_0), we can show that $D_{\mathbf{u}}f(x_0, y_0)$ does exist, and we can obtain a useful computational formula for the directional derivative. By Definition 14.1, we can write $\Delta f/\Delta s$ in the form

$$\frac{\Delta f}{\Delta s} = f_x(x_0, y_0)\frac{\Delta x}{\Delta s} + f_y(x_0, y_0)\frac{\Delta y}{\Delta s} + \varepsilon_1\frac{\Delta x}{\Delta s} + \varepsilon_2\frac{\Delta y}{\Delta s}$$

But $\Delta x = u_1\Delta s$ and $\Delta y = u_2\Delta s$, so $\Delta x/\Delta s = u_1$ and $\Delta y/\Delta s = u_2$. Furthermore, as $\Delta s \to 0$, both Δx and Δy approach 0, and hence $\varepsilon_1 \to 0$ and $\varepsilon_2 \to 0$. Thus,

$$D_{\mathbf{u}}f(x_0, y_0) = \lim_{\Delta s \to 0}\frac{\Delta f}{\Delta s} = f_x(x_0, y_0)u_1 + f_y(x_0, y_0)u_2$$

We restate this result for emphasis.

The Directional Derivative of a Differentiable Function

If f is differentiable at (x_0, y_0), then for any unit vector $\mathbf{u} = \langle u_1, u_2 \rangle$, $D_{\mathbf{u}}f(x_0, y_0)$ exists and is given by the formula

$$D_{\mathbf{u}}f(x_0, y_0) = f_x(x_0, y_0)u_1 + f_y(x_0, y_0)u_2 \qquad (14.17)$$

REMARK ────────────────────────────────────

■ For Equation 14.17 to be valid, it is essential that $\mathbf{u}$ be a *unit* vector. If the direction is given by some non-unit-vector $\mathbf{v}$, then we replace $\mathbf{v}$ with the unit vector $\mathbf{u} = \mathbf{v}/|\mathbf{v}|$.

EXAMPLE 14.13   Find the directional derivative of the function $f(x, y) = x^2 - 2xy^3$ at $(-2, 1)$ in the direction of the vector $\mathbf{v} = \langle 3, 4 \rangle$.

Solution   First, we obtain a unit direction vector:

$$\mathbf{u} = \frac{\mathbf{v}}{|\mathbf{v}|} = \left\langle \frac{3}{5}, \frac{4}{5} \right\rangle$$

Now we use Equation 14.17:

$$D_{\mathbf{u}}f(x, y) = f_x(x, y) \cdot \left(\frac{3}{5}\right) + f_y(x, y) \cdot \left(\frac{4}{5}\right)$$

$$= (2x - 2y^3)\left(\frac{3}{5}\right) + (-6xy^2)\left(\frac{4}{5}\right)$$

Setting $(x, y) = (-2, 1)$, we get

$$D_{\mathbf{u}}f(-2, 1) = (-6)\left(\frac{3}{5}\right) + (12)\left(\frac{4}{5}\right) = 6$$

■

EXAMPLE 14.14   A coordinate system is established on a flat metal plate, and it is determined that at a point (x, y) on the plate, other than the origin, the temperature $T(x, y)$ in degrees Celsius is given by $T(x, y) = 100(x^2 + y^2)^{-1/2}$. Find the instantaneous rate of change of temperature at the point $P(2, 6)$ in the direction from P toward $Q(4, 2)$.

Solution   We show the graph of the temperature function $z = T(x, y)$ in Figure 14.16. The instantaneous rate of change of T in the direction of $\overrightarrow{PQ}$ is the directional derivative at P in the direction of the unit vector

$$\mathbf{u} = \frac{\overrightarrow{PQ}}{|\overrightarrow{PQ}|} = \frac{\langle 2, -4 \rangle}{\sqrt{20}} = \left\langle \frac{1}{\sqrt{5}}, \frac{-2}{\sqrt{5}} \right\rangle$$

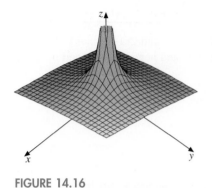

FIGURE 14.16

So we have, by Equation 14.17,

$$D_{\mathbf{u}}T(x, y) = \left[\frac{-100x}{(x^2 + y^2)^{3/2}} \right] \left(\frac{1}{\sqrt{5}} \right) + \left[\frac{-100y}{(x^2 + y^2)^{3/2}} \right] \left(\frac{-2}{\sqrt{5}} \right)$$

and after simplification, we obtain

$$D_{\mathbf{u}}T(2, 6) = \left(\frac{-200}{80\sqrt{10}} \right) \left(\frac{1}{\sqrt{5}} \right) + \left(\frac{-600}{80\sqrt{10}} \right) \left(\frac{-2}{\sqrt{5}} \right) = \frac{5\sqrt{2}}{4} \approx 1.768$$

Thus, as a point moves from P one unit on a line toward Q, one would expect an increase in temperature of approximately $1.768°$. (Remember, though, that this rate of change is instantaneous, so after moving away from P the rate changes.)   ■

The Gradient of a Function

The right-hand side of Equation 14.17 has the appearance of the dot product of two vectors. In fact,

$$\begin{aligned}
D_{\mathbf{u}}f(x, y) &= f_x(x, y)u_1 + f_y(x, y)u_2 \\
&= \langle f_x(x, y), f_y(x, y) \rangle \cdot \langle u_1, u_2 \rangle \\
&= \langle f_x(x, y), f_y(x, y) \rangle \cdot \mathbf{u}
\end{aligned}$$

The vector $\langle f_x(x, y), f_y(x, y) \rangle$ occurs in a number of contexts and is given the special name, the *gradient* of f.

Definition 14.4 The Gradient	The **gradient** of f at a point $P(x, y)$ is denoted by $\nabla f(P)$ or $\nabla f(x, y)$ and is defined as the vector $$\nabla f(x, y) = \langle f_x(x, y), f_y(x, y) \rangle \qquad (14.18)$$ provided these partials exist.

REMARK
■ The symbol ∇ is read "del." Another notation for the gradient of f is **Grad**. It is an operator, in the same sense that d/dx is an operator. So ∇f and **Grad** f mean the same thing. We will use ∇f exclusively in this book.

Using Definition 14.4, we can rewrite Equation 14.17 in the following form:

$$D_{\mathbf{u}} f(x, y) = \nabla f(x, y) \cdot \mathbf{u} \qquad (14.19)$$

EXAMPLE 14.15   Let $f(x, y) = \ln(x/y)$. Find ∇f and use it to find the directional derivative of f at $(1, 2)$ in the direction of $\mathbf{u} = \langle 1/2, -\sqrt{3}/2 \rangle$.

Solution   To simplify the computation of the partials f_x and f_y, we write $\ln \frac{x}{y}$ in the equivalent form $\ln x - \ln y$, so that $f(x, y) = \ln x - \ln y$. Then we see that

$$\nabla f = \langle f_x, f_y \rangle = \left\langle \frac{1}{x}, -\frac{1}{y} \right\rangle$$

At the point $(1, 2)$, we therefore have

$$\nabla f(1, 2) = \left\langle 1, -\frac{1}{2} \right\rangle$$

By Equation 14.19,

$$D_{\mathbf{u}} f(1, 2) = \left\langle 1, -\frac{1}{2} \right\rangle \cdot \left\langle \frac{1}{2}, -\frac{\sqrt{3}}{2} \right\rangle = \frac{1}{2} + \frac{\sqrt{3}}{4} = \frac{2 + \sqrt{3}}{4} \qquad ■$$

Extreme Values of the Directional Derivative

Sometimes we want to know the direction in which a function changes most rapidly and what this maximum rate of change is. The answer is provided in the following theorem.

THEOREM 14.3

Let f have continuous partial derivatives f_x and f_y in a region containing a point (x_0, y_0) at which $\nabla f \neq 0$. Then the maximum value of $D_{\mathbf{u}} f(x_0, y_0)$ is $|\nabla f(x_0, y_0)|$, which occurs when $\mathbf{u}$ is in the direction of $\nabla f(x_0, y_0)$. The minimum value of $D_{\mathbf{u}} f(x_0, y_0)$ is $-|\nabla f(x_0, y_0)|$ and occurs when $\mathbf{u}$ is in the direction of $-\nabla f(x_0, y_0)$.

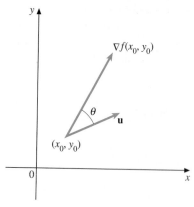

FIGURE 14.17

Proof   In Figure 14.17 we show the gradient vector $\nabla f(x_0, y_0)$ and a unit vector $\mathbf{u}$, both drawn from the point (x_0, y_0). If θ denotes the angle between these two vectors, then

$$D_\mathbf{u} f(x_0, y_0) = \nabla f(x_0, y_0) \cdot \mathbf{u} \quad \text{By Equation 14.19}$$

$$= |\nabla f(x_0, y_0)||\mathbf{u}| \cos \theta \quad \text{By Equation 11.4}$$

$$= |\nabla f(x_0, y_0)| \cos \theta \quad \text{Since } |\mathbf{u}| = 1 \qquad (14.20)$$

Now the maximum value $\cos \theta$ can assume is 1, which occurs when $\theta = 0$. Thus, to obtain the maximum value of $D_\mathbf{u} f(x_0, y_0)$, we take the angle between $\mathbf{u}$ and $\nabla f(x_0, y_0)$ to be 0 (the cosine is 1 at 0 and is never greater than 1); that is, we take $\mathbf{u}$ in the same direction as $\nabla f(x_0, y_0)$, namely,

$$\mathbf{u} = \frac{\nabla f(x_0, y_0)}{|\nabla f(x_0, y_0)|}$$

Furthermore, when we use this value of $\mathbf{u}$, we see from Equation 14.20 that

$$\max D_\mathbf{u} f(x_0, y_0) = |\nabla f(x_0, y_0)|$$

since $\cos 0 = 1$.

Similarly, the minimum value of $\cos \theta$ is -1, which occurs when $\theta = \pi$. In this case $\mathbf{u}$ is the opposite of $\nabla f(x_0, y_0)$. That is,

$$\mathbf{u} = -\frac{\nabla f(x_0, y_0)}{|\nabla f(x_0, y_0)|}$$

and from Equation 14.20 we see that

$$\min D_\mathbf{u} f(x_0, y_0) = -|\nabla f(x_0, y_0)| \qquad \blacksquare$$

EXAMPLE 14.16   For the function $f(x, y) = \ln(x/y)$ of Example 14.15, find the maximum value of the directional derivative at the point $(1, 2)$, and give the direction $\mathbf{u}$ that produces this maximum value.

Solution   In Example 14.15, we found that

$$\nabla f(1, 2) = \left\langle 1, -\frac{1}{2} \right\rangle$$

Thus, by Theorem 14.3,

$$\max D_\mathbf{u} f(1, 2) = \left| \left\langle 1, -\frac{1}{2} \right\rangle \right| = \sqrt{1 + \frac{1}{4}} = \frac{\sqrt{5}}{2}$$

This maximum value occurs in the direction of the vector

$$\mathbf{u} = \frac{\nabla f(1, 2)}{|\nabla f(1, 2)|} = \frac{2}{\sqrt{5}} \left\langle 1, -\frac{1}{2} \right\rangle = \left\langle \frac{2}{\sqrt{5}}, \frac{-1}{\sqrt{5}} \right\rangle \qquad \blacksquare$$

EXAMPLE 14.17   Refer to Example 14.14. In what direction from the point $P(2, 6)$ does the temperature $T(x, y) = 100(x^2 + y^2)^{-1/2}$ change most rapidly, and what is this maximum rate of change?

Solution   In Example 14.14, the temperature function $T(x, y)$ is given as

$$T(x, y) = 100(x^2 + y^2)^{-1/2}$$

The gradient vector $\nabla T = \langle T_x(x, y), T_y(x, y)\rangle$ is given by

$$\nabla T = 100\left\langle \frac{-x}{(x^2 + y^2)^{3/2}}, \frac{-y}{(x^2 + y^2)^{3/2}}\right\rangle$$

Evaluating ∇T at $(2, 6)$, we have

$$\nabla T(2, 6) = 100\left\langle \frac{-2}{(40)^{3/2}}, \frac{-6}{(40)^{3/2}}\right\rangle$$

and

$$|\nabla T(2, 6)| = 100\sqrt{\frac{4 + 36}{(40)^3}} = \frac{100}{40} = \frac{5}{2}$$

We therefore have, by Theorem 14.3,

$$\max D_{\mathbf{u}} T(2, 6) = \frac{5}{2}$$

and the maximum occurs in the direction of the vector

$$\mathbf{u} = \frac{\nabla T(2, 6)}{|\nabla T(2, 6)|} = \frac{100\left\langle -\frac{2}{(40)^{3/2}}, -\frac{6}{(40)^{3/2}}\right\rangle}{5/2} = 40\left\langle -\frac{2}{(40)^{3/2}}, -\frac{6}{(40)^{3/2}}\right\rangle$$

$$= \left\langle -\frac{2}{(40)^{1/2}}, -\frac{6}{(40)^{1/2}}\right\rangle = \left\langle -\frac{1}{\sqrt{10}}, -\frac{3}{\sqrt{10}}\right\rangle \qquad \blacksquare$$

Relation of ∇f to Level Curves

The relationship between the gradient vector $\nabla f(x_0, y_0)$ and the level curve to the surface $z = f(x, y)$ that passes through (x_0, y_0) is helpful in understanding the maximum value of the directional derivative $D_{\mathbf{u}} f(x_0, y_0)$. We suppose f is differentiable at (x_0, y_0) and that $\nabla f(x_0, y_0) \neq 0$. If $c = f(x_0, y_0)$, then the level curve $f(x, y) = c$ passes through (x_0, y_0). Suppose this level curve is parameterized by the differentiable functions $x = x(t), y = y(t)$. Let $x_0 = x(t_0)$ and $y_0 = y(t_0)$. Then, since $f(x(t), y(t)) = c$, we have, by the Chain Rule (Equation 14.10),

$$f_x(x_0, y_0)x'(t_0) + f_y(x_0, y_0)y'(t_0) = 0$$

or, equivalently,

$$\nabla f(x_0, y_0) \cdot \langle x'(t_0), y'(t_0)\rangle = 0$$

The vector $\langle x'(t_0), y'(t_0)\rangle$ is a tangent vector to the level curve $f(x, y) = c$ at the point (x_0, y_0). We conclude, therefore, that the gradient vector $\nabla f(x_0, y_0)$ is perpendicular to this tangent vector at (x_0, y_0). We illustrate this result in Figure 14.18. We summarize our results as follows.

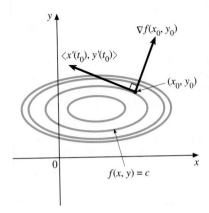

FIGURE 14.18

Orthogonality of the Gradient Vector and Level Curves

$\nabla f(x_0, y_0)$ is orthogonal to the level curve $f(x, y) = c$ passing through (x_0, y_0).

Combining this result with Theorem 14.3, we conclude that from any point in its domain a function $f(x, y)$ increases (or decreases) most rapidly in a direction perpendicular to the level curve through that point. For example, on a weather map, to move in the direction of the most rapid change in temperature, we would move in a direction perpendicular to the isotherms.

EXAMPLE 14.18   Let $f(x, y) = x^2/4 + y^2$. Sketch the level curve through the point $(3, 2)$, and show the gradient vector at that point.

Solution   Since $f(3, 2) = 9/4 + 4 = 25/4$, the level curve has the equation $x^2/4 + y^2 = 25/4$, or in standard form,

$$\frac{x^2}{25} + \frac{y^2}{25/4} = 1$$

The graph is an ellipse with semimajor axis 5 and semiminor axis 5/2. We also have

$$\nabla f = \left\langle \frac{x}{2}, 2y \right\rangle$$

so $\nabla f(3, 2) = \langle 3/2, 4 \rangle$. We show the level curve and the gradient orthogonal to it in Figure 14.19. ∎

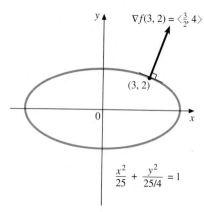

FIGURE 14.19

Extension to Three Variables

All of the concepts in this section can be extended in natural ways to functions of three (or more) variables. We summarize the results as follows.

Directional Derivative for a Function of Three Variables

If f is a differentiable function of three variables at the point $P_0 = (x_0, y_0, z_0)$ and $\mathbf{u}$ is a unit vector, then the directional derivative of f at P_0 in the direction $\mathbf{u}$ is given by

$$D_{\mathbf{u}} f(P_0) = \nabla f(P_0) \cdot \mathbf{u}$$

where $\nabla f = \langle f_x, f_y, f_z \rangle$. The maximum value of this directional derivative, which is $|\nabla f(P_0)|$, occurs when $\mathbf{u}$ is in the direction of $\nabla f(P_0)$. Furthermore, the direction of maximum change in f is orthogonal to the level surface $f(x, y, z) = c$ passing through P_0. The minimum value, which is $-|\nabla f(P_0)|$, occurs when $\mathbf{u}$ is directed opposite to $\nabla f(P_0)$.

Exercise Set 14.3

In Exercises 1–14, find the directional derivative of f at the given point in the direction of the given vector.

1. $f(x, y) = 2x^2 y - 3xy^2$ at $(3, 1)$; $\mathbf{u} = \langle \frac{3}{5}, \frac{4}{5} \rangle$

2. $f(x, y) = x^3 - 2xy + y^3$ at $(1, 0)$; $\mathbf{u} = \langle 1/\sqrt{10}, 3/\sqrt{10} \rangle$

3. $f(x, y) = \ln[xy/(x^2 + y^2)]$ at $(1, 2)$; $\mathbf{v} = \langle -2, 2 \rangle$

4. $f(x, y) = xe^y - ye^x$ at $(0, 0)$; $\mathbf{v} = \langle 8, -1 \rangle$

5. $f(x, y) = x(\cos y - \sin y)$ at $\left(-1, \dfrac{\pi}{2}\right)$; $\mathbf{v} = 4\mathbf{i} - 3\mathbf{j}$

6. $f(x, y) = \tan^{-1} \dfrac{y}{x}$ at $(2, 1)$; $\mathbf{v} = \mathbf{i} + \mathbf{j}$

7. $f(x, y) = x \cosh y + y \cosh x$ at $(0, 0)$; $\mathbf{v} = -2\mathbf{i} + 4\mathbf{j}$

8. $f(x, y) = y^2 \ln \sqrt{x}$ at $(1, -2)$; $\mathbf{v} = 3\mathbf{i} + 2\mathbf{j}$

9. $f(x, y) = \sqrt{9 - x^2 - y^2}$ at $(2, -1)$; $\mathbf{v} = \langle 1, -1 \rangle$

10. $f(x, y) = e^{(x-y)/(x+y)}$ at $(1, 1)$; $\mathbf{v} = \langle 5, 12 \rangle$

11. $f(x, y, z) = x^2 - 2y^2 + 3z^2$ at $(2, 0, -1)$;
$\mathbf{u} = \langle \frac{1}{3}, \frac{2}{3}, -\frac{2}{3} \rangle$

12. $f(x, y, z) = \dfrac{x + y}{x + z}$ at $(3, -2, -1)$;
$\mathbf{u} = \left\langle \dfrac{1}{\sqrt{6}}, -\dfrac{1}{\sqrt{6}}, \dfrac{2}{\sqrt{6}} \right\rangle$

13. $f(x, y, z) = \ln[x^2/(yz^3)]$ at $(1, 1, 3)$; $\mathbf{v} = 3\mathbf{i} - 4\mathbf{j} + 5\mathbf{k}$

14. $f(x, y, z) = x \cosh(y + z)$ at $(3, 2, -1)$; $\mathbf{v} = \mathbf{i} + \mathbf{j} - \mathbf{k}$

15. Find the directional derivative of $f(x, y) = 3x^2 - 2xy^2$ at $(3, -2)$ in the direction from $(3, -2)$ toward $(5, 4)$.

16. Find the directional derivative of $f(x, y) = \ln \sqrt{x^2 - 2y^2}$ at $(2, 1)$ in the direction from $(2, 1)$ toward $(5, -3)$.

In Exercises 17–20, find the unit vector $\mathbf{u}$ for which $D_{\mathbf{u}} f(P_0)$ is a maximum, and give this maximum value.

17. $f(x, y) = \sqrt{\dfrac{x - y}{x + y}}$; $P_0 = (5, 4)$

18. $f(x, y) = \ln \cos(x + 2y)$; $P_0 = \left(\dfrac{\pi}{4}, \dfrac{\pi}{4}\right)$

19. $f(x, y) = y^2 + e^{(\sin x)/y}$; $P_0 = (0, -1)$

20. $f(x, y, z) = \ln \dfrac{x + 2y}{z^3}$; $P_0 = (5, -2, 3)$

21. In what direction from the point $(1, -1)$ is the instantaneous rate of change of $f(x, y) = 2x^2 + 2xy - 3y^2$ equal to 2? (There are two solutions.) In what direction from $(1, -1)$ does this function increase most rapidly? What is this most rapid rate of change?

22. In what direction from the point $(4, 1)$ is the function $f(x, y) = x/(y + 1)$ stationary? From the same point, in what direction is the instantaneous rate of change of this function equal to 1? (There are two solutions.) Can the rate of change from the point $(4, 1)$ in any direction ever equal 2? Explain.

23. Two adjacent edges of a flat, rectangular, metal plate coincide, respectively, with the positive x- and y-axes. For points other than the origin, the temperature $T(x, y)$ at an arbitrary point (x, y) is inversely proportional to the distance from P to the origin. At the point $P(8, 6)$, the temperature is $10°$C. How rapidly is the temperature changing at P in the direction from P toward $Q(6, 10)$? In what direction from P does the temperature decrease most rapidly, and what is this rate of decrease?

24. Two adjacent edges of a large, square, metal plate are kept at temperatures $T = 0$ and $T = 100$, respectively, and the flat surfaces are well insulated. By taking the positive x- and y-axes along the edges held at $T = 0$ and $T = 100$, respectively, it can be shown that the temperature $T(x, y)$ at an arbitrary point in the plate is approximated by

$$T(x, y) = \frac{200}{\pi} \tan^{-1} \frac{y}{x}$$

How rapidly is the temperature changing at the point $(2, 4)$ in the direction of the vector $\mathbf{v} = 3\mathbf{i} - 4\mathbf{j}$? In what direction from $(2, 4)$ is the temperature increasing most rapidly, and what is this most rapid change?

25. A cross section of two long, coaxial, conducting cylindrical surfaces consists of the circles $x^2 + y^2 = 1$ and $x^2 + y^2 = 4$. If the smaller cylinder is held at electrostatic potential $V = 0$ and the larger at $V = 1$, then it can be shown that in the annular ring between the two, the potential $V(x, y)$ is given by

$$V(x, y) = \frac{\ln(x^2 + y^2)}{\ln 4}$$

Find the rate of change in potential at $P(\frac{3}{2}, \frac{1}{2})$ in the direction from P toward $Q(\frac{3}{4}, 1)$. In what direction from P does V increase most rapidly, and what is this rate of change?

26. In Exercise 24 find the equation of the isotherm $T(x, y) = c$ for c between 0 and 100 and identify the graph. Show that $\nabla T(x_0, y_0)$ is orthogonal to the isotherm through (x_0, y_0).

27. In Exercise 25 find the equation of the equipotential $V(x, y) = c$ for c between 0 and 1 and identify the graph. Show that $\nabla V(x_0, y_0)$ is orthogonal to the equipotential curve through (x_0, y_0).

28. Find a function $f(x, y)$ for which $\nabla f = \langle xe^x, e^{-y} \rangle$. Is this function unique? Explain.

29. Find the function f for which $\nabla f = (x\sin x)\mathbf{i} + (\cos y)\mathbf{j}$ and $f(\frac{\pi}{2}, 0) = 3$.

30. Show that if α, β, and γ are direction angles of the unit vector $\mathbf{u}$, and if f is differentiable at $P = (x, y, z)$, then

$$D_{\mathbf{u}}f(P) = f_x(P)\cos\alpha + f_y(P)\cos\beta + f_z(P)\cos\gamma$$

In Exercises 31–36, $u = f(x, y)$ and $v = g(x, y)$ are differentiable functions, and c and α are arbitrary real numbers. Prove each statement.

31. $\nabla(cu) = c\nabla u$

32. $\nabla(u + v) = \nabla u + \nabla v$

33. $\nabla(uv) = u\nabla v + v\,\nabla u$

34. $\nabla\left(\dfrac{u}{v}\right) = \dfrac{v\nabla u - u\nabla v}{v^2}$ if $v \neq 0$

35. $\nabla u^\alpha = \alpha u^{\alpha-1}\,\nabla u$

36. If $w = h(u, v)$, and h is differentiable, then

$$\nabla w = \frac{\partial w}{\partial u}\nabla u + \frac{\partial w}{\partial v}\nabla v$$

CAS *Use a CAS for Exercises 37 and 38.*

37. Let $f(x, y) = 2x - y^2$. Plot the surface and a contour map for the surface. Find the gradient of f and use it to compute the gradient at the point $(2, -1)$ and the tangent line to the level curve $f(x, y) = 3$. Plot the tangent line and gradient vectors on the contour map.

38. Let $f(x, y) = x^2 + y^2$. Plot the surface and a contour map for the surface. Find the gradient of f and use it to compute the gradient at the point $(1, -1)$ and the tangent line to the level curve $f(x, y) = 2$. Plot the tangent line and gradient vectors on the contour map.

14.4 TANGENT PLANES AND NORMAL LINES

Denote by S the surface that is the graph of $F(x, y, z) = 0$. For example, S might be the ellipsoid whose equation is $x^2 + 2y^2 + 4z^2 = 16$. In this case, $F(x, y, z) = x^2 + 2y^2 + 4z^2 - 16$. Assume that F is differentiable at the point (x_0, y_0, z_0) on the surface S, with $\nabla F(x_0, y_0, z_0) \neq \mathbf{0}$. Let C be a curve on the surface S passing through (x_0, y_0, z_0) and defined by the vector function $\mathbf{r}(t) = \langle f(t), g(t), h(t)\rangle$. If $\mathbf{r}(t_0) = \langle x_0, y_0, z_0\rangle$, then from Section 12.2, we know that $\mathbf{r}'(t_0)$ is a tangent vector to C at (x_0, y_0, z_0). Because C lies on the surface S, all points $(f(t), g(t), h(t))$ on C satisfy the equation of the surface; that is, $F(f(t), g(t), h(t)) = 0$, and so $dF/dt = 0$ wherever this derivative exists. It does exist at t_0 and is found by the Chain Rule:

$$\frac{dF}{dt}(f(t_0), g(t_0), h(t_0)) = F_x(x_0, y_0, z_0)f'(t_0) + F_y(x_0, y_0, z_0)g'(t_0)$$

$$+ F_z(x_0, y_0, z_0)h'(t_0)$$

$$= 0$$

or, in vector form,

$$\nabla F(x_0, y_0, z_0) \cdot \mathbf{r}'(t_0) = 0$$

Since $\mathbf{r}'(t_0)$ is tangent to C, it follows that $\nabla F(x_0, y_0, z_0)$ is orthogonal to the tangent line to C at (x_0, y_0, z_0).

The argument just given applies to *every* curve C on S that passes through (x_0, y_0, z_0) and has a tangent line there. Thus, the plane through (x_0, y_0, z_0) that is perpendicular to $\nabla F(x_0, y_0, z_0)$ must contain all the tangent lines to such curves (see Figure 14.20). It is natural to call this plane the **tangent plane** to S at (x_0, y_0, z_0).

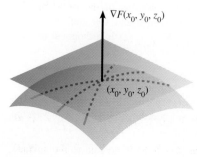

FIGURE 14.20

Definition 14.5 Tangent Plane and Normal Line	Let S be the surface that is the graph of $F(x, y, z) = 0$. Let (x_0, y_0, z_0) be a point on S at which F is differentiable, with $\nabla F(x_0, y_0, z_0) \neq \mathbf{0}$. Then the plane through (x_0, y_0, z_0) that has normal vector $\nabla F(x_0, y_0, z_0)$ is called the **tangent plane** to S at (x_0, y_0, z_0). The line through (x_0, y_0, z_0) with direction vector $\nabla F(x_0, y_0, z_0)$ is called the **normal line** to S at (x_0, y_0, z_0).

Since $\nabla F(x_0, y_0, z_0) = \langle F_x(x_0, y_0, z_0), F_y(x_0, y_0, z_0), F_z(x_0, y_0, z_0) \rangle$ is a normal vector to the tangent plane, we know from Section 11.6 that the equation of the tangent plane can be written in the following form:

Tangent Plane to the Surface $F(x, y, z) = 0$ at $(x_0,\ y_0,\ z_0)$

$$F_x(x_0, y_0, z_0)(x - x_0) + F_y(x_0, y_0, z_0)(y - y_0) + F_z(x_0, y_0, z_0)(z - z_0) = 0 \tag{14.21}$$

Also, the normal line has the following parametric equations:

The Normal Line to the Surface $F(x, y, z) = 0$ at $(x_0,\ y_0,\ z_0)$

$$\begin{cases} x &= x_0 + F_x(x_0, y_0, z_0)\, t \\ y &= y_0 + F_y(x_0, y_0, z_0)\, t \\ z &= z_0 + F_z(x_0, y_0, z_0)\, t \end{cases} \tag{14.22}$$

EXAMPLE 14.19   Find equations of the tangent plane and normal line to the ellipsoid $x^2 + 2y^2 + 4z^2 = 16$ at the point $(2, -2, 1)$.

Solution   We let $F(x, y, z) = x^2 + 2y^2 + 4z^2 - 16$, so that $\nabla F = \langle 2x, 4y, 8z \rangle$. Since F_x, F_y, and F_z are continuous everywhere, F is differentiable. Also, $\nabla F \neq \mathbf{0}$ for all points on the surface, since $\nabla F = \mathbf{0}$ only at the origin, which does not lie on the ellipsoid. Thus, a tangent plane exists everywhere. At $(2, -2, 1)$ the vector $\nabla F(2, -2, 1) = \langle 4, -8, 8 \rangle$ is normal to the tangent plane. So the equation of the tangent plane is

$$4(x - 2) - 8(y + 2) + 8(z - 1) = 0$$

which, on simplification, becomes

$$x - 2y + 2z = 8$$

Parametric equations of the normal line at $(2, -1, 1)$ are $x = 2 + 4t$, $y = -2 - 8t$, and $z = 1 + 8t$.

We might note that since $\nabla F(2, -2, 1) = \langle 4, -8, 8 \rangle$ is normal to the surface, so is $\frac{1}{4} \nabla F(2, -2, 1) = \langle 1, -2, 2 \rangle$, and this simpler vector could have been used to get the equations of both the tangent plane and the normal line.   ∎

Tangent Plane for $z = f(x, y)$

An equation in the form $z = f(x, y)$ can be written in the form $F(x, y, z) = 0$, where $F(x, y, z) = f(x, y) - z$. For example, $z = x^2 + y^2$ would be written as $x^2 + y^2 - z = 0$. So Equation 14.21 can be used to find the equation of the tangent plane. It is useful to obtain the general result for surfaces with equations in this form. We assume f is a differentiable function of two variables at (x_0, y_0). It follows that $F(x, y, z) = f(x, y) - z$ is a differentiable function of three variables at (x_0, y_0, z_0) (see Exercise 22 in Exercise Set 14.4). Furthermore, $\nabla F = \langle f_x, f_y, -1 \rangle$ is never $\mathbf{0}$. So the tangent plane at (x_0, y_0, z_0) exists and has $\langle f_x(x_0, y_0), f_y(x_0, y_0), -1 \rangle$ as a normal vector.

By Equation 14.21, the equation of the tangent plane is

$$f_x(x_0, y_0)(x - x_0) + f_y(x_0, y_0)(y - y_0) - (z - z_0) = 0$$

which can be written in the following equivalent form.

The Tangent Plane to the Surface $z = f(x, y)$ at (x_0, y_0, z_0)

$$z - z_0 = f_x(x_0, y_0)(x - x_0) + f_y(x_0, y_0)(y - y_0) \qquad (14.23)$$

You may use this result to get the tangent plane when $z = f(x, y)$, or you may use Equation 14.21 with $F(x, y, z) = f(x, y) - z$. The answers will be equivalent.

In employing Equation 14.23, it is important to remember the hypothesis that f is differentiable at (x_0, y_0), since otherwise there is no tangent plane. For example, it can be shown (see Exercise 37 in Exercise Set 14.1) that the function $f(x, y) = -\sqrt{|xy|}$ is continuous at the origin and that $f_x(0, 0) = 0$ and $f_y(0, 0) = 0$, yet f is not differentiable at $(0, 0)$. The nonexistence of a tangent plane to the surface $z = f(x, y)$ in this case can be seen clearly from the computer-generated graph in Figure 14.21.

REMARK
■ Equation 14.23 for the equation of the tangent plane to the surface $z = f(x, y)$ confirms our observation in Section 14.1 that the right-hand side of the linear approximation formula

$$f(x, y) \approx f(x_0, y_0) + f_x(x_0, y_0)(x - x_0) + f_y(x_0, y_0)(y - y_0)$$

is the vertical distance from the point (x, y) to the tangent plane to the surface at (x_0, y_0, z_0). (See Figure 14.2.)

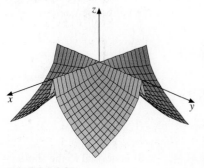

FIGURE 14.21

THE CHAIN RULE AND TANGENT PLANES USING COMPUTER ALGEBRA SYSTEMS

CAS 54

Use Maple and Mathematica to express the Chain Rule for a differentiable function $z = f(x, y)$ of x and y, where x and y are differentiable functions of t (see Equation 14.10). Then find dz/dt, where $z = f(x, y) = x^3 y + x \sin y$ and where $x = x(t) = t^2 + 1$ and $y = y(t) = 2t - 3$.

Maple:

z:=f(x,y);

Output: $z := f(x, y)$;

x:=X(t);y:=Y(t);

Here we use capital X and Y to avoid redefining the symbols x and y.

diff(z,t);

Output:

$$D_{[1]}(f)(X(t), Y(t)) \left(\frac{\partial}{\partial t} X(t) \right) + D_{[2]}(f)(X(t), Y(t)) \left(\frac{\partial}{\partial t} Y(t) \right)^*$$

Compare this with Equation 14.10.

f:=(x,y)–>x^3*y+x*sin(y);

Output: $f := (x, y) \to x^3 y + x \sin(y)$

z:=f(x,y);

Output: $z := x^3 y + x \sin(y)$

x:=t^2+1;y:=2*t–1;
diff(z,t);

Output:
$$6(t^2 + 1)^2(2t - 1)t + 2(t^2 + 1)^3 + 2t \sin(2t - 1)$$
$$+ 2(t^2 + 1) \cos(2t - 1)$$

Mathematica:

z=f[x,y]
x=X[t]
y=Y[t]
D[z,t]
Clear[x,y,z,t]
z=x^3*y+x*Sin[y]
x=t^2+1
y=2*t–1
D[z,t]

CAS 55

Find the tangent plane to $z = f(x, y) = 5e^{-x^2 - y^2}$ at the point $\left(0, 1/4, 5e^{-1/16} \right)$.

---

*In Maple, $D_{[1]}$ and $D_{[2]}$ represent the first partial derivatives with respect to x and y, respectively.

Maple:

f:=(x,y)–>5*exp(–x^2–y^2);

To use Equation 14.21, we first compute the partial derivatives.

fx:=diff(f(x,y),x);

Output: $fx := -10xe^{-x^2-y^2}$

fy:=diff(f(x,y),y);

Output: $fy := -10ye^{-x^2-y^2}$

a:=0;b:=0.25;
plane:=f(a,b)+subs(x=a,y=b,fx)*(x–a)+subs(x=a,y=b,fy)*(y–b);

Output: $plane := 4.697065314 - 2.50e^{-.0625}(y - .25)$

Now we plot the surface and the tangent plane.

p1:=plot3d(f(x,y),x=–5..5,y=–5..5,scaling=unconstrained,
    numpoints=2000);
p2:=plot3d(plane,x=–1..1,y=–0.5..0.5,scaling=unconstrained,
    numpoints=2000);
display3d({p1,p2},style=patch,axes=boxed,orientation=[40,77]);

Mathematica:

Clear[x,y,z,t]
f[x_, y_]=5*Exp[–x^2–y^2]
fx=D[f[x,y],x]
fy=D[f[x,y],y]
a=0
b=0.25
plane=f[a,b]+ReplaceAll[fx,{x–>a,y–>b}]*(x–a)
    +ReplaceAll[fy,{x–>a},y–>b}]*(y–b)
P1=Plot3D[f[x,y],{x,–5,5},{y,–5,5},
    DisplayFunction–>Identity]
P2=Plot3D[plane,{x,–1,1},{y,–0.5,0.5},
    DisplayFunction–>Identity]
Show[P1,P2,DisplayFunction–>$DisplayFunction]

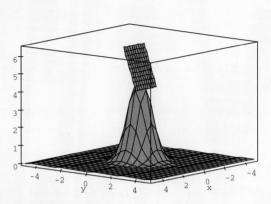

FIGURE 14.4.1

Exercise Set 14.4

In Exercises 1–14, find equations of the tangent plane and normal line to the given surface at the specified point.

1. $2x^2 + 3y^2 - z^2 = 5$; (3, −2, 5)

2. $z^2 = 3x^2 + 4y^2$; (−2, 1, −4)

3. $4x^2 + 3y^2 + 2z^2 = 12$; (1, 0, −2)

4. $xy + 2yz + 3xz = 16$; (4, −2, 3)

5. $(x + y)^2 + (y + z)^2 + (x + z)^2 = 10$; (1, −1, 2)

6. $xe^{2y-z} - 3 = 0$; (3, 1, 2)

7. $y = \ln\left(\dfrac{x + 2y}{y + 2z}\right) - 1$; (3, −1, 1)

8. $\sin\left(\dfrac{x}{y}\right) + \cos\left(\dfrac{y}{z}\right) = 0$; $\left(\pi, 1, \dfrac{2}{\pi}\right)$

9. $z = x^2 - 2y^2$; $(5, 4, -7)$

10. $z = \ln \sqrt{x + y}$; $(3, -2, 0)$

11. $z = \tan^{-1} \dfrac{y}{x}$; $\left(1, -1, -\dfrac{\pi}{4}\right)$

12. $z = \dfrac{x + y}{x - y}$; $(4, 3, 7)$

13. $z = e^{2x} \sin 3y$; $\left(0, \dfrac{\pi}{2}, -1\right)$

14. $z = \sqrt{\dfrac{x - 2y}{x + 2y}}$; $(5, -2, 3)$

15. Find the point on the hyperbolic paraboloid $z = 2x^2 - 3y^2$ at which the tangent plane is parallel to the plane $4x + 9y - 2z = 11$.

16. Find the point on the elliptic paraboloid $z = 3x^2 + 4y^2$ at which the tangent plane is perpendicular to the line through the points $(1, -2, 4)$ and $(-2, 0, 3)$.

In Exercises 17 and 18, assume F and G are differentiable functions at $P_0 = (x_0, y_0, z_0)$ and have nonzero gradients there.

17. The surfaces defined by $F(x, y, z) = 0$ and $G(x, y, z) = 0$ are said to be *tangent* at P_0 if they have the same tangent plane there.
    (a) Prove that the surfaces are tangent at P_0 if and only if $\nabla F(P_0) = k \nabla G(P_0)$ for some nonzero scalar k.
    (b) Find all points P_0 at which the surfaces $x^2 + 2y^2 - 2z^2 = 20$ and $xy - yz + 2xz = 5$ are tangent to each other.

18. The surfaces defined by $F(x, y, z) = 0$ and $G(x, y, z) = 0$ are said to be *orthogonal* at P_0 if their normal lines at P_0 are perpendicular to each other.
    (a) Prove that the surfaces are orthogonal at P_0 if and only if $\nabla F(P_0) \cdot \nabla G(P_0) = 0$.
    (b) Find two points P_0 at which the surfaces defined by $2x^2 - 3y^2 + 4z^2 = 10$ and $2x^2 + y^2 - 4z^2 + 2y - z = 21$ are orthogonal to each other.

19. Find all points on the surface $z = 2x^3 - 6x^2y + 9y^2 + 2y^3$ at which the tangent plane is horizontal.

20. The angle between a line l and a surface S is defined as the complement of the angle between l and the normal line to S at the point where l pierces S. Find the angle between the line $\mathbf{r}(t) = \langle t, 2t, 2 - t \rangle$ and the elliptic cone $2z^2 = 4x^2 + y^2$ at each point of intersection.

21. The angle between a curve C and a surface S is defined as the angle between the tangent line to C and the surface at each point of intersection. (See Exercise 20.) Find the angle between the curve $\mathbf{r}(t) = \langle 1 - t, 2 + t, t^2 \rangle$ and the paraboloid $9z = 4x^2 + y^2$ at each of their points of intersection.

22. Prove that if f is a function of two variables that is differentiable at (x_0, y_0), then $F(x, y, z) = f(x, y) - z$ is differentiable at (x_0, y_0, z_0), where $z_0 = f(x_0, y_0)$.

[CAS] *Use a CAS for Exercises 23 through 25.*

23. Find the equation of the tangent plane to $f(x, y) = \dfrac{x^2}{2} + \dfrac{y^2}{3}$ at the point $\left(1, 1, \dfrac{5}{6}\right)$. Plot the surface and the tangent plane to the surface at the point.

24. Find the equation of the tangent plane to $f(x, y) = xe^{-x^2 - y^2}$ at the point $(1, 1, 1)$. Also find the normal line at this point. Plot the surface, the tangent plane, and the normal line to the surface at the point.

25. The angle of inclination of a plane is defined to be the angle θ, $0 \le \theta \le \pi/2$, between the given plane and the xy-plane as shown in the figure. Find the angle of inclination of the tangent plane found in Exercise 24.

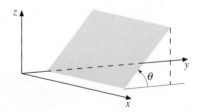

14.5 EXTREME VALUES

Just as in the case of one variable (see Chapter 4), some of the most important applications of multivariable differential calculus involve finding maximum and minimum values of functions. We will concentrate in this section on the two-variable case, but much of the theory can be extended to functions of three or more variables.

The definitions of local and absolute maxima and minima (which are referred to collectively as extreme values, or extrema) parallel those for functions of one variable.

Definition 14.6 Local and Absolute Extrema for Functions of Two Variables	Let f be a function of two variables with domain D. Then f is said to have a **local maximum** at a point (x_0, y_0) in D if there exists some circle centered at (x_0, y_0) such that $f(x_0, y_0) \geq f(x, y)$ for all points (x, y) of D that lie inside this circle. If this inequality holds true for all points (x, y) in D, then f is said to have an **absolute maximum** at (x_0, y_0). When the reverse inequality holds, f has a **local minimum** at (x_0, y_0) in the first instance and an **absolute minimum** in the second.

Relative to the surface of the Earth, Mt. Everest's peak is an absolute maximum.

The terms *relative maximum* and *relative minimum* are often used instead of *local* maximum and minimum. If f has a local maximum at (x_0, y_0), then $f(x_0, y_0)$ is called a **local maximum value** of f, and the point $(x_0, y_0, f(x_0, y_0))$ is a **local maximum point** on the graph of f. This point is the highest point on the graph in its immediate vicinity. Similar remarks apply for a local minimum. The absolute maximum value of f, if it exists, is the largest of its local maximum values, and the absolute minimum value is the smallest of its local minimum values.

Figure 14.22 shows the computer-generated graph of the function

$$f(x, y) = 4e^{-\sqrt{x^2+y^2}/4} \sin x \sin y$$

having many local maxima and minima.

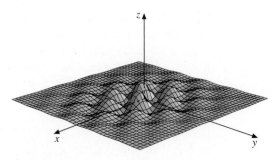

FIGURE 14.22

Sometimes it is possible to determine maximum and minimum values without using calculus. (In Chapter 1 we saw how we could find maximum or minimum values of quadratic functions of one variable without calculus.) For polynomial functions of degree 2, the technique of completing the square is especially useful in this regard, as we illustrate in the following example.

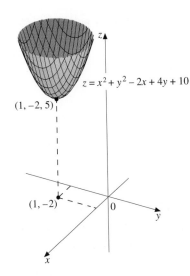

$z = x^2 + y^2 - 2x + 4y + 10$

$(1, -2, 5)$

$(1, -2)$

FIGURE 14.23

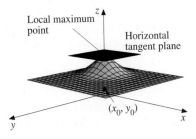

Local maximum point

Horizontal tangent plane

(x_0, y_0)

FIGURE 14.24

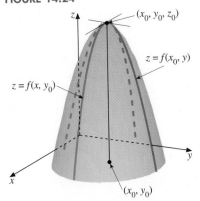

(x_0, y_0, z_0)

$z = f(x_0, y)$

$z = f(x, y_0)$

(x_0, y_0)

FIGURE 14.25

EXAMPLE 14.20   Find the absolute extrema and where they occur for the function

$$f(x, y) = x^2 + y^2 - 2x + 4y + 10$$

Solution   We complete the squares on x and y, getting

$$f(x, y) = (x^2 - 2x + 1) + (y^2 + 4y + 4) + 10 - 1 - 4$$
$$= (x - 1)^2 + (y + 2)^2 + 5$$

The two squared terms are positive except when $x = 1$ and $y = -2$, when each is 0. So the absolute minimum value of f occurs at $(1, -2)$, and this minimum value is 5. There is no maximum value. We show the graph in Figure 14.23. It is a paraboloid with vertex $(1, -2, 5)$. ∎

Since problems of this type are rather specialized, we need to develop methods for handling a wider class of problems. The first task is to find a systematic way of determining points in the domain at which maximum or minimum values *might* occur. Then we must test to see the actual nature of the function at these points.

If $z = f(x, y)$, and f is differentiable at a point (x_0, y_0) of its domain where f attains a local maximum or minimum value, then it is geometrically evident that the graph of f has a horizontal tangent plane at the point (x_0, y_0, z_0), where $z_0 = f(x_0, y_0)$. We illustrate this fact for a local maximum in Figure 14.24.

A horizontal tangent plane through the point (x_0, y_0, z_0) has the equation $z = z_0$. It follows, therefore, from Equation 14.23 that $f_x(x_0, y_0) = 0$ and $f_y(x_0, y_0) = 0$. That is, $\nabla f(x_0, y_0) = \mathbf{0}$. As the following argument shows, the condition $\nabla f(x_0, y_0) = \mathbf{0}$ holds true at a local maximum or minimum even if f is not differentiable, provided $f_x(x_0, y_0)$ and $f_y(x_0, y_0)$ both exist.

Suppose $\nabla f(x_0, y_0)$ exists—that is, $f_x(x_0, y_0)$ and $f_y(x_0, y_0)$ both exist—and suppose for definiteness that f has a local maximum at (x_0, y_0). Then, as we see in Figure 14.25, the curves $z = f(x, y_0)$ and $z = f(x_0, y)$ formed by the intersection of the surface $z = f(x, y)$ with the planes $y = y_0$ and $x = x_0$, respectively, have maximum points at (x_0, y_0, z_0), where $z_0 = f(x_0, y_0)$. Their slopes at (x_0, y_0, z_0) therefore both equal 0. But these slopes are $f_x(x_0, y_0)$ and $f_y(x_0, y_0)$, respectively. So $\nabla f(x_0, y_0) = \mathbf{0}$. A similar argument can be given when f has a local minimum at (x_0, y_0).

We see, then, that if $\nabla f(x_0, y_0)$ exists at a local maximum or minimum point, it must be $\mathbf{0}$. The only other possibility is that $\nabla f(x_0, y_0)$ does not exist. We therefore have the following theorem.

THEOREM 14.4

If f is a function of two variables that has a local maximum or local minimum at (x_0, y_0), then either $\nabla f(x_0, y_0) = \mathbf{0}$ or $\nabla f(x_0, y_0)$ does not exist. That is, either $f_x(x_0, y_0)$ and $f_y(x_0, y_0)$ both equal 0, or at least one of these partials does not exist.

Points in the domain of f where either the partial derivatives f_x and f_y both equal 0 or at least one of these partials fails to exist are *candidates* for points where f has a local maximum or minimum. We give such points a name in the following definition.

Definition 14.7 Critical Point	A point (x_0, y_0) in the domain of f for which $\nabla f(x_0, y_0) = \mathbf{0}$ or $\nabla f(x_0, y_0)$ fails to exist is called a **critical point** of f.

Note that this definition is analogous to that of a critical point for a function of one variable, with ∇f replacing f'.

EXAMPLE 14.21   Find all critical points of

$$f(x, y) = x^3 - 3x^2 y + 6y^2 + 24y$$

Solution   To calculate the gradient, ∇f, we calculate the first partials.

$$\frac{\partial f}{\partial x} = 3x^2 - 6xy \quad \text{and} \quad \frac{\partial f}{\partial y} = -3x^2 + 12y + 24$$

Thus, $\nabla f = \langle 3x^2 - 6xy, \ -3x^2 + 12y + 24 \rangle$. Since ∇f exists everywhere, the only critical points are those for which $\nabla f = \mathbf{0}$. We set $\partial f/\partial x = 0$ and $\partial f/\partial y = 0$ and solve simultaneously. Setting $\partial f/\partial x = 0$ gives

$$3x^2 - 6xy = 0 \quad \text{or} \quad 3x(x - 2y) = 0$$

Thus, $x = 0$ or $x = 2y$. Setting $\partial f/\partial y = 0$ gives

$$-3x^2 + 12y + 24 = 0$$

or, equivalently (dividing by -3),

$$x^2 - 4y - 8 = 0$$

Substituting $x = 0$ gives $-4y - 8 = 0$, so that $y = -2$. Thus, $(0, -2)$ is a critical point. Substituting $x = 2y$ gives $4y^2 - 4y - 8 = 0$, whose solutions are readily found to be $y = 2$ and $y = -1$. The corresponding x values are 4 and -2. The critical points of f are therefore $(0, -2)$, $(4, 2)$, and $(-2, -1)$. ∎

EXAMPLE 14.22   Find all critical points of

$$f(x, y) = (3x^2 + 4y^2)^{1/2}$$

and determine the nature of the function at each point.

Solution   The gradient in this case is

$$\nabla f = \left\langle \frac{3x}{\sqrt{3x^2 + 4y^2}}, \ \frac{4y}{\sqrt{3x^2 + 4y^2}} \right\rangle$$

which is never $\mathbf{0}$. It is undefined only when $x = 0$ and $y = 0$. So $(0, 0)$ is the only critical point. We see that $f(0, 0) = 0$, and for all other points (x, y), $f(x, y) > 0$. Thus, $f(0, 0)$ is the absolute minimum value. The graph of f is the upper nappe of the elliptical cone $z^2 = 3x^2 + 4y^2$, pictured in Figure 14.26. At the minimum point the graph comes to a sharp point, and there is no tangent plane there. ∎

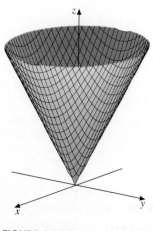

FIGURE 14.26

According to Theorem 14.4, a function can have a local maximum or minimum at a point of its domain only if that point is a critical point. Note carefully, however, it *does not* say that f *will* have a local maximum or minimum at each critical point. The next example illustrates this fact.

EXAMPLE 14.23   Show that for the function $f(x, y) = xy$ the point $(0, 0)$ is a critical point but $f(0, 0)$ is neither a local maximum nor a local minimum value of f.

Solution   Since $\nabla f = \langle y, x \rangle$, we see that $\nabla f(0, 0) = \mathbf{0}$. So $(0, 0)$ is a critical point. Also, $f(0, 0) = 0$.

If x and y are both nonzero and they are like in sign (first or third quadrants), then $f(x, y) > 0$. If both x and y are nonzero and unlike in sign (second and fourth quadrants), then $f(x, y) < 0$. So $f(0, 0)$ is neither larger nor smaller than all other values of f for points in the immediate vicinity of $(0, 0)$. That is, $f(0, 0)$ is neither a local maximum nor a local minimum value. (This situation is similar to $f(x) = x^3$ in the single-variable case, in which $f'(0) = 0$, but the origin is neither a maximum nor a minimum point.)

We show the graph of f in Figure 14.27. It is a hyperbolic paraboloid (rotated $45°$ from the position we showed in Chapter 13). The origin is a saddle point.   ■

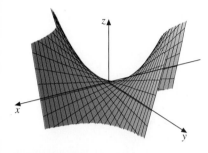

FIGURE 14.27

A Test for Local Extrema

Just as in the single variable case, finding that a function of two variables has a critical point does not guarantee that the function will have a maximum or minimum value there. We need a test that will enable us to determine the nature of the function at its critical points. The test we give is analogous to the Second Derivative Test for a function of one variable. It applies only to critical points for which the gradient exists and equals $\mathbf{0}$. The proof can be found in most advanced calculus texts.

THEOREM 14.5

A Test for Local Extrema

Let (x_0, y_0) be a critical point of the function f for which $\nabla f(x_0, y_0) = \mathbf{0}$, and let f have continuous second partial derivatives at all points inside some circle centered at (x_0, y_0). Define

$$D(x, y) = f_{xx}(x, y) f_{yy}(x, y) - \left(f_{xy}(x, y) \right)^2$$

1. If $D(x_0, y_0) > 0$ and $f_{xx}(x_0, y_0) < 0$, then $f(x_0, y_0)$ is a local maximum.

2. If $D(x_0, y_0) > 0$ and $f_{xx}(x_0, y_0) > 0$, then $f(x_0, y_0)$ is a local minimum.

3. If $D(x_0, y_0) < 0$, then f has a saddle point at (x_0, y_0). (That is, $f(x_0, y_0)$ is neither a local maximum nor a local minimum.)

4. If $D(x_0, y_0) = 0$, then the test is inconclusive.

If the test is inconclusive or not applicable, then it may be possible to determine the nature of f by examining its values near the critical point.

EXAMPLE 14.24   Find all local maximum and minimum values of the function f defined by

$$f(x, y) = x^2 - 2xy + 4y^2 - 2x - 4y + 1$$

Solution   First, we find the critical points by setting $\nabla f(x, y) = \mathbf{0}$.

$$\nabla f = \langle 2x - 2y - 2, \; -2x + 8y - 4 \rangle$$

By setting each component equal to 0 and dividing both sides of each equation by 2, we obtain the two equations

$$x - y = 1 \quad \text{and} \quad -x + 4y = 2$$

The simultaneous solution is $(2, 1)$. To test this point by Theorem 14.5, we need the second partials:

$$f_{xx} = 2 \quad f_{xy} = -2 \quad f_{yy} = 8$$

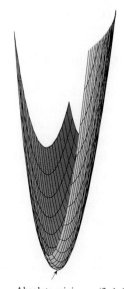

Absolute minimum $(2, 1, 3)$

FIGURE 14.28

So $D(x, y) = 2(8) - (-2)^2 = 16 - 4 = 12$. Since $D(x, y)$ is constant in this case, its value at the critical point $(2, 1)$ is 12 also, which is positive. So we see that $D(2, 1) > 0$ and $f_{xx}(2, 1) = 2 > 0$. By part 2 of Theorem 14.5, we conclude that $f(2, 1)$ is a local minimum value. To find this minimum value, we substitute $x = 2$ and $y = 1$ into the formula for $f(x, y)$ and obtain $f(2, 1) = 3$. A computer-generated graph of f is shown in Figure 14.28. ∎

EXAMPLE 14.25   Find and classify all extrema of $f(x, y) = x^3 - 3x^2y + 6y^2 + 24y$.

Solution   This is the function from Example 14.21, and we found the critical points to be $(0, -2)$, $(4, 2)$, and $(-2, -1)$. To test them we need the second partials f_{xx}, f_{xy}, and f_{yy}:

$$f_x = 3x^2 - 6xy \qquad f_y = -3x^2 + 12y + 24$$
$$f_{xx} = 6x - 6y \qquad f_{xy} = -6x \qquad f_{yy} = 12$$

The following table helps to keep track of things.

(x_0, y_0)	$f_{xx}(x_0, y_0)$	$f_{xy}(x_0, y_0)$	$f_{yy}(x_0, y_0)$	$D(x_0, y_0)$	Test result
$(0, -2)$	12	0	12	144	Minimum
$(4, 2)$	12	-24	12	-432	Saddle point
$(-2, -1)$	-6	12	12	-216	Saddle point

So the only local extremum is $(0, -2)$, where f has a minimum value. The minimum value is $f(0, -2) = -24$. Figure 14.29 shows a computer-generated graph of f, where the vertical scale has been compressed. We also show a contour map for f (Figure 14.30). ∎

EXAMPLE 14.26   A crate in the shape of a rectangular box is to be constructed so that its volume is 270 ft^3. The sides and top each cost \$1/ft^2 to construct, and the bottom, which must be stronger, costs \$1.50/ft^2. What are the dimensions of the crate that will yield the minimum cost? What is the minimum cost?

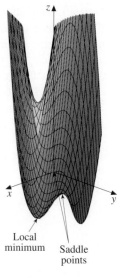

Local
minimum   Saddle
points

FIGURE 14.29

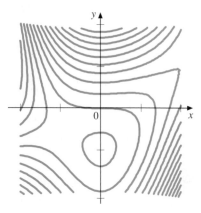

FIGURE 14.30
A contour map for
$f(x, y) = x^3 - 3x^2y + 6y^2 + 24y$

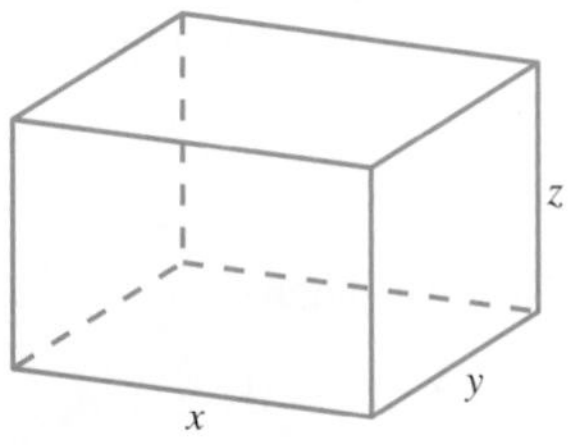

Solution   Denote the dimensions of the crate by x, y, and z, as shown in Figure 14.31. Then, since there are two sides of area xz and two ends of area yz, the cost function C is given by

$$C = \overbrace{2xz}^{\text{sides}} + \overbrace{2yz}^{\text{ends}} + \overbrace{1.5xy}^{\text{bottom}} + \overbrace{xy}^{\text{top}} = 2(xz + yz) + 2.5xy$$

We are also given that the volume must be 270. So the variables are related by the equation

$$xyz = 270$$

This relationship enables us to eliminate one of the variables. We solve for z and substitute in the cost function, reducing C to a function of two variables only:

$$z = \frac{270}{xy}$$

$$C(x, y) = 2\left(\frac{270}{y} + \frac{270}{x}\right) + \frac{5xy}{2} \qquad \text{Write 2.5 as } \tfrac{5}{2}.$$

To apply Theorem 14.5, we need the gradient of C:

$$\nabla C = \left\langle -\frac{540}{x^2} + \frac{5y}{2}, -\frac{540}{y^2} + \frac{5x}{2} \right\rangle$$

Setting the first component of ∇C equal to 0 gives

$$-\frac{540}{x^2} + \frac{5y}{2} = 0 \qquad \text{or} \qquad y = \frac{216}{x^2}$$

Setting the second component equal to 0 gives

$$-\frac{540}{y^2} + \frac{5x}{2} = 0 \qquad \text{or} \qquad x = \frac{216}{y^2}$$

Now we substitute $x = 216/y^2$ into the equation $y = 216/x^2$ to obtain (after simplification)

$$y = \frac{y^4}{216} \qquad \text{or} \qquad y\left(216 - y^3\right) = 0$$

FIGURE 14.31

Clearly, $y \neq 0$ (since we are talking about a real crate), so $y^3 = 216$, or $y = 6$. Thus, $x = 216/36 = 6$ also. Since $(6, 6)$ is the only critical value for the cost function C, and we know from the nature of the problem that C does assume a minimum value, we can conclude that this minimum occurs at $(6, 6)$. So it is not essential in this case to apply Theorem 14.5.

With $x = 6$ and $y = 6$, we get $C(6, 6) = 2\left(\frac{270}{6} + \frac{270}{6}\right) + \frac{5(36)}{2} = 270$. Thus, since $z = 270/xy = 7.5$, the dimensions for minimum cost are 6 ft $\times$ 6 ft $\times$ 7.5 ft, and the minimum cost is \$270. ∎

In the next two examples we show how to determine absolute extrema of a function on a closed and bounded set (a closed set is one that includes its boundary). The technique works when the boundary consists of a finite number of curves on each of which the function can be expressed in terms of one variable.

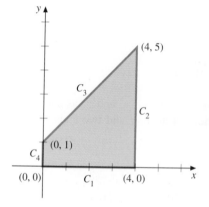

FIGURE 14.32

EXAMPLE 14.27   Find the absolute maximum and minimum values of

$$f(x, y) = x^2 - 2xy + 3y^2 - 4x$$

on the closed trapezoidal region pictured in Figure 14.32.

Solution   Name the boundary segments C_1, C_2, C_3, and C_4, as shown. First, we look for points in the interior where extrema occur. To find the critical points, we see where the gradient either is $\langle 0, 0 \rangle$ or is undefined.

$$\nabla f = \langle 2x - 2y - 4, -2x + 6y \rangle$$

In this case, ∇f is defined for all points. It is $\langle 0, 0 \rangle$ when

$$2x - 2y - 4 = 0 \quad \text{and} \quad -x + 3y = 0 \tag{14.24}$$

Thus,

$$x = y + 2 \quad \text{and} \quad x = 3y \tag{14.25}$$

Equating the two expressions for x, we get $3y = y + 2$, or $y = 1$. Thus $x = 3$. To test whether $(3, 1)$ gives a local extremum, we need the second partials $f_{xx} = 2$, $f_{xy} = -2$, and $f_{yy} = 6$. So $D(x, y) = 12 - (-2)^2 = 8 > 0$, and since $f_{xx} > 0$, we conclude from Theorem 14.5 that f has a local minimum at $(3, 1)$. The value there is found to be $f(3, 1) = -6$.

Now we consider each boundary segment. On C_1 we have $y = 0$. So, the function becomes $f(x, 0) = x^2 - 4x$, which is a function of one variable on $0 \leq x \leq 4$. By setting the derivative equal to 0 and testing, we find that $x = 2$ yields a minimum value—namely, $f(2, 0) = -4$. The endpoint values are $f(0, 0) = 0$ and $f(4, 0) = 0$.

We sketch briefly the results along C_2, C_3, and C_4. You should verify these. In each case we are working with one variable only.

C_2: $\underline{x = 4}$. $f(4, y) = 3y^2 - 8y$ on $0 \leq y \leq 5$. Minimum at $y = \frac{4}{3}$, $f\left(4, \frac{4}{3}\right) = -\frac{16}{3}$. Endpoint values: $f(4, 0) = 0$, $f(4, 5) = 35$.

C_3: $\underline{y = x + 1}$. $f(x, x+1) = 2x^2 + 3$ on $0 \leq x \leq 4$. No interior critical values. Endpoint values: $f(4, 5) = 35$, $f(0, 1) = 3$.

C_4: $\underline{x = 0}$. $f(0, y) = 3y^2$ on $0 \leq y \leq 1$. No interior critical values. Endpoint values already found.

The extreme values are to be found among the following: $f(3, 1) = -6$, $f(2, 0) = -4$, $f(4, \frac{4}{3}) = -\frac{16}{3}$, $f(0, 0) = 0$, $f(4, 0) = 0$, $f(4, 5) = 35$, and $f(0, 1) = 3$. So f assumes the absolute maximum value of 35 at $(4, 5)$ and the absolute minimum value of -6 at $(3, 1)$. ∎

EXAMPLE 14.28   Find the absolute maximum and minimum values of

$$f(x, y) = x^2 - 4xy - 2y^2$$

on the closed disk $x^2 + y^2 \le 5$.

Solution   The critical points in the interior of the disk, if any exist, occur where $\nabla f = \langle 0, 0 \rangle$. Since $\nabla f = \langle 2x - 4y, -4x - 4y \rangle$, the coordinates of such critical points satisfy $x = 2y$ and $x = -y$. The simultaneous solution gives the point $(0, 0)$.

Rather than use the rectangular equation $x^2 + y^2 = 5$ of the boundary, it is easier to use the parametric equations

$$\begin{cases} x = \sqrt{5} \cos \theta \\ y = \sqrt{5} \sin \theta \end{cases} \quad 0 \le \theta < 2\pi$$

On this boundary, we have

$$f(x, y) = f\left(\sqrt{5} \cos \theta, \sqrt{5} \sin \theta\right) = 5 \cos^2 \theta - 20 \cos \theta \sin \theta - 10 \sin^2 \theta$$

giving f as a function of the single variable θ.

We find critical values of this function of θ by setting $df/d\theta = 0$.

$$\begin{aligned} \frac{df}{d\theta} &= -10 \cos \theta \sin \theta - 20 \cos^2 \theta + 20 \sin^2 \theta - 20 \sin \theta \cos \theta \\ &= 10(2 \sin^2 \theta - 3 \sin \theta \cos \theta - 2 \cos^2 \theta) \\ &= 10(2 \sin \theta + \cos \theta)(\sin \theta - 2 \cos \theta) \end{aligned}$$

Thus, $df/d\theta = 0$ when

$$2 \sin \theta + \cos \theta = 0 \qquad \text{or} \qquad \sin \theta - 2 \cos \theta = 0$$

From the first of these equations we get, on dividing by $\cos \theta$,

$$\tan \theta = -\frac{1}{2}$$

and from the second,

$$\tan \theta = 2$$

In Figure 14.33, we show the primary angles determined by these two equations. Since $x = \sqrt{5} \cos \theta$ and $y = \sqrt{5} \sin \theta$, we obtain the four critical points on the boundary: $(2, -1)$, $(-2, 1)$, $(1, 2)$, and $(-1, -2)$.

Now we calculate the value of $f(x, y)$ at the interior critical point and each critical point on the boundary. We show the results in the margin.

We conclude that the absolute maximum value of f is 10, occurring at $(2, -1)$ and $(-2, 1)$, and the absolute minimum is -15, occurring at $(1, 2)$ and $(-1, -2)$. ∎

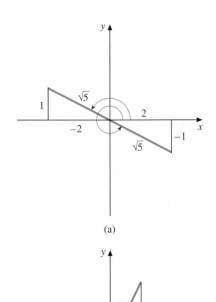

(a)

(b)

FIGURE 14.33

(x, y)	$f(x, y)$
$(0, 0)$	0
$(2, -1)$	10
$(-2, 1)$	10
$(1, 2)$	-15
$(-1, -2)$	-15

We can summarize the procedure illustrated in Examples 14.27 and 14.28 as follows.

A Procedure for Finding Absolute Extrema on a Closed and Bounded Set

Let f be a continuous function of two variables on a closed and bounded region R, with the boundary consisting of finitely many curves where f can be expressed as a function of one variable.

Step 1. Find all critical points in the interior of R and calculate $f(x, y)$ at each such point.

Step 2. Find the local and endpoint extrema of f along each boundary curve.

Step 3. Among all values of f found in Steps 1 and 2, select the largest and smallest values. The largest is the absolute maximum, and the smallest is the absolute minimum of f on all of R.

FINDING MAXIMUM AND MINIMUM VALUES OF FUNCTIONS OF SEVERAL VARIABLES USING COMPUTER ALGEBRA SYSTEMS

CAS 56

Find the extreme values for $f(x, y) = \dfrac{y^2}{3} - \dfrac{x^2}{4}$.

Since the function is differentiable and the domain does not contain boundary points, the function can have extreme values only where the first partials with respect to x and y are both 0.

Maple:

```
f:=(x,y)->y^2/3-x^2/4;
```

Output: $f := (x, y) \rightarrow \dfrac{1}{3}y^2 - \dfrac{1}{4}x^2$

Compute the first partials.

```
fx:=diff(f(x,y),x);
```

Output: $fx := -\dfrac{1}{2}x$

```
fy:=diff(f(x,y),y);
```

Output: $fy := \dfrac{2}{3}y$

Solve the partials simultaneously equal to 0.

```
solve({fx,fy},{x,y});
```

Mathematica:

```
f[x_, y_]=y^2/3-x^2/4
fx=D[f[x,y],x]
fy=D[f[x,y],y]
Solve[{fx==0, fy==0}, {x,y}]
Plot3D[f[x,y], {x,-5,5}, {y,-4,4},PlotRange->{-1,2},
    AspectRatio->Automatic]
D[f[x,y], {x,2}]*D[f[x,y], {y,2}]-(D[f[x,y], {x,1},{y,1}])^2
```

Maple:

Output: $\{y = 0, x = 0\}$

So the only possible critical value is the origin $(0, 0)$. A sketch will reveal that the origin is in fact a saddle point and not an extreme point. See Figure 14.5.1.

```
plot3d(f(x,y),x=-5..5,y=-4..4,scaling=constrained,axes=normal,
   tickmarks=[0,0,0],orientation=[10,80],view=-1..2);
```

Finally, we check algebraically that the origin is a saddle point. That is, we show that the discriminant function of f is negative at the origin (see Theorem 14.5). In fact, we see that the discriminant function is always negative in this case.

```
diff(f(x,y),x,x)*diff(f(x,y),y,y)-(diff(f(x,y),y,x))^2;
```

Output: $\dfrac{-1}{3}$

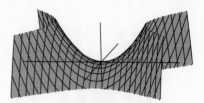

FIGURE 14.5.1

⊡ CAS 57

A thin metal plate is in the shape of the disk $x^2 + y^2 \le 1$. The plate is heated in such a way that the temperature at the point (x, y) on the surface of the plate is given by $T(x, y) = 2x^2 + 3y^2 - 2x$. Find the temperatures of the hottest and coldest points on the plate.

Maple:

```
f:=(x,y)->2*x^2+3*y^2-x;
```

Plot the temperature along with the region in the xy-plane corresponding to the plate.

```
p1:=plot3d(2*x^2+3*y^2-x,x=-1..1,y=2..2,view=-1..4,
   scaling=constrained):
p2:=plot3d({[t,sqrt(1-t^2),0],[t,-sqrt(1-t^2),0]},t=-1..1,
   scaling=constrained):
display3d({p1,p2},axes=boxed);
```

Mathematica:

```
f[x_,y_]=2*x^2+3*y^2-x

P1=Plot3D[f[x,y], {x, -2,2}, {y,-2,2},
   PlotRange->{-1,4},DisplayFunction->Identity]

P2=ParametricPlot3D[{t, Sqrt[1-t^2],0},{t,-1,1},
   DisplayFunction->Identity]

P3=ParametricPlot3D[{t, -Sqrt[1-t^2],0},{t,-1,1},
   DisplayFunction->Identity]

Show[P1,P2,P3,DisplayFunction->$DisplayFunction]

fx=D[f[x,y],x]

fy=D[f[x,y],y]

Solve[{fx==0,fy==0}, {x,y}]

D[f[x,y],{x,2}]*D[f[x,y],{y,2}]-(D[f[x,y],{x,1},{y,1}])^2

s=-x^2+1

g=Simplify[ReplaceAll[f[x,y],{x->x,y->Sqrt[s]}]]

dg=D[g,x]

Solve[dg==0,x]

ReplaceAll[s,{x->-1/2}]
```

Maple:

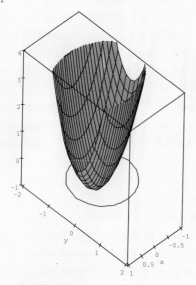

FIGURE 14.5.2

Mathematica:

h=Solve[y^2==3/4,y]

f[1/4,0]

f[3/4,Sqrt[3]/2]

f[3/4,–Sqrt[3]/2]

ContourPlot[f[x,y],{x,–2,2},{y,–2,2}]

First, locate any interior extreme values, as we did in CAS 56.

fx:=diff(f(x,y),x);

Output: $fx := 4x - 1$

fy:=diff(f(x,y),y);

Output: $fy := 6y$

solve({fx,fy},{x,y});

Output: $\left\{ y = 0, x = \dfrac{1}{4} \right\}$

Now check, using the discriminant function for f, if the point (1/4, 0) is an extreme value.

diff(f(x,y),x,x)*diff(f(x,y),y,y)–(diff(f(x,y),y,x))^2;

Output: 24

Since the value of the discriminant function is greater than 0, the point is a local maximum. Now analyze the boundary points. First solve for y^2 in terms of x and then substitute into $f(x, y)$.

s:=solve(x^2+y^2=1,y^2);

Output: $s := -x^2 + 1$

g:=subs(x=x,y^2=s,f(x,y));

Output: $g := -x^2 + 3 - x$

Maple:

Notice g is now a function of only one variable, x. Apply the critical point analysis for functions of one variable.

dg:=diff(g,x);

Output: $dg := -2x - 1$

solve(dg=0,x);

Output: $\dfrac{-1}{2}$

subs(x=–1/2,s);

Output: $\dfrac{3}{4}$

h:=solve(y^2=3/4,y);

Output: $h := \dfrac{1}{2}\sqrt{3}, -\dfrac{1}{2}\sqrt{3}$

f(1/4,0);

Output: $\dfrac{-1}{8}$

f(3/4,h[1]);

Output: $\dfrac{21}{8}$

f(–3/4,h[2]);

Output: $\dfrac{33}{8}$

Thus, we see that the coldest spot on the plate is at the point $(1/4,\ 0)$ and has temperature of $-1/8$ degree, and the hottest spot on the plate occurs at the point $(-\frac{3}{4}, \frac{\sqrt{3}}{2})$ with a temperature of 33/8 degrees.

Finally, we plot a contour map of the temperature surface showing curves of constant temperature. See Figure 14.5.3.

contourplot(2*x^2+3*y^2–x,x=–2..2,y=–2..2,view=–1..4,
    numpoints=1500, scaling=constrained,axes=normal,tickmarks=[0,0,0]);

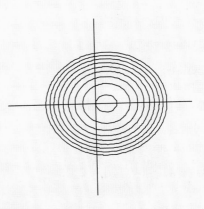

FIGURE 14.5.3

Exercise Set 14.5

In Exercises 1–4, find the extreme values of f and where they occur by completing the square.

1. $f(x, y) = x^2 + y^2 + 2x - 4y + 3$

2. $f(x, y) = 2x^2 + 3y^2 - 4x + 6y - 9$

3. $f(x, y) = x^4 + y^2 - 8y + 13$

4. $f(x, y) = x^2 + y^4 - 2y^2 + 4x + 1$

In Exercises 5–26, locate all critical points, and at each such point determine whether f has a local maximum, a local minimum, or a saddle point.

5. $f(x, y) = x^2 - 6xy + 2y^3 - 8x - 16$

6. $f(x, y) = x^2 + y^3 - 4xy - 8x + 13y + 1$

7. $f(x, y) = 2x^3 - 6xy + y^2 + 30$

8. $f(x, y) = x^3 + 3xy^2 - 3x^2 - 3y^2 + 4$

9. $f(x, y) = 2x^2 + y^4 - 4xy + 2$

10. $f(x, y) = x^4 + 2y^2 + 8xy - 7$

11. $f(x, y) = 6xy - x^2 - y^3$

12. $f(x, y) = 4xy - 2x^2 - y^3 + 3$

13. $f(x, y) = x^4 - 2x^2y + y^3 - y$

14. $f(x, y) = x^4 - y^4 - 4x^2y^2 + 20y^2$

15. $f(x, y) = xy + \dfrac{1}{x} + \dfrac{2}{y}$

16. $f(x, y) = 4 - \dfrac{2}{x} - \dfrac{1}{y} - x^2y$

17. $f(x, y) = 8x^2 - \dfrac{1}{y} + 2x - y$

18. $f(x, y) = \dfrac{8}{xy} + \dfrac{2}{x} - \dfrac{4}{y}$

19. $f(x, y) = x^3 + 2x^2y + y^3 + x$

20. $f(x, y) = 2x^3 + 3y^3 + xy^2 + 2y$

21. $f(x, y) = e^x(x^2 - y^2)$

22. $f(x, y) = e^{-y}(x^2 - 3x + 3y)$

23. $f(x, y) = \sin x \sin y, \quad -\pi < x < \pi, \quad -\pi < y < \pi$

24. $f(x, y) = \sin^2 x - 2\cos^2 y, \quad -\dfrac{\pi}{4} < x < \dfrac{3\pi}{4},$
    $-\dfrac{\pi}{4} < y < \dfrac{3\pi}{4}$

25. $f(x, y) = x^2 - 2x \cos y + 1, \quad 0 \le y \le 2\pi$

26. $f(x, y) = y^2 - 4y(\sin x + \cos x), \quad -\pi < x < \pi$

27. Show that $f(x, y) = 4 - x^{2/3} + 2x^{1/3}y^{1/3} - y^{2/3}$ has a critical point at $(0, 0)$ for which ∇f does not exist and that f has a local (and absolute) maximum value there.

In Exercises 28–31, find the absolute maximum and absolute minimum values of f on the closed domain bounded by the given curves.

28. $f(x, y) = x^2 + 2y^3$; the line segments joining $(0, 0)$, $(2, 0)$, and $(0, 1)$

29. $f(x, y) = x^2y - xy^2 - y$; x-axis, y-axis, $x = 1$, $y = 1$

30. $f(x, y) = 2x^3 - 3x^2y + 2y^3 - 3y$; $x + y = \pm 1$, $x - y = \pm 1$

31. $f(x, y) = x^2 - xy - x + y$; $y = 5 - x^2$, $y = 0$

32. Find the absolute maximum and minimum values of $f(x, y) = x^3 - y^3 - 3x$ on the closed unit disk $x^2 + y^2 \le 1$. At what points do these extrema occur? (*Hint:* Use the parameterization $x = \cos t$ and $y = \sin t$ for $0 \le t \le 2\pi$.)

33. An open-top rectangular box is to have a volume of 256 ft^3. What dimensions will require the least amount of material?

34. Find the point on the plane $3x + 2y - z = 4$ that is nearest the origin. (*Hint:* Minimize the *square* of the distance of the point (x, y, z) from the origin, using the fact that z satisfies the given equation.)

35. The temperature on the surface of the hemisphere $z = \sqrt{1 - x^2 - y^2}$ is given by $T(x, y, z) = 400xyz^2$. Find the hottest and coldest temperatures on the hemisphere and the points where these extrema occur.

36. A company makes two types of automatic ice cream freezers, Type A and Type B. The cost C of producing x Type-A and y Type-B machines per day is

$$C(x, y) = x^2 + xy + y^2 + 20x - 20y$$

and the revenue from selling x Type A and y Type B machines per day is $R(x, y) = 100x + 80y$. How many machines of each type should be manufactured and sold each day to maximize profit? What is the maximum profit?

37. An open-top rectangular box is to be constructed with a divider in the middle. The unit cost of the divider is half that of the bottom and sides. If the volume is to be 320 in.^{3}, find the dimensions that minimize the cost.

38. A common problem in experimental work is to find the line $y = mx + b$ that "fits" a set of data points $(x_1, y_1), (x_2, y_2), \ldots, (x_n, y_n)$ best in the sense that the sum of the squares of the vertical deviations of the data points from the line is minimum. This line is said to fit the data best in the sense of *least squares*. So the problem is to find m and b such that

$$F(m, b) = \sum_{k=1}^{n} (y_k - mx_k - b)^2$$

is a minimum. Determine the values of m and b.

39. In a chemistry experiment, the density (in gr/mL) of potassium chloride in a solution with water was measured for various solutions, with known weights of potassium chloride as a percentage of the weight of water. The

results were as shown.

weight %	23.04	17.73	15.57	4.78
Density	1.170	1.163	1.115	1.078

weight %	1.80	11.21	14.36	22.75
Density	1.058	1.107	1.135	1.165

(a) Fit a straight line to the data points using the results of Exercise 38.
(b) Draw a graph showing the line and the data points.
(c) Predict the density for a solution with a weight percentage of 20.

CAS *Use a CAS for Exercises 40 and 41.*

40. Find the maximum value of the function $f(x, y) = xy^2$ on the circle $x^2 + y^2 = 1$. Parameterize the circle by $x = \cos t$ and $y = \sin t$ for $-\pi \le t \le \pi$. This converts f into a function of t only—say, g—which can then be maximized. Plot the surface $z = f(x, y)$ and also the function g.

41. A thin metal plate is in the shape of a rectangle in the xy-plane with vertices $(1, 1)$, $(1, -1)$, $(-1, 1)$, and $(-1, -1)$. The plate is heated in such a way that the temperature of the plate is given by $T(x, y) = x^2 + 3y^2 + 3x$. Find the temperatures of the hottest and coldest points on the plate. Plot the temperature surface and the region in the xy-plane corresponding to the plate. Finally, plot a contour map of the temperature surface showing curves of constant temperature.

14.6 CONSTRAINED EXTREMUM PROBLEMS

In extremum problems we often seek to maximize or minimize some function subject to a *constraint* on the variables, as was the case in Example 14.26. There we wanted to find the dimensions x, y, and z of a crate that minimized the cost, subject to the constraint that the volume had to be constant at 270; that is, the constraint on the variables was that $xyz = 270$. In that example we solved the constraint equation for z and substituted into the cost function, thereby reducing it to a function of two variables. In this section we consider an alternative method of solving such problems that in many cases is easier. In fact, depending on the nature of the constraint equation, it may be difficult or impossible to use the substitution method.

The method we describe is called the **Method of Lagrange Multipliers**, after the French-Italian mathematician Joseph Louis Lagrange (1736–1813). We consider first the simplest case in which a function of two variables, say $f(x, y)$, is to be maximized or minimized subject to a constraint on x and y. We assume this constraint to be expressed as an equation, say $g(x, y) = k$, whose graph is a curve C in the xy-plane. We say that f has a *constrained local maximum* at a point (x_0, y_0) on C provided that $f(x, y) \le f(x_0, y_0)$ for all points near (x_0, y_0) that *lie on* C and are in the domain of f. A similar definition applies for a *constrained local minimum*.

Figure 14.34 illustrates a constrained local maximum. The function whose graph we have shown (a paraboloid) clearly has an absolute maximum when $x = 0$ and $y = 0$. But if (x, y) is constrained to lie on the curve C, then the maximum occurs at (x_0, y_0), as shown.

To be specific, suppose the paraboloid in Figure 14.34 is the graph of

$$f(x, y) = 6 - 2x^2 - y^2$$

and the curve C is the first-quadrant branch of the hyperbola $xy = 2$, so that $g(x, y) = xy$. Thus, we want to find the maximum value of the function f, where (x, y) is restricted to lie on C. In Figure 14.35 we show the graph of C, along with several level curves $f(x, y) = k$ of the function f. These level curves are the ellipses

$$2x^2 + y^2 = 6 - k \qquad (k > 0)$$

For increasing values of k the ellipses get smaller. We want the largest value of k (this value will be the maximum of the function f) for which (x, y) will be on the level curve $f(x, y) = k$ and also be on the constraint curve $g(x, y) = 2$. This value of k will be the one for which the level curve $f(x, y) = k$ and the constraint curve $g(x, y) = 2$ just touch; that is, where they are tangent to one another.

At such a point of tangency, the gradient vectors ∇f and ∇g must be parallel, since these gradient vectors are normal to the curves. Thus, if (x_0, y_0) is the point of tangency,

$$\nabla f(x_0, y_0) = \lambda \nabla g(x_0, y_0)$$

for some constant λ. The constant λ is called a **Lagrange Multiplier**.

This geometric discussion is intended to help you understand the next theorem. We will return to our example after the theorem and find the constrained maximum using the Lagrange Multiplier Method.

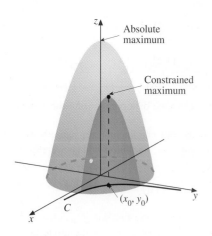

FIGURE 14.34

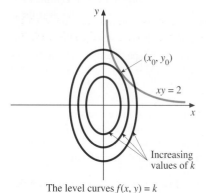

The level curves $f(x, y) = k$

FIGURE 14.35

THEOREM 14.6

Lagrange's Theorem

Let $f(x, y)$ have a constrained local maximum or minimum at (x_0, y_0), with the constraint curve given by $g(x, y) = k$. If f and g are differentiable in some circle centered at (x_0, y_0) with $\nabla g(x_0, y_0) \ne \mathbf{0}$, then there exists a constant λ such that

$$\nabla f(x_0, y_0) = \lambda \nabla g(x_0, y_0) \qquad (14.26)$$

Proof   Let the curve $g(x, y) = k$ be given by the vector equation

$$\mathbf{r}(t) = \langle x(t), y(t) \rangle$$

and let t_0 be the value of t for which

$$(x_0, y_0) = \big(x(t_0), y(t_0)\big)$$

Then, by hypothesis, $\mathbf{r}'(t)$ exists for values of t in some open interval about t_0, and $\mathbf{r}'(t_0) \ne \mathbf{0}$. Define the function F of the single variable t by

$$F(t) = f\big(x(t), y(t)\big)$$

Then F has a local maximum or minimum at t_0, so $F'(t_0) = 0$. Using the Chain Rule, we have

$$F'(t_0) = f_x(x_0, y_0)x'(t_0) + f_y(x_0, y_0)y'(t_0) = 0$$

We can write $F'(t_0)$ as the dot product

$$\nabla f(x_0, y_0) \cdot \mathbf{r}'(t_0) = 0$$

Since $\mathbf{r}'(t_0)$ is a tangent vector to the curve $g(x, y) = k$, it follows that ∇f is orthogonal to this curve at (x_0, y_0).

Since $g(x, y) = k$ can be interpreted as a level curve of $z = g(x, y)$, we know (see Section 14.3) that ∇g is orthogonal to this level curve at each point on the curve, and in particular, at the point (x_0, y_0). We have just shown that $\nabla f(x_0, y_0)$ is also orthogonal to $g(x, y) = k$. Hence, $\nabla f(x_0, y_0)$ and $g(x_0, y_0)$ are parallel vectors. So

$$\nabla f(x_0, y_0) = \lambda \nabla g(x_0, y_0)$$

for some constant λ.   ■

To make use of the theorem to find where f can assume a constrained local extreme value, we write the component equations that arise from Equation 14.26, namely,

$$f_x(x_0, y_0) = \lambda g_x(x_0, y_0)$$

$$f_y(x_0, y_0) = \lambda g_y(x_0, y_0)$$

These, together with the constraint equation

$$g(x_0, y_0) = k$$

constitute a system of three equations in the three unknowns x_0, y_0, and λ, which we solve simultaneously. Our objective is to find x_0 and y_0, and the multiplier λ is just a means to an end. So we might attempt to eliminate λ from the three equations as a first step. However, there are times when this approach is not feasible. It might be best, in fact, in some cases to solve first for λ and then find x_0 and y_0. The examples that follow illustrate some possible strategies.

EXAMPLE 14.29   Find the maximum value of the function $f(x, y) = 6 - 2x^2 - y^2$ subject to the constraint $xy = 2$.

Solution   This problem is the one we discussed just prior to Theorem 14.6. We set $g(x, y) = xy$. Then the constraint equation is $g(x, y) = 2$. The gradient vectors are

$$\nabla f(x, y) = \langle -4x, -2y \rangle$$

$$\nabla g(x, y) = \langle y, x \rangle$$

By Theorem 14.6, the maximum occurs when

$$\nabla f = \lambda \nabla g$$

or

$$\langle -4x, -2y \rangle = \lambda \langle y, x \rangle$$

Equating components, we get

$$-4x = \lambda y \qquad \text{and} \qquad -2y = \lambda x$$

Solving the second of these equations for λ, we find that $\lambda = -2y/x$, and when this value is substituted in the first equation, we obtain

$$-4x = \left(-\frac{2y}{x}\right) y$$

or, after simplification,

$$y^2 = 2x^2$$

From the constraint equation $xy = 2$, we have $x = 2/y$. So

$$y^2 = 2\left(\frac{4}{y^2}\right)$$

or

$$y^4 = 8$$

So $y = \sqrt[4]{8} \approx 1.68$ and $x = 2/\sqrt[4]{8} \approx 1.19$. Thus, the maximum value of $f(x, y) = 6 - 2x^2 - y^2$ subject to the given constraint is

$$f\left(\frac{2}{\sqrt[4]{8}}, \sqrt[4]{8}\right) = 6 - 2\left(\frac{4}{\sqrt{8}}\right) - \sqrt{8}$$

$$= 6 - \frac{8}{\sqrt{8}} - \sqrt{8}$$

$$= 6 - 2\sqrt{8} \approx 0.343 \qquad \blacksquare$$

EXAMPLE 14.30   Find the points on the ellipse $x^2 + 2y^2 = 6$ at which the function $f(x, y) = x^2 y$ assumes its largest and smallest values. What are these values?

Solution   The constraint on points (x, y) is that they lie on the ellipse $x^2 + 2y^2 = 6$. Let $g(x, y) = x^2 + 2y^2$. Now $\nabla f(x, y) = \langle 2xy, x^2 \rangle$ and $\nabla g(x, y) = \langle 2x, 4y \rangle$. So from $\nabla f = \lambda \nabla g$ and $g(x, y) = 6$, we get the three equations

$$\begin{cases} 2xy = 2\lambda x \\ x^2 = 4\lambda y \\ x^2 + 2y^2 = 6 \end{cases}$$

From the first of these equations we have $2x(y - \lambda) = 0$, so either $x = 0$ or $\lambda = y$. If $x = 0$, then from the third equation $2y^2 = 6$, or $y = \pm\sqrt{3}$. When $\lambda = y$, we substitute for λ in the second equation and get $x^2 = 4y^2$, or $x = \pm 2y$. Then replacing x^2 by $4y^2$ in the third equation gives $6y^2 = 6$, or $y = \pm 1$, and so $x = \pm 2$.

Summarizing, we have found the points $(0, \pm\sqrt{3})$ and $(\pm 2, \pm 1)$ as candidates for places where f reaches extreme values. Next, we calculate $f(x, y)$ at each point:

(x_0, y_0)	$(0, \sqrt{3})$	$(0, -\sqrt{3})$	$(2, 1)$	$(-2, 1)$	$(2, -1)$	$(-2, -1)$
$f(x_0, y_0)$	0	0	4	4	-4	-4

Clearly, f has an absolute maximum of 4 at $(2, 1)$ and $(-2, 1)$ and an absolute minimum of -4 at $(2, -1)$ and $(-2, -1)$. Now consider points (x, y) on the ellipse close to $(0, \sqrt{3})$. Since $y > 0$ and $x^2 > 0$, $f(x, y) > 0$. So $f(0, \sqrt{3}) = 0$ is a local minimum value. By similar reasoning, we see that $f(0, -\sqrt{3}) = 0$ is a local maximum.

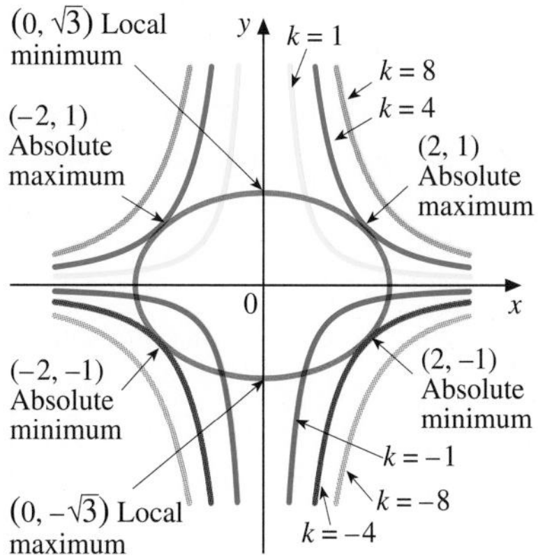

In Figure 14.36, we show the constraint curve $x^2 + 2y^2 = 6$ along with the level curves $f(x, y) = k$ for $k = 1, 4, 8, -1, -4,$ and -8. Notice that as k increases through positive values, the level curves are above the y-axis and move outward from the origin. The one farthest from the origin that touches the constraint curve is for $k = 4$. That is, the largest value of $f(x, y)$ for which (x, y) lies on the constraint curve is $f(x, y) = 4$, and as we have seen, this contact occurs at $(2, 1)$ and $(-2, 1)$. At these points the level curve $f(x, y) = 4$ and the constraint curve $x^2 + 2y^2 = 6$ have a common tangent line, since their normals are parallel. A similar analysis can be given for k negative, with $f(x, y) = -4$ being the minimum value f can be with (x, y) on the constraint curve, occurring at $(-2, -1)$ and $(2, -1)$. ∎

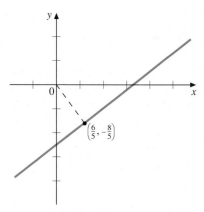

$(0, \sqrt{3})$ Local minimum

$(-2, 1)$ Absolute maximum

$k = 1$

$k = 8$

$k = 4$

$(2, 1)$ Absolute maximum

$(-2, -1)$ Absolute minimum

$(2, -1)$ Absolute minimum

$k = -1$

$(0, -\sqrt{3})$ Local maximum

$k = -8$

$k = -4$

FIGURE 14.36

EXAMPLE 14.31   Find the point on the line $3x - 4y = 10$ that is nearest the origin.

Solution   The distance from a point (x, y) to the origin is $\sqrt{x^2 + y^2}$, and this distance is a minimum if and only if its square is a minimum. So to simplify calculations, we use the square of the distance. We therefore want to minimize $f(x, y) = x^2 + y^2$ subject to the constraint that $3x - 4y = 10$. The constraint equation can be written as $g(x, y) = 10$, where $g(x, y) = 3x - 4y$. Proceeding as before, we have

$$\nabla f(x, y) = \lambda \nabla g(x, y)$$

$$\langle 2x, 2y \rangle = \lambda \langle 3, -4 \rangle$$

So we have three equations to solve:

$$\begin{cases} 2x = 3\lambda \\ 2y = -4\lambda \\ 3x - 4y = 10 \end{cases}$$

This time we solve for x and y in terms of λ from the first two equations and substitute into the third:

$$x = \frac{3\lambda}{2} \qquad y = -2\lambda$$

$$3\left(\frac{3\lambda}{2}\right) - 4(-2\lambda) = 10$$

Solving for λ, we get $\lambda = \frac{4}{5}$. So

$$x = \frac{3}{2}\left(\frac{4}{5}\right) = \frac{6}{5} \qquad \text{and} \qquad y = -2\left(\frac{4}{5}\right) = -\frac{8}{5}$$

Thus, the point to be tested is $\left(\frac{6}{5}, -\frac{8}{5}\right)$. The geometry of the situation tells us that there is some minimum distance and no maximum distance (see Figure 14.37), and since there is only one critical point, it must be the point at which the distance is minimum. ∎

$\left(\frac{6}{5}, -\frac{8}{5}\right)$

FIGURE 14.37

Let us summarize our discussion of the Method of Lagrange Multipliers for a constrained extremum of a function of two variables.

The Method of Lagrange Multipliers

Let f be a function of two variables that is to be maximized or minimized subject to the constraint $g(x, y) = k$, where f and g satisfy the conditions stated in Theorem 14.6.

Step 1. Set $\nabla f(x, y) = \lambda \nabla g(x, y)$, and write the equations obtained by equating corresponding components.

Step 2. Combine the equations from step 1 with the constraint equation $g(x, y) = k$ to obtain the system

$$\begin{cases} f_x(x, y) = \lambda g_x(x, y) \\ f_y(x, y) = \lambda g_y(x, y) \\ g(x, y) = k \end{cases}$$

of three equations with the unknowns x, y, and λ. Solve this system simultaneously.

Step 3. For each solution (x, y, λ) obtained in step 2, calculate $f(x, y)$. If f has a constrained maximum, it will be the largest of these values, and if f has a constrained minimum, it will be the smallest.

More Than Two Variables

The analogue of Theorem 14.6 for f and g functions of three (or more) variables also holds true, with the proof being virtually the same, and the procedure given above requires only slight modifications. We illustrate the procedure for the three-variable case in the next two examples.

EXAMPLE 14.32   Rework Example 14.26 using the Method of Lagrange Multipliers.

Solution   The problem can be phrased as follows:
Minimize

$$C(x, y, z) = 2(xz + yz) + \frac{5}{2}xy$$

subject to the constraint $g(x, y, z) = 270$, where

$$g(x, y, z) = xyz$$

We seek solutions to the system $\nabla C = \lambda \nabla g$ and $g(x, y, z) = 270$:

$$\begin{cases} 2z + \frac{5}{2}y = \lambda yz \\ 2z + \frac{5}{2}x = \lambda xz \\ 2(x + y) = \lambda xy \\ xyz = 270 \end{cases} \quad \text{or} \quad \begin{cases} 5y + 4z = 2\lambda yz \\ 5x + 4z = 2\lambda xz \\ 2x + 2y = \lambda xy \\ xyz = 270 \end{cases}$$

This system requires a little more ingenuity to solve than those of the preceding examples. One approach is to subtract the first equation from the second:

$$5x - 5y = 2\lambda xz - 2\lambda yz$$

$$5(x - y) = 2\lambda z(x - y)$$

$$(x - y)(5 - 2\lambda z) = 0$$

So either $x = y$ or $2\lambda z = 5$. The constraint equation ensures that $z \neq 0$, so from $2\lambda z = 5$ we get $\lambda = \frac{5}{2z}$. When this value of λ is substituted into the first equation, we get

$$5y + 4z = 2\left(\frac{5}{2z}\right) yz$$

$$5y + 4z = 5y$$

$$4z = 0$$

giving $z = 0$, which is not possible. Thus, the only feasible solution is $x = y$. The third equation then gives $4x = \lambda x^2$, or $\lambda = 4/x$, since $x \neq 0$. Substituting this value of λ into the second equation of our system gives

$$5x + 4z = 2\left(\frac{4}{x}\right) xz$$

$$= 8z$$

Thus,

$$5x - 4z = 0$$

$$z = \frac{5x}{4}$$

Now that we have $y = x$ and $z = (5x)/4$, the constraint equation $xyz = 270$ gives $5x^3/4 = 270$, or $x^3 = 216$. Finally, $x = 6$, $y = 6$, and $z = \frac{15}{2}$. That a minimum value of C exists is evident from physical considerations, so $(6, 6, \frac{15}{2})$ must yield the minimum—namely,

$$C\left(6, 6, \frac{15}{2}\right) = 2(45 + 45) + \frac{5}{2}(36) = 270 \qquad \blacksquare$$

EXAMPLE 14.33 The largest box the United Parcel Service will accept is one for which the length plus the girth (distance around) is 108 in. What are the dimensions of the box of maximum volume that can be sent by UPS?

Solution Let the dimensions be x, y, and z, as shown in Figure 14.38. Then we want to maximize $V = xyz$ subject to the constraint $x + 2(y + z) = 108$. Taking $g(x, y, z) = x + 2(y + z)$, we must have for the constrained maximum $\nabla V = \lambda \nabla g$ and $g(x, y, z) = 108$. These equations give

$$\begin{cases} yz = \lambda \\ xz = 2\lambda \\ xy = 2\lambda \\ x + 2(y + z) = 108 \end{cases}$$

Eliminating λ from the first two equations yields $2yz = xz$. Since $z = 0$ is not a feasible solution, we have $x = 2y$. Again eliminating λ from the second

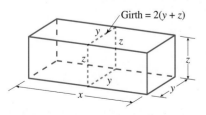

Girth $= 2(y + z)$

FIGURE 14.38

and third equations of our system, we obtain $xz = xy$, or $z = y$ (since $x \neq 0$). Substituting $x = 2y$ and $z = y$ into the constraint equation gives

$$2y + 2(2y) = 108$$

$$6y = 108$$

$$y = 18$$

So the dimensions that give the maximum volume are $x = 36$, $y = 18$, and $z = 18$. ∎

Two or More Constraints

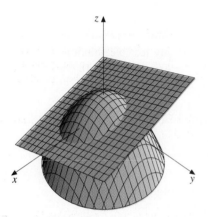

FIGURE 14.39

The Method of Lagrange Multipliers can also be applied to problems involving more than one constraint equation. We then have a multiplier for each constraint. For example, all local extrema of a function $f(x, y, z)$ subject to the constraints $g(x, y, z) = k_1$ and $h(x, y, z) = k_2$ will occur at points for which

$$\nabla f = \lambda_1 \nabla g + \lambda_2 \nabla h \tag{14.27}$$

EXAMPLE 14.34　　Find the points on the curve of intersection of the paraboloid of revolution $x^2 + y^2 + 2z = 4$ and the plane $x - y + 2z = 0$ that are closest to and farthest from the origin.

Solution　　We show the two surfaces and the curve of intersection in Figure 14.39. As in Example 14.31, we find the minimum and maximum values of the square of the distance from the origin. Thus, we take

$$f(x, y, z) = x^2 + y^2 + z^2$$

where the points (x, y, z) are constrained to lie on the given paraboloid and the given plane. We write the constraints in the form $g(x, y, z) = 4$ and $h(x, y, z) = 0$ by letting

$$g(x, y, z) = x^2 + y^2 + 2z$$

and

$$h(x, y, z) = x - y + 2z$$

From Equation 14.27, we have

$$\langle 2x, 2y, 2z \rangle = \lambda_1 \langle 2x, 2y, 2 \rangle + \lambda_2 \langle 1, -1, 2 \rangle$$

or, in terms of components,

$$\begin{cases} 2x = 2x\lambda_1 + \lambda_2 \\ 2y = 2y\lambda_1 - \lambda_2 \\ 2z = 2\lambda_1 + 2\lambda_2 \end{cases}$$

These three equations, together with the two constraint equations, constitute a system of five equations involving the five unknowns x, y, z, λ_1, and λ_2. By eliminating λ_2 from the first two equations, we get

$$(x + y)(1 - \lambda_1) = 0$$

so that $y = -x$ or $\lambda_1 = 1$. We leave it as an exercise to show that $y = -x$ yields the points $(2, -2, -2)$ and $(-1, 1, 1)$, and that $\lambda_1 = 1$ also gives the point $(-1, 1, 1)$. Since $f(2, -2, -2) = 8$ and $f(-1, 1, 1) = 3$, we conclude that the maximum distance from the origin is $\sqrt{8} = 2\sqrt{2}$ and the minimum distance is $\sqrt{3}$.   ∎

Exercise Set 14.6

In Exercises 1–8, find the local maxima and minima of f subject to the given constraint, using the Method of Lagrange Multipliers.

1. $f(x, y) = x^2 - y^2$; $x - 2y = 4$

2. $f(x, y) = 2x^2 + 3y^2$; $2x - 6y = 7$

3. $f(x, y) = x^2 + 2xy$; $y = x^2 - 2$

4. $f(x, y) = 2x - 4y$; $x^2 + y^2 = 5$

5. $f(x, y) = x^3 + 3y^2$; $xy + 4 = 0$

6. $f(x, y, z) = x^2 + 2y^2 - 2z^2$; $z = x^2 y$

7. $f(x, y, z) = x - 2y - 3z$; $xyz = 36$

8. $f(x, y, z) = 6x^2 + 3y^2 + 4z^2$; $3x^2 y + 2z^2 = 4$

In Exercises 9–16, re-do Exercises 1–8, without Lagrange Multipliers, by making a substitution to reduce the number of variables.

17. Use Lagrange Multipliers to find the distance from the line $2x + y = 3$ to the point $(1, -1)$.

18. Find the point on the plane $3x - y - 3z = 6$ that is nearest the origin, using Lagrange Multipliers.

In Exercises 19 and 20, use Lagrange Multipliers.

19. A coordinate system is set up on a flat metal plate so that the temperature $T(x, y)$ in degrees Celsius at the point (x, y) is $T(x, y) = 10x^2 y + 50$. Find the hottest and coldest spots at points on the ellipse $2x^2 + 3y^2 = 9$. What are these extreme temperatures?

20. A company determines from experience that its monthly revenue from the sale of a certain product is

$$R(x, y) = y^2 + 5xy + 20x$$

where x is the amount spent on magazine ads and y is the amount spent on television commercials, both in thousands of dollars. If the company plans to spend a total of \$60,000 per month on advertising, how should it be divided to maximize R?

In Exercises 21–24, rework the specified exercises from Exercise Set 14.5 using the Method of Lagrange Multipliers.

21. Exercise 33   **22.** Exercise 34

23. Exercise 35   **24.** Exercise 37

25. Supply the details for the solution of Example 14.34.

26. Prove that a function can have a constrained local extremum at a point but not have a local extremum there. (*Hint:* Consider $f(x, y) = xy$ with an appropriate constraint.)

27. Find the dimensions of the rectangular box of greatest volume that can be inscribed in the ellipsoid $2x^2 + y^2 + 4z^2 = 12$.

28. Find the dimensions of the cone of maximum volume that can be inscribed in a sphere of radius a.

29. Find the maximum and minimum values of $f(x, y, z) = x^2 + 2y^2 - 3z^2$ subject to the two constraints $2x^2 - 3y^2 = 8$ and $y^2 - 2z = 3$.

30. Use Lagrange Multipliers to derive the formula

$$d = \frac{|Ax_0 + By_0 + Cz_0 + D|}{\sqrt{A^2 + B^2 + C^2}}$$

for the distance d from the plane $Ax + By + Cz + D = 0$ to the point (x_0, y_0, z_0).

31. Find the points on the curve of intersection of the ellipsoid $2x^2 + 3y^2 + 4z^2 = 6$ and the paraboloid $z = 4 - x^2 - 2y^2$ that are closest to the origin and farthest from the origin.

Chapter 14 Review Exercises

In Exercises 1 and 2, find df.

1. (a) $f(x, y) = \ln \dfrac{x^2}{\sqrt{1 - y^2}}$

   (b) $f(x, y) = \dfrac{\sin x}{\cosh y}$

2. (a) $f(x, y) = \sin^{-1} \dfrac{x}{y}, \quad y > 0$

   (b) $f(x, y, z) = \dfrac{4x - 2z}{z + 3y}$

3. Approximate Δf using df.

   (a) $f(x, y) = \dfrac{(x - y)^2}{x^2 + y^2}$ from $(2, -4)$ to $(2.02, -3.97)$

   (b) $f(x, y) = \ln \sqrt{9 - x^2 - y^2}$ from $(-2, 1)$ to $(-1.99, 0.98)$

4. Suppose the electrostatic potential at a point in $\mathbb{R}^3$ is given by

   $$V(x, y, z) = \dfrac{140z}{\sqrt{x^2 + y^2 + z^2}}$$

   Find the approximate change in potential from $(3, -2, 6)$ to $(2.6, -1.8, 6.5)$.

5. A company makes two types of toasters, Models A and B. It costs \$15 to produce each unit of Model A and \$21 to produce each unit of Model B. The revenue from producing and selling x Model-A units and y Model-B units is

   $$R(x, y) = 42x + 56y - 0.02xy - 0.01x^2 - 0.03y^2$$

   The current weekly production level is 150 Model-A and 100 Model-B units. Find the approximate increase in profit if 5 more Model-A units and 8 more Model-B units are produced each week. Approximately what is the profit at this new level?

6. Use Theorem 14.2 to show that f is differentiable except at $(0, 0)$, where

   $$f(x, y) = \tan^{-1} \dfrac{x}{y}$$

7. Let

   $$f(x, y) = \begin{cases} \dfrac{x^2 y}{x^4 + y^2} & \text{if } (x, y) \neq (0, 0) \\ 0 & \text{if } (x, y) = (0, 0) \end{cases}$$

Show that f_x and f_y both exist at $(0, 0)$ but that f is not differentiable there.

8. Use Definition 14.1 to show that $f(x, y) = x^2 - 2xy + 3y$ is differentiable throughout $\mathbb{R}^2$.

In Exercises 9–13, use an appropriate Chain Rule.

9. Find dz/dt at $t = 5\pi/6$ if $z = x^2 - 2xy - y^3$ and $x = \cos 2t, \ y = \sin 2t$.

10. Find $\dfrac{dz}{dt}$ at $t = 2$ if $z = \ln \sqrt{\dfrac{x + y}{x - y}}$ and $x = t + \dfrac{1}{t}$, $y = t - \dfrac{1}{t}$.

11. Find dz/dt if $z = e^{-x^2/y}$ and $x = \sinh t, \ y = 1 + \cosh t$.

12. Find $\dfrac{\partial z}{\partial u}$ and $\dfrac{\partial z}{\partial v}$ if $z = \dfrac{x^2 - y^2}{xy}$ and $x = \dfrac{u}{v}, \ y = \dfrac{1}{u}$.

13. Find $\dfrac{\partial z}{\partial t}$ at $(s, t) = \left(1, \dfrac{1}{2}\right)$ if $z = x \sin \pi y - y \cos \pi x, \ x = s^2 - t^2$, and $y = 2st$.

14. Find the equation of the tangent line to the graph of

    $$2x^4 - 3x^2 y + 4xy^2 + y^3 + 4 = 0$$

    at $(-1, 1)$.

15. Find $\partial z/\partial x$ and $\partial z/\partial y$ if

    $$y \ln(\cos xz) + xyz = 3$$

16. Let $z = f\left(\dfrac{x - y}{x + y}\right)$. Show that

    $$x \dfrac{\partial z}{\partial x} + y \dfrac{\partial z}{\partial y} = 0$$

17. A water tank is in the form of a frustum of a cone, as shown in the figure, with bottom radius 3 ft. Water is being drained from the tank at the constant rate of

10π ft^3/min. When 210π ft^3 of water remain in the tank, the radius r of the upper surface of the water is 6 ft and is decreasing at the rate of 1 in./min. Find how fast the water level is falling at that instant.

In Exercises 18 and 19, find the directional derivatives of f in the direction indicated.

18. $f(x, y) = \ln(x^2/\sqrt{x - y})$ at $(1, -3)$, in the direction $6\mathbf{i} + 8\mathbf{j}$

19. $f(x, y, z) = x^2 y \cos \pi z$ at $(3, -1, 1)$, toward $(4, 1, -1)$

20. For the function

$$f(x, y, z) = \sqrt{\frac{x - 2y}{y - z}}$$

in what direction from $P_0 = (0, -2, -3)$ is $D_{\mathbf{u}} f(P_0)$ a maximum, and what is this maximum value?

21. Find $f(x, y)$ if $f(1, 1) = 1$ and

$$\nabla f = \left(\frac{1 - 2x^2}{x}, \frac{2y^2 - 1}{y} \right)$$

In Exercises 22 and 23, find the equation of the tangent plane and normal line at the indicated point.

22. $z = x^2 - xy - 2y^2$ at $(2, -1, 4)$

23. $z = 2e^{(x-y)/z}$ at $(1, 1, 2)$

24. The accompanying figure shows a cross section of a long semicircular cylinder with a flat base. The curved surface is kept at electrostatic potential $V = 1$ and the base at $V = 0$. It can be shown that the potential $V(x, y)$ at points inside the region is

$$V(x, y) = \frac{2}{\pi} \tan^{-1} \left(\frac{2y}{1 - x^2 - y^2} \right)$$

(a) Draw several equipotential curves $V(x, y) = c$ for $0 < c < 1$.

(b) Find the rate of change of V at $(\frac{1}{2}, \frac{1}{2})$ in the direction toward $(0, 1)$.

(c) Show that $\nabla V(\frac{1}{2}, \frac{1}{2})$ is orthogonal to the level curve through $(\frac{1}{2}, \frac{1}{2})$.

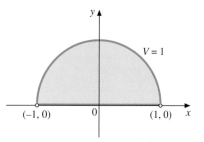

25. Find the angle between the hyperbolic paraboloid $z = x^2 - y^2$ and the line $\mathbf{r}(t) = \langle -t, 1 + 2t, 5(1 + t) \rangle$ at each of their points of intersection. (See Exercise 20 in Exercise Set 14.4.)

26. Find all maximum, minimum, and saddle points on the surface $z = x^3 - 2xy^2 - 9x + 4y^3$.

27. Find the absolute maximum and minimum values of the function

$$f(x, y) = \frac{4}{x} + \frac{1}{y} + 2xy$$

over the closed region bounded by $x = 1$, $x = 2y$, and $y = 2$, and give the points at which these extreme values occur.

28. Suppose the temperature in a three-dimensional region that contains the ellipsoid $x^2 + y^2 + 4z^2 = 12$ is given by $T(x, y, z) = xyz$. Find the hottest and coldest temperatures on the ellipsoid, and identify where they occur.

29. Use the Method of Lagrange Multipliers to find the dimensions of the right circular cylinder inscribed in a sphere of radius a that has a maximum (a) volume and (b) lateral surface area.

30. A company produces two types of electric brooms, the standard and deluxe models. Suppose the monthly profit (in thousands of dollars) from producing and selling x thousand standard models and y thousand deluxe models is given by

$$P(x, y) = 2x + 3y - 0.1x^2 - 0.2y^2 - 0.5xy - 4$$

If the combined production of the two models is to be 6000 per month, how many of each model should be produced to maximize profit?

Chapter 14 Concept Quiz

1. Define each of the following:
   (a) A differentiable function at a point (x, y)
   (b) The differential of a function f of two variables
   (c) The gradient of f
   (d) The tangent plane to the surface $F(x, y, z) = 0$ at a point P_0 where $\nabla F \neq \mathbf{0}$.
   (e) A critical point for a function of two variables

2. State the following:
   (a) A sufficient condition (other than the definition) for f to be differentiable at (x_0, y_0)
   (b) The Chain Rule for $f(x, y, z)$, where $x = x(u, v)$, $y = y(u, v)$, and $z = z(u, v)$, showing also an appropriate tree diagram
   (c) The formula for dy/dx if $F(x, y) = 0$, assuming appropriate hypotheses
   (d) A formula for the directional derivative $D_{\mathbf{u}} f(x_0, y_0)$
   (e) A test for determining the nature of a function f at a critical point (x_0, y_0), assuming appropriate hypotheses

3. (a) Give the linear approximation formula for a function f near (x_0, y_0) at which it is differentiable.
   (b) Let $z = f(x, y)$, with f differentiable at (x_0, y_0). Write the equation of the tangent plane and normal line to the graph at (x_0, y_0, z_0), where $z_0 = f(x_0, y_0)$.
   (c) Give the direction and the value of the maximum rate of change of a differentiable function f at a point (x_0, y_0).
   (d) Explain in your own words how to apply the Lagrange Multiplier Method.

4. Fill in the blanks.
   (a) If $z = f(x, y)$, f is differentiable at (x, y), and x and y are changed by small amounts dx and dy, respectively, then the approximate change in z is _____, and the percentage change in z is approximately _____.
   (b) If $w = g(s, t)$ and $s = s(x, y)$, $t = t(x, y)$, then $\dfrac{\partial w}{\partial x} = $ _____ and $\dfrac{\partial w}{\partial y} = $ _____
   (c) The gradient of f is orthogonal to the _____ _____ of f.
   (d) The normal line to the surface $F(x, y, z) = 0$ at a point (x_0, y_0, z_0) for which $\nabla F \neq \mathbf{0}$ has direction vector _____.

5. Indicate which of the following statements are true and which are false.
   (a) If f is continuous at (x_0, y_0) and $f_x(x_0, y_0)$ and $f_y(x_0, y_0)$ both exist, then f is differentiable at (x_0, y_0).
   (b) If $\nabla f(x_0, y_0) = \mathbf{0}$, then f has either a local maximum or a local minimum at (x_0, y_0).
   (c) If $f(x, y)$ has a constrained maximum or minimum value at (x_0, y_0), with constraint $g(x, y) = k$, then the gradient vector of f is parallel to the gradient vector of g at (x_0, y_0) (assuming appropriate differentiability conditions).
   (d) If f_x and f_y are continuous at (x_0, y_0), then f is differentiable at (x_0, y_0).
   (e) The differentiable function f changes most rapidly at (x_0, y_0) in the direction of $\nabla f(x_0, y_0)$.

APPLYING CALCULUS

1. In economics, the quotient

$$\frac{f'(t)}{f(t)} = \frac{\text{marginal function}}{\text{function}}$$

is called the *relative growth rate of f*. Show that it can also be calculated as

$$\frac{d}{dt}[\ln(f(t))]$$

(a) If the relative growth rate of the total consumption C of the population is c and the relative growth rate of the population P is p, show that the relative growth rate of per capita consumption C/P is $c - p$.

(b) If C is the total amount of cash on deposit in banks and D is the total of all demand deposits, then the money supply M is given by $M = C + D$. Let m, c, and d denote the relative growth rates of M, C, and D, respectively. Show that

$$m = \frac{cC + dD}{C + D}$$

2. As a rocket lifts off from the earth, its mass decreases as fuel is burned. Suppose at liftoff the mass is 80,000 kg and that for the time interval in question, the mass is decreasing at the constant rate of 72 kg/s. The rocket reaches an altitude of 30 km after 45 s, and at that time it is rising at the rate of 90 km/s. How fast is the magnitude F of the force of gravity decreasing at that instant? Use Newton's Law of Gravitation

$$F = -\frac{GMm}{r^2}$$

where G is the universal gravitational constant, M is the mass of the earth, and r is the distance of the rocket from the center of the earth. (The radius of the earth is approximately 6,370 km.)

A Saturn V rocket launches Apollo 17 to the Moon.

3. A mountain has the approximate shape of the surface $2x^2 + 3y^2 - 4x + z = 1$, for $x \geq 0$ and $y \geq 0$ and units are in kilometers. Suppose a climber is at the point $P(0.6, -0.4, 2.2)$. Assume that the positive y-axis points north and the positive x-axis points east.

(a) How high is the mountain?

(b) What is the climber's rate of ascent when moving from P in a northeasterly direction (i.e., in the direction $(\mathbf{i} + \mathbf{j})/\sqrt{2}$)?

(c) What direction from P should the climber take to ascend most rapidly? (Express as a unit vector.)

(d) What direction from P should the climber take in order not to change level?

4. The work W done during a reversible adiabatic process is given by

$$W = nC_v(T_1 - T_2)$$

where n is the number of moles of gas, C_v is molar specific heat, and T_1 and T_2 are the initial and final temperatures. Show that if both T_1, and T_2 are increased by $\alpha\%$, then the work also is increased by $\alpha\%$.

5. A storage tank for propane is in the form of a circular cylinder 6 m long and 3m in diameter, with hemispherical caps at both ends (see the figure). There is a possible

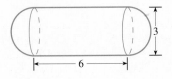

1% error in each of the measurements. Find the maximum percentage error in (a) the volume and (b) the surface area.

6.   A builder is in the process of designing low-cost, one-story, tract homes. Excluding the roof, each home will have the basic shape of a rectangular box of length x, width y, and height z, and will enclose 16,000 cubic feet. The builder is concerned about heat loss through the walls, floor, and ceiling during the winter months when the temperature averages $20°$F. The loss of heat through a surface is proportional to the surface area.

(a)   Based on some experiments, the builder has determined that the heat loss through the ceiling will be 5 times as great as the loss through the floor and the heat loss through the walls will be 3 times as great as the heat loss through the floor. What dimensions will minimize the heat loss? Are these dimensions reasonable for the type of house being considered?

(b)   The builder has decided that the height z of the walls of the house should be 10 feet. If the heat loss through the ceiling can be made arbitrarily low by adding insulation, is it possible to minimize the heat loss with $z = 10$? If so, what is the ratio of the heat loss through the ceiling to the heat loss through the floor?

7.   Let L represent the number of units of labor and K the number of units of capital in the manufacture of $P(L, K)$ units of production. Suppose that labor costs a dollars per unit and capital costs b dollars per unit, and suppose that there is a total of c dollars available for production. Then $aL + bK = c$.

(a)   Use Lagrange Multipliers to show that production is a maximum at the point (L_0, K_0) where

$$\frac{P_L(L_0, K_0)}{a} = \frac{P_K(L_0, K_0)}{b} = \lambda$$

Here the Lagrange Multiplier λ is called the *equimarginal productivity* of the production function P.

(b)   Consider the Cobb-Douglas Production Function given by

$$P(L, K) = mL^{\alpha} K^{\beta}$$

where α and β are positive constants and $\alpha + \beta = 1$. Using the result in part (a), show that at the point of maximum production (L_0, K_0)

$$\frac{L_0}{K_0} = \frac{\alpha b}{\beta a}$$

What can you conclude from this result?

N. Scott Urquhart

What do you remember of your first calculus course?

I remember lots about my first calculus course. I worked hard and learned the difference between limits and continuity, a distinction which escaped me in an earlier brief encounter with calculus in high school. I enjoyed the course—the instructor made us work hard, but we learned a lot. It was just the starting point; over the years, I went on to three higher levels of calculus after the first sequence, which lasted four quarters.

Did taking calculus (and later mathematics) courses in any way influence your career choice?

The calculus was an absolutely required starting point on my path to becoming a consulting statistician. My skills and insights in this area were essential to subsequent opportunities even appearing. Without the calculus, the opportunities would never have appeared.

How did your interests in statistics develop?

I grew up around agriculture, but always was good at math and science. Statistics provided a way to work in areas which interested me and to use mathematics on real and important problems. I never made a conscious choice to become a statistician, it just happened. Given my interests in real problems and association with consulting statisticians who worked on relevant problems, it was natural for me to move in that direction. Early on in college, I got a job running (old!) rotary calculators doing statistical analyses, and learned to use the old electronic accounting equipment (which was punch-card based) to accumulate results for these analyses. This job paid nearly twice as well as mopping floors in the dormitories and gym so I was very willing to learn more and become more valuable to the project as a result. The computer I had access to was in Los Angeles; we mailed card decks from Fort Collins, Colorado, to Los Angeles so we could do batch jobs there. Turnaround was a week. How is that for "Remote Job Entry"?

How do biologists and statisticians typically use calculus?

Biologists and statisticians use the calculus in a variety of ways, but the most fundamental aspect is to understand what derivatives and integrals are. Occasionally I do integrations or differentiations to compute some complex areas or probability functions. For that, the concepts of calculus are absolutely critical; experience with integration across complexly shaped regions was essential. Although I rarely use the machinery of calculus at the level of this text, an understanding of that material is critical to my present job.

Can you tell us about a particular problem that you solved for which calculus was useful?

I presently spend about half of my time collaborating with researchers in EPA's Environmental Monitoring and Assessment Program. I am heavily involved in developing sampling plans for inland aquatic ecological resources such as lakes, streams and wetlands. As a part of the comparison and use of spatially distributed sampling plans, I need to evaluate individual and joint inclusion probabilities for points that might be selected by a first-stage area sample. The points could be the "label points"— a geographic coordinate associated with an individual body of water as assigned by a geographic information system. The required probabilities depend on the sampling plan, the shape of the partitions across which the area sample is taken, and, in a major way, on the distance between the points.

Specifically, consider a shape formed by combining seven hexagons of the same size. Now evaluate the area of the intersection of two such shapes as a function of the location of their centers. The resulting values can be thought of as a surface over two-dimensional space; this surface rises to the area of the shaped area as the two points converge. This surface falls off quickly as the points are moved apart, going to zero when the points are as far apart as the maximum width of the sampling shape. Near its top, its cross sections have moderately pronounced corners, but they smooth out as the points pull apart. This problem has made it obvious how important it is for a calculus student to become as competent with complex regions of integration as he or she is with complex integrands.

How is multivariable calculus useful in biological analysis?

Most interesting biological quantities are functions that depend on several independent variables. For example, fisheries biologists sometimes want to know how body weight of a particular species of fish depends on other identifiable variables. Length would be an important variable, but age, water temperature, food availability, predatory pressure, etc., might all be significant in the problem. Biologists do not usually apply multiple integrals to problems involving specific functions; however, multiple integrals of general functions are sometimes used to communicate the workings of a biological system without ever explicitly stating an integrand to integrate. The function becomes specific only for a particular point in time, or space, and only for a given set of conditions. Even then, it is frequently evaluated only in a discrete manner using numerical approximations. It's more typical to use calculus to describe a kind of general representation of an accumulative biological process.

For example, the photosynthesis of carbohydrates in a plant depends on the variables of light, nutrients, temperature, carbon dioxide concentration, and leaf surface area. Photosynthesis uses solar radiation to transform inorganic compounds into carbohydrates. If this occurs in an aquatic environment where there are phytoplankton floating at all depths to about several hundred feet, the light intensity diminishes with depth, and net photosynthetic production also declines with depth. These changes occur continuously and so general integral formulas are used to describe the total photosynthetic production. The light intensity is a function of both water depth (or height) and time, because the amount and direction of incoming radiation depends on time, and its penetration depends on depth. Under a fixed nutrient supply, the instantaneous rate of net photosynthesis depends primarily on the energy available, that is, light intensity. Therefore, if q represents the functional dependence of instantaneous net photosynthesis on intensity, $q(t) = q(I(h, t)) = p(h, t)$, so p has an equivalent representation as a function of time and depth. If p has been evaluated on a per unit area basis, then the total net production between times t_1 and t_2 occurring from the surface down to depth h_l (where l stands for light) can be written as $\iint_R p(h, t) dA$ where $R = \{(h, t): 0 \leq h \leq h_l, t_1 \leq t \leq t_2\}$. This double integral can be written as an iterated integral, $\iint_R p(h, t) dA = \int_{t_1}^{t_2} \int_0^{h_l} p(h, t) \, dh \, dt$, which indicates the involvement of time and depth. Functions of *three* or more variables present only a modest kind of notational change mathematically, but are radically different when we try to give them a visual representation simply because figures cannot be drawn in four or more dimensions. A thorough treatment of these extensions to three or more variables requires linear algebra and students who are interested in serious mathematical biology will need to take linear algebra right after calculus. Biology abounds with situations that can be described by some form of mathematical transformation, especially linear transformations as a first approximation. For example, an organism's physical condition might be described in terms of a number of variables whose values at a given point in time would be a point in a vector space. A certain influence such as stress or feeding the organism would change these variables, that is, transform one point into another. A linear transformation approximates this very well in many cases.

Do you use alternative coordinate systems to model other biological phenomena?

Yes. Polar coordinates, for instance, are useful in describing certain regions bounded by curves, more easily than a rectangular system. We use double integration in applications such as studies of populations introduced to a particular location. When animals are introduced to a particular location, for example, they frequently disperse without any particular directional orientation. One well-known study of muskrats introduced to Central Europe in 1900 used a model of the time-dependent density of the population in a polar coordinate system. This was effectively a probability density function that described a dispersal whose contours of equal density spread out from the release point in ever-expanding circles, like the ripples created by tossing a pebble into calm water. The model suggests that in the absence of some kind of force encouraging directional dispersion, the square root of the area occupied should have a linear relation to time. This same model was used by the author of the muskrat study to show that oak trees in Great Britain after the last ice age had been aided in their dispersal by animals or some other extrinsic factor.

You were coauthor of a textbook on Mathematics in Biology. Do students get the proper exposure to mathematical methods in the biology curriculum?

My past experience suggests that biology students see far too few relevant examples because few mathematicians teaching such courses have enough relevant experience with biology. It can be difficult to convince biology students that the precision of mathematics and mathematical operations is needed by biologists. The most relevant experiences of a biology student in a calculus course may have nothing to do with any of the specific functions to which they are exposed. Biology students need to learn to be precise; at the local level, because relative to the quantitative sciences, biology has lots of "fuzz"—imprecision in what is really happening or being seen. Secondly, fruitful development of computational procedures relevant to biology require an assimilation of many parts of mathematics. The value of repeated exposure to the concepts of rates of change (derivatives), accumulation processes (integration), and their relation (differential equations) cannot be emphasized enough.

Just as with other disciplines, the precision and conciseness of mathematics provides biologists with compact ways of expressing complex ideas. The tools of mathematics open up new ways of examining these ideas and their implications. Some biological problems, even apparently simple ones, require a mathematical approach for their understanding. Since more and more biologists are using mathematics to express their ideas, it's likely that understanding the basic literature of biology will require an increased understanding of basic mathematics. The trend toward more use of mathematics has occurred for a variety of reasons. One is that biologists are drawing more on ideas from chemistry and physics, which have long adopted mathematics as a tool and a language; another is the growing interface between biology and technology, particularly in the area of environmental studies.

N. Scott Urquhart, professor of statistics at Oregon State University, grew up in the agricultural country of western Colorado and studied statistics, mathematics, and biology at Colorado State University, leading to a Ph.D. in 1965. He has taught, been an internal consultant, and done research affiliated with statistical aspects of agriculture at Cornell University, New Mexico State University, and Oregon State University. He has received several awards for excellence in teaching. His research interests span many areas of statistics in biology, including determining the factors affecting beef quality and associated consumer acceptance; using waste radioisotopes to convert sewage sludge to beneficial uses; developing resistance to root-rot diseases; projecting the effects of hunting regulations on deer populations; modeling plant growth in Alaska to minimize the environmental effects of energy development in fragile ecosystems; and developing a national sampling program for ecological resources, in cooperation with USEPA.

15

MULTIPLE INTEGRALS

In this chapter we consider the integral calculus of functions of more than one variable. We begin by defining a *double integral*, in which a function of two variables is integrated over a region in the xy-plane. Just as the integral of a nonnegative function of one variable can be interpreted geometrically as the area under its graph, the double integral of a nonnegative function of two variables can be interpreted as the volume under its graph.

After defining the double integral and listing some of its properties, we will give an important theorem that will let us evaluate many double integrals by evaluating two single integrals in succession. (By a single integral we mean an integral of a function of one variable—the type we have studied up to now.) This theorem plays a role for double integrals similar to that of the Second Fundamental Theorem of Calculus for single integrals.

The region in the xy-plane over which a function of two variables is integrated sometimes can be better described with polar coordinates than with rectangular coordinates. For this reason, we devote a section to studying double integrals in polar coordinates.

As one application of double integrals, we revisit moments and centers of mass of two-dimensional laminas. Recall that we studied these concepts in Chapter 6 as an application of the integral. There we were restricted to homogeneous laminas (ones with constant density). With double integration, we can remove this restriction.

As another application, we develop a formula for the area of a surface. This result is more general than the formulas we developed for surface area in Chapter 6, where we considered surfaces of revolution only.

Next, we extend our results for double integrals to *triple integrals*—that is, integrals of functions of three variables over suitably restricted three-dimensional regions. There are two three-dimensional analogues of polar coordinates—cylindrical coordinates and spherical coordinates. We will study triple integrals in each of these systems.

We conclude the chapter with a section on changing variables in double and triple integrals.

15.1 DOUBLE INTEGRALS

We begin by defining the double integral of a function of two variables over a region in the xy-plane. Since the definition has much in common with the definition of the single integral $\int_a^b f(x)dx$, you may find it helpful to review this latter definition, found in Section 5.2, before proceeding.

We begin with the simplest case, in which the region of integration is rectangular. Then we extend the definition to more general regions.

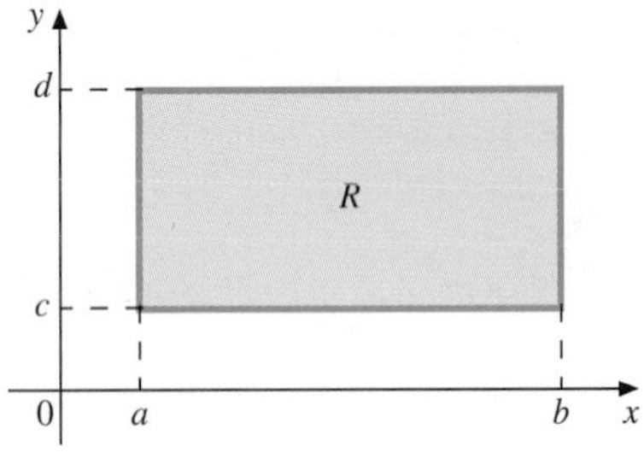

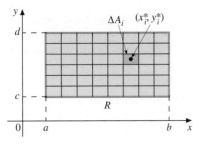

FIGURE 15.1
The rectangular region
$R = \{(x, y): a \le x \le b, c \le y \le d\}$

FIGURE 15.2
A partition of the region R

Double Integrals Over Rectangular Regions

Let f be a function of two variables that is defined on the rectangular region

$$R = \{(x, y): a \le x \le b, c \le y \le d\}$$

We have pictured such a region in Figure 15.1.

We form a rectangular grid over R as in Figure 15.2 by means of horizontal and vertical lines drawn at arbitrary points on the boundary of R. These lines divide R into finitely many smaller rectangles, which we call *subrectangles*. We call this collection of subrectangles a *partition* of R and designate the partition by P. The subrectangles need not all be of the same size. We call the length of the longest diagonal of all of them the *norm* of P and denote it by $\|P\|$.

We number the subrectangles of the partition P consecutively in any manner, starting with 1. Suppose there are n of them. In each subrectangle we select an arbitrary point. We designate the point selected in the ith subrectangle by (x_i^*, y_i^*) and the area of the ith subrectangle by ΔA_i, for $i = 1$ to $i = n$.

Now we form the sum

$$\sum_{i=1}^{n} f(x_i^*, y_i^*)\Delta A_i$$

This sum is called a *Riemann sum* for f corresponding to the partition P. If we take finer and finer partitions (that is, partitions whose norms approach 0), the corresponding Riemann sums may approach some limit L. If so, we call L the double integral of f over R and denote it by

$$\iint\limits_R f(x, y)dA$$

In summary, we have the following definition.

Definition 15.1
The Double Integral Over a Rectangular Region R

The **double integral** of a function f of two variables over the rectangular region R is

$$\iint\limits_R f(x, y)dA = \lim_{\|P\| \to 0} \sum_{i=1}^{n} f(x_i^*, y_i^*)\Delta A_i \qquad (15.1)$$

provided the limit on the right exists. When the limit does exist, we say that f is *integrable* on R.

REMARK
■ The precise meaning of the limit on the right-hand side of Equation 15.1 is as follows:

$$\lim_{\|P\|\to 0} \sum_{i=1}^{n} f(x_i^*, y_i^*)\Delta A_i = L$$

provided that, corresponding to each positive number ε, there is a positive number δ such that for all partitions P of R with $\|P\| < \delta$,

$$\left| \sum_{i=1}^{n} f(x_i^*, y_i^*)\Delta A_i - L \right| < \varepsilon$$

independently of how the point (x_i^*, y_i^*) is chosen in the ith subrectangle for $1 \le i \le n$.

Although we will not prove it here, it can be shown that a *continuous* function is always integrable over any rectangular region. Most of the functions we work with will be continuous. In the next section, we will see a way of evaluating double integrals that does not require us to calculate the limit in Equation 15.1 directly. In fact, evaluating most integrals by calculating this limit would be a hopeless task. However, we can approximate the integral by means of Riemann sums for particular partitions. The next example illustrates this type of approximation.

EXAMPLE 15.1   Approximate $\iint_R (x + 2y)\,dA$, where R is the rectangular region bounded by the lines $x = 0$, $x = 4$, $y = 0$, and $y = 1$. Use a partition formed by vertical lines at $x = 1, 2,$ and 3, and a horizontal line at $y = \frac{1}{2}$. Take (x_i^*, y_i^*) as the lower left-hand corner of the ith subrectangle.

Solution   Figure 15.3 shows the region R and the given partition of it. We have numbered the eight subrectangles. In each case, $\Delta A_i = (1)(\frac{1}{2}) = \frac{1}{2}$. We calculate the Riemann sum using the following table.

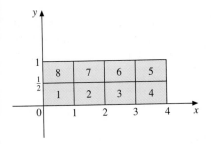

FIGURE 15.3

i	(x_i^*, y_i^*)	$f(x_i^*, y_i^*)$	$f(x_i^*, y_i^*)\,\Delta A_i$
1	$(0, 0)$	0	0
2	$(1, 0)$	1	$\frac{1}{2}$
3	$(2, 0)$	2	1
4	$(3, 0)$	3	$\frac{3}{2}$
5	$(3, \frac{1}{2})$	4	2
6	$(2, \frac{1}{2})$	3	$\frac{3}{2}$
7	$(1, \frac{1}{2})$	2	1
8	$(0, \frac{1}{2})$	1	$\frac{1}{2}$

$$\sum_{i=1}^{8} f(x_i^*, y_i^*)\Delta A_i = 8$$

By using the technique we will learn in the next section, we can show that the exact value of the integral is 12, so our approximation is not very good. We could improve it by using a partition with a smaller norm.   ■

Double Integrals Over General Regions

If the region R of integration is not rectangular but can be contained within some rectangle, then we partition this rectangle as before but count only those subrectangles of the partition that are completely contained within R. We illustrate a typical situation in Figure 15.4. Then we define the double integral of f over R by

$$\iint\limits_{R} f(x, y)\, dA = \lim_{\|P\| \to 0} \sum_{i=1}^{n} f(x_i^*, y_i^*) \Delta A_i$$

provided the limit exists. So the integral is defined exactly as in Definition 15.1, with the understanding that for each partition P of the outer rectangle, we count only those subrectangles of the partition that are contained in the region R.

When f is continuous on R, the double integral will exist provided the boundary of R is not too complicated. We will not state the most general conditions on the boundary, but in the next section we will describe two commonly occurring types of regions for which the integral will exist.

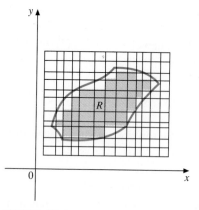

FIGURE 15.4

A Geometric Interpretation of the Double Integral

If f is a nonnegative continuous function over the region R, we can interpret the double integral of f over R as follows: Let S denote that part of the surface $z = f(x, y)$ that lies over R. Partition a rectangle containing R as before, numbering those subrectangles lying completely within R. For each i from 1 to n, build a rectangular box, called a rectangular *prism*, with the ith subrectangle as a base and having height $f(x_i^*, y_i^*)$, where (x_i^*, y_i^*) is any point in the ith subrectangle. We show a typical prism in Figure 15.5. The Riemann sum

$$\sum_{i=1}^{n} f(x_i^*, y_i^*) \Delta A_i$$

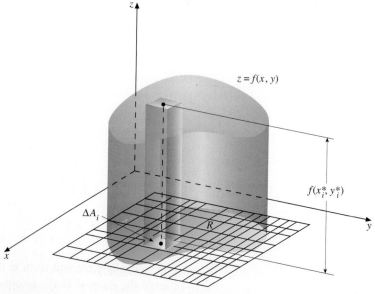

FIGURE 15.5

is the sum of the volumes of all such prisms. It is reasonable to expect that this sum approximates the volume under the surface S and above the region R. The sum may be an approximation only, since the prisms have flat tops, whereas the surface S may be curved. Also, the subrectangles inside R may not fill out all of R. Nevertheless, it seems reasonable to assume that as we take finer and finer partitions, these deviations become negligible. Since the Riemann sums approach the double integral as the norms of the partitions approach 0, we expect the double integral to give the exact volume under the surface. In fact, we *define* the volume by the double integral.

Definition 15.2
Volume Defined by a
Double Integral

Let f be a nonnegative continuous function on the closed and bounded region R. If the double integral of f over R exists, then the volume V under the graph of f and above the region R is equal to this double integral:

$$V = \iint\limits_{R} f(x, y)\, dA \tag{15.2}$$

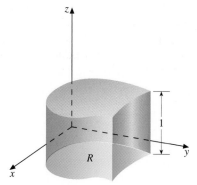

FIGURE 15.6

The special case of Equation 15.2 in which $f(x, y) = 1$ for all (x, y) in R is of particular interest. A typical volume of this type is shown in Figure 15.6. By Equation 15.2, the volume is

$$V = \iint\limits_{R} 1\, dA = \iint\limits_{R} dA$$

But we can also calculate the volume of such a slab by multiplying the area of the base by the altitude. So if we denote the area of R by A, we have $V = (A)(1) = A$. If we equate these two expressions for V, we get the following very useful formula for the area of R.

The Area of a Plane Region R

$$A = \iint\limits_{R} dA \tag{15.3}$$

Although our discussion showed the plausibility of this result, we can take Equation 15.3 as the *definition* of the area of the region R, when the integral exists. In Exercise 54 of Exercise Set 15.2 you will be asked to show that when R is of a particular type, this definition is consistent with that of the area between two curves, which we studied in Chapter 6.

Properties of the Double Integral

A number of properties of the double integral can be proved based on Definition 15.1 and its extension to general regions. Some of the more important ones are stated here. We assume that each integral exists.

Properties of the Double Integral

1. $$\iint_R c f(x, y)\, dA = c \iint_R f(x, y)\, dA \text{ for any constant } c$$

2. $$\iint_R [f(x, y) \pm g(x, y)]\, dA = \iint_R f(x, y)\, dA \pm \iint_R g(x, y)\, dA$$

3. If $R = R_1 \cup R_2$, where R_1 and R_2 are nonoverlapping, then
$$\iint_R f(x, y)\, dA = \iint_{R_1} f(x, y)\, dA + \iint_{R_2} f(x, y)\, dA$$

4. If $f(x, y) \geq 0$ for all (x, y) in R, then
$$\iint_R f(x, y)\, dA \geq 0$$

5. If f is integrable over R, then so is $|f|$, and
$$\left| \iint_R f(x, y)\, dA \right| \leq \iint_R |f(x, y)|\, dA$$

Exercise Set 15.1

1. Redo Example 15.1, taking (x_i^*, y_i^*) as the midpoint of the ith subrectangle.

2. Redo Example 15.1 with a partition that divides R into 16 squares $\frac{1}{2}$ unit on each side.

In Exercises 3–6, calculate the Riemann sum for f over R where P is the partition formed by vertical lines at the specified x values and horizontal lines at the specified y values. Take (x_i^*, y_i^*) as the lower right-hand corner of the ith subrectangle.

3. $f(x, y) = 2xy$; $R = \{(x, y) : 1 \leq x \leq 3, 0 \leq y \leq 4\}$; $x = 2$; $y = 1, 2, 3$

4. $f(x, y) = x^2 - y^2$; $R = \{(x, y) : -1 \leq x \leq 2, 1 \leq y \leq 3\}$; $x = 0$; $y = 2$

5. $f(x, y) = x^2 - 2xy$; $R = \{(x, y) : 2 \leq x \leq 6, -2 \leq y \leq 4\}$; $x = 3, 4$; $y = 0, 2$

6. $f(x, y) = 2x^2 + 3y^2$; $R = \{(x, y) : -2 \leq x \leq 2, 0 \leq y \leq 3\}$; $x = 0, 1$; $y = \frac{1}{2}, \frac{3}{2}$

In Exercises 7–10, approximate the volume under the surface $z = f(x, y)$ and above the region R, using a Riemann sum for the partition formed by the specified vertical and horizontal lines. Take (x_i^*, y_i^*) as the center of the ith subrectangle.

7. $f(x, y) = 2x + 3y$; R is bounded by the triangle with vertices $(0, 0)$, $(4, 0)$, and $(0, 4)$; $x = 1, 2, 3$; $y = 1, 2, 3$

8. $f(x, y) = \sqrt{x^2 + y^2}$; $R = \{(x, y) : 0 \leq x \leq 5, -x \leq y \leq x\}$; $x = 1, 2, 3, 4$; $y = \pm 1, \pm 2, \pm 3, \pm 4$

9. $f(x, y) = 2x^2 + 4y^2$; $R = \{(x, y) : |x| + |y| \leq 2\}$; $x = 0$, $\pm\frac{1}{2}, \pm 1, \pm\frac{3}{2}$; $y = 0, \pm\frac{1}{2}, \pm 1, \pm\frac{3}{2}$

10. $f(x, y) = x^2 + 2xy + 3y^2$; $R = \{(x, y) : y - 4 \leq x \leq 4 - y, 0 \leq y \leq 2\}$; $x = \pm 1, \pm 2, \pm 3$; $y = \frac{1}{2}, 1, \frac{3}{2}$

11. Make use of one or more of the properties of the double integral to show that if f and g are integrable over R and $f(x, y) \leq g(x, y)$ for all (x, y) in R, then

$$\iint_R f(x, y)\, dA \leq \iint_R g(x, y)\, dA$$

12. If f is integrable over R, and m and M are numbers such that $m \leq f(x, y) \leq M$ for all (x, y) in R, prove that

$$mA \leq \iint_R f(x, y)\, dA \leq MA$$

where A is the area of R.

In Exercises 13 and 14, use a calculator, a CAS, or a spreadsheet to approximate $\iint_R f(x, y)\, dA$, where R is the rectangular region specified. Use partitions formed by m equally spaced vertical lines and n equally spaced horizontal lines. Take the point (x_i^*, y_i^*) as the center of the ith rectangle.

13. $f(x, y) = e^x \sin y$; $R = \{(x, y) : 0 \leq x \leq 2, 0 \leq y \leq 4\}$
 (a) $m = 4, n = 4$
 (b) $m = 6, n = 10$
 (c) $m = 10, n = 8$
 (d) $m = 40, n = 20$

14. $f(x, y) = \sqrt[3]{2x^2 y - 3y^2}$;
 $R = \{(x, y) : -4 \leq x \leq 6, -1 \leq y \leq 4\}$
 (a) $m = 5, n = 10$
 (b) $m = 20, n = 40$
 (c) $m = 30, n = 20$
 (d) $m = 50, n = 100$

15.2 EVALUATING DOUBLE INTEGRALS BY ITERATED INTEGRALS

Certain double integrals can be evaluated by means of *iterated integrals*, in which two single integrals are evaluated in succession. To explain what an iterated integral is and its relationship to a double integral, let us return to the simplest case, in which the region R of integration is rectangular.

One of the single integrals in an iterated integral in this case will either be of the form

$$\int_a^b f(x, y)\, dx$$

or of the form

$$\int_c^d f(x, y)\, dy$$

In the first of these integrals, the integration is with respect to x, *holding y fixed*, whereas in the second, the integration is with respect to y, *holding x fixed*. For example, consider the two integrals

$$\int_0^2 x \cos y\, dx \quad \text{and} \quad \int_0^{\pi/2} x \cos y\, dy$$

For the first integral, we have

$$\int_0^2 x \cos y\, dx = \frac{x^2}{2} \cos y \bigg]_0^2 = 2 \cos y$$

Notice that $\cos y$ was treated as a constant, since we were holding y fixed. For the second integral, we hold x fixed to get

$$\int_0^{\pi/2} x \cos y\, dy = x \sin y \bigg]_0^{\pi/2} = x$$

since $\sin(\pi/2) = 1$ and $\sin 0 = 0$.

Notice the similarity between integrals of these two types and partial differentiation. In fact, the integration process we have illustrated is sometimes called *partial integration*.

As our example shows, an integral of the form

$$\int_a^b f(x, y)\, dx$$

is a function of y, since after integrating with respect to x, we substituted the upper and lower limits for x, leaving the variable y only. In our example we found that

$$\int_0^2 x \cos y\, dx = 2 \cos y$$

confirming that the result is a function of y. Suppose now that we integrate this function of y over the interval $[c, d]$. Then we have

$$\int_c^d \left[\int_a^b f(x, y)\, dx \right] dy$$

where the bracket on the inside indicates the (partial) integration with respect to x is to be performed first. For example,

$$\int_0^{\pi/2} \left[\int_0^2 x \cos y\, dx \right] dy = \int_0^{\pi/2} 2 \cos y\, dy$$
$$= 2 \sin y \Big]_0^{\pi/2} = 2$$

Similarly, the result of the integration

$$\int_c^d f(x, y)\, dy$$

is a function of x, which can then be integrated over the x interval to get

$$\int_a^b \left[\int_c^d f(x, y)\, dy \right] dx$$

For example,

$$\int_0^2 \left[\int_0^{\pi/2} x \cos y\, dy \right] dx = \int_0^2 x\, dx = x^2 \Big]_0^2 = 2$$

Notice that we got the same answer as when we integrated first with respect to x and then y. This equivalence is no accident, as we will soon see.

Each of the integrals

$$\int_c^d \left[\int_a^b f(x, y)\, dx \right] dy \quad \text{and} \quad \int_a^b \left[\int_c^d f(x, y)\, dy \right] dx$$

is called an *iterated integral*. It is customary to delete the brackets and write simply

$$\int_c^d \int_a^b f(x, y)\, dx\, dy \quad \text{and} \quad \int_a^b \int_c^d f(x, y)\, dy\, dx$$

The order in which the integration should be performed is indicated by the order of the dx and dy, from left to right.

REMARK ———

■ An iterated integral has much in common with a composite function. In a composite function $f(g(x))$, the value of the inner function is obtained first and then the result is substituted into the outer function. In an iterated integral, the inner integral is evaluated first and then the result becomes the integrand for the outer integral.

———

EXAMPLE 15.2   Evaluate the iterated integral

$$\int_0^3 \int_{-1}^1 \frac{x^2}{1+y^2}\, dy\, dx$$

Solution   An antiderivative of $1/(1+y^2)$ is $\tan^{-1} y$. So on integrating first with respect to y, holding x fixed, we get

$$\int_0^3 \left[x^2 \tan^{-1} y\right]_{-1}^1 dx = \int_0^3 x^2 \left[\tan^{-1} 1 - \tan^{-1}(-1)\right] dx$$

$$= \int_0^3 x^2 \left[\frac{\pi}{4} - \left(-\frac{\pi}{4}\right)\right] dx$$

$$= \frac{\pi}{2} \int_0^3 x^2\, dx$$

Now we integrate with respect to x to get

$$\frac{\pi}{2}\left[\frac{x^3}{3}\right]_0^3 = \frac{\pi}{2}\left(\frac{27}{3}\right) = \frac{9\pi}{2}$$

So we have the final result

$$\int_0^3 \int_{-1}^1 \frac{x^2}{1+y^2}\, dy\, dx = \frac{9\pi}{2} \qquad\blacksquare$$

We are ready now for the important theorem relating a double integral over a rectangular region and two associated iterated integrals. It is named for the Italian mathematician Guido Fubini (1879–1943).

THEOREM 15.1

Fubini's Theorem

If f is continuous on the rectangular region $R = \{(x, y) : a \leq x \leq b,\ c \leq y \leq d\}$, then

$$\iint\limits_R f(x, y)\, dA = \int_a^b \int_c^d f(x, y)\, dy\, dx = \int_c^d \int_a^b f(x, y)\, dx\, dy$$

REMARK ———

■ Fubini's Theorem tells us that the iterated integrals taken in either order are equal and that their common value equals the double integral over R.

———

Although we will not give a proof of Fubini's Theorem, we can show the result is plausible at least for the case in which $f(x, y)$ is nonnegative and continuous on R. We know in this case that the volume V under the surface $z = f(x, y)$ and above the region R is, according to Definition 15.2,

$$V = \iint_R f(x, y)\,dA \tag{15.4}$$

But we can also calculate the volume in another way. As indicated in Figure 15.7, we select an arbitrary x value between a and b and hold it fixed temporarily. To emphasize that it is held fixed, we designate it by x_0 (but we will soon remove the subscript). The plane $x = x_0$ intersects the given surface in a curve $z = f(x_0, y)$. Here y is the independent variable and z is the dependent variable. We are interested in that portion of this curve from $y = c$ to $y = d$. Denote the area under its graph by $A(x_0)$. Then from Chapter 5 we know that

$$A(x_0) = \int_c^d f(x_0, y)\,dy$$

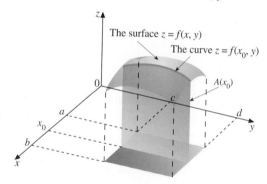

The surface $z = f(x, y)$

The curve $z = f(x_0, y)$

$A(x_0)$

FIGURE 15.7

Observe that $A(x_0)$ is the area of a typical cross section of the solid under the surface, taken perpendicular to the x-axis. So by Definition 6.1, we get the total volume by integrating this cross-sectional area over the interval $[a, b]$. Since we now want x_0 to vary over this interval, we drop the subscript and write

$$V = \int_a^b A(x)\,dx$$

That is,

$$V = \int_a^b \left[\int_c^d f(x, y)\,dy \right] dx \tag{15.5}$$

Equating the two expressions for volume (Equations 15.4 and 15.5) gives

$$\iint_R f(x, y)\,dA = \int_a^b \int_c^d f(x, y)\,dy\,dx$$

A similar argument can be given when we pass a plane perpendicular to the y-axis at $y = y_0$ to obtain the result

$$\iint_R f(x, y)\,dA = \int_c^d \int_a^b f(x, y)\,dx\,dy$$

(See Exercise 53 in Exercise Set 15.2.)

EXAMPLE 15.3   Use Fubini's Theorem to evaluate the double integral

$$\iint_R (x + 2y)\, dA$$

where R is the rectangular region $\{(x, y) : 0 \leq x \leq 4, 0 \leq y \leq 1\}$.

Solution   Note that the integral in question is the one that we estimated in Example 15.1.

By Fubini's Theorem, we can write

$$\iint_R (x + 2y)\, dA = \int_0^4 \int_0^1 (x + 2y)\, dy\, dx$$

$$= \int_0^4 \left[xy + y^2 \right]_0^1 dx$$

$$= \int_0^4 (x + 1)\, dx$$

$$= \frac{(x + 1)^2}{2} \Bigg]_0^4 = \frac{25}{2} - \frac{1}{2} = \frac{24}{2} = 12$$

We chose to integrate first with respect to y, then x. If we use the other order, we get

$$\iint_R (x + 2y)\, dA = \int_0^1 \int_0^4 (x + 2y)\, dx\, dy$$

$$= \int_0^1 \left[\frac{x^2}{2} + 2xy \right]_0^4 dy$$

$$= \int_0^1 (8 + 8y)\, dy$$

$$= 8y + 4y^2 \Big]_0^1 = 8 + 4 = 12$$

Of course, we expected that the results would be the same, as guaranteed by Fubini's Theorem.   ∎

Extension of Fubini's Theorem to Nonrectangular Regions

In order to extend Fubini's Theorem to more general regions, we first identify two special types of regions, which we call **type I** and **type II**. A type I region is shown in Figure 15.8 and a type II region in Figure 15.9.

More precisely, we say a region R is of type I if

$$R = \{(x, y) : a \leq x \leq b, \ g_1(x) \leq y \leq g_2(x)\}$$

where g_1 and g_2 are continuous functions on the closed interval $[a, b]$. Similarly, R is a type II region if

$$R = \{(x, y) : h_1(y) \leq x \leq h_2(y), c \leq y \leq d\}$$

where h_1 and h_2 are continuous functions on the closed interval $[c, d]$.

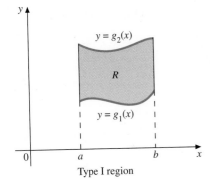

Type I region

FIGURE 15.8

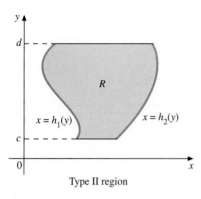

x = h_1(y)   x = h_2(y)

Type II region

FIGURE 15.9

When we are working with a type I region, an iterated integral will be of the form

$$\int_a^b \int_{g_1(x)}^{g_2(x)} f(x, y)\, dy\, dx$$

REMARK ───────────────────────────────

■ To help determine when a region is of type I, consider vertical lines drawn at x values between the left and right extremities of the region. Each such line must intersect the boundary of R in exactly two points—once on the lower bounding curve and once on the upper bounding curve. Similarly, for a type II region each horizontal line between the lower and upper extremities will intersect the boundary exactly twice—once on the left bounding curve and once on the right bounding curve.

───────────────────────────────────────

When the region is type II, we will have an iterated integral of the form

$$\int_c^d \int_{h_1(y)}^{h_2(y)} f(x, y)\, dx\, dy$$

Note that these two types of iterated integrals differ from those we considered previously only in the limits of integration on the inner integral. Previously, all the limits were constants. Now we have variables as limits on the inner integral. In the next example, we illustrate how to evaluate iterated integrals of these two types.

EXAMPLE 15.4   Evaluate the following iterated integrals.

(a) $\displaystyle \int_0^2 \int_{x^2}^{4-x} (x - 2y)\, dy\, dx$ (b) $\displaystyle \int_0^1 \int_{y-1}^{\sqrt{1-y^2}} (xy + y)\, dx\, dy$

Solution

(a) We proceed just as before, integrating first with respect to y, holding x fixed.

$$\int_0^2 \int_{x^2}^{4-x} (x - 2y)\, dy\, dx = \int_0^2 \left[xy - y^2 \right]_{x^2}^{4-x} dx$$

$$= \int_0^2 \left[x(4 - x) - (4 - x)^2 - x^3 + x^4 \right] dx$$

$$= \int_0^2 \left[4x - x^2 - (4 - x)^2 - x^3 + x^4 \right] dx$$

$$= 2x^2 - \frac{x^3}{3} + \frac{(4 - x)^3}{3} - \frac{x^4}{4} + \frac{x^5}{5} \Bigg]_0^2 = -\frac{164}{15}$$

(You should check the algebra.)

(b) $\displaystyle\int_0^1 \int_{y-1}^{\sqrt{1-y^2}} (xy + y)\,dx\,dy$

$$= \int_0^1 \left[\frac{x^2}{2}y + xy\right]_{y-1}^{\sqrt{1-y^2}} dy$$

$$= \int_0^1 \left[\frac{(1-y^2)y}{2} + y\sqrt{1-y^2} - \frac{(y-1)^2}{2}y - y(y-1)\right] dy$$

$$= \int_0^1 \left[\frac{1}{2}(y - y^3) + y\sqrt{1-y^2} - \frac{1}{2}(y^3 - 2y^2 + y) - y^2 + y\right] dy$$

$$= \int_0^1 (-y^3 + y + y\sqrt{1-y^2})\,dy$$

$$= -\frac{y^4}{4} + \frac{y^2}{2} - \frac{1}{2} \cdot \frac{2}{3}(1-y^2)^{3/2}\Big]_0^1$$

$$= -\frac{1}{4} + \frac{1}{2} + \frac{1}{3} = \frac{7}{12}$$

■

We can now state the following stronger form of Fubini's Theorem.

THEOREM 15.2

Fubini's Theorem (Stronger Form)

1. If f is continuous on the type I region
$$R = \{(x, y) : a \le x \le b, \ g_1(x) \le y \le g_2(x)\}$$
then
$$\iint_R f(x, y)\,dA = \int_a^b \int_{g_1(x)}^{g_2(x)} f(x, y)\,dy\,dx$$

2. If f is continuous on the type II region
$$R = \{(x, y) : h_1(y) \le x \le h_2(y), \ c \le y \le d\}$$
then
$$\iint_R f(x, y)\,dA = \int_c^d \int_{h_1(y)}^{h_2(y)} f(x, y)\,dx\,dy$$

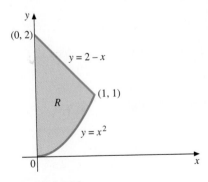

FIGURE 15.10

EXAMPLE 15.5   Evaluate the double integral
$$\iint_R (x + \sqrt{y})\,dA$$
where R is the region shown in Figure 15.10.

Solution   Observe that since R is bounded below and above by the continuous functions $y = x^2$ and $y = 2 - x$, respectively, for $0 \le x \le 1$, R is a type I region. It is also of type II, since it is bounded on the left by the line $x = 0$

and on the right by the continuous function consisting of the line $x = 2 - y$ when $1 \leq y \leq 2$ and the parabolic arc $x = \sqrt{y}$ when $0 \leq y \leq 1$. We choose to treat it as a type I region, however, to avoid the necessity of having to use one integral for $0 \leq y \leq 1$ and another for $1 \leq y \leq 2$, as we would have to do if we treated it as a type II region.

By part 1 of Theorem 15.2,

$$\iint_R (x + \sqrt{y})\, dA = \int_0^1 \int_{x^2}^{2-x} (x + \sqrt{y})\, dy\, dx$$

$$= \int_0^1 \left[xy + \frac{2}{3} y^{3/2} \right]_{x^2}^{2-x} dx$$

$$= \int_0^1 \left[x(2 - x) + \frac{2}{3}(2 - x)^{3/2} - x^3 - \frac{2}{3} x^3 \right] dx$$

$$= \int_0^1 \left[2x - x^2 + \frac{2}{3}(2 - x)^{3/2} - \frac{5}{3} x^3 \right] dx$$

$$= x^2 - \frac{x^3}{3} - \frac{2}{3} \cdot \frac{2}{5} (2 - x)^{5/2} - \frac{5}{12} x^4 \Big]_0^1$$

$$= 1 - \frac{1}{3} - \frac{4}{15} - \frac{5}{12} + \frac{4}{15}(2)^{5/2} = \frac{-1 + 64\sqrt{2}}{60} \approx 1.492 \quad \blacksquare$$

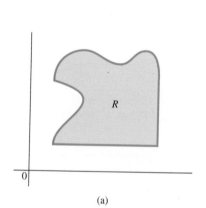

(a)

R is of type I or type II

REMARKS
- If, as in Example 15.5, a region can be treated either as type I or type II but treating it one way requires splitting an integral into two or more parts whereas treating it the other way does not, we usually choose to treat it as the type that does not require splitting up the integral.
- When a region is neither type I nor type II, it may be possible to divide it into two or more regions, each of which is either type I or type II. We show such a region in Figure 15.11. By dividing it as shown, we see that R_1 is of type II and R_2 is of type I. We first write

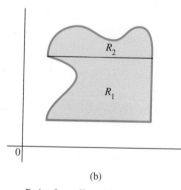

(b)

R_1 is of type II, and R_2 is of type I

FIGURE 15.11

$$\iint_R f(x, y)\, dA = \iint_{R_1} f(x, y)\, dA + \iint_{R_2} f(x, y)\, dA \qquad \text{By Property 3}$$

Then we apply Theorem 15.2 to each integral on the right.

The next three examples further illustrate the use of Theorem 15.2.

EXAMPLE 15.6   Evaluate the integral $\iint_R x^2 y\, dA$, where R is bounded by $x = 0$, $x = 1$, $y = \sqrt{x}$, and $y = 2$. Draw the region.

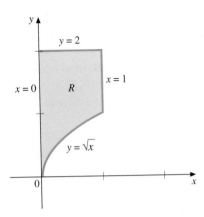

FIGURE 15.12

Solution   The region is pictured in Figure 15.12. It is a type I region (also type II, but it is better to treat it as type I). So by Theorem 15.2,

$$\iint_R x^2 y \, dA = \int_0^1 \int_{\sqrt{x}}^2 x^2 y \, dy \, dx$$

$$= \int_0^1 \left[\frac{x^2 y^2}{2} \right]_{\sqrt{x}}^2 dx$$

$$= \int_0^1 \left(2x^3 - \frac{x^3}{2} \right) dx = \frac{2x^3}{3} - \frac{x^4}{8} \Big]_0^1 = \frac{13}{24}$$ ∎

EXAMPLE 15.7   Find the volume under the paraboloid $z = x^2 + 3y^2$ and above the region bounded by the lines $x = 0$, $y = 0$, and $x + y = 1$.

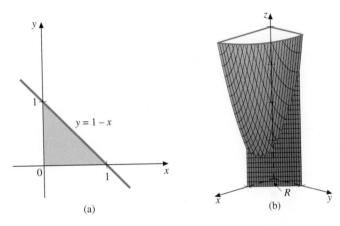

FIGURE 15.13

Solution   The region R in the xy-plane is shown in Figure 15.13(a) and the three-dimensional solid is shown in Figure 15.13(b). The region R is both a type I and a type II region. Treating it as type I and using Definition 15.2 and Theorem 15.2, we have

$$V = \iint_R (x^2 + 3y^2) \, dA = \int_0^1 \int_0^{1-x} (x^2 + 3y^2) \, dy \, dx$$

$$= \int_0^1 \left[x^2 y + y^3 \right]_0^{1-x} dx$$

$$= \int_0^1 \left[x^2 - x^3 + (1 - x)^3 \right] dx$$

$$= \frac{x^3}{3} - \frac{x^4}{4} - \frac{(1 - x)^4}{4} \Big]_0^1$$

$$= \left(\frac{1}{3} - \frac{1}{4} \right) + \frac{1}{4} = \frac{1}{3}$$ ∎

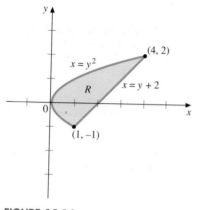

EXAMPLE 15.8   Using double integration, find the area of the region R bounded by the curves $y^2 = x$ and $x - y = 2$. Sketch the region.

Solution   The region is shown in Figure 15.14. Solving the two equations simultaneously, we find the points of intersection $(1, -1)$ and $(4, 2)$ as shown. (Verify.) The region is of type II. So we have from Equation 15.3 and Theorem 15.2,

$$A = \iint_R dA = \int_{-1}^{2} \int_{y^2}^{y+2} dx\,dy = \int_{-1}^{2} \left[x \right]_{y^2}^{y+2} dy = \int_{-1}^{2} (y + 2 - y^2)\,dy$$

$$= \left. \frac{y^2}{2} + 2y - \frac{y^3}{3} \right]_{-1}^{2} = \left(2 + 4 - \frac{8}{3} \right) - \left(\frac{1}{2} - 2 + \frac{1}{3} \right) = \frac{9}{2}$$

■

FIGURE 15.14

Interchange of Order of Integration

Sometimes an iterated integral that is difficult (or impossible) to evaluate can be evaluated by interchanging the order of integration. Such an interchange is possible when the region described by the limits is both type I and type II. We illustrate this procedure in the next example.

EXAMPLE 15.9   Evaluate the iterated integral

$$\int_0^1 \int_x^1 \sin y^2 \, dy \, dx$$

Solution   Since $\sin y^2$ has no elementary antiderivative, it seems we are stymied. But the given iterated integral, according to Theorem 15.2, equals the double integral $\iint_R \sin y^2 \, dA$, where R is the type I region $\{(x, y) : 0 \le x \le 1, x \le y \le 1\}$ pictured in Figure 15.15. But R can be viewed equally well as a type II region with left boundary $x = 0$ and right boundary $x = y$. Thus, the double integral, and hence the original iterated integral, equal

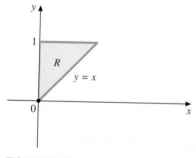

$$\int_0^1 \int_0^y \sin y^2 \, dx \, dy = \int_0^1 \left[x \sin y^2 \right]_0^y dy \qquad \text{Remember that } y \text{ is constant at this stage.}$$

$$= \int_0^1 y \sin y^2 \, dy$$

$$= \left. -\frac{1}{2} \cos y^2 \right]_0^1 = \frac{1}{2}(1 - \cos 1)$$

FIGURE 15.15

By treating this region R as type II, thereby changing both the order of integration and the limits on the integrals, we arrived at the integral of $y \sin y^2$. The presence of the factor y enabled us to make the (mental) substitution $u = y^2$. All that we needed for the differential $du = 2y\,dy$ was the constant factor 2, which we supplied, compensating on the outside with the factor $1/2$.   ■

Exercise Set 15.2

In Exercises 1–10, evaluate the iterated integral. Sketch the region R determined by the limits.

1. $\int_0^1 \int_1^2 (2x - 3y)\, dy\, dx$ **2.** $\int_0^2 \int_{-1}^1 (3x^2 - 2xy)\, dx\, dy$

3. $\int_1^2 \int_0^x xy\, dy\, dx$ **4.** $\int_0^1 \int_x^{4-x} (2x - 1)\, dy\, dx$

5. $\int_0^4 \int_{-2y}^{\sqrt{y}} y(2x - 1)\, dx\, dy$

6. $\int_{-2}^2 \int_{-y}^{\sqrt{2-y}} (x^3 + 2xy)\, dx\, dy$

7. $\int_0^{\sqrt{3}} \int_{2-\sqrt{4-x^2}}^{\sqrt{4-x^2}} x\, dy\, dx$ **8.** $\int_0^{\sqrt{5}} \int_{y^2-2}^{\sqrt{y^2+4}} y\, dx\, dy$

9. $\int_1^2 \int_{-y}^{4-y} \frac{y-4}{(x+4)^2}\, dx\, dy$

10. $\int_1^4 \int_{x-2}^{\sqrt{x}} (2xy + 3)\, dy\, dx$

In Exercises 11–21, evaluate the double integral of f over R, making use of Theorem 15.1 or 15.2. Sketch the region and identify its type.

11. $f(x, y) = 4x - 3y$; $R = \{(x, y) : 0 \le x \le 4, 0 \le y \le 2\}$

12. $f(x, y) = x^2 - y^2$; $R = \{(x, y) : 1 \le x \le 2, -1 \le y \le 3\}$

13. $f(x, y) = y/(1 + x^3)$; $R = \{(x, y) : 0 \le x \le 1, 0 \le y \le 2x\}$

14. $f(x, y) = x + y$; $R = \{(x, y) : y - 1 \le x \le \sqrt{1 - y^2}, 0 \le y \le 1\}$

15. $f(x, y) = 2x - y$; $R = \{(x, y) : y \le x \le \sqrt{4 - y^2}, 0 \le y \le \sqrt{2}\}$

16. $f(x, y) = xe^y$; $R = \{(x, y) : 0 \le x \le 1, x^2 \le y \le 2 - x^2\}$

17. $f(x, y) = \sqrt{(2x^2 + 7)/y}$; $R = \{(x, y) : 1 \le x \le 3, x^2/4 \le y \le x^2\}$

18. $f(x, y) = e^{x+2y}$; R is bounded by the triangle with vertices $(0, 0)$, $(1, 1)$, and $(3, 0)$.

19. $f(x, y) = 10xy^3$; R is bounded by the parallelogram with vertices $(-1, 0)$, $(0, 1)$, $(1, 0)$, and $(2, 1)$.

20. $f(x, y) = 4xy$; R is bounded by the triangle with vertices $(-1, 1)$, $(0, 2)$, and $(1, 0)$.

21. $f(x, y) = y^2$; R is bounded by the triangle with vertices $(0, 0)$, $(2, 2)$, and $(3, -1)$.

In Exercises 22–31, make use of Equation 15.3 and Theorem 15.2 to find the area of the region bounded by the given curves. Sketch the region and identify its type.

22. $y = x^2$, $y = 2 - x^2$ **23.** $y = x$, $y = 3x - x^2$

24. $y^2 = 4x$, $y = 2x - 4$ **25.** $y = \sqrt{x}$, $x = 0$, $y = 4$

26. $y = e^x$, $y = e^{-x}$, $x = \ln 3$

27. $y = 1 - x^2$, $y = \ln x$, $y = 1$

28. $y = \cos \pi x$, $4x^2 + 4y = 1$, $-\frac{1}{2} \le x \le \frac{1}{2}$

29. $y = \cos^{-1}\frac{x}{2}$, $y = \tan^{-1}\frac{x}{3}$, $x = 0$

30. $y = \sqrt{x}$, $x + y = 0$, $x - y = 2$

31. Below $y = 3$ and $y = 3(x + 1)$, and above $y = x^2 - 1$.

In Exercises 32–39, find the volume under the surface $z = f(x, y)$ that is above the region R bounded by the given curves.

32. $f(x, y) = 2 - x^2 - y^2$; $x = 0$, $x = 1$, $y = 0$, $y = 1$. Sketch the solid.

33. $f(x, y) = xy$; $y = \sqrt{8 - x^2}$, $y = x$, $y = 0$. Sketch the solid.

34. $f(x, y) = y\sqrt{1 + x^3}$; $y = x$, $x = 2$, $y = 0$

35. $f(x, y) = x + y$; $x = \sqrt{2 - y^2}$, $x = 0$, $x = y$. Sketch the solid.

36. $f(x, y) = x^2 e^{xy}$; $xy = 1$, $x = 2$, $y = 2$

37. $f(x, y) = 1 + x^2 + y^2$; $x = 0$, $y = x$, $y = 1$, $y = 2$

38. $f(x, y) = e^{-y}$; $x = 0$, $x = 4$, $y = 0$, $y = \ln 2$. Sketch the solid.

39. $f(x, y) = 4 - x^2$; $y = 0$, $x = 2$, $y = x$. Sketch the solid.

40. Find the volume of the tetrahedron formed by the planes $3x + 4y + 2z = 12$, $x = 0$, $y = 0$, and $z = 0$.

41. Find the volume of the tetrahedron with vertices $(0, 0, 0)$, $(2, 0, 0)$, $(0, 4, 0)$, and $(0, 0, 4)$.

42. Find the volume inside the paraboloid $z = 4 - x^2 - y^2$ and above the xy-plane.

In Exercises 43–46, evaluate the iterated integral by changing the order of integration.

43. $\displaystyle\int_0^1 \int_x^1 e^{-y^2}\, dy\, dx$

44. $\displaystyle\int_0^1 \int_{\sqrt{y}}^1 \sqrt{1 + x^3}\, dx\, dy$

45. $\displaystyle\int_0^2 \int_y^2 \frac{y}{(1 + x^3)^2}\, dx\, dy$

46. $\displaystyle\int_0^2 \int_{x^2}^4 \frac{1}{1 + y^{3/2}}\, dy\, dx$

In Exercises 47–52, give an equivalent integral with the order of integration reversed.

47. $\displaystyle\int_0^4 \int_{\sqrt{x}}^2 f(x, y)\, dy\, dx$

48. $\displaystyle\int_0^{\pi/2} \int_0^{\sin x} f(x, y)\, dy\, dx$

49. $\displaystyle\int_1^3 \int_{y+1}^4 f(x, y)\, dx\, dy$

50. $\displaystyle\int_0^4 \int_{y^2/4}^{2\sqrt{y}} f(x, y)\, dx\, dy$

51. $\displaystyle\int_0^1 \int_{1-\sqrt{1-x^2}}^{\sqrt{2x-x^2}} f(x, y)\, dy\, dx$

52. $\displaystyle\int_1^2 \int_0^{\ln y} f(x, y)\, dx\, dy$

53. Use an argument similar to that used in the text as a partial justification of Fubini's Theorem to show that if f is continuous and nonnegative on the rectangular region $R = \{(x, y) : a \le x \le b, \ c \le y \le d\}$, then

$$\iint\limits_R f(x, y)\, dA = \int_c^d \int_a^b f(x, y)\, dx\, dy$$

54. Show that if R is a type I region, then Equation 15.3 for the area of R is consistent with Equation 6.1 for the area between the graphs of two functions.

In Exercises 55–57, evaluate the double integral.

55. $\iint_R x^2 e^y dA$; $R = \{(x, y) : 0 \le x \le 1, \ -x \le y \le x\}$

56. $\iint_R (4xy + 4)\, dA$; R is bounded by $y = 0$ and $y = \sin x$ between $x = 0$ and $x = \pi$.

57. $\iint_R \sqrt{y}\,(x^2 + 1)\, dA$; R is the first-quadrant region bounded by $x^2 y = 4$, $x = 0$, $x = 2$, $y = 0$, and $y = 4$.

In Exercises 58–60, find the area of the region bounded by the given curves, using double integration.

58. Below $y = \sqrt{2x}$ and $x + 2y = 6$, and above $x - 4y = 0$

59. $y = x^2$, $y = x^3 - 2x$

60. $y^2 = x$, $y^2 = 8x$, $x + y = 6$, $5x + y = 48$ (first quadrant)

In Exercises 61–64, find the volume of the solid.

61. Under the surface $z = 2x/y^2$ and above the region bounded by $xy = 6$ and $x^2 + y = 7$

62. Bounded by the surface $z = (1 - \sqrt{x} - \sqrt{y})^2$ and the coordinate planes

63. Bounded by the surface $z = (a^{2/3} - x^{2/3} - y^{2/3})^{3/2}$ and the xy-plane (*Hint:* To evaluate the inner integral use a trigonometric substitution.)

64. Common to the two cylinders $x^2 + y^2 = a^2$ and $x^2 + z^2 = a^2$

[CAS] *Use a CAS for Exercises 65 through 69.*

65. Find the volume of the region under the paraboloid $f(x, y) = x^2 + 2y^2 + 3$ and above the square in the xy-plane $\{(x, y) : -1 \le x \le 1, -1 \le y \le 1\}$. Plot the solid in question.

66. Find the volume of the region under the hyperbolic paraboloid $f(x, y) = y^2 - x^2 + 2$ and above the circle in the xy-plane $x^2 + y^2 = 1$. Plot the solid in question.

67. Find the volume of the region under the plane $f(x, y) = 2 - 2(x - 1) + y$ and above the region in the xy-plane bounded by the curves $y = (x - 2)^2 + 1$ and $y = -(x - 2)^2 + 3$. Plot the solid in question.

68. Find the volume of the region under the graph of $f(x, y) = 4 - x^2 - y^2$ and above the graph of $x^2 + 3y^2 - 2$. Plot the region in question.

69. Find the area of the ellipse $\dfrac{x^2}{a^2} + \dfrac{y^2}{b^2} = 1$.

15.3 DOUBLE INTEGRALS IN POLAR COORDINATES

It is frequently more convenient to describe regions in the xy-plane using polar coordinates rather than rectangular coordinates. For example, the region R shown in Figure 15.16 can be expressed in polar coordinates by

$$R = \{(r, \theta) : 0 \leq r \leq 2, \ 0 \leq \theta \leq \pi/4\}$$

Expressing R in rectangular coordinates is more difficult, as you can see by trying it.

The integrand also can often be simplified by changing to polar coordinates. For example, if the integrand were $\sin\sqrt{x^2 + y^2}$, it would become $\sin r$ in polar coordinates.

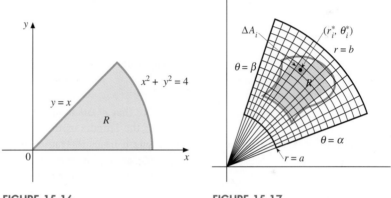

FIGURE 15.16

FIGURE 15.17

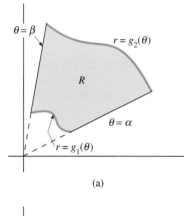

(a)

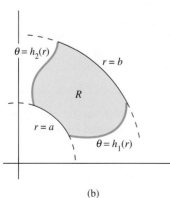

(b)

FIGURE 15.18

Suppose, then, that we want to integrate a function f of the polar variables r and θ over a bounded region R that is given in polar coordinates. We need to define the double integral in this case. We follow a procedure similar to that used in Section 15.1. First, we enclose R in a *polar rectangle* of the form $\{(r, \theta) : a \leq r \leq b, \ \alpha \leq \theta \leq \beta\}$, as shown in Figure 15.17. Then we partition this polar rectangle by rays and circular arcs as shown, thus forming finitely many polar subrectangles. We denote the partition by P. The norm of P, again denoted by $\|P\|$, is the length of the longest diagonal of the polar subrectangles formed by P (just as with a rectangle, the diagonal of a polar rectangle is a line segment joining opposite vertices). We number the polar subrectangles that lie entirely in R (shaded in Figure 15.17) from 1 to n. We let ΔA_i denote the area of the ith subrectangle and (r_i^*, θ_i^*) be polar coordinates of an arbitrary point in the ith polar subrectangle. Then we define the double integral of f over R by

$$\iint\limits_R f(r, \theta)\, dA = \lim_{\|P\| \to 0} \sum_{i=1}^{n} f(r_i^*, \theta_i^*)\, \Delta A_i \qquad (15.6)$$

provided this limit exists, independently of the choices of the points (r_i^*, θ_i^*). Again, it can be shown that the integral will exist when f is continuous on R.

To evaluate the double integral of a continuous function by iterated integrals when R is given in polar coordinates, we consider two types of regions only, analogous to type I and type II for rectangular coordinates. Figure 15.18 illustrates these two types. In part (a) of Figure 15.18, R is of the form $\{(r, \theta) : g_1(\theta) \leq r \leq g_2(\theta), \alpha \leq \theta \leq \beta\}$, where g_1 and g_2 are continuous functions on

$[\alpha, \beta]$. In part (b), R is of the form $\{(r, \theta) : a \leq r \leq b, \ h_1(r) \leq \theta \leq h_2(r)\}$, where h_1 and h_2 are continuous functions on $[a, b]$. Because the first type is the more common, we will concentrate our attention on it.

To help you understand the form of the iterated integral in this case, we consider a typical polar subrectangle formed by a partition P. We have enlarged such a polar subrectangle in Figure 15.19. To simplify our notation we have deleted all subscripts. As shown, let $\Delta\theta$ be the angle between the two rays that form the polar subrectangle, and let Δr be the change in radial distance between the two bounding arcs. In the definition of the double integral, we are free to choose an arbitrary point in each subinterval at which the function is evaluated. We now specify this point (r^*, θ^*) as the center of the subinterval. Then the inner radius is $r^* - \Delta r/2$ and the outer radius is $r^* + \Delta r/2$. The area ΔA is the difference in areas of two circular sectors. Since the area of a circular sector is one-half its angle (in radians) times its radius, we have

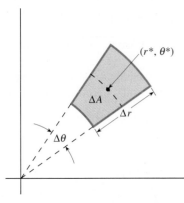

$$\Delta A = \frac{1}{2}\left(r^* + \frac{\Delta r}{2}\right)^2 \Delta\theta - \frac{1}{2}\left(r^* - \frac{\Delta r}{2}\right)^2 \Delta\theta$$

Squaring and collecting terms, we get

$$\Delta A = r^* \, \Delta r \, \Delta\theta \tag{15.7}$$

You might note that $r^*\Delta\theta$ is the length of the arc through the center, which is the average of the lengths of the inner and outer bounding arcs. So ΔA is the average arc length times the radial distance Δr. Notice the similarity with the formula for the area of a trapezoid. The Riemann sum in Equation 15.6 can now be written

$$\sum_{i=1}^{n} f(r_i^*, \theta_i^*) \, \Delta A_i = \sum_{i=1}^{n} f(r_i^*, \theta_i^*) r_i^* (\Delta r \, \Delta\theta)_i$$

where by $(\Delta r \, \Delta\theta)_i$ we mean the product of Δr and $\Delta\theta$ for the ith polar subrectangle.

By taking partitions with norms approaching 0, we are led to the following result, analogous to the stronger form of Fubini's Theorem.

> If f is continuous on the polar region
> $$R = \{(r, \theta) : g_1(\theta) \leq r \leq g_2(\theta), \ \alpha \leq \theta \leq \beta\}$$
> then
> $$\iint\limits_{R} f(r, \theta) \, dA = \int_{\alpha}^{\beta} \int_{g_1(\theta)}^{g_2(\theta)} f(r, \theta) r \, dr \, d\theta \tag{15.8}$$

Observe the extra factor r in the integrand of the iterated integral on the right. This discussion is intended to suggest to you where it comes from, but it is not a proof. (In Section 15.8, we will see that this factor is a special case of a more general result having to do with changing variables in double integrals.) It is useful to write

$$dA = r \, dr \, d\theta$$

and to call this form the *differential of area in polar coordinates*. For rectangular coordinates, the analogous formula is $dA = dy\,dx$. So in going from a double integral in polar coordinates to an iterated integral, dA is replaced by $r\,dr\,d\theta$, whereas in rectangular coordinates dA is replaced by $dy\,dx$. (In each case the order of the differentials may be reversed, depending on the order of integration.)

When the region is of the type in Figure 15.18(b), we have the following result, analogous to that given in Equation 15.8.

If f is continuous on the polar region

$$R = \{(r, \theta) : a \le r \le b, \; h_1(r) \le \theta \le h_2(r)\}$$

then

$$\iint\limits_R f(r, \theta)\, dA = \int_a^b \int_{h_1(r)}^{h_2(r)} f(r, \theta)r\, d\theta\, dr \qquad (15.9)$$

EXAMPLE 15.10   Evaluate the double integral $\iint_R r^2 \sin\theta\, dA$, where R is the region bounded by the polar axis and the upper half of the cardioid $r = 1 + \cos\theta$.

Solution   The region R is pictured in Figure 15.20. We have shown a typical small sector of the type used in a partition along with a typical "differential element" that we imagine to have area $dA = r\,dr\,d\theta$. This shorthand technique is an aid in setting up the integral. The upper half of the cardioid is traced out as θ varies from 0 to π. We can therefore describe the region R by $\{(r, \theta) : 0 \le r \le 1 + \cos\theta, \; 0 \le \theta \le \pi\}$. From Equation 15.8, we therefore have

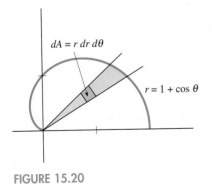

$dA = r\,dr\,d\theta$

$r = 1 + \cos\theta$

FIGURE 15.20

$$\iint\limits_R r^2 \sin\theta\, dA = \int_0^\pi \int_0^{1+\cos\theta} (r^2 \sin\theta)\, \overbrace{r\,dr\,d\theta}^{dA}$$

$$= \int_0^\pi \left[\int_0^{1+\cos\theta} r^3 \sin\theta\, dr \right] d\theta$$

$$= \frac{1}{4} \int_0^\pi \left[r^4 \right]_0^{1+\cos\theta} \sin\theta\, d\theta$$

$$= \frac{1}{4} \int_0^\pi (1 + \cos\theta)^4 \sin\theta\, d\theta \qquad \text{Let } u = 1 + \cos\theta.$$

$$= -\frac{1}{4} \left[\frac{(1 + \cos\theta)^5}{5} \right]_0^\pi = \frac{8}{5} \qquad\blacksquare$$

When f is continuous and nonnegative on R, $\iint_R f(r, \theta)\, dA$ again represents the volume above R and under the surface $z = f(r, \theta)$, and in the special case $f(r, \theta) = 1$ for all points (r, θ) in R, as before, the resulting volume is numerically the same as the area of R; that is, $A = \iint_R dA$. The next three examples illustrate these results.

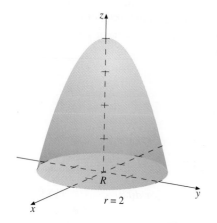

FIGURE 15.21

EXAMPLE 15.11   Use polar coordinates to find the volume below the paraboloid $z = 4 - x^2 - y^2$ and above the xy-plane.

Solution   In polar coordinates, $x^2 + y^2 = r^2$, so we can write the equation of the paraboloid as $z = 4 - r^2$. Thus, we take $f(r, \theta) = 4 - r^2$. We find the boundary of the region R by setting $z = 0$, giving $r^2 = 4$ or $r = 2$ (see Figure 15.21). So we have

$$V = \iint_R (4 - r^2)\, dA = \int_0^{2\pi} \int_0^2 (4 - r^2) r\, dr\, d\theta$$

$$= \int_0^{2\pi} \left[2r^2 - \frac{r^4}{4} \right]_0^2 d\theta = 4 \int_0^{2\pi} d\theta = 8\pi$$ ∎

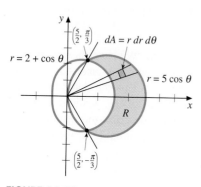

FIGURE 15.22

EXAMPLE 15.12   Find the area inside the circle $r = 5\cos\theta$ and outside the limaçon $r = 2 + \cos\theta$.

Solution   We show the curves in Figure 15.22. To find the points of intersection we solve the equations simultaneously:

$$5\cos\theta = 2 + \cos\theta$$

$$\cos\theta = \frac{1}{2}$$

$$\theta = \pm\frac{\pi}{3}, \qquad r = \frac{5}{2}$$

By symmetry, the total area is twice that for $0 \le \theta \le \pi/3$. By Equation 15.8,

$$A = \iint_R dA = 2 \int_0^{\pi/3} \int_{2+\cos\theta}^{5\cos\theta} r\, dr\, d\theta = 2 \int_0^{\pi/3} \left[\frac{r^2}{2} \right]_{2+\cos\theta}^{5\cos\theta} d\theta$$

$$= \int_0^{\pi/3} [25\cos^2\theta - (4 + 4\cos\theta + \cos^2\theta)]\, d\theta$$

$$= 4 \int_0^{\pi/3} (6\cos^2\theta - \cos\theta - 1]\, d\theta$$

$$= 4 \int_0^{\pi/3} [3(1 + \cos 2\theta) - \cos\theta - 1]\, d\theta$$

$$= 4 \left[2\theta + \frac{3\sin 2\theta}{2} - \sin\theta \right]_0^{\pi/3} = \frac{8\pi}{3} + \sqrt{3}$$ ∎

EXAMPLE 15.13   Find the area outside the circle $r = \sqrt{2}$ and inside the lemniscate $r^2 = 4\cos 2\theta$.

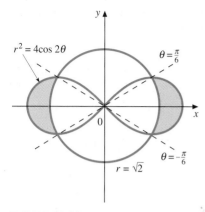

$r^2 = 4\cos 2\theta$

$\theta = \frac{\pi}{6}$

$r = \sqrt{2}$

$\theta = -\frac{\pi}{6}$

FIGURE 15.23

Solution   We show the two curves and the area in question in Figure 15.23. To find the points of intersection, we again solve the two equations simultaneously.

$$4\cos 2\theta = (\sqrt{2})^2$$

$$\cos 2\theta = \frac{1}{2}$$

$$2\theta = \pm\frac{\pi}{3}$$

$$\theta = \pm\frac{\pi}{6}$$

By symmetry, we can find the area of the first-quadrant region given by

$$\left\{(r, \theta) : \sqrt{2} \le r \le 2\sqrt{\cos 2\theta}, \ 0 \le \theta \le \frac{\pi}{6}\right\}$$

and multiply the result by 4. By Equation 15.8, then, we have

$$A = \iint\limits_{R} dA = 4\int_{0}^{\pi/6} \int_{\sqrt{2}}^{2\sqrt{\cos 2\theta}} r\, dr\, d\theta$$

$$= 4\int_{0}^{\pi/6} \left[\frac{r^2}{2}\right]_{\sqrt{2}}^{2\sqrt{\cos 2\theta}} d\theta$$

$$= 4\int_{0}^{\pi/6} (2\cos 2\theta - 1)\, d\theta$$

$$= 4\left[\sin 2\theta - \theta\right]_{0}^{\pi/6} = 2\left(\sqrt{3} - \frac{\pi}{3}\right) \qquad \blacksquare$$

Changing from Rectangular to Polar Coordinates

In the next example, we illustrate how an iterated integral in rectangular coordinates can sometimes be evaluated more easily by changing to polar coordinates.

EXAMPLE 15.14   Evaluate the integral

$$\int_{0}^{2} \int_{0}^{\sqrt{2x-x^2}} \sqrt{x^2 + y^2}\, dy\, dx$$

by changing to polar coordinates.

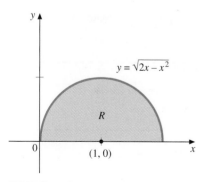

$y = \sqrt{2x - x^2}$

R

$(1, 0)$

FIGURE 15.24

Solution   First, we find the region R determined by the limits of integration. The limits tell us that R is of type I with lower bounding curve $y = 0$ and upper bounding curve $y = \sqrt{2x - x^2}$. Squaring and collecting terms, we find that the upper boundary is that part of the circle $x^2 + y^2 - 2x = 0$ for which $y \ge 0$. So the region R is that shown in Figure 15.24. In polar coordinates, the circle $x^2 + y^2 - 2x = 0$ becomes $r = 2\cos\theta$. (Verify.) So in polar coordinates the region of integration is

$$\left\{(r, \theta) : 0 \le r \le 2\cos\theta, \ 0 \le \theta \le \frac{\pi}{2}\right\}$$

The integrand, $\sqrt{x^2 + y^2}$, changes to r in polar coordinates, and so the double integral and hence the original iterated integral are both equal to

$$\int_0^{\pi/2} \int_0^{2\cos\theta} r(r\,dr\,d\theta) = \int_0^{\pi/2} \left[\frac{r^3}{3}\right]_0^{2\cos\theta} d\theta$$

$$= \frac{8}{3} \int_0^{\pi/2} \cos^3\theta\,d\theta = \frac{8}{3} \int_0^{\pi/2} (1 - \sin^2\theta)\cos\theta\,d\theta$$

$$= \frac{8}{3} \left[\sin\theta - \frac{\sin^3\theta}{3}\right]_0^{\pi/2} = \frac{8}{3}\left[1 - \frac{1}{3}\right] = \frac{16}{9} \qquad \blacksquare$$

The procedure suggested by this example for changing an iterated integral in rectangular coordinates to one in polar coordinates can be summarized as follows:

Procedure for Changing from Rectangular to Polar Coordinates

1. Sketch the region R determined by the limits of integration, and write the equations of the curves that form the boundary of R in polar coordinates.

2. Determine the new limits of integration as in Equation 15.8 or 15.9, depending on whether R is of the type shown in Figure 15.18(a) or (b). (In some cases it may be necessary to express R as the union of two or more such regions.)

3. In the integrand, replace x by $r\cos\theta$ and y by $r\sin\theta$. (Note that $x^2 + y^2$ can be changed immediately to r^2.)

4. Replace $dy\,dx$ or $dx\,dy$, whichever occurs, by $r\,dr\,d\theta$ or by $r\,d\theta\,dr$, depending on the order of integration to be used.

Don't forget the r in step 4. Remember that in rectangular coordinates the differential of area is $dy\,dx$ (or $dx\,dy$), but in polar coordinates it is $r\,dr\,d\theta$ (or $r\,d\theta\,dr$).

If we denote by S the region R described in polar coordinates, then steps 2, 3, and 4 of the procedure can be given by the equation

---

$$\iint_R f(x, y)\,dA = \iint_S f(r\cos\theta, r\sin\theta)\,r\,dr\,d\theta$$

Guidelines for Changing to Polar Coordinates

Changing from rectangular to polar coordinates is desirable if carrying out the integration in rectangular coordinates is more difficult than doing so in polar coordinates. If the curves that make up the boundary of R have simpler polar equations than rectangular equations, a change is indicated. In this connection, be on the lookout especially for bounding curves that are circles centered at the origin ($r = a$), circles centered on an axis and passing through the origin ($r = a\cos\theta$ or $r = a\sin\theta$), and lines through the origin ($\theta = \alpha$). The nature of the integrand also is a factor to be considered. If the original integrand is $f(x, y)$, we see from applying steps 3 and 4 that the new integrand is $rf(r\cos\theta, r\sin\theta)$. Whether the antiderivatives needed in the iterated integration can be found clearly affects the decision. As you practice problems such as Exercises 20–31, you should begin to develop a feel for when a change is worthwhile.

REMARK ─────────────────────────────────────

■ We investigate changing variables in double integrals more fully in Section 15.8.

Exercise Set 15.3

In Exercises 1–8, evaluate the double integral $\iint_R f(r, \theta)\, dA$, where R is the region described. Draw the region.

1. $f(r, \theta) = r\theta$; $|r| \le 2, 0 \le \theta \le 2\pi$

2. $f(r, \theta) = r(\sin\theta - \cos\theta)$; $\{(r, \theta) : 0 \le r \le 1, \pi/4 \le \theta \le \pi\}$

3. $f(r, \theta) = 2r + 1$; $1 \le r \le 4, 0 \le \theta \le 2\pi$

4. $f(r, \theta) = \sqrt{r}\sin\theta$; $\{(r, \theta) : 0 \le r \le 1 - \cos\theta, 0 \le \theta \le \pi\}$

5. $f(r, \theta) = r^2\cos\theta$; $\{(r, \theta) : 0 \le r \le 2\sin\theta, 0 \le \theta \le \pi/2\}$

6. $f(r, \theta) = r(\sin(\theta/2) - \cos(\theta/2))$; $1 \le r \le 2, 0 \le \theta \le 2\pi$

7. $f(r, \theta) = r^2\sin^2 2\theta$; $\{(r, \theta) : 0 \le r \le 2\sqrt{\cos 2\theta}, 0 \le \theta \le \pi/4\}$

8. $f(r, \theta) = \sqrt{1 + \theta^3}$; $\{(r, \theta) : 0 \le r \le \theta, 0 \le \theta \le 2\}$

In Exercises 9–19, use double integration and polar coordinates to find the area of the region described. Draw the region.

9. Inside $r = 2 + 2\cos\theta$

10. Inside $r^2 = 4\cos 2\theta$

11. Inside $r = 2\cos 2\theta$

12. Inside $r = 2 + \sin\theta$

13. Inside $r = 2(1 + \cos\theta)$ and outside $r = 1$

14. Inside $r = 8\sin\theta$ and outside $r = 3 + 2\sin\theta$

15. Inside $r^2 = 2\cos 2\theta$ and outside $r = 1$

16. Inside $r = 1 + 2\cos\theta$ and to the right of $r\cos\theta = 1$

17. Inside both $r = 2$ and $r = 3 - 2\cos\theta$

18. Inside the small loop of the limaçon $r = 1 + 2\cos\theta$

19. Inside both $r = 6\sin\theta$ and $r = 2(1 + \sin\theta)$

In Exercises 20–31, evaluate the iterated integral by changing to polar coordinates.

20. $\displaystyle\int_0^2 \int_0^{\sqrt{4-x^2}} x^2 y\, dy\, dx$

21. $\displaystyle\int_0^1 \int_x^{\sqrt{2-x^2}} (x^2 + y^2)^2\, dy\, dx$

22. $\displaystyle\int_{-1}^1 \int_{-\sqrt{1-x^2}}^{\sqrt{1-x^2}} (x^2 + y^2)^{3/2}\, dy\, dx$

23. $\displaystyle\int_0^2 \int_{-\sqrt{8-y^2}}^{-y} (x + y)\, dx\, dy$

24. $\displaystyle\int_0^4 \int_0^{\sqrt{4y-y^2}} y\,dx\,dy$ **25.** $\displaystyle\int_0^1 \int_x^{\sqrt{2x-x^2}} x\,dy\,dx$

26. $\displaystyle\int_{-\sqrt{3}}^{\sqrt{3}} \int_1^{\sqrt{4-y^2}} \frac{1}{(x^2+y^2)^{3/2}}\,dx\,dy$

27. $\displaystyle\int_0^{\sqrt{3}} \int_1^{\sqrt{4-x^2}} \frac{x}{y}\,dy\,dx$ **28.** $\displaystyle\int_1^{\sqrt{3}} \int_1^x \frac{x^2-y^2}{x^2+y^2}\,dy\,dx$

29. $\displaystyle\int_0^1 \int_y^{1+\sqrt{1-y^2}} \sqrt{4-x^2-y^2}\,dx\,dy$

30. $\displaystyle\int_{-2}^2 \int_2^{2+\sqrt{4-x^2}} \frac{x+y}{y}\,dy\,dx$

31. $\displaystyle\int_{-a}^a \int_{-\sqrt{a^2-y^2}}^{\sqrt{a^2-y^2}} e^{-(x^2+y^2)}\,dx\,dy$

In Exercises 32–39, use polar coordinates to find the indicated volume.

32. Inside the ellipsoid $x^2+y^2+4z^2=4$ and above the xy-plane

33. Inside the cone $z = 2 - \sqrt{x^2+y^2}$ and above the xy-plane

34. A sphere of radius a

35. Inside the sphere $x^2+y^2+z^2=a^2$ and outside the cylinder $x^2+y^2=b^2$, where $0<b<a$

36. Inside the cylinder $x^2+y^2=a^2$ between the upper and lower sheets of the hyperboloid $z^2-x^2-y^2=a^2$

37. Under the cone $z = \sqrt{x^2+y^2}$ and above the region R in the xy-plane inside the circle $x^2+y^2=2x$

38. Under the cylindrical surface $z = y^2$ and above the region $R = \{(x,y) : 0 \le x \le 1,\ x \le y \le \sqrt{2-x^2}\}$

39. Between the surfaces $z = 6 - x^2 - y^2$ and $z = 2$. (*Hint:* Use the difference of two volumes.)

40. Evaluate by changing to polar coordinates:

$$\int_{2-2\sqrt{2}}^{2+2\sqrt{2}} \int_{-y}^{(4-y^2)/4} (x^2 y + y^3)\,dx\,dy$$

41. Write as a single iterated integral in polar coordinates and evaluate:

$$\int_0^{3/2} \int_{\sqrt{2y-y^2}}^{\sqrt{4y-y^2}} xy\,dx\,dy + \int_{3/2}^3 \int_{y/\sqrt{3}}^{\sqrt{4y-y^2}} xy\,dx\,dy$$

42. Find the area bounded by the curves $r = 1$, $r = 2$, $r = 2\ln\theta$, and $r\theta = 1$.

43. Find the area between the y-axis and the parabola $y^2 = 9 - 6x$ that lies outside the circle $x^2 + y^2 - 4x = 0$.

44. **The Normal Probability Density Function**   Carry out the following steps to verify that the area under the normal probability density function is 1:
(a) Let $I = \int_0^\infty e^{-x^2}dx$. Write I^2 as follows:

$$I^2 = \left(\int_0^\infty e^{-x^2}dx\right)\left(\int_0^\infty e^{-y^2}dy\right)$$

$$= \int_0^\infty \int_0^\infty e^{-(x^2+y^2)}dy\,dx$$

(b) Change the iterated integral in part (a) to polar coordinates and evaluate it. Find the value of I.
(c) The normal probability density function with mean μ and standard deviation σ is

$$f(x) = \frac{1}{\sqrt{2\pi}\,\sigma} e^{-(x-\mu)^2/2\sigma^2}, \quad -\infty < x < \infty$$

Use the result of part (b) to show that $\int_{-\infty}^\infty f(x)\,dx = 1$. (*Hint:* Make a substitution, and use symmetry.)

15.4 MOMENTS, CENTROIDS, AND CENTERS OF MASS USING DOUBLE INTEGRATION

In Chapter 6 we saw how to find the center of mass of a homogeneous lamina (the density is constant) using single integrals. With the aid of double integration we can remove the restriction that the lamina be homogeneous. Suppose the lamina occupies a region R in the xy-plane, and its density at the point (x, y) is given by $\rho(x, y)$, where ρ is a continuous function on R. To explain what is meant by the density at the point (x, y), suppose that ΔA is the area of a small

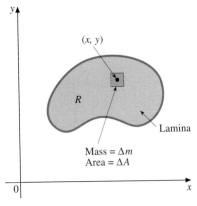

FIGURE 15.25

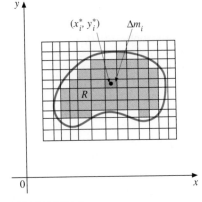

FIGURE 15.26

rectangular element of the lamina, centered at (x, y), as in Figure 15.25. If the mass of this element is Δm, then its average density is $\Delta m / \Delta A$. We define the density $\rho(x, y)$ at the point (x, y) by

$$\rho(x, y) = \lim_{d \to 0} \frac{\Delta m}{\Delta A}$$

where d is the length of the diagonal of the rectangular element.

Now we proceed exactly as in defining the double integral over R, enclosing R in some rectangle and partitioning this rectangle with horizontal and vertical lines, as in Figure 15.26. We number the subrectangles entirely contained within R. Denote the center of the ith subrectangle by (x_i^*, y_i^*). If the norm $\|P\|$ of the partition is small (so that all subrectangles are small), we would expect that the density throughout that part of the lamina in the ith subrectangle would differ very little from $\rho(x_i^*, y_i^*)$. Thus, its mass Δm_i should be approximately $\rho(x_i^*, y_i^*) \Delta A_i$.

We can think of each little rectangular piece of the lamina formed by the partition as if it were a "point mass" concentrated at its center. The total mass m of the lamina is approximately the sum of these point masses:

$$m \approx \sum_{i=1}^{n} \Delta m_i = \sum_{i=1}^{n} \rho(x_i^*, y_i^*) \Delta A_i$$

This last sum is a Riemann sum for the density function ρ over the region R. Thus, as we let $\|P\| \to 0$, we arrive at the following result.

Total Mass of a Lamina of Continuous Density $\rho(x, y)$, Occupying the Region R

$$m = \iint\limits_{R} \rho(x, y)\, dA \qquad\qquad (15.10)$$

Similarly, the moments M_x and M_y about the y-axis and x-axis, respectively, are approximated by

$$M_x \approx \sum_{i=1}^{n} y_i^* \Delta m_i = \sum_{i=1}^{n} y_i^* \rho(x_i^*, y_i^*) \Delta A_i$$

and

$$M_y = \sum_{i=1}^{n} x_i^* \Delta m_i = \sum_{i=1}^{n} x_i^* \rho(x_i^*, y_i^*) \Delta A_i$$

As we take partitions with norm approaching 0 we arrive at the following formulas.

The Moments M_x and M_y About the x- and y-Axes

$$M_x = \iint\limits_R y\rho(x, y)\, dA \qquad (15.11)$$

$$M_y = \iint\limits_R x\rho(x, y)\, dA \qquad (15.12)$$

Recall from Chapter 6 that the center of mass $(\bar{x}, \bar{y})$ satisfies

$$\bar{x} = \frac{M_y}{m} \quad \text{and} \quad \bar{y} = \frac{M_x}{m} \qquad (15.13)$$

If we write $dm = \rho(x, y)dA$, we can combine Equations 15.10, 15.11, 15.12, and 15.13 to get the following formulas.

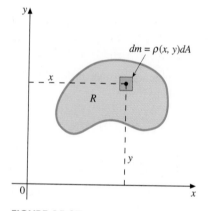

FIGURE 15.27

The Center of Mass

$$\bar{x} = \frac{\displaystyle\iint\limits_R x\, dm}{\displaystyle\iint\limits_R dm} \quad \text{and} \quad \bar{y} = \frac{\displaystyle\iint\limits_R y\, dm}{\displaystyle\iint\limits_R dm} \qquad (15.14)$$

where $dm = \rho(x, y)dA$.

REMARK ─────────────

■ To simplify the process of arriving at Equations 15.14, we can use the shorthand technique of designating a typical element of the lamina by $dm = \rho(x, y)\, dA$, as shown in Figure 15.27. Treating this element as a point mass, we get the moments $x\, dm$ and $y\, dm$ about the y-axis and x-axis, respectively. When we "sum" these moments, in the sense of integration, and divide by the "sum" of the masses, we arrive at Equation 15.14.

EXAMPLE 15.15   A lamina in the shape of the triangular region in Figure 15.28 has density $\rho(x, y) = \sqrt{xy}$. Find the center of mass.

Solution   First, let us calculate the mass m of the lamina. By Equation 15.10,

$$m = \iint\limits_R dm = \iint\limits_R \rho(x, y)dA = \int_0^2 \int_0^x \sqrt{xy}\, dy\, dx = \frac{16}{9}$$

(The integration here is straightforward, and we have omitted the details.)

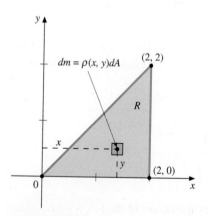

FIGURE 15.28

Next, we calculate the numerators for $\bar{x}$ and $\bar{y}$, which are the moments M_y and M_x about the y-axis and x-axis, respectively. By Equations 15.11 and 15.12,

$$M_y = \iint_R x \, dm = \iint_R x\rho(x, y) \, dA = \int_0^2 \int_0^x x\sqrt{xy} \, dy \, dx = \frac{8}{3}$$

$$M_x = \iint_R y \, dm = \iint_R y\rho(x, y) \, dA = \int_0^2 \int_0^x y\sqrt{xy} \, dy \, dx = \frac{8}{5}$$

(again, we omit the details of the integration). Thus, by Equation 15.13,

$$\bar{x} = \frac{M_y}{m} = \frac{\frac{8}{3}}{\frac{16}{9}} = \frac{3}{2} \quad \text{and} \quad \bar{y} = \frac{M_x}{m} = \frac{\frac{8}{5}}{\frac{16}{9}} = \frac{9}{10}$$

So the center of mass is the point $(\frac{3}{2}, \frac{9}{10})$. Notice that because of the variable density, the center of mass does not coincide with the centroid of the region, which is two-thirds of the horizontal distance from $(0, 0)$ to $(0, 2)$ and one-third of the vertical distance from $(2, 0)$ to $(2, 2)$—namely, $(\frac{4}{3}, \frac{2}{3})$—as we saw in Section 6.8.   ■

If the density is constant, then the center of mass coincides with the centroid of the region R. In this case, the constant density, say $\rho(x, y) = k$, can be taken outside the integrals for $\bar{x}$ and $\bar{y}$ in Equations 15.14 and then divided out. We thus have the following equations for the coordinates of the centroid.

The Centroid of a Region R in $\mathbb{R}^2$

$$\bar{x} = \frac{\displaystyle\iint_R x \, dA}{\displaystyle\iint_R dA} \qquad \bar{y} = \frac{\displaystyle\iint_R y \, dA}{\displaystyle\iint_R dA} \qquad (15.15)$$

EXAMPLE 15.16   Find the centroid of the region bounded by the parabola $y = 4 - x^2$ and the x-axis.

Solution   We show the region in Figure 15.29. By symmetry, the centroid lies on the y-axis, so $\bar{x} = 0$. (Note that if we were working with a nonhomogeneous lamina, this conclusion would hold only if the mass were symmetrically distributed with respect to the y-axis.) From Equations 15.15, we have

$$\bar{y} = \frac{\displaystyle\iint_R y \, dA}{\displaystyle\iint_R dA} = \frac{\displaystyle\int_{-2}^2 \int_0^{4-x^2} y \, dy \, dx}{\displaystyle\int_{-2}^2 \int_0^{4-x^2} dy \, dx} = \frac{8}{5}$$

(You should verify this result by supplying the details of the integration.) Thus, the centroid is at the point $(0, \frac{8}{5})$.   ■

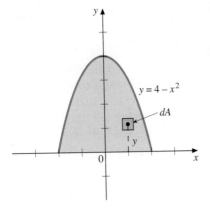

$y = 4 - x^2$

dA

FIGURE 15.29

EXAMPLE 15.17   A homogeneous lamina occupies the region outside the circle $x^2 + y^2 = 1$ and inside the circle $x^2 + y^2 - 2x = 0$. Find its center of mass.

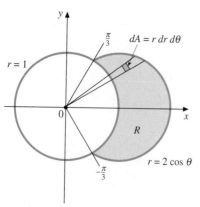

Solution   The region R in this case is more easily described in polar coordinates (see Figure 15.30). The bounding curves are $r = 1$ and $r = 2\cos\theta$, as you can verify. These intersect at $(1, \pm\frac{\pi}{3})$. By symmetry, $\bar{y} = 0$. For $\bar{x}$, Equations 15.15 give

$$\bar{x} = \frac{\iint\limits_R x \, dA}{\iint\limits_R dA} = \frac{\int_{-\pi/3}^{\pi/3} \int_1^{2\cos\theta} (r\cos\theta)(r \, dr \, d\theta)}{\int_{-\pi/3}^{\pi/3} \int_1^{2\cos\theta} r \, dr \, d\theta} = \frac{8\pi + 3\sqrt{3}}{4\pi + 6\sqrt{3}} \approx 1.321$$

Again, you should supply the missing steps.   ∎

FIGURE 15.30

Higher-Order Moments; Moment of Inertia

FIGURE 15.31

Consider again a system of point masses in the xy-plane. Let m be such a point mass and l a line in the plane (see Figure 15.31). Then, as we have seen, the moment of m about l is the product md, where d is the distance from l to m. To distinguish such a moment from those we are about to define, we sometimes refer to md as the **first moment**. Higher-order moments of m about l are defined by

$$md^2, \text{ second moment}$$
$$md^3, \text{ third moment}$$
$$\vdots$$
$$md^k, k\text{th moment}$$

We concentrate on the second moment because it is especially important in physics and engineering. Moments of orders 3 and 4 have applications in fields such as structural design, but the primary use of higher-order moments is in probability theory. We will say more about this use later. In a physical system, the second moment is also called the **moment of inertia**, and the distance d is called the **radius of gyration**. This terminology suggests a rotation, and we will explain shortly the sense in which the moment of inertia is related to a rotating mass.

For a lamina of the type we have been considering, with density function ρ and occupying a region R, we can use the same partition process as before to define the moments of inertia about the x-axis and y-axis, respectively, as follows.

Moments of Inertia About the x- and y-Axes

$$I_x = \lim_{\|P\| \to 0} \sum_{i=1}^{n} (y_i^*)^2 \Delta m_i = \iint_R y^2 \, dm \qquad (15.16)$$

$$I_y = \lim_{\|P\| \to 0} \sum_{i=1}^{n} (x_i^*)^2 \Delta m_i = \iint_R x^2 \, dm \qquad (15.17)$$

where $dm = \rho(x, y) \, dA$.

We also define the *polar moment of inertia*, I_0, as follows.

Polar Moment of Inertia

$$I_0 = \iint_R (x^2 + y^2) \, dm \qquad (15.18)$$

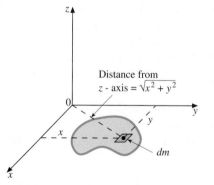

Distance from
z-axis $= \sqrt{x^2 + y^2}$

FIGURE 15.32

It follows that $I_0 = I_x + I_y$. This polar moment of inertia is sometimes said to be the moment of inertia about the origin. Perhaps a better way to describe it is as the moment of inertia about a line through the origin perpendicular to the xy-plane—that is, about the z-axis (see Figure 15.32).

If a lamina of mass m has moment of inertia I with respect to a line, then the positive number r that satisfies $I = mr^2$ is called the *radius of gyration* of the lamina with respect to that line. This terminology is consistent with the notion of radius of gyration for a point mass if we consider that the entire mass of the lamina is concentrated at a distance r from the line. In particular, if r_x is the radius of gyration with respect to the y-axis, and r_y is that with respect to the x-axis, we have

$$r_x = \sqrt{\frac{I_y}{m}} \quad \text{and} \quad r_y = \sqrt{\frac{I_x}{m}} \qquad (15.19)$$

The point (r_x, r_y) plays a role for second moments analogous to that of $(\bar{x}, \bar{y})$ for first moments. In general, these points are not the same.

The polar moment of inertia is helpful in getting an intuitive feeling for the role moments of inertia play in a dynamic system. Consider a point mass m in the xy-plane located a distance r from the origin. Suppose it is rotating about the z-axis at the constant angular velocity ω (the number of radians it turns through per unit of time). Its kinetic energy is $\frac{1}{2}mv^2$, where v is the linear velocity. By differentiating both sides of the arc length formula $s = r\theta$ (see Appendix 2) with respect to time, we get $v = r\omega$, since $v = ds/dt$ and $\omega = d\theta/dt$. The kinetic energy can thus be written as

$$KE = \frac{1}{2}m(r\omega)^2 = \frac{1}{2}(mr^2)\omega^2 = \frac{1}{2}I_0\omega^2$$

since $I_0 = m(x^2 + y^2) = mr^2$. Comparing this result with the formula $KE = \frac{1}{2}mv^2$, we can see that I_0 plays a role analogous to the mass m. The same result holds if we have a lamina in the xy-plane rotating about the z-axis, as the usual partitioning, summing, and passing to the limit would show. For example, the total kinetic energy of a rotating wheel is $\frac{1}{2}I_0\omega^2$. Now, kinetic energy is equal to the work required to bring the object to rest. So for a constant angular velocity, the moment of inertia is a measure of the work required to bring the rotating wheel to a stop.

EXAMPLE 15.18   A lamina is bounded by the curves $y^2 = x$ and $x - y = 2$, and its density is $\rho(x, y) = 2x$. Find I_x, I_y, r_x, and r_y.

Solution   By Equations 15.16 and 15.17, since $dm = \rho(x, y)dA$,

$$I_x = \iint_R y^2 \, dm = \iint_R 2xy^2 \, dA$$

$$I_y = \iint_R x^2 \, dm = \iint_R 2x^3 \, dA$$

The region occupied by the lamina is the type II region pictured in Figure 15.33. The points of intersection $(4, 2)$ and $(1, -1)$ are found by solving the equations simultaneously. Using iterated integrals, we obtain

$$I_x = \int_{-1}^{2} \int_{y^2}^{y+2} 2xy^2 \, dx \, dy \quad \text{and} \quad I_y = \int_{-1}^{2} \int_{y^2}^{y+2} 2x^3 \, dx \, dy$$

The results, found after some computation, are

$$I_x = \frac{531}{35} \quad \text{and} \quad I_y = \frac{369}{5}$$

Thus, $I_0 = I_x + I_y = 3114/35$.

To find r_x and r_y we need the mass m. By Equation 15.10,

$$m = \iint_R \rho \, dA = \int_{-1}^{2} \int_{y^2}^{y+2} 2x \, dx \, dy = \frac{72}{5}$$

So from Equations 15.19 we get, after some simplification,

$$r_x = \sqrt{\frac{I_y}{m}} = \sqrt{\frac{41}{8}} \approx 2.26 \qquad r_y = \sqrt{\frac{I_x}{m}} = \sqrt{\frac{59}{56}} \approx 1.03 \qquad \blacksquare$$

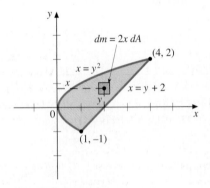

$dm = 2x \, dA$

$(4, 2)$

$x = y^2$

$x = y + 2$

$(1, -1)$

FIGURE 15.33

EXAMPLE 15.19   Consider a lamina that is the annular ring between the circles $x^2 + y^2 = 1$ and $x^2 + y^2 = 4$. The density at any point (x, y) is the reciprocal of its distance from the origin. Find the polar moment of inertia.

Solution   The nature of the region (Figure 15.34) suggests that it would be convenient to use polar coordinates. In polar coordinates, Equation 15.18 becomes

$$I_0 = \iint_R r^2 \, dm$$

and $dm = \rho(r, \theta) \, dA = \frac{1}{r}(r \, dr \, d\theta) = dr \, d\theta$. Thus,

$$I_0 = \int_0^{2\pi} \int_1^2 r^2 \, dr \, d\theta = \frac{14\pi}{3} \qquad \blacksquare$$

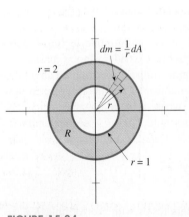

$dm = \frac{1}{r} \, dA$

$r = 2$

r

R

$r = 1$

FIGURE 15.34

Moments in Probability Theory

The ideas studied in this section are important in probability theory. Consider first the one-dimensional case in which $f(x)$ is a continuous probability density function on the interval $[a, b]$. The *kth moment about the origin M_k* is

$$M_k = \int_a^b x^k f(x) \, dx$$

This integral is what we referred to in Section 6.8 as the expected value of x^k and denoted by $E[x^k]$. In particular, for $k = 1$, $M_1 = E[x] = \mu$, the mean of the distribution. Since f is a probability density function, $\int_a^b f(x) \, dx = 1$. So we can write

$$\mu = \frac{\int_a^b x f(x) \, dx}{\int_a^b f(x) \, dx}$$

showing that μ is analogous to the center of mass. In fact, we can think of the "probability" as a continuously distributed mass on the interval $[a, b]$, where the "density" is $f(x)$. Indeed, the name *probability density* comes from this analogy.

The *kth moment about the mean* is defined by

$$M_k' = \int_a^b (x - \mu)^k f(x) \, dx$$

which is the same as the expected value $E[(x - \mu)^k]$. In particular, for $k = 2$, we get the variance σ^2:

$$\sigma^2 = E[(x - \mu)^2] = \int_a^b (x - \mu)^2 f(x) \, dx$$

These ideas can be extended to two dimensions. If $f(x, y) \geq 0$ and $\iint_R f(x, y) \, dA = 1$, then f can serve as a probability density function of two random variables. We define the x-mean and y-mean by

$$\mu_x = \iint_R x f(x, y) \, dA \quad \text{and} \quad \mu_y = \iint_R y f(x, y) \, dA \tag{15.20}$$

These means are analogous to $\bar{x}$ and $\bar{y}$ for a lamina with density $f(x, y)$. Similarly,

$$\sigma_x^2 = \iint_R (x - \mu_x)^2 f(x, y) \, dA \quad \text{and} \quad \sigma_y^2 = \iint_R (y - \mu_y)^2 f(x, y) \, dA \tag{15.21}$$

Moments of higher order are also sometimes useful, but we will not consider them here.

We illustrate these ideas in the next two examples.

EXAMPLE 15.20   Let $f(x, y) = k(2x + y)$ and $R = \{(x, y) : 0 \leq x \leq 1, 0 \leq y \leq 2x\}$.

(a) Find k so that $f(x, y)$ will be a probability density function on R.

(b) Find μ_x and μ_y.

Solution

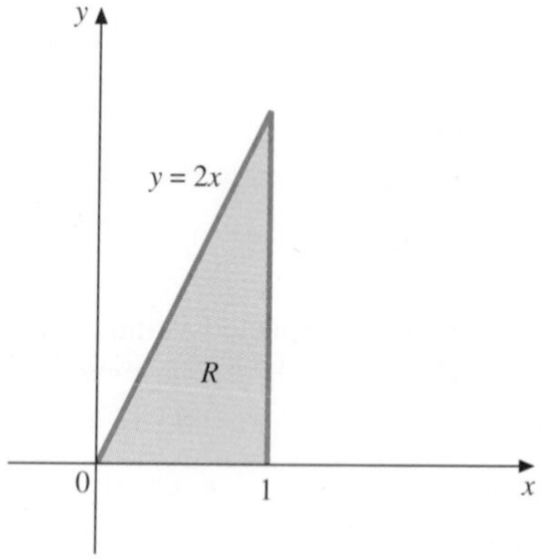

FIGURE 15.35

(a) We show the region R in Figure 15.35. Note first that in this region both x and y are nonnegative, so $f(x, y) = k(2x + y)$ will be nonnegative if $k > 0$. For $f(x, y)$ to be a probability density function, we must also have $\iint_R f(x, y)\, dA = 1$. So we calculate the integral, set it equal to 1, and solve for k.

$$\iint_R f(x, y)\, dA = k \int_0^1 \int_0^{2x} (2x + y)\, dy\, dx$$

$$= k \int_0^1 \left[2xy + \frac{y^2}{2} \right]_0^{2x} dx$$

$$= k \int_0^1 (4x^2 + 2x^2)\, dx$$

$$= k \left[2x^3 \right]_0^1 = 2k$$

Thus, we must have $2k = 1$, so $k = \frac{1}{2}$.

(b) Using $f(x, y) = \frac{1}{2}(2x + y)$, we have for μ_x, from Equations 15.20,

$$\mu_x = \frac{1}{2} \int_0^1 \int_0^{2x} x(2x + y)\, dy\, dx = \frac{1}{2} \int_0^1 \left[2x^2 y + x\frac{y^2}{2} \right]_0^{2x} dx$$

$$= \frac{1}{2} \int_0^1 (4x^3 + 2x^3)\, dx$$

$$= \frac{1}{2} \left[\frac{6x^4}{4} \right]_0^1 = \frac{3}{4}$$

Similarly, from Equations 15.20, for μ_y, we have

$$\mu_y = \frac{1}{2} \int_0^1 \int_0^{2x} y(2x + y)\, dy\, dx = \frac{1}{2} \int_0^1 \left[y^2 x + \frac{y^3}{3} \right]_0^{2x} dx$$

$$= \frac{1}{2} \int_0^1 \left(4x^3 + \frac{8x^3}{3} \right) dx$$

$$= \frac{1}{2} \int_0^1 \frac{20x^3}{3}\, dx$$

$$= \frac{1}{2} \left[\frac{5x^4}{3} \right]_0^1 = \frac{5}{6}$$ ∎

We can write Equations 15.21 for the variances σ_x^2 and σ_y^2 in a more useful form as follows, making use of Properties 1 and 2 of double integrals given in Section 15.1.

$$\sigma_x^2 = \iint_R (x - \mu_x)^2 f(x, y)\, dA = \iint_R (x^2 - 2x\mu_x + \mu_x^2) f(x, y)\, dA$$

$$= \iint_R x^2 f(x, y)\, dA - 2\mu_x \iint_R x f(x, y)\, dA + \mu_x^2 \iint_R f(x, y)\, dA$$

$$= \iint_R x^2 f(x, y)\, dA - 2\mu_x^2 + \mu_x^2$$

since $\mu_x = \iint_R x f(x, y) \, dA$ and $\iint_R f(x, y) \, dA = 1$. Thus,

$$\sigma_x^2 = \iint_R x^2 f(x, y) \, dA - \mu_x^2 \tag{15.22}$$

A similar calculation shows that

$$\sigma_y^2 = \iint_R y^2 f(x, y) \, dA - \mu_y^2 \tag{15.23}$$

EXAMPLE 15.21   Find σ_x^2 and σ_y^2 for the probability density function in Example 15.20.

Solution   In Example 15.20, we found that $f(x, y) = \frac{1}{2}(2x + y)$, $\mu_x = \frac{3}{4}$, and $\mu_y = \frac{5}{6}$. So, from Equation 15.22, we have

$$\sigma_x^2 = \frac{1}{2} \int_0^1 \int_0^{2x} x^2 (2x + y) \, dy \, dx - \left(\frac{3}{4}\right)^2$$

$$= \frac{1}{2} \int_0^1 \left[2x^3 y + \frac{x^2 y^2}{2} \right]_0^{2x} dx - \frac{9}{16}$$

$$= \frac{1}{2} \int_0^1 \left(4x^4 + 2x^4 \right) dx - \frac{9}{16}$$

$$= \frac{1}{2} \left[\frac{6x^5}{5} \right]_0^1 - \frac{9}{16} = \frac{3}{5} - \frac{9}{16} = \frac{3}{80}$$

Similarly, by Equation 15.23,

$$\sigma_y^2 = \frac{1}{2} \int_0^1 \int_0^{2x} y^2 (2x + y) \, dy \, dx - \left(\frac{5}{6}\right)^2$$

$$= \frac{1}{2} \int_0^1 \left[\frac{2x y^3}{3} + \frac{y^4}{4} \right]_0^{2x} dx - \frac{25}{36}$$

$$= \frac{1}{2} \int_0^1 \left(\frac{16 x^4}{3} + 4x^4 \right) dx - \frac{25}{36}$$

$$= \frac{1}{2} \left[\frac{28 x^5}{15} \right]_0^1 - \frac{25}{36} = \frac{14}{15} - \frac{25}{36} = \frac{43}{180} \qquad \blacksquare$$

Exercise Set 15.4

In Exercises 1–8, find the center of mass of the lamina that occupies the region R and has density $\rho(x, y)$.

1. $R = \{(x, y) : 0 \le x \le 1, \ x \le y \le 1\}$; $\rho(x, y) = 2y$

2. $R = \{(x, y) : 0 \le x \le 2, \ 0 \le y \le 1\}$; $\rho(x, y) = x + 2y$

3. $R = \{(x, y) : -y \le x \le y, \ 0 \le y \le 1\}$; $\rho(x, y) = y^2$

4. $R = \{(x, y) : 0 \le x \le 4, \ 0 \le y \le \sqrt{x}\}$; $\rho(x, y) = xy$

5. R is bounded by $y = x$ and $y = 2 - x^2$; $\rho(x, y) = 2$

6. R is bounded by $y = 2x - x^2$ and $y = 0$; $\rho(x, y) = 1$

7. R is bounded by $y = \sqrt{4 - x^2}$ and $y = 0$; $\rho(x, y) = \sqrt{x^2 + y^2}$

8. R is the region inside $x^2 + y^2 = 4x$ for which $x \geq 1$; $\rho(x, y) = y^2/x$

In Exercises 9–16, find the centroid of the region R.

9. $R = \{(x, y) : -1 \leq x \leq 1, \ 0 \leq y \leq x^2\}$

10. $R = \{(x, y) : -1 \leq x \leq 1, \ 0 \leq y \leq 2 - x^2\}$

11. $R = \{(r, \theta) : 0 \leq r \leq a, \ 0 \leq \theta \leq \pi\}$

12. $R = \{(r, \theta) : 0 \leq r \leq 2\cos\theta, \ 0 \leq \theta \leq \pi/4\}$

13. R is the region bounded by $y = x^3 - x$ and $y = 7x$ for which $x \geq 0$.

14. $R = \{(x, y) : -\pi/2 \leq x \leq \pi/2, \ 0 \leq y \leq \cos x\}$

15. $R = \{(x, y) : 0 \leq x \leq 2, \ 0 \leq y \leq e^x\}$

16. R is the region above the x-axis and between $x^2 + y^2 = 1$ and $x^2 + y^2 = 16$.

In Exercises 17–24, find I_x, I_y, I_0, r_x, and r_y for the lamina that occupies the region R and has density $\rho(x, y)$.

17. The region in Exercise 1

18. The region in Exercise 2

19. The region in Exercise 3

20. The region in Exercise 4

21. $R = \{(x, y) : 0 \leq x \leq 2, \ 0 \leq y \leq 2-x\}$; $\rho(x, y) = x+y$

22. $R = \{(r, \theta) : 0 \leq r \leq a, \ 0 \leq \theta \leq \pi/2\}$; $\rho(r, \theta) = k$ (constant)

23. R is the region under $y = \sin x$ and above the x-axis, from $x = 0$ to $x = \pi$; $\rho(x, y) = k$ (constant)

24. $R = \{(x, y) : 0 \leq x \leq \sqrt{3}, \ 1 \leq y \leq \sqrt{4 - x^2}\}$; $\rho(x, y) = x/y$

25. Prove that the moment of inertia of a homogeneous lamina of density ρ in the shape of a rectangle with base b and altitude h about its base is $I = (bh^3/3)\rho$.

26. Find the centroid of the region bounded by $y^2 = 4x$ and $2x - y = 4$.

27. Suppose a homogeneous lamina of density ρ occupies the region described in Exercise 26. Set up, but do not evaluate iterated integrals for the moments of inertia about the lines $x + 2 = 0$ and $y - 4 = 0$.

28. Consider the limaçon $r = 3 + 2\cos\theta$ and the circle $r = 3$. Set up, but do not evaluate, iterated integrals for $\bar{x}$ and I_0 for homogeneous laminas of density ρ that occupy each of the following regions:
  (a) outside the circle and inside the limaçon;
  (b) outside the limaçon and inside the circle;
  (c) inside both the limaçon and the circle.

29. Let $f(x, y) = kxy$ on the region $R = \{(x, y): 0 \leq x \leq 1, \ 0 \leq y \leq 1\}$.
  (a) Find k so that $f(x, y)$ will be a probability density function.
  (b) Find μ_x and μ_y.
  (c) Find σ_x^2 and σ_y^2.

30. Let $f(x, y) = k$ on the unit circular disk $\{(x, y) : x^2 + y^2 \leq 1\}$.
  (a) Find k so that $f(x, y)$ is a probability density function.
  (b) Find σ_x^2 and σ_y^2.

31. Let $f(x, y) = k(1 - y)$ on the region $R = \{(x, y): 0 \leq x \leq y \leq 1\}$.
  (a) Find k so that $f(x, y)$ is a probability density function.
  (b) Find μ_x and μ_y.
  (c) Find σ_x^2 and σ_y^2.

32. Prove that the first moment of a lamina about each of the lines $x = \bar{x}$ and $y = \bar{y}$ equals 0.

33. (a) Prove that the moment of inertia $I_{\bar{x}}$ of a lamina about the line $x = \bar{x}$ is

$$I_{\bar{x}} = I_y - m\bar{x}^2$$

  (b) State and prove an analogous result for the moment of inertia $I_{\bar{y}}$ about the line $y = \bar{y}$.

34. Prove that the medians of a triangle intersect at the centroid of the region enclosed by the triangle.

35. If a homogeneous lamina of density ρ is bounded by a triangle of altitude h and base b, prove that the moment of inertia about its base is

$$I_b = \frac{bh^3\rho}{12}$$

36. A lamina of density $\rho(x, y) = y^2$ is in the shape of the circular disk $x^2 + y^2 \leq 1$. Find its moment of inertia with respect to the line $x = 2$.

37. Use double integration to prove the following *First Theorem of Pappus*: If R is a plane region and l is a line in the plane of R that does not intersect R, then the volume of the solid formed by revolving R about l is the area of R times the distance traveled by the centroid of R. (See also Exercise 28 in Exercise Set 6.8, where a proof using single integration was called for.)

38. It can be shown that the function

$$f(x, y) = \frac{1}{8}xe^{-(x+y)/2}$$

defined on the infinite region $\{(x, y) : 0 \leq x \leq \infty,$
$0 \leq y \leq \infty\}$ is a probability density function. Assuming

that the double integrals over this unbounded region exist whenever the corresponding improper iterated integrals converge, find the following:

(a) μ_x (b) μ_y

(c) σ_x^2 (d) σ_y^2

15.5 SURFACE AREA

We learned in Chapter 6 how to find areas of surfaces of revolution using techniques of single-variable calculus. We now show how to use double integration to find areas of more general surfaces.

Let S be the surface that is the graph of a function f of two variables over a region R, and suppose f has continuous first partial derivatives on R. We partition a rectangle containing R in the usual way and count those subrectangles entirely contained in R. In the ith such subrectangle, whose area is ΔA_i, we select an arbitrary point (x_i^*, y_i^*). Because we have assumed that f_x and f_y are continuous, we know that f is differentiable in R, and so S has a tangent plane at each point. Let T_i denote its tangent plane at $P_i = (x_i^*, y_i^*, f(x_i^*, y_i^*))$. Now project the ith subrectangle vertically. The resulting prism cuts out a patch of the surface S, whose area we denote by $\Delta\sigma_i$, and a corresponding patch of the tangent plane T_i, whose area we denote by ΔT_i. Figure 15.36(a) illustrates this construction.

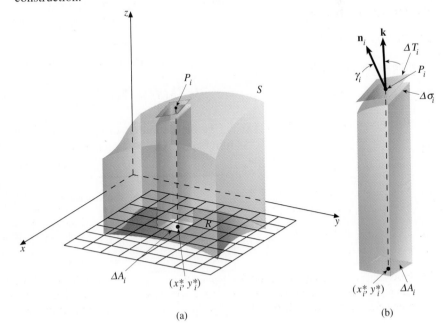

(a)   (b)

FIGURE 15.36

If the norm $\|P\|$ of the partition is small, we would expect the area of the patch of the surface, $\Delta\sigma_i$, to be approximated by the area of the patch of the tangent plane ΔT_i. Hence, denoting the area of S by $A(S)$, it is reasonable to suppose that

$$A(S) \approx \sum_{i=1}^{n} \Delta\sigma_i \approx \sum_{i=1}^{n} \Delta T_i$$

and that the approximation becomes better and better as we take partitions with norms approaching 0. Our next task is to express ΔT_i in terms of the function f.

As shown in the enlarged view in Figure 15.36(b), let $\mathbf{n}_i$ be a normal vector to the tangent plane at P_i, and let γ_i be the angle between $\mathbf{n}_i$ and the unit vertical vector $\mathbf{k} = \langle 0, 0, 1 \rangle$. Then it can be shown (see Exercise 20, in Exercise Set 15.5) that

$$\Delta T_i \, |\cos\gamma_i| = \Delta A_i \tag{15.24}$$

We learned in Section 14.4 that the vector $\langle f_x, f_y, -1 \rangle$ is normal to the surface S, and so we may take

$$\mathbf{n}_i = \langle f_x(x_i^*, y_i^*), f_y(x_i^*, y_i^*), -1 \rangle$$

Thus,

$$\cos\gamma_i = \frac{\mathbf{n}_i \cdot \mathbf{k}}{|\mathbf{n}_i||\mathbf{k}|} = \frac{-1}{\sqrt{f_x^2 + f_y^2 + 1}}$$

where f_x and f_y are evaluated at (x_i^*, y_i^*). Solving for ΔT_i from Equation 15.24 and substituting for $\cos\gamma_i$, we have

$$\Delta T_i = \sqrt{1 + f_x^2 + f_y^2} \, \Delta A_i$$

Thus,

$$A(S) \approx \sum_{i=1}^{n} \sqrt{1 + [f_x(x_i^*, y_i^*)]^2 + [f_y(x_i^*, y_i^*)]^2} \, \Delta A_i$$

We *define* $A(S)$ by the limit of this sum as $\|P\| \to 0$:

$$A(S) = \lim_{\|P\| \to 0} \sum_{i=1}^{n} \sqrt{1 + [f_x(x_i^*, y_i^*)]^2 + [f_y(x_i^*, y_i^*)]^2} \, \Delta A_i$$

By the definition of the double integral, we therefore have the following formula.

Area of the Surface $z = f(x, y)$ Over the Region R

$$A(S) = \iint\limits_{R} \sqrt{1 + [f_x(x, y)]^2 + [f_y(x, y)]^2} \, dA \tag{15.25}$$

REMARK
- In Section 6.4, we developed the formula for arc length

$$L = \int_a^b \sqrt{1 + [f'(x)]^2} \, dx$$

Note the similarity between this formula and Formula 15.25 for surface area.

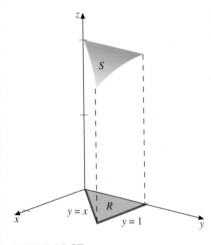

FIGURE 15.37

EXAMPLE 15.22 Find the area of that part of the cylinder $z = \sqrt{4 - x^2}$ above the region R bounded by $y = x$, $x = 0$, and $y = 1$. Sketch the surface.

Solution Figure 15.37 shows the surface. Writing $f(x, y) = \sqrt{4 - x^2}$, we have

$$f_x(x, y) = \frac{-x}{\sqrt{4 - x^2}} \qquad f_y(x, y) = 0$$

So, by Equation 15.25,

$$A(S) = \iint_R \sqrt{1 + \frac{x^2}{4 - x^2}}\, dA = \iint_R \frac{2}{\sqrt{4 - x^2}}\, dA$$

$$= \int_0^1 \int_x^1 \frac{2}{\sqrt{4 - x^2}}\, dy\, dx = \int_0^1 \left(\frac{2}{\sqrt{4 - x^2}} - \frac{2x}{\sqrt{4 - x^2}} \right) dx$$

$$= \left[2\sin^{-1} \frac{x}{2} + 2\sqrt{4 - x^2} \right]_0^1 = \frac{\pi}{3} + 2\sqrt{3} - 4 \qquad \blacksquare$$

EXAMPLE 15.23 Find the area of that portion of the paraboloid $z = 2 - x^2 - y^2$ that lies above the xy-plane.

Solution The surface S in question is shown in Figure 15.38. Writing $z = f(x, y)$, we have $f_x = -2x$ and $f_y = -2y$, so by Equation 15.25,

$$A(S) = \iint_R \sqrt{1 + 4x^2 + 4y^2}\, dA$$

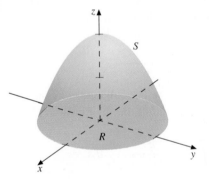

FIGURE 15.38

The region R is bounded by the xy-trace of the paraboloid—namely, the circle $x^2 + y^2 = 2$. Both the nature of R and the integrand suggest using polar coordinates to evaluate the integral. Since in polar coordinates $r^2 = x^2 + y^2$ and $dA = r\, dr\, d\theta$, we have

$$A(S) = \int_0^{2\pi} \int_0^{\sqrt{2}} \sqrt{1 + 4r^2}\, r\, dr\, d\theta$$

$$= \int_0^{2\pi} \left[\frac{1}{8} \cdot \frac{2}{3}(1 + 4r^2)^{3/2} \right]_0^{\sqrt{2}} d\theta = \frac{13\pi}{3} \qquad \blacksquare$$

Surfaces Defined by an Equation of the Form $F(x, y, z) = 0$

If the equation for the surface S has the form $F(x, y, z) = 0$, where F_x, F_y, and F_z are continuous and $F_z \neq 0$, then we know from Section 14.2 that z is implicitly a function of x and y, say $z = f(x, y)$, and

$$f_x = \frac{\partial z}{\partial x} = -\frac{F_x}{F_z} \qquad f_y = \frac{\partial z}{\partial y} = -\frac{F_y}{F_z}$$

When these values are substituted for f_x and f_y in Equation 15.25, we get the alternative formula

$$A(S) = \iint_R \frac{\sqrt{F_x^2 + F_y^2 + F_z^2}}{|F_z|} \, dA \qquad (15.26)$$

By observing that the numerator is the length of the gradient of F, $|\nabla F|$, and that $\nabla F \cdot \mathbf{k} = F_z$, we can put Equation 15.26 in the compact form

$$A(S) = \iint_R \frac{|\nabla F|}{|\nabla F \cdot \mathbf{k}|} \, dA \qquad (15.27)$$

In both Equation 15.26 and Equation 15.27 it must be understood that z is a function of x and y defined implicitly by $F(x, y, z) = 0$.

EXAMPLE 15.24   Find the area of that portion of the sphere $x^2 + y^2 + z^2 = a^2$ that is inside the cylinder $x^2 + y^2 = b^2$, where $0 < b < a$.

Solution   Because both surfaces are symmetric with respect to the xy-plane, we can consider the upper hemisphere only and double the result (see Figure 15.39). Taking $F(x, y, z) = x^2 + y^2 + z^2 - a^2$, we have $\nabla F = \langle 2x, 2y, 2z \rangle$, so that

$$|\nabla F| = \sqrt{4x^2 + 4y^2 + 4z^2} = 2\sqrt{x^2 + y^2 + z^2} = 2a$$

The region R in the xy-plane over which the surface lies is bounded by the xy-trace of the cylinder. Thus, $R = \{(x, y) : x^2 + y^2 \le b^2\}$. So, by Equation 15.26, we have

$$A(S) = 2 \iint_R \frac{2a}{2z} \, dA = 2 \iint_R \frac{a}{\sqrt{a^2 - (x^2 + y^2)}} \, dA$$

Using polar coordinates, we obtain

$$A(S) = 2a \int_0^{2\pi} \int_0^b \frac{r \, dr \, d\theta}{\sqrt{a^2 - r^2}} = 2a \int_0^{2\pi} \left[-\sqrt{a^2 - r^2} \right]_0^b \, d\theta$$

$$= 4\pi a \left[a - \sqrt{a^2 - b^2} \right] \qquad \blacksquare$$

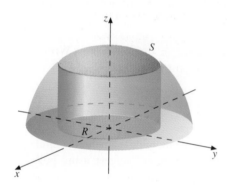

FIGURE 15.39

DOUBLE INTEGRATION USING COMPUTER ALGEBRA SYSTEMS

As we have seen, computing double integrals can be very messy and tedious. Computer algebra systems can, in many cases, compute double integrals very quickly. In this section, we present several examples that take advantage of the algebraic as well as the graphical capabilities of the CAS in applying double integration.

⊞ CAS 58

Find the volume of the region under the paraboloid $z = x^2 + 3y^2 + 4$ and above the unit disk $x^2 + y^2 \leq 1$ in the xy-plane.

The iterated integral that we need to compute for the volume is given by

$$\int_0^1 \int_{-\sqrt{1-x^2}}^{\sqrt{1-x^2}} \left(x^2 + 3y^2 + 2\right) \, dy \, dx$$

(See Theorem 15.2.)

Maple:

First, we present a plot to get an idea of the region in space. The first plot command plots the paraboloid and the second plots a cylinder with the unit disk as base and that cuts up through the paraboloid. Maple needs several libraries.

with(plots):with(student):
p:=plot3d(x^2+3*y^2+4,x=–2..2,y=–2..2,view=0..10,scaling=
  constrained,axes=boxed):
cylinder:=plot3d([cos(t),sin(t),s],t=0..2*Pi,s=0..27,
  scaling=constrained):
display3d({cylinder,p},orientation=[57,50]);

Mathematica:

P = Plot3D[x^2+3*y^2+4,{x,–2,2},{y,–2,2},
  DisplayFunction–>Identity]

cylinder = ParametricPlot3D[{Cos[t],Sin[t],s},
  {t,0,2*Pi},{s,0,27},DisplayFunction–>Identity]

Show[P,cylinder,DisplayFunction–>$DisplayFunction,
  ViewPoint–>{–2,–2,0},PlotRange–>{0,10}]

Integrate[x^2+3*y^2+4,{x,–1,1},{y,–Sqrt[1–x^2],
  Sqrt[1–x^2]}]

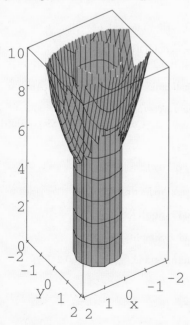

FIGURE 15.5.1

Now, to compute the volume is an easy matter using Maple. You may want to try this integral by hand to appreciate the algebra involved.

volume:=int(int(x^2+3*y^2+4,y=–sqrt(1–x^2)..sqrt(1–x^2)),x=–1..1);

Output: volume := 5π

CAS 59

The formula given for the surface area of $z = f(x, y)$ over some region R in Equation 15.25 is a special case of a more general formula for the surface area when the surface is given in parametric form, say,

$$\mathbf{r}(s, t) = x(s, t)\mathbf{i} + y(s, t)\mathbf{j} + z(s, t)\mathbf{k}$$

where (s, t) are in R. The formula in this case is

$$\iint_R |r_s(s, t) \times r_t(s, t)| \, dA$$

where $\mathbf{r}_s = x_s\mathbf{i} + y_s\mathbf{j} + z_s\mathbf{k}$ and $\mathbf{r}_t = x_t\mathbf{i} + y_t\mathbf{j} + z_t\mathbf{k}$. (Here the parametric surface must be smooth and the surface must be covered only once as (s, t) ranges over R.)

Use this formula to compute the surface area of the torus generated by rotating the circle $(x - 3)^2 + z^2 = 1$ in the xz-plane about the z-axis.

The parametric equations for the circle in the xz-plane are $x = \cos t + 3$ and $z = \sin t$ for $0 \leq t \leq 2\pi$.

Maple:

To rotate the circle around the z-axis we use:

```
plot3d([(cos(t)+3)*cos(s),(cos(t)+3)*sin(s),sin(t)],
    t=0..2*Pi,s=0..2*Pi,scaling=constrained);
```

Maple will need the linear algebra library:

```
with(linalg);
X:=(cos(t)+3)*cos(s);
Y:=(cos(t)+3)*sin(s);
Z:=sin(t);
r:=[X,Y,Z];
```

Mathematica:

```
ParametricPlot3D[{(Cos[t]+3)*Cos[s],(Cos[t]+3)*Sin[s],
    Sin[t]},{t,0,2*Pi},{s,0,2*Pi}]
```

Mathematica needs the library VectorAnalysis.

```
<<Calculus`VectorAnalysis`
r = {(Cos[t]+3)*Cos[s],(Cos[t]+3)*Sin[s],Sin[t]}

rs = D[r,s]

rt = D[r,t]

cp = CrossProduct[rs,rt]

integrand = Sqrt[cp.cp]

integrand = Simplify[%]

Integrate[integrand,{s,0,2*Pi},{t,0,2*Pi}]
```

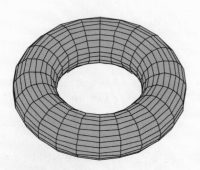

FIGURE 15.5.2

Maple:

Output: $r := [(\cos(t) + 3)\cos(s), (\cos(t) + 3)\sin(s), \sin(t)]$

rs:=diff(r,s);

Output: $rs := [-(\cos(t) + 3)\sin(s), (\cos(t) + 3)\cos(s), 0]$

rt:=diff(r,t);

Output: $rt := [-\sin(t)\cos(s), -\sin(t)\sin(s), \cos(t)]$

cp:=crossprod(rs,rt);

Output:

$cp := [(\cos(t) + 3)\cos(s)\cos(t)(\cos(t) + 3)\sin(s)\cos(t)$
$\qquad (\cos(t) + 3)\sin(s)^2\sin(t) + (\cos(t) + 3)\cos(s)^2\sin(t)]$

integrand:=sqrt(dotprod(cp,cp));

Output: $integrand := \Big((\cos(t) + 3)^2\cos(s)^2\cos(t)^2$
$\qquad + (\cos(t) + 3)^2\sin(s)^2\cos(t)^2 + \big((\cos(t) + 3)\sin(s)^2\sin(t)$
$\qquad + (\cos(t) + 3)\cos(s)^2\sin(t)\big)^2\Big)^{1/2}$

We better try to simplify this expression before asking Maple to integrate, because Maple will have trouble with the expression in its current form.

factor(integrand);

Output: $\dfrac{\sqrt{\sin(s)^2 + \cos(s)^2}\,(\cos(t) + 3)}{\sqrt{\cos(t)^2 + \sin(t)^2\cos(s)^2 + \sin(s)^2\sin(t)^2}}$

integrand:=simplify(");

Output: $integrand := \cos(t) + 3$
This is better and we can now integrate.

int(int(integrand,t=0..2*Pi),s=0..2*Pi);

Output: $12\pi^2$

Exercise Set 15.5

1. Find the area of the portion of the plane $z + 2x + 3y = 6$ that lies above the rectangular region R bounded by $x = 0$, $y = 0$, $x = 2$, and $y = 1$.

2. Find the area of that portion of the plane $4x - 3y - 6z + 12 = 0$ that lies above the triangular region R with vertices $(3, 0)$, $(0, 0)$, and $(0, 4)$.

3. Find the area of the first-octant portion of the cylinder $z = 2 - y^2$ that is cut off by the planes $x = 0$, $y = x$, and $z = 0$.

4. Find the area of the portion of the cylinder $z = \sqrt{4 - y^2}$ above the region $R = \{(x, y) : 0 \le x \le 3,\ 0 \le y \le 1\}$.

5. Find the area of the paraboloid $z = x^2 + y^2$ that lies inside the cylinder $x^2 + y^2 = 2$.

6. Find the area of the part of the sphere $x^2 + y^2 + z^2 = 4$ that is inside the cylinder $x^2 + y^2 - 2x = 0$.

7. Find the area of the part of the upper nappe of the cone $z^2 = x^2 + y^2$ that lies inside the cylinder $x^2 + y^2 - 4y = 0$.

8. Find the area of that portion of the cylinder $x^2 + z^2 = a^2$ that lies inside the cylinder $x^2 + y^2 = a^2$.

9. Find the area of the paraboloid $z = 1 + x^2 + y^2$ that is between the planes $z = 2$ and $z = 5$.

10. Find the area of the part of the sphere $x^2 + y^2 + z^2 = 25$ that is between the planes $z = 3$ and $z = 4$.

In Exercises 11–16, find $A(S)$, where S is the portion of the graph $z = f(x, y)$ that lies above the region R.

11. $f(x, y) = 3x + y^2$; $R = \{(x, y) : 0 \le x \le y, \ 0 \le y \le 2\}$

12. $f(x, y) = \frac{2}{3}(x^{3/2} + y^{3/2})$; $R = \{(x, y) : 0 \le x \le 3,$
$0 \le y \le 3 - x\}$

13. $f(x, y) = \ln \sec x$; $R = \{(x, y) : -\pi/4 \le x \le \pi/4,$
$0 \le y \le \sec x\}$

14. $f(x, y) = (1 - y^{2/3})^{3/2}$; $R = \{(x, y) : 0 \le x \le 1,$
$0 \le y \le (1 - x^{2/3})^{3/2}\}$

15. $f(x, y) = 2 - (x^2 + y^2)/2$; R is the region inside the lemniscate $r^2 = \cos 2\theta$.

16. $f(x, y) = x + y + 2$; R is the region outside $r = 1$ and inside $r = 2 \cos \theta$.

17. Use the methods of this section to find the area of the surface of a sphere of radius a.

18. Use the methods of this section to find the lateral surface area of a right circular cone (one nappe) of base radius a and height h.

19. Using the accompanying figure, verify that when a nonnegative smooth function f of one variable, defined on $[a, b]$, is rotated about the x-axis, the equation of the resulting surface of revolution is $y^2 + z^2 = [f(x)]^2$. Using this and the methods of this section, show that its

total surface area is

$$2\pi \int_a^b f(x)\sqrt{1 + [f'(x)]^2}\, dx$$

and so is in agreement with Definition 6.3.

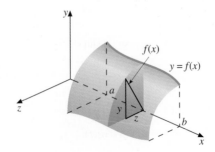

20. In the accompanying figure, $PQRS$ is a parallelogram of area ΔT formed by intersecting a plane with the prism that has the rectangular base $P'Q'R'S'$. The normal **n** to the plane makes an angle γ with the unit vertical vector **k**. If ΔA is the area of the rectangle $P'Q'R'S'$, show that

$$(\overrightarrow{PQ} \times \overrightarrow{PS}) \cdot \mathbf{k} = (\overrightarrow{P'Q'} \times \overrightarrow{P'S'}) \cdot \mathbf{k}$$

and then explain how it follows from this that

$$\Delta T \, |\cos \gamma| = \Delta A$$

(Hint: Write $\overrightarrow{PQ} = \overrightarrow{PP'} + \overrightarrow{P'Q'} + \overrightarrow{Q'Q}$, and write $\overrightarrow{PS}$ in a similar way.)

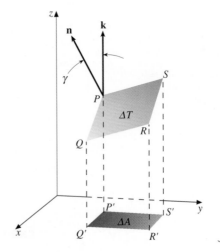

15.6 TRIPLE INTEGRALS

The ideas behind the triple integral of a function of three variables over a region in $\mathbb{R}^3$ are similar to those for a double integral in $\mathbb{R}^2$, so we will be briefer in

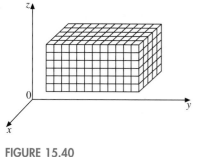

FIGURE 15.40

our treatment. We will designate the three-dimensional region of integration by Q. We begin with the simplest case, in which Q is a rectangular box of the form

$$Q = \{(x, y, z) : a_1 \le x \le a_2, b_1 \le y \le b_2, c_1 \le z \le c_2\}$$

Using planes parallel to each of the coordinate planes, we partition Q into finitely many small boxes that we will call *cells*. We show a typical partition in Figure 15.40. We denote the partition by P and its norm (the length of the longest diagonal of all the cells in P) by $\|P\|$.

We number the n cells of P from 1 to n, in any manner. Let (x_i^*, y_i^*, z_i^*) be any point in the ith cell, and denote the volume of the ith cell by ΔV_i. We define the triple integral of a function f of three variables over the region Q as follows.

Definition 15.3
The Triple Integral

$$\iiint\limits_{Q} f(x, y, z)\, dV = \lim_{\|P\| \to 0} \sum_{i=1}^{n} f(x_i^*, y_i^*, z_i^*) \Delta V_i \qquad (15.28)$$

provided the limit on the right-hand side exists.

It can be shown that the triple integral of a continuous function over a rectangular box Q does exist.

Evaluating Triple Integrals by Iterated Integrals

There is a version of Fubini's Theorem for triple integrals that enables us to evaluate the integral by a triple iterated integral. We can state the theorem as follows.

THEOREM 15.3

Fubini's Theorem for Triple Integrals

If f is continuous on the rectangular box $Q = \{(x, y, z) : a_1 \le x \le a_2, b_1 \le y \le b_2, c_1 \le z \le c_2\}$ then

$$\iiint\limits_{Q} f(x, y, z)\, dV = \int_{a_1}^{a_2} \int_{b_1}^{b_2} \int_{c_1}^{c_2} f(x, y, z)\, dz\, dy\, dx \qquad (15.29)$$

There are five other iterated integrals that could be used in place of the one on the right-hand side of Equation 15.29, obtained by changing the order of integration.

EXAMPLE 15.25   Evaluate the triple integral

$$\iiint\limits_{Q} (2xy + 3yz^2)\, dV$$

where Q is the rectangular box $\{(x, y, z) : 0 \le x \le 1, -1 \le y \le 2, 1 \le z \le 3\}$.

Solution   By Equation 15.29,

$$\iiint\limits_{Q} (2xy + 3yz^2)\, dV = \int_0^1 \int_{-1}^2 \int_1^3 (2xy + 3yz^2)\, dz\, dy\, dx$$

$$= \int_0^1 \int_{-1}^2 \left[2xyz + yz^3\right]_1^3 dy\, dx \qquad \text{We held } x \text{ and } y \text{ fixed and integrated with respect to } z.$$

$$= \int_0^1 \int_{-1}^2 (6xy + 27y - 2xy - y)dy\, dx$$

$$= \int_0^1 \int_{-1}^2 (4xy + 26y)\, dy\, dx$$

$$= \int_0^1 \left[2xy^2 + 13y^2\right]_{-1}^2 dx \qquad \text{We held } x \text{ fixed and integrated with respect to } y.$$

$$= \int_0^1 (8x + 52 - 2x - 13)\, dx$$

$$= \int_0^1 (6x + 39)\, dx$$

$$= 3x^2 + 39x\Big]_0^1 = 3 + 39 = 42 \qquad \blacksquare$$

In Exercise 7 of Exercise Set 15.6 we will ask you to redo Example 15.25, using two of the other five iterated integrals, to show they give the same result.

Triple Integrals Over General Regions

If the region Q of integration is not a rectangular box but can be enclosed in a rectangular box, then we partition the rectangular box that encloses Q but count only those cells that are completely contained within Q. We show such a partition in Figure 15.41. With the understanding that in Equation 15.28 the n cells are those contained within Q, Definition 15.3 also applies to this more general case.

It is beyond the scope of this book to describe the most general type of three-dimensional region Q for which the triple integral of a continuous function exists. Instead, we will limit consideration to regions of the following type:

$$Q = \{(x, y, z) : (x, y) \in R, \varphi_1(x, y) \le z \le \varphi_2(x, y)\}$$

where φ_1 and φ_2 are continuous and the projection R of Q onto the xy-plane is

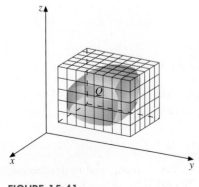

FIGURE 15.41

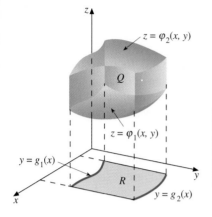

FIGURE 15.42

a two-dimensional region of type I or type II. In Figure 15.42, we have shown such a region Q in which the projection R is a type I region.

It can be shown that if f is continuous on a region Q as described, then the triple integral of f over Q is given by the following formula.

$$\iiint\limits_{Q} f(x, y, z)\, dA = \iint\limits_{R} \left[\int_{\varphi_1(x,y)}^{\varphi_2(x,y)} f(x, y, z)\, dz \right] dA \qquad (15.30)$$

To evaluate the integral on the right, first hold x and y fixed and integrate with respect to z. Then the double integral of the result is taken over R.

If R is the type I region shown in Figure 15.42, then Equation 15.30 can be rewritten in the form

$$\iiint\limits_{Q} f(x, y, z)\, dA = \int_{a}^{b} \int_{g_1(x)}^{g_2(x)} \int_{\varphi_1(x,y)}^{\varphi_2(x,y)} f(x, y, z)\, dz\, dy\, dx \qquad (15.31)$$

Similarly, if R is the type II region

$$R = \{(x, y) : h_1(y) \le x \le h_2(y), c \le y \le d\}$$

then Equation 15.30 can be written as

$$\iiint\limits_{Q} f(x, y, z)\, dA = \int_{c}^{d} \int_{h_1(y)}^{h_2(y)} \int_{\varphi_1(x,y)}^{\varphi_2(x,y)} f(x, y, z)\, dz\, dx\, dy \qquad (15.32)$$

REMARK

■ Formulas analogous to Equation 15.30 exist for regions of the form

$$Q = \{(x, y, z) : (y, z) \in R, \varphi_1(y, z) \le x \le \varphi_2(y, z)\}$$

or

$$Q = \{(x, y, z) : (x, z) \in R, \varphi_1(x, z) \le y \le \varphi_2(x, z)\}$$

In the first case, R is the projection of Q onto the yz-plane, and in the second, R is the projection of Q onto the xz-plane. In each case, R can be of type I or type II in the variables in question.

EXAMPLE 15.26   Evaluate the integral $\iiint_Q 2xz\,dV$, where Q is the region enclosed by the planes $x + y + z = 4$, $y = 3x$, $x = 0$, and $z = 0$.

Solution   The region Q is shown in Figure 15.43. Its projection onto the xy-plane is a type I region, so we will use Equation 15.31. To find the limits on z, we consider a point (x, y, z) in Q and find the smallest and largest values of z so that the point remains in Q. In this case, z varies from the horizontal plane $z = 0$ to the inclined plane $z = 4 - x - y$ that forms the top of the region. The limits on x and y are determined from the region R that is the xy-projection of Q, just as with double integrals. In this case, the limits on y are the traces $y = 3x$ and $y = 4 - x$. They intersect when $x = 1$, so x varies from 0 to 1. Thus,

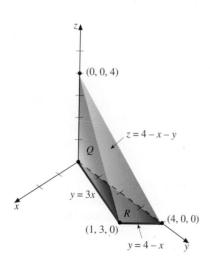

FIGURE 15.43

$$\iiint_Q 2xz\,dV = \int_0^1 \int_{3x}^{4-x} \int_0^{4-x-y} 2xz\,dz\,dy\,dx$$

$$= \int_0^1 \int_{3x}^{4-x} x(4 - x - y)^2\,dy\,dx$$

$$= \int_0^1 \left[\frac{-x(4 - x - y)^3}{3}\right]_{3x}^{4-x} dx = \frac{1}{3}\int_0^1 x(4 - 4x)^3\,dx$$

$$= \frac{64}{3}\int_0^1 x(1 - x)^3\,dx \qquad \begin{array}{l} \text{Integrate by parts} \\ \text{with } u = x, dv = (1 - x)^3\,dx \end{array}$$

$$= \frac{64}{3}\left[-\frac{x(1 - x)^4}{4} - \frac{(1 - x)^5}{20}\right]_0^1$$

$$= \frac{64}{3}\left(\frac{1}{20}\right) = \frac{16}{15} \qquad \blacksquare$$

Applications of Triple Integrals

We can use the triple integral to find volumes and centroids of regions in $\mathbb{R}^3$, masses, centers of mass, and moments of inertia of solids. The ideas parallel those for the two-dimensional case, so we omit the details and simply give the results.

Volume

By taking $f(x, y, z) = 1$, we get the volume of Q:

$$V = \iiint_Q dV \qquad (15.33)$$

Mass

If a solid is in the shape of a three-dimensional region Q and has density $\rho(x, y, z)$, then its mass m is

$$m = \iiint_Q dm \tag{15.34}$$

where $dm = \rho(x, y, z) \, dV$.

Moments

The (first) moments with respect to the xy-plane, the xz-plane, and the yz-plane are

$$M_{xy} = \iiint_Q z \, dm, \quad M_{xz} = \iiint_Q y \, dm, \quad M_{yz} = \iiint_Q x \, dm \tag{15.35}$$

Center of Mass

$$\bar{x} = \frac{\displaystyle\iiint_Q x \, dm}{\displaystyle\iiint_Q dm}, \quad \bar{y} = \frac{\displaystyle\iiint_Q y \, dm}{\displaystyle\iiint_Q dm}, \quad \bar{z} = \frac{\displaystyle\iiint_Q z \, dm}{\displaystyle\iiint_Q dm} \tag{15.36}$$

Centroid

The centroid of a region is the same as the center of mass of a homogeneous solid that occupies that region:

$$\bar{x} = \frac{\displaystyle\iiint_Q x \, dV}{V}, \quad \bar{y} = \frac{\displaystyle\iiint_Q y \, dV}{V}, \quad \bar{z} = \frac{\displaystyle\iiint_Q z \, dV}{V} \tag{15.37}$$

Moments of Inertia

With respect to the coordinate axes, the moments of inertia of a region Q are given by

$$I_x = \iiint_Q (y^2 + z^2) \, dm \quad I_y = \iiint_Q (x^2 + z^2) \, dm \quad I_z = \iiint_Q (x^2 + y^2) \, dm \tag{15.38}$$

EXAMPLE 15.27   A solid of density $\rho(x, y, z) = \sqrt{xyz}$ is in the shape of the region Q enclosed by the surfaces $z = 4 - x^2$, $z = x^2 + y^2$, $x = 0$, and $y = 0$. Set up, but do not evaluate, iterated integrals for (a) the mass m, (b) the x-coordinate of the center of mass, $\bar{x}$, and (c) the moment of inertia with respect to the z-axis, I_z.

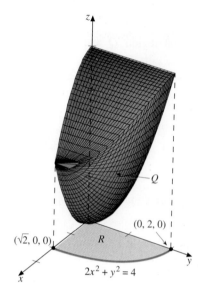

FIGURE 15.44

Solution   The region Q is shown in Figure 15.44. To find the projection of Q onto the xy-plane, we solve the equations $z = 4 - x^2$ and $z = x^2 + y^2$ simultaneously. Setting the two values equal to one another yields the equation $2x^2 + y^2 = 4$. Thus, the xy-projection R of Q is given by

$$R = \left\{ (x, y) : 0 \le x \le \sqrt{2},\ 0 \le y \le \sqrt{4 - 2x^2} \right\}$$

The z limits are from the lower bounding surface $z = x^2 + y^2$ to the upper bounding surface $z = 4 - x^2$.

(a) $m = \displaystyle\iiint_Q dm = \int_0^{\sqrt{2}} \int_0^{\sqrt{4-2x^2}} \int_{x^2+y^2}^{4-x^2} \sqrt{xyz}\ dz\, dy\, dx$

(b) $\bar{x} = \dfrac{M_{yz}}{m} = \dfrac{1}{m} \displaystyle\int_0^{\sqrt{2}} \int_0^{\sqrt{4-2x^2}} \int_{x^2+y^2}^{4-x^2} x\sqrt{xyz}\ dz\, dy\, dx$

(c) $I_z = \displaystyle\iiint_Q (x^2 + y^2)\, dm = \int_0^{\sqrt{2}} \int_0^{\sqrt{4-2x^2}} \int_{x^2+y^2}^{4-x^2} (x^2 + y^2)\sqrt{xyz}\ dz\, dy\, dx$

Note that we could have integrated first with respect to y, with limits from 0 to $\sqrt{z - x^2}$, since the xz-projection of Q is the type I region described by $\{(x, z) : 0 \le x \le \sqrt{2},\ x^2 \le z \le 4 - x^2\}$. Then for part (a),

$$m = \int_0^{\sqrt{2}} \int_{x^2}^{4-x^2} \int_0^{\sqrt{z-x^2}} \sqrt{xyz}\ dy\, dz\, dx$$

and the limits for parts (b) and (c) would have been the same as those for m. ∎

Exercise Set 15.6

In Exercises 1–6, evaluate the iterated integrals. Sketch the region Q determined by the limits of integration.

1. $\displaystyle\int_0^1 \int_1^2 \int_{-1}^1 (xy - 2yz)\, dy\, dz\, dx$

2. $\displaystyle\int_2^4 \int_0^1 \int_0^2 \frac{2x + y}{z^2}\, dx\, dy\, dz$

3. $\displaystyle\int_0^1 \int_0^{1-x} \int_0^2 x\, dz\, dy\, dx$

4. $\displaystyle\int_0^4 \int_0^{\sqrt{z}} \int_0^2 (2xy^2 - 1)\, dy\, dx\, dz$

5. $\displaystyle\int_0^4 \int_0^2 \int_{\sqrt{y}}^{6-y} 4xz\, dx\, dz\, dy$

6. $\displaystyle\int_0^1 \int_0^{2-2y} \int_2^{4-x-2y} z\, dz\, dx\, dy$

7. Redo Example 15.25 using an iterated integral in which
   (a) the first integration is with respect to y and the second is with respect to x;
   (b) the first integration is with respect to x and the second is with respect to z.

In Exercises 8–11, for the given region Q write an iterated integral whose value equals $\iiint_Q f(x, y, z)\, dV$.

8. Q is bounded by $z = 4 - x^2$, $y = x$, $y = 0$, and $z = 0$.

9. Q is the first-octant portion of the region inside the ellipsoid $4x^2 + y^2 + z^2 = 4$.

10. Q is bounded by $2x + 3y = 6$, $x + z = 3$, and the coordinate planes.

11. $Q = \{(x, y, z) : \sqrt{4 - z^2} \le x \le \sqrt{3},\ 0 \le y \le 4,$
$1 \le z \le 2\}$

12. Evaluate the integral in Exercise 8 for $f(x, y, z) = \sqrt{z}$.

13. Evaluate the integral in Exercise 11 for $f(x, y, z) = x(3 - y)z^2$.

In Exercises 14–17, find the volume of the region Q bounded by the given surfaces.

14. $x + 2y + z = 4$ and the coordinate planes

15. $z = x^2 + y^2,\ z = 2,\ y = x,\ y = 0$, in the first octant

16. $x^2 + z^2 = 4,\ y = x,\ y = 0,\ z = 0$, in the first octant

17. $x + z = 1,\ 4x + y + z = 4$, and the coordinate planes

In Exercises 18 and 19, write five different iterated integrals equal to the given integral.

18. $\displaystyle\int_0^1 \int_0^x \int_0^{1-x} f(x, y, z)\, dz\, dy\, dx$

19. $\displaystyle\int_0^4 \int_0^{4-z} \int_0^{\sqrt{y}} f(x, y, z)\, dx\, dy\, dz$

20. Find the centroid of the region bounded by $y^2 = 2x$, $2x + z = 4$, and $z = 0$.

21. For the solid of density $\rho(x, y, z) = 2x$ bounded by the planes $x + y = 1,\ y + z = 1$, and the coordinate planes, find m and $\bar{x}$.

In Exercises 22–28, set up iterated integrals for the specified quantities for the solid that has density $\rho(x, y, z)$ and occupies the region Q.

22. Q is the region above $z = x^2 + y^2$ and below $z = 8 - x^2 - y^2$, $\rho(x, y, z) =$ distance from the xy-plane to the point (x, y, z); $m, \bar{z}, I_z$.

23. The solid of Exercise 21; $\bar{z}, I_y$

24. Q is bounded by $x + 2y + 3z = 6$ and the coordinate planes, $\rho(x, y, z) = x^2 yz$; center of mass.

25. Q is the region inside both $x^2 + y^2 = 4$ and $x^2 + y^2 + z^2 = 16$ above the xy-plane, $\rho(x, y, z) =$ distance from the z-axis to the point (x, y, z); $m, \bar{z}, I_z$.

26. Q is the first-octant portion of the region bounded by $y^2 + z^2 = 4,\ z = x - 2y$, and the coordinate planes, $\rho(x, y, z) = \sqrt{x + y};\ \bar{x}, I_x$.

27. Q is the first-octant portion of the region inside both $x^2 + y^2 = 1$ and $x^2 + z^2 = 1$, $\rho(x, y, z) = xyz$; center of mass.

28. Q is the first-octant portion of the region inside the ellipsoid $2x^2 + y^2 + z^2 = 4$, $\rho(x, y, z) =$ distance from the y-axis to the point (x, y, z); $\bar{y}, I_y$.

29. Find the volume and the location of the centroid of a pyramid of height h and with a base that is a square of side a.

30. Evaluate the integral
$$\iiint_Q y^2 \sin xy\, dV$$
where Q is bounded by $z = x,\ 2xy = \pi,\ y = \pi/2,\ y = \pi$, and $z = 0$. Give two possible physical interpretations of this integral.

31. A solid of density $\rho(x, y, z) = (x+y)^2 e^{(x+y)z}$ occupies the region $Q = \{(x, y, z) : 0 \le x \le 2 - y,\ 0 \le y \le 1,\ 0 \le z \le 1\}$. Find $\bar{z}$.

32. Prove that the first moment of a solid with respect to any plane through its center of mass is 0. (*Hint:* If $\langle a, b, c \rangle$ is a unit normal vector to the plane, the distance d from the plane to a point (x, y, z) is
$$d = \pm[a(x - \bar{x}) + b(y - \bar{y}) + c(z - \bar{z})]$$
which is positive on one side of the plane and negative on the other. (Verify.))

33. The **Parallel Axis Theorem** states that for a solid of mass m, the moment of inertia I_l about any line l is
$$I_l = I_{l'} + md^2$$
where l' is a line through the center of mass parallel to l and at a distance d from it. Prove this theorem.

15.7 TRIPLE INTEGRALS IN CYLINDRICAL AND SPHERICAL COORDINATES

As we saw in Section 15.3, polar coordinates often can be used to simplify double integrals. In a similar way, triple integrals often can be more readily evaluated by using one of two alternatives to rectangular coordinates, called **cylindrical coordinates** and **spherical coordinates**.

Cylindrical Coordinates

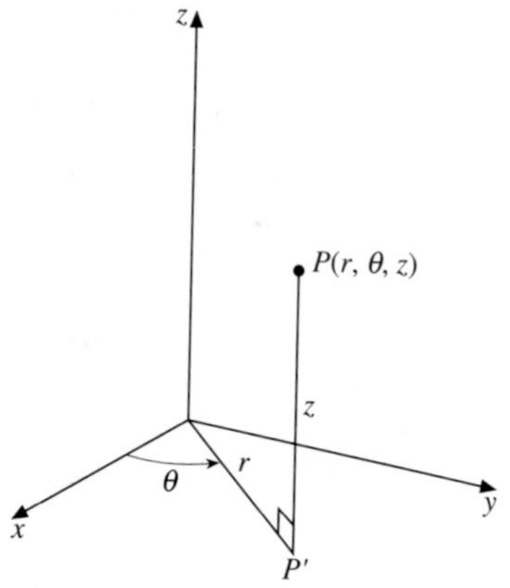

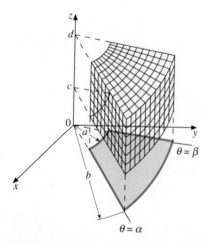

FIGURE 15.45

In a cylindrical coordinate system, polar coordinates are used for two variables and the rectangular coordinate for the third variable. For example, in Figure 15.45 the point P has coordinates (r, θ, z). The projection P' of P onto the xy-plane has polar coordinates r and θ, and z is the usual rectangular z-coordinate. We could equally well use polar coordinates in the xz- or yz-plane, along with the appropriate third rectangular coordinate, but we will concentrate on the situation illustrated in Figure 15.45.

Rectangular and cylindrical coordinates are related by the equations

$$\begin{cases} x = r\cos\theta \\ y = r\sin\theta \\ z = z \end{cases}$$

Some common equations of surfaces in rectangular coordinates, along with the corresponding cylindrical equations, are given below:

	Circular cylinder	Circular cone	Sphere	Paraboloid
Rectangular	$x^2 + y^2 = a^2$	$z^2 = a^2(x^2 + y^2)$	$x^2 + y^2 + z^2 = a^2$	$z = a(x^2 + y^2)$
Cylindrical	$r = a$	$z = ar$	$r^2 + z^2 = a^2$	$z = ar^2$

Triple Integrals in Cylindrical Coordinates

The simplest type of closed region in $\mathbb{R}^3$ to describe in cylindrical coordinates is a set of the form

$$Q = \{(r, \theta, z) : a \le r \le b,\ \alpha \le \theta \le \beta,\ c \le z \le d\}$$

Such a region is shown in Figure 15.46. We will call this region a *cylindrical box*. Notice that its projection onto the xy-plane is a polar rectangle.

Suppose now that f is a function of the cylindrical variables r, θ, and z that is defined on a cylindrical box of the type shown in Figure 15.46. We partition the box by means of horizontal planes, planes containing the z-axis, and circular cylinders centered on the z-axis. These surfaces correspond, respectively, to the equations $z = $ constant, $\theta = $ constant, and $r = $ constant. They divide the box into smaller boxes that we again call *cells*. We denote the partition by P and again call the length of the longest diagonal of all of the cells its norm, which we designate by $\|P\|$. If there are n cells, we number them from 1 to n in any way, and we denote their volumes by $\Delta V_1,\ \Delta V_2,\ \ldots,\ \Delta V_n$. Finally, we choose an arbitrary point $(r_i^*, \theta_i^*, z_i^*)$ in the ith cell, for $i = 1, 2, \ldots, n$. Then we define the triple integral in cylindrical coordinates of f over Q as follows.

FIGURE 15.46
The cylindrical box $Q = \{(r, \theta, z) :$ $a \le r \le b, \alpha \le \theta \le \beta, c \le z \le d\}$ and a partition of it.

Definition 15.4
The Triple Integral in
Cylindrical Coordinates

$$\iiint_Q f(r, \theta, z) \, dV = \lim_{\|P\| \to 0} \sum_{i=1}^{n} f(r_i^*, \theta_i^*, z_i^*) \Delta V_i \qquad (15.39)$$

provided the limit on the right-hand side exists independently of the choices of the points $(r_i^*, \theta_i^*, z_i^*)$.

Fubini's Theorem takes the following form for a cylindrical box such as that shown in Figure 15.46.

If f is continuous on the cylindrical box $Q = \{(r, \theta, z) : a \le r \le b, \; \alpha \le \theta \le \beta, c \le z \le d\}$, then

$$\iiint_Q f(r, \theta, z) \, dV = \int_\alpha^\beta \int_a^b \int_c^d f(r, \theta, z) r \, dz \, dr \, d\theta \qquad (15.40)$$

Equivalently, the iterated integral on the right-hand side may be replaced by any one of the other five iterated integrals obtained by integrating with respect to the three variables in other orders.

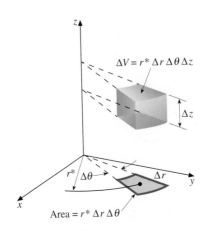

$\Delta V = r^* \Delta r \, \Delta \theta \, \Delta z$

Δz

$r^* \Delta \theta$ Δr

Area $= r^* \Delta r \, \Delta \theta$

FIGURE 15.47

The factor r in the integrand on the right-hand side of Equation 15.40 is suggested by Figure 15.47, in which we show a typical cell in the partition of Q. The area of the base of this cell is $r^* \Delta r \, \Delta \theta$ (we have omitted subscripts for simplicity), just as in the case of polar coordinates. Its height is Δz. So the volume is $\Delta V = r^* \Delta r \, \Delta \theta \, \Delta z$.

We write

$$dV = r \, dz \, dr \, d\theta$$

(or some permutation of the differentials) and call dV in this case the *differential of volume* in cylindrical coordinates.

EXAMPLE 15.28 Evaluate the integral

$$\iiint_Q (zr \sin \theta) \, dV$$

where $Q = \{(r, \theta, z) : 0 \le r \le 2, \; 0 \le \theta \le \pi/2, 0 \le z \le 4\}$.

Solution By Equation 15.40,

$$\iiint_Q (zr \sin \theta) \, dV = \int_0^{\pi/2} \int_0^2 \int_0^4 (zr \sin \theta) \overbrace{r \, dz \, dr \, d\theta}^{dV}$$

$$= \int_0^{\pi/2} \int_0^2 \left[\frac{z^2}{2} r^2 \sin \theta \right]_0^4 dr \, d\theta \qquad \begin{array}{l} \text{We held } r \text{ and } \theta \text{ constant and} \\ \text{integrated with respect to } z. \end{array}$$

$$= \int_0^{\pi/2} \int_0^2 (8r^2 \sin\theta) \, dr \, d\theta$$

$$= \int_0^{\pi/2} \left[\frac{8r^3}{3} \sin\theta\right]_0^2 d\theta \quad \text{We held } \theta \text{ constant and integrated with respect to } r.$$

$$= \int_0^{\pi/2} \frac{64}{3} \sin\theta \, d\theta$$

$$= -\frac{64}{3} \cos\theta\Big]_0^{\pi/2} = -\frac{64}{3}(0-1) = \frac{64}{3} \qquad \blacksquare$$

If the region Q is not a cylindrical box but can be enclosed in such a box, we proceed in the usual way by partitioning a cylindrical box that encloses Q and counting only those cells that lie completely inside Q. If there are n such cells in Q, then Definition 15.4 applies to this case also.

If Q is of the form

$$Q = \{(r, \theta, z) : (r, \theta) \in R, \varphi_1(r, \theta) \le z \le \varphi_2(r, \theta)\}$$

where R is a polar region of one of the types we considered in Section 15.3, then it can be shown that for f continuous on Q,

$$\iiint_Q f(r, \theta, z) \, dV = \iint_R \left[\int_{\varphi_1(r,\theta)}^{\varphi_2(r,\theta)} f(r, \theta, z) \, dz\right] dA$$

In particular, if $R = \{(r, \theta) : g_1(\theta) \le r \le g_2(\theta), \alpha \le \theta \le \beta\}$, then

$$\iiint_Q f(r, \theta, z) \, dV = \int_\alpha^\beta \int_{g_1(\theta)}^{g_2(\theta)} \int_{\varphi_1(r,\theta)}^{\varphi_2(r,\theta)} f(r, \theta, z) \, r \, dz \, dr \, d\theta \qquad (15.41)$$

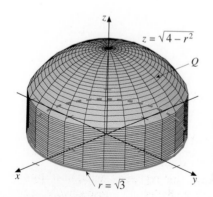

$z = \sqrt{4 - r^2}$

Q

$r = \sqrt{3}$

FIGURE 15.48

Analogous formulas exist for the cases in which R is the projection of Q onto the yz-plane or the xz-plane. In these cases, polar coordinates would be used in the yz-plane, or xz-plane, respectively.

EXAMPLE 15.29   A homogeneous solid is bounded laterally by the circular cylinder $x^2 + y^2 = 3$, on the top by the sphere $x^2 + y^2 + z^2 = 4$, and on the bottom by the xy-plane. Find its center of mass and the moment of inertia with respect to the z-axis.

Solution   The region Q occupied by the solid is shown in Figure 15.48. In cylindrical coordinates the bounding surfaces are $r = \sqrt{3}$, $z = \sqrt{4 - r^2}$, and

$z = 0$. Denote the density by ρ. First, we calculate the mass. By Equations 15.34 and 15.41, we have

$$m = \iiint_Q dm = \iiint_Q \rho \, dV = \int_0^{2\pi} \int_0^{\sqrt{3}} \int_0^{\sqrt{4-r^2}} \rho r \, dz \, dr \, d\theta$$

$$= \rho \int_0^{2\pi} \int_0^{\sqrt{3}} r\sqrt{4-r^2} \, dr \, d\theta = \rho \int_0^{2\pi} \left[-\frac{1}{2} \cdot \frac{2}{3}(4-r^2)^{3/2} \right]_0^{\sqrt{3}} d\theta = \frac{14\pi}{3}\rho$$

By symmetry, we see that $\bar{x} = \bar{y} = 0$, so we need only calculate $\bar{z} = M_{xy}/m$. By Equations 15.35 and 15.41,

$$M_{xy} = \iiint_Q z \, dm = \int_0^{2\pi} \int_0^{\sqrt{3}} \int_0^{\sqrt{4-r^2}} z\rho r \, dz \, dr \, d\theta$$

$$= \rho \int_0^{2\pi} \int_0^{\sqrt{3}} r\left(\frac{4-r^2}{2}\right) dr \, d\theta$$

$$= \rho \int_0^{2\pi} \left[r^2 - \frac{r^4}{8} \right]_0^{\sqrt{3}} d\theta = \frac{15\pi}{4}\rho$$

So

$$\bar{z} = \frac{M_{xy}}{m} = \frac{15\pi\rho}{4} \cdot \frac{3}{14\pi\rho} = \frac{45}{56}$$

From Equations 15.38 and 15.41, we have, for the moment of inertia with respect to the z-axis,

$$I_z = \iiint_Q (x^2 + y^2) \, dm = \iiint_Q r^2 \, dm = \int_0^{2\pi} \int_0^{\sqrt{3}} \int_0^{\sqrt{4-r^2}} \rho r^3 \, dz \, dr \, d\theta$$

$$= \rho \int_0^{2\pi} \int_0^{\sqrt{3}} r^3 \sqrt{4-r^2} \, dr \, d\theta = \frac{94\pi}{15}\rho$$

(You should check the integration on r using either trigonometric substitution, integration by parts, or the substitution $u = \sqrt{4-r^2}$.) ■

Changing from Rectangular to Cylindrical Coordinates

EXAMPLE 15.30   Write an integral in cylindrical coordinates equivalent to

$$\int_0^1 \int_0^{\sqrt{4-x^2}} \int_{\sqrt{x^2+y^2}}^{6-x^2-y^2} f(x, y, z) \, dz \, dy \, dx$$

Solution   The first thing to do is to sketch the region Q determined by the limits of integration. The lower and upper boundaries on z, respectively, are

$$z = \sqrt{x^2 + y^2} \quad \text{and} \quad z = 6 - x^2 - y^2$$

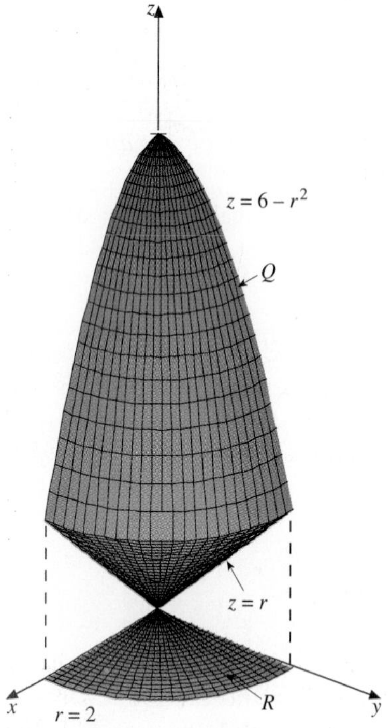

In cylindrical coordinates these equations become $z = r$ and $z = 6 - r^2$. They intersect when $r = 6 - r^2$, with the positive solution $r = 2$. The limits on x and y show that the projection R of Q onto the xy-plane is one-quarter of the circular disk bounded by $r = 2$, as shown in Figure 15.49. So an equivalent integral in cylindrical coordinates is

$$\int_0^{\pi/2} \int_0^2 \int_r^{6-r^2} f(r\cos\theta, r\sin\theta, z) r \, dz \, dr \, d\theta \qquad \blacksquare$$

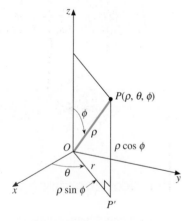

$z = 6 - r^2$

Q

$z = r$

x $r = 2$ R y

FIGURE 15.49

As we illustrated in Example 15.30, when we are given an iterated integral in rectangular coordinates and wish to convert it to cylindrical coordinates, we first find the region Q of integration determined by the rectangular limits. Then we write Q in cylindrical coordinates in order to determine the limits for r, θ, and z. The original integrand, $f(x, y, z)$, becomes $f(r\cos\theta, r\sin\theta, z)$, and the differential of volume $dV = dz\,dy\,dx$ (or some permutation) is replaced by $r\,dz\,dr\,d\theta$ (or the appropriate permutation).

If we denote by T the region R described in cylindrical coordinates, then the transformation equation can be written as

$$\iiint_Q f(x, y, z) \, dA = \iiint_T f(r\cos\theta, r\sin\theta, z) r \, dz \, dr \, d\theta$$

Spherical Coordinates

In the spherical coordinate system, a point P in space is located by its distance ρ from the origin, the polar angle θ from the positive x-axis to the projection OP' of OP onto the xy-plane, and the angle ϕ (phi) from the z-axis to OP. The angles θ and ϕ and the distance ρ are illustrated in Figure 15.50. We restrict ρ and ϕ so that $\rho \geq 0$ and $0 \leq \phi \leq \pi$. From the right triangle $OP'P$, we see that

$$\overline{OP'} = \rho \sin\phi \quad \text{and} \quad \overline{P'P} = \rho \cos\phi$$

FIGURE 15.50

REMARK ─────────────────────────────────

■ We have previously used ρ to designate density. Now we are using it as one of the three spherical coordinates. When we are using spherical coordinates, we will designate density by δ.

───

Thus, since $x = \overline{OP'}\cos\theta$, $y = \overline{OP'}\sin\theta$, $z = \overline{P'P}$, we have

$$\begin{cases} x = \rho\cos\theta\sin\phi \\ y = \rho\sin\theta\sin\phi \\ z = \rho\cos\phi \end{cases} \qquad (15.42)$$

as the equations relating rectangular coordinates to spherical coordinates. On squaring each of these and adding, we get (verify it)

$$x^2 + y^2 + z^2 = \rho^2 \qquad (15.43)$$

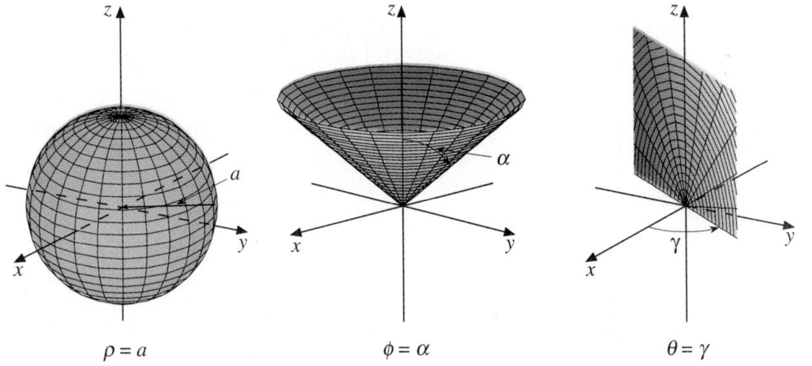

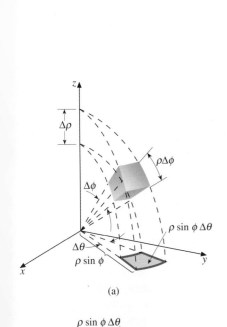

$\rho = a$ $\qquad\qquad$ $\phi = \alpha$ $\qquad\qquad$ $\theta = \gamma$

FIGURE 15.51

The simplest surfaces to represent in spherical coordinates are those with equations of the form $\rho = a$, $\phi = \alpha$, and $\theta = \gamma$, where a, α, and γ are constants. These are, respectively, a sphere of radius a centered at the origin, a half-cone with vertex at the origin and axis along the z-axis, and a plane that contains the z-axis. These surfaces are illustrated in Figure 15.51.

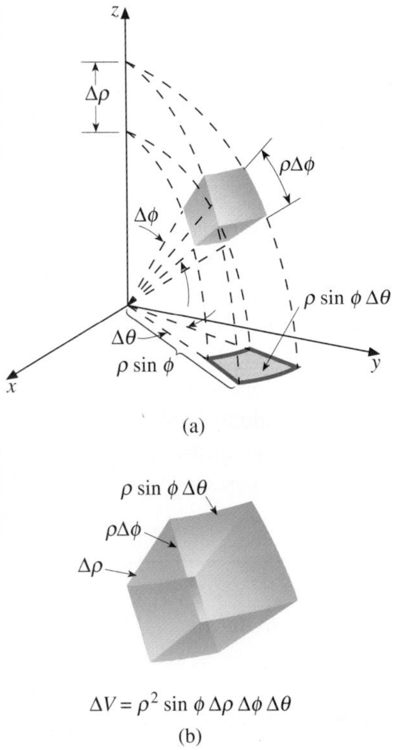

(a)

$\Delta V = \rho^2 \sin \phi \, \Delta \rho \, \Delta \phi \, \Delta \theta$

(b)

FIGURE 15.52

Triple Integrals in Spherical Coordinates

To define the triple integral of a bounded function f over an appropriately restricted closed and bounded region Q, we proceed in a familiar way. Suppose first that Q is of the form $\{(\rho, \theta, \phi) : a \le \rho \le b, \ \alpha \le \theta \le \beta, \ \gamma \le \phi \le \psi\}$. We will call such a region a *spherical box*. Then we partition Q by means of spheres ($\rho = $ constant), half-cones ($\phi = $ constant), and planes ($\theta = $ constant). A typical cell into which this partition subdivides Q is shown in Figure 15.52. If these cells are numbered from 1 to n, we denote their volumes by $\Delta V_1, \Delta V_2, \ldots, \Delta V_n$. In the ith such cell we choose an arbitrary point $(\rho_i^*, \theta_i^*, \phi_i^*)$. Then we define the triple integral of f over Q as follows.

Definition 15.5 The Triple Integral in Spherical Coordinates	$$\iiint\limits_{Q} f(\rho, \theta, \phi) \, dV = \lim_{\|P\| \to 0} \sum_{i=1}^{n} f(\rho_i^*, \theta_i^*, \phi_i^*) \, \Delta V_i \qquad (15.44)$$ provided the limit on the right exists independently of the choices of the points $(\rho_i^*, \theta_i^*, \phi_i^*)$.

To help you understand the appropriate form of dV when going to an iterated integral, let us consider the enlarged cell shown in Figure 15.52(b). Its volume ΔV is approximately the same as that of a rectangular box of dimensions $\Delta \rho$ by $\rho \Delta \phi$ by $\rho \sin \phi \, \Delta \theta$. So

$$\Delta V \approx \rho^2 \sin \phi \, \Delta \rho \, \Delta \phi \, \Delta \theta$$

The smaller the norm of the partition—that is, the smaller $\Delta \rho$, $\Delta \theta$, and $\Delta \phi$ are—the better the approximation. It can be proved that when f is continuous over the spherical box Q, the triple integral can be evaluated as the iterated integral in the following equation.

$$\iiint\limits_{Q} f(\rho, \theta, \phi) dV = \int_{\alpha}^{\beta} \int_{\gamma}^{\psi} \int_{a}^{b} f(r, \theta, \phi) \rho^2 \sin \phi \, d\rho \, d\phi \, d\theta \quad (15.45)$$

We write

$$dV = \rho^2 \sin \phi \, d\rho \, d\phi \, d\theta$$

and call this form the *differential of volume in spherical coordinates*. The factor $\rho^2 \sin \phi$ plays the analogous role to the factor r in the differential of volume in cylindrical coordinates.

If the region Q is not a spherical box but can be enclosed in such a box, we partition a spherical box that encloses Q and count only those cells that lie inside Q, as in previous cases. Then Definition 15.5 continues to hold true, with this restriction on the cells that are counted. The evaluation of the integral of a continuous function f over Q by an iterated integral in this case can be indicated by the following equation.

$$\iiint\limits_{Q} f(\rho, \theta, \phi) dV = \iiint\limits_{\substack{(appropriate \\ limits)}} f(r, \theta, \phi) \rho^2 \sin \phi \, d\rho \, d\phi \, d\theta \quad (15.46)$$

The limits of integration depend on the nature of the bounding surfaces. We will illustrate some common types in the examples.

EXAMPLE 15.31   Find the volume and the centroid of the region shaped like an ice cream cone (see Figure 15.53) bounded by the cone $z = \sqrt{3(x^2 + y^2)}$ and the hemisphere $z = \sqrt{4 - x^2 - y^2}$.

Solution   For the cone, the spherical equation is

$$\rho \cos \phi = \sqrt{3} \, \rho \sin \phi \quad \text{or} \quad \tan \phi = \frac{1}{\sqrt{3}}$$

So $\phi = \pi/6$. The sphere has the equation $\rho = 2$. Thus,

$$V = \iiint\limits_{Q} dV = \int_{0}^{2\pi} \int_{0}^{\pi/6} \int_{0}^{2} \rho^2 \sin \phi \, d\rho \, d\phi \, d\theta$$

$$= \int_{0}^{2\pi} \int_{0}^{\pi/6} \frac{8}{3} \sin \phi \, d\phi \, d\theta = \frac{8}{3} \int_{0}^{2\pi} \left[-\cos \phi \right]_{0}^{\pi/6} d\theta$$

$$= \frac{8}{3} \left[-\frac{\sqrt{3}}{2} + 1 \right] \cdot 2\pi = \frac{8\pi}{3} (2 - \sqrt{3})$$

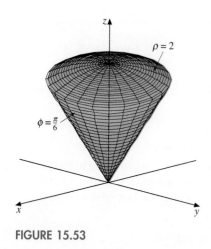

FIGURE 15.53

By symmetry, $\bar{x} = \bar{y} = 0$, so we need only calculate $\bar{z}$. First, we calculate M_{xy}:

$$M_{xy} = \iiint_Q z\, dV = \iiint_Q (\rho \cos \phi)\, dV$$

$$= \int_0^{2\pi} \int_0^{\pi/6} \int_0^2 \rho^3 \sin \phi \cos \phi \, d\rho \, d\phi \, d\theta$$

$$= \int_0^{2\pi} \int_0^{\pi/6} 4 \sin \phi \cos \phi \, d\phi \, d\theta = 4 \int_0^{2\pi} \left[\frac{\sin^2 \phi}{2} \right]_0^{\pi/6} d\theta = \pi$$

So

$$\bar{z} = \frac{M_{xy}}{V} = \frac{\pi}{\dfrac{8\pi}{3}(2 - \sqrt{3})} = \frac{3}{8(2 - \sqrt{3})} \cdot \frac{2 + \sqrt{3}}{2 + \sqrt{3}} = \frac{3(2 + \sqrt{3})}{8} \qquad \blacksquare$$

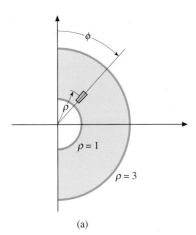

ϕ

ρ

$\rho = 1$

$\rho = 3$

(a)

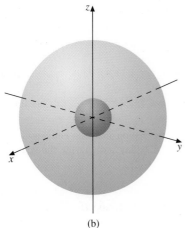

(b)

FIGURE 15.54

EXAMPLE 15.32 Find the mass of the solid between the two spheres $x^2 + y^2 + z^2 = 1$ and $x^2 + y^2 + z^2 = 9$ if the density is inversely proportional to the distance from the origin.

Solution The spheres have equations $\rho = 1$ and $\rho = 3$, and these are the limits of integration on ρ. If we next integrate with respect to ϕ, from $\phi = 0$ to $\phi = \pi$, we will then have integrated over the semiannular region shown in Figure 15.54(a). Then, allowing θ to vary from 0 to 2π in effect rotates this semiannular region around the z-axis to give the entire region shown in Figure 15.54(b). The density at the point (ρ, θ, ϕ) is of the form k/ρ for some constant k, so we have

$$m = \iiint_Q dm = \int_0^{2\pi} \int_0^{\pi} \int_1^3 \frac{k}{\rho}(\rho^2 \sin \phi \, d\rho \, d\phi \, d\theta)$$

$$= k \int_0^{2\pi} \int_0^{\pi} \left[\frac{\rho^2}{2} \right]_1^3 \sin \phi \, d\phi \, d\theta = \frac{8k}{2} \int_0^{2\pi} \int_0^{\pi} \sin \phi \, d\phi \, d\theta = 16k\pi \qquad \blacksquare$$

Exercise Set 15.7

In Exercises 1–6, evaluate the iterated integrals.

1. $\displaystyle\int_0^{\pi} \int_0^1 \int_0^4 rz \sin \theta \, dz \, dr \, d\theta$

2. $\displaystyle\int_0^{\pi/2} \int_1^2 \int_0^{4-r^2} \frac{\cos \theta}{r^2} \, dz \, dr \, d\theta$

3. $\displaystyle\int_0^{2\pi} \int_0^1 \int_0^r zr \, dz \, dr \, d\theta$

4. $\displaystyle\int_0^{\pi} \int_0^{2\sin\theta} \int_0^{\sqrt{4-r^2}} r \, dz \, dr \, d\theta$

5. $\displaystyle\int_0^{2\pi} \int_0^{\pi/2} \int_0^1 \rho^2 \sin \phi \, d\rho \, d\phi \, d\theta$

6. $\displaystyle\int_0^{2\pi} \int_{\pi/6}^{\pi/4} \int_0^{1/\cos\phi} \rho^3 \sin \phi \cos \phi \, d\rho \, d\phi \, d\theta$

In Exercises 7–13, use cylindrical coordinates.

7. Find the volume of the region enclosed by the cylinder $x^2 + y^2 = 4$ and the paraboloid $z = 8 - x^2 - y^2$ that lies above the xy-plane.

8. Find the centroid of the region inside the hemisphere $z = \sqrt{4 - x^2 - y^2}$ and outside the cylinder $x^2 + y^2 = 1$.

9. A homogeneous solid of density δ is bounded by $z = \sqrt{x^2 + y^2}$, $x^2 + y^2 = 1$, and $z = 0$. Find I_z.

10. Find the mass of the solid of constant density δ bounded by the surfaces $z = r$, $z = 0$, and $r = 2\cos\theta$.

11. Find the center of mass of the solid in Exercise 10.

12. Find the center of mass of the homogeneous solid of density δ bounded above by $x^2 + y^2 + z^2 = 4$ and below by $3z = x^2 + y^2$.

13. A solid occupies the region
$$Q = \left\{(r, \theta, z) : 1 \le r \le 3, \ 0 \le \theta \le \frac{\pi}{3}, \ 0 \le z \le 9 - r^2\right\}$$

Its density at any point is inversely proportional to the distance of the point from the z-axis. Find its center of mass.

In Exercises 14–18, use spherical coordinates.

14. Find the volume of a sphere of radius a.

15. Find the center of mass of a solid that occupies the first-octant region inside a sphere of radius a if the density is proportional to the distance from the z-axis.

16. Find the centroid of the region below the hemisphere $z = \sqrt{4 - x^2 - y^2}$ that lies between the upper nappes of the cones $z^2 = 3(x^2 + y^2)$ and $z^2 = x^2 + y^2$.

17. A solid that occupies the region above the xy-plane and between $z = \sqrt{9 - x^2 - y^2}$ and $z = \sqrt{1 - x^2 - y^2}$ has density proportional to the distance from the origin. Find its mass and its moment of inertia with respect to the z-axis.

18. A solid that occupies the region between the spheres $\rho = 1$ and $\rho = 2$ and inside the cone $\rho = \pi/3$ has

density inversely proportional to the distance above the xy-plane. Find its center of mass.

In Exercises 19–23, evaluate the integral by changing to cylindrical or spherical coordinates.

19. $\displaystyle\int_0^2 \int_0^{\sqrt{4-x^2}} \int_0^{\sqrt{x^2+y^2}} \left(z + \sqrt{x^2 + y^2}\right) dz\,dy\,dx$

20. $\displaystyle\int_{-1}^1 \int_{-\sqrt{1-y^2}}^{\sqrt{1-y^2}} \int_1^{2-x^2-y^2} \frac{1}{z^2}\,dz\,dx\,dy$

21. $\displaystyle\int_0^{\sqrt{2}} \int_0^{\sqrt{2-x^2}} \int_{\sqrt{x^2+y^2}}^{\sqrt{4-x^2-y^2}} \sqrt{x^2 + y^2 + z^2}\ dz\,dy\,dx$

22. $\displaystyle\int_0^1 \int_{-\sqrt{1-z^2}}^{\sqrt{1-z^2}} \int_{-\sqrt{1-y^2-z^2}}^{\sqrt{1-y^2-z^2}} x^2\,dx\,dy\,dz$

23. $\displaystyle\int_0^2 \int_{-\sqrt{2x-x^2}}^{\sqrt{2x-x^2}} \int_0^{\sqrt{4-x^2-y^2}} dz\,dy\,dx$

24. Find the centroid of the region that lies both inside the sphere $x^2 + y^2 + z^2 = 2az$ and outside the sphere $x^2 + y^2 + z^2 = a^2$.

25. A solid spherical ball $x^2 + y^2 + z^2 \le a^2$ has density equal to the distance from the xy-plane. The ball is cut by planes $z = h_1$ and $z = h_2$, where $0 < h_1 < h_2 < a$. Find the center of mass of the portion of the ball between these two planes. (*Hint:* Use cylindrical coordinates and integrate first with respect to r.)

26. Find the centroid of the region enclosed by the frustum of the cone shown in the accompanying figure.

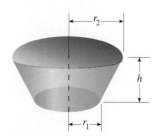

15.8 CHANGING VARIABLES IN MULTIPLE INTEGRALS

We have seen how changing variables from rectangular to polar in double integrals and from rectangular to cylindrical or spherical in triple integrals can often simplify the integration. In this section, we consider the general question of changing variables in double or triple integrals. Changing variables is essentially the same as making a substitution. Just as with substitutions in single integrals, substitutions in multiple integrals require changing not only the integrand, which is usually straightforward, but also the limits of integration and the differential (of area or of volume). These latter two types of changes are generally more difficult, and we concentrate on them.

First, let us review the substitution process in a single integral. Suppose that in the integral $\int_a^b f(x)\,dx$ we make the substitution $x = g(u)$. Then, under suitable restrictions on the functions f and g, we obtain

$$\int_a^b f(x)\,dx = \int_c^d f(g(u))\,\underbrace{g'(u)du}_{dx}$$

where the new limits $u = c$ and $u = d$ satisfy $c = g(a)$ and $d = g(b)$. Notice that the differential dx is replaced by a multiple of du, namely, $g'(u)\,du$.

A similar result holds true when we make the substitution $x = r\cos\theta$, $y = r\sin\theta$ in a double integral. Then, as we have seen, we get

$$\iint_R f(x, y)dA = \iint_S f(r\cos\theta, r\sin\theta)r\,dr\,d\theta$$

where S denotes the domain of integration described in polar coordinates. Note that the differential of area dA in the variables x and y is replaced by a multiple of $dr\,d\theta$, namely, $r\,dr\,d\theta$. In the case of cylindrical coordinates, the differential of volume dV is replaced again by r times the product $dr\,d\theta\,dz$. In spherical coordinates, the multiple is $\rho^2\sin\phi$; that is, dV is replaced by $\rho^2\sin\phi\,d\rho\,d\theta\,d\phi$.

In view of this discussion, it should not be surprising to find that changing from rectangular coordinates to any other set of coordinates causes the differential of area or volume to be replaced by some multiple of the differential of area or volume in the new variables. It turns out that there is a general formula for this multiple, as we now show.

Changing Variables in Double Integrals

Suppose we are given the double integral

$$\iint_R f(x, y)\,dA$$

of the continuous function f over the region R and we want to substitute

$$\begin{cases} x = g(u, v) \\ y = h(u, v) \end{cases} \tag{15.47}$$

Equations 15.47 can be thought of as a function from $\mathbb{R}^2$ to $\mathbb{R}^2$. That is, corresponding to a point (u, v), the equations determine a point (x, y). A function of this type is called a **transformation**, or a **mapping**, from the uv-plane to the xy-plane. Let us denote this transformation by T. Then we can write Equations 15.47 more compactly as

$$T(u, v) = \big(g(u, v), h(u, v)\big)$$

and we can indicate this relationship graphically as in Figure 15.55.

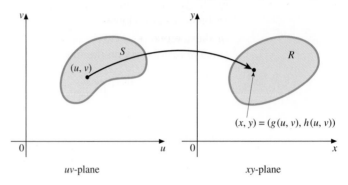

FIGURE 15.55

If (u, v) is a point in the uv-plane, then $T(u, v)$ is called the **image** of (u, v). The **image of a set of points S** in the uv-plane is the set **R** of all images of points in **S**.

EXAMPLE 15.33   Let $T(u, v) = (u^2 - v^2, uv)$. Find the image of the triangle in the uv-plane having vertices $(0, 0)$, $(1, 0)$, and $(1, 1)$.

Solution   Along the base of the triangle from $(0, 0)$ to $(1, 0)$, we see that $v = 0$ and u varies from 0 to 1. So its image under T is the set $\{(u^2, 0) : 0 \le u \le 1\}$ in the xy-plane. That is, $x = u^2$, $y = 0$ are parametric equations of the image. Since $0 \le u \le 1$, it follows also that $0 \le x \le 1$. Thus, the image is the line segment from $(0, 0)$ to $(1, 0)$ on the x-axis.

The image of the vertical side of the given triangle from $(1, 0)$ to $(1, 1)$ is the set $\{(1 - v^2, v) : 0 \le v \le 1\}$ since u is always 1 on this side. Parametric equations of the image are therefore $x = 1 - v^2$, $y = v$, with $0 \le v \le 1$. Eliminating the parameter gives $x = 1 - y^2$, with $0 \le y \le 1$. The graph is thus a parabolic arc from $(1, 0)$ to $(0, 1)$.

The image of the third side of the triangle, joining $(0, 0)$ and $(1, 1)$, is the set $\{(0, u^2) : 0 \le u \le 1\}$, since on this side $v = u$. The parametric equations $x = 0$, $y = u^2$, with $0 \le u \le 1$, define the line segment on the y-axis from $(0, 0)$ to $(0, 1)$.

We show the original triangle in the uv-plane and its image in the xy-plane in Figure 15.56. It is not difficult to show that the entire region S consisting of all points inside and on the triangle is mapped by T onto the region R in the xy-plane.   ■

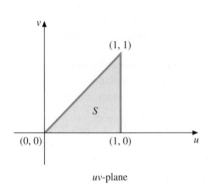

uv-plane

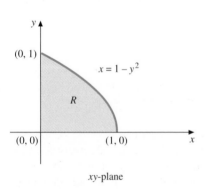

xy-plane

FIGURE 15.56

The transformations that are of most interest to us for our present purposes are one-to-one; that is, they map distinct points in the domain onto distinct points in the range. If T is such a one-to-one transformation given by $T(u, v) = (x, y)$, where $x = g(u, v)$ and $y = h(u, v)$, then there is an inverse transformation T^{-1} such that $T^{-1}(x, y) = (u, v)$. That is, we can solve uniquely for u and v in terms of x and y. We illustrate such a one-to-one mapping and its inverse in the next example.

EXAMPLE 15.34   Let $T(u, v) = (2u - v, \; u + 3v)$, and show that T is a one-to-one mapping from the entire uv-plane onto the entire xy-plane. Find $T^{-1}(x, y)$.

Solution   In order to show that T is one-to-one, we assume that $T(u_1, v_1) = T(u_2, v_2)$ and show as a consequence that $(u_1, v_1) = (u_2, v_2)$. It will follow, then, that no two distinct points in the uv-plane have the same image, or equivalently, if two points (u_1, v_1) and (u_2, v_2) are distinct, their images are distinct.

Assume, then, that $T(u_1, v_1) = T(u_2, v_2)$. Then

$$(2u_1 - v_1, u_1 + 3v_1) = (2u_2 - v_2, u_2 + 3v_2)$$

Thus,

$$\begin{cases} 2u_1 - v_1 = 2u_2 - v_2 \\ u_1 + 3v_1 = u_2 + 3v_2 \end{cases}$$

If we multiply both sides of the top equation by 3 and add the result to the bottom equation, we get $7u_1 = 7u_2$, so $u_1 = u_2$. By substituting u_1 for u_2 in the second equation and simplifying, we get $3v_1 = 3v_2$, so $v_1 = v_2$. Thus, whenever $T(u_1, v_1) = T(u_2, v_2)$, we have $(u_1, v_1) = (u_2, v_2)$, proving that T is one-to-one.

To find T^{-1}, let us set $(x, y) = T(u, v)$. Then

$$x = 2u - v \quad \text{and} \quad y = u + 3v$$

What we want to do is to solve this system of two equations for u and v in terms of x and y. We find that

$$u = \frac{3x + y}{7} \quad \text{and} \quad v = -\frac{x - 2y}{7}$$

or we can write

$$T^{-1}(x, y) = \left(\frac{3x + y}{7}, \; -\frac{x - 2y}{7} \right)$$

No restriction was placed on either (u, v) or (x, y), so the domain of T is the entire uv-plane, and its range is the entire xy-plane. Similarly, the domain of T^{-1} is the entire xy-plane, and its range is the entire uv-plane.   ∎

Another requirement we will place on the transformations we consider is that the component functions have continuous first partial derivatives. That is, if $T(u, v) = (g(u, v), h(u, v))$, we want g_u, g_v, h_u, h_v all to be continuous functions. If we write $x = g(u, v)$ and $y = h(u, v)$, these partials can be denoted by $\partial x / \partial u$, $\partial x / \partial v$, $\partial y / \partial u$, and $\partial y / \partial v$. A transformation having this continuity property is called a C^1 **transformation** (here C denotes "continuous" and the superscript 1 denotes first partial derivatives; a C^2 transformation has continuous second-order partials).

Let $T(u, v) = (g(u, v), h(u, v))$ denote a one-to-one C^1 transformation. We want to see how it transforms a small rectangular element of area, Δu units by Δv units, in the uv-plane. Such an element is of the type that would occur in a partition of a region. As we show in Figure 15.57(a), let (u_0, v_0) denote the lower left-hand corner of the rectangle. On the lower boundary of the rectangle,

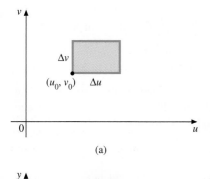

(a)

(b)

FIGURE 15.57

v is constant, namely, $v = v_0$. So its image is the set $\{(g(u, v_0), h(u, v_0):$ $u_0 \leq u \leq u_0 + \Delta u\}$. Parametric equations of this image are

$$\begin{cases} x = g(u, v_0) \\ y = h(u, v_0) \end{cases} \quad u_0 \leq u \leq u_0 + \Delta u$$

Equivalently, we can write its vector equation as

$$\mathbf{r}_1(u) = g(u, v_0)\mathbf{i} + h(u, v_0)\mathbf{k}, \quad u_0 \leq u \leq u_0 + \Delta u$$

Similarly, the image of the left boundary of the rectangle can be given by the vector equation

$$\mathbf{r}_2(v) = g(u_0, v)\mathbf{i} + h(u_0, v)\mathbf{j}, \quad v_0 \leq v \leq v_0 + \Delta v$$

We show these images, along with those of the other two sides of the rectangle, in Figure 15.57(b), where we have denoted by (x_0, y_0) the image of (u_0, v_0).

The images of the left and right endpoints of the lower boundary of the rectangle are $\mathbf{r}_1(u_0)$ and $\mathbf{r}_1(u_0 + \Delta u)$. If Δu is small, the secant vector $\mathbf{r}_1(u_0 + \Delta u) - \mathbf{r}_1(u_0)$ approximates the curved image of this lower boundary. Since

$$\mathbf{r}_1'(u_0) = \lim_{\Delta u \to 0} \frac{\mathbf{r}_1(u_0 + \Delta u) - \mathbf{r}_1(u_0)}{\Delta u}$$

it follows that

$$\mathbf{r}_1(u_0 + \Delta u) - \mathbf{r}_1(u_0) \approx \mathbf{r}_1'(u_0)\Delta u$$

if Δu is small. Consequently, the lower boundary image is approximated by the vector $\mathbf{r}_1'(u_0)\Delta u$. Similarly, the image of the left side of the rectangle is approximated by the vector $\mathbf{r}_2'(v_0)\Delta v$. We show these two vectors in red, drawn from the point (x_0, y_0) in Figure 15.57(b).

Let ΔA denote the area of the image of the rectangular element. Then ΔA is approximately equal to the area of the parallelogram having the two tangent vectors $\mathbf{r}_1'(u_0)\Delta u$ and $\mathbf{r}_2'(v_0)\Delta v$ as adjacent sides. We learned in Section 11.4 that this area is the magnitude of the cross product of the two vectors:

$$\Delta A \approx |\mathbf{r}_1'(u_0)\Delta u \times \mathbf{r}_2'(v_0)\Delta v| = |\mathbf{r}_1'(u_0) \times \mathbf{r}_2'(v_0)|\Delta u \Delta v$$

(Here, we must treat the vectors as three-dimensional, with third component 0, in order for the cross product to be defined.) If we write $x = g(u, v)$ and $y = h(u, v)$, we can calculate the cross product as follows:

$$\mathbf{r}_1'(u_0) \times \mathbf{r}_2'(v_0) = \begin{vmatrix} \mathbf{i} & \mathbf{j} & \mathbf{k} \\ \dfrac{\partial x}{\partial u} & \dfrac{\partial y}{\partial u} & 0 \\ \dfrac{\partial x}{\partial v} & \dfrac{\partial y}{\partial v} & 0 \end{vmatrix} = \begin{vmatrix} \dfrac{\partial x}{\partial u} & \dfrac{\partial y}{\partial u} \\ \dfrac{\partial x}{\partial v} & \dfrac{\partial y}{\partial v} \end{vmatrix}\mathbf{k}$$

where it is understood that the partial derivatives are evaluated at the point (u_0, v_0). Since $|\mathbf{k}| = 1$, we conclude that

$$\Delta A \approx \left| \dfrac{\partial x}{\partial u} \dfrac{\partial y}{\partial v} - \dfrac{\partial x}{\partial v} \dfrac{\partial y}{\partial u} \right| \Delta u \Delta v$$

The second-order determinant involved here is given a special name and symbol, after the German mathematician Carl Jacobi (1804–1851).

Definition 15.6 The Jacobian of a Transformation	The **Jacobian** of the C^1 transformation $T(u, v) = (x, y)$, where $x = g(u, v)$ and $y = h(u, v)$, is denoted by $J(u, v)$ and has the value $$J(u, v) = \begin{vmatrix} \dfrac{\partial x}{\partial u} & \dfrac{\partial y}{\partial u} \\ \dfrac{\partial x}{\partial v} & \dfrac{\partial y}{\partial v} \end{vmatrix} = \dfrac{\partial x}{\partial u}\dfrac{\partial y}{\partial v} - \dfrac{\partial x}{\partial v}\dfrac{\partial y}{\partial u} \qquad (15.48)$$

REMARKS

■ The Jacobian is also frequently designated by $\partial(x, y)/\partial(u, v)$. Thus,

$$\frac{\partial(x, y)}{\partial(u, v)} = J(u, v)$$

■ Note also that $J(u, v)$ can be written as

$$J(u, v) = \begin{vmatrix} \dfrac{\partial x}{\partial u} & \dfrac{\partial x}{\partial v} \\ \dfrac{\partial y}{\partial u} & \dfrac{\partial y}{\partial v} \end{vmatrix} \qquad (15.49)$$

since the value of this determinant is the same as the value given by Equation 15.48. Using the alternative notation

$$\frac{\partial(x, y)}{\partial(u, v)}$$

suggests writing the Jacobian with partials of x on the first row and partials of y on the second, as above.

EXAMPLE 15.35   Find the Jacobian of the transformation given in Example 15.33.

Solution   The transformation is $T(u, v) = (u^2 - v^2, uv)$. That is,

$$x = u^2 - v^2 \qquad \text{and} \qquad y = uv$$

Thus, by Equation 15.48,

$$J(u, v) = \begin{vmatrix} \dfrac{\partial x}{\partial u} & \dfrac{\partial y}{\partial u} \\ \dfrac{\partial x}{\partial v} & \dfrac{\partial y}{\partial v} \end{vmatrix} = \begin{vmatrix} 2u & v \\ -2v & u \end{vmatrix} = 2u^2 + 2v^2 \qquad ■$$

We have seen that under the change of variables $(x, y) = T(u, v)$, a small element of area ΔA in the xy-plane is related to the area $\Delta u \Delta v$ of a rectangular element of area in the uv-plane by the approximation

$$\Delta A \approx |J(u, v)|\Delta u \Delta v$$

It should come as no surprise, then, that when we make this change of variables in a double integral, the differential of area dA in the original integral is replaced by $|J(u, v)|\, du\, dv$ in the new integral.

Our discussion should make the following theorem seem plausible, but it is not a proof. A complete proof can be found in most advanced calculus textbooks. We assume that each of the regions R and S consists of all points inside or on a piecewise-smooth, simple, closed curve.

THEOREM 15.4

Change of Variables in a Double Integral

Let $T(u, v) = (g(u, v), h(u, v))$ be a one-to-one C^1 transformation, with nonzero Jacobian, that maps the region S in the uv-plane onto the region R in the xy-plane. If f is continuous on R, then

$$\iint_R f(x, y) \, dA = \iint_S f(g(u, v), h(u, v)) |J(u, v)| \, du \, dv \qquad (15.50)$$

REMARK

■ We are using $du \, dv$ here for the differential of area in the uv coordinates instead of the usual dA. We are reserving dA for the differential of area in the xy coordinates. So in the context of the double integral on the right-hand side of Equation 15.50, the order of writing the differentials du and dv is not significant. The order becomes significant only when we evaluate the integral by means of iterated integrals.

EXAMPLE 15.36   Find the result of changing from rectangular to polar coordinates in the integral $\iint_R f(x, y) \, dA$.

Solution   The transformation equations are

$$\begin{cases} x = r \cos \theta \\ y = r \sin \theta \end{cases}$$

or equivalently,

$$T(r, \theta) = (r \cos \theta, r \sin \theta)$$

Then (replacing u and v by r and θ, respectively), we have, by Equation 15.49,

$$J(r, \theta) = \begin{vmatrix} \cos \theta & \sin \theta \\ -r \sin \theta & r \cos \theta \end{vmatrix} = r \cos^2 \theta + r \sin^2 \theta = r$$

By restricting r to be the nonnegative value $r = \sqrt{x^2 + y^2}$ and θ to satisfy $0 \le \theta \le 2\pi$, the transformation is one-to-one and $J(r, \theta) \ne 0$ except at the origin, where $r = 0$. (It can be shown that Theorem 15.4 continues to hold true if the Jacobian is zero at finitely many points.) Let S be the region in the (r, θ)-plane having image R in the xy-plane. Then, by Equation 15.50

$$\iint_R f(x, y) dA = \iint_S f(r \cos \theta, r \sin \theta) r \, dr \, d\theta$$

■

Example 15.36 agrees with the result arrived at in Section 15.3, which we have used repeatedly.

Sometimes the nature of the region R suggests a suitable transformation, as the next example shows.

EXAMPLE 15.37   Make a suitable change of variables to evaluate the integral

$$\iint_R (3x - y)\, dA$$

where R is the parallelogram with vertices $(1, 2)$, $(3, 4)$, $(4, 3)$, and $(6, 5)$.

Solution   We show the region R in Figure 15.58. To evaluate the integral without changing variables, we would need to divide the region into three parts in order to have each part either type I or type II. The equations of the sides of the parallelogram are

$$x - y = -1, \ x - y = 1, \ x - 3y = -5, \ \text{and} \ x - 3y = -9$$

Suppose we make the substitution

$$u = x - y \quad \text{and} \quad v = x - 3y$$

Then, the region S in the uv-plane corresponding to R is the rectangle with sides $u = -1$, $u = 1$, $v = -5$, and $v = -9$, shown in Figure 15.59. Our change of variables has therefore simplified the region of integration.

In substituting for u and v in terms of x and y, we have given the inverse of the transformation T described in Theorem 15.4. To find T, we solve for x and y in terms of u and v:

$$
\begin{array}{rl}
x - y & = u \\
x - 3y & = v \\
\hline
2y & = u - v \qquad \text{Subtract.} \\
y & = \dfrac{u - v}{2}
\end{array}
$$

Thus,

$$x = u + y = u + \frac{u - v}{2} = \frac{3u - v}{2}$$

We can therefore write the transformation T as

$$T(u, v) = \left(\frac{3u - v}{2}, \frac{u - v}{2} \right)$$

The Jacobian of the transformation is

$$J(u, v) = \begin{vmatrix} \frac{3}{2} & \frac{1}{2} \\ -\frac{1}{2} & -\frac{1}{2} \end{vmatrix} = -\frac{3}{4} + \frac{1}{4} = -\frac{2}{4} = -\frac{1}{2}$$

Under the transformation, the original integrand becomes

$$3x - y = 3\left(\frac{3u - v}{2} \right) - \left(\frac{u - v}{2} \right) = \frac{9u - 3v - u + v}{2}$$

$$= \frac{8u - 2v}{2} = 4u - v$$

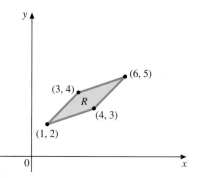

FIGURE 15.58

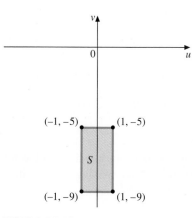

FIGURE 15.59

All of the hypotheses of Theorem 15.4 are readily seen to be satisfied. Thus, by that theorem, we have

$$\iint_R (3x - y)\, dA = \iint_S (4u - v) \left| -\frac{1}{2} \right| du\, dv$$

$$= \frac{1}{2} \int_{-9}^{-5} \int_{-1}^{1} (4u - v)\, du\, dv$$

$$= \frac{1}{2} \int_{-9}^{-5} \left[2u^2 - uv \right]_{-1}^{1} dv$$

$$= \frac{1}{2} \int_{-9}^{-5} [(2 - v) - (2 + v)]\, dv$$

$$= \frac{1}{2} \int_{-9}^{-5} (-2v)\, dv = -\frac{1}{2} \left[v^2 \right]_{-9}^{-5} = -\frac{25}{2} + \frac{81}{2} = 28 \quad \blacksquare$$

In the next example, the substitution is motivated by the form of the integrand.

EXAMPLE 15.38   Evaluate the integral

$$\iint_R (x + y) e^{x^2 - y^2}\, dA$$

by substituting $u = x - y$ and $v = x + y$, where R is the region shown in Figure 15.60.

Solution   Notice that as it stands, there is no elementary antiderivative for the integrand regardless of the order of integration. By making the specified change of variables, the integrand becomes

$$(x + y) e^{x^2 - y^2} = (x + y) e^{(x-y)(x+y)} = v e^{uv}$$

To find the region S in the uv-plane corresponding to R, we note that since $x^2 - y^2 = uv$, the boundaries $x^2 - y^2 = 1$ and $x^2 - y^2 = -1$ transform to $uv = 1$ and $uv = -1$, respectively. The boundaries $x + y = 1$ and $x + y = 2$ transform to $v = 1$ and $v = 2$, respectively. The region S is therefore as shown in Figure 15.61.

By solving for x and y in terms of u and v, we find that

$$(x, y) = T(u, v) = \left(\frac{u + v}{2}, \frac{v - u}{2} \right)$$

So

$$J(u, v) = \begin{vmatrix} \frac{1}{2} & -\frac{1}{2} \\ \frac{1}{2} & \frac{1}{2} \end{vmatrix} = \frac{1}{2}$$

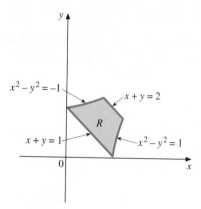

FIGURE 15.60

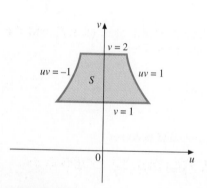

FIGURE 15.61

Consequently, T is a one-to-one C^1 transformation, and $J(u, v) \neq 0$. So by Theorem 15.4,

$$\iint\limits_{R} (x + y)e^{x^2 - y^2}\, dA = \iint\limits_{S} v e^{uv} \left| \frac{1}{2} \right| du\, dv$$

$$= \frac{1}{2} \int_{1}^{2} \int_{-1/v}^{1/v} v e^{uv}\, du\, dv$$

$$= \frac{1}{2} \int_{1}^{2} \left[e^{uv} \right]_{-1/v}^{1/v}\, dv$$

$$= \frac{1}{2} \int_{1}^{2} (e - e^{-1})\, dv = \frac{e - e^{-1}}{2} = \sinh 1 \qquad \blacksquare$$

Changing Variables in Triple Integrals

A result analogous to Theorem 15.4 holds true for triple integrals. The transformation T in this case maps a point (u, v, w) onto a point (x, y, z) by equations of the form

$$x = g(u, v, w), \quad y = h(u, v, w), \quad z = k(u, v, w)$$

We again assume that T is a C^1 transformation—that is, the first partial derivatives of g, h, and k are all continuous on the domain in question—and that T is one-to-one. The Jacobian is the third-order determinant

$$J(u, v, w) = \begin{vmatrix} \dfrac{\partial x}{\partial u} & \dfrac{\partial y}{\partial u} & \dfrac{\partial z}{\partial u} \\[2mm] \dfrac{\partial x}{\partial v} & \dfrac{\partial y}{\partial v} & \dfrac{\partial z}{\partial v} \\[2mm] \dfrac{\partial x}{\partial w} & \dfrac{\partial y}{\partial w} & \dfrac{\partial z}{\partial w} \end{vmatrix}$$

which can also be denoted by

$$\frac{\partial(x, y, z)}{\partial(u, v, w)}$$

If T has nonzero Jacobian and maps the region G onto the region Q, and f is continuous on Q, then the following equation holds true:

$$\iiint\limits_{Q} f(x, y, z)\, dV$$

$$= \iiint\limits_{G} f\left(g(u, v, w),\ h(u, v, w),\ k(u, v, w)\right) |J(u, v, w)|\, du\, dv\, dw$$

$$\text{(15.51)}$$

We leave it as an exercise (Exercise 11 in Exercise Set 15.8) for you to show that the Jacobian of the transformation $T(\rho, \theta, \phi) = (x, y, z)$, where

$$x = \rho \sin \theta \cos \phi, \quad y = \rho \sin \theta \sin \phi, \quad z = \rho \cos \phi$$

is

$$J(\rho, \theta, \phi) = \rho^2 \sin \phi$$

This transformation effects a change of variables from rectangular to spherical coordinates.

Exercise Set 15.8

In Exercises 1–4, find the image R of the region S under the given transformation.

1. $S = \{(u, v) : 1 \le u \le 2, \ 0 \le v \le 3\}$; $T(u, v) = (2u - 3v, \ 3u + 2v)$

2. $S = \{(u, v) : 0 \le u \le 2, \ 0 \le v \le u\}$; $T(u, v) = (2uv, \ v^2 - u^2)$

3. $S = \{(u, v) : 0 \le u \le 1, \ -u^2 \le v \le u^2\}$; $T(u, v) = (u^2 + v, u^2 - v)$

4. $S = \{(u, v) : |u| + |v| \le 1\}$; $T(u, v) = (u - v, u + v)$

In Exercises 5–10, find the Jacobian of the transformation.

5. $x = u + 3v, \ y = 3u - v$

6. $x = u/v, \ y = uv$

7. $x = e^u \cos v, \ y = e^u \sin v$

8. $x = u^2 + v^2, \ y = 2uv$

9. $x = 2u - v + w, \ y = u + v - 2w, \ z = u - 2v + 3w$

10. $x = uv, \ y = vw, \ z = uw$

11. Verify that the Jacobian of the transformation $x = \rho \cos \theta \sin \phi, \ y = \rho \sin \theta \sin \phi, \ z = \rho \cos \phi$ is $\rho^2 \sin \phi$.

In Exercises 12–16, evaluate the given integral by making the indicated change of variables.

12. $\displaystyle\iint\limits_R (x + 2y) \, dA$, where R is the square with vertices $(1, 0), (0, 1), (1, 2),$ and $(2, 1)$; $x = (u + v)/2, \ y = (u - v)/2$

13. $\displaystyle\iint\limits_R (2x - y) \, dA$, where R is the triangle with vertices

$(0, 0), (1, 2),$ and $(3, 3)$; $x = u - v, \ y = u - 2v$

14. $\displaystyle\iint\limits_R y e^{xy} \, dA$, where R is the region bounded by the hyperbolas $xy = 1$ and $xy = 3$ and the lines $y = 1$ and $y = 3$; $x = u/v, \ y = v$

15. $\displaystyle\iint\limits_R y \sin(y^2 - x) \, dA$, where R is the region bounded by $y = \sqrt{x}, x = 2,$ and $y = 0$; $x = v, \ y = \sqrt{u + v}$

16. $\displaystyle\iint\limits_R xy \, dA$, where R is the region bounded by the ellipse $x^2 + 4y^2 = 4$; $x = 2u, \ y = v$

17. Set up a triple integral for the volume enclosed by the ellipsoid

$$\frac{x^2}{a^2} + \frac{y^2}{b^2} + \frac{z^2}{c^2} = 1$$

and evaluate it by making the change of variables $x = au, \ y = bv, \ z = cw$.

In Exercises 18–22, evaluate the integral by making an appropriate change of variables.

18. $\displaystyle\iint\limits_R \cos(x - 2y) \, dA$, where R is the region bounded by the lines $y = 2x, \ y = 2x - 4, \ 2y - x = 0,$ and $2y - x = 3$

19. $\displaystyle\iint\limits_R \frac{4}{(x - y)^2} \, dA$, where R is the region bounded by the lines $x - y = 2, \ x - y = 4, \ x = 0,$ and $y = 0$

20. $\displaystyle\iint\limits_R (x - 2y)^2 \, dA$, where R is the region enclosed by the triangle with vertices $(0, 0), (1, -1),$ and $(2, 1)$

21. $\iint\limits_{R} \dfrac{1}{x}\,dA$, where R is the region bounded by the curves

$xy = 1,\ xy = 4,\ y = 1,$ and $y = 2$

22. $\iint\limits_{R} 6xy\,dA$, where R is the region enclosed by the ellipse

$x^2/9 + y^2/4 = 1$

In Exercises 23 and 24, evaluate the iterated integral by making a suitable change of variables.

23. $\displaystyle\int_{0}^{1/2}\int_{y}^{1-y} \dfrac{\sin(x-y)}{\cos(x+y)}\,dx\,dy$

24. $\displaystyle\int_{1}^{2}\int_{0}^{x-1} (x-y)^{-3}e^{(x+y)/(x-y)}\,dy\,dx$

Chapter 15 Review Exercises

In Exercises 1 and 2, evaluate the integrals. Sketch the region of integration.

1. $\displaystyle\int_{1}^{5}\int_{x}^{2x+1} \dfrac{6x}{(x+y)^2}\,dy\,dx$

2. $\displaystyle\int_{0}^{1}\int_{-\sqrt{y}}^{2-y} xy\sqrt[3]{x^2+4y}\,dx\,dy$

In Exercises 3 and 4, evaluate $\iint_{R} f(x,y)\,dA$.

3. $f(x,y) = (x+y)/y^2;\ R = \{(x,y) : y^2 \le x \le y+2, 1 \le y \le 2\}$

4. $f(x,y) = xe^y;\ R$ is the region bounded by $y = x^2$ and $x - y + 2 = 0$.

In Exercises 5–9, use double integration.

5. Find the area of the region bounded by $y = \ln x$ and $y = (x-1)/(e-1)$.

6. Find the area outside the circle $r = 1$ and inside the limaçon $r = 3 + 4\cos\theta$.

7. Find the volume of the region under the graph of $f(x,y) = x + y$ that lies above $R = \{(x,y) : 0 \le x \le 2,\ 0 \le y \le \sqrt{4-x^2}\}$.

8. Find the volume of the region under the graph of

$$f(x,y) = \dfrac{xy}{x^2+y^2}$$

above the region inside $r = 3\cos\theta$ that is outside $r = 1 + \cos\theta$ and in the first quadrant.

9. Find the volume enclosed by the tetrahedron with vertices $(0,0,0),\ (0,2,0),\ (0,2,4),$ and $(1,2,0)$.

10. Give an equivalent integral with the order of integration reversed:

$$\int_{0}^{4}\int_{1}^{\sqrt{2x+1}} f(x,y)\,dy\,dx$$

11. Evaluate the integral

$$\int_{0}^{1}\int_{x^2}^{1} x\cos^2(y^2)\,dy\,dx$$

12. Change to polar coordinates and evaluate:

$$\int_{0}^{9}\int_{y/3}^{\sqrt{10y-y^2}} (x+y)\,dx\,dy$$

13. A lamina occupies the region bounded by $2y = x^2$, $x = 2$, and $y = 0$. Its density is $\rho(x,y) = 5(x+y)$. Find $(\bar{x}, \bar{y})$ and (r_x, r_y).

14. Find the centroid of the leaf of the rose curve $r = 4\sin 2\theta$ that lies in the first quadrant.

15. A lamina occupies the region $R = \{(x,y) : 0 \le x \le \pi, 0 \le y \le \sin x\}$ and has density $\rho(x,y) = y$. Find $(\bar{x}, \bar{y})$, I_x, and r_y.

In Exercises 16–21, find the volume described using triple integration in rectangular, cylindrical, or spherical coordinates, whichever seems most appropriate.

16. Inside $z = 4 - \sqrt{x^2+y^2}$ and outside $x^2 + y^2 = 1$, in the first octant

17. Bounded by $z = \sqrt{x},\ x + z = 2,\ y = 0,$ and $y = 4$, in the first octant

18. Inside $x^2 + y^2 = 4$, above $z = 0$, and below $2x + y + z = 8$

19. Bounded by $z = 12 - x^2 - y^2,\ y = x^2,$ and $x = 0$, in the first octant

20. Inside $x^2 + y^2 + z^2 = a^2$, between $z^2 = x^2 + y^2$ and $z^2 = 3(x^2 + y^2)$

21. Bounded by $y = x^2 + z^2$, $z = \sqrt{3}\,x$, $z = 0$, and $y = 4$, in the first octant. (*Hint:* Use cylindrical coordinates having the polar variables in the xz-plane.)

22. A solid occupies the first-octant region bounded by the surfaces $y = x^2$, $y = z$, $y = 1$, $y = 4$, and $z = 0$. If its density is $\rho(x, y) = (x + z)/\sqrt{y}$, find its center of mass.

23. For the region that is common to the two half-cones $z = \sqrt{x^2 + y^2}$ and $z = 3 - 2\sqrt{x^2 + y^2}$, find (a) the centroid and (b) I_z.

24. For the solid described in Example 15.32, find the moment of inertia with respect to the z-axis.

25. Find the moment of inertia of the homogeneous solid inside $z = \sqrt{2 - x^2 - y^2}$, below $z = \sqrt{x^2 + y^2}$, and above $z = 0$, with respect to the line $\mathbf{r}(t) = 2\mathbf{j} + t\mathbf{k}$. (*Hint:* Use the law of cosines.)

In Exercises 26–28, find the area of the surface described.

26. The portion of the paraboloid $z = 4 - x^2 - y^2$ above the plane $z = 4$

27. The portion of the surface $z = 2e^{x/2}\sin(y/2)$ that lies above the region

$$R = \{(x, y) : 0 \le x \le \ln 3,\ 0 \le y \le e^x\}$$

28. The band on the sphere $x^2 + y^2 + z^2 = a^2$ cut off by the planes $y = b_1$ and $y = b_2$, where $0 < b_1 < b_2 < a$

29. Let $f(x, y) = (kx)/y$ on the region bounded by $x - 2y = 0$, $y = 1$, $y = 3$, and $x = 0$. Find the following:
(a) k so that f will be a probability density function;
(b) $P(x > 2)$;
(c) μ_x and μ_y;
(d) σ_x and σ_y.

30. Rewrite each integral in either cylindrical or spherical coordinates, as seems most appropriate, but do not evaluate.

(a) $\displaystyle\int_0^2 \int_0^{\sqrt{2y - y^2}} \int_0^{2 - \sqrt{x^2 + y^2}} (xy - y^2)\,dz\,dx\,dy$

(b) $\displaystyle\int_{-1}^1 \int_{-\sqrt{1-x^2}}^{\sqrt{1-x^2}} \int_{\sqrt{x^2+y^2}}^{\sqrt{2-x^2-y^2}} \frac{1}{x^2 + y^2 + z^2}\,dz\,dy\,dx$

(c) $\displaystyle\int_0^2 \int_0^{\sqrt{2x - x^2}} \int_0^{x^2+y^2} \frac{1}{1 + x^2 + y^2}\,dz\,dy\,dx$

(d) $\displaystyle\int_{-1}^1 \int_{-\sqrt{1-z^2}}^{\sqrt{1-z^2}} \int_{-\sqrt{5-x^2-z^2}}^{\sqrt{5-x^2-z^2}} \frac{z^2 - x^2}{z^2 + x^2}\,dy\,dx\,dz$

(*Hint:* Permute the variables in the transformation equations.)

In Exercises 31 and 32, find the Jacobian of the given transformation.

31. (a) $T(u, v) = (u^2 - v, u + 2v^2)$
(b) $x = ue^v,\ y = ve^u$

32. (a) $T(u, v, w) = (2u - 3v + w,\ u - 2w,\ 4u + v - w)$
(b) $x = uv + w,\ y = 2vw,\ z = uvw$

In Exercise 33–35, evaluate the double integral by making the given change of variables.

33. $\displaystyle\iint_R (x - 4y)\sin\frac{\pi(x - y)}{2}\,dA$, where R is the region enclosed by the parallelogram with vertices $(-2, -2)$, $(2, -1)$, $(2, 2)$, and $(6, 3)$; $x = 4u - v,\ y = u - v$

34. $\displaystyle\iint_R \frac{(x + y)^2}{\sqrt{xy}}\,dA$, where R is the region between the curves $y = \sqrt{x}$ and $y = x^2$; $x = u^2,\ y = v^2$

35. $\iint_R xy(\ln x)^2\,dA$, where R is the region bounded by the graphs of $y = 1/x$, $x = 1$, $x = 2$, and $y = 0$; $x = v,\ y = u/v$

In Exercises 36–38, evaluate the integral by making a suitable change of variables.

36. $\iint_R (2x - 3y)\,dA$, where R is the region enclosed by the triangle with vertices $(0, 0)$, $(-1, 1)$, and $(2, 2)$

37. $\iint_R (x - y)e^{x+y}\,dA$, where R is the trapezoid bounded by the lines $y = x + 2$, $y = x + 4$, $x = 0$, and $y = 0$

38. $\int_0^2 \int_0^{2-y} (x^2 - y^2)\sin(x + y)^2\,dx\,dy$

Chapter 15 Concept Quiz

1. Define each of the following:
   (a) The double integral of a bounded function f over a region R in $\mathbb{R}^2$
   (b) A type I region
   (c) A type II region
   (d) The triple integral of a bounded function f over a region Q in $\mathbb{R}^3$
   (e) The Jacobian of the transformation $x = g(u, v)$, $y = h(u, v)$
   (f) A C^1 transformation

2. In each of the following cases, express the double integral $\iint_R f(x, y) \, dA$ as an iterated integral.
   (a) $R = \{(x, y) : a \leq x \leq b, \ g_1(x) \leq y \leq g_2(x)\}$
   (b) $R = \{(x, y) : h_1(y) \leq x \leq h_2(y), \ c \leq y \leq d\}$

3. (a) Give equations relating rectangular coordinates x and y to polar coordinates r and θ. Also give the formula for dA in polar coordinates.
   (b) Give equations relating rectangular coordinates x, y, and z to cylindrical coordinates r, θ, and z. Also give the formula for dV in terms of cylindrical coordinates.

   (c) Repeat part (b), replacing cylindrical coordinates with spherical coordinates ρ, θ, and ϕ.

4. For a lamina of density $\rho(x, y)$ occupying a region R, give formulas in terms of double integrals for each of the following:
   (a) the mass of the lamina
   (b) the coordinates $\bar{x}$ and $\bar{y}$ of the center of mass;
   (c) the moments of inertia I_x, I_y, and I_0.

5. Give a formula for the area of a surface S over a region R in the xy-plane if the equation of S is of the form
   (a) $z = f(x, y)$
   (b) $F(x, y, z) = 0$

6. Suppose the change of variables $x = g(u, v)$, $y = h(u, v)$ is made in the double integral

$$\iint_R f(x, y) \, dA.$$

Make appropriate assumptions about the transformation, and write the equivalent double integral in terms of the new variables.

APPLYING CALCULUS

1. Epidemiologists often model the spread of a disease in a population by assuming that the probability an infected individual will spread the disease to a healthy individual is a function of the distance between them. Consider a circular city, of radius 10 mi. Assume that the population is uniformly distributed and that the individuals with a certain disease are also uniformly distributed throughout the city, with k such individuals per square mile. Suppose that the probability, $f(P)$, that an infected individual at location $P(x, y)$ will infect a healthy person at location $P_0(x_0, y_0)$ is given by

$$f(P) = 0.05[20 - d(P, P_0)]$$

where $d(P, P_0)$ is the distance between P and P_0. Define the total exposure, $E(P_0)$, of a healthy person at P_0 to be the sum of all probabilities of catching the disease from all infected persons.

(a) Write a double integral for $E(P_0)$.

(b) Evaluate the integral in part (a) for P_0 at the center of the city.

(c) Evaluate $E(P_0)$ for P_0 on the edge of the city.

(d) For minimum exposure, is it better to live at the center or on the edge?

2. Assume that the population density (number of people per unit of area) for a certain city is given by

$$N(r) = N_0 e^{-kr}$$

where N_0 is the population density at the center of the city, r is the radial distance from the center, and k is a constant. Let D be a circular disk of radius R, centered at the city center.

(a) Find an expression for the total number of people living in the region.

(b) Evaluate the quotient

$$\frac{\iint_D r N(r) \, dA}{\iint_D N(r) \, dA}$$

and explain what its value gives in practical terms.

(c) In parts (a) and (b), let $R \to \infty$ and explain the meaning of the result in each case.

3. A lamina of constant density $\rho = k$ occupies the circular region described in polar coordinates by $R = \{(r, \theta) : 0 \le r \le a, 0 \le \theta \le 2\pi\}$. A particle of mass m_1 is located on the z-axis at the point $(0, 0, h)$, where $h > 0$.

(a) Show that the magnitude of the force exerted by the laminar mass on the point mass m_1 is given by

$$\iint_R \frac{Gm_1hk}{(r^2 + h^2)^{3/2}} \, dA$$

and evaluate the integral. (*Hint:* Use Newton's Law of Universal Gravitation and the fact that the relevant component of the force is the z component.)

(b) Suppose the lamina occupies the entire xy-plane. By letting $a \to \infty$ in part (a) show that the total force is independent of the height h.

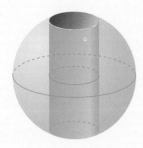

4. A cylindrical hole of radius b is drilled through the center of a solid sphere of radius a, where $0 < b < a$.

(a) Determine the volume of the material that is removed. (*Hint:* Set up a double integral and use polar coordinates.)

(b) What is the volume of the "ring" that remains after the hole has been drilled?

(c) Assume that the sphere has uniform density k. Find the moment of inertia of the "ring" about its axis of symmetry.

5. Schrödinger's Equation in two dimensions has the form

$$-\frac{h^2}{8\pi^2 m}\left(\frac{\partial^2 \psi(x, y)}{\partial x^2} + \frac{\partial^2 \psi(x, y)}{\partial y^2}\right) + V(x, y)\psi(x, y) = E\psi(x, y)$$

where h, m, ψ, V, and E have the meanings noted in Exercise 6 of the Applying Calculus Exercises for Chapter 13. Consider an atomic particle in the two-dimensional "box" $B = \{(x, y) : 0 \le x \le L, 0 \le y \le L\}$, and suppose that the potential energy V is zero inside the box. Since ψ^2 is the probability density function for the position of the particle, and the particle is confined to the box, we must have

$$\iint_B \psi^2(x, y)\,dA = 1$$

and outside the box $\psi^2(x, y) = 0$, reflecting the fact that the probability of the particle's being outside the box is zero. It follows also that $\psi(x, y) = 0$ for points outside the box.

Assume that $\psi(x, y)$ can be written in the form $f(x)g(y)$ where

$$f(x) = A\sin k_x x + B\cos k_x x$$

and

$$g(y) = C\sin k_y y + D\cos k_y y$$

Impose the continuity requirement on ψ on the boundary of the box to conclude that $B = 0$ and $D = 0$. Also show that k_x and k_y must be integral multiples of π/L. That is, $k_x = m_x\pi/L$ and $k_y = m_y\pi/L$, for integers m_x and m_y. Then use the condition

$$\iint_B \psi^2(x, y)\,dA = 1$$

to find the constant $C_1 = AC$ for which

$$\psi(x, y) = C_1 \sin\frac{m_x\pi x}{L}\sin\frac{m_y\pi y}{L}$$

6. A spherical planet of radius R has an atmosphere in which the density decreases exponentially with the altitude above the surface. That is, the density δ at height h above the surface is $\delta = \delta_0 e^{-kh}$, where δ_0 is the density at the surface and k is the constant of proportionality.

(a) Determine the total mass of the planet's atmosphere.

(b) For the planet Earth, half of the atmosphere lies below 5,500 m. Determine the mass of the Earth's atmosphere if the density is 1.20 kg/m^3 at sea level. Use 6,370 km for the radius of the Earth.

(c) Determine the average density of the Earth's atmosphere for that part that lies between 0 and 5,500 m.

Alfred L. Goldberg

Do you use calculus in your medical research?

We use the concepts that we learned in calculus in our analyses of physiological and biochemical processes all the time. Even though in our type of research we seldom use precise mathematical formulations, the concepts of the differential and integral calculus, of differential equations, and of multivariate calculus are nevertheless implicit in our thinking about biochemical and physiological phenomena. For example, if one analyzes a biochemical process, which in living cells is catalyzed by enzymes, the rates of product formation will depend on the concentrations of each of the precursor molecules and on the concentrations and properties of the enzyme. Therefore, we must analyze these processes experimentally as if we were solving multivariate equations. In our laboratory research, we handle logarithmic functions all the time in our analysis of bacterial growth. Bacteria grow exponentially at rates that depend on nutrient supply, temperature, and their genetic makeup, and the analysis of exponential curves can be very informative. Similarly, we use radioactive tracers in analyzing metabolism of cells and tissues, and therefore have to take into account the exponential decay of the radioisotopes when we design such experiments.

In my own research, we are very interested in how cells adapt to new conditions or to hormonal signals by increasing the levels of a specific enzyme. These responses follow a classic differential equation commonly seen in elementary calculus and commonly used in physics or engineering to describe the charging and discharging of an electrical capacitor. The levels of an enzyme depend on the rate at which the cell synthesizes it, which is a linear function of time, and the rate at which the cells destroy that enzyme, which turns out to be an exponential process proportional to the enzyme's concentration. Therefore, the rate of biochemical adaptation depends on an exponential rate constant, and one can predict how quickly the system adapts by knowing the intracellular half-life of the protein.

Is multivariable calculus of use in your field?

We very often encounter multivariable problems, and the researcher has to define the critical variable; that is, to identify the rate-limiting parameter in a process or on an experimental system. Our approach in the laboratory amounts to the experimental solution of a multivariate differential equation in which we control independently each parameter and define how the outcome depends on different variables. We are using calculus subconsciously.

Some areas of medical research are much more quantitative and really involve explicit solutions of multivariate differential equations. For example, in the development of new drugs, pharmacologists have to do research called *pharmacokinetics*, in order to understand the dynamic properties and the stability of a drug in the body. If you were trying to develop a new medication for the treatment of arthritis (disease of the joints), you would want to know how rapidly after ingestion the pill goes into solution in bodily fluids, how rapidly the resulting soluble drug is transported across the intestines into the blood stream, how rapidly it reaches the painful joints, how rapidly it acts once there, and how rapidly the body destroys the drug in the liver or secretes it through the kidneys. Each of these processes is described by rate equations. Moreover, each of these functions may differ with sex, age, or disease, and can determine the important result: how long the drug will be maintained in the blood in an active form. It is then the task of the chemist to synthesize different variant forms of the drug that optimize its pharmacokinetic properties.

These examples from our types of research involve only limited use of the calculus. However, there are very important areas of biochemistry and molecular biology that depend totally on advanced calculus and computer technology for rapid solutions of complex equations. Specifically, the recent revolutionary advances in our knowledge about DNA, enzyme, and protein function have come about from the elucidation of their molecular architectures by X ray diffraction analysis, in which crystals of these large molecules are bombarded with X rays. From the pattern of scattering of the X rays, it is possible through a complex Fourier analysis to determine how the protein or DNA is folded and where individual atoms are located. By this Fourier analysis, very exciting and enlightening images can be obtained of the structure of a virus or of how big protein molecules fold in space, interact with drugs, or are altered in disease states. For example, present efforts to develop drugs to combat AIDS rely on such a biophysical analysis of the critical enzymes of the virus. On this basis, small molecular drugs are being synthesized, and their interactions with targets in the virus are being analyzed by similar biochemical and mathematical approaches.

A somewhat similar mathematical analysis, which also involves Fourier transforms, is extremely useful in studies of large cellular structure. From tomographic analysis of electron-microscopic images, it is possible to obtain precise structural information and to construct three-dimensional images. At the very gross anatomical level, it is now routine in medical diagnosis to use similar mathematical approaches to construct images from nuclear magnetic resonance spectroscopy of the human body. In this case, temporary changes in the physical state of nuclei of the atoms within the body are induced by short exposures to magnetic fields. This perturbation leads to quantitative data that, if analyzed from all angles surrounding the body, can be converted with computers to useful images that indicate whether the structure is normal or affected by disease.

Do you always realize when you are using the concepts of calculus?

No, in fact, when you asked me initially, "Do I use calculus in my work?" my reaction was to deny it, because I always associated calculus with college courses, painful homework, and challenging exams. In fact, although I have forgotten most of the details, those couple of years of calculus have subconsciously permeated my thinking, so that nowadays I practice calculus without a license. In other words, learning how to describe variables precisely, to understand complex functions, focus on one parameter at a time (as occurs in multivariate calculus) was really an exercise in clear, rigorous thinking, from which I have greatly benefited and subconsciously use all the time in research. Also, I have often had the impression that some of my colleagues have contributed so much to our knowledge of biology because of their extensive studies in physics or mathematics, which trained them to think in rigorous, quantitative terms.

Calculus is a requirement for premedical students. Why?

Our medical school requires all entering students to have had a year of calculus and college-level physics, because many of the most common physiological phenomena that doctors have to think about involve the principles of rates or forces, integration over time, electrical events, and so on. For example, the performance of the heart is analyzed in terms of cardiac output, which is the amount of blood pumped per minute, and depends on the heart rate, the force generation during each contraction, the time the valves of the heart are open, and the back pressure and capacitance of the arteries. All these variables are taken into account by the cardiologist in analyzing the nature of cardiac disease and choosing the appropriate therapy. Thus, the cardiologist, like us researchers, is thinking in the rigorous terms of the calculus, though not explicitly formulating or solving mathematical equations. However, the machinery being used in diagnosis, or even in our medical student physiological laboratory exercises, automatically integrates or differentiates the measured parameters to convert them into medically important variables.

Do you use concepts of gradients and flows, which are discussed in multivariate calculus courses?

Absolutely. The concepts of gradients and flows permeates all our analyses of living systems. In evolution, when biological systems could not function adequately simply through diffusion, then active-transport systems or bulk-flow systems evolved, such as the cardiovascular and respiratory systems. Many of these different physiological systems have been analyzed rigorously by multivariate equations. A specific area of physiology and biophysics, where concepts of fields and potentials dominate our thinking, is in understanding the electrical events underlying communication in the nervous system. Signals along nerves and muscles travel as action potentials, which are electrical events whose generation and termination depend on the electrical potential, the ionic fluxes across the cell membrane, and the permeability of the cell membrane to charged ions, such as sodium, potassium, or chloride. Furthermore, these possibilities in turn are nonlinear functions of the electrical potential.

The analysis of these events by Hodgkin and Huxley in the 1950s was a classic achievement that led to our modern understanding of the mechanisms of nerve conduction. At present, many scientists are studying the integration of these ionic events at the synapses between neurons in order to understand how drugs or disease processes alter the actual potentials across the membrane, the molecular channels for different ions, and the resulting fluxes of ions across membranes. We now understand various disease processes in man and animals that lead to changes in ionic flux across membranes and influence whether signals are carried along neurons and muscles. None of these phenomena could be analyzed without knowledge of the mathematics of fields and potentials.

What advice do you have for people in biological studies?

I continue personally to find a life in scientific research exciting and fulfilling, as do my colleagues. My own ability to think rigorously about biological phenomena and to analyze data, I am sure, would be greater were I more adept in use of sophisticated mathematical tools, such as calculus and statistics. Also, I would hope that young people who are excellent in mathematics, physics, and engineering would greatly become more knowledgeable about biology and medicine, where their training and quantitative strengths would enable them to make valuable practical contributions.

Alfred L. Goldberg is professor of cell biology at the Harvard Medical School. He teaches human physiology to medical students and cell biology to graduate students, and trains postdoctoral fellows. A native of Providence, Rhode Island, Prof. Goldberg graduated from Harvard College, was a Churchill Scholar at the University of Cambridge, attended Harvard Medical School, and received his Ph.D. in physiology.

16

VECTOR FIELD THEORY

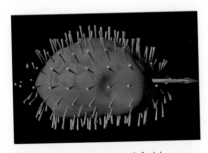

The vector gravitational field on Mars's moon Phobos is shown as blue spikes. The direction to Mars itself is shown with a long, blue arrow to the right.

We introduce the concept of a vector field in this chapter, and we study the calculus associated with vector fields. We will give a formal definition in the first section, but we can indicate the idea of a vector field by considering air currents in some region of space, say in a football stadium. At each point think of a vector giving the magnitude and direction of the air velocity at that point. The totality of such vectors is a vector field that is called a *velocity field*. Or consider the electric force exerted by an electric charge at a fixed point on other charges at various distances from it. At a given location of another charge, there is a force, which can be represented by a vector. The force is attractive if the charges are unlike (one positive and one negative) and repulsive if they are like charges (both positive or both negative). Again, the totality of such vectors is a vector field. It is called an electric *force field*. Another example of a force field is the gravitational field of the Earth.

Vector field theory is of fundamental importance in the physical sciences. For example, it is used in the study of ocean currents, in meteorology, in aerodynamics (such as the flow of air around an airplane or a car), in electricity and magnetism, in heat flow, and in the analysis of effects on objects moving through force fields (such as gravitational forces on a space vehicle).

The two most important tools of calculus used in the study of vector fields are *line integrals* and *surface integrals*. We define both of these types of integrals in this chapter and investigate their properties. We also relate them to integrals we have previously studied. We show applications to vector fields in calculating such things as *work*, *circulation*, and *flux* (all of which we will define). Along the way, we will give some important theorems concerning the various types of integrals we have previously encountered and those we introduce in this chapter.

16.1 VECTOR FIELDS

In Chapter 12 we studied vector-valued functions, which assign either two-dimensional or three-dimensional vectors to real numbers in their domains. Now we consider vector-valued functions whose domains are subsets of $\mathbb{R}^2$ or $\mathbb{R}^3$. We give such functions a special name.

Definition 16.1 A Vector Field	If D is a subset of $\mathbb{R}^2$, a **vector field** on D is a function $\mathbf{F}$ that assigns a two-dimensional vector $\mathbf{F}(x, y)$ to each point (x, y) in D. If D is a subset of $\mathbb{R}^3$, a vector field $\mathbf{F}$ on D assigns a three-dimensional vector $\mathbf{F}(x, y, z)$ to each point (x, y, z) in D.

FIGURE 16.1

As one illustration of a vector field, consider the flow of air over a moving car. We have shown a photograph simulating such airflow in Figure 16.1, with several vectors in the vector field of velocities indicated. Each vector represents the direction and magnitude of the wind velocity at the initial point of the vector. We call the vector field a **velocity field** in this case.

In Figure 16.2 we show the velocity field of the flow of water around a submerged object (such as a half-buried pipe). The lines shown in each of these figures are called *streamlines*. A streamline is the path followed by a particle of fluid (air in Figure 16.1, water in Figure 16.2). The velocity vectors are tangent to the streamlines.

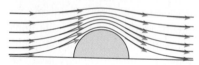

FIGURE 16.2

If $\mathbf{F}$ is a two-dimensional vector field, we can express $\mathbf{F}(x, y)$ in terms of its component functions, say f and g, as

$$\mathbf{F}(x, y) = \langle f(x, y), g(x, y) \rangle$$

or, alternatively, as

$$\mathbf{F}(x, y) = f(x, y)\mathbf{i} + g(x, y)\mathbf{j}$$

The component functions f and g are scalar functions; that is, their values are real numbers, rather than vectors.

Similarly, if $\mathbf{F}$ is three-dimensional, with component functions f, g, and h, we can write

$$\mathbf{F}(x, y, z) = \langle f(x, y, z), g(x, y, z), h(x, y, z) \rangle$$
$$= f(x, y, z)\mathbf{i} + g(x, y, z)\mathbf{j} + h(x, y, z)\mathbf{k}$$

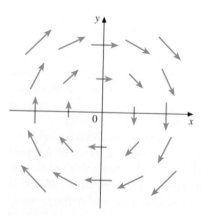

FIGURE 16.3
The vector field $\mathbf{F}(x, y) = \langle y, -x \rangle$

EXAMPLE 16.1   Let $\mathbf{F}(x, y) = \langle y, -x \rangle$. Sketch several members of the vector field defined by $\mathbf{F}$.

Solution   In Figure 16.3, we show vectors $\mathbf{F}(x, y)$ drawn at selected points (x, y). Notice that this vector field suggests a clockwise rotation, with increasing magnitudes as the distance from the origin increases.

We can show that $\mathbf{F}(x, y)$ is tangent to the circle about the origin that passes through the point (x, y) as follows. Let $\mathbf{r} = \langle x, y \rangle$ be the position vector of this point. Then

$$\mathbf{r} \cdot \mathbf{F}(x, y) = \langle x, y \rangle \cdot \langle y, -x \rangle = xy - xy = 0$$

So $\mathbf{F}(x, y)$ is perpendicular to the position vector $\mathbf{r}$, and since $\mathbf{r}$ is a radius vector of the circle, $\mathbf{F}(x, y)$ is tangent to the circle at (x, y). ∎

EXAMPLE 16.2   Sketch several vectors in the three-dimensional vector field $\mathbf{F}(x, y, z) = \langle 0, y, 0 \rangle$.

Solution   We show several vectors in Figure 16.4. Note that all the vectors are parallel to the y-axis. For points to the right of the xz-plane, the vectors point in the direction of the positive y-axis; for points to the left of the xz-plane, they point in the opposite direction. The farther a point is from the xz-plane, the greater the magnitude of the corresponding vector.   ■

The next example illustrates the gravitational field associated with a body such as the earth.

EXAMPLE 16.3   Newton's Law of Gravitation states that the magnitude of the gravitational force between two bodies is

$$|\mathbf{F}| = \frac{GmM}{|\mathbf{r}|^2}$$

where G is the universal gravitational constant (approximately 6.67×10^{-11} N·m²/kg²), m and M are the respective masses of the two bodies, and $\mathbf{r}$ is the vector from the center of mass of one of the bodies to the other. Suppose the body of mass M has its center of mass at the origin. Find the gravitational force field produced by the mass M on a mass m located at a point (x, y, z) in space.

Solution   Let $\mathbf{r}$ denote the position vector $\langle x, y, z \rangle$ of mass m. Then the gravitational force of mass M acting on mass m is in the direction from m toward M, that is, in the direction $-\mathbf{r}$. To find the force $\mathbf{F}$, we multiply its magnitude by the unit vector $-\mathbf{r}/|\mathbf{r}|$ in the direction of the force. Thus,

$$\mathbf{F}(\mathbf{r}) = \frac{-GmM}{|\mathbf{r}|^3}\mathbf{r} \tag{16.1}$$

In terms of components, since $\mathbf{r} = \langle x, y, z \rangle$, Equation 16.1 can be written in the form

$$\mathbf{F}(x, y, z) = \frac{-GmM}{\left(x^2 + y^2 + z^2\right)^{3/2}} \left(x\mathbf{i} + y\mathbf{j} + z\mathbf{k}\right)$$

We show some of the vectors in this gravitational field in Figure 16.5. Notice that all vectors point toward the origin and increase in magnitude as points come closer to the origin.   ■

The electric force field between an electric charge Q and other electric charges is similar to the gravitational field in Example 16.3. According to Coulomb's Law, if Q is located at the origin and q is another electric charge, located at the point (x, y, z) whose position vector is $\mathbf{r}$, then the force $\mathbf{F}(\mathbf{r})$ between the two charges is

$$\mathbf{F}(\mathbf{r}) = \frac{EqQ}{|\mathbf{r}|^3}\mathbf{r} \tag{16.2}$$

where E is a constant. When Q and q are like charges, $qQ > 0$, and the force is repulsive, and when Q and q are unlike, $qQ < 0$, and the force is attractive. Equations 16.1 and 16.2 have essentially the same form.

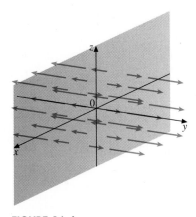

FIGURE 16.4
The vector field $\mathbf{F}(x, y, z) = \langle 0, y, 0 \rangle$

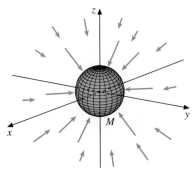

FIGURE 16.5
The gravitational field
$\mathbf{F}(\mathbf{r}) = \dfrac{-GmM}{|\mathbf{r}|^3}\mathbf{r}$

Gradient Fields

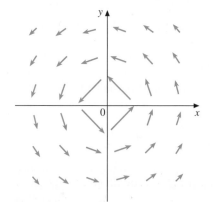

A false-color map of the gravitational potential of the Earth, which is known as the *geoid*.

If f is a scalar function of two variables, then its gradient

$$\nabla f(x, y) = \langle f_x(x, y), f_y(x, y) \rangle$$

is a vector field, called a **gradient field**. The function f is called a **potential function** for the gradient field.

Similarly, if f is a scalar function of three variables, then it is a potential function for the three-dimensional gradient field

$$\nabla f(x, y, z) = \langle f_x(x, y, z), f_y(x, y, z), f_z(x, y, z) \rangle$$

Finding the gradient field for a given potential function is straightforward, as the next example illustrates.

EXAMPLE 16.4   Find the gradient field of the potential function

$$f(x, y) = \tan^{-1} \frac{y}{x}$$

Sketch several vectors in the field.

Solution   First, we calculate the partial derivatives of f.

$$\frac{\partial f}{\partial x} = \frac{1}{1 + \frac{y^2}{x^2}} \cdot \left(-\frac{y}{x^2} \right) = \frac{-y}{x^2 + y^2}$$

$$\frac{\partial f}{\partial y} = \frac{1}{1 + \frac{y^2}{x^2}} \cdot \left(\frac{1}{x} \right) = \frac{x}{x^2 + y^2}$$

Thus, the gradient field is

$$\nabla f(x, y) = \left\langle \frac{-y}{x^2 + y^2}, \frac{x}{x^2 + y^2} \right\rangle$$

FIGURE 16.6
The gradient field
$$\nabla f = \left\langle -\frac{y}{x^2 + y^2}, \frac{x}{x^2 + y^2} \right\rangle$$

We show some of the vectors in Figure 16.6. As in Example 16.1, the vector $\nabla f(x, y)$ is tangent to the circle centered at the origin and passes through the point (x, y).   ∎

Gradient fields are special types of vector fields that have many interesting and useful properties, as we will show in Section 16.3. Such fields are said to be **conservative**. Finding the gradient field of a given potential is relatively easy, as we have seen. Determining whether a given vector field is conservative and finding a potential function for it when it is conservative is, in general, more difficult. We will take up this question in Section 16.3 also.

Exercise Set 16.1

*In Exercises 1–10, draw several members of the vector field **F**.*

1. $\mathbf{F}(x, y) = \langle -y, x \rangle$

2. $\mathbf{F}(x, y) = \langle 2x, y \rangle$

3. $\mathbf{F}(x, y) = \dfrac{x\mathbf{i} + y\mathbf{j}}{\sqrt{x^2 + y^2}}$

4. $\mathbf{F}(x, y) = -x\mathbf{i} - y\mathbf{j}$

5. $\mathbf{F}(x, y) = \dfrac{y\mathbf{i}}{|y|}$

6. $\mathbf{F}(x, y) = \langle x + y, y - x \rangle$

7. $\mathbf{F}(x, y, z) = \langle 1, 1, 0 \rangle$

8. $\mathbf{F}(x, y, z) = \dfrac{x\mathbf{i} + y\mathbf{j} + z\mathbf{k}}{\sqrt{x^2 + y^2 + z^2}}$

9. $\mathbf{F}(x, y, z) = z\mathbf{k}$

10. $\mathbf{F}(x, y, z) = \langle 0, -z, y \rangle$

In Exercises 11–14, find the gradient field of the given potential function, and sketch several of its vectors.

11. $f(x, y) = x^2 - 2y^2$

12. $f(x, y) = \sqrt{x^2 + y^2}$

13. $f(x, y) = \ln \sqrt{x^2 + y^2}$

14. $f(x, y) = \tan^{-1} \dfrac{x}{y}$

In Exercises 15–20, find the gradient field for the given potential function f.

15. $f(x, y) = e^{-x} \sin 2y$

16. $f(x, y) = \ln(\cos xy)$

17. $f(x, y, z) = xy - xz + yz$

18. $f(x, y, z) = \sqrt{x^2 + y^2 + z^2}$

19. $f(x, y, z) = x^2 y^3 z$

20. $f(x, y, z) = \ln\left(x^2 + y^2 + z^2\right)$

21. Show that the gravitational field in Example 16.3 is conservative by showing it is the gradient field of the potential function

$$f(x, y, z) = \dfrac{GMm}{\sqrt{x^2 + y^2 + z^2}}$$

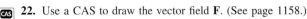 **22.** Use a CAS to draw the vector field **F**. (See page 1158.)

(a) $\mathbf{F}(x, y) = \dfrac{(y\mathbf{i} - x\mathbf{j})}{\sqrt{x^2 + y^2}}$

(b) $\mathbf{F}(x, y) = 2x\mathbf{i} + y\mathbf{j}$

(c) $\mathbf{F}(x, y) = -y\mathbf{i} + x\mathbf{j}$

(d) $\mathbf{F}(x, y, z) = x\mathbf{i} + y\mathbf{j} + z\mathbf{k}$

16.2   LINE INTEGRALS AND WORK

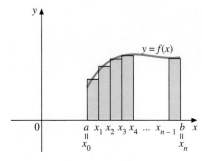

FIGURE 16.7

$$\int_a^b f(x)\, dx = \lim_{\max \Delta x_k \to 0} \sum_{k=1}^{n} f(c_k) \Delta x_k$$

In this section we introduce a new type of integral, called a **line integral**, that is especially useful in studying properties of vector fields. As one application, we will show how a line integral can be used to find the work done by a force field in moving a particle along a curve.

In Chapter 5, we defined the integral $\int_a^b f(x)\, dx$ of a function of one variable as a limit of Riemann sums formed by partitioning the interval $[a, b]$ on the x-axis. (See Figure 16.7.) For the line integral, we replace the interval $[a, b]$ by a curve C in the plane, and we replace the function $f(x)$ of one variable by a function $f(x, y)$ of two variables.

We begin with a curve C in the xy-plane given by the parametric equations

$$x = x(t),\ y = y(t),\quad a \le t \le b$$

and we assume that C is rectifiable (has finite length). We partition the parameter interval $[a, b]$ by points

$$a = t_0 < t_1 < t_2 < \cdots < t_n = b$$

which induces a partition of the curve C by the points

$$P_0 = (x(t_0), y(t_0)), \; P_1 = (x(t_1), y(t_1)), \ldots, P_n = (x(t_n), y(t_n))$$

as shown in Figure 16.8(a). Let Δs_k denote the length of the arc of C from P_{k-1} to P_k and let $P_k^* = (x_k^*, y_k^*)$ denote any point on C between P_{k-1} and P_k, for $k = 1, 2, \ldots, n$. In Figure 16.8(b) we show an enlarged view of a typical subarc from P_{k-1} to P_k, along with the point P_k^*. Now we form the sum

$$\sum_{k=1}^{n} f(x_k^*, y_k^*) \, \Delta s_k$$

where f is a function of two variables whose domain contains the curve C. Note the similarity between this sum and a Riemann sum for the integral $\int_a^b f(x) \, dx$. If this sum approaches a finite limit as we take partitions such that the lengths Δs_k of the subarcs all approach zero, then this limit is the line integral of f along C.

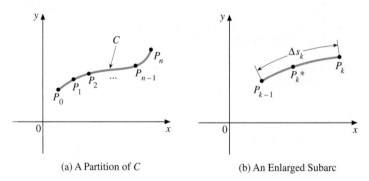

(a) A Partition of C (b) An Enlarged Subarc

FIGURE 16.8

Definition 16.2 The Line Integral	If f is a function of two variables defined on the rectifiable curve C, then the **line integral of f along C** is $$\int_C f(x, y) \, ds = \lim_{\max \Delta s_k \to 0} \sum_{k=1}^{n} f(x_k^*, y_k^*) \Delta s_k \qquad (16.3)$$ provided this limit exists independently of the choices of the points (x_k^*, y_k^*).

REMARK ————————————————————

■ If f is continuous on C, and C is rectifiable, then it can be shown that the line integral of f along C exists.

————————————————————————————————

Note that if f is the constant function $f(x, y) = 1$, then Equation 16.3 gives the length of C, since the sum on the right-hand side simply adds up the lengths of the subarcs.

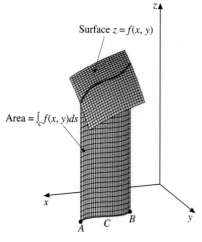

FIGURE 16.9

$$\text{Length of } C = \int_C ds$$

The name *line integral* is unfortunate, since we normally interpret "line" to mean "straight line." *Curvilinear integral* would be a better name. We will use *line integral*, however, since it occurs so widely in the literature.

Geometric Interpretation

When f is a nonnegative continuous function on C, we can give a geometric interpretation to $\int_C f(x, y) \, ds$ as the area of the cylindrical surface obtained by projecting C upward until it meets the surface $z = f(x, y)$. (See Figure 16.9.) To see why, observe that each term $f(x_k^*, y_k^*) \, \Delta s_k$ of the sum on the right-hand side of Equation 16.3 gives the area of a rectangle that approximates the area of a vertical strip with curved base Δs_i. Figure 16.10(a) shows one such strip. In Figure 16.10(b) we show a rectangle whose area approximates the area of the strip. In the limit we get the exact area. Note the similarity with the interpretation of $\int_a^b f(x) \, dx$ as the area above the interval $[a, b]$ and below the graph of f, when f is nonnegative.

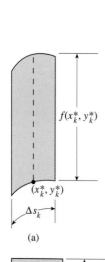

(a)

The Area of a Cylindrical Surface

If f is nonnegative and continuous on the rectifiable curve C, then the area A of the cylindrical surface formed by projecting C upward until it meets the surface $z = f(x, y)$ is

$$A = \int_C f(x, y) \, ds$$

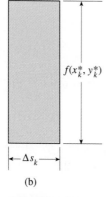

(b)

FIGURE 16.10

A Computational Formula for the Line Integral

Definition 16.2 does not provide a very useful means of calculating the values of line integrals. The following theorem, which we state without proof, provides a more efficient computational method. Recall that an arc C is *smooth* if it has a parameterization $x = x(t)$, $y = y(t)$, for $a \le t \le b$, where x' and y' are continuous on $[a, b]$ and not simultaneously 0 there. We saw in Chapter 9 that smooth curves are rectifiable.

THEOREM 16.1

If f is continuous on the smooth curve C, with parameterization $x = x(t), y = y(t)$ for $a \le t \le b$, then

$$\int_C f(x, y) \, ds = \int_a^b f(x(t), y(t)) \sqrt{\left(\frac{dx}{dt}\right)^2 + \left(\frac{dy}{dt}\right)^2} \, dt \qquad (16.4)$$

The result seems plausible since we saw in Chapter 9 that

$$ds = \sqrt{\left(\frac{dx}{dt}\right)^2 + \left(\frac{dy}{dt}\right)^2} \, dt$$

According to Theorem 16.1 we can evaluate a line integral as an ordinary integral over the parameter interval $[a, b]$ by writing both the integrand and the differential in terms of the parameter.

REMARK
■ Recall that a curve C can have more than one parameterization. It can be shown that the value of the integral is independent of the parameterization used. This fact is important because some choices of parameterization simplify computation of the integral.

EXAMPLE 16.5   Find the area of the portion of the parabolic cylinder $y = x^2/2$ from $x = 0$ to $x = 2$ that lies in the first octant and is bounded above by the cone $2z = \sqrt{x^2 + 4y^2}$.

Solution   The surfaces in question are shown in Figure 16.11. The desired area is given by the line integral

$$\text{Area} = \int_C f(x, y) \, ds$$

where C is the arc of the parabola $y = x^2/2$ from $(0, 0)$ to $(2, 2)$ and $f(x, y) = \sqrt{x^2 + 4y^2}/2$. A parameterization of C can be obtained by setting $x = t$:

$$\begin{cases} x = t \\ y = \dfrac{t^2}{2} \end{cases} \quad 0 \le t \le 2$$

With this parameterization, we have

$$ds = \sqrt{\left(\frac{dx}{dt}\right)^2 + \left(\frac{dy}{dt}\right)^2} \, dt = \sqrt{1 + t^2} \, dt$$

By Equation 16.4, we therefore have

$$\text{Area} = \int_C \frac{\sqrt{x^2 + 4y^2}}{2} \, ds = \int_0^2 \frac{\sqrt{t^2 + t^4}}{2} \sqrt{1 + t^2} \, dt$$

$$= \frac{1}{2} \int_0^2 t(1 + t^2) \, dt = 3$$

(You should supply the details of the integration.)   ■

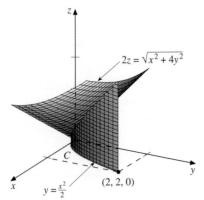

$2z = \sqrt{x^2 + 4y^2}$

$(2, 2, 0)$

$y = \frac{x^2}{2}$

FIGURE 16.11

Another parameterization of the curve C in Example 16.5 is $x = e^t$, $y = e^{2t}/2$, for $-\infty < t < \ln 2$, as you can verify by eliminating the parameter. In the exercises (Exercise 11 in Exercise Set 16.2) we will ask you to show that with this parameterization, the result found in the example is unchanged.

Sectionally Smooth Curves

If C is composed of finitely many smooth curves joined end-to-end, then C is said to be **sectionally smooth**. Suppose, for example, that the smooth curves $C_1, C_2, \ldots, C_n$ are joined end-to-end to form the curve C. Then we write $C = C_1 \cup C_2 \cup C_3 \cup \cdots \cup C_n$ and define the line integral of f over C by

$$\int_C f(x, y)\, ds = \int_{C_1} f(x, y)\, ds + \int_{C_2} f(x, y)\, ds + \cdots + \int_{C_n} f(x, y)\, ds$$

EXAMPLE 16.6   Find

$$\int_C xy^2\, ds$$

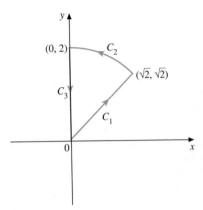

FIGURE 16.12

where C is the sectionally smooth closed curve shown in Figure 16.12, consisting of the line segment $y = x$ from $(0, 0)$ to $\left(\sqrt{2}, \sqrt{2}\right)$, the circular arc $x^2 + y^2 = 4$ from $\left(\sqrt{2}, \sqrt{2}\right)$ to $(0, 2)$, and the y-axis from $(0, 2)$ to $(0, 0)$.

Solution   Denote the three parts of C by C_1, C_2, and C_3, respectively, as shown. We can parameterize these curves as follows:

$$C_1 : \begin{cases} x = t \\ y = t \end{cases} \quad 0 \le t \le \sqrt{2} \qquad C_2 : \begin{cases} x = 2\cos t \\ y = 2\sin t \end{cases} \quad \pi/4 \le t \le \pi/2$$

$$C_3 : \begin{cases} x = 0 \\ y = 2 - t \end{cases} \quad 0 \le t \le 2$$

Then on C_1, $ds = \sqrt{1 + 1}\, dt = \sqrt{2}\, dt$. So by Equation 16.4,

$$\int_{C_1} xy^2\, ds = \int_0^{\sqrt{2}} t^3 \left(\sqrt{2}\, dt\right) = \sqrt{2} \left[\frac{t^4}{4} \right]_0^{\sqrt{2}} = \sqrt{2}$$

On C_2, $ds = \sqrt{(-2\sin t)^2 + (2\cos t)^2}\, dt = \sqrt{4\left(\sin^2 t + \cos^2 t\right)}\, dt = 2\, dt$, so

$$\int_{C_2} xy^2\, ds = \int_{\pi/4}^{\pi/2} (2\cos t)\left(4\sin^2 t\right)(2\, dt) = 16 \int_{\pi/4}^{\pi/2} \sin^2 t \cos t\, dt$$

$$= \frac{16}{3} \sin^3 t \Bigg]_{\pi/4}^{\pi/2}$$

$$= \frac{16}{3} - \frac{4}{3}\sqrt{2}$$

On C_3 the integrand $xy^2 = (0)(2 - t)^2 = 0$, so the integral is also 0.

Combining the results, we have

$$\int_C xy^2\, ds = \int_{C_1} xy^2\, ds + \int_{C_2} xy^2\, ds + \int_{C_3} xy^2\, ds$$

$$= \sqrt{2} + \frac{16}{3} - \frac{4\sqrt{2}}{3} = \frac{16}{3} - \frac{\sqrt{2}}{3} \qquad \blacksquare$$

Parameterizing Curves

We have seen that if a curve is given in the form $y = f(x)$ from (x_1, y_1) to (x_2, y_2), then we can take x as the parameter and write

$$\begin{cases} x = t \\ y = f(t) \end{cases} \quad x_1 \leq t \leq x_2$$

(If $x_2 < x_1$, then the t limits are from x_2 to x_1.) Similarly, if $x = g(y)$ from (x_1, y_1) to (x_2, y_2), we can use y as the parameter to get

$$\begin{cases} x = g(t) \\ y = t \end{cases} \quad y_1 \leq t \leq y_2$$

(where, again, if $y_2 < y_1$, the lower t limit is y_2 and the upper limit is y_1).

For a circle $x^2 + y^2 = a^2$, using the polar angle θ (see Figure 16.13) as the parameter gives $x = a \cos \theta$ and $y = a \sin \theta$, or on replacing θ by t,

$$\begin{cases} x = a \cos t \\ y = a \sin t \end{cases}$$

The parameter interval in this case depends on the orientation. Starting from $(a, 0)$, if the orientation is counterclockwise, then t goes from 0 to 2π. If only a part of the circle is to be used, the limits have to be adjusted accordingly.

To parameterize a straight line segment from $P_1(x_1, y_1)$ to $P_2(x_2, y_2)$, recall that a direction vector for the line through P_1 and P_2 is $\langle x_2 - x_1, y_2 - y_1 \rangle$. So a vector equation of the entire line is

$$\mathbf{r}(t) = \langle x_1, y_1 \rangle + \langle x_2 - x_1, y_2 - y_1 \rangle t \tag{16.5}$$

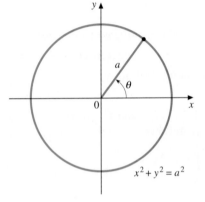

FIGURE 16.13
$x = a \cos \theta, \; y = a \sin \theta$

By limiting t to the interval $0 \leq t \leq 1$, we get the line segment from P_1 to P_2. Thus, the parametric equations are

$$\begin{cases} x = x_1 + (x_2 - x_1)t \\ y = y_1 + (y_2 - y_1)t \end{cases} \quad 0 \leq t \leq 1 \tag{16.6}$$

Although Equations 16.6 always work, certain line segments, such as parts of horizontal or vertical lines, can usually be parameterized more easily. For example, we could use $x = t, y = b$ for a horizontal line segment along $y = b$. The t limits are determined by the x-coordinates of the endpoints (taking direction into account) of the line segment. Similarly, $x = a, y = t$ is a parameterization of a vertical line segment along $x = a$, with the t limits determined by the endpoint y values.

Work Done by a Variable Force

An important application of line integrals is in calculating the work done by a variable force $\mathbf{F}$ that acts on a particle as it moves along a curve C. This application extends the definition given in Chapter 11 of the work done by a variable force as the particle moves along a straight line. The force $\mathbf{F}$ varies from point to point and so is a vector-valued function of x and y. That is, $\mathbf{F}$ is a force field, which as we saw in Section 16.1 is a particular kind of vector field.

To motivate the definition, let $\mathbf{F}(x, y) = \langle f(x, y), g(x, y) \rangle$ be a continuous force field whose domain includes the smooth curve C with vector equation $\mathbf{r}(t) = \langle x(t), y(t) \rangle$ for $a \leq t \leq b$. Partition the interval $[a, b]$, thereby obtaining a partition of the curve C, as in Definition 16.2, by the points $P_k = (x_k, y_k)$,

$k = 0, 1, 2, \ldots, n$. If the subarc of length Δs_k is sufficiently small, the force can be considered approximately equal to the constant value $\mathbf{F}(x_k^*, y_k^*)$ on that subarc, where (x_k^*, y_k^*) is any point on the subarc. Thus, the work ΔW_k in moving the particle along the kth subarc is approximately equal to the tangential component of $\mathbf{F}(x_k^*, y_k^*)$ multiplied by the length of the subarc, Δs_k (see Figure 16.14). That is,

$$\Delta W_k \approx \text{Comp}_{\mathbf{T}}\, \mathbf{F}(x_k^*, y_k^*)\Delta s_k \qquad \text{See Section 11.2}$$

$$= \mathbf{F}(x_k^*, y_k^*) \cdot \mathbf{T}(x_k^*, y_k^*)\Delta s_k \qquad \text{By Equation 11.9}$$

where $\mathbf{T}(x_k^*, y_k^*)$ is the unit tangent vector to the curve C at (x_k^*, y_k^*). (See Equation 11.7.) The total work W is therefore

$$W = \sum_{k=1}^{n} \Delta W_k \approx \sum_{k=1}^{n} \mathbf{F}(x_k^*, y_k^*) \cdot \mathbf{T}(x_k^*, y_k^*)\Delta s_k$$

By taking the limit as max $\Delta s_k \to 0$, we are led to the following definition.

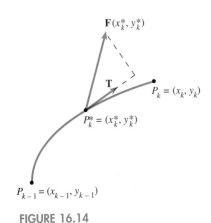

$\mathbf{F}(x_k^*, y_k^*)$

$\mathbf{T}$

$P_k = (x_k, y_k)$

$P_k^* = (x_k^*, y_k^*)$

$P_{k-1} = (x_{k-1}, y_{k-1})$

FIGURE 16.14

Definition 16.3
Work Done by a Variable Force Along a Curve

The work W done by a continuous force $\mathbf{F}$ in moving a particle along a smooth curve C in the domain of $\mathbf{F}$ is

$$W = \int_C \mathbf{F} \cdot \mathbf{T}\, ds \tag{16.7}$$

where $\mathbf{T}$ is the unit tangent vector to the curve C.

Before giving an example, we obtain a convenient alternative formulation of the integral in Equation 16.7. Using the vector equation $\mathbf{r}(t) = \langle x(t), y(t) \rangle$ of C, we can write the unit tangent vector $\mathbf{T}(t)$ as

$$\mathbf{T}(t) = \frac{\mathbf{r}'(t)}{|\mathbf{r}'(t)|} \qquad \text{By Equation 12.10}$$

So from Equation 16.7 and Equation 16.4,

$$W = \int_a^b \left[\mathbf{F}(\mathbf{r}(t)) \cdot \frac{\mathbf{r}'(t)}{|\mathbf{r}'(t)|} \right] |\mathbf{r}'(t)|\, dt \qquad \text{Since } ds = |\mathbf{r}'(t)|\, dt$$

That is,

$$W = \int_a^b \mathbf{F}(\mathbf{r}(t)) \cdot \mathbf{r}'(t)\, dt \tag{16.8}$$

We can also write the unit tangent vector $\mathbf{T}$ as

$$\mathbf{T} = \frac{d\mathbf{r}}{ds} \qquad \text{By Equation 12.6}$$

Formally, then

$$\mathbf{F} \cdot \mathbf{T}\, ds = \mathbf{F} \cdot \frac{d\mathbf{r}}{ds}\, ds = \mathbf{F} \cdot d\mathbf{r}$$

So, another way to write Equation 16.7 is

$$W = \int_C \mathbf{F} \cdot d\mathbf{r} \tag{16.9}$$

Combining Equations 16.7 and 16.9, we have the following alternative ways of calculating work.

The work W done by the continuous force field $\mathbf{F}$ in moving a particle along the smooth curve C given by the vector-valued function $\mathbf{r}(t)$, $a \leq t \leq b$, is

$$W = \int_C \mathbf{F} \cdot d\mathbf{r} = \int_a^b \mathbf{F}(\mathbf{r}(t)) \cdot \mathbf{r}'(t)\, dt \qquad (16.10)$$

EXAMPLE 16.7   Find the work done by the force field $\mathbf{F}(x, y) = x^2\mathbf{i} + 2xy\mathbf{j}$ on a particle as it moves along the curve C defined by $\mathbf{r}(t) = (\cos t)\mathbf{i} + (\sin t)\mathbf{j}$, for $0 \leq t \leq \pi$.

Solution   Note that C is the upper half of the circle $x^2 + y^2 = 1$, with a counterclockwise orientation. Since $\mathbf{r}'(t) = (-\sin t)\mathbf{i} + (\cos t)\mathbf{j}$, we have, by Equation 16.10,

$$W = \int_C \mathbf{F} \cdot d\mathbf{r} = \int_0^\pi [(\cos^2 t)\mathbf{i} + 2(\cos t \sin t)\mathbf{j}] \cdot [(-\sin t)\mathbf{i} + (\cos t)\mathbf{j}]\, dt$$

$$= \int_0^\pi (-\cos^2 t \sin t + 2\cos^2 t \sin t)\, dt$$

$$= \int_0^\pi \cos^2 t \sin t\, dt = -\frac{\cos^3 t}{3}\Big]_0^\pi = \frac{2}{3}$$

The units depend on those for force and distance. If, as is usual, distance is in meters and force is in newtons, then work is in joules. ■

EXAMPLE 16.8   Find the work done by the force $\mathbf{F}(x, y) = \langle x - y, xy \rangle$ that acts on a particle in moving it from point A to point B along the curve $C = C_1 \cup C_2$ shown in Figure 16.15. Assume force is measured in newtons and distance in meters.

Solution   First, we obtain parameterizations for C_1 and C_2. For the circular arc C_1 we have $\mathbf{r}(t) = \langle \cos t, \sin t \rangle$, where $0 \leq t \leq \pi/2$. For C_2 we use the vector equation of a line segment given in Equation 16.5:

$$\mathbf{r}(t) = \langle 0, 1 \rangle + t\langle -2, -2 \rangle = \langle -2t, 1 - 2t \rangle, \qquad 0 \leq t \leq 1$$

The work done is given by

$$W = \int_C \mathbf{F} \cdot d\mathbf{r} = \int_{C_1} \mathbf{F} \cdot d\mathbf{r} + \int_{C_2} \mathbf{F} \cdot d\mathbf{r}$$

We evaluate each integral on the right separately and then add the results:

$$\int_{C_1} \mathbf{F} \cdot d\mathbf{r} = \int_0^{\pi/2} \langle \cos t - \sin t, \cos t \sin t \rangle \cdot \langle -\sin t, \cos t \rangle\, dt$$

$$= \int_0^{\pi/2} (-\sin t \cos t + \sin^2 t + \cos^2 t \sin t)\, dt = \frac{\pi}{4} - \frac{1}{6}$$

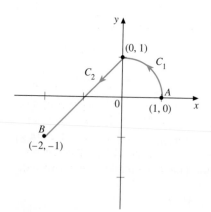

FIGURE 16.15

(Supply the missing steps.)

$$\int_{C_2} \mathbf{F} \cdot d\mathbf{r} = \int_0^1 \langle -2t - (1-2t), -2t(1-2t) \rangle \cdot \langle -2, -2 \rangle \, dt$$

$$= \int_0^1 (2 + 4t - 8t^2) \, dt = \frac{4}{3}$$

So $W = (\pi/4 - 1/6) + 4/3 = (\pi/4 + 7/6)$ joules. ∎

The integral $\int_C \mathbf{F} \cdot d\mathbf{r}$ can be expressed in yet another way by setting $d\mathbf{r} = \langle dx, dy \rangle$ and evaluating the dot product $\mathbf{F} \cdot d\mathbf{r}$. If $\mathbf{F}(x, y) = \langle f(x, y), g(x, y) \rangle$, then $\mathbf{F} \cdot d\mathbf{r} = f(x, y)dx + g(x, y)dy$. So we can write

$$\int_C \mathbf{F} \cdot d\mathbf{r} = \int_C f(x, y) \, dx + g(x, y) \, dy \qquad (16.11)$$

The form on the right is sometimes called the *differential form* of the line integral $\int_C \mathbf{F} \cdot d\mathbf{r}$.

EXAMPLE 16.9   Evaluate the line integral

$$\int_C (2x + y) \, dx + x^2 \, dy$$

where C is the curve

$$\begin{cases} x = t \\ y = 2t^2 \end{cases} \qquad 0 \le t \le 2$$

Solution   The given integral is the differential form of $\int_C \mathbf{F} \cdot d\mathbf{r}$, where $\mathbf{F}(x, y) = \langle 2x + y, x^2 \rangle$ and $\mathbf{r}(t) = \langle t, 2t^2 \rangle$. We can evaluate it by writing everything in terms of the parameter t. We have $dx = dt$ and $dy = 4t \, dt$. So, on substituting for x, y, dx, and dy, we get

$$\int_C (2x + y) \, dx + x^2 \, dy = \int_0^2 \left[(2t + 2t^2) \, dt + t^2 (4t \, dt) \right]$$

$$= 2 \int_0^2 \left(t + t^2 + 2t^3 \right) dt = \frac{76}{3}$$ ∎

REMARK ─────────────────────────────────

■ Although the line integral $\int_C f(x, y) \, ds$ is unchanged if C is given a new parameterization, the integral $\int_C \mathbf{F} \cdot \mathbf{T} \, ds$, or equivalently, $\int_C \mathbf{F} \cdot d\mathbf{r}$, changes in sign if the orientation of C is reversed. The sign changes because the unit tangent vector $\mathbf{T}$ reverses direction when the orientation is reversed. So the function $\mathbf{F} \cdot \mathbf{T}$ changes in sign. Thus, if $-C$ denotes the curve C with orientation reversed, we have

$$\int_{-C} \mathbf{F} \cdot d\mathbf{r} = -\int_C \mathbf{F} \cdot d\mathbf{r}$$

─────────────────────────────────

Extension to Three Dimensions

The concept of the line integral extends in a natural way to functions of three variables defined on space curves, and we will use such integrals freely from now on. For example, if the smooth curve C has vector representation $\mathbf{r}(t) = \langle x(t), y(t), z(t) \rangle$, and if $\mathbf{F}(x, y, z) = \langle f(x, y, z), g(x, y, z), h(x, y, z) \rangle$ is continuous on C, then

$$\int_C \mathbf{F} \cdot \mathbf{T} \, ds = \int_C \mathbf{F} \cdot d\mathbf{r} = \int_C \mathbf{F}(x(t), y(t), z(t)) \cdot \mathbf{r}'(t) \, dt$$

and in differential form

$$\int_C \mathbf{F} \cdot d\mathbf{r} = \int_C f(x, y, z) \, dx + g(x, y, z) \, dy + h(x, y, z) \, dz$$

VECTOR FIELDS AND WORK USING COMPUTER ALGEBRA SYSTEMS

A vector field is a function $\mathbf{F}$ that maps points (x, y) in the plane (or space) to a vector $\mathbf{F}(x, y)$ in the plane (or space). One way of describing a vector field is to draw a collection of arrows in the plane (or in space), representing the vector starting at the point (x, y). Sketching vector fields can be done easily and quickly using a CAS. The example in this section examines the work done by a force field in moving an object along a curve in the plane.

CAS 60

Find the work done by the force field $\mathbf{F}(x, y) = -2x\mathbf{i} + (y - 1)\mathbf{j}$ in moving an object along one arch of the cycloid $\mathbf{r}(t) = (t - \sin t)\mathbf{i} + (1 - \cos t)\mathbf{j}$, for $0 \le t \le 2\pi$. The formula we will need, given in Equation 16.8, is

$$W = \int_0^{2\pi} \mathbf{F}(\mathbf{r}(t)) \cdot \mathbf{r}'(t) \, dt$$

Maple:

First we show a sketch of the cycloid and the vector field $\mathbf{F}$. See Figures 16.2.1 and 16.2.2.

with(plots):with(linalg):
plot([t–sin(t),1–cos(t),t=–4*Pi..4*Pi],scaling=constrained);
fieldplot([–2*x,y–1],x=–4..4,y=–4..4);

Mathematica:

<<Graphics`Graphics`

ParametricPlot[{t–Sin[t],1–Cos[t]},{t,–4*Pi,4*Pi},
    AspectRatio–>Automatic]

PlotVectorField[{–2*x,y–1},{x=–4,4},{y,–4,4}]

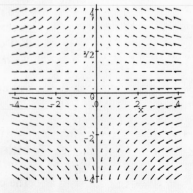

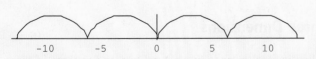

FIGURE 16.2.1

FIGURE 16.2.2

Maple:

Now we prepare to compute the integral.

F:=(x,y)->[-2*x,y-1];
R:=[t-sin(t),1-cos(t)];
F(R[1],R[2]);

Output: $[-2t + 2\sin(t), -\cos(t)]$

diff(R,t);

Output: $[1 - \cos(t), \sin(t)]$

integrand:=dotprod(F(R[1],R[2]),diff(R,t));

Output: $integrand := (-2t+2\sin(t))(1-\cos(t))-\cos(t)\sin(t)$

int(integrand,t=0..2*Pi);

Output: $-4\pi^2$

Mathematica:

F[x_,y_] = {-2*x,y-1}

R = {t-Sin[t],1-Cos[t]}

D[R,t]

integrand = F[t-Sin[t],1-Cos[t]].D[R,t]

Integrate[integrand,{t,0,2*Pi}]

Exercise Set 16.2

In Exercises 1–6, evaluate the integral $\int_C f(x, y)\,ds$, where C has the given parameterization.

1. $f(x, y) = x^2y;\ x = 2t - 1,\ y = 3t,\ -1 \le t \le 2$

2. $f(x, y) = x - 2y;\ x = \cos 2t,\ y = \sin 2t,\ 0 \le t \le \pi/2$

3. $f(x, y) = x/y;\ x = 2t,\ y = t^2 + 1,\ 0 \le t \le 2$

4. $f(x, y) = ye^x;\ x = t,\ y = 2t - 3,\ 0 \le t \le 1$

5. $f(x, y) = x - y;\ x = 2\cos^2 t,\ y = \sin 2t,\ 0 \le t \le \pi/2$

6. $f(x, y) = (x^3 + 9x)/y;\ x = 3t,\ y = t^2 + 1,\ 0 \le t \le 2$

In Exercises 7–10, find the area of the vertical cylindrical surface that has xy-trace C and is bounded above by $z = f(x, y)$ and below by the xy-plane.

7. C is the circle $x^2 + y^2 = 1;\ f(x, y) = 4 - x - 2y$.

8. C is the line segment from $(1, 0)$ to $(0, 2);\ f(x, y) = x^2 + y^2$.

9. C is the curve $y = \sin x$ from $x = 0$ to $x = \pi/2;\ f(x, y) = \sin 2x$.

10. C is the parabolic arc $y = x^2$ from $(0, 0)$ to $(1, 1);\ f(x, y) = \sqrt{1 + 4x^2}$.

11. Redo Example 16.5 using the parameterization $x = e^t,\ y = e^{2t}/2$, for $-\infty < t < \ln 2$, and show that the answer is unchanged.

12. Redo Exercise 10 using the parameterization $x = \sin t,\ y = \sin^2 t$, for $0 \le t \le \pi/2$, and show that the answer is unchanged.

In Exercises 13–16, find the work done by the force $\mathbf{F}$ acting on a particle moving in the positive direction along C.

13. $\mathbf{F}(x, y) = 16xy^2\mathbf{i} - (y/x^2)\mathbf{j};\ C$ is given by $x = \cosh t,\ y = \sinh t,\ 0 \le t \le \ln 2$.

14. $\mathbf{F}(x, y) = x\mathbf{i} + y\mathbf{j};\ C$ is the arc of the parabola $y = 2x^2 - 1$ from $(0, -1)$ to $(1, 1)$.

15. $\mathbf{F}(x, y, z) = \left\langle \dfrac{x + y}{z}, xyz, \dfrac{1}{z^2} \right\rangle;\ C$ has the vector equation $\mathbf{r}(t) = \left\langle t, t^2, \dfrac{1}{t} \right\rangle,\ 1 \le t \le 3$.

16. $\mathbf{F}(x, y, z) = xy\mathbf{i} + xz\mathbf{j} + yz\mathbf{k};\ C$ is the arc of the circular helix, $x = \cos t,\ y = \sin t,\ z = t$ for $0 \le t \le \pi/2$.

In Exercises 17–22, evaluate the line integral $\int_C \mathbf{F} \cdot d\mathbf{r}$, where C is the graph of $\mathbf{r}(t)$.

17. $\mathbf{F}(x, y) = \langle xy, x - y \rangle$; $\mathbf{r}(t) = \langle 2t, 1 - t \rangle$, $1 \le t \le 2$

18. $\mathbf{F}(x, y) = \langle x^2 - y^2, 2xy \rangle$; $\mathbf{r}(t) = \langle \cos t, \sin t \rangle$, $0 \le t \le \pi$

19. $\mathbf{F}(x, y) = (x + y)\mathbf{i} + (3x - 2y)\mathbf{j}$; $\mathbf{r}(t) = t^2\mathbf{i} - 2t\mathbf{j}$, $-1 \le t \le 1$

20. $\mathbf{F}(x, y) = xy\mathbf{i} - x^2\mathbf{j}$; $\mathbf{r}(t) = e^t\mathbf{i} + e^{-t}\mathbf{j}$, $0 \le t \le 1$

21. $\mathbf{F}(x, y, z) = \langle x+y-z, x-2z, y+z \rangle$; $\mathbf{r}(t) = \langle t+1, t-1, t \rangle$, $1 \le t \le 2$

22. $\mathbf{F}(x, y, z) = xyz^2\mathbf{i} + (x/y)\mathbf{j} + xz\mathbf{k}$; $\mathbf{r}(t) = t\mathbf{i} + (1/t^2)\mathbf{j} + \sqrt{t}\mathbf{k}$, $1 \le t \le 4$

In Exercises 23–28, evaluate the line integral along the given curve C.

23. $\int_C y\,dx - x\,dy$; C is the curve $y = e^x$, $0 \le x \le \ln 2$.

24. $\int_C 2xy\,dx + x^2\,dy$; C is the arc of the parabola $y = x^2$ from $(-1, 1)$ to $(2, 4)$.

25. $\int_C (x^2 - y^2)\,dx + xy\,dy$; C is the line segment from $(0, 1)$ to $(1, 2)$.

26. $\int_C (2x - 3y)\,dx + (4x + 2y)\,dy$; $C = C_1 \cup C_2$, where C_1 is the line segment from $(0, 0)$ to $(2, 0)$ and C_2 is the line segment from $(2, 0)$ to $(2, 4)$.

27. $\int_C x\,dx + y\,dy$; $C = C_1 \cup C_2 \cup C_3$, where C_1 is the arc of the parabola $y = x^2$ from $(0, 0)$ to $(1, 1)$, C_2 is the line segment from $(1, 1)$ to $(0, 1)$, and C_3 is the line segment from $(0, 1)$ to $(0, 0)$.

28. $\int_C (x^2 + y^2)\,dx + (1 - x^2)\,dy$; $C = C_1 \cup C_2$, where C_1 is the semicircle $y = \sqrt{1 - x^2}$ from $(1, 0)$ to $(-1, 0)$, and C_2 is the line segment from $(-1, 0)$ to $(1, 0)$.

29. For a certain vector field $\mathbf{F}$, defined on all of $\mathbb{R}^2$ except the origin, the vector at each point has length inversely proportional to the distance of that point from the origin, and it is directed toward the origin.
  (a) Draw several typical vectors of the field $\mathbf{F}$.
  (b) Find, without integration, the work done by $\mathbf{F}$ on a particle moving on a circle $x^2 + y^2 = a^2$. Explain your reasoning.
  (c) Find a formula for $\mathbf{F}(x, y)$.

30. Let $\mathbf{F}(x, y) = -y\mathbf{i} + x\mathbf{j}$. Find and show geometrically the vectors in this field at $45°$ intervals around the circle

C: $x^2 + y^2 = a^2$ directed in a counterclockwise direction. Show that for this curve C, $\mathbf{F} \cdot \mathbf{T} = |\mathbf{F}| = a$, and use this to evaluate $\int_C \mathbf{F} \cdot \mathbf{T}\,ds$ by inspection.

31. Evaluate $\int_C xyz\,ds$ along the circular helix $x = 2\cos t$, $y = 2\sin t$, $z = t$, for $0 \le t \le \pi$.

32. Evaluate $\int_C (x + z)\,dx + (x - y)\,dy + (2y - z)\,dz$, where C is the line segment from $(2, -1, 3)$ to $(3, 0, 4)$.

33. Evaluate $\int_C (2x + y)\,dx + (x - y)\,dy$ where C is the path from A to B shown in the accompanying figure.

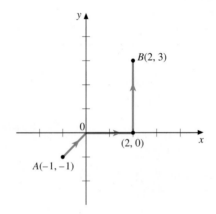

34. Evaluate $\int_C (x^2 + y^2)\,dx + 2xy\,dy$, where C is
  (a) the path shown in the figure for Exercise 33
  (b) the straight line segment from A to B.

35. Evaluate $\int_C x^2y\,dx + (x - 2y)\,dy$, where C is the closed curve shown in the accompanying figure.

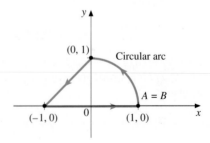

36. Show that

$$\int_C (2x + 3y)\,dx + (3x - 4y)\,dy = 0$$

for the curve C in the figure for Exercise 35.

37. Evaluate

$$\int_C (x+z)\,dx + (y-z)\,dy + (x-y)\,dz$$

where C is the path from A to B shown in the accompanying figure.

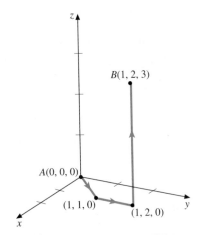

38. Evaluate $\int_C (x^2 - 4y)\,ds$, where C is defined by $x = 2t$, $y = t^2/2 - \ln t$, with $1 \le t \le e$.

39. Evaluate $\int_C \mathbf{F} \cdot d\mathbf{r}$ if C is given by $\mathbf{r}(t) = (e^t \cos t)\mathbf{i} + (e^t \sin t)\mathbf{j}$ for $0 \le t \le \pi$, and

$$\mathbf{F}(x,y) = \frac{-x}{\sqrt{x^2+y^2}}\mathbf{i} + \frac{y}{\sqrt{x^2+y^2}}\mathbf{j}$$

40. Evaluate $\int_C (x + y - z)\,ds$, where C is defined by $x = 4\cos^3 t$, $y = 4\sin^3 t$, $z = 3\sin 2t$, with $0 \le t \le \pi/2$.

41. Find the work done by the force field $\mathbf{F}(x,y,z) = \langle xz, xy^2, -yz \rangle$ on a unit mass moving along the elliptical helix $\mathbf{r}(t) = \langle 2\cos t, \sin t, 3t \rangle$, where $0 \le t \le 2\pi$.

Exercises 42 and 43 refer to a thin wire of density $\rho(x, y)$ (mass per unit length) in the shape of a smooth, plane curve C.

42. By partitioning C, explain the plausibility of defining the mass m of the wire by $m = \int_C dm$, where $dm = \rho(x, y)\,ds$.

43. (a) By using ideas from Section 15.4, show that the natural definition of the center of mass $(\bar{x}, \bar{y})$ of the wire is

$$\bar{x} = \frac{\int_C x\,dm}{m} \qquad \bar{y} = \frac{\int_C y\,dm}{m}$$

(b) Similarly, show that the natural definition of the moments of inertia I_x and I_y are

$$I_x = \int_C y^2\,dm \qquad I_y = \int_C x^2\,dm$$

44. A wire is in the shape of the catenary $y = a\cosh(x/a)$, where $-a \le x \le a$, and its density at any point is inversely proportional to the distance from the x-axis to the point. Use the results of Exercises 42 and 43 to find (a) its center of mass and (b) I_x and I_y.

45. Using results analogous to those in Exercises 42 and 43, find the center of mass of a homogeneous wire of density ρ in the shape of the circular helix

$$\mathbf{r}(t) = (2\cos t)\mathbf{i} + (2\sin t)\mathbf{j} + (3t)\mathbf{k}$$

from $t = 0$ to $t = 3\pi/2$. Also find I_x, I_y, and I_z.

46. An *inverse-square force field* is of the form

$$\mathbf{F}(x,y,z) = \frac{k}{|\mathbf{r}|^2}\mathbf{u}$$

where $\mathbf{u} = \mathbf{r}/|\mathbf{r}|$. Find the work done by such a force field in moving a particle of unit mass from $(1, 1, 1)$ to $(2, 1, 3)$.

47. Use Newton's Second Law of Motion, $\mathbf{F} = m\mathbf{a}$, to prove that the work done by a force $\mathbf{F}$ in moving a particle of mass m along a curve C: $\mathbf{r}(t) = \langle x(t), y(t), z(t) \rangle$, from $A = (x(a), y(a), z(a))$ to $B = (x(b), y(b), z(b))$, equals the change in kinetic energy $K(B) - K(A)$, where $K = \frac{1}{2}mv^2$. *Hint:* Show that

$$\int_C \mathbf{F} \cdot d\mathbf{r} = \int_a^b m\mathbf{r}'' \cdot \mathbf{r}'\,dt = \frac{m}{2}\int_a^b \frac{d}{dt}(\mathbf{r}' \cdot \mathbf{r}')\,dt$$

$$= \frac{m}{2}\int_a^b \frac{d}{dt}|\mathbf{r}'|^2\,dt$$

16.3 GRADIENT FIELDS AND PATH INDEPENDENCE

Recall from Section 16.1 that the gradient of a scalar function is a particular type of vector field, called a **gradient field**. For example, if ϕ is a scalar function of two variables, then $\nabla\phi = \langle \phi_x, \phi_y \rangle$ is a gradient field. The function ϕ is called a **potential function** for this gradient field. Note also that for any constant k, $\phi + k$ is another potential function for this field since the gradient is still $\langle \phi_x, \phi_y \rangle$.

If we begin with a scalar function (whose partial derivatives exist), it is easy to find the gradient field $\nabla\phi$. A more difficult problem in general is to determine if a given vector field $\mathbf{F}$ is a gradient field and if so, to find a potential function for it. In certain special cases this determination may not be difficult. For example, if $\mathbf{F}(x, y) = \langle 2x, 2y \rangle$, then it is easy to see that $\mathbf{F}(x, y) = \nabla\phi(x, y)$, where $\phi(x, y) = x^2 + y^2$. Or we could take $\phi(x, y) = x^2 + y^2 + 2$ or $x^2 + y^2 - 3$, or $x^2 + y^2 + k$ for any constant k. In each case, we would have $\nabla\phi(x, y) = \langle 2x, 2y \rangle = \mathbf{F}(x, y)$.

Later in this section, we will show a general procedure for determining whether a vector field is a gradient field and for finding a potential function for it when it is. For now we concentrate on some special properties of gradient fields not shared by other vector fields. The most important property is given in the next theorem, sometimes called the *fundamental theorem for line integrals*. The theorem refers to a function defined on an **open** region in $\mathbb{R}^2$, which means a set consisting only of interior points (no boundary points). That is, for every point (x_0, y_0) in the set, there is some circle centered at (x_0, y_0) enclosing only points of the set. For example, the region enclosed by a rectangle (excluding the rectangle itself) is an open region.

THEOREM 16.2

Fundamental Theorem for Line Integrals

Let $\mathbf{F}$ be a continuous gradient field in an open region R of $\mathbb{R}^2$, and let ϕ be any potential function for $\mathbf{F}$. If $A = (x_1, y_1)$ and $B = (x_2, y_2)$ are any two points in R and C is any piecewise-smooth curve from A to B lying entirely in R, then

$$\int_C \mathbf{F} \cdot d\mathbf{r} = \int_C \nabla\phi \cdot d\mathbf{r} = \phi(x_2, y_2) - \phi(x_1, y_1) \qquad (16.12)$$

Proof   We prove the theorem when C is a smooth curve only. The result for piecewise-smooth curves can be obtained by adding the results for each of the smooth component curves of C. (See Exercise 38 of Exercise Set 16.3.)

Let $\mathbf{F}(x, y) = \langle f(x, y), g(x, y) \rangle$. Then, since $\mathbf{F} = \nabla\phi = \langle \phi_x, \phi_y \rangle$, it follows that $f(x, y) = \phi_x(x, y)$ and $g(x, y) = \phi_y(x, y)$. So

$$\int_C \mathbf{F} \cdot d\mathbf{r} = \int_C f(x, y)\,dx + g(x, y)\,dy = \int_C \phi_x(x, y)\,dx + \phi_y(x, y)\,dy$$

If C has the vector equation $\mathbf{r}(t) = \langle x(t), y(t) \rangle$ for $a \leq t \leq b$, such that $\mathbf{r}(a) = (x_1, y_1)$ and $\mathbf{r}(b) = (x_2, y_2)$, we can evaluate the last integral in terms of the parameter t:

$$\int_C \phi_x(x, y)\,dx + \phi_y(x, y)\,dy$$

$$= \int_a^b \left[\phi_x(x(t), y(t))\,x'(t) + \phi_y(x(t), y(t))\,y'(t) \right] dt$$

$$= \int_a^b \left[\frac{\partial\phi}{\partial x}\frac{dx}{dt} + \frac{\partial\phi}{\partial y}\frac{dy}{dt} \right] dt \qquad \text{Using Leibniz notation}$$

By the Chain Rule given in Equation 14.10, the integrand is just $d\phi/dt$. So we have

$$\int_C \mathbf{F} \cdot d\mathbf{r} = \int_a^b \left[\frac{d}{dt} \phi(x(t), y(t)) \right] dt$$

$$= \phi(x(b), y(b)) - \phi(x(a), y(a)) \qquad \text{By the Second Fundamental Theorem of Calculus (Theorem 5.3)}$$

$$= \phi(x_2, y_2) - \phi(x_1, y_1) \qquad \qquad \blacksquare$$

The striking feature of Theorem 16.2 is that the value of the line integral $\int_C \mathbf{F} \cdot d\mathbf{r}$ depends only on the initial and terminal points of C and not on C itself. Thus, *any* piecewise-smooth curve C from (x_1, y_1) to (x_2, y_2) in the domain of $\mathbf{F}$ will give the same value of the line integral. Because of this fact, the line integral $\int_C \mathbf{F} \cdot d\mathbf{r}$ is said to be *independent of the path*. The term *path* is used here to mean a piecewise-smooth curve (which may, in particular, be simply a smooth curve). Our theorem can therefore be restated as follows:

Path Independence

If $\mathbf{F}$ is a continuous gradient field in an open region R, then $\displaystyle\int_C \mathbf{F} \cdot d\mathbf{r}$ is independent of the path C in R.

Because of this path independence, the integral $\int_C \mathbf{F} \cdot d\mathbf{r}$ is sometimes written as $\int_{(x_1, y_1)}^{(x_2, y_2)} \mathbf{F} \cdot d\mathbf{r}$ or even as $\int_A^B \mathbf{F} \cdot d\mathbf{r}$, where $A = (x_1, y_1)$ and $B = (x_2, y_2)$. Using the latter notation, the result of Theorem 16.2 assumes the form

$$\int_A^B \mathbf{F} \cdot d\mathbf{r} = \phi(B) - \phi(A)$$

which looks very much like the result of the Second Fundamental Theorem of Calculus.

Another result is immediately apparent. If C is a *closed* curve, then the initial point $A = (x_1, y_1)$ and the terminal point $B = (x_2, y_2)$ coincide. Thus,

$$\int_C \mathbf{F} \cdot d\mathbf{r} = \phi(x_2, y_2) - \phi(x_1, y_1) = 0$$

So we have the following additional result.

COROLLARY 16.2

If $\mathbf{F}$ is a continuous gradient field in an open region R, and C is any piecewise-smooth *closed* curve in R, then

$$\int_C \mathbf{F} \cdot d\mathbf{r} = 0$$

EXAMPLE 16.10   Evaluate the integral $\int_C 2x \, dx + 2y \, dy$, where C is a piecewise-smooth curve from $A = (1, -2)$ to $B = (3, 6)$.

Solution   As we observed earlier, the vector field $\mathbf{F} = \langle 2x, 2y \rangle$ is a gradient field with potential $\phi(x, y) = x^2 + y^2$, since $\nabla\phi = \langle 2x, 2y \rangle = \mathbf{F}$, valid throughout all of $\mathbb{R}^2$. Thus, by Theorem 16.2, the integral

$$\int_C \mathbf{F} \cdot d\mathbf{r} = \int_C 2x\,dx + 2y\,dy$$

is independent of the path from A to B, and

$$\int_C 2x\,dx + 2y\,dy = \phi(3,6) - \phi(1,-2)$$

$$= (9+36) - (1+4) = 40 \qquad \blacksquare$$

In Exercise 29 of Exercise Set 16.3, you will be asked to evaluate the integral in Example 16.10 along various paths from A to B by parameterizing the curves to verify that the answer is always the same.

EXAMPLE 16.11   Evaluate the integral $\int_C 2x\,dx + 2y\,dy$, where C is the path shown in Figure 16.16.

Solution   The hard way to do this problem is to parameterize each part of C and evaluate the integral on each part. The easy way is to use the fact that $\mathbf{F}(x,y) = \langle 2x, 2y \rangle$ is a gradient field, and so by Corollary 16.2, $\int_C \mathbf{F} \cdot d\mathbf{r} = 0$ for every piecewise-smooth closed path. Thus, $\int_C 2x\,dx + 2y\,dy = 0$. $\blacksquare$

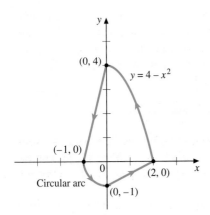

FIGURE 16.16

Recall that a force field that is a gradient field is called *conservative*. This term comes from the law of conservation of energy in physics, stating that the sum of the kinetic energy and potential energy of a particle moving through a conservative force field is constant.

EXAMPLE 16.12   Let $\mathbf{F}$ be the conservative force field $\mathbf{F}(x,y) = \nabla\phi(x,y)$, where $\phi(x,y) = x^2 - 2xy + y^3$. Find the work required to move a particle through this force field from the point $(-1,1)$ to $(3,-2)$.

Solution   By Equation 16.10, the work W is

$$W = \int_C \mathbf{F} \cdot d\mathbf{r}$$

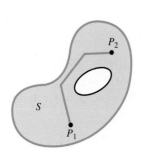

(a) S is connected

where the curve C joins the two given points. But since $\mathbf{F}$ is conservative, the integral is independent of the path, and its value is, by Theorem 16.2,

$$W = \int_C \nabla\phi \cdot d\mathbf{r} = \phi(3,-2) - \phi(-1,1)$$

$$= \left[3^2 - 2(3)(-2) + (-2)^3\right] - \left[(-1)^2 - 2(-1)(1) + 1^3\right]$$

$$= 13 - 4 = 9$$

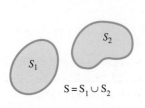

$S = S_1 \cup S_2$

(b) S is disconnected

FIGURE 16.17

Assuming the magnitude of the force $\mathbf{F}$ is in meters, the work required is 9 joules. $\blacksquare$

The converse of Theorem 16.2 is also true, provided we assume that R is **connected**, meaning that any two points in R can be joined by a sectionally smooth curve lying entirely in R. (See Figure 16.17.)

THEOREM 16.3

If $\mathbf{F}(x, y) = \langle f(x, y), g(x, y) \rangle$ is a continuous vector field in an open connected region R of $\mathbb{R}^2$ and $\int_C \mathbf{F} \cdot d\mathbf{r}$ is independent of the path C in R, then $\mathbf{F}$ is a gradient field.

Proof   To prove the theorem, it is sufficient to construct a function ϕ such that $\mathbf{F} = \nabla \phi$. To do so, we begin with an arbitrary fixed point (a, b) in R. Define $\phi(x, y)$ by

$$\phi(x, y) = \int_C \mathbf{F} \cdot d\mathbf{r} = \int_{(a,b)}^{(x,y)} \mathbf{F} \cdot d\mathbf{r}$$

where C is any piecewise-smooth curve in R from (a, b) to (x, y). Since by hypothesis $\int_C \mathbf{F} \cdot d\mathbf{r}$ is independent of the path, $\phi(x, y)$ depends only on the point (x, y) and not on the curve C connecting (a, b) to (x, y). Consequently, ϕ is a valid scalar-valued function of x and y. We now show that $\phi_x(x, y) = f(x, y)$.

We fix (x, y) temporarily. By Definition 13.7,

$$\phi_x(x, y) = \lim_{h \to 0} \frac{\phi(x + h, y) - \phi(x, y)}{h}$$

Since R is open, $|h|$ can be chosen small enough that the line segment from (x, y) to $(x + h, y)$ lies in R. Let C_1 be any path from (a, b) to (x, y), and let C_2 be the line segment from (x, y) to $(x + h, y)$ (see Figure 16.18). Then we have

$$\phi(x + h, y) - \phi(x, y) = \int_{C_1 \cup C_2} \mathbf{F} \cdot d\mathbf{r} - \int_{C_1} \mathbf{F} \cdot d\mathbf{r} = \int_{C_2} \mathbf{F} \cdot d\mathbf{r}$$

A parameterization of C_2 is $\mathbf{r}(t) = \langle t, y \rangle$, where $x \leq t \leq x + h$, so that $d\mathbf{r} = \langle dt, 0 \rangle$. We now have

$$\int_{C_2} \mathbf{F} \cdot d\mathbf{r} = \int_x^{x+h} \mathbf{F}(\mathbf{r}(t)) \cdot \mathbf{r}'(t) \, dt = \int_x^{x+h} \langle f(t, y), g(t, y) \rangle \cdot \langle 1, 0 \rangle \, dt$$

$$= \int_x^{x+h} f(t, y) \, dt$$

By the Mean-Value Theorem for Integrals (Theorem 5.1), the last integral can be written as

$$\int_x^{x+h} f(t, y) \, dt = f(c, y)h$$

where c is between x and $x + h$. Finally, then,

$$\phi_x(x, y) = \lim_{h \to 0} \frac{\phi(x + h, y) - \phi(x, y)}{h} = \lim_{h \to 0} \frac{f(c, y)h}{h} = f(x, y)$$

The last equality follows from the continuity of f, since as $h \to 0$, $c \to x$, so that $f(c, y) \to f(x, y)$.

A similar argument in which C_2 is the vertical segment from (x, y) to $(x, y + k)$ would show that $\phi_y(x, y) = g(x, y)$ (see Exercise 40 in Exercise Set 16.3). It follows that $\mathbf{F} = \nabla \phi$, and so $\mathbf{F}$ is a gradient field. ∎

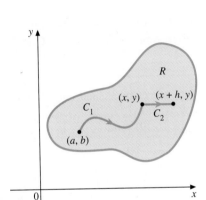

FIGURE 16.18

REMARK
- The way the function ϕ was defined in the proof of this theorem provides one means of obtaining a potential function for $\mathbf{F}$, once we know that $\int_C \mathbf{F} \cdot d\mathbf{r}$

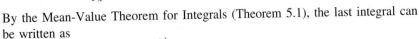

is independent of the path in a region R. We define ϕ by

$$\phi(x, y) = \int_{(a,b)}^{(x,y)} \mathbf{F} \cdot d\mathbf{r}$$

and evaluate the integral by selecting a convenient initial point (a, b) and a convenient path from (a, b) to (x, y). Often a path that consists of vertical and horizontal line segments is useful, or the line segment that goes from (a, b) to (x, y) might work well. The complete path, however, must lie in R. We will illustrate this technique after we have developed a test for determining path independence (see Example 16.14).

We have seen that when $\mathbf{F}$ is a continuous gradient field in an open region R, and C is any piecewise-smooth *closed* curve in R, $\int_C \mathbf{F} \cdot d\mathbf{r} = 0$. We now use Theorem 16.3 to show that the converse of this result is also true, providing another way to show when $\mathbf{F}$ is a gradient field.

THEOREM 16.4

If $\mathbf{F}$ is a continuous vector field in an open connected region R, with the property that $\int_C \mathbf{F} \cdot d\mathbf{r} = 0$ for *every* piecewise-smooth closed curve C in R, then $\mathbf{F}$ is a gradient field.

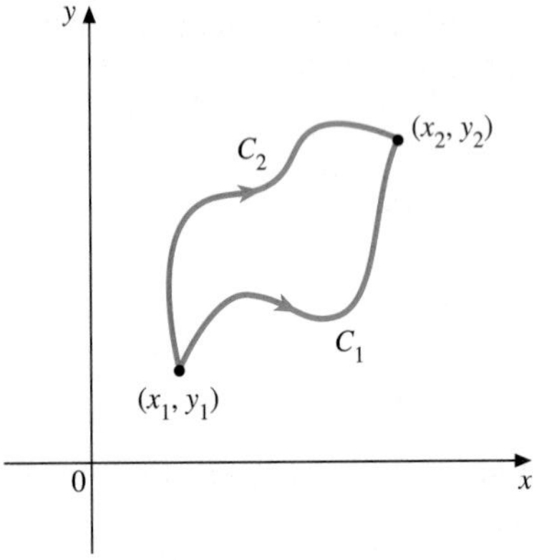

Proof   Let (x_1, y_1) and (x_2, y_2) be any two points in R, and consider any two piecewise-smooth curves C_1 and C_2 in R from (x_1, y_1) to (x_2, y_2). Then $C_1 \cup (-C_2)$ is a closed path in R (see Figure 16.19), and so by hypothesis, $\int_{C_1 \cup (-C_2)} \mathbf{F} \cdot d\mathbf{r} = 0$. But then we have

$$\int_{C_1 \cup (-C_2)} \mathbf{F} \cdot d\mathbf{r} = \int_{C_1} \mathbf{F} \cdot d\mathbf{r} + \int_{-C_2} \mathbf{F} \cdot d\mathbf{r} = \int_{C_1} \mathbf{F} \cdot d\mathbf{r} - \int_{C_2} \mathbf{F} \cdot d\mathbf{r} = 0$$

so that

$$\int_{C_1} \mathbf{F} \cdot d\mathbf{r} = \int_{C_2} \mathbf{F} \cdot d\mathbf{r}$$

This proves that $\int_C \mathbf{F} \cdot d\mathbf{r}$ is independent of the path. Thus, by Theorem 16.3, $\mathbf{F}$ is a gradient field.   ∎

FIGURE 16.19

REMARK

■ Recall from Section 14.1 that the differential of a differentiable function ϕ of two variables is given by

$$d\phi = \frac{\partial \phi}{\partial x} dx + \frac{\partial \phi}{\partial y} dy$$

When ϕ is a potential function for the continuous gradient field $\mathbf{F} = \langle f, g \rangle$, then $\phi_x = f$ and $\phi_y = g$, so that

$$d\phi = f(x, y) dx + g(x, y) dy$$

For this reason we call $f(x, y) dx + g(x, y) dy$ an *exact differential*, since it is exactly equal to the differential of ϕ. Using this terminology, we can say that *the continuous vector field $\mathbf{F} = \langle f, g \rangle$ in an open connected region R is a gradient field if and only if $f \, dx + g \, dy$ is an exact differential.*

Testing for Gradient Fields and Finding Potential Functions

In the study of gradient fields two crucial questions remain: (1) How can one tell whether a vector field is a gradient field? and (2) When it is a gradient field, how can a potential function for it be found? We discuss these questions now.

A simple test for determining whether **F** is a gradient field is given in the next theorem. First, though, we must distinguish between two types of open connected regions. We say that R is **simply connected** if, for every simple closed curve C in R, all points inside C are also in R. If R is not simply connected, it is said to be **multiply connected**. Figure 16.20 illustrates both types. Intuitively you can think of simple connectedness as meaning "no holes."

THEOREM 16.5

> If f and g have continuous first partial derivatives in the simply connected open region R, then $\mathbf{F} = \langle f, g \rangle$ is a gradient field if and only if
>
> $$\frac{\partial f}{\partial y} = \frac{\partial g}{\partial x} \qquad (16.13)$$

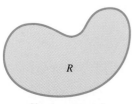

R

Simply connected

(a)

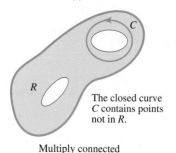

R

C

The closed curve
C contains points
not in *R*.

Multiply connected

(b)

FIGURE 16.20

Partial Proof   We will prove only that when **F** is a gradient field, Equation 16.13 is true. The converse is more difficult and will be omitted. Suppose then that **F** is a gradient field with potential function ϕ. Then $\mathbf{F} = \nabla \phi$ and so by definition

$$f(x, y) = \frac{\partial \phi}{\partial x} \quad \text{and} \quad g(x, y) = \frac{\partial \phi}{\partial y}$$

Thus,

$$\frac{\partial f}{\partial y} = \frac{\partial^2 \phi}{\partial y \, \partial x} \quad \text{and} \quad \frac{\partial g}{\partial x} = \frac{\partial^2 \phi}{\partial x \, \partial y}$$

In view of the continuity of f_y and g_x, it follows from Theorem 13.1 that the two second-order mixed partials of ϕ are equal; that is,

$$\frac{\partial^2 \phi}{\partial y \, \partial x} = \frac{\partial^2 \phi}{\partial x \, \partial y}$$

Consequently, we have the result that

$$\frac{\partial f}{\partial y} = \frac{\partial g}{\partial x}$$

This part of the proof did not require simple connectedness. It is only in proving the converse that this property comes into play.   ∎

Theorem 16.5 provides the key for determining when a vector field is a gradient field, at least in a simply connected region. In the next example we use this test and then illustrate a general technique for finding a potential function.

EXAMPLE 16.13   Show that

$$\mathbf{F}(x, y) = \langle e^y - 2xy, \, xe^y - x^2 + 2y \rangle$$

is a gradient field throughout $\mathbb{R}^2$, and find a potential function for **F**.

Solution   The functions $f(x, y) = e^y - 2xy$ and $g(x, y) = xe^y - x^2 + 2y$ are each continuous, and they have continuous partial derivatives in all of $\mathbb{R}^2$. The set $\mathbb{R}^2$, which is the entire xy-plane, is a simply connected open region. Furthermore,

$$\frac{\partial f}{\partial y} = e^y - 2x \quad \text{and} \quad \frac{\partial g}{\partial x} = e^y - 2x$$

so that the condition in Equation 16.13 is met. Thus, **F** is a gradient field.

We know, then, that a function ϕ exists for which

$$\frac{\partial \phi}{\partial x} = e^y - 2xy \quad \text{and} \quad \frac{\partial \phi}{\partial y} = xe^y - x^2 + 2y$$

To find ϕ, we can proceed in either of two ways: integrate $e^y - 2xy$ with respect to x while holding y fixed, or integrate $xe^y - x^2 + 2y$ with respect to y while holding x fixed. Let us choose the first:

$$\phi(x, y) = \int (e^y - 2xy)\, dx = xe^y - x^2 y + C(y) \qquad \text{\small y is held fixed.}$$

Note that we have written the general antiderivative with the "constant" of integration as $C(y)$ to allow for the fact that it may be a function of y. Since $\partial C(y)/\partial x = 0$, we have

$$\frac{\partial}{\partial x}\left[xe^y - x^2 y + C(y) \right] = e^y - 2xy$$

as required. To find $C(y)$, we force $\partial \phi/\partial y$ to equal $xe^y - x^2 + 2y$:

$$\frac{\partial \phi}{\partial y} = xe^y - x^2 + C'(y) = xe^y - x^2 + 2y$$

$$C'(y) = 2y$$

$$C(y) = y^2 + C_1$$

where C_1 is an arbitrary (numerical) constant. Since we are seeking any potential function for **F**, we might as well let $C_1 = 0$. Then

$$\phi(x, y) = xe^y - x^2 y + y^2$$

You can verify that $\mathbf{F} = \boldsymbol{\nabla}\phi$.   ∎

As we indicated in this example, once we have determined that the function $\mathbf{F} = \langle f, g \rangle$ is a gradient field, we know it has a potential function ϕ for which $\phi_x = f$ and $\phi_y = g$. So we can obtain $\phi(x, y)$ by integrating $f(x, y)$ with respect to x or $g(x, y)$ with respect to y. In the first case, the integration constant is a function $C(y)$ of y whose value can be determined by forcing $\phi_y = g$, as we did in the example. If the second approach is used, the integration constant is of the form $C(x)$, and its value is determined by forcing $\phi_x = f$. Although either way will work, sometimes one of the integrations may be simpler than the other, and so you should be on the lookout for this possibility.

An alternative way of finding a potential function when you know that **F** is a gradient field was suggested in the remark following Theorem 16.3. We illustrate this method in the next example.

EXAMPLE 16.14   Find a potential function for the gradient field **F** of Example 16.13 using the method suggested in the remark following Theorem 16.3.

Solution   The function **F** is

$$\mathbf{F}(x, y) = \langle e^y - 2xy, xe^y - x^2 + 2y \rangle$$

which was shown in Example 16.13 to be a gradient field throughout $\mathbb{R}^2$. Define ϕ by $\phi(x, y) = \int_{(0,0)}^{(x,y)} \mathbf{F} \cdot d\mathbf{r}$, where the path taken from $(0, 0)$ to (x, y) is the polygonal path $C_1 \cup C_2$ shown in Figure 16.21. We can parameterize C_1 by $\mathbf{r}(t) = \langle t, 0 \rangle$, $0 \le t \le x$, and C_2 by $\mathbf{r}(t) = \langle x, t \rangle$, $0 \le t \le y$. So

$$\phi(x, y) = \int_{C_1} \mathbf{F} \cdot d\mathbf{r} + \int_{C_2} \mathbf{F} \cdot d\mathbf{r} = \int_0^x \mathbf{F}(t, 0) \cdot \langle dt, 0 \rangle + \int_0^y \mathbf{F}(x, t) \cdot \langle 0, dt \rangle$$

$$= \int_0^x dt + \int_0^y (xe^t - x^2 + 2t) \, dt = x + \left[xe^t - x^2 t + t^2 \right]_0^y$$

$$= x + [(xe^y - x^2 y + y^2) - x] = xe^y - x^2 y + y^2 \qquad \blacksquare$$

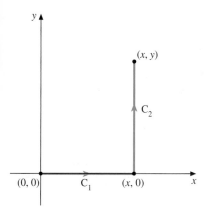

FIGURE 16.21

Summary of Results

The main results of this section can be summarized by the following list of equivalent statements for a continuous vector field $\mathbf{F} = \langle f, g \rangle$ in an open connected region R. They are equivalent in the sense that if any one of the statements is true, each of the others is also true.

1.   **F** is a gradient field.
2. $\int_C \mathbf{F} \cdot d\mathbf{r}$ is independent of the path.
3. $f \, dx + g \, dy$ is an exact differential.
4. $\int_C \mathbf{F} \cdot d\mathbf{r} = 0$ for every piecewise-smooth closed curve C in R.

Figure 16.22 illustrates these equivalencies. If, in addition, R is simply connected and f and g have continuous first partial derivatives in R, we can add a fifth statement equivalent to each of these four:

5. $\dfrac{\partial f}{\partial y} = \dfrac{\partial g}{\partial x}$

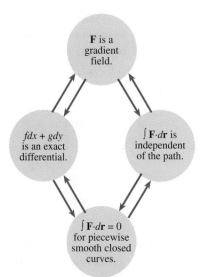

FIGURE 16.22

Extension to Three Dimensions

Most of the ideas of this section have natural extensions to three dimensions. In particular, the equivalence of the first four conditions given above remains true, where $\mathbf{F} = \langle f, g, h \rangle$. In Condition 3, the differential form becomes $f \, dx + g \, dy + h \, dz$. We will show in Section 16.7 that the appropriate form of Condition 5 in the three-dimensional case is the following:

$$\frac{\partial f}{\partial y} = \frac{\partial g}{\partial x} \qquad \frac{\partial g}{\partial z} = \frac{\partial h}{\partial y} \qquad \frac{\partial h}{\partial x} = \frac{\partial f}{\partial z} \qquad (16.14)$$

Conservation of Energy

We conclude this section by verifying the *Law of Conservation of Energy*. Let **F** be a three-dimensional conservative force field with potential function ϕ, so that $\mathbf{F} = \nabla\phi$. If the position vector of a particle of mass m moving through the field is $\mathbf{r}(t)$, then the **kinetic energy** of the particle is defined by

$$K(\mathbf{r}(t)) = \frac{1}{2} m \left|\mathbf{r}'(t)\right|^2$$

(Recall that $\left|\mathbf{r}'(t)\right|$ is the speed of the particle.) Its **potential energy** is defined by

$$P(\mathbf{r}(t)) = -\phi(\mathbf{r}(t))$$

Thus, $\mathbf{F} = -\nabla P$. Note that both kinetic energy and potential energy are scalars. The Conservation Law can be stated as follows.

Law of Conservation of Energy

In a conservative force field, the sum of the potential energy and kinetic energy remains constant as the object moves between any two points in the field.

We prove this law by making use of the ideas of this section. Our approach is to find the work done in moving the particle between two points in two different ways and then to equate the results. Let $A = \mathbf{r}(a)$ and $B = \mathbf{r}(b)$ be any two points in the vector field **F**. By the three-dimensional analogue of Theorem 16.2,

$$W = \int_C \mathbf{F} \cdot d\mathbf{r} = -\int_C \nabla P \cdot d\mathbf{r} = -[P(B) - P(A)] = P(A) - P(B)$$

We also can find the work by using Newton's Second Law of Motion, which states that $\mathbf{F} = m\mathbf{a}$, where $\mathbf{a}$ is the acceleration of the particle. Since $\mathbf{a} = \mathbf{r}''$, we have

$$W = \int_C \mathbf{F} \cdot d\mathbf{r} = \int_a^b m\mathbf{r}''(t) \cdot \mathbf{r}'(t)\, dt$$

$$= \frac{m}{2} \int_a^b \frac{d}{dt}\left[\mathbf{r}'(t) \cdot \mathbf{r}'(t)\right] dt$$

$$= \frac{m}{2} \int_a^b \frac{d}{dt}\left|\mathbf{r}'(t)\right|^2 dt$$

$$= \frac{m}{2}\left[\left|\mathbf{r}'(t)\right|^2\right]_a^b$$

$$= \frac{m}{2}\left[\left|\mathbf{r}'(b)\right|^2 - \left|\mathbf{r}'(a)\right|^2\right]$$

$$= \frac{1}{2} m \left|\mathbf{r}'(b)\right|^2 - \frac{1}{2} m \left|\mathbf{r}'(a)\right|^2$$

$$= K(B) - K(A)$$

The two values we have found for the work must be equal:

$$P(A) - P(B) = K(B) - K(A)$$

That is,

$$P(A) + K(A) = P(B) + K(B)$$

So the sum of the potential energy and kinetic energy remains constant from one point to another.

Exercise Set 16.3

In Exercises 1–12, determine whether $\mathbf{F}$ is a gradient field. If so, find a potential function for $\mathbf{F}$ using the method of Example 16.13. Assume the domain is $\mathbb{R}^2$ unless otherwise specified.

1. $\mathbf{F}(x, y) = \langle 2x + 3y, 3x - 2y + 1 \rangle$

2. $\mathbf{F}(x, y) = (3x^2 - 4xy)\mathbf{i} + (6y^2 - 2x^2)\mathbf{j}$

3. $\mathbf{F}(x, y) = \langle x^2 - y^2, 2xy \rangle$

4. $\mathbf{F}(x, y) = \langle y \sin x, \cos x + y^2 \rangle$

5. $\mathbf{F}(x, y) = x^2(3y^2 + 2)\mathbf{i} + 2(x^3y - 1)\mathbf{j}$

6. $\mathbf{F}(x, y) = \left\langle \dfrac{xy}{\sqrt{x^2 + 1}}, \sqrt{x^2 + 1} \right\rangle$

7. $\mathbf{F}(x, y) = \langle xe^{xy}(xy + 2), x^3 e^{xy} \rangle$

8. $\mathbf{F}(x, y) = \dfrac{x\mathbf{i} + y\mathbf{j}}{x^2 + y^2}$; $\mathbb{R}^2 - \{(0, 0)\}$

9. $\mathbf{F}(x, y) = \left(\dfrac{y}{\sqrt{x^2 + y^2}} \right)\mathbf{i} + \left(\tan^{-1}\dfrac{y}{x} \right)\mathbf{j}$

10. $\mathbf{F}(x, y) = \langle \sin^2 xy, 1 - \cos xy \rangle$

11. $\mathbf{F}(x, y) = \langle y^2 \cos xy, xy \cos xy + \sin xy - 2y \rangle$

12. $\mathbf{F}(x, y) = (2xy \sec^2 xy + 2 \tan xy + 1)\mathbf{i} + (2x^2 \sec^2 xy + 4y)\mathbf{j}$; in $\{(x, y) : 0 < x < \infty, -\pi/(2x) < y < \pi/(2x)\}$

In Exercises 13–20, use results from Exercises 1–12 to evaluate the given integrals.

13. $\displaystyle\int_{(-1,-2)}^{(3,5)} (2x + 3y)\, dx + (3x - 2y + 1)\, dy$

14. $\displaystyle\int_{(0,2)}^{(1,4)} (3x^2 - 4xy)\, dx + (6y^2 - 2x^2)\, dy$

15. $\displaystyle\int_{(-1,2)}^{(3,1)} (3x^2y^2 + 2x^2)\, dx + (2x^3y - 2)\, dy$

16. $\displaystyle\int_{(0,0)}^{(2,5)} \dfrac{xy\, dx}{\sqrt{x^2 + 1}} + \sqrt{x^2 + 1}\, dy$

17. $\displaystyle\int_{(1,0)}^{(2,\ln 2)} xe^{xy}(xy + 2)\, dx + x^3 e^{xy}\, dy$

18. $\displaystyle\int_{(1,1)}^{(3,-4)} \dfrac{x\, dx + y\, dy}{x^2 + y^2}$

19. $\displaystyle\int_{(1,0)}^{(5\pi/9,3/2)} \mathbf{F} \cdot d\mathbf{r}$, where $\mathbf{F}$ is given in Exercise 11

20. $\displaystyle\int_{(0,0)}^{(\pi/2,1/2)} \mathbf{F} \cdot d\mathbf{r}$, where $\mathbf{F}$ is given in Exercise 12

In Exercises 21–24, show that $\mathbf{F}$ is independent of the path in the given region, and use the method shown in Example 16.14 to find a potential function for $\mathbf{F}$.

21. $\mathbf{F}(x, y) = \langle 2x - 2y^2, 3y^2 - 4xy \rangle$; all of $\mathbb{R}^2$

22. $\mathbf{F}(x, y) = \left\langle \dfrac{1}{x - y}, \dfrac{-1}{x - y} \right\rangle$; $\{(x, y) : x > y\}$ [Hint: Use the straight-line path from $(1, 0)$ to (x, y).]

23. $\mathbf{F}(x, y) = \langle \sin y + y \sin x, x \cos y - \cos x \rangle$; all of $\mathbb{R}^2$

24. $\mathbf{F}(x, y) = \left\langle \ln \sqrt{x^2 + y^2}, \tan^{-1} xy \right\rangle$; $\mathbb{R}^2 - \{(0, 0)\}$. [*Hint:* Use $C_1 \cup C_2$, where C_1 is the vertical line segment from $(1, 0)$ to $(1, y)$ and C_2 is the horizontal line segment from $(1, y)$ to (x, y).]

In Exercises 25–28, prove that the given integral is independent of the path and then evaluate it in two ways: (a) using Theorem 16.2, where C is any piecewise-smooth curve from A to B, and (b) integrating along the specified curve.

25. $\int_C (2x - 2y) \, dx + (6y - 2x) \, dy$ from $A = (-1, 2)$ to $B = (3, -2)$; C is the straight line segment from A to B.

26. $\displaystyle\int_C \frac{x \, dx + y \, dy}{\sqrt{x^2 + y^2}}$ from $A = (-3, 4)$ to $B = (5, 12)$, where C does not pass through the origin; C is the horizontal line segment from A to $(5, 4)$ followed by the vertical line segment from $(5, 4)$ to B.

27. $\displaystyle\int_C \frac{-y \, dx + x \, dy}{x^2 + y^2}$ from $A = \left(\sqrt{6}, -\sqrt{6}\right)$ to $B = \left(3, \sqrt{3}\right)$, where C does not pass through the origin; C is the smaller arc of the circle $x^2 + y^2 = 12$ from A to B.

28. $\int_C e^{xy}(y \cos x - \sin x) \, dx + xe^{xy} \cos x \, dy$ from $A = (0, 4)$ to $B = (\pi/3, 0)$; C is the vertical line segment from A to $(0, 0)$ followed by the horizontal line segment from $(0, 0)$ to B.

29. Evaluate the integral in Example 16.10 by integrating along each of the following paths:
  (a) C is the line segment from A to B.
  (b) C is the horizontal line segment from A to $(3, -2)$ followed by the vertical line segment from $(3, -2)$ to B.
  (c) C is the arc of the parabola $y = 2x^2 - 4x$ from A to B.

30. Evaluate the integral in Example 16.11 by integrating along each of the component curves of C.

31. Show that the force field
$$\mathbf{F}(x, y) = (2x - 3y)\mathbf{i} + 3(y^2 - x)\mathbf{j}$$
is conservative throughout $\mathbb{R}^2$, and find the work done by $\mathbf{F}$ on a particle moving from $(2, 1)$ to $(5, -3)$.

32. Show that the force field
$$\mathbf{F}(x, y) = \frac{y}{x}\mathbf{i} + \left(\ln x - \frac{2}{y}\right)\mathbf{j}$$
is conservative in the region
$$R = \{(x, y) : x > 0, y > 0\}$$

and find the work done by $\mathbf{F}$ on a particle as it goes once around the ellipse
$$4x^2 + y^2 - 16x - 6y + 21 = 0$$

33. Evaluate the integral
$$\int_C \cos x \cos y \, dx - \sin x \sin y \, dy$$
along each path in the accompanying figure. (*Hint:* There is an easy way and a hard way.)

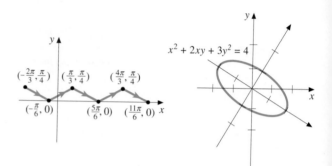

In Exercises 34–37, show that the given expression is an exact differential, and find ϕ so that the expression equals $d\phi$.

34. $(2x - 3y) \, dx + (8y - 3x) \, dy$

35. $(2x + y^2 \cos xy) \, dx + (\sin xy + xy \cos xy) \, dy$

36. $\dfrac{xy^2}{(1 + x^2)^2} \, dx + \dfrac{x^2 y}{1 + x^2} \, dy$

37. $(\ln \cosh y) \, dx + (x \tanh y - 1) \, dy$

38. Using the proof of Theorem 16.2 for a smooth curve C, show that the theorem is also true for any piecewise-smooth curve.

39. By inspection determine a potential function for $\mathbf{F}(x, y, z) = \langle 2x, 2y, 2z \rangle$ and use the result to evaluate
$$\int_{(1, -1, 2)}^{(3, 2, -1)} 2x \, dx + 2y \, dy + 2z \, dz$$

40. Complete the proof of Theorem 16.3 by showing that $\phi_y(x, y) = g(x, y)$, taking C_1 as any path from (a, b) to (x, y) and C_2 as the vertical line segment from (x, y) to $(x, y + k)$ for $|k|$ sufficiently small.

41. Using Equations 16.14, show that

$$\mathbf{F}(x, y, z) = \langle 2xyz - 3z^2, x^2z + 8yz^3, x^2y - 6xz + 12y^2z^2 \rangle$$

is a gradient field in $\mathbb{R}^3$, and follow a procedure similar to that used in Example 16.7 to find a potential function for **F**.

42. Prove that the gravitational field

$$\mathbf{F} = -\frac{C}{|\mathbf{r}|^3}\mathbf{r}$$

of a point mass is conservative, where **r** is the vector $\overrightarrow{OP}$ from the point located at O to a point P in space. Find the work done by **F** on an object as it moves from A to B, where $|\overrightarrow{OA}| = r_1$ and $|\overrightarrow{OB}| = r_2$.

16.4 GREEN'S THEOREM

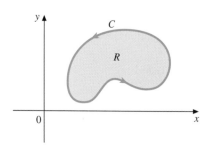

FIGURE 16.23

We now introduce Green's Theorem, which has far-reaching consequences in both pure and applied mathematics. The theorem provides a relationship between a line integral around a simple closed curve C and a double integral over the region R enclosed by C. (See Figure 16.23.) Recall that the direction, or *orientation* of a curve is determined by the parameterization, with the positive direction corresponding to increasing parameter values. In Green's Theorem the positive direction of the closed curve C is counterclockwise, as indicated in Figure 16.23.

THEOREM 16.6

> **Green's Theorem in the Plane**
>
> Let the simple closed curve C be oriented in the counterclockwise direction and let R be the region enclosed by C. If f and g have continuous first partial derivatives in an open set containing R, then
>
> $$\int_C f(x, y)\, dx + g(x, y)\, dy = \iint_R \left(\frac{\partial g}{\partial x} - \frac{\partial f}{\partial y} \right) dA \qquad (16.15)$$

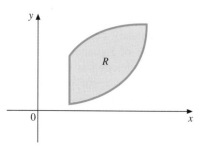

FIGURE 16.24
A simple region R

Recall that the left-hand side of Equation 16.15 is equivalent to $\int \mathbf{F} \cdot \mathbf{T}\, ds$ and also to $\int \mathbf{F} \cdot d\mathbf{r}$. We will prove Green's Theorem only for a region R that is simultaneously of type I and type II, as defined in Section 15.2. We call a region of this type a **simple region**. Figure 16.24 shows an example of a simple region. In Exercise 37 of Exercise Set 16.4 you will be asked to show that Green's Theorem is also true when R can be divided by horizontal and vertical line segments into a finite number of simple regions, as in Figure 16.25.

Partial Proof of Green's Theorem   Let R be a simple region with boundary C oriented counterclockwise. Figure 16.26(a) shows such a region. In Figure 16.26(b) it is viewed as type I and in Figure 16.26(c) as type II. The proof of Equation 16.15 is accomplished by showing that

$$\int_C f(x, y)\, dx = - \iint_R \frac{\partial f}{\partial y}\, dA \qquad (16.16)$$

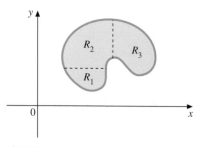

FIGURE 16.25
$R = R_1 \cup R_2 \cup R_3$ is the union of three simple regions

and

$$\int_C g(x, y)\, dy = \iint_R \frac{\partial g}{\partial x}\, dA \qquad (16.17)$$

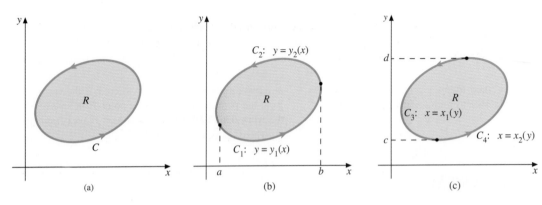

FIGURE 16.26

George Green (1793–1841) was the leading English mathematician at a time when most mathematical advances were taking place on the European continent. Green had little formal education and worked as a miller in Nottingham until he was more than 30 years old. At the age of 35, without prior history of publication, he wrote his "Essay on the Application of Mathematical Analysis to the Theories of Electricity and Magnetism." What we now call Green's Theorem was included. But the essay was not widely known. Seven years later, when William Thompson (who later became Lord Kelvin) was a senior at Cambridge University, he saw a reference to "the ingenious Essay by Mr. Green of Nottingham" and wound up introducing the work to a wider audience. Recently a memorial was dedicated to Green in Westminster Abbey, as was a stained glass window (photo below) in his old Cambridge college, where he began studying for a mathematics degree at age 40.

By adding the corresponding sides of these two equations, we get the desired result.

To prove Equation 16.16, we express R as the type I region

$$R = \{(x, y) : a \leq x \leq b, \; y_1(x) \leq y \leq y_2(x)\}$$

(See Figure 16.26(b).) Then the right-hand side of Equation 16.16 becomes

$$-\iint_R \frac{\partial f}{\partial y} \, dA = \int_a^b \int_{y_1(x)}^{y_2(x)} \frac{\partial f(x, y)}{\partial y} \, dy \, dx$$

$$= -\int_a^b \left[f(x, y_2(x)) - f(x, y_1(x)) \right] dx \quad \begin{array}{l}\text{By the Second}\\\text{Fundamental}\\\text{Theorem}\end{array} \quad (16.18)$$

In Figure 16.26(b), the lower boundary curve for the region R is denoted by C_1, with equation $y = y_1(x)$, and the upper boundary curve by C_2, with equation $y = y_2(x)$. To compute the left-hand side of Equation 16.16, we consider the curve C oriented in the counterclockwise direction as the union of the curves C_1 and C_2. To parameterize C_1, we can use x as the parameter to get $x = x$, $y = y_1(x)$, with $a \leq x \leq b$. For C_2, the parameter x will vary from b to a, since we want to traverse C_2 from right to left. Thus, $-C_2$ is traversed from left to right, and its parameterization is $x = x$, $y = y_2(x)$, with $a \leq x \leq b$. We can therefore write

$$\int_{C_2} f(x, y) \, dx = -\int_{-C_2} f(x, y) \, dx$$

$$= -\int_a^b f(x, y_2(x)) \, dx$$

Since $C = C_1 \cup C_2$, we have

$$\int_C f(x, y) \, dx = \int_{C_1} f(x, y) \, dx + \int_{C_2} f(x, y) \, dx$$

$$= \int_a^b f(x, y_1(x)) \, dx - \int_a^b f(x, y_2(x)) \, dx$$

$$= -\int_a^b \left[f(x, y_2(x)) - f(x, y_1(x)) \right] dx$$

When we compare this result with Equation 16.18, we see that Equation 16.16 is true. The proof of Equation 16.17 is similar, using Figure 16.26(c). You will be asked to carry out the details in Exercise 36 of Exercise Set 16.4. Equation 16.15 follows, as previously indicated, by adding the results of Equations 16.16 and 16.17.   ■

REMARK ───────────────────────────────────
■ There is a version of Green's Theorem for three dimensions that we will consider in Section 16.7, which explains the phrase "in the plane" in the name of the theorem as stated here.

We can view Equation 16.15 in either of two ways: (a) as a means of evaluating a line integral using a double integral, or (b) as a means of evaluating a double integral using a line integral. Both points of view are important, and we illustrate them in the examples that follow.

EXAMPLE 16.15   Use Green's Theorem to evaluate the line integral

$$\int_C \left(e^{-x^2} + xy^2 \right) dx + \left(x^3 + \sqrt{1 + y^3} \right) dy$$

where C is the path shown in Figure 16.27.

Solution   Observe first that evaluating the integral by parameterizing the component curves of C would lead to integrals that cannot be evaluated by elementary means. (Try it.) Let us see, then, if we can use Green's Theorem to replace the line integral with a double integral that we can evaluate. Note that the region R bounded by C is a simple region (both type I and type II). To apply Green's Theorem, we take

$$f(x, y) = e^{-x^2} + xy^2 \qquad \text{and} \qquad g(x, y) = x^3 + \sqrt{1 + y^3}$$

Since

$$\frac{\partial g}{\partial x} = 3x^2 \qquad \text{and} \qquad \frac{\partial f}{\partial y} = 2xy$$

we have, by Green's Theorem,

$$\int_C \left(e^{-x^2} + xy^2 \right) dx + \left(x^3 + \sqrt{1 + y^3} \right) dy = \iint_R (3x^2 - 2xy) \, dA$$

$$= \int_0^2 \int_0^x (3x^2 - 2xy) \, dy \, dx$$

$$= \int_0^2 \left[3x^2 y - xy^2 \right]_0^x dx$$

$$= \int_0^2 2x^3 \, dx$$

$$= 8 \qquad\qquad ■$$

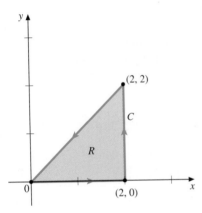

FIGURE 16.27

In Example 16.15, Green's Theorem was useful in evaluating a difficult line integral by means of a double integral. In the next example, the situation is reversed.

EXAMPLE 16.16   Use Green's Theorem to evaluate the integral $\iint_R y^2 \, dA$, where R is the elliptical region shown in Figure 16.28.

Solution   Direct evaluation by iterated integrals results in the integral

$$\int_{-3}^{3} \int_{-(2/3)\sqrt{9-x^2}}^{(2/3)\sqrt{9-x^2}} y^2 \, dy \, dx$$

which is difficult to evaluate. By Green's Theorem, we can replace this integral with a line integral that may be easier to evaluate. To use Green's Theorem, we must find functions f and g that satisfy the continuity requirements over R for which

$$\iint_R y^2 \, dA = \iint_R \left(\frac{\partial g}{\partial x} - \frac{\partial f}{\partial y} \right) dA$$

Many choices are possible, but a particularly simple one is $g(x, y) = xy^2$ and $f(x, y) = 0$. A parameterization of the ellipse C that gives a counterclockwise orientation is $x = 3\cos t$, $y = 2\sin t$, where $0 \le t \le 2\pi$. So we have, by Equation 16.15,

$$\iint_R y^2 \, dA = \iint_R \left(\frac{\partial g}{\partial x} - \frac{\partial f}{\partial y} \right) dA = \int_C f(x, y) \, dx + g(x, y) \, dy$$

$$= \int_C xy^2 \, dy = \int_0^{2\pi} (3\cos t)(4\sin^2 t)(2\cos t \, dt)$$

$$= 6 \int_0^{2\pi} \sin^2 2t \, dt = 3 \int_0^{2\pi} (1 - \cos 4t) \, dt \quad \text{Since } \sin^2 \theta = (1 - \cos 2\theta)/2$$

$$= 6\pi$$

∎

In the next example, we use Green's Theorem to evaluate a line integral around the boundary of a region that is the union of two simple regions.

EXAMPLE 16.17   Evaluate the line integral

$$\int_C (e^x - x^2 y) \, dx + (xy^2 + y^3) \, dy$$

using Green's Theorem, where C is the boundary of the semiannular region shown in Figure 16.29, oriented in a counterclockwise direction.

Solution   Let R denote the region enclosed by C. Although R is not simple, it is the union of the two simple regions into which the y-axis divides R. To apply Green's Theorem, we let $f(x, y) = e^x - x^2 y$ and $g(x, y) = xy^2 + y^3$. Then

$$\frac{\partial g}{\partial x} - \frac{\partial f}{\partial y} = y^2 - (-x^2) = x^2 + y^2$$

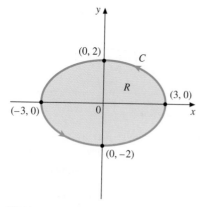

FIGURE 16.28

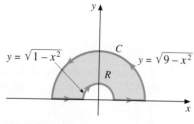

FIGURE 16.29

So we have, by Equation 16.15,

$$\int_C (e^x - x^2 y)\, dx + (xy^2 + y^3)\, dy = \iint_R (x^2 + y^2)\, dA$$

It is easier to use polar coordinates in this case. The bounding curves are $r = 1$ and $r = 3$, with θ going from 0 to π. Thus,

$$\iint_R (x^2 + y^2)\, dA = \int_0^\pi \int_1^3 r^2 (r\, dr\, d\theta) = \int_0^\pi \int_1^3 r^3 dr\, d\theta$$

$$= \int_0^\pi \frac{r^4}{4}\Big]_1^3 d\theta = \left(\frac{81}{4} - \frac{1}{4}\right)\pi = 20\pi \qquad \blacksquare$$

Areas by Line Integration

We can use Green's Theorem to find the area of a region as a line integral around its boundary, as we show in the following theorem.

THEOREM 16.7

Let R be the region enclosed by a piecewise-smooth, simple, closed curve C, oriented in a counterclockwise direction. Then the area A of the region R is given by any one of the following formulas:

$$A = \int_C x\, dy \qquad (16.19)$$

$$A = -\int_C y\, dx \qquad (16.20)$$

$$A = \frac{1}{2}\int_C x\, dy - y\, dx \qquad (16.21)$$

Proof   To prove Equation 16.19, we let $f(x, y) = 0$ and $g(x, y) = x$. Then, by Green's Theorem, we have

$$\int_C x\, dy = \iint_R 1\, dA = A$$

To prove Equation 16.20, we let $f(x, y) = y$ and $g(x, y) = 0$. By Green's Theorem

$$\int_C y\, dx = \iint_R (-1)\, dA = -A$$

So $A = -\int_C y\, dx$. To prove Equation 16.21, we add corresponding sides of Equations 16.19 and 16.20, getting $2A = \int_C x\, dy - \int_C y\, dx$. Then we divide by 2 and combine the two integrals. $\blacksquare$

EXAMPLE 16.18   Find the area of the region R enclosed by the ellipse

$$\frac{x^2}{a^2} + \frac{y^2}{b^2} = 1$$

(See Figure 16.30.)

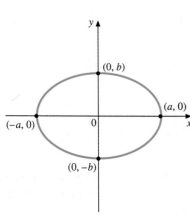

FIGURE 16.30

The ellipse $\dfrac{x^2}{a^2} + \dfrac{y^2}{b^2} = 1$

Solution   Parametric equations of the ellipse that give it the correct orientation are $x = a\cos t$, $y = b\sin t$, where $0 \le t \le 2\pi$. Although each of the equations in Theorem 16.7 will give the same result, we choose to use Equation 16.21. Using it, we get

$$
A = \frac{1}{2}\int_C x\,dy - y\,dx = \frac{1}{2}\int_0^{2\pi}[(a\cos t)(b\cos t\,dt) - (b\sin t)(-a\sin t\,dt)]
$$

$$
= \frac{1}{2}\int_0^{2\pi} ab(\cos^2 t + \sin^2 t)\,dt
$$

$$
= \pi ab \qquad\blacksquare
$$

Multiply-Connected Regions

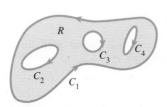

FIGURE 16.31
$C = C_1 \cup C_2 \cup C_3 \cup C_4$ is positively oriented with respect to R.

We now discuss how to extend Green's Theorem to certain multiply-connected regions (regions that are not simply connected). In particular, suppose R is a closed and bounded region with a boundary that consists of finitely many simple closed curves that are sectionally smooth and do not intersect one another. We illustrate such a region in Figure 16.31, with boundary curves C_1, C_2, C_3, and C_4. Let each of these boundary curves be oriented so that when it is traversed in its positive direction, the region R lies on the *left*, and let C be the union of all of these boundary curves. When the component curves of C are oriented according to this left-hand rule, we say that C is *positively oriented with respect to R*.

Green's Theorem holds for regions of the type described; that is, if f and g have continuous first partial derivatives in an open set that contains R, then

$$
\int_C f\,dx + g\,dy = \iint\limits_R \left(\frac{\partial g}{\partial x} - \frac{\partial f}{\partial y}\right)dA
$$

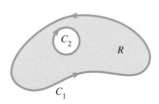

FIGURE 16.32

We indicate the idea of the proof for the simple case shown in Figure 16.32 in which the multiply-connected region R has one hole. The outer boundary of R is C_1 and the inner boundary C_2, both positively oriented with respect to R. We denote the total boundary of R by C. So $C = C_1 \cup C_2$. In Figure 16.33, we show R cut into two simply-connected regions, R_1 and R_2. Let us denote the boundary of R_1 by B_1 and the boundary of R_2 by B_2. Since Green's Theorem applies to R_1 and R_2 individually, we have

$$
\iint\limits_R \left(\frac{\partial g}{\partial x} - \frac{\partial f}{\partial y}\right)dA = \iint\limits_{R_1} \left(\frac{\partial g}{\partial x} - \frac{\partial f}{\partial y}\right)dA + \iint\limits_{R_2} \left(\frac{\partial g}{\partial x} - \frac{\partial f}{\partial y}\right)dA
$$

$$
= \int_{B_1} f\,dx + g\,dy + \int_{B_2} f\,dx + g\,dy \qquad \begin{array}{l}\text{By Equation 16.10}\\ \text{applied to both } R_1\\ \text{and } R_2\end{array}
$$

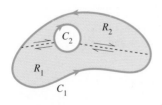

FIGURE 16.33

The boundaries B_1 and B_2 have the two line segments that we introduced in common, but oriented in opposite directions. So when we add the integrals along these line segments, they cancel, and we are left with the integrals around C_1 and C_2:

$$
\iint\limits_R \left(\frac{\partial g}{\partial x} - \frac{\partial f}{\partial y}\right)dA = \int_{C_1} f\,dx + g\,dy + \int_{C_2} f\,dx + g\,dy = \int_C f\,dx + g\,dy
$$

Thus, Green's Theorem holds true for the multiply-connected region R.

A very useful consequence of Green's Theorem for multiply connected regions is the following. It shows that in certain cases line integrals around different closed paths are equal.

THEOREM 16.8

Let C_1 and C_2 be any two nonintersecting piecewise-smooth simple closed curves, and let R be the closed annular region bounded by C_1 and C_2. If f and g have continuous partial derivatives and

$$\frac{\partial f}{\partial y} = \frac{\partial g}{\partial x}$$

throughout some open set that contains R, then

$$\int_{C_1} f\,dx + g\,dy = \int_{C_2} f\,dx + g\,dy$$

provided C_1 and C_2 are similarly oriented.

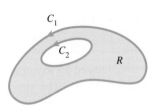

C_1

C_2 R

FIGURE 16.34

Proof   Suppose C_1 and C_2 are both oriented in the counterclockwise direction and C_2 is interior to C_1, as shown in Figure 16.34. Then Green's Theorem for the multiply-connected region R applies, where the boundary of R is $C = C_1 \cup (-C_2)$, since C is positively oriented with respect to R. Thus,

$$\int_C f\,dx + g\,dy = \iint_R \left(\frac{\partial g}{\partial x} - \frac{\partial f}{\partial y} \right) dA = 0 \quad \text{Since } \frac{\partial f}{\partial y} = \frac{\partial g}{\partial x}$$

But

$$\int_C f\,dx + g\,dy = \int_{C_1} f\,dx + g\,dy + \int_{-C_2} f\,dx + g\,dy$$

$$= \int_{C_1} f\,dx + g\,dy - \int_{C_2} f\,dx + g\,dy$$

Since we just showed that this difference equals 0, the result follows. If C_1 and C_2 are oriented in the clockwise direction, we let $C = (-C_1) \cup C_2$, and the reasoning follows the same lines.   ■

This theorem sometimes enables us to replace a complicated closed path with a simpler one. The next example illustrates this approach.

EXAMPLE 16.19   Evaluate the integral

$$\int_C \frac{-y}{x^2 + y^2}\,dx + \frac{x}{x^2 + y^2}\,dy$$

where C is the ellipse $x^2 - 2xy + 3y^2 = 4$.

Solution   The ellipse is pictured in Figure 16.35. Let

$$f(x, y) = \frac{-y}{x^2 + y^2} \quad \text{and} \quad g(x, y) = \frac{x}{x^2 + y^2}$$

It is easily verified that for all points except $(0, 0)$, $\partial g/\partial x = \partial f/\partial y$. If it were not for the exceptional point $(0, 0)$, we could conclude that $\langle f, g \rangle$ is a gradient field (why?) and therefore that the integral around the closed path C is 0. But this reasoning breaks down, since f and g fail even to exist at the origin.

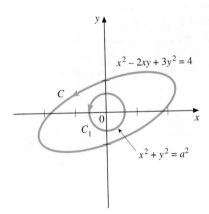

$x^2 - 2xy + 3y^2 = 4$

C

C_1

$x^2 + y^2 = a^2$

FIGURE 16.35

It is rather difficult to parameterize this ellipse, so we introduce a new curve C_1 inside C and with the origin in its interior. Because of its simplicity, we might as well take C_1 to be a circle centered at the origin. Let its radius be a. It does not matter what a is, as long as it is small enough that C_1 lies inside C. The conditions of Theorem 16.8 are now met, so

$$\int_C \frac{-y}{x^2 + y^2}\, dx + \frac{x}{x^2 + y^2}\, dy = \int_{C_1} \frac{-y}{x^2 + y^2}\, dx + \frac{x}{x^2 + y^2}\, dy$$

A parameterization of C_1 is $x = a\cos t$, $y = a\sin t$, where $0 \le t \le 2\pi$. So $dx = -a\sin t\, dt$ and $dy = a\cos t\, dt$. Also, $x^2 + y^2 = a^2\cos^2 t + a^2\sin^2 t = a^2$. So we have

$$\int_{C_1} \frac{-y}{x^2 + y^2} dx + \frac{x}{x^2 + y^2} dy = \int_0^{2\pi} \frac{-a\sin t}{a^2}(-a\sin t\, dt) + \frac{a\cos t}{a^2}(a\cos t\, dt)$$

$$= \int_0^{2\pi} (\sin^2 t + \cos^2 t)\, dt = \int_0^{2\pi} dt = 2\pi$$

The fact that the integral around the closed path is nonzero again shows that $\langle f, g \rangle$ is not a gradient field in the region enclosed by C. ■

REMARK ――――――――――――――――――――――――――――――

■ The ellipse in Example 16.19 could have been replaced by any other sectionally smooth simple closed curve with the origin inside it. The answer would have been the same.

―――――――――――――――――――――――――――――――――――――

Green's Theorem and Two-Dimensional Fluid Flow

We conclude this section with a brief discussion of how Green's Theorem can be used to determine certain characteristics of the flow of a fluid. We consider here a two-dimensional "laminar" flow (think of a thin sheet of water flowing over a flat surface), and in Sections 16.6 and 16.7 we will extend the ideas to three dimensions.

Let $\mathbf{F} = \langle f, g \rangle$ be the velocity field of the fluid, and let a region R bounded by the curve C, as in Green's Theorem, be in the field of flow. We assume the flow is in *steady state*, meaning that $\mathbf{F}$ does not change with time. At each point P on C, we wish to consider the component $\mathbf{F} \cdot \mathbf{T}$ of $\mathbf{F}$ in the direction of the unit tangent vector $\mathbf{T}$ and the component $\mathbf{F} \cdot \mathbf{n}$ in the direction of the *outer* unit normal $\mathbf{n}$, as shown in Figure 16.36. The integral $\int_C \mathbf{F} \cdot \mathbf{T}\, ds$ gives the amount of fluid per unit of time flowing tangentially around C, whereas $\int_C \mathbf{F} \cdot \mathbf{n}\, ds$ gives the amount of fluid per unit of time flowing out of R orthogonally across C. These integrals are given names as follows:

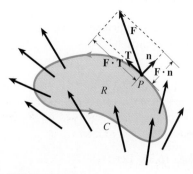

FIGURE 16.36

$$\int_C \mathbf{F} \cdot \mathbf{T}\, ds = \text{The \textbf{Circulation} of } \mathbf{F} \text{ Around } C$$

$$\int_C \mathbf{F} \cdot \mathbf{n}\, ds = \text{The \textbf{Flux} of } \mathbf{F} \text{ Through } C$$

Although the circulation and flux can be evaluated directly as line integrals, we wish to show how Green's Theorem can be used to find them. For the circulation, this result is immediate, since $\mathbf{F} \cdot \mathbf{T} \, ds = \mathbf{F} \cdot d\mathbf{r} = f \, dx + g \, dy$, so that Green's Theorem gives

$$\text{Circulation of } \mathbf{F} \text{ Around } C = \int_C \mathbf{F} \cdot \mathbf{T} \, ds = \iint_R \left(\frac{\partial g}{\partial x} - \frac{\partial f}{\partial y} \right) dA \qquad (16.22)$$

To obtain a similar representation of the flux, we can show (see Exercise 33 of Section 16.4) that the outer unit normal vector $\mathbf{n}$ is given by

$$\mathbf{n} = \left\langle \frac{dy}{ds}, \frac{-dx}{ds} \right\rangle$$

We can therefore write

$$\mathbf{F} \cdot \mathbf{n} \, ds = \langle f, g \rangle \cdot \left\langle \frac{dy}{ds}, \frac{-dx}{ds} \right\rangle ds = -g \, dx + f \, dy$$

Now we can apply Green's Theorem, replacing f with $-g$ and g with f in Equation 16.15, to get

$$\text{Flux of } \mathbf{F} \text{ Through } C = \int_C \mathbf{F} \cdot \mathbf{n} \, ds = \iint_R \left(\frac{\partial f}{\partial x} + \frac{\partial g}{\partial y} \right) dA \qquad (16.23)$$

At each point P of R, the integrands of the double integrals in Equations 16.22 and 16.23 are called, respectively, the **rotation of F** at P and the **divergence of F** at P. These are abbreviated rot $\mathbf{F}$ and div $\mathbf{F}$, respectively. So we have

$$\text{rot } \mathbf{F} = \frac{\partial g}{\partial x} - \frac{\partial f}{\partial y}$$

$$\text{div } \mathbf{F} = \frac{\partial f}{\partial x} + \frac{\partial g}{\partial y}$$

Each of these is a scalar function. Some authors use *scalar curl* or *two-dimensional curl* for rotation. (We will study the curl of a vector field in Section 16.7 and see how it is related to rotation.) With these definitions, we can rewrite Equations 16.22 and 16.23 as

$$\int_C \mathbf{F} \cdot \mathbf{T} \, ds = \iint_R (\text{rot } \mathbf{F}) \, dA \qquad (16.24)$$

$$\int_C \mathbf{F} \cdot \mathbf{n} \, ds = \iint_R (\text{div } \mathbf{F}) \, dA \qquad (16.25)$$

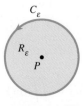

FIGURE 16.37

To get an intuitive understanding of the physical significance of the rotation and circulation of $\mathbf{F}$ at a point P in the field of flow, consider a small circle C_ε, of radius ε, centered at P, and denote the region it encloses by R_ε (see Figure 16.37). Then, by Equation 16.24, the circulation around C_ε is

$$\int_{C_\varepsilon} \mathbf{F} \cdot \mathbf{T} \, ds = \iint_{R_\varepsilon} (\text{rot } \mathbf{F}) \, dA$$

Now there is a mean-value theorem for double integrals analogous to the one for single integrals that enables us to write (when rot $\mathbf{F}$ is continuous)

$$\iint_{R_\varepsilon} (\text{rot } \mathbf{F}) \, dA = [\text{rot } \mathbf{F}(Q)](\text{Area of } R_\varepsilon)$$

where Q is some point in R_ε. Solving for $\operatorname{rot}\mathbf{F}(Q)$, we have

$$\operatorname{rot}\mathbf{F}(Q) = \frac{\displaystyle\iint_{R_\varepsilon}(\operatorname{rot}\mathbf{F})\,dA}{\text{Area of } R_\varepsilon} = \frac{\displaystyle\int_{C_\varepsilon}\mathbf{F}\cdot\mathbf{T}\,ds}{\text{Area of } R_\varepsilon}$$

If we let $\varepsilon \to 0$, then $Q \to P$, and we get

$$\operatorname{rot}\mathbf{F}(P) = \lim_{\varepsilon\to 0}\frac{\text{Circulation of }\mathbf{F}\text{ Around } C_\varepsilon}{\text{Area of } R_\varepsilon}$$

So $\operatorname{rot}\mathbf{F}$ at a point P is the *circulation per unit area at P*. If $\operatorname{rot}\mathbf{F} \neq 0$ at P, the fluid forms a whirlpool, called a **vortex**, at P. If $\operatorname{rot}\mathbf{F} = 0$ for all points of a region, then $\mathbf{F}$ is said to be **irrotational** in that region.

In an exactly analogous way, we can show that

$$\operatorname{div}\mathbf{F}(P) = \lim_{\varepsilon\to 0}\frac{\text{Flux of }\mathbf{F}\text{ Through } C_\varepsilon}{\text{Area of } R_\varepsilon}$$

so that $\operatorname{div}\mathbf{F}$ at a point P is the *flux per unit area* at that point. If $\operatorname{div}\mathbf{F}(P) > 0$, fluid is emerging from P, and we say that P is a **source**. If $\operatorname{div}\mathbf{F}(P) < 0$, fluid is flowing into P, and we say that P is a **sink**. If $\operatorname{div}\mathbf{F} = 0$ for all points of a region, we say the fluid is **incompressible**.

REMARK

■ Although the concepts of circulation, flux, rotation, and divergence were introduced for fluid flow, the terms are frequently used for other types of vector fields as well.

EXAMPLE 16.20   Let $\mathbf{F} = -2xy\mathbf{i} + x^2\mathbf{j}$ be the velocity field of a two-dimensional fluid flow, and let R be the region enclosed by the triangle C that has vertices $(0, 0)$, $(2, 0)$, and $(2, 4)$, oriented counterclockwise. Use Green's Theorem to find the circulation of $\mathbf{F}$ around C and the flux of $\mathbf{F}$ through C.

Solution   The region R is shown in Figure 16.38. First we calculate $\operatorname{rot}\mathbf{F}$ and $\operatorname{div}\mathbf{F}$. We write $f(x, y) = -2xy$ and $g(x, y) = x^2$, so that $\mathbf{F} = \langle f, g\rangle$. So

$$\operatorname{rot}\mathbf{F} = \frac{\partial g}{\partial x} - \frac{\partial f}{\partial y} = 2x - (-2x) = 4x$$

and

$$\operatorname{div}\mathbf{F} = \frac{\partial f}{\partial x} + \frac{\partial g}{\partial y} = -2y + 0 = -2y$$

Thus, by Equations 16.24 and 16.25,

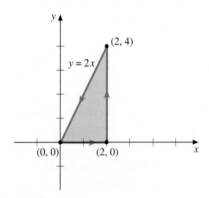

FIGURE 16.38

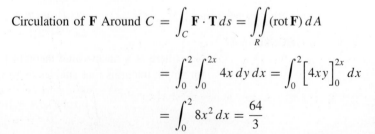

$$\text{Circulation of }\mathbf{F}\text{ Around } C = \int_C \mathbf{F}\cdot\mathbf{T}\,ds = \iint_R(\operatorname{rot}\mathbf{F})\,dA$$

$$= \int_0^2\int_0^{2x} 4x\,dy\,dx = \int_0^2\Big[4xy\Big]_0^{2x}dx$$

$$= \int_0^2 8x^2\,dx = \frac{64}{3}$$

$$\text{Flux of } \mathbf{F} \text{ Through } C = \int_C \mathbf{F} \cdot \mathbf{n}\, ds = \iint_R (\text{div } \mathbf{F})\, dA$$

$$= \int_0^2 \int_0^{2x} (-2y)\, dy\, dx = -\int_0^2 \left[y^2 \right]_0^{2x} dx$$

$$= -\int_0^2 4x^2\, dx = -\frac{32}{3}$$

Since the circulation is positive, the net flow of fluid around C is in the counterclockwise direction, and since the flux is negative, there is a net inflow of fluid into R. ■

REMARK ───────────────────────────────

■ Since we are dealing with two-dimensional flow, the "amount" of fluid is given by area rather than volume. Thus, for example, if $|\mathbf{F}|$ is in centimeters per second and distance is in centimeters, both circulation and flux will be in square centimeters per second.

Exercise Set 16.4

In Exercises 1–14, use Green's Theorem to evaluate the line integral. In each case, C has a counterclockwise orientation. For Exercises 1–4, take C to be the rectangle with vertices $(0, 0)$, $(2, 0)$, $(2, 1)$, *and* $(0, 1)$.

1. $\displaystyle \int_C \left(x^2 y - 2y^2 \right) dx + \left(x^3 - 2y^2 \right) dy$

2. $\displaystyle \int_C \left(\ln \sqrt{x+1} + xy \right) dx + \left(x^2 y + e^{y^2} \right) dy$

3. $\displaystyle \int_C y \cos \frac{\pi x}{2}\, dx - x \sin \frac{\pi y}{2}\, dy$

4. $\displaystyle \int_C \frac{x}{1+y^2}\, dx + \frac{y}{1+x^2}\, dy$

5. $\int_C e^{x+2y}\, dx$, C is the triangle with vertices $(0, 0)$, $(1, 1)$, and $(0, 1)$.

6. $\int_C (\tan^{-1} x)\, dy$, C is the boundary of the region between $y = 2 - x^2$ and $y = x$.

7. $\int_C y \sin 2x\, dx + \sin^2 x\, dy$, C is the ellipse $2x^2 + 3y^2 = 6$.

8. $\int_C (x^2 + y^2)\, dx + (x^2 - y^2)\, dy$, C is the boundary of the region determined by $y = x^2$, $y = 0$, and $x = 1$.

9. $\displaystyle \int_C \left(x^2 y + \frac{y^3}{3} \right) dx + (2x - y^5)\, dy$, C is the circle $x^2 + y^2 = 4$.

10. $\int_C (e^x + y^3)\, dx + (x^2 - \sqrt{y})\, dy$, C is the boundary of the region determined by $y = \sqrt{x}$, $y = 0$, and $x - y = 2$.

11. $\int_C y^2\, dx + x^2\, dy$, C is the boundary of the region determined by $x = -\sqrt{9 - y^2}$, $x + y = 3$, $y = 0$.

12. $\int_C (2y^3 - 3x^2)\, dx + (2x^3 + 5y^2)\, dy$, C is the boundary of the region determined by $y = \sqrt{4 - x^2}$ and $y = 0$.

13. $\int_C y^2\, dx + 3xy\, dy$, C is the cardioid $r = 1 + \cos \theta$.

14. $\int_C \sqrt{x^2 + 1}\, dx + x(1 + y)\, dy$, C is the boundary of the region outside the circle $r = 1$ and inside the cardioid $r = 2(1 + \cos \theta)$.

In Exercises 15–18, make use of Green's Theorem to evaluate each double integral by means of a line integral.

15. $\displaystyle \iint_R x\, dA$, R is the triangle with vertices $(0, 0)$, $(1, 1)$, and $(-1, 2)$.

16. $\displaystyle \iint_R [2(x - 1) - 2y]\, dA$, R is the circle $x^2 + y^2 = 2x$.

17. $\displaystyle \iint_R \left(x\sqrt{1 - y^2} - 4x^2 y \right) dA$, R is the ellipse $x^2 + 4y^2 = 4$.

18. $\displaystyle\iint_R y \, dA$, R is the parallelogram with vertices $(0, 0)$, $(4, 0)$, $(5, 2)$, and $(1, 2)$.

In Exercises 19–25, find the area of the specified region using line integration.

19. Bounded by the parallelogram with vertices $(0, 0)$, $(3, 1)$, $(4, 3)$, and $(1, 2)$

20. Between $y = x^2$ and $y = 2x$

21. Bounded by the triangle with vertices $(0, 2)$, $(1, 1)$, and $(2, 3)$

22. Inside the loop of $x = t^2 - 1$, $y = t^3 - t$, where $-\infty < t < \infty$

23. Inside the four-cusp hypocycloid $x = a\cos^3 t$, $y = a\sin^3 t$, where $0 \le t \le 2\pi$

24. Bounded by $x^{1/2} + y^{1/2} = 1$, $x = 0$, and $y = 0$. (*Hint:* Take $t = x^{1/2}$.)

25. Above the x-axis and under one arch of the cycloid $x = a(t - \sin t)$, $y = a(1 - \cos t)$

26. Let R be the region inside the circle C, having equation $x^2 + y^2 = a^2$, oriented in a counterclockwise direction. For each of the following vector fields find the circulation of $\mathbf{F}$ around C and the flux of $\mathbf{F}$ through C:
(a) $\mathbf{F}(x, y) = x\mathbf{i} + y\mathbf{j}$ (b) $\mathbf{F}(x, y) = -y\mathbf{i} + x\mathbf{j}$

In Exercises 27–30, $\mathbf{F}$ is the velocity field of two-dimensional fluid flow and R is the region enclosed by the curve C oriented in a counterclockwise direction. Find the circulation of $\mathbf{F}$ around C and the flux of $\mathbf{F}$ through C.

27. $\mathbf{F}(x, y) = \langle x^2 - y^2, 2xy \rangle$; R is the region in the first quadrant bounded by $y = \sqrt{1 - x^2}$, $x = 0$, and $y = 0$.

28. $\mathbf{F}(x, y) = \langle xy^2 - 3y, 2x + x^2 y \rangle$; C is the circle $x^2 + y^2 = 9$.

29. $\mathbf{F}(x, y) = -y^3\mathbf{i} + x^3\mathbf{j}$; C is the circle $x^2 + y^2 = 4$.

30. $\mathbf{F}(x, y) = (2x^2 - 3y^2)\mathbf{i} + (4y^2 - x^2)\mathbf{j}$; R is the region enclosed by the lines $y = x$, $y = 2 - x$, and $y = 0$.

In Exercises 31 and 32, C is the ellipse $x = 3\cos t$, $y = 2\sin t$, where $0 \le t \le 2\pi$.

31. Let $\mathbf{F} = \left\langle \sqrt{1 + x^4} - 4xy, x^3 - e^{y^2} \right\rangle$. Find the circulation of $\mathbf{F}$ around C.

32. Let $\mathbf{F} = \langle 2x + \cosh y^2, 4y - \sinh x^2 \rangle$. Find the flux of $\mathbf{F}$ through C.

33. Let C be a smooth simple closed curve, oriented in a counterclockwise direction, and let C be parameterized by arc length. Prove that the outer unit normal $\mathbf{n}$ is given by

$$\mathbf{n} = \left\langle \frac{dy}{ds}, -\frac{dx}{ds} \right\rangle$$

(*Hint:* Let θ denote the smallest positive angle from the unit vector $\mathbf{i}$ to the unit tangent vector $\mathbf{T}$. Then

$$\mathbf{T} = \left\langle \frac{dx}{ds}, \frac{dy}{ds} \right\rangle = \langle \cos\theta, \sin\theta \rangle$$

Now show that $\mathbf{n} = \langle \cos(\theta - \pi/2), \sin(\theta - \pi/2) \rangle$.)

34. Make use of Theorem 16.8 to evaluate the integral

$$\int_C \ln\sqrt{x^2 + y^2} \, dx - \left(\tan^{-1} \frac{y}{x} \right) dy$$

where C is the limaçon $r = 4 + 2\cos\theta$ (see Section 9.5). Explain why Green's Theorem is not applicable in this case.

35. Prove that for

$$F(x, y) = \frac{x\mathbf{i} + y\mathbf{j}}{\sqrt{x^2 + y^2}}$$

and for any piecewise-smooth simple closed curve C that does not pass through the origin,

$$\int_C \mathbf{F} \cdot d\mathbf{r} = 0$$

Consider separately the cases in which the origin is interior to C and exterior to C.

36. Complete the proof of Theorem 16.6 for a simple region by showing that Equation 16.17 is true.

37. Prove Green's Theorem for a region that can be divided by a horizontal or a vertical line segment into two simple regions. Extend this result by mathematical induction to a region that can be divided in this manner into finitely many simple regions.

16.5 SURFACE INTEGRALS

In this section we define a type of integral called a **surface integral** that is particularly useful in the study of three-dimensional vector fields. For example, surface integrals are used in the study of fluid dynamics, heat transfer, and electric and magnetic field theory. Surface integrals can also be used to find the mass, center of gravity, and moments of inertia of such objects as sheets of metal in various shapes.

The term *surface integral* comes from the fact that the domain of integration is a surface S in space, in contrast to a double integral, which involves integration over a region R in the xy-plane. In what follows, we will make use of results obtained in Section 15.5 on areas of surfaces, so you may need to review that section.

Let S be the surface defined by $z = f(x, y)$, where the domain of f is a region R in the xy-plane. Figure 16.39 illustrates such a surface. We will refer to the region R as the *xy projection* of S. Our primary concern will be with regions R of type I or type II, or that can be divided into a finite number of subregions of these two types. We assume that the function f and its first partial derivatives f_x and f_y are continuous on R. This assumption ensures that S has a tangent plane at each of its points.

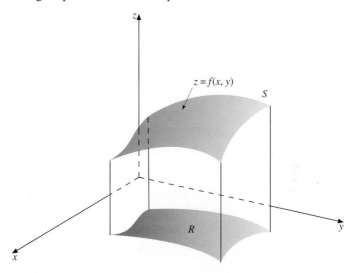

$z = f(x, y)$

S

R

FIGURE 16.39

Following exactly the same procedure as in Section 15.5, we partition a rectangle containing R and count only the subrectangles completely contained in R. When we project the ith subrectangle vertically, the resulting prism cuts out a patch $\Delta\sigma_i$ (called *a cell*) on the surface S that is approximated by the patch ΔT_i on the tangent plane, drawn at a point (x_i^*, y_i^*, z_i^*) on S that is the vertical projection on S of a point (x_i^*, y_i^*) in the subrectangle. We denote the area of the ith subrectangle by ΔA_i. The relationships among $\Delta\sigma_i$, ΔT_i, and ΔA_i are shown in Figure 16.40. We define the **surface integral of g over S** by

$$\iint\limits_{S} g(x, y, z)d\sigma = \lim_{\|P\|\to 0} \sum_{i=1}^{n} g(x_i^*, y_i^*, z_i^*)\Delta\sigma_i \qquad (16.26)$$

where as usual, $\|P\|$ is the norm of the partition of the region R, that is, the length of the longest diagonal of all the subrectangles in the partition. It can be shown that when g is continuous on S, and the function f and the region R satisfy the conditions stated, then the limit in Equation 16.26 exists, so the surface integral exists.

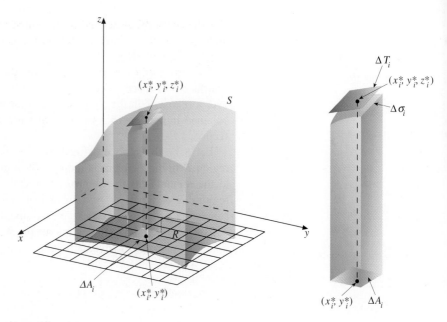

FIGURE 16.40

In Section 15.5, we found that the area $\Delta\sigma_i$ of the ith cell on the surface is approximated by ΔT_i, given by the following formula:

$$\Delta T_i \approx \sqrt{1 + \left[f_x(x_i^*, y_i^*)\right]^2 + \left[f_y(x_i^*, y_i^*)\right]^2}\, \Delta A_i$$

Thus, we can obtain the computational formula for the surface integral, given in the following theorem.

THEOREM 16.9

> Let the surface S be the graph of $z = f(x, y)$, and let R be the xy projection of S. If f_x and f_y are continuous on R, and g is a function of three variables that is continuous on S, then
>
> $$\iint\limits_{S} g(x, y, z)\, d\sigma$$
>
> $$= \iint\limits_{R} g\left(x, y, f(x, y)\right) \sqrt{1 + \left[f_x(x, y)\right]^2 + \left[f_y(x, y)\right]^2}\, dA \qquad (16.27)$$

REMARK ─────────────────────────

■ Equation 16.27 enables us to evaluate a surface integral as an ordinary double integral. Compare this result with that of Equation 16.4 in Theorem 16.1, in which a line integral is given as an ordinary integral of a function of one variable.

─────────────────────────

If the surface S is defined by the equation $x = f(y, z)$ and its yz projection is R, then by permuting variables in Equation 16.27, we obtain

$$\iint_S g(x, y, z)\, d\sigma = \iint_R g(f(y, z), y, z) \sqrt{1 + \left[f_y(y, z)\right]^2 + \left[f_z(y, z)\right]^2}\, dA$$

(16.28)

Similarly, if S is defined by $y = f(x, z)$ and R is the xz projection of S, we have

$$\iint_S g(x, y, z)\, d\sigma = \iint_R g(x, f(x, z), z) \sqrt{1 + \left[f_x(x, z)\right]^2 + \left[f_z(x, z)\right]^2}\, dA$$

(16.29)

In Equation 16.27, if we let $g(x, y, z) = 1$ for all points (x, y, z) on the surface S, then we see that the double integral on the right-hand side is exactly the area of the surface S, as given by Equation 15.25. Thus, we have the following result:

Let $A(S)$ denote the area of the surface S. Then

$$A(S) = \iint_S d\sigma \qquad (16.30)$$

EXAMPLE 16.21   Evaluate the surface integral $\iint_S (2xy + xz)\,d\sigma$, where S is the portion of the plane $3x + 2y + z = 6$ in the first octant.

Solution   We will do this problem in two ways, first using R as the xy projection of S and second as the yz projection, to illustrate that we get the same result.

In the first method, $z = f(x, y)$ (xy projection). Solving for z, we get $f(x, y) = 6 - 3x - 2y$. The region R is the triangular region in the xy-plane bounded by the x- and y-axes and the xy trace of S, $3x + 2y = 6$, as shown in Figure 16.41. Thus, $f_x = -3$ and $f_y = -2$, so $\sqrt{1 + f_x^2 + f_y^2} = \sqrt{14}$. Using Equation 16.27, we get (omitting some details)

$$\iint_S (2xy + xz)\, d\sigma = \iint_R [2xy + x(6 - 3x - 2y)]\sqrt{14}\, dA$$

$$= \sqrt{14} \iint_R (6x - 3x^2)\, dA$$

$$= 3\sqrt{14} \int_0^2 \int_0^{(6-3x)/2} (2x - x^2)\, dy\, dx = 6\sqrt{14}$$

In the second method, $x = f(y, z)$ (yz projection). Solving for x, we get $f(y, z) = (6 - 2y - z)/3$. This time, the region R is the yz projection of S bounded by the y- and z-axes and the line $2y + z = 6$. Since $f_y = -2/3$ and

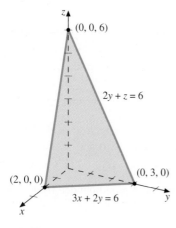

FIGURE 16.41

$f_z = -1/3$, we get $\sqrt{1 + f_y^2 + f_z^2} = \sqrt{14}/3$. Thus, by Equation 16.28

$$\iint\limits_{S} (2xy + xz)\, d\sigma = \iint\limits_{R} \left[2\left(\frac{6 - 2y - z}{3}\right)y + \left(\frac{6 - 2y - z}{3}\right)z\right]\frac{\sqrt{14}}{3}\, dA$$

$$= \frac{\sqrt{14}}{9} \int_0^3 \int_0^{6-2y} (12y - 4y^2 - 4yz + 6z - z^2)\, dz\, dy$$

$$= 6\sqrt{14}$$

Note that in this problem the integration is simpler using the first method. When there is a choice of methods, it pays to look ahead to anticipate which method may result in the easiest integration.   ■

Moments, Mass, and Center of Mass of a Lamina

Suppose a thin sheet (a lamina) of some material (such as metal) is in the shape of a surface S. If the density is $\rho(x, y, z)$ at the point (x, y, z) on S, then by the usual reasoning we can obtain the following formulas for the mass m, the center of mass $(\bar{x}, \bar{y}, \bar{z})$, and the moments of inertia I_x, I_y, and I_z with respect to the coordinate axes.

Mass, Center of Mass, and Moments of Inertia of a Lamina in the Shape of a Surface S

$$m = \iint\limits_{S} \rho(x, y, z)\, d\sigma \qquad (16.31)$$

$$\bar{x} = \frac{1}{m} \iint\limits_{S} x\, dm, \qquad \bar{y} = \frac{1}{m} \iint\limits_{S} y\, dm, \qquad \bar{z} = \frac{1}{m} \iint\limits_{S} z\, dm \quad (16.32)$$

$$I_x = \iint\limits_{S} (y^2 + z^2)\, dm, \quad I_y = \iint\limits_{S} (x^2 + z^2)\, dm, \quad I_z = \iint\limits_{S} (x^2 + y^2)\, dm$$

$$(16.33)$$

where $dm = \rho(x, y, z)\, d\sigma$.

EXAMPLE 16.22   A homogeneous lamina of density ρ is in the shape of the portion of the paraboloid $z = x^2 + y^2$ between $z = 0$ and $z = 4$. Find the center of mass.

Solution   The xy projection of S is the circular region bounded by $x^2 + y^2 = 4$, as shown in Figure 16.42. With $f(x, y) = x^2 + y^2$, we get

$$\sqrt{1 + f_x^2 + f_y^2} = \sqrt{1 + 4(x^2 + y^2)}$$

We first calculate the mass. By Equation 16.31,

$$m = \iint\limits_{S} \rho\, d\sigma = \iint\limits_{R} \rho\sqrt{1 + 4(x^2 + y^2)}\, dA = \rho \int_0^{2\pi} \int_0^2 \sqrt{1 + 4r^2}\, r\, dr\, d\theta$$

$$= \frac{\pi\rho}{6}\left[(17)^{3/2} - 1\right]$$

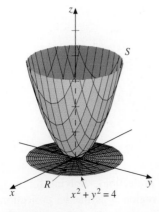

FIGURE 16.42

$x^2 + y^2 = 4$

where again we have omitted details of integration. By symmetry, $\bar{x} = \bar{y} = 0$. For $\bar{z}$ we have, by the third of Equations 16.32,

$$m\bar{z} = \iint_S z \, dm = \iint_S z\rho \, d\sigma = \rho \iint_R (x^2 + y^2)\sqrt{1 + 4(x^2 + y^2)} \, dA$$

$$= \rho \int_0^{2\pi} \int_0^2 r^3\sqrt{1 + 4r^2} \, dr \, d\theta = \frac{\pi\rho}{60}\left[23\,(17)^{3/2} + 1\right]$$

(The integration with respect to r can be accomplished by the substitution $u = \sqrt{1 + 4r^2}$. You should supply the details.) Thus,

$$\bar{z} = \frac{1}{10}\left[\frac{23(17)^{3/2} + 1}{(17)^{3/2} - 1}\right] \approx 2.335 \qquad \blacksquare$$

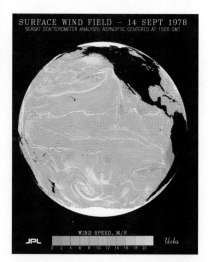

Windspeed and direction over the Pacific Ocean

Three-Dimensional Fluid Flow

Suppose now that $\mathbf{F}(x, y, z)$ is the velocity field of some fluid in steady state. The flow of water in a stream, ocean currents, wind flow, radiation in a star, and the flow of blood in the vascular system can be assumed to have an approximate steady-state motion, at least over relatively short periods of time. Let S be a surface in the given vector field through which the fluid can flow unimpeded. You can think of S as being a screen or netting (or even an imaginary surface). Let $\Delta\sigma$ be the area of one of the cells on S that results from a partition of S, and let $\mathbf{n}$ be a unit normal vector to S at an arbitrary point in this cell. If the cell is small, we can assume the velocity is approximately constant throughout the cell. The dot product $\mathbf{F} \cdot \mathbf{n}$ is the component of $\mathbf{F}$ in the direction of $\mathbf{n}$, and if we multiply this component by $\Delta\sigma$ we get the approximate volume of fluid that flows orthogonally through this cell per unit of time. This volume is represented in Figure 16.43 by the prism with height $\mathbf{F} \cdot \mathbf{n}$. By summing over all such cells and passing to the limit as the norms of partitions approach 0, we

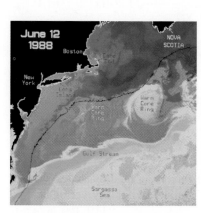

Ocean currents

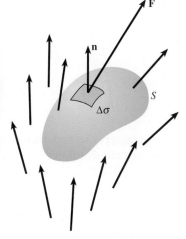
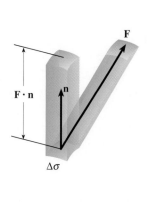

FIGURE 16.43

obtain the *flux of* **F** *through S:*

$$\text{Flux of } \mathbf{F} \text{ Through } S = \iint_S \mathbf{F} \cdot \mathbf{n} \, d\sigma \qquad (16.34)$$

The flux is the total net volume of the fluid that flows through S per unit of time. This volume flux is analogous to flux in two dimensions. If S is a closed surface (such as an ellipsoid), we typically take **n** to be directed toward the exterior to S, called the *outer unit normal*. Then if the flux is positive, there is a net outflow of fluid through S, and we say there is a *source* inside S. If the flux is negative, there is a net inflow of fluid, and we say there is a *sink* inside S. If the flux is 0, the flow is said to be *incompressible*. As we have defined it, flux is the net volume that passes through S per unit of time. If the fluid has density $\rho(x, y, z)$, then the integral $\iint_S \rho\mathbf{F} \cdot \mathbf{n} \, d\sigma$ is the *mass flux*—that is, the net mass of fluid that passes through S per unit of time.

EXAMPLE 16.23   Find the flux of a fluid that has velocity field $\mathbf{F} = 4x\mathbf{i} + 4y\mathbf{j} + 3z\mathbf{k}$ through the parabolic surface S defined by $z = 4 - x^2 - y^2$ for $z \geq 0$ in the direction of outer unit normals.

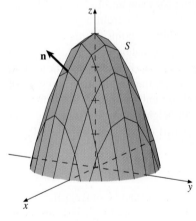

FIGURE 16.44

Solution   As shown in Figure 16.44, the outer normals are also the upward normals. In Section 14.4, we showed that a normal vector to the surface $z = f(x, y)$ is $f_x\mathbf{i} + f_y\mathbf{j} - \mathbf{k}$. But we take the negative of this normal so that the z component is positive (upward-directed). We make it into a unit vector by dividing by its length. Thus,

$$\mathbf{n} = \frac{-f_x\mathbf{i} - f_y\mathbf{j} + \mathbf{k}}{\sqrt{1 + f_x^2 + f_y^2}} = \frac{2x\mathbf{i} + 2y\mathbf{j} + \mathbf{k}}{\sqrt{1 + 4(x^2 + y^2)}}$$

The flux is therefore

$$\iint_S \mathbf{F} \cdot \mathbf{n} \, d\sigma = \iint_R \frac{8x^2 + 8y^2 + 3(4 - x^2 - y^2)}{\sqrt{1 + 4(x^2 + y^2)}} \sqrt{1 + 4(x^2 + y^2)} \, dA$$

$$= \int_0^{2\pi} \int_0^2 (5r^2 + 12)r \, dr \, d\theta = 88\pi$$

If velocity is in meters per second, for example, and area is in square meters, then the net amount of fluid that flows out of the surface each second is 88π cubic meters.   ∎

Although we have illustrated the idea of flux using fluid dynamics, we still call the integral $\iint_S \mathbf{F} \cdot \mathbf{n} \, d\sigma$ the flux of **F** through S for any vector field **F**, whether or not it represents velocity. Some other areas in which this notion is useful are heat flow, electricity, magnetism, and gravitational fields.

Exercise Set 16.5

In Exercises 1–8, evaluate the surface integral $\iint_S g(x, y, z)\, d\sigma$.

1. $g(x, y, z) = 2x - y + z$; S is the first-octant portion of the plane $x + y + z = 2$.

2. $g(x, y, z) = x^2 y - 2z$; S is the first-octant portion of the plane $z = x + 2y$ that lies below $z = 4$.

3. $g(x, y, z) = xz$; S is the portion of the plane $z = 2x - 3y$ inside the cylinder $x^2 + y^2 = 9$.

4. $g(x, y, z) = xy$; S is the first-octant portion of the cylinder $x^2 + z^2 = 4$ between $y = 0$ and $y = 4$ and above $z = 1$.

5. $g(x, y, z) = x^2 + y^2 - z$; S is the portion of the paraboloid $z = 2 - x^2 - y^2$ above the xy-plane.

6. $g(x, y, z) = xyz$; S is the portion of the cone $z^2 = x^2 + y^2$ between $z = 1$ and $z = 2$.

7. $g(x, y, z) = 8/z^2$; S is the portion of the sphere $x^2 + y^2 + z^2 = 25$ above $z = 3$.

8. $g(x, y, z) = xz^2$; S is the portion of the parabolic cylinder $y = x^2$ in the first octant bounded by $y = 2$, $y = 6$, $z = 0$, and $z = 4$. (*Hint:* Use a yz projection.)

In Exercises 9–12, find the mass and center of mass of the lamina in the shape of the surface S with density $\rho(x, y, z)$.

9. S is the first-octant portion of the plane $x + 2y + 4z = 8$; $\rho(x, y, z) = z$.

10. S is the portion of the cylinder $3z = x^2$ lying above the region $R = \{(x, y) : 0 \le x \le 2, 0 \le y \le 4\}$; $\rho = $ constant.

11. S is the portion of the paraboloid $z = 4 - x^2 - y^2$ that is inside the cylinder $x^2 + y^2 = 2$; $\rho = $ constant.

12. S is the upper portion of the sphere $x^2 + y^2 + z^2 = 16$ that is inside the cylinder $x^2 + y^2 = 8$; $\rho = 1/\sqrt{z}$.

In Exercises 13 and 14, set up, but do not evaluate, iterated integrals for I_x, I_y, and I_z for the specified laminas.

13. The lamina of Exercise 10

14. The lamina of Exercise 12

In Exercises 15 and 16, set up, but do not evaluate, iterated integrals for calculating the given surface integrals using (a) xy projections, (b) yz projections, and (c) xz projections.

15. $\iint_S x^2 yz^3\, d\sigma$; S is the first-octant portion of the plane

$2x + y - z = 0$ that is below the plane $z = 4$

16. $\iint_S (x + 2yz)\, d\sigma$; S is the first-octant portion of the elliptic paraboloid $z = 12 - 3x^2 - 4y^2$

In Exercises 17 and 18, set up, but do not evaluate, two iterated integrals for calculating the given surface integrals using projections on two different planes.

17. $\iint_S xyz\, d\sigma$; S is the portion of the cylinder $z = 2 - x^2$ in the first octant below $z = 1$ and between $y = 0$ and $y = 5$.

18. $\iint_S (xz/y)\, d\sigma$; S is the first-octant portion of the cylinder $y = e^x$ bounded by the planes $y = 2$ and $z = 3$.

In Exercises 19–26, find the flux of $\mathbf{F}$ through S. Take $\mathbf{n}$ as the upward normal unless otherwise specified.

19. $\mathbf{F} = \langle 3xy, yz, 2z \rangle$; S is the portion of the plane $x + 3y + z = 5$ lying above the region $R = \{(x, y) : 0 \le x \le 2, 0 \le y \le 1\}$.

20. $\mathbf{F} = \langle x, y, z \rangle$; S is the upper portion of the sphere $x^2 + y^2 + z^2 = 25$ that is inside the cylinder $x^2 + y^2 = 16$.

21. $\mathbf{F} = 2x\mathbf{i} + 2y\mathbf{j} + 3z\mathbf{k}$; S is the portion of the paraboloid $z = 4 - x^2 - y^2$ above the xy-plane.

22. $\mathbf{F} = yz\mathbf{i} + (x^3/z)\mathbf{j} + 2z^2\mathbf{k}$; S is the first-octant portion of the cylinder $z = e^x$ bounded above by $z = 2$ and on the right by $y = 2$. Use downward-directed normals.

23. $\mathbf{F} = \langle -x^3, y^3, -z \rangle$; S is the portion of the cone $z^2 = x^2 + y^2$ between $z = 1$ and $z = 2$. Use downward-directed normals.

24. $\mathbf{F} = 3x\mathbf{i} + 3y\mathbf{j} - z\mathbf{k}$; S is the portion of the hemisphere $z = \sqrt{8 - x^2 - y^2}$ that is inside the cone $z^2 = x^2 + y^2$.

25. $\mathbf{F} = (x^3 yz)\mathbf{i} + (x - y^2)\mathbf{j} + z^2\mathbf{k}$; S is the surface $z = 1 - |y|$ above the xy-plane and between $x = 0$ and $x = 4$. (*Hint:* Divide S into two parts and add the integrals over the separate parts.)

26. $\mathbf{F} = \langle xyz, x^2 - y^2, xz - yz \rangle$; S is the cube that is one unit on an edge that has one vertex at the origin and three of its edges on the positive coordinate axes. Use outward-directed normals. Is there a source or a sink within S? (*Hint:* Find the flux through each face and add the results.)

27. Find the mass and center of mass of a homogeneous hemispherical shell of radius a and density $\rho = k$. (*Hint:* To evaluate the improper integral, first integrate over the circular region R_b with radius $b < a$, and after integration let $b \to a^-$.)

28. The velocity field for a certain liquid is $\mathbf{F} = x\mathbf{i} - y\mathbf{j} + z\mathbf{k}$. In it is submerged a closed surface that forms the boundary of the region below the hemisphere $z = \sqrt{2a^2 - x^2 - y^2}$ and above the paraboloid $az = x^2 + y^2$. Find the total flux through S, using outward-directed normals.

29. Let $\mathbf{F}$ be the inverse-square field

$$\mathbf{F} = \frac{k\mathbf{u}}{|\mathbf{r}|^2}$$

where $\mathbf{u}$ is the unit vector in the direction of $\mathbf{r} = x\mathbf{i} + y\mathbf{j} + z\mathbf{k}$. Show that the flux $\mathbf{F}$ through a sphere S centered at the origin is independent of the radius of S.

30. If $\mathbf{F} = \langle L, M, N \rangle$ and S is the graph of $z = f(x, y)$ that has projection R in the xy-plane, then

$$\iint_S \mathbf{F} \cdot \mathbf{n} \, d\sigma = \iint_R (-Lf_x - Mf_y + N) \, dA$$

assuming the appropriate hypotheses. Prove this result.

31. Give a derivation to justify Equations 16.31, 16.32, and 16.33.

16.6 THE DIVERGENCE THEOREM

FIGURE 16.45
A Möbius strip

In this section and the next, we will be making use of surface integrals of vector fields, and for this purpose we limit consideration to surfaces that have two distinct sides, called **orientable surfaces**. It may surprise you to learn that some surfaces have only one side. The best known example is the *Möbius strip*, shown in Figure 16.45. You can easily construct such a surface by taking a strip of paper, giving one end a half twist, and pasting the ends together. To convince yourself that this surface is one-sided, take a crayon and start coloring the "top" side and continue until you return to the starting point. You will discover you have colored the entire surface without ever lifting the crayon from the paper!

Another way to describe an orientable surface S is by means of its unit normal vectors. If a direction for a unit normal can be chosen in such a way that, starting from any point on S and going around any closed curve C on S, the unit normal returns to its original direction at the starting point, then S is orientable. In Figure 16.46, we indicate how the Möbius strip fails this test. We call whichever unit normal we have selected positive, and then we say that S *is oriented* with respect to that normal. In effect, we have designated a positive side to the surface. For example, a surface may be oriented by upward normals. Then we are calling the top side positive.

FIGURE 16.46
Normals along C do not return to their original position.

Green's Theorem enabled us to express a line integral around a closed path C as a double integral over the region enclosed by C. The theorem we consider in this section enables us to express a surface integral over a closed surface S as a triple integral over the region enclosed by S. Before stating the theorem, we introduce the notion of the *divergence* of a vector field in $\mathbb{R}^3$, which is the natural extension of divergence in $\mathbb{R}^2$.

Definition 16.4	Let $\mathbf{F} = \langle f, g, h \rangle$ be a vector field in $\mathbb{R}^3$ for which the partial derivatives
The Divergence of a Vector Field	of f, g, and h exist. The **divergence** of $\mathbf{F}$, written $\operatorname{div} \mathbf{F}$, is the scalar field defined by

$$\operatorname{div} \mathbf{F} = \frac{\partial f}{\partial x} + \frac{\partial g}{\partial y} + \frac{\partial h}{\partial z}$$

The next theorem is known as both the *Divergence Theorem* and *Gauss's Theorem*. A precise formulation of the hypotheses of the theorem would require a deeper background in three-dimensional regions and their boundaries than we have presented. Our formulation is sufficient for most applications.

THEOREM 16.10

> **The Divergence Theorem (Gauss's Theorem)**
>
> Let G be a closed and bounded three-dimensional region with boundary S that is a piecewise-smooth closed surface. If $\mathbf{F}$ is a continuously differentiable vector field on some open set that contains G, then
>
> $$\iint_S \mathbf{F} \cdot \mathbf{n}\, d\sigma = \iiint_G \operatorname{div} \mathbf{F}\, dV \qquad (16.35)$$
>
> where $\mathbf{n}$ is the outer unit normal to S.

REMARK ────────────────────────────────────

■ Observe the similarity between Equations 16.35 and 16.25 for two dimensions.

────────────────────────────────────

Proof   We will give a proof of the theorem when G is a special (but commonly occurring) type of region that we call a **simple** three-dimensional region. By a simple region, we mean that G is bounded above and below by the graphs of two smooth functions $z = \phi_1(x, y)$ and $z = \phi_2(x, y)$ with $\phi_1(x, y) \le \phi_2(x, y)$ for all points (x, y) in the projection R of G onto the xy-plane. Similarly, for G to be simple, we require that the lateral bounding surfaces of G be graphs of smooth functions of y and z in the x direction and of x and z in the y direction. A rectangular box is an example of a simple region, as is an ellipsoid.

Let $\mathbf{F}(x, y, z) = f(x, y, z)\mathbf{i} + g(x, y, z)\mathbf{j} + h(x, y, z)\mathbf{k}$. Then

$$\iint_S \mathbf{F} \cdot \mathbf{n}\, d\sigma = \iint_S f\mathbf{i} \cdot \mathbf{n}\, d\sigma + \iint_S g\mathbf{j} \cdot \mathbf{n}\, d\sigma + \iint_S h\mathbf{k} \cdot \mathbf{n}\, d\sigma$$

Also,

$$\iiint_G \operatorname{div} \mathbf{F}\, dV = \iiint_G \frac{\partial f}{\partial x}\, dV + \iiint_G \frac{\partial g}{\partial y}\, dV + \iiint_G \frac{\partial h}{\partial z}\, dV$$

Thus, to prove Equation 16.35, it is sufficient to prove that

$$\iint_S f\mathbf{i} \cdot \mathbf{n}\, d\sigma = \iiint_G \frac{\partial f}{\partial x}\, dV \qquad (16.36)$$

$$\iint_S g\mathbf{j} \cdot \mathbf{n}\, d\sigma = \iiint_G \frac{\partial g}{\partial y}\, dV \qquad (16.37)$$

$$\iint_S h\mathbf{k} \cdot \mathbf{n}\, d\sigma = \iiint_G \frac{\partial h}{\partial z}\, dV \qquad (16.38)$$

We will prove Equation 16.38 only and leave the proofs of the other two equations as an exercise. (See Exercise 19 in Exercise Set 16.6.)

We are assuming G is a simple region. So we can write

$$G = \{(x, y, z): (x, y) \in R, \phi_1(x, y) \leq z \leq \phi_2(x, y)\}$$

where R is the xy projection of G. As we show in Figure 16.47, let S_1 be the lower bounding surface (the graph of $z = \phi_1(x, y)$) and S_2 the upper bounding surface (the graph of $z = \phi_2(x, y)$). Also, let S_3, S_4, S_5, and S_6 be the vertical lateral bounding surfaces, as shown. (In some cases, such as an ellipsoid, there may not be any vertical surfaces.)

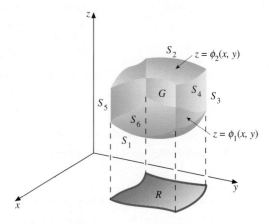

FIGURE 16.47

On each of the lateral surfaces S_3, S_4, S_5, and S_6, the unit normal vector $\mathbf{n}$ is horizontal. Thus, $\mathbf{k} \cdot \mathbf{n} = 0$ on these lateral surfaces. It follows that

$$\iint_S h\mathbf{k} \cdot \mathbf{n} \, d\sigma = \iint_{S_1} h\mathbf{k} \cdot \mathbf{n} \, d\sigma + \iint_{S_2} h\mathbf{k} \cdot \mathbf{n} \, d\sigma$$

The outer unit normal on S_1 is directed downward. So on S_1

$$\mathbf{n} = \frac{\dfrac{\partial \phi_1}{\partial x}\mathbf{i} + \dfrac{\partial \phi_1}{\partial y}\mathbf{j} - \mathbf{k}}{\sqrt{1 + \left(\dfrac{\partial \phi_1}{\partial x}\right)^2 + \left(\dfrac{\partial \phi_1}{\partial y}\right)^2}}$$

Thus,

$$\mathbf{k} \cdot \mathbf{n} = \frac{-1}{\sqrt{1 + \left(\dfrac{\partial \phi_1}{\partial x}\right)^2 + \left(\dfrac{\partial \phi_1}{\partial y}\right)^2}}$$

Similarly, on the upper boundary, S_2,

$$\mathbf{n} = \frac{-\dfrac{\partial \phi_2}{\partial x}\mathbf{i} - \dfrac{\partial \phi_2}{\partial y}\mathbf{j} + \mathbf{k}}{\sqrt{1 + \left(\dfrac{\partial \phi_2}{\partial x}\right)^2 + \left(\dfrac{\partial \phi_2}{\partial y}\right)^2}}$$

so that

$$\mathbf{k} \cdot \mathbf{n} = \frac{1}{\sqrt{1 + \left(\dfrac{\partial \phi_2}{\partial x}\right)^2 + \left(\dfrac{\partial \phi_2}{\partial y}\right)^2}}$$

We therefore have

$$\iint\limits_{S} h\mathbf{k} \cdot \mathbf{n}\, d\sigma = \iint\limits_{S_1} h\mathbf{k} \cdot \mathbf{n}\, d\sigma + \iint\limits_{S_2} h\mathbf{k} \cdot \mathbf{n}\, d\sigma$$

$$= \iint\limits_{S_1} \frac{-h(x, y, z)d\sigma}{\sqrt{1 + \left(\dfrac{\partial \phi_1}{\partial x}\right)^2 + \left(\dfrac{\partial \phi_1}{\partial y}\right)^2}}$$

$$+ \iint\limits_{S_2} \frac{h(x, y, z)d\sigma}{\sqrt{1 + \left(\dfrac{\partial \phi_2}{\partial x}\right)^2 + \left(\dfrac{\partial \phi_2}{\partial y}\right)^2}}$$

We evaluate the two surface integrals on the right, by Equation 16.27, as double integrals over the xy projection R:

$$\iint\limits_{S_1} \frac{-h(x, y, z)d\sigma}{\sqrt{1 + \left(\dfrac{\partial \phi_1}{\partial x}\right)^2 + \left(\dfrac{\partial \phi_1}{\partial y}\right)^2}}$$

$$= \iint\limits_{R} \frac{-h(x, y, \phi_1(x, y))}{\sqrt{1 + \left(\dfrac{\partial \phi_1}{\partial x}\right)^2 + \left(\dfrac{\partial \phi_1}{\partial y}\right)^2}} \sqrt{1 + \left(\dfrac{\partial \phi_1}{\partial x}\right)^2 + \left(\dfrac{\partial \phi_1}{\partial y}\right)^2}\, dA$$

$$= \iint\limits_{R} -h(x, y, \phi_1(x, y))\, dA$$

Similarly,

$$\iint\limits_{S_2} \frac{h(x, y, z)d\sigma}{\sqrt{1 + \left(\dfrac{\partial \phi_2}{\partial x}\right)^2 + \left(\dfrac{\partial \phi_2}{\partial y}\right)^2}}$$

$$= \iint\limits_{R} \frac{h(x, y, \phi_2(x, y))}{\sqrt{1 + \left(\dfrac{\partial \phi_2}{\partial x}\right)^2 + \left(\dfrac{\partial \phi_2}{\partial y}\right)^2}} \sqrt{1 + \left(\dfrac{\partial \phi_2}{\partial x}\right)^2 + \left(\dfrac{\partial \phi_2}{\partial y}\right)^2}\, dA$$

$$= \iint\limits_{R} h(x, y, \phi_2(x, y))\, dA$$

Combining the integrals over S_1 and S_2 gives

$$\iint\limits_{S} h\mathbf{k} \cdot \mathbf{n}\, d\sigma = \iint\limits_{R} \left[h(x, y, \phi_2(x, y)) - h(x, y, \phi_1(x, y))\right] dA \qquad (16.39)$$

The right-hand side of Equation 16.38 can be written as

$$\iiint_G \frac{\partial h}{\partial z} dV = \iint_R \left[\int_{\phi_1(x,y)}^{\phi_2(x,y)} \frac{\partial h}{\partial z} dz \right] dA$$

$$= \iint_R \left[h(x, y, \phi_2(x, y)) - h(x, y, \phi_1(x, y)) \right] dA \qquad \begin{array}{l} \text{By the Second} \\ \text{Fundamental} \\ \text{Theorem of} \\ \text{Calculus} \end{array}$$

The equality of this last integral and the integral on the right-hand side of Equation 16.39 establishes the result given in Equation 16.38:

$$\iint_S h\mathbf{k} \cdot \mathbf{n} \, d\sigma = \iiint_G \frac{\partial h}{\partial z} \, dV$$

Equations 16.36 and 16.37 are proved in a similar manner. Thus, Equation 16.35 is true when G is simple.   ■

The proof we have given can be extended to a region $G = G_1 \cup G_2$, where G_1 and G_2 are simple regions with an intersection that is a sectionally smooth surface, and by induction to any finite union of such simple regions. You will be asked to show this extension in Exercise 20 of Exercise Set 16.6.

EXAMPLE 16.24   Use the Divergence Theorem to evaluate $\iint_S \mathbf{F} \cdot \mathbf{n} \, d\sigma$, where S is the sphere $x^2 + y^2 + z^2 = a^2$ and $\mathbf{F} = xy^2\mathbf{i} + yz^2\mathbf{j} + x^2z\mathbf{k}$. Assume $\mathbf{n}$ is the outer unit normal.

Solution   Let G be the sphere together with its interior. Then, by Equation 16.35,

$$\iint_S \mathbf{F} \cdot \mathbf{n} \, d\sigma = \iiint_G \operatorname{div} \mathbf{F} \, dV$$

$$= \iiint_G (y^2 + z^2 + x^2) \, dV$$

Because G is a sphere and because of the nature of the integrand, it is convenient to change to spherical coordinates. The sphere has equation $\rho = a$, and the integrand becomes ρ^2. Thus,

$$\iint_S \mathbf{F} \cdot \mathbf{n} \, d\sigma = \int_0^\pi \int_0^{2\pi} \int_0^a \rho^2 (\rho^2 \sin \phi \, d\rho \, d\theta \, d\phi) = \frac{4\pi a^5}{5}$$

(You may wish to try evaluating the surface integral in this problem *without* using the Divergence Theorem, to compare the difficulty.)   ■

EXAMPLE 16.25   Evaluate the integral $\iint_S \mathbf{F} \cdot \mathbf{n} \, d\sigma$, where S is the boundary of the region G below the paraboloid $z = 4 - x^2 - y^2$, inside the cylinder $x^2 + y^2 = 1$, and above the xy-plane, and where

$$\mathbf{F} = \left\langle 2x + \sqrt{z^3}, 3y - e^{z^2}, (x^3 + y^3)^{4/3} \right\rangle$$

Use outer unit normals. (See Figure 16.48.)

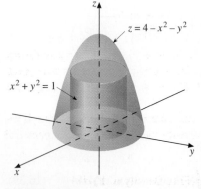

FIGURE 16.48

Solution   Since div $\mathbf{F} = 2+3+0 = 5$, we have, by the Divergence Theorem,

$$\iint\limits_{S} \mathbf{F} \cdot \mathbf{n} \, d\sigma = \iiint\limits_{G} 5 \, dV$$

This time we will use cylindrical coordinates. The paraboloid has the cylindrical equation $z = 4 - r^2$, and the cylinder has the equation $r = 1$. So we have

$$\iint\limits_{S} \mathbf{F} \cdot \mathbf{n} \, d\sigma = 5 \int_{0}^{2\pi} \int_{0}^{1} \int_{0}^{4-r^2} r \, dz \, dr \, d\theta$$

$$= 5 \int_{0}^{2\pi} \int_{0}^{1} (4r - r^3) \, dr \, d\theta = \frac{35\pi}{2}$$

Without the Divergence Theorem this problem would be virtually impossible to solve. (Try it!)   ■

Relationship Between Flux and Divergence

If $\mathbf{F}$ is the velocity field of a fluid, we know that $\iint_{S} \mathbf{F} \cdot \mathbf{n} \, d\sigma$ is the flux of $\mathbf{F}$ through S. So, assuming appropriate conditions on $\mathbf{F}$, S, and G, the Divergence Theorem says that

$$\text{Flux of } \mathbf{F} \text{ Through } S = \iiint\limits_{G} \text{div} \, \mathbf{F} \, dV$$

The physical interpretation of div $\mathbf{F}$ at a point P is analogous to that in two dimensions. We let S_ε be a sphere of radius ε centered at P, and let G_ε be the region enclosed by S_ε. Then, using a Mean-Value Theorem for triple integrals, we can write

$$\iiint\limits_{G_\varepsilon} \text{div} \, \mathbf{F} \, dV = [\text{div} \, \mathbf{F}(Q)] \, (\text{Volume of } G_\varepsilon)$$

where Q is some point in G_ε. Thus,

$$\text{div} \, \mathbf{F}(Q) = \frac{\text{Flux of } \mathbf{F} \text{ Through } S_\varepsilon}{\text{Volume of } G_\varepsilon}$$

Now we let $\varepsilon \to 0$, so that $Q \to P$, and if div $\mathbf{F}$ is continuous, div $\mathbf{F}(Q) \to$ div $\mathbf{F}(P)$. Thus,

$$\text{div} \, \mathbf{F}(P) = \lim_{\varepsilon \to 0} \frac{\text{Flux of } \mathbf{F} \text{ Through } S_\varepsilon}{\text{Volume of } G_\varepsilon}$$

The limit on the right is the flux per unit volume at P, called the *flux density* of $\mathbf{F}$ at P. So the divergence of $\mathbf{F}$ at a point is the flux density at that point. Just as with two-dimensional flow, we call P a *source* if div $P > 0$ and a *sink* if div $P < 0$.

The Divergence Theorem says that we can obtain the flux of $\mathbf{F}$ through the closed surface S by integrating the flux density over the volume G enclosed by S:

$$\text{Flux of } \mathbf{F} \text{ Through } S = \iiint\limits_{G} (\text{Flux Density of } \mathbf{F}) \, dV$$

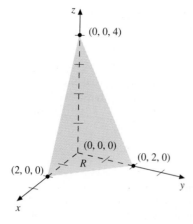

FIGURE 16.49

EXAMPLE 16.26   Use the Divergence Theorem to find the flux of the velocity field $\mathbf{F}(x, y, z) = 2xy\mathbf{i} + 3yz\mathbf{j} + xz\mathbf{k}$ through the tetrahedron with vertices $(0, 0, 0)$, $(2, 0, 0)$, $(0, 2, 0)$, and $(0, 0, 4)$.

Solution   Let S denote the surface of the tetrahedron. We show it in Figure 16.49. Let G denote the region enclosed by S. Its xy projection is the triangular region R, as shown, bounded by the lines $x = 0$, $y = 0$, and $x + y = 2$. The inclined plane that forms the upper bounding surface of G has the equation $2x + 2y + z = 4$, so $z = 4 - 2x - 2y$. Applying the Divergence Theorem, we have

$$\text{Flux of } \mathbf{F} \text{ Through } S = \iiint_G (\text{div } \mathbf{F})\, dV$$

$$= \int_0^2 \int_0^{2-x} \int_0^{4-2x-2y} (2y + 3z + x)\, dz\, dy\, dx = 12 \quad \blacksquare$$

Exercise Set 16.6

In Exercises 1–6, use the Divergence Theorem to evaluate $\iint_S \mathbf{F} \cdot \mathbf{n}\, d\sigma$, where $\mathbf{n}$ is the outer unit normal.

1. $\mathbf{F}(x, y, z) = x^2\mathbf{i} + y^2\mathbf{j} + z^2\mathbf{k}$; S is the rectangular parallelepiped formed by the planes $x = 1$, $x = 3$, $y = 2$, $y = 6$, $z = 0$, and $z = 4$.

2. $\mathbf{F}(x, y, z) = (2x + y)\mathbf{i} + (x - y)\mathbf{j} + x^2 y^3\mathbf{k}$; S is the tetrahedron formed by the planes $3x + 2y + z = 6$, $x = 0$, $y = 0$, and $z = 0$.

3. $\mathbf{F}(x, y, z) = \langle x + z, y^2, yz \rangle$; S is the boundary of the first-octant region enclosed by the cylinder $z = 4 - x^2$ and the planes $y = 0$, $z = 0$, and $y = x$.

4. $\mathbf{F}(x, y, z) = \langle xz, 2xy, 4yz \rangle$; S is the tetrahedron formed by the planes $x + y = 2$, $y + z = 2$, $x = 0$, $y = 0$, and $z = 0$.

5. $\mathbf{F}(x, y, z) = (e^z \sin y)\mathbf{i} + (e^z \cos x)\mathbf{j} + z\mathbf{k}$; S is the boundary of the region inside the cylinder $x^2 + y^2 = 4$ between the planes $z = 0$ and $z = 6 + x + 2y$.

6. $\mathbf{F}(x, y, z) = \langle x^3, y^3, \cosh x^3 \rangle$; S is the boundary of the region inside the cone $z = \sqrt{x^2 + y^2}$ between $z = 1$ and $z = 2$.

In Exercises 7–12, use the Divergence Theorem to find the flux of $\mathbf{F}$ through S in the direction of outer unit normals.

7. $\mathbf{F}(x, y, z) = (2x - 3y)\mathbf{i} + (4y + 2z)\mathbf{j} + (x + z)\mathbf{k}$; S is the sphere $x^2 + y^2 + z^2 = 16$.

8. $\mathbf{F}(x, y, z) = (e^y \cos z)\mathbf{i} + (e^z \sin x)\mathbf{j} + (e^{x^2 + y^2})\mathbf{k}$; S is the ellipsoid $3x^2 + 7y^2 + 12z^2 = 84$.

9. $\mathbf{F}(x, y, z) = \langle y/x, x/y, 1/z^2 \rangle$; S is the cube formed by the planes $x = 1$, $x = 2$, $y = 1$, $y = 2$, $z = 2$, and $z = 3$.

10. $\mathbf{F}(x, y, z) = \langle xy^2, yz^2, zx^2 \rangle$; S is the "ice cream cone" formed by the cone $z = \sqrt{x^2 + y^2}$ and the hemisphere $z = \sqrt{4 - x^2 - y^2}$.

11. $\mathbf{F}(x, y, z) = x^2 y\mathbf{i} + xy^2\mathbf{j} + z^2\mathbf{k}$; S is the boundary of the region inside the paraboloid $z = x^2 + y^2$ and below the hemisphere $z = \sqrt{2 - x^2 - y^2}$.

12. $\mathbf{F}(x, y, z) = xy\mathbf{i} + y^2\mathbf{j} + yz\mathbf{k}$; S is the boundary of the first-octant region inside the cylinder $x^2 + z^2 = 4$ between the planes $y = 0$ and $y = 2x$.

In Exercises 13–16, verify the Divergence Theorem by calculating $\iint_S \mathbf{F} \cdot \mathbf{n}\, d\sigma$ and $\iiint_G \text{div } \mathbf{F}\, dV$.

13. $\mathbf{F}(x, y, z) = 2x\mathbf{i} - 3y\mathbf{j} + 4z\mathbf{k}$; G is the region inside the paraboloid $z = x^2 + y^2$ and below the plane $z = 4$.

14. $\mathbf{F}(x, y, z) = \langle x - y, x + y, 2x \rangle$; G is the region enclosed by the tetrahedron formed by the coordinate planes and the plane $x + y + z = 2$.

15. $\mathbf{F}(x, y, z) = \langle 2x, 3y, z \rangle$; G is the first-octant region inside both of the cylinders $x^2 + y^2 = a^2$ and $x^2 + z^2 = a^2$.

16. $\mathbf{F}(x, y, z) = x^{3/2}\mathbf{i} + y^{3/2}\mathbf{j} + z^{3/2}\mathbf{k}$; G is the region enclosed by the planes $x + y = 4$, $z = 4$, and the coordinate planes.

17. Show that if a region G and its boundary S satisfy the conditions of the Divergence Theorem, then for $\mathbf{F}(x, y, z) = x\mathbf{i} + y\mathbf{j} + z\mathbf{k}$,

$$\iint_S \mathbf{F} \cdot \mathbf{n}\, d\sigma = 3V$$

where V is the volume of G.

18. Let G satisfy the Divergence Theorem hypotheses. Prove that the flux of any constant vector field through S is 0.

19. Verify Equations 16.36 and 16.37 for a simple region.

20. Let $G = G_1 \cup G_2$, where G_1 and G_2 are simple regions with an intersection that is a sectionally smooth surface T. Prove that the Divergence Theorem holds true for G. Extend the result by induction to a finite union of simple regions. (*Hint:* The outer unit normals for G_1 and G_2 across their common boundary T are oppositely directed.)

21. Let S be the sphere $x^2 + y^2 + (z - a)^2 = a^2$ and $\mathbf{F} = \langle x^2, y^2, z^2 \rangle$. Show that $\iint_S \mathbf{F} \cdot \mathbf{n}\, d\sigma = 8\pi a^4 / 3$. (*Hint:* Use the Divergence Theorem and spherical coordinates.)

22. By Coulomb's Law the force field $\mathbf{F}$ of a point charge of q coulombs located at the origin is

$$\mathbf{F} = \frac{cq\mathbf{u}}{|\mathbf{r}|^2}$$

where $\mathbf{r}$ is the position vector of a point P in space, $\mathbf{u}$ is a unit vector in the direction of $\mathbf{r}$, and c is a constant. Prove that the flux of $\mathbf{F}$ through any sectionally smooth closed surface S with the origin in its interior is $4\pi qc$. (*Hint:* Use the Divergence Theorem to show that if S_1

is a sphere centered at the origin lying inside S, then $\iint_S \mathbf{F} \cdot \mathbf{n}\, d\sigma = -\iint_{S_1} \mathbf{F} \cdot \mathbf{n}\, d\sigma$ with $\mathbf{n}$ directed toward the origin for S_1.)

Exercises 23–26 are to be done in sequence. The formulas in Exercises 23 and 24 are known as **Green's Identities**.

23. If f is a scalar function whose second partials exist, the **Laplacian** of f, denoted by $\nabla^2 f$, is defined by

$$\nabla^2 f = \frac{\partial^2 f}{\partial x^2} + \frac{\partial^2 f}{\partial y^2} + \frac{\partial^2 f}{\partial z^2}$$

Prove that if u and v are scalar functions that satisfy appropriate continuity requirements and G and S are as in the Divergence Theorem, then

$$\iiint_G (u\nabla^2 v + \nabla u \cdot \nabla v)\, dV = \iint_S u\nabla v \cdot \mathbf{n}\, d\sigma$$

Hint: Take $\mathbf{F} = \left\langle u\dfrac{\partial v}{\partial x}, u\dfrac{\partial v}{\partial y}, u\dfrac{\partial v}{\partial z} \right\rangle$.

24. Using Exercise 23, prove that

$$\iiint_G (u\nabla^2 v - v\nabla^2 u)\, dV = \iint_S (u\nabla v - v\nabla u) \cdot \mathbf{n}\, d\sigma$$

(*Hint:* Make use of Exercise 23 twice, once as it stands and once with u and v interchanged.)

25. Prove that

$$\iiint_G \nabla^2 u\, dV = \iint_S \nabla u \cdot \mathbf{n}\, d\sigma$$

26. Let $\mathbf{F}$ be a gradient field with potential function ϕ. Prove that if $\operatorname{div} \mathbf{F} = 0$,

$$\iiint_G |\mathbf{F}|^2\, dV = \iint_S \phi \mathbf{F} \cdot \mathbf{n}\, d\sigma$$

16.7 STOKES'S THEOREM

The main result of this section is a generalization of Green's Theorem to three dimensions, called Stokes's Theorem, named for the English mathematical physicist George Stokes (1819–1903). Green's Theorem relates a line integral around a closed curve C in the plane to the double integral over the region enclosed by C. In a similar way, Stokes's Theorem relates the line integral around a closed curve C in space to the surface integral over a surface that has C as its boundary.

Definition 16.5
Curl of a Vector Field

Let $\mathbf{F} = \langle f, g, h \rangle$ be a vector field for which the first partial derivatives of f, g, and h exist in some open region of $\mathbb{R}^3$. Then the **curl of F** is the vector

$$\text{curl } \mathbf{F} = \left\langle \frac{\partial h}{\partial y} - \frac{\partial g}{\partial z}, \frac{\partial f}{\partial z} - \frac{\partial h}{\partial x}, \frac{\partial g}{\partial x} - \frac{\partial f}{\partial y} \right\rangle \tag{16.40}$$

We can compute curl $\mathbf{F}$ using the symbolic determinant

$$\text{curl } \mathbf{F} = \begin{vmatrix} \mathbf{i} & \mathbf{j} & \mathbf{k} \\ \dfrac{\partial}{\partial x} & \dfrac{\partial}{\partial y} & \dfrac{\partial}{\partial z} \\ f & g & h \end{vmatrix} \tag{16.41}$$

EXAMPLE 16.27   Find curl $\mathbf{F}$, where

$$\mathbf{F}(x, y, z) = x^2 y \mathbf{i} + (2y - z)\mathbf{j} + xyz\mathbf{k}$$

Solution   Using Equation 16.41, we get

$$\text{curl } \mathbf{F} = \begin{vmatrix} \mathbf{i} & \mathbf{j} & \mathbf{k} \\ \dfrac{\partial}{\partial x} & \dfrac{\partial}{\partial y} & \dfrac{\partial}{\partial z} \\ x^2 y & 2y - z & xyz \end{vmatrix} = (xz + 1)\mathbf{i} - yz\mathbf{j} - x^2\mathbf{k} \qquad\blacksquare$$

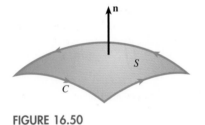

FIGURE 16.50

Now let S be a smooth oriented surface, and let its boundary be a sectionally smooth simple closed curve C. We will say that C is *positively oriented with respect to S* if, when viewed from the tip of a positive unit normal $\mathbf{n}$ to S, C is oriented in a counterclockwise direction. This orientation is illustrated in Figure 16.50. The direction of C then is such that if you walked around C in its positive direction, with your head in the direction of $\mathbf{n}$, the surface S would always be on your left.

THEOREM 16.11

Stokes's Theorem

Let C be a piecewise-smooth simple closed curve that forms the boundary of a smooth oriented surface S, and let C be positively oriented with respect to S. Then if $\mathbf{F}$ is a continuously differentiable vector field on some open set that contains both S and C,

$$\int_C \mathbf{F} \cdot d\mathbf{r} = \iint_S (\text{curl } \mathbf{F}) \cdot \mathbf{n} \, d\sigma \tag{16.42}$$

We will not give a proof here, but we illustrate the result and see some of its consequences. First, let us show that the theorem does generalize Green's Theorem in the plane. We can think of the two-dimensional vector field

$\mathbf{F} = \langle f, g \rangle$ as being three-dimensional, with $h = 0$; that is, we can write $\mathbf{F} = \langle f, g, 0 \rangle$. Then it is easy to verify that

$$\operatorname{curl} \mathbf{F} = \left(\frac{\partial g}{\partial x} - \frac{\partial f}{\partial y} \right) \mathbf{k}$$

A region R in the plane with boundary C oriented counterclockwise is a smooth surface in $\mathbb{R}^3$ oriented by upward unit normals; that is, $\mathbf{n} = \mathbf{k}$. Hence,

$$(\operatorname{curl} \mathbf{F}) \cdot \mathbf{n} = \frac{\partial g}{\partial x} - \frac{\partial f}{\partial y}$$

and so Equation 16.42 reduces to

$$\int_C f \, dx + g \, dy = \iint_R \left(\frac{\partial g}{\partial x} - \frac{\partial f}{\partial y} \right) dA$$

which is the conclusion in Green's Theorem.

EXAMPLE 16.28   Use Stokes's Theorem to evaluate the integral

$$\int_C (x^2 + y^2) \, dx + xy^2 \, dy + xyz \, dz$$

where C is the boundary of the surface S consisting of the first-octant portion of the cylinder $z = 4 - x^2$ between the planes $y = 0$ and $y = 2x$. Orient S with upward unit normals and orient C positively with respect to S.

Solution   In Figure 16.51, we show the surface S and its boundary $C = C_1 \cup C_2 \cup C_3$. We could parameterize each of these component curves and evaluate the integral directly, but it is easier to use Stokes's Theorem. Let $\mathbf{F} = (x^2 + y^2)\mathbf{i} + xy^2\mathbf{j} + xyz\mathbf{k}$. By Equation 16.40 or 16.41, we find that $\operatorname{curl} \mathbf{F} = xz\mathbf{i} - yz\mathbf{j} + (y^2 - 2y)\mathbf{k}$. The upward unit normal $\mathbf{n}$ to $z = 4 - x^2$ is

$$\mathbf{n} = \frac{-\dfrac{\partial z}{\partial x}\mathbf{i} - \dfrac{\partial z}{\partial y}\mathbf{j} + \mathbf{k}}{\sqrt{1 + \left(\dfrac{\partial z}{\partial x}\right)^2 + \left(\dfrac{\partial z}{\partial y}\right)^2}} = \frac{2x\mathbf{i} + \mathbf{k}}{\sqrt{1 + 4x^2}}$$

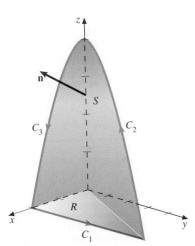

FIGURE 16.51

So we have

$$\int_C \mathbf{F} \cdot d\mathbf{r} = \iint_S (\operatorname{curl} \mathbf{F}) \cdot \mathbf{n} \, d\sigma = \iint_S \frac{2x^2 z + y^2 - 2y}{\sqrt{1 + 4x^2}} \, d\sigma$$

Using Equation 16.27, this surface integral can be evaluated as the following double integral over the xy projection R of the surface S.

$$\iint_R \frac{2x^2(4 - x^2) + y^2 - 2y}{\sqrt{1 + 4x^2}} \sqrt{1 + 4x^2} \, dA$$

$$= \int_0^2 \int_0^{2x} (8x^2 - 2x^4 + y^2 - 2y) \, dy \, dx$$

$$= \frac{64}{3}$$

EXAMPLE 16.29   Verify Stokes's Theorem for $\mathbf{F}(x, y, z) = \langle x+y, y-x, z \rangle$, where S is the portion of the paraboloid $z = x^2 + y^2$ below $z = 4$, oriented by downward unit normals.

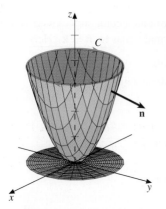

FIGURE 16.52

Solution   The surface S and its boundary C are shown in Figure 16.52. Note that with the unit normal $\mathbf{n}$ directed downward, the positive orientation for C is clockwise, as shown. We will verify Equation 16.42 by calculating each side separately.

The curve C can be parameterized by

$$\begin{cases} x = 2\cos(-t) = 2\cos t \\ y = 2\sin(-t) = -2\sin t, \qquad 0 \le t \le 2\pi \\ z = 4 \end{cases}$$

So $dx = -2\sin t\, dt$, $dy = -2\cos t\, dt$, and $dz = 0$. Thus,

$$\begin{aligned} \int_C \mathbf{F} \cdot d\mathbf{r} &= \int_C (x + y)\, dx + (y - x)\, dy + z\, dz \\ &= \int_0^{2\pi} [(2\cos t - 2\sin t)(-2\sin t) \\ &\qquad\qquad + (-2\sin t - 2\cos t)(-2\cos t)]\, dt \\ &= 4\int_0^{2\pi} (\sin^2 t + \cos^2 t)\, dt = 8\pi \end{aligned}$$

Next, we calculate the integral $\iint_S (\operatorname{curl} \mathbf{F}) \cdot \mathbf{n}\, d\sigma$ on the right-hand side of Equation 16.42. We find that $\operatorname{curl} \mathbf{F} = -2\mathbf{k}$. For $z = x^2 + y^2$, the downward unit normal is

$$\mathbf{n} = \frac{\dfrac{\partial z}{\partial x}\mathbf{i} + \dfrac{\partial z}{\partial y}\mathbf{j} - \mathbf{k}}{\sqrt{1 + \left(\dfrac{\partial z}{\partial x}\right)^2 + \left(\dfrac{\partial z}{\partial y}\right)^2}} = \frac{2x\mathbf{i} + 2y\mathbf{j} - \mathbf{k}}{\sqrt{1 + 4x^2 + 4y^2}}$$

So, by Equation 16.27,

$$\begin{aligned} \iint_S (\operatorname{curl} \mathbf{F}) \cdot \mathbf{n}\, d\sigma &= \iint_S \frac{2}{\sqrt{1 + 4x^2 + 4y^2}}\, d\sigma \\ &= \iint_R \frac{2}{\sqrt{1 + 4x^2 + 4y^2}}\sqrt{1 + 4x^2 + 4y^2}\, dA \\ &= 2\iint_R dA = 2(\text{Area of } R) = 2(4\pi) = 8\pi \end{aligned}$$

Our answers for the integrals on the two sides of Equation 16.42 agree, so we have verified the truth of Stokes's Theorem in this case. ∎

REMARK

■ Using ideas similar to those for extending Green's Theorem to multiply connected regions, we can also extend Stokes's Theorem to surfaces with holes, whose boundaries therefore consist of unions of two or more disjoint closed curves. We will not pursue these ideas here, however.

Application of Stokes's Theorem to Fluid Flow

To gain some insight into the physical interpretation of Stokes's Theorem, we return to the notion of fluid flow in which $\mathbf{F}$ is the velocity field of the fluid. For a closed curve C in the region of flow, then, just as in two-dimensional flow, the integral

$$\int_C \mathbf{F} \cdot d\mathbf{r} = \int_C \mathbf{F} \cdot \mathbf{T}\, ds \qquad \text{Where } \mathbf{T} \text{ is the unit}$$
$$\text{tangent vector to } C$$

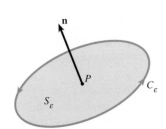

FIGURE 16.53

is called the **circulation of F around** C. Suppose now that S_ε is a small disk of radius ε, centered at a point P in the velocity field. Let $\mathbf{n}$ be the upward unit normal to S_ε, and let C_ε be the positively oriented boundary of S_ε, as in Figure 16.53. Then, by Stokes's Theorem,

$$\int_{C_\varepsilon} \mathbf{F} \cdot d\mathbf{r} = \iint_{S_\varepsilon} (\operatorname{curl} \mathbf{F}) \cdot \mathbf{n}\, d\sigma$$

A mean-value theorem for surface integrals enables us to write

$$\iint_{S_\varepsilon} (\operatorname{curl} \mathbf{F}) \cdot \mathbf{n}\, d\sigma = [\operatorname{curl} \mathbf{F}(Q) \cdot \mathbf{n}](\text{Area of } S_\varepsilon)$$

where Q is some point in S_ε. Thus,

$$\operatorname{curl} \mathbf{F}(Q) \cdot \mathbf{n} = \frac{\displaystyle\iint_{S_\varepsilon} (\operatorname{curl} \mathbf{F}) \cdot \mathbf{n}\, d\sigma}{\text{Area of } S_\varepsilon}$$

$$= \frac{\displaystyle\int_{C_\varepsilon} \mathbf{F} \cdot d\mathbf{r}}{\text{Area of } S_\varepsilon}$$

$$= \frac{\text{Circulation of } \mathbf{F} \text{ Around } C_\varepsilon}{\text{Area of } S_\varepsilon}$$

Now we let $\varepsilon \to 0$, so that $Q \to P$, and assuming the continuity of curl $\mathbf{F}$, curl $\mathbf{F}(Q) \to$ curl $\mathbf{F}(P)$. So

$$\operatorname{curl} \mathbf{F}(P) \cdot \mathbf{n} = \lim_{\varepsilon \to 0} \frac{\text{Circulation of } \mathbf{F} \text{ Around } C_\varepsilon}{\text{Area of } S_\varepsilon}$$

The limit on the right is called the **rotation of F around n** at P. So we can write

$$\text{curl } \mathbf{F}(P) \cdot \mathbf{n} = \text{Rotation of } \mathbf{F} \text{ around } \mathbf{n} \text{ at } P \qquad (16.43)$$

As we saw earlier in this section, when we interpret the two-dimensional vector field $\mathbf{F} = \langle f, g \rangle$ as being the same as the three-dimensional field $\mathbf{F} = \langle f, g, 0 \rangle$,

$$(\text{curl } \mathbf{F}) \cdot \mathbf{n} = \left(\frac{\partial g}{\partial x} - \frac{\partial f}{\partial y} \right) \mathbf{k} \cdot \mathbf{k} = \frac{\partial g}{\partial x} - \frac{\partial f}{\partial y}$$

and this scalar function is what we called the rotation of $\mathbf{F}$ at P in Section 16.4. So, in view of Equation 16.43, we see that rot $\mathbf{F}$ for two dimensions is the same as the rotation of $\mathbf{F}$ about $\mathbf{k}$ in three dimensions.

From Equation 16.43, if curl $\mathbf{F}(P) = 0$ or if $\mathbf{n}$ is perpendicular to curl $\mathbf{F}(P)$, then the rotation of $\mathbf{F}$ around $\mathbf{n}$ will be 0 at P. If curl $\mathbf{F} \neq \mathbf{0}$ at P, there is a circular motion **vortex** at P. The flow is said to be **irrotational** in a region if curl $\mathbf{F} = \mathbf{0}$ for all points in that region.

From Equation 16.43, we can see that the rotation at a point will be a maximum when $\mathbf{n}$ is in the direction of curl $\mathbf{F}$. The maximum value is $|\text{curl } \mathbf{F}|$ evaluated at the point. Suppose, for example, that curl $\mathbf{F}(P) \neq \mathbf{0}$ and a paddle wheel is submerged in the fluid at P, as in Figure 16.54. Then the paddle wheel will rotate as long as its axis is not perpendicular to curl $\mathbf{F}(P)$ (in which case curl $\mathbf{F} \cdot \mathbf{n} = 0$). It will rotate most rapidly when its axis is in the same direction as curl $\mathbf{F}(P)$.

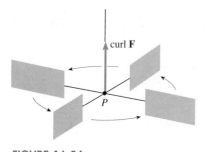

FIGURE 16.54

Three-Dimensional Conservative Vector Fields

The following list of equivalent conditions for conservative vector fields is analogous to those in Section 16.3 for two-dimensional conservative fields. We omit the proofs. We assume $\mathbf{F} = \langle f, g, h \rangle$ is a continuous vector field in an open region Q of $\mathbb{R}^3$.

1. $\mathbf{F}$ is a gradient field.
2. $\int_C \mathbf{F} \cdot d\mathbf{r}$ is independent of the path.
3. $f \, dx + g \, dy + h \, dz$ is an exact differential.
4. $\int_C \mathbf{F} \cdot d\mathbf{r} = 0$ for every piecewise-smooth simple closed curve C in Q.

If Q is simply connected and $\mathbf{F}$ is continuously differentiable, we can add a fifth condition equivalent to these four:

5. curl $\mathbf{F} = \mathbf{0}$ in Q.

Summary of Main Theorems

We conclude with a brief summary of the main theorems of this chapter.

CHART 16.1

**Fundamental Theorem
for Line Integrals**

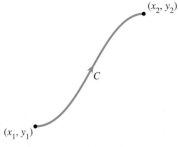

(x_1, y_1)

C is any path from (x_1, y_1)
to (x_2, y_2),

$$\int_{(x_1,y_1)}^{(x_2,y_2)} \nabla \phi \cdot d\mathbf{r}$$

$$= \phi(x_2, y_2) - \phi(x_1, y_1)$$

Green's Theorem

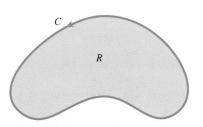

R is a plane region bounded
by the plane curve C.

$$\int_C \mathbf{F} \cdot d\mathbf{r} = \iint_R \left(\frac{\partial g}{\partial x} - \frac{\partial f}{\partial y} \right) dA$$

where $\mathbf{F} = \langle f, g \rangle$

**Divergence Theorem
(Gauss's Theorem)**

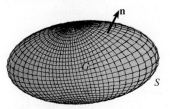

G is a three-dimensional
region bounded by the
surface S.

$$\iint_S \mathbf{F} \cdot \mathbf{n} \, d\sigma = \iiint_G \operatorname{div} \mathbf{F} \, dV$$

Stokes's Theorem

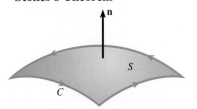

S is a three-dimensional surface
bounded by the space curve C.

$$\int_C \mathbf{F} \cdot d\mathbf{r} = \iint_S (\operatorname{curl} \mathbf{F}) \cdot \mathbf{n} \, d\sigma$$

Exercise Set 16.7

In Exercises 1–6, find curl **F**.

1. $\mathbf{F}(x, y, z) = \langle 2xyz, x - y, y + 2z \rangle$

2. $\mathbf{F}(x, y, z) = \langle x^2 y^2, x^2 z^2, y^2 z^2 \rangle$

3. $\mathbf{F}(x, y, z) = e^x yz\mathbf{i} + e^y xz\mathbf{j} + e^z xy\mathbf{k}$

4. $\mathbf{F}(x, y, z) = (y \ln xz)\mathbf{i} + (x \ln yz)\mathbf{j} + (z \ln xy)\mathbf{k}$

5. $\mathbf{F}(x, y, z) = \langle \cos xy, \sin xy, e^{-z^2} \rangle$

6. $\mathbf{F}(x, y, z) = \left\langle \ln \sqrt{x^2 + y^2}, \tan^{-1} \frac{y}{x}, \frac{z}{xy} \right\rangle$

In Exercises 7–10, use Stokes's Theorem to evaluate the line integral, where C is oriented in a counterclockwise direction when viewed from above.

7. $\displaystyle\int_C xy \, dx + (y + z) \, dy + (x - yz) \, dz$; *C is the triangle formed by the traces of the plane $3x + 2y + z = 6$ on the coordinate planes.*

8. $\displaystyle\int_C x^2 yz \, dx + xy^2 z^3 \, dy + x^4 y^3 z^2 \, dz$; *C is the curve* $\mathbf{r}(t) = \langle \cos t, \sin t, 2 \rangle$, *where $0 \le t \le 2\pi$.*

9. $\displaystyle\int_C \mathbf{F} \cdot d\mathbf{r}$; $\mathbf{F}(x, y, z) = \langle x + yz, 2yz, x - y \rangle$; *C is the intersection of the cylinder $x^2 + y^2 = 4$ and the plane $x + y + z = 1$.*

10. $\displaystyle\int_C \mathbf{F} \cdot d\mathbf{r}$; $\mathbf{F}(x, y, z) = (x + y)\mathbf{i} + (x - z)\mathbf{j} + (y + z)\mathbf{k}$; *C is the intersection of the hemisphere $z = \sqrt{1 - x^2 - y^2}$ and the cylinder $x^2 + y^2 = x$.*

In Exercises 11–16, verify Stokes's Theorem for the given surface S and vector field F. Assume S is oriented by upward unit normals unless otherwise specified.

11. *S is the part of the surface $z = 4 - x^2 - y^2$ above the xy-plane;* $\mathbf{F}(x, y, z) = \langle y - z, x - z, x - y \rangle$.

12. *S is the part of the plane $x + z = 2$ in the first octant between $y = 0$ and $y = 4$;* $\mathbf{F}(x, y, z) = \langle x^2 z, yz^2, x^2 + z^2 \rangle$.

13. *S is the triangular surface with vertices $(2, 0, 0)$, $(0, 1, 0)$, and $(0, 0, 3)$;* $\mathbf{F} = (1 - xy)\mathbf{i} + (y + z)\mathbf{j} + (2x + 3z)\mathbf{k}$.

14. *S is the hemisphere $z = \sqrt{4 - x^2 - y^2}$;* $\mathbf{F}(x, y, z) = xyz\mathbf{i} + (x + 1)\mathbf{j} + xz^2\mathbf{k}$.

15. *S is the part of the cone $z = \sqrt{x^2 + y^2}$ below $z = 1$, oriented by downward unit normals;* $\mathbf{F}(x, y, z) = \langle 2x - 3z, xy + 2z, y - xz \rangle$.

16. *S is the first-octant portion of the surface $z = e^x$ between $y = 0$ and $y = 3$, below $z = 2$, oriented by downward unit normals;* $\mathbf{F}(x, y, z) = \langle ye^x, ze^x, e^x \rangle$.

In Exercises 17 and 18, show that F is a gradient field, and find a potential function for F.

17. $\mathbf{F}(x, y, z) = \langle 2xyz, x^2 z, x^2 y \rangle$

18. $\mathbf{F}(x, y, z) = \langle 2y - 4z, 2x + 5z, 5y - 4x \rangle$

In Exercises 19 and 20, show that the given force field F is conservative, and find the work done by F on a particle moving from A to B.

19. $\mathbf{F}(x, y, z) = (y - 2z)e^x\mathbf{i} + e^x\mathbf{j} - 2e^x\mathbf{k}$; $A = (0, 0, 0)$, $B = (0, -3, -5)$

20. $\mathbf{F}(x, y, z) = 3xz\sqrt{x^2 + y^2}\mathbf{i} + 3yz\sqrt{x^2 + y^2}\mathbf{j} + [(x^2 + y^2)^{3/2} + 2]\mathbf{k}$; $A = (-2, 0, 3)$, $B = (3, 4, -1)$

In Exercises 21 and 22, show that the given expression is an exact differential and find a function for which it is the differential.

21. $e^z \cos x \cos y \, dx - e^z \sin x \sin y \, dy + e^z \sin x \cos y \, dz$

22. $(\ln y + zx) \, dx + (xy - \ln z) \, dy + (\ln x - yz) \, dz$

23. Show that $\text{div}(\text{curl } \mathbf{F}) = 0$.

24. Show that $\text{curl}(\nabla \phi) = 0$.

25. Let S be a closed surface that satisfies the hypotheses of the Divergence Theorem, and let $\mathbf{F}$ be a continuously differentiable vector field on some open region that contains S. Prove that

$$\iint_S (\text{curl } \mathbf{F}) \cdot \mathbf{n} \, d\sigma = 0$$

(*Hint:* Use the result of Exercise 23.)

26. A fluid has velocity field $\mathbf{F}(x, y, z) = y^2\mathbf{i} + z^2\mathbf{j} + x^2\mathbf{k}$. Find its circulation around the curve of intersection of the surfaces $z = x^2 + y^2$ and $z = 2(x + y + 1)$, oriented counterclockwise when viewed from above.

27. Let $\mathbf{a} = \langle a_1, a_2, a_3 \rangle$ be any constant vector and $\mathbf{r} = \langle x, y, z \rangle$ be the position vector of a point in $\mathbb{R}^3$.
(a) Prove that $\text{curl}(\mathbf{a} \times \mathbf{r}) = 2\mathbf{a}$.
(b) Show that $\int_C (\mathbf{a} \times \mathbf{r}) \cdot d\mathbf{r} = 2 \iint_S \mathbf{a} \cdot \mathbf{n}\, d\sigma$, where S and C satisfy the hypotheses of Stokes's Theorem.

28. Prove that if u has continuous first partials and v has continuous second partials in an open region Q of $\mathbb{R}^3$, then

$$\text{curl}(u \nabla v) = \nabla u \times \nabla v$$

29. Under the assumptions of Exercise 28, prove the following:
(a) $\iint_S (\nabla u \times \nabla v) \cdot \mathbf{n}\, d\sigma = \int_C u \nabla v \cdot d\mathbf{r}$
(b) $\int_C (u \nabla v - v \nabla u) \cdot d\mathbf{r} = 2 \iint_S (\nabla u \times \nabla v) \cdot \mathbf{n}\, d\sigma$, where S and C are in Q and satisfy the hypotheses of Stokes's Theorem.

Chapter 16 Review Exercises

1. Let C_1 be the quarter of the unit circle from $(1, 0)$ to $(0, 1)$, let C_2 be the line segment from $(0, 1)$ to $(-1, 0)$, and let $C = C_1 \cup C_2$. Find $\int_C f(x, y)\, ds$ for $f(x, y) = (x + y)^2$.

2. Let $\mathbf{F}(x, y, z) = \langle xy + z, x - y, z^2 - y^2 \rangle$ and C be the arc of the helix $\mathbf{r}(t) = \langle 2\cos t, \sin t, 3t \rangle$ from $(2, 0, 0)$ to $(0, 1, 3\pi/2)$. Find $\int_C \mathbf{F} \cdot d\mathbf{r}$.

3. Find the work done by the force field $\mathbf{F}(x, y, z) = x^2\mathbf{i} - y^2\mathbf{j} + xyz\mathbf{k}$ acting on a unit mass along the curve $\mathbf{r}(t) = \sqrt{t}\mathbf{i} + t^{3/2}\mathbf{j} + (\ln t)\mathbf{k}$ from $(1, 1, 0)$ to $(\sqrt{e}, e^{3/2}, 1)$.

4. Evaluate $\int_C e^{-x} \cos y\, dx + e^{-x} \sin y\, dy$ for each path in the accompanying figure. (*Hint:* There is an easy way.)

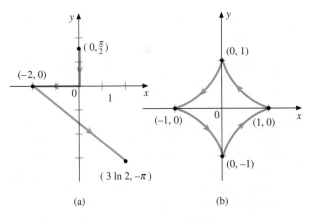

(a)   (b)

5. For each of the following, determine whether $\mathbf{F}$ is a gradient field in the specified domain. If so, find a potential function for $\mathbf{F}$.

(a) $\mathbf{F}(x, y) = \left\langle \dfrac{y^2}{(x^2 + y^2)^{3/2}} - 2, \dfrac{-xy}{(x^2 + y^2)^{3/2}} + 3 \right\rangle$; $\mathbb{R}^2 - \{(0, 0)\}$

(b) $\mathbf{F}(x, y) = \left\langle \dfrac{y}{x(x + y)}, \dfrac{x}{y(x + y)} \right\rangle$; $x > 0$, $y > 0$
(c) $\mathbf{F}(x, y) = y(2x + \tan xy)\mathbf{i} + x(x + \sec^2 xy)\mathbf{j}$; $\mathbb{R}^2$
(d) $\mathbf{F}(x, y) = \langle 2x \tanh x^2 + \tanh y, x \, \text{sech}^2 y - 2y \rangle$; $\mathbb{R}^2$

In Exercises 6 and 7, use Green's Theorem to evaluate $\int_C \mathbf{F} \cdot d\mathbf{r}$, where C has a counterclockwise orientation.

6. $\mathbf{F}(x, y) = \langle x^2 y, x/y^2 \rangle$; C is the boundary of the region enclosed by $xy = 2$, $y = x + 1$, and $y = 1$.

7. $\mathbf{F}(x, y) = (y^2 - \cos x^3)\mathbf{i} + (2xy + e^{y^2})\mathbf{j}$; $C = C_1 \cup C_2$, where C_1 is the arc of the parabola $y = x^2$ from $(-1, 1)$ to $(2, 4)$ and C_2 is the line segment from $(2, 4)$ to $(-1, 1)$.

In Exercises 8 and 9, evaluate the double integral using Green's Theorem.

8. $\iint_R (3x^2 - 2y)\, dA$, where R is the region bounded by the triangle with vertices $(0, 0)$, $(1, 0)$, and $(1, 1)$.

9. $\iint_R (y^2/4 - x^3/3)\, dA$, where R is the region enclosed by the ellipse $4x^2 + 9y^2 = 36$.

In Exercises 10–12, use Green's Theorem to find the area of R.

10. R is the region bounded by the quadrilateral with vertices $(0, 0)$, $(1, 2)$, $(3, 4)$, and $(-1, 3)$.

11. R is the region inside the loop of the *strophoid* $x = \cos 2t$, $y = \tan t \cos 2t$. (See the figure.)

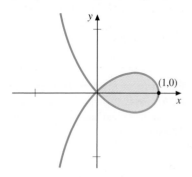

12. R is the region inside $\mathbf{r}(t) = \langle \cos t, \sin 2t \rangle$.

13. Evaluate the integral

$$\int_C \frac{xy^2\,dy - y^3\,dx}{x^2 + y^2}$$

where C is the star-shaped path that has vertices $(\pm 3, 0)$, $(\pm 1, 1)$, $(\pm 1, -1)$, and $(0, \pm 3)$ with a counterclockwise orientation. (*Hint:* Use Theorem 16.8.)

In Exercises 14 and 15, evaluate the surface integral $\iint_S g(x, y, z)\,d\sigma$.

14. $g(x, y, z) = 3y - x^2 - z$; S is the first-octant portion of the plane $z = 2x + y$ below $z = 4$.

15. $g(x, y, z) = 2x^2 + y^2 + z^2$; S is the portion of the sphere $x^2 + y^2 + z^2 = 5$ above $z = 1$.

16. A homogeneous lamina is in the shape of that portion of the hemisphere $z = \sqrt{4 - x^2 - y^2}$ that is inside the cylinder $(x - 1)^2 + y^2 = 1$.
 (a) Find its mass and center of mass.
 (b) Set up, but do not evaluate, iterated integrals for I_x, I_y, and I_z.

17. Set up, but do not evaluate, two iterated integrals for calculating $\iint_S x^2 yz^3\,d\sigma$, using projections on two different planes, where S is the first-octant portion of the cylinder $z = 4 - x^2$ bounded by $y = 0$, $y = x$, and $z = 0$.

18. Let S be the portion of the upper half of the hyperboloid $x^2 + y^2 - z^2 = 3$ below $z = 2$, oriented by downward normals. Find the flux of

$$\mathbf{F}(x, y, z) = \langle xy^2, -x^2 y, -2 \rangle$$

through S.

19. Use the Divergence Theorem to find the outward flux of $\mathbf{F}(x, y, z) = 3x\mathbf{i} + 2y\mathbf{j} + z\mathbf{k}$ through the boundary of the region G between the paraboloids $z = 2 - x^2 - y^2$ and $z = x^2 + y^2$.

In Exercises 20 and 21, use the Divergence Theorem to evaluate $\iint_S \mathbf{F} \cdot \mathbf{n}\,d\sigma$, where $\mathbf{n}$ is the outer unit normal to S.

20. $\mathbf{F}(x, y, z) = \langle x + \sqrt{1 + y^3}, e^{x^3} + \ln(1 + z^2), xz \rangle$; S is the boundary of the region G between $z = 0$, $z = y$, and $y = 2x - x^2$.

21. $\mathbf{F}(x, y, z) = 2xz\mathbf{i} - 2yz\mathbf{j} + z^2\mathbf{k}$; S is the boundary of the region G enclosed by the hemisphere $z = \sqrt{4 - x^2 - y^2}$ the plane $z = 0$, and the cylinder $x^2 + y^2 = x + \sqrt{x^2 + y^2}$. (*Hint:* You will recognize the xy trace of the cylinder when you change to cylindrical coordinates.)

22. A function $f(x, y, z)$ is said to be **harmonic** in a region G if it satisfies the **Laplace Equation**

$$\frac{\partial^2 f}{\partial x^2} + \frac{\partial^2 f}{\partial y^2} + \frac{\partial^2 f}{\partial z^2} = 0$$

for all (x, y, z) in G. Show that if u and v are harmonic in a bounded region G, with boundary S, where G and S satisfy the conditions of the Divergence Theorem, then

$$\iint_S u\nabla v \cdot \mathbf{n}\,d\sigma = \iint_S v\nabla u \cdot \mathbf{n}\,d\sigma$$

(See Exercise 24 in Exercise Set 16.6.)

In Exercises 23–25, use Stokes's Theorem. In each case C is oriented counterclockwise when viewed from above.

23. Evaluate the integral $\int_C 2yz\,dx + 2xz\,dy + 3xy\,dz$, where C is the triangle with vertices $(3, 0, 0)$, $(0, 2, 0)$, and $(0, 0, 6)$.

24. Evaluate $\int_C \mathbf{F} \cdot d\mathbf{r}$, where $\mathbf{F}(x, y, z) = \langle xz, y + 1, y^2 \rangle$ and C is the union of the traces of the ellipsoid $2x^2 + 4y^2 + z^2 = 8$ on the first-octant parts of the coordinate planes.

25. Find the circulation of the velocity field $\mathbf{F}(x, y, z) = yz\mathbf{i} + 8z\mathbf{j} + xy\mathbf{k}$ around the curve of the intersection of the cylinder $x^2 + y^2 = 4y$ and the plane $3x + 2y + z = 12$.

26. Verify Stokes's Theorem for $\mathbf{F}(x, y, z) = \langle 2yz, y^2, xy \rangle$ and S the portion of the hemisphere $z = \sqrt{5 - x^2 - y^2}$ inside the cylinder $x^2 + y^2 = 1$, oriented by upward normals.

27. Make use of the Divergence Theorem to find the volume of the region bounded by the elliptic paraboloids $z = 2x^2 + 3y^2$ and $z = 4 - 2x^2 - y^2$. (*Hint:* Choose $\mathbf{F}$ so that $\operatorname{div} \mathbf{F} = 1$.)

28. Show that the force field $\mathbf{F}(x, y, z) = (2xy - 1)\mathbf{i} + (x^2 + 2z^2)\mathbf{j} + (4yz + 3x^2)\mathbf{k}$ is conservative in $\mathbb{R}^3$ and find the work done by $\mathbf{F}$ on a particle moving from $(0, 1, -1)$ to $(2, 1, 3)$.

29. Let u and v have continuous second partial derivatives in an open region G of $\mathbb{R}^3$, and let C be a sectionally smooth simple closed curve in G. Prove that

$$\int_C (u\nabla v + v\nabla u) \cdot d\mathbf{r} = 0$$

(*Hint:* Let S be a surface in G with C as its boundary such that S and C satisfy the conditions of Stokes's Theorem. Then use Exercises 28 and 29 in Exercise Set 16.7.)

Chapter 16 Concept Quiz

1. Define each of the following.
   (a) A vector field
   (b) A gradient field
   (c) A potential function for a gradient field
   (d) A conservative field
   (e) The line integral of a function f along a curve C
   (f) The work done by a force field in moving a particle along a curve C
   (g) The divergence of a vector field $\mathbf{F} = \langle f, g, h \rangle$
   (h) The curl of a vector field $\mathbf{F} = \langle f, g, h \rangle$
   (i) The flux of a vector field through a surface S
   (j) The circulation of a vector field around a closed curve C
   (k) A simply-connected region

2. State the following.
   (a) The Fundamental Theorem for Line Integrals
   (b) Green's Theorem
   (c) The Divergence Theorem
   (d) Stokes's Theorem

3. (a) Give a formula for evaluating the line integral
$$\int_C f(x, y)ds,$$
   where $C = \{(x, y) : x = x(t), y = y(t) \text{ for } a \le t \le b\}$.

   (b) Give a formula for evaluating the surface integral $\iint_S f(x, y, z)d\sigma$, where S is the graph of a smooth function $z = \phi(x, y)$ and the xy projection of S is R.

4. State a test for determining whether a vector field $\mathbf{F} = f\mathbf{i} + g\mathbf{j}$ is conservative. If it is conservative, explain how to find a potential function for $\mathbf{F}$.

5. Let R be a simply-connected region in $\mathbb{R}^2$. State four conditions that are equivalent to the condition that $\mathbf{F} = \langle f, g \rangle$ is a gradient field in R.

6. Indicate which of the following statements are true and which are false.
   (a) In a conservative force field, the work done in moving a particle along a straight line from (x_1, y_1), to (x_2, y_2) is the same as moving it along the arc of a curve between these points.
   (b) If $\int_C \mathbf{F} \cdot d\mathbf{r} \ne 0$ for a closed path C, then there is no function ϕ for which $\nabla\phi = \mathbf{F}$.
   (c) If $\int_C (\mathbf{F} - \mathbf{G}) \cdot d\mathbf{r} = 0$ for every simple closed curve C in the common domain of $\mathbf{F}$ and $\mathbf{G}$, then $\mathbf{F} = \mathbf{G}$.
   (d) The value of $\int_C \mathbf{F} \cdot d\mathbf{r}$ depends on the parameterization of C.
   (e) If f is a scalar function with continuous partial derivatives in a region R, then $\int_C \nabla f \cdot d\mathbf{r}$ is independent of the path in R.

APPLYING CALCULUS

The Hubble Space Telescope after its 1993 repair

1. When the space shuttle went into orbit to service the Hubble Space Telescope, they had a combined mass of 20,000 kg and orbited the Earth in a circular orbit with radius 8,850 km from the center of the Earth.
    (a) Find the work W done by gravity on the space shuttle during one-half a revolution.

    (b) Find the work done by gravity on the shuttle during one full revolution.

    (c) In general, suppose that a continuous force acts on an object in a direction normal to its path. Show that the work done by the force on the object is zero. How does this result compare with your answers in parts (a) and (b)?

2. A vector field $\mathbf{F}$ in space (or in the plane) is a *central force field* if $\mathbf{F}(P)$ is parallel to the position vector $\overrightarrow{OP}$ from the origin to P for all points P for which $\mathbf{F}$ is defined. A central force field is *radially symmetric* if $|\mathbf{F}(P)| = |\mathbf{F}(Q)|$ whenever P and Q are points on the same circle centered at the origin.
    (a) Show that a radially symmetric vector field $\mathbf{F}$ has the form

    $$\mathbf{F}(P) = f(|\mathbf{r}|)\mathbf{r}$$

    where $\mathbf{r} = \overrightarrow{OP}$ and f is a scalar function.

    (b) Show that a radially symmetric field is conservative.

    (c) Assume that $\mathbf{F}$ is a radially symmetric vector field that is defined everywhere in space except at the origin and that $\mathbf{F}$ is differentiable. Show that if div $\mathbf{F} = 0$, then $f(|\mathbf{r}|) = k/|\mathbf{r}|^3$. Thus, a gravitational field is a radially symmetric vector field with divergence equal to 0.

3. A particle of mass m moves along a plane curve C, with position vector $\mathbf{r}(t)$ for $t_1 \le t \le t_2$. It is acted on by a force field given by $\mathbf{F}(\mathbf{r}) = -k\mathbf{r}$, where k is a positive constant.
    (a) Show that $\mathbf{F}$ is a conservative force field, and find a potential function for it.

    (b) Find the work done by $\mathbf{F}$ in moving the particle from $\mathbf{r}(t_1)$ to $\mathbf{r}(t_2)$.

    (c) Suppose that the force field is modified to include a damping force proportional to the velocity of the particle and opposite to the direction of motion. Then $\mathbf{F}(\mathbf{r}) = -k\mathbf{r} - c\mathbf{v}$, where $\mathbf{v} = \mathbf{r}'(t)$ and c is a positive constant. Give an integral in terms of the parameter t for the work done in this case in moving the particle from $\mathbf{r}(t_1)$ to $\mathbf{r}(t_2)$. Show that the force field in this case is not conservative. (*Hint:* By considering two different paths from $\mathbf{r}(t_1)$ to $\mathbf{r}(t_2)$, show that the integral is not independent of the path.)

4. In 1864, the Scottish physicist James Clerk Maxwell unified electricity and magnetism with the following set of equations, now known as **Maxwell's Equations**:

    (1) div $\mathbf{E} = \dfrac{\rho}{\varepsilon}$ (3) curl $\mathbf{E} = -\dfrac{\partial \mathbf{B}}{\partial t}$

    (2) div $\mathbf{B} = 0$ (4) curl $\mathbf{B} = \varepsilon\mu\dfrac{\partial \mathbf{E}}{\partial t} + \mu\mathbf{J}$

In these equations $\mathbf{E}$ represents an electric field and $\mathbf{B}$ a magnetic field. The constant ε is an electrical constant called the *permittivity* and μ is a magnetic constant called the *permeability*. The charge density is ρ and the current density is $\mathbf{J}$, where $\mathbf{J} = \rho\mathbf{v}$, for charge velocity $\mathbf{v}$. Note that a changing electric field results in a magnetic field, and vice versa. Maxwell's Equations govern all electromagnetic waves, including light, and among many other things led to the discoveries of radio and television.

(a) Show that Equation 3 implies Equation 2.

(b) Show that Equation 3 implies **Faraday's Induction Law**, that the electromotive force around any closed contour C that is sectionally smooth and forms the boundary of an orientable surface S in the electric field $\mathbf{E}$ is the negative of the time rate of change of the magnetic flux through S:

$$\int_C \mathbf{E} \cdot \mathbf{T} \, ds = -\frac{\partial}{\partial t} \iint_S \mathbf{B} \cdot \mathbf{n} \, d\sigma$$

(c) Use the special case of Equation 4 in which $\partial \mathbf{E}/\partial t = 0$ to prove **Ampere's Law** for a steady current, that the circulation of the magnetic field $\mathbf{B}$ induced by $\mathbf{J}$ around the boundary C of an orientable surface S is proportional to the total current flowing through S:

$$\int_C \mathbf{B} \cdot \mathbf{T} \, ds = \mu \iint_S \mathbf{J} \cdot \mathbf{n} \, d\sigma$$

5. Refer to Maxwell's Equations in Exercise 4 to derive two forms of **Gauss's Law** as follows:

(a) Use Equation 1 to derive Gauss's Law for the electric field, that the net outward flux is equal to the electric charge q inside, divided by the permittivity:

$$\iint_S \mathbf{E} \cdot \mathbf{n} \, d\sigma = \frac{q}{\varepsilon}$$

(b) Use Equation 2 to derive Gauss's Law for the magnetic field, that the net outward flux is equal to 0:

$$\iint_S \mathbf{B} \cdot \mathbf{n} \, d\sigma = 0$$

6. Consider an incompressible fluid (no sources or sinks) flowing in space with variable density $\rho = \rho(x, y, z, t)$ and velocity $\mathbf{v} = \mathbf{v}(x, y, z, t)$. Let S_ε be a sphere of radius ε centered at a point (x, y, z) in the velocity field, and let G_ε be the spherical region enclosed by S_ε.

(a) Write a triple integral for the mass $m(t)$ of fluid inside S_ε at time t, and show that

$$m'(t) = \iiint_{G_\varepsilon} \frac{\partial \rho}{\partial t} \, dV$$

(Assume differentiation inside the integral is valid.)

(b) Explain why $m'(t)$ is also given by

$$m'(t) = -\iint_{S_\varepsilon} \rho \mathbf{v} \cdot \mathbf{n} \, d\sigma$$

where $\mathbf{n}$ is the outer unit normal to S_ε.

(c) Use the Divergence Theorem to write the result in part (b) as a triple integral.

(d) Equate the triple integrals for $m'(t)$ in parts (a) and (c), and combine the integrals. By using a mean-value theorem for triple integrals (as in Section 16.6) and then letting $\varepsilon \to 0$, obtain the **continuity equation**

$$\frac{\partial \rho}{\partial t} + \text{div}(\rho \mathbf{v}) = 0$$

Frank McGrath

Your Ph.D. is in partial differential equations. How have you found the mathematics you learned in university useful in your professional life?

It's impossible to solve most real-world problems without mathematics. When I was a student solving problems like "how long a board can be carried around this corner?" I sometimes wondered whether calculus had any real-world use. It turned out that many of the problems people were willing to pay me to solve were much more interesting calculus problems than those I solved in class. So market forces are not necessarily bad when it comes to science. Partial differential equations is calculus and most of the engineering problems that one runs into involve some form of differential equations.

What role do computers play in these kinds of problems?

What computers are doing is numerically integrating differential equations, which involve derivatives. Often one has to numerically integrate. That is, you have to solve most of the problems that are of practical value on a computer. Very few real-world problems are amenable to closed-form solution and so you have to numerically integrate them—which, I contend, can often require more understanding of the nature of the differential equations than merely working with or solving the analytical forms of the equations.

Can you give us an example?

Sure. There are the problems of computing an orbit to a planet, which is relatively straightforward on the surface—just applying Newton's laws. They make a very simple differential equation, force equals mass times acceleration, except you need to account for solar wind, gravitational fields, and thrust that burns fuel which changes mass and the center of gravity. Force and mass become complex functions in the differential equation. It's a hard problem to arrive at a planet that itself is a moving target.

I was involved in tracking and intercepting satellites. Here there are a multiplicity of problems, all of which involve differential equations. Imagine that you have a satellite in orbit and you want to know its position with great precision. If you're going to intercept something and you can't see it until you are almost there, it's kind of nice to know where it is in advance. If you're going to arrive at a point in space at the same time the satellite arrives at the same point when the closing velocity may be 20,000 feet per second or more, you have to know the satellite's orbit with great precision. So where's the satellite? It's found by measuring the changing position of the satellite at earlier times, usually with radars, and then using mathematics to project forward in time both the position of the satellite and the interceptor's position to the point of rendezvous. Precise determination of position as a function of time is very, very important in this kind of problem. You have to consider things like solar wind and irregularities in the gravitational field. You might not think that the solar wind pushing on a satellite in an orbit of 300-mile altitude would have much effect on the orbit. In fact, its effect on the position in orbit must be taken into account whenever high accuracy is required, especially in applications such as navigation and measurement of gravitational fields. Also, you have to predict the gravitational field with great precision because, unlike rocket thrust, the acceleration due to gravity cannot be measured during powered flight.

To intercept your target, you have to know where you are at all times while you are still controlling the vehicle with rocket thrust. The cycle is: Where am I now? Exactly what velocity vector do I need to reach my target from this position? Do I have this velocity? If not, in what direction do I need to point the rocket thrust to get this velocity? When you arrive at the velocity that will create the trajectory to hit your target you shut the motor off. Every step in this cycle involves calculus. When you use polar and Cartesian coordinates in calculus you will see that some problems are better solved in polar coordinates and others in Cartesian. It turns out that the best coordinate system for the differential equations that are used in the cycle is a coordinate system called *inertial coordinates*. Devices that measure forces in inertial coordinates are based on gyroscopes, and the design of these gyros is a problem solved with mathematics based on calculus. When you shut the rocket motors off because you have reached the desired velocity, you still have to be able to predict the effect of terminating thrust. This involves the noninstantaneous termination of thrust and the relaxation of the metal of the final stage of the booster, which can add additional thrust. The mathematics needed to solve the problems of each of these steps involves differential equations, which are studied in advanced courses like applied mechanics, control theory, and continuum mechanics.

Is this is a problem for the space shuttle as well?

Yes. It is one very large mathematical problem after another to be able to fly the shuttle from Cape Kennedy to intercept the Hubble Space Telescope. But these mathematical problems are also relevant to the design of the shuttle. For instance, Thiokol, the company that builds the solid fuel rocket booster that unfortunately blew up the Challenger space shuttle because of a design flaw, had already solved a very difficult problem for solid rocket motors. When you ignite the solid fuel, huge pressures build up inside the rocket. If not properly controlled, they can oscillate back and forth and you can get a resonance that can destroy the rocket. So how do you prevent this resonance? The first step is to write down the applicable laws of physics in mathematical notation. You end up with equations that involve the same integrals and differentials that you learn about in calculus class. In some sense, this is the easy part of the problem. Then some person has the very interesting challenge of solving these equations. The difference being that if your solution is not correct the rocket might blow up, which is much more serious than getting a "C" in your calculus class. All of this very complex engineering gets back to calculus.

What other mathematics courses are particularly important for practical applications after you take your calculus?

I've already said that if you want to work on engineering problems in industry, differential equations are extremely important, both ordinary and partial. Real and complex variable analysis is also important, as is linear algebra. You need to understand how all this fits together. Another very useful mathematical area in industry that applied math majors often miss is that of Fourier transforms and spectral analysis. These too involve integrals—you just can't seem to get away from calculus.

For example, you may have noticed that there are several antenna towers at a radio or television transmitter. These antennas combine into something called a phased array. Basically, the engineering problem is to receive or transmit waves, sound or electromagnetic, on each of a number of omnidirectional receivers or transmitters and then combine all of these omnidirectional waves in such a way that the result is a beam that points in one direction. The more omnidirectional receivers or transmitters you have, the narrower you can make the beam. When the phased array is an antenna receiving sound, a narrow beam excludes sound from all directions except in the direction of the beam. The sound in one beam is then decomposed into its spectral components, or tones, by numerically integrating the integral, that is, the Fourier transform. Each sound source has a unique signature of tones that is used to distinguish the sound source from all other sound sources.

There is not only the Fourier transform involved in forming a beam and then producing the spectrum, but if your real-world problem is to simultaneously form a large number of beams, say 500, and then break the sound in each beam into its spectral components, you have one whale of a computation problem. You might be able to write the analytical equations and then find out that computers large enough to solve the equations within the necessary time *do not exist*! You didn't give enough thought to computational efficiency. In these types of problems, you have to worry about saving a microsecond or less in certain intermediate computations. For instance, a method of numerically integrating the Fourier Transform integral, called the Fast Fourier Transform, makes an enormous improvement in computational efficiency for beam forming and spectral decomposition by replacing very computationally expensive multiplications by right or left shifts of the string of 1s and 0s in the binary representation of numbers. Shifts are extremely efficient on a computer. So you not only need to know the mathematics, you also need to understand computation both at the level of computer functions and numerical algorithms. Be wary of computing a lot of trigonometric functions. Don't divide when you can multiply. Addition is better yet and shifts can be even better.

What do you look for in terms of mathematical aptitude when you hire employees?

Many people who think they're trained in mathematics have learned how to turn the crank and do computation but don't really understand mathematics. There is an enormous value for people to understand mathematics as opposed to merely learning

that the derivative of x^2 is $2x$. I really need people who understand a derivative and what a differential is, and the difference between the two, and how does this apply to solving a practical problem? Similarly, if you don't understand the physics behind an engineering problem you're going to be very limited in what you can do. You just can't work on real-world engineering problems without a broad base of knowledge to supplement your mathematics. You've got to understand the physics; you've got to understand the math; you've got to understand how they relate to each other. This is necessary, but not sufficient, for you to have the two most in-demand skills: understanding thought processes of mathematics and having mathematical creativity. You can't just be a worker of algorithms because computers can work the algorithms better than you can. You have to truly understand the problems and this involves the ability to reason quantitatively.

Mathematicians working in industry also have excellent opportunities to be part of a team developing large-scale software. The creative thought process of mathematics is the key skill you can bring to the team. One of the problems of great economic importance is how to improve productivity and quality in very large computer programs, say half a million instructions or more. The central problem is how to decompose this problem that can't be understood by one person into a large number of problems, each of which can be understood. The best current and future (potential) technology for doing this is called object-oriented design, which decomposes into levels of abstraction. Does this sound like mathematics? It may not at your stage of undergraduate calculus, but it will if you do a Ph.D. in applied mathematics. You are currently moving up the levels of abstraction in mathematics. Calculus generalized many of the more concrete properties of numbers. When you complete your calculus and then eventually get up the ladder of abstraction to abstract algebra and point set topology, you will appreciate why a mathematics education gives the thought process necessary to pull off the design of large-scale software.

Frank McGrath is a native of Ohio and is Vice-President for Information Technology at Logicon, Inc., in Arlington, VA. He received his B.A. in Mathematics from Iowa State University, and his Ph.D. in Applied Mathematics from the University of California at Berkeley. His speciality—both as a research and occupational interest—is developing extremely large computer systems involving both software and hardware. These projects have tackled a wide variety of problems involving direct applications of mathematics and physics.

DIFFERENTIAL EQUATIONS

We introduced the notion of a differential equation in Chapter 4, during our discussion of antiderivatives. At that stage, we considered only simple differential equations of the form

$$\frac{dy}{dx} = f(x)$$

having the general solution

$$y = F(x) + C$$

where F is an antiderivative of f. In Chapter 7, in our study of exponential growth and decay, we described the solution technique called *separation of variables*. This technique enabled us to solve differential equations that were more complex than those we had previously studied.

In this chapter we will review and expand on the basic idea of a differential equation and the nature of solutions. We will review the technique of separation of variables and introduce other methods for solving first-order differential equations (those involving only first derivatives). We will also study two techniques for solving second-order equations. We will conclude the chapter by indicating how to use power series to solve equations that otherwise would be difficult or impossible to solve.

The subject of differential equations is almost as old as calculus itself; both Newton and Leibniz originated many of the concepts involved. A tremendous body of knowledge on the subject exists, but much is still not known, and the subject continues to be an area of intensive research. In this single chapter we can only scratch the surface of this vast field, but the techniques we will present are nevertheless important. We leave to a later course the deeper theoretical aspects, in particular those having to do with existence and uniqueness of solutions.

The importance of differential equations in applications lies in their use as mathematical models to describe a wide variety of phenomena having to do with rates of change. Such phenomena occur in biology, chemistry, economics, engineering, physics, and psychology, to name only a few areas of study. For example, velocities, accelerations, chemical reaction rates, rates of spread of diseases, learning rates, and growth rates of investments all enter into models that are expressed by differential equations. We will illustrate a variety of such applications in this chapter.

17.1 BASIC CONCEPTS; GENERAL SOLUTIONS AND PARTICULAR SOLUTIONS

Recall that a differential equation is an equation that involves a function and one or more of its derivatives. If there is only one independent variable, so that ordinary derivatives occur, the equation is called an **ordinary differential equation**. If there are two or more independent variables, then partial derivatives occur, and the equation is called a **partial differential equation**. For example, the equation

$$Q'(t) = kQ(t)$$

of exponential growth or decay that we studied in Chapter 7 is an ordinary differential equation. The equation

$$\frac{\partial u}{\partial t} = c\frac{\partial^2 u}{\partial x^2}$$

is a partial differentiation equation, called the *heat equation*, that we discussed briefly in Chapter 13 in connection with our study of partial derivatives. In this chapter we limit our consideration to ordinary differential equations. When we use the term *differential equation*, we will be referring to an ordinary differential equation.

One way of classifying differential equations is by their order. The **order** of a differential equation is defined to be the order of the highest derivative that occurs. For example, the differential equations

$$\frac{dy}{dx} = 3x^2$$

$$y'' - 2y' - 3y = e^x$$

$$\frac{d^4x}{dt^4} - (\sin t)\left(\frac{dx}{dt}\right)^2 = t^3 + 1$$

have orders 1, 2, and 4, respectively.

REMARK ———————————————————————————————————
■ When we write y', it will be understood to mean $y'(x)$, or equivalently, dy/dx, unless a different independent variable is specified.
——

A function $y = f(x)$ is a **solution** of a differential equation if when y and its derivatives are substituted into the equation, the equation becomes an identity. For example, we know that $y = x^3$ is a solution of the differential equation $dy/dx = 3x^2$. Also, $y = x^3 + C$, where C is any constant, is a solution.

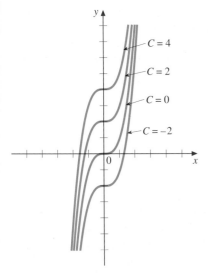

FIGURE 17.1
The family of solutions
$y = x^3 + C$

We call $x^3 + C$ the **general solution**. All solutions to the equation can be obtained from the general solution by appropriate choices of the constant C. This general solution is really a **family** of solutions. We refer to it as a **one-parameter** family, since there is only one arbitrary constant (the parameter C) involved. Several members of the family are sketched in Figure 17.1.

In solving differential equations, we often seek such a general solution from which all solutions can be obtained.* For first-order equations, the general solution will always be a one-parameter family. This result seems plausible, since you would expect to have to carry out only one integration, giving rise to one arbitrary constant. Similarly, a second-order equation will have a two-parameter family as its general solution, and in general, an nth-order equation will have an n-parameter family as its general solution.

REMARK
■ It is important to point out that, when speaking of the number of parameters in the general solution, we mean the number of *essential* constants. For example, the equation $y + C_1 = x^2 + C_2$ may appear to define a two-parameter family, but by writing it as $y = x^2 + (C_2 - C_1)$ and defining $C = C_2 - C_1$, we see that it is really the one-parameter family $y = x^2 + C$.

When specific values are given to the arbitrary constants in the general solution, we obtain a **particular** solution. In a first-order equation, specifying the value of y for one value of x is sufficient to determine the constant. For example, suppose we want the particular solution to $y' = 3x^2$ that satisfies $y(1) = 3$. Graphically, we want the curve of the family that passes through the point $(1, 3)$. To find it, we substitute $x = 1$ and $y = 3$ into the general solution $y = x^3 + C$. This substitution gives $3 = (1)^3 + C$, or $C = 2$. Thus, $y = x^3 + 2$ is the desired particular solution.

More generally, for a first-order differential equation, if we want the solution $y = y(x)$ that satisfies $y(x_0) = y_0$, we substitute $x = x_0$ and $y = y_0$ into the general solution to find C. The condition $y(x_0) = y_0$ is called an **initial condition**, and the differential equation with such a specified initial condition is called an **initial-value problem**. For a second-order equation, two initial conditions are needed to determine a particular solution, as the next example illustrates.

EXAMPLE 17.1   Verify that $y = C_1 e^{3x} + C_2 e^{-x}$ is a two-parameter family of solutions of the differential equation

$$y'' - 2y' - 3y = 0$$

and find the particular solution in this family that satisfies $y(0) = 2$ and $y'(0) = -1$. (*Note:* In Section 17.5 we will see that the given family is the general solution of this equation and see how it was obtained.)

Solution   We first calculate y' and y'' for the given family:

$$y' = 3C_1 e^{3x} - C_2 e^{-x}$$

$$y'' = 9C_1 e^{3x} + C_2 e^{-x}$$

---

*Sometimes there are additional solutions not obtainable from the general solution. These are called **singular** solutions, but we will not consider them here.

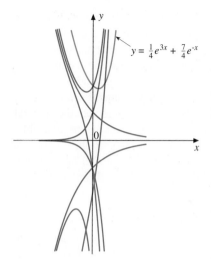

$$y = \tfrac{1}{4}e^{3x} + \tfrac{7}{4}e^{-x}$$

FIGURE 17.2

So

$$y'' - 2y' - 3y = 9C_1e^{3x} + C_2e^{-x} - 2(3C_1e^{3x} - C_2e^{-x}) - 3(C_1e^{3x} + C_2e^{-x})$$

$$= (9C_1 - 6C_1 - 3C_1)e^{3x} + (C_2 + 2C_2 - 3C_2)e^{-x} = 0$$

Thus, the differential equation is identically satisfied by the given function, so we have a two-parameter family of solutions. To find C_1 and C_2 that satisfy the given initial conditions, we impose those conditions on y and y' to get

$$C_1 + C_2 = 2$$

$$3C_1 - C_2 = -1$$

The simultaneous solution is easily found to be $C_1 = \frac{1}{4}$, $C_2 = \frac{7}{4}$. So the particular solution in question is

$$y = \frac{1}{4}e^{3x} + \frac{7}{4}e^{-x}$$

In Figure 17.2 we show several members of the family given by the general solution, along with the particular solution. ∎

In the next example, a particular solution of a second-order equation is obtained from its general solution by specifying values of y at two different values of x. Such values of y are called **boundary values** and are of the general form $y(x_0) = y_0$ and $y(x_1) = y_1$. The terminology reflects the fact that in applications the two known values of y occur at the endpoints (that is, on the boundary) of the interval under consideration. A differential equation to be solved, subject to such boundary conditions, is known as a **boundary-value problem**.

EXAMPLE 17.2   Show that for all choices of the parameters C_1 and C_2, the function $y = C_1 \cos x + C_2 \sin x$ is a solution (it is the general solution) of the equation $y'' + y = 0$, and determine C_1 and C_2 such that the boundary conditions $y(0) = 2$ and $y(\pi/2) = 1$ are satisfied.

Solution   Since for the given family $y' = -C_1 \sin x + C_2 \cos x$, and $y'' = -C_1 \cos x - C_2 \sin x$, we see that

$$y'' + y = (-C_1 \cos x - C_2 \sin x) + (C_1 \cos x + C_2 \sin x) = 0$$

So the equation is satisfied. From $y(0) = 2$, we get $C_1 = 2$, and from $y(\frac{\pi}{2}) = 1$, we get $C_2 = 1$. Thus, the solution to the boundary-value problem is

$$y = 2 \cos x + \sin x$$

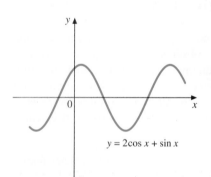

$$y = 2\cos x + \sin x$$

FIGURE 17.3

We show its graph in Figure 17.3. ∎

It is desirable to express a solution to a differential equation as an explicit function, $y = f(x)$, but sometimes doing so is difficult or even impossible and we settle for an **implicit solution** in the form $F(x, y) = 0$. The next example illustrates this situation.

EXAMPLE 17.3   Show that if y is defined as a function of x by the equation

$$x^3 - 2xy^2 + y^3 - 5 = C$$

then y is a solution to the differential equation

$$\frac{dy}{dx} = \frac{3x^2 - 2y^2}{4xy - 3y^2}$$

Solution   Let $F(x, y) = x^3 - 2xy^2 + y^3 - 5 - C$. Then from Chapter 14 we know that

$$\frac{dy}{dx} = -\frac{F_x}{F_y} = -\frac{3x^2 - 2y^2}{-4xy + 3y^2} = \frac{3x^2 - 2y^2}{4xy - 3y^2}$$

as long as the denominator is nonzero. So the function y defined implicitly by $F(x, y) = 0$ satisfies the given differential equation.   ∎

Sometimes we are interested in finding the differential equation of a given family. That is, we want to find the differential equation, given its general solution. We illustrate this procedure in the next example.

EXAMPLE 17.4   In each of the following, find the differential equation having the given equation as its general solution.
(a) $x^2 + y^2 = Cx$ (b) $y = C_1 e^{-x} + C_2 e^{2x}$

Solution

(a) Since there is one arbitrary constant, we expect the differential equation to be of first order. By differentiating both sides with respect to x, we obtain

$$2x + 2yy' = C$$

The differential equation cannot contain the constant of integration, C, so we want to eliminate this constant. We can do so by replacing C in the original equation by the value just found for C, namely, $2x + 2yy'$. Thus,

$$x^2 + y^2 = (2x + 2yy')x$$

or, on simplification,

$$2xyy' = y^2 - x^2$$

which is the desired differential equation.

(b) This time we differentiate twice, since there are two arbitrary constants:

$$y' = -C_1 e^{-x} + 2C_2 e^{2x}$$
$$y'' = C_1 e^{-x} + 4C_2 e^{2x}$$

By using these two equations, together with the original equation, we can eliminate the two constants. Adding the equations for y and y' gives

$$y + y' = 3C_2 e^{2x}$$

Similarly, adding y' and y'' gives

$$y' + y'' = 6C_2 e^{2x}$$

Thus,

$$2(y + y') = y' + y''$$

or equivalently,

$$y'' - y' - 2y = 0$$

This equation is the one we are seeking.   ∎

Slope Fields

The techniques we will study in subsequent sections will make it possible to solve a wide variety of differential equations, but just as there are functions that cannot be integrated in an exact form, there are differential equations for which exact solutions cannot be found. There are various approximation techniques studied in more advanced courses, but we will not consider them here. We will, however, mention one geometric approach that is sometimes useful in obtaining approximate graphs of the solution curves. Suppose the differential equation is of first order. By solving for dy/dx, we can write the equation in the form

$$\frac{dy}{dx} = F(x, y)$$

Since dy/dx is the slope of a given solution curve at a given point on the graph, we can obtain slopes at various points in the plane (in the domain of F). If at a large number of points we draw short line segments with slope $F(x, y)$, we call the result a **slope field** (also called a **direction field**) for the given differential equation. We illustrate such a slope field in Figure 17.4. A slope field is similar to a vector field except that instead of vectors we draw short line segments (parts of tangent lines). Once we have a slope field, we can sketch solution curves by following the tangent line segments. By starting at a particular point, we obtain a sketch of the particular solution passing through that point. We illustrate this technique in the next example.

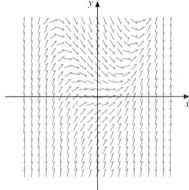

FIGURE 17.4
A slope field

EXAMPLE 17.5   Sketch the slope field for the differential equation

$$\frac{dy}{dx} = 2x - y$$

Use the result to sketch the graph of the particular solution that passes through the point $(1, 2)$.

Solution   We obtain the slope field in Figure 17.5 by substituting for x and y to calculate the slope $2x - y$ at various points (x, y). For example, the slopes at $(-1, 2)$, $(0, 1)$, and $(1, 2)$ are -4, -1, and 0, respectively. So at these points we draw short segments with the given slopes.

To obtain the particular solution passing through the point $(1, 2)$, we start our sketch at that point and then move to the right and left, using the tangent line segments as a guide. (See Figure 17.5)   ∎

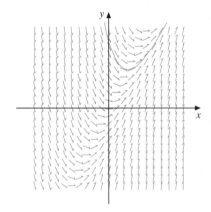

FIGURE 17.5

SOLVING DIFFERENTIAL EQUATIONS USING COMPUTER ALGEBRA SYSTEMS

The theory of differential equations is one of the oldest and richest in mathematics. There exist many techniques for finding explicit solutions for particular classes of differential equations as well as a multitude of approximation techniques for those that cannot be solved explicitly. In practice, the latter occur more frequently. A CAS can find explicit solutions for only a small collection of types of differential equations and can be used effectively in finding numerical approximations to solutions. In this section we give two examples to describe several ways a CAS can be useful in analyzing differential equations. The methods for solving the equations considered will be developed in the chapter.

CAS 61

Consider the first-order linear differential equation

$$\frac{dy}{dx} = x - y$$

Use Maple and Mathematica to first sketch a slope field for the differential equation that will describe the nature of the solutions. Then use the CAS to find the general solution and sketch several specific solutions along with the slope field.

Maple:

```
with(DEtools):with(plots):
dfieldplot(x–y,[x,y],x=–3..3,y=–3..3,scaling=constrained);
dsolve(diff(y(x),x)+y(x)–x=0,y(x));
```

Output: $y(x) = x - 1 + e^{-x}\_C1$

In returning the general solution, Maple indicates an arbitrary constant by _C1. Next, we set in turn the constant to 1, 0, and $4/e^3$ to get three particular solutions from the family of solutions and at the same time create plots that can then be generated together.

```
sol1:=plot(x–1+exp(–x),x=–3..3,y=–3..3):
sol2:=plot(x–1,x=–3..3,y=–3..3):
sol3:=plot(x–1+4/E^3*exp(–x),x=–3..3,y=–3..3):
slopefield:=dfieldplot(x–y,[x,y],x=–3..3,y=–3..3,scaling=constrained):
display({slopefield,sol1,sol2,sol3});
```

Mathematica:

```
DSolve[y′[x]==x–y[x],y[x],x]

Plot[{x–1+Exp[–x],x–1,x–1+4/E^3*Exp[–x]},{x,–3,3},
    PlotRange–>{–3,3}]
```

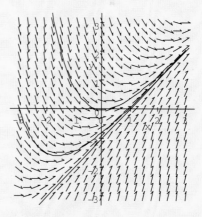

FIGURE 17.1.2
Slope field and several solutions

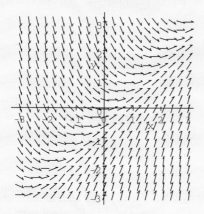

FIGURE 17.1.1
Slope field

CAS 62

Solve the initial-value problem

$$y'' + y = e^x, \ y(0) = 1, \ \text{and} \ y'(0) = 1$$

Maple:*

de:=diff(y(x),x\$2)+y−exp(x);

Output: $de := \left(\dfrac{\partial^2}{\partial x^2} y(x)\right) + y - e^x$

The particular solution to the initial-value problem can be solved by:

dsolve({de=0,y(0)=1,D(y)(0)=1},y(x));

Output: $y(x) = \dfrac{1}{2}e^x + \dfrac{1}{2}\sin(x) + \dfrac{1}{2}\cos(x)$

plot(1/2*exp(x)+1/2*sin(x)+1/2*cos(x),x=−10..10,y=−10..10,
    scaling=constrained);

Mathematica:

DSolve[y''[x]+y[x]==Exp[x],y[x],x]

y[x_] = Exp[x]/2+A*Cos[x]+B*Sin[x]

Solve[{y[0]==0,y'[0]==1},{A,B}]

Plot[Exp[x]/2−1/2*Cos[x]+1/2*Sin[x],{x,−10,10},
    PlotRange−>{−10,10}]

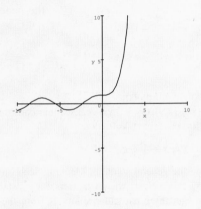

FIGURE 17.1.3

*In the Maple command **diff**, the use of x\$2 means the second derivative with respect to x.

Exercise Set 17.1

In Exercises 1–6, solve the given initial-value problem.

1. $\dfrac{dy}{dx} = x - 2; \ y(1) = 4$

2. $\dfrac{dy}{dx} = xe^{-x}; \ y(0) = -2$

3. $y' = \tan x$ on $[0, \pi/2); \ y(0) = 3$

4. $y' - x/\sqrt{9 + x^2} = 0; \ y(-4) = 3$

5. $y'' = 2; \ y(1) = 2, \ y'(1) = 3$

6. $d^2x/dt^2 = 3t; \ x(0) = 0, \ x'(0) = 2$

In Exercises 7–12, verify that for all choices of C_1 and C_2 the given function is a solution of the differential equation.

7. $y = C_1 e^{-x} + C_2 e^{-2x}; \ d^2y/dx^2 + 3\,dy/dx + 2y = 0$

8. $y = C_1e^{-2x} + C_2e^{4x}$; $y'' - 2y' - 8y = 0$

9. $y = C_1 \cos ax + C_2 \sin ax$; $y'' + a^2y = 0$

10. $y = C_1 \cosh ax + C_2 \sinh ax$; $y'' - a^2y = 0$

11. $x = e^{-t}(C_1 \cos t + C_2 \sin t)$; $d^2x/dt^2 + 2\,dx/dt + 2x = 0$

12. $y = C_1x^{-1} + C_2x^{-2}$; $x^2y'' + 4xy' + 2y = 0$ $(x > 0)$

In Exercises 13–18, find the particular solution of the problem specified that satisfies the given initial conditions or boundary conditions.

13. Exercise 7; $y(0) = 0$, $y'(0) = -2$

14. Exercise 8; $y(0) = 1$, $y'(0) = 0$

15. Exercise 9; $y(0) = 2$, $y\left(\dfrac{\pi}{2a}\right) = 1$

16. Exercise 10; $y(0) = 1$, $y'(0) = 4a$

17. Exercise 11; $x(0) = 0$, $x'(0) = 1$

18. Exercise 12; $y(1) = 2$, $y(2) = 3$

In Exercises 19–21, verify that any differentiable function $y = y(x)$ defined implicitly by the given equation is a solution of the differential equation.

19. $2x^3y - 3xy^2 + y^4 = 7$; $\dfrac{dy}{dx} = \dfrac{3y(y - 2x^2)}{2x^3 - 6xy + 4y^3}$

20. $y \tan xy - 2x = 1$; $\dfrac{dy}{dx} = \dfrac{2 - y^2 \sec^2 xy}{xy \sec^2 xy + \tan xy}$

21. $\tan^{-1}\dfrac{y}{x} + \ln\sqrt{x^2 + y^2} = 1$; $\dfrac{dy}{dx} = \dfrac{y - x}{y + x}$

In Exercises 22 and 23, solve the initial-value problem by first showing that the given family is a solution and then determining the proper constants.

22. $y''' + 3y'' - 4y = e^{-2x}$; $y(0) = 1$, $y'(0) = -2$, $y''(0) = 3$;
$y = C_1e^x + C_2e^{-2x} + C_3xe^{-2x} - (1/6)x^2e^{-2x}$

23. $y''' + y' - 10y = 10x^2 - 2x + 5$; $y(0) = 2$, $y'(0) = 1$,
$y''(0) = 0$; $y = e^{-x}(C_1 \cos 2x + C_2 \sin 2x) + C_3e^{2x} - x^2 - \frac{1}{2}$

In Exercises 24–29, find the differential equation that has the given family as a general solution.

24. $y = Cx^2$

25. $xy - y^2 = Cx$

26. $y = C_1e^x + C_2e^{-x}$

27. $y = C_1 \cos x + C_2 \sin x$

28. $y = C_1x + C_2x^2$

29. $y = C_1x + C_2x \ln x$

30. The general solutions of the equations

$$\frac{dy}{dx} = F(x, y) \qquad \text{and} \qquad \frac{dy}{dx} = -\frac{1}{F(x, y)}$$

are called **orthogonal trajectories** of each other. Explain the graphical significance of such orthogonal trajectories. Find the family of orthogonal trajectories of the family $y = x^3 + C$, and draw several members of each family. (*Hint:* First find the differential equation that has $y = x^3 + C$ as its general solution.)

In Exercises 31–34, sketch the slope field of the given differential equations. Then use your result to sketch the graph of the solution curve passing through the given point.

31. $\dfrac{dy}{dx} = x + y$; $(-1, 0)$ **32.** $\dfrac{dy}{dx} = x^2 - y$; $(0, 1)$

33. $\dfrac{dy}{dx} = xy$; $(1, 1)$ **34.** $\dfrac{dy}{dx} = x^2 + y^2$; $(2, 0)$

35. If $y' = F(x, y)$, then on each of the level curves $F(x, y) = c$, the slopes of solution curves all have the value c. These level curves are called **isoclines**. For each of the following, draw several isoclines and use them as an aid in making a sketch of the slope field.
(a) $y' = x - y^2$ (b) $y' = x^2/y$

CAS *Use a CAS for Exercises 36 through 38.*

36. In each of the following, first sketch a slope field for the differential equation. Then solve the equation and plot several solution curves on the same plot as the slope field.
(a) $dy/dx = x - y$
(b) $dy/dx - 4y = e^x$
(c) $dy/dx - \sin(x)y = \sin x$
(d) $dy/dx + x^2 - y = 0$

37. In each of the following, solve the initial-value problem and plot the solution curve.
(a) $y'' - y' - 6x = 0$, $y(0) = 1$, $y'(0) = 0$
(b) $(x + y)y' = x - y$, $y(0) = 1$
(c) $y'' + y' = \sin x$, $y(0) = 0$, $y'(0) = 1$

38. Solve the differential equation $y' + x^2y - y^2 = 0$.

17.2 FIRST-ORDER SEPARABLE AND FIRST-ORDER HOMOGENEOUS DIFFERENTIAL EQUATIONS

Separable Equations

A first-order differential equation is one that can be written in the form

$$\frac{dy}{dx} = F(x, y)$$

In this section and the next, we introduce solution techniques for cases in which $F(x, y)$ is of a special form. In the first case, $F(x, y)$ is the quotient of a function of x only and a function of y only.

Definition 17.1 Separable Equations	A first-order differential equation that can be written in the form $$\frac{dy}{dx} = \frac{f(x)}{g(y)} \quad (g(y) \neq 0) \qquad (17.1)$$ is said to be *separable*.

In order to explain the meaning of *separable*, let $y = y(x)$ be a solution of Equation 17.1, and write the equation in the form

$$g(y(x)) \frac{dy}{dx} = f(x) \qquad (17.2)$$

If G is an antiderivative of g, by the Chain Rule, we have

$$\frac{d}{dx} G(y(x)) = G'(y(x)) \frac{dy}{dx} = g(y(x)) \frac{dy}{dx}$$

Thus, we can write Equation 17.2 as

$$\frac{d}{dx} G(y(x)) = \frac{d}{dx} F(x)$$

where F is an antiderivative of f. The equality of these two derivatives implies that the functions F and G differ by a constant:

$$G(y(x)) = F(x) + C$$

Equivalently, we have

$$\int g(y) \, dy = \int f(x) \, dx \qquad (17.3)$$

where we understand that y is a function of x.

In Equation 17.3, the variables x and y are separated. We integrate the left-hand side with respect to y and the right-hand side with respect to x.

In practice, we usually shorten the process leading to Equation 17.3 by the following steps. First, we write Equation 17.1 in the *differential form*

$$g(y) \, dy = f(x) \, dx$$

(thus separating the variables). Then we integrate both sides to get Equation 17.3.

EXAMPLE 17.6   Find the general solution of the differential equation

$$\frac{dy}{dx} = \frac{2x}{y-1}, \qquad y > 1$$

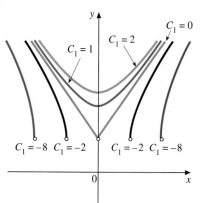

Solution   Separating variables, we get $(y-1)\,dy = 2x\,dx$. Thus, integrating both sides, we have

$$\int (y-1)\,dy = \int 2x\,dx$$

$$\frac{(y-1)^2}{2} = x^2 + C$$

Since $y > 1$, we can solve explicitly for y by taking the positive square root:

$$y - 1 = \sqrt{2(x^2 + C)}$$

or, on writing $C_1 = 2C$,

$$y = 1 + \sqrt{2x^2 + C_1}$$

FIGURE 17.6
The family $y = 1 + \sqrt{2x^2 + C_1}$

We show several members of this family in Figure 17.6.   ∎

EXAMPLE 17.7   Find the general solution of the differential equation

$$(x^2 + 1)\,dy = xy\,dx$$

Solution   Assume for the moment that $y \neq 0$. Then we can separate variables by dividing both sides of the equation by $(x^2 + 1)y$:

$$\frac{dy}{y} = \frac{x}{x^2 + 1}\,dx$$

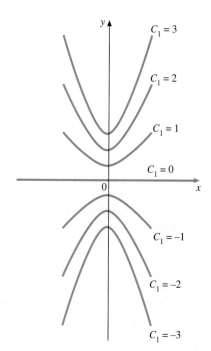

Thus,

$$\int \frac{dy}{y} = \int \frac{x}{x^2 + 1}\,dx \qquad \begin{array}{l}\text{Substitute } u = x^2 + 1 \\ \text{on the right-hand side.}\end{array}$$

$$\ln |y| = \frac{1}{2}\ln (x^2 + 1) + C$$

We can solve for y more readily by writing the constant C in the form $\ln |C_1|$, where $C_1 \neq 0$. Then we have

$$\ln |y| = \ln \sqrt{x^2 + 1} + \ln |C_1|$$

$$= \ln \left(|C_1|\sqrt{x^2 + 1} \right)$$

$$|y| = |C_1|\sqrt{x^2 + 1}$$

or equivalently, since C_1 can be either positive or negative,

$$y = C_1\sqrt{x^2 + 1}$$

Now let us return to the original equation and consider the case $y = 0$. The equation will be satisfied in this case if $dy = 0$ also—that is, if y is the zero function. But if we permit C_1 to be 0, the solution $y = C_1\sqrt{x^2 + 1}$ includes $y = 0$ as a particular case. Thus, $y = C_1\sqrt{x^2 + 1}$ is the general solution. (See Figure 17.7.)   ∎

FIGURE 17.7
The family $y = C_1\sqrt{x^2 + 1}$

EXAMPLE 17.8   Solve the initial-value problem

$$y' = xy - y + x - 1; \qquad y(1) = 3$$

Solution   We factor the right-hand side by grouping, and then we separate variables:

$$\frac{dy}{dx} = y(x - 1) + (x - 1)$$

$$\frac{dy}{dx} = (y + 1)(x - 1)$$

$$\frac{dy}{y + 1} = (x - 1)\,dx \qquad (y \neq -1)$$

$$\ln|y + 1| = \frac{(x - 1)^2}{2} + \ln|C| \qquad \text{Write the constant as } \ln|C|.$$

$$\ln\left|\frac{y + 1}{C}\right| = \frac{(x - 1)^2}{2} \qquad \text{Since } \ln|y + 1| - \ln|C| = \ln\left|\frac{y + 1}{C}\right|$$

$$\frac{y + 1}{C} = e^{(x-1)^2/2}$$

$$y = Ce^{(x-1)^2/2} - 1$$

Now we substitute $x = 1$ and $y = 3$ to get $3 = C - 1$, or $C = 4$. The desired particular solution is therefore

$$y = 4e^{(x-1)^2/2} - 1$$

We show the graph in Figure 17.8.   ■

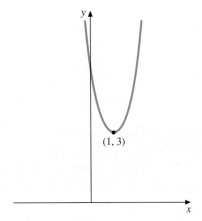

(1, 3)

FIGURE 17.8

Homogeneous Equations

Sometimes a differential equation in which the variables are not separable can be transformed into one in which they are. This transformation is especially applicable to so-called **homogeneous** first-order differential equations. To explain the procedure, we first define what is meant by a homogeneous function. Then we use this definition to define a first-order homogeneous differential equation.

Definition 17.2
A Homogeneous Function

A function $f(x, y)$ is said to be **homogeneous of degree** n if, for all $t > 0$ for which (tx, ty) is in the domain of f, the equation

$$f(tx, ty) = t^n f(x, y)$$

holds true.

EXAMPLE 17.9   Show that each of the following functions is homogeneous, and give the degree of homogeneity:

(a) $f(x, y) = x^3 - 2x^2y + xy^2$

(b) $g(x, y) = \dfrac{\sqrt{x^2 + y^2}}{x + y}$

Solution

(a) $f(tx, ty) = t^3x^3 - 2(t^2x^2)(ty) + (tx)(t^2y^2)$

$$= t^3(x^3 - 2x^2y + xy^2) = t^3 f(x, y)$$

Thus, f is homogeneous of degree 3.

(b) $g(tx, ty) = \dfrac{\sqrt{t^2x^2 + t^2y^2}}{tx + ty} = \dfrac{t\sqrt{x^2 + y^2}}{t(x + y)} = \dfrac{\sqrt{x^2 + y^2}}{x + y} = t^0 g(x, y)$

So g is homogeneous of degree 0. Note that we used the fact that $t > 0$ when we wrote $\sqrt{t^2x^2 + t^2y^2} = t\sqrt{x^2 + y^2}$.   ∎

Definition 17.3
Homogeneous Differential
Equation

An equation that can be expressed in the form

$$\frac{dy}{dx} = F(x, y)$$

where F is homogeneous of degree 0 is said to be a first-order **homogeneous differential equation**.

EXAMPLE 17.10   Show that each of the following differential equations is homogeneous:

(a) $\dfrac{dy}{dx} = \dfrac{\sqrt{x^2 + y^2}}{x + y}$

(b) $\left(y + x \sin \dfrac{x}{y}\right) dx + \left(y \cos \dfrac{x}{y} - 2x\right) dy = 0$

Solution

(a) We saw in Example 17.9, part (b), that $\sqrt{x^2 + y^2}/(x + y)$ is homogeneous of degree 0. So, by Definition 17.3, the given differential equation is homogeneous.

(b) Solving for dy/dx, we get

$$\frac{dy}{dx} = \frac{y + x \sin \dfrac{x}{y}}{2x - y \cos \dfrac{x}{y}}$$

Denote the function on the right by $F(x, y)$. Then

$$F(tx, ty) = \frac{ty + tx \sin \dfrac{tx}{ty}}{2tx - ty \cos \dfrac{tx}{ty}} = \frac{t\left(y + x \sin \dfrac{x}{y}\right)}{t\left(2x - y \cos \dfrac{x}{y}\right)} = F(x, y)$$

So F is homogeneous of degree 0, which means that the differential equation is homogeneous. ∎

If a differential equation is written in the differential form

$$f(x, y)\, dx + g(x, y)\, dy = 0$$

then the homogeneity requirement is satisfied provided f and g are homogeneous functions of the *same* degree. We can see this result by writing

$$\frac{dy}{dx} = -\frac{f(x, y)}{g(x, y)}$$

and observing that $F = -f/g$ will be homogeneous of degree 0 when f and g are homogeneous of the same degree. In Example 17.10(b), we see this result, where both $f(x, y) = y + x \sin(x/y)$ and $g(x, y) = y \cos(x/y) - 2x$ are homogeneous functions of degree 1.

The important and useful property common to all first-order homogeneous differential equations is that either of the substitutions

$$v = \frac{y}{x} \qquad \text{or} \qquad v = \frac{x}{y}$$

will invariably transform the equation to one in which the variables are separable. We will illustrate this result using $v = y/x$, but you should keep in mind that the second substitution $v = x/y$ also works and in some cases may involve simpler calculations.

Suppose, then, that $dy/dx = F(x, y)$ is a homogeneous differential equation. Setting $v = y/x$, we obtain

$$y = vx \qquad \text{and} \qquad \frac{dy}{dx} = v + x\frac{dv}{dx} \tag{17.4}$$

On substitution in the original differential equation, we get

$$v + x\frac{dv}{dx} = F(x, vx) \tag{17.5}$$

By our assumption that the original differential equation is homogeneous, we know by Definition 17.3 that F is a homogeneous function of degree 0. Thus, by Definition 17.2 (with x replacing t and v replacing y),

$$F(x, vx) = x^0 F(1, v) = F(1, v)$$

From Equation 17.5, we therefore have

$$v + x\frac{dv}{dx} = F(1, v)$$

Solving for dv/dx, we get

$$\frac{dv}{dx} = \frac{F(1, v) - v}{x}$$

The variables x and v can now be separated:

$$\frac{dv}{F(1, v) - v} = \frac{dx}{x} \qquad (17.6)$$

If F were known, we could solve this equation (assuming we can integrate the left-hand side) and then replace v by y/x to get the final result.

REMARK ————————————————————————————

■ There is no need to memorize Equation 17.6. The important thing to remember is that once you have identified a differential equation as being homogeneous, the substitution in Equations 17.4 should be made (or the similar substitution corresponding to $v = x/y$). Then in each individual case the variables can be separated and the resulting equation solved.

EXAMPLE 17.11   Find the general solution of the differential equation

$$\frac{dy}{dx} = \frac{x + y}{x - y}$$

Solution   If we write

$$F(x, y) = \frac{x + y}{x - y}$$

we see that F is homogeneous of degree 0, since

$$F(tx, ty) = \frac{tx + ty}{tx - ty} = \frac{x + y}{x - y} = F(x, y)$$

Letting $y = vx$, we have

$$F(x, vx) = \frac{x + vx}{x - vx} = \frac{1 + v}{1 - v}$$

By Equation 17.5,

$$v + x \frac{dv}{dx} = \frac{1 + v}{1 - v}$$

Simplifying, we obtain

$$x \frac{dv}{dx} = \frac{1 + v}{1 - v} - v$$

$$= \frac{1 + v - v(1 - v)}{1 - v}$$

$$= \frac{1 + v^2}{1 - v}$$

Now we can separate variables to get

$$\frac{1 - v}{1 + v^2} \, dv = \frac{dx}{x}$$

Thus,

$$\int \frac{1-v}{1+v^2}\,dv = \int \frac{dx}{x}$$

We divide the integral on the left into two integrals, obtaining

$$\int \frac{dv}{1+v^2} - \int \frac{v\,dv}{1+v^2} = \int \frac{dx}{x}$$

Integrating, we get

$$\tan^{-1} v - \frac{1}{2}\ln(1+v^2) = \ln|x| + C$$

Finally, we replace v by y/x:

$$\tan^{-1}\frac{y}{x} - \frac{1}{2}\ln\left(1 + \frac{y^2}{x^2}\right) = \ln|x| + C$$

Since $\ln\left[1 + (y^2/x^2)\right] = \ln\left[(x^2 + y^2)/x^2\right] = \ln(x^2 + y^2) - \ln x^2$ and $\ln x^2 = 2\ln|x|$, we can simplify our answer to get

$$\tan^{-1}\frac{y}{x} - \frac{1}{2}\ln(x^2 + y^2) = C$$

Solving for y would be difficult, so we leave the answer in implicit form.   ■

When you make the substitution $y = vx$ to solve a homogeneous differential equation, you will first get an answer in terms of x and v. Don't forget then to replace v by y/x to get the final answer. A similar remark applies when you substitute $x = vy$.

EXAMPLE 17.12   Solve the initial-value problem

$$(2x - 3y)\,dx + y\,dy = 0; \qquad y(1) = 3$$

Solution   The coefficients of dx and dy are both homogeneous functions of degree 1, so the differential equation is homogeneous. Setting $v = y/x$ gives $y = vx$ and $dy = v\,dx + x\,dv$. Thus, we get

$$(2x - 3vx)\,dx + vx(v\,dx + x\,dv) = 0$$

On dividing the preceding equation by x and simplifying, we get

$$(v^2 - 3v + 2)\,dx + xv\,dv = 0$$

We now separate variables and use partial fractions:

$$\frac{v\,dv}{(v-2)(v-1)} = -\frac{dx}{x}$$

$$\left(\frac{2}{v-2} - \frac{1}{v-1}\right)dv = -\frac{dx}{x}$$

Thus, on integrating both sides, we get

$$2 \ln |v - 2| - \ln |v - 1| = -\ln |x| + \ln |C|$$

Write $\ln |C|$ for the constant of integration.

$$\ln \frac{(v - 2)^2}{|v - 1|} = \ln \left| \frac{C}{x} \right|$$

$$(v - 2)^2 = (v - 1) \frac{C}{x}$$

Now we replace v with y/x and simplify to get $(y - 2x)^2 = C(y - x)$. Imposing the initial condition $y(1) = 3$ gives $C = \frac{1}{2}$ (found by substituting $x = 1$ and $y = 3$). Thus, the solution in question is given implicitly by

$$y = x + 2(y - 2x)^2$$

We show its graph in Figure 17.9.   ■

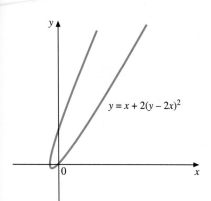

$y = x + 2(y - 2x)^2$

FIGURE 17.9

Exercise Set 17.2

In Exercises 1–5, use separation of variables to find the general solution.

1. $\dfrac{dy}{dx} = \dfrac{y}{x + 1}$

2. $xy\, dy = (y^2 + 1)\, dx$

3. $y' = \dfrac{x + 1}{y - 2}$

4. $y' = \dfrac{y}{x^2 y + x^2}$

5. $e^{-x}\, dy - (1 + y^2)\, dx = 0$

In Exercises 6–10, show that the differential equation is homogeneous, and find its general solution.

6. $y' = \dfrac{x + y}{x}$

7. $\dfrac{dy}{dx} = \dfrac{y(x + y)}{x(x - y)}$

8. $2x^2\, dx = (x^2 + y^2)\, dy$

9. $\dfrac{dy}{dx} = \dfrac{x}{x + 2y}$

10. $xy\, dy = y\left(y + x \cos^2 \dfrac{y}{x} \right) dx$

In Exercises 11–20, use whatever method is appropriate to find the general solution. If an initial value is given, find the particular solution that satisfies it.

11. $y' = x\sqrt{4 - y^2}$

12. $y' = \dfrac{\ln x}{2xy - 3x};\ y(1) = 1$

13. $x\, dy = \left(\sqrt{x^2 - y^2} + y \right) dx$

14. $\tan x \sec y\, dx = \sin^2 y\, dy$

15. $y' = \dfrac{2xy}{x^2 - y^2};\ y(1) = 2$

16. $\sqrt{y^2 + 4}\, dx + (2y - xy)\, dy = 0$

17. $ye^{x/y}\, dx - (y + xe^{x/y})\, dy = 0;\ y(0) = e$

18. $\dfrac{dy}{dx} = \dfrac{3x - y}{x + y}$

19. $(xy - x + y - 1)\, dy = (xy + x)\, dx$

20. $xy' = y + x \tan y/x;\ y(2) = \pi/3$

21. Show that if $y' = F(x, y)$ is homogeneous, then

$$x = C \exp \left(\int \frac{dv}{F(1, v) - v} \right)$$

where $v = y/x$. (Recall that $\exp x$ means e^x.)

22. Show that with the substitution $v = x/y$, the homogeneous differential equation $dy/dx = F(x, y)$ is transformed into the separable equation

$$y \frac{dv}{dy} = \frac{1}{F(v, 1)} - v$$

23. Use the method of Exercise 22 to solve the differential equation

$$\frac{dy}{dx} = \frac{y^3 + x^2 y}{x^3 + y^3}$$

24. Find the general solution of the equation

$$3(1 + y^2)\,dx + (y^2 - x^3y^2 - 2x^3y + 2y)\,dy = 0$$

25. Solve the initial-value problem

$$y' = \frac{5xy + y - 5x - 1}{x^3 - 3x + 2}; \qquad y(0) = 2$$

26. Use the idea of Exercise 30 in Exercise Set 17.1 to find the family of orthogonal trajectories of the family $x^2 - xy + y^2 = C$.

27. Prove that $dy/dx = F(x, y)$ is homogeneous if and only if there is a function G of one variable such that

$$F(x, y) = G\left(\frac{y}{x}\right)$$

17.3 FIRST-ORDER EXACT AND FIRST-ORDER LINEAR DIFFERENTIAL EQUATIONS

Exact Equations

Consider a first-order differential equation in differential form

$$f(x, y)\,dx + g(x, y)\,dy = 0$$

with f and g having continuous first partial derivatives in a region R. Suppose further that there is a function $\phi(x, y)$ such that

$$\frac{\partial \phi(x, y)}{\partial x} = f(x, y) \quad \text{and} \quad \frac{\partial \phi(x, y)}{\partial y} = g(x, y)$$

For example, the differential equation

$$(3x^2 - y)\,dx + (2y - x)\,dy = 0$$

can be written as

$$\frac{\partial \phi(x, y)}{\partial x}\,dx + \frac{\partial \phi(x, y)}{\partial y}\,dy = 0 \tag{17.7}$$

with $\phi(x, y) = x^3 - xy + y^2$, as you can readily verify. In Chapter 16, the expression on the left-hand side of Equation 17.7 was defined as the *exact differential* of ϕ. Differential equations that can be written in the form of Equation 17.7 are called *exact differential equations*.

Definition 17.4
Exact Differential Equation

The first-order differential equation

$$f(x, y)\,dx + g(x, y)\,dy = 0 \tag{17.8}$$

is said to be **exact** in a region R if there exists a function ϕ such that

$$f(x, y) = \frac{\partial \phi(x, y)}{\partial x} \quad \text{and} \quad g(x, y) = \frac{\partial \phi(x, y)}{\partial y}$$

in R.

If Equation 17.8 is exact, then by Definition 14.2, we know that $f(x, y)\,dx + g(x, y)\,dy$ is the exact differential of ϕ. That is, $d\phi(x, y) = f(x, y)\,dx + g(x, y)\,dy$.

From Section 16.2, we know that Equation 17.8 is exact if

$$\frac{\partial g}{\partial x} = \frac{\partial f}{\partial y} \qquad (17.9)$$

throughout R. In this case, Equation 17.8 is equivalent to

$$d\phi(x, y) = 0$$

whose solution is

$$\phi(x, y) = C \qquad (17.10)$$

(See Exercise 35 in Exercise Set 17.3.) Thus, Equation 17.10 is an implicit solution to Equation 17.8.

In the next example, we review the procedure for finding the function ϕ when Equation 17.9 is satisfied.

EXAMPLE 17.13   Show that the differential equation

$$(3x^2 - y)\,dx + (2y - x)\,dy = 0$$

is exact, and find its general solution.

Solution   We have already seen that the left-hand side of this equation is the exact differential of the function $\phi(x) = x^3 - xy + y^2$, but for purposes of this example, we will assume this result is not known.

Equation 17.9 is satisfied, since with $f(x, y) = 3x^2 - y$ and $g(x, y) = 2y - x$, we have

$$\frac{\partial g}{\partial x} = -1 \quad \text{and} \quad \frac{\partial f}{\partial y} = -1$$

So we know the equation is exact. Let ϕ denote a function for which

$$\frac{\partial \phi}{\partial x} = 3x^2 - y \quad \text{and} \quad \frac{\partial \phi}{\partial y} = 2y - x$$

Integrating both sides of the first of these equations with respect to x gives

$$\phi(x, y) = \int (3x^2 - y)\,dx = x^3 - xy + C_1(y) \qquad \text{The "constant" of integration can be a function of } y.$$

Now, imposing the condition $\partial \phi / \partial y = 2y - x$ gives

$$-x + C_1'(y) = 2y - x$$
$$C_1'(y) = 2y$$
$$C_1(y) = y^2$$

(We do not need to include a constant of integration at this stage.) Thus, we have verified that

$$\phi(x, y) = x^3 - xy + y^2$$

The original differential equation is $d\phi(x, y) = 0$, so its solution is $\phi(x, y) = C$. That is,

$$x^3 - xy + y^2 = C \qquad \blacksquare$$

EXAMPLE 17.14   Solve the initial-value problem

$$\frac{dy}{dx} = \frac{xe^{-x} + \cos y}{x \sin y - \cos y}, \qquad y(0) = \pi$$

Solution   The equation can be written in the differential form

$$(xe^{-x} + \cos y)\, dx + (\cos y - x \sin y)\, dy = 0$$

The variables cannot be separated, and the equation is not homogeneous. Testing for exactness, we have

$$\frac{\partial}{\partial y}(xe^{-x} + \cos y) = -\sin y, \qquad \frac{\partial}{\partial x}(\cos y - x \sin y) = -\sin y$$

Thus, the equation is exact. So we know that a function ϕ exists for which $d\phi = (xe^{-x} + \cos y)\, dx + (\cos y - x \sin y)\, dy$. To find ϕ, we choose this time to begin by integrating the coefficient of dy with respect to y, since it looks relatively easy to do.

$$\phi(x, y) = \int (\cos y - x \sin y)\, dy = \sin y + x \cos y + C_1(x)$$

Now we must also have $\partial \phi / \partial x = xe^{-x} + \cos y$. So

$$\cos y + C_1'(x) = xe^{-x} + \cos y$$

$$C_1'(x) = xe^{-x}$$

$$C_1(x) = -xe^{-x} - e^{-x} \qquad \text{Obtained by integrating by parts}$$

Thus

$$\phi(x, y) = \sin y + x \cos y - xe^{-x} - e^{-x}$$

and since the original equation can be written as $d\phi = 0$, its general solution is $\phi(x, y) = C$; that is,

$$\sin y + x \cos y - xe^{-x} - e^{-x} = C$$

Imposing the initial condition $y(0) = \pi$ gives $-1 = C$. The particular solution in question can therefore be written in the form

$$\sin y + x \cos y - e^{-x}(x + 1) + 1 = 0 \qquad \blacksquare$$

Linear Equations

Another class of differential equations that in many cases can be solved explicitly consists of equations involving y and its derivatives to the first power only, with coefficients that are functions of x only. Such equations are called *linear differential equations*. The formal definition is as follows:

Definition 17.5 Linear Differential Equation	An nth-order differential equation is said to be **linear** if it can be written in the form $$a_n(x)y^{(n)} + a_{n-1}(x)y^{(n-1)} + \cdots + a_1(x)y' + a_0(x)y = g(x) \qquad (17.11)$$ with $a_n(x) \neq 0$.

The term *linear* is used here because each term on the left-hand side of Equation 17.11 involves y or one of its derivatives raised to the first power only. To emphasize this reference to y and its derivatives, we sometimes say that the equation is *linear in y*. The coefficient functions $a_0(x), a_1(x), \ldots, a_n(x)$ may be highly nonlinear. For example, both

$$\frac{dy}{dx} + (\tan x)y = \sin x \quad \text{and} \quad x^2 y'' + 2xy' - 3y = \ln x$$

are linear in y, whereas

$$\left(\frac{dy}{dx}\right)^2 + 2xy = 0 \quad \text{and} \quad y'y'' - 3y = x$$

are nonlinear.

If $n = 1$ in Equation 17.11, we have the first-order linear equation

$$a_1(x)y' + a_0(x)y = g(x), \quad a_1(x) \neq 0$$

If we divide both sides by $a_1(x)$ and write $P(x) = a_0(x)/a_1(x)$ and $Q(x) = g(x)/a_1(x)$, we obtain the following, called the **standard form** of a first-order linear differential equation.

Standard Form of a First-Order Linear Differential Equation

$$y' + P(x)y = Q(x) \tag{17.12}$$

In order to develop a general procedure for solving first-order linear differential equations, let us rewrite Equation 17.12 in the differential form

$$\left[P(x)y - Q(x)\right]dx + dy = 0 \tag{17.13}$$

Our goal is to transform this equation into an exact differential equation, which we can then solve by the technique we have already developed.

Sometimes equations that are not exact can be transformed to exact equations by multiplying both sides by a function of x. For example, the linear equation

$$(5x + 2y)\,dx + x\,dy = 0$$

is not exact, but if we multiply both sides by x, we get

$$(5x^2 + 2xy)\,dx + x^2\,dy = 0$$

which is exact, since $\partial(x^2)/\partial x = 2x$ and $\partial(5x^2 + 2xy)/\partial y = 2x$ are equal. The factor x that we multiplied by is called an **integrating factor** for the differential equation.

We now generalize the idea illustrated in this example by attempting to find an integrating factor, which we denote by $\mu(x)$, for Equation 17.13. That is, we try to find a function $\mu(x)$ such that

$$\left[\mu(x)P(x)y - \mu(x)Q(x)\right]dx + \mu(x)\,dy = 0 \tag{17.14}$$

is an exact equation. The requirement for exactness is

$$\frac{\partial}{\partial y}\left[\mu(x)P(x)y - \mu(x)Q(x)\right] = \frac{\partial}{\partial x}\left[\mu(x)\right]$$

That is,

$$\mu'(x) = \mu(x)P(x) \tag{17.15}$$

We can separate variables in Equation 17.15 to get

$$\frac{d\mu(x)}{\mu(x)} = P(x)\,dx$$

When we integrate both sides, we obtain

$$\ln\mu(x) = \int P(x)\,dx$$

from which we get the following formula for $\mu(x)$.

An Integrating Factor for Linear First-Order Differential Equations

$$\mu(x) = e^{\int P(x)\,dx} \tag{17.16}$$

Note that $\mu(x)$ is positive, since e raised to any power is positive. We can take the constant of integration as 0, since we want *any* integrating factor, so we might as well use the simplest one.

We could proceed now to find a function ϕ for which the left-hand side of Equation 17.14 is the exact differential, using the technique we have developed. However, in this case there is an easier way. If we multiply both sides of Equation 17.12 by $\mu(x)$, we have

$$\mu(x)y' + \mu(x)P(x)y = \mu(x)Q(x)$$

But by Equation 17.15 this equation can be written as (suppressing the parentheses and the independent variable x)

$$\mu y' + \mu'y = \mu Q$$

The left-hand side is precisely the derivative of the product μy. Thus,

$$\frac{d}{dx}[\mu y] = \mu Q \tag{17.17}$$

So, on integrating both sides,

$$\mu y = \int \mu Q\,dx + C$$

and

$$y = \frac{1}{\mu}\int \mu Q\,dx + \frac{C}{\mu} \tag{17.18}$$

Direct substitution will show that the value of y given by Equation 17.18 is the solution we were seeking. There is no need to memorize the solution given by Equation 17.18. You can in each individual case carry out the following procedure.

Procedure for Solving a First-Order Linear Differential Equation

1. If the equation is linear, write it in the standard form

$$y' + P(x)y = Q(x)$$

2. Find the integrating factor

$$\mu(x) = e^{\int P(x)\,dx}$$

3. Multiply both sides of the standard form by $\mu(x)$, thereby obtaining

$$\frac{d}{dx}[\mu y] = \mu Q$$

(It is a good idea to check at this stage that the left-hand side is indeed the derivative of the product of μ and y.)

4. Take antiderivatives and solve for y.

EXAMPLE 17.15   Solve the differential equation

$$xy' + 2y = 2x^2 + 3x, \qquad x > 0$$

Solution   The equation is a first-order linear differential equation in y. To put it in standard form, we divide by x:

$$y' + \frac{2}{x}y = 2x + 3$$

So $P(x) = \dfrac{2}{x}$ and $Q(x) = 2x + 3$. The integrating factor is therefore

$$\mu(x) = e^{\int P(x)\,dx} = e^{\int (2/x)\,dx} = e^{2\ln x} = e^{\ln x^2} = x^2$$

(Note the use of the properties of logarithms and exponentials and the stipulation that $x > 0$.) We next multiply both sides of the standard form by $\mu(x)$:

$$x^2 y' + 2xy = 2x^3 + 3x^2$$

If our theory is correct, the left-hand side should be the derivative of μy—that is, of $x^2 y$—and a quick check using the Product Rule shows that this is correct. So we have

$$\frac{d}{dx}(x^2 y) = 2x^3 + 3x^2$$

Integrating both sides gives

$$x^2 y = \frac{x^4}{2} + x^3 + C$$

$$y = \frac{x^2}{2} + x + \frac{C}{x^2}$$

We show several solution curves in Figure 17.10. ∎

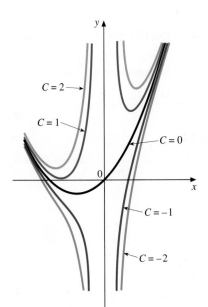

$C = 2$
$C = 1$
$C = 0$
$C = -1$
$C = -2$

FIGURE 17.10

The family $y = \dfrac{x^2}{x} + x + \dfrac{C}{x^2}$

A common mistake in this procedure is to forget the constant of integration in the next-to-last step. The constant is essential, however; otherwise we have not found the general solution.

EXAMPLE 17.16   Solve the initial-value problem

$$\frac{dy}{dx} = \frac{x+y}{x-1}; \qquad y(2) = 3$$

Solution   We see that the equation is linear in y, since both y and y' appear to the first degree. In standard form the equation is

$$y' + \left(\frac{1}{1-x}\right)y = \frac{x}{x-1} \qquad \text{Since } \frac{-y}{x-1} = \frac{y}{1-x}$$

Since we want the solution passing through $(2, 3)$, we restrict x so that $x > 1$. Thus,

$$\mu(x) = e^{\int dx/(1-x)} = e^{-\ln|1-x|} = e^{\ln(1/|1-x|)} = \frac{1}{|1-x|} = \frac{1}{x-1} \qquad \begin{array}{l} \text{Since } x > 1, \\ |1-x| = x-1. \end{array}$$

We multiply both sides by $\mu(x)$ to get

$$\frac{d}{dx}\left[\left(\frac{1}{x-1}\right)y\right] = \frac{x}{(x-1)^2} \qquad \text{Verify.}$$

Hence,

$$\frac{1}{x-1}y = \int \frac{x}{(x-1)^2}\,dx$$

Using partial fractions, we can write the integral on the right-hand side as

$$\int \left[\frac{1}{x-1} - \frac{1}{(x-1)^2}\right]dx$$

Thus,

$$\frac{1}{x-1}y = \ln(x-1) - \frac{1}{x-1} + C$$

$$y = (x-1)\ln(x-1) - 1 + C(x-1)$$

This is the general solution. Now we set $x = 2$ and $y = 3$, getting $C = 4$. Substituting this constant and simplifying gives the particular solution

$$y = (x-1)\ln(x-1) + 4x - 5$$

We show its graph in Figure 17.11.   ∎

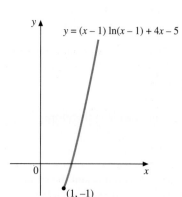

$y = (x - 1)\ln(x - 1) + 4x - 5$

$(1, -1)$

FIGURE 17.11

Exercise Set 17.3

In Exercises 1–6, show that the given equation is exact, and find its general solution. If an initial condition is given, find the corresponding particular solution.

1. $\dfrac{dy}{dx} = \dfrac{x - y + 1}{x - 2}$

2. $(e^y + 2x)\,dx + (xe^y - 2y)\,dy = 0$

3. $2xyy' = 3x^2 - y^2;\ y(1) = 2$

4. $\dfrac{dy}{dx} = \dfrac{y \sin x - \cos y}{\cos x - x \sin y}$

5. $\dfrac{y}{x}\,dx + (y + \ln x)\,dy = 0;\ y(1) = 4$

6. $y' = \dfrac{x \ln \cos^2 y}{x^2 \tan y - 3y^2};\ y(0) = 2$

In Exercises 7–12, show that the given equation is linear, and find its general solution. If an initial condition is given, find the corresponding particular solution.

7. $y' + y = x$

8. $y' - 2xy = x$

9. $x\dfrac{dy}{dx} = x^2 - y$; $y(2) = 1$

10. $dy = (\sin x + y \tan x)\, dx$ $\left(-\dfrac{\pi}{2} < x < \dfrac{\pi}{2} \right)$

11. $\dfrac{dy}{dx} + \dfrac{y}{1-x} = 1 - x^2$; $y(0) = 0$

12. $xy' = x^4 e^x + 2y$; $y(1) = 0$

Solve Exercises 13–27 by any method.

13. $(x - 2y)\, dx + (3y - 2x)\, dy = 0$

14. $y' = x - xy$

15. $(y + 1)\, dx + (x - 1)\, dy = 0$; $y(0) = 2$

16. $y' + y \cot x = 0$ $(0 < x < \pi)$

17. $y' - \dfrac{2xy}{x^2 + 1} = 1$; $y(1) = \pi$

18. $\dfrac{dy}{dx} = \dfrac{x^2 - y^2}{y(2x - y)}$

19. $\left(\dfrac{y}{x} - e^x + 2 \right) dx + (\ln xy + 3y^2 - 1)\, dy = 0$ $(x > 0)$

20. $(\cosh x)\, dy = (\cosh^3 x + y \sinh x)\, dx$

21. $\left(\dfrac{x}{\sqrt{x^2 + y^2}} + 1 \right) dx + \left(\dfrac{y}{\sqrt{x^2 + y^2}} - 2 \right) dy = 0$; $y(3) = 4$

22. $x\, dy + y(x + 1)\, dx = x^2\, dx$ $(x > 0)$

23. $\left(\dfrac{x - y}{x} \right) dx + (y - \ln x)\, dy = 0$ $(x > 0)$

24. $(e^x + 1)\, dy = e^x(x - y)\, dx$

25. $\dfrac{dy}{dx} = \dfrac{1 + y^2}{y(1 + x + y^2)}$ (*Hint:* Show it is linear in x.)

26. $y' + y \sin x = \sin 2x$; $y\left(\dfrac{\pi}{2} \right) = 3$

27. $[\ln(x^2 + y^2)]\, dx = 2 \left(\tan^{-1} \dfrac{y}{x} \right) dy$

28. An equation of the form
$$y' + P(x)y = y^n Q(x) \qquad (n \neq 1)$$
is called a **Bernoulli** equation. Show that the substitution $v = y^{1-n}$ transforms it into an equation that is linear in v. (*Hint:* First divide both sides by y^n.)

In Exercises 29 and 30, use the method of Exercise 28 to find the solution.

29. $y' + 2xy = xy^2$

30. $y' + \dfrac{2y}{x} = \sqrt{y} \sin x$; $y\left(\dfrac{\pi}{2} \right) = 4$

31. Show that a differential equation of the form
$$f(x, y)\, dx + g(x, y)\, dy = 0$$
if not already exact, can be made into an exact equation by multiplying both sides by an integrating factor of the form $\mu(x)$ provided $(f_y - g_x)/g$ is a function of x only. What is $\mu(x)$ in this case?

32. Find a condition analogous to that in Exercise 31 that will ensure the existence of an integrating factor of the form $\mu(y)$.

In Exercises 33 and 34, use the method of Exercise 31 or 32 to find an integrating factor, and solve the equation.

33. $(y^2 + 4x)\, dx + y\, dy = 0$

34. $2 \sin x \cos^2 y\, dx + \cos x \sin 2y\, dy = 0$

35. Prove that if ϕ is differentiable in a region R of P^2, then $y = y(x)$ is a solution to the differential equation $d\phi(x, y) = 0$ if and only if $\phi(x, y(x)) = C$.

17.4 APPLICATIONS OF FIRST-ORDER DIFFERENTIAL EQUATIONS

Differential equations are particularly useful in modeling real-world phenomena involving rates of change (derivatives). The following diagram illustrates the usual stages involved in mathematical modeling.

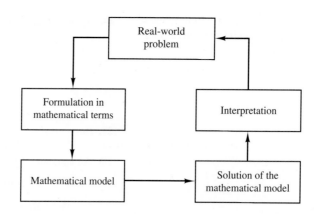

In this section we show how first-order differential equations can be used as models in a variety of applications. The first example is familiar from Section 7.5.

Radioactive Decay

Mushroom cloud from the explosion of a plutonium bomb

EXAMPLE 17.17   The isotope plutonium-241 produced in a nuclear explosion has a half-life of approximately 13.2 yr. Assume that the substance decays at a rate proportional to the amount present. Of a given initial amount, how much will remain after 5 yr? How long will it take for 90% of the original amount to decay?

Solution   Let $Q(t)$ denote the quantity present at time t (in years). Then since the rate of change dQ/dt is proportional to Q, we have $dQ/dt = kQ$. Denote the initial amount $Q(0)$ by Q_0. Since the half-life is 13.2 yr, we have

$$Q(13.2) = 0.5Q_0$$

Our problem is thus modeled by the differential equation $dQ/dt = kQ$, together with the known value $Q(13.2) = 0.5Q_0$. We can solve the differential equation by separating variables (note that it can also be solved as a linear equation).

$$\frac{dQ}{Q} = k\,dt$$

Integrating both sides gives

$$\ln Q = kt + \ln C \quad \text{\small Write the constant of integration as } \ln C.$$

$$\ln Q - \ln C = kt$$

$$\ln \frac{Q}{C} = kt$$

$$Q(t) = Ce^{kt}$$

Letting $t = 0$ gives $Q_0 = C$. So the solution is of the form

$$Q(t) = Q_0 e^{kt}$$

To find k, we use the fact that $Q(13.2) = 0.5Q_0$. Thus,

$$0.5Q_0 = Q_0e^{13.2k}$$

from which we find k by dividing by Q_0 and taking the natural logarithm of each side:

$$k = \frac{\ln 0.5}{13.2} \approx -0.0525$$

and the solution becomes

$$Q(t) = Q_0e^{-0.0525t}$$

We show the graph of Q in Figure 17.12.

For $t = 5$, we have

$$Q(5) = Q_0e^{(-0.0525)5} \approx 0.769Q_0$$

So after 5 yr, about 77% of the original amount remains.

To find how long it takes for 90% to decay, we set $Q(t) = 0.1Q_0$ and solve for t, since if 90% has decayed, 10% remains:

$$0.1Q_0 = Q_0e^{-0.0525t}$$

$$t = \frac{\ln 0.1}{-0.0525} \approx 43.7 \text{ yr} \qquad \blacksquare$$

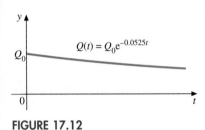

FIGURE 17.12

Population Growth

As our second example, we solve a model for population growth that in general gives a more realistic estimate than the Malthus Model that we considered in Section 7.4. In most models of population growth, the instantaneous rate of change of the size Q of the population is assumed to be some function of Q:

$$\frac{dQ}{dt} = f(Q) \qquad (17.19)$$

For the Malthus Model, $f(Q)$ is taken to be kQ for $k > 0$, and just as in the previous example, the solution is $Q(t) = Q_0e^{kt}$. In 1837, the Belgian mathematician Pierre-François Verhulst (1804–1849) proposed modifying the Malthus Model by taking $f(Q) = \alpha Q - \beta Q^2$, in Equation 17.19, where α and β are positive constants. Subtracting the term βQ^2 was Verhulst's way of accounting for competition among members of the population. The next example shows one way to solve Verhulst's equation.

EXAMPLE 17.18   Solve the Verhulst equation for population growth,

$$\frac{dQ}{dt} = \alpha Q - \beta Q^2 \qquad (\alpha > 0, \beta > 0)$$

where the initial size $Q(0)$ of the population is Q_0. Analyze the result.

Solution   One way to solve the problem is by separating variables, making use of partial fractions in the integration. We choose to illustrate a method suggested in Exercise 28 of Exercise Set 17.3 whereby we make a substitution that transforms the equation into a linear one. Equations of this type are called **Bernoulli equations**. We first rewrite the equation as

The Bernoullis were the greatest family of mathematicians in history. Jakob Bernoulli (1654–1705) was born and educated in Switzerland. After extending Descartes's book on geometry, he showed how to divide a triangle into four equal parts with two perpendicular straight lines.

Jakob elaborated on Leibniz's ideas and considered the problem of a body descending in a curve under the force of gravity. In this analysis the word *integral* first came to be used. He also worked out the parabolic and logarithmic spirals.

Jakob taught mathematics to his younger brother, Johann I (1667–1748), who became professor of mathematics in Basel, and extended the ideas of infinitesimals. He developed important ideas about series, about large numbers in probability theory, and on the method we call mathematical induction.

(continued next page)

Jakob and Johann together became the first to fully understand the differential calculus as expressed by Leibniz. In 1696, Johann proposed the problem of determining the curve by which a bead would descend most swiftly, the famous Brachistochrone Problem bested by Newton.

Johann taught calculus to L'Hôpital, the greatest French mathematician of the time. Johann's subsequent letters, after he left Paris, were the basis of the first textbook on differential calculus, published under the name of L'Hôpital.

Nikolaus I (1647–1759), the nephew of Jakob and Johann, studied with his uncles. Among his best known results is the proof that the binomial expansion $(1+x)^n$ diverges for $x > 1$.

Daniel (1700–1782), one of Johann I's sons worked on differential equations and series. He is also noted for his work in physics. In fact, we credit "Bernoulli's Principle" for explaining why airplanes fly—that pressure is lowest where speed is highest.

$$\frac{dQ}{dt} - \alpha Q = -\beta Q^2$$

and then divide through by Q^2:

$$Q^{-2}\frac{dQ}{dt} - \alpha Q^{-1} = -\beta$$

Now let $v = Q^{-1}$ and note that the derivative of v with respect to time t is

$$\frac{dv}{dt} = -Q^{-2}\frac{dQ}{dt}$$

When we substitute and multiply both sides by -1, we get

$$\frac{dv}{dt} + \alpha v = \beta$$

This equation is linear in v. The integrating factor is $\mu(t) = e^{\alpha t}$, and on multiplying by this value of $\mu(t)$ and simplifying (by the Product Rule), we get

$$\frac{d}{dt}[ve^{\alpha t}] = \beta e^{\alpha t}$$

Thus, on integrating both sides, we get

$$ve^{\alpha t} = \frac{\beta}{\alpha}e^{\alpha t} + C$$

The constant C can be found by setting $t = 0$, observing that $v(0) = \dfrac{1}{Q(0)} = \dfrac{1}{Q_0}$. Thus,

$$C = \frac{1}{Q_0} - \frac{\beta}{\alpha} = \frac{\alpha - \beta Q_0}{\alpha Q_0}$$

Substituting this constant, we can solve for v and then invert to obtain Q, since $Q = 1/v$. The details of the algebra are omitted, but the result is

$$Q(t) = \frac{\alpha Q_0}{\beta Q_0 + (\alpha - \beta Q_0)e^{-\alpha t}} \tag{17.20}$$

We make several observations about this solution. First, if β is small relative to α, the solution differs little from the Malthus solution $Q(t) = Q_0 e^{\alpha t}$, as you can see by neglecting the terms involving β. Second, if the denominator is written in the form

$$\beta Q_0(1 - e^{-\alpha t}) + \alpha e^{-\alpha t}$$

we see that for t near 0, the factor $1 - e^{-\alpha t} \approx 0$, so that again the solution is approximately $Q_0 e^{\alpha t}$. Finally, let us see what happens as $t \to \infty$:

$$\lim_{t \to \infty} Q(t) = \frac{\alpha Q_0}{\beta Q_0 + 0} = \frac{\alpha}{\beta}$$

Here, the result is strikingly different from that of Malthus. There is a limit to the size of the population, whereas in the Malthus model the growth is unlimited.

The Verhulst differential equation is sometimes referred to as a **logistic equation**, and its solution as given in Equation 17.20 is called the **Law of Logistic Growth**. The graph of Q is shown in Figure 17.13. ∎

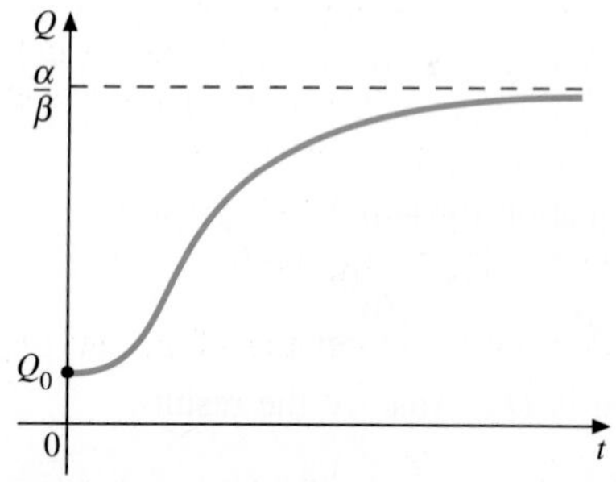

FIGURE 17.13
The logistic curve
$$Q(t) = \frac{\alpha Q_0}{\beta Q_0 + (\alpha - \beta Q_0)e^{-\alpha t}}$$

REMARK

■ If we put $m = \alpha/\beta$, the Verhulst equation can be written in the form

$$\frac{dQ}{dt} = \beta Q(m - Q)$$

and since m is the limiting maximum size of the population, this equation can be interpreted as saying that the rate of change of the population at any time is jointly proportional to the population at that time and the remaining capacity to expand. Equation 17.20 can then be written as

$$Q(t) = \frac{mQ_0}{Q_0 + (m - Q_0)e^{-m\beta t}}$$

(Compare this with Exercise 24 in Exercise Set 7.5.)

Mixing Problems

The third example is typical of *mixing problems*.

EXAMPLE 17.19   A fish tank with a capacity of 12,000 L has 6,000 L of pure water in it initially, and holds tilapia, which can live in both fresh and salt water. To convert the tank to salt water, brine containing $\frac{1}{4}$ kg of salt per liter is then fed into the tank at the rate of 40 L/min, and the well-stirred mixture is allowed to drain out at the rate of 20 L/min. Find how much salt is in the tank just as it begins to overflow.

A fish tank containing tilapia

Solution   Let $Q(t)$ be the number of kilograms of salt in the tank t minutes after the brine begins to enter. Because initially there was fresh water in the tank, we see that $Q(0) = 0$. The basic idea of this and other similar mixture problems is that the rate of change of Q is the rate at which salt (or some other substance) enters the tank minus the rate at which it leaves:

$$\frac{dQ}{dt} = (\text{Rate Salt Comes In}) - (\text{Rate Salt Goes Out})$$

Since 40 L of brine enters the tank each minute, and each liter contains $\frac{1}{4}$ kg of salt, it follows that salt enters the tank at the rate of 10 kg/min. To determine how much salt leaves the tank each minute, it is necessary to find the *concentration* (number of kilograms per liter) of salt in the tank at any given time. The net total increase in liquid in the tank is 20 L/min (40 L comes in and 20 goes out). So after t minutes it contains $6,000 + 20t$ liters of solution (see Figure 17.14). Contained within this solution are $Q(t)$ kg of salt. So we have at time t

$$\text{Concentration of Salt in Tank} = \frac{Q(t)}{6,000 + 20t} \text{ kg/L}$$

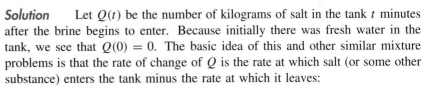

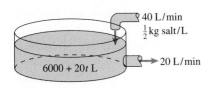

FIGURE 17.14

Since 20 L of solution leaves the tank each minute, the amount of salt leaving is

$$(20 \text{ L/min}) \left(\frac{Q}{6,000 + 20t} \text{ kg/L} \right) = \frac{Q}{300 + t} \text{ kg/min}$$

The differential equation is

$$\frac{dQ}{dt} = 10 - \frac{Q}{300 + t} \qquad \text{or} \qquad \frac{dQ}{dt} + \left(\frac{1}{300 + t} \right) Q = 10$$

This equation is linear in Q with integrating factor

$$\mu(t) = e^{\int dt/(300+t)} = e^{\ln(300+t)} = 300 + t$$

Multiplying the equation by this integrating factor, we get

$$\frac{d}{dt}[(300 + t)Q] = 10(300 + t)$$

$$Q(t) = 5(300 + t) + \frac{C}{300 + t} \qquad \text{Before the tank overflows.}$$

Since the initial amount of salt in the tank is $Q(0) = 0$, we find that $C = -5(300)^2$. The tank will overflow when $6{,}000 + 20t = 12{,}000$, or $t = 300$. The amount of salt in the tank at that instant is

$$Q(300) = 5(600) - \frac{5(300)^2}{2(300)}$$

$$= 3{,}000 - 750 = 2{,}250 \text{ kg} \qquad \blacksquare$$

Falling Body Subject to Air Resistance

In the next example we consider a falling body problem similar to ones we studied in Chapter 4 (see Example 4.26) but with the important difference that we now take air resistance into account.

EXAMPLE 17.20   From a height of s_0 meters above the ground an object of mass m is given an upward initial velocity of v_0 meters per second. Assume that air resistance has magnitude proportional to the speed. Find the height above the ground and the velocity of the object at time t. If the object could continue indefinitely without hitting the ground, what would be its limiting velocity?

Solution   As shown in Figure 17.15, let the distance $s(t)$ be measured positively upward from the ground, and let the velocity be $v(t)$. Then $s(0) = s_0$ and $v(0) = v_0$. The forces that act on the body are the pull of gravity, with magnitude mg, acting downward, and air resistance acting opposite to the direction of motion. (See Figure 17.15.)

Let $k > 0$ be the proportionality constant for the air resistance. When the object is moving upward, $v > 0$, and the resisting force is downward and so equals $-kv$. When the object is moving downward, $v < 0$, and the resisting force is upward, but again $-kv$ is the correct value, since this product is positive when v is negative. Thus, in all cases the resisting force is $-kv$.

Now we use Newton's Second Law of Motion, which for bodies of constant mass can be written as $m(dv/dt) = F$, where F is the magnitude of the net force acting on the body. In this case, then, we have

$$m\frac{dv}{dt} = -kv - mg$$

This differential equation can be written in the standard linear form

$$\frac{dv}{dt} + \frac{k}{m}v = -g$$

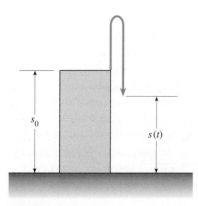

FIGURE 17.15

An integrating factor is $\mu(t) = e^{kt/m}$, and if we multiply both sides of this equation by $\mu(t)$ and simplify, we get

$$\frac{d}{dt}[ve^{kt/m}] = -ge^{kt/m}$$

Integrating both sides, we obtain

$$ve^{kt/m} = -\frac{gm}{k}e^{kt/m} + C_1$$

Setting $t = 0$ and $v = v_0$ gives

$$C_1 = v_0 + \frac{gm}{k}$$

Thus,

$$v(t) = \left(v_0 + \frac{gm}{k}\right)e^{-kt/m} - \frac{gm}{k}$$

To determine the limiting velocity (also called the *terminal* velocity), if the object could continue indefinitely, we let $t \to \infty$ to obtain

$$\lim_{t \to \infty} v(t) = -\frac{gm}{k}$$

Observe that this value is independent of the initial velocity.

To find $s(t)$, we integrate $v(t)$:

$$s(t) = -\frac{m}{k}\left(v_0 + \frac{gm}{k}\right)e^{-kt/m} - \frac{gmt}{k} + C_2$$

Since $s(0) = s_0$, we find that

$$C_2 = s_0 + \frac{m}{k}\left(v_0 + \frac{gm}{k}\right)$$

and $s(t)$ can be written as

$$s(t) = s_0 + \frac{m}{k}\left(v_0 + \frac{gm}{k}\right)(1 - e^{-kt/m}) - \frac{gmt}{k}$$ ∎

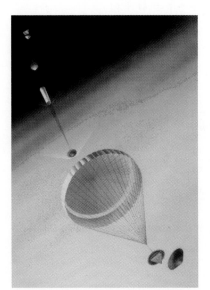

NASA's Pathfinder mission to Mars uses a parachute to slow the spacecraft during its descent through the Martian atmosphere.

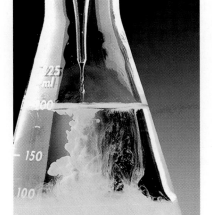

Chemical Reactions

In certain types of chemical reactions, called *second-order reactions*, two substances, say A and B, react with each other to form a third substance C. This reaction is written symbolically as $A + B \to C$. Suppose that initially the concentration of substance A is a (usually given in moles per liter) and of substance B is b. If, after t seconds, x moles per liter of A and of B have decomposed, the concentrations of what is left of A and of B are $a - x$ and $b - x$, respectively, and the concentration of C is x. For a second-order reaction, the rate of change of x is jointly proportional to $a - x$ and $b - x$:

$$\frac{dx}{dt} = k(a - x)(b - x)$$

In the next example, we solve this equation.

EXAMPLE 17.21   For the second-order chemical reaction described above, find the concentration $x(t)$ of the substance C after time t.

Solution   The initial-value problem to be solved is

$$\frac{dx}{dt} = k(a - x)(b - x), \qquad x(0) = 0$$

We separate variables and make use of partial fractions to perform the integration:

$$\frac{dx}{(a - x)(b - x)} = k\,dt$$

$$\frac{1}{a - b} \int \left[\frac{1}{b - x} - \frac{1}{a - x} \right] dx = kt + C$$

$$\frac{1}{a - b} \ln \left(\frac{a - x}{b - x} \right) = kt + C$$

When $t = 0$, $x = 0$. So

$$C = \frac{1}{a - b} \ln \frac{a}{b}$$

and we obtain

$$\frac{1}{a - b} \left[\ln \frac{a - x}{b - x} - \ln \frac{a}{b} \right] = kt$$

$$\ln \frac{b(a - x)}{a(b - x)} = (a - b)kt$$

$$\frac{b(a - x)}{a(b - x)} = e^{(a-b)kt}$$

Now we solve for x. You should verify that the answer can be put in the form

$$x(t) = \frac{ab(e^{akt} - e^{bkt})}{ae^{akt} - be^{bkt}} \qquad \blacksquare$$

Electrical Circuits

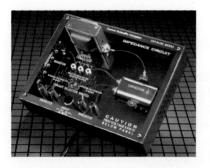

In the theory of the flow of electricity through an electrical circuit, it is known that the algebraic sum of all voltage drops is 0. That is, a voltage must be supplied by a battery or a generator to compensate for voltage losses elsewhere in the circuit. This result is known as **Kirchhoff's Second Law**. We will apply this to a circuit of the type shown in Figure 17.16, known as an **RL series circuit** since it involves resistance (R) and inductance (L). The customary units used are $R = $ ohms, $L = $ henrys, $I(t) = $ amperes, and $E(t) = $ volts. It can be shown that the voltage drops are RI across the resistor and $L(dI/dt)$ across

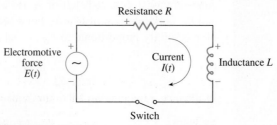

Resistance R

Electromotive force $E(t)$

Current $I(t)$

Inductance L

Switch

FIGURE 17.16

the inductor. The electromotive force (for example, a generator) provides the only voltage increase. Thus, by Kirchhoff's Second Law, we have

$$L\frac{dI}{dt} + RI = E(t) \tag{17.21}$$

We make use of this equation in the next example.

EXAMPLE 17.22 Find the current $I = I(t)$ in an RL circuit in which the electromotive force $E(t) = V$, a constant. Assume $I(0) = 0$. Draw the graph of I.

Solution From Equation 17.21, we have

$$L\frac{dI}{dt} + RI = V$$

In standard form, this linear equation is

$$\frac{dI}{dt} + \frac{R}{L}I = \frac{V}{L}$$

and an integrating factor is $e^{(R/L)t}$. So we have

$$\frac{d}{dt}\left[Ie^{(R/L)t}\right] = \frac{V}{L}e^{(R/L)t}$$

$$Ie^{(R/L)t} = \frac{V}{R}e^{(R/L)t} + C$$

From the initial condition $I(0) = 0$, we get $C = -V/R$. Thus,

$$I(t) = \frac{V}{R}\left(1 - e^{-(R/L)t}\right)$$

The graph is shown in Figure 17.17. ∎

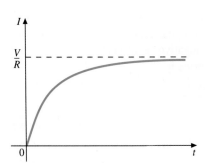

FIGURE 17.17

Graph of $I(t) = \dfrac{V}{R}\left(1 - e^{-(R/L)t}\right)$

Orthogonal Trajectories

For our final example, we give a geometric application. Suppose two one-parameter families of curves $F(x, y) = C_1$ and $G(x, y) = C_2$ have the property that whenever a curve of one family intersects a curve of the second family, it does so orthogonally (that is, the tangent lines are perpendicular). Then we say that the families are **orthogonal trajectories** of each other (see Exercise 30 in Exercise Set 17.1 and Exercise 26 in Exercise Set 17.2). For example, the family of circles $x^2 + y^2 = C_1$ and the family of lines $y = C_2x$ are easily seen to be orthogonal trajectories of each other (see Figure 17.18).

If the differential equation of one family is

$$\frac{dy}{dx} = \frac{f(x, y)}{g(x, y)}$$

then, since slopes of orthogonal curves are negative reciprocals of one another, the differential equation of the family of orthogonal trajectories is

$$\frac{dy}{dx} = -\frac{g(x, y)}{f(x, y)}$$

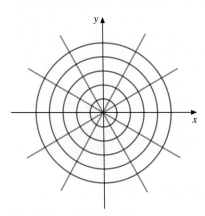

FIGURE 17.18

Orthogonal trajectories of circles $x^2 + y^2 = C_1$ are the lines $y = C_2x$.

This relation is the key to finding orthogonal trajectories of a given family, as we show in the example that follows.

EXAMPLE 17.23   Find the family of orthogonal trajectories of the family of curves $3x^2 + y^2 = Cx$.

Solution   To find the differential equation that has the given family as its general solution, we proceed, as in Example 17.4(a), by first differentiating both sides of the given equation:

$$6x + 2yy' = C$$

Now we substitute $C = 6x + 2yy'$ into the original equation:

$$3x^2 + y^2 = 6x^2 + 2xyy'$$

So, if $x \neq 0$ and $y \neq 0$,

$$y' = \frac{y^2 - 3x^2}{2xy}$$

Since at each point of intersection of a curve of the original family with any curve of the family of orthogonal trajectories, the tangents to the two curves are perpendicular, so that their slopes are negative reciprocals, the differential equation satisfied by the orthogonal trajectories is

$$y' = \frac{2xy}{3x^2 - y^2}$$

To solve this differential equation, we observe that it is homogeneous, so we substitute $y = vx$:

$$v + x\frac{dv}{dx} = \frac{2x^2v}{3x^2 - v^2x^2}$$

$$x\frac{dv}{dx} = \frac{2v}{3 - v^2} - v$$

$$x\frac{dv}{dx} = \frac{v^3 - v}{3 - v^2}$$

We now separate variables and make use of partial fractions to perform the integration:

$$\int \left[\frac{(3 - v^2)}{v(v + 1)(v - 1)}\right] dv = \int \frac{dx}{x}$$

$$\int \left(-\frac{3}{v} + \frac{1}{v + 1} + \frac{1}{v - 1}\right) dv = \int \frac{dx}{x}$$

$$-3\ln|v| + \ln|v + 1| + \ln|v - 1| = \ln|x| + \ln|C_1|$$

Write the constant of integration as $\ln|C_1|$.

$$\ln\left|\frac{v^2 - 1}{v^3}\right| = \ln|C_1 x|$$

Thus,

$$\frac{v^2 - 1}{v^3} = C_1 x$$

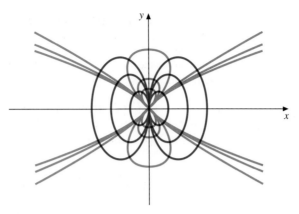

FIGURE 17.19

On replacing v by y/x and simplifying, we get

$$x^2 = y^2(1 - C_1 y)$$

as the equation of the orthogonal trajectories. Figure 17.19 shows several curves of the two families. ∎

Exercise Set 17.4

1. Assume that the rate of growth of a culture of bacteria is proportional to the number present. If a culture of 100 bacteria grows to a size of 150 after 2 hr, how long will it take for the size to double? How many will be present after 5 hr?

2. When interest on an investment is compounded continuously, the rate of growth of the amount of money in the account is proportional to the amount present, where the proportionality constant is the annual interest rate (expressed as a decimal). If an initial amount of P dollars is invested at $r\%$ compounded continuously, find a formula for the amount in the account at time t yr after it is invested. What value of r would cause the amount to double after 7 yr?

3. Use the result of Exercise 2 to determine the amount after 30 yr that results from an initial amount of $1000 invested at 6% compounded continuously. How many years will it take for the amount in the account to be $5000?

4. If 10% of a radioactive substance decays after 33 yr, how much will remain after 200 yr? What is the half-life of the substance?

5. Radium-226 has a half-life of 1620 yr. How long will it take for 80% of a given amount to decay?

6. If 75% of a quantity of the radioactive isotope uranium-232 remains after 30 yr, how much will be present after 100 yr? What is its half-life?

7. The half-life of thorium-228 is approximately 1.913 yr. If 100 g are on hand, how much will remain after 3 yr? How long will it take for the amount left to be 10 g?

8. Newton's Law of Cooling states that the surface temperature of an object changes at a rate proportional to the difference between the temperature of the object and that of the surrounding medium. Let $T(t)$ be the temperature of the object at time t, and let T_m be the temperature of the surrounding medium. If $T(0) = T_0$ (where $T_0 > T_m$), find the formula for $T(t)$.

9. Use the result of Exercise 8 to find the temperature of a body that was initially at 30°C, 30 min after it is placed in a medium of constant temperature 5°C if it cools to 20°C after 5 min. According to the model, will the object ever cool to 5°C? What does your answer tell about the model?

10. A thermometer registering 70°F is taken outside where the temperature is 28°F. After 5 min the thermometer registers 55°F. When will it register 30°F? (See Exercise 8.)

11. A vat contains 200 L of a 20% dye solution. A 40% solution of the same dye is then fed into the tank at the rate of 10 L/min, and the well-mixed solution is drained off at the same rate. Find an expression for the amount of pure dye in the tank at any time t. What is the concentration of dye in the tank after 30 min?

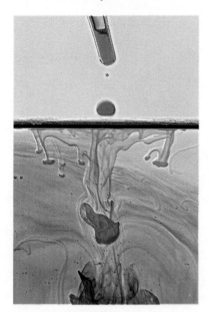

12. A tank with a 500-L capacity initially contains 100 L of brine with a salt concentration of 0.2 kg/L. Fresh water is allowed to enter the tank at the rate of 3 L/min, and the well-stirred mixture is drained from the tank at the rate of 1 L/min. How much salt does the tank contain after 20 min? How long will it take for the concentration to be reduced to 0.1 kg/L? Will this happen before the tank overflows? What is the salt concentration just as the tank overflows?

13. Solve the Verhulst model (Example 17.18) by separating variables.

14. Supply all the missing details in the solution of the Verhulst model in Example 17.18. Show that for $Q_0 < \alpha/\beta$, $Q(t)$ is an increasing function, and for $Q_0 > \alpha/\beta$, it is decreasing.

15. A body of mass m is dropped from a balloon high above the earth. Assume air resistance is proportional to the square of the velocity. Use Newton's Second Law of Motion to find a formula for the velocity at time t. What is the theoretical limiting velocity as $t \to \infty$? (Take the downward direction as positive.)

16. In Example 17.21, if the concentrations a and b are the same, then the solution given is not valid. (Why?) Solve the equation under this assumption, where $x(0) = 0$.

17. A 12-V battery is connected to an RL series circuit with a 6-ohm resistance and an inductance of 1 henry. If $I(0) = 0$, find $I(t)$.

18. Solve Equation 17.21 if $E(t) = E_0 \cos \omega t$ and $I(0) = I_0$.

19. A reasonably accurate model for the rate of dissemination of a drug injected into the bloodstream is given by

$$\frac{dy}{dt} = a - by, \qquad a > 0, \ b > 0$$

where $y = y(t)$ is the concentration of the drug at time t. Find $y(t)$ if the initial concentration of the drug is y_0.

20. Find the orthogonal trajectories of the family $x^2 + y^2 = Cx$. Sketch several members of each family.

21. Third-order chemical reactions are rare in the gaseous state, and those that do occur are almost always of the form $2A + B \to C$. For example, $2NO + O_2 \to 2NO_2$, a reaction that occurs in combustion of fossil fuels. If x is the concentration of C at time t, and a and b are the initial concentrations of A and B, respectively, then the rate of change of x is given by

$$\frac{dx}{dt} = k(a - 2x)^2(b - x)$$

Solve this equation with the initial condition $x(0) = 0$.

22. One model for the spread of an infectious disease in a community of N individuals is that the rate of change of the number $x(t)$ infected is jointly proportional to the number infected and the number uninfected. Set up and solve the relevant differential equation. Suppose $x(0) = 1$. Determine the limiting value of $x(t)$ as $t \to \infty$. (Observe that this model is mathematically the same as the Verhulst population growth model.)

23. The air in a room with dimensions $12 \times 20 \times 8$ ft initially contains 1% carbon dioxide. Air that contains 0.02% carbon dioxide is then forced into the room at the rate of 200 cu ft/min, and the well-circulated air leaves the room at the same rate. What will be the concentration of carbon dioxide after 5 min? After how long a time will the concentration be reduced to 0.05%?

24. A problem that is important in ecology has to do with the rise and decline of two species, where one species is a predator and the other is its prey. This problem is called the **predator–prey problem**. The following model is known as the **Lotka–Volterra model**, proposed by the American mathematician and biologist A. J. Lotka (1880–1949) and the Italian mathematician Vito Volterra (1860–1940) in the study of the interaction between the Alaskan snowshoe hare and the lynx:

$$\begin{cases} \dfrac{dx}{dt} = x(a - by) \\[2mm] \dfrac{dy}{dt} = y(-c + dx) \end{cases}$$

where a, b, c, and d are positive constants. Here, $x(t)$ is the population of the predator at time t and $y(t)$ is the population of the prey. By dividing the second equation by the first, find y as a function of x.

In Exercises 25–27, different models for population growth are given by specifying $f(Q)$ in Equation 17.19. Solve each model under the assumption that $Q(0) = Q_0$. The constants α and β are positive unless otherwise indicated.

25. $f(Q) = \alpha Q \cos \beta t$. This model is useful in describing seasonal growth, in which the population periodically increases and decreases. Sketch the graph of the solution.

26. $f(Q) = \alpha Q - \beta Q \ln Q$. This equation, called the **Gompertz model**, has applications in economic theory as well as population growth. For $\alpha > \beta \ln Q_0$, analyze and draw the graph of Q under each of the following circumstances:

(a) $\alpha > 0, \beta > 0$ (b) $\alpha > 0, \beta < 0$

27. $f(Q) = (Q - m)(\alpha - \beta Q)$. This equation is an appropriate model for the growth of a population that becomes extinct when its numbers are too small. Show that when $Q_0 < m$, $Q(t_1) = 0$ for some t_1. Find t_1.

17.5 SECOND-ORDER LINEAR DIFFERENTIAL EQUATIONS WITH CONSTANT COEFFICIENTS: THE HOMOGENEOUS CASE

We have seen a variety of applications of first-order differential equations. Second-order equations also are important in applications because they can serve as models for physical problems involving second derivatives, such as those dealing with acceleration. Oscillatory motion, such as that of a vibrating spring, can be modeled by a second-order differential equation, as we will show in Section 17.7. Many applications involve only second-order linear equations, and we will concentrate on these.

From Definition 17.5, we know that a second-order linear differential equation is of the form

$$a_2(x)y'' + a_1(x)y' + a_0(x)y = g(x)$$

with $a_2(x) \neq 0$. We limit our consideration to the case where the coefficient functions are constant, say $a_2(x) = a$, $a_1(x) = b$, and $a_0(x) = c$. When $g(x) = 0$ for all x on some interval, the equation is said to be **homogeneous**. Here we are using *homogeneous* in a different sense from that in Section 17.2. We will study the homogeneous case in this section and the nonhomogeneous case in the next section.

The homogeneous equation with constant coefficients has the form

$$ay'' + by' + cy = 0 \quad (a \neq 0) \tag{17.22}$$

Our main objective in this section is to find the general solution of the differential equation given by Equation 17.22. It will simplify our notation if we introduce the symbol $L[y]$ for the left-hand side of Equation 17.22. That is,

$$L[y] = ay'' + by' + cy$$

Then Equation 17.22 is equivalent to $L[y] = 0$. We call L a **linear differential operator**. The word *linear* is used here to mean that

$$L[ky] = kL[y] \quad \text{for any constant } k \tag{17.23}$$

and

$$L[y_1 + y_2] = L[y_1] + L[y_2] \tag{17.24}$$

We will ask you to prove these two properties in Exercise 37 of Exercise Set 17.5. (These properties are similar to properties of the linear *function* $f(x) = mx + b$.)

Suppose we have found two solutions, say $y_1 = y_1(x)$ and $y_2 = y_2(x)$, of Equation 17.22. Then $L[y_1] = 0$ and $L[y_2] = 0$. We can now show that any **linear combination**

$$C_1 y_1 + C_2 y_2$$

(C_1 and C_2 are constants) is also a solution. To see why, note that by Equation 17.23,

$$L[C_1 y_1] = C_1 L[y_1] = 0 \quad \text{and} \quad L[C_2 y_2] = C_2 L[y_2] = 0$$

Then, by Equation 17.24,

$$L[C_1 y_1 + C_2 y_2] = L[C_1 y_1] + L[C_2 y_2] = 0$$

Clearly, if C_1 and C_2 are both 0, then $C_1 y_1 + C_2 y_2 = 0$. If the *only* way that the linear combination $C_1 y_1 + C_2 y_2$ can be identically 0 (0 for all x) is for both C_1 and C_2 to be 0, then the functions y_1 and y_2 are said to be **linearly independent**. Otherwise, they are **linearly dependent**. For example, $y_1 = e^x$ and $y_2 = e^{-2x}$ are linearly independent, since no linear combination $C_1 e^x + C_2 e^{-2x}$ can be identically 0 except when both C_1 and C_2 are 0. On the other hand, $y_1 = 2x - 1$ and $y_2 = \frac{1}{2} - x$ are linearly dependent, since with $C_1 = 1$ and $C_2 = 2$,

$$C_1 y_1 + C_2 y_2 = 1(2x - 1) + 2\left(\frac{1}{2} - x\right) = 0$$

The importance of linear independence is that when y_1 and y_2 are linearly independent solutions of Equation 17.22, then $y = C_1 y_1 + C_2 y_2$ is the *general* solution of Equation 17.22. We leave a proof of this result to a course in differential equations.

REMARK ——

■ The definition we have given for the linear independence of two functions extends in a natural way to more than two. In the particular case of two functions, however, it is readily seen (see Exercise 38 of this section) that they are linearly independent if and only if *neither function is a multiple of the other*.

——

The Auxiliary Equation

With this background, we can obtain the general solution of Equation 17.22 if we can find two linearly independent solutions. For the remainder of this section we will concentrate on how to find two such solutions. To find a solution to Equation 17.22, we need to find a function y so that a constant times its second derivative y'', plus a constant times its first derivative y', plus a constant times the function y itself, add to give 0. Since an exponential function of the form $y = e^{mx}$ has the property that each successive derivative is a multiple of the function itself, this function seems as though it would be a natural candidate for a solution. So let us try $y = e^{mx}$. We have $y' = me^{mx}$ and $y'' = m^2 e^{mx}$, so that

$$L[y] = ay'' + by' + cy = am^2 e^{mx} + bm e^{mx} + c e^{mx}$$
$$= (am^2 + bm + c)e^{mx}$$

Since e^{mx} is always positive, it follows that $L[y]$ will be 0 if and only if

$$am^2 + bm + c = 0 \qquad\qquad (17.25)$$

Equation 17.25 is called the **auxiliary equation** for Equation 17.22. (It is also referred to as the **characteristic equation**.) Let m_1 and m_2 denote the roots of Equation 17.25. Then by the Quadratic Formula, we have

$$m_1 = \frac{-b + \sqrt{b^2 - 4ac}}{2a} \quad \text{and} \quad m_2 = \frac{-b - \sqrt{b^2 - 4ac}}{2a}$$

We distinguish three cases, according to the nature of the discriminant $b^2 - 4ac$.

Case 1: $b^2 - 4ac > 0$

In this case m_1 and m_2 are distinct real roots. Since $e^{m_1 x}$ and $e^{m_2 x}$ are linearly independent, the general solution is $y = C_1 e^{m_1 x} + C_2 e^{m_2 x}$.

Distinct Real Roots

If the auxiliary equation $am^2 + bm + c = 0$ has two distinct real roots m_1 and m_2, then the general solution of the equation

$$ay'' + by' + cy = 0$$

is

$$y = C_1 e^{m_1 x} + C_2 e^{m_2 x}$$

EXAMPLE 17.24   Find the general solution of the equation $y'' - 3y' - 4y = 0$.

Solution   The auxiliary equation is $m^2 - 3m - 4 = 0$. (Notice how the pattern of the auxiliary equation resembles the pattern of the differential equation.) By factoring, we have

$$(m - 4)(m + 1) = 0$$

So the two solutions are $m_1 = 4$ and $m_2 = -1$. These roots are real and distinct, so the general solution is

$$y = C_1 e^{4x} + C_2 e^{-x}$$ ∎

Case 2: $b^2 - 4ac = 0$

In this case, we have

$$m_1 = \frac{-b + \sqrt{0}}{2a} \quad \text{and} \quad m_2 = \frac{-b - \sqrt{0}}{2a}$$

That is, $m_1 = m_2 = -b/2a$. So there is only one solution of the form e^{mx}, namely,

$$y_1 = e^{-bx/2a}$$

By direct substitution into Equation 17.22, it can be shown that a second solution, independent from y_1, is

$$y_2 = xe^{-bx/2a}$$

(See Exercise 39 in Exercise Set 17.5.) Thus, writing $m = -b/2a$, the general solution is

$$y = C_1 e^{mx} + C_2 x e^{mx}$$

Equal Real Roots

If the roots of the auxiliary equation $am^2 + bm + c = 0$ are equal, with the common value $m = -b/2a$, then the general solution of the equation

$$ay'' + by' + cy = 0$$

is

$$y = C_1 e^{mx} + C_2 x e^{mx}$$

EXAMPLE 17.25   Find the general solution of the equation $y'' - 4y' + 4y = 0$.

Solution   The auxiliary equation is $m^2 - 4m - 4 = 0$, or equivalently $(m - 2)^2 = 0$, which has the double root $m = 2$. So the general solution is

$$y = C_1 e^{2x} + C_2 x e^{2x}$$ ■

Case 3: $b^2 - 4ac < 0$

In this case, the roots m_1 and m_2 of the auxiliary equation are the *complex conjugates*

$$m_1 = \alpha + i\beta \qquad \text{and} \qquad m_2 = \alpha - i\beta$$

where

$$\alpha = -\frac{b}{2a} \qquad \text{and} \qquad \beta = \frac{\sqrt{4ac - b^2}}{2a}$$

The complex number i satisfies $i^2 = -1$. Although the general solution can be written as

$$y = C_1 e^{(\alpha + i\beta)x} + C_2 e^{(\alpha - i\beta)x}$$

it is possible to rewrite the general solution in a form involving functions of real variables only, as we now show.

We will make use of the following formula, known as **Euler's Formula**:

$$e^{i\theta} = \cos\theta + i\sin\theta \tag{17.26}$$

To see why this formula is true, we begin by defining $e^{i\theta}$ as the result of replacing x with $i\theta$ in the Maclaurin series for e^x. (See Section 10.8.) When we do so and use $i^2 = -1$, so that $i^3 = -i$, $i^4 = 1$, and so on, we obtain

$$e^{i\theta} = 1 + (i\theta) + \frac{(i\theta)^2}{2!} + \frac{(i\theta)^3}{3!} + \frac{(i\theta)^4}{4!} + \frac{(i\theta)^5}{5!} + \cdots$$

$$= \left[1 - \frac{\theta^2}{2!} + \frac{\theta^4}{4!} - \cdots\right] + i\left[\theta - \frac{\theta^3}{3!} + \frac{\theta^5}{5!} - \cdots\right]$$

where we have rearranged the terms into the so-called real and imaginary parts. But from Section 10.8 we recognize that the bracketed series are the Maclaurin series for $\cos\theta$ and $\sin\theta$, respectively. Thus, $e^{i\theta} = \cos\theta + i\sin\theta$.

REMARK ————————————————————————————

■ If we set $\theta = \pi$ in Euler's Formula, we get $e^{i\pi} = -1$, or equivalently

$$e^{i\pi} + 1 = 0 \tag{17.27}$$

Equation 17.27 is one of the most remarkable equations in all of mathematics. Here, in one equation, appear the five special constants $0, 1, \pi, e,$ and i.

Returning now to the solution of Equation 17.22, we write

$$y = C_1 e^{(\alpha+i\beta)x} + C_2 e^{(\alpha-i\beta)x}$$
$$= e^{\alpha x}[C_1 e^{i\beta x} + C_2 e^{-i\beta x}]$$
$$= e^{\alpha x}[C_1(\cos \beta x + i \sin \beta x) + C_2(\cos \beta x - i \sin \beta x)]$$
$$= e^{\alpha x}[(C_1 + C_2)\cos \beta x + i(C_1 - C_2)\sin \beta x]$$
$$= e^{\alpha x}[C_3 \cos \beta x + C_4 \sin \beta x]$$

where C_3 and C_4 are new arbitrary constants, namely, $C_3 = C_1 + C_2$ and $C_4 = i(C_1 - C_2)$. Note that in applying Euler's Formula to $e^{-i\beta x}$, we used $\cos(-\beta x) = \cos \beta x$ and $\sin(-\beta x) = -\sin \beta x$. When the roots of the auxiliary equation are imaginary, we will use the result just obtained to write the general solution; that is, if the roots are $\alpha + i\beta$ and $\alpha - i\beta$, then we will write the general solution as

$$y = e^{\alpha x}(C_1 \cos \beta x + C_2 \sin \beta x) \tag{17.28}$$

(There is no longer any need to call the constants C_3 and C_4.)

Imaginary Roots

If the roots of the auxiliary equation $am^2 + bm + c = 0$ are the complex conjugates $m_1 = \alpha + i\beta$ and $m_2 = \alpha - i\beta$, then the general solution of the equation

$$ay'' + by' + cy = 0$$

is

$$y = e^{\alpha x}(C_1 \cos \beta x + C_2 \sin \beta x)$$

EXAMPLE 17.26   Find the general solution of each of the following:

(a) $y'' + 4y = 0$ (b) $y'' - 2y' + 3y = 0$

Solution

(a) The auxiliary equation is $m^2 + 4 = 0$, with roots $m_1 = 2i$ and $m_2 = -2i$. So we have $\alpha = 0$ and $\beta = 2$. Thus, since $e^{\alpha x} = e^0 = 1$, the general solution given by Equation 17.28 is

$$y = C_1 \cos 2x + C_2 \sin 2x$$

(You can check to see that this value of y does satisfy the differential equation.)

(b) The auxiliary equation is $m^2 - 2m + 3 = 0$, with the solutions

$$m = \frac{2 \pm \sqrt{4 - 12}}{2} = 1 \pm i\sqrt{2}$$

Thus, $\alpha = 1$ and $\beta = \sqrt{2}$. So, by Equation 17.28, we can write the general solution in the form

$$y = e^x(C_1 \cos \sqrt{2}x + C_2 \sin \sqrt{2}x) \qquad \blacksquare$$

Initial-Value and Boundary-Value Problems

The next two examples illustrate an initial-value problem and a boundary-value problem that involve second-order equations.

EXAMPLE 17.27   Solve the initial-value problem

$$y'' - 2y' - 8y = 0; \qquad y(0) = 1, \ y'(0) = -4$$

Solution   The auxiliary equation is

$$m^2 - 2m - 8 = 0$$
$$(m - 4)(m + 2) = 0$$
$$m = 4, \ -2$$

Thus, the general solution is

$$y = C_1 e^{4x} + C_2 e^{-2x}$$

From $y(0) = 1$, we get $1 = C_1 + C_2$. Now we calculate y' and then apply the second initial condition:

$$y' = 4C_1 e^{4x} - 2C_2 e^{-2x}$$
$$-4 = 4C_1 - 2C_2 \qquad \text{or} \qquad 2C_1 - C_2 = -2$$

To find C_1 and C_2, we solve the system

$$\begin{cases} C_1 + C_2 = 1 \\ 2C_1 - C_2 = -2 \end{cases}$$

simultaneously. The solution is found to be $C_1 = -\frac{1}{3}$, $C_2 = \frac{4}{3}$. Thus, the desired particular solution is

$$y = -\frac{1}{3} e^{4x} + \frac{4}{3} e^{-2x} \qquad \blacksquare$$

EXAMPLE 17.28   Find the solution of the equation $x''(t) + x(t) = 0$ that satisfies the boundary conditions $x(0) = 3$ and $x(\pi/2) = 5$.

Solution   The auxiliary equation $m^2 + 1 = 0$ has roots $\pm i$, and so by Equation 17.28 the general solution is

$$x(t) = C_1 \cos t + C_2 \sin t$$

Since $x(0) = C_1$ and $x(\pi/2) = C_2$, we see immediately that $C_1 = 3$ and $C_2 = 5$. Thus, the particular solution in question is

$$x(t) = 3 \cos t + 5 \sin t \qquad \blacksquare$$

We summarize below the results of this section.

The General Solution of a Second-Order Linear Homogeneous Differential Equation with Constant Coefficients

To find the general solution of

$$ay'' + by' + cy = 0$$

first solve the auxiliary equation

$$am^2 + bm + c = 0$$

Denote the roots by m_1 and m_2.

Case 1. If m_1 and m_2 are real and unequal, write the general solution as

$$y = C_1 e^{m_1 x} + C_2 e^{m_2 x}$$

Case 2. If m_1 and m_2 are real and equal, with common value m, write the general solution as

$$y = C_1 e^{mx} + C_2 x e^{mx}$$

Case 3. If m_1 and m_2 are the complex conjugates $m_1 = \alpha + i\beta$ and $m_2 = \alpha - i\beta$, with $\beta \neq 0$, write the general solution as

$$y = e^{\alpha x}(C_1 \cos \beta x + C_2 \sin \beta x)$$

Exercise Set 17.5

In Exercises 1–16, find the general solution of the differential equation. Unless otherwise indicated, the independent variable is x.

1. $y'' + 3y' + 2y = 0$

2. $y'' - y' - 2y = 0$

3. $y'' + 8y' + 16y = 0$

4. $y'' - 2y' + y = 0$

5. $y'' + 9y = 0$

6. $y'' + 9y' = 0$

7. $y'' - 2y' + 5y = 0$

8. $y'' + y' + 2y = 0$

9. $2y'' - 3y' - 5y = 0$

10. $4y'' + 5y' - 6y = 0$

11. $\dfrac{d^2 y}{dt^2} - 3\dfrac{dy}{dt} - 4y = 0$

12. $2x''(t) + 6x'(t) + 5x(t) = 0$

13. $2\dfrac{d^2 u}{dx^2} - 3\dfrac{du}{dx} = 0$

14. $\dfrac{d^2 v}{dx^2} - 9v = 0$

15. $4s''(t) - 12s'(t) + 9s(t) = 0$

16. $\dfrac{d^2 s}{dt^2} + 3\dfrac{ds}{dt} + 4s = 0$

In Exercises 17–24, solve the given initial-value problem or boundary-value problem.

17. $y'' + 2y' - 15y = 0$; $y(0) = 2$, $y'(0) = 3$

18. $2y'' - 3y' - 9y = 0$; $y(0) = -2$, $y'(0) = -6$

19. $y'' - 3y' = 0$; $y(0) = 2$, $y'(0) = 9$

20. $y'' - 9y = 0$; $y(0) = 0$, $y'(0) = 1$

21. $y'' - 4y' + 4y = 0$; $y(0) = 2$, $y(1) = e^2$

22. $y'' + 4y = 0$; $y(0) = 5$, $y(\pi/4) = 3$

23. $y'' + y = 0$; $y(\pi/6) = 0, y(\pi/3) = 1$

24. $y'' - 4y' + 5y = 0$; $y(0) = \frac{2}{3}$, $y'(0) = -\frac{1}{3}$

25. Show that when the roots m_1 and m_2 of the auxiliary equation are real and unequal, the general solution of Equation 17.22 can be written in the form

$$y = e^{ux}(C_1 \cosh vx + C_2 \sinh vx)$$

where $u = -b/2a$ and $v = \sqrt{b^2 - 4ac}/2a$.

26. Find the general solution of $y'' + 3y' + y = 0$ in the form given in Exercise 25. Find the particular solution that satisfies $y(0) = 4$, $y'(0) = -3$.

27. An equation of the form

$$ax^2 y'' + bxy' + cy = 0 \qquad (a \neq 0, \; x > 0)$$

is called an **Euler Equation**. Show that the substitution $t = \ln x$ changes it into a linear equation with constant coefficients.

In Exercises 28–30, use the result of Exercise 27 to find the general solution of the differential equation.

28. $x^2 y'' + 3xy' + 3y = 0$, $\quad x > 0$

29. $\dfrac{d^2 s}{dt^2} + \dfrac{1}{t}\dfrac{ds}{dt} = 0$, $\quad t > 0$ (*Hint:* Multiply by t^2.)

30. $v\dfrac{d^2 \psi}{dv^2} - \dfrac{1}{v}\psi = 0$, $\quad v > 0$ (*Hint:* Multiply by v.)

In Exercises 31–36, assume that the ideas of this section extend to higher-order linear homogeneous equations with constant coefficients, and find the general solution.

31. $y''' - 2y'' - 3y' = 0$ **32.** $y^{(4)} - 16y = 0$

33. $y^{(4)} + 5y'' + 4y = 0$ **34.** $y''' - 3y' + 2y = 0$

35. $y''' - y'' + 4y' - 4y = 0$. Also find the particular solution that satisfies $y(0) = 0$, $y'(0) = 1$, $y''(0) = 2$.

36. $y''' - 3y'' + 3y' - y = 0$

37. Prove that $L[ky] = kL[y]$ and that $L[y_1 + y_2] = L[y_1] + L[y_2]$.

38. Prove that two functions $y_1(x)$ and $y_2(x)$ are linearly independent on an interval if and only if neither is a multiple of the other.

39. Prove that if the solutions of the auxiliary equation of $ay'' + by' + cy = 0$ are equal, then $y_2 = xe^{-bx/2a}$ is a solution of the differential equation.

17.6 SECOND-ORDER LINEAR DIFFERENTIAL EQUATIONS WITH CONSTANT COEFFICIENTS: THE NONHOMOGENEOUS CASE

In this section, we give methods for solving second-order linear differential equations of the form

$$ay'' + by' + cy = g(x) \tag{17.29}$$

where g is a continuous function on some interval I and is not identically 0 on I. We refer to $g(x)$ as the *nonhomogeneous* term. As we will show, the corresponding homogeneous equation

$$ay'' + by' + cy = 0 \tag{17.30}$$

plays an important role in obtaining the general solution of Equation 17.29. We call Equation 17.30 the **complementary equation** to Equation 17.29, and its general solution, which we denote by y_c, is called the **complementary solution**.

The next theorem shows the relationship between the general solution of the nonhomogeneous equation (Equation 17.29) and the general solution y_c of the complementary equation (Equation 17.30).

THEOREM 17.1

If y_p is any particular solution of the nonhomogeneous equation

$$ay'' + by' + cy = g(x)$$

then its general solution is

$$y = y_c + y_p \qquad (17.31)$$

where y_c is the general solution of the homogeneous complementary equation $ay'' + by' + cy = 0$.

Proof   We must show that every solution of Equation 17.29 can be put in the form of Equation 17.31 for appropriate choices of the constants that occur in the complementary solution y_c. To simplify the proof, we make use of the linear differential operator, $L[y] = ay'' + by' + c$, that we introduced in Section 17.5. With this notation, Equation 17.29 can be written as $L[y] = g(x)$ and the complementary equation (Equation 17.30) as $L[y] = 0$. Now by hypothesis, $L[y_p] = g(x)$ and $L[y_c] = 0$. Suppose $y = Y(x)$ is any other solution of Equation 17.29; that is, $L[Y] = g(x)$. Then, by the properties of L,

$$L[Y - y_p] = L[Y] - L[y_p] = g(x) - g(x) = 0$$

Thus, $Y - y_p$ is a solution of the complementary equation $L[y] = 0$, whose general solution we know is y_c. This means, then, that for appropriate choices of the constants in y_c, $Y - y_p = y_c$ or $Y = y_c + y_p$. We have therefore shown that every solution of Equation 17.29 is a member of the family $y_c + y_p$. Thus, $y_c + y_p$ is the general solution of Equation 17.29.   ∎

REMARK ──

■ The remarkable thing about this theorem is that y_p can be *any* particular solution of Equation 17.29. So, for example, if y_{p_1} and y_{p_2} are two different particular solutions, then, even though the solutions $y_c + y_{p_1}$ and $y_c + y_{p_2}$ would *look* different, they would, in fact, describe the same family.

──

EXAMPLE 17.29   Show that $y_p = x + 2$ is a particular solution of

$$y'' - 3y' + 2y = 2x + 1$$

and find the general solution.

Solution   For $y_p = x + 2$, we have $y_p' = 1$ and $y_p'' = 0$. So

$$y_p'' - 3y_p' + 2y_p = 0 - 3(1) + 2(x + 2) = 2x + 1$$

Thus, $y_p = x + 2$ is a particular solution. According to Theorem 17.1, the general solution is $y_c + y_p$, where y_c is the complementary solution; that is, y_c is the general solution of the homogeneous equation

$$y'' - 3y' + 2y = 0$$

The auxiliary equation is $m^2 - 3m + 2 = 0$ and has roots $m_1 = 1$, $m_2 = 2$.

Thus,

$$y_c = C_1 e^x + C_2 e^{2x}$$

The general solution of the nonhomogeneous equation is therefore

$$y = \underbrace{C_1 e^x + C_2 e^{2x}}_{y_c} + \underbrace{x + 2}_{y_p}$$ ■

The crucial question, clearly, is how to find y_p. Sometimes we can find a solution by inspection. For example, the equation $y'' + 2y' = 4$ has $y_p = 2x$ as a particular solution, since $y'_p = 2$ and $y''_p = 0$. Usually, however, more work is involved. We will describe two methods. The first, called the **method of undetermined coefficients**, works when the nonhomogeneous term belongs to a certain class of functions. Although the restriction to this class limits the applicability of the method, it turns out that many of the functions that occur in common applications fall in this class. The second method, called **variation of parameters**, is more general in its applicability but is often more difficult to apply.

Method of Undetermined Coefficients

Before describing the method of undetermined coefficients in general, let us consider an example.

EXAMPLE 17.30   Find a particular solution of the equation

$$y'' + 2y' - 8y = g(x)$$

where

(a) $g(x) = 5e^{3x}$ (b) $g(x) = 5e^{2x}$

Solution

(a) Since $g(x)$ is an exponential function, it is reasonable to suppose that a solution of the form

$$y_p = Ae^{3x}$$

exists, since derivatives of y_p will all be multiples of e^{3x}. The coefficient A is yet to be determined (hence the name *undetermined* coefficients for this method). Trying this solution, we have $y'_p = 3Ae^{3x}$ and $y''_p = 9Ae^{3x}$. Substituting into the original equation gives

$$9Ae^{3x} + 2(3Ae^{3x}) - 8(Ae^{3x}) = 5e^{3x}$$

$$7Ae^{3x} = 5e^{3x}$$

$$A = \frac{5}{7}$$

So our trial solution works, with $A = \frac{5}{7}$; that is, a particular solution is $y_p = \frac{5}{7}e^{3x}$.

(b) Proceeding as in part (a), we try

$$y_p = Ae^{2x}$$

Then $y_p' = 2Ae^{2x}$ and $y_p'' = 4Ae^{2x}$. So, on substitution, we get

$$4Ae^{2x} + 2(2Ae^{2x}) - 8(Ae^{2x}) = 5e^{2x}$$
$$0 = 5e^{2x}$$

Clearly, something has gone wrong, since we have arrived at an impossibility. A look at the complementary solution reveals the problem. You can verify that

$$y_c = C_1 e^{2x} + C_2 e^{-4x}$$

Taking $C_2 = 0$, we see that any function of the form $C_1 e^{2x}$ satisfies the homogeneous equation. Since our trial solution $y_p = Ae^{2x}$ is of this form, it has no chance of satisfying the nonhomogeneous equation.

In this situation we alter our initial trial solution by multiplying it by x to obtain a new trial solution. Multiplying by x is analogous to the repeated root situation for the auxiliary equation. Thus, we try

$$y_p = Axe^{2x}$$

Then

$$y_p' = Ae^{2x} + 2Axe^{2x}$$
$$y_p'' = 2Ae^{2x} + 2Ae^{2x} + 4Axe^{2x}$$
$$= 4Ae^{2x} + 4Axe^{2x}$$

Now we substitute into the original differential equation:

$$4Ae^{2x} + 4Axe^{2x} + 2(Ae^{2x} + 2Axe^{2x}) - 8Axe^{2x} = 5e^{2x}$$

After collecting terms, we get $6Ae^{2x} = 5e^{2x}$, which is true if $A = \frac{5}{6}$. Our desired particular solution is therefore

$$y_p = \frac{5}{6}xe^{2x} \qquad \blacksquare$$

The method of undetermined coefficients works when the nonhomogeneous term $g(x)$ is one of the following types:

1. an exponential function
2. a polynomial function
3. a sine or cosine function

or else is a finite product of functions of one of these three types. In each case, our initial trial for y_p is a generalized function of the same type as $g(x)$, where unknown coefficients are used and where all *derived* terms (that is, terms obtained by differentiation) are included. Here are some examples to help guide your strategy:

$g(x)$	Initial Trial for y_p
$3x^2$	$Ax^2 + Bx + C$
$5 \sin 2x$	$A \sin 2x + B \cos 2x$
$2e^{-x} \cos 3x$	$e^{-x}(A \cos 3x + B \sin 3x)$
$(3x + 4)e^{2x}$	$(Ax + B)e^{2x}$
$x^2 \cos 3x$	$(Ax^2 + Bx + C) \cos 3x + (Dx^2 + Ex + F) \sin 3x$

Finally, suppose $g(x)$ is the sum of two functions of the type described above—say, $g(x) = g_1(x) + g_2(x)$. Then we find particular solutions y_{p_1} and y_{p_2} that satisfy

$$L[y_{p_1}] = g_1(x) \qquad \text{and} \qquad L[y_{p_2}] = g_2(x)$$

and set $y_p = y_{p_1} + y_{p_2}$. Since

$$L[y_p] = L[y_{p_1} + y_{p_2}] = L[y_{p_1}] + L[y_{p_2}] = g_1(x) + g_2(x)$$

it follows that y_p is a particular solution of the original equation. This procedure can be extended in a natural way to any finite sum.

Example 17.30 suggests the following algorithm for finding a particular solution by the method of undetermined coefficients:

An Algorithm for Finding y_p by the Method of Undetermined Coefficients

1. Find y_c.
2. Determine a trial solution that has the same general form as the non-homogeneous term $g(x)$, together with terms of the form obtained from it by differentiation.
3. If the trial solution has no term in common with y_c, substitute it into the original equation to find the unknown coefficients.
4. If the trial solution does have a term in common with y_c, multiply the trial solution by x. Use this product as a new trial solution, and return to step 3.

Notice that when step 4 applies, the new trial solution may again have a term in common with y_c, in which case we have to multiply by x again. For example, in the equation

$$y'' - 4y' + 4y = 5e^{2x}$$

we would determine $y_c = C_1 e^{2x} + C_2 x e^{2x}$. (Verify.) Since both trial solutions Ae^{2x} and Axe^{2x} occur in y_c, we would use

$$y_p = Ax^2 e^{2x}$$

EXAMPLE 17.31   Solve the initial-value problem

$$y'' - 2y' = 3x + 2e^{2x} \cos x; \qquad y(0) = 0, \; y'(0) = 1$$

Solution   We find from the auxiliary equation $m^2 - 2m = 0$ that

$$y_c = C_1 + C_2 e^{2x}$$

Since $g(x)$ is the sum of two functions of the types we have described, we write $g(x) = g_1(x) + g_2(x)$, where $g_1(x) = 3x$ and $g_2(x) = 2e^{2x} \cos x$. Our initial trial solution for $L[y] = g_1(x)$ is $y_{p_1} = Ax + B$, but this function contains a constant term (B) that duplicates a term (C_1) in y_c. Thus, we modify our initial trial and use

$$y_{p_1} = Ax + Bx^2$$

Substituting this value of y_{p_1} into $L[y] = 3x$ enables us to find A and B. You should verify the results: $A = -\frac{3}{4}$, $B = -\frac{3}{4}$.

For the equation $L[y] = g_2(x)$, the initial trial is

$$y_{p_2} = e^{2x}(C \cos x + D \sin x)$$

and since this form does not duplicate any term in y_c, we use it. (Even though e^{2x} occurs, it occurs only in combination with a sine or cosine and so does not duplicate the term $C_2 e^{2x}$ that occurs in y_c.) Calculating y'_{p_2} and y''_{p_2} and substituting into $L[y] = 2e^{2x} \cos x$ give

$$e^{2x}[(-C + 2D) \cos x + (-2C - D) \sin x] = 2e^{2x} \cos x$$

(Supply the missing steps.) For this equation to be an identity (like terms on left and right have the same coefficient), we must have

$$\begin{cases} -C + 2D & = 2 \\ -2C - D & = 0 \end{cases}$$

The simultaneous solution is $C = -\frac{2}{5}$, $D = \frac{4}{5}$.

A particular solution of the original differential equation is therefore

$$y_p = y_{p_1} + y_{p_2} = -\frac{3}{4}x - \frac{3}{4}x^2 + e^{2x}\left(-\frac{2}{5} \cos x + \frac{4}{5} \sin x\right)$$

and the general solution $y_c + y_p$, is

$$y = C_1 + C_2 e^{2x} - \frac{3}{4}x - \frac{3}{4}x^2 + e^{2x}\left(-\frac{2}{5} \cos x + \frac{4}{5} \sin x\right)$$

The determination of the constants C_1 and C_2 so that the two initial conditions are satisfied is a bit messy. The result is

$$y = -\frac{19}{40} + \frac{7}{8}e^{2x} - \frac{3}{4}x - \frac{3}{4}x^2 + e^{2x}\left(-\frac{2}{5} \cos x + \frac{4}{5} \sin x\right) \qquad \blacksquare$$

Variation of Parameters

Suppose we have already solved the homogeneous equation $ay'' + by' + cy = 0$ to obtain the complementary solution

$$y_c = C_1 y_1 + C_2 y_2.$$

In the method of **variation of parameters**, we replace the constants C_1 and C_2 (the parameters) in this complementary solution by variables (hence the name *variation* of parameters). That is, we try a particular solution y_p of the form

$$y_p(x) = u_1(x) y_1(x) + u_2(x) y_2(x)$$

where u_1 and u_2 are functions yet to be determined. We will be able to find u_1 and u_2 if we can obtain two equations relating them that we can solve simultaneously. One equation results from the requirement that y_p must satisfy the original differential equation. We are free to choose a second equation however we want. We do so in a way that simplifies the derivative y_p'. First, we calculate this derivative:

$$y_p' = (u_1'y_1 + u_1y_1') + (u_2'y_2 + u_2y_2')$$
$$= (u_1y_1' + u_2y_2') + (u_1'y_1 + u_2'y_2)$$

Since we are free to impose one condition on u_1 and u_2, we require them to satisfy

$$u_1'y_1 + u_2'y_2 = 0 \qquad (17.32)$$

Then y_p' becomes

$$y_p' = u_1y_1' + u_2y_2'$$

So

$$y_p'' = \left(u_1'y_1' + u_1y_1''\right) + \left(u_2'y_2' + u_2y_2''\right)$$

Now we substitute y_p and its derivatives into the original differential equation

$$ay'' + by' + cy = g(x)$$

After rearranging terms, we get (you should verify it)

$$u_1\left(ay_1'' + by_1' + cy_1\right) + u_2\left(ay_2'' + by_2' + cy_2\right) + a\left(u_1'y_1' + u_2'y_2'\right) = g(x)$$

But y_1 and y_2 are solutions of the complementary equation, so $L[y_1]$ and $L[y_2]$ both are 0. That is, the quantities inside the first two sets of parentheses vanish, and we have

$$a(u_1'y_1' + u_2'y_2') = g(x) \qquad (17.33)$$

Equations 17.32 and 17.33 can now be solved simultaneously for u_1' and u_2', from which u_1 and u_2 can be obtained by integration. (Of course, it is not always easy or even always possible to carry out this last step.)

We summarize our procedure below.

An Algorithm for Finding y_p by the Method of Variation of Parameters

1. Find the complementary solution $y_c = C_1y_1 + C_2y_2$.
2. Solve the system of equations

$$\begin{cases} u_1'y_1 + u_2'y_2 & = 0 \\ a(u_1'y_1' + u_2'y_2') & = g(x) \end{cases}$$

   simultaneously for u_1' and u_2'.
3. Integrate to find u_1 and u_2.
4. Determine y_p from

$$y_p = u_1y_1 + u_2y_2$$

REMARK ───────────────────────────────

■ In carrying out step 3, the constants of integration can be omitted, since we want *any* two functions u_1 and u_2 that work, and we might as well choose the simplest ones.

───────────────────────────────

EXAMPLE 17.32   Solve the differential equation

$$y'' + y = \sec^3 x \qquad 0 < x < \frac{\pi}{2}$$

by the method of variation of parameters.

Solution   From the auxiliary equation $m^2 + 1 = 0$, whose solutions are $m_1 = i$ and $m_2 = -i$, we know that the complementary solution can be written in the form

$$y_c = C_1 \cos x + C_2 \sin x$$

Thus, $y_1 = \cos x$ and $y_2 = \sin x$. To carry out step 2 of our algorithm, we solve the system

$$\begin{cases} u_1' \cos x + u_2' \sin x & = 0 \\ -u_1' \sin x + u_2' \cos x & = \sec^3 x \end{cases}$$

for u_1' and u_2'. The result is (which you will be asked to verify in Exercise 19 of Exercise Set 17.6)

$$u_1' = -\tan x \sec^2 x \qquad \text{and} \qquad u_2' = \sec^2 x$$

Thus, on integration, we obtain

$$u_1 = -\frac{1}{2} \tan^2 x \qquad \text{and} \qquad u_2 = \tan x$$

A particular solution is therefore

$$y_p = \left(-\frac{1}{2} \tan^2 x \right) \cos x + \tan x \sin x$$

The general solution is $y = y_c + y_p$:

$$y = C_1 \cos x + C_2 \sin x - \frac{1}{2} \tan^2 x \cos x + \tan x \sin x \qquad \blacksquare$$

REMARK ───
■ Although the method of variation of parameters will work for a broader class of functions $g(x)$ than the method of undetermined coefficients, it is often the more difficult of the two methods to apply since it involves integration, whereas undetermined coefficients involves only differentiation. So when there is a choice, you will probably find it easier to use the method of undetermined coefficients.

Exercise Set 17.6

In Exercises 1–4, find a particular solution by inspection. Then give the general solution.

1. $y'' + y = 2x$

2. $y'' - 3y' = 1$

3. $y'' - y' - 2y = 3$

4. $y'' - 4y = 3x + 5$

In Exercises 5–18, find a particular solution using the method of undetermined coefficients.

5. $y'' + y' - 6y = 3e^x$

6. $2y'' - 3y' - 5y = 3x - 1$

7. $y'' + 4y = \sin x$

8. $y'' - 4y = 2x^2$

9. $y'' - y' - 2y = e^{2x}$

10. $y'' - 2y' = x - 2$

11. $y'' + y = 3 \cos x$

12. $y'' - y = 3e^x$

13. $y'' - 2y' + y = e^x$

14. $y'' + 4y' + 4y = 2e^{-2x}$

15. $y'' - 2y' = xe^{2x}$

16. $y'' + y = e^x \sin x$

17. $2y'' + y' + y = 2e^{-x} \cos x$

18. $3y'' - 2y' + 5y = (2x - 1)e^x$

19. Supply the missing details in Example 17.32 for calculating u'_1 and u'_2.

20. Show that the simultaneous solution of Equations 17.32 and 17.33 is

$$u'_1 = -\frac{1}{a} \frac{y_2 g(x)}{W}, \qquad u'_2 = \frac{1}{a} \frac{y_1 g(x)}{W}$$

where

$$W = \begin{vmatrix} y_1 & y_2 \\ y'_1 & y'_2 \end{vmatrix}$$

[*Note:* W is called the *Wronskian* of y_1 and y_2, after the Polish mathematician Josef Wronski (1778–1853).]

In Exercises 21–30, use the method of variation of parameters to find a particular solution. Then write the general solution.

21. $y'' + y = \sec x \tan x, \quad 0 < x < \pi/2$

22. $y'' + y = \tan x, \quad 0 < x < \pi/2$

23. $\dfrac{d^2 x}{dt^2} + x = \sec t, \quad 0 < t < \pi/2$

24. $\dfrac{d^2 y}{dx^2} + y = \sec^2 x, \quad 0 < x < \pi/2$

25. $y'' - y = xe^x$ **26.** $y'' + y' - 2y = e^x$

27. $y'' + 3y' + 2y = \cos(e^x)$

28. $y'' - 2y' - 3y = \cosh 2x$

29. $y'' + 2y' + 2y = e^x \sin x$

30. $y'' - y' = \dfrac{e^x}{1 + e^x}$

In Exercises 31–38, determine an appropriate form for a particular solution using undetermined coefficients, after finding the complementary solution. Do not calculate the coefficients.

31. $y'' + 3y' - 4y = x^3 e^x$

32. $y'' - 4y' + 4y = 3e^{2x} + x \cos x$

33. $y'' - 2y' = 2x^2 - 1 + xe^{2x}$

34. $y'' - 2y' + 5y = 3e^x \sin 2x$

35. $y'' - 6y' + 25y = e^{3x} \cos 4x - 1$

36. $3y'' + 4y' - 7y = x^2 \sin 2x$

37. $2y'' + 3y' - 9y = xe^{-x} \sin x$

38. $y'' + y' + 2y = 2 \sin^2 x$

In Exercises 39–46, solve the initial-value problem or boundary-value problem. Use either undetermined coefficients or variation of parameters to find a particular solution.

39. $y'' - 4y = x + 3; \ y(0) = 0, \ y'(0) = 1$

40. $y'' + 2y' = 2e^{-x}; \ y(0) = 1, \ y'(0) = -1$

41. $y'' + y = \cos 2x; \ y(0) = 2, \ y(\pi/2) = 3$

42. $y'' + 3y' - 10y = x^2 + 2x; \ y(0) = 1, \ y'(0) = 0$

43. $y'' - 4y' + 5y = 2e^{-2x}; \ y(0) = -1, \ y'(0) = 2$

44. $y'' + 4y = 3 \sin x; \ y(0) = 4, \ y(\frac{\pi}{4}) = 0$

45. $y'' - y' - 2y = e^{2x}; \ y(0) = 2, \ y'(0) = 3$

46. $y'' + 4y' - 5y = xe^x; \ y(0) = 0, \ y'(0) = 1$

In Exercises 47–51, find the general solution.

47. $y'' - y = \cosh x$

48. $y'' + 4y = 2 \cos^2 x - 1$

49. $y'' + 4y = 2 \sin x \cos x$

50. $d^2 x/dt^2 + \omega_0^2 x = k \cos \omega t$, where (a) $\omega \neq \omega_0$ and (b) $\omega = \omega_0$

51. $d^2 x/dt^2 - 6(dx/dt) + 9x = 2e^{3t} + t^2$. Also find the particular solution that satisfies $x(0) = 4, \ x'(0) = -2$.

52. Prove that

$$ay'' + by' + cy = x^n \qquad (a \neq 0, \ n \geq 0)$$

has a particular solution of the form

$$y_p = A_n x^n + A_{n-1} x^{n-1} + \cdots + A_1 x + A_0$$

if and only if $c \neq 0$.

In Exercises 53–56, use the same undetermined coefficient procedure as for second-order equations to find a particular solution. Also give the general solution.

53. $y''' - 3y'' = x^2$

54. $y''' - y'' + 4y' - 4y = \sin 2x$

55. $y''' - 3y' + 2y = 2e^x + 3x$

56. $d^4y/dx^4 - 16y = \cosh 2x + \cos 2x$

17.7 THE VIBRATING SPRING

An important application of second-order differential equations is in modeling problems dealing with oscillatory motion. We restrict our attention here to a discussion of a vibrating spring, although the ideas presented are applicable to a broad range of problems involving oscillatory motion, extending into quantum mechanics.

Consider a spring of natural length l attached to a support, as shown in Figure 17.20(a). Suppose a weight w is attached, causing the spring to stretch a distance Δl as it comes to an equilibrium position, as shown in Figure 17.20(b). We assume the weight is not great enough to stretch the spring beyond its elastic limit. According to Hooke's Law, the force exerted by the spring is proportional to the elongation, within this elastic limit. When the spring and weight are in equilibrium and the force exerted by the spring is w, we have

$$w = k\,\Delta l,$$

where k is the constant of proportionality (the *spring constant*). Since $w = mg$, where m is the mass of the spring, we have $mg = k\,\Delta l$, or

$$mg - k\,\Delta l = 0 \qquad\qquad (17.34)$$

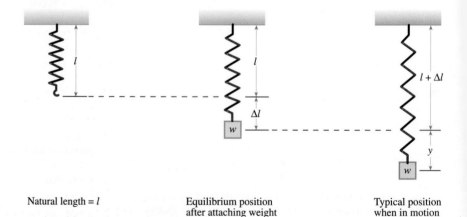

Natural length $= l$

(a)

Equilibrium position
after attaching weight

(b)

Typical position
when in motion

(c)

FIGURE 17.20

Now suppose the spring with attached weight is set in motion in some way. For example, it might be pulled below the equilibrium position and then released. As shown in Figure 17.20(c), we let $y = y(t)$ be the distance of the weight from the equilibrium position at time t. We take y as positive downward and try to determine y as a function of t. By Newton's Second Law of Motion,

$$m\frac{d^2y}{dt^2} = F$$

where F is the summation of all the forces acting on the weight. The force of gravity is mg, acting downward. The force exerted by the spring is of magnitude $k(\Delta l + y)$, since $\Delta l + y$ is the total displacement from the spring's natural length. When $\Delta l + y > 0$, the force of the spring is upward and so equals $-k(\Delta l + y)$. But this value is also correct when $\Delta l + y < 0$, since the spring is then compressed and its force is downward, in agreement with $-k(\Delta l + y) > 0$.

Two other forces may need to be considered. If the action takes place in air, we might reasonably neglect air resistance, but often vibrations occur in some viscous medium, such as oil, in which case the resistance of the medium cannot be neglected. Experimentally, it can be shown that for relatively small velocities, it is reasonable to assume that the resisting force is proportional to the velocity. Since this force is always opposite to the direction of motion, it is of the form $-c(dy/dt)$ for $c > 0$. Finally, we allow for some external force $f(t)$. For example, $f(t)$ might be a force applied to the entire support system.

We can now sum all the forces and obtain the following result from Newton's Second Law:

$$m\frac{d^2y}{dt^2} = \underbrace{mg}_{\substack{\text{Force} \\ \text{of} \\ \text{Gravity}}} - \underbrace{k(\Delta l + y)}_{\substack{\text{Force of} \\ \text{Spring}}} - \underbrace{c\frac{dy}{dt}}_{\substack{\text{Resisting} \\ \text{Force}}} + \underbrace{f(t)}_{\substack{\text{External} \\ \text{Force}}}$$

Since by Equation 17.34, $mg - k\Delta l = 0$, we can collect terms to obtain

---

Differential Equation for the Vibrating Spring

$$m\frac{d^2y}{dt^2} + c\frac{dy}{dt} + ky = f(t) \qquad (17.35)$$

This equation is the basic differential equation for the vibrating spring. We now consider certain special cases.

Undamped Free Vibrations

The simplest case is the one in which the resisting force is so small that it can be neglected, and for which there is no external force. We describe this situation by saying the motion is **undamped** (no resisting force) and **free** (no external force). Equation 17.35 then reduces to

$$m\frac{d^2y}{dt^2} + ky = 0$$

or, on dividing by m and writing $\omega^2 = k/m$ (ω is the Greek letter omega), we have the following formula.

Undamped Free Vibrations

$$\frac{d^2y}{dt^2} + \omega^2 y = 0 \qquad (17.36)$$

Equation 17.36 is a second-order linear homogeneous equation that has auxiliary equation $m^2 + \omega^2 = 0$, with roots $m = \pm i\omega$. So the general solution is

$$y = C_1 \cos \omega t + C_2 \sin \omega t \qquad (17.37)$$

The motion described by Equation 17.37 is called **simple harmonic motion**. Its **period** T, the time required to go through one complete cycle and return to the original position, is $2\pi/\omega$. **Frequency**, the number of cycles completed per unit of time, is $1/T = \omega/2\pi$. In Exercise 15 of this section you will be asked to show that Equation 17.37 can be written in the form

$$y = C \sin(\omega t + \alpha) \qquad (17.38)$$

where $C = \sqrt{C_1^2 + C_2^2}$ and α satisfies $\sin \alpha = C_1 / \sqrt{C_1^2 + C_2^2}$, $\cos \alpha = C_2 / \sqrt{C_1^2 + C_2^2}$. In this form we see that the motion is sinusoidal, with amplitude C. Its graph is shown in Figure 17.21.

EXAMPLE 17.33   A spring is attached to a rigid overhead support. A 4-lb weight is attached to the end of the spring, causing it to stretch 6 in. Then the weight is pulled down 3 in. below the equilibrium position and released. Air resistance is negligible. Describe the resulting motion.

Solution   For consistency, we will use feet instead of inches as the unit of distance. Then we have $y(0) = \frac{1}{4}$ and $y'(0) = 0$, since the weight was released without imparting any velocity. Since the 4-lb weight stretched the spring $\frac{1}{2}$ ft, we get the spring constant k from Hooke's Law:

$$4 = k \cdot \frac{1}{2}$$

$$k = 8 \text{ lb/ft}$$

Approximating g as 32 ft/s², we have

$$m = \frac{w}{g} = \frac{4}{32} = \frac{1}{8}$$

So using Equation 17.36 we have the initial-value problem

$$\frac{d^2y}{dt^2} + 64y = 0; \qquad y(0) = \frac{1}{4}, \ y'(0) = 0$$

Here we have used the fact that $\omega^2 = k/m = 8 \cdot 8 = 64$. The general solution is

$$y = C_1 \cos 8t + C_2 \sin 8t$$

From the initial conditions, we get $C_1 = \frac{1}{4}$ and $8C_2 = 0$, so that $C_2 = 0$. Thus, the equation of motion is

$$y = \frac{1}{4} \cos 8t$$

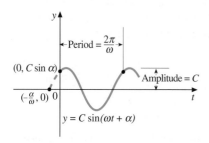

FIGURE 17.21
Simple harmonic motion

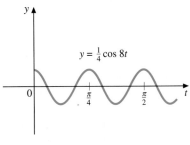

FIGURE 17.22

The period is $2\pi/\omega = \pi/4$, and the frequency is $4/\pi$ cycles per second. The amplitude is $1/4$ ft. We show the graph of the solution in Figure 17.22. ∎

Damped Free Vibrations

According to the undamped model, the motion of the spring continues forever without diminishing. This solution does not accurately describe reality, although it does give reasonable accuracy over a relatively short time span when the medium in which the motion takes place offers little resistance. When this resistance is taken into consideration and there is no external force, Equation 17.35 takes the following form.

Damped Free Vibrations

$$m\frac{d^2y}{dt^2} + c\frac{dy}{dt} + ky = 0 \qquad\qquad (17.39)$$

It is important to remember in Equation 17.39 that all the constants m, c, and k are positive. The nature of the motion in this case depends on the relative sizes of the resistance constant c and the spring constant k. On solving the auxiliary equation, we get the roots

$$r_1 = \frac{-c + \sqrt{c^2 - 4km}}{2m} \qquad \text{and} \qquad r_2 = \frac{-c - \sqrt{c^2 - 4km}}{2m}$$

There are three cases to consider.

1. $c^2 > 4km$: the roots are real and unequal, so the solution is of the form

$$y = C_1 e^{r_1 t} + C_2 e^{r_2 t}$$

Observe that since $\sqrt{c^2 - 4km} < c$, it follows that $r_1 < 0$ and $r_2 < 0$.

2. $c^2 = 4km$: the roots are real and equal. Denote the common value by r; that is, $r = -c/2m < 0$. The solution is therefore of the form

$$y = (C_1 + C_2 t)e^{rt}$$

3. $c^2 < 4km$: the roots are imaginary—say, $r_1 = \alpha + i\beta$ and $r_2 = \alpha - i\beta$, where $\alpha = -c/2m < 0$ and $\beta = \sqrt{4km - c^2}/2m > 0$. The solution then is of the form

$$y = e^{\alpha t}(C_1 \cos \beta t + C_2 \sin \beta t)$$

REMARK
■ In Case 3 we wrote

$$\sqrt{c^2 - 4km} = \sqrt{(-1)(4km - c^2)}$$
$$= i\sqrt{4km - c^2}$$

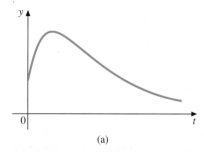

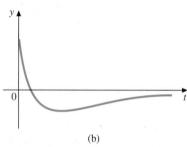

(a)

(b)

FIGURE 17.23
Overdamped motion $c^2 > 4km$

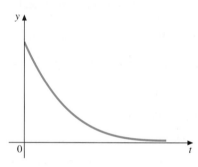

FIGURE 17.24
Critically damped motion $c^2 = 4km$

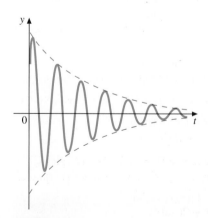

FIGURE 17.25
Underdamped motion $c^2 < 4km$

We analyze the motion for each case. For Case 1, $c^2 > 4km$, the solution is of the form

$$y = C_1 e^{r_1 t} + C_2 e^{r_2 t}$$

where both r_1 and r_2 are negative. As t increases, y approaches 0 and there is no oscillation. This type of motion is said to be **overdamped**. The resistance is so strong that the motion rapidly dies out. The form of the graph of y versus t depends on the initial conditions that determine C_1 and C_2. Figure 17.23 shows two possibilities.

In Case 2, $c^2 = 4km$, the solution is of the form

$$y = (C_1 + C_2 t)e^{rt}, \qquad r < 0$$

Again there is no oscillation, and the motion tends to die out as t increases because of the factor e^{rt}, where $r < 0$. The slightest change in the resisting force changes the situation to either Case 1 or Case 3. We describe this as **critically damped** motion. The graph is similar to that for overdamping, with the initial conditions determining the exact form. Figure 17.24 shows one possibility.

For Case 3, $c^2 < 4km$, the solution is of the form

$$y = e^{\alpha t}(C_1 \cos \beta t + C_2 \sin \beta t), \qquad \alpha < 0, \ \beta > 0$$

The motion in this case is oscillatory but with decreasing "amplitude" of the oscillations because of the $e^{\alpha t}$ factor (where $\alpha < 0$). The motion is said to be **underdamped**. A typical situation is shown in Figure 17.25.

EXAMPLE 17.34   An 8-lb weight stretches a spring 1.6 ft beyond its natural length. A damping force equal in magnitude to that of the velocity is present. If the spring is pushed upward 4 in. above the equilibrium position and then given a downward velocity of 6 ft/sec, find the equation of motion.

Solution   Using $g \approx 32$, we get $m = w/g = 1/4$, and by Hooke's Law $8 = k(1.6)$, so that $k = 5$. The resistance constant is $c = 1$. So the differential equation is

$$\frac{1}{4} \cdot \frac{d^2 y}{dt^2} + \frac{dy}{dt} + 5y = 0$$

or equivalently,

$$\frac{d^2 y}{dt^2} + 4\frac{dy}{dt} + 20y = 0$$

and the initial conditions are $y(0) = -\frac{1}{3}$, $y'(0) = 6$.

The auxiliary equation has roots $-2 \pm 4i$, so the general solution is

$$y = e^{-2t}(C_1 \cos 4t + C_2 \sin 4t)$$

To find C_1 and C_2, we impose the two initial conditions. From $y(0) = -\frac{1}{3}$, we have immediately that $C_1 = -\frac{1}{3}$. We find y' to be

$$y' = e^{-2t}[(4C_2 - 2C_1)\cos 4t + (-4C_1 - 2C_2)\sin 4t] \quad \text{Verify.}$$

So from $y'(0) = 6$, we get

$$4C_2 - 2C_1 = 6$$

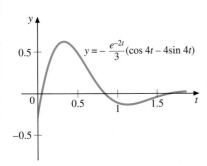

FIGURE 17.26

Substituting for C_1, we find $C_2 = \frac{4}{3}$. Thus, the solution is

$$y = -\frac{e^{-2t}}{3}(\cos 4t - 4\sin 4t)$$

The motion is underdamped in this case. (See Figure 17.26.)   ∎

Forced Vibrations

When an external force $f(t)$ is present, we describe the motion as having **forced vibrations**, and $f(t)$ is called the **forcing function**. Frequently the forcing function is sinusoidal, as in the example that follows.

EXAMPLE 17.35   A 4-lb weight stretches a spring 2 ft beyond its natural length. Air resistance is negligible, but the system is subjected to an external force $f(t) = 2\cos\lambda t$. The weight is pulled down 6 in. and released. Describe the motion for each of the following values of λ: (a) $\lambda = 3$ and (b) $\lambda = 4$.

Solution   In the usual way we find $m = \frac{1}{8}$ and $k = 2$. We assume $c = 0$. Thus, Equation 17.35 becomes

$$\frac{1}{8} \cdot \frac{d^2y}{dt^2} + 2y = 2\cos\lambda t$$

or equivalently,

$$\frac{d^2y}{dt^2} + 16y = 16\cos\lambda t$$

and the initial conditions are $y(0) = \frac{1}{2}$, $y'(0) = 0$. The complementary solution y_c is

$$y_c = C_1 \cos 4t + C_2 \sin 4t$$

(a) For $\lambda = 3$, we expect a particular solution of the form

$$y_p = A\cos 3t + B\sin 3t$$

By substituting this value into the equation and comparing coefficients, we get $A = \frac{16}{7}$, $B = 0$. The general solution is therefore

$$y = C_1 \cos 4t + C_2 \sin 4t + \frac{16}{7}\cos 3t$$

Imposing the initial conditions, we find $C_1 = -\frac{25}{14}$ and $C_2 = 0$, giving the equation of motion as

$$y = -\frac{25}{14}\cos 4t + \frac{16}{7}\cos 3t$$

(b) For $\lambda = 4$, the forcing function $f(t) = \cos 4t$ is a solution of the homogeneous equation, and so our trial solution y_p is of the form

$$y_p = t(A\cos 4t + B\sin 4t)$$

A straightforward calculation shows that $A = 0$, $B = 2$. Then applying the initial conditions to the general solution $y = y_c + y_p$, we obtain the equation of motion

$$y = \frac{1}{2}\cos 4t + 2t\sin 4t$$

We show its graph in Figure 17.27.   ∎

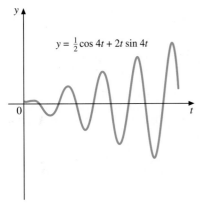

FIGURE 17.27

It is instructive to analyze the motion in part (b) of the preceding example. The presence of the factor t in the term $2t \sin 4t$ means that as t increases, the values of y become unbounded. For example, when $t = (2n + 1)\pi/8$, we get

$$y\left(\frac{2n + 1}{8}\pi\right) = \frac{(2n + 1)\pi}{4}(-1)^n$$

Tacoma Narrows Bridge as it collapsed

so that as $n \to \infty$, $|y| \to \infty$, with y alternating between positive and negative values. This phenomenon is known as **resonance**. It occurs when the forcing function is periodic and has the same period as the complementary solution (called the natural period of the system). The consequences of resonance to a physical system can be catastrophic. The presence of the factor t indicates large oscillations. Physical structures usually break, or the system changes so that the differential equation no longer describes it.

Exercise Set 17.7

In Exercises 1–10, find the equation of motion. Neglect the resisting force unless otherwise indicated. When the resisting force is considered, identify the motion as underdamped, overdamped, or critically damped.

1. A 4-lb weight stretches a spring 6 in. beyond its natural length. After the system comes to equilibrium, the weight is pulled down 6 more inches and released. Give the amplitude and period.

2. In Exercise 1, instead of the weight being pulled down, it is given an upward push from the equilibrium position of 4 ft/s.

3. A 16-lb weight stretches a spring 2 ft beyond its natural length. The weight is then pulled down 8 in. below the equilibrium position and given an upward velocity of 12 ft/s.

4. Repeat Exercise 3 under the assumption that there is a resisting force equal to 4 times the velocity.

5. A mass of 50 g stretches a spring 5 cm beyond its natural length. The mass is pushed upward 10 cm and released. (Take $g = 980$ cm/s^2.)

6. A 32-lb weight is attached to a spring with a natural length of 3 ft. When the system comes to equilibrium, the

spring is 4 ft long. The system is in a viscous medium that produces a resisting force equal to the velocity and opposite in direction. The weight is started in motion with a downward velocity from the equilibrium position of 6 ft/s.

7. A 20-g mass attached to a spring stretches it 10 cm beyond its natural length. The mass is attached to a mechanism that produces a resisting force of magnitude $420\,|v|$ dynes, where v is the velocity. The mass is pulled down 4 cm and also given an initial downward velocity of 7 cm/s.

8. An 8-lb weight stretches a spring from its natural length of 20 in. to a length of 28 in. The weight is then pulled down an additional 4 in. and given a downward velocity of 4 ft/s. Find (a) the time when the weight first passes through the equilibrium position and (b) its maximum displacement from the equilibrium position.

9. Repeat Exercise 1 if there is an external force $f(t) = 3 \cos 4t$ pounds acting on the system.

10. An 8-lb weight stretches a spring 1 ft beyond its natural length. The system is subjected to an external force of $f(t) = 2 \cos t + 3 \sin t$ pounds.

11. A weight of 8 lb is attached to a spring with a spring constant of 5 lb/ft. Assume a resisting force equal to $-c(dy/dt)$, where $c > 0$. Find c such that the motion is (a) overdamped, (b) critically damped, and (c) underdamped.

12. The angle θ from the vertical to a pendulum (see the figure) can be shown to satisfy the nonlinear second-order differential equation

$$\frac{d^2\theta}{dt^2} + \frac{g}{l}\sin\theta = 0$$

where l is the length of the pendulum rod. For small values of θ, $\sin\theta \approx \theta$. Using this approximation, determine the equation of motion of a pendulum 2 ft long, where θ is initially $\frac{1}{4}$ radian (with the positive direction for θ taken as counterclockwise, measured from the vertical), if the initial angular velocity is $d\theta/dt = \sqrt{3}$ radians per second.

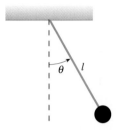

13. For the pendulum of Exercise 12, find the following:
    (a) the maximum angle from the vertical through which the pendulum will swing;
    (b) the time to complete one back-and-forth swing;
    (c) the instant when the pendulum will be vertical for the first time;
    (d) the magnitude of its velocity when it is in the vertical position.

14. Prove that for free damped vibrations of a spring, in the overdamped and critically damped cases, the weight will pass through the equilibrium position at most one time.

15. Show that the equation

$$y = C_1 \cos \omega t + C_2 \sin \omega t$$

of simple harmonic motion can be expressed in either of the following ways:
(a) $y = C \sin(\omega t + \alpha)$ (b) $y = C \cos(\omega t - \beta)$
where $C = \sqrt{C_1^2 + C_2^2}$. Describe the angles α and β.

16. Consider the following equation for damped forced vibrations:

$$m\frac{d^2 y}{dt^2} + c\frac{dy}{dt} + ky = F_0 \cos \lambda t$$

Find the general solution. Show that if $c \neq 0$, there can be no resonance. Find $\lim_{t\to\infty} y(t)$. (This is called the *steady-state* solution.)

17. Consider a simple series circuit that contains a resistance of R ohms, an inductance of L henrys, a capacitance of C farads, and an electromotive force of $E(t)$ volts (see the figure). By one of Kirchhoff's laws, the current I, in amperes, satisfies

$$L\frac{dI}{dt} + RI + \frac{Q}{C} = E(t)$$

where Q is the electric charge, related to I by $I = dQ/dt$. By differentiating both sides of the differential equation, we obtain

$$L\frac{d^2 I}{dt^2} + R\frac{dI}{dt} + \frac{I}{C} = E'(t)$$

Find I if $L = 10$, $R = 20$, $C = 0.02$, and $E(t) = 200 \sin t$. What is the steady-state current? (Let $t \to \infty$.)

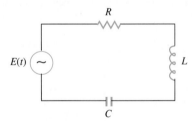

17.8 SERIES SOLUTIONS

Using methods we have studied, we can solve a large number of differential equations, but there are many others for which none of the methods works. For example, we have not yet shown a means of finding the general solution of

$$y'' + xy' + y = 0$$

This equation is a homogeneous linear equation, but one of the coefficients is a variable, so the theory of Section 17.5 is not applicable. For this problem and others it is often possible to find a solution in the form of a power series,

$$y = \sum_{n=0}^{\infty} a_n x^n \qquad (17.40)$$

Recall from Chapter 10 that every such power series defines a function within its interval of convergence, $|x| < R$. We assume $R > 0$. The series can be differentiated term by term, and the resulting series converges to y' in the same interval, $|x| < R$. Similarly, differentiating again, we obtain a series that converges to y'' for $|x| < R$, and so on. Thus, if y is given by Equation 17.40, then

$$y' = \sum_{n=1}^{\infty} n a_n x^{n-1}$$

and

$$y'' = \sum_{n=2}^{\infty} n(n-1) a_n x^{n-2}$$

Notice we started the index of summation with 1 for y' and with 2 for y'', since the constant term in each case drops out on differentiation.

It is useful to shift the index of summation at times. For example, we can see by expanding the terms that

$$\sum_{n=0}^{\infty} a_n x^n = \sum_{n=1}^{\infty} a_{n-1} x^{n-1}$$

The key to handling such shifts is that when the initial value of n on the summation sign is increased, n in the summand must be decreased by the same amount, and vice versa. (More formally, in the example shown, we could make the change of index $n = m - 1$ and write the second sum in terms of m, but the way we have suggested is faster and is easy to apply.) Test yourself by seeing whether you agree that

$$\sum_{n=2}^{\infty} n(n-1) a_n x^{n-2} = \sum_{n=1}^{\infty} (n+1)(n) a_{n+1} x^{n-1} = \sum_{n=0}^{\infty} (n+2)(n+1) a_{n+2} x^n$$

We give two examples to illustrate solution by series. The first problem could be done more easily by separating variables, but we include it to illustrate the technique first on a simple problem and also to provide a means of comparing our answer with that obtained by separating variables.

EXAMPLE 17.36   Find the general solution of the equation $y' = 2xy$ using infinite series.

Solution   We try a solution of the form given in Equation 17.40: $y = \sum_{n=0}^{\infty} a_n x^n$. The equation can then be written as

$$\sum_{n=1}^{\infty} n a_n x^{n-1} - 2 \sum_{n=0}^{\infty} a_n x^{n+1} = 0 \quad \text{Since } x \sum_{n=0}^{\infty} a_n x^n = \sum_{n=0}^{\infty} a_n x^{n+1}$$

To have the same power of x appearing in the two summations, we shift the first index down by 1 and the second one up by 1:

$$\sum_{n=0}^{\infty} (n+1) a_{n+1} x^n - 2 \sum_{n=1}^{\infty} a_{n-1} x^n = 0$$

After separating the $n = 0$ term of the first summation, both sums will begin with $n = 1$, and so they can be brought together, giving

$$a_1 + \sum_{n=1}^{\infty} [(n+1) a_{n+1} - 2 a_{n-1}] x^n = 0$$

We want this equation to be an identity in x. So we must have $a_1 = 0$, and every coefficient of x^n for $n = 1, 2, 3, \ldots$ equal to 0:

$$(n + 1)a_{n+1} - 2a_{n-1} = 0, \qquad n \geq 1$$

The last equation can be written in the form

$$a_{n+1} = \frac{2a_{n-1}}{n + 1}, \qquad n \geq 1$$

This equation is a **recursion formula** that enables us to find the coefficients $a_2, a_3, \ldots$ in sequence:

$$\underline{n = 1}: \quad a_2 = \frac{2a_0}{2} = a_0$$

$$\underline{n = 2}: \quad a_3 = \frac{2a_1}{3} = 0 \quad \text{Since } a_1 = 0$$

$$\underline{n = 3}: \quad a_4 = \frac{2a_2}{4} = \frac{a_2}{2} = \frac{a_0}{2} \quad \text{Since } a_2 = a_0$$

$$\underline{n = 4}: \quad a_5 = \frac{2a_3}{5} = 0 \quad \text{Since } a_3 = 0$$

$$\underline{n = 5}: \quad a_6 = \frac{2a_4}{6} = \frac{a_4}{3} = \frac{a_0}{3 \cdot 2} \quad \text{Since } a_4 = \frac{a_0}{2}$$

$$\vdots \qquad\qquad \vdots$$

Continuing in this way, we see that

$$\begin{cases} a_{2k-1} = 0 & \text{for } k = 1, 2, 3, \ldots \\ a_{2k} = \dfrac{a_0}{k!} & \text{for } k = 1, 2, 3, \ldots \end{cases}$$

Since no condition is placed on a_0, it is arbitrary. Thus, our solution $y = \sum_{n=0}^{\infty} a_n x^n$ becomes

$$y = a_0 + \sum_{k=1}^{\infty} a_{2k} x^{2k}$$

$$= a_0 + a_0 \sum_{k=1}^{\infty} \frac{x^{2k}}{k!}$$

Since $0! = 1$, we can combine the first term with the summation by starting k with 0:

$$y = a_0 \sum_{k=0}^{\infty} \frac{x^{2k}}{k!}$$

By the Ratio Test, we can determine that the series converges for all values of x.

Now let us compare this solution with that obtained by separating variables. We have

$$\frac{dy}{y} = 2x\, dx$$

$$\ln|y| = x^2 + \ln|C| \quad \text{Write the constant as } \ln|C|.$$

$$\ln\left|\frac{y}{C}\right| = x^2$$

$$y = Ce^{x^2}$$

To see that our series solution agrees with this result, recall from Section 10.8 that the Maclaurin series for e^{x^2} is

$$e^{x^2} = \sum_{k=0}^{\infty} \frac{x^{2k}}{k!}$$

Thus, the solution found by separating variables agrees with the series solution, with the arbitrary constant called C in one case and a_0 in the other.   ∎

EXAMPLE 17.37   Find the general solution of the equation $y'' + xy' + y = 0$ using infinite series.

Solution   This problem is the one we posed at the beginning of this section. Substituting $y = \sum_{n=0}^{\infty} a_n x^n$, we get

$$\sum_{n=2}^{\infty} n(n-1)a_n x^{n-2} + \sum_{n=1}^{\infty} n a_n x^n + \sum_{n=0}^{\infty} a_n x^n = 0$$

We shift the index down by 2 on the first sum, and by observing that the second sum is unaltered by starting with $n = 0$ (since the term corresponding to $n = 0$ will be 0), we obtain

$$\sum_{n=0}^{\infty} (n+2)(n+1)a_{n+2} x^n + \sum_{n=0}^{\infty} n a_n x^n + \sum_{n=0}^{\infty} a_n x^n = 0$$

The terms can be brought together now in one summation:

$$\sum_{n=0}^{\infty} [(n+2)(n+1)a_{n+2} + (n+1)a_n]x^n = 0$$

To be an identity in x, every coefficient must be 0. Thus,

$$(n+2)(n+1)a_{n+2} + (n+1)a_n = 0$$

or

$$a_{n+2} = -\frac{a_n}{n+2}, \qquad n \geq 0$$

We consider even subscripts and odd subscripts separately:

n Even	n Odd
$a_2 = -\dfrac{a_0}{2}$	$a_3 = -\dfrac{a_1}{3}$
$a_4 = -\dfrac{a_2}{4} = \dfrac{a_0}{2 \cdot 4}$	$a_5 = -\dfrac{a_3}{5} = \dfrac{a_1}{3 \cdot 5}$
$a_6 = -\dfrac{a_4}{6} = -\dfrac{a_0}{2 \cdot 4 \cdot 6}$	$a_7 = -\dfrac{a_5}{7} = -\dfrac{a_1}{3 \cdot 5 \cdot 7}$
$\vdots$	$\vdots$
$a_{2k} = \dfrac{(-1)^k a_0}{2 \cdot 4 \cdot 6 \cdot \ \cdots \ \cdot (2k)}$	$a_{2k+1} = \dfrac{(-1)^k a_1}{3 \cdot 5 \cdot 7 \cdot \ \cdots \ \cdot (2k+1)}$

Since a_0 and a_1 have no restrictions, they are arbitrary. The solution can now be written as

$$y = a_0 \left[1 + \sum_{k=1}^{\infty} \frac{(-1)^k}{2 \cdot 4 \cdot 6 \cdot \ \cdots \ \cdot (2k)} x^{2k} \right]$$

$$+ a_1 \left[x + \sum_{k=1}^{\infty} \frac{(-1)^k}{3 \cdot 5 \cdot 7 \cdot \ \cdots \ \cdot (2k+1)} x^{2k+1} \right]$$

The expression in the first bracket can be simplified somewhat by observing that $2 \cdot 4 \cdot 6 \cdot \ \cdots \ \cdot (2k) = 2^k \cdot k!$, and since $0! = 1$, we can combine the first term with the summation by starting k with 0. Similarly, we can include the first term of the second bracket with the summation by starting with $n = 0$. We can therefore write

$$y = a_0 \sum_{k=0}^{\infty} \frac{(-1)^k x^{2k}}{2^k k!} + a_1 \sum_{k=0}^{\infty} \frac{(-1)^k}{1 \cdot 3 \cdot 5 \cdot \ \cdots \ \cdot (2k+1)} x^{2k+1}$$

Both series can be shown to converge for all x. If we let y_1 be the first sum and y_2 the second, and write $C_1 = a_0$ and $C_2 = a_1$, then the solution is in the familiar form $y = C_1 y_1 + C_2 y_2$. ∎

REMARK ───────────────────────────────────────

■ In the two preceding examples, the recursion formulas were such that we could find explicit formulas for the coefficients. In some cases finding such explicit formulas is not possible, and the best we can do is calculate as many coefficients as we need in sequential order.

───

The procedure we have illustrated will also work for nonhomogeneous equations provided that the nonhomogeneous term can be expanded in a power series. In particular, if the nonhomogeneous term is a polynomial, it is already a (finite) power series. By comparing coefficients of like powers of x, we can find the unknown coefficients a_n in the series $\sum a_n x^n$.

As you might expect, there is a good deal more to solution by power series than we have gone into here, but the deeper aspects will have to be deferred to a course on differential equations.

Exercise Set 17.8

In Exercises 1–13, find the general solution in terms of power series in x. Where possible, solve the equation by other means and compare answers. If initial values are given, find the particular solution.

1. $y' = x^2 y$

2. $y' - 2xy = x$

3. $y'' = y$

4. $y'' - xy' - y = 0$

5. $y'' = xy$

6. $y' - y = 2x; \ y(0) = 3$

7. $y'' + y = 0; \ y(0) = 1, \ y'(0) = 0$

8. $(1 - x)y'' = y'$

9. $y'' + x^2 y = 0$

10. $(1 + x^2)y'' + 2xy' = 0$

11. $(1 - x^2)y'' - 5xy' - 3y = 0$

12. $(1 + x^2)y'' - 3xy' - 5y = 0$

13. $y'' + xy' + 2y = x^2 + 1; \ y(0) = 2, \ y'(0) = 1$

14. Obtain the general solution of the equation

$$y'' + (x - 2)y = 0$$

as a power series about $x = 2$—that is, in the form $\sum a_n (x - 2)^n$.

15. Find two linearly independent solutions in the form of power series for the equation

$$y'' + (1 - x)y = 0$$

Then write the general solution. (*Hint:* For one solution take $a_0 = 1$ and $a_1 = 0$; then do the opposite.)

16. Obtain the first 10 nonzero terms of the Maclaurin series solution

$$y(x) = y(0) + y'(0)x + \frac{y''(0)}{2!}x^2 + \frac{y'''(0)}{3!}x^3 + \cdots$$

for the initial-value problem

$$y'' - xy' + 2y = 0; \qquad y(0) = 1, \ y'(0) = -1$$

by carrying out the following steps:

(a) Solve the equation for y''.
(b) Put $x = 0$ and substitute the given values of $y(0)$ and $y'(0)$ to get $y''(0)$.
(c) Differentiate the equation in step (a) to get y'''.
(d) Put $x = 0$ and substitute known values for $y(0)$, $y'(0)$, and $y''(0)$ to get $y'''(0)$.
(e) Differentiate the equation in step (c) and continue in this manner.

17. Follow the procedure in Exercise 16 to obtain the first 10 nonzero terms of the Maclaurin series solution of the following initial-value problem:

$$y'' + (x - 1)y = 0; \qquad y(0) = 2, \ y'(0) = 1$$

Chapter 17 Review Exercises

In Exercises 1–18, find the general solution of the differential equation. If initial or boundary conditions are given, find the particular solution that satisfies them.

1. $(2 + e^{-x} \cos y) \, dx + (3 + e^{-x} \sin y) \, dy = 0$

2. $xy' = y + x \cot \frac{y}{x}; \ y(1) = 0$

3. $y' = x\sqrt{1 - y^2}; \ y(0) = 1$

4. $(\cosh x)y' + (\sinh x)y = \cosh^2 x$

5. $\dfrac{dy}{dx} = \dfrac{y(2x + y)}{x(x + y)}$

6. $xy' - 1 = e^{-y}; \ y(1) = 0$

7. $(y - 3x) \, dy + (3y + 4x) \, dx = 0; \ y(1) = 0$

8. $\left(\dfrac{2x^2 + y^2}{\sqrt{x^2 + y^2}} \right) dx + \left(\dfrac{xy}{\sqrt{x^2 + y^2}} - 3 \right) dy = 0; \ y(3) = 4$

9. $\dfrac{dy}{dx} = 1 + \dfrac{y(1 - x)}{x(2 - x)}; \ y(1) = 2$

10. $\dfrac{dy}{\tan y} = \dfrac{\sin 2x}{1 + \sin^2 x} \, dx$

11. $(\cos xy - xy \sin xy) \, dx = (x^2 \sin xy) \, dy$

12. $y' + \dfrac{3y}{x} = \dfrac{x}{y^2}$ (See Exercise 28 in Exercise Set 17.3.)

13. $e^{2x^2} \, dy + x(2ye^{2x^2} - 1) \, dx = 0; \ y(0) = 1$

14. $2y'' + 5y' - 3y = 0; \ y(0) = 1, \ y'(0) = 0$

15. $y'' + 6y' + 9y = 0; \ y(0) = 3, \ y'(0) = -4$

16. $y'' + 4y' + 5y = 0$

17. $\dfrac{d^2 s}{dt^2} + 2\dfrac{ds}{dt} = 0; \ s(0) = 0, \ s'(0) = 4$

18. $y'' + 9y = 0; \ y(0) = 3, \ y(\pi/2) = -2$

In Exercises 19–26, find a particular solution by: (a) undetermined coefficients, (b) variation of parameters.

19. $\dfrac{d^2 x}{dt^2} - \dfrac{dx}{dt} - 6x = 2e^t$

20. $y'' + 2y' - 3y = 2e^{-x} - 1$

21. $y'' - y' = x + 3e^x$

22. $y'' + 4y = 2\cos^2 x$

23. $2y'' + 3y' - 5y = e^x \cos x$

24. $y'' + y = xe^{-x}$

25. $y'' - y' = \sinh x; \ y(0) = 0, \ y'(0) = 1$

26. $y'' - 4y' + 4y = 8x^2; \ y(0) = 1, \ y'(0) = -1$

*In Exercises 27 and 28, give the appropriate form for a partic-
ular solution using undetermined coefficients. Do not calculate
the coefficients.*

27. $y'' - 3y' - 4y = xe^{-x} + \cos 2x$

28. $y'' + 2y' + 2y = x^2 e^{-x} \sin x$

29. If 10% of a certain radioactive substance decays in 2 yr,
find its half-life. How long will it take for 90% to decay?

30. A 50% dye solution is fed at the rate of 8 L/min into a
vat that originally contained 200 L of pure water, and the
well-mixed solution is drained off at the same rate. How
long will it take for the mixture to become a 25% dye
solution?

31. A modification of the Malthus model that accounts for
the culling of a population $Q(t)$ at a constant rate H (for
example, the culling of a deer population by hunters or
of some other animal species by predators) is given by

$$Q'(t) = \alpha Q(t) - H \quad (\alpha > 0)$$

with $Q(0) = Q_0$. Solve this equation and show that the
model predicts three possible outcomes, depending on
the relative sizes of H and α: (1) the population dies out
in a finite time, (2) the population grows without limit,
or (3) the population size stays constant. Determine the
relationship between H and α that produces each result.

32. In an *autocatalytic* chemical reaction, a substance A
is converted to a substance B in such a way that the
reaction is stimulated by the substance being produced.
If the original concentration of A is a and at time t the
concentration of B is $x = x(t)$, then the reaction rate is
modeled by

$$\frac{dx}{dt} = kx(a - x)$$

Find the solution if $x(0) = x_0$.

33. A 4-lb weight stretches a spring 0.64 ft beyond its natural
length. The system is pushed up 4 in. from equilibrium
and then given a downward velocity of 5 ft/sec. There is
a damping force present of 0.25 v. Find the equation of
motion.

34. In the equation

$$L\frac{d^2 I}{dt^2} + R\frac{dI}{dt} + \frac{I}{C} = E'(t)$$

obtained for the RLC circuit in Exercise 17 in Exercise
Set 17.7, suppose $E(t) = E \sin \omega t$ and $R^2 C = 4L$. Find
$I(t)$. To simplify notation, write $\alpha = (R/2L)$.

*In Exercises 35 and 36, find the general solution using power
series. Find the largest interval in which the solution is valid.*

35. $y'' - 2xy' - 4y = 0$

36. $(1 - x^2)y'' - 6xy' - 4y = 0$

Chapter 17 Concept Quiz

1. Define each of the following:
   (a) The order of a differential equation
   (b) A first-order separable differential equation
   (c) An exact differential equation
   (d) A homogeneous first-order differential equation
   (e) A linear differential equation of order n
   (f) A homogeneous function of degree k in x and y
   (g) Linearly independent functions $f_1, f_2, \ldots, f_n$

2. Explain in your own words how to solve a first-order
differential equation of each of the following types:
   (a) Separable   (b) Homogeneous
   (c) Exact   (d) Linear

3. Consider the equation $ay'' + by' + cy = 0$.
   (a) Write the auxiliary equation and express its solutions
   m_1 and m_2 in terms of a, b, and c.

   (b) Give the general solution in each of the following
   cases:
   (i) $b^2 - 4ac > 0$
   (ii) $b^2 - 4ac = 0$
   (iii) $b^2 - 4ac < 0$

4. Explain how to use the method of variation of parameters.

5. Consider the equation $ay'' + by' + cy = g(x)$. For each
of the following values of $g(x)$, tell what your initial trial
would be for a particular solution, y_p. Also, tell under
what circumstances you would modify your initial trial
and in what way you would modify it.
   (a) $g(x) = e^{kx}$ (b) $g(x) = kx^2$
   (c) $g(x) = \cos kx$ (d) $g(x) = xe^{kx}$

6. Consider the equation

$$m\frac{d^2y}{dt^2} + c\frac{dy}{dt} + ky = 0$$

for the motion of a mass m attached to a spring having spring constant k, if there is a damping force proportional to velocity in which the constant of proportionality is c. Describe the motion for each of the following cases:
(a) $c^2 - 4km > 0$,   (b) $c^2 - 4km = 0$,   (c) $c^2 - 4km < 0$

7. Fill in the blanks.
(a) A test for exactness of the differential equation $f(x, y)dx + g(x, y)dy$ is that _____ = _____ .

(b) An integrating factor, μ, for the equation $y' + P(x)y = Q(x)$ is _____ .

(c) If $f(x, y)dx + g(x, y)dy = 0$ and f and g are homogeneous functions of the same degree, then either of the substitutions _____ or _____ will result in a _____ equation.

(d) If $dy/dx = f(x, y)$, the differential equation of the orthogonal trajectories of its family of solutions is _____ .

(e) In forced, undamped vibrations, if the forcing function is periodic and has the same period as the complementary solution, the phenomenon of _____ occurs.

(f) If y_1 and y_2 are solutions of the homogeneous differential equation $ay'' + by' + cy = 0$, then _____ is also a solution.

8. Which of the following statements are true and which are false?
(a) It is impossible to have three linearly independent solutions of the equation $ay'' + by' + cy = 0$.
(b) In the general solution $y = y_c + y_p$ of a nonhomogeneous linear differential equation with constant coefficients, both y_c and y_p are uniquely determined.
(c) If a first-order differential equation is homogeneous, it cannot be exact.
(d) If y_1, y_2, and y_3 are all solutions of $ay'' + by' + cy = 0$, and y_2 is not a multiple of y_1, then there must be constants C_1 and C_2 such that $y_3 = C_1y_1 + C_2y_2$.
(e) The sum of two solutions of a nonhomogeneous equation $ay'' + by' + cy = g(x)$ is also a solution.

APPLYING CALCULUS

1. A skydiver and his parachute weigh 175 pounds. He free-falls from rest (i.e., his initial velocity is 0) from a height of 10,000 feet for 10 seconds at which time his parachute opens. Assume that the air resistance during the free fall is $(1/5)v$ and that the air resistance after his chute opens is $5v$, where $v = v(t)$ is his velocity at time t.

   (a) Determine expressions for the skydiver's velocity v and position s at any time t. (*Hint*: the motion is described by two differential equations.)

   (b) How long will it take for the skydiver to hit the ground?

   (c) What is the skydiver's terminal velocity and how does this compare with the velocity with which he hits the ground?

2. A dog sees a rabbit running in a straight line across an open field and immediately begins to chase it. In a rectangular coordinate system, assume:

   (i) The rabbit is at the origin and the dog is at the point $(L, 0)$ at the instant the dog sees the rabbit.

   (ii) The rabbit runs up the y-axis and the dog always runs toward the rabbit.

   (iii) The rabbit runs with speed u and the dog runs with speed v.

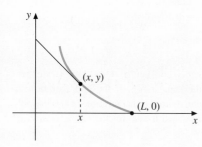

   Suppose that the dog's path is the graph of a function $y = y(x)$.

   (a) Show that the function y must satisfy the differential equation

   $$x\frac{d^2y}{dx^2} = \frac{u}{v}\sqrt{1 + \left(\frac{dy}{dx}\right)^2}$$

   (b) Determine $y(t)$ such that $y(L) = 0$ and $y'(L) = 0$.

   (c) Show that the dog will catch the rabbit only if $v > u$. When this condition is satisfied, where will the dog catch the rabbit?

3.

   (a) If an object of mass m is projected vertically upward from the ground with an initial velocity v_0, and if air resistance is neglected, then it follows from Newton's Second Law that

   $$m\frac{d^2y}{dt^2} = -mg$$

   where $y = y(t)$ represents the position of the object at time t, g is the acceleration due to gravity and $y(0) = 0$, $y'(0) = v_0$. Let t_a denote the time for the object to reach its highest point, called the *ascent time*, and let t_d denote the time that it takes for the object to descend from its highest point to the ground—the *descent time*. Let v_f denote the velocity with which the object hits the ground. Show that $t_a = t_d$ and $|v_0| = |v_f|$.

   (b) Now suppose that the object is projected vertically upward from the ground with initial velocity v_0 and that air resistance exerts a force proportional to the velocity. Then

   $$m\frac{d^2y}{dt^2} = -k\frac{dy}{dt} - mg$$

   subject to the initial conditions $y(0) = 0$, $y'(0) = v_0$.

   (i) Determine the solution of this initial-value problem.

(ii) Determine the ascent time t_a. (*Note*: It can be shown, with some difficulty, that $t_a < t_d$.)

(iii) Let $E(t) = (1/2)mv^2 + mgy$ be the total energy of the object at time t. Show that $E(t)$ is strictly decreasing on the interval $[0, T]$, where $T = t_a + t_d$ is the total time of flight. Conclude that $|v_0| > |v_f|$.

4. If glucose is given intravenously at a constant rate, the change in the overall concentration $c(t)$ of glucose in the blood with respect to time may be described by the differential equation

$$\frac{dc}{dt} = \frac{G}{100V} - kc$$

where G, V, and k are positive constants, G being the rate at which glucose is administered, in milligrams per minute, and V the volume of blood in the body, in liters (around 5 L for an adult). The concentration $c(t)$ is measured in milligrams per centiliter. The term $-kc$ is included because the glucose is assumed to be changing continually into other molecules at a rate proportional to its concentration.

(a) Solve the equation above for $c(t)$, using c_0 to denote $c(0)$.

(b) Find the so-called steady-state concentration, $\lim_{t \to \infty} c(t)$.

5.

(a) The volume of Lake Michigan is approximately constant at 4,900 km³ (water enters and leaves the lake at approximately the same rate). If the concentration of pollutants in the lake is now 0.20%, and water with a concentration of 0.05% pollutants enters the lake at 158 km³/yr, how long will it take for the concentration of pollutants in the lake to be reduced to 0.10%?

(b) Lake Superior has a volume of approximately 12,200 km³, and the rate of inflow and outflow is about 65.2 km³/yr. Redo part (a) for Lake Superior, assuming the same concentrations of pollutants.

6.

(a) Show that if a radioactive substance A decomposes into a second radioactive substance B, then the amount $y(t)$ of substance B present at time t satisfies the differential equation

$$\frac{dy}{dt} = -k_2 y + k_1 x_0 e^{-k_1 t}$$

where x_0 is the initial amount of substance A. Solve for $y(t)$, with the initial condition $y(0) = 0$.

(b) A serious problem in many homes worldwide is that radioactive radon-222 enters the basements through cracks. Radon results from the decomposition of radium-226 in the soil. The half-life of radon-222 is 3.82 days, and the half-life of radium-226 is 1620 yr. Suppose a new home is built in an area with a radon problem and that cracks in the basement cause infiltration of radon. Use the result of part (a) to determine how long it will take for the amount of radon in the basement to be a maximum.

7. For a series circuit containing a resistance of R ohms, an inductance of L henrys, a capacitance of C farads, and an electromotive force of $E(t)$ volts, one of Kirchhoff's laws yields the differential equation

$$L\frac{dI}{dt} + RI + \frac{Q}{C} = E(t) \tag{1}$$

where $I = I(t)$ is the current, in amperes, and $Q = Q(t)$ is the electric charge. The current and charge are related by the equation $I = dQ/dt$. Thus, from Equation 1, Q satisfies the second-order differential equation

$$L\frac{d^2Q}{dt^2} + R\frac{dQ}{dt} + \frac{Q}{C} = E(t) \tag{2}$$

It is common practice in electrical engineering to express $E(t)$ in the complex exponential form

$$E(t) = E_0 e^{i\omega t} \tag{3}$$

Recall that by Euler's Formula (Equation 17.26)

$$e^{i\omega t} = \cos \omega t + i \sin \omega t$$

(a) Find the general solution of Equation 2, with $E(t)$ given by Equation 3. Assume that $R^2 C < 4L$, and to simplify notation, let $\alpha = R/(2L)$ and $\beta = \sqrt{1/(LC) - \alpha^2}$. (*Hint:* For a particular solution try $Q_p = A e^{i\omega t}$.)

(b) Find the steady-state current. (*Hint:* Let $t \to \infty$ and use the fact that $I = dQ/dt$.)

(c) By finding the real and imaginary parts of the solution in part (b), determine the steady-state current when the impressed voltage $E(t)$ is of the form $E_0 \cos \omega t$ and when it is of the form $E_0 \sin \omega t$.

ANSWERS

CHAPTER 1

Exercise Set 1.1

1. (a) 0 **(b)** 0 **(c)** $-\frac{1}{2}$ **(d)** $\frac{a^2-1}{a+2}$ **(e)** $\frac{x^2+2x}{x+3}$ **3. (a)** 1 **(b)** 2 **(c)** 3 **(d)** $|t|$ **(e)** x^2 since $x^2 \geq 0$

5. (a) domain of $f = \mathbb{R}$, range of $f = -2 \cup (2, \infty)$. **(b)** domain of $f = (-\infty, 3)$, range of $f = (0, \infty)$

7. domain of $f = \mathbb{R}$, range of $f = [-4, \infty)$ **9.** domain $f = (-\infty, -1]$, range $f = [0, \infty)$. **11.** domain $f = \mathbb{R}$, range $f = \mathbb{R}$.

13. **15.** **17.** **19.** **21.**

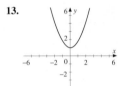 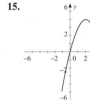 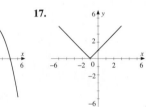 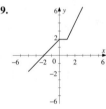 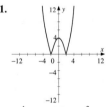

23. Assume that $0 < w \leq 12$. The postage is 0.29 for the first ounce or fraction thereof plus 0.23 for each additional ounce or fraction thereof. This is equivalent to **25.** $-\frac{1}{2x}$ **27.** $-\frac{3}{1+h}$

$$P(w) = \begin{cases} 0.29 + 0.23\,[w] & \text{if } w \text{ is not an integer} \\ 0.29 + 0.23\,[w - 1] & \text{if } w \text{ is an integer} \end{cases}$$

If $w = 1.5$, for example, $P(w) = 0.29 + 0.23\,[1.5] = 0.52$. Similarly, if $w = 2.0$, $P(w) = 0.29 + 0.23\,[w - 1] = 0.52$

29. $f(x) = x^n$ is even if n is even and odd if n is odd. **31. (a)** odd **(b)** odd **(c)** neither even nor odd **(d)** even **(e)** even

33. (a) odd
(b) even
(c) even
(d) neither even nor odd
(e) odd

35. (a) **(b)**

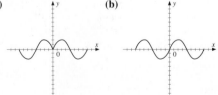

37. (a) $f(x) = 1000x - 25x^2$
(b) domain $= [x : 0 \leq x \leq 40]$

39. (a) $V = 2\pi r^3$
(b) $L = 4\pi r^2$
(c) $S = 6\pi r^2$

41. $\frac{12b - 2b^2}{3}$ **43.** domain $f = [-1, 3]$ **45.** domain $g = \{x : x \geq 1\}$. In Exercise 44, the square root is applied to the entire fraction, not the numerator and denominator individually.

47. domain $h = (-3, 2)$; range $h = \mathbb{R}$. **49.** $V = 2\pi r^2 \sqrt{a^2 - r^2}$ **51.** Area $= \frac{4}{25}(10 - h)^2$

53. $f(x) = \begin{cases} 0.015x & \text{if } x \leq 1000 \\ 5 + 0.01x & \text{if } x > 1000 \end{cases}$

55. range $= \{y : |y| \leq 1\}$.   **57.** The parameter a is the slope of the line. If $a > 0$, the slope is positive, and $x_2 > x_1 \Rightarrow y_2 > y_1$. If $a > 0$, the slope is positive, and $x_2 > x_1 \Rightarrow y_2 > y_1$. If $a < 0$, the slope is negative, and $x_2 > x_1 \Rightarrow y_2 < y_1$. The parameter b represents the y-intercept, i.e., $y(0)$.   **59. (a)** the graph is shifted vertically upward a units.   **(b)** the graph is shifted horizontally left a units.   **(c)** $a = 1$ gives the original graph, $a = -1$ reflects that graph about the x-axis, $a = 2$ stretches the original graph vertically, and $a = -2$ stretches the graph and reflects it about the x-axis.   **(d)** $a = -1$ reflects the original graph about the y-axis, $a = 2$ shrinks the graph horizontally to half its original width (for a particular value of y), and $a = -2$ shrinks the graph and reflects it about the y-axis.

61. $\lim\limits_{x \to -3/2} f(x) = 0.2$ **63.** $\lim\limits_{x \to 8} f(x) = -12$

Exercise Set 1.2

1. (a) $y = f(x) + 2$ **(b)** $y = f(x - 3)$ **(c)** $y = 2f(x)$ **(d)** $y = f(2x)$ **(e)** $y = -f(x)$

(f) $y = f(-x)$ **3. (a)** $y = f(x) + 2$ **(b)** $y = f(x - 3)$ **(c)** $y = 2f(x)$ **(d)** $y = f(2x)$

(e) $y = -f(x)$ **(f)** $y = f(-x)$ **5. (a)** $y = f(x) + 2$ **(b)** $y = f(x - 3)$ **(c)** $y = 2f(x)$

(d) $y = f(2x)$ **(e)** $y = -f(x)$ **(f)** $y = f(-x)$

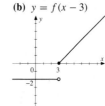

7. $y = \frac{1}{2}(x - 1)^3 + 2$
Move the graph 1 unit to the right, 2 units upward, and then compress it by a factor of $\frac{1}{2}$ in the y-direction.

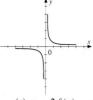
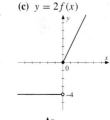
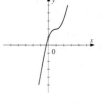

9. $y = g(x) = 3 - 2f(\frac{1}{2}x - 1)$

11. domain $f = \mathbb{R}$, domain $g = \mathbb{R}$
domain $f + g = \mathbb{R}$
domain $f - g = \mathbb{R}$
domain $fg = \mathbb{R}$
domain $f/g = \{x : x \neq -1\}$

13. domain $f = [-4, \infty)$
domain $g = (-\infty, 4]$
domain $f + g = [-4, 4]$
domain $f - g = [-4, 4]$
domain $fg = [-4, 4]$
domain $f/g = [-4, 4)$

15. domain $f = \{x : x \neq 0\}$
domain $g = \{x : x \neq 3\}$
domain $f + g = \{x : x \neq 0, 3\}$
domain $f - g = \{x : x \neq 0, 3\}$
domain $fg = \{x : x \neq 0, 3\}$
domain $f/g = \{x : x \neq 0, 3\}$

17. Domain of f and g = $\mathbb{R}$
domain $f + g = \mathbb{R}$
domain $f - g = \mathbb{R}$
domain $fg = \mathbb{R}$
domain $f/g = \{x : x < 0\}$

19.

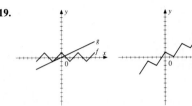

21. (a) g is even. h is odd   **(b)** $g(x) + h(x) = \frac{1}{2}[f(x) + f(-x) + f(x) - f(-x)] = f(x)$

Exercise Set 1.3

1. domain f and domain $g = \mathbb{R}$, domain $f \circ g = \mathbb{R}$, domain $g \circ f = \mathbb{R}$,   **3.** domain $f = \{x : x \neq 1\}$, domain $g = \{x : x \neq 0\}$, domain
$f \circ g = \{x : x \neq 0, 3\}$, domain $g \circ f = \{x : x \neq 1, 0\}$ **5.** $F(x) = (4 - 3x^2)^6 = (f \circ g)(x)$, where $f(x) = x^6$ and $g(x) = 4 - 3x^2$
7. Let $f(x) = x^{3/2}$, $g(x) = x^2 - 1$.   **11.** $[f \circ (g \circ h)](x) = f \circ (g \circ h)(x) = f \circ (g(h(x)) = f(g(h(x)))$
$= (f \circ g)(h(x)) = [(f \circ g) \circ h](x)$. Hence, $f \circ (g \circ h) = (f \circ g) \circ h$ whenever both sides are defined.   **13.** $f^{-1}(x) = (3x + 2)/5$
domain $f^{-1} = \mathbb{R}$ **15.** $F^{-1}(x) = (x^2 + 3)/2$, domain F^{-1} must be restricted to range $F = \{x : x \geq 0\}$.
17. $\varphi^{-1}(t) = \sqrt{t + 1}$; domain $\varphi^{-1} = \{t : t \geq -1\}$ **19. (a)**   **(b)**

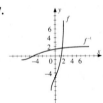

21. To make $f 1 - 1$, restrict the domain of f to $\{x : x \geq 0\}$. $f^{-1}(x) = \sqrt{\frac{1-2x}{x}}$; domain f^{-1} = range $f = \left(0, \frac{1}{2}\right)$.   **23.** To make $F 1 - 1$,
restrict its domain to $\left[\frac{3}{2}, \infty\right)$. $F^{-1}(x) = \frac{3+x}{2}$; domain F^{-1} = range $F = [0, \infty)$. Alternatively, one can restrict domain F to $\left(-\infty, \frac{3}{2}\right]$ to
get $F^{-1}(x) = \frac{3-x}{2}$ and domain $F^{-1} = [0, \infty)$.
25. $f^{-1}(x) = \frac{a-bx}{x-1}$; domain $f^{-1} = \{x : x \neq 1\}$ range $f^{-1} = \{y : y \neq -b\}$ **27.**

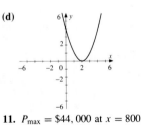

Exercise Set 1.4

1. $f(x) = -x + 2$ $f(10) = -10 + 2 = -8$ **3.** A: $21 + 0.21$/mile $21 + 0.21(320) = \$88.20$, B: $32 + 0.18$/mile $32 + 0.18(320) = 89.60. The better rate is given by agency A. For equal charges, $x \approx 367$ miles   If the mileage is greater than 367, agency B has the lower
rate.   **5. (a)** $f(x) = 3x + 4$ or $f(x) = -3x - 8$ **(b)** $f(x) = d, d \in \mathbb{R}$ or $f(x) = x$

7. (a)

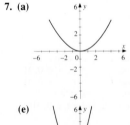

(b)

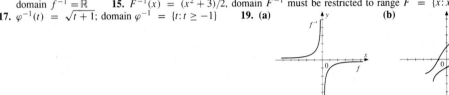

(c)

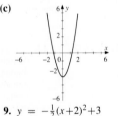

(d)

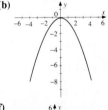

(e)

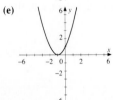

(f)

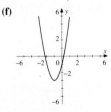

9. $y = -\frac{1}{2}(x+2)^2 + 3$ **11.** $P_{\max} = \$44,000$ at $x = 800$

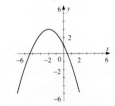

13. $R_{\max} = \$3{,}698$ at $x = 7$ **15. (a)** $(f \circ g)(x) = a(cx + d) + b = acx + (ad + b)$, which is linear.
(b) $(f \circ g)(x) = a_2(b_2x^2 + b_1x + b_0)^2 + a_1(b_2x^2 + b_1x + b_0) + a_0$ which is a quartic function. Therefore, the composition of two quadratic functions is not quadratic.   **(c)** $p \circ q$ is of degree mn.   **17.** If $f_n(x) = x^n$ and n is an even positive integer, then increasing n causes the graph to lie closer to the x-axis on $(0,1)$ and to rise more steeply for $x > 1$. The functions are even. If n is an odd positive integer, the function is odd. As n increases, the graph lies closer to the x-axis on $(0,1)$ and rises more steeply for $x > 1$. **(a)**

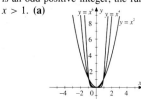

(b)

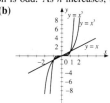

19. $f(x) = 3x^3 - 9x^2 + 12$

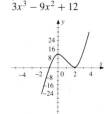

21. lowest possible value of $n = 5$, since the concavity changes at three points

23.

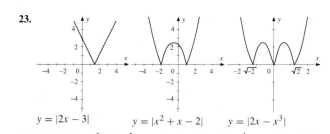

$$y = |2x - 3| \qquad y = |x^2 + x - 2| \qquad y = |2x - x^3|$$

25. (a) $a + b + c = 1$.
(b) vertex is at $\left(-\dfrac{b}{2a}, \dfrac{4ac - b^2}{4a}\right)$.
(c) $c = 6$.   **(d)** $c = 6$. $a = 5$, $b = -10$, domain $f = \mathbb{R}$; range $f = [1, \infty)$

27. (a) $f_1(x) = x^3 + b_1x^2 + c_1x + d_1$, where $b_1 = \dfrac{b}{a}$, $c_1 = \dfrac{c}{a}$, and $d_1 = \dfrac{d}{a}$. The graph of f is a contraction of the graph of f_1 if $|a| < 1$ and an expansion if $|a| > 1$.   **(b)** $f_2(x) = x^3 + c_2x + d_2$, where $c_2 = c_1 - \dfrac{b_1^2}{3}$ and $d_2 = \dfrac{2b_1^3}{27} - \dfrac{b_1c_1}{3} + d_1$. The graph of f_1 is the graph of f_2 shifted $\dfrac{b_1}{3}$ units to the left.   **(c)** $f_3(x) = f_2(x) - d_2 = x^3 + c_2x$ The graph of f_2 is displaced d_2 units upward from the graph of f_3.   **(d)** Let $k = c_2$, and write $g(x) = x^3 + kx$. Given the graph of g, find the graph of f_2 by displacing the graph d_2 units upward. The graph of f_1 is obtained by shifting the graph of f_2 to the left by $\dfrac{b_1}{3}$ units. The graph of f is now a contraction or expansion of the graph of f_1, as explained in part **(a)**.   **(e)** If $k = 0$, the graph of g is simply that of the cubic x^3. If $k < 0$, the graph has zeros at 0, $\sqrt{k}$, and $-\sqrt{k}$. It has a local maximum at $x = -\sqrt{k/3}$ and a local minimum at $x = \sqrt{k/3}$. If $k > 0$, the graph has no local maxima or minima. It rises more rapidly than the graph of x^3.

29. (a) local maxima at $\left(-3, \dfrac{72}{5}\right)$ and $\left(1, \dfrac{88}{15}\right)$. Local minima at $\left(-1, -\dfrac{88}{15}\right)$ and $\left(3, -\dfrac{72}{5}\right)$.   **(b)** increasing on $(-\infty, -3) \cup (-1, 1) \cup (3, \infty)$. decreasing on $(-3, -1) \cup (1, -3)$.   **(c)** concave up on $(-\sqrt{5}, 0) \cup (\sqrt{5}, \infty)$. concave down on $(-\infty, -\sqrt{5}) \cup (0, \sqrt{5})$.

Exercise Set 1.5

1. (a) Vertical asymptote: $x = 2$. Horizontal asymptote: $y = 2$ **(b)** Vertical asymptotes: $x = -2, 2$. Horizontal asymptote: $y = 0$

3. Vertical asymptote:
$x = 3$
Horizontal asymptote:
$y = 0$
y-intercept: $-\dfrac{1}{3}$

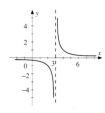

5. Vertical asymptotes:
$x = -1, 1$
Horizontal asymptote:
$y = 1$
x-intercepts: $-2, 2$
y-intercept: 4

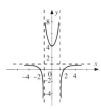

7.

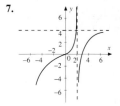

11. (a) oblique asymptote is $y = x$.   **(b)** oblique asymptote is $y = x - 4$.

13. (a) shift the graph of $y = \sqrt{x}$ two units to the left

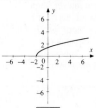

$y = \sqrt{x+2}$

(b) shift the graph of $y = \sqrt{x}$ one unit to the right.

$y = \sqrt{x-1}$

15. (a) $y = \sqrt{3}\sqrt{-(x-4/3)} - 2$; shift the graph of $y = \sqrt{x}$ to the right $\frac{4}{3}$ unit. Reverse the direction of the graph along the x-axis, and expand the graph in the y-direction by a factor of $\sqrt{3}$. Drop the resulting graph down 2 units.

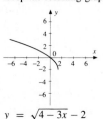

$y = \sqrt{4-3x} - 2$

(b) expand the graph of $y = \sqrt{x}$ in the y-direction by a factor of 3. Raise the graph 1 unit.

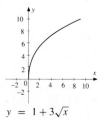

$y = 1 + 3\sqrt{x}$

17. (a) upper half of an ellipse

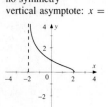

$y = \sqrt{9 - 4x^2}$

(b) upper half of an ellipse

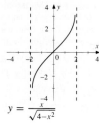

$y = \frac{3}{2}\sqrt{4 - x^2}$

19. (a) the graph opens upward like that of $y = x^2$, but does not rise so rapidly. The function is even.

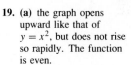

$y = x^{3/2}$

(b) the graph is a displacement two units to the right of the graph of $y = x^{2/3}$.

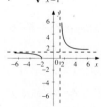

$y = (x-2)^{2/3}$

21. domain $f = \{x : -2 < x \le 2\}$
x-intercept: $x = 2$
y-intercept: $y = 1$
no symmetry
vertical asymptote: $x = -2$

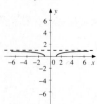

23. domain $f = \{x : |x| < 2\}$, x-intercept: $x = 0$, y-intercept: $y = 0$, symmetry: f is symmetric about the origin, vertical asymptotes: $x = \pm 2$

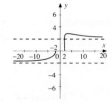

$y = \dfrac{x}{\sqrt{4-x^2}}$

25. $y = \sqrt{\dfrac{x+2}{x-1}}$

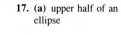

27. $y = \sqrt{\dfrac{x^2-1}{x^2+1}}$

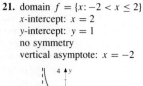

29. $y = \dfrac{2\sqrt{x^2-4}}{x-1}$

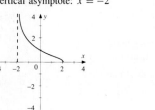

Exercise Set 1.6

1.

x	0	$\pi/6$	$\pi/4$	$\pi/3$	$\pi/2$
$\sin x$	0	$1/2$	$\sqrt{2}/2$	$\sqrt{3}/2$	1
$\cos x$	1	$\sqrt{3}/2$	$\sqrt{2}/2$	$1/2$	0
$\tan x$	0	$\sqrt{3}/3$	1	$\sqrt{3}$	DNE
$\cot x$	DNE	$\sqrt{3}$	1	$\sqrt{3}/3$	0
$\sec x$	1	$2\sqrt{3}/3$	$\sqrt{2}$	2	DNE
$\csc x$	DNE	2	$\sqrt{2}$	$2\sqrt{3}/3$	1

x	$2\pi/3$	$3\pi/4$	$5\pi/6$	π
$\sin x$	$\sqrt{3}/2$	$\sqrt{2}/2$	$1/2$	0
$\cos x$	$-1/2$	$-\sqrt{2}/2$	$-\sqrt{3}/2$	-1
$\tan x$	$-\sqrt{3}$	-1	$-\sqrt{3}/3$	0
$\cot x$	$-\sqrt{3}/3$	-1	$-\sqrt{3}$	DNE
$\sec x$	-2	$-\sqrt{2}$	$-2\sqrt{3}/3$	-1
$\csc x$	$2\sqrt{3}/3$	$\sqrt{2}$	2	DNE

x	$7\pi/6$	$5\pi/4$	$4\pi/3$	$3\pi/2$	$5\pi/3$	$7\pi/4$	$11\pi/6$
$\sin x$	$-1/2$	$-\sqrt{2}/2$	$-\sqrt{3}/2$	-1	$-\sqrt{3}/2$	$-\sqrt{2}/2$	$-1/2$
$\cos x$	$-\sqrt{3}/2$	$-\sqrt{2}/2$	$-1/2$	0	$1/2$	$\sqrt{2}/2$	$\sqrt{3}/2$
$\tan x$	$\sqrt{3}/3$	1	$\sqrt{3}$	DNE	$-\sqrt{3}$	-1	$-\sqrt{3}/3$
$\cot x$	$\sqrt{3}$	1	$\sqrt{3}/3$	0	$-\sqrt{3}/3$	-1	$-\sqrt{3}$
$\sec x$	$-2\sqrt{3}/3$	$-\sqrt{2}$	-2	DNE	2	$\sqrt{2}$	$2\sqrt{3}/3$
$\csc x$	-2	$-\sqrt{2}$	$-2\sqrt{3}/3$	-1	$-2\sqrt{3}/3$	$-\sqrt{2}$	-2

3. amplitude = 2, period = 2π

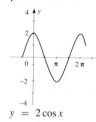

$y = 2\cos x$

5. amplitude = 3, period = 4π

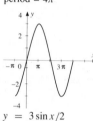

$y = 3\sin x/2$

7. amplitude = 1, period = $\frac{2\pi}{\pi} = 2$

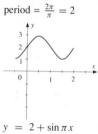

$y = 2 + \sin \pi x$

9. amplitude = $\frac{1}{2}$, period = $\frac{2\pi}{2} = \pi$

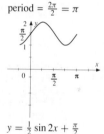

$y = \frac{1}{2}\sin 2x + \frac{\pi}{2}$

11. amplitude = 3, period = $\frac{2\pi}{2} = \pi$

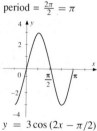

$y = 3\cos(2x - \pi/2)$

13. amplitude = $\frac{3}{2}$, period = $\frac{2\pi}{2} = \pi$

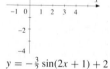

$y = -\frac{3}{2}\sin(2x + 1) + 2$

15.

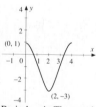

Period = 4. The graph of $\cos x$ has been expanded by 2 and shifted down 1 unit. Thus,

$$y = 2\cos\left(\frac{\pi x}{2}\right) - 1.$$

17.

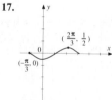

Period = $4\left(\frac{\pi}{3}\right)$; graph of $\cos x$ is reversed about the x-axis and has a contraction of $\frac{1}{2}$. Thus,

$$y = -\frac{1}{2}\cos\left(\frac{3}{2}x\right).$$

19. Period = $(4 - 1)(2) = 6$ (y attains minimum and maximum values at $x = 1$ and $x = 4$, respectively). $ax = 2\pi$ when $x = 6$; $6a = 2\pi$; $a = \frac{\pi}{3}$ The graph of $\cos x$ has been shifted 4 units to the right and raised 2 units on the y-axis. Thus,

$$y = \cos\left[\frac{\pi}{3}(x - 4)\right] + 2.$$

21. (a) Smallest positive value is $\frac{\pi}{3}$.
(b) $x = \frac{7\pi}{12}$ (c) $x = \frac{\pi}{12}$

23.

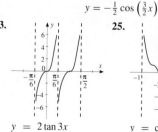

$y = 2\tan 3x$

25.

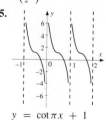

$y = \cot \pi x + 1$

27. Asymptotes $x = \frac{1}{b}(2n+1)\frac{\pi}{2} - \frac{c}{b}$. The factor a causes expansion of the graph in the y-direction if $|a| > 1$ and contraction if $0 < |a| < 1$. If $a < 0$, the graph is reversed about the x-axis.

29.

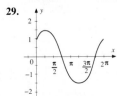

The graph of $y = \sin x + \cos x$ is periodic with period 2π. maximum of $\sqrt{2}$ at $x = \frac{\pi}{4}$; minimum of $\sqrt{2}$ at $x = \frac{5\pi}{4}$; zeros occur at $\frac{3\pi}{4}$ and $\frac{7\pi}{4}$. In terms of the sine function alone, $y = \sqrt{2}\sin\left(x + \frac{\pi}{4}\right)$.

31.

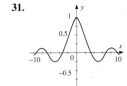

(a) The graph appears to approach 1 as $x \to 0$. The function is periodic with period 2π. Its amplitude $\to 0$ as $x \to \pm\infty$.

(b)

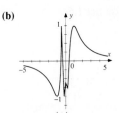

$y = \sin\left(\frac{1}{x}\right) (x \neq 0)$

The graph oscillates more and more rapidly as $x \to 0$. $f(x) \to 0$ as $x \to \pm\infty$.

(c)

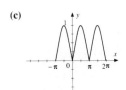

$y = \sin^2 x$
The function is periodic with period π. Its range is $[0,1]$.

(d)

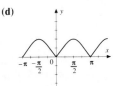

$y = |\sin x|$
The graph is the same as that of $y = \sin x$ wherever $\sin x \geq 0$ and is its reflection across the x-axis wherever $\sin x < 0$.

(e)

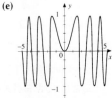

$y = \sin x^2$
The graph oscillates more and more rapidly as $x \to \pm\infty$. Since $x^2 \geq 0$, the function is even.

(f)

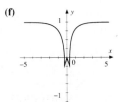

$y = x \sin\frac{1}{x}$ $(x \neq 0)$
The graph appears to approach 1 as $x \to \pm\infty$. It appears to approach 0 as $x \to 0$. The function is even.

33. **(a)** $x = \frac{2}{(2n+1)\pi}$, $n = 0, \pm 1, \pm 2, \ldots$

(b) $x = \frac{2}{(4n+3)\pi}$, $n = 0, \pm 1, \pm 2, \ldots$

35.

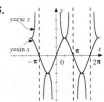

Chapter Review

1. **(a)** domain $f = \left(-\infty, -\frac{7}{3}\right] \cup [5, \infty)$

(b) domain $g = (-\infty, -1) \cup (-1, 1] \cup (3, \infty)$

3. $(f \circ g)(t) = f(g(t)) = -t - 2$
domain $f \circ g = (-\infty, 1]$

$(g \circ f)(t) = \sqrt{4 - t^2}$
domain $g \circ f = [-2, 2]$

5. **(a)** $f(-3) = 3$
(b) $f(0) = \frac{1}{2}$
(c) $f(1) = 1$
(d) $f(1.8) = 5$
(e) vertical asymptote: $x = 2$

7. $(g \circ h)(t) = -\frac{t+2}{t+1}$
domain $g \circ h = \{t : t \neq -2, -1\}$
$(h \circ g)(t) = \frac{t-1}{2t-1}$

domain $h \circ g = \left\{t : t \neq \frac{1}{2}, 1\right\}$

9. **(a)** $(f - g)(3) = 11$
(b) $\left(\frac{f}{g}\right)(-1) = -2$
(c) $(f \circ g)(x) = \sec^2\frac{\pi x}{4}$

11. **(a)** $y = f(x) + 1$

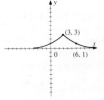

(b) $y = f(x+1)$ **(c)** $y = 2f(x)$ **(d)** $y = f(-x)$ **(e)** $y = f\left(\frac{x}{2}\right)$

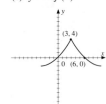

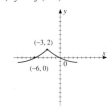

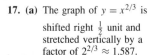

13. (a) $f^{-1}(x) = \frac{1}{(1-x)^2}$ **(b)** $f^{-1}(x) = \frac{3x+2}{x-1}$

15. zeros of f are $0, 0, -2, 4$

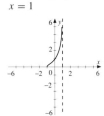

(a) increasing on $(-\infty, -1.4) \cup (0, 2.9)$
(b) decreasing on $(-1.4, 0) \cup (2.9, \infty)$
(c) concave up on $(-0.76, 1.76)$
(d) concave down on $(-\infty, -0.76) \cup (1.76, \infty)$

17. (a) The graph of $y = x^{2/3}$ is shifted right $\frac{1}{2}$ unit and stretched vertically by a factor of $2^{2/3} \approx 1.587$.

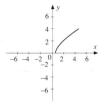

(b) domain $f = [-1, 1)$
vertical asymptote: $x = 1$

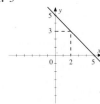

19. (a) period $= 9 - 1 = 8$,
amplitude $= 2$,
horizontal shift $= 3$,
vertical shift $= -2$

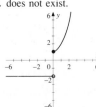

(b) period $= 2\pi$
amplitude $= 2$
horizontal shift $= -\frac{5}{6\pi}$,
vertical shift $= 1$
$y = 2\sin\left(x + \frac{5\pi}{6}\right) + 1$

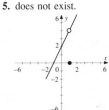

CHAPTER 2

Exercise Set 2.1

1. 3

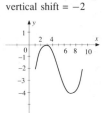

3. does not exist.

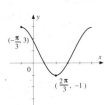

5. does not exist.

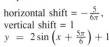

7. $\lim_{x \to 2^+} f(x) = \lim_{x \to 2^-} f(x) = 2$; $\lim_{x \to 2} f(x) = 2$

9. $\lim_{x \to 2^+} f(x)$ does not exist; $\lim_{x \to 2^-} f(x) = 0$; $\lim_{x \to 2} f(x)$ does not exist.

11. $\lim_{x \to 2^+} f(x)$ does not exist; $\lim_{x \to 2^-} f(x)$ does not exist; $\lim_{x \to 2} f(x)$ does not exist.

13. $\lim_{x \to 2^+} f(x) = 4$; $\lim_{x \to 2^-} f(x) = 4$; $\lim_{x \to 2} f(x)$ exists; its value is 4.

15. $\lim_{x \to 1} f(x) = 2$; $\lim_{x \to -1^-} f(x) = 2$; $\lim_{x \to 1} f(x)$ exists; its value is 2.

17. $\lim_{x \to 1^+} f(x) = 1$; $\lim_{x \to 1^-} f(x) = -1$; $\lim_{x \to 1} f(x)$ does not exist.

19.

x	-2.1	-2.01	-2.001	-1.9	-1.99	-1.999
$f(x)$	12.61	12.06	12.006	11.41	11.94	11.994

Conjecture: $\lim_{x \to -2} f(x) = 12$

21.

t	1.1	1.01	1.001	0.9	0.99	0.999
$f(t)$	-1.281	-1.328	-1.333	-1.393	-1.339	-1.334

Conjecture: $\lim_{t \to 1} f(t) = -\frac{4}{3}$

23.

x	4.1	4.01	4.001	3.9	3.99	3.999
$f(x)$	0.3315	0.3331	0.3333	0.3352	0.3335	0.3334

Conjecture: $\lim\limits_{x\to 4} f(t) = \frac{1}{3}$

25.

x	0.1	0.01	0.001
$f(x)$	0.4996	0.499995	0.50000

Conjecture: $\lim\limits_{x\to 0} f(x) = \frac{1}{2}$

27.

x	$\frac{\pi}{2} - 0.1$	$\frac{\pi}{2} - 0.01$	$\frac{\pi}{2} - 0.001$	$\frac{\pi}{2} - 0.0001$
$f(x)$	1.2580	1.0471	1.00693	1.00092

Conjecture: $\lim\limits_{x\to(\pi/2)^-} f(x) = 1$

29. $\lim\limits_{x\to 2}(3x-1) = 5$, $x \in \left(\frac{5.99}{3}, \frac{6.01}{3}\right)$, $x \in \left(\frac{5.999}{3}, \frac{6.001}{3}\right)$

31. $\lim\limits_{x\to 0} x\sin\left(\frac{1}{x}\right) = 0$, $x \in (-0.01, 0.01)$, $x \in (-0.001, 0.001)$.

Exercise Set 2.2

1. (a) 0 (b) -2 **3.** (a) 1 (b) $\sqrt{2}$ **5.** -4 **7.** 5   **9.** -7 **11.** $\frac{1}{2}$ **13.** $\frac{3}{8}$ **15.** $\frac{1}{2}$

17. $f(x) = x - 2$
domain
$f = \{x : x \neq -1\}$
$\lim\limits_{x\to-1^+} f(x) = -3$
$\lim\limits_{x\to-1^-} f(x) = -3$
$\lim\limits_{x\to-1} f(x) = -3$

19. $f(x) = 2x - 1$
domain $f = \{x : x \neq \frac{1}{2}\}$
$\lim\limits_{x\to-\frac{2}{3}^+} f(x) = -\frac{7}{3}$
$\lim\limits_{x\to-\frac{2}{3}^-} f(x) = -\frac{7}{3}$
$\lim\limits_{x\to-\frac{2}{3}} f(x) = -\frac{7}{3}$

21. $f(x) = x^2$
domain $f = \{x : x \neq -2\}$
$\lim\limits_{x\to 2^+} f(x) = \lim\limits_{x\to-2^-} f(x) = 4$
$\lim\limits_{x\to-2} f(x) = 4$

23. $f(x) = \frac{1}{x-1}$
domain $f = \{x : x \neq -1, 1\}$
$\lim\limits_{x\to-1^+} f(x) = \lim\limits_{x\to-1^-} f(x) = -\frac{1}{2}$
$\lim\limits_{x\to-1} f(x) = -\frac{1}{2}$

25. 2   **27.** -1 **29.** 1   **31.** (a) Let $f(x) = \begin{cases} 1 & \text{if } x \geq 0 \\ 0 & \text{if } x < 0 \end{cases}$ $g(x) = \begin{cases} 0 & \text{if } x \geq 0 \\ -1 & \text{if } x < 0 \end{cases}$ **(b)** Let $f(x) = \begin{cases} 1 & \text{if } x \geq 0 \\ 0 & \text{if } x < 0 \end{cases}$

$g(x) = \begin{cases} 0 & \text{if } x \geq 0 \\ -1 & \text{if } x < 0 \end{cases}$ **33.** for $-\frac{\pi}{4} < x < \frac{\pi}{4}$, $0 \leq \lim\limits_{x\to 0} x^2\cos 2x \leq \lim\limits_{x\to 0} x^2 = 0$ therefore $\lim\limits_{x\to 0} x^2\cos 2x = 0$

Exercise Set 2.3

1. continuous except at $x = 0$ **3.** continuous except at $x = 2$ **5.** continuous except at $x = 0$ **7.** continuous except at $x = 3$
9. continuous except at $x = (2n+1)\frac{\pi}{2}$ **11.** continuous at $x = 2$ **13.** continuous at $x = 3$ **15.** not continuous at $x = -1$
17. continuous at $x = 2$ **19.** continuous at $x = 0$ **21.** not continuous at $x = -3$ **23.** not continuous at $x = 0$
25. continuous on $(-\infty, \infty)$.   **27.** continuous on $(-\infty, 0] \cup (0, \infty)$.   **29.** G is continuous on $(-\infty, 0) \cup [0, \infty)$
31. continuous on all of $\mathbb{R}$.   **33.** discontinuous at $x = 2$ only.   **35.** define $f(3) = \frac{6}{5}$ **37.** ϕ is discontinuous at all points

39. $f(x) = \frac{1}{x}$ on $(0, 1)$ **41.** Yes, $\phi(x)$ of Exercise 37 is discontinuous at every point. Yes, $\psi(x)$ of Exercise 38 is continuous only at

$x = 0$.   **43.** (a) False. Let $f(x) = \begin{cases} 1 & \text{if } x \geq 0 \\ 0 & \text{if } x < 0 \end{cases}$ $g(x) = \begin{cases} 0 & \text{if } x \geq 0 \\ 1 & \text{if } x < 0 \end{cases}$ **(b)** False. Use same two functions as in part **(a)**.

45. Show $P_n(a)$ and $P_n(-a)$ have opposite signs for a large. Then use the intermediate value theorem.   **47.** continuous for all real x when x is not an integer and is discontinuous when x is an integer value.

Exercise Set 2.4

1. (a) $-\infty$ (b) ∞ **3.** (a) ∞ (b) $-\infty$ **5.** (a) $-\infty$ (b) ∞ **7.** 0   **9.** 1   **11.** 2   **13.** does not exist.
15. (a) 2 (b) -2 **17.** (a) $\lim\limits_{x\to+\infty} f(x) = 1$ (b) $\lim\limits_{x\to+\infty} f(x) = 0$ **19.** (a) $\lim\limits_{x\to\infty} f(x) = 0$ (b) $\lim\limits_{x\to\infty} f(x)$ does not exist

$\lim\limits_{x\to-\infty} f(x) = -1$ $\lim\limits_{x\to-\infty} f(x) = 0$ $\lim\limits_{x\to-\infty} f(x) = 0$ $\lim\limits_{x\to-\infty} f(x) = -1$

21. vertical: $x = 2, x = -2$ **23.** vertical: $x = 3, x = -2$ **25.** vertical: $x = -1$ no horizontal asymptote
horizontal: $y = 0$ horizontal: $y = 2$

27. vertical: $x = 3$, horizontal: $y = 0$ **29.** vertical: $x = -1$, horizontal: $y = 2$ **31.** horizontal: $y = 0$, vertical: $x = 1$

33. horizontal: $y = 0$ (on right), no vertical   **35.** vertical: $\begin{cases} x = \pm\frac{\pi}{6} + 2n\pi \\ x = \pm\frac{5\pi}{6} + 2n\pi \end{cases}$ $n = 0, \pm 1, \pm 2, \ldots$, no horizontal

37. Conjecture: $\lim\limits_{x\to+\infty}\left(1 + \frac{1}{x}\right)^x \approx 2.7182$ **39.** Property 2: Consider $\lim\limits_{x\to-1}\left[\frac{1}{x+1} + \frac{x}{x+1}\right] = \lim\limits_{x\to-1}\frac{1+x}{x+1} = 1$.

Property 3: Consider $\lim\limits_{x\to 1}\left[\frac{1}{x-1} - \frac{x}{x-1}\right] = \lim\limits_{x\to 1}\frac{1-x}{x-1} = -1$.

Property 4: Consider $\lim\limits_{x\to 0}\left[x \cdot \frac{1}{x}\right]$.

Property 5: Consider $\lim\limits_{x\to 0}\frac{x}{1/x}$.

41. (a) Find $P(1)$, $P(2)$, and extend to $P(n)$.

(b)

How Compounded	Number of Times Per Year, n	$P(n)$
Annually	1	$1100.00
Semiannually	2	$1102.50
Quarterly	4	$1103.81
Monthly	12	$1104.71
Weekly	52	$1105.06
Daily	365	$1105.16
Hourly	8760	$1105.17

(c) Conjecture: $\lim\limits_{n \to \infty} P(n) = 1000e^{0.10} = \$1105.17$

43. 20,000   **45.** $\approx 2.39 \times 10^3$ m/sec

Exercise Set 2.5

1. Take $\delta = \frac{\epsilon}{2}$.   **3.** Take $\delta = 3\epsilon$.   **5.** Take $\delta = \frac{\epsilon}{4}$.   **7.** Take $\delta = \frac{\epsilon}{2}$.   **9.** Take $\delta = 5\epsilon$.   **11.** Choose $\delta = \epsilon$.
13. Choose $\delta = \frac{\epsilon}{2}$.   **15.** Choose $\delta = \epsilon$.   **17.** There is a $\delta_1 > 0$ such that if $0 < |x - a| < \delta_1, |f(x) - 0| < 1$.   **19.** By

Problem 15 and Problem 18, we know that $\lim\limits_{x \to a} |x| = |a|$. $|x|$ is continuous at a, hence on all of $\mathbb{R}$.   **21.** Choose $\delta^+ = \sqrt{\frac{2}{M}}$.

Choose $\delta^- = \frac{3}{M}$, and let $\delta = \frac{3}{M}$.   **23.** Choose $\delta < \frac{2}{M}$ **25.** $\frac{1}{2}$ Choose $N = \frac{3 - 2\epsilon}{4\epsilon}$ **27.** 0   Choose $N = \frac{2}{\epsilon}$
29. (a) $\lim\limits_{x \to \infty} f(x) = \infty$ if $f(x) > M$ for all $x > N$ in the domain of f where M is an arbitrary positive constant.   **(b)** $\lim\limits_{x \to \infty} f(x) = -\infty$
if $f(x) < -M$ for all $x > N$ in the domain of f where M is an arbitrary positive constant.   **(c)** $\lim\limits_{x \to -\infty} f(x) = \infty$ if $f(x) > M$ for
all $x < -N$ in the domain of f where M is an arbitrary positive constant and $N > 0$.   **(d)** $\lim\limits_{x \to -\infty} f(x) = -\infty$ if $f(x) < -M$ for
all $x < -N$ in the domain of f where M is an arbitrary positive constant and $N > 0$.   **31.** Choose $\delta = \frac{\epsilon}{2a}$ **33.** Choose
$\delta = \min\left\{\frac{1}{2}, \frac{\epsilon}{2}\right\}$.   **35.** Choose $\delta = \min\left\{1, \frac{\epsilon}{3}\right\}$.   **37.** For $a = 0$, let $\delta = \epsilon$. For $a \neq 0$, assuming continuity leads to a
contradiction.   **39.** By continuity of f at a, $f(a) = \lim\limits_{x \to a} f(x)$; so let $f(a) = L > 0$ in Exercise 38. This holds for $x = a$ also.
If $f(a) = L < 0$, the proof is analogous.

Chapter Review

1. (a) ∞ **(b)** $-\infty$ **(c)** $-\infty$ **(d)** ∞ **(e)** 1   **(f)** 1   **(g)** 0 and 1.   **3. (a)** 0   **(b)** $\frac{1}{4}$ **5. (a)** does not exist.   **(b)** does not
exist.   **7. (a)** $-\frac{1}{4}$ **(b)** -1 **9. (a)** 0   **(b)** 0   **11.** $\frac{1}{\sqrt{2x+3}}, \frac{1}{\sqrt{x}}$ **13. (a)** Discontinuous at $x = 2$.   **(b)** Discontinuous at
$x = \pm\sqrt{2}$ **15.** Take $\delta = \frac{\epsilon}{2}$.   **17.** Take $\delta = \min\{2, \epsilon\}$.   **19.** Choose $\delta = \frac{\epsilon}{|m|}$.   **21.** Choose $M = \frac{3}{\sqrt{\epsilon}}$

CHAPTER 3

Exercise Set 3.1

1. (a)

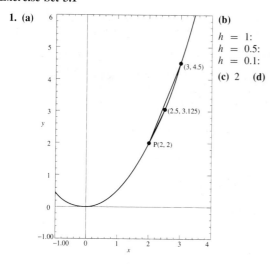

(b)

$h = 1$:   slope = 2.5
$h = 0.5$:   slope = 2.25
$h = 0.1$:   slope = 2.05

(c) 2   **(d)** $y = 2x - 2$

3. $2x + y = 2$ **5.** $x + 2y = -3$

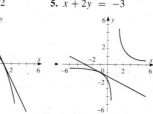

7. $2x - y = 5$

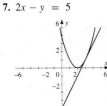

9. $4x + 2y + 1 = 0$

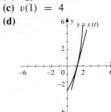

11. (a) 6; 5; 4.2; 4.02
(b) $v_{av} = 4 + 2h$
(c) $v(1) = 4$
(d)

13. (a) $v(t) > 0$
(b) $v(t) < 0$
(c) $v(t)$ increasing
(d) $v(t)$ decreasing
(e) $v(t_0) = 0$

15. (a) -8 m/sec   **(b)** $v(5) = 0$ **17.** 2   **19.** $\frac{1}{2}$ **21.** 5   **23.** $-\frac{4}{9}$ **25.** $\frac{9}{2}$ **27.** $2(x - 1)$

29. $\frac{1}{(x+1)^2}$ **31.** $x - 3y + 5 = 0$ **33.** Increasing on $(-\infty, 1) \cup (1, \infty)$ and decreasing on $(-1, 1)$.

35. (a) Increasing on $(0.45, 3.72) \cup (5.42, 6.86) \cup (5.42 + n\pi, 6.86 + n\pi), n = 1, 2, \ldots$ Also on
$(-0.45, -3.72) \cup (-5.42, -6.86) \cup (-5.42 - n\pi, -6.86 - n\pi), n = 1, 2, \ldots$ Decreasing on $(-0.45, 0.45) \cup (3.72, 5.42)$
$\cup(3.72 + n\pi, 5.42 + n\pi), n = 1, 2, \ldots$ Also on $(-5.42, -3.72) \cup (-5.42 - n\pi, -3.72 - n\pi), n = 1, 2, \ldots$ **(b)** ≈ -0.10 m/s

Exercise Set 3.2

1. $3; 3x - y + 4 = 0$ **3.** $-4; 4x + y + 8 = 0$ **5.** $1; x - y - 1 = 0$ **7.** $\frac{1}{6}; \frac{1}{6}x - 6y + 9 = 0$ **9.** $-2; 2x + y + 3 = 0$

11. $3; 3x - y + 4 = 0$ **13.** $-4; 4x + y + 8 = 0$ **15.** $1; x - y - 1 = 0$ **17.** $\frac{1}{6}; \frac{1}{6}x - 6y + 9 = 0$ **19.** $-2; 2x + y + 3 = 0$

21. 2   **23.** $\frac{-2}{x^2}$ **25.** $\frac{1}{(x+1)^2}$ **27. (a)**

(b)

29. (a)

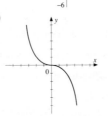

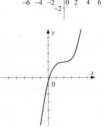

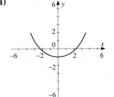

(b)

31. (a)

(b)

33. $\frac{-1}{(2x-3)^{3/2}}$

domain $f' = \left(\frac{3}{2}, \infty\right)$

35. $\frac{3}{2}x^{1/2}$ domain $f' = [0, \infty)$ **37.** 0   **39.**

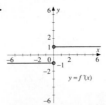

$y = f'(x)$

41. $\frac{1}{(t+1)^2}$ **43.** $-2\frac{Gm_1 m_2}{r^3}$

domain $f' = (-\infty, 0) \cup (0, \infty)$
f' and g are the same.

45. (a) The graph of $y = f'(x)$ appears to be that of $y = \cos x$.

(b) The graph of $y = g'(x)$ appears to be that of $y = -\sin x$.

47. Conjecture: $g'(x) = e^x$

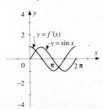

$y = f'(x)$
$y = \sin x$

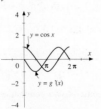

$y = \cos x$
$y = g'(x)$

Exercise Set 3.3

1. 0 **3.** $4x^3$ **5.** -2 **7.** a **9.** $12x^2 - 14x + 8$ **11.** $5x^4 - 24x^3$ **13.** $2ax + b$ **15.** $4x + 3$ **17.** $18x + 24$
19. $15x^2 - 6x - 6$ **21.** $y = 5x - 4$ **23.** $k = \frac{4}{3}$ **25.** tangent line: $2x - y - 9 = 0$; normal line: $x + 2y + 3 = 0$ **27.** 20
29. -10 **31.** $a = \frac{5}{4}, b = -4$ **33.** $(2, 4), (-2, 4)$; (1) $y = 4x - 4$; (2) $y = -4x - 4$
35. $v(t) = 48t^5 - 144t^3 + 108t$; $a(t) = 240t^4 - 432t^2 + 108$

Exercise Set 3.4

1. $f'(x) = 12x - 1$ **3.** $-20x^3 + 6x$ **5.** $f'(x) = 14x^6 + 4x^3 - 1$ **7.** $-12x^3 + 18x^2 - 10x - 8$
9. $7x^6 - 15x^4 + 12x^3 + 9x^2 - 12x - 1$ **11.** $70x^6 + 72x^5 - 75x^4 - 52x^3 + 60x^2 - 21$ **13.** $\frac{-2}{(x-2)^2}$ **15.** $\frac{4x}{(x^2+1)^2}$ **17.** $\frac{-8x}{(x^2-4)^2}$
19. $\frac{-2x^2+6x+6}{(x^2+2x)^2}$ **21.** $\frac{-27x^2}{(x^3-1)^2}$ **23.** $-6x^{-4}$ **25.** $2x - 2x^{-3}$ **27.** $1 - 4x^{-2}$ **29.** $\frac{8}{x^3}$ **31.** $4x + y = 13$ **33.** $\frac{27}{16}$
35. If g is continuous at $x = a$, then $\lim\limits_{x \to a} g(x) = g(a)$. Let $x = a + h$ and consider $\lim\limits_{h \to 0} g(x + h)$. **37.** $-\frac{2(9x^4-18x^3+14x^2-5x+5)}{[(3x^2+5)(x^2-2x)]^2}$
39. $1 - \frac{2x}{(x^2+1)^2}$ **41.** $2x(2x+1)(3x^2-4x) + (2x+1)(6x-4)(x^2+4) + 2(3x^2-4x)(x^2+4)$

Exercise Set 3.5

1. $y' = 8x^3$, $y'' = 24x^2$, $y''' = 48x$ **3.** $y' = 12x^2 - 4x$, $y'' = 24x - 4$, $y''' = 24$ **5.** $y' = 2x + 2x^{-3}$, $y'' = 2 - 6x^{-4}$,
$y''' = 24x^{-5}$ **7.** $12x + 7$; 12 **9.** $4x^3 - 6x - \frac{5}{x^2}$; $12x^2 - 6 + \frac{10}{x^3}$ **11.** $\frac{x^2-3}{x^2}$; $\frac{6}{x^3}$ **13.** $3x^2 - 6x$ **15.** $\frac{5}{(x+3)^2}$
17. $6x - 6x^{-4}$ **19.** $2 + \frac{3}{2}x^{-4}$ **21.** $n! a_n$; 0 **23.** $f'(x) = \frac{1}{2}x^{-1/2}$; $f''(x) = -\frac{1}{2^2}x^{-3/2}$,
$f'''(x) = \frac{1 \cdot 3}{2^3}x^{-5/2}$; $f^{(4)}(x) = -\frac{1 \cdot 1 \cdot 5}{2^4}x^{-7/2}$, $f^{(n)}(x) = (-1)^n \frac{(-1)1 \cdot 3 \cdot 5 \cdots (2n-3)}{2^n}x^{-(2n-1)/2}$, $(n = 1, 2, \ldots)$
25. $v(t) = 4t - \frac{t^2+6t+1}{t^2+6t+9}$, $a(t) = 4 - \frac{16}{(t+3)^3}$, $v(1) = \frac{7}{2}, a(1) = \frac{15}{4}$
27. $y' = 2 - \frac{1}{x^2}$; $y'' = \frac{2}{x^3}$ Substitute into the differential equation **29.** (a) $6(3x - 2)$ (b) $8(x + 2)$ (c) $4(x - 1)^3$ (d) $6x^2(x^3 + 1)$

Exercise Set 3.6

1. 2 **3.** 1 **5.** 0 **7.** $\frac{3}{2}$ **9.** $\frac{2}{5}$ **11.** $2x \sin x + x^2 \cos x$ **13.** $\frac{-2 \sin x \cos x}{(1+\sin x)^2}$ **15.** $\frac{x \cos x - \sin x}{x^2}$
17. $-2 \sin x + \cos x$ **19.** $\frac{1}{1-\sin 2x}$ **21.** (a) $2 \sin x \cos x$ (b) $-2 \sin x \cos x$ **23.** $y = \frac{\pi}{2}$ **25.** $-x \cos x - \sin x$
27. $\frac{x(1+\sin^2 x)+2\sin x \cos x}{\cos^3 x}$ **29.** $v(t) = (t+1)\cos t + \sin t$; $a(t) = -(t+1)\sin t + 2\cos t$ **31.** $y' = x(-2\sin x + \cos x) + (2\cos x + \sin x)$;
$y'' = -x(2\cos x + \sin x) + 2(\cos x - 2\sin x)$. Substitute into equation.
33. Let $\sec x = \frac{1}{\cos x}$ and use quotient rule. **35.** (a) $s(0) = 3$; $v(0) = 2$ (b) $s\left(\frac{2\pi}{3}\right) \approx 0.2321$, $v\left(\frac{2\pi}{3}\right) \approx -3.5981$, $a\left(\frac{2\pi}{3}\right) \approx -0.2321$,
moving to the left and speeding up. (c) $t \approx 2.1588$ **37.** No $\lim\limits_{x \to 0} \frac{\sin x}{x} = \frac{\pi}{180} \approx 0.01745$.
39. $\frac{d}{dx}(2\cos^2 x - 1) = -4\sin x \cos x = -2\sin 2x$ **41.** (a) odd (b) $\frac{1+\cos x+x+\sin x}{\cos x}$ (c) $y \approx 6.826x - 3.322$

Exercise Set 3.7

1. $8x(x^2 + 2)^3$ **3.** $-20x(1 - 2x^2)^4$ **5.** $120x^3(3x^4 - 2)^9$ **7.** $3(x^2 - 3x + 4)^2(2x - 3)$ **9.** $-8x(3x^2 - 1)(3x^4 - 2x^2 + 1)^{-3}$
11. $2(2x + 1)(x^2 + 2)^2(8x^2 + 3x + 4)$ **13.** $\frac{-12x}{(x^2+1)^4}$ **15.** $\frac{(3x-4)^2(6x+13)}{(x+1)^2}$ **17.** $\frac{-2x^3-6x}{(x^2-1)^3}$ **19.** $\frac{6(x-1)^2}{(x+1)^4}$ **21.** $3\cos 3x$
23. $2\sin(1 - 2x)$ **25.** $6x\sec^2(3x^2)$ **27.** $-6\csc 3x \cot 3x$ **29.** $10\sin x \cos x$ **31.** $4\tan(2x + 3)\sec^2(2x + 3)$
33. $6x\sec^2 x^2 \tan^2 x^2$ **35.** $-2\cos 2x$ **37.** $-3(3x + 2)\csc 3x \cot 3x + 3\csc 3x$ **39.** $-\sin x \cos(\cos x)$ **41.** $6(2x + 3)^2$
43. $-2(x^2 + \sin x)^{-3}(2x + \cos x)$ **45.** $18x^2(x^3 - 8)$ **47.** $12x(x^2 - 3)^2$ **49.** $-\frac{18x^2}{(x^3+4)^3}$ **51.** $2(\sin x + \cos x - \cos 2x)$
53. $8(x^2 + 1)^2(7x^2 + 1)$ **55.** $\frac{2}{(1-x)^3}$ **57.** $2[(1 - 2x^2)\cos 2x - 4x\sin 2x]$ **59.** $48\sec^5 2x - 36\sec^3 2x$
61. $(h \circ g)'(t) = \frac{-4t}{(t^2-1)^2}$, $(g \circ h)'(t) = \frac{-4(t+1)}{(t-1)^3}$ **63.** $3(y^2 + 3y - 1)(y^2 + 3y - 3)(2y + 3)$ **65.** $\frac{4(6x^3-5x^2-6x+1)}{(x^2+1)^4}$
67. $36x^2(2x^3 - 3)[(2x^3 - 3)^2 + 4]^2$ $f(x) = x^3, g(x) = x^2 + 4, h(x) = 2x^3 - 3$ **69.** $-2\sqrt{3} + 80\pi \frac{\sqrt{3}}{3}$, $32 + 16\sqrt{3} - \frac{1280\pi}{3}$
71. $\left[-\sin(y^2 + 2xy)\right]\left[2y\frac{dy}{dx} + 2x\frac{dy}{dx} + 2y\right]$ **73.** $f'(f_{n-1}(x))f'(f_{n-2}(x)) \ldots f'(f_2(x))f'(f(x))f'(x)$ **75.** (a) $-\frac{1}{x}\cos\frac{1}{x} + \sin\frac{1}{x}$
(b) $f'(0) = \lim\limits_{h \to 0} \sin\frac{1}{h}$, which does not exist. (c) If $x \neq 0$, f is the product of two continuous functions. If $x = 0$,
$\lim\limits_{x \to 0} x \sin\frac{1}{x} = f(0) = 0$

Exercise Set 3.8

1. $\frac{x}{y}$ **3.** $\frac{2(1-xy^3)}{1+3x^2y^2}$ **5.** $\frac{3x^2}{1+3y^2}$ **7.** $\frac{2xy}{4y-x^2}$ **9.** $\frac{1}{2y-1}$ **11.** $-\frac{\sin y+y\cos x}{\sin x+x\cos y}$ **13.** $\frac{3-y\sec(xy)\tan(xy)}{x\sec(xy)\tan(xy)}$ **15.** $\frac{1-\cos x\cos y}{1-\sin x \ \sin y}$

17. $\frac{x(2-x^2-y^2)}{y(2+x^2+y^2)}$ **19.** $\frac{x}{\sqrt{x^2+1}}$ **21.** $\frac{x}{(4-x^2)^{3/2}}$ **23.** $\frac{4x^2}{\sqrt[3]{2x^3+5}}$ **25.** $\frac{1}{(1-x^2)^{3/2}}$ **27.** $\frac{\cos x}{2\sqrt{\sin x}}$ **29.** $-\frac{1}{2x^{3/2}}\sec^2\left(\frac{1}{\sqrt{x}}\right)$

31. $-\frac{2\sec(2x+3)}{\tan^2(2x+3)}=-2\csc(2x+3)\cot(2x+3)$ **33.** $2\cos x$ **35.** $\frac{-4}{x^2\sqrt{4-x^2}}$ **37.** $-\frac{1}{y^3}$ **39.** $\frac{-4}{x^{1/2}y^2}$ **41.** $-\frac{1}{(1-x^2)^{3/2}}$

43. $-\left[\frac{1}{2\sqrt{x}}\right]^2\sin\sqrt{x}-\frac{1}{4x^{3/2}}\cos\sqrt{x}$ **45.** $y=y_0-\frac{b^2x_0}{a^2y_0}(x-x_0)$ **47.** $-\sin^2(x+y);\ -2\sin(x+y)\cos^3(x+y)$

49. (a) $f_1(x)=x+\sqrt{4-x^2};\ f_2(x)=x-\sqrt{4-x^2}$ **(b)** $f_1'(x)=\frac{\sqrt{4-x^2}-x}{\sqrt{4-x^2}};\ f_2'(x)=\frac{\sqrt{4-x^2}+x}{\sqrt{4-x^2}}$ **(c)** $y'=\frac{y-2x}{y-x}$

51. $v(t)=\frac{4t}{3(t^2-1)^{1/3}};\ a(t)=\frac{4(t^2-3)}{9(t^2-1)^{4/3}};\ v(3)=2;\ a(3)=\frac{1}{6}$ **53.** Product of the derivatives is -1

Exercise Set 3.9

1. $(2x+2)dx$ **3.** $-\frac{x\,dx}{\sqrt{4-x^2}}$ **5.** $\frac{2-x}{2(1-x)^{3/2}}dx$ **7.** $(x\sin x+\sin x)dx$ **9.** $(\sec x\tan x+\sec^2 x)dx$ **11.** $\frac{4x^5dx}{(x^6+1)^{1/3}}$

13. $\left(\frac{x\sec^2 x-\tan x}{x^2}\right)dx$ **15.** $(2x\sin 2x^2)dx$ **17.** $\frac{3x^2-2y^2}{4xy}$ **19.** $\frac{3y-4x}{8y-3x}$ **21.** $-\frac{y^{1/3}}{x^{1/3}}$ **23.** $\frac{y\cos x-\sin y}{x\cos y-\sin x}$

25. (a) $1.8\text{ cm}^2\approx\Delta A$ **(b)** $0.2\text{ cm}\approx\Delta P$ **27.** $0.001;\ 0.1\%;\ 0.0005;\ 0.05\%$ **29.** $\approx 1.958;\ 0.289\%$ **31.** $0.099;\ 0.015\%$

33. $1+2\left(x-\frac{\pi}{4}\right)$ **35.** $3x-4$ **37.** $4x-\frac{25}{3}$ **39.** $1+3x$ **41.** $1-2x$ **43.** $1-3x$

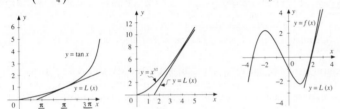

45. $V_{\max}\approx 16\frac{1}{4};\ V_{\min}\approx 15\frac{3}{4}$; Actual $V_{\max}\approx 16.2092$; Actual $V_{\min}\approx 15.7925$ **47.** \$3.   **49. (a)** $4\cos 2t\ dt$

51. $\frac{z(1+z^2)\sec^2 z+\tan z}{(1+z^2)^{3/2}}dz$ **53.** ≈ 26.74

Chapter Review

1. 9 **3.** $\frac{1}{9}$ **5.** $60x(3x^2-2)^9$ **7.** $\frac{1-x}{(1-2x)^{3/2}}$ **9.** $\frac{1}{1+\cos x}$ **11.** $-\frac{(a^{2/3}-x^{2/3})^{1/2}}{x^{1/3}}$ **13.** $\sqrt{2}\sec^2 x$

15. tangent line: $3x+y=4$; normal line: $x-3y=8$ **17.** $v(3)=12;\ a(3)=-\frac{52}{5}$ **19.** $\frac{4x^3+2y^2}{3y^2-4xy}$ **21.** $\frac{\sec y-y\sec^2 x}{\tan x-x\sec y\ \tan y}$

23. $1-\frac{a^{1/2}}{x^{1/2}},\ y''=\frac{a^{1/2}}{x^{3/2}}$ **25. (a)** $=33.9\text{ cm}\approx 50\%$ **(b)** $\approx 11.92\text{ cm}^2\approx 3.32\%$ Uncertainty in calorie count $\approx 5.0\%$

27. Substitute $y,\ y'=x^{-1/2}\cos x-\frac{1}{2}x^{-3/2}(1+\sin x)\ y''=-x^{-1/2}\sin x-\frac{1}{2}x^{-3/2}\ \cos x-\frac{1}{2x^{-3/2}}\cos x+\frac{3}{4}x^{-5/2}(1+\sin x)$ into the differential equation.

CHAPTER 4

Exercise Set 4.1

1. increasing on $(-\infty,-1)$; decreasing on $(-1,\infty)$ **3.** increasing on $\left(-\infty,-\frac{4}{3}\right)\cup(2,\infty)$; decreasing on $\left(-\frac{4}{3},2\right)$ **5.** increasing on $\left(-\infty,-\frac{2}{3}\right)\cup(2,\infty)$; decreasing on $\left(-\frac{2}{3},2\right)$ **7.** decreasing on $(-\infty,0)\cup(0,1)$; increasing on $(1,\infty)$

9. increasing on $(-\infty,2)$, decreasing on $(2,\infty)$ **11.** increasing on $(-\infty,-1)\cup(3,\infty)$; decreasing on $(-1,3)$ **13.** minimum at $(4,-9)$ **15.** maximum at $(-1,8)$, minimum at $(1,4)$ **17.** $(4,-76)$ is a minimum point; $(-2,32)$ is a maximum point.

19. $(0,12)$ is a maximum point; $(-2,-4)$ and $(2,-4)$ are minimum points.   **21.** $(3,-19)$ is a minimum point.   **23.** $(3,1)$ is a minimum point.   **25.** local minimum at $f(2)=-1$; local maxima: $f(0)=3,\ f(5)=8$; absolute minimum $=-1$ at $x=2$; absolute maximum $=8$ at $x=5$ **27.** local maximum at $\left(-\frac{2}{3},\frac{256}{27}\right)$, local minimum at $(2,0)$; absolute maximum at $\left(-\frac{2}{3},\frac{256}{27}\right)$, absolute minimum at $(-2,0)$ and $(2,0)$ **29.** local minima at $\left(-\frac{3}{2},\frac{297}{16}\right)$, $(3,-27)$; local maximum at $(0,27)$; absolute minimum at $(0,-27)$; absolute maximum at $(0,27)$ **31.** local and absolute maximum at $\left(\frac{\pi}{4},\sqrt{2}\right)$; absolute minimum at $(\pi,-1)$

33. local and absolute maximum at $\left(-\frac{\pi}{4}, \frac{\pi}{2}\right)$; local and absolute minimum at $\left(\frac{\pi}{4}, 2 - \frac{\pi}{2}\right)$ **35.** local maximum at $\left(\frac{\pi}{6}, \frac{5\sqrt{3}}{2} + \frac{\pi}{6} - 1\right)$;

local minimum at $\left(\frac{5\pi}{6} - \frac{5\sqrt{3}}{2} + \frac{5\pi}{6} - 1\right)$; absolute max at 2π; 9.28 $\left(\frac{5\pi}{6}, -2.71\right)$

37. absolute maximum $= 24$ at $x = 3$; absolute minimum $= -12$ at $x = -3$ **39.** absolute maximum $= 32$ at $x = -2$; absolute minimum
$= -76$ at $x = 4$ **41.** absolute maximum $= 21$ at $x = 3$ and $x = -3$; absolute minimum $= -4$ at $x = 2$ and $x = -2$
43. absolute maximum $= 8$ at $x = -4$; absolute minimum $= -19$ at $x = 3$ **45.** absolute maximum $= 9$ at $x = -5$; absolute minimum $= 1$
at $x = 3$ **47.** only critical number is $x = 0$. $f'(x)$ does not change sign at $x = 0$. If n is an odd positive integer, then f does not have a
local extremum at $x = 0$. $f(x) = (x - a)^n + b$ changes sign at $x = a$ only if n is an even positive integer   **49.** If $f(a)$ is a local
maximum, then $f(a) > f(a - h)$ and $f(a) > f(a + h)$. If $g(x) = -f(x)$, then $-f(a) < -f(a - h)$ and $g(a) < g(a - h)$. Similarly,
$g(a) < g(a + h)$.
51. $f(x) = 0$ can have two solutions or no solutions.   **53.** g is increasing on $(0,1)$.

55. 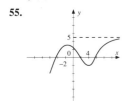   **57.** Absolute minimum $= 0$ at $x = 1$;   **59.** on $\left(-\infty, \frac{1}{2}\right]$; absolute maximum $= 2$ at
no absolute maximum.

$x = \frac{1}{2}$; local and absolute minimum
$= -\frac{1}{4}$ at $x = -1$. On $(-\infty, 1)$ f has no
maximum. Absolute minimum remains $-\frac{1}{4}$
at $x = -1$. On $(1, \infty)$ f has no maximum
or minimum.

61. local minimum at $\left(4, \frac{1}{3}\right)$; absolute maximum $= \frac{1}{3}$ at $x = 3$ and $x = 12$; absolute minimum $= 0$ at $x = 4$ **63.** local minimum at
$(-3, -177)$; absolute max $= 255$ at $x = -5$; absolute min $= -177$ at $x = -3$ **65.** absolute max $= 13$ at $x = 3$; local and absolute min
$= -19$ at $x = -1$ **67.** local and absolute max $= 51.3$ at -1.92; local and absolute min $= 25.6$ at 0.97. (All results are approximate.)
69. Assume that $\lim\limits_{x \to a} f(x) = L < 0$. Show that this contradicts the given information. Consider $f(x) = x^2$. In every deleted neighborhood of

$0, f(x) > 0$. But $\lim\limits_{x \to 0} f(x) = 0$.   **71.** As $h \to 0$, $\sin \frac{1}{h}$, used in the definition, does not approach a limit.

73. Let $f(x) = \begin{cases} x & \text{if } x \in (0, 1) \\ \frac{1}{2} & \text{if } x = 0 \text{ or } 1 \end{cases}$ **75.** max $(f + g) \le$ max $f +$ max g. If max $f \ge 0$ and max $g \ge 0$, then max (fg)
$\le$ (max f)(max g). If max $f \ge 0$ and max $g \le 0$, then max $(fg) \le 0$. If max $f \le 0$ and
max $g \le 0$, then we can draw no conclusion about max fg, unless we know that min f
and min g exist, in which case max $fg \le$ (min f)(min g).

Exercise Set 4.2

1. f is always concave up.   **3.** f is always concave up if $a > 0$ and always concave down if $a < 0$.   **5.** f is concave up for $x > -2$
and concave down for $x < -2$.
7. G is concave up on $(-\infty, 0) \cup (1, \infty)$ and concave down on $(0,1)$.   **9.** f is concave up on $(-\infty, -1) \cup (1, \infty)$ and concave down on

$(-1, 1)$.   **11. (a)** $f'(x) > 0$ at points B and C   **(b)** $f''(x) > 0$ at points A and B   **13.** $(-2, 13)$ is a local maximum; $\left(\frac{2}{3}, \frac{95}{27}\right)$ is

a local minimum   **15.** $(3, 11)$ is a local maximum; $\left(\frac{1}{3}, \frac{41}{27}\right)$ is a local minimum   **17.** $(-2, 16)$ and $(1, 38)$ are local maximum points;

$(-1, -38)$ and $(2, 16)$ are local minimum points   **19.** $(-1, 2)$ is a local minimum point   **21.** $(-2, 71)$ is a local maximum point;
$(2, -57)$ is a local minimum point   **23.** decreasing on $(-2, 2)$
increasing on $(-\infty, -2) \cup (2, \infty)$
concave down on $(-\infty, 0)$
concave up on $(0, \infty)$
point of inflection: $(0, 0)$
intercepts: $x = 0, \; x = \pm 2\sqrt{3}$
local maximum at $x = -2$
$f(-2) = 16$
local minimum at $x = 2$
$f(2) = -16$

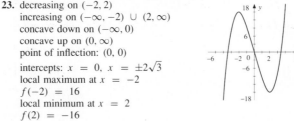

25. increasing on $\left(-\infty, -\frac{1}{3}\right) \cup (1, \infty)$; decreasing on $\left(-\frac{1}{3}, 1\right)$;

concave down on $\left(-\infty, \frac{1}{3}\right)$; concave up on $\left(\frac{1}{3}, \infty\right)$; point of

inflection $\left(\frac{1}{3}, \frac{16}{27}\right)$; intercepts: $(-1, 0)$, $(1, 0)$ and $(0, 1)$; local

maximum at $x = -\frac{1}{3}$; $f\left(-\frac{1}{3}\right) = \frac{32}{27}$; local minimum at $x = 1$;

$f(1) = 0$

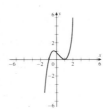

27. increasing on $(-\infty, -2) \cup \left(\frac{2}{3}, \infty\right)$

decreasing on $\left(-2, \frac{2}{3}\right)$

concave down on $\left(-\infty, -\frac{2}{3}\right)$

concave up on $\left(-\frac{2}{3}, \infty\right)$

point of inflection: $\left(-\frac{2}{3}, \frac{230}{27}\right)$

local maximum at $x = -2$
$f(-2) = 18$
local minimum at $x = \frac{2}{3}$

$f\left(\frac{2}{3}\right) = -\frac{26}{27}$

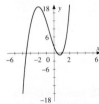

29. increasing on $(-\infty, 2) \cup (2, \infty)$
concave down on $(-\infty, 2)$
concave up on $(2, \infty)$
point of inflection $(2, 5)$
no local extrema
horizontal tangent at $(2, 5)$

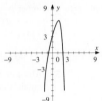

31. increasing on $(-\infty, 1)$
decreasing on $(1, \infty)$
no point of inflection
concave down on
$(-\infty, 0) \cup (0, \infty)$
local maximum at $x = 1$
$f(1) = 6$

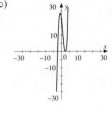

33. increasing on $\left(-\infty, -\frac{2}{3}\right) \cup (2, \infty)$

decreasing on $\left(-\frac{2}{3}, 2\right)$

concave down on $\left(-\infty, \frac{2}{3}\right)$

concave up on $\left(\frac{2}{3}, \infty\right)$

point of inflection: $\left(\frac{2}{3}, \frac{128}{27}\right)$

intercept: $x = 2$ or $x = -2$

local maximum at $x = -\frac{2}{3}$

$f\left(-\frac{2}{3}\right) = \frac{256}{27}$

local minimum at $x = 2$
$f(2) = 0$

35. increasing on $(-\infty, -1) \cup (2, \infty)$
decreasing on $(-1, 2)$

concave down on $\left(-\infty, \frac{1}{2}\right)$

concave up on $\left(\frac{1}{2}, \infty\right)$

point of inflection: $\left(\frac{1}{2}, \frac{27}{2}\right)$

intercepts: $x = 2$ or $x = -\frac{5}{2}$
local maximum $(-1, 27)$
local minimum $(2, 0)$

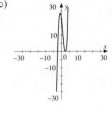

37. increasing on $\left(0, \frac{\pi}{4}\right) \cup \left(\frac{5\pi}{4}, 2\pi\right)$

decreasing on $\left(\frac{\pi}{4}, \frac{5\pi}{4}\right)$

concave down on $\left(0, \frac{3\pi}{4}\right) \cup \left(\frac{7\pi}{4}, 2\pi\right)$

concave up on $\left(3\frac{\pi}{4}, \frac{7\pi}{4}\right)$

point of inflection: $\left(\frac{3\pi}{4}, 0\right), \left(\frac{7\pi}{4}, 0\right)$

local maximum $\left(\frac{\pi}{4}, \sqrt{2}\right)$

local minimum $\left(\frac{5\pi}{4}, -\sqrt{2}\right)$

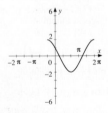

39. decreasing on $\left(-\frac{\pi}{3}, \frac{2\pi}{3}\right)$,

increasing on $\left(\frac{2\pi}{3}, \frac{5\pi}{3}\right)$

concave down on $\left(-\frac{\pi}{3}, \frac{\pi}{6}\right) \cup \left(\frac{7\pi}{6}, \frac{5\pi}{3}\right)$,

concave up on $\left(\frac{\pi}{6}, \frac{7\pi}{6}\right)$

local minimum at $\left(\frac{2\pi}{3}, -2\right)$

endpoint maximum at $\left(-\frac{\pi}{3}, 2\right)$ and $\left(\frac{5\pi}{3}, 2\right)$

points of inflection: $\left(\frac{\pi}{6}, 0\right)$ and $\left(\frac{7\pi}{6}, 0\right)$

41. increasing on $\left(-\pi, -\frac{7\pi}{8}\right) \cup \left(-\frac{3\pi}{8}, \frac{\pi}{8}\right) \cup \left(\frac{5\pi}{8}, \pi\right)$

decreasing on $\left(-\frac{7\pi}{8}, -\frac{3\pi}{8}\right) \cup \left(\frac{\pi}{8}, \frac{5\pi}{8}\right)$

concave up on $\left(-\frac{5\pi}{8}, -\frac{\pi}{8}\right) \cup \left(\frac{3\pi}{8}, \frac{7\pi}{8}\right)$

concave down on $\left(-\pi, -\frac{5\pi}{8}\right) \cup \left(-\frac{\pi}{8}, \frac{3\pi}{8}\right) \cup \left(\frac{7\pi}{8}, \pi\right)$

local minima at $x = -\frac{3\pi}{8}, \frac{5\pi}{8}$

local maxima at $x = -\frac{7\pi}{8}, \frac{\pi}{8}$

$f\left(-\frac{3\pi}{8}\right) = -1 - \sqrt{2} = f\left(\frac{5\pi}{8}\right)$;

$f\left(-\frac{7\pi}{8}\right) = f\left(\frac{\pi}{8}\right) = \sqrt{2} - 1$

absolute maximum $= \sqrt{2} - 1$ at $\frac{\pi}{8}$

absolute minimum $= -1 - \sqrt{2}$ at $x = -\frac{3\pi}{8}$
points of inflection:

$\left(-\frac{\pi}{8}, -1\right), \left(\frac{3\pi}{8}, -1\right), \left(\frac{7\pi}{8}, -1\right), \left(-\frac{5\pi}{8}, -1\right)$

43. intercept: $(0, 0)$
symmetric with respect to
the origin
vertical asymptotes:
$x = -1, x = 1$
horizontal asymptote: $y = 0$
inflection point: $(0,0)$
f decreasing on
$(-\infty, -1) \cup (-1, 1) \cup (1, \infty)$
f concave down on
$(-\infty, -1) \cup (0, 1)$
f concave up on
$(-1, 0) \cup (1, \infty)$

45. intercept: $(0,0)$
symmetric with respect to the origin
vertical asymptotes: none
horizontal asymptote: $y = 0$
f decreasing on $(-\infty, -1) \cup (1, \infty)$
f increasing on $(-1, 1)$
f concave down on $(-\infty, -\sqrt{3}) \cup (0, \sqrt{3})$
f concave up on $(-\sqrt{3}, 0) \cup (\sqrt{3}, \infty)$
inflection points: $\left(-\sqrt{3}, \frac{-\sqrt{3}}{4}\right), (0, 0),$

$\left(\sqrt{3}, \frac{\sqrt{3}}{4}\right)$

local minimum at $\left(-1, -\frac{1}{2}\right)$

local maximum at $\left(1, \frac{1}{2}\right)$

47. intercepts: $(-2, 0), (2, 0), (0, 4)$
symmetric with respect to the y-axis
vertical asymptotes:
$x = -1, x = 1$
horizontal asymptote: $y = 1$
f decreasing on
$(-\infty, -1) \cup (-1, 0)$
f increasing on $(0, 1) \cup (1, \infty)$
f concave down on
$(-\infty, -1) \cup (1, \infty)$
f concave up on $(-1, 1)$
local minimum at $(0, 4)$
inflection points: none

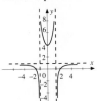

49. intercepts: $(-1, 0), (1, 0), \left(0, -\frac{1}{4}\right)$
symmetric with respect to the y-axis
vertical asymptotes:
$x = -2, x = 2$
horizontal asymptote: $y = -1$
f decreasing on
$(-\infty, -2) \cup (-2, 0)$
f increasing on $(0, 2) \cup (2, \infty)$
f concave down on
$(-\infty, -2) \cup (2, \infty)$;
f concave up on $(-2, 2)$
local minimum at $\left(0, -\frac{1}{4}\right)$;
inflection points: none

51. intercepts: none, vertical asymptotes: $x = -1, x = 0$, horizontal asymptote: $y = 0$,
symmetry: none (Actually, symmetric about $x = -\frac{1}{2}$.) f decreasing on $\left(-\frac{1}{2}, 0\right) \cup (0, \infty)$,
f increasing on $(-\infty, -1) \cup \left(-1, -\frac{1}{2}\right)$, f concave down on $(-1, 0)$; f concave up on
$(-\infty, -1) \cup (0, \infty)$, local maximum at $\left(-\frac{1}{2}, -4\right)$; inflection points: none

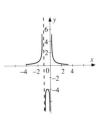

53. $\lim\limits_{x\to 0^-} f'(x) = -\infty;\ \lim\limits_{x\to 0^+} f'(x) = +\infty;\ f$ has a vertical tangent line at $x = 0$. **55.** $\lim\limits_{x\to 1^-} f'(x) = -\infty,;\ \lim\limits_{x\to 1^+} f'(x) = -\infty;$

point of inflection at $(1, 0)$, with vertical tangent **57.** $\lim\limits_{x\to 0^-} f'(x) = +\infty;\ \lim\limits_{x\to 0^+} f'(x) = -\infty;\ f$ has a vertical tangent line at $x = 0$.

59. $\lim\limits_{x\to 0^+} f'(x) = -\infty;\ f(0) = 1;\ f$ has a vertical tangent line at $x = 0$.

61. $\left(-2, \frac{-19}{3}\right)$ is a local minimum; $\left(1, \frac{59}{12}\right)$ is a local maximum; $\left(2, \frac{13}{3}\right)$ is a local maximum **63.** $\left(-1, \frac{7}{2}\right)$ is a local minimum

65. $(3, 11)$ is a local minimum

67. decreasing on $(-\infty, -2)$,
increasing on $(-2, 2) \cup (2, \infty)$
concave up on
$\left(-\infty, -\frac{2}{3}\right) \cup (2, \infty)$
concave down on $\left(-\frac{2}{3}, 2\right)$
points of inflection:
$\left(-\frac{2}{3}, -\frac{121}{27}\right)$ and $(2, 5)$
local minimum at $(-2, -11)$
horizontal tangent line at $(2, 5)$

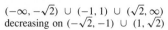

69. increasing on
$(-\infty, -\sqrt{2}) \cup (-1, 1) \cup (\sqrt{2}, \infty)$
decreasing on $(-\sqrt{2}, -1) \cup (1, \sqrt{2})$
concave up on $\left(-\sqrt{\frac{3}{2}}, 0\right) \cup \left(\sqrt{\frac{3}{2}}, \infty\right)$
concave down on $\left(-\infty, -\sqrt{\frac{3}{2}}\right) \cup \left(0, \sqrt{\frac{3}{2}}\right)$
points of inflection:
$\left(-\sqrt{\frac{3}{2}}, (-\frac{19}{4})\sqrt{\frac{3}{2}} - 3\right),\ (0, -3),$
$\left(\sqrt{\frac{3}{2}}, (\frac{19}{4})\sqrt{\frac{3}{2}} - 3\right)$
maximum points: $(-\sqrt{2}, -4\sqrt{2} - 3)$,
$(1, 3) \approx (-1.41, -8.66),\ (1, 3)$
minimum points:
$(-1, -9),\ (\sqrt{2}, 4\sqrt{2} - 3)$
$\approx (-1, -9),\ (1.41, 2.66)$

71. increasing on $(-1, 2) \cup (2, \infty)$
decreasing on $(-\infty, -1)$
concave up on $(-\infty, 0) \cup (2, \infty)$
concave down on $(0, 2)$
points of inflection: $(0, -4),\ (2, 12)$
local minimum at $(-1, -15)$
horizontal tangent at $(2, 12)$

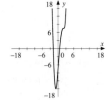

73. increasing on $\left(-\pi, \frac{\pi}{6}\right) \cup \left(\frac{5\pi}{6}, \pi\right)$
decreasing on $\left(\frac{\pi}{6}, \frac{5\pi}{6}\right)$
concave up on $\left(-\pi, -\frac{\pi}{2}\right) \cup \left(\frac{\pi}{2}, \pi\right)$
concave down on $\left(-\frac{\pi}{2}, \frac{\pi}{2}\right)$
points of inflection: $\left(-\frac{\pi}{2}, 0\right),\ \left(\frac{\pi}{2}, 0\right)$
local maximum: $\left(\frac{\pi}{6}, \sqrt{3}\right)$
local minimum: $\left(\frac{5\pi}{6}, -\sqrt{3}\right)$
endpoints: $\left(-\pi, -\frac{3}{2}\right),\ \left(\pi, -\frac{3}{2}\right)$

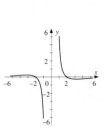

75. symmetry about y-axis
increasing on $\left(-\frac{3\pi}{2}, -\pi\right) \cup (0, \pi)$
decreasing on $(-\pi, 0) \cup \left(\pi, \frac{3\pi}{2}\right)$
concave down on
$\left(-\frac{3\pi}{2}, -\frac{4\pi}{3}\right) \cup \left(-\frac{4\pi}{3}, -\frac{\pi}{2}\right) \cup \left(\frac{\pi}{2}, \frac{4\pi}{3}\right) \cup \left(\frac{4\pi}{3}, \frac{3\pi}{2}\right)$
concave up on $\left(-\frac{\pi}{2}, 0\right) \cup \left(0, \frac{\pi}{2}\right)$
points of inflection: $\left(-\frac{\pi}{2}, 3\right),\ \left(\frac{\pi}{2}, 3\right)$

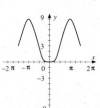

77. intercepts: $(\pm\sqrt{3}, 0)$
symmetry: $f(-x) = -f(x)$;
symmetry about origin
asymptotes: $x = 0,\ y = 0$
maxima, minima: $f'(x) = 0$ if
$x = \pm 3$
increasing on $(-\infty, -3) \cup (3, \infty)$,
decreasing on $(-3, 0) \cup (0, 3)$
local maximum at $\left(-3, \frac{2}{9}\right)$,
local minimum at $\left(3, -\frac{2}{9}\right)$
concave down on
$(-\infty, -3\sqrt{2}) \cup (3\sqrt{2}, \infty)$
concave up on $(-3\sqrt{2}, 3\sqrt{2})$
points of inflection:
$\left(-3\sqrt{2}, \frac{5}{18\sqrt{2}}\right),\ \left(3\sqrt{2}, -\frac{5}{18\sqrt{2}}\right)$

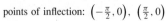

79. intercepts: $(0, 0)$
symmetry: none
asymptotes: $x = 2, y = 0$
local minimum at $(-1, -1)$
vertical tangent at $x = 0$.
concave down on
$(-\infty, -2.3416) \cup$
$(0, 0.3416), \cup(2, \infty)$
concave up on
$(-2.3416, 0) \cup (0.3416, 2)$
points of inflection:
$(-2.3416, -0.9176), (0, 0), (0.3416, 1.2646)$

81. intercepts: $(-1, 0)$
symmetry: none
asymptotes: $x = 0, \ y = x$
increasing on
$(-\infty, 0) \cup (\sqrt[3]{2}, \infty)$
decreasing on $(0, \sqrt[3]{2})$
local minimum at
$\left(\sqrt[3]{2}, \frac{3}{\sqrt[3]{4}}\right)$
point of inflection: none.
Always concave up.

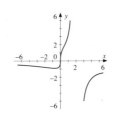

83. $a = 0$: $f(x) = x^3$ is symmetric about the origin, concave down on $(-\infty, 0)$, concave up on $(0, \infty)$, has one critical point at $(0, 0)$, and is increasing on $(-\infty, \infty)$.

$a > 0$: The graph is symmetric about the origin, concave down on $(-\infty, 0)$, concave up on $(0, \infty)$, has no critical points, and is increasing on $(-\infty, \infty)$.

$a < 0$: The graph is symmetric about the origin, concave down on $(-\infty, 0)$, concave up on $(0, \infty)$, has two critical points, has a local maximum at $\left[-\sqrt{\frac{-a}{3}}, \ f\left(-\sqrt{\frac{-a}{3}}\right)\right]$ and a local minimum at $\left[\sqrt{\frac{-a}{3}}, \ f\left(\sqrt{\frac{-a}{3}}\right)\right]$. The general cubic equation, $g(x) = ax^3 + bx^2 + cx + d$, can be converted to the simpler form $f(x) = a_1 x^3 + c_1 x$. The graph of the general equation is related to the graph of f by horizontal and vertical shifts and compression or expansion in the y-direction.

85. (a) $|c| > 0.648$ (approximate). (b) inflection points at $x = \pm\sqrt{\frac{1}{3}}; \left(-\sqrt{\frac{1}{3}}, \frac{3}{4} - c\sqrt{\frac{1}{3}}\right), \left(\sqrt{\frac{1}{3}}, \frac{3}{4} + c\sqrt{\frac{1}{3}}\right)$ (c) $c \approx -0.6495, \ 0.6495$

Exercise Set 4.3

1. 12.5 ft $\times$ 25 ft. **3.** Dimensions are 100 m $\times$ 120 m. Minimum cost is $3,600. **5.** 1.5 ft $\times$ 3 ft $\times$ 2 ft. **7.** Dimensions are 2×3. **9.** Construct a rectangle of sides x and y and hypotenuse $2a$. Show that $x = y$ for maximum area. **11.** $r = \frac{a\sqrt{6}}{3}, \ h = \frac{2a\sqrt{3}}{3}$

13. 80 **15.** 210 **17.** 70 **19.** 5 **21.** dimensions are $\sqrt{3}$ ft $\times \sqrt{6}$ ft. **23.** dimensions are 10 in $\times$ 12 in. **25.** $12\frac{1}{2}$ in wide and 10 in high. **27.** base radius $= \frac{2a\sqrt{2}}{3}$, height $= \frac{4a}{3}$ **29.** $x = 8\sqrt{5}$. **31.** $x = \frac{24}{\pi+4}, \ y = \frac{12}{\pi+4}$ **33.** 30 kilometers

35. $\left(3^{2/3} + 5^{2/3}\right)^{3/2} \approx 11.19$ feet **37.** (a) $x = 4$ (b) $x = \frac{9}{2}$

Exercise Set 4.4

1. 5 m/s **3.** 1 ft/s **5.** 10π m^2/s **7.** $\frac{2,000\pi}{3}$ m/s **9.** $\frac{48}{125}$ m/s **11.** $\frac{2}{5}$ ft/min **13.** $-\frac{5}{2}$ m/s **15.** -0.0114 m^3/s

17. -20 N/s **19.** $72,000 \, \pi$ cm^3/min **21.** $\frac{17}{4}$ m/s **23.** $\frac{63\pi}{10}$ mm^3/day **25.** 10 cm/s **27.** $\frac{5}{3}$ cm/min **29.** ≈ 0.1548 m/min

Exercise Set 4.5

1. 2.2247 **3.** -0.8850 **5.** 0.4429 **7.** $r_1 \approx 0.7140$; $r_2 \approx 1.6449$ **9.** 1.4142 **11.** -2.0801 **13.** $x_0 = 6.5$
15. $x_0 = 1.2$ **17.** $f'(x) > 0$ for all x and $f''(x) > 0$ for $x > \frac{1}{2}$ $f(x_0)$ agrees in sign with $f''(x)$ on I. On $I = (1.7, 1.8)$, take $M = 18$ and $m = 4$. $E_3 \leq 0.0000029$ **19.** f is monotone in I and is concave down. $f(2.1) \approx -0.4786 < 0$. Let $I = (1.9, 2.1)$, $m \geq \frac{\pi}{2}, M \leq 3$. $E_3 \leq 0.000002$ **21.** $z_3 = 1.703928, x_3 = 1.7039397$ **23.** $z_3 = 1.9333131, x_3 = 1.933754$ **25.** 2.44
27. (i) $f(0.2594) \approx -19.42$, local maximum; (ii) $f(2.2994) \approx = -50.63$; (iii) $f(-0.5588) \approx -22.11$ **29.** Use Eq. 4.7 with $E_n \leq 5 \times 10^{-(k+1)}$

Exercise Set 4.6

1. -6 **3.** $\frac{1}{2}$ **5.** 1 **7.** limit does not exist **9.** 6 **11.** 4 **13.** $\frac{1}{4}$ **15.** $\frac{1}{2}$ **17.** 1 **19.** 2

21. 0 **23.** $-\frac{1}{3}$ **25.** $\frac{1}{2}$ **27.** (a) (b) **29.** (a) $y = 0$ (b) $x \approx 0.4534$

(a)

x	$f(x)$
0.1	-0.05449
0.01	-0.005064
0.001	-0.000827
0.0001	-0.0000306

The limit appears to be 0.

(b)

x	$f(x)$
0.1	0.9161
0.01	1.0086
0.001	1.0006
0.0001	0.9999

The limit appears to be 1.

(c)

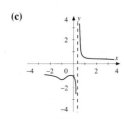

(d) local maximum at $(0, -1)$
local minimum at
$(-0.9180, -1.2526)$

Exercise Set 4.7

1. $c = \frac{3}{2}$ **3.** $c = \frac{3}{4}$ **5.** $c = \sqrt{3}$ **7.** $c = 1$ **9.** $c = \frac{1}{2}$ **11.** $c = \frac{7}{3}$ **13.** $c = \frac{2}{3}$ **15.** $c = \frac{3}{2}$

17. f' does not exist at 0.   **19.** f is discontinuous at $x = \frac{\pi}{2}$ **21.** f is discontinuous at $x = 0$.   **23.** f' does not exist at $x = 1$

25. $c = \frac{2}{3}$ satisfies
conclusion of Rolle's
Theorem. No
contradiction, since
theorem gives sufficient
conditions only.

27. $\left(-\frac{1}{3}, \frac{37}{27}\right)$, $(2, -9)$

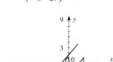

29. Use the result of
Ex. 28 and let
$h = x - a$ and
$\theta = \frac{(c-a)}{h}$

31. Use $\frac{f(x)-f(0)}{x-0} = $
$\frac{\tan x}{x} = \sec^2 c > 0$

33. Suppose $f'(a) > 0$ and $f'(b) < 0$. Since f is continuous on $[a, b]$, it must attain a maximum at some number c in (a, b), By Theorem 4.2, $f'(c) = 0$. A similar argument applies if $f'(a) < 0$ and $f'(b) > 0$.

Exercise Set 4.8

1. $\frac{x^6}{6} + C$ **3.** $\frac{2x^3}{3} - 2x^2 + 5x + C$ **5.** $\frac{3x^4}{4} - \frac{5x^3}{3} + x^2 - 3x + C$ **7.** $-\frac{1}{2}\left[\frac{1}{x} + \frac{x^3}{3}\right] + C$ **9.** $\frac{4}{3}x^{3/2} - 6x^{1/2} + C$

11. $\frac{3x^2}{2} - \frac{8}{3}x^{3/2} + 7x + C$ **13.** $\frac{2}{5}x^{5/2} + \frac{3}{2}x^2 + C$ **15.** $-2\cos x - 3\sin x + C$ **17.** $\frac{x^2}{2} + \cot x + C$ **19.** $x + \csc x + C$

21. $\frac{2}{5}t^{5/2} - \frac{2}{3}t^{3/2} + C$ **23.** $-\frac{1}{2x^2} + \frac{1}{x} + C$ **25.** $\sec x + \tan x + C$ **27.** $v = t^2 - 3t + 2$; $v(t) = \frac{1}{3}t^3 - \frac{3}{2}t^2 + 2t + C$

29. $f(x) = x^2 - 3x + 5$ **31.** $g(x) = x - \frac{x^2}{2} - \frac{4}{3}x^3 + \frac{79}{2}$ **33.** $y = x^2 - \frac{x^3}{3} - \frac{8}{3}$ **35.** $f(x) = \frac{1}{2}x^3 + 2$

37. $f(x) = \frac{x^4}{12} - \frac{x^2}{2} - \frac{2x}{3} + 2$ **39.** $y = \frac{x^2}{2} - \cos x + 4$ **41.** 64 feet = maximum height; Strikes ground when $t = 4$ sec $v(4) = -64$ ft/sec

43. Maximum height = 78.4 meters above canyon floor, or 19.6 meters above top of cliff. Rock strikes canyon floor at $t = 6$ sec

Chapter Review

1. (a) increasing on $(-1, 1) \cup (3, \infty)$; decreasing on $(-\infty, -1) \cup (1, 3)$ **(b)** increasing on $(-\infty, -2) \cup \left(-2, -\frac{1}{2}\right) \cup \left(\frac{1}{2}, \infty\right)$; decreasing on $\left(-\frac{1}{2}, \frac{1}{2}\right)$ **3.** local maximum at $(2, 13)$; local minimum at $\left(-\frac{2}{3}, \frac{95}{27}\right)$ **5.** local maximum at $(0, 0)$; local minimum at $(2, 4)$

7. local minimum at $(2, -19)$; absolute maximum at $(-2, 13)$; absolute minimum at $(2, -19)$ **9.** local and absolute maximum at $\left(\frac{\pi}{12}, \frac{\sqrt{3}}{2} + \frac{\pi}{12}\right)$; local minimum at $\left(\frac{5\pi}{12}, -\frac{\sqrt{3}}{2} + \frac{5\pi}{12}\right)$; absolute minimum at $\left(-\frac{\pi}{2}, -1 - \frac{\pi}{2}\right)$ **11. (a)** concave up on $\left(-\infty, -\frac{1}{2}\right) \cup (1, \infty)$ concave down on $\left(-\frac{1}{2}, 1\right)$ **(b)** concave up on $(3, \infty) \cup (-1, 1)$ concave down on $(-\infty, -1) \cup (1, 3)$

13. intercepts:
$(0, -1), (1, 0), (-1, 0)$
symmetry: none
asymptotes: none
local maximum at $(-1, 0)$
local minimum at $\left(\frac{1}{3}, -\frac{32}{27}\right)$
point of inflection: $x = -\frac{1}{3}$

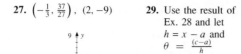

15. intercept: $(0, 60)$
$\approx (\pm 3.7, 0)$, $(\pm 2.1, 0)$
symmetry: y-axis; asymptotes:
none; maxima, minima:
$(\pm 3, -21)$ local minimum, $(0, 60)$ local maximum; points of
inflection: $(\pm\sqrt{3}, 15)$

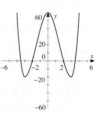

17. intercept: $(0, 0)$
symmetry: origin
asymptote: $y = 0$
maxima, minima: max $(2, 2)$,
min $(-2, -2)$
points of inflection:
$(-2\sqrt{3}, -\sqrt{3})$, $(0, 0)$, $(2\sqrt{3}, \sqrt{3})$

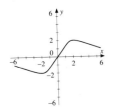

19. intercepts: $(0, 15)$
x intercepts between 1 and 2
and between 3 and 4
symmetry: none
asymptotes: none
maxima, minima: $(3, -12)$ local
minimum
point of inflection: $(2, -1)$;
horizontal tangent at $(0, 15)$

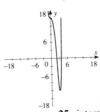

21. intercepts: $(2, 0)$, $(-1, 0)$,
$(0, 2)$
symmetry: none
asymptotes: $x = 1$, $y = x$
maxima, minima: none
points of inflection: none

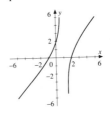

23. intercepts: $(0, -1)$
x-intercepts between 0 and $\frac{1}{3}$, between $\frac{1}{3}$ and 1,
and between 8 and 9
symmetry: none
asymptotes: none
maxima, minima: $(3, -1)$ local minimum
$\left(\frac{1}{3}, \frac{1}{3}\right)$ local maximum,
cusp points of inflection: none

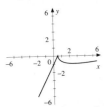

25. intercept: $(0, 0)$
symmetry: none
asymptotes: $x = 1$, $y = 0$
maximum, minimum: $(-1, -1)$
is a local min
point of inflection: $\left(-2, -\frac{8}{9}\right)$

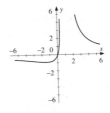

27. intercepts: $\left(\pm\frac{1}{\sqrt{3}}, 0\right)$; symmetry: origin; asymptotes:
$x = 0$, $y = 0$; maxima, minima: $(1, 2)$ local
maximum; $(-1, -2)$ local minimum; points of
inflection: $\left(\sqrt{2}, \frac{5}{2\sqrt{2}}\right)$, $\left(-\sqrt{2}, -\frac{5}{2\sqrt{2}}\right)$

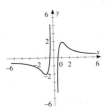

29. intercept: $(0, -1)$; symmetry: y-axis; asymptotes: $x = \pm\frac{\pi}{4}$
in $\left[-\frac{\pi}{2}, \frac{\pi}{2}\right]$; maxima, minima: $f(0) = -1$ (local
maximum); $f\left(\pm\frac{\pi}{2}\right) = 3$ (local minimum); no inflection
points; f is periodic with period π

31. 6 ft $\times$ 6 ft $\times$ 9 ft.   **33.** $x = 18, y = 36$ **35.** $h = 10$ in, $r = 5$ **37.** $x = \frac{m}{2}$ **39.** $5\sqrt{5}$ ft   **41.** 17 mph

43. (a) $\frac{2}{\pi}$ ft/min (b) $\frac{1}{\pi\sqrt{2}}$ ft/min   **45.** $r \approx 0.7430$ **47.** $r_1 \approx 2.1149; r_2 \approx -0.2541; r_3 \approx -1.8608$

49. $f(0.8603336) \approx 0.561096$ **51.** 4   **53.** $f\left(\frac{\pi}{6}\right) \approx 0.142$, $f\left(\frac{\pi}{3}\right) \approx -0.523$, f' and f'' do not change sign on

$\left(\frac{\pi}{6}, \frac{\pi}{3}\right)$, $E_{n+1} \leq 0.113E_n$ **55.** (a) 1 (b) 0   **57.** (a) 0 (b) $\frac{3}{2}$ **59.** (a) f and f' are continuous everywhere $c = 1$

(b) $f \& f'$ are continuous everywhere. $c = 0$ **61.** (a) f and f' are continuous everywhere $c = \frac{1}{2}$ (b) f and f' are continuous

everywhere   **63.** (a) 7 (b) $g(x) = x^2 + 1$ **65.** (a) $-4\cos x - 3\sin x + C$ (b) $\tan x + 2\sec x + C$

67. (a) $f(x) = x^3 + \frac{2}{x} + 1$ (b) $y = -2\cos x - \frac{x^2}{2} + 7$

69. (a) $f(x) =$
$-\frac{1}{6x} - \frac{x^2}{6} + \frac{x}{2} + \frac{5}{2}$
(b) $y =$
$\cos x + x^2 - x + 1$

71. (a) $v(t) = 3t^2 - 12$ $s(t) = t^3 - 12t$ **(b)** $s(0) = 0$ when $t = 2$, 16 units to the left of the origin.   **(c)** increasing speed for $t > 2$; decreasing speed for $0 < t < 2$
(d)

CHAPTER 5

Exercise Set 5.1

1. $L_5 = 12.5$, $U_5 = 15.0$, $L_{10} = 13.125$, $U_{10} = 14.375$, $L_{20} = 13.4375$, $U_{20} = 14.0625$, $L_{100} = 13.6875$, $U_{100} = 13.8125$.
3. $L_5 = 4.48$, $U_5 = 6.08$, $L_{10} = 4.92$, $U_{10} = 5.72$, $L_{20} = 5.13$, $U_{20} = 5.53$, $L_{100} = 5.2932$, $U_{100} = 5.3732$ **5.** $L_5 = 4.704$,
$U_5 = 7.656$, $L_{10} = 5.298$, $U_{10} = 6.792$, $L_{20} = 5.635125$, $U_{20} = 6.385875$, $L_{100} = 5.925453$, $U_{100} = 6.075447$
7. $L = 141.5$, $U = 181.5$ **9.** Add $1 + n, 2 + (n-1), \ldots, n + 1$. Sum and divide by 2.   **11.** $P(1)\ 1^2 = \frac{1(1+1)(2+1)}{6}$. Assume $P(k)$ is true, and add $(k+1)^2$ to both sides.   **13.** $P(1)\ 1^3 = \left[\frac{1(1+1)}{2}\right]^2$. Assume $P(k)$ is true and add $(k+1)^3$ to both sides.

15. $L_n = \frac{b-a}{n}\sum_{k=1}^{n} m_k$, $U_n = \frac{b-a}{n}\sum_{k=1}^{n} M_k$, error $\le |U_n - L_n| = \frac{b-a}{n}\left|\sum_{k=1}^{n}[f(x_{k-1}) - f(x_k)]\right| = \frac{b-a}{n}|f(x_n) - f(x_0)| = \frac{b-a}{n}|f(b) - f(a)|$

17. (a) 240 **(b)** 640, with $\Delta x = 0.0125$ **19. (a)** Area $= \frac{1}{2}\left(2r\tan\left(\frac{\pi}{n}\right)\right)(r)$ **(b)** Let $n = \frac{\pi}{t}$, $\lim_{t \to 0} \pi r^2 \frac{\tan t}{t} = \lim_{t \to 0} \frac{\pi r^2}{\cos t}\frac{\sin t}{t}$

Exercise Set 5.2

1. -19 **3.** -5.025 **5.** $\sum_{k=1}^{n}\frac{1}{n+k}$ **7.** $\frac{3}{m}\sum_{k=1}^{m}\sqrt[3]{3\left(1 + \frac{3k}{m}\right)} - 4$ **9.** $\sum_{k=1}^{n}\frac{1}{n+k-1}$ **11.** $\frac{3}{m}\sum_{k=1}^{m}\sqrt[3]{3\left(1 + \frac{3(k-1)}{m}\right)} - 4$

13. $\int_0^2 \sqrt{1+x^3}\,dx$ **15.** $\int_0^2 x^2\,dx$ **17.** $\int_{-2}^3 \sin x\,dx$ **19.** $\int_0^3 (1+x)\,dx$ **21.** $\int_0^2 x\,dx + \int_2^4 (4-x)\,dx$ **23.** 9

25. 2π **27.** $6 + \pi$ **29.** $9\left(\frac{\pi}{4} + 1\right)$ **31.** 10   **33.** 10   **35.** $3 < \int_1^4 f(x)\,dx < 7$, $11 < \int_{-3}^2 f(x)\,dx < 17$

37.

n	2	8	16	32	100	200	≈ 0.3333
LHS	0.125	0.2734	0.3027	0.3179	0.3284	0.3308	
RHS	0.625	0.3984	0.3652	0.3491	0.3384	0.3358	

39.

n	2	8	16	32	100	200	≈ 1.532
LHS	-0.523	1.221	1.418	1.479	1.515	1.523	
RHS	1.277	1.672	1.643	1.592	1.551	1.541	

41. $f'(x) = 2x \ge 0$
$n \ge 80$
Integral ≈ 2.7
43. 0.41

45. 4.97   **47. (a)**

(b) $\int_0^{2\pi} \cos^5 x\,dx = 0$
by symmetry with respect to $x = \pi$ and
$\int_0^{2\pi} \sin^5 x\,dx = 0$ by symmetry about the point $x = \pi$.

49. $\lim_{n \to \infty} \Delta x \sum_{k=1}^{n} (c) = \lim_{n \to \infty} \frac{b-a}{n} \cdot nc = c(b-a)$ **51.** $\lim_{n \to \infty} \Delta x \sum_{k=1}^{n} [f(x_k) + g(x_k)] = \lim_{n \to \infty} \Delta x \sum_{k=1}^{n} f(x_k) + \lim_{n \to \infty} \Delta x \sum_{k=1}^{n} g(x_k)$
$= \int_a^b f(x)\,dx + \int_a^b g(x)\,dx$ **53.** Let $h(x) = g(x) - f(x)$, $\int_a^b h(x)\,dx = \int_a^b g(x)\,dx - \int_a^b f(x)\,dx \ge 0$, or $\int_a^b f(x)\,dx \le \int_a^b g(x)\,dx$

55. $P(1)$ is true for $m = 2$ by Property 4. Assume for $n = m$ and add $\int_a^b f_{m+1}(x)\,dx$ to both sides.   **57.** $c_k = \frac{kx}{n}$, $\frac{x}{n}\sum_{k=1}^{n} c_k = \frac{x^2}{n^2}\sum_{k=1}^{n} k$
$= \frac{x^2}{n^2}\frac{n(n+1)}{2} = \frac{x^2}{2} + \frac{x^2}{2n} = \frac{x^2}{2}$ as $n \to \infty$ **59.** $c_k = \frac{kx}{n}$, $\frac{x}{n}\sum_{k=1}^{n} c_k^3 = \frac{x^4}{n^4}\sum_{k=1}^{n} k^3 = \frac{x^4}{n^4}\left[\frac{n(n+1)}{2}\right]^2 = \frac{x^4}{4} + \frac{x^4}{2n} + \frac{x^4}{4n^2} = \frac{x^4}{4}$ as
$n \to \infty$

Exercise Set 5.3

1. $\frac{44}{3}$ **3.** 9   **5.** 16   **7.** $\frac{17}{6}$ **9.** $\frac{84}{5}$ **11.** -2 **13.** 4   **15.** $\sqrt{3}$ **17.** 2   **19.** $\frac{961}{15}$ **21.** $\frac{48}{5}$ **23.** $\frac{1}{3}\pi^2 + \sqrt{3}$

25. 1   **27.** $\sqrt{3} - 1$ **29.** $\frac{28}{3}$ **31.** -1 **33.** $-\frac{15}{4}$ **35. (a)** $V_{ave} = \frac{s(b)-s(a)}{b-a}$ **(b)** $V_{ave} = \frac{\int_a^b s'(t)\,dt}{b-a}$ **37.** 2

39. $1 - \sqrt{3}$ **41.** $\frac{x}{x-1}$ for all $x > 1$ **43.** $\frac{\sin x}{2+\cos x}$ for all x **45.**   **47.** $\approx 5{,}692{,}100$ kg; $\approx 58{,}706{,}700$ kg

49. $\approx 116{,}000{,}000$ kwh.   **51.** $-\frac{2}{3}$ **53.** $-\frac{\sin x}{x}$ **55.** $2(x-1)\sin^3(x^2 - 2x)$ **57.** $2 - \frac{\sqrt{2}}{2}$ **59.** $\frac{4}{3}$ **61.** $\frac{1}{6}$

63. 3   **65.** $\frac{8}{\sqrt{3}}$ **67.** $2(\sqrt{3} + \sqrt{2} - 1) - \frac{7\pi}{12}$ **69.** $\int_a^{x+h} f(t)\,dt - \int_a^x f(t)\,dt = -\int_{x+h}^x f(t)\,dt = -f(c)[x - (x+h)] = f(c)h$

Thus, $F'(x) = \lim\limits_{h \to 0} \frac{f(c)h}{h} = f(x)$, by the Squeezing Theorem, Property 3, and the mean value theorem for integrals.

71. If $x = 0$, $f'(0) = \lim\limits_{h \to 0} \frac{h^2 \sin\frac{1}{h^2}}{h} = \lim\limits_{h \to 0} h\left(\sin\frac{1}{h^2}\right) = 0$. For all other x, $f'(x) = \frac{-2}{x}\cos\frac{1}{x^2} + 2x\sin\frac{1}{x^2}$ which is unbounded as $x \to 0$.

f' is not integrable.   **73.** Let $F' = f$. Then, $\int_a^b f(x)\,dx = \int_a^b F'(x)\,dx = F(b) - F(a) = F'(c)(b-a)$ **75.** Let x

and $x_0 \in [a,b]$, with x_0 fixed. Since f is integrable, it is bounded, so there is an $M > 0$ such that $|f(t)| \le M$ for all $t \in [a,b]$.

$|F(x) - F(x_0)| = \left|\int_x^b f(t)\,dt - \int_{x_0}^b f(t)\,dt\right| = \left|\int_x^{x_0} f(t)\,dt\right| = \int_x^{x_0} |f(t)|dt \le M\int_x^{x_0} dt = M(x_0 - x)$ when $x < x_0$,

and when $x > x_0 |F(x) - F(x_0)| = |\int_{x_0}^x f(t)\,dt| \le \int_{x_0}^x |f(t)|dt \le M(x - x_0)$. Let $\delta = \frac{\epsilon}{M}$.   **77.** Define $g(x) = f(x)$ on

(a,b) and $g(a) = \lim\limits_{x \to a^+} f(x)$, $g(b) = \lim\limits_{x \to b^-} f(x)$. Then g is continuous on $[a,b]$. Let c be any number in (a,b). Then, by

Exercise 74, $\int_c^b f(x)\,dx$ exists and equals $\int_c^b g(x)\,dx$ and by Exercise 76, $\int_a^c f(x)\,dx$ exists and equals $\int_a^c g(x)\,dx$. Therefore,

$\int_a^b f(x)\,dx = \int_a^c f(x)\,dx + \int_c^b f(x)\,dx$ exists and equals $\int_a^b g(x)\,dx$.

Exercise Set 5.4

1. $\frac{1}{8}(1+x^2)^4 + C$ **3.** $\frac{1}{18}(x^3 - 1)^6 + C$ **5.** $\frac{1}{9}(3x^2 + 4)^{3/2} + C$ **7.** $\frac{2}{9}(1+3x)^{3/2} + C$ **9.** $\frac{1}{3}(x^2 - 2x + 3)^{3/2} + C$

11. $-\frac{1}{2}\cos 2x + C$ **13.** $-\frac{1}{k}\cos kx + C$ **15.** $\sin^2 x + C$ or $-\cos^2 x + C_1$ or $-\frac{1}{2}\cos 2x + C_2$ **17.** $\frac{1}{2}\tan x^2 + C$

19. $\frac{2}{3}(\tan x)^{3/2} + C$ **21.** $-\frac{1}{1-\cos x} + C$ **23.** $-2\cot\sqrt{x} + C$ **25.** $\frac{1}{6}\left(\frac{x-1}{x}\right)^6 + C$ **27.** $\frac{2}{5}(x^{3/2} - 1)^{5/3} + C$ **29.** $\frac{121}{5}$

31. $\frac{33}{10}$ **33.** $\frac{254}{7}$ **35.** 60   **37.** $\frac{1}{3}$ **39.** $\frac{1}{8}$ **41.** -1 **43.** $\frac{7}{3}$ **45.** $\frac{2}{9}$ **47. (a)** $\frac{1}{2}\tan^2 x + C$ **(b)** $\frac{1}{2}\sec^2 x + C_1$

Both are correct; they differ in the constant.   **49.** $g(x) = \frac{1}{6}(4x + 2)^{3/2} - \frac{1}{3}$ **51.** \$90,990   **53.** $\sqrt{x^2 + 1} + C$

55. $\frac{1}{2}(x^2 + \sin 2x) + C$ **57.** $\frac{2}{5}(1 + \tan t)^{5/2} + C$ **59.** $y = \frac{2}{3}\sqrt{x^3 + 1} + 3$ **61.** $y = \frac{2}{3}\left[(1 + \sin^2 x)^{3/2} + 2\right]$

63. $y = -\frac{1}{2}\left(\frac{1}{x^2+1} + 1\right)$ **65. (a)** ≈ 413.1 ft   **(b)** ≈ 128.6 m   **67. (a)** $-9.8t - 39.2$ **(b)** $s(t) = -4.9t^2 - 39.2t + 1176$; 12 sec

69. $-\frac{1}{\tan x + 1} + C$ **71.** $-4\sqrt{1 - \sqrt{x}} + C$ **73.** $\sin x - \frac{2}{3}\sin^3 x + C$ **75.** $\frac{d}{dx} - 2\left[\frac{1}{\sqrt{x}}\sin\frac{1}{\sqrt{x}} + \cos\frac{1}{\sqrt{x}}\right] + C = \frac{1}{x^2}\cos\frac{1}{\sqrt{x}}$

77. Property 1: $\frac{d}{dx}[F(x) + C] = F'(x) = f(x)$. Property 2: If f is differentiable on I, f is then an antiderivative

of f'. Property 3: Let $u(x) = cF(x)$ where F is an antiderivative of f. Then, $du = cF'(x)\,dx = cf(x)\,dx$.

$\int cf(x)\,dx = \int du = u + C = cF(x) + C = c\int f(x)\,dx$. Property 4: Let $u(x) = F(x) \pm G(x)$ Then,

$du = [F'(x) \pm G'(x)]dx = [f(x) \pm g(x)]dx$. $\int[f(x) + -g(x)]dx = \int du = u + C = F(x) + -G(x) + C$

$= \int f(x)\,dx + -\int g(x)\,dx$.

Exercise Set 5.5

1. (a) ≈ 0.50899, $E_4 \le 0.03125$ **(b)** ≈ 0.4955, $E_4 \le 0.015625$ **(c)** ≈ 0.5004, $E_4 \le 0.002604$, $\int_0^1 \frac{dx}{(x+1)^2} = \frac{1}{2}$ **3. (a)** ≈ 0.62955,

$E_8 \le 1.3125$.   **(b)** ≈ 0.77489, $E_8 \le 0.65625$ **(c)** ≈ 0.73782, $E_8 \le 0.2703$, $\int_1^3 x\sin x^2\,dx = -\frac{1}{2}(\cos 9 - \cos 1) \approx 0.72572$

5. (a) ≈ -0.9375, $E_6 \le 0.75$ **(b)** ≈ -0.6563, $E_6 \le 0.375$ **(c)** -0.7500, $E_6 = 0$, $\int_{-2}^{1}(1+x^3)\,dx = -0.7500$ **7. (a)** $n = 30$
(b) $n = 22$ **(c)** $n = 5$ **9. (a)** $n = 116$ **(b)** $n = 82$ **(c)** $n = 11$ **11. (a)** $n = 142$ **(b)** $n = 100$ **(c)** $n = 14$
13. (a) ≈ 2.29 **(b)** 2.30 **15.** ≈ 243 cu meters **17. (a)** ≈ 0.6592, $E_{10} \le 0.00833\ldots$ **(b)** ≈ 0.6594, $E_{10} \le 0.00417$
(c) ≈ 0.6593, $E_{10} \le 0.0000933$ **19.** 2.6368, $M_2 = 2.00$. $E_n \le 0.00667$ **21.** $n = 10$: ≈ 0.785398153, $n = 20$: ≈ 0.785398163,
$n = 100$: ≈ 0.785398163, $n = 1000$: ≈ 0.785398163, $\pi/4 \approx 0.785398163$,

Chapter Review

1. $U_n = \frac{14}{3} + \frac{4}{n} + \frac{4}{3n^2}$, $L_n = \frac{14}{3} - \frac{4}{n} + \frac{4}{3n^2}$, $\lim_{n \to \infty} U_n = \lim_{n \to \infty} L_n = \frac{14}{3}$ **3. (a)** $-\frac{8}{3}$ **5.** Overestimate by Trapezoidal rule: $= 17.8$

liters. Underestimate by Midpoint rule: $= 17.2$ liters **7. (a)** 10.47 **(b)** 28.07 **9.** $-\frac{55}{6}$ **11.** $\sqrt{3} - 1$ **13.** $\frac{1}{16}$

15. $\frac{9}{2}$ **17.** 21 **19.** $\frac{3072}{5}$ **21.** $\frac{\tan^4 \theta}{4} - \sec \theta + C$ **23.** $\frac{(x^2+2)^6}{12} + C$ **25.** $-\cot x^2 - \frac{x^2}{2} + C$ **27.** $-\frac{(1+\cos^2 t)^4}{4} + C$

29. 28 **31. (a)** 0. **(b)** $-\frac{3}{\pi}$ **33.** $\frac{1}{2}$ **35.** $y = x^2 + \frac{1}{2}\cos 2x + C$ **37.** $\frac{1}{3}(t^2 + 4)^{3/2} + C$ **39.** $\frac{1}{5}(t^2 + 4)^{5/2} - \frac{22}{5}$

41. $\frac{1}{2}(\sqrt{x} + 2)^4 - \frac{71}{2}$ **43. (a)** ≈ 0.4137, $E_{10} \le 0.000716$ **(b)** ≈ 0.4145, $E_{10} \le 0.000358$ **(c)** ≈ 0.41422, $E_{10} \le 4.48 \times 10^{-6}$

$\int_0^1 \frac{x}{\sqrt{x^2+1}}\,dx = \sqrt{2} - 1 \approx 0.414214$ **45. (a)** ≈ 15.7536, $E_8 \le 0.0992$ **(b)** ≈ 15.7484, $E_4 \le 0.0496$ **(c)** ≈ 15.7511,

$E_8 \le 0.0563$, $\int_1^8 (x^{1/3} + x^{-1/3})\,dx = 15.75$ **47. (a)** $n = 260$ **(b)** $n = 184$ **(c)** $n = 24$ **49.** 0; $E_{20} \le 0.00889$

51. ≈ 34.2 m **53. (a)** ≈ -0.0073115 **(b)** 0 **(c)** ≈ 46.7097

CHAPTER 6

Exercise Set 6.1

1. $\frac{14}{3}$ **3.** $\frac{22}{3}$ **5.** $\frac{20}{3}$ **7.** 3 **9.** $\frac{8\pi}{3}$

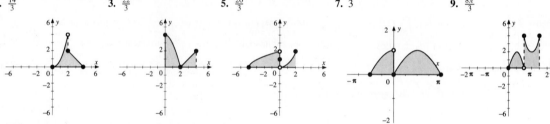

11. $\frac{32}{3}$ **13.** 4 **15.** $\frac{9}{2}$ **17.** 4 **19.** $\frac{5}{6}$ **21.** $\frac{80}{3}$ **23.** $4\frac{1}{2}$

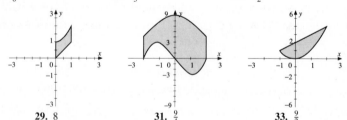

25. $\frac{4}{3}$ **27.** $\frac{1}{3}$ **29.** 8 **31.** $\frac{9}{4}$ **33.** $\frac{9}{2}$

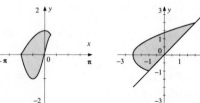

35. 8 **37.** $\frac{256}{27}$ **39.** $\frac{8}{3}$ **41.** 3 **43.** 8 **45.** 3

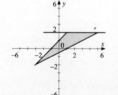
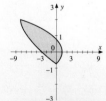

47. (a) For $\int_0^{1/2} f(2x)\,dx$, let $u = 2x$, $\frac{1}{2}\int_0^1 f(u)\,du = \frac{1}{2}A$ (b) If $f(x) = c$, $f(2x) = c$ then $\int_0^1 f(x)\,dx = \int_0^1 f(2x)\,dx = c$. If f
is periodic, with period 1, then $\int_0^1 f(x)\,dx = \int_0^1 f(2x)\,dx = 0$.   (c) A

Exercise Set 6.2

1. $\frac{37\pi}{12}$ **3.** $\frac{256\pi}{15}$ **5.** 8π **7.** $\frac{16\pi}{7}$ **9.** $\frac{434\pi}{3}$ **11.** $\frac{92\pi}{15}$ **13.** $\frac{477\pi}{4}$ **15.** $\frac{72\pi}{5}$ **17.** $\frac{16\pi}{3}$ **19.** $\frac{32\pi}{3}$ **21.** $\frac{32\pi}{3}$

23. $\frac{99\pi}{5}$ **25.** 10π **27.** $\frac{277\pi}{3}$ **29.** $\frac{120\pi}{7}$ **31.** $\frac{4}{3}\pi a^3$ **33.** $\frac{28\pi}{3}$ **35.** (a) $\frac{88\pi}{3}$ (b) 24π **37.** $\frac{\pi}{3}$ **39.** $\frac{2}{3}\pi$

41. $\frac{\pi}{3}$ **43.** $\frac{\pi}{6}(3\pi - 8)$ **45.** $\frac{\pi}{3}$ **47.** $\frac{\pi}{6}(10 - 3\pi)$ **49.** $\frac{\pi}{2}(\pi - 1)$ **51.** 8π

53. $\frac{16}{3}$ **55.** 2π **57.** $\frac{1612\pi}{15}$ **59.** $\frac{5000\pi}{3}$ liters   **61.** $16\sqrt{3}$ **63.** $\frac{256\pi}{3}$ **65.** (a) ≈ 15.4282 (b) ≈ 15.3319

Exercise Set 6.3

1. $\frac{16\pi}{3}$ **3.** 8π **5.** $\frac{32\pi}{3}$ **7.** $\frac{110\pi}{3}$ **9.** $\frac{381\pi}{14}$ **11.** $\frac{8\pi}{3}$ **13.** $\frac{12\pi}{5}$ **15.** $\frac{4\pi}{27}$ **17.** $\frac{16\pi}{5}$ **19.** $\frac{5\pi}{6}$

21. $\frac{16\pi}{3}$ **23.** $2\pi^2 a^2 b$ **25.** $\frac{1024\pi}{3}$ **27.** (a) $\frac{875\pi}{48}$ (b) $\frac{2125\pi}{96}$ **29.** (a) $2\pi \int_0^\pi x \sin^2 x\,dx$; $\frac{\pi^3}{2} \approx 15.503$

(b) $2\pi \int_0^\pi (x + \pi)\sin^2 x\,dx$; $\frac{3\pi^3}{2} \approx 46.509$

Exercise Set 6.4

1. $3\sqrt{10}$ **3.** $\sqrt{13}$ **5.** $\frac{14}{3}$ **7.** $\frac{335}{27}$ **9.** $\frac{22}{3}$ **11.** $\frac{1}{54\sqrt{2}}\left[(77)^{3/2} - (32)^{3/2}\right]$ **13.** $\frac{53}{6}$ **15.** $6a$

17. $x = \pm \left(\frac{2y}{3}\right)^{3/2} - 1$. The point $\left(-2, \frac{3}{2}\right)$ is on $x = -\left(\frac{2y}{3}\right)^{3/2} - 1$ and $(7, 6)$ is on $\left(\frac{2y}{3}\right)^{3/2} - 1$. Thus, there are two functions of y. Each
must be dealt with independently of the other. $L = 5\sqrt{5} + 2\sqrt{2} - 2$ **19.** ≈ 0.22 **21.** ≈ 4.6468 **23.** ≈ 17.261

Exercise Set 6.5

1. $24\pi\sqrt{5}$ **3.** $\frac{2\pi}{3}(5\sqrt{5} - 1)$ **5.** 2π **7.** $\frac{\pi}{27}[(145)^{3/2} - (10)^{3/2}]$ **9.** $\frac{\pi}{6}[(17)^{3/2} - (5)^{3/2}]$ **11.** $4\pi a^2$ **13.** $\frac{12}{5}\pi a^2$

15. Partition $[a,b]$ so that the zeros of f occur at partition points. For those intervals on which $f(x) < 0$, the band obtained has radius $|f(x_k^{**})|$,
where $x_k^{**} \in (x_{k-1}, x_k)$, and so its area is $2\pi|f(x_k^{**})|\sqrt{1 + [f'(x_k^*)]^2}\Delta x_k$. If $f(x) > 0$, the same result holds. Sum and let $||P|| \to 0$ to
get the result.   **17.** $\frac{\pi}{2}$ **19.** $\frac{448\pi}{9}$ **21.** (a) $\approx 1.341 \times 10^8$ km^2 **23.** (a) 14.42
                                                                                       (b) $\approx 9.824 \times 10^{-5}$ (b) 53.23

Exercise Set 6.6

1. $\frac{4}{3}$ joules   **3.** 24 inch-pounds   **5.** $\frac{45}{4}$ newton-cm; 30 newton-cm   **7.** $k = 0.04$ N/m   **9.** 1350 ft-lb   **11.** 1930 joules

13. $\approx 6.904 \times 10^6$ joules   **15.** $\frac{k}{3}$ dyne-cm   **17.** $\approx 17,200$ joules   **19.** $\approx 6.87 \times 10^6$ joules   **21.** $\approx 17,020$ joules

Exercise Set 6.7

1. $56\rho g$ **3.** $135\rho g$ **5.** $\frac{184\rho g}{3}$ **7.** ≈ 124.8 lb; ≈ 62.4 lb

9. (a) $\approx 49,050$N; $\approx 784,800$N   (b) $\approx 515,000$N   **11.** $\frac{52\rho g}{3}$ **13.** $\frac{304\rho g}{15}$

15. The kth strip, cut off by horizontal   **17.** $\approx 285,700$N
planes $x = k_{k-1}$ and $x = x_k$
is inclined and so its dimensions are
$w(x*)$ by $\Delta x \sec\theta$ (see sketch).

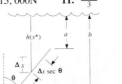

$$F \approx \sum_{k=1}^n \rho g_{\text{depth}} h(x*) w(x*) \Delta x_{\text{area}} \sec\theta$$

Take limits to get the integral.

Exercise Set 6.8

1. $\frac{27}{59}$ **3.** $\left(-\frac{33}{32}, \frac{63}{32}\right)$ **5.** $\frac{3}{2}$ **7.** $\frac{633}{95} \approx 6.663$ **9.** $\left(\frac{4}{3}, \frac{4}{3}\right)$ **11.** $\left(-\frac{2}{3(4+\pi)}, 0\right)$ **13.** $\left(\frac{8}{3}, \frac{10}{3}\right)$ **15.** $\left(\frac{8}{3\pi}, \frac{8}{3\pi}\right)$

17. $\left(-\frac{1}{2}, \frac{2}{5}\right)$ **19.** $\left(\frac{8}{5}, 0\right)$ **21.** $(\bar{x}, \bar{y}) = \left(0, \frac{3}{5}b\right)$, which is independent of a.   **23.** $\bar{y} = \frac{2}{\pi a^2} \int_{-a}^{a} \left(\frac{1}{2}\sqrt{a^2 - x^2}\right)$

$\left(\sqrt{a^2 - x^2}\right) dx = \frac{4a}{3\pi}$ **25.** $\left(\frac{23}{8}, \frac{5}{8}\right)$ **27.** $\left(\frac{192-7\pi}{48-\pi}, \frac{144-4\pi}{48-\pi}\right)$ **29.** Rotate a semicircular region of radius a about the y-axis.

Centroid of region is $\left(\frac{4a}{3\pi}, a\right)$. Area $= \frac{\pi a^2}{2}$. Volume $= \frac{\pi a^2}{2} \cdot 2\pi\left(\frac{4a}{3\pi}\right) = \frac{4}{3}\pi a^3$ **31. (a)** 90π **(b)** 36π **(c)** 72π

Exercise Set 6.9

1. $\frac{1}{28}$ **3.** $\frac{3}{38}$ **5.** $\frac{4}{3}$ **7.** 60   **9.** 1   **11.** $\frac{1}{2}$ **13.** $\frac{1}{4}$ **15.** $\frac{37}{125}$ **17.** $\mu = 1$ **19.** $\mu = \frac{1}{2}$
$\sigma^2 = \frac{1}{2}$ $\sigma^2 = \frac{1}{20}$

21. $\mu = \frac{4}{5}$ **23.** $\mu = \frac{1}{2}(a + b)$ **25.** $\frac{3}{4}$ **27. (a)** $\frac{1}{8}$ **(b)** μ = about 333,000,000 barrels   **29.** ≈ 1.58 hundred
$\sigma^2 = \frac{2}{75}$ $\sigma^2 = \frac{1}{12}(a^2 - 2ab + b^2)$

hours   **31.** $\sigma^2 = E[(x - \mu)^2] = E[x^2 - 2\mu x + \mu^2] = E[x^2] - 2\mu E[x] + E[\mu^2] = E[x^2] - 2\mu^2 + \mu^2 = E[x^2] - \mu^2$ **33.** $\frac{\pi-2}{8}$

Exercise Set 6.10

1. $\frac{x^4}{4} - x^2 + 200$ **3.** $\approx \$2092$ **5.** Consumers' surplus $= \$15,340$
Producers' surplus $= \$66,120$

7. (a) equilibrium price $= \$700$;
equilibrium quantity $= 10,000$ units
(b) $\approx \$1867$
(c) $\approx \$3467$
(d)

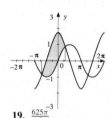

9. $C = 0.1I + 0.8\sqrt{I + 1} + 3.1$; $C(20) \approx \$8.77$ billion

Chapter Review

1. $\frac{46}{3}$ **3.** 24   **5.** 16   **7.** $\frac{2}{\pi} + \frac{1}{6}$ **9.** 4

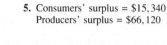

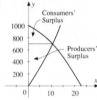

11. $\frac{\pi}{2}$ **13.** $\frac{176\pi}{3}$ **15.** $\frac{500\pi}{3}$ **17.** $\frac{28\pi}{3}$ **19.** $\frac{625\pi}{6}$

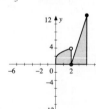

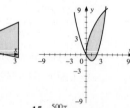

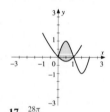

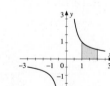

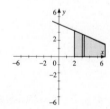

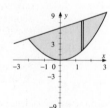

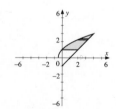

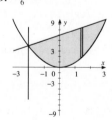

21. $\frac{32}{3}$ **23.** $\frac{1}{27}\left[(40)^{3/2} - 8\right]$ **25.** $\frac{22}{3}$ **27.** $\frac{49\pi}{9}$ **29.** (a) $\frac{1744}{45}\pi$ **31.** 37,500 ft-lb

(b) $\frac{173\pi}{3}$

33. (a) ≈ 266 lb   **35.** $\frac{22}{5}$ **37.** $\left(\frac{1}{2}, \frac{8}{5}\right)$ **39.** $\left(-\frac{8}{5}, 2\right)$ **41.** $\left(\frac{19}{12(3+2\pi)}, \frac{1+8\pi}{6+4\pi}\right)$ **43.** (a) $\frac{1}{2}$ (b) $\frac{17}{8}$ (c) $\frac{139}{320}$ (d) $\frac{2}{27}$

(b) ≈ 165 lb

(e) $\frac{17}{27}$ **45.** $C(x) = 4x - \frac{x^2}{2000} + 5400,\ R(x) = 40x - \frac{x^2}{200},\ P(x) = 36x - \frac{9x^2}{2000} - 5400;\ P'(x) = 0 \to x = 4000$, where

$M_C(x) = M_R(x)$; Maximum profit $\approx \$66,600$

CHAPTER 7

Exercise Set 7.1

1. $\{x: x > 0\}; -\frac{1}{x}$ **3.** $\{x: x > 0\}; \frac{3}{x}$ **5.** $\{x: 0 < x < 1\}; \frac{1}{x} - \frac{1}{2(1-x)}$ **7.** $\{x: x > 0\}; 1 + \ln x$

9. $\{x: -1 < x < 1\}; \frac{1}{x^2 - 1}$ **11.** $\{x: x > 1\}; \frac{1}{x \ln x}$ **13.** $\{x: -\frac{\pi}{2} + 2n\pi < x < \frac{\pi}{2} + 2n\pi\}; \tan x$

15. $\{x: x > 0\}; \ln \frac{x-2}{x^2}$ **17.** $\{x: x < -1\} \cup \{x: 1 < x < 2\}; \frac{x}{x^2-1} + \frac{1}{2-x}$ **19.** $\{x: -3 < x < 0\} \cup \{x: 2 < x\}; \frac{1}{x+3} - \frac{1}{x} - \frac{1}{x-2}$ **21.** 1

23. $-\infty$ **25.** 0   **27.** (a) Let $f(x) = \frac{\ln x}{\sqrt{x}}$. For $0 < x \le 1$, $\ln x < 0$ and $\ln x < \sqrt{x}$; $f'(x) = \frac{2 - \ln x}{2x^{3/2}}$ has critical number: $x = e^2$

$f(e^2) \approx 0.736$, and $\ln x < \sqrt{x}$, $\lim\limits_{x \to +\infty} \frac{\ln x}{\sqrt{x}} = 0$. Thus, $\ln x < \sqrt{x}$ for $x > 0$.   (b) Let $f(x) = x - 1 - \ln x$, For $x > 1$, $f'(x) = 1 - \frac{1}{x} > 0$.

Therefore, f is an increasing function, and $f(x) > 0$ for $x > 1$.

29. $\ln(1 - \cos x) + C,\ x \ne 2n\pi$ **31.** $\frac{1}{3}\ln|x^3 + 1| + C,\ x \ne -1$ **33.** $\ln|\sin x| + C,\ x \ne n\pi$

35. $\ln|\ln x| + C,\ x > 0,\ x \ne 1$ **37.** $\frac{1}{3}\ln|x(2\sqrt{x} - 3)| + C,\ x > 0,\ x \ne \frac{9}{4}$ **39.** $\frac{x\sqrt{1-x}}{(x+2)(x-3)}\left[\frac{1}{x} - \frac{1}{2(1-x)} - \frac{1}{x+2} - \frac{1}{x-3}\right]$

41. $\frac{\sqrt[3]{(x+1)^2}}{x^4(3-4x)^2}\left[\frac{2}{3(x+1)} - \frac{4}{x} + \frac{8}{3-4x}\right]$ **43.** $\left(\frac{1}{2}, -\frac{1}{2}\ln 2 - \frac{1}{4}\right)$ is a local maximum.

45. $(1,1)$ is a local minimum, $\left(-\frac{1}{2}, -\frac{1}{2} - \ln 4\right)$ is a local maximum.   **47.** $\left(\frac{3}{2}, -\ln 3\right)$ is a local maximum.

49. domain $= \{x: x \ne 0\}$
intercepts: $(\pm 1, 0)$
symmetry: to y-axis
asymptote: $x = 0$
maxima, minima: none
concavity: always down
inflection points: none

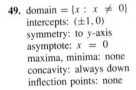

51. domain $= (-1, 1)$
intercepts: $x = 0,\ y = 0$
symmetry: y-axis
asymptotes: $x = \pm 1$
maxima, minima: $(0,0)$
local maximum.
concavity: always down
inflection points: none

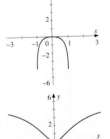

53. domain $= \{x: x > 0\}$
intercepts: none
symmetry: none
asymptote: $x = 0$
maxima, minima: $(1, 1)$
local minimum
concavity: always up
inflection points: none

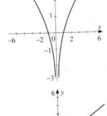

55. domain $= \mathbb{R}$
intercepts: $(0, 0)$
symmetry: y-axis
asymptotes: none
maxima, minima: $(0, 0)$
local minimum
concavity: down on
$(-\infty, -1) \cup (1, \infty)$
up on $(-1, 1)$
inflection points:
$(-1, \ln 2),\ (1, \ln 2)$

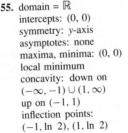

57. Let $x = n + 1$. Left-hand Riemann sum is $\sum\limits_{k=1}^{n} \frac{1}{k}$. For a decreasing function, the area of the kth rectangle $\ge$ area under the graph, and

$\sum\limits_{k=1}^{n} \frac{1}{k} \ge \int_1^x \frac{1}{t}\,dt = \ln x$. If $\ln x = 100$, then $100 \le \sum\limits_{k=1}^{n} \frac{1}{k}$, or n is the smallest integer satisfying the given inequality. x must be greater than m.

59. $2\ln x + (\ln x)^2;\ \frac{2}{x}(1 + \ln x)$ **61.** $\frac{1}{x\sqrt{\ln x^2}};\ -\frac{1 + \ln x^2}{x^2(\ln x^2)^{3/2}}$ **63.** $\frac{y(2x + \ln y)}{x(1-x)}$ **65.** $-\ln(1 + \cos^2 x) + C$

67. $x - 2\ln|x - 1| - \frac{1}{x-1} + C$

69. domain: $x \neq 1, -1$
intercept: $(0, 0)$
asymptotes: $y = 0$, $x = \pm 1$
symmetry: origin
maxima, minima: none
concavity: down if $x > 0$
up if $x < 0$
inflection points: $(0, 0)$

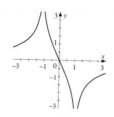

71. domain $= (-\infty, 0) \cup (3, \infty)$
$\left(-3, \ln \frac{3}{2}\right)$ is a local maximum.
intercept: $(-1, 0)$
symmetry: none
asymptotes: $x = 0$, $x = 3$
$y = 0$
Inflection point: $(-4.91, .36)$
Concave up on: $(-\infty, -4.91)$
Concave down on:
$(-4.91, 0) \cup (0, \infty)$

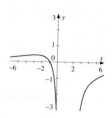

73. Let $\ln n = \int_1^n \frac{1}{x} dx$, and consider the Riemann sums L_{n-1} and U_{n-1} with $\Delta x = 1$. Then, $L_{n-1} \le \ln n \le U_{n-1}$. $\sum_{k=1}^{n-1} \frac{1}{k+1} < \ln n$
$< \sum_{k=1}^{n-1} \frac{1}{k}$. Hence, $\frac{1}{2} + \frac{1}{3} + \ldots + \frac{1}{n} < \ln n < 1 + \frac{1}{2} + \frac{1}{3} + \ldots + \frac{1}{n-1}$.   **75.** If $0 < a < 1$, it shrinks the graph of $\ln(bx)$ in the vertical

direction. If $1 < a < \infty$, it stretches the graph. If $a < 0$, it reflects the graph about the x-axis and shrinks or stretches the graph depending on $|a|$. Let $y = a(\ln b + \ln x) = a \ln b + a \ln x$. For $b > 1$, the graph is moved up if $a > 0$ and down if $a < 0$. For $0 < b < 1$, the graph is moved down if $a > 0$ and up if $a < 0$.

Exercise Set 7.2

1. (a) $\sqrt{3}$ (b) $\frac{1}{16}$ **3.** (a) $\ln 5 - 1$ (b) $1 - \ln 2$ **5.** (a) $\frac{1}{e}$ (b) $e^2 - 1$ '**7.** (a) $\frac{1}{\sqrt[3]{e}}$ (b) $1 + \frac{1}{e^3}$ **9.** $2xe^{x^2}$

11. $3x^2$ **13.** $\frac{1}{2\sqrt{x}} e^{\sqrt{x}}$ **15.** $e^x e^{e^x} = e^{x+e^x}$ **17.** $e^x(x+1)$ **19.** $e^{2x}(3\cos 3x + 2\sin 3x)$ **21.** $\frac{e^x + e^{-x}}{2}$ **23.** $\frac{4}{(e^x + e^{-x})^2}$

25. $2e^x(\sin e^x \cos e^x) = e^x \sin 2e^x$ **27.** $\frac{-e^x}{(e^x-1)^2}$ **29.** $\frac{y - e^{x+y}}{e^{x+y} - x}$ **31.** $\frac{e^{-x}\sin y - e^{-y}\sin x}{e^{-x}\cos y + e^{-y}\cos x}$ **33.** $-\frac{1}{2}e^{(1-x^2)} + C$

35. $e^{-\cos x} + C$ **37.** $\ln(e^x + e^{-x}) + C$ **39.** $\frac{7}{3}$ **41.** $-\frac{3}{8}(1 - e^6)^{4/3}$ **43.** $\ln \frac{3}{2}$ **45.** $3 - \sqrt{3}$ **47.** $\frac{1}{1-e^x} + C$

49. intercept: $(0, 1)$
symmetry: to y-axis
asymptote: $y = 0$
maxima, minima: local max $(0, 1)$
inflection points: $\left(2, \frac{1}{e}\right)$, $\left(-2, \frac{1}{e}\right)$
concave up on $(-\infty, -2) \cup (2, \infty)$
down on $(-2, 2)$

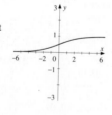

51. intercept: $(0, 0)$
symmetry: none
asymptote: $y = 0$ is an
asymptote on the right
maxima, minima: $\left(2, \frac{2}{e}\right)$ is a
local max
inflection points: $\left(4, \frac{4}{e^2}\right)$ is
an inflection point
concave up on: $(4, \infty)$
concave down on: $(-\infty, 4)$

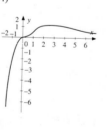

53. intercept: $\left(0, \frac{1}{2}\right)$
symmetry: none
asymptotes: $y = 1$ on right
$y = 0$ on left
maxima, minima: none
inflection point: $\left(0, \frac{1}{2}\right)$
concave up on: $(-\infty, 0)$
concave down on: $(0, \infty)$

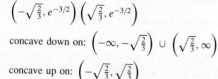

55. intercept: $(0, 0)$
symmetry: none
asymptote: $y = 0$
maxima, minima: $(3, 27e^{-3})$
is a local maximum
inflection points: $(0, 0)$,
$(3 - \sqrt{3}, 0.574)$,
$(3 + 3\sqrt{3}, 0.9334)$
concave down on:
$(-\infty, 0) \cup (3 - \sqrt{3}, 3 + \sqrt{3})$
concave up on:
$(0, 3 - \sqrt{3}) \cup (3 + \sqrt{3}, \infty)$

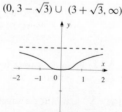

57. intercept: $(0, 0)$; symmetry: about y-axis; asymptote: $y = 1$;
maxima, minima: $(0, 0)$ is a local minimum inflection points:
$\left(-\sqrt{\frac{2}{3}}, e^{-3/2}\right) \left(\sqrt{\frac{2}{3}}, e^{-3/2}\right)$

concave down on: $\left(-\infty, -\sqrt{\frac{2}{3}}\right) \cup \left(\sqrt{\frac{2}{3}}, \infty\right)$

concave up on: $\left(-\sqrt{\frac{2}{3}}, \sqrt{\frac{2}{3}}\right)$

59. $\ln e^{-x} = -x$ and $\ln \frac{1}{e^x} = \ln 1 - \ln e^x = 0 - x = -x$. So $\ln e^{-x} = \ln \frac{1}{e^x}$ and, therefore, $e^{-x} = \frac{1}{e^x}$.   **61.** $\frac{1}{2}$ **63.** $\frac{3}{4}$

65. $f'(x) = \frac{e^x + e^{-x}}{2} > 0$ for all x. Therefore, monotone increasing; hence $1-1$ on $\mathbb{R}$. $f^{-1}(x) = \ln(x + \sqrt{x^2+1})$

67. intercepts: none
symmetry: none
asymptotes: $x = 0$, $y = 1$
maxima, minima: no local extrema
inflection point: $\left(-\frac{1}{2}, e^{-2}\right)$
concave up on: $\left(-\frac{1}{2}, 0\right) \cup (0, \infty)$
concave down on: $\left(-\infty, -\frac{1}{2}\right)$

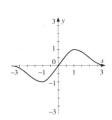

$$\begin{array}{c} - \qquad + \\ \hline \\ -\frac{1}{2} \end{array}$$

69. intercept: $(0, 0)$
symmetry: origin
asymptote: $y = 0$
maxima, minima: $(1,1)$ local max, $(-1, -1)$ local min
inflection points:
$(0,0)$, $\left(\sqrt{3}, \frac{1}{e}\right)$, $\left(-\sqrt{3}, -\frac{1}{e}\right)$
concave up on: $(-\sqrt{3}, 0) \cup (\sqrt{3}, \infty)$
concave down on: $(-\infty, -\sqrt{3}) \cup (0, \sqrt{3})$

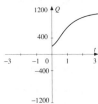

71. $k = -\frac{1}{500} \ln \frac{3}{8}$; $\left(\frac{\ln 2}{\ln \frac{8}{3}}, 500\right)$ is where rate of growth is fastest

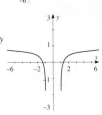

73. (a) $\lim\limits_{x \to \infty} \frac{h(x)}{f(x)} = \lim\limits_{x \to \infty} \frac{x^c}{e^{ax}} = 0$. Use L'Hôpital's rule until the exponent of x in the numerator ≤ 0. (b) $\lim\limits_{x \to \infty} \frac{g(x)}{h(x)} = \lim\limits_{x \to \infty} \frac{(\ln x)^b}{x^c} = 0$.
Use L'Hôpital's rule until the exponent of $\ln x \leq 0$.

Exercise Set 7.3

1. (a) $e^{\sqrt{3} \ln \pi} \approx 7.26255$ (b) $e^{\pi \ln \pi} \approx 36.46216$ (c) $e^{e \ln \pi} \approx 22.45916$ (d) $e^{\sqrt{2} \ln \sqrt{2}} \approx 1.63253$

(e) $e^{\pi \ln(\sin 2)} \approx 0.74177$ **3.** $2^x \ln 2$ **5.** $4^{-x} x(2 - x \ln 4)$ **7.** $\frac{1}{\ln 3}\left[\frac{1}{x} - \frac{3x^2}{1+x^3}\right]$ **9.** $\frac{1}{x(x^2-1)\ln 10}$ **11.** $\frac{1}{x(\ln 2)(\ln x)}$

13. $x^x(1 + \ln x)$ **15.** $x^{1/(1-x)}\left[\frac{1}{x(1-x)} - \frac{\ln x}{(1-x)^2}\right]$ **17.** $(1-x^2)^{1/x}\left[\frac{2}{x^2-1} - \frac{1}{x^2}\ln(1-x^2)\right]$ **19.** (a) 2   (b) ∞

21. (a) $-\frac{1}{2}$ (b) -1 **23.** (a) e (b) $\frac{1}{e}$ **25.** (a) 1   (b) e^2 **27.** $-\frac{1}{2x^2+1 \, \ln 2} + C$ **29.** $\frac{2^{\sin x}}{\ln 2} + C$

31. $\frac{1}{\ln 2}\ln(2^t + 1) + C$ **33.** $\frac{8}{\ln 3}$ **35.** $-\frac{1}{2 \ln 2}\frac{1}{(2^{2t}-1)} + C = \frac{1}{(\ln 4)(1-4^{-t})} + C$ **37.**

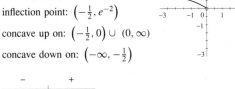

$y = \log_2 x$
$y = \log_e x$
$y = \log_{10} x$
$y = \log_{1/2} x$

39. (a) $\left(\frac{1}{x-1} - \frac{1}{x+2}\right)\frac{1}{\ln 10} = \frac{3}{(x-1)(x+2)\ln 10}$
(b) $\frac{1}{\ln 2}\left[\frac{1}{x} - \frac{1}{2(1-x)} - \frac{6}{2x-3}\right]$

41. intercepts: $(\pm\sqrt{2}, 0)$
symmetry: y-axis
asymptotes: $x = \pm 1$ for negative y
maximum, minimum: none
inflection points: none
concave down on
$(-\infty, -1), \cup (1, \infty)$

43. $\frac{\ln 2}{\ln 6} \approx 0.3869$ **45.** $\frac{\ln \frac{5}{3}}{\ln \frac{4}{3}} \approx 1.7757$ **47.** (a) Let $u = \log_a x$ and $v = \log_a y$. Then
$x = a^u$ and $y = a^v$; $xy = a^u \cdot a^v = a^{u+v}$ $\log_a xy = u + v = \log_a x + \log_a y$
(b) Let $u = \log_a x$ and $v = \log_a y$ $x = a^u$, $y = a^v$; $\frac{x}{y} = \frac{a^u}{a^v} = a^{u-v}$ $\log_a \frac{x}{y} = u - v = \log_a x - \log_a y$
(c) $\log_a x^\alpha = \alpha \log_a x$ Let $u = \log_a x$ $x = a^u$; $x^\alpha = (a^u)^\alpha = a^{\alpha u}$ $\log_a x^\alpha = \alpha u = \alpha \log_a x$
49. (a) $\approx \$16,288.95$ (b) $\approx \$16,386.16$ (c) $\approx \$16,486.65$ (d) $\approx \$16,487.21$ **51.** $\frac{\pi}{\ln 2}$

53. (a) Since for $1 \leq t \leq 1 + \frac{1}{n}$, $\frac{1}{1+\frac{1}{n}} \leq \frac{1}{t} \leq \frac{1}{1}$, it follows that $\int_1^{1+1/n} \frac{1}{1+\frac{1}{n}} dt \leq \int_1^{1+1/n} \frac{1}{t} dt \leq \int_1^{1+1/n} 1 \, dt$ **(b)** Left integral:

$\int_1^{1+1/n} \frac{1}{1+\frac{1}{n}} dt = \frac{1}{n+1}$. Middle integral: $\int_1^{1+1/n} \frac{1}{t} dt = \ln\left(1+\frac{1}{n}\right)$. Right integral: $\int_1^{1+1/n} 1 \, dt = \frac{1}{n}$. So $\frac{1}{n+1} \leq \ln\left(1+\frac{1}{n}\right) \leq \frac{1}{n}$

(c) $\frac{n}{n+1} \leq n \ln\left(1+\frac{1}{n}\right) \leq 1$ or $\frac{1}{1+\frac{1}{n}} \leq \ln\left(1+\frac{1}{n}\right)^n \leq 1$ **(d)** Let $n \to \infty$, $1 \leq \lim_{n \to \infty} \ln\left(1+\frac{1}{n}\right)^n \leq 1$, and thus,

$\lim_{n \to \infty} \ln\left(1+\frac{1}{n}\right)^n = 1$. Since the logarithm is continuous, $\lim(\ln) = \ln(\lim)$, and so $\ln\left[\lim_{n \to \infty}\left(1+\frac{1}{n}\right)^n\right] = 1$. Thus,

$\lim_{n \to \infty}\left(1+\frac{1}{n}\right)^n = e$ **55.** Let $h(x) = \frac{f(x)}{g(x)} = \frac{x^n}{a^x}$; if $n > 0$ and $a > 1$, $\lim_{x \to \infty} \frac{x^n}{a^x} = \lim_{x \to \infty} \frac{nx^{n-1}}{a^x \ln a} = 0$. f grows more slowly than g.

If $n < 0$ and $a > 0$, $f(x)$ decreases and $g(x)$ increases as $x \to \infty$. If $n > 0$ and $0 < a < 1$, $f(x)$ increases and $g(x)$ decreases as $x \to \infty$. If $n < 0$ and $0 < a < 1$, both $f(x)$ and $g(x) \to 0$ as $x \to \infty$. Let $h(x) = a^x x^n$; $h'(x) = na^x x^{n-1} + a^x(\ln a) x^n$. $h'(x) < 0$, since $n < 0$ and $\ln a < 0$. Hence, h is a decreasing function. Because $\lim_{x \to \infty} h(x) = 0$, we can conclude that a^x decreases

more rapidly than x^n.   **57. (a)** If $x^e < e^x$, then $\ln x^e < \ln e^x$, since $\ln x$ is an increasing function. Or, $e(\ln x) < x$, and

$\frac{\ln x}{x} < \frac{1}{e}$, since $x > 0$. If $\frac{\ln x}{x} < \frac{1}{e}$, then $e \ln x < x$, $\ln x^e < x$, and $e^{\ln x^e} = x^e < e^x$, since the exponential function is

increasing.   **(b)** $f'(x) = \frac{1-\ln x}{x^2}$. The only critical value is at $\ln x = 1$, or $x = e$. $f''(e) = -\frac{1}{e^3} < 0 \Rightarrow$ local maximum

$\lim_{x \to 0^+} f(x) = -\infty$; $\lim_{x \to \infty} f(x) = 0$. Therefore, $\left(e, \frac{1}{e}\right)$ is the absolute maximum of f.   **(c)** $x \in (0, e) \cup (e, \infty)$

59. $a^x = x^b \Rightarrow x \ln a = b \ln x, a > 0$; $x = \frac{b}{\ln a} \ln x$. We are looking for intersections of $y = x$ and $y = \frac{b}{\ln a} \ln x$. There can be 0, 1, or 2

intersections. If $\frac{b}{\ln a} > 0$, the graph will intersect the line $y = x$ zero or two times. If $\frac{b}{\ln a} < 0$, the graph of $\ln x$ will intersect the graph of $y = x$ at one point.

61. (a) $f(x) = e^x$, $g(x) = x^\alpha$, $h(x) = \ln x$; $\alpha > 0$ and real. Suppose that n is an integer such that $n \leq \alpha \leq n+1$ $n = [\alpha]$, the

greatest integer in α. $\lim_{x \to \infty} \frac{g(x)}{f(x)} = \lim_{x \to \infty} \frac{x^\alpha}{e^x} = \lim_{x \to \infty} \frac{\alpha(\alpha-1)\cdots(\alpha-n+1)x^{\alpha-n}}{e^x}$. If $\alpha = n$, we are through, since the numerator is a constant,

and in this case, the limit is 0. If $\alpha > n$, apply L'Hôpital's rule once more, and the limit is 0. e^x grows more rapidly than x^α, regardless

of how large α is. Similarly, $\lim_{x \to \infty} \frac{h(x)}{g(x)} = \lim_{x \to \infty} \frac{\ln x}{x^\alpha} = \lim_{x \to \infty} \frac{1/x}{\alpha x^{\alpha-1}} = \lim_{x \to \infty} \frac{1}{\alpha x^\alpha} = 0$. Thus, x^α grows more rapidly than $\ln x$,

regardless of how small the positive exponent α is.   **(b)** Let $f(x) = e^{ax}$, $g(x) = (\ln x)^\beta$, and $h(x) = x^\gamma$; $a > 0, \beta, \gamma > 0$ and

real. $\lim_{x \to \infty} \frac{h(x)}{f(x)} = \lim_{x \to \infty} \frac{x^\gamma}{e^{ax}} = \lim_{x \to \infty} \frac{\gamma(\gamma-1)\cdots(\gamma-n+1)x^{\gamma-n}}{a^n e^{ax}}$, where $n = [\gamma]$. If $\gamma = n$, we are through, since the numerator is a constant.

If $\gamma > n$, apply L'Hôpital's rule once more. In either case, the limit of the quotient is 0. Therefore, e^{ax} grows more rapidly than x^γ. Next

consider, $\lim_{x \to \infty} \frac{g(x)}{h(x)} = \lim_{x \to \infty} \frac{(\ln x)^\beta}{x^\gamma} = \lim_{x \to \infty} \frac{(\beta/x)(\ln x)^{\beta-1}}{\gamma x^{\gamma-1}}$. Continue applying L'Hôpital's rule until the exponent on $\ln x$ is ≤ 0. The limit

will then be zero. Thus, x^α grows more rapidly than $(\ln x)^\beta$.

Exercise Set 7.4

1. 13,500   **3.** 10 hours   **5.** 16 kilograms   **7.** $\frac{\ln 9}{\ln 2} \approx 3.17$ hours   **9.** ≈ 33.87 hours   **11.** $\approx 55, \; 902$

13. ≈ 70.7 grams; ≈ 26.6 years   **15.** $-6.25°$ **17.** Suppose $T_m > T_0$, and assume $\frac{dT}{dt} = k(T - T_m)$. Then, $\frac{dT}{T-T_m} = k \, dt$,

$\ln |T - T_m| = kt + C$; $|T - T_m| = e^{kt+C} = C_1 e^{kt}$, $T - T_m < 0$, so $|T - T_m| = T_m - T$. $T = T_m - C_1 e^{kt}$. When $t = 0, T = T_0; C_1 = T_m - T_0$ and $T = T_m - (T_m - T_0)e^{kt}$ or $T(t) = T_m + (T_0 - T_m)e^{kt}$ $t \approx 32.1$ minutes   **19.** 9.6 psi

21. ≈ 9.97 m   **23. (a)** $N(t) = N_1 e^{k(t-1)}$; $N(1) = N_1$.   **(b)** 5   **25.** $\approx 19,035$ years   **27. (a)** $t = -\frac{1}{k} \ln(1 + \lambda)$

(b) $\approx 7.18 \times 10^9$ years

Exercise Set 7.5

1. (a) $\frac{\pi}{6}$ **(b)** $\frac{\pi}{3}$ **(c)** $\frac{\pi}{4}$ **(d)** $\frac{\pi}{3}$ **(e)** $\frac{\pi}{2}$ **3. (a)** $-\frac{\pi}{2}$ **(b)** π **(c)** 0   **(d)** $\frac{2\pi}{3}$ **(e)** $\frac{3\pi}{4}$

5. (a) $\frac{4}{3}$ **(b)** $\frac{3}{5}$ **7. (a)** $\frac{7}{9}$ **(b)** $-\frac{24}{25}$

9. (a) $\frac{\pi}{5}$ **(b)** $-\frac{\pi}{8}$ **11.** $\frac{33}{65}$ **13. (a)** 2.412   **(b)** 4.965   **15.** $\frac{6}{4x^2+9}$ **17.** $\frac{1}{2\sqrt{x-x^2}}$

19. $1 + 2x \tan^{-1} x$ **21.** $\frac{x \cos^{-1} x - \sqrt{1-x^2}}{(1-x^2)^{3/2}}$ **23.** $\frac{2x\sqrt{1-y^2}}{\sqrt{1-y^2}-1}$ **25.** $\frac{\pi}{2}$ **27.** $\frac{\pi}{12}$ **29.** $\frac{\pi}{16}$ **31.** $\frac{1}{3} \sin^{-1} \frac{3x}{4} + C$

33. $\frac{1}{3} \sec^{-1} \frac{4x}{3} + C$ **35.** $\frac{1}{3} \tan^{-1}(3x - 2) + C$ **37.** $\frac{1}{2} \sin^{-1} x^2 + C$ **39.** $-\frac{\pi}{6\sqrt{3}}$ **41.** $\frac{\pi}{3}$ **43.** $\frac{\pi^2}{2\sqrt{3}}$

45. $\int \frac{du}{\sqrt{a^2-u^2}} = \int \frac{du}{a\sqrt{1-\frac{u^2}{a^2}}} = \int \frac{dv}{\sqrt{1-v^2}} = \sin^{-1} v + C$ Let $v = \frac{u}{a}$, $dv = \frac{1}{a} du = \sin^{-1} \frac{u}{a} + C$

47. Let $y = \tan^{-1} x$, $-\infty < x < \infty$, $-\frac{\pi}{2} < y < \frac{\pi}{2}$, $x = \tan y$; $1 = (\sec^2 y)y'$, $y' = \frac{1}{\sec^2 y} = \frac{1}{1+\tan^2 y} = \frac{1}{1+x^2}$

49. Let $y = \sec^{-1} x$, $|x| > 1$ $0 < y < \frac{\pi}{2}$ if $x > 1$, $\pi < y < \frac{3\pi}{2}$ if $x < -1$, $x = \sec y$; $1 = (\sec y \tan y)y'$,

$y' = \frac{1}{\sec y \tan y} = \frac{1}{\sec y \sqrt{\sec^2 y - 1}}$ (since $\tan y > 0$ for y in given range) $= \frac{1}{x\sqrt{x^2-1}}$

51. $\int \frac{du}{1+u^2} = \tan^{-1} u + C$, since $\frac{d}{du}(\tan^{-1} u + C) = \frac{1}{1+u^2}$

53. $\int \frac{du}{u\sqrt{u^2-a^2}} = \frac{1}{a}\int \frac{du}{u\sqrt{\frac{u^2}{a^2}-1}} = \frac{1}{a}\int \frac{\frac{1}{a}du}{\frac{u}{a}\sqrt{\frac{u^2}{a^2}-1}} = \frac{1}{a}\int \frac{dv}{v\sqrt{v^2-1}}$

Let $v = \frac{u}{a}$ $= \frac{1}{a}\sec^{-1} v + C$
$dv = \frac{du}{a}$ $= \frac{1}{a}\sec^{-1}\frac{u}{a} + C$

55. $\frac{d}{dx}\left[\sin^{-1} x + \sin^{-1}\sqrt{1-x^2}\right] = \frac{1}{\sqrt{1-x^2}} - \frac{x}{\sqrt{x^2}\sqrt{1-x^2}} = 0$. Therefore, $\sin^{-1} x + \sin^{-1}\sqrt{1-x^2} = C$. Let $x = \frac{1}{2}$ and

$\sin^{-1} x + \sin^{-1}\sqrt{1-x^2} = \frac{\pi}{2}$ which holds for all $x \in (0, 1)$.   **57.** $\frac{\pi}{6}$ **59.** $\frac{\pi}{4}$ **61.** $\frac{\pi^2}{4}$ **63.** $-\frac{6}{425}$ rad/sec

Exercise Set 7.6

1. $\sinh x$; Domain $= \mathbb{R}$, Range $= \mathbb{R}$, $\cosh x$; Domain $= \mathbb{R}$, Range $= [1, \infty)$, $\tanh x$; Domain $= \mathbb{R}$, Range $= (-1, 1)$, $\coth x$; Domain $= \{x : x \neq 0\}$, Range $= (-\infty, -1) \cup (1, \infty)$, $\operatorname{sech} x$; Domain $= \mathbb{R}$, Range $= (0, 1]$, $\operatorname{csch} x$; Domain $= \{x : x \neq 0\}$, Range $= (-\infty, 0) \cup (0, \infty)$

3. (a)

x	0.5	1	2	4
$\tanh x$	0.462	0.762	0.964	0.9993

(b)

x	0.5	1	2	4
$\coth x$	2.16	1.31	1.04	1.00

(c)

x	0.5	1	2	4
$\operatorname{sech} x$	0.887	0.648	0.266	0.0366

(d)

x	0.5	1	2	4
$\operatorname{csch} x$	1.92	0.851	0.276	0.0366

5. $\cosh x \cosh y + \sinh x \sinh y$

$= \left(\frac{e^x+e^{-x}}{2}\right)\left(\frac{e^y+e^{-y}}{2}\right) + \left(\frac{e^x-e^{-x}}{2}\right)\left(\frac{e^y-e^{-y}}{2}\right) = \frac{1}{4}[e^{x+y} + e^{-x+y} + e^{x-y} + e^{-(x+y)} + e^{(x+y)} - e^{-x+y} - e^{x-y} + e^{-(x+y)}]$

$= \frac{1}{4}[2e^{x+y} + 2e^{-(x+y)}] = \frac{e^{x+y}+e^{-(x+y)}}{2} = \cosh(x+y)$ **7.** $\cosh(x+x) = \cosh 2x = \cosh x \cosh x + \sinh x \sinh x$

$= \cosh^2 x + \sinh^2 x$, which $= \cosh^2 x + \cosh^2 x - 1 = 2\cosh^2 x - 1$ or $= \sinh^2 x + 1 + \sinh^2 x = 2\sinh^2 x + 1$

9. In $\cosh 2x = 2\cosh^2 x - 1$, replace x by $\frac{x}{2}$ and solve for $\cosh\frac{x}{2}$: $\cosh x = 2\cosh^2\frac{x}{2} - 1$; $\cosh^2\frac{x}{2} = \frac{\cosh x+1}{2}$ $\cosh\frac{x}{2} = \frac{\sqrt{\cosh x+1}}{\sqrt 2}$

since $\cosh\frac{x}{2} > 0$ **11.** $\cosh^2 x - \sinh^2 x = 1$. Divide through by $\sinh^2 x$: $\coth^2 x - 1 = \operatorname{csch}^2 x$

13. $6\tanh^2 2x \operatorname{sech}^2 2x$ **15.** $\coth x$ **17.** $\frac{1}{\cosh x\sqrt{\cosh^2 x-1}} = \frac{1}{\cosh x\,|\sinh x|}$ **19.** $-x\sqrt{\operatorname{sech} x^2}\tanh x^2$ **21.** $\frac{y\,\cosh x-\cosh y}{x\,\sinh y-\sinh x}$

23. $(\sinh x)^x[x\coth x + \ln\sinh x]$ **25.** $\frac{\sinh^2 x}{2} + C$ **27.** $\ln(\cosh x) + C$ **29.** $\ln|x + \sinh x| + C$ **31.** $\ln(\ln\cosh x) + C$

33. $\frac{\cosh^3 x}{3} - \cosh x + C$ **35.** $\sinh x + \frac{\sinh^3 x}{3} + C$ **37.** $\frac{\tanh^2 x}{2} - \frac{\tanh^4 x}{4} + C$ **39.** $\ln\frac{16}{9}$ **41.** $\frac{\pi}{12}$ **43. (a)** $-\frac{1}{2}$

(b) $-\frac{1}{2}$ **45. (a)** 1   **(b)** 1   **47.** $\pi[2 - \tanh 2]$ **49.** $\sinh x(\coth x - \tanh x) = \sinh x\left(\frac{\cosh x}{\sinh x} - \frac{\sinh x}{\cosh x}\right)$

$= \cosh x - \frac{\sinh^2 x}{\cosh x} = \frac{\cosh^2 x-\sinh^2 x}{\cosh x} = \frac{1}{\cosh x} = \operatorname{sech} x$ **51.** $\frac{2}{\coth x-\tanh x} = \frac{2}{\frac{\cosh x}{\sinh x} - \frac{\sinh x}{\cosh x}} = \frac{2\sinh x\cosh x}{\cosh^2 x-\sinh^2 x} = \frac{\sinh 2x}{1} = \sinh 2x$

53. $\tanh\frac{x}{2} = \frac{\sinh\frac{x}{2}}{\cosh\frac{x}{2}} \cdot \frac{2\cosh\frac{x}{2}}{2\cosh\frac{x}{2}} = \frac{2\sinh\frac{x}{2}\cosh\frac{x}{2}}{2\cosh^2\frac{x}{2}} = \frac{\sinh x}{2\left(\frac{1+\cosh x}{2}\right)} = \frac{\sinh x}{1+\cosh x}$ Also, $\frac{\sinh x}{\cosh x+1} \cdot \frac{\cosh x-1}{\cosh x-1} = \frac{\sinh x(\cosh x-1)}{\cosh^2 x-1} = \frac{\sinh x(\cosh x-1)}{\sinh^2 x}$

$= \frac{\cosh x-1}{\sinh x}$ **55. (a)** $\frac{1}{2}(\cosh t)(\sinh t)$ **(b)** $\frac{1}{2}\sinh t\cosh t - \frac{t}{2}$ **(c)** $\frac{t}{2}$ **57.** $\ln\cosh x - \frac{\tanh^2 x}{2} - \frac{\tanh^4 x}{4} + C$

59. $-\operatorname{csch} x - \coth x + C$ **61.** $\int \frac{\sinh x}{\sinh^2 x}dx = \int \frac{\sinh x}{\cosh^2 x-1}dx$ Let $u = \cosh x$, $du = \sinh x\,dx$ $\int \frac{du}{u^2-1} = \frac{1}{2}\int\left[\frac{1}{u-1} - \frac{1}{u+1}\right]du$

$= \frac{1}{2}\ln\left|\frac{u-1}{u+1}\right| + C = \frac{1}{2}\ln\left|\frac{\cosh x-1}{\cosh x+1}\right| + C$ and $\sqrt{\frac{\cosh x-1}{\cosh x+1} \cdot \frac{\cosh x+1}{\cosh x+1}} = \left|\frac{\sinh x}{\cosh x+1}\right| = \left|\tanh\frac{x}{2}\right|$

63. $\frac{\sinh 3x}{\sinh x} - \frac{\cosh 3x}{\cosh x} = \frac{\sinh 3x\cosh x-\cosh 3x\sinh x}{\sinh x\cosh x} = \frac{\sinh(3x-x)}{\sinh x\cosh x} = \frac{2\sinh x\cosh x}{\sinh x\cosh x} = 2$

65. (a) 30 m   **(b)** ≈ 57.05 m   **(c)** ≈ 85.17 m   **67.** domain $= (-1, 1)$ **69.** domain $= (0, 1]$
range $= \mathbb{R}$ range $= [0, \infty)$

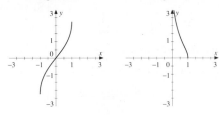

71. (a) Let $y = \operatorname{sech}^{-1} x, x > 0$. Then $x = \operatorname{sech} y$. So $\cosh^{-1}\left(\frac{1}{x}\right) = \cosh^{-1}\left(\frac{1}{\operatorname{sech} y}\right) = \cosh^{-1}(\cosh y) = y = \operatorname{sech}^{-1} x$ **(b)** Let

$y = \operatorname{csch} x, x \neq 0$; $\operatorname{csch} y = x$. So, $\sinh^{-1}\frac{1}{x} = \sinh^{-1}\left(\frac{1}{\operatorname{csch} y}\right) = \sinh^{-1}(\sinh y) = y = \operatorname{csch} x$ **73.** Let $y = \coth^{-1} x, |x| > 1$,

$x = \coth y$ $y' = -\frac{1}{\operatorname{csch}^2 y} = -\frac{1}{\coth^2 y - 1}$, $1 = (-\operatorname{csch}^2 y)y' = \frac{1}{1-x^2}$

75. Let $y = \operatorname{csch}^{-1} x$, $x \neq 0$; $x = \operatorname{csch} y$ $1 = -(\operatorname{csch} y \coth y)y'$; $y' = -\frac{1}{\operatorname{csch} y \coth y}$. Since $\coth^2 y = 1 + \operatorname{csch}^2 y$,

$\coth y = \begin{cases} \sqrt{1 + \operatorname{csch}^2 y} & \text{if } \coth y > 0 \\ -\sqrt{1 + \operatorname{csch}^2 y} & \text{if } \coth y < 0 \end{cases}$. That is, $\coth y = \begin{cases} \sqrt{1+x^2} & \text{if } x > 0 \\ -\sqrt{1+x^2} & \text{if } x < 0 \end{cases}$. Thus, $y' = -\frac{1}{|x|\sqrt{1+x^2}}$

77. $\frac{2x}{\sqrt{x^4-1}}$ **79.** $\sec x$ **81.** $\operatorname{csch}^{-1} x - \frac{1}{\sqrt{1+x^2}}$ **83.** $-\frac{1}{\sqrt{-(x+x^2)}}$ **85.** $\frac{e^x}{1+2e^x}$ **87.** $\frac{1}{2}\cosh^{-1} 2x + C$

89. $-\operatorname{sech}^{-1} 2x + C$ **91.** $\cosh^{-1}(\tan x) + C$ **93.** $\frac{1}{2}\tanh^{-1}\frac{1}{4} = \frac{1}{4}\ln\frac{5}{3}$ **95.** $\sinh^{-1}(x+1) + C$

97. $\int \frac{du}{a\sqrt{\left(\frac{u}{a}\right)^2-1}} = \int \frac{\frac{1}{a}du}{\sqrt{\left(\frac{u}{a}\right)^2-1}} = \cosh^{-1}\frac{u}{a} + C, \frac{u}{a} > 1$ **99.** $\int \frac{du}{au\sqrt{1-\left(\frac{u}{a}\right)^2}} = \frac{1}{a}\int \frac{\frac{1}{a}du}{\frac{u}{a}\sqrt{1-\left(\frac{u}{a}\right)^2}} = -\frac{1}{a}\operatorname{sech}^{-1}\frac{u}{a} + C$ if $0 < \frac{u}{a} < 1$

101. $\frac{1}{3}\sinh^{-1}\frac{3x}{2} + C$ **103.** $\frac{1}{2} \cdot \frac{1}{3}\tanh^{-1}\frac{2}{3} = \frac{1}{12}\ln 5$ **105.** $-\frac{1}{2}\operatorname{csch}^{-1}\frac{e^x}{2} + C$

107. (a) If $\cosh^{-1} x = y$, $x \geq 1$, $y \geq 0$ then $x = \cosh y = \frac{e^y + e^{-y}}{2}$; $e^y - 2x + e^{-y} = 0$; $e^{2y} - 2xe^y + 1 = 0$. $e^y = x \pm \sqrt{x^2 - 1}$.

$e^y > 1$. The solution $x - \sqrt{x^2 - 1}$ is not possible, therefore, $y = \ln(x + \sqrt{x^2 - 1})$ **(b)** Let $y = \tanh^{-1} x$. $x = \tanh y = \frac{e^y - e^{-y}}{e^y + e^{-y}}$.

$e^{2y}(1-x) = 1 + x$; $y = \frac{1}{2}\ln\left(\frac{1+x}{1-x}\right)$ **109. (a)** Let $y = \operatorname{csch}^{-1} x$. $x = \operatorname{csch} y = \frac{2}{e^y - e^{-y}}$; $xe^{2y} - 2e^y - x = 0$;

$e^y = \frac{1+\sqrt{1+x^2}}{x}$ if $x > 0$ and $e^y = \frac{1-\sqrt{1-x^2}}{x}$ if $x < 0$. Thus $y = \ln\left[\frac{1}{x} + \frac{\sqrt{1+x^2}}{|x|}\right]$.   **(b)** $\int \frac{du}{(1+u)(1-u)} = \frac{1}{2}\int\left(\frac{1}{1+u} + \frac{1}{1-u}\right)du$

$= \frac{1}{2}(\ln|1+u| - \ln|1-u|) + C = \frac{1}{2}\ln\left|\frac{1+u}{1-u}\right| + C$

111. If $x > 0$, $\frac{d}{dx}\operatorname{csch}^{-1}|x| = -\frac{1}{x\sqrt{1+x^2}}$ If $x < 0$, $\frac{d}{dx}\operatorname{csch}^{-1}|x| = \frac{d}{dx}\operatorname{csch}^{-1}(-x) = -\frac{1}{(-x)\sqrt{1+x^2}}(-1) = -\frac{1}{x\sqrt{1+x^2}}$. So, for all

$x \neq 0$, $\frac{d}{dx}\operatorname{csch}^{-1}|x| = -\frac{1}{x\sqrt{1+x^2}}$ Thus, $\int \frac{du}{u\sqrt{a^2+u^2}} = \frac{1}{a}\int \frac{\frac{1}{a}du}{\frac{u}{a}\sqrt{1+\left(\frac{u}{a}\right)^2}} = -\frac{1}{a}\operatorname{csch}^{-1}\frac{|u|}{a} + C$, $u \neq 0$, $a > 0$.

113. (a) $v(t) = \sqrt{\frac{mg}{k}}\tanh\left(\sqrt{\frac{kg}{m}}t\right)$ **(b)** $s(t) = \frac{m}{k}\ln\cosh\sqrt{\frac{kg}{m}}t$ **(c)** $\lim_{t\to\infty} v(t) = \sqrt{\frac{mg}{k}}$

Chapter Review

1. (a) $f'(x) = \frac{1}{x} + \frac{1}{2(x-2)}$, domain: $x > 2$ **(b)** $f'(x) = \frac{1}{x} + \frac{1}{x-1} - \frac{1}{x-4} - \frac{1}{x+2}$, domain: $(-\infty, -2) \cup (0, 1) \cup (4, \infty)$

3. (a) $f'(x) = xe^{-x}(2-x)$, domain: $\mathbb{R}$ **(b)** $f'(x) = -e^{-x}$, domain: $\mathbb{R}$ **5. (a)** $f'(x) = e^{-x}(\sin e^{-x} + \cos e^{-x})$,

domain: $\mathbb{R}$ **(b)** $f'(x) = \frac{1}{2}\frac{e^x - e^{-x}}{e^x + e^{-x}} \cdot \ln(e^x + e^{-x})$, domain: $\mathbb{R}$ **7. (a)** $f'(x) = 2(\ln x)x^{\ln x - 1}$, domain: $x > 0$

(b) $f'(x) = (\sec x)^{\tan x}(\tan^2 x + \sec^2 x \ln \sec x)$, domain: $\left(-\frac{\pi}{2}, \frac{\pi}{2}\right)$, $\left(\frac{3\pi}{2}, \frac{5\pi}{2}\right)$, ... or $\{x : \sec x > 0\}$ **9. (a)** $\frac{y}{xye^y + x}$

(b) $\frac{x^2 - y^2 + 2y}{2x}$ **11. (a)** $-\frac{1}{x} + C$ **(b)** $\frac{1}{x - e^x} + C$

13. (a) $\frac{1}{2}\ln\frac{9}{8}$ **(b)** 2   **15. (a)** $\frac{1}{\ln 4}\ln(4^x + 1) + C$ **(b)** $\frac{\sqrt{3}}{4\ln 2}$

17. intercepts: $(2, 0)$, $(0, 0)$, symmetry: to $x = 1$, asymptote: $x = 1$, maxima, minima: none, inflection points: none, concave down on: $(-\infty, 1) \cup (1, \infty)$

19. intercept: $(0, 0)$, symmetry: origin, asymptote: $y = 0$, maxima, minima: $(\sqrt{3}, 3\sqrt{3}e^{-3/2})$ local maximum, $(-\sqrt{3}, -3\sqrt{3}e^{-3/2})$ local minimum, inflection points: $(0, 0)$, $(\pm 1, \pm e^{-1/2})$, $(\pm\sqrt{6}, \pm 6\sqrt{6}e^{-3})$, concave up on: $(-\sqrt{6}, -1) \cup (0, 1) \cup (\sqrt{6}, \infty)$, concave down on: $(-\infty, -\sqrt{6}) \cup (-1, 0) \cup (1, \sqrt{6})$

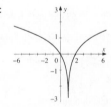

21. $\pi\left(\frac{16}{3} + 4\ln 3\right)$ **23. (a)** $x = 2$ **(b)** $x = 3$ **25.** ≈ 20.6 hours   **27.** $Q(t) = m\left(\frac{Q_0}{m}\right)^{e^{-kt}}$. As $t \to \infty$, $Q(t) \to m$. Thus, m is

the limiting size of the population.   **29. (a)** $-\frac{2}{3}$ **(b)** $-\frac{4}{5}$

31. (a) $-\frac{56}{65}$ **(b)** $-\frac{1+4\sqrt{2}}{3\sqrt{5}}$

33. (a) $\frac{2}{\sqrt{1-x^4}} - \frac{1}{x^2}\sin^{-1} x^2$ **(b)** $\frac{1}{x\sqrt{x^2-1}}$ **35. (a)** $\frac{\sin^{-1}\sqrt{x} - \cos^{-1}\sqrt{x}}{\sqrt{x-x^2}}$ **(b)** $\frac{y-x}{y+x}$ **37. (a)** $\frac{1}{1+\cosh x}$ **(b)** $-\frac{\cosh y}{\cosh x}$

39. (a) $-\frac{1}{2x}$ (b) $\frac{1}{\sqrt{x+2}} + \frac{1}{2\sqrt{x}}\cosh^{-1}(x+1)$ **41.** $\frac{\pi}{3\sqrt{3}}$ **43.** $\frac{\pi}{8}$ **45.** $\frac{\pi}{6}$ **47.** $\frac{1}{2}\tan^{-1}x^2 + C$ **49.** $-\frac{1}{e^{\sinh^2 x}} + C$

51. $\frac{\pi}{4}$ **53.** $\cosh^{-1}e^x + C$ **55.** $\frac{1}{2}\ln\frac{3}{2}$ **57.** (a) $\frac{1}{2}$ (b) 0 **59.** (a) $-\frac{1}{2}$ (b) $-\frac{1}{2}$ **61.** (a) 1 (b) $\sqrt{e}$

63. $2\ln\frac{(1+\sqrt{3})}{2} - \frac{\pi}{6}$ **65.** ≈ -0.04937 radian/sec

67. $\frac{1}{1-\tanh x} - \frac{1}{1+\tanh x}$

$= \frac{1+\tanh x-1+\tanh x}{1-\tanh^2 x}$

$= \frac{2\tanh x}{\operatorname{sech}^2 x} = \frac{2\sinh x}{\cosh x}$

$\cdot \cosh^2 x = 2\sinh x\cosh x = \sinh 2x$

69. $\frac{\cosh x-\sinh x}{\cosh 2x-\sinh 2x} = \frac{\cosh x-\sinh x}{\cosh^2 x+\sinh^2 x-2\sinh x\cosh x}$

$= \frac{\cosh x-\sinh x}{(\cosh x-\sinh x)^2} = \frac{1}{\cosh x-\sinh x}$

$= \frac{1}{\cosh x-\sinh x} \cdot \frac{\cosh x+\sinh x}{\cosh x+\sinh x}$

$= \frac{\cosh x+\sinh x}{\cosh^2 x-\sinh^2 x} = \cosh x + \sinh x$

71. (a) $8\sinh\frac{1}{2}$

CHAPTER 8

Exercise Set 8.1

1. $-xe^{-x} - e^{-x} + C$ **3.** $\frac{\pi}{8} - \frac{1}{4}$ **5.** $e^x(x^4 - 4x^3 + 12x^2 - 24x + 24) + C$ **7.** $x\sec x - \ln|\sec x + \tan x| + C$

9. $\frac{3}{4}\left[(\ln 2)^2 + 2\right] - \frac{5}{2}\ln 2$ **11.** $-x\cot x + \ln|\sin x| + C$ **13.** $-\frac{x^2\ln x}{2} + \frac{x^2}{4} + C$ **15.** $\sin x\ln(\sin x) - \sin x + C$

17. $\frac{1}{6} + \frac{1}{4}\ln 3$ **19.** $\frac{-e^{-2x}}{8}\left[4x^3 + 6x^2 + 6x + 3\right] + C$ **21.** $x\tan x + \ln|\cos x| - \frac{x^2}{2} + C$ **23.** $\frac{e^{-x}}{5}(2\sin 2x - \cos 2x) + C$

25. $\frac{1}{4}(e^{2x} + 2x) + C$ **27.** $\frac{1}{2}\left[\operatorname{sech} x\tanh x + \tan^{-1}(\sinh x)\right] + C$ **29.** $\frac{x^3}{3}\cot^{-1}x + \frac{x^2}{6} - \frac{1}{6}\ln(1 + x^2) + C$ **31.** $\frac{3\pi}{2}$

33. $4\pi(\pi - 2)$ **35.** (a) $\left(1, \frac{\pi}{8}\right)$ (b) $\left(\frac{\pi-2}{2}, \frac{\pi}{8}\right)$ **37.** (a) $\frac{1}{2}\sec^2 x + C$ (b) $\frac{1}{2}\tan^2 x + C_1$ **39.** $\frac{x}{2}[\sin(\ln x) + \cos(\ln x)] + C$

41. $\frac{x^2}{4} + \frac{x\sin 2x}{4} + \frac{\cos 2x}{8} + C$ **43.** $\frac{(x^2-4)^{5/2}}{35}(5x^2 + 8) + C$ **45.** $\frac{e^{ax}(a\cos bx+b\sin bx)}{a^2+b^2} + C$ **47.** (a) $\begin{aligned}\text{Let } u &= x^n \quad du = nx^{n-1}\,dx \\ dv &= \sin x\,dx \quad v = -\cos x\end{aligned}$

(b) $\begin{aligned}\text{Let } u &= x^n \quad du = nx^{n-1}\,dx \\ dv &= \cos x\,dx \quad v = \sin x\end{aligned}$ **49.** By Exercise 48, $\int_0^{\pi/2}\sin^n x\,dx = -\frac{\sin^{n-1}x\cos x}{n}\Big]_0^{\pi/2} + \frac{n-1}{n}\int_0^{\pi/2}\sin^{n-2}x\,dx$

$= \frac{n-1}{n}\int_0^{\pi/2}\sin^{n-2}x\,dx$; apply this same result again so that: $\int_0^{\pi/2}\sin^n x\,dx = \frac{n-1}{n}\cdot\frac{n-3}{n-2}\int_0^{\pi/2}\sin^{n-4}x\,dx$. Continue. If n is odd,

$\int_0^{\pi/2}\sin^n x\,dx = \frac{n-1}{n}\cdot\frac{n-3}{n-2}\cdot\frac{n-5}{n-4}\cdots\frac{2}{3}\int_0^{\pi/2}\sin x\,dx$ and $\int_0^{\pi/2}\cos^n x\,dx = \frac{n-1}{n}\cdot\frac{n-3}{n-2}\cdot\frac{n-5}{n-4}\cdots\frac{2}{3}\int_0^{\pi/2}\cos x\,dx$. Since

$\int_0^{\pi/2}\sin x\,dx = 1$ and $\int_0^{\pi/2}\cos x\,dx = 1$, $\int_0^{\pi/2}\sin^n x\,dx = \int_0^{\pi/2}\cos^n x\,dx = \frac{2\cdot 4\cdot 6\cdots (n-1)}{1\cdot 3\cdot 5\cdots n}$. If n is even the result in each case

is: $\frac{n-1}{n}\cdot\frac{n-3}{n-2}\cdot\frac{n-5}{n-4}\cdots\frac{1}{2}\int_0^{\pi/2}dx = \frac{1}{2}\cdot\frac{3}{4}\cdot\frac{5}{6}\cdots\frac{n-1}{n}\frac{\pi}{2}$

51. (a) Let $u = \sec^{n-2}x \quad\Big|\quad du = (n-2)\sec^{n-3}x\sec x\tan x\,dx$

$dv = \sec^2 x\,dx \quad\Big|\quad v = \tan x$.

Then $I = \sec^{n-2}x\tan x - (n-2)\int\sec^{n-2}x(\sec^2 x - 1)dx = \sec^{n-2}x\tan x - (n-2)\left[I - \int\sec^{n-2}x\,dx\right]$.

$I(n-1) = \sec^{n-2}x\tan x + (n-2)\int\sec^{n-2}x\,dx$; and $I = \frac{\sec^{n-2}x\tan x}{n-1} + \frac{n-2}{n-1}\int\sec^{n-2}x\,dx$.

(b) Let $u = \csc^{n-2}x \quad\Big|\quad du = -(n-2)\csc^{n-3}x\csc x\cot x\,dx$

$dv = \csc^2 x\,dx \quad\Big|\quad v = -\cot x$.

Then $I = -\csc^{n-2}x\cot x - (n-2)\int\csc^{n-2}x(\csc^2 x - 1)\,dx = -\csc^{n-2}x\cot x - (n-2)\left[I - \int\csc^{n-2}x\,dx\right]$;

$I(n-1) = -\csc^{n-2}x\cot x + (n-2)\int\csc^{n-2}x\,dx$ and $I = \frac{-\csc^{n-2}x\cot x}{n-1} + \frac{n-2}{n-1}\int\csc^{n-2}x\,dx$

Exercise Set 8.2

1. $\sin x - \frac{\sin^3 x}{3} + C$ **3.** $\frac{3x}{8} - \frac{\sin 2x}{4} + \frac{\sin 4x}{32} + C$ **5.** $\frac{16-7\sqrt{2}}{120}$ **7.** $\frac{x}{8} - \frac{\sin 4x}{32} + C$ **9.** $\frac{1}{4}$ **11.** $\frac{\sec^3 x}{3} - \sec x + C$

13. $\frac{1}{2}(\sec x\tan x - \ln|\sec x + \tan x|) + C$ **15.** $\ln 2 - \frac{3}{8}$ **17.** $\tan x - \cot x + C$ **19.** $\frac{3}{2} - \ln 2$ **21.** $\frac{\sin^3 3x}{9} - \frac{\sin^5 3x}{15} + C$

23. $\frac{10}{21}$ **25.** $\frac{1}{2}\left[\frac{\sin^3 x^2}{3} - \frac{2\sin^5 x^2}{5} + \frac{\sin^7 x^2}{7}\right] + C$ **27.** $-\frac{2\cot^3(x/2)}{3} + C$ **29.** $\frac{\sinh^2 x}{2} + \frac{\sinh^4 x}{4} + C = \frac{\cosh^4 x}{4} + C_1$

31. $\frac{\tanh^4 x}{4} - \frac{\tanh^6 x}{6} + C$ **33.** $-\frac{\cos 5x}{10} - \frac{\cos x}{2} + C$ **35.** $\frac{\sin 5x}{10} + \frac{\sin x}{2} + C$ **37.** $\frac{1}{8}\left[\frac{5x}{2} - 2\sin 2x + \frac{3\sin 4x}{8} + \frac{\sin^3 2x}{6}\right] + C$

39. $\frac{\sec^3 x}{3} + 2\ln|\cos x| + \cos x + C$ **41.** $\int\frac{\sin^3 x}{\cos^8 x}\,dx = \frac{1}{7}\sec^7 x - \frac{1}{5}\sec^5 x + C$

43. $\int_{-\pi}^{\pi} \sin mx \, \sin nx \, dx = \frac{1}{2} \int_{-\pi}^{\pi} [\cos(m-n)x - \cos(m+n)x] \, dx = \frac{1}{2} \left[\frac{\sin(m-n)x}{m-n} - \frac{\sin(m+n)x}{m+n} \right]_{-\pi}^{\pi} = 0$ if $m \neq n$;

$\int_{-\pi}^{\pi} \cos mx \, \cos nx \, dx = \frac{1}{2} \int_{-\pi}^{\pi} [\cos(m+n)x + \cos(m-n)x] \, dx = \frac{1}{2} \left[\frac{\sin(m+n)x}{m+n} + \frac{\sin(m-n)x}{m-n} \right]_{-\pi}^{\pi} = 0$ if $m \neq n$ and

$\int_{-\pi}^{\pi} \sin mx \, \cos nx \, dx = \frac{1}{2} \int_{-\pi}^{\pi} [\sin(m+n)x + \sin(m-n)x] \, dx = 0$. Therefore, the set is orthogonal on $[-\pi, \pi]$

Exercise Set 8.3

1. $\ln|\sec\theta + \tan\theta| + C$ **3.** 1 **5.** $\frac{2511}{20\sqrt{3}}$ **7.** $\frac{1}{27} + \frac{5}{81\sqrt{3}}$ **9.** $\frac{1}{4(1-x^2)^2} - \frac{1}{1-x^2} - \ln\sqrt{1-x^2} + C$ **11.** $\frac{31}{15}$

13. (a) $\frac{x^2}{2(1+x^2)} + C$ **(b)** $-\frac{1}{2(1+x^2)} + C$ **15.** $\frac{\sqrt{1+x^2}}{3}(x^2-2) + C$ **17.** Let $u = a\sin\theta, \theta = \sin^{-1}\frac{u}{a}, du = a\cos\theta d\theta$

19. Let $u = a\sec\theta, du = a\sec\theta\tan\theta d\theta, \theta = \sec^{-1}\frac{u}{a}$ **21.** Let $u = a\sec\theta, du = a\sec\theta\tan\theta d\theta, \theta = \sec^{-1}\frac{u}{a}$,

$\int \frac{a\sec\theta\tan\theta \, d\theta}{a^2\tan^2\theta} = \frac{1}{a}\int\csc\theta \, d\theta, = \frac{1}{a}\left[\ln|u-a| - \frac{1}{2}\ln|u+a| - \frac{1}{2}\ln|u-a|\right] + C, = \frac{1}{2}a \, \ln\left|\frac{u-a}{u+a}\right| + C$ **23.** Let

$u = a\tan\theta, du = a\sec^2\theta \, d\theta, \theta = \tan^{-1}\frac{u}{a}, \int\frac{a\sec^2\theta \, d\theta}{a\tan\theta \, a\sec\theta} = \frac{1}{a}\int\csc\theta \, d\theta, = \frac{1}{a}\ln|\csc\theta - \cot\theta| + C,$

$= \frac{1}{a}\ln\left|\frac{\sqrt{a^2+u^2}-a}{u} \cdot \frac{\sqrt{a^2+u^2}+a}{\sqrt{a^2+u^2}+a}\right| + C$ **25.** $3\ln\left|\frac{3-\sqrt{9-x^2}}{x}\right| + \sqrt{9-x^2} + C$ **27.** $\frac{2511}{20\sqrt{3}}$ **29.** $3\sqrt{2} + \frac{1}{2}\ln(3+2\sqrt{2})$

31. $\frac{3}{8}\sin^{-1}(x-1) + \frac{3}{8}(x-1)\sqrt{2x-x^2} + \frac{1}{4}(x-1)(2x-x^2)^{3/2} + C$ **33.** $\frac{1}{3}\left[\frac{3x+2}{\sqrt{5-12x-9x^2}} - \sin^{-1}\frac{3x+2}{3}\right] + C$

35. $\frac{\sqrt{\tan^2 t - 1}}{3}(\tan^2 t + 2) + C$ **37.** $\frac{1}{2}\sin^{-1}e^x - \frac{e^x\sqrt{1-e^{2x}}}{2} + C$ **39.** $V = 2(2\pi)\int_{-r}^{r}(R-x)\sqrt{r^2-x^2} \, dx$. Let

$x = r\sin\theta, dx = r\cos\theta \, d\theta$. Then $4\pi R\int_{-\pi/2}^{\pi/2}r^2\sqrt{1-\sin^2\theta}\cos\theta \, d\theta = 4\pi r^2 R\int_{-\pi/2}^{\pi/2}\cos^2\theta \, d\theta = 2\pi^2 r^2 R$

Exercise Set 8.4

1. $\frac{1}{2}\left[\frac{1}{x} - \frac{1}{x+2}\right]$ **3.** $\frac{2}{5}\ln|x+2| + \frac{3}{5}\ln|x-3| + C$ **5.** $5\ln 6$ **7.** $\ln\frac{(x+2)^4}{|x+3|} + C$ **9.** $\ln\frac{\sqrt{|x^2-1|}}{|x|} + C$

11. $x + 2\ln|x+2| + \ln|x-1| + C$ **13.** $5\ln|x+3| - 6\ln|x+4| + 4\ln|x-2| + C$

15. $x^2 - x - 3\ln|x+1| + \ln|x-1| + C$ **17.** $\ln 12 + 2$ **19.** $\ln\left|\frac{x^3}{(x-2)}\right| - \frac{4}{x-2} + C$ **21.** $\ln\frac{(x-2)^2}{x^2+4} + \frac{1}{2}\tan^{-1}\frac{x}{2} + C$

23. $\frac{\pi}{3\sqrt{3}} - \frac{4}{3} + \ln\sqrt{3}$ **25.** $\frac{1}{6}\ln\frac{(x-1)^2}{x^2+x+1} - \frac{1}{\sqrt{3}}\tan^{-1}\frac{2x+1}{\sqrt{3}} + C$ **27. (a)** $\ln\frac{4}{3}$ **(b)** $\pi\left(\frac{2}{3} + \ln\frac{9}{16}\right)$

29. $\int\frac{du}{u(au+b)} = \frac{1}{b}\int\left(\frac{1}{u} - \frac{a}{au+b}\right)du = \frac{1}{b}\ln\left|\frac{u}{au+b}\right| + C$ **31.** $\int\frac{du}{u(au+b)^2} = \frac{1}{b^2}\int\left(\frac{1}{u} - \frac{a}{au+b}\right)du - \frac{1}{b}\int\frac{a}{(au+b)^2}du$

$= \frac{1}{b^2}\ln\left|\frac{u}{au+b}\right| + \frac{1}{b(au+b)} + C$ **33.** $\int\frac{du}{u^3-a^3} = \frac{1}{3a^2}\int\left[\frac{1}{u-a} - \frac{u+2a}{u^2+au+a^2}\right]du = \frac{1}{3a^2}\left[\ln|u-a| - \frac{1}{2}\int\frac{(2u+a)+3a}{u^2+au+a^2}du\right]$

$= \frac{1}{3a^2}\left[\ln|u-a| - \frac{1}{2}\ln(u^2+au+a^2) - \frac{3a}{2}\int\frac{du}{(u+a/2)^2+3a^2/4}\right] = \frac{1}{3a^2}\left[\ln\frac{|u-a|}{\sqrt{u^2+au+a^2}} - \sqrt{3}\tan^{-1}\frac{2u+a}{\sqrt{3}a}\right] + C$

35. $\ln\left|\frac{x-1}{x+1}\right| - \frac{2}{x-1} + C$ **37.** $-\ln(x^2+1) + \tan^{-1}x + \frac{3}{2}\ln(x^2+4) - 2\tan^{-1}\frac{x}{2} + C$ **39.** $Q(t) = \frac{mQ_0}{Q_0+(m-Q_0)e^{-mkt}}$

Exercise Set 8.5

1. $\frac{1}{2(2+3x)} - \frac{1}{4}\ln\left|\frac{2+3x}{x}\right| + C$ (#22)   **3.** $\frac{4x^2+1}{8}\tan^{-1}2x - \frac{x}{4} + C$ (#97)   **5.** $-\frac{\sqrt{4x-x^2}}{2x} + C$ (#66)   **7.** $\frac{1}{2x} + \frac{1}{4}\ln\left|\frac{-2+x}{x}\right| + C$ (#23)

9. $\frac{x}{9\sqrt{x^2+9}} + C$ (#49)   **11.** $\frac{x-1}{2}\sqrt{2x-x^2} + \frac{1}{2}\cos^{-1}(1-x) + C$ (#59)   **13.** $\frac{1}{3}x^4e^{3x} - \frac{4}{9}x^3e^{3x} + \frac{4}{9}x^2e^{3x} - \frac{8}{27}xe^{3x} + \frac{8}{81}e^{3x} + C$ (#102)

15. $\frac{1}{5}\ln\left|\frac{x}{3x+5}\right| + C$ (#26)   **17.** $\frac{-3}{16(-3+4x)} + \frac{1}{16}\ln|-3+4x| + C$ (#20)   **19.** $\frac{e^{-2x}}{13}(-2\cos 3x + 3\sin 3x) + C$ (#112)

21. $-x - \frac{17}{5}\ln|x-4| + \frac{2}{5}\ln|x+1| + C$ (#26, 27)   **23.** $\frac{1}{3}\left[x^3\sin^{-1}x + \sqrt{1-x^2} - \frac{(1-x^2)^{3/2}}{3}\right] + C$ (#98)

25. $-\frac{1}{2}\left[\ln\left(\cos\frac{x}{2} - \sin\frac{x}{2}\right)\right] + \frac{1}{2}\left[\ln\left(\cos\frac{x}{2} + \sin\frac{x}{2}\right)\right] + \frac{\sec x\tan x}{2}$ (Mathematica result) $= \frac{1}{2}\ln\left[\frac{\cos^2\left(\frac{x}{2}\right)+2\sin\left(\frac{x}{2}\right)\cos\left(\frac{x}{2}\right)+\sin^2\left(\frac{x}{2}\right)}{\cos^2\left(\frac{x}{2}\right)-\sin^2\left(\frac{x}{2}\right)}\right] + \frac{1}{2}\frac{\sin x}{\cos^2 x}$

$= \frac{1}{2}\ln\left[\frac{1+\sin x}{\cos x}\right] + \frac{1}{2}\frac{\sin x}{\cos^2 x}$ (Maple result)   **27.** $\sqrt{-9+4x^2} + 3\tan^{-1}\frac{3}{\sqrt{-9+4x^2}}$ (Mathematica result) $= 2\sqrt{x^2 - \frac{9}{4}} + 3\left(\frac{\pi}{2} - \sec^{-1}\frac{2x}{3}\right)$

(Maple result, except for a constant)

Exercise Set 8.6

1. $\frac{1}{2}$ **3.** 1   **5.** $\frac{\pi}{2}$ **7.** ln 2   **9.** ∞ diverges   **11.** ∞ diverges   **13.** 0   **15.** ln 3   **17.** does not exist

19. diverges   **21.** 2   **23.** 2   **25.** $\frac{\pi}{3}$ **27.** diverges   **29.** diverges   **31.** diverges   **33.** 2   **35.** 0

37. diverges   **39.** diverges   **41.** diverges   **43.** diverges   **45.** diverges   **47.** ln $(2+\sqrt{3})$ **49.** diverges   **51.** converges

53. converges   **55.** $k = \frac{1}{3}; \frac{1}{\sqrt[3]{e}} \approx 0.72$ **57.** $k = 2; \frac{e^3-1}{e^4} \approx 0.35$ **59.** $\frac{\pi}{4}$ **61.** $\frac{\pi}{\ln 2}$ **63.** 2

65. (a) f even:, Consider $\lim_{t\to-\infty} \int_t^0 f(x)dx$. Let $u = -x$, $x = -u$, $dx = -du$. $\int_t^0 f(x)dx = \int_{-t}^0 f(-u)(-du) = \int_0^{-t} f(u)\,du$,

since $f(u) = f(-u)$. Write $\tau = -t$. Then $\lim_{t\to-\infty} \int_t^0 f(x)\,dx = \lim_{\tau\to-\infty} \int_0^\tau f(u)\,du = \int_0^\infty f(u)\,du$, and this converges,

by hypothesis. Since, $\int_0^\infty f(u)\,du = \int_0^\infty f(x)dx$, $\int_{-\infty}^\infty = \int_0^\infty + \int_{-\infty}^0 = \int_0^\infty + \int_0^\infty = 2\int_0^\infty f(x)\,dx$,   (b) f odd:

Make substitutions as above. Then, since $f(-u) = -f(u)$, $\int_{-\infty}^0 f(x)\,dx = \lim_{t\to-\infty} \int_t^0 f(x)\,dx = \lim_{t\to-\infty} \int_0^{-t} -f(u)\,du$

$= -\lim_{\tau\to\infty} \int_0^\tau f(u)\,du = -\int_0^\infty f(u)\,du = -\int_0^\infty f(x)\,dx$; $\int_{-\infty}^\infty f(x)\,dx = 0$ **67.** (a) $0 + e^{-1.5} \approx 0.223$ (b) 3930; 950

69. π **71.** diverges   **73.** diverges   **75.** Let $u = \frac{1}{x}$; $\int_0^1 \cos\frac{1}{x}dx = \int_\infty^1 \cos u\left(-\frac{1}{u^2}\right) du =$

$\int_1^\infty \frac{\cos u}{u^2}du$; $\left|\cos\frac{u}{u^2}\right| \leq \frac{1}{u^2}$

and $\int_1^\infty \frac{du}{u^2}$ converges. Thus,

$\int_1^\infty \cos\frac{u}{u^2}du$, converges.

77. $\int_0^1 \frac{dx}{x^p} = \lim_{t\to0^+} \int_t^1 \frac{dx}{x^p} = \lim_{t\to0^+} \frac{x^{1-p}}{1-p}\Big]_t^1$

$= \frac{1}{1-p}\left(1 - \lim_{t\to0^+} t^{1-p}\right) = \begin{cases} 1/(1-p) & \text{if } p < 1 \\ \infty & \text{if } p > 1 \end{cases}$

If $p = 1$, $\int_0^1 \frac{dx}{x} = \ln|x|\Big]_0^1$, which diverges.

79. (a) Converges; compare with $\int_0^{\pi/2} \frac{dx}{x^{2/3}}$

(b) Diverges; compare with $\int_0^1 \frac{dx}{x}$

81. (a) $\frac{\pi}{2}$

(b) $2\pi^2 - 4\pi$

(c) $f'(x) = \frac{x}{(4-x^2)^{3/2}}$, $ds = \sqrt{1 + \frac{x^2}{(4-x^2)^3}}\ dx$;

$S = \int_0^2 2\pi \frac{\sqrt{(4-x^2)^3+x^2}}{(2+x)^{3/2}(2-x)^{1/2}}\ dx$; $0 < \frac{2\pi\sqrt{(4-x^2)^3+x^2}}{(2+x)^{3/2}(2-x)^{1/2}} \leq \frac{2\pi\sqrt{4^3+2^2}}{2^{3/2}(2-x)^{1/2}}$

$= \pi\sqrt{34}\frac{1}{(2-x)^{1/2}}$ on $[0,2)$, $\int_0^2 \frac{dx}{(2-x)^{1/2}} = 2\sqrt{2}$. Thus, by the

comparison test, S is finite.

83. (a) $-|f(x)| \leq f(x) \leq |f(x)|$.   So, $0 \leq f(x) + |f(x)| \leq 2|f(x)|$ (b) By comparison test, $\int_1^\infty (f(x) + |f(x)|)\ dx$

converges, since $2\int_a^\infty |f(x)|\ dx$ converges, by hypothesis.   (c) $\int_0^t f(x)\ dx = \int_0^t [f(x) + |f(x)|]\ dx - \int_0^t |f(x)|\ dx$ Now let

$t \to \infty$. Both integrals on the right approach finite limits. The integral on the left approaches a finite limit also (the difference of

the limits on the right). That is, $\int_a^\infty f(x)\ dx$ converges.   **85.** (a) $\frac{1}{s}$, $s > 0$ (b) $\frac{1}{s^2}$, $s > 0$ (c) $-\frac{1}{1-s}$, $(s > 1)$

(d) $\frac{1}{1+s^2}$, $(s > 0)$ **87.** $\mu = E[x] = \int_{-\infty}^\infty x\ f(x)\ dx = \int_{-\infty}^\infty x\frac{1}{\beta\sqrt{2\pi}}e^{-(x-\alpha)^2/2\beta^2}\ dx$. Write $x = (x-\alpha)+\alpha$,

and get $\mu = \frac{1}{\beta\sqrt{2\pi}}\int_{-\infty}^\infty (x-\alpha)e^{-(x-\alpha)^2/2\beta^2}\ dx + \alpha\int_{-\infty}^\infty \frac{1}{\beta\sqrt{2\pi}}e^{-(x-\alpha)^2/2\beta^2}\ dx = \left[\frac{-\beta}{\sqrt{2\pi}}e^{-(x-\alpha)^2/2\beta^2}\right]_{-\infty}^\infty + \alpha = \alpha$,

$\sigma^2 = E[(x-\mu)^2] = \int_{-\infty}^\infty (x-\alpha)^2\frac{1}{\beta\sqrt{2\pi}}e^{-(x-\alpha)^2/2\beta^2}\ dx = \frac{-\beta^2}{\beta\sqrt{2\pi}}(x-\alpha)e^{-(x-\alpha)^2/2\beta^2}\Big]_{-\infty}^\infty + \beta^2\underbrace{\int_{-\infty}^\infty \frac{1}{\beta\sqrt{2\pi}}e^{-(x-\alpha)^2/2\beta^2}\ dx}_{=1} = \beta^2.$

So, $\sigma = \beta$

Chapter Review

1. $-\frac{x^2 e^{-2x}}{2} - \frac{xe^{-2x}}{2} - \frac{1}{4}e^{-2x} + C$ **3.** $\frac{\pi}{8} - \frac{\ln 2}{4}$ **5.** $2\left[\sqrt{x}\tan\sqrt{x} + \ln\,|\cos\sqrt{x}|\right] + C$ **7.** $\frac{\sin^6 x}{6} - \frac{\sin^8 x}{8} + C$

9. $\frac{1}{3}(1 + 3\sqrt{3})$ **11.** $-\frac{1}{4}x\cos^4 x + \frac{3x}{32} + \frac{\sin 2x}{16} + \frac{\sin 4x}{128} + C$ **13.** $\frac{1}{2}\left[\frac{\cos 2x}{2} - \frac{\cos 6x}{6}\right] + C$ **15.** $\frac{2\sqrt{3}}{3} - \frac{\pi}{6}$

17. $\frac{\pi}{3}$ **19.** $\frac{\sqrt{1+e^{2x}}}{3}(e^{2x} - 2) + C$ **21.** $6\sin^{-1}\frac{x-2}{2} - 4\sqrt{4x - x^2} - \frac{1}{2}(x - 2)\sqrt{4x - x^2} + C$ **23.** $1 + \ln\frac{3}{8}$

25. $\ln\frac{|x|}{\sqrt{x^2+1}} + \frac{1}{2}\left[\tan^{-1} x + \frac{x}{x^2+1}\right] + C$ **27.** $\left(-\frac{\pi}{2\sqrt{3}} - \frac{\ln 3}{2}\right)$ **29.** $\frac{\sqrt{x^2-4}}{3}(x^2 + 8) + C$ **31. (a)** $\frac{\pi^2}{2}$ **(b)** $2\pi^2$

33. (a) $\frac{2\pi}{5}\ln 4$ **(b)** $10\pi\sin^{-1}\frac{3}{5}$ **35.** $y = \frac{1}{2}\left[1 - \frac{kt+C}{\sqrt{16+(kt+C)^2}}\right]$ **37.** diverges   **39.** $\frac{243}{10}$

41. diverges   **43.** $\ln 4$ **45.** $-\frac{1}{4}$ **47.** 2 **49.** $\frac{1}{2}$ **51.** $\frac{1}{\ln 2}$ **53.** Let $u = \frac{1}{x^\alpha}$, $x^\alpha = \frac{1}{u}$, $x = \frac{1}{u^{1/\alpha}}$,

$dx = -\frac{1}{\alpha}(u^{-(1/\alpha)-1})\,du$. $\int_0^1 \sin\frac{1}{x^\alpha}\,dx = \int_\infty^1 (\sin u)\left(-\frac{1}{\alpha}u^{-(1/\alpha)-1}\,du\right) = -\frac{1}{\alpha}\int_1^\infty \frac{\sin u}{u^{1+1/\alpha}}\,du \left|\frac{\sin u}{u^{1+1/\alpha}}\right| \le \frac{1}{u^{1+1/\alpha}}$, and

$\int_1^\infty \frac{du}{u^{1+1/\alpha}}$ converges. $\left(p = 1 + \frac{1}{\alpha} > 1\right)$. Therefore, $\int_2^\infty \frac{\sin u}{u^{1+1/\alpha}}\,du$ converges.   **55. (a)** $\int_e^\infty f(x)\,dx = -\frac{k\theta^k}{kx^k}\Big]_\theta^\infty = 1$

(b) $k > 1; \mu = \frac{k\theta}{k-1}$.   **(c)** $k > 2; \sigma^2 = \frac{k\theta^2}{(k-1)^2(k-2)}$

57. (a) $\Gamma(\alpha + 1) = \int_0^\infty x^\alpha e^{-x}dx = -x^\alpha e^{-x}\Big]_0^\infty +$

$\alpha\int_0^\infty x^{\alpha-1}e^{-x}\,dx$, and $\lim_{x\to\infty}\frac{x^\alpha}{e^x} = 0$. Hence,

$\Gamma(\alpha + 1) = \alpha\int_0^\infty x^{\alpha-1}e^{-x}dx = \alpha\Gamma(\alpha)$.

(b) By repeated applications of the result in (a) we get: $\Gamma(\alpha + n) = (\alpha + n - 1)\Gamma(\alpha + n - 1) = (\alpha + n - 2)\Gamma(\alpha + n - 2) = \ldots = (\alpha + n - 1)(\alpha + n - 2)\ldots(\alpha + 1)\Gamma(\alpha)$ **(c)** In (b), let $\alpha = 1$. $\Gamma(n + 1) = n(n - 1)\ldots(2)\Gamma(1)$. But we have shown $\Gamma(1) = 1$. So, $\Gamma(n + 1) = n!$ If we take $n = 0$, it is natural to define $0! = \Gamma(1) = 1$.

59. Right-hand derivative at 0 equals 0. $x = 0$ is a vertical asymptote. Left-hand derivative does not exist at 0. $y = e$ is a horizontal asymptote. minimum at (0, 0) f is always increasing for $x \ne 0$.

Inflection point: $\left(\frac{1}{2}, \frac{1}{e}\right)$ concave up on:

$(-\infty, 0) \cup (0, \frac{1}{2})$ concave down on: $(\frac{1}{2}, \infty)$

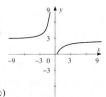

CHAPTER 9

Exercise Set 9.1

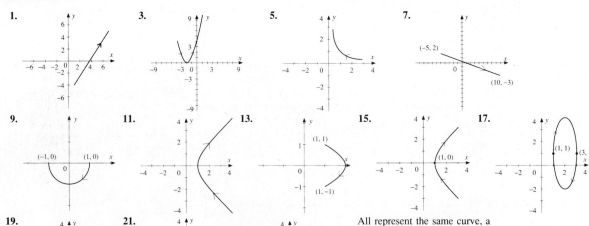

1.   **3.**   **5.**   **7.** (-5, 2) (10, -3)

9. (-1, 0) (1, 0)   **11.**   **13.** (1, 1) (1, -1)   **15.** (1, 0)   **17.** (1, 1) (3, 1)

19. (0, 1) (1, -1)   **21.** (3, 2) (1, -2) C_1, C_3, and C_4 (3, 2) (1, -2) C_2

All represent the same curve, a straight line segment with endpoints (1, −2) and (3,2). C_1, C_3, and C_4 have the same orientation; C_2 has the reverse orientation. If t represents time, we can say that the point (x,y) covers the same path in each case, but it reaches points along the way at different times.

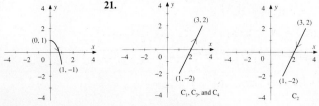

23. 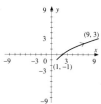 If t_1 and t_2 produce the same point (x,y), then $t_1^3 + 1 = t_2^3 + 1$ and $t_1 = t_2$. Thus there is no double point.

25.

27.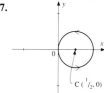

29. $x = r\left[\cos^{-1}\left(\frac{r-y}{r}\right) - \frac{\sqrt{2ry-y^2}}{r}\right]$ for $0 \le x \le \pi r$; $x = r\left[2\pi - \cos^{-1}\left(\frac{r-y}{r}\right) + \frac{\sqrt{2ry-y^2}}{r}\right]$ for $\pi r < x \le 2\pi r$. The graph is periodic of period $2\pi r$.   **31.**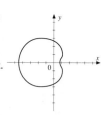

33. $x = (a+b)\cos\theta + b\sin\left(\phi + \theta - \frac{\pi}{2}\right) = (a+b)\cos\theta - b\cos(\phi+\theta)$; $a\theta = B\phi$, so $\phi + \theta = \left(\frac{a+b}{b}\right)\theta$ $\therefore x = (a+b)\cos\theta - b\cos\left(\frac{a+b}{b}\right)\theta$

$y = (a+b)\sin\theta - b\cos\left(\phi+\theta-\frac{\pi}{2}\right) = (a+b)\sin\theta - b\sin\left(\frac{a+b}{b}\right)\theta$

35. In each case, parameter θ is restricted to the interval $[0, 2\pi]$ and θ increases. **(a)** $x = r\cos\theta$, $y = r\sin\theta$ **(b)** $x = r\cos 2\theta$, $y = r\sin 2\theta$ **(c)** $x = r\cos 2\theta$, $y = -r\sin 2\theta$ **(d)** $x = r\sin 3\theta$, $y = r\cos 3\theta$ **(e)** $x = -r\cos 3\theta$, $y = -r\sin 3\theta$

37. When $t = 0$, $x = x_1$, and $y = y_1$. When $t \ne 0$, $\frac{x-x_1}{y-y_1} = \frac{(x_2-x_1)t}{(y_2-y_1)t}$, is the equation of a line passing through (x_1, y_1) and (x_2, y_2). When $t = 1, x = x_2$, and $y = y_2$.   **39.** Both are circles, but the first starts at $(-1, 0)$ and is traced counterclockwise, while the second starts at $(0, -1)$ and is traced clockwise.   **41.** 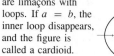 If $a < b$, the curves are limaçons with loops. If $a = b$, the inner loop disappears, and the figure is called a cardioid.

If $a > b$, the figure is that of a limaçon without loop. If $a > 2b$, the dimple on the right is absent.

43. $x = h + r\cos t$, $y = k + r\sin t$, $0 \le t \le 2\pi$.

Exercise Set 9.2

1. 1   **3.** $-\frac{1}{3}$ **5.** $-\frac{1}{2}$ **7.** $-\frac{1}{4}$ **9.** $\frac{1}{20}$ **11.** $x - 8y = 13$.   **13.** $4x - 3y = 11$.   **15.** horizontal tangent at $\left(\frac{-27}{64}, \frac{-49}{8}\right)$, vertical tangents at $\left(\frac{14}{27}, -\frac{52}{9}\right)$ and $(-18, 22)$ **17. (a)** Both $x'(t)$ and $y'(t)$ are rational functions with nonzero denominators in their domain; hence both are continuous. Neither is ever zero. So C is smooth, and the equations define y as a differentiable function of x in $(-\infty, 2)$ and $(2, \infty)$.   **(b)** $\frac{-1}{t^2-4t+5}$ **19.** $-\frac{3}{4}\csc^3 t$ **21.** -2 **23.** $\frac{d^2y}{dx^2} = \frac{d}{dt}\left(\frac{\dot{y}}{\dot{x}}\right)/\frac{dx}{dt} = \frac{\dot{x}\ddot{y}-\dot{y}\ddot{x}}{\dot{x}^3}$ $\frac{d^3y}{dx^3} = \frac{\dot{x}^2\dddot{y}-\dot{x}\dot{y}\dddot{x}-3\dot{x}\ddot{x}\ddot{y}+3\dot{y}\ddot{x}^2}{\dot{x}^5}$

Exercise Set 9.3

1. $4\sqrt{13}$ **3.** $\frac{335}{27}$ **5.** $\frac{488}{27}$ **7.** $2(5^{3/2} - 8) \approx 6.36$ **9.** $2\sqrt{2}$ **11.** $\frac{3}{2}$ **13.** $\frac{9}{2}$

15. $\sqrt{3} + \frac{1}{2}\ln(\sqrt{3} + 2)$ **17.** $ds = \sqrt{r^2(\sin^2 t + \cos^2 t)}\, dt = r\, dt$; $L = \int_0^{2\pi} r\, dt = 2\pi r$

19. $\frac{4088\pi}{27}$ **21.** $2\pi\,(\pi + 1)$ **23.** $\frac{8\pi}{3}(5\sqrt{5} - 2\sqrt{2})$ **25.** 70π **27.** $\frac{64}{3}\pi a^2$ **29. (a)** $\frac{\sqrt{6}-2}{\sqrt{3}} + \ln\frac{\sqrt{3}+2}{\sqrt{2}+1}$

(b) $\pi\left[\sqrt{2} - \frac{2}{3} - \ln(\sqrt{6} - \sqrt{3})\right]$ **31. (a)** $L = 4\int_0^{\pi/2}\sqrt{(a^2 - b^2)\sin^2 t + b^2}\, dt = 4a\int_0^{\pi/2}\sqrt{e^2\sin^2 t - e^2 + 1}\, dt$ $= 4a\int_0^{\pi/2}\sqrt{1 - e^2(1 - \sin^2 t)}\, dt = 4a\int_0^{\pi/2}\sqrt{1 - e^2\sin^2 t}\, dt$, by symmetry.   **(b)** ≈ 15.87

33.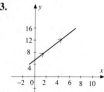

35.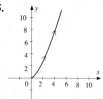

37.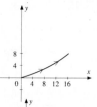

39.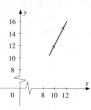

41.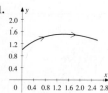

43.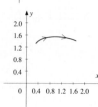

45.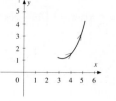

47.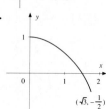

$\left(\sqrt{3}, -\frac{1}{2}\right)$

Exercise Set 9.4

1. (a)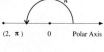

$(2, \pi)$　0　Polar Axis

(b)

$\left(3, -\frac{\pi}{2}\right)$

(c)

$\left(-1, \frac{\pi}{4}\right)$

(d)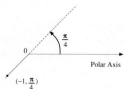

$\left(0, \frac{5\pi}{9}\right)$　Polar Axis

3. (a)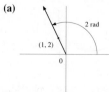

$(1, 2)$　2 rad

(b)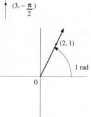

$(2, 1)$　1 rad

(c)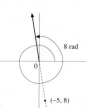

8 rad　$(-5, 8)$

(d)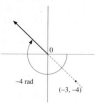

-4 rad　$(-3, -4)$

5. (a) $(-\sqrt{3}, 1)$
(b) $(0,3)$
(c) $(0,0)$
(d) $(\sqrt{2}, \sqrt{2})$

7. (a) $(4, 0)$
(b) $\left(4, \frac{\pi}{2}\right)$
(c) $(4, \pi)$
(d) $\left(4, \frac{3\pi}{2}\right)$

9. (a) $x^2 + y^2 = 4$
(b) $y = x$

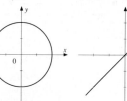

$y = x$

11. (a) $x = 3$
(b) $y = -1$

13. (a) $2x - 3y = 4$
(b) $3x + 2y = 5$

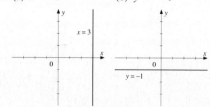

$x = 3$　　$y = -1$

$2x - 3y = 4$

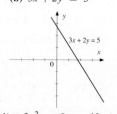

$3x + 2y = 5$

15. $(x^2 + y^2)^2 - 2ax^3 - ay^2(2x + a) = 0$　**17.** $(x^2 + y^2)^2 + 4x^3 + y^2(4x - 1) + 3x^2 = 0$　**19.** $(x^2 + y^2)^{3/2} = a(x^2 - y^2)$
21. $y^2 = 4(x + 1)$　**23. (a)** $r = a$　**(b)** $\theta = \frac{\pi}{3}$　**25. (a)** $r = -\csc\theta$　**(b)** $r = 3\sec\theta$　**27.** $r = \sec\theta \tan\theta$

29. $r = \frac{4\cos\theta}{\cos 2\theta}$　**31.** $r^2 = \tan 2\theta$　**33.** If P is in quadrant I or IV, $\theta = \tan^{-1}\frac{y}{x}$ and $r = \sqrt{x^2 + y^2}$. If P is in quadrant II or III,

$\theta = \tan^{-1}\frac{y}{x}$ if we choose $r = -\sqrt{x^2 + y^2}$.　**35.** $e = 1/2$, $d = 2$

37. One example $r = 2\cos\theta$. The pole lies on the graph, since the equation is satisfied by $\left(0, \frac{\pi}{2}\right)$. But the equation is not satisfied by $(0,0)$; that is $r = 0$, $\theta = 0$.

Exercise Set 9.5

1. (a) circle   **(b)** circle   **3. (a)** line   **(b)** line   **5.** circle

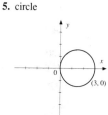

7. cardioid   **9.** cardioid   **11.** Limaçon without loop.   **13.** Limaçon with loop.

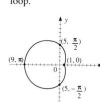

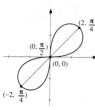

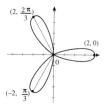

15. Limaçon without loop, no dimple.   **17.** Lemniscate.   **19.** Lemniscate.   **21.** Three-Leaf Rose.

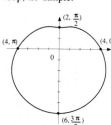

23. Four-Leaf Rose.

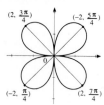

25. $r_1 = a + b\cos\theta$ and $r_2 = -a + b\cos\theta$, where $a > 0$ and $b > 0$, have the same graphs, although θ is shifted by π radians for a given value of the radius. $r_2(\theta) = -r_1(\theta + \pi)$.

27.

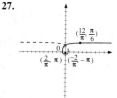

29.

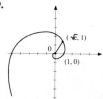

31.

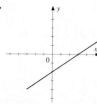

33.

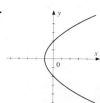

35. (a) The graph is symmetric to the line $\theta = 0$ if, and only if, when (r, θ) satisfies the equation, either $(r, 2n\pi - \theta)$ or $(-r, (2n+1)\pi - \theta)$ satisfies the equation for some integer n.   **(b)** The graph is symmetric to the line $\theta = \pi/2$ if, and only if, when (r, θ) satisfies the equation, either $(r, (2n+1)\pi - \theta)$ or $(-r, 2n\pi - \theta)$ satisfies the equation for some integer n.   **(c)** The graph is symmetric to the pole if, and only if, when (r, θ) satisfies the equation, either $(r, (2n+1)\pi + \theta)$ or $(-r, 2n\pi + \theta)$ satisfies the equation for some integer n.   **37. (a)** Let $F(r, \theta) = r - \tan\theta$. Then, $F(-r, \pi - \theta) = -F(r, \theta)$. $\therefore$ the graph is symmetric to the polar axis. $F(r, -\theta) = r + \tan\theta$, however, and this does not necessarily equal 0 when $r - \tan\theta = 0$. Thus, the test given in the text fails. Similarly, $F(-r, -\theta) = -F(r, \theta) = 0$ if $F(r, \theta) = 0$. But, $F(r, \pi - \theta) = r + \tan\theta$, and so the test for symmetry to $\theta = \pi/2$ fails. The graph is, nevertheless, symmetric to $\theta = \pi/2$.   **(b)** Let $F(r, \theta) = r - \sin 2\theta \cos\theta$ Then $F(-r, -\theta) = -F(r, \theta)$. But $F(r, \pi - \theta) \neq \pm F(r, \theta)$, and so the test for symmetry to $\theta = \pi/2$, as given in the text, fails.   **(c)** Let $F(r, \theta) = r - 2\csc 2\theta + 1$ Then $F(r, \pi + \theta) = F(r, \theta)$. But $F(-r, \theta) \neq \pm F(r, \theta)$, and so the test for symmetry to the pole fails.   **(d)** Let $F(r, \theta) = r - \sin 2\theta - \cos\theta$. But $F(r, \theta)$ Then

$F(-r, \pi - \theta) = -F(r, \theta) \neq \pm F(r, \theta)$ and so the test for symmetry to $\theta = 0$ fails.   **39. (a)** $\sin\theta = \cos\left(\theta - \frac{\pi}{2}\right)$. Hence,

$F(r, \sin\theta) = F\left(r, \cos\left(\theta - \frac{\pi}{2}\right)\right)$, or the graph of $F(r, \cos\theta) = 0$ is rotated $\frac{\pi}{2}$ radians.   **(b)** $F(r, -\cos\theta) = F(r, \cos(\theta - \pi))$

(c) $F(r, -\sin\theta) = F\left(r, \cos\left(\theta - \frac{3\pi}{2}\right)\right)$ **41.** Let $\sin\alpha = \dfrac{a}{\sqrt{a^2+b^2}}$ and $\cos\alpha = \dfrac{b}{\sqrt{a^2+b^2}}$. Then, $r = \sqrt{a^2 + b^2}\,[\cos(\theta - \alpha)]$, which

is the equation of the circle $r = \sqrt{a^2 + b^2}\,\cos\alpha$, rotated through the angle α.

43.

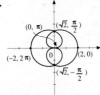

45.

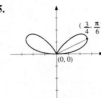

47.

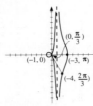

49. $\left(1, \frac{\pi}{3}\right), \left(1, \frac{5\pi}{3}\right)$

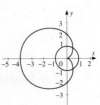

51. $\left(1, \frac{\pi}{3}\right), \left(1, \frac{5\pi}{3}\right)$, and $(0, 0)$

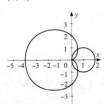

53. $\left(1, \frac{\pi}{2}\right), \left(1, \frac{3\pi}{2}\right), (-1, \pi)$

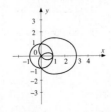

55. $\left(2, \frac{\pi}{12}\right), \left(2, \frac{5\pi}{12}\right), \left(2, \frac{13\pi}{12}\right)$, and $\left(2, \frac{17\pi}{12}\right)$

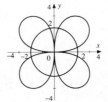

57.

Exercise Set 9.6

1. 6π

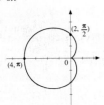

3. 9

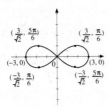

5. $\frac{3\pi}{2}$

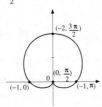

7. π

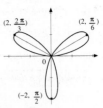

9. 4π

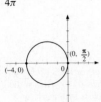

11. $8 + \pi$

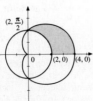

13. $2 - \frac{\pi}{2}$

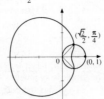

15. $\sqrt{3} - \frac{\pi}{3}$

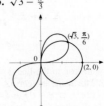

17. $8\sqrt{3} - \frac{8\pi}{3}$

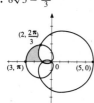

19. $9\sqrt{3} - 4\pi$

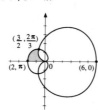

21. $\pi - \frac{3\sqrt{3}}{2}$

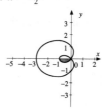

23. $\frac{11\pi}{3} + \frac{5\sqrt{3}}{2}$

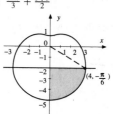

25. $2\pi - 4$ **27.** $\pi + 3$ **29.** $1 + \frac{3\sqrt{3}}{2} - \pi$ **31.** $\frac{27\pi^3}{2}$ **33.** $r = \cos n\theta$ n odd:

$$A = \frac{1}{4} \int_0^\pi (1 + \cos 2n\theta)d\theta = \frac{\pi}{4}$$

n even:

$$A = \frac{1}{2} \int_0^\pi (1 + \cos 2n\theta)d\theta = \frac{\pi}{2}$$

$r = \sin n\theta$; n odd:

$$A = \frac{1}{4} \int_0^\pi (1 - \cos 2n\theta)d\theta = \frac{\pi}{2}$$

n even:

$$A = \frac{1}{2} \int_0^\pi (1 - \cos 2n\theta)d\theta = \frac{\pi}{2}$$

Exercise Set 9.7

1. $-\frac{1}{\sqrt{3}}$ **3.** $-\frac{1}{2}$ **5.** $\sqrt{3}$ **7.** $-\sqrt{3}$ **9.** $-\frac{\sqrt{3}}{2}$ **11.** 0   **13.** $\frac{1}{\sqrt{3}}$ **15.** 1   **17.** $\frac{1}{2}$ **19.** $-\frac{3}{2}$ **21.** $\tan\phi = 0$

23. $\cot\phi = 0$ **25.** $\tan\psi = \frac{1}{b} = \frac{\pi}{4}$ **27.** $\tan\alpha = -\sqrt{3}$ at $\theta = \frac{2\pi}{3}$ and $\sqrt{3}$ at $\theta = -\frac{2\pi}{3}$ **29.** At $\theta = \frac{\pi}{2}$, $\tan\alpha = 0$. At

$\theta = \frac{3\pi}{2}$, $\tan\alpha = 0$. At $\theta = \frac{\pi}{3}$, $\tan\alpha = -\frac{\sqrt{3}}{9}$. At $\theta = \frac{2\pi}{3}$, $\tan\alpha = \frac{\sqrt{3}}{9}$ **31.** $\cot\alpha = \frac{1}{\tan(\psi_2 - \psi_1)} = \frac{1 + \tan\psi_2 \psi_1}{\tan\psi_2 - \tan\psi_1} = 0$ if and only

if $\tan\psi_1 \tan\psi_2 = -1$ for ψ_1 and $\psi_2 \neq \frac{\pi}{2}$. This condition is satisfied if and only if $\alpha = \frac{\pi}{2}$.   **33.** For $\theta = \frac{\pi}{3}$, $\tan\psi_1 = -\frac{1}{\sqrt{3}}$, and

$\tan\psi_2 = \frac{2 - 1/2}{\sqrt{3}/2} = \sqrt{3}$ $\tan\psi_1 \tan\psi_2 = -1$. For $\theta = -\frac{\pi}{3}$, $\tan\psi_1 = \frac{1}{\sqrt{3}}$, and $\tan\psi_2 = \frac{2 - 1/2}{-\sqrt{3}/2} = -\sqrt{3}$ $\tan\psi_1 \tan\psi_2 = -1$

35. $\tan\psi_1 = -\cot\theta$, $\tan\psi_2 = \tan\theta$, $\tan\psi_2 = -1$ **37.** $\tan\psi_1 = \frac{1 - \sin\theta}{-\cos\theta}$, $\tan\psi_2 = \frac{1 + \sin\theta}{\cos\theta}$, $\tan\psi_1 \tan\psi_2 = -1$ **39.** $\frac{7\sqrt{10}}{3}$

41. 4   **43.** 12   **45.** $\frac{\tan\phi - \tan\theta}{1 + \tan\phi \ \tan\theta} = \frac{\frac{f'(\theta)\sin\theta + f(\theta)\cos\theta}{f'(\theta)\cos\theta - f(\theta)\sin\theta} - \tan\theta}{1 + \left(\frac{f'(\theta)\sin\theta + f(\theta)\cos\theta}{f'(\theta)\cos\theta - f(\theta)\sin\theta}\right)\tan\theta} = \frac{f(\theta)[\sin^2\theta + \cos^2\theta]}{f'(\theta)[\sin^2\theta + \cos^2\theta]} = \frac{f(\theta)}{f'(\theta)}$

47. $\alpha = \phi_2 - \phi_1 = [(\theta + \psi_2) + n\pi] - [(\theta + \psi_1) + m\pi] = (\psi_2 - \psi_1) + (n - m)\pi$ (m, n integers). Hence, $\tan\alpha = \tan(\psi_2 - \psi_1)$, since $\tan$

has period π.   **49.** $1 + \frac{1}{2\sqrt{3}} \ln(2 + \sqrt{3})$

51. (a) $S = 2\pi \int y \, ds = 2\pi \int_\alpha^\beta r \sin\theta \sqrt{r^2 + (dr/d\theta)^2} \, d\theta = 2\pi \int_\alpha^\beta f(\theta) \sin\theta \sqrt{[f(\theta)]^2 + [f'(\theta)]^2} \, d\theta$ **(b)** Assume that $\cos\theta \geq 0$

for $\alpha \leq \theta \leq \beta$ and that f' is continuous. Then, about $\theta = \frac{\pi}{2}$, $S = 2\pi \int x \, ds = 2\pi \int_\alpha^\beta r \cos\theta \sqrt{r^2 + (dr/d\theta)^2} \, d\theta$

$= 2\pi \int_\alpha^\beta f(\theta) \cos\theta \sqrt{[f(\theta)]^2 + [f'(\theta)]^2} \, d\theta$ **53.** 2π **55. (a)** $\frac{4\pi\sqrt{2}}{5}$ **(b)** $\frac{8\pi}{5}(3\sqrt{2} - 4)$

Exercise Set 9.8

1. parabola

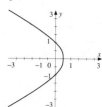

3. hyperbola

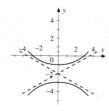

5. ellipse

7. hyperbola

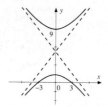

9. hyperbola

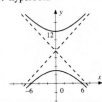

11. $r = \frac{5}{3+\sin\theta}$ **13.** $r = \frac{3}{2-3\cos\theta}$ **15.** $y^2 = 4(x+1)$ vertex: $(-1, 0)$; focus: $(0, 0)$

17. $\frac{(y+24/7)^2}{(18/7)^2} - \frac{x^2}{\left(6/\sqrt{7}\right)^2} = 1$, center: $\left(0, -\frac{24}{7}\right)$; vertices: $\left(0, -\frac{6}{7}\right)$, $(0, -6)$; foci: $(0, 0)$, $\left(0, -\frac{48}{7}\right)$; asymptotes: $y = \frac{3}{\sqrt{7}}x - \frac{24}{7}$ and

$y = -\frac{3}{\sqrt{7}}x - \frac{24}{7}$ **19.** Let $r = \frac{de}{1-e\,\cos\theta}$, $e < 1$; center $= \frac{de^2}{1-e^2}$; $a = \frac{de}{1-e^2}$. Thus, $c = ae$. The distance from the directrix to the

center $= \frac{a}{e}$. **21.** Let $b^2 = a^2 - c^2$; $(a^2 - c^2)x^2 + a^2y^2 = a^2(a^2 - c^2)$. Let $e = c/a$; $a^2(1 - e^2)x^2 + a^2y^2 = a^4(1 - e^2)$ and

$(1 - e^2)r^2\cos^2\theta + r^2\sin^2\theta = a^2(1 - e^2)$. Thus $r^2 = \frac{a^2(1-e^2)}{(1-e^2\cos^2\theta)}$.

23. $r \approx \frac{9.481}{1-0.0484\cos\theta}$; $r_a \approx 9.963$ A.U. $\approx 1.490 \times 10^{12}m$; $r_p \approx 9.043$ A.U. $\approx 1.353 \times 10^{12}m$ **25.** $r \approx \frac{2.42}{1-0.135\,\cos\theta}$, $r_a \approx 2.80$ A.U.,

$r_p \approx 2.14$ A.U. **27.** From Definition 9.3, $\frac{\text{distance between } P \text{ and } F}{\text{distance between } P \text{ and } \ell} = e$,
where P is a point on the graph, F is a focus, and
ℓ is the corresponding directrix. Let $F =$ the pole
$= (0, 0)$. The ratio becomes $\frac{\sqrt{x^2+y^2}}{d-y} = e$ if the
directrix is d units above the pole and is horizontal.

Thus, $\frac{r}{d-r\sin\theta} = e$; $r = \frac{ed}{1+e\sin\theta}$. If the directrix is
d units below the pole and is horizontal, then:

$r(1 - e\sin\theta) = ed$; and $r = \frac{ed}{1-e\sin\theta}$

Chapter Review

1. $y^2 = \frac{1}{2}(1 - x)$, $y \geq 0$

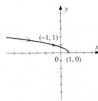

3. $y = 1 - e^{-x}$, $-\infty < x < \infty$

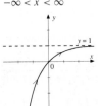

5. $(y - 1)^2 = x + 2$, $-1 \leq x \leq \infty$

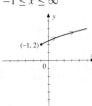

7. $\frac{dy}{dx} = 3t$; $\frac{d^2y}{dx^2} = \frac{3}{2t}$

9. $\frac{dy}{dx} = -\frac{1}{2}\sec^2 t$; $\frac{d^2y}{dx^2} = -\frac{1}{2}\sec^4 t$ **11.** tangent line:
$3x + 4y = 8$
normal line:
$4x - 3y + 6 = 0$

13. $\frac{20}{9} + \ln 3$

15. (a) $4 - 2\sqrt{2} - \ln(9 - 6\sqrt{2})$
(b) $8\pi(\sqrt{3} - \frac{\sqrt{2}}{2} + \frac{1}{2}\ln\frac{\sqrt{3}+2}{\sqrt{2}+1})$

17. (a) $(x - 1)^2 + (y - 1)^2 = 2$, circle

(b) $\frac{x^3}{1/3} + \frac{(y+1/3)^2}{4/9} = 1$, ellipse

19. (a) cardioid

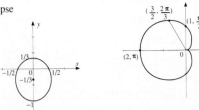

(b) lemniscate   **21. (a)** four-leaf rose   **(b)** hyperbolic spiral   **23.** $\frac{5\pi}{2} - 3\sqrt{3}$

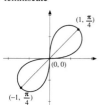

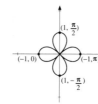

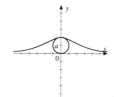

25. (a) $\frac{4}{3}\pi + 2\sqrt{3}$ **(b)** $\frac{8\pi}{3} - 2\sqrt{3}$ **(c)** $\frac{4\pi}{3} + 2\sqrt{3}$ **27.** At $\theta = \frac{\pi}{4}, \alpha = \frac{\pi}{4}$. At $\theta = -\frac{\pi}{4}, \alpha = \frac{3\pi}{4}$. At $\theta = -\frac{3\pi}{4}, \alpha = \frac{\pi}{4}$

29. At $\theta = \pm\frac{\pi}{3}$, $\tan\psi_1 \tan\psi_2 = $ **31.** $2\sqrt{2} + 7\ln(3 + 2\sqrt{2})$ **33.** $x = 2a\cot\theta, \ y = 2a\sin^2\theta, \ y > 0$

$-\frac{(1/2)}{(1-1/2)} = -1$

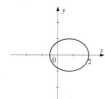

35. parabola   **37.** ellipse   **39.** ellipse   **41.** $\tan\psi_1 = \frac{1+\cos\theta}{\sin\theta}$; $\tan\psi_2 = -\frac{(1-\cos\theta)}{\sin\theta}$;

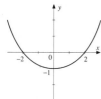

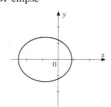

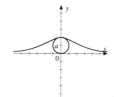

$\tan\psi_1 \tan\psi_2 = -1$

CHAPTER 10

Exercise Set 10.1

1. $1 - x + \frac{x^2}{2!} - \frac{x^3}{3!} + \frac{x^4}{4!} - \frac{x^5}{5!}$ **3.** $1 - x + x^2 - x^3 + x^4 - x^5 + x^6$ **5.** $1 + \frac{x^2}{2!} + \frac{x^4}{4!} + \frac{x^6}{6!} + \frac{x^8}{8!}$

7. $1 - (x-1) + (x-1)^2 - (x-1)^3 + (x-1)^4 - \ldots + (-1)^n(x-1)^n$.   **9.** $x - \frac{x^3}{3!} + \frac{x^5}{5!} - \ldots + \frac{(-1)^{k+1}x^{2k-1}}{(2k-1)!}$, $n = 2k-1$ or $n = 2k$

$k = 1, 2, 3, \ldots$ **11.** $-(x-1) - \frac{(x-1)^2}{2} - \frac{(x-1)^3}{3} - \frac{(x-1)^4}{4} - \ldots - \frac{(x-1)^n}{n}$ **13.** $R_n(x) = \frac{(-1)^{n+1}(x-1)^{n+1}}{c^{n+2}}$; c is between 1 and x,

where $x \in (0,2)$ **15.** $R_{2k}(x) = \frac{(-1)^k \cos c}{(2k+1)!}x^{2k+1}$, $k = 1, 2, \ldots$; c is between 0 and x; $x \in (-\infty, \infty)$

17. $R_n(x) = \frac{-(x-1)^{n+1}}{(2-c)^{n+1}(n+1)}$, c is between 0 and x; $x \in (0, 2)$ **19.** $P_5(1) \approx 0.36667$, $|R_5(1)| \leq \frac{1}{6!} \approx 0.0013889$, $P_5(1)$ has at least a

2-decimal place accuracy. The most we can say, then, is $e^{-1} \approx 0.37$.   **21.** $P_7(1.5) \approx 0.99739$, $|R_8(1.5)| \leq \frac{(1.5)^9}{9!} \approx 0.0001059$,

$\sin 1.5 \approx 0.997$, correct to 3 places.   **23.** $P_8(2) \approx -0.415873$, $|R_9(2)| \leq \frac{2^{10}}{(10)!} \approx 0.0002822$, $\cos 2 \approx -0.416$, correct to 3 places.

25. $P_5(5) \approx 2.236076$, $|R_5(5)| \leq \frac{1 \cdot 3 \cdot 5 \cdot 7 \cdot 9}{2^6 \cdot 2^{11} \cdot 6!} \approx 0.000010014$. $\sqrt{5} \approx 2.2361$ to 4 places.

27. $P_{12}(3) \approx 20.0852$; $|R_{12}(3)| < \frac{3^{16}}{(13)!} \approx 0.0069$, $e^3 \approx 20.1$, with accuracy assured only to 1 place.   **29.** $P_6(1) \approx 1.54306$,

$R_7(1) < \frac{2}{8!} \approx 0.0000496$, $\cosh 1 \approx 1.5431$ to 4 places.   **31.** $P_4\left(\frac{\pi}{3} - \frac{\pi}{90}\right) \approx 0.5299193$; $\left|R_4\left(\frac{\pi}{3} - \frac{\pi}{90}\right)\right| \leq \frac{\left(\frac{\pi}{90}\right)^5}{5!} \approx 4.3 \times 10^{-10}$,

$\cos 58° = 0.5299193$ is accurate to 7 places, and we would have accuracy to 9 places if the calculator went that far.

33. $k = 4$, $\cos 72° = \cos\frac{2\pi}{5} \approx 0.309017$. By calculator, $\cos 72° = 0.309016994$.   **35.** Use $k = 7$. $\sinh 2 \approx 3.626860$

37. (a) $P_n(x) = 1 + \alpha x + \frac{\alpha(\alpha-1)}{2!}x^2 + \frac{\alpha(\alpha-1)(\alpha-2)}{3!}x^3 + \ldots + \frac{\alpha(\alpha-1)(\alpha-2)\ldots(\alpha-n+1)}{n!}x^n$; $R_n(x) = \frac{\alpha(\alpha-1)(\alpha-2)\ldots(\alpha-n)}{(n+1)!\ (1+c)^{n+1-\alpha}}x^{n+1}$; c is between 0 and x

(b) If $\alpha = m$, then $R_m(x) = 0$, since $(\alpha - m) = m - m = 0$. Then, $f(x) = P_m(x) + R_m(x) = P_m(x)$ **(c)** $f^{(n+1)}(x)$ fails

to exist for $x = -1$; the neighborhood of 0 in which $f(x) = P_n(x) + R_n(x)$ is $(-1, 1)$.   **(d)** Let $x = \frac{1}{2}$, $P_4\left(\frac{1}{2}\right) \approx 1.1440$;

$\left|R_4\left(\frac{1}{2}\right)\right| \leq 0.000943$; $\sqrt[3]{1.5} \approx 1.14$, is correct to 2 places   **39. (a)** If $\frac{1+x}{1-x} = u$, then $x = \frac{u-1}{u+1} = 1 - \frac{2}{u+1} < 1$. Also

$1 - \frac{2}{u+1} > -1$. For any positive number u, there is an x in $(-1, 1)$ such that $\ln u = \ln \frac{1+x}{1-x}$.   (b) $P_{2k-1}(x) = P_{2k}(x) =$

$2\left[x + \frac{x^3}{3} + \frac{x^5}{5} + \frac{x^7}{7} + \ldots + \frac{x^{2k-1}}{(2k-1)}\right]$, $k = 1, 2, 3, \ldots$, $R_{2k}(x) = \left[\frac{1}{(1+c)^{2k+1}} + \frac{1}{(1-c)^{2k+1}}\right]\frac{x^{2k+1}}{2k+1}$, $k = 1, 2, \ldots$ (c) Let $u = 2$.

Then, $x = \frac{1}{3}$, $\left|R_8\left(\frac{1}{3}\right)\right| \leq 0.000223$, $P_8\left(\frac{1}{3}\right) = 2\left[\frac{1}{3} + \frac{1}{3 \cdot 3^3} + \frac{1}{5 \cdot 3^5} + \frac{1}{7 \cdot 3^7}\right] \approx 0.693$.

41. (a) $P_n(x) = x - \frac{x^2}{2} + \frac{x^3}{3} - \frac{x^4}{4}$
     $+ \ldots + (-1)^{n-1}\frac{x^n}{n}$ **(b)**   **(c)** The graphs of $P_n(x)$ approach the graph of $\ln(1 + x)$

on $(-1, 1)$, but diverge outside that interval.   **(d)** Conjecture: $R_n(x) \to 0$ as $n \to \infty$ for $x \in (-1, 1)$. The actual interval is

$(-1, 1]$.   **43.** $-0.33098 \leq x \leq 0.33098$.   **45. (a)** $P_n(x) = P_{2k}(x) = x - \frac{x^3}{3!} +$ **(b)**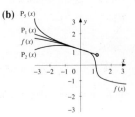

$\frac{x^5}{5!} + \ldots + (-1)^k \frac{x^{2k+1}}{(2k+1)!}$

(c) $P_3(x): (-1.04, 1.04)$.
     $P_5(x): (-1.75, 1.75)$.
     $P_{10}(x): (-3.25, 3.25)$.

47. (a) $P_n(x) = 1 - \frac{1}{3}x - \frac{2}{9}\frac{x^2}{2!} - \frac{10}{27}\frac{x^3}{3!} - \ldots - \frac{1\cdot2\cdot5\ldots(3n-4)}{3^n}\frac{x^n}{n!}$ **(b)**

(c) $P_3(x): (-0.78, 0.60)$.
     $P_5(x): (-0.94, 0.74)$.
     $P_{10}(x): (-0.995, 0.875)$.

49. (a) About $x = 0$: $P_n(x) = 1 + \frac{1}{2}x - \frac{1}{4}\frac{x^2}{2!} + \frac{3}{8}\frac{x^3}{3!} - \frac{15}{16}\frac{x^4}{4!} +$ **(b)**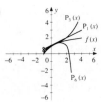

$\ldots + (-1)^{n+1}\frac{1 \cdot 3 \cdot 5 \ldots (2n-3)}{2^n}\frac{x^n}{n!}$

(c) $P_3(x): (-0.611, 0.792)$.
     $P_5(x): (-0.755, 0.974)$.
     $P_{10}(x): (-0.883, 1.083)$.

(a) About $x = 2$: $P_n(x) = \sqrt{3} + \frac{1}{2\sqrt{3}}(x - 2) -$ **(b)**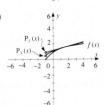

$\frac{1}{4 \cdot 3^{3/2}}\frac{(x-2)^2}{2!} + \frac{3}{8 \cdot 3^{5/2}}\frac{(x-2)^3}{3!}$

$+ \ldots + (-1)^{n+1}\frac{1 \cdot 3 \cdot 5 \ldots (2n-3)}{2^n 3^{(2n-1)/2}}\frac{(x-2)^n}{n!}$

(c) $P_3(x): (0.361, 4.046)$.
     $P_5(x): (-0.107, 4.647)$.
     $P_{10}(x): (-0.554, 5.084)$.

Exercise Set 10.2

1. (a) 1, $\frac{-2}{3}$, $\frac{3}{5}$, $\frac{-4}{7}$, $\frac{5}{9}$ **(b)** 2, $\frac{4}{2!}$, $\frac{8}{3!}$, $\frac{16}{4!}$, $\frac{32}{5!}$ **3. (a)** $1, -2, 6, -24, 120$ **(b)** $1, 2, 2, 1, \frac{1}{2}$ **5. (a)** $a_n = \frac{2^n-1}{2^n}$

(b) $a_n = \frac{(-1)^{n-1}(2n)}{2n+3}$ **7.** $\frac{1}{3}$ **9.** diverges   **11.** 0 **13.** Does not exist. $a_{2k} = 0$, $a_{2k+1} = 1$ **15.** 1 **17.** 0
19. 0 **21.** 0 **23.** Does not exist.   **25.** 0 **27.** 0 **29.** 0 **31.** monotone increasing.   **33.** monotone decreasing for
$n \geq n_0 = 2$.   **35.** monotone decreasing.   **37.** monotone decreasing.   **39.** monotone increasing.

41. $\frac{a_{n+1}+1}{a_n} = \frac{e}{n+1} < 1$ for $n \geq 2$ $\therefore$ $\{a_n\}$ is decreasing. Also, $a_n \geq 0$. $\therefore$ convergent.   **43.** Let $y = x^{1/x}$.

$y' < 0$ for $x \geq e$. $\therefore x^{1/x}$ is decreasing for $x \geq e$, and $\{n^{1/n}\}$ is decreasing. Since it is bounded below by 0, the sequence converges.
45. Show that $\{a_n\}$ is monotone increasing and bounded above by 2.   **47.** Suppose that L_1 and L_2 are both limits of $\{a_n\}$, $L_1 \neq L_2$,

say $L_2 > L_1$. Let $\varepsilon = \frac{L_2 - L_1}{2}$. Then for all n large enough, $|a_n - L_1| < \varepsilon$, so that $a_n < L_1 + \varepsilon = \frac{L_1 + L_2}{2}$. Similarly, $a_n > L_2 - \varepsilon = \frac{L_1 + L_2}{2}$. The two inequalities cannot both be true.   **49.** Since $|a_n - 0| = ||a_n| - 0|$, it follows that for any $\varepsilon > 0$, there is an N such that, for all $n > N$, $|a_n - 0| < \varepsilon$ if, and only if, there is an N such that, for all $n > N$, $||a_n| - 0| < \varepsilon$. Or $\lim_{n \to \infty} a_n = 0$ if, and only if, $\lim_{n \to \infty} |a_n| = 0$.   **51.** $a_{n+1} - a_n > 0$. $a_n < n\left(\frac{1}{n+1}\right) < 1$. $\therefore$ $\{a_n\}$ is monotone increasing and bounded above; it is convergent.   **53.** $\frac{1 + \sqrt{5}}{2}$ **55. (a)** The limit is 1 if $0 < x_1 < \infty$ and 0 if $x_1 = 0$. Proof: Let $\lim_{n \to \infty} x_n = L$. Then, $L = \sqrt{L}$; $L^2 - L = 0$; $L(L - 1) = 0$; $L = 0, 1$. The answer depends on x_1 if x_1 may be zero; otherwise, it does not. **(b)** $x_{x+1} = x_n^2 - 2$. Let $\lim_{x \to \infty} x_n = L$. Then $L = L^2 - 2$; $L^2 - L - 2 = 0$; $L = -1, 2$. The numbers -1 and 2 are fixed points of the function $f(x) = x^2 - 2$. If some other number is substituted into the function, the interation may or may not converge. If we let $x_1 = -1 + \delta$ or $x_1 = 2 + \delta$, both sequences diverge.

57.

(a) upper bound of 2 **(b)** Assume that $a_1 = 1$, which is ≤ 2. Assume that $a_n \leq 2$. Then $a_{n+1} = \sqrt{2 + a_n} = \sqrt{b}$, where $b \leq 4$. Hence, $a_{n+1} \leq 2$. By induction, $a_n \leq 2$ for all n. **(c)** $a_{n+1}^2 - a_n^2 = 2 + a_n - (2 + a_{n+1}) = 2 + a_n - a_n^2 = (2 - a_n)(1 + a_n) > 0$, since $2 - a_n$ and $1 + a_n$ are > 0. Then, $a_{n+1}^2 - a_n^2 > 0$, which shows that $\{a_n\}$ is monotone increasing $(a_n > 0)$. **(d)** $L = 2$.

59. (a) $\left\{11 + \frac{n-1}{n}\right\}$; **(b)** $\{n\}$ **(c)** $\{\sin n\}$ **(d)** $\left\{\frac{\sin n}{n}\right\}$ **(e)** $\{n\}$ **(f)** $\left\{7\left(1 + \frac{\sin(n\pi/2)}{n}\right)\right\}$ **(g)** Such a sequence is not possible because if it converges to 7, there exists a finite number N such that $|a_n - 7| < \epsilon$ for $n > N$. **(h)** Not possible, since every convergent sequence is bounded. **(i)** $\left\{1 + \frac{1}{n}\right\}$ and $\left\{1 - \frac{1}{n}\right\}$ **(j)** Not possible, by the same reasoning as in part **(g)**, with 7 replaced by 2.

Exercise Set 10.3

1. $\frac{1}{2} + \frac{2}{5} + \frac{3}{10} + \frac{4}{17} + \ldots + \frac{n}{n^2 + 1} + \ldots$ $S_1 = \frac{1}{2}$, $S_2 = \frac{9}{10}$, $S_3 = \frac{6}{5}$, $S_4 = \frac{122}{85}$ **3.** $\frac{\ln 2}{1!} + \frac{\ln 3}{2!} + \frac{\ln 4}{3!} + \frac{\ln 5}{4!} + \ldots + \frac{\ln(n+1)}{n!} + \ldots$

$S_1 = \ln 2$, $S_2 = \ln 2 + \frac{\ln 3}{2}$, $S_3 = \ln 2 + \frac{\ln 3}{2} + \frac{\ln 4}{6}$, $S_4 = \ln 2 + \frac{\ln 3}{2} + \frac{\ln 4}{6} + \frac{\ln 5}{24}$ **5.** $\displaystyle\sum_{n=1}^{\infty} \frac{(-1)^{n-1}}{3^{n-1}}$; $S_n = \displaystyle\sum_{k=1}^{n} \frac{(-1)^{k-1}}{3^{k-1}}$

7. $\displaystyle\sum_{k=1}^{\infty} \frac{1}{(k+1)\ln(k+1)}$; $S_n = \displaystyle\sum_{k=1}^{n} \frac{1}{(k+1)\ln(k+1)}$ **9.** converges to 1.   **11.** converges to 2.   **13.** $a = 2$, $r = -\frac{1}{2}$; $S = \frac{4}{3}$

15. $a = 3$, $r = \frac{1}{2}$; $S = 6$ **17.** $a = \frac{2}{e}$, $r = \frac{-2}{e}$; $S = \frac{2}{e+2}$ **19.** $a = 1, r = 0.99; S = 100$ **21.** $\frac{5}{33}$ **23.** $\frac{4}{27}$

25. $\frac{1}{2}$ **27.** 1   **29.** 2   **31.** $-\frac{1}{2}$ **33.** 50 m   **35. (a)** $\lim_{n \to \infty} \frac{n}{100n+1} = \frac{1}{100} \neq 0$ **(b)** $\lim_{x \to \infty} \frac{x}{(\ln x)^2} = \infty \neq 0$

37. (a) False **(b)** False **(c)** True **(d)** False **(e)** True   **39. (a)** Suppose that $\sum (a_n - b_n)$ converges. Then, $-\sum b_n = \sum -b_n = \sum [(a_n - b_n) - a_n] = \sum (a_n - b_n) - \sum a_n$ converges also, contrary to the statement that $\sum b_n$ diverges.

Thus, $\sum (a_n - b_n)$ diverges. **(b)** Let $A = \displaystyle\sum_{n=1}^{\infty} \frac{1}{2m-1}$ and $B = \displaystyle\sum_{n=1}^{\infty} \frac{1}{2n}$; both diverge but $A - B$ converges.   **41.** Since $a_n \geq 0$ and $S_{n+1} = S_n + a_n$, it follows that $\{S_n\}$ is monotone increasing. Also, $S_n = a_1 + a_2 + \ldots + a_n \leq k$ for all n. Thus $\{S_n\}$ is bounded above. $\therefore$ the sequence $\{S_n\}$ converges. Call its limit S. Then, by definition, $\displaystyle\sum_{n=1}^{\infty} a_n = S$.   **43. (a)** $x < -5$ or $x > 1$

(b) $e^{-1} < x < e$; $\frac{1}{1 - \ln x}$; $\frac{1}{x+5}$ **45.** For a convergent series, $|S_n - L| < \epsilon$ whenever $n > N$, a finite number. The partial sum S_n is the sum of the first N terms, and since addition is associative, these terms may be arranged in any order whatsoever.   **47.** ≈ 5.1873775

49. (a) $s_n = 3 \cdot 4^{n-1}$, $l_n = \frac{a}{3^{n-1}}$, $P_n = 3a\left(\frac{4}{3}\right)^{n-1}$ **(b)** $\lim_{n \to \infty} P_n = \infty$, since $\lim_{n \to \infty} r^n = \infty$ if $r > 1$, and $r = \frac{4}{3}$ **(c)** After the first term the series is geometric, with $r = \frac{4}{9}$. Area $= \frac{2\sqrt{3}}{5}a^2$

Exercise Set 10.4

1. diverges   **3.** converges   **5.** diverges   **7.** diverges   **9.** diverges   **11.** converges   **13.** diverges   **15.** diverges
17. converges   **19.** converges   **21.** diverges   **23.** converges   **25.** converges   **27.** diverges   **29.** converges
31. converges   **33.** converges   **35.** converges   **37.** converges   **39.** converges   **41.** diverges   **43.** converges
45. diverges   **47.** converges   **49.** converges if, and only if, $p > 1$ **51. (a)** diverges **(b)** diverges **(c)** converges   **53.** Let $\lim_{n \to \infty} na_n = L > 0$. Then, there is a natural number N such that, for all $n > N$, $na_n > \frac{L}{2}$; that is, $a_n > \frac{L}{2} \cdot \frac{1}{n}$ for all $n > N$. But

$$\sum_{n=N+1}^{\infty} \frac{L}{2} \cdot \frac{1}{n} = \frac{L}{2} \sum_{n=N+1}^{\infty} \frac{1}{n} \text{ diverges, since } \sum_{n=1}^{\infty} \frac{1}{n} \text{ diverges. } \therefore \sum a_n \text{ diverges.}$$ **55. (a)** The first series behaves like $\sum \frac{1}{n}$ and

diverges   **(b)** The second series behaves like $\sum \frac{1}{n^2}$ and converges

Exercise Set 10.5

1. converges   **3.** diverges   **5.** converges   **7.** diverges   **9.** converges   **11.** converges   **13.** converges   **15.** converges
17. converges   **19.** converges   **21.** diverges   **23.** converges   **25.** diverges   **27. (a)** $\lim\limits_{n\to\infty} \frac{a_{n+1}}{a_n} = \lim\limits_{n\to\infty} \frac{n}{n+1} = 1$, No

conclusion   **(b)** $\lim\limits_{n\to\infty} \frac{n^2}{(n+1)^2} = 1$, No conclusion   **29.** If $0 \le L < 1$, choose r such that $L < r < 1$. Then, $\sqrt[n]{a_n} < ra_n < r^n$;

$\sum\limits_{k=N}^{\infty} a_k$ converges as a geometric series. If $1 < L$, choose r such that $1 < r < L$ and $\sum\limits_{k=N}^{\infty} a_k$ diverges. If $L = 1$, no conclusion

can be drawn.   **31. (a)** $\lim\limits_{n\to\infty} \sqrt[n]{a_n} = \frac{1}{2} < 1$ **(b)** n even: 2^{-3}; n odd: 2^1 $\lim\limits_{n\to\infty} \frac{a_{n+1}}{a_n}$ does not exist   **33.** It appears that $n^{1/n}$

passes through a maximum at $n = 3$ and then decreases toward its limit of 1.

Exercise Set 10.6

1. converges   **3.** converges   **5.** converges   **7.** diverges   **9.** converges   **11.** Error $\le a_{10} = \frac{1}{10^3} \approx 0.001$ **13.** Error

$\le a_6 = \frac{1}{6^6} \approx 0.0000214.$ **15.** $n \ge 39,998$ **17.** $n \ge 101$ **19.** ≈ 0.531 **21.** converges conditionally   **23.** converges

absolutely   **25.** converges absolutely   **27.** converges absolutely   **29.** converges absolutely   **31.** diverges.   **33.** Alternate

the 2's from 2^n with the elements of $n!$; group factors to get form for a_n; compare to $\sum\limits_{n=0}^{\infty} \frac{1}{2n+1}$ which diverges. $0 < a_n < \frac{1}{\sqrt{n+1}}$ implies

$\lim\limits_{n\to\infty} a_n = 0$ and the alternating series converges conditionally.   **35.** $S_{2n+1} = (a_1 - a_2) + (a_3 - a_4) + \ldots + (a_{2n-1} - a_{2n}) + a_{2n+1} \ge 0$

for all n. Also, $S_{2n+1} - S_{2n-1} = a_{2n+1} - a_{2n} \le 0$, since $a_{2n+1} \le a_{2n}$. Hence, $\{S_{2n+1}\}$ is a decreasing sequence and is bounded
below by 0.   **37. (a)** $p_n - q_n = a_n$, since when $a_n \ge 0$, $p_n = a_n$ and $q_n = 0$, and when $a_n < 0$, $-q_n = a_n$ and

$p_n = 0$. Also, $p_n + q_n = \begin{cases} a_n & \text{if } a_n \ge 0 \\ -a_n & \text{if } a_n < 0 \end{cases}$ or $p_n + q_n = |a_n|$.   **(b)** The series could not both converge, since if they did,

$\sum (p_n + q_n) = \sum |a_n|$ would also converge and $\sum a_n$ would be absolutely convergent. Assume one series converges and compare S_n
of each to get a contradiction.   **39.** If the process is continued, the sum can be made arbitrarily close to 2. At any stage the maximum
difference between the partial sum and 2 is q_k, for the last q_k subtracted. Since $\lim\limits_{k\to\infty} Q_k = 0$, the partial sums approach 2.

41. $1 - \frac{1}{2} + \left(\frac{1}{3} + \frac{1}{5} + \frac{1}{7}\right) - \frac{1}{4} + \left(\frac{1}{9} + \frac{1}{11} + \ldots + \frac{1}{19}\right) - \frac{1}{6} + \ldots$ **43.** Add terms of $\sum p_k$ until the sum first exceeds S. Then subtract

terms of $\sum q_k$ until the sum less than S.

Exercise Set 10.7

1. $-1 < x < 1$ **3.** $-\infty < x < \infty$ **5.** $-1 \le x \le 1$ **7.** $0 \le x < 2$ **9.** $-5 \le x < 1$ **11.** $-1 < x \le 1$
13. $-\frac{1}{2} \le x \le \frac{1}{2}$ **15.** $-1 \le x < 1$ **17.** $-\infty < x < \infty$ **19.** $\frac{5}{2} \le x \le \frac{7}{2}$ **21.** $-\frac{7}{3} < x < \frac{1}{3}$ **23.** $-\frac{1}{3} < x < \frac{1}{3}$
25. $4 \le x \le 6$ **27.** $-\infty < x < \infty$ **29.** $\frac{1}{3} \le x < 1$ **31.** converges only if $x = -4$.   **33.** $-1 \le x < 1$
35. $-2 < x < 2$ **37.** $-\infty < x < \infty$ **39.** $R = 1$ **41. (a)** Choose x_1 such that $|x_0| < |x_1| \le R$ and such that $\sum\limits_{a^n} |x_1|^n$

converges. If no such x_1 existed, then $|x_0|$ would be an upper bound to the set for which the series converges absolutely, contradicting the
definition of R.   **(b)** Thus, by Theorem 10.18, $\sum a_n x^n$ converges absolutely for all x such that $|x| < |x_0|$.   **(c)** If $|x| > R$ and

$\sum a_n x^n$ converged, then we could find an x_1 such that $R < |x_1| < |x|$ and conclude by Theorem 10.18 that $\sum a_n x_1^n$ converges
absolutely, contrary to the definition of R. Hence, $\sum a_n x^n$ diverges.

Exercise Set 10.8

1. $\sum\limits_{n=1}^{\infty} x^{n-1}$, $|x| < 1$.   **3.** $\sum\limits_{n=1}^{\infty} \frac{(-1)^n x^{2n-1}}{(2n-1)!}$, $|x| < \infty$.   **5.** $\sum\limits_{n=1}^{\infty} \frac{nx^{n-1}}{2^{n+1}}$, $|x| < 2$ **7.** $\sum\limits_{n=0}^{\infty} \frac{x^{n+1}}{(n+1)^2} + C$, $|x| < 1$.

9. $\sum\limits_{n=0}^{\infty} \frac{(-1)^n x^{2n+1}}{(2n+1)!} + C$, $|x| < \infty$.   **11.** $\sum\limits_{n=0}^{\infty} \frac{(-1)^n x^{2n+1}}{2^n} + C$, $|x| < \sqrt{2}$.   **13.** $\sum\limits_{n=0}^{\infty} \left(\frac{1}{2}\right)^{n+1} = 1$

15. $\sum\limits_{n=0}^{\infty} \frac{2^{n+1}}{(n+1)!} - \sum\limits_{n=0}^{\infty} \frac{1}{(n+1)!}, e^2 - e.$ **17.** $\sum\limits_{n=0}^{\infty} \frac{(-1)^n}{2^n(2n+1)^2}$ **19.** $\sum\limits_{n=0}^{\infty} (-1)^{n-1} nx^{n-1}, R = 1.$ **21.** $-\sum\limits_{n=0}^{\infty} \frac{x^{n+1}}{n+1}, R = 1$

23. $2\sum_{n=1}^{\infty} nx^{2n-1}$, $R = 1$,   **25.** $\frac{1}{2}\sum_{n=0}^{\infty} \frac{(-1)^n x^{n+1}}{n+1}$, $R = 1$.   **27.** $\sum_{n=0}^{\infty} \frac{(-1)^n x^{2n+1}}{2^{2n+1}(2n+1)}$, $R = \sqrt{2}$

29. $\frac{1}{2}\sum_{n=2}^{\infty} n(n-1)x^{n-2}$, $R = 1$ **31.** $\sum_{k=1}^{\infty} kx^{k+1}$, $R = 1$ **33.** $\int_0^1 f(x)dx = \frac{1}{2}(e^2 - 1) = \sum_{n=0}^{\infty} \frac{2^n}{(n+1)!}$ **35.** $\frac{x^2+x}{(1-x)^3}$

37. $\frac{3}{2}$ **39.** $\sum_{n=1}^{\infty} \frac{x^{n-1}}{n!}$ **41.** ≈ 0.74682413 **43. (a)** $\frac{10^{50}}{50!} \approx 3.288 \times 10^{-15}$ **(b)** $\frac{20^{100}}{100!} \approx 1.3583 \times 10^{-28}$

(c) $\frac{100^{1000}}{1000!} \approx 10^{-567.6}$

Exercise Set 10.9

1. $\sum_{k=0}^{\infty} \frac{(-1)^k x^{2k+1}}{(2k+1)!}$, $-\infty < x < \infty$.   **3.** $\sum_{k=0}^{\infty} \frac{x^{2k}}{(2k)!}$, $-\infty < x < \infty$ **5.** $\sum_{k=0}^{\infty} \frac{x^{2k+1}}{2k+1}$, $-1 < x < 1$

7. $\sum_{k=1}^{\infty} \frac{(-1)^{k-1} 2^{2k-1} x^{2k}}{(2k)!}$, $-\infty < x < \infty$.   **9.** $\sum_{k=0}^{\infty} \left(\binom{-3}{k}\right)(-1)^k x^k$, $-1 < x < 1$ **11.** $\sum_{k=0}^{\infty} \frac{(\ln 2)^k x^k}{k!}$, $-\infty < x < \infty$.

13. $\sum_{k=1}^{\infty} \frac{(-1)^{k+1} x^{2k-2}}{(2k)!}$, $-\infty < x < \infty$ **15.** $\sum_{k=0}^{\infty} (-1)^k \left(\binom{1/3}{k}\right) \frac{x^k}{2^{3k+1}}$, $-8 < x < 8$ **17.** $\sum_{k=0}^{\infty} \frac{(-1)^k x^k}{(2k)!}$, $-\infty < x < \infty$

19. $e \sum_{k=0}^{\infty} \frac{(x-1)^k}{k!}$, $-\infty < x < \infty$ **21.** $-\sum_{k=0}^{\infty} (x+1)^k$, $-2 < x < 0$.

23. $\frac{1}{2}\left[1 + \sqrt{3}\left(\frac{\pi}{6}\right)(x-1) - \left(\frac{\pi}{6}\right)^2 \frac{(x-1)^2}{2!} - \sqrt{3}\left(\frac{\pi}{6}\right)^3 \frac{(x-1)}{3!} + \left(\frac{\pi}{6}\right)^4 \frac{(x-1)^4}{4!} + --++\dots\right]$, $-\infty < x < \infty$

25. (a) $|R_{2k+1}(x)| = |R_{2k+2}(x)| = \frac{|\cos c| \, |x|^{2k+2}}{(2k+2)!} \le \frac{|x|^{2k+2}}{(2k+2)!}$, and $\lim_{k\to\infty} |R_{2k+2}(x)| = 0$ **(b)** $|R_{2k+1}(x)| = |R_{2k+2}(x)| = \frac{(\cosh c) \, |x|^{2k+2}}{(2k+2)!}$,

and $\lim_{k\to\infty} R_{2k+2}(x) = 0$.   **(c)** $|R_{2k}(x)| = |R_{2k+1}(x)| = \frac{(\cosh c) \, |x|^{2k+1}}{(2k+1)!} \le \frac{(\cosh x) \, |x|^{2k+1}}{(2k+1)!}$ and $\lim_{k\to\infty} R_{2k+1}(x) = 0$.

27. $R = \lim_{n\to\infty} \left|\frac{a_n}{a_{n+1}}\right| = \lim_{n\to\infty} \left|\frac{\alpha(\alpha-1)(\alpha-2)\dots(\alpha-n+1)}{n!} \cdot \frac{(n+1)!}{\alpha(\alpha-1)(\alpha-2)\dots(\alpha-n)}\right| = \lim_{n\to\infty} \left|\frac{n+1}{\alpha-n}\right| = 1$ **29.** The series converges to 0 for all x

and so represents $f(x)$ only at $x = 0$. By using the definition of the derivative $f'(0) = 0$. Also $f^{(n)}(0) = 0$ for all n.

31. ≈ 0.5077. error ≤ 0.000006.   **33.** ≈ 0.19737, error ≤ 0.0000003.   **35.** $1 + x^2 + \frac{2}{3}x^4$

37. $\frac{2}{\sqrt{\pi}} \sum_{n=0}^{\infty} \frac{(-1)^n}{(2n+1)n!}$ erf $(1) \approx 0.843$ **39. (a)** $|\sin^{-1}(\sin x)| = |x|$; but with a period of π **(b)** The graph of the sum of the first three terms is

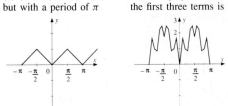

Exercise Set 10.10

1. $\frac{a_0}{2} = \frac{\int_{-\pi}^{\pi} f(x)dx}{\pi - (-\pi)} = f_{ave}$ on $[-\pi, \pi]$ **3.** $\int_{-\pi}^{\pi} f(x) \sin mx \, dx = \frac{a_0}{2}\int_{-\pi}^{\pi} \sin mx \, dx + \sum_{n=1}^{\infty} \left[a_n \int_{-\pi}^{\pi} \cos nx \sin mx \, dx\right.$

$\left. + b_n \int_{-\pi}^{\pi} \sin nx \, \sin mx \, dx\right]$. Apply equations 10.37, 10.34 and 10.35. $\int_{-\pi}^{\pi} f(x) \sin mx \, dx = b_m \int_{-\pi}^{\pi} \sin^2 mx \, dx = b_m \pi$ (equation

10.26). Hence, $b_m = \frac{1}{\pi}\int_{-\pi}^{\pi} f(x) \sin mx \, dx$.   **5.** $\frac{\pi}{4} - \frac{2}{\pi}\left(\cos x + \frac{\cos 3x}{9} + \frac{\cos 5x}{25} + \dots\right) + \sin x - \frac{\sin 2x}{2} + \frac{\sin 3x}{3} - \dots$

7. $a_n = 0$, except for $a_2 = 1$, and $b_n = 0$.   **9.** $a_0 = 0$, $a_n = \frac{1}{n}\sin\frac{n\pi}{2}$, $b_n = 0$, So $f(x) = \cos x - \frac{\cos 3x}{3} + \frac{\cos 5x}{5} - \dots$

11. For 10 terms

13. Graphs for $n = 1$ and 3

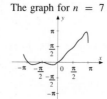

The graph for $n = 7$

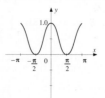

15. The graphs are identical for $n = 2, 5, 8, 11$.

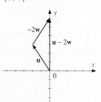

Chapter Review

1. $P_5(x) = -x - \frac{x^2}{2} - \frac{x^3}{3} - \frac{x^4}{4} - \frac{x^5}{5}$; error ≤ 0.0000407; actual error ≈ 0.000013

3. $P_5(x) = 1 + 2\left(x - \frac{\pi}{4}\right) + 2\left(x - \frac{\pi}{4}\right)^2 + \frac{8}{3}\left(x - \frac{\pi}{4}\right)^3 + \frac{10}{3}\left(x - \frac{\pi}{4}\right)^4 + \frac{64}{15}\left(x - \frac{\pi}{4}\right)^5$

5. $\sqrt{2}\sin\left(x - \frac{\pi}{4}\right) = \sqrt{2}\left[x - \frac{\pi}{4} - \frac{(x-\pi/4)^3}{3!} + \frac{(x-\pi/4)^5}{5!} - \cdots + (-1)^{2n+1}\frac{(x-\pi/4)^{2n+1}}{(2n+1)!} + \cdots\right]$

7. $P_3(x) = \frac{\pi}{4} + \frac{1}{2}(x-1) - \frac{1}{4}(x-1)^2 + \frac{1}{12}(x-1)^3$, $R_3(x) = \frac{c(1-c)^2}{(1+c^2)^4}(x-1)^4$

9.

x	$P_1(x)$	$P_2(x)$	$P_3(x)$	$P_4(x)$	e^x
0.25	1.25	1.2813	1.2839	1.2840	1.2840
−0.25	0.75	0.7813	0.7786	0.7788	0.7788
0.50	1.50	1.6250	1.6458	1.6484	1.6487
−0.50	0.50	0.6250	0.6042	0.6068	0.6065
0.75	1.75	2.0313	2.1016	2.1147	2.1170
−0.75	0.25	0.5313	0.4609	0.4741	0.4724
1.00	2.00	2.5000	2.6667	2.7083	2.7183
−1.00	0.00	0.5000	0.3333	0.3750	0.3779

11. (a) 2   **(b)** 0

13. (a) Increasing, bounded above by e, limit is e.   **(b)** Increasing, 2   **15.** For large n, $|a_n^2 - L^2| = |a_n + L| \, |a_n - L| \leq (|a_n| + |L|) \, |a_n - L| < (2|L| + 1) \, |a_n - L|$. For $\epsilon > 0$ and large n, $|a_n - L| < \frac{\varepsilon}{2|L|+1}$, and $|a_n^2 - L^2| < \varepsilon$. The converse is not true. Consider $a_n = \frac{(-1)^n n}{n+1}$.

17. $S_n = 2 - \frac{2}{n+1}$; $\lim\limits_{n \to \infty} S_n = 2$ **19. (a)** $\frac{3}{10}$ **(b)** $\frac{3}{4}$ **21.** diverges   **23.** converges   **25.** diverges.

27. converges.   **29.** converges conditionally   **31.** converges.   **33.** converges   **35.** converges   **37.** conditionally convergent.

39. converges.   **41.** diverges   **43.** converges conditionally.   **45.** $-\frac{3}{2} < x < \frac{3}{2}$ **47.** $1 \leq x < 3$.

49. $-1 < x < 1$ **51.** $-\frac{1}{2} < x \leq \frac{3}{2}$.   **53.** $-2\sum\limits_{n=0}^{\infty}\frac{x^{2n+2}}{2n+2}$, $R = 1$.   **55.** $\sum\limits_{n=0}^{\infty}\binom{-1/2}{n}\frac{x^{2n+1}}{2n+1}$, $R = 1$.

57. (a) $\lim\limits_{x \to 0}\left(1 - \frac{x^2}{3!} + \frac{x^4}{5!} - \cdots\right) = 1$ **(b)** $\lim\limits_{x \to 0}\left(\frac{1}{2!} - \frac{x^2}{4!} + \frac{x^4}{6!} - \cdots\right) = \frac{1}{2}$ **(c)** $\lim\limits_{x \to 0}\left(1 + \frac{x}{2!} + \frac{x^2}{3!} + \cdots\right) = 1$

59. $\frac{1}{2}\sum\limits_{n=1}^{\infty}\frac{(-1)^{n-1}(x+1)^n}{n}$ **61.** ≈ 0.444, $|\text{error}| \leq 0.000052$ **63.** $\cos^2 x - \sin^2 x = \cos 2x$; $\lim\limits_{n \to \infty} R_n(x) = 0$;

$$\sum_{k=0}^{\infty}\frac{(-1)^{k+1}2^{2k+1}(x-\pi/4)^{2k+1}}{(2k+1)!}$$

CHAPTER 11

Exercise Set 11.1

1. $\langle 2, 5 \rangle$ **3.** $\langle -3, 1 \rangle$ **5.** $\langle -6, 1 \rangle$ **7.** $\langle 1, -5 \rangle$ **9.** $\langle 0, 7 \rangle$

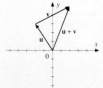

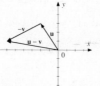

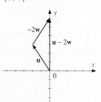

11. $\langle 11, 2 \rangle$ **13.** $\langle -4, -2 \rangle$ **15.** $\langle -3, -4 \rangle$ **17.** $(3, -2)$ **19.** $(-2, 0)$

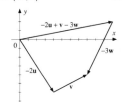

21. 5   **23.** 17   **25.** $\sqrt{5}$ **27.** (a) $\sqrt{5}$ **29.** $\frac{5}{13}\mathbf{i} + \frac{12}{13}\mathbf{j}$ **31.** $\left(-\frac{2}{\sqrt{13}}, -\frac{3}{\sqrt{13}} \right)$ **33.** $\langle 2\sqrt{5}, -4\sqrt{5} \rangle$
      (b) $\sqrt{41}$
      (c) 7
      (d) $5\sqrt{10}$

35. $\frac{12}{5}\mathbf{i} + \frac{16}{5}\mathbf{j}$ **37.** $3\sqrt{5}\mathbf{i} - 6\sqrt{5}\mathbf{j}$ **39.** Inclination of $\mathbf{R} \approx 46.6°$ **41.** Speed $\approx$ 294.62 kph. Heading of plane $\approx 240.16°$.

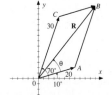

43. Let $\mathbf{u} = \langle a, b \rangle$ and $\mathbf{v} = \langle c, d \rangle$, and consider geometric representatives of $\mathbf{u}$ and $\mathbf{v}$ with initial points at the origin.

45. (1) $\mathbf{u} + \mathbf{0} = \langle a, b \rangle + \langle 0, 0 \rangle = \langle a + 0, b + 0 \rangle = \langle a, b \rangle = \mathbf{u}$ (2) $\mathbf{u} + (-\mathbf{u}) = \langle a, b \rangle + \langle -a, -b \rangle = \langle 0, 0 \rangle = \mathbf{0}$
(3) $(1)\mathbf{u} = (1)\langle a, b \rangle = \langle a, b \rangle = \mathbf{u}$ (4) $(-1)\mathbf{u} = (-1)\langle a, b \rangle = \langle -a, -b \rangle = -\mathbf{u}$ (5) $\mathbf{u} = 0\langle a, b \rangle = \langle 0, 0 \rangle = \mathbf{0}$
(6) $k\mathbf{0} = k\langle 0, 0 \rangle = \langle 0, 0 \rangle = \mathbf{0}$

47. $a = 2, b = -1$ **49.** Let $\mathbf{w} = \langle w_1, w_2 \rangle = a\langle u_1, u_2 \rangle + b\langle v_1, v_2 \rangle = \langle au_1 + bv_1, au_2 + bv_2 \rangle$ and solve the simultaneous equations
$u_1 a + v_1 b = w_1$; $u_2 a + v_2 b = w_2$ for $a = (v_2 w_1 - v_1 w_2)/(u_1 v_2 - u_2 v_1)$ and $b = (u_1 w_2 - u_2 w_1)/(u_1 v_2 - u_2 v_1)$, provided $u_1 v_2 - u_2 v_1 \neq 0$,
which is equivalent to $\mathbf{u} \neq k\mathbf{v}$ for all scalars k.   **51.** Let M_1 and M_2 be the midpoints of $\overrightarrow{AC}$ and $\overrightarrow{BD}$, respectively. Then,

$\overrightarrow{AM_1} = \frac{1}{2}(\overrightarrow{AB} + \overrightarrow{AD})$) and $\overrightarrow{AM_2} = \overrightarrow{AD}\frac{1}{2}(\overrightarrow{DB})\overrightarrow{AD} + \frac{1}{2}(\overrightarrow{AB} - \overrightarrow{AD}) = \frac{1}{2}(\overrightarrow{AB} + \overrightarrow{AD})$. Hence, $\overrightarrow{AM_1} = \overrightarrow{AM_2}$; that is, the midpoints coincide.

55. Heading = 348.43°. On the return trip, heading = 191.57°. Total time = 3.50 hrs

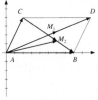

Exercise Set 11.2

1. -6 **3.** -4 **5.** $\mathbf{u} \cdot \mathbf{v} = 0$ **7.** $\mathbf{u} \cdot \mathbf{v} = 0$ **9.** $-\frac{3}{5}$ **11.** $-\frac{1}{5\sqrt{2}}$ **13.** $\frac{7}{25}$ **15.** $\frac{1}{2}; \theta = \frac{\pi}{3}$ **17.** $-\frac{1}{7}, 7$

19. $-\frac{5}{\sqrt{2}}$ **21.** $-\frac{5}{\sqrt{2}}$ **23.** 28 joules   **25.** 30 joules   **27.** $\text{Proj}_{\mathbf{v}}\mathbf{u} = \left\langle -\frac{1}{5}, -\frac{2}{5} \right\rangle$; $\text{Proj}_{\mathbf{v}}^{\perp}\mathbf{u} = \left\langle \frac{16}{5}, -\frac{8}{5} \right\rangle$

29. $\text{Proj}_{\mathbf{v}}\mathbf{u} = \frac{6}{5}\mathbf{i} - \frac{8}{5}\mathbf{j}$; $\text{Proj}_{\mathbf{v}}^{\perp}\mathbf{u} = \frac{24}{5}\mathbf{i} + \frac{18}{5}\mathbf{j}$ **31.** 500 lb   **33.** Let $A = P(2, 1), B = P(6, 9), C = P(-2, 3)$ Let

$\mathbf{u} = \overrightarrow{AB} = \langle 4, 8 \rangle, \mathbf{v} = \overrightarrow{AC} = \langle -4, 2 \rangle$ $\mathbf{u} \cdot \mathbf{v} = -16 + 16 = 0$. $\therefore \angle BAC = 90°$; Area = 20 square units

35. Let $P_1 = (-5, 2), P_2 = (-3, -2), P_3 = (6, 1), P_4 = (4, 5)$ $\overrightarrow{P_1 P_2} = \langle 2, -4 \rangle, \overrightarrow{P_3 P_4} = \langle -2, 4 \rangle$. $\therefore \overrightarrow{P_1 P_2} \| \overrightarrow{P_3 P_4}$.

$\overrightarrow{P_1 P_4} = \langle 9, 3 \rangle, \overrightarrow{P_2 P_3} = \langle 9, 3 \rangle$. $\therefore \overrightarrow{P_1 P_4} \| \overrightarrow{P_2 P_3}$; angle at $P_1 \approx 81.9°$; angle at $P_2 \approx 98.1°$

37. (a) $|\mathbf{u} + \mathbf{v}|^2 + |\mathbf{u} - \mathbf{v}|^2 = (\mathbf{u} + \mathbf{v}) \cdot (\mathbf{u} + \mathbf{v}) + (\mathbf{u} - \mathbf{v}) \cdot (\mathbf{u} - \mathbf{v})$ (Theorem 11.4)

$= |\mathbf{u}|^2 + 2\mathbf{u} \cdot \mathbf{v} + |\mathbf{v}|^2 + |\mathbf{u}|^2 - 2\mathbf{u} \cdot \mathbf{v} + |\mathbf{v}|^2 = 2\left(|\mathbf{u}|^2 + |\mathbf{v}|^2 \right)$ (b) $|\mathbf{u} + \mathbf{v}|^2 - |\mathbf{u} - \mathbf{v}|^2 = 2\mathbf{u} \cdot \mathbf{v} - (-2\mathbf{u} \cdot \mathbf{v}) = 4\mathbf{u} \cdot \mathbf{v}$ (following the method used in part a)

39. $(\text{Proj}_{\mathbf{v}}\mathbf{u}) \cdot (\text{Proj}_{\mathbf{v}}^{\perp}\mathbf{u}) = \left(\frac{\mathbf{u} \cdot \mathbf{v}}{|\mathbf{v}|^2} \right) \mathbf{v} \cdot \left[\mathbf{u} - \left(\frac{\mathbf{u} \cdot \mathbf{v}}{|\mathbf{v}|^2} \right) \mathbf{v} \right] = \frac{(\mathbf{u} \cdot \mathbf{v})(\mathbf{v} \cdot \mathbf{u})}{|\mathbf{v}|^2} - \frac{(\mathbf{u} \cdot \mathbf{v})^2(\mathbf{v} \cdot \mathbf{v})}{|\mathbf{v}|^4} = \frac{(\mathbf{u} \cdot \mathbf{v})^2}{|\mathbf{v}|^2}\left[1 - \frac{|\mathbf{v}|^2}{|\mathbf{v}|^2} \right] = 0$

41. From Exercise 40, $\mathbf{n} = \langle a, b \rangle$ is perpendicular to the line $ax + by + c = 0$. Also, $\overrightarrow{P_0 P_1} = \langle x_1 - x_0, y_1 - y_0 \rangle$. Then,

$d = |\text{Comp}_{\mathbf{n}} \overrightarrow{P_0 P_1}| = \frac{|\langle x_1 - x_0, y_1 - y_0 \rangle \cdot \langle a, b \rangle|}{|\langle a, b \rangle|} = \frac{|ax_1 + by_1 - ax_0 - by_0|}{\sqrt{a^2 + b^2}} = \frac{|ax_1 + by_1 + c|}{\sqrt{a^2 + b^2}}$

Exercise Set 11.3

1. (a)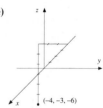

(3, 2, 4)

(b)   (4, −2, 1)

(c)   (−3, 2, 4)

(d)   (0, −5, −2)

(e)   (−4, −3, −6)

3. (a) xy-plane   **(b)** yz-plane   **(c)** xz-plane   **(d)** x-axis   **(e)** a point on a coordinate axis   **5. (a)** $\overrightarrow{PQ} = \langle -2, -4, 3 \rangle$

(b) $\overrightarrow{PQ} = \langle -3, 1, 4 \rangle$ **(c)** $\overrightarrow{PQ} = \langle 3, 2, -2 \rangle$ **(d)** $\overrightarrow{PQ} = \langle 6, -2, -2 \rangle$ **7. (a)** -14 **(b)** -27 **(c)** $3\sqrt{14}$ **(d)** $\langle 3, 2, -3 \rangle$

9. (a) $\left\langle \frac{2}{3}, -\frac{1}{3}, \frac{2}{3} \right\rangle$ **(b)** $\left\langle \frac{4}{5\sqrt{2}}, \frac{3}{5\sqrt{2}}, -\frac{1}{\sqrt{2}} \right\rangle$ **(c)** $\frac{1}{\sqrt{3}}\mathbf{i} - \frac{1}{\sqrt{3}}\mathbf{j} + \frac{1}{\sqrt{3}}\mathbf{k}$ **(d)** $\frac{2}{3\sqrt{5}}\mathbf{i} - \frac{4}{3\sqrt{5}}\mathbf{j} + \frac{5}{3\sqrt{5}}\mathbf{k}$

11. (a) $\mathbf{u} \cdot \mathbf{v} = 8 - 6 - 2 = 0$ **(b)** $x = \frac{1}{2}$ **13. (a)** $\cos\alpha = \frac{2}{3}, \cos\beta = -\frac{1}{3}, \cos\gamma = \frac{2}{3}$ **(b)** $\cos\alpha = \frac{3}{5\sqrt{2}}, \cos\beta = -\frac{1}{\sqrt{2}}, \cos\gamma = \frac{4}{5\sqrt{2}}$

15. $\left\langle -\frac{1}{2}, \frac{1}{2}, \frac{1}{\sqrt{2}} \right\rangle$ **17.** $\cos\alpha \le \frac{1}{2}; \cos\beta \le \frac{1}{2}; \cos^2\gamma = 1 - \cos^2\alpha - \cos^2\beta \ge \frac{1}{2} \Rightarrow \cos\gamma \ge \frac{1}{\sqrt{2}}$, and $\gamma \le \frac{\pi}{4}$.   **19.** $\frac{31}{5}$ **21.** $-\frac{2}{\sqrt{3}}$

23. 184 ergs   **25.** $\text{Proj}_{\mathbf{v}}\mathbf{u} = \left\langle -\frac{2}{7}, -\frac{6}{7}, \frac{4}{7} \right\rangle$; $\text{Proj}_{\mathbf{v}}^{\perp}\mathbf{u} = \left\langle \frac{37}{7}, -\frac{1}{7}, \frac{17}{7} \right\rangle$ **27.** $\text{Proj}_{\mathbf{v}}\mathbf{u} = \left(\frac{11}{14}\mathbf{i} + \frac{11}{7}\mathbf{j} - \frac{33}{14}\mathbf{k} \right)$; $\text{Proj}_{\mathbf{v}}^{\perp}\mathbf{u} = \left(\frac{17}{14}\mathbf{i} - \frac{32}{7}\mathbf{j} - \frac{37}{14}\mathbf{k} \right)$

29. Applying the rules for the addition and multiplication of real numbers, we have: **(1)** $\mathbf{u} + \mathbf{v} = \langle u_1, u_2, u_3 \rangle + \langle v_1, v_2, v_3 \rangle =$
$\langle u_1 + v_1, u_2 + v_2, u_3 + v_3 \rangle = \langle v_1 + u_1, v_2 + u_2, v_3 + u_3 \rangle = \langle v_1, v_2, v_3 \rangle + \langle u_1, u_2, u_3 \rangle = \mathbf{v} + \mathbf{u}$ **(2)** $\mathbf{u} + (\mathbf{v} + \mathbf{w}) =$
$\langle u_1, u_2, u_3 \rangle + (\langle v_1, v_2, v_3 \rangle + \langle w_1, w_2, w_3 \rangle) = \langle u_1 + v_1 + w_1, u_2 + v_2 + w_2, u_3 + v_3 + w_3 \rangle = \langle u_1 + v_1, u_2 + v_2, u_3 + v_3 \rangle + \langle w_1, w_2, w_3 \rangle = (\mathbf{u} + \mathbf{v}) + \mathbf{w}$
(3) $k(\mathbf{u} + \mathbf{v}) = k\langle u_1 + v_1, u_2 + v_2, u_3 + v_3 \rangle = \langle ku_1 + kv_1, ku_2 + kv_2, ku_3 + kv_3 \rangle = \langle ku_1, ku_2, ku_3 \rangle + \langle kv_1, kv_2, kv_3 \rangle = k\mathbf{u} + k\mathbf{v}$
(4) $(k+l)\mathbf{u} = (k+l)\langle u_1, u_2, u_3 \rangle = \langle (k+l)u_1, (k+l)u_2, (k+l)u_3 \rangle = \langle ku_1 + lu_1, ku_2 + lu_2, ku_3 + lu_3 \rangle = \langle ku_1, ku_2, ku_3 \rangle + \langle lu_1, lu_2, lu_3 \rangle = k\mathbf{u} + l\mathbf{u}$
(5) $k(l\mathbf{u}) = k(l\langle u_1, u_2, u_3 \rangle) = k\langle lu_1, lu_2, lu_3 \rangle = \langle klu_1, klu_2, klu_3 \rangle = (kl)\langle u_1, u_2, u_3 \rangle = (kl)\mathbf{u}$

31. (1) $|\mathbf{u}| = \sqrt{u_1^2 + u_2^2 + u_3^2} \ge 0$; $|\mathbf{u}| = 0$ if and only if u_1, u_2, and u_3 are all zero and $\mathbf{u} = \mathbf{0}$.

(2) $|-\mathbf{u}| = |\langle -u_1, -u_2, -u_3 \rangle| = \sqrt{u_1^2 + u_2^2 + u_3^2} = |\mathbf{u}|$ **(3)** $|k\mathbf{u}| = |\langle ku_1, ku_2, ku_3 \rangle| = |k||\langle u_1, u_2, u_3 \rangle| = |k||\mathbf{u}|$

(4) $(\mathbf{u} + \mathbf{v}) \cdot (\mathbf{u} + \mathbf{v}) = |\mathbf{u} + \mathbf{v}|^2 = |\mathbf{u}|^2 + 2\mathbf{u} \cdot \mathbf{v} + |\mathbf{v}|^2$; $|\mathbf{u}|^2 + 2\mathbf{u} \cdot \mathbf{v} + |\mathbf{v}|^2 \le |\mathbf{u}|^2 + 2|\mathbf{v}||\mathbf{v}| + |\mathbf{v}|^2 = (|\mathbf{u}| + |\mathbf{v}|)^2$; Take the square root of both
sides of the equation to get the desired result $|\mathbf{u} + \mathbf{v}| \le |\mathbf{u}| + |\mathbf{v}|$.   **35.** $\frac{2}{\sqrt{93}}\mathbf{i} - \frac{8}{\sqrt{93}}\mathbf{j} - \frac{5}{\sqrt{93}}\mathbf{k}$ **37.** $a = \frac{5}{3}, b = \frac{7}{3}, c = \frac{4}{3}$

39. Suppose that $\alpha + \beta < \frac{\pi}{2} \Rightarrow \alpha < \left(\frac{\pi}{2} - \beta \right)$. Then, $\cos^2\alpha + \cos^2\beta > (\cos\pi/2 \cos\beta + \sin\pi/2 \sin\beta)^2 + \cos^2\beta = \sin^2\beta + \cos^2\beta = 1 \Rightarrow$
$\cos^2\alpha + \cos^2\beta > 1$. Now, since $\cos^2\alpha + \cos^2\beta + \cos^2\gamma = 1$, we have $\cos^2\gamma = 1 - (\cos^2\alpha + \cos^2\beta) < 0$, which is impossible.
$\alpha + \beta \ge \frac{\pi}{2}$. Since α and β were chosen arbitrarily from the set $\{\alpha, \beta, \gamma\}$, the conclusion that the sum of any two of the three
direction angles must be $> \frac{\pi}{2}$ follows.   **41.** $\cos\alpha = .486, \cos\beta = .741, \cos\gamma = .463; \alpha = 60.904°, \beta = 42.184°, \gamma = 117.588°$;
$\alpha = 1.063$ radians, $\beta = .736$ radians, $\gamma = 2.052$ radians   **43.** $\cos\alpha = \frac{1}{165}\sqrt{165}, \cos\beta = -\frac{8}{165}\sqrt{165}, \cos\gamma = \frac{2}{33}\sqrt{165}$;
$\alpha = 85.535°, \beta = 128.521°, \gamma = 38.877°; \alpha = 1.493$ radians, $\beta = 2.243$ radians, $\gamma = .679$ radians   **45.** The two vectors
are $\left\langle -\frac{3}{4}, -\frac{9}{8}, 1 \right\rangle$ and $\left\langle -\frac{3}{2}, -\frac{9}{4}, 2 \right\rangle$.   **51. (a)** $\mathbf{u} + \mathbf{v} = \langle -2, 1, 12 \rangle$; $\mathbf{u} - \mathbf{v} = \langle 6, -3, -2 \rangle$; $-2\mathbf{u} = \langle -4, 2, -10 \rangle$ **(b)** $\sqrt{69}$
(c) $\left\langle -\frac{4}{69}\sqrt{69}, \frac{2}{69}\sqrt{69}, \frac{7}{69}\sqrt{69} \right\rangle$ **(d)** 25   **(e)** .989 radians or 56.668°   **(f)** $25\sqrt{69}$ **(g)** $\left\langle -\frac{100}{69}, \frac{50}{69}, \frac{175}{69} \right\rangle$

Exercise Set 11.4

1. $\langle 3, -1, 4 \rangle$ **3.** $3\mathbf{i} + 9\mathbf{j} + 6\mathbf{k}$ **5.** $\langle -9, -1, -21 \rangle$ **7.** $19\mathbf{i} + 13\mathbf{j} + 5\mathbf{k}$ **9.** $\langle 11, -6, -2 \rangle$ **11.** $14\mathbf{i} - 14\mathbf{j} + 14\mathbf{k}$ or $\mathbf{i} - \mathbf{j} + \mathbf{k}$
13. 12   **15.** $12\mathbf{i} + 12\mathbf{j} - 12\mathbf{k}$ **17.** 69   **19.** $\mathbf{v} \cdot (\mathbf{u} \times \mathbf{v}) = \langle v_1, v_2, v_3 \rangle \cdot \langle (u_2v_3 - u_3v_2), (u_3v_1 - u_1v_3), (u_1v_2 - u_2v_1) \rangle =$
$v_1(u_1v_3 - u_3v_2) + v_2(u_3v_1 - u_1v_3) + v_3(u_1v_2 - u_2v_1) = u_1v_1v_3 - u_3v_1v_2 + u_3v_1v_2 - u_1v_2v_3 + u_1v_2v_3 - u_2v_1v_3 = 0$, or $\mathbf{v} \perp (\mathbf{u} \times \mathbf{v})$

21. $\sqrt{146}$ **23.** $3\sqrt{78}$ **25.** $\frac{3\sqrt{5}}{2}$ **27.** $\left\langle \frac{4}{\sqrt{89}}, -\frac{3}{\sqrt{89}}, -\frac{8}{\sqrt{89}} \right\rangle$ **29.** 23   **31.** $\overrightarrow{AB} \cdot (\overrightarrow{AC} \times \overrightarrow{AD}) =$
$\langle 2, -3, -1 \rangle \cdot (\langle -1, 2, 0 \rangle \times \langle 0, 1, -1 \rangle) = 0$

33. Assume that the second and third rows of the determinant are identical. If this is not the case, then
interchange row 1 with either row 2 or row 3, as needed. This will change the sign of the determinant only.

$$\begin{vmatrix} a_1 & a_2 & a_3 \\ b_1 & b_2 & b_3 \\ b_1 & b_2 & b_3 \end{vmatrix} = a_1 \begin{vmatrix} b_2 & b_3 \\ b_2 & b_3 \end{vmatrix} - a_2 \begin{vmatrix} b_1 & b_3 \\ b_1 & b_3 \end{vmatrix} + a_3 \begin{vmatrix} b_1 & b_2 \\ b_1 & b_2 \end{vmatrix} = 0$$

35. Property 1. $\mathbf{u} \times \mathbf{v} = \langle u_1, u_2, u_3 \rangle \times \langle v_1, v_2, v_3 \rangle = \langle u_2v_3 - u_3v_2, u_3v_1 - u_1v_3, u_1v_2 - u_2v_1 \rangle = -(v_2u_3 - v_3u_2, v_3u_1 - v_1u_3, v_1u_2 - v_2u_1) = -\mathbf{v} \times \mathbf{u}$
Property 2. $k(\mathbf{u} \times \mathbf{v}) = k\langle u_2v_3 - u_3v_2, u_3v_1 - u_1v_3, u_1v_2 - u_2v_1 \rangle = \langle (ku_2)v_3 - (ku_3)v_2, (ku_3)v_1 - (ku_1)v_3, (ku_1)v_2 - (ku_2)v_1 \rangle = (k\mathbf{u}) \times \mathbf{v}$
Property 3. $\mathbf{u} \times \mathbf{0} = \langle u_1, u_2, u_3 \rangle \times \langle 0, 0, 0 \rangle = \langle 0, 0, 0 \rangle = \mathbf{0}$ **37.** Use components to show in each case that the two sides are equal.
39. $(\mathbf{u} + \mathbf{v}) \times (\mathbf{u} - \mathbf{v}) = \mathbf{u} \times (\mathbf{u} - \mathbf{v}) + \mathbf{v} \times (\mathbf{u} - \mathbf{v})$ (Property 6) $= \mathbf{u} \times \mathbf{u} - \mathbf{u} \times \mathbf{v} + \mathbf{v} \times \mathbf{u} - \mathbf{v} \times \mathbf{v}$ (Property 5) $= \mathbf{0} - \mathbf{u} \times \mathbf{v} - \mathbf{u} \times \mathbf{v} - \mathbf{0} = -2(\mathbf{u} \times \mathbf{v}) = 2(\mathbf{v} \times \mathbf{u})$

41. $\mathbf{u} \times (\mathbf{v} + \mathbf{w}) + \mathbf{v} \times (\mathbf{w} + \mathbf{u}) + \mathbf{w} \times (\mathbf{u} + \mathbf{v}) = \mathbf{u} \times \mathbf{v} + \mathbf{u} \times \mathbf{w} + \mathbf{v} \times \mathbf{w} + \mathbf{v} \times \mathbf{u} + \mathbf{w} \times \mathbf{u} + \mathbf{w} \times \mathbf{v}$ (Property 5) $=$
$\mathbf{u} \times \mathbf{v} + \mathbf{u} \times \mathbf{w} + \mathbf{v} \times \mathbf{w} - \mathbf{u} \times \mathbf{v} - \mathbf{u} \times \mathbf{w} - \mathbf{v} \times \mathbf{w}$ (Property 1) $= \mathbf{0}$

43. (a) The area of the given triangle is one-half the area of the parallelogram with adjacent sides $\overrightarrow{P_1P_2}$ and $\overrightarrow{P_1P_3}$. $\therefore$ the area of the triangle

$= \frac{1}{2}\left| \overrightarrow{P_1P_2} \times \overrightarrow{P_1P_3} \right|$ **(b)** Evaluate the cross product in part (a).   **45.** $\langle 17, -4, -5 \rangle$ **47.** Volume of the
parallelepiped = 44 cubic
units

Exercise Set 11.5

1. $\langle 2, 5, -1 \rangle + t\langle -3, 1, 2 \rangle$; $x = 2 - 3t, y = 5 + t, z = -1 + 2t$ **3.** $\langle 5, 8, -6 \rangle + t\langle 2, -3, 4 \rangle$; $x = 5 + 2t, y = 8 - 3t, z = -6 + 4t$
5. $\langle 4, -1, 8 \rangle + t\langle -1, 3, -3 \rangle$ **7.** $\langle 7, -2, -4 \rangle + t\langle -4, 3, 2 \rangle$ **9.** $\theta \approx 52.5°$ **11.** $\theta \approx 92.8°$ **13.** $\langle 3, -2, 1 \rangle \cdot \langle 7, 8, -5 \rangle = 0$
15. $x = 3 - 3t, y = -1 + t, z = 2 + 5t$ **17.** $\langle 5 + 2t, 2 + t, -3 \rangle$ **19.** The point of intersection is $(-1, 2, 1)$. A vector orthogonal is
$\langle -1 - 7t, 2 - 9t, 1 + 3t \rangle$ or $x = -1 - 7t, y = 2 - 9t, z = 1 + 3t$ **21.** The lines are skew.

23. The point of intersection is $(-1, 3, 2)$ **25.** Let $x = x_0 + at \Rightarrow t = \frac{x - x_0}{a}, a \neq 0; y = y_0 + bt \Rightarrow t = \frac{y - y_0}{b}, b \neq 0; z = z_0 + ct \Rightarrow t =$
$\frac{z - z_0}{c}, c \neq 0$. Equate the expressions for t: $\frac{x - x_0}{a} = \frac{y - y_0}{b} = \frac{z - z_0}{c}$ **27. (a)** $\frac{x-3}{-2} = \frac{y+1}{4} = \frac{z-4}{3}$ **(b)** $x = 3 - 2t, y = -1 + 4t, z = 4 + 3t$
(c) $\langle 3, -1, 4 \rangle + t\langle -2, 4, 3 \rangle$ **(d)** $\left(0, 5, \frac{17}{2}\right); \left(\frac{5}{2}, 0, \frac{19}{4}\right); \left(\frac{17}{3}, -\frac{19}{3}, 0\right)$

29. $t = -1 : \mathbf{r} = \langle -1 + 3s, 2 + 4s, 8 - 5s \rangle; t = \frac{7}{3} : \mathbf{r} = \langle -1 + 19s, 2 - 18s, 8 + 5s \rangle$ **31.** $|\overrightarrow{PQ} \times \mathbf{v}| = |\overrightarrow{PQ}| \cdot |\mathbf{v}| \cdot \sin\theta$;

$d = |\overrightarrow{PQ}| \cdot \sin\theta = \frac{|\overrightarrow{PQ} \times \mathbf{v}|}{|\overrightarrow{PQ}| \cdot |\mathbf{v}|} \cdot |\overrightarrow{PQ}| = \frac{|\overrightarrow{PQ} \times \mathbf{v}|}{|\mathbf{v}|}$ **33.** The point of intersection is
$(3, 0, 1)$; $10.025°$

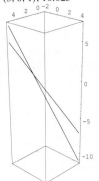

35. Solution of parametric
equations is $x = 1 - 3t, y = -2 + 7t, z = 4 + 2t$

Exercise Set 11.6

1. (a) $\langle x - 4, y - 2, z - 6 \rangle \cdot \langle 3, 2, -1 \rangle = 0$ **(b)** $3(x - 4) + 2(y - 2) - (z - 6) = 0$ **(c)** $3x + 2y - z - 10 = 0$
3. (a) $\langle x - 5, y + 3, z + 4 \rangle \cdot \langle 2, 3, -4 \rangle = 0$ **(b)** $2(x - 5) + 3(y + 3) - 4(z + 4) = 0$ **(c)** $2x + 3y - 4z - 17 = 0$ **5.** $x - 2y - 4z + 8 = 0$
7. $3x - 2y + 4z + 25 = 0$ **9.** $z = 4$ **11.** $3x - 4y + 5z = 0$ **13.** $x - 4y - 3z - 1 = 0$ **15.** $5x + y - 4z - 9 = 0$

17. $\theta \approx 78.55°$ **19.** $\frac{\mathbf{n}_1 \cdot \mathbf{n}_2}{|\mathbf{n}_1||\mathbf{n}_2|} = \frac{\langle 3, -4, 2 \rangle \cdot \langle 2, 3, 3 \rangle}{|\mathbf{n}_1||\mathbf{n}_2|} = 0$ **21.** $3y - z - 2 = 0$ **23.** $17x + 5y + 19z - 55 = 0$

25.

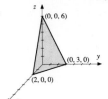

27.

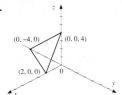

29.

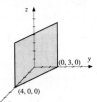

31.

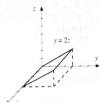

33.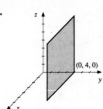

35. $x = t, y = -13t, z = -2 + 8t.$

37. $\frac{22}{5\sqrt{5}}$

39. $\mathbf{n}_2 = \langle 3, -6, 6 \rangle = 3\mathbf{n}_1; D = \frac{14}{9}$

41. The intersection occurs at $(-1, -4, 3)$; $x - y - z = 0.$

43. $(-1, 9, 4)$

45. Parametric equations for the line of intersection are $x = 1 + 5t, y = -1 + 10t, z = 5t$

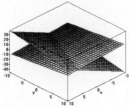

47. Direction vectors for the lines are $\mathbf{u} = \langle 1, -1, 1 \rangle$ and $\mathbf{v} = \langle 2, 3, 1 \rangle$ respectively, and $\mathbf{u} \cdot \mathbf{v} = \mathbf{0}$.

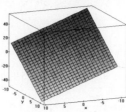

49. $\frac{1}{6}\sqrt{6}$ **51.** 72 units

Chapter Review

1. (a) 201   (b) $\sqrt{5}\mathbf{i} + 2\sqrt{5}\mathbf{j}$ **3.** (a) -82 (b) $\theta = 45°$ **5.** (a) $3\sqrt{30}$ (b) $\left\langle -\frac{20}{9}, \frac{10}{9}, \frac{20}{9} \right\rangle$

7. If two sides of the triangle are the vectors $\mathbf{u}$ and $\mathbf{v}$, then the third side is $\mathbf{u} - \mathbf{v}$. In the smaller triangle, the sides are $\mathbf{v}/2$ and $(\mathbf{u} - \mathbf{v})/2$ and $\mathbf{x}$; we have $\mathbf{x} = \frac{\mathbf{v}}{2} + \frac{\mathbf{u} - \mathbf{v}}{2} = \frac{\mathbf{u}}{2}$. Hence, $\mathbf{x}$ is parallel to $\mathbf{u}$, and its length is one-half that of $\mathbf{u}$.

9. 253.74 mph; Heading = 186.79°

11. $\langle 3, 1, -2 \rangle + t\langle -1, 2, 5 \rangle; x = 3 - t, y = 1 + 2t, z = -2 + 5t$ **13.** $\langle 4, 2, -1 \rangle + t\langle 5, -3, 4 \rangle; x = 4 + 5t, y = 2 - 3t, z = -1 + 4t$

15. Lines do not intersect.   **17.** $x - y + z - 1 = 0$ **19.** (a) $\alpha \approx \alpha 107.3°, \beta \approx 53.4°, \gamma \approx 41.8°$ (b) $\langle -\frac{16}{3}, \frac{28}{3}, \frac{16}{3} \rangle$

21. (a) $\frac{16}{5}$ (b) $\frac{5}{3}$ **23.** Let $a\mathbf{u} + b\mathbf{v} + c\mathbf{w} = 0$. Then $a(\mathbf{u} \cdot \mathbf{u}) + b(\mathbf{u} \cdot \mathbf{v}) + c(\mathbf{u} \cdot \mathbf{w}) = a|\mathbf{u}|^2 = 0 \Rightarrow a = 0$, since $|\mathbf{u}| \neq 0$;

$a(\mathbf{u} \cdot \mathbf{v}) + b(\mathbf{v} \cdot \mathbf{v}) + c(\mathbf{v} \cdot \mathbf{w}) = b|\mathbf{v}|^2 = 0 \Rightarrow b = 0$, since $|\mathbf{v}| \neq 0$; $a(\mathbf{u} \cdot \mathbf{w}) + b(\mathbf{v} \cdot \mathbf{w}) + c(\mathbf{w} \cdot \mathbf{w}) = a|\mathbf{w}|^2 = 0 \Rightarrow c = 0$, since $|\mathbf{w}| \neq 0$

CHAPTER 12

Exercise Set 12.1

1. $\{t : t < 1\}$, or $-\infty < t < 1$ **3.** $\{t : 1 \leq t < \infty\}$ **5.** $\{t : t > 0, t \neq \frac{n\pi}{2}, n = 1, 2, 3, \ldots\}$ **7.**

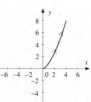

9.   **11.** 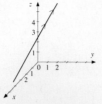   **13.** 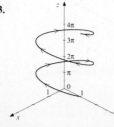   **15.**

17.

19.

 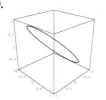 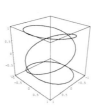

21.

Exercise Set 12.2

1. $\langle 1, 2, 3 \rangle$ **3.** $\mathbf{i} + 2\mathbf{j}$ **5.** $\{t : t < 1\}$ **7.** $\{t : -1 < t < 1\}$ **9. (a)** $\langle 3t - 1, t - 2, 3 - t \rangle$ **(b)** $\langle t + 3, 2t + 6, 3t + 6 \rangle$

11. (a) $\langle t^2 - 6, 3 - 3t - 2t^2, 3t + t^2 \rangle$ **(b)** $\langle t, 2, 1 - t \rangle$ **13. (a)** $(2t \cos t)\mathbf{i} + (2t \sin t)\mathbf{j} + (2t \sin t)\mathbf{k}$

  (b) $\cos t \sin t + \sin t \cos t + \sin t = \sin 2t + \sin t$ **15.** $\left\langle 2t, 2e^{2t}, -\frac{1}{t^2} \right\rangle$ **17.** $\langle t \cos t + \sin t, 1 + \ln t \rangle$ **19.** $e^{-t}\mathbf{i} + e^{2t}(2t + 1)\mathbf{j}$

21. $\frac{1}{(1-t^2)^{3/2}}\mathbf{i} + \frac{1}{\sqrt{1-t^2}}\mathbf{j} - \frac{t}{\sqrt{1-t^2}}\mathbf{k}$ **23.** $x = 2 + 2u, y = 3 + 6u, z = 1 + 3u$ **25.** $x = 4 - 16u, y = \frac{1}{2} + u; z = \frac{1}{4} + 2u$

27. $x = -1 + u, y = 4 + \frac{5}{2}u, z = 1 - u$ **29.** $\mathbf{r}'\left(\frac{\pi}{3}\right) = \langle -2\sqrt{3}, 1 \rangle$ **31.** $\mathbf{r}'(-1) = -2\mathbf{i} + 3\mathbf{j}$ **33.** $\mathbf{r}'(1) = \langle 2, -3, 3 \rangle$

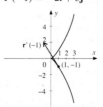

35. $\left\langle -\cos t, t - \sin t, \frac{t^2}{2} \right\rangle + \langle C_1, C_2, C_3 \rangle$ **37.** $\langle \frac{\pi}{4}, \frac{2}{3}, \frac{1}{3} \rangle$ **39.** $\frac{\pi}{4}\mathbf{i} + \frac{1}{3}\mathbf{j}$ **41.** $\frac{-3\sqrt{3}}{8}\mathbf{i} + \left(\sqrt{3} - \frac{\pi}{3}\right)\mathbf{j} + \frac{\pi}{3}\mathbf{k}$

43. $\frac{1}{3}(32 - \sqrt{2}) \approx 10.195$ **45.** $\left| \int_a^b \mathbf{r}(t)dt \right| = |\langle \cos a - \cos b, \sin b - \sin a, b - a \rangle| = [2 - 2\cos(a - b) + (b - a)^2]^{1/2}$ and

  $\int_a^b |\mathbf{r}(t)|dt = \sqrt{2}(b - a)$. Let $(b - a) = \alpha \geq 0$. $\sqrt{\alpha^2 - 2\cos\alpha + 2} \leq \sqrt{2}\alpha$ since $\alpha^2 + 2\cos\alpha - 2$ is nondecreasing.

47. $\frac{d}{dt}[c\mathbf{u}(t)] = \frac{d}{dt}[\langle cu_1(t), cu_2(t), \ldots, cu_n(t) \rangle] = \langle cu_1'(t), cu_2'(t), \ldots, cu_n'(t) \rangle = c\mathbf{u}'(t)$

49. $\frac{d}{dt}[\mathbf{u}(t) \cdot \mathbf{v}(t)] = \frac{d}{dt}[u_1(t)v_1(t) + u_2(t)v_2(t) + \ldots + u_n(t)v_n(t)] = u_1(t)v_1'(t) + u_1'(t)v_1(t) + u_2(t)v_2'(t) + u_2'(t)v_2(t) + \ldots + u_n(t)v_n'(t) + u_n'(t)v_n(t) = $
  $u_1(t)v_1'(t) + u_2(t)v_2'(t) + \ldots + u_n(t)v_n'(t) + u_1'(t)v_1(t) + u_2'(t)v_2(t) + \ldots + u_n'(t)v_n(t) = \mathbf{u}(t) \cdot \mathbf{v}'(t) + \mathbf{u}'(t) \cdot \mathbf{v}(t)$

51. $\frac{d}{dt}[\mathbf{u}(f(t))] = \frac{d}{dt}[\langle u_1(f(t)), u_2(f(t)), \ldots, u_n(f(t)) \rangle] = \langle u_1'(f(t))f'(t), u_2'(f(t))f'(t), \ldots, u_n'(f(t))f'(t) \rangle = \mathbf{u}'(f(t))f'(t)$

53. $c \int_a^b \mathbf{u}(t)dt = c \int_a^b \langle u_1(t), u_2(t), \ldots, u_n(t) \rangle dt = \int_a^b \langle cu_1(t), cu_2(t), \ldots, cu_n(t) \rangle dt = \int_a^b c\mathbf{u}(t)dt$ **55.** $\mathbf{K} \cdot \int_a^b \mathbf{u}(t)dt = $

  $\langle k_1, k_2, \ldots, k_n \rangle \cdot \langle \int_a^b u_1(t)dt, \int_a^b u_2(t)dt, \ldots, \int_a^b u_n(t)dt \rangle = k_1 \int_a^b u_1(t)dt + k_2 \int_a^b u_2(t)dt + \ldots + k_n \int_a^b u_n(t)dt = \int_a^b \mathbf{K} \cdot \mathbf{u}(t)dt$

57. $\mathbf{u}'(t) = \mathbf{v}'(t) \Rightarrow u_i'(t) = v_i'(t), i = 1, 2, \ldots, n$ Since the derivatives of $u_i(t)$ and $v_i(t)$ are equal, $u_i(t)$ and $v_i(t)$ can differ only by a constant,
  say C_i. Thus, $\langle u_1(t), u_2(t), \ldots, u_n(t) \rangle = \langle v_1(t) + C_1, v_2(t) + C_2, \ldots, v_n(t) + C_n \rangle$. $\mathbf{u}(t) = \mathbf{v}(t) + \mathbf{C}$.

59. Parametric equations of the
  tangent line are $x = t + 2$,
  $y = 4t + 4, y = 12t + 8$

61. $\left(\frac{3}{2}, 2\sqrt{3}, 0 \right)$

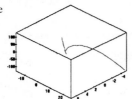

Exercise Set 12.3

1. $2\sqrt{14}$ **3.** $4(3\sqrt{6}-\sqrt{2})$ **5.** $\sqrt{5}\pi$ **7.** $\sqrt{65}(t+1)$ **9.** $\frac{3}{2}(t^2+2t)$ **11.** $\frac{1}{27}[(9t^2+4)^{3/2}-8]$

13. $\mathbf{R}(u)=\mathbf{r}(\alpha(u))=\left\langle\sqrt{u},2\sqrt{u},\frac{1}{\sqrt{u}}\right\rangle$, $u>0$; The orientation is unchanged.   **15.** $\mathbf{R}(u)=(1-u)\mathbf{i}+u\mathbf{j}+(1+u)\mathbf{k}$, $\frac{1}{2}\leq u\leq 1$; The orientation is reversed.

17. $\mathbf{R}(s)=\mathbf{r}\left(\frac{s}{a}\right)=\langle a\cos(s/a),a\sin(s/a)\rangle$ **19.** $\mathbf{R}(s)=\mathbf{r}\left(\frac{s}{5}\right)=\left(4\cos\frac{s}{5}\right)\mathbf{i}+3\frac{s}{5}\mathbf{j}+\left(4\sin\frac{s}{5}\right)\mathbf{k}$

21. $L=\frac{1}{27}(44^{3/2}-17^{3/2})\approx 8.2137$; domain of $\mathbf{R}$ is $0\leq u\leq\ln 2$ **23.** $\frac{2}{27}(343-80\sqrt{10})\approx 6.6680$ **25.** $s(t)=\frac{5}{2}(t^2-1)=u$; so $\mathbf{R}(u)=\left\langle\frac{3u}{5}+\frac{3}{2},6+\frac{4u}{5},3\right\rangle$ and $\mathbf{R}'(u)=\left\langle\frac{3}{5},\frac{4}{5},0\right\rangle$ **27.** $\mathbf{R}'(u)=\mathbf{r}'(\alpha(u))\alpha'(u)$ Assume that $\alpha'(u)>0$. Let $I=[a,b]=[\alpha(c),\alpha(d)]$, where $J=[c,d]$. $L=\int_c^d|\mathbf{r}'(\alpha(u))|\,|\alpha'(u)|\,du=$ length of graph of $\mathbf{R}$ over J. Let $t=\alpha(u)$, $dt=\alpha'(u)du$. Then, $L=\int_a^b|\mathbf{r}'(t)|\,dt=$ length of graph of $\mathbf{r}$ over I. If $\alpha'(u)<0$, the orientation of the curve is reversed. Then $I=[\alpha(d),\alpha(c)]=[a,b]$, and the conclusion follows.

29. 8.151998483

Exercise Set 12.4

1. $\mathbf{T}(2)=\left\langle\frac{1}{\sqrt{17}},\frac{4}{\sqrt{17}}\right\rangle$; $\mathbf{N}(2)=\left\langle-\frac{4}{\sqrt{17}},\frac{1}{\sqrt{17}}\right\rangle$ **3.** $\mathbf{T}\left(\frac{\pi}{4}\right)=-\frac{1}{\sqrt{2}}\mathbf{i}+\frac{1}{\sqrt{2}}\mathbf{j}$; $\mathbf{N}\left(\frac{\pi}{4}\right)=-\frac{1}{\sqrt{2}}\mathbf{i}-\frac{1}{\sqrt{2}}\mathbf{j}$

5. $\mathbf{T}\left(\frac{3\pi}{4}\right)=\left\langle-\frac{1}{\sqrt{10}},-\frac{1}{\sqrt{10}},\frac{2}{\sqrt{5}}\right\rangle$; $\mathbf{N}\left(\frac{3\pi}{4}\right)=\left\langle\frac{1}{\sqrt{2}},-\frac{1}{\sqrt{2}},0\right\rangle$ **7.** $\mathbf{T}(1)=\frac{4}{5}\mathbf{i}+\frac{3}{5}\mathbf{j}+0\mathbf{k}$; $\mathbf{N}(1)=-\frac{3}{5}\mathbf{i}+\frac{4}{5}\mathbf{j}+0\mathbf{k}$

9. $\mathbf{T}(0)=\left\langle\frac{1}{3},-\frac{2}{3},\frac{2}{3}\right\rangle$; $\mathbf{N}(0)=\left\langle\frac{2}{3},\frac{2}{3},\frac{1}{3}\right\rangle$ **11.** $\frac{6}{13\sqrt{13}}$ **13.** $\frac{3}{2\sqrt{2}}$ **15.** $\frac{32}{343}$ **17.** $\frac{2}{5\sqrt{5}}$ **19.** $\frac{4}{5\sqrt{5}}$ **21.** $\frac{14}{17\sqrt{17}}$ **23.** $\frac{2}{9}$

25. $\frac{\sqrt{2}}{3}$ **27.** $\left(\frac{7}{4},0\right)$ **29.** $Q=(-2,3)$ **31. (a)** $\mathbf{T}=\langle|\mathbf{T}|\cos\theta,|\mathbf{T}|\sin\theta\rangle=\langle\cos\theta,\sin\theta\rangle$, since $|\mathbf{T}|=1$

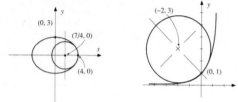

(b) $\kappa=\left|\frac{d\mathbf{T}}{ds}\right|=\left|\frac{d\mathbf{T}}{d\theta}\right|\left|\frac{d\theta}{ds}\right|=|\langle-\sin\theta,\cos\theta\rangle|\left|\frac{d\theta}{ds}\right|=\left|\frac{d\theta}{ds}\right|$ **(c)** $\mathbf{N}=\frac{1}{\kappa}\frac{d\mathbf{T}}{ds}=\frac{1}{|d\theta/ds|}\langle-\sin\theta,\cos\theta\rangle\frac{d\theta}{ds}=\langle-\sin\theta,\cos\theta\rangle\frac{d\theta}{ds}/\left|\frac{d\theta}{ds}\right|=$ $\begin{cases}\langle-\sin\theta,\cos\theta\rangle & \text{if }d\theta/ds>0 \\ \langle\sin\theta,-\cos\theta\rangle & \text{if }d\theta/ds<0\end{cases}$; If $d\theta/ds>0$, C is concave toward the left, and the x-component of $\mathbf{N}$ is negative $(0<\theta<\pi)$. If $d\theta/ds<0$, C is concave toward the right, and again the normal is directed toward the concave side of C.

33. $\mathbf{T}(t)=\left\langle\frac{-ab\sin bt}{\sqrt{a^2b^2+c^2}},\frac{ab\cos bt}{\sqrt{a^2b^2+c^2}},\frac{c}{\sqrt{a^2b^2+c^2}}\right\rangle$; $\mathbf{N}(t)=\frac{\mathbf{T}'(t)}{|\mathbf{T}'(t)|}=\langle-\cos bt,-\sin bt,0\rangle$; $\mathbf{B}=\left\langle\frac{c\sin bt}{\sqrt{a^2b^2+c^2}},\frac{-c\cos bt}{\sqrt{a^2b^2+c^2}},\frac{ab}{\sqrt{a^2b^2+c^2}}\right\rangle$; $\kappa=\frac{|a|b^2}{a^2b^2+c^2}$

35. $\mathbf{r}'\times\mathbf{r}''=|\mathbf{r}'|\kappa\left(\frac{ds}{dt}\right)^2\mathbf{B}$; $\mathbf{r}'=|\mathbf{r}'|\mathbf{Tj}$; So $(\mathbf{r}'\times\mathbf{r}'')\times\mathbf{r}'=|\mathbf{r}'|^2\left(\frac{ds}{dt}\right)^2\kappa\mathbf{N}$; (since $\mathbf{B}\times\mathbf{T}=\mathbf{N}$) $=|\mathbf{r}'|^4\kappa\mathbf{N}$ $\therefore$ $\mathbf{N}=\frac{(\mathbf{r}'\times\mathbf{r}'')\times\mathbf{r}'}{\kappa|\mathbf{r}'|^4}$

37. $\mathbf{B}\times\mathbf{T}=(\mathbf{T}\times\mathbf{N})\times\mathbf{T}=-\mathbf{T}\times(\mathbf{T}\times\mathbf{N})=-(\mathbf{T}\cdot\mathbf{N})\mathbf{T}+(\mathbf{T}\cdot\mathbf{T})\mathbf{N}=0\mathbf{T}+\mathbf{N}=\mathbf{N}$ (Theorem 11.8). $\therefore$ $\mathbf{B}\times\frac{d\mathbf{T}}{ds}+\frac{d\mathbf{B}}{ds}\times\mathbf{T}=\frac{d\mathbf{N}}{ds}$ Since $\mathbf{N}=\frac{1}{\kappa}\frac{d\mathbf{T}}{ds}$, $\frac{d\mathbf{T}}{ds}=\kappa\mathbf{N}$. Also, $\frac{d\mathbf{B}}{ds}=-\tau\mathbf{N}$ Thus, $\frac{d\mathbf{N}}{ds}=\kappa(\mathbf{B}\times\mathbf{N})-\tau(\mathbf{N}\times\mathbf{T})=-\kappa\mathbf{T}-\tau(-\mathbf{B})=\tau\mathbf{B}-\kappa\mathbf{T}$

39. (a)   **(b)** velocity $=\langle\cos(t)-t\sin(t),\sin(t)+t\cos(t),1\rangle$

speed $=\sqrt{(\cos(t)-t\sin(t))^2+(\sin(t)+t\cos(t))^2+1}$; accel $=\langle-2\sin(t)-t\cos(t),2\cos(t)-t\sin(t),0\rangle$

(c) $\kappa=\frac{\sqrt{5t^2+8+t^4}}{(2+t^2)^{3/2}}$ **(d)** $\mathbf{T}=\frac{\langle\cos(t)-t\sin(t),\sin(t)+t\cos(t),1\rangle}{\sqrt{2+t^2}}$;

$\mathbf{N}=(2\langle-2\sin(t)-t\cos(t),2\cos(t)-t\sin(t),0\rangle+\langle-2\sin(t)-t\cos(t),$

$2\cos(t)-t\sin(t),0\rangle t^2-\langle\cos(t)-t\sin(t),\sin(t)+t\cos(t),1\rangle t)/\left(\sqrt{2+t^2}\sqrt{t^4+5t^2+8}\right)$ **(e)** $T_a=\frac{t}{\sqrt{4+t^2}}$; $N_a=\frac{\sqrt{5t^2+8+t^4}}{\sqrt{2+t^2}}$

41. $\mathbf{T} = 2\dfrac{\langle \frac{3}{2}t^2, 2t \rangle}{t\sqrt{9t^2+6}}$; $\mathbf{N} = -\dfrac{1}{6}\dfrac{-9\langle 3t,2\rangle t^3 - 16\langle 3t,2\rangle t + 18\langle \frac{3}{2}t^2, 2t\rangle t^2 + 16\langle \frac{3}{2}t^2, 2t\rangle}{t^2\sqrt{9t^2+16}}$; $\kappa = 3\dfrac{t^2}{\left(\frac{9}{4}t^4+4t^2\right)^{3/2}} \approx .0320019$ at $t = 2$

Exercise Set 12.5

1. $\mathbf{v}\left(\frac{\pi}{4}\right) = \langle -\sqrt{2}, \frac{3\sqrt{2}}{2} \rangle$;

$\mathbf{a}\left(\frac{\pi}{4}\right) = \langle -\sqrt{2}, -\frac{3\sqrt{2}}{2} \rangle$;

$\left|\mathbf{v}\left(\frac{\pi}{4}\right)\right| = \frac{\sqrt{26}}{2}$

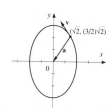

3. $\mathbf{v}(0) = 2\mathbf{i} - 3\mathbf{j}$; $\mathbf{a}(0) = 2\mathbf{i} + 3\mathbf{j}$; $|\mathbf{v}(0)| = \sqrt{13}$

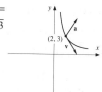

5. $\mathbf{v}(2) = \langle 2, 4 \rangle$; $\mathbf{a}(2) = \langle 0, 2 \rangle$; $|\mathbf{v}(2)| = 2\sqrt{5}$

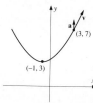

7. $\mathbf{v}(1) = 2\mathbf{i} + 2\mathbf{j} + 3\mathbf{k}$; $\mathbf{a}(1) = 0\mathbf{i} + 2\mathbf{j} + 6\mathbf{k}$; $|\mathbf{v}(1)| = \sqrt{17}$

9. $\mathbf{v}(0) = \langle 1, 2, -1 \rangle$; $\mathbf{a}(0) = \langle 1, 0, 1 \rangle$; $|\mathbf{v}(0)| = \sqrt{6}$

11. $\mathbf{v}(t) = -3(5 - 2t)^{1/2}\mathbf{i} + (t + 2)\mathbf{j}$; $\mathbf{a}(t) = 3(5 - 2t)^{-1/2}\mathbf{i} + \mathbf{j}$; $|\mathbf{v}(t)| = 7 - t$

13. $\mathbf{v}(t) = \langle -t\sin t + \cos t, t\cos t + \sin t, (2t)^{1/2} \rangle$; $\mathbf{a}(t) = \langle -t\cos t - 2\sin t, -t\sin t + 2\cos t, (2t)^{-1/2} \rangle$; $|\mathbf{v}(t)| = t + 1$

15. $\mathbf{v}(t) = \langle -e^t\sin t + e^t\cos t, e^t\cos t + e^t\sin t, e^t \rangle$; $\mathbf{a}(t) = \langle -2e^t\sin t, 2e^t\cos t, e^t \rangle$; $|\mathbf{v}(t)| = \sqrt{3}e^t$ **17. (a)** Force will become $\sqrt{840} \approx 29.0$ N   **(b)** Force will become $\sqrt{612} \approx 24.7$ N   **19.** $h \approx 284$ km   **21.** $y_{max} \approx 4592$ m = 4.60 km; $x_{max} \approx 31,800$ m = 31.8 km

23. $v_0 \approx 256$ m/s   **25.** $\mathbf{v} \cdot \mathbf{a} + \mathbf{a} \cdot \mathbf{v} = 0 \Rightarrow 2\mathbf{a} \cdot \mathbf{v} = 0 \Rightarrow \mathbf{a} \cdot \mathbf{v} = 0 \Rightarrow \mathbf{a} \perp \mathbf{v}$ **27.** Suppose that $\mathbf{a}(t) = \mathbf{0}$. Then, $\mathbf{v}(t) = \mathbf{C}_1$, and $\mathbf{r}(t) = \mathbf{C}_1 t + \mathbf{C}_2$ (let $\mathbf{C}_1 = \mathbf{A}$ and $\mathbf{C}_2 = \mathbf{B}$). Now suppose that $\mathbf{r}(t) = \mathbf{A}t + \mathbf{B}$. Then $\mathbf{v}(t) = \mathbf{A}$, and $\mathbf{a}(t) = \mathbf{0}$. The motion is along a line with direction vector $\mathbf{A}$, and the speed is constant $= |\mathbf{A}|$.   **29.** $x_{max} = (500)(\cos 25°)(47.4) \approx 21,500$ m; $|\mathbf{r}'(t)| \approx 519$ m/s at impact

31. $\mathbf{v}(t) = \langle 2, -3, 1 - \frac{t^2}{2} \rangle$, or $2\mathbf{i} - 3\mathbf{j} + \left(1 - \frac{t^2}{2}\right)\mathbf{k}$; $\mathbf{r}(t) = \langle 2t + 4, -3t + 2, t - \frac{t^3}{6} \rangle$, or $(2t + 4)\mathbf{i} + (-3t + 2)\mathbf{j} + \left(t - \frac{t^3}{6}\right)\mathbf{k}$

33. $\boldsymbol{\omega} \times \mathbf{r}(t) = \begin{vmatrix} \mathbf{i} & \mathbf{j} & \mathbf{k} \\ 0 & 0 & \omega \\ r\cos\omega t & r\sin\omega t & 0 \end{vmatrix} = \langle -r\omega\sin\omega t, r\omega\cos\omega t, 0 \rangle$

$= \mathbf{r}'(t) = \mathbf{v}(t)$

The same result holds if $\mathbf{r}(t) = \langle r\cos\omega t, r\sin\omega t, C \rangle$

35. $a_{\mathbf{T}}(1) = \frac{22}{13}$; $a_{\mathbf{N}} = \frac{6}{\sqrt{13}}$

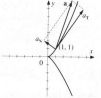

37. $a_{\mathbf{T}}(\ln 2) = \frac{15}{2\sqrt{34}}$; $a_{\mathbf{N}} = \frac{4}{\sqrt{34}}$

39. $a_{\mathbf{T}}(0) = \frac{9}{\sqrt{5}}$; $a_{\mathbf{N}} = \frac{2}{\sqrt{5}}$

41. $\frac{6}{13\sqrt{13}}$ **43.** $\frac{32}{17\sqrt{34}}$ **45.** $\frac{2}{5\sqrt{5}}$ **47.** $a_{\mathbf{T}} = -\frac{2}{t^3}$; $a_{\mathbf{N}} = \frac{2}{t^2}$

49. $a_{\mathbf{T}} = 3$; $a_{\mathbf{N}} = 3\sqrt{t^2 + 3 + \frac{1}{2t}}$ **51.** $a_{\mathbf{T}} = \sqrt{5}$; $a_{\mathbf{N}} = t$ **53.** $\mathbf{T} = \left\langle \frac{2t}{1+2t^2}, \frac{1}{1+2t^2}, \frac{2t^2}{1+2t^2} \right\rangle$; $\mathbf{N} = \left\langle \frac{1-2t^2}{1+2t^2}, \frac{-2t}{1+2t^2}, \frac{2t}{1+2t^2} \right\rangle$; $\kappa = \frac{2t^2}{(1+2t^2)^2}$

55. $\mathbf{T} = \frac{t\cos t + \sin t}{t+1}\mathbf{i} + \frac{\cos t - t\sin t}{t+1}\mathbf{j} + \frac{(2t)^{1/2}}{t+1}\mathbf{k}$; $\mathbf{N} = \frac{-t^2\sin t + t(\cos t - \sin t) + 2\cos t - \sin t}{d}\mathbf{i} - \frac{t^2\cos t + t(\sin t + \cos t) + 2\sin t + \cos t}{d}\mathbf{j} + \frac{(1-t)/(2t)^{1/2}}{d}\mathbf{k}$, where $d = (t + 1)\sqrt{t^2 + 3 + 1/(2t)}$; $\kappa = \frac{\sqrt{2t^3 + 6t + 1}}{3(t+1)^2\sqrt{2t}}$ **57.** $\mathbf{T} = \left\langle \frac{\sin t}{\sqrt{5}}, \frac{\cos t}{\sqrt{5}}, \frac{2}{\sqrt{5}} \right\rangle$; $\mathbf{N} = \langle \cos t, -\sin t, 0 \rangle$; $\kappa = \frac{1}{5t}$

59. $a_N = \kappa \left(\frac{ds}{dt}\right)^2 = \frac{|\mathbf{v} \times \mathbf{a}|}{|\mathbf{v}|}$. Then, $\kappa = \frac{a_N}{(ds/dt)^2} = \frac{|\mathbf{v} \times \mathbf{a}|}{|\mathbf{v}||\mathbf{v}|^2} = \frac{|\mathbf{v} \times \mathbf{a}|}{|\mathbf{v}|^3}$ **61.** $\mathbf{F} = k\mathbf{N} = m\mathbf{a}$, or $\mathbf{a} = \frac{k}{m}\mathbf{N}$ $\therefore a_T = \frac{d^2s}{dt^2} = 0$, and

$a_N = \frac{k}{m}$; $\frac{d^2s}{dt^2} = 0 \Rightarrow \frac{ds}{dt} = v = C$ **63. (a)** $\mathbf{v} = \frac{d\mathbf{r}}{dt} = r\frac{d\mathbf{u}_r}{dt} + \frac{dr}{dt}\mathbf{u}_r = r\mathbf{u}_\theta\frac{d\theta}{dt} + \frac{dr}{dt}\mathbf{u}_r$ **(b)** $|\mathbf{v}| = \left|\frac{d\mathbf{r}}{dt}\right| = \sqrt{r^2\left(\frac{d\theta}{dt}\right)^2 + \left(\frac{dr}{dt}\right)^2}$, since

$|\mathbf{u}_\theta| = |\mathbf{u}_r| = 1$ **65. (a)** $\text{Comp}_{\mathbf{u}_r}\mathbf{a} = \frac{\mathbf{a} \cdot \mathbf{u}_r}{|\mathbf{u}_r|} = \mathbf{a} \cdot \mathbf{u}_r$ From Exercise 64, since $\mathbf{u}_r \cdot \mathbf{u}_r = 1$ and $\mathbf{u}_r \cdot \mathbf{u}_\theta = 0$, $\mathbf{a} \cdot \mathbf{u}_r = \frac{d^2\mathbf{r}}{dt^2} - \mathbf{r}\left(\frac{d\theta}{dt}\right)^2$

(b) Similarly, $\text{Comp}_{\mathbf{u}_\theta}\mathbf{a} = \mathbf{a} \cdot \mathbf{u}_\theta = r\frac{d^2\theta}{dt^2} + 2\frac{dr}{dt}\frac{d\theta}{dt}$

67. $\mathbf{v} = (-2t\sin t^2)\mathbf{u}_r + (2t + 2t\cos t^2)\mathbf{u}_\theta$; $\mathbf{a} = -2(\sin t^2 + 2t^2 + 4t^2\cos t^2)\mathbf{u}_r + 2(1 + \cos t^2 - 4t^2\sin t^2)\mathbf{u}_\theta$

69. Maximum height = 2040.816327; Range =   **71.** Maximum height = 7485.001135; Range =
14139.19027 The speed at impact is equal to   25122.64675
the initial speed.

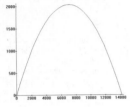

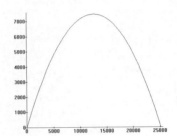

Exercise Set 12.6

1. $\dfrac{\left(x + \frac{pe^2}{1-e^2}\right)^2}{\left[\frac{e^2p^2}{(1-e^2)^2}\right]} + \dfrac{y^2}{\left(\frac{e^2p^2}{1-e^2}\right)} = 1$ with $a^2 = \frac{e^2p^2}{(1-e^2)^2}$ and $b^2 = \frac{e^2p^2}{1-e^2}$. **3. (a)** $\mathbf{u}_r \cdot \mathbf{u}_\theta = \langle\cos\theta, \sin\theta, 0\rangle \cdot \langle-\sin\theta, \cos\theta, 0\rangle = 0$

(b) $\mathbf{u}_r \times \mathbf{u}_\theta = \langle 0, 0, 1\rangle = \mathbf{k}$ **(c)** $\mathbf{r} = r\mathbf{u}_r$; $\mathbf{v} = \frac{d\mathbf{r}}{dt} = r\frac{d\mathbf{u}_r}{dt} + \frac{dr}{dt}\mathbf{u}_r$; $\frac{d\mathbf{u}_r}{dt} = \mathbf{u}_\theta\frac{d\theta}{dt}$ and $\frac{d\mathbf{u}_\theta}{dt} = -\mathbf{u}_r\frac{d\theta}{dt}$ $\therefore$ $\mathbf{v} = \frac{dr}{dt}\mathbf{u}_r + r\frac{d\theta}{dt}\mathbf{u}_\theta$

5. (a) We know that $\mathbf{c} = \mathbf{r} \times \mathbf{v}$ is directed along the positive z-axis (by choice). Hence, $\mathbf{v} \times \mathbf{c}$ is perpendicular to the z-axis. $\therefore$ $\mathbf{v} \times \mathbf{c}$ lies in the xy-plane. **(b)** Since $\mathbf{d} = \mathbf{v} \times \mathbf{c} - GM\mathbf{u}_r$ (by Equation 12.37) and both $\mathbf{v} \times \mathbf{c}$ and $\mathbf{u}_r$ lie in the xy-plane, so also does $\mathbf{d}$.

7. (a) $\frac{dA}{dt} = \frac{c}{2}$; $\therefore A(t) = \frac{c}{2}t + C_1$. Since $A(0) = 0, C_1 = 0$. Let $A = A(T)$ be the area swept out in one revolution. Then,

$A = \frac{1}{2}cT$. **(b)** $b = a\sqrt{1 - e^2}$; hence, $\pi a^2\sqrt{1 - e^2} = \frac{1}{2}cT$. $T = \frac{2\pi a^2}{c}\sqrt{1 - e^2}$. **(c)** $a = \frac{ep}{1 - e^2} \Rightarrow 1 - e^2 = \frac{ep}{a}$; thus,

$T^2 = \frac{4\pi^2 a^4}{c^2}\left(\frac{ep}{a}\right) = \frac{4\pi^2 ep}{c^2}a^3$. Since $ep = \frac{c^2}{GM}$, we can also write $T^2 = \frac{4\pi^2}{GM}a^3$ **9. (a)** $r_0 = $ minimum distance of planet from

the sun when $\theta = 0$; $r_0 = \frac{ep}{1 + e\cos 0} = \frac{ep}{1+e}$ **(b)** $|\mathbf{v} \times \mathbf{c}| = |\mathbf{v}||\mathbf{c}|\sin\frac{\pi}{2} = |\mathbf{v}||\mathbf{c}| = vc$, since the angle between $\mathbf{v}$ and $\mathbf{c}$ is $\frac{\pi}{2}$. Also,

$\mathbf{v} \times \mathbf{c} = GM\mathbf{u}_r + \mathbf{d}$. At $\theta = 0$, $\mathbf{u}_r$ and $\mathbf{d}$ are collinear, and $|GM\mathbf{u}_r + \mathbf{d}| = GM + d$. Thus, $v_0 = \frac{(GM+d)}{c}$. **(c)** Since $d = GMe$, we can

write $v_0 = \frac{GM(1+e)}{c} = c\left(\frac{GM}{c^2}\right)(1+e) = \frac{c(1+e)}{ep} = \frac{c}{r_0}$ $\left(\text{since } ep = \frac{c^2}{GM}\right)$

Chapter Review

1. (a) $t^3\ln t - 2t^3 + 3 - 2t - t^2$ **(b)** $\langle e^{2t}, e^{3t}, 1 - e^t\rangle$ **(c)** $\langle t^2 - t\ln t, t^3 + 2, -2 - 2t\rangle$

(d) $\langle 0, -2, 4\rangle$ **(e)** $\langle 3, -3, -3\rangle$ **3. (a)** $\left\langle\sin^{-1}t, -\sqrt{1 - t^2}, \ln\frac{\sqrt{1+t}}{1-t}\right\rangle + \langle C_1, C_2, C_3\rangle$

(b) $\left(\frac{\pi}{6} + \frac{\sqrt{3}}{8}\right)\mathbf{i} - \frac{5}{24}\mathbf{j} + \left(\sqrt{3} - \frac{\pi}{3}\right)\mathbf{k}$ **5. (a)** $4(\ln 2)\mathbf{i} + \mathbf{j} + 4\mathbf{k}$ **(b)** 5 **7.** 30 **9. (a)** 6

(b) $\mathbf{R}(u) = \left\langle\ln\frac{1+u^2}{2u}, 2\tan^{-1}\frac{1}{u}, -\sqrt{3}\ln u\right\rangle$; $e \geq u \geq e^{-2}$ **(c)** 6 **11.** $\mathbf{T} = \left\langle\frac{2}{3}[\cos u(s) + \sin u(s)], \frac{2}{3}[-\sin u(s) + \cos u(s)], \frac{1}{3}\right\rangle$ with

$u(s) = \ln\left(\frac{1+s}{3}\right)$; $\mathbf{N} = \left\langle\frac{\cos u(s) - \sin u(s)}{\sqrt{2}}, \frac{-\cos u(s) - \sin u(s)}{\sqrt{2}}, 0\right\rangle$; $\kappa = \frac{2\sqrt{2}}{3(s+3)}$; $\mathbf{T}(0) = \left\langle\frac{2}{3}, \frac{2}{3}, \frac{1}{3}\right\rangle$, $\mathbf{N}(0) = \left\langle\frac{1}{\sqrt{2}}, -\frac{1}{\sqrt{2}}, 0\right\rangle$, $\kappa(0) = \frac{2\sqrt{2}}{9}$

13. $\mathbf{v}(t) = \left\langle e^{-t}(2\cos 2t - \sin 2t), e^{-t}(-2\sin 2t - \cos 2t), -2e^{-t}\right\rangle$; $\mathbf{a}(t) = \left\langle e^{-t}(-3\sin 2t - 4\cos 2t), e^{-t}(-3\cos 2t + 4\sin 2t), 2e^{-t}\right\rangle$; $\frac{ds}{dt} = 3e^{-t}$

15. (a) $v \approx 28, 100$ km/hr **(b)** $h \approx 1670$ km; $v = \frac{2\pi(8110)}{2} \approx 25, 500$ km/hr

17. (a) $\kappa = |\sin x|$ **(b)** $\kappa = \frac{8}{5\sqrt{10}}$ **19.** $\mathbf{a}(t) = \left\langle\frac{-1}{(t+1)^2}, 0, 2\right\rangle$; $a_T = \frac{-1}{(t+1)^2} + 2$; $a_N = \frac{2}{t+1}$ **21.** The greatest distance occurs when

$\theta = \pi$. Then, $\mathbf{u}_r = -\mathbf{i}$ and $\mathbf{d} = d\mathbf{i}$. $r_m = \frac{ep}{1-e}$; $v_m = \frac{|\mathbf{v} \times \mathbf{c}|}{c} = \frac{|GM\mathbf{u}_r + \mathbf{d}|}{c} = \frac{|(-GM+d)\mathbf{i}|}{c} = \frac{GM-d}{c} = \frac{GM-GMe}{c} = \frac{GM}{c}(1-e)$; $c = \sqrt{GMep} = $

$\sqrt{GMa(1 - e^2)}$; and $v_m = \frac{GM(1-e)}{\sqrt{GMa(1-e^2)}} = \sqrt{\frac{GM}{a}}\sqrt{\frac{1-e}{1+e}} = \sqrt{\frac{4\pi^2 a^3}{T^2 a}}\sqrt{\frac{1-e}{1+e}} = \frac{2\pi a}{T}\sqrt{\frac{1-e}{1+e}}$; $v_m \approx 65, 620$ mi/hr.

CHAPTER 13

Exercise Set 13.1

1. (a) $\frac{3}{2}$
 (b) 0
 (c) $\frac{3}{22}$
 (d) $\frac{1}{4}$

3. (a) 1
 (b) 3
 (c) 2
 (d) 0

5. (a) 0   **(b)** a
 (c) $\frac{x^2+2x\Delta x+(\Delta x)^2-y}{x+\Delta x-y^2}$
 (d) $\frac{x^2-y-\Delta y}{x-y^2-2y\Delta y-(\Delta y)^2}$

7. (a) $\frac{2}{3(-3+h)}$
 (b) $\frac{-4}{3(3+k)}$

9. $1+\tan 2t$

11. (a) domain $=\{(x,y):y\neq x^2\}$

(b) domain $=\{(x,y):y\neq \pm x\}$

13. domain $=\{(x,y):xy\geq 0\}$

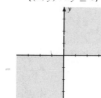

15. domain $=\{(x,y):x^2+y^2\neq 0\}$

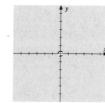

17. domain $=\{(x,y):x^2>y^2\}$

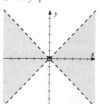

19. domain $=\{(x,y):x^2>2y\}$

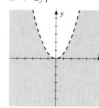

21. domain $=\{(x,y):xy>0\}$

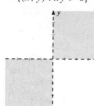

23. domain $=\{(x,y,z):x\neq 1,y\neq -2,z\neq 3\}=$ all of $\mathbb{R}^3$ except the planes $x=1,y=-2,$ and $z=3$

25. domain $=\{(x,y,z);x+y-z>0\}=$ all points in $\mathbb{R}^3$ below the plane $z=x+y$

27. $x^2+2x\Delta x+(\Delta x)^2-3(xy+x\Delta y+y\Delta x+\Delta x\Delta y)+2(y^2+2y\Delta y+(\Delta y)^2)=x^2-3xy+2y^2+(2x-3y)\Delta x+(4y-3x)\Delta y+(\Delta x-3\Delta y)\Delta x+(2\Delta y)\Delta y$
Then, $\varepsilon_1=\Delta x-3\Delta y$ and $\varepsilon_2=2\Delta y.$ We can take $g(x,y)=x-3y,h(x,y)=2y$

29. domain $=\{(x,y):|x|+|y|<1\}$

31. $C(r,h,p)=\pi rp(5r+2h)$

33.

35.

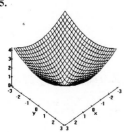

37.

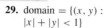

39.

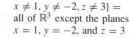

41.

43.

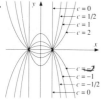

45.

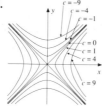

47.

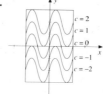

49. Circles of radius $\frac{k}{c}$. $x^2 + y^2 = \frac{625}{9}$ or circles of radius $\frac{25}{3}$.

51. $p = \frac{dv^2}{k}$.

53.

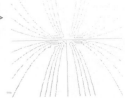

55. (a)

(b)

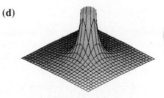

(c)

(d)

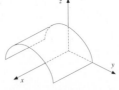

(e)

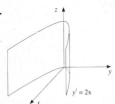

Exercise Set 13.2

1.

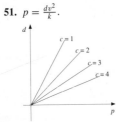

$y^2 = 2x$

3.

$y = 4 - x^2$

5.

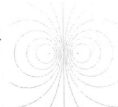

7.

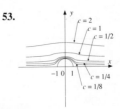

9.

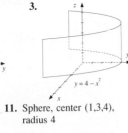

11. Sphere, center $(1,3,4)$, radius 4

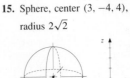

$(1, 3, 4)$

13. Degenerate sphere: the point $(-5, 1, -3)$

15. Sphere, center $(3, -4, 4)$, radius $2\sqrt{2}$

$(3, -4, 4)$

17. Elliptic hyperboloid of one sheet

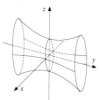

19. Elliptic cone

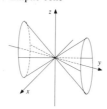

21. Elliptic paraboloid

23. One nappe of an elliptic cone

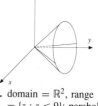

25. domain $= \mathbb{R}^2$, range $= \mathbb{R}$; plane

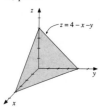

27. domain $= \mathbb{R}^2$, range $= \mathbb{R}$; plane

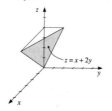

29. domain $= \{(x, y) : x \geq 0\}$, range $= \{z : z \geq 0\}$; half a parabolic cylinder

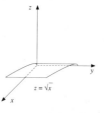

31. domain $= \mathbb{R}^2$, range $= \{z : z \leq 9\}$; parabolic cylinder

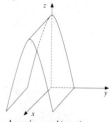

33. domain $= \mathbb{R}^2$, range $= \{z : z \geq 0\}$; paraboloid

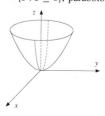

35. domain $= \{(x, y) : 4x^2 - 3y^2 \leq 12\}$, range $= \{z : z \geq 2\sqrt{3}\}$; upper half of a hyperboloid of one sheet

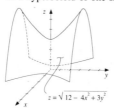

37. domain $= \mathbb{R}^2$, range $= \{z : z \geq 1\}$; paraboloid of revolution

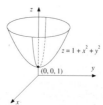

39. domain $= \{(x, y) : x^2 + y^2 \geq 4\}$, range $= \{z : z \geq 0\}$; upper half of a hyperboloid of one sheet

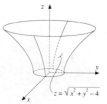

41. domain $= \{(x, y) : x^2 \leq 4y\}$, range $= \{z : z \geq 0\}$; elliptic paraboloid, upper half

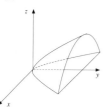

43. domain $= \mathbb{R}^2$, range $= \mathbb{R}$; hyperbolic paraboloid

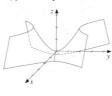

45. $\dfrac{(z+2)^2}{4} - \dfrac{(x-1)^2}{2} - \dfrac{(y-2)^2}{8} = 1$; hyperboloid of two sheets

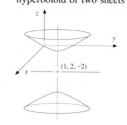

47. $4(x-3)^2 + (y-2)^2 = 4(z+5)$; elliptic paraboloid

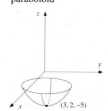

49. $z - 4 = \dfrac{(y+2)^2}{4} - \dfrac{(x-2)^2}{6}$; hyperbolic paraboloid

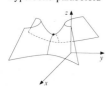

51.

53.

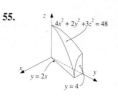

55.

Exercise Set 13.3

1. 13   **3.** -8 **5.** 3   **7.** 1   **9.** 0   **11.** 0   **13.** If $y = x$, the limit is $-\frac{1}{2}$, but if $y = 2x$, the limit is -2 **15.** Along the x-axis $f(x, y) = f(x, 0) = \frac{x}{x^2} = \frac{1}{x}$ if $x \neq 0$, and $\lim\limits_{x \to 0} f(x, 0)$ does not exist.   **17.** Consider $x = 0$ and $x = y^2$. The respective limits are 0 and $\frac{1}{2}$.   **19.** Continuous on all of $\mathbb{R}^2$ **21.** Continuous on $\mathbb{R}^2 - \{(x, y) : y = x\}$ **23.** Continuous on $\mathbb{R}^2 - \{(x, y) : y = -x\}$

25. Continuous on $\mathbb{R}^2 - \{(x, y) : |y| = |x|\}$ **27.** Continuous on $\mathbb{R}^2$ **29.** Continuous on $\mathbb{R}^2 - \{(0, 0)\}$

31. f is continuous at $(0, 0)$ **33.** f is continuous at $(0, 0)$ **35.** f is continuous at $(0, 0)$ **37.** Let $\varepsilon > 0$. Let $P = (x, y)$.

Choose δ_1 and δ_2 such that $|f(P) - L| < \frac{\varepsilon}{2}$ if $0 < |\overrightarrow{P_0 P}| < \delta_1$ and $|g(P) - M| < \frac{\varepsilon}{2}$ if $0 < |\overrightarrow{P_0 P}| < \delta_2$. Let $\delta = \min(\delta_1, \delta_2)$. Then,

if $0 < |\overrightarrow{P_0 P}| < \delta, |(f + g)(P) - (L + M)| = |(f(P) - L) - (g(P) - M)| \le |f(P) - L| + |g(P) - M| < \frac{\varepsilon}{2} + \frac{\varepsilon}{2} = \varepsilon$. Thus,

$\lim_{P \to P_0} (f + g)(P) = L + M$.   **39.**

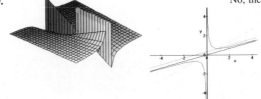

No, the limit does not exist.

Exercise Set 13.4

1. $f_x = 2x, f_y = 2y$ **3.** $f_x = y \cos xy, f_y = x \cos xy$ **5.** $f_x = \frac{-2y}{(x-y)^2}, f_y = \frac{2x}{(x-y)^2}$ **7.** $f_x = \ln y - \frac{y}{x}, f_y = \frac{x}{y} - \ln x$

9. $f_x = \frac{-y^3}{(x^2-y^2)^{3/2}}, f_y = \frac{x^3}{(x^2-y^2)^{3/2}}$ **11.** $\frac{\partial z}{\partial x} = \frac{x}{(1-x^2-y^2)^{3/2}}, \frac{\partial z}{\partial y} = \frac{y}{(1-x^2-y^2)^{3/2}}$ **13.** $\frac{\partial z}{\partial x} = -\tan(x - y), \frac{\partial z}{\partial y} = \tan(x - y)$

15. $\frac{\partial z}{\partial x} = \frac{-y}{\sqrt{1-x^2 y^2}}, \frac{\partial z}{\partial y} = \frac{-x}{\sqrt{1-x^2 y^2}}$ **17.** $f_x(3, -2) = -12, f_y(3, -2) = 17$ **19.** $f_r\left(0, \frac{\pi}{3}\right) = \frac{1}{2}, f_\theta\left(0, \frac{\pi}{3}\right) = -\frac{\sqrt{3}}{2}$

21. $\frac{\partial w}{\partial s}\Big|_{(3,-2)} = -\frac{1}{2}, \frac{\partial w}{\partial t}\Big|_{(3,-2)} = -1$ **23.** $\frac{\partial z}{\partial x} = \frac{-y}{x^2+y^2}, \frac{\partial z}{\partial y} = \frac{x}{x^2+y^2}; x\frac{\partial z}{\partial y} - y\frac{\partial z}{\partial x} = \frac{x^2+y^2}{x^2+y^2} = 1$ **25.** $f_x = 4x - 3y^2, f_y = -6xy + 9y^2;$

$f_x(3, 2) = 12 - 12 = 0 =$ slope of tangent to C_1 at $(3, 2, 6); f_y(3, 2) = -6(6) + 9(4) = 0 =$ slope of tangent to C_2 at $(3, 2, 6)$

27. $f_{xx} = \frac{2(3y^2-x^2)}{(x^2+3y^2)^2}, f_{yy} = \frac{6(x^2-3y^2)}{(x^2+3y^2)^2}, f_{xy} = f_{yx} = \frac{-12xy}{(x^2+3y^2)^2}$ **29.** $f_{xx} = 2ye^{x^2 y}(2x^2 y + 1), f_{yy} = x^4 e^{x^2 y}, f_{xy} = f_{yx} = 2xe^{x^2 y}(x^2 y + 1)$

31. $f_{xx} = e^{-x} \sin y - e^{-y} \cos x, f_{yy} = -e^{-x} \sin y + e^{-y} \cos x, f_{xy} = f_{yx} = -e^{-x} \cos y + e^{-y} \sin x$

33. $u_{xx} = -18(\cos 3x)e^{-9kt}, u_t = -18k(\cos 3x)e^{-9kt} = ku_{xx}$ **35.** $y_{xx} = -\sin x \cos at, y_{tt} = -a^2 \sin x \cos at = a^2 y_{xx}$

37. $u_{xx} = -\cos x \cosh y, u_{yy} = \cos x \cosh y, u_{xx} + u_{yy} = 0, u$ is harmonic   **39.** $u_{xx} = \frac{2xy}{(x^2+y^2)^2}, u_{yy} = \frac{-2xy}{(x^2+y^2)^2}, u_{xx} + u_{yy} = 0 \Rightarrow u$

harmonic   **41.** v is harmonic since $v_{xx} = \sin x \sinh y, v_{yy} = -\sin x \sinh y; v_{xx} + v_{yy} = 0. u_x = -\sin x \cosh y = v_y$ and

$u_y = \cos x \sinh y = -v_x; u$ and v are harmonic conjugates.   **43.** In Exercise 42, $\tan^{-1} \frac{y}{x}$ and $\ln \sqrt{x^2 + y^2}$ were proved to be harmonic. conjugates. Here, the sign of one of the functions has been changed. Suppose that $u_x = v_y$ and $u_y = -v_x$, with the original function v. If v is replaced by $w = -v$, we have $u_x = v_y = -w_y$ and $u_y = -v_x = w_x$. But, these equations can be written as $w_x = u_y$ and $w_y = -u_x. \therefore -v$ and u are harmonic conjugates if u and v are.

45. (a) $u_{xx} = 2; u_{yy} = -2; u_{xx} + u_{yy} = 0 \Rightarrow u$ harmonic.

$v_{xx} = 0; v_{yy} = 0; v_{xx} + v_{yy} = 0 \Rightarrow v$ harmonic.

(b) $u_x = v_y$ and $u_y = -v_x \Rightarrow u$ and v are harmonic conjugates

(c) Level curves for u: $x^2 - y^2 + 2x + 1 = c_1$;

Level curves for v: $2xy + 2y = c_2$;

slope of $u(x, y) = m_1 = \frac{x+1}{y} (y \ne 0)$,

slope of $v(x, y) = m_2 = \frac{-y}{x+1} (x \ne -1)$

$m_1 m_2 = -1 \Rightarrow$ level curves of $u \perp$ level curves of v (excluding the lines $x = -1$ and $y = 0$)

(d)

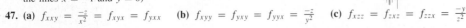

47. (a) $f_{xxy} = \frac{-z}{x^2} = f_{xyx} = f_{yxx}$ **(b)** $f_{xyy} = f_{yxy} = f_{yyx} = \frac{-z}{y^2}$ **(c)** $f_{xzz} = f_{zxz} = f_{zzx} = \frac{-y}{z^2}$

(d) $f_{xyz} = f_{yxz} = f_{xzy} = f_{zxy} = f_{yzx} = f_{zyx} = \frac{1}{z} + \frac{1}{y} + \frac{1}{x}$ **49. (a)** \$20   **(b)** \$30   **(c)** \$2

51. $Q_{KK} = \alpha(\alpha - 1)AK^{\alpha-2}L^{1-\alpha}; Q_{LL} = (1 - \alpha)(-\alpha)AK^\alpha L^{-\alpha-1}; K^2 Q_{KK} = \alpha(\alpha - 1)AK^\alpha L^{1-\alpha} = L^2 Q_{LL}$

53. $f_x(0, 0) = \lim_{h \to 0} \frac{f(h,0)-f(0,0)}{h} = \lim_{h \to 0} \frac{h^3/h^2-0}{h} = 1; f_y(0, 0) = \lim_{k \to 0} \frac{f(0,k)-f(0,0)}{k} = \lim_{k \to 0} \frac{-k^3/k^2-0}{k} = -1$

55. $f_x = \lim_{h \to 0} \frac{\frac{hy(h^2-y^2)}{h^2+y^2}}{h} = \lim_{h \to 0} \frac{y(h^2-y^2)}{h^2+y^2} = -y; f_{xy}(0, 0) = -1; f_y = \lim_{k \to 0} \frac{\frac{xk(x^2-k^2)}{x^2+k^2}}{k} = \lim_{k \to 0} \frac{x(x^2-k^2)}{x^2+k^2} = x; f_{yx}(0, 0) = 1$

57. (a) slope is 2   **(b)** slope is 2

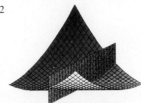

Chapter Review

1. (a) domain $= \{(x, y) : x^2 + y^2 > 1\}$ **(b)** domain $= \{(x, y): x^2 > y^2\}$

3. (a) parabolic cylinder   **(b)** sphere

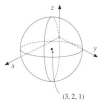

$(3, 2, 1)$

5. (a) hyperbolic paraboloid   **(b)** cylinder

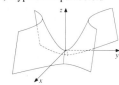

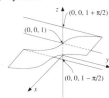

$(0, 0, 1 + \pi/2)$

$(0, 0, 1)$

$(0, 0, 1 - \pi/2)$

7. (a) domain $= \mathbb{R}^2$, range $= \{z : z \geq 2\}$, half a hyperboloid of two sheets

$(0, 0, 2)$ y

(b) domain $= \mathbb{R}^2$, range $= \{z : z \leq 4\}$, paraboloid of revolution

$(0, 0, 2)$ y

9. $T_y(1, 3) = -300e^{-5}$, decreasing

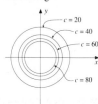

$c = 20$
$c = 40$
$c = 60$
$c = 80$

11. (a) $f_x = \frac{1}{x}\sinh\frac{2x}{y} - \frac{y}{x^2}\cosh^2\frac{x}{y}$,

$f_y = -\frac{1}{y}\sinh\frac{2x}{y} + \frac{1}{x}\cosh^2\frac{x}{y}$

(b) $f_x = \frac{1}{\sqrt{2x(y^2 - 2x)}}$, $f_y =$

$\frac{2x}{y\sqrt{2x(y^2 - 2x)}}$

13. $u_{xx} = -\sin x \cosh y$, $u_{yy} = \sin x \cosh y$, and $u_{xx} + u_{yy} = 0 \Rightarrow u$ harmonic
$v_{xx} = -\cos x \sinh y$, $v_{yy} =$ $\cos x \sinh y$, $v_{xx} + v_{yy} = 0 \Rightarrow v$ harmonic $u_x = \cos x \cosh y = v_y$ and $u_y = \sin x \sinh y = -v_x \Rightarrow u$ and v harmonic conjugates

15. (a) $\frac{\partial^2 w}{\partial z^2} = \frac{-x^2 - y^2}{(z^2 - x^2 - y^2)^{3/2}}$ **(b)** $\frac{\partial^2 w}{\partial x \partial z} = \frac{xz}{(z^2 - x^2 - y^2)^{3/2}}$ **(c)** $\frac{\partial^2 w}{\partial y \partial z} = \frac{yz}{(z^2 - x^2 - y^2)^{3/2}}$ **(d)** $\frac{\partial^2 w}{\partial y \partial x} = \frac{-xy}{(z^2 - x^2 - y^2)^{3/2}}$

17. $|f(x, y) - 0| \leq \frac{x^2|x + 3y|}{x^2 + 2y^2} \leq \frac{(x^2 + y^2)(|x| + 3|y|)}{(x^2 + 2y^2)} \leq 4\sqrt{x^2 + y^2} < \varepsilon$ if $\sqrt{x^2 + y^2} < \frac{\varepsilon}{4}$. Thus, $\lim\limits_{(x, y) \to (0,0)} f(x, y) = 0$. Assigning 0 to $f(0, 0)$.

19. (a) $f(x, 0) = \frac{x^3}{3x^3} = \frac{1}{3}$ if $x \neq 0$; $f(0, y) = \frac{-2y^3}{4y^3} = -\frac{1}{2}$ if $y \neq 0$ **(b)** $f(x, 0) = 0$, $f(y^2, y) = \frac{3}{7}$ if $y \neq 0$

21. $\frac{\partial P}{\partial V} = -\frac{nRT}{V^2}$; $\frac{\partial V}{\partial T} = \frac{nR}{P}$; $\frac{\partial T}{\partial P} = \frac{v}{nR}$; $\frac{\partial P}{\partial V}\frac{\partial V}{\partial T}\frac{\partial T}{\partial P} = -1$ **23. (a)** $y_{xx} = -\frac{2a^2t^2 + 2x^2}{(x^2 - a^2t^2)^2}$, $y_{tt} = \frac{-2a^2(x^2 - a^2t^2) - 4a^4t^2}{(x^2 - a^2t^2)^2} = a^2 y_{xx}$

(b) $y_{xx} = Ae^x e^{at} + Be^x e^{-at}$; $y_{tt} = Aa^2 e^x e^{at} + Ba^2 e^x e^{-at} = a^2 y_{xx}$ **25.** $Q_K(240, 150) = 18\left(\frac{150}{240}\right)^{0.4} \approx 14.92$, $Q_L(240, 150) =$

$12\left(\frac{240}{150}\right)^{0.6} \approx 15.91$; Q_K represents the marginal change in Q for a unit change in K. Q_L represents the marginal change in Q

for a unit change in L.   **27. (a)** $f_x(0, 0) = \lim\limits_{h \to 0} \frac{2h^3/h^2}{h} = 2$ **(b)** $f_x(x, y) = \frac{(x^2 + 2y^2)(6x^2) - (2x^3 - y^3)(2x)}{(x^2 + 2y^2)^2} = \frac{2x^4 + 12x^2y^2 + 2xy^3}{(x^2 + 2y^2)^2}$

for $(x, y) \neq (0, 0)$ **(c)** $\lim\limits_{(x, 0) \to (0,0)} f_x(x, 0) = 2$, but $\lim\limits_{(x, x) \to (0,0)} f_x(x, x) = \frac{16}{9}$ ∴ $\lim\limits_{(x, y) \to (0,0)} f_x(x, y)$ does not exist.

(d) $f_y(0, 0) = \lim\limits_{k \to 0} \frac{-k^3/(2k^2)}{k} = -\frac{1}{2}$; $f_y(x, y) = \frac{(x^2 + 2y^2)(-3y^2) - (2x^3 - y^3)(4y)}{(x^2 + 2y^2)^2} = \frac{-2y^4 - 3x^2y^2 - 8x^3y}{(x^2 + 2y^2)^2}$; $\lim\limits_{(0, y) \to (0,0)} f_y(0, y) = -\frac{1}{2}$ but

$\lim\limits_{(y, y) \to (0,0)} f_y(y, y) = \frac{-13}{9}$ ∴ $\lim\limits_{(x, y) \to (0,0)} f_y(x, y)$ does not exist.   **29.** Suppose that S is open, and let $P \in S$. Then P is not a

boundary point of S, and so some neighborhood of P contains only points of S. Conversely, let S be any nonempty set with the given property. Then no point of S is a boundary point of S, and so S is open.

CHAPTER 14

Exercise Set 14.1

1. $f_x = 3x^2 + 6xy$; $f_y = 3x^2 - 8y^3$; these are continuous on $\mathbb{R}^2$ **3.** $f_x = e^x \sin y$; $f_y = e^x \cos y$; these are continuous on all of $\mathbb{R}^2$

5. $f_x = \frac{1}{x} - \frac{2x}{x^2 + y^2}$; $f_y = \frac{1}{y} - \frac{2y}{x^2 + y^2}$. For $xy > 0$, both f_x and f_y are continuous.   **7.** $(4x - 3y)dx + (3y^3 - 3x)dy$

9. $ye^{xy}dx + xe^{xy}dy$ **11.** $\frac{2y}{(x+y)^2}dx - \frac{2x}{(x+y)^2}dy$ **13.** $\frac{y}{x^2+y^2}dx - \frac{x}{x^2+y^2}dy$ **15.** $\frac{x}{\sqrt{x^2+y^2+z^2}}dx + \frac{y}{\sqrt{x^2+y^2+z^2}}dy + \frac{z}{\sqrt{x^2+y^2+z^2}}dz$

17. $df = 0.58$ **19.** $df = 0.012$ **21.** $df = -0.0044$ **23.** $\Delta V \approx dV = 51.6$ cu ft; $\approx 2.56\%$ **25.** $\Delta T \approx dT = 0.20$; Actual change $= 0.2059$; Approximate percentage change $\approx 25.7\%$ **27.** $\Delta C \approx dC = \$54.10$; Approximate percentage increase in cost $\approx 8.78\%$ **29.** Approximate percentage change $\approx 1.60\%$ **31.** $\Delta f = y\Delta x + x\Delta y + \Delta x\Delta y = f_x\Delta x + f_y\Delta y + E_1\Delta x + E_2\Delta y$, where $E_1 = \Delta y, E_2 = 0$ Then, $E_1 \to 0$ and $E_2 \to 0$ as $(\Delta x, \Delta y) \to (0,0)$ **33.** $\Delta f = [(x+\Delta x) - 3y(y+\Delta y)]^2 - (x-3y)^2$ $= 2(x-3y)\Delta x - 6(x-3y)\Delta y + (\Delta x - 6\Delta y)\Delta x + 9\Delta y(\Delta y) = f_x\Delta x + f_y\Delta y + E_1\Delta x + E_2\Delta y$, where $E_1 = \Delta x - 6\Delta y, E_2 = 9\Delta y$

Then, $E_1 \to 0$ and $E_2 \to 0$ as $(\Delta x, \Delta y) \to (0,0)$ **35.** $f_x(0,0) = \lim\limits_{h\to 0}\frac{f(h,0)-f(0,0)}{h} = \lim\limits_{h\to 0}\frac{0}{h} = 0$ Similarly, $f_y(0,0) = 0$. For

$(x,y) \neq (0,0), f_x(x,y) = \frac{2xy^4}{(x^2+y^2)^2}$ and $f_y(x,y) = \frac{2x^4y}{(x^2+y^2)^2}$. If $(x,y) \neq (0,0)$, then f_x and f_y are continuous. To show continuity at

$(0,0)$, let $\varepsilon > 0$ be given. $|f_x(x,y) - f_x(0,0)| = \frac{|2xy^4|}{(x^2+y^2)^2} \leq \frac{2\sqrt{x^2+y^2}(x^2+y^2)^2}{(x^2+y^2)^2} < \varepsilon$ if $\sqrt{x^2+y^2} < \frac{\varepsilon}{2}$. Hence, f_x is continuous at $(0,0)$; so

is f_y. Thus, f_x and f_y are continuous on all of $\mathbb{R}^2$, and so, by Theorem 14.2, f is differentiable on $\mathbb{R}^2$.

37. (a) $\lim\limits_{(x,y)\to(0,0)} f(x,y) = \lim\limits_{(x,y)\to(0,0)} |\sqrt{|xy|} - 0| = \sqrt{|x||y|} \leq \sqrt{x^2+y^2} < \varepsilon$ if $\sqrt{x^2+y^2} < \delta = \varepsilon$. Since $\lim\limits_{(x,y)\to(0,0)} f(x,y)$ exists and equals

$f(0,0), f$ is continuous at $(0,0)$.   **(b)** $f_x(0,0) = \lim\limits_{h\to 0}\frac{f(h,0)-f(0,0)}{h} = \lim\limits_{h\to 0}\frac{0-0}{h} = 0$ By symmetry $f_y(0,0) = 0$ **(c)** $\Delta f = \sqrt{|\Delta x\Delta y|} - 0$.

Take $\Delta x = \Delta y$, to get $\Delta f = |\Delta x| = 0 + 0 + (E_1 + E_2)\Delta x$. If $\Delta x > 0, E_1 + E_2 = 1$; if $\Delta x < 0, E_1 + E_2 = -1$. Since E_1 and E_2 do not go to zero as $(\Delta x, \Delta y) \to (0,0)$, f is not differentiable at $(0,0)$.

Exercise Set 14.2

1. 56   **3.** 2   **5.** $-(\tan t + \cot t)$ **7.** $1 - \cos t^2 + 2t^2\sin t^2$ **9.** $\frac{\partial z}{\partial u}\big|_{(2,-1)} = -4; \frac{\partial z}{\partial v}\big|_{(2,-1)} = 8$

11. $\frac{\partial z}{\partial u}\big|_{(1,1/2)} = 2e^2; \frac{\partial z}{\partial v}\big|_{(1,1/2)} = 0$ **13.** $\frac{\partial z}{\partial u} = 2v\tan 2uv; \frac{\partial z}{\partial v} = 2u\tan 2uv$ **15.** $\frac{\partial z}{\partial u} = -4u; \frac{\partial z}{\partial v} = 6v$ **17.** $\frac{3x^2-2y^2}{4y(x-y^2)}$

19. $\frac{\sin y - y\cos x}{\sin x - x\cos y}$ **21.** $\frac{2x-x^2y+y^3}{2y-xy^2+x^3}$ **23.** $\frac{\partial z}{\partial x} = \frac{-2xyz}{x^2y+6z}; \frac{\partial z}{\partial y} = \frac{4y-x^2z}{6z+x^2y}$ **25.** $\frac{\partial z}{\partial x} = \frac{2z(1-2xyz-2y^2z)}{(x+y)(1+4xyz)}; \frac{\partial z}{\partial y} = \frac{2z(1-2x^2z-2xyz)}{(x+y)(1+4xyz)}$

27. $\frac{\partial z}{\partial x} = \frac{-z\cos xz}{x\cos xz - y\tan xz}; \frac{\partial z}{\partial y} = \frac{z\tan yz}{x\cos xz - y\tan xz}$ **29.** $\frac{\partial w}{\partial x} = \frac{2yz(y^2+z^2)}{(x^2+y^2+z^2)^{3/2}}; \frac{\partial w}{\partial y} = \frac{2xz(x^2+z^2)}{(x^2+y^2+z^2)^{3/2}}; \frac{\partial w}{\partial z} = \frac{2xy(x^2+y^2)}{(x^2+y^2+z^2)^{3/2}}$

31. $\frac{\partial z}{\partial r} = \frac{\partial z}{\partial x}\cos\theta + \frac{\partial z}{\partial y}\sin\theta; \frac{\partial z}{\partial\theta} = \frac{\partial z}{\partial x}(-r\sin\theta) + \frac{\partial z}{\partial y}(r\cos\theta); \left(\frac{\partial z}{\partial r}\right)^2 + \frac{1}{r^2}\left(\frac{\partial z}{\partial\theta}\right)^2 = \left(\frac{\partial z}{\partial x}\right)^2\cos^2\theta + 2\frac{\partial z}{\partial x}\frac{\partial z}{\partial y}\sin\theta\cos\theta$

$+ \left(\frac{\partial z}{\partial y}\right)^2\sin^2\theta + \left(\frac{\partial z}{\partial x}\right)^2\sin^2\theta - 2\frac{\partial z}{\partial x}\frac{\partial z}{\partial y}\sin\theta\cos\theta + \left(\frac{\partial z}{\partial y}\right)^2\cos^2\theta = \left(\frac{\partial z}{\partial x}\right)^2 + \left(\frac{\partial z}{\partial y}\right)^2$ **33.** 0.68 km^2/hr   **35.** Let

$u = \frac{x}{y}; \frac{\partial z}{\partial x} = f'(u)\left(\frac{1}{y}\right), \frac{\partial z}{\partial y} = f'(u)\left(\frac{-x}{y^2}\right)$, and $x\frac{\partial z}{\partial x} + y\frac{\partial z}{\partial y} = \frac{x}{y}f'(u) - \frac{x}{y}f'(u) = 0$

37. $\frac{\partial z}{\partial u} = f_x g_u + f_y h_u; \frac{\partial z}{\partial v} = f_x g_v + f_y h_v; \frac{\partial^2 z}{\partial u^2} = g_u(f_{xx}g_u + f_{xy}h_u) + f_x g_{uu} + h_u(f_{yx}g_u + f_{yy}h_u) + f_y h_{uu}$

$= f_{xx}g_u^2 + 2f_{xy}g_u h_u + f_{yy}h_u^2 + f_x g_{uu} + f_y h_{uu}$ (assuming that $f_{xy} = f_{yx}$). Similarly, $\frac{\partial^2 z}{\partial v^2} = f_{xx}g_v^2 + 2f_{xy}g_v h_v) + f_{yy}h_v^2 + f_x g_{vv} + f_y h_{vv}$

39. $\frac{\partial^2 z}{\partial\theta\partial r} = \left(\frac{\partial^2 z}{\partial y^2} - \frac{\partial^2 z}{\partial x^2}\right)(r\sin\theta\cos\theta) + \frac{\partial^2 z}{\partial y\partial x}(r\cos^2\theta - r\sin^2\theta) - \frac{\partial z}{\partial x}\sin\theta + \frac{\partial z}{\partial y}\cos\theta$ **41.** $\frac{dh}{dt} = \frac{9}{10\pi}$ ft/min

43. Let $w = f(x,y,z)$, where $x = x(t), y = y(t)$, and $z = z(t)$ Then the chain rule can be written as

$D_1(f)(x(t), y(t), z(t))\left(\frac{\partial}{\partial t}x(t)\right) + D_2(f)(x(t), y(t), z(t))\left(\frac{\partial}{\partial t}y(t)\right) + D_3(f)(x(t), y(t), z(t))\left(\frac{\partial}{\partial t}z(t)\right)$

45. $\frac{\partial f}{\partial u} = D_1(f)(x(u,v), y(u,v))\left(\frac{\partial}{\partial u}\right) + D_2(f)(x(u,v), y(u,v))\left(\frac{\partial}{\partial u}y(u,v)\right) \frac{\partial f}{\partial v} = D_1(f)(x(u,v), y(u,v))\left(\frac{\partial}{\partial v}x(u,v)\right) +$

$D_2(f)(x(u,v), y(u,v))\left(\frac{\partial}{\partial v}y(u,v)\right)$ For the special case: $\frac{\partial f}{\partial u} = e^{(u-3uv+\sin(v))}(u + e^{u^2v}) \frac{\partial f}{\partial v} = e^{(u-3uv+\sin(v))}(u + e^{u^2v})$

Exercise Set 14.3

1. $\frac{27}{5}$ **3.** $-\frac{9}{10\sqrt{2}}$ **5.** $-\frac{7}{5}$ **7.** $\frac{1}{\sqrt{5}}$ **9.** $-\frac{3}{2\sqrt{2}}$ **11.** $\frac{16}{3}$ **13.** $\frac{1}{\sqrt{2}}$ **15.** $\frac{82}{\sqrt{10}}$ **17.** $\mathbf{u} = \left\langle\frac{4}{\sqrt{41}}, -\frac{5}{\sqrt{41}}\right\rangle; \frac{\sqrt{41}}{27}$

19. $\mathbf{u} = \left\langle-\frac{1}{\sqrt{5}}, -\frac{2}{\sqrt{5}}\right\rangle; \sqrt{5}$ **21.** $\mathbf{u} = \langle 1,0\rangle$ or $\mathbf{u} = \left\langle-\frac{15}{17}, \frac{8}{17}\right\rangle$. f increases most rapidly if $\mathbf{u} = \left\langle\frac{1}{\sqrt{17}}, \frac{4}{\sqrt{17}}\right\rangle$. Maximum rate of change

$= 2\sqrt{17}$.   **23.** $-\frac{2}{5\sqrt{5}}; \left\langle\frac{4}{5}, \frac{3}{5}\right\rangle; -1$ **25.** $-\frac{14}{5\sqrt{13}\ln 4}; \mathbf{u} = \left\langle\frac{3}{\sqrt{10}}, \frac{1}{\sqrt{10}}\right\rangle; \frac{2\sqrt{10}}{5\ln 4}$ **27.** $x^2 + y^2 = 4^c$, circles of radii between 1 and 2. y'

at $(x_0, y_0) = \frac{-x_0}{y_0}$, and a vector tangent to the curve is $\mathbf{u} = \left\langle 1, -\frac{x_0}{y_0}\right\rangle$. $\nabla V(x_0, y_0) = \frac{1}{\ln 4}\left\langle\frac{2x_0}{4^c}, \frac{2y_0}{4^c}\right\rangle$, and $\mathbf{u}\cdot\nabla V(x_0, y_0) = \frac{1}{\ln 4}\left(\frac{2x_0-2x_0}{4^c}\right) = 0$.

Hence, the gradient vector is orthogonal to the equipotential curve through (x_0, y_0).   **29.** $f(x,y) = \sin x - x\cos x + \sin y + 2$

31. $\nabla(cu) = \left\langle\frac{\partial}{\partial x}cu, \frac{\partial}{\partial y}cu\right\rangle = \left\langle c\frac{\partial u}{\partial x}, c\frac{\partial u}{\partial y}\right\rangle = c\left\langle\frac{\partial u}{\partial x}, \frac{\partial u}{\partial y}\right\rangle = c\nabla u$ **33.** $\nabla(uv) = \left\langle u\frac{\partial v}{\partial x} + v\frac{\partial u}{\partial x}, u\frac{\partial v}{\partial y} + v\frac{\partial u}{\partial y}\right\rangle = \left\langle u\frac{\partial v}{\partial x}, u\frac{\partial v}{\partial y}\right\rangle + \left\langle v\frac{\partial u}{\partial x}, v\frac{\partial u}{\partial y}\right\rangle =$

$$u\left\langle \tfrac{\partial v}{\partial x}, \tfrac{\partial v}{\partial y}\right\rangle + v\left\langle \tfrac{\partial u}{\partial x}, \tfrac{\partial u}{\partial y}\right\rangle = u\nabla v + v\nabla u \qquad \textbf{35. } \nabla u^{\alpha} = \left\langle \tfrac{\partial}{\partial x}u^{\alpha}, \tfrac{\partial}{\partial y}u^{\alpha}\right\rangle = \left\langle \alpha u^{\alpha-1}\tfrac{\partial u}{\partial x}, \alpha u^{\alpha-1}\tfrac{\partial u}{\partial y}\right\rangle = \alpha u^{\alpha-1}\nabla u$$

37. $\nabla f = \langle 2, -2y\rangle$; the gradient at $(2, -1)$ is $\langle 2, 2\rangle$
The vector equation of the tangent line at $(1, -1)$

is $\mathbf{r}(t) = \left\langle 2 + \tfrac{1}{2}t\sqrt{2}, -1 - \tfrac{1}{2}t\sqrt{2}\right\rangle$

Exercise Set 14.4

1. Tangent plane: $\langle 6, -6, -5\rangle \cdot \langle \mathbf{x} - (3, -2, 5)\rangle = 0$ or $6x - 5y - 5z = 5$; Normal line: $\mathbf{r}(t) = \langle 3 + 6t, -2 - 6t, 5 - 5t\rangle$.
3. Tangent Plane: $\langle 1, 0, -1\rangle \cdot \langle \mathbf{x} - (1, 0, -2)\rangle = 0$, or $x - z = 3$; Normal Line: $\mathbf{r}(t) = \langle 1 + t, 0, -2 - t\rangle$
5. Tangent Plane: $\langle 3, 1, 4\rangle \cdot \langle \mathbf{x} - (1, -1, 2)\rangle = 0$, or $3x + y + 4z = 10$; Normal Line: $\mathbf{r}(t) = \langle 1 + 3t, -1 + t, 2 + 4t\rangle$
7. Tangent Plane: $\langle -1, 0, 2\rangle \cdot \langle \mathbf{x} - (3, -1, 1)\rangle = 0$, or $x - 2z = 1$; Normal Line: $\mathbf{r}(t) = \langle 3 - t, -1, 1 + 2t\rangle$
9. Tangent Plane: $\langle 10, -16, -1\rangle \cdot \langle \mathbf{x} - (5, 4, -7)\rangle = 0$ or $10x - 16y - z + 7 = 0$; Normal Line: $\mathbf{r}(t) = \langle 5 + 10t, 4 - 16t, -7 - t\rangle$
11. Tangent Plane: $\langle 1, 1, -2\rangle \cdot \left\langle \mathbf{x} - \left(1, -1, -\tfrac{\pi}{4}\right)\right\rangle = 0$ or $x + y - 2z = \tfrac{\pi}{2}$; Normal Line: $\mathbf{r}(t) = \langle 1 + t, -1 + t, -\tfrac{\pi}{4} - 2t\rangle$
13. Tangent Plane: $\langle 2, 0, 1\rangle \cdot \left\langle \mathbf{x} - \left(0, \tfrac{\pi}{2}, -1\right)\right\rangle = 0$, or $2x + z + 1 = 0$; Normal Line: $\mathbf{r}(t) = \langle 2t, \tfrac{\pi}{2}, -1 + t\rangle$ **15.** $\left(\tfrac{1}{2}, -\tfrac{3}{4}, -\tfrac{19}{16}\right)$
17. (a) The surfaces have the same tangent plane at $P_0 = (x_0, y_0, z_0)$ if and only if their normal vectors are parallel at P_0, which is equivalent to $\nabla F(P_0) = k\nabla G(P_0)$. Note that neither vector is the zero vector, by assumption.   **(b)** $(2, 3, -1)$ and $(-2, -3, 1)$
19. $(0, 0, 0)$ and $(2, 1, 3)$ **21.** $\approx 6.34°$; $\approx 26.93°$ **23.** $-\tfrac{5}{6} + x + \tfrac{2}{3}y$ **25.** .2938577115 radians
16.83680664°

Exercise Set 14.5

1. Absolute minimum of -2 at $(-1, 2)$ **3.** Absolute minimum of -3 at $(0, 4)$ **5.** Local minimum at $(16, 4)$; saddle point at $(1, -1)$
7. Saddle point at $(0, 0)$; local minimum at $(3, 9)$ **9.** Saddle point at $(0, 0)$; local minimum at $(-1, -1)$; local minimum at $(1, 1)$
11. Saddle point at $(0, 0)$; local maximum at $(18, 6)$ **13.** Saddle point at $(0, 1/\sqrt{3})$; saddle point at $(0, -1/\sqrt{3})$; local minimum at $(-1, 1)$;
local minimum at $(1, 1)$.   **15.** Local minimum at $\left(\tfrac{1}{\sqrt[3]{2}}, \sqrt[3]{4}\right)$ **17.** Local minimum at $\left(-\tfrac{1}{8}, -1\right)$; saddle point at $\left(-\tfrac{1}{8}, 1\right)$
19. No local extrema or saddle points   **21.** Saddle point at $(0, 0)$; local maximum at $(-2, 0)$ **23.** Saddle point at $(0, 0)$; local minima at
$\left(\tfrac{\pi}{2}, -\tfrac{\pi}{2}\right), \left(-\tfrac{\pi}{2}, \tfrac{\pi}{2}\right)$; local maxima at $\left(\tfrac{\pi}{2}, \tfrac{\pi}{2}\right)$ and $\left(-\tfrac{\pi}{2}, -\tfrac{\pi}{2}\right)$ **25.** Local minimum at $(1, 0)$; saddle point at $\left(0, \tfrac{\pi}{2}\right)$; local maximum
at $(-1, \pi)$; saddle point at $\left(0, \tfrac{3\pi}{2}\right)$ **27.** $\nabla f = \left\langle -\tfrac{2}{3}x^{-1/3} + \tfrac{2}{3}x^{-2/3}y^{1/3}, \tfrac{2}{3}x^{1/3}y^{-2/3} - \tfrac{2}{3}y^{-1/3}\right\rangle$ which does not exist at $(0, 0)$.
$f(0, 0) = 4$; for $(x, y) \neq (0, 0)$, $f(x, y) < 4$. $\therefore f$ has an absolute maximum (and local maximum) at $(0, 0)$.   **29.** Absolute maximum of
0; absolute minimum of $-\tfrac{5}{4}$.   **31.** $5 + 4\sqrt{2} =$ absolute maximum; $5 - 4\sqrt{2} =$ absolute minimum   **33.** $l = 8$; $w = 8$; $h = 4$
35. The hottest spots are $\left(\tfrac{1}{2}, \tfrac{1}{2}, \tfrac{1}{\sqrt{2}}\right)$ and $\left(-\tfrac{1}{2}, -\tfrac{1}{2}, \tfrac{1}{\sqrt{2}}\right)$, where the temperature is $50°$; the coldest spots are $\left(\tfrac{1}{2}, -\tfrac{1}{2}, \tfrac{1}{\sqrt{2}}\right)$ and $\left(-\tfrac{1}{2}, \tfrac{1}{2}, \tfrac{1}{\sqrt{2}}\right)$,
where the temperature is $-50°$.   **37.** $l = 10$, $w = 8$, and   **39. (a)** $y = 1.051 + 0.00525x$ **(b)**
$h = 4$

(c) ≈ 1.156 **41.** So the hottest point is $(-3/2, 0)$ at a temp of 12
and coldest is -2 at $(-1, 0)$.

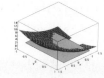

Exercise Set 14.6

1. $f\left(-\frac{4}{3}, -\frac{8}{3}\right) = -\frac{16}{3}$ is a constrained local minimum.   **3.** $f(-1, -1) = 3$ is a constrained local maximum; $f\left(\frac{2}{3}, -\frac{14}{9}\right) = -\frac{44}{27}$

is a constrained local minimum.   **5.** $f(2, -2) = 20$ is a constrained local minimum.   **7.** $f(6, -3, -2) = 18$ is a

constrained local minimum.   **9.** Let $g(x) = x^2 - \left(\frac{x-4}{2}\right)^2$; $f\left(-\frac{4}{3}, -\frac{8}{3}\right) = -\frac{16}{3}$, a constrained local minimum.   **11.** Let

$g(x) = x^2 + 2x(x^2 - 2)$; $f\left(\frac{2}{3}, -\frac{14}{9}\right) = -\frac{44}{27}$, a constrained local minimum; $f(-1, -1) = 3$, a constrained local maximum.   **13.** Let

$g(x) = x^3 + 3\left(\frac{16}{x^2}\right)$; $f(2, -2) = 20$, a constrained local minimum.   **15.** Let $g(x, y) = x - 2y - 3\left(\frac{36}{xy}\right)$; $f(6, -3, -2) = 18$, a

constrained local minimum   **17.** $\frac{2\sqrt{5}}{5}$ **19.** $T(\pm\sqrt{3}, 1) = 80 = $ maximum of T; $T(\pm\sqrt{3}, -1) = 20 = $ minimum of T.

21. $x = 8, y = 8, z = 4$ **23.** Maximum of $50°$ at $\left(\frac{1}{2}, \frac{1}{2}, \frac{1}{\sqrt{2}}\right)$ and $\left(-\frac{1}{2}, -\frac{1}{2}, \frac{1}{\sqrt{2}}\right)$ and minimum of $-50°$ at $\left(-\frac{1}{2}, \frac{1}{2}, \frac{1}{\sqrt{2}}\right)$ and

$\left(\frac{1}{2}, -\frac{1}{2}, \frac{1}{\sqrt{2}}\right)$.   **25.** $y = -x$: $2x^2 + 2z - 4 = 0, 2x + 2z = 0$, Eliminate z to get $x^2 - x - 2 = (x - 2)(x + 1) = 0$. If

$x = 2, y = -2, z = -2$; if $x = -1, y = 1, z = 1$. $\therefore$ $(2, -2, -2)$ and $(-1, 1, 1)$ are critical points. $\lambda_1 = 1$: $2y - 2y = 0 = -\lambda_2 \rightarrow \lambda_2 = 0, z = \lambda_1 = 1$$x^2 + y^2 - 2 = 0, x - y + 2 = 0 \rightarrow y = x + 2, x^2 + (x + 2)^2 - 2 = 2(x^2 + 2x + 1) = 0 \rightarrow x = -1$ $\therefore$ $(-1, 1, 1)$ is a critical

point.   **27.** $2\sqrt{2}$ by 4 by 2   **29.** $f\left(\pm 2\sqrt{3}, \frac{\pm 4}{\sqrt{3}}, \frac{7}{6}\right) = \frac{223}{12}$ are constrained local maxima. $-\frac{11}{4} = f\left(\pm 2, 0, \frac{-3}{2}\right)$ are constrained

local minima.   **31.** $\left(0, \pm\frac{\sqrt{29}}{4}, \frac{3}{8}\right) = $ Closest point; $\left(\pm\frac{\sqrt{10}}{6}, \pm\frac{4}{3}, \frac{1}{6}\right) = $ Farthest point

Chapter Review

1. (a) $\frac{2}{x}dx + \frac{y}{1-y^2}dy$ (b) $\frac{\cos x}{\cosh y}dx - \frac{\sin x \sinh y}{\cosh^2 y}dy$ **3.** (a) $\Delta f \approx df = 0.0084$ (b) $\Delta f \approx df = 0.01$ **5.** $\Delta P \approx \$318$ increase in

profit; \$7,043   **7.** $f_x(0, 0) = \lim\limits_{h\to 0} \frac{f(h,0)-f(0,0)}{h} = \lim\limits_{h\to 0} \frac{0}{h} = 0$; $f_y(0, 0) = \lim\limits_{k\to 0} \frac{f(0,k)-f(0,0)}{k} = \lim\limits_{k\to 0} \frac{0}{k} = 0$. But f is not continuous at

$(0, 0)$, since $\lim\limits_{(x,y)\to(0,0)} f(x, y) = 0$ along the x-axis, but $\lim\limits_{(x,y)\to(0,0)} f(x, y)$ along $y = x^2 = \lim\limits_{x\to 0} \frac{x^4}{x^4+x^4} = \frac{1}{2}$. Thus, f is not differentiable at

$(0, 0)$.

9. $\left.\frac{dz}{dt}\right|_{t=5\pi/6} = \sqrt{3} - \frac{1}{4}$ **11.** $\frac{dz}{dt} = -(\sinh t) e^{(1-\cosh t)}$ **13.** $\left.\frac{\partial z}{\partial t}\right|_{(1,1/2)} = \sqrt{2} - \frac{\pi}{2}(3 + \sqrt{2})$ **15.** $\frac{\partial z}{\partial x} = -\frac{z}{x}$; $\frac{\partial z}{\partial y} = \frac{-[\ln(\cos xz)+xz]}{xy(1-\tan xz)}$

17. $\frac{dh}{dt} = -\frac{5}{18}$ ft/min $= -\frac{10}{3}$ in./min   **19.** -4 **21.** $y^2 - x^2 + \ln\frac{x}{y} + 1$ **23.** Tangent plane: $\langle 1, -1, -1\rangle \cdot \langle \mathbf{x} - (1, 1, 2)\rangle = 0$ or

$x - y - z + 2 = 0$; Normal line: $\mathbf{r}(t) = \langle 1, 1, 2\rangle + t\langle 1, -1, -1\rangle$ **25.** $4.3°$ **27.** Absolute maximum of $f = \frac{35}{2}$, at $(4, 2)$; absolute

minimum of $f = \frac{9}{\sqrt[3]{3}}$, at $\left(\sqrt[3]{3}, \frac{\sqrt[3]{3}}{2}\right)$ **29.** (a) $h = \frac{2a}{\sqrt{3}}, r = \frac{2a}{\sqrt{6}}$ (b) $r = \frac{a}{\sqrt{2}}, h = 2r = \sqrt{2}a$

CHAPTER 15

Exercise Set 15.1

1.

i	(x_i^*, y_i^*)	$f(x_i^*, y_i^*)$	$f(x_i^*, y_i^*)\Delta A_i$
1	(1/2, 1/4)	1	1/2
2	(3/2, 1/4)	2	1
3	(5/2, 1/4)	3	3/2
4	(7/2, 1/4)	4	2
5	(7/2, 3/4)	5	5/2
6	(5/2, 3/4)	4	2
7	(3/2, 3/4)	3	3/2
8	(1/2, 3/4)	2	1

$\sum\limits_{i=1}^{8} f\left(x_i^*, y_i^*\right) \Delta A_i = 12$ **3.** $\sum\limits_{i=1}^{8} f\left(x_i^*, y_i^*\right) \Delta A_i = 60$

5. $\sum\limits_{i=1}^{9} f\left(x_i^*, y_i^*\right) \Delta A_i = 582$ **7.** $\sum\limits_{i=1}^{6} f\left(x_i^*, y_i^*\right) \Delta A_i = 35 \approx$ volume   **9.** $4\sum\limits_{i=1}^{6} f\left(x_i^*, y_i^*\right) \Delta A_i = \frac{69}{4} \approx$ volume

11. Since $g(x, y) \geq f(x, y), g(x, y) - f(x, y) \geq 0$. By Property 4, $\iint\limits_{R}[g(x, y) - f(x, y)]dA \geq 0$. By Property 2,

$\iint\limits_{R} g(x, y)dA - \iint\limits_{R} f(x, y)dA \geq 0$ and $\iint\limits_{R} f(x, y)dA \leq \iint\limits_{R} g(x, y)dA$.   **13.** (a) 10.905   (b) 10.587   (c) 10.658   (d) 10.580

Exercise Set 15.2

1. $-\frac{7}{2}$

3. $\frac{15}{8}$

5. $-\frac{4352}{15}$

7. $\frac{5}{3}$

9. $4 \ln \frac{6}{7}$

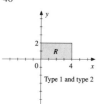

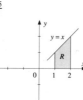

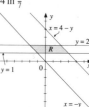

11. 40

13. $\frac{2}{3} \ln 2$

15. $4\sqrt{2} - \frac{8}{3}$

17. $\frac{49}{3}$

19. 4

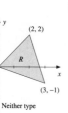

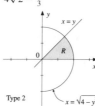

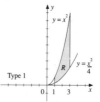

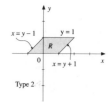

21. 2

23. $\frac{4}{3}$

25. $\frac{64}{3}$

27. $e - \frac{5}{3}$

29. $1 + \frac{3}{2} \ln \frac{4}{3}$

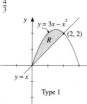

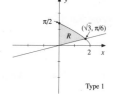

31. Either two type II regions or a type I and a type II region.

33. 4

35. $\frac{2\sqrt{2}}{3}$

37. $\frac{13}{2}$

39. 4

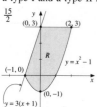

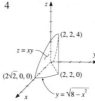

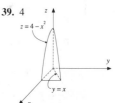

41. $\frac{16}{3}$

43. $\frac{e-1}{2e}$

45. $\frac{4}{27}$

47. $\int_0^2 \int_0^{y^2} f(x, y)\,dx\,dy$

49. $\int_2^4 \int_1^{x-1} f(x, y)\,dy\,dx$

51. $\int_0^1 \int_{1-\sqrt{1-y^2}}^{\sqrt{2y-y^2}} f(x, y)\,dx\,dy$

53. By Definition 15.2, the volume V under the surface $z = f(x, y)$ and above R is the double integral $V = \iint\limits_R f(x, y)\,dA$. Let y_0 be an arbitrary value of y, such that $c \le y_0 \le d$. The plane $y = y_0$ intersects the given surface in a curve $z = f(x, y_0)$. Consider the portion from $x = a$ to $x = b$, and the area under its graph is $A(y_0) = \int_a^b f(x, y_0)\,dx$, the area of a typical cross-section of the volume taken perpendicular to the y-axis. Let y_0 now be variable; drop the subscript and write $V = \int_c^d A(y)\,dy = \int_c^d \left[\int_a^b f(x, y)\,dx \right] dy = \iint\limits_R f(x, y)\,dA$.

55. $e + \frac{5}{e} - 4$

57. $\frac{82}{9} + \frac{16}{3} \ln 2$

59. $\frac{37}{12}$

61. $\frac{7}{9} - \ln 2$

63. $\frac{2\pi a^3}{35}$

65. 16

67. $\frac{16}{3}$

69. $a\pi b$

Exercise Set 15.3

1. $\frac{16}{3}\pi^2$ **3.** 99π **5.** $\frac{4}{5}$ **7.** $\frac{\pi}{8}$ **9.** 6π **11.** 2π

13. $\frac{10\pi}{3} + \frac{7\sqrt{3}}{2}$ **15.** $\sqrt{3} - \frac{\pi}{3}$ **17.** $\frac{19\pi}{3} - \frac{11\sqrt{3}}{2}$ **19.** 5π **21.** $\frac{\pi}{3}$ **23.** $\frac{16\sqrt{2}}{3}(1 - \sqrt{2})$

25. $\frac{\pi}{4} - \frac{2}{3}$ **27.** $\ln 4 - \frac{3}{4}$ **29.** $\frac{2\pi}{3} - \frac{10\sqrt{2}}{9}$ **31.** $\pi\left(1 - \frac{1}{e^{a^2}}\right)$ **33.** $\frac{8\pi}{3}$ **35.** $\frac{4\pi}{3}(a^2 - b^2)^{3/2}$ **37.** $\frac{32}{9}$ **39.** 8π

41. $\int_0^{\pi/3} \int_{2\sin\theta}^{4\sin\theta} (r^2\cos\theta\sin\theta)r\,dr\,d\theta = \frac{135}{32}$ **43.** $6 - \frac{4\pi}{3} + \frac{\sqrt{3}}{3}$

Exercise Set 15.4

1. $\left(\frac{3}{8}, \frac{3}{4}\right)$ **3.** $\left(0, \frac{4}{5}\right)$ **5.** $\left(-\frac{1}{2}, \frac{2}{5}\right)$ **7.** $\left(0, \frac{3}{\pi}\right)$ **9.** $\left(0, \frac{3}{10}\right)$ **11.** $\left(0, \frac{4a}{3\pi}\right)$ **13.** $\left(\frac{16\sqrt{2}}{15}, \frac{176\sqrt{2}}{35}\right)$

15. $\left(\frac{e^2+1}{e^2-1}, \frac{e^2+1}{4}\right)$ **17.** $I_x = \frac{2}{3}; I_y = \frac{2}{15}; I_0 = \frac{8}{15}; r_x = \frac{1}{\sqrt{5}}; r_y = \sqrt{\frac{3}{5}}$ **19.** $I_x = \frac{1}{3}; I_y = \frac{1}{9}; I_0 = \frac{4}{9}; r_x = \frac{\sqrt{2}}{3}; r_y = \frac{\sqrt{6}}{3}$

21. $I_x = \frac{32}{15}; I_y = \frac{32}{15}; I_0 = \frac{64}{15}; r_x = \frac{2}{\sqrt{5}}; r_y = \frac{2}{\sqrt{5}}$ **23.** $I_x = \frac{4\rho}{9}; I_y = \rho(\pi^2 - 4); I_0 = \rho\left(\pi^2 - \frac{32}{9}\right); r_x = \sqrt{\frac{\pi^2-4}{2}}; r_y = \frac{\sqrt{2}}{3}$

25. Let $R = \{(x, y) : 0 \le x \le b, 0 \le y \le h\}$; $I_x = \iint_R y^2 dm = \int_0^h \int_0^b y^2\rho\,dx\,dy = \rho \int_0^h \left[y^2 x\right]_0^b dy = \rho b \int_0^h y^2 dy = \frac{\rho b h^3}{3}$

27. $I_{x=-2} = \rho \int_{-2}^4 \int_{y^2/4}^{(y+4)/2} (x + 2)^2 dx\,dy; I_{y=4} = \rho \int_{-2}^4 \int_{y^2/4}^{(y+4)/2} (4 - y)^2 dx\,dy$ **29.** (a) $k = 4$ (b) $\mu_x = \frac{2}{3}; \mu_y = \frac{2}{3}$

(c) $\sigma_x^2 = \frac{1}{18}; \sigma_y^2 = \frac{1}{18}$ **31.** (a) $k = 6$ (b) $\mu_x = \frac{1}{4}; \mu_y = \frac{1}{2}$ (c) $\sigma_x^2 = \frac{3}{80}; \sigma_y^2 = \frac{1}{20}$

33. (a) $I_{x=\bar{x}} = \iint_R (x - \bar{x})^2 dm = \iint_R x^2 dm - 2\bar{x}\iint_R x\,dm + \bar{x}^2\iint_R dm = I_y - 2\bar{x}^2 m + \bar{x}^2 m = I_y - m\bar{x}^2$ (b) $I_{y=\bar{y}} = I_x - m\bar{y}^2$

Proof: $I_{y=\bar{y}} = \iint_R (y - \bar{y})^2 dm = \iint_R y^2 dm - 2\bar{y}\iint_R y\,dm + \bar{y}^2\iint_R dm = I_x - 2\bar{y}^2 m + \bar{y}^2 m = I_x - m\bar{y}^2$ **35.** Take the vertices as $(0, 0)$,

$(b, 0)$, and (a, h). $I_b = \iint_R y^2 dm = \int_0^h \int_{ay/h}^{b-(b-a)y/h} y^2\rho\,dx\,dy = \rho \int_0^h \left(b - \frac{by}{h}\right) y^2 dy = b\rho \left[\frac{y^3}{3} - \frac{y^4}{4h}\right]_0^h = \frac{bh^3\rho}{12}$ **37.** Orient the

coordinate axes so that l coincides with the y-axis. Then, we can compute the volume of the solid of revolution obtained by revolving R

about the y-axis by using the method of cylindrical shells: $V = 2\pi \int_a^b x \int_{g_1(x)}^{g_2(x)} dy\,dx = 2\pi A\bar{x}$, where $\bar{x}$ is the x-coordinate of the centroid

of R, and A is the area of R. $2\pi\bar{x}$ = distance traveled by the centroid of R. A similar formula can be derived to compute the volume of the

solid of revolution about the x-axis. The result is $V = 2\pi A\bar{y}$, where $\bar{y}$ is the y-coordinate of the centroid of R.

Exercise Set 15.5

1. $2\sqrt{14}$ **3.** $\frac{13}{6}$ **5.** $\frac{13}{3}\pi$ **7.** $4\pi\sqrt{2}$ **9.** $\frac{\pi}{6}(17\sqrt{17} - 5\sqrt{5})$ **11.** $\frac{1}{6}(13\sqrt{26} - 5\sqrt{10})$ **13.** 2 **15.** $\frac{20}{9} - \frac{\pi}{3}$

17. $A(S) = \iint_R \frac{2\sqrt{x^2+y^2+z^2}}{2|z|} dA = 8\int_0^{\pi/2}\int_0^a \frac{a}{\sqrt{a^2-r^2}} r\,dr\,d\theta = 4\pi a^2$ **19.** At a point (x, y, z) on the surface, y and z form the

legs of a right triangle, and $f(x)$ is the hypotenuse. Hence, $y^2 + z^2 = [f(x)]^2$. By equation (15.26), with $y^2 + z^2 - [f(x)]^2 =$

$F(x, y, z) = 0; F_x = -2f(x)f'(x), F_y = -2y$, and $F_z = -2z$, $A(S) = \iint_R \frac{2\sqrt{[f(x)]^2[f'(x)]^2+y^2+z^2}}{2z} dA = \iint_R \frac{f(x)\sqrt{1+[f'(x)]^2}}{\sqrt{[f(x)]^2-y^2}} dy\,dx$.

Thus, $4\int_a^b \int_0^{f(x)} \frac{f(x)\sqrt{1+[f'(x)]^2}}{\sqrt{[f(x)]^2-y^2}} dy\,dx = 4\int_a^b \left[f(x)\sqrt{1 + [f'(x)]^2}\sin^{-1}\frac{y}{f(x)}\right]_0^{f(x)} dx = 2\pi\int_a^b f(x)\sqrt{1 + [f'(x)]^2}\,dx$

Exercise Set 15.6

1. 0

3. $\frac{1}{3}$

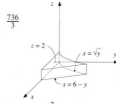

5. $\frac{736}{3}$

7. (a) $\int_1^3 \int_0^1 \int_{-1}^2 (2xy + 3yz^2)dy\,dx\,dz =$

$\int_1^3 \int_0^1 \left(3x + \frac{9}{2}z^2\right)dx\,dz =$

$\int_1^3 \left(\frac{3}{2} + \frac{9}{2}z^2\right)dz = 42$

(b) $\int_{-1}^2 \int_1^3 \int_0^1 (2xy + 3yz^2)dx\,dz\,dy = \int_{-1}^2 \int_1^3 (y + 3yz^2)dz\,dy = \int_{-1}^2 y(3 + 27 - 1 - 1)dy = 42$

9. $\int_0^1 \int_0^{2\sqrt{1-x^2}} \int_0^{\sqrt{4-4x^2-y^2}} f(x, y, z)dz\,dy\,dx$

11. $\int_1^2 \int_0^4 \int_{\sqrt{4-z^2}}^{\sqrt{3}} f(x, y, z)dx\,dy\,dz$

13. $\frac{116}{15}$

15. $\frac{2}{3}$

17. $\frac{7}{6}$

19. $\int_0^4 \int_0^{\sqrt{4-z}} \int_{x^2}^{4-z} dy\,dx\,dz$; $\int_0^4 \int_0^{4-y} \int_0^{\sqrt{y}} dx\,dz\,dy$; $\int_0^4 \int_0^{\sqrt{y}} \int_0^{4-y} dz\,dx\,dy$; $\int_0^4 \int_{x^2}^4 \int_0^{4-y} dz\,dy\,dx$; and $\int_0^2 \int_0^{4-x^2} \int_{x^2}^{4-z} dy\,dz\,dx$

21. $m = \frac{1}{4}$; $\bar{x} = \frac{8}{15}$

23. $\bar{z} = 4 \int_0^1 \int_0^{1-x} \int_0^{1-y} 2xz\,dz\,dy\,dx$; $I_y = \int_0^1 \int_0^{1-x} \int_0^{1-y} 2x(x^2 + z^2)dz\,dy\,dx$

25. $m = 4 \int_0^2 \int_0^{\sqrt{4-x^2}} \int_0^{\sqrt{16-x^2-y^2}} \sqrt{x^2+y^2}dz\,dy\,dx$; $\bar{z} = \frac{4}{m} \int_0^2 \int_0^{\sqrt{4-x^2}} \int_0^{\sqrt{16-x^2-y^2}} \sqrt{x^2+y^2}\,z\,dz\,dy\,dx$;

$I_z = 4 \int_0^2 \int_0^{\sqrt{4-x^2}} \int_0^{\sqrt{16-x^2-y^2}} (x^2 + y^2)^{3/2}dz\,dy\,dx$

27. $\bar{x} = \frac{1}{m} \int_0^1 \int_0^{\sqrt{1-x^2}} \int_0^{\sqrt{1-x^2}} x^2yz\,dz\,dy\,dx$; $\bar{y} = \frac{1}{m} \int_0^1 \int_0^{\sqrt{1-x^2}} \int_0^{\sqrt{1-x^2}} xy^2z\,dz\,dy\,dx$; $\bar{z} = \frac{1}{m} \int_0^1 \int_0^{\sqrt{1-x^2}} \int_0^{\sqrt{1-x^2}} x\,yz^2\,dz\,dy\,dx$

29. $\bar{x} = \bar{y} = 0, \bar{z} = \frac{h}{4}$

31. $\bar{z} = \frac{3(4e-3)}{6e^2+6e-23} \approx 0.627$

33. Select the coordinate system so l' coincides with the z-axis and l

passes through the point $(d, 0, 0)$. The distance from l to an element of mass located at (x, y, z) is $\sqrt{(d-x)^2 + y^2}$. Then

$I_l = \iiint_Q [(d-x)^2 + y^2]dm = \iiint_Q (x^2 + y^2)dm + \iiint_Q d^2\,dm - 2d \iiint_Q x\,dm$. But, since l' goes through the center of mass, $\bar{x} = 0$, and

$\iiint_Q x\,dm = m\bar{x}$, and so $\iiint_Q x\,dm = 0$. Thus, $I_l = \iiint_Q (x^2 + y^2)dm + d^2 \iiint_Q dm = I_{l'} + md^2$.

Exercise Set 15.7

1. 8

3. $\frac{\pi}{4}$

5. $\frac{2\pi}{3}$

7. 24π

9. $\frac{2\pi\delta}{5}$

11. $\left(\frac{6}{5}, 0, \frac{27\pi}{128}\right)$

13. $\left(\frac{18\sqrt{3}}{7\pi}, \frac{18}{7\pi}, \frac{102}{35}\right)$

15. $\left(\frac{64a}{15\pi^2}, \frac{64a}{15\pi^2}, \frac{16a}{15\pi}\right)$

17. $m = 40\pi$; $I_z = \frac{1456\pi}{9}$

19. 3π

21. $(2 - \sqrt{2})\pi$

23. $\frac{8}{9}(3\pi - 4)$

25. $\bar{z} = \frac{4}{15}\left[\frac{5a^2(h_2^3-h_1^3)-3(h_2^5-h_1^5)}{2a^2(h_2^2-h_1^2)-(h_2^4-h_1^4)}\right], \bar{x} = \bar{y} = 0$

Exercise Set 15.8

1. R is bounded by a parallelogram with vertices $(2, 3), (4, 6), (-7, 9), (-5, 12)$.

3. R is bounded by a triangle, with vertices

$(0, 0), (0, 2),$ and $(2, 0)$.

5. -10

7. e^u

9. 0

11. $\frac{\partial(x,y,z)}{\partial(\rho,\theta,\phi)} = \begin{vmatrix} \cos\theta\sin\phi & -\rho\sin\theta\sin\phi & \rho\cos\theta\cos\phi \\ \sin\theta\sin\phi & \rho\cos\theta\sin\phi & \rho\sin\theta\cos\phi \\ \cos\phi & 0 & -\rho\sin\phi \end{vmatrix} =$

$-\rho^2(\cos^2\theta\sin^3\phi + \sin^2\theta\sin\phi\cos^2\phi + \cos^2\theta\sin\phi\cos^2\phi + \sin^2\theta\sin^3\phi) = -\rho^2[\sin\phi(\sin^2\phi + \cos^2\phi)] = -\rho^2\sin\phi$. If we take the absolute

value, we get $\rho^2\sin\phi$.

13. $\frac{3}{2}$

15. $\frac{1}{2}(\sin 2 - 2) \approx -0.5454$

17. $V = 8 \int_0^a \int_0^{b\sqrt{1-x^2/a^2}} \int_0^{c\sqrt{1-x^2/a^2-y^2/b^2}} dz\,dy\,dx$; $V = \frac{4\pi}{3}abc$

19. Let $u = x - y$ and $v = x + y$. $4\ln 2 \approx 2.773$

21. Let $x = u/v$ and $y = v$. $\ln 4$

23. Let $u = x + y$ and $v = x - y$. $\frac{1}{2}[\ln(\sec 1 + \tan 1) - 1] \approx 0.1131$

Chapter Review

1. $4 + \frac{2}{3}\ln 4$

3. $4\ln 2 - \frac{1}{6}$

5. $\frac{3-e}{2}$

7. $\frac{16}{3}$

9. $\frac{4}{3}$

11. $\frac{1}{16}(2 + \sin 2)$

13. $(\bar{x}, \bar{y}) = \left(\frac{34}{21}, \frac{110}{147}\right)$; $(r_x, r_y) = \left(\frac{20\sqrt{3}}{21}, \frac{5\sqrt{14}}{21}\right)$

15. $(\bar{x}, \bar{y}) = \left(\frac{\pi}{2}, \frac{16}{9\pi}\right)$; $I_x = \frac{3\pi}{32}$; $r_y = \frac{\sqrt{6}}{4}$

17. $\frac{14}{3}$

19. $\frac{321\sqrt{3}}{70} + 6\pi$

21. $\frac{4\pi}{3}$

23. (a) centroid: $\left(0, 0, \frac{5}{4}\right)$ **(b)** $I_z = \frac{3\pi}{10}$

25. $\frac{46\pi}{15}$

27. $\frac{4}{3}(4 - \sqrt{2})$

29. (a) $k = \frac{1}{8}$ **(b)** $1 - \frac{1}{4}\ln 3$ **(c)** $\mu_x = \frac{26}{9}, \mu_y = \frac{13}{6}$ **(d)** $\sigma_x = \frac{\sqrt{134}}{9}, \sigma_y = \frac{\sqrt{11}}{6}$

31. (a) $8uv + 1$ **(b)** $e^{u+v}(1 - uv)$ **33.** 0 **35.** $\frac{1}{6}(\ln 2)^3 \approx 0.05550$

37. Let $u = x - y$ and $v = x + y$; $-4\cosh 4 + \sinh 4 + 2\cosh 2 - \sinh 2 \approx -78.045$

CHAPTER 16

Exercise Set 16.1

1. 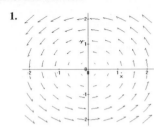   **3.** 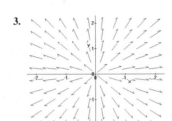   **5.**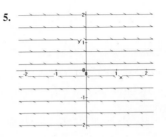

7. 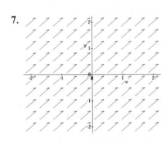   **9.** 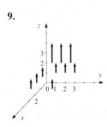   **11.** $\nabla f(x, y) = \langle 2x, -4y \rangle$ **13.** $\nabla f(x, y) = \left\langle \dfrac{x}{x^2+y^2}, \dfrac{y}{x^2+y^2} \right\rangle$

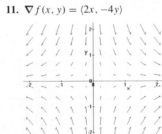   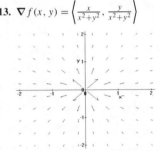

15. $\nabla f(x, y) = \langle -e^{-x} \sin 2y, 2e^{-x} \cos 2y \rangle$ **17.** $\nabla f(x, y, z) = \langle y - z, x + z, -x + y \rangle$ **19.** $\nabla f(x, y, z) = \langle 2xy^3z, 3x^2y^2z, x^2y^3 \rangle$

21. $\nabla f(x, y, z) = \dfrac{-GMm}{(x^2+y^2+z^2)^{3/2}} \langle x, y, z \rangle$

Exercise Set 16.2

1. $\dfrac{27\sqrt{13}}{2}$ **3.** $4(\sqrt{5} - 1)$ **5.** $\pi - 2$ **7.** 8π **9.** $\dfrac{2}{3}(2\sqrt{2} - 1)$ **11.** $A = \int_{-\infty}^{\ln 2} \dfrac{1}{2}\sqrt{e^{2t} + e^{4t}} \left(e^t \sqrt{1 + e^{2t}} \right) dt = 3$

13. $\dfrac{81}{64} - \ln\dfrac{5}{4}$ **15.** $\dfrac{200}{3}$ **17.** $-\dfrac{41}{6}$ **19.** $-\dfrac{20}{3}$ **21.** 3 **23.** $2 - \ln 4$ **25.** $-\dfrac{7}{6}$ **27.** 0 **29. (a)**

(b) $W = 0$ **(c)** $\mathbf{F}(x, y) = \left\langle -\dfrac{kx}{x^2+y^2}, -\dfrac{ky}{x^2+y^2} \right\rangle$ **31.** $-\pi\sqrt{5}$ **33.** 4 **35.** $\dfrac{3\pi}{16} + \dfrac{5}{12}$ **37.** $-\dfrac{1}{2}$ **39.** $\dfrac{3}{5}(1 - e^{\pi})$

41. $\dfrac{49\pi}{2}$ **43. (a)** Let C be partitioned into the subarcs $P_0P_1, P_1P_2, \cdots, P_{n-1}P_n$, and let Δs_i be the length of $P_{i-1}P_i$. If Δs_i is small, approximate $m_i\bar{x}_i$ for the subarc by $x_i^* \rho(x_i^*, y_i^*)\Delta s_i$, where (x_i^*, y_i^*) is some point on $P_{i-1}P_i$.

Let $\bar{x} = \dfrac{\lim\limits_{\|P\|\to 0} \sum\limits_{i=1}^{n} x_i^* \rho(x_i^*, y_i^*)\Delta s_i}{\lim\limits_{\|P\|\to 0} \sum\limits_{i=1}^{n} \rho(x_i^*, y_i^*)\Delta s_i} = \dfrac{\int_C x\rho(x, y)ds}{\int_C \rho(x, y)ds} = \dfrac{\int_C x\,dm}{m}$; Similarly, $\bar{y} = \dfrac{\lim\limits_{\|P\|\to 0} \sum\limits_{i=1}^{n} y_i^* \rho(x_i^*, y_i^*)\Delta s_i}{\lim\limits_{\|P\|\to 0} \sum\limits_{i=1}^{n} \rho(x_i^*, y_i^*)\Delta s_i} = \dfrac{\int_C y\,dm}{m}$ **(b)** Use the same partition of

C as in part (a). Then, by the definitions of I_x and I_y, $I_x = \lim\limits_{\|P\|\to 0} \sum\limits_{i=1}^{n} (y_i^*)^2 \rho(x_i^*, y_i^*)\Delta s_i = \int_C y^2dm$; $I_y = \lim\limits_{\|P\|\to 0} \sum\limits_{i=1}^{n} (x_i^*)^2 \rho(x_i^*, y_i^*)\Delta s_i =$

$\int_C x^2dm$

45. Center of mass: $\left(-\dfrac{4}{3\pi}, \dfrac{4}{3\pi}, \dfrac{9\pi}{4} \right)$; $I_x = \dfrac{3\pi\sqrt{13}\rho}{8}(8 + 27\pi^2)$; $I_y = \dfrac{3\pi\sqrt{13}\rho}{8}(8 + 27\pi^2)$; $I_z = 6\pi\sqrt{13}\rho$ **47.** $\int_C \mathbf{F} \cdot d\mathbf{r} = \int_a^b m\mathbf{r}'' \cdot \mathbf{r}'dt$ by

Newton's Second Law of Motion, since $\mathbf{F} = m\mathbf{a} = m\mathbf{r}''(t)$. Next, we note that $\dfrac{d}{dt}(\mathbf{r}' \cdot \mathbf{r}') = \mathbf{r}'' \cdot \mathbf{r}' + \mathbf{r}' \cdot \mathbf{r}'' = 2\mathbf{r}'' \cdot \mathbf{r}'$, which allows us to

rewrite the integral as $\dfrac{m}{2} \int_a^b \dfrac{d}{dt}(\mathbf{r}' \cdot \mathbf{r}')dt = \dfrac{m}{2} \int_a^b \dfrac{d}{dt}|\mathbf{r}'|^2dt$.

Exercise Set 16.3

1. $\phi(x, y) = x^2 + 3xy - y^2 + y$ **3.** not a gradient field   **5.** $\phi(x, y) = x^3y^2 + \dfrac{2}{3}x^3 - 2y$ **7.** $\phi(x, y) = x^2e^{xy}$

9. not a gradient field   **11.** $\phi(x, y) = y \sin xy - y^2$ **13.** 33   **15.** $\frac{155}{3}$ **17.** 15   **19.** $-\frac{3}{2}$

21. $\phi(x, y) = x^2 + y^3 - 2xy^2$ **23.** $\phi(x, y) = x \sin y - y \cos x$ **25.** $\frac{\partial f}{\partial y} = \frac{\partial g}{\partial x} = -2$; (a) $\int_C (2x - 2y)dx + (6y - 2x)dy = 16$

(b) $\int_C \mathbf{F} \cdot d\mathbf{r} = \int_0^1 \langle 2x - 2y, 6y - 2x \rangle \cdot \langle 4, -4 \rangle\, dt = 16$ **27.** Let $\mathbf{F} = \langle f, g \rangle = \left(\frac{-y}{x^2+y^2}, \frac{x}{x^2+y^2} \right)$. Then $\frac{\partial f}{\partial y} = \frac{\partial g}{\partial x} = \frac{y^2-x^2}{(x^2+y^2)^2}$;

(a) $\int_C \frac{-y\,dx+x\,dy}{x^2+y^2} = \frac{5\pi}{12}$ **(b)** $\int_C \mathbf{F} \cdot d\mathbf{r} = \int_{-\pi/4}^{\pi/6} \left[\frac{(-2\sqrt{3}\sin t)(-2\sqrt{3}\sin t)}{12} + \frac{(2\sqrt{3}\cos t)(2\sqrt{3}\cos t)}{12} \right] dt = \frac{5\pi}{12}$

29. (a) $\int_C \mathbf{F} \cdot d\mathbf{r} = \int_0^1 \langle 2(1 + 2t), 2(-2 + 8t) \rangle \cdot \langle 2, 8 \rangle\, dt = 40$ **(b)** $\int_C \mathbf{F} \cdot d\mathbf{r} = \int_0^1 \langle 2(1 + 2t), 2(-2) \rangle \cdot \langle 2, 0 \rangle\, dt + \int_0^1 \langle 2(3), 2(-2 + 8t) \rangle \cdot$

$\langle 0, 8 \rangle\, dt = 40$ **(c)** $\int_C \mathbf{F} \cdot d\mathbf{r} = \int_1^3 \left\langle 2t, 4(t^2 - 2t) \right\rangle \cdot \langle 1, 4t - 4 \rangle\, dt = 40$ **31.** $\frac{\partial f}{\partial y} = \frac{\partial g}{\partial x} = -3$; $W = 44$ **33. (a)** $\frac{\sqrt{6}-2}{4}$ **(b)** 0

35. $\frac{\partial f}{\partial y} = \frac{\partial g}{\partial x} = y \cos xy + y \cos xy - xy^2 \sin xy$; $\phi(x, y) = x^2 + y \sin xy$ **37.** $\frac{\partial f}{\partial y} = \frac{\partial g}{\partial x} = \tanh y$; $\phi(x, y) = x \ln \cosh y - y$

39. $\phi(x, y, z) = x^2 + y^2 + z^2$; 8   **41.** $\frac{\partial f}{\partial y} = \frac{\partial g}{\partial x} = 2xz$; $\frac{\partial f}{\partial z} = \frac{\partial h}{\partial x} = 2xy - 6z$; $\frac{\partial g}{\partial z} = \frac{\partial h}{\partial y} = x^2 + 24yz^2$; $\mathbf{F}$ is a gradient field

$\phi(x, y, z) = x^2yz - 3xz^2 + 4y^2z^3$

Exercise Set 16.4

1. $\frac{28}{3}$ **3.** $-\frac{4}{\pi}$ **5.** $\frac{3e^2 - 2e^3 - 1}{3}$ **7.** 0   **9.** 0   **11.** -36 **13.** 0   **15.** 0   **17.** $\frac{3\pi}{2}$ **19.** 5   **21.** $\frac{3}{2}$

23. $\frac{3}{8}\pi a^2$ **25.** $3\pi a^2$ **27.** Circulation of $\mathbf{F}$ around $C = \frac{4}{3}$. Flux of $\mathbf{F}$ through $C = \frac{4}{3}$ **29.** Circulation of $\mathbf{F}$ around $C = 24\pi$.

Flux of $\mathbf{F}$ through $C = 0$ **31.** Circulation of $\mathbf{F}$ around $C = \frac{81\pi}{2}$.   **33.** Suppose $\mathbf{T}$ is the unit tangent vector at point $P(x, y)$

on the curve C and C is parameterized by arc length, then $\mathbf{T} = \left(\frac{dx}{ds}, \frac{dy}{ds} \right)$. Let θ denote the smallest positive angle from the unit

vector $\mathbf{i}$ to $\mathbf{T}$. Then $\overrightarrow{PA} = \frac{dx}{ds}$, $\overrightarrow{AC} = \frac{dy}{ds}$. Consider the outer normal $\mathbf{n}$. The angle between $\mathbf{T}$ and $\mathbf{n}$ is $\frac{\pi}{2}$, measured clockwise. Thus,

$\angle BPD = \theta - \frac{\pi}{2}$, and $\mathbf{n} = \left(\cos \left(\theta - \frac{\pi}{2} \right), \sin \left(\theta - \frac{\pi}{2} \right) \right)$.   **35.** Suppose that C contains the origin. By Theorem 16.8, replace C by

the circle $C_1 : \mathbf{r}(t) = \langle a \cos t, a \sin t \rangle$, $0 \le t \le 2\pi$, where a is small enough that C_1 lies inside C. Then, $\int_C \mathbf{F} \cdot d\mathbf{r} = \int_{C_1} \mathbf{F} \cdot d\mathbf{r} =$

$\int_0^{2\pi} \langle \cos t, \sin t \rangle \cdot \langle -a \sin t, a \cos t \rangle\, dt = \int_0^{2\pi} (-a \cos t \sin t + a \sin t \cos t)dt = 0$; If the origin is exterior to C, Green's theorem can be used:

$\int_C \mathbf{F} \cdot d\mathbf{r} = \iint_R \left(\frac{\partial g}{\partial x} - \frac{\partial f}{\partial y} \right) dA = \iint_R \left[-\frac{xy}{(x^2+y^2)^{3/2}} + \frac{xy}{(x^2+y^2)^{3/2}} \right] dA = 0$ **37.** Assume that R can be divided by a vertical line segment into

two simple regions R_1 and R_2, as shown. Then, $\iint_R \left(\frac{\partial g}{\partial x} - \frac{\partial f}{\partial y} \right) dA = \iint_{R_1} \left(\frac{\partial g}{\partial x} - \frac{\partial f}{\partial y} \right) dA + \iint_{R_2} \left(\frac{\partial g}{\partial x} - \frac{\partial f}{\partial y} \right) dA = \left(\int_{C_1} \mathbf{F} \cdot d\mathbf{r} + \int_{C_2} \mathbf{F} \cdot d\mathbf{r} \right) +$

$\left(\int_{C_3} \mathbf{F} \cdot d\mathbf{r} + \int_{C_4} \mathbf{F} \cdot d\mathbf{r} \right) = \int_{C_1} \mathbf{F} \cdot d\mathbf{r} + \int_{C_4} \mathbf{F}\, d\mathbf{r} = \int_C \mathbf{F} \cdot d\mathbf{r}$, since $\int_{C_2} \mathbf{F} \cdot d\mathbf{r} = -\int_{C_3} \mathbf{F} \cdot d\mathbf{r}$. In a similar fashion, R_1 and R_2 can be subdivided

into two regions each and the above proof used to show that Green's theorem applies to R. The process may be continued for a finite number of such divisions; each time the new integrals that are introduced along the common boundary of the subregions cancel.

Exercise Set 16.5

1. $\frac{8\sqrt{3}}{3}$ **3.** $\frac{81\sqrt{14}\pi}{2}$ **5.** $\frac{19\pi}{15}$ **7.** $\frac{32\pi}{3}$ **9.** $m = \frac{8\sqrt{21}}{3}$; $(\bar{x}, \bar{y}, \bar{z}) = (2, 1, 1)$ **11.** $m = \frac{13\pi k}{3}$; $(\bar{x}, \bar{y}, \bar{z}) = (0, 0, 2.854)$

13. $I_x = \frac{k}{3} \int_0^2 \int_0^4 \left(y^2 + \frac{x^4}{9} \right) \sqrt{9 + 4x^2}\, dy\, dx$; $I_y = \frac{k}{3} \int_0^2 \int_0^4 \left(x^2 + \frac{x^4}{9} \right) \sqrt{9 + 4x^2}\, dy\, dx$; $I_z = \frac{k}{3} \int_0^2 \int_0^4 (x^2 + y^2) \sqrt{9 + 4x^2}\, dy\, dx$

15. (a) $\sqrt{6} \int_0^2 \int_0^{4-2x} x^2 y(2x + y)^3 dy\, dx$ **(b)** $\sqrt{\frac{6}{8}} \int_0^4 \int_0^z (z - y)^2 yz^3\, dy\, dz$ **(c)** $\sqrt{6} \int_0^4 \int_0^{z/2} x^2(z - 2x)z^3\, dx\, dz$ **17.** Projection onto the

xy-plane, $\int_1^{\sqrt{2}} \int_0^5 xy(2 - x^2)\sqrt{1 + 4x^2}\, dy\, dx$; projection onto the yz-plane, $\frac{1}{2} \int_0^1 \int_0^5 yz\sqrt{9 - 4z}\, dy\, dz$ **19.** Flux $= 19$

21. Flux $= 56\pi$ **23.** Flux $= \frac{14\pi}{3}$ **25.** Flux $= \frac{8}{3}$ **27.** $m = 2\pi ka^2$; center of mass $= \left(0, 0, \frac{a}{2} \right)$

29. $\mathbf{F} = \frac{k\mathbf{u}}{|\mathbf{r}|^2} = \frac{k\mathbf{r}}{|\mathbf{r}|^3}$; $S : x^2 + y^2 + z^2 = a^2$, or $z = \sqrt{a^2 - x^2 - y^2}$; $\mathbf{n} = \frac{\langle 2x, 2y, 2z \rangle}{2\sqrt{x^2+y^2+z^2}} = \frac{\mathbf{r}}{a}$; $\mathbf{F} \cdot \mathbf{n}\, d\sigma = \left(\frac{k\mathbf{r}}{|\mathbf{r}|^3} \cdot \frac{\mathbf{r}}{a} \right) = \frac{k}{a^2}$; flux

$= \iint_S \mathbf{F} \cdot \mathbf{n}\, d\sigma = \iint_S \frac{k}{a^2}\, d\sigma = 4\pi a^2 \left(\frac{k}{a^2} \right) = 4\pi k$ **31.** Let S denote the surface of a lamina having continuous density $\rho(x, y, z)$. Partition S

into finitely many small surface elements having areas $\Delta\sigma_1, \Delta\sigma_2, \cdots, \Delta\sigma_n$. Denote the partition by P and define its norm $\|P\|$ as the length of the longest diagonal of all the cells. Choose a point (x_i^*, y_i^*, z_i^*) arbitrarily in the ith element. The mass of the ith cell is approximately $\rho(x_i^*, y_i^*, z_i^*)\Delta\sigma_i$ and the mass of the entire lamina is $m = \lim_{||P||\to 0}\sum_{i=1}^{n}\rho(x_i^*, y_i^*, z_i^*)\Delta\sigma_i = \iint_S \rho(x, y, z)d\sigma$, provided the

limit exists. The coordinates of the center of mass are approximated as follows: $\bar{x} \approx \dfrac{\sum_{i=1}^{n}(\Delta m_i)x_i^*}{\sum_{i=1}^{n}\Delta m_i}$, $\bar{y} = \dfrac{\sum_{i=1}^{n}(\Delta m_i)y_i^*}{\sum_{i=1}^{n}\Delta m_i}$, $\bar{z} = \dfrac{\sum_{i=1}^{n}(\Delta m_i)z_i^*}{\sum_{i=1}^{n}\Delta m_i}$

Taking the limits as $||P|| \Rightarrow 0$, provided they exist, gives $\bar{x} = \dfrac{\iint_S x\,dm}{m}$, $\bar{y} = \dfrac{\iint_S y\,dm}{m}$, $\bar{z} = \dfrac{\iint_S z\,dm}{m}$, where $dm = \rho\, d\sigma$. Similarly,
approximations for the moments of inertia are $I_x \approx \sum_{i=1}^{n}(\Delta m_i)(y_i^2 + z_i^2)$, $I_y \approx \sum_{i=1}^{n}(\Delta m_i)(x_i^2 + z_i^2)$, $I_z \approx \sum_{i=1}^{n}(\Delta m_i)(x_i^2 + y_i^2)$,
where $(x_i, y_i, z_i) = (x_i^*, y_i^*, z_i^*)$ and $\Delta m_i = \rho(x_i^*, y_i^*, z_i^*)\Delta\sigma_i$. Taking the limits as $||P|| = 0$, provided they exist, gives
$I_x = \iint_S (y^2 + z^2)dm$, $I_y = \iint_S (x^2 + z^2)dm$, $I_z = \iint_S (x^2 + y^2)dm$.

Exercise Set 16.6

1. 512 **3.** $\frac{52}{5}$ **5.** 24π **7.** Flux $= \frac{1792\pi}{3}$ **9.** Flux $= -\frac{59}{36}$ **11.** Flux $= \frac{7\pi}{6}$ **13.** $\iint_S \mathbf{F}\cdot\mathbf{n}\,d\sigma = \iiint_G \text{div } \mathbf{F}\,dV = 24\pi$

15. $\iint_S \mathbf{F}\cdot\mathbf{n}\,d\sigma = \iiint_G \text{div } \mathbf{F}\,dV = 4a^3$ **17.** $\mathbf{F} = x\mathbf{i} + y\mathbf{j} + z\mathbf{k}$; div $\mathbf{F} = 1 + 1 + 1 = 3$ $\iint_S \mathbf{F}\cdot\mathbf{n}\,d\sigma = \iiint_G \text{div } \mathbf{F}\,dV = 3\iiint_G dV = 3V$,

if G and its boundary S satisfy the conditions of the Divergence Theorem. **19.** Equation (16.36): let G be a region that is
yz-simple. Its boundary S can be decomposed into the surfaces S_1, S_2, and S_3 as in Figure 16.46 except that the projection of G is
onto the yz-plane. If S_3 is nonempty, on it the outer unit normal vectors are parallel to the yz-plane, so that $\alpha = \pi/2$ and $\cos\alpha = 0$.
Thus, $f\cos\alpha = 0$, and $\iint_S f\cos\alpha\,d\sigma = 0$. Whether or not S_3 is empty, $\iint_S f\cos\alpha\,d\sigma = \iint_{S_1} f\cos\alpha\,d\sigma + \iint_{S_2} f\cos\alpha\,d\sigma$. Let S_1 be the
graph of $x = x_1(y, z)$ and S_2 the graph of $x = x_2(y, z)$. On S_2 the outer normal is directed in the positive x-direction, so that
α is acute. Thus, by equation (16.26) $\iint_{S_2} f\cos\alpha\,d\sigma = \iint_R f(x_2(y, z), y, z)dA$ On S_1 the outer normal is directed in the negative
x-direction, so that α is obtuse, and $|\sec\alpha| = -\sec\alpha$. Equation (16.26) then gives $\iint_{S_1} f\cos\alpha\,d\sigma = -\iint_R f(x_1(y, z), y, z)dA$

Combining, we get $\iint_S f\cos\alpha\,d\sigma = \iint_R [f(x_2(y, z), y, z) - f(x_1(y, z), y, z)]dA = \iiint_G \frac{\partial f}{\partial x}dV = \iint_R \left[\int_{x_1(y,z)}^{x_2(y,z)}\left(\frac{\partial f}{\partial x}\right)dx\right]dA$; it

follows that equation (16.36) is true. The proof of equation (16.37) is similar to the foregoing. The projection of G is onto
the xz-plane, and the outward normals to S_3 are parallel to the xz-plane, so that $\beta = \pi/2$ and $\cos\beta = 0$. Thus, $g\cos\beta = 0$,
and $\iint_{S_3} g\cos\beta\,d\sigma = 0$. Hence, $\iint_S g\cos\beta\,d\sigma = \iint_{S_1} g\cos\beta\,d\sigma + \iint_{S_2} g\cos\beta\,d\sigma$; Let S_1 be the graph of $y = y_1(x, z)$ and S_2
the graph of $y = y_2(x, z)$. Following the earlier steps, we get $\iint_S g\cos\beta\,d\sigma = \iint_R [g(x, y_2(x, z), z) - g(x, y_1(x, z), z)]dA$ and

$\iiint_G \frac{\partial g}{\partial y}dV = \iint_R \left[\int_{y_1(x,z)}^{y_2(x,z)}\left(\frac{\partial g}{\partial y}\right)dy\right]dA = \iint_R [g(x, y_2(x, z), z) - g(x, y_1(x, z), z)]dA$ from which it follows that equation (16.37) is true.

21. In spherical coordinates, $x^2 + y^2 + (z - a)^2 = a^2$ becomes $\rho = 2a\cos\phi$. $\iiint_G \text{div } \mathbf{F}\,dV$

$= 2\int_0^{2\pi}\int_0^{\pi/2}\int_0^{2a\cos\phi}[\rho\sin\phi(\cos\theta + \sin\theta) + \rho\cos\phi]\rho^2\sin\phi\,d\rho\,d\phi\,d\theta = 8a^4\int_0^{2\pi}\int_0^{\pi/2}[(\cos\theta + \sin\theta)\cos^4\phi\sin^2\phi + \cos^5\phi\sin\phi]d\phi\,d\theta$

Note that the terms $\int_0^{2\pi}(\cos\theta + \sin\theta)d\theta = 0$ (or we can reverse the order of integration). We now have $16\pi a^4\int_0^{\pi/2}\cos^5\phi\sin\phi\,d\phi = \frac{8}{3}\pi a^4$

23. $\mathbf{F} = \left(u\frac{\partial v}{\partial x}, u\frac{\partial v}{\partial y}, u\frac{\partial v}{\partial z}\right)$; div $\mathbf{F} = u\frac{\partial^2 v}{\partial x^2} + \frac{\partial u}{\partial x}\frac{\partial v}{\partial x} + u\frac{\partial^2 v}{\partial y^2} + \frac{\partial u}{\partial y}\frac{\partial v}{\partial y} + u\frac{\partial^2 v}{\partial z^2} + \frac{\partial u}{\partial z}\frac{\partial v}{\partial z} = u\,\nabla^2 v + \nabla u\cdot\nabla v$. $\iint_S \mathbf{F}\cdot\mathbf{n}\,d\sigma = \iint_S u\,\nabla v\cdot\mathbf{n}\,d\sigma = \iiint_G \text{div }$

$\mathbf{F}\,dV = \iiint_G (u\nabla^2 v + \nabla u\cdot\nabla v)dV$ **25.** From Exercise 23, $\iiint_G (u\nabla^2 v + \nabla u\cdot\nabla v)dV = \iint_S u\nabla v\cdot\mathbf{n}\,d\sigma$. Replace v by u and u by

$u(x, y, z) = 1$; $\iiint_G (\nabla^2 u + 0)dV = \iint_S \nabla u\cdot\mathbf{n}\,d\sigma$; $\therefore \iiint_G \nabla d^2 u\,dV = \iint_S \nabla u\cdot\mathbf{n}\,d\sigma$

Exercise Set 16.7

1. curl $\mathbf{F} = \langle 1, 2xy, 1 - 2xz\rangle$ **3.** curl $\mathbf{F} = x(e^z - e^y)\mathbf{i} + y(e^x - e^z)\mathbf{j} + z(e^y - e^x)\mathbf{k}$ **5.** curl $\mathbf{F} = \langle 0, 0, y\cos xy - x\sin xy\rangle$ **7.** -35

9. -12π **11.** $\int_C \mathbf{F}\cdot d\mathbf{r} = \iint_S (\text{curl } \mathbf{F})\cdot\mathbf{n}\,d\sigma = 0$ **13.** $\int_C \mathbf{F}\cdot d\mathbf{r} = \iint_S (\text{curl } \mathbf{F})\cdot\mathbf{n}\,d\sigma = -\frac{41}{6}$ **15.** $\int_C \mathbf{F}\cdot d\mathbf{r} = \iint_S (\text{curl } \mathbf{F})\cdot\mathbf{n}\,d\sigma = 0$

17. curl $\mathbf{F} = \langle 0,0,0 \rangle$; $\phi(x,y,z) = x^2 yz$ **19.** curl $\mathbf{F} = \langle 0,0,0 \rangle$; work $= 7$ **21.** curl $\mathbf{F} = \langle 0,0,0 \rangle$; $\phi(x,y,z) = e^z \sin x \cos y$

23. $\operatorname{div}(\operatorname{curl} \mathbf{F}) = \frac{\partial}{\partial x}\left(\frac{\partial h}{\partial y} - \frac{\partial g}{\partial z}\right) + \frac{\partial}{\partial y}\left(\frac{\partial f}{\partial z} - \frac{\partial h}{\partial x}\right) + \frac{\partial}{\partial z}\left(\frac{\partial g}{\partial x} - \frac{\partial f}{\partial y}\right) = \frac{\partial^2 h}{\partial x \partial y} - \frac{\partial^2 g}{\partial x \partial z} + \frac{\partial^2 f}{\partial y \partial z} - \frac{\partial^2 h}{\partial y \partial x} + \frac{\partial^2 g}{\partial z \partial x} - \frac{\partial^2 f}{\partial z \partial y} = 0$

25. Let $\mathbf{H} = \operatorname{curl} \mathbf{F} = \nabla \times \mathbf{F}$. By the divergence theorem, $\iint_S \mathbf{H} \cdot \mathbf{n}\, d\sigma = \iiint_G \operatorname{div} \mathbf{H}\, dV = \iiint_G \nabla \cdot (\nabla \times \mathbf{F}) dV = 0$, from the result of Exercise 23.

$\therefore \iint_S (\operatorname{curl} \mathbf{F}) \cdot \mathbf{n}\, d\sigma = 0$ **27. (a)** $\mathbf{a} \times \mathbf{r} = \langle a_2 z - a_3 y, a_3 x - a_1 z, a_1 y - a_2 x \rangle$; $\operatorname{curl}(\mathbf{a} \times \mathbf{r}) = \langle a_1 + a_1, a_2 + a_2, a_3 + a_3 \rangle = 2\mathbf{a}$

(b) $\int_C (\mathbf{a} \times \mathbf{r}) \cdot d\mathbf{r} = \iint_S \operatorname{curl}(\mathbf{a} \times \mathbf{r}) \cdot \mathbf{n}\, d\sigma = 2 \iint_S \mathbf{a} \cdot \mathbf{n}\, d\sigma$ **29. (a)** $\iint_S (\nabla u \times \nabla v) \cdot \mathbf{n}\, d\sigma = \iint_S \operatorname{curl}(u \nabla v) \cdot \mathbf{n}\, d\sigma$ (by Exercise 28)

$= \int_C (u \nabla v) \cdot d\mathbf{r}$ (by Stokes's theorem)   **(b)** $\int_C (u \nabla v - v \nabla u) \cdot d\mathbf{r} = \int_C u \nabla v \cdot d\mathbf{r} - \int_C v \nabla u \cdot d\mathbf{r} =$

$\iint_S (\nabla u \times \nabla v) \cdot \mathbf{n}\, d\sigma - \iint_S (\nabla v \times \nabla u) \cdot \mathbf{n}\, d\sigma$ (by part a) $= \iint_S (\nabla u \times \nabla v) \cdot \mathbf{n}\, d\sigma + \iint_S (\nabla u \times \nabla v) \cdot \mathbf{n}\, d\sigma = 2 \iint_S (\nabla u \times \nabla v) \cdot \mathbf{n}\, d\sigma$

Chapter Review

1. $\frac{\pi}{2} + 1 + \frac{\sqrt{2}}{3}$ **3.** $\frac{e^{3/2}(1 - e^3)}{3} + \frac{e^2 + 1}{4}$ **5. (a)** $\phi(x,y) = \frac{x}{\sqrt{x^2 + y^2}} - 2x + 3y$ **(b)** $\phi(x,y) = \ln \frac{xy}{x+y}$ **(c)** not a gradient field

(d) $\phi(x,y) = x \tanh y - y^2 + \ln \cosh x^2$ **7.** 0 **9.** 3π **11.** $2\left(1 - \frac{\pi}{4}\right)$ **13.** π **15.** $\frac{4\pi}{3}(50 - 11\sqrt{5})$

17. xy-projection: $\int_0^2 \int_0^x x^2 y(4 - x^2)^3 \sqrt{1 + 4x^2} dy\, dx$; yz-projection: $\frac{1}{2} \int_0^2 \int_0^{4-y^2} yz^3 \sqrt{(4-z)(17-4z)} dz\, dy$ **19.** Flux $= 6\pi$

21. $\frac{61\pi}{16}$ **23.** 0 **25.** Circulation $= -128\pi$ **27.** 2π **29.** $\int_C u \nabla v\, d\mathbf{r} = \iint_S (\nabla u \times \nabla v) \cdot \mathbf{n}\, d\sigma$ (from Exercise 29 of Exercise

Set 16.7). Then, $\int_C (u \nabla v + v \nabla u) \cdot d\mathbf{r} = \iint_S (\nabla u \times \nabla v + \nabla v \times \nabla u) \cdot \mathbf{n}\, d\sigma = \iint_S (\nabla u \times \nabla v - \nabla u \times \nabla v) \cdot \mathbf{n}\, d\sigma = \iint_S (0) \cdot \mathbf{n}\, d\sigma = 0$

CHAPTER 17

Exercise Set 17.1

1. $y = \frac{x^2}{2} - 2x + \frac{11}{2}$ **3.** $y = \ln \sec x + 3$ **5.** $y = x^2 + x$ **7.** $C_1 e^{-x} + 4C_2 e^{-2x} + 3(-C_1 e^{-x} - 2C_2 e^{-2x}) + 2(C_1 e^{-x} + C_2 e^{-2x}) = 0$

9. $-C_1 a^2 \cos ax - C_2 a^2 \sin ax + a^2 (C_1 \cos ax + C_2 \sin ax) = 0$ **11.** $e^{-t}(-2C_2 \cos t + 2C_1 \sin t) + 2e^{-t}[(C_2 - C_1) \cos t -$
$(C_2 + C_1) \sin t] + 2e^{-t}(C_1 \cos t + C_2 \sin t) = 0$ **13.** $y = -2e^{-x} + 2e^{-2x}$ **15.** $y = 2 \cos ax + \sin ax$

17. $x = e^{-t} \sin t$ **19.** Let $F(x,y) = 2x^3 y - 3xy^2 + y^4 - 7 = 0$; $\frac{dy}{dx} = -\frac{F_x}{F_y} = \frac{-(6x^2 y - 3y^2)}{2x^3 - 6xy + 4y^3} = \frac{3y(y - 2x^2)}{2x^3 - 6xy + 4y^3}$ **21.** Let

$F(x,y) = \tan^{-1} \frac{y}{x} + \ln \sqrt{x^2 + y^2} - 1 = 0$; $\frac{dy}{dx} = -\frac{F_x}{F_y} = \frac{-\left(\frac{-y}{x^2 + y^2} + \frac{x}{x^2 + y^2}\right)}{\frac{x}{x^2 + y^2} + \frac{y}{x^2 + y^2}} = \frac{y - x}{y + x}$

23. $y''' + y' - 10y = e^{-x} \cos 2x(11C_1 - 2C_2 - C_1 + 2C_2 - 10C_1) + e^{-x} \sin 2x(2C_1 + 11C_2 - 2C_1 - C_2 - 10C_2) + e^{2x}(8C_3 + 2C_3 - 10C_3) - 2x + 10x^2 + 5 =$
$10x^2 - 2x + 5$; $C_3 = \frac{33}{26}, C_2 = -\frac{4}{26}, C_1 = \frac{32}{26}$.   **25.** $y' = \frac{y^2}{x(2y - x)}$ **27.** $y'' = -y$ **29.** $x^2 y'' - xy' + y = 0$

31. 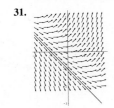   **33.** 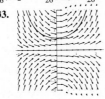   **35. (a)** 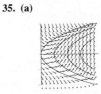   **(b)**   **37. (a)** $y(x) = -3x^2 - 6x - 5 + 6e^x$

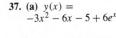

(b) $y(x) = -x - \sqrt{2}\sqrt{x^2 + \frac{1}{2}}, y(x) = -x + \sqrt{2}\sqrt{x^2 + \frac{1}{2}}$ **(c)** $y(x) = -\frac{1}{2}\cos(x) - \frac{1}{2}\sin(x) + 2 - \frac{3}{2}e^{-x}$

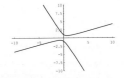

Exercise Set 17.2

1. $y = C(x+1)$ 　　**3.** $(y-2)^2 = (x+1)^2 + C$ 　　**5.** $y = \tan(e^x + C)$ 　　**7.** $Ce^{-x/y} = xy$

9. $(2y-x)^2(y+x) = C$ 　 $(Cx > 0)$ 　　**11.** $y = 2\sin\left(\frac{x^2}{2} + C\right)$ 　　**13.** $y = x\sin(\ln Cx)$ 　 $(Cx > 0)$

15. $2(x^2 + y^2) = 5y$; $y = \frac{5 + \sqrt{25 - 16x^2}}{4}$ 　　**17.** $e^{x/y} = \ln Cy$; $e^{x/y} = \ln y$ 　　**19.** $e^x(y+1)^2 = Ce^y(x+1)$ 　　**21.** $y' = F(x,y)$; let

$y = vx$; $v + x\frac{dv}{dx} = F(1, v)$; $x\frac{dv}{dx} = F(1, v) - v$; $\frac{dv}{(1,v)-v} = \frac{dx}{x}$; $\int (dv)/(F(1, v) - v) = \ln C_1|x|$; $C_1 x = e^{\int (dv)/(F(1,v)-v)}$; $x = Ce^{\int dv/F(1,v)-v}$

23. $\frac{(x-y)^2}{y} = Ce^{-(x+y)^2/(2y^2)}$ 　　**25.** $y = 1 + 2\left(\frac{1-x}{2+x}\right)e^{-2x/(x-1)}$ 　　**27.** If the differential equation is homogeneous, then F is homogeneous of degree 0, and so substituting $v = y/x$ gives $F(x, y) = F(x, vx) = x^0 F(1, v) = F(1, v)$. Define $G(v) = F(1, v)$, so that $F(x, y) = G(y/x)$. If, conversely, $F(x, y) = G(y/x)$, then $F(tx, ty) = G(ty/tx) = G(y/x)$. Thus, $F(tx, ty) = F(x, y) = t^0 F(x, y)$.

Exercise Set 17.3

1. $y = \frac{(x+1)^2}{2(x-2)} + \frac{C}{x-2}$ 　　**3.** $x^3 - xy^2 = C$; $xy^2 - x^3 = 3$ 　　**5.** $y\ln x + \frac{y^2}{2} = C$; $y\ln x + \frac{y^2}{2} = 8$ 　　**7.** $y = x - 1 + Ce^{-x}$

9. $y = \frac{x^2}{3} + \frac{C}{x}$; $y = \frac{x^2}{3} - \frac{2}{3x}$ 　　**11.** $\frac{y}{1-x} = \frac{(x+1)^2}{2} + C$; $y = \frac{2x - x^2 - x^3}{2}$ 　　**13.** exact; $x^2 - 4xy + 3y^2 = C$ 　　**15.** exact; also

variables separate; $x(y+1) - y = C$; $y = \frac{2+x}{1-x}$ 　　**17.** linear; $\frac{y}{x^2+1} = \tan^{-1}x + C$; $y = (x^2 + 1)(\tan^{-1}x + \frac{\pi}{4})$ 　　**19.** exact;

$y\ln xy - e^x + 2(x-y) + y^3 = C$ 　　**21.** exact; $\sqrt{x^2 + y^2} + x - 2y = C$; $\sqrt{x^2 + y^2} + x - 2y = 0$ 　　**23.** exact; $x - y\ln x + \frac{y^2}{2} = C$

25. linear; $x = 1 + y^2 + C\sqrt{1 + y^2}$ 　　**27.** exact; $x\ln(x^2 + y^2) - 2x - 2y\tan^{-1}\frac{y}{x} = C$ 　　**29.** let $v = y^{-1} = \frac{1}{y}$; $y = \frac{2}{1+Ce^{x^2}}$

31. $f(x, y)dx + g(x, y)dy = 0$; multiply both sides by $\mu(x)$ and $[\mu(x)f(x, y)]dx + [\mu(x)g(x, y)]dy = 0$ is exact if and only if

$\frac{\partial}{\partial y}[\mu(x)f(x, y)] = \frac{\partial}{\partial x}[\mu(x)g(x, y)]$, or $\mu f_y = \mu g_x + g\mu'$; $\frac{du}{\mu} = \frac{f_y - g_x}{g}dx$. If $\frac{f_y - g_x}{g}$ is a function only of x, then $\ln\mu = \int \frac{f_y - g_x}{g}dx$, or

$\mu(x) = e^{\int (f_y - g_x/g)dx}$ 　　**33.** $\mu(x) = e^{2x}$; $e^{2x}\left(\frac{y^2}{2} + 2x - 1\right) = C$ 　　**35.** If $\phi(x, y(x)) = C$, then $d\phi = \frac{\partial\phi}{\partial x} + \frac{\partial\phi}{\partial y}\frac{dy}{dx} = 0$. Conversely,

if $d\phi(x, y) = 0$ and $y = y(x)$ solves $d\phi(x, y) = 0$, then $\phi(x, y(x)) = C$

Exercise Set 17.4

1. ≈ 3.42 hrs; ≈ 276 　　**3.** $A(30) \approx \$6049.65$; $t \approx 26.82$ yrs 　　**5.** $t \approx 3760$ yrs 　　**7.** $Q(3) \approx 33.7g$; $t \approx 6.35$ yrs
9. $T(30) \approx 6.17°C$; $T(t) \neq 5°$ C for any finite value of t, showing that the mathematical model is only an approximation of the physical system.

11. $Q(t) = 80 - 40e^{-t/20}$; concentration ≈ 0.355, or 35.5% 　　**13.** $\frac{dQ}{dt} = \alpha Q - \beta Q^2$, $(\alpha > 0, \beta > 0)$; separating variables, $\frac{dQ}{Q(\alpha - \beta Q)} = dt$; by

partial fractions, $\frac{1}{Q(\alpha - \beta Q)} = \frac{1}{\alpha Q} + \frac{\beta}{\alpha(\alpha - \beta Q)}$; $\frac{1}{\alpha}\ln Q - \frac{1}{\alpha}\ln(\alpha - \beta Q) = t + \ln C_1$; $\frac{Q}{\alpha - \beta Q} = Ce^{\alpha t}$; $\frac{Q_0}{\alpha - \beta Q_0} = C$; $Q(t) = \frac{\alpha Q_0}{\beta Q_0 + (\alpha - \beta Q_0)e^{-\alpha t}}$

15. $v(t) = \sqrt{\frac{mg}{k}}\tanh\sqrt{\frac{kg}{m}}t$; $\lim\limits_{t\to\infty} v(t) = \sqrt{\frac{mg}{k}}$ 　　**17.** $I(t) = 2(1 - e^{-6t})$ 　　**19.** $y(t) = \frac{a}{b}(1 - e^{-bt}) + y_0e^{-bt}$

21. $\ln\frac{b(a-2x)}{a(b-x)} - \frac{2x(a-2b)}{a(a-2x)} = (a - 2b)^2 kt$ 　　**23.** concentration $\approx 0.602\%$; $t \approx 33.47$ min

25. $Q(t) = Q_0 e^{(\alpha/\beta)\sin\beta t}$ 　　　　　　　　**27.** $Q(t) = \frac{\alpha(Q_0 - m) + m(\alpha - \beta Q_0)e^{-(\alpha - m\beta)t}}{\beta(Q_0 - m) + (\alpha - \beta Q_0)e^{-(\alpha - m\beta)t}}$. If $Q_0 < m$, then $Q_0 - m < 0$ and $Q(t_1) = 0$ when

$m(\alpha - \beta Q_0)e^{-(\alpha - m\beta)t_1} = \alpha(m - Q_0)$, or $e^{-(\alpha - m\beta)t_1} = \frac{\alpha(m - Q_0)}{m(\alpha - \beta Q_0)}$; $t_1 = \frac{1}{\alpha - m\beta}\ln\frac{m(\alpha - \beta Q_0)}{\alpha(m - Q_0)}$

Exercise Set 17.5

1. $y = C_1 e^{-x} + C_2 e^{-2x}$ 　　**3.** $y = C_1 e^{-4x} + C_2 x e^{-4x}$ 　　**5.** $y = C_1\cos 3x + C_2\sin 3x$ 　　**7.** $y = e^x(C_1\cos 2x + C_2\sin 2x)$
9. $y = C_1 e^{-x} + C_2 e^{5x/2}$ 　　**11.** $y = C_1 e^{4t} + C_2 e^{-t}$ 　　**13.** $u = C_1 + C_2 e^{3x/2}$ 　　**15.** $s = e^{3t/2}(C_1 + C_2 t)$

17. $y = \frac{3}{8}e^{-5x} + \frac{13}{8}e^{3x}$ 　　**19.** $y = 3e^{3x} - 1$ 　　**21.** $y = e^{2x}(2 - x)$ 　　**23.** $y = \sqrt{3}\sin x - \cos x$ 　　**25.** For real

and unequal $m_1 = \frac{-b + \sqrt{b^2 - 4ac}}{2a}$, $m_2 = \frac{-b - \sqrt{b^2 - 4ac}}{2a}$, $y = C_1 e^{m_1 x} + C_2 e^{m_2 x}$. Let $C_1 = \frac{C_3 + C_4}{2}$, $C_2 = \frac{C_3 - C_4}{2}$, and then

$y = \frac{C_3}{2}(e^{m_1 x} + e^{m_2 x}) + \frac{C_4}{2}(e^{m_1 x} - e^{m_2 x})$. Since $u = \frac{-b}{2a}$ and $v = \frac{\sqrt{b^2 - 4ac}}{2a}$, we have $m_1 = u + v$, $m_2 = u - v$.

$y = \frac{C_3}{2}(e^{(u+v)x} + e^{(u-v)x}) + \frac{C_4}{2}(e^{(u+v)x} - e^{(u-v)x}) = e^{ux}(C_3\cosh vx + C_4\sinh vx)$ (Now rename the constants.)

27. $t = \ln x \Rightarrow \frac{dt}{dx} = \frac{1}{x}$; $\frac{dy}{dx} = \frac{dy}{dt}\frac{dt}{dx} = \frac{1}{x}\frac{dy}{dt}$; $\frac{d^2y}{dx^2} = -\frac{1}{x^2}\frac{dy}{dt} + \frac{1}{x^2}\frac{d^2y}{dt^2}$. Substitute into the differential equation:

$ax^2\left(-\frac{1}{x^2}\frac{dy}{dt} + \frac{1}{x^2}\frac{d^2y}{dt^2}\right) + bx\left(\frac{1}{x}\frac{dy}{dt}\right) + cy = 0$. Then, $a\frac{d^2y}{dt^2} + (b - a)\frac{dy}{dt} + cy = 0$, which is linear with constant coefficients.

29. $s = C_1 \ln t + C_2$ **31.** $y = C_1 + C_2 e^{3x} + C_3 e^{-x}$ **33.** $y = C_1 \cos x + C_2 \sin x + C_3 \cos 2x + C_4 \sin 2x$

35. $y = C_1 e^x + C_2 \cos 2x + C_3 \sin 2x; \ y = \frac{2}{5} e^x - \frac{2}{5} \cos 2x + \frac{3}{10} \sin 2x$ **37.** $L[ky] = aky'' + bky' + cky = k(ay'' + by' + cy) =$

$kL[y]L[y_1 + y_2] = a(y_1'' + y_2'') + b(y_1' + y_2') + c(y_1 + y_2) = (ay_1'' + by_1' + cy_1) + (ay_2'' + by_2' + cy_2) = L[y_1] + L[y_2]$ **39.** If

$am^2 + bm + c = 0$ has equal solutions, then they must be $-b/2a$, and $\sqrt{b^2 - 4ac}$ must be zero, $y = C_1 e^{-bx/2a} + C_2 x e^{-bx/2a}$, or

$y_2 = xe^{-bx/2a}$. Using $y_2' = e^{-bx/2a}\left(1 - \frac{bx}{2a}\right)$ and $y_2'' = e^{-bx/2a}\left(-\frac{2b}{2a} + \frac{b^2 x}{4a^2}\right)$, we get $e^{-bx/2a}\left(-b + \frac{b^2 x}{4a} + b - \frac{b^2 x}{2a} + cx\right) = 0$.

Exercise Set 17.6

1. $y_p = 2x; \ y = C_1 \cos x + C_2 \sin x + 2x$ **3.** $y_p = -\frac{3}{2}; \ y = C_1 e^{2x} + C_2 e^{-x} - \frac{3}{2}$ **5.** $y_p = Ae^x = -\frac{3}{4}e^x$

7. $y_p = A \cos x + B \sin x = \frac{1}{3} \sin x$ **9.** $y_p = Axe^{2x} = \frac{1}{3}xe^{2x}$ **11.** $y_p = x(A \cos x + B \sin x) = \frac{3}{2}x \sin x$

13. $y_p = Ax^2 e^x = \frac{1}{2}x^2 e^x$ **15.** $y_p = (Ax^2 + Bx)e^{2x} = \frac{1}{4}x(x-1)e^{2x}$ **17.** $y_p = e^{-x}(A \cos x + B \sin x) = -\frac{2}{3}e^{-x} \sin x$

19. $u_1' \cos x + u_2' \sin x = 0 \, (1) \ -u_1' \sin x + u_2' \cos x = \sec^3 x \, (2)$ Multiply (1) by $\sin x$ and (2) by $\cos x$. Add the resulting equations to get

$u_2'(\sin^2 x + \cos^2 x) = \sec^3 x \cos x = \sec^2 x; \ u_2' = \sec^2 x$. Substitute into (1) and solve for $u_1' = -\sec^2 x \tan x$.

21. $y = C_1 \cos x + C_2 \sin x + (x - \tan x) \cos x + (\sin x) \ln(\sec x)$

23. $x = C_1 \cos t + C_2 \sin t - \cos t \ln(\sec t) + t \sin t$ **25.** $y = C_1 e^x + C_2 e^{-x} + \frac{x^2 e^x}{4} - \frac{1}{4}xe^x + \frac{1}{8}e^x$

27. $y = C_1 e^{-x} + C_2 e^{-2x} - e^{-2x} \cos(e^x)$ **29.** $y = e^{-x}(C_1 \cos x + C_2 \sin x) + \frac{e^x}{8}(\sin x - \cos x)$ **31.** $y_p = (Ax^4 + Bx^3 + Cx^2 + Dx)e^x$

33. $y_p = Ax^3 + Bx^2 + Cx + (Dx^2 + Ex)e^{2x}$ **35.** $y_p = e^{3x}(Ax \cos 4x + Bx \sin 4x) + C$ **37.** $y_p = e^{-x}[(Ax+B)\cos x + (Cx+D)\sin x]$

39. $y = \frac{11}{16}e^{2x} + \frac{1}{16}e^{-2x} - \frac{1}{4}(x+3)$ **41.** $y = \frac{7}{3}\cos x + \frac{8}{3}\sin x - \frac{1}{3}\cos 2x$ **43.** $y = e^{2x}\left(-\frac{19}{17}\cos x + \frac{76}{17}\sin x\right) + \frac{2}{17}e^{-2x}$

45. $y = \frac{14}{9}e^{2x} + \frac{4}{9}e^{-x} + \frac{1}{3}xe^{2x}$ **47.** $y = C_1 e^x + C_2 e^{-x} + \frac{x}{2}\sinh x$ **49.** $y = C_1 \cos 2x + C_2 \sin 2x - \frac{1}{4}x \cos 2x$

51. $x(t) = C_1 e^{3t} + C_2 t e^{3t} + t^2 e^{3t} + \frac{1}{9}t^2 + \frac{4}{27}t + \frac{2}{27}; \ x(t) = \left(\frac{106}{27} - \frac{376}{27}t + t^2\right)e^{3t} + \frac{1}{27}(3t^2 + 4t + 2)$

53. $y_p = -\frac{x^4}{36} - \frac{x^3}{27} - \frac{x^2}{27}; \ y = C_1 + C_2 x + C_3 e^{3x} - \frac{x^4}{36} - \frac{x^3}{27} - \frac{x^2}{27}$ **55.** $y_p = \frac{1}{3}x^2 e^x + \frac{3}{2}x + \frac{9}{4}; \ y = C_1 e^x + C_2 x e^x + C_3 e^{-2x} + \frac{1}{3}x^2 e^x + \frac{3}{2}x + \frac{9}{4}$

Exercise Set 17.7

1. $y = \frac{1}{2}\cos 8t$ **3.** $y = \frac{2}{3}\cos 4t - 3 \sin 4t$ **5.** $y = -10 \cos 14t$ **7.** $y = 9e^{-7t} - 5e^{-14t}$; overdamped   **9.** $y = \frac{1}{2}\cos 4t$;

underdamped   **11.** (a) $c > \sqrt{5}$;   (b) $c = \sqrt{5}$;   (c) $c < \sqrt{5}$ **13.** (a) $\frac{1}{2}$ radian;   (b) $\frac{\pi}{2}$;   (c) $\frac{5\pi}{24}$;   (d) 2 ft/sec

15. $C_1 \cos \omega t + C_2 \sin \omega t = \sqrt{C_1^2 + C_2^2}\left(\frac{C_1}{\sqrt{C_1^2+C_2^2}}\cos \omega t + \frac{C_2}{\sqrt{C_1^2 C_2^2}}\sin \omega t\right)$. Let α be such that $\sin \alpha = \frac{C_1}{\sqrt{C_1^2+C_2^2}}$ and $\cos \alpha = \frac{C_2}{\sqrt{C_1^2+C_2^2}}$. Let

β be such that $\cos \beta = \frac{C_1}{\sqrt{C_1^2+C_2^2}}$ and $\sin \beta = \frac{C_2}{\sqrt{C_1^2+C_2^2}}$. Let $C = \sqrt{C_1^2 + C_2^2}$. Then, (a) $C(\sin \alpha \cos \omega t + \cos \alpha \sin \omega t) = C \sin(\omega t + \alpha)$

(b) $C(\cos \beta \cos \omega t + \sin \beta \sin \omega t) = C \cos(\omega t - \beta)$ **17.** $I(t) = e^{-t}(C_1 \cos 2t + C_2 \sin 2t) + 4 \cos t + 2 \sin t; \ \lim\limits_{t \to \infty} I(t) = 4 \cos t + 2 \sin t$

Exercise Set 17.8

1. $y = a_0 \sum_{k=0}^{\infty} \frac{x^{3k}}{3^k k!} = Ce^{x^3/3}$ **3.** $y = a_0 \sum_{k=0}^{\infty} \frac{x^{2k}}{(2k)!} + a_1 \sum_{k=0}^{\infty} \frac{x^{2k+1}}{(2k+1)!} = a_0 \cosh x + a_1 \sinh x = C_1 e^x + C_2 e^{-x}$

5. $y = a_0 \left[1 + \sum_{k=1}^{\infty} \frac{1 \cdot 4 \cdot 7 \cdots (3k-2)}{(3k)!} x^{3k}\right] + a_1 \left[x + \sum_{k=1}^{\infty} \frac{2 \cdot 5 \cdot 8 \cdots (2k-1)}{(3k+1)!} x^{3k+1}\right]$ **7.** $y = a_0 \sum_{k=0}^{\infty} \frac{(-1)^k}{(2k)!} x^{2k} + a_1 \sum_{k=0}^{\infty} \frac{(-1)^k}{(2k+1)!} x^{2k+1}$;

$y = \sum_{k=0}^{\infty} \frac{(-1)^k}{(2k)!} x^{2k} = \cos x$ **9.** $y = a_0 \left(1 - \frac{x^4}{3 \cdot 4} + \frac{x^8}{3 \cdot 5 \cdot 7 \cdot 8} - \frac{x^{12}}{3 \cdot 4 \cdot 7 \cdot 8 \cdot 11 \cdot 12} + \cdots\right) + a_1 \left(x - \frac{x^5}{4 \cdot 5} + \frac{x^9}{4 \cdot 5 \cdot 8 \cdot 9} - \frac{x^{13}}{4 \cdot 5 \cdot 8 \cdot 9 \cdot 12 \cdot 13} + \cdots\right)$

11. $y = a_0 \sum_{k=1}^{\infty} \frac{1 \cdot 3 \cdot 5 \cdots (2k+1)}{2^k k!} x^{2k} + a_1 \sum_{k=1}^{\infty} \frac{2^k(k+1)!}{1 \cdot 3 \cdot 5 \cdots (2k+1)} x^{2k+1}$ **13.** $y = 2 + x + \frac{x^2}{4} + \frac{7}{4}\sum_{k=1}^{\infty} \frac{(-1)^k x^{2k}}{1 \cdot 3 \cdot 5 \cdots (2k-1)} + \sum_{k=1}^{\infty} \frac{(-1)^k x^{2k+1}}{2 \cdot 4 \cdot 6 \cdots (2k)}$

15. $y_1 = 1 - \frac{x^2}{2!} + \frac{x^3}{3!} + \frac{x^4}{4!} - \frac{4x^5}{5!} + \frac{3x^6}{6!} + \frac{9x^7}{7!} - \frac{27x^8}{8!} + \frac{12x^9}{9!} + \cdots; \ y_2 = x - \frac{x^3}{3!} + \frac{2x^4}{4!} + \frac{x^5}{5!} - \frac{6x^6}{6!} + \frac{9x^7}{7!} + \frac{12x^8}{8!} - \frac{51x^9}{9!} + \frac{60x^{10}}{10!} + \cdots; \ y = C_1 y_1 + C_2 y_2$

17. $y(x) = 2 + x + \frac{2x^2}{2!} - \frac{x^3}{3!} - \frac{7x^5}{5!} + \frac{4x^6}{6!}x - \frac{7x^7}{7!} + \frac{46x^8}{8!} - \frac{35x^9}{9!} + \frac{102x^{10}}{10!}$

Chapter Review

1. $2x + 3y - e^{-x} \cos y = C$ **3.** $y = \sin\left(\frac{x^2}{2} + \frac{\pi}{2}\right)$ **5.** $Cx^2 = ye^{y/x}$ **7.** $\frac{3}{2}\tan^{-1}\left(\frac{y}{2x}\right) - \frac{1}{2}\ln\left[\left(\frac{y}{x}\right)^2 + 4\right] = \ln x + C; \ \tan^{-1}\frac{y}{2x} =$

$\frac{2}{3}\ln\frac{\sqrt{y^2+4x^2}}{2}$ **9.** $y = -\sqrt{x(2-x)}\sin^{-1}(1 - x) + C\sqrt{x(2-x)}; \ y = \sqrt{2x - x^2}[2 - \sin^{-1}(1-x)]$

11. $x \cos xy = C$ **13.** $ye^{x^2} = -\frac{1}{2}e^{-x^2} + C; \ y = -\frac{1}{2}e^{-2x^2} + \frac{3}{2}e^{-x^2}$ **15.** $y = C_1 e^{-3x} + C_2 x e^{-3x}; \ y = 3e^{-3x} + 5xe^{-3x}$

17. $s(t) = C_1 + C_2 e^{-2t}$; $s(t) = 2(1 - e^{-2t})$ **19.** $x_p = -\frac{1}{3}e^t$ **21.** $y_p = -\frac{x^2}{2} - x + 3xe^x$ **23.** $y_p = \frac{e^x}{53}(7\sin x - 2\cos x)$

25. $y_p = \frac{xe^x}{2} - \frac{1}{4}e^{-x}$ **27.** $y_p = x(Ax + B)e^{-x} + C\cos 2x + D\sin 2x$ **29.** ≈ 43.71 years   **31.** $Q(t) = e^{\alpha t}\left(Q_0 - \frac{H}{\alpha}\right) + \frac{H}{\alpha}$; (1) If $\frac{H}{\alpha} > Q_0$, then $Q(t) = 0$ when $\frac{H}{\alpha} = e^{\alpha t}\left(\frac{H}{\alpha} - Q_0\right)$ and $t = \frac{1}{\alpha}\ln\frac{H}{H - \alpha Q_0}$. Thus, the population dies out in a finite time. (2) If $\frac{H}{\alpha} < Q_0$, then the population grows without limit. (3) If $\frac{H}{\alpha} = Q_0$, then $Q(t) = \frac{H}{\alpha}$, a constant.   **33.** $y(t) = \frac{e^{-t}}{3}(2\sin 7t - \cos 7t)$

35. $y = a_0 \sum_{k=0}^{\infty} \frac{2^{2k}k!}{(2k)!}x^{2k} + a_1 \sum_{k=0}^{\infty} \frac{x^{2k+1}}{k!}$; series converges for $-\infty < x < \infty$.

Photo Credits

PROLOGUE: ix, top, Kathleen Olson; bottom, courtesy of Texas Instruments.
CHAPTER 1: 4, Fig. 1.1, Keeling, C. D. et al., in *Aspects of Climate Variability in the Pacific and the Western Americas*, Part 1: Atmospheric CO_2 Data and Analysis, Geophysical Monograph **55**, pp. 165–236, American Geophysical Union, Washington, D.C., 1989 (updated, private communication); **33**, Kathleen Olson; **51**, Ken Whitmore/Tony Stone Images; **67**, TIB/West Jay Brousseau © 1994; **69**, Loren M. Winters; **79**, courtesy, Central Scientific Company; **80**, top, Art Resource, NY; center, Erich Lessing/Art Resource, NY; bottom, Bettmann Archive.
CHAPTER 2: 82, photo by Jay M. Pasachoff.
CHAPTER 3: 139; © Chad Slattery/Tony Stone Images; **142**, © Loubat/Vandystadt/Photo Researchers; **173**, courtesy, Central Scientific Company; **176**, © Gary Buss 1990/FPG Selects; **192**, © Jim Goodwin/Photo Researchers; **203**, © 1990 Smithsonian Institution and Lockheed Corporation; **204**, Mick Roessler/Superstock.
CHAPTER 4: 207, Ralph Cowan/Tony Stone Worldwide; **242**, photo by Jay M. Pasachoff; **247**, © 1992, 1993, PhotoDisc, Inc.; **251**, left, © 1992, 1993, PhotoDisc, Inc.; right, © Day Williams/Photo Researchers; **257**, top left, Uniphoto/Pictor; top right, William Strode/Superstock; bottom, © 1992, 1993, PhotoDisc, Inc.; **259**, Westlight/Orion; **289**, Exercise 4, reprinted by permission from Mathematical Association of America Notes, Vol. 4 (by John Ramsay, College of Wooster); **298**, courtesy, Central Scientific Company.
CHAPTER 5: 302, photo by Jay M. Pasachoff; **304**, Steve Lipofsky; **311**, Globus Brothers/The StockMarket; **319**, photo by Jay M. Pasachoff; **339**, ARS-USDA; **373**, TIB West/Harald Sund © 1994; **383**, Exercise 3, reprinted by permission of Dr. Judith H. Morrel, Department of Mathematics and Computer Science, Butler University; **384**, top, © J. Fields/Science Source/Photo Researchers; bottom, courtesy, Central Scientific Company.
CHAPTER 6: 388, photo by Jay M. Pasachoff; **399**, photo by Jay M. Pasachoff; **403**, Kathleen Olson; **405**, Kathleen Olson; **412**, © 1992, 1993, PhotoDisc, Inc.; **420**, photo by Jay M. Pasachoff; **427**, Robert Llewellyn/Superstock; **430**, Charles Orrica/Superstock; **435**, © 1990 Tom Van Sant/The GeoSphere™ Project, Santa Monica, CA; **438**, courtesy, Central Scientific Company; **441**, NASA Photo/Lyndon B. Johnson Space Center; **442**, Grant Heilman/Grant Heilman Photography; **447**, Billy E. Barnes/PhotoEdit; **462**, photo by Jay M. Pasachoff; **479**, photo by Jay M. Pasachoff; **480**, top photo, Eloise Pasachoff; left, Dennis Di Cicco/Peter Arnold, Inc; right, photo by Jay M. Pasachoff.
CHAPTER 7: 525, Jet Propulsion Laboratory; **527**, M.G. Giles/Pix* Elation/Fran Heyl Associates; **530**, © 1992, 1993, PhotoDisc, Inc.; **534**, courtesy, Central Scientific Company; **548**, left, © Garry D. McMichael/Photo Researchers; center, Jeremy Hayhurst; right, © 1992, 1993, PhotoDisc, Inc.; **552**, Larry Chiger/ Superstock; **568**, photo by Jay M. Pasachoff/courtesy the Sterling and Francine Clark Art Institute; **570**, © Dr. R. Clark and M. Goff/Science Photo Library/Photo Researchers.
CHAPTER 8: 572, photo by Jay M. Pasachoff; **612**, © Yoav Levy/ Phototake NYC; 644. © Paul Hanny/Gamma Liaison; **644**, courtesy Jay M. Pasachoff.
CHAPTER 9: 646, Musées de Narbonne; **652**, Fundamental Photographs, NY; **655**, The Bettmann Archive; **667**, courtesy, Kimbell Art Museum; **676**, Kathleen Olson; **683**, photo by Jay M. Pasachoff; **700**, Lee Hocker; **701**, Lee Hocker;

INDEX

Page numbers followed by a "d" refer to definitions on those pages.

Forms Containing $\sqrt{u^2 \pm a^2}$ (continued)

41. $\displaystyle\int \frac{\sqrt{u^2 - a^2}}{u}\,du = \sqrt{u^2 - a^2} - a\sec^{-1}\left|\frac{u}{a}\right| + C$

42. $\displaystyle\int \frac{\sqrt{u^2 \pm a^2}}{u^2}\,du = -\frac{\sqrt{u^2 \pm a^2}}{u} + \ln\left|u + \sqrt{u^2 \pm a^2}\right| + C$

43. $\displaystyle\int \frac{du}{\sqrt{u^2 \pm a^2}} = \ln\left|u + \sqrt{u^2 \pm a^2}\right| + C$

44. $\displaystyle\int \frac{u^2\,du}{\sqrt{u^2 \pm a^2}} = \frac{u}{2}\sqrt{u^2 \pm a^2} \mp \frac{a^2}{2}\ln\left|u + \sqrt{u^2 \pm a^2}\right| + C$

45. $\displaystyle\int \frac{du}{u\sqrt{u^2 + a^2}} = -\frac{1}{a}\ln\left|\frac{a + \sqrt{u^2 + a^2}}{u}\right| + C$

46. $\displaystyle\int \frac{du}{u\sqrt{u^2 - a^2}} = \frac{1}{a}\sec^{-1}\left|\frac{u}{a}\right| + C$

47. $\displaystyle\int \frac{du}{u^2\sqrt{u^2 \pm a^2}} = \mp\frac{\sqrt{u^2 \pm a^2}}{a^2 u} + C$

48. $\displaystyle\int (u^2 \pm a^2)^{3/2}\,du = \frac{u}{8}(2u^2 \pm 5a^2)\sqrt{u^2 \pm a^2} + \frac{3a^4}{8}\ln\left|u + \sqrt{u^2 \pm a^2}\right| + C$

49. $\displaystyle\int \frac{du}{(u^2 \pm a^2)^{3/2}} = \pm\frac{u}{a^2\sqrt{u^2 \pm a^2}} + C$

Forms Containing $\sqrt{a^2 - u^2}$

50. $\displaystyle\int \sqrt{a^2 - u^2}\,du = \frac{u}{2}\sqrt{a^2 - u^2} + \frac{a^2}{2}\sin^{-1}\frac{u}{a} + C$

51. $\displaystyle\int u^2\sqrt{a^2 - u^2}\,du = \frac{u}{8}(2u^2 - a^2)\sqrt{a^2 - u^2} + \frac{a^4}{8}\sin^{-1}\left(\frac{u}{a}\right) +$

52. $\displaystyle\int \frac{\sqrt{a^2 - u^2}}{u}\,du = \sqrt{a^2 - u^2} - a\ln\left|\frac{a + \sqrt{a^2 - u^2}}{u}\right| + C$

53. $\displaystyle\int \frac{\sqrt{a^2 - u^2}}{u^2}\,du = -\frac{\sqrt{a^2 - u^2}}{u} - \sin^{-1}\frac{u}{a} + C$

54. $\displaystyle\int \frac{u^2}{\sqrt{a^2 - u^2}}\,du = -\frac{u}{2}\sqrt{a^2 - u^2} + \frac{a^2}{2}\sin^{-1}\left(\frac{u}{a}\right) + C$

55. $\displaystyle\int \frac{du}{u\sqrt{a^2 - u^2}} = -\frac{1}{a}\ln\left|\frac{a + \sqrt{a^2 - u^2}}{u}\right| + C$

56. $\displaystyle\int \frac{du}{u^2\sqrt{a^2 - u^2}} = -\frac{\sqrt{a^2 - u^2}}{a^2 u} + C$

57. $\displaystyle\int (a^2 - u^2)^{3/2}\,du = \frac{u}{4}\left(a^2 - u^2\right)^{3/2} + \frac{3a^2 u}{8}\sqrt{a^2 - u^2} + \frac{3a^4}{8}\sin^{-1}\frac{u}{a} + C$

58. $\displaystyle\int \frac{du}{(a^2 - u^2)^{3/2}} = \frac{u}{a^2\sqrt{a^2 - u^2}} + C$

Forms Involving $\sqrt{2au - u^2}$

59. $\displaystyle\int \sqrt{2au - u^2}\,du = \frac{u - a}{2}\sqrt{2au - u^2} + \frac{a^2}{2}\cos^{-1}\left(\frac{a - u}{a}\right) + C$

60. $\displaystyle\int u\sqrt{2au - u^2}\,du = \frac{2u^2 - au - 3a^2}{6}\sqrt{2au - u^2} + \frac{a^3}{2}\cos^{-1}\left(\frac{a - u}{a}\right)$

61. $\displaystyle\int \frac{\sqrt{2au - u^2}}{u}\,du = \sqrt{2au - u^2} + a\cos^{-1}\left(\frac{a - u}{a}\right) + C$

62. $\displaystyle\int \frac{\sqrt{2au - u^2}}{u^2}\,du = -\frac{2\sqrt{2au - u^2}}{u} - \cos^{-1}\left(\frac{a - u}{a}\right) + C$

63. $\displaystyle\int \frac{du}{\sqrt{2au - u^2}} = \cos^{-1}\left(\frac{a - u}{a}\right) + C$

64. $\displaystyle\int \frac{u\,du}{\sqrt{2au - u^2}} = -\sqrt{2au - u^2} + a\cos^{-1}\left(\frac{a - u}{a}\right) + C$

65. $\displaystyle\int \frac{u^2\,du}{\sqrt{2au - u^2}} = -\frac{(u + 3a)}{2}\sqrt{2au - u^2} + \frac{3a^2}{2}\cos^{-1}\left(\frac{a - u}{a}\right) + C$

66. $\displaystyle\int \frac{du}{u\sqrt{2au - u^2}} = -\frac{\sqrt{2au - u^2}}{au} + C$

67. $\displaystyle\int \frac{\sqrt{2au - u^2}}{u^n}\,du = \frac{(2au - u^2)^{3/2}}{(3 - 2n)au^n} + \frac{n - 3}{(2n - 3)a}\int \frac{\sqrt{2au - u^2}}{u^{n-1}}\,du, \quad n \neq \frac{3}{2}$

68. $\displaystyle\int \frac{u^n\,du}{\sqrt{2au - u^2}} = -\frac{u^{n-1}\sqrt{2au - u^2}}{n} + \frac{a(2n - 1)}{n}\int \frac{u^{n-1}}{\sqrt{2au - u}}$

69. $\displaystyle\int \frac{du}{u^n\sqrt{2au - u^2}} = \frac{\sqrt{2au - u^2}}{a(1 - 2n)u^n} + \frac{n - 1}{(2n - 1)a}\int \frac{du}{u^{n-1}\sqrt{2au - u^2}}$

70. $\displaystyle\int \frac{du}{(2au - u^2)^{3/2}} = \frac{u - a}{a^2\sqrt{2au - u^2}} + C$

71. $\displaystyle\int \frac{u\,du}{(2au - u^2)^{3/2}} = \frac{u}{a\sqrt{2au - u^2}} + C$

Forms Containing Trigonometric Functions

72. $\displaystyle\int \sin^2 u\,du = \frac{u}{2} - \frac{\sin 2u}{4} + C$

73. $\displaystyle\int \cos^2 u\,du = \frac{u}{2} + \frac{\sin 2u}{4} + C$